Diesel Engines Part I

Kees Kuiken

This book is dedicated to my wife Truus,
who has supported me for over forty years
and without whose faith, help and
encouragement this book would never
have been written.

DIESEL ENGINES

for ship propulsion and power plants

FROM 0 TO 100,000 kW

2nd revised edition

Kees Kuiken

VIV More than engines

Target is member of the Association of Importers of Combustion Engines

Second revised edition, July 2012

Author:
Kees Kuiken

Technical advice:
Johan Heuff, Chief Engineer ret. P&O Nedlloyd.
Jan Leekstra, Chief Engineer ret. P&O Nedlloyd.
Christiaan Kuiken, Maritime Officer, Maersk Ship Management.

Translation:
AAG Translations, A.A. Geerts MA, Almelo, The Netherlands.

Textual corrections and proof reading:
S.E. Brent BSc, Almelo, The Netherlands.
Mimi Kuijper-Heeres, Senior Lecturer ret. Maritime Institute Willem Barentsz, Terschelling, The Netherlands.

Lay-out:
Little Shop of Graphics, Almelo, The Netherlands.

Technical drawings:
Niels Kuiken.
Maarten Bosch (LSoG).

Printing:
Epospress, Zwolle, The Netherlands.

Finish:
Stronkhorst Van der Esch/Embé, Groningen, The Netherlands.

A publication of:
Target Global Energy Training
Noorderhooidijk 2
9755 PJ Onnen, The Netherlands

Website: www.targettrainingcentre.nl
www.dieselengines.nl
E-mail: info@targettrainingcentre.nl

Target Global Energy Training.
Onnen, The Netherlands.

ISBN/EAN: 978-90-79104-00-0

Information
In order to gather sufficient information for this book, many companies both in The Netherlands and abroad have been visited in the past five years.
These companies, each in their own way, have contributed to the creation of this book by participating in interviews, allowing photographs and providing technical information.
Nevertheless, special thanks go to MAN Diesel and Wärtsilä AG, for their cooperation in making a great deal of material available.

Preface

Preface to the first edition

Diesel engines play an important role in today's society: we are quite dependent on them.
Over 100 years after Rudolf Diesel developed a working diesel engine, there is still no real alternative for ship propulsion and electric generators in tropical and/or remote areas. The diesel engine is indispensable for road haulage, inland shipping, aquatics, electric power emergency systems, agriculture, and passenger transport by road or rail, oil and gas industry and various other industries. We have chosen to make use of many pictures accompanied by a written explanation.
Much highly in-depth technical theory has been omitted as these topics are covered by specialist books available in the market; these topics include thermodynamics, vibrations, materials, and electronics.
We at Target Global Energy Training, have opted for a more practical approach. This includes ample information with respect to the construction of engines, use of materials, various engine categories, maintenance, repairs, and the use of engines.
Much attention has been paid to the choice of proper graphic material. This, in our opinion, is helpful for the reader to gain insight in the various subjects. This publication is indispensable for every person who has dealings with the diesel-engine industry, from the smallest engine to 'The Cathedrals of the Oceans'.

Kees Kuiken, Onnen, July 2008

Preface to the second completely revised edition

Based on feedback from various users of the first edition of this book, including companies in the diesel engine field, maritime institutes and shipping companies, we decided to add a number of topics to the second edition.
It became apparent that many users had an interest in certain subjects, but were not in the position to visit these particular businesses to learn how certain components are manufactured or repaired. Therefore, the following topics have been added:

- The manufacture of pistons, valves, cylinder liners and gear drives for the larger four-stroke and two-stroke crosshead engines.
- A damage report from a four-stroke medium-speed diesel engine.
- A report on the manufacture of crankshafts.
- The excessive wear on cylinder liners in two-stroke crosshead engines, called 'scuffing'.
- Slower sailing larger vessels to save fuel, the so-called 'slow steaming'.
- Future improvements in propulsion efficiency and the reduction of CO_2 emissions by ships.
- A third example in ship propulsion; a vibration analysis for two engines driving a single propeller.
- An overview of the market share 2010 of diesel engine manufacturers worldwide.
- The Hercules Project, a collaboration between various parties, including the engine manufacturers for the purpose of improving efficiency and reducing emissions.
- A new thrust measurement system.
- The theoretical background of and damage to bearings.
- A new method for the alignment of amongst others, engines and shafts.
- Additional theoretical calculations for the aeration of engines.
- Additional theoretical calculations of the efficiencies of engines.
- An improvement to both to the many images and the text.

Kees Kuiken, Onnen, July 2012

◄◄ Maastricht, May 2007.

At the special general meeting of the 60th anniversary of the VIV, de Vereniging van Importeurs van Verbrandingsmotoren (Association of Importers of Combustion Engines) at the 'Theater aan het Vrijthof' in Maastricht, author Kees Kuiken presents the first proof print to the chairman of the Association FME-CWM, Mr. Jan Kamminga.

◄ Dordrecht, November 2008.

During the general meeting of V.I.V., the Association of Importers of combustion engines in November 2008, Kees Kuiken presents the first proof print of the English edition to Mr. Jules L. Th. Schöningh, chairman of the V.I.V.

People and engines

For engineering, operating and the maintenance of diesel engines, enthusiastic, motivated and well-trained technicians are essential for manufacturers, proprietors and users of diesel engines.

Contents

>CH 1

The use of industrial diesel engines

Diesel engines are used for a variety of applications.

- A Propulsion and generator sets for yachts.
- B Generating electricity in the tropics.
- C Propulsion of sea-going ships.
- D Propulsion of inland shipping.
- E Generating the electric energy required on large passenger ships.
- F Generating electrical energy on board ships.

A

B

C

E

D

F

1.1 Introduction

Throughout the last century the internal combustion engine has become increasingly important in a society that has come to rely heavily on machinery. The engine is the elementary chain in, for instance, the transportation of goods and persons by road and water, propelling various machines and generating electrical energy. The use of combustion-engine equipment such as chain saws, water pumps, concrete mixers and lawn mowers has seen an explosive increase.

Modern agriculture is completely dependent on internal combustion engines for propulsion of tractors, combines, and other farming equipment.

The industrial diesel engines mentioned in this book are almost all used for propelling an enormous diversity of ships. Also in generating electricity the diesel-power plant plays an important role.
In the internal combustion engine one distinguishes two principles: the **Otto cycle** and the **Diesel cycle**.

▶
The largest diesel engines are used in the marine industry; shown here, part of a twelve-cylinder two-stroke crosshead engine on a container ship.

Shaft power 68,000 kW.
The cylinders are in-line and the engine is equipped with a supercharger.

▶
A twenty-cylinder four-stroke piston engine with supercharger, and cylinders in V-position.

▲
A six-cylinder in-line engine for the generation of electrical energy on board a ship.

▶
A harbour towboat needs power of several thousands kilowatts to tow large sea vessels.

Container feeders with a carrying capacity of 9000 tons have an engine capacity of approximately 6000 kilowatts at 18 nautical miles per hour.

A common diesel-power plant in the tropics with diesel-engine driven electric generators.

Five S.E.M.T. Pielstick (now part of MAN Diesel SA) heavy-oil four-stroke diesel engines.

1.2 Otto cycle

In 1876, after years of experimentation **Nicolaas August Otto** develops the first four-stroke engine (four strokes of the piston - two rotations of the crankshaft) which compresses a mixture of air and fuel. A spark plug provides an ignition spark at the right moment, which ignites and then combusts the mixture.
This cycle will not be discussed any further in this book.

1.3 Diesel cycle

In 1897 Rudolf Diesel markets the first diesel engine. In these diesel engines, the air is compressed to such a degree that the high-end temperature of the air effects a very swift ignition and combustion of the injected fuel. The efficiency of this diesel cycle is higher than that of the Otto cycle, which resulted in a huge expansion of the diesel-engine industry.

Total efficiency of the internal combustion engine = the ratio of the produced shaft power and the supplied fuel power. This ranges from between 25% and 50%, depending on the engine size.

▶ Rudolf Diesel's first diesel engine at the test bed of the Augsburg Machine factory; MAN.

▶ Rudolf Diesel's first diesel engine built in Augsburg, Germany.

This engine never ran independently and looked more like a canon barrel than an internal combustion engine. The design pressure was 150 bar! Rudolf Diesel was nearly killed when one of his test models exploded.

PATENT-URKUNDE

№ 67207.

AUF GRUND DER ANGEHEFTETEN BESCHREIBUNG UND ZEICHNUNG IST DURCH BESCHLUSS DES KAISERLICHEN PATENTAMTES

an Rudolf Diesel, Ingenieur, in Berlin

EIN PATENT ERTHEILT WORDEN.

GEGENSTAND DES PATENTES IST:

Arbeitsverfahren und Ausführungsart für Verbrennungskraftmaschinen.

ANFANG DES PATENTES: 28. Februar 1892.

DIE RECHTE UND PFLICHTEN DES PATENTINHABERS SIND DURCH DAS PATENTGESETZ VOM 7. APRIL 1891 (REICHS-GESETZBLATT FÜR 1891 SEITE 79) BESTIMMT.

ZU URKUNDE DER ERTHEILUNG DES PATENTES IST DIESE AUSFERTIGUNG ERFOLGT.

Berlin, den 23. Februar 1893.

KAISERLICHES PATENTAMT.

Beglaubigt durch Frank

Bureau-Vorsteher des Kaiserlichen Patentamtes.

▶ The patent for the engine of Rudolf Diesel, the Diesel engine!

▼ Diesel-engine losses.

Left figure: The fuel supply of a line engine. This is always set at 100%.

Right figure: Shaft power of 50%. So half of the supplied energy (fuel) is converted into mechanical power on the crankshaft.

The remainder of the supplied energy is waste, such as cooling-water and exhaust-gas losses.

Far right figure: in order to use at least part of the 35% exhaust-gas losses, steam is generated in an exhaust-gas boiler. This steam drives a steam turbine, which subsequently generates electricity.

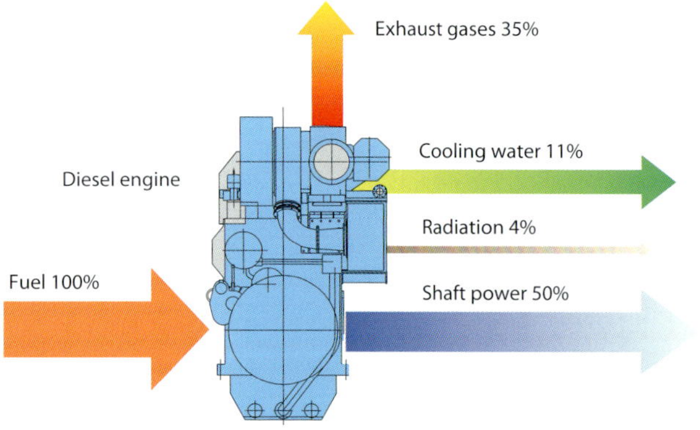

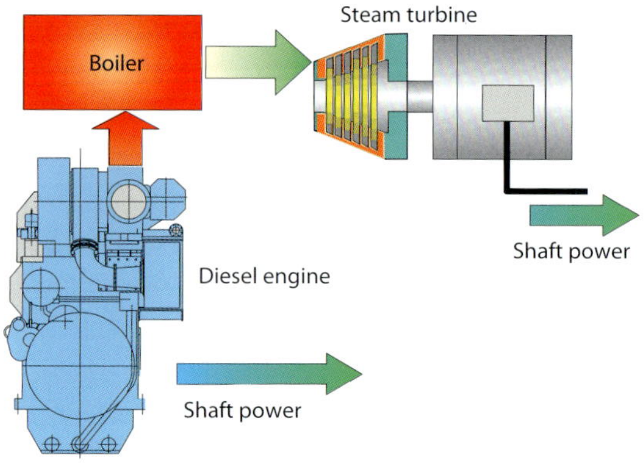

1.4 The use of Otto engines

They are often used in:
- (hand) tools;
- garden equipment;
- automotive industry;
- outboard engines;
- gas engines;
- small aviation craft.

1.5 The use of Diesel engines

They are often used in:
- ship propulsion;
- diesel-power plants;
- agriculture;
- back-up generator sets;
- lorries;
- earth-moving machines;
- military vehicles such as tanks.

Generally, diesel engines are more often used when the operating hours increase and fuel costs become important. The required shaft power is often much higher than in Otto engines.

▲
A seven-cylinder diesel engine at the test bed; a Wärtsilä Sulzer 7RT-flex60C, a typical propulsion engine for large, high-speed container ships.

1.6 Properties of both principles

	Otto engine	Diesel engine
Principle	mixture compression	air compression
Ignition	with spark plug ignition spark	by hot-air combustion
Fuels	gases and petrol	diesel oil, heavy oil and bio-fuels
Engine weight	light	heavier
Efficiency	low: 15–45%	high: 25–55%
Capacity	maximum 10 MW (gas engines)	maximum 100 MW (marine engines)

◀
The properties of both principles.

In this book, we will only discuss diesel engines. These are commonly used in navigation and industry.

◀
A small three-cylinder Volvo Penta diesel engine for yachts.

>CH 2

Classification of diesel engines

The classification of engines into four engine categories.

Category I: High-speed four-stroke engines with up to 100 kW shaft power.
The number of revolutions (rpm) normally lies between 1500 and 3000 revolutions per minute (rpm).
The fuel is diesel oil.

Category II: High-speed four-stroke engines with a shaft power between 100 and 10,000 kW.
The rpm normally lies between 960 and 2100 revolutions per minute.
The fuel is diesel oil.

Category III: Medium-speed four-stroke engines with a shaft power between 500 and 30,000 kW.
The rpm normally lies between 400 and 1000 revolutions per minute.
The fuel is heavy fuel oil or diesel oil.

Category IV: Low-speed two-stroke crosshead engines with a shaft power of 1500 up to 100,000 kW and rpm between 50 and 250 revolutions per minute.
The fuel is heavy fueloil.

The categories could of course overlap in some cases.

III IV

Classification of engines

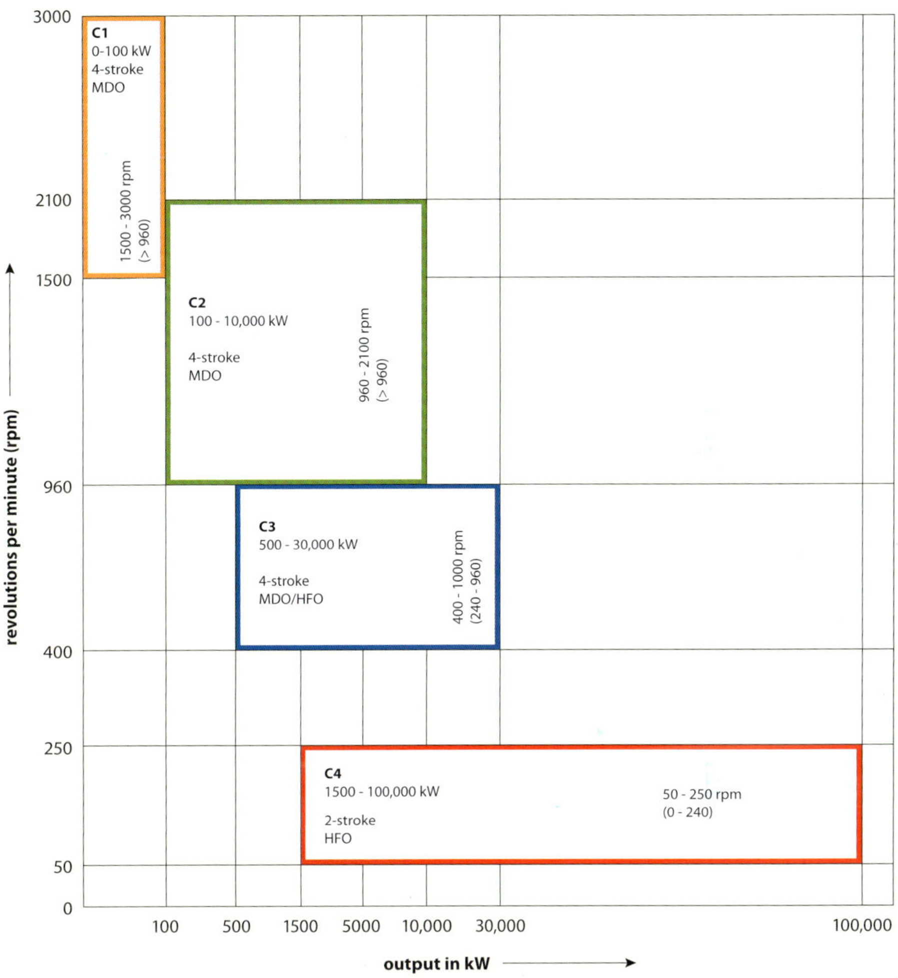

II I

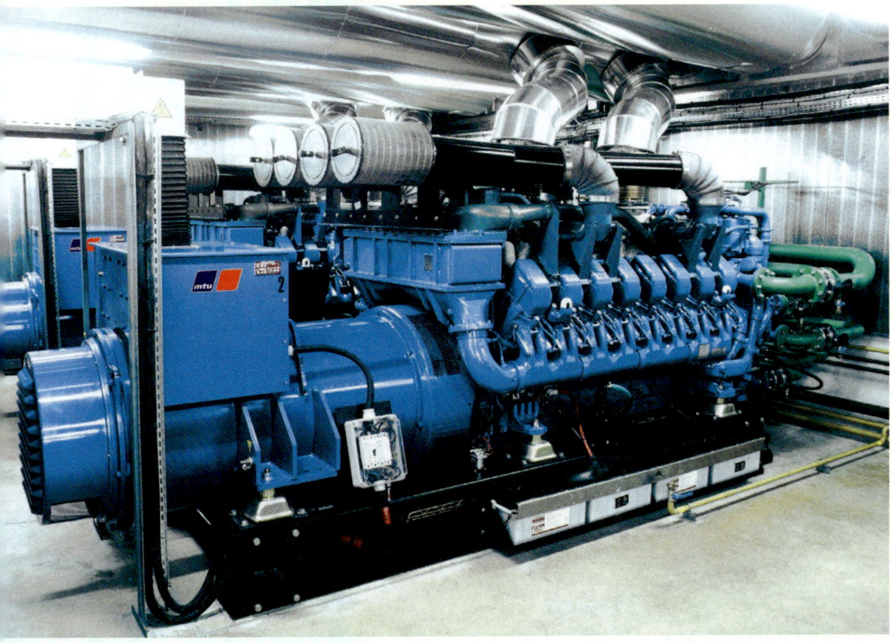

2.1 Introduction

Diesel engines are categorised by working principle, design, rotational speed, power output, fuel used, usage and other characteristics.

2.2 Working principle

Two-stroke principle

The entire combustion cycle is completed in one revolution of the crankshaft, which means in two piston strokes.
This principle is most often applied in small- and very large-sized diesel engines. Small diesel engines are often used to drive a generator or a pump. The very large diesel engines are without exception crosshead engines, which are predominantly used for ship propulsion and diesel-power plants.

Four-stroke principle

The entire combustion cycle is completed in two revolutions of the crankshaft, which means in four piston strokes. This principle is generally applied in high-speed and medium-speed diesel engines. Smaller propulsion engines, engines for driving generator sets, back-up generators, and also larger propulsion engines of up to ± 30 MW shaft power all work according to the four-stroke principle.
Where industrial diesel engines are concerned, the number of four-stroke engines that are built annually far outnumber that of two-stroke engines.

▶

A large container ship with feeder-container ships both to the front and rear of the ship, used for taking cargo to smaller ports.

The largest two-stroke engines for these ships have a cylinder bore of 96, 98 and 108 cm with a fourteen-cylinder in-line arrangement and a shaft power of 100,000 kW at a hundred revolutions per minute.

▲

Cylinder heads, with on top the hydraulically driven exhaust valves of a MAN Diesel two-stroke crosshead engine with a cylinder bore of 500 millimetres.

▶

Installation of a four-stroke trunk piston V-engine in Augsburg, Germany, in the MAN Diesel factories.

2.3 Design

We distinguish between trunk piston engines and crosshead engines.

Trunk piston engines

In these four-stroke engines, the connecting rod is hinged on the piston by a gudgeon pin. The engine is low in comparison to crosshead engines. The pistons have to absorb the lateral forces and transfer them to the cylinder liner and engine frame.

Crosshead engines

In these engines, the piston rod is bolted to the piston. The crosshead is situated below the piston rod with which it is hinged to the shaft. The lateral forces of the crankshaft mechanism are transferred to guide shoes, which are fixed, rigidly to the engine frame via crosshead guides.

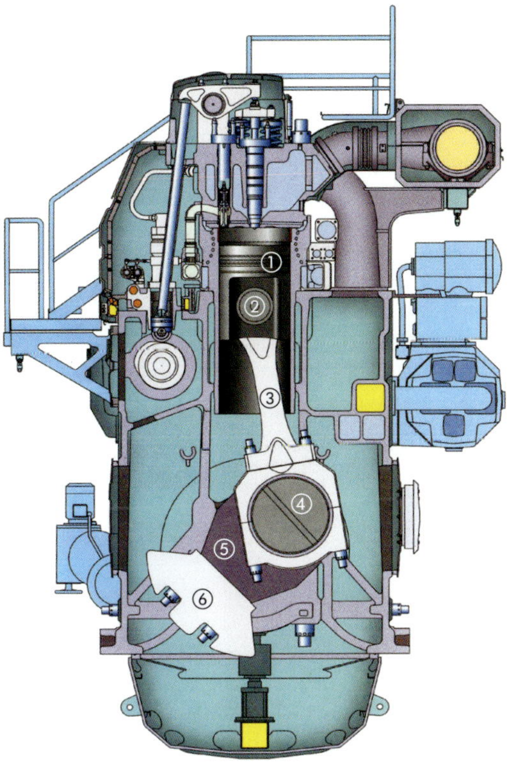

◄

Cross view of an in-line engine; the cylinders are positioned upright and in a straight row. This is a typical trunk piston engine.

1 *piston*
2 *piston pin*
3 *connecting rod*
4 *crankpin*
5 *crankweb*
6 *counterweight*

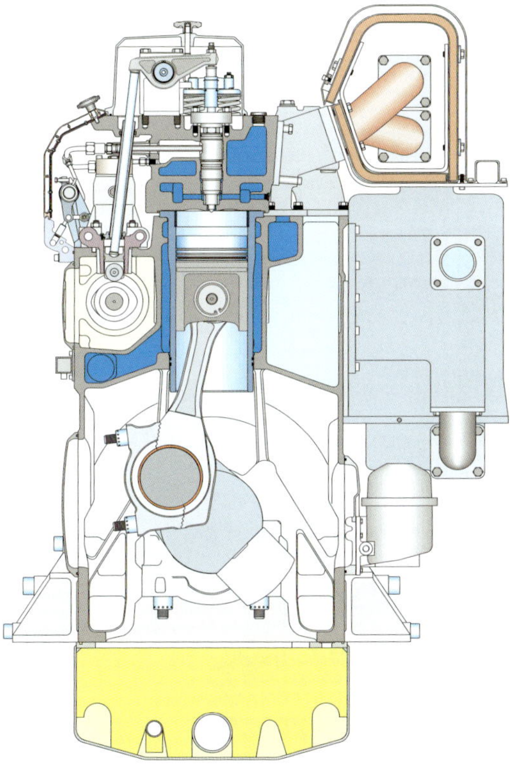

▲

Cross view of a trunk piston engine.

►

The two-stroke crosshead engine.

1 *piston*
2 *fixed piston rod*
3 *crosshead*
4 *guide shoes*
5 *crosshead guides*
6 *connecting rod*
7 *crankshaft*
8 *crankweb*
9 *crankpin*
10 *A-frame*

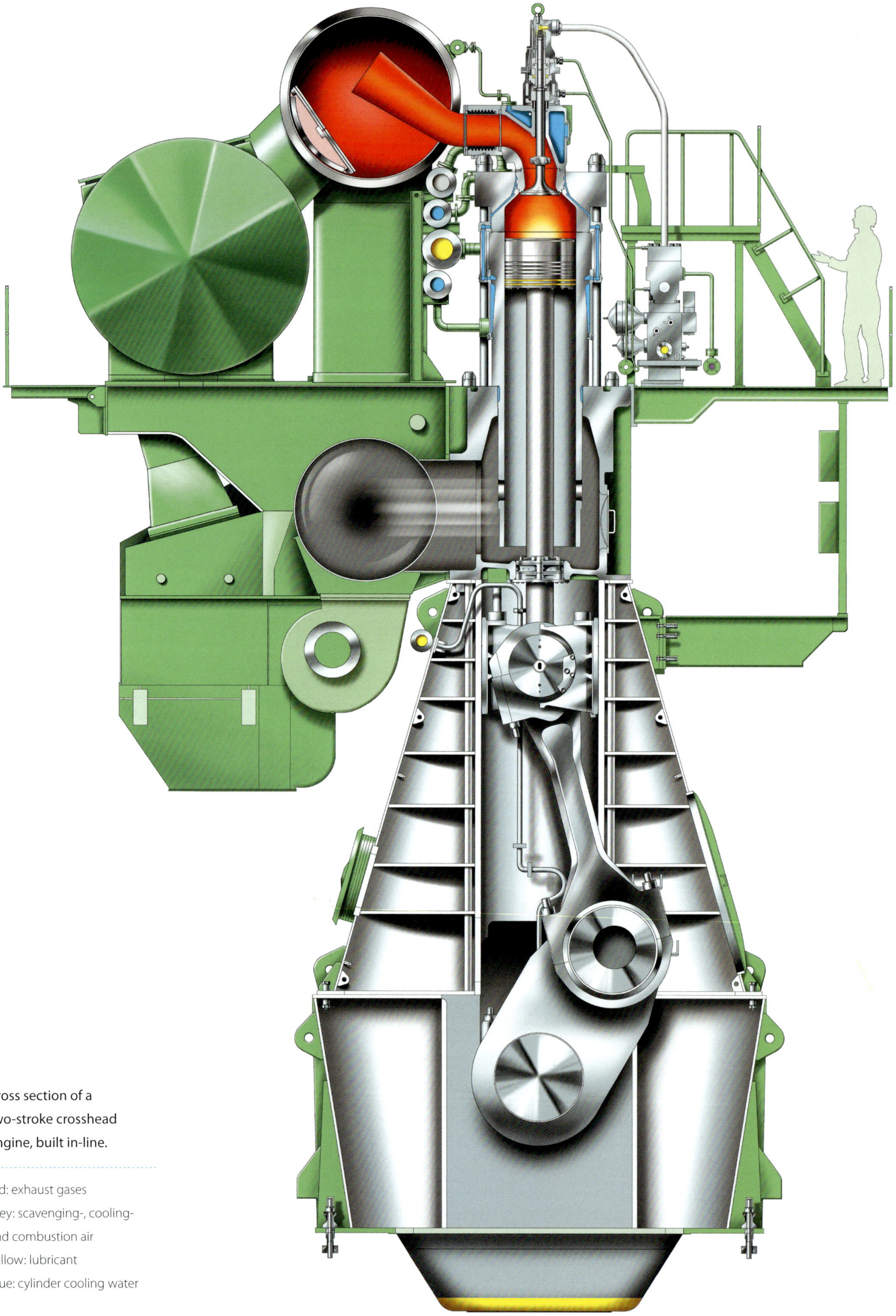

▶
Cross section of a two-stroke crosshead engine, built in-line.

red: exhaust gases
grey: scavenging-, cooling- and combustion air
yellow: lubricant
blue: cylinder cooling water

2.4 Rotational speed

In the engine industry one still often speaks in terms of revolutions per minute, whereas in today's technological industry one indicates the maximum permissible operational speed by means of frequency: the number of revolutions per second.

Low-speed engines

Low-speed engines have a maximum of 240 revolutions per minute or a frequency of 4 revolutions per second max.
These are always very large-sized and heavy crosshead engines, which mainly serve to propel large oil tankers, ore tankers and container ships. Some diesel-power plants are also equipped with these engines.
Low-speed engines are marketed by just three engine manufacturers. Those actually building the engines are also called 'cathedral builders'.
The cylinder bore varies from 260 to 1080 millimetres. For the largest diameters and a fourteen-cylinder version, the shaft power constitutes 97,300 kW. The piston stroke for this engine is 2660 millimetres.
This MAN diesel engine has a dry weight of 2300 tons and is 28 metres long and 14 metres high.

▲

View from above of part of a two-stroke crosshead engine. The diameter of the exhaust gases pipe is approximately 3.5 metres!

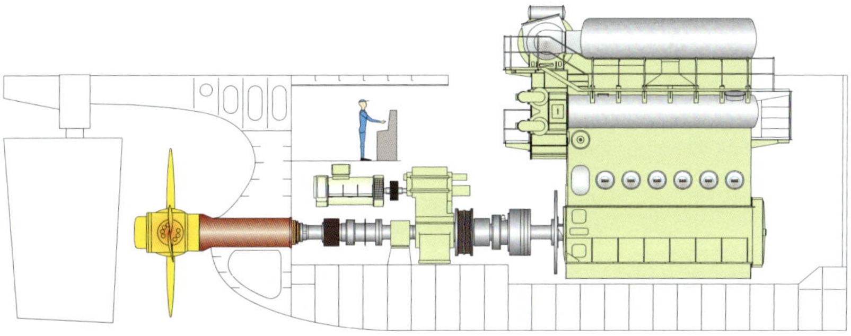

▲

A directly driven propeller is the standard arrangement for two-stroke crosshead engines in ship propulsion.

◀

View of a small two-stroke crosshead engine.

Medium-speed engines

Medium-speed engines have an rpm of between 240 and 960 or a frequency of between 4 and 16 revolutions per second (Hz). They are usually trunk piston engines using the four-stroke principle. However, one does find medium-speed two-stroke engines, and apart from the trunk piston design, one occasionally finds small crosshead engines, which are (older) existing designs.

The medium-speed engines are employed in very divergent applications: for ship propulsion, generating electricity and driving various kinds of machinery.

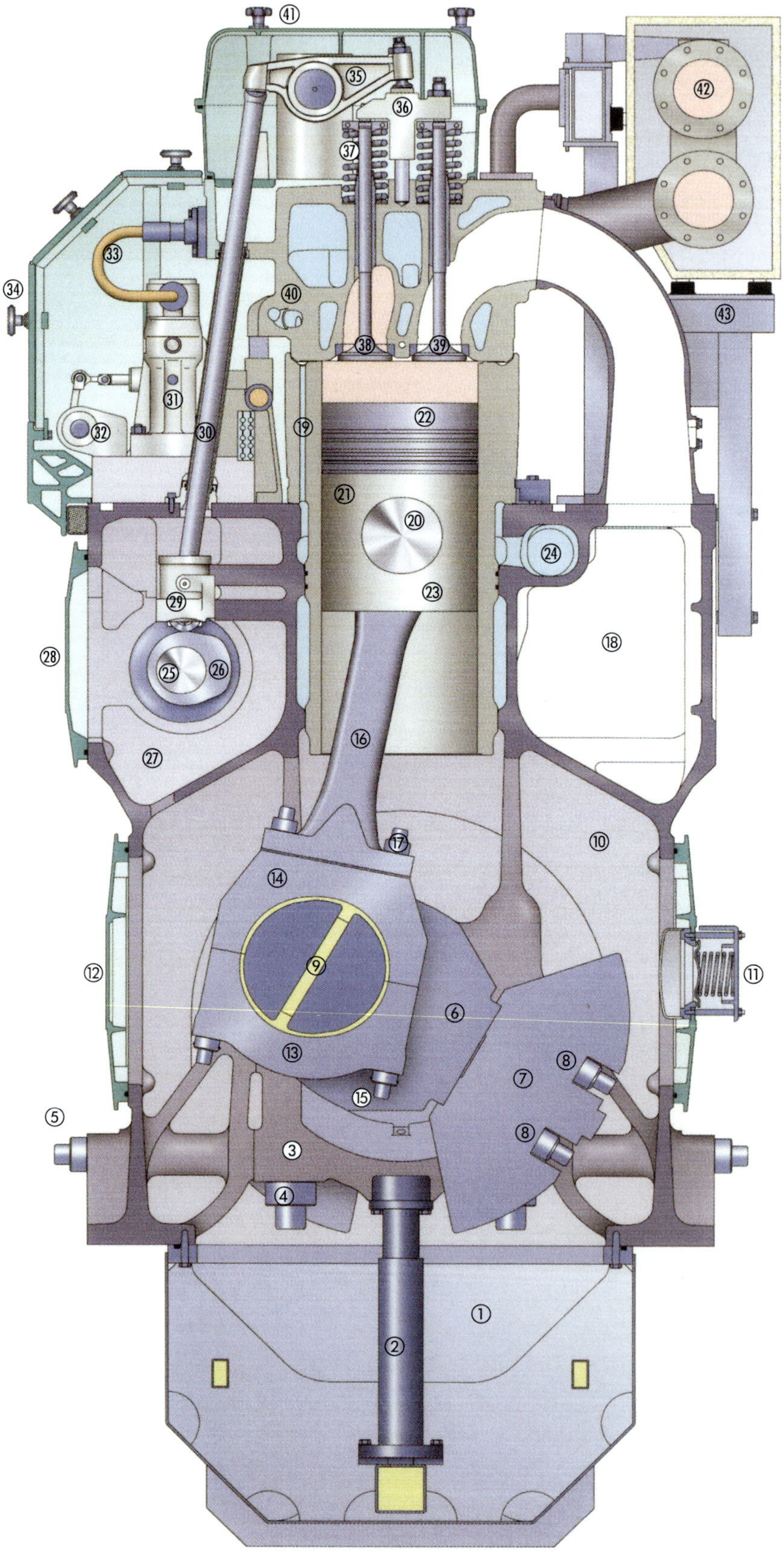

►

Cross view of a four-stroke medium-speed trunk piston engine with a cylinder bore of 380 millimetres, a Wärtsilä 38.

1 *oil sump*
2 *stay bolt/ lubricant supply pipe*
3 *lower bearing cap of the crankshaft bearing*
4 *vertical bolt for the lower bearing cap of the crankshaft bearing*
5 *horizontal bolt for the lower bearing cap of the crankshaft bearing*
6 *crankshaft web*
7 *counterweight*
8 *counterweight bolts*
9 *crankpin*
10 *engine block*
11 *crankcase relief valve*
12 *crankcase cover*
13 *lower cap of the crankpin bearing*
14 *upper cap of the crankpin bearing*
15 *crankpin-bearing bolt*
16 *connecting rod type 'marine'*
17 *connecting rod bolt*
18 *scavenging-air space*
19 *cylinder line with coolant*
20 *gudgeon pin*
21 *piston*
22 *piston gasket with piston rings*
23 *piston skirt*
24 *coolant supply*
25 *camshaft*
26 *cam*
27 *cam box*
28 *camshaft cover*
29 *guide sleeve*
30 *push rod*
31 *fuel pump*
32 *fuel rack*
33 *high-pressure fuel pipe*
34 *hot-box lid*
35 *rocker arm*
36 *spreader/bridge*
37 *valve spring*
38 *exhaust valve*
39 *inlet valve*
40 *cylinder head*
41 *cylinder cover*
42 *exhaust-gases duct*
43 *turbocharger frame*

yellow: lubricating oil
light blue: jacket water
brown: fuel
white: air
pink: exhaust gases

Medium-speed engines are manufactured by a limited number of engine builders. There are a large number of types when taking into account the cylinder bore, the piston stroke and the number of cylinders. The cylinder bore varies from approximately 200 to 640 millimetres.
The maximum shaft power is approximately 30,000 kW.

High-speed engines

High-speed engines have an rpm in excess of 960 or a frequency of more than 16 revolutions per second (Hz). Industrial engines often have a maximum rpm of just over 2000.

High-speed engines are built by many engine manufacturers and there are a large number of types when the cylinder bore, the piston stroke and the number of possible cylinders are taken into account. The cylinder bore varies from 40 millimetres to approximately 200 to 300 millimetres.
The maximum shaft power is approximately 10,000 kW.

Of course, it is possible to have engines delivered with a divergent rpm, which are on the fringe of these categories. They are then usually placed into a higher rpm category.

▲
A high-speed four-stroke V-engine for a generator.

▼
A high-speed four-stroke in-line engine with radiator cooling for a generator.

2.5 Power output or shaft power

Category subdivisions

The following categories are used:

I	0 – 100 kW,	fuel MDO,	four-stroke high-speed
II	100 – 10,000 kW,	fuel MDO,	four-stroke high-speed
III	500 – 30,000 kW,	fuel HFO,	four-stroke medium-speed
IV	1,500 – 100,000 kW,	fuel HFO,	two-stroke low-speed

These subdivisions are discussed elaborately in Chapter 5, *Standard figures of various types of diesel engines.*

2.6 Fuel used

Use of a particular type of fuel is related to the fuel costs, which is often expressed in dollars per metric ton.

▲

The price of heavy-fuel oil is about half of that of diesel oil.

The price of heavy oil ranged from 480 to 510 USD per metric ton in December 2011.
In ship propulsion the fuel- and lubrication costs are often more than 50% of the ship's running costs and one therefore tries to keep these costs as low as possible.

A diesel engine that has low lubricating-oil consumption and high fuel efficiency, is absolute key in the running of a ship. Diesel engines that are suitable for Heavy Fuel Oil (HFO), a waste product

◄

An illustration of the large range of different engine types.

Far right: a high-speed engine with a cylinder bore of 170 millimetres; to its left a medium-speed engine with a cylinder bore of 320 millimetres; far left: a very large medium-speed engine with a cylinder bore of 640 millimetres; in the background the only two-stroke crosshead engine with a cylinder bore of 960 millimetres. They are all Wärtsilä-engines.
The only diesel-engine type missing here is a small high-speed engine suitable for, for instance, yachts.

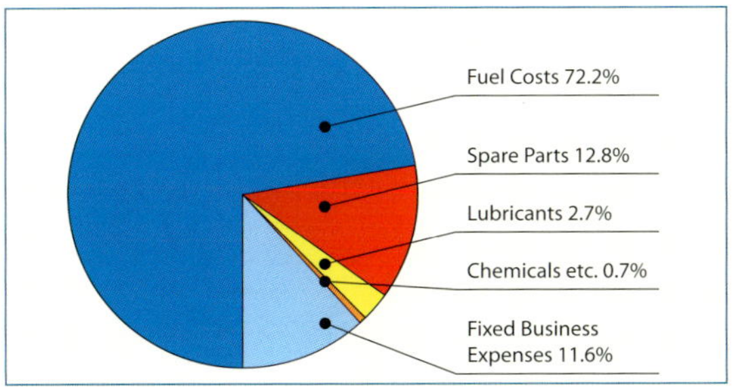

▲
The fuel and lubricating-oil costs often constitute three-quarters of the running costs of diesel engines.

from oil refineries, have lower fuel costs than diesel engines that only consume distillate fuels, for example Marine Diesel Oil (MDO).
MDO costs are roughly twice those of HFO.

Considerations for the use of MDO/HFO

This is a complex matter. Investment and exploitation play an important role in the development of a diesel engine that can use both kinds of fuel. Heavy-fuel oil requires expensive purifying and treatment equipment, which has to be recouped within a reasonable time span. Furthermore, legislation on toxic emissions such as nitrogen, sulphur dioxides and carbon dust particles is becoming increasingly important.

See also Chapter 22, *Diesel-engine emissions.*

Generally, all large two-stroke crosshead engines, as well as the large four-stroke trunk piston engines use HFO. The decision point between the use of MDO and HFO depends largely on the required power, engine operating hours per year, local environmental legislation and numerous other factors.
In the navigation sector, ships with a propulsion power of over 3000 kW are often equipped with HFO-engines.

380 IFO		Feb-04 2-6	Feb-04 9-13	Feb-04 16-20	Feb-04 23-27	Mar-04 1-5	Mar-04 8-12	Mar-04 15-19	Mar-04 22-26
Rotterdam	d	135	134	142	144	149	147	142	144
Gibraltar	d	160	157	156	156	158	155	149	152
Piraeus	d	148	154	145	146	150	148	144	147
Suez	d	158	156	159	159	161	161	159	164
Fujairah	d	174	168	167	165	164	162	162	162
Durban	w	n/a	n/a	n/a	n/a	n/a	n/a	n/a	n/a
Tokyo	d	201	198	195	196	197	198	197	196
Busan	d	184	181	182	183	186	185	182	179
Hong Kong	d	183	185	184	184	181	182	181	180
Singapore	d	171	172	170	167	166	164	165	166
Los Angeles	w	168	163	161	155	155	155	157	164
Houston	w	153	162	152	154	154	154	153	154
New York	w	163	163	162	158	161	161	157	158
Panama	w	160	160	160	160	161	164	168	168
Santos	d	156	142	144	146	156	156	154	153
Buenos Aires	d	n/a	n/a	n/a	n/a	n/a	n/a	n/a	n/a

180 IFO		Feb-04 2-6	Feb-04 9-13	Feb-04 16-20	Feb-04 23-27	Mar-04 1-5	Mar-04 8-12	Mar-04 15-19	Mar-04 22-26
Rotterdam	d	144	144	152	154	159	158	152	154
Gibraltar	d	169	166	164	164	167	164	157	161
Piraeus	d	160	155	159	160	165	164	161	164
Suez	d	164	162	165	166	167	167	166	171
Fujairah	d	181	174	173	172	171	169	169	169
Durban	w	170	167	169	166	165	163	161	159
Tokyo	d	205	202	200	200	201	202	201	200
Busan	d	192	191	192	192	195	194	192	189
Hong Kong	d	186	187	187	188	187	186	185	184
Singapore	d	175	172	175	170	171	168	170	170
Los Angeles	w	193	186	182	169	172	174	173	178
Houston	w	161	162	160	160	160	164	157	159
New York	w	183	182	177	177	179	178	172	173
Panama	w	170	170	169	171	171	174	178	179
Santos	d	159	146	147	150	160	160	158	157
Buenos Aires	d	171	170	167	165	166	164	165	165

MDO		Feb-04 1-5	Feb-04 8-12	Feb-04 15-19	Feb-04 22-26	Mar-04 5-9	Mar-04 12-16	Mar-04 19-23	Mar-04 26-30
Rotterdam	d	243	236	242	238	254	253	268	266
Gibraltar	d	284	281	300	306	322	319	328	327
Piraeus	d	264	261	277	282	300	298	307	307
Suez	d	321	321	321	321	325	336	335	335
Fujairah	d	310	310	310	310	309	309	309	309
Durban	w	267	263	274	285	301	299	308	311
Tokyo	d	287	288	291	288	287	292	292	290
Busan	d	332	325	322	322	321	321	319	318
Hong Kong	d	274	274	282	283	282	281	279	273
Singapore	d	277	278	282	282	284	281	279	277
Los Angeles	w	317	320	344	358	345	342	342	340
Houston	w	281	280	278	278	279	282	285	282
New York	w	345	343	335	327	325	330	331	330
Panama	w	317	317	318	321	321	323	326	327
Santos	d	350	343	342	344	347	350	345	344
Buenos Aires	d	317	311	302	306	295	303	303	302

▲
Fuel prices of diesel oil (MDO), heavy oil (HFO) with a viscosity of 180 cSt and heavy oil with a viscosity of 380 cSt in USD/ton (2004).

MDO: Marine Diesel Oil;
IFO: Intermediate Fuel Oil/ Heavy Fuel Oil.

Prices March 2012:
MDO: 1005-1050 dollar
HFO: 690-750 dollar

◄
Most large ships with a lot of propulsion power use HFO. This is less suitable from an environmental viewpoint.

Diesel-power plants, which generally apply medium-speed four-stroke engines, usually use HFO. A decisive factor is the high number of operating hours per year. Here fuel costs may be exorbitant.
HFO is seldom used for high-speed diesel engines, because the real combustion time is too short to fully combust the fuel.

See also Chapter 8, *Fuels, fuel-line systems and fuel cleaning.*

2.7 Use of engines

The distinguishing characteristics are:
- the drive type, such as propulsion, power generation, pumps, compressors, traction or different tool applications;
- the number of operating hours on a yearly basis;
- the degree of load variations;
- the atmospheric conditions such as air pressure, humidity, air pollution, minimum and maximum air temperatures.

▶
An offshore supply ship.

▶▶
A floating diesel power plant in a port.

▲
An ore mine far from civilisation.

The required electrical power is generated with diesel engines. The fuel, heavy-fuel oil, is supplied by a tank lorry. In the background, two white storage tanks containing heavy-fuel oil.

▲
Fast ships require a great deal of power; a high-speed four-stroke V-engine using diesel oil (MDO) is the most common.

◀
A relatively small container ship, a so-called 'feeder'.

Propulsion by means of a medium-speed diesel engine with in-line cylinders, fuel is heavy-fuel oil.

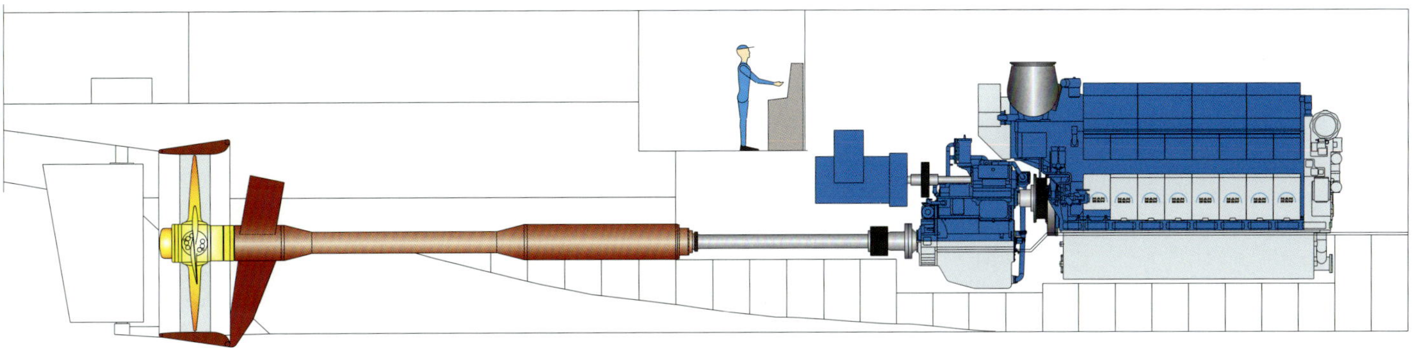

A four-stroke in-line engine with reduction gearing, intermediate shaft, propeller shaft, and propeller arrangement in a ship.

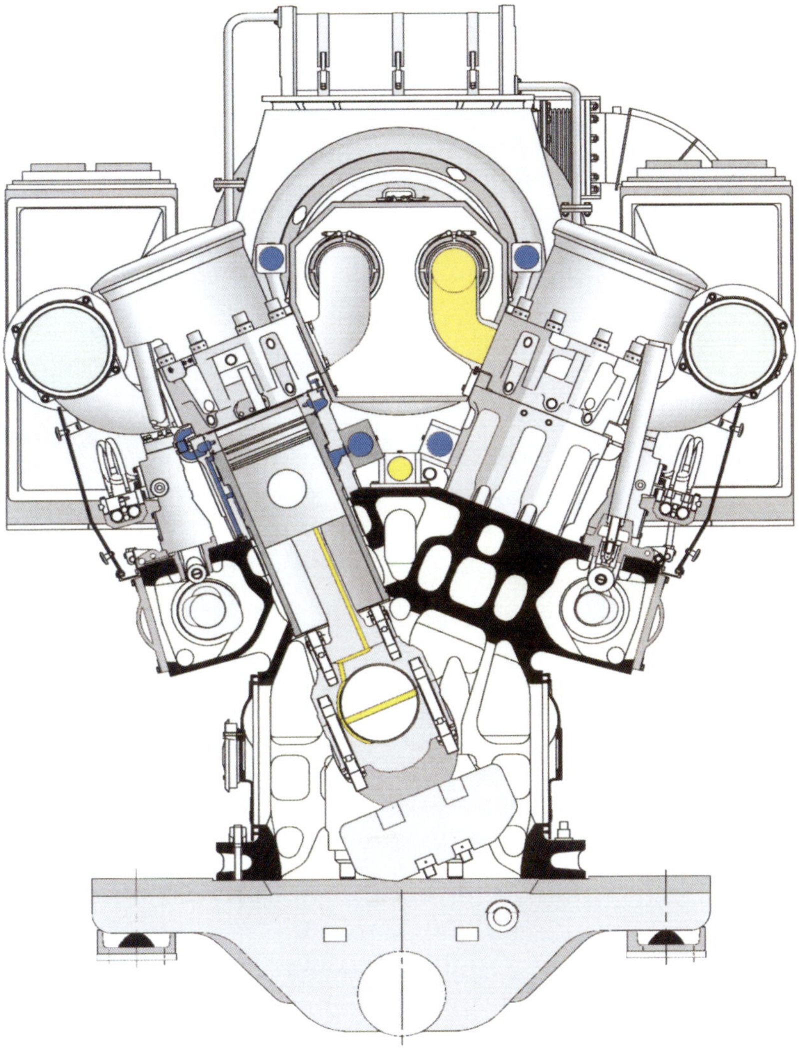

A large four-stroke medium-speed diesel engine with cylinders in V-arrangement. Make MAN Diesel.

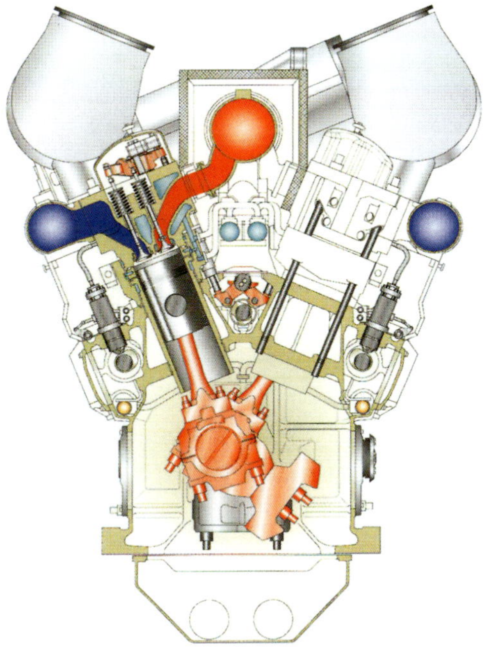

A medium-speed V-engine. Note the large turbochargers.

2.8 Other characteristics of diesel engines

In-line and V-engines

In in-line engines, the cylinder axis lines are planar. The maximum number of cylinders in a two-stroke crosshead engine is usually twelve. As the size of container ships increases, fourteen- or sixteen-cylinder in-line engines are also possible. The total length of the engine then increases to 25 and even 35 metres! The number of cylinders of four-stroke trunk piston engines does not normally exceed ten to twelve. Nine cylinders generally being the largest in-line engine.

In V-engines the cylinder axis lines are biplanar at an angle of 45° to 120° from each other.
They normally have a maximum of twenty cylinders. Sixteen to eighteen cylinders in V-designs in are often found in large diesel-power plants.

2.9 The use of in-line and V-engines

V-engines are most often used as they have the advantage of being shorter in length in comparison to an in-line engine with the same power output. Obviously, an eighteen-cylinder V-engine takes up far less space than two nine-cylinder in-line engines. In the engine industry, everything revolves around economics. The capital costs of V-engines are considerably lower then those of the same number of cylinders in-line.

V-engines are often used in diesel-power plants. In the shipping industry, the preference is for in-line engines for propulsion. Larger vessels usually employ auxiliary in-line engines, whereas smaller vessels use V-engines.

For maintenance, in-line engines have the advantage of having vertically dismountable and mountable parts, such as the cylinder head, piston, connecting rod and cylinder liner. For V-engines with a somewhat larger cylinder bore,

▶
Lifting a large cylinder head of a two-stroke crosshead engine.

▶▶
A nine cylinder Caterpillar MaK 43 four-stroke in-line engine with turbocharger.

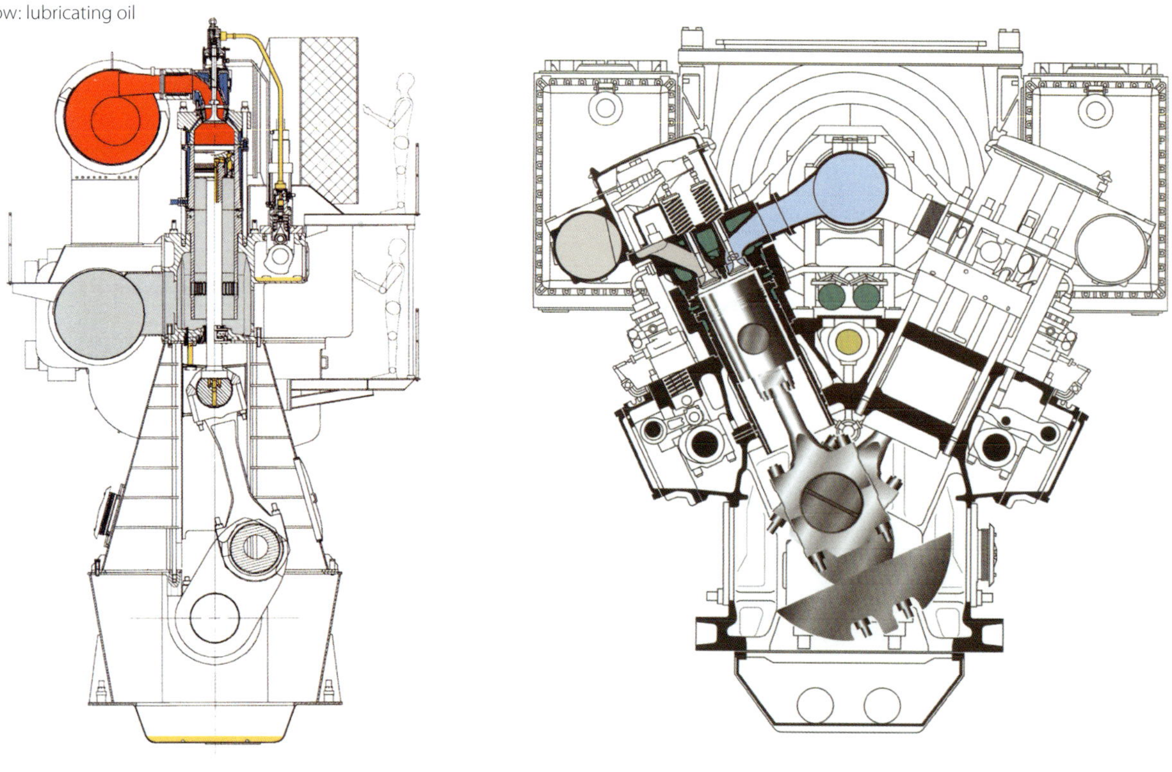

▼
An in-line and a V-engine next to one another.

left: an in-line engine
right: a V-engine
red: exhaust gas
blue: scavenging air
blue: cooling water, left
green: cooling water, right
yellow: lubricating oil

auxiliary tools and adapted hoisting machinery are indispensable.

V-engines and in-line engines seem identical when considering their components. Cylinder heads, pistons and liners are often the same. The piston stroke of some V-type engines is shorter than that of in-line engines, as there is not enough space in the V-engine block to make the same stroke length as in the in-line engine. Since the stroke is shorter, the number of revolutions can be increased; the mean piston speed will then remain the same.

2.10 Direction of rotation of the diesel engine

If one were to stand at the far end of the crankshaft where the power output is delivered and look at the engine, it turns clockwise. A large number of engines turn clockwise. In large in-line engines, the exhaust gas ducts are situated on the right-hand side, seen from the outgoing engine shaft.

◄

The rotation direction of a diesel engine.

Clockwise: the side of the engine where a propeller or generator is driven.

▼

A regular clockwise-turning engine.

This picture shows a large two-stroke crosshead engine with a cylinder bore of 960 millimetres and the common-rail fuel system by engine manufacturer Wärtsilä Sulzer. The exiting part of the crankshaft can still be seen. For a regular clockwise-turning engine, cylinder 1 is the cylinder furthest removed from the exiting crankshaft. So, in this twelve-cylinder in-line engine cylinder 12 is closest to the exiting crankshaft.

The central exhaust-gas duct of a two-stroke crosshead engine is located on the right-hand side of the ship (starboard) with the turbochargers, air coolers, exhaust-gas boiler, silencers and finally the exhaust.
Because of this, there is enough space at the left-hand side of the ship (port) for the driving gear of the exhaust valves, the fuel supply, a floor for ship repairs and a hoisting space for an overhead crane.

▼

Description of the direction of rotation and cylinder numbering.

Driving end: the part of the engine where the exiting crankshaft is located.
Free end: the 'blind' part of the engine, so without exiting crankshaft.
Control side: this is the side of the engine with the camshaft to drive the fuel pump and the valves.
Exhaust side: this is the side of the engine where the exhaust-gas duct is located.

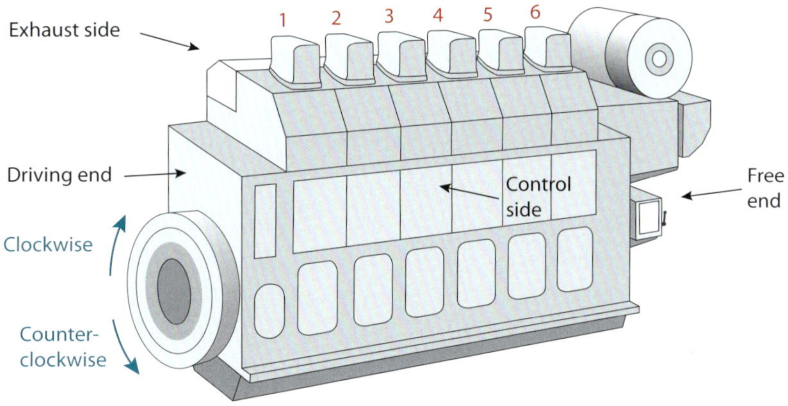

▶

Cylinder numbering and direction of rotation.

For the in-line engine of this engine manufacterer, numbering of the cylinders starts at the driving end. In V-engines one speaks of banks, the left bank ranging from A1 up to and including A8 and the right bank ranging from B1 up to and including B8.

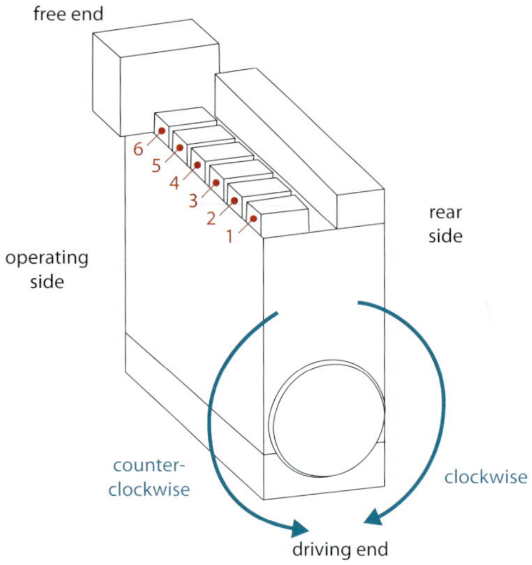

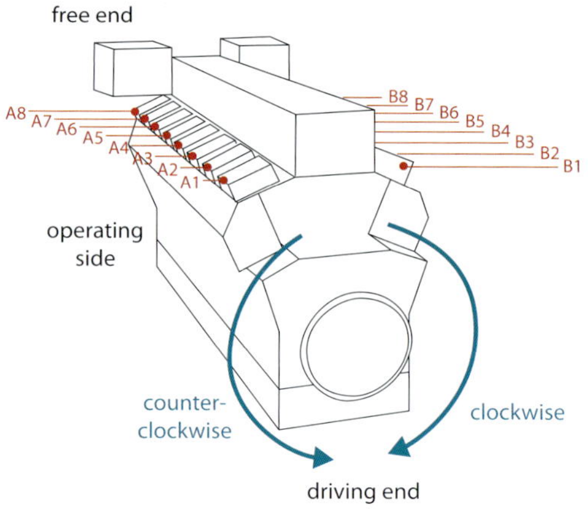

▼

V-engine numbering with a left and right bank.

1 *driving end*
2 *vibration damper*

2.11 Cylinder number

Cylinder 1 is often the cylinder furthest removed from the crankshaft side. So in traditional propulsion engines cylinder 1 is situated towards the bow of the ship and the cylinder with the highest number towards the stern of the ship.

Note

It is recommended that the manual be consulted in order to check the correct numbering. This also applies to V-engines!

2.12 Natural aspiration and turbocharging

Natural aspiration

When the engine draws in air by means of the piston movement, this is called natural aspiration. The air enters the cylinder with a pressure slightly lower than atmospheric pressure. This can only be achieved in four-stroke engines.

Supercharger or Turbocharger

When the air supply to the cylinder takes place under a pressure that is higher than the atmospheric pressure, it is referred to as turbo- or super-charging. Here air is compressed and transported to the engine by means of an exhaust-gas driven turbocharger. This process is always applied in two-stroke engines.

◀

Virtually every sizeable industrial diesel engine is equipped with a turbocharger.

1 *turbocharger*
2 *exhaust-gas manifold*
3 *inlet-air manifold*
4 *cylinders*

▶

The turbocharger principle.

The turbocharger, driven by exhaust gases draws in ambient air by means of a centrifugal pump and forces the air under overpressure into the air-inlet manifold of the cylinders. As soon as the inlet valves open, the air flows into the combustion chamber. Of course, the downward piston movement also has an inductive effect on the airflow.

1 *turbocharger, air section with a centrifugal pump*
2 *turbocharger, exhaust-gas section with an impeller for a gas turbine*
3 *air-inlet duct*
4 *exhaust-gas manifold*

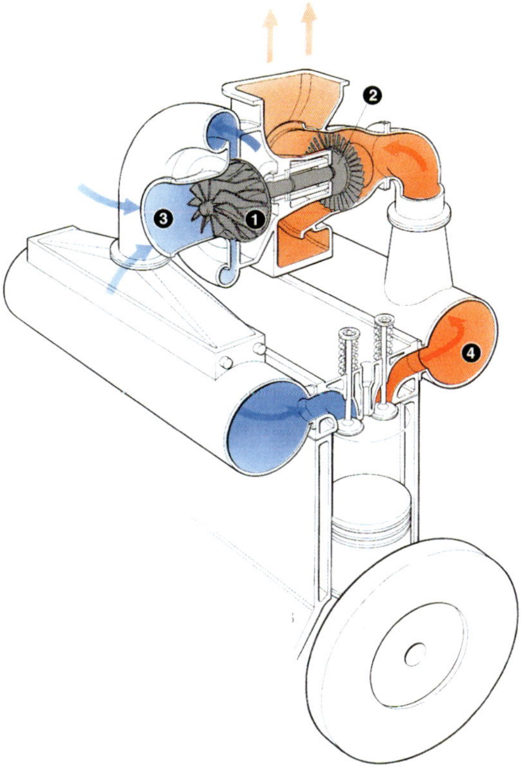

Today many engines are equipped with turbochargers. One of the advantages is that with an identical cylinder volume, the power output can be considerably increased.
With an increased air intake, using the same stroke volume, more fuel is supplied to the cylinder for combustion, which increases the power output.

Also, see Chapter 12, *Air supply.*

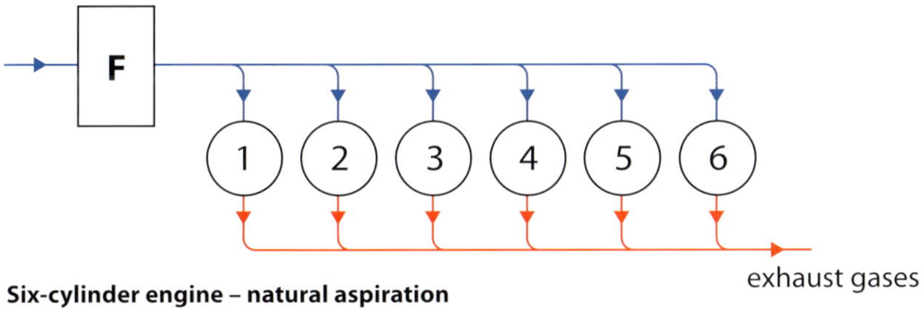

Six-cylinder engine – natural aspiration

F - air filter

◀

The principle of natural aspiration and turbocharging

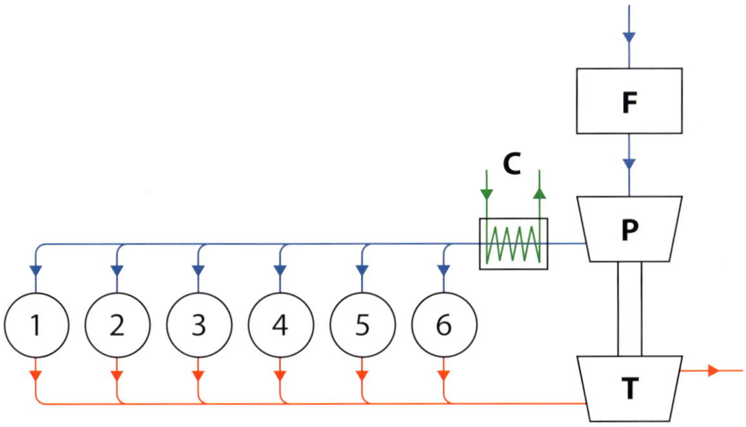

Six-cylinder engine – turbocharging

F - air filter
P - turbocharger - air section
T - turbocharger - exhaust-gas section
C - air-cooler

▶

Cross view of a turbocharger group.

1 *impeller charger*
2 *exhaust-gas turbine*
3 *bearings*
4 *turbine shaft*
5 *lubricating-oil inlet*
6 *lubricating-oil outlet*
7 *casing*
8 *exhaust gas gasket*
9 *air gasket*

▶

An exhaust-gas turbine of a large two-stroke crosshead engine.

1 *cylinder*
2 *cylinder head*
3 *exhaust valve*
4 *exhaust-gas manifold of the cylinder*
5 *central exhaust-gas manifold*
6 *auxiliary chargers*
7 *gas turbine*
8 *exhaust lead towards exhaust-gas boiler and outside air*

>CH 3

Working principles of diesel engines

The working principles of diesel engines have remained unchanged for over a hundred years.
Improvements in certain aspects of the process have enhanced the performance throughout the years, for instance, efficiency, power output per stroke volume and, also of significance today, reduced toxic emissions.

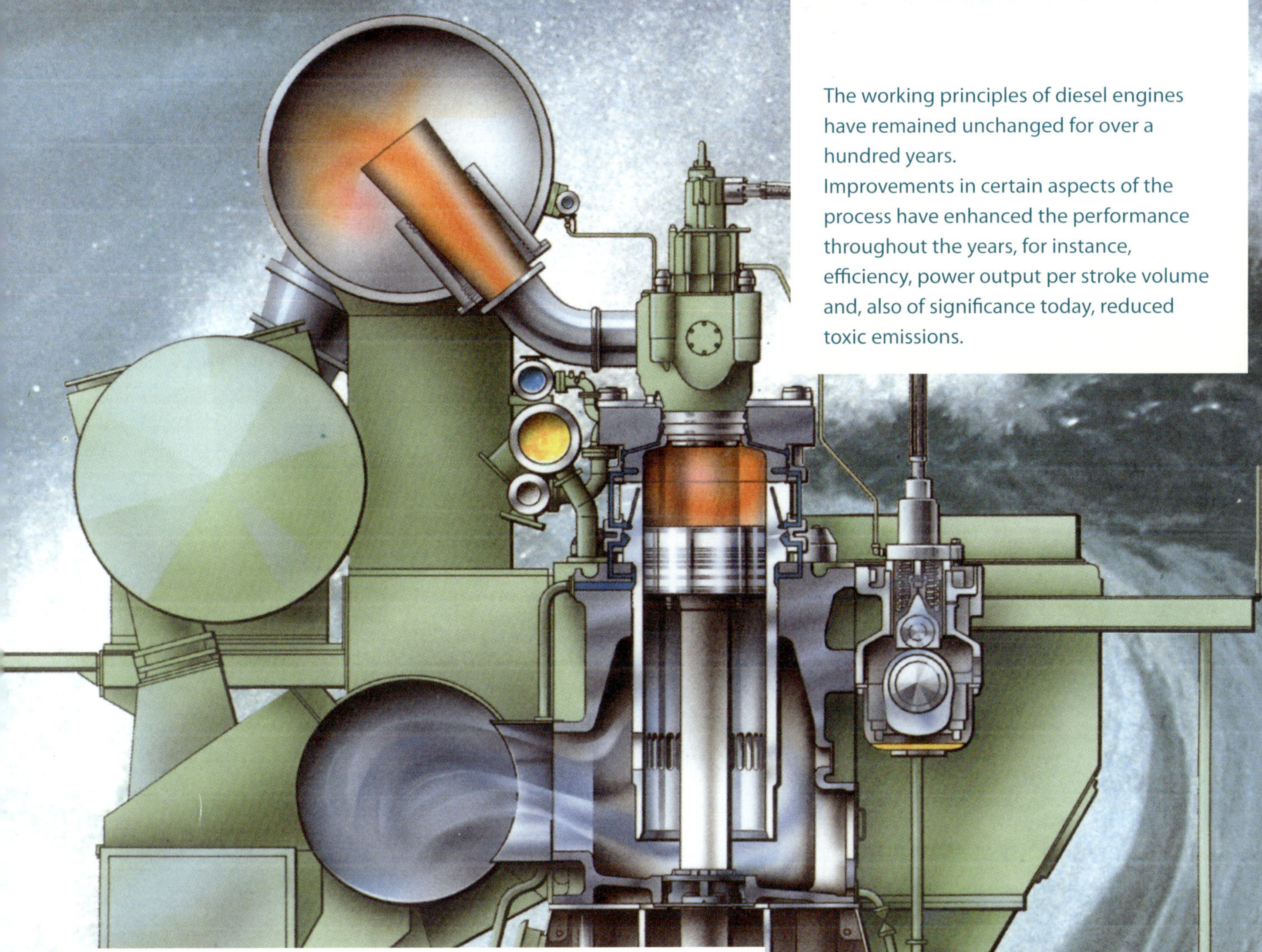

2-stroke cycle diesel engine
2 piston strokes
1 revolution of the crankshaft (360°)
1 power stroke

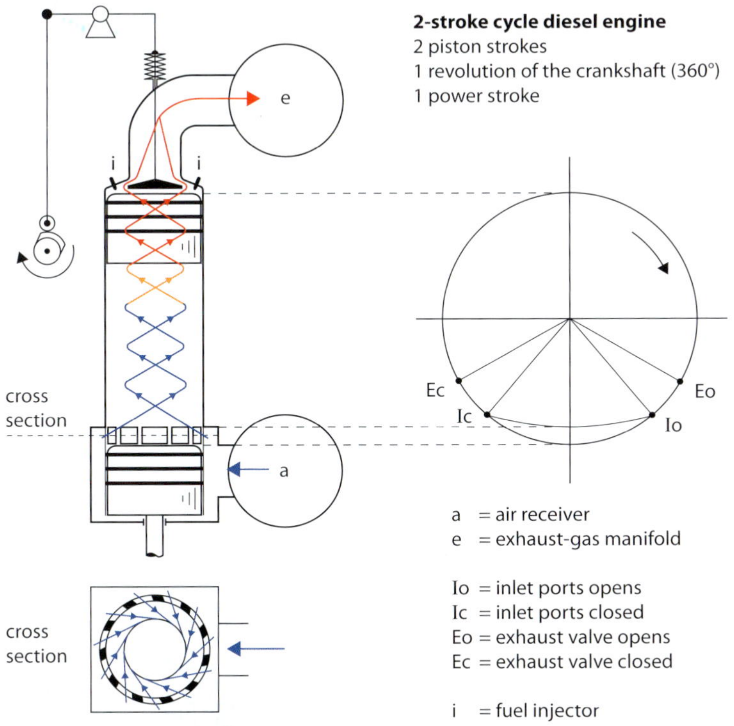

a = air receiver
e = exhaust-gas manifold

Io = inlet ports opens
Ic = inlet ports closed
Eo = exhaust valve opens
Ec = exhaust valve closed

i = fuel injector

3.1 Working principles

In the internal combustion engine, there are two combustion cycles, namely:

1. **the four-stroke cycle**, which is four piston strokes during two crankshaft revolutions;
2. **the two-stroke cycle,** which is two piston strokes during one crankshaft revolutions.

Four-stroke engine

A **four-stroke engine** requires at least one inlet valve and one exhaust valve in the cylinder. Today most four-stroke engines have two inlet and two exhaust valves in the cylinder head. This has the added advantage of improved gas exchange and that the open valves intrude less far into the cylinder, thus allowing for a higher compression ratio.

3.2 Two-stroke engine build

Since industrial two-stroke diesel engines of the trunk piston type are very rare today, we will only discuss the two-stroke crosshead engine principle. In a **two-stroke**-diesel engine, there are two types of gas exchange:

1. **with inlet and exhaust ports in the cylinder liner;**
2. **with inlet ports in the lower cylinder liner and a central exhaust valve in the cylinder head, the so called uniflow scavenging.**

▶ The four-stroke cycle.

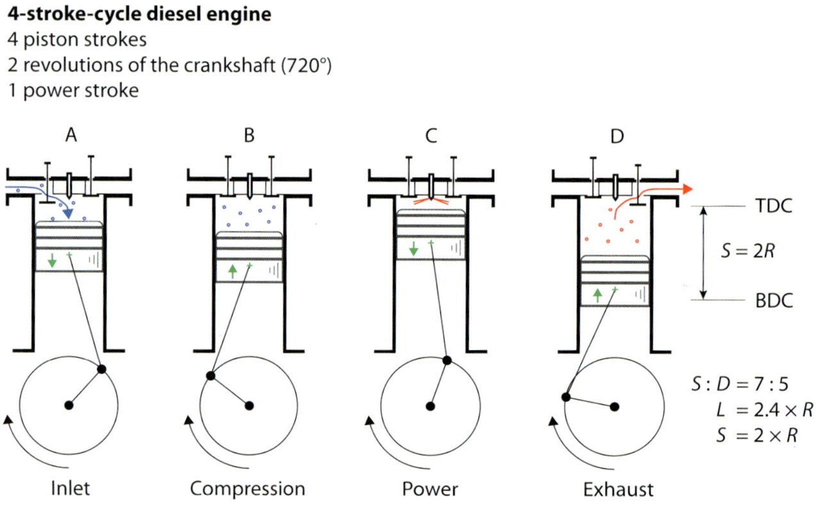

▶ The two-stroke cycle.

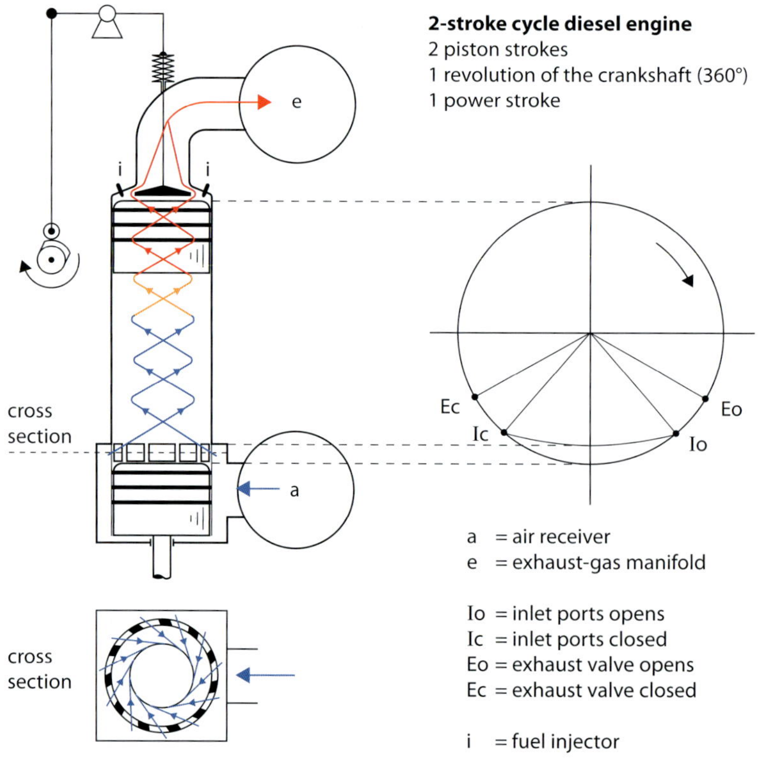

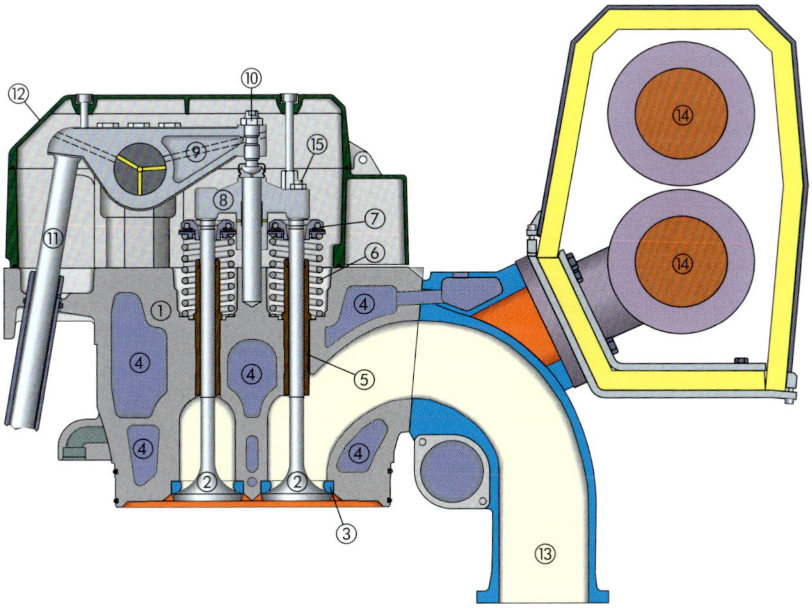

◄

Sectional view of a cylinder head of a four-stroke engine.

1 *cylinder head*
2 *inlet valves (behind which are the exhaust valves, not visible)*
3 *valve seats*
4 *coolant passages*
5 *valve guides*
6 *valve springs*
7 *valve disc/rotor caps*
8 *spreader/bridge*
9 *valve lever/rocker arm*
10 *second adjusting bolt*
11 *push rod*
12 *valve cap*
13 *inlet-air manifold*
14 *exhaust manifold*
15 *first adjusting bolt/clearance of the valve*

3.2.1 Inlet and exhaust ports in the cylinder liner.

This is often the case with small engines such as those for hand tools, outboard engines etc. These have the advantage of having a simple and lightweight construction. Most of these engines work according to the Otto-process. Port-type diesel engines are now only produced by some American manufacturers. Traditionally, the old two-stroke crosshead engines were also equipped with only ports. These engines are still in use, but are no longer built.

▲

A cylinder liner of an early type two-stroke crosshead engine.

1 *inlet port*
2 *exhaust port*
3 *grooves for the sealing O-rings of the coolant passages above and below the ports*
4 *holes for cylinder lubrication*
5 *liner with bore cooling*

◄

A typical four-stroke high-speed diesel engine by engine manufacturer Cummins.

A cylinder head contains two inlet and two exhaust valves.

3.2.2 Inlet ports in the lower cylinder liner and one central exhaust valve in the cylinder head

All modern two-stroke crosshead engines are built with this configuration.

Note

The two-stroke cycle is characterised by one rotation of the crankshaft and two piston strokes in one combustion cycle. Inlet and exhaust ports are located in the cylinder liner (small two-stroke diesel engines) or the inlet ports are situated around the cylinder liner with one central exhaust valve (large two-stroke diesel engines).

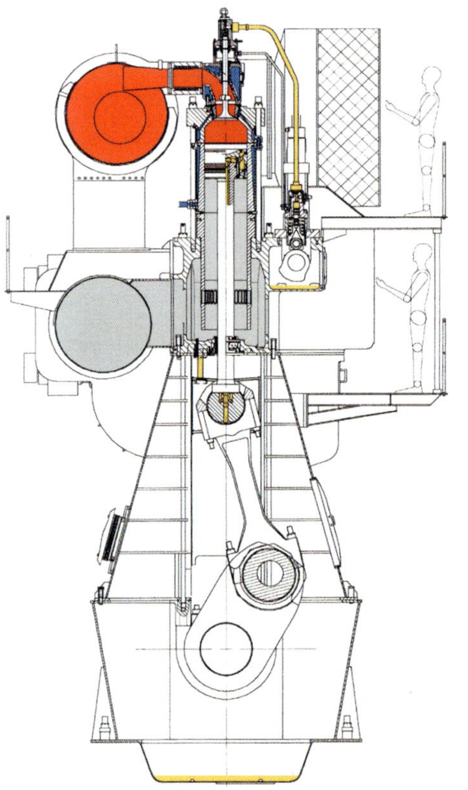

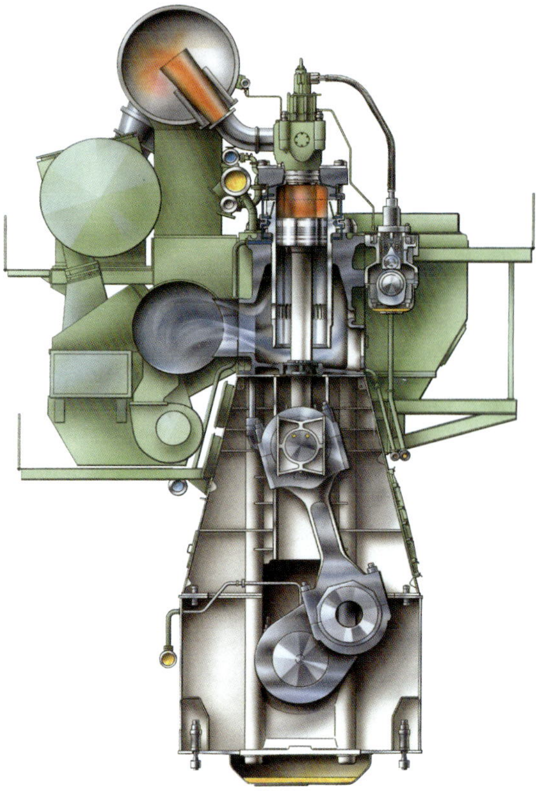

▶

A typical two-stroke crosshead engine by engine manufacturer MAN Diesel.

The engine is equipped with inlet ports and a central exhaust valve.

▶▶

A large two-stroke crosshead engine which is almost always used for large propulsion plants.

On the left the air-inlet manifold, the exhaust manifold, the turbocharger and the air coolers. On the right the exhaust-valve drive and the fuel pumps (not visible).

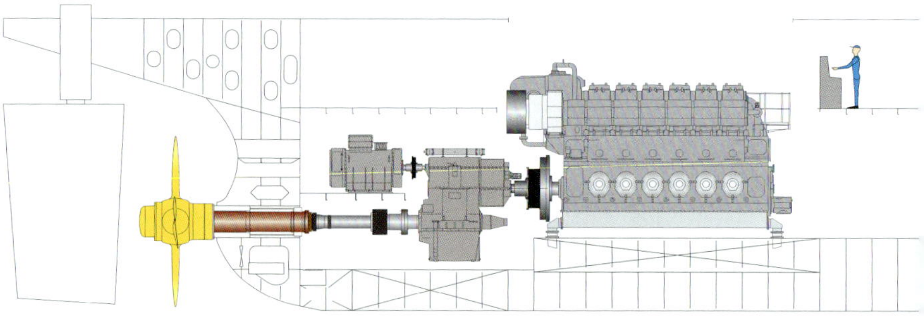

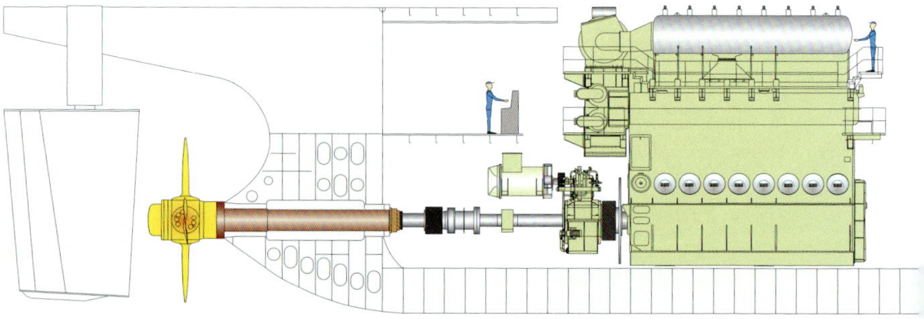

▶

Arrangement of two propulsion systems.

Above: four-stroke trunk-piston engine with gear box to an adjustable-pitch propeller with a shaft generator on the gear box.
Propeller: rpm much lower than the engine rpm.
Shaft generator: rpm much higher than the engine rpm.

Comment

The built-in height of this propulsion installation is low.

Below: two-stroke crosshead engine with a direct-driven controllable-pitch propeller. The shaft generator, driven by the engine shaft via a reduction-gearing box.
Propeller: rpm identical to that of the engine.
Shaft generator: rpm much higher than that of engine.

Comment

The built-in height of this propulsion installation is high. In both of these systems, the shaft generators and controllable-pitch propellers can be disconnected from the engine.

3.3 Four-stroke engine build/set-up

There is a cylinder liner in the engine block, in which a piston makes upwards and downwards movements known as a linear motion. By means of a gudgeon pin, the piston is hinged to the connecting rod that is fixed rotationally to the crankpin of the crankshaft.

When the piston has reached its highest point it stops for a moment: this is called 'top dead centre' (TDC).
The bottom dead point is called 'bottom dead centre' (BDC).

Intake stroke

During the downward stroke of the piston both inlet valves are open and the air flows into the cylinder under an over-pressure. Both inlet valves close when the piston is almost at the bottom of its downward stroke (BDC).

Compression stroke

During the upward stroke of the piston the air is compressed and the pressure and the temperature of the compressed air rise considerably.

Combustion stroke

Just before the piston has reached the top of its upward stroke, an atomiser injects fuel under high pressure into the cylinder filled with hot air and the fuel/air mixture subsequently combusts swiftly. This is called the 'self combustion' of the diesel engine. The fuel/air mixture combusts rapidly, causing an increase in pressure and temperature.
During the downward piston stroke the gases formed in the cylinder expand and exert force on the piston. This force is transferred to the crankshaft via the piston pin, connecting rod and the crank.

Exhaust stroke

Before the end of the combustion stroke the exhaust valves open and exhaust gases flow out of the cylinder. During the upward exhaust stroke most of the exhaust gases are expelled from the cylinder. Subsequently, the four-stroke cycle starts anew with the intake stroke.
At the top of the upward piston stroke the inlet and exhaust valves remain slightly open. The residual gases are removed by the scavenging air via the exhaust valves and the hot components of the combustion space, such as the exhaust valves, the cylinder head and the piston crown, are cooled.

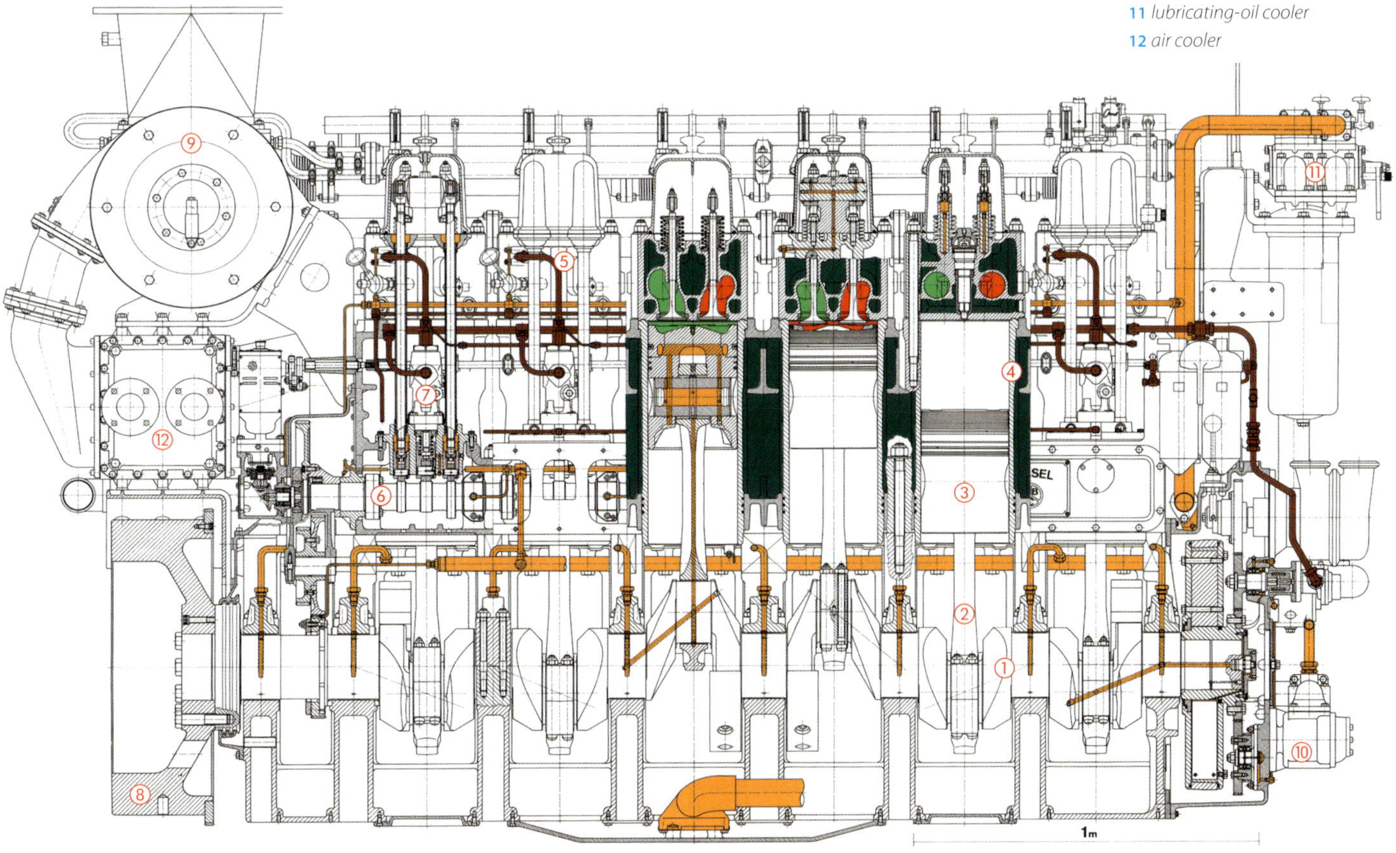

▼
A traditional four-stroke in-line engine. ABC, Ghent, Belgium.

1 *crankshaft*
2 *connecting rod*
3 *piston*
4 *cylinder liner*
5 *cylinder head*
6 *camshaft*
7 *fuel pump*
8 *flywheel and turning wheel*
9 *turbocharger*
10 *lubricating-oil pump*
11 *lubricating-oil cooler*
12 *air cooler*

Note

The four-stroke cycle is characterised by the fact that one combustion cycle is completed in two revolutions of the crankshaft, so four piston strokes. There are two inlet and two exhaust valves in the cylinder head.

The four-stroke cycle is often applied to small, medium and large diesel engines. These are the high-speed and medium-speed engines.

It is only in cases requiring very high power outputs that one resorts to engines with the two-stroke principle. These are without exception crosshead engines that directly drive propellers or generators. These are always low-speed engines.

In crosshead engines, the forces moving the pistons are transferred to the crosshead via a piston rod that is fixed to the piston. The crosshead functions as a pivotal point between the fixed piston rod and the connecting rod fixed to the crank. Lateral forces are absorbed in the crosshead and are transferred to the rod via the crosshead guides, which are fixed to the frame.

3.4 A few significant differences between the two-stroke and four-stroke cycles

In the two-stroke cycle there is a combustion process for every revolution, whereas in the four-stroke cycle there is one for every two revolutions.

Theoretically, a two-stroke engine with a similar cylinder capacity and rpm should have twice the power output.

In the two-stroke cycle the exhaust gases flow out of cylinder under an over-pressure, at the same moment that the piston releases the inlet port. The air supplied must have an over-pressure in relation to the exhaust-gas pressure, otherwise the exhaust gases will flow back into the cylinder! The two-stroke engine requires an air supply that has a pressure exceeding that of the exhaust-gas pressure and therefore far higher than atmospheric pressure (over-pressure).

In very small two-stroke engines the bottom part of the piston in conjunction with the smallest possible crankcase functions as a scavenging pump.

In slightly larger two-stroke engines, one finds a scavenging pump driven by the crankshaft for the air supply. In industrial engines the scavenging air is always supplied by a turbocharger.

In the **four-stroke cycle** all stages are executed by the piston:

- intake;
- compression;
- power;
- exhaust.

This is a simpler process. Additionally, the exhaust gases are far better expelled; the process is cleaner! It does not require a separate air-supply system (over-pressure); even four-stroke engines without turbochargers have good quality charging of the fresh-air cylinder.

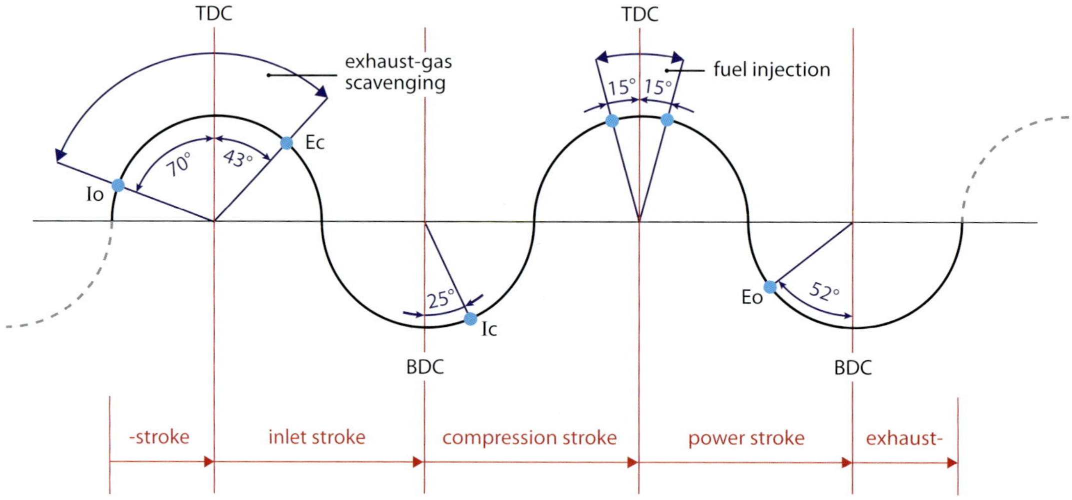

► The four-stroke cycle, the continual valve diagram.

Inlet stroke	70 + 180 + 25 = 275°	Low pressures: long processes (inlet/exhaust)
Compression stroke	65 + 90 = 155°	High pressures: short processes (compression/power)
Power stroke	90 + 38 = 128°	
Exhaust stroke	52 + 180 + 43 = 275°	
Fuel injection	15 + 15 = 30°	
Exhaust-gas scavenging	70 + 43 = 113°	

◀
A ten-cylinder two-stroke crosshead engine on a large container ship.

In this case a Wärtsilä Sulzer 10RTA96C, Southampton class (6800 TEU) of shipping company P&O Nedlloyd, now part of Maersk.

Even so, today most four-stroke engines have turbocharging with a turbocharger driven by the exhaust gases. This produces a considerable increase in the power output of the engine. An increase in the kilograms of air that can be fed into the cylinder means an increase in the amount of fuel combustion, thus creating a proportional increase in power output.

Note
Despite the fact that, theoretically, a two-stroke engine has twice the power output with an identical cylinder capacity, one generally uses a four-stroke engine when a high power output in conjunction with a higher rpm is required. Only for a low-power output are two-stroke trunk piston engines still used. Also in case of a very high output, one generally uses two-stroke crosshead engines. In the latter case, propellers or generators are directly driven by the engine and therefore have the same rpm.

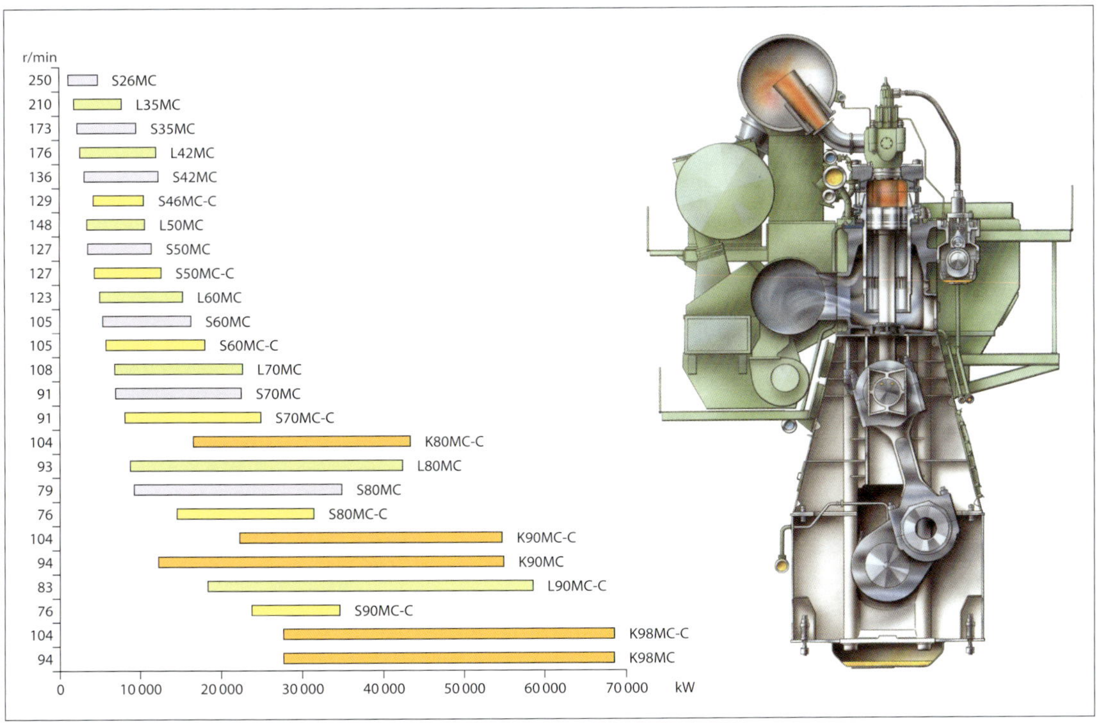

◀
The engine programme of the two-stroke crosshead engines of MAN Diesel type MC.

Here the smallest engine has a cylinder bore of 260 millimetres and the largest has a bore of 980 millimetres. Today there are also 1080 millimetre engines in the market.

3.5 Examples of supply programmes of engine manufacturers

▶

The engine programme of manufacturer MTU.

These are high-speed four-stroke diesel engines running on MDO.
The power output varies from 298 to 8200 kW.
Rpm varies from 1150 to 2100.
The stroke volume varies from 14 to 347.4 litres.
The weight varies from 1840 to 43,000 kg.

S60

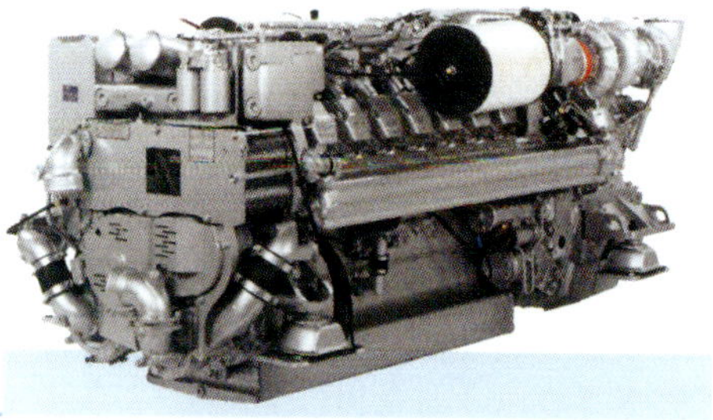

Baureihe **2000**

Baureihe **396**

Baureihe **4000**

Baureihe **595**

Baureihe **8000**

Baureihe **1163**

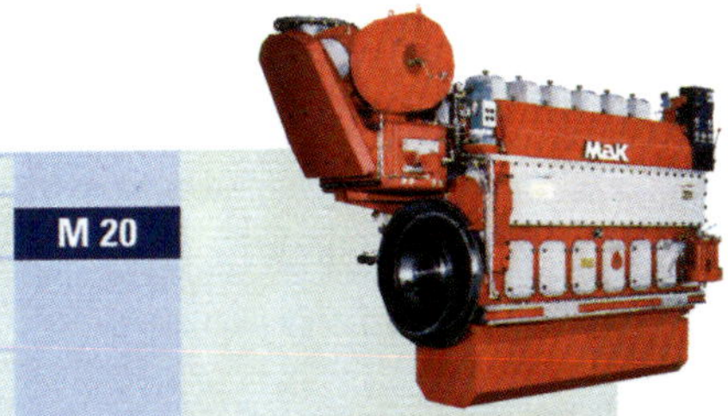

M 20

Bore	200 mm ∅
Stroke	300 mm
Cylinders	6, 8, 9
Output	1,020–1,710 kW
Rated speed	900–1000 rpm

M 25

Bore	255 mm ∅
Stroke	400 mm
Cylinders	6, 8, 9
Output	1,800–2,700 kW
Rated speed	720–750 rpm

M 32 C

Bore	320 mm ∅
Stroke	480/420 mm
Cylinders	6, 8, 9, 12, 16
Output	2,880–8,000 kW
Rated speed	600–750 rpm

M 43

Bore	430 mm ∅
Stroke	610 mm
Cylinders	6, 7, 8, 9, 12, 16, 18
Output	5,400–16,200 kW
Rated speed	500–514 rpm

◄ The engine programme of manufacturer Caterpillar MaK, four-stroke medium-speed diesel engines for engines using HFO.

The engines' power output ranges from 1020 to 16,200 kW. Rpm varies from 500 to 1000.

3.6 Important terms and definitions

3.6.1 The piston stroke (*S*)

The piston moves between the bottom dead centre and the top dead centre; one could also say that the piston is either BDC or TDC. This means that the crank is positioned vertically upwards or vertically downwards, respectively.
The length the piston travels is the engine stroke. This varies in industrial engines from approximately 60 millimetres to 680 millimetres (four-stroke trunk piston engines) to about 3200 millimetres (two-stroke crosshead engines).

3.6.2 Cylinder bore (*D*)

In order to determine the cylinder bore, one takes the diameter of the cylinder liner. In industrial engines this ranges from 40 millimetres to 640 millimetres (four-stroke trunk piston engines) to 1080 millimetres (two-stroke crosshead engines).

3.6.3 The stroke/bore ratio (*S*/*D*)

The ratio varies from ± 0.8 to 1.5 for four-stroke trunk piston engines and 2 to 4 for two-stroke crosshead engines.

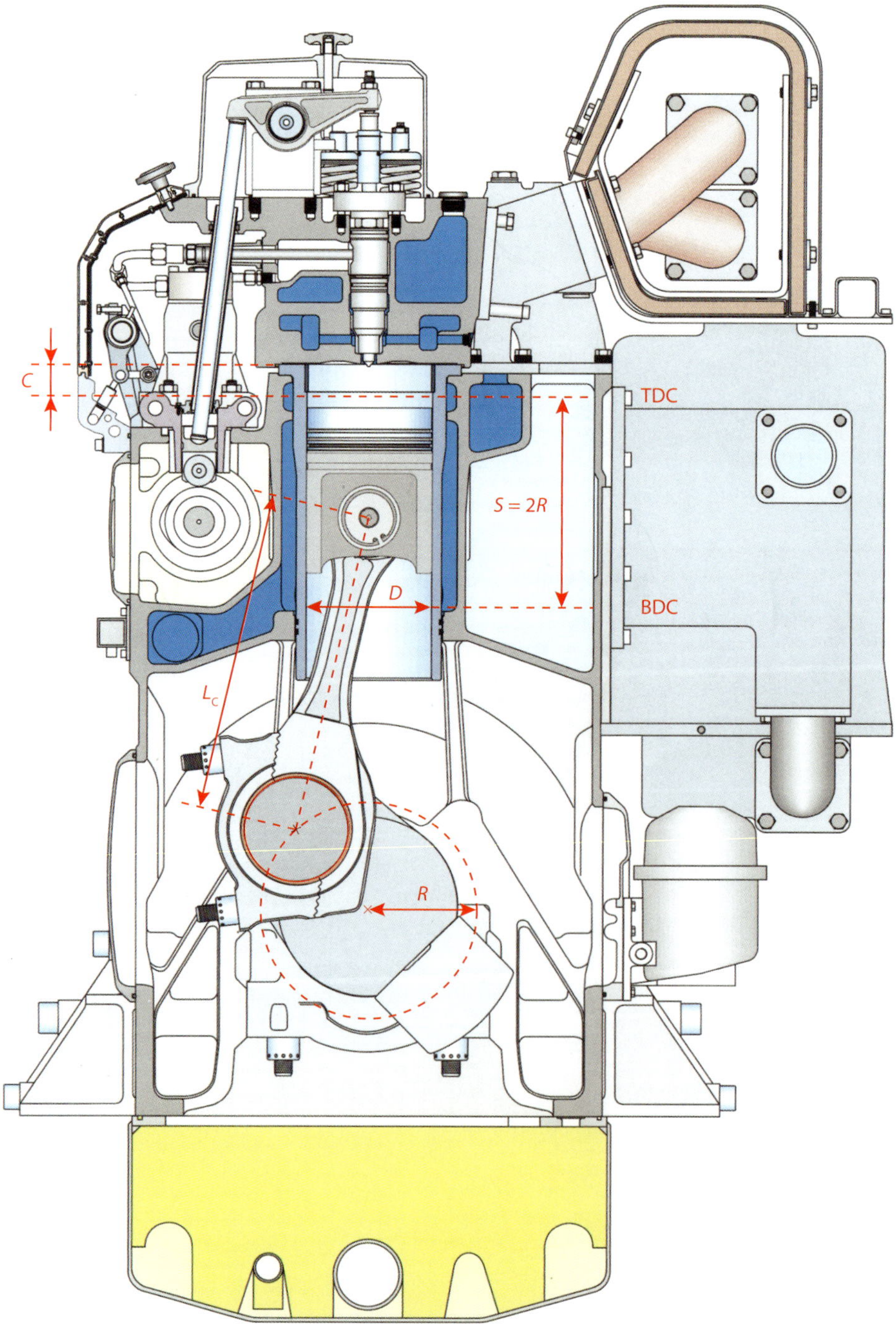

▶ The cylinder bore, the stroke, the connecting-rod length and the crank circle.

3.6.4 Mean piston speed in metres per second ($C_{P\ mean}$)

This is directly related to the rpm and the stroke.

$C_{P\ mean} = 2 \times S \times n$

Two-stroke crosshead engines have a maximum piston speed of 6 to 8 m/sec. and four-stroke engines have a maximum piston speed varying from 8 to 12 m/sec.
This is determined by the acceleration and deceleration forces of the gearing and the lubrication of the pistons, rings and cylinder liners.

3.6.5 Crank length of the crankshaft (R)

The crank forms the link between the connecting rod and the crankshaft and ensures that the up- and downwards movement of the piston is converted to crankshaft rotation (R) via the piston pin, connecting rod and crankpin (four-stroke) or via the fixed piston rod, crosshead, connecting rod and crankpin (two-stroke).

3.6.6 Crank circle (S)

In a frontal view of the engine the centre axis of the crankpin makes a circular movement with at its centre the centre axis of the crankshaft, which is denoted by the radius R and the diameter $2R = S$.

3.6.7 Connecting rod length (L_C)

This is the distance between the centre axes of the piston pin and the crankpin. This is usually 3.5 to 4.5 times the length of the crank, so:
$D = (3.5 - 4.5)\ R$.

3.6.8 Stroke volume (V_S)

This is the cylinder volume displacement between bottom- and top dead centre.

$$V_S = \frac{\pi}{4} \times D^2 \times S$$

In smaller diesel engines the total cylinder volume is indicated in cm^3 (cc) or in litres.
This is less common in larger engines. Here one speaks of output per cylinder, total output, the stroke, cylinder diameter and the number of cylinders.

3.6.9 Distance (C)

This is the distance between the cylinder head and the piston in top dead centre (TDC).

3.7 Some engine names

MAN Diesel 12K80ME-C

This is a twelve-cylinder two-stroke crosshead in-line engine with a cylinder bore of 800 millimetres (or 80 centimetres).
The stroke is 2300 millimetres (or 230 centimetres) (project guide), the rpm 104 and the maximum output 43,320 kW. Electronic fuel injection.

MAN Diesel 9L 16/24

This is a nine-cylinder four-stroke trunk piston in-line engine with a cylinder bore of 160 millimetres and a 240 millimetre stroke.
Here the power output is a maximum of 900 kW at 1200 rpm (project guide).

Caterpillar 3516 B

This is a sixteen-cylinder four-stroke trunk piston V-engine with a cylinder bore of 170 millimetres and a 215 millimetre stroke. The maximum output is 2237 kW at 1800 rpm.

S.E.M.T. Pielstick PC4.2 B-20V

This is a twenty-cylinder four-stroke trunk piston V-engine with a cylinder bore of 570 millimetres and a 660 millimetre stroke. The maximum output is 28,000 kW at 430 rpm.

◄ The type of main engine on a container ship, a twelve-cylinder Wärtsilä Sulzer two-stroke crosshead engine with a shaft power of 61,900 kW at a speed of 94 rpm. The cylinder diameter is 960 mm and the piston stroke is 2600 mm. Flex means a common-rail fuel injection system. Above in the picture is the firing order. This engine was built under license at Diesel United, Aioi, Japan in June 2004.

>CH 4

Efficiency and losses of diesel engines

A medium-speed four-stroke Caterpillar MaK 43 propulsion engine in the container-feeder 'Elite' of J.R. Shipping, Harlingen, the Netherlands. These engines have a high fuel efficiency of 45%, the fuel used is HFO. Nevertheless, 55% is wasted as unused heat in the form of exhaust gases and engine cooling.

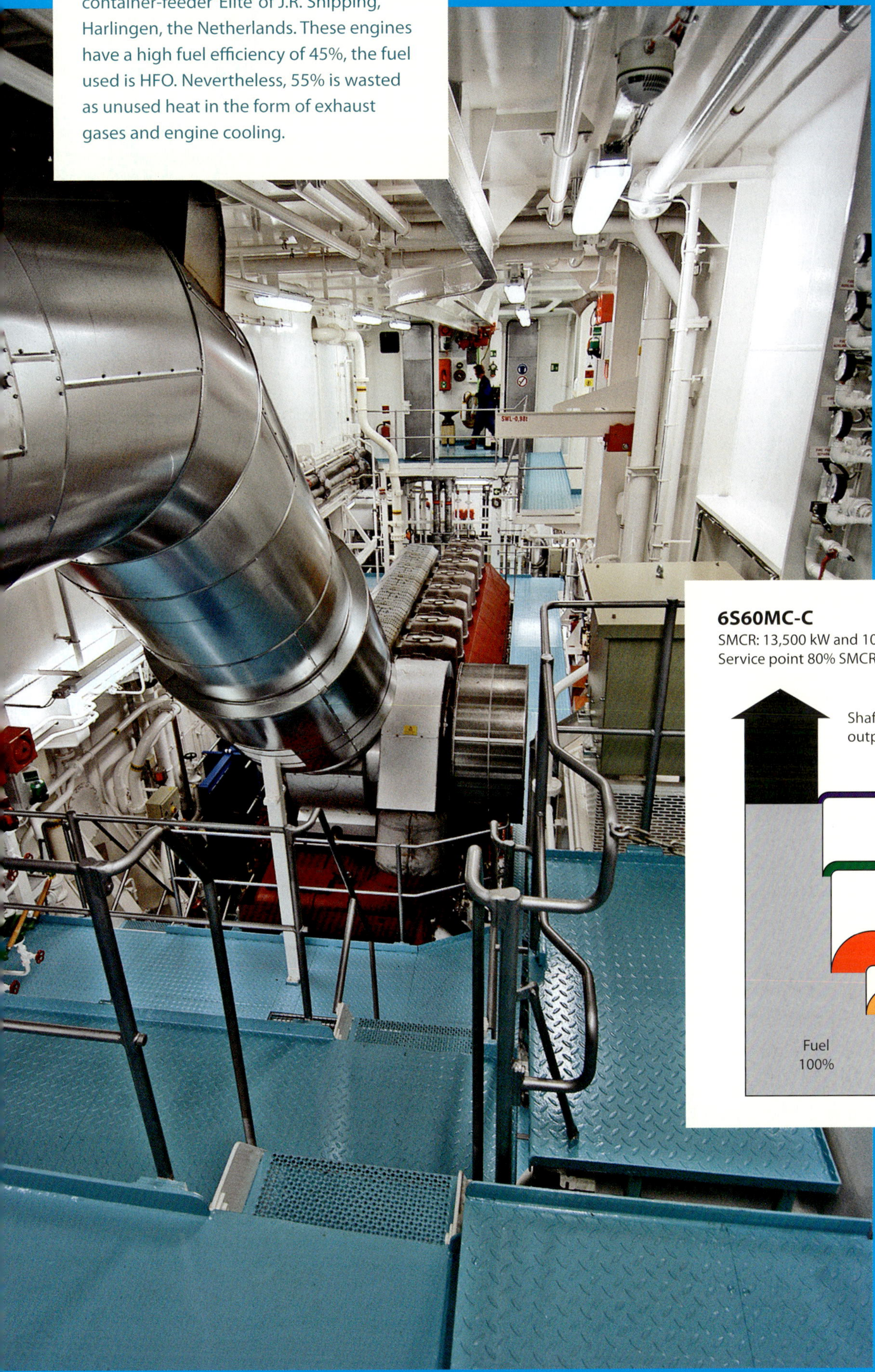

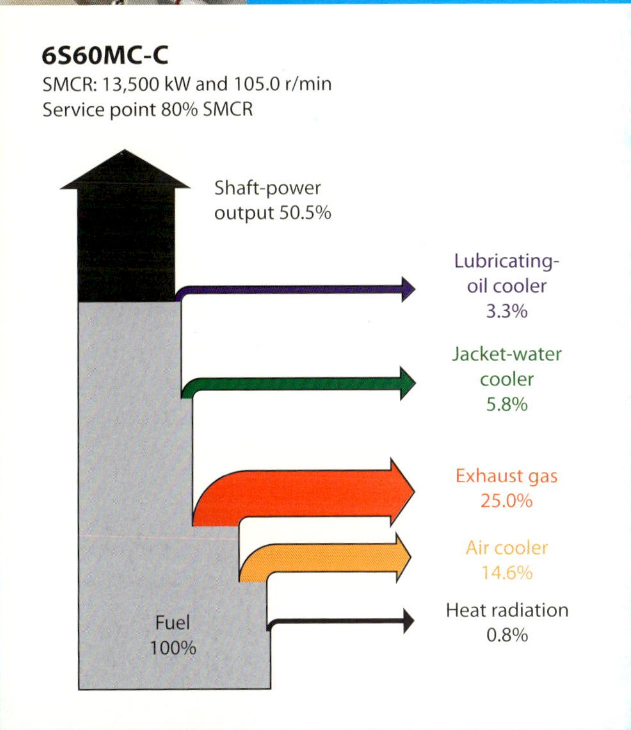

The Sankey diagram of a two-stroke crosshead engine. Efficiency is 50.5%.

▶
The total propulsion efficiency, from the engine fuel supply to the available ship thrust is approximately a third, or 33 %.

4.1 Efficiency and losses

In diesel engines only a small part of the fuel is converted into mechanical work.
The ratio of energy supplied to energy delivered to the outgoing shaft is referred to as the total engine output.

$$\text{Engine Efficiency} = \frac{\text{Shaft power (effective power)}}{\text{Energy in supplied fuel}}$$

Formulated form:

$$\eta_{tot} = \frac{P_E}{F_C \times H_L}$$

η_{tot} = efficiency as a ratio of 1
P_E = effective or shaft power in MW
F_C = fuel consumption in kg/sec.
H_L = calorific value in MJ/kg

Small diesel engines have an efficiency of 25% whereas that of very large engines exceeds 50%.
All other fuel energy is eventually lost in the form of heat, namely:

▶
The larger the cylinder bore, the higher the total efficiency of a diesel engine.

This example by Wärtsilä shows that the efficiency of an engine with a cylinder bore of 640 millimetres is 7% higher than with an engine with a 200 millimetre cylinder bore.
This is mainly due to the fact that a larger cylinder volume produces an increasingly smaller radiant surface – liners, cylinder heads and piston crowns – so more heat remains for power output.

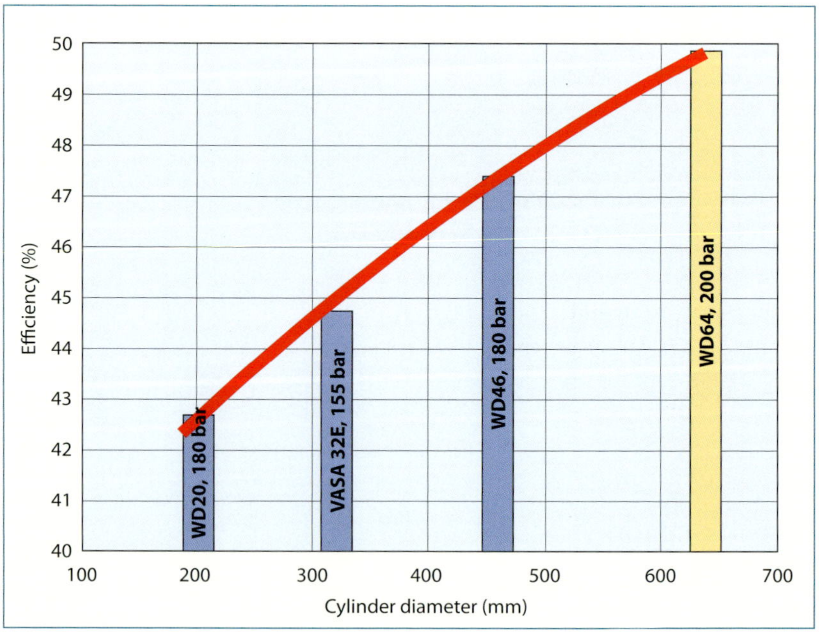

- the exhaust gases;
- the cooling water;
- the lubricating oil;
- the intercooler or air cooler;
- the heat radiation.

Additionally, energy is used to drive the valves, the fuel pumps and the engine-driven cooling-water pumps.
We will return to this subject later in this chapter in the **section; thermal energy balance or Sankey diagram.**

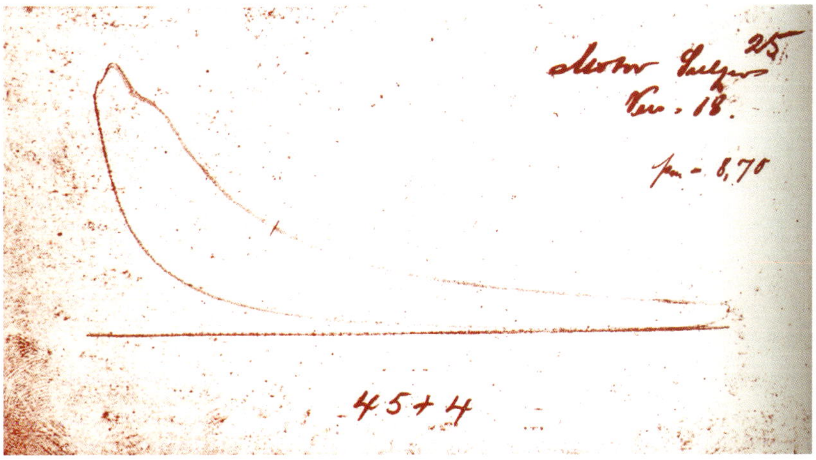

▲ Rudolf Diesel's very first cycle diagram.

Essentially, very little has changed over the past one hundred years. The power output, the enclosed surface of the banana-curve, was much higher than that of the Otto cycle at an identical energy supply.

4.2 Indicator diagram

In order to obtain a good insight into the working of the diesel engine, the Pressure-Volume or *pV*-diagram is fundamental. From the development of the very first **Diesel** and **Otto** internal combustion engines, the compression in the engine cylinder has been indicated as a function of the continuous stroke volume of the piston.
This can be measured by means of an indicator gauge. This results in an **indicator diagram**, popularly known as the 'banana-curve'. This provides a good insight into the compression in the cylinder and the work done.

The following pictures show *pV*-diagrams for both two- and four-stroke engines. Shown below in both diagrams are the most important aspects of the crank circles, such as the opening and closing of the inlet- and exhaust valves as well as the start and end of the fuel injection.

4.2.1 Four-stroke cycle

Let us assume that de piston is at bottom-dead-centre position (BDC) and commencing its upward stroke. From the moment that the inlet valves are closed the pressure mounts significantly from about 2.5 bar over-pressure (supercharging with turbocharger) to about 150 bar (final compression pressure). The internal air temperature has increased from ± 50 °C (supercharged air temperature after cooler) to ± 750 °C (final compression temperature). Just before the piston reaches top-dead-centre position (TDC) an exceedingly fine fuel spray is rapidly injected into the cylinder and ignites.

During combustion the pressure climbs to ± 200 bar (maximum combustion pressure) and the temperature rises to ± 1600 °C (maximum combustion temperature).

In the course of the combustion- or power stroke the pressure decreases and the exhaust valves subsequently open at 6 bar, the temperature of the exhaust gases are now about 700 °C.
At this time, the pressure in the exhaust pipe amounts to ± 2 bar and the temperature of the exhaust gases 450 °C.

Of course, each modern four-stroke diesel engine motor has its own cycle and accompanying pressures and temperatures that deviate slightly from the aforementioned.

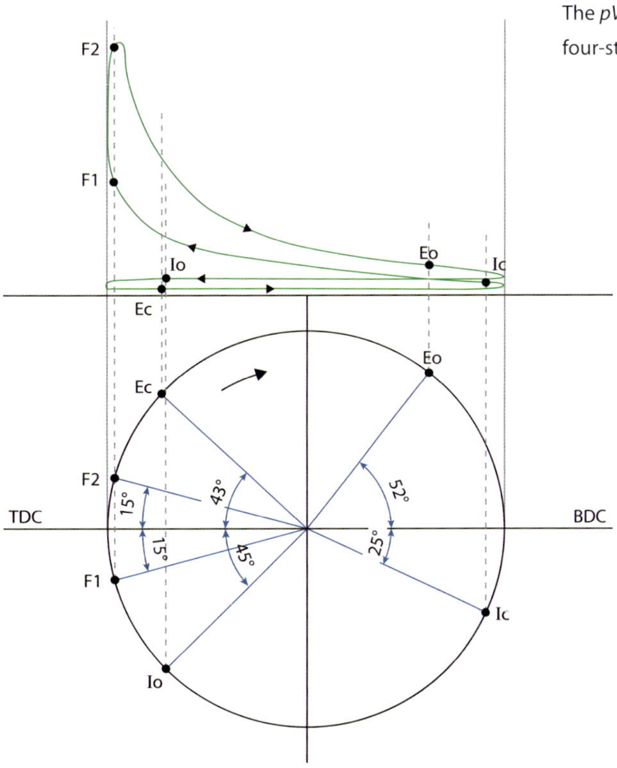

In diagram: starting on the right side
After the inlet valve is closed, the pressure increases (Ic)
Fuel is injected at (F1) until the power stroke commences at (F2)
The exhaust valve opens at (Eo) until the exhaust stroke commences at (Ec)
Exhaust-gas scavenging from (Io) to (Ec), followed by the inlet stroke

◄ The *pV*-diagram for a four-stroke diesel engine.

The mechanical work done is positive (a clockwise-working cycle) and is the internal power of the cylinder (heat engine).

Three strokes in this process use energy from the crankshaft: the inlet stroke, the compression stroke and the exhaust stroke. The mechanical work done is negative.
Only the power stroke supplies energy to the crankshaft. This is positive mechanical work.

4.2.2 Two-stroke process

Let us assume that the piston is in BDC and commencing its upward stroke. From the moment that the piston passes the inlet ports and that the central exhaust valve is shut, compression begins. Pressure increases from approximately 2 bar over-pressure (supercharging with turbocharger) to about 100 bar (final compression pressure). The air temperature has risen from ± 50 °C (scavenging-air temperature after the cooler) to ± 650 °C (final compression temperature).
Just before TDC is reached, the finely atomised fuel mixture is rapidly injected into the cylinder and ignites.
During combustion the pressure climbs to ± 150 bar (maximum combustion pressure) and the temperature to 1500 °C (maximum combustion temperature).

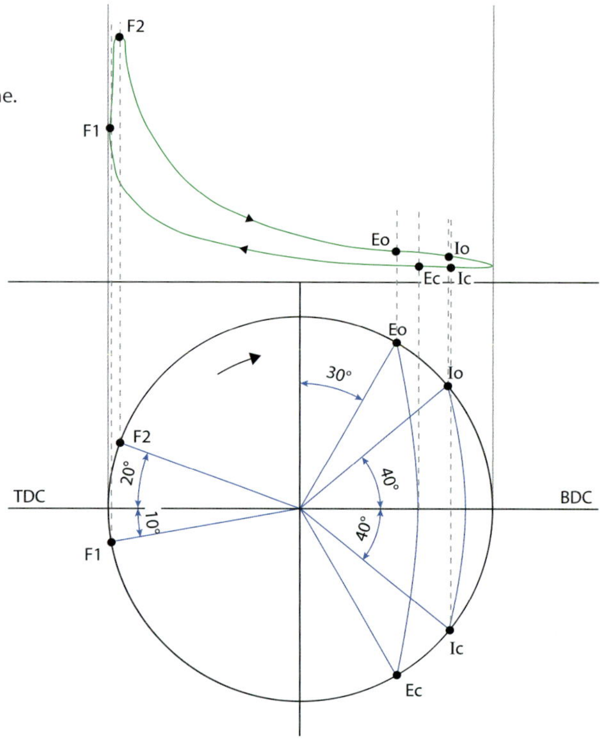

▶ The *pV*-diagram for a two-stroke diesel engine.

In the course of the combustion- or power stroke, the pressure further decreases and the exhaust valves subsequently open at approximately 4 bar. The exhaust-gas temperature is then about 600 °C.
At this time, the pressure in the exhaust pipe is ± 2 bar and the exhaust-gas temperature ± 375 °C.

Of course, each modern two-stroke engine has its own cycle and accompanying pressures and temperatures that deviate slightly from the aforementioned.

The mechanical work done is positive (a clockwise-working cycle) and is the internal power of the cylinder (heat engine).

In this process one stroke uses energy (the compression stroke = negative mechanical work) and one stroke supplies energy (combustion stroke = positive mechanical work) to the crankshaft.

Scavenging process

When the piston releases the inlet ports, the scavenging process takes place, at which point the scavenging-air pressure should exceed the pressure of the exhaust gases in order to prevent reflux of the exhaust gases.

Note

A diagram of an engine with inlet ports and one central exhaust valve has been made for the two-stroke process. Two-stroke industrial crosshead diesel engines with inlet and exhaust ports still exist, but are no longer manufactured.

4.2.3 A comparison of both processes

At an identical rpm of the crankshaft, a two-stroke engine of the same cylinder size delivers twice the amount of power strokes than that of a four-stroke engine.
Theoretically, a two-stroke engine should have twice the power output than that of a four-stroke engine at the same rpm and an identical cylinder volume.
However, scavenging and charging for a two-stroke engine are less efficient than that of a four-stroke engine, which means less fuel is burnt.
The power-output ratio of four-stroke versus two-stroke engines at the same rpm and cylinder volume lies just below 2, namely 1.6 to 1.7.
The thermal load of the material of the components of the two-stroke engine is higher than that of the four-stroke engine, so the time available between two power strokes, the hottest part of the process, is reduced by half.

Moreover, cylinder scavenging in a two-stroke engine is more complex, since it is only the over-pressure of the scavenging air in relation to the exhaust gases that determines the scavenging, as opposed to the four-stroke engine where a piston draws in air and forcibly expels the exhaust gases. Therefore **the mean piston speed** for a two-stroke engine is set slightly lower (6 to 8 m/sec. for two-stroke and 8 to 11 m/sec. for four-stroke) in order to avoid scavenging problems.

Generally, all medium- and high-speed diesel engines are four-stroke engines. These can effortlessly achieve a high mean piston speed as well as a high mean effective pressure without having the problems that gas exchange or thermal load of the components may cause. Moreover, the lubricative oil film on the piston and the cylinder liner is better maintained in a four-stroke engine.

4.3 Parameters of both working principles

The parameters of both working principles differ considerably.

4.3.1 Mean effective pressure: $p_{\text{effective mean}}$

For the latest two-stroke crosshead engines, this is approximately 20 bar and for the latest four-stroke trunk piston engines about 28 bar. This disparity is predominantly caused by the fact that there is less fresh air for combustion in the cylinder; as a consequence less fuel can be injected per combustion stroke. This ultimately results in a lower mean effective pressure for two-stroke engines.

4.3.2 Mean piston speed: $C_{\text{P mean}}$

For large two-stroke engines this is around 8.5 m/sec. and for the four-stroke engine 10 m/sec. The mean piston speed is an important factor with regard to the thermal and mechanical load of the various engine components.
The lubrication of, among other things, the piston rings is slightly lower in two-stroke engines as they constantly pass the ports. this is not conducive to cylinder lubrication. Therefore, the piston speed is lower and of course in order to obtain optimum scavenging, the piston speed is also set slightly lower.

For each engine this is connected to a maximum value.

$$C_{\text{P mean}} = 2 \times S \times n$$

$C_{\text{P mean}}$ = mean piston speed in m/sec.
S = piston stroke in m
n = number of revolutions per second (rps)

Conclusion
The shorter the stroke of the engine, the higher the achievable rpm!

4.3.3 Load parameters L_p, product of the mean effective pressure and the mean piston speed

See also 4.3.1 and 4.3.2.

$$L_P = p_E \times C_P$$

Example:
For the two-stroke cycle: $L_P = 19 \times 8.5 = 161.5$ bar/m/sec.
For the four-stroke cycle: $L_P = 25 \times 10 = 250$ bar/m/sec.

4.3.4 Exhaust gas temperatures

For the two-stroke cycle: $T_{gas} = 325\ °C$ to $375\ °C$
For the four-stroke cycle: $T_{gas} = 400\ °C$ to $500\ °C$

▼
A large two-stroke engine. Next to it a four-stroke trunk piston engine.

Example 1
Set stroke = 0.2 m.
Mean piston speed 10 m/sec.

$$C_{P\ mean} = 2 \times S \times n$$
$$10 = 2 \times 0.2 \times n$$
$$0{,}4\ n = 10$$
$$n = 25 \text{ rps} \rightarrow 25 \times 60 = 1500 \text{ rpm}$$

This is a high-speed four-stroke engine.

◄ Fast-running four-stroke Caterpillar diesel engines ready to be sent to genset manufacturers or companies for assembly in propulsion systems.

Example 2
Set stroke = 0.6 m.
Mean piston speed 9.6 m/sec.

$$C_{P\ mean} = 2 \times S \times n$$
$$9.6 = 2 \times 0.6 \text{ x } n$$
$$1.2\ n = 9.6$$
$$n = 8 \text{ rps} \rightarrow 8 \times 60 = 480 \text{ rpm}$$

This is a medium-speed four-stroke engine.

◄ A medium-speed four-stroke HFO diesel engine driving a generator on a large container ship.

Example 3
Set stroke = 2 m.
Mean piston speed 8 m/sec.

$$C_{P\ mean} = 2 \times S \times n$$
$$8 = 2 \times 2 \times n$$
$$4\ n = 8$$
$$n = 2 \text{ rps} \rightarrow 2 \times 60 = 120 \text{ rpm}$$

This is a low-speed two-stroke engine.

◄ Cylinder heads of a large two-stroke crosshead engine for ship propulsion.

Note the hydraulically tightened bolts for the exhaust valves and the cylinder covers. Here the cylindrical nuts are covered with protective caps.

4.4 Determining cylinder output using an indicator diagram and the mean induced pressure

This is still done in the traditional way, but today usually by means of an electronic pressure gauge on the cylinder which computes all the necessary values using a dedicated software program.

4.4.1 Using the indicator gauge

◀

An indicator gauge.

1 *sleeve nut on indicator cock*
2 *piston casing*
3 *indicator spring (pressure spring)*
4 *lock pin*
5 *lever system with needle*
6 *indicator drum with indicator paper*
7 *indicator drive belt for drum*

▲

Part of the cylinder cover of a two-stroke crosshead engine.

1 *cylinder cover*
2 *one of the two fuel valves*
3 *indicator cock for making a pV-diagram*
4 *safety valve*
5 *exhaust-valve casing*

▼▶

An indicator.

1 *pressure-measuring staff*
2 *extra spring*

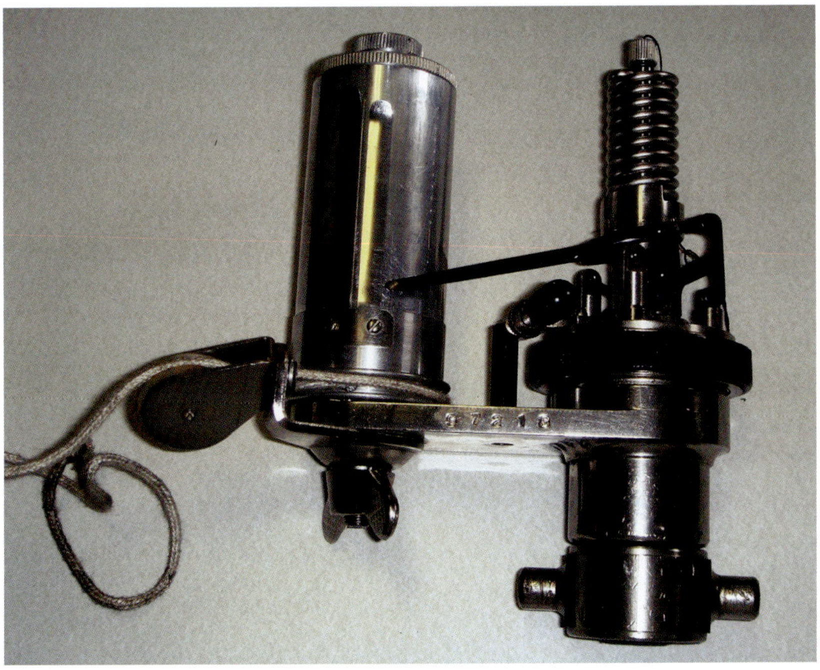

▼

A pressure gauge for measuring the maximum combustion pressure in the cylinder.

The gauge stays at peak pressure

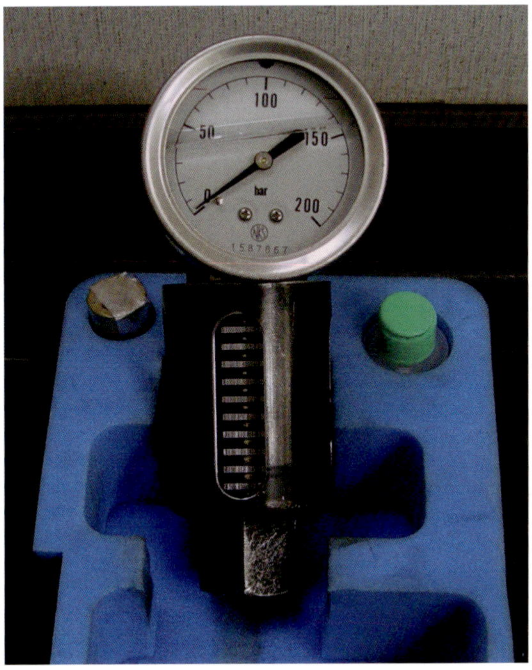

4.4.2 Determining cylinder output using an indicator diagram

A pV-diagram represents the working cycle of an internal combustion engine; in this case a diesel engine.

Theoretical diagram

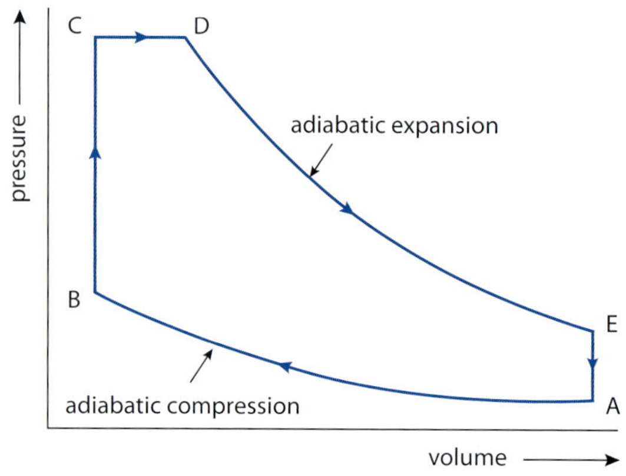

▲

The extent of the clockwise-turning process is indicative of the amount of positive power exerted on the piston.

The enclosed surface A–B–C–D–E–A is the resulting positive work.

Practical diagram

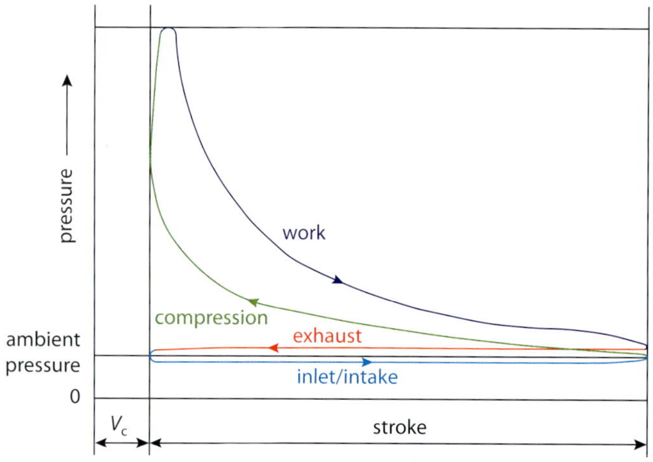

▲

The actual pressure development in the cylinder of a four-stroke engine as measured by an indicator.

Horizontal axis, the piston position and the vertical axis, the pressure development. Here the exhaust-gas pressure slightly exceeds the inlet-air pressure at the exhaust and inlet stroke respectively.
This is a diesel engine without turbocharger.

Theoretical and practical diagram

The mean induced pressure on the piston is not different from the measured (or calculated) pressure throughout **the entire process**.

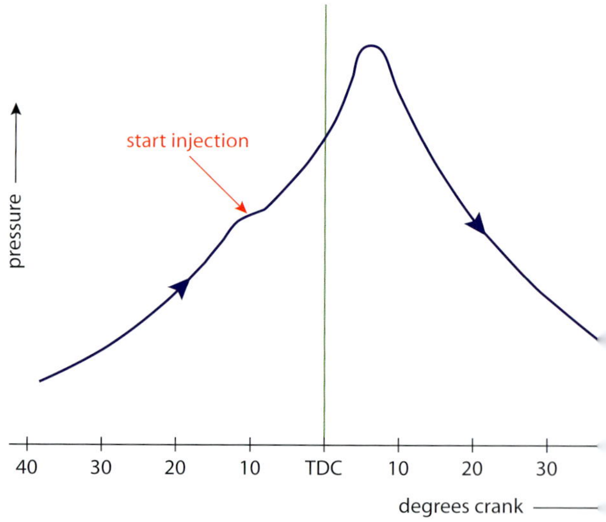

▲

The 90° shifted diagram.

This clearly shows the initial moment of fuel injection.

4.5 Determining the mean induced pressure

There are three ways in which to determine the mean induced pressure in a pV-diagram.

4.5.1 Measuring the diagram with a planimeter

With a measuring instrument, the planimeter, the diagram-surface area is meticulously calculated in mm² and divided by the length of the diagram in mm.
Mean height is calculated using the formula below.

$$\text{mean height in mm} = \frac{\text{diagram-surface area in mm}^2}{\text{length diagram in mm}}$$

The calculated mean height in mm is divided by the linear spring scale and this provides the mean induced pressure..

$$\begin{array}{l}\text{mean induced}\\ \text{pressure in bar}\end{array} = \frac{\text{mean height in mm}}{\text{spring scale mm/bar}}$$

To warrant accuracy one takes three planimeter measurements.

4.5.2 Oldest method, the division of the acquired diagram into ten equal-sized sections

The classical method in determining the mean induced pressure from a pV-diagram: split the diagram into ten pieces of the same size. Determine the mean height of each part. This is roughly the height in the middle of one part.
The sum of the average heights (ten parts) divided by 10 gives the mean height in mm.
By dividing this number by the spring scale the mean induced pressure in bar is obtained.

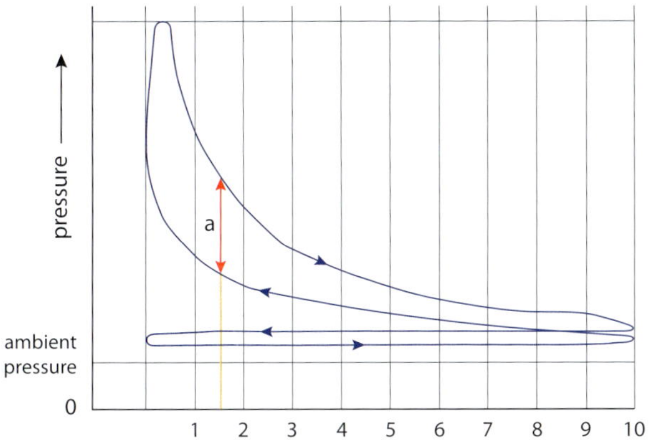

▲
Determining the mean induced pressure from an indicator diagram.

4.5.3 Modern method, electronic

Here a pressure sensor is connected to the indicator. All the cylinder data are passed on to a collecting unit. This processes all the information such as pressure development, crankshaft position, rpm, etc. All the data is then processed by a central computer.

The possibilities are extensive. Here some examples of the values that can be established:
- mean induced pressure;
- maximum cylinder pressure;
- induced power output;
- torsional vibration;
- exhaust-gas pressure;
- general trends.

In this manner, all the important data of the engine are available at any given moment.

See also Chapter 25, *Operational management and automation*.

4.6 Engine formula

The internal and external power output of the engine can be calculated with the aid of the engine formula.
The following applies to a combustion engine:

$$P_i = \frac{\pi}{4} \times D^2 \times S \times \frac{n}{a} \times Z \times p_i$$

P_i = induced power output in MW (Megawatt)
D = piston bore in m (metres)
S = piston stroke in m (metres)
n = number of revolutions of the crankshaft per second
a = 1 for two-stroke, a = 2 for four-stroke
Z = number of cylinders
p_i = induced mean pressure in MPa (megapascal)

As a rule, pressure in engine technology is expressed in bar: **1 MPa = 10 bar**.

The engine power output is usually often expressed in kW: **1 MW = 1000 kW**.

▼
An overview of a modern measuring system mounted on a two-stroke crosshead engine.

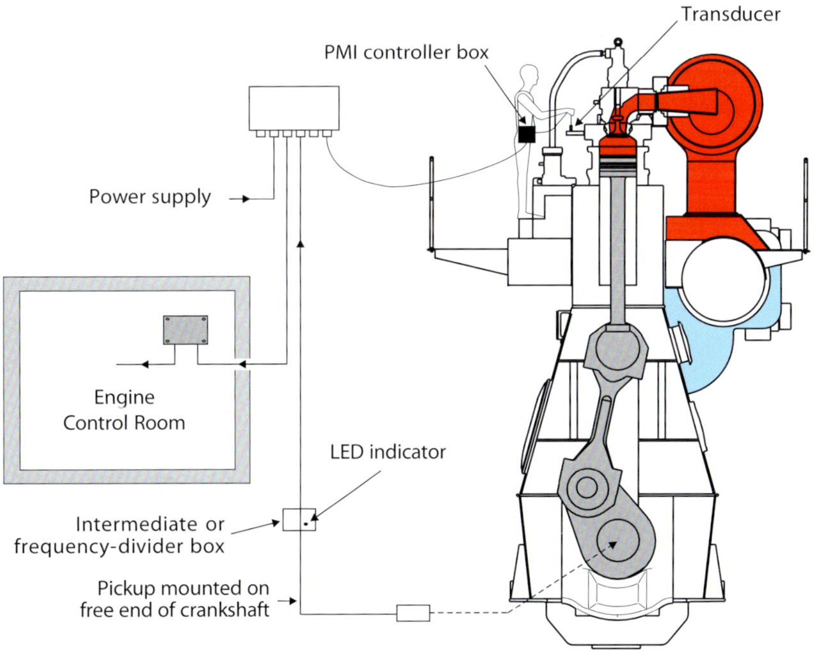

The PMI System as applied on a two-stroke diesel engine

Example

How to calculate the induced (power) output of an engine by using an indicator gauge and a planimeter.

Two-stroke crosshead engine

number of cylinders	8
cylinder bore D	= 700 mm
piston stroke S	= 2360 mm
number of revolutions	
108 per minute	= 1.8 rps
Spring scale 1 bar	= 0.3 mm
Surface-area diagram	= 399 mm^2
length diagram	= 70 mm

Calculate:

a The mean effective pressure in bar and MPa.

b The mean piston speed in m/sec.

c The mean induced (power) output provided that the induced mean pressure of all cylinders is identical.

Solution

a The mean height of the indicator diagram is:

$$h_M = \frac{399}{70} = 5.7 \text{ mm}$$

For a spring scale of 1 bar = 0.3 mm, the mean induced pressure is now $\frac{5.7}{0.3} = 19$ bar or 1.9 MPa.

b The mean piston speed is:

$$C_{P\,mean} = 2 \times S \times n = 2 \times 2.36 \times 1.8 = 8.5 \text{ m/s}$$

c The induced power output:

$$P_i = \frac{\pi}{4} \times D^2 \times S \times \frac{n}{a} \times Z \times p_i$$

$$P_i = \frac{\pi}{4} \times 0.7^2 \times 2.36 \times \frac{1.8}{1} \times 8 \times 1.9$$

$$P_i = 24.840 \text{ MW or } 24{,}840 \text{ kW}$$

This is an example of a MAN Diesel two-stroke crosshead engine type 8L70ME-C.

Additional information:

length	11.35 metres
height	10.25 metres
width	4 tot 6.5 metres
weight (dry)	642 ton

Fuel consumption per day 101,350 kg or 101.4 tons weight.

Note

It is not possible to use the indicator gauge for high-speed diesel engines. It is generally used for low-speed two-stroke crosshead engines in order to calculate the output compression and combustion pressures.

In higher rotating four-stroke diesel engines, it is/was merely used to measure the compression- and combustion pressure.

4.7 Induced thermal efficiency

Induced thermal efficiency is also called 'the useful effect'.

The formula in words is:

$$\text{Induced thermal effect} = \frac{\text{Induced engine output}}{\text{Fuel power input}}$$

Induced engine power	P_i (MW)
Fuel consumption	F_C (kg/sec.)
Fuel energy density (also heat value)	H_L (MJ/kg)
Fuel power input	P_F (MW)

$$\eta_i = \frac{P_i}{F_C \times H_L}$$

η_i = eta internal = the internal efficiency.

This ranges from about 25% to 55% for a four-stroke engine and for a two-stroke engine is slightly over 55%. The remainder of the fuel energy is lost in the form of heat.

▲
The cylinder and shaft power.

1 *shaft- or effective power (P_{eff})*
2 *cylinder or internal power (P_i)*

4.8 Mechanical and total efficiency

In practice, one prefers to work with a slightly more practical efficiency, namely the total efficiency or shaft efficiency of an engine.
Engine manufacturers always present the maximum engine power output as the maximum continuous capacity or shaft power.

A percentage of the induced power is not transferred to the crankshaft as useful shaft power, as it is required to overcome frictional forces. Naturally, this is just a very small percentage. A far larger percentage of the induced power is required to drive the high-pressure fuel pumps, the inlet- and exhaust valves and the lubricating- and cooling-water pumps. For a modern diesel engine this amounts to about 8 to 10% of the induced power in total. Therefore, the mechanical efficiency is approximately 90 to 92%.

The mechanical efficiency equals:

$$\frac{\text{the shaft power}}{\text{the cylinder output}} \text{ or } \frac{P_{eff}}{P_i}$$

The mechanical efficiency $\eta_{mech.}$ therefore is:

$$\eta_{mech.} = \frac{\text{effective power or shaft power on the crankshaft}}{\text{induced power}}$$

$$\eta_{mech.} = \frac{P_{eff}}{P_i}$$

Also: $\eta_t = \eta_m \times \eta_i = \frac{P_i}{F_C \times H_L} \times \frac{P_{eff}}{P_i}$ or $\eta_t = \frac{P_{eff}}{F_C \times H_L}$

4.9 Specific fuel consumption

This is the amount of fuel consumed per power unit and per time unit. It is normally indicated in grams per kilowatt.
The lower the fuel consumption, the less fuel the engine uses and the higher the engine's 'return efficiency' is. This way, one can easily compare different engines.

Lubricating oil- and fuel consumption constitute by far the highest costs in ship exploitation.

Of course the measurement of fuel consumption should take place under identical conditions; these have been laid down in the ISO norm 3046-1:1995. This stipulates amongst others:
- the inlet-air temperature of 25 °C;
- the temperature after the air cooler of 25 °C;
- the ambient pressure 1000 millibar;
- relative humidity of 30%;
- the calorific value of the fuel.

All engine manufacturers provide these details when stating fuel consumption.
The fuel consumption in conjunction with the lubricating-oil consumption (in grams per kWh) forms a major consideration in deciding which engine to chose.
The fuel consumption varies from 167 gr/kWh to 220 gr/kWh and depends for instance on the size of the diesel engine. The larger the engine the higher the efficiency and the lower the fuel consumption.

4.10 Mean effective pressure

We already know:

$$\text{mechanical efficiency} = \frac{\text{the effective power}}{\text{the induced/internal power}}$$

$$\eta_{mech.} = \frac{P_{eff.}}{P_i}$$

Thus follows: $\eta_{mech.} = \frac{p_{eff.\,mean}}{p_{i\,mean}}$;

And also: $P_{eff.} = \frac{\pi}{4} \times D^2 \times S \times \frac{n}{a} \times Z \times p_{eff.\,mean}$

This is the more practical formula with regard to diesel engines.
Engine manufacturers also always provide the effective power output.
The difference between the induced and the effective mean pressure is called the mean frictional pressure p_f.

So: $p_f = p_i - p_{eff.}$

Furthermore : $P_f = \frac{\pi}{4} \times D^2 \times S \times \frac{n}{a} \times Z \times p_f$

Therefore the frictional work of the engine is also:

$$P_f = P_i - P_{eff}$$

Measurements show that the (frictional) torque and therefore also the frictional pressure are barely influenced by the number of revolutions (rpm) and engine load. Consequently, the mean frictional pressure for an engine can be taken as a constant.

Overall, engine load is proportional to $P_{eff.}$, so to the generated power per cycle. This means that the engine output is only proportional to the load at a constant rpm.
It is also given that:

$$\eta_{mech.} = \frac{P_{eff.}}{P_{eff.} + P_w} = \frac{p_{eff.}}{p_{eff.} + p_w} = \frac{1}{1 + \frac{p_w}{p_{eff.}}}$$

Example

At full load an engine with a constant rpm runs with a mean effective pressure of 25 bar and a mean induced pressure of 28 bar.
The mean frictional pressure is:

$p_f = p_i - p_{eff.} = 28 - 25 = 3$ bar.

The mechanical efficiency at full load of this engine is:

$$\eta_{mech.} = \frac{1}{1 + \frac{3}{25}} = 0.8928 \text{ of } 89.3\%$$

The mechanical efficiency drops slowly if this is calculated for a partial load.

At 75% power output $p_{eff.}$ is= 0.75 × 25 = 18.75 bar and the mechanical efficiency is 86.2%.
At 50% power output $p_{eff.}$ is = 0.50 × 25 = 12.0 bar and the mechanical efficiency is 80%.
At 25% power output $p_{eff.}$ is = 0.25 × 25 = 6.25 bar and the mechanical efficiency is 67.6%.
Therefore, the mechanical efficiency drops considerably at a reduced engine-power output.

4.11 Thermal energy balances or Sankey-diagrams

An energy balance or Sankey-diagram indicates how the heat supplied by the fuel next to the delivered shaft power is distributed over various systems. Exhaust-gas loss is always the greatest loss, followed by the air-cooler loss, or jacket-coolant loss and lubricating-oil loss. Furthermore, there is a radiation loss and residual loss. The following examples are two overviews of a four-stroke trunk piston engine and a two-stroke crosshead engine.

▼
One of the two Wärtsilä 9L46B four-stroke diesel engines on the 'Oranjeborg', Wagenborg Shipping.

An example of a thermal energy balance I

The Wärtsilä 9L46B has the following values as detailed in the Project Guide Marine Applications:

four-stroke cylinder bore	460 mm
stroke	580 mm
cylinder output at 514 revolutions	975 kW
mean effective pressure	23.6 bar
mean piston speed	9.9 m/sec.
number of revolutions	514 rpm
shaft power	8775 kW
amount of combustion air	15.8 kg/sec.
exhaust-gas temperature after turbocharger	375 °C (CCP)
amount of exhaust gases	16.2 kg/sec.
lubricating-oil cooling	1150 kW
cylinder cooling-water cooling	950 kW
air cooler high temperature (HT)	1500 kW
air cooler low temperature (LT)	990 kW
radiation heat	340 kW
fuel consumption (CCP)	173 gram per KWh

CPP = controllable pitch propeller installations

Heat value fuel (calorific value) 42.777 MJ/kg (ISO-norm). Inlet-air temperature to the engine 25 °C (ISO-norm).

Specific heat of exhaust gases	1.04 kJ/Kg K
Specific heat of the air	1.0 kJ/Kg K

This information enables us to set up a thermal energy balance.

Calculation

The fuel consumption is 0.173 × 8,775 =

1,518 kilogram per hour or $\frac{1518}{3600} = 0.4216$ kg/sec.

It follows that the supplied fuel output is:
0.4216 × 42,700 = 18,006 kW. In a thermal-energy balance this normally comprises the total energy supply, combustion-air thermal energy supply is not included.

The volume flow of the exhaust gases is 16.2 kg/sec.
The exhaust-gas thermal energy after the turbocharger is m_{gas} (kg/sec.) × q. spec. gas (kJ.kg.K) × t_{gas} (°C) and subsequently: 16.2 × 1.04 × 375 = 6318 kW.

The combustion-air thermal energy to the engine is 15.8 × 1.0 × 25 = 395 kW.

The net loss with the exhaust gases is 6318 – 395 = 5923 kW.

In summary

supplied by fuel		18,006 kW
thermal energy of the combustion air		395 kW
total supplied	18,006 + 395 =	18,401 kW
exhaust-gas loss	5923 kW	32,18%
absorbed by lubricating oil	1150 kW	6.24%
absorbed by cylinder-cooling water	950 kW	5.16%
air cooler low temperature	990 kW	5.38%
radiation heat	340 kW	1.84%
effective power	8775 kW	47.68%
residual loss	273 kW	1.48%
total	**18,401kW**	**100%**

The high exhaust-gas losses (approximately a third) are remarkable together with the high total efficiency of this relatively large four-stroke trunk piston engine.
This engine weighs 134 tons and is 10.3 metres long, 3.3 metres wide and 4 metres high.

An example of a heat balance II

MAN Diesel K98MC twelve cylinders

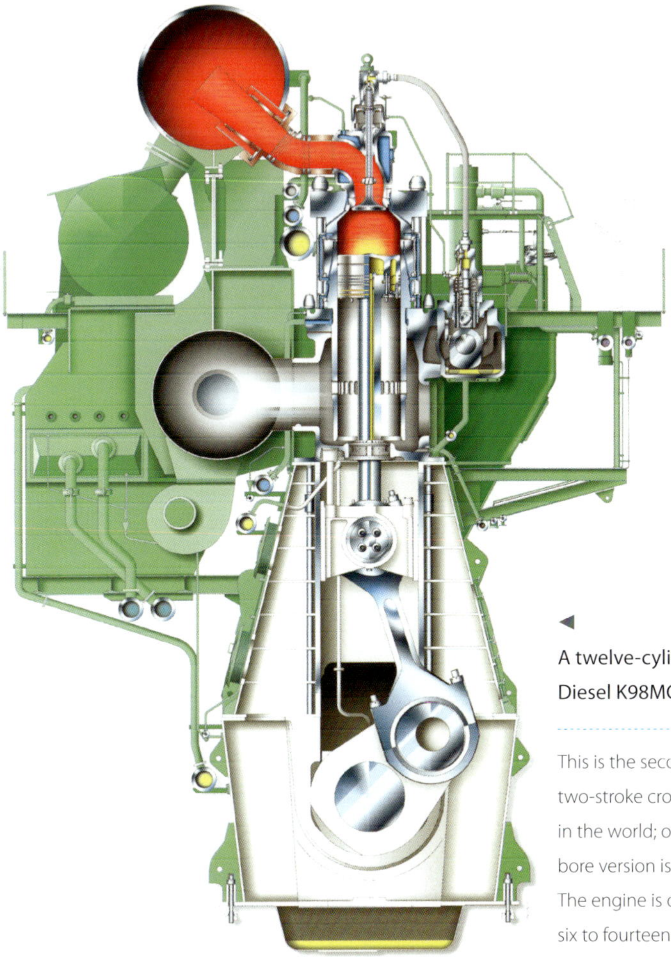

◄ A twelve-cylinder MAN Diesel K98MC Engine.

This is the second largest two-stroke crosshead engine in the world; only the1080 mm bore version is larger.
The engine is delivered with six to fourteen cylinders.

Some data

cylinder bore	980 mm
stroke MC-version	2660 mm
total power output	80,080 kW
number of revolutions	94 rpm
mean effective pressure	18.2 bar
mean piston speed	8.3 m/sec.
effective fuel consumption	171 g/kWh
air cooler	19,416 kW
lubricating-oil cooler	6770 kW
cylinder cooling-water cooler	11,600 kW
amount of exhaust gases	
754,600 kg/h =	209.6 kg/sec.
amount of combustion air	205.8 kg/sec.
fuel consumption	3.8 kg/sec.
exhaust gas temperature after turbocharger	245 °C

heat value fuel (calorific value) 42,700 MJ/kg (ISO-norm). Inlet-air temperature to the engine 25 °C (ISO-norm)

specific heat of the exhaust gases	1.04 kJ/kg K
specific heat of the air	1.0 kJ/kg K

With this information we can set up a thermal-energy balance.

Calculation

The fuel consumption per second is:

$$\frac{0.171 \times 80{,}080}{3600} = 3.8038 \text{ kg/sec.}$$

The supplied fuel output is:

3.8038 × 42,700 = 162,422 kW

In a thermal-energy balance this normally comprises the total energy supply, combustion air thermal-energy supply not included.

The volume flow of the exhaust gases is: 205.8 kg/sec.

The heat of the exhaust gases after the turbocharger is: m_{gas} (kg/sec.) × q.spec.gas (kJ.kg.K) × t_{gas} (°C).

Therefore 205.8 × 1.04 × 245 = 52,437 kW.

The heat in the combustion air to the engine is: 205.8 × 1.0 × 25 = 5145 kW.

Therefore, the net loss with the exhaust gases is: 53,406 – 5145 = **48,261 kW**.

Summary:

supplied by fuel	162,422 kW
heat of the combustion air	5,145 kW
total supplied to the engine	**167,567 kW**

exhaust-gas loss	48,261 kW	28.2 %
absorbed by lubricating oil	6,770 kW	4.0 %
absorbed by cylinder-cooling water	11,600 kW	6.9 %
air cooler	19,416 kW	12.6 %
radiation heat (1.1% of P_e)	880 kW	0.5 %
effective power	80,080 kW	47.8 %
residual loss	560 kW	0.03%
total	**167,567 kW**	**100.0%**

▶ Ship propulsion by diesel engines has been the most effective propulsion method for over a hundred years.

4.12 Efficiencies of diesel-engine driven power plants

4.12.1 Ship propulsion

Most ships are propelled by propellers. The propulsion efficiency of a propeller is not 100%, as a propeller 'slips' in water. For this reason the actual shaft power made available for ship propulsion amounts to 60 to 70% of the power supplied to the propeller shaft by the diesel engine. Therefore, the power available for ship propulsion is only part of the power supplied to the engine in the form of fuel.

$$P_{propeller} = P_{fuel} \times \eta_{engine} \times \eta_{propeller}$$

Let's assume that the total efficiency of the engine is 45% and propeller efficiency 60%, then the propeller output is a mere 27% (0.60 times 0.45) of the fuel-power output.

Where are the losses?

1 Heat losses in the engine; 55% of P_{fuel}.
2 Propeller slippage; 40% of the shaft power delivered by the engine ($P_{eff.}$).

See also chapter 19, *Ship propulsion* and Chapter 31, *Propellers.*

4.12.2 Gensets for ships and diesel-power plants

Apart from the engine losses, in diesel-engine power plants or in ordinary diesel gensets, there are generator losses. In general these are very small. The efficiencies of generators vary from 95 to 98% (100 to 10,000 kW), which means:

$$P_{elec.\ generator} = P_{fuel} \times \eta_{engine} \times \eta_{gen.}$$

Let us assume that the total efficiency of a low-speed crosshead engine in a diesel-power plant is 48% and the generator efficiency 96%. Then the generated electric efficiency from the supplied fuel is:

0.48 × 0.96 = 0.46 or 46%.

Clearly this is much higher than the ship propulsion efficiency!

▲
Ships for container cargo.

The power required for power generation and propulsion is virtually always supplied by diesel engines.

Where are the losses?

1 The total engine losses; 52% of the fuel power output.
2 The copper losses in the generator; these are heat losses, a mere 4% of the diesel engine's shaft power. The losses due to the friction of the bearings and the power required for driving the generator's scavenging-air ventilator are minor.

▲
A modern genset with a high-speed four-stroke V-diesel engine, category II.

4.13 More complex ship propulsion

4.13.1 Diesel-electric propulsion

▼
A diagram of diesel-electric propulsion with two rotary-driven electric motors driving the propeller via the gear box.

Diesel-electric propulsion is a technology that was developed many years ago.

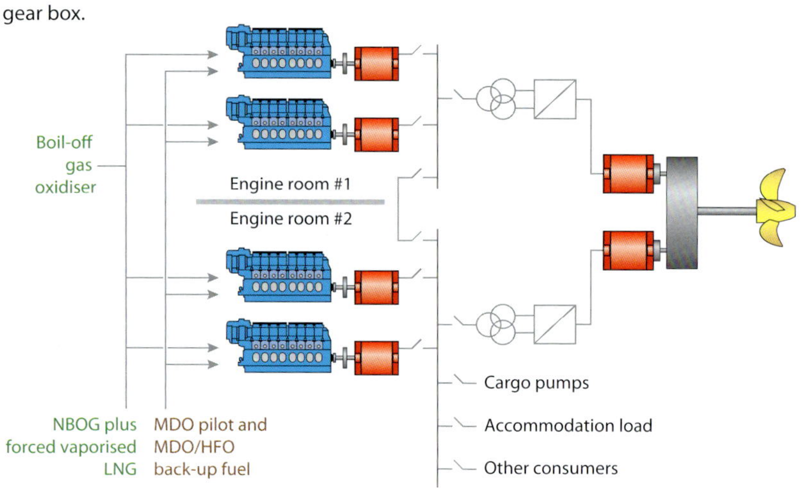

This technology uses diesel-generator sets to provide electric power to a distributing panel (rail). The propeller or propellers are therefore driven by an electromotor. These generators often generate a medium voltage of, for instance, 6.6 kilovolt, which is why the copper diameter or the cable diameter between the various components is acceptable.
The total propulsion efficiency from the supplied fuel is :

$$\eta_{propulsion} = \eta_{diesel\ engine} \times \eta_{generator} \times \eta_{electromotor} \times \eta_{propeller}$$

Example

$\eta_{propulsion} = 0.45 \times 0.95 \times 0.95 \times 0.65$
$\boldsymbol{\eta_{propulsion}} = \mathbf{0.2639\ or\ 26.4\%}$

4.13.2 Electric propulsion by means of 'PODS' or electric blade propellers

In this case the generator is driven by diesel engines or diesel engines and gas turbines, occasionally these include a steam turbine.
The generated electric power is then made available to frequency-controlled and reversible rotary-current motors, which drive the 'PODS', the fixed propeller or propellers, which are attached to the keel of the ship. These are for example used on the latest passenger ships **'The Jewel of Norway'** and **'The Queen Mary II'**.

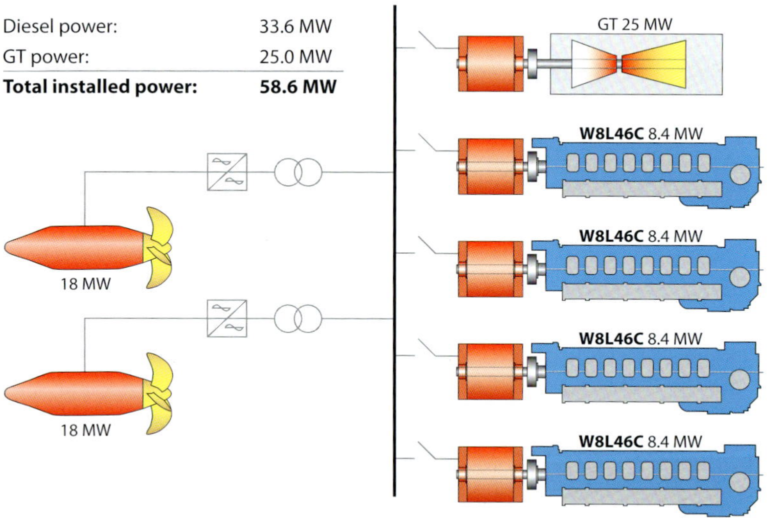

▲
Passenger liner 'The Queen Mary' is powered by two 360° rotatable PODS, each with 18 MW output.
PODS – podded propeller.

The required electric power is generated by four medium-speed Wärtsilä 8L46C HFO. in-line engines of 8.4 MW each.
Gas-turbine genset with a 25 MW capacity using MDO fuel can also be used for the ship's electricity network.

▼
Passenger liner 'The Queen Mary II' is a good example of a diesel-/gas-turbine combination used for electric propulsion.

Total output 117 MW with four Wärtsilä 46V-engines and one gas-turbine. Propulsion by two fixed PODS and two azimuth thrusters.

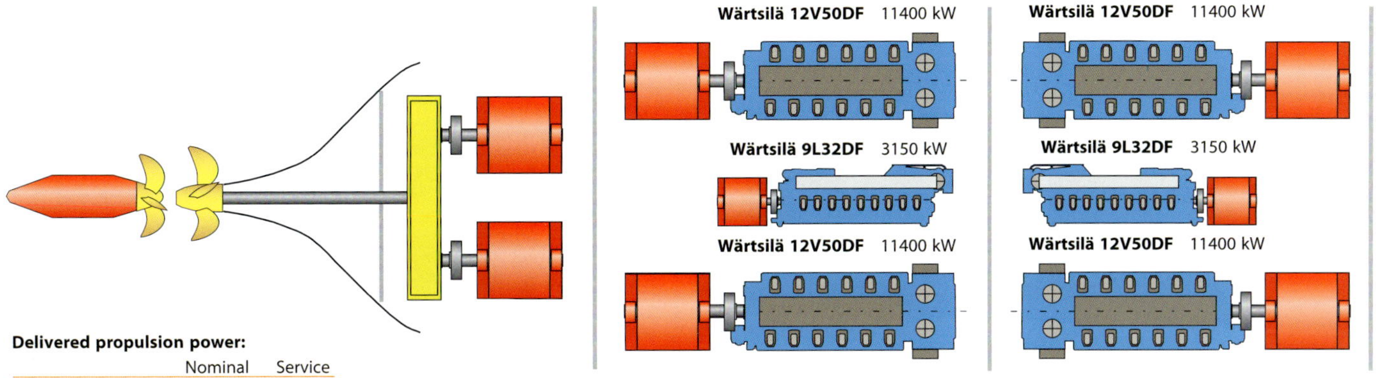

Delivered propulsion power:

	Nominal	Service
Electric pod	17 MW 40 %	17 MW 41 %
LIPS CP propeller	25 MW 65 %	24 MW 62 %
Total shaft power	**42 MW**	**41 MW**

Installed engine power:

Mechanical transmission	0 MW
Electrical power generation	51.9 MW
Total installed power	**51.9 MW**

▲
Diesel-electric propulsion with four large and two smaller gensets for a controllable-pitch propeller and a contra propeller.

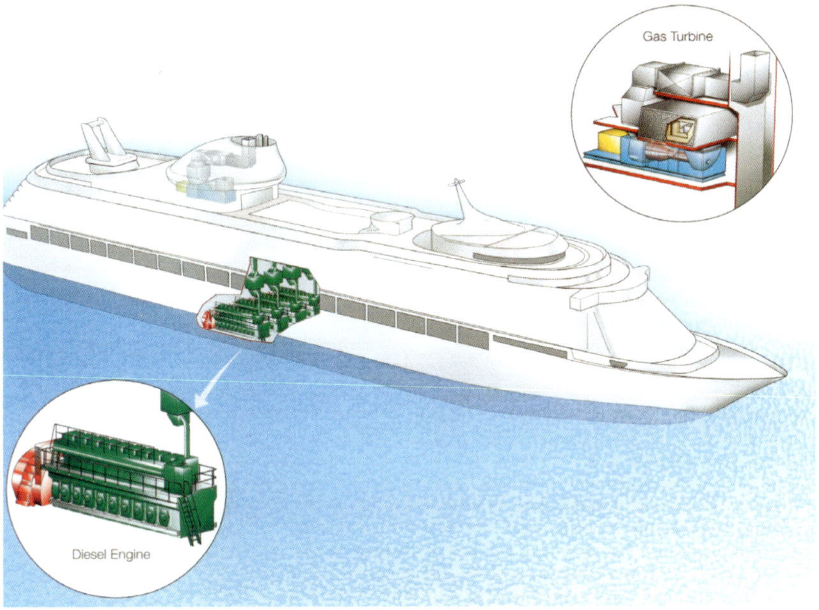

◄
A diesel-engine and gas-turbine combination used for electric propulsion.

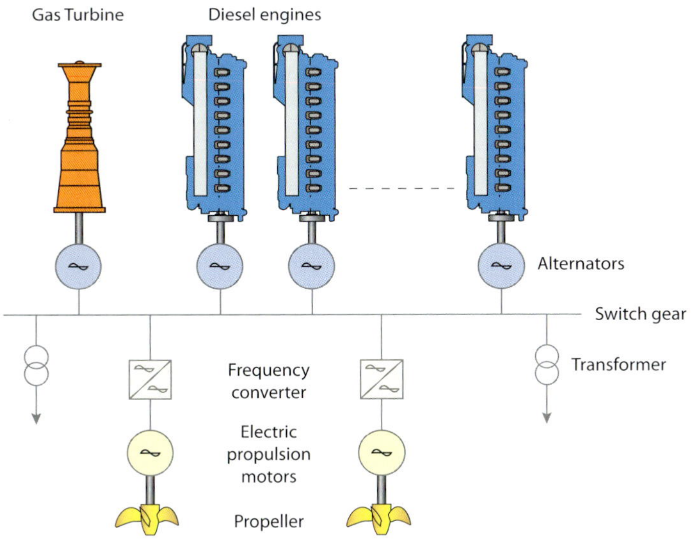

◄
A propulsion system with three diesel engines and one gas turbine with two electrically driven propellers.

The gas-turbine genset has been mounted on the upper deck, so it can be replaced as a whole by an overhauled set during major repairs. The elevated position does not pose a problem for the ship's stability, since the gas-turbines' weight, unlike the diesel-engines', is low in relation to the generated power output.

4.13.3 Propulsion by diesel-engine driven water jets

This allows for several types of propulsion.
A fast-running four-stroke diesel engine drives a waterpump that provides the thrust for the ship's propulsion via a nozzle. This nozzle is adjustable and rotatable. It is often used for light ships such as fast passenger liners, catamarans and yachts.

For additional information on propellers, see Chapter 19, *Ship propulsion* and Chapter 31, *Propellers.*

▲
A fast motor yacht with propeller propulsion.

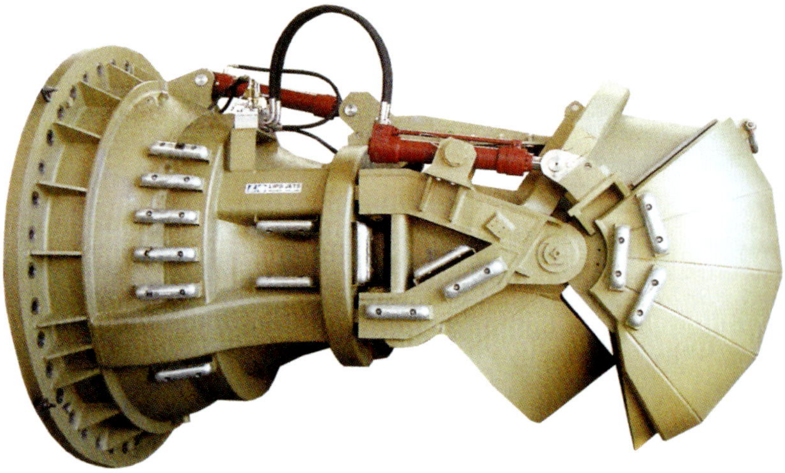

►
A water jet with stream controls.

4.14 Propulsion of water pumps, dredging pumps, crude-oil pumps or compressor drives

The most common drives are used in dredging pumps and other machines on large dredgers. They can be diesel-electrical driven and have a diesel-engine driven pump. Since the dredging-pump output varies significantly, the diesel-engine load may change markedly and is often quite low.

►
On dredgers, diesel engines are not only used for propulsion, but also for driving dredging pumps and water-jet systems for unloading cargo.

▲
The propulsion engine on a dredger in the foreground and back right; the gear box for driving the propeller and the generator.

1 *engine*
2 *gear box*
3 *propeller shaft*
4 *generator*

▲
The other side of the same propulsion engine with a manually controlled gear box for driving the dredging pump on the other side of the engine-room bulkhead, the pump room.

1 *engine*
2 *manually controlled gear box for on/off/two-speed*
3 *engine-room bulkhead*

◀
On this suction-cutter dredger, two eight-cylinder Wärtsilä 26 diesel engines drive two dredge pumps via a switchable gearbox (in the background) through the pump chamber wall.

In the foreground; both adjustable propellers and two shaft generators are also driven via a gearbox. Fuel MDO. On top of both gearboxes, the lubrication system with pumps and coolers.

▶

Diesel process efficiency.

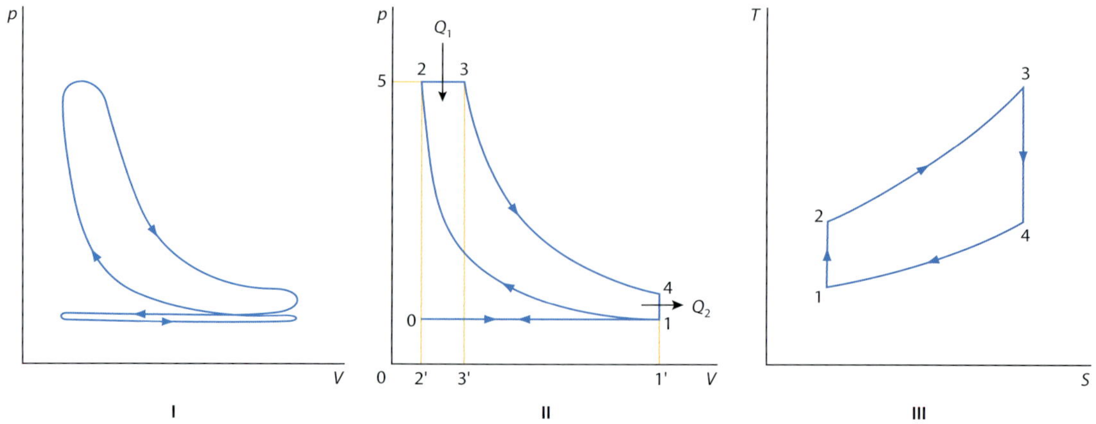

The left diagram represents the standard *pV*-diagram for intake, compression, power and exhaust.

The middle/centre diagram II represents a *pV*-diagram of the same process: with adiabatic compression, an isobar for heat supply, isentropic expansion (power) and an isochoric discharge of heat (exhaust).

The right diagram III represents a *TS*-diagram for the same process: isentropic compression, isobaric expansion, isentropic expansion and an isochoric exhaust.

4.15 Theoretical background of the processes in an internal combustion engine - the diesel engine

4.15.1 Diesel-process efficiency

Figure I:
The standard indicator process of a diesel engine consists of inlet, compression, power and outlet.

Figure II:
The standard diesel process can be approximated by drawing a cycle consisting of 2 isentropes, no heat exchange with the environment, an isobar, a line of constant pressure and an isochore, a line of constant volume.

1 – 2 compression, an isentrope.
2 – 3 expansion, an isobar.
3 – 4 expansion, an isentrope.
4 – 1 exhaust, an isochore.

Note:
Fuel injection occurs between 2 - 3 during which the pressure still remains more or less constant. Subsequently, isentropic expansion takes place between 3 - 4.

Figure III:
Here the process is shown again on a Temperature–Entropy, or *TS*-graph.

1 – 2 compression, an isentrope (isotropic line).
2 – 3 expansion, an isobar (isobaric line).
3 – 4 expansion, an isentrope (isotropic line).
4 – 1 exhaust, an isochore (isochoric line).

The above graph should in fact be an 'open' rather than closed graph as air is continually drawn in and fuel continuously injected. Therefore line 4 - 1 is often depicted as a dotted line.

1 – 2 compression: energy consuming, compression work.
2 – 3 expansion: energy supply (fuel) and energy supply to the piston.
3 – 4 expansion: energy supply to the piston.
4 – 1 exhaust: energy discharge to ambient air.

Diesel-engine process efficiency

Here we assume that the cycle is reversible and that we are dealing with an ideal gas.

Cycle efficiency:

$$\eta_{\text{cycle}} = \frac{A}{Q_1} = \frac{Q_1 - Q_2}{Q_1} = 1 - \frac{Q_2}{Q_1}$$

A = work done
Q_1 = energy supply
Q_2 = energy discharge

$Q_1 = c_p \times (T_3 - T_2)$ en $Q_2 = c_V \times (T_4 - T_1)$

$$\eta_{\text{cycle}} = 1 - \frac{c_V \times (T_4 - T_1)}{c_p \times (T_3 - T_2)} = 1 - \frac{1}{k} \times \frac{(T_4 - T_1)}{(T_3 - T_2)} \qquad (1)$$

According to the law of Boyle-Gay Lussac, the following applies to ideal gases:

$$T_4 = \frac{p_4 \times V_4}{p_1 \times V_1} \times T_1 = \frac{p_4}{p_1} \times T_1 \text{ and}$$

$$T_3 = \frac{p_3 \times V_3}{p_2 \times V_2} \times T_2 = \frac{V_3}{V_2} \times T_2$$

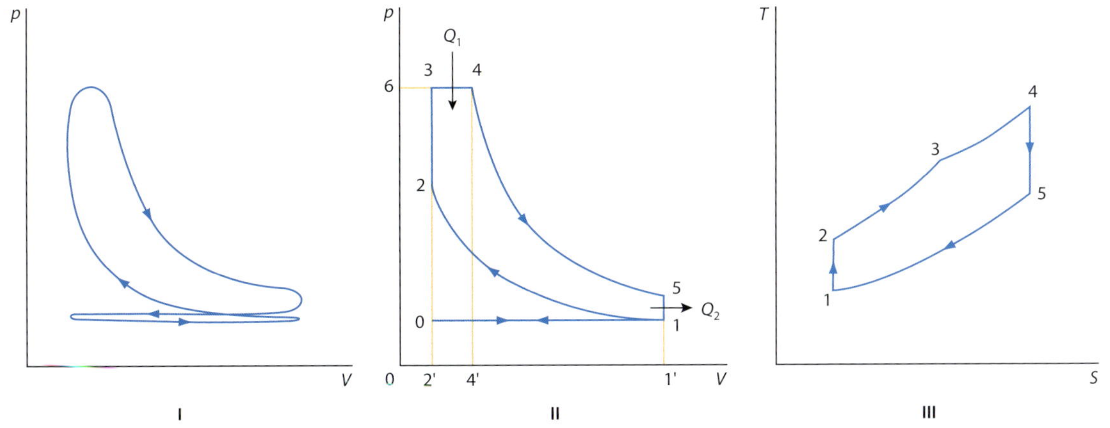

The left diagram I represents the standard pV-diagram for intake, compression, power and exhaust.

The middle diagram II represents a pV-diagram for the same process: with adiabatic compression, an isochore and isobar for heat supply, an adiabatic expansion and an isochoric heat discharge.

The right diagram III represents a TS-diagram with adiabates 1-2 and 4-5, isochores 2-3 and 5-1 and an isobar 3-4.

◄ The semi-diesel of the Otto-Diesel standard cycle.

Filling in the formula gives:

$$\eta_{cycle} = 1 - \frac{1}{k} \times \frac{T_1}{T_2} \times \frac{\frac{p_4}{p_1} - 1}{\frac{V_3}{V_2} - 1} \qquad (2)$$

As lines 3 – 4 and 1 – 2 are adiabatic, the law of Poisson states the following:

$$p_4 = \left(\frac{V_3}{V_4}\right)^k \times p_3 \text{ and } p_4 = \left(\frac{V_3}{V_4}\right)^k \times p_3$$

As $p_3 = p_2$ and $V_1 = V_4$ it follows that:

$$\frac{p_4}{p_1} = \left(\frac{V_3}{V_2}\right)^k$$

Integrating this into (2) gives:

$$\eta_{cycle} = 1 - \frac{1}{k} \times \frac{T_1}{T_2} \times \frac{\left(\frac{V_3}{V_2}\right)^k - 1}{\frac{V_3}{V_2} - 1} \qquad (3)$$

According to Poisson:

$$\frac{T_1}{T_2} = \left(\frac{V_2}{V_1}\right)^{k-1}$$

Integrating this into (3) gives:

$$\eta_{cycle} = 1 - \frac{1}{k} \times \left(\frac{V_2}{V_1}\right)^{k-1} \times \frac{\left(\frac{V_3}{V_2}\right)^k - 1}{\frac{V_3}{V_2} - 1}$$

Assuming that:

$\frac{V_1}{V_2} = \varepsilon$ the compression ratio and $\frac{V_3}{V_2} = \nu$ the turbocharger ratio, the cycle efficiency of a typical diesel engine is:

$$\eta_{cycle} = 1 - \frac{1}{k} \times \frac{1}{\varepsilon^k - 1} \times \frac{\nu^k - 1}{\nu - 1}$$

Conclusion:

In order to achieve high efficiency:

1 The compression ratio should be high.
2 The turbocharger ratio should be low.

Work done for each cycle:
$A = Q_1 - Q_2 = c_p\,(T_3 - T_2) - c_V\,(T_4 - T_1)$ kJ/kg

4.15.2 The semi-diesel or standard Otto-Diesel cycle

Obviously, the pressure in today's diesel engines does not remain constant throughout the entire combustion process.
Fuel injection is started towards the end of the compression stroke and also after TDC at the beginning of the power- or expansion stroke, fuel continues to be injected. Together this is the heat supply, Q_1.

Next we will make a calculation based on the cycle in figures 2 and 3.

This consists of:

1 – 2 adiabatic compression.

2 – 3 isochore:beginning of the heat supply with the fuel, section Q_1.

3 – 4, isobar: residual heat supply from the fuel, remainder Q_2.

4 – 5, adiabatic expansion.

5 – 1, isochore: heat supply from exhaust gases Q_2.

This process is referred to as the semi-diesel- or the combined Otto-Diesel process.

Efficiency of the standard semi-diesel cycle

Again we assume a reversible cycle and an ideal gas.

$$\eta_{\text{cycle}} = \frac{A}{Q_1} = \frac{Q_1 - Q_2}{Q_1} = 1 - \frac{Q_2}{Q_1}$$

$$Q_1 = c_V \times (T_3 - T_2) + c_p\,(T_4 - T_1)$$

and

$$Q_2 = c_V \times (T_5 - T_1)$$

Substituting this gives:

$$\eta_{\text{cycle}} = 1 - \frac{(T_5 - T_1)}{(T_3 - T_2) + k \times (T_4 - T_3)}$$

Poisson's law for adiabatics reads:

$\frac{V_1}{V_2} = \varepsilon$ the compression ratio, $\frac{V_4}{V_3} = v$ the turbocharger ratio and $\frac{T_3 = p_3}{T_3 = p_3} = d$ the pressure ratio before and after combustion.

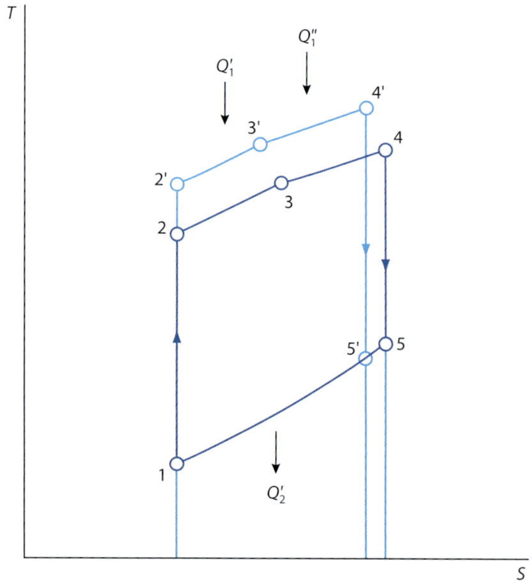

Filling in the formula gives:

$$\eta_{\text{cycle}} = 1 - \frac{1}{\varepsilon^{k-1}} \times \frac{d \times v^k - 1}{(d - 1) + k \times d(v - 1)}$$

Here the same applies:

For optimum engine efficiency:

1 the compression ratio ε should be high;

2 the turbocharger ratio v should be small;

3 the pressure ratio d should have the highest possible value.

Work done per cycle:

$$A = Q_1 - Q_2$$

$$A = c_V \times (T_3 - T_2) + c_p \times (T_4 - T_3) - c_V \times (T_5 - T_1) \text{ kJ/kg}$$

4.15.3 How to increase diesel-engine efficiency

We now know that in order to achieve maximum efficiency:

1 the compression ratio ε should have the highest possible value.

2 the volume ratio v during heat supply should have the lowest possible value.

3 the ratio between the maximum pressure and compression pressure d should have the highest possible value.

The effects of the compression ratio

◀

The effects of the compression ratio.
A *TS*-diagram ('Seiliger-process').

The surface 1-2-3-4-5, clockwise cycle.
Heat addition occurs between 2 and 3 at a constant volume Q_1'.
Heat supply occurs between 3 and 4 at a constant pressure Q_1''.
The surfaces below both lines represent heat. The heat is discharged between 5 and 1 (exhaust gases) Q_2.
If the compression ratio ε is increased, points 2, 3 and 4 move upward, as the increased compression ratio produces higher temperatures! Hence, the new points 2', 3' and 4'. The heat supply (fuel) remains unchanged. The surfaces below the new points 2', 3' and 4' represent heat.
From 2' to 3': the heat supply remains the same.
As a result 3' has shifted slightly to the top left of 3 in order to maintain equal surfaces in both conditions. The same applies to point 4' in relation to 4. The heat supply Q_2 exhaust gas heat, occurs at a constant volume which is why 5' is directly situated below 5. Therefore, point 5' lies to the right of point 5.
This means that the heat supply is reduced as the surface below line 5'-1 is smaller than that below line 5-1.
The thermal output is consequently higher!

The effects of adjusting the moment of heat supply, fuel injection

▶

In the *TS*-diagram the (Seilinger) process is represented moving clockwise through surface 1-2-3-4.
The effect of adjusting the time of heat supply, or fuel injection.

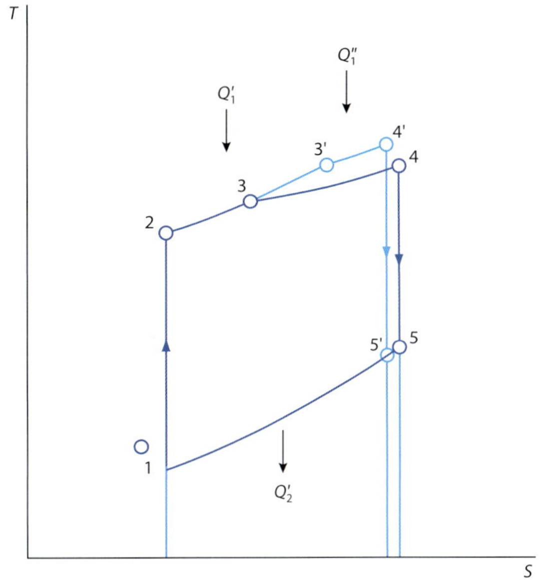

If the heat supply by means of fuel injection is accelerated and occurs at a constant pressure rather than a constant volume, the volume ratio *v* is reduced and the thermal efficiency increased. Point 3 moves along the constant volume line to point 3' and point 4 shifts to 4', vertically above and left of point 4. That is: the pressure is higher and the volume smaller.
The total heat supply (the fuel injected) remains constant. The new surface directly below the line 2-3'-4' must be equal to the surface directly below the 2-3-4 line.
Here point 4' should be situated left of point 4. Directly below 5' can be found on the 5-1 line (heat discharge, exhaust gases).
As a consequence, the maximum pressure and temperature rise.
This is a typical example of how to sustain (temperature and pressure) a proper combustion process at a reduced power output by accelerating the fuel injection. In this instance the heat discharge is reduced!

The effect of turbo charging in relation to natural aspiration on diesel-engine efficiency

▶

Once again the clockwise cycle for natural aspiration in the *TS*-diagram is represented by points 1-2-3-4-5.
The effects of turbo charging as opposed to natural aspiration on engine efficiency.

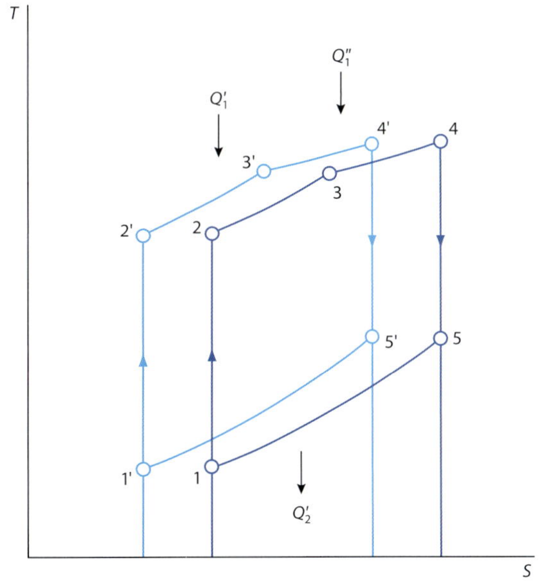

Air is compressed and cooled to equal the temperature of the ambient air drawn in by natural aspiration.
So when using turbo charging with an initial compression temperature identical to that of natural aspiration point 1 becomes point 1' and point 2 becomes point 2'. Here pressure increases during compression.
This means that if the compression ratio of the diesel engine has remained constant, the final compression temperature also remains the same, so point 2 shifts horizontally to point 2'.
Seeing that the heat supply by fuel injection is not altered, point 3 and 4 move horizontally to the left.
This means that the shape of the diagram itself has not changed, it has simply moved to the far left in the TS-diagram.
Here the thermal efficiency remains unchanged provided that:
the compression ratio ε, the pressure ratio δ and the volume ratio *v* remain the same.

>CH 5

Standard figures of various types of diesel engines

The compression ratio, efficiency and the maximum pressure.

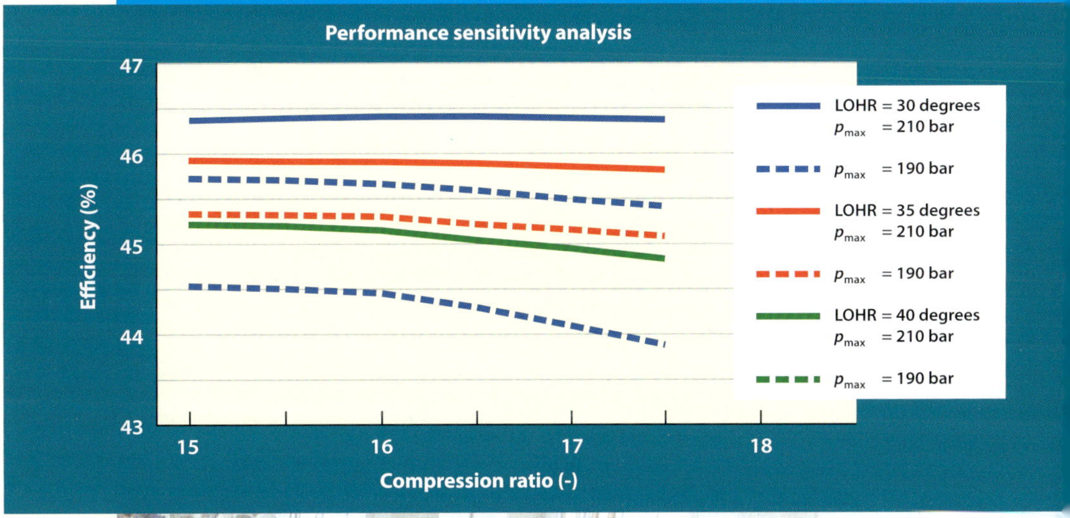

In order to have a firm grasp of the principles of diesel engines, it is essential to have a basic knowledge of the operating systems and their corresponding parameters.
Shown in the inset, an excerpt from the MTU training centre for diesel engines in Friedrichshafen, Germany.

5.1 Mean effective pressure

$p_{\text{mean effective pressure}}$; expressed in bar or megapascal.

This is the pressure that mathematically is effective throughout the entire process and is available to the crankshaft. It is, of course a fictional pressure as in reality for example in the four-stroke cycle, the pressure fluctuates considerably from the intake pressure to the compression pressure, then to maximum combustion pressure and subsequently to the exhaust pressure.
The magnitude of the mean effective pressure is important when considering the diesel engine's performance and has shown a slow and gradual development over the past one hundred years. Often new inventions such as the turbocharger, lubrication of piston rings under pressure and the anti-polishing ring have resulted in an increase of the mean effective pressure.

An improved turbocharger, that is an increased air supply with a higher charging pressure, allows more kilograms of air to flow into the cylinders during the very brief moment that the inlet valves are opened. More air means extra fuel injection, thus increasing the combustion pressure. An augmentation of the compression ratio has a similar effect. The compression ratio is the ratio of the volume when the piston is in its bottom position to the volume when it is in its top position.

		Mk	mep bar	Cm m/s
1981	L35MC introduced			
1982	Full L-MC programme	1	15.0	7.2
1984	L-MC upgraded	2	16.2	
1985	L42MC introduced	2	16.2	7.2
1986	K-MC introduced		16.2	
	S-MC introduced		17.0	
	L-MC upgraded	3	16.2	7.6
1987	S26MC introduced		16.8	8.2
1988	K-MC-C introduced		16.2	8.0
1991	MC programme upgraded			8.0
	K and L-MC	5	18.0	
	S-MC	6	18.0	
1992	S26MC and L35MC upgraded		18,5	8.2
1993	S35MC and S90MC introduced			
	K90MC/MC-C upgraded	6	18.0	8.0
1994	S42MC introduced	6	18.5	8.0
1994	K98MC-C introduced	6	18.2	8.3
1995	K80MC-C upgraded		18.0	8.0
1996	L70MC upgraded	6	18.0	8.2
1996	S70MC-C, S60MC-C, S50MC-C		19.0	8.5
	and S46MC-C introduced		19.0	8.3
1996	S80MC upgraded		19.0	8.0
1997	L80MC upgraded		18.0	8.0
	K98MC introduced		18.2	8.3
1998	S80MC-C, S90MC-C and			
	L90MC-C introduced		19.0	8.1
	S35MC upgraded	7	19.1	8.1
1999	S42MC upgraded	7	19.5	8.0
2001	L70MC-C introduced		19.0	8.5
2001	L60MC-C introduced		19.0	8.3

mep = mean effectieve pressure   Cm = mean piston speed

▲

The development of the mean effective pressure over twenty years. Shown here a two-stroke crosshead engine by MAN Diesel.

An increase of the mean effective pressure by 3.5 bar and the mean piston speed by a metre per second.

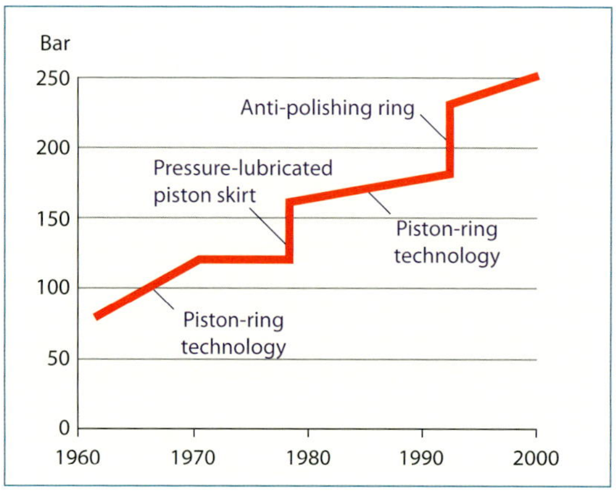

▲

The development of maximum combustion pressure during the last forty years.

Shown here medium-speed four-stroke diesel engines running on HFO, by Wärtsilä. At this time, the piston-ring packets were improved. The pressure-lubricated piston skirt was introduced and the greatest improvement, the anti-polishing ring.

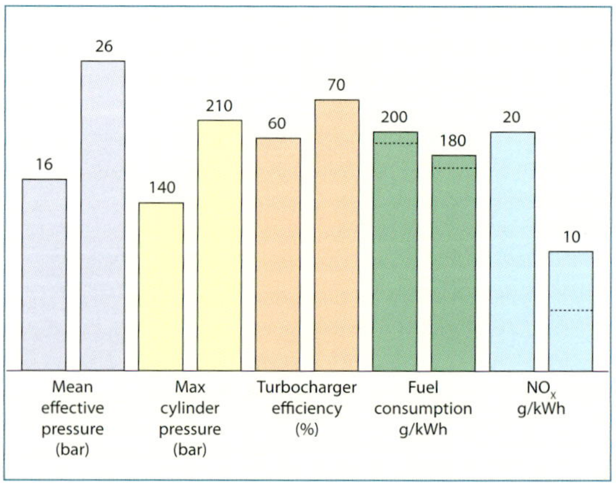

▲

When the compression rate is increased, both the effective mean pressure and the maximum combustion pressure increase. The specific fuel consumption decreases, and the efficiency increases.

Note: Future diesel engines with a compression rate of 18 and a maximum combustion pressure of 250 bar will have an efficiency increase of 2.5%. Approximately 1% higher than that of current engines in 2011.

A higher effective mean pressure means that engines of a certain stroke volume can produce more power. The so-called power density increases. Due to severe competition, various engine manufacturers try to achieve increasingly more power per engine weight and volume. For a certain engine type, this can be achieved by increasing the mean effective pressure and the number of revolutions.

The mean effective pressure for two-stroke crosshead diesel engines is slowly but surely moving towards 20 bar. Most types achieve up to 18.5-20 bar (2012).

The mean effective pressures for four-stroke medium-speed diesel engines have reached circa 26 bar. Only a few very heavily loaded diesel engines achieve 32 bar (2012).

For four-stroke high-speed diesel engines these pressures lie somewhat lower but still easily exceed the 20 bar limit. Very heavily loaded engines can register pressures of over 30 bar.

The future: as far as future development in flow (in m³/sec) and suction lift (in bar or metres water column) of turbochargers is concerned, much depends on the development of turbocharger efficiency.

5.2 Mean piston speed

$C_{P\ mean}$; expressed in metres per second.

When accelerating the mean piston speed there are two important factors:

1. **The magnitude of the accelerating and decelerating forces that affect the up- and downward moving piston.**
2. **The degree of lubrication of the piston rings and cylinder liner.**

The magnitude of the accelerating and decelerating forces that affect the up- and downward moving piston.
When increasing the revolutions of an engine with the same piston stroke, the accelerating and decelerating forces develop steadily as the piston mass must accelerate and decelerate in a shorter period. In TDC and BDC the piston comes to a brief stop.

The forces on the piston increase according to the formula $F = m \times a$, in which:

F = the forces exerted on the piston in Newton
m = piston mass in kg
a = acceleration/deceleration in m/sec.²

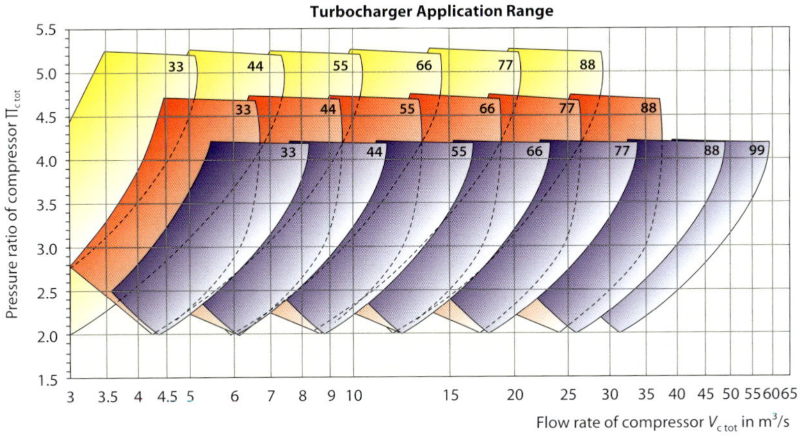

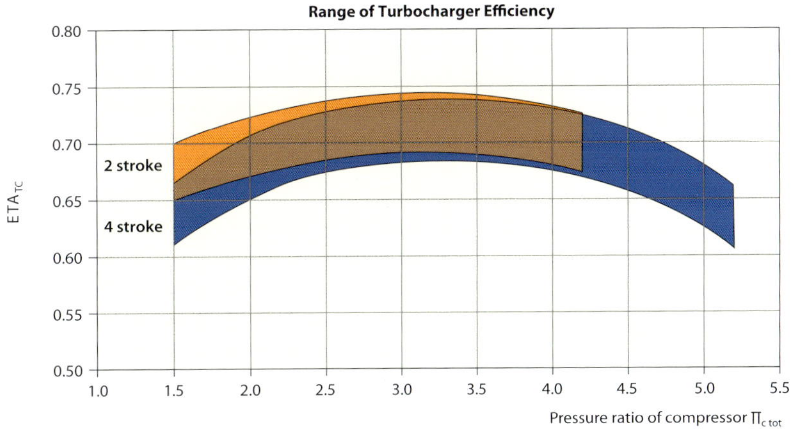

Supercharged engine output [kW]

Type/size	2-stroke engines (l_{spec} = 8.5*)	4-stroke engines (l_{spec} = 6.5*)
TCA 33	2,800 – 4,300	
TCA 44	4,100 – 6,200	
TCA 55	4,400 – 8,200	5,800 – 10,800
TCA 66	6,200 – 11,600	8,200 – 15,200
TCA 77	8,800 – 16,400	11,500 – 21,500
TCA 88	12,400 – 23,300	16,300 – 30,300
TCA 99	18,700 – 31,300	

* specific air consumption in kg/kWh

▲
Increasing the compression rate and the total efficiency of the turbochargers is decisive in improving the total efficiency of the diesel engine.

An increase in the number of kilograms of air supplied over a certain period means that more fuel can be combusted, and therefore more power can be generated.

Top figure
Horizontal: the capacity in m³/s.
Vertical: the compression rate.
In the figure: the different types of compressors.

Middle figure
Horizontal: the compression rate.
Vertical: the efficiency.
In the figure: yellow and brown for two-stroke engines, blue for four-stroke engines.

Bottom figure
Left: types of turbochargers.
Middle and right: power of two-stroke (middle) and four-stroke (right) diesel engines.
Specific air usage two-stroke l_{spec} = 8.5 kg/kWh
four-stroke l_{spec} = 6.5 kg/kWh

In practice one tries to make the piston mass as light as possible. Therefore, low-weight light-metal pistons are used for high-speed diesel engines. These diesel engines use Marine Diesel Oil (MDO) and exert lower pressures on the piston, which can then withstand the compression pressures produced in the process.

▶
Low mass, especially of the piston is important in reducing the accelerating and decelerating forces.

Light metal pistons of a high-speed four-stroke diesel engine.

Medium-speed diesel engines often use Heavy Fuel Oil (HFO). The pressures and the temperatures are higher and light-metal pistons are therefore not suitable.
For these types, a piston is used with a cast steel piston crown that can withstand the higher load forces. The piston skirt is manufactured of a light metal or cast iron.
Since the latter is much heavier, it is often milled in places where this is possible.

▶
The piston mass in medium-speed engines is reduced by keeping the skirt a light as possible.

The material surrounding the piston pin is removed where possible (1).

Degree of lubrication of piston rings and cylinder liner

Proper lubrication is of the utmost importance in maintaining a good spring-pack seal, situated between the piston and the cylinder liner, for a considerable time (for instance 15,000 operating hours). The functions of lubricating oil are among others, the reduction of friction, assisting the discharge of combustion by-products and the discharge of heat that cause problems when a certain piston speed is exceeded.
Today an average piston speed of 8 to 11 m/sec. is standard for four-stroke engines. Anything beyond this poses a potential problem. There are diesel engines available that have a mean piston speed of 12 m/sec. (2012).
Two-stroke crosshead engines usually have a slightly lower speed. This is related to the time required to scavenge with fresh air and the interruptions in the cylinder liner because of the surrounding inlet port: all of which interfere with lubrication.
Here mean piston speeds average between 6 and 8.5 m/sec.

5.3 Load parameters

The product of the mean effective pressure and the mean piston speed is a measurement of diesel engine load. Indicated by L_P – the load parameter of the diesel engine.

$$L_P = p_{\text{eff mean}} \times C_{\text{P mean}} \text{ (bar/m/sec.)}$$

The product of this equation gives a good indication of the degree of engine load.

Examples

Diesel engine I $p_{\text{eff mean}} = 15$ bar
$C_{\text{P mean}} = 8$ m/sec.
$L_P = 15 \times 8 = 120$ bar/m/sec.
This is low. Probably from an early diesel engine.

Diesel engine II $p_{\text{eff mean}} = 25$ bar
$C_{\text{P mean}} = 10$ m/sec.
$L_P = 25 \times 10 = 250$ bar/m/sec.
This is high. An ultra-modern diesel engine.

Diesel engine III $p_{\text{eff mean}} = 27$ bar
$C_{\text{P mean}} = 11$ m/sec.
$L_P = 27 \text{ x } 11 = 297$ bar/m/sec.

This is extremely high and is only achievable in few finely tuned engines. The thermal and mechanical load in this diesel engine is also very high.

Rule of thumb

L_p	=	100 – 150	**low load, earlier models**
L_p	=	150 – 250	**normal load, contemporary engines**
L_p	=	250 – 300	**heavy load, most modern engines**

5.4 Compression ratio

Also known as compression rate. This is the ratio of the cylinder volume above the piston at the beginning of the compression stroke (so the piston is at BDC) to the cylinder volume above the piston at the end of the compression stroke (so the piston is at TDC).
The compression ratio is indicated by ε = epsilon.

So, the theoretical compression ratio is:

$$\varepsilon = \frac{V_s + V_c}{V_c}$$

In which:
V_s = Volume moved by the piston, the swept volume.
V_c = Volume above the piston in TDC or volume compression space. This is the combustion space above the piston in TDC.

In reality, compression only starts when, for a four-stroke engine the inlet valves and for a two-stroke engine the exhaust ports or valves, are closed.
This is called the effective compression ratio. The distance which the piston has travelled from BDC prior to the beginning of the effective compression is denoted by *k*.
The volume above the piston is then:

$$V_{eff} = (1 - k) \times V_s + V_c.$$

The effective compression ratio is indicated by $\varepsilon_{effective}$.

The formula:

$$\varepsilon_{effective} = \frac{(1 - k)\, V_s + V_c}{V_c}$$

The effective compression ratio varies from approximately 6 to 20. In Otto-engines where the compressed mixture is not self-igniting the compression ratio varies from 6 to 15, dependent on the air volume of the mixture. The greater the volume the higher the compression ratio.
For diesel engines used in industrial applications, the compression ratio ranges from 10 to 20. Most diesel engines have ratios of between 12 and 16.

This is always the theoretical compression ratio, because the effective (practical) compression ratio is difficult to determine. Therefore, the effective compression ratio is always slightly lower than the theoretical compression ratio.

Significance of the compression ratio

The compression ratio is of great importance for the final compression pressure and the final compression temperature in the engine and is therefore key in the working of the process and consequently that of the engine.
Therefore, dependent on the inlet-air pressure and the inlet-air temperature the compression ratio determines the pressure and the temperature at the end of the compression stroke.
In order to control this somewhat, the compression ratio is often changed slightly. In earlier times, a compression plate was fitted between the connecting-rod foot and the top crankpin-bearing cap.

▼
The theoretical compression rate is the ratio of the cylinder volume over the piston in BDC to the cylinder volume over the piston in TDC.

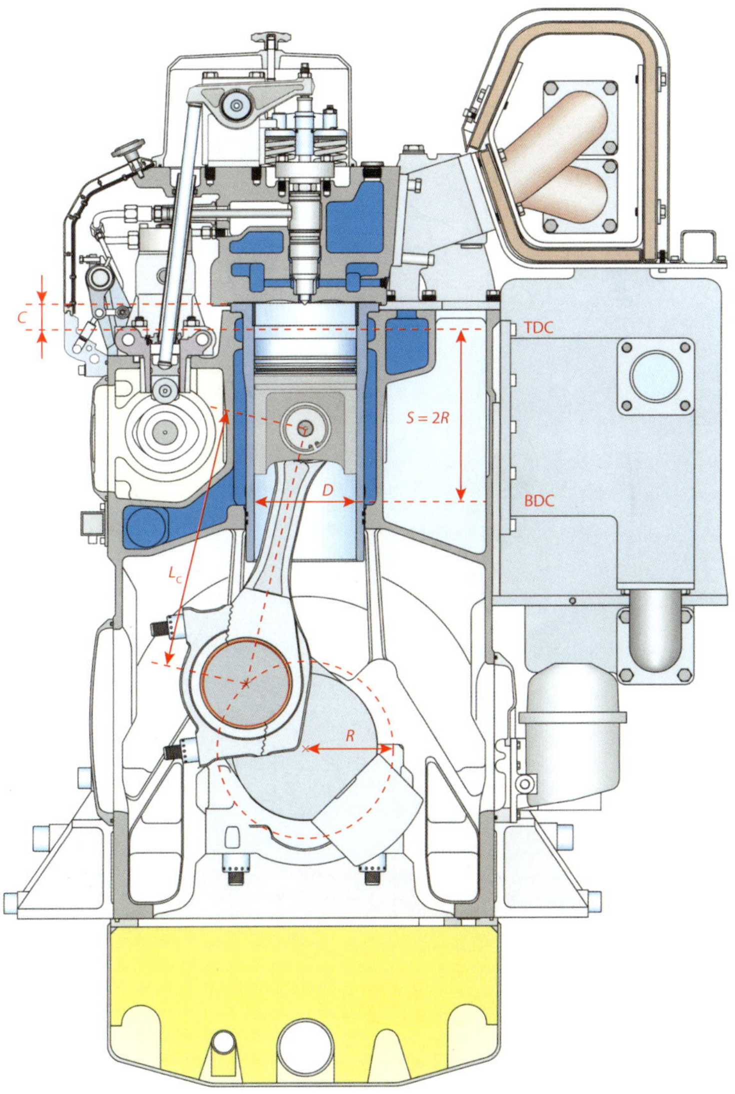

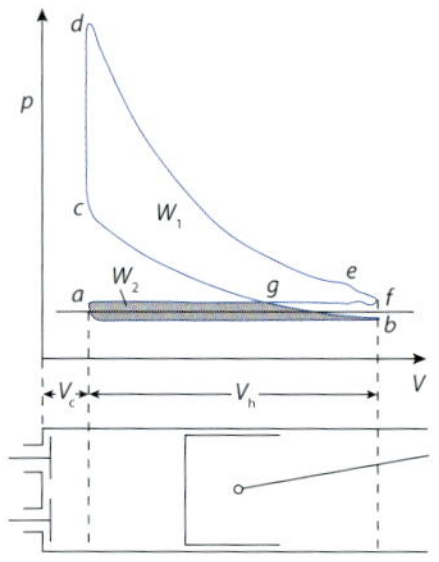

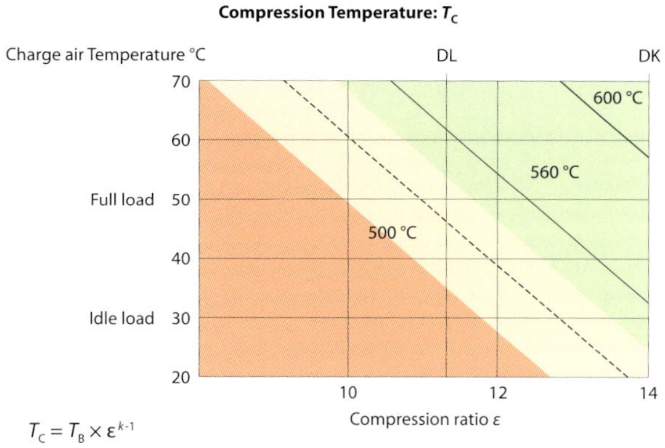

$$T_C = T_B \times \varepsilon^{k-1}$$
$$= T_B \times \left[\frac{V_h + V_c}{V_c}\right]^{k-1}$$

▲

The final compression temperature in relation to the compression rate, the engine load and the air temperature after the inter cooler.

Shown here a Daihatsu medium-speed four-stroke engine Type DL and DK.

In small engines, the compression ratio can be increased by slightly milling out the top of the engine block. Additionally, the piston height above the piston pin may be adjusted by the placement of a 'higher' (higher compression ratio) or a 'lower' (lower compression ratio) piston. Although this does not alter the stroke volume, it does change the final compression space.

Influence of the compression ratio

Theoretically, the thermal efficiency of the combustion cycle of an internal combustion engine is:

$$\eta_{th} = 1 - \left(\frac{1}{\varepsilon}\right)^{k-1}$$

The efficiency is therefore solely dependent on the compression ratio ε and the adiabatic component *k*. The *k* value, for air under standard conditions, is 1.4. Theoretically, inlet temperature, initial pressure and the calorific value of the fuel-air mixture is of no consequence.
At infinitely high compression ratios, the efficiency approximates up to 100%.

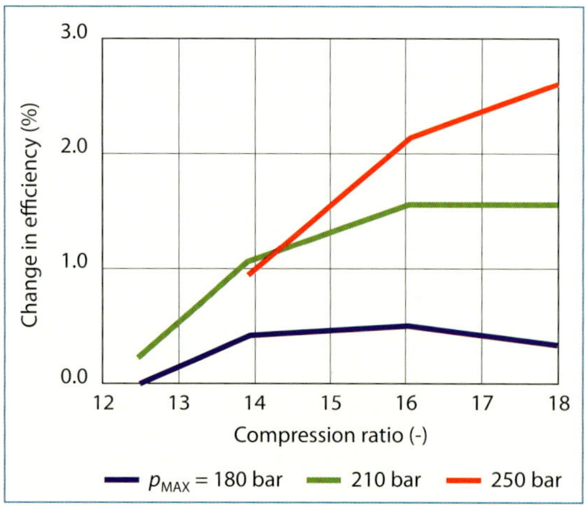

▶

The increase in engine efficiency when both compression rate and combustion pressures are increased.

The example above shows that the theoretical process nowhere reflects the actual process.

In approximation, the total efficiency of diesel engines is as follows:
high-speed runners (small diesel engines, cat. 1):
0 – 100 kW 25 – 30%
high-speed runners (larger diesel engines, cat. 2):
100 – 10,000 kW 30 – 40%
medium-speed runners (large diesel engines, cat. 3):
500 – 30,000 kW 40 – 45%
Slow-speed runners (very large diesel engines, cat. 4):
1500 – 100,000 kW 48 – 53%

Only the low-speed runners are two-stroke engines, all others are four-stroke engines.

Compression ratios very seldom exceed ± 15 to 20, as the top pressures and temperatures would get too high, the mechanical and thermal load on the components would be too high. The friction losses would also rise significantly.
Large medium-speed diesel engines have a compression ratio of 13 to 15.
In smaller diesel engines sometimes 16 to 24. Large low-speed engines average around 11.5 to 12.5.
At high compression ratios, the final compression pressures and temperatures are significantly higher.

5.5 Power density

The term power density has come into being as a result of the continual development of industrial diesel engines and growing competition in the diesel-engine industry.
Smaller engines sizes are required as well as a reduction in engine weight. Furthermore, one coninues to look to increase power output for a specific stroke volume.
This is why there are generally three types of power density, namely:
- the shaft power produced in kW in relation to the weight of the engine;
- the shaft power produced in kW in relation to the total stroke volume of the engine ;
- the shaft power produced in kW in relation to the total volume of the engine.

5.5.1 Shaft power produced in kW in relation to the weight of the engine

Power density per weight =

$$\frac{\text{Total shaft power}}{\text{Total engine weight}}$$

or: kW shaft power produced per kilogram engine weight.

$P_{density}$ **= kW/kg engine weight.**

Examples

Perkins 1103C-33
Propulsion, diesel-fuel oil, three-cylinder in-line engine, four-stroke, natural aspiration, cylinder bore 105 mm, stroke 127 mm, compression ratio 18.2 : 1, stroke volume 3.3 litres, maximum continuous power at 2200 revolutions per minute 41.5 kW.
Weight 249 kg.
Category I
per litre stroke volume 12.6 kW
per kW power 6 kg weight
mean effective pressure 6.9 bar
mean piston speed 7.7 m/sec.

Cummins KT19-M
Propulsion, diesel-fuel oil, six-cylinder in-line engine, four-stroke, turbocharger, cylinder bore 159 mm, stroke 159 mm, compression ratio 15.5 : 1, stroke volume 19 litres, maximum continuous power at 1800 revolutions per minute 317 kW.
Weight 1937 kg.
Category II
per litre stroke volume 16 kW
per kW power 6.1 kg weight
mean effective pressure 11.2 bar
mean piston speed 9.54 m/sec.

Caterpillar MaK M25
Propulsion, heavy fuel oil, six-cylinder in-line engine, four-stroke, turbocharger, cylinder bore 255 mm, stroke 400 mm, stroke volume 122.5 litres, maximum continuous power at 720 revolutions per minute 1800 kW.
Weight 22,000 kg.
Category III
per litre stroke volume 14.7 kW
per kW power 12.2 kg weight
mean effective pressure 23.7 bar
mean piston speed 9.6 m/sec.

Wärtsilä 18V32B2 output
Propulsion, heavy fuel oil, eighteen-cylinder V-engine, four-stroke, turbocharger, cylinder bore 320 mm, stroke 400 mm, stroke volume 578.8 litres, maximum continuous power at 750 revolutions per minute 9000 kW.
Weight 82,500 kg.
Category III
per litre stroke volume 15.55 kW
per kW power 9.2 kg weight
mean effective pressure 24.9 bar
mean piston speed 10.0 m/sec.

MAN Diesel 14K108ME-C
Propulsion, heavy fuel oil, fourteen-cylinder in-line engine, two-stroke, crosshead, turbocharger, cylinder bore 960 mm, stroke 2500 mm, stroke volume 25,320 litres, maximum continuous power at 102 revolutions per minute 80,080 kW.
Weight 2,300,000 kg.
Category IV
per litre stroke volume 3.16 kW
per kW power 28.7 kg weight
mean effective pressure 18.2 bar
mean piston speed 8.3 m/sec.

Wärtsilä 12RTA96C
Propulsion, heavy fuel oil, fourteen-cylinder in-line engine, two-stroke, crosshead, turbocharger, cylinder bore 960 mm, stroke 2500 mm, stroke volume 25,320 litres, maximum continuous power at 102 revolutions per minute 80,080 kW.
Weight 2,300,000 kg.
Category IV
per litre stroke volume 3.16 kW
per kW power 28.7 kg weight
mean effective pressure 18.6 bar
mean piston speed 8.5 m/sec.

Remarks

Today the total load is often expressed in the load parameter, L_p.

$L_p = p_{eff}$ mean × C_p mean. Unit in bars/m/sec.

In all five examples, the following applies:

Perkins 1103C-33
L_p = 6.9 × 7.7 = 53.1 bar/m/sec.
Cummins KT19-M
L_p = 11.2 × 9.5 = 106.4 bar/m/sec.
Caterpillar MaK M25
L_p = 23.7 × 9.6 = 227.5 bar/m/sec.
Wärtsilä 18V32B2
L_p = 24.9 × 10.0 = 249.0 bar/m/sec.
MAN Diesel 14 K108ME-C
L_p = 18.2 × 8.3 = 151.1 bar/m/sec.
Wärtsilä 12RTA96C
L_p = 18.6 × 8.5 = 158.1 bar/m/sec.

Clearly, the small Perkins engine has the lowest load. The medium-speed runners (Caterpillar MaK and Wärtsilä) have the highest load.
The two-stroke engines by MAN Diesel and Sulzer have a far lower load.
It is much more practical to compare engines that have an identical cylinder bore, similar operating processes and revolutions per minute. In these cases, one often finds small, but significant differences. Well-known examples are medium-speed HFO engines with a cylinder bore of 320 mm.

MAN Diesel L 32/40
$L_p = 24.9 \times 10 = 249$ bar/m/sec.
Wärtsilä 32
$L_p = 23.3 \times 9.6 = 223.7$ bar/m/sec.
MaK 32 C
$L_p = 24.9 \times 9.6 = 239.0$ bar/m/sec.

It is obvious that the differences in engines in the same category are far smaller.
Caterpillar Mak for instance, has a stroke – bore ratio of $\frac{480}{320}$, while Wärtsilä and MAN Diesel have a stroke – bore ratio of $\frac{400}{320}$.

When comparing a specific category of engines the following aspects are important:
- **specific fuel consumption;**
- **specific lubricating-oil consumption;**
- **the purchase price;**
- **the delivery time;**
- **the maximum continuous power output;**
- **the torque;**
- **the service;**
- **the warranties.**

Past experiences with the engine in question

For ships, and in particular ships of a certain design, the propulsion engine of choice is often pre-determined. Changing the engine often means additional costs.

5.5.2 Shaft power produced in kW in relation to the total stroke volume of the engine

Power density per stroke volume =

$$\frac{\text{Total shaft power}}{\text{Total stroke volume}}$$

Or: kW shaft power produced per litre stroke volume.

$$P_{density} = \textbf{kW/dm}^3 \textbf{ stroke volume}$$

5.5.3 Shaft power produced in kW in relation to the total volume of the engine

Power density per total engine volume =

$$\frac{\text{Total shaft power}}{\text{Total engine volume}}$$

$$P_{density} = \textbf{kW/dm}^3 \textbf{ engine volume}$$

This is, of course, far more difficult to determine and so is apparently used less frequently. Even so, in reality the size of a particular engine type is certainly taken into account. For example, the built-in height, the free height required to enable piston withdrawal, the width of a V-engine.
The decision-making process in choosing a certain engine automatically includes choosing the space in which to fit in the engine. For fast compact ships, the engine's measurements most certainly play an important role. The size of engine space is often very limited.
For larger engine sizes, this is also of importance. Often large car ferries have a limited height in their engine rooms. Sometimes a hatch in the ferry's deck must be opened in order to withdraw a piston from the engine ('piston drawing').

In feeder-container ships, the engine room is kept as small as possible in order to load as many containers as possible in the holds. This means that the propulsion engine is fitted into the small space between the propeller shaft and the engine room bulk-head. In this case, the gear box used to reduce the engine speed to the most effective number of revolutions for the propeller also needs to be as short as possible.

▲
Car ferries require that the entire main deck is utilized for rolling material and therefore the engine-room height is limited. In this case, four-stroke engines are the only option.

▼

Abundant data is important in running a propulsion installation.

The shaft power, as well as the number of revolutions, the fuel consumption, the weight and size are required to make a well-informed informed decision.

MAN B&W Four-stroke Propulsion Systems

L23/30A-KV
800-1280kW
(1090-1740 BHP)

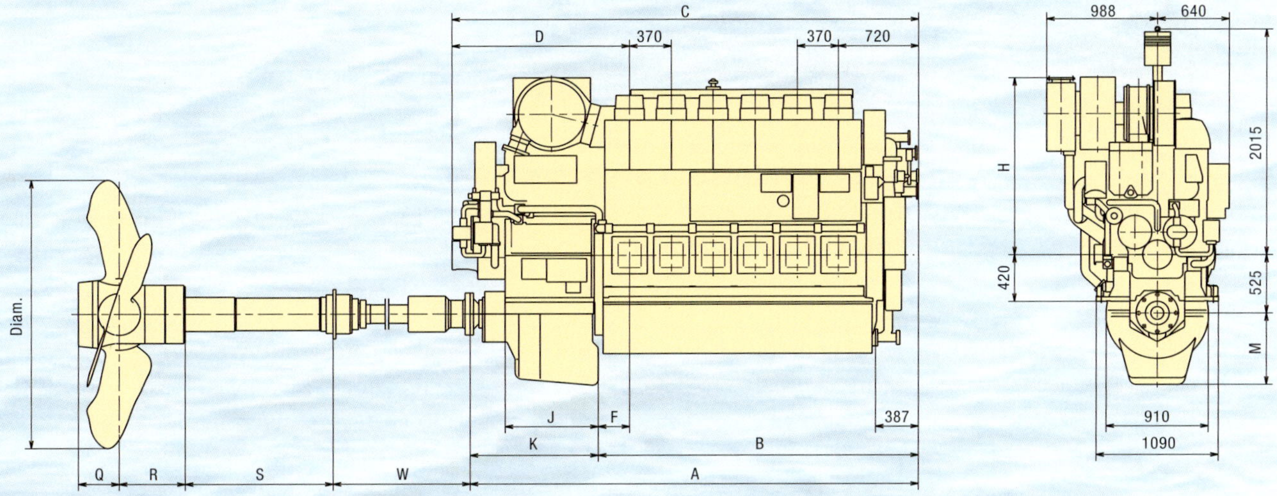

Standard programme

ENGINE	REDUCTION GEAR		PROPELLER			DIMENSIONS IN MM											
Type.	Series	Type	Type	Speed	Diam	A	B	C	D	F	H	J	K	M	Q	R	W
Output mcr				rpm				mm									min
6L23/30A-E	AMG 8	31KV8	VB 560	268	2200	3975	2898	4136	1566	328	1587	770	1077	636	293	553	900
800 kW	AMG 8	39KV8	VB 640	214	2450	3975	2898	4136	1566	328	1587	770	1077	636	360	595	900
1090 BHP	AMG 8	44KV9	VB 640	190	2600	3975	2898	4136	1566	328	1587	770	1077	636	360	595	900
	AMG 16	52KV13	VB 740	159	2850	4550	2898	4660	2090	318	1587	1282	1662	505	445	655	1200
6L23/30A	AMG 8	31KV8	VB 560	292	2250	3975	2898	4136	1566	328	1587	770	1077	636	293	553	900
960 kW	AMG 8	39KV8	VB 640	233	2450	3975	2898	4136	1566	328	1587	770	1077	636	360	595	900
1305 BHP	AMG 8	44KV9	VB 640	207	2600	3975	2898	4136	1566	328	1587	770	1077	636	360	595	900
	AMG 16	52KV13	VB 740	173	2850	4550	2898	4660	2090	318	1587	1282	1662	505	445	655	1200
8L23/30A	AMG 11	31KV11	VB 640	292	2350	4953	3628	5064	1754	318	1587	950	1325	639	415	595	900
1280 kW	AMG 11	39KV11	VB 640	233	2600	4953	3628	5064	1754	318	1587	950	1325	639	415	655	1200
1740 BHP	AMG 11	44KV13	VB 740	207	2800	4953	3628	5064	1754	318	1587	950	1325	639	445	655	1200
	AMG 16	52KV13	VB 860	173	3100	5290	3628	5400	2090	318	1587	1282	1662	505	445	745	1350

The propeller is calculated according to DnV, No Ice with high skew

Main data

ENGINE	BORE	STROKE	SPEED	MEP	PISTON SPEED	OUTPUT/CYLINDER	
Type	mm	mm	rpm	bar	m/s	kW	BHP
L23/30A-E	225	300	825	16,3	8,25	133	181
L23/30A	225	300	900	17,9	9,00	160	217,5

Specific consumption

ENGINE	FUEL OIL				LUBRICATING OIL	
Type	mcr		85% mcr		mcr	
	g/kWh	g/BHPh	g/kWh	g/BHPh	g/kWh	g/BHPh
L23/30A-E	188	138	187	137	1.0	0.7
L23/30A	190	140	189	139	1.0	0.7

Weight

ENGINE	GEAR	PROP	Dry weight in tons (approx)	
Type	Type	Type	Engine/Gear	Prop*
6L23/30A-E	31KV8	VB 560	15.0	2.7
	39KV8	VB 640	15.0	3.4
	44KV9	VB 640	15.0	3.5
	52KV13	VB 740	17.6	4.4
6L23/30A	31KV8	VB 560	15.0	2.7
	39KV8	VB 640	15.0	3.4
	44KV9	VB 640	15.0	3.5
	52KV13	VB 740	17.6	4.4
8L23/30A	31KV11	VB 640	17.9	3.4
	39KV11	VB 640	17.9	3.8
	44KV13	VB 740	17.9	4.5
	52KV13	VB 860	20.1	6.1

** Weight incl 4.0 m shaft and 2.0 m stern tube*

▲
A medium-speed four-stroke in-line engine used for propulsion in a 'feeder-container ship'.

Shown here an eight-cylinder Caterpillar MaK M43-C heavy fuel-oil engine.

5.6 Number of revolutions of diesel engines in relation to the size of the stroke of an engine

Explanation

Many characteristics of the diesel engine can be established once the number of revolutions is determined, for example:

thc maximum allowable mean piston speeds:
- ± 12 m/sec. for four-stroke trunk piston engines;
- ± 8.5 m/sec. for two-stroke crosshead engines.

The stroke/diameter ratio for four-stroke as well as two-stroke engines in practice often lies within certain values.

Four-stroke	normally	1.1 – 1.3
	long stroke	1.4 – 1.6
two-stroke	normally	1.8 – 2.2
	long, ultra-long stroke	3 – 4

▼
A two-stroke crosshead engine on a large container ship, a twelve-cylinder Wärtsilä Sulzer RT-FLEX, with the common-rail system.

In smaller diesel engines one also sees S/D ratios of 1 or less. This is, for instance, the case in very small diesel engines with a high number of revolutions.
A **relatively long stroke** means that the engine will be tall with a long cylinder liner and a crankshaft with a large turning circle (especially for two-stroke crosshead engines).
The world's biggest engine, a fourteen-cylinder MAN Diesel K108ME-C has a 1080 mm bore with a 2660 mm stroke and has therefore a limited stroke/diameter ratio of 2.5 at an rpm of 94 and a shaft power of 97,300 kW.
A Wärtsilä Sulzer, the nine-cylinder RTA84F-B has an 840 mm bore at 3150 mm stroke and so a high stroke/diameter ratio of 3.75 at an rpm of 74 and a shaft power of 34,920 kW.
In the formula C_p mean = $2 \times S \times n$, if the stroke has been established, the number of revolutions can also be determined.
Conversely: at a certain number of revolutions the stroke and, to a lesser extent, the stroke volume can be established.

A stroke/diameter ratio of 1 can be seen in category I and II of this subdivision; these are the relatively smaller engines running on MDO.
This relatively short stroke runs at a slightly higher rpm of often over 1500 revolutions, for instance, 1800 and 2100 rpm.

5.7 Rpm of generators

Diesel engine gensets ought to provide a **constant voltage** in conjunction with a **constant frequency** under all load variations of the electrical net.
The European net frequency along with most parts of the world is 50 Hz. The United States of America and other countries related to the USA have a frequency of 60 Hz. In general, the frequency on board ships is 60 Hz.
The **frequency** generated by a rotary generator is directly dependent on the **rpm** and the **number of polar pairs** (*pp*).
A polar pair consists of a north and a south pole.

◄ A four-stroke medium-speed Deutz 640 diesel engine for HFO in a diesel-power plant in the Gambia.

Electrically generated power 5,2 MW

Formula

Frequency = number of polar pairs × number of revolutions

Frequency in Hz (Hertz)
Number of polar pairs is an undefined number.
1 polar pair – bi-polar. A north and a south pole.
Number of revolutions in rps.

Example

Number of revolutions of diesel engines for 50 Hz gensets.

1 polar pair	bi-polar	50 rps	=	3000 rpm
2 polar pairs	4-polar	25 rps	=	1500 rpm
3 polar pairs	6-polar	16.67 rps	=	1000 rpm
4 polar pairs	8-polar	12.5 rps	=	750 rpm
5 polar pairs	10-polar	10 rps	=	600 rpm
6 polar pairs	12-polar	8.33 rps	=	500 rpm

At 60 Hz this is then multiplied by 1.2, so 3600, 1800, 1200, 900, 720 and 600 revolutions.

▼ A diesel genset on a large container ship.

The diesel engines for gensets are frequently delivered for two different engine revolutions, for example 1500/1800 or 1000/1200 revolutions, especially by American companies that supply markets outside the Americas.

Large power plants predominantly use 3000 rpm generators. These are directly driven by a gas- and steam-turbine combination or in smaller plants, using gear boxes with 1500 rpm generators.

Extremely small emergency gensets have an rpm of 3000. The slightly larger ones, 1500 rpm and diesel-power plants equipped with medium-speed diesel engines using (HFO) often have an rpm of 600, 750 or 1000.

Large two-stroke crosshead engines for diesel power plants have an even smaller rpm. With eighteen polar pairs and therefore a 36-polar generator, the number of revolutions is 166.67 rpm. In order to install a 36-polar generator, a generator diameter of between 6 to 12 metres is required. These are the so-called **disc generators**.

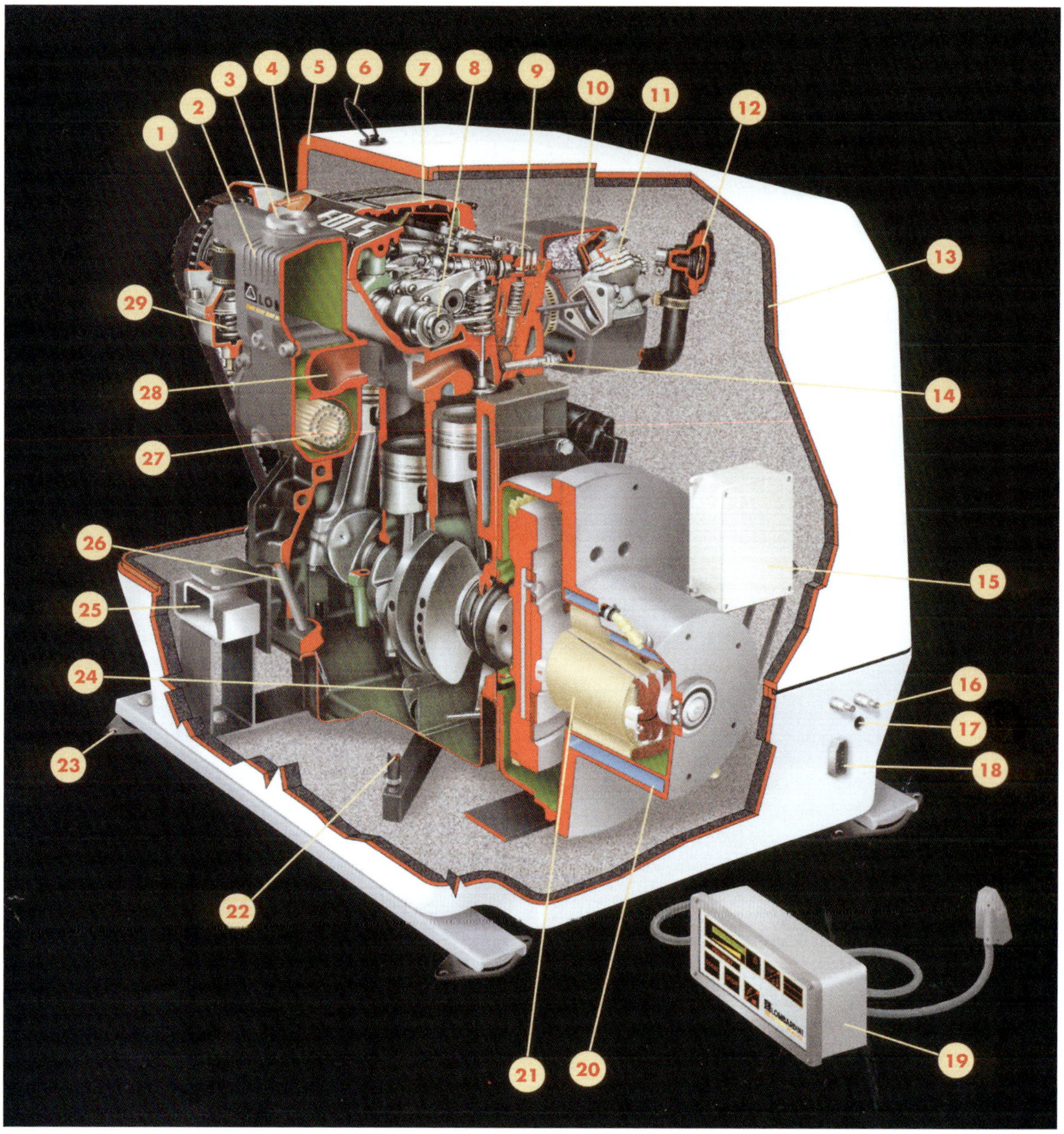

▶
A small sound-proof genset for yachts from Mastervolt with 3000 rpm. Therefore, one polar pair in de generator (21).

1 *timing belt*
2 *heat exchanger*
3 *coolant filler cap*
4 *oil filler cap*
5 *sound damping casing*
6 *rubber valves*
7 *valve cap*
8 *camshaft*
9 *fuel injector*
10 *air filter*
11 *fuel-supply pump*
12 *bleeder valve*
13 *sound insulation*
14 *glow plug*
15 *electric junction box*
16 *coupling for fuel line*
17 *sea water inlet*
18 *multi connector for remote control*
19 *remote control panel*
20 *generator jacket*
21 *generator*
22 *oil cooler*
23 *external vibration dampers*
24 *sump partition wall*
25 *internal vibration dampers*
26 *dip stick*
27 *cooling element of the heat exchanger*
28 *water-cooled exhaust manifold*
29 *thermostat*

▶
A high-speed diesel power unit.

1 *air filter*
2 *turbocharger air section*
3 *turbocharger exhaust gas section*
4 *compressed air t -air- cooled intercooler*
5 *radiator ventilator*
6 *radiator with cooling section for combustion air, fresh water coolant and lubricating oil*
7 *engine block*
8 *lubricating-oil filter*
9 *crankpan*

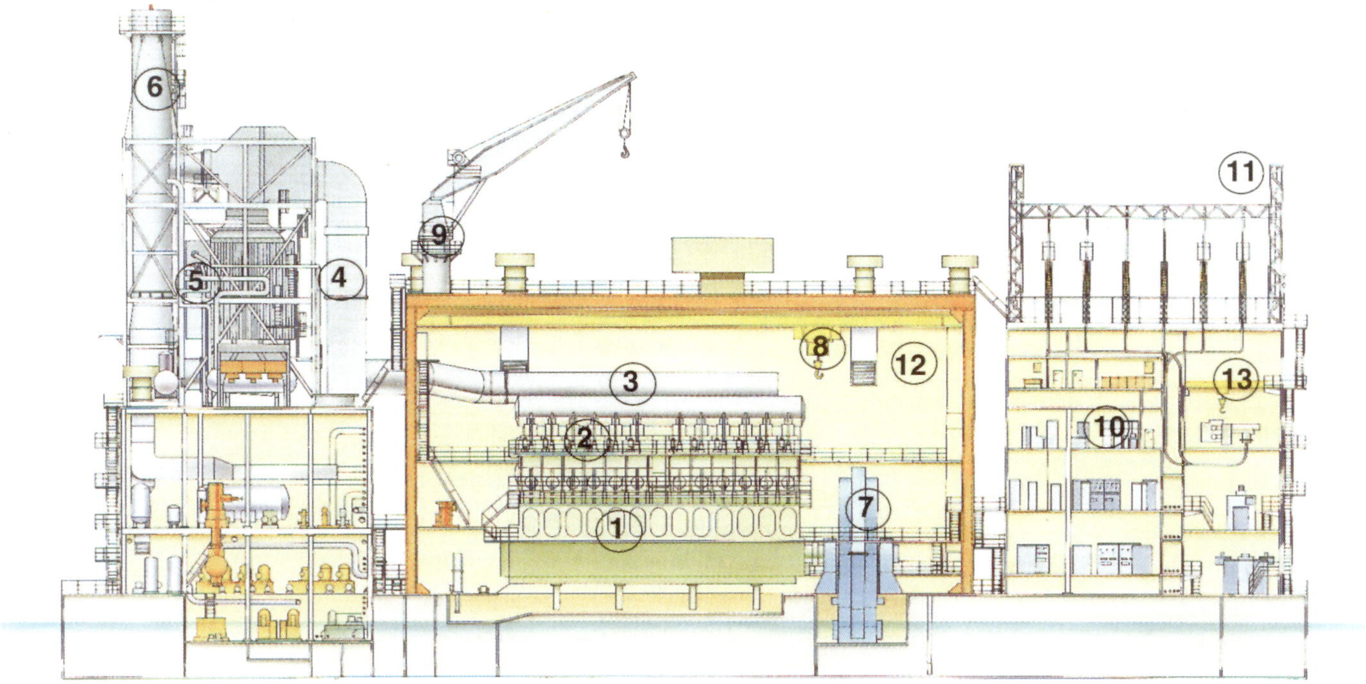

▲
A floating diesel-power plant.

The diesel engine has ± 84 rpm. Therefore 36 polar pairs in the disc generator (7).

1 *two-stroke crosshead engine*
2 *cylinder head plateau*
3 *exhaust-gas manifold*
4 *sound damper*
5 *exhaust-gas boiler*
6 *funnel*
7 *disc generator*
8 *overhead crane*
9 *crane*
10 *electric control box*
11 *high-voltage connection to the shore*
12 *ventilation shafts*
13 *spare parts and maintenance area*

◄
The generator poles are clearly visible. The north pole is followed by a south pole, which in turn is followed by a north pole and so on.

>CH 6

Construction of various types of diesel engines

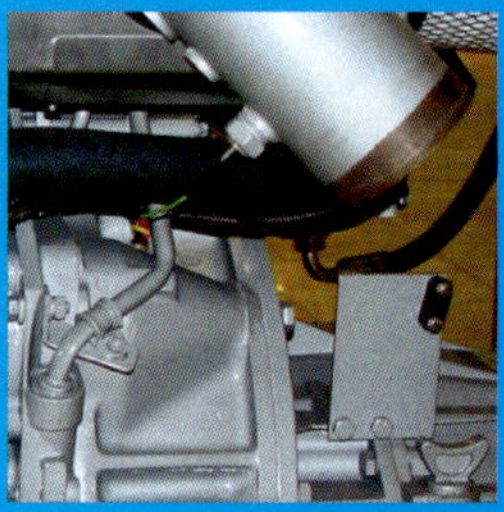

Mounting a fast running MTU diesel engine.

Modern diesel-engine construction is a time-consuming, slow process requiring a great deal of research. Strict consumer requirements with regards to diesel engines, such as a high power density, low oil and fuel consumption, low toxic emissions, lengthy operating-time of the components and simple mounting and dismounting methods make it difficult for engine manufacturers to continue to compete in this market segment.

6.1 Category I: Industrial diesel engines from 0 to 100 kW shaft power, fuel MDO, four-stroke, high-speed engines

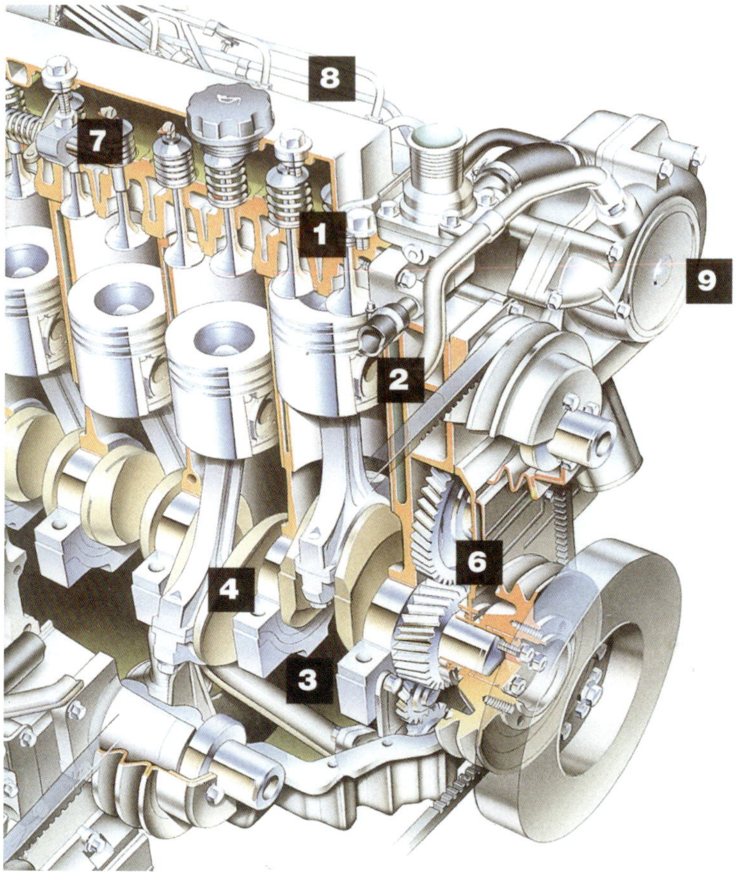

▶ High-speed four-stroke diesel engine.

1 *cylinder head with two inlet valves and two exhaust valves*
2 *piston*
3 *crankcase or carter*
4 *crank web*
6 *timing gear between the crankshaft and camshaft*
7 *valve motion*
8 *high-pressure fuel lines*
9 *coolant pump*

These relatively small industrial diesel engines are used in a large number of applications, amongst others:

- **propulsion;**
- **gensets;**
- **pump geardrives;**
- **traction transport and machinery**
- **numerous areas** which will not be discussed any further, such as transportation, earth moving, cranes, etc.

Within the scope of this book, they are mainly found:

- in propulsion of smaller ships and yachts;
- for generating electrical energy, such as in gensets or as back-up generators;
- for driving pumps and other equipment.

In general, they are based on the four-stroke principle, are a trunk-piston type and today more and more often fitted with two inlet- and exhaust valves.

Here one also finds that the modern systems such as fuel injection with, among other things, a common-rail system are increasingly applied.
The engine consists of a light-metal or cast-iron block. The smallest types do not have separate cylinder liners.
This means that with excessive wear and tear of the liner the cylinder bore needs to be rebored and then fitted with an oversized (larger piston and piston rings) piston.

See Chapter 26, *Overhauling diesel engines and their parts.*

The crankshaft is fixed to the block by means of bearing caps and vertical tap bolts. In this case, a four-cylinder diesel engine has five bearings (popularly said: the engine is five times bearinged).
At the bottom of the frame is a freely suspended oil sump that contains the oil pump, oil filter, and the lubricating oil.
The engine is not directly placed on the sump but on the engine mountings that are attached to the block.

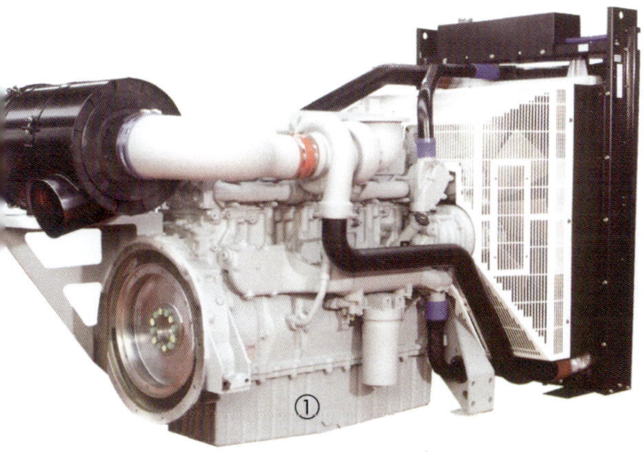

▲

Diesel engine with suspended oil sump (1).

Traditional designs have a closed frame under the block onto which the engine is mounted.

▲

A diesel engine adapted for maritime use.

Often the original uncooled hot exhaust pipes are water cooled to reduce the fire risk. The engine is mounted on vibration dampers.

Often vibration dampers are placed underneath the frame or the engine mountings in order to stop vibrations from the engine being transmitted to its surroundings.

▲

Vibration dampers.

▲

A small high-speed four-stroke diesel engine for yacht propulsion.

This so-called 'z-drive' is fairly common: there is no need for a long propeller shaft, so the propulsion space is kept to a minimum. The propeller is a folding propeller; at sea the two propeller blades fold back for minimum drag. At the large horizontal flange the gear drive passes through the ship's hull.

1 *diesel engine*
2 *bevel gear in the yacht*
3 *bevel gear in the water*
4 *flange in the hull*
5 *folding propeller*
6 *oil sump*

Often the **cylinder head** mounted on top of the engine block already comprises all the cylinders, so the engine does not have separate cylinder heads as is the case in larger engines. Of course, the weight of such a multi-cylinder head in larger engines is impractical.

▲

A complete multi-cylinder head of an full-sized engine.

▶
A small diesel generator with a crank handle on the camshaft.

▶
A decompression handle to keep the exhaust valve open during cranking.

1 *compression handle for releasing the exhaust valve*
2 *cylinder head*
3 *air filter*
4 *fuel tank*
5 *lubricating-oil filler cap*
6 *engine block*

Usually one or two **inlet**- and **exhaust valves** are placed in the head for the combustion-air supply and the discharge of exhaust gases.
These are mechanically opened and shut at the right time by means of the camshaft, guide pulleys, push rods and rockers. Rings around the valve seal ensure the valves close swiftly. For smaller engines which are manually started it is possible to keep the exhaust valve open with a **decompression handle** until a certain number of revolutions is achieved with the crank handle. By pulling the handle again, the exhaust valve(s) are closed at the right moment and compression, combustion etc. take place: the engine is running.

Also see Chapter 14, *Starting systems of diesel engines.*

Usually the **fuel** is injected into the combustion chamber in the centre of the cylinder head. This is achieved with a high-pressure fuel pump, a high-pressure fuel-supply line and a fuel atomiser. Often one central 'block-fuel pump' is installed.
The engine speed is restricted by, for instance, a mechanical speed regulator. The regulator directly controls the amount of fuel supplied to the engine by adjusting the effective capacity of the high-pressure fuel pumps.
With a regulator it is possible to maintain the number of revolutions and therefore keeping the frequency and the voltage of the generator constant during varying generator loads.

Engine lubrication is done by a crankshaft driven gear pump which is positioned in the crankcase/sump. Via a suction-filter whichh protects the pump, the lubricating oil is transported through lines and bores to the various engine parts that require lubrication and cooling. A lubricating-oil fine filter ensures that the lubricating oil remains clean.
An oil cooler keeps the lubricating-oil temperature at a normal operating temperature. For water-cooled engines, the oil is cooled by the coolant circuit. A thermostatic control valve sends enough cooling water through the cooler to ensure that the lubricating-oil temperature remains constant.
Most engines are cooled by using a closed cooling-water system, an open cooling-water system with for instance canal water, or by fuel-oil or air. An

▶
A four-cylinder diesel genset with an rpm of 1500. The engine is equipped with a turbocharger without an air cooler.

1 *diesel engine*
2 *starting motor*
3 *alternator*
4 *radiator cooling*
5 *three-phase generator*
6 *generator ventilation*
7 *air-inlet filter*
8 *turbocharger*
9 *inlet manifold*
10 *oil-filler cap*

increasingly large number of these engines are equipped with a **turbocharger** and an intercooler in order to cool the air.

The number of cylinders varies from one to six in in-line-engines or four to eight in V-engines, depending on the cylinder-stroke volume and the number of revolutions. Cylinder bores vary from ± 50 to 150 mm.

The engines are **started either manually or electrically**. There are also other types of ignition systems, **hydraulics**, **spring pressure** or **compressed air** that drive the ring gear.

Also see Chapter 14, *Starting systems of diesel engines*.

The number of manufacturers and types is large and diverse.

Also see Chapter 21, *Diesel-engine manufacturers*.

▲
A diesel generator.

◄
A very small, compact and silent diesel genset which can be placed in the most suitable position in a sailing- or motor-yacht. Manufacturer Mastervolt, Drachten, The Netherlands.

6.2 Category II: Industrial diesel engines from 100 to 10,000 kW shaft power, fuel MDO, four-stroke, high-speed engines

The slightly larger industrial diesel engines ranging from 100 to 10,000 kW are also used for very divergent purposes. A 2000 kW engine is a large propulsion engine for the Rhine, inland waterways and coastal navigation and has a considerable capacity for driving the generator. Applications are among others:

- propulsion of numerous ships especially in inland navigation or for pleasure craft;
- generation of electricity;
- engine drives for multi-purpose machines such as pumps and compressors.

▼
A Deutz propulsion engine with reduction drive to the controllable pitch propeller.

▶

Sailing yachts also have a diesel engine for auxiliary propulsion. Larger yachts also have a diesel genset on board to generate the required electricity.

Important points of interest:

- vibration-free set-up;
- low-noise production;
- compact and lightweight;
- easy access for repairs;
- elaborate service network;
- exhaust-gas emissions below legal requirements;
- low fuel consumption;
- reasonable purchase price;
- easy to operate;
- reliable.

▶

Dredger pumps on dredgers are usually driven by diesel engines.

▶

Rhine and (other) inland navigation ships are becoming increasingly larger in order to be cost-effective.

Ships of 2000 to 8000 tons carrying capacity are no exception. Propulsion is usually done by high-speed four-stroke diesel engines with diesel oil as fuel. Heavy oil is not allowed in these navigational areas due to the increasingly strict regulations with regard to exhaust-gas emissions.
The auxiliary genset for generating electricity or direct pump drive is also becoming considerably larger. Here one uses a similar engine with a lower capacity.

▲
One of two propulsion engines on an inland tanker.

Dozens of engine manufacturers are active in this 'power category', many of whom cover a global market.

These diesel engines still use MDO, because heavy fuel is not suitable for these fast running engines (over 960 rpm). The number of cylinders for a V-engine can amount to sixteen, eighteen or even twenty cylinders. Cylinder bore ± 80 tot 200 mm.

Construction

The block or frame is constructed from a single piece of cast iron and in smaller engines sometimes of light metal (a cast alloy with a high aluminium content) and generally has separate interchangeable cylinder liners which are usually cooled on all sides by cooling water (or coolant).
In engines with higher power outputs each cylinder has its own fuel pump, however, block-fuel pumps are also often used.

These engines are increasingly often equipped with common-rail fuel systems, which enhance the potential of the diesel engine with regard to process control, fuel consumption and toxic emissions.

▲
A cross section of the cylinder of a Caterpillar diesel engine with the mechanically driven fuel injector positioned centrally.

▲
An independent fuel pump for medium-speed larger engines running on HFO.

▼
A block fuel pump on a high-speed in-line engine.

1 *fuel filter*
2 *low-pressure transfer pump*
3 *block fuel pump*
4 *propeller shaft from crankshaft*
5 *manual fuel pump for deaeration*
6 *high-pressure pipe to the cylinder*

▲
A block fuel pump with six independent high-pressure in-line plunger pumps.

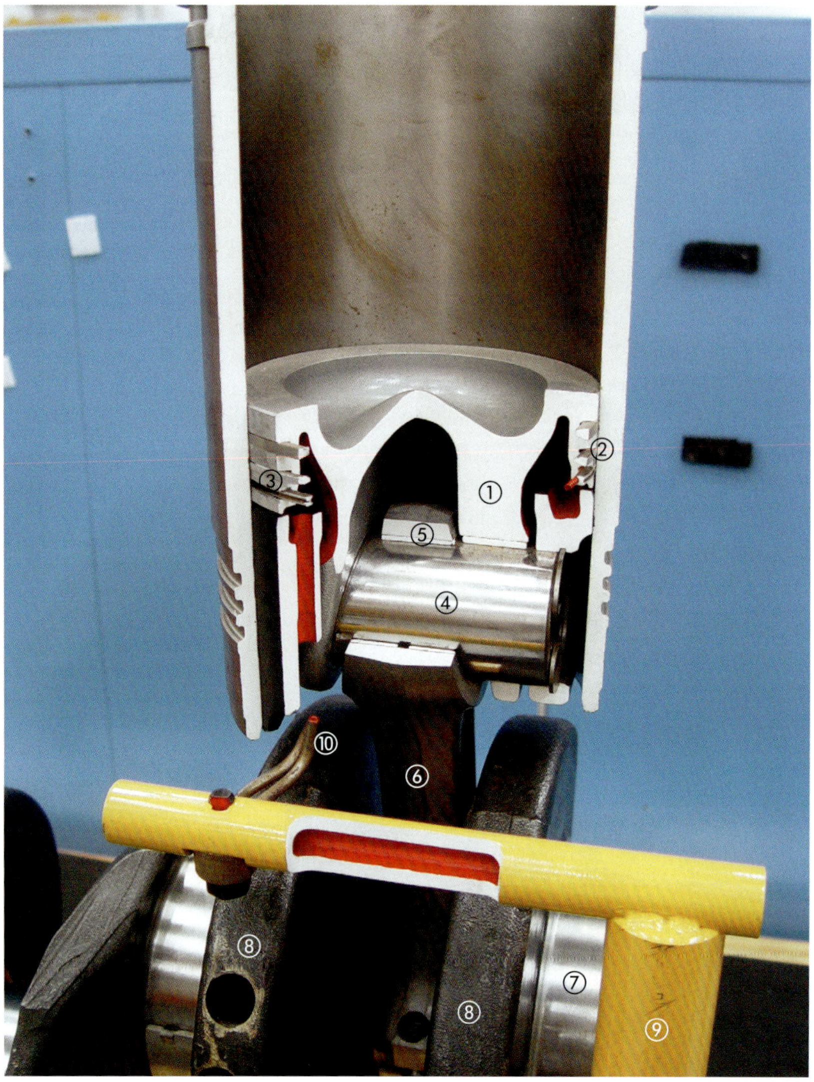

▲

In smaller engines the piston is cooled internally with an oil jet which is vertically directed through a nozzle.

1 *light-metal piston*
2 *piston rings*
3 *scraper and divider ring*
4 *gudgeon pin*
5 *gudgeon-pin bearing*
6 *connecting rod*
7 *crankshaft*
8 *crank webs*
9 *main lubricating-oil duct*
10 *nozzle*

◀

A twelve-cylinder Caterpillar diesel engine for gensets.

Note the double electric starter motors in the fly wheel, right (1).

Normally these engines are cooled by means of a closed cooling-water- or coolant system. The heat in the cooling water can be lost naturally by contact with the surface water or air.
The lubrication of theses engines is becoming increasingly important. The heavy thermal and mechanical load of these engines require thorough lubrication and cooling all around the combustion space (piston, piston rings, cylinder liner and exhaust valves), as well as proper lubrication of the highly loaded mechanical parts such as the piston pin, crankpin and the crankshaft.

Further noticeable is an extremely high power density: high charging with turbocharger groups, a high compression ratio and therefore a high mean effective pressure. What is striking is that the number of V-type versions is extensive for these engines. Arranging the cylinders in a V-shape makes the engine construction short and compact, which is especially important when there is limited space. Parts that need to be dismounted when overhauling the engine, such as the cylinder head, piston, connecting rod and cylinder liner have reasonable sizes and weights, thus restricting the number of hoists and other auxiliary tools required.

Number of revolutions

The number of revolutions of these diesel engines is generally over 1000 rpm, which puts them into **category II**, high-speed engines (number of revolutions > 960 rpm).
Almost without exception, MDO is used as a fuel, since the process time of HFO is too short to combust properly.
There are, of course, HFO engines on the market with an rpm of 1000 to 1200, but these are exceptions. The group of engines we are presently discussing often has 1500, 1800 or even 2100 revolutions per minute.

Also see Chapter 21, *Diesel-engine manufacturers.*

The mean piston speeds are approximately 10 to 11 m/sec. and the mean effective pressures can go up to 25 bar.

6.3 Category III: Industrial diesel engines from 500 to 30,000 kW shaft power, fuel HFO, four-stroke, medium-speed engines

Almost all of this relatively large group of industrial diesel engines uses HFO. This means that the maximum number of revolutions is limited from 1000 up to 1200 rpm. According to engine classification these are really high-speed engines. Some engine manufacturers call these engines 'high medium-speed four-stroke diesel engines'. The cylinder bore varies from 160 to 640 mm. A fair number of engine manufacturers, about a dozen, manufacture diesel engines that can run on HFO under all loads and under varying operating conditions.

Definition of HFO diesel engine

This is a diesel engine that can start and run on HFO at any load. The heavy oil then has a viscosity of at least 50 cSt at 50 °C.

Also see Chapter 8, *Fuels, fuel-line systems and fuel cleaning.*

▲
A medium-speed Caterpillar MaK in-line engine.

▲
A medium-speed V-engine.

Here an 18-cylinder MAN Diesel 48/60 B diesel engine capable of generating an electrical power output of 18.4 MW

◄
A cut-away section of a Caterpillar MaK diesel engine with mechanically adjustable camshaft. A typical example of a medium-speed diesel engine running on heavy oil.

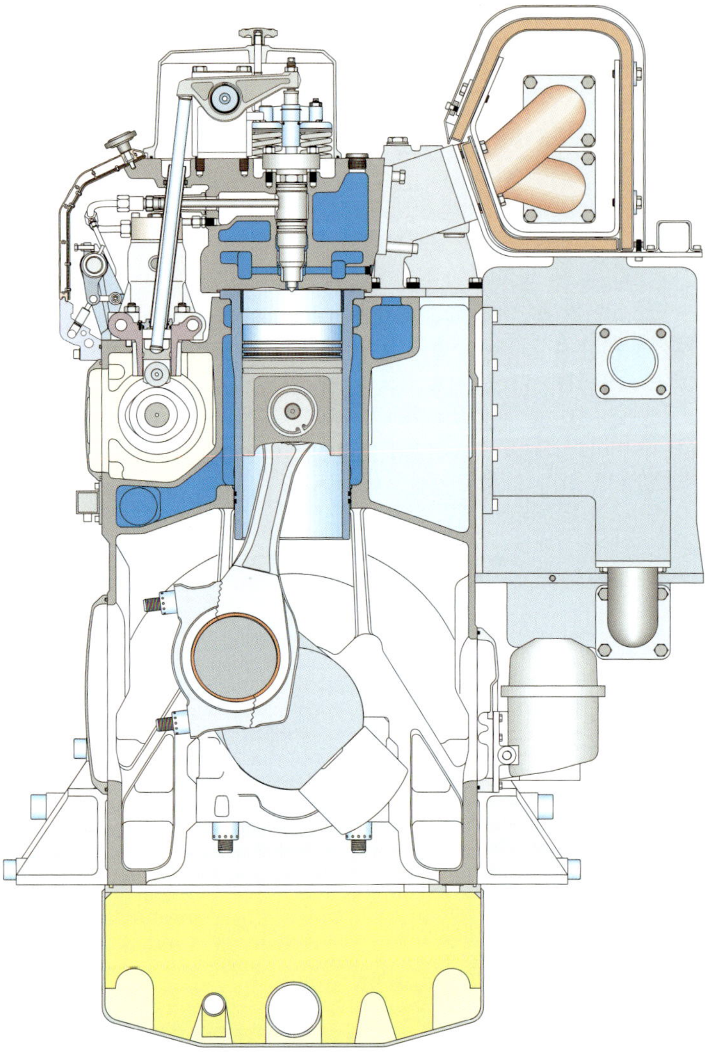

▲

A modern four-stroke heavy-fuel oil engine, a Wärtsilä 20.

All cylinder-block frames are manufactured from a single metal piece. The casting is usually done with grey cast iron, nodular cast iron or cast iron with different additives used to enhance the material's properties. Due to the strong increase in power densities, the mechanical and thermal loads have increased considerably.

Where possible, the lines for lubricating oil, cooling water and air are incorporated in the cast engine block. This way a very robust entity is obtained with a minimum of extra piping and other connectors (potential faults).

The one-piece forced-steel cast crankshaft is incorporated in the bottom of the frame and the bottom bearing caps are fixed with two vertical bolts and laterally supported by two horizontal pressure bolts. In this way the bottom bearing cap becomes more or less one with the frame.

▲

Crankshaft with mounted counterweights for correct balancing.

▲

The cylinder block of a Caterpillar MaK 43 in-line engine.

The air supply to the cylinders is integrated in the solid block. Here only the crank shaft bearing caps have been fitted.

1 *engine block*
2 *position crankshaft*
3 *scavenging-air space*
4 *position camshaft*

▲

The engine block of a V-engine with fitted cylinder liner and cylinder-head bolts. In the middle the air-inlet manifold.

The cylinder liners can be fitted level in the frame, but with the larger engines in this group, a substantial part of these cylinders sticks out of the block.
In the first instance, they are conventionally cooled with cooling water and in the second the liners have been furnished with bore holes through which the cooling water flows. This principle is called 'bore cooling'

◄

A cylinder block of a Caterpillar MaK 25 in-line engine.

It is clearly visible how the cylinder liners protrude from the block.

▲

An in-line crankshaft.

The crankshaft at final inspection. Tiny burrs are also manually removed.

◄

An in-line cylinder block with mounted fuel pumps, camshaft and cylinder-head bolts.

▲

The camshaft part of one cylinder.

▲

The cylinder liner of a modern highly loaded diesel engine.

1 *upper collar, for absorbing the high compression and combustion pressures, equipped with bore-cooling*
2 *grooves for coolant tubes, above these the traditional coolant areas*
3 *edge support in engine block*
4 *top of the engine block*

▶

A medium-speed diesel-engine piston skirt.

One can clearly see the reduction of the material around the gudgeon pin. Part of the skirt has been removed under the pin. Here the forces exerted on the skirt are very small in comparison to the forces exerted on the skirt above the gudgeon pin.

▶

A piston in a cylinder with bore-cooling, anti-polishing ring and a piston skirt lubricated under pressure in a medium-speed diesel engine.

The piston rings have a chromium-ceramic coating.

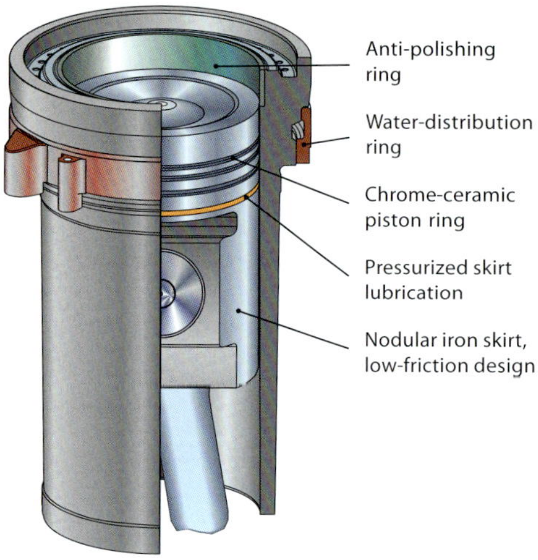

The piston is a two-part construction and consists of a wrought iron- or cast steel piston crown, which can absorb the high thermal and mechanical loads and contains the piston rings and a skirt.
The skirt is made of a light metal or cast iron and is used for the piston conveyance and the absorption of the lateral forces.
In order to keep the piston as light as possible considering the acceleration and deceleration forces transmitted to the piston, all material not construction-technically required is removed, especially when the piston skirt is manufactured of cast iron.
The piston pin must have a large diameter in relation to its length in order to be rigid and barely flexible.
The piston-ring package for larger cylinder bores is lubricated under pressure and also cooling of the piston crown receives special attention.

The connecting rod has to transfer the extremely high forces produced by the piston to the crankshaft. Numerous possible types can achieve this.

Also see Chapter 13, *Driving gears.*

In the upper part of the **cylinder liner** special so-called **'anti-polishing rings'** are placed to ensure that the mean effective pressures are significantly increased and that the wear and tear of the piston, piston rings and cylinder liner is diminished.

Also see Chapter 13, *Driving gears.*

▼

The 'Marinehead' version of a connecting rod.

The piston can be removed without the crankpin bearing via the cylinder liner by unbolting the eight studs.

1 *connecting rod*
2 *big end bearing box*
3 *studs*
4 *connecting-rod eye*

▼

An anti-polishing ring.

This can be easily removed and plays no part in the construction or sealing of the combustion chamber.

Four-valve cylinder heads in a medium-speed Caterpillar MaK 43 in-line diesel engine on the feeder-container ship 'Endurance' of JR Shipping.

In the foreground three high-pressure fuel pumps with fuel-supply lines.

1 *high-pressure fuel pump*
2a *fuel-supply line*
2b *fuel-return line*
3 *cylinder head*
4 *cylinder-head bolt*
5 *safety valve*
6 *indicator cock*
7 *valve cover*
8 *push rod*
9 *rocker*
10 *rocker shaft*
11 *exhaust valve*
12 *inlet valve*
13 *set screw-valve clearance*
14 *exhaust-gas line*

The **cylinder heads** are made of cast iron in which alloys are increasingly added in order to improve thermal and mechanical properties.
The **gas exchange** occurs **with two inlet valves and two exhaust valves**. The exhaust valves in larger engines are placed in an interchangeable casing.

The exhaust valves and valve seats are manufactured using a special high performance alloyed steel to ensure that wear and corrosion are reduced to a minimum.

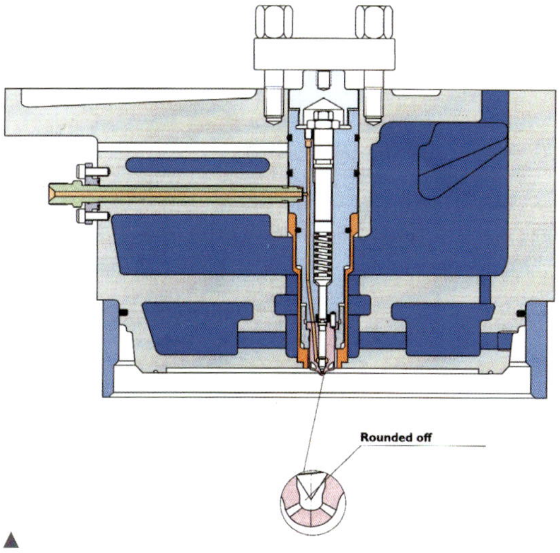

Sectional view of a cylinder head with in the centre the injector.

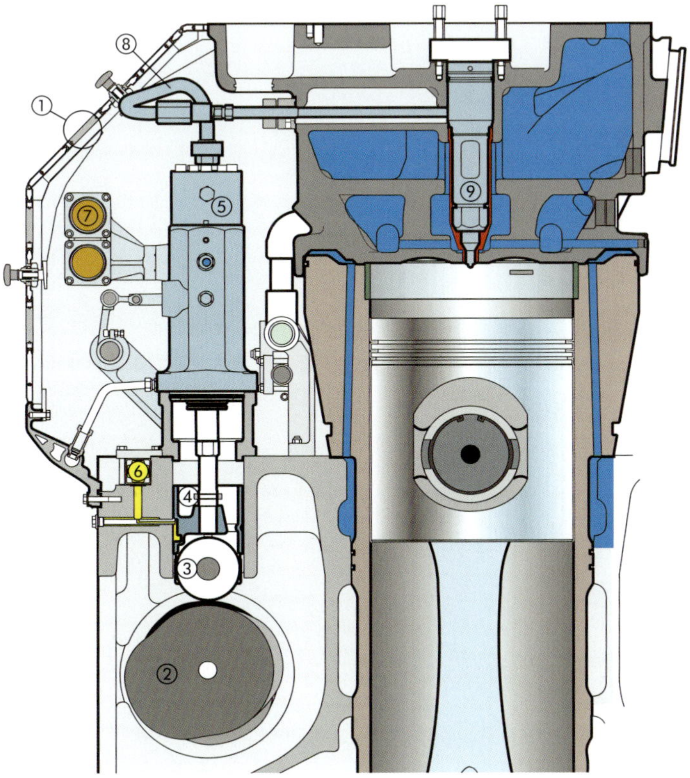

A camshaft-driven high-pressure pump and the fuel injector.

Note the so-called 'hotbox' for maintaining the temperature of the fuel section.

1 *'hotbox' cover*
2 *camshaft*
3 *roll*
4 *roller guide*
5 *high- pressure fuel pump*
6 *lubricating-oil supply*
7 *fuel supply*
8 *high-pressure delivery pipe to injector*
9 *fuel injector*

▶
View of a cylinder head with the fuel pump (1).

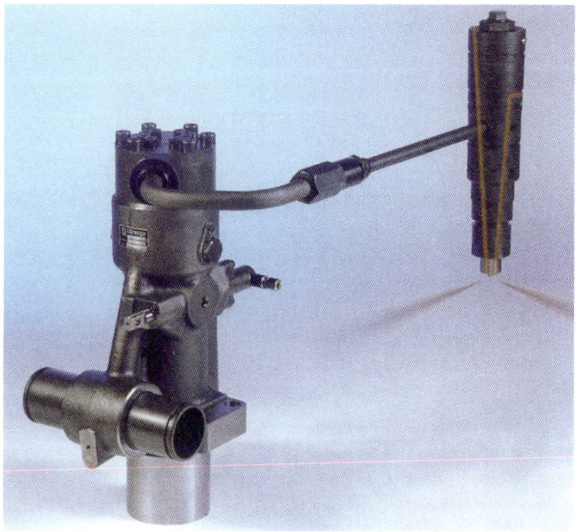

▲
The high-pressure fuel pump, a very important part of the diesel engine.

The centrally placed fuel atomiser in smaller engines often has its own double-walled fuel line to prevent fuel leakage; the larger engines have a fuel supply through the cylinder head itself.

All engines, without exception, are equipped with a **turbocharger** by means of (a) turbocharger group(s) and fitted with an air cooler, which ensures that the compressed air is adequately cooled, thus increasing the specific mass of the air and consequently the number of kilograms of air per time unit which are added to the engine. This way more fuel can be combusted in the cylinder, which allows for extra power generation.

▶
Sectional view of a modern MAN Diesel four-stroke diesel engine with required details.

Note the separate camshaft for the inlet- and exhaust valves (left) and the separate camshaft for the high-pressure fuel pump (right).

MCR = Maximum Continuous Rating

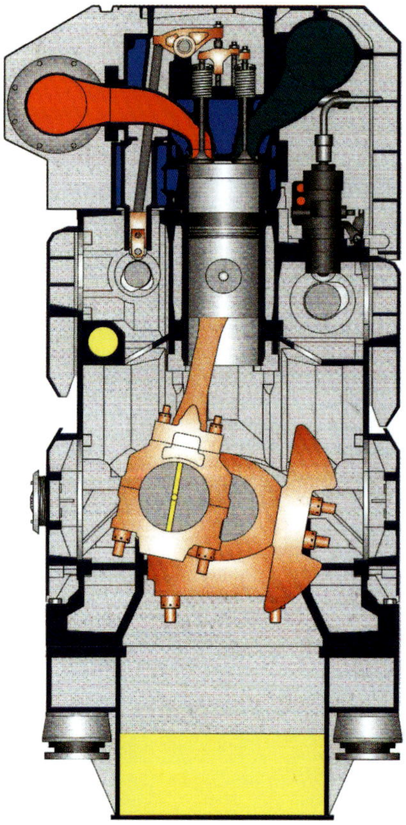

Main particulars

Cycle	:	4-stroke
Configuration	:	In-line
Cyl. Nos. available	:	5-6-7-8-9
Power range	:	950-1935 kW
Speed	:	900/1000 rpm
Bore	:	210 mm
Stroke	:	310 mm
Stroke/bore ratio	:	1.48:1
Piston area per cyl.	:	346 cm^2
Swept volume per cyl.	:	10.7 ltr.
Compression ratio	:	15.5:1
Max. combustion pressure	:	210 bar
Turbocharging principle	:	Constant pressure system and intercooling
Fuel quality acceptance	:	HFO up to 700 cSt/50 °C (BSMA 100-M9)

Power lay-out		MCR version	
Speed	rpm	900	1000
Mean piston speed	m/sec.	9.3	10.3
Mean effective pressure			
5 cylinder engine	bar	23.6	22.4
6, 7, 8, 9 cylinder engine	bar	24.8	24.0
Power per cylinder:			
5 cylinder engine	kW/cyl.	190	200
6, 7, 8, 9 cylinder engine	kW/cyl.	200	215

In large four-stroke engines the turbocharger groups with the air coolers placed below take up a large amount of space.

Especially in V-engines the overall width exceeds the total height of the engine. The exhaust-gas line can be seen in the background.

1 *air-inlet manifold*
2 *cylinder-head cover*
3 *exhaust-gas line*
4 *turbocharger*
5 *air coolers*
6 *central exhaust-gas line*

Lubrication is performed by the displacement pump, which is driven by the engine. Prior to the initial start, the engine can be pre-lubricated by an external, electrically driven pre-lube pump. There are also electrically driven main-lubricating oil pumps.

The lubricating oil is not only purified in fine filters on the feed side of the lubricating-oil system but also in centrifuges or separators that clean the oil thoroughly by removing water and regular contaminants such as metal particles, combustion products and dust.

Also see Chapter 11, *Lubrication of engines*.

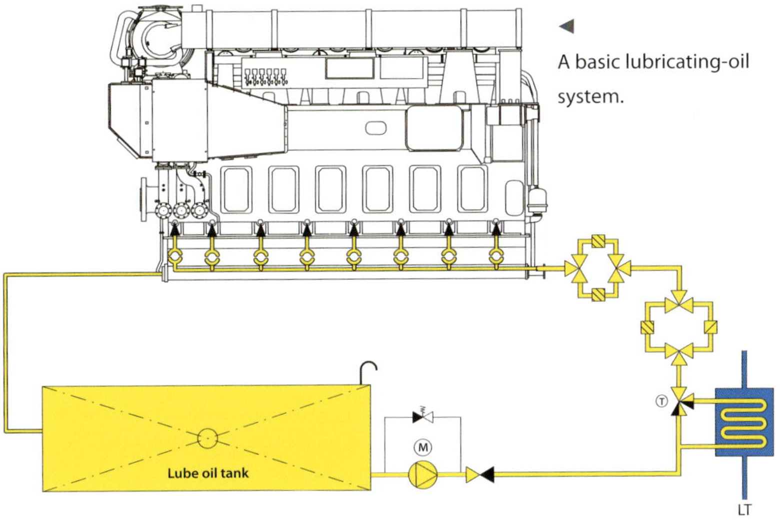

A basic lubricating-oil system.

Cooling takes place by an engine-driven cooling-water pump in a standard closed system with pressure build-up obtained by using a buffer tank or an open expansion tank.

The temperature of the cooling water is between 80 and 90 °C.

Also see Chapter 10, *Cooling diesel engines*.

This group is frequently used on various ships for propulsion, either as an auxiliary engine for generating electrical energy or for directly driving the pumps.

An overview of a coolant system with a high- and low-temperature section.

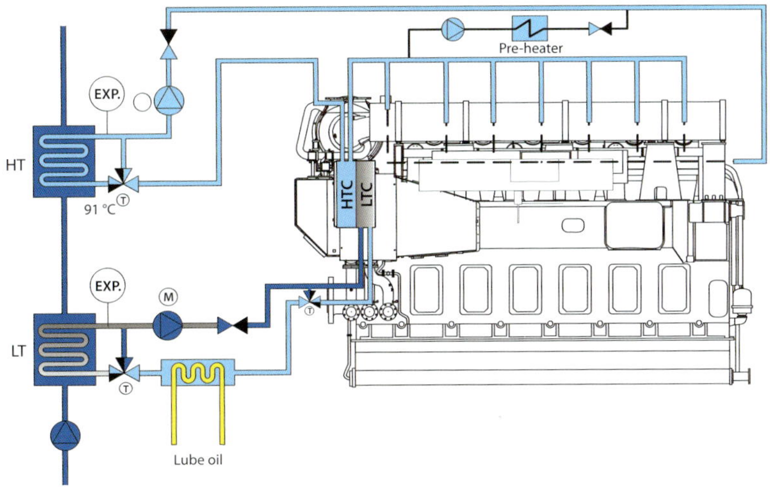

▲
Building a diesel-generator set at Caterpillar MaK in Kiel, Germany.

▶
A Caterpillar MaK four-stroke diesel engine for the genset in the engine room of a large container ship.

▲
Fourteen eighteen-cylinder MAN V48/60 type A&B diesel engines in a diesel-electric power plant in San Pedro Sula, Honduras, Central America.

In the merchant navy with a total line-up capacity from approximately 15 MW, the main engines as well as the auxiliary engines use the same heavy fuel. This cuts the fuel costs considerably.

These engines are very often used in diesel plants. The latest developments (2011) show that in particular the fuel-injection systems continue to be further optimised with among others common-rail injection systems, in which the injection period, the moment of injection and the amount of fuel is electronically regulated.

Also see Chapter 9, *Fuel-injection systems*.

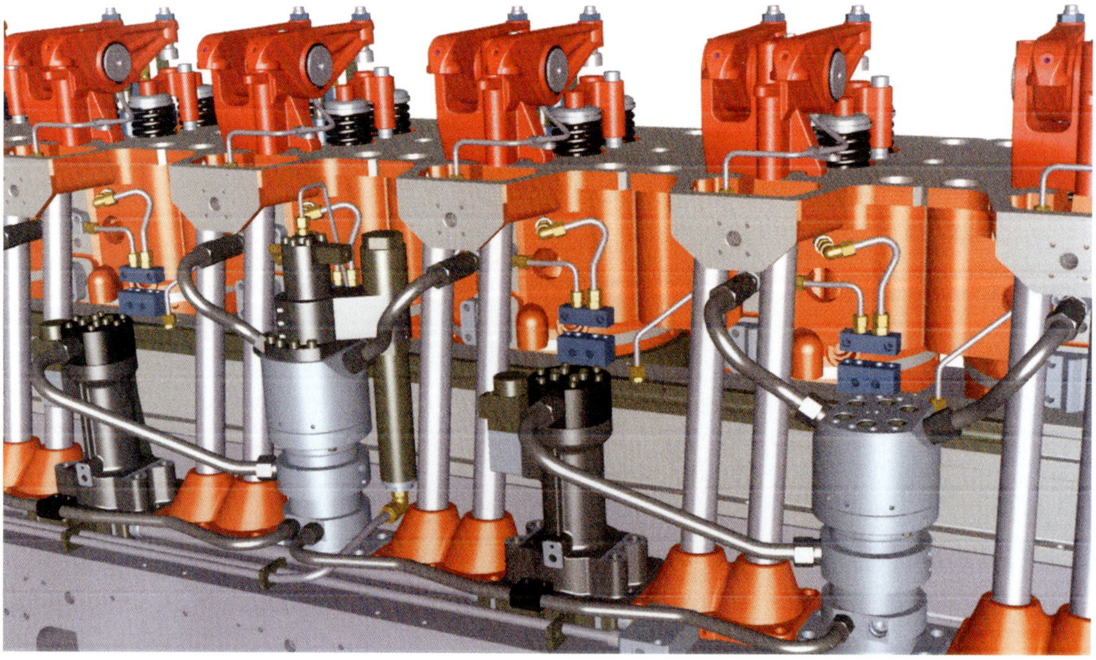

▶
A modern fuel-injection system, the Common-Rail System for Wärtsilä four-stroke engines.

6.4 Category IV: Industrial diesel engines of 1500 to 100,000 kW shaft power, fuel HFO, two-stroke with crosshead, low-speed

▶

All tankers, bulk carriers and large container ships use the bulky low-speed two-stroke crosshead HFO engine for propulsion.

These are the immensely large and also remarkably high (due to the crosshead construction) diesel engines also referred to as the **Cathedrals** among engines. They can be as high as 14 metres with a length of 30 metres and a weight of 2300 tons.

▲

On the cylinder-head mounting floor.

The curved lines show the lubricating-oil supply ducting for the operation of the exhaust valves. The exhaust valves (1) with valve casing exceed two metres.

▼

Two-stroke crosshead engine construction.

1 *crankcase bedplate*
2 *crankcase*
3 *A-frame*
4 *fuel- and lubricating-oil pumps for exhaust-valve actuation*
5 *cylinder block*
6 *cylinder liners*
7 *cylinder covers*
8 *exhaust-gas line*
9 *scavenging-air line*

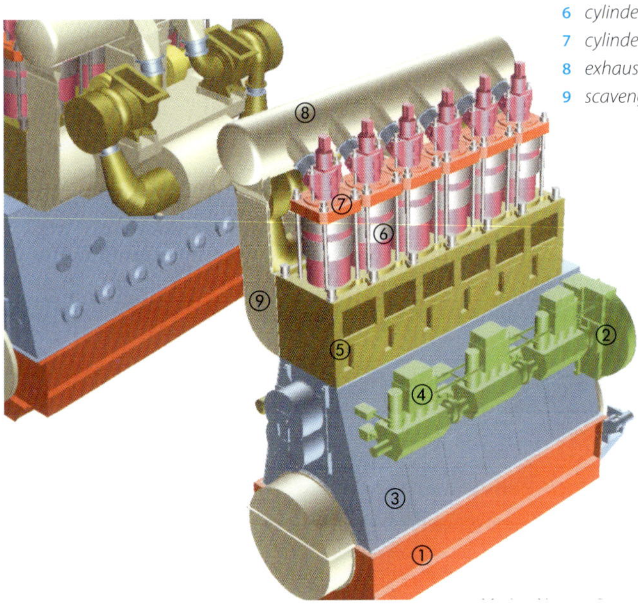

These very large, low-speed engines with a number of revolutions from 60 to 250 rpm with a cylinder bore of 260 to 1080 mm are mainly used for propulsion of very large container ships, oil- and ore tankers and sporadically in diesel-power plants for the generation of electricity.

▼

Large high-speed container ships are equipped with the largest two-stroke crosshead engines in the world with a shaft power of up to 100,000 kW.

▲
A large container ship leaves the port of Rotterdam.

They always directly drive a propeller or generator, so the number of revolutions is always equal to that of the propeller or the generator. This makes their installation transparent and straightforward.
All these engines run on HFO and are for the greater part built by license holders in South Korea, Japan and China, as the shipbuilding industry is situated in this region. The dry weight of these engines range from 100 to 3,300 tons.

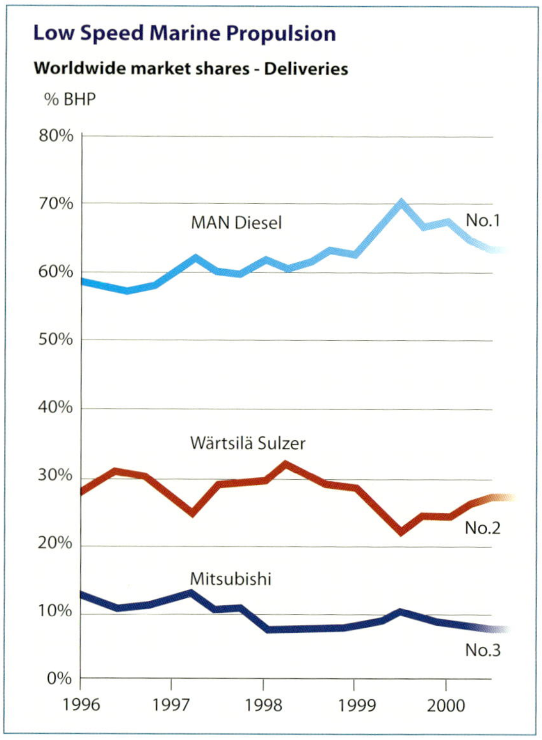

▲
There are only three two-stroke crosshead-engine manufacturers.

Engine builders

There are only three manufacturers that build these big two-stroke crosshead engines:

MAN Diesel	a German-Danish factory;
Wärtsilä Sulzer	a Finnish factory which manufactures the traditional Swiss two-stroke engines;
Mitsubishi	a Japanese factory.

▼
Globally, there are a considerable number of engine factories which build engines of a certain design 'under license'.

▼
The engine program of the Wärtsilä Sulzer RTA two-stroke crosshead engines with a cylinder bore of 380 to 960 millimetres. The number of revolutions varies from approximately 50 to 200 per minute.
See also Chapter 21, *Diesel-engine manufacturers.*

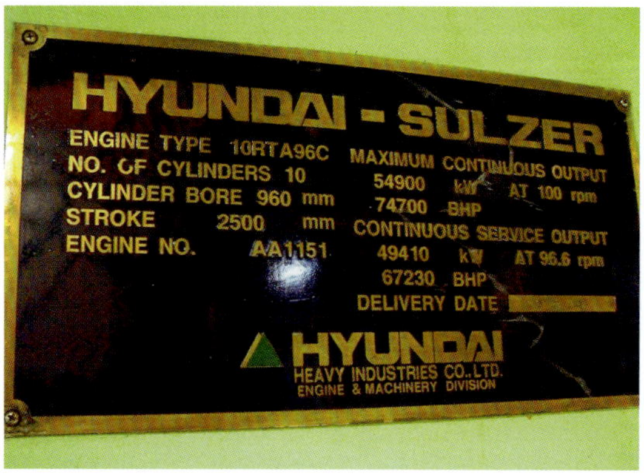

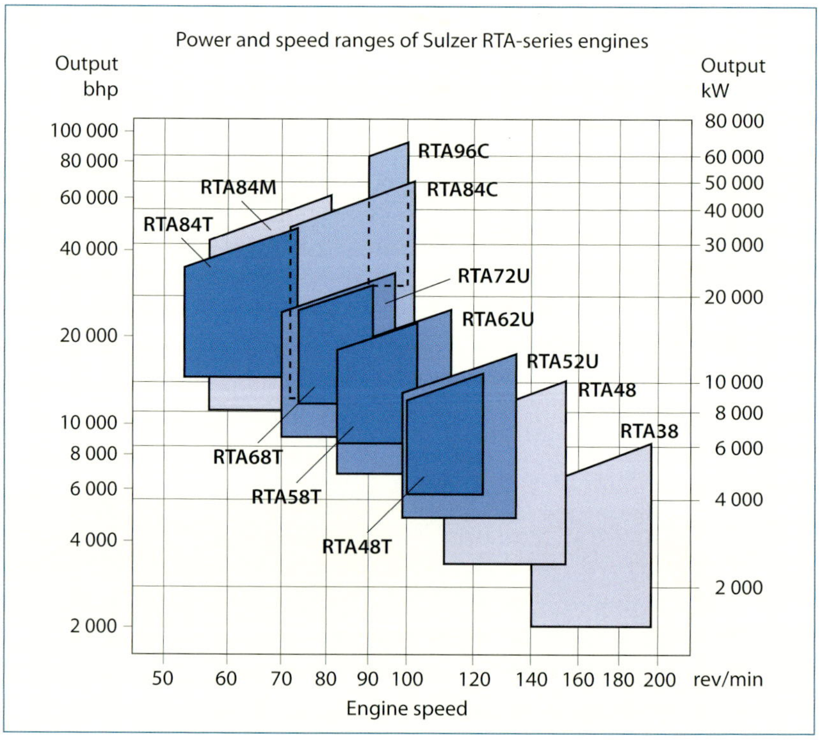

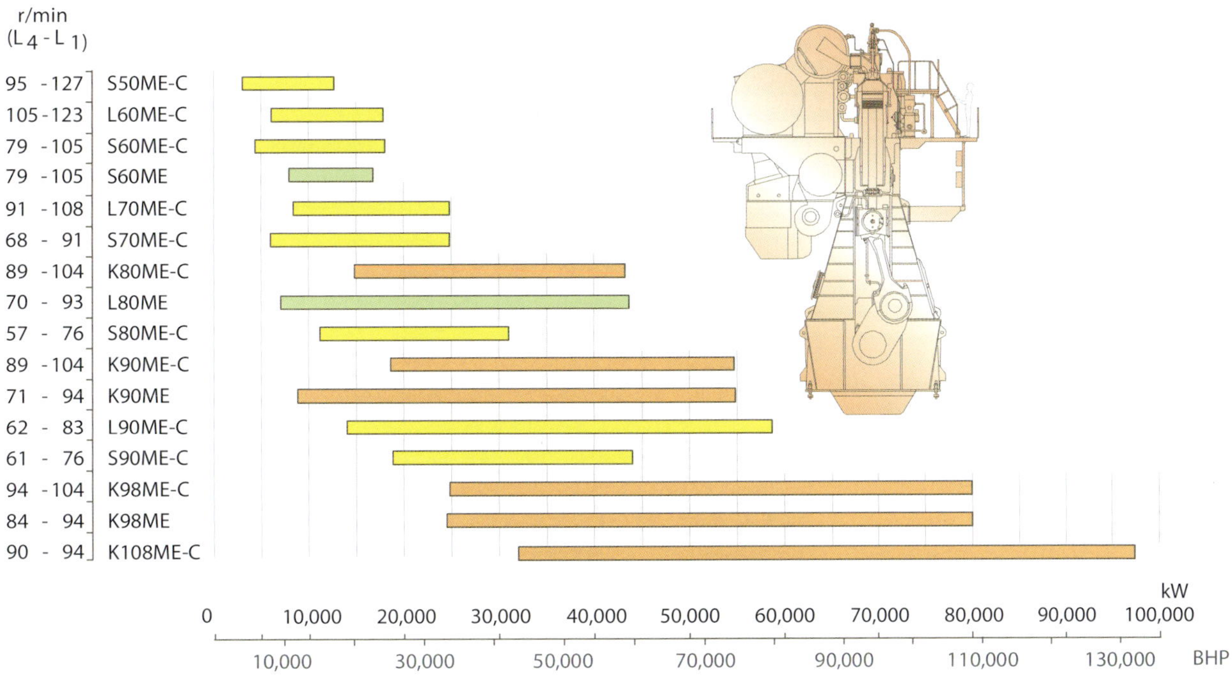

▲

The engine program of the MAN Diesel ME-series.

The ME is the version with electronically controlled fuel injection. Today, 2012, even smaller cylinder bores are delivered in the ME-version.

See also Chapter 21, *Diesel-engine manufacturers*.

Construction

These engines are entirely manufactured from a partially welded construction, such as the engine foundation and A-columns including cast parts such as the cylinder axis and cylinder liner. The cylinder head is made using forged steel as is the crankshaft.

▼

A cylinder block with liners.

▼

The welded A-frame.

▶

Building a crosshead engine.

1 *crankshaft with flywheel*
2 *A-frame*
3 *cylinder block*
4 *cylinder liners*
5 *cylinder-head bolts*
6 *crankshaft location*

▼

The camshaft.

1 *cylinder block*
2 *cylinder liner*
3 *cylinder-lubricating point*
4 *cylinder-head bolt*
5 *camshaft*
6 *cam*
7 *camshaft bearing*
8 *sleeve-coupling bearing, hydraulic*

On the specially adjusted ship foundation plate the engine bedplate is affixed using many holding-down bolts. The forged crankshaft is situated inside the engine bedplate. It can be as long as 25 metres with a weight of 150 tons!

The welded-steel A-columns of, for example, two, three or four cylinders are then placed on top of the bedplate.

These are joined together with tie bolts. The cast iron cylinder beams are placed on the columns with very long drawing rods that join the three parts together.

▲

A large crankshaft in a two-stroke crosshead engine with a turning circle of 1680 mm.

The forged-steel cylinder heads are attached to the cylinder axis and they contain the hydraulically driven central exhaust valve, two or three fuel valves, a relief valve, an indicator cock and a starting-air valve.

The **driving gear** comprises a **piston** to which a **piston rod** is connected, a **crosshead** with **guide shoes** attached to it that transfer the lateral forces of the crankshaft mechanism to the **crosshead guide** fixed to the A-frame.

◀

One spare crosshead.

1 crosshead pin
2 crosshead guide shoe

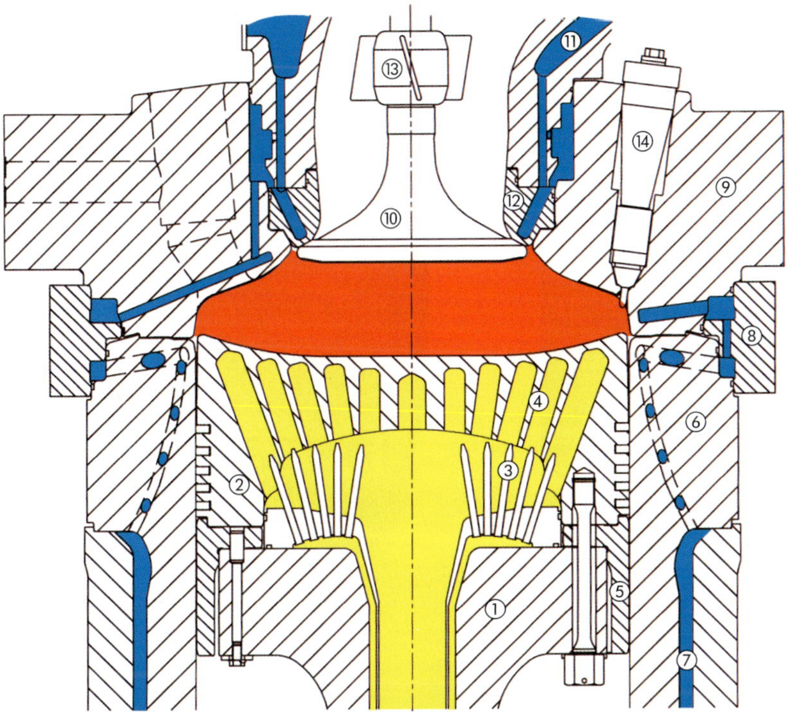

◀

The combustion chamber.

1 piston rod
2 piston
3 inlet drillings for lubricating oil
4 honeycomb holes
5 guide ring
6 cylinder liner
7 bore cooling
8 coolant circulation
9 cylinder head
10 exhaust valve
11 exhaust-valve line
12 valve seat
13 valve rotator
14 fuel injector

blue: coolant

yellow: lubricating-oil for cooling

red: exhaust gases

▲

A piston with piston rod.

The flanged piston rod has almost the same diameter as the piston.

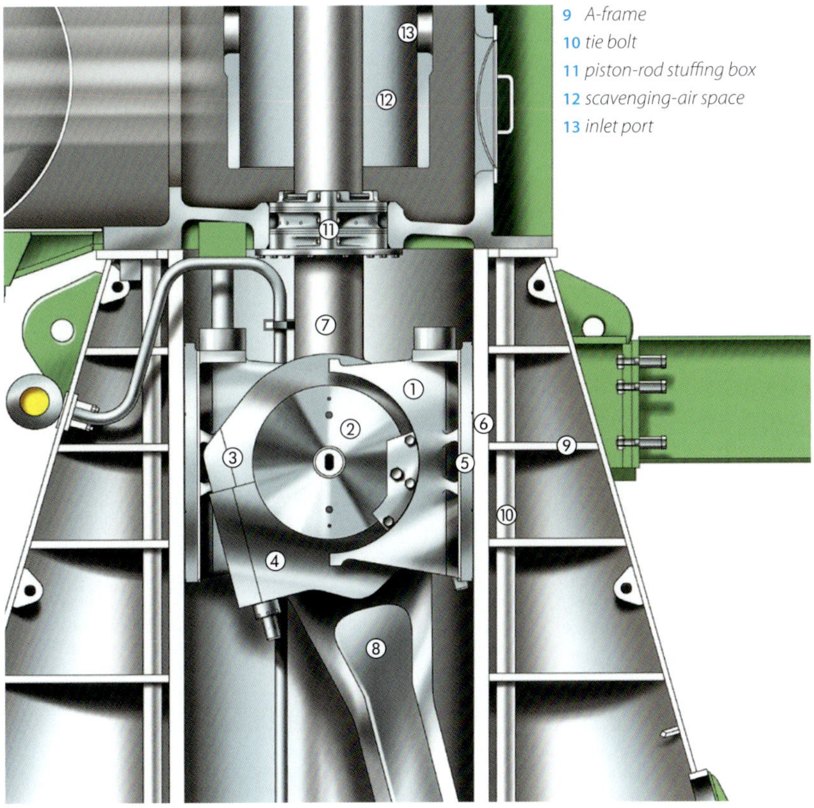

▼

The crosshead.

1 crosshead
2 crosshead pin
3 upper bearing cap
4 lower bearing cap
5 crosshead guide shoe
6 crosshead guide
7 piston rod
8 connecting rod
9 A-frame
10 tie bolt
11 piston-rod stuffing box
12 scavenging-air space
13 inlet port

▶ A piston-rod stuffing box for the separation of the scavenging-air space above the sealing rings and the crankcase below the sealing rings.

The forged **connecting rod** is fixed to the crosshead and forms the link to the crankshaft with a **crankpin**. The **piston-rod stuffing box** ensures an excellent seal between the **scavenging space** surrounding the inlet ports in the cylinder liner and the **crankcase**.

▲ A cylinder liner.

1 *inlet ports*
2 *support collar with the cylinder block*
3 *cylinder lubricating-oil inlet holes*
4 *thickened upper collar*
5 *bore cooling*

▼ Exhaust-valves at the factory for repair.

▲ High-pressure fuel pumps (1).

For most existing engines the camshaft for driving the fuel pump and the hydraulic pump driving the exhaust valves are driven from the crankshaft either using cogwheels or a chain.
The latest types have a very advanced fuel system such as the common-rail system and no longer have their own camshaft.

Also see Chapter 9, *Fuel-injection systems and Chapter 21, Diesel-engine manufacturers.*

The lubrication of these crosshead engines consists of two separate systems:

- the cylinder lubrication system;
- the crankcase lubrication system.

▶
A cylinder lubrication unit for a cylinder with sixteen small independent plunger pumps. The lubricating-oil is pre-warmed.

Cylinder lubrication system

A series of small plunger pumps driven either independently by the engine itself, or by a frequency-controlled electromotor that forces special cylinder lubricant through the drillings in the liner and is spread around the cylinder via grooves machined in the cylinder liners. This is done to ensure that the lubricant is injected at the right time and at the most effective piston height. The cylinder lubricant lubricates the piston, the piston rings and the cylinder liner. A certain amount of the lubricating oil should combust odourlessly and ashlessly, another part seeps along the liner and drips to the bottom of the scavenging-air space from which it is manually tapped.

Also see Chapter 11, *Lubrication of engines.*

Crankcase lubricating system

This system is operated by electromotor-driven displacement pumps.
The lubricating oil is pumped past the main- and crankpin bearings, the crosshead and the camshaft. In addition, the piston is often internally cooled using this circuit. Furthermore, the exhaust-valve drive and the reserve gear use this system.

Also see Chapter 11, *Lubrication of engines.*

▲
A crosshead lubricating-oil pump.

This is a displacement pump and provides the crossheads, the directional change of the camshaft and the exhaust-valve actuators with lubricating-oil

Cooling system

Large two-stroke engines have two cooling-water systems with two temperature levels, the HT-(high temperature) and the LT-(low temperature) system. **The high-temperature system** is used to cool the cylinder liner, the cylinder head and the exhaust valve casing. Pistons are also sometimes cooled using this system.

The low-temperature system cools the scavenging air and the lubricating oil.

Also see Chapter 10, *Cooling diesel engines.*

▶
Two heat exchangers of the lubricating-oil system.

Air supply

The **air supply** is usually provided by several turbocharger groups that compress the combustion air; the air is subsequently cooled and led to the scavenging-air pipe which runs along the length of the engine. The air flows into the cylinder via the scavenging-air space surrounding the cylinder liner when the piston is in BDC and releases the inlet ports.

In case of a **low load** the turbocharger groups provide too little air for an optimal combustion cycle and an electrically driven auxiliary charger is automatically switched on.

Also see Chapter 12, *Air supply*.

▲

A cut-away diagram of a turbocharger group.

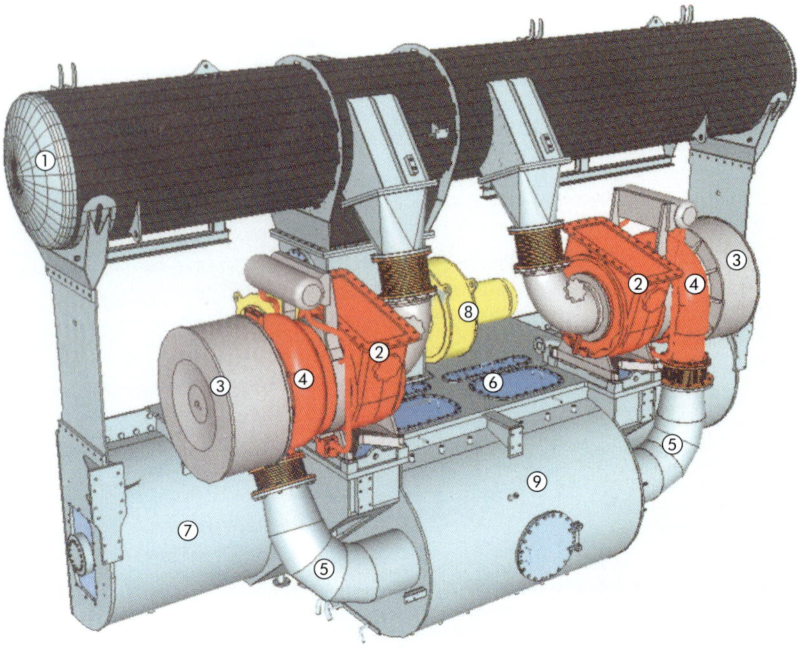

▼

An overview of the turbocharger of a two-stroke crosshead engine.

1 *exhaust-gas ducts*
2 *exhaust-gas turbine*
3 *air filter*
4 *air-compressor space*
5 *compressed air*
6 *air cooler*
7 *air inlet*
8 *auxiliary turbocharger*
9 *mist catcher*

Fuel

For conventional fuel pumps the fuel supply is provided by a camshaft-driven plunger pump. The cylinders are fitted with two or three fuel atomisers surrounding the centrally positioned exhaust valve and injecting the fuel almost horizontally into the cylinder.

Both plunger as well as valve-controlled fuel pumps are used. The recently (2012) marketed electronically controlled common-rail systems have numerous advantages and are increasingly applied.

Also see Chapter 9, *Fuel-injection systems*.

▼

The gear drives from the camshaft of:

1 *the fuel pump;*
2 *the hydraulic-oil pump for the exhaust valve.*

▲

One of the fuel injectors.

There are usually two or three assembled around the central exhaust valve.

Reversing the direction of rotation

Two-stroke engines up to a cylinder bore of about 600 mm are often equipped with an adjustable-pitch propeller which allows the engine to operate in only one direction of rotation.
For larger engines and therefore larger propellers one opts for a fixed-pitch propeller, which requires the rotation direction to be reversible. This is made possible by moving or turning the camshaft under a lubricating-oil pressure causing the starting-air valves and fuel injection to the various cylinders to obtain a different operating order.

Also see Chapter 14, *Starting systems of diesel engines.*

Number of cylinders

This can vary from four to fourteen cylinders and the capacity from approximately 1500 to 100,000 kW. All these engines have in-line cylinders.

Comment

The capacity of the small two-stroke crosshead engines overlaps the power supply of the 'larger' four-stroke trunk piston engines. For the latter the power can reach approximately 30,000 kW.

The choice of a two-stroke crosshead engine or a four-stroke trunk piston engine depends on many factors such as:

- desired power output;
- desired number of revolutions;
- available space;
- weight of the engine;
- specific fuel consumption;
- specific lubricating-oil consumption;
- economic aspects;
- ship yard lay-out for certain types of ships;
- delivery time;
- service package.

The client's preference for, for instance, other shipbuilding companies and/or investors' wishes often play an important role as well.

The 'Berlin Express' of Hapag-Lloyd has a capacity of 8600 TEU and an engine output of 68,000 kW. It is one of the largest container ships launched in 2004.

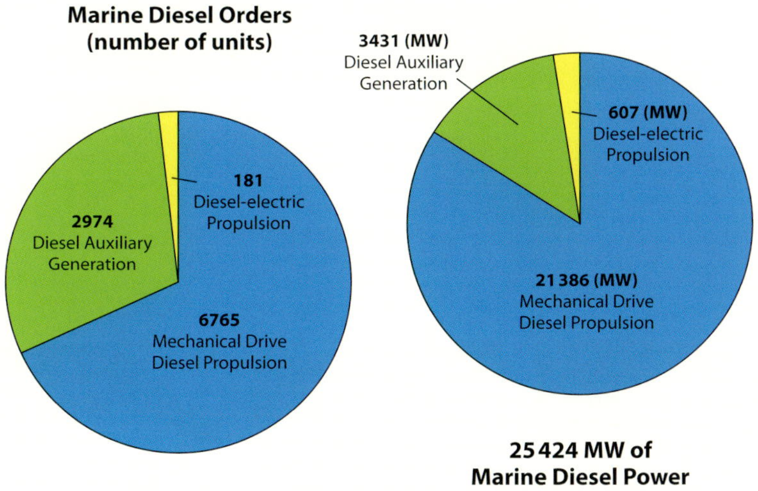

A overview of the orders per year for diesel engines with a shaft power exceeding one Megawatt.

blue: left – the number of propulsion installations
green: left – the number of generator sets
yellow: left – the number of engines used for diesel-electric propulsion
blue: right – the total power output
green: right – the total power output
yellow: right – the total power output

>CH 7

Use of materials for diesel engines

The basis of every engine since they first appeared in the market at the end of the nineteenth century is a top-quality cast engine block. Apart from cast iron, steel and light metals are the most important materials used in engines.

A cast engine block aligned for the initial machining process.

The following sub-divisions can be made:
- general material usage;
- specific materials used in the four engine categories according to their classification;
- special materials and manufacturing processes.

7.1 General use of materials

Cast iron was the material used in the very first internal combustion engines and is still often used for many engine parts. Forged steel and later **welded steel sheeting** are also generally used. The properties of these materials are often enhanced by adding different metals, thereby creating an alloy with improved properties.
Materials used in the manufacture of high-load parts such as exhaust valves, piston crowns, bearing caps and fuel pumps have also further been developed, enabling them to resist the increasingly higher **mechanical, thermal** and **chemical** loads found in modern industrial engines.
Firstly, we will study the various materials, their properties and general use in the engine manufacturing industry.

7.2 Cast iron

Cast iron is an iron-carbon alloy with a minimum 2.5% carbon content.
The carbon percentage determines the iron alloy properties.

Properties:
- thin liquid, good pouring characteristics for casts (+);
- hard (+);
- good machining properties such as planing, milling and drilling (+);
- relatively inexpensive (+);
- rigid(+);
- good running characteristics in, for instance, cylinder liners (+);
- sound-dampening (+);
- corrosion resistant (+);
- low shrinkage (+);
- forms an alloy with the addition of other metals, for example, with magnesium to give improved characteristics; the so-called nodular cast iron (+);
- heavy and often thick-walled (–);
- low melting point (± 1150 °C) (+);
- low tensile strength (–);
- brittle (–);
- not malleable (–);
- possible crack formation with sudden temperature variation (–);
- not weldable (–).

Cast iron is often used for engine blocks, cylinder liners, cylinder heads, exhaust-valve casings, inlet lines and piston rings.
It is further used in components with comparatively low mechanical load requirements such as supports, air ducts, covers and cooler casings.

▶
Casting an engine block in the Caterpillar MaK foundry, Kiel, Germany.

▲

The casting of a large block for a four-stroke diesel engine.

The casting time is very important for the mould quality. A large block with a weight of 115 tons must be cast within 100 seconds. If this time is exceeded, the quality is such that the block is rejected. The costs, approximately € 100,000.- for manufacturing the mould and the melting-down of the base materials, are lost.

▲

A cast engine block just out of the mould.

The deburring and cleaning of the rough block has yet to take place.

◀

A cast block of an in-line engine in manufacture.

7.3 Steel

This is an iron-carbon alloy with a maximum carbon content of 1.7%.

Properties:

- malleable (+);
- ductile (+);
- weldable (+);
- light in welded constructions (+), for example the A-frame for crosshead engines;
- good machinability, such as planing, milling, drilling, grinding and polishing (+);
- high melting point of approximately 1450 °C (+);
- **over 0.3% carbon hardness**, surface hardening possible (+);
- soft (–).

Steel in a steel plate form is often used in welded constructions such as bedplates, crosshead-engine A-frames, air inlets in crosshead engines, air-cooler casings, various supports and exhaust-gas manifolds.

▲

The manufacture of two-stroke crosshead engines at MAN Diesel, Frederikshavn, Denmark (2005).

Foreground: a welded steel crankshaft bedding or lower crankcase of a two-stroke crosshead engine.
Background: a welded steel A-frame.

7.4 Cast steel

This is an iron-carbon alloy with a carbon content of between 0.5 and 2%. The base material is white pig iron.

Properties:
- can be cast in crude moulds(+);
- extremely high tensile strength (+);
- malleable (+);
- high melting point (+);
- possibility of strong yet lightweight constructions (+);
- viscous (–);
- shrinkage is twice as high as that of cast iron (–);
- due to its viscosity, only used in simple casting moulds (–);
- requires annealing to resist high casting pressures (–).

Casting properties are improved by adding manganese and silicon. By adding 0.5 to 1.0% molybdenum, a higher tensile strength at higher temperatures is achieved. In older and larger crosshead engines, it is used in, for instance, cylinder heads, high-pressure stop-valve housing, piston crowns and cylinder heads in four-stroke engines.

7.5 Forged steel

This is an iron-carbon alloy with a carbon content of between 0.15 to 0.5%.

Properties:
- malleable (+);
- high tensile strength also at high temperatures (+);
- ductile (+);
- machinable (+);
- coarse moulding which requires machining operation after casting (–).

Today generally used for the cylinder heads of crosshead engines due to the mechanical and thermal load increases.
Crankshafts for all types of engines, connecting rods, piston rods and camshaft components also are made of this material.

7.6 Steel alloys

Alloys are added to steel, an iron-carbon alloy, in order to enhance certain physical properties, or to create new properties.

Alloy elements are metals, such as chromium, nickel, molybdenum, tungsten, manganese, vanadium and boron.

The various alloy properties provide steel with particular characteristics, for example:

chromium	– increases the hardness and toughness
nickel	– increases the tensile strength and is anticorrosive
molybdenum	– increases the tensile strength and maintains the hardness up to 600 °C
tungsten	– makes steel heat resistant
manganese	– increases the tensile strength and reduces wear and corrosion
vanadium	– increases ductility
boron	– increases wear resistance of cast iron cylinder liners

▼
A cast-steel cylinder head of a two-stroke crosshead engine.

▲
The shaft and the blades in the exhaust-gas turbine of the turbocharger are manufactured from high-alloyed steel for resistance to high temperatures and corrosion caused by chemicals.

The crankshafts of high-load fast-running engines are often made of **chromium nickel- or chromium molybdenum steel.**
The bearing surface of the crankpins and journal bearings are then hardened, thus increasing their resistance to wear.
Steel used in exhaust valves is often alloyed with 8 to 12% chromium and, for example, silicon. 'Nimonac 80A', an alloy of 80% nickel, chromium, titanium and aluminium is resistant to the high temperatures and the corrosive products contained in exhaust gases. This is, however, an expensive material and today frequently used in the exhaust valves of two-stroke engines. The cost of this material has decreased considerably in recent years.

◄
Steel Alloys.

These are used for the manufacture of parts for diesel engines, for example:
1 *exhaust valves*
2 *exhaust-valve seats*

◄
Crankshafts are the most heavily loaded parts in a diesel engine; they are manufactured from high-alloy forged steels.

Steel alloys are used in a large number of diesel-engine parts, such as toothed wheels, fuel pumps, shafts, cams, atomisers, valve springs and valves. Each component has different material requirements, this is the reason that there are so many different alloyed steel types.

7.7 Aluminium

Properties:
- high fatigue resistance (+);
- good bearing properties (+);
- light-weight (+);
- good heat conductor (+);
- good strength properties at high temperatures (+);
- high wear rate (–);
- high coefficient of expansion (–).

By forming aluminium alloys, certain properties are enhanced.

Copper, nickel and magnesium ensure good heat conduction and favourable strength properties at high temperatures, the so-called Y-gamma alloy. This is called an **aluminium-copper alloy**.
The 'light metal' alloy formed with **silicon, copper, nickel** and **magnesium** has a low coefficient of expansion and a reasonable wear rate. This is called an **aluminium-silicon alloy.**

Use
'Light metals' with aluminium as the main component are used for engine blocks of small diesel engines, the pistons of fast-running diesel engines, the piston skirts of medium-speed diesel engines, tri-metal axle bearings in all engines and, for example, for impellers in the air inlets of the turbocharger groups. Today, many crankcase and camshaft

covers are also made of a light metal. The 'hot box' of Wärtsilä engines, which maintains the high temperature found in the common-rail fuel injection system, is made entirely of light metals.

▶
The impeller of the air blower in a turbocharger is manufactured from an aluminium alloy, the so-called light metal.

7.8 Ceramic materials

Technical ceramic has unique properties such as extreme hardness and high wear rates. It can consist of the following compounds:
Al_2O_3, ZrO_2, TiC, Si_3N_4 en $ALTiO_2$.
These are:

- aluminium-, silicon-, zirconium-, titanium-, beryllium- en magnesium oxides;
- silicon-, boron-, titanium- en aluminium nitrides;
- silicon-, boron- en titanium carbides;
- composites.

They are easily applied as a thin coating on, for instance, piston rings.

▼
Light-metal pistons are generally used in high-speed diesel engines.

The main advantage is their light weight which ensures that the acceleration and deceleration forces of the piston remain within the maximum limits.

7.9 Specific materials for engine parts; engine classification according to the four categories

7.9.1 Engines from 0 to 100 kW, high-speed, four-stroke, fuel MDO: category I, small industrial diesel engines

The engine blocks are often constructed of cast light metal or cast iron. An added advantage of a light metal is, of course, its weight.
The **cylinder liners** or bushings of a light-metal block need to be fitted separately, while a cast-iron block does not require an separate cylinder liner. Moreover, a separate cylinder liner or 'bushing' is regularly applied to the somewhat larger engines of this category.

The pistons are often made of light metal and the entire cylinder head can be constructed from both light metal and cast iron.

The crankshaft of high-load engines is often made of high-alloyed forged steel, but for standard engines normal unalloyed carbon steel suffices.
However, the crankpin and the shaft journals are hardened.

▼
A diesel-engine series for pleasure yachts with from front to back increasing power outputs.

All diesel engines are standard diesel engines and later modified for use on ships. This is known as 'marinisation'.

7.9.2 Engines of 100 to 10,000 kW, high-speed, four-stroke, fuel MDO: category II, larger industrial diesel engines

The greater the power output, the more often cast iron is used for the block. In this case, the crankshaft is also made of forged iron that, depending on the load, may or may not be alloyed. The **cylinder heads** are usually manufactured of cast iron. Pistons are often made of **light metal.**
The piston mass is very important when the number of revolutions is high because of the acceleration and deceleration forces working on the moving parts. The lighter the piston mass, the lower the acceleration and deceleration forces.
The piston of high-load high-speed engines is often constructed from **two components**, the **cast-iron piston crown for** the high thermal- and mechanical load, and the **light-metal piston skirt enabling a** good piston conduction and absorption of the lateral forces exerted on the piston by the crank-connecting-rod mechanism.

With a large power output cylinder liners are of cast iron and water cooled.

7.9.3 Engines of 500 to 30,000 kW, medium-speed, four-stroke, fuel HFO: category III, large industrial diesel engines

Generally, these engines are highly powered and are both mechanically and thermally highly loaded. HFO is also paramount in the decisions taken in, for instance, the choice of material for the exhaust valves.
The **engine block** is made of alloyed cast iron. This is often **nodular cast iron** compounded with magnesium creating a compound with a high tensile strength.

Cylinder liners are made of **special cast iron** with superior running properties and often have a built-up upper collar for the absorption of the high compression and combustion pressures. The upper collar contains bore-cooling and protrudes from the engine block. This has the added advantage of reducing the size and weight of the engine block.

▲
A diesel engine with a large power output such as this sixteen-cylinder MTU engine which has a high thermal- and mechanical load uses many high-alloyed parts, for example, the crankshaft, the cylinder block and the exhaust valves.

◄
A high-speed Caterpillar diesel engine for the propulsion of an inland-navigation tanker.

▼
A block of an in-line engine rotated 180° for the mounting of the underslung crankshaft.

1 *engine block*
2 *crankpin*
3 *counterweight*
4 *lower bearing cap*
5 *bearing-cap bolts*
6 *hydraulic nuts*
7 *joint interface between bearing cap and block*

▶
A cylinder liner with a built-up top collar containing drilled-out cooling ducts.

The liner is conserved after repair and wrapped partially to prevent corrosion.

▼
Assembling the forged-steel crankshaft must be done precisely and carefully.

The engine block is rotated 180° for assembling the crankshaft with bearings and bearing caps.

The crankshaft is of forged steel and as in all four-stroke engines 'underslung'. The **piston consists** of a **cast-iron crown** and a **cast-iron skirt.** Tri-metal bearings are used for the crankshaft and the crankpin **bearings**. Today, this is the case in virtually all engines.

The **cylinder heads** are made of **cast iron** or **cast steel.** The **exhaust valves** in particular are heavily loaded. Sodium and vanadium compounds in the fuel can create a sticky deposit on the valve seat which is a very aggressive corrosive. This is called high temperature corrosion, or abbreviated, HTC. Low temperature corrosion, or abbreviated, LTC, is caused by sulphur in the fuel found at much lower temperatures. This occurs, for instance, on the cylinder liners at extremely low engine loads.

Also see Chapter 8, *Fuels, fuel-line systems and fuel cleaning.*

Inlet valves are often made of high-quality **carbon steel**. This certainly does not suffice for the exhaust valves and therefore high-chromium steel with, for example, 8 to 12% chromium content and also silicon is used.
The **exhaust valve** is often rewelded with so-called armour steel, stellite. This is a very hard and wear-resistant alloy containing **chromium, tungsten, cobalt** and **carbon**.
Sometimes the entire valve or the valve disc and part of the valve stem are made of '**Nimonac 80A**', an **alloy with 80% nickel, chromium, titanium** and **aluminium**. This material is costly, but gives a tremendous increase in the operating lifetime of the valve.

▲
Cylinder heads during maintenance at engine works Bolier, Dordrecht, The Netherlands.

▲
The exhaust-valve seat is ground.

To improve the resistance of exhaust valves against high-temperature corrosion, they are often completely or partially manufactured from a high-alloyed steel type, Nimonac.

7.9.4 Engines of 1500 to 100,000 kW, low-speed, two-stroke crosshead engines, fuel HFO: category IV, extremely large industrial diesel engines

These are exceptionally large and heavy engines. Today, they are the only engines that are constructed from parts.

▲

The cylinder head of a crosshead engine with in the background the turbocharger.

◀

The welded A-frame of a crosshead engine under construction at MAN Diesel, Frederikshavn, Denmark.

A large number of these engines are built from **welded steel-plate constructions** and a smaller number from **castings.**

The crankshaft of these large engines rests on the engine bedplate, a welded construction with incorporated crossbeams in which the forged-steel supports are integrated below the crankshaft bearings.

The A-shaped engine frame constructed from welded steel plates is mounted on the bedplate**.** The crosshead guides, used to convey the crosshead and absorb the shearing forces of the crosshead, are welded or bolted against the columns found in the A-frame.

The cylinder beam rests on the columns which in larger crosshead type engines consist of different parts bolted together.

Bedplates, columns and the **cylinder beam** are linked by very long vertical **alloyed steel** tension **bolts**, which run through the columns in ducts. The bolts in the ducts are fastened in some places with stay-bolts in order to prevent shearing vibrations of the tie rods. These vibrations could cause cracks in the long term. These are as in all other large linkages hydraulically pre-stressed.

▲

The welded steel crankshaft bed of a crosshead engine; it is of utmost importance that the crankshaft is placed in line.

▼

Parts of crosshead engines under construction.

1 *forged crankshaft*
2 *cylinder block with cylinder liners and camshaft*
3 *crankshaft bedplate*

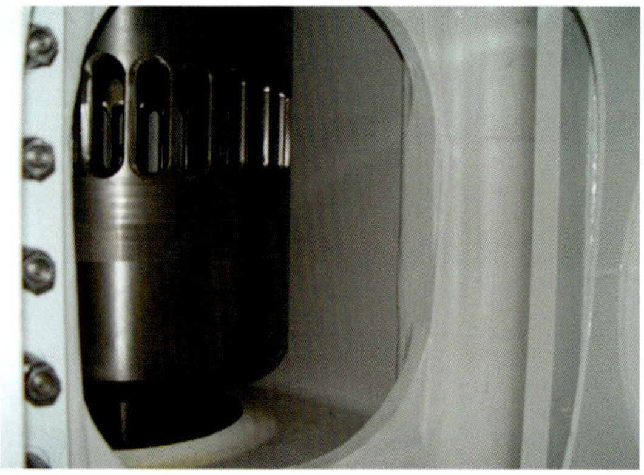

▲

The manufacture of a crosshead engine.

1 *camshaft*
2 *cylinder liners*
3 *cylinder head*
4 *exhaust-valve casing*
5 *cylinder-head bolts*
6 *exhaust-valve casing bolts*
7 *fuel injector*

▲▲

The scavenging-air space at the cylinder liner with the inlet ports.

▲

Forged steel connecting rods with mounted crankpin bearings and crosshead-pin bearings of a two-stroke crosshead engine.

The **cylinder beams** used to be made of cast iron. Today however, they are generally made of cast steel, which has superior mechanical properties. The bottom part of the cast-iron **cylinder liners** is situated in the cylinder beam. The **forged-steel cylinder heads** are fixed to the cylinder beam by means of long bolts.

The cooling of the cylinder liner and cylinder head occurs through drilled canals, the so-called 'bore-cooling'.

The pistons are made of forged steel or cast steel in view of their high thermal and mechanical loads.

The piston rod, connecting rod and the crankshaft are made of top quality malleable steel; special attention is paid to the crankshaft.
For crosshead engines unalloyed carbon steel usually provides sufficient strength, which is in contrast to engines with a higher load, such as medium- and high-speed engines.

▶

The forged crankshaft; manufactured in sections, and fixed together using shrinkage joints.

1 *crankshaft*
2 *crankweb*
3 *crankpin*
4 *weight economy*
5 *roundings*

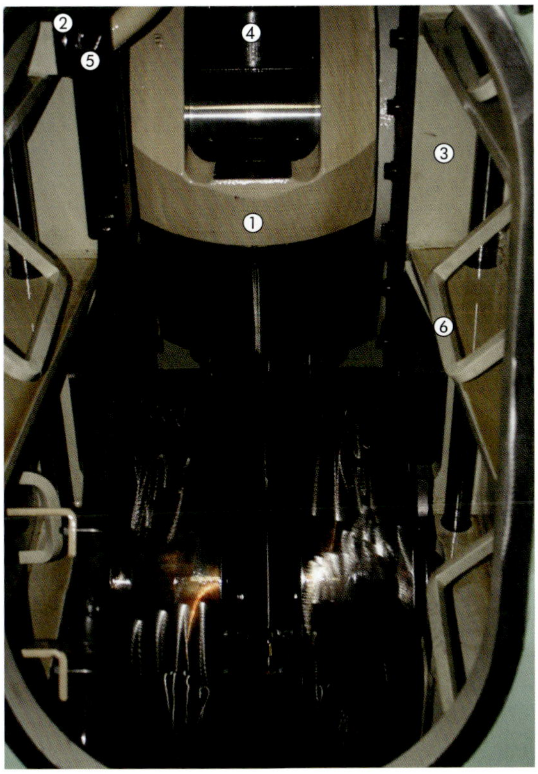

◄
The mounted crosshead, seen from above.

1 *crosshead*
2 *guide shoes – four pieces*
3 *crosshead guides – four pieces*
4 *place for the fixed piston rod; not mounted here*
5 *A-frame*
6 *step*

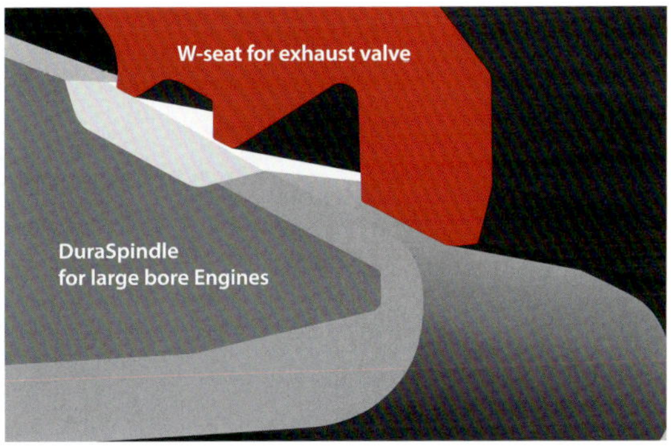

▲
The valve seat also has a 'Wave Cut' profile as that of the cylinder liner in a MAN Diesel engine.

Lubrication here requires special attention. Unlike four-stroke engines, two-stroke engines have no 'rest' revolution. Therefore the crosshead bearings are always subjected to great forces from the piston. Without a lubricating-oil film, wear and tear of the bearings increases.
One of the measures taken to prevent premature wear and tear at the boundary lubricating conditions is a surface treatment of the steel crosshead pin. **Hard chromium plating, a grind or super-finishes are often applied**.

Also see Chapter 13, *Driving gears*.

Piston rings

These usually consist of **fine-grained perlitic cast iron** with **flake graphite** which has excellent running properties in the cast-iron cylinder liner.
The piston ring hardness lies between 2000 and 2400 N/mm², hardness Brinell.
The **cylinder liners** are also made of **perlitic cast iron** with flake graphite and therefore also have good running properties. Often alloys are added in order to improve wear and tear resistance. The **hardness** is considerably lower than that of the piston rings, namely 800 N/mm², hardness Brinell.

Exhaust valve

All modern crosshead engines are equipped with one centrally placed hydraulically controlled exhaust valve. In large engines the valve diameter is approximately 40 centimetres.
All material consists of a high-alloy type of steel usually covered with a rewelded stellite layer.
Alloys such as '**Nimonac 80A**' are also often used.

▼
A series of exhaust valves manufactured from Nimonac, a high-grade material.

The total life of these valves given by the factory is up to 96,000 hours!

▼
The finished surface of a cylinder liner. Shown here is a special finishing that deviates from the normal honing cross-hatching, the 'ribbed' surface of diesel engines of MAN Diesel. Also called 'wave cut'.

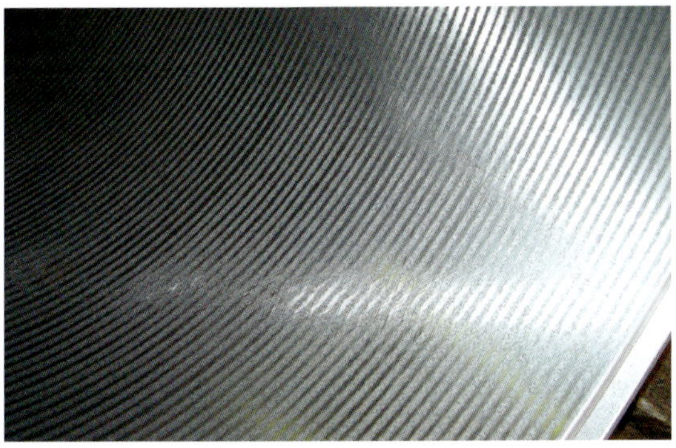

▲

The cast-iron exhaust-valve casing.

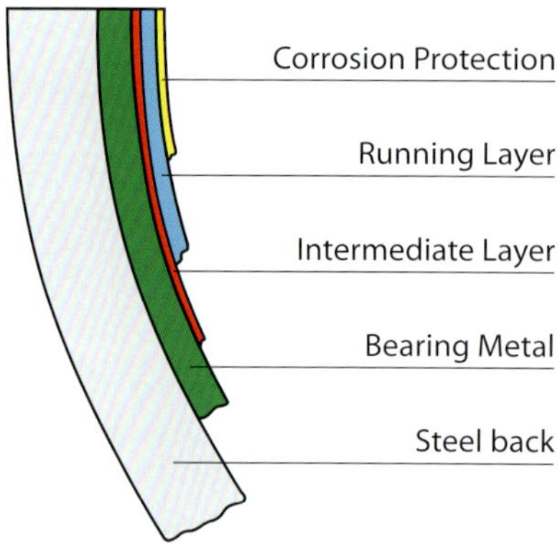

▲

The manufacture of a tri-metal bearing. Most of the layers are very thin.

The exhaust-valve casing is made of high quality cast iron and is today the only remaining engine component with cooling water ducting in the casting. The remainder of the cooled components such as the cylinder liner, piston and the cylinder head are of course cooled using 'bore-cooling' technology.

Bearing materials

Due to the heavy load exerted on the **bearings** today one often uses **tri-metal** types. These can bear high loads and do not require custom-manufacturing unlike the previously used 'soft' white-metal bearings.

They generally consist of a steel bearing cap with a lead-bronze layer, a very thin nickel barrier and the actual sliding material, a 'white metal'. 'The white metal often comprises mainly aluminium and tin in an 80% aluminium and 20% tin ratio. This material has excellent properties, such as high fatigue strength and good running properties. High loading can be achieved with an **aluminium alloy where either a zinc or silicon component is added**.

▶

A complete exhaust valve (1) right and two spare cylinder liners (2). Far left cylinder-head bolts (3).

7.10 Special finishes and heat treatments

7.10.1 Nitration

Nitration is thermal treatment in which the surface layer of virtually all iron parts can be nitrogen enriched at temperature ranges of 500 °C to 600 °C. The nitrogen enrichment increases the surface tensile strength thus achieving a higher corrosion and wear resistance as well as a higher oscillatory resistance. The nitrate layer consists of a very thin outerlayer of some thousandths of a millimetre (micro millimetre) with a hardness of about 800 to 1200 N/mm², hardness Brinell.

Application

It is often applied in overhauling cylinder liners and crankshafts.

7.10.2 Annealing

This occurs in temperature ranges of 450 °C to 650 °C. It reduces the internal stresses in the constructions caused by the electric welding processes.
After a certain 'glow' time the work piece or the construction needs to be cooled slowly in order to prevent development of new stresses.

Application

Welded-steel constructions in engines such as bedplates, A-frames and air-inlet manifolds.
Overhauled engine parts such as welded cylinder covers, pistons and exhaust valves.

7.10.3 Surface hardening

During this process the parts are only heated and hardened at the surface, thus causing only minor shape and size changes. Heating takes place by means of induction or with a gas jet, the so-called 'flame hardening'. The latter is sometimes applied when overhauling small crankshafts. After grinding the valves to a smaller size, the ground surfaces must be hardened.
There are many other surface treatments. Surface hardening causes tensile stress on the surface, which subsequently enables a higher load.

Application

Crankshafts, piston pins and crosshead pins.
Steel crankshafts of a 0.5 to 0.6% carbon content are heated to approximately 720 °C and then subsequently water cooled. The core of the crankshaft remains fairly soft, but on the outside a hard layer of approximately 0.9 to 2.0 mm is formed.

7.10.4 Applying surface layers

Some engine components' surfaces run either continuously or discontinuously across another surface, such as:

1 piston rings on the cylinder liner;
2 the stuffing box on the connecting rod of two-crosshead engines;
3 the bearings of crankshafts, camshafts and piston pins.

◄ The honing pattern on the running surface of a cylinder liner.

▼ Shown is a very hard, wear-resistant layer sprayed on the valve stem, followed by very fine sandpapering.

◄◄ A cylinder cover in an electric oven for annealing.

The temperature and the timing of this treatment are very important.

▲
Forged-steel connecting rods with above the crosshead and guide shoes left and right the crosshead guides.

1 *A-frame*
2 *step*
3 *crosshead guides*
4 *guide shoes*
5 *crosshead*
6 *connecting rod*

▲
A chromed valve stem that is very wear-resistant and resistant to chemical corrosion.

Chromium

In order to make the surfaces **harder** and more **wear resistant** they are often furnished with a thin **chromium** coating, which increases the operating life of the components considerably. There are, for instance, chromium-plated piston rings, connecting rods and crosshead pins.
Chromium has also good resistance properties with respect to the corrosive substances in the combustion products found on the piston rings. Furthermore, chromium has an extremely low friction coefficient on the generally cast-iron cylinder liners.

Hard chromium plating or chromium hardening is the electrolytic plating of chromium on the components. The so-called plating is done with a chromium-salt solution.

Plasma layer

A ceramic coating is applied to chromium-plated parts such as the top piston rings. Ceramic has very good wear resistance properties under heavy operating conditions.
Both the running-in of a piston in the cylinder liner, as well as the great many operating hours that follow, show that the wear and tear of the cylinder liner and the rings is much lower.

The **rewelding of special material** that has good corrosion and erosion resistance is called component cladding. This is mainly applied in combustion areas with excessively high temperatures such as those found in RTA-C Wärtsilä Sulzer crosshead engines. Here the material is attached to the inner wall of the cylinder head, in the direction of the fuel jet. In the latest designs the material temperature of the cylinder head has been reduced to such an extent that this surface treatment method is no longer required.

▼
This cylinder head is waiting outside for repair.

The high-grade material is clearly seen near the injector nozzle; shown here a very strong welded layer of material that is resistant to the high temperature at the injector nozzle.

▶
Reconditioning the injector nozzle by rewelding.

In a later phase, the liner is machined resulting in the original dimensions and the injector nozzle is drilled out and finished.

◄

The honing of a cylinder liner.

1 *honing machine*
2 *cylinder liner*
3 *honing-stone holder*
4 *honing stone*
5 *driving shaft*
6 *oil supply*
7 *clamping strips*

7.10.5 Grinding, polishing and lapping

Grinding, **polishing** and **lapping** are standard expressions used in the finishing of metal surfaces. They are general terms which give an impression of the surface finishes. However, they do not provide a surface-roughness value.
A **ground** and polished crankshaft journal can have a very smooth surface.
Polishing is usually a finish which removes the final roughness.
Lapping involves the use of two surfaces which are rubbed together such as seal surfaces of the different parts of a fuel atomiser.

7.10.6 Honing

This surface finish is applied to the running surface of cylinder liners in various engines. The objective of honing is to improve the running surface for the reciprocating piston. The honing cross-hatch pattern is used to retain a certain amount of lubricating oil which reduces the wear and tear of piston rings, piston and liner.
Honing is done with a pivoting shaft head on which four honing stones are held under pressure against the cylinder liner using springs. With a specified pressure, peripheral speed and longitudinal movement of the liner a rough kind of cross-hatch pattern is applied to the liner in which the scratches are at an angle of about 60°. There are a great many types of honing stones with a specified roughness. The honing oil that is used is also important. Honing stones, pressure, speed and the process length are crucial in order to achieve optimum results.

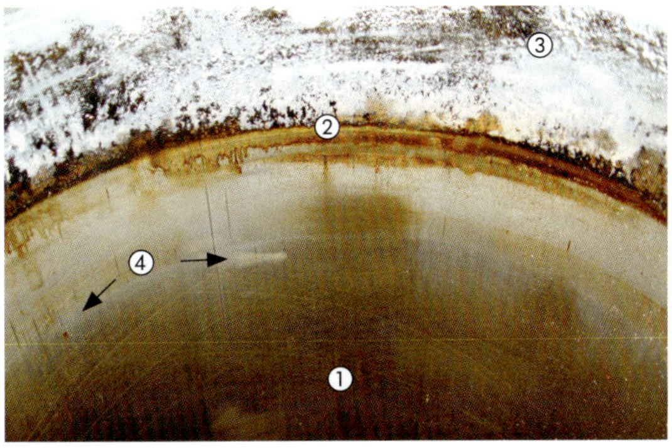

◄

A worn cylinder liner after thousands of operating hours.

Stop shoulder for the upper piston ring is clearly seen.

1 *worn part of the cylinder liner with still visible honing cross-hatching*
2 *wear edge*
3 *non-worn part of the cylinder liner with incrustation*
4 *vertical scratches in cylinder liner caused by damage, dirt or metal particles (catfines)*

▼

Honing.

The use of the correct honing stones, turning tension on the stones, the number of revolutions and the horizontal movement of the stones through the liner are important points of interest. The choice of honing oil also plays a role.

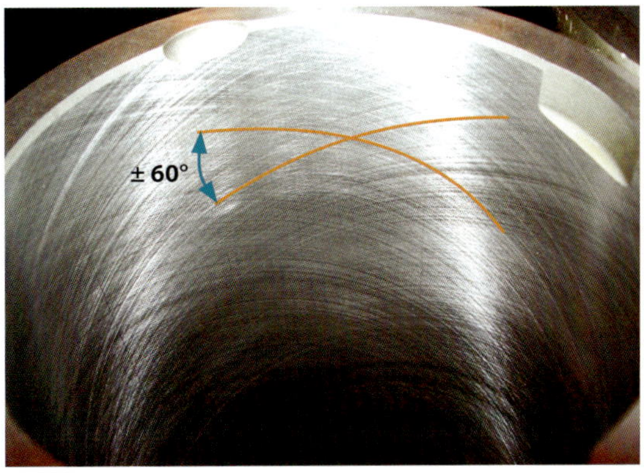

◄

With a good honing cross-hatching, the scratches have an angle of roughly 50° tot 60°.

▶
A honing machine with a rotating and a forward and backward movement, both are adjustable.

Plateau honing

This type of processing is used by some engine manufacturers and reconditioning companies. It takes place in two steps, namely the **pre-honing** and **fine-honing**.
With pre-honing, coarse-grained honing stones are used resulting in a surface roughness of 8 to 10 microns.
With **fine-honing,** fine-grained honing stones are used where the roughness 'peaks' are ground off. Between the deeper scratches resulting from the pre-honing, peaks are obtained with finer structured profiles.
Using this method, the bearing capacity of a thus processed cylinder liner achieves the same level as conventionally honed cylinder liners after very few operating hours.
Simply put: plateau honing reduces the running-in time. This process also reduces the lubricating oil consumption without having a negative effect on the lubrication.

▶
Plateau honing.

Figure 1: Shows an almost smooth worn cylinder liner. This is known as 'glazing'.
Figure 2: the surface profile after a standard honing process. The surface is still quite rough with deep grooves and high peaks.
Figure 3: the surface profile after plateau-honing. Very fine honing stones remove the high peaks resulting in a flat running surface being created for the piston rings.

Surface Condition

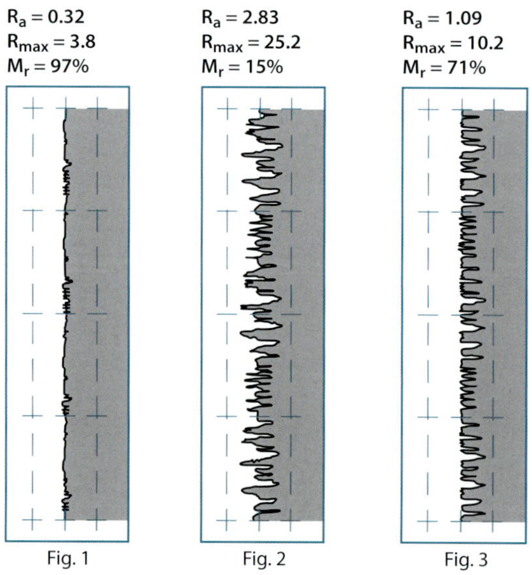

7.11 Examples of modern material usage

Competition in the engine-building industry is fierce. Customers' requirements are always increasing, therefore engine designers (Research and Development) are continuously looking for materials that have better resistance to the thermal and mechanical loads and the effects of aggressive compounds.

7.11.1 Example 1: Design requirements for Wärtsilä Sulzer RTA-C crosshead engines, fuel HFO, Engine category IV

Construction
- stiff frame.
- gondola-type engine bedplate.
- cast-iron cylinder blocks.

Bearings
- lower crankshaft bearing: thin white-metal bearings.
- upper crankshaft bearing: thick cast white-metal bearings.
- crosshead bearings: thin white-metal bearings.

Combustion chambers
- special forged steel with high resistance to corrosion fatigue.
- rewelding cylinder cover with heat resistant material at the fuel nozzle tips to prevent burning.

Piston
- forged-steel piston crown with chromium-plated piston-ring grooves.

Liner
- cylinder liner of high wear and tear resistant rigid cast iron.
- honed cylinder liner.
- piston rings with a plasma coating.

Furthermore, all measures are taken to planish/smooth bore the combustion room, as summed up in the so-called '**Tribo Pack Concept**'. This entails the following:
- isolated 'bore'-cooling holes' throughout the combustion space in piston, cylinder liner and cylinder head;
- multilevel cylinder lubrication;
- mid-stroke cylinder-liner insulation;
- a 'deeply' honed liner;

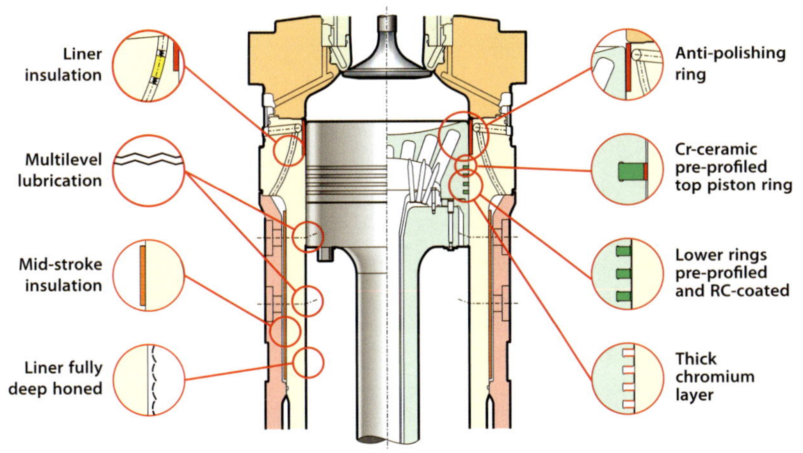

▲
Eight improvements in the combustion space of a Wärtsilä Sulzer RTA84 crosshead engine.

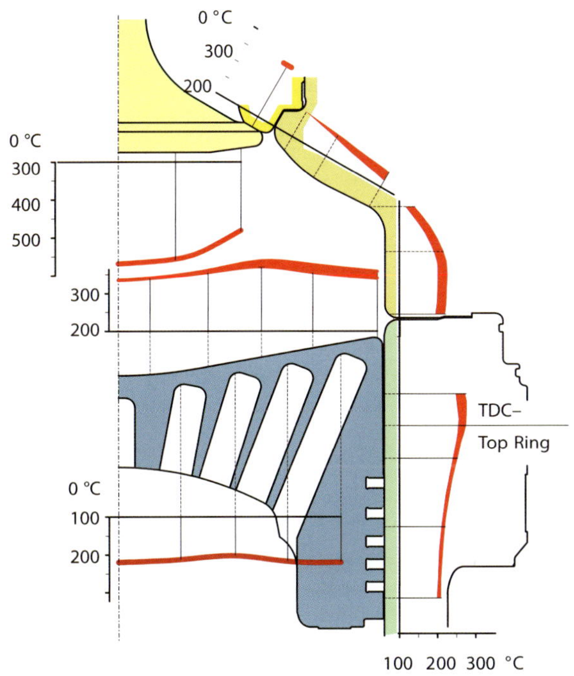

▲
The material temperatures in an improved combustion space of a Wärtsilä Sulzer RTA84 crosshead engine.

Lower material temperatures means in principle a lower load on the parts around the cylinder.

- an anti-polishing ring which prevents the combustion residue from settling in the ring packing;
- a chromium- and ceramic-treated pre-profiled top piston ring;
- applying a thin run-in layer to the other piston rings;
- thick chromium-coated piston-ring grooves to prevent pitting and wear and tear.

7.11.2 Example 2: Wärtsilä four-stroke engines, medium-speed, fuel HFO, Engine category III

Wärtsilä has a programme with cylinder diameters of 20, 26, 32, 38, 46 and 64 centimetres for this type of engine. The latter is also the largest four-stroke engine presently available in the market, with a cylinder capacity of 2000 kW.

To designers the following elements are essential:
- reliability;
- low user costs;
- low exhaust-gas emissions;
- simple installation at low cost;
- applied mounting methods need to be reliable;
- low, simple maintenance.

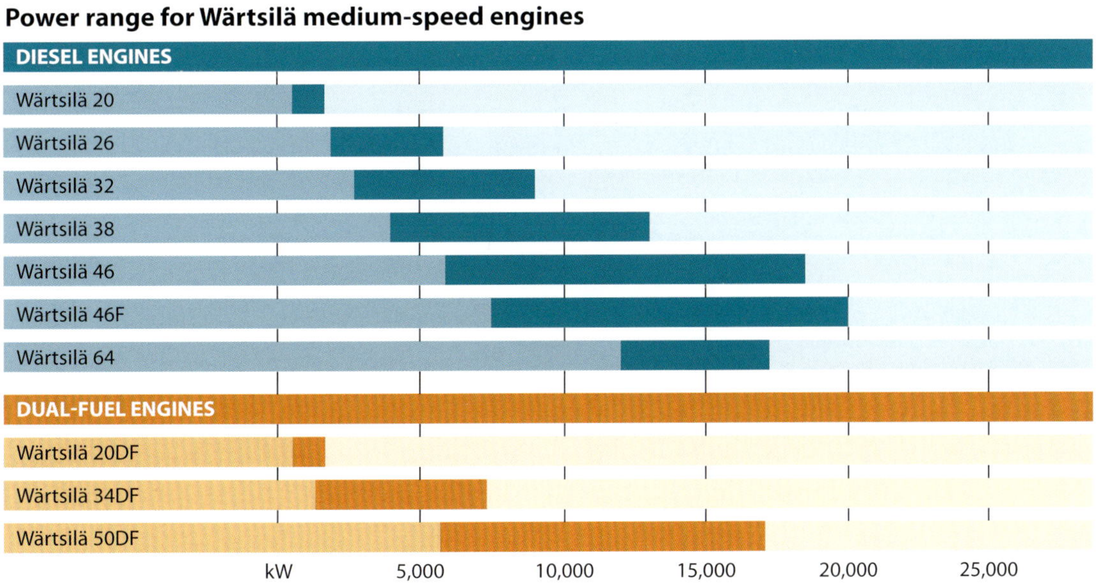

◀
The different cylinder diameters of four-stroke diesel engines at Wärtsilä.

Wärtsilä 20DF (IMO Tier II)

Cylinder bore	200 mm
Piston stroke	280 mm
Cylinder output	146/176 kW/cyl
Engine speed	1000/1200 rpm
Mean effective pressure	20.0 bar
Piston speed	9.3/11.2 m/s
Generator voltage	0.4–13.8 kV
Generator efficiency	0.95–0.96

Fuel oil specification:
ISO-F-DMX, DMA and DMB
ISO 8217:2005(E)

SFOC 194/199 g/kWh at ISO conditions
Natural gas
Methane number: 80

Rated power				
Engine type	60Hz 176 kW/cyl, 1200 rpm		50Hz 146 kW/cyl, 1000 rpm	
	Engine kW	Gen. kW	Engine kW	Gen. kW
6L20DF	1056	1014	876	841
8L20DF	1408	1352	1168	1121
9L20DF	1584	1521	1314	1261

Genset dimensions (mm) and weights (tonnes)						
Engine type	A*	E	I*	K	L*	weight
6L20DF	5325	2070	900/980/1030	1800	2688	17
8L20DF	6030	2070	1030/1080	1800	2824	20.9
9L20DF	6535	2300	1080/1130	1800	2874	24

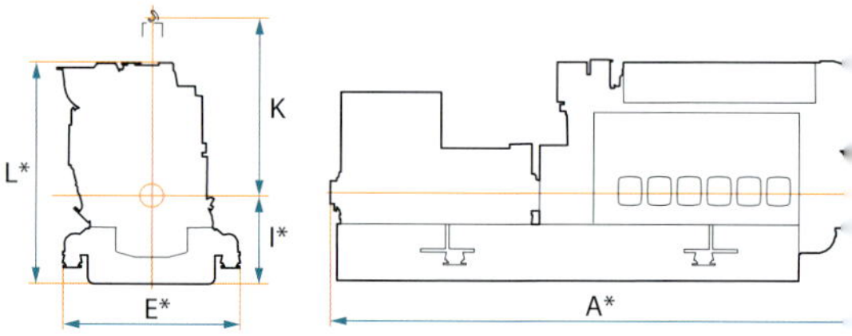

Wärtsilä 50DF (IMO Tier II)

Cylinder bore	500 mm
Piston stroke	580 mm
Cylinder output	950/975 kW/cyl
Engine speed	500/514 rpm
Mean effective pressure	20.0 bar
Piston speed	9.7/9.9 m/s

Fuel oil specification:
Fuel oil 700 cSt/50 °C
7200 sR1/100 °F
ISO 8217, category ISO-F-DMX, DMA and DMB

Natural gas
Methane number: 80
LHV: min. 28 MJ/Nm^3, 5.5 bar
BSEC 7410 kJ/kWh

Rated power				
Engine type	50Hz		60Hz	
	Engine kW	Gen. kW	Engine kW	Gen. kW
6L50DF	5700	5500	5850	5650
8L50DF	7600	7330	7800	7530
9L50DF	8550	8250	8775	8470
12V50DF	11400	11000	11700	11290
16V50DF	15200	14670	15600	15050
18V50DF	17100	16500	17550	16940

Generator output based on a generator efficiency of 96.5%.

Genset dimensions (mm) and weights (tonnes)						
Engine type	A	B	C	D	F	weight
6L50DF	8115	3580	2850	3820	1455	96
8L50DF	9950	3600	3100	3820	1455	128
9L50DF	10800	3600	3100	3820	1455	148
12V50DF	10465	4055	3810	3600	1500	175
16V50DF	12665	4055	4530	3600	1500	220
18V50DF	13725	4280	4530	3600	1500	240

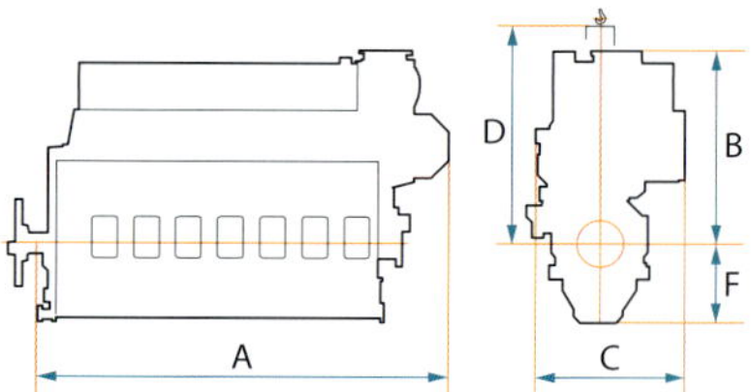

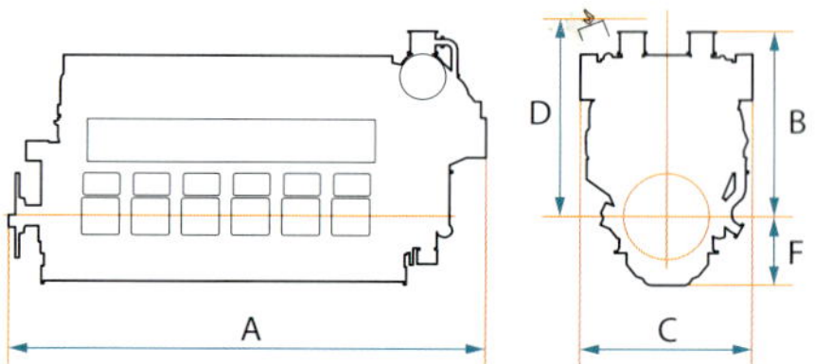

Wärtsilä 34DF (IMO Tier II)

Cylinder bore	340 mm
Piston stroke	400 mm
Cylinder output	435/450 kW/cyl
Engine speed	720/750 rpm
Mean effective pressure	20.0/19.8 bar
Piston speed	9.6/10.0 m/s

Fuel oil specification:
Fuel oil 700 cSt/50 °C
7200 sR1/100 °F
ISO 8217, category ISO-F-DMX, DMA and DMB

Natural gas
Methane number: 80
LHV: min. 24 MJ/Nm^3, 5.5 bar
BSEC 7700 kJ/kWh

Rated power			
Engine type	Main engine kW	60 Hz Gen. kW	50 Hz Gen. kW
6L34DF	2700	2510	2600
9L34DF	4050	3760	3890
12V34DF	5400	5010	5190
16V34DF	7200	6680	6920

Engine dimensions (mm) and weights (tonnes)						
Engine type	A	B	C	D	F	weight
6L34DF	5280	2550	2385	2345	1155	34
9L34DF	6750	2550	2385	2345	1155	47
12V34DF	6615	2665	3020	2120	1475	59
16V34DF	7735	2430	3020	2120	1475	75

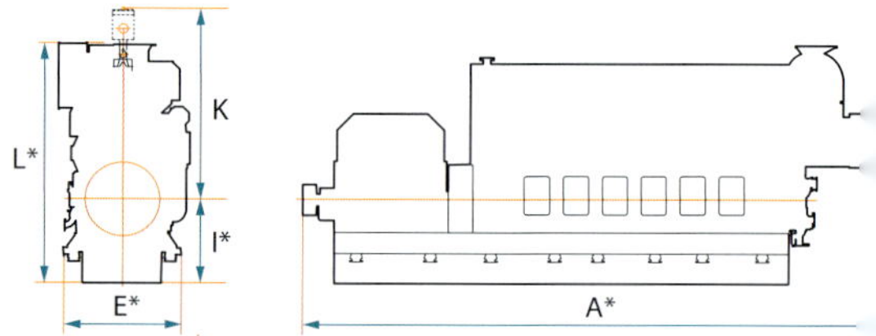

Genset dimensions (mm) and weights (tonnes)						
Engine type	A*	E*	I*	K	L*	weight
6L34DF	8700	2290	1450	2345	4000	55
9L34DF	10475	2890	1630	2345	4180	84
12V34DF	10075	3060	1700	2120	4365	96
16V34DF	11175	3060	1850	2120	4280	121

* Dependent on generator type.
Generator output based on a generator efficiency of 96%.

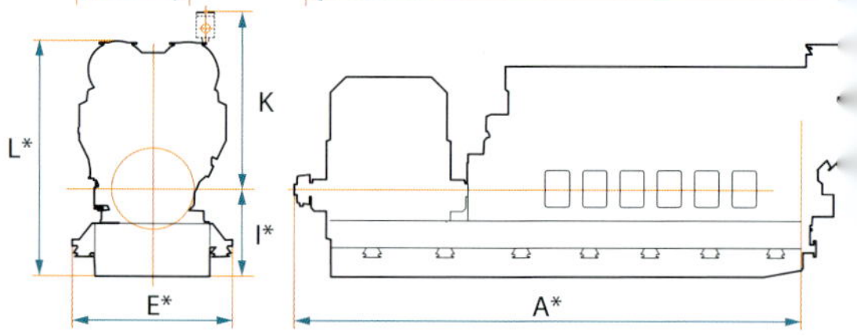

▲
The DF Dual Fuel engines, 20DF, 50DF and 34DF. This is a new development to reduce emissions significantly.

Engine block

All lubricating-oil and cooling-water ducting is integrated in the engine block, which results in compact and safe operating systems. The camshafts have also been fully incorporated in the block which creates a very solid entity.
The engine block is made from special nodular cast iron, which is very strong and rigid and especially suitable for casting complicated forms.

Crankshaft and crankshaft bearings

Crankshafts are increasingly heavily loaded. When increasing the cylinder capacity the pressure in the cylinder rises, which results in higher forces on the crankshaft. This means the size of the crankshaft and the bearings must be increased. An optimum design of the crankshaft such as a reduction of the cylinder centre radius and expansion of the bearing surface reduces the surface pressure in the bearings.

The crankshafts are fabricated from high-quality forged steel. The crankpin and the crankshaft are case-hardened and subsequently very finely ground.

The crankshaft and crankpin bearings are manufactured from a tri-metal type using an aluminium/tin or antimonium/tin basis and work according to the so-called 'Thick-Pad Bearing' principle.

The cylinder liner is made of fine perlitic cast iron with high wear resistance and excellent running properties.
The thick high-collar type cylinder liner is very stiff and therefore distortion is minimal. The collar has bore-cooling for an optimum liner temperature. The anti-polishing ring prevents carbon particles from ending up between the piston and cylinder liner, which diminishes the wear and tear of these parts considerably.

▲
A cast-iron engine block of an in-line engine after machining.

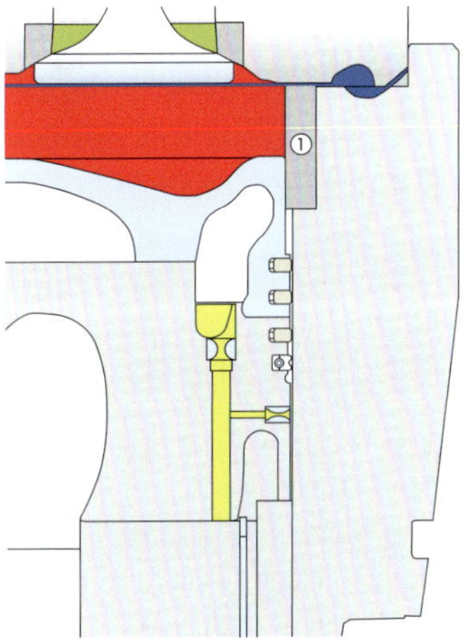

◄
Cross-section of a cylinder liner with bore-cooling in the thickened upper collar and an anti-polishing ring (1).

▼
A crankshaft of a V-engine.

Note the two lubricating-oil holes in the crankpin for both connecting rods. Crankshaft reconditioning at Mark van Schaick, Schiedam, The Netherlands.

►
A cylinder liner with a thickened upper collar, in the partial cutaway the antipolishing ring is visible.

Cylinder head

All modern cylinder heads have four valves: two inlet valves and two exhaust valves. This allows for the gas exchange to take place swiftly and completely. The cylinder head is usually affixed with four cylinder-head bolts.

The head is manufactured from special cast steel, very stiff and heat resistant.

The supply of combustion air, the discharge of exhaust gases and the supply and discharge of cooling water takes place via nodular cast iron 'multi-ducts' which remain in one place when the cylinder head is dismantled.

▲

A modern cylinder head with four valves.

Pistons

Throughout the years the thermal- and mechanical load on the pistons has increased. Therefore all pistons consist of two parts:

1. **a cast-steel piston crown** which contains the ring packing. This has good resistance against the high thermal and mechanical loads. This material has good resistance against pitting of the piston rings in the piston-ring grooves.
2. **a nodular cast-iron piston skirt for surveying the piston and absorbing the lateral forces caused by the connecting-rod mechanism.** The nodular cast iron has a very low expansion coefficient and retains sufficient tensile strength at high operating temperatures.

▲

A mounted piston with a cast-steel piston crown and cast-iron piston skirt.

Note the gaps for the valves in the piston crown for exhaust-gas scavenging.

Piston rings

A three-piston ring design has been made for the sealing package. These are made of fine perlitic cast iron and both sealing rings are chromium plated. On top of that the upper ring has an extra plasma coating and has high wear and tear resistance.

▼

The piston rings as shown here are chromed and the top ring has an extra ceramic layer.

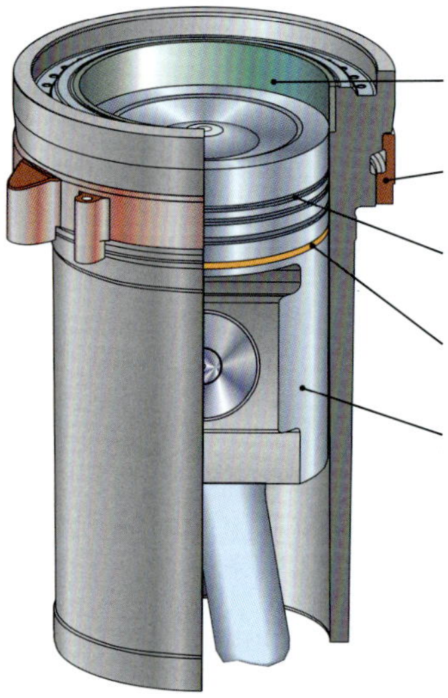

7.11.3 Example 3: Four-stroke high-speed engines with high capacity at a relatively high rpm, fuel MDO, Engine category II

Marine engines of this type are always subdivided into two types:

1. **High output,** often used in the military services. Additional characteristics are: complex designs, few operating hours between big overhauls, costly components, extremely high effective mean pressures and a low weight in relation to the output.

▲
A modern diesel engine, manufacturer Caterpillar.

2 **Engines for heavy operating conditions** such as: earth-moving equipment, gensets and other gear-driven devices. Characteristics are a simple and robust design with normal mean effective pressures and a fairly high weight in relation to the output.

The demand for the somewhat smaller high-speed diesel engines for propulsion and other gear driving has resulted in designs that can be easily integrated in high-volume serial production lines. The following manufacturers are market leaders in the manufacturing of engines with a cylinder bore of about 170 mm:
Caterpillar, Cummins, Deutz MWM, GMT, Isotta Franchini, MAN Diesel, Perkins, Mitsubishi, MTU, MTU/DDC, Niigata, Paxman, SEMT Pielstick, Wärtsilä and others. All manufacturers produce high-speed engine types.

Characteristics

Blocks of cast iron, with integrated cooling-water and lubricating-oil ducts. Nodular cast iron is often applied.

- **An underslung alloyed forged-iron crankshaft with tri-metal bearings**.
- **Cast fine perlitic cylinder liners**.
- **Light metal cast pieces** for pistons, covers, inlet manifold and other engine parts. This creates a considerable weight reduction.
- **Weight** in general plays a role in these engines in that they are often used in relatively light ships, such as catamarans, high-speed single-hull ships and mobile gensets.
- **Many V-engines of this type are built; the size is limited in relation to the power output.**
- **Pistons** often consist of two parts, a cast-steel piston crown and a light-metal piston skirt.
- **Cylinder heads** are made from cast iron and contain four valves.

Further technical details in this type of engine are

- high mean effective pressures > 25 bar;
- high rpm up to 2100 per minute;
- high mean piston speeds up to12 m/sec.;
- high power output per litre stroke volume;
- combustion pressures up to 200 bar;
- power output/weight ratio high;
- state-of-the-art fuel-injection systems;
- high air charging with turbocharger groups up to 4 bar;
- optimized coolant temperature at varying loads;
- improved lubricating-oil cooling for the piston.

▼
A high-speed heavy loaded twenty-cylinder V-engine for propelling a fast catamaran.

The engine rooms have been placed in both pontoons. The fuel is diesel oil. Power output and weight play an important role. The operating speed of this catamaran is 37.5 knots.

▼
Two diesel generator sets on board a container-feeder with a Scania diesel engine.

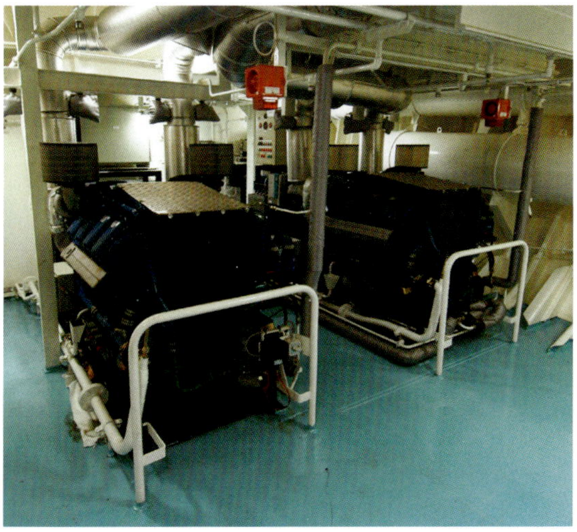

7.11.4 Example 4: Smaller engines actually in engine category I: 0 tot 100 kW but also higher in capacity, high speed, four-stroke fuel MDO

These engines are used for a large number of applications, such as the propulsion of smaller ships, gensets, pumps and traction. A noticeable difference is that, generally, many light metals are used. The main metal is aluminium with added alloys.
Properties of light metal
- low crystal density
- good heat conduction
- good corrosive properties
- many machining capabilities and especially low weight

Cylinder heads can also be made of light metal, but of course cast iron is also often used.
Generally, crankshafts are made of malleable steel, although cast crankshafts still exist, especially in older type engines.

Steel is often alloyed with **chromium** and **molybdenum.**
The journals are often hardened by means of nitration. Flame hardening is also applied.

Also see Chapter 26, *Overhauling diesel engines and their parts.*

Timing belts, (also known as toothed-, notch- or cog belts)

These are increasingly used in this engine category. They require no heavy cog wheels between the crankshaft and the camshaft. However, they must to be replaced after a certain operating time.

Bearings

Usually hard steel tri-metal sleeve bearings are applied. The friction metal most frequently used is an aluminium alloy.

▶ A modern high-speed four-stroke diesel engine in V-form (manufacturer MTU) ready to be placed in an inland navigation tanker.

These fairly large ships with a cargo capacity from 2000 to 4000 tons usually have two propulsion engines. In view of the strict emission requirements all inland vessels run on diesel oil.

▶ A Caterpillar high-speed four-stroke diesel engine with twelve cylinders in V-form.

1. *air filter*
2. *turbocharger (not visible, behind filter)*
3. *compressed air to intercooler*
4. *intercooler*
5. *cylinder head*
6. *control console*
7. *lubricating-oil filter*
8. *lubricating-oil cooler*

Category	Mahle 105	Mahle 109	Mahle 120	Mahle 121	Mahle 147	Mahle 226
Composition (%)						
Si	4.5 – 6.0	9.0 – 10.0	10.3 – 13.5	10.3 – 13.5	16.0 – 18.0	7,5 – 9,5
Mg	0.5 – 0.8	0.2 – 0.4	0.06 – 0.12	0.1 – 0.3	0.4 – 0.7	≤ 0,3
Cu	≤ 0.03	≤ 0.01	≤ 0.1	± 0.1	4.0 – 5.0	2,5 – 3,5
Fe	≤ 0.4	≤ 0.4	≤ 0.25	≤ 0.25	≤ 0.7	≤ 1,3
Mn	≤ 0.5	≤ 0.4				≤ 0,5
Zn						≤ 0,7
Al	rest	rest	rest	rest	rest	rest
Tensile strength Rm (N/mm²)	≤ 260	≤ 260	≤ 140	≤ 145	≤ 200	240 – 310
Yield strength Ro 0,2 (N/mm²)	≤ 240	≤ 200	≤ 67	≤ 80	≤ 160	160 – 240
Stretching at breaking load A (%)	≤ 1	≤ 4	≤ 5	≤ 3.5	≤ 0.5	0.5 – 3
Brinell hardness (HB 2,5/62,5)	≤ 90	≤ 80	48 – 60	50 – 70	90 – 120	80 – 110
Properties and applications	Rust proof; Alfin-cooling fin cylinder	Rust proof; cylinder block of water-cooled ship engines	Ductile, rust proof, not heat cured; motor-car wheel rims	Ductile, rust proof, not heat cured	high wear and tear resistance; cylinder without motor-car wheel rims	Die cast alloy for cooling fin cylinder running surface protection SILUMAL-cylinder

▲

A table for the light-metal blocks for diesel engines. The main component is aluminium.

The table shows an **aluminium cast alloy** for **cylinder blocks** of water-cooled marine engines, Mahle 109.

9 to 10% silicon and 0.2 to 0.4% magnesium have been added.

The table also provides the **aluminium alloy Mahle 147**, which is applied to unprotected bores in cylinder blocks without separate cylinder liners.

The material is highly wear-resistant due to the alloys:

silicon: 16 to 18%

magnesium: 0.4 to 0.7%

copper: 4.0 to 5.0%

◄

A Deutz high-speed propulsion engine for a small dredgingcraft.

In the foreground the gear box for the propeller.

◄

A sixteen-cylinder high-speed V-engine, make Mitsubishi.

1 *turbocharger*
2 *intercooler*
3 *cylinder head*
4 *block fuel pump*
5 *fuel filters*
6 *speed controller/governor*
7 *high-pressure fuel lines*
8 *lubricating-oil filters*
9 *engine supports*
10 *safety switch for lubricating-oil level in sump*
11 *emergency stop*

>CH 8

Fuel, fuel-line systems and fuel cleaning

The fuel quality injected into the combustion space is essential for optimal running of the diesel engine.
Cleaning the fuel, here performed by Alfa Laval centrifuges for HFO on a large container ship, is an important aspect of fuel treatment prior to injection.

Poor quality HFO, these filters are completely clogged.

▶

Diesel engines have been powered by liquid carbon compounds distilled from crude oil for over a hundred years.

The refined crude oil is used as gas oil, diesel oil and heavy oil for all four diesel-engine categories. From approximately 2000 onwards, other fuels such as natural gas and biogas are also successfully used. Biofuels are increasingly used, especially in smaller diesel engines in road transport and the agricultural sector. Today a few medium-speed engines are equipped to run on bio diesel.

8.1 Introduction

Since the first internal combustion engine was built based on the Otto principle around 1860 and on the Diesel principle in 1896, to the present day, 2012, mixtures of gaseous and liquid hydrocarbons have generally been used.
The gaseous hydrocarbons used in the Otto engine were predominantly a mixture of mining gases, process gases and land-fill gases. Much later, after gas- and oil supplies were discovered deep in the earth's crust, mostly natural gas and oil were used.

Today, natural-gas resources exploration continues with gas fields being discovered and exploited. Otto engines are also used frequently, for instance, in motor cars with liquid fuels such as petrol, a very light liquid hydrocarbon compound.
Of the gaseous hydrocarbons natural gas is generally used for stationary gas engines.
However, over the past five years, other gases such as land-fill gas, sewer gas, biogas and other waste gases are becoming increasingly important due to energy-saving and environmental issues.

Also see Chapter 29, *Recent developments in the fuel industry*.

The comparatively light liquid hydrocarbon compound diesel oil has been the most frequently used fuel since the introduction of the diesel engine. Around 1950 heavier fuels, initially used only in steam boilers, were used as a fuel in low-speed crosshead engines with an rpm of 80 to 120 revolutions per minute. As the fuel costs soared this so-called heavy oil was used more regularly.

From 1985 onwards, heavy oil has been increasingly used for four-stroke diesel engines with an rpm of about 1200 revolutions per minute. Above these rpm's, the combustion time is too short to achieve perfect combustion of heavy oil. The price of gas oil is presently twice that of heavy oil.
Today, heavy oil is becoming increasingly a residual product in oil refineries with a corresponding loss in quality.

▼

Diesel-oil requirements according to ISO 8217:2010

There is a subdivision consisting of four categories: DMX, DMA, DMZ and DMB.

ISO 8217:2010

MARINE DISTILLATE FUELS

Parameter	Unit	Limit	DMX	DMA	DMZ	DMB
Viscosity at 40 °C	mm^2/s	Max	5.500	6.000	6.000	11.00
Viscosity at 40 °C	mm^2/s	Min	1.400	2.000	3.000	2.000
Micro Carbon Residue						
at 10% Residue	% m/m	Max	0.30	0.30	0.30	–
Density at 15 °C	kg/m3	Max	–	890.0	890.0	900.0
Micro Carbon Residue	% m/m	Max	–	–	–	0.30
Sulphur[a]	% m/m	Max	1.00	1.50	1.50	2.00
Water	% V/V	Max	–	–	–	0.30[b]
Total sediment by hot filtration	% m/m	Max	–	–	–	0.10[b]
Ash	% m/m	Max	0.010	0.010	0.010	0.010
Flash point	0 °C	Min	43.0	60.0	60.0	60.0
Pour point, Summer	0 °C	Max	–	0	0	6
Pour point, Winter	°C	Max	–	-6	-6	0
Cloud point	°C	Max	-16	–	–	–
Calculated Cetane Index		Min	45	40	40	35
Acid Number	mgKOH/g	Max	0.5	0.5	0.5	0.5
Oxidation stability	g/m3	Max	25	25	25	25[c]
Lubricity, corrected wear scar diameter (wsd 1.4 at 60 °C)[d]	um	Max	520	520	520	520[c]
Hydrogen sulphide[e]	mg/kg	Max	2.00	2.00	2.00	2.00
Appearance			Clear & Bright[f]			b, c

[a] A sulphur limit of 1.00% m/m applies in the Emission Control Areas designated by the International Maritime Organization. As there may be local variations, the purchaser shall define the maximum sulphur content according to the relevant statutory requirements, notwithstanding the limits given in this table.
[b] If the sample is not clear and bright, total sediment by hot filtration and water test shall be required.
[c] Oxidation stability and lubricity tests are not applicable if the sample is not clear and bright.
[d] Applicable if sulphur is less than 0.050% m/m.
[e] Effective only from 1 July 2012.
[f] If the sample is dyed and not transparent, water test shall be required. The water content shall not exceed 200 mg/kg (0.02% m/m).

8.2 Composition of liquid fuels

Apart from carbon and hydrogen, many other elements are found in fuels, such as oxygen, nitrogen, sulphur, and various metals which are often bound to the hydrogen-carbon molecules.

8.3 Definition of heavy oil

This is a fuel which is manufactured from the remaining distillate of the crude-oil refining process and is characterised by its viscosity. Residual fuel is a general term for heavy fuels that are characterised by a high specific mass and a high viscosity. **Heavy oil** consists mainly of residues from the distillation and refining process.

Heavy-fuel values

For specific mass: between 950 and 1020 kilograms per m³.
For viscosity: over 30 cSt at 50 °C.

Diesel oil consists mostly of lighter fractions of crude oil.

Gas oil: specific mass 820 to 880 kilograms per m³.

Marine diesel oil: specific mass 840 to 920 kilograms per m³ for a viscosity to 30 cSt at 50 °C.

Definition heavy-oil diesel engine: a genuine heavy oil diesel engine can start, stop and run on heavy oil at every load.

8.4 Refining crude oil

Refining process

Step 1: Salt is removed by water washing.

Step 2: Atmospheric distillation. Crude oil is heated up to 375 °C under atmospheric pressure.
Products
- gases
- distillates
- vapours

Step 3: Vacuum distillation at an absolute pressure of 0.05 to 0.15 bar and a temperature of 400 °C. Here a further distillation of the atmospheric residue takes place.
Products
- vacuum gas oils
- vacuum residues

These residues already have a high viscosity and specific mass.

ISO 8217:2010

MARINE RESIDUAL FUELS

Parameter	Unit	Limit	RMA[a]	RMB	RMD	RME	RMG				RMK		
			10	30	80	180	180	380	500	700	380	500	700
Viscosity at 50 °C	mm²/s	Max	10.00	30.00	80.00	180.0	180.0	380.0	500.0	700.0	380.0	500.0	700.0
Density at 15 °C	kg/m³	Max	920.0	960.0	975.0	991.0	991.0				1010.0		
Micro Carbon Residue	% m/m	Max	2.50	10.00	14.00	15.00	18.00				20.00		
Aluminium + Silicon	mg/kg	Max	25	40		50	60						
Sodium	mg/kg	Max	50	100		50	100						
Ash	% m/m	Max	0.040	0.070			0.100				0.150		
Vanadium	mg/kg	Max	50	150			350				450		
CCAI	–	Max	850	860			870						
Water	% V/V	Max	0.30	0.50									
Pour point (upper)[b], Summer	°C	Max	6		30								
Pour point (upper)[b], Winter	°C	Max	0		30								
Flash point	°C	Min	60.0										
Sulphur[c]	% m/m	Max	Statutory requirements										
Total Sediment, aged	% m/m	Max	0.10										
Acid Number[e]	mgKOH/g	Max	2.5										
Used lubricating oils (ULO): Calcium and Zinc; or Calcium and Phosphorus	mg/kg	–	The fuel shall be free from ULO, and shall be considered to contain ULO when either one of the following conditions is met: Calcium > 30 and zinc >15; or Calcium > 30 and phosphorus > 15.										
Hydrogen sulphide[d]	mg/kg	Max	2.00										

a This residual marine fuel grade is formerly DMC distillate under ISO 8217:2005.
b Purchasers shall ensure that this pour point is suitable for the equipment on board, especially in cold climates.
c The purchaser shall define the maximum sulphur content according to the relevant statutory requirements.
d Effective only from 1 July 2012.
e Strong acids are not acceptable, even at levels not detectable by the standard test methods for SAN. As acid numbers below the values stated in the table do not guarantee that the fuels are free from problems associated with the presence of acidic compounds, it is the responsibility of the supplier and the purchaser to agree upon an acceptable acid number.
Source: ISO 8217 Fourth Edition 2010-06-15

◄ Heavy-oil requirements according to ISO 8217:2010

Fuels can be divided into ten different fuel classes. The viscosity at 50 °C ranges from 10 to 700 cSt. This is often expressed in centistokes; sometimes in mm² per second.

▶
A simple diagram of the processes that crude oil undergoes in order to obtain products such as gases, gas oil, diesel oil and lubricating oil.

The most popular heavy fuel is made from the waste product residue that remains after these four processes. Due to an increased demand for lighter fuels in for instance road transport, aviation and the chemical industry, refineries are increasingly looking for ways in which lighter carbon compounds can be refined from crude oil. This results in a poor quality residue, which forms the basic product of the heavy fuel and therefore poses problems for the heavy-fuel powered diesel engines.

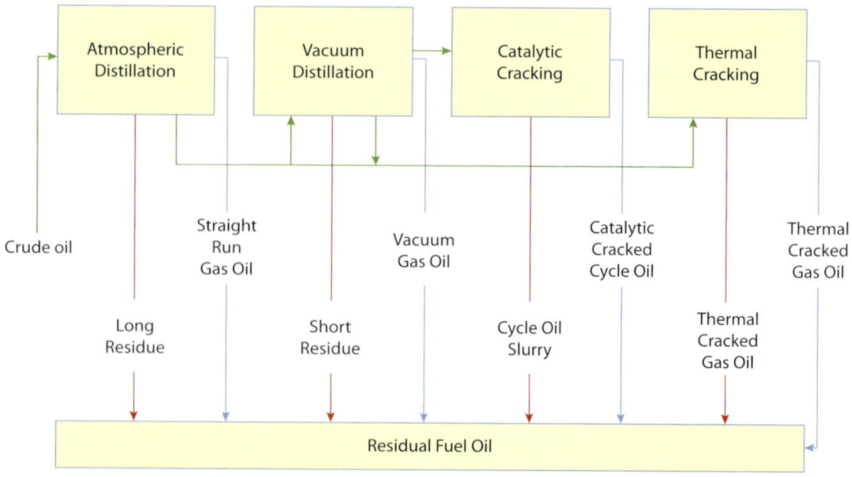

Step 4: Thermal-cracking; the vacuum residue is thermally treated with high temperatures and pressures to produce products with a lower viscosity.
Products
- gas
- distillates
- thermal-cracking residue

Step 5: Catalytic cracking: catalysts added to the process aiding the process of breaking down long hydrocarbon chains without requiring high temperatures.
Products
- petrol
- light fuels
- heavy fuels

Note
Part of the catalysts, the so-called **'cat-fines'** can remain in the heavy oil. These aluminium compounds can seriously damage the fuel pumps and atomisers and must be removed by centrifuges.

Shell Hycon installation
About a dozen years ago Shell developed a special method to process crude oil, the so-called Hycon installation. During this process hydrocarbon is injected into the crude oil under a pressure of 200 bar and at 500 °C and heavy asphaltenes are broken down into smaller carbon chains which can then be distilled. Sulphur and vanadium are then removed and very little residue remains.
Due to the growing number of Hycon refineries that are being commissioned, the amount of residue from which heavy oil can be obtained is decreasing. This, along with its poor quality, makes it less suitable for diesel-engine combustion.

8.5 Chemical composition of hydrocarbon compounds

One distinguishes between **paraffins** and **aromats**.

Paraffins consist of long rigid chains.
As crystallisation occurs at relatively high temperatures, this fuel requires heating at low temperatures in order to prevent problems such as pipe blockage.
Properties: good ignition properties and high solidification point.

Aromats consist of conjugated planar ring systems. They have poor ignition properties and may contain asphaltenes that have a high specific mass. Asphaltenes are oil molecules that do not dissolve in light crude oils.
Properties: high specific mass and poor ignition properties.

Heavy oil is always a mixture of paraffins and aromats.

8.6 Standardisation of liquid fuels

It is imperative that guidelines be established for liquid fuels. We have already seen that fuels can roughly be classified as follows.

Gas oil
This product consists of distillates only and is used for small fast-running diesel engines with low-temperature operating conditions.
Examples are back-up generators and boat engines.
Category I engines.

Marine diesel oil

Often indicated as 'MDO'.
This product consists for a large part of distillates and contains a small percentage of residues. This percentage usually lies between 15 and 20 per cent. Generally, the fuel is not heated. There is risk of pipe blockage due to crystallisation of the paraffins at low temperatures. This is also called 'flocculation'.
This fuel is generally used in high-speed diesel engines. **Category I and II engines.**

Heavy oil

Often indicated as 'HFO'.
This product consists mainly of various residues mixed with lighter fuels so that the correct viscosity is obtained. Depending on the type of mixture, these fuels must be heated to a maximum of 150 °C for optimum combustion. HFO is used in large low-speed two-stroke crosshead engines and larger medium-speed four-stroke trunk-piston engines. **Category III en IV engines.**

Fuels must meet minimum requirements

These requirements have been laid down by ISO, the International Organization for Standardization, in collaboration with the association of diesel-engine manufacturers, the **Cimac**, the French '**Conseil International des Machines à Combustion**'.
Cimac is an interest group in the field of the 'non-automotive' combustion engines for gas- and diesel engines as well as gas turbines.

The requirements are stipulated in the standard specification ISO 8217:2010.
Each specification has a test method indicating a minimum or a maximum value acknowledged by the ISO.

The following table contains the specifications for heavy fuels. Shown here, the standards for diesel engines make MAN Diesel, four-stroke medium-speed type V40/50. **Engine category III.**

8.7 Fuel properties

8.7.1 Viscosity, in centistokes or mm²/second

This is a measure of the viscosity or 'resistance to flow' of fuel. It is measured by time taken for a fluid to drain from a graduated cylinder through a calibrated opening. The longer the discharge time the thicker the oil and the higher the viscosity.
The unit of measure is **centistokes indicated by cSt**. This can also be indicated by mm^2 per second. The viscosity is, of course, strongly influenced by the temperature; this is why the temperature at which the measurement is taken is always noted.

Examples

At -10 °C, gas oil still has good viscosity of ± 10 cSt.
Diesel oil still has good atomising qualities at 15 °C.
Heavy oil of 30 cSt. requires heating to approximately 75 °C to atomise properly.
Heavy oil of 380 cSt. requires heating to approximately 130 to 140 °C to obtain the correct viscosity.

Marine Residual Fuels

Parameter	Unit	Limit	RMA 30	RMB 30	RMD 80	RME 180	RMF 180	RMG 380	RMH 380	RMK 380	RMH 700	RMK 700
Density at 15 °C	kg/m^3	max.	960.0	975.0	980.0	991.0	991.0	991.0	991.0	1010.0	991.0	1010.0
Viscosity at 50 °C	mm^2/s	max.	30.0	30.0	80.0	180.0	180.0	380.0	380.0	380.0	700.0	700.00
Water	% (V/V)	min.	0.5	0.5	0.5	0.5	0.5	0.5	0.5	0.5	0.5	0.5
Sulfur[1]	% (m/m)	max.	3.5	3.5	4.0	4.5	4.5	4.5	4.5	4.5	4.5	4.5
Aluminium + Silicon[2]	mg/kg	max.	80	80	80	80	80	80	80	80	80	80
Flashpoint[3]	°C	max.	60	60	60	60	60	60	60	60	60	60
Pour point, summer	°C	max.	6	24	30	30	30	30	30	30	30	30
Pour point, winter	°C	max.	0	24	30	30	30	30	30	30	30	30

1) Max sulfur content is 1.5% in designated areas (since 1-07-2010 1% is max).
2) The aluminium + silicon value is used to check for remains of the catalyst after catalytic cracking.
Most catalysts contains asluminium or silicon and remains of catalyst can cause damage to the engine.
3) The flash point of all fuels used in the engine room should be at least 60 °C.

◄ The kinematic viscosity is specified at 50 °C. All fuel data is issued in English as a standard. The table shows the specifications that heavy oil must meet. Here, the viscosity is divided in ten classes.

▼

A viscosity-temperature table for fuels.

horizontal: the fuel temperature between –10 and + 170 °C

vertical: the fuel viscosity in various units.

The unit to the right is most frequently used: the kinematic viscosity in mm^2 per second of centistokes.

The light blue horizontal band indicates the optimum fuel-injection viscosity area. In this case 13 to 17 centistokes.

The horizontal red line at 1000 centistokes gives the maximum viscosity value at which the fuel is pumpable from, for instance, a storage tank to a settling tank.

The slanted lines are the lines for various fuels at a certain viscosity.

Bottom left: gas oil, subsequently diesel oil and then the heavy-fuel oils of 30, 60, 100, 180, 380, 600 and 700 centistokes.

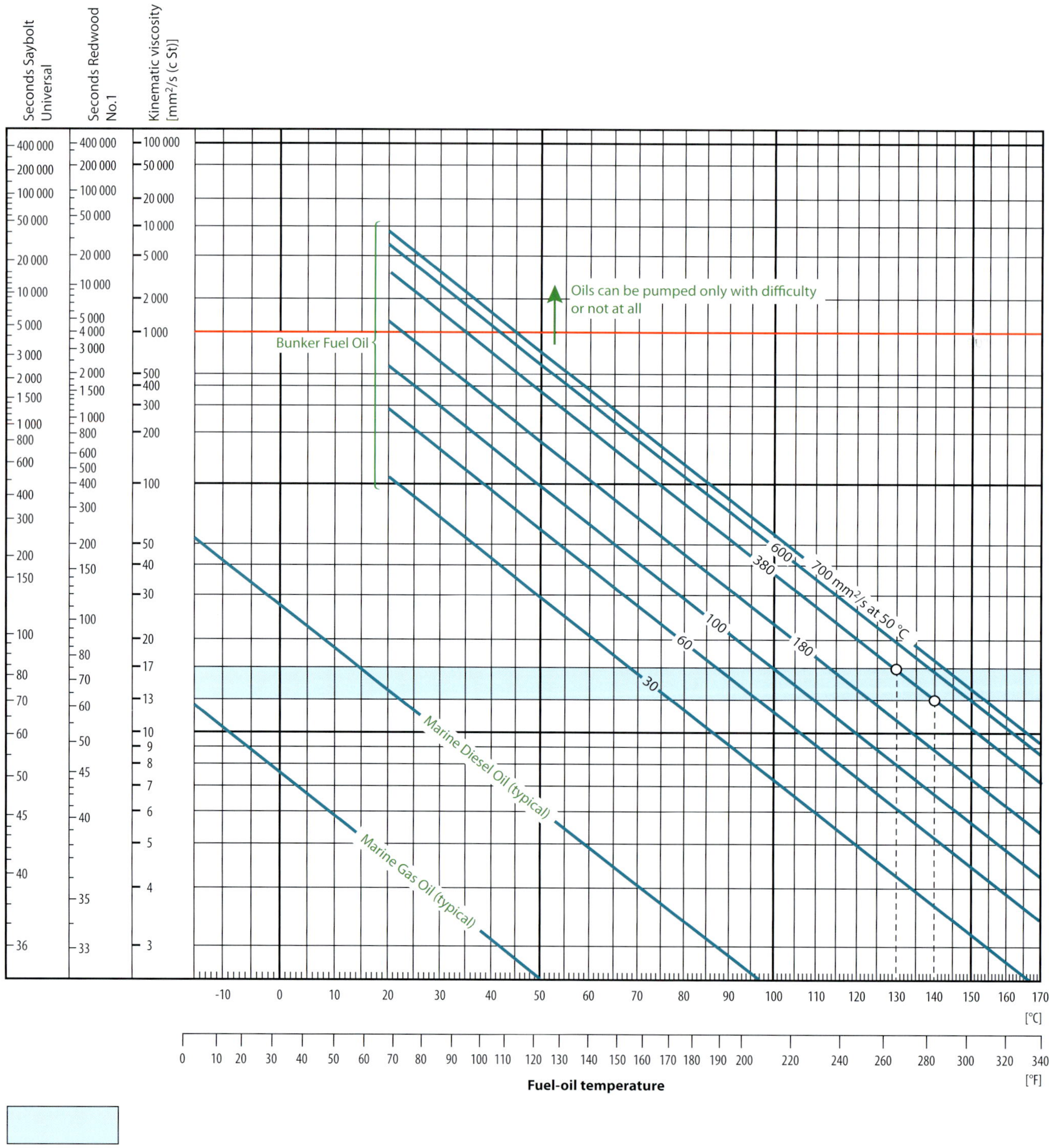

Example

To obtain the recommended viscosity before fuel injection pumps a fuel oil of 380 mm^2/s (cSt) at 50 °C must be heated to 130 to 140 °C

Comment

Naturally every engine manufacturer has his own prescribed viscosity range for optimum atomisation. **Medium-speed engines with a high number of revolutions** and therefore a shorter process time often have lower-viscosity fuels.
Two-stroke low-speed crosshead engines with a longer process time often have fuels with a viscosity that can increase up to 15 cSt.
The maximum allowable fuel temperature is often prescribed by manufacturers and amounts to 150 °C. This is mainly done to avoid problems with the fuel-pump seals. Fuels with a viscosity of 600 and 700 centistokes are rarely used as the price difference with fuels having a viscosity of 380 cSt. is negligible. The fuel temperature needs to be approximately 10 °C higher at 600 and 700 cSt.!

8.7.2 Density denoted by kilogram per m^3

This signifies the mass (m) in kilograms per m^3 at 15 °C. This is very important with regard to, for instance, the required storage space and the settings of the centrifuges. The most advanced centrifuges can clean fuel with a density of 1010 kg per m^3 at 15 °C.

Also see Chapter 24, *Auxiliary systems: Fuel- and lubricating-oil centrifugal separators.*

The rule of thumb for the density correction at an increased temperature:

$$m_{\text{actual}} = m_{15\,°\text{C}} - 0{,}64 \times (T_{\text{actual}} - 15)$$

8.7.3 Flash point, denoted by degrees Celsius

The flash point is the minimum temperature at which the fuel vapours formed over an oil sample that is warmed up in a special heating device, first burns after being ignited with a test flame.
The Pensky Martens closed-cup flash-point tester is a well-known and often applied instrument for determining the flash point. In the shipping industry the minimum flash point of 60 °C is enforced because of safety issues (explosion and fire hazards). Fuel for barge engines and back-up generators forms an exception. To ensure low-temperature ignition of these engines a flash point of 43 °C is allowed. Modified rules apply to the MOB (man-over-board boat) which is equipped with an outboard motor and where very light **petroleum** is used as a fuel.

8.7.4 Flow point, denoted by degrees Celsius

This is the lowest temperature expressed in degrees Celsius at which a fuel can be treated or still flows. This is about three degrees above the solidifying point. The solidifying point is of the utmost importance in the storage and processing of fuels. **Furthermore, the fuel must be pumpable. Once heavy oil has solidified, it is extremely difficult to liquify it by using heating coils. The oil often only liquifies around the coils as the oil no longer circulates. When using heavy oil heating is an absolute necessity.**

8.7.5 Carbon residue or Conradson-carbon number, also denoted by Micro Carbon Residue abbreviated as MCR, indicated in percentage mass

This is a measurement for the formation of carbon and coal deposits during combustion. This is important when considering the contamination of the combustion space in which the atomiser nozzles, the exhaust valves, the pistons, the piston rings and the piston-ring grooves are found and in the exhaust system containing among others the turbocharger, the exhaust-gas boiler and the silencers.
The carbon residue is determined in the Conradson apparatus. Fuel is heated in a crucible. At high temperatures the lighter hydrocarbons vaporise and the heavier hydrocarbons are cracked, breaking down the long chains into smaller ones. The pure carbon residue that remains is subsequently measured.

8.7.6 Ash content, denoted by percentage mass

This represents the inorganic material content that is left behind after combustion at extremely high temperatures when the carbon is removed. For ship fuels the ash content lies between 0.002% and 0.009% mass.
The inorganic metal oxides found are often those of vanadium, nickel, chromium and sometimes sodium. The latter mostly resulting from salt-water flooded oil fields or during transport. Sodium can also be introduced into the fuel during crude-oil processing such as washing. Aluminium and silicon are also found in fuel. They are introduced during the catalytic cracking process. Fuel also contains substances such as rust and dirt, mostly from storage tanks, bunkers or double-bottom tanks.

If the ash content in the fuel is considerably higher than the sum of the parts of sodium, aluminium and silicon, further analysis is required. Ashes may cause wear and engine corrosion.

8.7.7 Water content, denoted by percentage volume

Fuels always contain a small amount of water. This enters during transportation, processing in the refinery and storage in tanks, especially by air condensation.
A maximum of 1% of water is allowed in fuel. In reality 80% of the fuels contain less than 0.3% water. At high water content values the heat value of fuel decreases.

8.7.8 Sulphur content denoted by percentage mass

This is one of the most common elements in liquid fuels. The sulphur content in heavy oil varies from 1.5 to 3.5%. Sulphur poses problems in the form of low-temperature corrosion, LTC (Low Temperature Corrosion). Together with water and oxygen, which are always present in air, sulphur forms sulphuric acid. Sulphuric acid can damage the cylinder liner and, for instance, the exhaust systems at low engine loads.
In the future, IMO will check ship fuels in order to reduce air pollution. Worldwide the maximum sulphur content will be set at 4.5% and in vulnerable areas, a maximum of 1.5%. This measure will become effective before 2010.

8.7.9 Vanadium content denoted by milligrams per kilogram and also in parts per million (normally indicated by ppm)

This metal is found in all fuels. It is bound to hydrocarbons and can not normally be separated from the fuel. Vanadium contents can vary from 10 ppm to 1000 ppm, dependent on the oil field. After combustion of the fuel in the cylinder the vanadium remains in the ash. At certain temperatures and compositions it forms, when combined with sodium and sulphur oxides, a very aggressive precipitate especially on the exhaust valves.
Adequate cooling of the exhaust valve and valve seats and usage of, for instance, corrosion-resistant valve material such as 'Nimonac', this high temperature corrosion, (abbreviated HTC), can be limited.

8.7.10 Aluminium and silicon content, denoted by milligrams per kilogram and often ppm

This occurs mainly in the form of an aluminium silicate, a catalyst used in catalytic cracking of crude oil. These compound silicates are small (1 to 100 micromillimetres) and very hard and can cause serious wear to atomisers, fuel pumps, pistons, piston rings and cylinder liners.
A good centrifuge can remove the majority of these silicates as their specific mass far exceeds that of the fuel.
A maximum of 80 ppm total of the two metals is allowed to be present in the fuel and aluminium alone has a limit of 30 ppm.

8.7.11 Total sediment, indicated by percentage mass

These are ash-like substances or insoluble hydro-carbons. Heavy oil contains asphaltenes, which are usually soluble in the fuel. Fuels that contain insoluble asphaltenes are called unstable fuels.
These unstable fuels may contain large quantities of sediment, overloading the centrifuges and causing them to work poorly.
Fuel obtained from the catalytic-cracking process contains many aromats. This is one of the main reasons that suppliers accept it as a diluent.
Fuels acquired from either thermal cracking or from the 'Hycon'-process pose more problems concerning fuel stability.
Simply put: the latter techniques increase the likelihood of unstable fuels.

8.8 Additional fuel specifications

These are not incorporated in standard specifications, but are in practice important to the user and are often prescribed by engine builders.

8.8.1 Calculated Carbon Aromaticity Index, indicated by the abbreviation CCAI: indefinite number

This is a unit with which the ignition quality of heavy oil is expressed.
This CCAI has been introduced because the quality of heavy oil continues to deteriorate. Modern refineries can increasingly refine light fuels from various residues. As a result the last remnants are becoming less and less suitable for diesel engines.
The ignition quality is especially important.
The Shell system of quality indicators is used.
Fuels that have a high carbon percentage ('heavy oil') which are bound to aromatic compounds produce a longer ignition delay than fuels with fewer aromats.
Research has also shown that there is a relationship between the carbon aromaticity and the fuel properties of density and viscosity.
The CCAI value can be established by means of a formula or a nomogram.

Formule:

$$\text{CCAI} = \rho - 81 - 141 \times \lg \cdot [\lg \cdot (v + 0{,}85)] - 483 \times \lg \frac{T + 273}{323}$$

Hierbij is:
ρ = fuel density in kg/m^3 at 13 °C;
v = kinematic viscosity in mm^2/sec. or cSt at 50 °C;
T = temperature in °C when the viscosity measurement is taken.

One often uses the above nomogram in which the viscosity and the density of the fuel are known by drawing a straight line through these two known points which bisect the CCAI-line. The value at this intersection is the CCAI-value for the fuel.

CCAI-values

Normal heavy oil has CCAI-values of 770 to 840 and is easy to ignite. Cracked residues have values that can increase to 840 or even to values exceeding 900. Most heavy oils have maximum values of approximately 850 to 870.

In many engine manufacturers' product guides the maximum CCAI-value is mentioned in addition to all other (ISO) specifications.

8.8.2 Calorific value

This is the heat that is released during the combustion of 1 kilogram of fuel, expressed in kilojoules per kilogram of fuel.

We know the higher heating value. Here the combustion gases are cooled to the point that the water vapour in the flue gases condenses thereby releasing the condensation heat present in the water vapour. In reality this condensation very rarely occurs in diesel engines. Just think of the formation of sulphuric acid!
In gas engines one often uses this condensation heat by placing a flue-gas condenser in the exhaust pipe. This is manufactured from stainless steel and on top of this the gases contain very little sulphur.
We usually work with the lower heating value or net calorific value.

Lower heating value = higher heating value – heat of vaporisation.

The lower heating value is determined by, among other things, the following properties:
- the density;
- the water content;
- the ash content;
- the sulphur content.

Examples

Fuel with a specific mass of 800 kg per m^3 and a sulphur content of 0.5% has a lower heating value of 43.5 MJ/kg.
Fuel with a specific mass of 1050 kg per m^3 and a sulphur content of 4% has a lower heating value of 39.2 MJ/kg.

The lower heating value can also be established by using formulas and nomograms.
Below is an example of how to calculate the lower heating value or **'specific energy'**.
A higher density means that the fuel contains more carbon and fewer hydrogen atoms.
The combustion of:

Carbon	C	34,000 kJ/kg
Hydrogen	H	144,000 kJ/kg
Sulphur	S	10,500 kJ/kg

Lower heating values of common fuels

gas oil	43,000 kJ/kg
diesel oil	42,700 kJ/kg
heavy oil	40,500 kJ/kg

▼

A so-called nomogram to determine the CCAI-number of a heavy fuel.

The first vertical column represents the kinematic viscosity at 50 and 100 °C. Unit in mm² per second.
The middle vertical column represents the specific mass at 15 °C. Unit in kilograms per m³.

The vertical column to the right represents the CCAI-number. This is an abstract number.
The fuel RMH 55 is an example. It has a viscosity of ± 750 cSt and a specific mass of 990 kg per m² at 15 °C.
By drawing a line to the right through the points mentioned above, the line intersects the right column at a CCAI-number of approximately 852. This is fairly high, but nevertheless acceptable.

$$CCAI = \rho - 81 - 141 \log_{(10)} [\log_{(10)} (\nu + 0{,}85)] - 483 \log_{(10)} \left(\frac{T + 273}{323}\right)$$

ρ = density at 15 °C (kg/m³)
ν = kinematic viscosity (mm²/s) at T temperature
T = temperature corresponding to the viscosity measurement (°C)

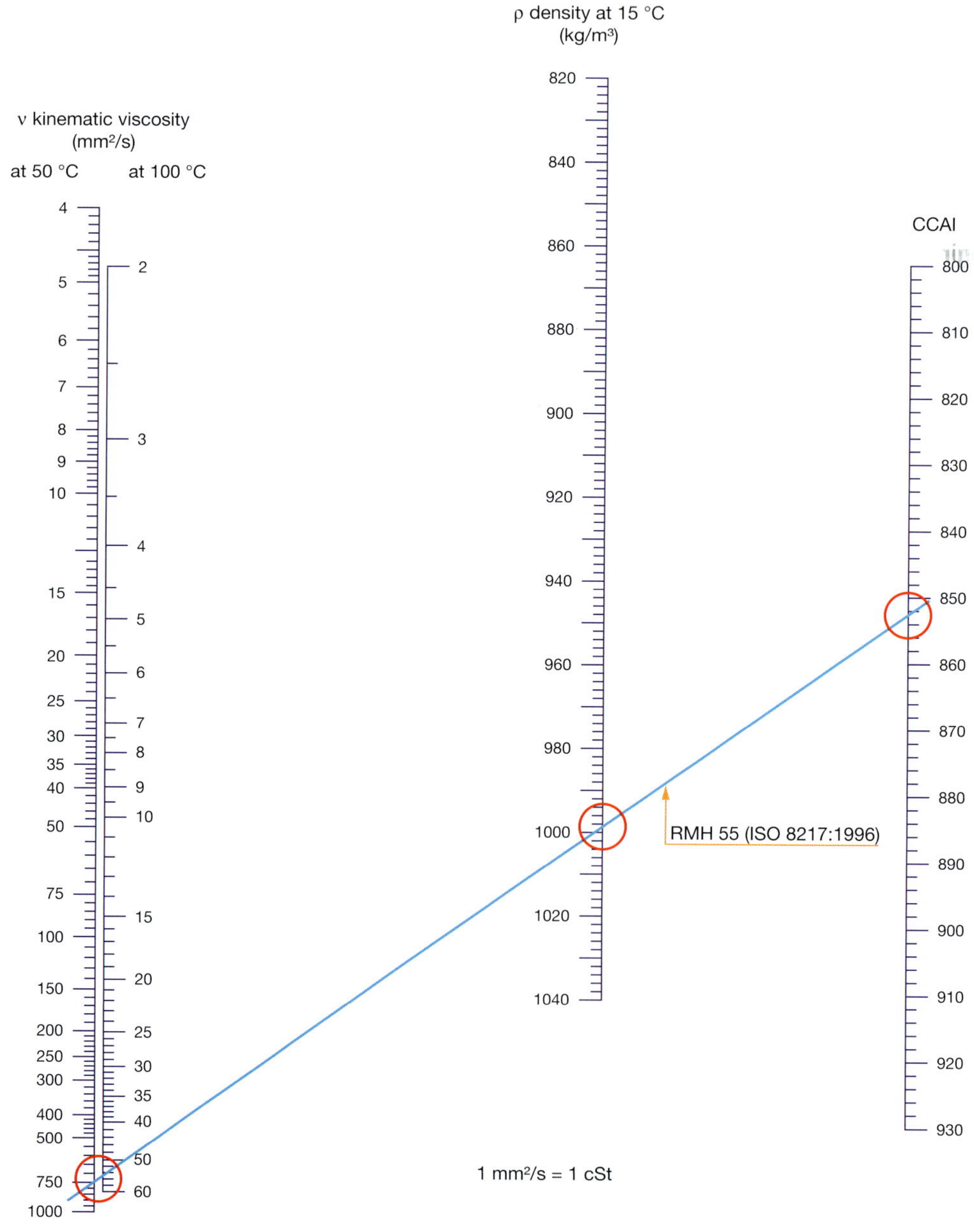

Nomogram to deduce CCAI

This diagram can be used to find the CCAI number of a fuel oil when density and viscosity are known.
The CCAI number can be determined by drawing a straight line through the viscosity and density values.
The line for the RMH 55 (ISO 8217:1996) is outlined as an example.

It is easy to calculate the energy supply to the diesel engine by using the lower heating value.

Examples

A six-cylinder Caterpillar MaK 32 C diesel engine has a power output of 500 kW per cylinder.
The specific fuel consumption for the propeller propulsion is 178 grams per kWh at full capacity (100%). The engine uses heavy oil with a lower heating value of 42,000 kJ per kg.
What is the total efficiency of this engine?

Energy supplied to the engine.
Fuel consumption = number of cylinders × power per cylinder × specific fuel consumption.

Fuel consumption = 6 × 500 × 0.178 = 534 kilograms per hour.

Supplied energy = mass fuel per second × lower heat value

$$P_F = \frac{kg}{sec} \times \frac{kJ}{kg} = \frac{kJ}{sec} = kW$$

$$P_F = \frac{534}{3600} \times 42{,}000 = 6230 \text{ kW}$$

$$\text{Total engine efficiency} = \frac{\text{Shaft power}}{\text{Energy input}}$$

$$\text{Total efficiency} = \frac{3000}{6230} = 0.4815 \text{ of } 48.15\%.$$

1. Engine description

The M 32 C is a four-stroke diesel engine, non-reversible, turbocharged with direct fuel injection.

V-engine M 32 C

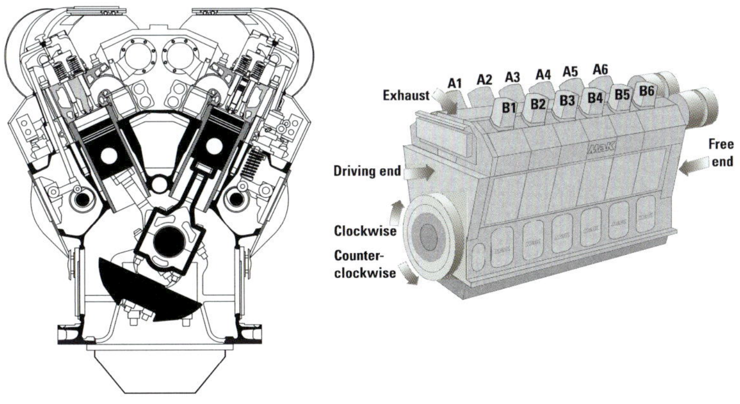

Cylinder configuration:	12, 16 V
Bore:	320 mm
Stroke:	420 mm
Stroke/Bore-Ratio:	1.3
Swept volume:	33.8 l/Cyl.
Output/cyl.:	480/500 kW
BMEP:	23.7/23.7 bar
Revolutions:	720/750 rpm
Mean piston speed:	10.1/10.5 m/s
Turbocharging:	single pipe system
Direction of rotation:	clockwise, option: counter-clockwise

Important data for a Caterpillar MaK 32 C diesel engine.

8.8.3 Fuel stability

Fuel is stable when during long-term storage in, for example, tanks, bunkers or a double bottom on ships, heavy components do not separate and gravitate to the bottom of the storage tanks. Especially in the thermal cracking process, the residue usually contains very long and heavy chain compunds which are difficult to dissolve in the surrounding (lighter) oil distillates.
When these residues are mixed with much lighter distillate oils in the manufacture of the heavy oil, it is a possibility that the heavy compounds slowly sink. This is called a volatile fuel. One often uses gas oil in the mixture.

Using these volatile fuels can create serious problems:
- **sludge forming in the fuel tanks;**
- **problems with the centrifuges during the fuel cleaning process. Here a large amount of sludge (sediment) is extracted, which often blocks the centrifuge entirely;**
- **fuel fine-filters can get completely obstructed.**

8.8.4 Miscible fuels

It should be stressed that mixing different fuels is highly inadvisable.
There is always the possibility that this can result in an unstable fuel. Tanks should preferably be drained completely before being filled with a 'new fuel'. In reality it is not always feasible to remove all the 'old' fuel from a tank.
To establish the fuel stability Shell has developed a test which can be performed on site. **Heavy-fuel instability is a common occurrence.**

8.8.5 Ignition quality, the Cetane number

This only applies to light distillates such as gas oil and diesel oil.
The Cetane number is a measurement of the ignition quality of fuel. Highly aromatic fuels have a lower cetane number. Pre-chamber diesel engines require fuel that has a cetane number of at least 35. Direct-injection engines require a cetane number of at least 40. The cetane number is established using a test engine.

8.8.6 Combustion quality

When a fuel ignites easily this means it will also combust well. The hydrocarbons with chain-like molecule structures (paraffins), burn faster than the hydro carbons with circular molecule structures (naphthenes). During combustion of the latter a large amount of free carbon is produced by the cracking reactions, which in turn translates into high carbon numbers. Today the carbon number is extremely high, over 20% to be precise. This causes slow combustion.

8.8.7 Hydrogen sulphide (H_2S)

This is a toxic gas which can be present in some crude oils and heavy-oil fuels. Damage can occur when present in high concentrations. Water, a combustion by-product, can form sulphuric acid (H_2SO_4) when mixed with hydrogen sulphide.

Inhaling hydrogen sulphide

This is an extremely dangerous health hazard. Death can occur rapidly at high concentrations.

8.9 Decreasing the sulphur content in fuels

Regulations with regard to sulphur content in fuels.

In order to reduce the negative environmental effects, the regulations concerning the maximum allowable sulphur content of diesel fuel or heavy oil are being continually intensified. Here one differentiates between:
- **inland waters;**
- **coastal waters;**
- **regional waters, for instance the European Union.**

These stringent regulations already apply to diesel engines used in the automotive industry or, for example, cranes and trains (traction). The regulations are also being tightened for other engine types in inland navigation such as for propulsion, gensets and bow-thruster propulsion.
Diesel engines require a certificate stating that they meet the stipulated emission values. Rules and regulations have been implemented and are becoming more stringent for the international shipping industry and diesel-power plants.

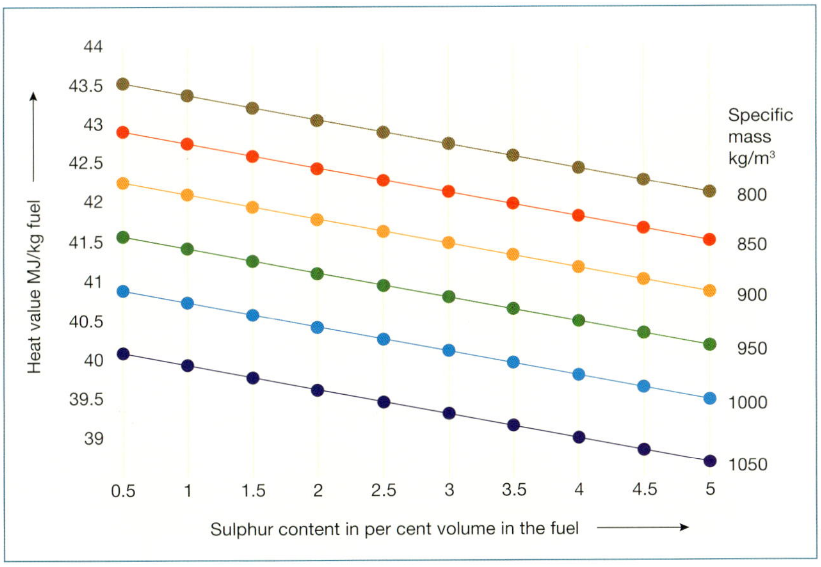

◀

A chart which clearly shows the effects of the sulphur content and the specific fuel mass on the heat value.

Horizontal: the sulphur content in per cent volume in the fuel.
Vertical: the heat value in MJ per kilograms fuel.
The slanted lines: the specific mass of the fuel in kg per m^3.
The bottom line has a specific mass of 1050 kg per m^3, the top line has a specific mass of 800 kg per m^3.

In relation to heavier fuels, light fuels contain less carbon atoms and more hydrogen atoms. Carbon has a relatively high atomic weight and hydrogen a relatively light atomic weight.
Hydrogen generates more energy during combustion than carbon.
A high sulphur content also inhibits the production of energy.
Fuel with a low specific mass and sulphur content has therefore the highest combustion value.

A table providing the concentration of H_2S in parts per million.

At a concentration of approximately 100 ppm, sulphur gas inhalation is life-threatening. Lower concentrations can also pose a health hazard.

H_2S concentrations and their effect on humans	
H_2S Concentration ppm	Physiological effect
0.2 – 0.5	First detectable by smell.
10	Limit for eight hours working, may cause some nausea, eye irritation.
15	Limit for 15 minutes working. Eye and respiratory tract irritation.
50 – 100	Eye and respiratory tract irritation after 1 hour. Longer exposures to concentrations at 100 PPM induce a gradual increase in the severity of these symptoms and death may occur after 4-48 hours exposure.
150	Loss of sense of smell.
350	Could be fatal after 30 minutes inhalation.
700	Rapidly (few minutes) induces unconsciousness and death. Causes seizures, loss of control of bowel and bladder. Breathing will stop and death will result if not rescued promptly.
700+	Immediately fatal.

It is technically fairly simple to further reduce the sulphur content of engine fuels. The rules regarding the maximum sulphur content of gas oil and diesel oil are already stringent.
In the territorial waters of Europe ships are required to prove that the sulphur content of their diesel oil amounts to less than 0.2%.
So ships not running on low-sulphuric diesel oil are obliged to reserve a separate tank suitable for fuels with low sulphur content.

Let's assume that the regulations for heavy fuel are as follows: the sulphur limit is 1.5% according to IMO Marpol Annex VI for regions within which the sulphur rules apply, or perhaps even 1% for certain areas inside the European Union.
This means that ships which normally use heavy oil with sulphur contents exceeding these values, which is generally the case, are required to have separate tanks for this low-sulphuric fuel. Many ships are already equipped with these special tanks, so they can meet the sulphur limit when the rules become effective.

Low-sulphur fuel costs

These are largely a matter of supply and demand, but the costs are considerably higher than those of regular heavy oil. Furthermore, these low-sulphur fuels are generally difficult to obtain.
One can expect to pay 15 to 20 US-dollar per ton surcharge.

At even higher surcharge costs on-shore companies, such as diesel-power plants, can consider using a different fuel, such as natural gas for Dual Fuel engines or bio-fuels, which are presently in great demand.

Also see Chapter 29, *New fuel developments*.

▼
Fuel price development of heavy oil with a viscosity of 380 centistokes and diesel oil (fourth quarter of 2010) at four bunker stations: Houston, Singapore, Fujairah and Rotterdam.

The prices are provided in USD per tonnage weight. For heavy oil the average price is 490 USD per ton and for diesel oil 785 USD per ton.

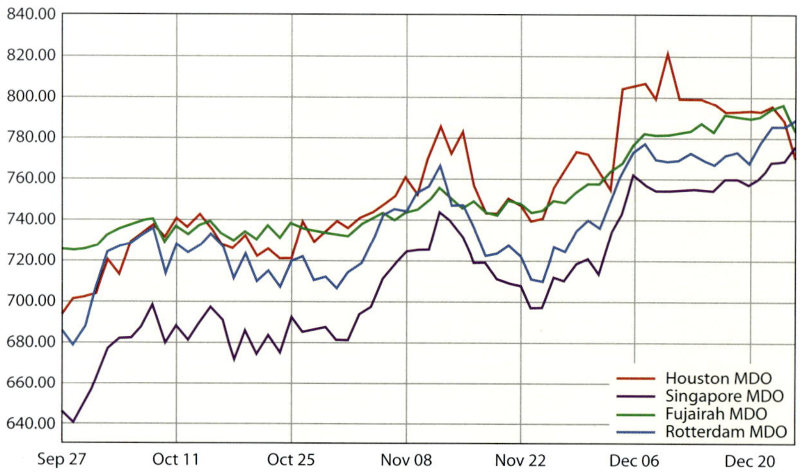

8.9.1 Sulphur-free diesel oil

On shore, sulphur-free fuel will be increasingly prescribed. In 'sulphur-free' diesel oil the sulphur content is less than 1%.

A maximum sulphur content of 0.2% applies to the territorial waters of the European Union for gas oil and diesel oil, including DMA-, DMB-, DMC- and DMX-specifications.
The latest IMO guideline and the European Commission state that the sulphur content of heavy oil on the North sea, the English channel and the Baltic sea may not exceed 1.5%. This new rule became effective on the 19th of May 2005!

Presently, heavy bunker oil has an average sulphur content of 2.7%. The price difference between 1.5% and 2.7% sulphur content bunker oil currently amounts to approximately 30 USD per ton. It is very probable that most ships will drain a fuel tank in order to use it for these low-sulphur content fuels and run on the 'low-sulphur tank' in said areas.

8.10 Bunkering

Everywhere diesel engines are used, fuel delivery is a periodic reoccurrence. Engines with high operating hours such as in navigation and in diesel-power plants use enormous amounts of fuel.
The specific fuel consumption of engines lies roughly between 170 and 220 grams per kWh, depending on the working principle, the engine size and the load.

Example 1

A large inland boat with a power output of 1600 kW and 5000 operating hours per annum has a specific fuel consumption of 200 gr/kWh:
Fuel consumption = 1600 × 0.200 × 5000 = 1,600,000 kg per annum.

This is approximately $\frac{1{,}600{,}000}{0.850} = 1{,}882{,}350$ litres or 1882 m^3!

Example 2

A container ship of 8600 TEU (TEU = twenty-feet equivalent unit = standard 20-feet container) with a power output of 72,000 kW and 6500 operating hours has at full load a specific fuel consumption of 175 grams per kWh:
Fuel consumption: 72,000 × 0.175 × 6500 = 81,900,000 kilograms

or $\frac{81{,}900{,}000}{0.950} = 86{,}210{,}526$ litres or 86,210 m^3!

380 IFO		Feb-04 2-6	Feb-04 9-13	Feb-04 16-20	Feb-04 23-27	Mar-04 1-5	Mar-04 8-12	Mar-04 15-19	Mar-04 22-26
Rotterdam	d	135	134	142	144	149	147	142	144
Gibraltar	d	160	157	156	156	158	155	149	152
Piraeus	d	148	154	145	146	150	148	144	147
Suez	d	158	156	159	159	161	161	159	164
Fujairah	d	174	168	167	165	164	162	162	162
Durban	w	n/a	n/a	n/a	n/a	n/a	n/a	n/a	n/a
Tokyo	d	201	198	195	196	197	198	197	196
Busan	d	184	181	182	183	186	185	182	179
Hong Kong	d	183	185	184	184	181	182	181	180
Singapore	d	171	172	170	167	166	164	165	166
Los Angeles	w	168	163	161	155	155	155	157	164
Houston	w	153	162	152	154	154	154	153	154
New York	w	163	163	162	158	161	161	157	158
Panama	w	160	160	160	160	161	164	168	168
Santos	d	156	142	144	146	156	156	154	153
Buenos Aires	d	n/a	n/a	n/a	n/a	n/a	n/a	n/a	n/a

180 IFO		Feb-04 2-6	Feb-04 9-13	Feb-04 16-20	Feb-04 23-27	Mar-04 1-5	Mar-04 8-12	Mar-04 15-19	Mar-04 22-26
Rotterdam	d	144	144	152	154	159	158	152	154
Gibraltar	d	169	166	164	164	167	164	157	161
Piraeus	d	160	155	159	160	165	164	161	164
Suez	d	164	162	165	166	167	167	166	171
Fujairah	d	181	174	173	172	171	169	169	169
Durban	w	170	167	169	166	165	163	161	159
Tokyo	d	205	202	200	200	201	202	201	200
Busan	d	192	191	192	192	195	194	192	189
Hong Kong	d	186	187	187	188	187	186	185	184
Singapore	d	175	172	175	170	171	168	170	170
Los Angeles	w	193	186	182	169	172	174	173	178
Houston	w	161	162	160	160	160	164	157	159
New York	w	183	182	177	177	179	178	172	173
Panama	w	170	170	169	171	171	174	178	179
Santos	d	159	146	147	150	160	160	158	157
Buenos Aires	d	171	170	167	165	166	164	165	165

MDO		Feb-04 1-5	Feb-04 8-12	Feb-04 15-19	Feb-04 22-26	Mar-04 5-9	Mar-04 12-16	Mar-04 19-23	Mar-04 26-30
Rotterdam	d	243	236	242	238	254	253	268	266
Gibraltar	d	284	281	300	306	322	319	328	327
Piraeus	d	264	261	277	282	300	298	307	307
Suez	d	321	321	321	321	325	336	335	335
Fujairah	d	310	310	310	310	309	309	309	309
Durban	w	267	263	274	285	301	299	308	311
Tokyo	d	287	288	291	288	287	292	292	290
Busan	d	332	325	322	322	321	321	319	318
Hong Kong	d	274	274	282	283	282	281	279	273
Singapore	d	277	278	282	282	284	281	279	277
Los Angeles	w	317	320	344	358	345	342	342	340
Houston	w	281	280	278	278	279	282	285	282
New York	w	345	343	335	327	325	330	331	330
Panama	w	317	317	318	321	321	323	326	327
Santos	d	350	343	342	344	347	350	345	344
Buenos Aires	d	317	311	302	306	295	303	303	302

◄

The prices of heavy oil and diesel oil in February and March 2004.

The price of fuels is affected by, amongst others, supply and demand, war, natural disasters and international politics.

d = day price

w = week price

Prices in March 2012

Rotterdam:

MDO/MGO 990 USD per ton

180 IFO 730 USD per ton

380 IFO 690 USD per ton

▼

A container ship for inland shipping.

Diesel oil is the fuel for propulsion engines and power aggregates.

On average 2000 tons of heavy oil is bunkered by these ships, but this can go op to as much as 8000 tons! Moreover, large amounts of lubricating oil are used. For the latter example, approximately 120,000 litres per annum!

►

A large sea-going container ship.

Heavy oil is the fuel for both propulsion engines as for diesel gensets.

▶
Emergency power aggregates usually run on diesel oil.

This type of engine should start within a few seconds and run at full load. With relatively light diesel oil this is not a problem.

Example 3
A back-up genset for a hospital with a power output of 1500 kW has at full load a specific fuel consumption of 190 grams per kWh:

$1500 \times 0.190 = 285$ kilograms of $\dfrac{285}{0.850} =$
335 litres per hour.

If, in case of an emergency situation or any other circumstances requiring the engine to run for two whole days, a fuel storage tank of 2 × 24 × 335 = 16,100 litres would be required.
It is for the diesel-engine user of the utmost importance that the quality of the fuel meets all requirements.

8.10.1 Bunkered fuel quality, who determines the quality?

Firstly, refineries adhere to the ISO 8217 norm. Additionally, there are the engine manufacturers' recommendations. The latter are far stricter than the former. The owners and users can also follow their own regulations.

Gas oil makes a good example. Apart from liquid it often contains minute metal particles that are so small that most fuel filters cannot filter them out. Especially aluminium silicate, the so-called 'cat-fines', that have a two to ten micron diameter and cause increased wear and tear to fuel pumps, atomisers, pistons, and cylinder liners. One also occasionally finds soot particles in filters.
According to ISO 8217 the maximum allowed amount of metal particles in gas oil is 80 ppm. Engine manufacturers of modern engines recommend 15 ppm as the maximum value.
Most metal particles can be removed by centrifuges. Of course one must have access to such a centrifuge on an inland boat.
Another solution would be to use very fine fuel filters for these installations, which are able to filter out most of the metal particles. The mesh of these filters is only two microns!

8.10.2 Fuel samples

It is always advisable to take a sample of the fuel supplied. Of course this is not practical for the small amounts purchased at petrol stations. For large to very large amounts the purchase price plays an important role, especially when one encounters problems with regard to storage, cleaning and usage of the bunkered fuel.

Recommendations for taking fuel samples
The sample needs to be representative. This means that it should represent the average value of the entire fuel intake. It is advisable to take various samples at regular intervals during the bunkering. There is special equipment available which automatically taps small amounts of fuel during bunkering.

▶
Testing fuel samples in a laboratory.

It is very important to identify the fuel composition prior to using it in the diesel engine.

▶▶
Testing fuel in the laboratory of Inpechem Inspectors B.V., Rotterdam, The Netherlands.

8.10.3 Bunker procedures

Ensure that:
- the fuel supplier knows exactly the type and amount of fuel oil required;
- both parties are in agreement on the arrangements, the pump capacity and manner of communication during the bunkering operation;
- both parties are aware of how to stop the bunkering operation in case of an emergency situation;
- both parties follow the bunkering process;
- both parties take all safety and anti-pollution measures prior to commencing the bunkering operation.

The following is important:
- discuss and make arrangements regarding the starting procedure of opening the valves and tanks;
- discuss and make arrangements regarding the sampling procedures that take place during the bunkering operation;
- take collective fuel samples during bunkering at an agreed location;
- discuss and make arrangements as to how the supply line will be closed when the bunkering operation is terminated;
- most shipping companies and manufacturers have stipulated strict procedures supported by checklists.

Make sure that both parties sign the following documents at the correct time:
- amount of fuel;
- type of fuel;
- the bunkering sequence;
- the pump capacity;
- safety and anti-pollution regulations;
- starting procedures for the pumps;
- terminating procedures for the pumps;
- method used for the verification of measurements;
- bunkering data;
- sampling data;
- an agreement stipulating that all complaints be submitted in writing and that the signature on the complaint form will only be used to acknowledge receipt.

◄ The bunkering station on a large container ship.

Apart from heavy oil and diesel oil, lubricating oil can also be bunkered. The picture also shows the pipes required for the discharge of contaminated lubricating oil and separator sludge.

◄ A sample-taking device.

A sample-taking device which is placed in the bunker line and periodically takes some fuel from the line in order to collect a representative sample of the bunkered fuel. This is often referred to as a 'drip sample'.

The sampling method and location, and a record of the witnesses attending the sampling is very important. This forms the basis for all the discussions concerning possible ensuing disagreements as to the quality of the fuel supplied.

8.10.4 Gas-oil shipping associations

The VOS (Vignet Olie Scheepvaart) foundation, for instance, is active in The Netherlands and monitors the quality of gas oil and issues a hall mark for shipping gas oil.
There are recurring signals of serious malfunctions in engines in the inland shipping branch. This apparently mainly concerns the decrease in the operating life time of fuel filters. This often causes engines to stall.

Possible problems

It is often suggested that paraffin could be one of the causes. Paraffin is an essential part of every (diesel) fuel. The type and the amount of paraffin in the fuel affect the **'Cloud Point'** (CP). The cloud point is the stage at which the paraffins start to 'flocculate' in the fuel, thus blocking the filters. In physics this is referred to as 'crystallizing out'.

A process during which the fuel is slightly heated and consequently prevents the paraffin from crystallizing out could form one of the solutions. This, of course, requires some technical adjustments. The **quality of the paper used in the filters** could also be one of the problems. Gas oil naturally absorbs 22 ppm of water before it separates from the water. This could pose problems if the filter paper is not manufactured for these conditions. There are also instances where the filters are obstructed by a black substance. The composition of this particular substance remains unknown.

Metallic pollution

According to the experts iron and silicon (sand) do not naturally occur in gas oil. Gas oil is a distillate and these elements do not vaporise. They must therefore originate from an external source.

Filter sizes

In the past filters up to 10 microns were used and today one uses even finer filters of less than 2 microns. In the automotive industry much coarser filters up to 25 microns are used. Tests have demonstrated that filters less than 2 microns get blocked after a period of time.

Diesel- and gas-oil storage

Generally, the maximum storage time of fuel is approximately one year. Contamination with water (liquid) and other substances has to be prevented.

Fuel costs

In exploiting engines in shipping, trade, industry and diesel-power plants, the fuel- and lubricating-oil consumption form, especially with the engines in continuous use, the greatest cost factor.

Example

A large container ship with a cargo capacity of 7200 TEU can transport about 10,000 tons of heavy oil of which it will consume approximately 250 tons per day. For instance: a price difference of 30 USD per ton between Rotterdam and Singapore amounts to a price advantage of 250,000 USD at full bunkering of the ship in Rotterdam. For the standard eight return trips to Asia that a container ship makes on a yearly basis, this amounts to nearly 2,000,000 USD!

Fuel-line systems

In using diesel engines the storage, cleaning and treatment of the fuel before it is ready for use in the engine is of the utmost importance.
Before the fuel can be used, the following applies:

1 Fuel storage
This should be such that the fuel can not be contaminated with corrosives, sand or other impurities. Any condensate should be drained.

2 The amount of fuel
This should be adequate for the planned voyage, that is: the number of operating hours.

3 Fuel quality
This should be established through sampling.

4 Cleaning fuel
The fuel should be cleaned sufficiently by means of filters and/or separators.

5 Engine supply
An adequate amount of fuel should be transported to the engine at a certain pressure and temperature.

6 Fuel specifications
The fuel needs to meet the relevant ISO 8217 specifications as well as the local and inter-local regulations such as maximum sulphur content.

7 Fuel specifications of engine suppliers
Of course, the fuel must comply with the specifications detailed for the engine make and type. This is also important in case there is damage, excessive wear and tear or other problems.

General storage of fuels

This can diverge tremendously according to the engine size, the fuel used, the application, such as a diesel-power plant or ship propulsion and the number of operating hours.
For small category I diesel engines a simple fuel tank may suffice, whereas category III and IV engines which use large amounts of heavy oil require large storage tanks equipped with heating systems.
There are building instructions in place for both fuel-storage tanks on shore and on ships. These serve to prevent fuels polluting the environment and unsafe situations such as fire and explosions. It is important for mechanical engineers to bear in mind the following points of interest:

- storage tanks must never be over-pressurised; therefore a pressure valve should be placed on top of the tank.
- storage tanks should be equipped with a drain valve which can remove water and dirt particles.
- a lid or manhole makes it possible to check for rust formation and dirt during an internal inspection.
- in order to prevent dirt particles, water and rust in the fuel system, the connection between fuel line and engine should preferably not be placed at the bottom of the tank but at a certain distance from the tank floor.
- gauge glasses on tanks should be closed at all times and only be opened for reading.

Of course, there are many specific rules and regulations with regard to the building of fuel tanks which are outside the scope of this book.

8.11 Fuel-line systems according to the engine classification

8.11.1 Category I: 0 to 100 kW, four-stroke, high-speed, fuel MDO

These are often very straightforward systems. From an elevated tank the diesel oil passes through a fuel filter towards the high-pressure fuel pump. The atomiser-drain or return pipe is also connected to the fuel tank. If the tank is not placed in an elevated position, the engine is equipped with a low-pressure fuel pump. It is important that the filter, transfer pumps and lines can be easily bled. It is also advisable to fit a water tap at the bottom of the fuel filter.

Examples

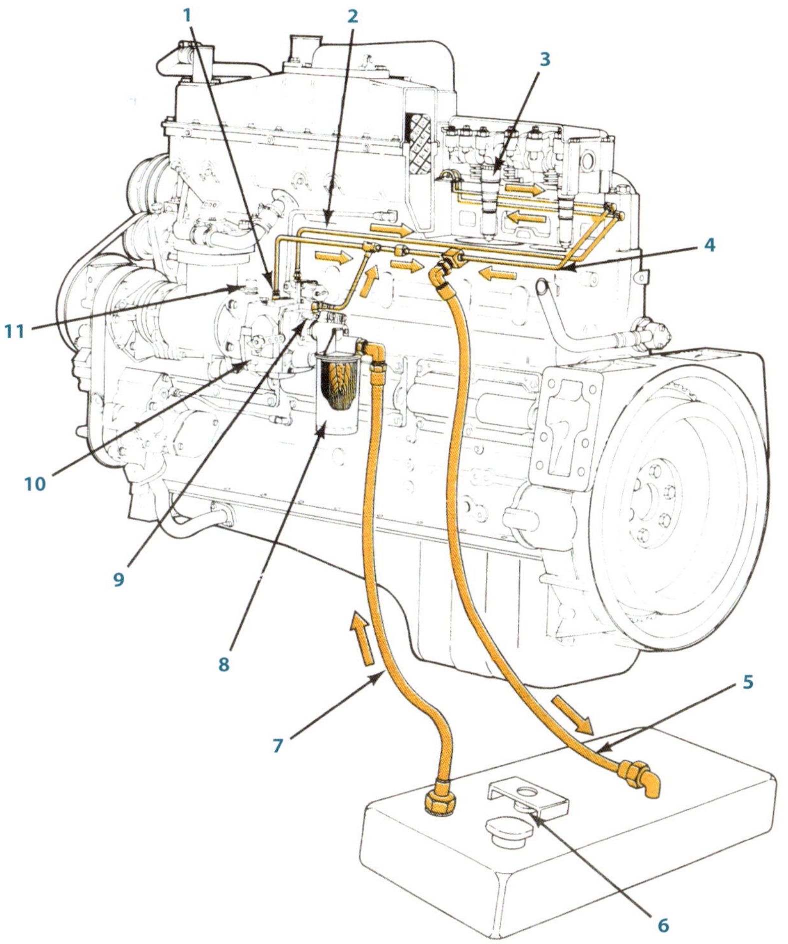

◀ Fuel system of a Cummins NTA 855 Big Cam III diesel engine.

This is an engine with a fuel injector mechanically driven from the camshaft in the cylinder head. This fuel injector is a combination high-pressure fuel pump and a fuel atomiser.
The storage tank can be placed anywhere and is often positioned lower than the engine. The tank is equipped with a filler cap, a vent 6; a suction line 7 and a return line 5.
An engine-driven low-pressure fuel pump 10 draws fuel from the storage tank via a filter 8 to the high-pressure line on the suction side of the injectors 3. The return fuel of the injectors flows from pipe 5 to the storage tank. Mounted on top of the low-pressure fuel pump is a vent valve 1.

1 *AFC cavity drain*
2 *fuel rail pressure line*
3 *injector*
4 *injector fuel drain return*
5 *fuel return to tank*
6 *fuel tank breather*
7 *fuel inlet supply*
8 *fuel filter*
9 *gear pump coolant drain*
10 *fuel pump*
11 *tachometer drive*

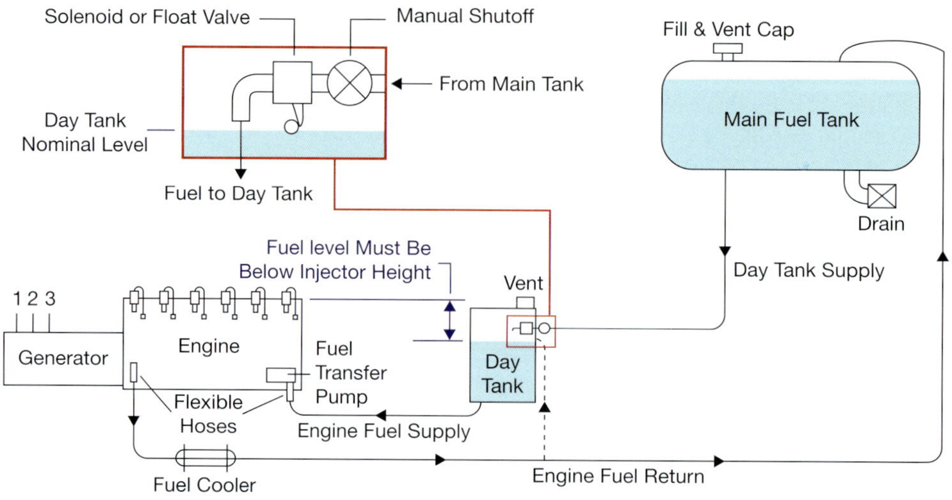

▲

Fuel system of a Caterpillar diesel-power unit, general.

This consists of an elevated storage tank with a fill-and bleed connection. A drain for water and dirt is fitted at the bottom of the tank. Via an automatic float system or a manual stop valve, the day tank can be maintained at the correct level. From the day tank the fuel flows to the low-pressure fuel-suction pump. This provides a certain fuel pre-pressure to the mechanically driven fuel injectors. A suction filter has been placed before the fuel-suction pump to protect the pump. After the fuel pump a duplex filter is installed to clean the fuel. In the engine the supply and return lines are fitted with a flexible section in order to avoid damage to the fuel lines as a result of vibration. The fuel return line has an integrated fuel cooler.

8.11.2 Category II: 100 to 8000 kW, four-stroke, high-speed, fuel MDO

Engines can be equipped with either a high-pressure fuel pump for each cylinder or one block-fuel pump. This also applies to category I engines. These larger engines usually have several clean-fuel tanks. Two duplex section filters are fitted in order to avoid problems with blocked filters, which could cause the engine to stall. The atomiser-drain or return pipe must be connected to the tank being used.

Examples

Some larger engines in this category, which have a power output easily exceeding 1000 kW, are generally equipped with a fuel separator which cleans the diesel oil pumped from the fuel tanks prior to storage in the clean-fuel tank or the daily supply tank.

►

A fuel system for diesel oil of a Caterpillar, Series 3500 B MEUI.

The diagram is self-explanatory. The green section of the line has no pressure build-up. The pressure governor above the fuel tank by-passes the fuel to the storage tank at high pressure. The suction line for both low-pressure fuel-suction pumps is located to the left of the tank. The filter in the suction line protects the pumps. Sometimes a water separator is also placed here. The grey casing left in the drawing is called the 'Engine Control Module' (ECM) and controls all the important engine functions.

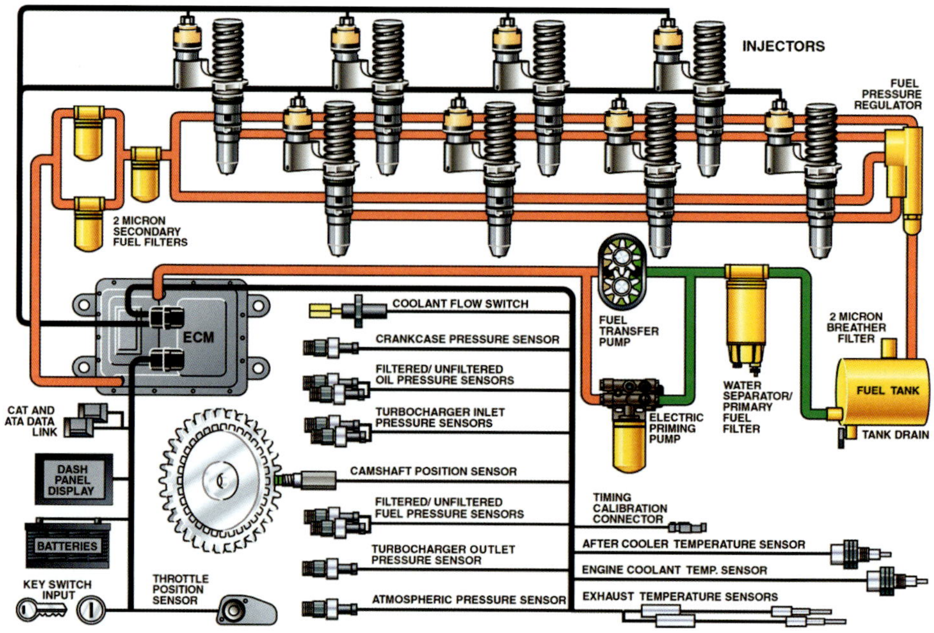

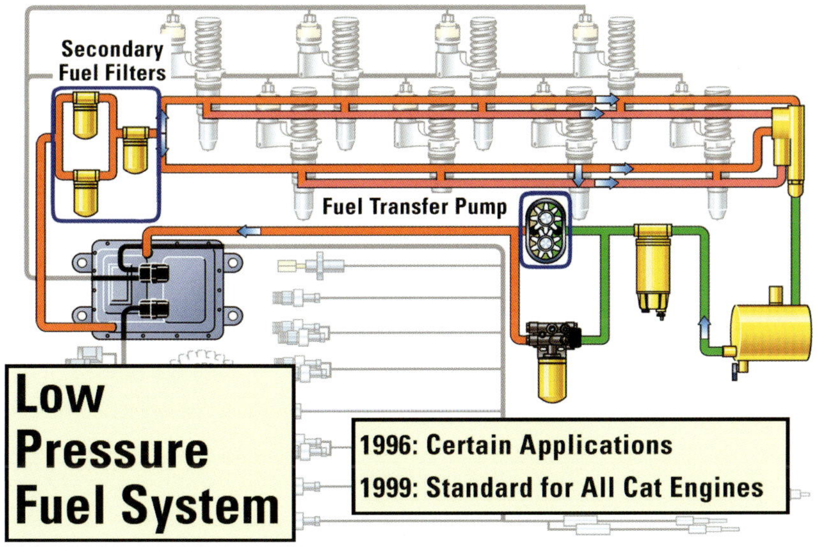

◄
The low-pressure fuel system of a Caterpillar diesel engine, category II.

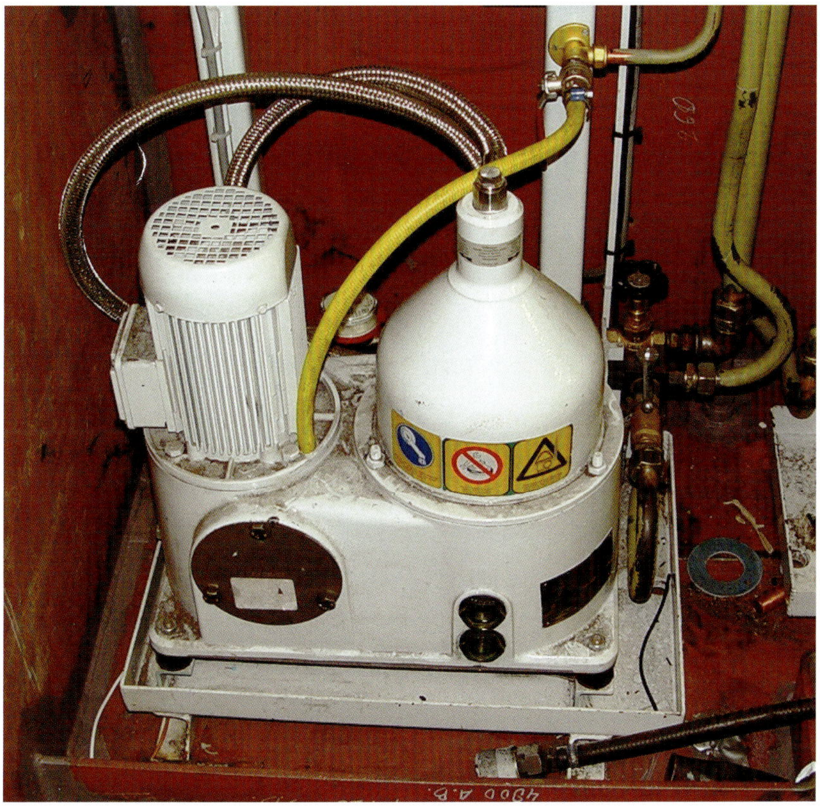

◄
Fuel separator.

If required, engines running on diesel oil are equipped with a small separator, where dirt and water are removed. This separator or centrifuge is located between the fuel-storage tank and the fuel-day tank.

8.11.3 Category III: 500 to 30,000 kW, four-stroke, medium-speed, fuel HFO

Engines in this category use very elaborate fuel systems. The heavy oil must be cleaned and attain the correct pressure and temperature before it reaches the high-pressure fuel pipes.

Examples

Fuel systems with heavy fuel

When using heavy oil it is of the utmost importance that the fuel temperatures during the various stages of the pre-treatment are read, such as:

– the sedimentation or settling tanks	± 60 to 80 °C
– the separator	± 95 to 98 °C
– the fine-filter to the engine (max. 150 °C)	± 125 to 135 °C

▶
Fuel system with diesel oil.
Shown, an example of the system of a Caterpillar MaK 43 C propulsion diesel engine.

Brief description.
Diesel oil can be pumped from storage tank DT 4 to the diesel oil tank DT 1 via the suction filter DF 3 and diesel-oil trimming pump DP 3. Additionally, from the storage tank DT 4, diesel oil can be pumped via an electric heater DH 2 using the diesel-oil trimming pump DP 5 to the diesel-oil separator, where additional cleansing (water and dirt) occurs. The fuel is then pumped to the diesel-day tank DT 1,

From the elevated diesel-oil tank, the diesel oil flows through the duplex filter DF 2, the flow meter FQ 1 and the diesel-oil pre-heater DH 1. Following, one of the two diesel-oil feed pumps pumps the fuel through a very fine filter to the high-pressure fuel pumps KP 1. Spilled fuel oil is collected in the spillage collection fuel-oil tank KT 1.
This fuel flows back into the day tank DT. The high-pressure fuel pump return line feeds the day tank DT via a flow meter FQ 1.
The difference between both flow meters is the actual fuel consumption of the engine.
After the high-pressure fuel pumps, the fuel pressure is set using the adjustable valve DR 2. As there are two pre-heaters and a day tank, which contains a heating element, the fuel can be maintained at the correct temperature in cold conditions. In this manner the injected fuel viscosity is acceptable, which in this engine running on diesel oil is a maximum of 14 cSt.

10. Fuel oil system
Gas oil / MDO operation

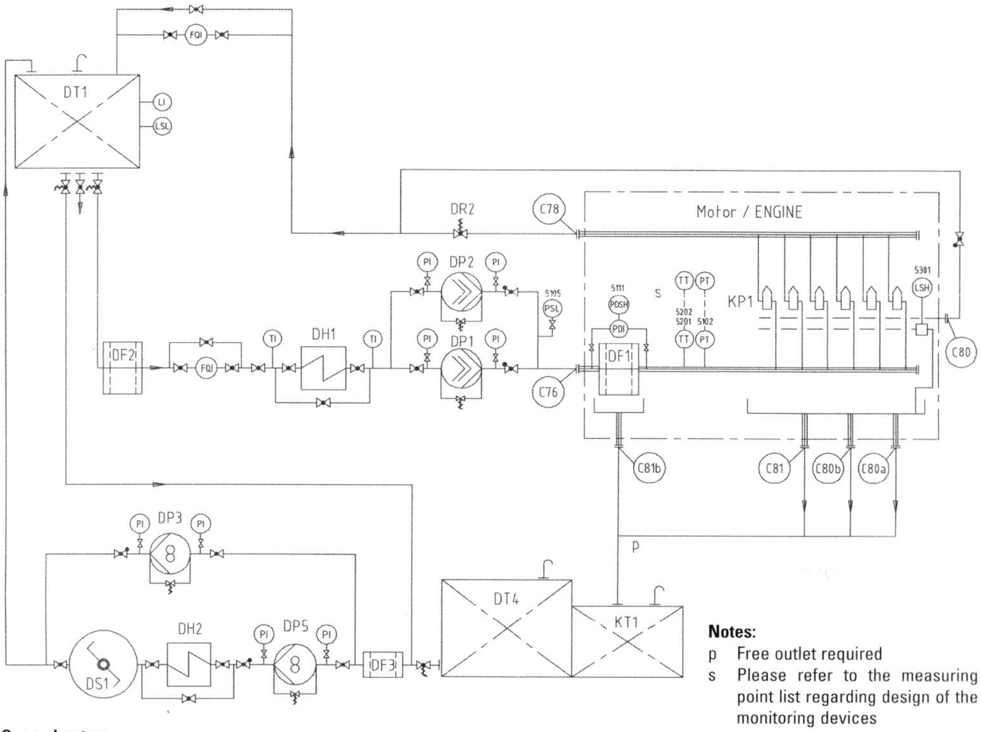

Notes:
p Free outlet required
s Please refer to the measuring point list regarding design of the monitoring devices

General notes:
For location, dimensions and design (e. g. flexible connection) of the disconnecting points see engine installation drawing.
DH1 not required with:
- Gas oil ≤ 7 cSt/40°
- heated diesel oil day tank DT1

Accessories and fittings:

DF1	Fuel fine filter (duplex filter)
DF2	Fuel primary filter (duplex filter)
DF3	Fuel coarse filter
DH1	Diesel oil preheater
DH2	Electrical preheater for diesel oil (separator)
DP1	Diesel oil feed pump
DP2	Diesel oil stand-by feed pump
DP3	Diesel oil transfer pump (to day tank)
DP5	Diesel oil transfer pump (separator)
DR2	Fuel pressure regulating valve
DS1	Diesel oil separator
DT1	Diesel oil day tank
DT4	Diesel oil storage tank
KP1	Fuel injection pump
KT1	Drip fuel tank
FQ1	Flow quantity indicator
LI	Level indicator
LSH	Level switch high
LSL	Level switch low
PDI	Diff. pressure indicator
PDSH	Diff. pressure switch high
PI	Pressure indicator
PSL	Pressure switch low
PT	Pressure transmitter
TI	Temperature indicator
TT	Temperature transmitter (PT 100)

Connecting points:

C76	Inlet duplex filter
C78	Fuel outlet
C80	Drip fuel
C80a	Drip fuel (injection pump)
C80b	Drip fuel (sealing oil injection pump)
C81	Drip fuel
C81b	Drip fuel (filter pan)

▶

Fuel system for HFO Caterpillar MaK 43 C propulsion diesel engine.

Brief description.

The fuel must be maintained at the correct temperature throughout the system. This entails adequate insulation of the tanks and pipes and that pre-heaters are mounted in the fuel tanks trimming section, the fuel separators, the fine filters and in the day tank. Moreover, some pipes are heated with warm water, steam or electricity. This is usually referred to as 'tracing'.

The fuel trimming pump HP 5 or HP 6 can draw fuel up from the settling tanks HT 5 and HT 6. In these tanks, it will take at least 24 hours for the heavy oil to achieve the correct temperature and separate from the water and dirt. This sinks and can be drained (water). Via heater HH 3 the heavy oil is cleaned in separators HS 1 and HS 2. The waste product, the sludge, is stored in the sludge tank KT 2. The heavy oil is subsequently pumped from the separator to the heated day tank HT 1 by its own pump. From the day tank the fuel passes first through the filter HF2 and is then forced through a self-cleaning automatic fine filter HF 4 by feed pump HPO 1 or HP 2. Via a flow meter FQ 1 the fuel flows through a mixing tank GT 2 to the circulating pumps HPO 3 and HP 4. Both pre heaters HH 1 and HH 2 ensure that the fuel reaches the correct temperature. This is controlled by the viscosity gauge HR 2. Via a final Duplex fine filter HF 1 the fuel flows to the high-pressure fuel pumps. At extremely low fuel consumption the cooler DH 3 can ensure that the maximum fuel temperature in the day tank HT 1 is not exceeded.

In fine filters pressure-difference gauges indicate if a filter is blocked. Fine-filter HF 4 is automatically flushed clean when certain pressure differences are exceeded. The mixing tank HT 2 ensures an even transition from diesel oil to heavy oil and vice versa.

MaK

10. Fuel oil system

Heavy fuel operation

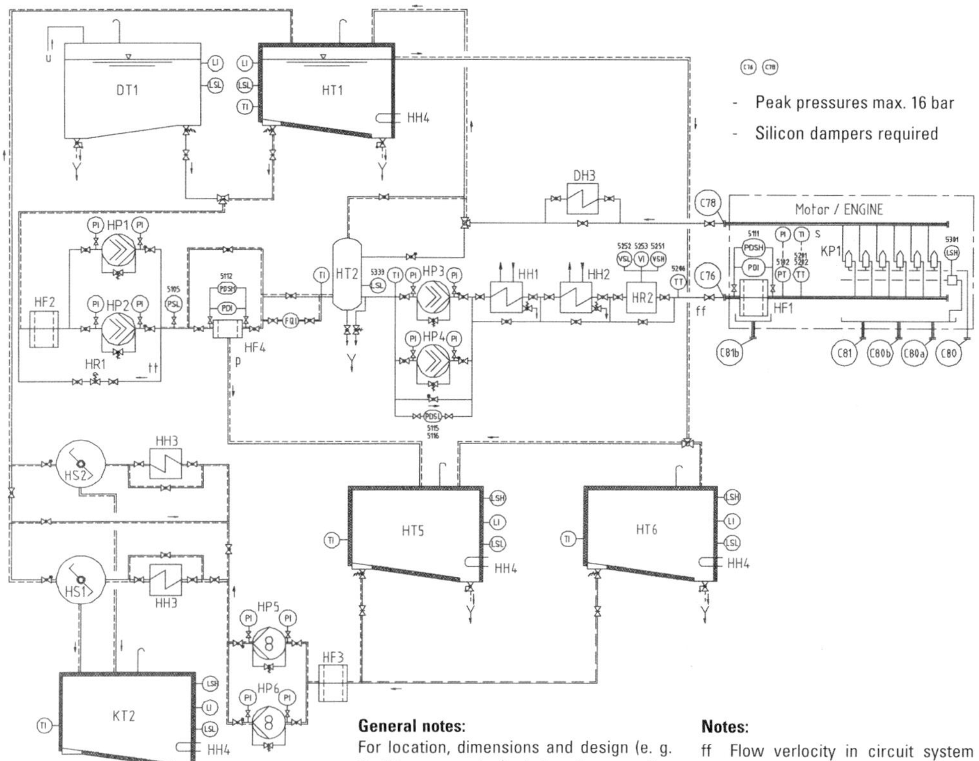

Accessories and fittings:

DH3	Gas oil cooler
DT1	Diesel oil day tank
HF1	Fine filter (duplex filter)
HF2	Primary filter
HF3	Coarse filter
HF4	Self cleaning fuel filter
HH1	Heavy fuel final preheater
HH2	Stand-by final preheater
HH3	Heavy fuel preheater (separator)
HH4	Heating coil
HP1/HP2	Pressure pump
HP3/HP4	Circulating Pump
HP5/HP6	Heavy fuel transfer pump (separator)
HR1	Pressure regulating valve
HR2	Viscometer
HS1/HS2	Heavy fuel separator
HT1	Heavy fuel day tank
HT2	Mixing tank
HT5/HT6	Settling tank

General notes:

For location, dimensions and design (e. g. flexible connection) of the disconnecting points see engine installation drawing. Valve fittings with loose cone are not accepted in the admission and return lines.

KP1	Injection pump
KT2	Sludge tank
FQ1	Flow quantity indicator
LI	Level indicator
LSH	Level switch high
LSL	Level switch low
PDI	Diff. pressure indicator
PDSH	Diff. pressure switch high
PDSL	Diff. pressure switch low
PI	Pressure indicator
PSL	Pressure switch low
PT	Pressure transmitter
TI	Temperature indicator
TT	Temperature transmitter (PT 100)
VI	Viscosity indicator
VSH	Viscosity Control switch high
VSL	Viscosity Control switch low

Notes:

ff	Flow verlocity in circuit system ≤ 0,5 m/s
p	Free outlet required
s	Please refer to the measuring point list regarding design of the monitoring devices
tt	not insulated nor heated pipe
u	From diesel oil separator or diesel oil transfer pump

All heavy fuel pipes have to be insulated.

---- heated pipe

Connecting points:

C76	Inlet duplex filter
C78	Fuel outlet
C80	Drip fuel
C80a	Drip fuel (injection pump)
C80b	Drip fuel (sealing oil injection pump)
C81	Drip fuel connection
C81b	Drip fuel (filter pan)

8.11.4 Category IV: 1500 to 100,000 kW, two-stroke, low-speed, fuel HFO

These resemble the system described above in many ways.

Example

The fuel systems of a Wärtsilä Sulzer RTA48T and a RTA58T are used as an example.

▶

Description of a complete fuel-treatment system.

Via a three-way valve 002, the diesel oil or heavy oil can be pumped from the day tanks with the low-pressure feed pump 004. Suction filter 003 protects these pumps. Pressure regulating valve 005 ensures that the correct pressure is found in the fuel lines leading to the mixing tank 006, where the engine's fuel-return pipe exits. The high-pressure circulating pumps 007 press the fuel through the final pre-heater 008, which, again, is controlled by the viscosity gauge V, via a flow meter. Subsequently, the fuel travels through a duplex fine-filter to the high-pressure fuel pumps 011 of the engine. A control valve 012, also referred to as the pressure retaining valve, ensures that the fuel pressure in the suction line 010 of the high-pressure fuel pumps is correct. Drain pipe 013 leads back to the mixing tank 006.

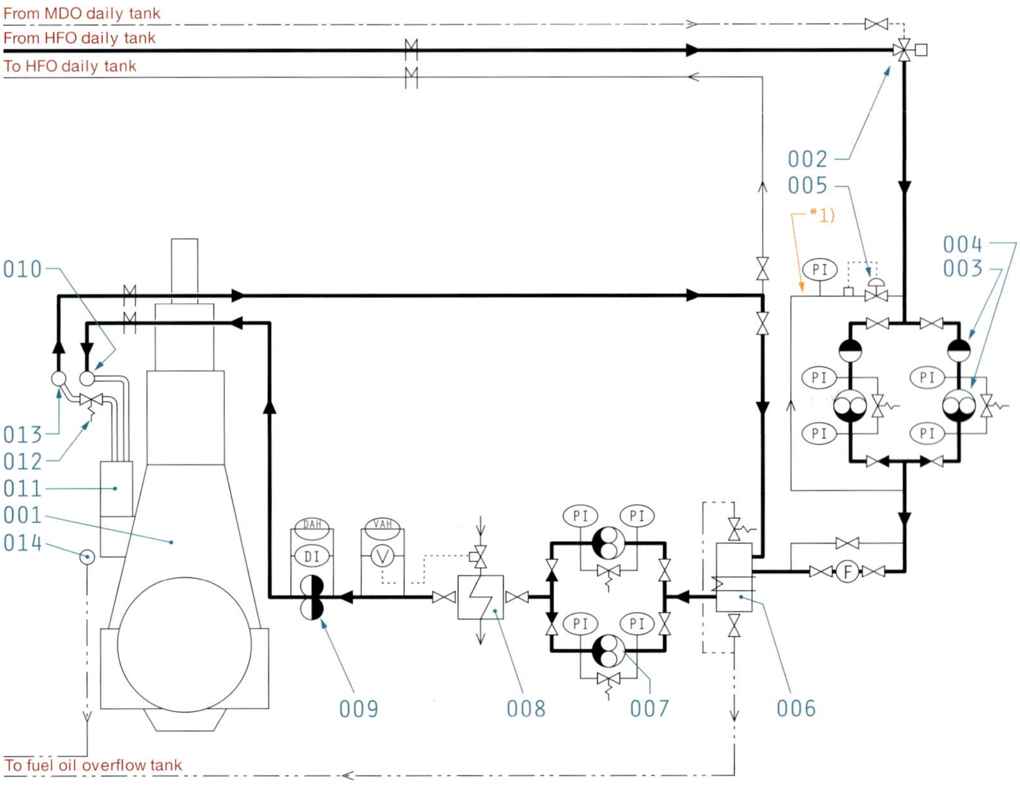

001 Main engine
002 Three-way valve, maually or remotely operated
003 Suction filter, heated (trace heating acceptable)
004 Low-pressure feed pump
005 Pressure regulating valve
006 Mixing unit, heated and insulated
007 High-pressure booster pump
008 Fuel-oil endheater
009 Fuel-oil filter, heated (trace heating acceptable)
010 Fuel-oil inlet
011 Fuel-injection pump
012 Pressure retaining valve
013 Fuel-oil outlet
014 Fuel-oil leakage from fuel pump

Remarks:

*1) The return pipe may also be led to the HFO daily tank.

– Air vent and drain pipes must be fully functional at all inclinational angles of the ship at which the engine must b operational.

HFO pipes, heated and insulated
MDO pipes
Drain and overflow pipes
Piping on engine/pipe connection

8.12 Modular fuel-treatment systems

Today complete modules mounted on frames that can be connected directly to the fuel lines, alarm systems and electricity grid are delivered to ships as well as diesel power plants.
The modules are mounted, tested and inspected in the engine factory and subsequently installed on board. The advantages are short delivery times and smaller fuel-treatment systems.
These fuel-treatment sets are usually called 'booster units'. Worldwide there are a dozen leading companies which specialise in these systems.

8.12.1 Standard form for fuel treatment system

Operating mode

- One supply pump.
- One circulating pump.
- One pre-heater.
- One filter of the duplex filter where the off-line filter is clean.
- The fine-filter is in automatic-reflush mode.
- The viscosity sensor controls the fuel temperature via the supply valve using steam or thermal oil and regulates the pre-set viscosity.
- Shock absorbers have been fitted in order to level off the pressure impact of the high-pressure fuel pumps in the suction line.
- The pressure in the suction line of the high-pressure fuel pumps must be checked at regular intervals.

▲
A modular fuel-treatment system by Alfa Laval.

The complete installation is mounted on a frame and can be installed between the day tanks and the diesel engines on location.

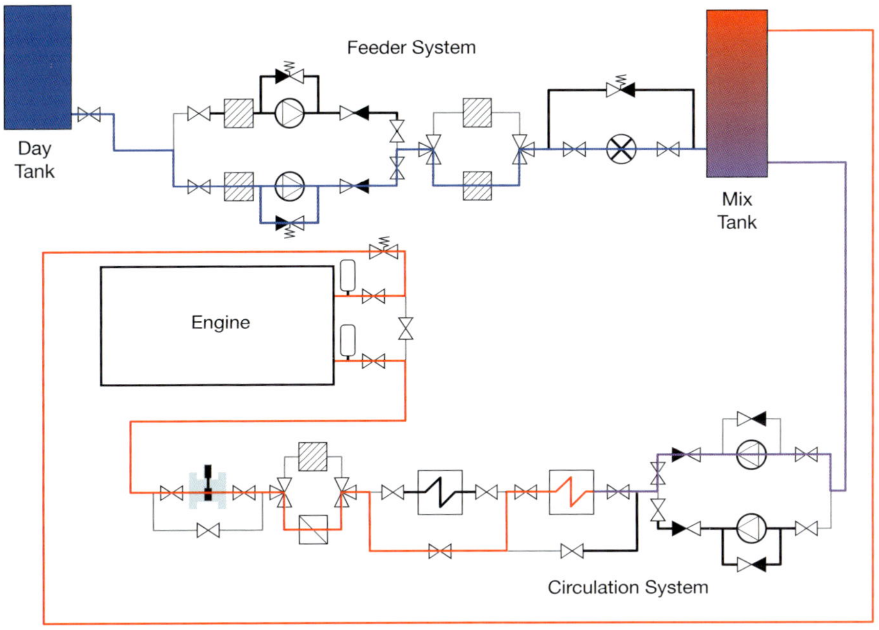

▶
A standard flow diagram for a fuel-treatment system.

One of the feed pumps draws fuel up from the clean-fuel day tank via a suction filter and pumps it to the circulating pumps via a mixing tank. A single feed pump always has a larger capacity than the maximum fuel consumption of the diesel engine. A fraction of the pumped fuel flows back into the suction pipe via an adjustable by-pass valve.

The fuel then passes through an automatic fine-filter, a flow meter and is subsequently pumped through a pre-heater and a viscosity gauge by one of the two circulating pumps. The fuel reaches the suction pipe of the high-pressure fuel pumps via a duplex fine-filter. The pressure in the suction pipe is such that the fuel cannot boil. Gas pockets can cause serious damage to the fuel pumps. This pressure can be set by means of a spring-loaded overflow valve where the excess fuel flows back into the mixing tank. The mixing tank ensures an even fuel temperature before the circulating pumps.
Furthermore, on top of the tank an automatic vent is fitted.
The capacity of each circulating pump is considerably higher than the maximum fuel consumption of the engine.

The viscosity sensor is placed in the full fuel flow just before the engine. In order to maintain the viscosity set on the controller, a control valve regulates the steam or thermal oil flow to the fuel pre-heater. During normal operation both viscosense by-pass valve and the control valve must remain shut. Only during repairs or damage to the sensor or the control valve is the viscosity regulated by the by-pass valve. When the control valve is fully opened while the pre-set viscosity has not yet been achieved it is possible that the steam pressure and/or the temperature will remain too low. If thermal oil is used for heating, only the temperature will be too low.

Operating principle of viscosity control, specifically, the VAF Instrument 'ViscoSense'

The sensor consists of a stainless steel pendulum which is attached to a base plate by a torsion tube. A pair of piezo-elements is integrated in the pendulum and they are actuated by an alternating signal thus causing the pendulum to move by torsional vibration. A second pair of piezo-elements, (receiver piezo-elements), measures the torsional movement and sends this signal to the processor which also ensures that the first pair of piezo-elements provides a continual torsional vibration.

Special details for the standard form

Data

Fuel velocity in booster unit 0.6 to	0.8 m/sec.
Circulation factor supply pumps	1.5 to 2.0
Circulation factor circulating pumps	1.5 to 4.0

$$\text{circulation factor} = \frac{\text{max. capacity pump}}{\text{max. engine fuel consumption}}$$

Fuel pressure of the engine (suction line) is between 4 and 8 bar depending on the engine type and the fuel used.
Maximum fuel temperature of the engine is between 125 and 150 °C, dependent on the engine type and the fuel used.

Location of the automatic fine filter

The most favourable location is the 'cold part' of the installation directly behind the booster pumps. Here the liquid flow is much lower than past the circulating pumps; only the fuel used is filtered. This allows a considerably smaller filter (smaller filter capacity is sufficient).
The fuel is more stable at these lower temperatures; there is no separation of hydrocarbon compounds.

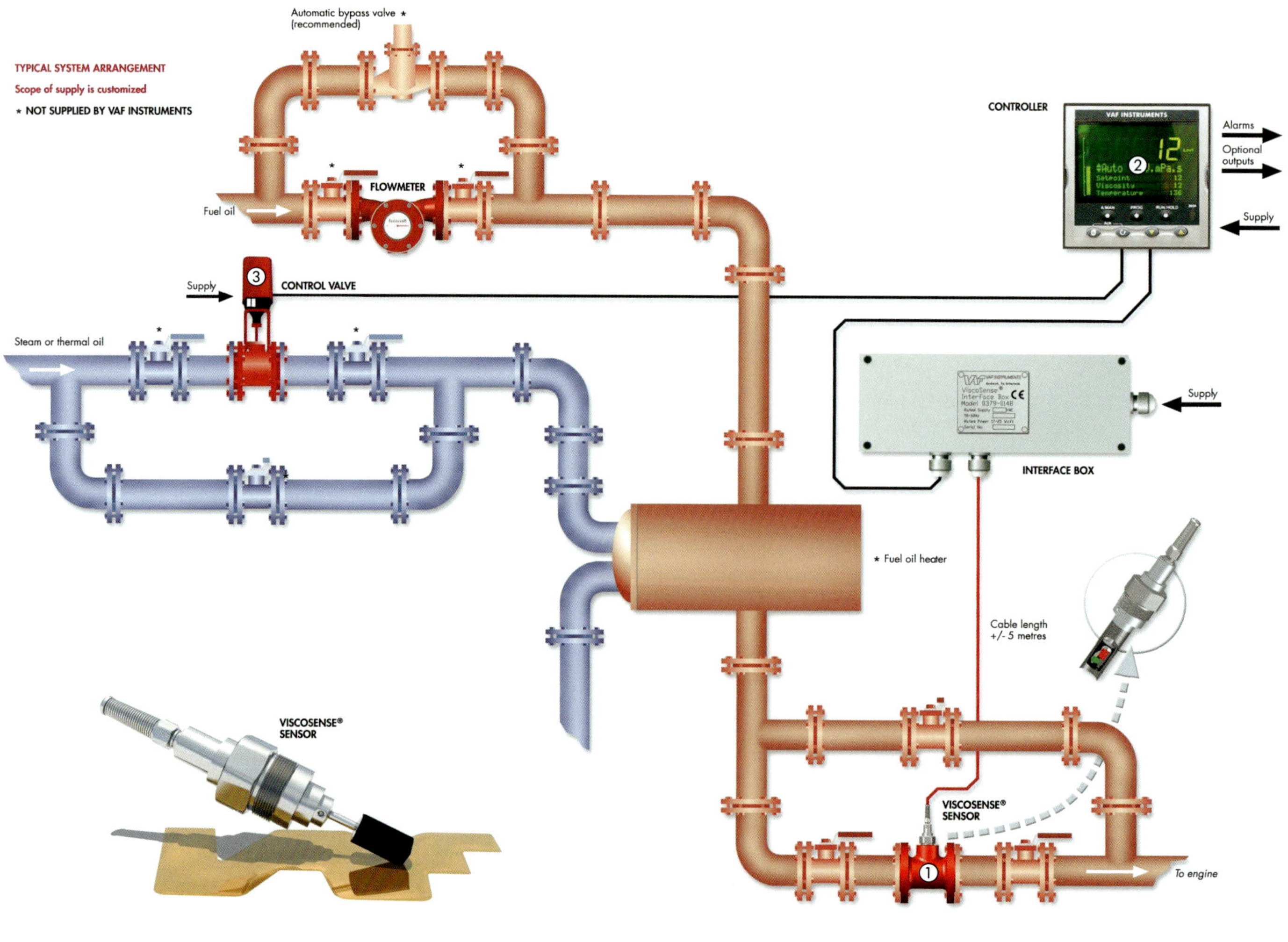

▼
Fuel viscosity indicator. Shown, a system by VAF-Instruments with the ViscoSense system.

1 *The ViscoSense sensor. This measures fuel viscosity.*
2 *The ViscoSense sensor is connected to an electronic control unit which sets and controls the viscosity.*
3 *A control valve is regulated by a control unit which deals with the steam or thermal oil flow to a fuel pre-heater.*

Regulations with regard to fuel systems

These are clearly described in the 'Project guide' of the engine concerned. Engine manufacturers require diesel engines with a certain fuel consumption, a certain viscosity and a certain circulation factor. Furthermore, at full load of the diesel engine the fuel supply pressure and the fuel discharge pressure (regulated with the retaining valve or the overflow valve) require a certain value.

Viscosity sensor

The viscosity sensor is always set at the desired viscosity after which the fuel temperature is adjusted accordingly. **The correct viscosity, not the correct temperature, is important.** Therefore: do not adjust the temperature.
One fuel pre-heater with a shut by-pass valve should suffice in maintaining the right fuel temperature.
A processor measures the phase difference between the sent and received signal. This phase difference is processed and provides the vibration-damping value which is proportional to the viscosity of the liquid.

◄ The ViscoSense sensor; this measures the fuel viscosity and temperature.

◄ An automatic fuel fine-filter in the low-pressure heavy-fuel system.

The filter is cleaned by automatically back-flushing the fine-filter elements. This occurs automatically at a certain pressure difference in the filter.

◄ A complete fuel-treatment system with:

1. *steam control valve*
2. *fuel pre-heaters*
3. *circulating pumps*
4. *mixing tank*

◄ A retaining-valve measurement in a large two-stroke crosshead engine. Also known as pressure difference.

The pressure difference between the supply low-pressure fuel line to the high-pressure fuel pumps and the fuel return pipe to the fuel treatment. This pressure difference should not exceed a certain value. If the fuel treatment unit does not supply sufficient fuel, the fuel pressure will drop below the green sector, preventing the engine from achieving the required rpm.

The Project guide also often provides the specifications of the:

- fuel velocity at the suction side of the pumps;
- fuel velocity in the fuel booster unit;
- fuel velocity in the supply line to the engine;
- all the pressures and temperatures in the system;
- the fuel viscosity between the minimum and maximum value.

8.12.2 General comments with regard to 'booster units'

- **Ship propulsion plants** often have loads that are much lower than those of diesel-power plants.
- **A propulsion engine** usually operates at **± 85%** of its capacity due to the sea, current and draught.
- **A diesel-power plant engine** usually runs at the maximum allowed operating speed (Maximum Continuous Rating or MCR) of **100%** and therefore has a heavier but more consistent load.
- **Many two-stroke propulsion engines** have a booster system developed by the ship yard and it is normally built on a false floor in the engine room. Sizeable systems with few problems.
- **Some two-stroke propulsion engines** have a compact booster unit manufactured by a third party and delivered to the ship yard. Comparatively, more problems are to be expected. However, the components are often much smaller.
- **For four-stroke propulsion engines** compact booster units are generally placed with few problems.
- **Fuel consumption for four-stroke auxiliary engines on large seagoing vessels** can vary greatly and can range, depending on the electricity demand for, for instance, the number of reefers, bow-thruster manoeuvring etc. from 8 to 100%. The control valve of the heater is often fully closed at low load and the fuel flow extremely low. However, the fuel viscosity is usually stable.
- **For container ships** the number of refrigerated containers ('reefers') is important in order to determine the load of the auxiliary engines.
  It is also important to pay attention to the main-engine driven shaft generators. A large shaft generator has an electric power output of 2.5 to 3.5 MW and often 'replaces' one of the four auxiliary engines.
- **Main engine type:** The MAN Diesel crosshead engines are not usually equipped with mixing tanks and only have a raised tank venting system. MAN Diesel four-stroke auxiliary engines are also often equipped with this kind of system.
- **Make.** There are approximately fifteen booster manufacturers, five of whom are prominent. These are Alfa Laval, Aura Marine, Kupke and Wolf, Westfalia and Eefting. All the sets are fairly similar. The customers play an important role as they stipulate their specific requirements and will pay for them.
- **Most engine manufacturers** have regulations with regard to switching from **HFO to MDO**. Most of them always operate their engines on HFO, also when they are stationary. Others change to MDO under certain conditions, for instance, under a certain load or during repairs. Most fuels used are not heavier than **IF 380**, because heavier fuel does not offer any financial advantages. It is only just cheaper to purchase.
- The so-called **pressure compensators** are often not in place. This causes an enormous pressure pulse in the piping and consequently in the booster units.

8.12.3 Points of interest for booster units on location

1. Does the ViscoSense, including the software, work properly?
2. Ensure that the viscosity is regulated as opposed to the temperature. People often erroneously think that the temperature is more important than viscosity!
3. Is there one heater? Two are not normally required!
4. Is the ViscoSense bypass shut?
5. For four-stroke auxiliary engines on large ships, the fuel temperature of (both) auxiliary engines often diverges. The fuel temperature closest to the booster unit is often higher than the fuel temperature further away from the booster unit (usually the other side of the ship).
   The difference is often 5 °C or more. This can generally only be avoided by placing a separate ViscoSense/heater near the engine 'furthest away'.
6. Is there a quick conversion from MDO to HFO and vice versa?! The ViscoSense will react immediately!
7. Does one not resort too quickly to adjusting the temperature when the viscosity does not reach the correct value fast enough?
8. Does one use the correct fuel? Sampling is important!

8.12.4 Plastics in Heavy Fuel Oil (HFO)

The presence of plastics in **cleaned** heavy oil is a common occurrence.
These plastics are predominantly:

polypropylene	– melting point approximately 120 °C
polystyrene atyrene monomer	– melting point approximately 100 °C

If filters are placed in a system where the fuel temperature is higher than the melting point of these contaminants, they filter may clog up. The temperature of the duplex fine-filter just before the diesel engine ranges from 125 to 150 °C and is therefore prone to blockage when these substances are present.

▼
Heavily contaminated fuel fine-filters.

8.13 Bunkering fuels

This does not only apply to shipping but also to diesel-power plants.
One can generally say that much attention should be paid to fuel intake, among other things:

- to the amount of fuel;
- to the available storage capacity;
- to the manner in which fuel is taken in and the supplier's capacity;
- in case of large quantities the sampling procedure;
- to prevent environmental pollution due to leakage or tank overflow.

It is self-evident that 'filling up' 200 litres of diesel oil for a small motor yacht is much easier than 'bunkering' 8000 tons of heavy oil for a large container ship.
Especially in the second example it is imperative that the entire procedure is meticulously followed. In particular the quality of the fuel delivered is often a moot point.

8.13.1 Points of interest for heavy fuel

Mixing two types of fuels that were bunkered at different times should be avoided if possible. Before bunkering, the 'old' fuel should preferably be stored in tanks that are filled to capacity so the 'new' fuel can be bunkered when the storage capacity is full. The 'old' fuel should be used first. If the fuel remains in the storage tanks too long it can separate into light and heavier components.

◀
A bunker boat alongside a container ship for inland shipping.

A practical way to refuel swiftly and efficiently. Operation does not have to cease for refuelling, so no time is wasted. Moreover, there are often no suitable berthing facilities.

When fuels are mixed, asphaltenes can separate out of the fuel and produce sludge. One is generally advised not to mix heavy-fuel oil with gas oil or diesel oil on location to reduce the heavy-oil viscosity. It is difficult to assess the consequences of such mixing on the viscosity and also any other potential side effects.

Volatile heavy oil or heavy oil mixed with lighter fuels often cause the following problems:
- sludge forming in storage- and sedimentation tanks;
- excessive sludge excretion by the centrifuges;
- rapid clogging of filters.

8.13.2 Cleaning of volatile fuels with centrifuges

Actions:
- the centrifuge should be set at the lowest possible capacity.
- avoid multiple centrifugation of heavy oil.
- centrifuges should be cleaned or shot every fifteen minutes.
- ensure that the highest fuel temperature is achieved in the centrifuge, namely 98 °C.

8.13.3 Heating double-bottom tanks on seagoing vessels

By heating the heavy oil to make it pumpable, the temperature in the cargo holds above the fuel tanks rises, which may cause problems for certain types of cargo.
The temperature of the cargo containers in these holds may rise as well.
Depending on the type of cargo that is exposed to this temperature increase, one usually maintains a maximum hold temperature of 40 °C.
In the tropics, where many reefers are in operation, this temperature is easily reached in the hold. It is advisable to temporarily reduce the tank heating when temperatures exceed 40 °C.

8.13.4 Fuel-oil additives

There is an enormous amount of additives available which are used for:
- preventing sludge and sludge forming in fuel tanks;
- separating water from fuel;
- improving combustion;
- increasing the melting point of vanadium;

▶
A completely soiled disc-stack package in a heavy-oil separator.

▶
A severely damaged separator casing as a result of the imbalanced overly full separator bowl.

It is absolutely necessary to regularly discharge the sludge in the separator. The backwash frequency is dependent on the amount of dirt, which in this case, is removed from the fuel.

▲
Heavy-oil separators in the engine room of a large container ship.

They are crucial in cleaning the fuel and the lubricating oil for engines using heavy oil as a fuel.

▶

Valve damage caused by poor-quality heavy fuel.

The vanadium and sodium in the fuel bind with the valve material above 550 °C and forms a very aggressive deposit on the valve dish. This is known as high temperature corrosion, HTC. Clearly shown here are the results of sintering: pits and hairline cracks in the valve material. Both elements are always present in the heavy fuel. Keeping the valve material temperature lower than 550 °C can limit HTC.

- preventing sedimentation in the engine, the exhaust-gas turbines and the exhaust-gas boilers;
- preventing high- and low-temperature corrosion.

There are no general rules for the use of these additives. Some ship yards and owners of diesel-power plants use one or multiple additives and give positive feedback.

▼

A 'drip sample' of Marine Diesel Oil (MDO).

Check the label detailing all the required information and the tamper-proof seal.

▼

Two 'drip samples' of Heavy Fuel Oil (HFO).

In this instance 380 cSt with a low sulphur content.

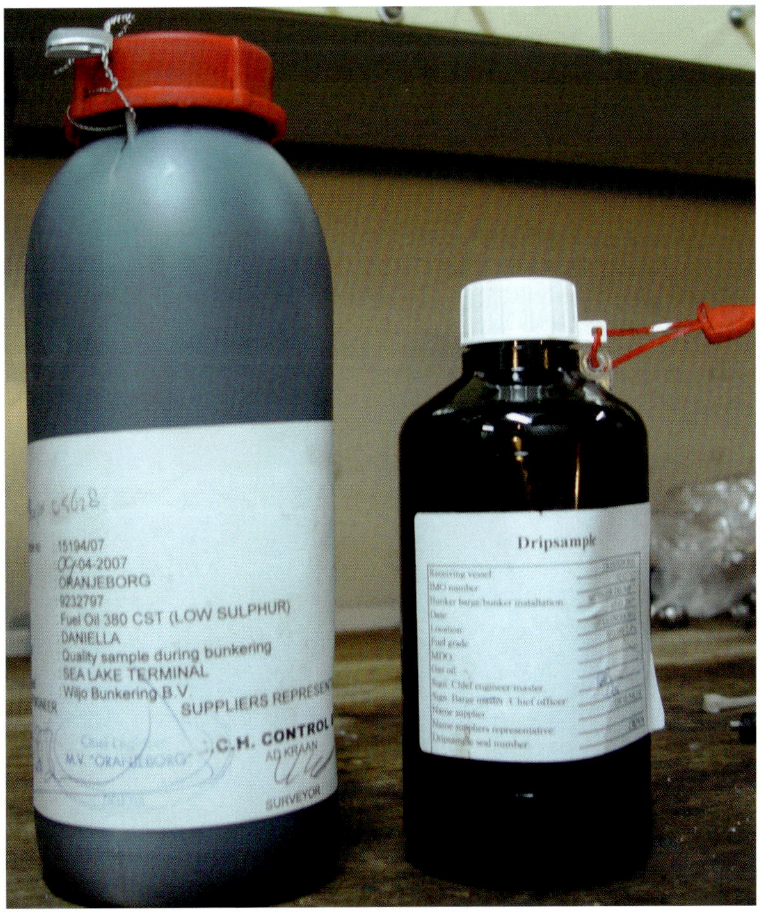

>CH 9

Fuel-injection systems

The most important factor in the combustion process in a diesel-engine cylinder is an optimum fuel injection.
The fuel-injection velocity, the droplet size of the fuel spray and the compressed-air temperature to which the droplets are exposed are crucial for perfect combustion.

The injected fuel must fully combust in the combustion space within a fraction of a second.

9.1 Introduction

The fuel-injection process has been the principal focal point from the first working diesel engine. Swift and complete combustion of the fuel is absolutely imperative for the diesel engine to work optimally. In addition, low specific fuel consumption and a limited emission of pollutants in the exhaust gases have become increasingly important.
One should also bear in mind that the materials used in the diesel-engine components are subjected to an increasingly heavy mechanical, thermal and chemical load.
The rising pressures especially in the cylinder exert very strong forces in the engine.
Fuel-injection systems have continually been adjusted, modified and at times essentially re-designed during the century that the diesel engine has been in use.
One of the most recent developments is the so-called 'common-rail' injection system for four-stroke engines and the 'flexible' injection system for two-stroke crosshead engines.
These electronically controlled injection techniques have been applied to small diesel engines for a dozen years in mainly the automotive industry: motor cars and lorries.

However, the basics of these injection techniques are the same; in a short time span, they accurately regulate the amount of fuel that is injected into the cylinder in such a way that a very rapid and perfect combustion follows.

The combustion process in a diesel engine occurs in a fraction of a second.

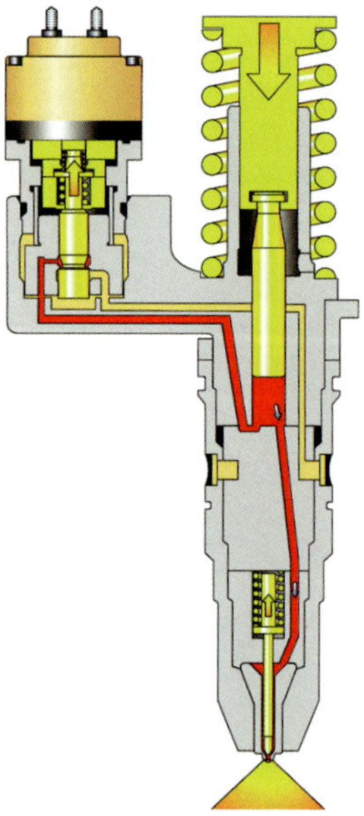

▶ Fuel injectors ensure that fuel is swiftly injected into the combustion chamber as a finely distributed spray. Here accurate timing is key.

9.2 Examples of injection times

9.2.1 A small high-speed four-stroke diesel engine (category I)

Number of revolutions	3000 rev/min.
Power stroke	140 crank degrees.
Fuel injection	30 crank degrees.

Therefore, 50 × 360 = 18,000 crank degrees per second.
Power stroke takes 140 crank degrees.

In time $= \frac{140}{18{,}000} = 0.007$ seconds.

Injection time 30 crank degrees; in time: $\frac{30}{18{,}000} =$ 0.0016 seconds.

Number of processes per second: $\frac{18{,}000}{720} = 25.$

Number of injections is therefore 25 per second.

9.2.2 A high-speed larger four-stroke diesel engine (category II)

Number of revolutions	1500 rev/min.
Power stroke	130 crank degrees.
Fuel injection	30 crank degrees.

Revolutions per second: $\frac{1500}{60} = 25$ rev /sec.

So per second 25 × 360 = 9000 crank degrees.

Then the power stroke takes: $\frac{130}{9000} = 0.014$ seconds.

The injection takes: $\frac{30}{9000} = 0.0033$ seconds.

Number of power strokes: $\frac{9000}{720} = 12.5$ per second.

Number of injections is therefore 12.5 per second.

9.2.3 A medium-speed four-stroke diesel engine (category III)

Number of revolutions	600 rev /min.
Power stroke	150 crank degrees.
Fuel injection	30 crank degrees.

Number of revolutions: $\frac{600}{60} = 10$ rev/sec.

So per second 10 × 360 = 3,600 crank degrees.

Then the power stroke takes: $\frac{150}{3600} = 0.041$ seconds.

The injection time takes: $\frac{30}{3600} = 0.0083$ seconds.

Number of processes per second: $\frac{3600}{720} = 5.$

Number of injections is therefore 5 per second.

9.2.4 A low-speed two-stroke crosshead diesel engine (category IV)

Number of revolutions	94 rev/min.
Power stroke	120 crank degrees.
Fuel injection	25 crank degrees.

Number of revolutions: $\frac{94}{60} = 1.566$ rev/sec.

So per second: $1.566 \times 360 = 564$ crank degrees.

Then the power stroke takes: $\frac{120}{564} = 0.212$ seconds.

Then the injection takes: $\frac{25}{564} = 0.044$ seconds.

Number of power strokes: $\frac{564}{360} = 1.566$ seconds.

Number of injections is 1.566 per second.

Comments

For both high-speed engines (3000 and 1500 rev/min.), the injection times (0.0016 and 0.0033 sec.) and the power strokes (0.007 and 0.014 sec.) are extremely short.
For these types of engines, only gas oil and diesel oil are considered. Heavier fuels such as heavy oil combust too slowly.
For medium- and low-speed engines, the injection times are short (0.0083 and 0.044 sec.) and the power strokes are considerably longer (0.041 and 0.212 sec.).
These engines can also use a cheaper heavy fuel as the combustion time is much longer. Heavy fuels can be used for four-stroke engines up to 1000 or sometimes 1200 revolutions per minute. For all rpm's above this value, the combustion time is too short and after-burning will take place whilst the exhaust valves are open. Not only does this cause power loss, but also the exhaust valves can reach too high a temperature and ignite.
Therefore, mixing fuel and air has to occur as quickly as possible and the fuel mixture should be homogeneous for the fuel particles to react rapidly with the oxygen.
In order to achieve a swift reaction the mixture has to have a sufficiently high temperature. This usually means a final compression temperature between 650 °C and 850 °C. This temperature is achieved by compression of the combustion air.
The required compression ratio is between 10 and 20, dependent on the engine type. The combustion process of a diesel engine is based on the self-ignition principle. This means that the engine will not work if the fuel is compressed when the final compression temperature is too low. This is mostly the case during a 'cold' start, when the engine is not warm. Therefore, many engines are equipped with devices that prevent a cold start such as hot-wire filaments (small engines category I) or heating in the coolant system (larger engines).

9.3 Ignition delay

In practice, the mixing rate of fuel and air is lower than the chemical reaction time during ignition and combustion, which means that there is always a delay: the so-called ignition delay.
The combustion process, of course, will be longer as the number of revolutions decreases, so ensuring that the larger droplets will still fully combust within the limited combustion time. This is why there are very few problems with regard to the combustion time in low-speed two-stroke crosshead engines running on heavy oil. The shorter the combustion time, the greater the potential problems, for instance, medium-speed four-stroke engines with a comparatively high rpm.

9.4 Partial-load conditions

The injection pressures typically drop at a lower rpm. This will negatively affect the combustion process. Therefore, one must take into account these partial-load conditions.
Only by means of special mechanical, and today, electronically controlled injection systems can the full-load injection pressures be maintained.
Virtually all new diesel-engine designs have advanced, electronically controlled injection systems which, among others, achieve lower pollutant emissions in exhaust gases at partial-load conditions. Generally, the fuel consumption is also reduced.
At partial-load conditions, the final compression pressure and temperature both drop as the increased compression time allows heat to escape in the cylinder. This of course has a negative effect on the ignition and combustion time.
Medium-speed diesel engines, which usually run on heavy oil, are converted to diesel oil below a 25% load. Increasing the inlet air- and cooling-water temperatures also positively affects the final compression temperature.

9.5 Processes in the cylinder; injection, ignition and combustion

The fuel is injected into the cylinder by a camshaft-driven high-pressure fuel pump. The correct amount of fuel is forced into the injector or the fuel valve at a meticulously timed moment. At the start of the compression stroke of the pump, the fuel is compressed in the pump cylinder, the high-pressure fuel line and the injector. An injection delay lasts until the moment that the needle valve opens because of the increasing pressure in the injector. **This is therefore a mechanical delay that is dependent on the fuel volume in the pump, the high-pressure line and the injector. It is also dependent on the construction of the high-pressure fuel pump and the injector. So this injection delay cannot be manipulated in an existing engine.**

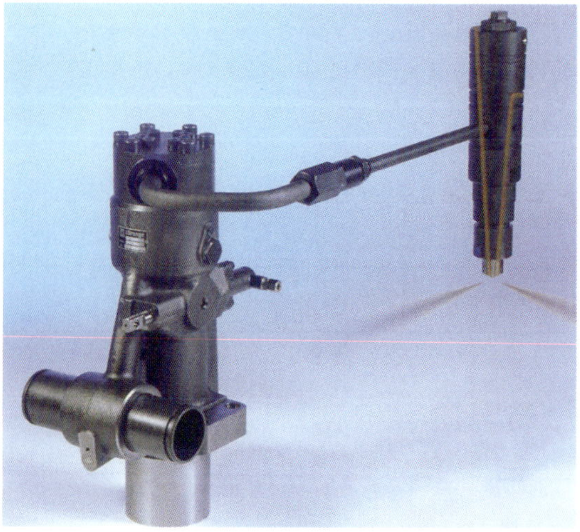

▲

On the left a high-pressure fuel pump driven by the camshaft. Middle the high-pressure fuel line and the fuel injector on the right.

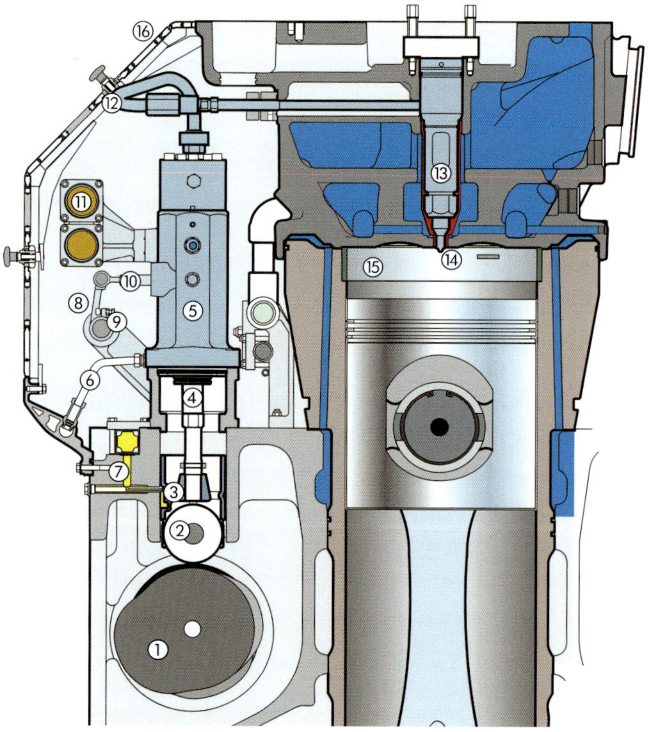

►

A complete high-pressure fuel-injection system.

1 *fuel cam on the camshaft*
2 *cam rollers*
3 *cam roll guides*
4 *pump plunger*
5 *fuel-pump casing*
6 *fuel-return line*
7 *lubricating-oil supply to the cam roll guides*
8 *fuel-pump control rod*
9 *central adjusting spindle*
10 *fuel-pump gear rack*
11 *fuel-pump suction line*
12 *fuel-pump high-pressure line*
13 *fuel injector*
14 *injector tip*
15 *combustion space with the piston in the top position*
16 *cover over the fuel pumps, the so-called 'hot box'*

9.6 The four phases by Ricardo

From the moment that the initial fuel droplets are sprayed into the cylinder, the chemical process of ignition and combustion begins. This has been described by various engine builders, among whom the Englishman Ricardo. He put together a four-phase classification.

9.6.1 Phase 1: Ignition delay

The injected fuel droplets are heated by hot compressed air. They then vaporise and mix with the hot compressed air. After several thousandths of a second, the mixture ignites. The time that elapses between the fuel entering the cylinder and the ignition of the mixture is called the ignition delay. The ignition delay is affected by:

- the compression pressure and temperature;
- the shape of the combustion space at the moment of injection;
- the type of fuel.

The heavier the fuel the longer the ignition delay.
For gas oil and diesel oil this delay is very short, just a few thousandths of a second. For heavy fuel, it can be considerably longer.
When the ignition delay is represented by the number of crank degrees, the high-speed four-stroke engines obviously have the greatest value.

9.6.2 Phase II: Uncontrolled combustion

From the moment that the fuel is ignited, there is a rapid pressure and temperature increase of the combustion gases in the cylinder. The rate at which the pressure rises in comparison to the crank degrees is called the pressure gradient. Sometimes, too steep a pressure gradient manifests itself as a knocking sound due to the pressure impact on the piston that travels to the driving mechanism.
The injection time in high-speed engines is very brief. This is particularly noticeable during the ignition delay. This is not the case in low-speed engines running on heavy oil. Here a minimum value has been set due to the increasingly poor ignition quality of heavy fuel.

Also see Chapter 8, *Fuels, fuel-line systems and fuel cleaning.*

9.6.3 Phase III: Partially controlled combustion

After the ignition delay, the fuel injection and combustion take place in a relatively controlled manner. The heat development increases synchronously with the fuel supply. The greater the volume of air consumed, the smaller the amount of heat generated.

9.6.4 Phase IV: After burning

The fuel injection terminates when the effective compression stroke of the fuel-pump plunger ends. The needle valve shuts as a result of the rapidly decreasing pressure in the high-pressure pipeline. At this moment, the last of the fuel injected combusts. By the time that the exhaust valves open all the required fuel must have combusted in order to avoid overheating of the valves and valve seats.

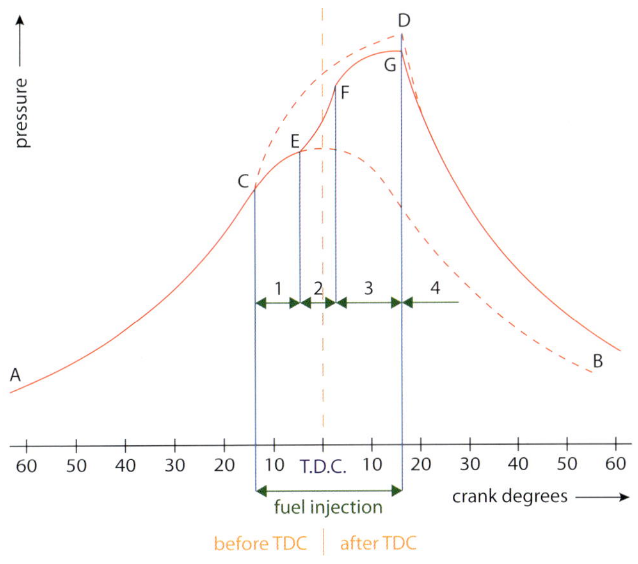

▲
The pressure curve in the cylinder during the four phases.

Line AB shows the compression pressure without fuel injection. Line 2, starting at point E, shows the combustion pressure of the fuel injection. The maximum combustion pressure is at point G, just after the top position of the piston. Point B is the time at which the exhaust valve or valves are open. Most of the energy formed in the combustion process is converted to pressure exerted on the piston.

The four phases of the combustion process are:
1 the ignition delay;
2 the uncontrolled combustion;
3 the partially controlled combustion;
4 the after burning.

The fuel is injected at approximately 15 degrees before top dead centre and terminated at 15 degrees past the top dead centre. Consequently, the entire injection angle constitutes 30 crank degrees. The higher the number of revolutions, the earlier the fuel is injected. The maximum combustion pressure should be achieved just after the TDC of the piston in order to obtain maximum pressure on the piston as it passes TDC. If the maximum combustion pressure is reached too early, for instance, with the piston in TDC position, this can cause considerable damage to the driving mechanism because of the extremely high maximum combustion pressure at the TDC position. Conversely, a delayed maximum combustion pressure produces a much lower pressure as the piston has long passed TDC and the piston speed is increasing rapidly. The engine generates less power and the energy created by the high pressure and temperature of the exhaust gases is lost.

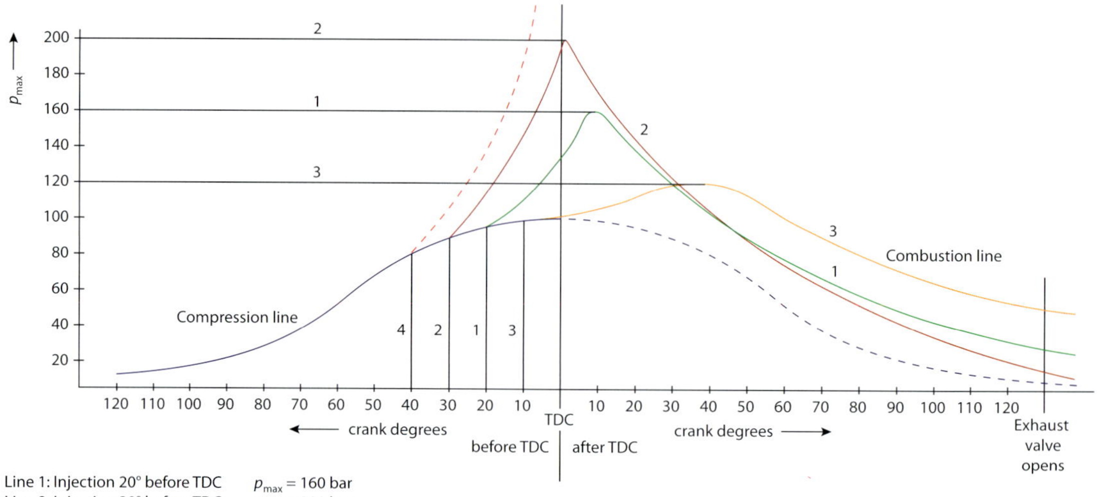

Line 1: Injection 20° before TDC p_{max} = 160 bar
Line 2: Injection 30° before TDC p_{max} = 200 bar
Line 3: Injection 10° before TDC p_{max} = 120 bar
Line 4: Injection 40° before TDC p_{max} = extreme high

▲

The pressure curve in the cylinder at varying fuel injection times.

Line 1 (green) is the normal pressure curve in the cylinder at an optimally tuned fuel injection. The maximum combustion pressure is approximately 15 to 20 degrees after the TDC position.
Line 3 (orange) is the pressure curve in the cylinder with a later fuel injection where less power is transferred to the piston and the exhaust-gas pressure and temperature increases.
Line 2 (red) is the pressure curve in the cylinder at a slightly early fuel injection. Here the maximum combustion pressure increase occurs too early to be effective on the piston that is almost at its top position. Less power is produced and the exhaust-gas pressure and temperature are approximately equal to line 1.
The dotted line shows the pressure curve at an early fuel injection. The piston is going through the upward stroke and the combustions process has started. Slightly before the top position and in the top position, the combustion pressure is extremely high and causes damage to the piston and the rest of the gearing. Often this results in irreversible damage. The power output is minimal.

Altering the fuel injection seriously affects the combustion process and the power output of the diesel engine. Meticulously tuned fuel pumps and injectors are a prerequisite for optimal operation of the diesel engine.

As aforementioned, it is fairly simple to calculate the injection and the total combustion time.

$$\text{Injection time in seconds} = \frac{\text{injection time in travelled crank degrees}}{\text{360 times the number of revolutions per second}}$$

$$\text{Total combustion time} = \frac{\text{combustion in travelled crank degrees}}{\text{total combustion process in crank degrees times the number of revolutions per second}}$$

9.7 Ignition delay; causes

Ignition delay is dependent on the shape of the combustion space, the pressure and temperature at the time of the fuel injection, as well as the nature of the atomisation and the ignition quality of the fuel.

Further explanation

Shape of the combustion space

From the time of the fuel injection just prior to the TDC position until the end of the fuel injection just after the TDC position, the combustion space is a cylindrically shaped space with little headspace. It resembles a flat disc. The higher the compression rate the flatter the disc. This means that the fuel is injected almost horizontally in most four-stroke and two-stroke diesel engines.

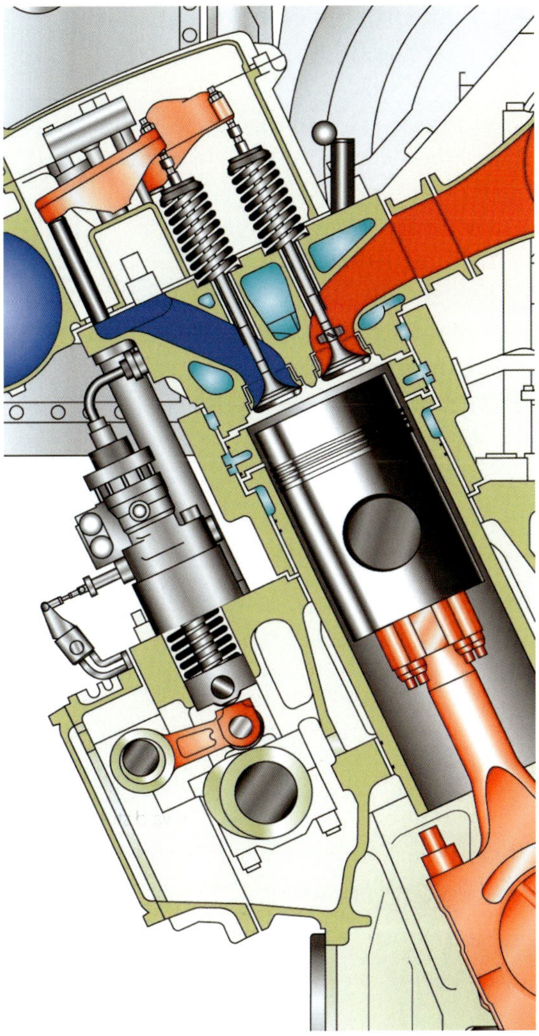

▲
Fuel injection is an extremely precise process; the injection angle and the fuel-injection radius are continually adjusted in order to achieve optimal injection.

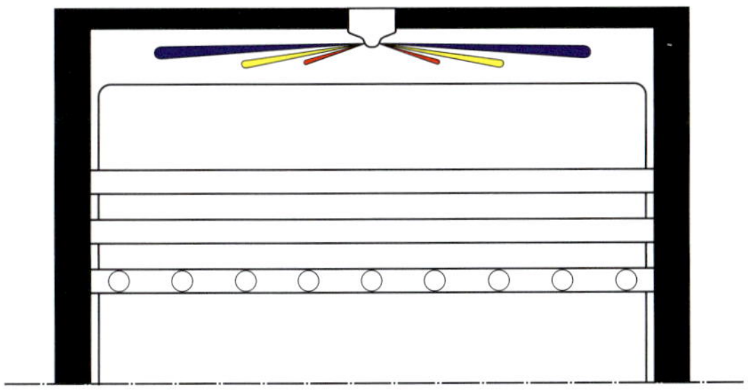

▲
Various fuel-injection possibilities.

Experimentally, the number and the diameter of injector apertures are adjusted, as well as the fuel-injection angle and the distance that the fuel droplets travel in the compressed hot air. This is done in order to obtain optimal combustion.

The crank angle at which the fuel is injected varies, but averages between 30 and 45 crank degrees for medium- and high-speed four-stroke diesel engines, and approximately 20 crank degrees for large low-speed two-stroke engines.

Pressure and temperature of the air

The pressure and temperature of the compressed air can also greatly influence the ignition delay. The high final compression pressure is a result of a high turbocharger supply pressure in conjunction with a high compression rate. The turbocharger temperature is also of importance. Depending on the engine category, one finds the following values:

turbocharger fill pressures	between 1.0 and 3.5 bar over-pressure
turbocharger temperature	between 25 °C and 65 °C
compression ratios	approximately between 10 and 20
final compression temperatures	from 650 °C to 850 °C
final compression pressures	from 20 to 200 bar
final combustion pressures	from 30 bar tot 250 bar
final combustion temperatures	from 1400 °C tot 2000 °C

9.8 Nature of atomisation

The quality of the atomisation is dependent on the injection pressure of the injector, the diameter of the injector apertures, so therefore the velocity at which the fuel droplets travel through the 'thick' compressed air.
The droplets must move through the cylinder in just a few milliseconds.

9.9 Ignition quality of the fuel

The ignition quality is dependent on the type of fuel. Light distillate fuels such as gas oil and diesel oil ignite much quicker than heavy oils. The ignition quality of heavy oils can also diverge considerably. Aromats and asphaltenes (cyclic compounds) are more difficult to crack than aliphatic hydrocarbons (chain compounds) and thus produce a longer ignition delay.
For heavy oil the CCAI-value is used: a measure for the ignition quality of heavy oil.

The ignition delay of heavy oil can sometimes be improved by mixing it with a fuel that has a better ignition quality (usually a lighter fuel). There is also a (theoretical) formula for the ignition quality.

$$t_v = \frac{k}{p^{1.19}} \times e^{\frac{4650}{T}}$$

Where:

t_v = the ignition delay in milliseconds;

p = the absolute pressure in the chamber during injection;

T = the absolute temperature in the chamber during injection;

e = radix for natural logarithms;

k = the constant; this is dependent on the ignition quality of the fuel and is established for a particular fuel.

This laboratory test shows that the higher the pressure and the temperature, the shorter the ignition delay.

9.10 Examples of combustion processes

Engine manufacturers often provide interesting graphs when introducing a new or improved diesel engine:

- the pressure build-up in the cylinder;
- the pressure build-up in the high-pressure fuel line;
- the lift of the atomising needle;
- the injection angle of the fuel.

This is in relation to the position of the crankshaft.

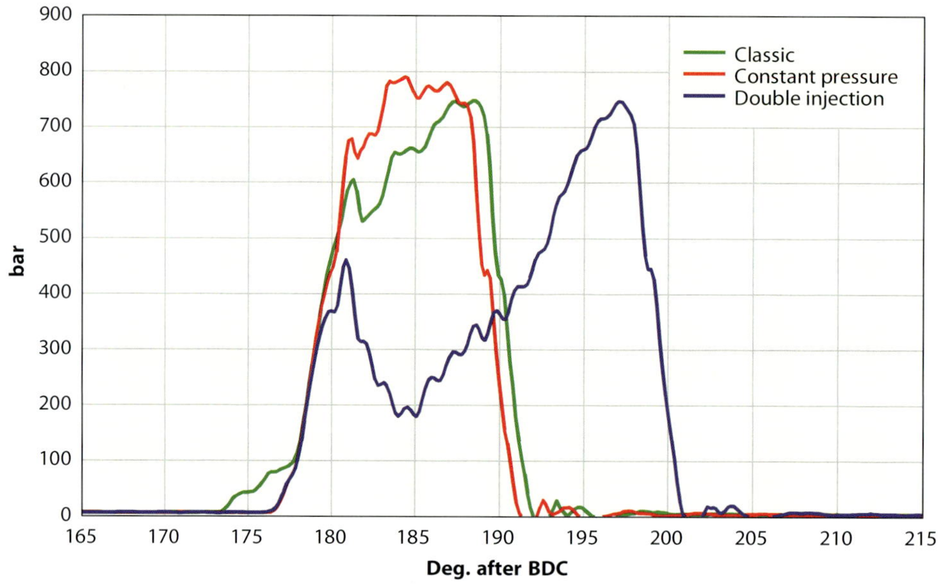

▶ Fuel-injection profile for the MAN Diesel ME series, two stroke crosshead engines.

Horizontally: the position of the crankshaft in crank degrees.
At 180 crank degrees, the piston is in top 'deg. after BDC' = crank degrees after the bottom dead point (Bottom Dead Centre).

Vertically: the fuel injection pressure in bars.

Green line: the pressure curve for normal conventional injection.
Start: 173 crank degrees, this is 7 crank degrees before the top.
Finish: 192 crank degrees, this is 12 crank degrees after the top.
Total injection time: 19 crank degrees.
Maximum fuel injection pressure: 740 bar.

Red line: the pressure curve according to the constant pressure system.
Start: 177 crank degrees after the bottom, this is 3 crank degrees before the top.
Finish: 191 crank degrees after the bottom, this is 11 crank degrees after the top.
Total injection time: 14 crank degrees.
Maximum fuel injection pressure: 795 bar.

Remark: the injection time is shorter and at a higher pressure.

Blue line: double injection.
Start: 177 crank degrees after the bottom, this is 3 crank degrees before the top.
Finish: 201 crank degrees after the bottom, this is 21 crank degrees after the top.
Total injection time: 24 crank degrees.
Maximum fuel injection pressure: 740 bar.
Remark: the fuel injection takes place in two phases and takes longer. Due to this, the maximum combustion pressures and maximum combustion temperatures are lower. This results in lower nitrous oxide emissions.

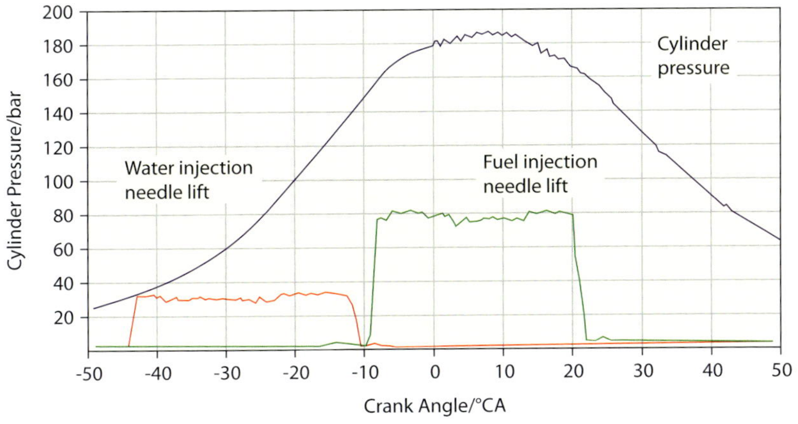

▲

The pressure curve in the cylinder of a four-stroke medium-speed diesel engine operating on heavy fuel oil.

Horizontal: the position of the crankshaft with the top dead point of the piston at zero.
Vertical: the cylinder pressure in bars.

The maximum combustion pressure in the cylinder is approximately 185 bar and occurs approximately 10 degrees after the top position of the piston. The fuel injection starts at the moment the injector needle is opened by the increased fuel pressure from the high-pressure fuel pump.
This occurs at 10 degrees before the top position to 20 degrees after the top position of the piston (green).
Injection takes place in total over 30 crank degrees. This engine is also equipped with water injection to limit nitrous oxide emissions. This takes place before the top position of the piston (red).

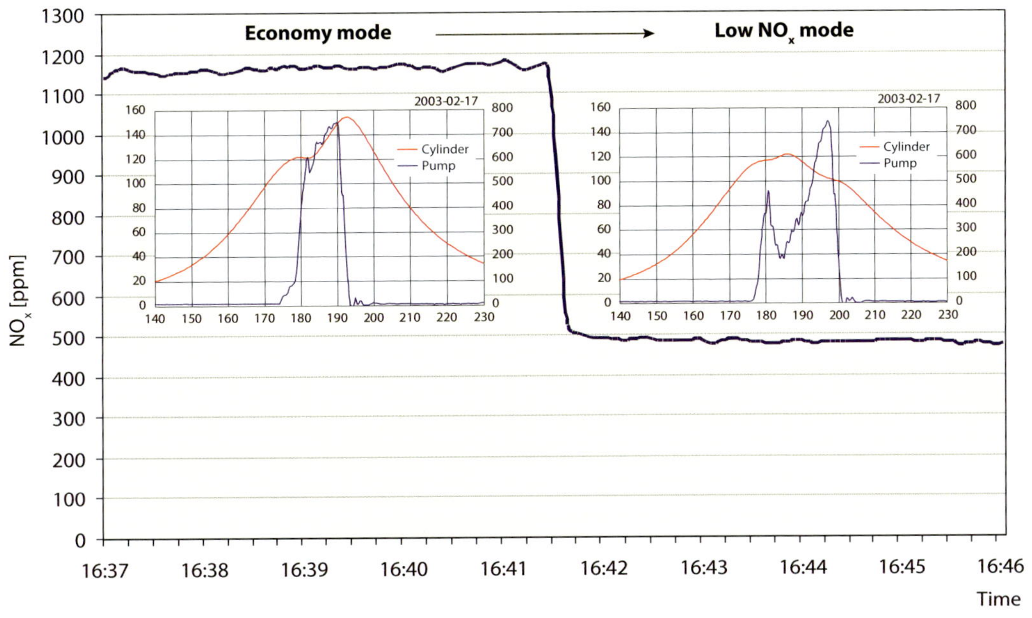

▲

This graph shows the pressure curve in the cylinder of a two-stroke crosshead engine operating on heavy fuel oil.

The graph shows a two-stroke crosshead engine operating at the lowest specific fuel usage.

Left graph:
Horizontally: the position of the crankshaft. At 180 crank degrees, the piston is in the top position.
Left vertically: the cylinder pressure in bars
Right vertically: the fuel injection pressure in bars
Red line: the cylinder pressure curve
Blue line: the fuel pressure curve at injection

Right graph:
This graph shows the same engine that has been tuned so that nitrous oxide content in the exhaust gasses is limited. The fuel injection takes place later and consists of pre- and main injection. Due to this, the combustion is less explosive and more gradual. This results in lower maximum combustion temperatures and pressures. Therefore less nitrous oxides are produced.

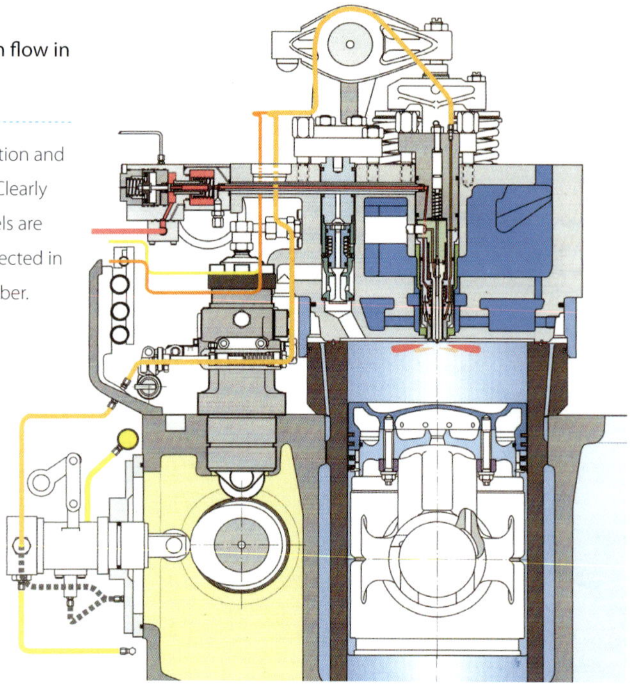

▶
Shown, the injection flow in a 'dual-fuel' engine.

Brown is the fuel injection and red the gas injection. Clearly shown is that both fuels are almost horizontally injected in the combustion chamber.

9.11 Injection pressure and droplet size

The ignition quality, including the ignition delay, is also determined by the size of the fuel droplets.
The smaller the droplet the quicker it heats up, vaporises and gasifies. In other words; the smaller the droplet, the higher the ratio of the surface area to the volume of the droplet. The smaller the droplet, the faster the fuel mass heats up!
The **droplet size** is dependent on the average injection pressure. For two-stroke crosshead engines, this lies between the 250 and 1000 bar and for four-stroke engines between 250 and 2000 bar. This, of course, largely depends on the engine type.
The injection pressure has increased considerably throughout the years. Due to the growing demand for an increase in cylinder capacity, the effective mean pressures have risen significantly. This has been facilitated by increasing the pressures, including the final compression pressure during the combustion process. In order to spray rapidly the fuel sufficiently far into the cylinder the fuel injection pressure must also increase, so that the increased resistance due to the final compression pressure can be overcome.
Electronically controlled modern fuelling systems for two-stroke crosshead engines currently (2012) use pressures of approximately 1000 bar and for four-stroke engines, approximately 2000 bar.
Here the droplet size varies from 15 to about 60 micro millimetres.

It is of the utmost importance that the fuel pumps and injectors are maintained and set correctly.

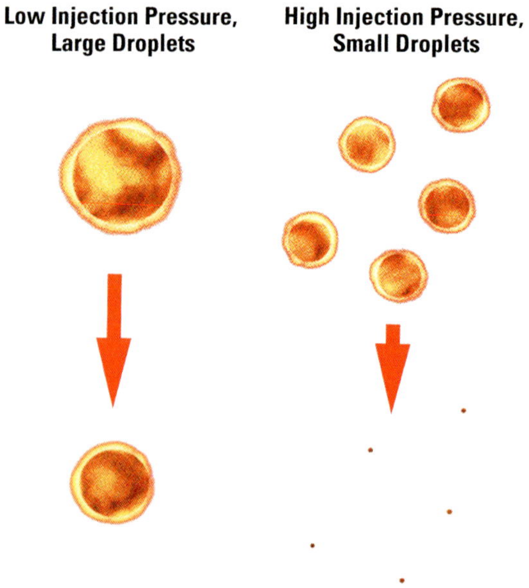

▶
The droplet size is important in achieving an ideal combustion process.

At a low fuel injection pressure, the droplets are too large and therefore incomplete combustion occurs. At higher fuel injection pressures, the droplet size is reduced and the fuel combusts faster and more completely.

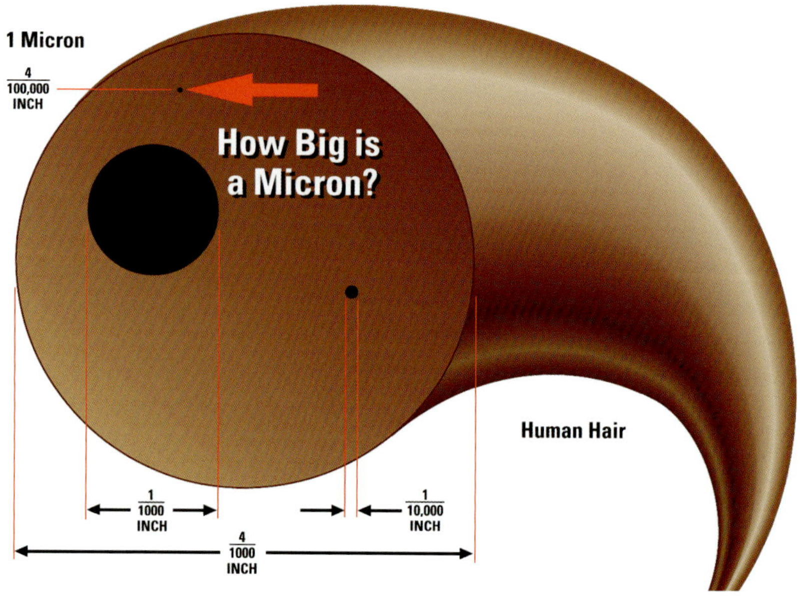

▶
The droplet size relative to a human hair.

A human hair has a diameter of approximately 0,004 inch, approximately 0.1 millimetres. A fuel droplet is approximately 0,001 inch in diameter, approximately 0,025 millimetres. This is 25 microns; the ideal droplet size!

This is ideally performed by specialised firms that have all the required maintenance- and test equipment in-house.

9.12 Injection principles

There are two systems: **direct** and **indirect**.

Direct injection

The fuel is injected into the single combustion space under pressures of 1000 to 2000 bar. The modern four stroke engines have one fuel injector in the centre of the cylinder head. They are usually multiple-aperture injectors that inject the fuel almost horizontally into the disc-shaped combustion chamber at a maximum pressure, the so-called peak pressure, of 2000 bar. The distance that is covered during the injection amounts to approximately 0.3 *D* from the centre axis. In fact, this means that the fuel is injected into the cylinder over a diameter of 0.6 *D*. So in most of today's engines the fuel is not sprayed against the piston crown surface.

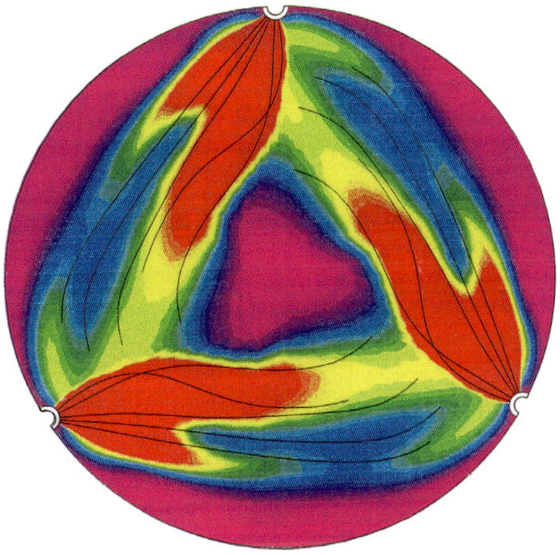

◄ The injection flow of the three injectors in a large two-stroke crosshead engine.

As the exhaust valve is centrally positioned, the injectors are positioned around the circumference. The various colours show the gas temperatures. Red is the highest temperature then decreasing yellow, blue and purple.

In two-stroke crosshead engines, two or three injectors inject the fuel almost horizontally into the engines. The fuel is injected into the cylinder at a certain angle in a circular spray form. The injectors in question are always multiple-aperture injectors.

▲ The combustion process above the piston is a very quick and complicated process. The fuel injection quality is essential for the good operation of the diesel engine.

◄ The three injectors per cylinder for a two-stroke crosshead engine.

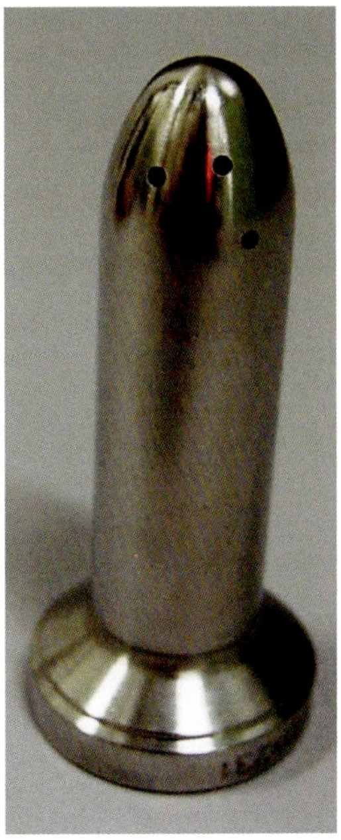

◄ The injector tip of a similar engine with multiple injector apertures.

Indirect injection

The fuel is injected into a separate pre-chamber or pre-heating chamber at a relatively low pressure. The system is only applied in four-stroke high-speed engines.

▶ A pre-chamber injection in an air-cooled Deutz diesel engine. Old design.

Left in the pre-chamber the single aperture pintle-type injector and right the glow plug for a cold start.

1 *pre-chamber*
2 *injector*
3 *glow plug*
4 *cylinder head*

There are many different types of pre-chambers. In pre-chamber injection, the fuel is injected into the pre-chamber via a pintle nozzle at a fairly low injection pressure of a 100 to 200 bar. There are numerous different pre-chamber designs. Essentially, they all serve a similar purpose.

- The walls are extremely hot and rapidly heat up the injected fuel.
- This allows the fuel to ignite swiftly.
- Due to the high temperature the fuel in the pre-chamber cracks and breaks down the long hydrocarbons molecules into smaller chains which are easier to combust.

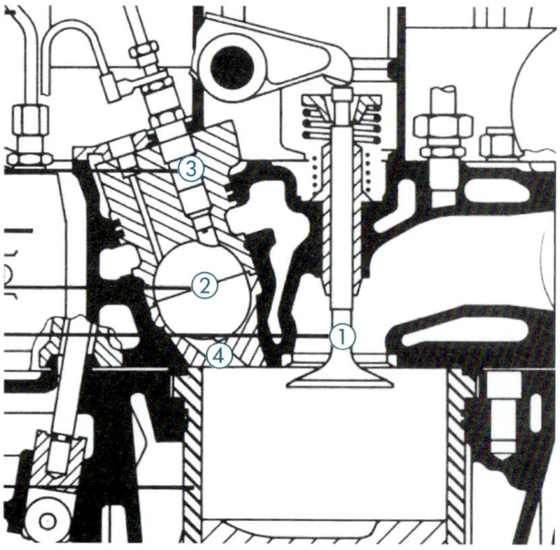

▶ A Ricardo-pre-chamber in a conventional four-stroke medium-speed diesel engine.

Cylinder diameter 210 mm. Fuel MDO. Note the position of the single aperture injector and the channel to the main combustion chamber.

1 *exhaust valve*
2 *pre-chamber*
3 *pintle- type injector*
4 *tangential channel*

- As the pressure in the relatively small pre-chamber rapidly increases, the mixture is pushed, at a great force, through a number of specially shaped ducts into the main combustion chamber. Here there is more than sufficient air to fully combust all the fuel in the pre-chamber.
- As a result of the special manner in which the pipes are connected to the main combustion chamber, there is increased air movement which means that remaining fuel combusts fully and swiftly.

Whirl-chamber engines have a conically shaped whirl chamber in their cylinder head. This is connected to the main combustion space by means of a tangential pipe. During the compression stroke, a part of the compressed air is fed into the whirl chamber, causing rapid air rotation. The fuel injected by the single-point injector is now intensively mixed with the rotating air. The walls of these whirl chambers are also very hot as they are only partially cooled. This causes the incoming air to achieve high temperatures very quickly and therefore the injected fuel will swiftly combust. Due to the high pressure rise in the relatively small space the mixture, which still contains a large amount of uncombusted fuel, is forced into the main space where the remaining fuel combusts fully and swiftly.

Some advantages of this system

- Swift ignition that makes the system suitable for high-speed engines.
- The system is insensitive to the ignition quality of the fuel.
- The maximum combustion pressures are relatively low.
- The single-point injector is relatively simple and pipe blockage will therefore seldom occur.
- The fuel pump is relatively simple due to the low pressures.

Some disadvantages of this system

- The efficiency is lower than that of a directly injected diesel engine; there are considerable flow-and whirl losses. The cooling-water losses are also slightly higher.
- Compared to a directly injected engine, starting is more difficult due to the low injection quality and the higher loss of compression warmth to the still cool ambient air. Therefore, the engine often requires aids such as hot wire filaments to assist starting.
- Pre-chambers tend to crack with a cold start or rapidly changing loads.

During the past years, larger numbers of high-speed engines have been built on the principle of direct injection to avoid the aforementioned disadvantages of indirect injection.
The volume of an pre-chamber in relation to the main chamber can range from 20 to 50%.

9.13 Shape of the combustion chamber

9.13.1 Four-stroke

The space at the end of the compression stroke, so at the beginning of the combustion process is increasingly 'flat' as a result of the increasing compression rate.

Let us assume the compression rate is 14 and the cylinder bore 200 millimetres.

$$\varepsilon = \text{compression rate} = \frac{V_s + V_c}{V_c}$$

$$14\ V_c = V_s + V_c$$
$$13\ V_c = V_s$$

The volume of the final compression space then is: $\frac{1}{13}$ part of the stroke volume.
Let us assume that the stroke at a cylinder bore of 200 millimetres constitutes 220 millimetres, which gives a height of the final compression space $\frac{220}{13}$ = 16.9 millimetres!

In this 'flat disc' the fuel must be injected almost horizontally.

▲
Left, the cylinder head of an older diesel engine type with above the pre-chamber, in the middle the inlet- and the exhaust valve; below the air-starting valve. Right a modern cylinder head with direct injection.

The small hole bottom left (see circle) of the inlet valve is the relief valve. Right, the cylinder head of a more recent diesel engine type with an injector placed centrally in the head for direct fuel injection and around this aperture, both inlet- and exhaust valves. Fuel is MDO and/or HFO.

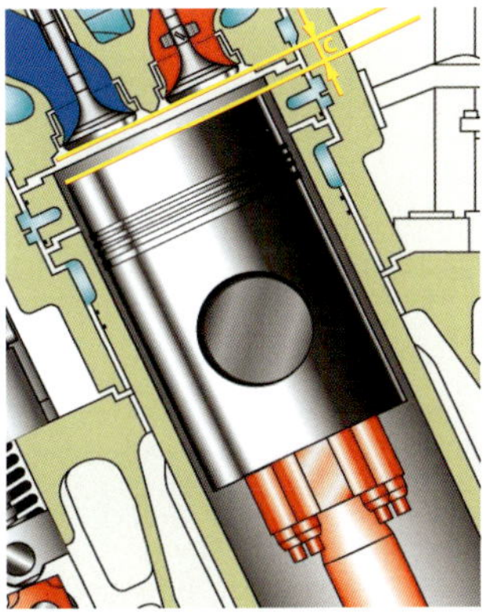

◄
The cross-section of a V-engine with the piston in top position and the both valves closed.

Clearly visible is how small the space is above the piston.

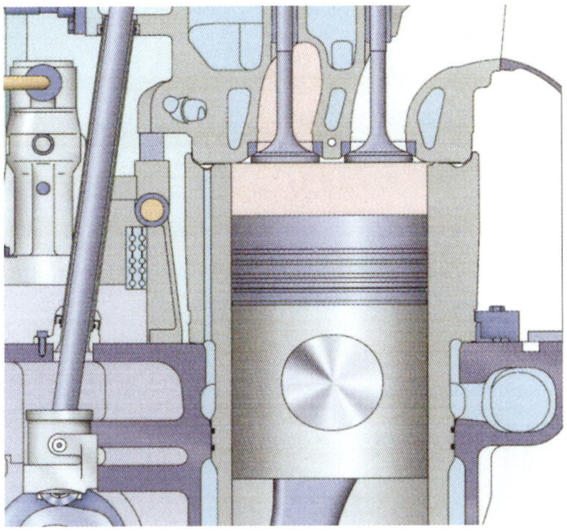

▲
In this line engine with the piston almost in the top position, the space above the piston is also only centimetres high.

◄
Recesses in the piston crown of a high-speed Cummins-diesel engine.

These serve to provide room for the inlet- and exhaust valves when the piston is in top position and the valves are still slightly opened. This means during exhaust air scavenging. Also called top-scavenging.

▲

Recesses in the piston crown of a MaK 453B diesel engine in order to create space for the valves.

In the background, the dismounted steel piston crowns. Note the bowl shaped combustion chamber in the piston crown.

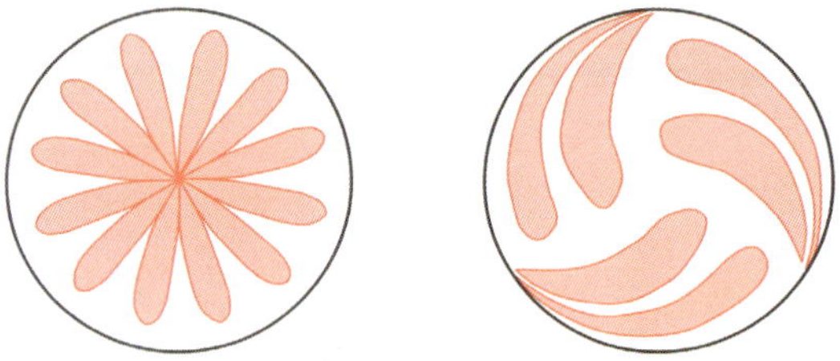

▲

The injection flow of a four-stroke trunk-piston engine and a two-stroke crosshead engine.

Left, the injection flow pattern from the cylinder heart when the fuel is centrally injected.
Right, the circular injection flow pattern that is slightly directed towards the inside when the fuel is injected from the circumference.

The **piston crown** is also often given a specific shape and the two inlet- and exhaust valves require some room.
Often there is so little space at the TDC position of the piston during exhaust air scavenging that recesses have been made in the upper rim of the cylinder liner and the piston crown in order to create room for the valves.

Injecting the fuel

In a test phase or in practice, the injector nozzle is often adjusted to such an extent that the injection angle and -jet are modified slightly. In this manner, the most ideal injection method is investigated. The number of injector apertures or the location of the apertures is regularly modified. One could obviously also modify the opening pressure of the injector, the maximum pressure of the fuel injection or the diameter of the injector apertures. The injection curve and the moment of injection are important values in finding the ideal injection.

9.13.2 Two-stroke crosshead engines

All two-stroke crosshead engines work with the uniflow- or direct flow principle. This means that the air is supplied via the scavenging ports located in the lower section of the cylinder liner and that the discharge of exhaust gases occurs by means of the hydraulically controlled exhaust valve centrally situated in the cylinder cover. Therefore, there is no space to place a fuel injector in the heart of the cylinder cover. The injectors are then mounted in a

▶

The combustion chamber of a Wärtsila Sulzer RTA60C two-stroke crosshead engine.

Note the rounded upper angle of the combustion chamber in the cylinder cover. Centrally the large exhaust valve. Two of three injectors are visible.

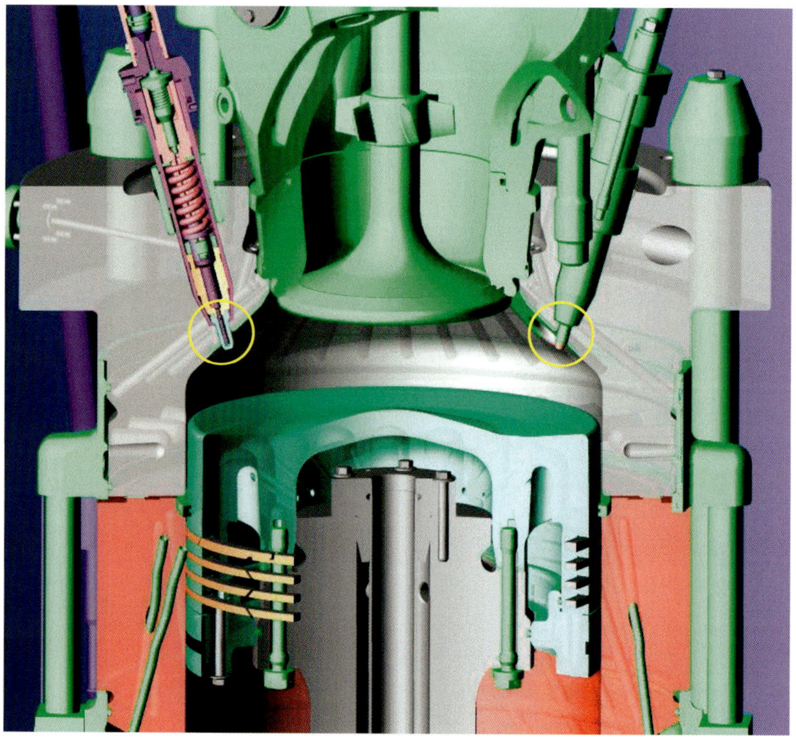

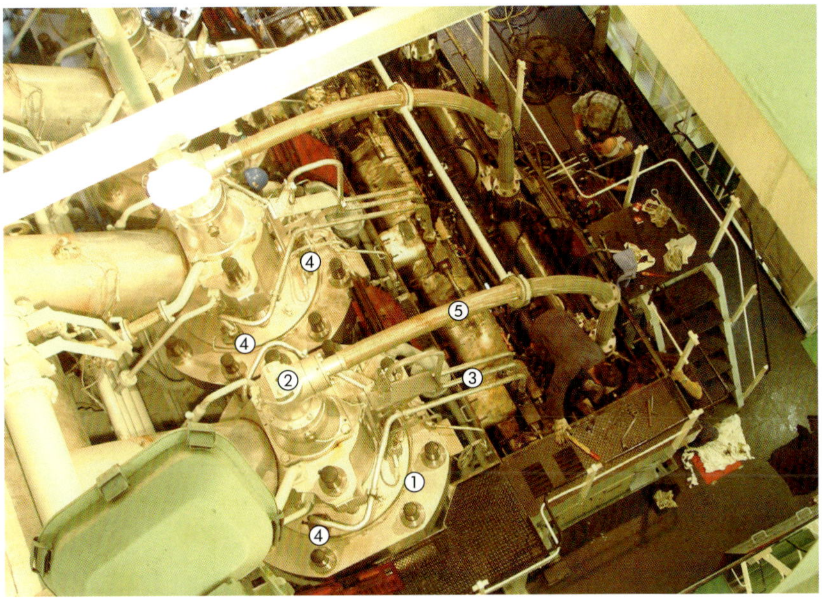

◀
A large two-stroke crosshead engine as seen from above.

Note the dimensions of the cylinder heads in relation to the engineers.

1 *cylinder cover*
2 *exhaust valve*
3 *high-pressure fuel lines to the three injectors*
4 *injectors*
5 *hydraulic pipe lines to open the exhaust valve*

circle in between the exhaust valve and the cylinder wall.

Two or three injectors, which inject the fuel more or less in the direction of this circle, are fitted.

The upper corners of the combustion chamber are rounded in order to avoid scavenging shadows. The shape of the piston bottom may vary, but is generally fairly flat and either slightly bowl-shaped or provided with an elevation in its heart ('island').

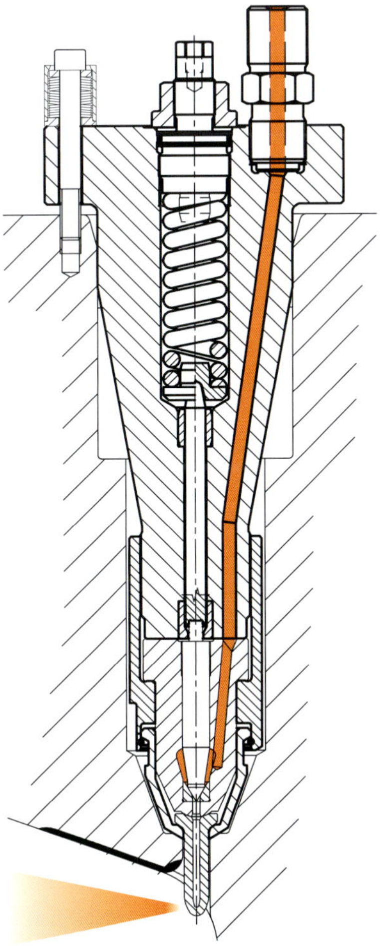

◀
The position of the injector at the edge of the combustion chamber in a two-stroke crosshead diesel engine.

The fuel is almost horizontally injected.

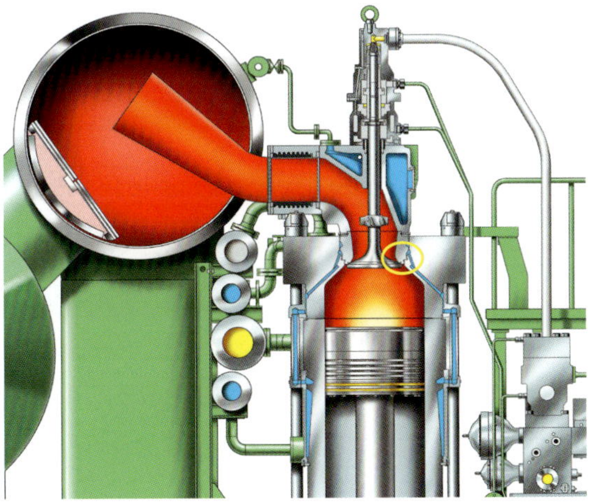

▲
The upper part of the combustion space in this two-stroke crosshead engine is rounded to avoid scavenging shadows.

Scavenging shadows are areas which are not (fully) reached by the scavenge air. Therefore exhaust gas-particles remain behind.

◀
An injector tip of a similar engine with multiple injector apertures.

Injecting fuel

Fuel injection in two-stroke crosshead engines is frequently adjusted in order to obtain optimum injection. The injector nozzles are often changed.

9.14 Fuel-injection mechanism: the fuel pump

Fuel pumps have to meet demanding requirements since the manner in which the injection takes place is crucial for the quality of the combustion process, and therefore the optimal operation of the diesel engine.

9.14.1 Fuel pump requirements

- **To put the fuel under high-pressure swiftly.** The pressure time varies from several thousandths of a second to several hundredths of a second. The required pressure ranges from approximately 200 to 2000 bar.
- **To pump a very meticulously regulated amount of fuel to the injector where the quantity should be continually controllable from zero output to maximum output.** The maximum amount varies from several tenths of a cubic millimetre for small engines to several dozen cubic centimetres for the largest engines. These amounts apply to each plunger stroke and each cylinder.
- **The delivery of the fuel must commence at the correct time and last for a certain time.** The complete injection curve for four-stroke high-speed and medium-speed diesel engines amounts to approximately 30 to 40 crank degrees and about 20 crank degrees for two-stroke crosshead engines.
- **Fuel pump leakage losses should be minimal.** For the most frequently used fuel pumps, the plunger pump, fuel leakage is negligible. Long plungers, which have very little clearance in relation to the plunger housing, guarantee minimal leakage.
- **The fuel pump should be able to cope with fuel temperatures of approximately 150 °C to ensure that the heavy fuel is atomised into the cylinder at the correct viscosity.**

Heavy fuel viscosity prior to injection must lie between 8 and 12 centistokes for four-stroke engines and between 10 and 15 centistokes for two-stroke crosshead engines. The fuel temperatures lie between 125 °C to 150 °C.

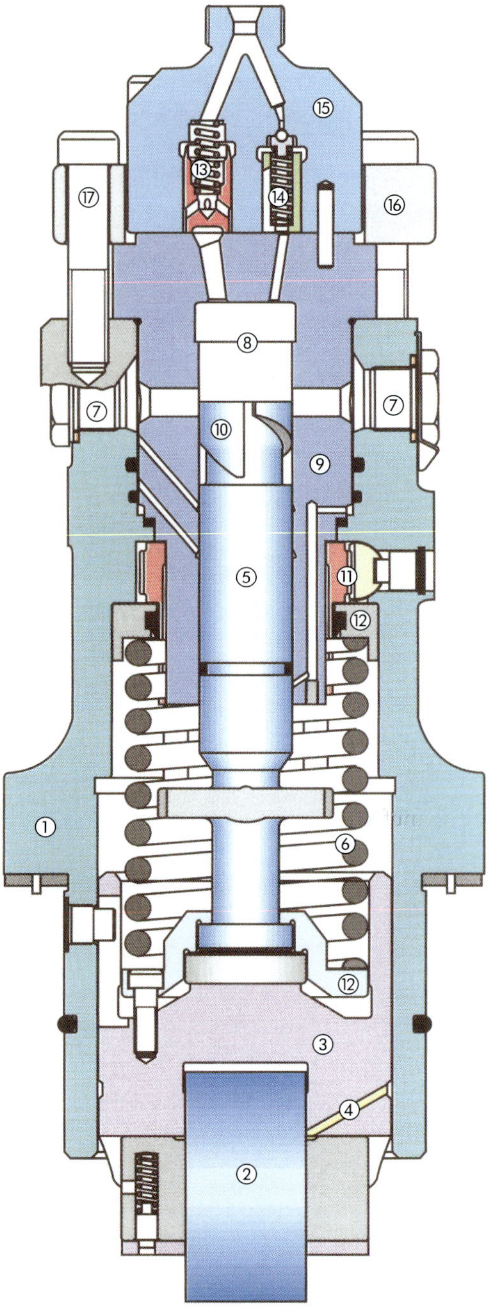

▲
A high-pressure fuel pump in a medium-speed four-stroke diesel engine operating on HFO.

1 *pump housing*
2 *cam rollers*
3 *roll guide*
4 *lubricating-oil supply*
5 *plunger*
6 *plunger spring*
7 *suction chamber*
8 *pressure chamber*
9 *pump cylinder*
10 *machined part of the plunger*
11 *gear rack for plunger rotation*
12 *spring seats*
13 *pressure valve*
14 *overflow valve, pressure release*
15 *pump cylinder head*
16 *pressure plate*
17 *bolt*

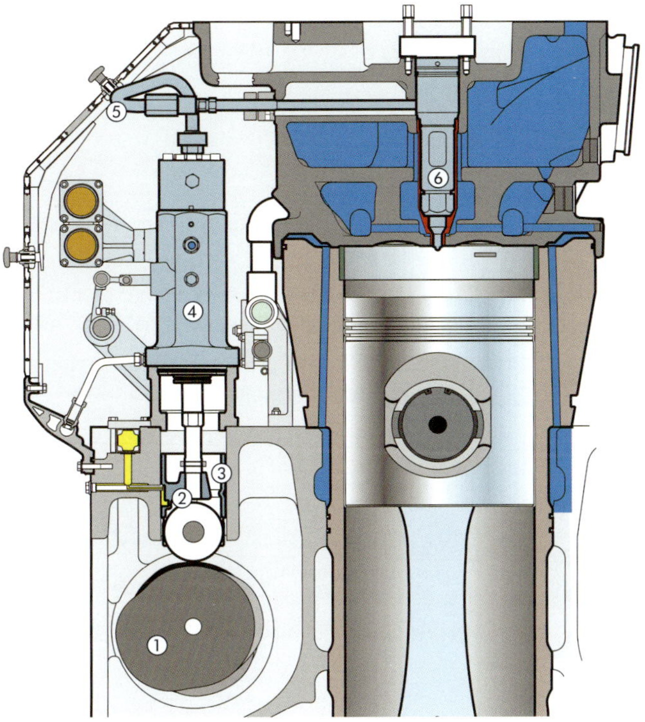

◀

A complete high-pressure fuel system of a four-stroke medium-speed diesel engine.

The last part of the high-pressure fuel line runs through the cylinder head. The injector tip is intensively cooled around its surface perimeter.

1 *camshaft*
2 *roll*
3 *guide*
4 *fuel pump*
5 *high-pressure fuel line*
6 *fuel injector*

Engine manufacturers often set an upper limit to the fuel temperature because the quality of the fuel-pump's seals deteriorates rapidly when exposed to higher temperatures.

9.14.2 Fuel pump drives

Traditionally, fuel pumps are driven from the camshaft by means of fuel cams.

Today there are also other drive systems such as the common rail systems. This will be discussed later in this chapter.

9.14.3 Types of injection systems

Six different fuel injection systems can be distinguished.

1 With plunger controlled pumps for each cylinder

Each pump is individually driven from the camshaft by a fuel cam.

This system can be applied to each type and size of engine.

▲

In this Caterpillar MaK diesel engine, every cylinder has its own high-pressure fuel pump.

▼

Two high-pressure plunger fuel pumps. In the centre the hydraulic pump for the exhaust valve. Engine type MAN Diesel 50MC.

◀

This two-stroke crosshead engine has an integrated high-pressure fuel pump block for every two cylinders comprising two fuel pumps.

2 With a block fuel pump with small plunger-controlled pumps for each cylinder

The block fuel pump has a small plunger-controlled pump for each cylinder. The block fuel pump is driven from the 'timing gear train' between the crankshaft and the camshaft.

This system is often used in high-speed four-stroke engines.

3 With valve controlled fuel pumps

This system is used in large two-stroke crosshead engines.

▶

A fuel pump block for an in line six-cylinder, four-stroke high-speed diesel engine.

▶

A fuel pump block for a three-cylinder, four-stroke high-speed diesel engine found in yachts.

▶

A horizontally positioned rotated block fuel pump for a small high-speed, six-cylinder, four-stroke diesel engine.

▲

A pair of valve regulated high-pressure fuel pumps for two cylinders of a two-stroke crosshead engine. Manufacturer Wärtsilä Sulzer RTA.

The pump is placed close to the low positioned camshaft, so not at the same height as the cylinder covers.

Left and right, high-pressure lubricating-oil pumps for operating the exhaust valves. In this engine type, two fuel pumps are assembled in one housing.

◀

A plunger regulated high-pressure fuel pump of the plunger type for MAN Diesel MC two-stroke crosshead engines.

1 *pump housing*
2 *high-pressure plunger*
3 *pump head*
4 *pressure valve*
5 *erosion/cavitation plug*
6 *overflow valve*
7 *fuel to injectors*
8 *fuel adjustment spindle, fuel quantity*
9 *fuel adjustment, timing*
10 *fuel supply*

4 With a common rail system

In this system, the fuel supply is regulated electronically. There are different systems for four-stroke trunk piston engines and two-stroke crosshead engines.

▶

A two-stroke crosshead engine from manufacturer Wärtsilä Sulzer 12RT-flex96C.

1 *fuel common rail 1000 bar*
2 *lubricating-oil common rail 200 bar*
3 *fuel return lines*
4 *high-pressure injector fuel lines*
5 *exhaust valve*
6 *high-pressure, lubricating-oil supply to the exhaust valve*

▶

The high-pressure injector of a Caterpillar high-speed, four-stroke diesel engine with the common rail system.

1 *fuel injector*
2 *high- pressure fuel supply line*
3 *fuel return lines*
4 *exhaust valve*
5 *inlet valve*
6 *combustion chamber*

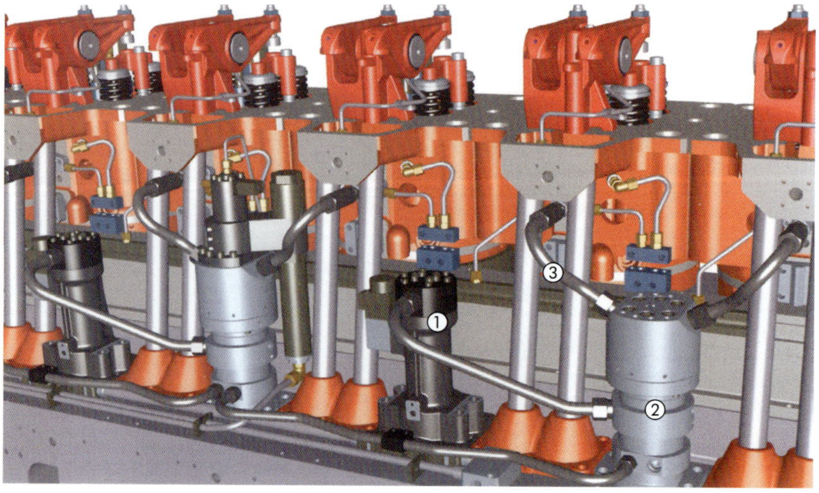

▶

The common rail system for the four-stroke medium-speed diesel engines from manufacturer Wärtsilä.

1 *high- pressure fuel pump for two cylinders*
2 *high-pressure fuel accumulator for two cylinders*
3 *high-pressure fuel line to the fuel injectors*

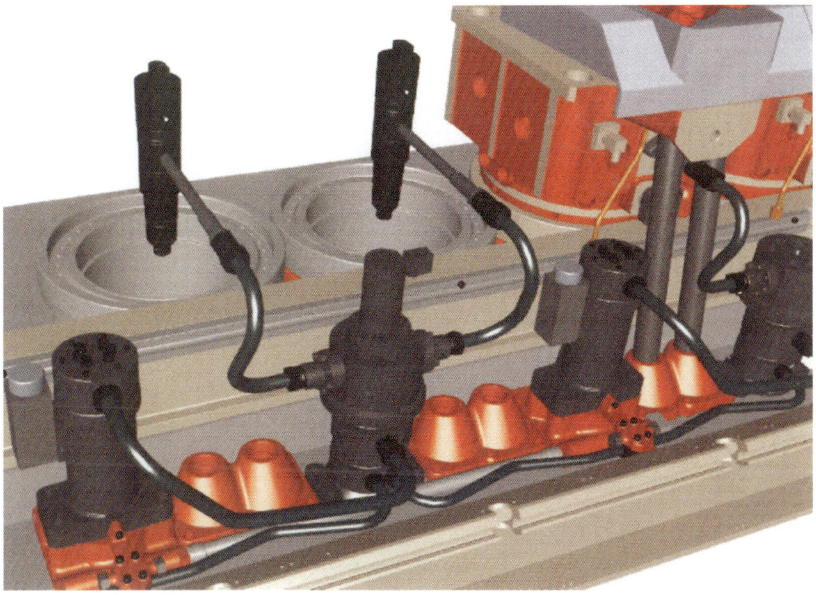

◀

Identical common rail system for Wärtsilä four-stroke engines where the cylinder heads of the two cylinders have been removed.

▼

A mechanically driven fuel injector from manufacturer Caterpillar.

1 *push rod*
2 *rocker arm*
3 *pressure spring*
4 *fuel injector*
5 *combustion chamber*

5 With an injector system

In this system, the high-pressure fuel pump and the atomiser are located in one casing, the so-called fuel injector. This fuel injector is placed in the cylinder head and is mechanically driven from the camshaft by means of a fuel cam, guide pulley, push rod and lever, just like the inlet- and exhaust valves. The system is used in high-speed four-stroke engines of, among others, Cummins and Caterpillar.

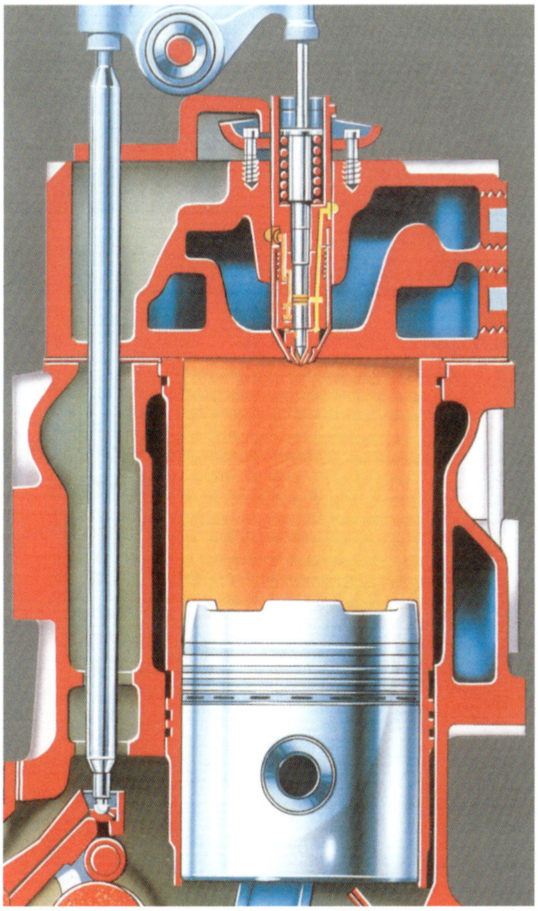

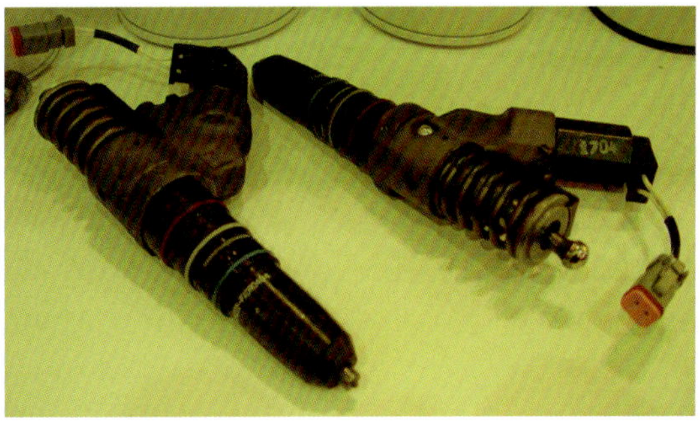

◀

Fuel injectors with mounted servo valves.

◀

A mechanically driven fuel injector from manufacturer Cummins.

▶

The mechanically adjustable camshaft in the medium-speed diesel engines from Caterpillar MaK.

Through rotation of the eccentricities, the timing of the valves and high-pressure fuel pump is modified.

6 With fuel pumps with adjustable cam drives

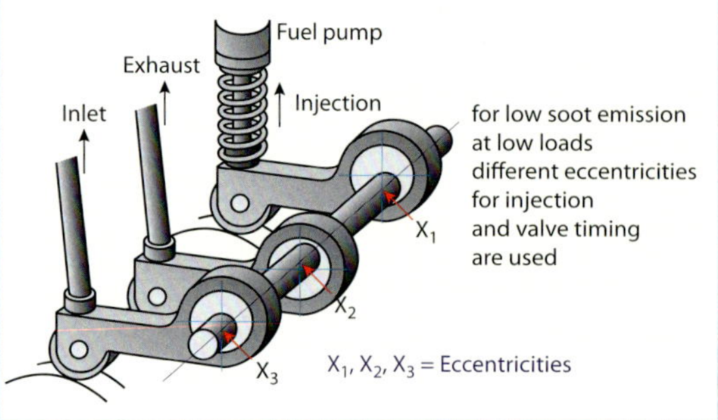

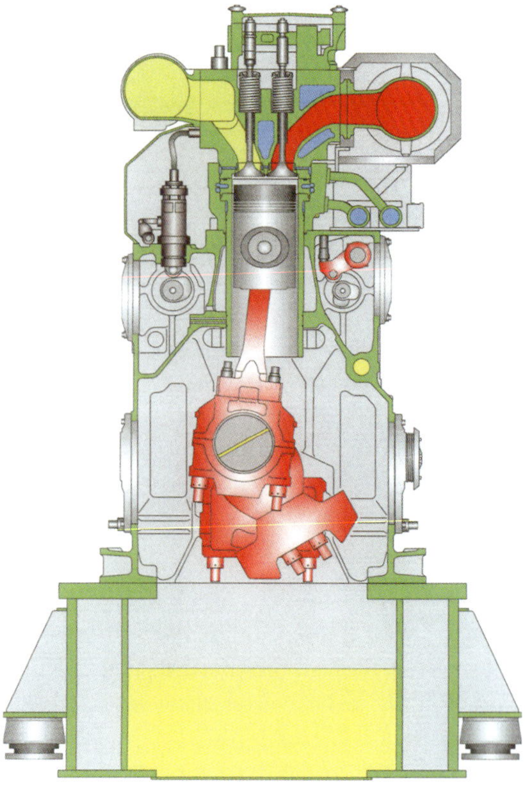

▶

A MAN Diesel four-stroke medium-speed diesel engine shown here is provided with a separate camshaft for the high-pressure fuel pump and the inlet- and exhaust valves. In this engine, the timing of the valves can also be adjusted.

9.14.4 Fuel pumps with modification for fuel quality, expressed in CCAI

This system is, among others, applied to Wärtsilä-Sulzer two-stroke crosshead engines. Two systems can either independently or concurrently be applied to the control system of the fuel pumps.

9.14.5 VIT-system

The Variable Injection Timing-system. This is used in order to advance the fuel injection at partial load. This way the maximum combustion pressure, and consequently the engine efficiency, remains at a high level. The system operates automatically and is dependent on the fuel control mechanism.
The VIT-system operates at an engine load scope of 100% to 85% and automatically ensures that the combustion pressure retains its maximum value. Maximum fuel economy is achieved at an engine load between 80 and 90% and amounts to two grams per kWh.
The automatic VIT-system is always combined with a manually adjusted FQS-system. Engines that are unsuitable for a VIT-system are equipped with a separate FQS-system.

9.14.6 FQS-system

The Fuel Quality Setting-system serves to ensure that the maximum allowable combustion pressure is not exceeded. The mechanism is manually adjusted and based on the ignition quality of the fuel.
The ignition quality is expressed in CCAI (Calculated Carbon Aromaticity Index). Both systems have limits in the degree of adjustment possible of the moment of injection. The maximum adjustment is three crank degrees for both advancing and retarding the injection.

Operation of both systems

Today these systems are electronically controlled, which allows for a reduction in the maximum combustion temperatures, which in turn facilitates a decreased emission of nitrous oxides in certain navigational areas.

9.14.7 VEC-system

The Variable Exhaust Control-system.
This controls the hydraulically driven plunger pump in the exhaust valve, which facilitates the timing of the opening and closing of the exhaust valve within certain restrictions.
So at present, the injection can be electronically controlled for:

Fuel regulation with 'VIT' combined with manual fuel quality setting 'FQS'

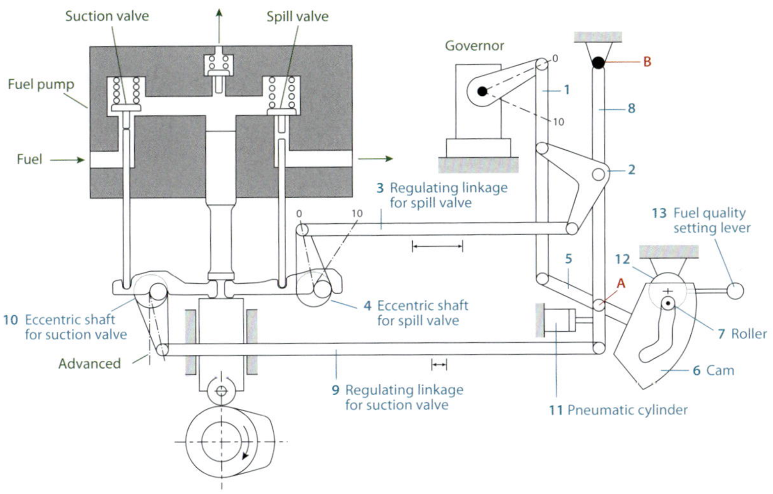

◀

Wärtsilä Sulzer two-stroke crosshead engines type RTA. The automatically regulated VIT-system in combination with the manually regulated FQS-system.

The FQS-system is operated with a fuel quality-setting lever 13 that can be fixed using a pin 12, an eccentric positioned roller 7 moves. This moves via a recess in the cam 6. Via levers 5, 1, and 2, the position of the overflow valve is regulated and via shaft 9, the suction valve.

The VIT-system is automatic and can at partial loads advance the fuel injection.

Variable beginning of injection

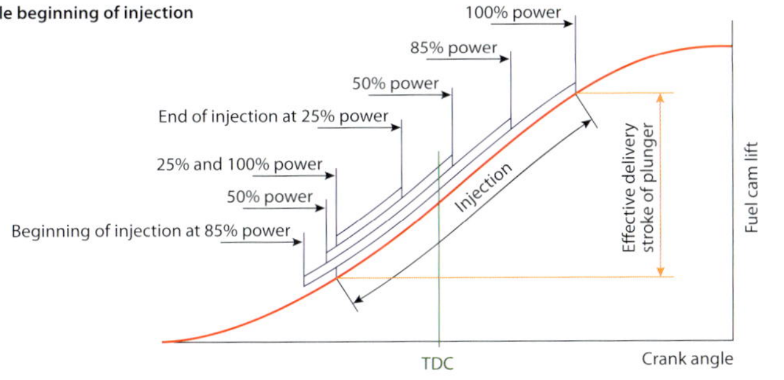

– minimum fuel consumption at partial load (VIT);
– an altered fuel quality (FQS);
– an optimum exhaust period of time (VEC).

Obviously, the development of common-rail engines is a continuation of the aforementioned techniques.

9.15 Fuel-capacity adjustments

Essentially, three fuel-capacity adjustment systems can be distinguished.

1 Initial adjustment
Here the power output of the engine is controlled by allowing an adjustment to the start of the injection.

2 Final adjustment
Here the power output of the engine is controlled by allowing an adjustment to the end of the injection.

3 Initial and final adjustment
Here the power output of the engine can be controlled. The initial adjustment ensures that the injection starts at the correct time at varying loads. A reduced power output and lower load will modify the ignition delay. With the final adjustment, one controls the power output.

The initial and final adjustment can also be applied to the electronically controlled common-rail systems. One is in this case no longer bound by mechanical limitations.

9.16 Working of a plunger pump

9.16.1 Plunger controlled fuel pumps

This type of pump is still often used. Not only in very small high-speed four-stroke diesel engines, but also in medium-speed four stroke engines. Two-stroke crosshead engines use both valve- and plunger controlled fuel pumps.

Principle

The principle of the plunger-controlled pumps is based on the rotation of the plunger in relation to the plunger barrel. Due to the rotation, the pressure is released either earlier or later in the fuel-suction chamber. The gear racks of all fuel pumps are connected to each other by pinions and form a fuel adjusting spindle, which can be moved several centimetres (depending on the size of the engine). Therefore, the fuel output of the pumps can vary from 0 to 100%.

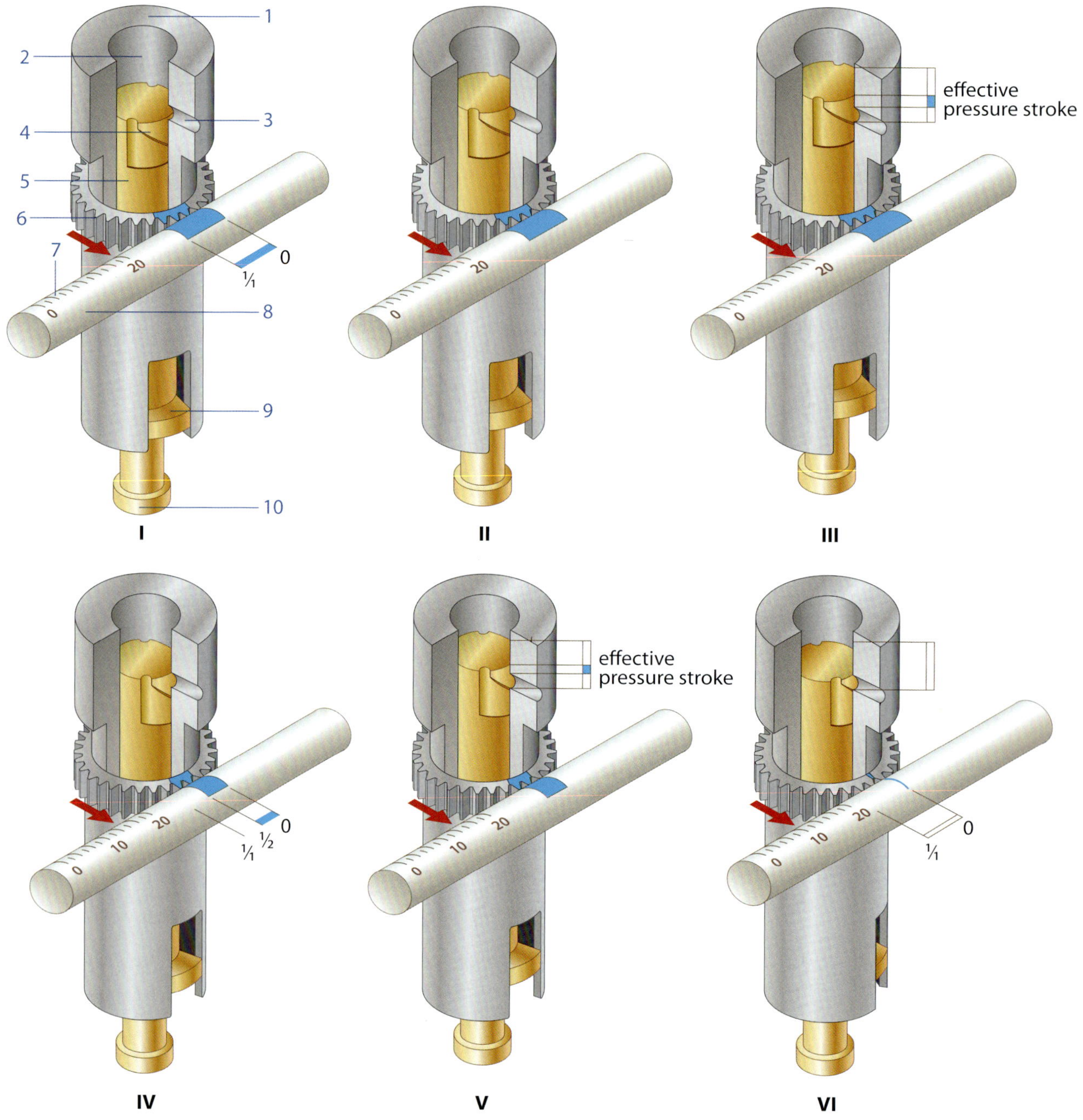

▲

The capacity adjustment of plunger controlled fuel pumps.

1 *pump casing*
2 *compression space above the plunger*
3 *fuel supply*
4 *control guide of the plunger*
5 *plunger*
6 *rotating the cog on the control guide carries the plunger to the milled groove*
7 *control position-capacity of the pump*
8 *rack with control position*
9 *cam on pump plunger*
10 *drive from the camshaft via guide pulley*

Operation:

In **position I**, the upwards stroke of the plunger starts, the connection between the fuel supply (3) and the compression space (2) above the plunger is open and there is no yield.

In **position II**, the fuel supply is closed by the un-milled section of the plunger (4) and the yield is 100%.

In **position III**, the plunger has reached the end of the compression stroke; a short while later there is once again an open connection between the fuel supply (3) and the compression space (2).

In **position IV**, the plunger is rotated by the adjustment of the rack (8) that takes the cam on the plunger along. De effective compression stroke is now much shorter than at the first position of the plunger, the yield is approximately 50%. **Example V**.

In **position VI**, the plunger with respect to the fuel supply (3) is further rotated and there is no effective compression stroke; the yield is null!

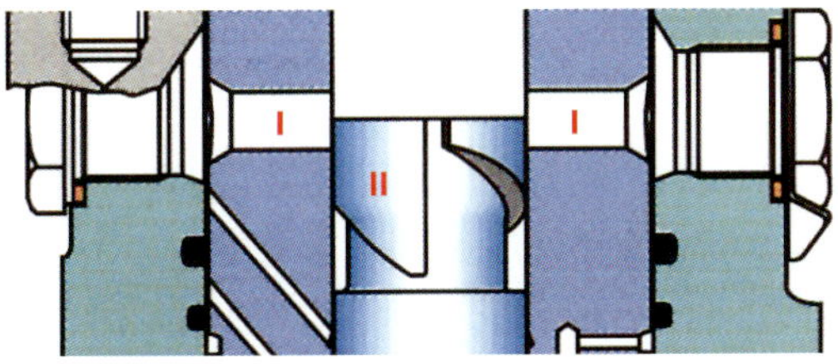

▲

In this position of the ascending plunger, the suction hole I is open. After closure of the suction hole, the plunger starts the compression stroke to the milled area. The yield is approximately 50%. If the plunger is rotated to position II, the yield is 100%.

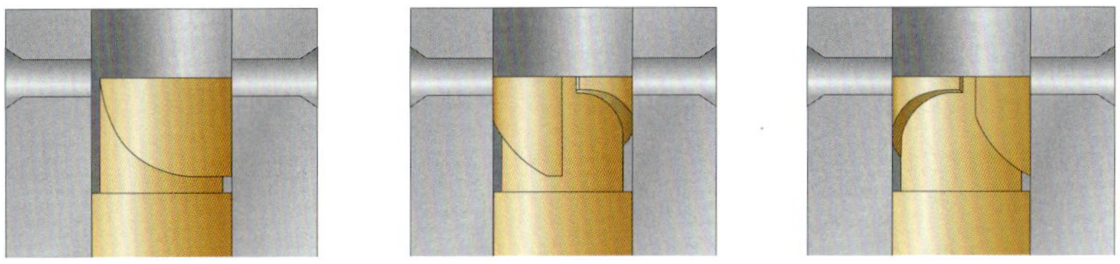

▲

Three positions with different yields. The plunger must slightly ascend before compression starts.

In positions 1, 2 en 3 of the pump plunger, the yield of the plunger is 0, 50 en 100%.
Material is milled from the plunger. The plunger has two opposing milled areas that are connected to the compression space.

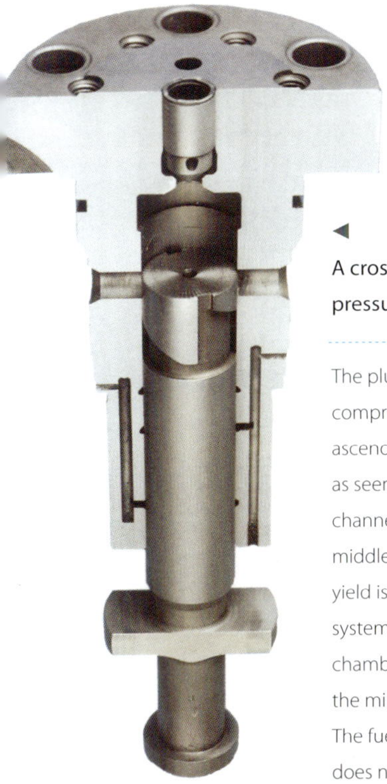

◀

A cross-section of a high-pressure fuel pump.

The plunger starts the compression stroke when it ascends in the cylinder. The yield as seen is below, 50%. If the channel is positioned in the middle of the suction hole, the yield is zero; there is an open system between the pressure chamber above the plunger and the milled area of the plunger. The fuel travels, but the plunger does not provide any capacity.

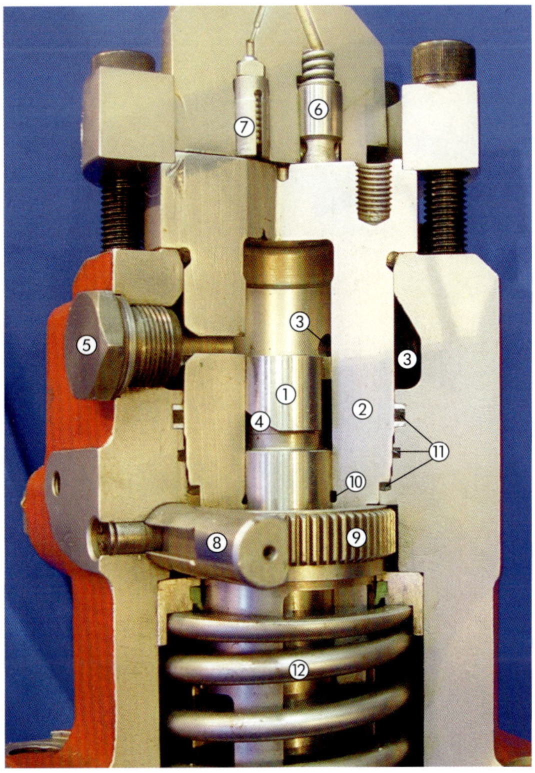

◀

A cross-section of a high-pressure fuel pump.

1 *plunger*
2 *housing*
3 *suction chamber (2x)*
4 *adjustable plunger*
5 *erosion plug*
6 *pressure valve*
7 *safety valve*
8 *gear rack for plunger adjustment – yield control*
9 *gear wheel*
10 *lubricating- oil channel for plunger lubrication*
11 *gaskets*
12 *spring for fast falling plunger when the fuel cam releases it for the descending stroke*

◀

A disassembled high-pressure fuel pump.

1 *pump housing*
2 *yield control rod*
3 *roller*
4 *guide*
5 *pump plunger*
6 *milled area*
7 *pump cylinder*
8 *high-pressure connection*
9 *spring*

◄

Two plungers of high-pressure fuel pumps with the associated high-pressure barrels.

There are two types: one with and one without lubricating-oil grooves in the plunger.

►

The high-pressure fuel pump of a four-stroke medium-speed diesel engine, manufacturer Caterpillar MaK 32.

1 *high-pressure fuel pump*
2 *fuel supply and discharge pipes*
3 *fuel-adjustment spindle*
4 *lever*
5 *gear rack*
6 *adjustable area*
7 *lubricating- pipe lines for fuel-pump drives*

►

The high-pressure fuel pump of a four-stroke medium-speed diesel engine, manufacturer MAN B&W 27-32.

1 *high-pressure fuel pump*
2 *fuel-supply pipe lines*
3 *high-pressure fuel pipe lines to the injector*
4 *fuel adjustment spindle*
5 *lever*
6 *gear rack*
7 *fuel return pipe lines*

◀

A high-pressure fuel pump positioned horizontally in a small four-stroke high-speed diesel engine.

1 *fuel pump*
2 *high-pressure fuel lines to the injectors*
3 *capacity adjusting fuel pump*
4 *gear drive*

9.16.2 Plunger pumps with initial- and final adjustments

The pump can also be fitted with initial-adjustment controls by machining the **top** of the plunger. Small plungers are milled on one side at the top of the plunger. In larger plungers, the lateral forces rise too significantly because of the uneven radial load. Therefore, two control guides are mounted opposite each other, which cancel out any radial forces. Of course, two spill ports are also installed.

▼

Description of the operation of the fuel pump of a Wärtsilä 64 four-stroke diesel engine.

With the left plunger, the start of the fuel injection is determined and with the right plunger, the duration of the injection and so the end of the injection.
Figure 1: Charging: both plungers are in the lowest position. Via the left and right suction lines, the spaces above the plungers are filled. The fuel cams are not touching the rollers.
Figure 2: Start of the ascending stroke. Both plungers are driven by the fuel cams. The right suction line is closed by the right plunger. Via the left suction line fuel can flow out.
Figure 3: Fuel injection. The speed of the plungers is maximal. The left plunger closes the left suction line and the fuel injection is optimal by the high plunger-speed.
Figure 4: the right plunger releases the right suction line and the fuel injection quickly stops. Both plungers are in their highest position.

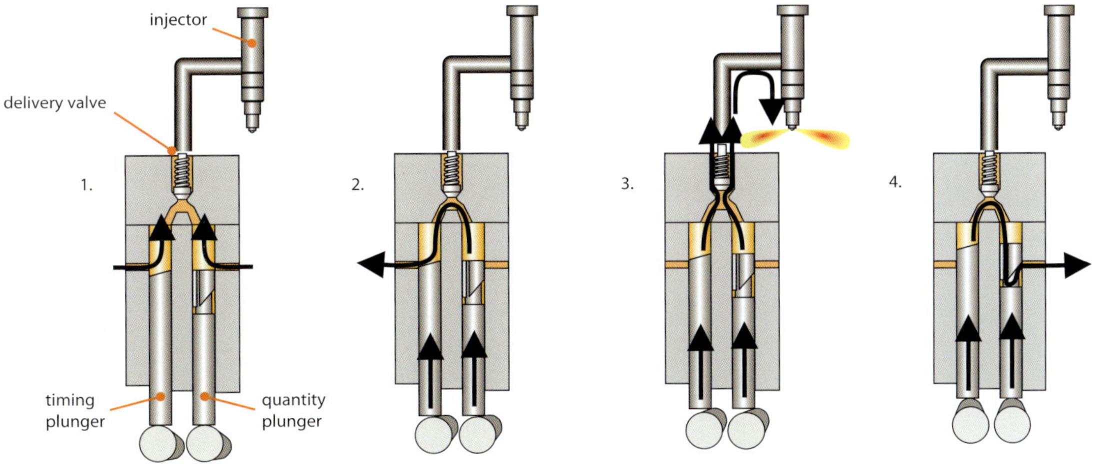

- tappets on cam base circle
- filling of injection pump

- quantity plunger shuts off spill port
- excessive fuel out to lowpressure side through filling port

- both ports are shut off
- delivery valve lifts
- start of injection

- spill port opens
- excessive fuel out to lowpressure side through spill port
- end of injection

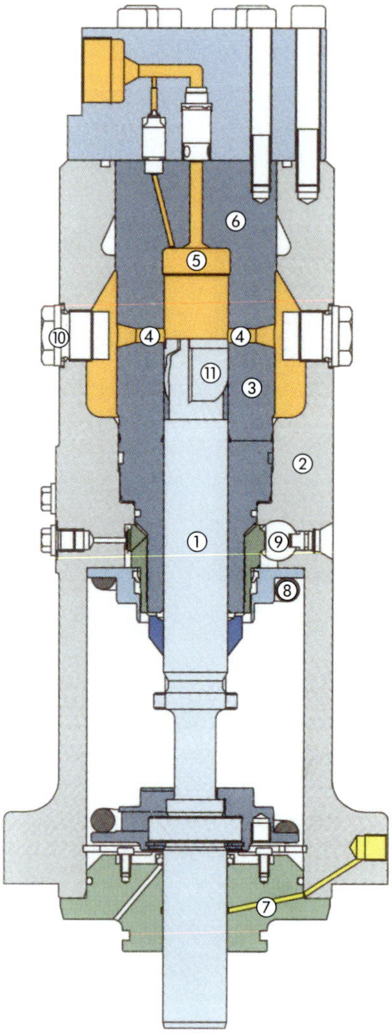

▶

A high-pressure fuel pump.

1 *plunger*
2 *pump housing*
3 *cylinder or barrel*
4 *suction chamber*
5 *pressure chamber*
6 *pump cover*
7 *lubricating-oil supply*
8 *spring*
9 *control rod*
10 *spreader bolt to avoid cavitation in the pump*
11 *milled area of the plunger*

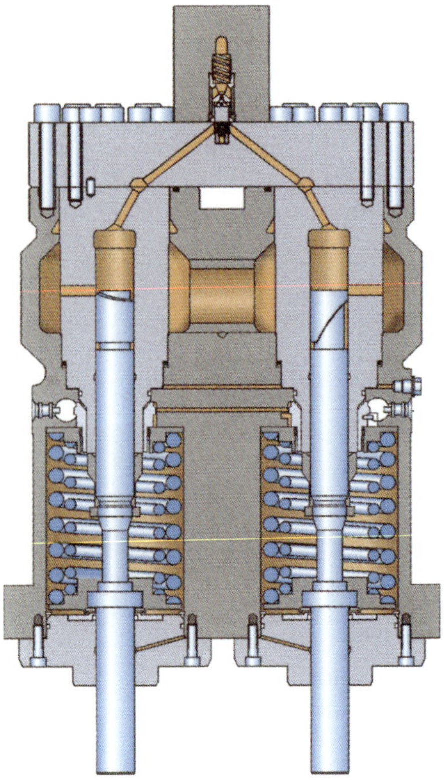

▲

A high-pressure fuel pump with two plungers for the four-stroke medium-speed Wärtsilä 64 diesel engine, fuel HFO. The pump has start and finish controls.

9.16.3 Plunger pumps with variable injection timing

This is the so-called VIT-system (Variable Injection Timing).
This is used for marine propulsion engines in ships. The system is used in low-speed two-stroke crosshead engines and medium-speed four-stroke trunk piston engines. Propulsion engines have widely varying loads at normal operating speed due to weather conditions, the load and hull fouling; these are in practice never the same.
Generally, a ship propulsion engine operates far below its maximum allowable capacity. This is usually referred to as the **'Maximum Continuous Rating' or MCR**. The engines' fuel control for this system has been modified by engine manufacturers for a load between 85 and 100%. The goal is to increase efficiency for all loads and therefore decrease the specific fuel consumption.

The system ensures that the maximum combustion pressure remains constant at engine loads between 85 to 100%.. MAN Diesel has plunger controlled fuel pumps for two-stroke crosshead engines.

This system is also used in four-stroke medium-speed trunk piston engines that run on heavy oil. This is the case with MAN B&W engines, in particular the largest series, the L 58/64.
Apart from the much discussed VIT-system, the fuel pumps have an adjustable plunger drive, which can change the beginning of the fuel injection to the ignition quality of the fuel over the entire engine load range. This system does not depend on the load. The objective is to keep the engine efficiency as high as possible at any speed. The pump settings are related to the fuel quality expressed in CCAI.

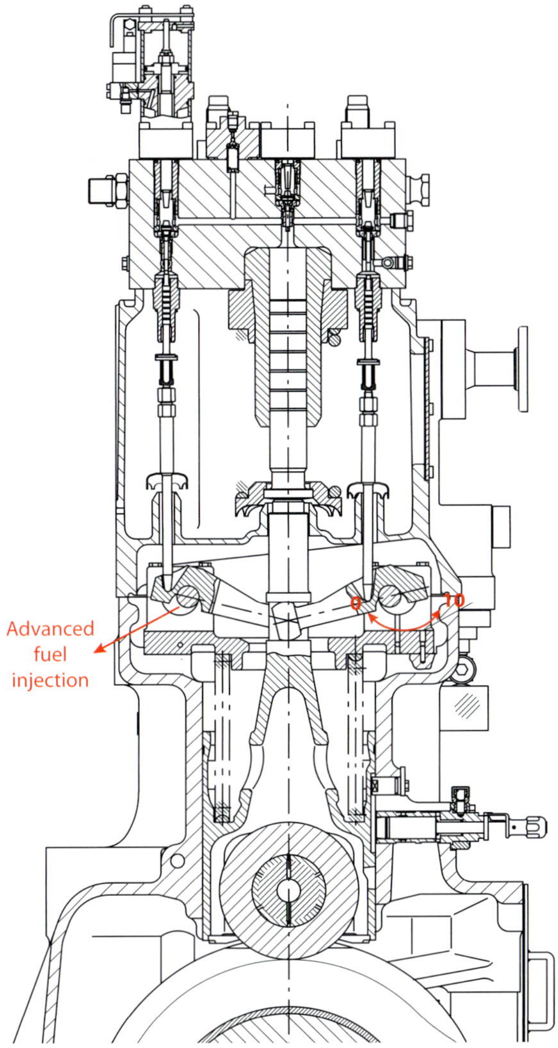

◀

An overview of the controls of the high-pressure plunger pump of the two-stroke crosshead engines from Wärtsilä Sulzer RTA with suction valve and overflow valve.

Under, a graph with horizontally, the crank angle and vertically, the plunger stroke of the high-pressure fuel pump. In addition, the fuel injection at different loads.
Clearly shown is that at a lower engine load, the fuel injection occurs earlier. Normally at decreased power output, the final compression pressure decreases and so the final combustion pressure. To retain the final combustion pressure at partial load, the fuel injection occurs earlier. This has a favourable influence on the efficiency and therefore the fuel consumption.

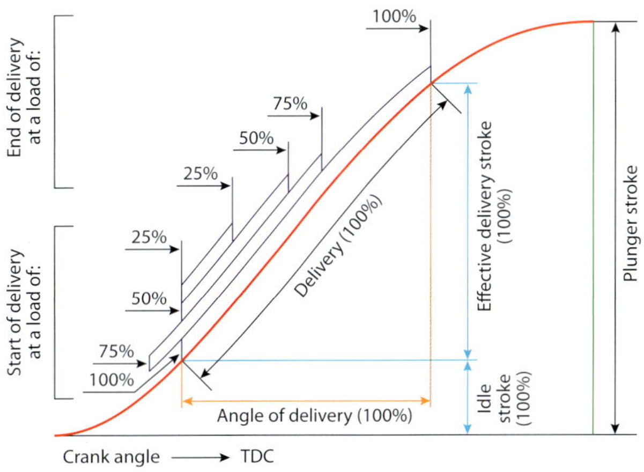

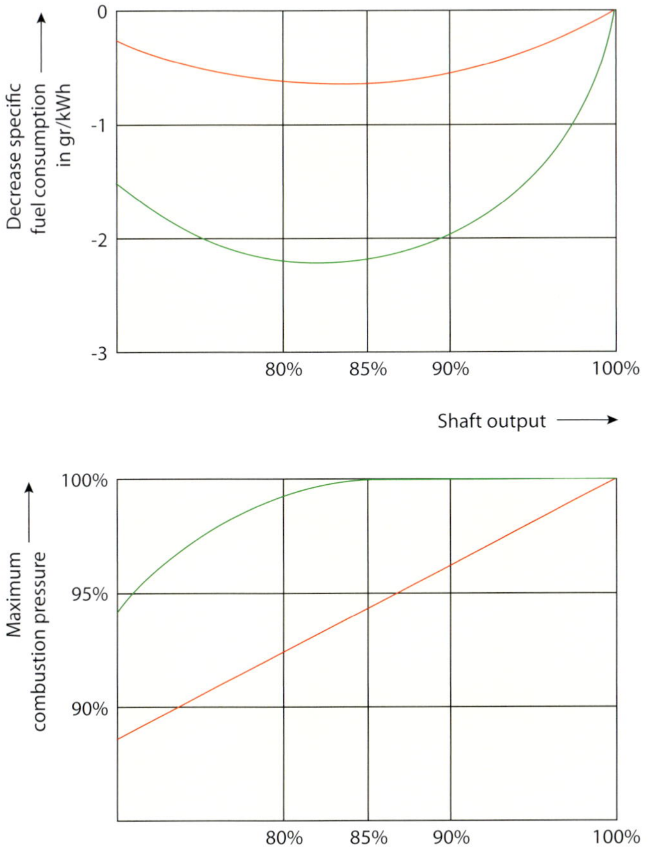

◀

Influence of timing advance at a partial load.

In the both graphs, the difference between a conventional injection with respect to a VIT-system it is clearly shown.
At partial load, the final combustion pressure is maintained and fuel usage decreases.

Red: conventional injection
Green: VIT-system

▶

The plunger surface of a fuel pump for a MAN Diesel two-stroke crosshead engine type MC.

The plungers are machined in such a way that the profile has five control areas, A, B, C, D and E.

A Injection starts later.
B Injection remains constant.
C Injection starts earlier.
D Final yield.
E Final yield.

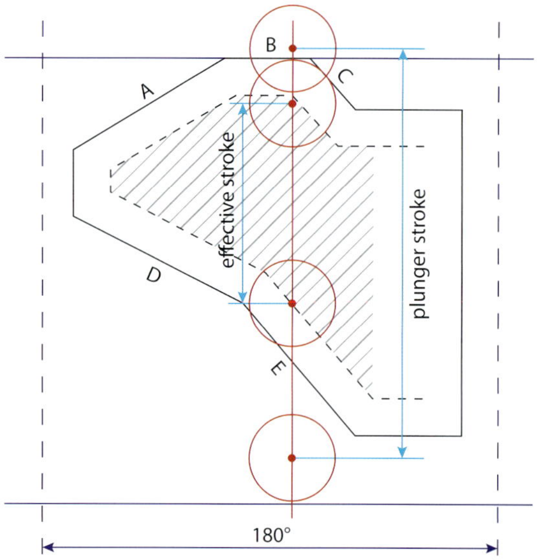

The start of the effective compression stroke is regulated by the above three control guides A, B and C. the end of the effective compression stroke is regulated by the lower control guides D and E.
At a low engine load, the overflow channels in the pump cylinder are opened and closed by control guides A and D in the left side of the figure.
In this area, the start and finish controls are applied. At increasing load, the fuel is injected earlier. From the start of control guide B, the start of the injection remains constant and at the end of the injection, the engine load increases. At the end of control guide B, an engine load of 85% of MCR is achieved. With a load increase to 100% of MCR, the start of the effective injection (combustion stroke) is determined by control guide C in such a way that the maximum combustion pressure remains constant.

9.16.4 With integrated block fuel pumps and plunger controlled pumps for each cylinder

This is often seen in small to medium-sized high-speed four-stroke diesel engines in the categories I and II (0 tot 5000 kW). One of the advantages is that, for instance in a sixteen cylinder V-engine, only one drive is required: the block fuel pump as opposed to sixteen separate fuel pumps with a camshaft drive. Another advantage is that a complete block pump can be changed and overhauled by a specialised company. To ensure that the injection delays are identical for each cylinder, the high-pressure fuel lines to the injector are of equal length. They are often expertly bent in order to make them fit between the block pump and the cylinder in question.

▼

A block fuel pump for a twelve-cylinder, four-stroke high-speed V-diesel engine.

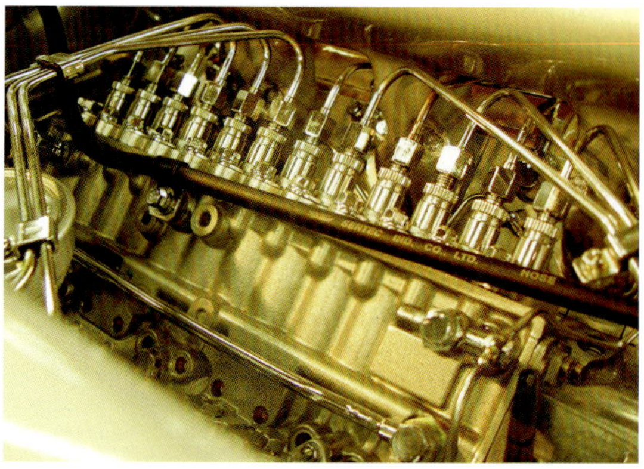

▶

Length of the high-pressure fuel lines.

To keep the injection delay of every cylinder equal, the pressure pipelines to every fuel injector are equal in length. The cylinders furthest from the pump have high-pressure fuel lines that run more or less directly to the cylinders.
The cylinders nearest to the pump have fuel lines of equal length that are so bent that they end at the relevant cylinder.

Depending on the engine's rpm, the moment of the initial fuel injection can be adjusted. The output of all small pumps is controlled by one gear rack that is horizontally mounted in the pump block. Due to increasingly strict emission regulations, special systems have been designed for this technology.

9.17 Valve-controlled fuel pumps

These are found, for instance, in low-speed two-stroke crosshead engines by Wärtsilä Sulzer RTA.

In these engines, the fuel pumps of two cylinders are located in one block. Each pump is driven from the camshaft with cam, cam rollers and cam roller guides.

▼

An exploded view of a fuel pump.

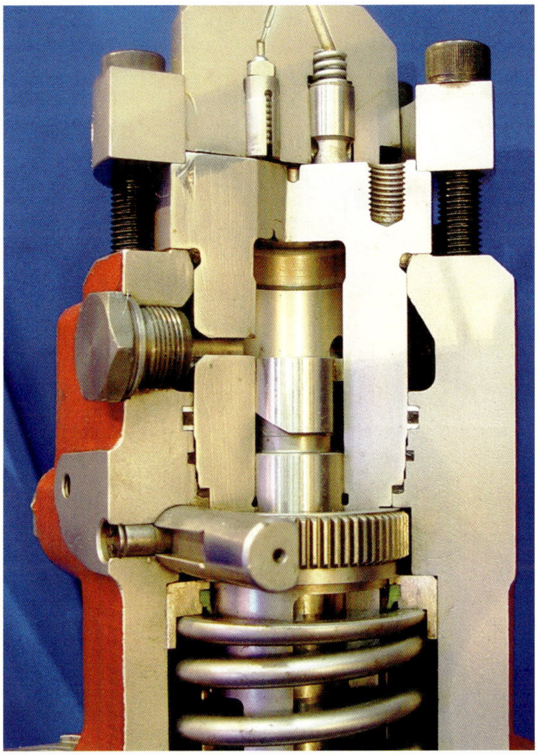

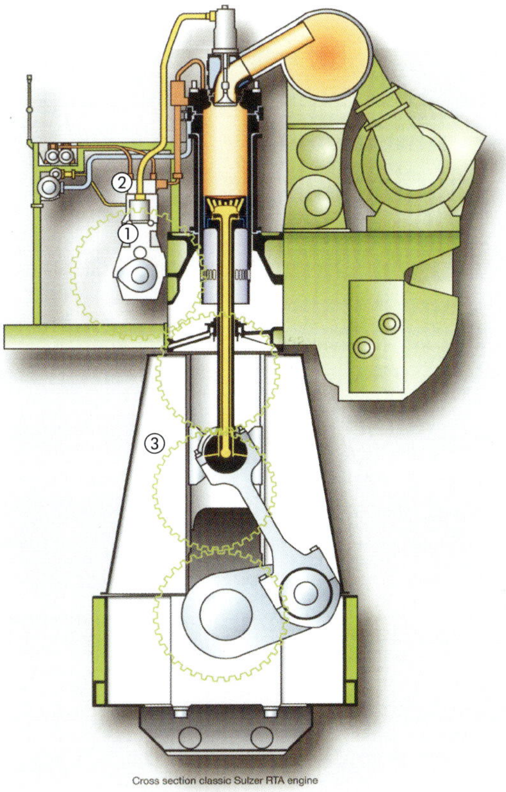

◀

A low-speed two-stroke crosshead engine from Wärtsilä Sulzer RTA with valve regulated fuel pumps.

They are driven by the camshaft. The transmission from the crankshaft and camshaft is driven with gear wheels.
The fuel pump positioned behind the camshaft driven high-pressure lubricating-oil pump for the operation of the exhaust valve.

1 *lubricating-oil pump for the operation of the exhaust valve*
2 *high-pressure fuel pump, behind lubricating-oil pump*
3 *gear wheels*

◀

Two high-pressure fuel pumps mounted in one housing in a Wärtsilä RTA 84 crosshead engine.

1 *pump housing with two high-pressure fuel pumps*
2 *hydraulic pump for an exhaust valve drive*

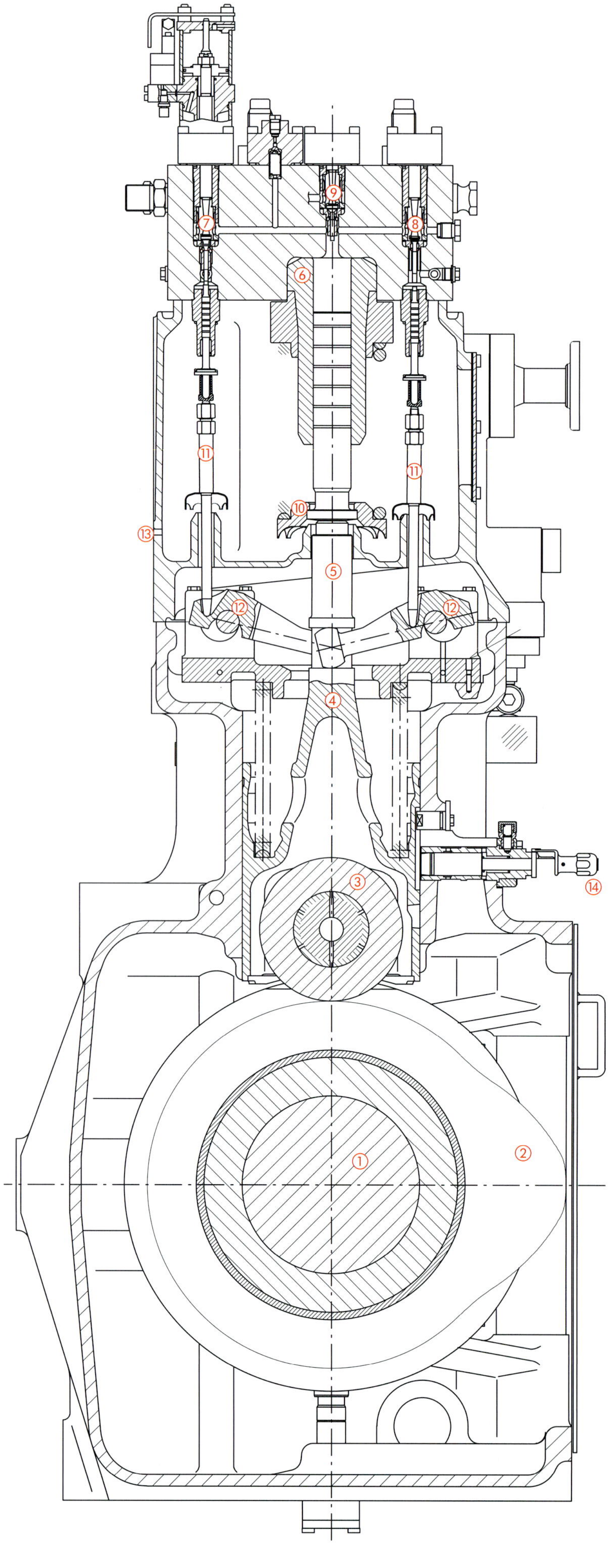

◄

A valve regulated high-pressure fuel pump for Wärtsilä RTA two-stroke crosshead engines.

1 *camshaft*
2 *cam*
3 *roller*
4 *roller guide*
5 *pump plunger*
6 *pump cylinder or barrel*
7 *suction valve*
8 *overflow valve*
9 *pressure valve*
10 *plunger springs with seats*
11 *adjustable valve stems*
12 *levers*
13 *discharge excess fuel*
14 *mechanism for stopping pump*

Fuel quantity is regulated using a **suction port** and a **spill port** as opposed to the rotation of the pump cylinder around a machined plunger. Therefore, the pump has initial and final adjustment controls.

Pump operation

At the downward stroke (overpressure) of the plunger, the piston valve is mechanically opened by a lever after which the fuel flows into the cylinder at a certain overpressure (5 to 10 bar). The pressure valve of the pump is still closed. At the upward piston compression stroke, the effective pressure stroke will commence as soon as the piston valve has been closed automatically by the lever. The overflow valve is closed. The pressure valve opens because of the rapidly increasing fuel pressure. At a certain moment, the overflow valve is mechanically opened by the lever.
The fuel pressure above the plunger drops, the pressure valve closes and the effective compression stroke is now terminated. During the remainder of the upward stroke of the plunger, the fuel flows back to the suction side of the pump.

Lever operation

The levers are moved up and down by the gearing mechanism of the plunger. Both sides of the lever are equipped with eccentric discs that are mounted on adjusting spindles. Rotation of the adjusting spindles alters the closure of the suction valve and the opening of the overflow valve.
If the plunger has no output, the overflow valve has opened before the suction valve has shut.
This Wärtsilä Sulzer system is also equipped with the VIT-system to maintain a constant maximum combustion pressure between 85 and 100% MCR.

Fuel regulation with 'VIT' combined with manual fuel quality setting 'FQS'

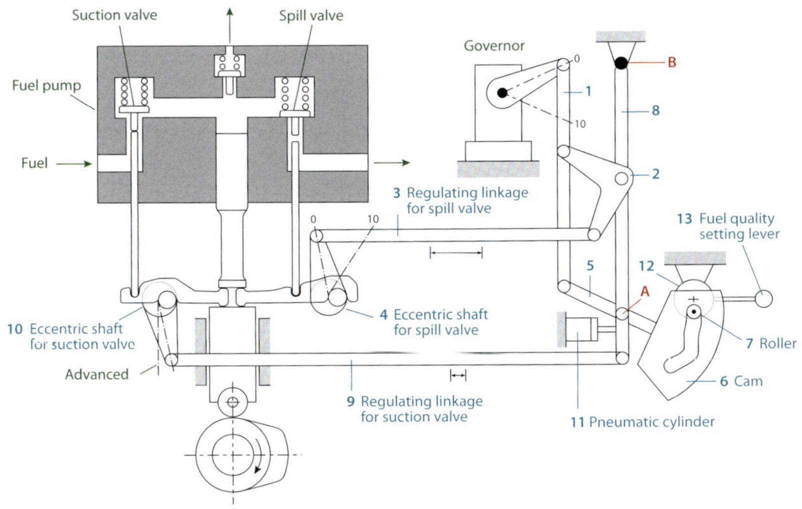

Variable beginning of injection

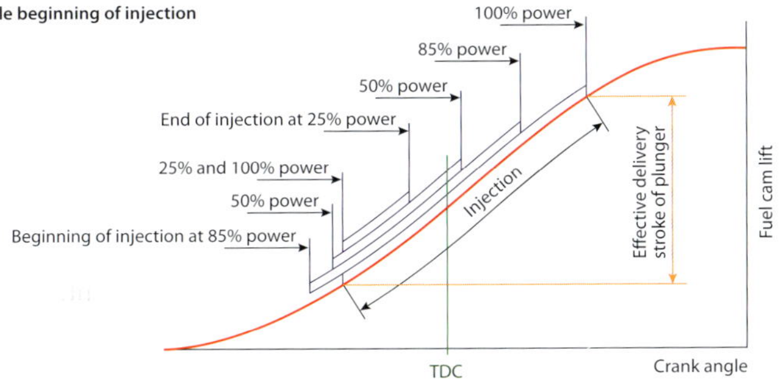

◀ A valve regulated high-pressure fuel pump for Wärtsilä RTA two-stroke crosshead engines.

The levers for the load- and VIT-controls are electronically controlled. Due to this, it is possible to, for example, retard the injection and so lower the maximum combustion pressures (and the maximum combustion temperatures) and reduce the nitrous oxide emissions.
Furthermore, in a similar fashion, the operation of the hydraulic driven exhaust valve is achieved. This is called at Wärtsilä Sulzer, the Variable Exhaust Control, VEC-system.

Furthermore, a system that allows the fuel pumps to start the effective compression stroke earlier in case of deteriorating fuel quality is included in the fuel system. This system is known as the Fuel Quality Setting, FQS. Once again, the fuel quality is expressed in CCAI-values. The higher this value, the earlier the fuel injection.

9.18 Common-rail system

In this system, the one-century-old system of the camshaft driving the high-pressure pump has been abandoned.
In the numerous systems that are currently applied, the fuel is brought to extremely high pressures of 1000 to 2000 bar in various ways. The fuel injection is electronically controlled.
The common rail system was introduced in the smaller diesel engines for motor cars and lorries towards the end of the last century.
The system has advanced so quickly over the past few years that it is now applied to all diesel engine categories. In particular, the increasingly strict emission regulations have required that engine manufacturers choose these very flexible fuelling systems. The common rail system allows for the reduction of the specific fuel consumption, especially at partial load.

General construction of a common rail system

The fuel supply pressure to the injectors can be attained with either one or several pump(s). The pumps can be driven traditionally, from the camshaft, but also directly from the crankshaft using gear transmissions or a gear belt (timing belt).

In larger four-stroke engines, every cylinder or every second cylinder has their own high-pressure fuel pump that is connected to a high-pressure collecting pipe. The electronically controlled and hydraulically operated supply valves to each injector ensure the fuel supply from the collecting pipe to the injectors.

Advantages of the common rail system

The injection curve of the fuel can be adjusted to guarantee optimal combustion at every engine load.

Explanatory note

Both the moment of the initial- and the final fuel injection, as well as the total injection time can be predetermined very accurately thus allowing for optimal combustion at all rpm's and every engine load. In this way the specific fuel consumption, the emission of nitrous oxides ('NOx') and fine soot particles are reduced to a minimum.

This has the following benefits:

- A lower specific fuel consumption especially at partial load.
- Lower emission of pollutants such as soot and nitrous oxides. At partial load, the emission of soot particles is considerably lower.
- A consistently high fuel-injection pressure, thus maintaining an optimum combustion process at every engine load. In existing diesel engines with a conventional fuel injection system, the combustion quality at partial load is mediocre. This is caused by a much lower fuel injection pressure as well as decreasing final compression pressures and -temperatures.
- Some adjustment, such as the moment of injection, the time and the pressure can be made during the process. In two-stroke crosshead engines, one can also modify the opening and closing of the exhaust valve, and in medium-speed four stroke engines the opening and closing of the inlet valves.
- Driving high-pressure fuel pump(s) is much simpler because not every cylinder requires a separate pump.
- In conventional fuel injection systems, high-pressure fuel pumps have frequently been in use for many years and therefore represent a proven technology.
- The opening and closing of the valves occurs more meticulously as a result of the electronic controlling unit, therefore the thermal load is more evenly distributed over the various cylinders.
- All data is stored and settings are improved where necessary, which increases the life-time of the engine components.
- Also, low rpm can be achieved. This is especially advantageous for large two-stroke crosshead engines, which directly drive a fixed propeller. The result is a rapid increase of the rpm and an improved stop- and turn ability in these engines. Cylinder lubrication dosage and control of the starting valves can now be regulated and adjusted electronically.
- The required software for electronically controlled systems can be upgraded throughout the entire lifetime of the engines.

Generally, one can say that in applying electronically controlled fuel systems, often referred to as common rail systems, it is easier to comply with present day requirements for modern diesel engines. These are lower specific fuel consumption and the reduction of pollutants in the exhaust gases. Engine manufacturers that produce for the international market certainly consider these. Most engine manufacturers either already have common rail systems or are in the process of developing one. A significant improvement is that at partial load there is a remarkable reduction of the so-called 'visible smoke'. This smoking used to be one of the main problems of diesel engines not equipped with common rail systems. It is presently also easier to control and reduce engine vibrations.

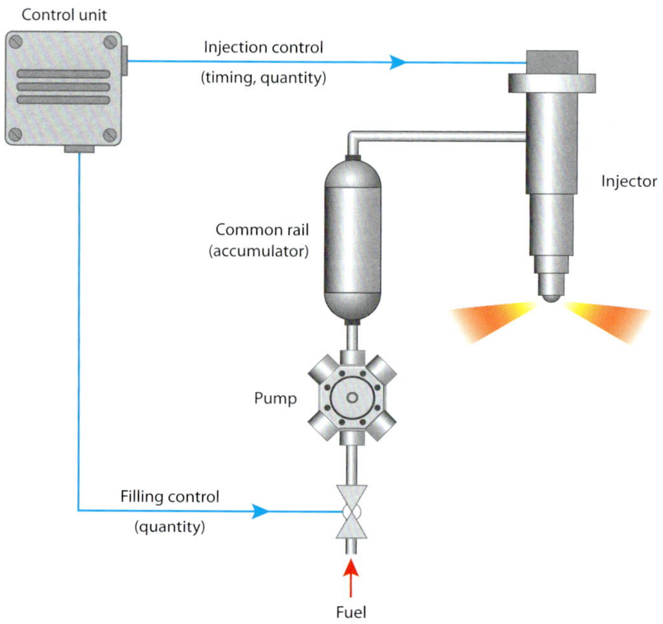

▶ The principal build of a common rail fuel system for diesel engines.

The fuel is drawn up by a high-pressure plunger pump and pumped into a spacious collection pipe; this pipe is known as the 'common rail'. The fuel pressure in this pipe is kept constant. A control-unit regulates the fuel injection with the injector. The greater the fuel demand, the greater the fuel supply. This is measured using a quantity measurement on the suction line of the pumps.

9.18.1 Example 1: common rail systems

Four-stroke medium-speed diesel engines by MAN Diesel, fuel HFO (engine category III): engine types 48/60 B en 32/40)

At the design stage, the following was established:

- The system must be able to inject fuel up to 700 cSt. at 50 °C.
- The fuel temperature can rise to 150 °C to achieve the correct injection viscosity of 10 to 12 cSt.
- An high-pressure supply rail of excessive length requires radial connections with the injector. This results in extremely high material stresses. In larger diesel engines, these problems increase as the internal inner diameter of the rail increases.
- In instances of a reduced accumulator volume, it is virtually impossible to reach equivalent injections for each cylinder. Considerable pressure fluctuations in the pipes may ensue.
- For larger engines, one would have to resort to individual rails for each cylinder.
- A very large pressure accumulator also proved to be insufficient.

These findings have been incorporated in the design and demonstrated that a six-cylinder engine requires at least two high-pressure pumps with two independent high-pressure supply lines.

Consequences for the diesel engine

- Increasingly strict regulations with regard to the emission of pollutants create high demands on the fuel injection system and the combustion.
- The common-rail system has obvious advantages when compared to conventional injection systems.

General comment

The common rail is particularly effective at partial load operation. At full load, the differences with the normal high-pressure fuel systems are negligible.

Results

- Smoke reduction over the entire load range.
- Nitrous oxides reduction, especially at partial loads.
- Lower specific fuel consumption especially at very low loads.

Electronic controlling system

Since there is no mechanical 'back up'-system present, the electronic control has to be of superior quality and safety.

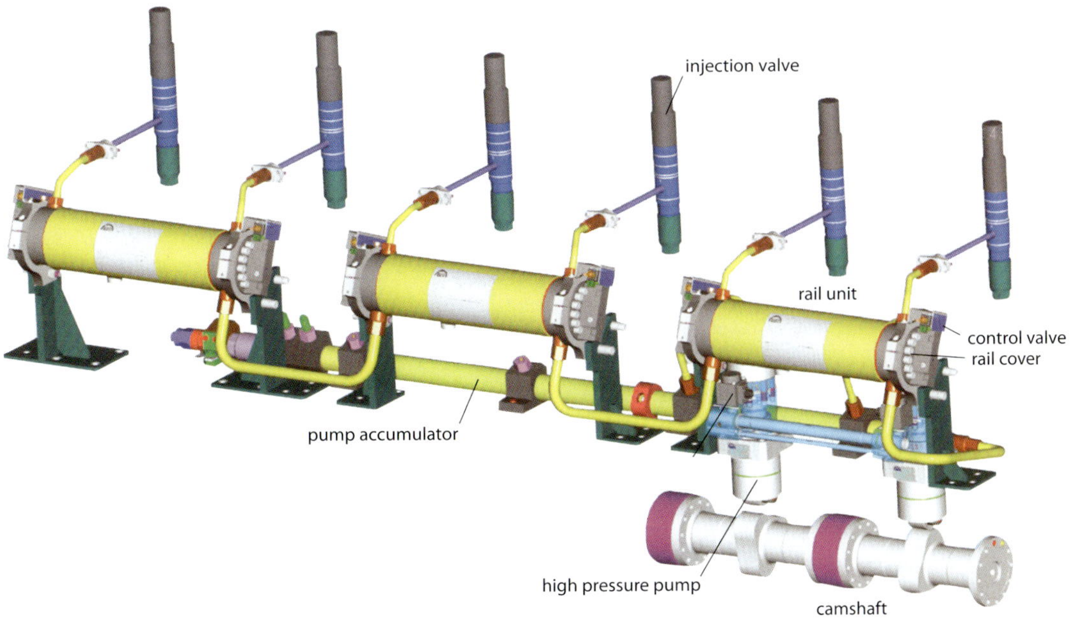

◀ The common rail system of a MAN Diesel four stroke medium speed engines operating on HFO.

Due to this, there is a complete common rail system for two cylinders. The camshaft now has cams with three sections, allowing the high-pressure fuel pumps to be operated three times per rotation of the camshaft. In this four-stroke engine, 1.5 times per revolution of the crankshaft.

▶

The common rail system of MAN Diesel for the four-stroke engines.

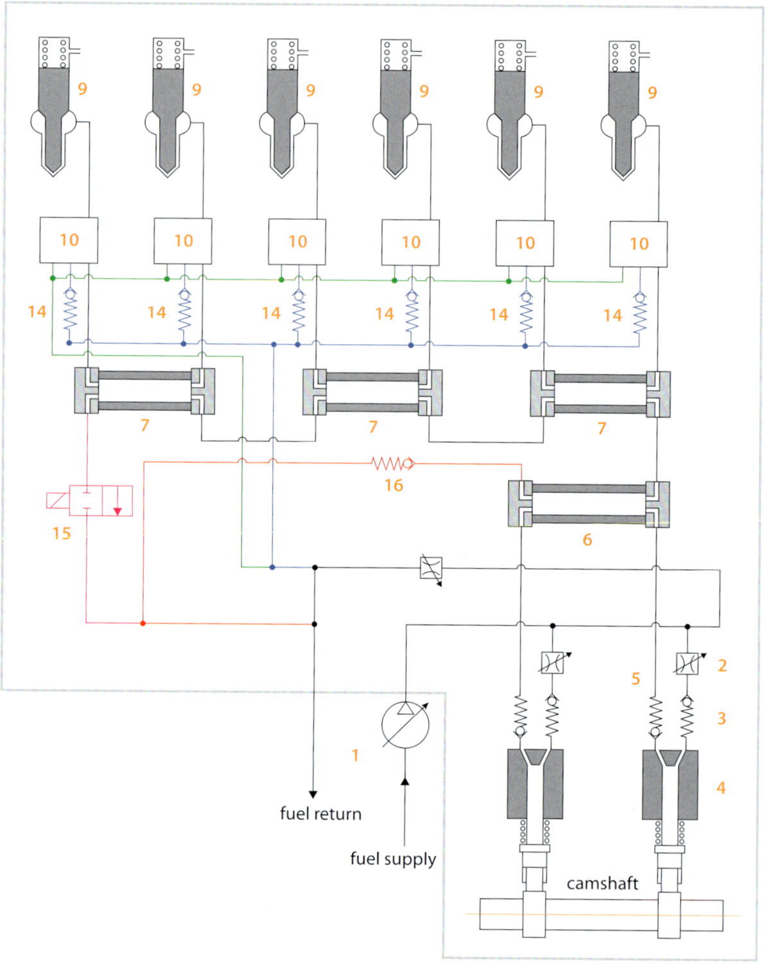

A low-pressure fuel pump (1) delivers the fuel via electromagnetic activated throttle valves (2) and suction valves (3) to the high-pressure pumps (4), which force the fuel into the pump accumulators (6) via pressure valves (5).

Each pump is connected to the pump accumulators (6) which collect the fuel delivered by the pumps.

From the accumulators, the fuel flow goes to the accumulator units (7), which are connected in series, the common rail. Both sides of the accumulator-unit cover have radial connections for the high-pressure pipes leading to the injectors (8) as well as for the connecting pipe to the next accumulator unit.

The unit itself does not contain any radial drillings and is therefore easy to produce and very resistant to high fuel pressures. The high-pressure pumps (4) are driven by a fuel cam on the camshaft. The delivery quantity of the high-pressure pumps is calculated by the electronic control system based on an evaluation of the fuel pressure indicated by the rail pressure sensor and the corresponding operating condition of the engine. The electro-magnetically activated throttle valve (2) in the low-pressure pipe regulates the fuel quantity supplied to the high-pressure pumps.

▶

The operation of the accumulator.

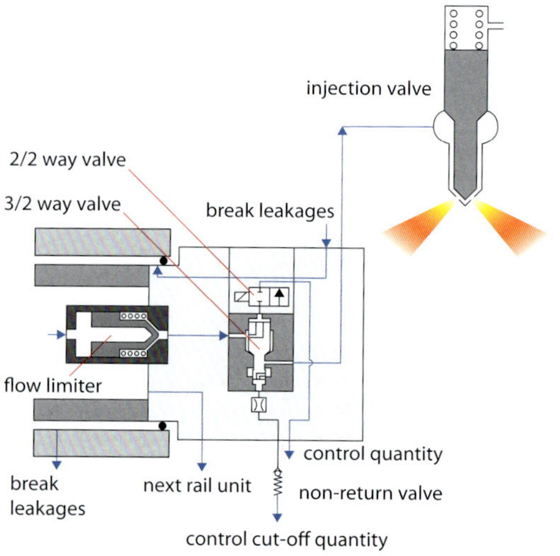

Every accumulator unit (7) contains in its cover components and connections, which serve for fuel supply as well as for fuel injection control.

On its way from the accumulator unit (7) to the 3/2-way valve and then to the injector, the fuel is passed through a flow limiter. A spring-loaded piston in this component carries out a stroke for each injection, which is proportional to the injection quantity and returns in its original position in the time between the injections. Should the injection quantity exceed a specified limit value, the piston in the 3-2 valve will be pressed to a sealing seat of the exit valve to the injector and thus avoid constant fuel flow to the injector.

The 3/2-way valve inside the accumulator cover is electro-magnetically activated by the control system. It permits the high-pressure fuel to be supplied from the accumulator unit, via the flow limiter, to the injector.

With this valve, it is possible to change pre- and post injection. When a specified pressure is exceeded in the system, a safety valve (16) opens. The high-pressure pipes and accumulator units are designed with double walls in order to prevent fuel leakage. In this case, the operating personnel will be warned by means of the float level switch. For start-up of the cold engine with heavy fuel oil, the high-pressure part of the injection system is heated by means of circulating hot heavy fuel oil. A flushing valve or circulation valve (15), can be opened pneumatically when the engine stalls or to keep the system at the correct temperature. The 3/2-way valve ensures that there is only pressure on the injector during injection.

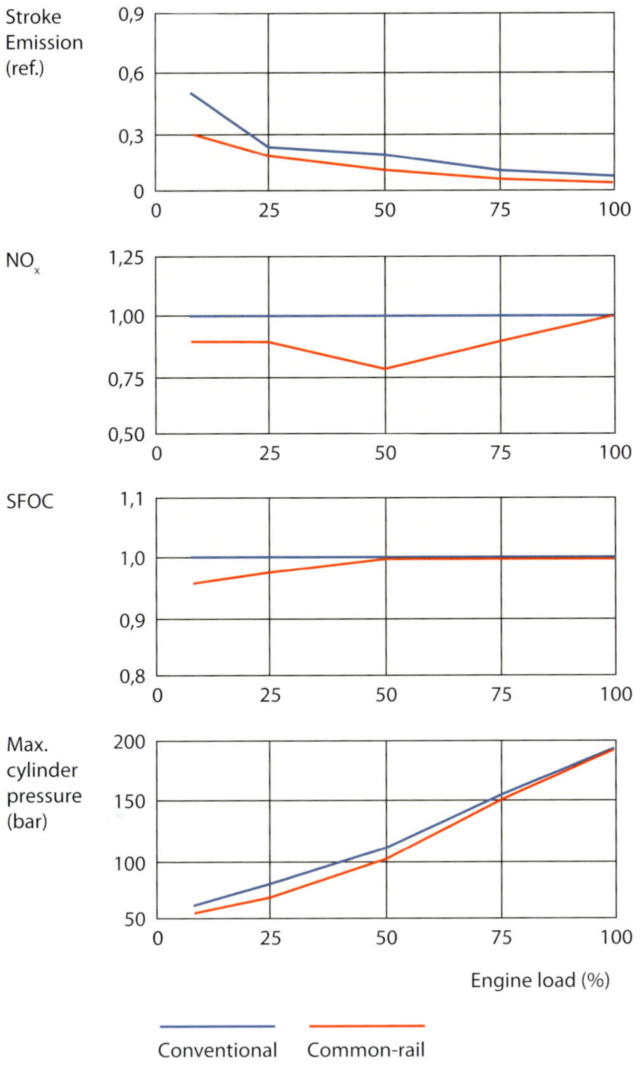

◀ Four graphs showing the changes with use of the common rail system in MAN Diesel four-stroke engines.

Horizontal: the engine load.

Blue line: conventional engines with a high-pressure fuel pump per cylinder.

Red line: engines with a common rail system.

Figure 1: vertical: the smoke opacity number, at partial load a great deal lower.

Figure 2: the nitrous oxide emissions in percent, at partial load much lower.

Figure 3: the specific fuel usage in percent, at partial load lower.

Figure 4: the maximum cylinder pressure in bars, at partial load slightly lower.

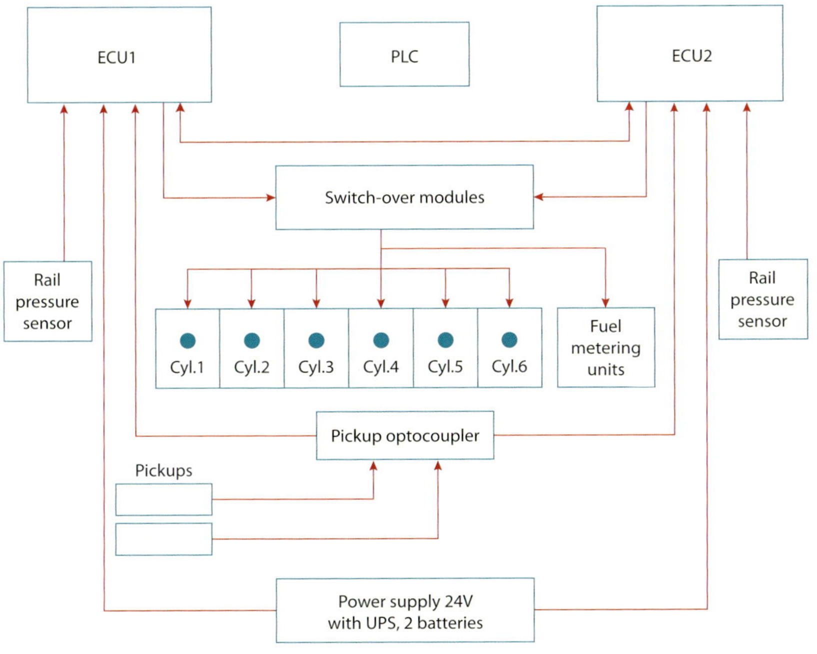

◀ The controls of the common rail system of MAN Diesel four-stroke engines.

The two electronic control units ECU1 and ECU2 are responsible for the solenoid valve control, the high-pressure pump control and therefore the speed governor. Each ECU controls half of the engine but is also able to control the complete engine. All necessary sensors, the power supply and all field bus connections are in double. Therefore, a single failure will never lead to an engine stop.

The PLC is only responsible for communication with the ship or diesel power plant control room, the alarm system and for the Man Machine Interface.

The electronic controls allow pre- and post-injection and of course control the fuel pressure of the common rail system. Due to this, optimal combustion occurs at every engine load.

9.18.2 Example 2: common rail systems

The Wärtsilä Sulzer RT-FLEX-systems for low-speed two-stroke crosshead engines, fuel HFO (engine category IV)

Note
There is no camshaft! There is a common rail system for lubrication of the exhaust valves. Two electrically driven pumps incorporated in the fuel pump's supply unit ensure that the system retains the correct pressure.

▶
The difference between the conventional high-pressure fuel system and the common rail system in Wärtsilä Sulzer two-stroke crosshead engines.

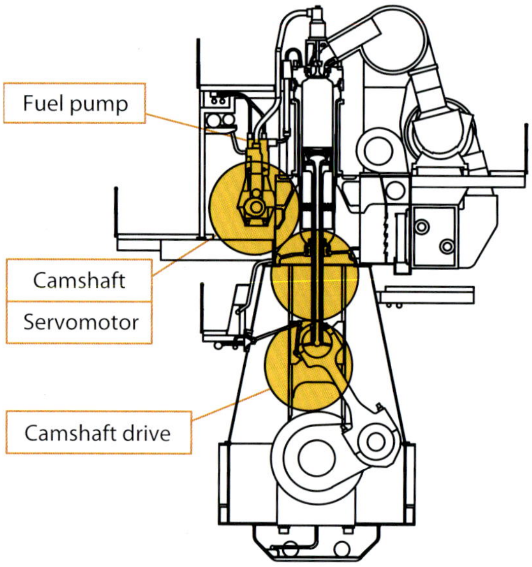

Sulzer RTA series.

Left: the conventional system with a camshaft driven high-pressure fuel pump on each cylinder. Between the crankshaft and the camshaft gear wheels.

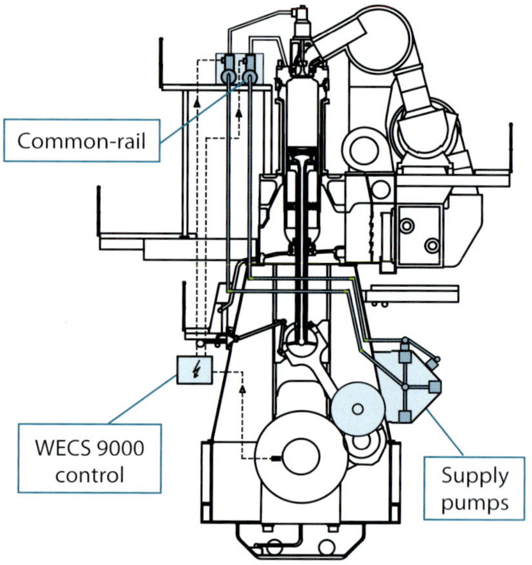

Sulzer RT-flex series.

Right: the common rail system with a crankshaft driven series of high-pressure fuel pumps in one block.
The fuel travels via the collection pipelines or the common rail to the injectors. An engine management system controls the supply of the fuel to the injectors. This system is known as the Wärtsilä Engine Control System (WECS) 9000, 9500 etcetera.

▶
A complete schematic diagram of the common rail system for the Wärtsilä RT-FLEX crosshead engines.

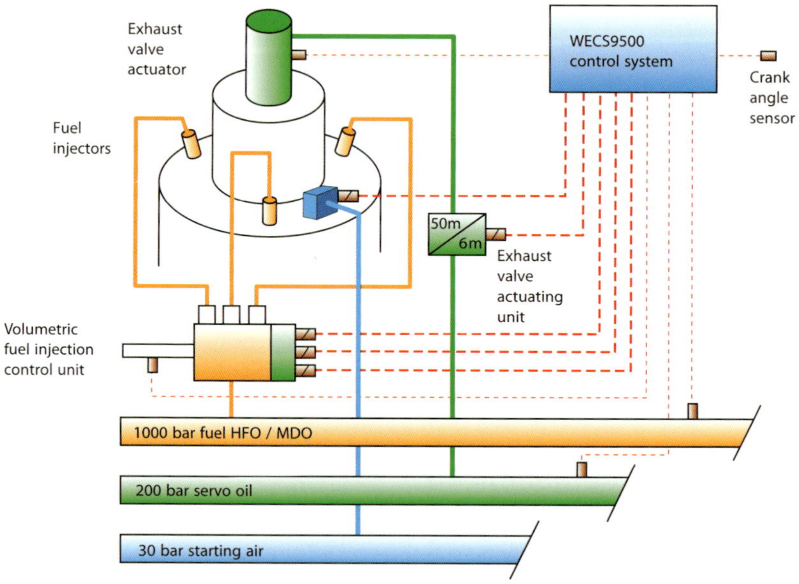

The system comprises the following components:
- A crankshaft driven high-pressure fuel unit and plunger pumps with adjustable yield.
- The camshaft for driving the individual fuel pumps and exhaust valve has been removed.
- The high-pressure fuel unit pumps into a large supply fuel line (common rail) that runs past all the cylinders. Pressure approximately 1000 bar.
- With a control unit, the fuel supply to the three injectors per cylinder is tuned to the requirements. The control unit is electronically driven by a WECS 9500-system (Wärtsilä Engine Control System) and hydraulically operated by a lubricating-oil system with a pressure of approximately 200 bar. This pressure is supplied by a separate pump system. The 200 bar line is also used for operating the central exhaust valve.

▲

The common rail system on a Wärtsilä Sulzer RT-FLEX96C two-stroke crosshead engine on board the containership 'P&O Nedlloyd Mondriaan'.

The space under the walkway for the cylinder heads is fitted out for the common rail system, the high-pressure lubricating-oil system for the operation of the exhaust valves, all appendages and control equipment.

◀

The common rail system on a Wärtsilä Sulzer RT FLEX-motor.

Advantages

- 'Smokeless' exhaust gases, no visible smoke.
- A potentially low rpm; the minimal rpm is approximately 10 to 12% of the nominal rpm. At this low rpm, one injector per cylinder is switched off.
- Lower specific fuel consumption at partial load and longer life of the parts (less soot formation at low loads).
- Easier engine-setting and thus reduced maintenance. The running settings determined at construction are automatically maintained.
- Reduced maintenance costs through precise fuel injection controls producing a more evenly balanced cylinder power output for the entire engine as well as a more balanced thermal load of all cylinders.
- The high-pressure fuel unit consists of a number fuel pumps for four-stroke engines that already have been in use for several years.
- High operational reliability, because even if one fuel pump were to fail the capacity of the remaining pumps can still easily supply a sufficient amount of fuel.
- The common rail is situated next to the cylinders on the upper level and is easily accessible for inspection and repair.
- The fuel supply to the three injectors is regulated at once by one control unit, but is separately checked per individual injector.
- The units regulate the time of injection, the amount of fuel and the type of injection.

Comments

- Free choice of the injection pressure.
- Evenly levelled pressure in the common rail and other high-pressure lines.
- The injector may be checked and if necessary switched off independently.
- The fuel system is completely separated from the lubricating-oil system of the engine.

Operation of the exhaust valves and starting air valves

Apart from regulating the fuel supply, the system also controls the opening and closing of the exhaust valves and the starting air valves. All valves can be controlled and adjusted individually.

▶

A high-pressure fuel division gate for the three injectors of one cylinder.

1 *common rail with a pressure of 1000 bar*
2 *division gate*
3 *high-pressure fuel line to the injectors*
4 *cylinder liner*
5 *valve control of the high-pressure fuel lines to the injector*

▶

The common rail system of a Wärtsilä Sulzer RT-FLEX engine.

1 *lubricating-oil filter*
2 *lubricating-oil pumps for the hydraulic system*
3 *high- pressure fuel pumps*
4 *local operation of the engine*

The high-pressure fuel unit and the high-pressure hydraulic system are positioned on the first floor of the engine. The high-pressure fuel unit is driven from the crankshaft, as is the high-pressure lubricating-oil system. In addition, the high-pressure part of the hydraulic system that with the help of electro motors is used in stationary engines to bring- and keep the system under pressure.

▶

A complete module for a Wärtsilä Sulzer RT-FLEX-motor.
Shown, the lubricating-oil pumps are built into the fuel module.

1 *high-pressure fuel pumps*
2 *high-pressure lubricating-oil pumps*
3 *adjustment spindle of the fuel pumps*

The fuel pumps originated from Wärtsilä four stroke engines.

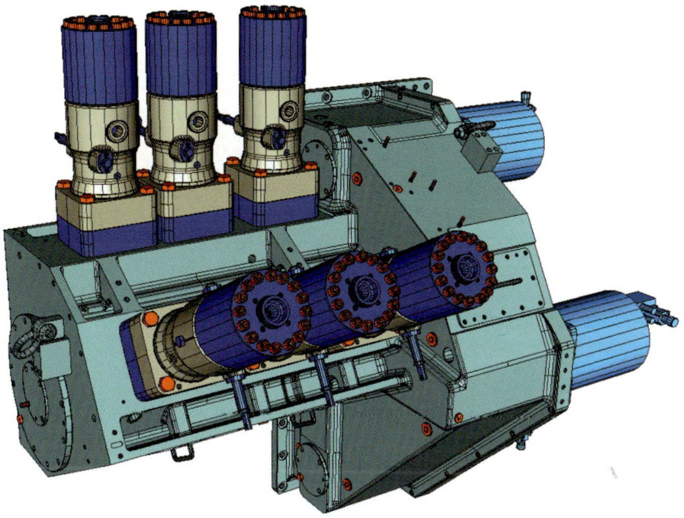

◀ A complete unit with left six high-pressure fuel pumps and right two high-pressure lubricating-oil pumps.

◀ The common rail system under the walkway of the cylinder heads.

1 *exhaust valve actuator*
2 *fuel injection controls*
3 *control pipe lines for the hydraulic system*
4 *discharge pipe lines for operating- and control oil*

◀ The common-rail system on a slow speed 12 cylinder Wärtsilä RTA-flex engine on the containership the 'P&O Nedlloyd Mondriaan' (now Maersk Line).

1 *high-pressure fuel supply line 1000 bar*
2 *high-pressure hydraulic-oil lines 200 bar*
3 *high-pressure hydraulic-oil lines for operation of the fuel to the injectors*
4 *valve block for injector fuel*
5 *high-pressure hydraulic-oil lines for operation of the exhaust valve*
6 *injector lines*

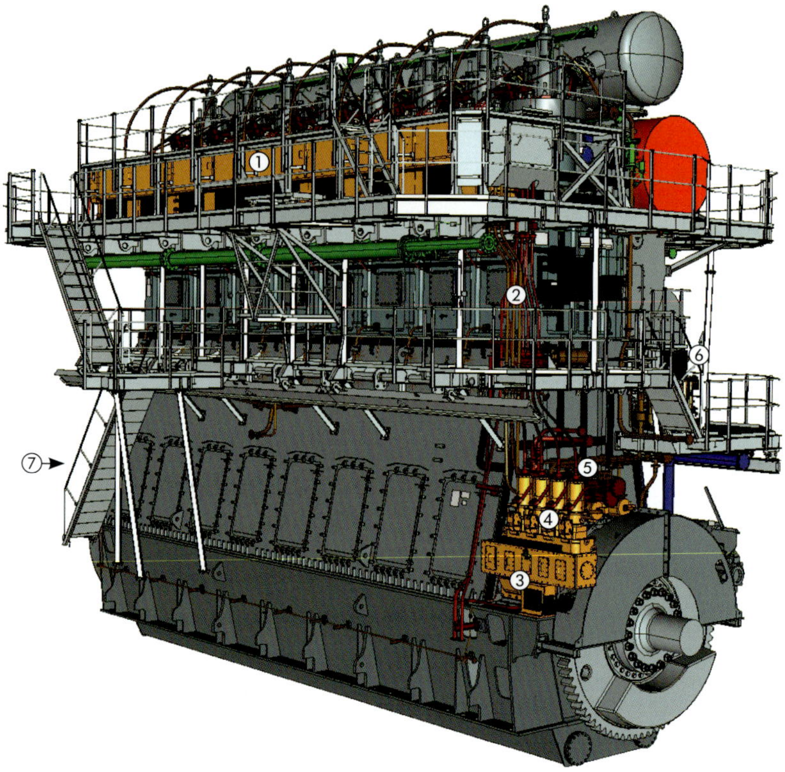

▶

A overview of the system.

1 *rail unit with electronic control units*
2 *supply lines for fuel and servo oil*
3 *pump block supply unit*
4 *fuel pumps*
5 *servo oil pumps*
6 *automatic filters for servo- and control oil*
7 *crank angle sensor at the free end of the crank shaft*

▶

The common rail system on a Wärtsilä Sulzer RT-FLEX two-stroke crosshead engine with twelve cylinders.

1 *high-pressure hydraulic servo oil lines (200 bar)*
2 *high-pressure fuel line (1000 bar)*
3 *fuel to three injectors*
4 *hydraulic servo oil lines for operation of the exhaust valve*

WECS 9500-system

The whole system is controlled by a Wärtsilä WECS 9500-system. It is a modular system with separate microprocessor control-units for each cylinder and a general control and management system with a second series of microprocessor control-units. The electronic speed regulation as well as the general monitoring system with alarm systems has also been incorporated in the WECS 9500-system.

▲

The air-start valves are electronically operated from the WECS 9500-controls system.

9.18.3 Example 3: new fuel systems

The high-pressure fuel system of low-speed two-stroke crosshead engines of the ME type by engine manufacturer MAN Diesel, fuel HFO (engine category IV)

This system deviates widely from the standard common-rail systems. The fuel pumps are hydraulically driven.

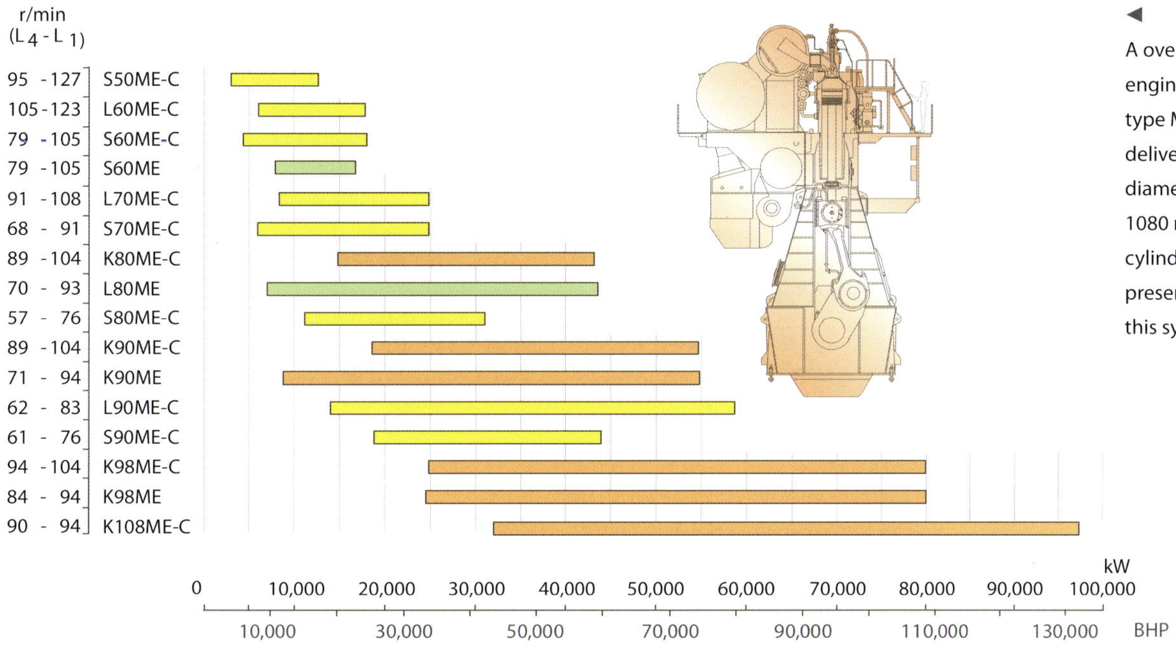

◄ A overview of the type of engines from MAN Diesel type ME-C that can be delivered. The cylinder diameter is from 500 to 1080 millimetre. Smaller cylinder diameters can at present be equipped with this system.

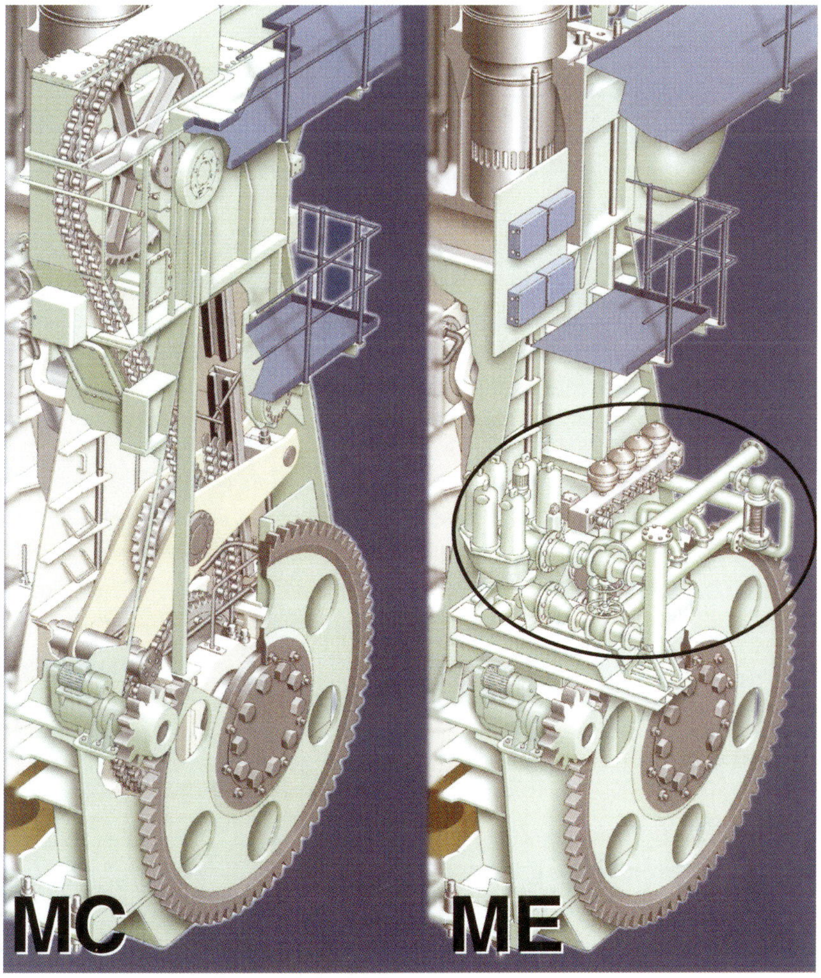

◄ The high-pressure fuel system of two-stroke crosshead engines type ME from engine manufacturer MAN Diesel.

Left: the normal version, with a chain driven camshaft and per cylinder a separate high-pressure plunger fuel pump (MC-version) Right: the ME-version, with electronic fuel adjustment. The camshaft with gear drives has been completely removed and in its place is a hydraulic high-pressure unit positioned above the crank shaft that supplies high-pressure lubricating oil for driving the fuel pumps and the operating the exhaust valves.

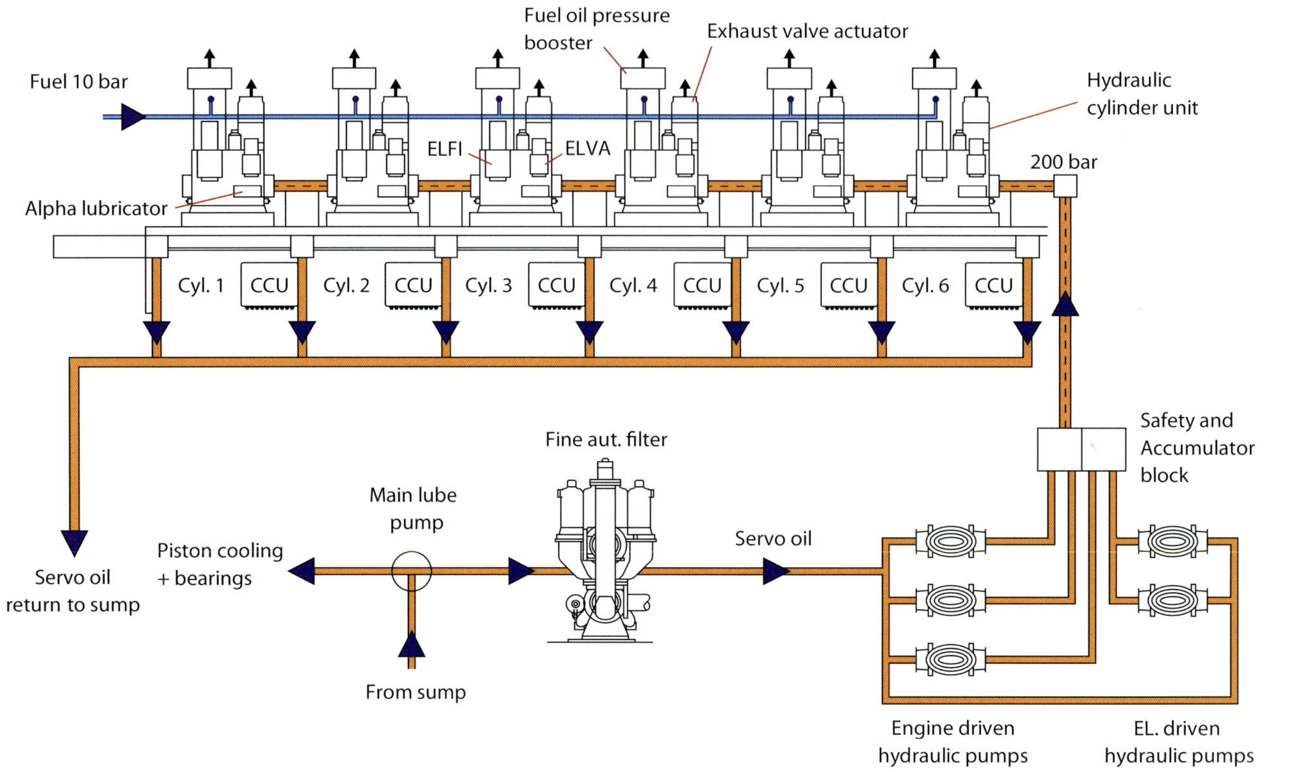

▼

The high-pressure fuel system of MAN Diesel type ME.

Left:: The is electronically driven fuel system where the high-pressure plunger pump is driven by high pressure lubricating oil. An accumulator prevents pressure pulses. Per cylinder, there are two injectors..

Right: the operation of the exhaust valve. The plunger pump of the exhaust valve is also driven with high pressure lubricating oil.

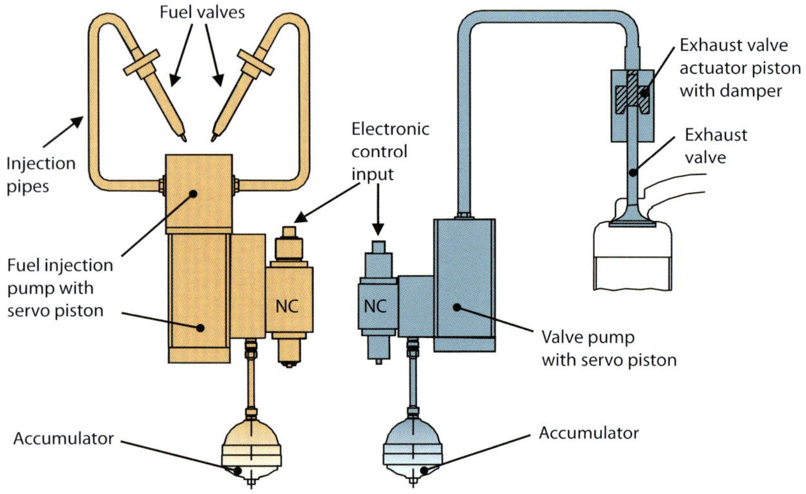

▲

The schematic diagram of the hydraulic system for the drive and controls of the high-pressure fuel pumps and the exhaust valve for MAN Diesel type ME two stroke crosshead engines.

The system comprises the following components.

- An engine driven set of hydraulic lubricating-oil pumps that supply a constant pressure of 200 bar for driving the fuel pumps and the operation of the exhaust valves. The pumps are furnished with lubricating oil from the main lubricating-oil system via an automatic fine filter. The discharge lines of the hydraulic system lead to the lubricating-oil discharge tank in the main engine.

  The fuel supply pressure is held at 10 bar.

ELVA Electronic Valve Actuator – an electronic exhaust valve operation

CCU Cylinder Control Unit – a cylinder control unit

ELVI Electronic Fuel Injection – an electronic fuel injection unit

Remark

The dosing of the cylinder lubrication is automatically adjusted to the engine's operating conditions.

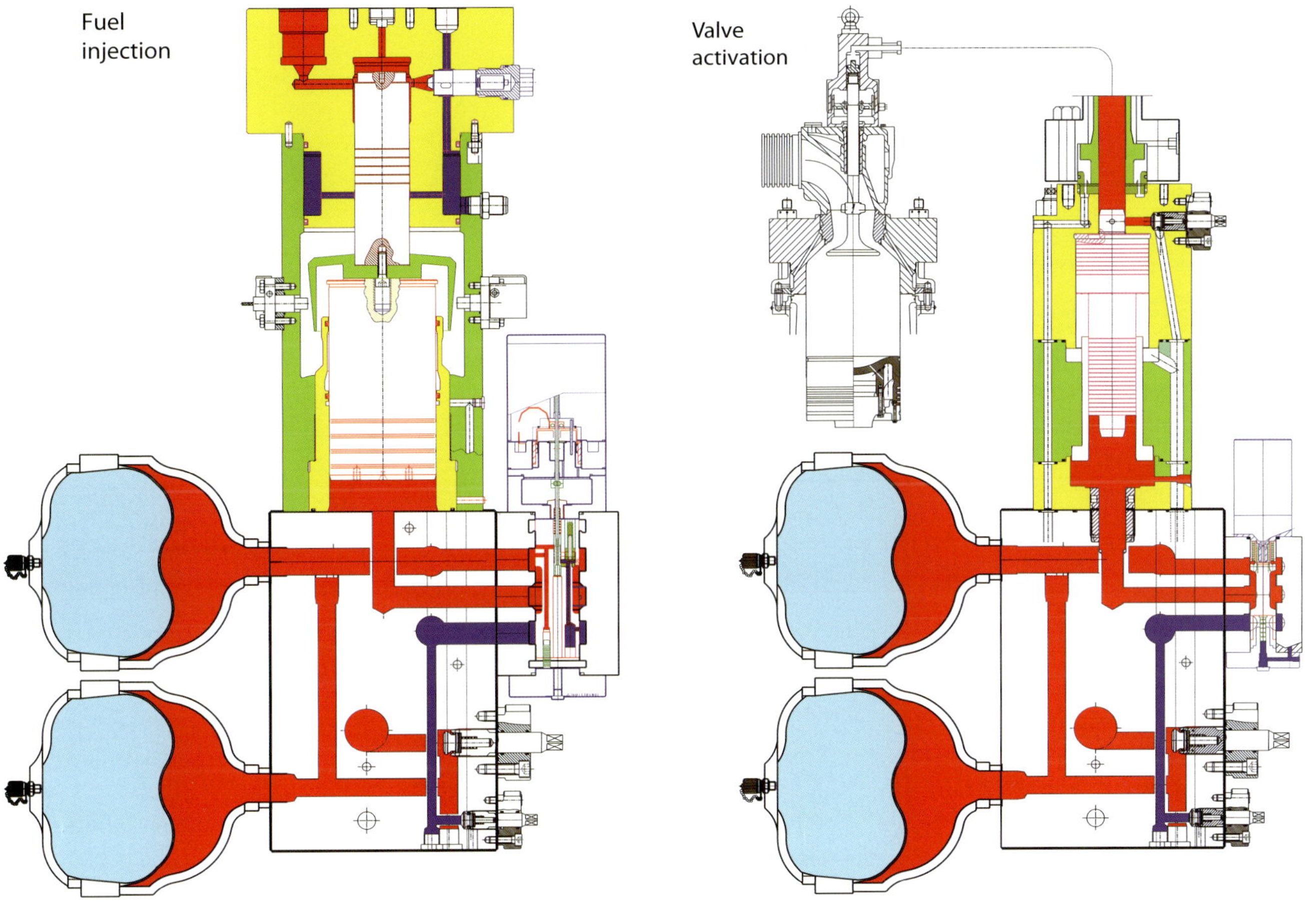

▲

The high-pressure fuel system of MAN Diesel.

Left: the fuel system.

Right: the operating system for the exhaust valves.

Advantages of the system

- Lower specific fuel consumption and performance improvement as a result of the adjustable electronic controlled fuel injection and exhaust valve timing at every load.
- The correct high fuel injection pressure and injection time at every load.
- An improved combustion process where the nitrous oxide emissions are decreased and the soot emissions fall dramatically, especially at a low engine load.
- It is simple to modify the various settings during operation with the process computer, the ECS, 'Engine Control System'.
- A simplified mechanical system with solid traditional fuel injection systems that are well known to engineers.
- A control system with a more accurate timing so there is a better balance between the cylinders and a more even thermal load.
- All engine data is displayed and saved and shows as to whether adjustments should be made. Through this good diagnose tool and can lead to the number of operating hours between maintenance being increased.
- Manoeuvring at lower revolutions is possible due to improved combustion processes in the cylinders at low loads.
- The advancing and retarding of the revolutions, reversing, starting and running astern allow for a faster change in the ship's speed in an emergency stop situation.
- The cylinder lubrication is incorporated in the electronic controls. This results in a reduction of the lubricating oil usage and therefore less contamination in the cylinder -and exhaust systems.
- It is simple to modify the software during the lifetime of the engine.
- The start air valves and auxiliary turbochargers are driven precisely at the correct moment.
- The total weight of the engine is decreased.
- Furthermore it is possible with this system to modify the timing of the opening and closing of the exhaust valve.

▶

Possibilities for modifying the timing of the exhaust valve.

Horizontal: crank degrees after top dead centre
Vertical: valve lift in millimetres
Purple: earlier closure
Yellow: later closure
Brown: earlier opening
Green: later opening
Black: normal line

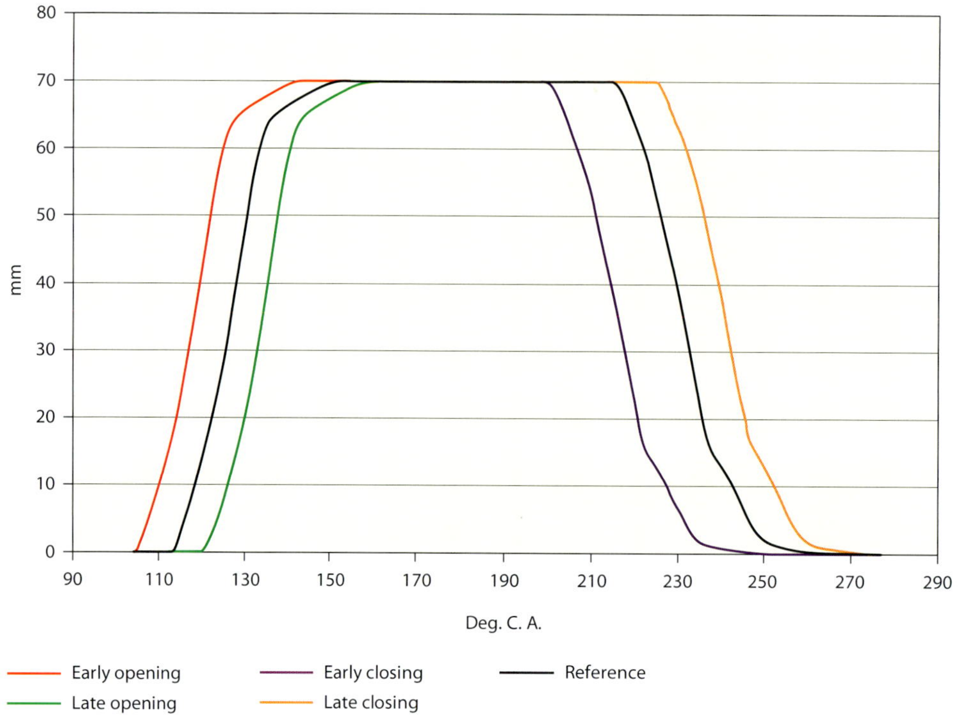

Remark

On average, the valve is for approximately 140 crank degrees open. An example of the possibilities of the ME-version.

▶

By modifying the fuel injection, the engine can be tuned to different requirements such as lower specific fuel consumption and lower nitrous oxide emissions. In the second case is the fuel injection later and in two phases.

Left figure: low specific fuel consumption
Red line: pressure curve in the cylinder
Blue line: injection pressure curve of the fuel
Right figure: the pressure in the cylinder during the entire process is smoother by allowing the fuel injection to take place in two phases. Due to this, the pressure curve is more gradual and lower in temperature.

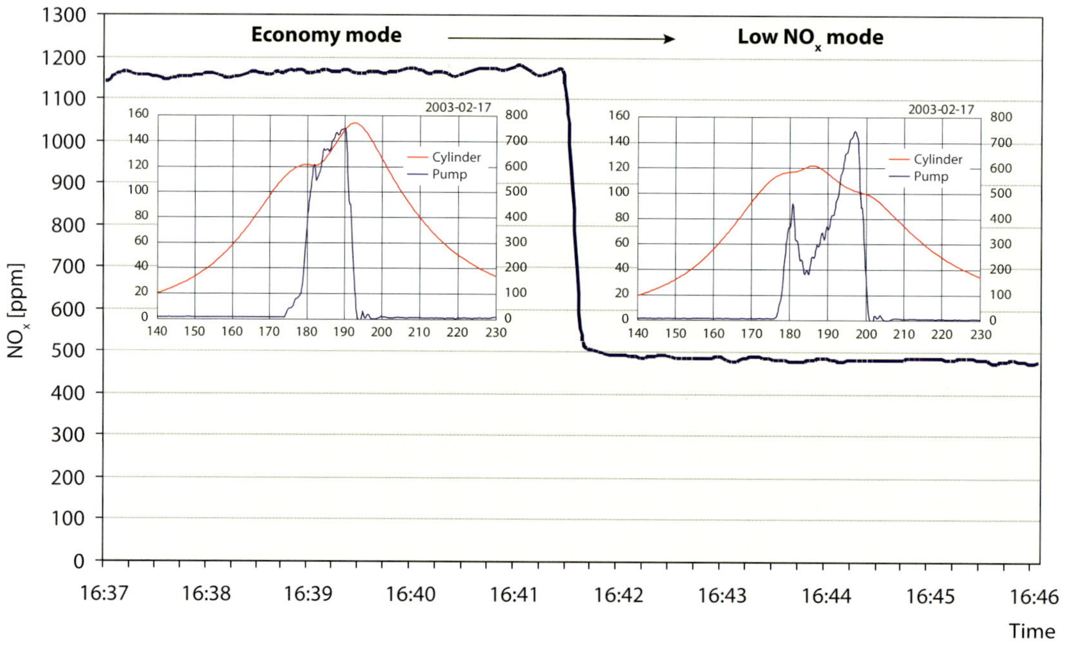

9.18.4 Example 4: common rail systems

Common rail systems for four-stroke medium-speed trunk piston engines by engine manufacturer Wärtsilä, fuel HFO (engine category III)

The principle is as follows: high-pressure fuel pumps force the fuel towards the accumulators that are connected to the electronically controlled fuel injectors of the two cylinders. The accumulators are connected with pipes to the common rail system. The fuel pumps are driven by the camshafts. As the timing of the fuel pumping is no longer tied to the timing of injection, it is now possible to realise **two pump cycles in one revolution of the crank shaft**. Consequently, the engine requires fewer fuel pumps: usually one high-pressure fuel pump for every two cylinders. A hydraulically driven control system is used to send the fuel to the injectors. The oil required is fed from the general lubricating-oil system of the engine and pressurised to 200 bar by a separate, pump driven by the engine.

The same oil is used for the start up and safety valve (SSV) that allows the fuel to circulate during preheating and as safety valve for the common-rail safety system. All functions are controlled by the engine management system. Apart from regulating and controlling engine speed, the system monitors the safety controls in the engine. The common rail system was developed for new engines, but the modular system can also be applied to engines that are still equipped with conventional fuel injection systems.

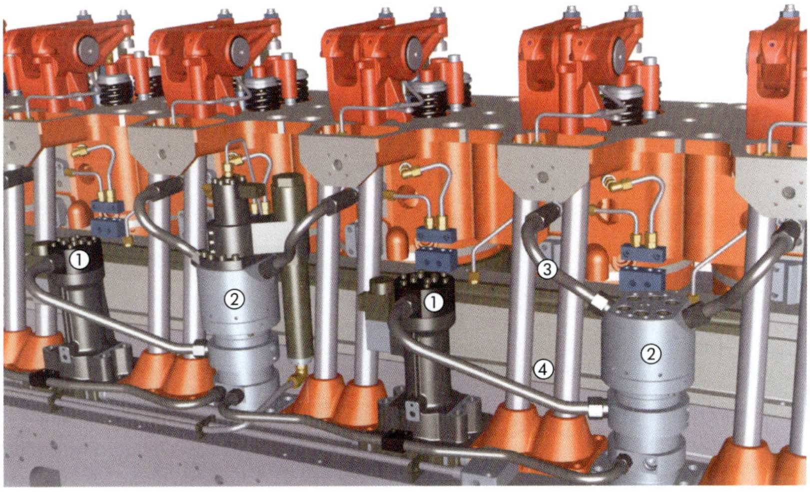

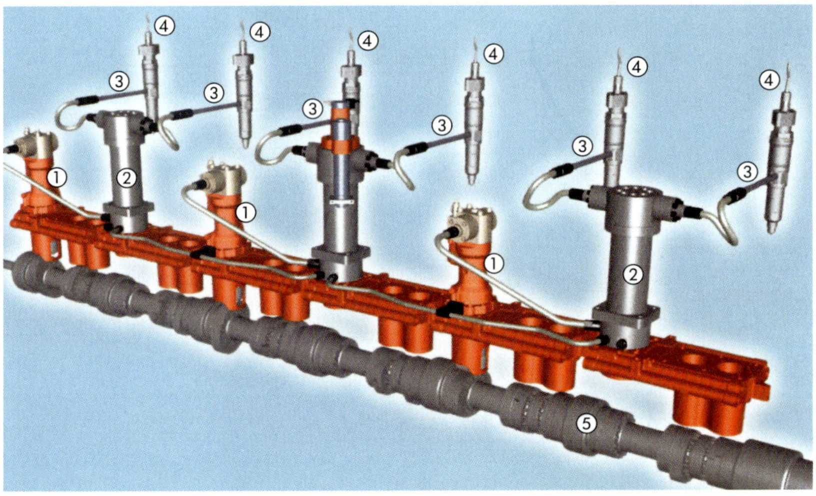

◀

The common rail system for Wärtsilä medium-speed, four-stroke engines, fuel HFO.

1 *high-pressure fuel pump*
2 *accumulator*
3 *high-pressure fuel line to the injector*
4 *injectors*
5 *camshaft*

▲

The common rail system for Wärtsilä medium-speed, four-stroke engines, fuel HFO.

1 *high-pressure fuel pump*
2 *accumulator*
3 *high-pressure fuel line to the injector*
4 *push rods for the inlet- and exhaust valves*

The high-pressure fuel pumps 1 are driven by the camshaft 5 and have sufficient capacity for two cylinders.
Every pump is linked with an accumulator 2 that levels the pressure spikes and supplies fuel to two cylinders.
The accumulators are linked together by a double walled pipe. Due to this, the pressure in all the accumulators is always equal and in emergencies, one or two high-pressure fuel pumps can be disconnected.

From the accumulators, the fuel with the required pressure is supplied to the injectors 4. The fuel valves for the injectors are hydraulically operated and electronically driven. Due to this, it is possible to change the timing and duration of the fuel injection in every cylinder. From a safety point of view, it is important that between two injections, the injector is unpressurized, so fuel leakage to the cylinders is not possible.
In a new injector design, the injector needle is opened and closed at full fuel pressure. Due to this, the atomisation of the fuel at every load and revolutions of the engine is optimal.
A consequence is a smokeless engine with low emissions of amongst other carbon monoxide.
Pressures: these can be adjusted between 900 and 1500 bar.

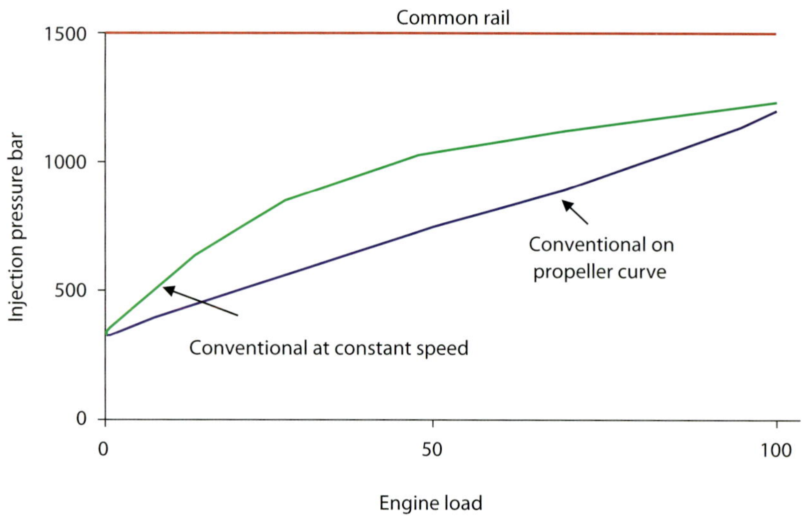

◀

The common rail system for Wärtsilä medium-speed, four-stroke engines, fuel HFO. Comparison to the conventional system with individual fuel pumps per cylinder.

Horizontal: the load of the engine.
Vertical: the fuel pressure in bars.
Above horizontal: the fuel pressure with a common rail system.
Curved green line: the fuel pressure at constant revolutions with a conventional system.
Straight blue line below: the fuel pressure at the propeller drive.

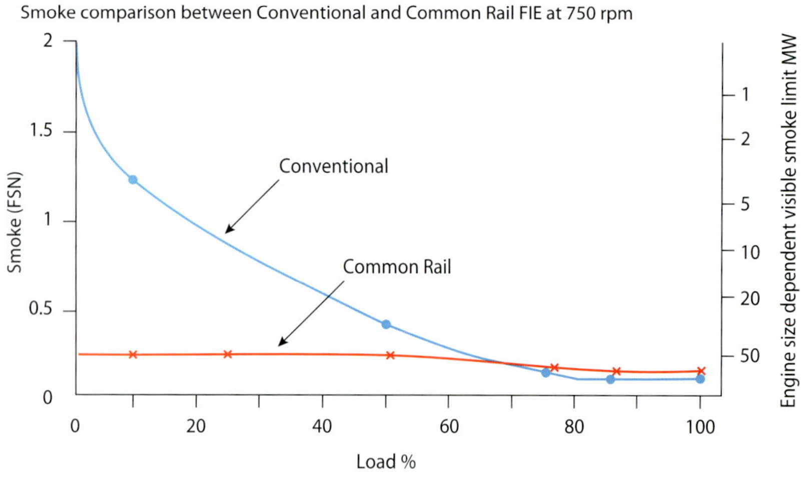

◀

The effect on the soot emissions of the common rail system from Wärtsilä for medium-speed, four-stroke engines, fuel HFO.

Horizontal: the load of the engine in percent.
Vertical left: the soot number.
Vertical right: the power output in MW of an engine, the soot number is dependant on this.
Red line: the soot emissions with the common rail system.
Blue line: the soot emissions with the conventional system.

From experience, it has been shown that the specific fuel consumption at partial load falls slightly.
Furthermore, with a common rail system every cylinder and so every individual combustion process can be adjusted for optimal performance. This limits the engine vibrations.

It is expected that these systems will remain on the market for at least another twenty years as the visible smoke can effectively be controlled with these new injection systems. It is most effective at partial load.

Regulating- and controlling system

The system has two functions:

1 to regulate the injection time and the amount of fuel injected;
2 to control the refilling and the pressure of the accumulators.

The control system also includes the monitoring of all the important data such as the pre-heating, the maximum allowable pressure in the system, the pressure in case of an emergency stop etcetera. So far, three types of four-stroke engines have been equipped with the common rail system, namely the Wärtsilä 32, 38 and 46 engines.

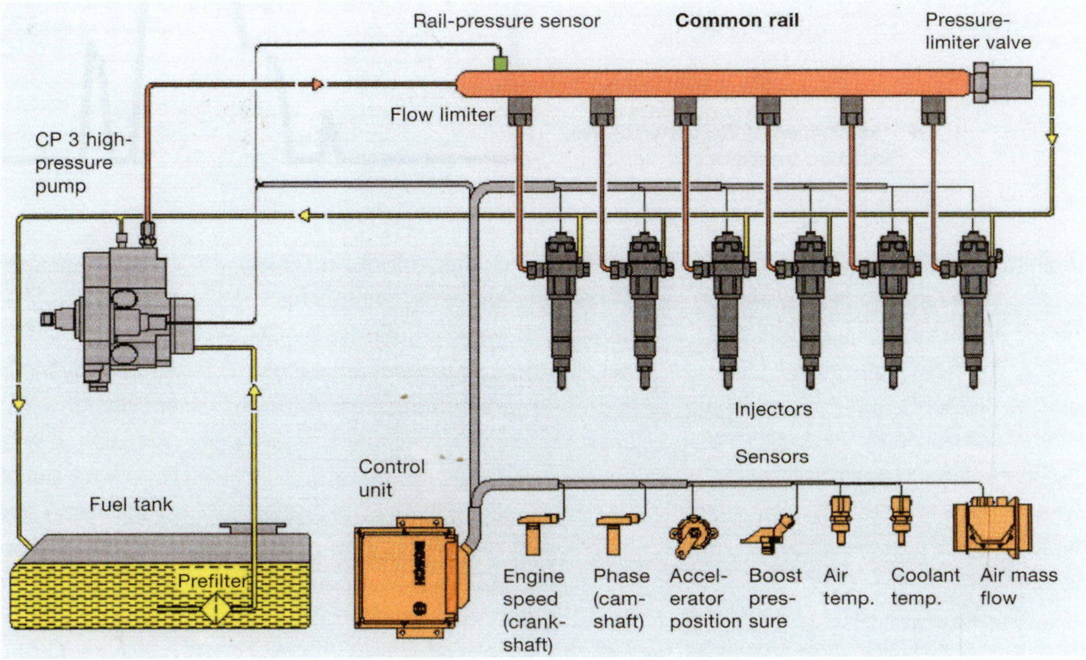

◀

For MAN trucks, the Bosch common rail (CR) system is used.

The system comprises of a fuel tank where the high-pressure, multi-plunger pump draws up the fuel and delivers it via a spacious fuel supply line (CR) to the injectors. The plunger pump is driven by the pressure sensor on the rail.

A safety valve ensures that the maximum allowed pressure is not exceeded.

There are restrictions in the supply lines to the injectors that guarantee the maximum flow capacity.

The actual fuel supply to the injectors is controlled with electronically controlled magneto valves. Here, there are also restrictions for the supply and discharge of fuel.

9.18.5 Example 5: common rail systems

Common rail systems for high-speed four-stroke diesel engines, fuel MDO (engine category II)

Many engines in this category have been equipped with common-rail systems. This modern fuel injection system was developed for the automotive industry and is in now applied in navigation and diesel power plants due to the increasingly strict requirements with regard to the emission of toxic substances in exhaust gases.

In principle, these systems are very similar and all have the same objective:

1 Reduce the emission of pollutants at varying loads and numbers of revolutions.
2 Reduce the specific fuel consumption at varying loads and numbers of revolutions.

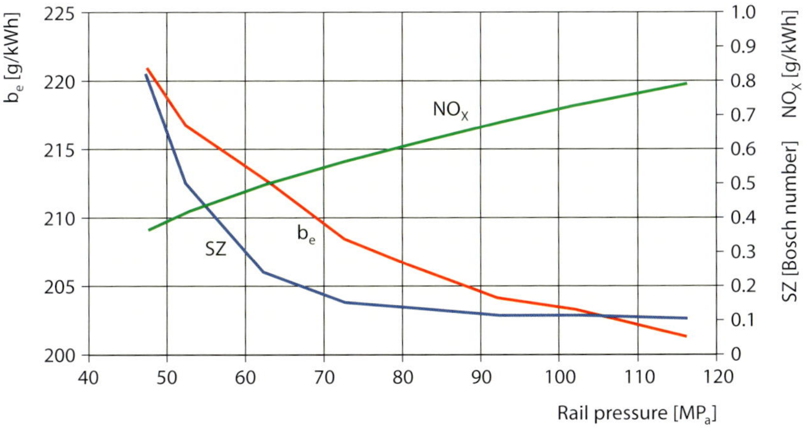

◀

The effect of the common rail pressure on the soot number SZ (Bosch), the specific fuel consumption b_e (in grams per kWh) and the nitrous oxide emissions NOx (in grams per kWh). The common rail pressure is in Mega Pascal.

Clearly shown is that at a fuel pressure increase, the soot number and the specific fuel consumption decreases to a large extent due to improved combustion and that the nitrous oxide content increases because of the higher maximum combustion temperatures in this improved combustion.

▶

The fuel pressure in the common rail system with respect to:
horizontal: the revolutions of the engine in revolutions per minute;
vertical: the torque of the engine in Newton meter.

The fuel pressure of the common rail system varies from 1150 to 1290 bar; this is a difference of approximately 11%.

In a conventional system, this varies from 1250 bar, dark blue to 110 bars, light blue to at the lowest load 70 bar, yellow. In the yellow area the engine is running at a very low partial load and the fuel pressure is so low that the atomisation and therefore combustion is poor.

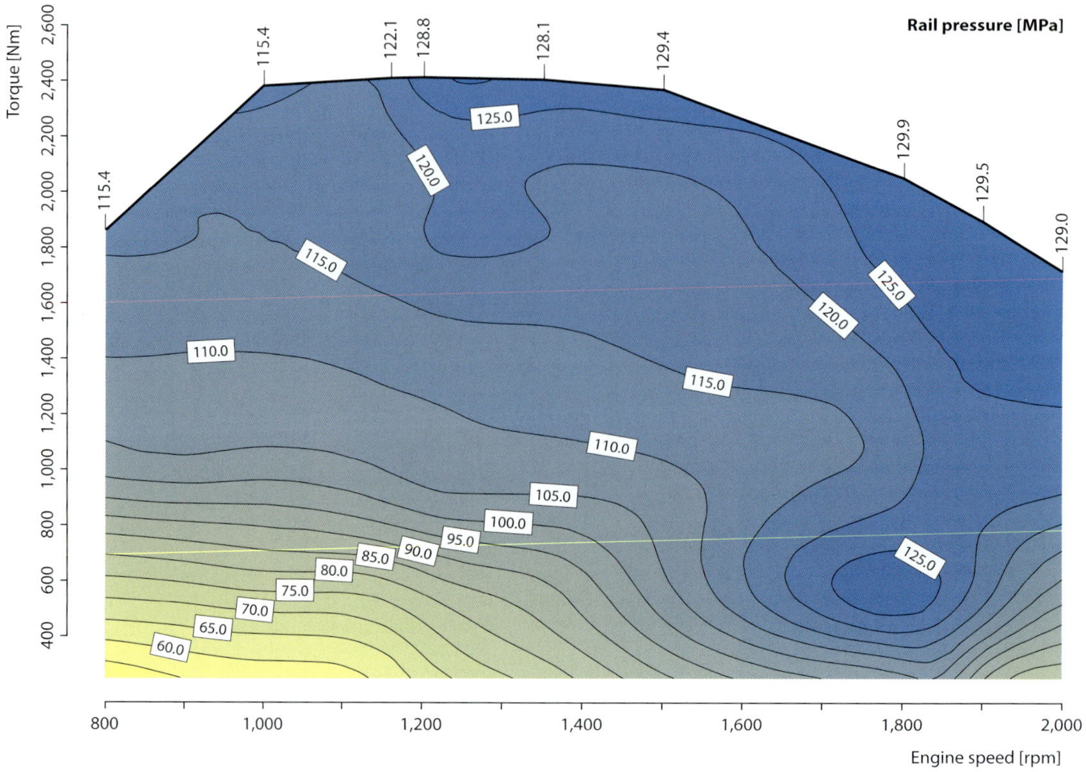

▼

In truck diesel engines, all the systems are used for recirculation of the cooled exhaust gases.

Due to this, the maximum combustion temperatures are disproportionately lower, so less nitrous oxide is produced. As from 2007, this system is an option for all engine categories. The normal maximum rail pressure for this system is 160 MPa or 1600 bar.
For the Euro 3 standards, this is sufficient. The Euro 4 standards can only be met with a pressure of approximately 180 MPa. Pressures of 200 MPa will be necessary in the future to meet the increasingly strict emission standards.

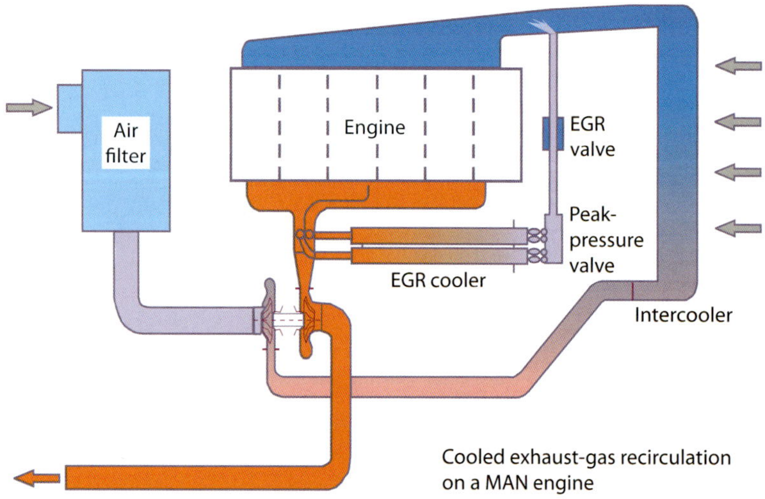

Cooled exhaust-gas recirculation on a MAN engine

There are a few important details:

- Electronic engine management systems regulate the time and duration of the fuel injection.
- The fuel injection pressures are extremely high and constant and ensure adequate atomisation under all circumstances.
- The high fuel pressure can be produced by numerous different pump systems such as in-line pumps and block pumps as well as injector pumps, mechanically driven from the camshaft by a push rod and a lever.
- The development started at the end of the last century and continues to develop rapidly for application in all diesel engine categories.
- By increasing the fuel pressure, the emission of soot particles diminishes noticeably, but due to the improved combustion at higher temperatures the production of nitrous oxides increases. With the application of exhaust gas recirculation with an inter cooler this effect is offset.

9.18.6 Example 6: common rail systems

The Cummins common rail system for high-speed four-stroke diesel engines, fuel MDO (engine category II)

This system also uses a high-pressure pump unit that maintains the fuel at the correct pressure in the widespread supply line to the injectors. The fuel supply to the injectors is electronically controlled from a central computer.

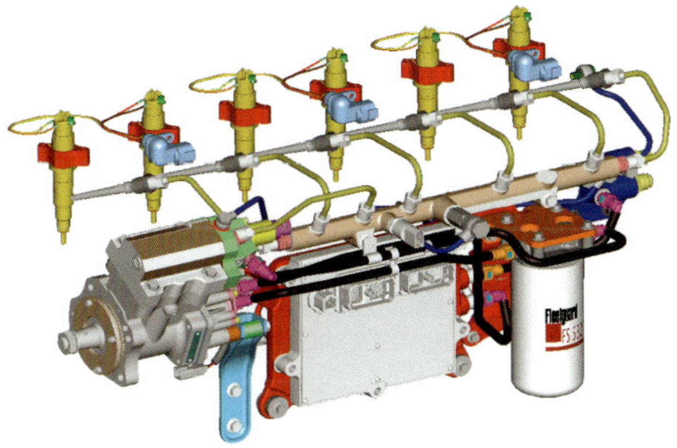

◀ The Cummins common rail system.

These are high speed, four-stroke engines running on MDO.

9.18.7 Example 7: common-rail systems

The Caterpillar common-rail system for high-speed four-stroke diesel engines, fuel MDO (engine category II)

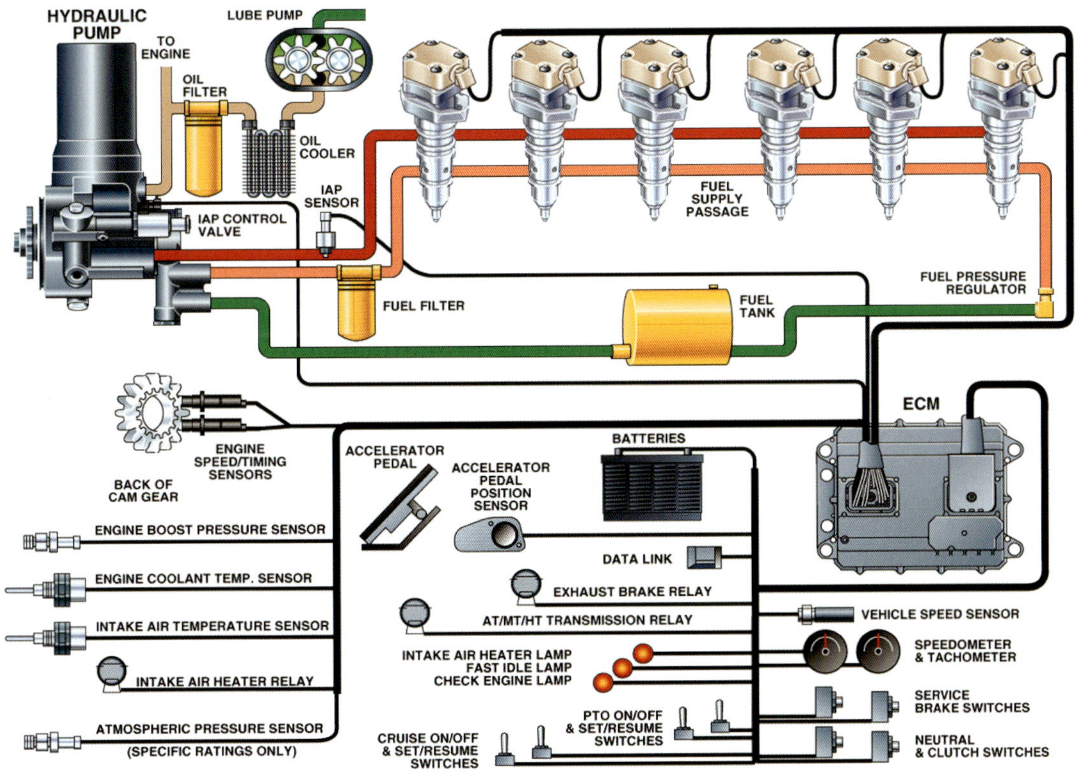

◀ The Caterpillar common rail system.

These are high-speed, four-stroke engines running on MDO. The high-pressure fuel pump is driven by high-pressure lubricating oil supplied by a separate pump from the main lubricating-oil system of the engine.
Green: low-pressure suction- and discharge lines.
Red: high-pressure supply and common rail lines.
Light red: discharge lines with an adjustable minimal pressure.
Brown: normal pressure lines of the regular lubricating-oil pump.
ECM: Engine Control Module.

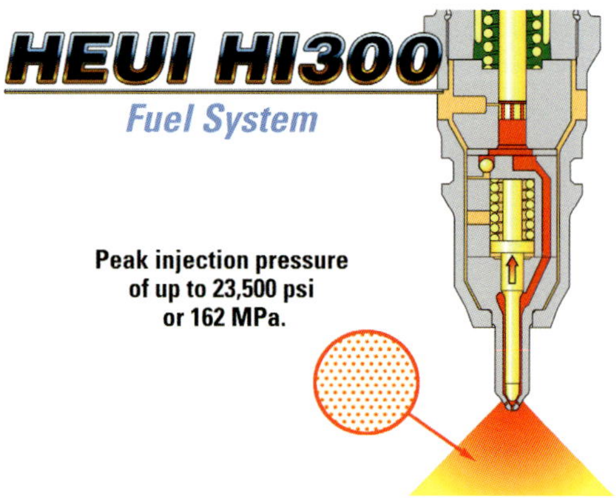

▲

The injectors ensure that the fuel is injected in a very fine spray into the cylinder.

In the common rail systems of Caterpillar, the fuel injection pressure can increase to 162 MPa or 1620 bar.

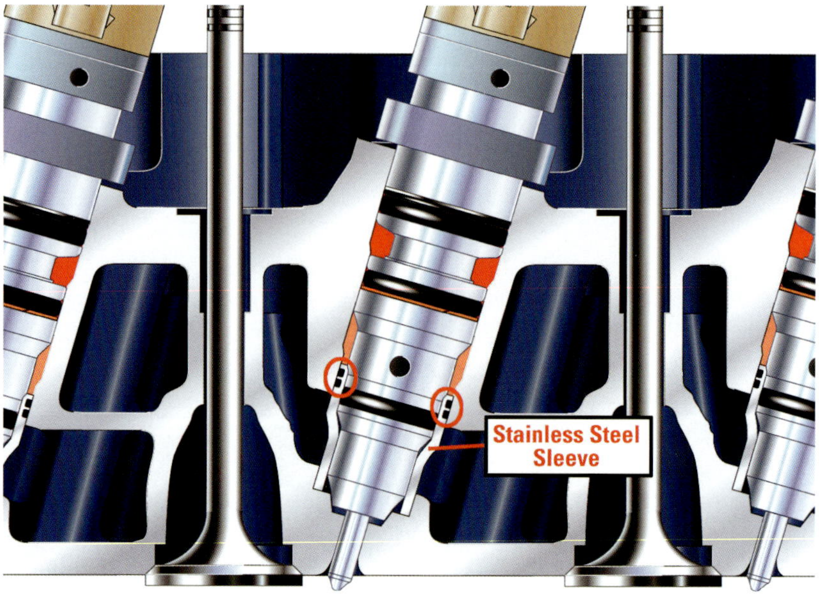

▲

The position of the injector in the cylinder head.

The lowest part is intensively cooled with coolant.

9.18.8 Example 8: common-rail systems

The common-rail system by MTU for high-speed four stroke diesel engines, fuel MDO (engine category II)

▶

The common rail system of MTU.

This common rail system has together with a suction filter for the engine driven high-pressure fuel pump, a low pressure-suction pump and a fuel cooler. The common rail pipelines have a diameter as such that pressure waves are limited.

▼

The fuel system for fast speed four stroke MTU-diesel engines, fuel MDO.

1 Non-return valve
2 Pressure relief valve

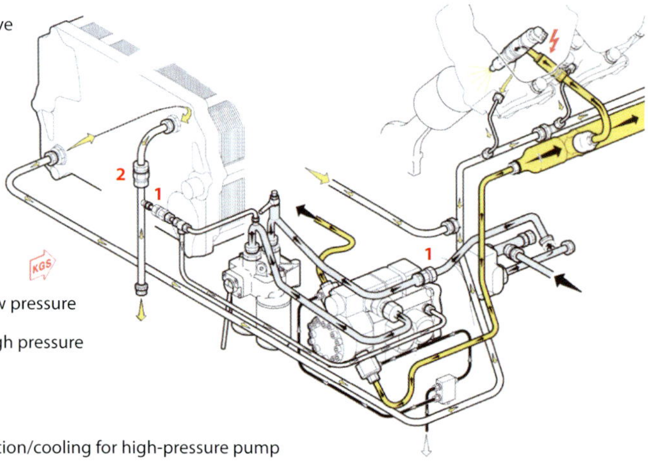

Fuel supply, low pressure
Fuel supply, high pressure
Fuel return
Leak-off fuel
Return, lubrication/cooling for high-pressure pump

▲

The fuel filters of a MTU-diesel engine.

1 *fuel filters*
2 *lubricating-oil filters*
3 *fuel pump*

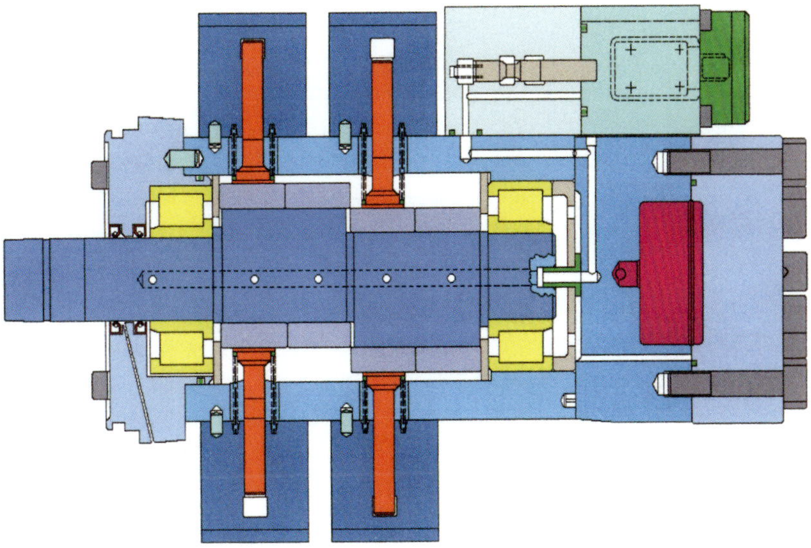

The multiple plunger pump for the MTU common rail engines.

The parts of the shaft that drive the plungers are asymmetrically placed with respect to of the shaft drive.

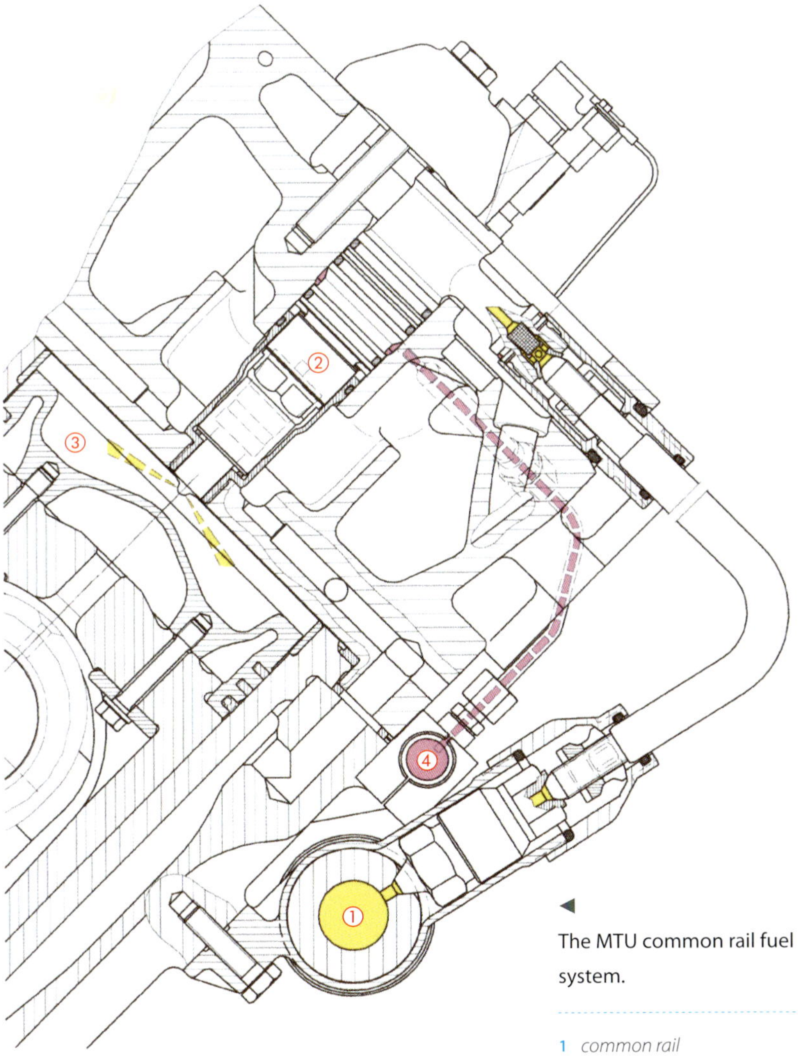

The MTU common rail fuel system.

1 *common rail*
2 *high-pressure fuel injector*
3 *combustion chamber*
4 *discharge fuel lines*

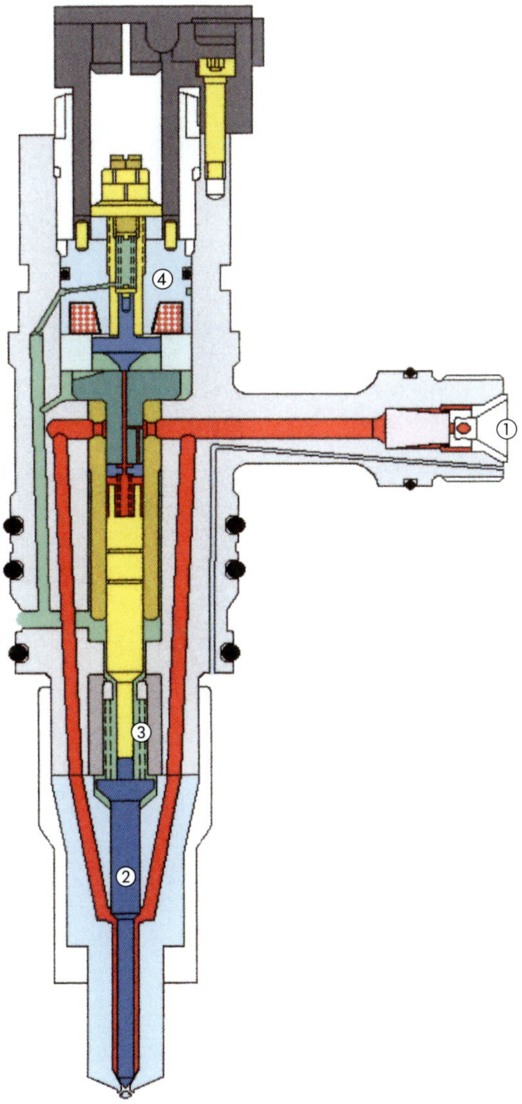

A cross-section of the injector.

1 *fuel supply*
2 *injector needle*
3 *injector needle spring*
4 *solenoid valve*

9.19 Injector system

In this system, the high-pressure fuel pump and the injector have been placed in one casing. The fuel injector is situated in the cylinder and is mechanically driven from the camshaft, as are the inlet- and exhaust valves by means of a fuel cam, guide pulley, push rod and lever.

9.19.1 Mechanically driven injector by Cummins-diesel engines

High-speed four-stroke engines, fuel MDO (engine category II)

The PT-pump is a low-pressure system, self-compensating by wear and tear, easy maintenance and repair. When compared to the conventional system there is relatively little risk of leakage in the PT-fuel system because the pressure difference between the pump and the injector is low. Cummins has designed the fuel supply in such way that it runs through internal ducts in the cylinder heads to the injectors, thus further reducing the risk of leakage. Although the fuel pressure in the PT-system is low, the injection pressure is much higher than in ordinary systems. This is achieved by the mechanical movements of the injectors produced by the push rod, rocker and a separate cam on the camshaft for each cylinder. The high injection pressure effects efficient fuel consumption that results in low exhaust gas emissions, without requiring a complicated combustion space with high whirling as other engines do.

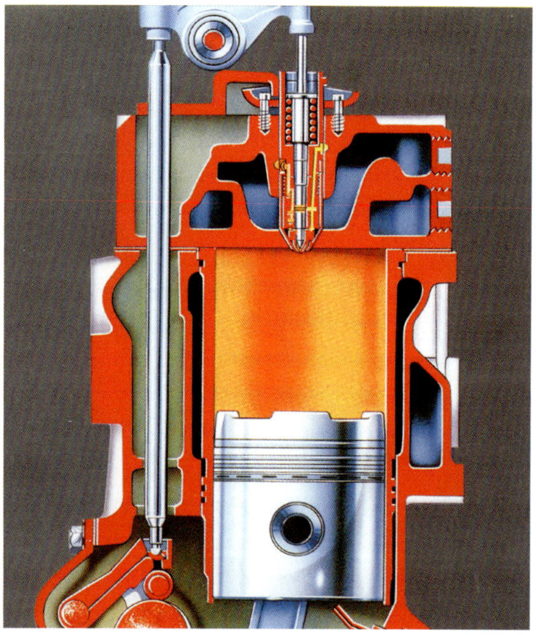

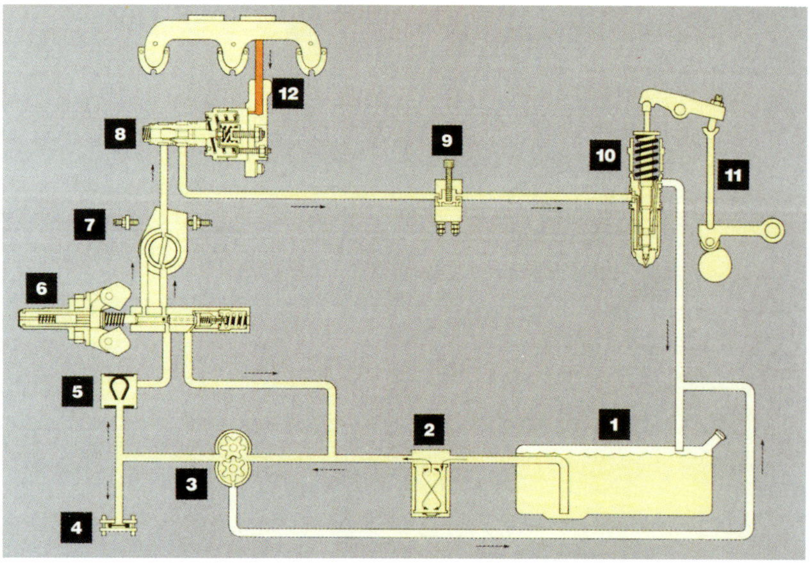

▶ The mechanically driven injector for Cummins-diesel engines.

1 *fuel tank*
2 *fuel filter*
3 *gear wheel pump*
4 *vibration damper*
5 *filter with magnet*
6 *governor en pressure control parts*
7 *throttle lever*
8 *air-fuel control*
9 *shut-off valve*
10 *injector*
11 *rocker arm, push rod and cam shaft follower*
12 *air from the exhaust manifold*

Fuel regulation

A gear pump in the fuel unit draws fuel from the tank through a filter and forces this into the pump. The fuel flows from the gear pump through a combined speed- and pressure regulator, then through a throttle valve and a shut off valve to the injectors. A central supply line transports the fuel to all the injectors of an in-line engine and to each bank in V-engines. This fuel line provides equal fuel pressure to each injector, and an evenly distributed amount of fuel and power output to cylinders. The fuel circulates through the injectors for the injection, the cooling and lubrication of the system as well as the bleeding. The superfluous fuel is returned to the tank by a return pipe. The warm return fuel sent back to the tank, improves the flow of the fuel, which is advantageous in a cold climate.

Fuel pumps function

Fuel consumption in individual injectors depends on the pressure in the supply line. This pressure is fully controlled by the fuel pump.
Therefore, it is not necessary to set the timing of the pump, since it only has to provide the specific pressure and the flow to the injectors as a function of revolutions, load and the position of the throttle control.

Operating under low pressure

Since the high injection pressure is built up mechanically in the injector, the fuel pump only has to ensure that fuel is supplied under relatively low pressure and in the correct quantity. This is the reason that the pressure in the fuel lines is lower than 17.5 bar as opposed to a pressure of 700 bar and higher required for systems with hydraulically functioning injectors.

Injector function

The Cummins-injector provides for both the dosage and the injection of the fuel. The fuel pressure at the supply port and the time the supply port is open is determined by the camshaft velocity, it measures the quantity of fuel that then flows to the so-called cup. The downward movement of the injector plunger subsequently injects the fuel from the cup into the cylinder of the engine.

Optimum timing in the STC-injection (Stepped Timing Control)

This facility is applied to special engines with a high power output in order to obtain optimum performance and economic fuel consumption.

Cummins-PT-fuel injector

Dosage and injection

The injector plunger movement is produced by the camshaft, a push rod and a rocker combination. Therefore, all that is required during regular maintenance inspections, is that the rocker be put under a certain torque in order to tune the injector, similar to tuning the valve setting.

Dosage operation

The working of the dosage is shown below. When the plunger is in top position the fuel flows from a calibrated opening into the cup under the injector plunger.

Injector calibration

The amount of fuel in the reservoir is dependent on the pressure at port B, the diameter of the port and the time that the calibrated port C is open. The diameter of port B can be increased so that each injector capacity can be correctly calibrated. In this manner, new or repaired injectors suitable for an engine can be installed in a cylinder, without having to replace or recalibrate the other injectors. Due to the numerous port variations, each basic fuel injector can be used for various engine types.

Two-part injector

The injector consists of two elements, the cup and the cup holder. The space between the cup holder and the cup is sufficiently large to centre the cup on the injector plunger. This self-centring ensures a meticulous alignment of the injector plunger in de cup.

Fuel preparation

The fuel intake (in the correct composition) in the cup occurs at the end of the inlet stroke. During the compression stroke, the fuel is put under a compression pressure by the plunger and heated air flows though the injector orifices in the cup.

▼ Operation of the Cummins injector.

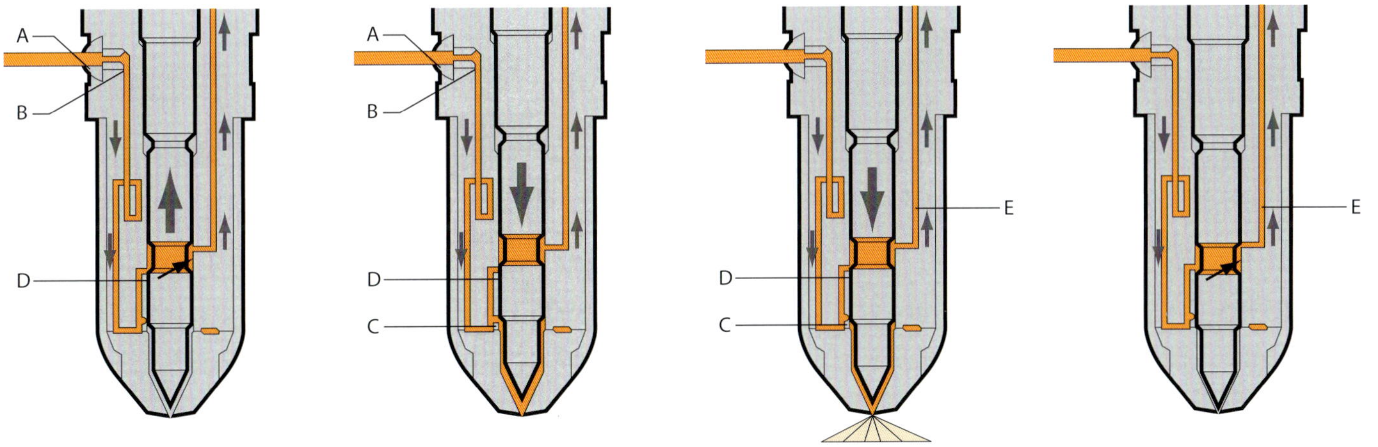

The start of the upwards stroke (the fuel circulates).
The fuel is supplied under low pressure to the injector (A) and flows through the inlet port (B) via internal channels through opening (D) back to the fuel tank. The amount of fuel that flows through injector depends on the fuel pressure before the inlet port (B). The fuel pressure is dependant on of the engine revolutions, the speed governor and the gas valve.

The finish of the upwards stroke (fuel enters the injector cup).
When the injector plunger rises, the calibrated opening (C) opens and fuel flows into the injector cup. The amount of fuel is determined by the fuel pressure. Passage (D) is blocked and temporarily stops the fuel circulation, so that the calibrated opening (C) in not influenced by pressure fluctuations.

The downwards stroke (the fuel injection).
When the plunger moves downwards and the calibrated opening (C) closes, the fuel supply to the cup is stopped. When the plunger travels further downwards the fuel present is compressed with a high pressure out of the cup through small apertures, so atomisation occurs. This leads to a complete combustion of the fuel. When the passage (D) opens again due to the constriction in the plunger, the fuel starts to flow through channel (E) back to the fuel tank.

The complete downwards stroke.
After injection, the plunger remains under pressure in the cup until the following fuel dosage and injection occurs. As no fuel can enter the injector cup, the fuel flows freely through the injector back to the tank through channel (E). This results in injector cooling and the warming of the fuel in the tank.

Injection

The injector plunger starts its downward movement before the compression stroke is completed, but injection only commences when the pressure of the air/fuel mixture in the reservoir exceeds the compression pressure in the combustion space. Then pre-injection takes place and subsequently the main injection. As the downward movement of the injector plunger hits the fuel, the fuel is forced into the cylinder under pressure. The PT-injector starts the combustion slowly after which the fuel is fully combusted with optimal economy and efficiency.

Highest injection pressure

The Cummins-injector develops an injection pressure of up to 1330 bar. This is twice as high as observed in other hydraulically controlled injectors. The higher the injection pressure, the better the fuel atomises, which in turn improves the combination with the air for an efficient combustion.

Corresponding timing, advancing or retarding

At lower power outputs, the injector cup is not completely filled. Therefore, the injector plunger must travel a longer distance to reach the fuel. This results in a delayed injection at low power outputs and an accelerated injection at high power outputs (as the reservoir is fuller).

Cummins BIG/CAM-concept

The PT-fuel system, combined with the Cummins BIG/CAM-technology produces engines with reduced fuel consumption.
These camshafts with large diameters produce shorter injection times, higher injection pressures, reduced emissions and an improved fuel consumption.

Top Stop injectors

Some engine types have a predetermined mechanical stop, which restricts the injector plunger stroke. This version is not shown in the diagram.

Cummins PT-fuel pump

Flexible design

The PTG-type fuel pump is flexible, so a basic fuel pump can be adapted to fit a large range of designs and capacities. With minor adjustments, the same fuel pump can be used for: tractors, buses, locomotives, ships, building machines and other industrial applications.
The PTG-fuel pump has a wide range of coupling characteristics to attune engine performances to requirements.

Fuel pump function

The PT-fuel pump manages the fuel pressure at the inlet of the injectors in order to obtain the rpm required for that particular engine type.

Fuel pump components

The basic components of the flow control system and its functions are:

- A self-priming gear pump for flow and pressure.
- A governor and plunger with ports and a groove in the plunger to allow the fuel to flow through the governor when the port and grooves are opposite each other.
- A gas valve flanked by a speed control and the fuel-pump exit-port controls the pressure from the fuel line in order to regulate the power output. In most instances, the gas valve is directly operated by an 'engineer'.
- A fuel line to the injectors.
- A fuel-bypass circuit to re-circulate the fuel from the control plunger back to the inlet side of the gear pump.

AFC-fuel pump

A fuel pump with AFC (Air Fuel Control) includes an extra valve to control the fuel pressure during acceleration in turbocharged engines. This valve is placed between the gas valve and the stop valve and is activated by the air pressure in the inlet manifold. During acceleration, the air supply lags behind the fuel injection due to the turbocharger. Without this control, there would be a fuel overdose in the cylinders. This AFC-valve reduces the fuel pressure to the point that the turbocharger can provide sufficient air for full combustion of all the injected fuel that is required to generate the correct power output.

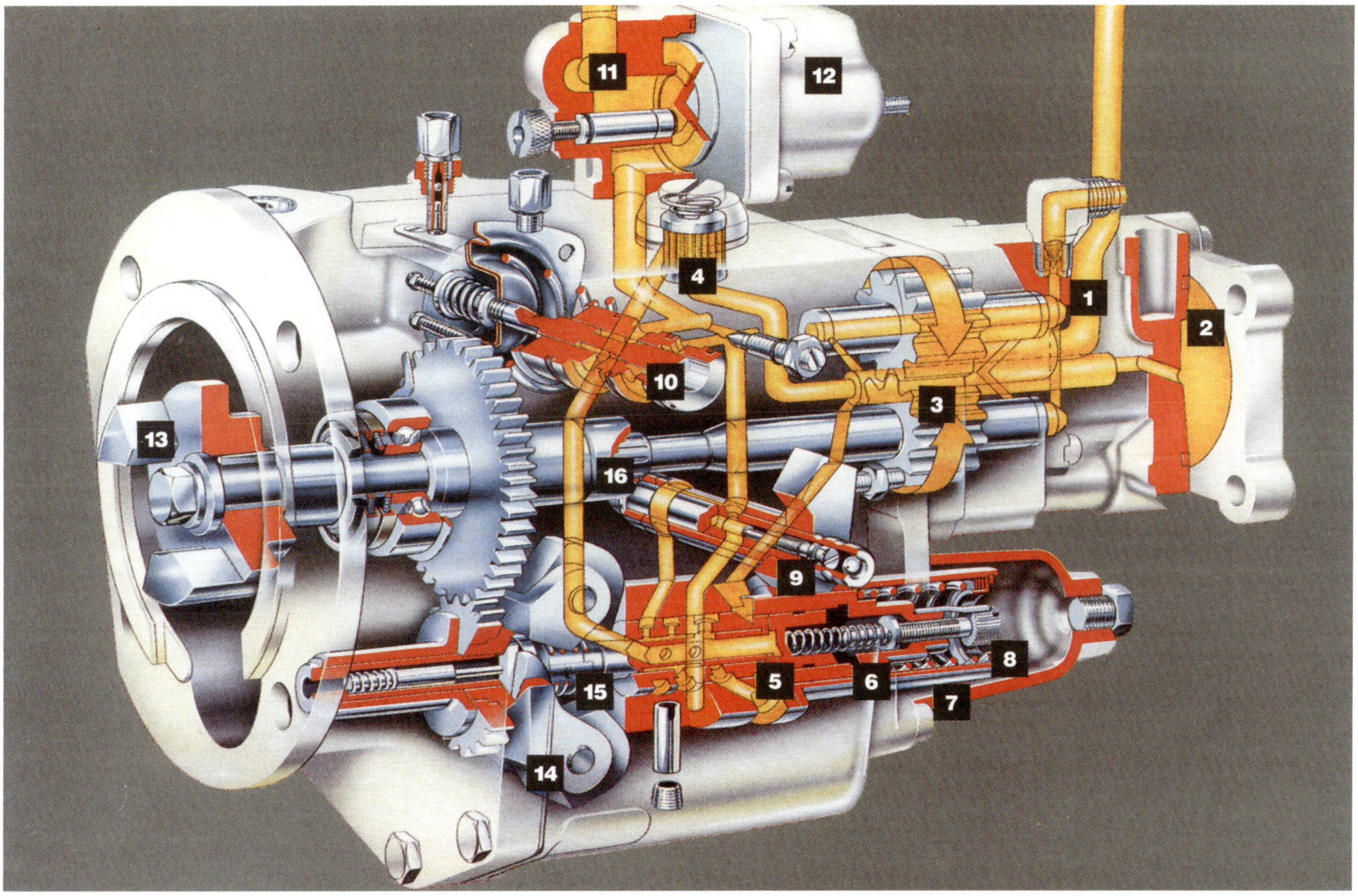

▲ The low-pressure fuel pump of Cummins.

1 fuel inlet
2 vibration damper
3 gear wheel pump
4 filter with magnet
5 governor plunger
6 by-pass plunger
7 stationary spring packet
8 stationary adjusting screw
9 gas valve
10 air-fuel control housing
11 fuel to injectors
12 shut-off valve
13 drive coupling
14 governor weight
15 link spring
16 main drive shaft

Other PT-fuel pump components

Other essential fuel pump components:

- A filter with a magnet between the gear pump and the speed regulator to filter out possible metal particles in the fuel, which could damage other parts of the system.
- A stop valve (electrically-, manually- or air controlled) mounted on top of the pump.
- A pulsation damper on the gear pump that regulates pulsations in the fuel flow in order to obtain a constant fuel pressure.
- A speed regulating drive shaft for installing a rev counter.
- A fuel filter directly mounted on the pump to prevent dirt and air from penetrating between the filter and the pump.

9.19.2 Mechanically driven injectors by Caterpillar-diesel engines

These are high-speed four-stroke engines; fuel MDO (engine category II)

The injector is equipped with a magnet valve to regulate the fuel injection. This fuel system comprises fuel injectors that are mechanically driven from the camshaft as well as a rail system with a separate high-pressure fuel pump.

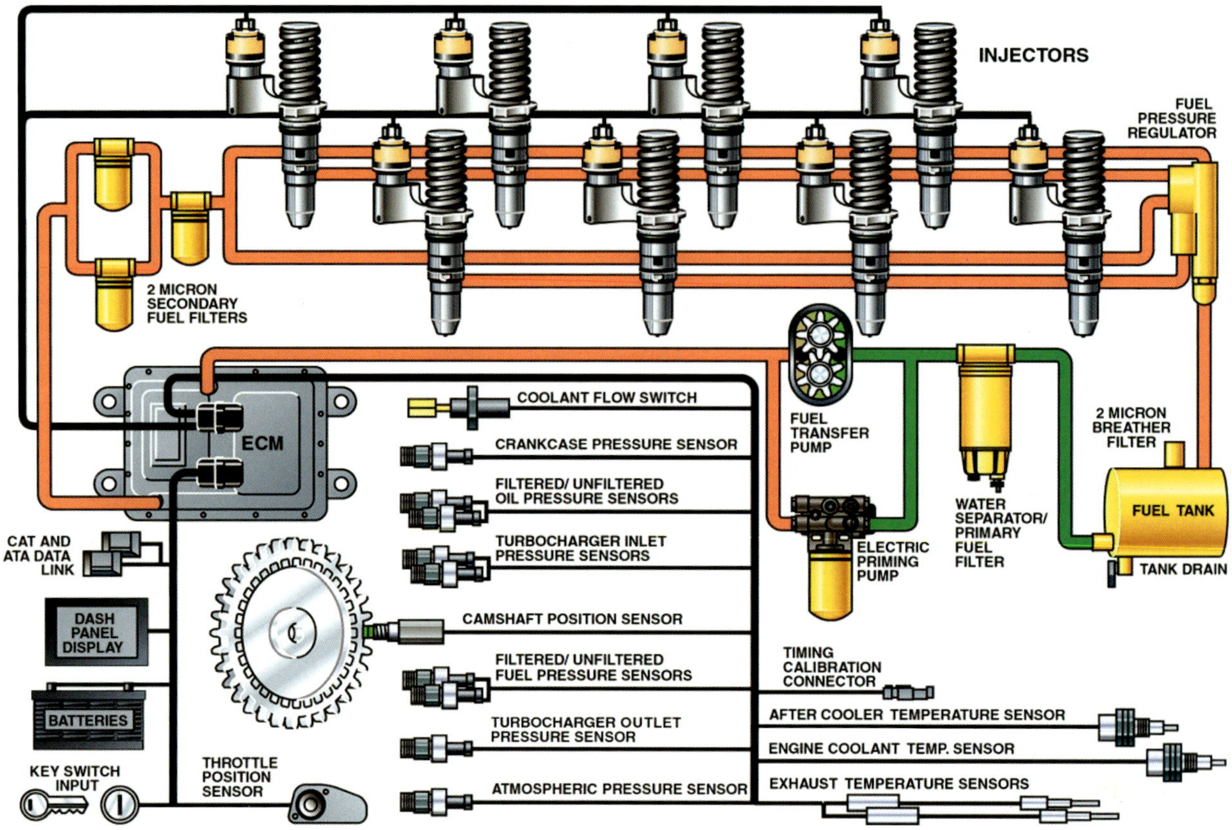

▲

The mechanically driven injectors in Caterpillar-engines.

Green: suction lines, low pressure

Red: pressure pipe lines to the fuel priming pump, low pressure

A low pressure -priming pump draws fuel via a fuel filter/water separator out of the day tank. The fuel flows through an engine management system (ECM Electronic Control Module) if certain requirements are met. When engine failures occur, the fuel is immediately cut off and the engine stops. The fuel flows through a very fine fuel filter with a mesh size of two micron and then flows to the electronically regulated fuel injectors. In the discharge lines after the injectors is an adjustable pressure control valve.

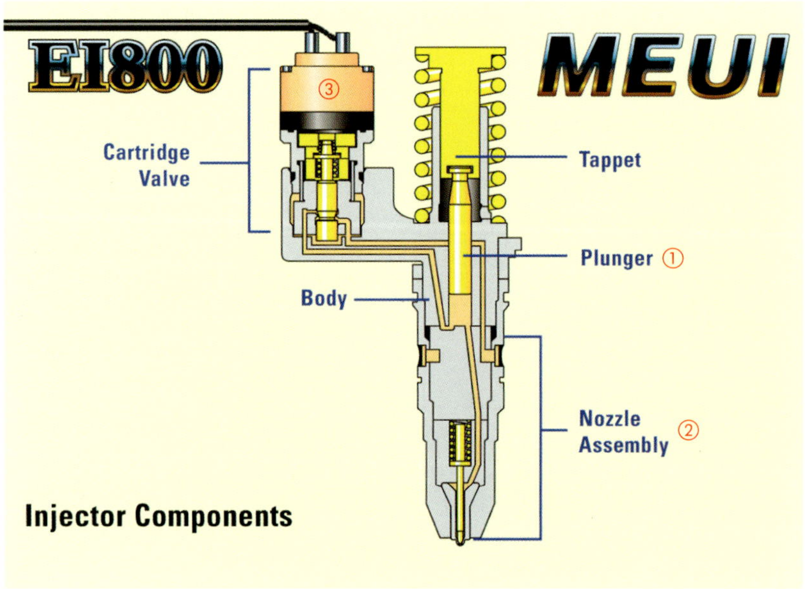

◄

The mechanical injector from Caterpillar. Outside the injector, the fuel pressure is very low so fire hazards are limited to a minimum.

This injector comprises four parts:

1 The high-pressure compression part with a mechanically driven pump plunger that draws fuel from a fuel supply line.
2 The injector part underneath the fuel plunger with a spring loaded injector needle.
3 An electronically operated solenoid valve that supplies the fuel supply to the high-pressure part.

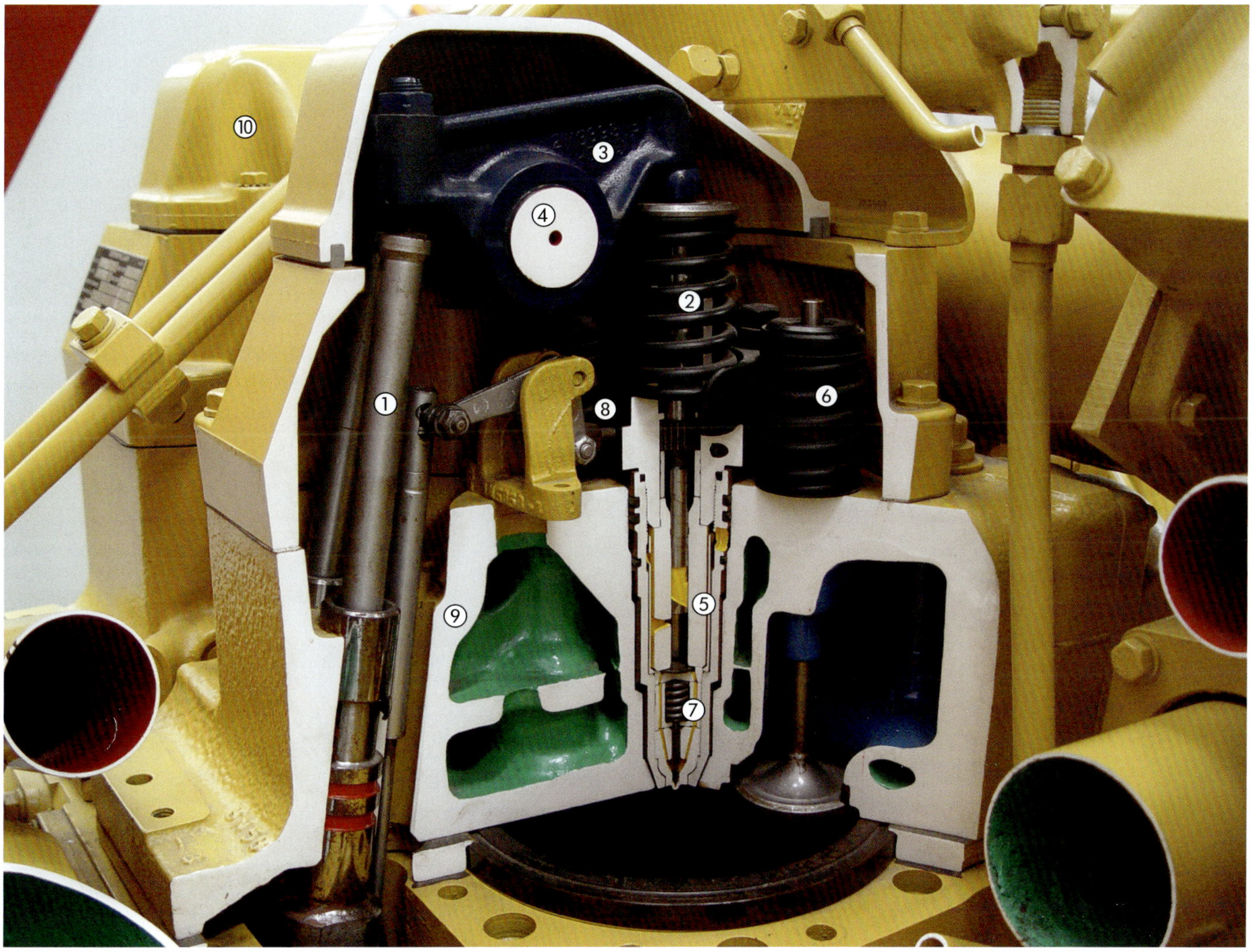

▲

The mechanical injector of Caterpillar.

1 *push rod*
2 *spring*
3 *rocker arm*
4 *rocker arm shaft*
5 *injector*
6 *valve spring*
7 *injector spring*
8 *solenoid valve (not visible)*
9 *cylinder head*
10 *cylinder head cover*

Operation

The pressure exerted by the injector spring results in **the upwards stroke of the high-pressure plunger** and therefore at a fully adjustable point the suction valve opens which in turn is operated by the solenoid valve and the pressure chamber is filled.

In the downwards stroke of the high-pressure plunger the pressure increases and the injector needle opens at approximately 38 MPa or 380 bar. Fuel still flows via the solenoid valve to the injector discharge lines. Then the solenoid valve shuts and due to the high speed of the plunger, the injection pressure quickly increases to 152 MPa or 1.520 bar. This is the main injection.
The end of the injection occurs at the moment that the solenoid valve opens and the injector needle caused by the spring pressure closes.

The supply and discharge of fuel occurs on the right side outside the injectors. All fuel lines outside the injectors have a harmless low pressure of approximately 3 bar.
With an electrical of hand driven priming pump the low pressure-system can be filled and vented.
Since 1999, this is the standard-fuel system for all Caterpillar-diesel engines.
The entire system operation can be controlled with a pc and contains all modern control - and failure programmes.
The system is abbreviated MEUI and means:
Mechanically Actuated – **E**lectronically Controlled – **U**nit – **I**njector.
As in all previous systems, this system has all the known advantages such as lower emissions, lower fuel consumption and better balancing and harmonisation of the different cylinders and therefore less vibration.

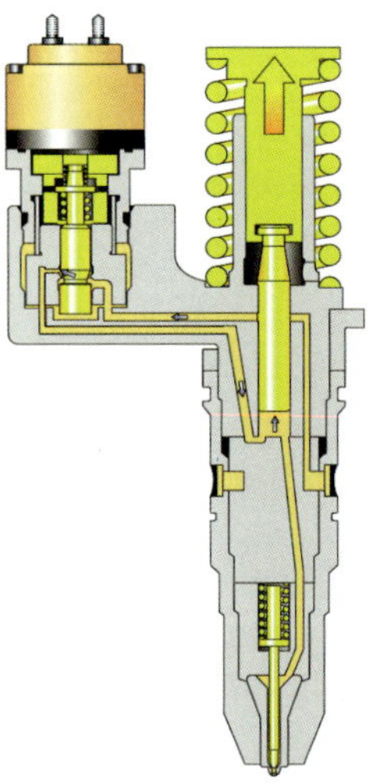

▲ Filling.

The injector is filled with fuel, the press plunger moves upwards. The solenoid valve opens and admits the fuel.

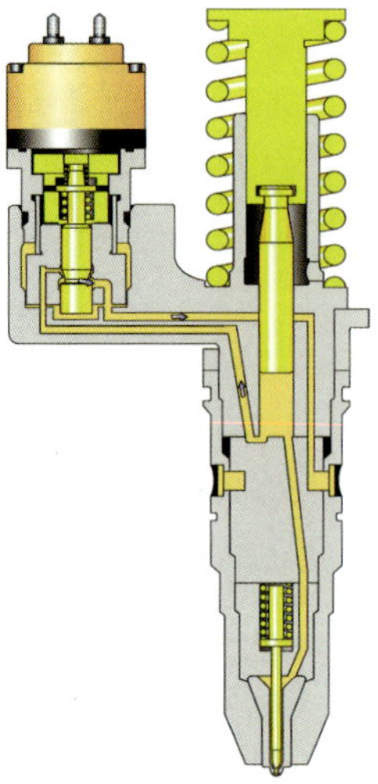

▲ Compression.

The solenoid valve is closed. The moment the press plunger moves downwards, the pressure in the injector increases. The needle is raised as soon as the force the fuel pressure exerts on the needle exceeds the force of the needle spring.

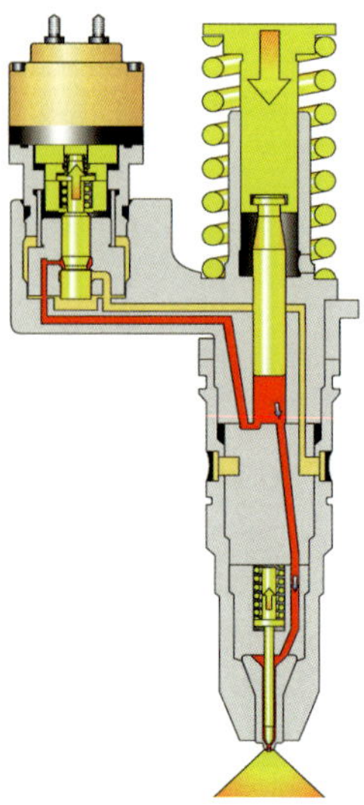

▲ Injection.

The injector needle moves upwards against the spring pressure. Then the fuel is finely distributed in the extremely hot, compressed air via the injector and ignited. The injection pressure subsequently increases drastically.

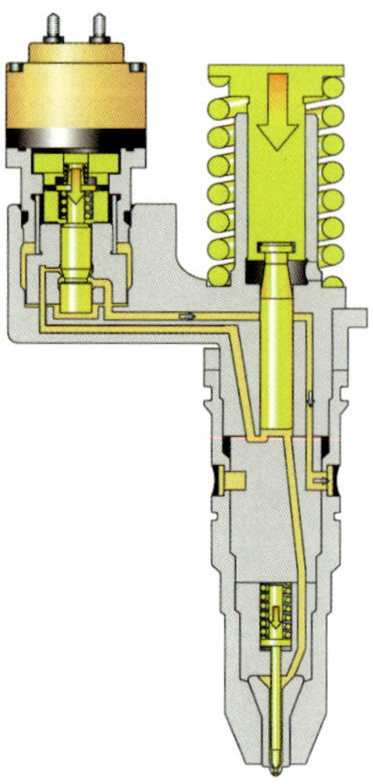

▲ The end of the injection.

The solenoid valve opens, the fuel pressure falls and the injector needle quickly closes. The injection is finished.

9.20 Fuel injectors

For large two-stroke crosshead engines, these are also called fuel valves. In small high-speed engines, the term injector is frequently used.

Function of the injector

To inject the fuel into the cylinder at such a speed that the extremely fine fuel droplets, in the little time available, are spread over the entire combustion space without coming into contact with the 'walls' such as the cylinder head, cylinder liner and piston crown.

Some technical data:

- Injection time dependent on engine type and rpm, several hundredths of a second.
- Injection speed over 300 metres per second.
- Injection length in the cylinder dependent on the cylinder content, several centimetres to decimetres.
- Droplet size of the fuel 20 to 60 micro millimetres (microns).
- Opening pressure of injectors 100 to 400 bar.
- Maximum injection pressure of injectors up to 2000 bar.
- Diameter injector holes 0.2 to 1.0 millimetres.

Number of injectors per cylinder

Virtually all four-stroke engines are equipped with an injector that is centrally placed in the cylinder head. The two-stroke crosshead engines have two or three injectors that are placed around the periphery of the central exhaust valve.

◄

The injector position in a small high-speed, four-stroke engine for yachts.

1 *fuel filter*
2 *high-pressure fuel pump*
3 *high-pressure fuel lines*
4 *fuel injectors*

◄

The three injectors on the cylinder head in a two-stroke crosshead engine. They are positioned around the central exhaust valve.

1 *common rail fuel supply lines*
2 *high-pressure fuel lines to injectors*
3 *fuel injectors*

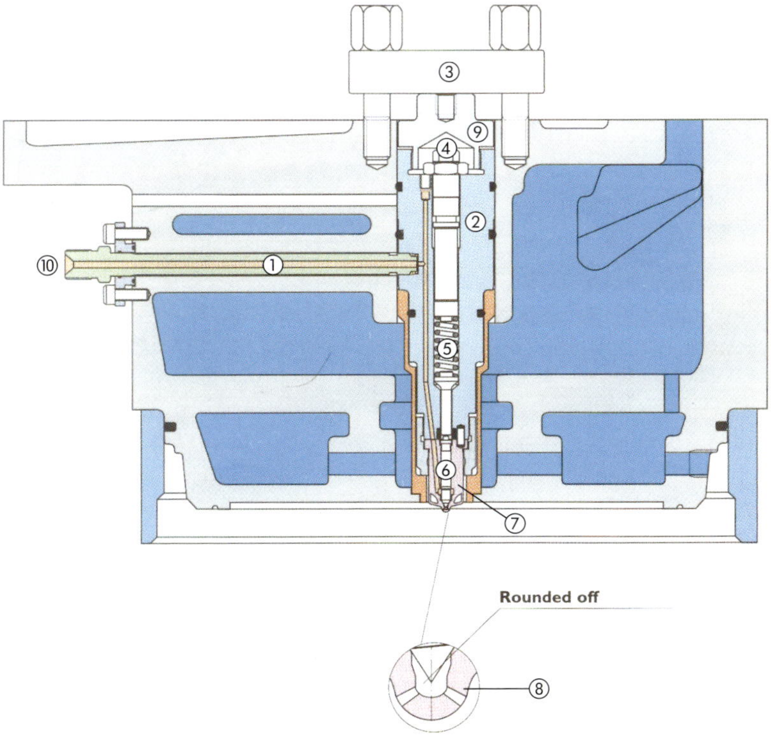

◄

A fuel injector in the cylinder head of a large four-stroke engine, category III.

1 *high-pressure fuel supply line*
2 *injector housing*
3 *gland*
4 *adjustable bolt for the opening pressure*
5 *injector spring*
6 *injector needle*
7 *injector tip*
8 *injector holes*
9 *cap*
10 *high-pressure fuel supply*

▶
A cross-section of an injector tip.

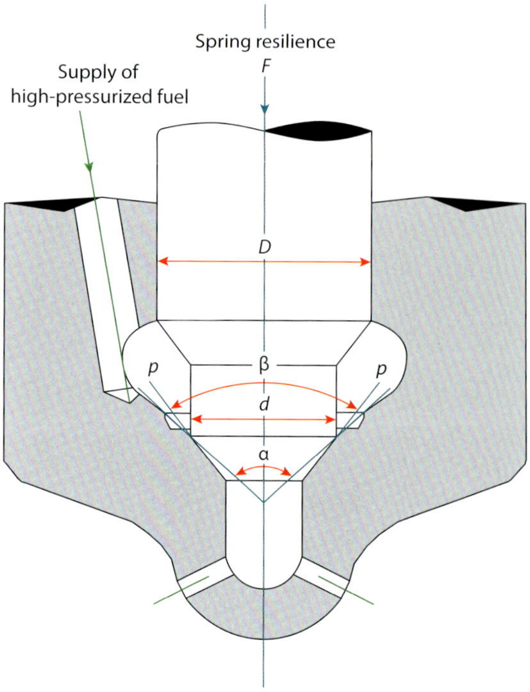

9.20.1 Construction of the injector

A simple injector consists of an injector casing, at the bottom a needle guide with a tension sleeve enabling the connection of the needle guide liquid with the injector housing. The needle valve sits in a tapered valve seat and held against the valve seat by a compression spring. The pressure pen bridges the distance between the compression spring and the atomising needle.
The spring tension of the compression spring can be adjusted with a screwed tensioner. The injector is equipped with a fuel supply connection and a spill fuel connection.

9.20.2 Working of the injector

At the moment that the high-pressure fuel pump starts to compress the fuel, there is still a remaining fuel pressure in the high-pressure fuel line, the so-called residual pressure. This is much lower than the closure pressure of the valve and avoids dripping from the injector tip. The supply line leads to a chamber in the injector just above the needle valve. This needle has two diameters, D large and d small. At the bottom, the needle valve has a milled, polished surface with an angle α that seals it against the valve seat of the needle guide; this seat is also milled and polished, at an angle β. The angle α is slightly larger than angle β causing the valve to close rapidly and the fuel flows to the cylinder. As the fuel-pressure rises, the force exerted on surface of the needle valve increases, which overcomes the spring pressure and opens the needle valve.

$$\frac{\pi}{4}(D^2 - d^2)$$

Now the fuel injection takes place. As the fuel pressure decreases, the needle closes because the force of the spring exceeds the upward force of the fuel on the circular surface of the needle valve. The opening and closing of the needle must occur very swiftly in order to obtain optimum atomising. Opening and closing the needle valve too slowly reduces the pressure near the injector apertures, thus abating the atomising process.

Opening and closing of the injector needle

Let's assume:

p_o = the fuel pressure at the time that the needle valve opens.
p_c = the fuel pressure at the time that the needle valve closes.
F_s = the force of the spring exerted on the needle valve via the needle guide.
D = the largest diameter of the needle.
d = the smallest diameter of the needle.

The following applies for the opening time of the injector needle valve:

$$p_o \times \frac{\pi}{4}(D^2 - d^2) = F_s.$$

As the needle valve closes, the fuel pressure pushes against the entire bottom surface of the needle valve with surface area $\frac{\pi}{4} \times D^2$.

$$F_s = F_c \times \frac{\pi}{4} \times D^2$$

Deducing from equation 1 and 2

$$p_o \times \frac{\pi}{4}(D^2 - d^2) = p_c \times \frac{\pi}{4} \times D^2$$

$$p_c = \frac{D^2 - d^2}{D^2} \times p_o$$

Therefore, the closing pressure is lower than the opening pressure. Often $p_c = 0{,}7 \times p_o$. The pressure that remains in the high-pressure fuel line after the injection is terminated, the residual pressure, is lower than the closing pressure in order to avoid dripping. This often amounts to 0.7 of the residual pressure.
So for example:

p_o = 400 bar
p_c = 0.7 x 400 bar = 280 bar
p_r = 0.7 x 280 bar = 196 bar

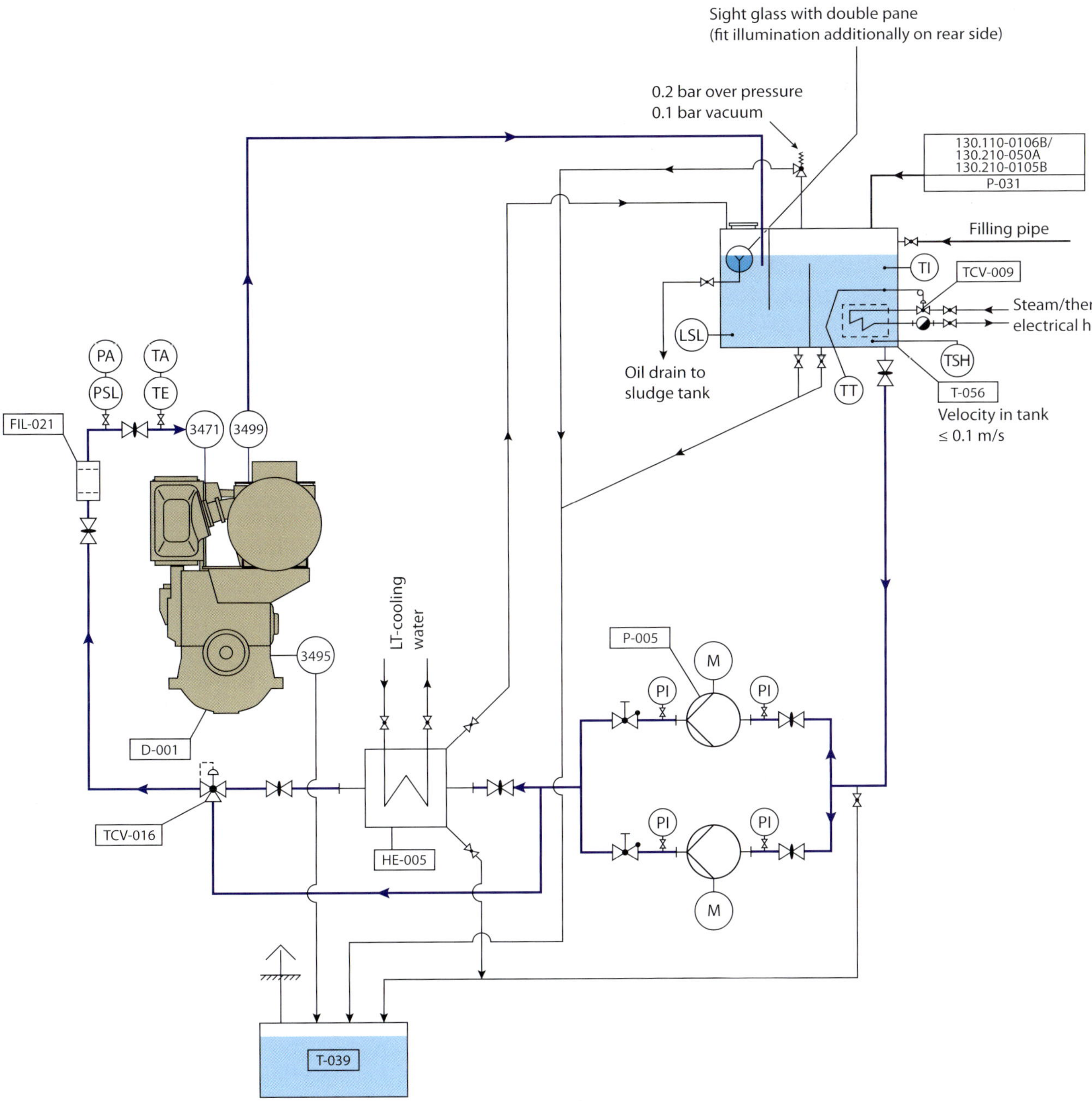

D-001	Diesel Engine
FIL-021	Strainer
HE-005	Nozzle cooling water cooler
P-005	Nozzle cooling water pump
T-039	Cooling water storage tank
T-056	Nozzle cooling water expansion tank
TVC-009	Temperature control valve for heating
TVC-016	Temperature control valve for nozzle cooling water

3471	Nozzle cooling water inlet
3495	Nozzle cooling water drain
3499	Nozzle cooling water outlet

▲
The injector cooling system of a MAN Diesel four-stroke diesel engine running on HFO.

Short description: With two circulation pumps, the coolant is supplied from a supply and expansion tank that can be kept at a certain temperature using steam, thermal oil of electricity. The temperature of the coolant keeps the heavy fuel oil in the injector at the correct temperature when the engine stops.

A cooler to the circulation pumps control the temperature during operation.

During repairs, the coolant can be drained to a storage tank.

When heavy fuel-oil leaks into the coolant in the injector this is visible in the expansion tank. The oil floating on the water surface can be skimmed to a sludge tank.

▲

Different injector tips of four-stroke and two-stroke engines.

Injector cooling

In larger engines, the injectors are cooled with coolant or fuel. This is necessary to prevent the fuel that remains in the injector space or cup to vaporise, crack and form a carbon residue on the outside of the injector nozzle, thus disrupting the atomising process. It is also possible to use the cooling system to pre heat the engine and in this instance ensure that the injector nozzle achieves the correct temperature for a smooth start. In this manner, the temperature is controlled, including that of the injector nozzle itself. Modern diesel engines do not normally have direct injector cooling. The reason is, among others, to avoid fuel contamination of the cooling circuit. One generally uses indirect cooling, which means that the coolant circuit of the cylinder head is close to the injector nozzle.

Details

- Number of injector apertures. This varies from one for the single hole injector, which is solely used for indirect injection in pre-chambers, to four. A maximum of sixteen apertures for multi-hole injectors for direct injection into the combustion space.
- The injector apertures have a diameter of approximately 0.2 to 1 millimetres. Apertures smaller than 0.2 millimetres are rare as they are difficult to drill and clog up easily. It is also possible for injector nozzles to contain apertures with varying diameters and for apertures to be placed above each other.

Aperture diameter

For two-stroke crosshead engines, approximately one thousandth of the cylinder diameter. For four-stroke engines slightly over one thousandth of the cylinder diameter. The length of the injector aperture is at least four times the aperture diameter in order to acquire a good fuel focus.

9.20.3 Various examples of injectors

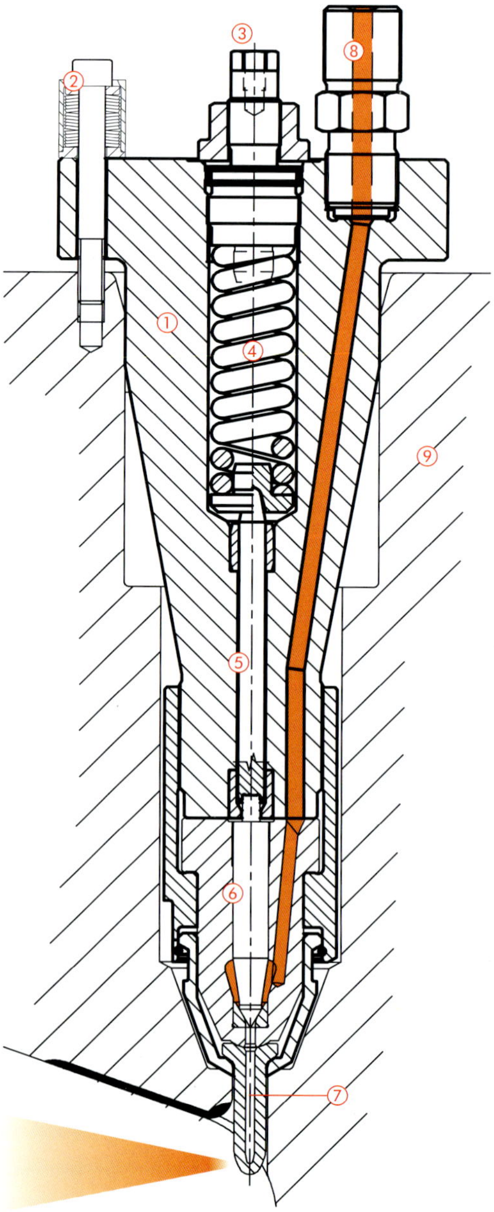

▶

An injector for a large two-stroke crosshead engine.

1 *injector housing*
2 *bolts with spring packet*
3 *adjustable bolts for the spring pressure*
4 *spring*
5 *spindle*
6 *injector needle*
7 *injector tip*
8 *fuel supply*
9 *cylinder cover*

◀ A modern injector in a four-stroke, high-speed Caterpillar diesel engine.

1. *high-pressure fuel supply of the common-rail system*
2. *solenoid valve for the control of the fuel injection*
3. *injector tip*

◀◀ Entire injector tips on a vibrating sanding-disc for the accurate planing of the packing face with the injector casing.

◀ The correct angle of the seat of the injector needle is important for the good operation of the injector.

◀ The position of the injectors in the cylinder head.

The injectors are not always placed vertically in the cylinder head. Shown here is a Deutz diesel engine for yachts with the injectors positioned in the side of the head. The engine distributor here is the Vetus Company in Schiedam, The Netherlands.

9.21 Residual-pressure valves

Most small plunger-controlled fuel pumps and some larger plunger-controlled and valve-controlled fuel pumps are equipped with **residual pressure valves. As the effective compression stroke of the pump plunger is terminated these residual-pressure valves reduce the fuel pressure in the high-pressure supply line to such an extent that post-injection of fuel is avoided.** The effective compression stroke of the high-pressure fuel pump is terminated when the pressure over the plunger drops due to the opening of the spill valve (other types of two-stroke crosshead engines) or spill ports. A pressure valve, if present, will close immediately. The injector needle could lift due to pressure waves that move up and down the high-pressure fuel line resulting in poor quality post-injection. This reduces the quality of the post-combustion process, which results in contamination as well as extra fuel consumption. The residual-pressure valves make sure that the pressure in the lines is kept sufficiently low to ensure that post-injection is avoided. This residual pressure should not be too low as this would increase the injection delay of the following injection.

There are two types of residual valves:
1 pressure relief;
2 volume relief.

9.21.1 Pressure relief

Operation

This can be fitted near or inside the pressure valve. The valve allows a certain amount of fuel to flow back into the pump cylinder after fuel injection and pressure-valve closure.
The residual pressure in the high-pressure fuel line is determined by the spring force of the residual-pressure valve. **The residual-pressure valve operates independently of engine load and fuel-pump capacity. The residual-pressure valve including pressure relief is predominantly used for larger four-stroke and two-stroke diesel engines that run on HFO. The residual pressure can be adjusted to prevent it from falling to the extent that vapour bubbles are produced in the extremely hot fuel.**
In fact, these residual-pressure valves work similarly to a safety valve on a pressure vessel; they open at a predetermined pressure and ensure that the desired temperature is not exceeded.

9.21.2 Volume relief

Operation of residual-pressure valves including volume relief

These are usually part of a normal pressure valve in the high-pressure fuel line. They are equipped with a collar or a piston body underneath the actual valve setting.

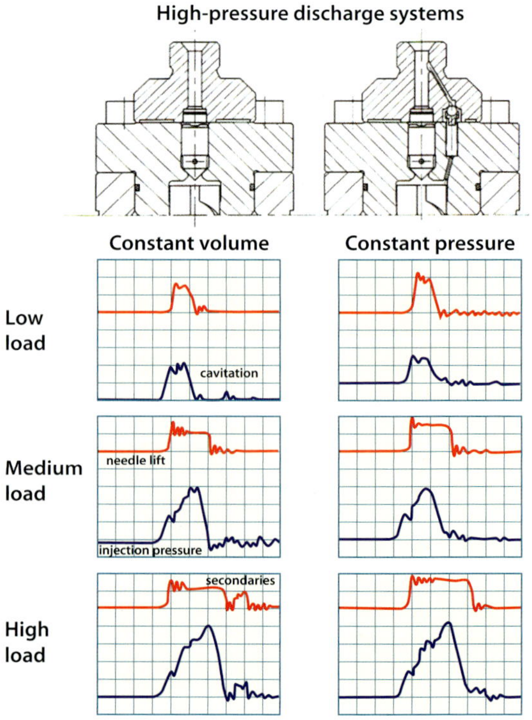

▶
Left the residual pressure with volume relief and right the residual-pressure valve with pressure relief.

With volume relief, cavitation can occur at partial engine loads because the residual pressure of the fuel can fall below the vapour pressure of the fuel. Residual-pressure valves with pressure relief can be used in fuel-injection systems for large engines that run on heavy fuel oil. Because of the very high fuel temperatures required for the correct injection viscosity of the fuel, there is no risk of formation of a vapour lock at partial loads.
Residual-pressure valves with volume relief have a pressure loss in the fuel-supply line that is independent of the engine load. The pressure loss must be low enough to prevent post-injection.

9.22 Cavitation

Opening the residual-pressure valve may generate pressure waves in the pump, supply lines or piston valves found, for instance, in valve-controlled pumps of some crosshead engines. An accurate plunger and valve design may prevent this. In plunger-controlled pumps, a spread bolt or a cavitation bolt at the spill port in the fuel pump ensures that this phenomenon is reduced to a minimum.

9.23 Fuel-injection characteristics

The pressure curve of the fuel in the high-pressure fuel line prior to, during and after the injection may vary considerably. Depending on the engine type, the engine load, the fuel type and the type of fuel system there are many different injection patterns.

Several examples

Two-stroke crosshead engines by MAN Diesel-ME

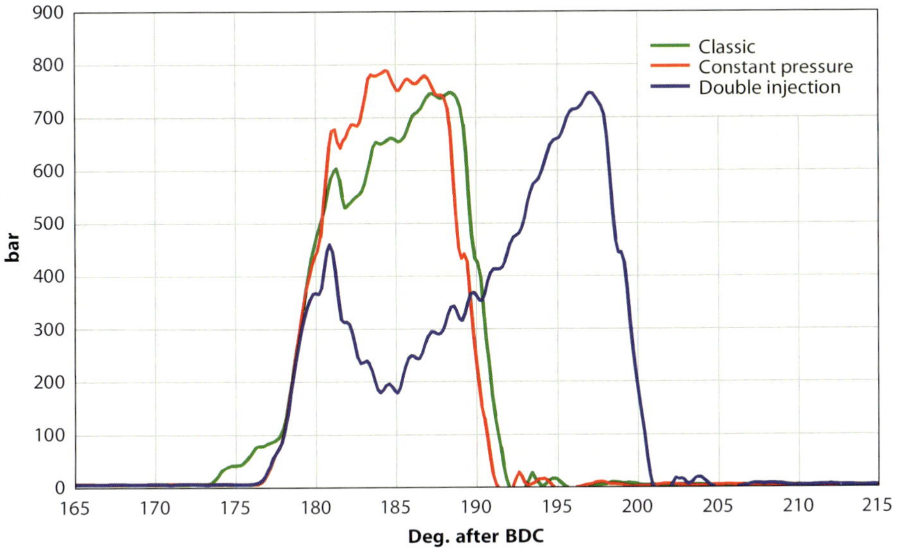

▲

Fuel-injection characteristics of the two-stroke crosshead engines of MAN Diesel-ME.

Horizontal: crank degrees after the bottom dead centre.
BDC 180 Crank degrees is the top dead centre.
Vertical: the fuel injection pressure in bar.
The pressure curve is shown here for a seven-cylinder ME-C-diesel engine with a cylinder diameter of 50 centimetres at a load of 75%.

Green pressure line

The conventional injection starts at 173 crank degrees after bottom and so 7 crank degrees before top and the injection ends at approximately 192 crank degrees after bottom and so 12 crank degrees after top.
Total injection time is 7 + 12 = 19 crank degrees.
The maximum injection pressure is approximately 750 bar.

Red pressure line

For the constant-pressure process.
The injection begins slightly later (3 crank degrees before top) and finishes slightly earlier (11 crank degrees after top). Total injection time 3 + 11 = 14 crank degrees.
The maximum injection pressure is approximately 780 bar. The pressure curve is at the injection smoother than the former conventional injection. The injection is slightly better over the entire injection area due to the slightly higher injection pressures.

Purple pressure line

Double injection: this is especially intended to lower the maximum combustion temperatures so the nitrous oxide content in the exhaust gases decreases. The first injection takes place approximately 3 crank degrees before top and the second injection finishes at approximately 21 crank degrees after top.
Total injection time is 3 + 21 = 24 crank degrees.
The first injection peak is approximately 450 bar and the second injection peak is approximately 740 bar.
The entire combustion process is less explosive and takes place later so the maximum combustion pressures and temperatures are lower that in the two former systems. With electronically regulated fuel injection in this engine type, every required injection profile can be obtained.
This engine does not have a camshaft and the exhaust valve is operated electronically.

Four-stroke trunk-piston engines by MAN Diesel with electronic injection

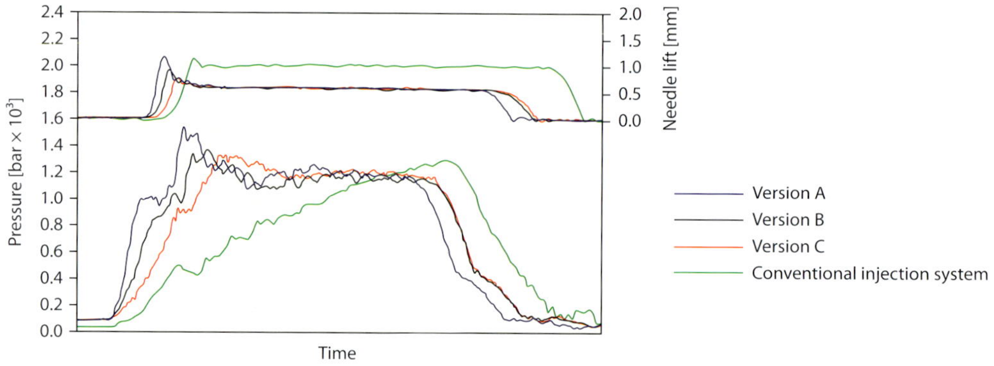

▲

The four-stroke trunk-piston engines of MAN Diesel with electronic injection.

In the figure:

Horizontal: the position of the crankshaft, not given in crank degrees

Vertical: the fuel injection pressure in bar

In the figure, three new injection versions are compared with a conventional fuel-injection system (green).

Above in the figure: the injector needle lift in millimetres

The most conspicuous difference between the three new versions and conventional injection is that the conventional system takes longer to reach the full injection pressure of ± 1250 bar.

The three new versions reach the maximum pressure faster and therefore have a better injection and so a better combustion.

Due to this, the formation of soot particle and nitrous oxide emissions are reduced.

At partial-load operation, the injection pressure can also be kept at high levels and the differences in exhaust-gas emissions between the electronic and the conventional mechanically regulated fuel injection are larger.

Two-stroke crosshead engine by Wärtsilä, the Sulzer RT-FLEX-design

▶

The two-stroke crosshead engine of Wärtsilä, the Sulzer RT-FLEX- version.

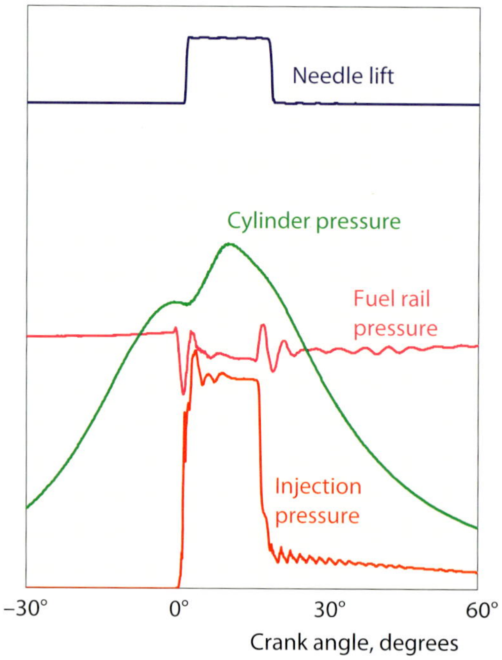

In the figure:

The **green line** shows the pressure curve in the cylinder.

Conspicuous is the 'depression' at the moment that the injection starts. The cause is that the injected fuel requires sufficient heat for vaporisation and gasification. Slightly later, combustion takes place.

The **purple line** shows the pressure curve in the common fuel line for all cylinders. At the opening and closing of the fuel injector needle a variation in the fuel pressure takes place.

The **red line** shows the pressure curve of the injected fuel. Note the very fast increase in the beginning and at the end of injection with respect to the crankshaft position. Due to the common-rail system with a large volume capacity, the steep pressure line is possible.

After the closure of the injector needle, the pressure 'ripples' for a time.

The blue line shows the injector-needle lift.

>CH 10

Cooling diesel engines

 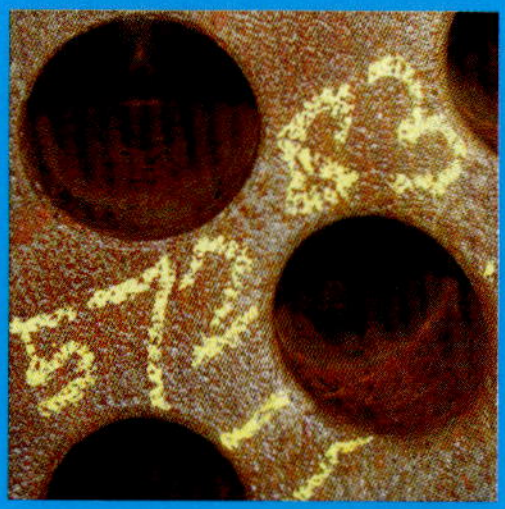 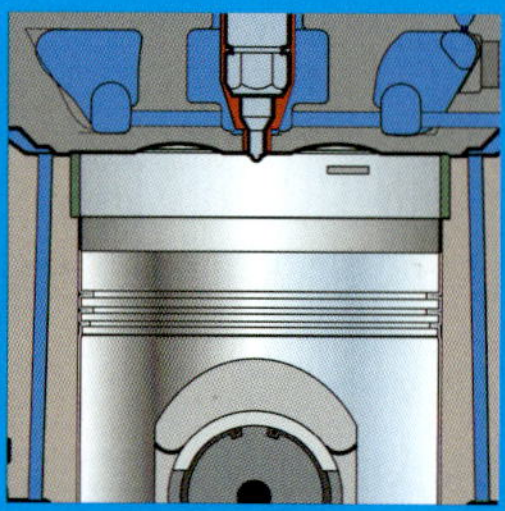 

Cooling engines is essential in order to avoid overheating of engine parts and subsequent damage. Approximately 30% of the heat present in the injected fuel is cooled by fresh-water coolant, lubricating oil and air coolers.
Shown, cooling-water pumps for cooling the lubricating-oil system for a two-stroke crosshead engine.

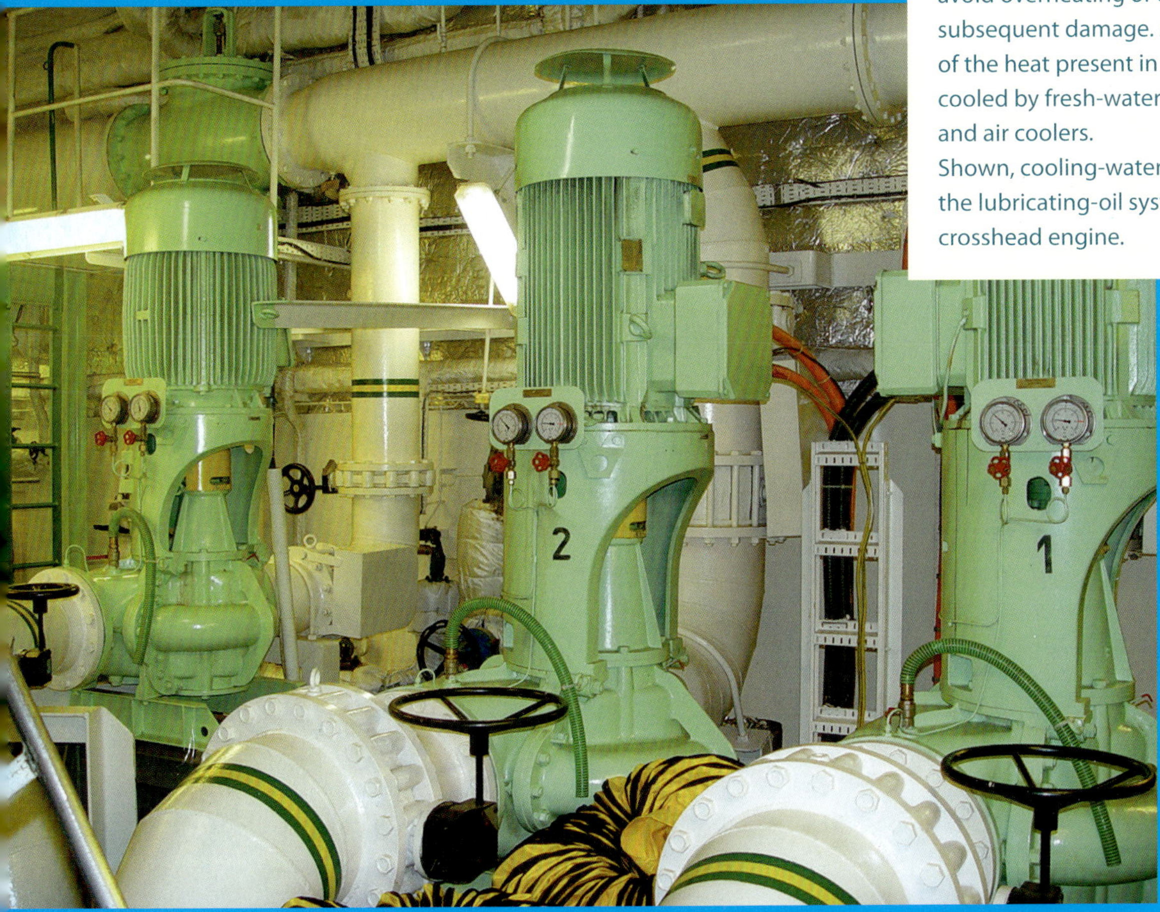

Coolers are essential for the discharge of heat.

10.1 Introduction

Internal combustion engines such as the diesel engine are provided with cooling systems. During the combustion process the temperatures in the cylinder can reach 1500 °C to 2000 °C, during the initial release of the combustion gases the temperatures can increase to 700 °C to 800 °C. Finally, the exhaust gases in the exhaust lines have temperatures of 325 °C to 600 °C and in fully loaded engines, these gases pass through the turbocharger. Without a cooling system, the material of the parts that are exposed to these gases would be damaged irreparably and undergo thermal distortion.
In practice, an uncooled engine would be seriously damaged in a very short period. Therefore cooling is an absolute necessity and a constant point of interest for engine designers.

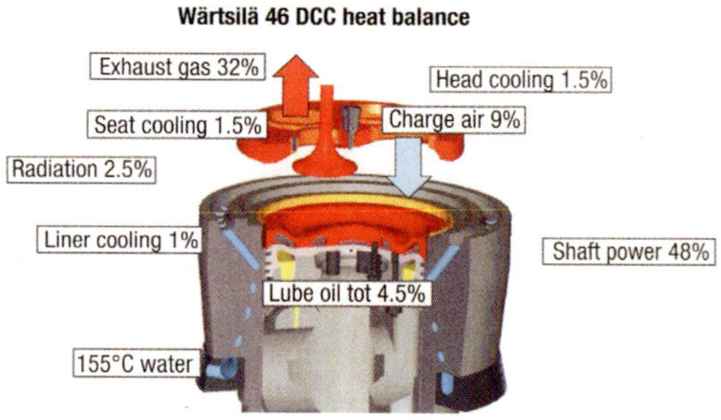

▲
An overview of the heat flow around the combustion space. Shown here a four-stroke medium-speed diesel engine by Wärtsilä 46 DCC – Category III.

Fuel energy supply is 100%
Shaft power is 48%

Losses:

exhaust gases	32.0 %
cylinder head cooling	1.5 %
valve seat cooling	1.5 %
cylinder liner cooling	1.0 %
piston cooling with lubricating oil	4.5 %
radiation losses	2.5 %
air cooling	9.0 %
Total losses	52.0 %

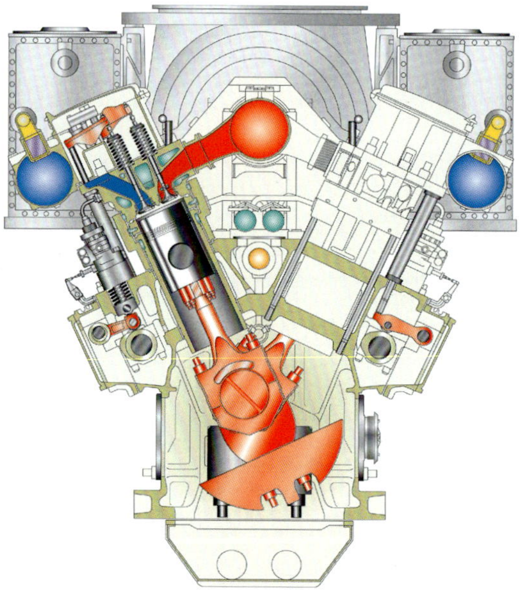

▶
Each diesel engine requires cooling in order to avoid overheating of the parts.

▼
The combustion chamber.

Engine parts in proximity to the combustion chamber, such as the piston, cylinder liner, cylinder cover and valves require intensive cooling. Shown here; the cooling water is light blue.

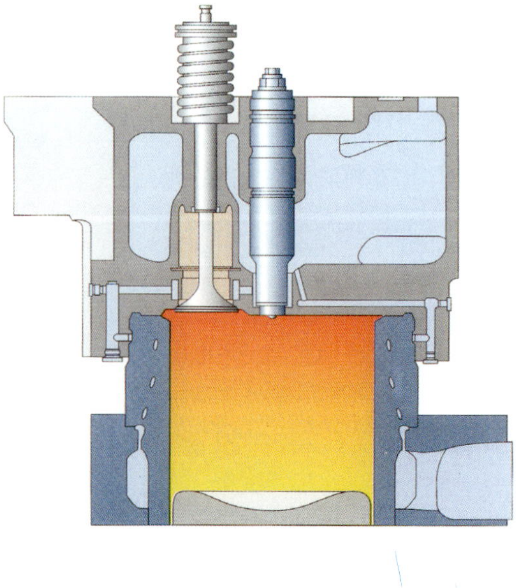

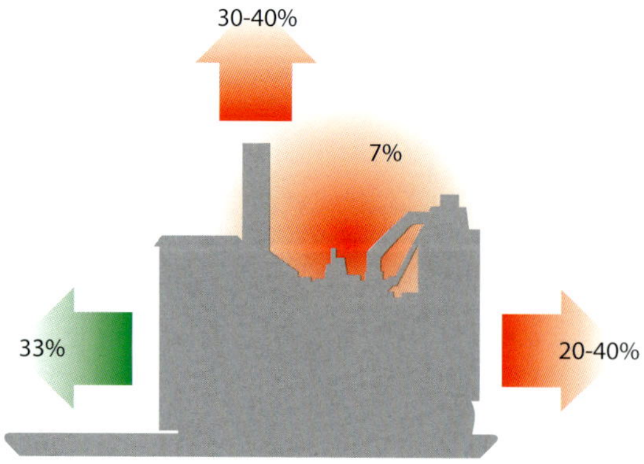

▶
An overview of the heat flow in a small diesel generator category I with radiator cooling. The radiator is cooled by means of a mechanically driven ventilator.

Approximately 33% of the heat supplied by the fuel is available as mechanical energy on the shaft, 30 to 40% is dispersed in the exhaust gases and 20 to 40% is utilised for cooling, the residual losses, among which radiation loss, constitute 7%.

▲

Guides of a two-stroke crosshead engine, viewed from above.

All modern crosshead engines use lubricating oil to cool the guides.

1 *crosshead pin*
2 *guide shoes*
3 *crosshead guides*
4 *A-frame*

▲

A cross-section of a four-stroke engine.

In this four-stroke diesel engine, the fresh water is indicated in blue. A conspicuous detail is that only the upper rim of the cylinder liner protruding from the block is cooled. So, the 'dry' engine block is not exposed to the coolant, which means that corrosion of the block by coolant no longer occurs.

The following engine parts are exposed to the hot combustion gases and therefore require cooling:

1 cylinder covers
2 piston crowns with the piston ring package
3 upper part of the cylinder liners
4 exhaust valves and the exhaust-valve casings
5 parts of the fuel injectors closest to the combustion chamber
6 parts of the turbochargers, such as the bearings and sometimes the casings

In two-stroke crosshead engines, the crosshead guides are also cooled because of frictional heat generated.

Moreover, the temperature of the compressed air in the air cooler after the turbocharger decreases considerably thereby cooling the combustion space internally. **Four-stroke trunk-piston engines** have in addition to the suction stroke which takes in fresh air an additional phase in the process: **exhaust-gas scavenging**. Here the piston is in the TDC position whilst both inlet valves as well as both exhaust valves are opened for a considerable part of the crank curve, thus cooling hot engine parts such as

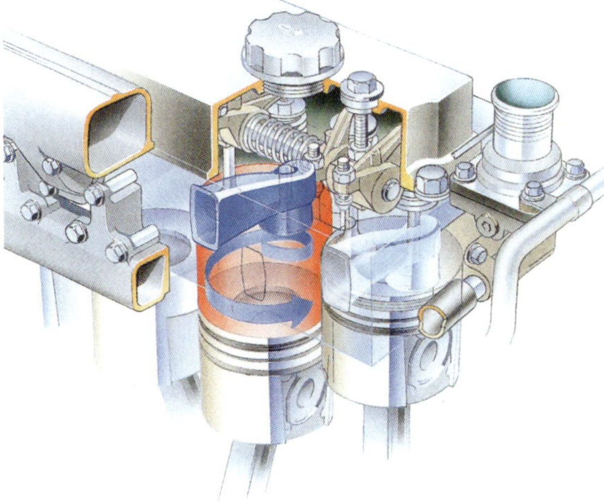

◀

The air supply in a four-stroke high-speed Perkins diesel engine category I.

The combustion space is intensively cooled by the rotational movement of the inlet air.

▶
There is an infinite amount of seawater for discharging heat generated by the diesel engines.

It is no longer directly used as a coolant. However, seawater is indirectly used in a heat exchanger.

the cylinder head, piston crown, cylinder liner and exhaust valves.

In two-stroke crosshead engines, the hot engine parts are cooled at the moment that the inlet ports are uncovered by the downward movement of the piston while the exhaust valve is already partially opened. At this point, exhaust-gas scavenging, cooling and supercharging of the cylinder takes place.

10.2 Cooling agents for diesel engines

The following liquids can be considered:

- seawater
- fresh water
- coolants
- lubricating oil
- fuel

Furthermore, air is used as a coolant.

10.2.1 Seawater

Clearly, in shipping this is in endless supply in seas and oceans.

Advantageous properties

High specific heat at a relatively high density. Therefore, small cooling-water pumps and coolers are required to discharge the unwanted heat. The heated seawater is directly discharged into the sea. This system concept is very simple.

Disadvantageous properties

Seawater has a fairly high mineral content, which can crystallise on the cooling surfaces after heating and produce scale. This is a hard layer that reduces the heat transmission and can cause blockages in the cooling ducts. Furthermore, the high chloride content of seawater may cause severe corrosion of the cooled parts.
This is why today seawater is only used for indirect cooling; it cools a closed fresh-water coolant system.

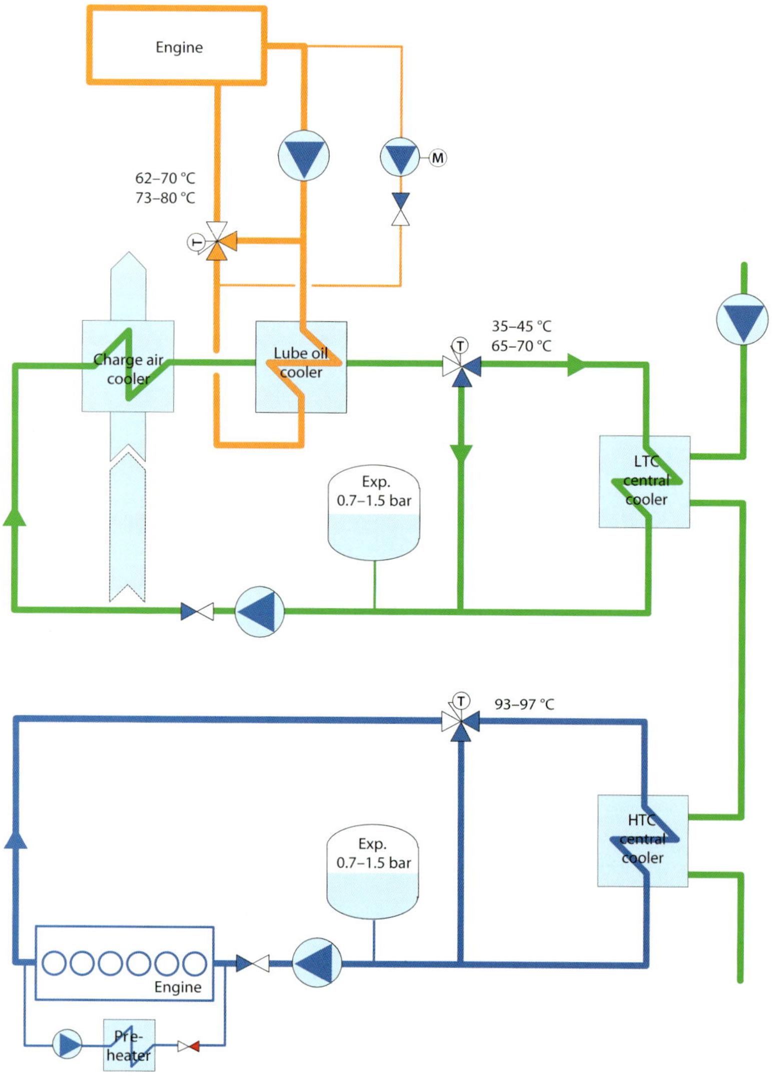

◀
A complete cooling-water system with a low-temperature system (LT colour, green) and a high-temperature system (HT colour, blue).

Both systems have two circulating pumps each, one of which is in operation and the other on stand-by. The LT-system cools the scavenging air and subsequently the lubricating oil. The HT-system is the standard cooling system for fresh-water coolant. This is also where the scavenging air is cooled for the second time. The central cooler is cooled with seawater (not shown here).

The seawater cooling section is designed to be as short as possible and the material used is seawater-proof, such as 'cunial' or stainless steel.

The cooling-, scavenging- and combustion air in the air cooler after the turbocharger is occasionally cooled by seawater; this has the advantage that the low temperature of the seawater can cool the scavenging air for the engine to the required 35 °C to 40 °C. Usually these coolers are finished with a special seawater-proof coating.

◄ A scavenging-air cooler of a two-stroke crosshead engine with the supply and discharge pipes for the LT coolant system.

10.2.2 Fresh water

This has the same positive properties as seawater. Additionally, there are no problems with regard to corrosion and scale forming. The fresh-water system requires proper deaerating and water softening. **Furthermore, the cooling-water is treated with chemicals.** Fresh water is always used in a closed circuit. In case of leakage, fresh water is supplied by the storage tanks. Most ships have the ability to produce fresh water from seawater, which enables them to maintain their fresh-water store for both drinking water as well as fresh-water coolant for the engines.

Regular drinking water is not suitable for diesel-engine cooling and first needs to be softened. In today's highly thermally loaded engines, small amounts of salt in the water would cause scale formation on the warm engine parts. This could cause cracks.

▲ A seawater evaporator.

The heat of the HT cooling-water system is also often used to distil pure fresh water from seawater by means of a vacuum seawater evaporator. Seawater reaches boiling point at ± 40 °C when the absolute pressure is as low as ± 0.05 bar, so the HT coolant temperature of 70 to 90 °C is sufficient to evaporate seawater.

◄ Fissures in the cooling channels of a piston crown in a two-stroke crosshead engine.

A lot of damage to cooled engine parts is caused by scale formation, which inhibits the heat convection of the cooled surfaces. The material becomes excessively hot and tears. Shown here is part of a piston crown in a two-stroke crosshead engine with bore cooling. The cracks are marked with chalk.

10.2.3 Lubricating oil

Apart from excellent lubricating properties, a large amount of heat is discharged with the lubricating oil. For instance, heat generated in the combustion process by the pistons and heat generated by friction in the bearings is removed. In small engines, the piston is simply cooled with a lubricating-oil nozzle that is attached to the cylinder liner. A return pipe is not required; the lubricating oil simply falls back into the sump.

In larger engines, the oil is transported to the piston via a perforation in the connecting rod and is sometimes discharged in a similar way. This is also the case for two-stroke crosshead engines.

Piston leakage in these types of cooling systems cause few problems as opposed to, for instance, leakages in the water-cooled pistons in crosshead engines.

Several disadvantages of lubricating oil as a coolant as compared with fresh water are:

Lubricating oil has a lower density and a lower specific heat. This is a considerable difference!

specific heat	fresh water	4.19 kJ/kg/K
	lubricating oil	1.68 kJ/kg/K
density	fresh water	1000 kg/m³
	lubricating oil	915 kg/m³

▶ The lubricating-oil lines in a connecting rod.

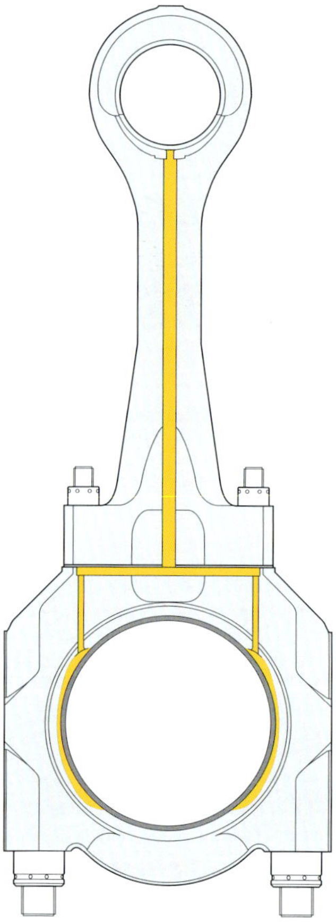

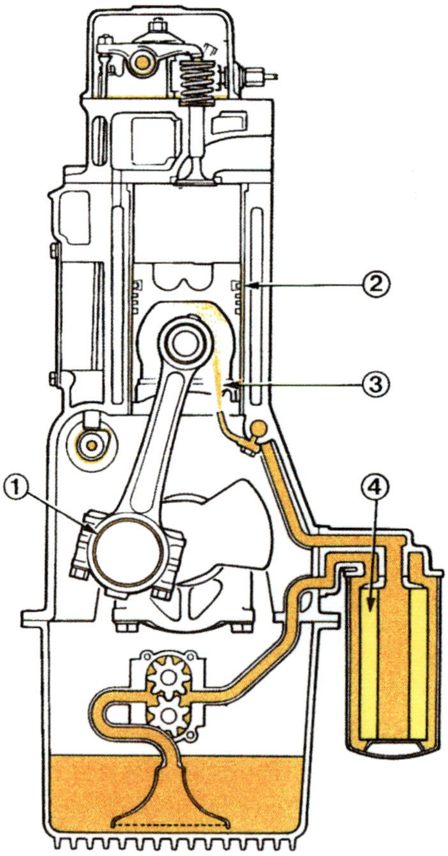

▲ Spray cooling in pistons using lubricating oil.

Lubricating oil absorbs a large amount of heat during the lubricating process. The use of spray cooling in the piston crown of this smaller diesel engine aids this process.

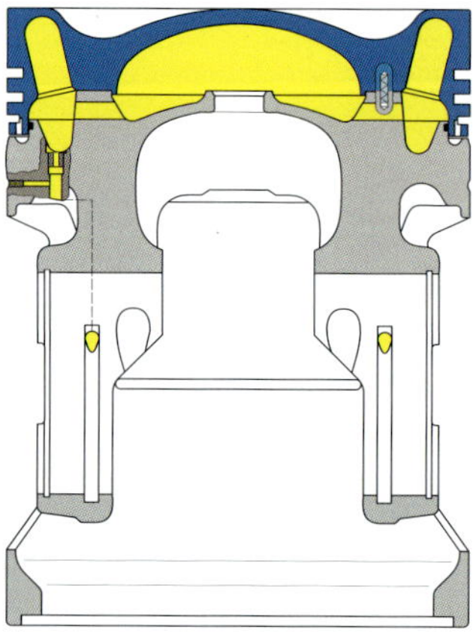

▶ Piston cooling.

In larger medium-speed diesel engines, the piston requires intensive cooling with lubricating oil via perforations in the connecting rod.

Heat absorption per m^3 at a temperature increase of one degree Kelvin or one degree Celsius:

heat = mass in kg × specific heat kJ/kg/K × temperature difference

The following applies for fresh water:
heat = 1000 × 4.19 × 1 = **4190 kJ**

The following applies for lubricating oil:
heat = 915 × 1.68 × 1 = **1537 kJ**

Therefore the cooling effect of fresh water compared to that of lubricating oil is $\frac{4190}{1537} =$ **2.7 times better!**

One must also take into account that lubricating oil oxidises at higher temperatures and forms a carbon layer on the surfaces of the cooled parts.

Note

Diesel engine parts cooled with lubricating oil, such as the inside of small pistons or the cooling pipes in larger pistons, must be checked for carbon formation.

In lubricating-oil coolers, problems may occur when the temperature of the lubricating oil is too high. Carbon layers form on the lubricating-oil side of the cooler. This is only visible when the entire pipe assembly is removed from the cooler casing.

10.2.4 Air

This coolant is also used in small high-speed diesel engines, category I, 0 to 100 kW shaft power.

Properties: Very low density and -specific heat; it is only suitable for small power outputs in engines with low thermal loads. A crankshaft-driven ventilator is required which via a ducting system forces cold air to the cooling fins on the cylinders.

Dust and other dirt particles can therefore penetrate these cooled engine parts.

▲
The cooling of piston crowns with lubricating oil in a MAN Diesel MC (left) and a Wärtsilä Sulzer RTA (right) two-stroke crosshead engine.

Disadvantages of air-cooling

- low cooling capacity, only suitable for small, diesel engines of low thermal loads;
- contamination of the cooling fins by dust, grease, oil and dirt;
- no sound damping by cooling water;
- requires air ducting to external air.

Advantages air-cooling

- no cooling water required; therefore no corrosion or frost risks;
- simple engine construction.

Note

Small stationary diesel engines are nevertheless still regularly provided with air-cooling. They are then not supercharged.

▼
Occasionally small and simple diesel engines without turbochargers still use cooling fins, along which air is forced, thus cooling the cylinder.

◄ A small diesel-generator set with air-cooling.

From the fan flywheel, the air is led to the engine via sheet-metal casing.

1 *flywheel casing with fan blades*
2 *air-inlet grid*
3 *sheet-steel casing*

10.3 Cooling-water treatment

The most suitable fresh-water coolant is distilled water that is obtained with an evaporation installation. Large seagoing ships have a seawater evaporator on board; this heats the seawater under vacuum and evaporates the seawater. The cooling systems of engines serve as a heating source. In diesel-power plants, one can distil drinking water in a similar manner with for instance heat generated by the fresh-water cooling system or with steam from the exhaust-gas boiler.

This distilled water contains dissolved air and therefore oxygen, which may cause corrosion of the various materials. **Proper deaeration of the cooling water is therefore absolutely vital.** This is the reason that a cooling-water treatment is always required. This takes place by adding chemicals from a small feed tank situated in the cooling-water system. Metal surfaces are protected by adding chemicals ('inhibitors'). These substances apply a very thin coating on the surfaces in the cooling-water system, which inhibits contact between the coolant and the parts. A nitrate/borate mixture is often used.

Cooling-water treatment

Cooling-water treatment is applied to numerous systems, amongst others:

- fresh-water cooling systems for diesels;
- fresh-water cooling systems for air compressors;
- central fresh-water cooling systems;
- central-heating systems;
- warm-water systems for tank heating;
- cold-water systems for air-conditioning systems.

We will focus on the first system.

Fresh-water cooling for diesel engines

The following problems may arise with inadequate cooling-water treatment:

- Corrosion due to a low pH-value of the cooling water or a combustion-gas leakage.
- Deposits on the cooling surfaces due to water hardness. This is usually the case with tap water.
- Oil leakage causing oil emulsification that forms a sludge layer, which absorbs dirt and consequently impedes heat conduction and causes overheating.
- Oxygen corrosion, resulting in pitting.
- Biological contamination, producing acid products that cause corrosion.
- Cavitation and erosion. Erosion results from hard particles that circulate and literally grind down the material.

▼ A complete auxiliary steam installation with an oil-fired boiler and exhaust-gas boiler.

The steam is utilised for various purposes, amongst others fresh-water distillation.

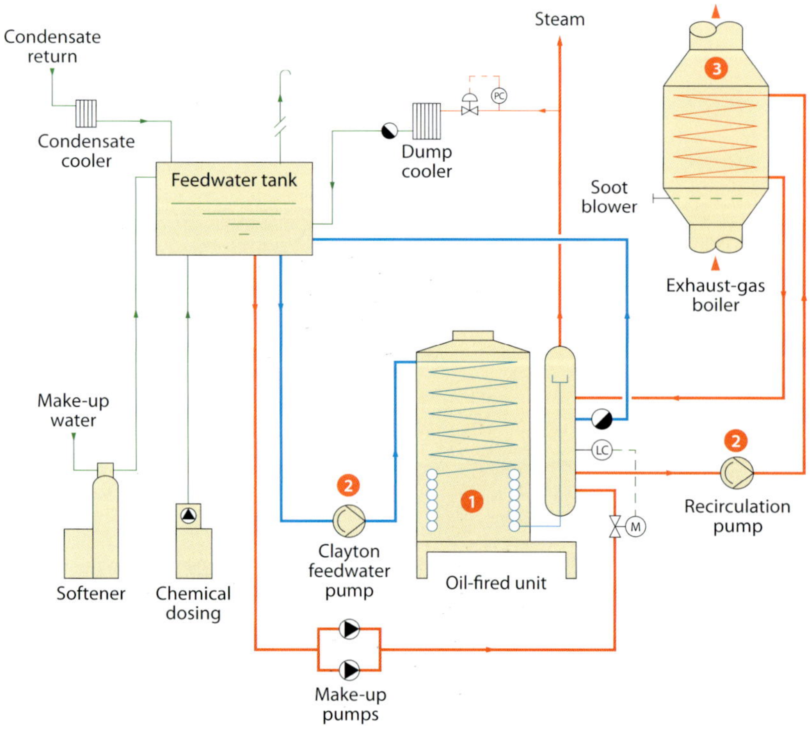

- Electrolysis between various materials. This phenomenon may develop randomly, depending on the nature of the water supplied, or close to two different materials.

So a **fresh-water evaporator** can be saturated with **oxygen** and **carbon dioxide** and therefore contain corrosive products. After supplying fresh water to the cooling-water tank, the supplemented water much reach the correct operating temperature, in order for oxygen and carbon dioxide to be bled off.

1 Cylinder cooling-water systems

A modern diesel-engine with a relatively high heat-transmission coefficient, a high cooling-water temperature and often narrow cooling-water ports demands a very effective cooling water treatment where corrosion can be reduced to a minimum and ensures that deposit formation on the cooling surfaces is avoided.

2 Injector cooling-water system

This system is particularly vulnerable to the coagulation of various substances due to the very narrow cooling-water passages, mainly fuel that leaks into the cooling system. In the injector-coolant expansion tank, corrosion products can collect and cause blockages further down in the system. With fluctuating levels in this expansion tank, the metal not in contact with the water corrodes. In order to avoid this, the expansion tank needs to be finished with a coating that is resistant to high temperatures and a low pH. Contamination by oil leakage is a common occurrence in these systems.

3 Piston cooling-water system for crosshead engines (old types of engines)

This system can be contaminated by lubricating-oil leakage. **The lubricating oil is predominantly found on the piston crown.** This layer reduces the heat transmission considerably, which can cause overheating of the piston crown. Corrosion is often the major point of attention in these types of cooling systems. The so-called galvanic corrosion takes place through a potential difference between two different materials. This is prevented by using, if possible, a limited number of materials in the piston design.

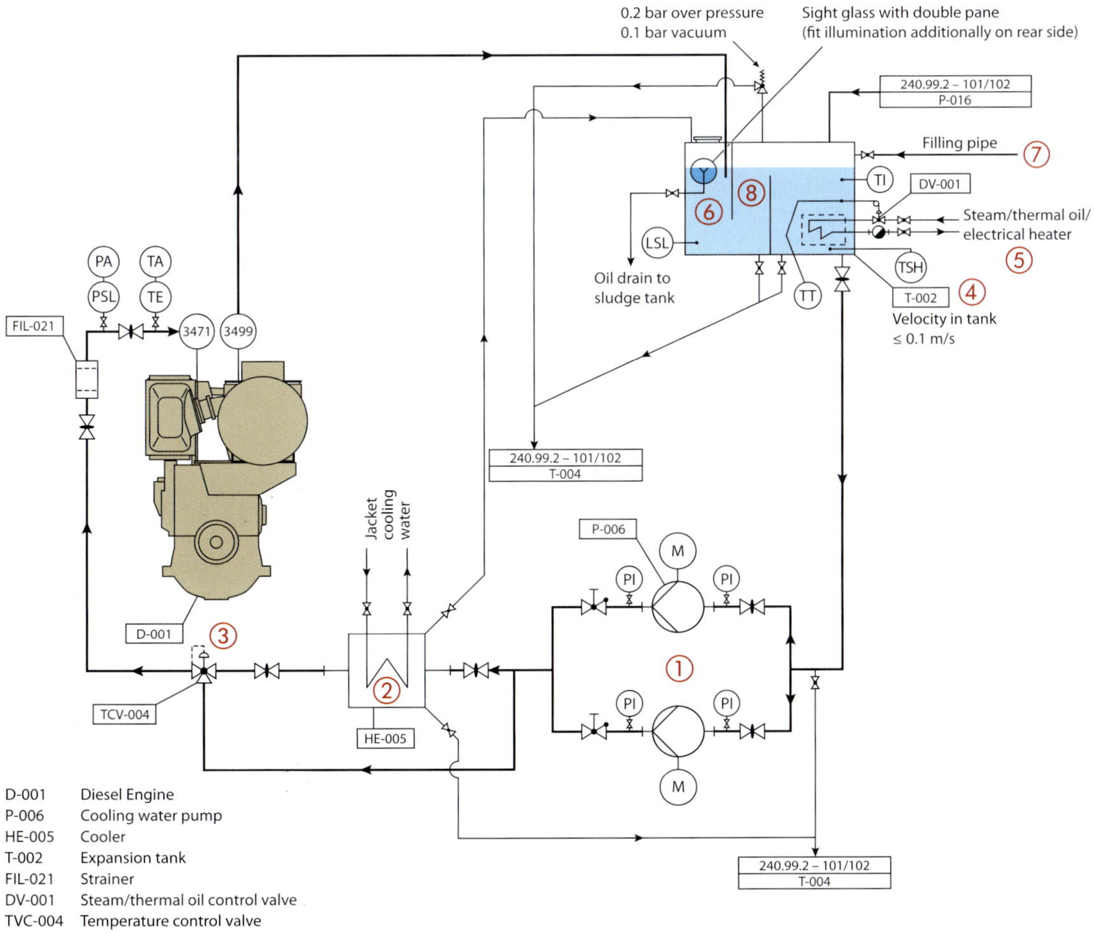

◄ A fuel-injector cooling system with two circulating pumps (1), a pre-heater fed by warm cylinder coolant (2), a temperature-controlled three-way valve (3) and an expansion tank (4).

The expansion tank is equipped with a heating system (5), an oil separator (6), a filling pipe (7) and, amongst others, a low-water warning system (8). The expansion tank is coated with a special finish to avoid corrosion. In a low-load engine, the injector-coolant temperature should remain high to prevent settling of sulphur on the injector-nozzle tip. The expansion tank is also an observation tank where potential leakage of fuel into the coolant water of the injectors can be detected.

▶
The exhaust-valve casing in a two-stroke crosshead engine.

This is the only part of this engine type that still contains a large amount of coolant. All other parts have been provided with bore-cooling. Note the low cooling-water volume.

10.4 Corrosion

One can distinguish the following types of corrosion:

- **Pitting**, a local corrosion on unprotected surfaces.
- **Galvanic corrosion**, with two or more different materials in the system.
- **Cavitation and erosion**, a combination of mechanical factors, such as oscillation, flow velocities and chemical damage.
- **Corrosion**, which is formed beneath the sediment or by micro-biological contact.

Protection against corrosion

The best protection is finishing all metal surfaces in the cooling system with a tough coating. Cathodic and/or anodic products have therefore been formed to prevent the aforementioned chemical or electrolytic reactions. High sulphur-chloride and sulphate concentrates must be avoided. Cooling water should not contain zinc. Zinc is corrosive at 60 °C!
At higher temperatures, zinc can become anodic and the iron atoms in metals will be exchanged for zinc atoms!

Deposits

These may consist of:

- salts from temporary hardeners, such as calcium and magnesium carbonates;
- sludge, formed by a chemical reaction when water is treated with alkaline or phosphate additives;
- sludge, formed by corrosion products of metals;
- sludge, formed by micro-biological contamination, such as oil products.

▶
Severe degradation because of poor cooling with lube oil.

Prevention of deposits

- Only use distilled water produced by an evaporation installation so that the water hardness due to dissolved salts is as low as possible. However, this kind of water has a low pH, is therefore corrosive, and requires an anti-corrosive additive.
- Use a biocide product to avoid micro-biological contamination.
- Use a dispersing agent to ensure that the sludge remains in solution and is not precipitated.

10.5 Products for cooling-water treatment

There are a considerable number of products available, which have the following advantages and disadvantages.
New products presuppose the usage of distilled water as a basis for cooling water, simple test methods, dosage systems and treatment.

10.5.1 Chromates

This is a good anti-corrosive, but is no longer commonly used. It is toxic and creates drainage- and treatment problems. Furthermore, it is **prohibited** when an evaporation installation is used for the production of drinking water.

10.5.2 Phosphates

One can distinguish between polyphosphates and orthophosphates.
Polyphosphates have properties that prevent corrosion and bind hardeners. In the course of time and at the present cooling-water temperatures they are converted into orthophosphates.
Orthophosphates such as sodium triphosphate do not have anti-corrosive properties. All phosphates form a basis for the generation of bacteria and micro-biological contamination. They are therefore less frequently used.

10.5.3 Emulsion oils

These are used with water in piston cooling for two-stroke crosshead engines.
There is limited protection against corrosion. In practice, hardness and pH-values can scarcely be influenced and testing is very difficult.
Bacteriological contamination occurs frequently.

10.5.4 A nitrite-borate organic mixture

Apart from chromate nitride, this is the most effective anti-corrosive compound in combating corrosion. This compound consists mainly of nitrite borate, and dependant on the producer, some additives. These cooling-water treatment agents are not toxic and can also be mixed with anti-freezing additives.
The compound forms a coating on all the materials, thus protecting it from oxygen corrosion and chemical corrosion.
A nitrite surplus maintains the coating and the borate increases the pH of the water to 9 or 10.
Formation of the protective layer requires alkaline water!
This agent also contains poly-acrylics that keep the sediment in solution.
In order to avoid micro-biological contamination, biocides are added.

10.5.5 Carbon-xylates

At present, these agents are rarely used. They also form a coating on the metal surfaces and neutralise the galvanic corrosion. Only in case of water leakage, an additional agent is required. It will be a good alternative for **'nitrite borate' in the near future**.

Some remarks

Treated coolant should be clear. Opaque coolant indicates the presence of corrosive products.
The only way to remove these is to drain, flush and re-fill the system. A temporary solution is the addition of a 'dip solution'. The nitrite value should be maintained at approximately 1500 parts per million (ppm). In case of a water leakage, nitrite must be added to the water supply to maintain this value. Whenever combustion gases leak into the cooling water the pH and the 'P' alkalinity will decrease to a stable nitrite content. In order to quickly resolve the problem and increase the 'P' alkalinity and the pH, caustic soda or soda is added.

Normal 'P' alkalinity at least 20 ppm, pH at least 8.5.

10.6 Cooling-water treatment products, brands

Nalfleet
Engine Water Treatment 9 – 108.
For aluminium protection 9 – 111 or Nalcool 2000, for severe combustion-gas leakage into the cooling water 9 – 131.

Drew
DEWT NC usually in solution, liquidewt.
For hard water: Maxiguard.

Vecom
QC – 2.

Unitor
Diesel guard NB or ROCOR NB liquid.

10.7 Cleaning cooling-water systems contaminated with oil

Contamination of cooling-water systems by oil or oil residues can produce severe overheating due to a decrease in heat conduction. In the long term it can cause metal cracking.

When cooling water has been contaminated by oil it is imperative that the system is cleaned immediately!

Procedure
- Reduce the engine capacity to 60 to 70% of the normal maximum capacity.
- Continuously add fat-dissolving chemicals to the cooling water to an approximately 3% content. Apply a non-emulsifying sort. Ideally, this agent is added at least four hours prior to arrival in port, at a diesel-power plant or another installation (four hours prior to stopping the engine). The agent may not circulate in a cooling system for over 24 hours.
- Keep the cooling-water pump running for at least 4 hours after stopping the engine.
- Any oil or grease on the surface of the liquid in the expansion tank needs to be 'skimmed'.
- Drain the entire cooling-water system immediately.
- Replenish the system with fresh distilled water and keep the circulating pump running for an hour.
- Drain the entire cooling-water system again.
- If the water remains clear after having replenished the system, the system is clean. If not: drain and re-fill!

Products

Nalfleet:	Clearbreak 79103
Drew:	Drew clean 2000, Ameroid OWS
Vecom:	Veclean clear break or Veclean Tank
Unitor:	Clean break

10.8 Bacteriological contamination

Although most chemicals for cooling-water treatment contain substances which suppress bacteria and fungi growth, these organisms can rapidly propagate in favourable conditions.
Oil-contaminated bilges and supply-water drain tanks are the main source of these contaminations. Central freshwater-cooling systems with their favourable temperatures can form an ideal environment for the rapid development of these organisms when air, combustion gases or oil enters the system. It even occurs in water used for piston cooling. Some organic cultures flourish at 70 °C.
This bacterial growth can be controlled by adding a biocide similar to the ones used in seawater systems.
Nalfleet advises a dosage of 0.2 litre agent 9-321 per ton of cooling water per week.
For severe contamination one may use Nalco 7.330 with a dosage of 1litre per ton of cooling water per week.

10.9 Testing cooling water

Every manufacturer of cooling-water treatment agents has their own test method. Overall, the following tests are important.

The 'P' alkalinity and the chloride percentages can be tested similarly as to those for boiler water.

For the nitrite surplus, reaction agent '1' is added to the sample to which reaction agent '2' is titrated until the colour changes. All manufacturers apply different test sets, so the instructions must be read carefully.
As with boiler water, the **pH-value** is tested with litmus test strips or an electronic pH test meter.

By performing this test at regular intervals and carefully following the instructions many problems with the cooling-water system can be avoided, which is vital in engine maintenance.

The Project Guide of a diesel engine describes the following:
- the requisite water quality;
- the permitted water-treatment products;

- the method in which the glycol or other anti-freeze products can be used.
- In addition, the instructions of the cooling-water treatment manufacturer and the water-testing method have to be read carefully.

10.10 Macro-biological growth in seawater-cooling systems

This is a common phenomenon in seawater cooling systems. The larvae that are found floating in seawater settle in the system and grow from 1 to 15 mm in a fortnight! The growth of these barnacle larvae can be such that the resistance in the system increases dramatically and the capacity of the seawater-cooling pumps is drastically reduced. In fresh water, the barnacles will die within a matter of days. However, the shell is left behind and can cause serious blockages if they come loose, especially in plate coolers. Mussels occur less frequently, but cause similar problems. In some instances, macro-biological control is an absolute necessity. There are special pesticides available. The current method includes adding copper to retard growth. Not all organisms respond to this. The recurring annual **polishing of propellers** is an example.

Biocide treatment of seawater systems

Biocides are delivered in liquid form in order to facilitate the dosage.

The following products are often used:

Nalfleet: 9 –321
Drew: Amersperse 280 of Drewsperse SWD
Vecom: SWT Vecosperse
Unitor: Biogard SW

They are used as anti-fouling and to fight actual fouling.
The **dosage** is product dependent and is generally from four to six grams per m^3 of seawater for one hour per day; in particular in coastal waters where the fouling is extensive. This includes diesel-power plants that are situated along the coast as well as floating diesel-power plants. The treatment often has to continue for an extended period, several days for instance, in order to be effective. Increased dosages have no effect. Lower dosages extend the period of treatment considerably. The liquid is injected by means of a special dosage pump which introduces the agent into the warmest section of the system.
At a dosage of 5 to 6 ppm the barnacles remain open and are promptly destroyed. At, for example, 10 ppm the barnacles shut instantly and therefore cannot be destroyed. By dosing, a thin coating is applied to all metal surfaces which inhibits the barnacles from adhering to them.

Important: When applying cooling-water treatment agents and biocides the instructions must be strictly followed.

Seawater evaporator

In cases when the seawater evaporator is operational during biocide treatment, health is not affected; only pure water evaporates and the biocides are discharged when flushing.
Even under extreme circumstances seawater can be converted into drinking water, and will not contain biocide levels that pose a heath hazard!

10.11 Design of cooling-water systems

General

In engine cooling-water systems, the pump pressure, pump capacity, fluid velocity in the pipes, cooler resistance, control system and temperature are meticulously designed for optimal engine cooling. Both insufficient cooling and overcooling of engine parts can be unfavourable for the life time of these components. Phenomena such as cavitation, pipe vibration and corrosion of materials should also be taken into account. It is important to maintain the cooling-water system at a constant temperature.

◄ A photograph of the weed strainer in a seawater inlet. Picture taken below the water surface during inspection. The shells gradually block the strainer.

▼

A complete fresh-water cooling system with a low temperature (LT) and a high temperature (HT) section for a four-stroke diesel engine.

01 High temperature (HT)-coolant pump

In four-stroke engines, most circulating pumps are driven by the engine itself. They take care of the closed coolant system of the engine block.

Temperatures between 80 °C and 95 °C.

02 Low temperature (LT)-coolant pump

These circulating pumps are usually driven by the engine. They take care of the closed coolant system for the air cooler and the lubricating oil cooler.

Temperatures between 40 °C and 50 °C.

03 Air cooler

Cools the compressed air using a turbocharger in combination with the LT system before the air is led into the inlet line of the engine.

04 Lubricating-oil cooler

Cools the lubricating oil heated by the engine.

05 HT thermostatic control valve

Keeps the HT-coolant temperature at a predetermined value by means of a three-way valve.

06 LT thermostatic control valve

Maintains the LT-coolant temperature at a predetermined value by means of a three-way valve.

07 Adjustable restriction, usually referred to as an orifice

In both systems, this type of adjustable restriction is applied. These orifices are set in such a way that when the engine is operated, the three-way controlling valve does not produce significant pressure divergences in the coolant system.

08 Safety valve

This is installed to avoid damage to the cooler casing when the supercharged-air pressure is too high.

Connection outside of the diagram

401 and 402: HT cooler in- and outlets
404: HT system air vent.
406: connection to the HT pre-heater system
408: HT coolant from the stand-by coolant pump
411: drainage HT cooling-water system
451 en 452: LT coolant in- and outlet from the LT seawater-cooler
454: LT-cooler air vent
457: LT-coolant from stand-by pump
464: LT-coolant drain

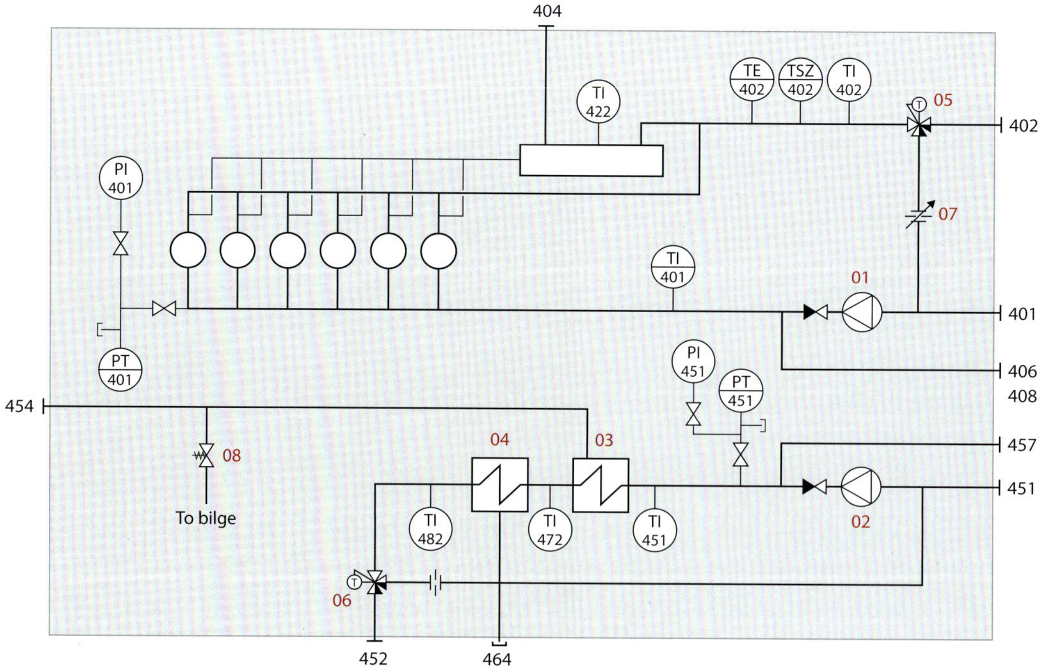

System components
01 HT-cooling water pump
02 LT-cooling water pump
03 Charge-air cooler
04 Lubricating-oil cooler
05 HT-thermostatic valve
06 LT-thermostatic valve
07 Orifice
08 Safety valve

Pipe connections
401 HT-water inlet
402 HT-water outlet
404 HT-water air vent
406 Water from preheater to HT-circuit
408 HT-water from stand-by pump
451 LT-water inlet
452 LT-water outlet
454 LT-water air vent
457 LT-water from stand-by pump
464 LT-water drain

▶
A corroded valve casing.

On board ship, attempts have been made to weld the leaking corroded cast-iron valve casing with steel-plate strips. This is obviously not very effective as cast iron cannot be welded! The welds will tear with temperature fluctuations.

▼
Centrifugal pumps are often used as coolant circulating pumps; they have a simple design, sufficient discharge head and a consistent capacity.

10.11.1 Pump capacities

Most engine-driven cooling-water pumps are centrifugal pumps.

Each pump has an accompanying discharge head and capacity curve generated at a certain rpm. Here the capacity is indicated in m^3 per hour, the discharge head in metres water column and the number of revolutions per minute.

A conspicuous detail here is that the discharge head drops considerably when capacity is increased. Cause: the resistances in the pump casing and the resistances of the whole system!
For certain engine types, the maximum capacity and discharge head are specified.

▶
A series of coolant circulating pumps in a propulsion engine.

▶

Two pump characteristics or Q-h curves. Vertically, the discharhe head is shown in metres water column. Horizontally, the capacity in cubic metres per hour. As shown the discharge head H decreases as the capacity Q increases.

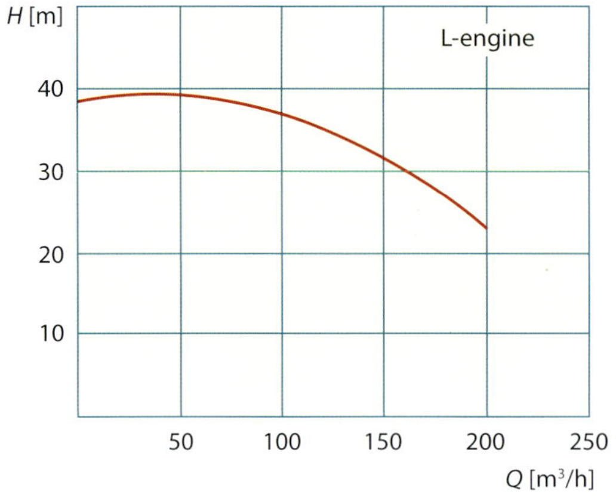

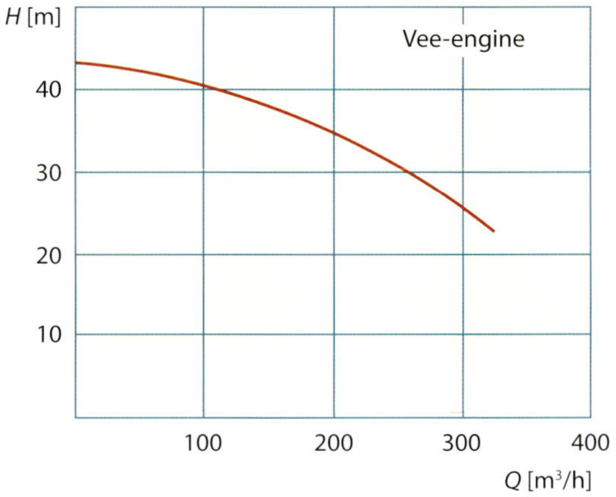

10.11.2 Pipe diameters

These are stated in a table for pipe diameters in which the velocities on the suction- and delivery side of the pump are provided.

Also see Chapter 8, *Fuels, fuel-line systems and fuel cleaning*.

10.11.3 Cooler capacities

A cooler has to be designed to have some overcapacity in view of the risk of contamination on **both** sides of the heat exchanging surface.

Pipe dimensions

Recommended maximum fluid velocities and flow rates for pipework*

Nominal pipe diameter ↓	Flow rate [m/sec] Flow amount [m³/h]									
Media →	Sea-water		Fresh water		Lubricating oil		Marine diesel oil		Heavy fuel oil	
Pipe material →	Steel galvanized		Mild steel		Mild steel		Mild steel		Mild steel	
Pump side →	suction	delivery	suction	delivery	suction	delivery	suction	delivery	suction	delivery
32	1.0 2.9	1.4 4.1	1.5 4.3	1.5 4.3	0.6 1.7	1.0 2.9	0.9 2.6	1.1 3.2	0.5 1.4	0.6 1.7
40	1.2 5.4	1.6 7.2	1.7 7.7	1.7 7.7	0.7 3.2	1.2 5.4	1.0 4.5	1.2 5.4	0.5 2.3	0.7 3.2
50	1.3 9.2	1.8 12.7	1.9 13.4	1.9 13.4	0.8 5.7	1.4 9.9	1.1 7.8	1.3 9.2	0.5 3.5	0.8 5.7
65	1.5 17.9	2.0 23.9	2.1 25.1	2.1 25.1	0.8 9.6	1.5 17.9	1.2 14.3	1.4 16.7	0.6 7.2	0.9 10.8
80	1.6 29.0	2.1 38.0	2.2 39.8	2.2 39.8	0.9 16.3	1.6 29.0	1.3 23.5	1.5 27.1	0.6 10.9	1.0 18.1
100	1.8 50.9	2.2 62.2	2.3 65.0	2.3 65.0	0.9 25.5	1.6 45.2	1.4 39.6	1.6 45.2	0.7 19.8	1.2 33.9
125	2.0 88.4	2.3 101.6	2.4 106.0	2.4 11.4	1.1 48.6	1.7 75.1	1.5 66.3	1.7 75.1	0.8 35.3	1.4 61.9
150	2.2 140.0	2.4 152.7	2.5 159.0	2.6 165.4	1.3 82.7	1.8 114.5	1.5 95.4	1.8 114.5	0.9 57.3	1.6 108.2
200	2.3 260.2	2.5 282.8	2.6 294.1	2.7 305.4	1.3 147.0	1.8 203.6	– –	– –	– –	– –
Aluminium brass	2.6 294.0									
250	2.5 441.8	2.6 459.5	2.7 477.2	2.7 477.2	1.3 229.8	1.9 335.8	– –	– –	– –	– –
Aluminium brass	2.7 447.2									
300	2.6 661.7	2.6 661.7	2,7 687.2	2,7 687.2	1.3 330.9	1.9 483.6	– –	– –	– –	– –
Aluminium brass	2.8 712.5									
350	2.6 900.5	2.6 900.5	2.7 935.2	2.7 935.2	1.4 484.9	2.0 692.7	– –	– –	– –	– –
Aluminium brass	2.8 969.8									
400	2.6 1176.2	2.7 1221.5	2.7 1221.5	2.7 1221.5	1.4 633.3	2.0 904.8	– –	– –	– –	– –
Aluminium brass	2.8 1266.7									
450	2.6 1488.6	2.7 1545.9	2.7 1545.9	2.7 1545.9	1.4 801.6	2.0 1145.1	– –	– –	– –	– –
Aluminium brass	2.9 1660.4									
500	2.6 1837.8	2.7 1908.5	2.7 1908.5	2.7 1908.5	1.5 1060.4	2.1 1484.6	– –	– –	– –	– –
Aluminium brass	2.9 2049.9									

* The velocities given in the above table are guidance figures only. National standards can also be applied.

◀

Pipe dimensions and recommended maximum flow velocities.

The velocities are denoted in metres per second.
For example, the coolant flow rate for pipe diameters of 32 to 200 millimetres lies between 1.5 and 2.5 metres per second on the suction side and between 1.5. and 2.7 metres per second on the delivery side. These data are shown in the column 'Fresh water – Mild Steel'.
The capacities on the suction side are 4.3 and 294 m³ per hour respectively and 4.3 and 305 m³ on the delivery side.
Too high a flow velocity generates too much resistance and consequently pressure losses in the pipes.
This table also provides a good guideline for other liquids, such as seawater, lubricating oil and fuels.
This is particularly significant for fuels when designing **'booster units'**.

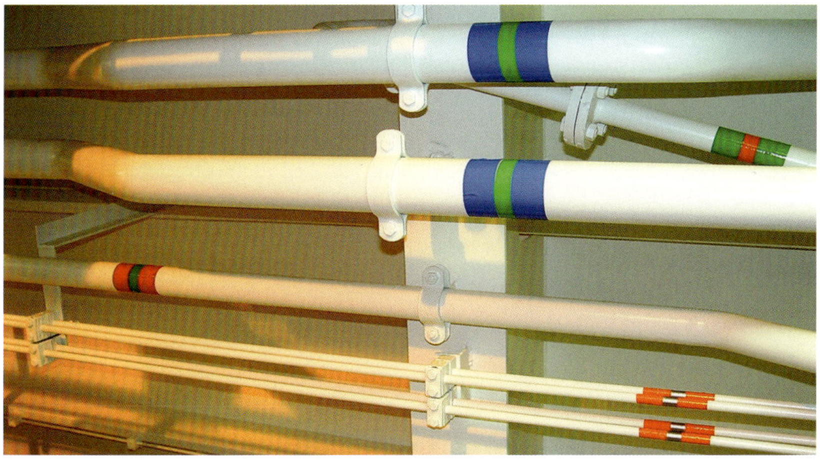

◄

An example of piping codes.

By means of a standard code, it is possible to indentify the gas or liquid flowing through the pipes at any given point in the installation.

Important data with regard to coolers

This concerns:

- the capacity in kW;
- the amount of flow in m^3 per hour;
- the temperature difference before and after the cooler in degrees Celsius;
- the maximum pressure drop over the cooler on the fresh-water side in kilopascal;
- the amount of flow for seawater or another liquid in m^3 per hour;
- the maximum pressure drop over the cooler on the seawater side in kilopascal;
- the maximum fresh-water temperature after the cooler;
- the contamination factor of the cooler. This constitutes approximately 15%.

◄

Plate coolers in a MAN Diesel four-stroke medium-speed diesel engine. Shown here the lubricating-oil and fresh-water cooler.

▲

The air coolers (1) after the turbochargers in this high-speed four-stroke MTU-diesel engine are plate coolers.

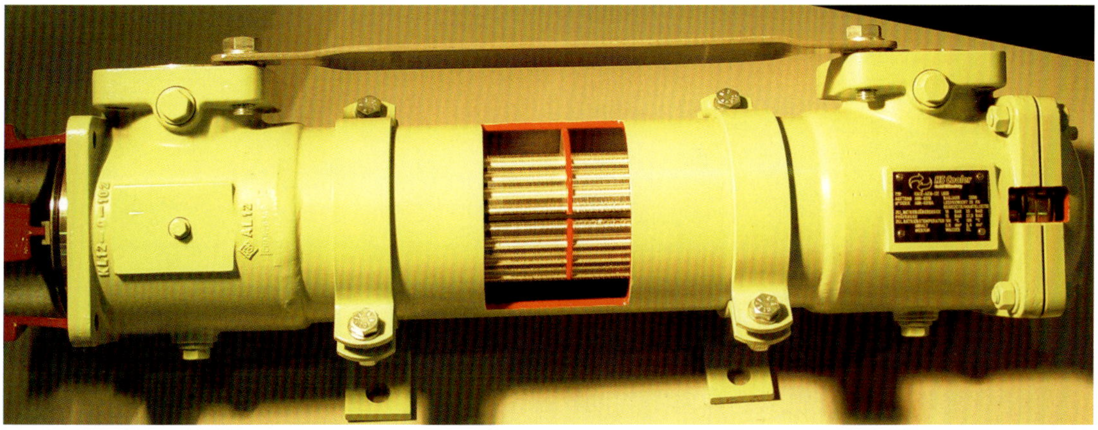

◄

When designing coolers, it should be taken into account that there may be some contamination of the cooling surfaces. Here a cutaway view of a pipe cooler.

▲

A plate-cooler detail.

On the right, the bolts are of sufficient length to increase the cooling capacity by fitting more plates.

10.11.4 Fresh-water expansion tank

This compensates volume fluctuations at different temperatures in the system. Furthermore, the fresh-water cooling expansion tank serves as an air-bleed system, storage tank and generates the required static pressure on the suction pipe of the cooling-water pumps. So, in stationary cooling-water pumps there is always pressure on the suction side of the pump which is why water is produced the instant the engine starts. **The centrifugal pumps are not self-priming and consequently experience problems with initial coolant flow when this tank is excluded.**

Static pressure is formed by placing the expansion tank at a certain height in relation to the pumps. This is often between 5 and 15 metres water column. The volume is usually 10 to 15% of the system content. The cooling-water treatment chemicals can be dosed in this tank. The individual bleeding pipes of each engine are fitted to the tank. **These pipes should never be placed horizontally in order to avoid air accumulation.**

10.11.5 Cooling-water drain tank

It is recommended in larger systems to drain the cooling water into a tank so it can be re-used after repairs to the cooling-water system.

10.11.6 Cooling-water pre heater

For the slightly larger engines it is advisable to either preheat the cooling water and therefore the engine block or maintain the correct temperature for an extended time before starting the engine. This is particularly important for back-up gensets, in order to achieve an easy start and to avoid heat tension in the engine. At sea, stopped diesel engines are generally kept at the correct temperature with preheating systems. This also applies to heavy-fuel systems.

As a result of short berthing times a large part of the machinery, such as the pumps, remain operational and the temperatures of fresh cooling water, lubricating oil and fuel remain at a constant operating temperature.

▶

An expansion tank in a small high-speed four-stroke diesel engine for coolant storage.

The system is under pressure. Opening the filler cap (1) can be very dangerous (burns).

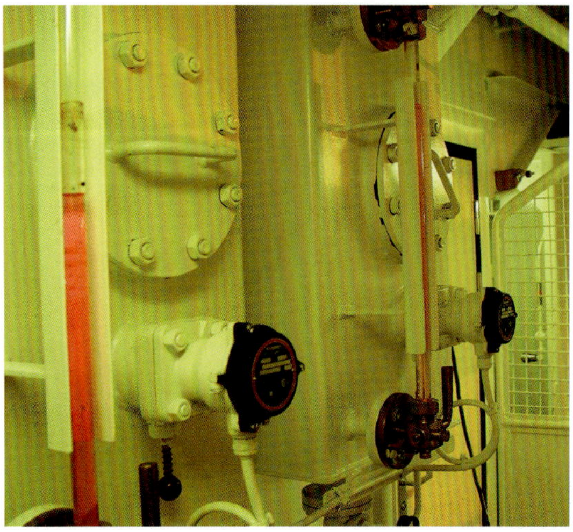

◀

Two fresh-water expansion tanks for high-speed four-stroke Deutz-diesel engines.

Shown here, the cooling-water treatment system colours the coolant pink. The tank is provided with a manhole for inspection purposes, a gauge glass and a low-water alarm. The system is not under pressure; the expansion tank is placed in an elevated position in the engine room to ensure a static pressure on the coolant pumps.

An electric heating element mounted on the coolant system of a stand-by diesel generator.

A stable coolant temperature guarantees a smooth quick start of the diesel engine. Obviously, this is imperative in emergencies.

This also applies to stopped diesel engines in diesel-power plants. Heated cooling water is also often used for emergency diesel engines or back-up gensets. Finally, in cold regions, small diesel engines are warmed or placed in heated spaces to achieve suitable temperatures prior to starting.

10.11.7 Deaerating systems

These are very important in cooling water systems. Air and other gases can cause corrosion of materials and cool the engine parts poorly causing severe damage. Pumps do not operate properly in systems containing air. Vents should be fitted on top of the systems or engines and be connected to a vent tank or expansion tank by pipes mounted vertically and preferably without bends and horizontal sections.

10.11.8 Adjustable restrictions or orifices

These have been fitted to adjust the cooling-water circulating pumps and the pressure drop when the water does not flow through the cooler.

Cooling-water vents.

Cooling-water vents are fitted on the top side of a system and should run upward to the cooling-water expansion tank. In horizontally placed vents, potential air bubbles can remain in the system and cause damage to the engine (tearing).

1 *cooling-water vents*

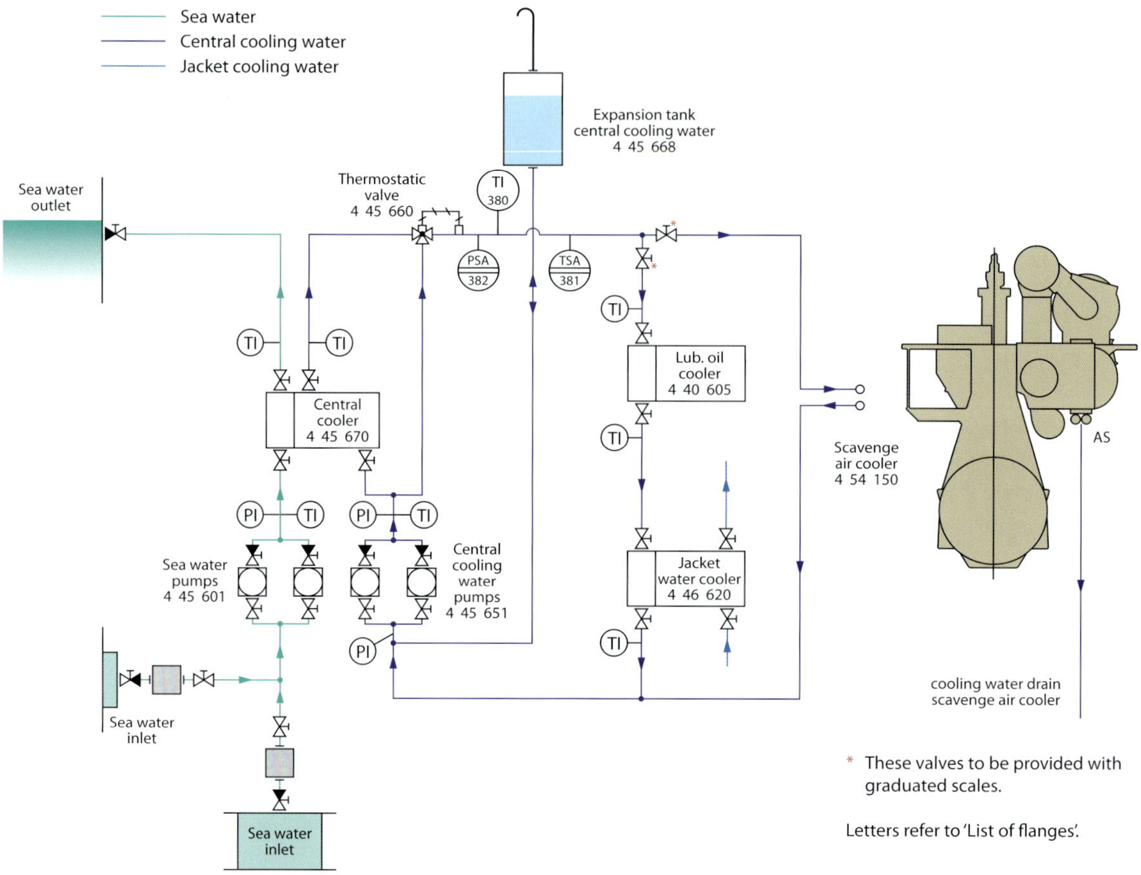

▲

A central cooling-water system for a large two-stroke crosshead engine.

The seawater-cooling system has been designed to be as short as possible and is manufactured from corrosion-resistant material. It consists of two sea inlets, two cooling-water circulating pumps and a central cooler. Further along in the ship, the seawater is discharged overboard. The closed cooling system comprises two fresh-water circulating pumps, the central cooler where the heat is discharged to the seawater, the scavenging-air cooler, the lubricating-oil cooler and the fresh-cooling water cooler (jacket-water cooler). At the top of the system, the coolant expansion tank has been installed. A thermostatic control valve regulates the amount of fresh-water coolant circulating through the central cooler. This is dependent on the required temperature of the fresh-water coolant system.

It is, of course, important that the scavenging air has a sufficiently low temperature of between 40 °C and 50 °C.

▶

A complete cooling-water system for a four-stroke medium-speed propulsion engine.

System parts

01 HT cooling-water pump
02 LT cooling-water pump
03 Scavenging-air cooler
04 Lubricating-oil cooler
05 Three-way valve, thermostatic HT
06 Three-way valve, thermostatic LT
07 HT level at cooling-water pump
08 LT level at cooling-water pump
09 Central cooler
10 Pre-heater pump
11 Pre-heater
12 Expansion tank
13 Seawater pump
14 Seawater inlet filter
15 Lubricating-oil cooler reduction gear
16 Seawater discharge valve
17 Vent
18 Drain tank
19 Transfer pump

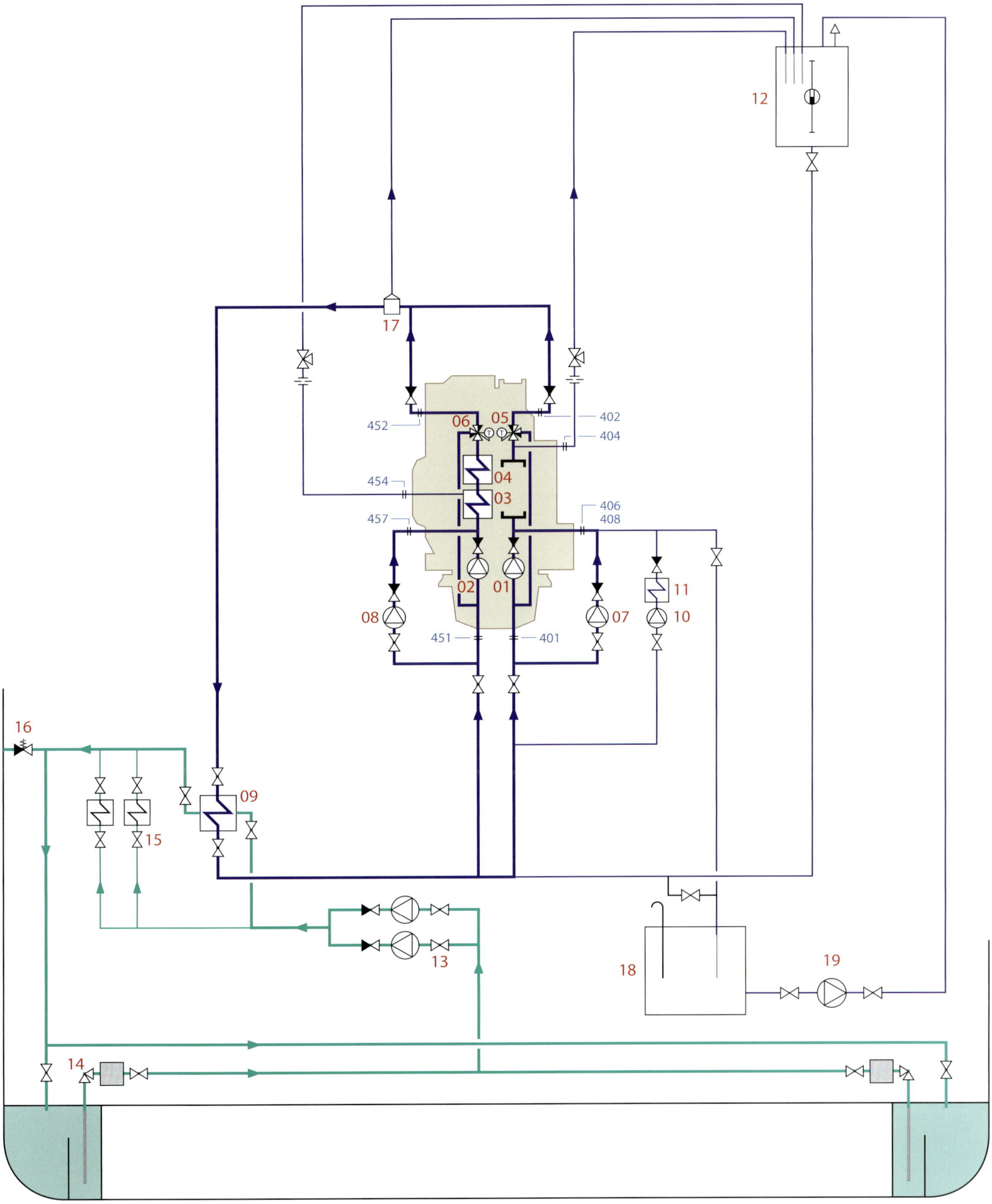
12
17
452
06
05
402
404
454
04
03
457
406
408
02
01
11
08
07
10
451
401
16
09
15
13
18
19
14

10.11.9 Cavitation

This is a phenomenon that may occur at low water pressures and relatively high cooling-water temperatures. At 1 bar, water boils at approximately 100 °C. For instance, when the HT cooling-water system on the suction side of the cooling-water pump has a temperature of 75 °C and the piston pressure drops below 0.4 bar absolute, which is 0.6 bar vacuum, the water will boil at the entrance of the pump; vapour bubbles will form in the inflowing liquid.
In the compression section of the pump the water pressure is increased considerably to, for instance, 3 bar and the water only achieves its boiling point at 133 °C. The vapour bubbles are then compressed: so-called **imploding**. **During this process huge forces are released.** Vapour bubbles which implode against the material surface cause severe damage to the material. Pump vanes begin to show an open structure and wear very rapidly. The pump capacity is then quickly reduced.

Note
Keep the pressure in the suction line at least 0.5 bar above the vapour pressure belonging to the boiling point of the cooling water related to temperature.

Example
- Cooling-water temperature 45 °C
- Boiling point at 0.1 bar absolute
- Safe suction pressure 0.1 + 0.5 = 0.6 bar absolute
- Boiling point at 0.6 bar absolute is 85 °C
- So range is 85 – 45 = 40 °C

10.12 Cooling-water system defects

10.12.1 Air in the cooling-water system

Air in the system inhibits the flow and results in little or no cooling of engine parts causing damage, such as tearing, cracking and piston seizure. Solution: ensure that the expansion tank is filled to the correct level and ensure that the air-bleeder pipes are open. The air deaerators should be fitted below the water surface in the expansion tank.

Make sure there are no flattened or bent pipes and/or incorrect packing between the flanges to under-sized holes. There should be a pipe running straight from the expansion tank to the suction side of the cooling-water pump to ensure that the pump is always filled with water.

10.12.2 Exhaust gases in the cooling-water system

In cooled exhaust-gas manifolds or double-walled casings of turbochargers, exhaust gas can enter the cooling water of a running engine through tearing. This can also occur as a result of a packing leakage under the cylinder head. **In stationary engines the cooling water can end up in the exhaust manifold, turbocharger or combustion space by tearing or a faulty packing and produce severe problems!** Controls and safety alarms for the water level in the expansion tank are therefore absolutely imperative. A **humming sound** in the expansion tank is usually indicative of gas leakage – often a leaking cylinder-head packing.

▼
Shown here each engine has an individual cooling-water expansion tank.

The connections of the cooling-water expansion pipes are found below the tanks.

1 *cooling-water expansion tank*
2 *manhole*
3 *liquid-level gauge*
4 *shut-off valve for the liquid-level gauge*
5 *float for low-water level indication*
6 *pipe to cooling-water system*

10.12.3 Cooling-water capacity is too low

Apart from (shut off) valves that are not entirely closed, the pump impeller can be affected by cavitation, thus reducing the capacity. Severe contamination in the system produces an increased discharge head due to the increased resistance in the system. Consequently, the heat transmission to the second cooling-water system is reduced.

10.12.4 Temperature control is inoperative or works poorly

This is probably produced by a defective temperature gauge, the control itself or a sticking three-way valve. In smaller engines the thermostatic control valve is often defective and needs to be replaced. It is easy to check if a control valve is functioning by dismantling it and placing it in a bucket of warm water with a temperature at least 10 °C higher than the normal cooling-water temperatures. The valve should open!

10.12.5 Contaminated coolers

This may occur at the fresh-water cooling side when there is insufficient cooling-water treatment; a thin corroded metal layer and other deposits on the material reduce the heat transmission. At the untreated water side or seawater side fine dirt particles that have passed the suction filter of the pump can form a thick layer of, for instance, fine silt, sand, vegetable- and animal remains as well as the growth of shell fish such as 'Balanus' barnacles, a mollusc.

10.13 Damaged engine parts

- When cooling is insufficient the material temperature will increase to the point where the material properties will show signs of deterioration; this could cause tearing. It could also cause irreparable damage, for instance, piston seizure due to inadequate cooling of the piston or the cylinder liner.

- Inadequate cooling of the lubricating-oil system may cause the lubricating oil to decompose leading to deposition of carbon particles on, for instance, the lubricating-oil cooler, the inside of the piston or behind the piston rings, so-called 'ring sticking'.

- When an air cooler is contaminated the fresh combustion-air temperature, and subsequently all temperatures in the combustion process increase. The temperature of the material near the combustion chamber could also rise.

- Special attention is required for: injectors, exhaust-valve seats and turbocharger bearings and -shafts. Components that are provided with bore-cooling, such as pistons, cylinder liners and cylinder heads in large engines.
  Lubricating oil that enters the pistons via a nozzle in small engines – a minimal temperature increase could cause complications.

- Material tension as a result of rapid temperature fluctuations can be fatal for an engine part. A stopped engine should not be preheated too quickly and increasing the shaft power of a running engine should also be bound to a certain period of time.

10.13.1 Pressures and temperatures in cooling-water systems

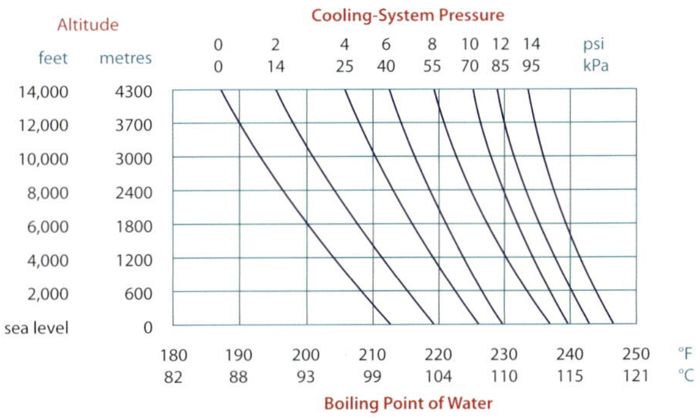

▲

A graph with on the vertical axis the seawater level in metres and feet, on the bottom horizontal axis the boiling point of water in degrees Celsius and Fahrenheit and the top horizontal axis, the cooling-water pressure in kPa and psi (pounds per square inch).

Example: At an altitude of 3000 metres above sea level (for instance, mining in the mountains) and a prescribed cooling-water pressure of 0.3 bar (30 kPa) over-pressure, water boils at 101 °C. At a cooling-water pressure of 0.7 bar (70 kPa) over-pressure, water boils at 115 °C. Without a cooling-water pressure, water boils at 89 °C at an altitude of 3000 metres.

The pressure in cooling-water systems should be sufficiently high to prevent the forming of vapour bubbles or boiling. If the cooling-water pressure is lower than that of the vapour pressure at the current temperature, the water will begin to boil. Cooling-water temperatures of 80 to 90 °C are fairly standard. In dual-fuel and gas engines, cylinder head cooling-water temperatures can reach 125 °C! The cooling-water pressure after pumping is usually 3 to 4 bar.

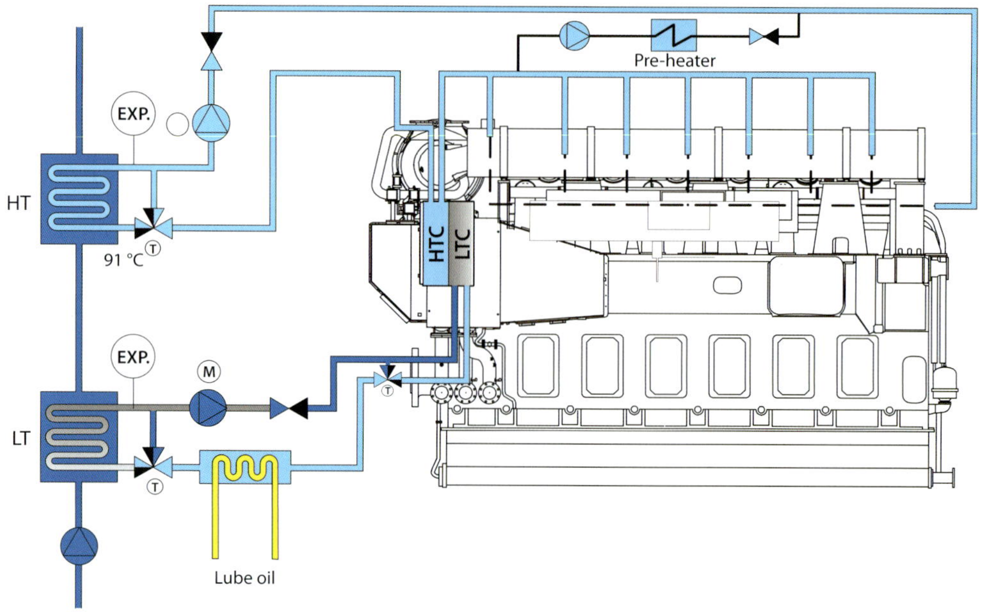

▶

A combined HT and LT cooling-water system for a four-stroke medium-speed diesel engine.

Here the lubricating-oil cooler is mounted on the LT system. The scavenging-air cooler consists of two heat exchangers; first, the air for the engine is cooled with the HT system and subsequently with the LT system. A pre-heater is placed in the HT system.

10.14 Standard cooling-water system

A simple closed fresh-water cooling system always consists of a fresh-water cooling circulating pump, the engine parts that must be cooled, an air-venting and expansion system, a fresh-water cooler and a temperature control valve, usually a three-way thermostatic control valve.

▲

Cooling fins on the cylinder liner of a small high-speed diesel engine.

10.15 Cooling systems according to the engine classification

10.15.1 Engines with a shaft power of 0 to 100 kW, four-stroke, high-speed, MDO (category I)

Small industrial diesel engines usually have a very simple but effective cooling system.
There a various cooling methods:

A with air;
B with untreated water or seawater;
C with a closed fresh-water cooling system;
D radiator cooling;
E with coolant in a closed system;
F with lubricating oil in a closed system.

▲

A small single cylinder air-cooled diesel-power unit (arrow = air supply).

A Air cooling

Engines with air cooling are still fairly common, especially in smaller power outputs.

Cooling method

The heat of the engine is discharged to the cooling fins on the cylinder. A ventilator driven from the crankshaft by a V-belt or an impeller integrated into the fly wheel, sends cool air through sheet-steel ducts along the fins. The heated air must be expelled far from the engine block to avoid it short circuiting with the cool air supply. For instance, in sailing- and motor boats or back-up gensets attention is given to the hot air discharge and cool air supply.
This system can only be applied to low-load engines.

Advantages

- Simple.
- No risk of frost damage.
- No cooling-water system required.
- Economical.

Disadvantages

- Dirt and dust settles on the cooling fins.
- Mediocre cooling capacity.
- No temperature control.
- No sound-proof cooling-water jacket.
- Supply- and discharge ducting require a great deal of space in built-in engines.
- The power output has to be reduced at high air temperatures.

In practice this cooling method is seldom used.

B With untreated water or seawater

This method is often applied. The canal-, river- or seawater is drawn in via an inlet or weed chest by a engine-driven cooling-water pump. The coolant then flows through the engine block and is discharged back into the untreated water header above the water level. The temperature is regulated by a temperature controlled valve placed in the discharge pipe after the engine. Since this system is often absent, the flow of the untreated water depends on the number of revolutions of the water pump and therefore those of the engine.

Advantage

- Simple system.

Disadvantages

- The cooling water is neither clean nor treated; dirt particles can settle on the cooling surfaces and heat transmission is reduced.
- Potential blocking of the weed chest.
- All cooling water needs to be drained during periods of frost.

In the direct seawater cooling system it is possible for salts to be deposited on the cooling surfaces when the cooling-water temperature exceeds approximately 50 °C. This will cause scaling. Due to the high chloride content of seawater, metal corrosion is severe.

C With a closed fresh-water cooling system

This is commonly applied. The closed cooling-water system is filled with distilled water and equipped with a cooling-water treatment. The system is often maintained at an over-pressure to prevent boiling and is provided with a closed expansion tank.
The heat of the engine parts is absorbed by this closed fresh-water cooling system and is transmitted to a raw water or seawater system through a heat exchanger (cooler). The system, including the cooler, is designed to be as small as possible, and consists mostly of seawater resistant materials, such as stainless steel, brass and cunifer. In smaller engines the pipes may be manufactured from synthetic materials. The temperature of the first cooling-water system is controlled with a thermostatic three-way valve. In an initial start in a cold engine the coolant circulates through the engine. Only when the water achieves the correct temperature in proportion to the engine load increase, does more water flow through the cooler.

Advantage

- Possibility of cooling-water treatment; no corrosion.

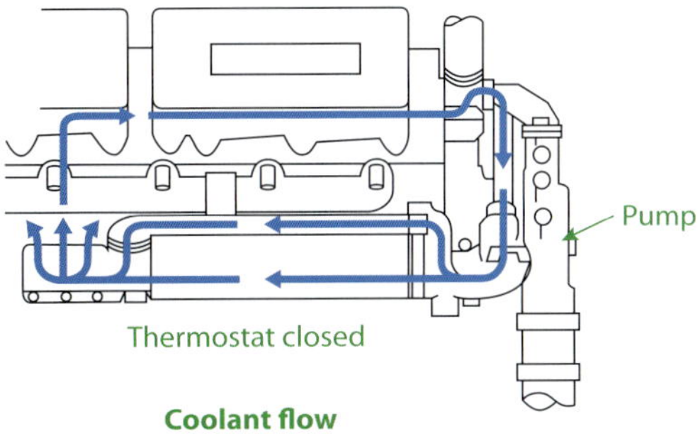

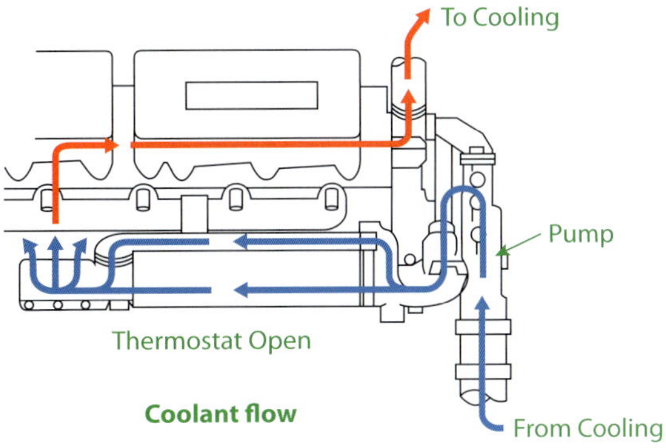

◄ A simple cooling-water system.

Top picture: The cooling system is cold. The water circulates through the engine.
Bottom picture: The cooling system has achieved the set temperature. The thermostatic control valve allows a certain amount of water through the heat exchanger, thus ensuring that the pre-set cooling-water temperature of the engine remains constant.

▶

Radiator cooling.

Whenever the heat in the cooling water cannot or only with great difficulty be emitted to the surface of the cooling water is radiator cooling used, the heat is released into the ambient air.

▲

Regular deaeration prevents problems.

Properly deaerated cooling water is imperative for adequate cooling of each part. Here a (manual) deaerator vent. Regular dearation prevents problems.

▶

Cooling-water pumps mounted on the engine.

Many modern four-stroke medium-speed diesel engines drive several cooling-water pumps. These are driven from the crankshaft using cogwheels.

▼

Radiator cooling.

A radiator is relatively large in comparison to the common fresh-water cooler. Cooling with ambient air temperature sometimes in excess of 40 °C requires a large cooling surface.
The ventilator is operated by the engine by means of a belt- or gear-belt drive from the crankshaft or the camshaft.

1 *heat exchanger*
2 *ventilator (behind grill guard)*
3 *cooling-water pump*
4 *thermostatic three-way valve*
5 *grill guard.*

▼

A modern cooling-water system.

D Radiator cooling

If there is no untreated water available or if one would prefer not to use untreated or salt water in the 'second' system, one could cool to the ambient air temperature with a radiator. A radiator resembles a heat exchanger which has warm fresh water on the one side and cool air on the other, and discharges heat with an engine-driven ventilator. This is often used for 'on-shore' -installations such as diesel gensets, mobile diesel power units in, for instance, large buildings such as hospitals, office buildings, factories and other facilities requiring electricity.

E With coolant in a closed system

There are engine manufacturers that prefer to have their engines filled with a ready-made coolant. This has the following properties:

- The agent contains an anti-corrosion additive; all the oxygen atoms are bound.
- The agent is resistant to below zero temperatures.
- The agent does not attack packings, rubber seals or metals.
- The agent forms a strong, thin coating on the metal parts preventing corrosion.
- This kind of cooling system often contains a liquid filter which filters out sand grains, metal particles and dirt and so avoiding abrasion.

Advantages

- No cooling-water treatment.
- Agent meets all the required properties.
- No problems at below zero temperatures.

Disadvantages

- Refilling after leakages is expensive.
- Prevent contact with eyes and skin. Follow the user instructions.

F With a closed system with lubricating oil

Some manufacturers build small engine series equipped with this cooling system.

▲
A diesel generator with on the left, the radiator cooler and on the right, the electric generator with its own closed-air ventilation system. The air for the electric generator is cooled by fresh water in a heat exchanger.

10.15.2 Engines with a shaft power of 100 to 10,000 kW, four-stroke, high-speed, MDO (category II)

These large industrial engines all have closed fresh-water cooling systems:

- **with a secondary raw water or seawater system;**
- **with a secondary system to ambient air cooled with a radiator.**

Both the fresh-water circulating pump and the raw cooling-water circulating pump are driven by the engine. Since most of these engines are provided with a turbocharger, an inter-cooler has been placed after the turbocharger cooled by the circulating water in the block. This is insufficient for highly loaded engines, and therefore the cooling system is provided with a high- and low-temperature system. This is often referred to as the HT- and LT-system.

▼
The Deutz D 4.29 with natural aspiration and the Deutz DT 4.29 with supercharged air; both diesel engines are cooled with lubricating oil.

A HT cooling-water system by Caterpillar.

This has a:

- low-water flow shut-down;
- exhaust-heat recovery device;
- two thermostatic control valves; a warm-up thermostatic valve and a load-balancing thermostatic valve.
- the expansion tank is under pressure.

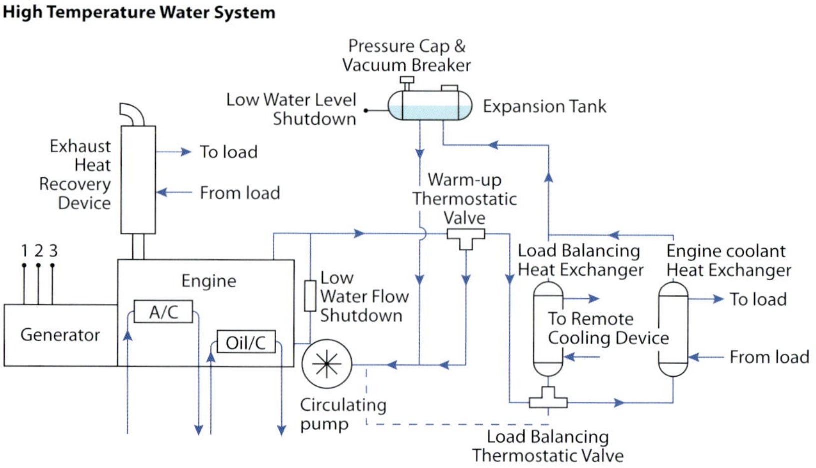

10.15.3 Engines with a shaft power of 500 to 30,000 kW, four-stroke, medium-speed, HFO (category III)

These large industrial engines have closed fresh-water cooling systems, often with a HT- and LT-system where cooling of both the air cooler and the lubricating-oil cooler is performed by the LT-system and the engine block by the HT-system.

At these high shaft outputs, a large amount of heat is released; the cooling systems increase in size proportionally .

Following is an example of a Caterpillar MaK 12 M 32 C diesel engine, a V-engine running on HFO.

shaft power P_{shaft} = 5760 kW

heat dissipation – engine block	800 kW
air cooler I	1880 kW
air cooler II	530 kW
lubricating-oil cooler	840 kW
gear box and generator coolers	223 kW
total	4273 kW

The HT-cooling absorbs heat from the engine block and air cooler I = 2680 kW.
The LT-cooling circuit absorbs heat from air cooler II, the lubricating-oil cooler and the gear box/generator coolers = 1593 kW.

▶
A cooling-water system of a Caterpillar MaK 16 M 32 C diesel engine.

Description

The engine has a two-stage air cooler with CH1 and CH2. The system consists of a HT system and fresh-water cooling pumps FP3 and FP5.

The HT cooling-water system cools air cooler CH1 and the engine block, it then flows to the thermostatic three-way valve FR1. In a cold system, it circulates back to the suction side of the cooling-water pumps and at the moment that the system achieves the correct temperature, the three-way valve opens and delivers, depending on the engine load, more of less water through the heat exchanger FH1. The expansion tank FT1 is connected to the suction pipe of pumps FP3 and FP5. It exerts a sufficient amount of static pressure on the pump to have it start effortlessly as the pump is always filled with water.

Pre-heating

The HT system has its own pre-heater pump FP7 and pre-heater FH5. The LT system cools the air cooler CH2 and lubricating-oil cooler LH1. The LT circulating pumps FP4 and FP6 first pump the water through the air cooler in order for the air temperature to be lowered as much as possible. This is usually to 40 °C.
The thermostatic three-way valve FR2 ensures that the LT temperature is constant and delivers more or less water through the LT cooler FH2.
This LT system also has its own expansion tank FT2 with a connection to the suction side of the pumps.

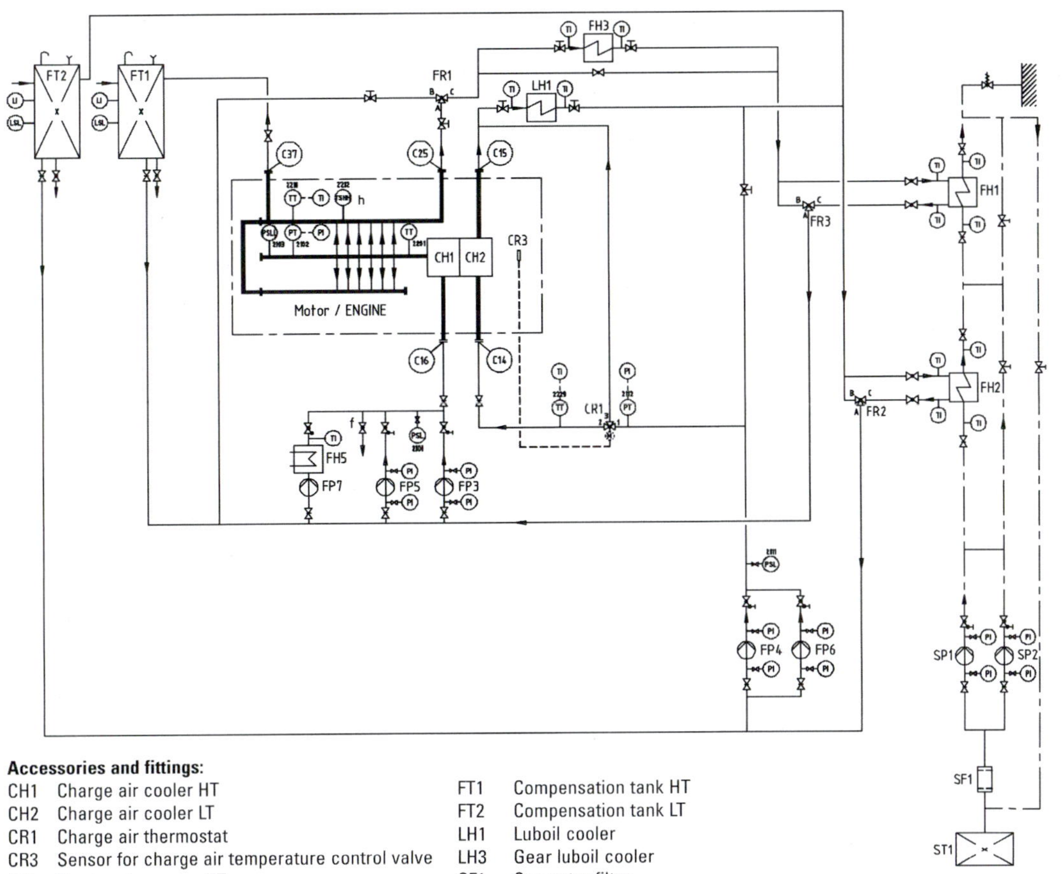

Accessories and fittings:

CH1 Charge air cooler HT
CH2 Charge air cooler LT
CR1 Charge air thermostat
CR3 Sensor for charge air temperature control valve
FH1 Freshwater cooler HT
FH2 Freshwater cooler LT
FH3 Heat Consumer
FH5 Freshwater preheater
FP3 Freshwater pump (separate) HT
FP4 Freshwater pump (separate) LT
FP5 Freshwater stand-by pump HT
FP6 Freshwater stand-by pump LT
FP7 Preheating pump
FR1 Temperature control valve HT
FR2 Temperature control valve LT
FR3 Temperature control valve HT

FT1 Compensation tank HT
FT2 Compensation tank LT
LH1 Luboil cooler
LH3 Gear luboil cooler
SF1 Seawater filter
SP1 Seawater pump
SP2 Seawater stand-by pump
ST1 Sea chest
LI Level indicator
LSL Level switch low
PI Pressure indicator
PSL Pressure switch low
PSLL Pressure switch low
PT Pressure transmitter
TI Temperature indicator
TSHH Temperature switch high
TT Temperature transmitter (PT 100)

Notes:

f Drain
h Please refer to the measuring point list regarding design of the monitoring devices

Connecting points:

C14 Charge air cooler LT, inlet
C15 Charge air cooler LT, outlet
C16 Charge air cooler HT, inlet
C25 Cooling water, engine outlet
C37 Vent

General notes:
For location, dimensions and design (e. g. flexible connection) of the disconnecting points see engine installation drawing.

With skin cooler not required:
- Seawater system (SP1, SP2, SF1, ST1) Temp. - control valve FR3 required, if heat recovery installed.

Raw-water cooling system

This system is designed to be as short as possible due to potential corrosion and contamination risks. Via the untreated or seawater inlet filter SF1 the seawater circulating pumps SP1 and SP2 draw in cooling water, pump it to the LT and HT coolers FH2 and FH1 and then the water is discharged overboard. Re-circulation is also possible using a manual valve. Furthermore, both coolers are equipped with a 'bypass' valve.

General

Pressure and temperature gauges are fitted throughout the system. Furthermore, the expansion tanks are provided with low-water alarms. Before and after each cooler, thermometers are installed to provide accurate readings of temperature differences in the cooler.

Today, this data is made available on a pc, where all required data is gathered in the 'cooling-water system' databank.

Some remarks

- The inlet temperature in the LT system may not exceed 38 °C. These temperatures are only achieved in tropical areas, shallow coastal waters and estuaries.
- Expansion tanks should be installed at least four metres above the centre line of the crankshaft of the engine.
- It is possible to insert an oil detector in the cooling-water system.
- Flow velocities in the fresh-water systems on both the suction and delivery sides are 2 to 3 m/sec.

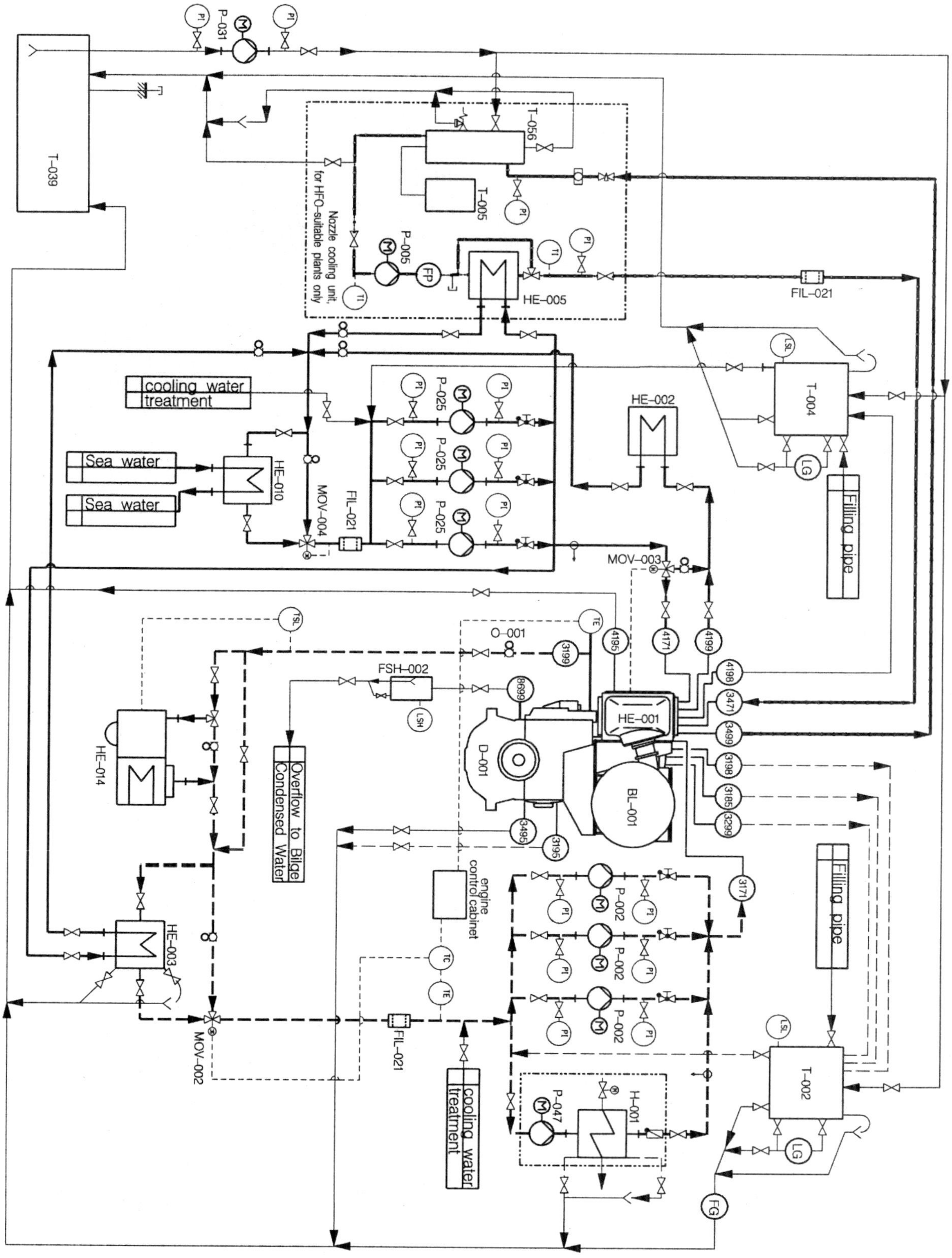
T-039
P-031
T-056
T-005
P-005
FP
HE-005
Nozzle cooling unit, for HFO-suitable plants only
FIL-021
T-004
LSL
LG
Filling pipe
cooling water treatment
P-025
HE-002
Sea water
Sea water
HE-010
MOV-004
MOV-003
O-001
FSH-002
LSH
TSL
TE
HE-001
D-001
BL-001
HE-014
Overflow to Bilge Condensed Water
engine control cabinet
P-002
HE-003
MOV-002
TC
H-001
P-047
T-002
FG

◄

A complete cooling-water system for a medium-speed four-stroke diesel engine by MAN Diesel.

Some remarks

- Here a HT and LT system are used.
- The cooling-water pumps are not driven by the engine but rather by an electromotor.
- A seawater evaporator HE-014 has been integrated into system.
- Engines running on heavy oil are equipped with separate nozzle cooling with a circulating pump P-005 and cooler HE-005.
- Additionally, cooling water that, for instance, is required during repairs, can be drained and stored in drain tank T-039 and pump P-031 can pump the drained water back into the system via expansion tank T-002.
- The seawater system only cools the LT system.
- Here the HT system is cooled by the LT system. Therefore, this seawater system is very compact.
- Air cooler HE-001 and the lubricating-oil cooler are cooled by the LT system.
- Seawater evaporator HE-014 is heated by the HT system.
- Cooling-water treatment takes place on the suction side of the HT and LT circulating pumps.
- Orifice 0-001 has been added to the outlet of the HT system to maintain sufficient cooling-water pressure in the engine block in order to avoid boiling the cooling water in the hottest sections.
- Potential fuel leakage to the cooling-water system can be clearly detected in expansion tank T-005 of the nozzle cooling-water system.
- Air coolers of turbochargers for these engines often comprise two parts: a section cooled by HT water and after this, a section cooled with LT water.
- Temperature LT system 36 °C -40 °C.
- Three-way valve MOV-003 regulates the following temperatures.

1 At a low engine load, the LT coolant is not sent through the cooler in order to prevent the air for the engine from becoming too cold. This has a negative effect on the combustion process and promotes corrosion due to the presence of sulphur in heavy oil.

2 It reduces the amount of condensation that is formed in the cooler in tropical conditions.
The air-inlet temperature is continuously compared with the outside-air temperature and if this rises, the air-inlet temperature is increased accordingly in order to avoid condensation of water vapour.

- Most coolers have an over-capacity of at least 15% with regards to contamination.
- When anti-freezing agents are used, the cooling-water system capacity is considerably reduced.
- In order to remain within the maximum NOx emissions set by the IMO (International Maritime Organisation) at a shaft power of 100% and seawater temperature of 25 °C, a LT system temperature for the LT cooler of 32 °C must be maintained. At a higher air-inlet temperature the process temperature also rises, which in turn results in higher nitrous-oxide (NOx) emissions.
- Small filters FIL-021 have been placed in the cooling-water system. These must be inspected after maintenance on the system.
- The injector-cooling water system includes a pre-heater HE-500 that heats up the injectors prior to starting the engine.
- If a central cooling-water system is used for several diesel engines or, for instance, main and auxiliary engines on a large ship, adjustable orifices should be installed in the cooling-water system so that the flow can be meticulously distributed throughout all the engine systems. These are generally not readjusted later!
- The outlet temperature of the HT system is approximately 90 °C.
- The engine is preheated and maintained at a temperature of up to 60 °C.
- Preheating a cold engine from 10 °C to 60 °C takes four hours!
- The pre-heater is also often switched on regularly in diesel-power plants which are operating at low load and when fresh-water generators run at maximum capacity.
- The seawater evaporator is switched off when the HT cooling water falls below 88 °C.
- Bureau Veritas requires that the condensed water in the air cooler be drained to a condensate observation tank. There should be a high-water level alarm installed. In this way, condensate reaching the combustion chamber is avoided.
- The seawater filter should have a mesh size of between 2 to 4 mm. The recommended mesh size for dredgers in waters containing a large amount of sand is 0.3 to 0.5 mm.

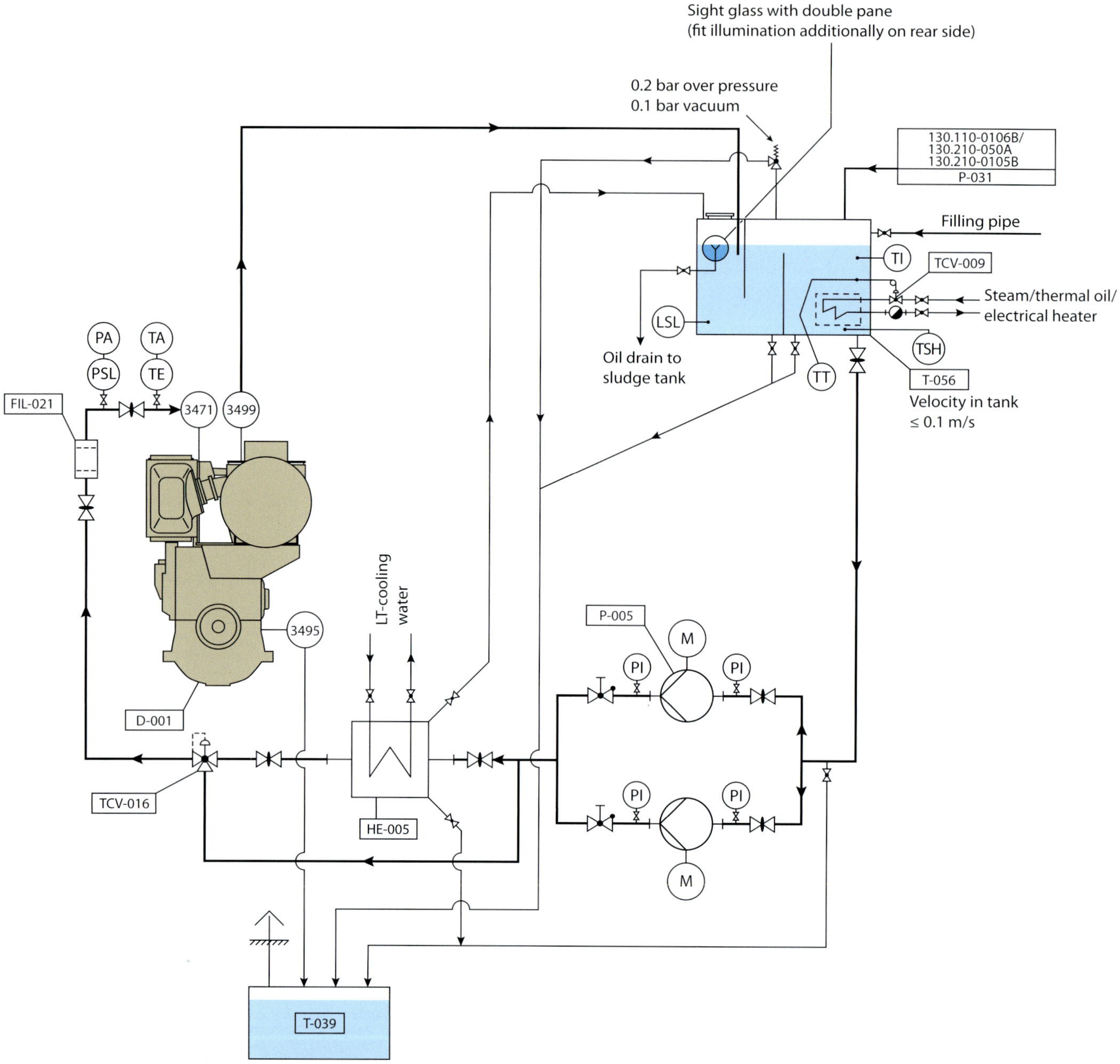

D-001	Diesel Engine	3471	Nozzle cooling water inlet
FIL-021	Strainer	3495	Nozzle cooling water drain
HE-005	Nozzle cooling water cooler	3499	Nozzle cooling water outlet
P-005	Nozzle cooling water pump		
T-039	Cooling water storage tank		
T-056	Nozzle cooling water expansion tank		
TVC-009	Temperature control valve for heating		
TVC-016	Temperature control valve for nozzle cooling water		

▲

The nozzle cooling-water system in MAN Diesel four-stroke diesel engines.

Some remarks

The expansion tank is fitted as an observation tank to detect potential leakage of fuel into the nozzle cooling water. Possible oil residue can be funnelled off. A fine-filter FIL-021 prevents dirt particles from obstructing the narrow passages in the nozzles.

- The cooling water is preheated to between 60 °C and 80 °C.
- Depending on the engine type the cooling-water inlet temperature for the nozzle is:

  type 40/54 and 58/64 80 to 85 °C

  type 32/40 and 48/60 55 to 60 °C

There are complete modules available for nozzle cooling-water systems. These modules are frame-mounted.

10.15.4 Engines with a shaft power output of 1500 to 100,000 kW, two-stroke, low-speed, HFO (category IV)

The large two-stroke crosshead engines have beside the standard cooled parts, two additional special systems:

1 **a separate piston-cooling system**; cooling with both water and lubricating oil;
2 **a cooling system for the crosshead guides**; this is connected to the lubricating-oil system.

General

The engines in this category with shaft power outputs up to 100,000 kW have to discharge enormous amounts of heat. For a MAN Diesel 12K98MC with a shaft power of 68,640 kW the following values can be read in the 'Project Guide':

air coolers	27,390 kW
lubricating-oil cooler	5,760 kW
cylinder cooling-water cooler	9,920 kW
	33,070 kW

A total of 33 MW of heat is discharged into the seawater!
The circulating pumps also have a high capacity:

- Fresh-water cooling pump 590 m^3 per hour
- Sea-water cooling pump 770 m^3 per hour

In these extremely large diesel engines the 'consumption figures' are also huge!

air consumption per hour	635,040 kg
fuel consumption per hour	11,737 kg
lubricating-oil consumption per hour	61 kg
shaft power output of this giant	68,640 kW

Details

Fresh-water cooling system. Standard system.

Cylinder cooling-water system

This system cools the cylinder liner, the cylinder cover, the exhaust-valve casing and heats the fuel drainage lines.

▼
Both plate coolers of a HT cooling-water system.

▼
Diesel generator sets on board large seagoing ships often share their cooling-water system with that of the main engine when at sea.

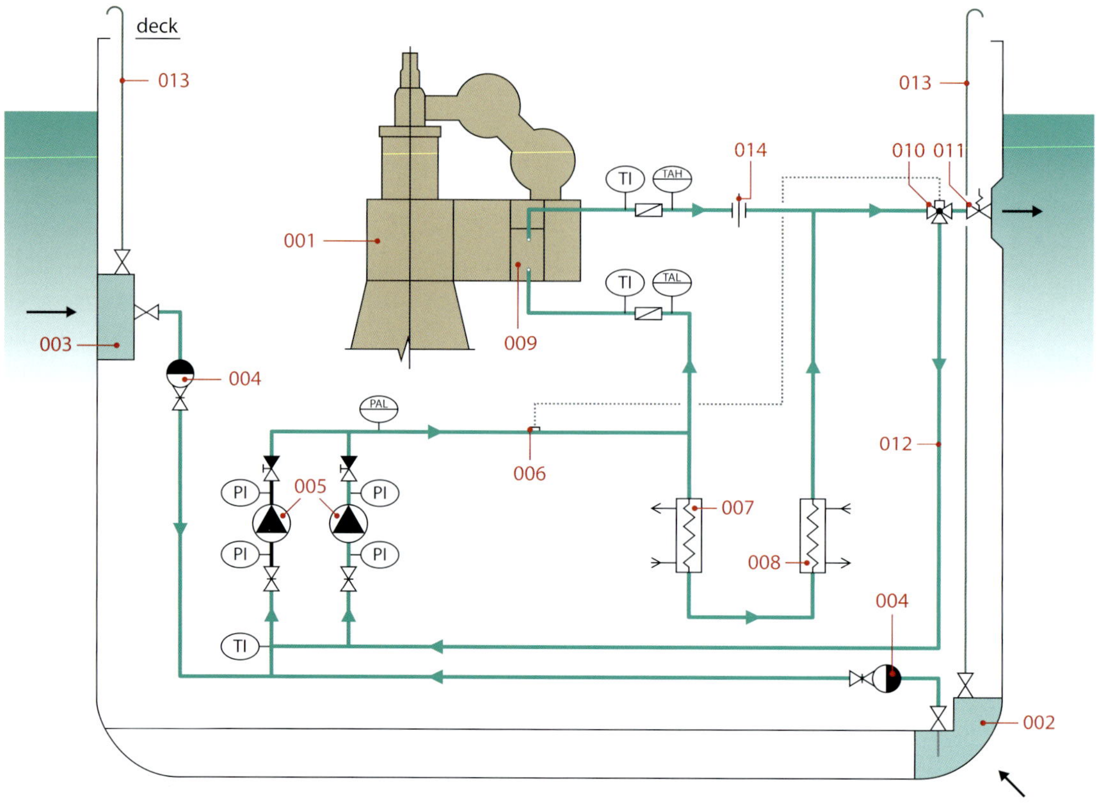

001 Main engine
002 Low sea chest[1]
003 High sea chest
004 Sea water filter screen max. mesh size 6 mm
005 Sea-water pump
006 Temperature sensor
007 Lubricating-oil cooler
008 Cylinder cooling-water cooler
009 Scavenge-air cooler
010 Automatic temperature control valve
011 Overboard discharge valve
012 Warm sea-water return line
013 Air vent[2]
007 Throttling disc[3]

Remarks:
1 If requested, two low sea chest are applicable.
2 Air vent (air vent pipe or equal system acc. to shipyard's design.
3 When using a valve, lock in proper position to avoid mishandling.
– Air vent and drain pipes must be fully functional at all inclination angels of the ship at which the engine must be operational.

▲
The cooling-water system of a two-stroke crosshead engine from MAN Diesel.

The cylinder cooling-water pumps draw the water from the cylinder cooling-water cooler and force the water through the engine.

- An orifice has been fitted to the cooling-water outlet to maintain sufficient cooling-water pressure in the engine.
- The preheated cooling water can also be used to heat the seawater evaporator and is then returned to the cylinder cooling-water cooler.
- The thermostatic valve sends more or less water to the cooler and responds to the temperature sensor placed in the engine cooling-water outlet.
- A deaerator tank prevents air from entering the system. The air rises to the cooling-water expansion tank. If air is in the system, this is indicated by an alarm.
- Several pipelines are connected to the expansion tank. They are all mounted on the top of objects, such as the:
- main engine;
- deaerator tank;
- cylinder cooling-water cooler;
- auxiliary engines.
- The heating for the seawater evaporator should be gradually increased in order to avoid fluctuations in the cooling-water temperature. The minimum is a period of three minutes.
- The expansion tank must be placed at least five metres higher than the highest point of the discharge cooling-water pipe of the engine. The discharge cooling-water temperature of the three generators is kept constant at 80 °C using three-way valves.

Port operations

A stopped main engine is heated by stand-by generators. As the preheating capacity of a low-load generator is probably too low, a pre-heater must be installed.

The pre-heater is on stand-by when the stop valves A are closed and the stop valve B is open. The engine-driven generator cooling-water pumps pump the heated water from the top to the bottom through the engine towards the thermostatic three-way valve.

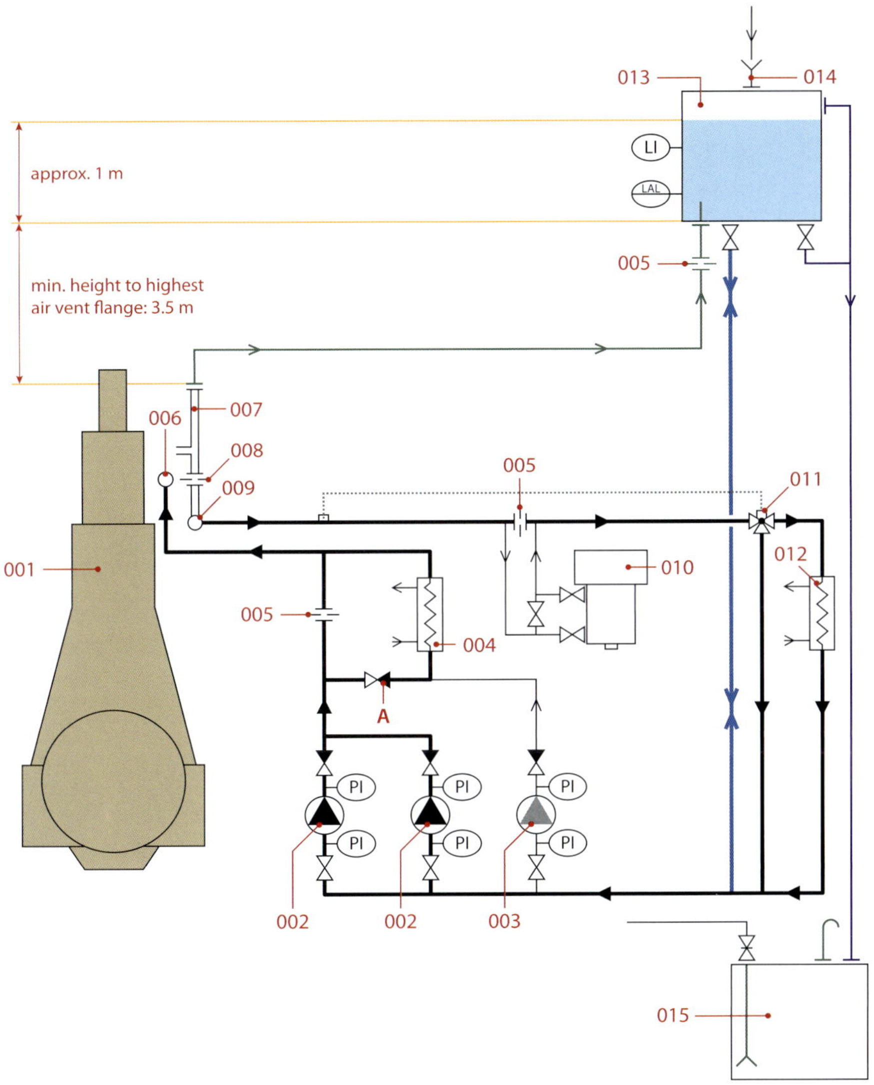

001 Main engine
002 Cylinder cooling water pump
003 Pre-heating circulating pump (optional)
004 Heater for cylinder cooling-water circuit
005 Throttling disc[1]
006 Cylinder cooling-water inlet
007 Air vent pipe
008 Throttling disc (adjustable, on engine)
009 Cylinder cooling-water outlet
010 Freshwater generator
011 Automatic temperature control valve
012 Cylinder cooling-water cooler
013 Cylinder cooling water expansion tank
014 Filling pipe/inlet chemical treatment
015 Cooling water drain tank

Remarks:

1 When using a valve, lock in proper position to avoid mishandling.

A Only when pos. 003 is installed.

– Air vent and drain pipes must be fully functional at all inclination angels of the ship at which the engine must be operational.

Fresh water pipes
Balance pipe
Water drain pipes
Air vent pipes
Pipes on engine/pipe connections

▲
A standard cooling-water system with a relatively large seawater-cooling system in a two-stroke crosshead engine.

Brief description

The two main seawater cooling-water pumps 005 draw the seawater in via a low-inlet chest 002 or in shallow water via the high-inlet chest 003. The seawater passes through the seawater filter 004 with a mesh size of between 2 to 4 mm. The water flows through the lubricating-oil cooler(s) 007 and subsequently through the cylinder cooling-water cooler 008.
Parallel to these two coolers the water flows through the scavenging- air coolers 009 on the engine. A temperature control valve 010 measures the inlet water temperature before the cooler and circulates the water or partially discharges the water overboard with discharge valve 011.

Temperature control conditions

1 The inlet seawater temperature before the lubricating-oil cooler should not drop below 10 °C as to prevent the lubricating oil from becoming viscous.
2 The lowest possible seawater inlet temperature for the air cooler must be applied to keep fuel consumption as low as possible.

All the parts of the seawater system are manufactured from seawater resistant materials.
The auxiliary engines are provided with seawater cooling for the lubricating-oil and air coolers via a non-return valve.

Central cooling-water system

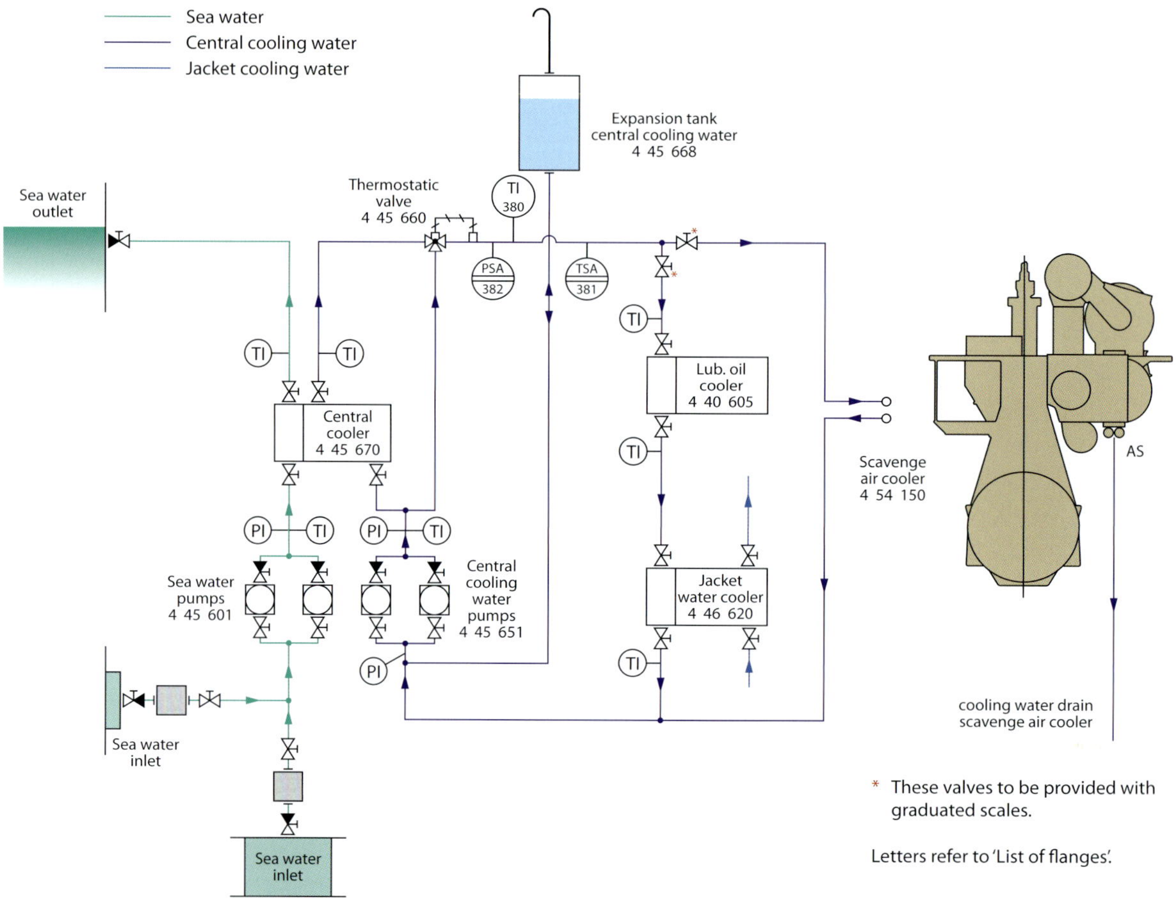

▲ A modern cooling-water system with a short seawater-cooling section.

In a short seawater cooling system, corrosion is kept to a minimum. Furthermore, the seawater section is manufactured from corrosion-resistant materials such as stainless steel and cunial, a copper nickel aluminium alloy.
The most remarkable feature of this system is that the seawater only passes through one cooler, the central cooler. The seawater system is, therefore, very short with high and low seawater inlets, a suction filter, two seawater circulating pumps, a central cooler after which the water is pumped overboard.
The fresh-water circulating system consists of two pumps that deliver the water through the central cooler and the air cooler, which is arranged parallel to the lubricating-oil cooler and the cylinder cooling water.

An elevated expansion tank has a vertical connecting pipe to the suction side of both fresh water-circulating pumps, so they can pump water immediately the engine is started.
A temperature-control valve ensures a constant supply temperature to the engine.
The cooling-water temperature to the air cooler must remain as low as possible in order to achieve optimum engine efficiency. Consequently, the temperature of the fresh-water cooling system is kept as low as possible by the thermostatic control valve. It follows the seawater temperature when this rises above 10 °C.

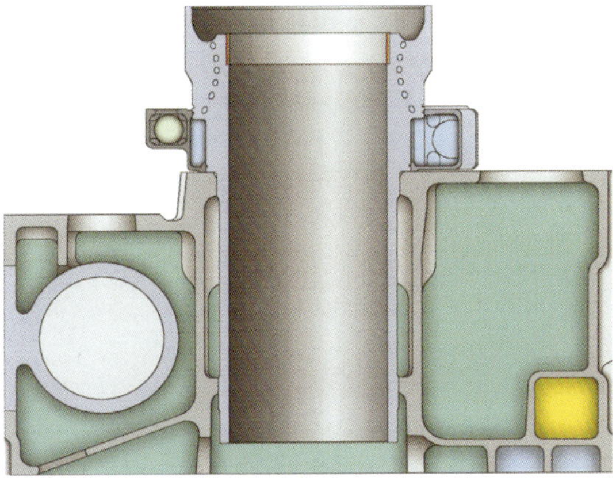

▲
A cylinder liner of a four-stroke medium-speed diesel engine of the Wärtsilä type 64 series.

Only the cylinder liner protruding from the block is cooled, the entire block is dry. Corrosion is therefore no longer possible.

10.16 Examples of cooling methods for engine parts

10.16.1 'Wet' and 'dry' cylinder liners

These terms are often applied in relation to smaller diesel engines. 'Wet' indicates that there is cooling-water directly aft of the cylinder liner. 'Dry' indicates that cooling-water spaces have been installed in the engine block, so there is no direct contact between the cooling-water and the cylinder liner.

Advantage 'wet'

– superior cooling.

Disadvantage 'wet'

– leakages along the liner for instance due to leaking rubber seals; this contaminates the lubricating system with water.

10.16.2 Bore cooling or cooling through drilled holes

With the 'power density' increase, or rather: with an increase in power per stroke volume, the mean effective pressure increases. This can only be accomplished when compression and combustion pressures increase significantly. Therefore, the forces on the cylinder head, the cylinder liner and the piston increase. Consequently, the wall thickness of the engine components must be increased, but at the same time the material must be adequately cooled. This is made possible by drilling holes fairly close to the surface of the combustion chamber walls through which the cooling-water flows. This

▲
A bored piston crown from a Wärtsilä Sulzer RTA two-stroke crosshead engine, oil cooled.

▲
Pistons are usually cooled with lubricating oil.

Shown here a piston crown of a two-stroke crosshead engine of MAN Diesel. Note the drilled holes in the circumference.

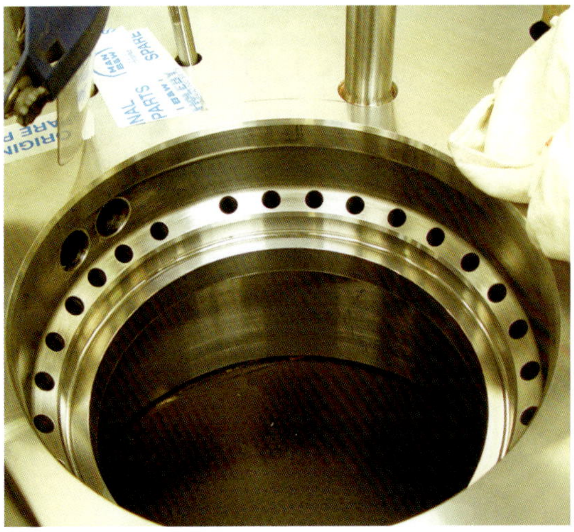

▲
A cylinder liner in a MAN Diesel four-stroke medium-speed diesel engine with bore cooling. Coolant is water.

way the various components remain sturdy and yet thoroughly cooled

Cylinder liners with a diameter in excess of 200 to 250 millimetres have a thicker upper collar and are provided with 'bore cooling'. In some of the current engine brands, this is the only part of the liner that is cooled. Therefore the engine block is manufactured 'dry' which prevents corrosion caused by cooling water.

Modern cylinder heads of four-stroke engines are very rigid in order to withstand combustion pressures of 200 bar or more without deformation; this can cause gas- or cooling-water leakages. The bottom part of the cylinder head, the so-called flame plate is a thick walled section of the cast iron cylinder head and is provided with 'bore-cooling'.

Modern cylinder heads of two-stroke-crosshead engines are manufactured from forged steel and completely solid. Centrally positioned is a hole for the exhaust valve including the casing, holes for the injectors, starting valve, discharge valve and an indicator cock.

This solid head is provided with bore cooling to ensure effective cooling.

In modern diesel engines pistons are both thermally and mechanically heavily loaded. This requires them to be both strong and adequately cooled. Here the forged- or cast-steel piston crown has also been provided with drilled holes in the crown.

▼

The thick flame plates of the cylinder heads in a Caterpillar MaK 32 four-stroke engine.

▼

Large cylinder covers in crosshead engines.

Large cylinder covers in crosshead engines are almost completely solid and provided with bore cooling. The water capacity of these large cylinder covers amounts to only tens of litres.

▶

A cross-section of the combustion space in a four-stroke diesel engine.

Note the thick flame plates are provided with bore cooling.

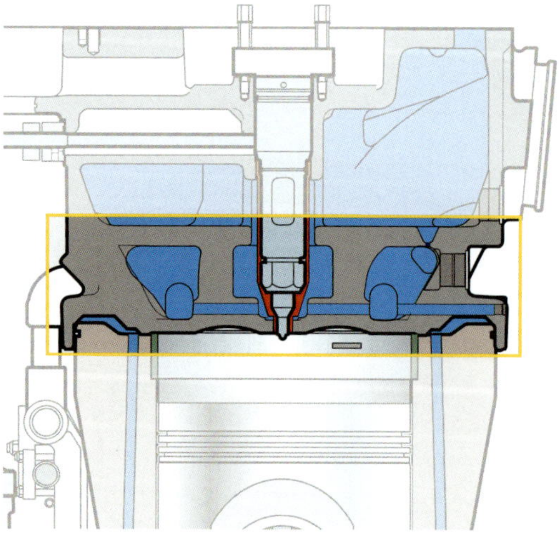

▶

Piston crowns in modern four-stroke medium-speed diesel engines are highly loaded and therefore thoroughly cooled with lubricating oil.

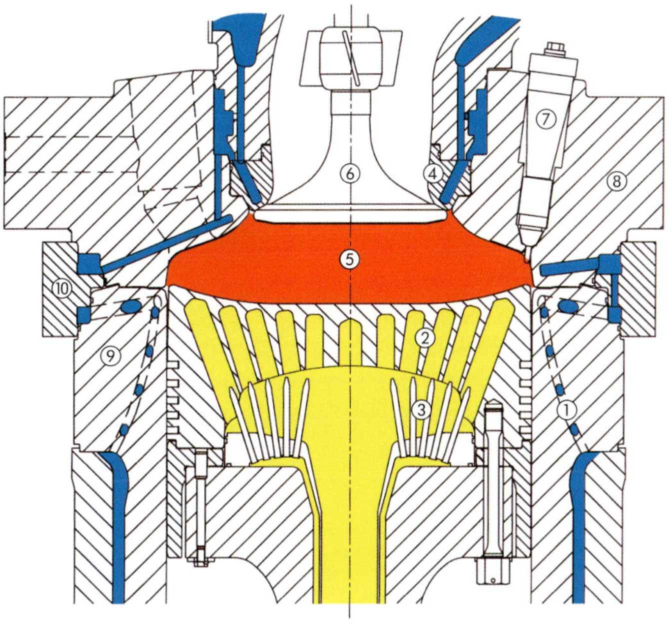

▶

A cross-section of a large Wärtsilä Sulzer RTA two-stroke crosshead engine.

The parts around the combustion space are solid and provided with bore cooling.

1 *drilled channels for fresh-water cooling.*
2 *drilled channels for lubricating-oil cooling*
3 *lubricating-oil spray lances*
4 *deep-cooled valve seat*
5 *combustion space*
6 *exhaust valve*
7 *fuel injector*
8 *cylinder head*
9 *cylinder liner*
10 *cooling-water ring*

▲

Nozzle cooling with lubricating oil in a two-stroke crosshead engine, a Wärtsilä Sulzer RTA.

The spray lances fit exactly in the drilled holes of the piston crown shown here.

▲

Nozzle cooling with lubricating oil in a two-stroke crosshead engine, a Wärtsilä Sulzer RTA.

The spray lances shown above fit exactly in the drilled holes of the piston crown.

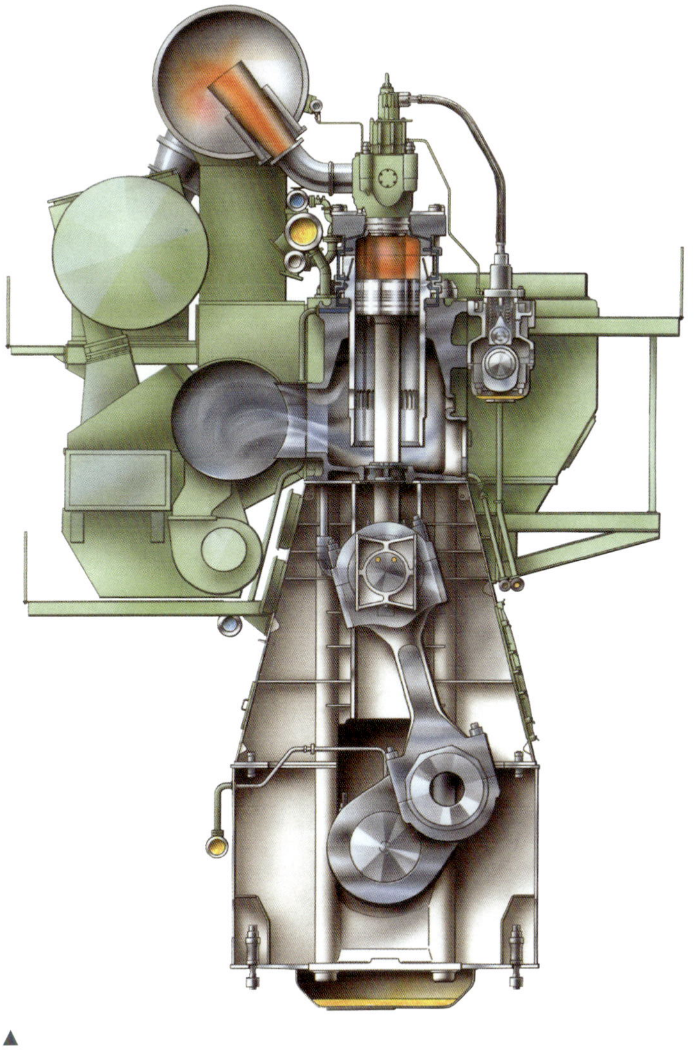

▲

The combustion space of this two-stroke crosshead engine from MAN Diesel type MC has a completely solid combustion space.

In **two-stroke crosshead engines** the piston crown is provided with drilled holes which are injected with lubricating oil by nozzles.

So all important components near the combustion space of a two-stroke crosshead engine are provided with bore cooling except for the central exhaust valve.

In **four-stroke trunk-piston engines** only parts of components have bore cooling, such as the flame plate of the cylinder head, the upper rim of the cylinder liner and the outer rim of the piston crown.

Piston cooling

Virtually all pistons are cooled with lubricating oil. Some older crosshead engine types are still equipped with a separate fresh water cooling system.

10.17 Examples according to the engine classification

10.17.1 Engine category I: 0 to 100 kW, four-stroke, high-speed, MDO

All pistons are cooled with lubricating oil. Lubricating-oil supply to the piston can occur in various ways.

A. Via a nozzle mounted on the bottom rim of the cylinder liner.

The oil jet sprays **continuously** and directly into the piston and cools the inside of the piston.

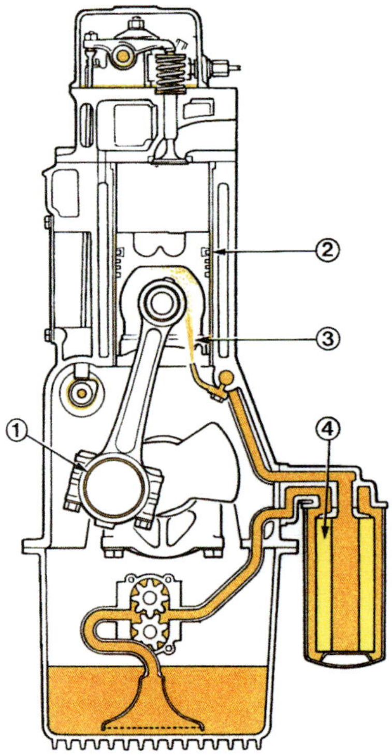

▶ Nozzle cooling of the piston in a small high-speed four-stroke diesel engine of DAF.

The heated lubricating oil falls back into the sump. Cooling is rather limited; a relatively large amount of lubricating oil must be supplied. In this simple system, this is referred to as the absorption gradient. This is the lubricating oil which is injected by the nozzle and is used as piston coolant.

It is very important that the lubricating-oil injection path is parallel to the heart of the cylinder.

If the nozzle including the supply line is only misaligned by a few degrees, the absorption gradient and consequently the piston cooling is adversely affected.

It is then possible that this so-called cooling-duct piston could seize.

▲ A completely seized piston.

The subsequent damage can be huge.

B. The lubricating oil is pumped under pressure from the crankshaft through a perforation in the connecting rod. Via the piston pin and a through perforation it is led into the light-metal cast piston-cooling spiral.

This system has a far better cooling effect. The lubricating-oil velocity and flow in the coil has to be reasonably high. After cooling, the oil falls back into the sump.

C. A spray nozzle in the small or large end of the connecting rod. This is also a form of spray cooling. Here the cooling effect is moderate.

10.17.2 Engine category II: 100 to 10,000 kW, four-stroke, high-speed, MDO

All the pistons are cooled with lubricating oil. In engines with a high load which require intensive cooling, the simple cooling systems with spray cooling are less prevalent. As aforementioned, the lubricating oil is pumped to the piston through the connecting rod and cools the piston via a cast-cooling spiral. For the somewhat larger cylinder bores, cast-iron pistons are sometimes still used. Here the stroke/bore ratio is normally slightly higher than that of the light-metal pistons and directly linked to the somewhat higher stroke, the rpm is lower so that the mean piston speed does not increase.
In this category II one also finds assembled pistons, but since the various categories have some overlaps, these will be discussed in category III.

It is evident that the construction of the piston and also the cooling of the larger engines is far more complex than that of smaller engines.
This is primarily due to the much larger power output that, for instance, is generated per litre stroke volume. The mechanical- and thermal loads are consequently also larger!

10.17.3 Engine category III: 500 to 30,000 kW, four-stroke, medium-speed, HFO

Piston construction

These pistons usually consist of two parts: the steel piston crown and the cast-iron or light-metal skirt. The oil supply to the inside of the piston can take place in various ways:

- via the connecting rod to a nozzle on the head of the small end where it sprays against the piston crown;
- via the connecting rod to the gudgeon pin and from here to the piston and the piston crown via bore holes.

The piston crown can be machined in various ways to ensure maximum cooling with the lubricating oil:

- to force the lubricating oil to flow past the thermally heavily loaded parts;
- manufacture the piston crown surface in an undulating wave pattern to expand the cooling surface;
- to apply bore cooling to the rims of the piston crown.

'Jet shaker'-cooling system

The reciprocating movement of the piston can be used to cool the piston lubricating oil. This is referred to as 'the cocktail shaker'-effect or the 'splash effect'.

Some engine manufacturers still use one-piece pistons manufactured from cast iron, such as in the Daihatsu DK-series with 20, 26, 28, 32 and 36 centimetre bores.

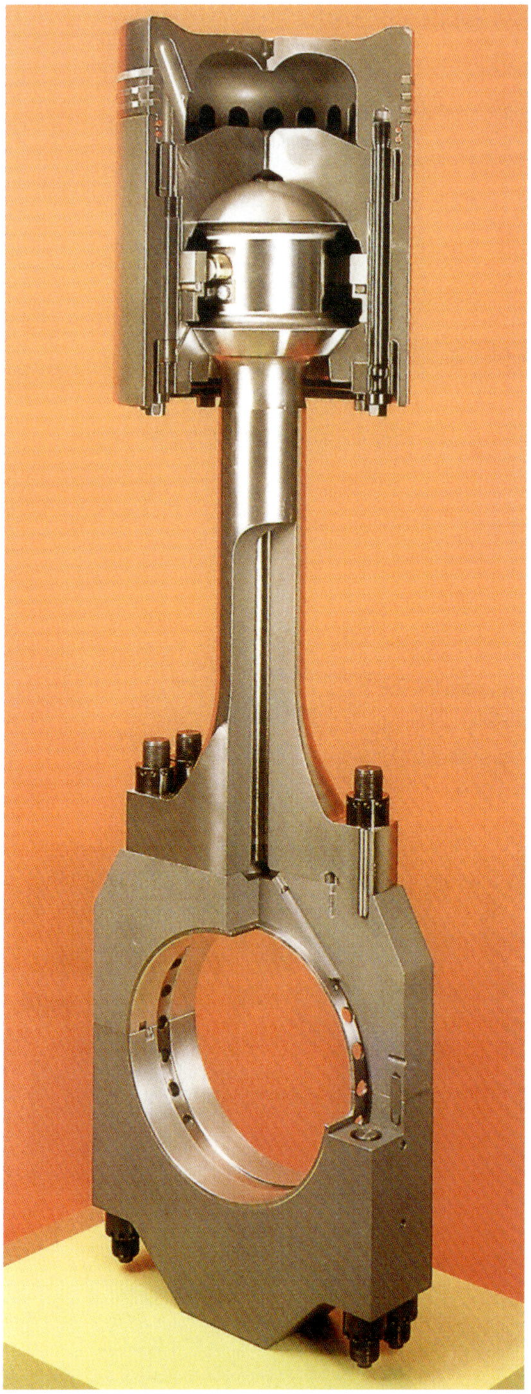

◄ A special piston in a Wärtsilä Sulzer Z-40 four-stroke medium-speed diesel engine.

The lubricating oil is pumped from the crankshaft via openings in the bearing caps, through the perforations in the connecting rod to the gudgeon pin and then to the piston-ring pack and the piston for cooling. The top of the connecting rod has a ratchet mechanism that during the up- and downward movement of the connecting rod allows the piston to rotate around the vertical axis. This improves the distribution of the heat flow through the piston, which provides a more evenly distributed temperature for the entire piston, preventing localised overheating. The rotating motion of the piston ensures that the cylinder liner is run in properly. The back- and forth-motion forces of the connecting rod are stored temporarily in a spring; the moment the gas and mass forces are balanced and the spheroid bearing is not under any load, the piston is turned.

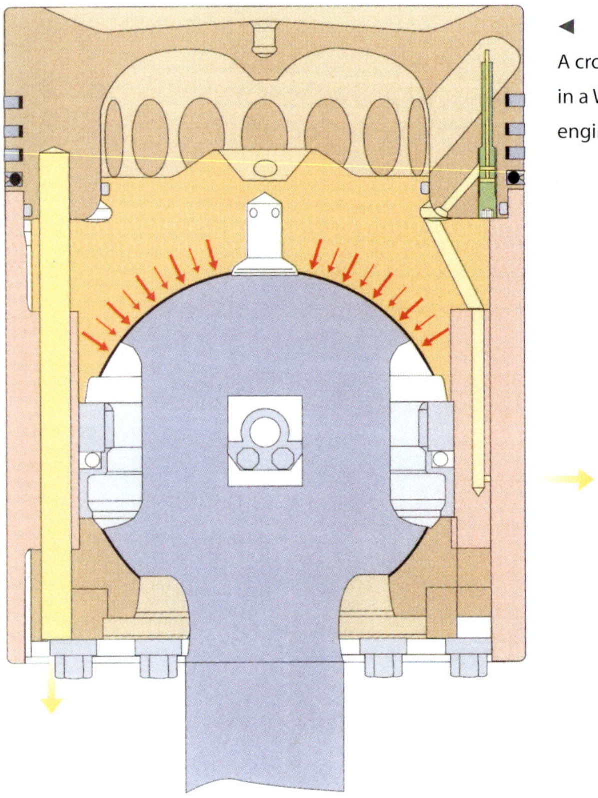

◀
A cross section of a piston in a Wärtsilä Sulzer Z-40 engine.

10.17.4 Engine category IV: 1500 to 100,000 kW, two-stroke crosshead engines, low-speed, HFO

These engines have a deviating process in the supply- and discharge system for piston cooling. Due to the crosshead construction it is possible to install the lubricating-oil supply and -discharge in the crosshead and subsequently via the hollow piston rod to and from the piston.

Today all piston cooling is performed via the piston rod. In the supply and discharge of the crosshead, telescopic pipes and swivel pipes for lubricating-oil or fresh-water cooling are used.

All modern type crosshead engines use lubricating oil for piston cooling. The main advantage is that, in case of leakages, there is no water contamination of the lubricating oil which may form an emulsion.

Disadvantages are the reduced heat absorption capacity and possibility of contamination of the metal surfaces with carbonised lubricating oil, which occurs at high material temperatures.

Therefore special attention is required to limit the amount of contamination in the piston.

▼
This Daihatsu diesel engine, series DK has one-piece cast iron pistons. They are cooled with lubricating oil.

▲
All the pistons of modern two-stroke crosshead engines are cooled with lubricating oil.

Extra attention should be paid to the contamination of the inner side of the piston. If the temperatures are too high, the lubricating oil will carbonise and the cooling of the piston will deteriorate. This could cause the piston to overheat, crack and/or seize.

▶

From the scavenging-air section of the turbocharger, the air channel widens towards the air cooler.

This produces a decrease in the air velocity that improves the air-cooling as the time available for exposure of the cold cooling-water pipes to the air in the cooler is lengthened.
After the cooling section is a tapered part or converging section, this increases the velocity as the air is forced towards the inlet valves or the inlet ports of the cylinders.

10.18 Combustion-air cooling

After the turbocharger, the air is compressed to a higher pressure and temperature. The compressor capacity generated by the exhaust-gas turbine converts this capacity to a higher air pressure and a higher air temperature.
Most chargers operate with pressures of 2 to 5 bar (over-pressure) and an air temperature of 100 °C to 160 °C. Generating ample power requires both large amounts of fuel and air. When the air is cooled, the specific mass of the air increases which allows an increased amount of air to be injected into the cylinder or engine, thereby enabling an additional fuel supply resulting in an increase in the engine capacity. The air coolers after the turbocharger are usually pipe coolers in which the low-temperature coolant flows, the so-called LT cooling-water system. The air circulates around the pipes which are often provided with cooling fins to increase the cooling surface. Low air velocity is critical for optimum heat transmission.
The air cooler has therefore a large volume and the turbocharger supply line is fitted with a divergent section to reduce the air velocity in the cooler and therefore time to intensify the air cooling.

◀

A highly loaded high-speed four-stroke diesel engine in V-shape.

Both air coolers are placed between the cylinder rows.

1 *air cooler*
2 *air section turbocharger*
3 *inlet-air manifold to the cylinders*
4 *exhaust-gas turbine*

10.18.1 Condensation

When air is cooled it can, dependent on vapour pressure, begin to condensate at a certain temperature.
When the water that is produced is carried to the cylinder with the air flow this can adversely affect the piston movement. This can also lead to an increase in wear and tear of the piston rings and the cylinder. The water droplets interfere with the lubricating-oil film on the cylinder liner. This is why the water is drained and two-stroke engines have a partition installed in the air supply to the cylinder, the so-called **'mist catcher'**.
In tropical climates where the air has high humidity, large quantities of water are drained from large crosshead engines: sometimes **several thousands of litres per hour!**

Standard air temperatures after the cooler are 40 °C to 45 °C for two-stroke engines. The four-stroke engine air temperatures are similar or slightly higher.

10.18.2 Bore cooling

This type of cooling has been applied for dozens of years in engine parts that are in proximity to the combustion space, such as the cylinder liner, cylinder head and the piston. They are now often applied to large highly loaded four-stroke engines running on heavy-fuel oil. Due to the increased mechanical load ensuing from combustion pressures that sometimes exceed 200 bar, the walls of the combustion space are becoming increasingly thicker. The piston crown and the cylinder head are steel: cast steel or forged steel. The cylinder liner is manufactured from cast iron in view of the

▶
Two turbochargers in a V-engine with both air coolers positioned below.

The air coolers are connected to avoid pressure differences between both banks.

1 *connecting pipe*
2 *turbocharger*
3 *air cooler*
4 *cylinder*
5 *air-inlet filter*
6 *exhaust flange*

▶
The scavenging-air condensate outlet in a large two-stroke crosshead engine.

1 *condensate sump*
2 *condensate-discharge pipe*
3 *collection tank with high-water level alarm*

advantageous running properties of, for instance, perlitic cast iron.
As opposed to the cast-iron parts, it is not possible to install a cooling-water space in the mould of the steel piston crown and the cylinder head. The former forged parts are today machined and the drilled cooling holes allow the materials near the high process temperatures of the combustion space to be cooled sufficiently without a decline in component strength.

Two-stroke crosshead engines

The cylinder liner designs of all three existing manufacturers have a thicker upper rim provided with bore cooling.

In Wärtsilä Sulzer RTA engines, the other parts of the cylinder liner are cooling-water free!
In MAN Diesel and Mitsubishi the middle section is cooled with water.
In all three engines the bottom section including the scavenging ports is cooled with scavenging air.

10.18.3 Turbocharger cooling

One speaks of uncooled and cooled turbochargers in order to indicate whether they have water cooling.
In smaller turbochargers the thick robust bearing casing is cooled and lubricated with lubricating oil.

▼

The thickened upper collar of a cylinder liner in this two-stroke crosshead engine is provided with bore-cooling channels.

This is known as 'bore cooling'.

1 *thickened upper collar for 'bore-cooling'*

▼

An ABB VTR turbocharger with water-cooling.

Modern versions no longer have water-cooling.
light yellow: lubricating-oil part
light blue: cooling-water part
dark blue: left - exhaust-gas part, right - scavenging-air part

◄

Drilled channels in a cylinder cover in a MAN Diesel two-stroke crosshead engine.

The large holes are for the cylinder-cover studs.

▶
A turbocharger with an uncooled casing.
Manufacturer ABB.

Yellow: exhaust gases.
Blue: scavenging air.
White: lubricating- and cooling oil for both journal bearings and the thrust bearing.
Blue canal: here the scavenging air flows from the delivery side of the air part of the charger to the shaft seals of the exhaust-gas part. The scavenging-air pressure is higher than that of the exhaust gases so the dirty hot exhaust gases do not flow near the right shaft bearing.

Small turbochargers often only have a combination of lubrication and cooling with lubricating oil. There are also turbochargers which have a water cooled casing. This ensures lower material temperatures. Furthermore, it reduces radiant heat emission into the environment. This is fitted to exhaust gas ducts for the same reason.

Large turbochargers were traditionally cooled with fresh water and manufactured with double walls. However, today their construction resembles that of smaller uncooled turbochargers and they are no longer cooled. This makes turbine design simpler, cooling-water related maintenance such as leakages and cleaning of the water spaces is no longer required. The material of the casing is now subject to much higher temperatures and its quality has consequently been improved.

10.18.4 Cooling crosshead engine guides

In older engine types the hollow cast guides used to be cooled with relatively cold seawater. Today, guides are usually welded into the frame and cooled and lubricated with lubricating oil.

10.19 Special cooling systems

10.19.1 Box cooling

This is a system in which the heat exchanger of the closed fresh-water cooling system and the raw- or seawater system has been integrated in a casing in the ship's hull along which the outboard water can flow freely during sailing. The box cooler can be mounted alongside the ship, but also in the bottom

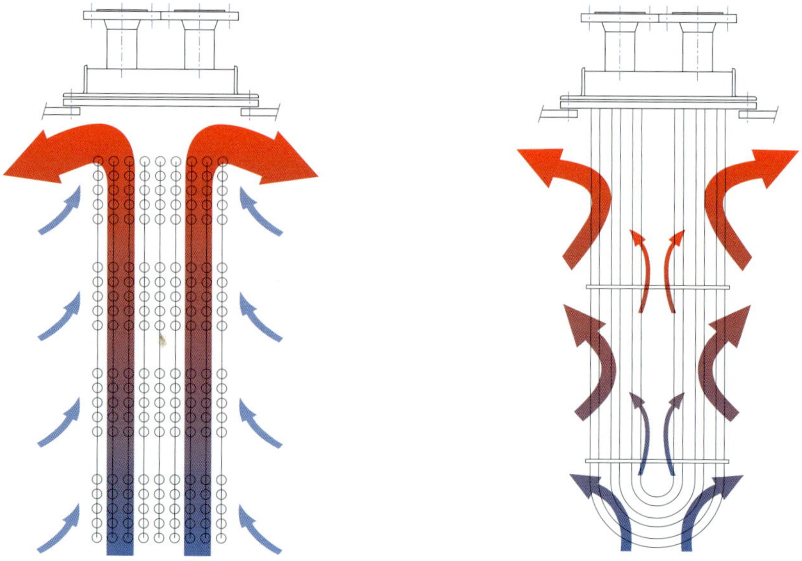

▶
Keel or box coolers have an open connection with water and therefore seawater-circulating pumps are not needed.

◀

The top view of a box-cooler series.

The connections with the various cooling systems are always made on the top side and provided with stop valves. Note the vent cocks at the highest point of the piping on the left hand side.

1 *vent cocks*

of the ship. The system is often applied in inland navigation, coastal navigation and for instance in dredgers. It has the advantage that any kind of dirt in the untreated cooling water cannot enter the pumps, and pipes do not come into contact with foul.

10.19.2 Bilge or keel cooling

This is frequently used in cabin cruisers and other comparatively small propulsion systems.
The heat exchanger is often simple: a long cooling pipe is mounted below the surface often next to the keel. Sometimes several pipes are fitted.

10.20 Pipe coolers and plate coolers

▶

Plate coolers are being more frequently used.

◀

A lubricating-oil pipe cooler in a four-stroke medium-speed diesel engine.

10.20.1 Pipe coolers

Principle: The heated water flows through pipes, and the cooling liquid or gas moves around those pipes. By means of guide panels the flow can be directed so that optimum heat transmission can take place. Traditionally, the **pipe cooler** was often used for lubricating oil, fresh-water cooling and raw-water cooling. They are arranged on a flow- or counter-flow principle.

Advantages and disadvantages of pipe coolers

The pipes can be cleaned on the inside by removing the caps. This is often the raw-water side. They can be easily cleaned with pipe brushes, compressed air or a high-pressure creeping hose. If the pipes need cleaning the entire assembly must be removed. Therefore, there always has to be ample room available and in today's smaller engine rooms there is insufficient space.

10.20.2 Plate coolers

The plate coolers are increasingly used for cooling liquids such as fresh water, lubricating oil and fuels.

Principle

The cooling water flows along one side of these sequentially arranged plates flowing over to the next plate via holes. The cooling liquid flows along the other side of the plates and flows over to the following plate through different holes. Between the plates, **packings** are fitted. Surface relief on the plates ensures that the plates are spaced correctly and can change the flow. They are very compact and have a relatively large capacity.

To clean, the plates are simply slid off the tie rods. Each plate can then be cleaned individually. **Packing seals are vulnerable.** However, cleaning the plates on location requires a lot of space. After mounting the clean plates there is a risk of leakage.
The capacity of the plate cooler can easily be modified by increasing or decreasing the number of plates. As a result of the compact construction of engine rooms there has been an increase in the number of plate coolers used. In smaller engines – category I and part of II – they are seldom used, the much smaller pipe coolers are used here.

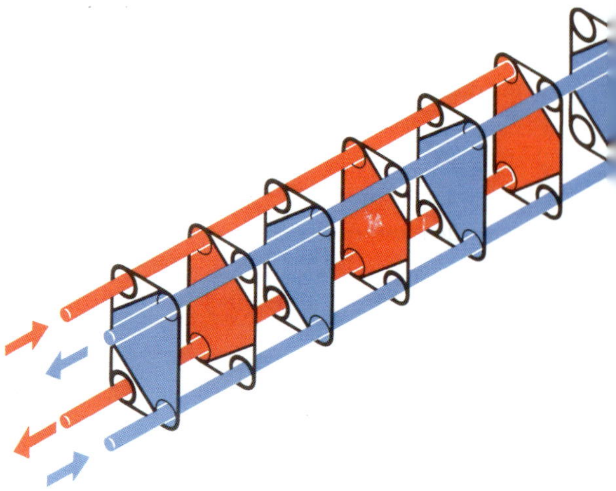

▲
The flow through a plate heat exchanger.

Red: the circulating cooling water.
Blue: the cooling water.

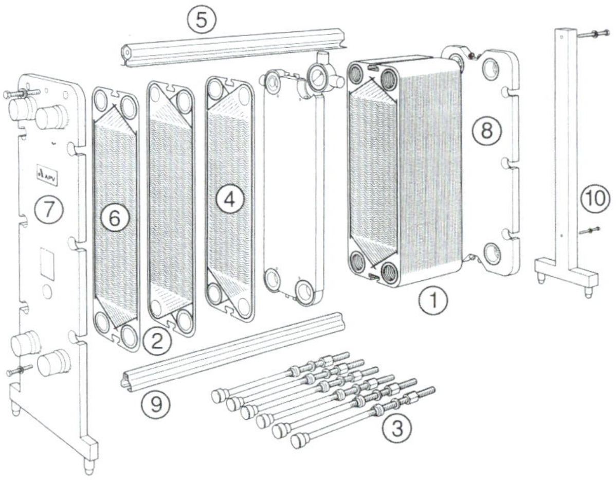

◀
The construction of a plate cooler.

1 *cooling-plates packet*
2 *ribbed cooling plates with an inlet and outlet*
3 *draw bolt*
4 *gasket with gasket groove*
5 *upper beam*
6 *first plate*
7 *stationary end cover with inlet and outlet connections*
8 *back plate*
9 *lower beam*
10 *column*

◀ Air coolers in a large diesel-power station in Honduras.

Approximately 25% of the heat supplied by the fuel is wasted to the environment.

10.21 Cooling systems in a diesel-power plant

Most diesel-power plants, particularly in tropical regions, do not use the exhaust gases. **The exhaust gas constitutes over 50% of the heat content of the fuel!** In cogeneration the total efficiency can amount to 80% and rise up to 90%. The remainder consists of flue-gas losses.
If we assume that that the diesel-genset efficiency is 45%, then at least 25% to 35% of the heat from the fuel must be discharged. The remainder are exhaust gases. In a diesel-power plant of 100 MW this is between 55 and 78 MW!
If there is river- or seawater available, the heat can simply be discharged with the normal and aforementioned cooling systems using HT-, LT- and raw-water systems.

If there is no or insufficient cooling water available the heat is discharged using a range of air coolers. These heat exchangers operate similarly to a radiator: a ventilator forces air through the heat exchanger.
In view of space, the radiators are usually arranged vertically with the ventilator placed at the opposite side to the output shaft of the diesel engine.
In diesel-power plants the heat exchanger is positioned horizontally and blows air upwards.

Air-cooler circuits

The number of heat exchangers in operation depends on the required power production and the ambient air temperature.
The cooling systems themselves are kept at a constant temperature by thermostatic three-way valves.

These air coolers are also often used in large back-up gensets. Smaller back-up gensets are usually provided with a radiator behind the engine. Here the radiator is normally mounted on the outside wall of the building and is in direct contact with outside air. If the ambient air temperatures fall below freezing point, it is important to add an anti-freezing agent to the cooling water.

◀ A large diesel-power station, Honduras.

1. *medium voltage substation, 13.8 kV, to high voltage, 110 kV.*
2. *plant with large four-stroke medium-speed diesel engines operating on HFO, category III.*
3. *exhaust-gas pipes with silencers and below, a building containing the auxiliaries for the diesel engines.*
4. *diesel-engine heat exchangers that discharge their heat to the outside environment.*
5. *drinking-water storage tanks.*

▶
Many diesel generators are provided with a radiator that in a simple manner discharges the engine heat to the outside environment.

10.22 Cogeneration systems

When the heat generated in engines is used for other purposes this is referred to as cogeneration. Heat emitted by exhaust gases, lubricating oil, cooling water and air-cooling is suitable for this purpose.

The heat is used for:
- room heating in residential areas, large buildings and companies;
- heating of swimming pools and factories;
- generation of electricity with an exhaust-gas steam boiler and a steam-turbine genset;
- heating auxiliary systems for engines such as fuel storage and cleaning, pre-heating lubricating oil and cooling water and the heating of fresh-water evaporators.

It should be noted that although the **total efficiency** of the installation increases, the **total shaft power** does not.

Comments:
- **In small cabin cruisers the engine heat is sometimes used to heat up water for showers and for cooling.**
- **In shipping distilled water** is made by means of a seawater evaporator that is heated with the HT fresh-water cooling system.
- **With an exhaust-gas boiler**, low pressure steam is formed which is used to heat the accommodation, the cargo and fuel. Exhaust-gas boilers are also often used in diesel-power plants.
- **In power plants** with diesel engines, heat is used to heat buildings, for city heating and for the process industry.

▶
The use of cooling-water and exhaust-gas heat is applied on a large scale.

This is the case in countries with a temperate climate. City heating, production processes, swimming pool heating and greenhouse/glasshouse cultivation projects are some of the applications. A problem with the use of diesel-engine exhaust gases is the high emission of harmful substances. Flue gas cleansing occurs, but gas engines are more often used as they produce less harmful emission. Shown here, a company in The Netherlands where peppers are cultivated. The heat generated by the gas engines is used in this greenhouse complex. An emergency cooler is used in case of very high external ambient air temperatures.

In countries where engines that run on fuels such as diesel oil or heavy oil are prohibited due to environmental regulations, gas engines are used.

Also see Chapter 29, *New fuel developments.*

Diesel engine flue gases are sometimes also cleaned.

Also see Chapter 22, *Engine emissions.*

10.23 Summary for cooling-water systems

- Only use pure water for cooling water.
- Cooling-water treatment is imperative. Follow manufacturer's instructions!
- Cooling-water tests must take place regularly. Test procedure: see manufacturer.
- Deaeration of cooling water is very important.
- In case of repairs, collect drained cooling water and re-use it.
- Perform surface inspections during the repair of cooled engine parts.
- Monitor cooling-water pressures and temperatures. Fluctuations are indicative of malfunctioning pumps and control systems.
- Pre- and post-heating of cooling water must take place gradually to minimise heat tension in the engine material.

◄

A cast-iron exhaust valve housing.

A very meticulous cooling water treatment with regular sampling is important for the often high thermally loaded parts in a diesel engine.
Here, the cast iron has been removed by the hot exhaust gases from a leaking exhaust valve.

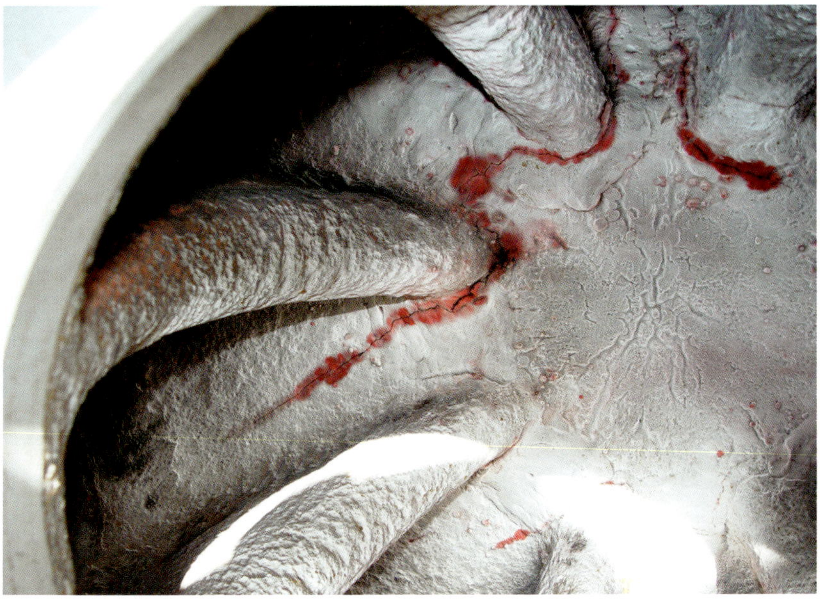

◄

This piston in an older model water-cooled two-stroke crosshead engine is very badly affected by pitting. This is clearly seen after blast cleaning. The thick deposit layer, which is removed by blasting, has reduced the cooling of the piston crown to such an extent that the piston material has become too hot causing thermal cracks.

>CH 11

Lubrication of engines

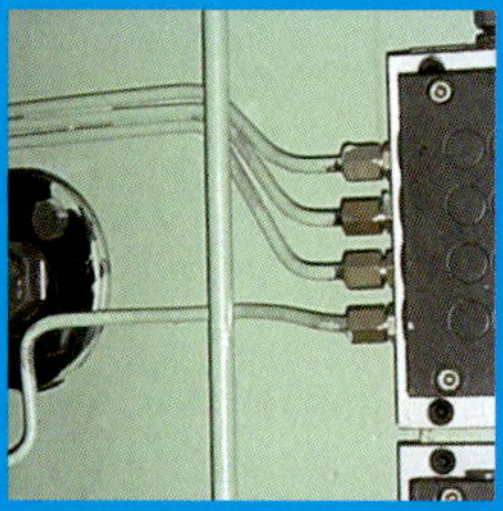

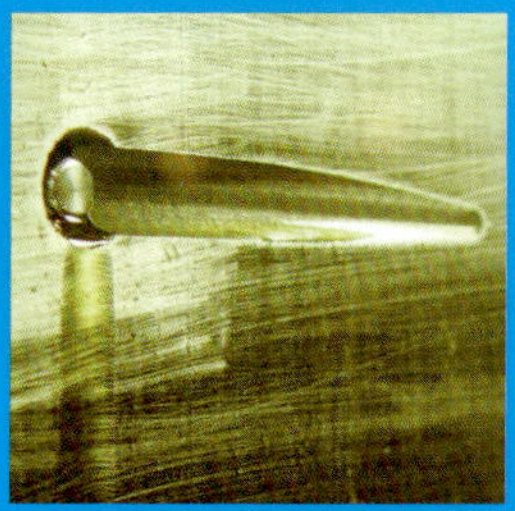

The piston, only just visible, in the cylinder liner of a Wärtsilä 38 medium-speed diesel engine.

The moving parts of the engine such as the crankshaft, the cam shaft, the connecting rods, the pistons, the valves, the cog wheels as well as components that are in contact with these parts, such as cylinder liners and bearings, require lubrication and constant cooling.
Adequate lubrication and proper cooling are crucial to keep maintenance costs as low as possible. Here a MAN Diesel 12 cylinder four-stroke diesel engine in V-alignment.

11.1 Introduction

Lubrication of engines is vital.

▶
Almost all the moving parts in a diesel engine require lubrication, such as, the cam of a fuel pump shown here.

11.2 The purpose of lubrication

11.2.1 To reduce friction between the moving parts and consequently limit wear and tear.

Obviously, the friction in engines should be minimal because friction causes wear and tear. Parts that are subject to friction are, amongst others: pistons, piston rings, cylinder liners, bearings, shafts, valve stems, cogwheels and fuel-pump plungers. They have an operating life ranging from 10,000 to 60,000 hours.

11.2.2 Discharge of frictional heat

The amount of frictional heat is negligible.

▼
Various engine parts must be lubricated. From left to right: a section of a cam shaft, a connecting rod, a gear wheel, a cylinder liner, valves, gudgeon pin bushings and a piston.

11.2.3 Cooling parts

Lubricating-oil also serves to cool parts and the amount of heat that must be discharged is considerable, particularly where piston cooling is concerned.

11.2.4 Protecting surfaces from corrosion

Many engine parts do not receive any special anti-corrosive treatment. Lubricating-oil is very suitable for the protection of the surfaces.

11.2.5 Lubrication as a sealant

Sealing the piston, piston rings and cylinder liner, the stem valves and, for instance, the piston-rod stuffing box in two-stroke crosshead engines.

11.2.6 Flushing and keeping various harmful substances in solution

These could be debris from wear, water, dirt, acids and/or corrosive particles.

11.2.7 Sound damping

Liquids have a sound damping effect; this also applies to lubricating-oil.

11.2.8 Neutralising acids to avoid severe damage to engine parts by chemical corrosion

Here one should think of sulphuric acid that is found in fuel. The **sulphur content** of heavy fuel is particularly high (1 to 4%). Therefore lubricating-oils have very specific properties and must meet strict requirements.

11.3 Three types of lubrication

11.3.1 Hydrodynamic or fluid film lubrication

This is obtained when two congruent surfaces are completely separated from each other by a thin layer of lubricating-oil, the so-called lubricating film. This film is several hundred microns to tens of microns thick. **The lubricating-oil pressure in for instance the crankshaft bearing carries the shaft and is itself generated by the rotating shaft.**

The pressure in the bearing can amount to 1000 bar in some places! The **lubricating-oil pressure** of lubricating systems provides **sufficient supply** to these places alone. Most rotating shafts, such as the crankshaft and the camshaft, work with this principle.

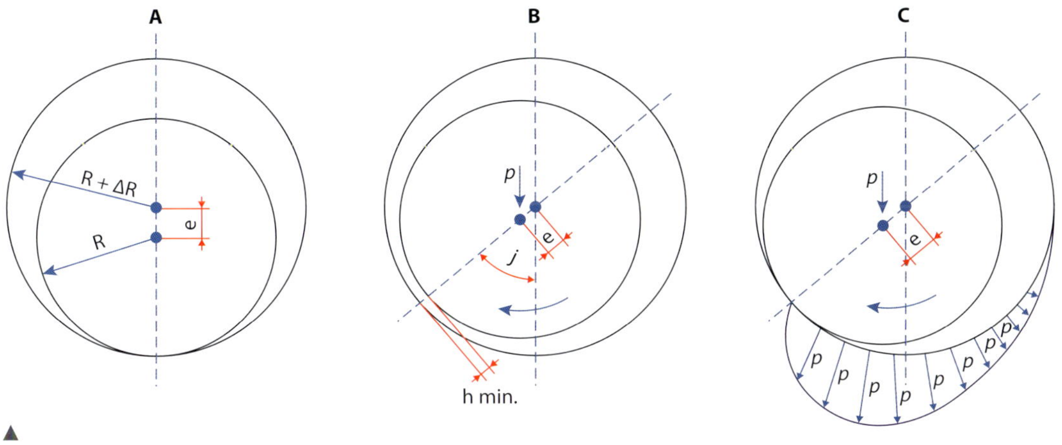

Pressure increase of a rotating shaft in a bearing

p = lubricating oil pressure in bearing

▲

The hydrodynamic lubrication of a crankshaft bearing.

Left figure: The stationary shaft is supplied with lubricating oil from the main lubricating-oil pump at a pressure of 2 to 4 bars.

Centre figure: The crankshaft rotates at a regular speed. There is some clearance between the shaft and the bearing preventing the metal parts from coming into contact. The minimum clearance is several tenths of a millimetre.

Right figure: The lubricating oil pressure increase in the bearing. The rotating shaft carries the lubricating oil. Due to the wedge-shape, the lubricating oil pressure increases. This lubricating oil pressure when distributed over part of the bearing surface must supply a counterforce to that exerted by the combustion process from above the piston downwards. **This pressure can vary from 100 to 1500 bars!**
Therefore, the bearing material must be adequately resistant to these high pressures.

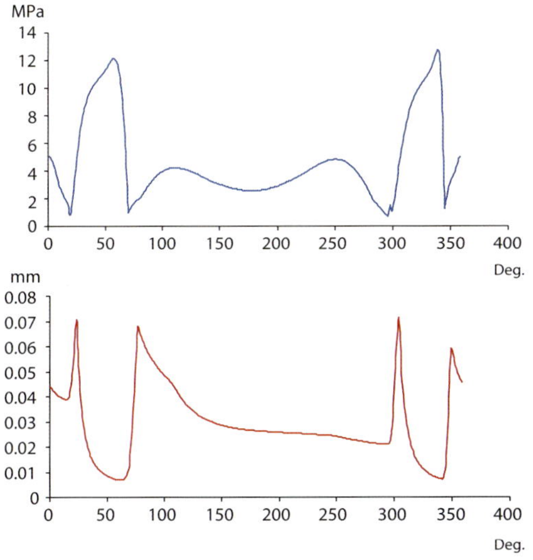

◀

The lubricating-oil pressure in the bearing and the minimum oil film thickness.

Top figure
horizontal axis: the position of the crank shaft in a four-stroke engine.
vertical axis: the lubricating oil pressure in the bearing in MPa.
The maximum shown here is approximately 13 MPa or 130 bar
Bottom figure
vertical axis: the minimum oil film thickness in hundredths of a millimetre. Minimum film thickness is approximately 0.01 mm, maximum thickness 0.07 mm

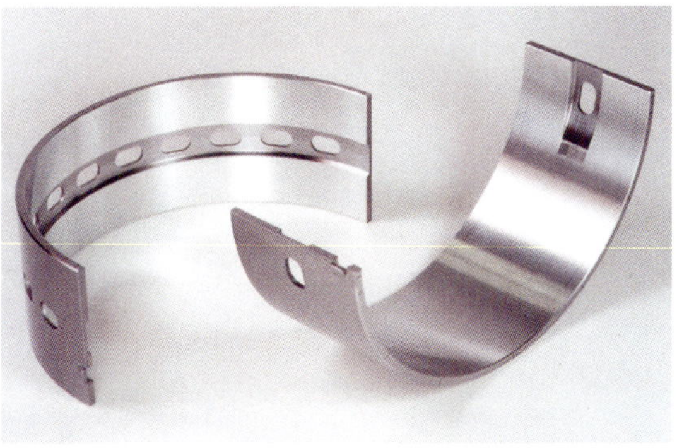

◀

Main bearings.

The clearance between the crank pins and the main bearing caps is so small that even the smallest dirt particles (several tenths of a millimetre) can cause severe damage to the bearing surface. This usually results in a damaged crank pin.

▲

A damaged crank pin.

Insufficient lubrication often causes such damage.

11.3.2 Hydrostatic lubrication

In this system, the lubricating-oil pressure is provided by a pump or an elevated tank. Here both congruent parts are also separated by a continuous lubricating film. It is often applied in the crossheads of two-stroke engines. The pressure can vary from 20 to 30 bar for the crosshead to over a 100 bar to **'lift and support'** a steam turbine's heavy shafts from their bearings during start-up.

11.3.3 Boundary lubrication

In boundary lubrication there is direct contact between both congruent surfaces. This is the case with a bearing and a rotating shaft. If boundary lubrication cannot be avoided, great care should be taken to reduce the surface roughness of both the shaft and the bearing. Chrome plating, grinding and polishing are common machining operations.

11.4 Engine parts that require lubrication and cooling

All driving mechanisms in diesel engines require lubrication. Congruent metal surfaces of all moving parts will wear rapidly without a lubricating-oil film to separate them. Furthermore, the frictional heat and heat generated during the combustion process must be discharged.
The following engine parts need to be lubricated and cooled:

- crankshaft with crankshaft bearings;
- pistons, piston pins and piston rings;
- connecting rods with connecting-rod bearings;
- camshaft with camshaft bearings;
- cams, guide pulleys, push rods, rockers, valve springs and valve stems;
- parts of the fuel pumps;
- cylinder liners via the pistons;
- gear transmissions between the crankshaft and the cam shaft;
- additional shafts such as balance shafts or separate cam shafts with bearings for fuel pumps as found in MAN Diesel four-stroke engines.

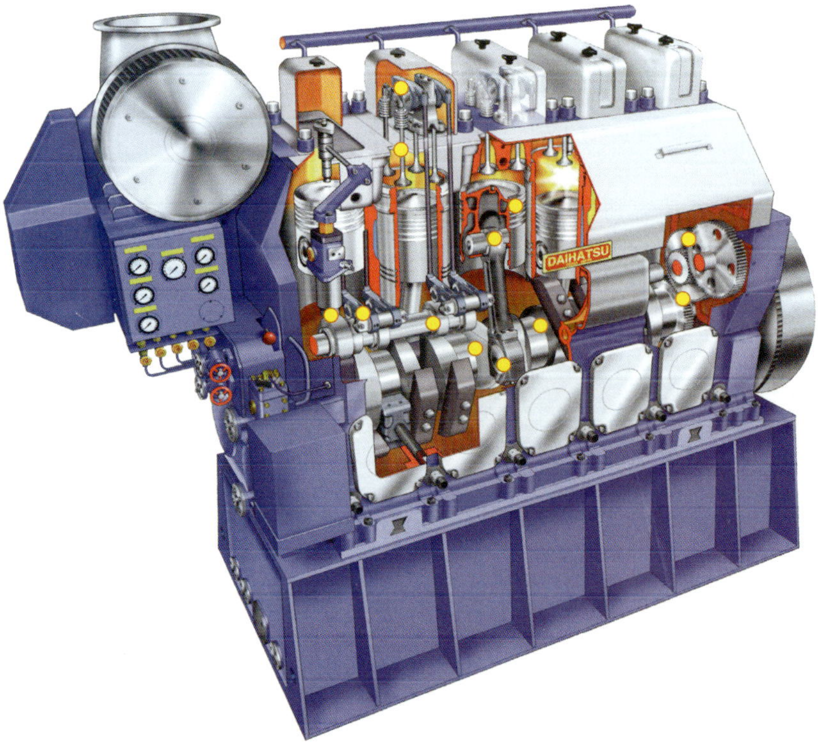

◄

Almost all moving parts in this four-stroke medium-speed diesel engine from Daihatsu require lubrication and/ or cooling.

Yellow dot: lubricated engine parts.

In **two-stroke crosshead engines**, the piston rod and the piston rod stuffing box is lubricated as well as the crosshead including all bearings, guides and guide blocks. There is separate lubrication for the cylinder liner.
All engine **governors** have independent lubricating systems.
Turbochargers' bearings and shafts are either equipped with their own lubricating-oil system or they are connected to the engine's main lubricating-oil system.
Thrust bearings and in larger engines thrust block bearings are also lubricated by the main lubricating-oil system.

Discharge of heat

Apart from discharging small amounts of frictional heat, the lubricating systems plays a vital part in cooling engine parts such as pistons and piston rings.

11.5 Common lubricating-oil system

A common lubrication system usually consists of the following components:
- lubricating-oil pump;
- lubricating-oil coolers;
- lubricating-oil fine filter;
- lubricating-oil pipes;
- through drilled holes;
- nozzles;
- splash lubrication;
- cylinder lubrication;
- special properties of the lubricating-oil
- other parts of the diesel engine.

11.5.1 Lubricating-oil pump

An engine-driven **lubricating-oil pump** has a coarse filter in the suction pipe, which protects the pump from dirt particles. These include metal, broken-off split pins, small nuts and packing material. The pump draws lubricating-oil from the sump or from a separate drain tank and forces the lubricating-oil into the central lubricating-oil line. The pump is generally a displacement pump. A gear pump is also often used.

11.5.2 Lubricating-oil cooler

A **lubricating-oil cooler is placed** in the main lubricating-oil line; this maintains the lubricating oil at a predetermined temperature using a thermostatic control valve.

▲
A lubricating-oil cooler positioned next to a diesel engine.

11.5.3 Lubricating-oil fine filter

The lubricating-oil passes through a **fine filter**, which removes all dirt and wear debris from the lubricating-oil. These are duplicated, the so-called **duplex** filter.
In larger diesel engines, self-cleaning fine filters make transferring, cleaning and replacing of the filter cartridges redundant.
Many types of fine filters are fitted with pressure-difference gauges that signal contamination of the filters and allow the necessary precautions to be taken.

11.5.4 Lubricating-oil lines

Lubrication-oil lines are installed throughout the engine. Today many lubricating-oil ducts are actually drilled in the engine block. The lubricating oil is delivered to the crankshaft bearings, camshaft bearings and cogwheels.

▼
Automatic fine-filters in a lubricating oil system of a large two-stroke-crosshead engine.

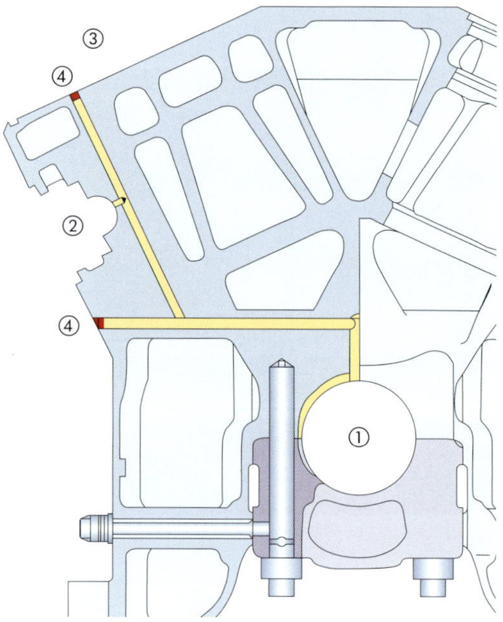

◄

At present, many engine blocks are provided with drilled lubricating-oil channels.

Shown here the lubricating-oil supply to the main bearings, the camshaft and the cylinder head with a valve timing drive.

1 *main bearing*
2 *cam shaft*
3 *cylinder head*
4 *packing plug*

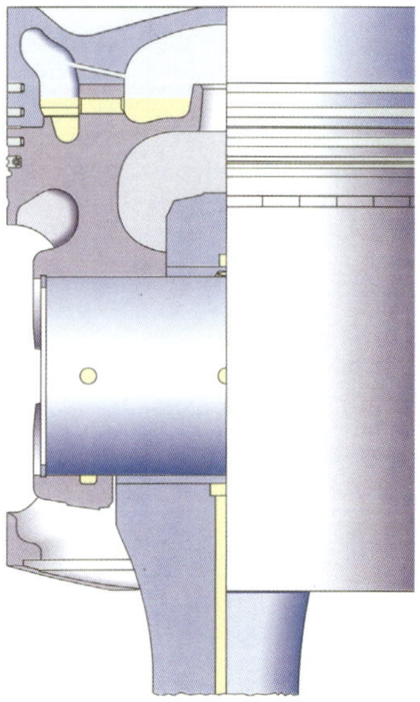

▲

This piston is cooled using the 'cocktail shaker' effect.

The piston is partially filled with lubricating oil. The piston crown is intensively cooled by the reciprocating piston motion.

11.5.5 Through-drilled holes

Through-drilled holes in the crankshaft, the connecting rod and the piston pin allow the lubricating-oil to be delivered to the crank pin bearings, connecting rod bearings, pistons and piston rings.

►

Cooling and lubricating a piston.

The entire piston crown is filled with circulating lubricating-oil. The piston-ring pack and the gudgeon pen are lubricated under pressure.

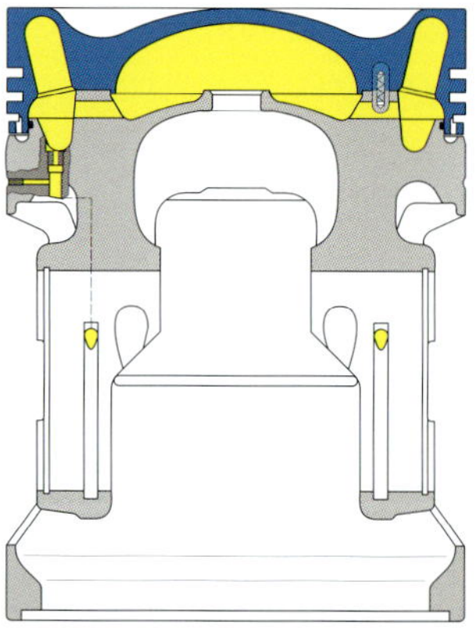

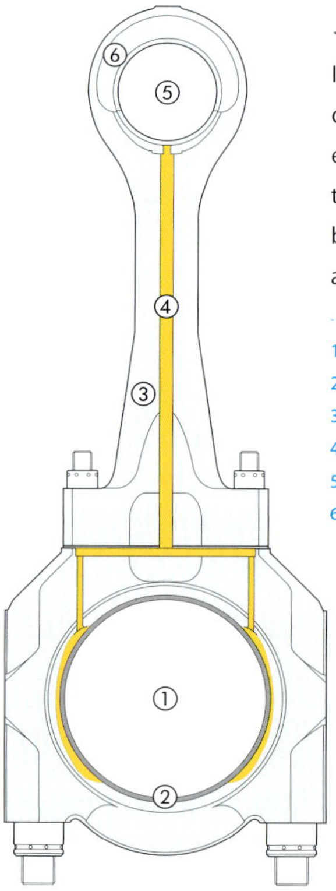

◄

In larger engines, a drilled duct in the connecting rod enables the lubricating oil to flow from the crank-pin bearing to the gudgeon pin and the piston.

1 *crank pin*
2 *crank-pin bearing*
3 *connecting rod*
4 *drilled duct*
5 *gudgeon pin*
6 *connecting-rod eye*

◄

Drilled ducts in the de crankshaft take the lubricating oil from the crankshaft to the crank pin.

The crankshaft bearing is supplied with lubricating oil from the lubricating-oil supply line. The lubricating oil flows to the crank-pin bearings via the ducts in the crankshaft and subsequently to the gudgeon pin and the piston rings via ducts in the connecting rod.

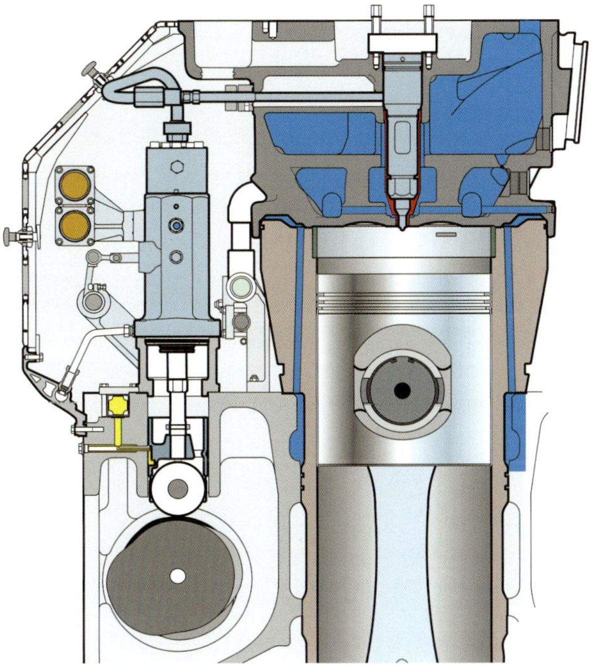

▲
Lubrication of the high-pressure fuel-pump drive.

It is important that fuel from the fuel-pump drive does not spill into the lubricating oil. Lubricating oil mixed with fuel does not lubricate adequately!

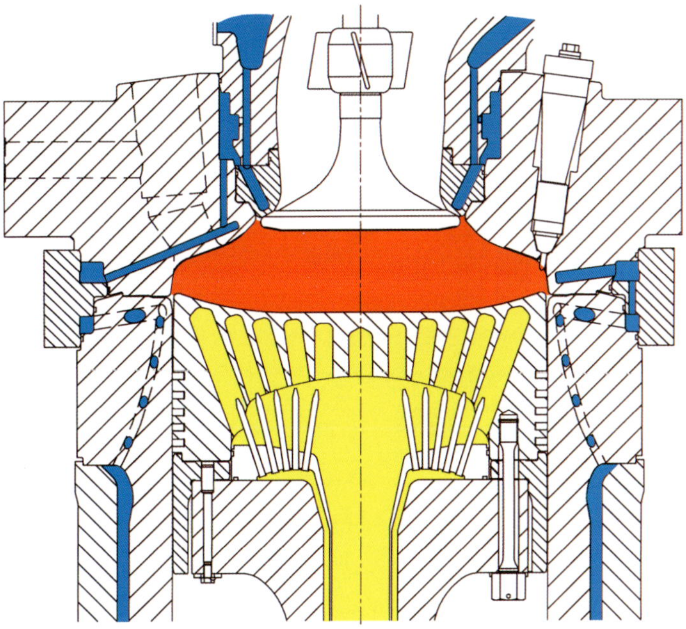

▲
Piston cooling using lubricating oil in two-stroke crosshead engines from Wärtsilä Sulzer; type RTA.

The nozzles are directed towards the matching drilled cooling channels in the forged iron piston crown. This allows each hot part to be intensively cooled.

11.5.6 Spray nozzles

Especially in small engines, lubricating-oil is sprayed on the cogwheels and pistons using **nozzles**. Sometimes pistons of large two-stroke crosshead engines are also cooled in this way.

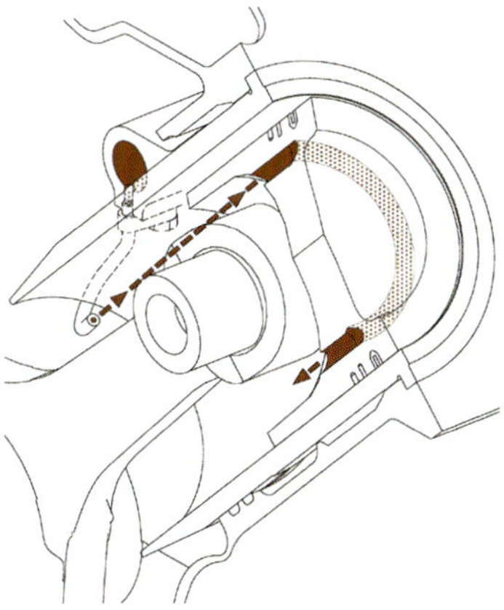

▲
Spray jet cooling of the pistons in a four-stroke high-speed MTU-diesel engine.

It is important that the jets be correctly directed. After maintenance, it is imperative to check that there is no deviation through incorrect mounting or damage.

11.5.7 Splash lubrication

Obviously, **splash lubrication** is mostly applied in small diesel engines. In every engine, the sump and the space directly below the piston are filled with swirling lubricating-oil droplets. It is therefore ill advised to open the crankcase covers while the engine is running!

11.5.8 Cylinder lubrication

In some large medium-speed four-stroke engines and in all two-stroke crosshead engines, the cylinder liners and consequently the pistons and piston rings are lubricated separately. Here the cylinder liners around the circumference of the cylinder are provided with a number of lubricating points. Using small plunger pumps, the lubricating oil is forced through the drilled holes onto the cylinder liner at the correct time. The plunger pump capacity can be modified by mechanically adjusting the engine gearing. At present, it is possible to alter electronically the rpm of the commonly used electro motor.

▶
Cylinder lubrication of a two-stroke crosshead engine.

right: the distribution blocks for various lubrication points around the cylinder liner.
left: two lubrication points which lead horizontally to the cylinder-liner bearing surface via a drilled duct.

11.5.9 Special properties of lubricating-oil

Reverse gear for large two-stroke cross head engines often operate on lubricating-oil pressure. In these engines, the lubricating-oil is also used for the high-pressure lubricating-oil pump. In some common-rail fuel injection systems, it provides power for the fuel pumps and exhaust valves.

11.5.10 Other diesel-engine parts

Other diesel-engine parts that require lubrication are:

- chain drives in traditional two-stroke crosshead engines;
- moment compensators;
- axial vibration dampers.

▶
A view from above of a Wärtsilä Sulzer two-stroke crosshead engine, type RT Flex with hydraulically driven exhaust valves and hydraulically operated fuel supply systems to the injectors.

▶
An automatic lubricating-oil fine filter with the lubricating-oil pumps for the high-pressure lubricating-oil system placed behind the filter.

This 200 bar hydraulic pressure system operates the exhaust valves and controls the fuel supply to a two-stroke crosshead engine from Wärtsilä Sulzer; type RT Flex.

1 *automatic lubricating-oil fine filter*
2 *high-pressure lubricating-oil pumps*
3 *high-pressure fuel pumps*

11.6 Examples of lubricating-oil systems according to their classification in the four categories

11.6.1 Category I: 0 to 100 kW, four-stroke, high-speed, MDO

In principle, these are simple lubricating systems with:
- an oil sump for collecting the lubricating oil;
- a lubricating-oil pump driven by the crankshaft;
- a simple lubricating-oil cooler;
- a duplex lubricating-oil fine-filter;
- a lubricating-oil supply system via the crankshaft bearings, through-drilled holes in the crankshaft and the connecting rod to the piston/piston rings and the cylinder liner;
- branching to the camshaft.

There is spray cooling as well as splash cooling of the piston.

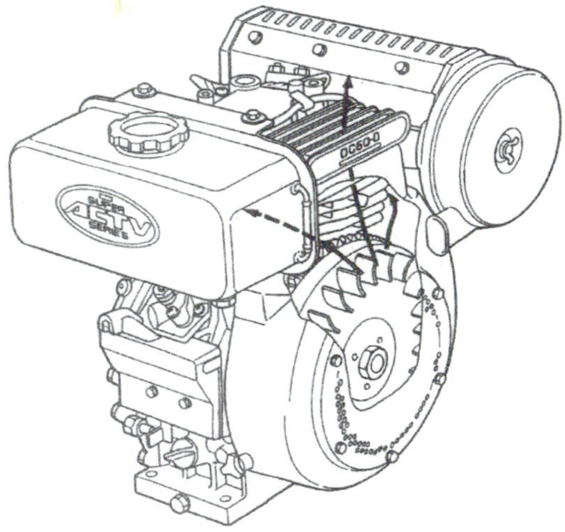

▲
The air-cooling of the lubricating-oil cooler using a flywheel in a small Kubota diesel engine.

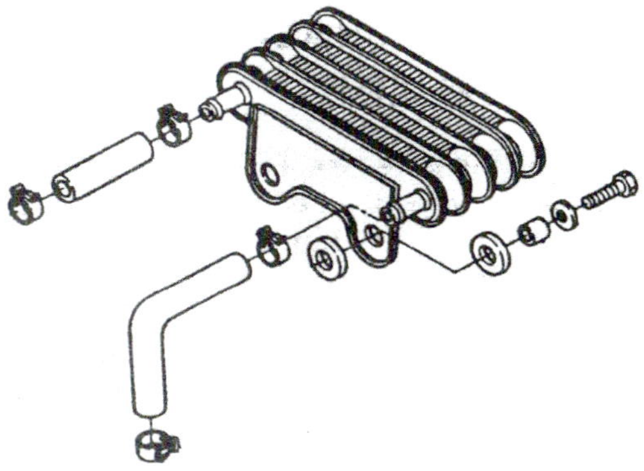

▲
A simple lubricating-oil cooler in a Kubota diesel engine.

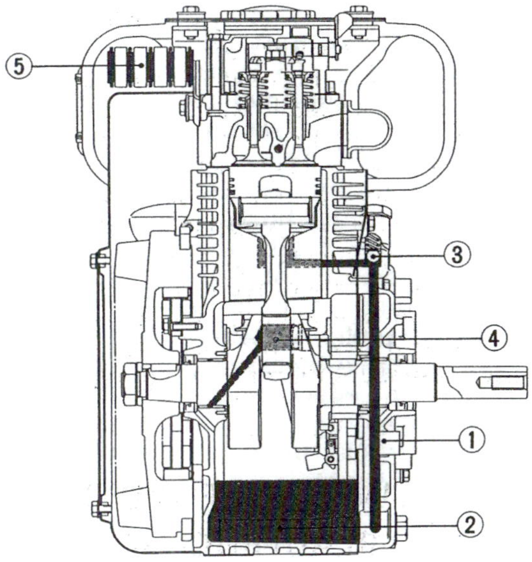

▲
The lubricating oil system in a diesel engine from Kubota, model OC 60 – I. This is a small high-speed four-stroke single-cylinder diesel engine with a power output of 4.2 kW.

The lubricating oil is drawn via a coarse filter (2) from the sump by the elevated lubricating-oil pump placed behind the pressure control valve (3). The lubricating-oil pressure is 1.5 to 5.0 bar, depending on the engine speed. The lubricating oil lubricates all the moving parts such as the crankshaft and main bearings, via drilled ducts in the crankshaft, the crank-pin bearings and with a spray nozzle, the piston. The cylinder liner is cooled with lubricating oil and the remainder of the engine block is air-cooled. Via a small interconnected pipe (not visible), the valve mechanism fitted on the cylinder head is lubricated. The lubricating oil passes through an oil cooler that in turn is cooled by the airflow generated by a flywheel placed on top of the diesel engine. The cooled lubricating oil is returned to the sump via a drilled duct. The suction filter in the sump is over dimensioned with a mesh size of 50 microns to avoid blockages. Three magnets are fitted to the filter to capture metal particles in the oil flow. Therefore, this small diesel engine does not have a fine filter in the pressure pipe of the lubricating-oil pump.

11.6.2 Category II: 100 to 10,000 kW, four-stroke, high-speed, MDO

Essentially, t this system is very similar to that of category I. It is often found in V-engines with various cylinders.

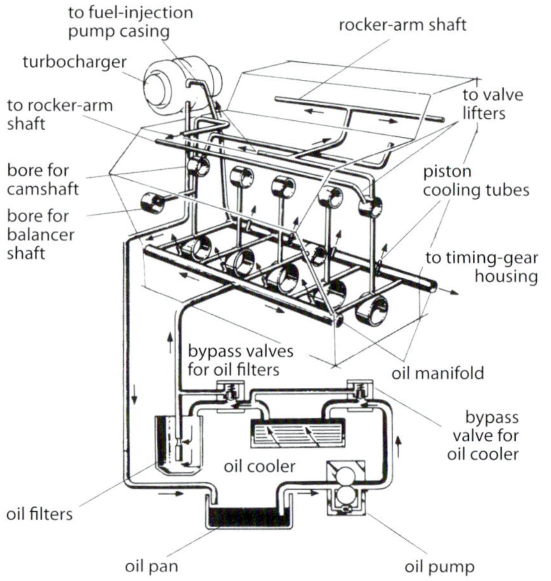

► The lubricating oil system of a high-speed four-stroke-diesel engine from Caterpillar, type 3408.

▼ Four lubricating-oil fine filters (1) and above a centrifugal lubricating-oil filter (2) in a four-stroke high-speed MTU diesel engine.

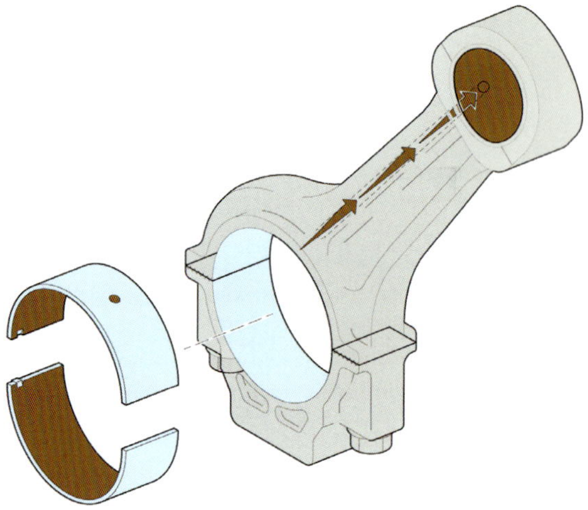

▲ The lubrication of the crank-pin- and gudgeon-pin bearings via drilled ducts in the connecting rod in MTU diesel engines.

► The lubricating oil system of a four-stroke high-speed MTU diesel engine.

The lubricating oil drains into the sump.

1 *main lubricating-oil pump*
2 *lubricating-oil fine filters*
3 *lubricating-oil cooler*
4 *crank shaft*
5 *cam shaft*
6 *sump*
7 *turbocharger*
8 *piston*
9 *cylinder head*

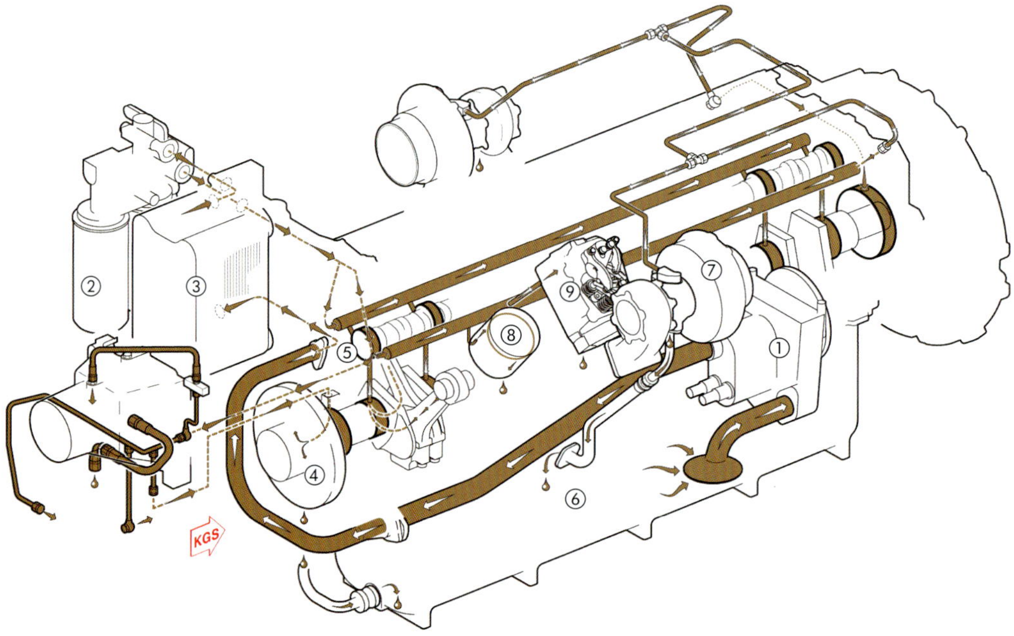

11.6.3 Category III: 500 tot 30,000 kW, four-stroke, medium-speed, HFO

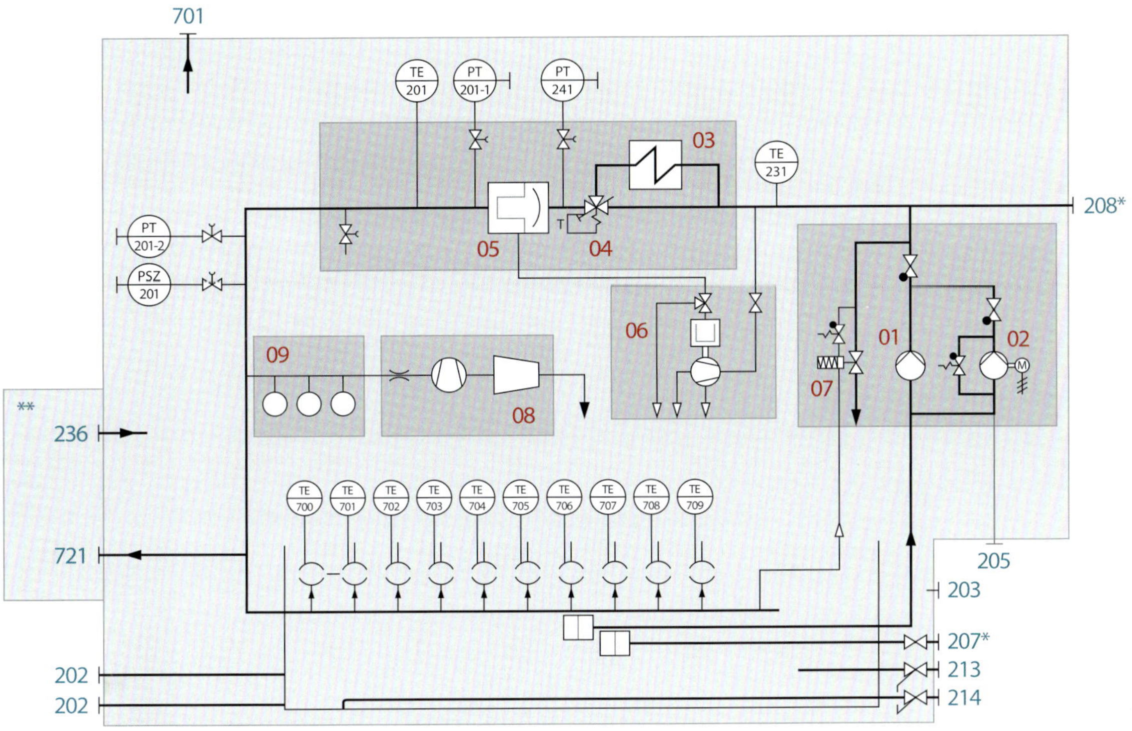

* If stand-by pump; ** If common rail

System components:
01 Lubricating oil main pump
02 Prelubricating oil pump
03 Lubricating oil cooler
04 Thermostatic valve
05 Automatic filter
06 Centrifugal filter
07 Pressure control valve
08 Turbocharger
09 Camshaft bearings and cylinder head lubrication

Pipe connections:

202	Lubricating oil outlet (if dry sump)	DN150
203	Lubricating oil to engine driven pump (if dry sump)	DN200
205	Lubricating oil to priming pump (if dry sump)	DN80
207	Lubricating oil to el. driven pump	DN150
208	Lubricating oil from el. driven pump	DN100
213	Lubricating oil from separator and drain (if wet sump)	DN40
214	Lubricating oil to separator and drain (if wet sump)	DN40
236	Sludge from external filter (common rail)	DN25
701	Crankcase ventilation	DN80
721	Control to external filter (common rail)	DN25

All connections DIN 2576, PN10

Sensors, transmitters and switches:

LS204	Lube oil level, wet sump, low
PSZ201	Lube oil pressure, engine inlet
PT201-1	Lube oil pressure, engine inlet
PT201-2	Lube oil pressure, engine inlet (back-up)
PT241	Lube oil pressure, filter inlet
TE201	Lube oil temperature, engine inlet
TE231	Lube oil temperature, LOC inlet
TE70x	Main bearing temperature

▲
The internal lubricating-oil system of a larger four-stroke diesel engine from MAN Diesel Type 32/40.

The main lubricating-oil pump 01 which is driven by the engine and pumps the pressurised lubricating oil to the main lubricating-oil line.

Right from 208, the lubricating oil can be supplied by an electrically driven lubricating-oil pump.

Pre-lubricating pump 02, which is electrically driven, switches on automatically when the lubricating-oil pressure is too low. For instance, when the engine is turned off. Pre-lubrication is generally used for pre-heating the engine prior to start and for cooling after turning-off the engine.

Both pumps are fitted with a prescribed overflow valve in the bypass to avoid excessive pressures in the lubricating-oil system.

In the main pipeline, the lubricating oil first travels through oil cooler 03. Thermostatic three-way valve 04 monitors, independent of the operating conditions, the predetermined lubricating-oil exit temperature. The lubricating oil then passes through the automatic filter 05 that discharges the contaminated lubricating oil to the centrifugal filter 06.

From the main lubricating-oil line, the lubricating oil flows to the camshaft bearings and cylinder head valves 09 to the turbocharger 08.

Underneath these camshaft bearings, the 10 main shaft bearing are situated. The lubricating oil continues through the drilled ducts in the crank shaft to the crank-pin bearings and via ducts in the connecting rod to the gudgeon pin, the piston rings and inside of the piston for cooling.

- **The lubricating-oil pumps** can be driven by the engine or by electric motors. In the latter, it is possible to switch on a lubricating-oil pump during warming up and cooling down of the engine.
- **The lubricating-oil fine filters** are often manufactured as automatic filters, which means that during operation one filter element is automatically back flushed and therefore cleaned. The contaminated lubricating oil is subsequently bled off outside the lubricating-oil system to a contamination tank. In this category, most turbochargers are lubricated by the main lubricating-oil system. Some engine types in large four-stroke engines operating on a low rpm (300 to 600 revolutions per minute) are equipped with a separate cylinder lubricating system. This creates a fresh lubricating-oil barrier between the combustion space and the crankcase preventing the drive gearing in the crankcase being contaminated by combustion products such as sulphur.

▶

The external lubricating-oil system from the same engine, a MAN Diesel type 32/40.

The **third** lubricating-oil pump in the system 2PO4 is electrically driven and provided with a suction filter 2F06.
The 'internal' main- and auxiliary oil pumps have a common suction filter 2F01.
Left of the engine is the separator module 2N01, which consists of an independent separator pump 2P03 with suction filter 2F03, the pre heater 2F02 and the separator 2S01.
The separator pump draws the dirty lubricating oil from the system oil tank 2T01. The pre-heater ensures that the lubricating-oil temperature is maintained in the separator, this is the temperature when the specific mass difference between water and lubricating oil is as high as possible ensuring optimal water separation. The cleaned lubricating oil drains to the lubricating-oil drain tank. In larger four-stroke- and two-stroke engines, this type of separator unit is constantly in stand-by mode (for the lubricating-oil system). Only when repairs to the centrifuge take place or when the engine is out of operation for an extended period is the separator switched off. The modern versions operate fully automatically and clean the separator bowl periodically. The dirt is discharged to the sludge tank 1T05.
The bleeder mounted on the engine contains an oil separator 2S01 which leads the oil back to the sludge tank. The gases are vented outside the engine room.

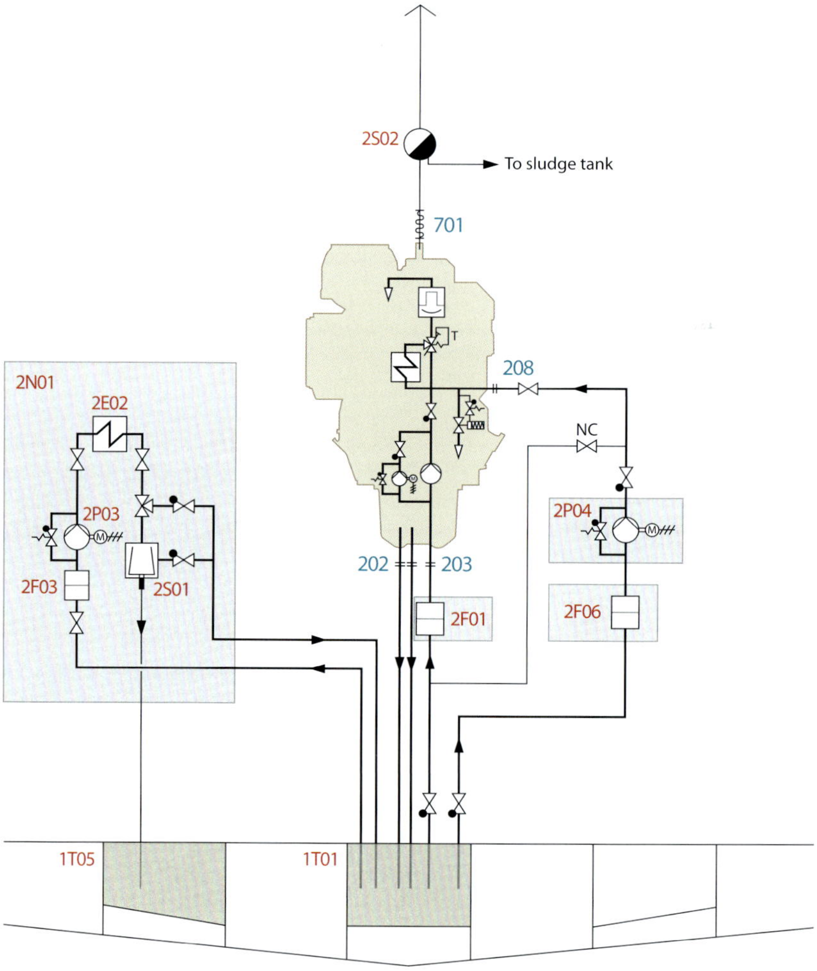

System components:
1T05 Sludge tank
2E02 Heater (separator unit)
2F01 Suction strainer (main lube oil pump)
2F03 Suction filter (separator unit)
2F06 Suction strainer (stand-by pump)
2N01 Separator unit
2P03 Separator pump (separator unit)
2P04 Stand-by pump
2S01 Separator
2S02 Condensate trap
2T01 System oil tank

Pipe connections:		L32	V32
202	Lubricating oil-outlet	DN150	DN150
203	Lubricating oil to engine driven pump	DN200	DN250
208	Lubricating oil from el. driven pump	DN100	DN125
701	Crankcase air vent	DN80	DN100

11.6.4 Category IV: 1500 tot 100,000 kW, two-stroke, low-speed, HFO

The size of the lubricating-oil system in these crosshead engines is substantial. A lubricating-oil drain tank can hold up to 40,000 litres of lubricating oil and the lubricating-oil pumps, coolers and filters are man size.

In this category, three types of lubricating-oil systems can be distinguished:

1 **lubrication of the crankshaft, crosshead with guides, thrust block, drive gearing of the fuel pumps and piston cooling;**

2 **cylinder lubrication,** a completely separate system for the lubrication of pistons, piston rings and cylinder liners;

3 **the oil system for controlling the exhaust valves and fuel pumps.** The latter refers to common rail systems.

Standard lubricating-oil system

In crosshead engines, there is a partition between the combustion space and the crankshaft including its gearing. The piston rod stuffing box in the scavenging-air space ensures that the contaminated lubricating oil that leaks down the piston cannot enter the crankcase. **In this way, the lubricating oil in the crankcase remains clean.**

The system is cleaned with a lubricating-oil separator and an automatic fine filter.

The lubricating oil- and cooling-oil system of the two-stroke crosshead engines of MAN Diesel type ME/ME-C have been used as an example.

▲ Maintenance on the drive gear in the crankcase of a two-stroke crosshead engine.

▼ The external lubricating-oil system in a MAN Diesel two-stroke crosshead engine. Type ME/ME-C/ME-GI.

Two main lubricating-oil pumps 4 40 601 draw the lubricating oil from the lubricating-oil drain tank and force the lubricating oil through the lubricating-oil cooler 4 40 610 and an automatic fine filter 4 40 615 to the main lubricating-oil line in the engine.

Connection RU provides lubrication and cooling of the main bearings, thrust block, axial vibration damper, piston cooling, crosshead bearings and connecting-rod bearings.

RU also provides the lubricating-oil supply to the hydraulic system for driving the exhaust valves, the fuel pumps, the moment compensators and the torsional vibration dampers.

Lubricating-oil separators that continuously clean the lubricating oil of dirt particles and water are also connected to the system.

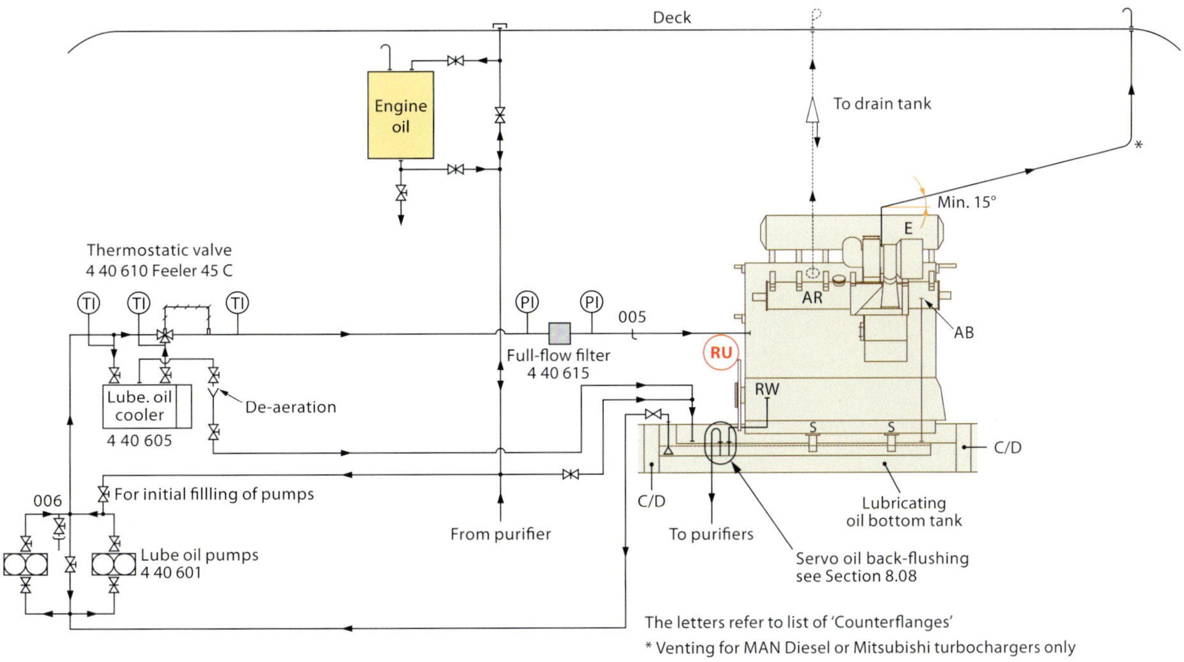

▶

The lubricating-oil system of two-stroke crosshead engines consists nowadays of three separate systems.

1. *lubrication of the crank-case drive gear and piston cooling*
2. *cylinder lubrication using special cylinder lubricating oil*
3. *the hydraulic section for driving the exhaust valves and operating the fuel injection system. The lubricating oil used here is the same as that used in the drive gear lubricating-oil system.*

▼

The hydraulic system of a MAN Diesel two-stroke crosshead engine type ME.

The hydraulic system consists of:

- an automatic fine filter and parallel to it a by-pass filter;
- two electrically driven lubricating-oil pumps;
- three engine driven lubricating-oil pumps;
- a safety- and accumulator unit.

At start-up, one of the two electrically driven lubricating-oil pumps is activated and is stopped as soon as the three engine driven lubricating-oil pumps have taken over the oil supply.

The high-pressure lubricating oil flows to the hydraulic units and is distributed to the electronically controlled fuel injection system and the electronically controlled exhaust-valve activation system. The exhaust valve is closed by a conventional 'air spring'.

Hydraulic Power Supply: HPS.

Hydraulic Cylinder Units: HCU.

Fuel Injection system and Valve Activation system: FIVA.

Engine Control System: ECS.

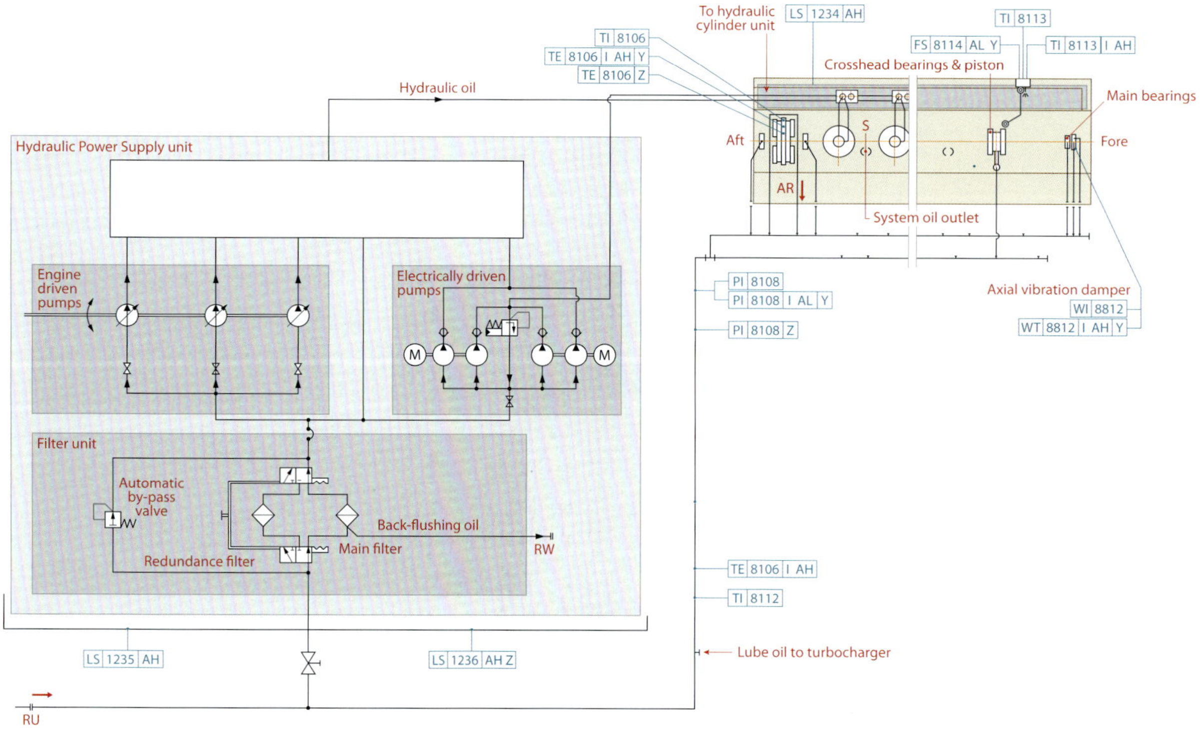

Turbocharger lubrication

As opposed to the older systems, the bearings are not lubricated with a separate system but by the main lubrication system.

Mesh of the automatic fine-filter of the turbocharger

Depending on the material used in the manufacture of the bearings:
white metal: 50 micro-millimetres;
aluminium tin: 40 micro-millimetres.

Comment

All metal particles larger than 25 micro millimetres, 85 to 90%, remain in a filter with a 40 micro millimetres mesh, and metal particles of 35 micro millimetres in a fine filter with a 50 micro millimetres mesh.

Cylinder lubricating-oil system

The cylinders are lubricated independently of the main system using a special cylinder lubricating oil. This cylinder lubricating-oil has to comply with strict requirements:

- the fuel has high sulphur content;
- a fraction of the lubricating oil combusts in the exhaust-gas flow, this must be ashless;
- the other fraction of the cylinder lubricating-oil that leaks into the scavenging air space is transported outside the engine from the top part of the piston rod-stuffing box;

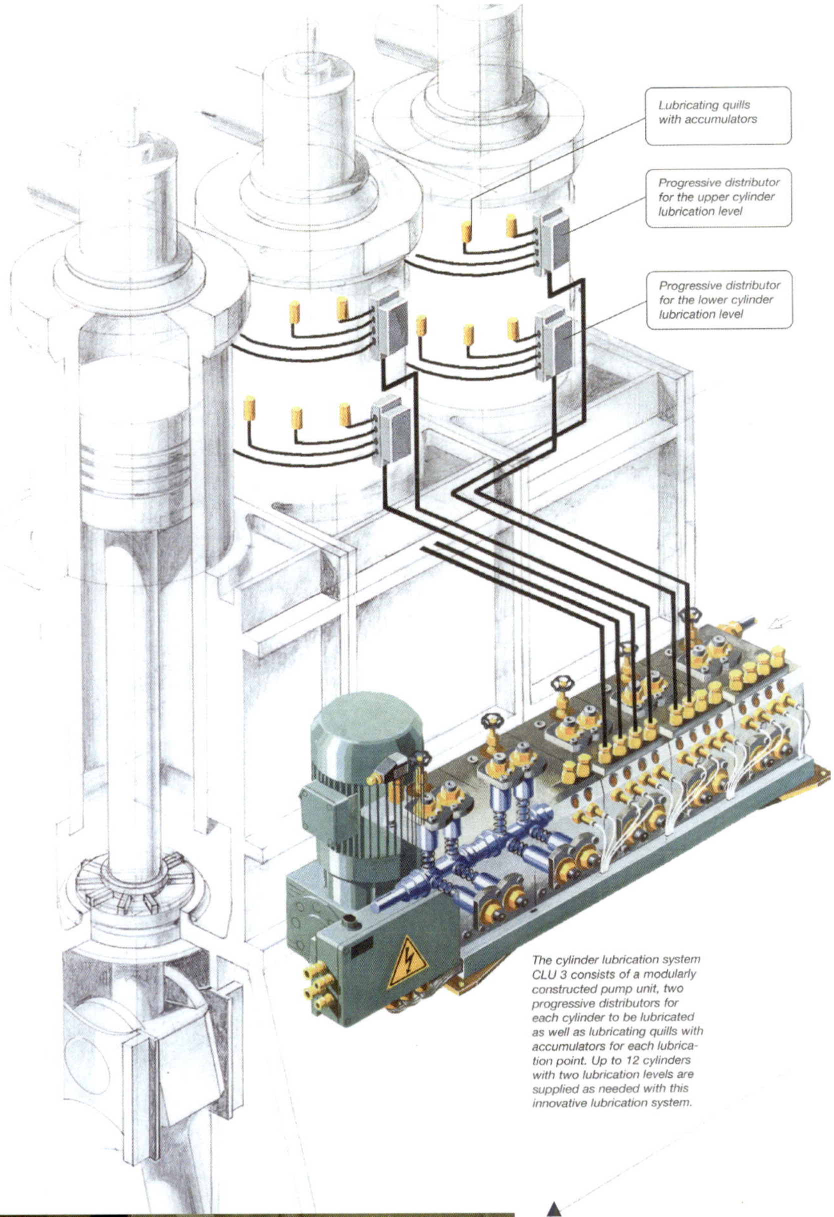

▲ The electrically driven cylinder lubricating-oil unit

Each lubrication point is provided with special cylinder lubricating oil by its own plunger pump. Lubrication of the cylinder liner occurs on two levels.

▲ Bearings of turbochargers or turbocharged groups are often directly lubricated from the main lubricating-oil system used for lubricating the crank-case drives. In this large turbocharger, lubrication takes place using its own lubricating-oil system.

◄ An automatic lubricating-oil fine filter.

A lubrication point with groove in the cylinder liner.

A large cylinder liner.

1 *inlet ports*
2 *grooves for cylinder lubrication*

– the scavenging-air spaces themselves are fitted with an oil drain, often set to drip in order to avoid blockage of the pipe and an accumulation of lubricating oil in the scavenging air spaces.

Next to fuel consumption, the consumption of cylinder lubricating-oil is a major cost factor in the exploitation of these engines.
Too high a dosage of cylinder lubricating-oil produces extra contamination of the piston-ring package, the combustion space, the exhaust-gas duct and the turbocharger. Too low a dosage causes additional wear of the piston, piston rings and cylinder liner.

Optimal cylinder lubrication

Precise dosing of cylinder lubricant can be achieved with the '**Alpha Adaptive Cylinder oil Control System, Alpha ACC-system**'.
This system takes into account engine load and fuel quality.
Sulphur causes wear and tear in the combustion space. Therefore, knowledge of the fuel's sulphur content is important in accurately determining the cylinder lubricant quantity.

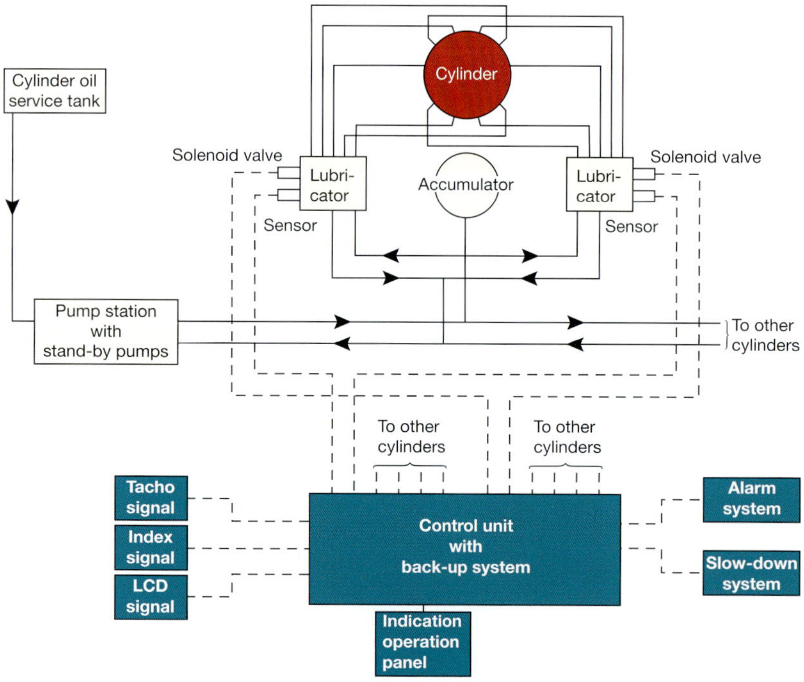

The cylinder lubricating system for MAN Diesel two-stroke crosshead engines type ME.

The cylinder lubricating oil is pumped from a storage tank to a cylinder lubricating-oil day tank and subsequently to the lubrication pump for each cylinder.
Each cylinder has eight lubricating points. Cylinder lubrication oil is supplied every fourth engine revolution at a constant pressure and with an electronically controlled timing and dosage control. The dosage of the lubricating oil can be tuned with an adjusting screw. The entire system is regulated by a cylinder control unit that calculates the fuel injections frequency based on the number of rotations and the set capacity (index) of the high-pressure pump. This is required to supply the correct dosage to the cylinder.

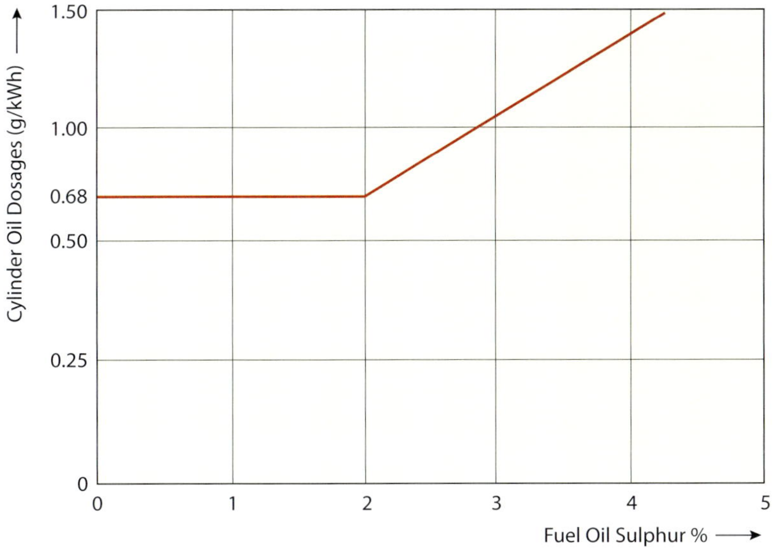

The dosage of the cylinder lubricating oil in the 'Alpha ACC' system from MAN Diesel.

With sulphur content in fuel in excess of 2%, the cylinder lubricating-oil dosage rises from 0.68 grams per kWh to 1.5 grams per kWh when the sulphur content is 4%

The dosage is dependent on the sulphur content of the fuel and the engine load.
The following rule of thumb applies:
Cylinder lubricating-oil dosage = 0.34 gram/kWh × S%
The minimal dosage is 0.68 gram per kWh at 2% sulphur.

Example
Cylinder lubricating-oil consumption: engine capacity 86,000 kW, sulphuric content of fuel 3.5%.
Consumption per hour:
86,000 × 0.34 × 3.5 = 102,340 grams or 102 kg
per day: 24 × 102 = 2448 kg
annually with an operating time of 6500 hours:

per annum $\frac{6500}{24} \times 2448 = 663{,}000$ kg or $\frac{663{,}000}{0.900}$

or 736.7 m^3.

11.7 Lubricating-oil properties

In order to acquire a good insight into engine lubrication, it is imperative to be familiar with the most important properties of lubricating oil.
The significance of these properties is shown each time problems occur with engines and in the assessment of lubricating-oil samples.

11.7.1 Viscosity expressed in centistokes (cSt) or mm^2/sec. at 40 °C

To ensure that, for instance, the lubricating film between the crankshaft and crankshaft bearing is sufficiently thick, it is imperative that the oil has a certain viscosity.

Thick lubricating oil, lubricating oil with too high a viscosity produces high frictional loss and therefore heat increase. Thin lubricating oil, lubricating oil with too low a viscosity seeps from the crankshaft, resulting in a pressure loss and allowing the bearing and shaft to come into contact with each other. This in turn causes wear.

Viscosity is divided into viscosity categories. This classification is incorporated in the standard specification 150–3348. Lubricating oil viscosity is divided into eighteen divisions. Each division encompasses the viscosities between two limits.

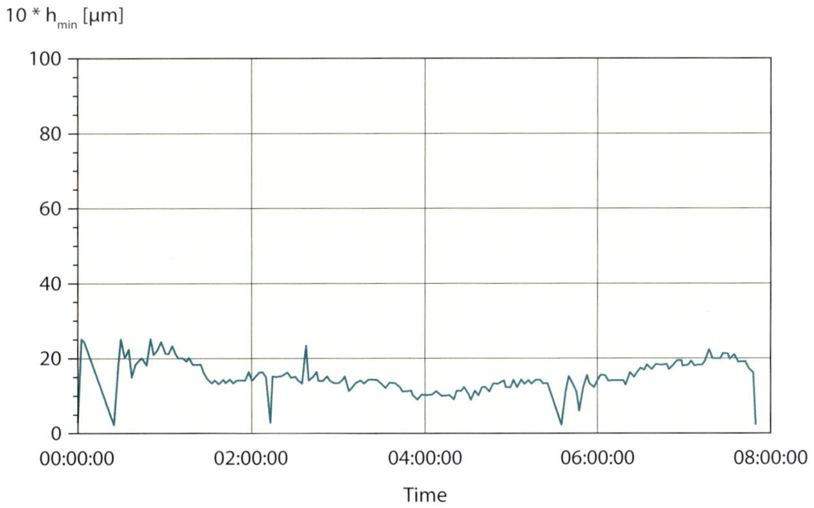

During lubrication of the engine parts, a very thin lubrication film separates the various parts.

The film thickness is usually no more than several tenths of a millimetre. Here the thickness is approximately 15 to 20 microns.

Lubricant properties

SAE Viscosity Grade	Viscosity at 40 °C (mPa/sec.)	Viscosity at 100 °C (mPa/sec.)	Viscosity at HTHS (mPa/sec.)
10W	31.87	5.18	2.20
20	61.02	7.36	2.79
30	85.72	9.70	3.54
40	139.72	12.81	4.31
50	189.74	16.34	5.30
5W-30(A)	47.87	8.90	3.24
5W-30(B)	59.79	9.99	3.24
10W-40 (A)	63.66	11.06	3.10
10W-40 (B)	92.69	11.22	4.19
10W-40 (C)	74.65	12.10	3.88
20W-40	89.40	11.00	3.89
10W-50 (A)	105.14	15.80	4.05
10W-50 (B)	134.19	15.92	4.80

▲
The lubricating-oil viscosity is divided into a number of grades. Each grade contains the viscosities between two limits.

11.7.2 Viscosity index

The viscosity index denotes the degree in which the viscosity changes in relation to the temperature. Lubricating oil with a viscosity barely susceptible to temperature differences is said to have a high viscosity index.

Viscosity index graph

A diesel engine that has a cold start and is subsequently fully loaded requires lubricating oil with a high viscosity index. The lubricating-oil temperature of these engine types varies from –10 °C to 100 °C!

The Society of Automotive Engineers, or SAE has a numerical code system for grading oils according to their viscosity. These 'Grades' are, for instance, SAE 10 W, SAE 30, SAE 15 W-40. The numbers represent the viscosity at a given temperature. The index W applies to lubricating oils which must have a low viscosity at low temperatures.

11.7.3 Neutralisation number

Usually referred to as the 'Total Base Number'. Fuels always contain sulphur. In light fuels, the sulphur content is extremely low, for instance, 0.15%. When using heavy fuels, this can rise to 4%. The acids produced in the combustion process can severely damage engine parts. These acids must be chemically neutralised by adding alkaline agents to the lubricating oil.

Calcium hydroxide is in this case commonly used. Therefore, the neutralisation value is a measure of the alkalinity of the lubricating oil. The higher the possibility of acid formation in combustion, the higher the required neutralisation value or the Total Base Number (**TBN**). Today one often speaks of the BN, Base Number.

Every engine manufacturer provides the required **TBN** values. They are denoted in mg KOH per gram lubricating oil.

▶
The viscosity grades of lubricating oil; in this instance for Caterpillar diesel engines.

The grades run from 0W to 60. 'W' indicates 'Winter'; that is low temperatures. At low temperatures, the viscosity is established for operation, shown here, cranking and pumping lubricating oil.

SAE Viscosity Grade	Low Temperature Viscosities		Low Temperature Viscosities		
	Cranking[b] (cP) max at temp °C	Pumping[c] (cP) max with no yield stress at temp °C	Low Shear rate Kinematic[d] (cSt) at 100 °C		High Shear rate (cP) at 150 °C min
			min	max	
0W	6200 at -35	60,000 at -40	3.8	–	–
5W	6600 at -30	60,000 at -35	3.8	–	–
10W	7000 at -25	60,000 at -30	4.1	–	–
15W	7000 at -20	60,000 at -25	5.6	–	–
20W	9500 at -15	60,000 at -20	5.6	–	–
25W	13,000 at -10	60,000 at -15	9.3	–	–
20	–	–	5.6	< 9.3	2.6
30	–	–	9.3	< 12.5	2.9
40	–	–	12.5	< 16.3	2.9 (0-40, 5W-40, 10W-40 grades)
40	–	–	12.5	< 16.3	3.7 (15-40, 20W-40, 25W-40 grades)
50	–	–	16.3	< 21.9	3.7
60	–	–	21.9	< 26.1	3.7

SAE: Society of Automotive Engineers
API: American Petroleum Institute
ASTM: American Society of Testing Materials

Examples

1 Caterpillar MaK 43
This is a four-stroke medium-speed diesel engine, fuel HFO.
TBN 30 to 40 mg KOH/g
Viscosity SAE 40

2 Wärtsilä 38 B
This is a four-stroke medium-speed diesel engine, fuel HFO.
TBN 30 to 55 mg KOH/g
Viscosity SAE 40
'At high sulphur contents adhere to the maximum TBN value.'

3 MAN Diesel K98MC
This is a two-stroke low-speed crosshead engine, fuel HFO.
Lubricating-oil circulation in the crankcase.
Viscosity SAE 30
Cylinder lubricating-oil TBN 70 mg KOH/g
Viscosity SAE 50

Generally, the following values are identical for all engine manufacturers:

- **for high-speed four-stroke engines with MDO as fuel (Category I + II) 4 to 6;**
- **for medium-speed four-stroke engines with HFO as fuel (Category III) 30 to 55;**
- **for two-stroke crosshead engines with HFO as fuel (Category IV):**
  - **in the crankcase 5 to 10;**
  - **cylinder lubricating-oil 70 to 90.**

11.7.4 Solidification point

This is the temperature at which lubricating oil solidifies. This property is only pertinent in extremely cold conditions. Lubricating oils that contain small amounts of paraffins have a lower solidification point.

11.7.5 Oxidation stability

There is a lot of air is in diesel engines, especially in the crank case. Lubricating oil is therefore exposed to air. The oil in the crank case splashes in all directions. If the lubricating oil oxidises, acids with a high viscosity are produced. These may cause blocking of filters and the small lubricating-oil pipes. Oxidation is accelerated at higher temperatures. By fine-tuning the refining process and using an anti-oxidation additive this can be averted. These additives comprise zinc or sulphur compounds.

11.7.6 Additives in lubricating oil

For each specific lubricating-oil application, chemicals may be added which improve its properties, create new properties, or compensate poor lubricating properties.

11.7.7 Detergents or 'solvents'

Combustion products, among which carbon, can be deposited on parts of the combustion space, such as pistons, piston rings, cylinder heads and scavenging ports in two-stroke engines. This may cause the piston rings to lodge in the grooves thus accelerating the wear and tear of the piston, piston rings and cylinder liner. Leakage of combustion gases along the ring packet will also increase. If carbon settles **behind** the piston rings, the piston rings will slowly begin to protrude through the lubricating film on the cylinder liner. This may result in a corroded or seized piston. This could in fact cause complete engine failure. The seizing of piston rings is referred to in engine technology as **'ring sticking'**. The detergents often consist of sulphur/phosphor compounds.

11.7.8 Dispersants

These lubricating-oil additives ensure that the combustion products do not settle and stick to engine parts but are kept in suspension. They can then be filtered in the fine lubricating-oil filter or in larger installations by a centrifugal separator.

▲
A seized piston in a cylinder liner.

The piston has been broken through the gudgeon pen. This usually results in total engine failure.

▲

A severely damaged piston in a two-stroke crosshead engine.

The piston rings are jammed in the piston-ring grooves and are barely recognisable. The piston and rings are heavily grooved.

11.7.9 Extreme Pressure (EP) additives or wear reducing compounds

These are added in exceptional instances where a lot of boundary lubrication occurs. This may occur in cogwheels and very highly loaded engine bearings. They form a film on the metal surfaces. They are mostly sulphur and phosphor compounds.

11.7.10 Viscosity index improving agents

These ensure that the lubricating oil does not become too thin at increasing temperatures. A whole range of chemicals can be used, such as:
- polyisobutenes;
- polymethacrylates;
- styrene;
- polyesters.

11.7.11 Ash content

This is a non-combustible lubricating-oil residue. It contains metal compounds such as barium, calcium and magnesium, which provide the correct alkalinity or TBN of the lubricating oil. High ash contents may affect the diesel engine's performance.

11.7.12 Biocides or bactericidal agents

In spite of all the precautionary measures taken, lubricating-oil will be exposed to water. Consider moist air in the crank case of a stationary diesel engine. In this kind of environment, certain bacteria develop very rapidly. The bacterial growth can become excessive and cause severe damage to engines.

11.8 Cleaning lubricating oil

Lubricating oil absorbs many substances in the engine that adversely affect the quality and reduce lubrication and cooling of the engine.
These substances are among other things:
- **water**, from cooling water leaks and water vapour precipitated from air in the engine;
- **fuel**, usually from leaking fuel pumps. The seals of the fuel pump and camshaft gear drives are not optimal;
- **metal particles**, from wear of the moving parts such as bearings, shafts, pistons, piston rings and cylinder liners. Due to corrosion or cavitation metal particles from pipe walls, heat exchangers and pumps can also be introduced.
- **combustion products**. Especially in four-stroke trunk-piston engines, there is always some 'deposition' of carbon particles along the piston rings; sulphuric acid, from the sulphur in the fuel, and other combustion by-products on the piston rings;
- **coolants**, similar to water leakage, due to cooling-system leaks;
- **carbon particles** from the lubricating oil itself. When lubricating oil is used for cooling, and when a certain temperature, approximately 400 °C, is exceeded, the lubricating oil begins to decompose and carbonise;
- **oxidation products** of the lubricating oil. Due to air exposure, lubricating-oil may oxidise. This produces long chained compounds that thicken the lubricating oil;
- **dirt particles** such as sealant materials, sand, welding beads and fine dust.

During engine repairs, sufficient attention should be paid to working 'hygienically'. Often engines are contaminated with grime ranging from fine dust to tools and cleaning cloths.

Note

In extreme conditions such as sand storms, for instance, in the Middle East, or the spread of pollen in large agricultural areas, one must take special precautionary measures to protect the combustion air from these dust particles.

Lubricating-oil storage should also be a focal point. **It is easy for moisture in the air to condensate when a storage tank is nearly empty.**

Lubricating-oil can be cleaned in various ways.

11.8.1 Fine lubricating-oil filters in the supply pipes of lubricating-oil pumps

These are used in all engine types. The mesh size determines the way in which filtering occurs.
In small diesel engines, filters are renewed. In larger engines, they are cleaned.

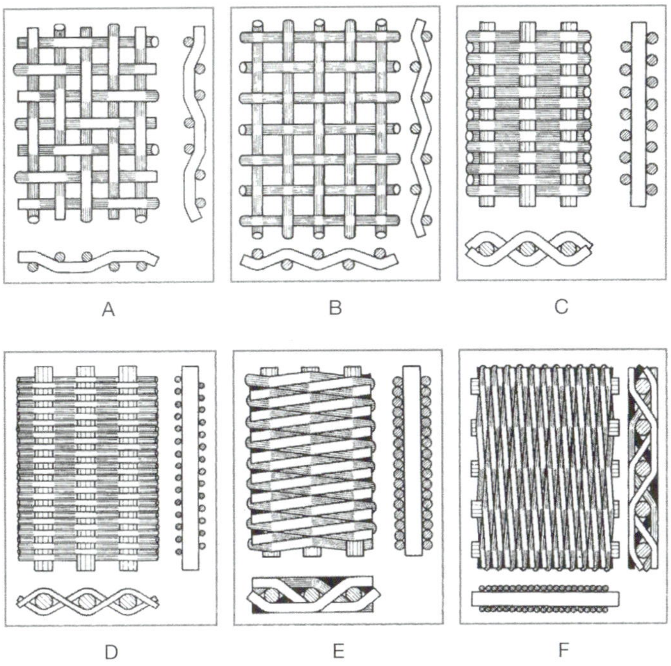

▶

Various metal filter constructions.

Depending on, amongst others, the required filter mesh size, the filter surface, the flow rate and the lubricating-oil temperature, a certain filter element is selected. Most mesh sizes vary from 10 to 25 microns, but there are also filters with a smaller mesh size, namely 2 to 4 microns.

▲

Four lubricating-oil fine filters in a diesel engine from MTU.

Left: two fuel filters (1). A centrifugal lubricating-oil filter (3) is just visible above the lubricating-oil filters (2).

▲

Two lubricating-oil fine filters in a diesel engine from Volvo Penta.

Left: a fuel filter.

11.8.2 Centrifugal filters

In increasingly more engines in categories II and III, centrifugal filters are used alongside the regular fine lubricating-oil filters. The centrifuge is driven by lubricating-oil pressure and dirt and water are thrown out along the outer surface. After a specified number of operating hours, the filters are manually cleaned.

◀

A centrifugal lube-oil filter on an ABC V-type diesel engine.

▼

A centrifugal filter driven by its own lubricating-oil pressure.

Contaminated lubricating oil enters from the back-flushed filter on the left. It is thrown against the filter walls by the centrifugal forces in the rotating filter element. At the nozzles the lubricating oil is sprayed onto the outside of the element and the filter begins to turn like a garden sprayer.
The dirt is thrown against the rotor wall and is periodically removed. The clean lubricating oil is discharged to the lubricating-oil sump at the bottom.

◄

Two centrifugal lubricating-oil filters mounted on a diesel engine.

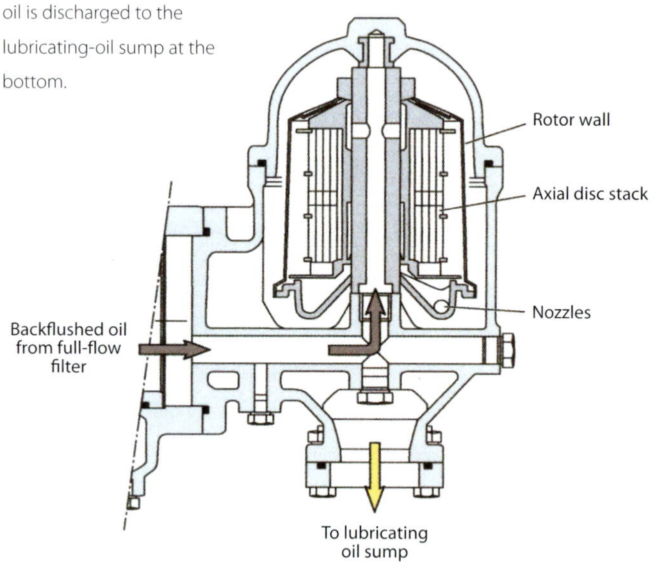

▼

A combination of an automatic full-flow filter and a centrifugal filter that periodically cleans the contaminated lubricating oil from the automatic full-flow filter.

The full-flow filter is driven by a hydraulic engine.
A distribution ring provides good lubricating oil supply distribution.
light blue: unfiltered lubricating oil to the full-flow filter.
yellow: filtered lubricating oil to the engine.
red: contaminated lubricating oil to the centrifugal filter.
blue: centrifugal filter.
yellow: cleaned lubricating oil returned to the lubricating-oil sump.

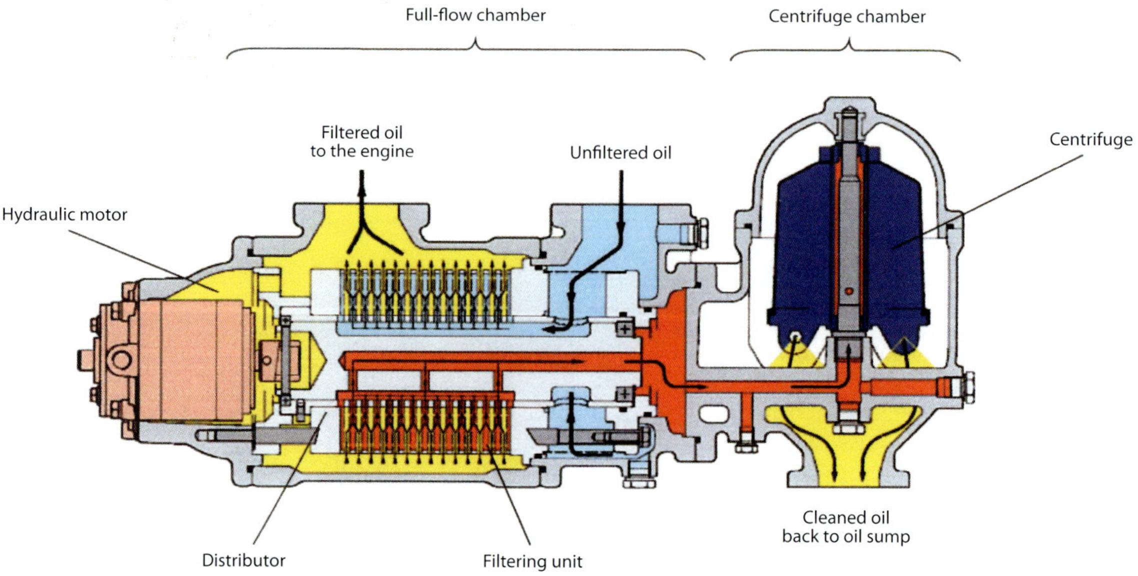

11.8.3 Automatic fine-filters

These filters are periodically cleaned by an automatic back-flushing system.
The filters are always used in category III and IV engines.

▼

The same type of automatic back-flow filter arranged vertically.

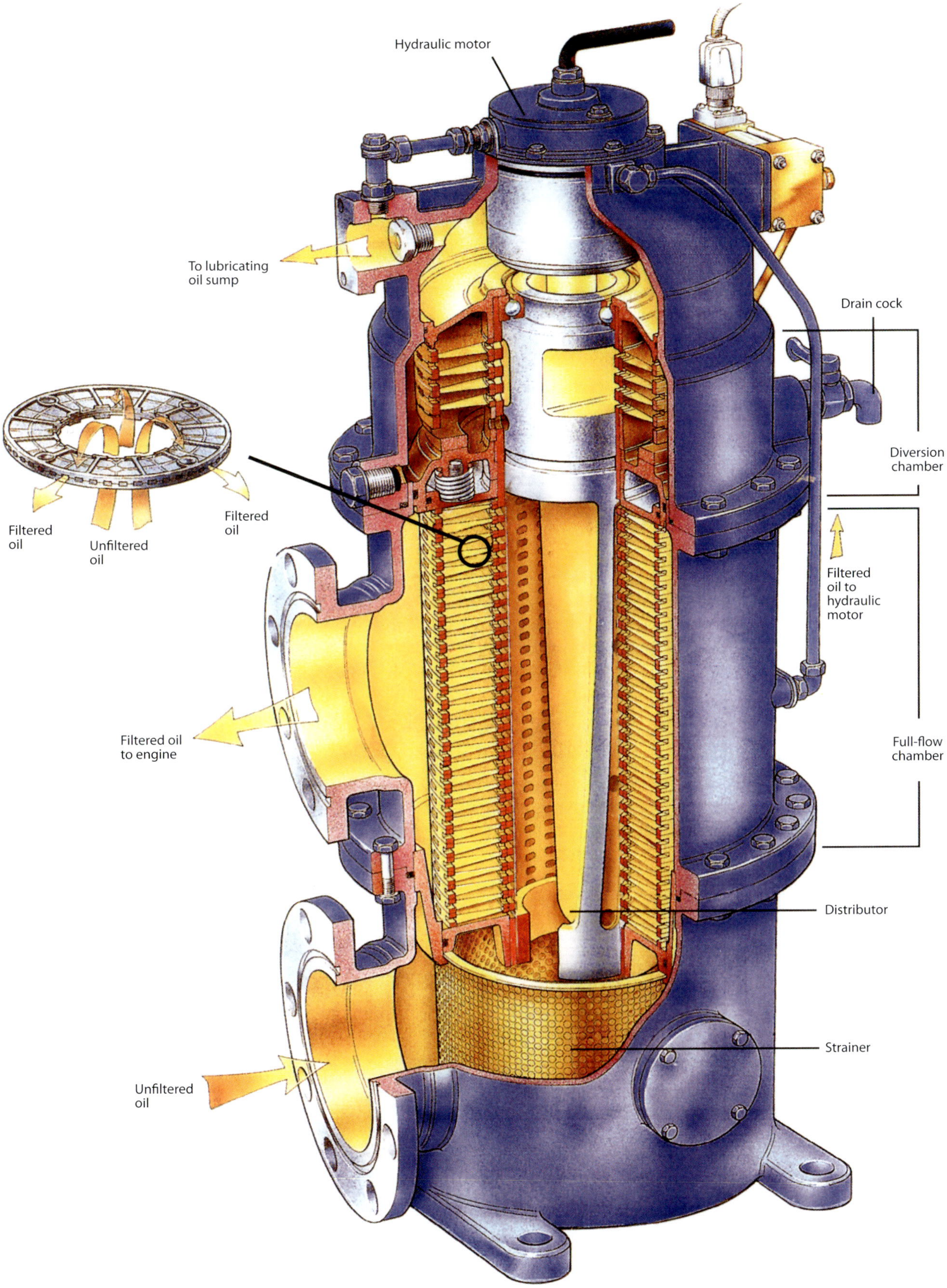

▲
An automatic lubricating-oil fine filter.

▼
The operating principle of a centrifuge or separator.

Contaminated lubricating oil is supplied at the top. The heavier dirt particles are flung outwards and removed, as shown in the lower figure. This is referred to as 'shooting' of the centrifuge. Prior to commencement of the 'shooting programme', the lubricating-oil supply is automatically shut and the entire rotating bowl filled with water. At the moment that the moveable bottom drops, the dirt is removed from the bowl by water pressure and discharged to the sludge tank via the centrifuge housing. It is possible to remove the water from the sludge and avoid the sludge tank from filling up too rapidly by using a separate sludge centrifuge.

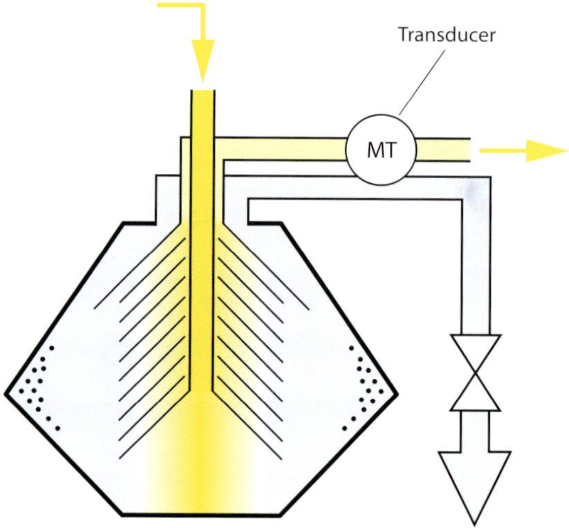

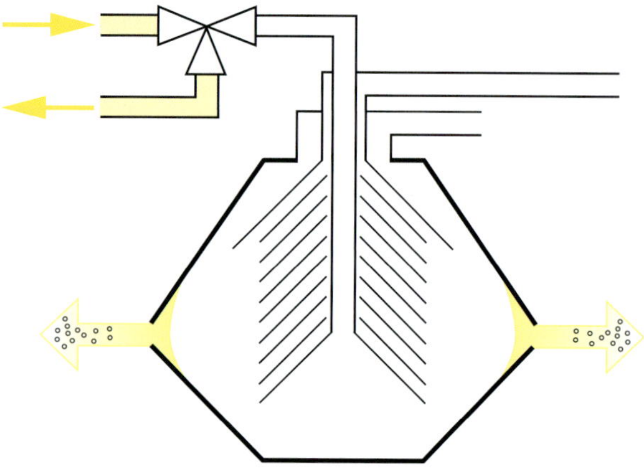

11.8.4 Magnetic elements or plugs

These plugs are mostly found in small engines. They capture mostly the iron particles. They require regular manual cleaning.

11.8.5 Lubricating-oil separators

The slightly larger diesel engines running on HFO often use centrifuges. They generally operate automatically and in the larger types cleaning occurs automatically on a periodic basis. This is called 'discharge'.

►
An automatic lubricating-oil centrifuge.

Water and dirt are separated and the dirt is automatically removed at regular intervals by 'shooting' of the centrifuge and is stored in a sludge tank.

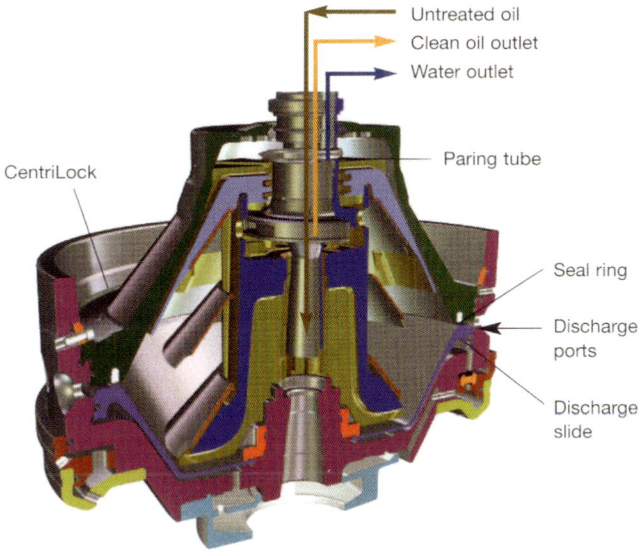

◄ The construction of a modern lubricating-oil centrifuge.

Brown: the unfiltered oil supply
Orange: the filtered lubricating-oil discharge
Blue: the water outlet

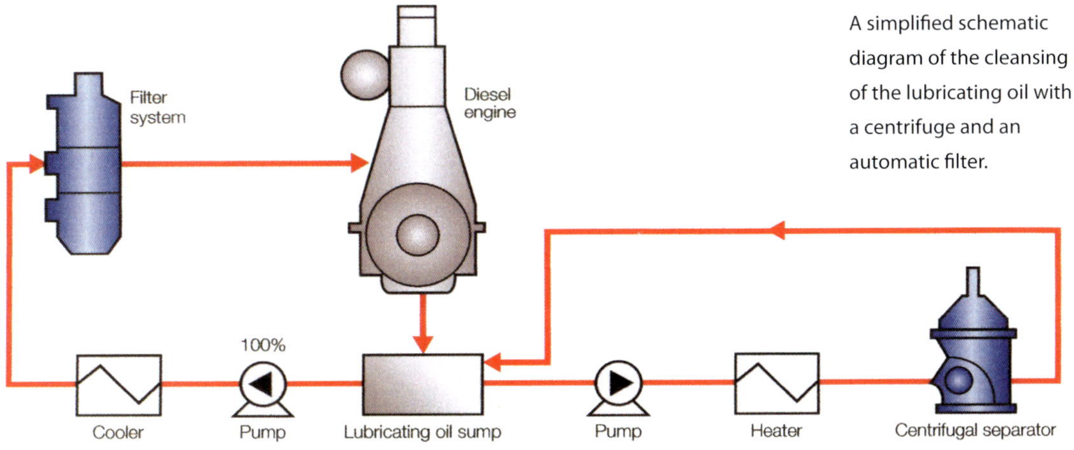

◄ A simplified schematic diagram of the cleansing of the lubricating oil with a centrifuge and an automatic filter.

▼ An automatic lubricating-oil separator in an Alfa Laval type S.

This type from 2005 operates with the principle Alcap. The automated module EPC 50 is a fully automatic operational management system. Inspection: every six months or 4000 operating hours. Maintenance: every eighteen months or 12,000 operating hours.

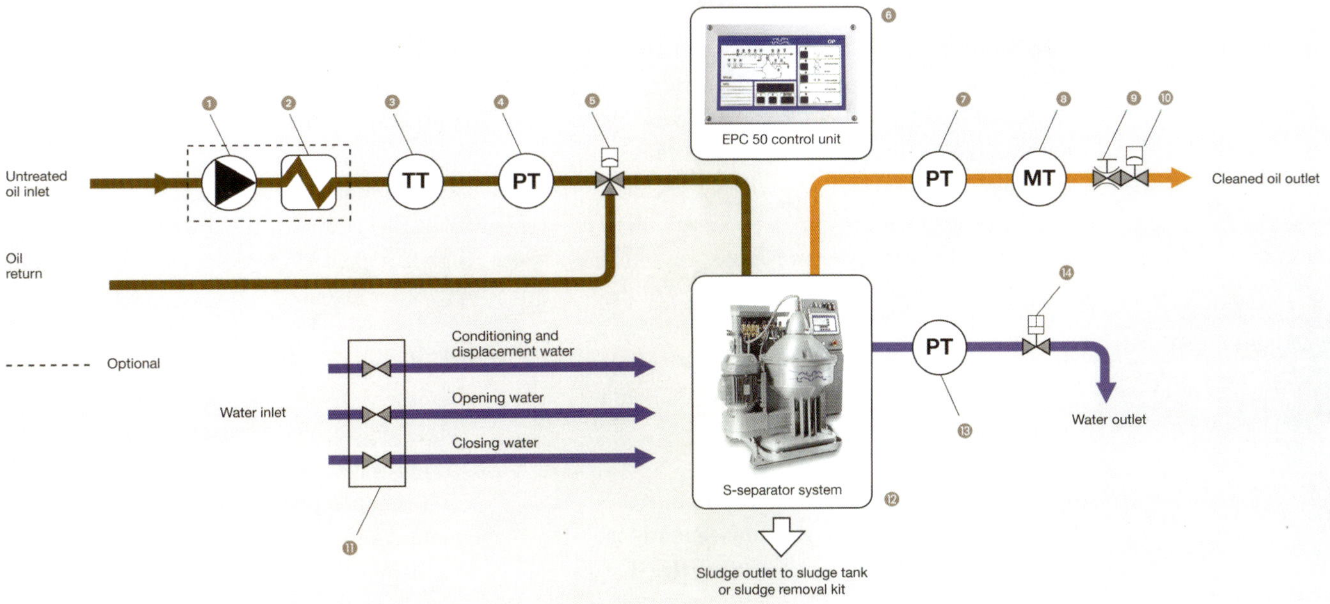

General comments

- In four-stroke trunk-piston engines running on HFO, lubricating-oil contamination is considerable.
- In four-stroke trunk-piston engines running on MDO, lubricating-oil contamination is noticeably less and cleaning the lubricating oil appreciably simpler.
- Due to rising fuel prices in recent years, many ships equipped with a modern propulsion engine and using MDO have been converted to run on HFO. Naturally, the engine must be suitable for HFO. These are engines with a 200 to 260 mm piston bore with a power output of 1500 to 3000 kW. The investment is recouped in **one year**. Adequate cleaning of the lubricating oil is now an important point of attention. Here automatic fine filters and centrifuges are indispensable.

Also see Chapter 24, *Auxiliary systems: Fuel and lubricating-oil centrifugal separators.*

11.8.6 Changing lubricating-oil

In smaller diesel engines that only have a fine filter behind the lubricating-oil pump, regularly renewing **all** lubricating oil and the fine-filter is simple and not expensive.
The lubricating-oil content often amounts to several litres to a maximum of 50 litres (Category I). In larger systems, the lubricating oil is cleaned more intensively and the operating lifetime is longer.
The lubricating-oil volume can amount to several thousands of litres (Category II and III).

In crosshead engines, the lubricating oil in the gear-drive section is very clean; this is due to the aforementioned partition between the combustion space and the gear drive by the piston rod-stuffing box.
The lubricating oil is cleaned with an automatic fine filter and a lubricating-oil separator.
The lubricating-oil volume is maintained at the correct level by refilling (leakages). If lubricating-oil samples indicate that the quality is substandard, the entire volume is replaced. This may amount to 10,000 litres or more.

In general: To acquire a good insight into the quality of lubricating oil and consequently the state of the diesel engine, it is imperative that samples be taken methodically to monitor the lubricating oil.

11.9 Lubricating-oil analysis

In the larger diesel engines in categories II, III and IV, it is customary to regularly take samples of the lubricating oil and have them checked in laboratories.
The analysis results of such samples provide information with regard to the lubricating-oil properties after a specified number of operating hours and therefore the state of maintenance of the diesel engine.
By consistently checking the lubricating oil, severe damage to the engine can be avoided.

Examples:

1 **The alkaline additive content decreases.**
The alkaline additive in the lubricating oil, which neutralises acids, indicated by BN or TBN value (BN – Base Number, TBN – Total Base Number) has been saturated by acids in the fuel.

2 **The metal-particle content increases.**
The content of, for instance, aluminium particles increases. This is found when bearing caps come into contact with an aluminium compound, this is indicative of worn bearing caps.
When there is a significant increase in the number of aluminium particles, the engine should be stopped to inspect the bearings.

3 **The viscosity decreases; the lubricating-oil is diluted.**
Lubricating oil has a certain viscosity, thickness. When the viscosity gradually increases, this is usually the result of the lubricating-oil ageing. The lubricating oil oxidises in air and thickens. If the viscosity decreases, this is usually due to dilution by the fuel.
It is, for instance, possible for fuel to leak near the fuel pumps into lubricating oil. The drainpipe of the fuel pump drain may be defective. The lubricating properties deteriorate, which result in engine corrosion.

4 **The water content in the lubricating oil increases.**
This is indicative of cooling water leakage. The lubricating properties decrease, which cause rust formation on the white metal parts in the engine.

>CH 12

Air supply

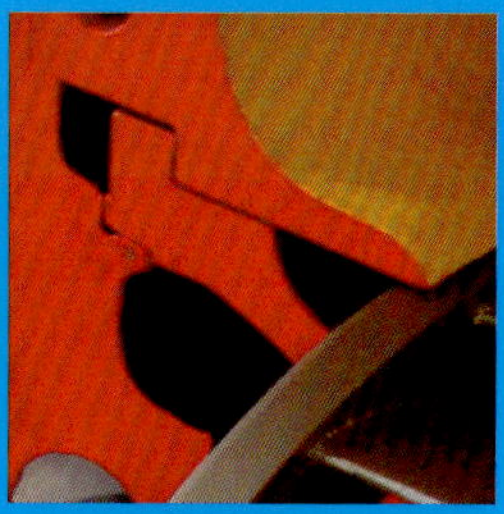
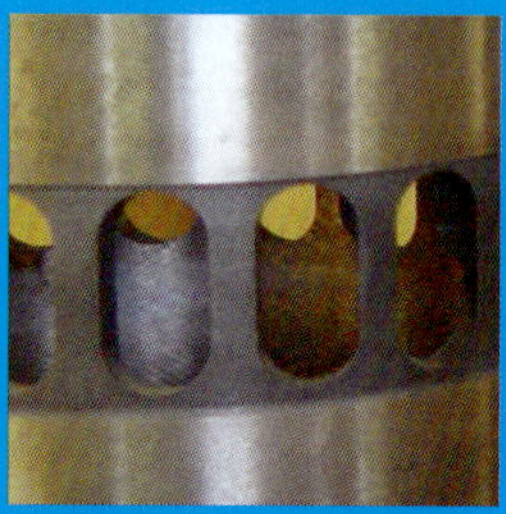

Diesel engines require large amounts of air for the combustion process. For example, this crosshead engine, a Wärtsilä-Sulzer 12RTA96C at full power output needs approximately 3 kg per second, 11,000 kg per hour and 201,000 kg per day.

The turbocharger, essential for the supply of air for modern diesel engines.

12.1 Introduction

Diesel engines use a considerable amount of air when operating. Not only do they require air for the combustion process, air is also indispensable in the scavenging process and cooling hot engine parts that are in close proximity to the combustion space.

The total amount of air required must be sufficient for:

- the combustion process;
- scavenging of the cylinder;
- cooling the parts surrounding the combustion space, such as the piston crown, the flame plate of the cylinder head and the exhaust valves.

12.2 The amount of air

Theoretically, the complete chemical combustion of 1 kg of fuel requires 14 to 14.5 kg of air. Lighter fuels with a high hydrogen number use approximately 14.5 kg of air and the heavier fuels with lower hydrogen number, approximately 14 kg.
In actuality, **twice** the amount of air required for the combustion process is supplied. Obviously, this depends on the type of diesel engine and the combustion system.

This is referred to as the practical amount of air that is available for combustion.

Combustion air factor

$$\text{combustion air factor} = \frac{\text{effective amount of air supplied for combustion}}{\text{theoretical amount of air supplied for combustion}}$$

λ = lambda = air factor

$$\lambda_{combustion} = \frac{\text{air effective}}{\text{air theoretical}}$$

The scavenging and cooling processes also require air. The total air consumption is denoted by the total air factor – λ_{total}. This is considerably higher than the combustion air factor, dependent on the type of engine and design, approximately 3.

▶

Diesel engines use large amounts of air in the combustion process.

Therefore, for 1 kg of fuel, approximately 3 x 14 = 42 kg of air is supplied!
In two-stroke engines approximately 3 x 16.5 = **51.5 kg air**.
The density of air at 0 °C and 1 bar is 1.276 kg per m^3.
This means that for each kilogram of fuel $\frac{42}{1.276}$ m^3 air supply, so each kilogram of fuel requires ± 33 m^3 of air!

Specific air consumption

Furthermore, there is the so-called **specific air consumption.** This is the mass of air that is supplied to the engine per power output and per time unit.

$$L_{specific} = \lambda_{total} \times L_{theoretical} \times b_e$$

$$\frac{L_{specific}}{L_{theoretical} \times b_e} = \frac{2{,}5}{14 \times \left(\frac{186}{3600}\right)} = 3{,}5$$

Specific fuel consumption

The **specific fuel consumption** is the amount of fuel that is supplied per power output and per time unit. **This is usually specified in grams per kilowatt per hour.**

$L_{specific}$ for four-stroke engines is approximately: 2 kg/MJ and for two-stroke crosshead engines approximately 2.5 kg/MJ.

$$\text{Theoretical air supply: } \frac{100}{23} \times \left(\frac{8}{3}C + 8H + S\right)$$

C = carbon content in mass percentage
H = hydrogen content in mass percentage
S = sulphur content in mass percentage

A well-ventilated engine room is the most important requirement. Shown here, an inlet grill of an emergency diesel generating set.

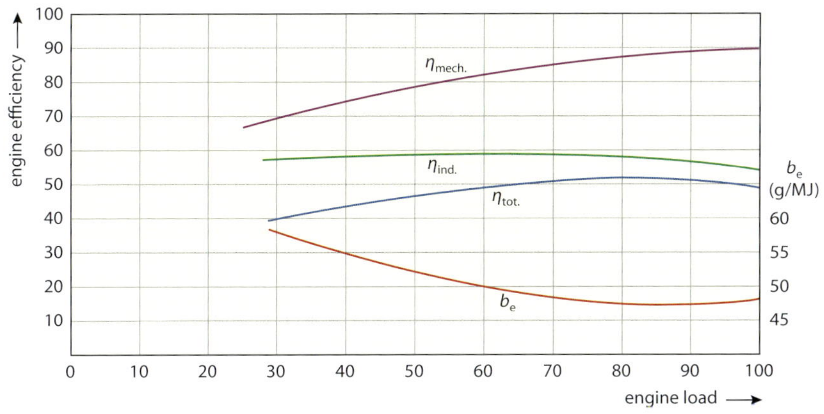

◄

The mechanical-, internal- and total efficiency and the specific fuel consumption in relation to the engine load.

Vertical left: the engine efficiency in %

Horizontal: the engine load in %

Vertical right: specific fuel consumption in grams per MJ.

The graph clearly shows that optimum efficiency, and therefore the lowest specific fuel consumption is achieved at an engine load of approximately 85%. This is generally also the maximum engine load for propulsion systems.

Examples

1 **Wärtsilä Vasaa 6 R 32:** four-stroke trunk-piston engine, six-cylinder in-line, fuel HFO.
Shaft power 2220 kW at 720 rev/min.
Engine category III.
Amount of air 4.6 kg/sec. The standard specific mass of air is taken as 1.276 kg per m³ at an atmospheric pressure of 1013 millibar at an air temperature of 15 °C.

This is therefore $\frac{4.6}{1.276}$ m³ per second or
3.6 × 3600 = 12,978 m³ per hour!

Fuel consumption 186 gram/kWh
So: 412.9 kg per hour
So: $\frac{186}{1000} \times 2220 = 412.9$ kg per hour

Per **kg fuel** is effectively
$\frac{4.6 \times 3600}{412.9} =$ **40.1 kg** of air supplied.
This is in accordance with the theory!

2 **MAN Diesel MC-C**: two-stroke crosshead engine, twelve cylinders in-line, fuel HFO.
Shaft power 68,520 kW at 104 rev/min. **Engine category IV.**
Amount of air 180.7 kg/sec.
So this is $\frac{180.7}{1.276} = 141.6$ m³ per second or
141.6 × 3600 = 50,976 m³ per hour!

Fuel consumption 171 g per kWh
So $\frac{171 \times 68{,}520}{1000} = 11{,}717$ kg per hour.

Per kilogram fuel $\frac{180.7 \times 3600}{11{,}717} =$ **55.6** kg of air is effectively supplied!
This is in accordance with the theory!

▼

A very small four-stroke single-cylinder diesel power-generating set with natural aspiration for the power process and air cooling by a ventilator(1) fitted on the fly-wheel for cooling the cylinder.

▼

A high-speed four-stroke V-engine with supercharging and an air cooler.

1 *air-inlet filter*
2 *air compressor*
3 *exhaust-gas turbine*
4 *exhaust-gas pipe from the cylinders*
5 *air cooler*
6 *exhaust-gas pipe to the exterior environment*

12.3 Air supply to the engine

Air may be supplied in two ways:

1 **Natural aspiration**
This can only be applied to four-stroke engines. The piston moves down during the inlet stroke, the inlet valves are opened and air flows into the cylinder because of the vacuum created.

2 **Turbo charging**
Here the air is compressed by a pump and subsequently flows into the cylinder:
- **in two-stroke engines via the inlet ports;**
- **in four-stroke engines via the inlet valves.**

Turbo charging generally occurs with a turbocharger, which is driven by the flow of exhaust gases. However, there are also mechanical scavenging pumps that are driven, for instance, by the crankshaft of the diesel engine.

▼
Small engines with a shaft power lower than 100 kW sometimes do not have supercharging.

1 *air-inlet filter*

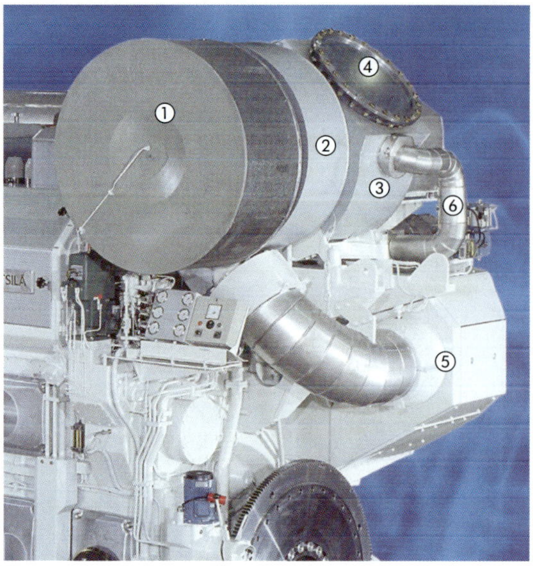

►
A turbocharger for a larger four-stroke engine.

In this case, a Wärtsilä W46, medium-speed in-line engine operating on heavy fuel oil.
1 *air-inlet filter*
2 *air compressor*
3 *exhaust-gas turbine*
4 *exhaust to funnel*
5 *air cooler*
6 *'waste-gate', a by-pass valve for exhaust gases*

12.3.1 Naturally aspirated air in four-stroke engines

The air is drawn in by the vacuum created during the inlet stroke.
The air flows through the air filter and via one or two inlet valves into the cylinder. Today, the system is only applied in small engines with a relatively low power output.
- The power output per litre stroke volume is low.
- The mean effective pressure is low.
- It follows that the mechanical load of these engines is also low.
- The engine has a simple construction; it has no turbocharger with piping and an intercooler.

12.3.2 Turbo charging for four-stroke and two-stroke engines

General

Today, almost all larger industrial diesel engines are equipped with a turbocharger.
Air flows into the cylinder due to over-pressure. The compressed air has a higher specific mass, which means that a larger air mass and consequently a larger fuel mass are inserted per fresh air charge. This allows for a capacity increase of as much as 200% to that of natural aspiration.

Advantages of turbo charging:
- **The power-to-weight ratio of the engine is increased significantly.**
- **The torque at partial load is increased significantly.**
- **The specific fuel consumption at partial load is reduced considerably.**
- **The power output is increased significantly.**
- **At high altitudes, both the power output and the specific fuel consumption stay almost constant.**
- **In natural aspiration at high altitudes, both the power output and the torque reduce significantly and fuel consumption increases.**
- **The total efficiency is higher at all loads.**
- **The fuel consumption is lower at all loads.**
- **The sound levels with turbo charging are lower over the entire frequency range.**
- **The power output to volume ratio of the engine is more favourable. An engine with turbo charging is much smaller than an engine with natural aspiration with a similar power output.**
- **The combustion process with turbo charging is cleaner and therefore the emission levels are lower.**

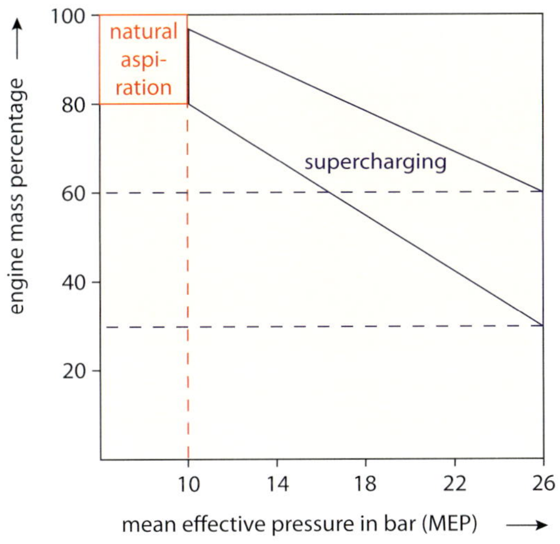

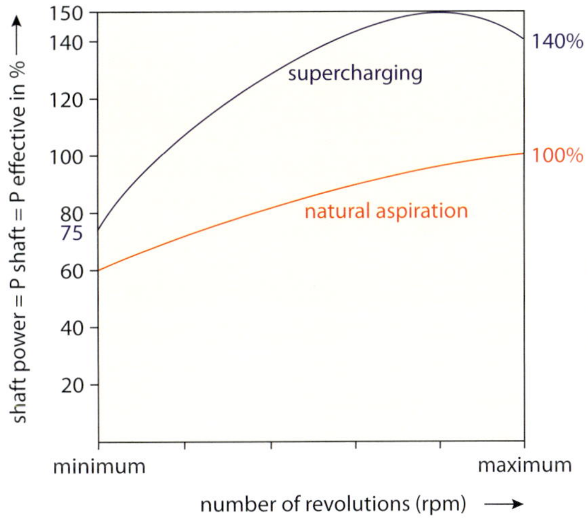

◄ Engine performances with and without supercharging based on an identical-stroke volume.

Left graph.

Dependent on the load the suction charging varies between 100 and 80%.

The maximum mean effective pressure is approximately 10 bar.

With supercharging, the mass varies between 60 and 80%.

The maximum mean effective pressure is approximately 26 bar.

Conclusion: an engine at full load with supercharging has less than half the mass!

Right graph.

The shaft power of supercharging shaft opposed to air charging based on the identical- stroke volume.

At the lowest rpm, an engine with supercharging produces more power.

At close to maximum revolutions per minute, this is 150%. 50% more shaft power compared to the same engine with air charging.

- **In combination with direct injection and the common-rail systems, the turbocharger is indispensable.**
- **Turbochargers can be fitted with 'waste-gates' and a variable geometry for the gas turbine with which the correct amount of air can be tuned to the engine load. In smaller engines with a fluctuating load, this can be matched very precisely.**

12.4 Principle of turbo charging

The exhaust gases of internal combustion engines retain a considerable amount of energy in the form of heat and pressure waves. A heat balance shows that, dependent on the engine type and size, approximately 25 to 40% of the generated fuel energy is wasted in the exhaust gases.

In essence, a turbocharger is a gas-turbine driven centrifugal compressor.

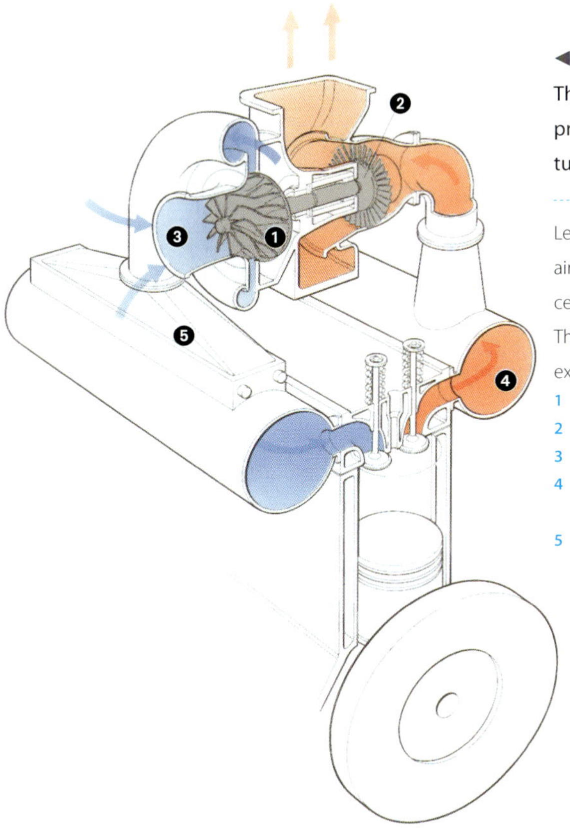

◄ The supercharging principle with a turbocharger.

Left; the air is drawn in via an air filter through the centrifugal compressor.

This is driven by a single wheel exhaust-gas turbine.

1 *air compressor*
2 *exhaust-gas turbine*
3 *inlet-air duct*
4 *exhaust-gas pipes in the diesel engine*
5 *air cooler*

▶
The turbocharger of a small four-stroke diesel engine.

From front to back: the exhaust-gas turbine (1), the air compressor (2) and the air filter (3)

▶▶
A cutaway view of a turbocharger from manufacturer ABB.

Left the air section and right the exhaust-gas section

Two-stage turbocharger

It is also possible to achieve the required inlet scavenging-air pressure in two stages; this can be accomplished by placing two relatively simple turbochargers in series.

Air-cooling

By compressing air, both the pressure and the temperature of the compressed air increase. Subsequently, the specific mass is significantly reduced. Cooling the air in an air cooler or intercooler increases the specific mass of the air, which means that in the same time, extra kilograms of air can be pushed into the cylinder. Therefore, more fuel can be injected into the cylinder, thus increasing the cylinder capacity. **Air mass plays a significant role in engine power output.**

▶
An air cooler for a large two-stroke crosshead engine.

Cooling the air significantly increases the specific mass of air. Due to this, more fuel can be injected in the cylinder and therefore engine power increases. Conclusion: lower air-inlet temperatures can lead to higher power outputs.

▶
A turbocharger of an eight-cylinder V-engine of MTU.

In this case, with a 'waste gate' on the exhaust-gas turbine and a 'by-pass' valve on the air compressor.

1 *'waste gate'*
2 *'by-pass' valve*

Air temperature directly influences the power output of the engine!
The air temperature after the turbocharger varies according to the engine type and load from 80 to 200 °C. After the intercooler, this range is 35 to 85 °C.

12.5 Turbocharger manufacturers

There are a large number of suppliers serving this market. They can be placed in two categories:
1 suppliers who both design and manufacture engines as well as turbochargers;
2 suppliers who only manufacture turbochargers, not engines.

▲

A cut-away view of a modern turbocharger.

1 *air-compressor fan*
2 *shaft with plain bearings*
3 *exhaust-gas turbine*
4 *blower casing*
5 *compressor section diffuser*
6 *exhaust-gas turbine nozzle ring*

▼

A high-speed four-stroke sixteen cylinder MTU-diesel engine with supercharging and intercooling – Series 2000.

Each 'bank' of this V-engine has its own turbocharger.
1 *location for the air filters*
2 *air compressor*
3 *pressure line from the air compressor to the air cooler*
4 *air cooler*
5 *cooled inlet air to the two inlet valves of every cylinder*
6 *cylinder*

Small turbochargers have an exhaust-gas turbine, which has an impeller with vanes, while **larger turbochargers** consist of an exhaust-gas turbine constructed of a rotor with separate vanes attached to it. They resemble the rotors of gas turbines.

12.5.1 Construction of turbochargers for small diesel engines (engine categories I and II)

Description of a small turbocharger

The rotor consists of an exhaust-gas turbine with a fan manufactured in one mass and a centrifugal air compressor. The shaft is short and thick and is supported by two plain bearings and an axial bearing on the air compressor side. Lubrication is realized by the engine's main lubricating system, even in the very large turbochargers found in two-stroke crosshead engines. The casings are not cooled and therefore made of high-grade heat resistant material. The turbine casings are insulated externally for fire safety.

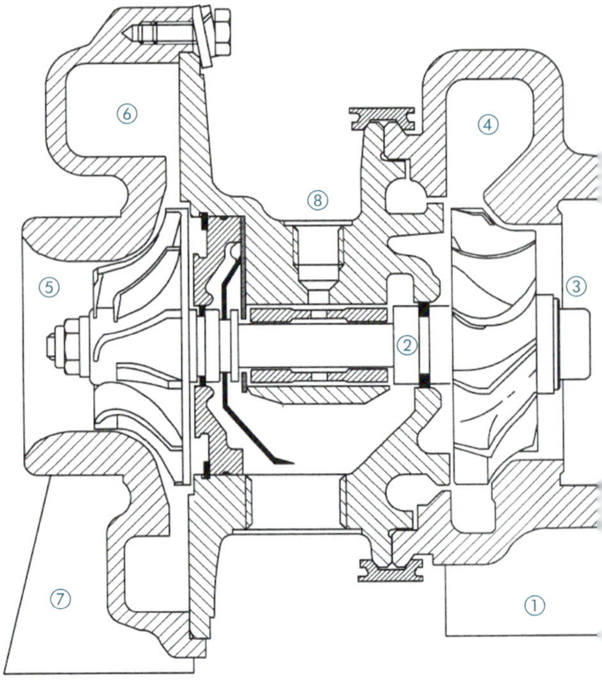

▲
Cross-section of a small turbocharger.

1 *inlet turbine*
2 *shaft*
3 *outlet turbine*
4 *turbine casing*
5 *inlet compressor*
6 *compressor casing*
7 *outlet compressor*
8 *oil supply*

▼
The rotor of a small turbocharger often includes a radial air compressor and radial exhaust-gas turbine.

The gas flows from the inside to the outside along the shaft.

1 *air compressor*
2 *turbine shaft*
3 *exhaust-gas turbine*
4 *lubricating-oil section*

▼
View of an air compressor.

1 *first row of blades of air compressor.*
2 *second row of blades of air compressor.*
3 *diffuser: this stationary section converts the high air velocity, generated in the rotating impeller into pressure. The output channels of the diffuser create a decrease in air velocity and an increase in air pressure.*

▲
A cut-away of a radial turbocharger.

The inlet segment of the exhaust-gas turbine is clearly visible in the foreground (1).

◀

A turbocharger of a small diesel engine.

There is no air cooler.

1 *air filter*
2 *turbocharger*
3 *engine block*

These small turbines rotate at a speed of 40,000 tot 80,000 revolutions per minute.

The exhaust gases are led to the turbine casing via the exhaust manifold. The diameter of the inlet duct inside the turbine casing becomes increasingly smaller which increases the speed of the exhaust gases. Due to its spiral shape the hot gases surrounding the turbine wheel are dispersed and hit the fan blades, subsequently the fan starts to rotate. The turbine casing clearances determine the rotation speed of the exhaust-gas turbine. The choice regarding turbine clearance dimension is dependent on the cylinder content, the number of rotations and the maximum power output of the engine.

▲

A very large turbocharger for a two-stroke crosshead engine.

1 *exhaust-gas pipes from the cylinders*
2 *exhaust-gas inlet to the turbine*
3 *exhaust-gas discharge to the outside air via an exhaust-gas boiler and silencer*
4 *inlet-air cooler*

◀

A relatively small diesel genset with turbo charging.

An oil separator is placed between the sump and the inlet-air filter for sump venting. The lubricating-oil fumes travel via the turbocharger back to the engine and are combusted.

1 *oil separator*
2 *connection to the air filter*
3 *liquid drain*

Large two-stroke crosshead engines have turbochargers with a rotor diameter of 60 to 100 cm and an rpm of 5000 to 6000.
Small four-stroke trunk-piston engines have turbochargers with a rotor diameter of 5 to 20 cm and an rpm of 80,000 to 120,000.
So large engines have large turbochargers with relatively low rpm's, whereas small engines have small turbochargers with relatively high rpm's.

The centrifugal compressor draws in fresh air via a large suction filter in order to reduce the pressure loss in the filter to a minimum. The air is accelerated to high speeds by the fan. As a result of the increasingly high velocity air passing through the blades in the diffuser, the velocity energy is converted into pressure. An over pressure of up to 5 bar can be achieved!
The short and thick shaft is equipped with two ball- or plain bearings and has seals on the exhaust-gas side and the airside.
This shaft is lubricated by the engine lubricating oil and is delivered to the top of the shaft; it then flows at the bottom of the blower back into engine oil sump.
Obviously, the air output depends on the amount exhaust gases supplied to the exhaust-gas turbine. There is a power balance between the gas turbine and the air compressor.
High capacity - plentiful exhaust gases – abundant air – high rpm.

12.5.2 Construction of a turbocharger for large diesel engines (engine categories III and IV)

Description of a large turbocharger

At present, there is a trend in the construction of large turbochargers where they now resemble the construction of much smaller turbochargers. The construction of the exhaust-gas turbine is the major difference. This consists of a wheel that has separate blades on all sides. The construction resembles that of gas- and steam turbines.
Here too the engine's lubricating system is employed to lubricate and cool the turbine shaft and bearings. The shaft's seals have a labyrinth construction. The lubricating oil is normally supplied from the main engine. Sleeve bearings are become increasingly used. Occasionally, the turbine-shaft casing is cooled with a separate cooling-water system.

▼

A cut-away view of a turbocharger.

Note the short thick turbine shaft between both bearings.

▶

Renewing the labyrinth seals on the exhaust-gas turbine side.

Several labyrinth segments have been placed close to the wheel.

▲

The blade wheel of the exhaust-gas turbine for a four-stroke engine.

The support wire between the blades shown here is necessary to contain the vibrations. These occur in four-stroke engines caused by pressure surges. Note the blade-foot fastening.

▲

With two-stroke crosshead engines, the pressure surges are levelled off in the large spacious exhaust-gas pipe and a support wire for the blades is not necessary.

12.6 Capacity curves

A number of well-known charts are used that apply to the compressor of a turbocharger.

12.6.1 Compressor efficiency in relation to the compression ratio

▼

The turbocharger efficiency with respect to the compression ratio for three types of turbochargers from manufacturer ABB.

Clearly shown, the following:

- VTR-4E (dark red) has the highest efficiency of 72% with a compression ratio of 2.7;
- VTR-4D (grey) has the highest efficiency of 71% with a compression ratio of 3.5;
- TPL-B (light red) has the highest efficiency of 74% with a compression ratio of 3.3.

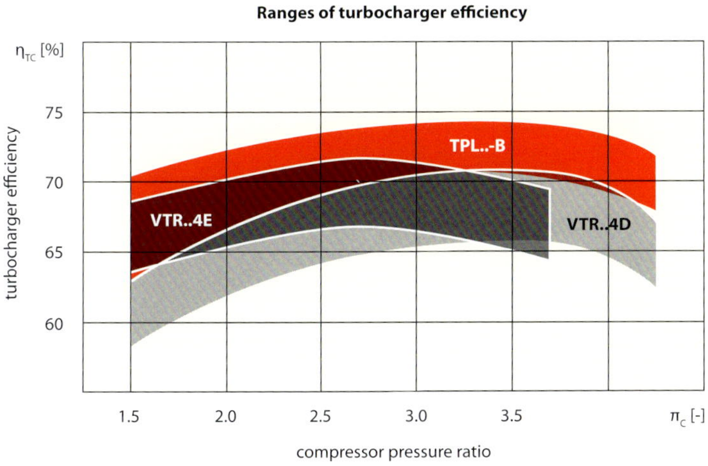

12.6.2 Turbocharger of the engine in relation to the engine load

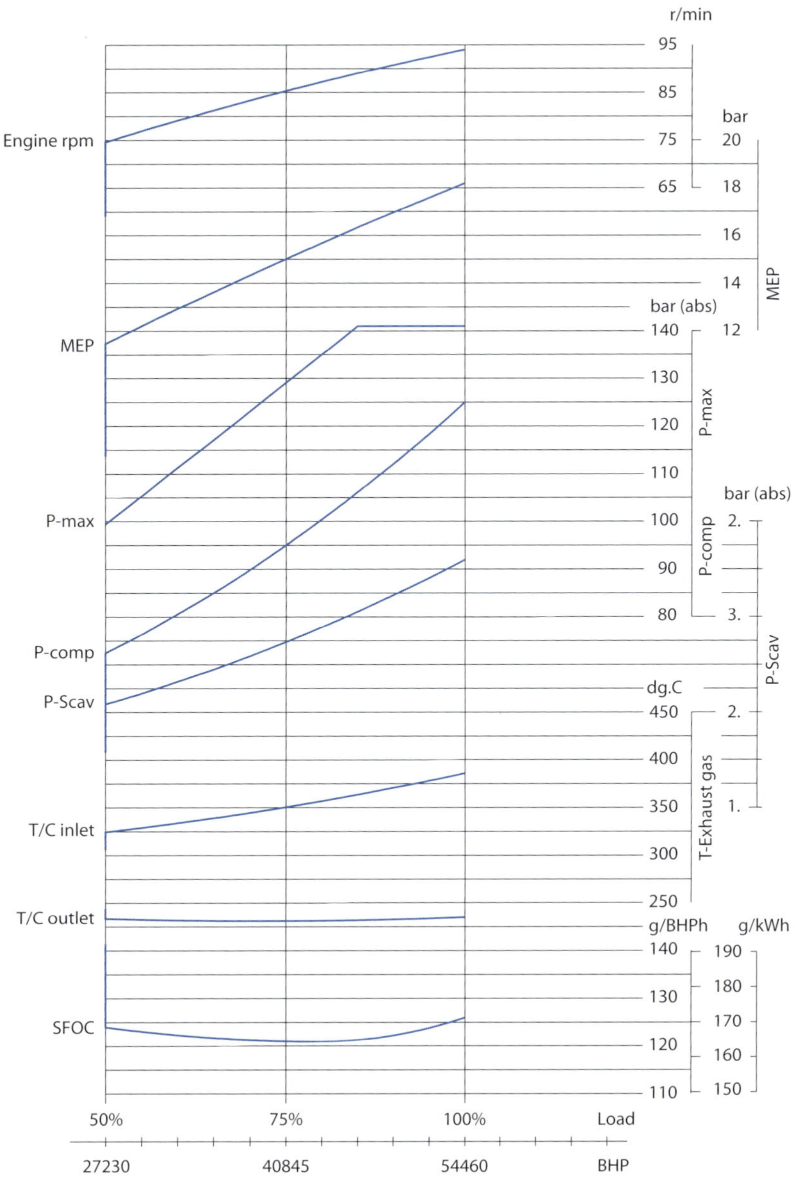

▲

The performance chart of a MAN Diesel 98MC two-stroke crosshead engine.

Between an engine load of 50 and 100%, all the data increase. Only the fuel consumption is at 85% the lowest.

The scavenging-air pressure increases from 2 to 3.5 bar.

The maximum combustion pressure increases to 85% of engine load and then remains equal.

12.6.3 Compression ratio in relation to engine power output

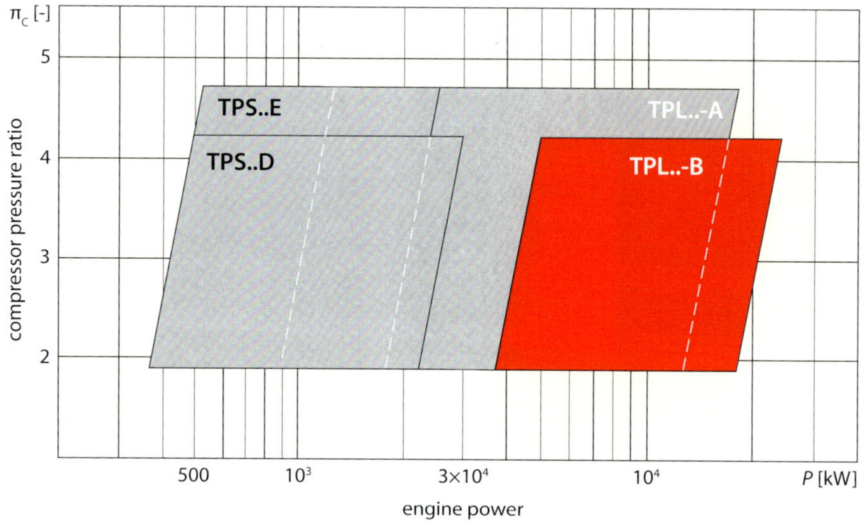

The compression ratio with respect to the engine power for four types of turbochargers from manufacturer ABB.

horizontal: engine power in kW.

vertical: compression ratio.

12.6.4 Performance graph of the compressor showing: the capacity, the efficiency, the rpm and the compression ratio of the compressor

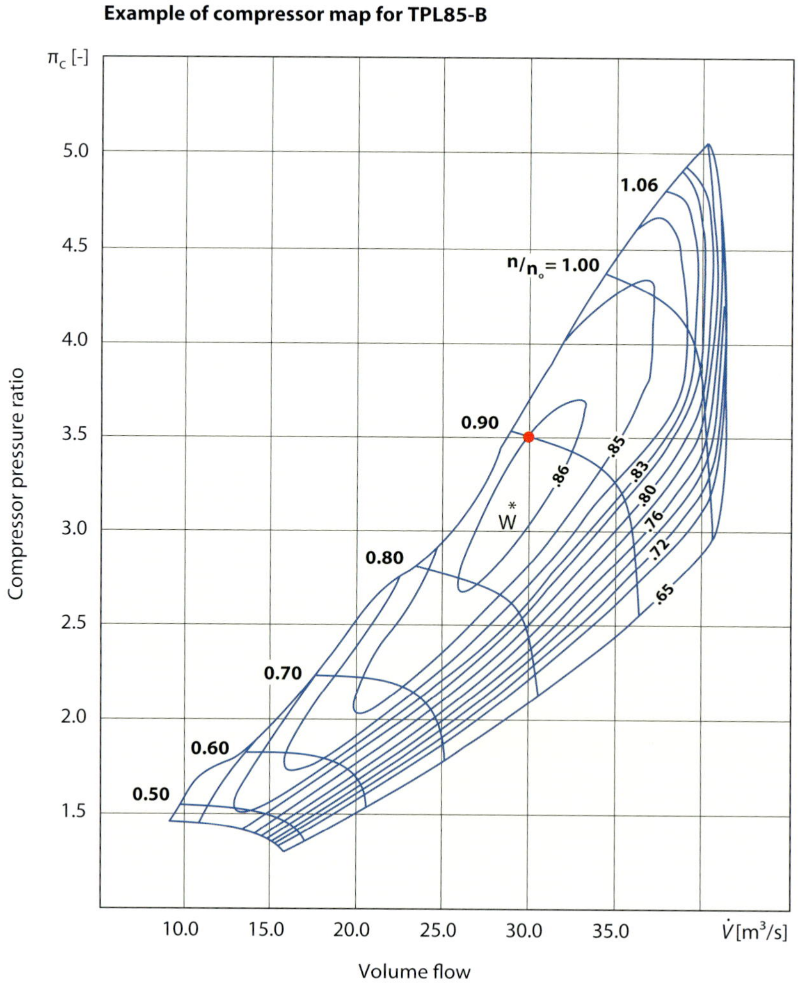

A delivery characteristic of a turbocharger.

Curved lines: the turbocharger speed.

Closed lines: the turbocharger efficiency.

The optimum operating area in this case is **point W**. The efficiency is 86%. The speed is then approximately 85%. The compression ratio is approximately 3.3 and the suction volume of the air is 30 m^3 per second. This graph is also known as a 'MAP'-graph.

12.7 Description of three turbocharger manufacturers – development of modern turbochargers

12.7.1 Mitsubishi

This Japanese manufacturer has the following principles:

- In two-stroke crosshead engines, the increasing cylinder capacity will produce a mean effective pressure of 2 MPa or 20 bar in the short term.
- In four-stroke trunk-piston engines, a standard mean effective pressure of 2.5 to 3.0 MPa or 25 to 30 bar will become standard.
- The pressure of the inlet/scavenging air is approximately 0.4 MPa or 4 bar. This pressure will have to increase.
- The efficiency and compression ratio of the compressor must therefore increase. This entails an increase in the air velocity in the compressor, which will increase the noise production and axial forces in the turbine shaft.

Clearly, turbocharger manufacturers have a large number of focal points in the development of new turbochargers!

1. Extremely high compressor output at low and high compression ratios.
2. Special attention to the design of the axial bearing.
3. Reduction of noise levels.
4. To increase the operating life of the compressor wheel.
5. Improve the protection of the exhaust-gas rotor, in particular against corrosion and high temperatures. Presently, the blades are manufactured from a 12% chromium alloy.
6. To improve the flow efficiency in the exhaust-gas turbine, the blades should be attached in such a way that no lacing wire is required and so preventing vibrations.
7. To optimise the air and gas flow in the turbocharger.

12.7.2 ABB

ABB – Four-stroke engines

This manufacturer in Switzerland has the following principles:

The importance of exhaust-gas emissions and the increase of mean effective pressure are constant focal points in diesel engines. **Due to present and future international regulations, a clean combustion process in internal combustion engines, such as the diesel engine, is crucial!**

Here the turbocharger has an important role to play.

Turbochargers have to meet the following requirements:

- higher compressor efficiency;
- higher compression ratios;
- higher capacities.

Mechanical reliability and a service-friendly design are also required.

Compressor cooling ensures a longer operating life of hot engine parts. Cooled air found after the intercooler is drawn in to cool the shaft seals and the compressor wheel. This considerably reduces the temperature of the materials and increases their operating life. The air-inlet filter is designed so that generated noise is optimally absorbed.

ABB – Two-stroke crosshead engines

Large diesel engines on board ships are often provided with several turbochargers. A twelve-cylinder two-stroke crosshead engine has a maximum of four turbochargers. The increase in the compression ratios and air- and gas flows produce a considerable increase in the noise pollution levels. **At a distance of one meter from the main engine, the sound levels generated are often 110 dB at a 100% load.**

The noise produced by turbochargers is generally lower, but nevertheless far too high.

In particular, in the pressurized part of the compressor, good results can be achieved by installing a damping system.

Alterations to the design of the compressor fan have also led to a reduction in noise levels.

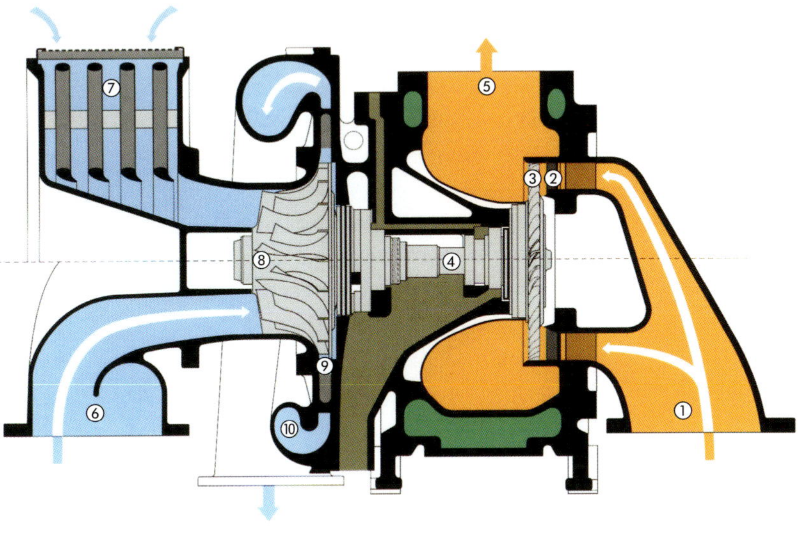

▼ The assembly of a turbocharger.

1. *exhaust-gas supply*
2. *nozzle segment*
3. *exhaust-gas turbine*
4. *turbine shaft*
5. *exhaust-gas discharge*
6. *air supply*
7. *air-inlet filter with sound damping*
8. *air-compressor rotor*
9. *diffuser*
10. *slug casing*

▶
A cross-section of an ABB TPL turbocharger with graphs of the compression ratio with respect to the air capacity of the various types.

D-version: large capacity, lower compression ratio
E-version: large capacity, higher compression ratio
F-version: smaller capacity with the highest compression ratio

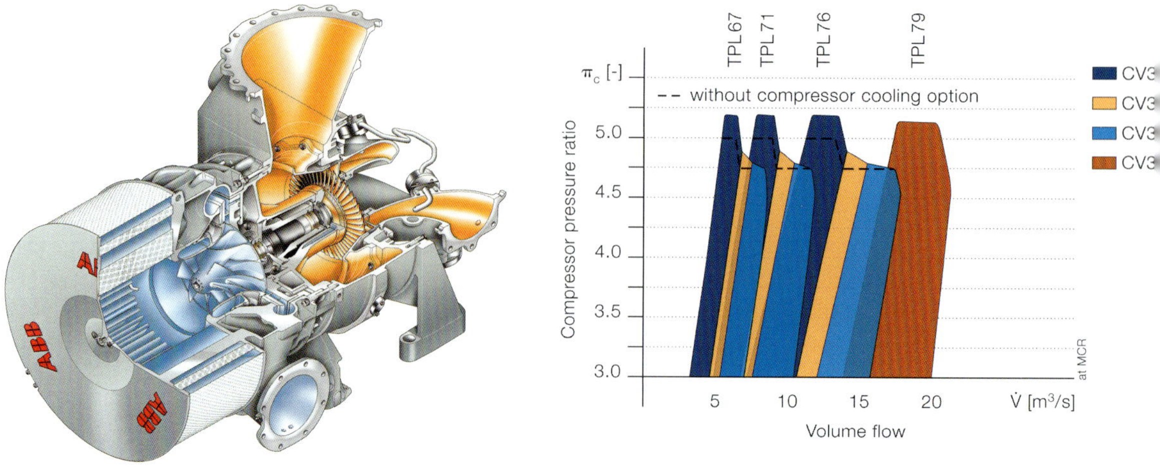

▶
A cross-section of an ABB TPS turbocharger with a graph of the compression ratio with respect to the air capacity for the various types.

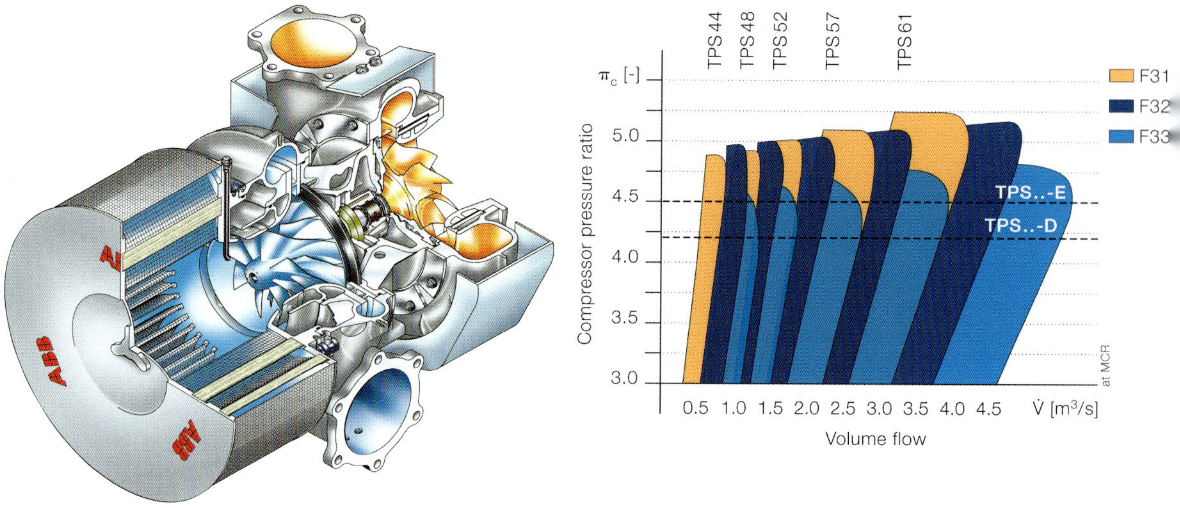

▶
A cross-section of an ABB VTC turbocharger with a graph of the compression ratio with respect to the air capacity for both types.

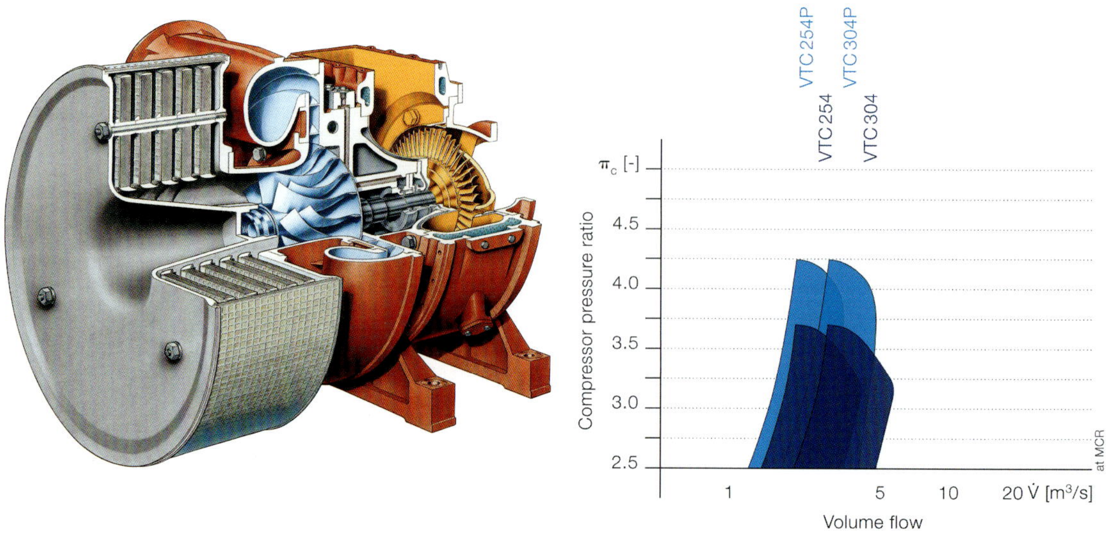

▲
A two-stroke crosshead engine always has a constant-pressure system.

At the end of the engine is a turbocharger. A backup lubricating-oil tank is placed above the compressor casing so that the rotating rotor of the turbocharger can always be lubricated even with engine failure.

▲
The manufacture of the casing of the exhaust-gas section of the largest turbocharger from ABB.

12.7.3 MAN Diesel, a German-Danish manufacturer

Here the same requirements apply as with the aforementioned manufacturers:
- Increase compressor efficiency;
- Reduce noise levels;
- Service friendly;
- Increase capacity.

12.8 Small turbochargers, engine categories I and II

Small engines are usually provided with **axial** turbines, large engines with **radial** turbines.

12.8.1 Small turbochargers ABB-TPS D/E series: 500 to 3200 kW diesel-engine shaft power (engine categories I and II)

Due to the strict regulations regarding exhaust-gas emissions, a growing number of diesel engines are designed according to the **Miller-principle**; in this system, the closing of the valves is either accelerated or retarded. This diminishes the pressures and the temperatures in the cylinder and causes a decrease in the NO_x emissions.

Also, see Chapter 22, *Diesel-engine emissions.*

With the use of this system, the scavenging-air pressure needs to be increased so the combustion process can be adequately cooled; lower process temperatures generate less NO_x.

Requirements
- Compression ratios of 4.7 and higher.
- Higher compressor efficiency, lower exhaust-gas temperatures and higher engine-shaft power.
- Wider capacity operating range.
- A simple, compact, modular and sturdy construction.
- No auxiliary systems, lubrication by the engine-lubricating system.
- Corrosion-resistant to sulphur found in the fuel.
- If the rotor shatters at unusually high speeds, the debris should remain inside the casing. To this end, special explosion-rings are fitted. **Generally, the so-called crack velocity lies at 60% over the maximum rpm**.
- A solid, vibration-damped foundation is vital to keep the vibration levels low.

▼
A cross-section of a small axial turbocharger.

▶ A 'waste-gate' of the turbocharger of an eight-cylinder diesel engine.

1 *'waste-gate'*
2 *turbine section*
3 *air section*

12.8.2 Special provisions in turbochargers

- waste-gate, an exhaust gas by-pass valve;
- air by-pass valve;
- compressor trims;
- ceramics in turbochargers;
- titanium for turbine wheels;
- re-circulation valve on compressor;
- water-cooled turbine casing;
- turbochargers are linked in parallel;
- serial connection of turbochargers.

12.8.3 Waste-gate; a by-pass valve for exhaust gases

This valve opens when a certain scavenging-air pressure is exceeded and the quantity of exhaust-gases flowing to the turbocharger is controlled and therefore the scavenging-air quantity, pressure and ultimately the engine capacity. Under normal conditions, this valve is not activated.

12.8.4 By-pass valve

The air by-pass valve is located in the air space and is a connection between the compression and suction sides of the compressor.
This amongst other effects ensures compressor stability at low loads. Effectively, when the by-pass is open, the air volume flow of the compressor exceeds the air volume flow required for the engine.

▼ Various designs of 'waste-gates'.

A **'waste-gate'** or by-pass valve in the exhaust-gas pipe limits the amount of air to the engine by reducing the flow of the exhaust gases to the turbine and consequently the engine power is limited. A by-pass valve in the air section after the air compressor can limit the amount of air to the engine.
This is also possible with the **'air waste-gate'**.

Schematic arrangement for charge-air by-pass

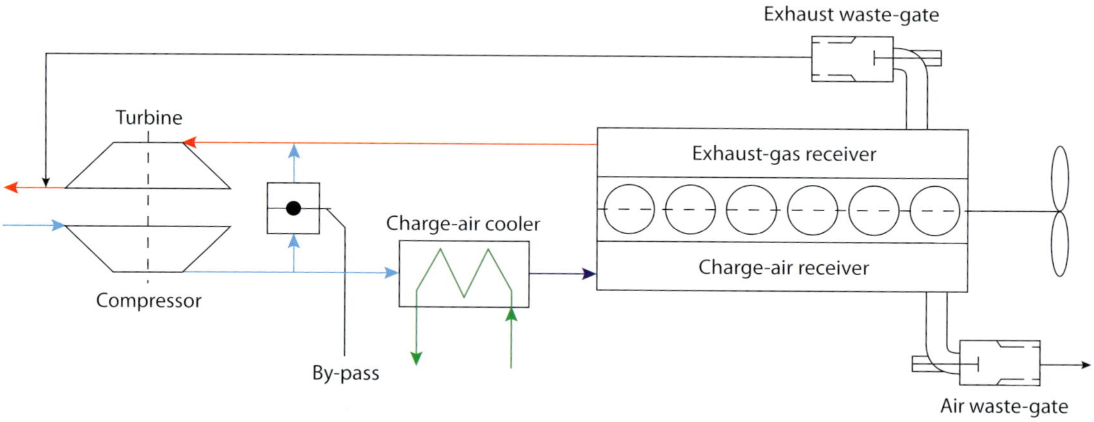

▶
The use of the 'by-pass' valve and the 'waste-gate' at different power outputs.

tEaT = temperature after the exhaust-gas turbine
BSEF = specific exhaust-gas mass
BSFC = specific fuel consumption

The specific fuel consumption BSFC is the lowest: approximately 75% power. This is in the region of the maximum power output for ship propulsion, viz. 80 to 85%. Due to the increased volume, the filling pressure to the engine increases at partial loads; this is good for scavenging!
By-pass: in air section.
Waste-gate: in exhaust-gas section.

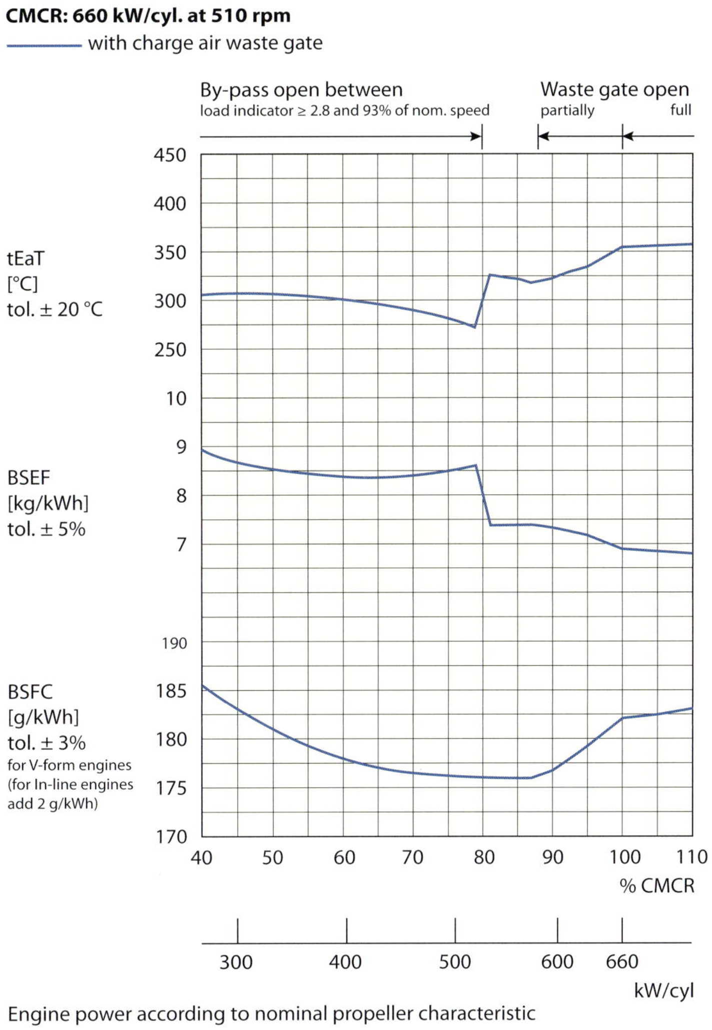

12.8.5 Compressor trims

Small turbochargers are typically cast with a single casing. By modifying the compressor wheel and applying various 'trims', the shape of the wheel is so that varying capacities can be generated.
The roughcast light-metal wheel has the identical basic measurements for an entire series. The finishing machining process gives it the distinctive properties. Generally, one roughcast wheel can produce twenty different 'trims'. In these radial exhaust-gas turbines, 'trims' are also employed.

12.8.6 Ceramics in turbochargers

Advantages of the use of ceramics for the turbine are:
- Resistant to high temperatures, approximately 1250 °C;
- Low coefficient of expansion; merely 1/5 that of metals;
- Low weight, rapid wheel acceleration;
- Thinner casing is possible. Less mass if, the light turbine wheel shatters.

The **disadvantages** are:
- Difficult to fix to the turbine shaft. Welding is not an option;
- Ceramic is brittle;
- Soot particles can damage ceramic;
- X-ray quality control is impossible.

12.8.7 Titanium for turbine wheels

Titanium has a higher resistance to the corrosive properties of soot particles than ceramic. The traditional nickel alloys that are applied in the small to very small turbochargers weigh twice as much as titanium. However, titanium is difficult to cast and very expensive. At present, it is not generally utilized.

12.8.8 Re-circulation valve on compressor

Serves primarily to keep a high rotor speed at various engine loads. Therefore, when extra power output is required, the rotor velocity is inside operating parameters. Opening the re-circulation valve reduces the turbocharger load and increases the number of revolutions.

12.8.9 Water-cooled turbine casing

These small turbochargers are also equipped with water-cooling that cools the casing around the rotor shaft and decreases the radiant heat in the vicinity of the turbocharger.

12.8.10 Turbochargers connected in parallel

In V-engines, it is common that the left- and right banks have individual turbochargers in which the individual compressors compress the air led from the joint air-inlet manifold. This inlet manifold is placed at the centre of the engine with the central exhaust pipes on either side. The advantage of this system is that the charge-air pressure of each cylinder and each turbo bank is equal and therefore the power output per cylinder and bank is distributed equally compared to separate turbocharger systems.

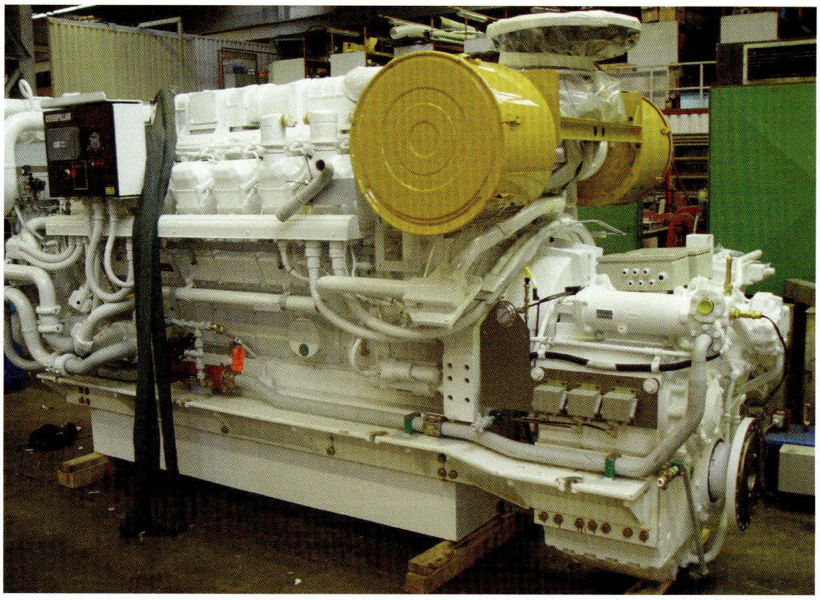

▶ A Caterpillar diesel engine in V-shape with two turbochargers.

Right the attached shaft reduction-gear of a luxury motor yacht.

▶ A V-type Caterpillar diesel engine with two turbochargers.

Generally, both turbochargers of V-type engines supply air to a common inlet-air space so that air-pressure differences between the left and right engine banks are not probable.

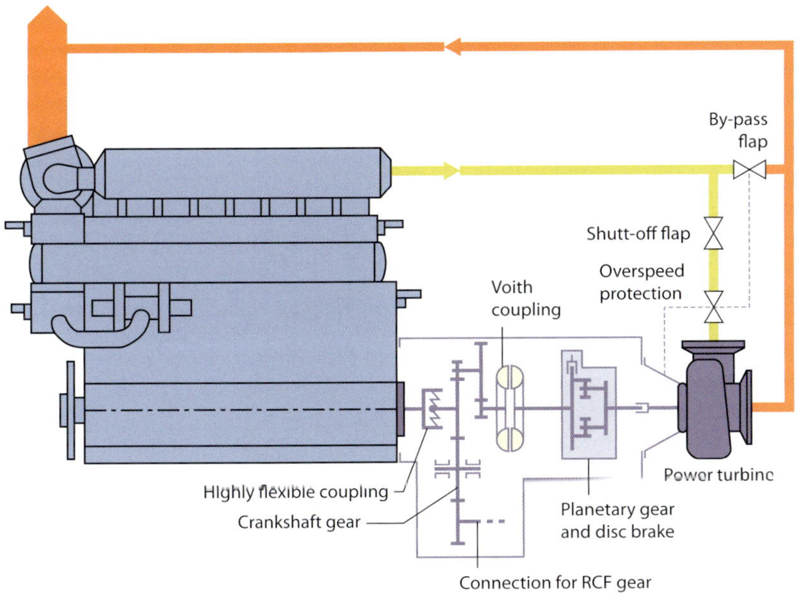

◄ A schematic diagram of a power turbine driven by surplus exhaust gases.

This power turbine drives the engine using a planetary gearbox and a coupling. This system are seldom used; the power gain of a few per cent does not compensate the high investment cost. The fuel savings at full load are approximately 3%. In reality these power turbines are seldom used.

12.8.11 Serially linked turbochargers

This is occasionally seen in category II diesel engines. The efficiency of standard turbochargers is too low when charged-air pressure is higher than 3 bar. Two standard turbochargers are then placed in series. The exhaust gases pass the high-pressure turbocharger followed by the low-pressure turbocharger. An intercooler is placed after the high-pressure turbocharger or there are two intercoolers, where the second intercooler is placed after the low-pressure turbocharger.

12.9 Supercharger with a separate power turbine

This has been applied sporadically to large two-stroke crosshead engines for years. At full load not all the kinetic energy present in the exhaust gases is required to drive the turbochargers. The residual energy is used for a power turbine. This energy is converted into mechanical power and supplied via special reduction gearing to the crankshaft. In this case, the power turbine is always connected in parallel.

This system has received a good deal of interest for small engines – **Categories I and II**. In these categories, both exhaust gas turbines are arranged in series.
Initially the turbocharger is driven and then the power turbine. At full load, this four-stroke engine can save approximately 10%.
It should be taken into account that this type of power-turbine driven supercharger requires a large investment cost!

12.10 Power turbines with generator

This connection has been applied sporadically for a number of years.
The financial implications are considerable:

- An exhaust-gas boiler that can generate sufficiently high steam pressure and overheated steam for the steam turbine.
- A steam-turbine driven generator that is connected to the on-board power system.
- A condensate system to feed the processed steam produced by the steam turbine as feed water to the exhaust-gas boiler.
- Reduction gear fitted between the steam turbine and the generator
- A complex control system for the entire mechanism/ installation
- The whole steam installation must comply with strict requirements and be inspected at regular intervals.

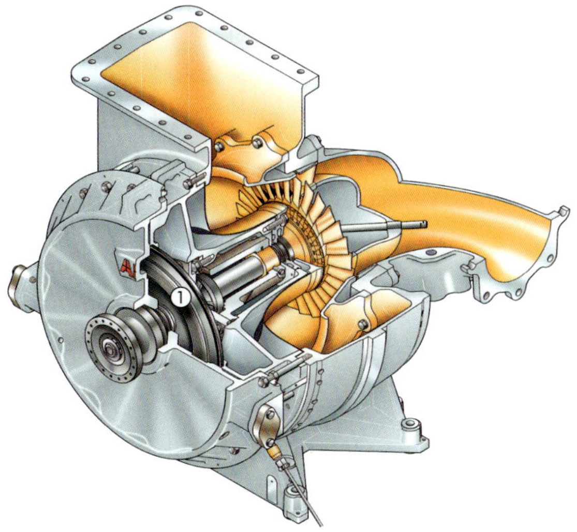

◄ An ABB power turbine, type TC.

The large disc (1) on the shaft is the reduction gear.

▶

An ABB installation with a power turbine, reduction gear, and alternator for generating electric energy from the exhaust gasses not used to drive the turbochargers.

The generator is air cooled in a closed circuit, using a heat exchanger in the LT cooling-water system.

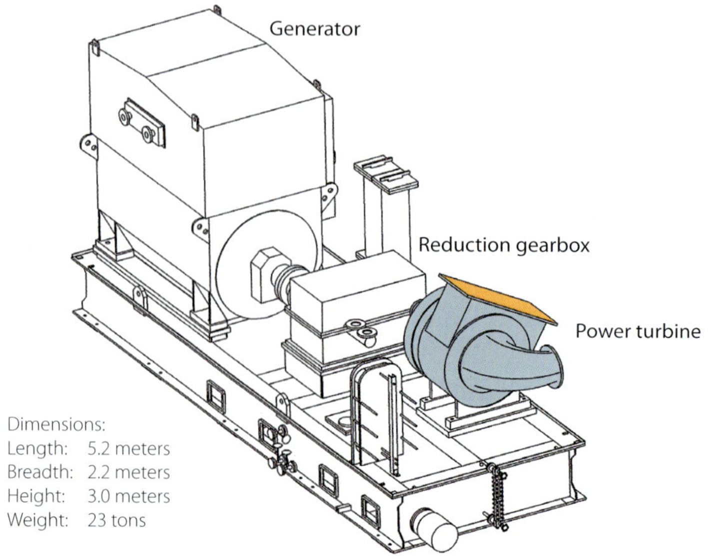

The system can be supplemented with a power turbine assisting in driving the generator. This requires fitting a reduction gearbox between the power turbine and the steam-turbine shaft. A shaft generator/motor is mounted on the propeller shaft. This produces electricity for the on-board power system and can serve as an electromotor driving the propeller using the power system.

This option could be used to get back to port in case of a break down or as auxiliary power to overhaul whilst in port. The costs and benefits are only favourable if the fuel costs are high and the engine runs on full load. In reality, these systems are rarely used: it appears that the costs are too exorbitant.

▶

A complete installation for maximum usage of a propulsion installation with a two-stroke crosshead engine with a fixed propeller.

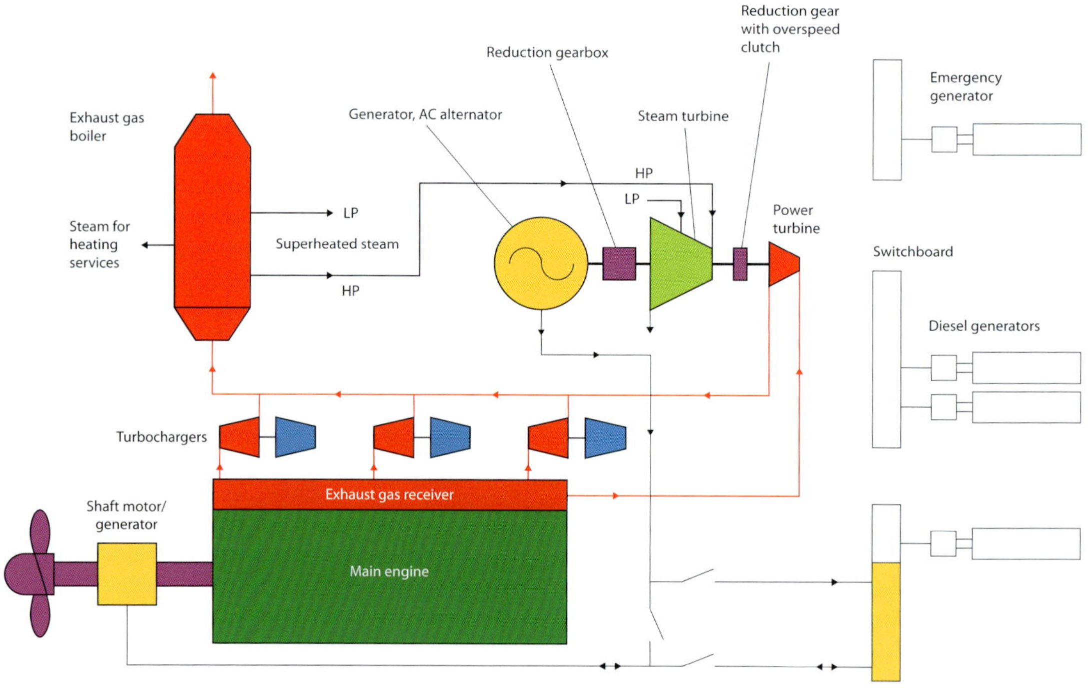

The exhaust-gas boiler provides superheated high- and lower pressure steam for a separately housed high- and lower-pressure steam turbine. These turbines drive the three-phase generator at a constant rpm via a reduction gearbox. A power turbine is also coupled to the turbine shaft by a reduction gearbox. The power turbine is driven by excess exhaust gases that are not required for driving the three turbochargers.

A shaft generator providing the electric energy for the on-board electricity system is mounted on the propeller shaft. This technique allows the number of auxiliary engines running while at sea to be kept to a minimum. The generator can be switched to the electromotor allowing the electricity generated by the combined steam/power-turbine installation to assist in driving the propeller and save fuel. This is a costly and complicated installation!

In small - Category I and II engines, the system is rather popular. Here both exhaust-gas turbines are arranged in series. First, the turbocharger is started followed by the power turbine. Cost reduction of approximately 10% can be achieved when this four-stroke engine runs at full load.
One should bear in mind that a power-turbine driven crankshaft requires a considerable investment

12.11 Air supply in four-stroke engines

In the four-stroke process, the exhaust gases must be expelled from the cylinder at the correct time.

- If the exhaust valves open too early, this results in power loss due to a reduced piston stroke.
- If the exhaust valves open too late, this results in an increase of exhaust-gas pressure on the piston during the upward stroke at the expense of power output.
- The inlet valves open before the end of the exhaust stroke so a suitably open passage for the incoming fresh air at the beginning of the inlet stroke is available.
- The exhaust valves must remain open to the end of the exhaust stroke and only close at the commencement of the inlet stroke.
- The time taken for the exhaust- and inlet strokes overlap. This overlap, the so-called exhaust-air scavenging, takes place from 25 to 75 crank degrees for naturally aspirated diesel engines and from 90 to 130 crank degrees for engines with a supercharger system.
- The higher the revolutions, the greater the overlap. Air needs sufficient time to adequately cool the hot parts and discharge the remaining exhaust gases.
- A naturally aspirated four-stroke engine can always maintain its own air supply using the piston, which functions as a scavenging pump.
- A disadvantage of naturally aspirated engines is the low initial compression pressure. This is found to be below the atmospheric pressure caused by the flow resistance of the air filter, the air ducting and the inlet valves.
- The cylinder capacity is limited as it is entirely dependent on the mass of air that is drawn into the cylinder per cycle.

Superchargers are categorised according to the manner in which the exhaust gases are expelled to the turbocharger.

There are roughly four systems.

1. Equal-pressure and constant-pressure systems.
2. Pulse system.
3. SPEX-system.
4. Pulse-Converter system.

12.11.1 Equal-pressure or constant-pressure systems

All exhaust gases are discharged into one single, capacious manifold. Pressure impacts when the exhaust valves open are levelled off in this large exhaust manifold and the exhaust-gas pressure before the turbocharger is virtually constant.
At low engine loads, it is possible that the pressure in the exhaust-gas manifold exceeds the pressure in the air-inlet manifold. This could cause exhaust gas flow into the air inlet manifold during scavenging.

The fresh air is drawn in by the compressor, compressed and cooled by an intercooler and subsequently channelled into a capacious scavenging-air inlet or receiver. In many four-stroke engines, the scavenging-air receiver is integrated in the engine block.

From the receiver the air flows to the cylinders via the inlet valve or valves.
The exhaust gases from the cylinder flow to an uncooled exhaust-gas manifold and then through a turbine.
At certain engine loads, the exhaust gases flow at a constant speed through the turbine, which maintains turbine-flow efficiency.

▼
Contemporary four-stroke in-line engines have one turbocharger.

A constant-pressure system is used for cylinder charging. See also, the engine manufacturer's guide for supercharging.

▶
The inlet manifold of a Caterpillar MaK medium-speed HFO-diesel engine.

1 *inlet manifold*

At an engine load below 25% of full load, the exhaust-gas scavenging may not function. The exhaust gases flow back into the cylinder as the scavenging-air pressure is lower than the exhaust-gas pressure. A moderate number of engines are equipped with this type of system.

12.11.2 Pulse system

All the cylinders are connected individually to short and relatively narrow exhaust-gas manifolds in which a pressure wave, the so-called pressure pulse is produced when the exhaust valves are opened.

Advantages

- Reduced exhaust-gas expansion losses when expelled from the cylinder.
- There is a sufficient over-pressure of scavenging air during the exhaust-air scavenging, so 'back flow' does not occur.

Disadvantages

- Due to the increase in the exhaust-gas pressure caused by the narrow exhaust-gas manifolds, the scavenging air over-pressure will be lower at full load than it would be in the equal-pressure system. This results in a shorter scavenging period.
- An exhaust-air scavenging occurring around piston TDC, when the inlet and exhaust valves are open, is selected for the pressure-pulse system. Scavenging time is extended to compensate for the reduced scavenging over-pressure and guarantees an adequate exhaust-air scavenging.

Ignition timing

In order to avoid a pressure-development distortion in the cylinder during the exhaust-air scavenging, the ignition times in crank degrees must be sufficient in magnitude. In an eight-cylinder four-stroke

▶
The inlet manifold of two medium-speed V-type engines.

Left an unmachined Wärtsilä W38 engine block and right the unmachined Caterpillar Mak M43 engine block.

1 *inlet manifold*

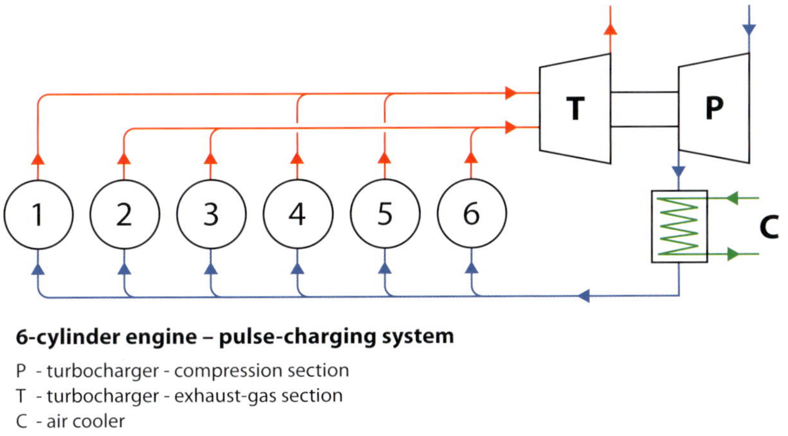

◀

The pulse-charging system.

The compressor draws ambient air via a filter and forces it through a cooler to the inlet-air manifold. All the pipes are connected to this manifold.

The exhaust of every cylinder is so chosen that the cylinders do not impede each other. In this example; the ignition sequence is 1–2–4–6–5–3, cylinders 1, 4 and 5 and 2, 3 and 6 are connected.

engine, two cylinders are connected to one exhaust manifold in four independent exhaust-gas manifolds. Six cylinder engines have two independent manifolds.

Example

An explanation based on a six-cylinder engine:

The four-stroke cycle is 720 crank degrees.

Ignition times $\frac{720}{3} = 240$ crank degrees.

Remember, there are **three** cylinders connected to one manifold. Therefore, the ignition times are 240 crank degrees and the pressure surges follow in succession after 240 crank degrees. There is an optimum exhaust-air scavenging between the pressure surges.

The crank position is determined by the required **equal** ignition times, which are:

$\frac{720}{6} = 120$ crank degrees.

Balancing the free forces and the engine moments is also pertinent.

Also, see Chapter 17, *Vibrations and Balancing*.

The manner in which the exhaust-gas manifolds of each cylinder are connected to a shared pipe is entirely dependent on the selected ignition sequence. Let us assume that the ignition sequence due to the desired crank degrees balancing conditions is as follows 1–5–3–6–2–4, the cylinders should be connected as follows: 1 to 3, 2 to 5 and 6 to 4. The crank degrees between the pressure surges are then 240°.

Stratagem

In an ignition sequence, one cylinder can be skipped each time. Consequently, for an ignition sequence of 1–2–4–6–5–3–1, 1 is connected to 4, 5 to 2 and 6 to 3.

▼

The ignition sequence of a six-cylinder four-stroke in-line engine.

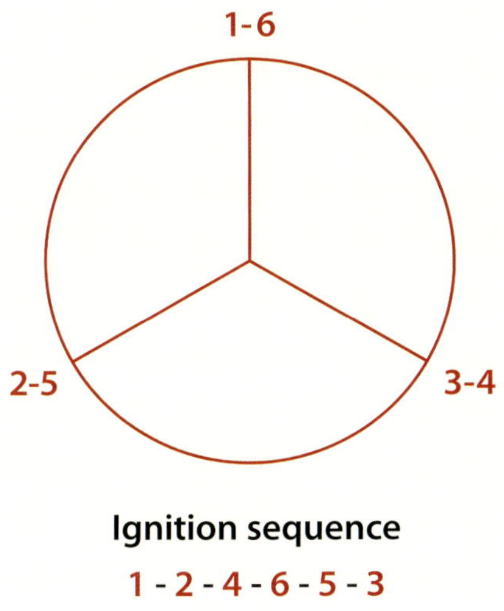

◀

A crankshaft with counterweights for balancing the moving parts.

The combustion sequence is very important when balancing the engine.

Also, see Chapter 17, *Vibrations and Balancing*.

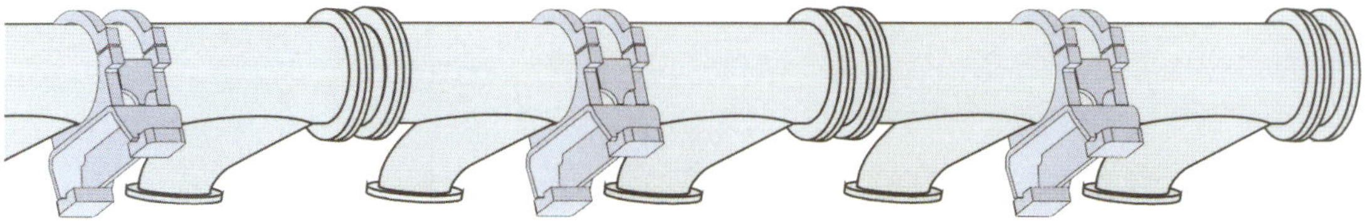

▶
The SPEX exhaust system from engine manufacturer, Wärtsilä.

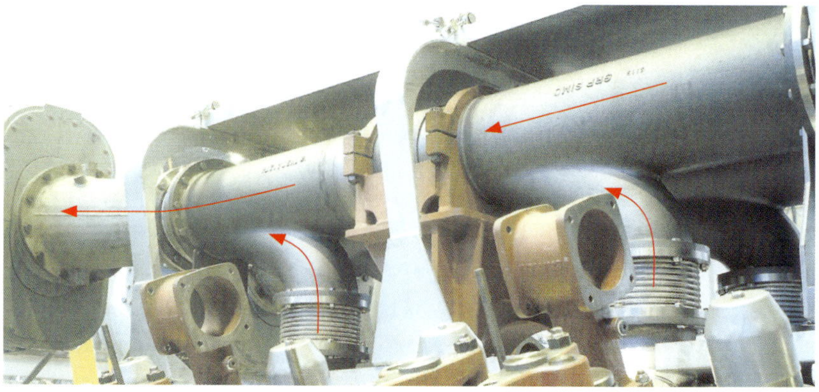

▶
The curved individual exhaust-gas pipes from each cylinder are attached to the common exhaust-gas receiver by a radial pipe.

12.11.3 SPEX-system

The single exhaust-pipe system comprises a single exhaust pipe into which the individual exhaust ducts from each cylinder are connected at right angles to the pipe in the exhaust-gas flow direction. In this approach, the exhaust gases from the cylinders do not disrupt one another and a very simple exhaust pipe is achieved.

12.11.4 Pulse-Converter system

Here, the advantages of both the pressure pulse and constant pressure systems are used. So both the pressure pulses, combined with the even flow in the gas turbine are used to increase the efficiency. The exhaust-gas manifolds of two cylinders where the ignition time is shorter than the time that both exhaust valves are open are connected to a diffuser via jet nozzles. The diffuser leads to a joint collection pipe for the exhaust-gas turbine. The cylinder exhaust-pressure pulse is converted into speed by the nozzle. As, the exhaust-gas from the other cylinders is not accelerated and flows to the joint collection pipe at a lower speed, there is an exchange of kinetic energy. The exhaust-gas speed from the first cylinder is slowed by the exhaust-gas speed from the second cylinder, which in turn is accelerated due to the velocity of the exhaust gas from the first cylinder.
In the diffuser, the collective speeds are converted to pressure and the exhaust gases then flow at a constant and increased speed to the turbine.

Advantages and disadvantages

Constant-pressure and pressure-pulse systems
In practice, one wants to see the advantages and the disadvantages of the constant-pressure and pulse-surge systems in operation (see picture below).

▶
A single turbocharger is sufficient for the air supply of this six-cylinder in-line engine, a Wärtsilä 46 medium-speed heavy-fuel oil engine.

The exhaust-gas receiver is of a simple construction and is not cooled.

1 *exhaust-gas pipe*
2 *turbocharger, exhaust section*
3 *turbocharger, air section*
4 *coolant discharge pipe*
5 *cylinder heads*

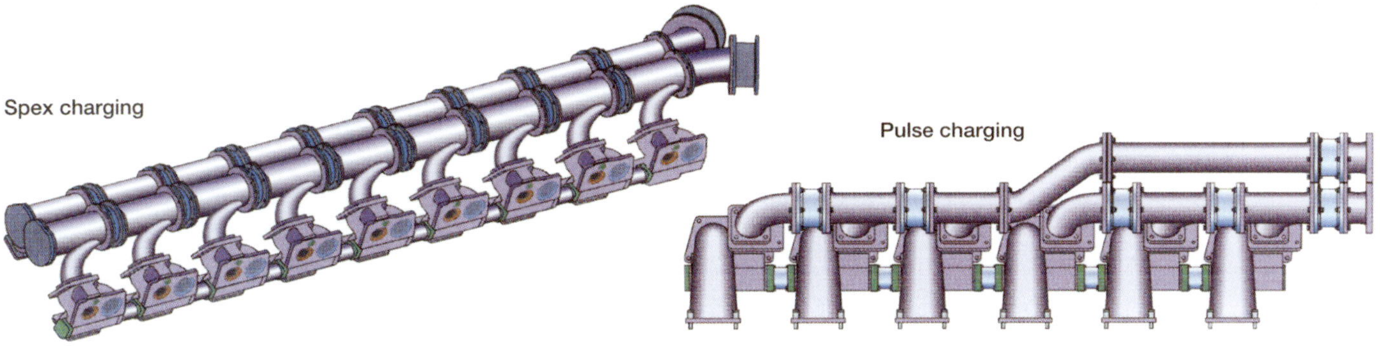

▲
The SPEX supercharging system of Wärtsilä with next to it a pulse-charging system.

Advantages constant-pressure system

- The quantity of turbochargers and pipes is not dependent on the number of cylinders. In general, one turbocharger positioned at the end of the central exhaust-gas pipe is sufficient.
- The mean exhaust-gas pressure is lower during the exhaust stroke in comparison to the pressure-pulse system because the gases are led into a more capacious manifold. This reduces flow resistance.
- As a result of the even flow of the exhaust gases, turbine efficiency is high and therefore the compressor can operate at maximum capacity.

Disadvantages constant-pressure system

- At low engine loads, the scavenging-air pressure may drop below that of the exhaust gases and the exhaust gases can flow back into the scavenging-air line. This negatively affects the efficiency of the exhaust-air scavenging process, the removal of the residual gases and cooling the hot parts in close proximity to the combustion space. Therefore, these parts can be poorly cooled, thus instigating various problems.
- Rapid power increase in the engine may cause temporarily, incomplete combustion. The large volume of the exhaust-gas manifold can result in a delayed reaction in increasing the power of the engine.

Time passage charts of a four-stroke diesel engine

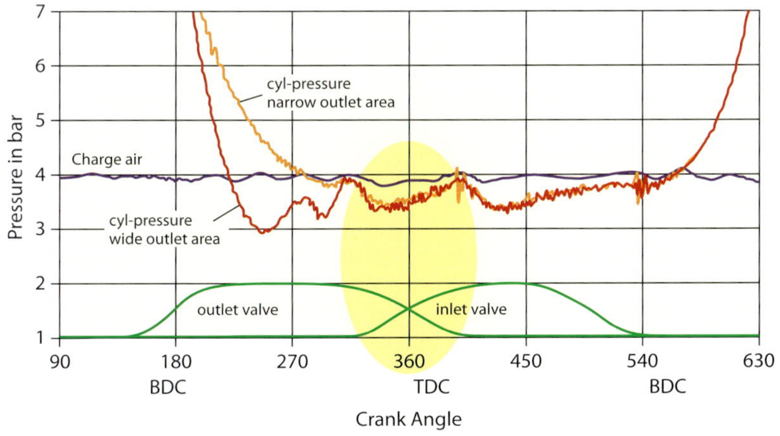

◄
In this line chart of a four-stroke diesel engine, the scavenging process is clearly shown.

At the top position of the piston, shown here at 360 crank degrees, the exhaust and inlet valves open momentarily to cool the hot engine parts. The scavenging-air pressure remains at approximately four bar during scavenging. In total, scavenging takes approximately 90 crank degrees. The entire area represents the scavenging process.

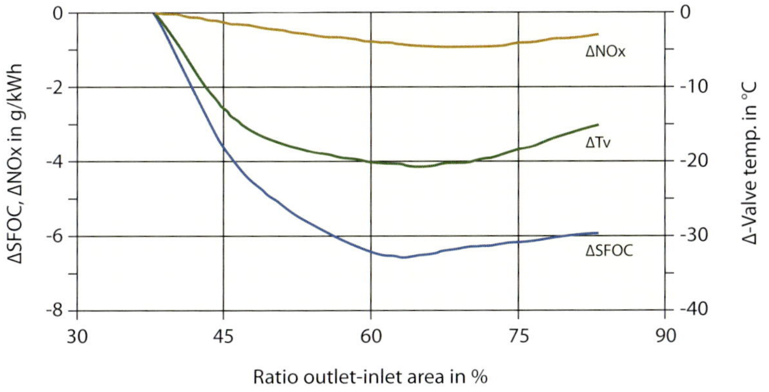

◄
This chart shows how the nitrogen oxide content, the temperature of the exhaust valve material and the specific fuel usage develops during different overlaps of open inlet- and exhaust valves.

Horizontal: the overlap in %.
Vertical left: the reduction in specific fuel consumption and nitrogen oxide concentrations in g/kWh.
Vertical right: the decrease in material temperature of the exhaust valve.

12.12 Air supply in two-stroke crosshead engines

This differs from the four-stroke engine in that the piston stroke of a two-stroke engine in no way contributes to the aspiration, as there is no separate suction stroke. Moreover, there is no actual exhaust stroke. Therefore, an independently driven scavenging pump is required. Initially, the supplied scavenging air should scavenge all the exhaust gases found in the cylinder and subsequently fill the entire cylinder with fresh air.

Presently, there are three manufacturers producing crosshead engines in the world market: **MAN Diesel, Wärtsilä Sulzer and Mitsubishi.**
Essentially, the scavenging principle in all three systems is identical, namely:
– uniflow scavenging.

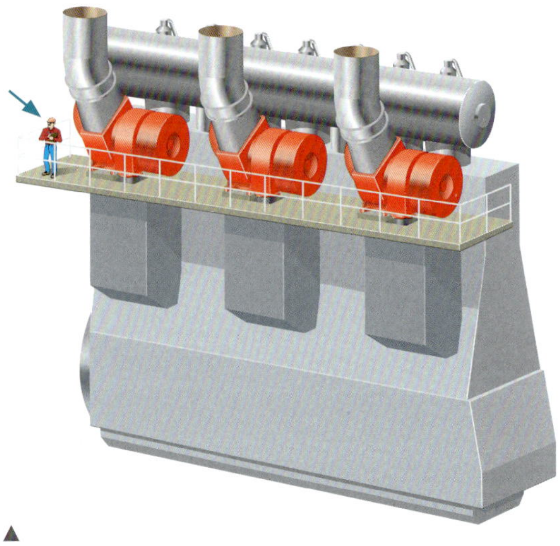

▲
Large two-stroke crosshead engines consume vast amounts of air.

The capacity of the largest turbochargers is limited and therefore two to four turbochargers are mounted on the central exhaust-gas receiver in order to supply sufficient air. Note the size of the engineer to the left on the platform!

12.12.1 Uniflow scavenging

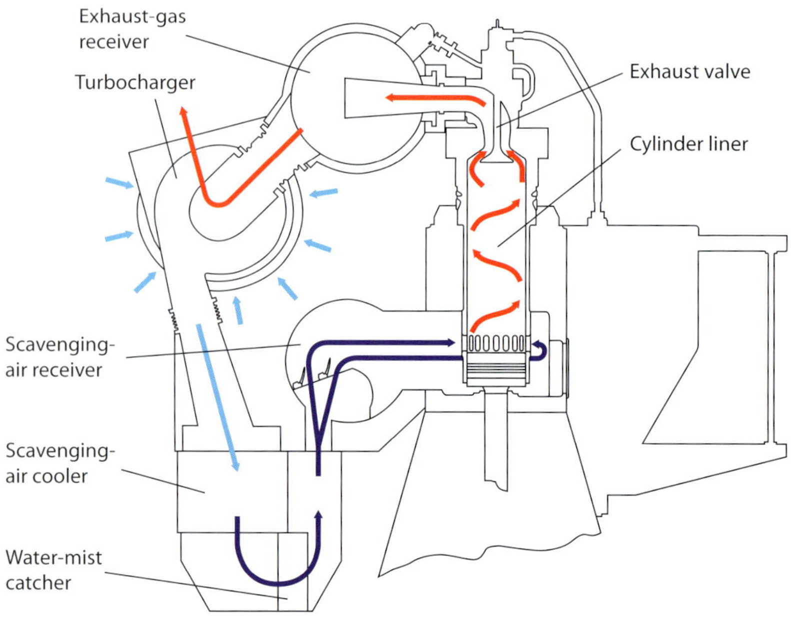

▲
The principle of uniflow scavenging in a two-stroke crosshead engine.

When the piston is in bottom position and the inlet ports are open, the scavenging air flows from the scavenging-air space through the inlet ports to the cylinder and in a twisting movement in the direction of the opened exhaust valve.
In this way, the air is supplied from scavenging ports mounted at the bottom of the cylinder liner and discharged via the exhaust valve mounted in the cylinder head. Scavenging occurs from the bottom to the top.

▲
Air enters the cylinder (see arrow) via the inlet ports at the bottom of the liner,

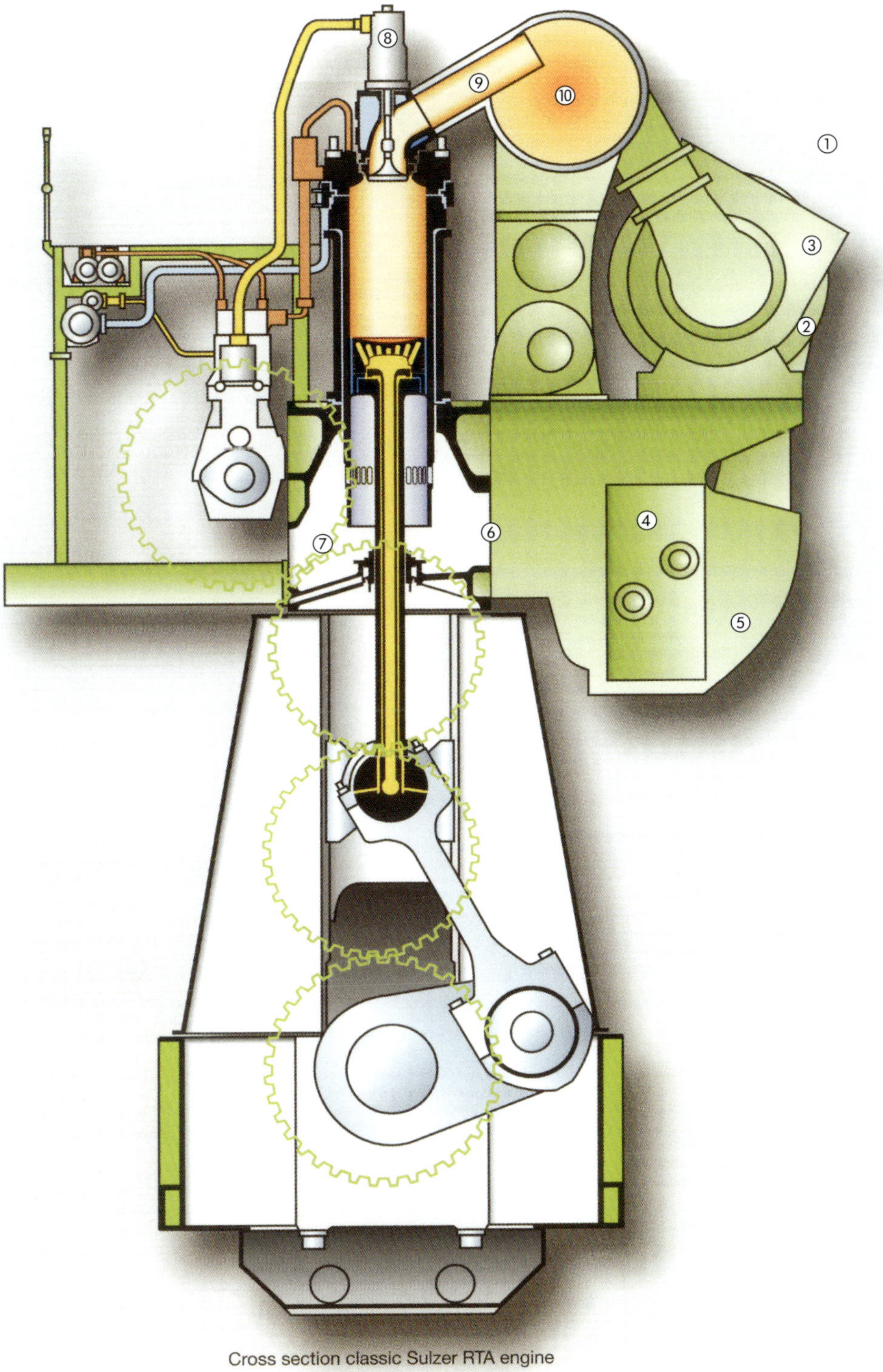

◄ The pressure-charging system of a two-stroke crosshead engine.

1 *ambient air*
2 *air-inlet filter*
3 *turbocharger*
4 *air cooler*
5 *moisture separator*
6 *flap valves (not visible)*
7 *scavenging-air space*
8 *exhaust valve*
9 *exhaust of one cylinder*
10 *central exhaust-gas receiver*

12.12.2 Scavenging effect or Scavenging ratio *R*

In this type of engine, one refers to scavenging effect or scavenging ratio.
This is the air volume added per cycle and per cylinder in relation to the stroke volume of the cylinder.

$$R = \frac{\text{Actual added air volume}}{\text{Stroke volume of the cylinder}}$$

During the uniflow process, *R* is approximately 1.2 to 1.3 and over 95% of the exhaust gases are removed. Five per cent is the residual gas.

$$\text{Scavenging effect} = \frac{\text{Volume of clean air in cylinder}}{\text{Stroke volume}}$$

$$= 90 \text{ to } 95\%$$

▶
The hydraulically operated central exhaust valve.

The hydraulically operated central exhaust valve of every cylinder opens during the power stroke at approximately 110 crank degrees after top to discharge the exhaust gases and scavenge the cylinder.

Description of the scavenging process

The entire circumference of the lower part of the cylinder is fitted with tangentially placed inlet ports connected to the scavenging-air space around the bottom part of the cylinder. The cylinder cover has one large, centrally positioned exhaust valve, which is opened hydraulically.

When the inlet ports are released by the piston during the downward stroke, the scavenging-air pressure is higher than the exhaust-gas pressure so scavenging occurs swiftly and effectively. This is the reason that the exhaust valve opens just prior to the release of the scavenging ports, the exhaust pressure is brought below that of the scavenging-air pressure.

▶
The principle of the two-stroke engine with right the crank circle.

The height of the scavenging ports varies between approximately 10 tot 15% of the piston stroke.
From the scavenging-air receiver, the air is supplied at the moment that the piston releases the inlet ports via the tangentially positioned scavenging ports. Due to this, the scavenging air moves in a rotating movement in the cylinder and removes all the residual gases via the exhaust pipe.

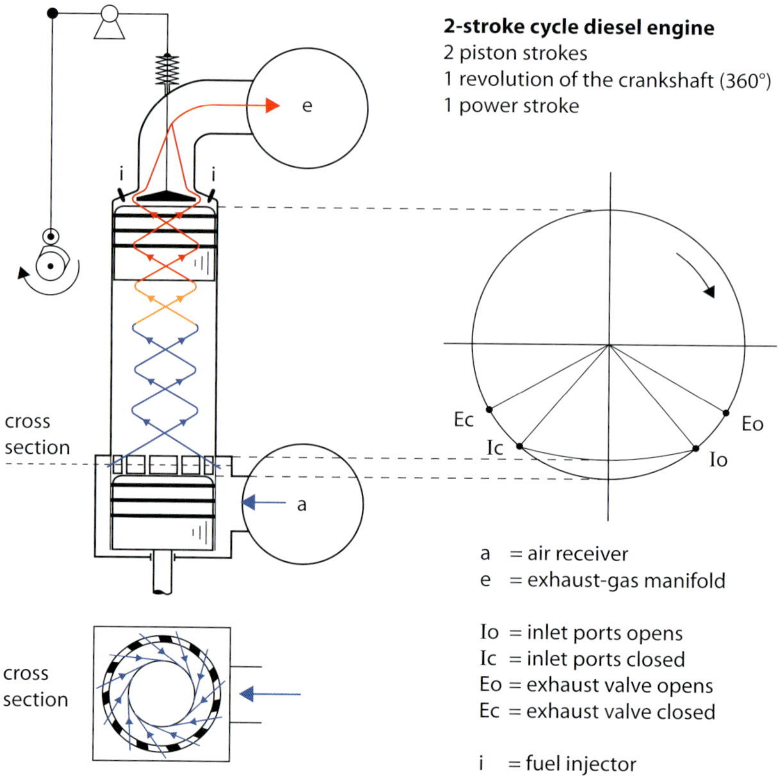

◀

Cylinder liners of a DETROIT high-speed two-stroke trunk-piston engine.

Left an indirectly cooled version and right a directly cooled version. Note: the difference in port height.

Heat transfer:

- indirectly via the material of engine block
- directly via coolant in the cylinder liner

◀

Cylinder liners in large two-stroke crosshead engines have a height of approximately 4 metres and a mass of 11,000 kg.

The tangentially placed scavenging ports guide the incoming air along the cylinder liners, therefore minimizing scavenging shadows. The ports themselves are often fitted with 'fingers': notches which further increase the turbulence of the air and assist the removal of the residual gases.

In this type of scavenging, one also refers to the 'swirl factor'.

This is the ratio $\frac{r}{R}$; r is the radius of the circle that the inlet air touches and R is the radius of the cylinder.

In modern engines the swirl factor is approximately 0.3. After the scavenging period, the inlet ports are closed by the upward piston stroke and shortly afterwards the exhaust valve.
Air compression can now start! At this stage, at least 20% of the upward stroke has been completed!

◀

The tangentially positioned inlet ports have a 'ribbed' profile to promote the scavenging-air turbulence and therefore improve the removal of the residual exhaust gasses.

12.13 Various approaches to supercharging in two-stroke crosshead engines

Equal- or constant-pressure systems

The exhaust gases from all the cylinders are discharged to a capacious, central exhaust-gas manifold or 'exhaust-gas receiver'.

Pressure pulses are equalized, therefore the gas flow to the exhaust-gas turbine(s) occurs at constant pressure and velocity.

Due to the constant gas flow, both turbine-flow efficiency and capacity are high. The number of turbochargers depends solely on the mass flow of the exhaust gases and the turbocharger capacities available.

Impact- or pulse system

Each cylinder is connected to a narrow, short exhaust-gas manifold, connected directly to the exhaust-gas turbine. The kinetic energy of the high-speed discharged gases and the expansion energy of the exhaust gases can be easily converted by turbochargers placed in close proximity to the cylinders concerned.

Disadvantages of this system

- The flow efficiency of the turbine is lower when the exhaust gases flow in pulses or bursts.
- The cylinder quantity is often limited to a maximum of **three** in order to avoid disturbing the scavenging of the adjacent cylinders.
- A **nine**-cylinder engine has therefore **three** turbochargers.

In practice, all three types of two-stroke crosshead engines are equipped with a constant pressure system. Experience has shown that a constant pressure system is the optimum system for high-cylinder capacities that have mean effective pressures between 1.6 and 1.9 MPa or between 16 and 19 bar. Diffusers are used to convert kinetic energy into pressure, for instance, the RTA-series by Wärtsilä Sulzer and the MC-series by MAN Diesel.

The third two-stroke crosshead engine manufacturer, Mitsubishi, also uses the constant pressure system, including a diffuser. This type has three cylinders connected to one diffuser, which lead to the central exhaust-gas receiver.

▶ The electrically driven auxiliary turbocharger of a two-stroke crosshead engine.

This starts automatically when the scavenging-air pressure is too low at reduced power.

▶ A ten-cylinder two-stroke crosshead engine with the exhaust pipes from every cylinder directly attached to the very spacious central exhaust-gas receiver.

Depending on the engine dimensions, one to four turbochargers can be installed.

12.13.1 Three engine manufacturers

The latest design by MAN Diesel is the MC-series, from Wärtsilä 'Sulzer', the RTA-series and the UEC-LS-series from Mitsubishi.
They all utilize the equal- or constant-pressure system with diffusers, uncooled turbochargers and non-return flap valves in the supply pipes to the scavenging-air spaces. In order to avoid scavenging problems in low-loaded engines that have an extremely low-scavenging air pressure in relation to the exhaust-gas pressure, all manufacturers make use of electro-driven auxiliary chargers that automatically switch on when the scavenging-air pressure drops below a certain point.

For an extensive engine description, see Chapter 21, *Diesel engine manufacturers*.

12.13.2 Absorption characteristics

The turbocharger compressor delivers a certain amount of air to the scavenging-air receiver where there is a certain scavenging-air pressure. The air then flows through the opened inlet valves into the cylinder (four-stroke engines) or through the inlet ports released by the piston into the cylinder (two-stroke crosshead engines).

The efficiency of the centrifugal compressor depends on the suction pressure.
Every altered rpm has a certain supply characteristic.

Turbocharger suppliers use these charts, often called maps instead of charts. The pressure ratio or the compression ratio lies on the vertical axis, instead of the compressor pressure.

▼
Absorption characteristic.

Description:
Vertical; compression ratio, MPa of bar or as shown in the compression ratio.
Speed: In the graph, four speeds are shown with their own delivery characteristic and the points where the compressor is set at these speeds, the operating points. Connecting these four points gives the **absorption characteristic**.
Left of the **boundary line of instability**, the compressor becomes unstable. The piping characteristics are also shown and in this case also the four points where the compressor map cuts the four speed lines.
Left of these four intersection points, the pressure in the scavenging-air piping is higher than the pressure supplied by the compressor so the air flows backwards in a pulsing motion. At the moment that the scavenging-air pressure falls below the compressor map, the flow reverses and flows in the correct direction. This is known as compressor 'surge'.
The sound produced is known as 'barking of the turbocharger'.
The vibrating rotor may be damaged by this.

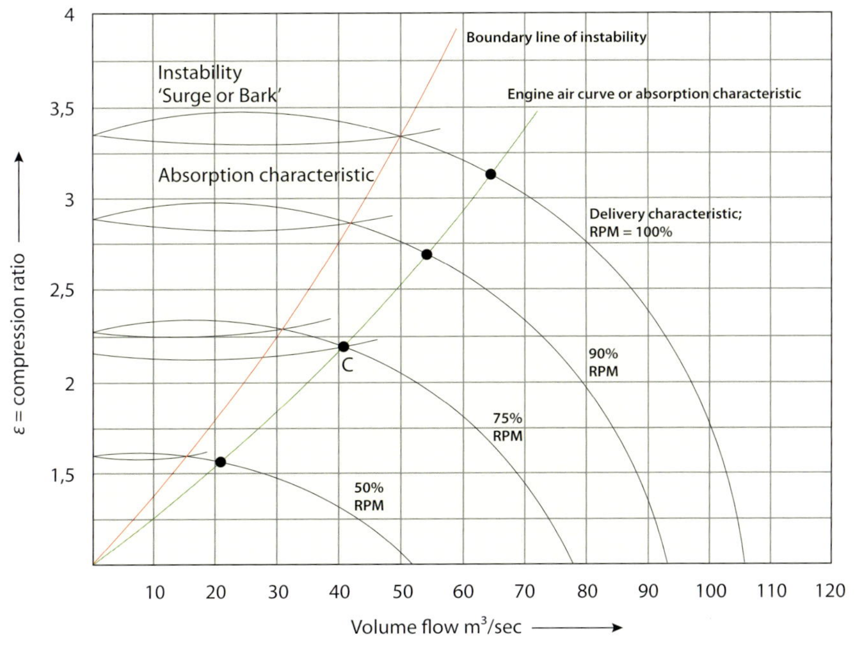

▶

A compressor map of an ABB TPL 85-B turbocharger.

Example: With an air output of 30 m³ per second and a compression ratio of 3.5, the speed is 90% of full load and the compressor yield 86%.

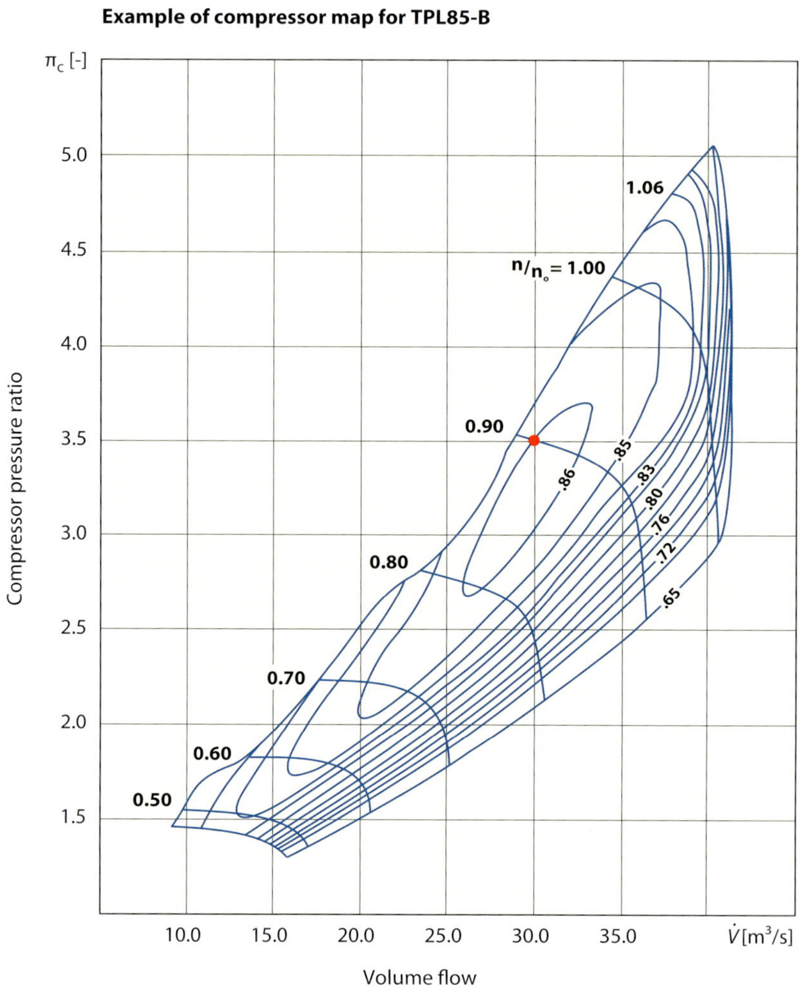

12.13.3 By-pass valve

To prevent compressor instability at reduced capacity in a four-stroke engine and stop engine surge or 'barking', a by-pass valve placed in the pressure supply line can be opened so that the compressed air on the suction side of the compressor can flow back into the non-pressurized side of the intake. This requires increased compressor output, the compressor remains to the right of the instability boundary line. This usually occurs in propulsion engines with a variable rpm.
Most four-stroke propulsion engines have a permanently high rpm, the speed and consequently the engine load is controlled by an adjustable-pitch propeller, therefore 'barking' very seldom occurs. Four-stroke engines for gensets should not experience any problems as they operate at a constant rpm.

In two-stroke crosshead engines with a fixed propeller used for propulsion and therefore reversible with a variable rpm, instability of the compressor usually does not occur. The compressor actually becomes more stable at a reduced output and/or rpm.

12.14 Some important points of interest with regard to the air supply in diesel engines

12.14.1 Air temperature

This is the temperature of the air after the intercooler taken the instant it passes the inlet valves (four-stroke engines) or the inlet ports (two-stroke engines).

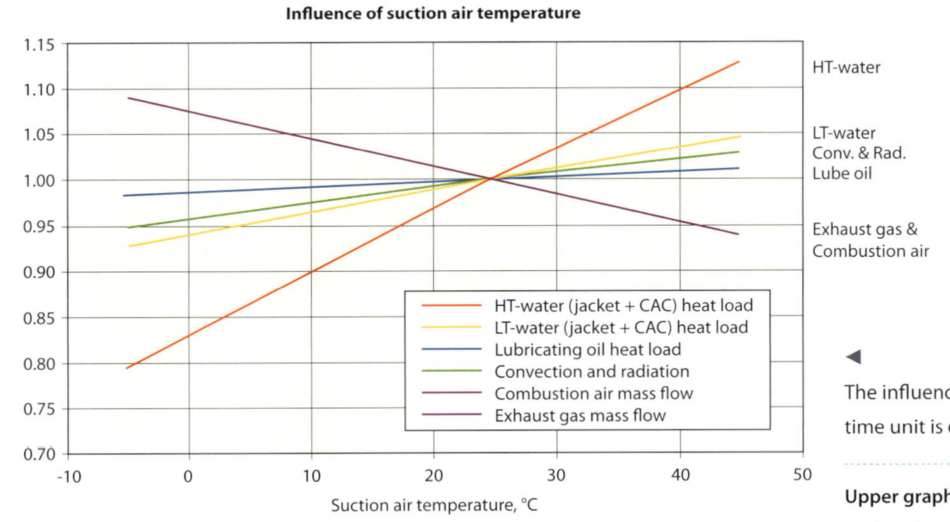

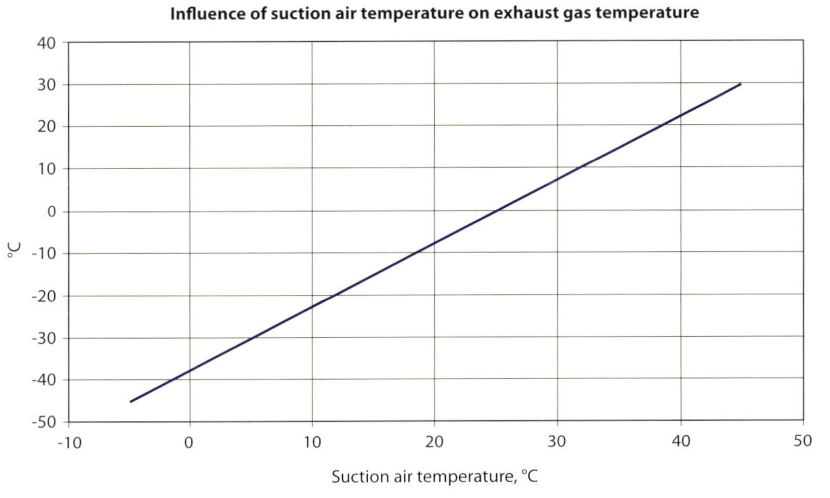

◄

The influence of air temperature on the air mass that per time unit is drawn into the engine.

Upper graph:

vertical: the correction factor

Clearly shown, the influence of air temperature on the air mass per time unit drawn into the engine. The difference between – 10 °C, correction factor 1.10 and 40 °C, correction factor 0.95 is large. The HT coolant correction factor changes a significantly from of 0.75 at – 10 °C air temperature to 1.10 at 40 °C.

Lower graph:

vertical: the influence of air temperature on the exhaust-gas temperature

Between a suction-air temperature of 0 °C and 30 °C, the exhaust-gas temperature increases approximately 44 °C.

The air temperature is important for the total process.

- The mass of air per time unit that can be brought into the engine depends directly on the air temperature.
- In essence, the capacity of the engine is reduced at higher air temperatures.

12.14.2 Effects of the air temperature on the temperature of exhaust gases, coolant, lubricating oil and the radiant heat

The a**ir temperature** has a direct effect on the engine output. At higher air temperatures, the power output decreases because the specific air mass is lower at higher temperatures and so the total air mass per time unit available to the engine. Less fuel can therefore be supplied to the engine, thus reducing the available shaft power.

◄

The cylinder of a large two-stroke crosshead engine.

1 *central exhaust-gas pipe*
2 *exhaust-gas pipe for one cylinder*
3 *exhaust valve*
4 *cylinder cover*
5 *cylinder liner*

12.14.3 Air pressure

Air pressure affects the engine's capacity. The higher the engine is above sea level, the further the drop in air pressure and so engine output.
At a higher atmospheric pressure, the specific air mass increases and therefore more power can be generated.
At a lower atmospheric pressure, the specific mass decreases and therefore less power can be generated.

12.14.4 Air moisture content

The moisture content of air is also an important factor. At too low a scavenging-air temperature, the water vapours can condensate. The water droplets will then precipitate out on the lubricating-oil film found on the cylinder liner, causing extra wear and tear in the cylinder.
For this reason, two-stroke crosshead engines are equipped with a water-mist catcher in the scavenging-air space, which collects the condensed water and discharges it, avoiding additional wear and tear of the cylinder liner.
Particularly in humid tropical areas, large amounts of water are drained in this way!

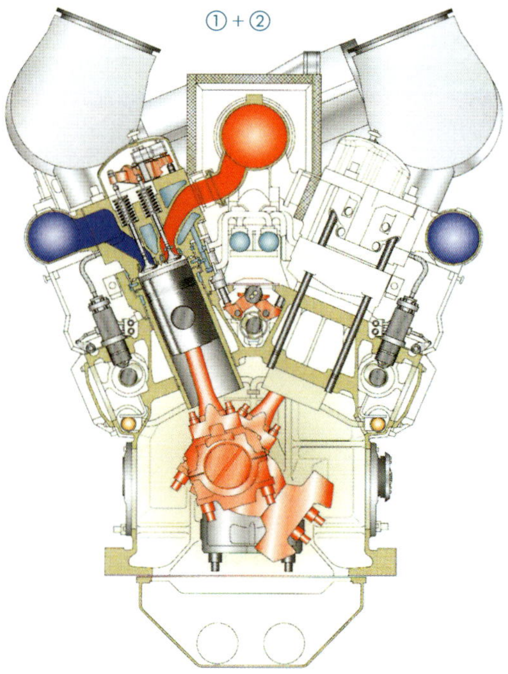

▲
The atmospheric conditions that influence diesel-engine performance.

1 *air temperature, T_{air}*
2 *air pressure, p_{air}*

▶
The schematic diagram shows a standard ventilation system in an engine room.

Sufficient air supply is very important not only for the combustion process in the engine, but also discharging the radiation heat around the engine and refreshing air in the room for personnel.
The lower graph shows the amount of condensed air in an air cooler after the turbocharger.
In this process, several litres of water per hour for small engines, and thousands of litres per hour for large engines are produced. The water drops have a negative effect on the lubricating-oil film on the cylinder liner. So in two-stroke crosshead engines, a water-mist catcher is installed after the air cooler that discharges the water outside the engine.

Engine room ventilation (4V69E8169)

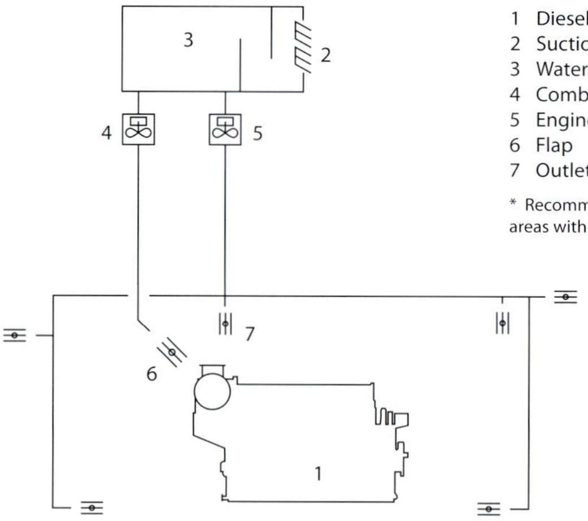

1 Diesel engine
2 Suction louver*
3 Water trap
4 Combustion air fan
5 Engine room ventilation fan
6 Flap
7 Outlets with flaps

* Recommended to be equipped with a filter for areas with dirty water (rivers, coastal areas. etc.)

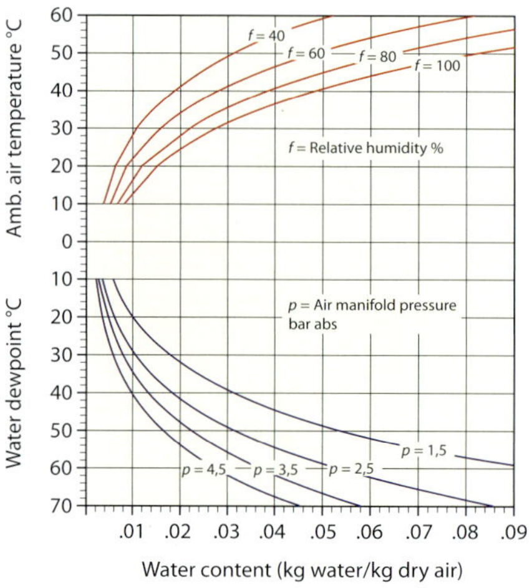

Condensation in charge air coolers
Example, according to the diagram:
At an ambient air temperature of 35 °C and a relative humidity of 80%, the content of water in the air is 0.029 kg water/kg dry air. If the air manifold pressure (receiver pressure) under these conditions is 2.5 bar (= 3.5 bar absolute), the dewpoint will be 55 °C. If the air temperature in the air manifold is only 45 °C, the air can only contain 0,018 kg/kg. The difference, 0.011 kg/kg (0.029 – 0.018) will appear as condensed water.

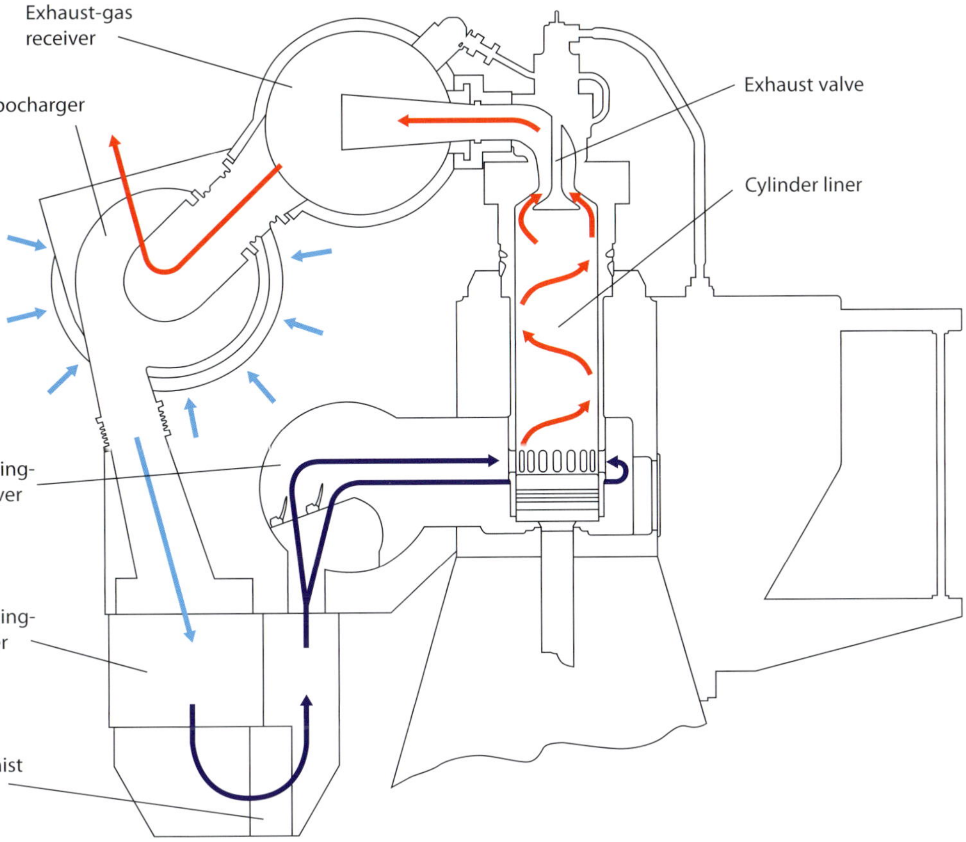

◄ The water-mist catcher at the bottom of the scavenging-air supply to the cylinder.

12.15 Maintenance of turbochargers

Proper maintenance of the turbochargers, including air filters and air coolers is essential for an optimum operation of the diesel engine.
If the air filter is contaminated, the air resistance on the suction side increases and the air capacity of the charger diminishes, so less air is supplied to the engine.
Contamination of the exhaust-gas turbine of the air compressor also causes a reduction in produced air. Therefore, turbochargers are supplied with a washing- or scavenging system to clean both wheels at regular intervals.

Detergents are clean water and solid substances such as burned walnut shells or other organic materials that burn virtually ash-free in the exhaust-gas turbine.

▲ The water-mist catcher is placed at the lowest point in the air-piping system so that the water droplets cannot reach the cylinder.

▶

Dirty filters impede the passage of fresh air to the turbochargers.

Shown here, the air filter is badly soiled. A pressure-difference meter will show that this filter should have been cleaned a while back!

▶

Damaged turbine blades cause a reduction in the air capacity of the turbochargers.

Regular cleaning and periodic maintenance can prevent this. Repair is essential.

▼

A water-rinsing system for the exhaust-gas turbine of the turbocharger.

The dirty water is discharged at the bottom.

1 *rinsing pipes*

▶

A water-wash system with the water supply above and both wastewater discharge pipes below.

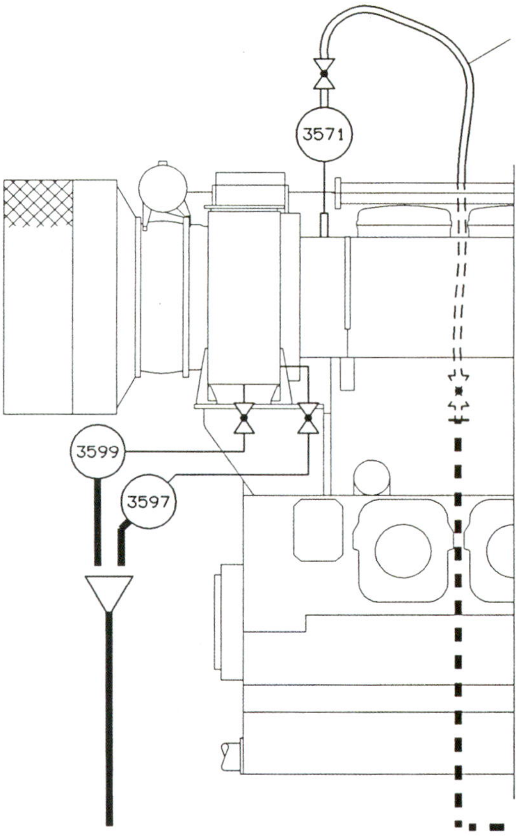

12.15.1 Air coolers

Contamination of air coolers reduces the heat transmission and increases the pressure drop over the air cooler.
Many engines are designed in such a way that the airside including the cooling fins can be cleaned without dismantling the cooler.
The application of a blind flange allows the cooler side to be filled with a solvent and cleaned using a circulating pump. Filter mats remove all the dirt particles from the solvent.

12.15.2 Arctic conditions

At extremely low air temperatures, it is imperative to surpass the minimum scavenging-air temperature.

- In order to start a diesel engine, a minimum air temperature of 5 °C in the engine is required.
- At low load, a minimum air temperature of – 5 °C.
- At full load, a minimum air temperature of – 20 °C.
- Air temperatures lower than the aforementioned require special heating to pre-heat the air.

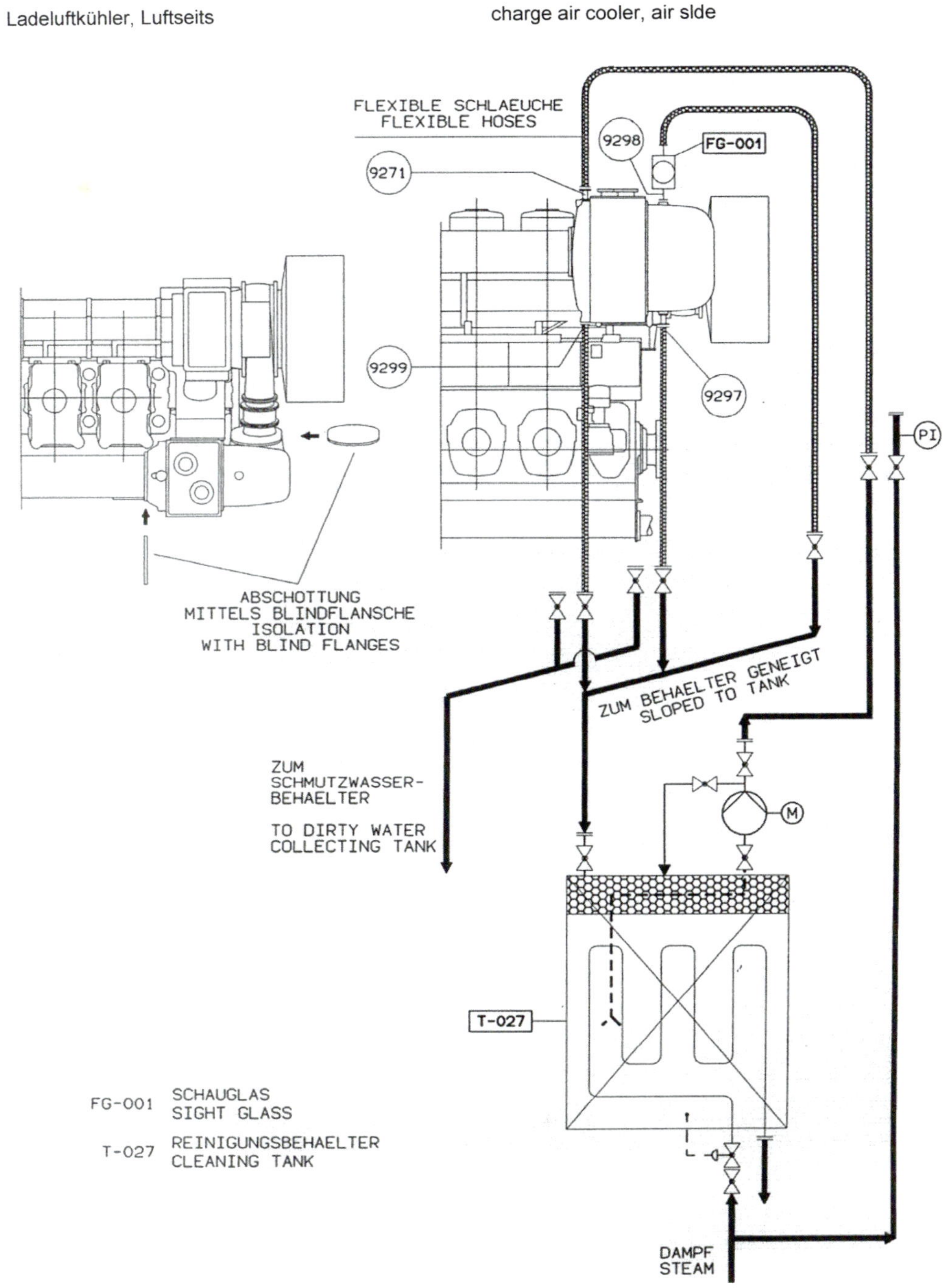

◄ A rinsing system is available for air coolers.

When the engine is stopped, the cooler can be cleaned without removing the air cooler. Blind flanges are installed in the air-cooler ducts and the airside is cleaned using a hot fat-soluble solution and a circulating pump.

▶
It is important to monitor the air temperature after the air cooler as this temperature affects the combustion process of the engine.

12.16 Problems with supercharging

The following problems may occur:
- contamination of the air filter;
- contamination, damage and corrosion of the gas turbines;
- contamination of the air compressor;
- leaking of the shaft seals;
- wear and tear of the shaft bearings;
- contamination and corrosion of coolant spaces;
- turbine-rotor imbalance;
- problems with the turbocharger-lubricating system or the supply and discharge of the lubricating oil from the diesel engine;
- air-compressor 'surge';
- contaminated air-cooler on the gas- or water side;
- too high or too low air suction temperatures when taking the ambient temperature into consideration;
- blockage of the condensate drain of the mist-/ condensate separator.

▶
Mechanical damage to the turbine blades of a four-stroke engine.

Probably caused by a fragment of a piston ring or an exhaust valve in the turbine.

◄

A protective screen is placed in the central exhaust-gas pipe of a two-stroke crosshead diesel engine.

This prevents the exhaust-gas turbine from being damaged by pieces of piston rings or parts of the exhaust valves.

◄

A badly contaminated exhaust-gas turbine.

Due to this, the exhaust-gas flow through the turbine wheel is reduced making it less effective and therefore the amount and pressure of the air diminishes. The result is that there is less air available for the engine.

12.17 Engine load

12.17.1 Explanatory notes

a = 1 for two-stroke, 2 for four-stroke
c = effective compression stroke
C = constant ($V_s \times \frac{n}{a} \times Z$)
$\dot{m}_b$ = fuel mass (kg/sec)
$\dot{m}_l$ = air mass (kg/sec)
n = rotations per second
p_a = initial pressure compression
p_n = pressure after compression
P_c = effective power output of compressor
V_s = stroke volume
W_e = effective work per process
Z = number of cylinders

Increasing the mean effective pressure generates more shaft power.

$$P_e = p_e \times V_s \times \frac{n}{a} \times z = P_e = p_e \times c$$

Furthermore: Increasing the mass of air supplied means that the mass of fuel supplied can increase and more power is generated.

Combustion

Air volume = stroke volume + volume combustion space – volume initial compression.

$$V_l = V_s + V_c - V_a$$

Volume to initial compression = V_a
$V_a = b \times V_s$ and $V_c = a \times V_s$
$V_l = (1 + a - b) \times V_s$

▶ Effective compression stroke.

The piston has travelled distance b before the inlet- and exhaust valves are completely closed.

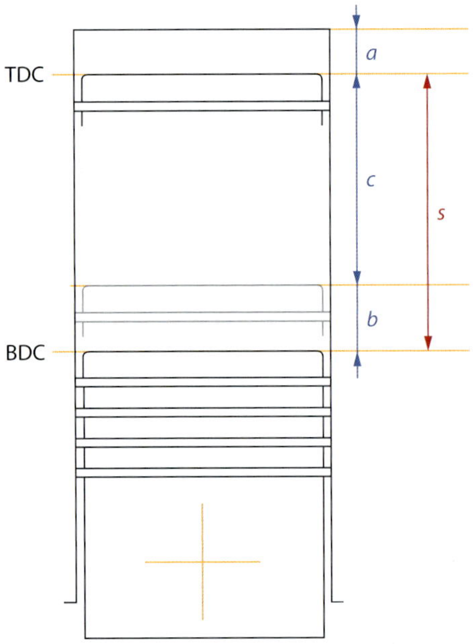

$c = s - b$ = effective compression stroke

Air volume that can be deduced to cylinder conditions is:

$$m = \rho \times V \text{ and } \frac{PV}{T} = c \quad (I)$$

ρ = specific mass in kg/m^3

Air mass in cylinder

Making use of both formulas in (I).

$$m_l = V_l \times \rho_n \times \eta_s \times \frac{p_a}{p_n} \times \frac{T_n}{T_a} \text{ (kg)}$$

Indication a, initial compression.
Indication n, at 1 bar and 273 °C
η_s = scavenging efficiency

$$\dot{m}_l = \dot{m}_b \times \lambda_v \times L_{th} \text{ (kg)}$$

T_a reduction is limited due to low temperature corrosion (LTC).

Effective work per process

$$W_e = \dot{m}_b \times H_o \times \eta_{tot} \text{ (kJ)}$$

$$W_e = V_s \times p_e \text{ (kJ)}$$

$$P_e = (1 + a - b) \times \eta_s \times \frac{P_a}{T_a} \times \frac{\eta_{tot}}{\lambda_v} \times \frac{273 \times H_o \times \rho_n}{L_{th}}$$

η_{tot} = total engine efficiency

To achieve a higher power output without increasing the rpm, the mean effective pressure must rise!

Effective work per process

$$W_e = \dot{m}_b \times H_o \times \eta_{tot} \text{ (kJ)}$$

$$W_e = V_s \times p_e \text{ (kJ)}$$

η_{tot} = total efficiency

$$P_e = \underbrace{(1 + a - b) \times \eta_s}_{I} \times \underbrace{\frac{p_a}{T_a} \times \frac{\eta_{tot}}{\lambda_v}}_{II} \times \underbrace{\frac{273 \times H_o \times \rho_n}{L_{th}}}_{III}$$

I = aspiration.
II = increase p_a by reducing turbocharger and T_a by air cooler.
III = is constant for a given fuel.

Conclusion

- At partial load the combustion-air factor decreases.
- At partial load the specific air consumption increases.
- At partial load the combustion- and scavenging-air ratio changes.
- As a result of optimising the turbocharger more air is available for the engine than previously.

◀
Adjustable guide blades in a turbocharger from ABB, type TC DE.

The guide blades are centrally controlled in a single motion; therefore the exhaust gas and air quantities can be perfectly controlled.

Higher turbocharger efficiency due to:
- Improved design of the compressor impeller.
- Compressor-housing enhancement.
- Improvement of the nozzle ring or diffuser.
- Omitting the mute lead of the turbine blades.
- Application of uncooled turbine housings.

The turbocharger can operate optimally at varying engine loads by:
- The waste-gate in the scavenging air or exhaust gas pipe.
- The by-pass, a connection between the air-inlet manifold and the exhaust gas pipe.
- The power turbine: part of the exhaust-gas flow is used for generating power.
- Further improvement of the adjustable deflector blades in the turbine section, allowing precision inlet-gas flow regulation.

12.17.2 Thermodynamics of the turbocharger

Effective compressor output:

$$P_c = \dot{m}_l \times \Delta \cdot h_l = \dot{m}_l \times c_{pl}\,(T_2 - T_1)$$

Isentropic compressor output:

$$P_{c_{isen}} = \dot{m}_l \times \Delta_{hl_{isen}} \cdot h_l = \dot{m}_l \times c_{pl}\,(T_{2_{isen}} - T_1)$$

Isentropic compressor efficiency:

$$\eta_c = \frac{P_{c_{isen}}}{P_c} = \frac{T_{2_{isen}} - T_1}{T_2 - T_1}$$

Effective turbine output:

$$P_T = \dot{m}_g \times \Delta_{h_{gas}} = \dot{m}_g \times c_{pg} \times (T_3 - T_4)$$

Isentropic turbine output:

$$P_{T_{isen}} = \dot{m}_g \times \Delta_{h_{isen}} = \dot{m}_g \times c_p \times (T_3 - T_{4_{isen}})$$

Isentropic turbine efficiency:

$$\eta_T = \frac{P_T}{P_{T_{isen}}} = \frac{T_3 - T_4}{T_3 - T_{4_{isen}}}$$

There is an energy balance between the compressor and the turbine!
Therefore:

$$P_T \times \eta_m = P_c \rightarrow P_{T_{isen}} \times \eta_T \times \eta_m = \frac{P_{c_{isen}}}{P_{T_{isen}}}$$

or

$$\eta_{Tc} = \eta_T \times \eta_c \times \eta_m = \frac{P_{c_{isen}}}{P_{T_{isen}}}$$

Or: The exhaust-gas temperatures are controlled by the required pressure ratios $\left(\frac{p_{end}}{p_{start}}\right)$

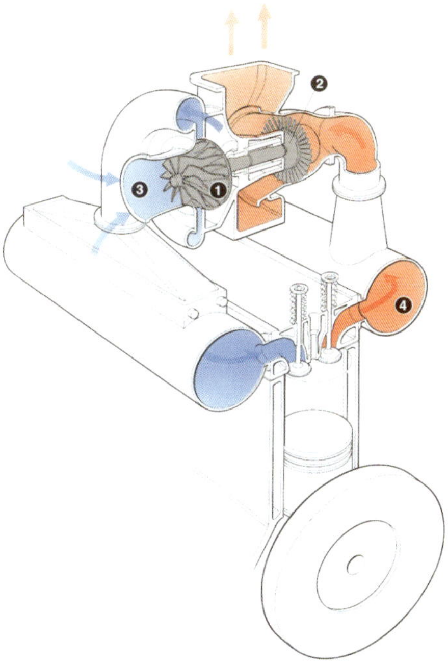

◀
A turbocharger generates its energy from the exhaust-gas flow and converts it into mechanical work using a gas turbine fitted with one wheel. The mechanical work is utilised to drive a centrifugal compressor that compresses aspirated air, which can then flow to the cylinders.

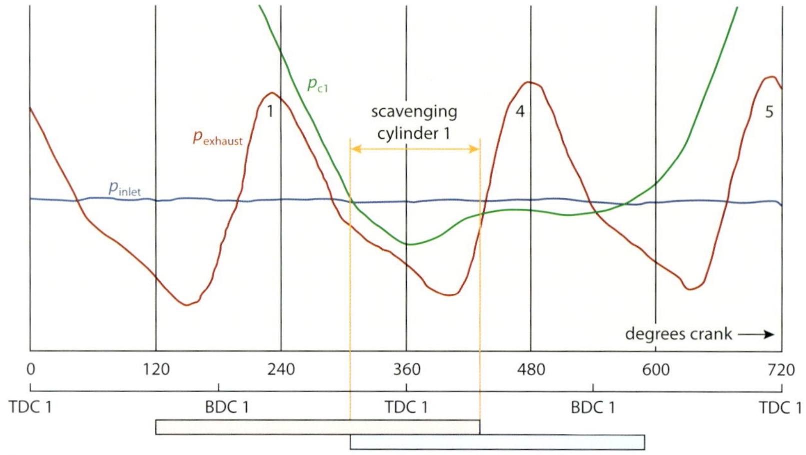

▲ Pressure gradient for a four-stroke engine with the pulse system.

Horizontal: the number of crank degrees up to a complete process of 720° crank.
Vertical: the pressure of the scavenging air and the exhaust gases.
Blue: the inlet pressure before the inlet valves.
Red: the exhaust pressure after the exhaust valves.
Green: the compression pressure.
Light red period: exhaust cylinder 1.
Light blue period: inlet cylinder 1.
Overlap angle scavenging period between the inlet- and exhaust valve of +/- 310° to +/- 440°, so +/-130° crank.
Combustion sequence: 1-2-4-6-5-3.

Note

The air compressor provides the exact amount of air required by the engine.

For an isentrope the following applies: $\frac{T^{k}}{P^{k=1}} = C$

Pressure ratio for the compressor:

$$\frac{p_2}{p_1} = \Pi_c$$

Pressure ratio for the turbine:

$$\frac{p_3}{p_4} = \Pi_T$$

Compressor output:

$$P_{c_{isen}} = \dot m_1 \times c_{pl} \times T_1 \left[\frac{k^1 - 1}{k_1} - 1 \right]$$

Turbine output:

$$P_{T_{isen}} = \dot m_g \times c_{pg} \times T_3 \left[1 - \varepsilon_T^{-\left(\frac{k^{11}-1}{k^{11}}\right)} \right]$$

Met: $k = \frac{c_p}{c_v}$ and $R = c_p - c_v$

$$c_{pl} = \frac{k^1}{k^1 - 1} \quad \text{and} \quad \frac{k^1 - 1}{k^1} = k$$

Compressor output:

$$P_{c_{isen}} = \dot m_1 \times \frac{R_{sl}}{k^{11}} \times T_1 \left(\varepsilon_c^{k^1} - 1 \right)$$

Turbine output:

$$P_{T_{isen}} = \dot m_g \times \frac{R_{sl}}{k''} \times T_3 \left(1 - \varepsilon_T^{k''} \right)$$

Fill in: $P_{T_{isen}} \times \eta_T \times \eta_m = \frac{P_{c_{isen}}}{\eta_c}$

$$\dot m_g \times \frac{R_{sg}}{k''} \times T_3 \left(1 - \varepsilon_T^{-k''} \right) \times \eta_T \times \eta_m \times \beta =$$

$$\dot m_1 \times \frac{R_{sl}}{k^1} \times T_3 \left(\varepsilon_T^{k'} - 1 \right) \times \frac{1}{\eta_c}$$

Suppose $\dot m_g = \dot m_1 \quad R_{sg} = R_{sl}{}^{k'} = k''$

$$T_3 = T_1 \frac{\left(\varepsilon_c^{k} - 1 \right)}{\left(1 - \varepsilon_T^{-k} \right)} \times \frac{1}{\beta \times \eta_T \times \eta_c \times \eta_m}$$

β = pulse factor.

Significance of the formula

To achieve scavenging over pressure, $\varepsilon_c > \varepsilon_T$. Exhaust-gas temperature T_3 should be sufficiently high, requiring the exhaust valves to be opened early in the process. The efficiency η_{engine} is lower.

High turbocharger efficiency may result in a lower T_3 value.

Note

The exhaust-gas turbine output is approximately 25% of the total power output!

12.18 Flow performance of scavenging- and combustion-air in four-stroke trunk piston engine and a two-stroke crosshead engine

12.18.1 Four-stroke trunk piston engine

The amount of aspirated air during the inlet stroke:
$V_{air} = V_{stroke} \times n$

Amount of air required for the scavenging process:

$$m_{air} = \mu \sqrt{2 \times \rho (p_i - p_u)} \, \frac{A \cdot \text{angle}}{720}$$

m_{air} = air mass.
μ = flow number.
ρ = air density.
p_i = inlet-air pressure before the valve.
p_u = exhaust-air pressure after the valve.
$A \cdot$ angle = angle of passage.

▶

Angle of passage diagram for a two-stroke crosshead engine with uniflow scavenging.

Inlet (blue): A_{inlet}, calculated.
A_{red}, actual.
Exhaust (red): A_{ex}, calculated.
$A_{u\ red}$, actual.

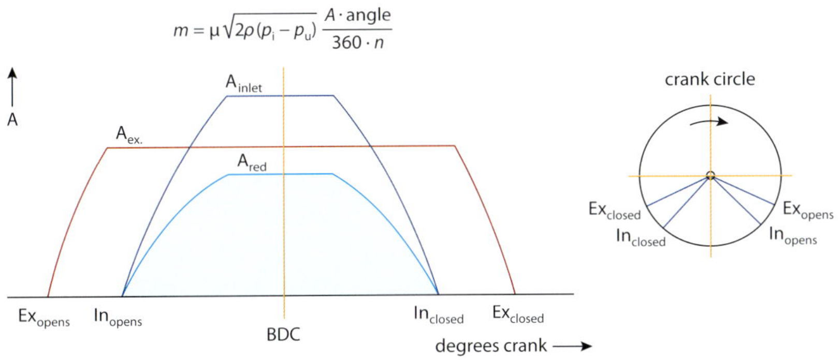

▶

The Angle of passage diagram for a four-stroke engine.

Horizontal: the position of the crankshaft.

Vertical: valve passage area in cm^2.

The light orange area represents the overlap of both valves. This Is the scavenging period. Here approximately 130 crank degrees.

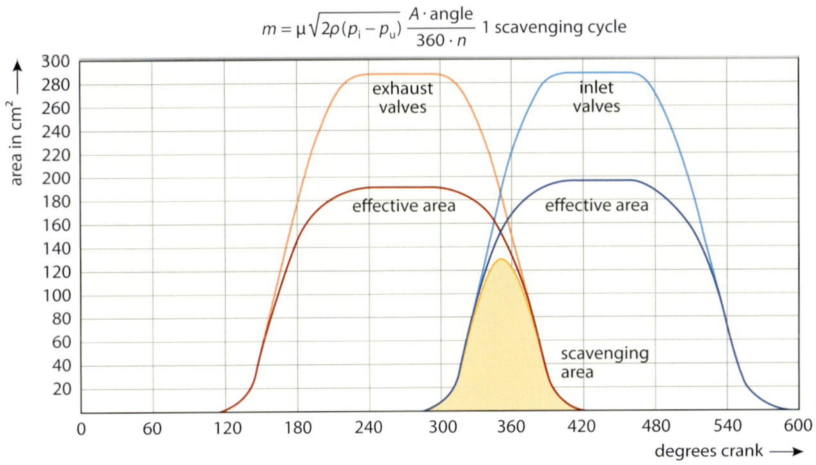

Exhaust valves: left red.

Light red: calculated area.

Red: effective area after calculation with contraction number, a loss.

Inlet valves: right blue.

Light blue: calculated area.

Blue: effective area after calculation with contraction number, a loss.

▶

The pressure supplied by the turbocharger and the pressure required for the engine.

Horizontal: the generated power output in %.

Vertical: the scavenging pressure supply in bar.

Vertical: the mean effective pressure in bar.

Blue: pressure supply

Red: mean effective pressure.

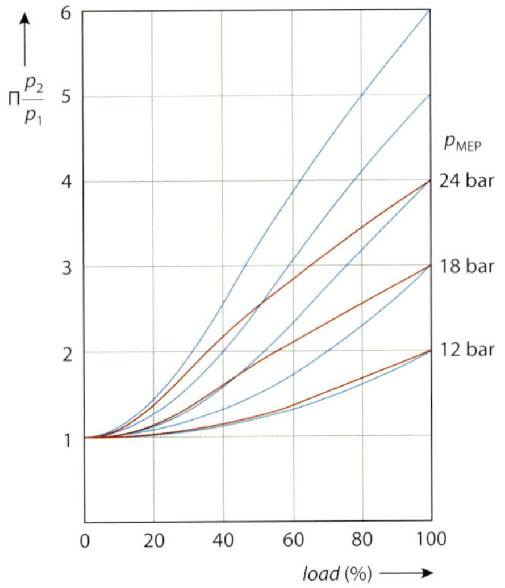

Note:

To increase the power generated by a diesel engine at a constant rpm, the mean effective pressure needs to increase.

This can be achieved by increasing the pressure generated by the turbocharger. At present, this is up to 5 bars to obtain a mean effective pressure of 30 bars!

The turbine output is roughly proportional to the power output.

This is irrespective of the fact that the output is modified according to the propeller law, at a constant rpm or at a constant torque.

In flow mechanisms such as the air- and exhaust-gas section of the turbocharger similar rules apply.

Volume: $V_{eff\,2} = \frac{n_2}{n_1} \times V_{eff\,1}$

Pressure: $p_{man\,2} = \left(\frac{n^2}{n^1}\right)^2 \times p_{man\,1}$

Power output: $P_{eff\,2} = \left(\frac{n^2}{n^1}\right)^3 \times P_{eff\,1}$

12.18.2 Air requirements for four-stroke piston engines

Generator operation

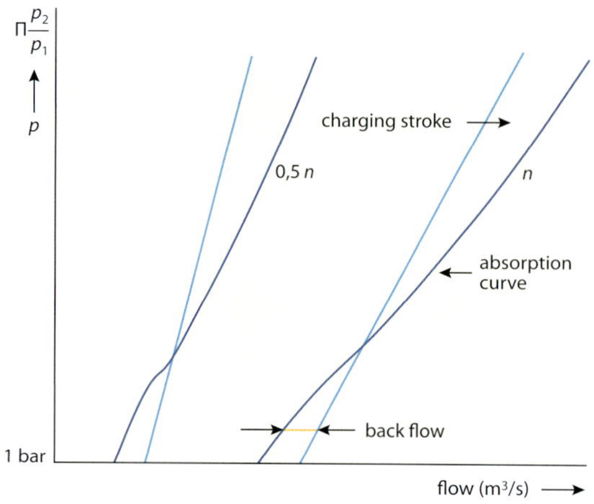

▲

Absorption characteristics of a four-stroke diesel engine with turbo charging.

Absorption: the amount of air required by the engine.

Horizontal: turbocharger capacity in m3/s.

Vertical: pressure ratio of the compressor: the pressure- and suction ratio.

Light blue lines: the amount of air drawn in by the turbocharger at 50% and an engine rpm of 100%.

Dark blue lines: the amount of air supplied by the turbocharger at 50% and an engine rpm of 100%

Note:

At low engines loads, the air will backflow. This means that the exhaust-gas pressure is higher than the inlet scavenging-air pressure.

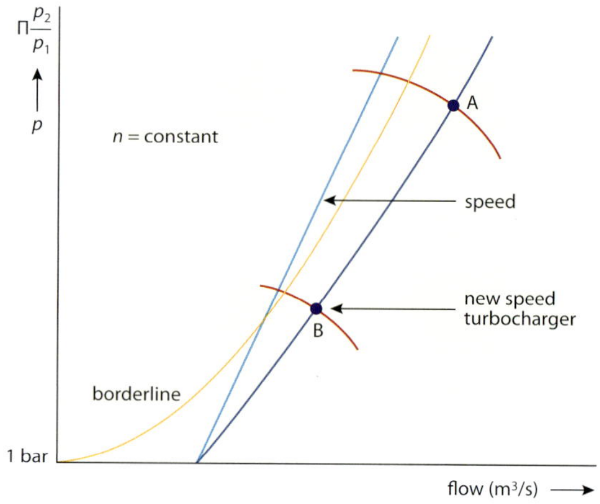

▲

The supply- and absorption characteristics for a four-stroke engine with turbocharger. Load: a generator.

The diesel engine as generator. This means that the diesel engine runs at a constant rpm at varying loads.

The output of the turbocharger P_T equals $P_T = c \times n^3$

c = a constant.

n^3 = rpm to the third power.

This shows that the rpm can vary considerably at different loads.

Conclusion: no 'barking' engine.

Horizontal: turbo-compressor capacity in m3/sec.

Vertical: compressor pressure ratio: the compression- and suction-pressure ratio.

Blue line: absorption characteristics of the turbocharger. A and B constitute the generator load. The pressure ratio for A and B is set out on the vertical line and the air flow on the horizontal line.

Light blue: represents the characteristics of the air drawn in by the engine during the intake stroke.

Yellow line: this is the borderline. If the blue absorption line is situated left of the borderline, the compressor provides insufficient air, which renders the compressor unstable. In shipping terms this is referred to as 'barking' and with reference to pumps this is called 'surging'. It can cause serious damage to the compressor.

▶

The Temperature-Entropy diagram of a turbocharger.

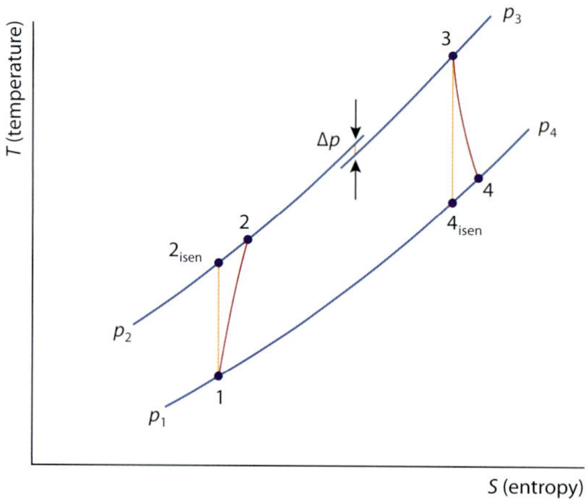

p_1 : suction pressure of the compressor.

p_2 : compression pressure of the compressor.

Theoretical process 1 - 2_{isen} - 3 - 4_{isen} (no losses).

Real process 1 - 2 - 3 - 4 (including losses).

Variable rpm according to propeller law

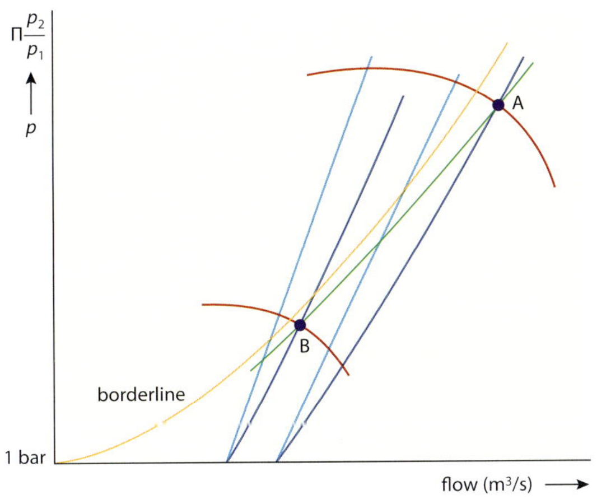

▲
Variable rpm according to the propeller law.

Horizontal: turbo-compressor capacity in m^3/sec.
Vertical: compression ratio of the compressor.

Yellow: borderline for unstable performance of the turbo-compressor. Left of the borderline unstable, right of the borderline stable.
Dark blue: supply characteristics of the turbocharger.
Light blue: absorption characteristics of the engine.

When the power supply of an the engine with a fixed propeller and a varying rpm is modified, the following applies:

$P_{engine} = c \times V^3$

V = speed
c = constant

So at a constant propeller efficiency: $P_{engine} = c \times n^3$
n = rpm of the propeller shaft.

Conclusion: The power output of a diesel engine is proportional to the cubic speed and the cubic rpm. New operating point is near the surge-line! Possible 'barking' due to instability.

Constant torque

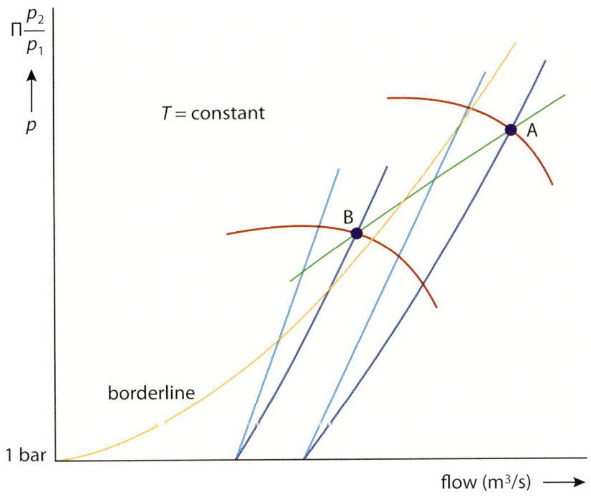

▲
A diesel engine with a constant torque.

A = intersection: absorption and supply characteristics at the maximum rpm of the propeller.

B = intersection: absorption and supply characteristics at a reduced rpm of the propeller.

Horizontal: the turbo-compressor capacity in m^3/sec.
Vertical: the compression pressure ratio of the compressor.

Yellow: borderline for unstable performance of the turbo-compressor. Left of the borderline unstable, right of the borderline stable.

The following applies for a diesel engine:

$P_{engine} = 2 \times \pi \times n \times T_{engine}$

P_{engine} = engine power.
T_{engine} = engine moment.
n_{engine} = engine speed.

It follows that the output is proportional to the rpm.

Note:
When the power output and consequently the rpm decrease, the absorption line will shift to the left into the unstable area. Point B is situated left of the borderline. Measures should be taken in order to prevent this from happening.

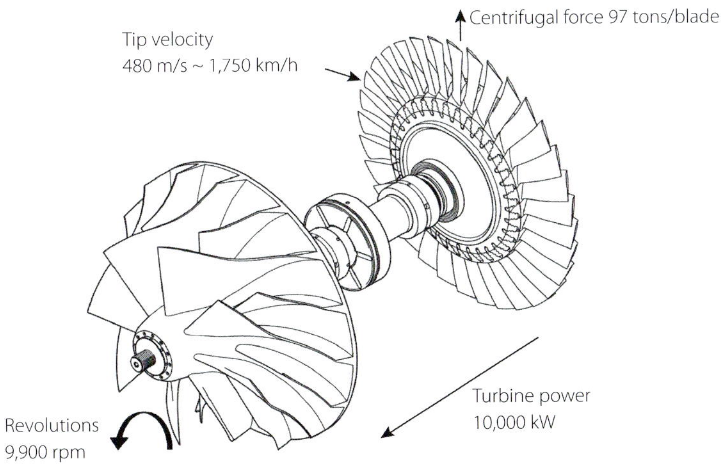

◄
Rotor of a turbocharger of ABB, type TC.

The centrifugal force exerted on the blades is enormous, in part due to a high rpm. This force amounts to as much as 97,000 kg for each blade! Here the peripheral speed at the tip of the blade is 480 m/s or 1750 km per hour. The shaft power generated by the turbine is 10,000 kW.

12.18.3 Air requirements for two-stroke crosshead engines

The amount of air supplied to the engine is solely dependent on the scavenging process.

$$m_{\text{air}} = \mu\sqrt{2 \times \rho(p_{\text{i}} - p_{\text{u}})}\,\frac{A \cdot \text{angle}}{720}$$

The amount of air is dependent on the rpm and proportional to the root of the scavenging overpressure.

Two-stroke crosshead engine performance in poor weather conditions such as swells, currents and wind

If the propeller periodically rises above the water surface during navigation, the rpm increases. The regulator ensures that less fuel is transported to the engine and therefore less power is generated.

When sailing in heavy head seas, the propeller is at times deeply submerged. The engine rpm decreases. The governer ensures that the fuel supply to the engine increases and more power is generated. So, the fuel supply continually reacts to a varying rpm and consequently a varying power output.

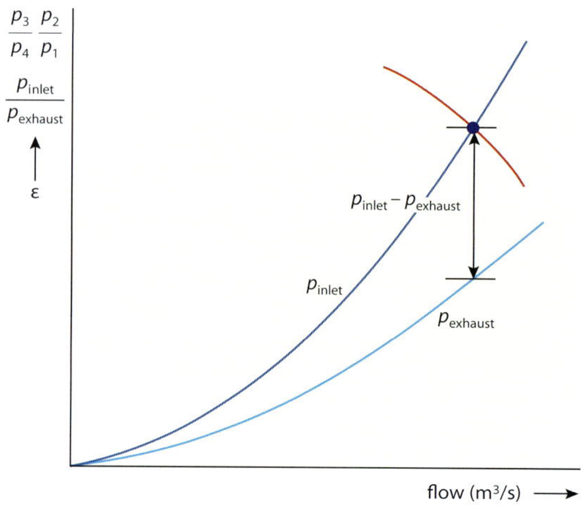

▲ The supply and absorption characteristics of a two-stroke crosshead diesel engine.

Horizontal: turbo-compressor capacity in m³/sec.
Vertical: pressure ratio of the pressure in the scavenging-air pipe (p_i) and the pressure in the exhaust-gas pipe (p_u).
Red: rpm

Here the pressure difference determines the air-flow volume.
Small pressure difference: little scavenging air.
Large pressure difference: substantial amount of scavenging air.

Conclusion: a turbocharger running at a varying rpm is less susceptible to unstable operation than a turbocharger for a four-stroke engine.

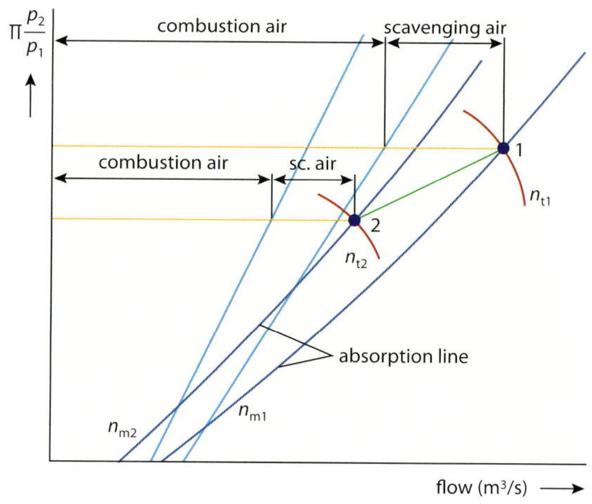

▲ Shifting of the operating point of a turbo-compressor at a reduced rpm in a four-stroke engine.

Horizontal: turbo-compressor capacity in m³/sec.
Vertical: compression pressure of the compressor.
Red: rpm.

Horizontally in the figure:
The arrows show the combustion- and scavenging-air flow at both speeds.
At the top at a high rpm, n_{t1} and at a lower rpm, n_{t2}.
Scavenging air supply is shown between the dark blue absorption lines and the light blue absorption characteristic lines.

Line 1: propeller curve.

Line 2: natural aspiration.

Line 3: one-step turbocharger, equal-pressure system.

p_{mep} = 20 bar.

Line 4: one-step turbocharger, impact system.

p_{mep} = 20 bar.

Line 5: modern turbocharger with extremely high flow efficiency.

p_{mep} = 30 bar.

Line 6: Multiple-step turbocharger.

The graph clearly shows that by applying six divergent methods for supplying the cylinder with fresh scavenging air, maximum torque is achieved at varying speeds.

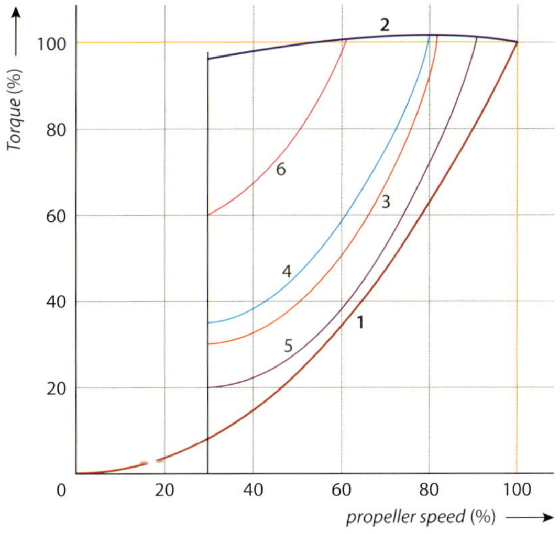

Example of a graph showing different methods of air supply varying from natural aspiration to a heavily loaded engine with a turbo-compressor with a mean effective pressure of 30 bars.

Example:

- When using method 6, the torque is 100% at a rpm of approximately 60%.
- When using method 5, the torque is 100% at a rpm of approximately 87.5%.
- When using natural aspiration, the torque is virtually constant for each load.

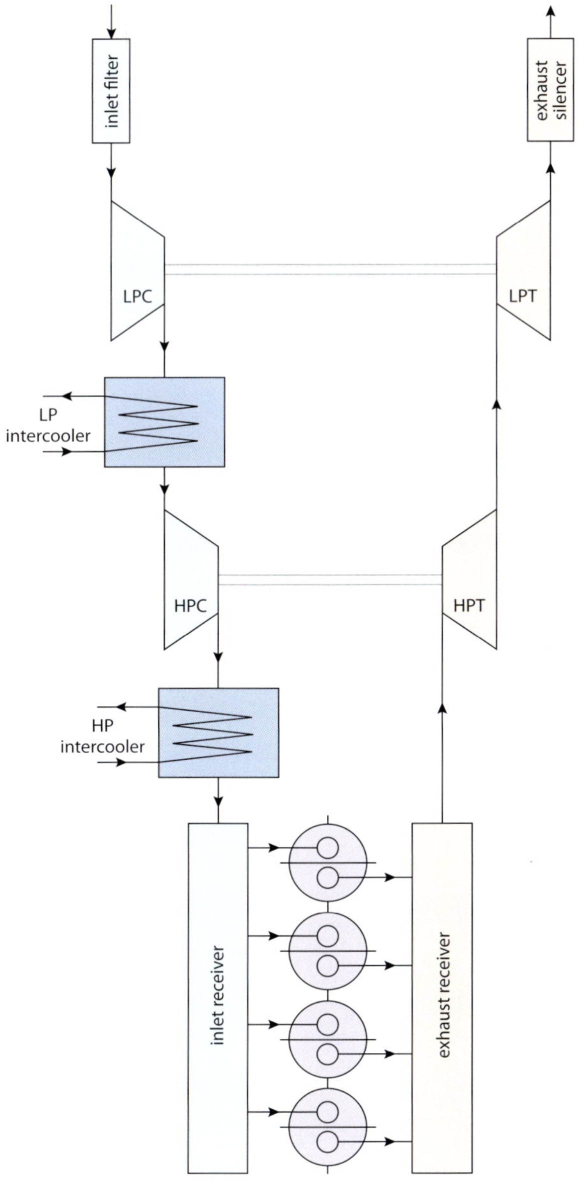

12.18.4 Examples of air-supply and exhaust-gas systems in four-stroke engines.

In recent years improved systems supplying the correct amount of air to the cylinders at each load and rpm have become available. By-pass valves and waste-gates are generally applied and the turbo-compressor is increasingly equipped with adjustable deflector plates. All these facilities have effected an optimum combustion process in all operation conditions resulting in low emission values in the flue gases and a cleaner engine.

Four-stroke in-line diesel engine equipped with two serial turbochargers, both fitted with an air cooler.

This arrangement was often used in high-speed diesel engines. Experiments are performed to meet the requirements for increasing compression ratios for turbo-compressors in order to achieve mean effective pressures of 30 bars and upwards, for medium-speed four-stroke engines. By placing two modern turbo-compressors together, compression ratios of 6 and upwards can be achieved. The air is cooled after each compressor.

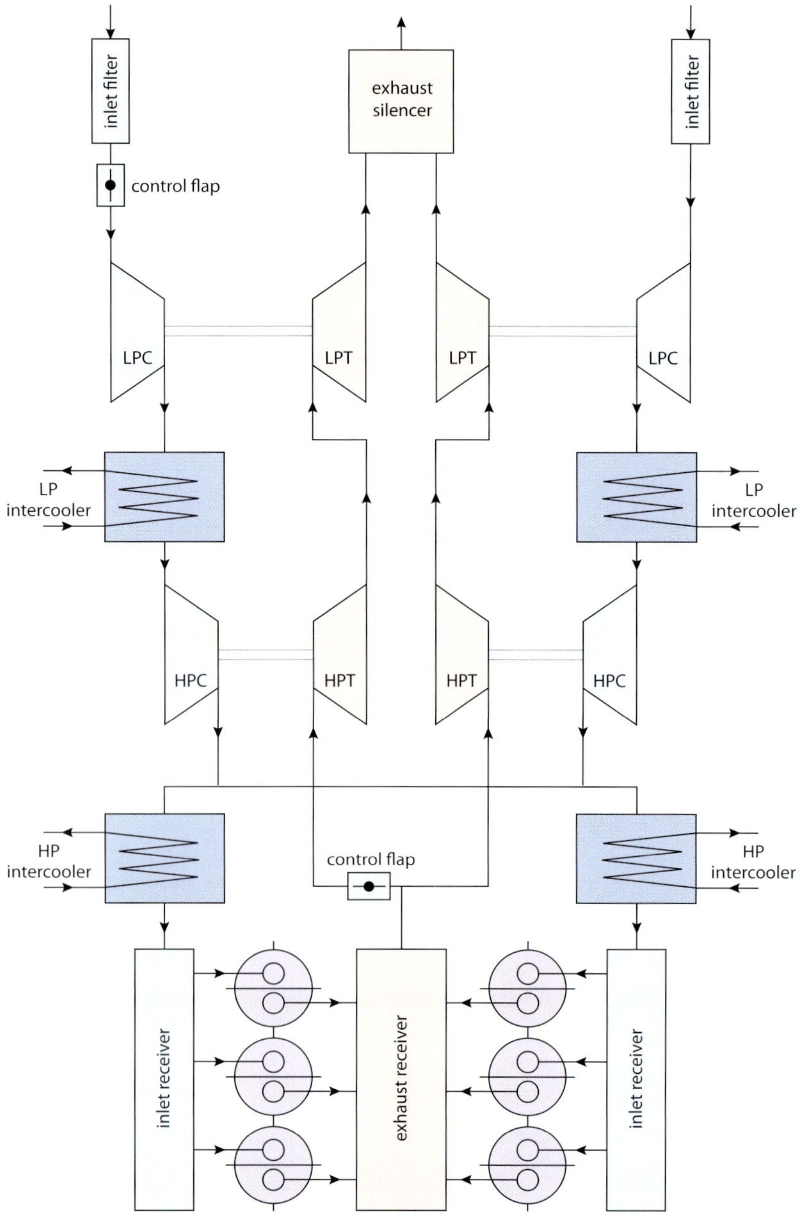

▲
Arrangement of a four-stroke V-diesel engine provided with a double set of serial turbochargers. Each turbocharger is fitted with an air cooler; therefore, each bank has two serial turbochargers.

A special control flap provides both gas turbines in series with equal amounts of exhaust gases.

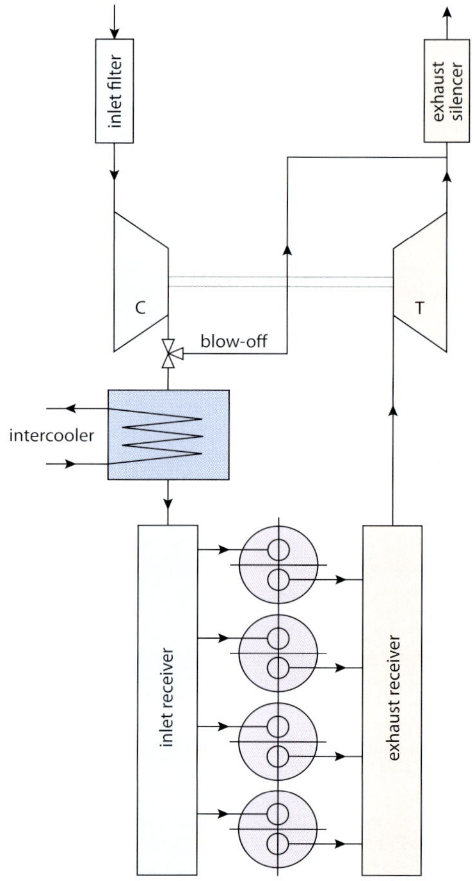

▲
Arrangement of a four-stroke in-line diesel engine with one turbocharger fitted with an air cooler and a by-pass valve in the air section.

Arrangement with a waste-gate on the compression side of the turbocharger. Here two things may happen. At full load the exhaust-gas flow is excessive, which means that the turbo-compressor supplies too much air, which is subsequently vented. Alternatively, at low loads, the air flow is too high and cools the combustion process, which results in poor combustion. This system is commonly used in the larger medium-speed gas engines. Excessive air effect gives a poor mixture - little fuel in relation to the amount of oxygen, which makes ignition with a spark plug problematic.

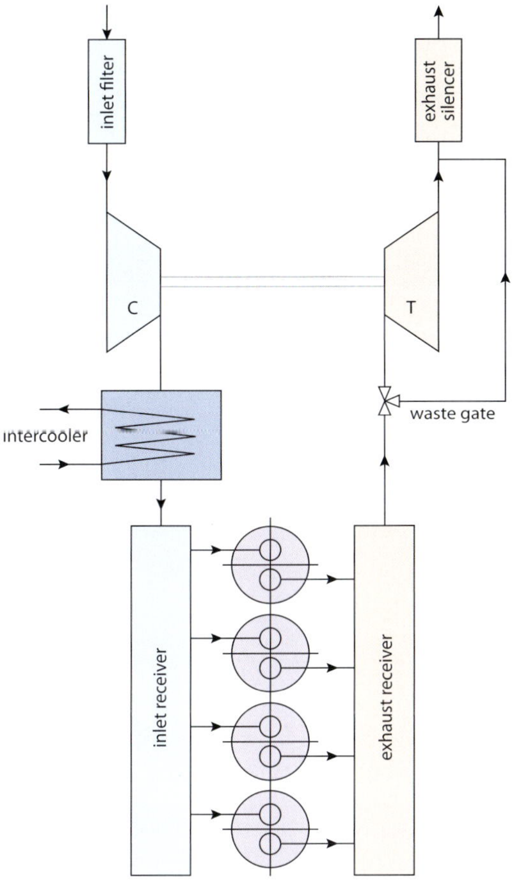

▲

Arrangement of a four-stroke in-line diesel engine with a single turbocharger with an air cooler.

For the proper control of the amount of air, part of the exhaust-gas flow can be diverted through a bypass valve, called a 'waste gate'. Therefore the supplied power to the turbine is reduced and consequently the quantity of air.

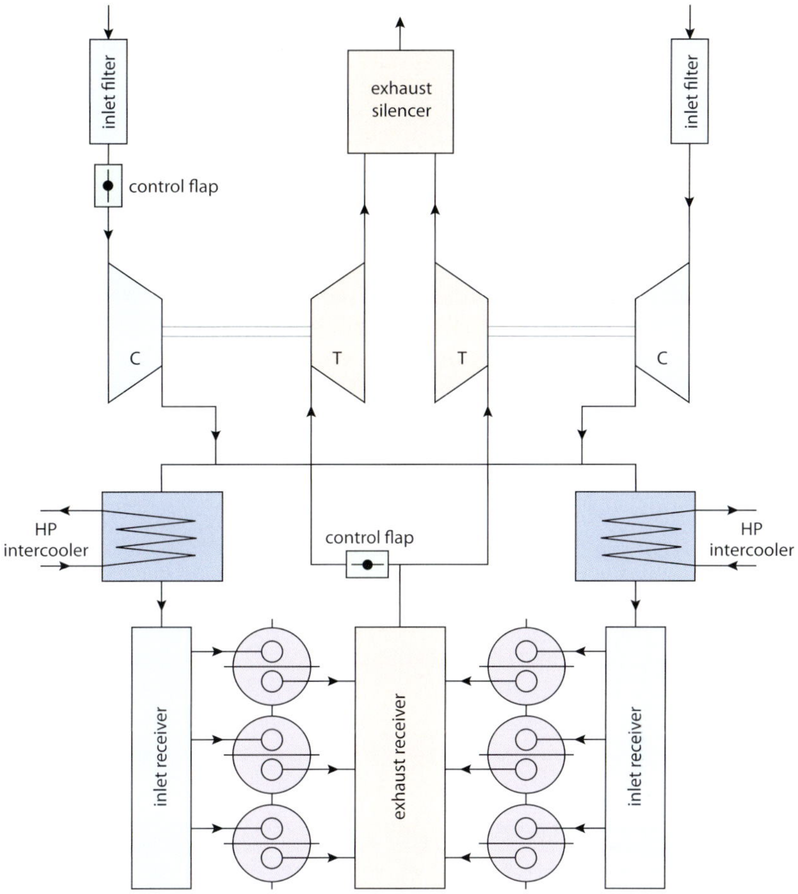

▲

Arrangement of a four-stroke V-diesel engine with a single set of turbochargers.

Apart from the special control flap to effect an equal exhaust-gas flow for the four gas turbines, a control valve has been installed after the air inlet filter for the left air duct. This method allows the air flow to both banks to be precisely controlled on the air side as well.

◄

Adjustable blades of a turbocharger by ABB, type TC.

Top picture: guide blades fully opened; maximum exhaust-gas flow and therefore maximum air efficiency.
Bottom picture: guide blades virtually shut; minimum exhaust-gas flow and therefore minimum air efficiency.

>CH 13

Driving gears

The piston and connecting rod assembly of a two-stroke crosshead engine. Piston, piston rod, crosshead, connecting rod and crankshaft.

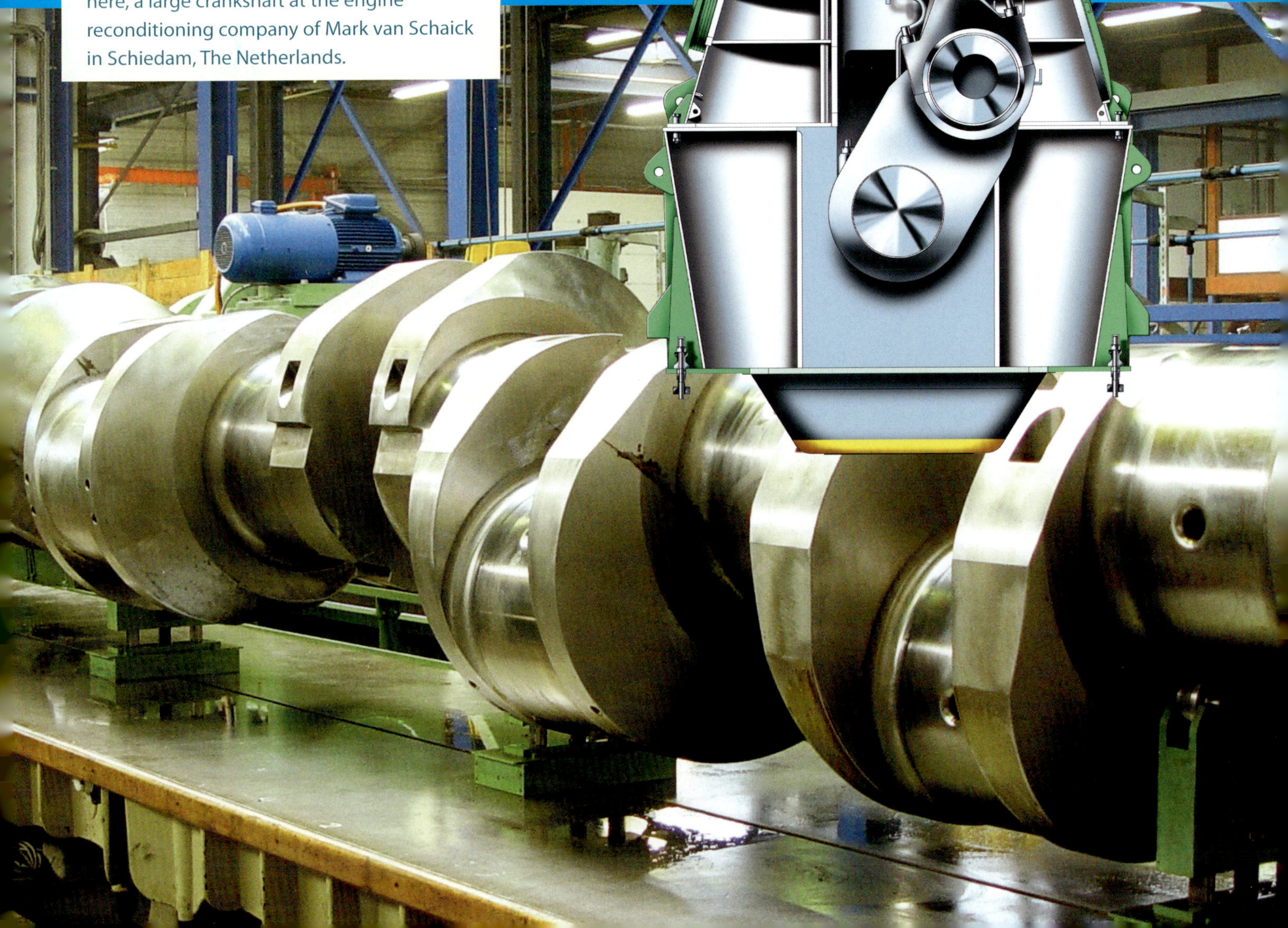

The purpose of the engine gearing is to convert the reciprocating up- and downward movement of the piston into a rotary movement of the shaft. This is done by the crankshaft, making it the most important component of the driving gear. Shown here, a large crankshaft at the engine reconditioning company of Mark van Schaick in Schiedam, The Netherlands.

13.1 Introduction

The engine gearing is subjected to huge forces. The pressures generated in the combustion space are transferred by the piston to the piston pin, the connecting rod and the crankshaft. In today's heavily charged diesel engines, the pressure exerted on the piston during the power stroke varies from 500 kilograms for a category I engine to 1100 tons for a category IV engine. Furthermore, these forces vary continuously and swiftly. Therefore, engine designers pay special attention to the construction of the gearing in order to control these enormous forces.

13.2 Driving gear of four-stroke diesel engines

The engine running gear consists of the following parts:

- piston, including the piston rings;
- piston pin;
- connecting rod and bearings;
- crankshaft and bearings;
- camshaft drive and pumps;
- camshaft;
- valve drive.

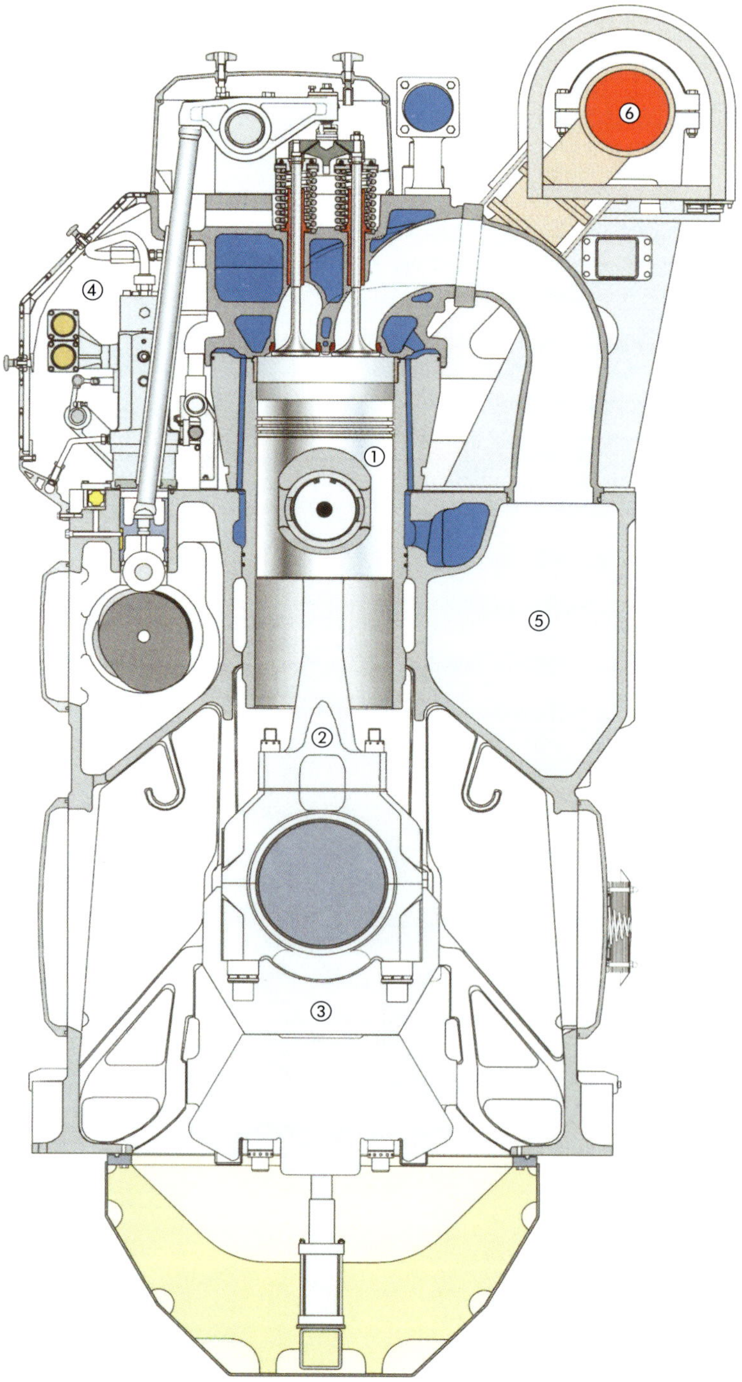

▶

A modern four-stroke medium-speed diesel engine (Wärtsilä 46) suitable for HFO.

Note:

1 *two-piece piston*
2 *big end type 'marine head'*
3 *under slung crankshaft*
4 *hot box for the fuel section*
5 *air-inlet manifold integrated in the engine block*
6 *uncooled exhaust with protective casing*

The piston skirt with its large surface area is suitable for the absorption and transmission of the lateral forces to the cylinder liner and the engine block.
During both the compression- and power strokes, the forces are considerable and constantly changing direction. During the compression stroke they are, as seen in a cross-section, directed to the left and during the combustion stroke to the right with a right-turning crankshaft.

13.2.1 Pistons

Pistons in four-stroke diesel engines not only transmit the gas forces to the crankshaft via the connecting rod, but also absorb the lateral forces that are produced by the crank-connecting rod mechanism.

Construction of the pistons

This is different for each engine category.
As the revolutions decrease and the cylinder bore increases, the pistons are increasingly manufactured in two parts, namely the steel piston crown that is resistant to the huge forces exerted on it with material temperatures at 300 to 400 °C, and the piston skirt for absorbing the lateral forces. The steel crown is provided with piston rings, often an oil scraper, and a distribution ring.
The light-metal- or cast-iron skirt absorbs the lateral forces. The skirts often have special finishes comprising a special coating that enhances the running properties.
In small engines, the pistons are usually manufactured from a single piece of high-grade light metal, mainly aluminium.

Accelerating- and decelerating forces

In order to restrict the accelerating- and decelerating forces, the pistons as generally kept as light as possible.

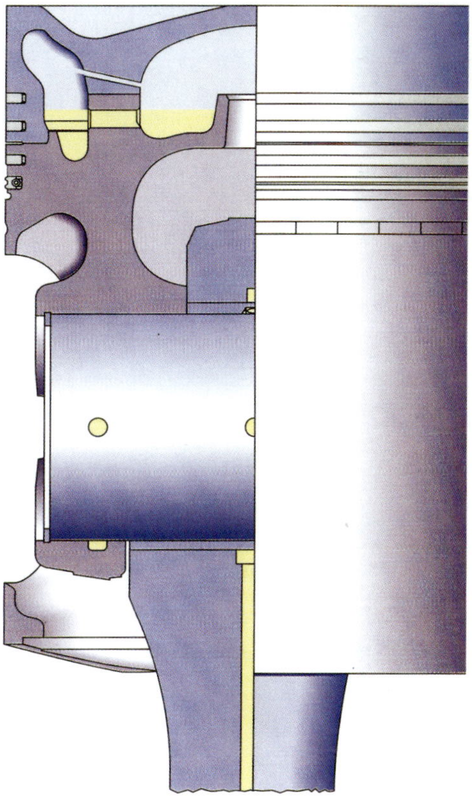

▲
A modern two-piece piston.

- The steel piston crown for transfer of the combustion forces via the gudgeon pin to the connecting rod. Heavy-loaded top rings are found in the piston crown.
  The material, cast steel or forged steel, has heat-resistant properties to maintain strength with the high piston-crown temperatures which rise to 350°C to 400 °C.
- The huge forces on the piston crown necessitate a stiff gudgeon pin to prevent bending of the pin and subsequent problems with the gudgeon-pin bush. The diameter of the pin is large in comparison to its length. This means that the perforation in the piston is also large! In fact, this weakens the piston skirt!

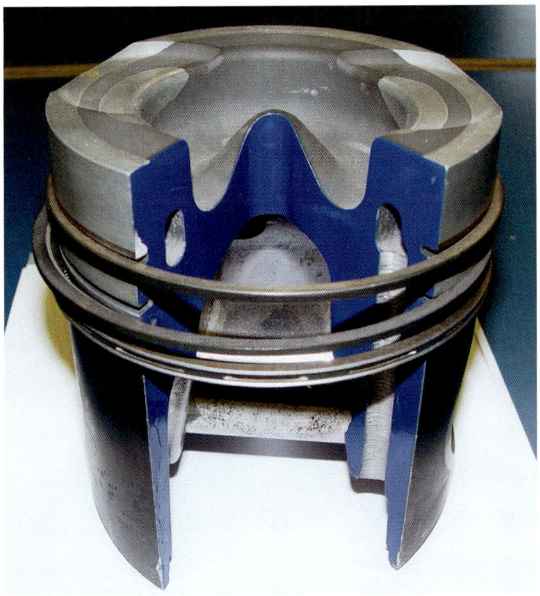

◄
A cut-away view of a piston from a Scania diesel engine.

Note the special shape of the piston crown, designed for optimal combustion. The piston consists of two materials clearly visible by the colour difference in the piston crown. The piston only has three rings: two compression rings and one scraper ring for oil distribution. The piston is cooled with lubricating oil from the gudgeon-pin via cooling ducts.

▲

A piston for a small engine with two compression rings and one oil-scraper ring.

A large section of the side where the gudgeon pin is normally positioned is missing. Only the section where the piston rings are placed is round and seals the combustion space. The skirt from the left to the right in the picture absorb the lateral forces.

▲

A set of spare parts for a modern highly charged four-stroke high-speed diesel engine (Cummins).

1 *camshaft section*
2 *connecting rod*
3 *cogwheel*
4 *valves*
5 *gudgeon-pin bushes*
6 *cylinder liner*
7 *piston, two-piece*

▼

An aluminium piston for engine categories I or II.

◄

New pistons with a steel crown and a cast-iron skirt in Caterpillar MaK 25 engines.

Note the sunken piston crown for optimal combustion space.

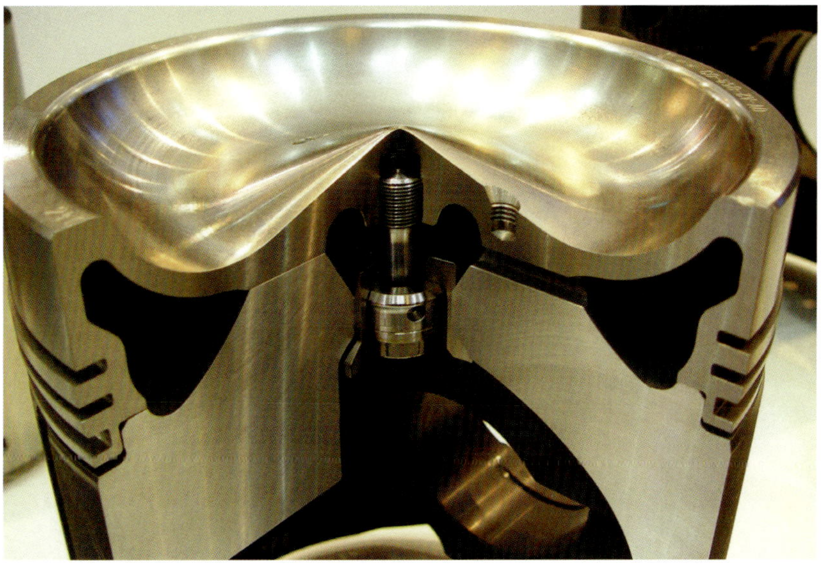

A 'cut-away' section of a two-piece piston in a four-stroke medium-speed diesel engine. In the centre; the lubricating-oil spray nozzle directed at the piston crown for cooling.

The piston rings are located in the cast-steel piston crown. Note the shape of the piston crown; the atomised fuel is injected towards the sides.

Piston rings

The piston rings serve to seal the combustion space and guide the piston in the cylinder liner. The number of rings varies from two to five, depending on the cylinder bore, the rpm and the forces exerted on the piston crown.

A large piston for a four-stroke medium-speed MAN diesel engine.

Note the amount of cast iron removed around and underneath the gudgeon pin, to reduce the weight of the piston.
At the bottom of the skirt the studs of the top section of the connecting rod and the eye, still attached to the gudgeon pin, are only just visible.

1 *piston crown*
2 *ring pack*
3 *piston skirt*
4 *gudgeon pin including seal*
5 *removed material*
6 *studs for the connecting-rod eye*
7 *cylinder-head studs*
8 *anti-polishing ring*

◀

A special aluminium piston for a Cummins NTA 855 four-stroke high-speed diesel engine.

Due to the temperature-expansion coefficient of the light-metal piston, the diameter of the piston-crown section is slightly decreased (extra clearance with the cylinder liner). The notches in the piston crown provide space for the valves, which are not entirely closed when the piston is in TDC.

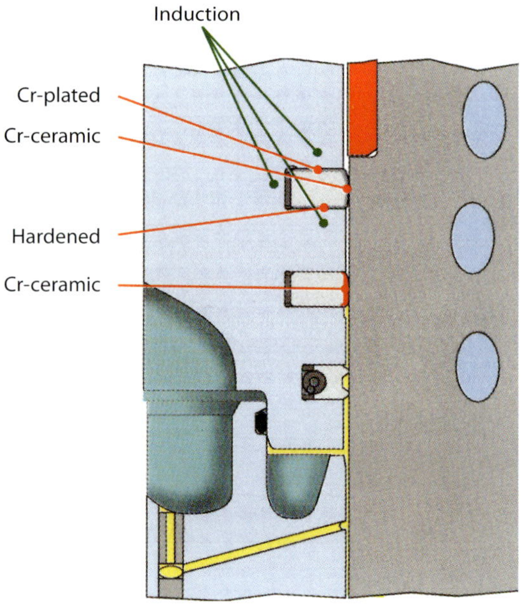

▲

Piston rings, in conjunction with an adequate lubricating-oil film between the cylinder liner and the piston/piston rings, provide a seal for the increasingly high final compression- and final combustion pressures.

Here the following aspects are important:

- heat absorption capacity of the lubricating oil, also expressed in μm (microns);
- adequate clearance between the piston and cylinder liner;
- correct spring and its specific properties such as tension on the liner, shape and size of the running surface area;
- the ring-groove dimensions.

In the drawing, the top ring has a thin chromium coating and the ring is hardened. The running surface is chromium plated and finished with a wear-resistant ceramic coating. Induction hardening has been applied to the circumference of the piston-ring groove. The cylinder liner has an anti-polishing ring and bore cooling. The piston lubricating-oil is pushed towards the oil-scraper rings under pressure.

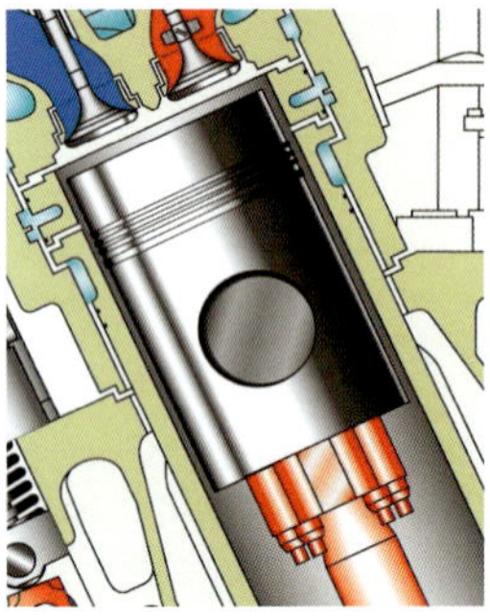

◀

A modern piston for a four-stroke medium-speed diesel engine from MAN Diesel, type 48/60.

In view of the high compression- and combustion pressures, the liner, with bore cooling, is very thick and placed partially above the block. In this approach, the engine block is smaller and lighter! The studs for the connecting-rod eye are visible at the bottom of the cylinder liner. The exhaust valves have flaps for valve rotation.

◀

Much attention is given to the piston-ring grooves.

A steel piston crown has better resistance to the 'beating' of the hard cast-iron rings than a light-metal one-piece piston. The materials used for the grooves are either hardened or chromium-plated to increase wear resistance.

▶

Piston rings are available in many variations.

Generally, each engine type and make has its own specific ring package. For high wear and corrosion as well as other piston-ring problems, modifications are made to improve ring-package properties.

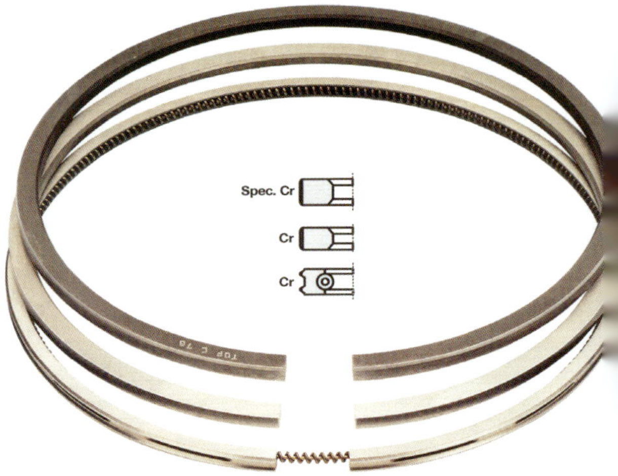

Piston pin

The piston pen transfers the forces exerted on the piston crown to the connecting rod. The connecting rod 'swings' around the piston pin so **impeding** hydrodynamic lubrication. The bearing surfaces, therefore the piston pin and the connecting-rod bearing must be large enough to ensure a low surface pressure. The diameter of the piston pin increases as the pressures exerted on the piston increase; this also applies to the perforations in the piston skirt. The piston pin must be rigid to avoid problems with the piston skirt and piston-pin bearing.

In order to lubricate sufficiently the connecting-rod bearings, lubricating oil is supplied under pressure via perforations in the connecting rod.
The piston pins are manufactured from a steel alloy and are often hardened or provided with a thin chromium coating.

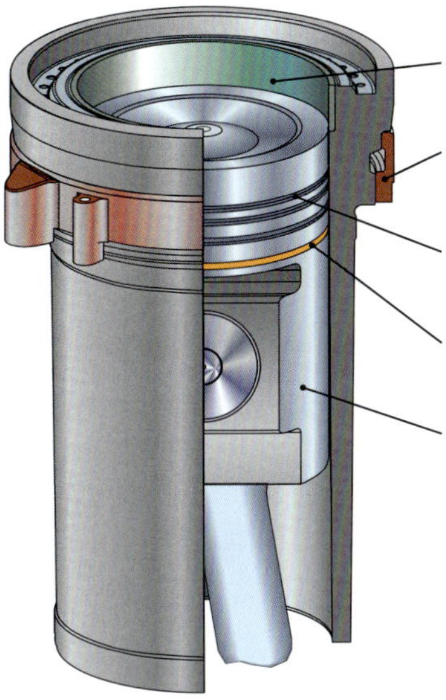

▲
An 'optimum package' of components in the combustion space. Shown, the Wärtsilä 32 four-stroke medium-speed diesel engine.

▼
Small engines often have lower loads (lower load numbers) and therefore have a less complicated construction.

Shown, a DAF-diesel engine with spray cooling for piston lubrication and -cooling.

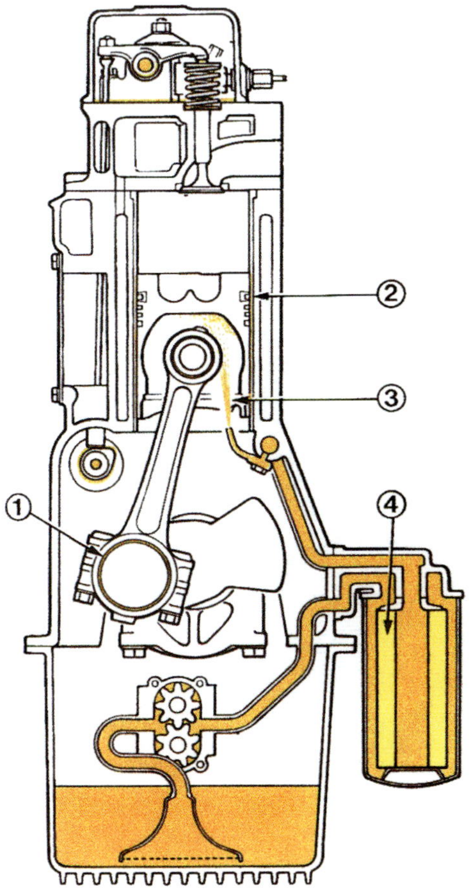

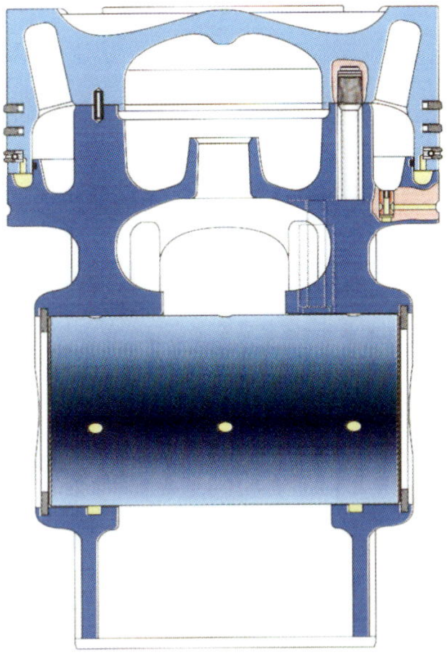

◀
The gudgeon pin placed in the cast-iron skirt of the piston.

The diameter of the gudgeon pin is such that bending of the pin is prevented. Note the lubrication of the connecting-rod eye and gudgeon-pin bushing. The lubricating-oil supply to the piston rings is not visible, however, the lubricating oil used for cooling the piston crown is.

◀
Installing the gudgeon pin in the world's largest four-stroke diesel engine, a Wärtsilä 64 medium-speed diesel engine with a cylinder output of over 2000 kW.

The pin is too heavy to be fitted manually.

13.2.2 Connecting rods

In modern heavily loaded diesel engines, connecting rods are manufactured from forged steel.
Only the smallest engines in category I still have connecting rods manufactured from special cast iron. Connecting rods are often manufactured as a single-drop forging in a mould from a white-hot piece of alloyed steel. It is later sawn through at the big end or at the flange connection of the connecting rod.

▼

Roughly forged steel connecting rods for Caterpillar MaK engines.

During machining the connecting-rod eye is severed from big end.

▲

Roughly forged steel connecting rods for Caterpillar MaK engines.

During the machining process, the big end is obliquely severed. The grain flow in the material, created during the forging process, produces strong components. This occurs to a lesser extent when the pieces are forged separately.

- - - - - = saw cut

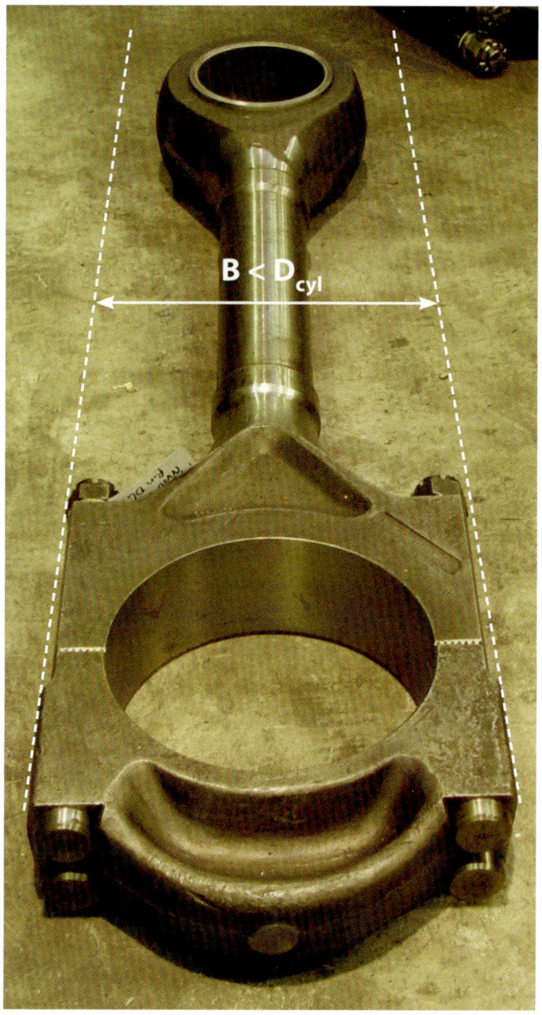

▲

A traditional connecting rod of an older engine type.

Partioning of the big end is horizontal to the rod. The cylinder-liner diameter is larger than the width of the big end; this means that the load parameter is low.

▶

Overhauled connecting rods with an angled big end, a common construction.

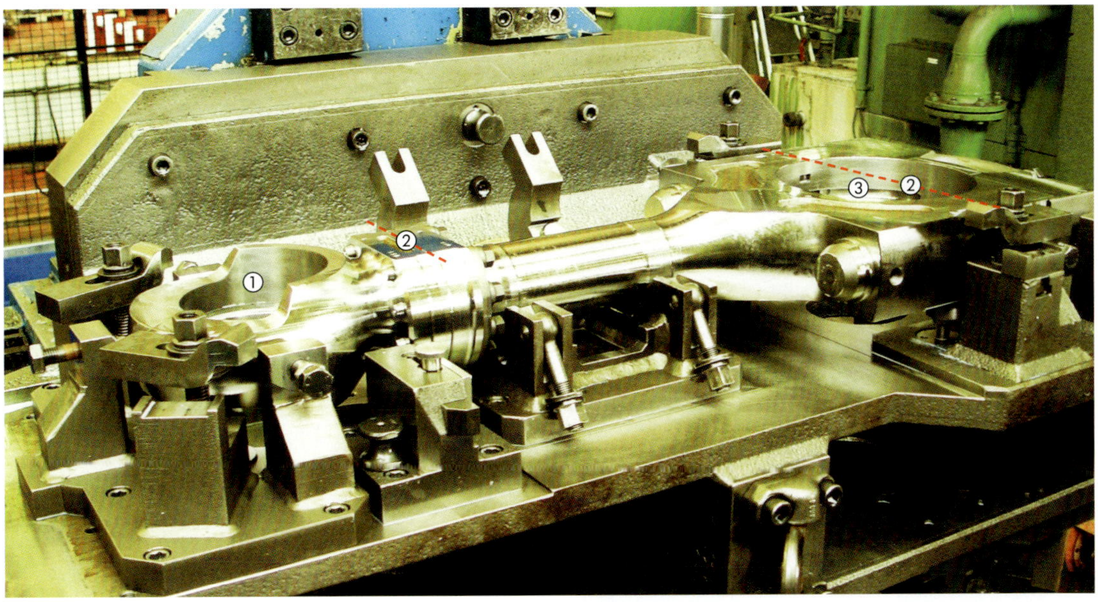

◄

The big end with a partition below the connecting-rod eye for a Caterpillar MaK diesel engine is clamped down for machining.

1 *position gudgeon pin*
2 *partition*
3 *position crankpin*

With an increase in cylinder capacity at an identical cylinder bore, an increase of the crankpin diameter is required to minimise the higher surface pressure on the crankpin bearings. Consequently, the dimensions of the crankpin bearings and hence the big end will increase. A horizontal split in the big end is no longer sufficient, as the width of the big end is too large for drawing the piston. An oblique division also proved to be problematic. The big end was therefore split into three sections (SWD 410) and today the marine version has a separate, small big end mounted on a separate upper- or lower cap or a division of the connecting rod close to the piston. Owing to this construction, dismantling remains possible.

◄

Connecting rod for a Caterpillar diesel engine. See the detail.

▼

A robust angled partition of the connecting-rod big end. An outdated model.

▼

Detail of a connecting rod for a Caterpillar diesel engine.

Big end and cover are fitted together at an angle (circle).

▶

'Connecting-rod eyes' for installing the gudgeon pin.

Note the large lubricating-oil grooves and the size of the bottom of the gudgeon-pin bushing; these are subjected to combustion forces of approximately 250,000 kg or 250 tons!

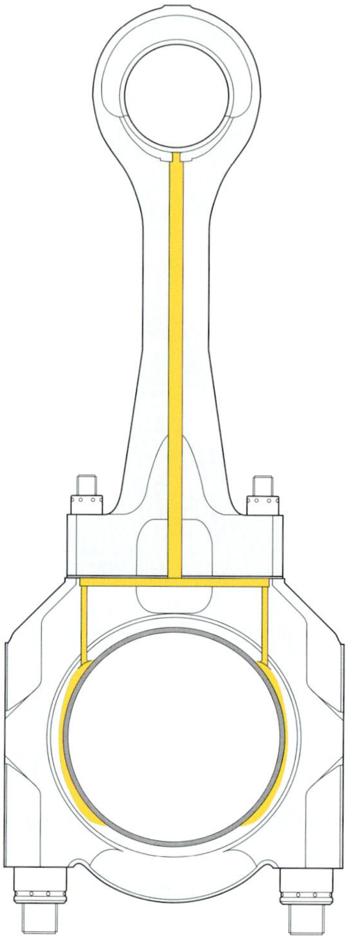

▶

The lubricating-oil channels in a modern connecting rod.

This connecting-rod version is known as a 'marine head'. The big end has been narrowed to such an extent that the piston, including the connecting rod, can be hoisted out of the engine. The lubricating oil is supplied through ducts in the crankshaft and flows to the gudgeon pin via the crankpin, the grooves in the bearing shell, both ducts in the upper crankpin cover and the duct in the connecting rod. Subsequently, the lubricating oil cools the piston and lubricates the piston rings.

13.2.3 Crankshaft

The crankshaft transforms the reciprocating up- and down movement of the piston to a rotary movement. Small category I crankshafts are manufactured from a special cast iron. However, all larger crankshafts are manufactured from one-piece forged iron, and are, dependent on the load, alloyed with chromium and nickel. The engines in the high-speed category usually have surface-hardened crankshafts for higher bearing loads.

The crankpins are often hollow and therefore restrict the centrifugal forces of high-speed crankshafts. The webs are relatively thin and so ensure that the centre-to-centre distance between the cylinders is as small as possible. In this way, the engine is shorter and deflections between the main bearings are limited.

◀

Three crankshafts from Detroit diesel engines.

Note the narrow webs and the large crankshaft- and crankpin diameter.
The drive gears for, amongst others, the camshaft is already installed.

▲

A crankshaft for a four-stroke high-speed engine.

The two lubricating-oil supply holes on each crankpin are indicative of a V-engine. The counterweights, which serve to balance the crankshaft, are an extension of the crank webs and not, as in larger engines, bolted to the crank webs. The crankshaft is forged in 'one throw'.

▲

The crankshaft is the most elementary component of the diesel engine driving gears.

Crankshaft damage often entails high costs. Replacing a crankshaft costs:

category I : ± € 1,000 – € 4,000
category II : ± € 50,000 – € 100,000
category III : ± € 200,000 – € 400,000
category IV : the overall damage can amount to millions of euros.

The main problem when a crankshaft is total loss is the removal of the crankshaft and installation of an overhauled or a new crankshaft.

▼

A roughly forged crankshaft.

The steel is heated to approximately 1100 degrees Celcius. In a systematic process, a hydraulic press forges the steel into a succession of moulds to reach its final rough form.

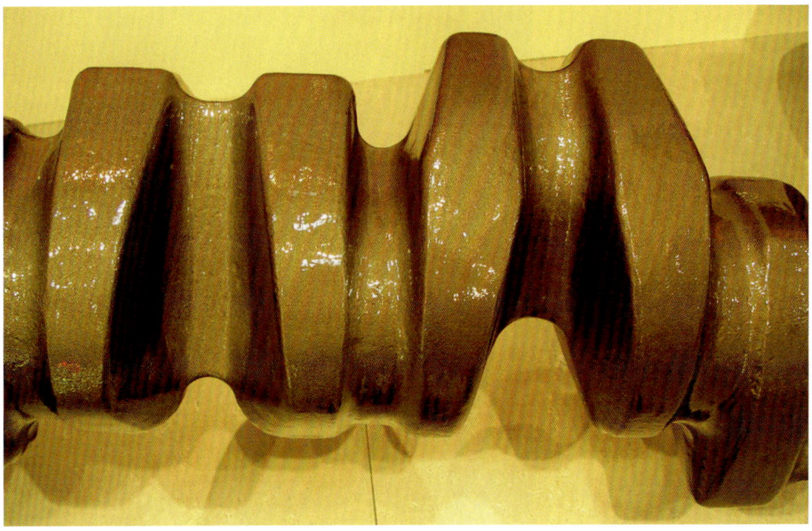

These crankshafts are forged using a crankshaft mould in which the hot steel is pressed under force into the mould. This is followed by a machining process.

▶
An extremely hot rough crankshaft is hoisted from a press mould.

Larger crankshafts are hydraulically pressed into the correct shape throw by throw. Each time, one crank throw and web is heated and formed. Alfing, Aalen, Germany.

▼
The final manual machining of the crankpin for a new crankshaft.

The surface is polished with a honing mould to remove the last surface irregularities.

The crankshafts of category III engines, the medium-speed diesel engines running on HFO are also forged from steel; however, the journals are not hardened. Crankshafts with a cylinder bore of 400mm to approximately 450 mm are forged in one throw over their entire length. Larger crankshafts in this category are generally forged crank by crank from a single piece of metal using a throw-by-throw forging process; this entails partially heating the crankshaft in the area where the crank is to be forged.
This is followed by a machining process and then the mounting of counterweights.

▶
Drilling holes for fixing the studs for the counter-weights.

Reconditioning workshop MAN Diesel, Hamburg, Germany. Shown, a new crankshaft.

13.2.4 Bearings

The bearing caps of the main shaft and the crank pin are the so-called tri-metal bearings or the 'hard bearing caps'. The steel caps are often coated with a thin copper layer, followed by a very thin running-surface layer, often an aluminium alloy.

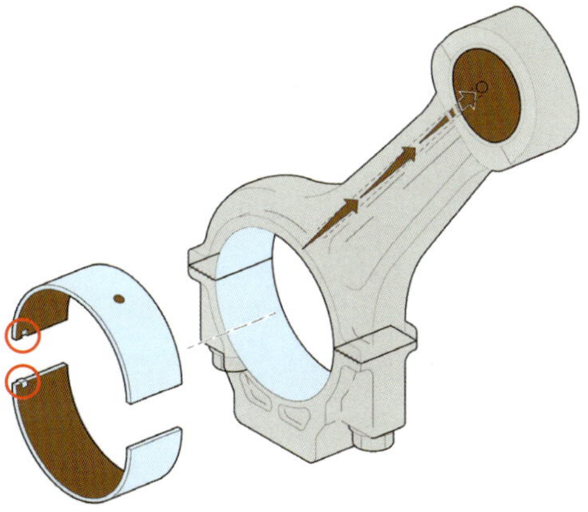

◄

A fitted bearing for the crankpin in an MTU diesel engine.

Note the notches on the left of the bearing shells and the lubricating-oil hole in the upper part of the bearing shell for the flow of lubricating-oil to the connecting-rod eye and gudgeon pin.

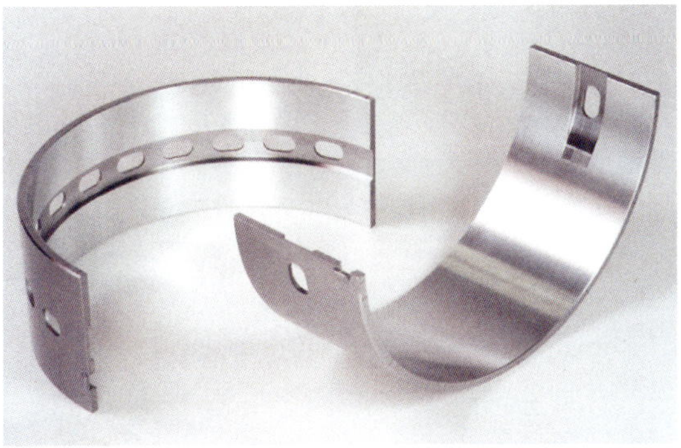

▲

Both main-bearing shells for a Wärtsilä 38 diesel engine.

left: the upper bearing with a lubricating-oil groove with holes for lubricating-oil supply to the shaft.
right: the bottom bearing with a large bearing surface for absorbing the large forces during the compression- and combustion strokes.

▲

Bearing shells for a Caterpillar diesel engine.

Bearing shells are usually replaced after disassembly. As maintenance intervals have become increasingly longer, new bearings are fitted, thus warranting an extended operating time and avoiding problems that could occur by re-using the bearings.

◄

White-metal bearing shells for a two-stroke crosshead engine by MAN Diesel.

These bearings have been re-cast, provided with a new white-metal coating and subsequently machined to the correct specifications and dimensions. Shown, traditional 'thick' bearings for the main bearings that are referred to as the 'thick shell' bearings. Modern bearings are 'thin shell' bearings and used in the crosshead. Due to increasing surface pressures, caused by the increasing gas forces on the piston, MAN Diesel also uses 'thin shell' bearings for the main bearings. The material is a tin-aluminium alloy, Sn40Al.

◀ The main journals and crankpins must be absolutely round and have a very smooth finish.

Naturally, the machining must occur when the crankshaft has been aligned.
Alignment is done with supports, the so-called 'rests'.
Shown: grinding a used crankshaft at Mark van Schaick, Schiedam, The Netherlands.

For the main bearings found in an engine, oversized thicker bearings are available thereby allowing the main journal to be polished and given an oversized bearing after damage repair. Depending on the degree of damage to the journal this can be repeated two or three times.

Also, see Chapter 26, *Overhauling diesel engines and their parts.*

The bearings of the connecting-rod small end and/ or the piston-pin bearings in smaller engines are made of a copper alloy (bronze) and in larger engines, of a hard tri-metal alloy.

◀ The bronze sleeve bearing of a connecting-rod small end.

Note the lubricating-oil groove and the ducts to the gudgeon pin.

◀ Overhauling a connecting rod.

Clearly visible is that one side (bottom part) of the piston-rod bushing of the small end has a larger surface area than the other side (top part). The larger surface area is required to absorb the high combustion forces.

13.2.5 Camshaft drive

In small category I and occasionally category II engines, this is sometimes performed by a chain drive. This is light, inexpensive and does not take up much space. Replacement takes place after a certain number of operating hours, thereby avoiding drive problems such as cracks in the shaft.

In most category II engines and certainly in category III engines, the camshaft is driven using toothed wheels that, as with chain drives, may also drive other engine components such as pumps. The steel-toothed wheels have polished and hardened teeth and run together very precisely. In these four-stroke engines, the number of revolutions of the camshaft is half the rpm of the crankshaft; a full four-stroke process lasts 720 crank degrees.

13.2.6 Camshaft

The camshaft carries the cams, which control the fuel pumps and the inlet- and exhaust valves. In small engines, the steel camshaft is manufactured from a single piece of metal with fixed cams. Here the 'bearing surfaces' are hardened with a surface finish. Larger engines, in particular those in category III, have built-up camshafts in which each cylinder has its own section that is connected to the following section by fitted bolts. This simplifies the removal of a damaged cam.

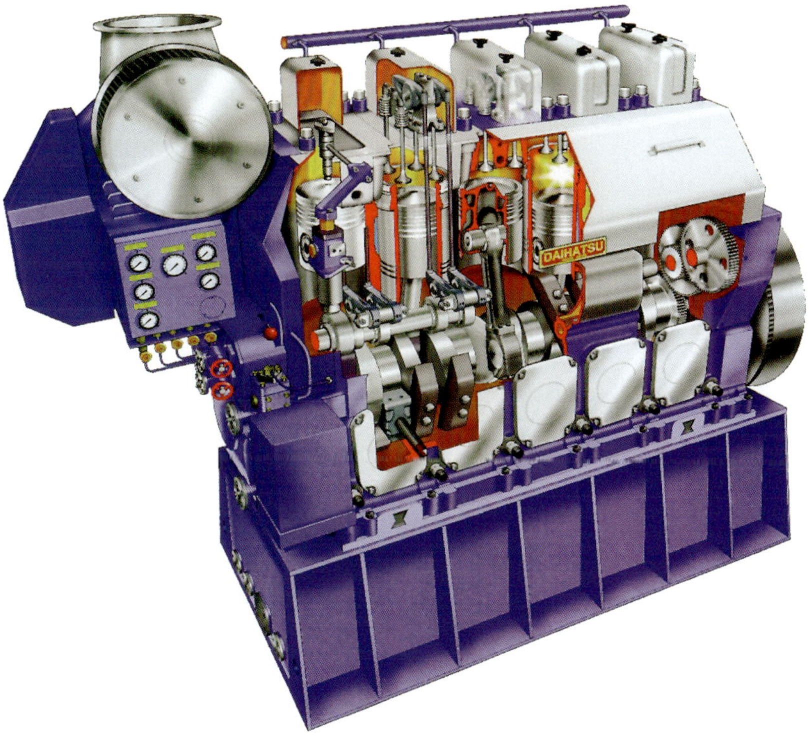

▲

A Daihatsu four-stroke medium-speed diesel engine operating on HFO.

The camshaft is driven by gear wheels connected to the crankshaft; on both shafts one gear wheel and one intermediate shaft are mounted, a standard version.
The four-stroke process has 720 crank degrees; therefore, the camshaft rotates at half the revolution rate of the crankshaft. This is clearly visible when comparing the diameter of the gear on the crankshaft and the diameter of the gear on the camshaft (twice as many teeth on the camshaft wheel).

◄

Camshaft for a MAN Diesel medium-speed diesel engine.

The high-pressure fuel pump and the inlet- and exhaust cams are operated by the camshaft, driven by gears or cog belts (smaller engines in category I) from the crankshaft.

▶

The camshaft drives for a high-speed four-stroke Cummins diesel engine.

The ratio between the cogwheel (gear) on the crankshaft and the camshaft (half the rpm of the crankshaft) is clearly visible.

▲

The drive of both camshafts in a V-engine from RK 280, now MAN Diesel.

1 *crankshaft*
2 *camshaft*
3 *intermediate wheel*

▶

A helical drive.

The objective is to reduce the pressure on the gear teeth. An angled tooth has a larger surface area with an identical cogwheel width, thereby reducing tooth pressure. The angled teeth generate an axial force on the shafts that must be absorbed by a thrust bearing.

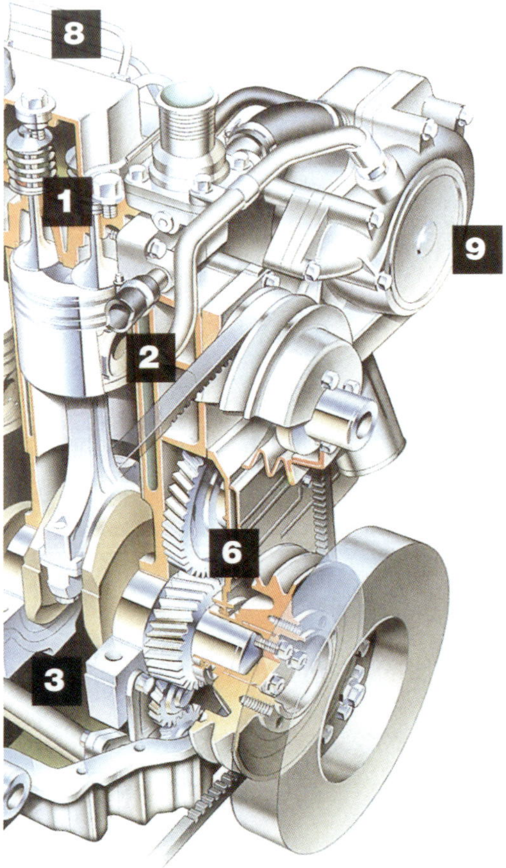

▲

A crankshaft and a camshaft; both forged from a single throw for a diesel engine, category II.

Camshafts with separate cams

In older engines, these cams were bolted to the camshaft. Today most engines have hydraulic cams. They are attached to a node on a finely polished shaft with hydraulic pressure. Therefore, it is possible to adjust slightly the position of the cam and therefore the timing of the exhaust valves or fuel pumps.

▲

Hydraulic cams.

They are mounted on the camshaft using oil pressure. They can be easily replaced when damaged.

◄

This version has three cams: for the inlet- ,exhaust valves and the high-pressure fuel pump.

13.2.7 Inlet- and exhaust-valve drives

These are driven from the camshaft by cams that force the roller upwards, thus moving the valve via the roller guide, the push rod and the lever system. As most of today's engines are equipped with two inlet- and two exhaust valves, the drives of these valves on the cylinder head can be very complex. The valve springs ensure that the valves close as soon as the position of the cam allows this.

►

A cylinder head with mounted valve drives.

1 *cylinder head*
2 *inlet port*
3 *exhaust port*
4 *location for cylinder-head bolts*
5 *inlet valve*
6 *exhaust valve*

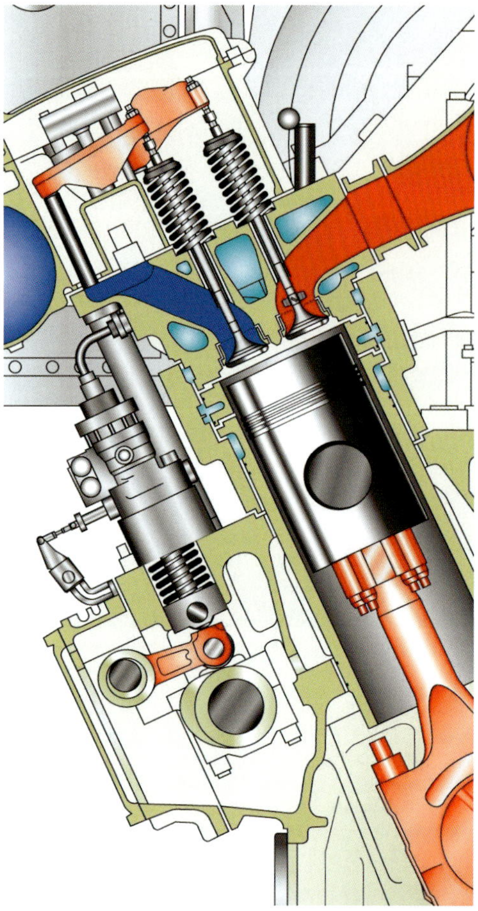

▶
A cross-section of a four-stroke medium-speed diesel engine from MAN Diesel type 51/60 V dual-fuel.

The valves are driven by two rocker arms, left for both inlet valves and right for both exhaust valves.

▲
Top view of a valve drive.

The valves are deflected in relation to the cylinder centre line. This approach allows the screw mechanism in the inlet-air cylinder to work optimally.

1 *rocker*
2 *spreader*
3 *valve spring*
4 *cylinder-head bolts*

▶
The drive for the inlet- and exhaust valves in medium-speed Caterpillar MaK diesel engines.

1 *location for the push rod*
2 *rocker*
3 *rocker shaft*
4 *adjusting bolt for the valve clearance*
5 *spreader*
6 *spreader guide*
7 *valve springs*
8 *cylinder head*
9 *inlet duct*
10 *exhaust duct*

◀

Construction of a Caterpillar MaK engine at Kiel, Germany.

The valve rockers are upright; the push rods are placed above the camshaft.

1 *valve rocker*
2 *push rods*
3 *cylinder head*
4 *high-pressure fuel pump*
5 *fuel- adjusting spindle*

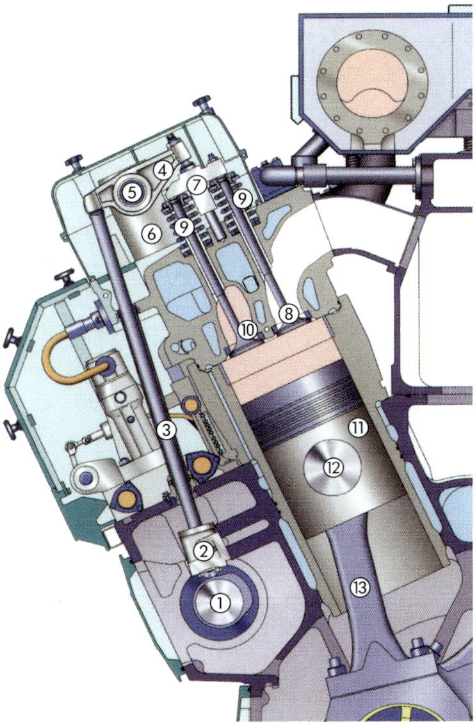

▲

A complete valve drive.

1 *camshaft with inlet cam*
2 *pulley with guide*
3 *push rod*
4 *rocker*
5 *rocker shaft*
6 *rocker-arm seat*
7 *spreader*
8 *inlet valve*
9 *valve spring*
10 *exhaust valve*
11 *piston*
12 *gudgeon pin*
13 *connecting rod*

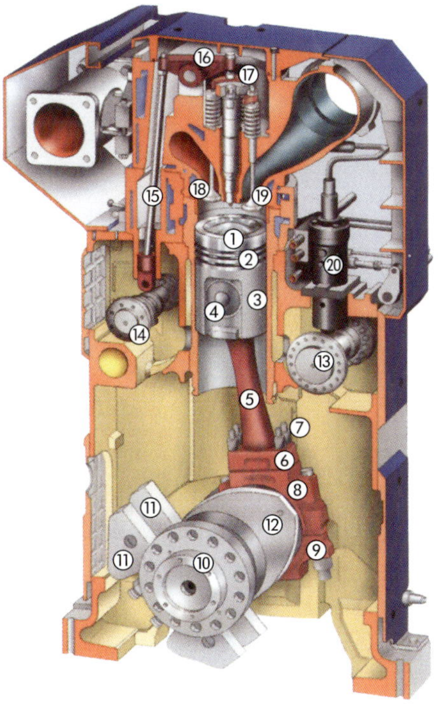

◀

The driving gear for a four-stroke MAN diesel engine.

Note that there are two camshafts in place; one for the valve movement and one for the high-pressure fuel pump.

1 *piston crown*
2 *piston rings*
3 *piston skirt*
4 *gudgeon pin*
5 *connecting rod*
6 *'Marine head big-end'*
7 *connecting-rod stud*
8 *top cover big-end*
9 *bottom cover big-end*
10 *crankshaft*
11 *counterweight*
12 *crank web*
13 *camshaft for high-pressure fuel pump*
14 *camshaft for the inlet- and exhaust valves*
15 *push rod*
16 *valve lever*
17 *spreader*
18 *exhaust valve*
19 *inlet valve*
20 *high-pressure fuel pump*

13.2.8 Fuel-pump drive

In four-stroke trunk piston engines, most fuel pumps are driven in the long-established manner:
- Individual high-pressure fuel pumps for each cylinder with camshaft.
- In diesel engines with a sturdy combined high-pressure fuel pump, the so-called block-fuel pump, driven by cogwheels.

Diesel engines with the latest common-rail system use the following systems:
- In high-speed diesel engines, a high-pressure fuel pump is often installed near the cogwheels.
- In medium-speed diesel engines the high-pressure fuel pump is driven by the camshaft; however, fewer high-pressure pumps are present, normally half the number of cylinders. These pumps operate in tandem on a mutual accumulator that leads the fuel to the cylinders.

▶
A sixteen-cylinder MTU-diesel engine category II with a common-rail fuel system.

The high-pressure fuel pump is driven directly by the cogwheel train.

1 *high-pressure fuel pump*
2 *vibration damper*

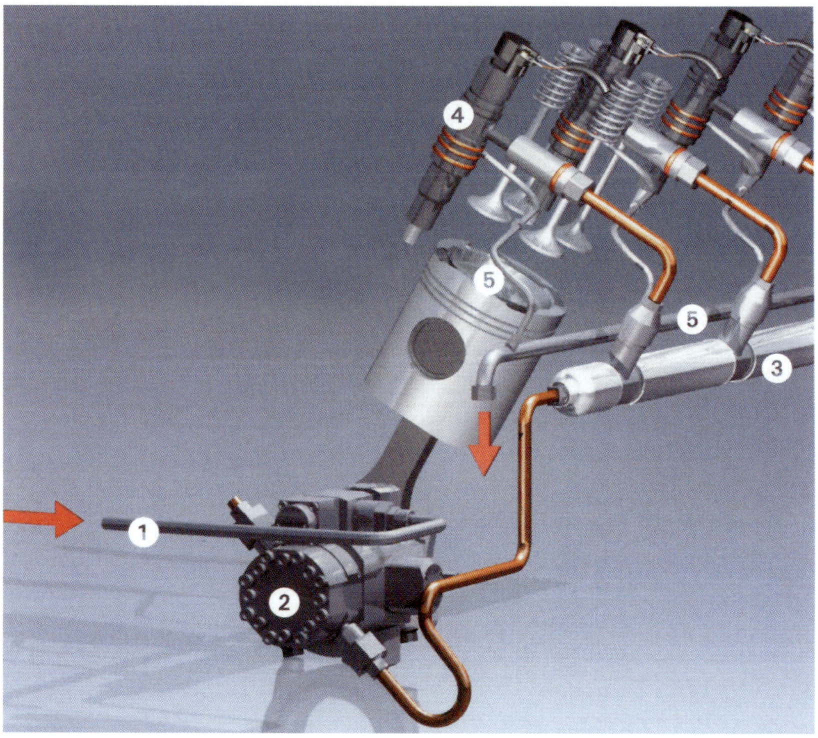

▶
The fuel system of an MTU-diesel engine, series 2000.

1 *fuel-supply line*
2 *fuel pump*
3 *common-rail high-pressure fuel line*
4 *fuel injectors*
5 *fuel-discharge line*

Four-stroke engines equipped with a common-rail fuel system need camshafts for driving the inlet- and exhaust valves. The main reason is that hydraulically driven valves at high engine speeds are complicated and therefore expensive.
The high-pressure fuel pump is driven by a cogwheel train connecting the crankshaft and camshaft.

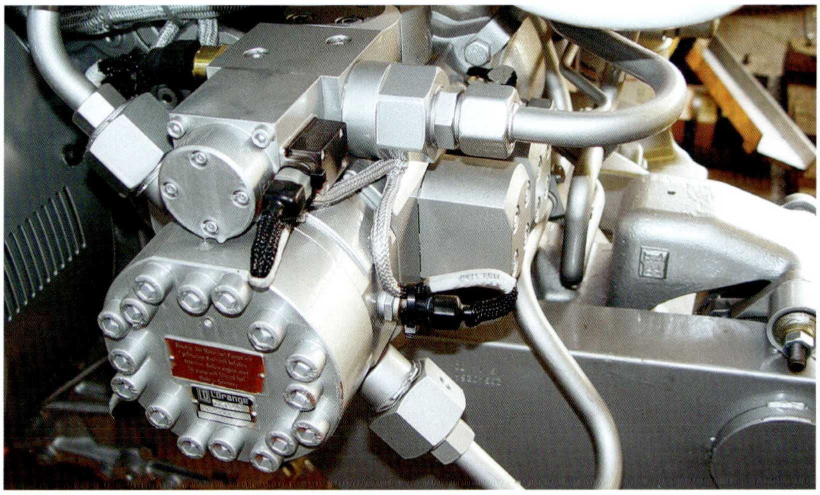

◀
The high-pressure fuel pump from l'Oranje for MTU engines with a common-rail system.

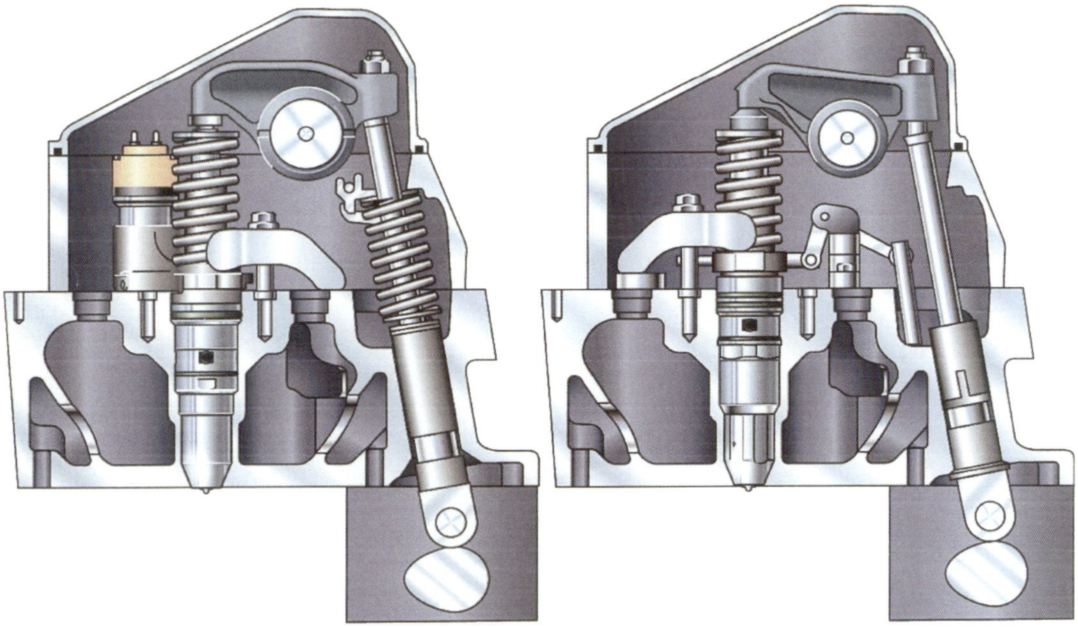

◀
The fuel-injector drive in four-stroke high-speed Caterpillar engines.

Left: electronic control. Right: mechanical control.

◀
The common-rail system from MAN Diesel for medium-speed engines, category III.

1 *accumulator*
2 *high-pressure fuel pump*
3 *inlet-valve control (VIT)*

13.3 Driving gears in two-stroke crosshead engines

Here the driving gear includes the following parts:

- piston with piston rings;
- piston rod;
- crosshead;
- crosshead guides and crosshead shoes;
- connecting rod with the crosshead bearings;
- crankshaft with the crankpin and main shaft bearing;
- camshaft drive;
- exhaust-valve drive;
- conventional fuel-pump drive;
- common-rail system drive for crosshead engines;
- reverse gears.

13.3.1 Pistons

The pistons of two-stroke crosshead engines serve only to transfer the gas forces to the piston rod, the crosshead and via the connecting rod to the crankshaft.

The lateral crank-connecting rod mechanism forces are entirely absorbed by the crosshead and via the crosshead-guides conveyed to the crosshead shoes that are fixed to the welded A-frame. The piston, together with the piston rings, should also secure a gas-proof sealing of the cylinder. When the piston is short, the cylinder diameter is often larger than the height of the piston and therefore the piston-ring package does not contain a separate scraper and distribution ring; the cylinder-lubricating oil is supplied through perforations at the beginning of the compression stroke.

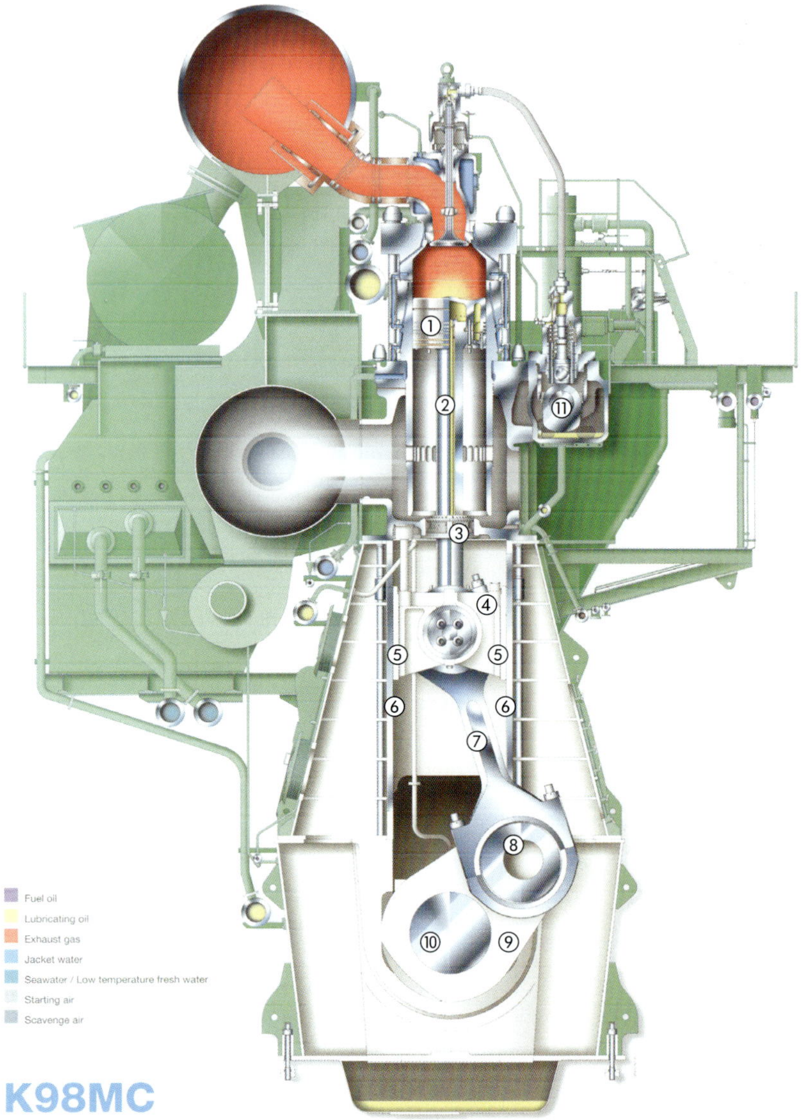

▼

Two-stroke low-speed crosshead engine from MAN Diesel, MC 98.

The reciprocating movement of the piston converts combustion pressure to a rotating motion of the crankshaft via the crank-connecting rod mechanism.

1 *piston*
2 *piston rod*
3 *piston-rod stuffing box*
4 *crosshead*
5 *guide shoe*
6 *crosshead guides*
7 *connecting rod*
8 *crankpin*
9 *crank web*
10 *crankshaft*
11 *high-pressure lubricating-oil pump for exhaust-valve control*

▲

A new piston for a two-stroke crosshead engine from MAN Diesel.

Note the large number of hydraulically tensioned studs used to connect the piston to the piston-rod flange and the bronze guide rings in the piston skirt.

▲

A dismounted piston for a large two-stroke crosshead engine.

Note the 'seized' top ring and the ribbed profile on the piston skirt, which serves as a lubricating-oil chamber.

▼

A cross-section of an earlier piston type and a more recent piston type from MAN Diesel MC.

The differences between the combustion chambers are remarkable. The shape is different and the mean material temperature of the 'piston crown' is approximately 100 °C lower. The piston is cooled with lubricating oil, which is circulated by pumps.

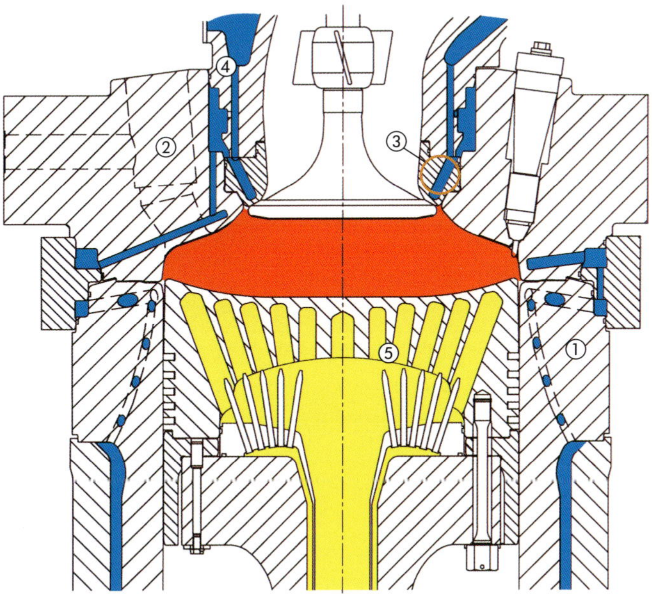

▲

A cross-section of the combustion space in Wärtsilä RTA58T-B and RTA68T-B engines.

The following parts around the combustion space are provided with ducts, also referred to as bore cooling.

1 *cylinder liner*
2 *cylinder head*
3 *exhaust-valve seat*
4 *exhaust-valve casing*
5 *piston crown*

Comment

In comparison to other two-stroke crosshead engines, these engines have very little cooling water! Therefore, draining the cooling water of a cylinder during repairs does not take time!

Features:
- High topland
- Oros piston top geometry
- CPR top ring
- Alu-coat piston rings
- Bore-cooled, forged piston of heat resistant steel
- Piston-cleaning ring

Improvements:
- Approx. 100 °C lower temperature on top compared to former type piston
- Elimination of Inconel coating on piston top
- Increased chrome-layer thickness in bottom of ring grooves
- Anti-erosion bushing in oil outlet in piston-rod foot

Verification:
- Extensive calculations
- Comprehensive tests on K90MC and K90MC-C
- Service test on K90MC

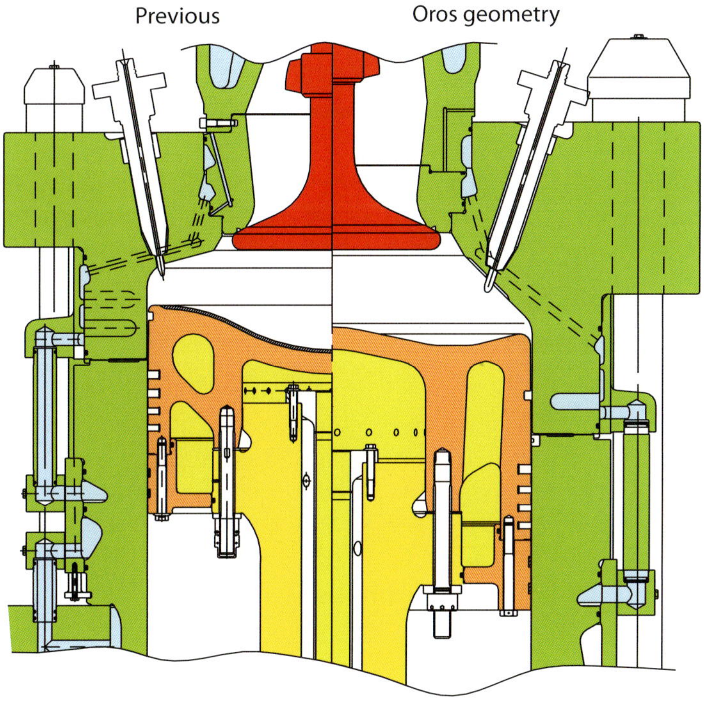

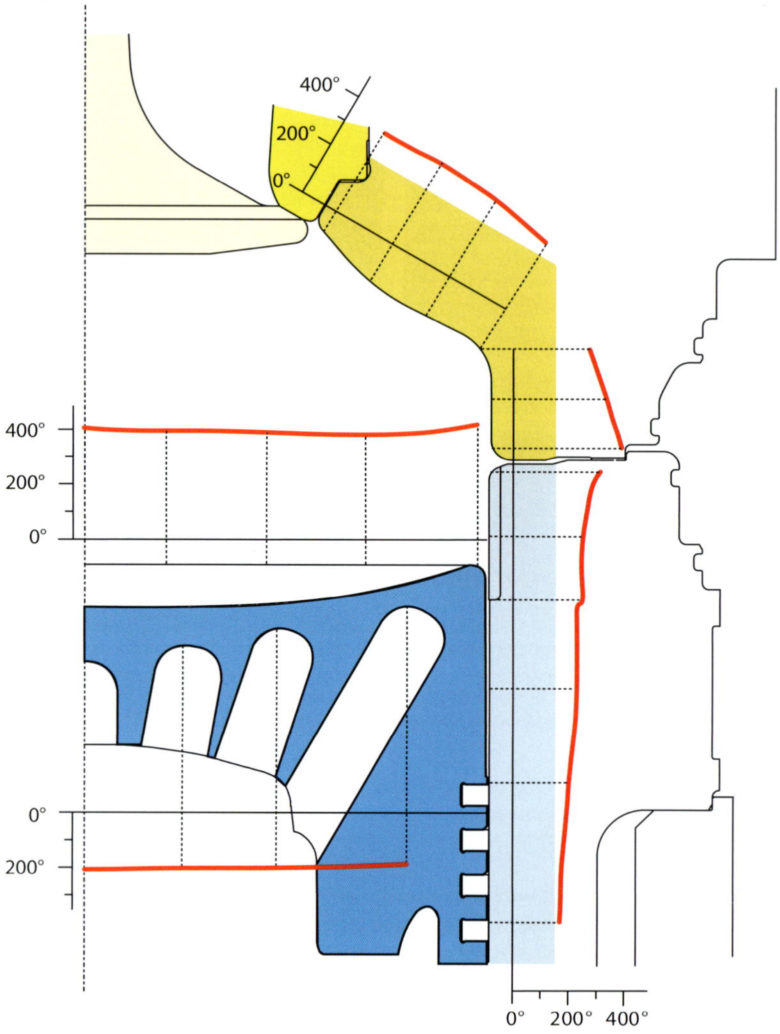

▶

The temperature developments in the various parts of the combustion space of the Wärtsilä RTA58T-B and RTA68T-B engines.

The engine design is continually being adapted to ensure that material temperatures of the parts do not dramatically increase when the cylinder output increases. In fact, temperatures of the parts are slightly lower in new designs.

Temperatures:

piston crown: ± 300 to 375 °C

piston crown, lubricating-oil cooled interior: ± 150 to 170 °C

cylinder liner: ± 200 to 260 °C

exhaust-valve disc: ± 450 to 550 °C (highest in the centre, lowest at the seat; here intensive cooling occurs when the valve is closed.)

exhaust-valve seat: ± 250 to 350 °C

Today, all pistons are cooled with lubricating oil via a separate piston cooling-oil system. In view of the high load, pistons are manufactured from forged steel so allowing repairs to take place using a rewelding process.

13.3.2 Piston rod

This is permanently bolted to both the piston and the crosshead using heavy flanges. The rod is manufactured from forged steel and has a very smooth finish. The piston-rod stuffing box forms the seal between the scavenging-air space and the crankcase.

▶

A piston rod in a lathe.

In this instance, the piston rod is to be cleaned (removal of the coating) and polished. The rectangular flange in the foreground is used to attach the rod to the crosshead.

▲
A piston, including piston rod, and a bare piston rod after repair.

▲
Dissembling a piston and piston rod.

The piston crown is grooved so the hoisting gear can be attached. When the piston is in its top position, it is of sufficient height above the cylinder liner so that the hoisting gear can be attached.

13.3.3 Piston-rod stuffing box

The piston-rod stuffing box separates the space above the sealing rings, the scavenging-air space, from the space below the sealing rings, the crankcase. The scavenging-air space always contains some form of dirt. Residues from the combustion process above the piston can reach this space through the piston-ring package, which is subject to wear and tear.
This, combined with cylinder lubricating oil and condensed water from the inlet air, can form a soiled layer on the bottom of the scavenging-air space. A scraper ring on the **upper part** of the stuffing box scrapes the dirt off the piston rod and transports it outside the engine.
The base plate of the scavenging-air space is provided with a drainpipe with ball valve that transports dirty oil substances to a foul-oil tank outside the engine. The amount of contaminated oil drained from the scavenging-air space is checked at regular intervals using a measuring cup. In this manner, the amount of cylinder lubricating oil burnt in the cylinder is measured. The amount of lubricating oil supplied to the cylinder is known and as the amount of lubricating oil drained from the dirty scavenging air is measured, the amount of combusted lubricating oil in the cylinder combustion space can be found.

There is a relatively clean space below the piston-rod stuffing box; the crankcase with driving gears, the crosshead guides, connecting rod and crankshaft. These are supplied with lubricating oil from the main lubricating-oil system, which is continuously circulating the lubricating oil in the engine. The pistons are cooled using this clean system. The **bottom part** of the piston-rod stuffing box scrapes the relatively clean lubricating oil from the piston rod and subsequently drains it to the crankcase.

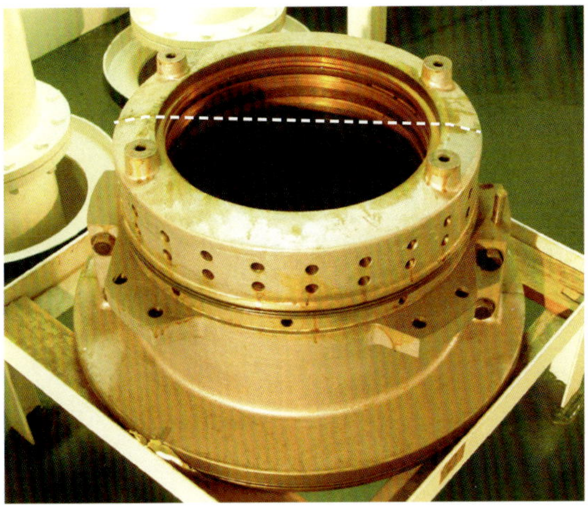

◄
A complete piston-rod stuffing box.

Note the large holes in the upper section for discharging contaminated cylinder lubricating oil. The packing is vertically partioned for assembly and disassembly.

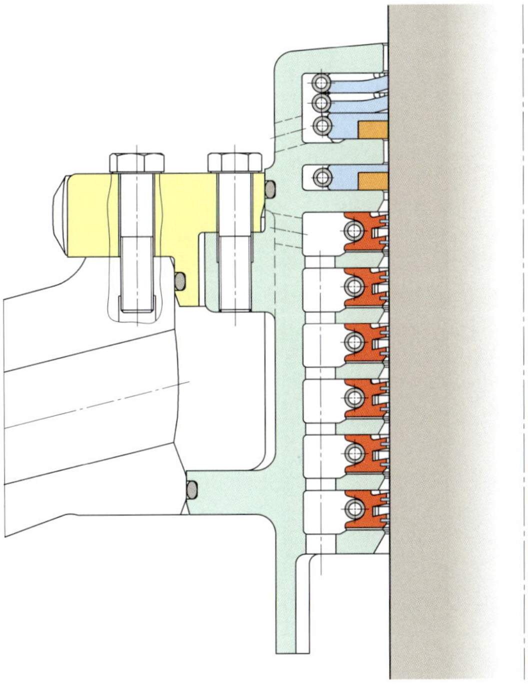

◄
A piston-rod stuffing box for sealing the scavenging-air space from the crankcase.

The top ring package scrapes the contaminated lubricating oil from the scavenging-air space and discharges it outside the engine. The bottom ring package scrapes the clean oil off the rod, after which it falls back in the crankcase sump.

▲

The piston with piston rod after removal from the engine. It is temporarily placed on the floor next to the cylinders. The piston-rod stuffing box (1) can now be renewed.

▲

A complete ring package for the stuffing box.

Each sealing ring contains three segments that are kept pressed against the piston rod by a spiral spring.

◀

A piston-rod stuffing box mounted on the piston rod.

On top the rectangular piston-rod foot for attachment to the crosshead.

▲

The two halves of the casing of a piston-rod stuffing box. Note the drainage holes for both the clean lubricating oil for the crankshaft as well as the contaminated cylinder lubricating oil.

13.3.4 Crosshead

The crosshead connects the piston rod and the connecting rod. The piston rod is fixed to the crosshead and the connecting rod can pivot around the crosshead.
The lateral forces acting on the connecting-rod mechanism are transferred to crosshead shoes fixed to the frame with crosshead guides.
The forged-iron connecting rod is attached to the crosshead with a forked connection, which means that there are two bearings on each side of the crosshead pin.
In the past, the bearing caps were manufactured from a ductile white metal. Today, modern engines have hard tri-metal caps that have better resistance to the significantly increased surface pressures produced by the higher combustion pressures exerted on the piston.

◄ A crosshead in an engine.

1 *welded A-frame*
2 *crosshead guide*
3 *connecting rod*
4 *crosshead*
5 *(crosshead) guide shoes*
6 *steps in crankcase*

◄ A complete crosshead with crosshead pin and guide shoes.

▲ Maintenance on the crosshead. Fitting hydraulic jackets to the studs for attachment of the connecting rod to the crosshead.

1 *hydraulic jack*
2 *connecting rod*
3 *crosshead*
4 *crosshead guide*
5 *hydraulic pressure bolts for attachment of the top cover of the main journal bearing*

▲ The crosshead pin for a Wärtsilä Sulzer RTA two-stroke diesel engine.

The piston rod is fixed to the top of the pin by four bolts. The centre of the pin is perforated for the supply and discharge of cooling lubricating-oil to and from the piston. The connecting-rod bottom bearing supports the entire bottom section of the crosshead pin. The top bearing comprises two parts; they are mounted on the left and right side of the piston rod.

13.3.5 Connecting rod

The connecting rod is bolted to the top of the crosshead by means of a 'fork' on two complete crosshead bearings. The bottom of the connecting rod is bolted to the crankpin. Connecting rods are forged from steel that can be an alloy. The forces in the drive gears for these category IV engines can be several hundred tons with a maximum of 1100 tons! **Therefore, all driving gears, including the bearings, have a heavy-duty finish. Today, tri-metal bearings are increasingly used as opposed to the soft white-metal bearings, as they have a better resistance to high engine loads.**

▶

The driving gears for a Wärtsilä Sulzer RTA engine.

Note the large dimensions of the driving-mechanism components in relation to the piston.
Piston diameter: 980mm
diameter piston rod: 375 mm
diameter crosshead pin: 980 mm
diameter connecting rod: 530 mm
diameter crankpin: 1350 mm
diameter crankshaft: 1200 mm

▶

The connecting rods for MAN Diesel 50MC engines.

1 *crankpin side*
2 *crosshead side*

◀

Connecting rods for MAN Diesel K50MC engines.

Left: the connection with the crosshead. Right: the connection with the crankpin.

◀

The connecting-rod head for a MAN Diesel K50MC engine.

Between the notches, the crosshead oscillates in relation to the reciprocating movement of the piston rod.

▲

The crosshead with on top the piston rod and below the connecting rod and the crankwebs. Left and right the crosshead guides.

▲

The crankshaft in the engine.

1 *connecting rod*
2 *big-end with connecting-rod bolts*
3 *crank webs*
4 *main journal bearing lubricating-oil supply*
5 *main journal bearing top cover*
6 *hydraulic bolt for attachment of the top cover*

13.3.6 Crankshaft

All crankshafts of two-stroke engines are constructed by joining individual forgings comprising one crankpin with two webs forged from a single piece of metal with main shafts shrunk into the webs.
Long crankshafts are manufactured in two parts connected by large flanges.
Increasing combustion forces have resulted in an increase in the crankpin and crankshaft diameters, whereas the distance between the crank webs has been decreased to avoid distortion of the crankpin. This also serves to shorten the engine despite increasing cylinder capacity; this translates into a smaller engine room, thus providing more cargo space. Crankshafts of large crosshead engines are 'soft'; they do not have a hardened finish. The crankpin- and main-shaft bearings are manufactured from a tri-metal.

13.3.7 Camshaft drive

There are two traditional methods for the driving the camshaft by the crankshaft:

1 **With cogwheels**
This method is still applied in the Wärtsilä Sulzer engines series RTA and the Mitsubishi UC engines.
2 **With chains**
This method is applied in the MAN Diesel M series MC.

◀

A partially built crankshaft for a low-speed two-stroke crosshead engine from MAN Diesel, type 50MC.

Crankpin and webs are made from a single throw. The main journals have been shrunk in the webs. The crankpins are hollow (weight saving) and the crank webs are positioned closely together in order to reduce the crankpin length. This results in a robust crank with little distortion.

Modern engines with the common-rail system from both engine manufacturers are no longer equipped with camshafts. New series: Wärtsilä Sulzer RT-FLEX and MAN Diesel K ME.

Advantages of cogwheel drives

- Finely-tuned drive in which slight wear and tear of the teeth does not affect the 'timing' of the camshaft.

Disadvantages of cogwheel drives

- Expensive, heavy.
- Requires meticulous alignment of the cogwheels.

Advantages of chain drives

- Relatively inexpensive.
- Light in weight.

Disadvantages of chain drives

- Requires periodic tensioning due to wear and tear of the links in the chain, otherwise the camshaft timing will lag.

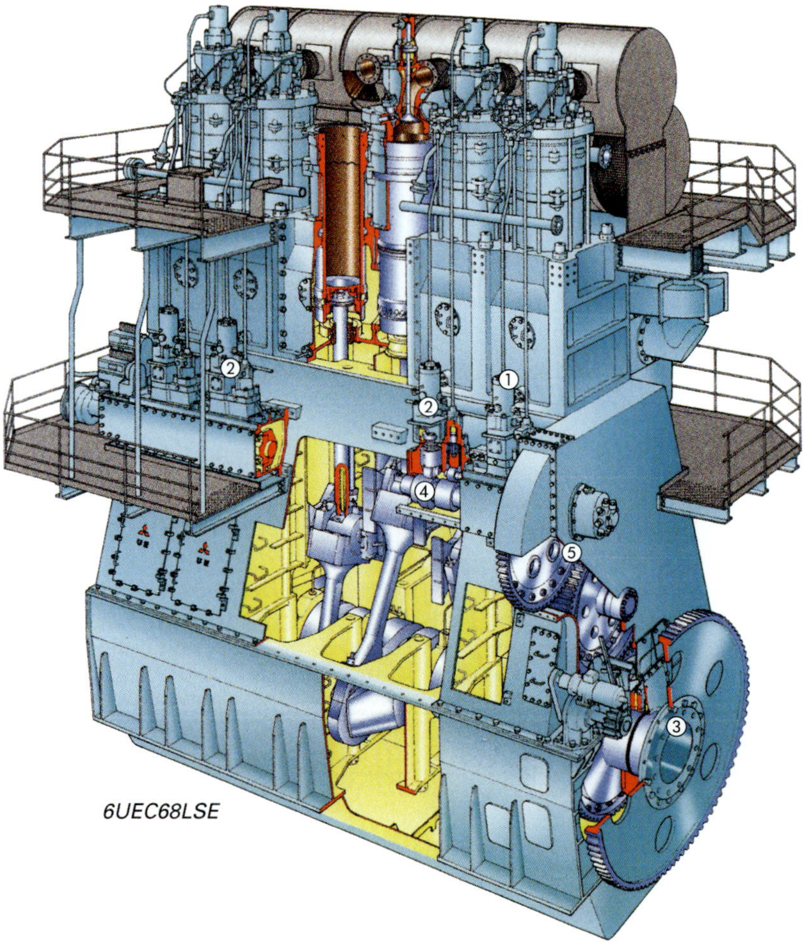

▼

A low-speed two-stroke crosshead engine, manufacturer Mitsubishi, type 6UEC68LSE.

Here the camshaft is driven by spur gears.
On the right, next to the high-pressure fuel pumps the hydraulic pump for the exhaust valve.

1 *high-pressure fuel pump*
2 *hydraulic pump for the exhaust valve*
3 *crankshaft*
4 *camshaft*
5 *intermediate wheel, two pieces*

▲

A chain and a gearwheel for driving the camshaft of a large two-stroke crosshead engine from MAN Diesel.

▲

A camshaft for a MAN Diesel K50MC two-stroke low-speed crosshead engine.

The camshaft is chain-driven.

▲

A section of the camshaft for a low-speed two-stroke crosshead engine, manufacturer Wärtsilä, type RTA96C.

The cams are hydraulically fitted. The section of the camshaft shown has been hoisted out of the bearing horizontally to the right.

13.3.8 Fuel-pump drives

Two-stroke crosshead engines: traditionally, fuel pumps were driven from the camshaft, with a high-pressure fuel pump placed directly above the cam. By rotating the cam or by shifting the camshaft, the fuel injection was adjusted for an opposite rotational direction.

▲

The chain-drive in a low-speed two-stroke crosshead engine from MAN Diesel, type MC.

The drive is installed with two chains and a tensioning device.

1 *crankshaft*
2 *camshaft*
3 *spanner device*
4 *turning wheel*
5 *turning drive*
6 *double chain*

◀

Traditional drive.

The traditional fuel-pump drive in a two-stroke crosshead engine from Wärtsilä Sulzer, type RTA96C. Two pumps (1) in one casing with the camshaft below. The pumps are valve-controlled. The exhaust valve is driven by the hydraulic pump (2).

◀

Two high-pressure fuel pumps (1) driven by the camshaft of a MAN Diesel 50MC crosshead engine.

In the centre the hydraulic exhaust-valve drive (2)..

▶

In the latest two-stroke crosshead engines with a common-rail system, the high-pressure fuel unit is driven directly from the crankshaft with gearwheels.

There are two camshafts in the V-shaped unit (1), which drive the original high-pressure fuel pumps of the four-stroke engine, the Wärtsilä 46. This engine has no camshaft.

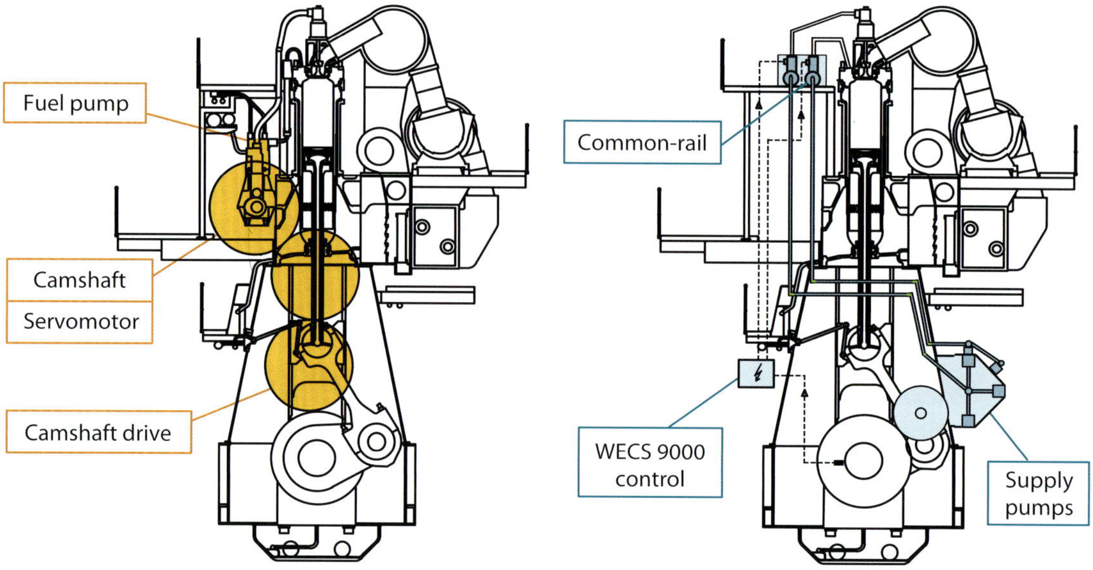

▶
Left: the traditional Wärtsilä Sulzer RTA series. Right: the common-rail version without camshaft, the Wärtsilä Sulzer RT-FLEX.

The hydraulic power-supply unit of the common-rail fuel system of the Wärtsilä series RTA engines (figure on the right) is placed at the lower section of the engine, shown here on the right.
The camshaft, driven from the crankshaft, in the figure on the left (orange), is not present in the right figure. The drive gearing for the engine has been simplified.

13.3.9 Exhaust-valve drive

Approximately 25 years ago, engine manufacturers changed over from a conventional mechanical exhaust-valve drive to a hydraulically driven exhaust valve in two-stroke crosshead engines.
The mechanically driven exhaust valves, driven from the camshaft with a cam lobe, guide pulley, push rod and rocker waned and never returned. Today, exhaust valves in two-stroke crosshead engines built by the three remaining engine manufacturers, MAN Diesel, Wärtsilä (Sulzer) and Mitsubishi are hydraulically operated. Closing the valve by a heavy valve (pressure) spring has also disappeared. Presently, the valve is swiftly closed by an air piston.

There are two techniques

1 A hydraulic pump is mounted on the camshaft. It opens the valve using lubricating-oil pressure obtained from the lubricating-oil circulation system. These engines have common high-pressure fuel pumps that are driven by the camshaft.
2 The latest common-rail fuel systems have no camshaft, but a fuel/lubricating-oil unit placed next to the crankshaft, from which it is driven. The part of the unit containing the lubricating oil forces the lubricating oil to the exhaust valves under great pressure. These valves can be opened and closed very precisely and valve settings can be easily adjusted. In this way, the valve movement can be optimised depending on, for instance, engine load, rpm and type of fuel.

▶
The hydraulic drive for the exhaust valves in a MAN Diesel 50MC two-stroke crosshead engine.

1 *high-pressure lubricating-oil line*
2 *servo piston for the exhaust valve*
3 *exhaust-valve casing*
4 *cylinder head*
5 *air piston*

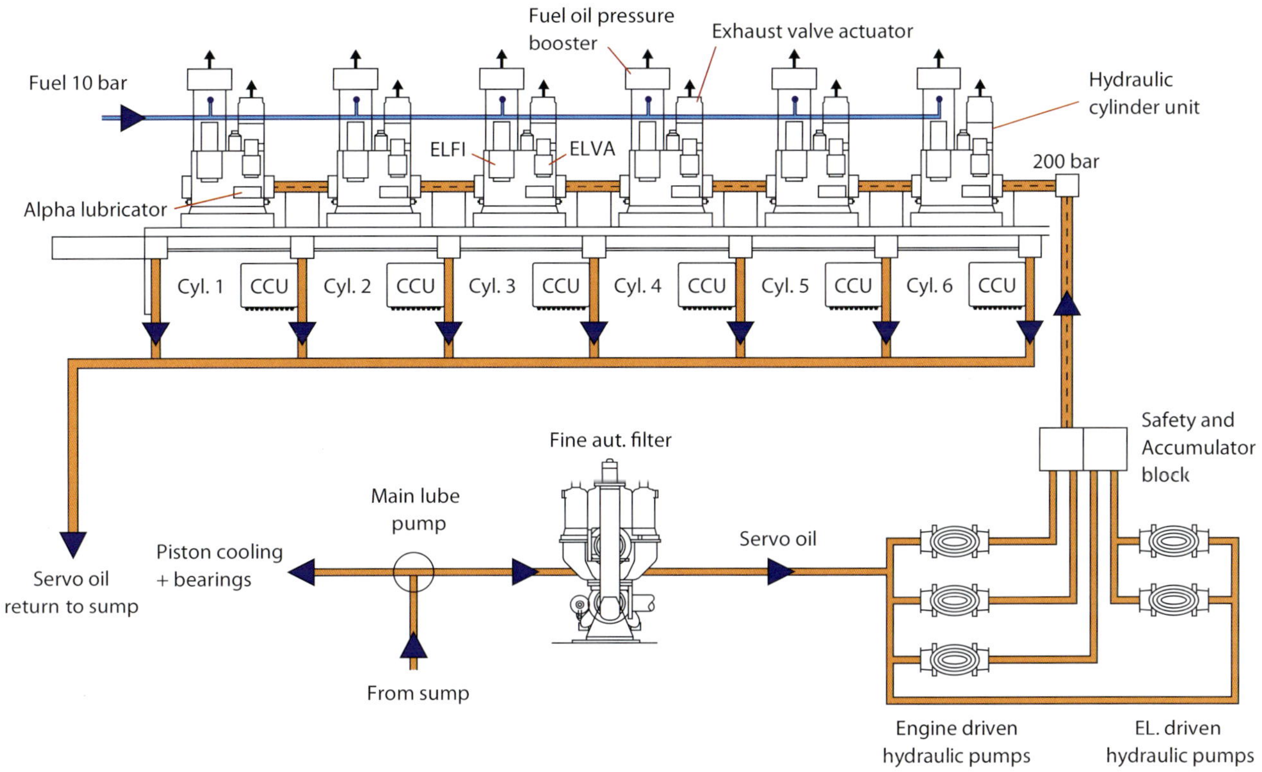

▲

The high-pressure lubricating-oil system for fuel injection and the exhaust-valve movement of MAN Diesel ME (E = 'electronic') two-stroke crosshead engines.

Operation: The main lubricating-oil pumps supply lubricating oil for the piston cooling, the bearings and other parts that require lubrication; they also feed the hydraulic system through a self-cleaning fine-filter. There are two types of hydraulic pumps:

1 engine-driven pumps during normal operation and:
2 electrically driven pumps when the engine has stopped, has not yet started, or is non-operational.

Via a distribution block fitted with safety valves (to prevent an excessive build-up of the lubricating-oil pressure), the hydraulic pumps supply the fuel pump hydraulic-drive unit with servo-oil at a pressure of 200 bars and the hydraulic pumps for the exhaust-valve movement.

ELFI: Electronic Fuel Injection control valve
ELVA: Electronic Exhaust Valve-Actuator
CCU: Cylinder Control Unit

▼

Via large ducts in the MAN Diesel ME engines, both the fuel pump, left picture, and the valve movement, right picture, are hydraulically driven under high pressure.

Central picture: both fuel injectors for each cylinder.

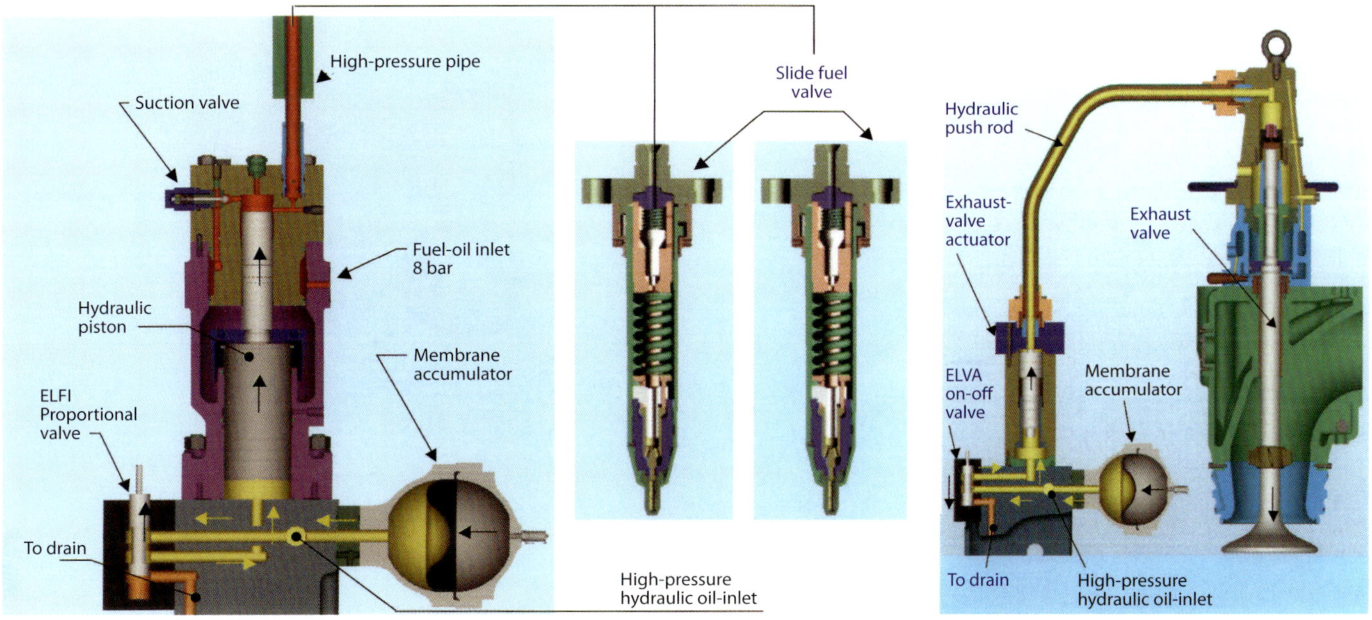

▶
Via large ducts in the MAN Diesel ME engines, both the fuel pump, left picture, and the valve movement, right picture are hydraulically driven under high pressure.

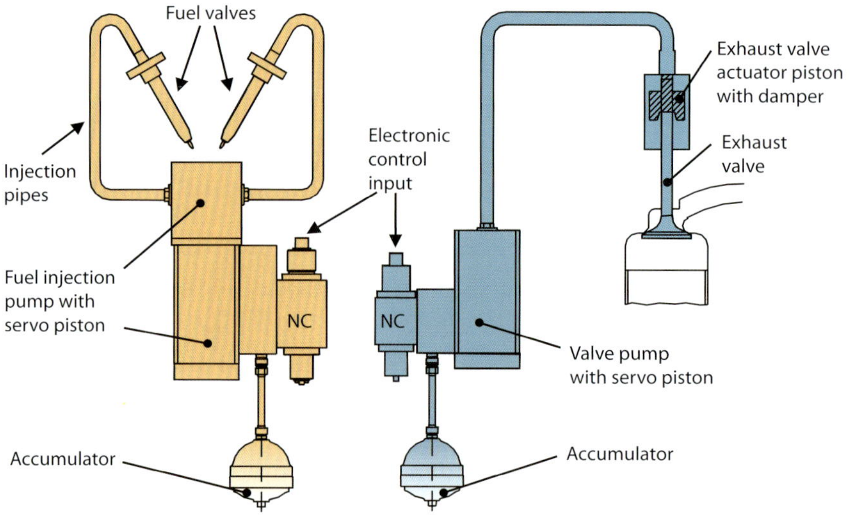

▶
The operation of the high-pressure fuel pump.
The intake stroke.

1 The plunger moves down to bottom position because the 'ELVI' valve moves downwards and the hydraulic-oil pressure underneath the piston moving the plunger, is bled off.
2 The hydraulic oil leaves the hydraulic-oil chamber via valve 3
3 Fuel oil flows to the top of the fuel-pump plunger.
4 'ELVI' valve is opened and the hydraulic oil discharged.

The accumulators are supplied by the hydraulic-oil pumps.

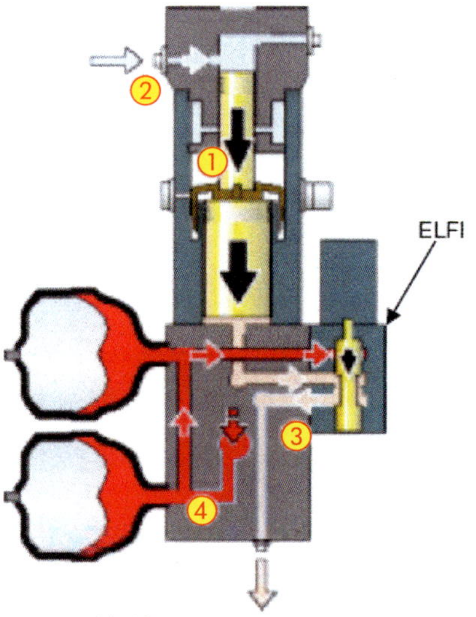

▶
The operation of the high-pressure fuel pump.
The compression stroke.

5 The ELVI-valve is closed and the hydraulic-oil pressure moves the piston and consequently the fuel plunger upwards.
6 The piston moves upwards.
7 The fuel-supply valve is closed.
8 The fuel is pushed towards the injectors under high pressure.

white: non-pressurised fuel (suction pressure)
light blue: pressurised fuel
red: pressurised hydraulic oil
pink: non-pressurised hydraulic oil (return pressure)

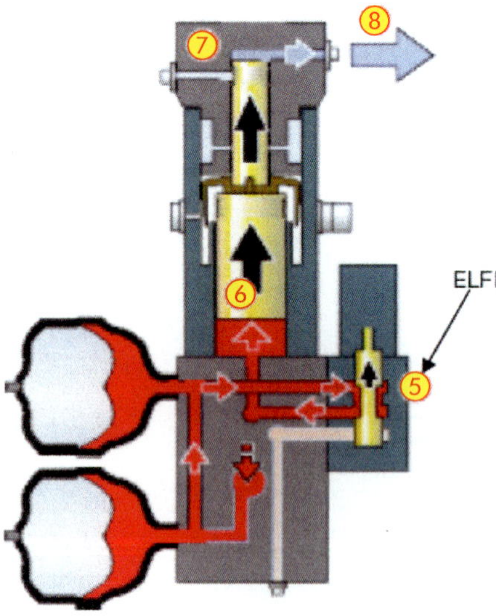

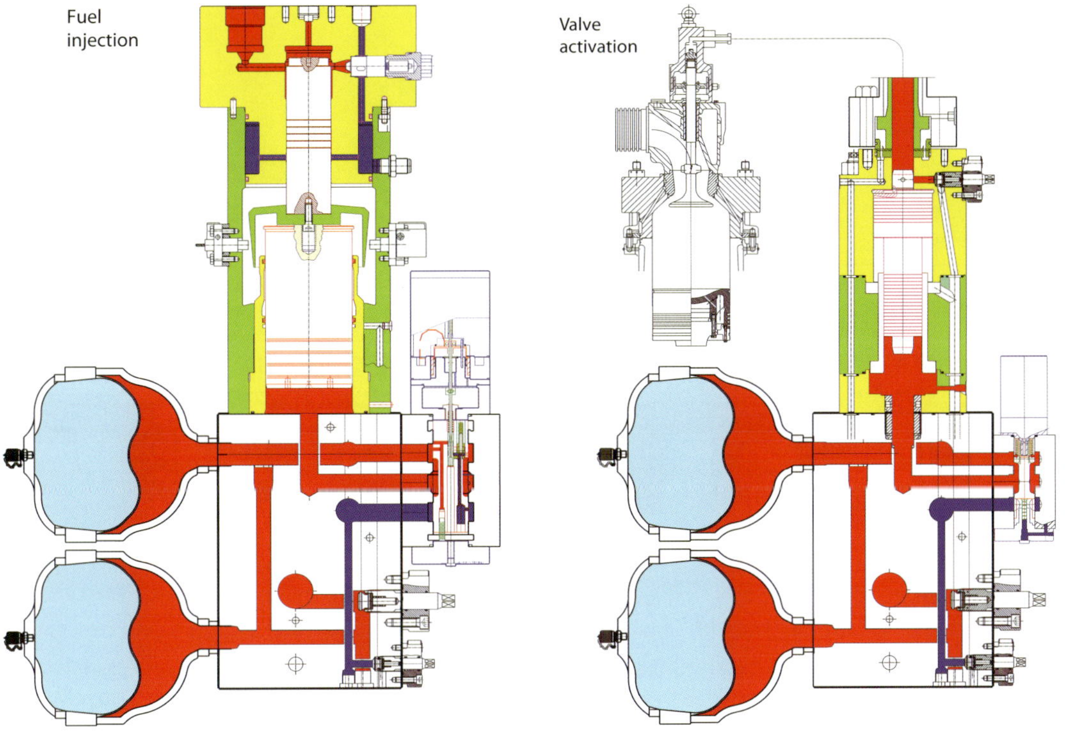

◄ The hydraulic drive for both the high-pressure fuel pump and the exhaust valve.

Red: hydraulic-oil supply, pressure 200 bar
Blue: hydraulic-oil return.

Red above the plunger:
Left: pressurised fuel to the injector.
Right: pressurised hydraulic oil for the exhaust-valve piston.

13.4 Thrust blocks and thrust bearings

In ship propulsion, propeller rotation generates a propulsion force on the water behind the ship. This thrust is transferred to the propeller shaft that is transmitted to the ship foundation by a thrust bearing.

In large two-stroke crosshead engines, the propeller shaft and the transmission shafts are directly connected to the diesel engine. The thrust bearing is placed either inside the engine or immediately behind it. This ensures that no axial forces can be exerted on the crankshaft.

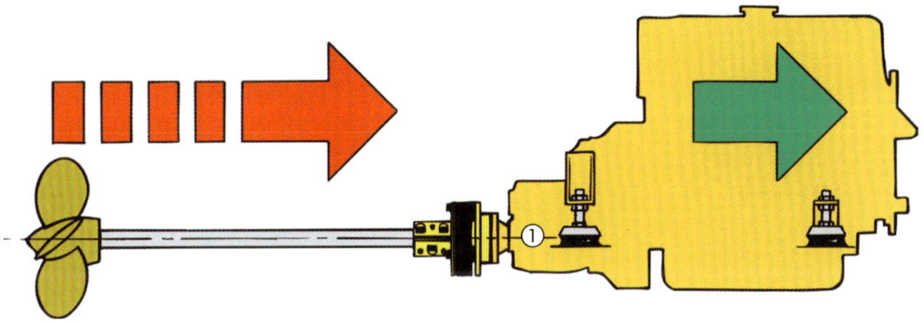

▲ The propeller thrust (red arrow) carries on in this flexibly arranged yacht engine.

The thrust block in the reduction gearing (1) mounted on the left of the engine absorbs this force.

◄ The thrust block mounted on the engine for a two-stroke low-speed crosshead engine, manufacturer Wärtsilä Sulzer, type RT-fLEX96C.

1 *crankcase engine*
2 *crankcase relief valve*
3 *thrust block*
4 *fly/turning wheel*
5 *intermediate shaft*

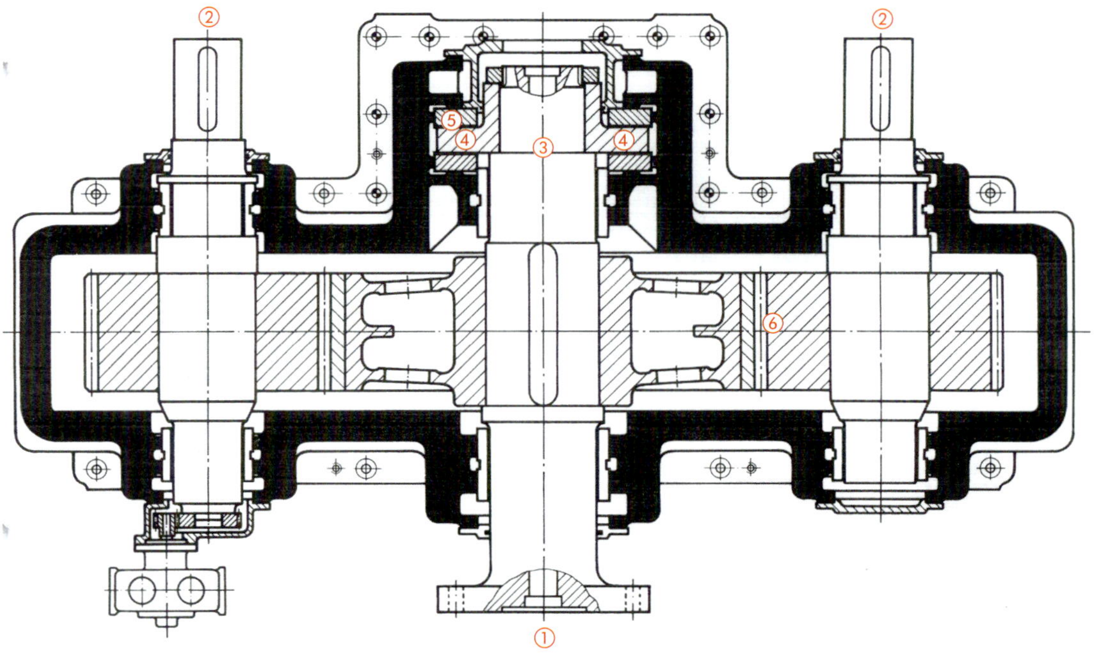

▶

Reduction gearing from Renk for driving a propeller with two engines.

The thrust block, located at the end of the intermediate shaft, absorbs the propeller thrust. The thrust block has been adapted for bi-directional thrust: small thrust blocks have been fitted on both sides of the thrust-block flange.

1 *propeller shaft*
2 *engine shaft*
3 *thrust block*
4 *thrust-block collar*
5 *small thrust blocks*
6 *reduction gearing*

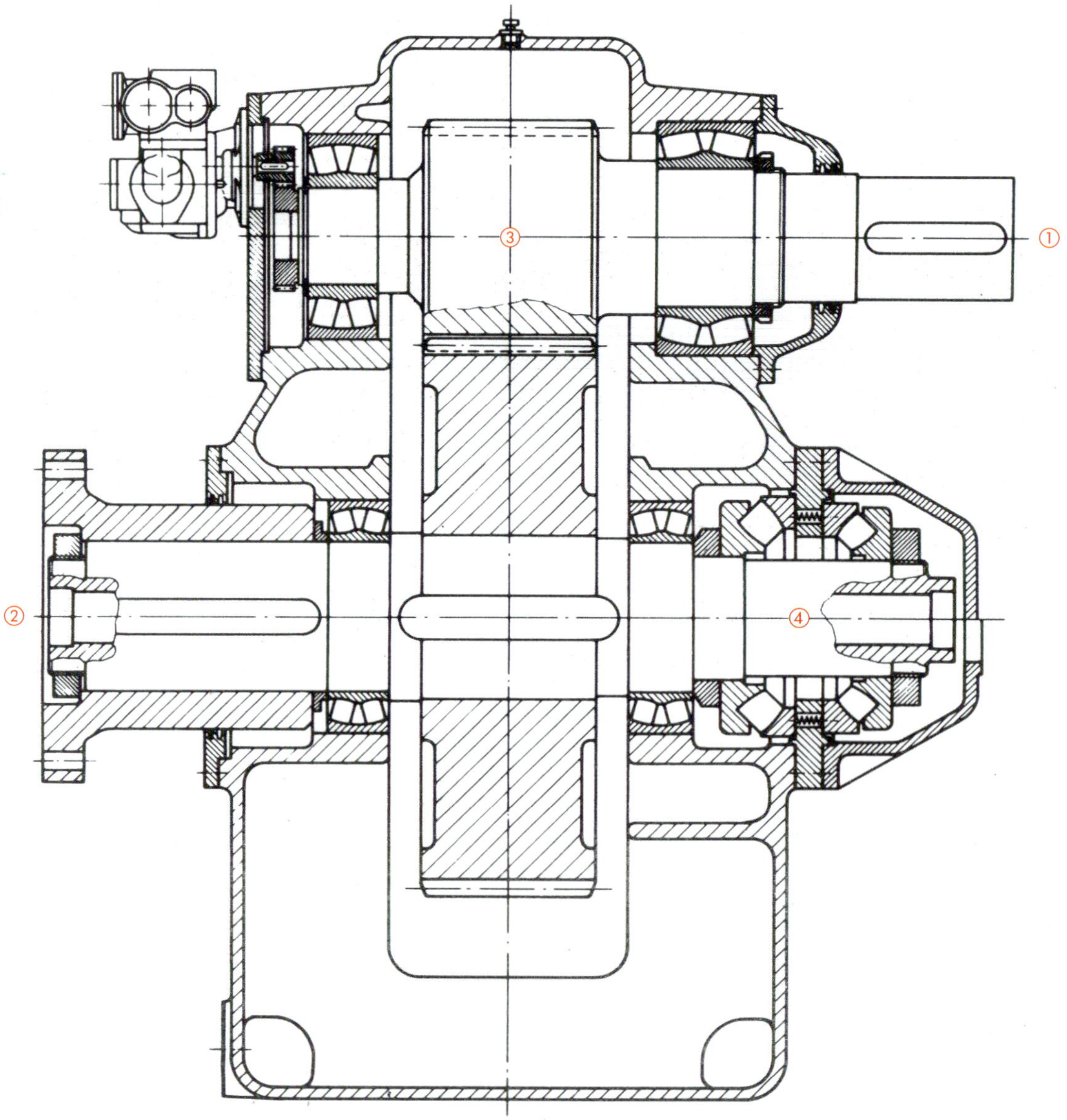

▶

Reduction gearing for a diesel engine with a lower positioned propeller shaft.

The thrust bearing is mounted to the right of the picture.

1 *engine-drive shaft*
2 *propeller shaft*
3 *reduction gearing*
4 *thrust bearing*

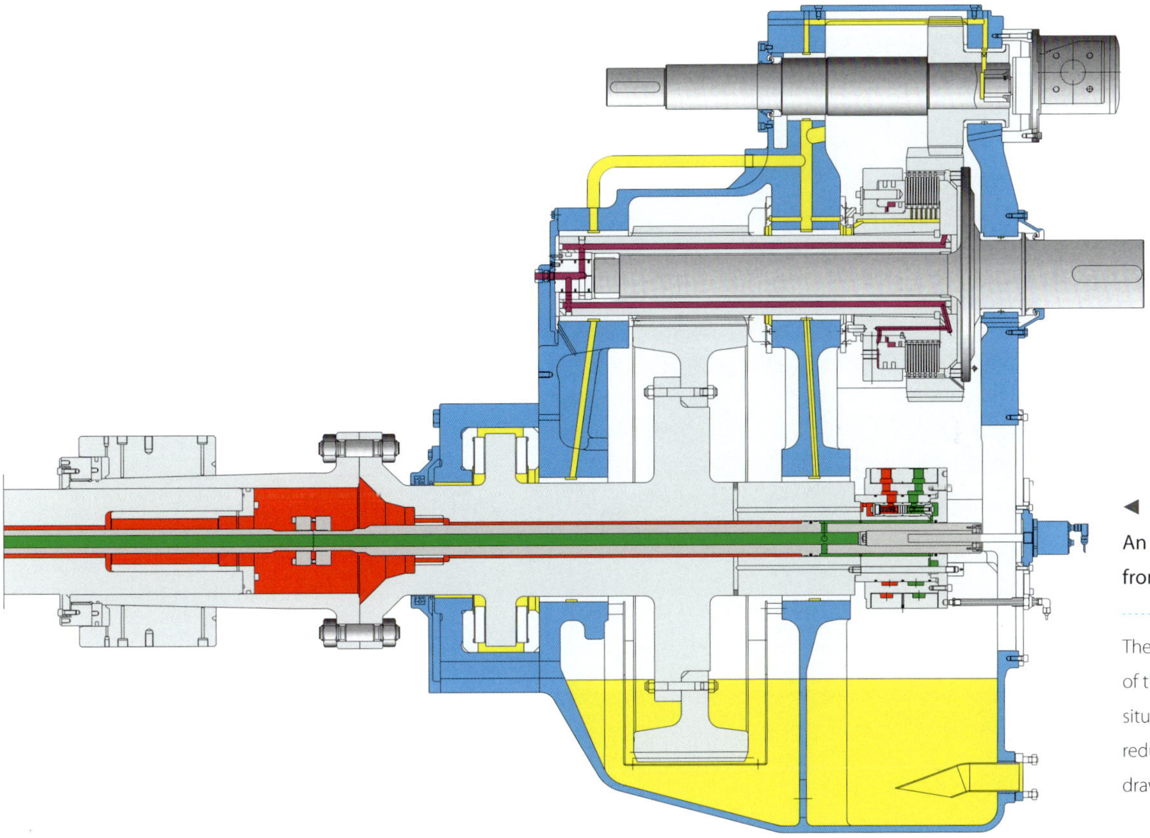

◀

An Alpha reduction gearing from MAN Diesel.

The thrust block for absorption of the propeller thrust is situated to the left in the reduction gearing. See detail drawing.

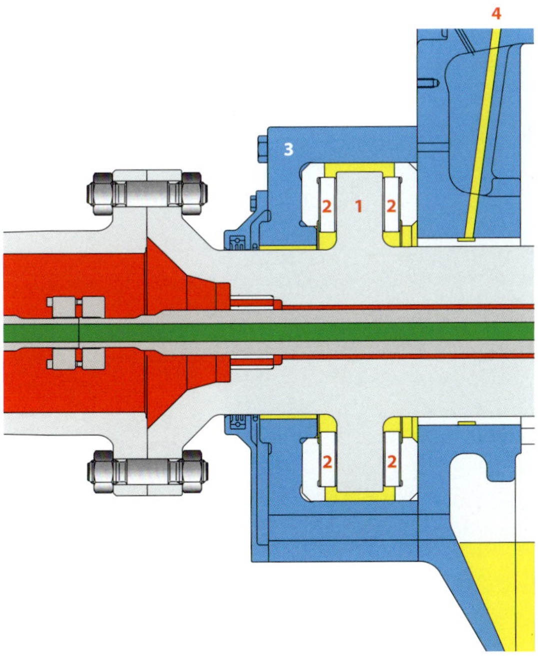

▲

Detail drawing of the thrust block.

1 *thrust-block flange*
2 *small thrust blocks*
3 *thrust-block casing*
4 *lubricating-oil supply from the lubricating-pump gearbox*

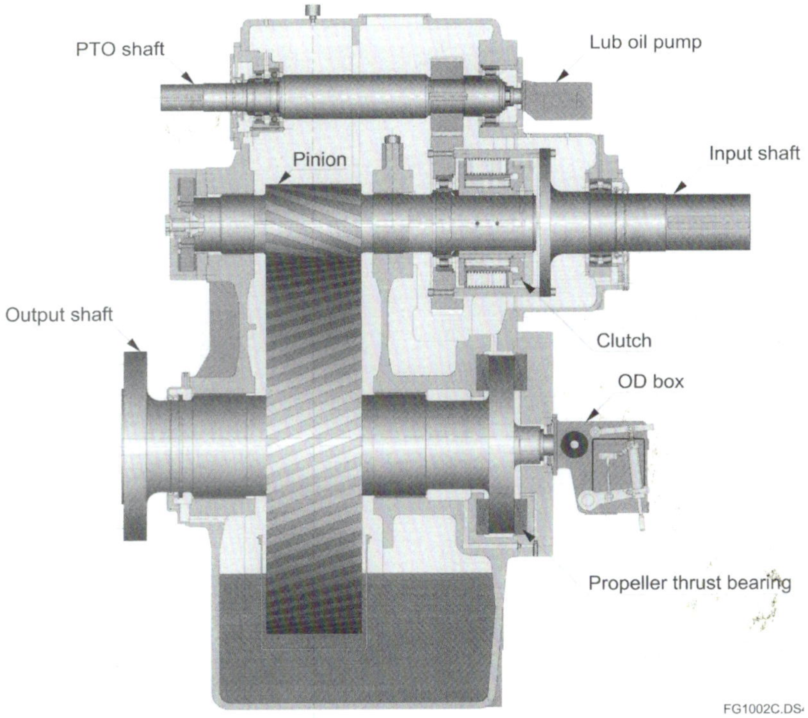

▲

A gearbox with a thrust block at the right side of the reduction gearing.

The engine shaft and the generator shaft have a thrust block/thrust bearing to maintain the shaft's position.

▶
A Renk thrust block, which is usually placed in the shaft between the propeller and the engine. The circular discs are the thrust blocks.

Pedestal bearing series H.

▲
A thrust block without top cover.

The round notch on the circular thrust block is positioned slightly off-centre; the block is tilted at the slightest movement and builds up a lubricating-oil pressure in the wedge-shaped gap between the thrust-block collar and the thrust block, which is sufficiently high to annul the thrust pressure.

▲
The thrust block in a large two-stroke crosshead engine from Wärtsilä, type RTA84C.

1 *crankcase*
2 *thrust-block location*
3 *fly/turning wheel*
4 *turning motor*

In four-stroke trunk piston engines that are normally provided with a propeller-shaft driven reduction gear, the thrust bearings have been integrated in the gearbox.

Principles of the thrust bearing

▶

The thrust block of a Wärtsilä Sulzer RTA96C two-stroke crosshead engine in the engine frame. The thrust block consists of a shaft with a large collar.

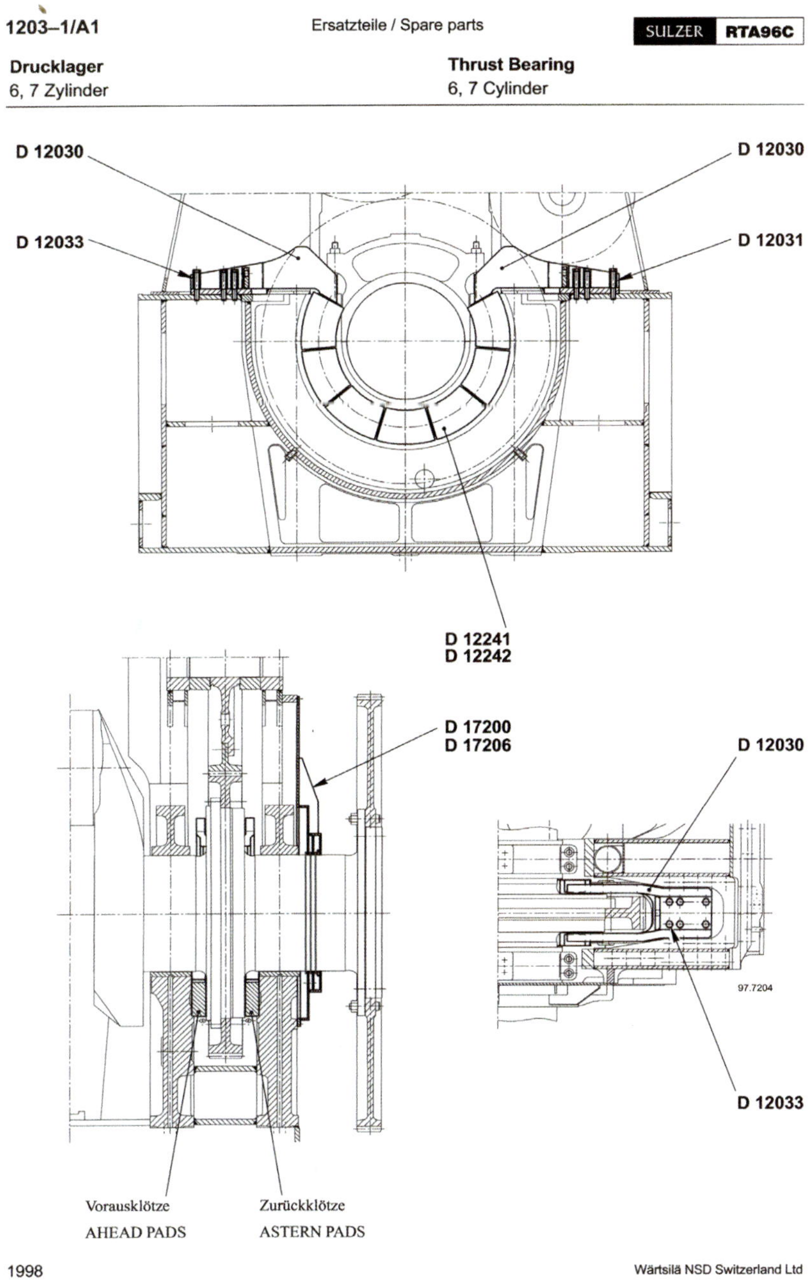

The casing consists of a number of tilting pads, which, in a directly reversible diesel engine, are fitted to both sides of the collar.
The pads have an elevation, the so-called pivot; this is placed off-centre of the pad.
The front of the pad has been rounded so the lubricating oil from the collar, which partially runs in lubricating oil, is carried between the collar and the pad.
The pad is pressed on the lubricating film by the pivot.
There is a pressure build-up in the lubricating film between both parts that causes the pad to tilt slightly.
A stable condition is then achieved as the product of the lubricating-oil forces are balanced with the force exerted on the pivot by the thrust force.
The effects of thrust fluctuation are neutralised as the equilibrium is constantly recreated. This allows the thrust to be transmitted to the foundation of the ship via the thrust collar, the lubricating-oil film, the pads and the thrust-block casing.
The angle of tilt is so small that it cannot be seen with the naked eye (1:1000).
The lubricating film in the system is not broken up, thus preventing metal-on-metal contact. Therefore, friction is negligible.
A number of tilting pads have been placed on the other side of the collar for engine reversal. Obviously, these tilt in the other direction due to the changed direction of rotation.
The top of the loose thrust blocks often lack tilting pads, so the total load on the bearing surface lies below the centre line. Therefore, the bending moment on the bedplate is smaller.
The large loose thrust blocks normally have a separate lubricating-oil system. If the thrust block is integrated in the engine or part of a reduction gearing, the lubricating-oil system is often used for the engine or the reduction gearing.

>CH 14

Starting systems of diesel engines

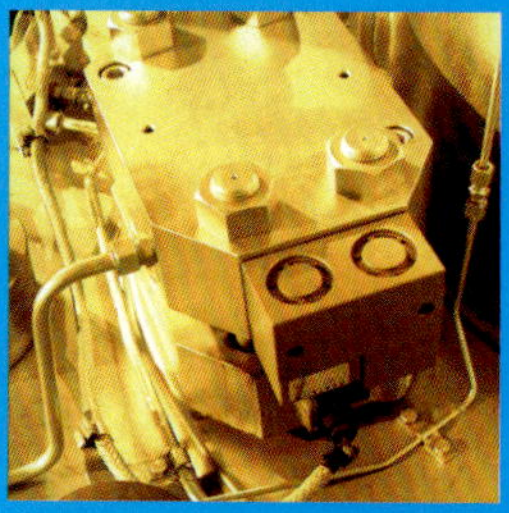

The only way in which a stationary engine can operate is by obtaining power from an external source. This is realised in various ways, such as by hand, with batteries or, for example, with starting air.
Shown here, a pair of starting-air valves for a large two-stroke crosshead engine.
The starting air has a pressure of 25 to 30 bar.

Many category I and II engines make use of electric starters to start the engine, like this diesel generator from Mastervolt.
Above left, the dynamo for charging the storage batteries, in the middle, the start motor.

14.1 Introduction

An internal combustion engine cannot be started from a standstill. There is no internal combustion process and consequently no working matter (combustion gases) present.
Therefore, the engine has to be started with an external energy source. The energy that is supplied when the engine is started is used by the piston to compress air, so fuel can be injected before the piston reaches top dead centre position and the combustion process can commence. The compression stroke requires a considerable amount of energy. Small single- or multiple-cylinder engines can be started manually. Larger engines, however, require far more energy.

14.2 Starting methods

The following starting methods are used:

A By hand combined with a decompression lever.
B Starting with a small energy source, such as a spring mechanism or a hydraulic-pressure vessel.
C Electrical start with a battery.
D Starting with starting rotors, operated by compressed air, mounted on the flywheel.
E Air start with compressed air fed to the starting-air valve, which is mounted directly on the cylinder.

Methods A, B, and C are only applied in relatively small four-stroke engines in categories I and II.

Method D is generally used in larger high-speed four-stroke engines. Method E is often applied in large four-stroke medium-speed engines and always in two-stroke crosshead engines.

14.2.1 Method A: Manual start in combination with a decompression lever

Turning a crank handle achieves a certain number of revolutions in an engine with an opened exhaust valve; due to this, the piston has no compression so the crankshaft can rapidly attain acceptable revolutions. The energy produced is 'stored' in the flywheel. When the decompression lever is pulled, the exhaust valve closes, allowing the piston to start its compression stroke and the injected fuel to ignite. The combustion stroke switches on the engine and the engine runs.
Obviously, hand-operated starting energy is limited, so this method is only suitable for small engines, with few cylinders.
The 'decompression lever' ensures that the exhaust valve remains open during hand cranking. However, there are also larger, electrically started engines provided with exhaust valves that initially remain open and only close after the crankshaft has rotated several times.

▼
The starting system of a single cylinder four-stroke Lister diesel engine category I.

1 *crank handle*
2 *decompression lever*

▼
The decompression lever on the cylinder head.

When starting the flywheel using a decompression lever, the exhaust valve is opened so compression cannot take place in the cylinder.

◀

An emergency generator built by Sandfirden Technics, Den Oever, The Netherlands.

Regulations stipulate that emergency generators must be operated by two independent starting systems, an electric starting system and a second, independent starting system. Shown, a second starting-system by means of a wind-up spring starter.

1 *electric starting-air motor using a battery*
2 *spring starter*
3 *spring-winder handle*
4 *red starting button*

14.2.2 Method B: Starting with a small energy source, such as a spring mechanism or an hydraulic-pressure vessel

This method is often used in lifeboats, as a manual start is not always possible in poor weather conditions. The starting energy is stored in a heavy spring or in a high-pressure lubricating-oil system stored in a starting reservoir.

When the engine is started, this energy is released and the engine starts. In this system, a decompression lever is also often used.

14.2.3 Method C: Electrical start with a battery

This method is often applied in high-speed four-stroke engines. Many engines with a shaft power well over 1000 kW are still started in this manner. It should be noted here that the size of the battery is increasingly large.

Description

The battery consists of a large number of cells, which generate a voltage of 12 or 24 Volt. When the engine is started with a key or button, the rotor of the starter rotor is shifted and the cogwheel on the starter shaft engages the gear ring on the flywheel.

◀

A different type of spring starter.

1 *spring-winder handle*
2 *trip lever*
3 *bendix, pinion*
4 *spring pack*

▶

The electric starting system on a Volvo Penta yacht engine category I.

1 *starting motor*
2 *generator*
3 *reduction gearing with shaft*

After this, the starter switches on automatically and the engine begins to run. When the engine itself operates at a certain rpm, the revolutions will increase and when they reach a certain rate the starter is switched off by resetting the rotor motion to nil. The engine is running. The battery is charged by a diesel-engine driven generator. In larger installations, this can also be performed by a charging system using mains electricity.

14.2.4 Method D: Starting with starting rotors mounted on the flywheel, which are operated by compressed air

This method is more and more used in the larger high-speed and medium-speed four-stroke diesel engines that were previously started with an extremely large battery (high-speed) or with starting air fed directly to the cylinder (medium-speed). In the first instance, the expensive high-maintenance batteries are no longer required, and in the latter, the construction of the cylinder head is simplified. Furthermore, while operating the engine when it is started for the second time, the relatively cold air no longer enters the cylinder. This cold wave often causes cracking of the cylinder head where the starting valve is mounted.
Larger engines with several cylinders often have two or more rotors in place to provide sufficient power for the crankshaft. The starter rotors resemble simple gas turbines.

▶

The electric starting motor of a 'Vetus' Deutz yacht engine category I.

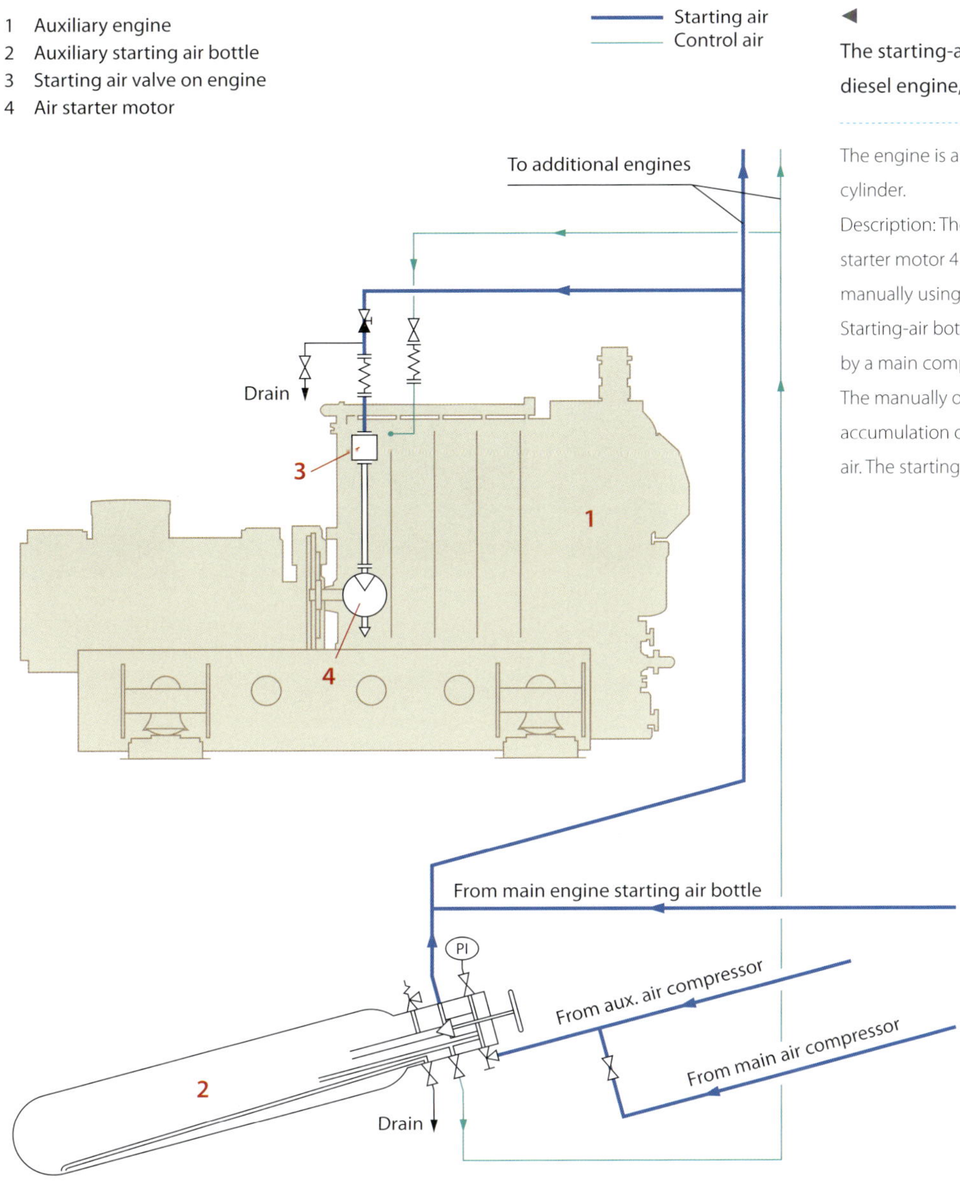

◄

The starting-air system of a Sulzer S20 medium-speed diesel engine, category III.

The engine is also available with a starting-air valve for every cylinder.

Description: The diesel generator set 1 is started using an air starter motor 4 that drives the flywheel. This can be done manually using starting valve 3 on the engine.

Starting-air bottle 2 is automatically kept at the correct pressure by a main compressor or an auxiliary/emergency compressor.

The manually operated water drain prevents the slow accumulation of condensed water vapour from the compressed air. The starting-air capacity slowly decreases!

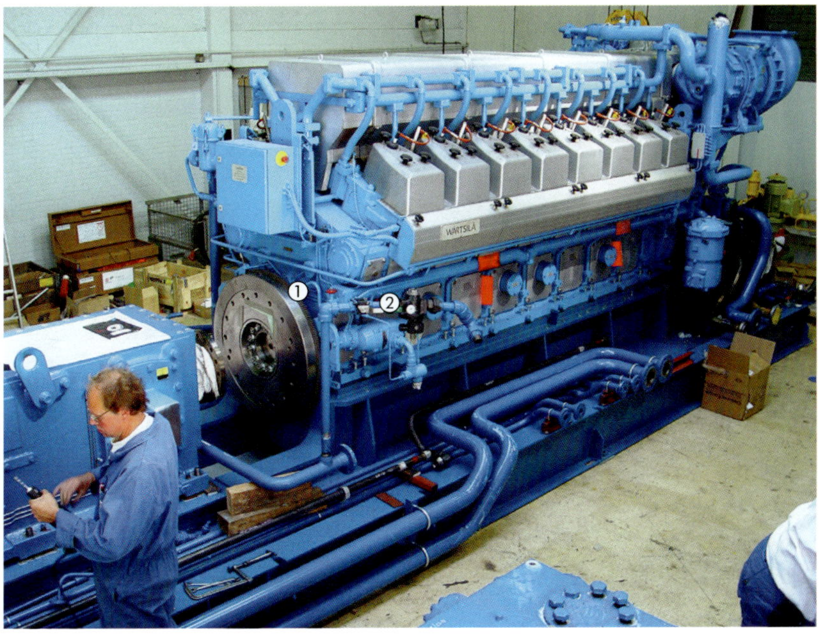

◄

The air-starter motor on the 'flywheel' of a sixteen-cylinder Wärtsilä 25 SG gas engine.

Mounting one or more air-starting motors that operate the flywheel is simpler than the traditional starting-air valves on each cylinder.

1 *fly-turning wheel*
2 *air-starter motor*

▲

The starting-air turbine/motor on a Kromhout diesel engine. Build 1956!

▲

A manual air-stop valve for an air-starting motor on a older Kromhout diesel engine, category I.

The lever is kept depressed until the engine takes on fuel.

In order to have an adequate amount of air available during the start, large starting-air reservoirs are used that are automatically kept under pressure by two starting-air compressors.
The pressure in the air reservoirs is approximately 30 bars. Classification Societies have laid down certain requirements for air-reservoir capacities.

- **One propulsion engine, directly reversible:** the capacity of the air receivers should be sufficient to start the engine at least **twelve** times in turn ahead and/or astern, without requiring the starting-air compressors.
- **Non-reversible engines:** the capacity of the air receivers should be sufficient to start the engine at least six times without assistance from the starting-air compressors.

▶

The starting-air turbine and the turning motor mounted near the flywheel on a four-stroke MAN Diesel engine, category III.

1 *starting-air turbine*
2 *electric-turning motor*

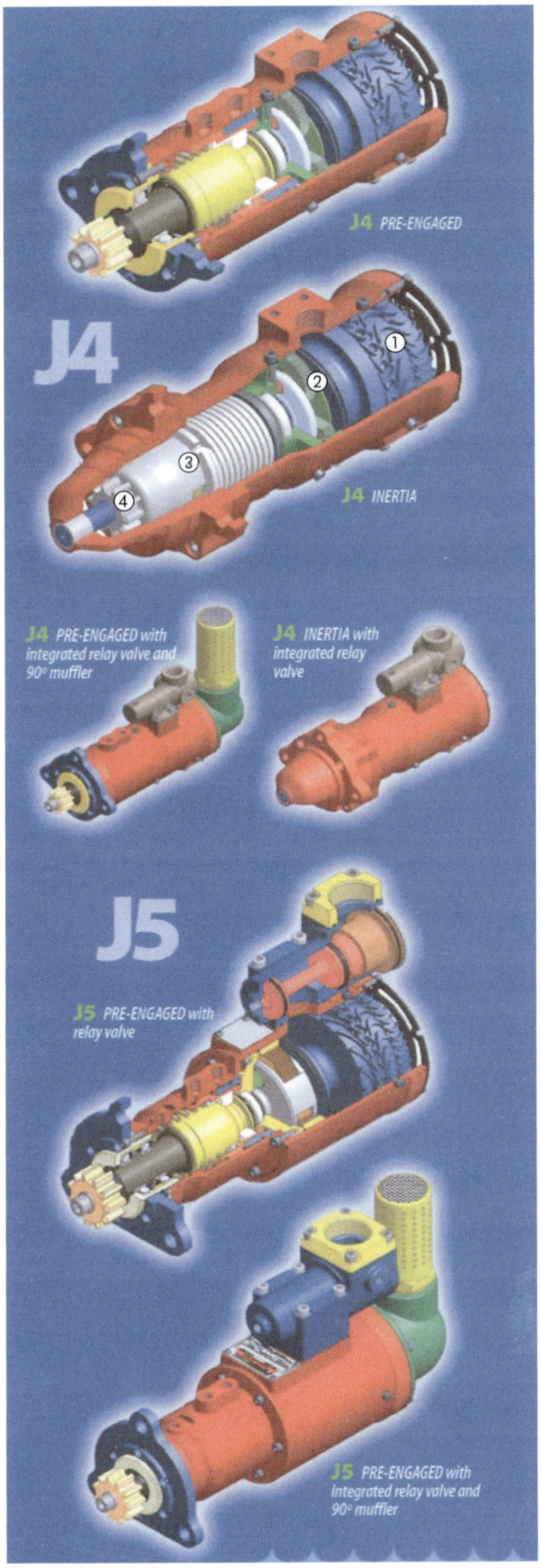

◄

A series of rotating starting-air turbines with different air capacities.

The blue section in three of the cut-away pictures is the air section.

1 *rotor with small blades on the outer surface*
2 *gear box*
3 *bearing section with space for positioning the gearwheel 4*
4 *gear drive*

There are many various types. Important: The air discharge of the starting motor is designed to reduce noise emissions.

14.2.5 Method E: Air Start with compressed air fed to the starting-air valve mounted directly on the cylinder

This system is applied in all two-stroke crosshead engines and frequently in large four-stroke medium-speed engines.
Each cylinder has a starting-air valve which feeds air into the cylinder just after the piston's TDC position and sets the piston in a downward motion.
The starting-air valves are opened by starting-air control valves. The starting-air valves open when the exhaust ports or valves are shut, as the starting air would escape without exerting any force on the piston. The starting-air valves remain open for a certain number of crank degrees. This is referred to as the starting arc. Therefore, in two-stroke engines the starting-air arc has a maximum of approximately 110 crank degrees and in four-stroke engines 130 to 140 crank degrees. As the combustion process of **two-stroke engines** is **360 crank degrees**, at least **four cylinders** are required in order to start the engine in any piston position. Four times 110 crank degrees = 440 crank degrees.
In **four-stroke engines**, the combustion process is **720 crank degrees**. This means that at least **six cylinders** are required to start the engine in any piston position.

▲

An air-starter for Yanmar diesel engines.

◄

A cut-away image of an air-starter with rotor blades.

▶

The starting-air valve mounted on the cylinder cover of a two-stroke crosshead engine from manufacturer Wärtsilä Sulzer, type RTA96C, category IV.

1 *starting-air valve*
2 *starting-air duct*

▶

The starting-air valve from a Wärtsilä Sulzer, type RT-FLEX96C, category IV.

This is an electronically controlled starting-air valve.

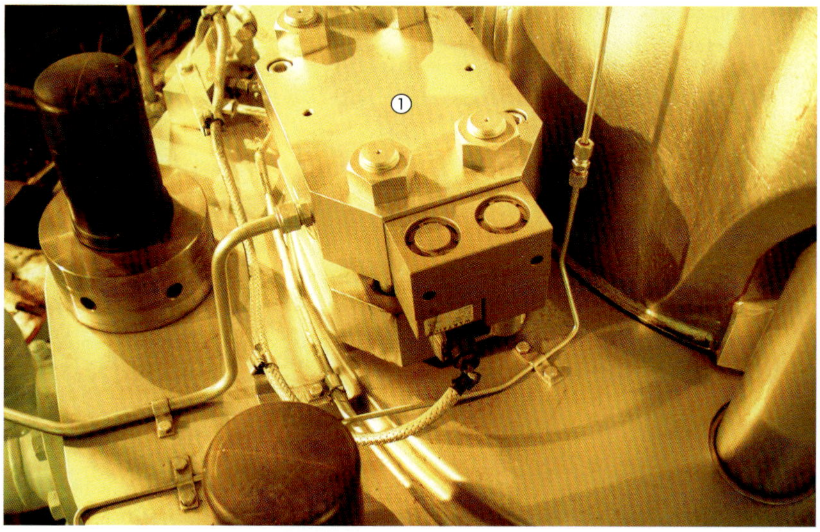

▶

The water-cooled two-stage high-pressure starting-air compressors for a large two-stroke crosshead engine.

They keep the large air receivers under pressure automatically.

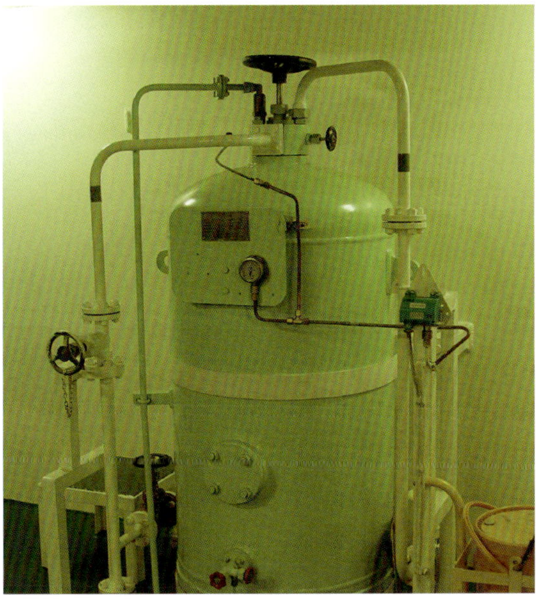

▲

The air receiver in four-stroke medium-speed diesel engines, category III.

▲

The air receivers for a large two-stroke crosshead engine, category IV.

The air receivers are manufactured and installed in accordance with regulations. Safety is a priority when working on these systems with pressures of 30 bars!

▼

The starting curve of a four-stroke medium-speed diesel engine, category III.

The starting curve is in this case 113 crank degrees from the start at A1 to the finish of the starting air at A2.

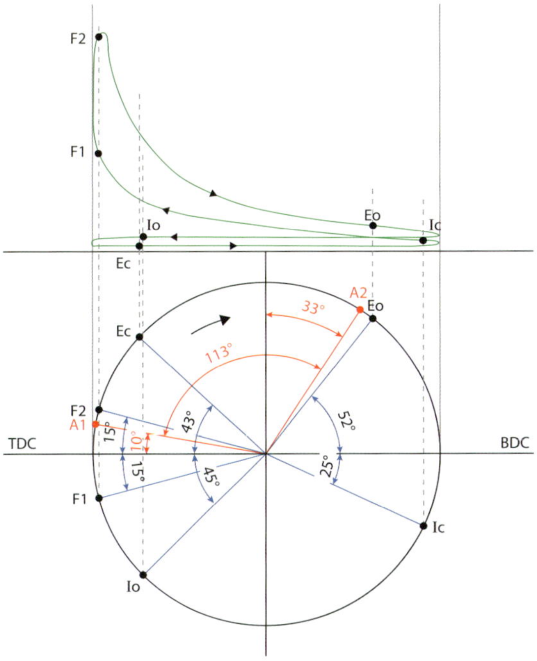

▼

The starting curve of a two-stroke low-speed crosshead engine, category IV.

The starting curve is 100 crank degrees, beginning at the start A1 and ending at A2.

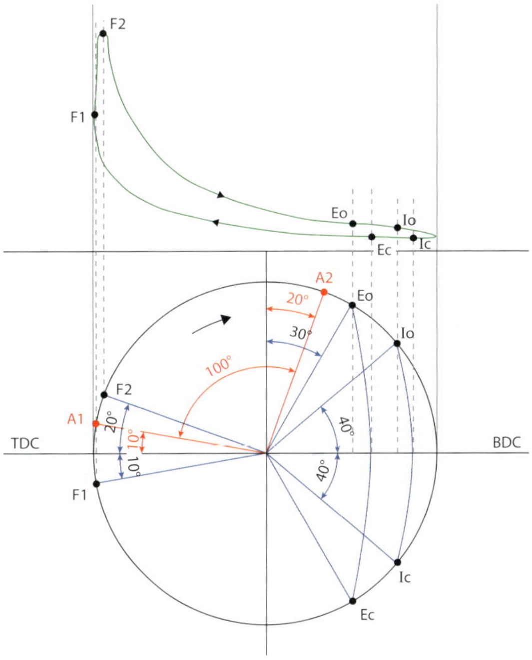

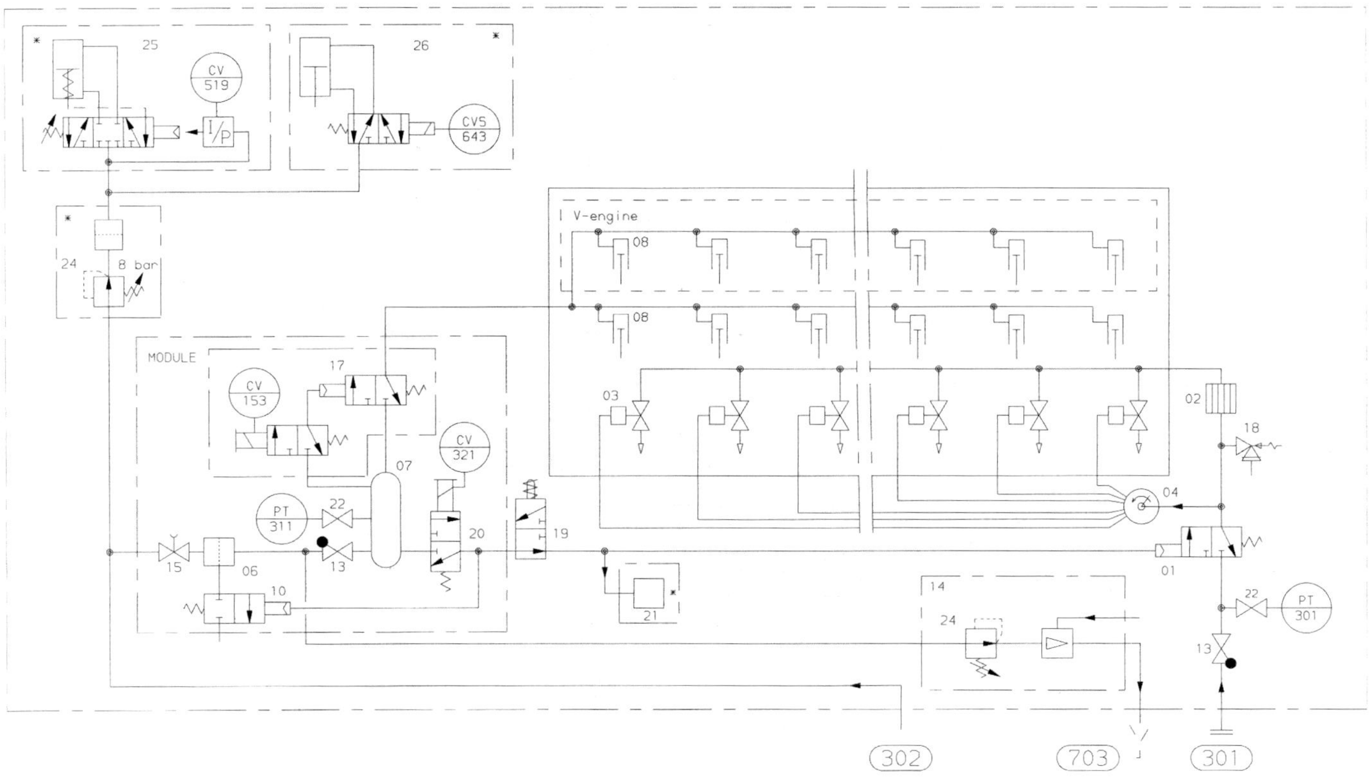

▲

The starting-air system of a four-stroke Wärtsilä 38 diesel engine, category III.

Description: a starting-air valve is mounted on every cylinder. In an in-line engine on all the cylinders and in a V- engine on a bank, therefore on half the number of cylinders.

At 301 the starting air under a pressure of maximum 30 bar enters the system via non-return valve 13, main starting-air valve 01, flame arrestors 02 and the six starting-air valves 003 on the cylinder heads of this six-cylinder engine. Activation of the starting-air valves on the cylinder heads is controlled by the starting-air distributor 04 mounted on the engine camshaft ensuring that the starting-air valves only open during the combustion stroke. If the electric turning motor has engaged the flywheel, safety valve 19 ensures no starting air enters the system via the main starting-air valve 01.

Via connection 302 the pneumatic control, air is supplied to the control unit, to operate the waste-gate 25 and the bypass valve 28.

Control unit: This comprises a ball valve 15, an air filter 06, an automatic drain 10, a non-return valve 13, a buffer header 07, a stop cylinder for the high-pressure fuel pump and a starting-air valve 20.

On every high-pressure fuel pump, a pneumatic stop cylinder is mounted ensuring that air and fuel are not supplied to the cylinder at the same time. The fuel-adjusting spindle on every pump is set to 0.

The module in a normal starting operation supplies control air to the main starting-air valve 01.

The oil-mist detector 14 is also supplied with control air. When mist is detected, the engine's speed and therefore load is decreased.

The entire system is electrically activated and electronically controlled.

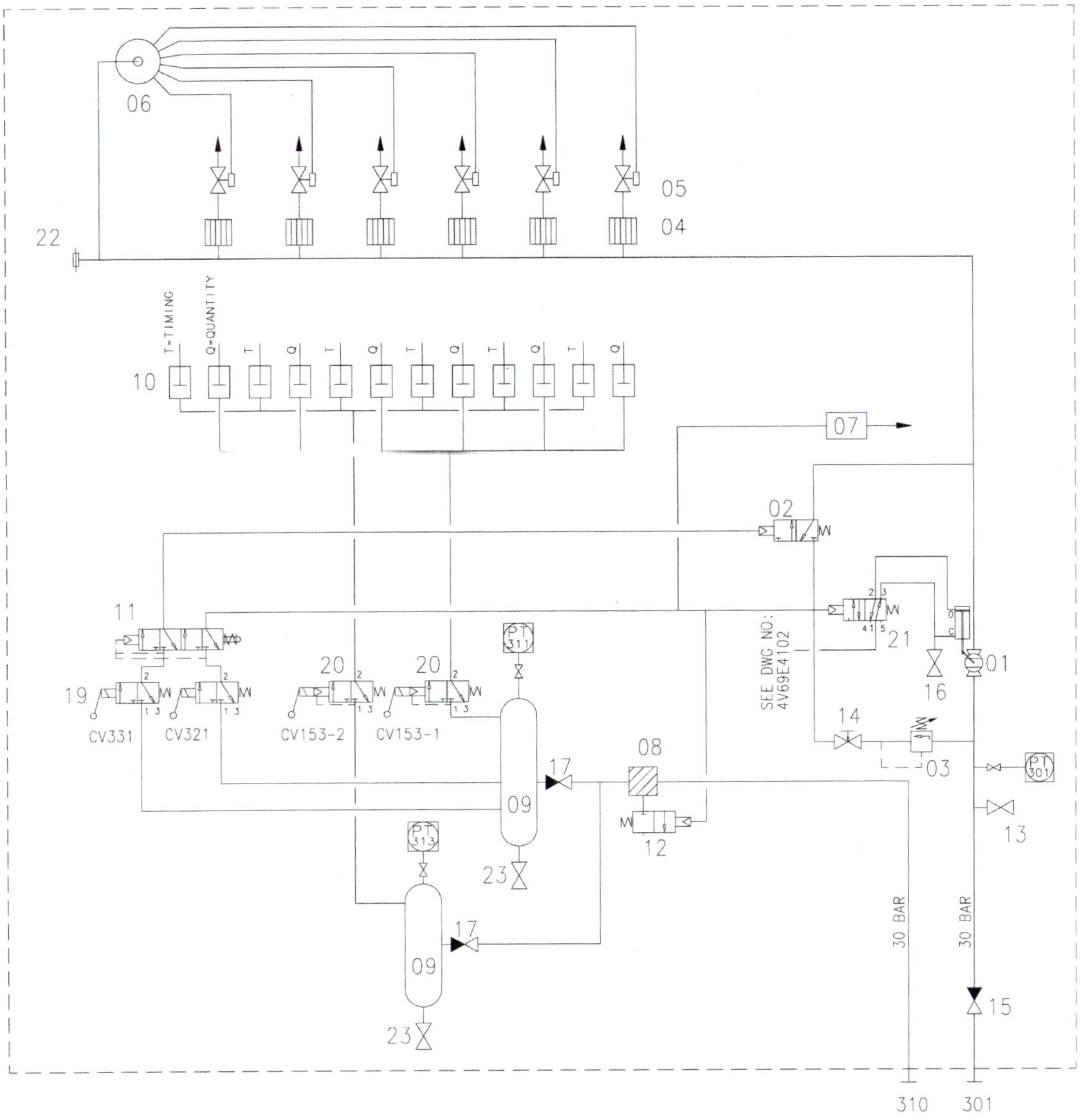

▲

The starting-air system of a four-stroke medium-speed Wärtsilä 64 diesel engine, category III.

01 *main starting valve*
02 *slow turning valve*
03 *pressure control for slow turning*
04 *flame arrestor*
05 *starting-air valve in the cylinder head*
06 *starting-air distributor*
07 *starting booster for governor*
08 *air filter*
09 *air receiver*
10 *pneumatic cylinder*
11 *blocking valve on turning gear*
12 *valve for automatic water draining*
13 *drain valve*
14 *throttle valve*
15 *non-return valve*
16 *ball valve*
17 *non-return valve*
18 *3/2 solenoid valve*
19 *3/2 valve, pilot-controlled*
20 *4/2 valve, pressure-controlled*
21 *bursting disc, break pressure 40 bar*
22 *ball valve*

Pipe connections
301 *starting-air inlet 30 bar*
310 *control-air inlet 30 bar (including emergency stop)*

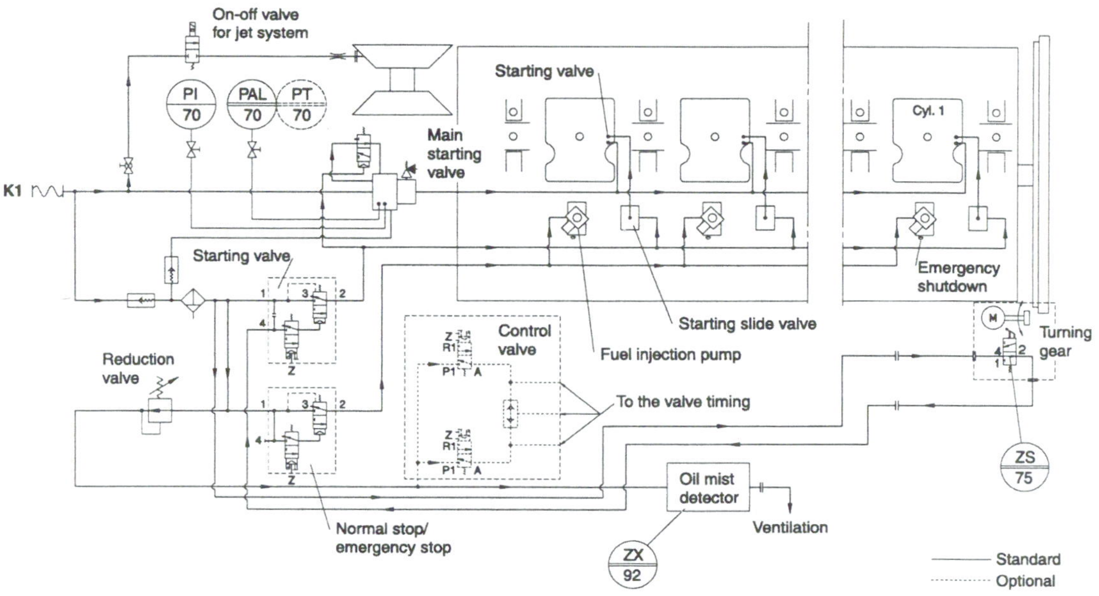

▶ The starting-air system of a medium-speed diesel engine of MAN Diesel, type L32/40.

Every cylinder has its own starting-air valve and control valve. The control valve opens the starting-air valve for the cylinder during the power stroke. The fuel-pump output is pneumatically set to zero, and the emergency stop works with this system.

14.3 Reversing the engine

This pertains mainly to large propulsion engines with a fixed propeller. A fixed screw does not have controllable blades and therefore has to be driven by a directly reversible engine.
In practice, this is often applied in large two-stroke crosshead engines and occasionally in the larger four-stroke engines. The smaller two-stroke crosshead engines are often connected to an adjustable propeller, also found with most four-stroke engines.

In directly reversible engines, the starting- and fuelling systems must comply with the following requirements.

- The engine can start on air in two rotational directions.
- The fuel injection must commence at the correct time in both directions.
- The opening and closing of the exhaust valves in two-stroke crosshead engines with uniflow scavenging must take place at the correct time.
- Safety devices must be installed in order to prevent dangerous situations ensuing from reversing the rotational direction.

A few examples of reversing devices in diesel engines.

14.3.1 Example 1: Large two-stroke crosshead engine, directly reversible

Series MC – MAN Diesel K98MC.

K = short piston stroke ratio ± 2.8 (stroke = 2660 mm)
98 = cylinder bore in centimetres
M = engine program
C = camshaft controlled

In the modern two-stroke crosshead engines of the three remaining manufacturers, the reversing devices are fairly simple.
The fuel cams remain in the same position as during the forward rotation.
Modification of the pressure stroke of the fuel cams occurs by reversing the roller-end of the plunger pump attached to a reversing lever, against the rotation direction of the new position (reverse rotation). **See the detailed drawing of the complete starting system.**
For safety reasons, the roller-end and the lever mechanism are locked in this position.

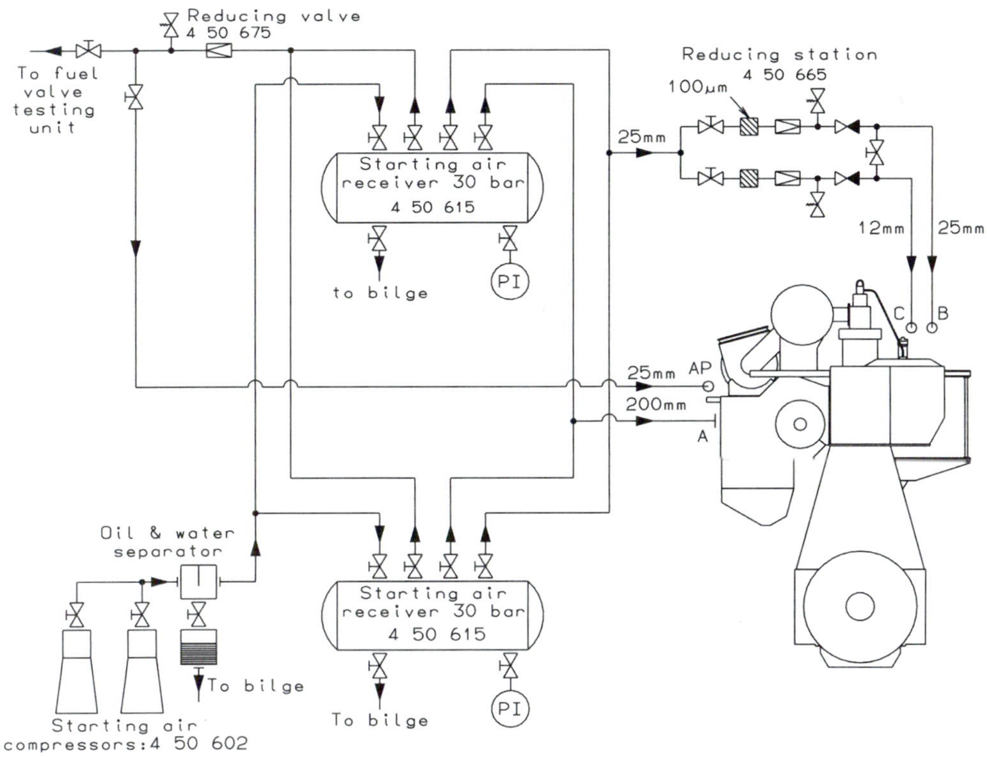

A: Valve "A" is supplied with the engine
AP: Air inlet for dry cleaning of turbocharger
The letters refer to "List of flanges"

▲
The diagram of the starting-air system of a MAN Diesel K98MC-C diesel engine.

Both starting-air compressors supply air at a pressure of maximum 30 bars via an oil- and water separator to two large air receivers. The receiver capacity of both is sufficient for twelve engine starts without starting the compressors.

The **connection A** with a diameter of 200 mm supplies the starting-air valve on every cylinder.

The **connection B** is fed with compressed air via a fine-filter (100µm) and reduction valve with an output pressure of 7 bars. This is control air for manoeuvring and operating the exhaust-valve air springs.

The **connection C** is fed with compressed air via a fine-filter (100µm) and reduction valve with an output pressure of 7 bars. This is safety air for the shutdown system of the main engine.

The **connection AP** is fed with compressed air via a reduction valve with an output pressure of 10 bar and is used for turbocharger cleaning (soft blast), and a negligible volume used for fuel-valve testing.

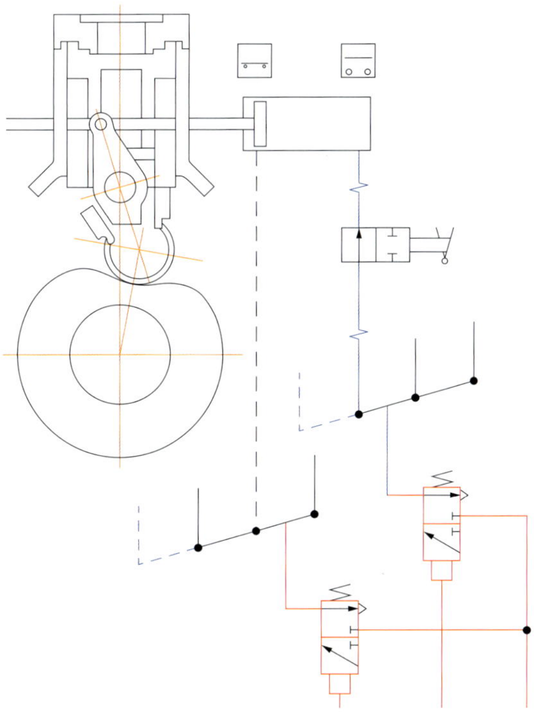

▲
For directly reversible engines, the camshaft can be moved axially allowing the fuel injection to take place at a different time. The axial movement of the camshaft occurs hydraulically.

▶

Starting-air system MAN Diesel 98MC.

The starting-air system consists of a manoeuvring console on the engine for emergencies, for example, if the remote control from the bridge does not work or after maintenance for testing purposes. The starting-air valve is opened using control air allowing air to enter the cylinder and the piston begins the power stroke. Furthermore, there is a 'slow-turning valve', which allows the engine to turn slowly using starting air for testing purposes after maintenance.

The starting air with a maximum pressure of 30 bars is supplied via A.

Main starting-air valve: main starting-air valve in the supply line to the engine.

Slow-turning valve: valve for slow engine turning.

Starting valve: starting-air valve on the cylinder.

Starting-air distributor: distributes the control air to every starting-air valve for opening of this valve.

Manoeuvring console: manual starting console on the engine.

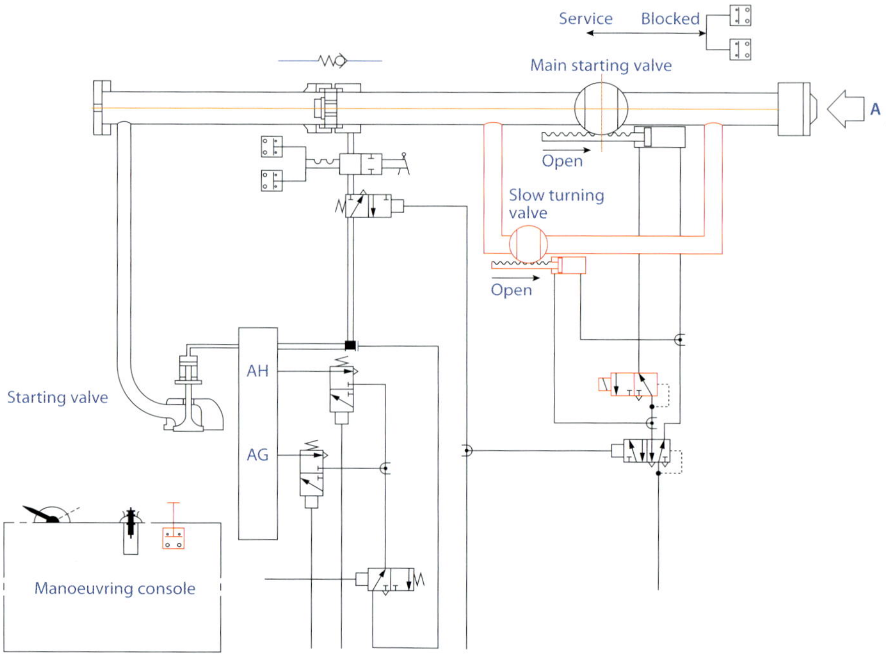

14.3.2 Example 2: Large two-stroke crosshead engine, directly reversible – Series RTA, RTA96C

In this engine series the fuel cams are arranged in such a way that they can be moved using the engine lubricating-oil system.

In these engines, a pump block consists of two fuel pumps. Each cam is served by one servomotor.

▼

An overview of the starting-air system of the Wärtsilä RTA96C diesel engine.

Above left: cylinder head with the starting-air valve

Above left middle: main starting-air valve

Above right middle: blocking valve on the turning gear

Above right: manual control on the engine

Below left: starting-air distributor to operate the starting-air valves in the cylinder head

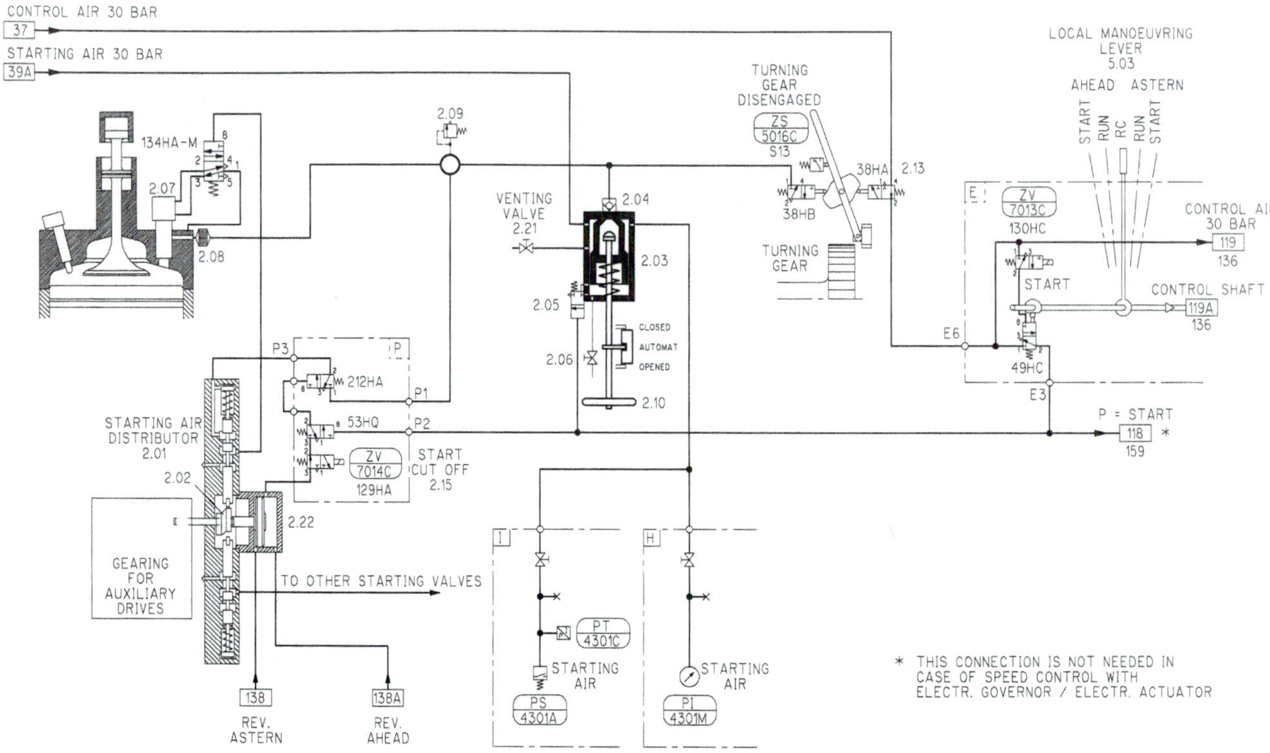

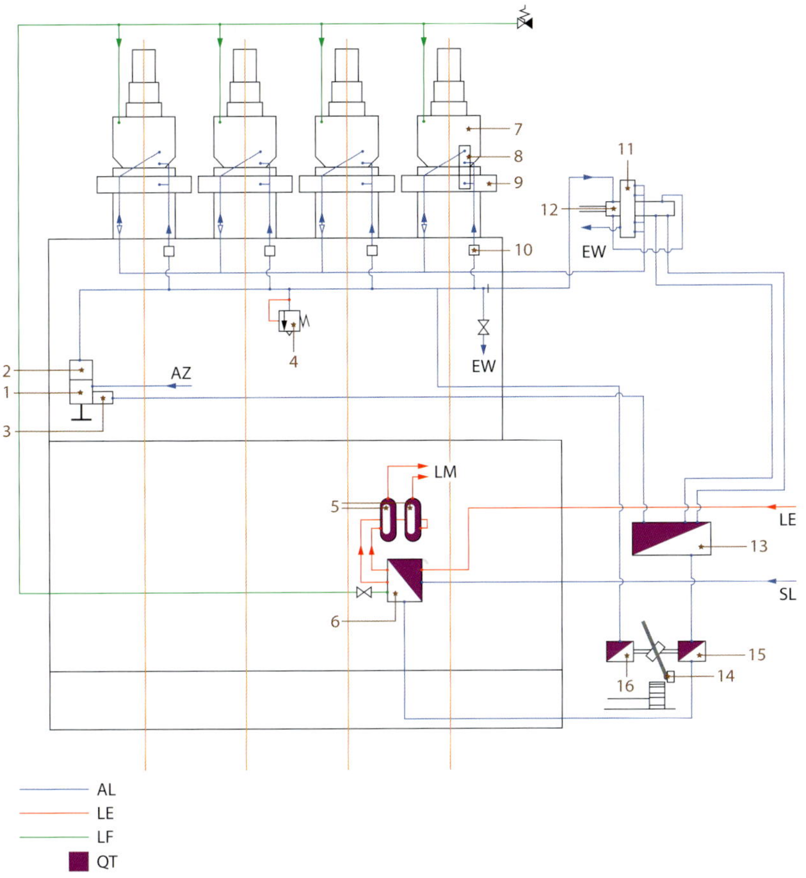

◀ Starting system of the Wärtsilä RTA96C diesel engine.

1 main starting-air valve
2 non-return valve
3 control valve
4 safety valve
5 air bottles for control air
6 control-air module
7 exhaust-valve casing
8 starting-air valve
9 cylinder head
10 flame plate
11 starting-air distributor
12 starting-air shut-off valve
13 pneumatic control unit
14 turning gear
15 shut-off valve on turning gear
16 blow-off valve on turning gear

AL starting-air connection
LE control-air connection
LF air-spring connection for the exhaust valve
QT components delivered by other suppliers
SL air connection for stand-by and control air
LM air for control and safety
EW blow-off and drain
AZ starting-air supply

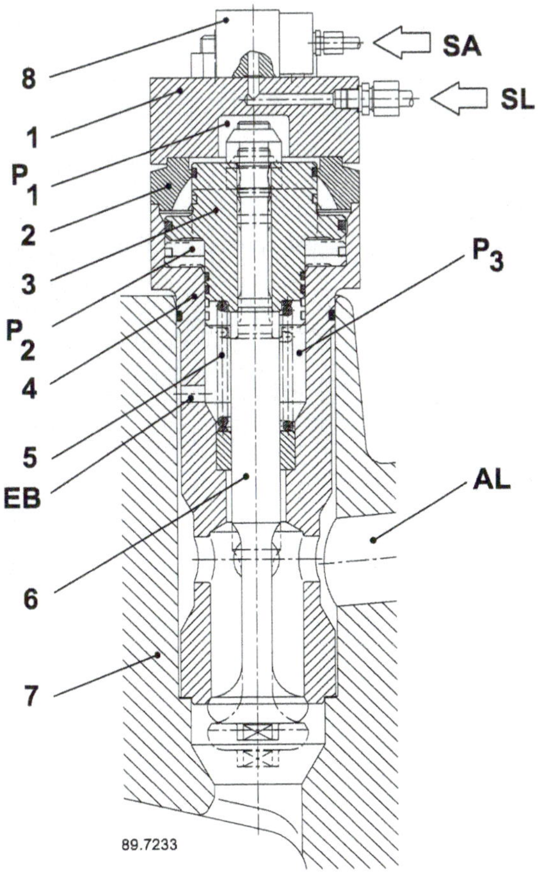

◀ A starting-air valve in the cylinder head of a Wärtsilä RTA96C low-speed two-stroke crosshead engine.

1 cover
2 ring
3 operational piston
4 casing
5 pressure spring
6 valve spindle
7 cylinder head
8 control valve

SA control-air connection from control-air distributor
SL control-air connection from control-airline
AL starting-air connection
EB connection duct for starting air
P1, P2 and P3 spaces

14.3.3 Example 3: Similar engines with the common-rail system – MAN Diesel ME (electronic), Wärtsilä Sulzer FLEX (electronic)

The fuel injection is controlled electronically in engines from both manufacturers.
As the gas exchange in modern engines occurs almost symmetrically for the current equal-pressure systems, adjustment of the exhaust valve in both mechanically- and electronically controlled engines is no longer required.
Controlling the starting air is simple as this can now be electronically activated.

Also, see Chapter 9, *Fuel-injection systems.*

▲

A cutaway of a starting rotor with on the left, the turbine and right, the coupling gear for driving the flywheel. This is mainly used for categories II and III diesel engines.

◄

A medium sized two-stroke crosshead engine during manufacture. An eight-cylinder MAN Diesel MC 50.

The starting-air valve is attached on the left side of the cylinder head (1).
All category IV engines use this system.

◄

Large engines in categories III en IV still have the starting-air valve in the cylinder head.
An electrically driven turning motor (1) is used to let the engine run at a very slow speed, for example, during maintenance. This is named 'cranking'.

>CH 15

Speed control

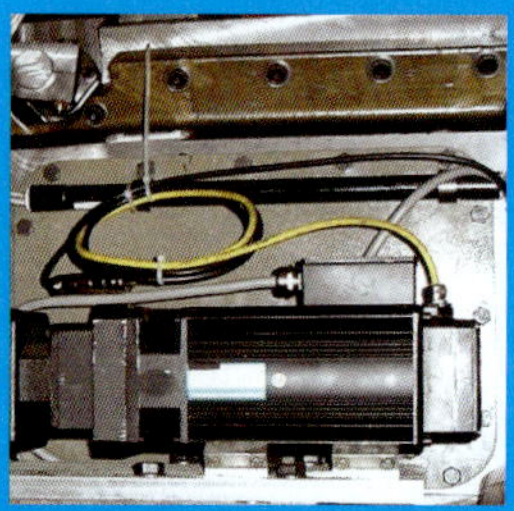

The governor is responsible for the correct and safe engine speed and is indispensible for every diesel engine.
Shown, a Woodward governor, type EGB, on a four-stroke medium-speed MAN Diesel propulsion diesel engine running on HFO on the passenger ship 'Maasdam' of the Holland America Line.

The passenger ship 'Maasdam'.

15.1 Introduction

Speed regulators or governors are an integral part of fuel-injection systems in diesel engines.
In general, the fuel-injection governor ensures that a diesel engine operates at a predetermined speed at varying loads.
The speed tolerances are very small, especially in gensets, which require a constant frequency.

15.2 Summary

Governors have the following functions:
- fuel-quantity control when the engine is started;
- fuel-quantity control during engine idling;
- maintain a constant speed at fluctuating loads, such as in gensets with alternating power decreases as well as in propulsion systems with varying cargo- and weather conditions;
- ensure that the maximum number of revolutions is not exceeded. Exceeding the maximum speed may cause damage.

15.2.1 Examples of damage

Drive-gear fractures due to impermissibly high acceleration and/or retardation forces.
Common damage: breakage of the connecting rod.
The piston can collide with the inlet and exhaust valves at too high a speed, as the valve springs cannot close the valves in time.

If the engine torque no longer matches the torque required by the propeller or generator, the result will be an angular acceleration or retardation: the speed alters. An adequately working governor will regulate the quantity of injected fuel via the (fuel) adjusting spindle in order to achieve the predetermined value. So the governor operates in a state of equilibrium.
The speed of the engine changes slightly continuously and the finely tuned governor intervenes and modifies the injected fuel quantity.

15.3 Types of governors

With regard to their operation, there are roughly three different types of governor:
1 mechanical governors;
2 hydraulic governors;
3 electronic governors.

There are often combinations of all three.

► The Woodward governors on three six-cylinder four-stroke medium-speed Wärtsilä 20 diesel engines used in gensets on the 'Oranjeborg' of Wagenborg Shipping, The Netherlands. The gensets run on heavy oil.

1 *camshaft drive*
2 *governor type 3161*

15.3.1 Mechanical governors

These have been in use since the first steam engines, steam turbines and internal combustion engines. Today, they are predominantly used in small engines in engine categories I and II. This is because they are reliable and relatively inexpensive. Their operation is based on a principle formulated over 200 years ago by the renowned physicist James Watt.
Here the centrifugal forces work on the rotating mass of the governor weights, in counter-resistance to the governor-spring thrust. The centrifugal force is squared in relation to the speed.
These systems have two different types:

1. The springs with which the speed can be set are embodied in the rotating mass.
2. The springs are placed outside the rotating mass.

In general, one can say that the greater the spring depression, the greater the resistance and the higher the speed before the injected fuel quantity is reduced by the governor.

Most mechanic governors operate within a small deviation. The deviation is between 2 and 10%, depending on the type of actuation and the type of engine.

The various parts of the governor have different functions:

- a speed-adjustment mode;
- an actual speed-measurement mode;
- a comparison of the actual speed and the desired speed mode;
- a means in which the governor modifies the fuel quantity supplied to the engine;
- an engine-stabilisation mode, that is, maintenance of a certain number of revolutions after the fuel quantity has been altered.

Setting the speed

The hydro-mechanical governors utilise a spring. The greater the pressure exerted on the spring, the higher the speed. The higher the spring resistance, the greater the quantity of fuel supplied to the engine.

Measuring the speed

The governor must receive a force proportional to the speed of the engine. In hydro-mechanical governors, this is supplied by the centrifugal forces of the flyweights, which are rotationally driven by a mechanical connection to the engine.
In electronic systems, this force originates from measuring the frequency of a magnetic pick-up and a converter, which are directly linked to the speed of the engine.

◄ A Bosch mechanical governor mounted directly on the block fuel pump rebuilt from an old, eight-cylinder Kromhout diesel engine.

1. *block fuel pump, in-line*
2. *manually operated deaerating pump*
3. *mechanical governor*
4. *speed-adjusting handle; the engine is switched off*
5. *minimum speed*
6. *normal speed*

Modifying the fuel supply to the engine

The hydro-mechanical governors or actuators (electronic speed measurement) usually have a rotating or a linear output shaft that is attached to the fuel-adjusting spindle of the engine. When the governor needs to make a modification to retain the speed or load, the output shaft is moved in the required direction. At a constant fuel supply, the engine generates a certain power output. Usually, the governor measures the speed and adjusts the fuel supply that in turn modifies the power output.

Flyweights

By increasing the speed, the flyweights spin further outwards and the fuel supply to the engine is reduced. the speed returns to its set value. Obviously, the reverse also occurs. At a decreased speed the weights rotate inwards, the fuel supply to the engine is increased, and the speed rises to its set value. Friction in the system produces a time delay before the speed is altered. This is called the 'dead band', when momentarily nothing happens.

Disadvantage of flyweights

In larger engines, the flyweights do not have sufficient force to serve the larger fuel systems. In order to operate these fuel systems, often the fuel-adjusting spindle, an auxiliary hydraulic system, is used. Hence the term: hydro-mechanical governor. Most hydro-mechanical governors and actuators use the governor drive to operate the hydraulic-oil pump. They often are gear-driven pumps.

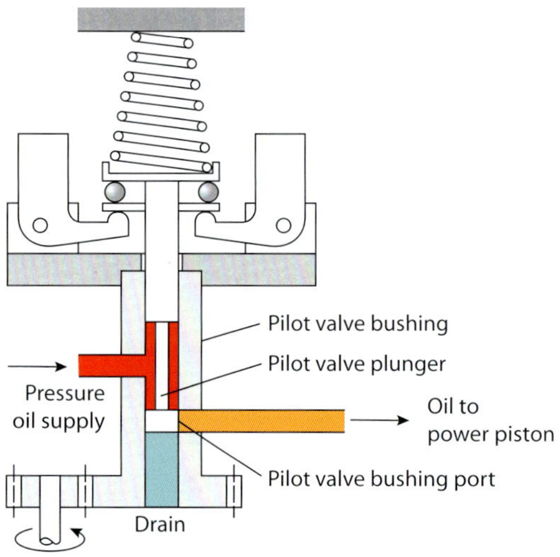

◀ Drawing showing the working principle of the mechanical governor.

The entire device is driven by the engine from below. The governor flyweights pivot outwards exerting pressure on the pilot-control valve, thus depressing the spring. The control valve, attached to the shaft regulates the oil flow to the fuel-control piston. Engine speed lower than the pre-set: the spring extends and forces open the pilot valve, thus causing the piston to allow more fuel to enter the engine. The engine accelerates to slightly over a pre-set speed and runs more or less at that speed.

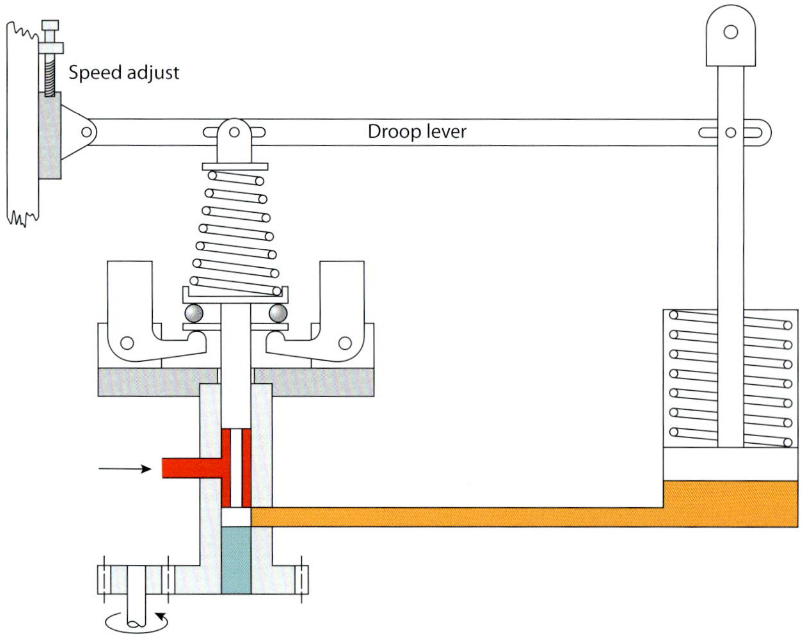

◀ The complete working principle showing the piston, spring mechanism and levers.

Top right: the governor is connected to the fuel-control system.

15.3.2 Hydro-mechanical governors

Here the adjustment forces used to control the fuel-pump capacity are not directly provided by the measuring device, but by a hydraulic system, which allows extra features to be added to the governor. One of the most frequently used hydraulic governors is the Woodward UG-8.
In the world of governors, one uses unusual terms such as 'speed droop'.

Speed droop

This is the governor function that reduces the governor reference speed at an increasing load.
In many governors, the extent of the droop can be set on the governor.

15.3.3 Electronic speed control

These governors are often applied in diesel engines in categories III and IV and increasingly often in category II.
Apart from controlling the set speed, electronic governors serve numerous purposes:

- avoidance of excessively high speeds;
- limiting the maximum fuel supply;
- limiting the maximum scavenging-air pressure;
- controlling the maximum fuel quantity available when the engine is started;
- controlling the actuation of a propeller screw with, for instance, two engines;
- controlling the delay of, for instance, switching on a coupling or in general alarms.

Operation of the electronic governor or digital speed governor (actuator)

The basic principle of the electronic governor is its comparison of the set speed and the actual speed.
At a certain discrepancy between these values, (the discrepancy can be configured), a signal is sent to the actuator fixed to the fuel-adjusting spindle and the spindle is adjusted by the actuator until the required speed is achieved.
The actual speed is measured by means of magnetic pick-ups that are often installed close to the flywheel. There are often several (two to four) pick-ups installed for various purposes. They send signals to the governor.
The number of signals is dependent on the speed and is in practice often equal to the number of teeth on the flywheel that pass the pick-up per time unit.
This system is often applied in high-speed and medium-speed four-stroke engines. For low-speed two-stroke crosshead engines, this system is too slow and other systems are used

15.4 Examples of engine configurations with different types of governors

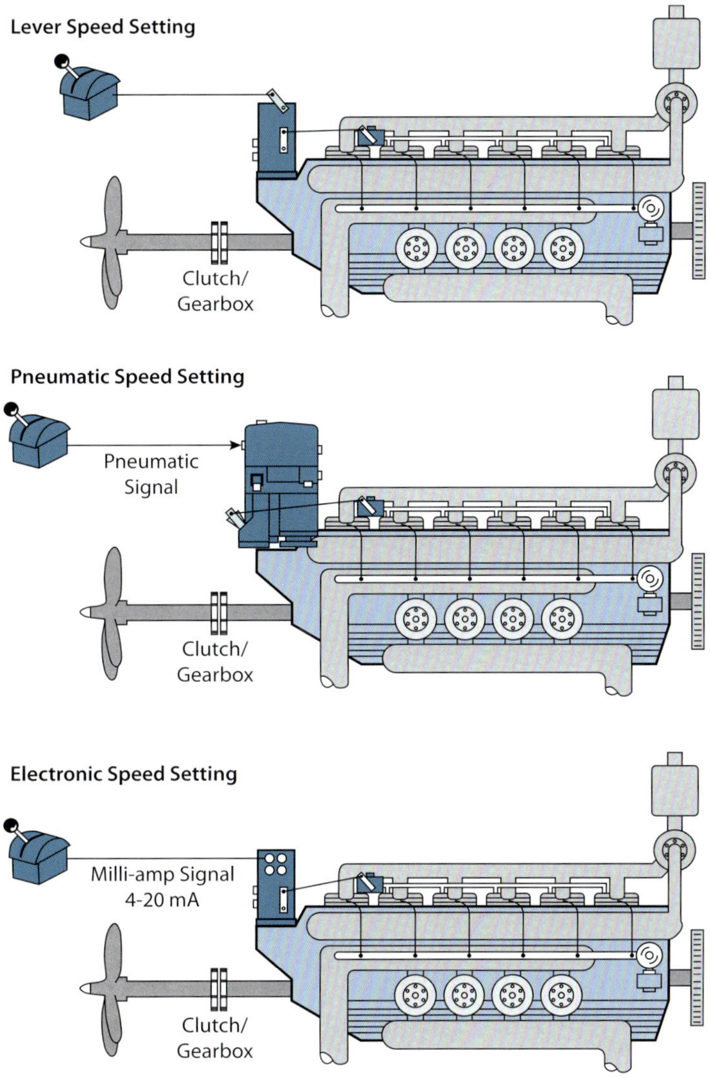

▲

Woodward governors, mechanical.

Various ways to set the speed of a propulsion engine.

First figure: mechanical transmission.

Second figure: pneumatic transmission.

Third figure: electronic control using a 4-20 milli-ampere control system.

▶

Woodward governors, electronic.

An example of a propulsion engine with an electronic governor. The engine speed is measured electronically with a tachometer attached to the gear ring on the turning/ flywheel (MPU).
The electronic control unit controls the actuator that mechanically controls the fuel-control shaft. The speed can be adjusted with an 'electronic' control handle.
MPU = Magnetic Pick Up

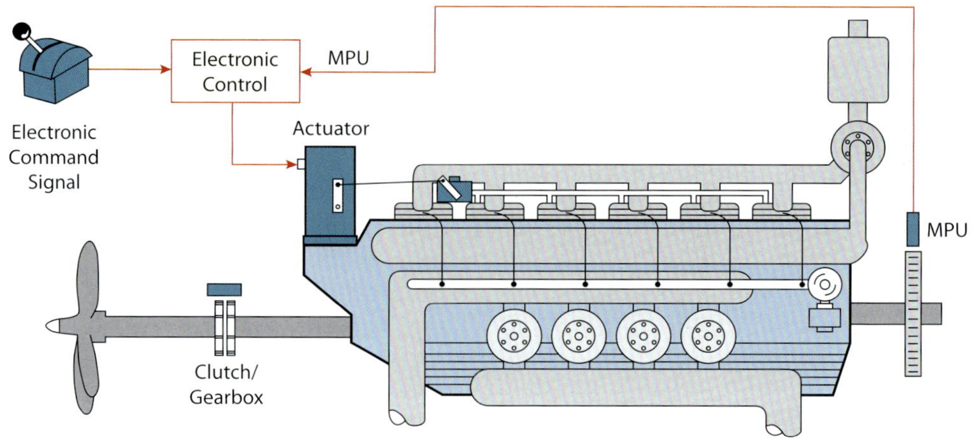

▶

Woodward governors, electronic.

Identical system for two separate propulsion engines, a single propeller each. The hydraulically controlled coupling in the reduction gearing is incorporated.

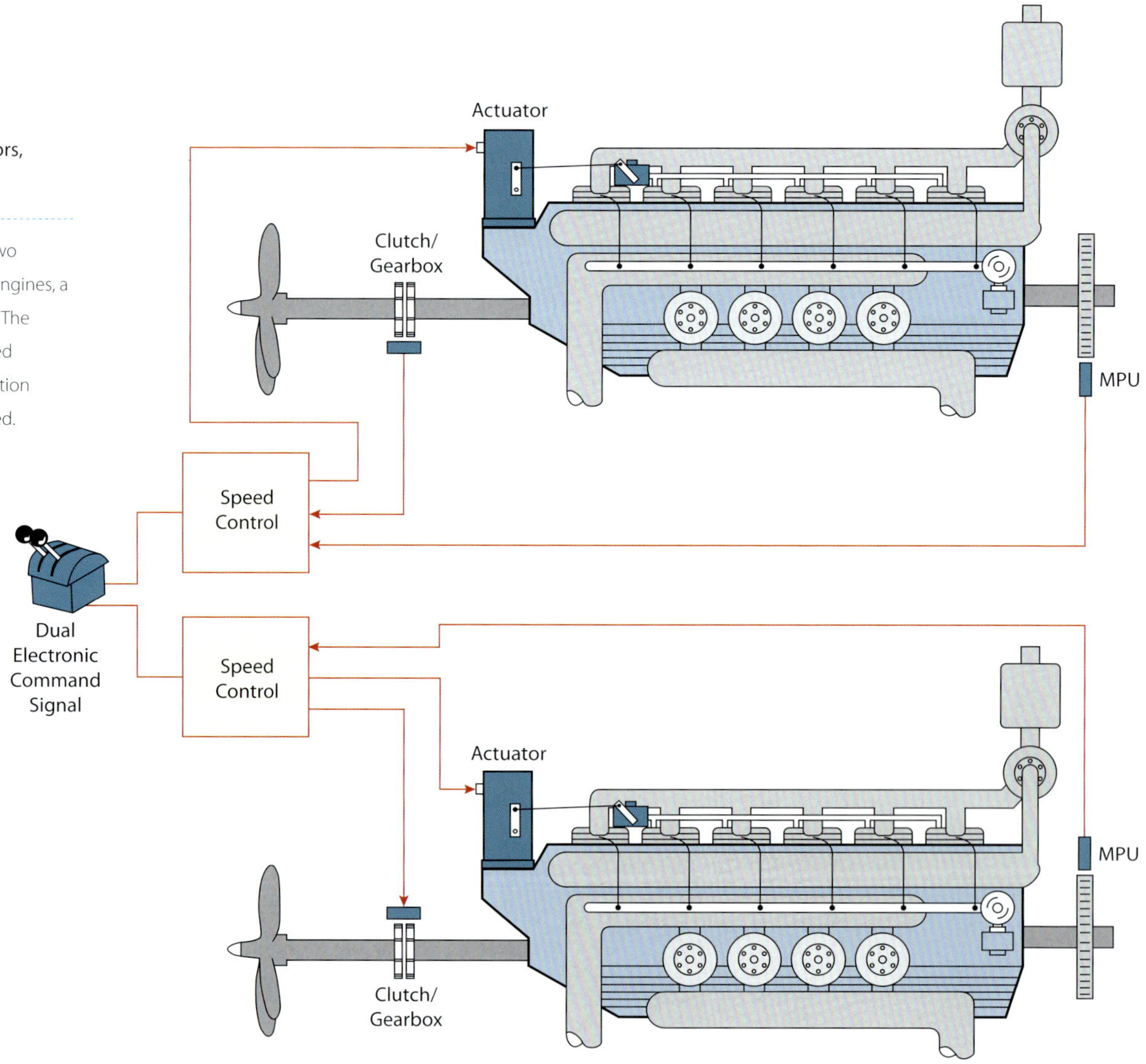

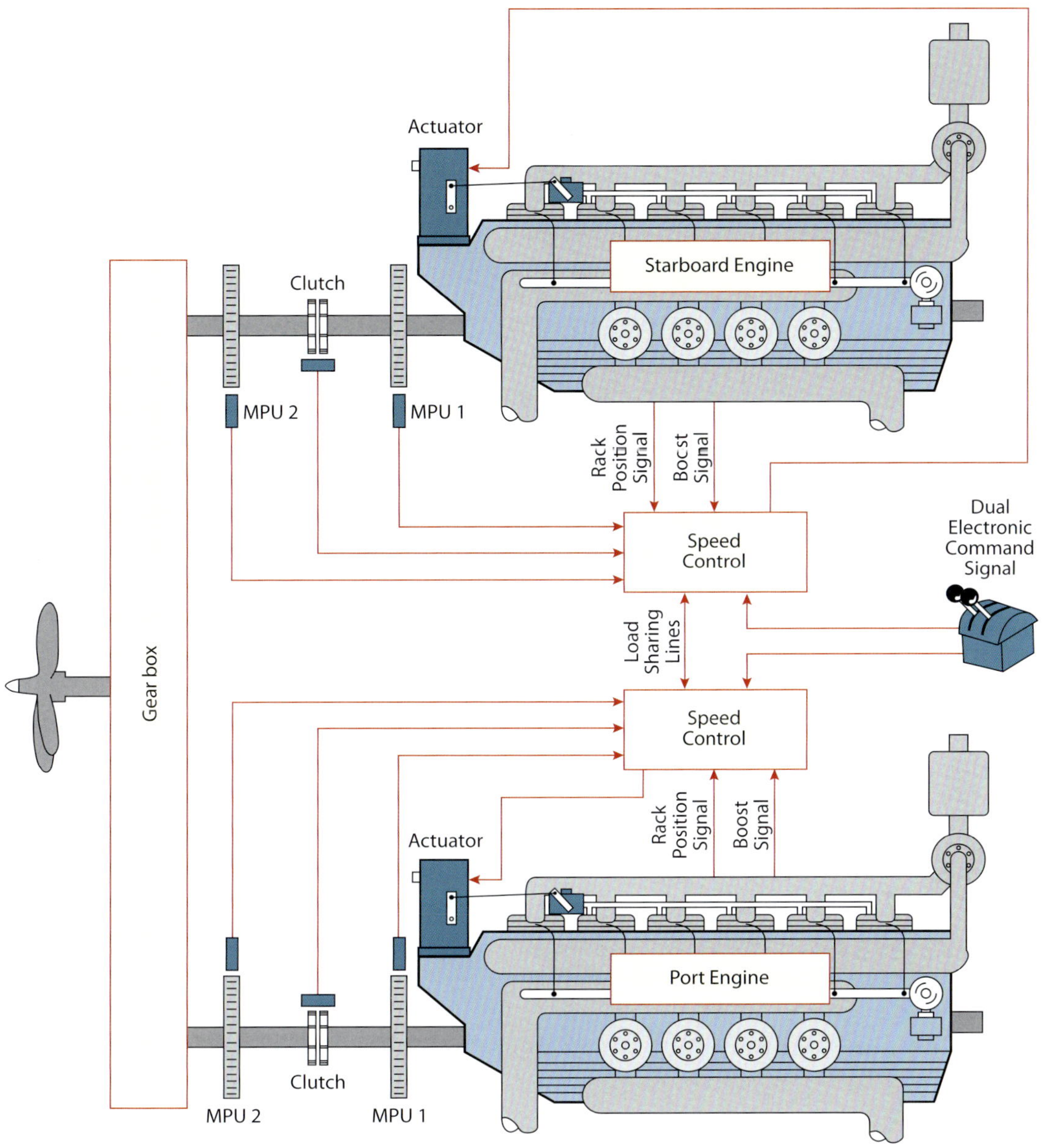

◀ Woodward governors, electronic.

A system using two propulsion engines for a single propeller. The master/slave system. Both engines are electronically controlled. The actuators are electronically controlled from the 'Speed Control' units. The 'Speed Control' units receive data from both, the position of the fuel-control shaft and the turbocharger pressures. This determines the engine load. Two revolution counters are mounted on the output shaft: MPU 1 before the coupling and MPU 2 after the coupling. This allows a comparison between the outgoing shaft speed of the propulsion engine and the incoming shaft speed of the reduction gearing when the connection is switched off. This is necessary in order to achieve the correct engine speed prior to the engagement of the coupling. The connection between both Speed Control units allows a comparison between both loads in relationship to both propulsion engines. At normal speed with two engines, the load is distributed as equally as possible to both engines. One engine has a certain speed to which the other engine adjusts its speed.

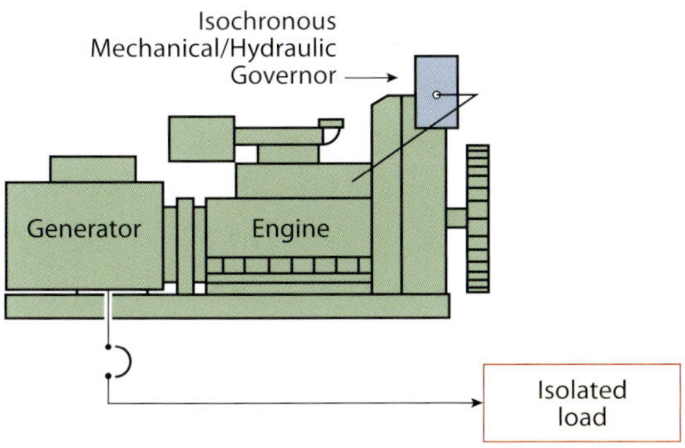

◀ Woodward governor, hydro-mechanical.

A single electric genset with an isolated load. The pre-set governor ensures a relatively constant rpm at various loads.

Woodward governors, electronic.

A system for single- or multiple diesel gensets with automatic synchronisation of the individual sets in the utilities network and separate load-control units.

Apart from the speed control with a magnetic 'pick-up' which controls the actuator and hence the fuel control, the electronic load of each set is separately measured.

AGLC: Automatic Generator Loading Control. This system serves to ensure that the genset operates safely at a certain pre-set load.

SPM-A: Speed and Phase Matching Synchronizer. This provides automatic frequency- and phase-control matching for the generator.

This system prevents undesirable loads.

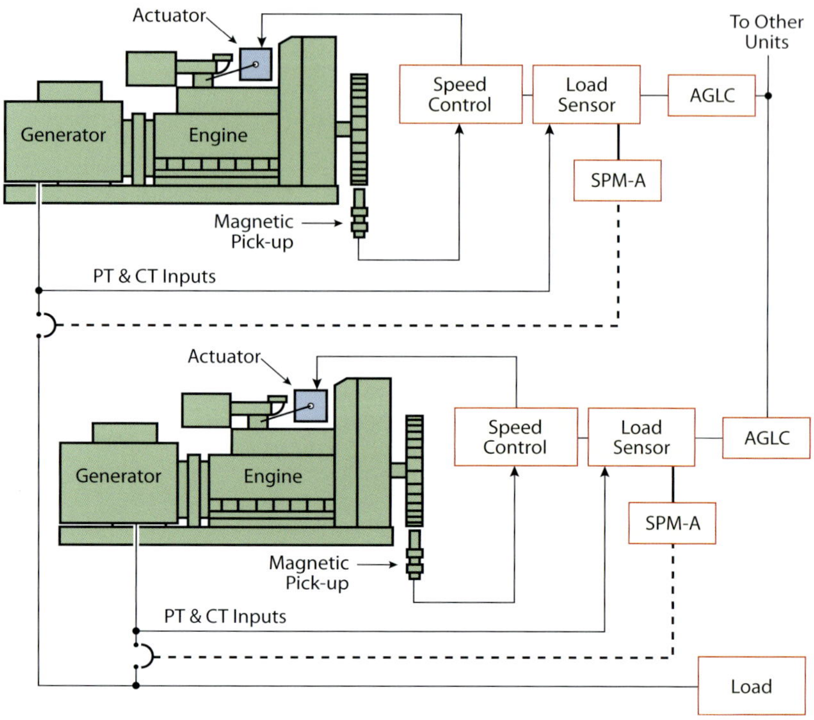

The governor in the complete monitoring and control system for a Wärtsilä 26 diesel engine, Category III.

The Main Control Unit can transmit a start/stop signal to the governor. The electronic governor has two 'pick-up' systems.

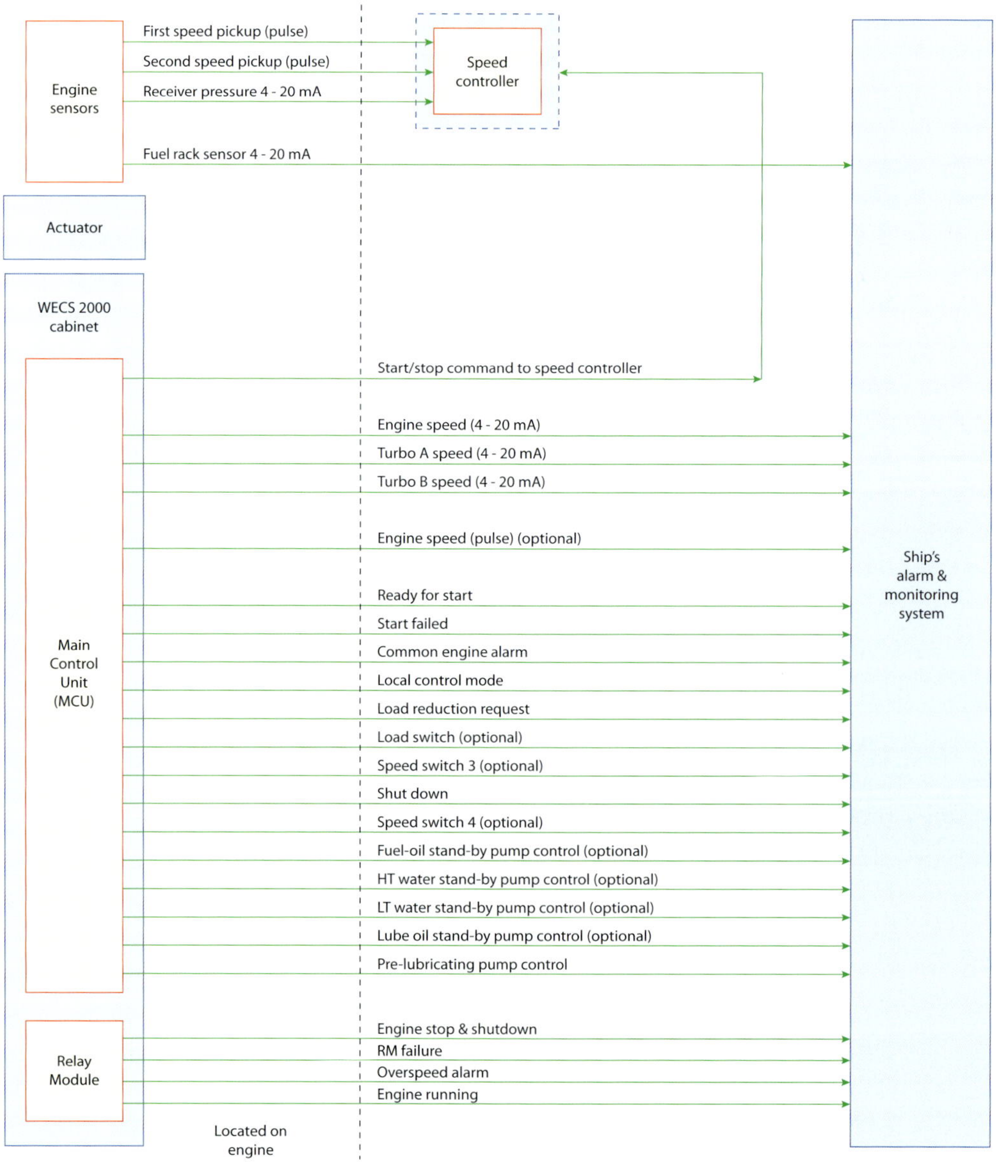

15.5 Theoretical background of governors

Governors are driven by the engine and control the injected fuel quantity in such a way that the predetermined diesel-engine speed remains within strict margins. One could say that a governor controls the energy quantity, in the form of fuel that is supplied to the engine.

In principle, each governor comprises a number of basic components:

1 a way to set the desired speed;
2 a way to obtain an actual speed reading;
3 a way of modifying the actual speed to the set speed;
4 a way of controlling the fuel supply quantity to the diesel engine;
5 a way of stabilising the speed after fuel supply alterations.

1: Setting the required speed
Modern governors have advanced systems to set the speed, which can compensate for varying operating conditions in order to maintain the desired the speed.

2: Governor must receive a signal of the diesel engine speed
In hydro-mechanical governors, this takes place by the centrifugal forces of the flyweights, which can be rotated by a system of drive shafts that are connected to the diesel engine.
In electronic governors, this force is obtained by measuring the frequency with a magnetic pick-up, converter or generator, which is directly related to the speed of the diesel engine.
The frequency is subsequently converted into an electrical force that is made available for the adjustable direction of the fuel. In both instances, the force increases as the engine speed increases.

3: Comparing the actual speed to the desired speed
The 'force 'of the required speed and the 'force' of the actual speed are totalled. The desired speed setting is a force in one direction and the actual speed is an opposing force. When these opposing forces are balanced, there is no change in the engine fuel supply and the speed remains constant.
If the force of the desired speed exceeds the force of the actual speed, the governor will increase the fuel supply to the engine. This is also applied when the force of the desired speed is lower than the force of the actual speed. In hydro-mechanical governors, this difference in forces is absorbed by the so-called thrust block or thrust bearing. In electronic governors, these forces are totalled in the so-called 'summing point'.

4: Ways of controlling the fuel supply to the diesel engine
The hydro-mechanical governors usually have a rotating linear output shaft, which is connected to the fuel system of the diesel engine. In order to alter the speed, the shaft, and therefore the control rod that is connected to the output shaft in the governor, is moved.

5: Ways of stabilising the diesel engine speed
Stabilisation can be achieved in various ways, but all systems have a feedback system that exerts pressure on the thrust bearing or summing point.

Speed, shaft power and load

These terms are often randomly used where governors are concerned.
The governor controls the fuel supply to the engine in order to maintain a constant speed or engine load.

▼
The Woodward governor in a MAN Diesel type-L9-58/64 inline engine for the propulsion of a container ship.

15.5.1 Basic principle of a hydro-mechanical governor

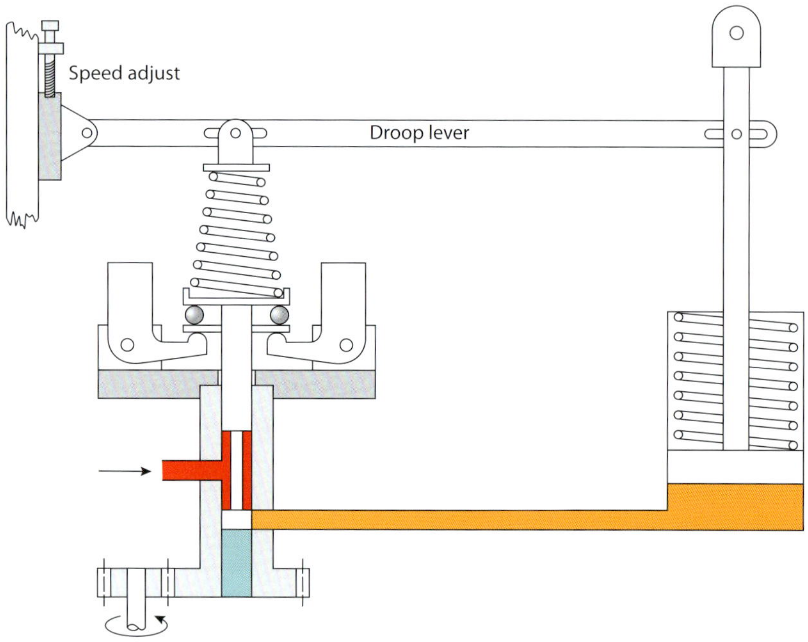

▲

Working principle of the hydro-mechanical governor.

Flyweights are mounted on the shaft, which is connected to the camshaft of the engine via a gear transmission. When the engine speed is increased, the flyweights pivot outwards around their shaft against a predetermined force set for the spring.
Next, a control valve moves downward and allows hydraulic oil to enter the piston under pressure. The top of the piston is provided with compression spring. The hydraulic-oil pressure causes piston movement against the compression spring that subsequently moves a fuel-control rod attached to the high-pressure fuel pumps.

15.5.2 Droop

This term is frequently applied in relation to governors. Droop is a system that prevents an excessive fluctuation of the speed around the set point. This is called 'hunting'.
Droop ensures that the engine maintains a stable speed in spite of fluctuating loads.

15.5.3 Actuators or operating mechanisms

Each governor must exert a certain pressure on the fuel-adjusting spindle of the high-pressure fuel pumps in order to control the fuel supply. The angle under which the actuation takes place is also important. Data with regard to governors always state for which type of engine they are suitable. Information such as output and shaft travel is also of significance.

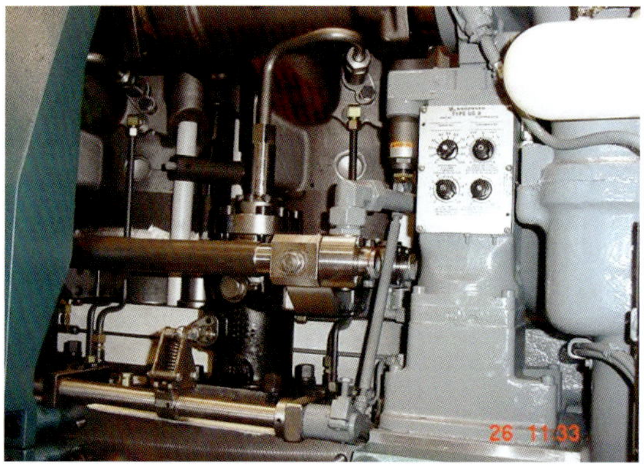

▲

An UG–3 Woodward governor on a medium-speed four-stroke engine of MAN Diesel type 27-38.

▶

An example of droop.

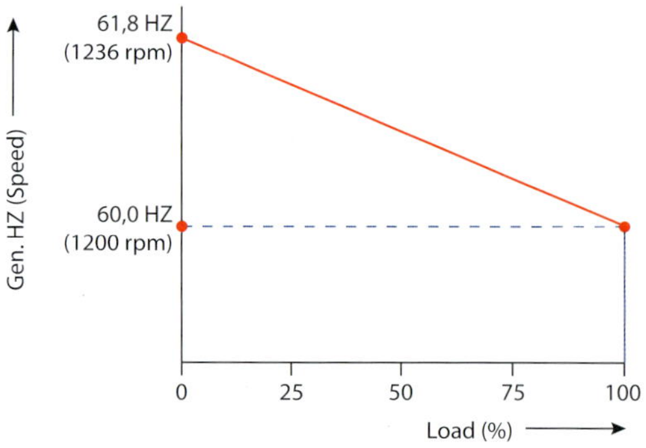

Here thee droop is $\frac{1236 - 1200}{1200} \times 100\% = 3\%$.

Droop formula:

$$\% \text{ droop} = \frac{\text{unloaded engine speed} - \text{full-loaded engine speed}}{\text{full-loaded engine speed}}$$

▲

Detail of the UG-8 Woodward governor with the settings for 'Speed droop' and 'Speed Setting'.

Specifications

Governor Drive	Standard drive is 5/8-36 (inch) serrated Drive shafts are available with 0.187 inch-wide keyways and threaded castle nut to carry gear
Drive Speeds and Requirements	Maximum speed range of 375 to 1500 rpm Recommended constant speed operating range of 1000 to 1500 rpm Requires 249 W (1/3 hp) at normal speed and operating temperature Rotation clockwise or counterclockwise
Work Output	7.1 N·m (5.2 lb-ft) for the UG-5.7 13.2 N·m (9.7 lb-ft) for the UG-8 15.9 N·m (11.7 lb-ft) for the UG-10

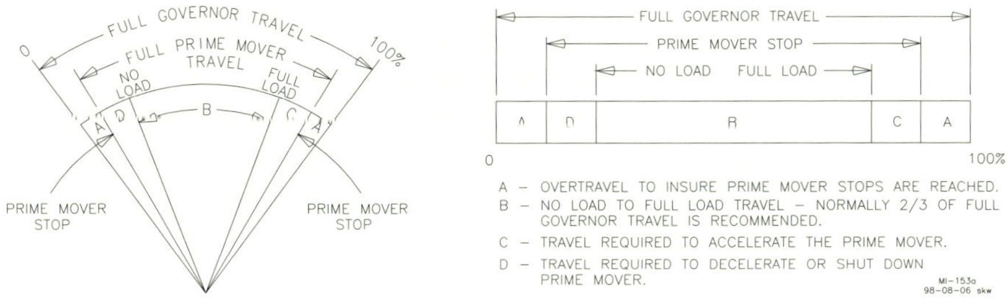

Terminal Shaft	12.70 mm (0.500 inch) diameter, SAE-36 serrations. May extend from either or both sides of the governor Shafts designed for specific applications are available
Linkage	The relationship between engine torque output and governor terminal shaft travel should be linear (very important for gas or dual fuel engines)
Steady State Speed Band	±0.25% of rated speed
Variable Speed Range	375 to 1500 rpm
Droop	Adjustable on the dial governor from 0 to 12.5% at 1500 rpm and from 0 to 19% at 1000 rpm Adjustable on the lever governor from 0 to 26.5% at 1500 rpm and from 0 to 40% at 1000 rpm All droop figures are based on 42° of terminal shaft travel. If less than full shaft travel is utilized, available droop will be decreased by the same percentage as is output shaft travel.
Ballhead/Drive Configuration	A spring-driven, oil-damped ballhead and flexible drive is often used to dampen the high-frequency, low-amplitude torsional vibration which may be present in the drive to the governor Ballheads are also available in undamped natural frequencies of: none, 50, 70, 100, or 150 cpm
Operating Temperature	Gas-fueled engines, particularly those used on gas pipelines, often have not operated at optimum efficiency because automated controls to maintain that peak condition have not been readily available
Pilot Valve Plunger Movement	Balanced between ballhead centrifugal force and speeder spring force
Pilot Valve Bushing	Rotated as an integral part of governor drive shaft
Pilot Valve Porting	A selection of chopper, 2 slotted or 8 round. Chopper gives slow response in acceleration

◄ Data of a Woodward governor type UG/UG-5.7, UG-8 and UG-9.

Apart from the speed range, the mechanical stops for the output-shaft travel to the fuel-control shaft and the forces they may exert on the control shaft are important. Droop is also provided and the operating principle of the governor.

◄◄ A Woodward UG–40 governor with an actuator 4 on a 16 VM 32 C Caterpillar MaK diesel engine.

◄ Installing a governor on the mechanical drive for a diesel engine.

▲ A Wärtsilä 20 medium-speed four-stroke engine running on HFO with a Woodward 3161 governor.

▲ A UG-40 Woodward governor on a Caterpillar MaK diesel engine.

▲ A Woodward PG governor on a Ruston 20 RK 270 high-speed four-stroke diesel engine running on MDO for a high-speed catamaran from Trasmediterranea.

▲ The fuel-control position. See arrow.

► A large MAN Diesel four-stroke medium-speed diesel engine in V-arrangement. The governor (to the left of the stairs) is significant in ensuring that large engines such as these operate safely. See arrow.

15.5.4 Examples of hydro-mechanical governors with 'droop'

Basic compensated mechanical governor with droop

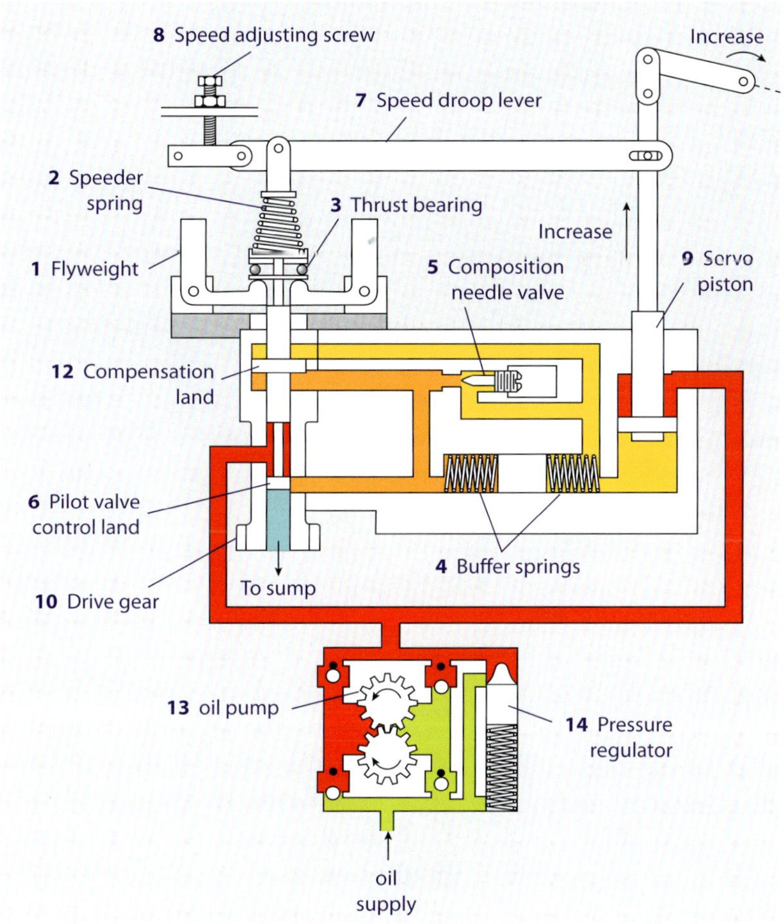

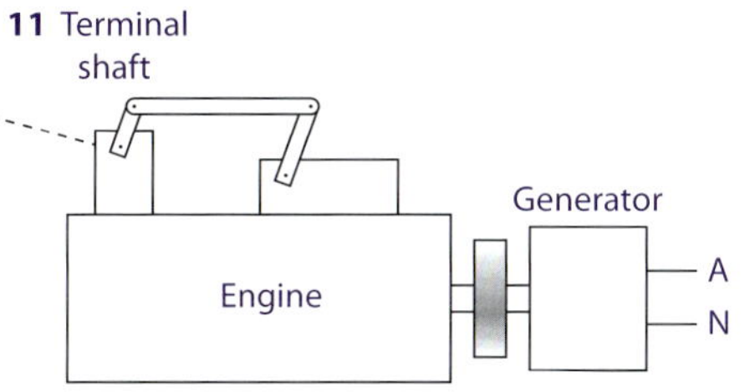

1	Flyweights	– Sense prime move speed.
2	Speeder spring	– Force for desired speed.
3	Thrust bearing	– Sum of actual speed and desired speed, droop and force from compensation land.
4	Buffer springs	– Sets gain of hydraulic circuits.
5	Needle valve	– Stabilize prime mover.
6	Pilot valve control land	– Controls flow of oil to servo piston.
7	Speed droop lever	– Feed back from servo piston to speeder spring for droop.
8	Speed adjusting screw	– Set desired speed.
9	Servo piston	– Force to move terminal shaft.
10	Drive gear	– Connected to engine to rotate pilot valve bushing and ballhead.
11	Terminal shaft	– Rotational output of actuator to move linkage.
12	Compensation land	– Transmits force diff. to thrust bearing to assist in return of pilot valve to center when fuel correction is made.
13	Oil pump	– Provides pressurized oil.
14	Pressure regulator	– Regulates pressure of oil from pump.

▲
The diagram above shows a mechanical governor for an electric generator driven by a diesel engine.

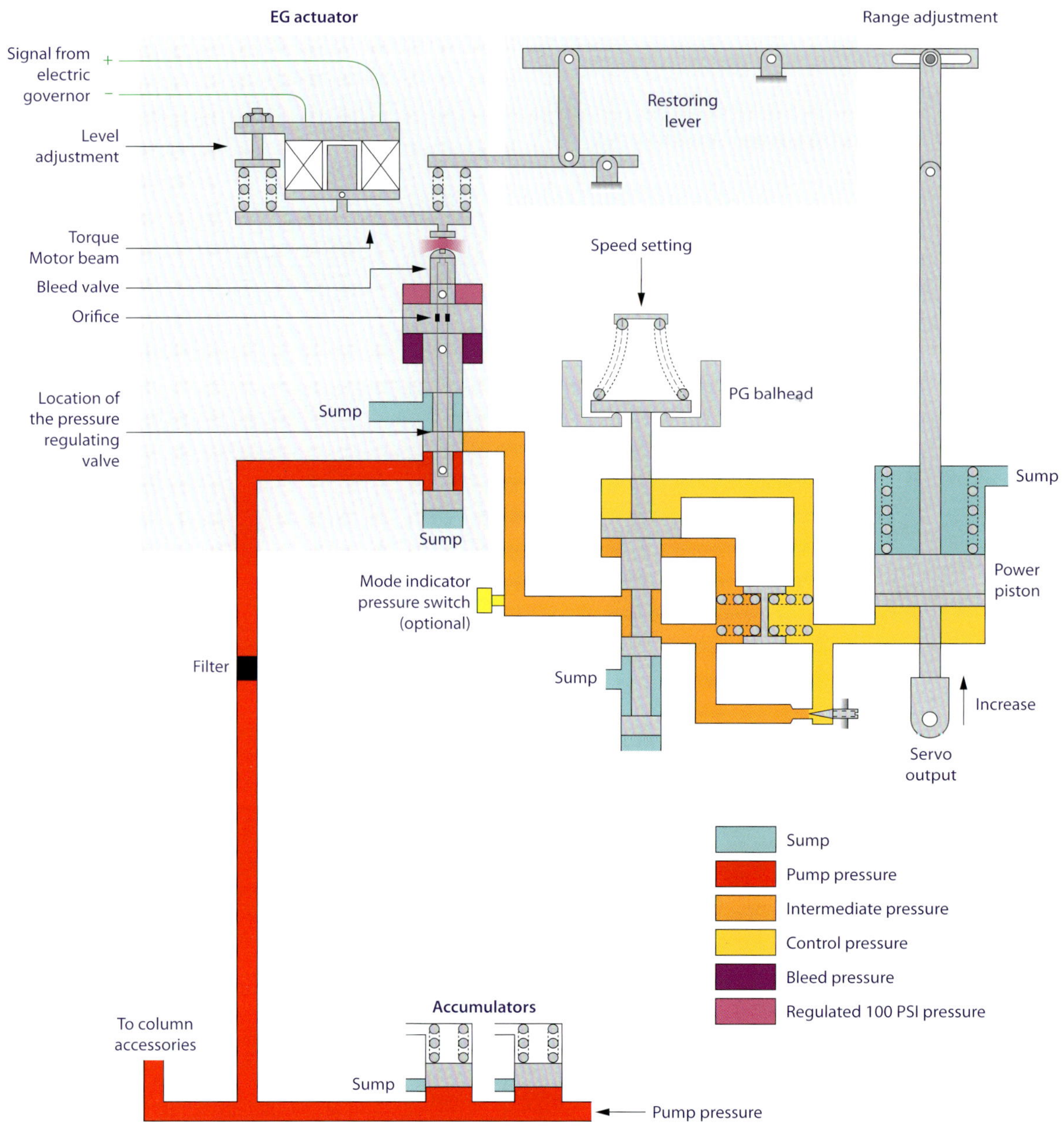

▲

A Woodward governor, PG-EG version with electronic governor/actuator.

Flyweight-control valves are opened. This governor has no 'select' setting, manually, electrically or pneumatically.

The flyweights (mechanical) and the electronic governors (with actuators) operate continually, each for their own required fuel supply.

▶

A Woodward governor on a Caterpillar MaK engine.

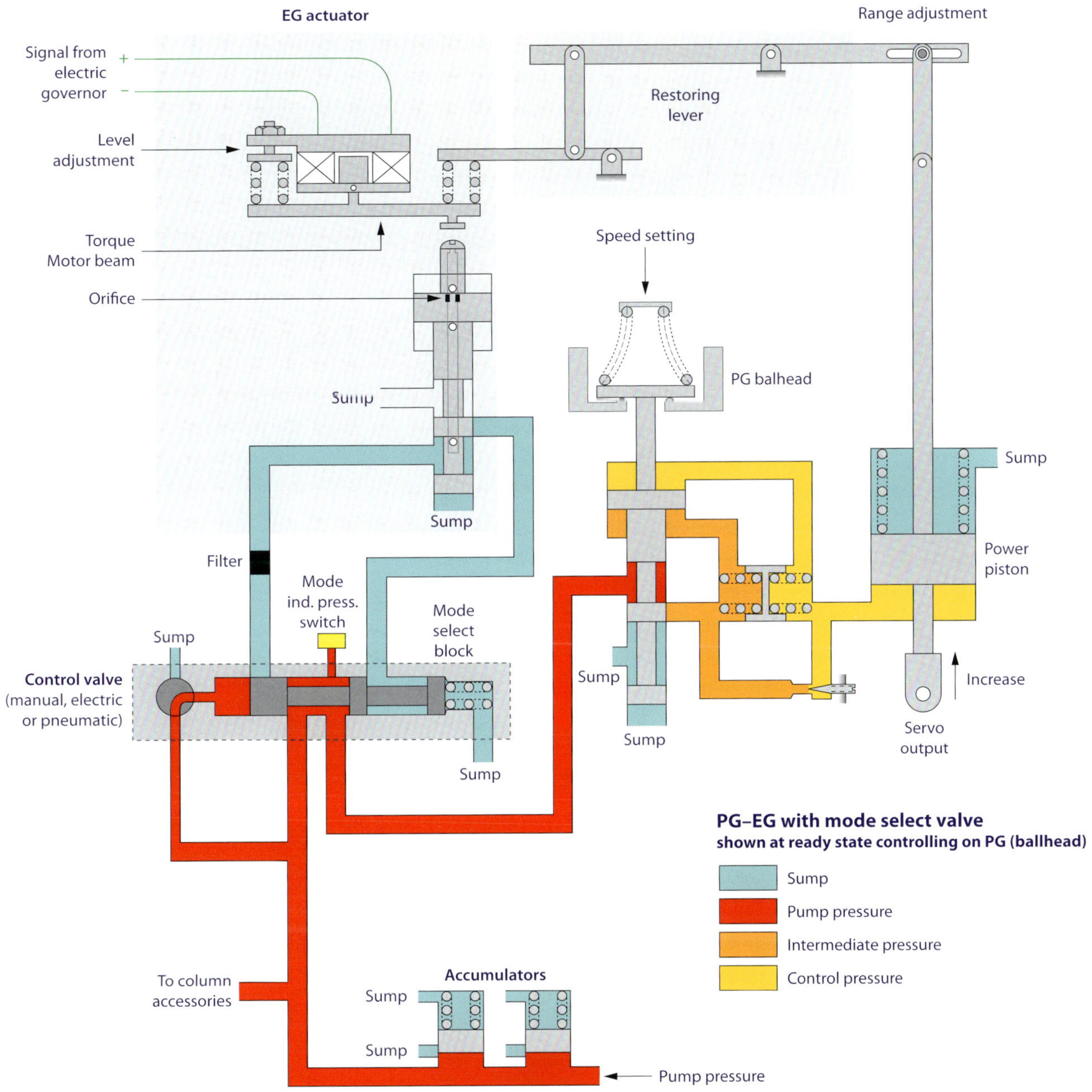

▲ A Woodward governor, PG-EG with mode-selection switch controlled with flyweights. The control valve can be set to manual, electrical or pneumatic control.

◄ A Woodward governor, PGG-EG 200 for a Man Diesel type 48/60 with actuator under the governor.

1. *Woodward PGG-EG 200.*
2. *Actuator.*
3. *Control shaft*
4. *Fuel-supply spindle*

>CH 16

Noise, origin and damping

One of the world's largest private sailing yachts 'the Athena' was built by Royal Huisman, Vollenhove, The Netherlands.

These enormous yachts have elaborate technical installations. In this instance two Caterpillar 3516 1472 kW propulsion engines and a large 870 kW genset.

Today, the noise (damping) requirements on ships are very strict as are those for reducing vibrations.

Owners of these large yachts are very demanding concerning noise and vibration reduction.

This creates a tremendous challenge for these specific companies!

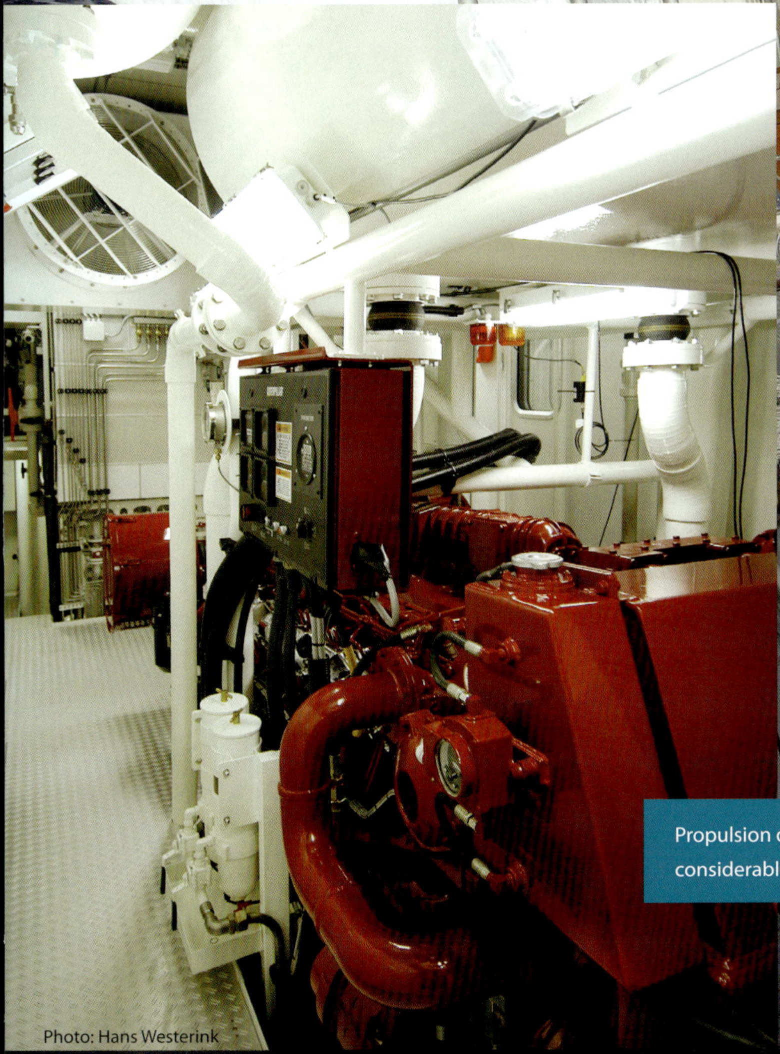

Photo: Hans Westerink

Propulsion of a yacht with a 982-ton displacement requires a considerable power output.

Photo: Franco Pace

16.1 Introduction

Diesel engines produce considerable noise levels. Noise pollution is a nuisance and can be damaging to health for both humans and animals. Excessive sound levels in a working environment can negatively affect an employee's well-being and result in productivity losses.
Techniques ensuing from guidelines pertaining to environmental and legal requirements reduce sound levels to acceptable levels. These are applicable for the surroundings of a diesel engine that is to say, in the immediate vicinity of a diesel plant.
Anyone working with diesel engines should wear proper hearing protection.
Working on or near extremely highly loaded modern diesel engines, with consequently high load numbers operating at full load can be very unpleasant even with the use of protective gear.

Working near or on an operational diesel engine should be avoided when possible.

16.2 Origin of noise in diesel engines

Diesel engines are operated by the principle of internal combustion. In the cylinder above the piston, combustion processes continuously take place. For instance, in a four-stroke engine the pressure development in these processes, the inlet/ intake-, compression-, power- and exhaust strokes, fluctuates rapidly. These pressure waves generate huge noise levels.

▼

Three sources of engine noise.

1 Exhaust-gas noises: this is especially important near the accommodation. Noise abatement: good silencers in the exhaust-gas ducting.
2 The air borne noise in the engine room, particularly important in the engine room itself. Noise abatement: good soundproofing of the engine room.
3 The structure-borne noise, particularly important with regard to the ship's accommodation. Noise abatement: vibration-proof arrangement of the diesel engine.

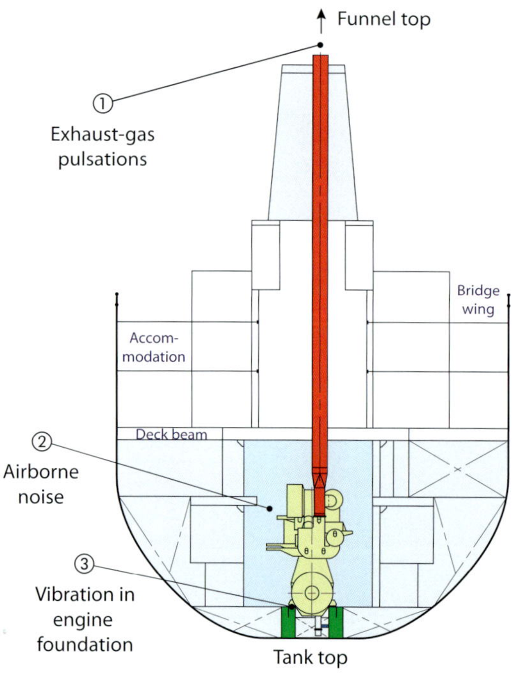

▼

An example of a MAN Diesel 6L80MC-Mk 5 two-stroke crosshead engine. MCR = 20,580 kW at 93 rev/min.

horizontally in the chart: frequency in Hertz
vertically in the chart: noise production in dB(A)

The highest noise production is airborne in the engine room. Obviously, adequate soundproofing is imperative!

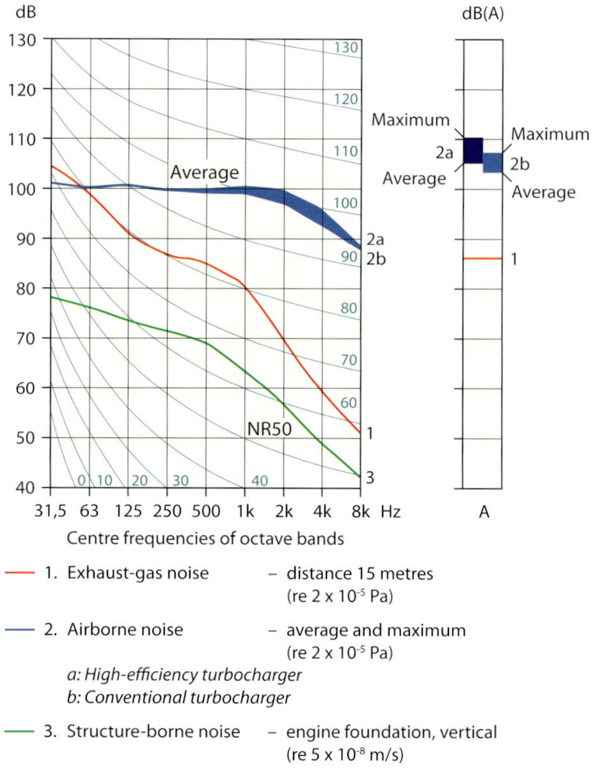

16.2.1 Sound

Sound is a physical phenomenon.
One can differentiate two aspects:

1. the pitch or frequency, denoted in Hertz;
2. the volume, the level.

- The sound output level is denoted in Watts.
- The sound pressure level is denoted in N per m^2 = Pa.
- The difference in sound levels is expressed in decibel, dB.

16.2.2 Weighting filters

The human ear with respect to frequency is 'non linear' and 'level dependent'. This means that the ear is not equally sensitive to both low and high frequencies. Generally, one can say that the human ear is less sensitive to tones below 1000 Hz and this sensitivity decreases as the frequency drops. The human ear becomes increasingly less sensitive to tones exceeding 6000 Hz.
Both these frequencies are affected by the sound level. The lower the sound level, the less sensitive the ear is to frequencies within the 1000 to 6000 Hz range and, obviously, outside this range. This is why so-called weighting filters have been chosen, for example, the A-, B-, C-, D- and NR-filters.

The A filter is the standard filter for industrial noise. If sound levels are measured by an operational A filter, the sound level is expressed in dB(A).

Possible effects during or after brief exposure to noise

These may vary considerably from, for instance, shock, increased heart rate and dizzy spells. Furthermore, speech impediments, decreased memory capacity, irritability and fatigue are possible side effects.

Long-term effects

Amongst others, high blood pressure, heart disease, ulcers and dizziness. Temporary and/or permanent hearing loss.

Legislation

The most important article in relation to noise issues is article 179a of the Safety Decree for factories and workplaces (1938).
Article 179a stipulates that machinery, the work place and the conducted activities should not produce harmful sound levels, unless such cannot reasonably be prevented. At present, the limit is the equivalent of 85 dB(A).
According to the law, **noise abatement** should occur as much as possible at its source.

Protection

From a sound level of 80 dB(A) hearing protection equipment should be provided to the employee and at sound levels above 90 dB(A), use is compulsory.

What is allowable during work hours?

For a working day of eight hours a maximum sound pressure level of Lp(A) = 85 dB(A) measured at the working place. **With every 3 dB increase, the sound level is doubled and the following rule applies:**

Sound pressure level	Exposure to sound
80 dB(A)	8 hour
83 dB(A)	4 hour
86 dB(A)	2 hour
89 dB(A)	1 hour
92 dB(A)	30 min
95 dB(A)	15 min
98 dB(A)	8 min
101 dB(A)	4 min
104 dB(A)	2 min
107 dB(A)	1 min
110 dB(A)	30 sec

Statement

Hearing loss is always permanent!

The 120 dB(A) level is known as the 'pain threshold'.

	dB(A)
Work spaces	
Machinery spaces (continuously manned*)	90
Machinery spaces (not continuously manned*)	**110**
Machinery control rooms	75
Workshops	85
Unspecified work spaces**	90
Navigation spaces	
Navigation bridge and chartrooms	65
Listening posts, including navigation bridge wings and windows	**70**
Radio rooms (with radio equipment operating but not producing audio signals)	60
Radar rooms	65
Accommodation spaces	
Cabins and hospital	60
Mess rooms	65
Recreation rooms	65
Open recreation areas	75
Offices	65

* Ear protectors should be used when the noise level is above 85 dB(A), and no individual's daily exposure duration should exceed four hours continuously or eight hours in total.

◄ An overview of areas on a ship indicating the maximum sound levels in dB(A).
Requirements by the IMO, the International Maritime Organisation.

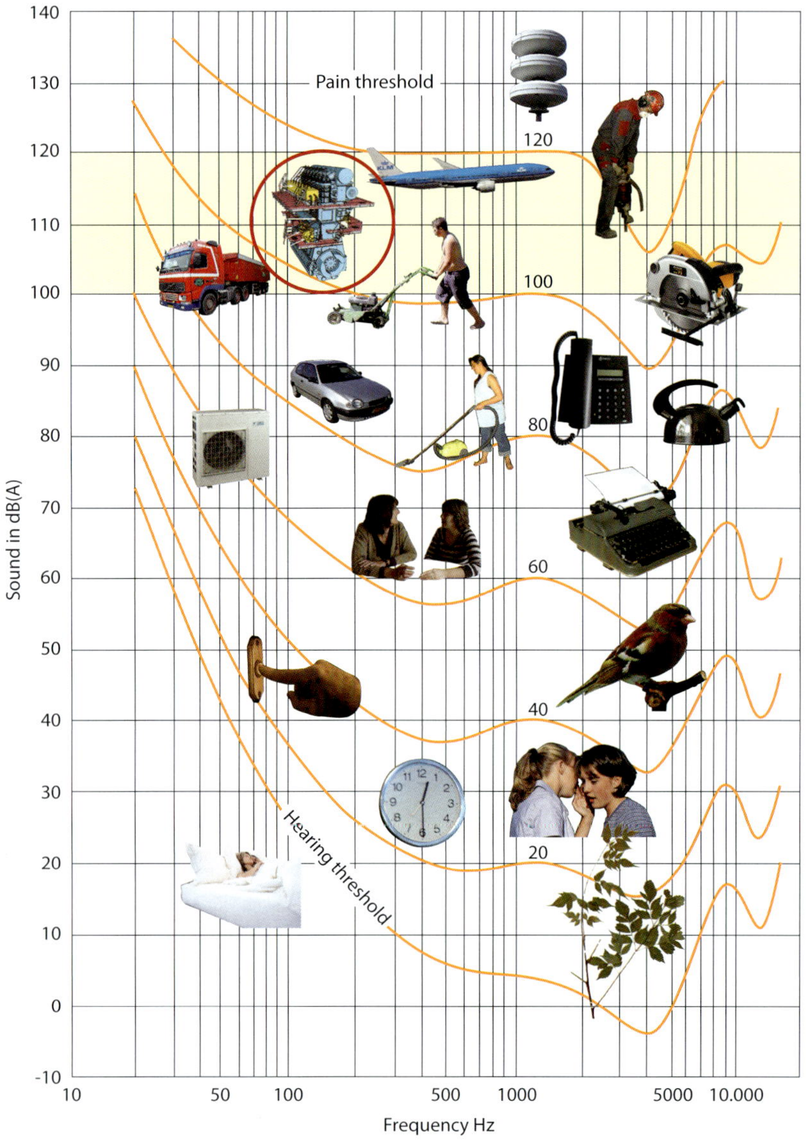

◀

Noise production in our environment at various frequencies.

horizontal: frequency in Hertz.

vertical: sound levels in dB(A).

Diesel engines are found in the upper section of the graph between 100 and 120 dB(A).

16.3 Sound-transmission paths

It is important to establish the paths that sound are transmitted to the receiver. There are two paths:

- **airborne noise**, sound transmitted through the air;
- **impact sound**, noise transmitted by constructions such as engine bed plates, pipes and casings.

Rigid 'couplings' between a source of noise (vibration) and, for instance, a casing may even amplify the noise; the construction behaves as an amplifier. These sound transmissions can be measured.

▶

A survey of the sound levels found a certain distance from its source. Shown here, the funnel, the diesel engine exhaust above the bridge on the deck.

This sound level reduction by 6 dB(A) is a noteworthy difference!

Sound level – far field law

Sound level at distance R_2 compared to distance R_1:

Reference sound intensity	$I_o = 10^{-12}\ W/m^2$
Sound intensity level (dB) at R_1	$L_1 = 10 \times \log_{10} (I_1/I_o)$
Area of sphere	$A = 4\,\pi R^2$
Sound intensity at R_2	$I_2 = I_1 \times (R_1/R_2)^2$
Sound level at R_2	$L_2 = L_1 - 20 \times \log_{10} (R_2/R_1)$

In general, the sound level will be reduced by 6 dB for each doubling of the distance from the noise source.

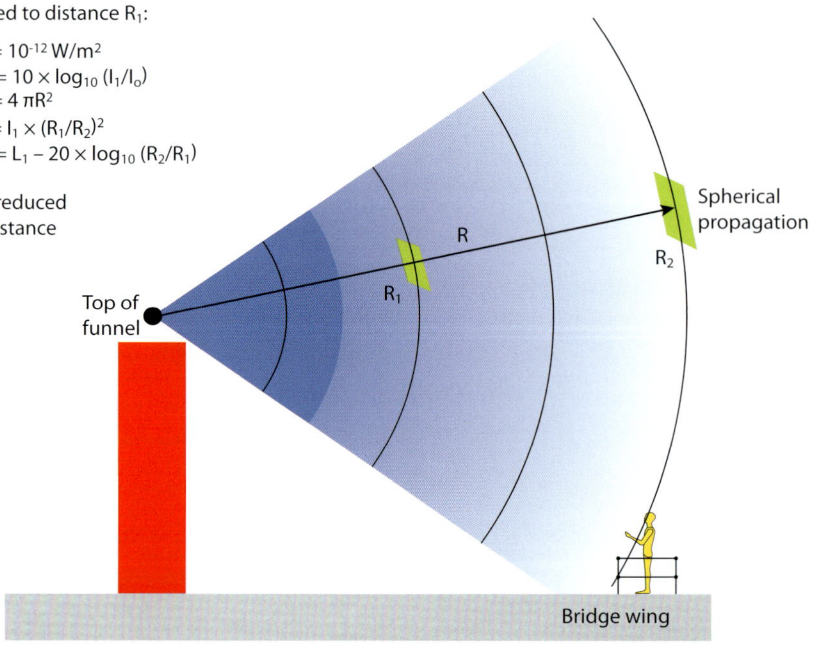

16.4 Silencers for diesel engines – choosing a silencer

There are diverse versions. For two- and four-stroke engines, these are explosion-proof silencers with spark arresters and baffles for the different frequencies generated by engines.

Note

Exhaust systems in combustion engines are subjected to pulsations and other vibrations. For this reason, the entire exhaust system is a vibration-free assembly because of the use of suitable vibration insulators.
How do silencers work?
The following principles apply:
Adsorption silencers and throttle silencers. The gases flow along sound absorbent materials, such as nonflammable mineral wool that significantly reduce the sound levels. These are suitable for frequencies between 150 and 5000 Hz but cannot be the first silencer in the exhaust system.
Reflection silencers. The silencer is constructed is such a way that the sound is reflected inside the silencer and not emitted via the exhaust. The sound waves created inside the silencer cancel out each other and therefore the sound levels decrease. It is often the first silencer in an exhaust system.
These various silencer types are often used in combinations.

Silencers that are a combination of both these principles are most often used.

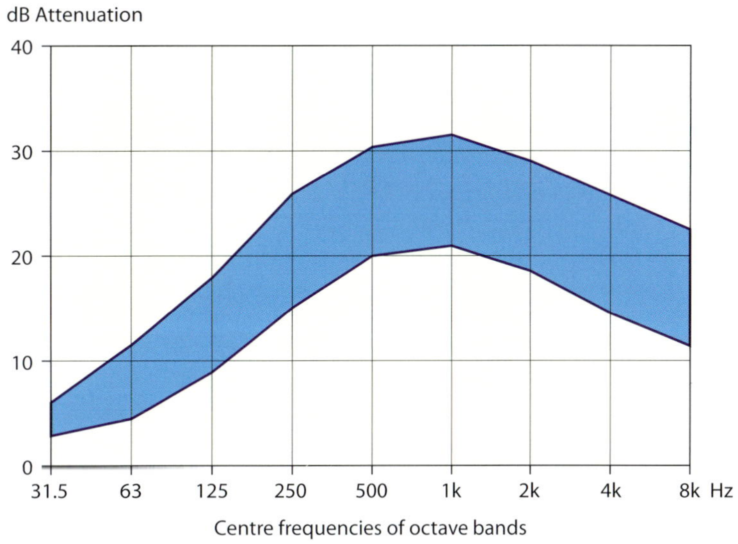

▲
A good example of noise reduction in a silencer at various frequencies. Approximately 25 to 30 dB (A) at 500 to 1000 Hz is viable.

Horizontal: Sound frequency Hertz.
Vertical: The decrease of sound in dB with a sound silencer.

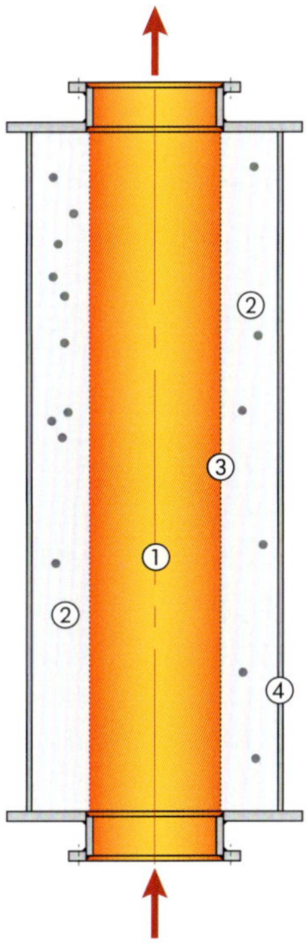

◀
A conventional absorption silencer for a two-stroke crosshead engine.

1 *exhaust-gas manifold*
2 *nonflammable mineral wool*
3 *'holes'-pipe*
4 *steel outer wall*

16.4.1 Interiors of sound silencers

Noise in a silencer is reduced to levels so that the noise produced by the diesel power plant or ship does not exceed the legally established level, as measured outside the engine.
The noise emitted by diesel engines travels into the exhaust-gas manifold and enters the silencer; it is first led through the expansion chamber, where low-frequency impulse sounds are eliminated. The gases then flow into the absorption section where the remaining medium- and high frequency sounds are dampened by special sound absorbent materials.

Materials

The casing and expansion section are often manufactured from carbon steel with a thickness of at least 3 mm. The absorption section contains sound-absorbing, nonflammable mineral wool of varying densities.

Model selection

Using a chart of the mass flow in kilograms per hour and the temperature of the exhaust gases, the volume flow in m^3 per hour is determined.
In the following chart, the correct model can be chosen by intersecting the pressure loss (horizontal line) and the discharge (vertical line).
In the last table the weights and measurements of the models are given.

▶ A graph showing Merford Industrial silencers.

horizontal: Y, the specific volume in kg per m³
vertical: exhaust-gas temperature in °C

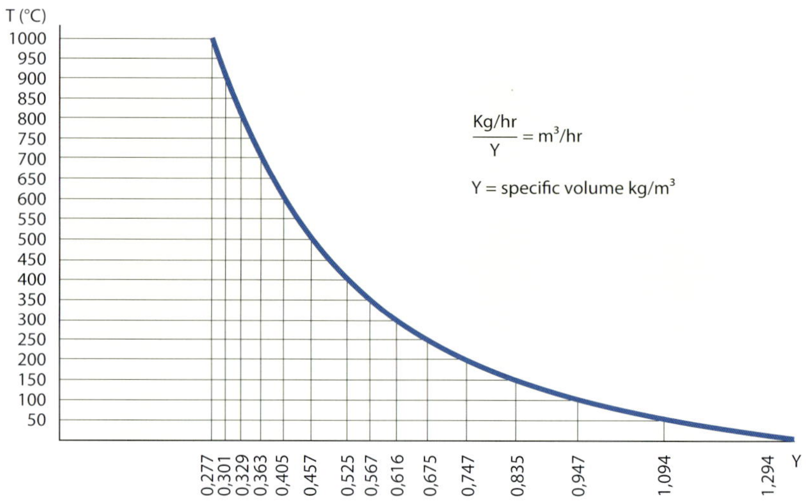

▶ A graph showing Merford Industrial silencers.

vertical: pressure difference of the silencer in millimetres water column.
slanted lines: various models of exhaust-gas silencers.

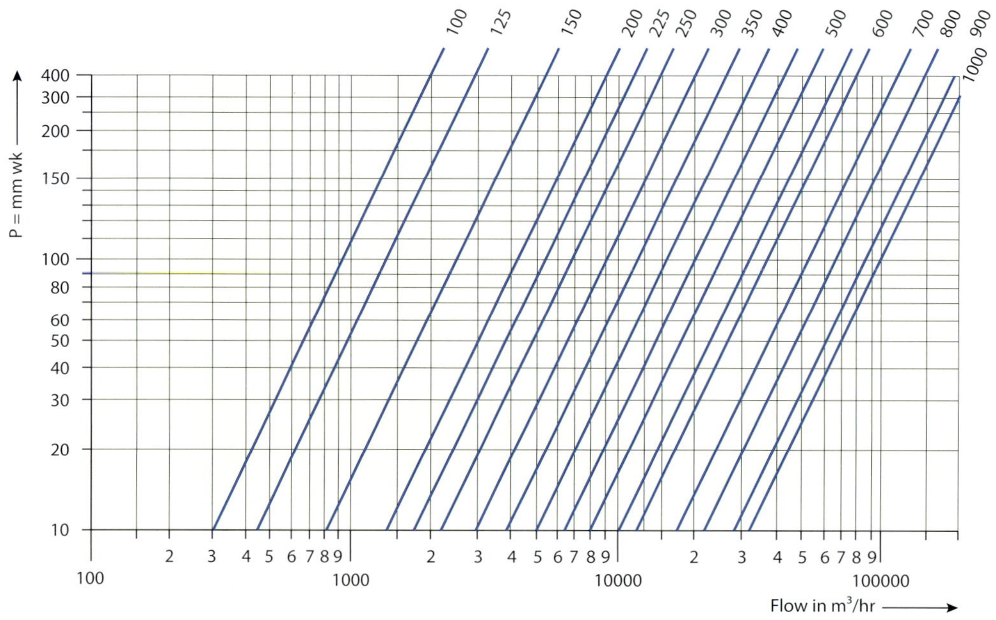

16.5 Noise reduction of diesel engines in categories I and II

The exhaust gases of diesel engines carry the noise generated by the engine to outside the engine. This noise can be reduced to acceptable levels for the immediate vicinity using silencers. Noise emitted directly from the engine block is more difficult to dampen; soundproofing of the engine has often been attempted, especially in small engines, but this has proved to be impractical.

16.5.1 Soundproofed rooms or containers

This is common and frequently used. The engine is mounted in a locked, well ventilated room or container with optimal insulation.
Air is supplied via a supply- and an exhaust channel. A ventilator provides a constant airflow sufficient for combustion as well as air to adequately vent and cool the room. The air-cooling ducts are often equipped with a labyrinth system to ensure that the noise escaping via the air ducts is kept to a minimum.
Soundproofed rooms or containers are often used for diesel gensets in category I and II engines.
Aboard ships, this is ideal when the ship is docked as maintenance or other work can take place without disturbance by the aggravating noise of 'auxiliary engines' (diesel gensets).

▲

A diesel genset at the Krone circus with a Scania diesel driven engine.

The hinged 'outer doors' are open; the triple glass soundproofing guarantees that the noise generated by the two engines is practically inaudible. The entire genset is manufactured by Genpower, Groot Ammers, The Netherlands.

▲

A small diesel genset by Vetus for yachts.

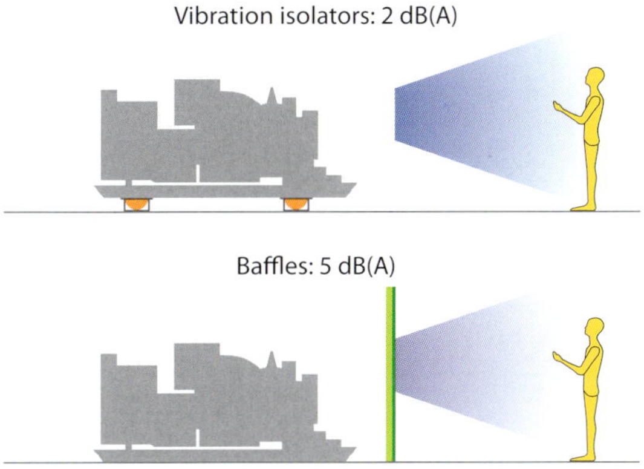

◀

An example from Caterpillar showing noise reduction.

Top figure: On vibration dampers a reduction of 2 dB(A).
Bottom figure: An insulated upright wall, noise reduction is 5 dB(A).

Unfortunately, port or auxiliary engines that are not soundproofed are still in operation, this cost reduction leads to the detriment of the working environment and consequently productivity.

During repairs, most panels can be easily removed. Larger engines in categories II and III and category IV are seldom placed in containers. They are on occasion mounted in a separate room, as are diesel gensets, so a non-operating engine can be repaired without the noise of an auxiliary engine. Obviously, spaces on ships as well as land-based engine rooms are soundproofed with respect to the external environment. Crew accommodation in ships is often built above and/or around the engine room.
A badly arranged propulsion engine and inferior soundproofing of the engine room can be very disturbing to the crew. The same applies to vibrations; these can negatively affect the crew.

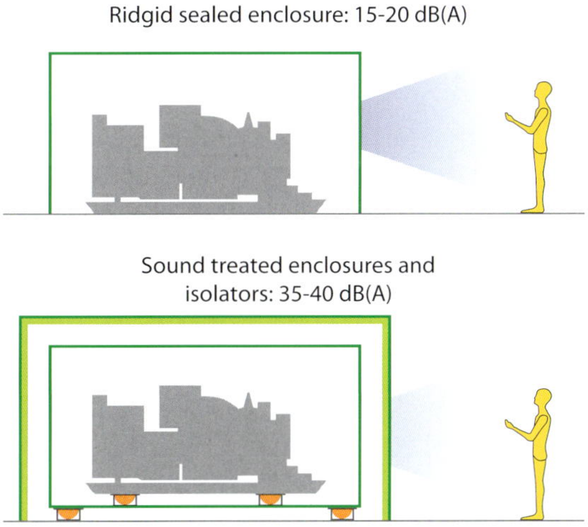

◀

An example from Caterpillar showing noise reduction.

Top figure: Arranged in an insulated room, noise reduction is 15 to 20 dB(A).
Bottom figure: Arranged in an insulated room and mounted on vibration dampers, 35 to 40 dB(A).

16.6 Turbocharger noise

The first prerequisite in increasing the power output of diesel engines is the increase of the air supplied for the combustion process. Increased air supply means more fuel can be injected and therefore more power can be generated. The supply of a sufficient quantity of air to the cylinder in a short time-span through the inlet valves, as in the case of a four-stroke engine, and the inlet ports in two-stroke engines, requires higher turbocharger air pressures. The compression ratio of the centrifugal compressor of the turbocharger has been increased throughout the years from 2 to approximately 4.5. The turbocharger capacity, the 'flow', has also increased significantly. Sound measurement around the turbochargers of large two-stroke crosshead engines show values of 110 dB(A) at full load; at a slightly lower power output they can still be as high as 105 dB(A).
On passenger ships, comfort is vital and therefore the noise levels of the large four-stroke propulsion engines are kept to a minimum. On cargo ships, the sound levels are kept as low as possible for the crew's comfort.

Improvements introduced by turbocharger manufacturers

Improving the compressor wheel and diffuser reduce noise at its source. A silencer consisting of absorption- and a resonance sections is fitted in the compressor exhaust-outlet and significantly reduces the sound output. In this way, noise levels have been reduced to approximately 95dB(A).

16.7 Sound levels in diesel engines

Examples from engine manufacturers.

16.7.1 Example 1: Wärtsilä 38A diesel engine, category III

Measurements are taken using a specific method; in this instance, one metre from the engine at the same height as the cylinder heads.

▶ The sound levels on the top floor near the cylinder heads and turbochargers of a two-stroke crosshead engine.

All measurements lie between 100 and 108 dB(A).

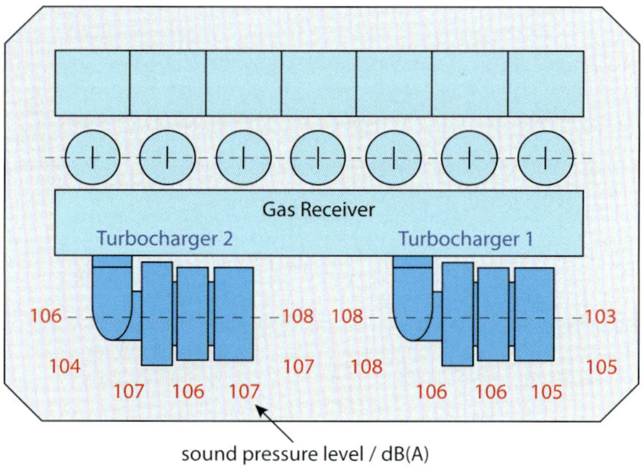

Measurements show that sound does not remain in the turbocharger only, but is carried to the inlet manifold and the air cooler. These sections may even reinforce the noise.

▶ The same engine. The noise levels are measured at the air-inlet manifold and at both air coolers.

This is the reason that the noise carried from the turbocharger is boosted to a maximum of 113 dB(A).

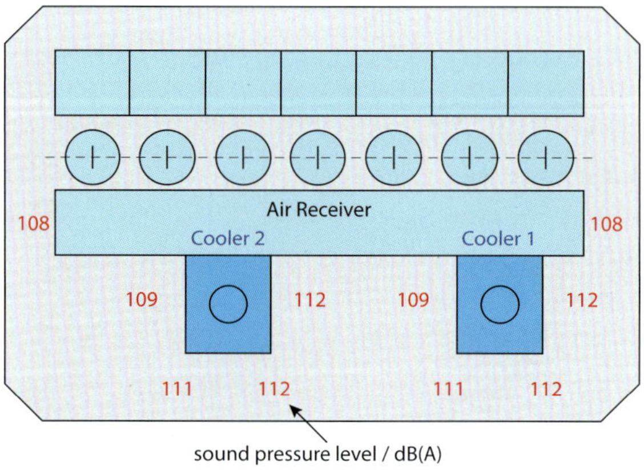

Air-borne sound

8 measuring points, dB(A) re 2.10^{-5} N/m^2 at 1 metre from the engine and at cylinder height.

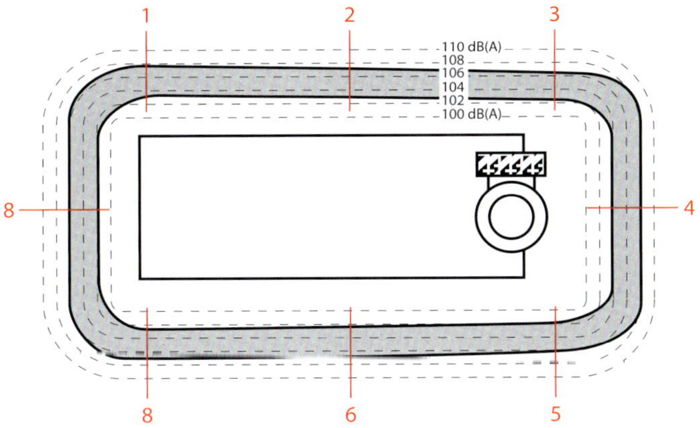

Air-borne sound power level L_w (dB re 1.10^{-12} W).

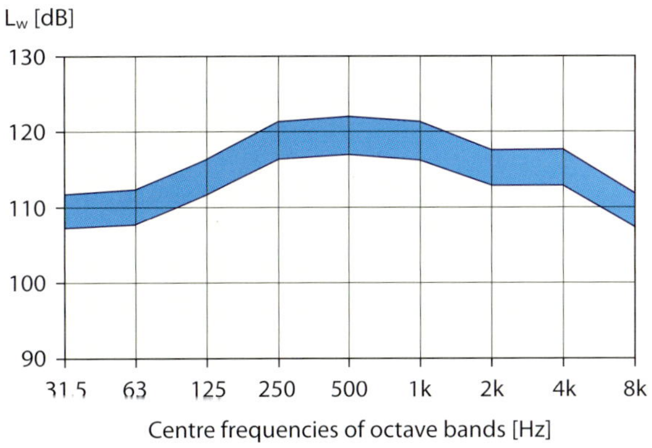

Inlet-air sound

Unsilenced inlet-air sound power level
$L_w = 10 \log P/Po$ dB (re $Po = 10^{-12}$w) after the turbocharger.

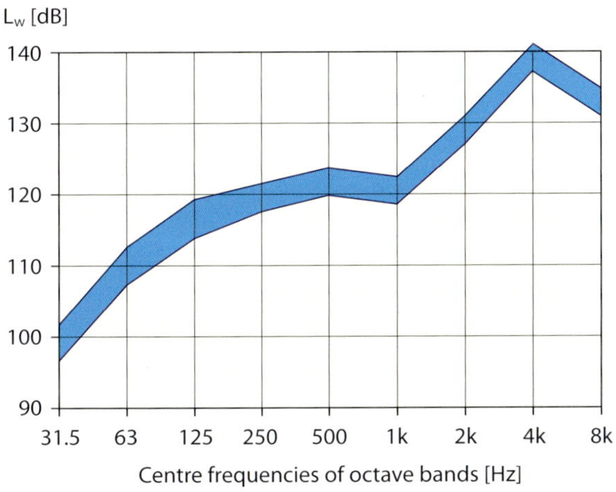

Exhaust sound

Unsilenced exhaust sound power level
$L_w = 10 \log P/Po$ dB (re $Po = 10^{-12}$w) after the turbocharger.

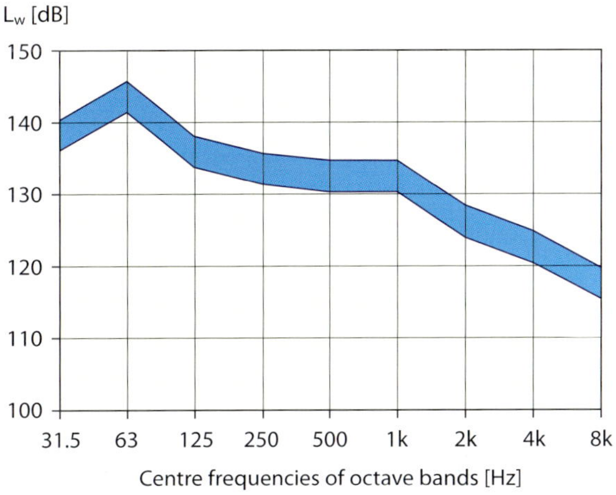

Structure-borne sound

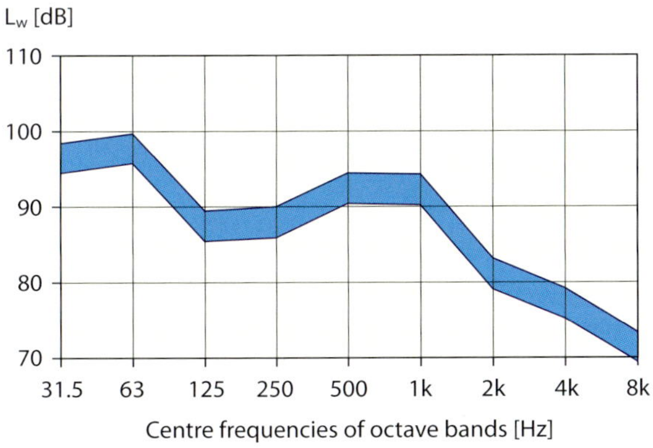

Average vibration velocity level L_v of the four corners of the engine foundation flange in three directions.
Reference level $v_0 = 5.10^{-8}$ m/s.

$$L_v = 10 \log \frac{\sum_{i=1}^{n} (v_i / v_0)^2}{n} \text{ [dB]}$$

▲

Example of a Wärtsilä 38A diesel engine, category III.

Top left: measuring points; shown here the noise production lies between 100 and110 dB(A).

Top right: noise production in dB(A) and frequency in Hz, k = kilo.

middle: noise production of the turbocharger, a notable noise producer!

Left: an air inlet that is not soundproofed.

Right: an exhaust that is not soundproofed. Both produce high noise levels.

Bottom right: noise produced in three directions towards the bedplate measured at the four corners of the bedplate.

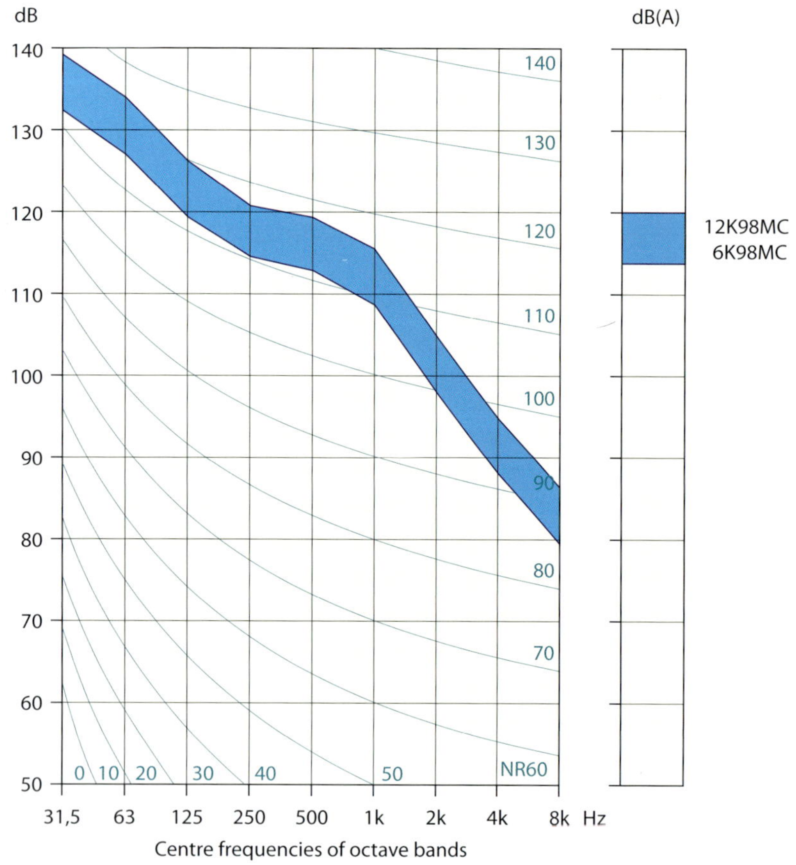

16.7.2 Example 2: MAN Diesel K98MC, the second largest crosshead engine, category IV

The silencers for these engines are typically based on the absorption principle and are installed after the exhaust-gas boiler. The gas velocity is approximately 35m/sec.

The sound levels of the exhaust gas discharged to the bridge wing may not exceed 60 to 70 dB(A).

This is achieved using a relatively simple silencer.

The pressure drop of this type of silencer is approximately 200 millimetres water column at MCR power output.

◀

The noise in the exhaust-gas system of a two-stroke crosshead engine from MAN Diesel for the six-and twelve-cylinder K98MC models.

Shown here, an average of 105 and120 dB(A). The frequency is approximately 500 Hz.

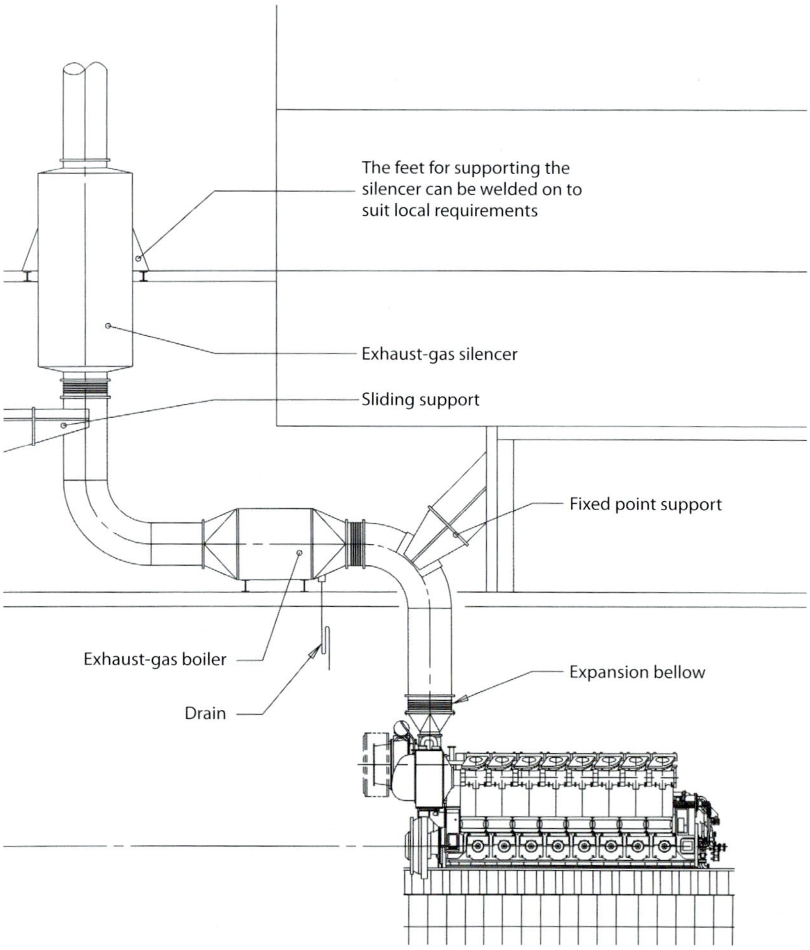

16.8 Examples of the engine arrangement with silencers

16.8.1 Example 1: MAN Diesel V 40/50

A medium-speed four-stroke diesel engine, engine category III.

◀

An engine arrangement with a silencer.

As shown here, one can appreciate the huge exhaust system's dimensions in comparison to the engine when an exhaust-gas boiler and a silencer are installed. Three exhaust-gas compensators absorb expansion and contraction of the pipes due to temperature fluctuations.

16.8.2 Example 2: Arrangement of a complete exhaust-gas system with two Discom silencers

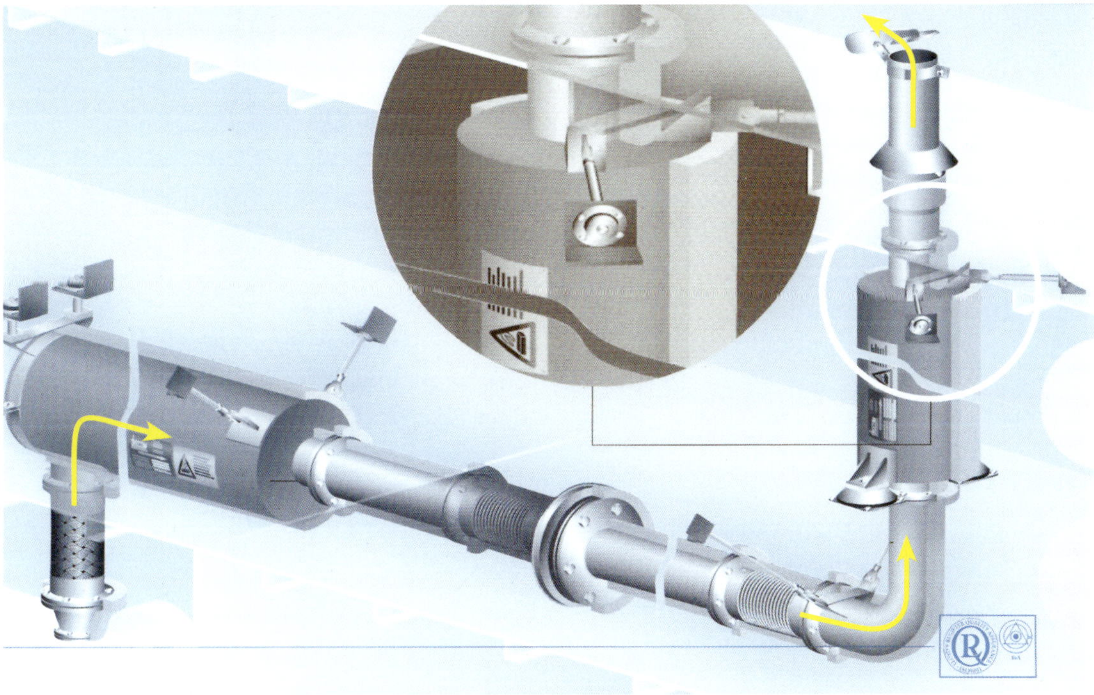

◄

The arrangement of a complete exhaust-gas system with two Discom silencers.

Apart from the correct type of silencers, the flanges, the flexible sections and the rain cover are essential in order to obtain an adequate overall system.

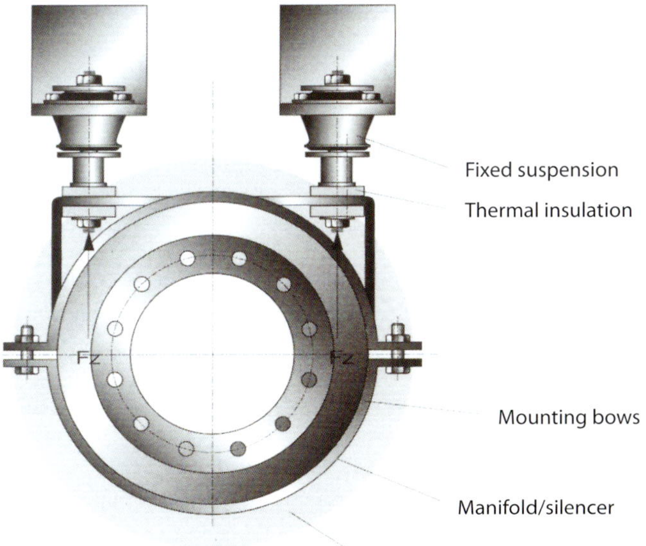

◄

A flexible suspension of the Discom silencer.

It is important that vibrations are not transmitted to the structures around the exhaust-gas manifolds.

►

The flexible suspension of the exhaust-gas manifold of a Wärtsilä 38A diesel engine, category III.

▶
The inventory of silencers at Discom, Alblasserdam, The Netherlands.

In the foreground small silencers with spark arresters (1).

▶
Part of the diesel genset at the water and electricity company on Aruba, the WEB.

In the engine hall, an eighteen-cylinder Wärtsilä 18V32 engine has been installed. The silencer has been placed outside.

▶
A genset in Africa with ten ABC engines, the silencers protrude vertically. In the foreground, the heat exchangers are facing outwards so that the heat of the lubricating oil and coolant can be discharged to the outside air.

ABC – Anglo Belgium Corporation, an engine manufacturer of medium-speed diesel engines in Ghent, Belgium.

>CH 17

Vibrations and balancing

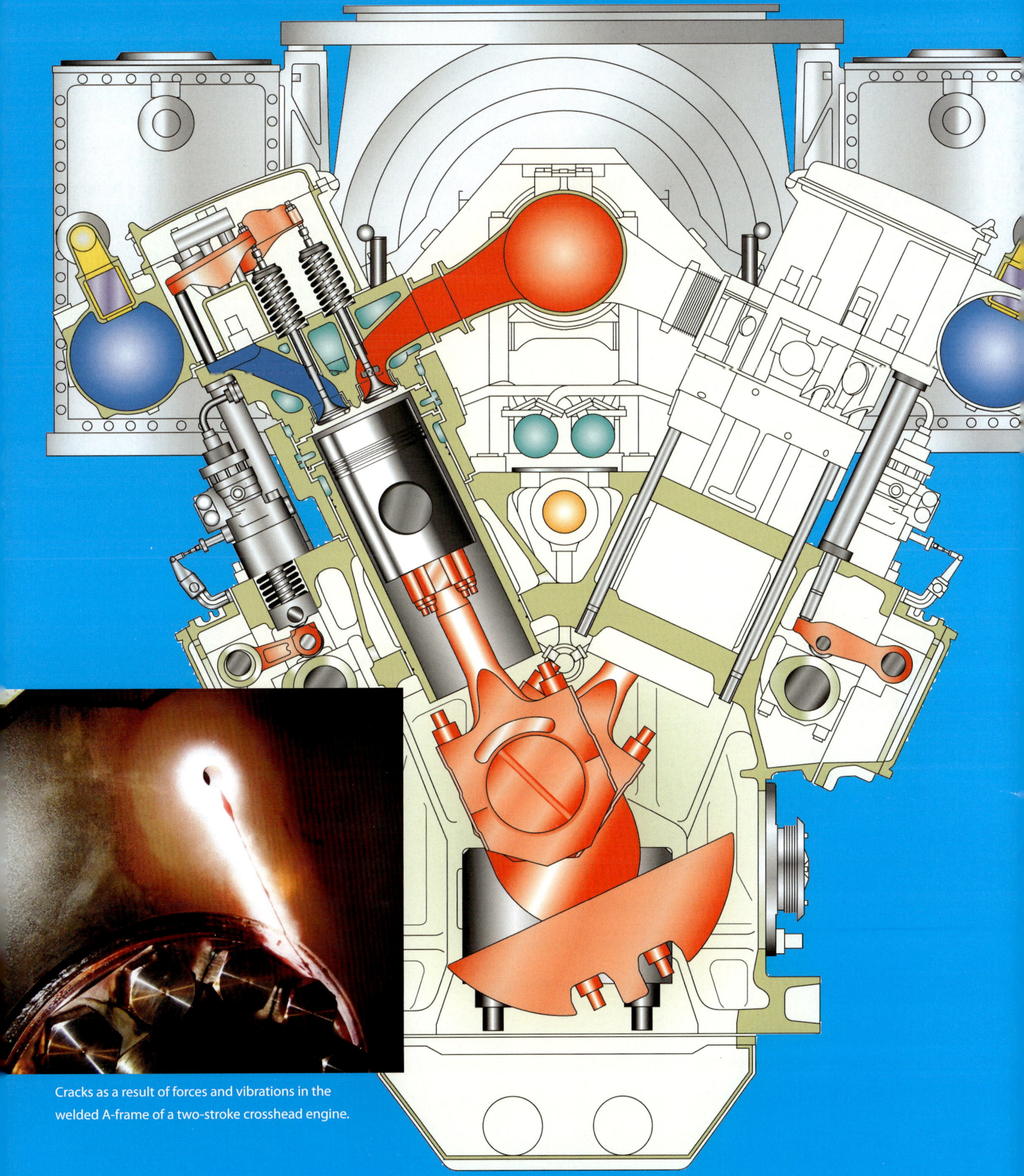

Cracks as a result of forces and vibrations in the welded A-frame of a two-stroke crosshead engine.

The combustion process above the piston produces vibrations that are difficult to combat. These vibrations are generated by pressures, which constantly fluctuate during the process.

17.1 Introduction

Running diesel engines are potential sources of vibrations.
Around an operating diesel engine, free forces and moments are generated that can produce vibrations. In ship propulsion, the ship is constructed in such a manner that vibrations travel to the ship's structure via the engine bedplate. In a diesel-power plant, the vibrations can cause the entire building to shake as vibrations travel from the engine via the bedplate to the whole structure. This should be avoided where possible.
Torsional vibrations are generated in the crankshaft and the output transmission shafting. The fluctuating gas forces over the piston can produce axial vibrations in the crankshaft and the outgoing shafts.
When designing a new diesel-engine type, much consideration is given to vibrations and their absorption by mounting vibration dampers on the crankshaft. In order to reduce vibrational transmittance to the bedplate and consequently the environment, engines in specifically categories I and II are mounted on flexible elements. This is also frequently used in category III engines.
All crosshead engines, category IV, are fixed to the bedplate.
Ship's propellers also generate vibrations and these can be reinforced in combination with certain types of diesel engines.
This could be due to the number of propeller blades and number of cylinders used in the propulsion engine.

17.2 Methods to reduce torsional vibration using dampers

If the generated vibrational moments can be compensated by damping moments, a more stable equilibrium is achieved.
The vibrational energy generated in the engine is then converted into damping energy.
This is usually frictional energy in the vibration dampers.
Using rule of thumb analyses, the damping factor or the relative damping can be deduced.

Also, see Chapter 20, *Transmission gears, flexible couplings, vibration dampers, shafting and shaft-generator drives.*

17.3 Main causes of vibration

There are three sources of vibrations.

1 **Torsional vibrations in the drive shafts**, especially in low-speed two-stroke crosshead engines.
The diesel-engine torque changes constantly during the combustion process for two-stroke and the four-stroke cycles.
As opposed to the steam and gas turbines that have a constant torque at constant loads.
The cranks of the crankshaft are also positioned at a certain angle to each other.
The position of these cranks is dependent on the combustion process and the number of engine cylinders.

▶
The main causes of vibrations.

As shown in the flow diagram, the vibrations found in the left side can be simply solved when evaluating the main causes of vibrations in an internal combustion engine.

- **The vibrations caused by the rotating masses** are vibrations to the first harmonic order. The symptoms are excitation/vibration and bending of the block/frame.
- **The vibrations of the oscillating masses**, such as the piston movement, are complex and to the lowest first and second harmonic order. The symptoms are vibration/excitation, bending and torsion of the block/frame.
- **The vibrations caused by the changing pressure in the cylinder** have all the harmonic orders, also the half orders. The symptoms are torsion generated by the deflections and excitations/vibrations in the engine. This is a complex and not easily solvable problem.

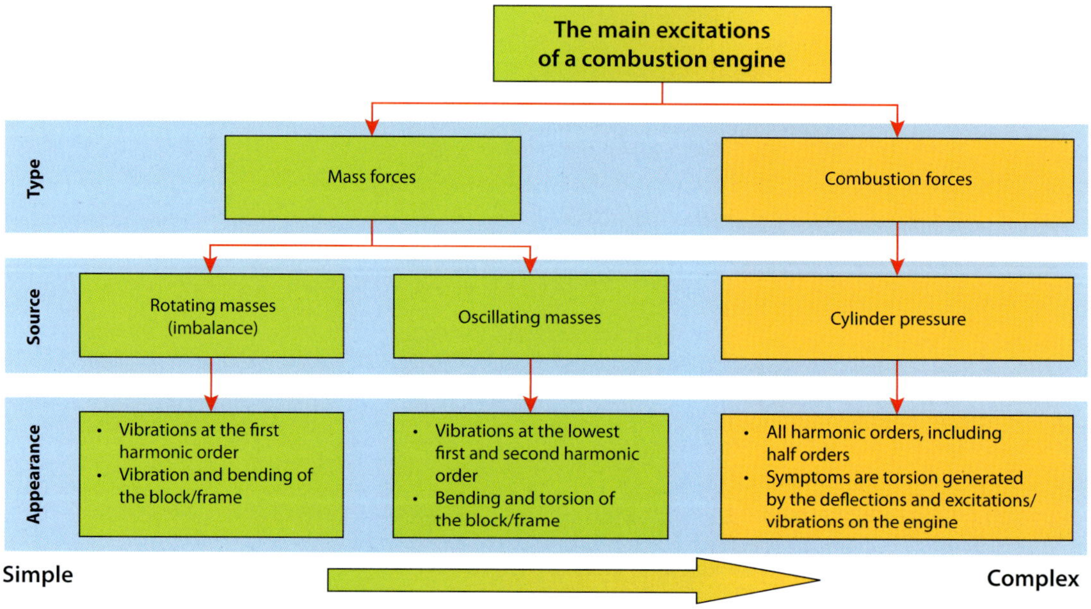

2 **Axial vibrations in the drive shafts**, such as the crankshaft, the thrust block, the intermediate shafts and the propeller shaft.
These vibrations originate from the radial loads generated from gas- and acceleration forces in the individual cylinders as well as the axial forces exerted on the propeller.
3 **Bending vibrations** caused by misalignment of the intermediate shafts and the propeller shaft. These deform due to compressional- and tensional forces and take on the shape of the ship. This is referred to as hogging or sagging.

17.4 Forces exerted on the driving gear and engine block

The following are important:
1 gas forces;
2 mass forces.

1 Gas forces

The combustion process produces constantly changing pressures over the piston, and in two-stroke crosshead engines, the pressures at the bottom of the piston in the scavenging-air space may also vary slightly.
This results in a downward force on the piston, which is equal to the pressure difference on the piston multiplied by the piston surface.

▼
A broken crankshaft.

The crank web is torn transversely. The lubricating-oil drilling is clearly visible. Vibrations can cause every shaft to break. The large vibrations cause material fatigue followed by breakage.

2 Mass forces

The rotating motion of the crankshaft and reciprocating movement of the piston with the connecting rod in four-stroke engines and the connecting rod, crosshead and piston rod in two-stroke crosshead engines, produce acceleration torque leading to mass forces.

Crankshaft

The rotating crankshaft generates centrifugal forces that are equal for every crank.

$$F_c = m_r \times \omega^2 \times r$$

F_c = centrifugal force
m_r = mass rotating crankshaft
ω = angular velocity of the crankshaft
r = distance from the centroid to the axis of rotation

The centroids of all these rotating engine parts are situated at varying distances from the axis of rotation, the centre line of the crankshaft. Simply, the rotating masses are replaced by imaginary masses with a crank radius equal to r from the crankshaft centre line on which inertia forces identical to the original masses are exerted.
The masses of the drive gears play a part in the tangential force diagram and the load of the crosshead, crankpin and the crankshaft bearings.

The value of these forces can be calculated as follows.

$$F = m \times g$$

F = mass force
m = mass of part
g = acceleration of gravity/gravitational constant

In low-speed two-stroke crosshead engines the acceleration forces which work on the rotating mass are between 10 and 25 times higher than the weight of the rotating mass.

In a medium-speed engine, these are 75 to 100 times higher and for a high-speed engine are negligible.
This is the reason that the loads of bearings in engine categories I, II and III are not included.

17.5 Resonance

The complete drive system consisting of the crankshaft, intermediate shafts and the propeller shaft form, together with the pistons and connecting rods, an elastic system subjected to torsional vibrations caused by the continuously changing torque exerted on the drive gears and cranks.
This can produce dangerous vibrations, which result in crankshaft breakage (torsional break).
This may occur due to **resonance**, when one of the elastic system's vibration frequencies matches the frequency of one of the moments that generates a vibration.

Resonance
If a system is subjected to an external frequency that is equal to its own frequency, resonance occurs. Vibrations will then occur that can lead to damage. In resonance, the vibrations are reinforced and can lead to metal fatigue en breaks.

17.6 Balancing diesel engines

Balancing engines is a matter of balancing the various forces and moments that are generated by an operating engine.

This is a complicated matter and balancing piston-driven engines can only be realised in part. It is simpler in a rotating machine such as a steam or gas turbine, an electromotor or the electrical genset for a centrifugal pump!

17.7 Principle of an internal combustion engine

The following can be stated for a crankshaft connected to a reciprocating piston in a cylinder:
- the crankshaft rotates, as do the camshaft(s) and the gear train connecting them.
- the piston moves between top dead centre (TDC), and bottom dead centre (BDC) where it comes to a complete standstill!

Between these two positions, the piston accelerates and decelerates continually, as do the connecting rod in a four-stroke engine and in a two-stroke crosshead engine, the connecting rod, the crosshead with guide shoes and piston rod.
The accelerating and decelerating forces can have a very large magnitude, depending on the number of crankshaft revolutions and the masses of the reciprocating parts.
Therefore, the piston is designed to be as light as possible.

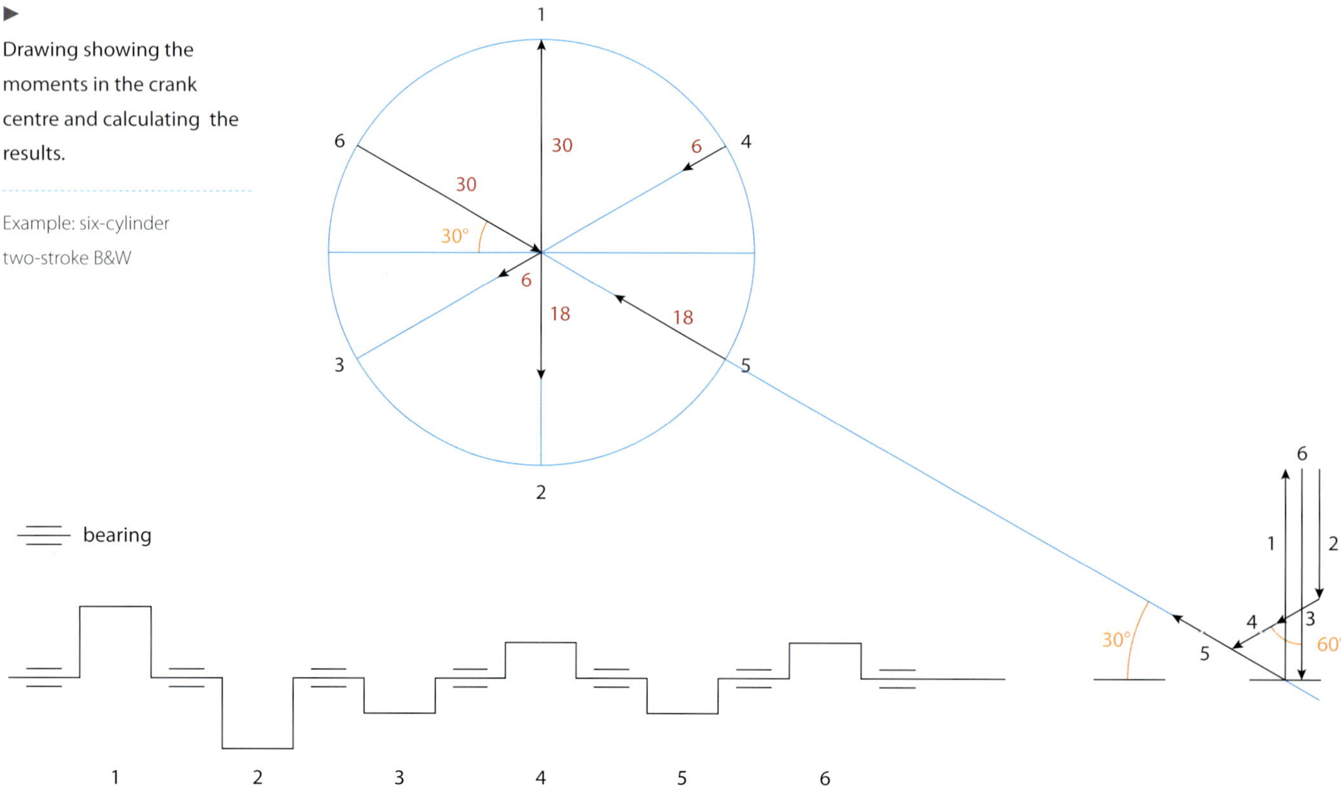

▶ Drawing showing the moments in the crank centre and calculating the results.

Example: six-cylinder two-stroke B&W

17.8 Crank - connecting rod mechanism

Derivation of the applied formulas

If the crank has turned an angle θ clockwise to the top point, the connecting rod has turned an angle φ counter-clockwise.

Here the following applies

$x = l + r - l\cos\varphi - r\cos\theta.$

To eliminate φ consider that

$BB' = r\sin\theta = l\sin\varphi$, so $\sin\frac{r}{l}\sin\theta.$

As generally $\sin^2\alpha + \cos^2\alpha = 1$ and therefore

$\cos\alpha = \sqrt{1-\sin^2\alpha}$ becomes $\cos\varphi = \sqrt{1-\frac{r^2}{l^2}\sin^2\theta}.$

The calculation is exact to this point. However, we will now apply an approach, which is generally used for standard ratios of r and l.

Let's assume $\frac{r}{l} = \lambda \equiv$ crank-connecting rod ratio

$x = r\cdot(1 + \frac{1}{\lambda} - \cos\theta - \frac{1}{\lambda}\cdot\sqrt{1-\lambda^2\cdot\sin^2\theta})$

$\sqrt{1-\lambda^2\cdot\sin^2\theta}$ is according to Mc Laurin resolved as:

$(1-\lambda^2\cdot\sin^2\theta)^{½} = 1 - \frac{1}{2}\lambda^2\sin^2\theta - \frac{1}{8}\lambda^4\cdot\sin^4\theta - \ldots$

The first terms usually suffice.

Therefore: $(1-\lambda^2\cdot\sin^2\theta)^{½} = 1 - \frac{1}{2}\lambda^2\sin^2\theta.$

As $\sin^2\theta = \frac{1}{2}(1-\cos 2\theta)$ it follows that:

$x = r\cdot(1 + \frac{\lambda}{4} - \cos\theta - \frac{\lambda}{4}\cdot\cos 2\theta)$

it then follows $\frac{d\theta}{dt} = \omega$ is the piston velocity $\underline{v}$

$v = \frac{dx}{dt} = \frac{dx}{d\theta}\cdot\frac{d\theta}{dt} = r\cdot\omega\,(\sin\theta + \frac{\lambda}{2}\sin 2\theta)$

and the acceleration of the piston

$a = \frac{dv}{dt} = r\cdot\omega^2\,r\cdot(\cos\theta + \lambda\cos 2\theta).$

The derived expression for a is significant in studying piston-engine dynamics.
Based on the equations for v and a, graphs can be drawn of v and a, as a function of time, or of the crank angle θ.

Crank/connecting rod mechanism

$Sx = r\,(1-\cos\alpha) + L\left(1-\sqrt{1-\frac{r^2}{L^2}\sin^2\alpha}\right)$

$\sqrt{1-\frac{r^2}{L^2}\sin^2\alpha}$ developed by means of

$\sqrt{1-x^2}$ Taylor or Mc Laurin?

$x = \frac{r}{L}\sin^2\alpha \rightarrow -\frac{r}{L} \le x \le \frac{r}{L} \rightarrow$ so around 0!

(λ is currently 0.3 à 0.4, depending on the engine stroke.)

Developed by applying Mc Laurin!

$f(x) = f(0) + \frac{f'(0)}{1!}x + \frac{f''(0)}{2!}x^2\ldots$

$f(x) = \sqrt{1-x^2} \rightarrow (1-x^2)^{½}$ $\qquad f(0) = 1$

$f'(x) = \frac{1}{2}(1-x^2)^{½}\cdot(-2x) = -x(1-x^2)^{½}$

$f'(0) = 0$

$f''(x) = \frac{1}{2}\cdot\frac{1}{2}(1-x^2)^{-⅔}\cdot 2x + (-2)\cdot\frac{1}{2}(1-x^2)^{-½}$

$f''(0) = -1$

Complete!

$f(x) = 1 + 0x - \frac{1}{2}x^2\ldots$ so $\sqrt{1-x^2} \approx 1 - \frac{1}{2}x^2$

In addition, $\sqrt{1-\frac{r^2}{L^2}\sin^2\alpha} \approx 1 - \frac{1}{2}\cdot\frac{r^2}{L^2}\sin^2\alpha.$

Therefore:

$S_x = r(1-\cos\alpha) + \frac{r^2}{2L}\sin^2\alpha.$

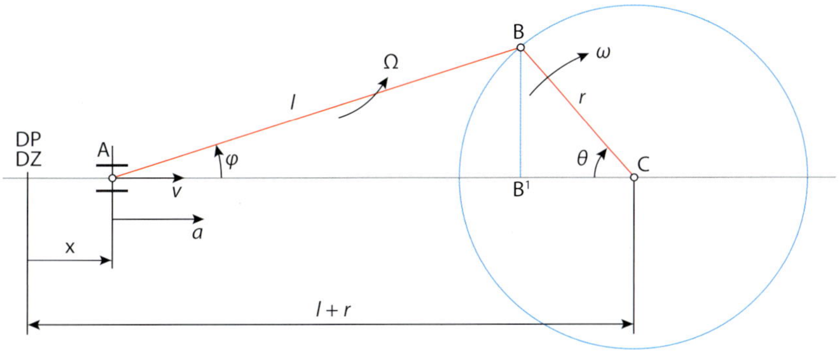

◀ The crank-connecting rod mechanism.

Forces exerted on a crankpin bearing.

In order to obtain a clear picture of the forces exerted on the crankpin bearing, it is assumed that the position of the piston during the power stroke is random.

The force on the piston is resolved into a force exerted on the connecting rod F_{CR} and a horizontal force on the piston skirt and therefore the cylinder liner.

The rotating part of the connecting rod is placed in the heart of the crankpin.

The rotating force on this point is located on a line drawn through the heart of the crankshaft and the heart of the crankpin.

As F_{CR} is known, F_R and F_{KPL} can be either compiled or calculated.

Therefore, the force on the crankpin bearing is opposite to the force on the crankpin and is at an angle γ with the connecting rod and this force is then placed at an angle γ as shown in the polar-load diagram for a crankpin bearing.

The force F_{CP} is subsequently used to establish the force in the main-shaft bearing. Here, it is of importance to remember that the force F_{CP} is distributed over the bearings on both sides of the crank webs! This should also be applied to the mass forces.

In order to determine the resultant moment, the forces are shifted to a so-called moment plane. Generally, the forces will not produce a resultant moment, as most engines are homogeneous with respect to the masses and the crankshaft is usually symmetric around its axis. The moments vary, but by drawing them in the centre of the crank and subsequently compiling them, their moments can be determined.

Here the following sign convention is used: for cylinders in front of the moment's plane, the moment vector is in line with the crank; for cylinders behind the moment plane, the moment vector is opposed to the crank. This is applied to both the primary and the secondary crank centre. If a considerable resultant moment remains, the reduction of the value of the resultant moment to a minimum can be attempted by adjusting the combustion sequence.

It should be remembered that optimum combustion sequence might give balancing problems regarding torsional vibrations, bearing load and X- and H moments that are exerted on the bedplate. The manufacturer should decide what is best given a certain situation.

Following are examples for establishing the resultant moment.

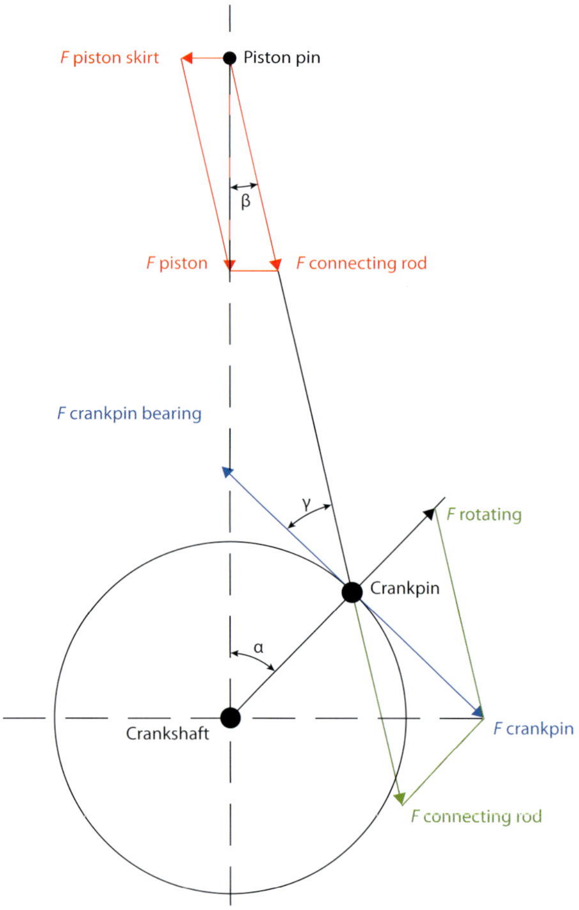

▲ Forces exerted on a crankpin bearing.

▲ The crankpin and crankpin bearing are the heaviest loaded parts of the drive gearing. Goltens, Dubai, UAE.

An overheated crankpin as shown in the picture always leads to buckling of the crankshaft. Shown here, one has tried to 'stretch' the crankshaft by hitting it with a hammer. It was not successful. The larger area is carved out material that had become harder due to overheating. Grinding the material away allows one to see how deeply the hardness has been increased in the material. This axis will probably be rejected. A threaded rod is placed between the webs to avoid flexing during transportation.

Polar-load diagrams

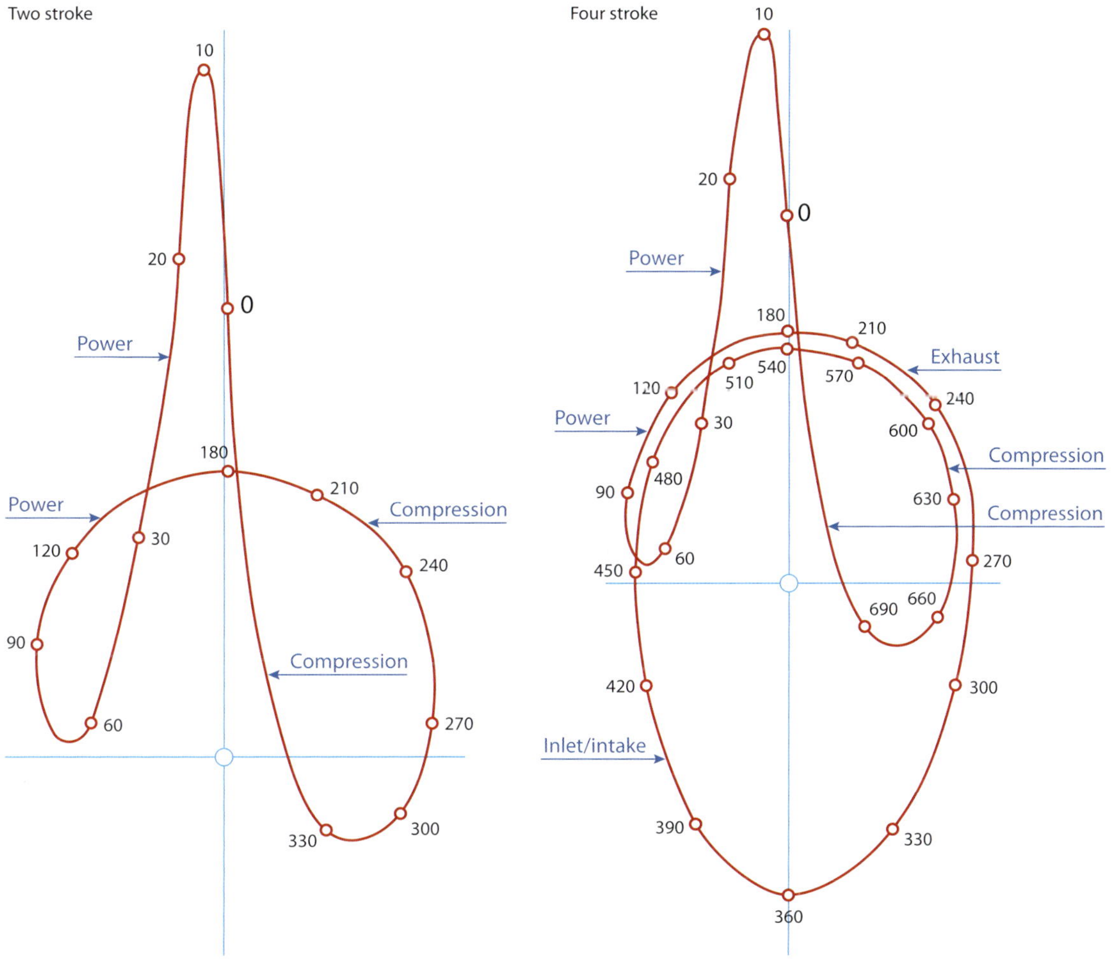

◀ Polar load charts of the crankpin bearings of a two-stroke (left) and a four-stroke engine (right).

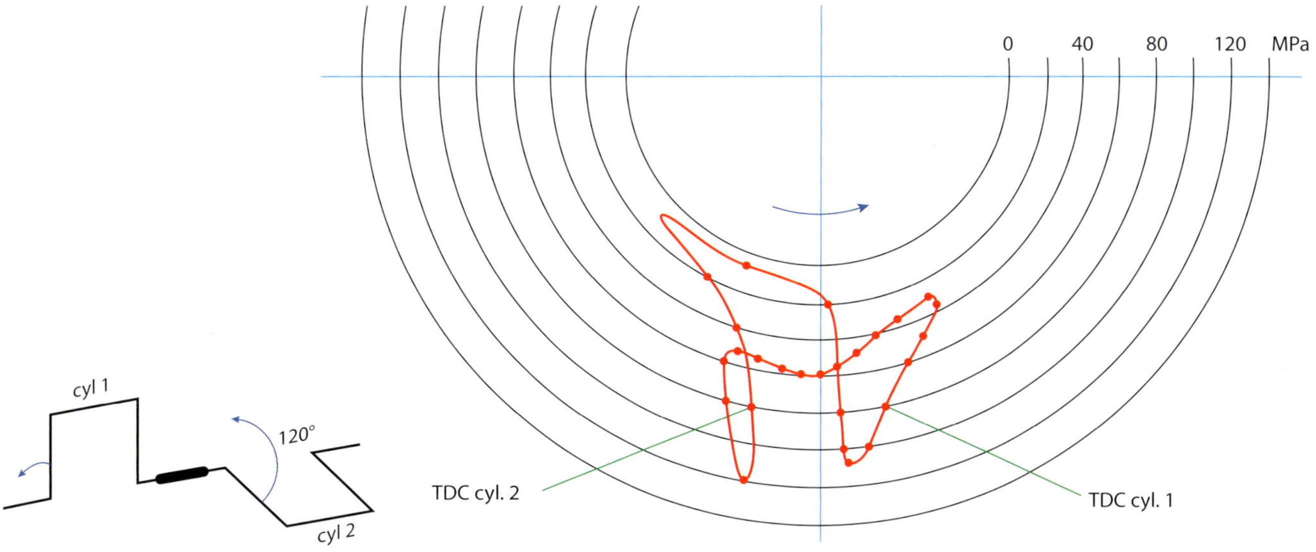

▲ A polar-load chart of a main-shaft bearing for a medium-speed four-stroke diesel engine from Wärtsilä, a Sulzer Z 40/48.

17.9 Forces in a two-stroke crosshead engine

The following forces and moments take place:

1. tilting moment $F_n \times a$
2. turning moment $F_t \times R \cos \alpha$
3. force exerted on the piston F_p
4. force exerted on the cylinder cover $F_h = F_p$
5. acceleration force F_a of the reciprocating parts
6. centrifugal force F_c of the rotating parts

Acceleration force F_a

The acceleration force F_a is linear and gives momentum to various parts in the engine, such as the piston, piston rod, crosshead and part of the connecting rod, from TDC to BDC.
This acceleration a, can be established with a formula (not deduced):

$$a = \omega^2 \times R \times \cos \alpha + \frac{R}{L} \times \omega^2 \times R \times \cos 2\alpha.$$

In which:
R = crank radius
L = connecting-rod length
α = travel through the crank angle

The acceleration force F_a can be established with the formula (not deduced):

$$F_a = \text{mass}_{\text{parts}} \times a = \text{mass}_{\text{parts}} \times \omega^2 \times R \times \cos \alpha + \text{mass}_{\text{parts}} \times \frac{R}{L} \times \omega^2 \times R \times \cos 2\alpha$$

Forces exerted on the piston and the cylinder cover

Usually, these are of the same magnitude and opposing. Resolution of the piston force in the centre of the crosshead produces a force in the connecting rod, F_c, and a horizontal force, F_n, on the crosshead guide.

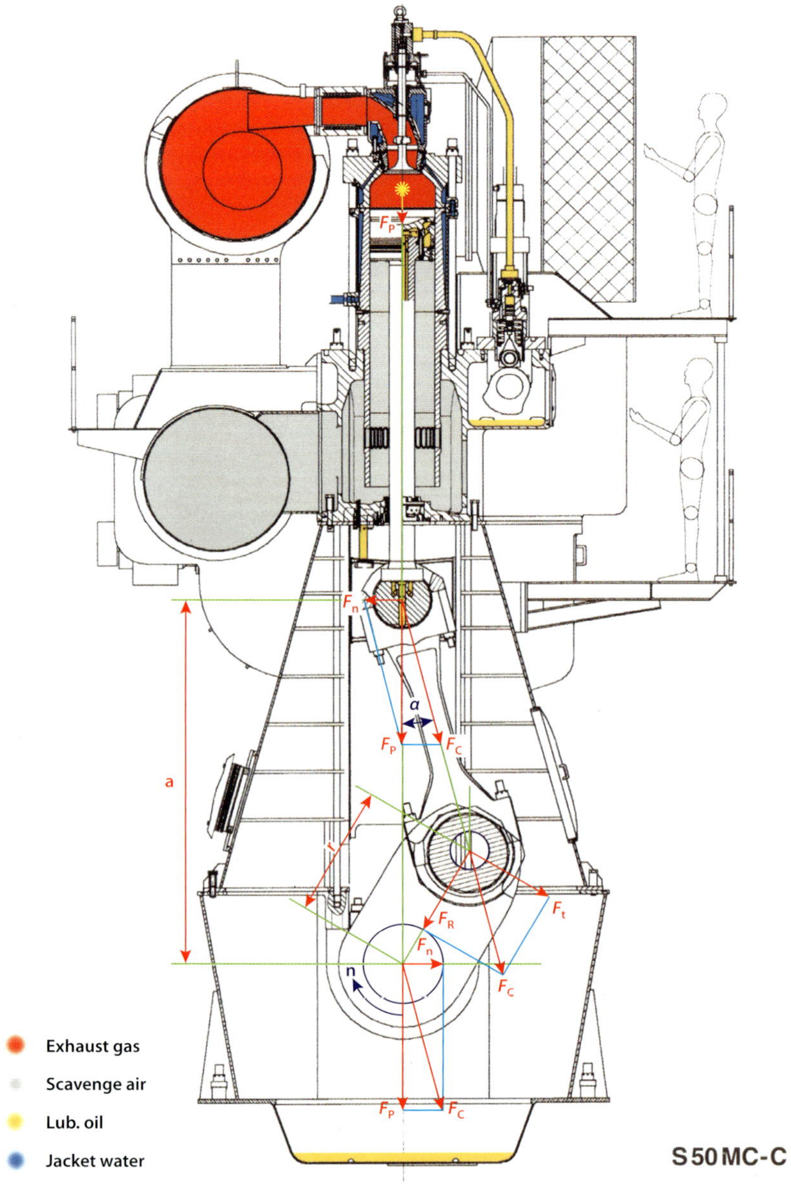

▶ The forces exerted by gas pressures in a running engine.

Designation
F_n = guide force
a = distance from the heart of the crosshead to the heart of the crankshaft
F_t = tangential force
r = crank radius
α = travelled crank angle

In a four-stroke trunk-piston engine, this piston force is resolved in the central axis of the gudgeon pin and F_n, becomes the force in the piston skirt. Vertical acceleration and inertia forces also work on the engine frame.
Part of the piston force is used for the acceleration force F_a to provide momentum for the moving parts. A force of $F_p - F_a$ remains as the tangential force and the guide force, which is exerted in the direction of the engine bedplate.
The force F_h is larger than the force $F_p - F_a$, so the engine has the tendency to move upwards. This force is not balanced in the engine and is therefore referred to as a **free force**.
So, during deceleration of the moving parts, a **free force**, directed downwards, is generated. As these free forces fluctuate periodically, the diesel engine tends to bounce up and down, thus causing vibrations in engine and ship via the bedplate.

17.10 Vibrational levels, acceptable values

There are two main points of consideration for assessing vibrational levels:

1. The vibrational levels should not produce material fatigue, thus causing damage to the engine or ship's construction.
2. Vibrations should not cause disturbance or discomfort for the personnel.

To comply with these criteria certain limitations for vibrational levels at various frequencies can be established.

a **Low frequencies** – transmission limitations. These are determined by the static stress levels.
b **Intermediate frequencies** – test limitations. Here the test limitation ensures that the kinetic energy remains constant, thus resulting in a reduction of the deviations.
c **High frequencies** – acceleration limitations. The acceleration limit for high frequencies reduces the admissible deviations even further to keep the noise production within certain margins.

The low-speed two-stroke crosshead engine is designed for high internal varying forces and high vibration levels in its frame.

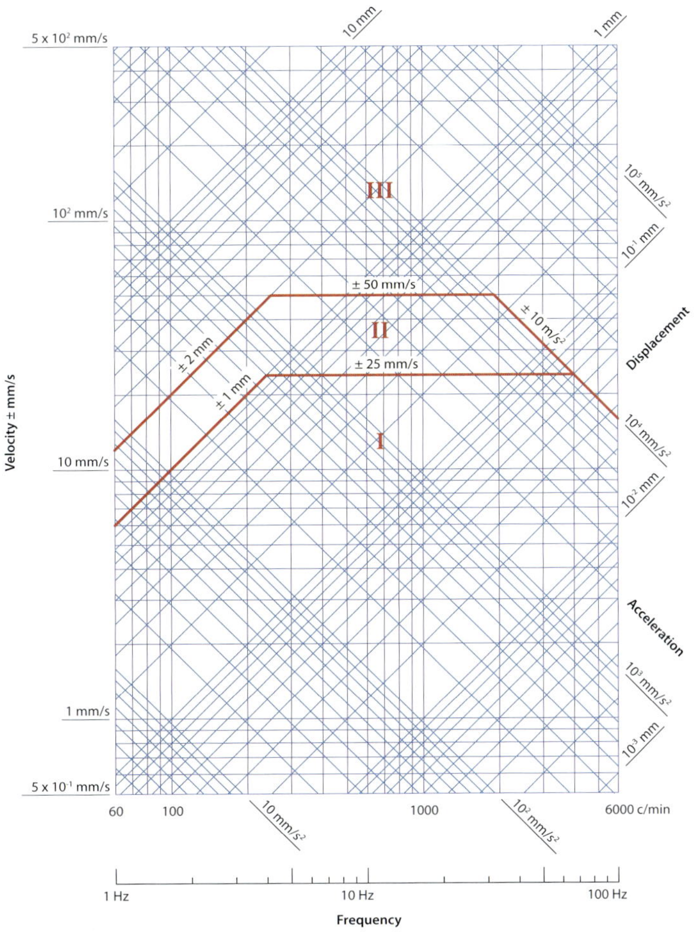

▶ Vibrational limits.

Zone I: Acceptable.
Zone II: Vibration will not damage the main engine. However, under adverse conditions, annoying/harmful vibrational responses may appear in the connected structures.
Zone III: Not acceptable: damage to the engine.

17.10.1 Conclusion

Mechanical vibrations in steel constructions are complex. In a particular steel construction in a ship driven by a low-speed two-stroke crosshead engine, the generated vibrations and the natural (vibration) frequencies of the construction may have unacceptable consequences, unless adequate action is taken.
In spite of modern technology with models and advanced software computations, there are still vibration-related problems. A familiar example is tearing of the welded plate-steel A-frame.

17.11 Tangential-force diagram

The crank-connecting rod mechanism is loaded with a force F_p on the gudgeon pin in a four-stroke engine and in a two-stroke engine, the piston rod.
The pressure on the piston varies considerably and consequently the force F_p changes constantly.

Calculations show that F tangential is approximately equal to F piston.

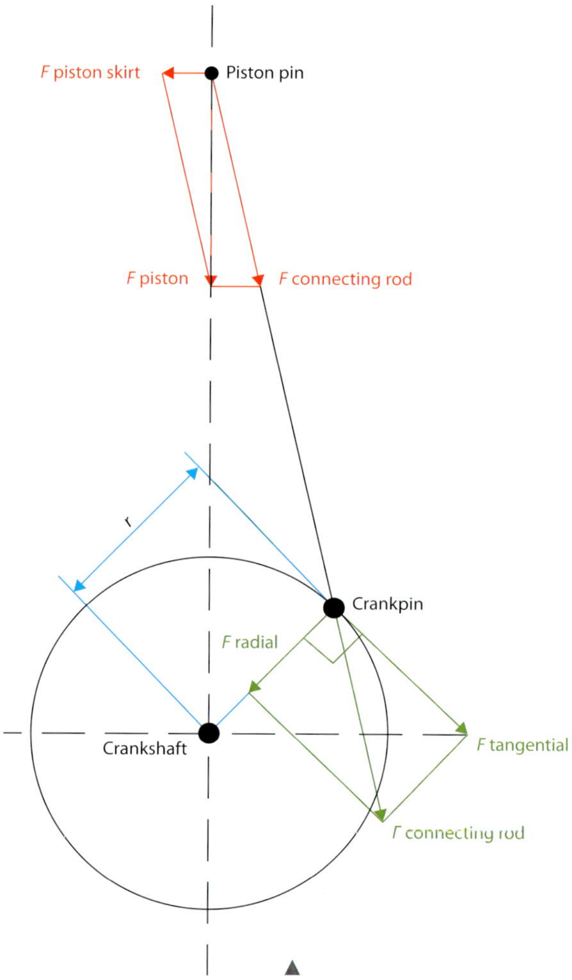

▲
The gas force F_p exerted on the piston results in force F_c acting on the connecting rod and force F_n, acting on the piston skirt.

This connecting-rod force F_c when exerted on the crankpin results in a tangential force F_t and a radial force F_r.
As the gas force exerted on the piston changes constantly, all the other forces also change in magnitude constantly.
With the ever-changing piston position and therefore the crank, the direction of the tangential force also changes continually.

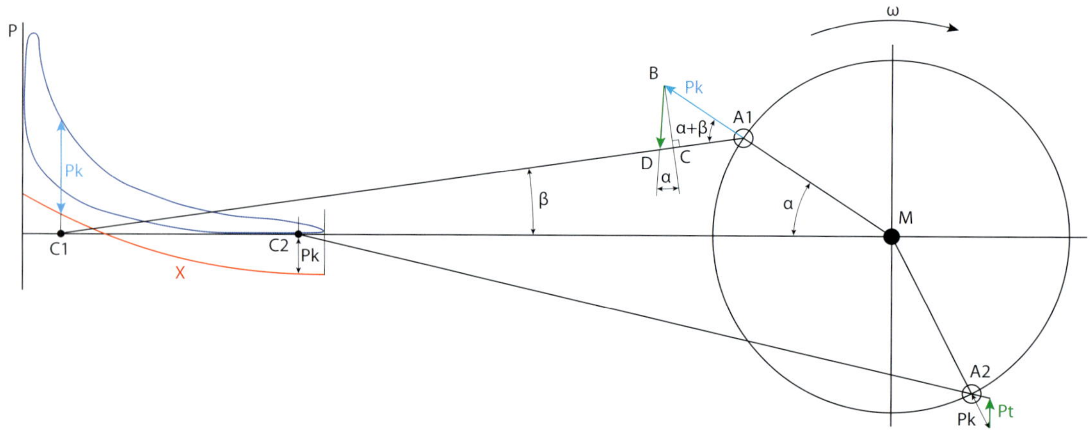

▲
A tangential-force diagram for a two-stroke crosshead engine.

Left, the indicator chart or banana and right the crank circle.
In these diagrams, one works with pressures not forces.
The crank MA is in a random position at an angle α and with the vertical axis (here, horizontal line MC).
AM = crank length
AC = connecting-rod length
At crank position A1, the connecting-rod length AC is circled and cuts the horizontal line of the diagram at C1.
At crank position A2, the connecting-rod length AC is circled and cuts the horizontal line of the diagram at C2.
Line x is the reaction of the mass forces translated to pressure.
Above the zero line of the operating diagram, the mass forces are assisting; below, they are counter-active.
Measuring the crosshead pressure from the working diagram at crank position A1/C1 and plotting them on the extension of MA1.
A line is dropped between point B and the connecting rod A1-C1.
Angle β, between the connecting rod A1-C1 and the 'vertical' MC is placed left of the perpendicular line BC and PT, the tangential pressure is then found. This can be found for every crankshaft position.
In crank position A2, the tangential pressure is directed upwards and therefore exerts a negative pressure on the shaft.

The development of these tangential forces can be observed in a diagram, with a length that is the circumference of the crank circle $2 \times \pi \times R$.

Magnitude of the decelerations and accelerations

The deceleration and acceleration rates and therefore the divergence from the mean angular velocity (read rpm) will decrease when the mass-inertia moment of the rotating parts, such as the crankshaft, camshaft, cog wheels, flywheel and coupling shafts, increases. This is why the flywheel is mounted on the crankshaft.

A single-cylinder diesel engine has a rather irregular angular velocity and therefore needs a 'thick' flywheel. The more cylinders engines have the thinner and lighter the flywheel.

In, for instance, an eighteen-cylinder V-engine it is a thin disc intended to start the engine and turn it during repairs.

The statement that an engine does not have a constant angular velocity ω is only partially correct.

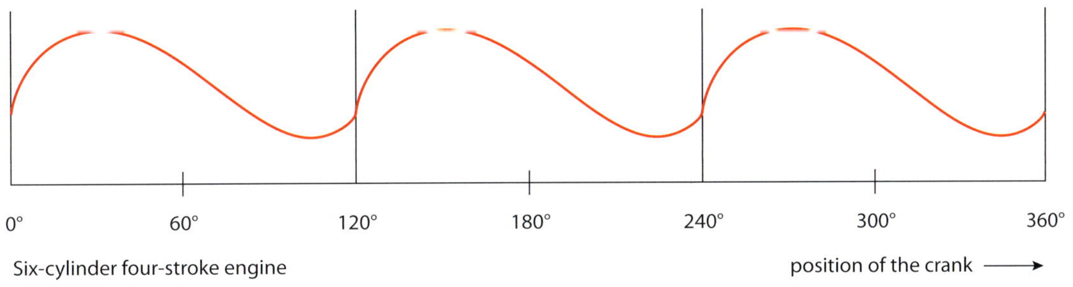

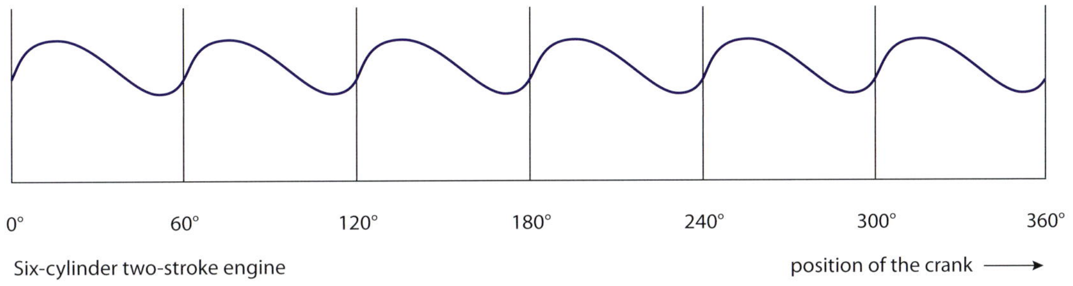

◄ Tangential diagrams.

above: tangential-force diagram for a six-cylinder four-stroke engine.
below: tangential-force diagram for a six-cylinder two-stroke engine
With both charts, only the positive tangential forces are shown.
Noticeable is that the tangential force in two-stroke engines have variation than those in four-stroke engines.

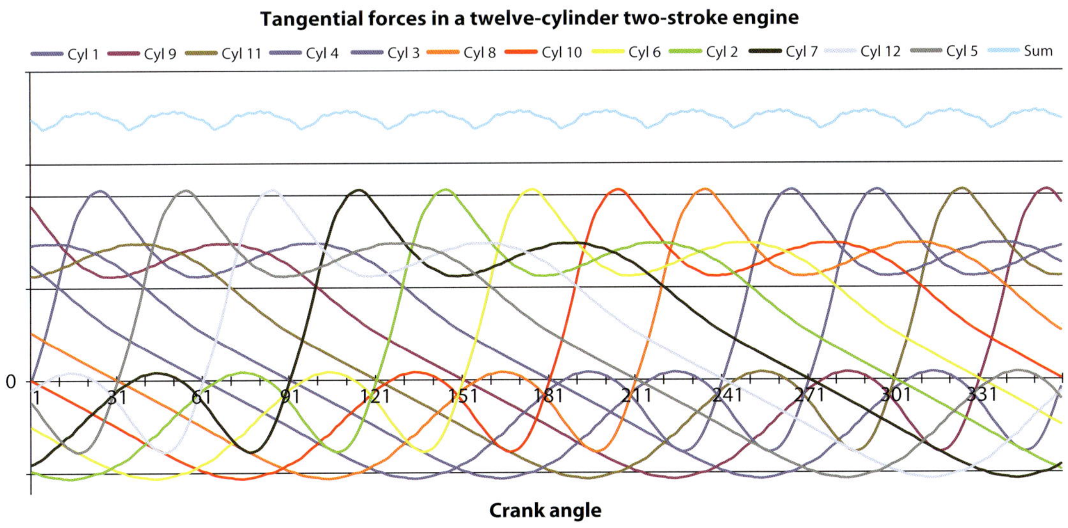

◄ The tangential forces in a twelve-cylinder two-stroke engine, a Wärtsilä RTA96C.

The ignition sequence is shown:
1-9-11-4-3-8-10-6-2-7-12-5

Horizontal:

One revolution of the crankshaft and so one process in the two-stroke engine.

Every cylinder has a different colour and one line is equal to one complete revolution and shows the tangential force at that moment.

The blue line above in the figure is the sum of all tangential forces at that specific point. Because of the large number of cylinders, this line could be called regular.

Vertical:

Above the zero line 'positive' tangential force – directed downwards.

Below the zero line 'negative' tangential force – directed upwards.

17.12 Degree of cyclic irregularity

Formule is:

$W_p = 0{,}5 \times J \times \omega^2$

W_p = power surplus provided by the combustion process
J = collective mass inertia moment of all the rotating parts
ω = angular velocity

From the formula, it follows that: The flywheel may get lighter as the angular velocity ω increases, the so-called power surplus W_p, or the power shortfall is smaller.
The power surplus or -shortfall can be reduced by distributing the capacity over more cylinders and ensuring that the angles between the corresponding cranks remain identical.

Example

Five-cylinder two-stroke engine:

angle $\frac{360°}{5} = 72°.$

Five-cylinder four-stroke engine:

angle $\frac{720°}{5} = 144°.$

The requirements with respect to the degree of cyclic irregularity are dependent on the apparatus driven by the engine.

Example

An alternator:

$\delta = \frac{1}{200} \text{ to } \frac{1}{300} = \frac{\omega_{max} - \omega_{min}}{\omega_{mean}}.$

A water pump:

$\delta = \frac{1}{50} \text{ to } \frac{1}{100}.$

17.13 Vibrations in engine frame and propeller shaft

Introduction

Construction of the crankshaft in a diesel engine is a complex matter.
The shaft has a complicated shape and it is therefore difficult to calculate meticulously the material stresses created by torque exerted on the cranks.
Moreover, additional material stresses may occur when the shaft vibrates.

Three types of vibrations can be distinguished:

1 Longitudinal vibrations, where the masses connected to the shaft periodically shift in relation to each other in the direction of the shaft.
2 Bending vibrations, where the masses connected to the shaft periodically shift in the direction perpendicular to the centre axis of the shaft.
3 Torsional vibrations, where the masses connected to the shaft periodically execute rotating movements around the centre axis of the shaft. This rotational vibration is superimposed on the normal rotation of the shaft. In this case, parts of the shaft positioned between the vibrating masses are torsion-loaded.

Each crankshaft, including the connecting shaft, is to some extent subject to these vibrations. Using suitable instruments, the different vibration forms can be distinguished and subsequently registered.

For most types of diesel engines, the **torsional vibrations** pose the most serious problems. Therefore, much attention is paid to this form of vibration.
Apart from the crankshaft, vibrations occur in the entire engine frame. These vibrations produced by crosshead-guide forces are divided into parallel or H-vibrations and X-vibrations.
Torque exerted on the various cranks in the engine is not constant, varies considerably in amplitude, and so creates torsional vibrations.
The assembly of drive shafts, for example, the crankshaft, intermediate shafts and the propeller shaft with the connected masses, such as the cranks, pistons with rods and crossheads, flywheel and propeller form an elastic torque system that can vibrate torsionally, because of the constantly changing torque on the cranks. Dangerous vibration conditions may ensue with resonance, which means that one of the natural vibrational frequencies of the elastic system corresponds with the frequency of one of the vibration-generating moments. The vibrational amplitudes can increase significantly resulting in a significant increase in the stresses in the shaft material, which they create. **A fatigue break can result from the extra load of the shaft.**

As will be demonstrated below, these resonance conditions occur at specific rotational velocities of the crankshaft. It is therefore sometimes necessary to avoid operating in one or more of these velocity areas.
Furthermore, investigations are ongoing to deduce which means and processes can avert the production of intensive torsional vibrations, and the resulting material stresses.

The natural vibrational frequency of a torsional-vibration system

The natural vibrational frequency of a torsional-vibration system is determined by different factors. The simplest system consists of one mass, with a low mass-inertia moment J, attached to the end of a shaft with length 1, and a diameter of which the polar-inertia moment is equal to I_p.

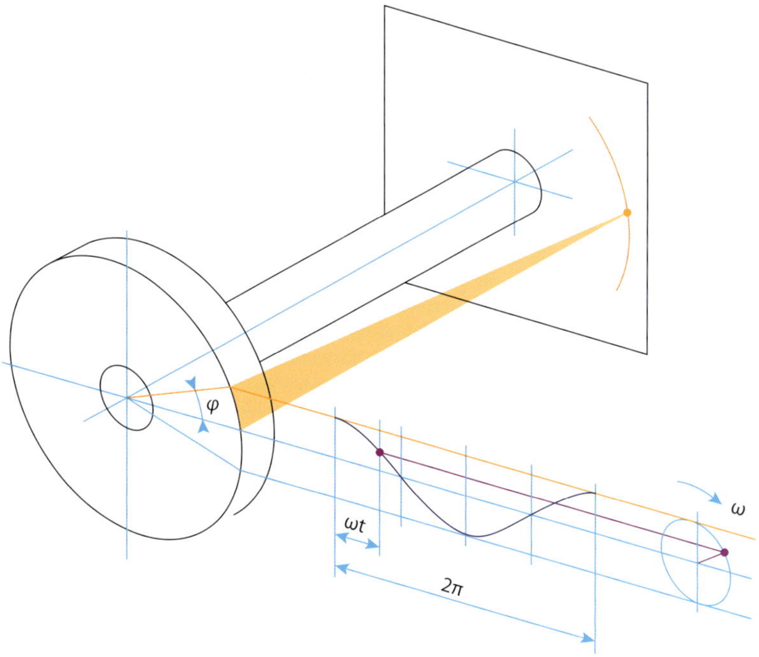

▶ The natural vibrational frequency of a torsional-vibration system.

An external moment T is exerted on the disk mass turning the disk through an angle φ (rad) from the middle position. The shaft is then loaded with a twisting stress and the following is valid:

$\varphi = T \cdot l \,/\, (G \cdot I_p)$

Where G is the shear modulus of the shaft material, the factor:
$c = G \cdot I_p \,/\, l$ is shaft rigidity.

The shaft exerts a counter-moment on the disk with a magnitude of:

$-T = -\varphi \cdot (G \cdot I_p \,/\, l)$

If the external moment suddenly ceases, only the counter-moment will exert a force on the disk. The disk then has an angle acceleration $\underline{\varphi}$.
The movement comparison is then:

$T = -\varphi \cdot (G \cdot I_p \,/\, l) = J \cdot \varphi$ or: $J \cdot \varphi + (G \cdot I_p \,/\, l) \cdot \varphi = 0$

divided by J gives: $\underline{\varphi} + \varphi \cdot (G \cdot I_p \,/\, l \cdot J)$

This differential comparison is solved using the Laplace equation:

$s^2 \cdot \varphi + \varphi \cdot (G \cdot I_p \,/\, l \cdot J) = 0$
$s^2 + (G \cdot I_p \,/\, l \cdot J) = 0$
$s^2 = -(G \cdot I_p \,/\, l \cdot J)$
$s = \pm j \cdot \sqrt{(G \cdot I_p \,/\, l \cdot J)}$

So valid for reverse transformation standard: $s = \pm j \cdot \omega$
So the solution is $\omega = \sqrt{(G \cdot I_p \,/\, l \cdot J)}$ and this is called the natural circle frequency ω_e.

It is often referred to as shaft natural vibration number and is:

$n_e = \omega_e \,/\, 2\pi$

The vibrational frequency is:

$n_e = 1 \,/(2\pi) \cdot \sqrt{(G \cdot I_p \,/\, l \cdot J)}$

From the last formula it follows that a larger rigidity of the shaft leads to a higher natural vibration frequency; increasing the mass-inertia moment leads to a decrease in frequency. The amplitude of the vibration is not important for this calculation.

The natural vibrational frequency of a shaft with two masses

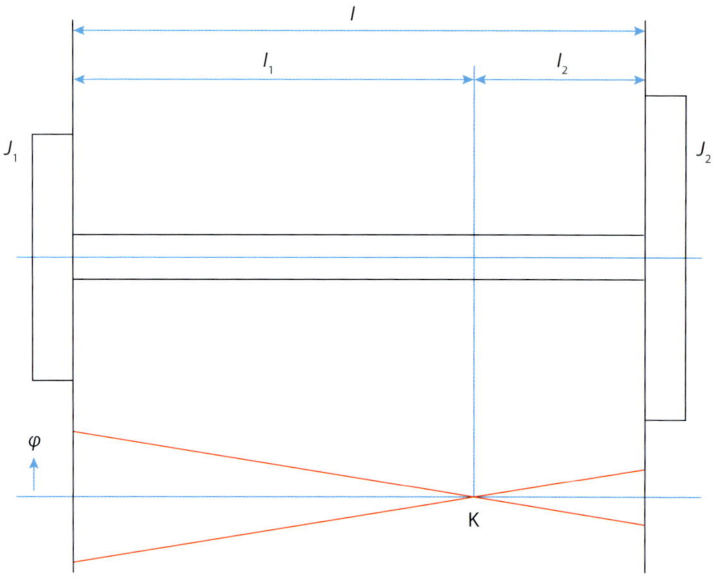

◀

Torsional-vibration system with two masses.

In the figure, a shaft with a mass mounted on both ends is shown. When these masses are twisted in relation to each other, torsion occurs in the shaft, and thereby a counter-moment. With the release of the masses a free vibration occurs. The deflections of both masses are constantly contrary to one another. One shaft section will remain static. This section is referred to as the node of the vibration system.

At the node position, the shaft can be clamped in such a way that it does not influence the vibration. For both shaft sections, the formula for a shaft with a single mass applies as both sections have a vibration with the same frequency. Therefore, the following is valid:

$1 / (2\pi) \cdot \sqrt{(G \cdot I_{p1} / l \cdot J_1)} = 1 / (2\pi) \cdot \sqrt{(G \cdot I_{p2} / l \cdot J_2)}$

The calculation gives:

$n_e = 1 / (2\pi) \cdot \sqrt{(G \cdot I_p / l \cdot J_v)}$ met $J_v = J_1 \cdot J_2 / (J_1 + J_2)$

The natural vibrational frequency for a shaft with three masses

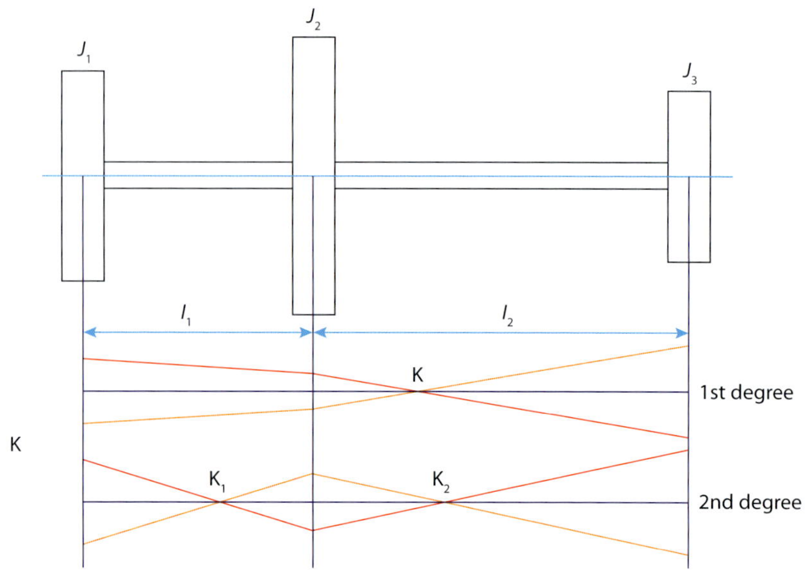

◀

Torsional-vibration system with three masses.

A similar system can take different vibrational modes:

a Mass 1 and 2 move opposite to mass 3; there is one shaft cross-section, the node. At rest, the node is positioned, in the example, between the masses 2 and 3. The natural vibrational frequency for this vibrational mode is referred to as being the 1st degree.

b Masses 1 and 3 move opposite to mass 2 and a node in both shaft sections occurs. The natural vibrational frequency belonging to these vibrational modes is higher than that of a vibration with one node and the frequency is referred to as being to the 2nd degree.

The following relationship is again valid:

$n_{e1} = \omega_{e1} / 2\omega$ en $n_{e2} = \omega_{e2} / 2\omega$

Therefore:

$\omega_{e1}^2 = (p + q)/2 - \sqrt{(((p + q) / 2)^2 - r)}$

$\omega_{e2}^2 = (p + q)/2 + \sqrt{(((p + q) / 2)^2 - r)}$

In these comparisons:

$p = (G \cdot I_p / l_1) \cdot ((J_1 + J_2) / (J_1 \cdot J_2))$

$q = (G \cdot I_p / l_2) \cdot ((J_2 + J_3) / (J_2 \cdot J_3))$

$r = (G \cdot I_p)^2 / (l_1 \cdot l_2) \cdot ((J_1 + J_2 + J_3)/(J_1 \cdot J_2 \cdot J_3))$

From this follows: $n_{e1} < n_{e2}$

The natural vibrational frequency of a shaft with a random quantity of masses

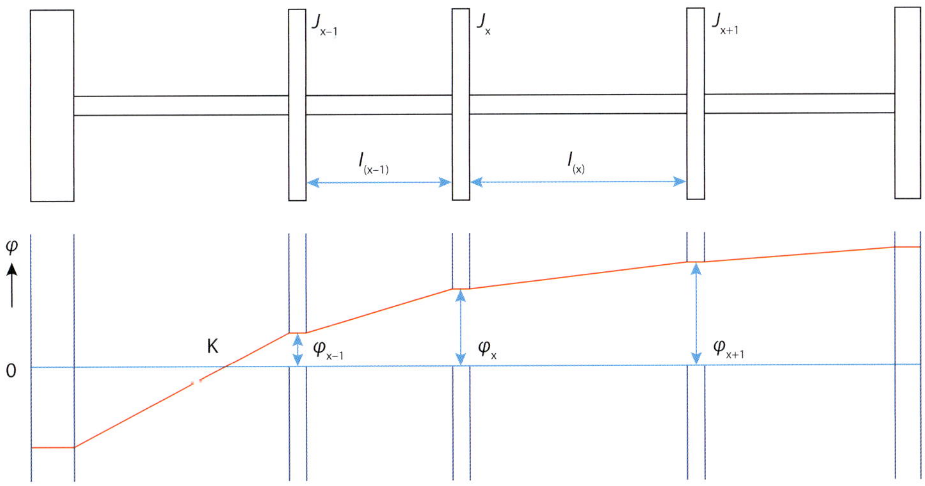

▲

Torsional-vibration system with a random number of masses.

Over the full length of a cylindrical shaft with an equal cross-section z, masses are mounted with mass-inertia moments of resp.: $J_1, J_2, J_3 \ldots\ldots J_z$.

When this system is in a torsional vibration, an isolated mass with mass-inertia moment J_x will move resulting in the following movement comparison:

$\varphi_x = \varphi_{0x} \cdot \cos \omega_e \cdot t.$

Here φ_{0x} is the vibrational amplitude of mass J_x. The angle acceleration of this mass is equal to:

$\varphi_x = - \omega_e^2 \cdot \varphi_{0x} \cdot \cos \omega_e \cdot t.$

The maximum angle acceleration $\underline{\varphi}$ will occur at the maximum mass deflection and therefore:

$\varphi_x = - \omega_e^2 \cdot \varphi_{0x}$

At the exact time of maximum deflection, the angle of shaft-section twist between the masses J_x and J_{x-1} will be equal to: $\varphi_{0x} - \varphi_{0x-1}$. The torsion in this shaft section l_{x-1} is then equal to:

$T_{x-1} = (G \cdot I_p / l_{x-1}) \cdot (\varphi_{0x} - \varphi_{0x-1})$

This moment acts on mass J_x.
With the same method, it can be shown that in the shaft section l_x a moment is exerted on mass J_x equal to:

$T_x = (G \cdot I_p / l_x) \cdot (\varphi_{0x-1} - \varphi_{0x})$

The moments T_{x-1} and T_x act on mass J_x and furthermore the acceleration moment $J_x \cdot \omega_e^2 \cdot \varphi_{0x}$.
The moments that try to increase the deflection of the mass are now positive. The moments that try to decrease the deflection are negative. This is because the sum of the moments acting on mass J_x must be equal to zero. This is valid at maximum deflection:

$T_x - T_{x-1} - J_x \cdot \omega_e^2 \cdot \varphi_{0x} = 0$ and so the moment on shaft section l_x is:

$T_x = T_{x-1} - J_x \cdot \omega_e^2 \cdot \varphi_{0x}$

This comparison is valid for all shaft parts, from l_1 tot l_z. It can now be stipulated:
On shaft section l_1 acts:
$T_1 = T_o + J_1 \cdot \omega_e^2 \cdot \varphi_{01}$ and because $T_o = 0$:
$T_1 = J_1 \cdot \omega_e^2 \cdot \varphi_{01}$
On shaft section, l_2 exerts a moment:
$T_2 = T_1 + J_2 \cdot \omega_e^2 \cdot \varphi_{02} = J_1 \cdot \omega_e^2 \cdot \varphi_{01} + J_2 \cdot \omega_e^2 \cdot \varphi_{02}$

This is consecutive for all shaft sections, until the moment T_z is exerted in the shaft section right of the last mass l_z. As this shaft section does not exist, $T_z = 0$, so:
$T_z = \Sigma J_x \cdot \omega_e^2 \cdot \varphi_{0x} = 0$ where x runs from 1 to z.
Alternatively, in words:
In a torsional-elastic system, in a free torsional-vibration state, the sum of the acceleration moments acting on the separate masses in every moment equals 0.

The conditions above are valid for a number of values of ω_e. It can be shown that for a system with z masses, a number of natural vibrational frequencies equal to z-1 can be found where this comparison is correct. The system has z-1 natural vibrational frequencies. Therefore it is valid that $\omega_e < \omega_{e+1}$, and that the vibrational frequencies are independent of the magnitude of φ_0.

The intrinsic values of the system can be found by estimating one of the values of ω_e, and then entering this value in $J_x \cdot \omega_e^2 \cdot \varphi_{0x}$.
If the estimation is correct, the sum of the acceleration moments is equal to zero.
This method is easily performed with the HOLZER algorithm; it is applied in the following easy example.

Example

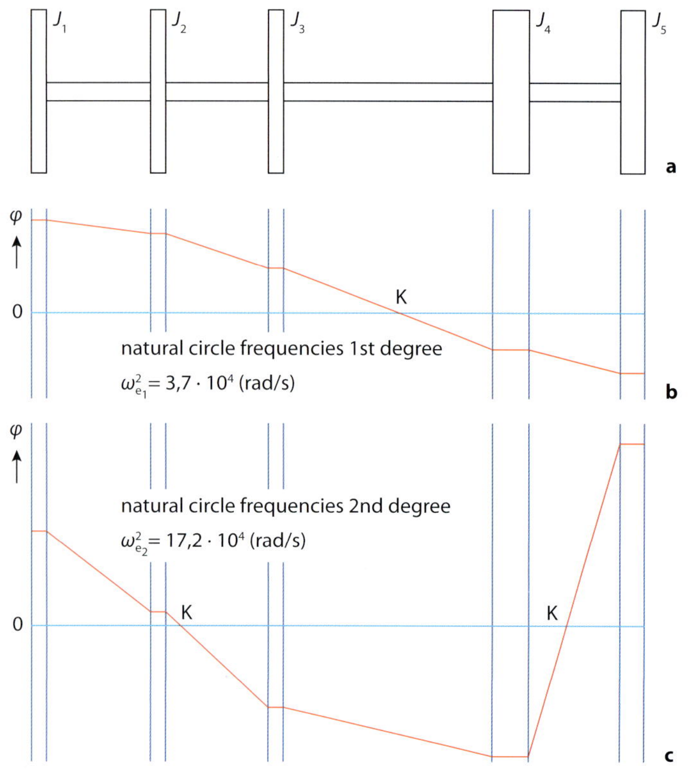

▲

Free vibrational modes of a system with five masses.

Five steel disk masses with a diameter of 0.223 metres are mounted on a steel shaft with a diameter of 1.067 metres. The diameter of the disks is from left to right:

$b_1 = b_2 = b_3 = 0{,}1$m, $b_4 = 0.3$ m and $b_5 = 0.2$ m.

The mass-inertia moments of the disks are resp.

$J_1 = J_2 = J_3 = 100$ kg/m², $J_4 = 300$ kg/m² and $J_5 = 200$ kg/m².

The lengths of the shaft sections between the disks are from left to right $L_1 = 1$ m, $L_2 = 1$m, $L_3 = 2$ m and $L_4 = 1$ m.

The torsional rigidity of a shaft section with a length of 1 metre is 2.10^7 N · m/m · rad.

An approximate value for the natural circle frequencies to the 1st and 2nd degree can be found by simplifying the system to a system with 3 masses. Here masses 1 and 3 are added to mass 2. From this, the natural circle frequencies can be calculated. So calculated:

$\omega_{e1}^2 = 3.34 \cdot 10^4$ (rad/s)² and $\omega_{e2}^2 = 17.77 \cdot 10^4$ (rad/s)²

These values are used as initial values in an algorithm (the HOLZER-algorithm)

The solution is as follows. In column 2, the mass-inertia moments of the different masses are inserted. In column 3, these mass-inertia moments are multiplied with the temporary value for ω_e^2 that was found by the calculation for the 3-mass system. The maximum deflections, amplitudes of the different masses φ_r are inserted in column 4.

As only the relative deflections are important for the amplitude of mass 1, the value $\varphi_{r1} = 1.000$ rad is chosen. The amplitude of mass 2 is then $\varphi_{r1} - \Delta\varphi$, the value of $\Delta\varphi$ is deduced from column 8.

The moments acting on the disk masses are inserted in column 5.

The sum of the moments on the masses is inserted in column 6.

In column 7, the lengths of the shaft sections are inserted. The angle of twist $\Delta\varphi$ in these shaft sections as a result of the operating moments acting on it are inserted in column 8. As the system must meet the comparison $T_z = \sum J_x \cdot \omega_e^2 \cdot \varphi_{0x} = 0$ the last value in column 6 is equal to 0, if the correct value for ω_e^2 has been estimated. This of course is mostly not the case, and a positive or negative value is found in column 6.

When in calculating the natural vibration frequency to the 1st degree, a positive value is found, the estimated value for ω was too low, and the algorithm must be recalculated for a corrected value.

The calculations are repeated until an adequate small value is found.

This method is used for the calculation of the natural circle frequency to the 2nd degree. However, with a positive value in column 6, a lower value for ω_2 must be inserted. Both algorithms occur only after several recalculations.

In the figures, the free-vibrational modes of the system to the 1st and 2nd degree are shown. Higher vibrational degrees may also be calculated using this method. The frequencies here are so high that resonance can be excluded.

In the figure, it is shown that with a vibration to the 1st degree, one node occurs in the shaft; with a vibration to the 2nd degree, 2 nodes. In the shaft sections with a node, the specific angle of twist is largest, and therefore the torsion occurring because of the torsional vibration in these shaft sections will be of the largest magnitude.

1	2	3	4	5	6	7	8
m	J	$J \cdot \omega_e^2 \times 10^{-6}$	φ_r	$J \cdot \omega_e^2 \cdot \varphi_r \times 10^{-6}$	$\sum J \cdot \omega_e^2 \cdot \varphi_r \times 10^{-6}$	1	$Dg = \frac{1}{G \cdot I_p} \sum J \cdot \omega_e^2 \cdot \varphi_r$
	$kg \cdot m^2$	$N \cdot m$	rad	$N \cdot m$	$N \cdot m$	m	rad
1	100	3,70	1,000	3,700	3,700	1	0,185
2	100	3,70	0,815	3,015	6,715	1	0,336
3	100	3,70	0,479	1,772	8,487	2	0,849
4	300	11,10	-0,370	-4,107	4,380	1	0,219
5	200	7,40	-0,589	-4,358	0,022	–	–

The Holzer algorithm.

1st degree natural vibrational frequency

$\omega_{e1}{}^2 = 3.7 \cdot 10^4 \ (rad/s)^2$

($n_{e1} = 30.6$ Hz)

1	2	3	4	5	6	7	8
m	J	$J \cdot \omega_e^2 \times 10^{-6}$	φ_r	$J \cdot \omega_e^2 \cdot \varphi_r \times 10^{-6}$	$\sum J \cdot \omega_e^2 \cdot \varphi_r \times 10^{-6}$	1	$Dg = \frac{1}{G \cdot I_p} \sum J \cdot \omega_e^2 \cdot \varphi_r$
	$kg \cdot m^2$	$N \cdot m$	rad	$N \cdot m$	$N \cdot m$	m	rad
1	100	17,2	1,000	17,200	17,200	1	0,860
2	100	17,2	0,140	2,408	19,608	1	0,980
3	100	17,2	-0,840	-14,455	5,153	2	0,515
4	300	51,6	-1,356	-69,954	-64,801	1	3,240
5	200	34,4	1,884	64,823	0,022	–	–

The Holzer algorithm.

2nd degree natural vibration frequency

$\omega_{e2}{}^2 = 17.2 \cdot 10^4 \ (rad/s)^2$

($n_{e2} = 66.0$ Hz)

The replacement system

In the abovementioned, the point of departure was a straight shaft with the same diameter along its entire length with a number of mounted masses.
An elastic torque system comprising a crankshaft with connected intermediate shafts, a propeller shaft or other gearing has however, a complex shape.
The mass-inertia moments of the various masses are difficult to calculate and the establishment of the rigidity of shaft parts, such as the cranks, often involves complicated calculations.
To obtain a well-ordered view when calculating the natural vibrational frequencies, the actual shaft is transformed to a shaft with a uniform shaft diameter, and the rigidity between the different masses will be identical to the actual shaft.
A transformed length is also deduced for the shaft flanges and other protuberances. If an elastic torque coupling is assembled in the propeller shaft, this is replaced by a deduced transformed shaft with an identical rigidity to the coupling.
The mass-inertia moments of the various engine parts should be established as meticulously as possible. The inertia moment of, for instance, a crank is obtained by determining the inertia moment of the different parts, such as the crank webs and crankpin, in relation to the crankshaft centre axis; these inertia moments are then totalled. The mass-inertia moment of a rotating part of the connecting rod in relation to the crankshaft centre axis is also calculated. The mass-inertia moments of unevenly shaped masses is found by dividing the mass in concentric rings and establishing the mass-inertia moments of these rings or parts of rings.
Whether the reciprocating moving parts of the drive gear participate in the torsional vibration is dependent on the position of the accompanying crank. If they are in one of the outer extreme positions, the connected reciprocating moving parts will hardly vibrate. If the crank is horizontal, the amplitude of the vibration of the crankpin and of the reciprocating moving parts is almost equal. An adequate approximation of the reciprocating moving parts is obtained by calculating half of their mass.
This is often prescribed by Classification Societies, amongst others, Lloyd's Rules.
When the propeller vibrates, part of the water it travels through will also vibrate. In order to correct this, it is assumed that the total mass-inertia moment of the water and screw is 1.25 times that of the individual propeller.

The free vibrational shape of a crankshaft in a diesel engine with the propeller and associated shaft

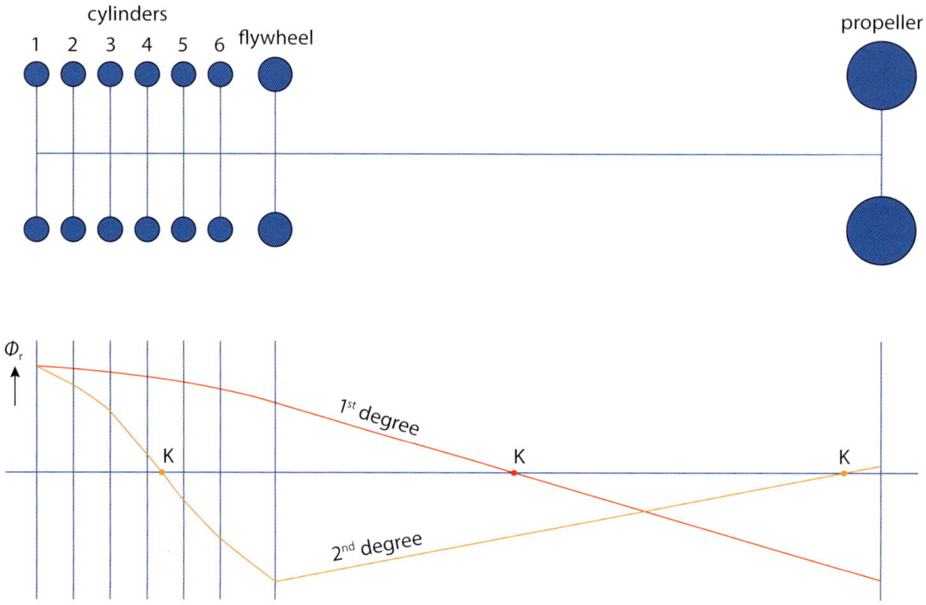

▲

Free-vibrational modes of a shaft in a six-cylinder diesel engine for propulsion.

In the figure: the system, with both free-vibrational modes, of a 6-cylinder 2-stroke diesel engine with intermediate shafts and a propeller. It is a SULZER engine, type RND 68. The cylinder diameter of this engine is 680 mm; the piston stroke is 1250 mm. At full load, rotational frequency 2.5 s^{-1} at an effective power output of 7280 kW. Both free-vibrational modes can be found using a HOLZER algorithm. The natural vibration frequencies are:

$\omega_{e1}^2 = 2.768 \cdot 10^3$ $(rad/s)^2$, $n_{e1} = 8.37$ Hz

$\omega_{e2}^2 = 30.480 \cdot 10^3$ $(rad/s)^2$, $n_{e2} = 27.79$ Hz

The node of 1st degree vibration is found in the connecting shaft. This is because 1st degree vibration is also the most heavily loaded. Notable is that a node of 2nd degree vibration is found between cranks 4 and 5. As the specific angle of the shaft twist in this section of the shaft is the greatest, this shaft section is because of 2nd degree vibration is also the most heavily loaded.

Vibration-generating moments, critical rotational velocities

As described previously, a torsion-elastic system can acquire torsional vibrations. However, there must be a cause for the generation of this type of vibration. In a diesel engine used for ship propulsion, the cause of vibrations in the propeller shaft is found in the irregular rotational moment supplied to the individual cranks by the pistons.
The magnitude of the load moment exerted on the propeller in the water varies because of the variations in the wake current. In the following example, the influence of these variations in the drive torque of the crankshaft will be considered.
In this chapter, the development of the tangential pressure, its multiplication by the piston surface and crank length is discussed, as is the torque development in one cylinder. A periodic phenomenon such as this can be analysed by using a harmonic analysis of one component of constant magnitude and an infinite number of components in a sinusoidal curves.

This kind of component has the following form:

$$p_{t(q)} = p_{Ot(q)} \cdot \sin(q \cdot \omega_a \cdot t + \beta_q)$$

Here ω_a is the angular velocity of the crankshaft, $p_{t(q)}$ is the q^e component of the tangential pressure. β_q is the phase shift in relation to the position of the crank by the vector through which the q^e harmonic component is proposed.
Furthermore, q is an integral number, which indicates the number of completed cycles during one combustion process in the cylinder.
However, the number of run-through periods by the harmonic component per crankshaft revolution is also of importance, and for a two-stroke engine this is equal to the number of periods per process, for a four-stroke engine it is equal to half the number of periods per process. If the number of periods per revolution is identified, the harmonic component can be presented as:

$$p_{t(q)} = p_{Ot(q)} \cdot \sin(p \cdot \omega_a \cdot t + \beta_q)$$

$p = q$ for a two-stroke engine, and has the consecutive values 1, 2, 3, 4, etcetera.
$p = q/2$ for a four-stroke engine, and has the consecutive values 0.5, 1, 1.5, 2, 2.5 etcetera. One refers to p as the order number of the harmonic component.
The alternating torque, which the harmonic component of the tangential pressure exerts on a crank, is therefore equal to:

$$T_{t(p)} = A_p \cdot r \cdot p_{t(p)}$$

Here A_p is the piston surface area and r the crank radius. $T_{t(p)}$ is called the harmonic moment of the p_e order.
At a given angular velocity of the crankshaft the frequency of a harmonic moment of the p_e order is then equal to $p \cdot \omega_a$. This applies to both two-stroke- and four-stroke engines. When this frequency matches the natural circle frequency of the torsional vibration ω_e of the shaft, then a moment $T_{t(p)}$, which is in phase with the crank vibration will work on the crank in question and the vibration will be amplified by the moment. A condition of resonance ensues.
The rotational frequency of the crankshaft in this condition is called a critical rotational frequency.

The following applies; there are a large number of potential critical rotational frequencies for a crankshaft with a propeller shaft:

$$P \cdot \omega_a = \omega_e \quad \text{or also: } p \cdot n_a = n_e$$

Obviously, resonance can occur with the natural vibrational frequencies to both the 1st and 2nd degree.
A number of critical rotational frequencies can fall inside the operational range of the engine. Examination will indicate if one or more of the critical rotational speeds may cause high torsional stress in parts of the propeller shaft.

Harmonic analysis of the tangential pressure in four-stroke- and two-stroke engines

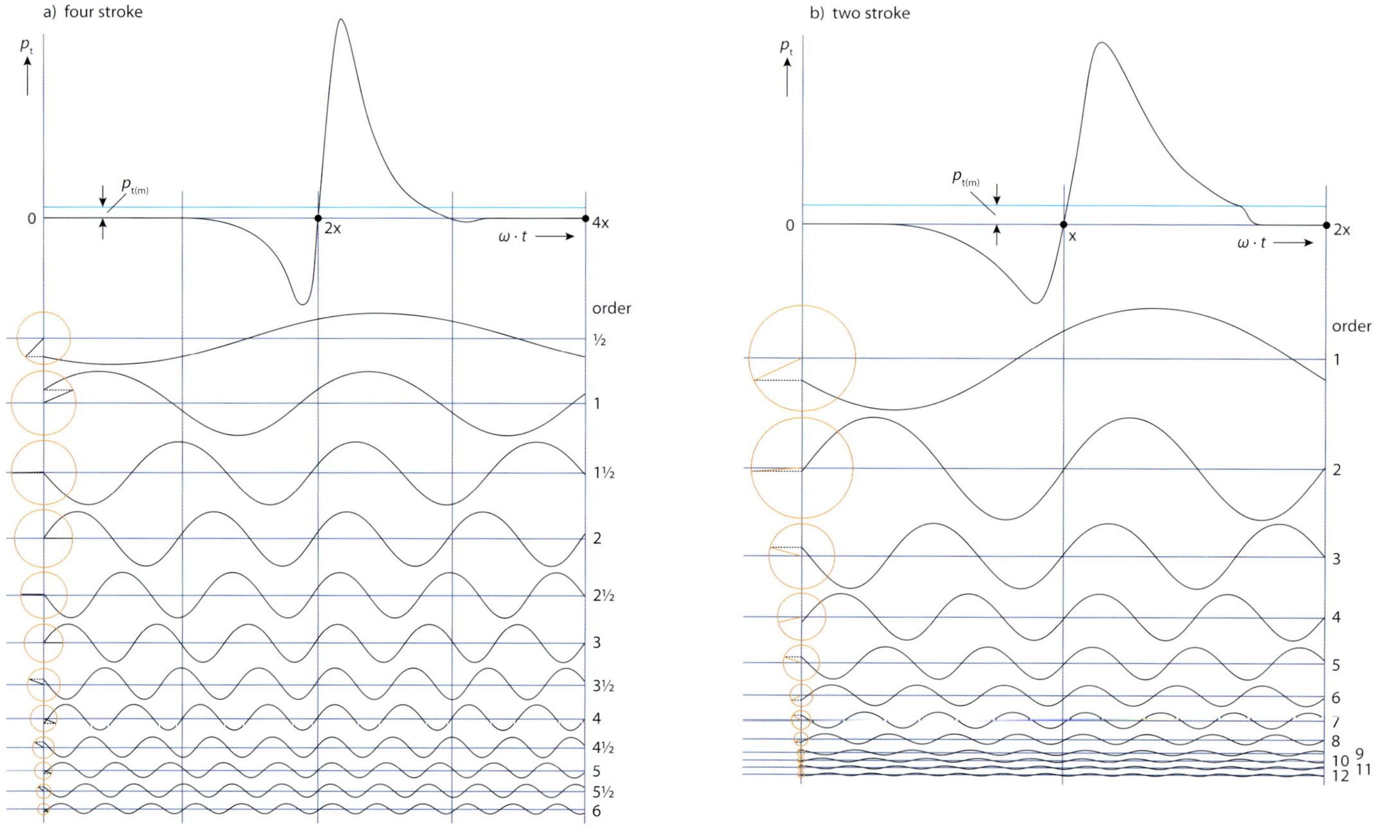

▲

In the figures a and b, the tangential pressure curve for a two-stroke and four-stroke diesel engine.

Below in the diagrams, a number of the harmonic components of the tangential pressure are shown. Noticeable is that in the four-stroke engine the harmonic components to the 1st and 1,5th order have a larger amplitude than that to the first order, but that the higher harmonic components rapidly decrease in amplitude. In two-stroke engines, the components to the increasing order are consistently smaller.

The tangential pressure curves are more or less calculated from representative indicator charts, when taking into account that the amplitude magnitudes of the harmonic components are a function of the mean indicated pressure. In, amongst others, 'Lloyd's Rules' are graphs showing certain mean indicated pressures where the harmonic components can be read. The values obtained are based on the tangential-pressure curve because of the gas pressure in the cylinder. The influences of the oscillating-mass forces have been omitted, as this would make it difficult to obtain a generally valid value. With harmonic analysis of the occurring tangential pressure because of the acceleration pressure on the piston, the following can be achieved:

$$P_{t(osc)} = m_o/A_z \cdot \omega_a^2 \cdot r \cdot (\lambda/4 \cdot \sin \omega_a \cdot t - ½ \sin 2\omega_a \cdot t - ¾\, \lambda \cdot \sin 3\omega_a \cdot t - \lambda/4 \cdot \sin 4\omega_a \cdot t)$$

waarin:

M_0 = oscillating mass

A_z = piston surface

$\lambda = \frac{R}{L}$

R = crank radius

L = length connecting rod

From this formula, it is shown that the components of the harmonic acceleration pressures comprise a 1st order component with small amplitude, a larger component to the 2nd order, a smaller to the 3rd order and a negligible component to the 4th order. Components to higher orders can be ignored. Resonances with components to low orders are mostly found at rotational speeds far above the operating area of the engine.

The question as to whether certain critical rotational speeds will damage the shaft is not dependent on the magnitude of the harmonic component concerned. More important is that the harmonic moments acting on the different cranks are in phase with each other, thereby strengthening or weakening each other. Assuming that the ignition distances of an engine with z cylinders are equal and the harmonic components of the tangential pressure to a certain order acting on the z cranks strengthen each other (in phase), the following is valid:

For two-stroke engines: $p = z; p = 2z; p = 3z$ etc.

for four-stroke engines: $p = \frac{z}{2}, p = 2 \times \frac{z}{2}, p = 3 \times \frac{z}{3}$ etc.

These rotational frequencies of the engine are known, as are the main critical rotational frequencies; the others are known as secondary critical rotational frequencies.

Example

The natural vibrational frequency of an eight-cylinder four-stroke engine is to the 1st degree (n_{el}) = 9.6 H. What are the main critical rotational frequencies to the 1st degree for this engine?
According to the above, it follows that: p = 8/2 = 4; p = 2 · 8/2 = 8; p = 3 · 8/2 = 12, etc.
So resonance occurs at a crankshaft-rotational frequency of $n_a = n_e/p$; so at: $n_a = 9.6/4 = 2.4\ s^{-1}$; $n_a = 9.6/8 = 1.2\ s^{-1}$; $n_a = 9.6/12 = 0.8\ s^{-1}$, as far as these rotational velocities fall inside the working range of the engine.

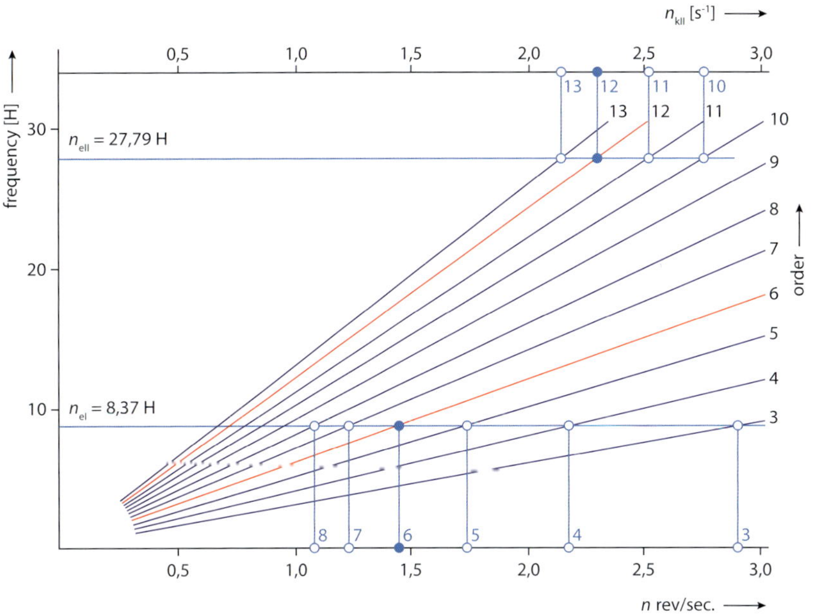

▲
For this two-stroke engine of SULZER, an RND 68, the critical rotational frequencies are easily read in the chart above.

For the frequency of a harmonic component to the pth order, the following is valid:
$n_p = p \cdot n_a$ the relationship between n and the rotational frequency of the shaft can be represented with a direction coefficient p. The points of intersection of these lines with the lines for n_{e1} and n_{e2} give the positions of the critical rotational frequencies inside the operating area of the engine.
A main critical rotational frequency (1/6) to the 1st degree and the 6th order occurs at rotational speeds of approximately 1.4 s^{-1}.
The main critical frequency to the 2nd degree, 12th order, is at approximately 2.3 s^{-1}. This last main critical frequency does not give any problems; the reason will be discussed later.

Transference of vibrational energy by harmonic moments

If the harmonic moment exerted on the crank is to the order p^e ,and has an amplitude equal to $T_{t(p)}$ and if the amplitude of the crank vibrations concerned is equal to φ_x, it can be proved that the power exerted on the crank for each vibrational period is equal to:

$$W_c = \pi \cdot A_p \cdot r \cdot p_{t(p)} \cdot \varphi_{1w}/\varphi_{1r} \cdot \varphi_x \cdot \sin\gamma$$

Here γ is the phase difference between the harmonic moment and the measurement of vibration of the crank and $\varphi_{1w}/\varphi_{1r}$ is the ratio between the actual measurement of vibration and the measurement according to the Holzer table.
Consequently, the energy transference will be at its highest level when $\sin\gamma = 1$, so when the phase difference between harmonic moment and the measurement of vibration is equal to 90°. If moment and the measurement of vibration are in phase, so when $\sin\gamma = 0$, the energy transference will also be equal to 0. For $\sin\gamma < 0$, as may occur on one or more cranks of a multiple-crank machine when a node of the free vibrational form falls between the cranks in a journal, the energy transference is also smaller than 0. This means that the vibration caused by the torque exerted on that crank is reduced.
At an angular velocity ω_a of the crankshaft, the vectors which represent the harmonic components of the tangential pressure have an angular velocity $p \cdot \omega_a$. Let us assume a crank position where the phase difference between the harmonic moment and the measurement of vibration for crank 1 equals 0 ($\sin\gamma = 0$). The harmonic moments to a given order working on all the cranks are equally high, but have shifted in phase in relation to the moment on crank 1, and at an angle of $p \cdot \delta$, when δ is the ignition distance between cylinder 1 and cylinder x. The measurements of vibration are, excepting special conditions, in phase with each other. This means that all cranks go through the state of equilibrium simultaneously.
Therefore, the total amount of energy transferred per period to the crankshaft by a harmonic moment to the order p amounts to:

$$W_{c(tot)} = \Sigma\, \pi \cdot A_z \cdot r \cdot p_{t(p)} \cdot \varphi_{1w}/\varphi_{1r} \cdot \varphi_x \cdot \sin\gamma \text{ with } x = 1 \text{ to } z$$

However, the actual extent of the measurement of vibration cannot yet be calculated based on this data. Only the relative size of the measurement of vibration can be inferred from the free-vibrational form. Nevertheless, a comparison can be made of the vibrational energy that is transferred during resonance by the various harmonic moments to the crankshaft, without having to know the absolute size of this energy.

As a certain order $p_{t(p)}$ has a certain constant value, therefore:

$$W_{c(tot)} = \pi \cdot A_p \cdot r \cdot p_{t(p)} \cdot \varphi_{1w}/\varphi_{1r} \cdot \Sigma\, \varphi_{r(x)} \cdot \sin \gamma_{(x)}$$
where x = 1 to z

The factor $\gamma_{(x)}$ which determines the actual magnitude of the harmonic moment on the crank x, is from a mathematical viewpoint detached from this moment, and connected to the measurement of vibration of crank x. Now an imaginary resultant measurement of vibration can be introduced with the magnitude:

$$\varphi_{r(res)} = \Sigma\, \varphi_{r(x)} \cdot \sin \gamma_{(x)} \text{ where x = 1 to z}$$

The maximum value of the magnitude of the harmonic moment to the p^e order of the vibrational energy transferred to the crankshaft then is:

$$W_{c(max)} = \pi \cdot A_p \cdot r \cdot p_{t(p)} \cdot \varphi_{1w}/\varphi_{1r} \cdot \varphi_{r(res)}$$

The $\varphi_{r(res)}$ can be established simply by construction, in a manner that resembles the construction of the secondary mass forces and moments. In the following example, the procedure will be clarified.

The magnitude of $W_{c(tot)}$ for the various critical rotational frequencies can be influenced by choosing a different ignition sequence. This has a particular influence on the critical rotational frequencies of half orders in a six-cylinder four-stroke engine.
Although the deduced values for $W_{c(tot)}$ provide a good idea of the transferred vibrational energy at varying critical velocities, they cannot be used to measure the actual measurements of vibration and ensuing torque-vibrational stresses. This is because the absolute value of the measurements of vibration is not only dependent on the vibrational energy conveyed by the harmonic moments, but also on the degree of damping. This is the measure in which the increase of the vibrational amplitudes by friction and otherwise is opposed.

Damping torsional vibrations and the produced stresses

If a torsion-elastic system vibrates because of external moments working on the system, the measurements of vibration can be infinitely high. This premise is based on a mathematical model for resonance. In practice, this means that the shaft will break because of excessive material stresses.

▼
A six-cylinder four-stroke engine. From top left to bottom right: the free-vibration s-shape, the crank middle point, the 5.5 middle point and the vibration spectrum.

Aimed to calculate the relative magnitude of the vibration-generated energy per vibration period by the harmonic component to the 5.5 order for a 6-cylinder four-stroke engine. The ignition sequence is 1 – 5 – 3 – 6 – 2 – 4. The free-vibration shape is found in the chart below, resonance with a natural vibration frequency to the 1st degree occurs at rotational speeds where the indicated mean pressure is equal to 12 bar.

Solution:

The crank middle point of the engine is shown in the figure. The ignitions follow each other every 120°. The total phase shift of the vibrational deflection is obtained from the position in the ignition sequence.
The positions of the vectors that the vibrational deflections demonstrate are:

$\gamma_{(1)} = 0°$
$\gamma_{(2)} = 5½ \cdot 4 \cdot 120 = 7 \cdot 360 + 120°$
$\gamma_{(3)} = 5½ \cdot 2 \cdot 120 = 3 \cdot 360 + 240°$
$\gamma_{(4)} = 5½ \cdot 5 \cdot 120 = 9 \cdot 360 + 60°$
$\gamma_{(5)} = 5½ \cdot 1 \cdot 120 = 1 \cdot 360 + 300°$
$\gamma_{(6)} = 5½ \cdot 3 \cdot 120 = 5 \cdot 360 + 180°$

From these positions, the 5.5 middle point can be calculated, such as shown in the figure.
With the help of the free-vibration shape where the magnitude of φ_r is given for every crank and to the 5.5-order middle point, the magnitude of $\varphi_{r\,(res)}$ can be calculated. Assuming a random scale, the product $W_{5\,½\,(rel)}$ can be calculated by multiplying $p_{t(5\,½\,)} = 0.8$ bar with the value obtained for $\varphi_{r(res)}$.
The value for $p_{0t(5½\,)}$ is obtained from graphs found in 'Lloyd's Rules'.
Using this process, the magnitude of W_p can be found for the critical rotational frequencies for every order. A vibration spectrum can be obtained by using a plot based on the rotational speeds of the engine.

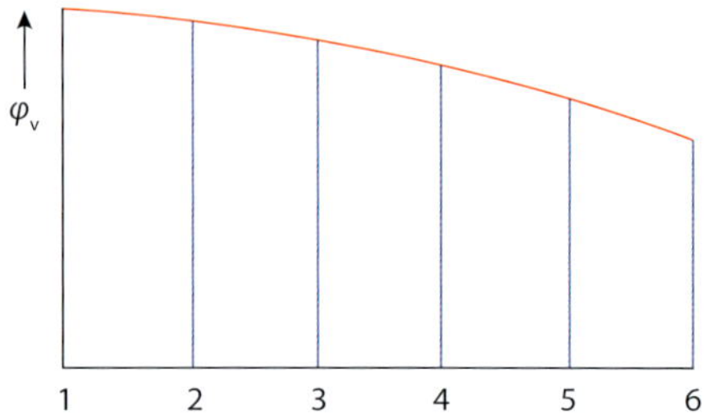

the free-vibration s-shape

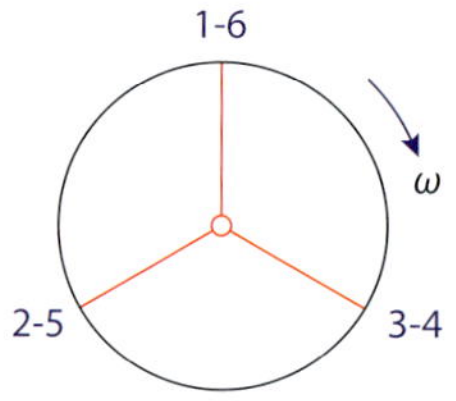

the crank middle point six-cylinder four-stroke

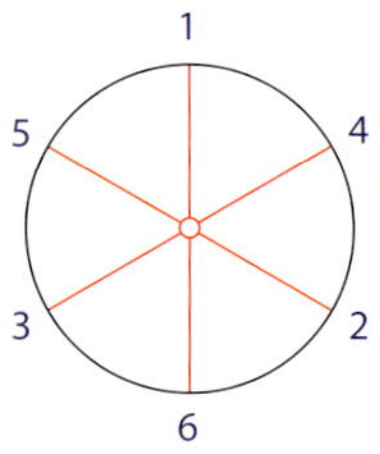

the 5.5 middle point

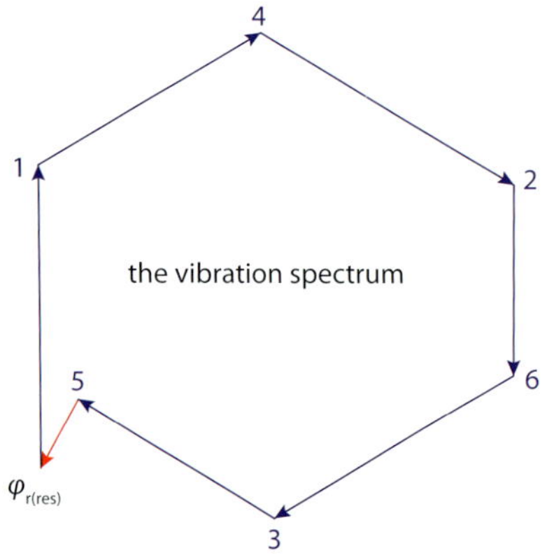

the vibration spectrum

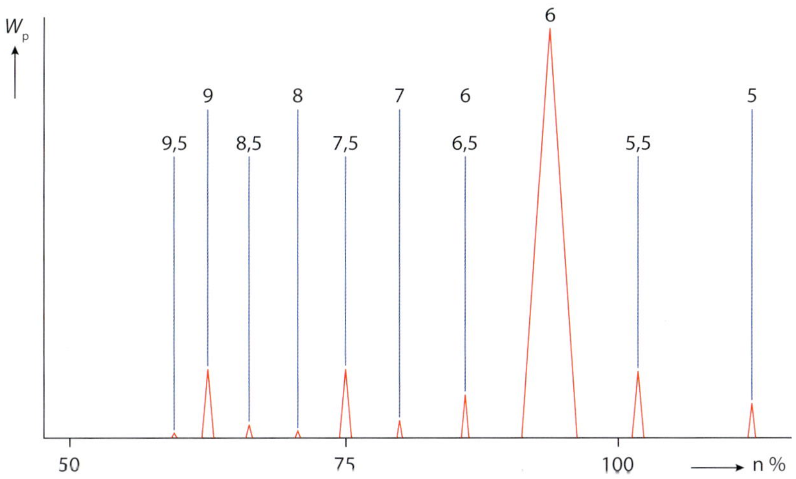

Vibration spectrum of a six-cylinder four-stroke engine.

When sufficient opposing or damping moments work on the system, a state of equilibrium will occur at certain amplitudes of vibration where the damping moments working on the system are equal to the vibration-generating moments. All the vibrational energy supplied to the system is subsequently converted into damping energy. This is usually frictional heat.

In a theoretical analysis of a vibration system, damping is expressed in the damping factor or in the relative damping. The difficulty with a complex system such as that of an engine crankshaft with connecting shafts lies in the fact that the measure of damping can only be approximated theoretically. Therefore, empirical formulas are applied that have been developed based on measuring and studying vibrations in a large number of installations.

The formula for calculating this damping power for all cylinders is:

$W_d = \pi \cdot A_p \cdot r^2 \cdot k \cdot \omega_e \cdot \varphi_{1w}{}^2/\varphi_{1r}{}^2 \cdot \Sigma\, \varphi_{r(x)}{}^2$ where x = 1 to z

Here the damping factor and the ratio between the actual measurement of vibration φ_{1w} and the measurement of vibration according to the Holzer-table φ_{1r} have been taken into account. The actual measurement of vibration can now be established as follows. If the displacement of the crank is at a maximum, then the maximum vibrational energy $W_{c(max)}$ is equal to the damping power W_d.

The actual measurement of vibration φ_{1w} is then (near cylinder 1):

$$\pi \cdot A_p \cdot r \cdot p_{t(p)} \cdot \varphi_{1w}/\varphi_{1r} \cdot \varphi_{r(res)} = \pi \cdot A_p \cdot r^2 \cdot k \cdot \omega_e \cdot \varphi_{1w}{}^2/\varphi_{1r}{}^2 \cdot \Sigma\, \varphi_{r(x)}{}^2$$

$$p_{t(p)} \cdot \varphi_{r(res)} = r \cdot k \cdot \omega_e \cdot \varphi_{1w}/\varphi_{1r} \cdot \Sigma\, \varphi_{r(x)}{}^2$$

$$\varphi_{1w} = \frac{p_{t(p)} \cdot \varphi_{r(res)} \cdot \varphi_{1r}}{r \cdot k \cdot \omega_e \cdot \Sigma\, \varphi_{r(x)}{}^2} \text{ [rad]}$$

Here $\varphi_{r(res)}$ is the vector sum of the vibration measurements, as calculated in the example; ω_e is the circle frequency of the natural vibration at which the harmonic moment concerned resonates; $\Sigma\, \varphi_{r(x)}{}^2$ is the sum of the square of the vibration measurements, found in the Holzer-table.

The actual measurement of vibration is now determined by the degree of damping. In Lloyd's Rules, damping is taken into account by the so-called Dynamic Magnifier or magnifier factor *M*. Total damping is obtained by calculating the magnifier factor from the damping factor of the engine, the propeller and a vibration damper, if present. The *k* applied in the formula now is $k = 1/M$

If the actual measurement of vibration at cylinder 1 is known, the measurements from the other masses can also be read from the free-vibrational mode shape.

It is now possible to calculate the torsional moments working in the shaft parts, and consequently the torsional stress, from column 6 of the Holzer-table. The highest stress level will occur at a node in that part of the shaft.

This stress can be computed by first calculating the maximum torsional moment in the shaft part containing the node. Here the following applies:

$$M_t = G \cdot I_p \cdot \varphi_{1w}/\varphi_{1r} \cdot (\varphi_{x(r)} - \varphi_{x+1(r)}) / l_x \text{ [Nm]}$$

Here the ratio of actual measurement of vibration and the measurement according to Holzer has been taken into account. Furthermore, $(\varphi_{x(r)} - \varphi_{x+1(r)})$ is the angular displacement in the shaft part containing the node according to the Holzer-table.

The extra torsional stress is now the torsional moment divided by the resistance moment opposing the torque, consequently

$$\tau_x = M_t / (\pi/16 \cdot d_{as}{}^3) \text{ [N/mm}^2\text{]}$$

Admissible torsional-vibration stress

In spite of damping the engine and the propeller themselves, unacceptable levels of alternating stresses can be generated in the propeller shaft at certain critical rotational frequencies. Lloyd's Rules provide formulas and graphs that give the admissible torsional-vibration stress as a function of the rotational frequency of the shaft and its diameter.

Genset and propeller drives via gear transmission

So far, this chapter has discussed the torsional vibrations in the propeller shaft of a propulsion engine directly connected to the propeller shaft. As diesel-engine driven electrical gensets operate at a constant rotational frequency, the regulations with respect to torsional vibrations are restricted to an area between 0.95 and 1.10 times the nominal rotational velocity of the engine. Stresses in the crankshaft and transmission shafts because of the critical speeds that are traversed by the engine during starting and stopping should not exceed 5.5 times the admissible stress inside aforementioned rotational area.

Propulsion installations in which the propeller is driven by one or multiple engines via gearing, and in which torsion-elastic couplings have been incorporated, may exhibit complex forms of vibration. The effects of a failure, for instance no combustion in one of the cylinders, must be investigated in these installations for torsional vibrations.

If any of the electrical gensets is driven by the so-called PTO (Power Take Off) gearing, the torsional vibrations of these branches should be carefully computed.

▼

Acceptable torsional-vibration stresses in a crankshaft.

In the figure, a graph for crankshafts with diameters of 100 mm and of 600 mm. In this graph, n are the rotational speeds of the crankshaft, n_s are the rotational speeds at nominal load.

The lines f shows the stress boundaries for continuous operation. If at a critical rotational frequency the torsional vibration stress obtains a value that lies between the lines f_c and f_t then continuous operation may not take place in this area. When increasing or decreasing the rotational speeds, the restricted area must be passed through quickly. The restricted area must be clearly marked on the tachometer.

Above a value $n/n_s = 0.8$, vibration stresses are allowed only if they lie below the line f. It is undesirable for a restricted area to lie too close to the maximum rotational speeds.

In the chart, it can be seen that as the value of n/n becomes larger, the acceptable torsional vibration stresses become smaller. This is obvious if one remembers that with a propulsion engine the moments of torsion increase with the rotational speeds and therefore the twist stress in the shaft because of this useful moment. The 'space' left for a torsional-vibration stress will then become smaller with increasing rotational speeds.

Above $n/n_s = 1$ a higher stress is allowed by preventing line f_c being intersected by the flank of a resonance peak of a critical rotational frequency that lies far above the nominal value. The temporary turning above the nominal frequency, for example the propeller coming out of the water is of such a short time-span that the vibration amplitudes do not reach their maximum value. Furthermore, in these cases, the normal shaft load is eliminated.

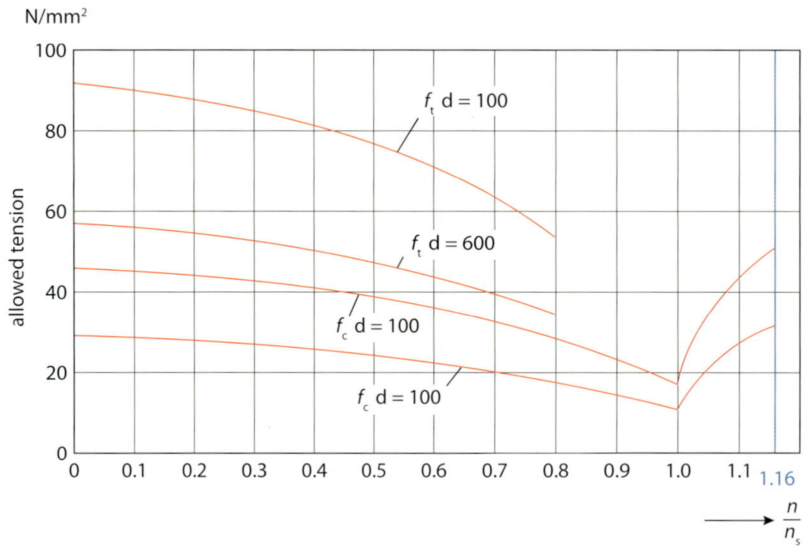

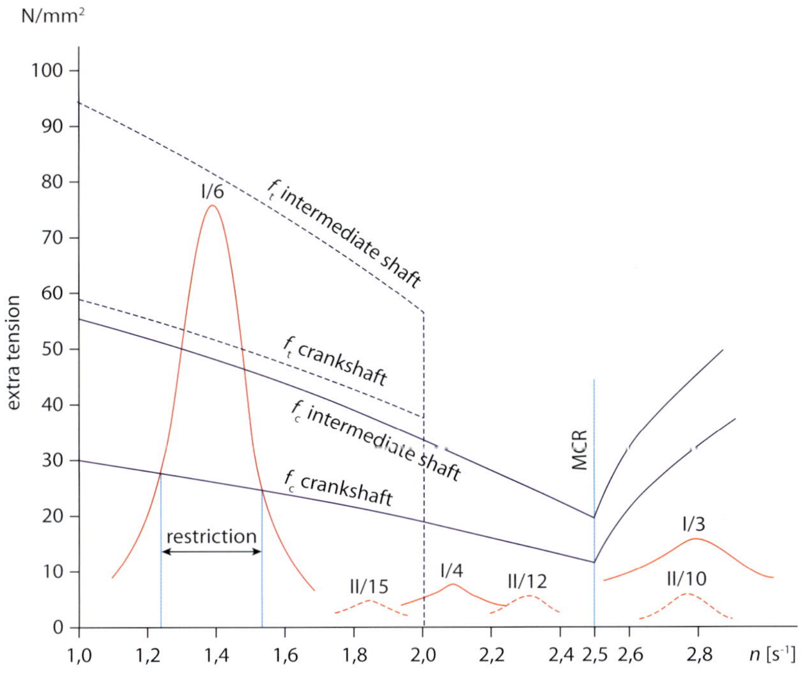

Resonance spectrum of a two-stroke propulsion engine.

In the chart, the resonance spectrum of the earlier mentioned 6-cylinder two-stroke SULZER type RND 68 engine. Clearly recognisable is the large stress peak at the critical rotational frequency to the 1st degree and 6th order. This crosses not only the f_c line, but also the f_t line of the crankshaft. As the node of the 1st degree vibration is found in the intermediate shaft, only the stress in this shaft needs to be taken into account. The resonance stress in the crankshaft is a fraction of that in the intermediate shaft.
To protect the intermediate shaft, a restricted area is marked. The magnitude of this area can be calculated with formulas.
Another critical rotational frequency is that to the 2nd degree/12th order. This has a limited effect. The reason is that the free-vibration shape to the 2nd degree for this system, where a node is found between the cylinders 4 and 5. Therefore, the vibration results for the middle crank are relatively small, and counteractive. The results of the sum of the vectors for the vibration results to the 12th order have a limited value. Simply: the harmonic moments acting on the cranks 1, 2, 3 and 4 are counteracted by the moments acting on the cranks 5 and 6.

17.14 Axial vibrations

These are vibrations in longitudinal direction of the crankshaft.

MAN Diesel MC engines with more than six cylinders have critical resonances just below the nominal engine speed. Vibration to the first degree is usually less important. The natural vibrational frequency has been established by the mass and the complete shaft system.

The rigidity of the thrust block and the bedplate is decisive.
The natural vibrational frequency is so high that there is no dynamic reinforcement to this order.
There are axial-vibration dampers for all cylinder numbers for MC engines.
It is important to keep the axial vibrations below a certain value to protect the crankshaft from high material stress caused by axial vibrations.
To this end, all MAN Diesel MC engines with six or more cylinders are equipped with an axial vibration damper.

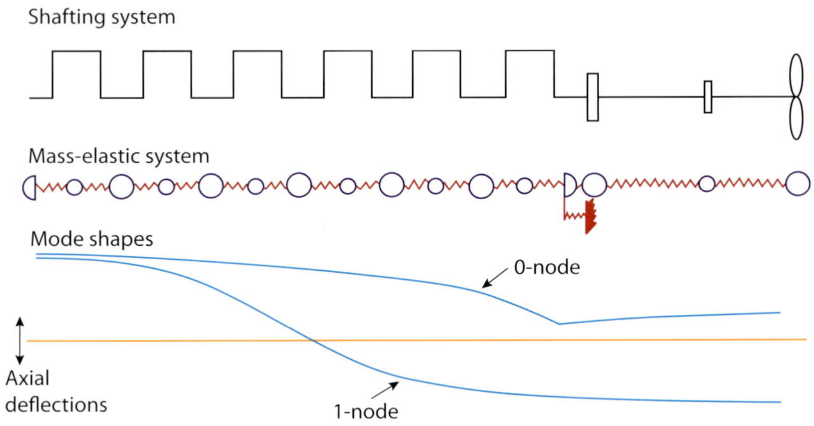

An example of a mass-elastic system for axial-vibration calculations.

Shown are the free-vibrational modes of the nominal speed and those to the first degree.

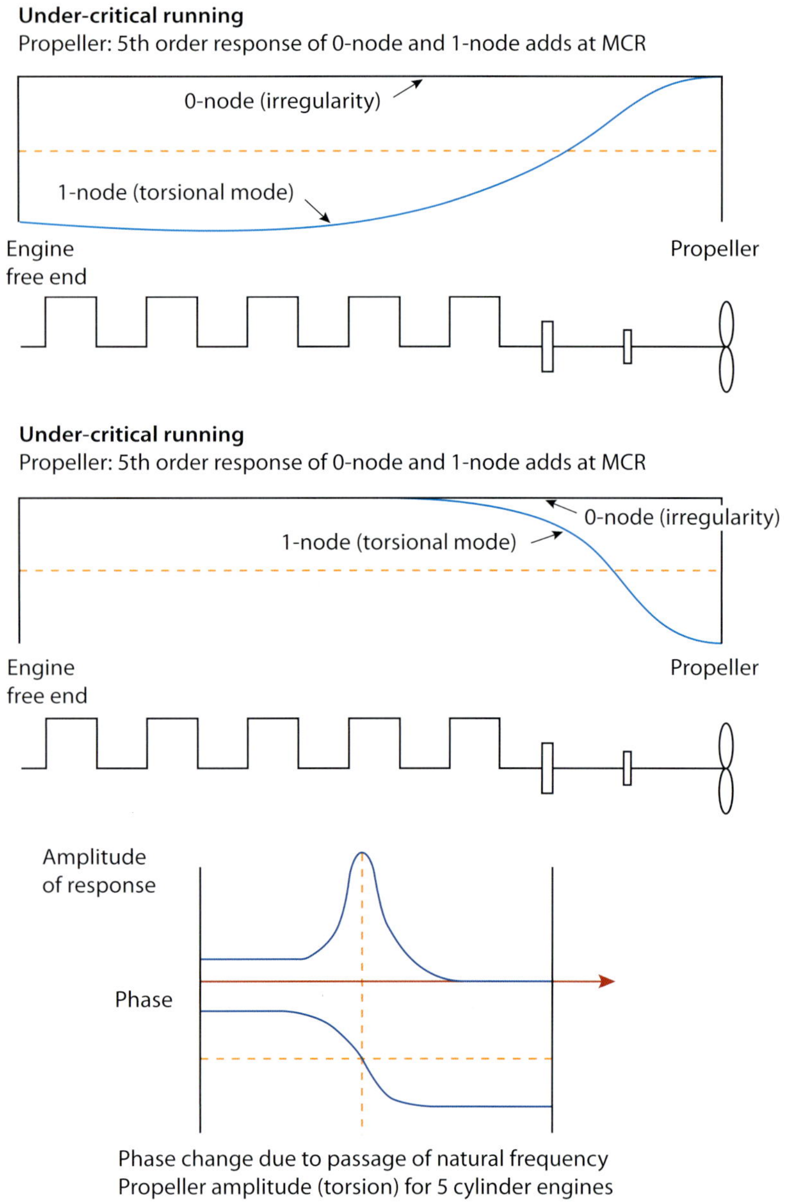

▲

An example of torsional response for five-cylinder two-stroke crosshead engines from MAN Diesel is shown to illustrate how the thrust pressure of the propeller induces torsional vibrations/deflection.

Upper: under-critical running.
Centre: over-critical running.
Lower: phase change when passing through a resonance of the crankshaft.

The second reason for installing an axial-vibration damper is to reduce the vibrations that occur in the ship's hull due to the fluctuating forces in the thrust block. For this reason, four- and five-cylinder engines are also fitted with an axial-vibration damper. An axial-vibration damper effectively reduces the vibrations generated by the fluctuating forces in the crankshaft and the thrust block.

17.15 Engine balancing: theoretical background

Free forces and moments originate in the engine but are important only if they remain in the engine. The ignition sequence of the engine is very important.

17.15.1 Free forces

The following free forces are found in engines:
Rotating forces: Horizontal and vertical forces.
Oscillating forces: Primary and secondary forces.

Rotating forces are generated by the rotating gearing in an engine.

The rotating force is equal to:

$$F_r = m_r \times \omega^2 \times r$$

F_r = the centrifugal force in N.
m_r = the rotating mass in kg.
ω = the angular velocity in rad/sec.
r = the crank radius in m.

The rotating force F_r is represented by a vector in line with the crank.

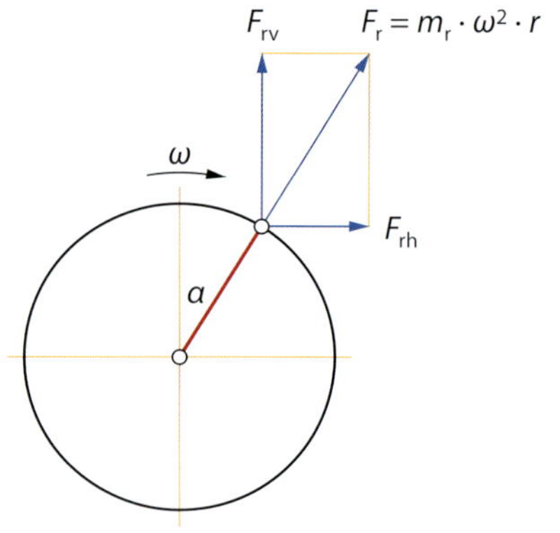

▲

Rotational free force.

The rotational force F_r, exerted on the engine is resolved in a vertical and horizontal force.

$F_{rv} = m_r \times \omega^2 \times r \times \cos\alpha$

and

$F_{rh} = m_r \times \omega^2 \times r \times \sin\alpha$

Both forces, known as primary free forces, have a rpm (frequency) equal to the rpm of the crankshaft. Oscillating forces are produced by the acceleration of the gearing, the reciprocating parts of the engine. This oscillating force is equal to:

$F_o = m_o \times \omega^2 \times r(\cos\alpha + \lambda \times \cos 2\alpha)$

Where λ is the ratio of crank radius and connecting rod length. This oscillating force can be divided into:

A primary oscillating force with a rpm/frequency equal to the rpm/frequency of the crankshaft.

A secondary oscillating force with a rpm/frequency equal to twice the rpm/frequency of the crankshaft.

The size of the primary oscillating force can be determined by a constructive proof.

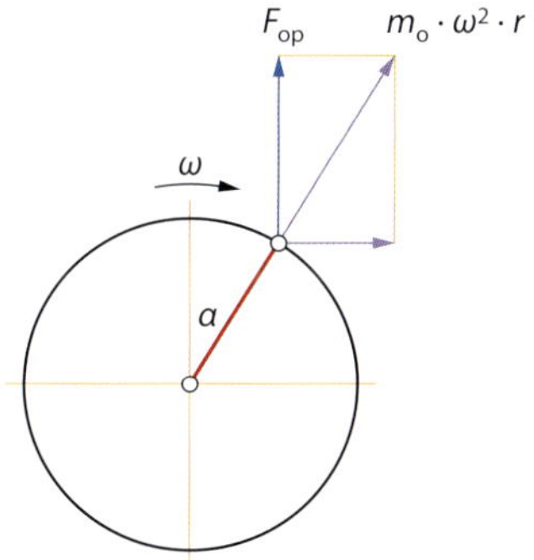

▲

Primary oscillating force.

F_{op} is a vector $m_o \times \omega^2 \times r$ in line with the crank resolved in a vertical direction.
In practice, F_o operates vertically in the centre line of the shaft. The horizontal resolution $m_o \times \omega^2 \times r$ has no meaning.

Therefore:

$F_{op} = m_o \times \omega^2 \times r \times \cos\alpha$

The secondary oscillating force can be similarly determined.

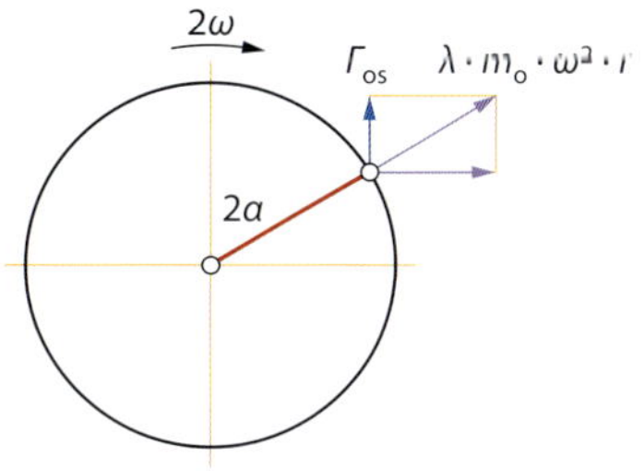

▲

Secondary oscillating force.

This is based on an imaginary crankshaft where the crank runs through an angle that is always twice as large as the actual crank angle.
A vector of magnitude inline with the crank is now:

$\lambda \times m_o \times \omega^2$ r

The vertical resolution is then:

$F_{os} = \lambda \times m_o \times \omega^2 \times r \times \cos 2\alpha$

In this single-cylinder engine (example), the following free forces are operational.

In a vertical plane through the crankshaft axis:

$F_{pv} = F_{rv} + F_{op} = (m_r + m_o) \times \omega^2 \times r \times \cos\alpha$

In the horizontal plane through the crankshaft axis, the primary force:

$F_{ph} = F_r = m_r \times \omega^2 \times r \times \sin\alpha$

In the vertical plane through the axis of the crankshaft, the secondary force:

$F_{sv} = F_{os} = \lambda \times m_o \times \omega^2 \times r \times \cos 2\alpha$

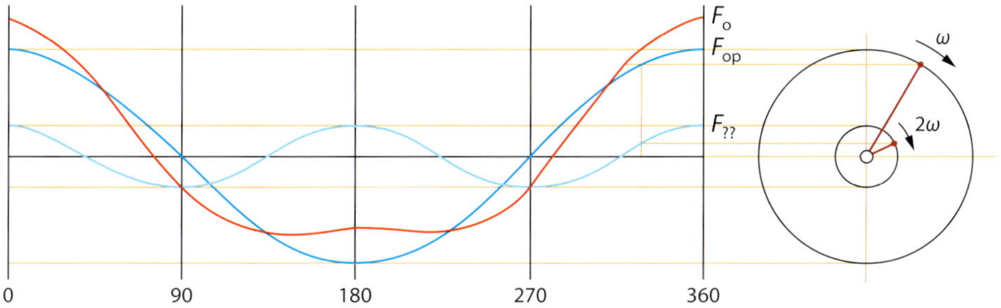

◀

Curve for oscillating forces.

17.15.2 Balancing the free forces in a single-cylinder engine: theory

The aim is to determine how these forces can be fully balanced.
As explained earlier, free forces in engines are not balanced.
We need to eliminate them by applying balancing forces.
These balancing forces must be of equal magnitude but opposite to the free forces in the engine.
This is achieved using rotating counterweights or oscillating weights.
Here, we will only consider that rotating counterweights, for example, the counterweights on crank shafts and counterweights on rotating shafts.

17.15.3 Balancing the rotating forces

A counterforce is applied to balance the free forces fully.

$F_{kr} = -F_r = m_r \times \omega^2 \times r$

A mass is mounted on both crank webs opposite to the crank.

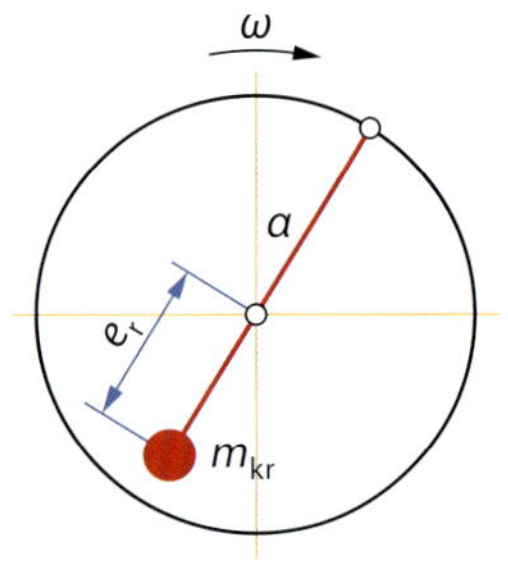

▲
Balancing a rotating force.

Assuming that at complete balancing, the mass m_{kr} and the rotational radius e_r of the common centre of gravity of both masses on the crank webs is:

$m_{kr} \times e_r = m_r \times r$

17.15.4 Complete balancing of the primary oscillating forces

In addition, here, an opposing primary oscillating force must be generated, because:

$$F_{kop} = -F_{op} = -m_o \times \omega^2 \times r \times \cos \alpha$$
$$= m_o \times \omega^2 \times r \times \cos \alpha \; (\alpha + 180°)$$

These opposing oscillating force can also be seen as the sum of the vectors of two opposing rotating forces, each with the magnitude: $½ \times m_o \times \omega^2 \times r$, with an angular velocity of ω and $-\omega$ rad/sec, that at any moment compared to the vertical centre line make an angle of 180° + α , and 180° – α.

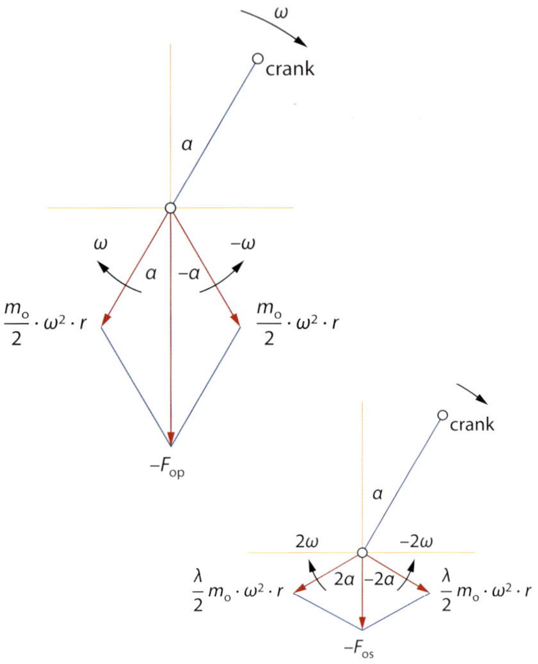

▲
Total balancing of the secondary oscillating forces.

These opposing forces are generated by attaching two counter-rotating weights m_{kop} in the plane through the axis of the cylinder that rotate in a radius e_r with velocities ω and $-\omega$.

The following applies for these weights:

$m_{kop} \times \omega^2 \times e_p = ½ \times m_o \times \omega^2 \times r$

or

$$m_{kop} = \frac{1}{2} \times m_o \times \frac{r}{e_p}$$

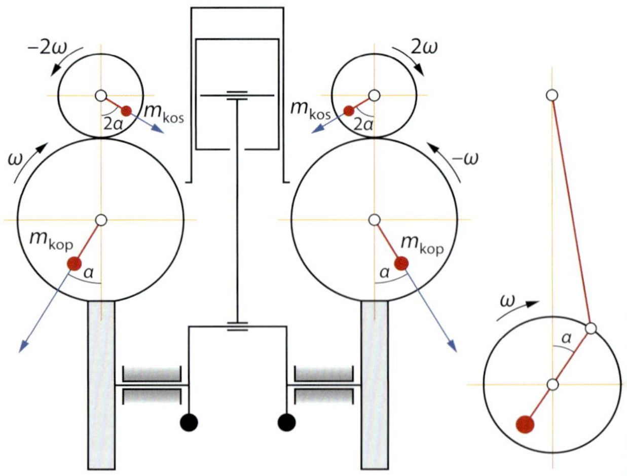

▲
The figure above shows a constructive execution of a total balancing system for a single-cylinder engine.

▲
Two balancing shafts in a Wärtsilä Vaasa 4R 32, this is a medium-speed four-stroke diesel engine, category III. Both shafts rotate at a speed twice that of the crankshaft.

Two counterweights are mounted on the crank webs to balance F_{rot}.

Two counterweights, m_{kop} are mounted on two auxiliary shafts driven by the crankshaft. The two auxiliary shafts have the same rotational velocity as that of the crankshaft. Both shafts have opposing rotations.

The secondary forces are balanced by two counterweights m_{kos}, mounted on a second set of shafts driven by the auxiliary shafts. The second set of shafts have twice the rotational velocity as that of the crankshaft, Both shafts also have opposing rotations.

17.15.5 The resulting free forces in a machine with more than one crank

The free forces generated in every cylinder by the drive gearing can be combined to a single resultant free force.
This is easy to see graphically.
The displacement of the free forces in a plane is allowed only when a torque is added to the displaced force.
The torque moment must be equal to the sum of the displaced force and the distance l, which is the distance that the force is displaced parallel to itself.

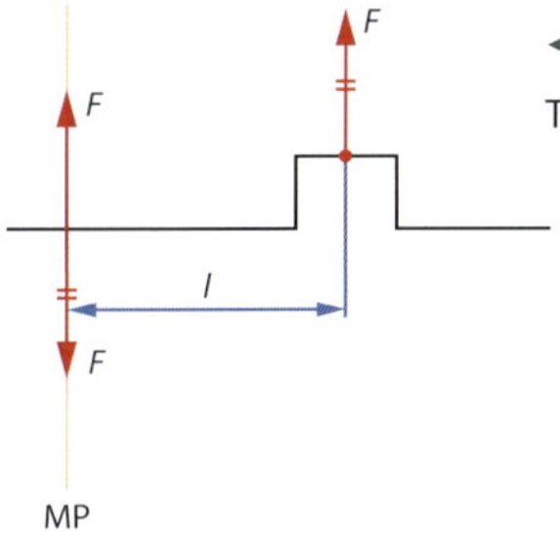

◀
Torque moment.

Let us assume that the original force F is engaged in the middle of the crank. Two forces parallel to the original force are generated in the chosen moment plane MP. that are opposing and therefore eliminating.
In this manner, a torque is generated by two forces F. An instance occurs when $F \times l$ while the force F is active in the moment plane.

Suppose we now continue with a three-cylinder engine.

The moment plane in an engine with three cylinders is taken through the middle of the crankshaft (2). From crank 1 to crank 3, the rotating forces are displaced to the centre line of the centre crank, (2). This is the moment plane.
A torque is added to these forces.

▼
Structure of a moment vector.

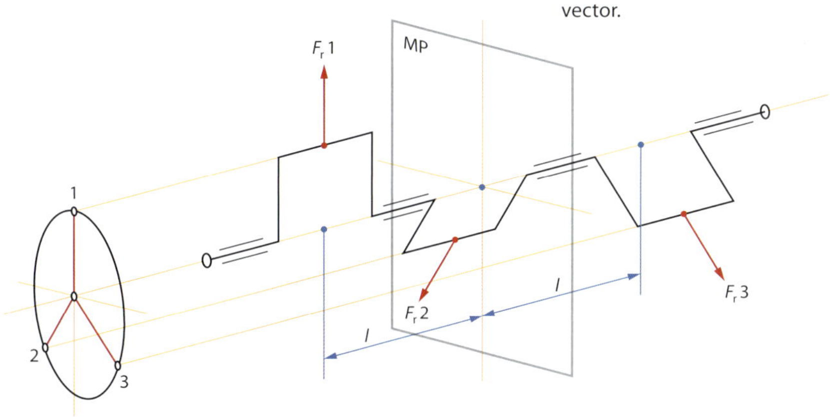

The moments are now:

$$T_{re} = F_{re} \times l \text{ en } T_{r3} = F_{r3} \times -l = -F_{r3} \times l$$

The distance l to the moment plane is positive when found on the same side as crank 1; the distances on the other sides of the moment plane are negative.
No torque has to be added to the rotating force on crank 2, as it is not displaced.

A moment can be represented by a vector. This vector is placed perpendicular to the plane where the torque is active. The length of the vector is a measure of the amplitude of the torque moment. The arrow indicates the direction.

▼
Structure of a moment vector.

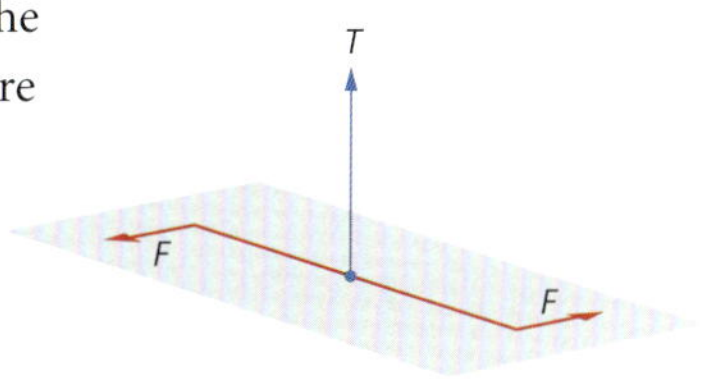

Vectors are also used for determining the resultant of the free forces.

Recapitulating some important issues.
Each moment vector is turned 90° clockwise.
Therefore, the vectors will point in the direction of the corresponding crank, away from the crankshaft centreline when it is situated on the positive side of the moment surface area, and toward the crankshaft centreline when the cranks are situated in the negative side of the moment surface area.

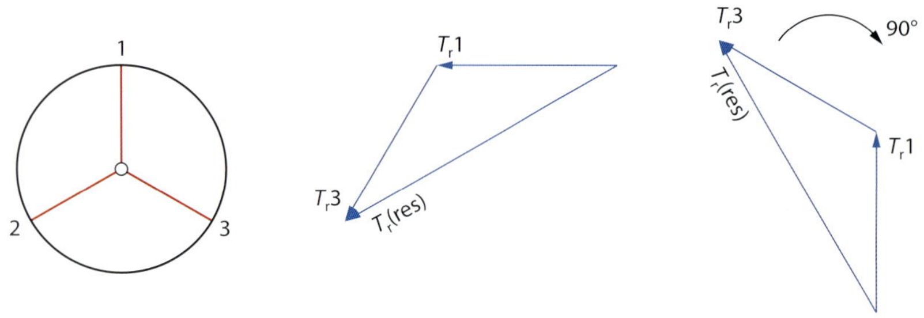

▲
Determining the rotating-moment resultant.
Left: the crank heart for the three cylinders.
In the middle: moment resultants 1 and 3, which are produced by the rotating forces.
Right: once again, moments 1 and 3, but here the moment vectors have rotated 90°.

In both cases, a resultant torque occurs. The plane in which the resultant torque is employed, goes through the axis of the crankshaft and is then perpendicular to crank 2. This is a rotating torque. This can generate horizontal and vertical vibrations in the ship.

Similarly, the resultant moment can be construed from the primary oscillating forces. The resultant vector represents the maximum value of the moment. The moment value can be established by projection to the vertical centreline.
This projection is turned 90 ° counter clockwise to touch the correct position.

▶
Structure of a secondary oscillating moment.

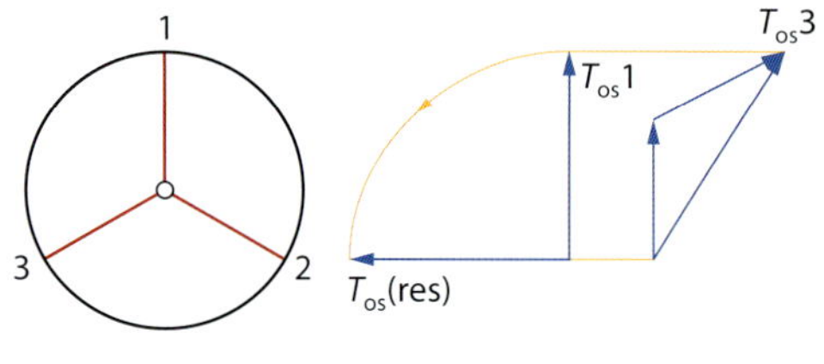

▶
Structure of a primary oscillating moment.

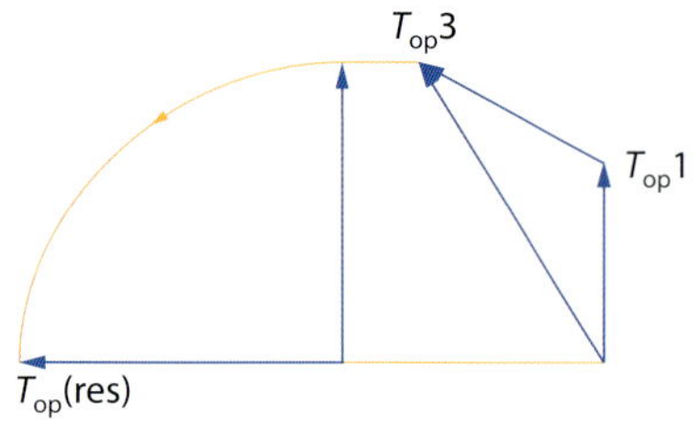

17.15.6 Secondary oscillating moments

When considering the secondary crankshaft, the structure is identical to that for primary oscillating moments.

17.15.7 Conclusion, three-cylinder engine

This engine is not balanced for either rotating-, or the primary- and secondary oscillating moments. The rotating forces can be balanced for each cylinder using counterweights mounted on the crank webs. Therefore negating the rotational forces.
Balancing primary moments, either rotating or oscillating, is possible by using a perpendicular crankshaft application.

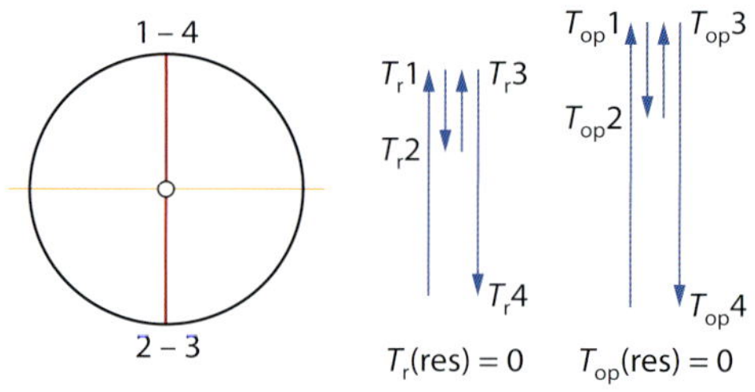

▲
Primary moments in a four-cylinder four-stroke engine.

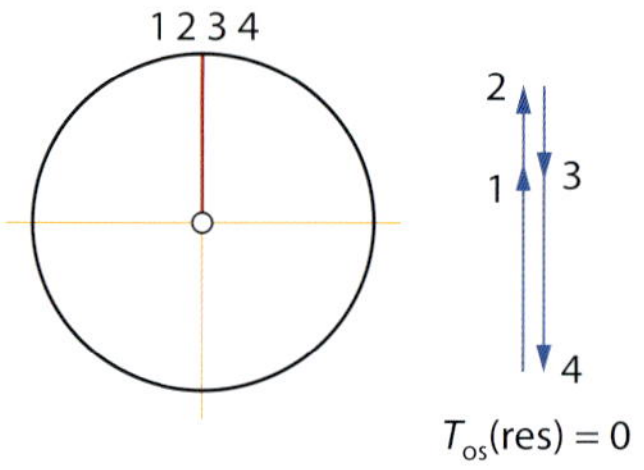

▲
Secondary moments in a four-cylinder four-stroke engine.

Example

To obtain equal ignition distances in a four-cylinder four-stroke engine, the angles between the cranks must be equal:
720° : 4 = 180°.

The vectors of the primary forces form a closed triangle so that the product of these forces is 0.

The primary moments, both rotating and oscillating negate each other. The moments produced in cylinders 1 and 2 are negated by the moments produced in cylinders 3 and 4.

The secondary shaft is not symmetrically centred as shown in figure below.
The secondary forces reinforce each other that result in a larger force.
The product of the secondary moments compared to the chosen moment plane MP is once again, 0. This occurs because the shaft is symmetrical in its length.

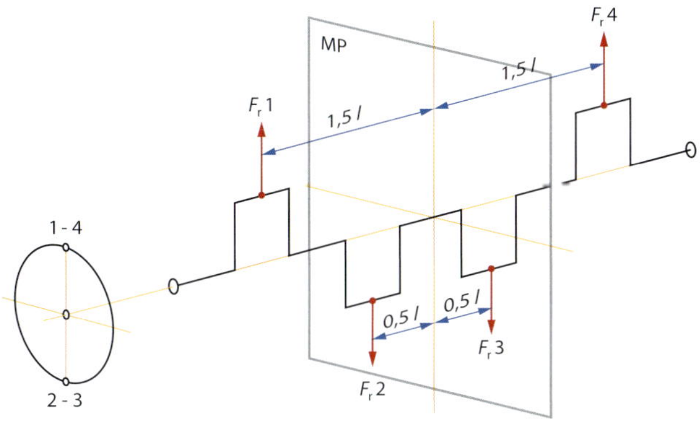

▲ Rotating moments in a four-cylinder four-stroke engine.

17.15.8 The mutual crank distances in engines with more than one cylinder

The mutual crank distances must be chosen in such a way that the resulting forces are kept as small as possible.

Example 1

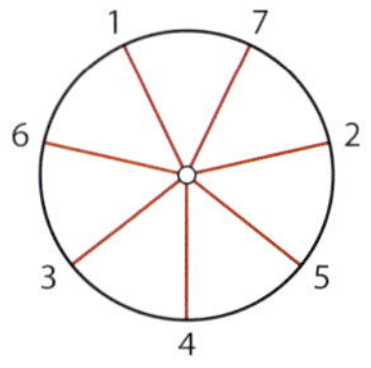

First to seven cylinders, odd number, for instance 7.
In the crank circle, the first cylinder that is drawn is the middle cylinder, cylinder 4 in a seven-cylinder engine. This crank is drawn straight down. The other six cranks are drawn in equidistant sections from each other symmetrical to the centre of the cylinder.
Numbering the crank lines starts left top with cylinder 1.
The crank that on the same horizontal line follows and is given a number that is the sum of the two following numbers equal to the number of cranks plus one. In this example 7.
Then, the second row is numbered in the opposite direction beginning with 2. The sum of the cylinders horizontally positioned must be 7 + 1 = 8; therefore, cylinder 6 is directly opposite to cylinder 2. Then, once again in the opposite direction, cylinder 3 and opposite, cylinder 5 (5 + 3 = 8).

Example 2

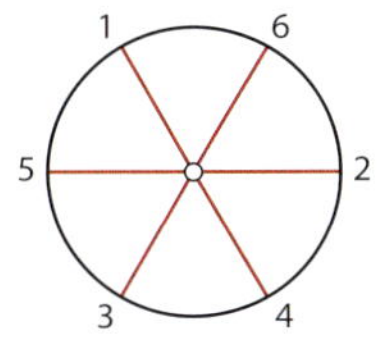

A six-cylinder two-stroke engine
Once more, the cranks are evenly divided over the crank circle.

Here: $\frac{360}{6} = 60°$

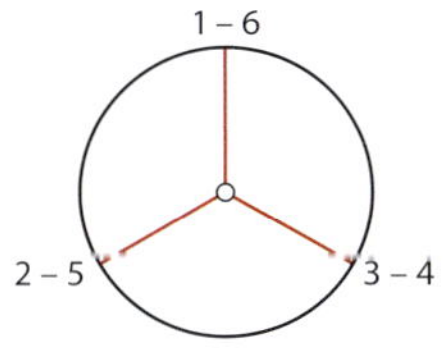

A six-cylinder four-stroke engine
Now the cranks are in the same position, two by two. Therefore, the angles between the cranks are

$\frac{360}{3} = 120°$

The calculations above are strictly theoretical. Apart from the magnitude of the free external moments, the following factors should also be taken into account:

– The internal bending moments in the engine frame.
– The longitudinal vibrations in the crankshaft.
– The transverse moments on the motor frame, caused by varying tipping moments of each cylinder, the so-called X- and H vibrations.
– The varying values of torsional-vibration generating moments.
– The various cylinders connected to the exhaust-gas pipes for turbo charging in four-stroke engines according the impact system.
– The bearing load of two simultaneously executed power strokes of two adjacent cylinders produces a heavy load on the intermediate main bearing.

There are a number of options for engines supplied with eight or more cylinders, each of which should be considered carefully in order to find the right compromise.

Example:
There are 2520 potential crankshaft sequences for eight-cylinder two-stroke crosshead engines.
For twelve-cylinder two-stroke crosshead engines, there are as many as 19,958,400! All of which are examined by engine manufacturers.

17.16 Forces and moments

Here we will treat the forces and moments as they are found in a crosshead engine. In a standard, vertically set up crosshead engine, the force that is exerted vertically downward on the crosshead equals: the gas force – the oscillating- or inertial force + the weight force of the parts.

$F_k = F_g - F_o + G_{o+r}$

Assuming that the total mass of the oscillating (reciprocating moving parts) is concentrated in the upper point (centreline of the connecting rod, small end).

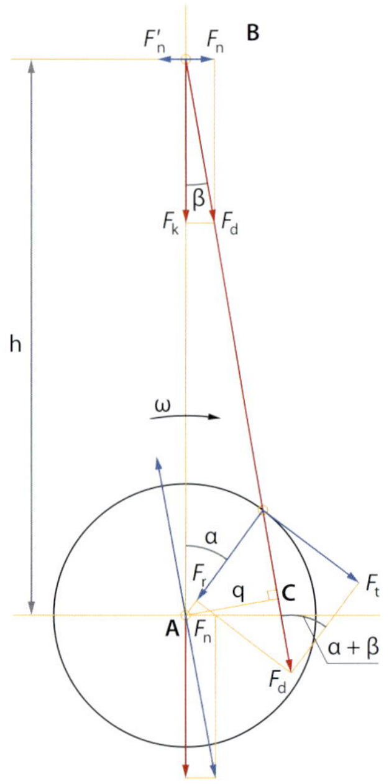

The inertial force equals:

$F_o = m_o \times a_z$

in addition, the weight force equals:

$G_o = m_{g+r} \times g$

Therefore, the weight force is found by totalling the gearing masses.

17.16.1 Forces in a two-stroke crosshead engine

The vertical force exerted on the crosshead is dissolved into a force perpendicular to the crosshead, F_n.
This is absorbed by the guide shoes. So when reversing the engine this force changes horizontally in the opposite direction. The other force is dissolved towards the connecting rod, F_d.

The value of $F_d = \frac{F_k}{\cos\beta}$ and $F_n = F_k \times \tan g \cdot \beta$

The connecting-rod force F_d is seized by the crankpin and dissolved into a force perpendicular to the crank, the tangential force F_t and a radial force toward the centreline of the crankshaft F_r.

$$F_t = F_d \times \sin(\alpha + \beta) = F_k \times \frac{\sin(\alpha + \beta)}{\cos\beta}$$

$$F_r = F_d \times \cos(\alpha + \beta) = F_h \times \frac{\cos(\alpha + \beta)}{\cos\beta}$$

17.16.2 Torque on the crankshaft

In order to calculate the torque, two opposing connecting-rod forces, F_d in the centreline of the crankshaft, A, are used: these forces cancel each other out. Next, two of these three parallel forces, F_d couple with a moment $T_d = F_d \times$ q. q is the perpendicular distance between A and C.
C is positioned on the F_d line.
At position A, a connecting rod force F_d must be absorbed by the crankshaft bearing. This force, F_d is decomposed in a vertical force equal to a horizontal force F_h equal to F_n, the force on the guide shoes.
Forces F_n deliver torque to the engine block with a moment called the turning moment because the engine may tilt.
In triangle ACB, B is the heart of the crosshead. This triangle is comparable to the triangle formed by F_d and F_h.

From this, we can also calculate:

$AB : AC = -F_d : F_n$

and

$F_n \times AB = -F_d \times q$

$F_n \times AB$ is the turning moment, this is equal but opposite to the torque.

17.16.3 Torque on the crankshaft

$$T_d = F_t \times r = F_k \times r \frac{\sin(\alpha + \beta)}{\cos\beta}$$

The power exerted on the crankshaft by the torque is:

$$P_{mom} = T_d \times \omega = F_k \times \omega \times r \frac{\sin(\alpha + \beta)}{\cos\beta}$$

17.17 Resultant forces and moments in the engine block

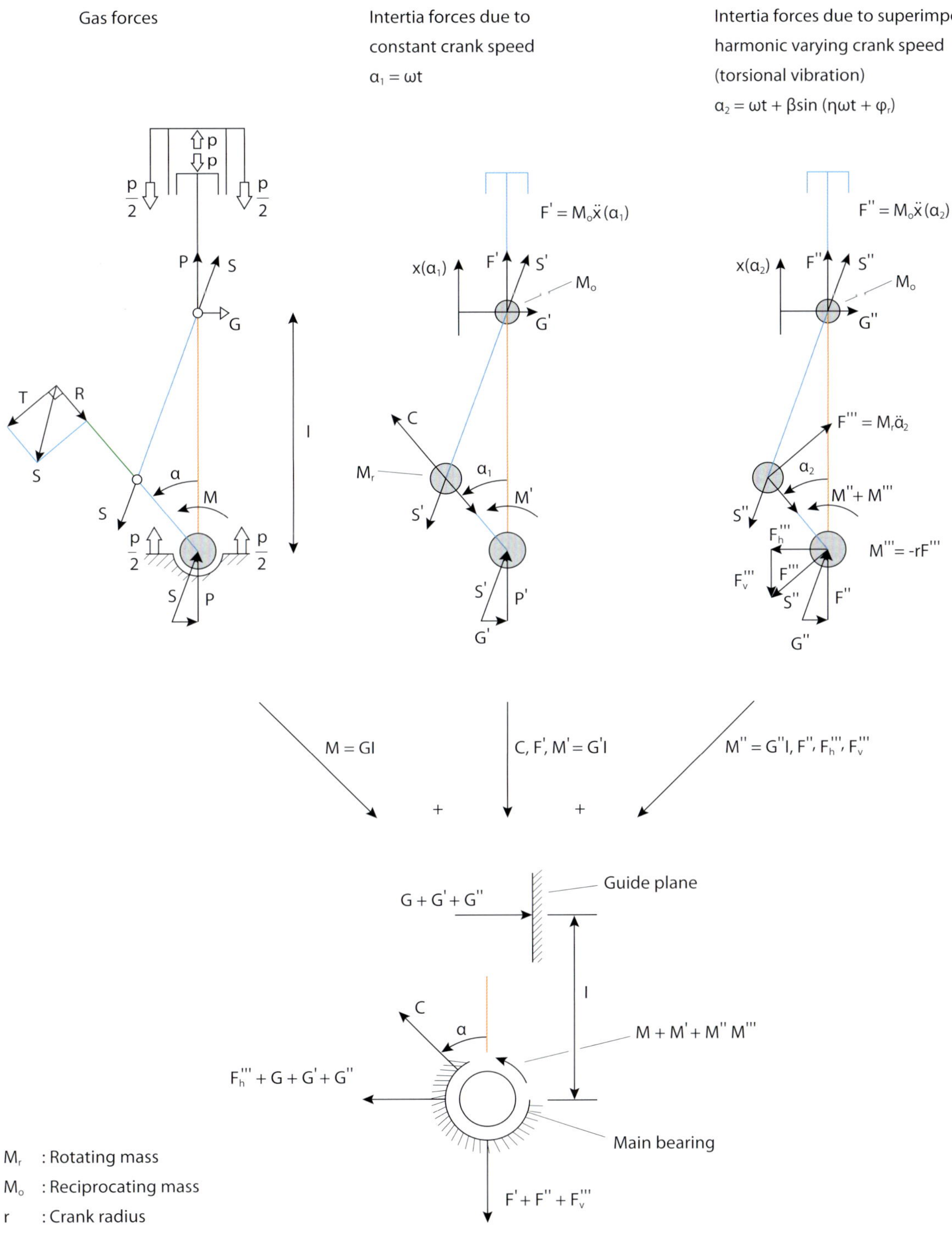

◀ The resulting forces and moments in the engine block of a two-stroke crosshead engine.

S, S' and S'': Connecting-rod forces acting on the crosshead, equal to the connecting-rod force acting on the crankpin, equal to the force acting on the main-shaft bearings.
M, M', M'' and M''': Torque on the main-shaft bearings from combustion-pressure forces and inertia forces.
T and R: S, S' and S'': At the crankpin, these are the sum of the radial component R and a tangential component T.
Resulting forces on the engine frame in vertical direction C, F', F'' and F_V'''.
Resulting forces on the engine frame in horizontal direction C and F_H'''.
Resulting moment on the engine frame M + M' + M'' = l × (G + G' + G'').

17.18 External forces and moments

The entire ship forms a mass-elastic system with a natural vibrational frequency and generated vibrations. The horizontal and vertical bending moments of the hull construction and the corresponding natural (vibration) frequencies can be calculated. Calculating vibrations with four, five or more orders requires complicated software. Vibrations with two or three orders are easier to establish. In practice, the external forces for all MAN Diesel engines are assumed zero, as their value is very low. Normally external moments to the first and second order are important.

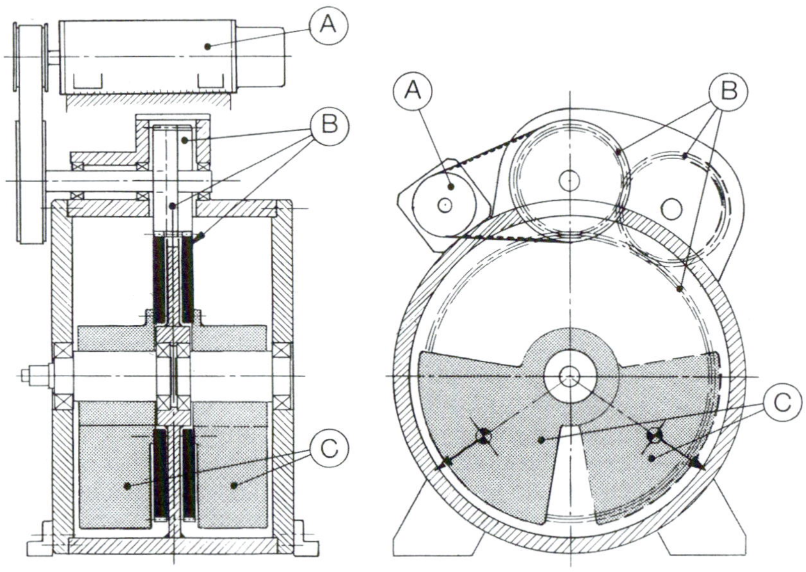

A servomotor-driven compensator.

A: AC-powered operating motor

B: Gearbox

C: Counterweights

This is a special vibration damper designed to reduce longitudinal ship vibrations in, amongst others, the superstructure.
The damper is designed to level out first- and second-order vibrations.
This damper is positioned aft, mostly in the steering-engine room as the amplitude of the vibrations are here the largest and so the effects are maximal.

17.18.1 Balancing the engine

Moments to the first order work in a vertical as well as a horizontal direction.
With (the application of) MAN Diesel standard balancing these moments are of equal magnitude.
Moments to the second order only work in the vertical plane.
Measures for four-, five- and six-cylinder engines should be considered.
In order to assess the extent of the external moments, the so-called Power Related Unbalance (PRU) system is applied.

The following applies to four-cylinder engines:

– **standard**, adjustable counterweights;
– **option**, a compensator for the moments to the first order.

Resonance between the vertical moment and the second order of the vertical hull construction is often critical, whilst the resonance between the horizontal moment and the second order of the horizontal hull construction normally occurs at engine speeds higher than average.

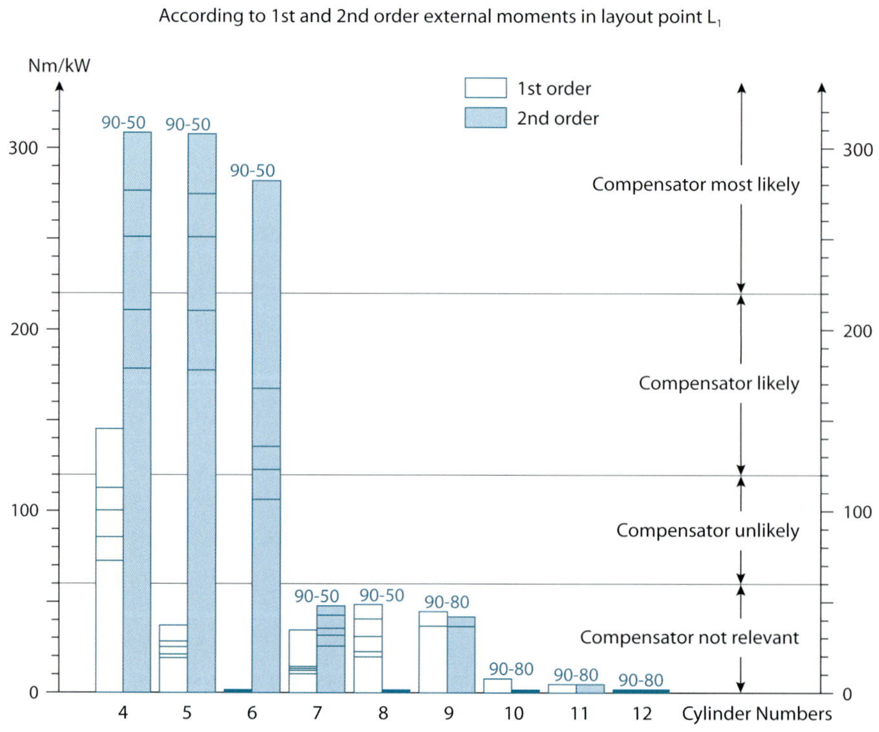

The graph shows the connection between the shaft power and the vibrations to the first and second order.

horizontal: number cylinders = shaft power

vertical: power-related unbalance in Nm/kW

Power-related unbalance =

$$\frac{\text{external moments in Nm}}{\text{shaft power in kW}} = \text{Nm/kW}$$

Numbers on the columns are the cylinder diameters.
Clearly shown is that mostly the four-, five- and six-cylinder engines are unbalanced.
The large moments are clearly present.

▲
Counterweights on the chain wheels in a MAN Diesel two-stroke crosshead engine.

1 *counterweights*

17.18.2 Adjustable counterweights

Four-cylinder engines are equipped with adjustable counterweights as a standard.

These counterweights considerably reduce the vertical moments. With the use of this method, no problems should be expected with the resonance to the second order of vertical hull constructions.
In exceptional cases, when the first-order moments generate a resonance at a normal engine speed in both the vertical and horizontal second order of the vertical hull construction, a first-order moment compensator can be installed in the chain-tensioning wheel. This reduces the horizontal first-order moment to an acceptable magnitude.
The compensator contains two counterweights that rotate at the same rate as the engine (two-stroke principle).
As resonance of the vertical as well as the horizontal hull shapes rarely occurs, standard engines are not designed do have these compensators installed.
Resonance between the vertical moment to the second-, third-, fourth- and fifth-degree vibrations of the hull construction occur at normal engine speeds.
A compensator to the second order can be fitted on four-, five-, and six-cylinder engines to reduce resonance.

There are various ways to reduce the vertical moment to the second order:

1 **No compensators** if it is established that this is not required because of its natural vibration number and the magnitude of the moment to the second order.
2 **A compensator mounted on the rear end of the engine** (crankshaft side) driven by the chain drive of the camshaft.
3 **A compensator on the front end of the engine** ('blind' side of the crankshaft) driven by a separate chain system from the crankshaft.
4 **An electrically driven compensator**, which runs in phase with the free moment.
This type of compensator requires an additional bedplate, preferably in the steering engine room where the deviations are highest and the compensator most effective.
5 **Compensators on both sides of the crankshaft,** which eliminate the external moments to the second order.

Solutions 2, 3 and 4 are effective when they are operational near the keel rather than the node of the ship's vibration.
Solutions 4 and 5 must be considered when the critical value of the hull construction is close to that of the engine's.
If the electrical compensator (solution 4) is placed in the steering engine room, it has the advantage that it is not as sensitive to the order position as those in solutions 2 and 3 are.

▼
Adjustable counterweights for compensating 1st order external moments.

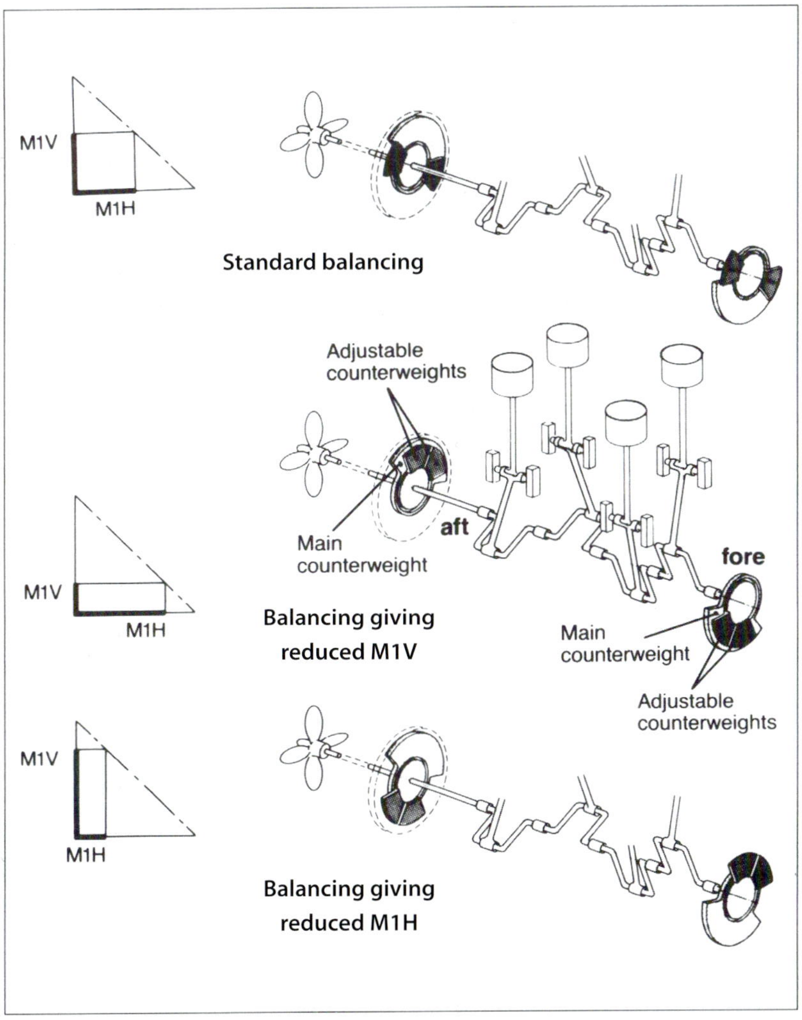

17.18.3 Vibrations in the bedplate/engine frame

Vibrations in the engine frame resulting from guide forces are part of the ship's vibrations.

There are three forms of vibration.

1. **H-mode**
   Works on the top side of the engine from the first to the last cylinder. It is also referred to as parallel vibration.

2. **X-mode**
   Works on the top side of the engine at a longitudinal angle. Affects the middle of the engine.

3. **L-mode**
   Works on the top side of the engine in the longitudinal direction of the engine.

The natural vibrational frequencies of these constructions are largely dependent on the rigidity of the bedplate and the double bottom of the ship below the engine and bedplate.

H- and X-type moments are generated by the guide forces of the H- and X-type.
The values of these guide forces can be calculated for each engine, based on the gas and mass forces.
These occur in all engines.

▶
An example of vibrations in the engine frame from the double bottom of the ship to above of the cylinder block, measured in a two-stroke MAN Diesel 5L80ME-C crosshead engine.

vertical: measuring points L1 t/m L7
horizontal: scale in tenths of millimetres

At the top of the cylinder block, the result is ± 0.18 millimetre.
On the flat bottom surface, the values V1, V12 and V2 are clearly smaller than L1.

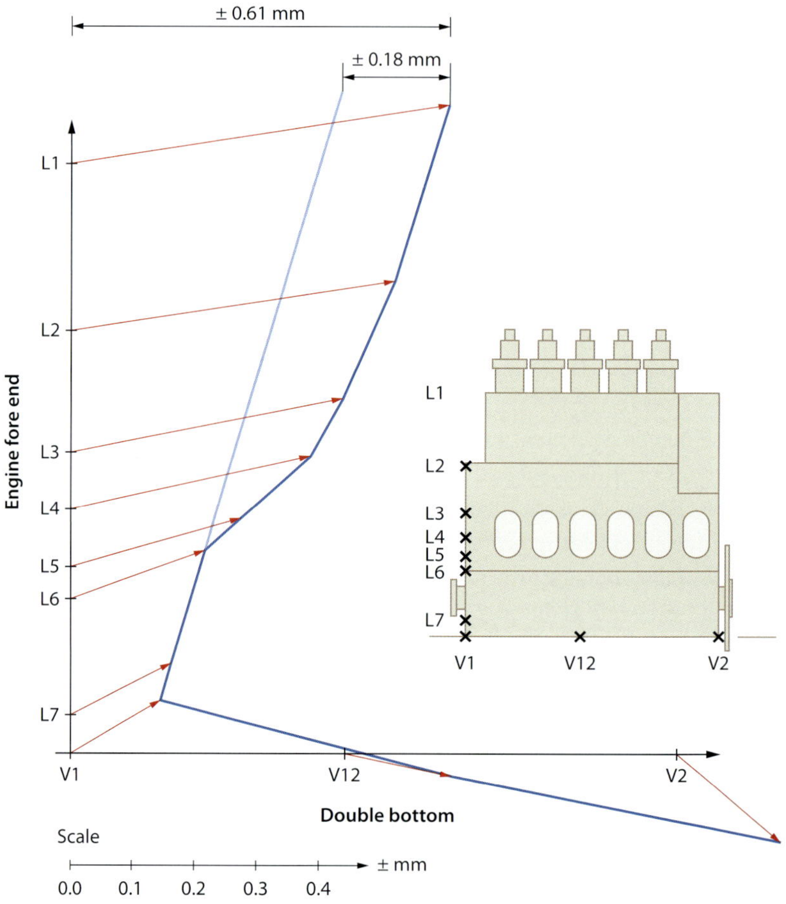

▶
Free moment effects on the hull.

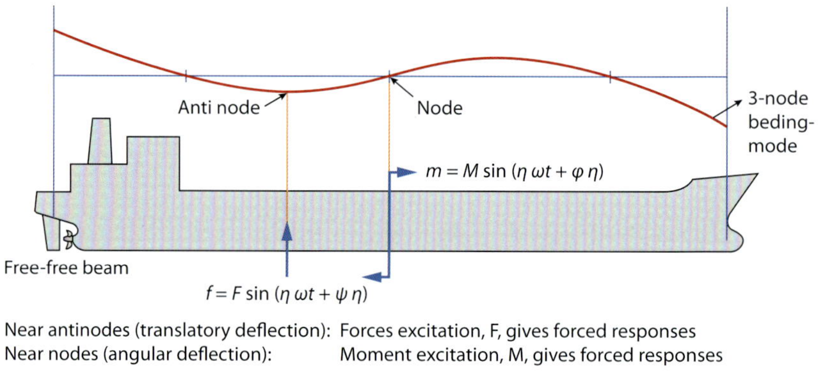

The secondary values of the guide moments originate from torsional vibrations and are therefore difficult to calculate.
Appendix B specifies the primary and secondary values of the guide moments for a MAN Diesel 5L70MC engine with two different shaft designs. The secondary values are low.

L-**type moments** are generated by installation-dependent factors.
Alternating forces in the thrust block, caused by the torsional vibrations of the propeller often cause these forces.
Furthermore, L-**type moments can** also be produced by axial vibrations of the crankshaft.

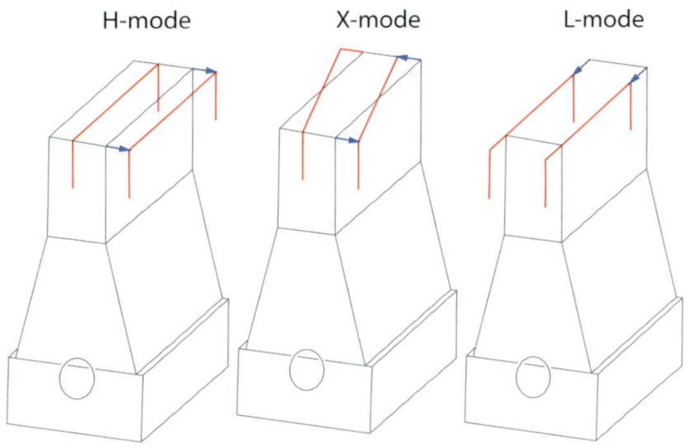

▲
Vibrational shapes of the motor frame of a cross-head engine as a result of the guide shoe forces, the measured H-, X- and L-shapes.

Appendix B

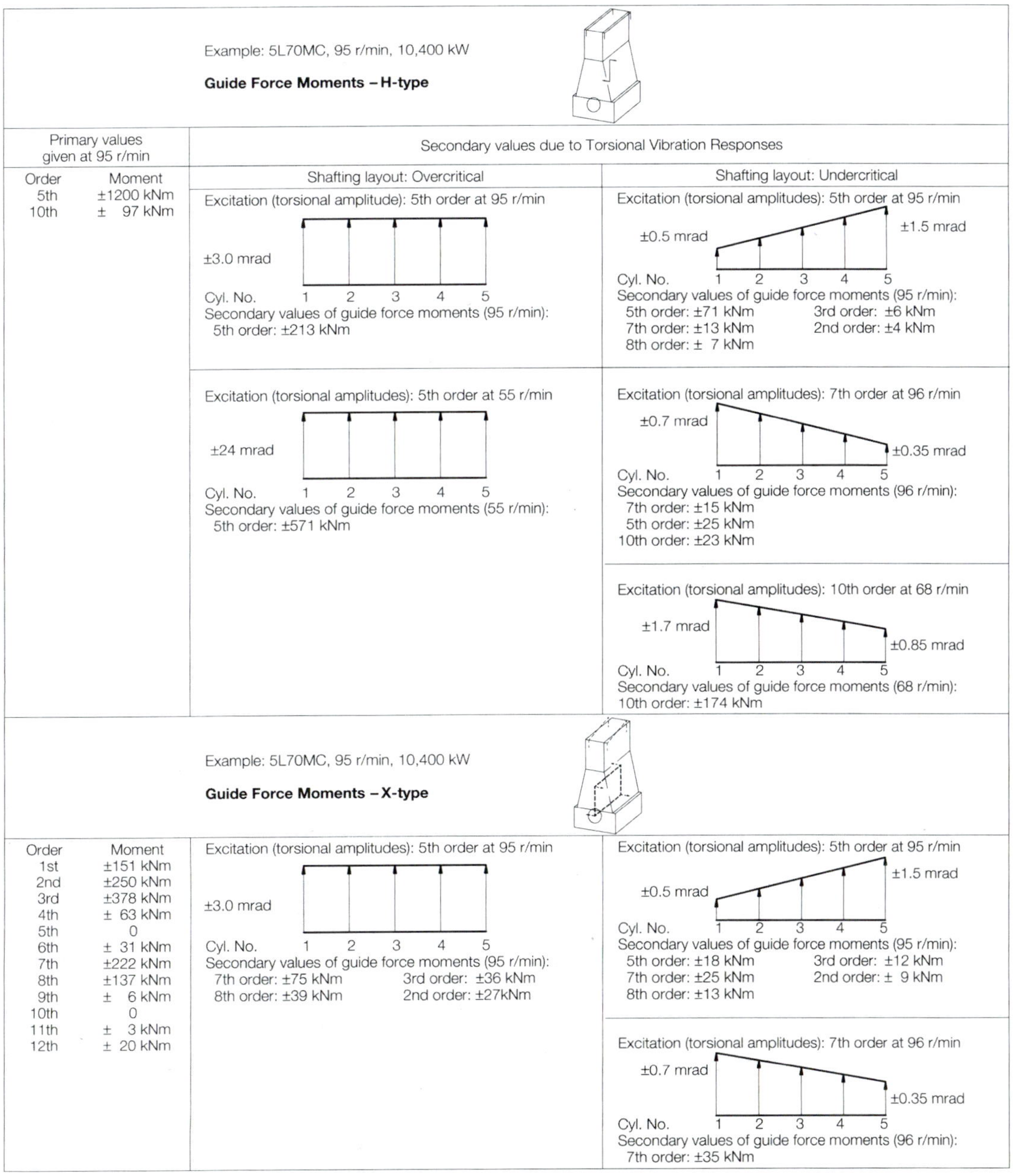

◄
The H- and X-type guide-force moments.

left: for shafts with over-critical speeds
right: for shafts with under-critical speeds

▶ Mechanical top bracing for a two-stroke eight-cylinder Wärtsilä Sulzer 8RTA84C crosshead engine.

17.18.4 Top Bracing – Braces on the top of the engine

H- and X-type vibration-mode shapes are traditionally controlled by providing the top of the engine with braces that rest on the hull construction and so obtain resonances with critical orders **in excess** of the normal engine speed. Here the in-situ rigidity of the construction is important.
Sometimes braces are placed in the longitudinal direction of the engine in order to reduce vibrations of the L-**type moments**.
If the crankshaft forms the main source of the vibrations, an axial vibration damper can reduce the vibration sufficiently.

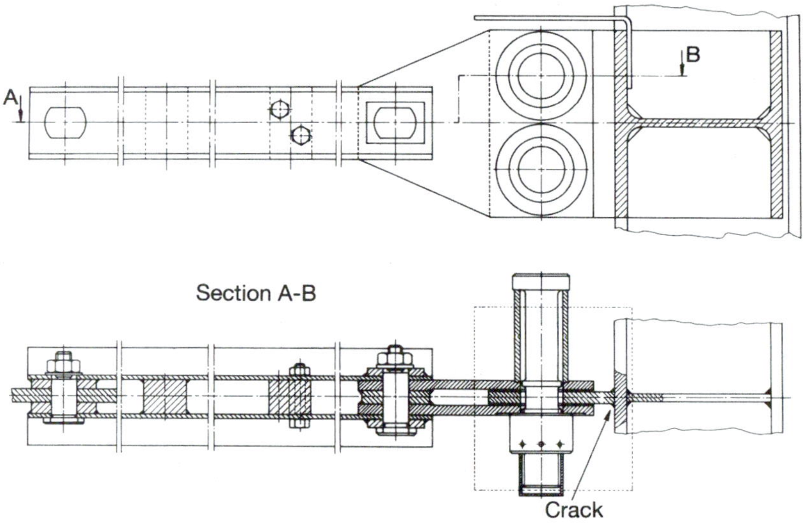

◀ Mechanical top bracing for MAN Diesel MC engines.

This is a standard support, mainly in modern ships. It is built of thin high tensile steel. Tearing occurs at the welds.
This type of ship has an elastic hull and so forces became too large for the welds.

▶ Hydraulic top bracing for a two-stroke twelve-cylinder Wärtsilä Sulzer RTA96C crosshead engine.

Hydraulic Top Bracing

This is capable of hydraulically compensating the brace distortions in ships that distort more than usual. This occurs, for instance, in heavy seas or loading and unloading.
This hydraulic top brace ensures that there is a constant force between the engine and the hull construction, independent of the distortion and functions as a damper.

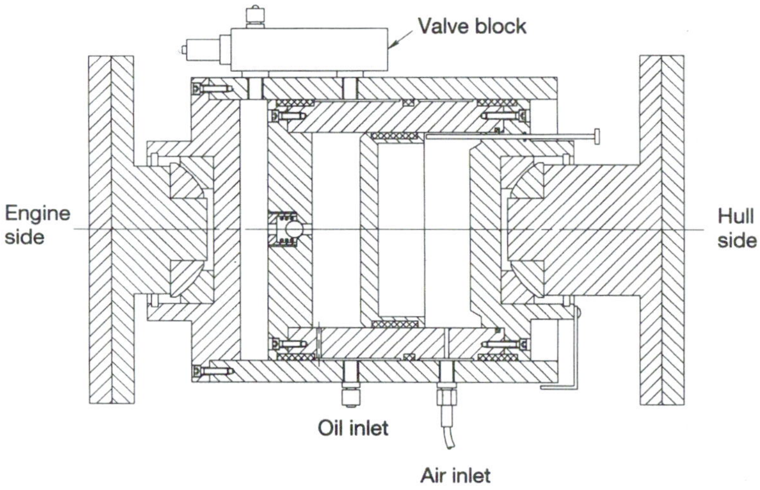

◀ Hydraulic top bracing for MAN Diesel MC engines.

This system includes a hydraulic cylinder and two spheroid bearings.
Oil is supplied from its own oil system and an overflow valve controls the hydraulic force.
In order to simplify the system, this does not have its own lubricating-oil system but an air-driven piston. The guide moments ensure that the oil is pressurised and therefore the vibration forces are transmitted to the ship's hull.

17.19 Example of the balancing used in a Wärtsilä 9L46 four-stroke engine, category III

This example demonstrates how balancing and vibrations are managed in a modern, heavily loaded, medium-speed four-stroke diesel engine.

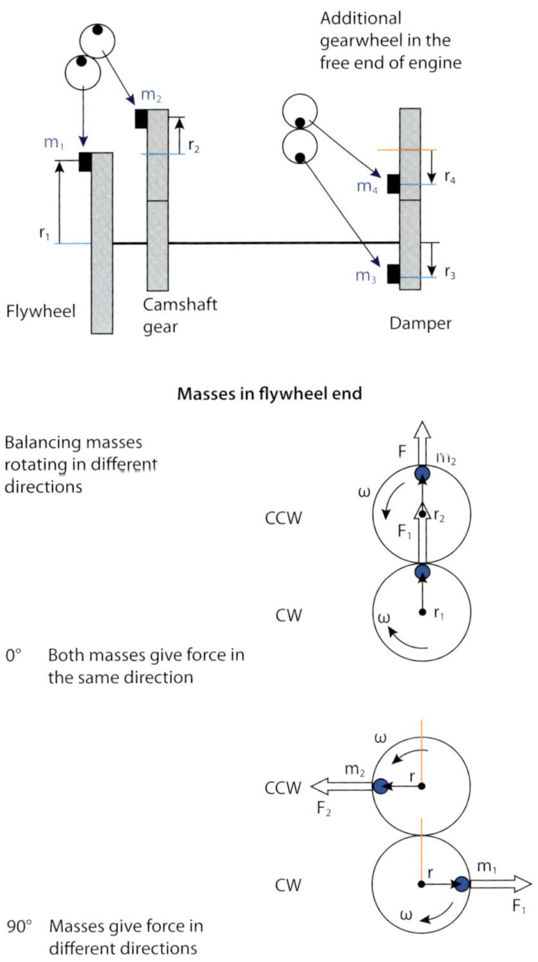

▶ An example of the balancing used in a Wärtsilä 9L46 four-stroke diesel engine, category III.

Four extra masses are mounted:

m_1 on the fly-turning wheel

m_2 on the camshaft gear wheel

m_3 and m_4 on an extra gear-wheel transmission mounted on the free end of the engine.

Every weight has its own radius; r_1, r_2, r_3 and r_4.

At position 0 of the crankshaft, both masses m_1 and m_2 provide a force in the same direction and at 90°; they provide a force in different directions.

ccw: counter-clockwise

cw: Clockwise

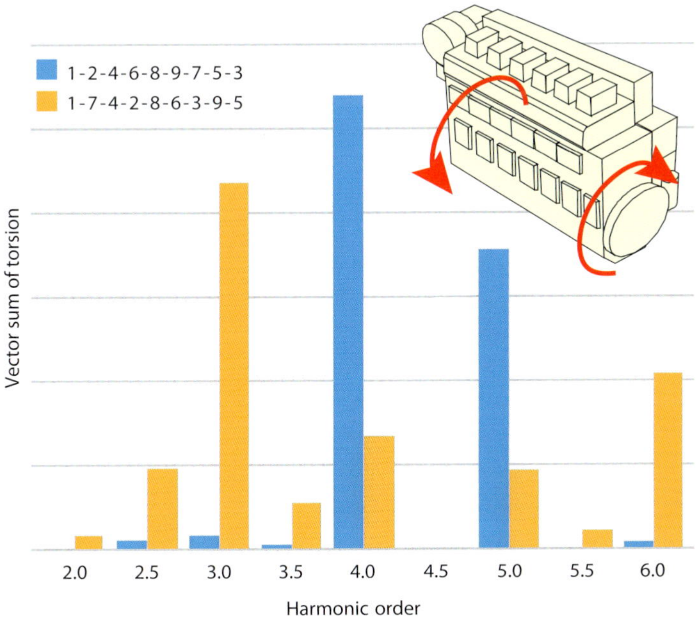

◀ Torsional forces in the engine block for two different ignition sequences. Shown here, a nine-cylinder four-stroke engine.

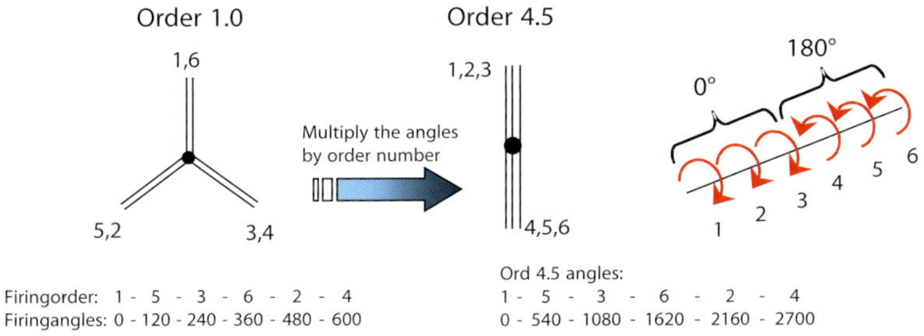

◀ An example of the composition of the torsional excitation-mode shape to order 4.5 for a six-cylinder four-stroke engine.

17.19.1 Wärtsilä four-stroke medium-speed diesel engines: Several examples

Wärtsilä roughly defines the relationship between the lowest harmonic orders and vibrational forms of flexibly arranged engines as follows:

Harmonic order 0.5
imbalance for gas forces – movement of the rigid frame/block
Harmonic order 1.0
mass forces – movement of the rigid frame/block
Harmonic order 2.0
mass forces – bend
Half orders 1.5, 2.5, ...
gas forces, torsion

If a local resonance frequency, such as the natural frequency of the turbocharger, is situated too close to one of these harmonic orders, the vibrations of this part dominates the produced form of this order.

17.19.2 Balancing

The standard manner to reduce the negative effects of mass forces is balancing.
Counterweights are used predominantly to balance the individual cranks mounted on the crankshaft to reduce local bending of the crankshaft as well as the bearing load.
More important, they can affect all the deviations in the engine block caused by bending.

Degree of balancing

The degree of balancing indicates which part of the rotating mass is balanced by counterweights. In many engines, including the Wärtsilä engines, the degree of balancing lies between 75 and 95%.
The present (2012) trend is to increase this percentage. In the modern Wärtsilä 46F diesel engine, in which the mass forces have increased due to a higher engine speed, the degree of balancing is 100%.

The Wärtsilä 12V46 diesel engine is an exception, the degree of balancing exceeds 100% so obtaining the lowest possible bending stresses and therefore deviations in the block in both vertical- and horizontal planes.

An even number of cranks in engines such as the 6 L, 8 L, 12 V or 16V are symmetrically aligned, enabling the moving masses to balance each other. This is referred to as an externally balanced engine. They do not have free forces or torques that can induce vibrations in the engine block.

17.19.3 Actual differences in the weight of engine parts

In practice, moving parts never have an identical weight as manufacturers work with tolerances. These produce free forces, so vibrations in permanently mounted engines can be experienced when these forces and torques are powerful enough when the engine passes through a certain speed range in a flexibly arranged installation.

Resonance

Resonance often cannot be avoided, especially in engines with ever-fluctuating engine speeds. The forces generated by the weight variances in the rotating parts near the crankshaft can be decreased by 80 to 90% if the components are manufactured to tighter tolerances. In connecting rods of high-speed engines, this can be measured in grams!

Balancing holes in the flywheel and other weights

It is not possible to arrange the cranks of seven-, nine- and eighteen cylinder engines symmetrically, which means that free forces are generated.
Special balancing holes in the flywheel and extra weights at the free end of the crankshaft ensure matching the vertical and horizontal torques.

New type of balancing

Since 2004, Wärtsilä has added additional balancing for the Wärtsilä 9L46. This has also been applied in the Wärtsilä 38 and 46 F diesel engines. Using this method, all free forces to the first order can be eliminated. Wärtsilä has fitted an extra balancing mass to the counter-clockwise turning intermediate camshaft drive and placed a counter-clockwise turning balance mass on a separate cogwheel at the free end of the crankshaft.
This, in conjunction with the clockwise turning balancing masses makes it possible to eliminate all torques!

A different method for tackling free torques is the realisation of unequal forces between the various cranks.
Irregularities in the combustion process can produce small tilting torques at low frequencies.
It can sometimes be accepted that some resonance will be present, if the resonance does not coincide with the natural vibrational frequencies, such as bending-generated torsional vibration.
In the severest case of resonance, the natural vibrational amplitudes together with the generated vibrational amplitudes have an identical frequency.
Comparing the generated vibrational frequencies and the natural vibrational frequencies provides information as to the expected degree of damage, at a certain harmonic order, to a construction.

17.19.4 Vector summation

The various kinds of torsional flexural-buckling vibrations are often analysed by summing the vectors. This is a simple method to compare generated and natural vibrations. The 'vector resultant' is the sum of the force and torque vectors for specific harmonic orders, and is calculated by:

$$V = \sqrt{\left(\sum_{i=1}^{c} f(i)\,A\sin\varphi_i\right)^2 + \left(\sum_{i=1}^{c} f(i)\,A\cos\varphi_i\right)^2}$$

▲

Here:

c = number cylinders

φ_i = ignition angle of cylinder 1 multiplied by the harmonic order

φ_i = order × $\propto i$

A = force of the torque amplitude

$f(i)$ = function defining a weighting factor depending on the estimated mode shape

17.19.5 Combustion sequence

The manner that the various harmonic orders are generated is mainly dependent on the chosen combustion sequence (the ignition sequence). Altering the combustion sequence has an important influence on the vibrational characteristic of an engine or genset. Today, the combustion sequence is defined in the early stages of the design process. This allows for an optimum design for engine performance. Special software has been designed to achieve these ends.

17.19.6 Vibration simulation

There are many software packages on the market to analyse the vibrational activity and accurately calculate the forces generated in the engine. The analyses can be displayed in time and frequency.

17.19.7 Finite-element method

Known as the FE model.
This is a three-dimensional computer model of, amongst others, the engine block. In this model, all possible bending and loads are visible and it is widely used in the engine industry. The load of a part of the block can be indicated, for instance, with a different colour.
Bending (or degrees of) can be grossly exaggerated so a visual picture is formed.

In the Wärtsilä 6L46F model, approximately 3.5 million 'degrees' must be simulated! In this type of FE model for an engine block, the working forces for each harmonic order can be calculated individually.
These forces are saved and may be re-used for calculations for components that are to be installed in the engine block, such as main bearings and cylinder liners.

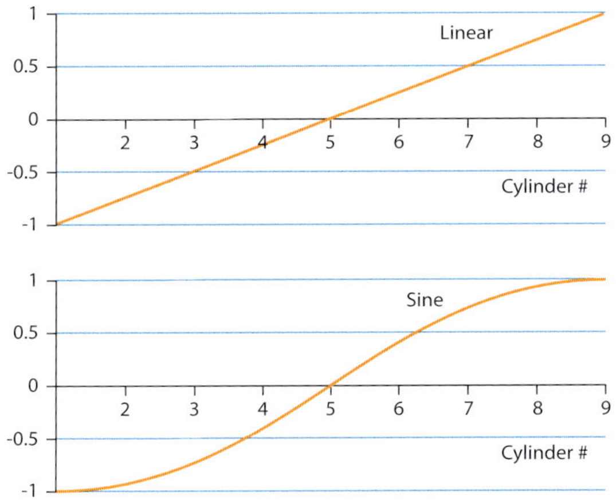

◀

When analyzing torsional excitations, a linear weighting function or a sinus wave can be used.

It is also possible to adjust the location of the mode's nodal point if the torsional mode shape is well known.

▶

A finite-element model showing the deformations in a pump assembly with a Wärtsilä V-engine mounted on a soft foundation. The colours show the degree of deformation and therefore load.

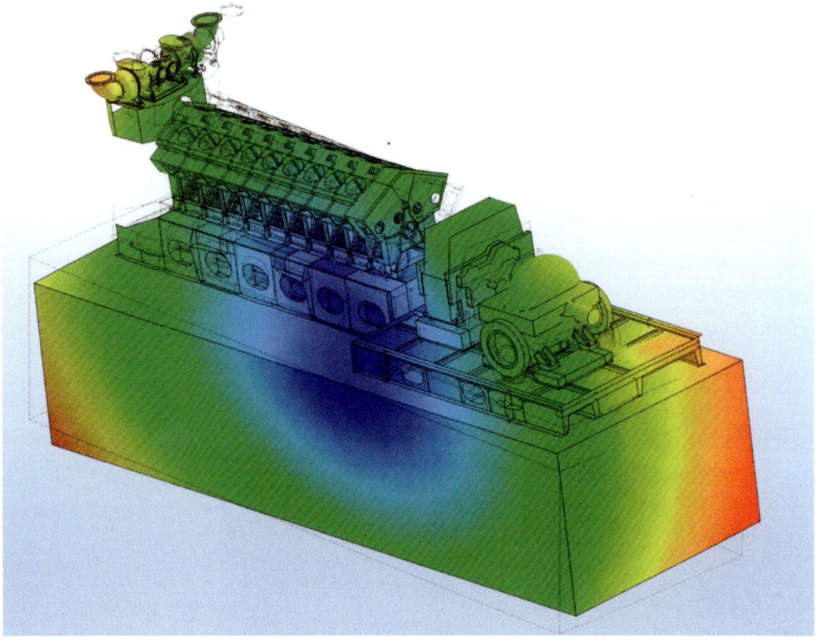

▶

The finite-element model applied to the crank for a two stroke Wärtsilä Sulzer RTA96C crosshead engine. Using this model, engine designers can perform simulations detailing exactly where problems could be expected.

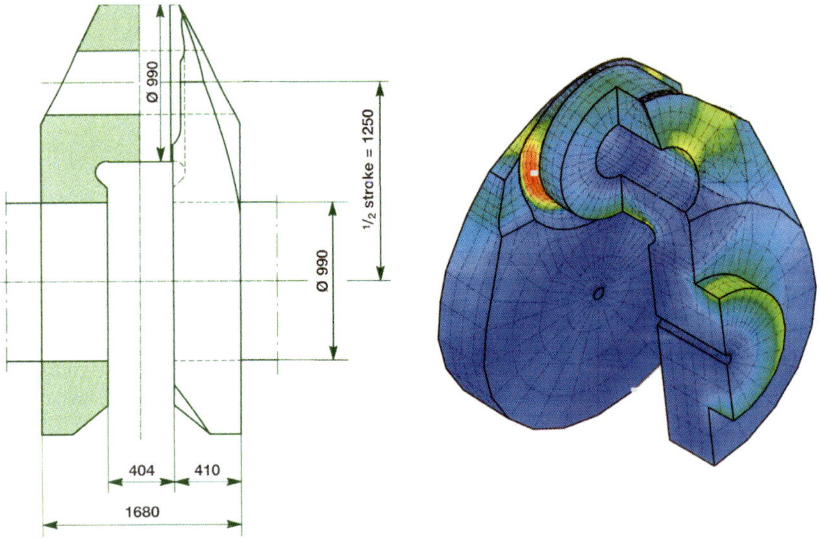

▲

The finite-element model applied to the crankshaft bedplate, the A-frame and cylinder beam for a two-stroke crosshead engine, a Wärtsilä Sulzer RTA96C.

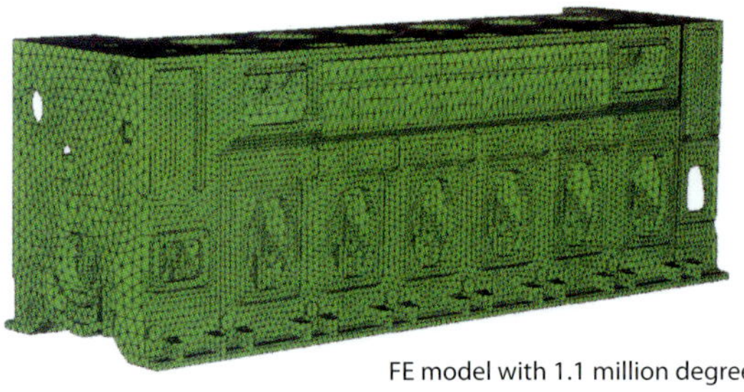

FE model with 1.1 million degrees of freedom

▶

The finite-element model applied to the block of a Wärtsilä 6L46F engine.

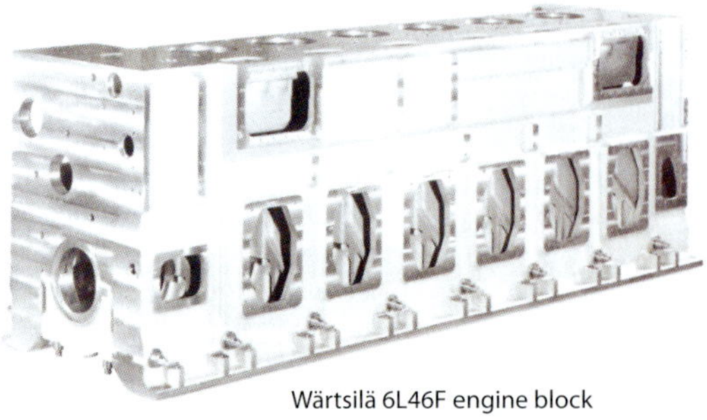

Wärtsilä 6L46F engine block

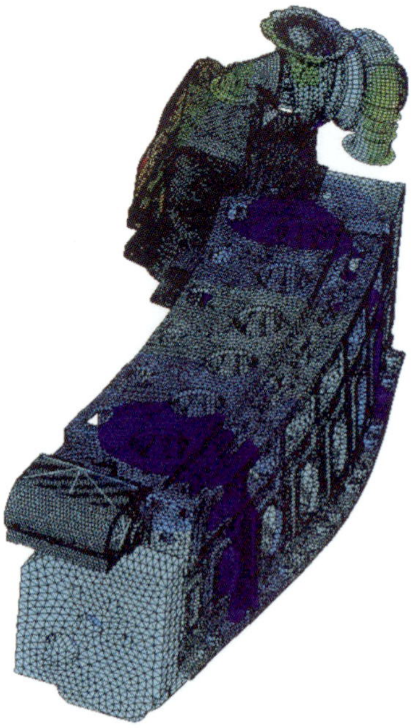

17.19.8 Simulating reality

This finite-element model simulates the entire propulsion system of an electricity generating plant. Therefore, the engine can operate with variable engine speeds in accordance with the propeller law or at a constant torque in an electricity generating plant.

17.19.9 Assessing vibrational levels

The vibrational levels of engines are usually assessed according to ISO standards:

- for piston engines: ISO 10816-1/ ISO 10816-3.
- for gensets: ISO 8528-9 and ISO 10816-3.

In addition to these regulations, various Classification Societies have their own regulations for finding vibrational levels in the different systems.

Today, diesel-engine manufacturers are often asked for their opinion about the vibrational levels in their engines and components.
Proper collaboration between all parties is the best way to achieve minimum vibrational levels in the engines.

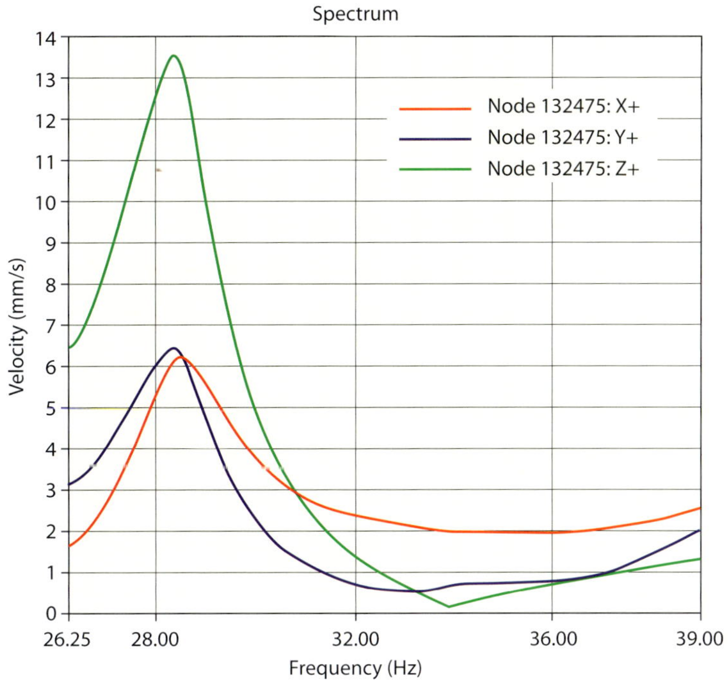

▲
An example of the torsional vibrations for the three axes.

17.19.10 Rule of thumb

Wärtsilä's rule of thumb is that the average vibrational level of an engine block should not exceed a velocity of 28 mm per second.
Limitations for standard engine blocks, added components and pipes are given separately.

It is not easy to indicate exact values, as the average vibrational level for each individual point constitutes the sum of their various characteristics, such as movement in a rigid frame, elastic deviations and local effects.
Engines with a diverging number of cylinders also behave differently and consequently some engines have above average vibrational levels.

17.19.11 Exceptions

It is possible for a vibrational level to be unacceptable even though the measured values are below the stipulated values. This may occur when average vibrational levels are dominated by the vibrational level in one component; this is often a sign of resonance. Simulation models and calculations which provide the maximum vibrational level for the particular component and can compare the values with the average vibrational level are necessary to eliminate this problem.
A good rule of thumb is that the contribution of one vibrational component should not constitute more than 80% of the average reasonable level.

▶
An example of the torque variation for the different Wärtsilä 46 medium-speed diesel engine types at speeds of 450, 500 and 514 revolutions per minute.

Torque variation, D-rating

Engine	Speed [RPM]	Frequency [Hz]	M_X [kNm]	Frequency [Hz]	M_X [kNm]	Frequency [Hz]	M_X [kNm]
6L46	450	22.5	132.5	45.0	48.1	67.5	6.8
	500	25.0	101.2	50.0	49.6	75.0	9.2
	514	25.7	94.0	51.4	50.0	77.1	9.6
6L46, idle	450	22.5	48.6	45.0	11.9	67.5	3.0
	500	25.0	69.8	50.0	12.0	75.0	3.0
	514	25.7	75.8	51.4	12.1	77.1	3.0
8L46	450	30.0	176.5	60.0	18.0	90.0	2.5
	500	33.3	167.9	66.7	21.6	100.0	3.6
	514	34.3	166.9	68.5	22.2	102.8	3.9
9L46	450	33.8	158.9	67.5	10.2	101.2	2.0
	500	37.5	156.0	75.0	13.9	112.5	2.8
	514	38.6	156.0	77.1	14.4	115.6	3.0
12V46	450	22.5	101.4	45.0	68.0	67.5	12.5
	500	25.0	77.4	50.0	70.2	75.0	17.1
	514	25.7	71.9	51.4	70.7	77.1	17.7
12V46, idle	450	22.5	37.2	45.0	16.9	67.5	5.5
	500	25.0	53.4	50.0	17.0	75.0	5.6
	514	25.7	58.0	51.4	17.1	77.1	5.6
16V46	450	–	–	60.0	36.0	120.0	1.3
	500	–	–	66.7	43.3	133.4	2.2
	514	–	–	68.5	44.4	137.0	2.5
18V46	450	33.8	311.8	67.5	18.8	101.2	3.3
	500	37.5	305.9	75.0	25.6	112.5	4.6
	514	38.6	306.1	77.1	26.6	115.6	5.0

The values are instructive and are to be calculated case by case.

– couples and forces = zero or insignificant

▶
The external forces and moments in a six-cylinder Wärtsilä 46 diesel engine.

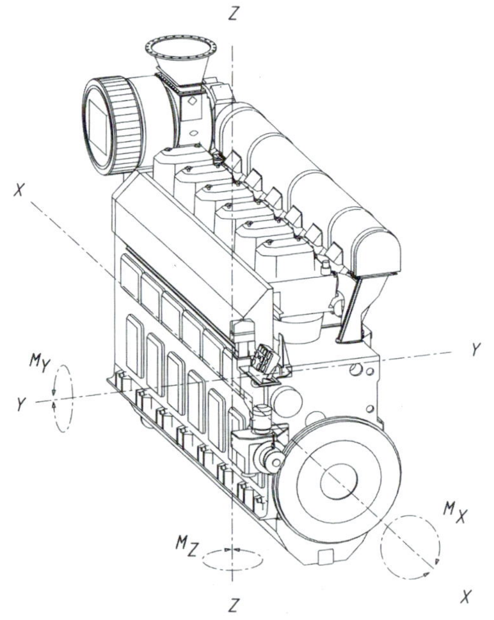

External forces F = 0 for all cylinder numbers

External couples (the values are instructive and to be calculated case by case)

Engine	Speed [RPM]	Frequency [Hz]	M_Y [kNm]	M_Z [kNm]	Frequency [Hz]	M_Y [kNm]	M_Z [kNm]	Frequency [Hz]	M_Y [kNm]	M_Z [kNm]
9L46 *)	450	7.5	25.5	25.5	15,0	30.8	–	30.0	10.5	–
	500	8.3	31.5	31.5	16.7	38.0	–	33.3	12.9	–
	514	8.6	33.3	33.3	17.1	40.2	–	34.4	13.6	–
18V46	500	8.3	283.8	283.8	16.7	135.1	55.9	33.3	–	4.0
	514	8.6	299.9	299.9	17.1	142.7	59.1	34.3	–	4.3

*) Subject to selected firing orders

– couples and forces = zero or insignificant

17.19.12 Summary for Wärtsilä four-stroke medium-speed diesel engines

The vibrational level in a diesel engine is a combination of structural properties, such as natural vibrational frequencies, mode shapes and damping. However, the vibrational level also includes the vibrations generated by gas forces, mass forces of the crank-connecting rod mechanism and moving parts in the running engine.
Using advanced software programmes allows first-rate calculations to be made for production series as well as for use in experimental tests for the probable vibrational performances of engines.
Optimising the vibrational characteristic of an engine usually consists of the analysis of the parts, such as engine block and its components, and analysing the generation of vibration.
An investigation can also be done to determine if and/or when resonance can occur between natural vibrational frequency and generated vibrational frequency. Naturally, this should be avoided as much as possible due to damaging resonances.
Sometimes resonances cannot be averted and must be kept as low as possible or compensated. Several manufacturers have developed good methods in this field. Available software and fast computers allow this data to be known quickly and can be used for combating disasters at sea.

Damage and its consequences in the engine with respect to allowable vibrational levels

It is common practice for a four-stroke engine with a below-average number of cylinders to continue to operate in the same fashion as two-stroke crosshead engines.
For the following assumption, a nine-cylinder four-stroke engine is used as the propulsion engine for a ship that is in the middle of the ocean. The combustion process is impeded due to problems with one cylinder. A high-pressure fuel pump could then be switched off, or the drive gear, piston and connecting rod removed.
The question is, at which capacity an engine can operate without causing damage to the engine. The damage can be caused by abnormally high torsional vibrations in the crankshaft.

General procedure for vibration- and balancing-related problems in four-stroke engines

Communicate with the engine manufacturer!
After mutual agreeing, an assessment can be made establishing at which speed the engine can be operated safely without damaging the engine.
Each case must be regarded individually as there is no general rule of thumb.
There are also engines, such as the Caterpillar MaK series, which have crankshafts that are so powerful that, for instance, a six-cylinder MaK 25 runs 'effortlessly' on five cylinders with 83,3 % of the power output capacity.
Consulting the manufacturer is always recommended!

Special operating conditions for four-stroke diesel engines

If damage is caused to any four-stroke engine and combustion no longer occurs in all the cylinders, measures must be taken with regard to vibrations and balancing.

- If one or more **fuel pumps must be switched off**, it is no longer possible to operate at full capacity.
- **During overheating of, for instance, the crankpin bearing**, the big end of the connecting rod, the corresponding piston and connecting rod must be removed.

Here the valve drive should also be dismantled in order to prevent exhaust gases entering the cylinder and subsequently the crankcase.
Apart from the 'Actions in case of damage and emergencies', it is of the utmost importance to determine at which capacity and revolution rate the engine can operate.
In principle, calculations are made for each type of engine. These meticulously indicate the allowable margins of vibration.
Individual Classification Societies also prescribe these margins.
The first course of action is to contact the engine supplier, who rapidly can provide appropriate advice regarding actions that should be taken.
It is possible to make a separate torsional vibration calculation for each situation, so damage to the crankshaft can be averted.

17.20 Balancing V-engines

17.20.1 Introduction

V-engines frequently used in engine categories I, II and III. They are supplied with from four- up to and including twenty cylinders. One of the advantages of V-engines compared to in-line engines is the short engine length in relation to the shaft power produced.
Furthermore, in-line engines in these three categories rarely have more than nine- or ten in-line cylinders. Only in engine category IV, the large two-stroke crosshead engines, deliver engine manufacturers as standard, the largest cylinder diameters with twelve cylinders.
Fourteen-cylinder engines manufactured with an option to increase this to sixteen- and eighteen cylinders, reach 30 metres in length!
In engine categories I, II and III, two ten-cylinder engines are installed for a twenty-cylinder V-engine. This takes up more space and is less interesting from a financial viewpoint. Diesel gensets in general are driven by V-engines, mainly with sixteen and eighteen cylinders. In engine category II, twenty-cylinder engines are common.

17.20.2 Arrangement and balancing

The engines are aligned in two rows at an angle varying from 60 to 120 degrees. An engine that must have equal-ignition distances has an angle between the cylinder rows identical to the angle between the cranks of the crankshaft. Generally, an angle of 45 to 60 degrees between the cylinders is used, making it impossible to obtain equally distributed ignitions for each cylinder number. For smaller engines in category II, angles of up to 120 degrees can be used.

17.20.3 Several examples of V-engines

▲
Mounting the complete cylinder unit with cylinder head, piston, connecting rod and cylinder liner in an eighteen-cylinder MTU V-engine block, type 8000, category II in Friedrichshafen, Germany.

The angle between the cylinders is small, approximately 45°.

▶
This twelve-cylinder V-engine block, category II is undergoing a revision. Seen here on its head.

The angle between the two cylinder rows is more than 90°.

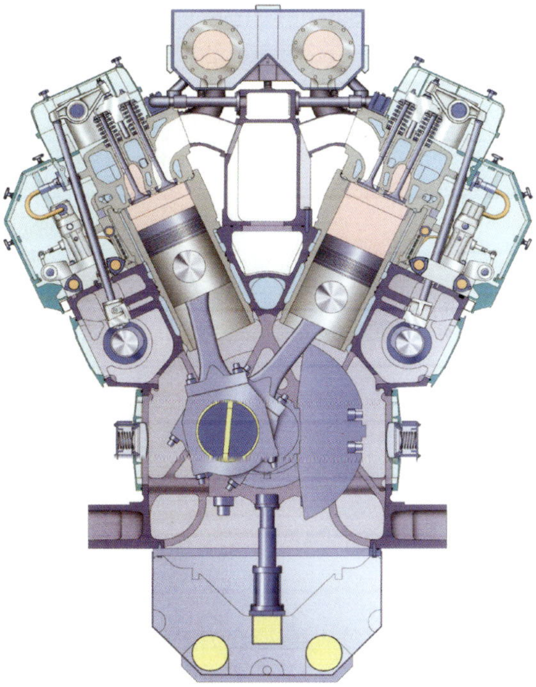

▲

A Wärtsilä 38 V-engine, category III, has the traditional angle between the cylinder rows for the large bores, between 45° and 60°.

The angle is 50°.

17.20.4 Arrangement of cylinders in the left- compared to the right banks

Today, in the majority of engines, both pistons of the opposing cylinders with their own connecting rods are fixed on the broad crankpin. Consequently, the cylinders are not directly across from each other but in an 'interleave' pattern, they are shifted a certain distance because of the two connecting rods next to each other attached to one crankpin.
Older engines often have one main connecting rod with a mounted second connecting rod, the so-called mother-daughter construction. In these engines, the cylinders are arranged directly opposite each other. The latter, complicated, construction is no longer used in new engines.

17.20.5 Rotating forces and moments

Essentially, these are mainly dependent on the crankshaft shape but also the fact that both bottom parts of the connecting rods are also part of the rotating forces.
The resultant rotating, primary and secondary oscillatory forces working on the engine can be found by summing the forces working on one cylinder pair.
The crankshaft is often constructed in central-symmetrical manner. If the engine is homogeneous (symmetrical), the forces and the resultant forces for the entire engine are zero.
In four-stroke V-engines from Wärtsilä, these forces are completely balanced.
Primary oscillating forces: the pistons, the piston coolant, the top parts of the connecting rods and the secondary oscillating forces.

17.21 Balancing examples for two-stroke crosshead engines, category IV

Example MAN Diesel

In general, there are relatively numerous two-stroke crosshead engines in use with four or five cylinders. This forms an ideal propulsion mechanism for the relatively low-speed large tankers and bulk carriers. They have relatively few moving parts, therefore low installation, and operational costs.
Due to the large number of engines with relatively few cylinders, a large amount of experience in the reduction of vibrations has been gained.
Requirements for the reduction of vibrations often originate from complaints in the accommodation areas aboard ship where vibrations are a nuisance factor.
However, when insufficient attention is paid to vibrations, the mechanical- and electronic parts of two-stroke crosshead engines and other parts of the ship could be damaged by for instance, tearing.
Internal forces and moments can distort/misalign the engine.
External forces and moments generated by the engine work on the lower bedplate under the engine and the upper bedplate above the engine through the fitted top braces.

17.21.1 External and internal forces and moments

An engine frame should be constructed with such rigidity that the internal moments and forces cannot cause excessive bending of the frame with its accompanying stresses.
If it can be presumed that the engine bedplate is sufficiently rigid, the internal moments and forces will not cause distortion in the ship's construction. The term 'rigid' is, of course, relative and there is always a certain amount of movement in the engine frame, the bedplate and ship's construction. Consider, for example, poor weather conditions. Information gained by experience at MAN Diesel has shown that the internal forces and moments to the first order caused by the mass-rotating forces and the reciprocating moving masses are not capable of generating vibrations in the ship.
X-type moments must be closely examined as they generate higher vibrational levels and work in the less rigid directions of the engine frame, especially in multiple-cylinder engines, such as the twelve- and the new fourteen-cylinder versions.

17.21.2 Torsional vibrations

The entire shaft system, the crankshaft, intermediate shaft and propeller shaft, is subject to vibrations that are predominantly caused by the tangential force F_t.
Torsional vibrations can cause vibrations throughout the entire ship because the engine-drive gearing is connected to the propeller.

17.21.3 Torsional vibrations generated by the thrust pressure of the propeller

The propeller is nothing but a part that with the power supplied by a standard propulsion force thrusts the ship forward. The variation in the rotating torque of the diesel engine continues in the propeller as a variation of the torque. Here the torque variation of the propeller produces vibrations, which are passed along the propeller shaft to the engine. Systems with flexible torques largely eliminate these vibrations.
Large two-stroke crosshead engines usually have rigid connections between the fixed-pitch propeller, the intermediate shaft and the crankshaft of the engine. The crankshaft is equipped with a vibration damper, if required.

17.21.4 Axial vibrations in the crankshaft

These are generated by the radial force F_r and by the tangential force F_t.
The vibrations generated by the propeller are a contributing factor to the axial vibrations in the crankshaft. Remember, large two-stroke crosshead engines have no flexible torque between the engine and the directly driven fixed-pitch propeller.
Axial vibrations produce a reaction force in the thrust block that then itself becomes a source of vibrations in the rest of the ship.

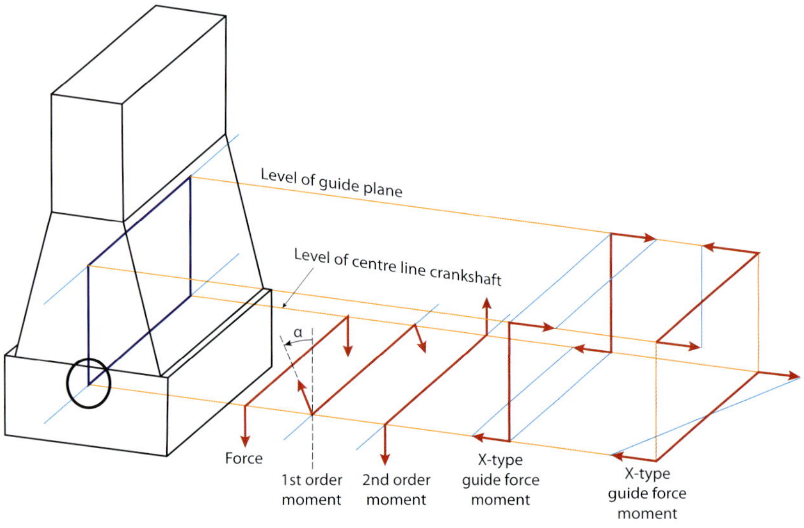

◄ External forces and moments.

- **first-order moments in vertical and horizontal directions.** They are of equal magnitude in MAN Diesel engines with standard balancing.
- **second-order moments in the vertical direction.** Moments to the fourth order and higher occur in engines with a certain number of cylinders, but are small and can be ignored.
- **H-type moments** are found between a stationary engine frame and the rotating and reciprocating moving parts of the engine. From a practical engineering point of view, it should be applied to the engine frame as an external moment.

17.21.5 Torsional vibrations and their limitations

Limiting the torsional vibrations is important for propulsion installations as excessive vibrations may cause damage to, or even break the crankshaft. Intermediate shafts, propeller shafts, gear transmissions and flexible couplings may also be damaged.
Therefore, Classification Societies have from the beginning of the use of propulsion engines required that calculations and control measurements of these vibrations take place.

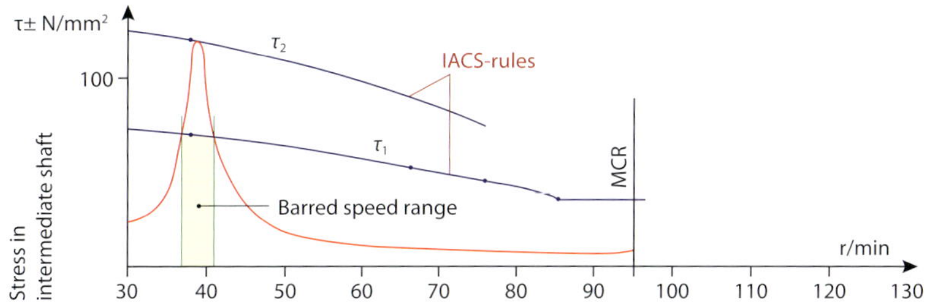

▲

The torsional stresses in the shaft between the engine and propeller shaft.

MCR = 95 revs/min.

yellow bar: barred speed range

In the graph, there are two torsional-stress limits. At engine speeds, where a lower-limit value of τ_1 is reached, the engine may not operate continuously, and the upper-limit value τ_2 may not be exceeded.

Above 80% speed, only limit τ_1 should be accepted.

17.21.6 Requirements of Classification Societies

The Classification Societies stipulate two limits, these being τ_1 and τ_2, for the torsion stresses at an engine speed of 80% of MCR.

17.21.7 Engines with four, five and six cylinders

As the engine room is situated aft, the torsional characteristics of the engine are dominated by the resonance of the torsional vibrations to the first degree, generated by the harmonic order, which is equal to the number of cylinders. So in a five-cylinder engine, a resonance to the fifth order and the first degree is generated.
Generally, this resonance occurs between the minimum and maximum engine speed, depending on the length and diameter of the shaft system or the total torsional flexibility of the shaft system between the propeller and the engine.
Resonance will produce torsional stresses in the shaft system and these must lie within the limitations stipulated by Classification Societies.
The degree of these resonance stresses is dependent on the sensitivity of the damping system. In general, the sensitivity increases as the engine velocity increases.
Damping of the system is dependent on, A, the ratio between the moments of **movement** of the propeller and the engine, B, the damping of the propeller and C, the existence of torsional dampers.

Solution C is only chosen when inevitable due to very short shaft systems or when a ship owner determines that the ship may not be bound to a certain number of revolutions.

Apart from making certain that, the ship is not bound to a certain number of revolutions, solution C allows a considerable variation in the shaft coupling, which, as previously explained, creates a reasonably high variation in thrust pressure so that torsional vibrations generated by the thrust pressure of the propeller are created.
Here it should be noted that under unfavourable conditions, the fluctuating thrust pressure could rise to 50% of the normal thrust pressure, which considerably exceeds the value that the designer of the propeller can accept for vibrations generated by travelling through a non-uniform wake.

Choice A, B or C

Of these three solutions, solution A usually costs the least. However, solutions neither A nor B nor C can be directly coupled to a certain engine. Only detailed torsional-vibration calculations in the design phase can provide the best solution.
Which solution is the most probable?

For four-cylinder engines
Solution A is used most often. C sporadically.

For six-cylinder engines
Solution A, without adjusting disk.

▶
'Lay-out' for a five-cylinder two-stroke crosshead engine, the MAN Diesel 5L70MC.

In the figures A, B and C, the 'lay-outs' for a five-cylinder two-stroke crosshead engine that is directly coupled to the propeller.
The three examples have differing circumstances as to how torsional vibrations occur.

Solution A: System with a relative flexible-shaft system
The material strength is increased to reduce the diameter of the shafts, thereby making them even more flexible. A turning wheel is mounted on the front end of the crankshaft to increase the ratio of **mass moments of inertia** between engine and propeller resulting in higher damping in the system.
The resonance will occur below engine MCR speed (over-critical).
The torsional stresses will be below the 2-limit, but above the 1-limit, with a **barred speed range** as a consequence.

Solution A

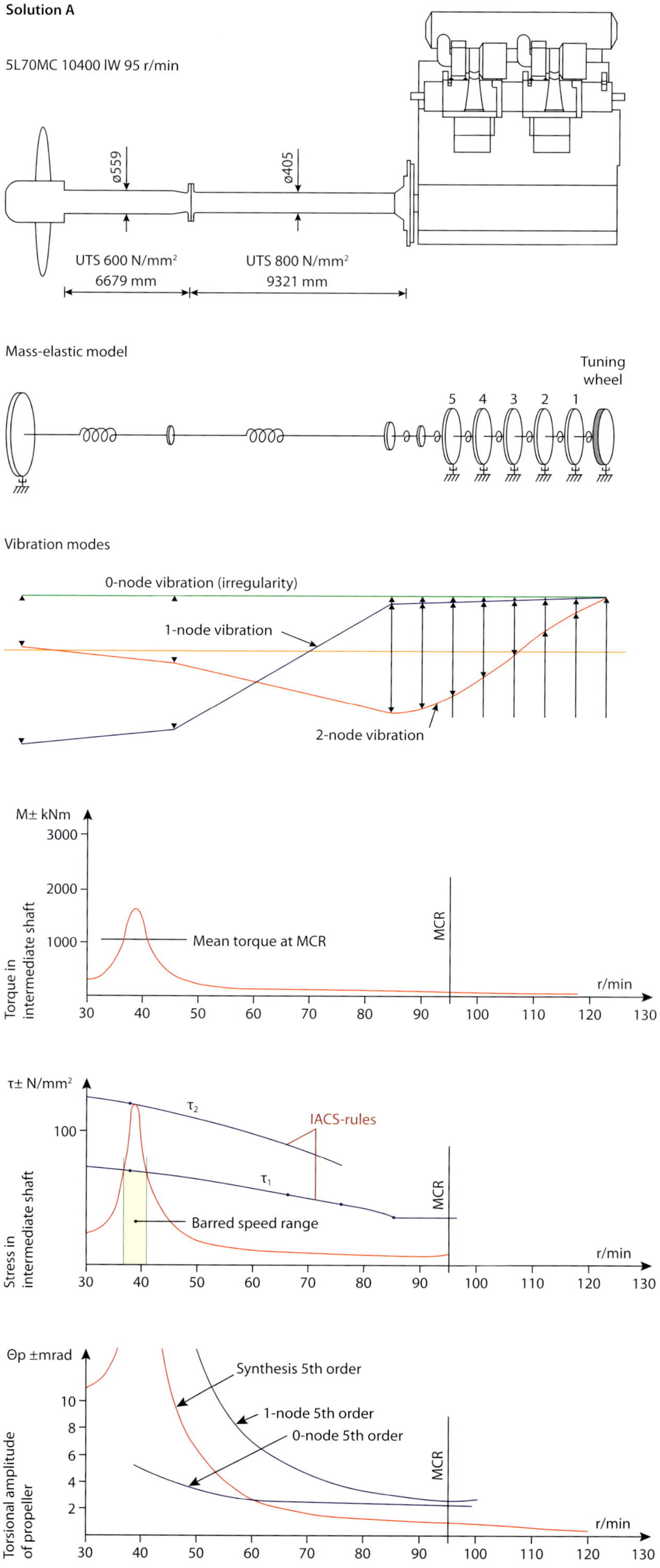

Solution B: System with a relatively stiff-shaft system

Due to the short distance between propeller and engine or implemented stiff shaft elements such as shrink-fit couplings, oil-distribution box to CP-propeller and ice-class requirements on shaft diameters.

This has brought the main critical resonance relatively close to MCR, and the resonance stresses will exceed the upper limit 2. In this case, there will be four possibilities:

1 Mounting a torsional-vibration damper of appropriate size to reduce stresses to below the 2-limit. The propulsion installation will have a barred speed range.
2 Mounting a torsional-vibration damper of appropriate size to reduce stresses to below the 1-limit. The propulsion installation will have no barred speed range.
3 Increasing shaft diameters in order to move the main critical resonance above MCR.
4 QPT – Quick Passage through a barred speed range Technique. The procedure is as follows;
Performing extensive torsional-vibration calculations in maximum and minimum pitch conditions using pessimistic propeller damping values. The establishment of a simulation model in the barred speed range to investigate the torsional vibrations. The evaluation of stresses in the barred speed range based on results found in the start and stop test simulations for rapid passage through the barred speed range.

Critical speed application

An engine is added to the system so that when operated via the bridge manoeuvring system, as well as when operating from the engine-control room, the engine operates on the speed-setting signal in such a way that automatic rapid passage through the barred speed range is obtained. MAN Diesel calls this system **'The critical speed unit'**.

Solution B

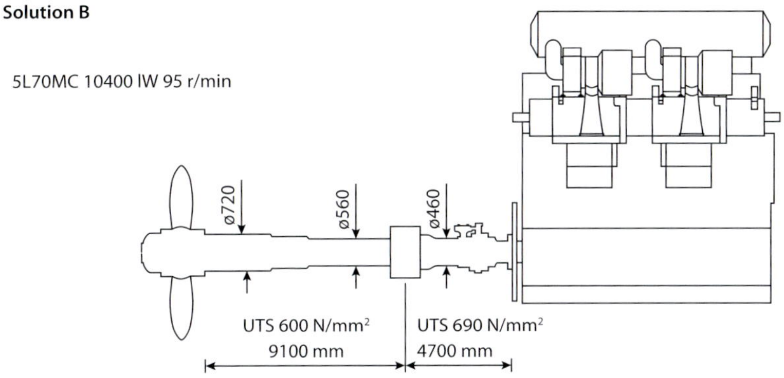

Mass-elastic model

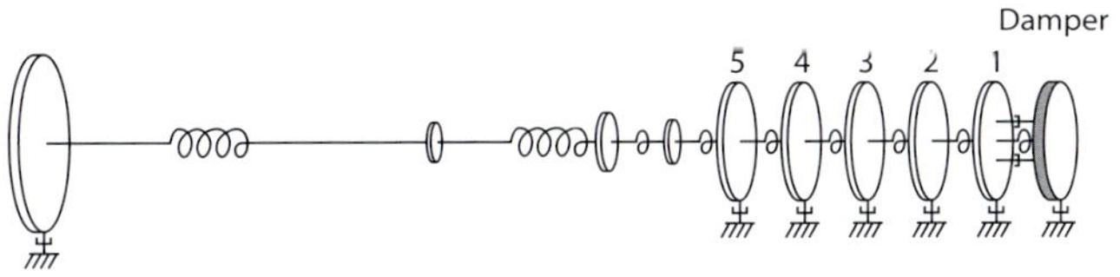

Vibration modes with damper

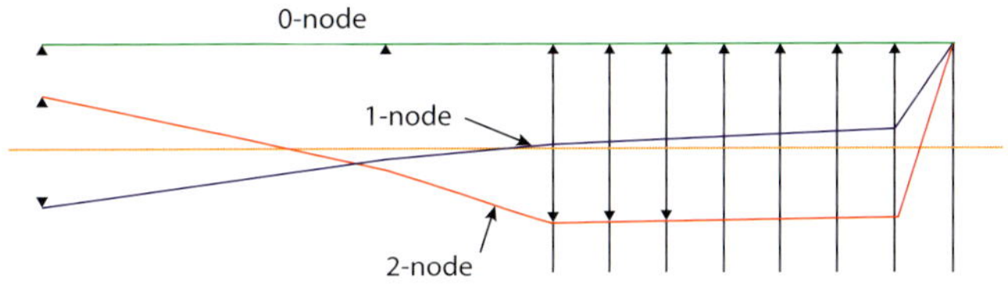

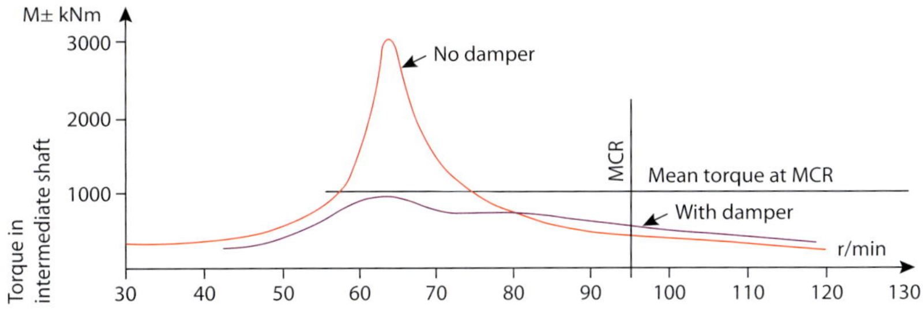

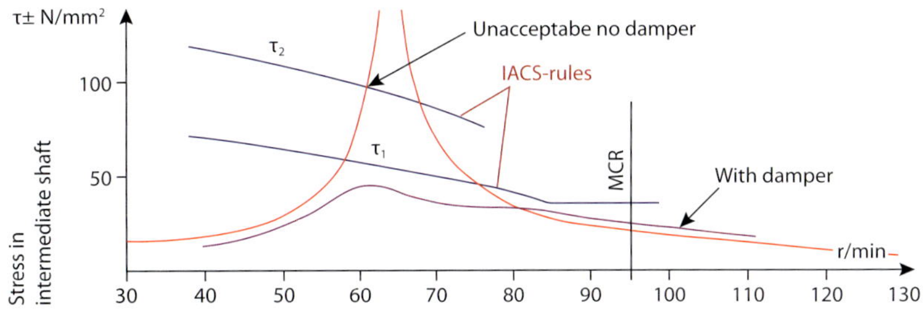

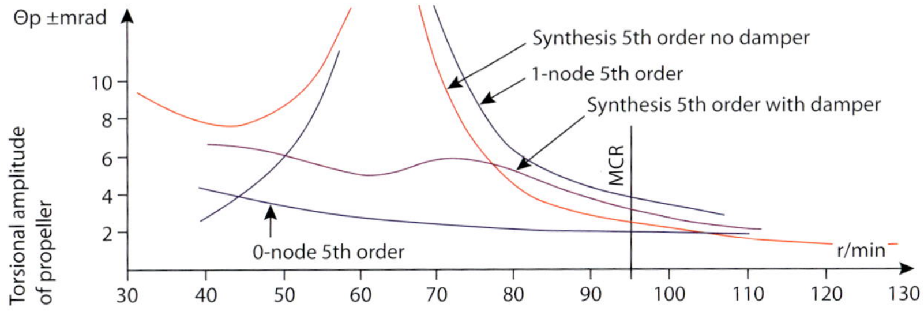

Solution C is obtained by increasing the diameter of shafts until the main critical resonance is positioned approximately 40 to 45 per cent above the nominal speed (under-critical). Due to the large shaft diameter (large moment of resistance), only moderate torsional stresses appear even though the varying torsional torque in the shaft is high.

Solution C

5L70MC 10400 IW 95 r/min

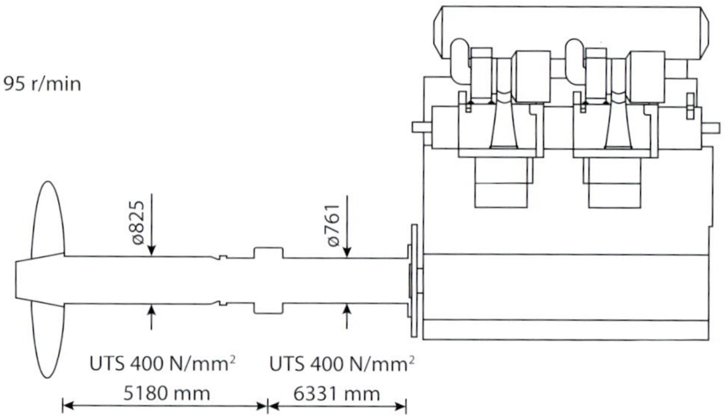

Mass-elastic model

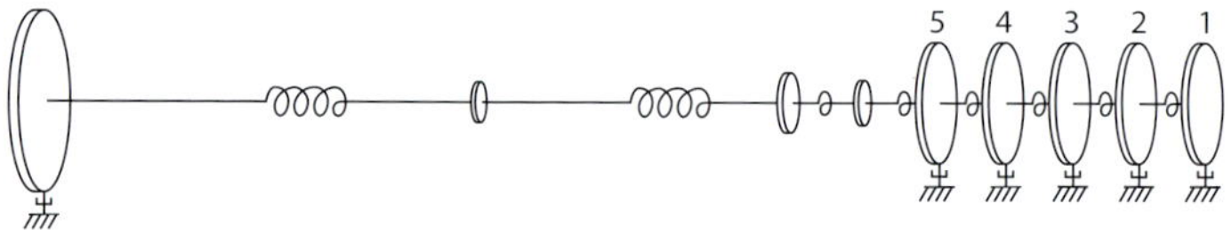

Vibration modes

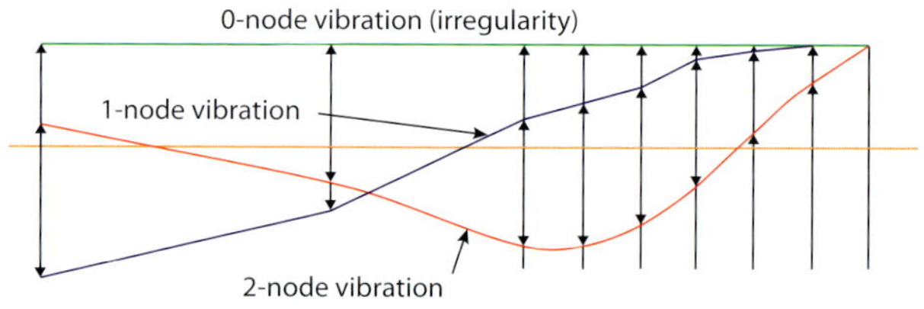

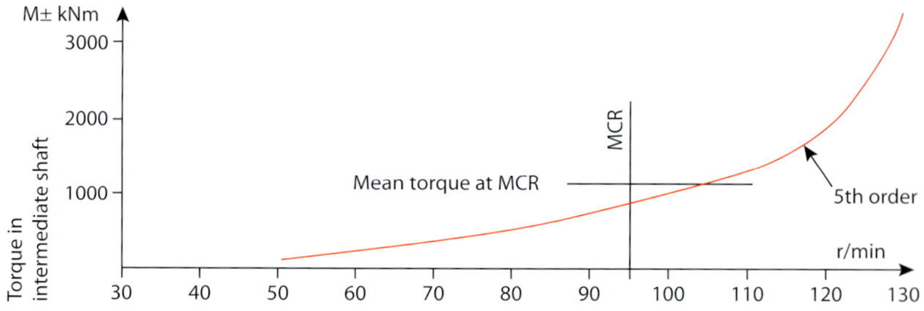

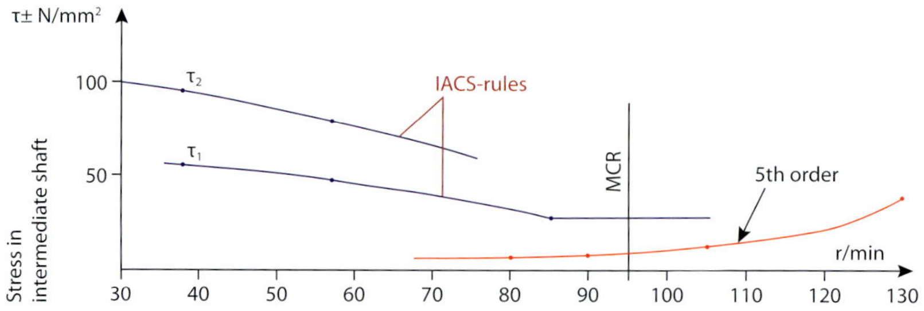

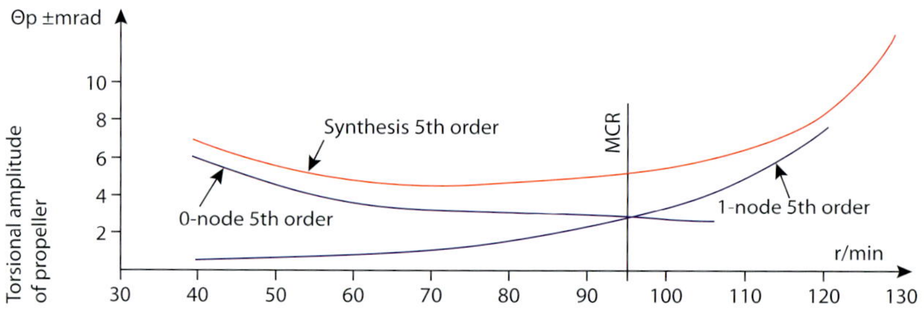

17.21.8 Engines with seven or more cylinders

The critical resonance to the first degree is not usually important for these engines, as it occurs close to or below the minimum engine speed.
Moreover, the torsional vibrations to the seven or more orders are lower than the fourth, fifth and sixth order, which means that the engine speeds are not bound to a range.
The second-degree torsional vibrations (with node in the crankshaft) become increasingly important and need attention. The main resonances of these vibrations should be avoided close to MCR.

17.22 Shaft generators in two-stroke crosshead engines

The MAN Diesel engines are equipped as standard with PTO's, Power Take Off, at the blind end of the crankshaft for driving the shaft generator, unlike the Wärtsilä Sulzer engines where it is part of the shaft between the crankshaft and propeller shaft.
PTI's, Propulsion Shafting Power Take In, the drive for exhaust-gas driven power turbines is also placed on this side.
Both systems require protection from the relatively high torsional vibrations of the crankshaft.
A flexible coupling connects the engine and the gearbox.
This is adequate for up to 10% shaft power for the PTO's or PTI's.

Large shaft generators in relation to the engine capacity

This is frequently used on special ships such as ferries, cement bulk carriers and shuttle tankers.
They often require a large electric capacity that is as high as the propulsion-power capacity.
The torsional vibrations in these large installations with mechanical couplings, elastic couplings, gearboxes, generators and/or hydraulic systems are complex.

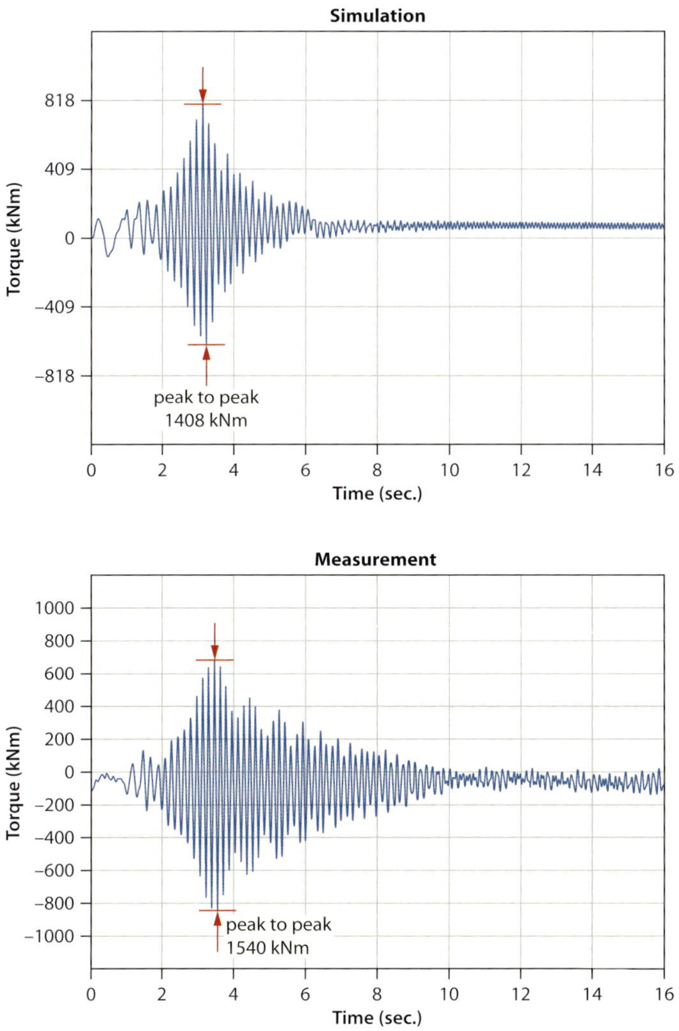

◄ A comparison between simulated and measured torque in the intermediate shaft during starting the main engine and passage through the barred speed range at a minimum propeller pitch.
Shown here, the MAN Diesel 5L50MC engine.

The measured values are minimally higher than the simulated values.

The following are significant:

- Flexible couplings are necessary for an optimal alignment (the alignment should of course always be as good as possible) and to protect the gear transmission from high torsional vibrations in the main engine.
- The elasticity of flexible couplings or couplings in general should therefore be such that the PTO-system has a natural frequency approximately 1.5 times the engine speed or lower than 0.75 times the engine speed.

The result is critical resonances in the PTO-system to the fourth, fifth and sixth orders at extremely low velocities or velocities below the minimum engine speed.

Furthermore, first- and second-order vibrations dominant during misfiring will generate resonances well outside the nominal number of revolutions.

17.23 Vibration numbers and orders

A propulsion installation operated by a diesel engine consists of a rotating crankshaft with flywheel, a propeller shaft and propeller.
The crankshaft normally has fitted counter- or balancing weights. The pistons are connected to the crankshaft by connecting rods. Behind the connecting rod, crosshead engines have the crosshead with guides, the piston rod and the piston.
The crank masses and items considered to be rotating parts are placed together at one point.
The shaft masses with the coupling flanges are not included as their influence is negligible in comparison to the other moving masses.
Simply, there is a shaft with three fixed masses: the rotating parts of the engine, the flywheel and the propeller.
Theoretically, a shaft with n masses (n – 1) has natural numbers to the first, second, third and to the (n – 1) degree, respectively.
The vibration numbers of the higher degrees are usually so large that they do not pose problems.
In standard propulsion installations, or when the engine is checked, the first-, second- and sometimes the third-degree vibrations must be examined.

When the engine supplies power, it provides a driving moment to the propeller or the generator, and the propeller or generator supplies a counter moment. This means that the shaft is twisted. If these moments were to cease, the rotating shaft could vibrate at one of the natural (vibration) frequencies. Resistances of the propeller or rotor of the generator and in the engine itself dampen this vibration considerably.
The driving- or the resistance moment in an engine is not constant, but changes frequently caused by the fluctuating combustion processes in each cylinder. Therefore, the twisting in the shaft varies significantly and produces forced vibrations in the (crank) shaft.
It can be deduced from the changing magnitudes of the tangential force calculated from the engine combustion process how the vibrations can be classified.

Orders

First-order moment:	at the same engine speed.
Second-order moment:	at twice the engine speed.

17.23.1 Example I

A moment in an eight-cylinder four-stroke diesel engine changes from a minimum to a maximum value **four times** per revolution**,** each time one cylinder makes a combustion stroke. Therefore, the twisting of the shaft and the stress in the shaft fluctuates around an average value **four times per revolution**.
The shaft acquires a forced stress that is referred to as a vibration to the **fourth order**.

17.23.2 Example II

In a six-cylinder two-stroke crosshead engine forced vibration **to the sixth order** are generated and in **a ten-cylinder two-stroke crosshead engine** a forced vibration to **the tenth order.**
The resistance moment of the driven propeller is not constant but depends on many factors, the wake fraction, the draught and the depth below the keel (under keel clearance).
When, for instance, two blades of a four-bladed propeller pass by the stern, the moment of resistance is temporarily lower. As this occurs **four times per revolution**, this is a forced vibration to the **fourth order**.

17.23.3 Example III

The vibrational frequency to the first degree of the shaft (the natural vibration) is 480 per minute.
In a **ten-cylinder four-stroke engine** a resonance will be produced at $\frac{480}{5} = 96$ revolutions per minute with a forced vibration of the engine and with a four-bladed propeller at $\frac{480}{4} = 120$ revolutions per minute.
Both 96 and 120 are critical revolutions per minute.
96: first degree and fifth order
120: first degree and fourth order

The shaft operates at a constant angular velocity and has a winding motion in relation to the angular velocity with these vibrational phenomena.
Due to this winding motion the material stress continuously increases and decreases at a constant angular velocity.

Eventually, the fluctuating stresses, if extensive, may lead to shaft fracturing.
If the critical area is **lower** than the number of revolutions then the critical engine speed must be increased **rapidly** by increasing the fuel supply to the engine.
The red-coloured section on the tachometer indicates this critical engine speed.

17.24 Vibrational frequencies

Every system has several natural frequency, which depends on the mass and rigidity of the system.
Only natural frequencies within the exitation frequency range have to be considered.
When a system is operational, there is a force or a generating frequency that can cause parts of the engine to vibrate. These vibrations can reach very high values when the vibrational frequency of the engine is equivalent to the natural vibrational frequency of the environment.
This is called **resonance**.
Vibrations pose problems as they create stress in the material and when these stresses reach inadmissibly high levels they can cause fatigue fractures.
It is impossible to prevent vibrations in a running diesel engine with, for instance, a propeller drive.
However, it is possible to reduce vibrations by, for example, balancing the engines.

◄
One of the two balancing shafts in a four-cylinder medium-speed diesel engine. Both shafts are driven by the crankshaft using gear wheels. Engine type, Wärtsilä 4 R 32.

17.25 Example 1: Adjusting the engine speed because of damaged cylinder liners

On a large container ship with a Wärtsilä Sulzer 12RTA96C propulsion engine and shaft power at MCR of 65,880 kW, horizontal tearing of two cylinder liners of cylinder eight and cylinder twelve was discovered during a stop at sea. After consulting Wärtsilä Switzerland, it was decided that the fuel pumps of both cylinders be switched off and to continue the voyage on ten cylinders with a reduced engine speed and shaft power. Cylinders eight and twelve could compress normally without a combustion process.

A calculation of the maximum value of the torsional vibrations allowed in an engine running on ten cylinders was sent by e-mail.

Standard engine speed at MCR = 100 rpm.
Mean effective pressure = 18.2 bar

Conclusion

The graphs show that the engine speed should not exceed 86 rpm. Here the power output is reduced by 36%, so 65,880 – 23,977 = 41,903 kW. The calculations indicate that the torsional stress with ten of the twelve cylinders in service does not pose any problems.

▶ First picture.

The engine data at MCR (Maximum Continuous Rating).

WÄRTSILÄ

TORSIONAL VIBRATION CALCULATION

Calculation No.	12RTA96C-TV-7C
Installation	9000 TEU Container Vessel
Plant Builder	
Engine Type	12RTA96C
Date	07.05.2007
Calculated by	R. Rusch
Company	Wärtsilä Switzerland Ltd
Address	CH-8401 Winterthur

Main installation data:

Engine type	: 12RTA96C (Eng.no. 1)
Type	: 2-stroke
Cyl.	: 12 In-line
Bore	: 960.0 mm
Stroke	: 2500.0 mm
Conn. rod ratio	: 0.434
Osc. mass	: 17834.0 kg
P(CMCR)	: 65880.0 kW
Speed	: 100.0 rpm
M.E.P.	: 18.20 bar
Excitation data	: rt_24a.tan
Cranks	: No. 1 to 12 at masses no. 14 to 3
Crank pin	: 990.0 mm (at # 3)
Sense of rotation	: Clockwise
Ahead firing order	: 1 - 9 -11 - 4 - 3 - 8 -10 - 6 - 2 - 7 -12 - 5 Regular

Minimum shafting diameters:

Intermediate shaft	: 800.0 mm (at # 17)
Propeller shaft	: 975.0 mm (at # 19)
Modulus of rigidity:	81.4 kN/mm2
Calculation option	: Including variable crank inertia
Diagram group no.	: 12RTA96C-TV-7C

Disclaimer note:

The enclosed calculation has been carried out by highly qualified specialists using thoroughly checked and well proven advanced computer programs.
All input data relating directly to Wärtsilä engines have been prepared with the utmost care and precision. However, for data related to the installation components such as dampers, elastic or hydraulic couplings, gears, generators, propellers, elastic mountings etc. WARTSILA SWITZERLAND LTD has to rely on input data furnished by the technical documentation of the different suppliers of the installation components. WARTSILA SWITZERLAND LTD cannot guarantee the correctness of these data and therefore excludes any warranty, liability or responsibility whatsoever as to damage, failure or malfunction of the installation and consequences thereof due to wrong input data furnished by or resulting from the above mentioned technical documentation of the suppliers or any estimated data.

Wärtsilä Switzerland Ltd   Visual TORSVIB V4.2 #7056-3082737   © WCH

▼

Second picture.

Sketch of torsinal vibrating system.

1 *outer ring of Geislinger vibration damper*
2 *inner ring of Geislinger vibration damper*
3 t/m 14 *cylinders*
15 *thrust bearing flange*
16 *turning wheel*
17 *internal shaft*
18 *shaft generator*
19 *propeller shaft*
20 *fixed pitch propeller*

The shaft diameters are in millimetres, the mass moments of inertia in kgm² and the flexibility in rad/Nm *e – 9.

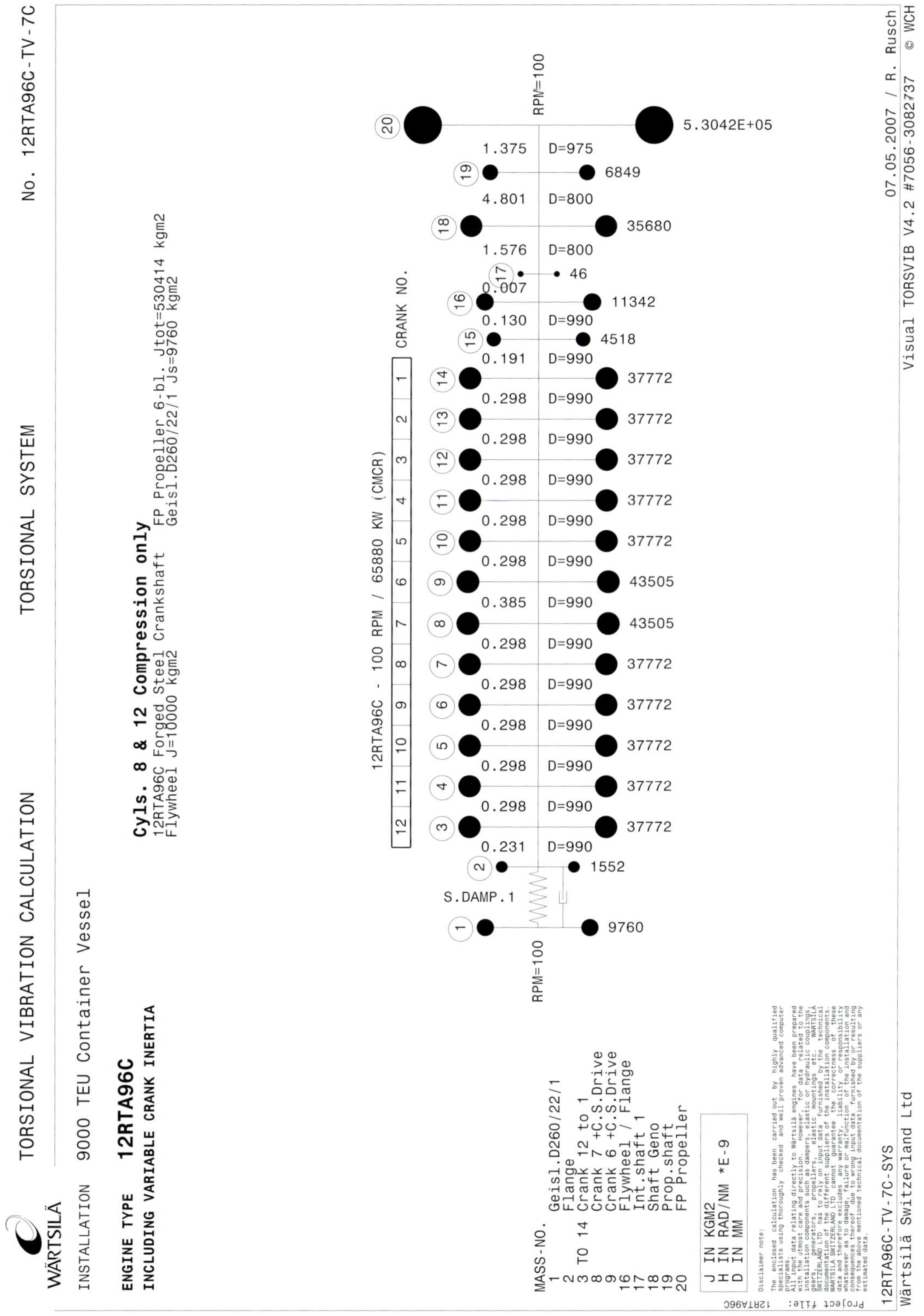

Third picture.

Horizontal: speed in revs/min.

Vertical: alternating torsional stress in N per mm².

First graph: for crankshaft D = 990 mm.

Second graph: for intermediate shaft D = 800 mm.

Third graph: for propeller shaft D = 975 mm

For all, the speed is 100 revolutions per minute.

First graph: crankshaft

At 86 revs/min., the actual torsional stress is ± 25 N/mm², the limit is 31.1 N/mm².

Second graph: for the intermediate shaft

At 86 revs/min., the actual torsional stress is just below 37.5 n/mm², the maximum value allowed.

Third graph: for the propeller shaft

At 86 revs/min., the actual torsional stress is just below the maximum allowed torsional stress of ± 24 N/mm².

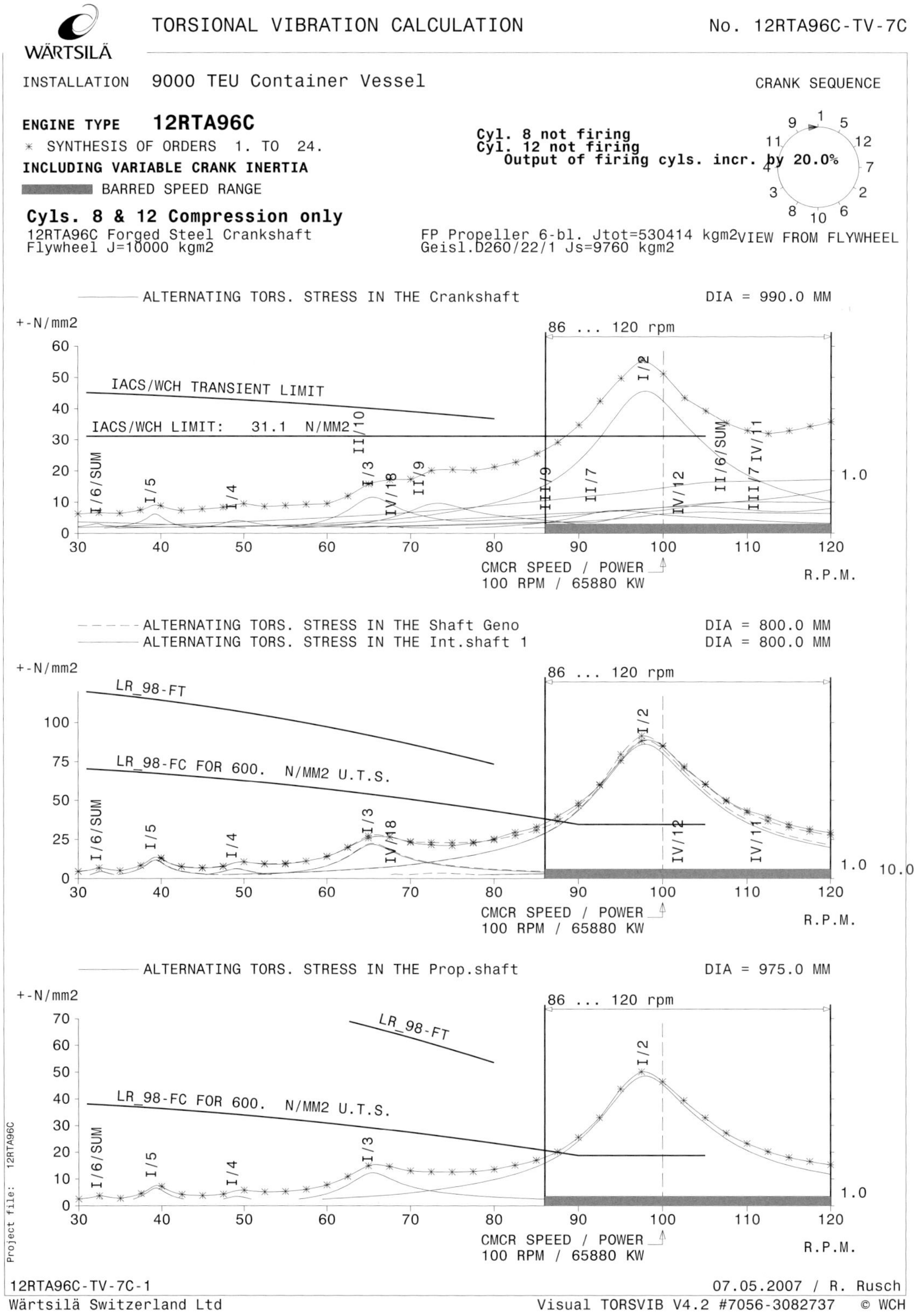

▼

Fourth picture: Graphs for the Geislinger vibration-damper, type: spring damper.

Horizontal: speed in revs/min.

First graph, vertical: torque in kNm.

The maximum elastic torque of ± 650 kNm is not reached (± 500 kNm).

Second graph, vertical: maximum allowed damping for torques at 2,0 bar absolute oil pressure.

This is also inside the 180 kNm limit.

Third graph, vertical: thermal load of the Geislinger damper in kW.

This is not reached. It can have a maximum of 160 kW and is only ± 15 kW!

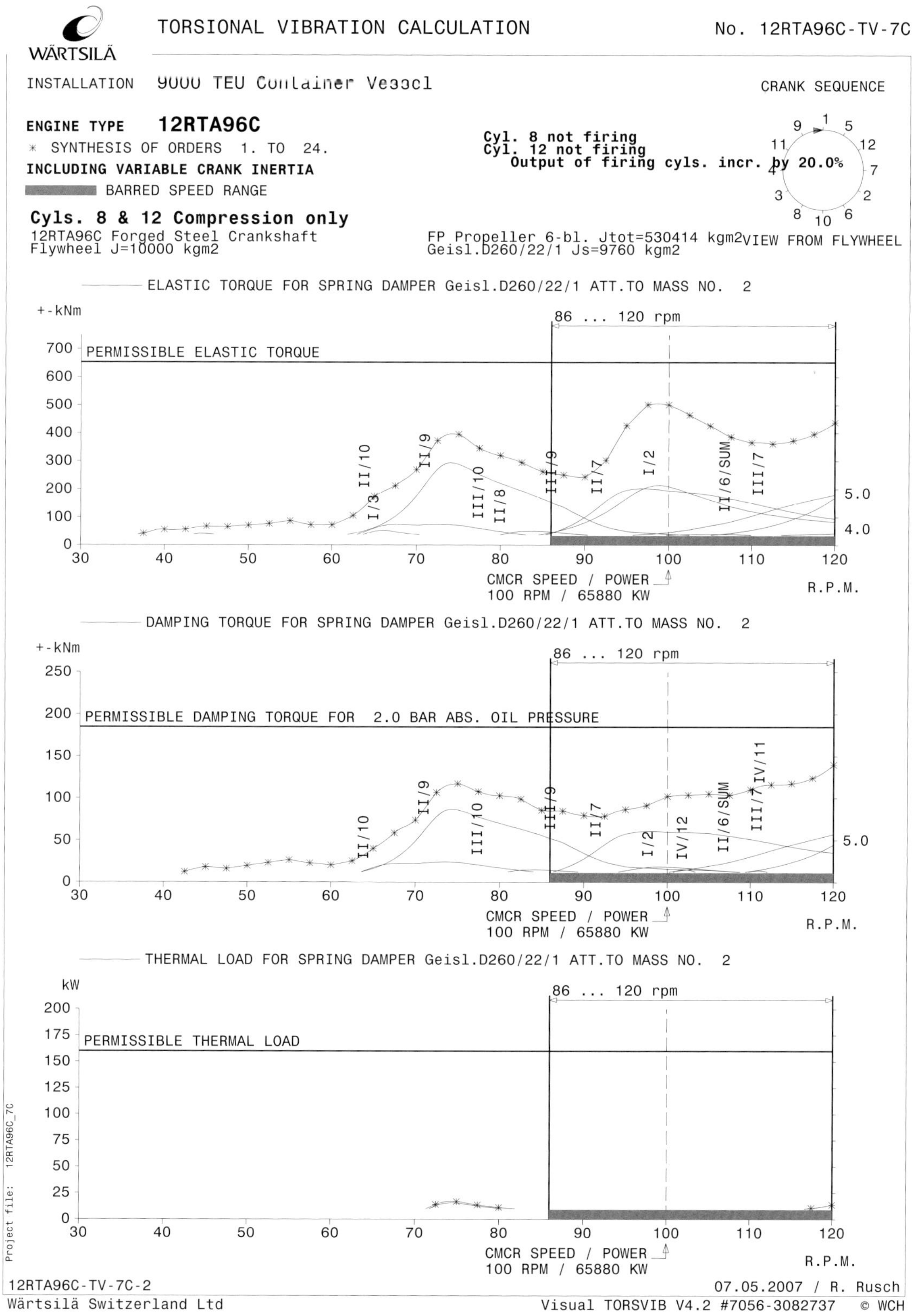

17.26 Example 2a: Simulation on six cylinders

This example is of a theoretical situation. In reality a cylinder does not have to be turned off by discontinuing the fuel supply.
A computer simulation for shutting down cylinder 3 was performed.

This means that this cylinder has no combustion process. That is to say, the fuel supply to this cylinder was stopped. The simulated calculations are then tested to verify that the calculation model is accurate and reliable.

This shows a two-stroke crosshead engine of the Wärtsilä Sulzer 6RT-Flex50 type at 124 revs/min. The shaft-power output is 9,720 kW.
This is a propulsion engine of a moderately sized chemical tanker.

▶ The engine data at Maximum Continuous Rating (MCR).

WÄRTSILÄ

TORSIONAL VIBRATION CALCULATION

Calculation No.	6rtflex50-tv-3e
Installation	37000 DWT Chemical Product Tanker
Plant Builder	
Engine Type	6RT-flex50
Date	10.04.2008
Calculated by	M. Barraud
Company	Wärtsilä Switzerland Ltd
Address	CH-8401 Winterthur

Main installation data:

Engine type	: 6RT-flex50 (Eng.no. 1)
Type	: 2-stroke
Cyl.	: 6 In-line
Bore	: 500.0 mm
Stroke	: 2050.0 mm
Conn. rod ratio	: 0.500
Osc. mass	: 3330.0 kg
P(CMCR)	: 9720.0 kW
Speed	: 124.0 rpm
M.E.P.	: 19.47 bar
Excitation data	: rt_49a.tan
Cranks	: No. 1 to 6 at masses no. 8 to 3
Crank pin	: 600.0 mm (at # 3)
Sense of rotation	: Clockwise
Ahead firing order	: 1 - 6 - 2 - 4 - 3 - 5 Regular

Minimum shafting diameters:

Intermediate shaft	: 410.0 mm (at # 11)
Propeller shaft	: 570.0 mm (at # 12)
Modulus of rigidity	: 81.4 kN/mm2
Calculation option	: Including variable crank inertia
Diagram group no.	: 6rtflex50-tv-3e

Disclaimer note:

The enclosed calculation has been carried out by highly qualified specialists using thoroughly checked and well proven advanced computer programs.
All input data relating directly to Wärtsilä engines have been prepared with the utmost care and precision. However, for data related to the installation components such as dampers, elastic or hydraulic couplings, gears, generators, propellers, elastic mountings etc. WARTSILA SWITZERLAND LTD has to rely on input data furnished by the technical documentation of the different suppliers of the installation components. WARTSILA SWITZERLAND LTD cannot guarantee the correctness of these data and therefore excludes any warranty, liability or responsibility whatsoever as to damage, failure or malfunction of the installation and consequences thereof due to wrong input data furnished by or resulting from the above mentioned technical documentation of the suppliers or any estimated data.

Visual TORSVIB © WCH V4.4 #7056-3082737 Wärtsilä Switzerland Ltd

▼

A schematic overview of the crankshaft with all the cylinders.

1 *left the Geislinger vibration damper*
2 *with crankshaft flange*
3 t/m 8 *cylinders*
9 *thrust block*
10 *flywheel/turning wheel*
11 *intermediate shaft*
12 *propeller shaft*
13 *Fixed Pitch propeller, four blades*

The shaft diameters are in millimetres, the mass-inertia moment *J* in kgm² and H in rad/Nm.

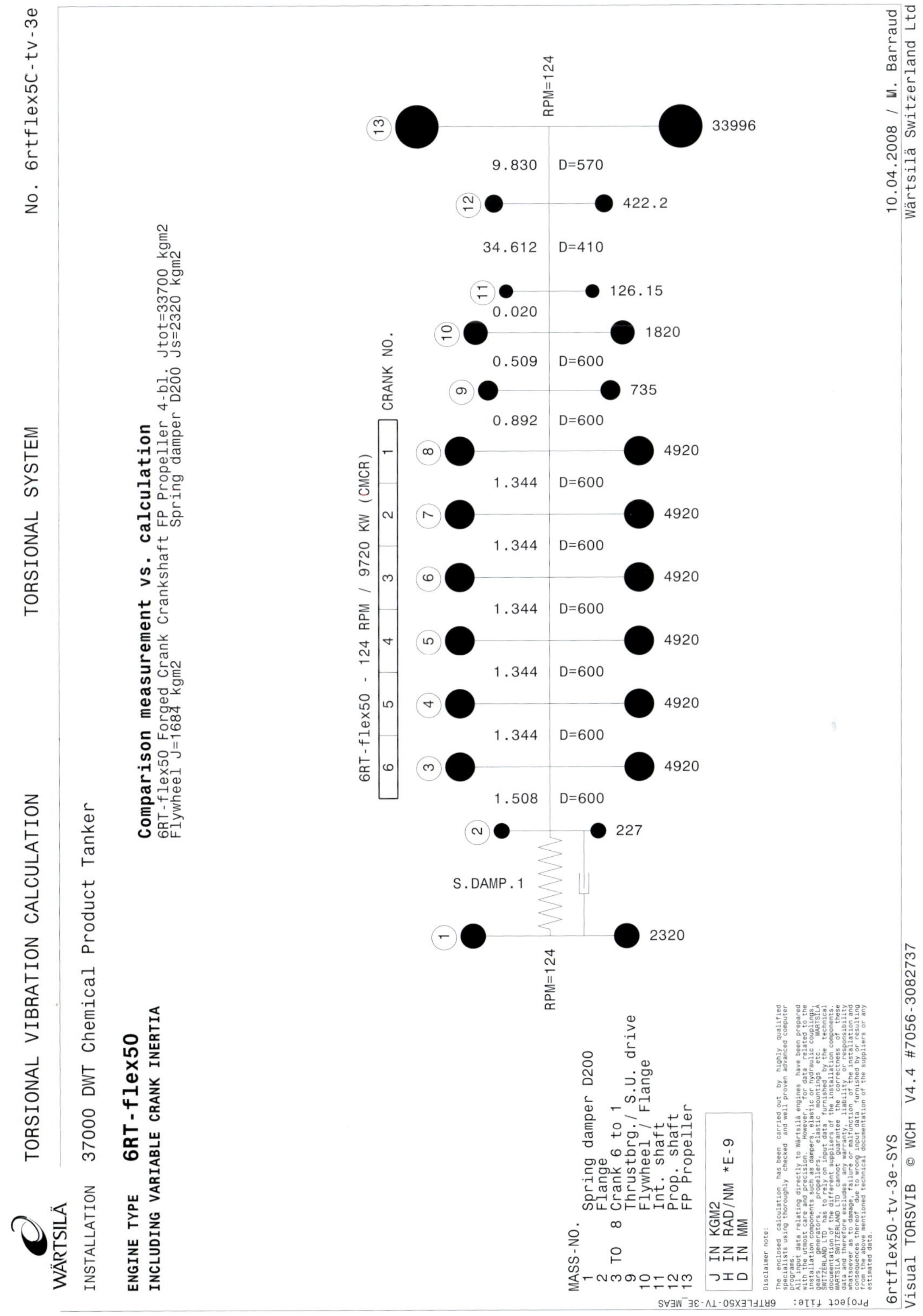

▼

A comparison between the actual measurements and simulated calculations.

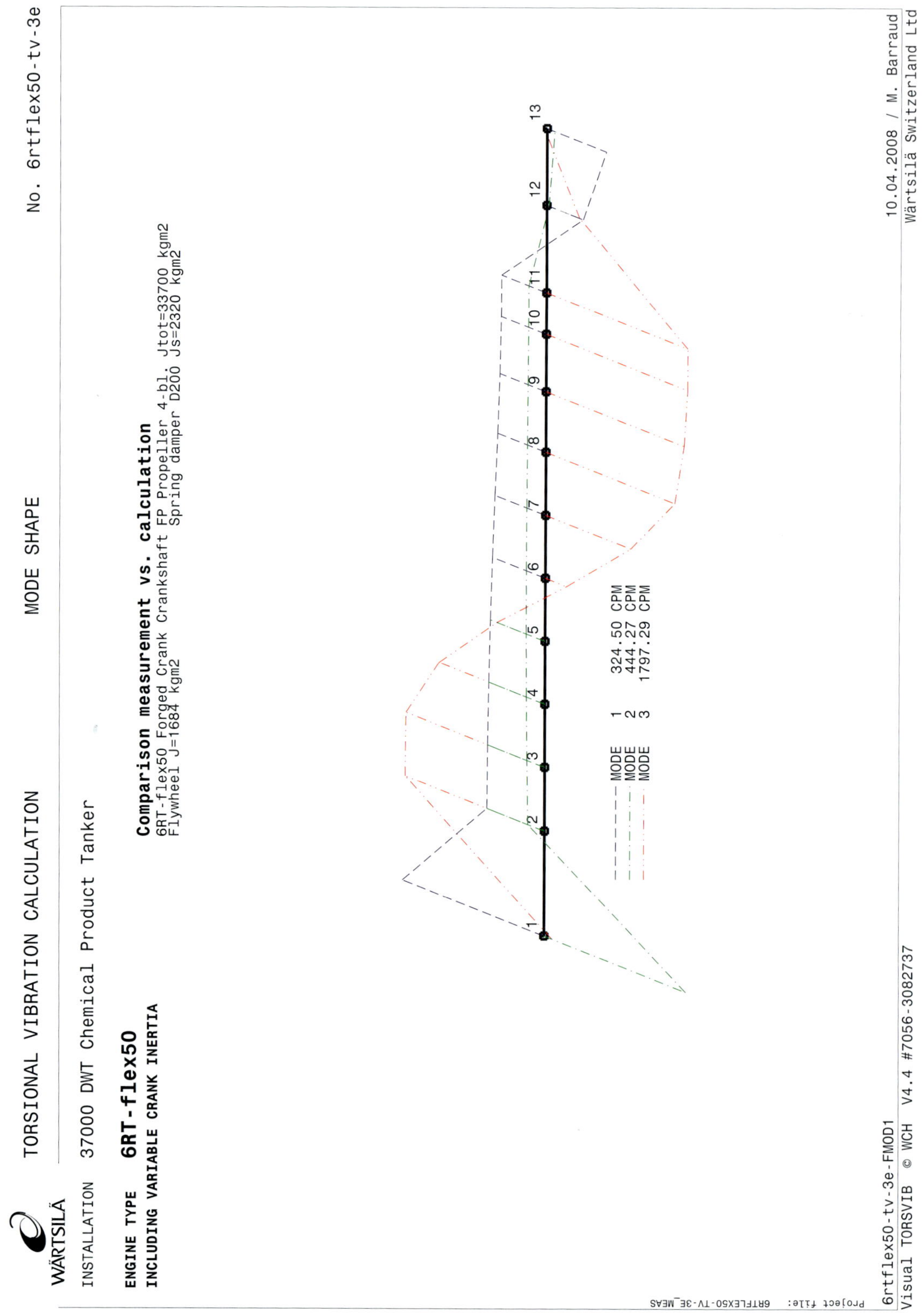

▼

Normal combustion of all six cylinders.

Horizontal: speed in revs/min.

Vertical: alternating torsional stress in N per mm².

First graph: crankshaft D = 600 mm.

Second graph: intermediate shaft D = 410 mm.

Third graph: propeller shaft D = 570 mm.

In the figures, the limits are set by classification societies, such as ABS (American Bureau of Shipping) and IACS (International Association of Classification Societies), the umbrella organisation of classification bureaus.

Only at a speed of 124 revs/min.

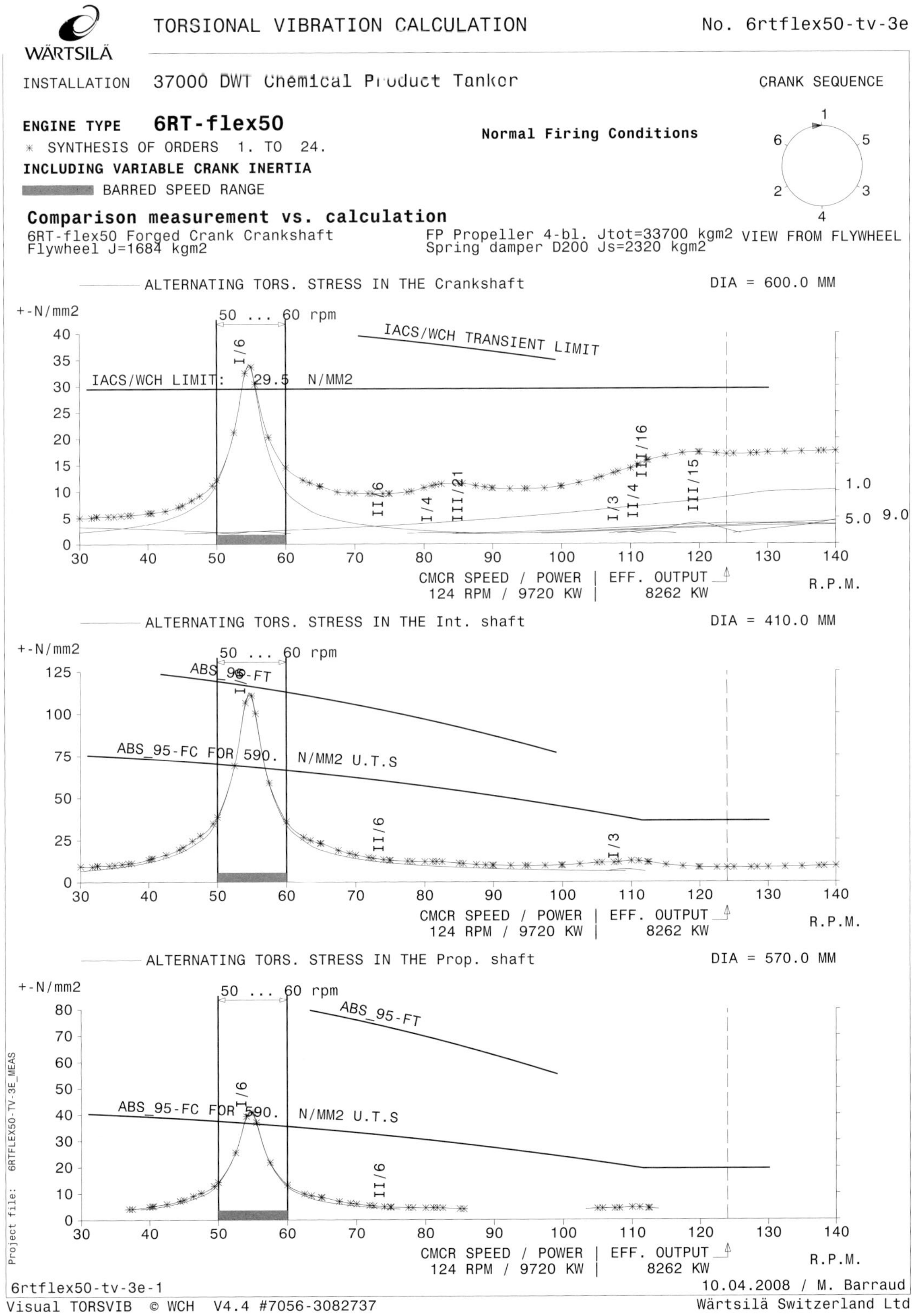

First graph: crankshaft.

At 124 revs/min., the torsional stress is ± 17.5 N/mm², the limit is ± 29.5 N/mm².

Second graph: intermediate shaft.

At 124 revs/min., the torsional stress is ± 8 N/mm, the limit is ± 37 N/mm².

Third graph: propeller shaft.

At 124 revs/min., the torsional stress is ± 5 N/mm², the limit is ± 20 N/mm².

The barred area lies between 50 and 60 revs/min for this engine.

In this area, the torsional stresses are too high.

▼

With 6 operating cylinders.

Horizontal: the speed in revs/min.

Vertical: the angular amplitude in crank degrees.

First picture:

The angular amplitude at the flange of mass 2.

First degree, sixth order.

Green and red: Measured values, different degrees and orders.

Second picture:

The dynamic angle of twist in the crankshaft between mass numbers 2 and 10.

Red: sixth order

Third picture:

The dynamic angle of twist in the crankshaft between mass numbers 2 and 10.

Red: third order

Green: fourth order

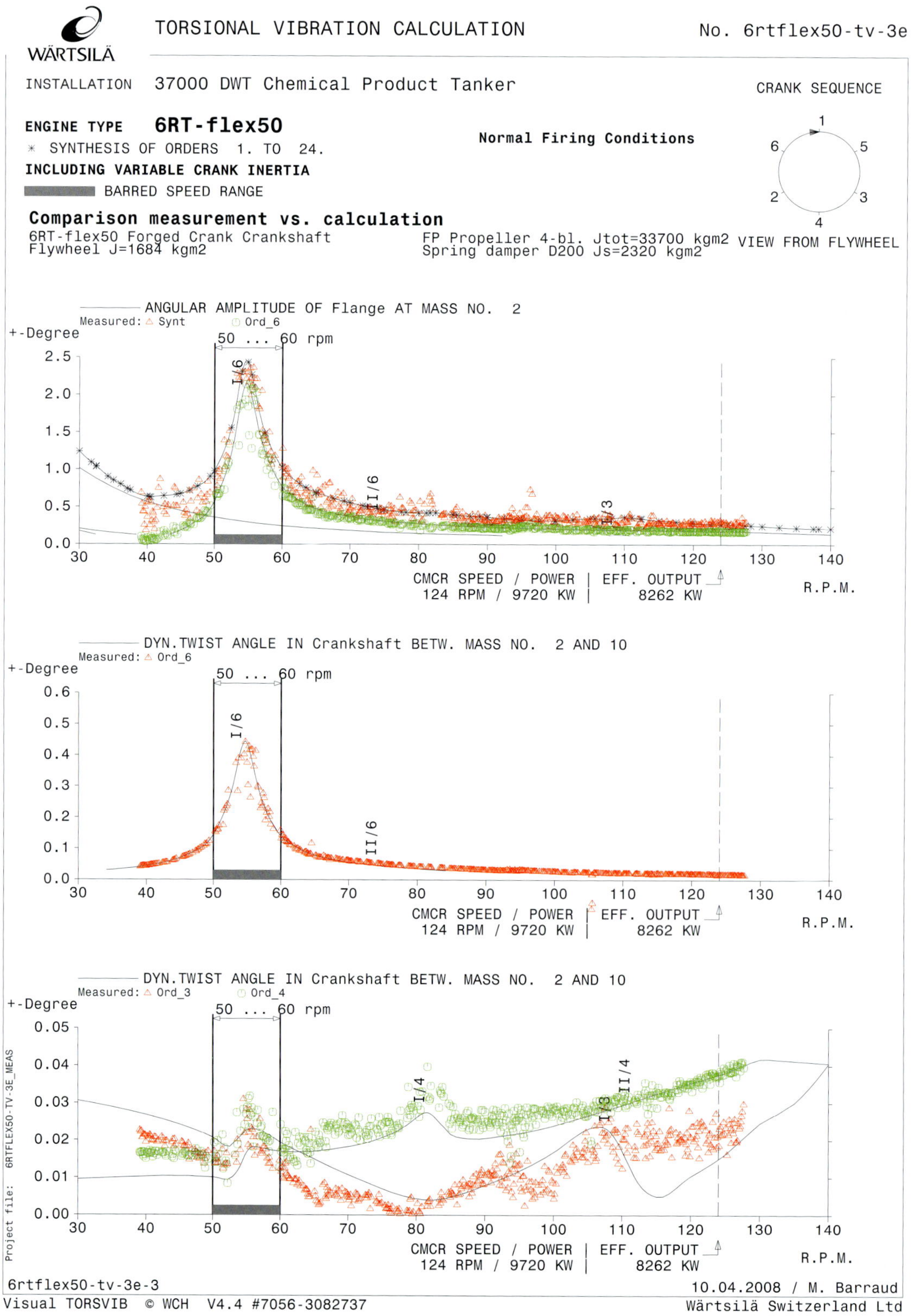

17.27 Example 2b: Simulation on five cylinders

▼

Simulation: In the simulation, one cylinder is switched off; the engine is now operating on five cylinders and the sixth cylinder bumps along. It has compression but no combustion. The other five cylinders are now producing the same power as in the first case. The total power remains the same.

With a FP propeller, the relationship between power and speed according to the propeller with P (kW) is equal to n^3 (rpm). The engine must supply the power demands of the propeller irrespective of the number of operating cylinders.

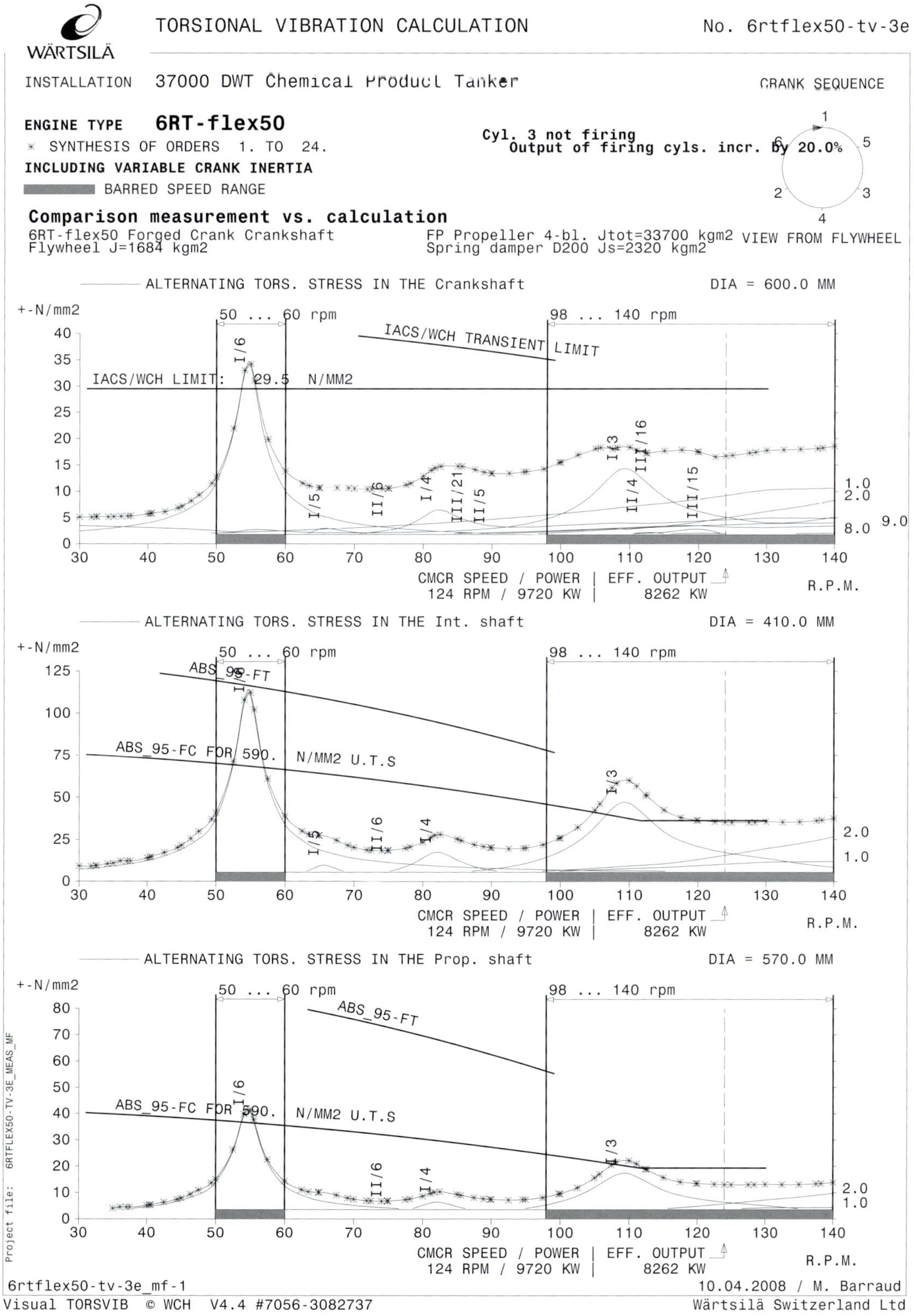

First picture: The generated torsional stresses in the crankshaft. A second barred speed area is present; above 98 revs/min. the engine may not be operated.

Second picture: The generated torsional stresses in the intermediate shaft. At approximately 110 revs/min., these are greatest and exceed the ABS limit.

Third picture: The generated torsional stresses in the propeller shaft. Here, at 110 revs/min. a maximum value is exceeded.

Conclusion: The engine can only operate at full power when the speed is reduced from 124 to 98 revs/min.

Conclusion

This engine can still supply full power to the propeller with five cylinders in service and a reduced engine speed.
The engine now has two barred engine speed ranges: of 50 to 60 revs/min and above 98 revs/min.

▶
Simulation (continued).

Fourth picture:
The angle of twist of the flange at mass 2.
Sixth order – green.

Fifth picture: coloured
The angle of twist in the crankshaft between mass 2 and mass 10.
Sixth order – red.
No problems at high speeds.

Sixth picture; coloured
The angle of twist in the crankshaft between mass 2 and mass 10.
Third order – red
Fourth order – green

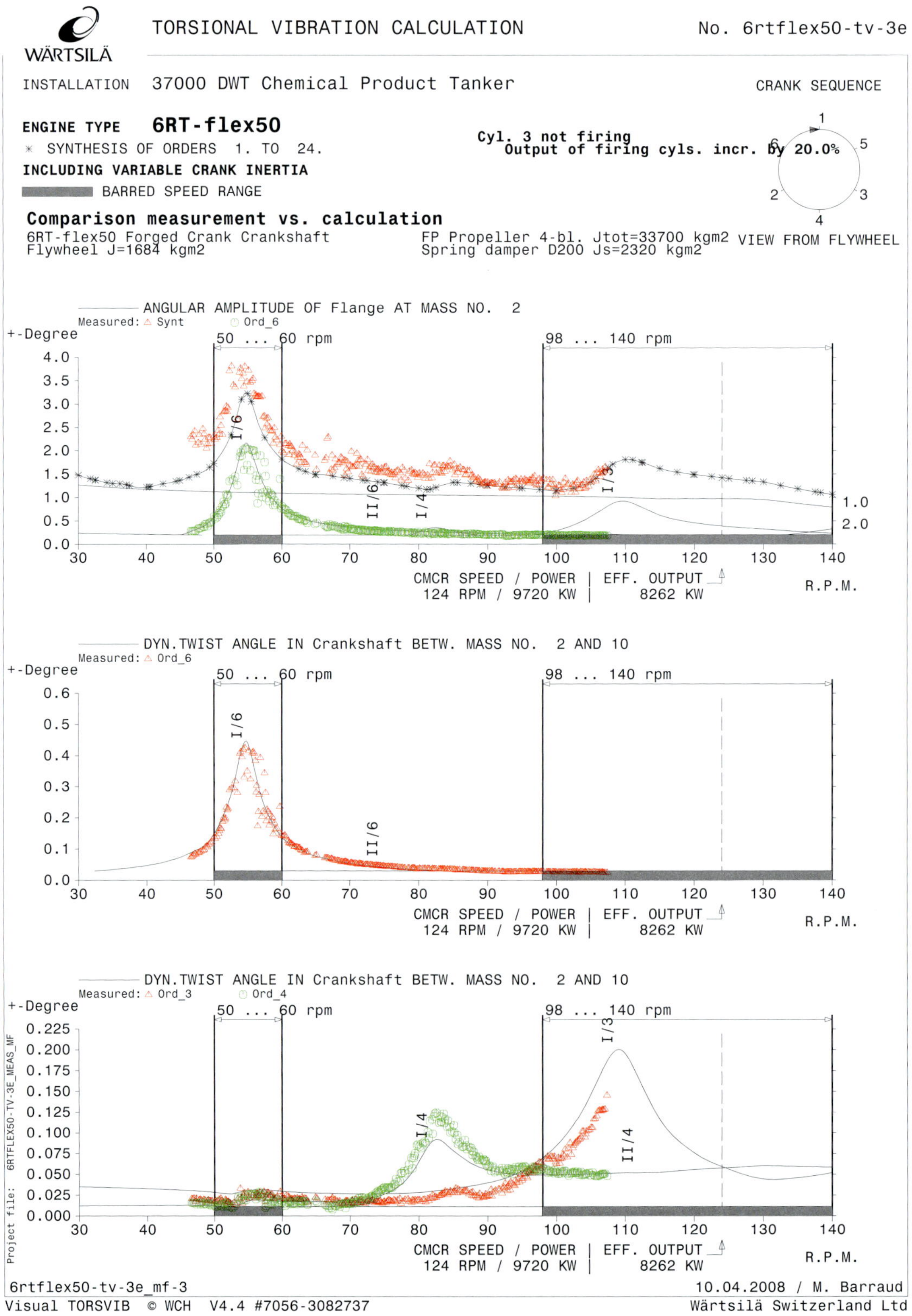

17.28 Example 3: Simulation of two engines and one propeller shaft

This involves a ferry with the following propulsion system, two medium speed diesel engines driving via pitch propeller via gearing.

Engine I:
A nine-cylinder four-stroke diesel engine.
Manufacturer, Wärtsilä Sulzer 9 ZA 40 S.
6480 kW shaft power at a speed of 510 rpm/min.
This is the starboard engine.

Engine II
A twelve-cylinder four-stroke diesel engine.
Manufacturer, Wärtsilä Sulzer 12 ZA 40 S.
8640 kW shaft power at a speed of 510 rpm/min.
This is the port engine.

WÄRTSILÄ

TORSIONAL VIBRATION CALCULATION

Calculation No.	TV-Twin_ZA40S
Installation	Twin engine propulsion for Ferry
Plant Builder	
Engine Type	9ZA40S + 12ZA40S
Date	10.05.1998
Calculated by	H. Keller
Company	Wärtsilä Switzerland Ltd
Address	CH-8401 Winterthur

Main installation data:

Engine type 1	: 9ZA40S (Eng.no. 1) STB Eng.
Type / Cyl.	: 4-stroke / 9 cyl. In-line
P(CMCR) / Speed	: 6480.0 kW / 510.0 rpm
Excitation data	: za40s_1a.tan
Cranks	: No. 1 to 9 at masses no. 11 to 3
Sense of rotation	: Clockwise
Ahead firing order	: 1 - 5 - 9 - 3 - 6 - 8 - 2 - 4 - 7 Regular

Engine type 2	: 12ZA40S (Eng.no. 2) PS Eng.
Type / Cyl.	: 4-stroke / 12 cyl. Vee
P(CMCR) / Speed	: 8640.0 kW / 510.0 rpm
Excitation data	: za40s_1a.tan
V-angle	: 50.00 degr.
Firing dist.	: 50.0 degr. (Consequtive)
Cranks	: No. 1 to 6 at masses no. 25 to 20
Sense of rotation	: Clockwise
Ahead firing order	: 1 - 4 - 5 - 6 - 3 - 2 Regular

Minimum shafting diameters:

Intermediate shaft	: 423.0 mm (at # 42)
Propeller shaft	: 491.0 mm (at # 44)

Modulus of rigidity: 81.4 kN/mm2

Diagram group no. : TV-Twin_ZA40S

Disclaimer note:

Visual TORSVIB © WCH V4.4 #7056-3082737 — Wärtsilä Switzerland Ltd

◄ Picture 1.

The data for both engines at nominal power. MCR: Maximum Continuous Rating.

▼

Picture 2.

A schematic overview of both engines, on top, the 9-cylinder engine and below, the 12-cylinder engine.
To the right, the reduction gearbox for lowering the speed of the adjustable-pitch propeller with underneath the coupling to the reduction gearbox for increasing the speed of the shaft generator.

The shaft diameters *D* are given in mm, the moments of mass inertia, *J* in kgm^2 and angle of twist, *H* in rad/Nm.

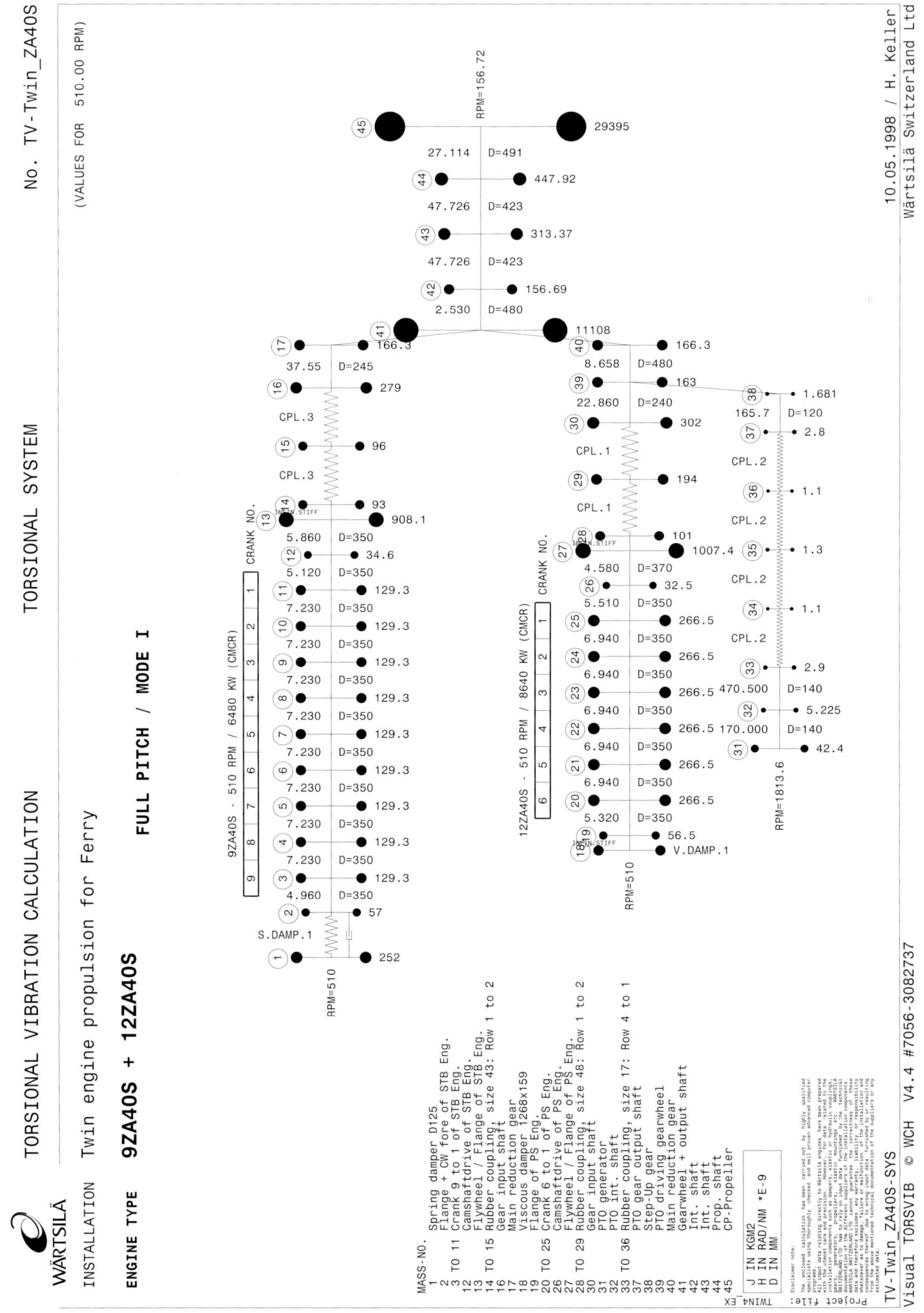

Picture 3.
A simulated malfunction in one of the cylinders.

A calculation is made based on the assumption that cylinder 9 in the starboard engine, the nine-cylinder engine, has no fuel injection and therefore no combustion- and power stroke. The engine runs on eight cylinders. The shaft generator is idling and at approximately 455 revs/min normally loaded.

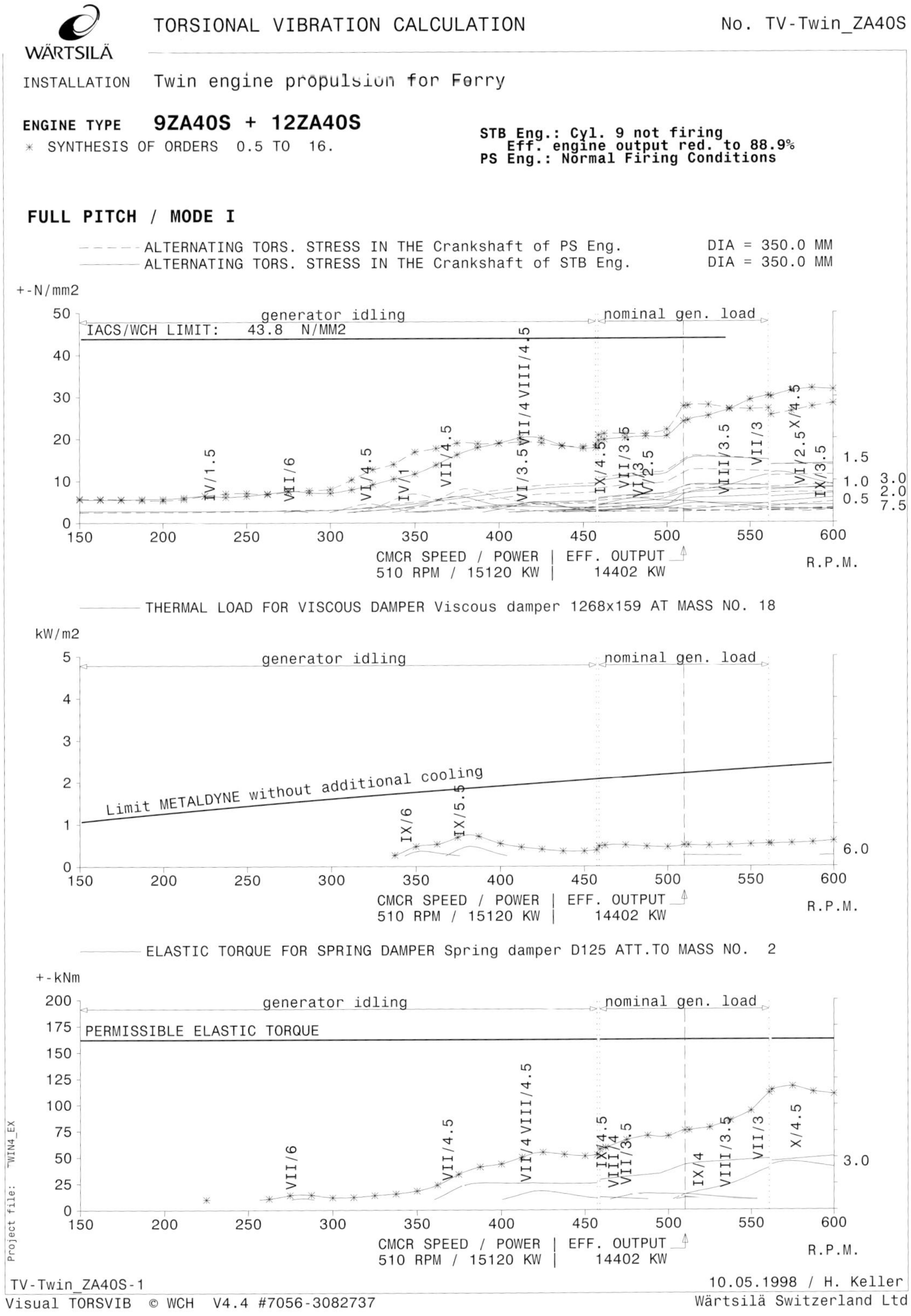

Top picture

First graph, the torsional tension for both crankshafts.

Horizontal: speed in revs/min.

Vertical: the torsional tension for the crankshaft of the port engine in N/mm2 (dotted line) and in the starboard engine (solid line).

The engines now have a total capacity of 14,402 kW at a load of 88.9% MCR.

The maximum permissible torsional tension for both crankshafts is 43.8 N/mm^2.

At 510 revs/min the torsional tension for the port engine is approximately 27 N/mm^2 and that for the starboard engine approximately 15 N/mm^2.

Both well below the maximum value.

Centre picture

The thermal load for the viscous damper.

Horizontal: speed in revs/min.

Vertical: heat development in kW/m^2.

The maximum value for the viscous liquid, here metadyne, may not exceed 4.8 kW/m^2 at 510 revs/min. Here it is considerably lower: +/- 2.2 kW/m^2.

Bottom picture

The permissible elastic torque for the spring damper.

Horizontal: speed in revs/min.

Vertical: torque in kNm.

Maximum permissible value +/- 190 kNm.

The true value is +/- 75 kNm at 510 revs/min and 14,402 kW, well below the maximum value.

Picture 4.

Top picture

Torque for spring damper no. 2.
Horizontal: speed in revs/min.
Vertical: torque in kNm.
The permissible torque is +/- 310 kNm.
The torque is a mere +/- 40 kNm at 510 revs/min.

Centre picture

The thermal load for spring damper no. 2
Horizontal: speed in revs/min.
Vertical: generated heat.
The generated heat is +/- 8 kW at 510 revs/min, well below the maximum value.

Bottom picture

The thermal stability for spring damper no. 2.
Horizontal speed in revs/min.
Vertical: generated heat in Nm/sec.
The maximum value for the thermal stability is - 0.45 W/Nm/sec. at 510 revs/min. The value is + 0.10 and therefore well below maximum value.

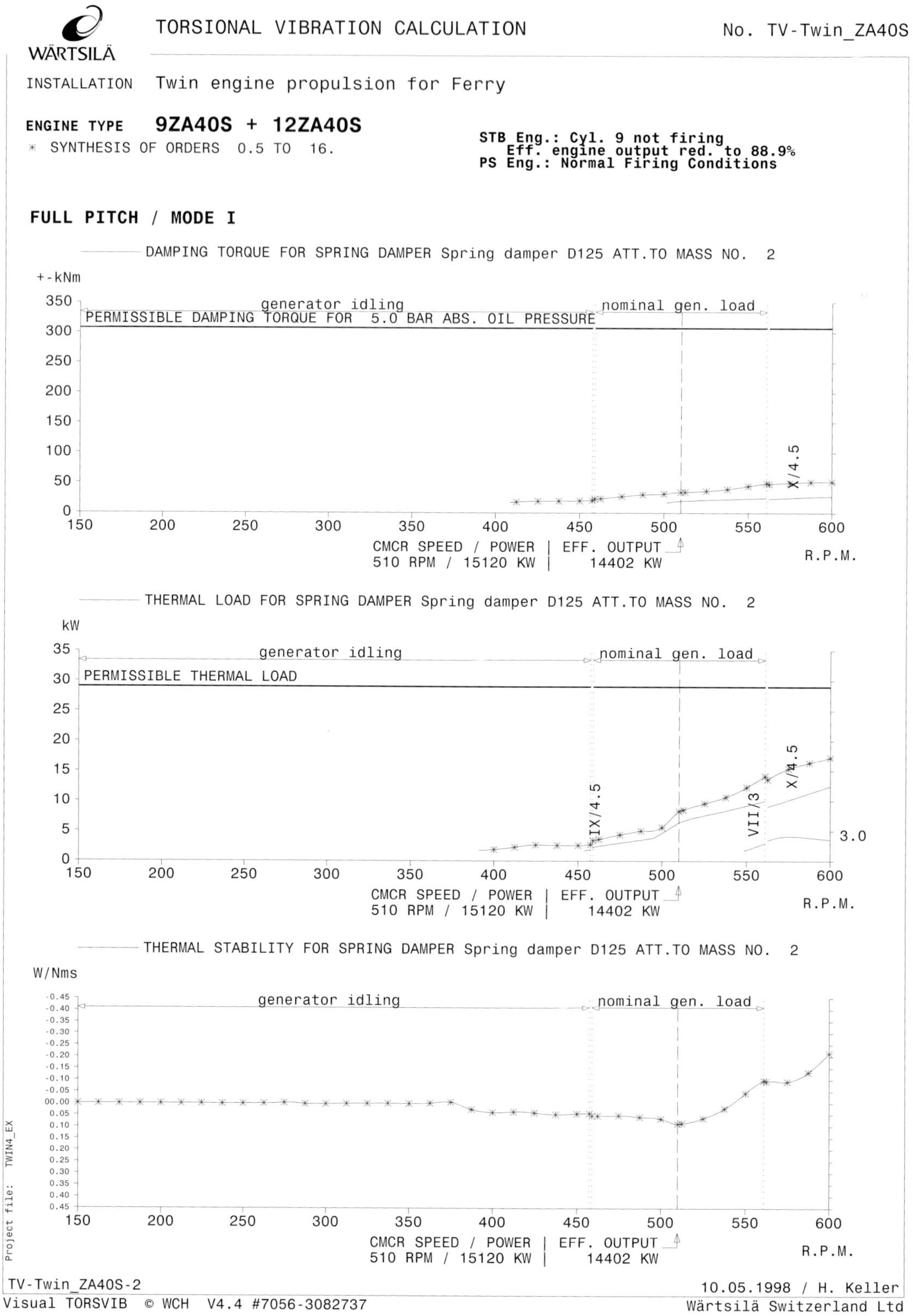

▼

Picture 5.

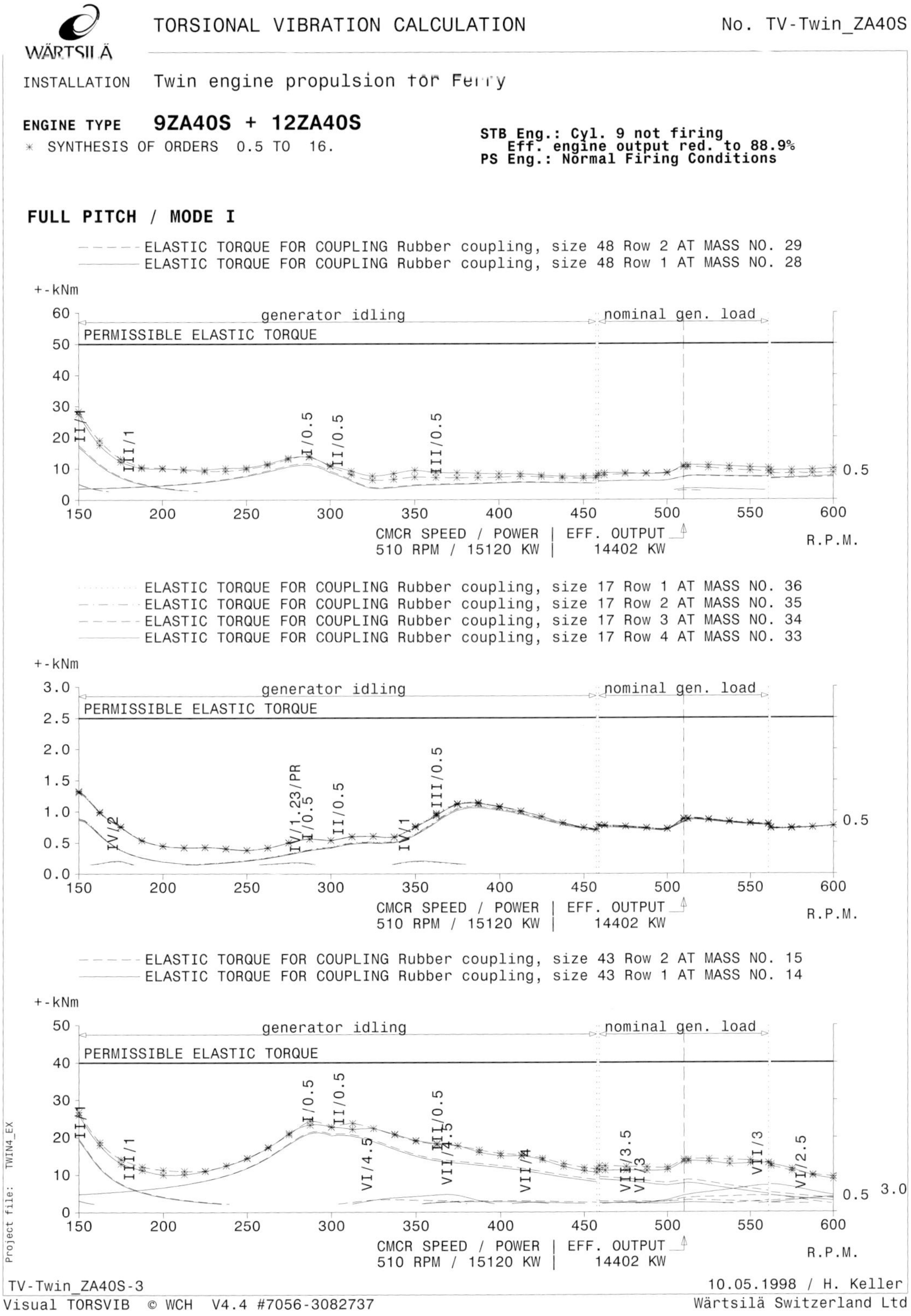

Top picture

The elastic torque for rubber couplings no. 28 and no. 29.
Horizontal: speed in revs/min,
Vertical: the elastic torque in kNm.
The maximum permissible torque is 50 kNm.
The elastic torque has a value well below the maximum value at 510 revs/min at +/- 10 kNm.

Centre picture

The elastic torque for rubber couplings no. 33, 34, 35 and 36.
Horizontal: speed in revs/min.
Vertical: the elastic torque in kNm.
The maximum permissible elastic torque is 2.5 kNm.
The elastic torque for four couplings is +/- 0.9 kNm at 510 revs/min, well below the maximum value.

Bottom picture

The elastic torque for rubber couplings no. 14 and no. 15.

Horizontal: speed in revs/min.
Vertical: the elastic torque in kNm.

The maximum permissible elastic torque is 40 kNm.
In the graph, it is +/- 13 kNm, well below the maximum permissible torque.

Picture 6

Conclusion: no special measures are required as at a normal speed of 510 revs/min, one malfunctioning cylinder only effects a power reduction of 6,480 + 8,640 = 15,120 kW. to 14,402 kW.

Top picture

The angle of twist for shaft generator no. 2.

Horizontal: the speed in revs/min.

Vertical: the angle of twist for generators in degrees. The maximum value for the generator shaft, so for the generator rotor, it is +/- 3.75 degrees.

At a speed of 510 revs/min this is minimally below the maximum value.

Centre picture

Torque in reduction gearbox no. 38, which increases the speed of the shaft generator.

Horizontal: speed in revs/min.

Vertical: torque- and gear hammering expressed in kNm for torque.

The maximum permissible value is +/- 1.75.

At 510 revs/min the value is +/- 1 kNm, thus dropping below the maximum permissible value.

Bottom picture

Torque in reduction-gear box no. 40.

Horizontal: speed in revs/min.

Vertical: torque in kNm.

The maximum permissible value varies from +/- 15 to +/- 48 kNm. At +/- 16 kNm, the values are much lower.

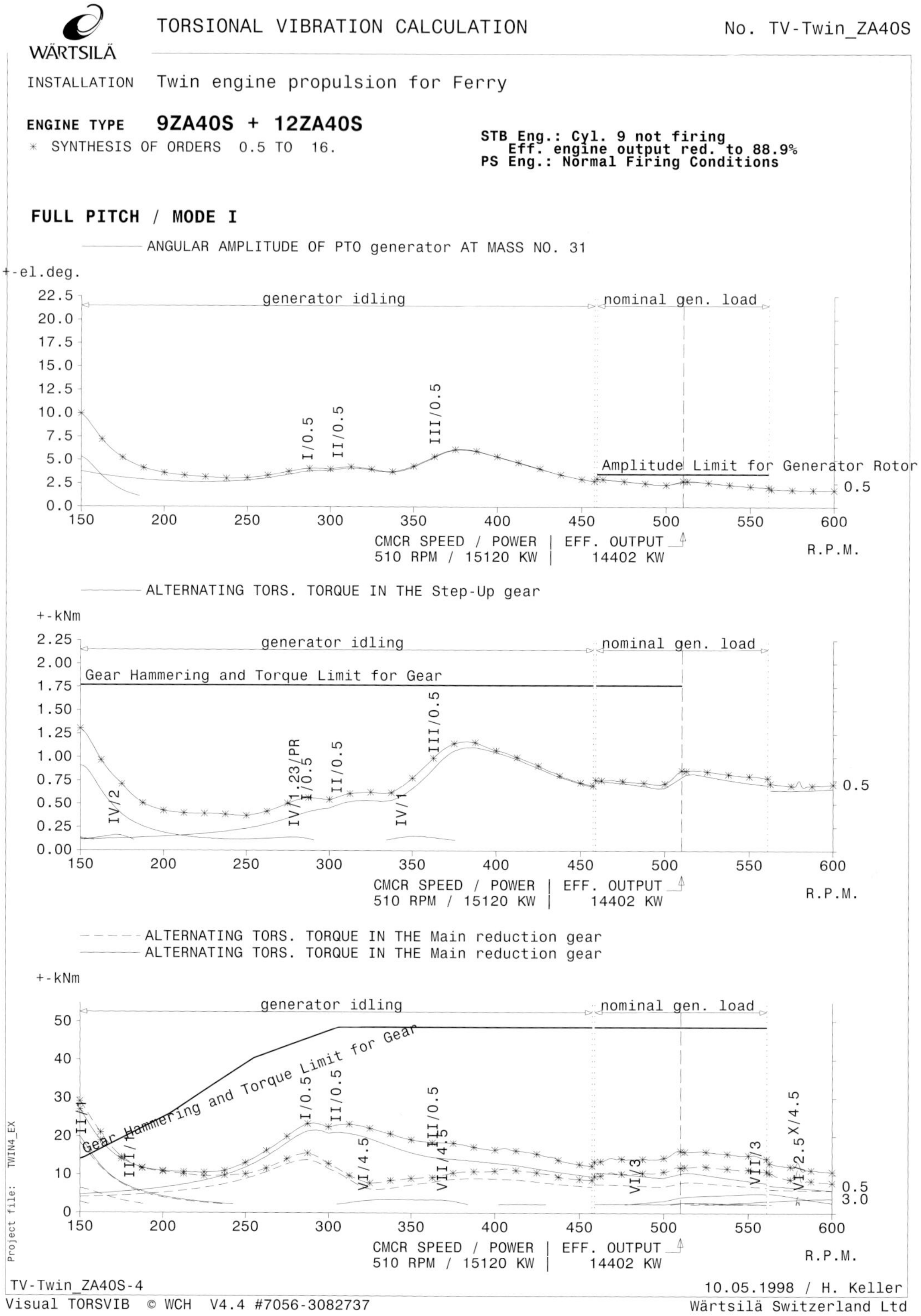

17.29 Examples of engine-frame tearing

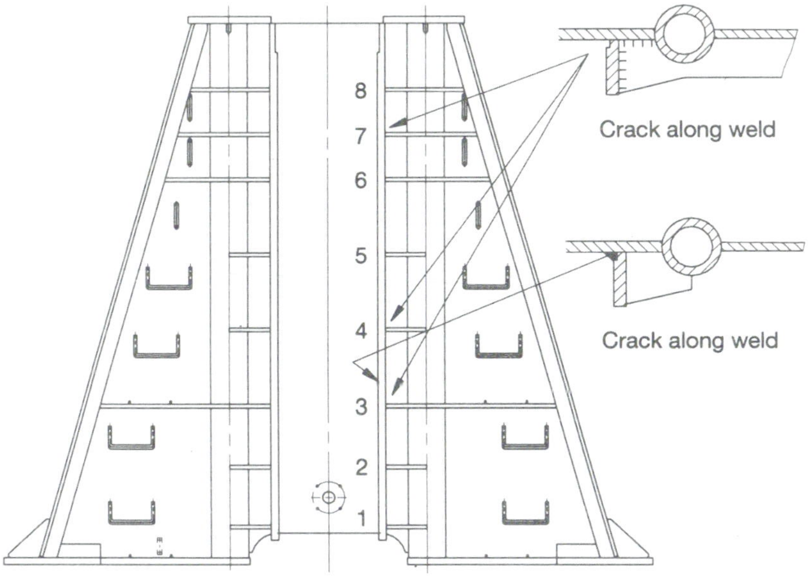

▲

Cracks in the welded plate-steel A-frame in MAN Diesel MC engines.

This problem can be solved in time.

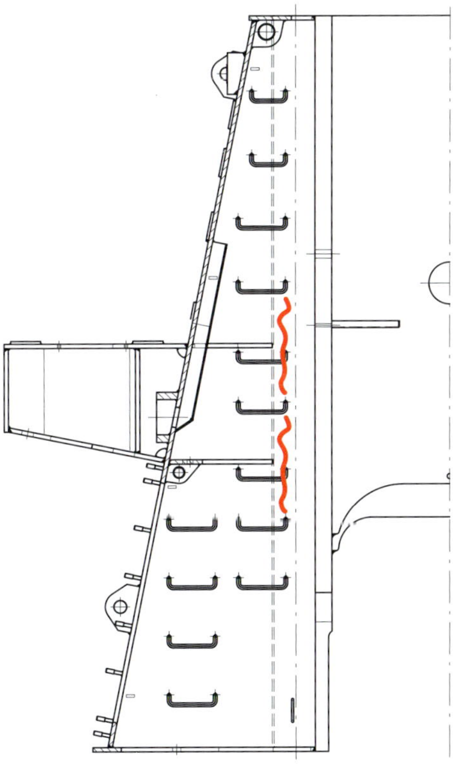

▲

Cracks in the welded plate-steel A-frame in a Wärtsilä Sulzer RTA96C two-stroke crosshead engine.

This is a familiar problem for all two-stroke engine manufacturers and is usually solved in time with the introduction of a new engine type that has small changes in the construction.

◄▼

Cracks in the A-frame of a two-stroke crosshead engine.

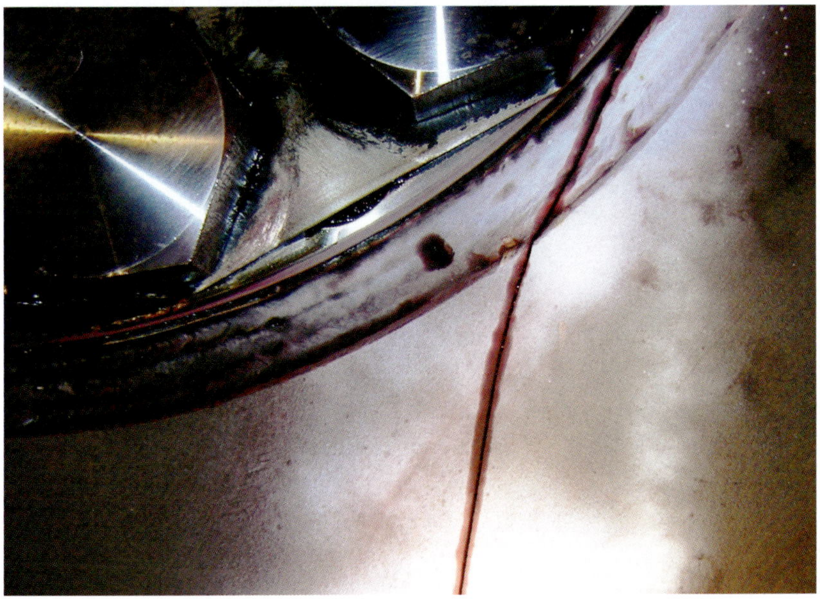

▶

Measuring the torsional vibration for a Geislinger vibration damper.

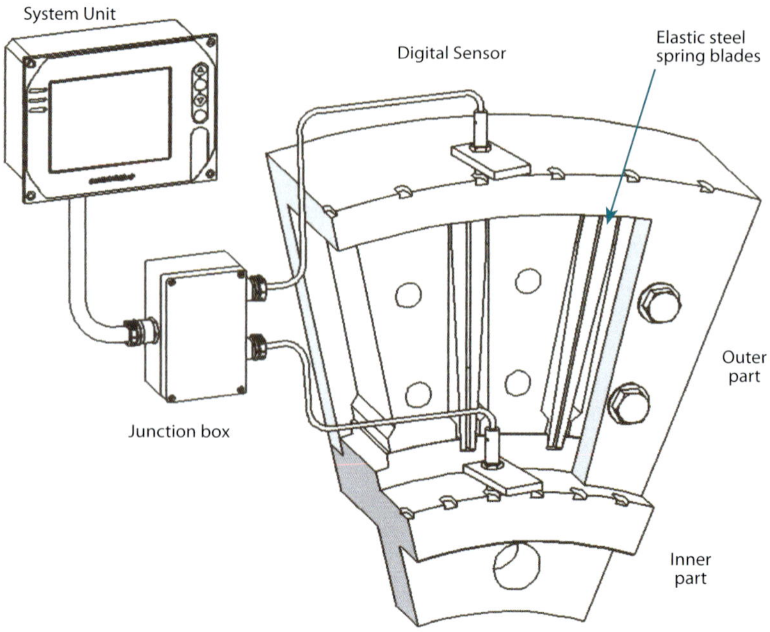

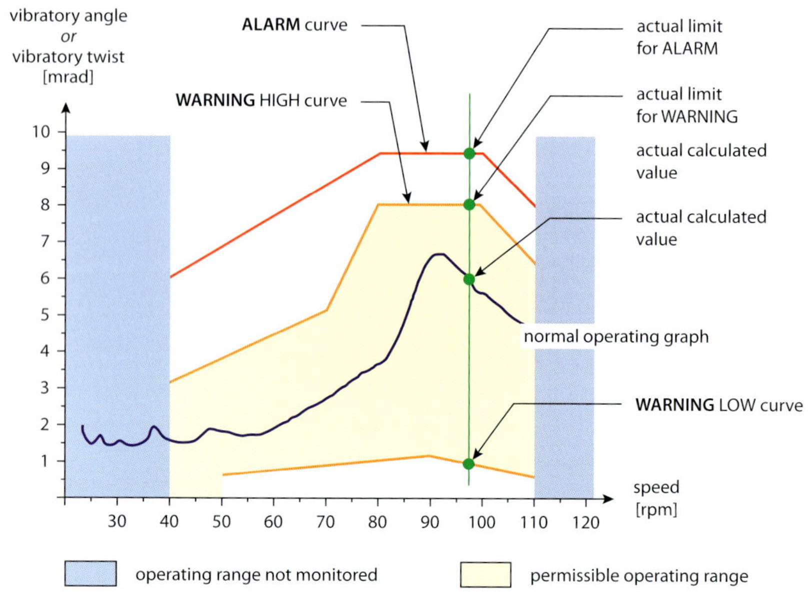

▲

Measurements for the Geislinger vibration dampers.

In modern large two-stroke crosshead engines that use Geislinger vibration dampers, an LCD display can be supplied that monitors and measures torsional-vibration amplitudes.

The system monitors and shows:

- the actual torsional vibration;
- an alarm signal when values are too low;
- the pre-alarm limit value for a value that is too high;
- the alarm limit value for a value that is too high;
- the actual calculated value;
- the normal operational value.

17.30 Measurements for vibration dampers

Vibration dampers serve to reduce the generated torsional vibrations of the engine to an acceptable level. By applying modern measuring techniques, it is possible to determine in which vibration area the damper is effective.

17.31 Measuring equipment

Today, measuring equipment that meticulously registers the engine vibrations of a complete system, the so-called torsion diagram, is available.

Most torsion diagrams show a waveform measurement, which is a result of the fact that the shaft does not rotate at a constant angular velocity, but continuously decelerates and accelerates. This is also referred to as the non-uniformity degree δ, of the engine.

δ = non-uniformity degree of the engine

$$\delta = \frac{n_{max} - n_{min}}{n_{mean}}$$

This expression is as old as the internal combustion engine.

Torsional vibrations are unavoidable; this is why many engines are equipped with vibration dampers.

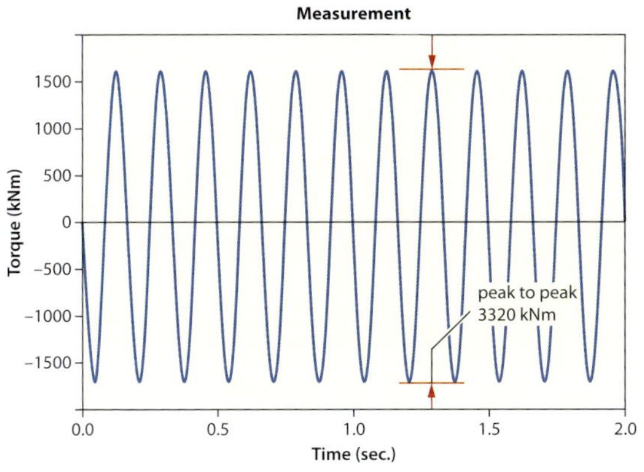

▲

A torsiogram for a five-cylinder two-stroke MAN Diesel crosshead engine, type 5L50MC, at approximately 72 revs/min.

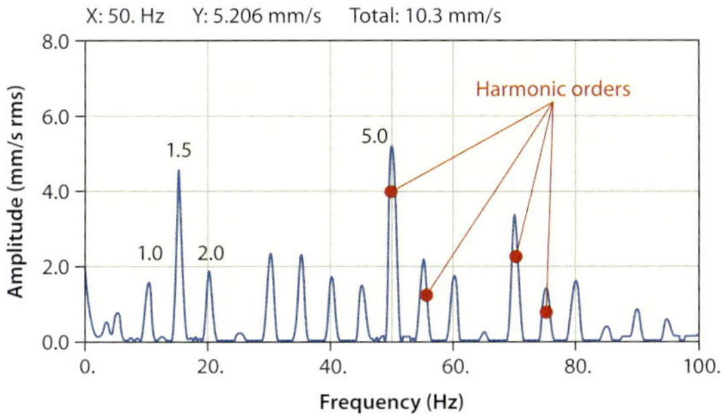

▲

A torsiogram for a four-stroke diesel engine.

Harmonic orders are shown: 1.0, 1.5, 2.0 and 5.0.

17.32 Vibrational energy

To create and maintain vibrations in a diesel engine, however negative, energy is required. This energy will then not be available for shaft power. Part of this energy is absorbed by the previously discussed vibration dampers.

In large engines, the lubricating oil from the main lubricating-oil system is used to cool the steel-spring type torsional vibration dampers, such as those of Geislinger.

In a large twelve-cylinder two-stroke crosshead engine type Wärtsilä Sulzer RTA84T, 80 kW is the maximum power requirement.

In comparison with the maximum shaft power of 45,000 kW, this is approximately 0.06%.

When taking into consideration that not all the vibrations are absorbed by the vibration damper, but only approximately 50%, the vibration energy amounts to a maximum of around 0.1% of the shaft power.

Therefore, small vibration dampers do not require cooling. A rapidly turning damper can easily discharge its heat to ambient air.

17.33 Design of a propulsion installation

The natural (vibration) frequency of the system is calculated to determine at which engine speeds the various critical areas lie and the magnitude of the created stresses.

If the exerted stresses are high, or if this area is too close to the regular engine speed, then the critical area is shifted by adjusting the natural vibration frequency of the engine or the system.

17.34 Effects of vibrational frequencies

The following formula can be used for the calculation of the vibrational frequency for a single-mass system.

$$n_{tr} = \frac{1}{2 \times \pi} \sqrt{\frac{G \times I_p}{l \times J}}$$

Where:

n_{tr} = the natural vibration frequency
G = shear modulus of the material
I_p = the polar inertia moment
l = the length of the shaft
J = the mass inertia moment

The following factors influence the vibration frequency:

- the strength of the material, G, the shear modulus;
- the shaft diameter, the polar inertia moment;
- the length of the shaft, l;
- the mass inertia moment, I_p of the rotating mass.

In practice, the following points may be adjusted:

- The mass inertia moment, J, of for instance, the propeller, flywheel and extra masses.
- The polar inertia moment, I_p.

In the formula of the inertia moment, the shaft diameter ($I_p = \frac{1}{32} \times \pi \times D^4$) is to the fourth power and the mass of the rotating parts to the first power.

Therefore, an increase in the shaft diameter is the easiest solution. If the shaft diameter is increased, the natural vibration frequency and the critical area of the system are moved to a higher engine speed.
In an existing propulsion installation, one of the few ways to reduce the vibrational frequency is to install a propeller with a smaller moment of inertia.
In this approach, the propeller has a slightly smaller diameter. In order to achieve an identical displacement, the propeller pitch must be increased.

17.35 Flywheel

17.35.1 Introduction

Virtually all engines are provided with a flywheel or the combination flywheel - turning wheel.
Engines with a small number of cylinders have in proportion a heavy 'thick' flywheel. Multiple-cylinder engines, such as V-engines usually have a light 'thin' flywheel.

17.35.2 Functions of the flywheel

1 Overcoming compression power

A single-cylinder four-stroke engine has two revolutions in one combustion process. In order to start this engine manually, a compression lever is used. This lever opens the exhaust valve, thus preventing compression so enabling 'elbow steam to achieve a certain speed of the moving engine parts, including the relatively heavy flywheel.
If the cranking is then discontinued and at the same time the compression lever is depressed, the exhaust valve is closed, the piston may compress the air. The fuel is injected at the end of the stroke and directly after TDC, the power stroke is initiated.
The energy required to compress the piston is generated by the moving parts of the engine, especially by the flywheel. Therefore, this flywheel serves as kinetic energy storage for starting the engine.

2 Levelling irregular engine speeds

In principle, a single-cylinder four-stroke engine has an irregular speed. There are three piston strokes that cost energy, and only one stroke that provides energy.
By placing a flywheel on the crankshaft, the fluctuating engine speed is levelled off.
Accelerating and decelerating the rotating mass takes time and the piston strokes follow each other rapidly at a normal engine speed, thus producing a more stable engine speed.
The flywheel, then, serves as energy storage.
The flywheel is no longer required to start the engine in multiple-cylinder engines. These are started electrically, mechanically or with air, but it assures a more regular engine speed.

3 Additional functions of the flywheel: use as a turning wheel

This can be manually performed in smaller engines, in which case the flywheel has holes that allow the turning of the crankshaft using a turning rod during inspections and repairs.
In larger engines, the flywheels are equipped with a gear ring to which a turning engine can be connected. This usually works on electricity or air.
Many starting engines working on compressed air have two circuits – one for extremely slow running or cranking, and one for rapid turning to start the engine.

4 Measuring crank positions

The final important function, of course, is to determine with the use of markers on the flywheel the TDC and BDC position of the cylinders.

17.35.3 Examples of types of flywheel versions

▲
An eight-cylinder four-stroke diesel engine from ABC in Ghent, Belgium, installed in a trawler in winter. The weight of the flywheel is found mainly in its circumference.

◄

A turning-wheel/flywheel in a two-stroke MAN Diesel 50MC crosshead engine.

The disk is actually more like a turning wheel with shown left, the electric turning motor. There are holes in the turning-wheel for balancing the crankshaft.

◄

A large flywheel for a six-cylinder four-stroke Caterpillar MaK diesel engine.

In the flywheel, the so-called turn holes are used to crank this category III engine manually.

◄

The flywheel of a four-stroke in-line engine, manufacturer MAN Diesel.

The flywheel is quite large. In the middle, the material has been removed as far as possible to ensure that the load on the crankshaft bearings is as low as possible. Right; next to the flywheel, the electric turning motor and below, acting on the same starting-ring gear, a starting engine operating on compressed air.

1 *flywheel mass*
2 *removed section*
3 *fixing bolts on crankshaft*
4 *turning motor*
5 *starting motor*

▶
A four-cylinder four-stroke Kromhout diesel engine in a pumping station.

As this engine has only four cylinders and a relatively low speed (600 revs/min), a large flywheel is necessary. The engine is started electrically with a starting motor attached to an accumulator battery.

▶
A small two-cylinder four-stroke diesel engine, a Volvo MD 2010, often used in yachts.

The large flywheel is mounted between the engine and the reduction gearing.

1 *engine block*
2 *flywheel casing*
3 *reduction gear*

17.35.4 The mass-inertia moment of flywheels

The formula for the mass-inertia moment of a disc:

$$J = \frac{1}{2} \times m_{disk} \times R^2$$

Where:
J = mass-inertia moment in kgm²
m_{disk} = mass of a disk in kg
R = radius from the centre of the mass

The mass-inertia moment is also expressed as $G \times D^2$. This gives exactly four times the value of formule in above.

Where:
G = weight of the flywheel in kg
D = diameter where the entire flywheel mass is concentrated, expressed in metres

17.35.5 Construction of flywheels

The flywheel is designed to be as light as possible to achieve a shaft load that is as low as possible. The further its mass centre is removed from the shaft, the higher the mass-inertia moment and the greater the effects on the engine speed.
Most flywheels have a rim-like construction: a thin centre area for attachment to the shaft and a very thick outer ring for the required mass.

17.36 Examples of crankshafts, either with or without counterweights

▶

Mounting a crankshaft in a reverse V-block of an MTU diesel engine.

Clearly shown is the circumference of the counter-/balancing weights with a radius from the centre of the crankshaft.

◀

Mounting a crankshaft in an MTU V diesel engine.

Every crank web has a counterweight.

◀

Three crankshafts in the Detroit Diesel factory in the United States.

This type of crankshaft has counterweights forged with the cranks.

◀

In this V-engine crankshaft, it can be seen that the counterweights are directly opposite the crankpin, where the two pistons will be fixed.

1 *crankshaft*
2 *crankpin*
3 *counterweights*

▶

This Caterpillar V-3408 crankshaft as well as that of the V-3508 type does not have one crankpin for the two pistons, but a crankpin in two 'staggered' sections.

The reason: In this manner, the ignition angle of all eight cylinders is identical and therefore the torsional vibrations lowest.

▶

This crank of a medium-speed four-stroke diesel engine is equipped with bolted counterweights.

1 *crank webs*
2 *connecting rod*
3 *crankpin bearing*
4 *counterweight*

▶

This MAN Diesel 8-50MC two-stroke crosshead engine crankshaft does not have counterweights.

17.37 Combustion forces exerted on the driving gear

In order to get an idea of the extent of the forces, a rough calculation of the moment the piston is in TDC position during the combustion stroke is made for all four engine categories.
The pressure and cylinder diameter are measured at the piston top position and put into the correct engine category.

Category I:
Pressure in top 30 bar
D = 50 mm.
Force = Surface × pressure

$$\text{Force} = \frac{\pi}{4} \times D_z^{\,2} \times P.$$

F = 5887,5 N or **0,58 ton.**

Category II:
Pressure in top 120 bar
D = 120 mm.
F = 135648 N or **13.5 tons.**

Category III:
Pressure in top 200 bar
D = 320 mm.
F = 1,607,680 N or **160.7 tons.**

Category IV:
Pressure in top 160 bar
D = 960 mm.
F = 11,575,296 N or **1157.5 tons.**

These are extremely high forces generated rapidly, which follow each other very quickly.
All parts of the drive gear, such as the piston, gudgeon pin, connecting rod, crankpin, crankshaft and the bearings must be strong enough to withstand these pulsating loads for tens of thousands of hours. Therefore, it is of great importance that the moving parts in the drive gears are separated sufficiently by a film of lubricating oil.
The pressure of the lubricating-oil film is often high and can rise from 100 to over 1.000 bars!
It is therefore important that the journals are absolutely round and that the bearing caps fit properly in order for the rotating crankshaft working as 'a high-pressure lubricating-oil pump' to separate the metal parts.

Lateral forces of the piston movement on the cylinder liner.

Consequently, these are also high. In four-stroke engines, the piston has to transfer these forces to the engine block via the cylinder liner by means of the piston skirt. In two-stroke crosshead engines, the crosshead ensures that the forces are transferred to the welded A-frame via the guides and guide shoes.
These lateral forces in the various engine categories vary roughly from 100 N to 2000,000 N or 0.1 to 200 tons!

>CH 18

Diesel-Power Plants

Caterpillar engines in the diesel power plant at Peng Hu II, China.

Diesel-power plants in most parts of the world generate electricity. This includes mobile floating power plants.
Shown here, discharging of a floating power plant with MAN Diesel engines from a semi-submersible vessel of 'Dockwise'.

◄ A large diesel power station with fourteen eighteen-cylinder V-engines running on HFO for driving the electric generators.

On the left: the gensets and on the right, the turbochargers and exhaust-gas pipes on the engines.

18.1 Introduction

For over fifty years, diesel-power plants have been used to generate electricity all over the world. They are located mainly on islands, in remote areas and often in regions with poor or non-existent infrastructure, for instance in the tropics or developing countries.

18.2 Classification of diesel-power plants

A Four-stroke high-speed diesel engines operating on MDO. Generates up to ± 5 MW power output per set.

B Four-stroke medium-speed diesel engines operating on HFO. Generates up to ± 20 MW power output per set.

C Two-stroke low-speed crosshead engines operating on HFO. Generates up to ± 50 MW power output per set.

D Floating diesel-power plants with four-stroke trunk piston- or two-stroke crosshead diesel engines operating on HFO. Delivers from ± 5 MW up to ± 20 MW power output per set.

E Mobile gensets installed in a container with four-stroke high-speed diesel engines, operating on MDO. Generates up to ± 5 MW power output per set.

18.3 Uses for diesel-power plants

Electricity is generated in various parts of the world by diesel engines for the following reasons:

1 **The required electric power capacity is not large enough to justify building large-scale power stations with gas and steam turbines.** The return on capital expenditure of these power stations is only acceptable when they operate at full load.

2 **The electric power capacity is required for a short time-span**, such as in projects for building large infrastructure, special events or following major natural disasters. In these situations, large emergency generators are used. These comprise diesel gensets placed in containers. These gensets

▼ Bottom left: Manufacture of a floating diesel-power plant. Manufacturer: Wärtsilä.

Bottom right: A floating diesel-power plant or power barge travelling to its final destination. Manufacturer: MAN Diesel.

At the front: the electric distribution section with the transformer, switches and pylons for connection to the high-voltage feeder line on-shore. Left of the engine room are the air-inlet filters, on the right, the exhaust-gas lines and sound dampers.

▲

A transformer station fed by diesel gensets.

Many countries with diesel-power plants, such as those found in developing countries, have an electricity grid in densely populated areas or areas with a high electricity demand. Here the transformers are fed by the diesel gensets increasing the voltage from 13,800 volt to 110,000 volt.

can also be used on a smaller scale with a single genset providing electricity.

3 **The local high-voltage electricity network is not able to transport large amounts of electric power.**

4 **The electricity requirement for an area is relatively small**, for instance only a few dozen megawatts. The efficiency at this power output is much higher (± 40%) compared to the efficiency when using a gas and/or steam turbine. A diesel-power plant of 100 MW is already substantial.

5 **The investment per kilowatt-hour is much lower than that of traditional power generators such as steam and gas turbines.**

6 **At partial load with an installation comprising several engines, a certain number can be switched off, so that the remainder operate at full load.** Thus, the total efficiency for the generation of electric power remains high. The efficiency for the generation of electric power using steam and gas turbines at partial loads is extremely low.

7 **Operating and maintaining gas and steam turbines requires highly skilled personnel.** This is to a lesser degree applicable for the maintenance of diesel engines.

8 **Maintaining diesel engines is relatively inexpensive.**

9 **Stand-by diesel gensets:** The stand-by genset is automatically started when there is a mains power failure. This system is often found in public buildings such as hospitals, institutions and large office blocks, but also as a 'black start' in power plants, in industry and other vital processes. These diesel gensets can be started within several seconds and directly operate at full load.

Some disadvantages of diesel-power plants

- Relatively high exhaust-gas emissions.
- Limited power output per engine:
  - four-stroke: maximum 40 MW;
  - two-stroke crosshead engine: maximum 100 MW;
  - gas turbine: maximum 400 MW;
  - steam turbine: maximum 1500 MW;
- Relatively heavy.
- Relatively high maintenance costs.

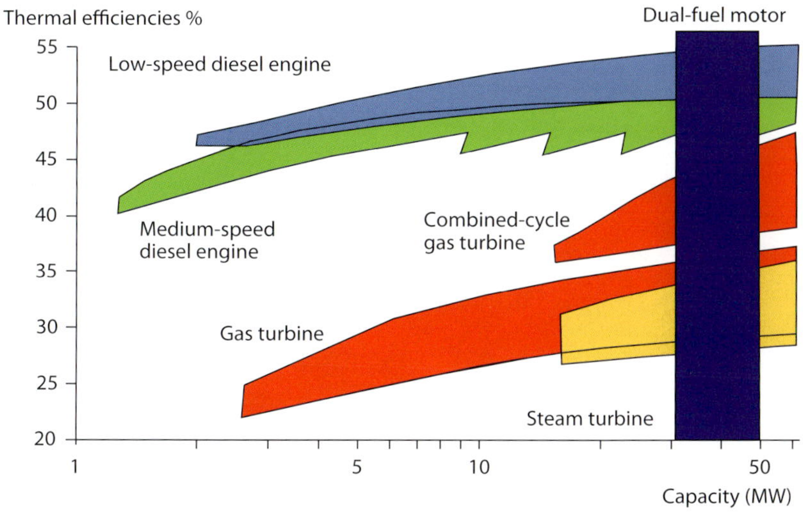

◀

An efficiency graph for the various engines and engine combinations up to 50 megawatt shaft power.

steam turbine	yellow	33%
gas turbine	red	36%
steam- and gas-turbine combination	red	46%
medium-speed engine	green	48%
low-speed engine	purple	53%
dual-fuel engine	blue	56%

Location

- Often in scarcely populated areas (with modest power requirements).
- Especially in the tropics, (poor infrastructure).
- Often in third-world countries (low investment levels).
- On islands and other isolated areas.
- In remote areas with large industrial activities, such as for mining, gas- and oil-drilling industries.

▲

In diesel-power plants, four-stroke medium-speed diesel engines are often used.

Shown, an eighteen-cylinder V-engine. Considering weight, size and capital cost, this is the best choice as two in-line nine-cylinder engines have a similar maximum output, but are heavier, take up more space and the capital cost is higher.

18.4 Types of diesel-power plants

18.4.1 Four-stroke high-speed diesel engines operating on MDO (engine category II)

Here the power output per diesel genset is limited. 1 MW to 5 MW due to the comparatively expensive diesel oil.
They are used to generate electricity in the vicinity for various purposes:

- lighting;
- infra-structure;
- production processes in manufacturing plants such as breweries and water treatment.

The number of revolutions per minute is usually 1500 and 1800, depending on the frequency (50 or 60 Hz). For the slightly larger engines, the number of revolutions is lower. However, at approximately 1200 revolutions or less, the use of heavy fuel oil is possible. This is cheaper than diesel oil. These engine types often fall in category II.
Familiar brands are Caterpillar, Cummins, MTU/DDC, GM amongst others.

▼

An artist's impression of a modularly built diesel-power plant for electric power generation.

1 *high-speed four-stroke diesel engine*
2 *electric generator*
3 *main switchboard*
4 *engine-room ventilation*
5 *space for distribution station and transformers*
6 *air coolers for discharging the heat from the fresh water-, lubricating oil- and combustion-air cooling systems.*

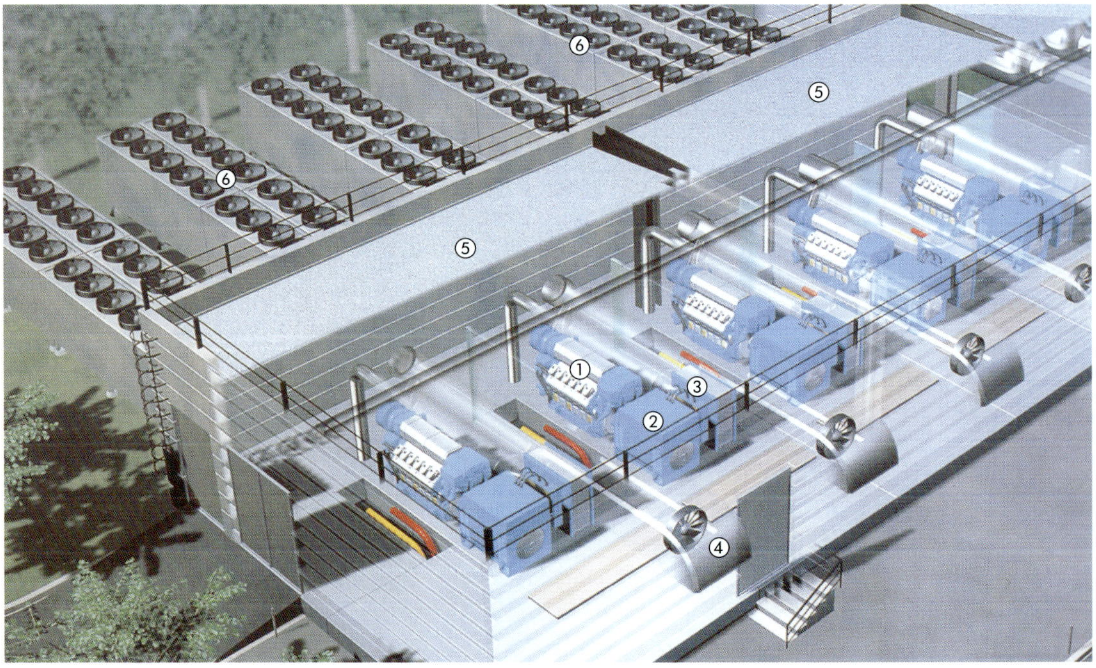

With the use of the light diesel oil (MDO), the fuel cleaning process is simple and there are fewer exhaust-gas emissions in comparison to those of heavy fuel oil (HFO).
The high-speed high-load diesel engines used for gensets are usually arranged in V-shape. They are compact and relatively light, are mounted either in a fixed or flexible manner on a steel frame and can be placed immediately on the factory floor.

In the gas and oil industry, many high-speed diesel gensets are set up on locations that lack (natural) gas for operating gas-engine gensets. The total power output varies from several megawatts to over ten megawatts.

◄

A high-speed four-stroke genset running on diesel oil used for the generation of electricity.

The frame can be easily positioned on the shop floor.

►

A high-speed four-stroke diesel genset with its own generator cooling.

In the foreground, the electric generator with its own air-cooling system. Offshore, for instance gas and oil platforms, complete sets are often substituted during major maintenance, so as little work as possible needs to be done on location. This is cost-effective.

18.4.2 Four-stroke medium-speed diesel engines operating on HFO (engine category III)

Most of the world's larger diesel-power plants are in this category. These plants have a total capacity of up to 150 MW-250 MW. To reach these capacities, several dozen sixteen- or eighteen-cylinder V-engines with large cylinder bores are installed. A fuel-treatment and supply system is an integral part of this installation. The large amounts of fuel are delivered by pipeline, ship or trucks and stored temporarily in storage tanks, enough to supply several days of consumption.

▲

The entrance to a large diesel-power plant in the tropics. Shown here, near Mombassa, Kenya.

Left the plant and right the electric distribution to the high-voltage grid.

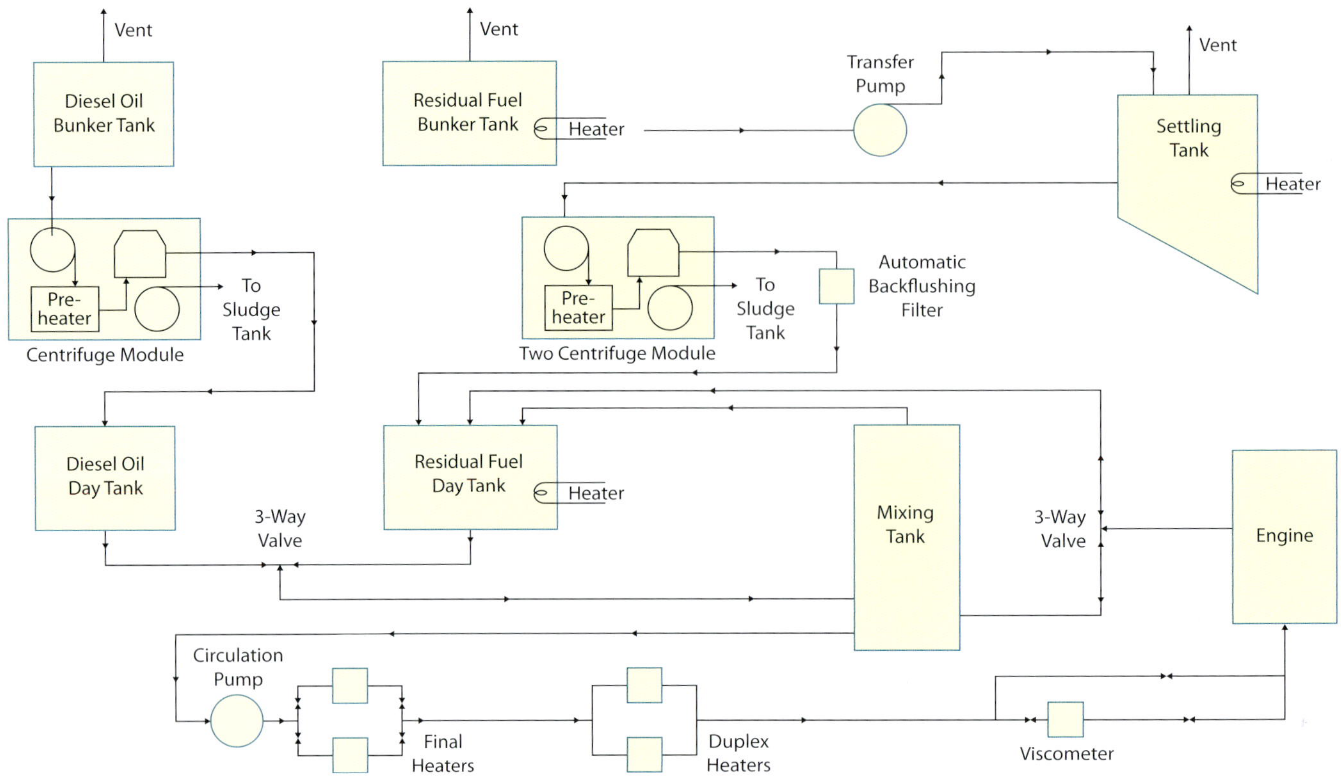

▲

A fuel system with storage, cleaning and treatment of diesel oil and HFO.

▶

A floating diesel-power plant, the 'Seaboard', in the Dominican Republic.

It is moored alongside a highway and close to a viaduct in a tidal estuary.

Electric power 72 MW.

▶

Arrangement of the diesel engines in the same floating plant.

Left, the turbochargers and exhaust-gas lines with right, the electric generator with air-cooling.

◀ Spare pistons requiring periodic overhaul after 12,000 operating hours.

In these relatively large plants, only a fraction of the total productivity is lost with one malfunctioning diesel genset.
If, for instance, ten engines of 8 MW each are assembled together, one non-operational engine would amount to a mere 10% of the total capacity and the genset will still produce 72 MW. The engine load in these types of plants is high, generally 95% to 98% of the nominal capacity using a three-phase generator.

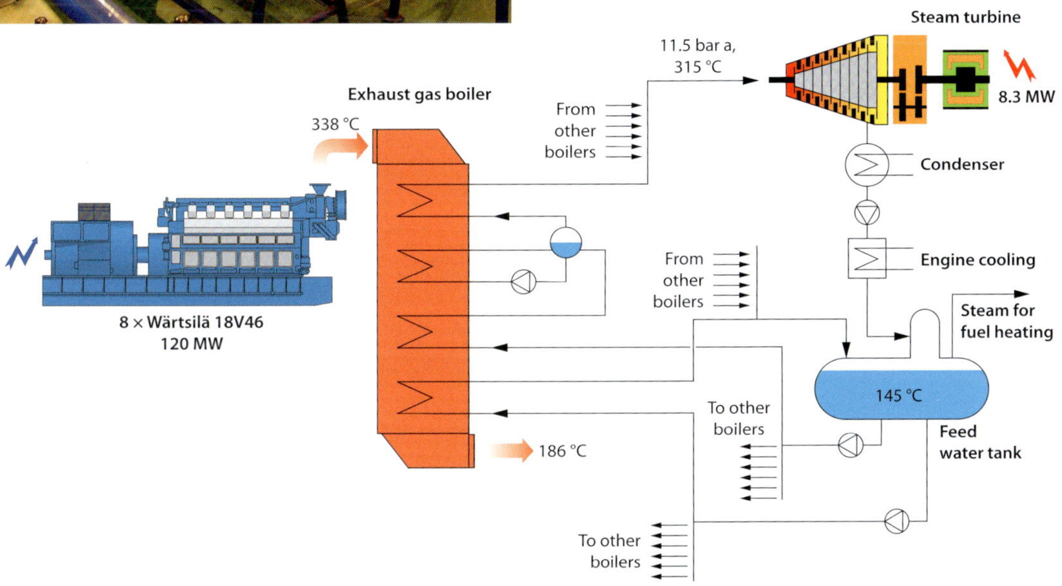

◀ A diesel genset producing steam using the exhaust gases and an exhaust-gas boiler.

The steam drives a turbine genset. This relatively high investment is only cost-effective in large installations with several engines connected to one exhaust-gas boiler. Shown here, eight Wärtsilä 18V46 engines. Total power output 120 MW. The steam turbine generates 8.3 MW at full load.

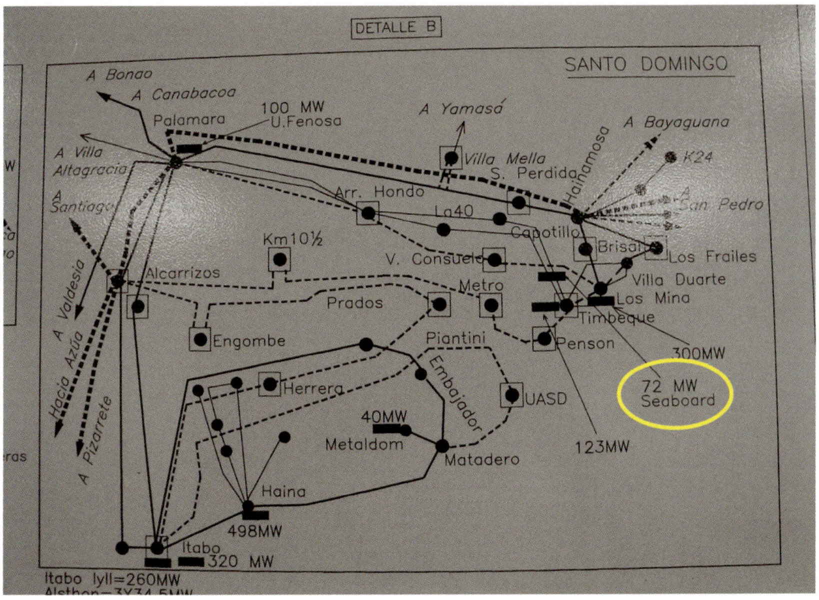

◀ The electricity network near Santo Domingo, the Dominican Republic.

The black horizontal rectangles are plants. The 'Power-barge Seaboard' on the right generates 72 MW.

◀ A large two-stroke crosshead engine from MAN Diesel for generating electricity.

In the background and right, a large disc generator with a diameter of 8 metres.

1 *disc generator*

18.4.3 Two-stroke low-speed crosshead engines operating on HFO (engine category IV)

Compared to engine category III, the exploitation of these engines is negligible. The power output for each plant varies from 20 to 240 MW. Obviously, this is dependent on the type and number of engine(s). Low engine revolutions result in low revolutions of the generator.

According to the formula:

$$f = p \times n$$

in which:

f = Hz;

p = number of pole pairs;

n = number of revolutions per second. The number of pole pairs at an engine speed of two revolutions per second (120 per minute) is 25 and the number of poles therefore: $25 \times 2 = 50$

Therefore: $50 = \frac{120}{60} \times p$.

Fifty poles at these large capacities take up a great deal of space, and consequently the generator has a very large diameter, the so-called **disc generator or disc dynamo**. The diameters of these generators can reach as much as 5 to 12 metres.

The size and weight of these sets are vast, but compared to the more prevalent four-stroke trunk-piston engines driven by gensets, the number of moving parts is significantly lower. This generally implies that the amount of malfunctions is also limited. However, the energy company's preferences play an important part.

From an economic view, diesel-power plants with four-stroke medium-speed V-engines using HFO, are the most cost-effective investment.

18.4.4 Floating diesel-power plants with both four-stroke trunk-piston- or two-stroke crosshead engines operating on HFO (engine categories III and IV)

The power output for each plant reaches approximately 200 MW. The plant consists of a large floating pontoon equipped with all the required facilities, including **diesel gensets**, **a control room**, **a workshop**, **fuel- and lubricating-oil storage tanks** and a **step-up transformer** for the medium-voltage gensets.
Floating diesel-power plants are completely self-sufficient; they can be supplied with fuel and spare parts when at sea and can be towed to most locations. The underlying concept is that of a mobile power plant, which can be deployed anywhere in the world.
Obviously, the floating plant must be connected to the medium- or high-voltage local network.

Advantages of floating plants

- The floating installation is anchored at the location where the energy is required.
- Expeditious delivery times of approximately twelve months.
- Modular construction, of both gensets and buildings.
- Very flexible deployment at the locations.
- Construction in boat yards by modular construction.
- Fuel storage in double-bottom tanks. Therefore, no separate tanks are required.
- Shore-power connection to the plant.

- Auxiliary systems are tested at the dockyard as opposed to on location.
- Possibility of using (natural) gas at a later phase.
- The engines are manufactured with two fuel systems (Dual-Fuel).
- When natural gas is available, the floating power plant can be converted to operate on natural gas.
- Fewer risks in cases of natural disasters, earthquakes and landslides.
- Sea water-cooling is possible, so a simple installation with low temperatures can be sufficient.
- Mobility, versatility and easy deployment of these plants represents a low-risk investment.

Several significant advantages of using natural gas as fuel

- Very low exhaust-gas emissions.
- Clean combustion process.
- No fuel storage required.
- No danger of fuel spillage, so reduced risk of environmental damage, such as ground or water contamination.

These dual-fuel systems are frequently used for land-based diesel-power plants.

Also, see Chapter 8, *Fuels, fuel-line systems and cleaning fuels.*

18.4.5 Mobile gensets arranged in a container with four-stroke high-speed diesel engines, fuel MDO (engine categories I and II)

Capacity per container is up to ± 5 MW per genset. These are often used for:

- **temporary electricity supply** to large events, building activities, military activities, fairs;
- **emergency power installations** for hospitals, public buildings, the IT-sector, the banking sector, pig farms and a variety of industrial processes where power-cuts cause damages.

Today, **legal regulations** stipulate the actions that should be taken to limit the damage caused by a power failure. Financiers, such as the **World Bank** also make demands, for instance, with respect to the pollutant emissions. There are many **rental companies** active in this sector that lease mobile gensets ranging from several kilowatts to several megawatts electrical capacity per set.
Some companies make use of these mobile gensets on a continual basis, for instance: **gas- and oil-drilling companies, construction firms, the army, circuses, the Red Cross and firms involved in constructing pipelines, mines, tunnels and waterpower stations**.

◀ The 'Powerbarge Seaboard' as seen from sea. Eight Wärtsilä 38 engines with eighteen cylinders in V-arrangement generate electricity.

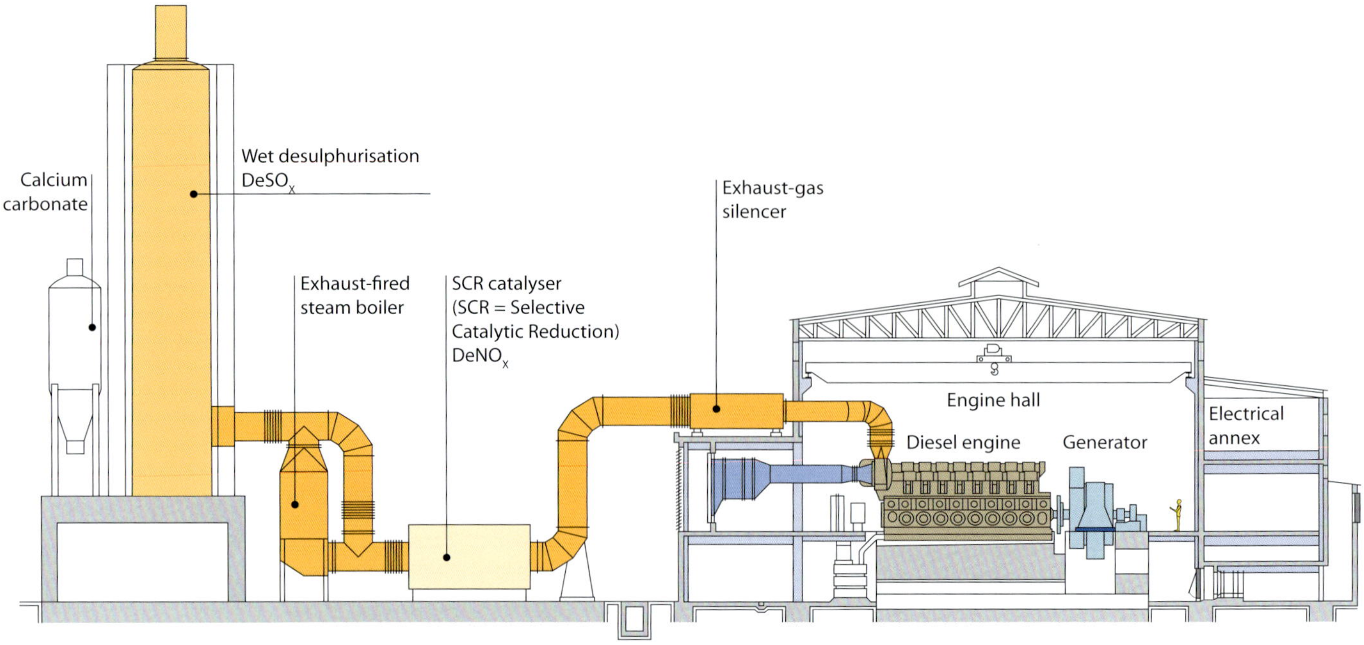

▲

A diesel-power plant with a two-stroke crosshead engine.

On the far right, the electrical switching section. Next to it, a large disc generator, crosshead engine and air inlet. Above, the sound dampers, the DeNO$_x$ installation for the removal of nitrogen oxides and to the far left, the DeSO$_x$ installation for the removal of sulphur oxides.

18.5 Special performances of diesel-power plants

18.5.1 Reduction of toxic flue-gas emissions

Reduction of nitrogen oxides, sulphur oxides and soot particles.

Also, see Chapter 22, *Engine emissions.*

18.5.2 Use of the residual heat

Heat generated by exhaust gases, coolant, lubricating oil and air-cooling.

This is also referred to as combined heat and power or CHP.

▼

A district heating project with diesel engines in Vaasa, Finland.

Both twelve-cylinder V-engines with bores of 46 and 64 centimetres drive generators that supply electricity to the public grid. The exhaust-gas heat is predominantly used for exhaust-gas boilers. The generated steam is used for driving a steam-turbine genset also connected to the network. Since the plant is located in the port near the city, the regulations with regard to the exhaust-gas emissions are very strict. Here both DeNO$_x$ and DeSO$_x$ installation are required (far right).

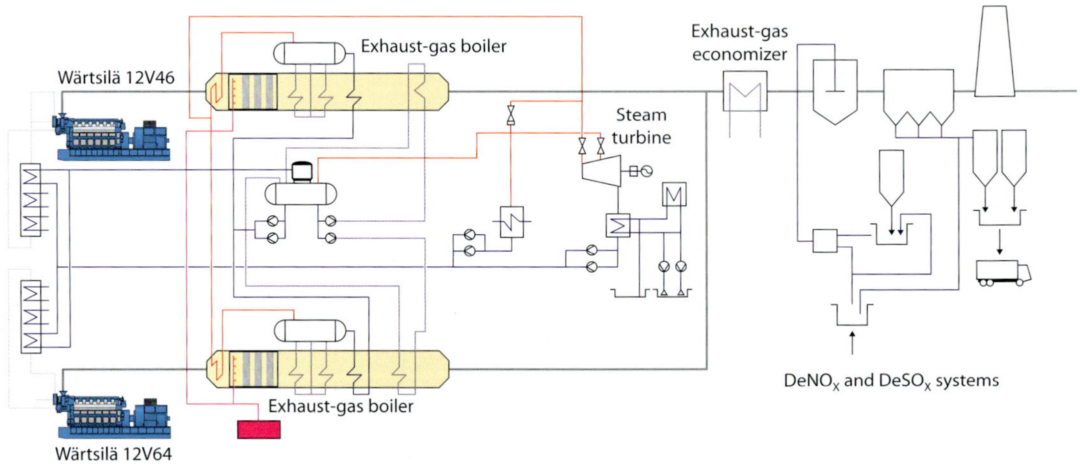

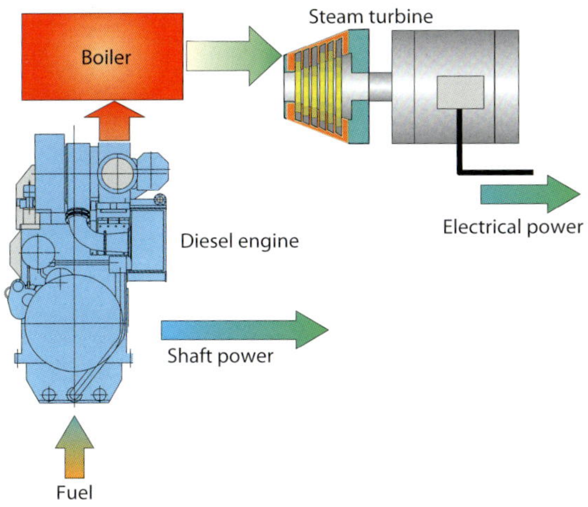

▲
The energy retained by the exhaust gases is approximately 25 to 45% of the delivered fuel energy and can be partially used for electricity generation using an exhaust-gas boiler combined with a steam-turbine genset.

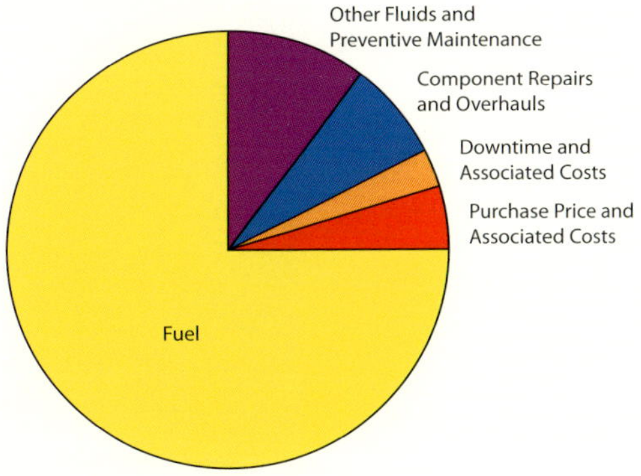

◄
Fuel costs are the largest cost driver in the exploitation of diesel engines used in shipping and diesel-power plants.

At high power outputs and many operating hours, HFO is often used as opposed to more expensive diesel oil.

Price indication March 2012, Rotterdam:

- Heavy oil 380 cSt.: 695 USD per tonnage weight
- Heavy oil 180 cSt.: 730 USD per tonnage weight
- Diesel oil: 1005 USD per tonnage weight

With the use of **heat exchangers**, water can be heated for use in city heating or manufacturing processes.
Using **exhaust-gas boilers**, steam for use in operating a steam-turbine genset is produced which subsequently **generates additional electric power**. The steam used in the steam turbine can then be re-used in processes that require a large amount of heat energy. Of course, much of the heat is used in the diesel-power plant itself for heating the fuel storage, heating the cleaned heavy oil and maintaining temperatures in stationary engines. Clearly, the extra energy 'gain' must be weighed against the extra investment.

General

Diesel-power plants often have a high operational reliability and capability. In the relatively smaller capacities of up to approximately 250 MW, higher efficiency in comparison to steam and/or gas turbines is usual. The fuel- and lubricating-oil costs form the highest expense posts, as is the case in the shipping industry. Diesel oil, which on average is twice as expensive, is only used for smaller capacities.

Legislation with regard to, amongst others, the pollutant emissions is also becoming increasingly strict in countries where diesel-power plants are prolific. Naturally, engine suppliers capitalise on these developments.

Also, see Chapter 22, *Engine emissions*.

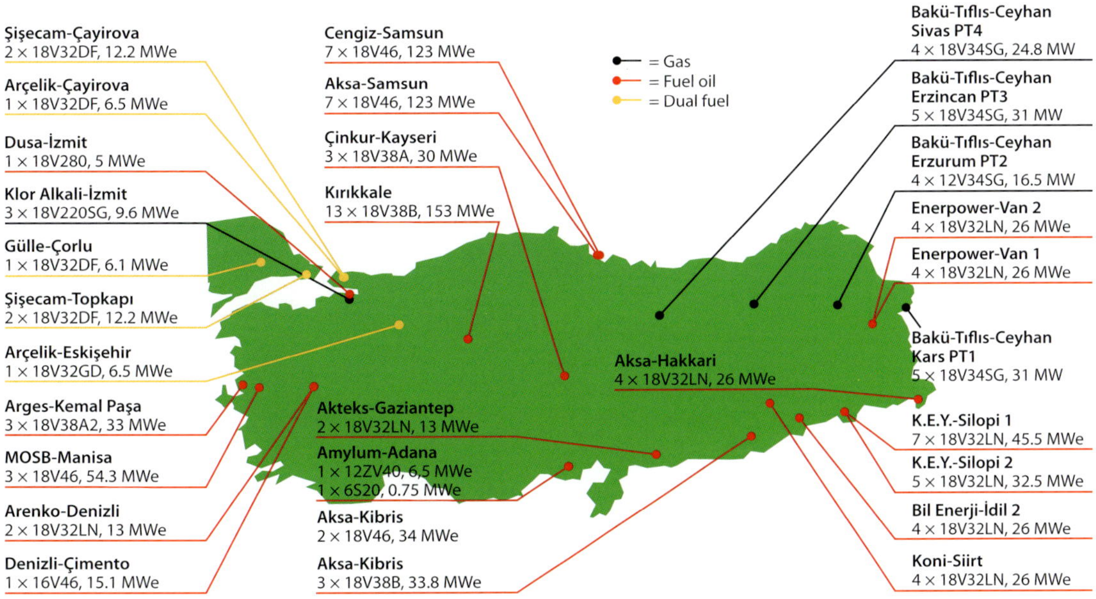

◄
Example: In a country like Turkey with huge distances and relatively few industries, many diesel-power plants are built locally. Shown here, plants from engine manufacturer, Wärtsilä.

GD = gas diesel. This is a diesel engine using a minimum of 5% diesel oil and a maximum of 95% (natural) gas.
SG = spark gas. This is an Otto engine running on gas. All the other engines are diesel engines.

▶

A diesel-power plant in China, also a country with huge distances and large areas with little industry.

▶

Modern diesel-power plants work fully automatically and are completely computer-controlled.

All technical data can be read by, for instance, engine manufacturers using modern communication systems.

▶

Two Caterpillar diesel-engine driven generators, engine category II, for electricity generation in China.

Both V-engines each have four turbochargers, two of which are mounted parallel to one suction-air filter.

◀
A large diesel-power plant in Kenya. On the left, an old plant and right, the new extension.

◀
A diesel-power plant far outside a city, China.

These plants are often built to support the extended electricity network and generate electricity for local industrial activities.

▼
Part of the 20 km long pipeline to the diesel power plant of the water and electricity of Aruba, the WEB.

The pipes for diesel, heavy oil and water run from the refinery in the east to the centre of the island, where the power plant is situated. Note the vertical expansion bends on the horizon for the free expansion of the pipes.

▲
The head platform in a power plant.

Each bank of these eighteen-cylinder medium-speed Wärtsilä V 38 diesel engines operating on HFO has its own turbocharger and inter-cooler.

▲

The first diesel-power plant at Choloma, with five MAN Diesel Pielstick medium-speed four-stroke engines running on HFO.

18.6 An example of a large diesel-power plant for the generation of electricity

20 February 2007 – Choloma

This power station is situated in the northwestern part of Honduras, near the village of Choloma, approximately 20 km outside the city of San Pedro de la Sula. The power company, Enersa, which has Egyptian investors, owns several diesel-power plants in Honduras and is the owner of the San Pedro de la Sula airport.

▼

The new diesel-power plant layout.

1 *fourteen MAN Diesel 48/60 four-stroke medium-speed HFO engines*
2 *cooling for the HT and LT cooling-water system to ambient air*
3 *HFO-, MDO- and lubricating-oil storage*
4 *loading station for tankers*
5 *centrifuge building for HFO*
6 *workshop*
7 *transformer station*

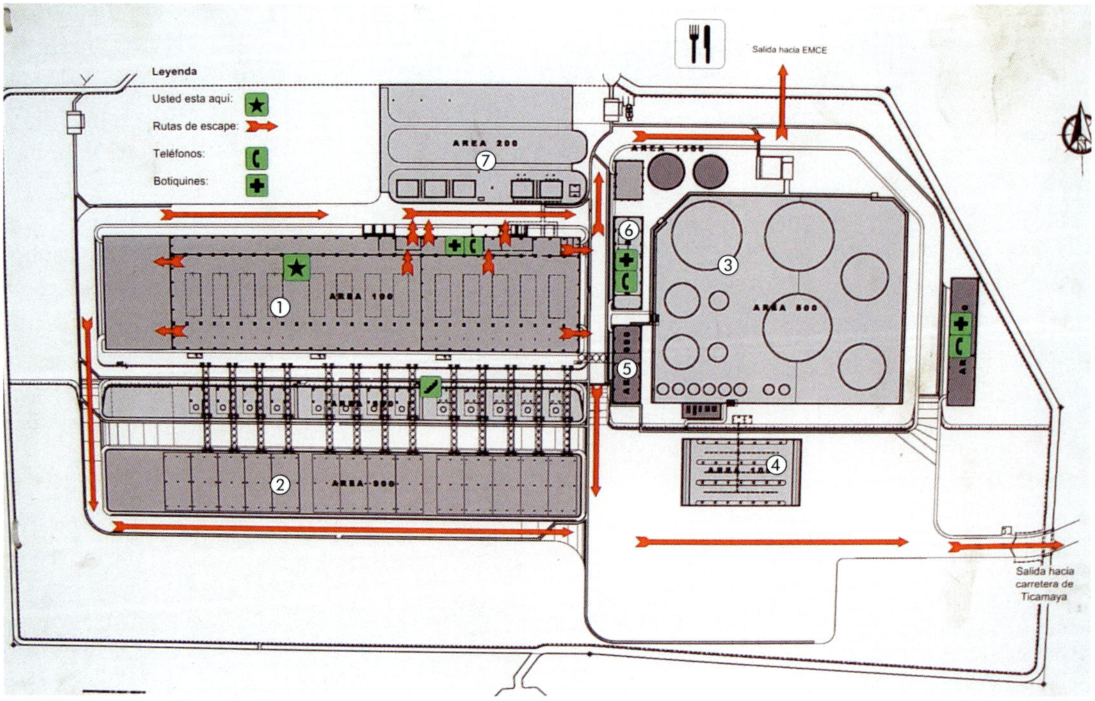

◄
The new diesel-power plant.

1 *engine hall*
2 *transformer station*
3 *high-voltage lines*
4 *HFO storage*
5 *contaminated-oil storage*
6 *workshops*

Using HFO has very noticeable consequences: high emissions of visible combustion products, amongst others, soot particles.

►
In the engine hall.

1 *electric 18.4 MW generator*
2 *MAN Diesel four-stroke medium-speed HFO diesel engines*
3 *exhaust-gas lines*

18.6.1 The Choloma III project

The Choloma III project comprises fourteen diesel gensets with the following data:
Diesel engines: Four-stroke medium-speed diesel engines operating on HFO, by MAN Diesel, Augsburg.
Type 18 V 48/60 of which three are type18 V 48/60 A and eleven type 18 V 48/60 B.

Electrical power output per genset 18.4 MW. Total power output 250 MW.
This diesel-power plant is in the top twenty largest plants in the world.

Generator: ABB

This genset generates a current of 13,800 volt, which is increased to 100,000 volt using a 'step-up' transformer.

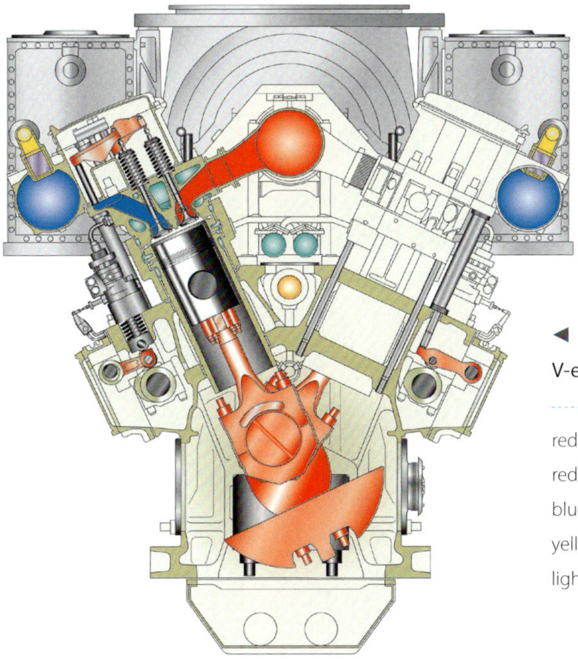

◄
V-engine cross-section.

red: exhaust gases
red: drive gearing
blue: inlet air
yellow: lubricating oil
light blue: HT cooling-water

▶ ABB three-phase gensets with in the foreground, the 'generator row'. Here large parts are moved by 10-ton overhead cranes and trucks.

Installation

The fourteen diesel-gensets are arranged alongside each other in one large hall with two 10-ton overhead cranes.

18.7 Auxiliary engines

The auxiliary engines are also situated in the hall.

18.7.1 Fuel-booster systems

The so-called 'booster units' heat the fuel to approximately 135 °C to achieve a viscosity of 10 to 12 cSt.

18.7.2 HT and LT coolant systems

The HT and LT coolant systems include two pumps for each engine. The HT system is used to cool the fresh-water- and lubricating-oil system of the LT system and cool the combustion air.

▼ A low-pressure fuel-supply system. Each engine has its own system, the so-called 'fuel boosters'.

▼ The pump row is almost 200 metres long.

left: HT and LT cooling-water pumps (yellow)
right: main lubricating-oil pump (grey) with heat exchangers (blue)

18.7.3 Main lubricating-oil pump

An electrically driven main lubricating-oil pump with an integrated lubricating-oil cooler and a lubricating-oil filter.

18.7.4 Valve-lubricating system

A valve-lubricating system for lubricating the inlet valves.

18.7.5 Fuel-nozzle coolant system

A nozzle-coolant system for cooling the fuel-injection nozzles. Pre-heating the nozzle coolant before starting the engine is possible.

18.7.6 Starting-air system

One system for starting the engines. The starting-air valves are mounted in the cylinder heads.

18.7.7 Lubricating-oil separator system

One lubricating-oil separator system for each engine, which automatically cleans the lubricating oil circulating in the engine.

18.7.8 Fire-extinguishing system

A fire-extinguishing system using foam as an extinguishing agent.

18.7.9 Ventilation system

A ventilation system for the engine hall to safeguard the combustion-air supply and simultaneously cool and ventilate the area.

18.7.10 Electrical generators

Each sleeve bearing in the electrical generator has a separate pre-lubricating pump, which is automatically switched on prior to starting and stopping.

18.8 Outside the engine hall

18.8.1 Waste-heat exhaust-gas steam boilers

Exhaust-gas boilers are connected to the exhaust-gas manifolds of all fourteen gensets. These produce steam for heating the fuel-booster units, the heating for the ten fuel oil- and fourteen lubricating-oil separators, the pre-heaters for the eight fuel-transfer pumps and storage-tank heating.

▼

Left in the picture, the HT- and LT system cooling-blocks.

Over 50% of the supplied fuel heat is wasted by the engine-cooling system and the exhaust gases seen to the right in the picture.

18.8.2 Silencers

All exhaust-gas systems have silencers (sound dampers).

18.8.3 Separator room

A dozen separators for cleaning the heavy-fuel oil are installed in a separate building. The diesel oil that is used during maintenance on the fuel systems or when there are problems with the viscosity of the heavy fuel oil is cleaned using fine-filters.

18.8.4 Fuel-receiving station

The fuel is supplied from the storage tanks in the port of Puerto Córtes and delivered using road-tankers. There is sufficient storage capacity for a two-week plant operation at the maximum fuel consumption of 1100 tons per day.

18.9 Fuel quality

Usually, a heavy fuel oil with a viscosity of 380 cSt at 50 °C and a sulphur content of 2.5% is used. The CCAI number fluctuates between 845 and 860. This means that the financial department rather than the plant operator determines the fuel quality. Obviously, the quality must conform to the specifications stipulated by the engine suppliers!

18.10 A few particulars

18.10.1 Operational management

Generally, the gensets operate at full load between 06:00 and 22:00 hours. All the gensets are automatically controlled proportional to the load. When the load of all the gensets is too low, with a limit of, for instance 90%, the first genset in the computer-controlled programme is automatically turned off, and vice versa.

▶ The separator room with ten separators for cleaning heavy-fuel oil. These Alfa Laval centrifuges work fully automatically.

▶ A fuel station.

North American influence is obvious: all the tankers are of an U.S.A. make! Fuel supply from the port at Puorto Córtes, 20 km away continues day and night,

Between 22:00 and 06:00 hours, the electricity requirement and therefore the loads are low. In this period, only two to three gensets are operational. The others remain on stand-by, as does the entire auxiliary plant. A genset can be started within a matter of minutes and is automatically loaded to a maximum of 98%. As fewer gensets are operational at night, the steam production of the exhaust-gas boiler is proportionally reduced. At low steam production a separate stand-by oil-fired auxiliary boiler is started.

18.10.2 'Turn-key' projects by MAN Diesel AG

The entire construction of the diesel-power plant including machinery, outbuildings and tank storage was outsourced to a special department at MAN Diesel AG.
Construction occurs in three phases:
Phase I: five sets (2003/2004)
Phase II: five sets (2004)
Phase III: four sets (2005)

In 2007, extra systems were delivered to the owner 'Enersa'. At this point approximately five hundred claims, standard for such an extensive installation, must be finalised.

▲
Exhaust-gas boilers for heat generation (March 2007).

18.10.3 Heat exploitation of exhaust gases

In addition to the small exhaust-gas boiler, fourteen high-pressure steam boilers have been installed to drive the steam-turbine genset that has a power output of 15 MW. Separate from the fourteen high-pressure exhaust-gas boilers with fittings and piping, a separate engine hall has been built. This hall contains the steam turbine with reduction gearing, the electrical generator, a high-pressure steam collection drum, and feed-water pumps.
A large air condenser condenses the steam expelled from the steam turbine into water and returns it to the exhaust-gas boiler. Unfortunately, there is no company in the vicinity that could use the low-pressure steam in a production process; this would lead to a considerable increase in the total fuel efficiency. With the present operation, the power efficiency amounts to approximately 54%, after the installation of the steam turbine this will increase to about 56%.

◄
Tank storage.

black: HFO
white: MDO
yellow: lubricating oil

18.10.4 Fuel supply

Diesel-power plants situated close to a suitable port usually receive the fuel directly from an oil tanker via a short pipeline installed in the harbour. Diesel-power plants located further inland receive fuel via either a pipeline or fuel-tanker. Diesel-power plants in third-world countries often opt for transportation by fuel-tanker, as building transport pipelines over long distances is costly and prone to sabotage and theft.

This plant requires 1100 tons of fuel per day when operating continuously at full load!

▶

The HT- and LT system for each engine consists of two pumps; one is operational and the other is on stand-by.

◀

Each engine has one large lubricating-oil pump with a plate heat exchanger. An automatic fine-filter is incorporated in the lubricating-oil system.

▼

Each engine has its own Alfa Laval automatic lubricating-oil separator.

▲

Engine maintenance. Shown here, the fuel injectors are exchanged.

◄

Engines weighing 110 tons are flexibly arranged on steel pressure-springs with, a centrally placed damper filled with a highly viscous liquid.

The engines oscillate approximately 2 millimetres. The flexible coupling between the diesel engine and generator must absorb this movement as the generator is mounted on a concrete foundation.

◄

The high-pressure fuel pump, one for each cylinder.

1 *fuel-supply line*
2 *fuel-return line*
3 *high-pressure supply line to the fuel injector*
4 *pump casing*
5 *supply-control mechanism*

◄

Starting-air compressors.

The engines are started with an air pressure of maximum 30 bars, supplied to the cylinder via a starting-air valve. The compressors maintain the required air-pressure in the start-air reservoirs.

►

A foam fire-extinguishing system.

Automatic 'sprinklers' are positioned near the engines. These operate when the ambient temperatures are too high. First, a water mist is ejected followed shortly afterwards by foam.

▼

Each engine has a crankcase mist detector that indicates when and where, if the bearing temperature is too high.

▲

A diesel engine including generator. Note the size!

◄

The transformer station 13,800/110,000 Volt.

◄ The electric main rail to which all engines are connected.

◄ The main switchboard where all fourteen generators can be switched into the network.

◄ The cylinder-head platform.

Right; the cylinder inlet-air manifold.

▶
Transfer pumps for the fuel-storage system.

▶
The air ventilators. The combustion process and building ventilation require large amounts of air.

▶
The camshaft 'doors' (1) with the crankcase covers (2) below containing several emergency vent covers (3).

▲

The valve drive of the inlet- and exhaust valves; a special construction.

▲

The fuel injector between two inlet- and exhaust valves. The high-pressure double-walled fuel-supply line has been removed.

◄

The valve-drive's yoke (upside down).

◄

Two transport pumps for draining tankers.

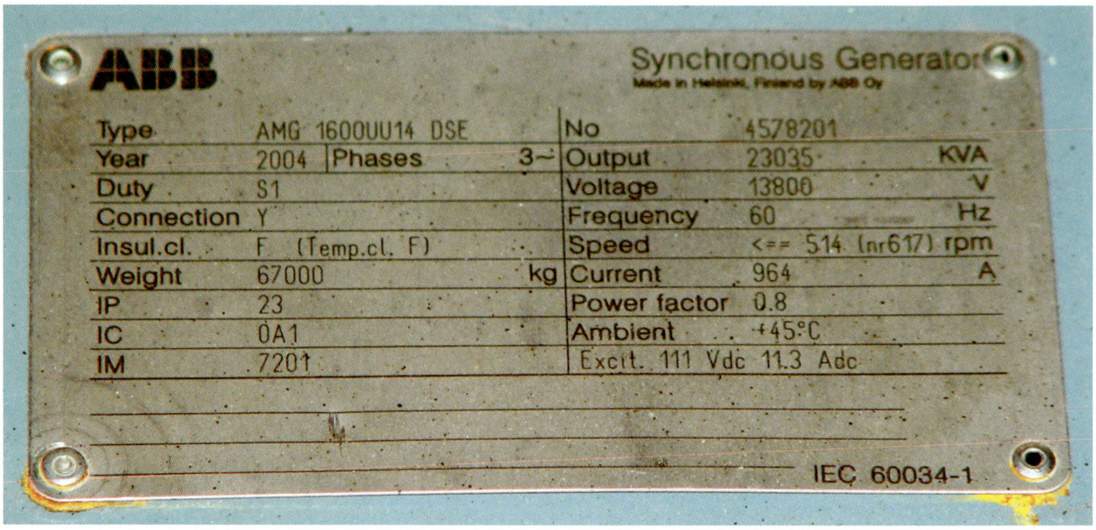

▶

Manufacturing data of the ABB three-phase generator.

The frequency is 60 Hz instead of 50 Hz, as in Europe. A sizable generator weighing 67,000 kilograms!

18.11 Examples of power plants

▶

Three complete diesel gensets with a power output of 5 MW.

Generators are driven by Caterpillar diesel engines category II. After manufacturing, testing and release for certification, the sets are disassembled and transported to Qatar where they will be re-assembled for a refinery. The test facility here on the premises of Topec/Pon Power in Papendrecht, The Netherlands.

A: One of the three Caterpillar V-engines, the largest from this manufacturer running on MDO, a 3616C 280-16 with a speed of 1000 revs/min.

B: The two starting rotors near the flywheel.

C: Both starting-air compressors, one is electric ally driven and the other is driven by a two-cylinder Hatz diesel engine. In the background, both starting-air reservoirs.

D: The air coolers used for testing. With these air coolers, the heat of the cooling-water system is discharged.

▲
A large floating power plant in a river mouth with Wärtsilä medium-speed diesel engines running on heavy-fuel oil.

▼
A power plant in a harbour.

The companies mentioned below all have contributed to the realisation of this book.
This has been made possible by allowing interviews, consenting to the use of photographs, providing abundant information or by correcting texts.
For this we would like to express our sincere gratitude!

Company	Location
A.M.W. Marine	Hendrik-Ido-Ambacht
ABB B.V.	Rotterdam
ABB Turbochargers	Baden, Switzerland
ABC, Anglo Belgium Corporation	Ghent, Belgium
Alfa Laval Benelux B.V.	Breda
Arhus Maskinmesterskole	Arhus, Denmark
Breko Reparatie B.V.	Papendrecht
Brinkman & Niemeijer Motoren B.V.	Twello
Caterpillar Mak	Kiel, Germany
Caterpillar Marine Power Systems	Hamburg, Germany
Caterpillar Motoren GmbH	Kiel, Germany
CENTA Nederland B.V.	Stellendam
Chalmers Maritime Academy	Kalmar, Sweden
Chris Marine	Malmö, Sweden
Cummins Holland B.V.	Dordrecht
DBR	Hardinxveld
Det Norske Veritas B.V.	Groningen
Diesel Engines Online B.V.	Barendrecht
Diesel Publications	Waukesha, USA
Discom B.V.	Alblasserdam
Easy Connections B.V.	Dordrecht
Eefting Engineering	Kolham
Fredericia Maskinmesterskole	Fredericia, Denmark
G.M.S. Instruments B.V.	Rotterdam
Geislinger GmbH	Salzburg, Austria
Genpower B.V.	Groot-Ammers
Germanischer Lloyd/Benelux	Schiedam
Goltens Dubai	Dubai, UAE
Goltens Rotterdam B.V.	Spijkenisse
Hatz Nederland B.V.	Culemborg
HRP Europe B.V.	Krimpen a/d Lek
IHC Holland	Kinderdijk
IHC Lagersmit B.V.	Kinderdijk
JR SHIPPING B.V.	Harlingen
Kemper en Van Twist Diesel B.V.	Dordrecht
Koedood Dieselservice B.V.	Hendrik-Ido-Ambacht
L.P. Koster & Zn. B.V.	Rotterdam
Ludwig Hunger	München, Germany
M. Jürgensen GmbH & Co KG	Sörup, Germany
Machinefabriek BOLIER	Dordrecht
Machinefabriek G. Olthof N.V.	Capelle a/d IJssel
Machinenfabrik Alfing Kessler GmbH	Aalen, Germany
Maersk Line	Copenhagen, Denmark
Maerkisches Werk GmbH	Halver, Duitsland
Mahle GmbH	Markgroningen, Germany
Maine Maritime Academy	Pleasant St. Castine, USA

MAK Nederland B.V.	Dordrecht
MAN Diesel A/G	Augsburg, Germany
MAN Diesel A/G	Hamburg, Germany
MAN Diesel A/S Alpha	Frederikshavn, Denmark
MAN Prime Serve	Schiedam
MAN Diesel Prime Serv Academy	Copenhagen, Denmark
Maritieme Academie Willem Barentsz	West-Terschelling
Mark van Schaick B.V.	Schiedam
Massachusetts Maritime Academy	Buzzards Bay, USA
Mastervolt B.V.	Drachten
Merford	Gorinchem
Miba Bearing Group	Laakirchen, Austria
Mobil Oil B.V.	Breda
M.R.C.	Schiedam
MTI Holland B.V.	Kinderdijk
MTU Detroit Diesel Benelux B.V.	Dordrecht
Nicoverken Holland B.V.	Schiedam
Noordhollandse Motorenrevisie B.V.	Winkel
Pasman Motoren & Aggregaten	Baak
Perkins	Peterborough, UK
Pon Power B.V.	Papendrecht
Rederij Doeksen	Harlingen
Royal Huisman Shipyard	Vollenhove
Rubber Design B.V.	Heerjansdam
Ruysch Technisch Handelsbureau	Zutphen
Sandfirden Technics	Den Oever
SKF Marine Service Centre	Ridderkerk
The Vestfold Academy	Göteborg, Sweden
Tognum, MTU Friedrickshafen GmbH	Friedrickshafen, Germany
Topec B.V.	Papendrecht
Total Nederland N.V.	's-Gravenhage
Turbo Ned	Zwijndrecht
Turbo's Hoet Nederland B.V.	Baarn
Turner B.V.	Hoofddorp
VAF Instruments B.V.	Dordrecht
Van Dinteren Technische Handelsonderneming	Tiel
Van West-Holland B.V.	IJmuiden
Vereniging Importeurs Verbrandingsmotoren	Zoetermeer
Vetus B.V.	Schiedam
Wagenborg Shipping B.V.	Delfzijl
Wärtsilä Nederland B.V.	Kruiningen
Wärtsilä Nederland B.V.	Schiedam
Wärtsilä Nederland B.V.	Zwolle
Wärtsilä Switzerland Ltd.	Winterthur, Switzerland
Waterschap Groot Salland	Zwolle
Westfalia Separator Nederland B.V.	Cuijk
Woodward Governor Nederland B.V.	Hoofddorp

Diesel Engines Part II

Kees Kuiken

This book is dedicated to my wife Truus, who has supported me for over forty years and without whose faith, help and encouragement this book would never have been written.

DIESEL II ENGINES

for ship propulsion and power plants

FROM 0 TO 100,000 kW

2nd revised edition

Kees Kuiken

VIV
More than engines

Target is member of the Association of Importers of Combustion Engines

Second revised edition, July 2012

Author:
Kees Kuiken

Technical advice:
Johan Heuff, Chief Engineer ret. P&O Nedlloyd.
Jan Leekstra, Chief Engineer ret. P&O Nedlloyd.
Christiaan Kuiken, Maritime Officer, Maersk Ship Management.

Translation:
AAG Translations, A.A. Geerts MA, Almelo, The Netherlands.

Textual corrections and proof reading:
S.E. Brent BSc, Almelo, The Netherlands.
Mimi Kuijper-Heeres, Senior Lecturer ret. Maritime Institute Willem Barentsz, Terschelling, The Netherlands.

Lay-out:
Little Shop of Graphics, Almelo, The Netherlands.

Technical drawings:
Niels Kuiken.
Maarten Bosch (LSoG).

Printing:
Epospress, Zwolle, The Netherlands.

Finish:
Stronkhorst Van der Esch/Embé, Groningen, The Netherlands.

A publication of:
Target Global Energy Training
Noorderhooidijk 2
9755 PJ Onnen, The Netherlands

Website: www.targettrainingcentre.nl
www.dieselengines.nl
E-mail: info@targettrainingcentre.nl

Target Global Energy Training.
Onnen, The Netherlands.

ISBN/EAN: 978-90-79104-00-0

Information
In order to gather sufficient information for this book, many companies both in The Netherlands and abroad have been visited in the past five years. These companies, each in their own way, have contributed to the creation of this book by participating in interviews, allowing photographs and providing technical information.
Nevertheless, special thanks go to MAN Diesel and Wärtsilä AG, for their cooperation in making a great deal of material available.

Preface

Preface to the first edition

Diesel engines play an important role in today's society: we are quite dependent on them.
Over 100 years after Rudolf Diesel developed a working diesel engine, there is still no real alternative for ship propulsion and electric generators in tropical and/or remote areas. The diesel engine is indispensable for road haulage, inland shipping, aquatics, electric power emergency systems, agriculture, and passenger transport by road or rail, oil and gas industry and various other industries. We have chosen to make use of many pictures accompanied by a written explanation.
Much highly in-depth technical theory has been omitted as these topics are covered by specialist books available in the market; these topics include thermodynamics, vibrations, materials, and electronics.
We at Target Global Energy Training, have opted for a more practical approach. This includes ample information with respect to the construction of engines, use of materials, various engine categories, maintenance, repairs, and the use of engines.
Much attention has been paid to the choice of proper graphic material. This, in our opinion, is helpful for the reader to gain insight in the various subjects. This publication is indispensable for every person who has dealings with the diesel-engine industry, from the smallest engine to 'The Cathedrals of the Oceans'.

Kees Kuiken, Onnen, July 2008

Preface to the second completely revised edition

Based on feedback from various users of the first edition of this book, including companies in the diesel engine field, maritime institutes and shipping companies, we decided to add a number of topics to the second edition.

It became apparent that many users had an interest in certain subjects, but were not in the position to visit these particular businesses to learn how certain components are manufactured or repaired. Therefore, the following topics have been added:

- The manufacture of pistons, valves, cylinder liners and gear drives for the larger four-stroke and two-stroke crosshead engines.
- A damage report from a four-stroke medium-speed diesel engine.
- A report on the manufacture of crankshafts.
- The excessive wear on cylinder liners in two-stroke crosshead engines, called 'scuffing'.
- Slower sailing larger vessels to save fuel, the so-called 'slow steaming'.
- Future improvements in propulsion efficiency and the reduction of CO_2 emissions by ships.
- A third example in ship propulsion; a vibration analysis for two engines driving a single propeller.
- An overview of the market share 2010 of diesel engine manufacturers worldwide.
- The Hercules Project, a collaboration between various parties, including the engine manufacturers for the purpose of improving efficiency and reducing emissions.
- A new thrust measurement system.
- The theoretical background of and damage to bearings.
- A new method for the alignment of amongst others, engines and shafts.
- Additional theoretical calculations for the aeration of engines.
- Additional theoretical calculations of the efficiencies of engines.
- An improvement to both to the many images and the text.

Kees Kuiken, Onnen, July 2012

◄◄ Maastricht, May 2007.

At the special general meeting of the 60th anniversary of the VIV, de Vereniging van Importeurs van Verbrandingsmotoren (Association of Importers of Combustion Engines) at the 'Theater aan het Vrijthof' in Maastricht, author Kees Kuiken presents the first proof print to the chairman of the Association FME-CWM, Mr. Jan Kamminga.

◄ Dordrecht, November 2008.

During the general meeting of V.I.V., the Association of Importers of combustion engines in November 2008, Kees Kuiken presents the first proof print of the English edition to Mr. Jules L. Th. Schöningh, chairman of the V.I.V.

Contents

>CH 19

Ship propulsion

Ever since the introduction of the reciprocating piston-driven steam engine in the 19th century propellers have been used for ship propulsion.
The diesel engine has been in use as the standard propulsion engine since the 1950's. The combination of a diesel engine and a propeller amounts to a propulsive efficiency of approximately 33%. So two-thirds of the energy generated by the fuel is wasted!

Marine craft often are powered by high-load, high-speed four-stroke diesel engines, category II.

19.1 Introduction

Source: A summery of a study by MAN Diesel, the largest engine manufacturer for medium- and low speed- diesel engines in the world.

Since the introduction of auxiliary power – a steam engine - on sailing vessels, ships have been fitted with a propeller shaft between the propulsion engine and the screw or propeller.
This is still the case today: nearly all seagoing vessels with a power output over 2000 kW, approximately 90,000 vessels, uses a propeller driven by a diesel engine for propulsion.
Numerous ships for inland-water navigation, including pleasure craft use propellers. Other propulsion systems such as jet-propulsion are only used for very specific craft.

Every vessel requires a different propeller to achieve maximum efficiency from the chemical (fuel) energy supplied to the diesel engine. This depends, among other things, on its dimensions, hull shape, the desired speed and available propulsion power.

To obtain a basic idea of marine propulsion, the subject is subdivided into three areas of interest.

1 **The most elementary issues, such as ship dimensions, hull shape, displacement, ship weight, design draught, length between perpendiculars and the block coefficient are discussed.**
Other topics are towing resistance, comprising frictional-, residual- and air resistance, and the effects of these resistances in real terms.

2 **Further elaboration on marine propulsion and the flow conditions near the propeller. The wake- and thrust coefficient will also be discussed.** The total engine output required for the propeller is computed from the aforementioned effective-towing resistance and the efficiencies depending on the various propeller- and hull shapes. Next, the operating conditions for a screw deduced from propeller laws for fixed-pitch propellers are discussed. These conditions vary from navigating in calm waters to conditions varying from light to heavy operating resistances, such as fouling and heavy swell.

3 **In this section the significance of choosing the correct Maximum Continuous Rating (MCR) and the optimal point for the propulsion engine will be discussed. Attention is paid to the load curve of the engine in relation to the design of the screw.** The compilation of the accompanying load diagrams is described in detail using a number of examples. It is important to be aware of how various ship resistances affect the continuous service rating of the engine.

▶
The combination of a diesel engine driving a propeller has been the most commonly used propulsion system for over 75 years. It is used by more than 90,000 large seagoing vessels, hundreds of thousands of inland vessels, and millions of yachts, cabin cruisers and other ships.

Category	Class	Type	abbrev.
Tanker	Oil tanker	Crude (oil) Carrier Very Large Crude Carrier Ultra Large Crude Carrier Product Tanker	CC VLCC ULCC
	Gas tanker Chemical tanker	Liquefied Natural Gas carrier Liquefied Petroleum Gas carrier	LNG LPG
	OBO	Oil/Bulk/Ore carrier	OBO
Bulk carrier	Bulk carrier		
Container ship	Container ship	Container carrier Roll On-Roll Off	Ro-Ro
General cargo ship	General cargo Coaster		
Reefer	Reefer	Refrigerated cargo vessel	
Passenger ship	Ferry Cruise vessel		

◄

The three largest ship categories.

The three main ship categories are; container ships, bulk carriers (grain, ore) and tankers that can be further subdivided into classes that are more specific, and types, such as oil, gas and chemical tankers.

There are often 'multi-purpose' ships, which are suitable for various tasks.

19.2 Ship types and hull resistance

Depending on the cargo and loading conditions, ships can be subdivided into different categories, classes and types.

19.3 Load lines of a ship

Every seagoing vessel has load-line marks amidships on each side of the hull; this is known as the 'Plimsoll Mark'. This mark usually consists of steel plates that are painted in a distinctive colour and have been welded on the port and starboard side of the hull. The lines and letters of the Plimsoll Mark correspond with the freeboard regulations as laid down by the IMO, the International Maritime Organisation, and indicate the maximum draft to which a ship can safely be loaded. This draft is dependent on sea, season and salinity. The salinity of the water affects the level; fresh water is less dense than seawater and therefore exerts an upward force on the ship.

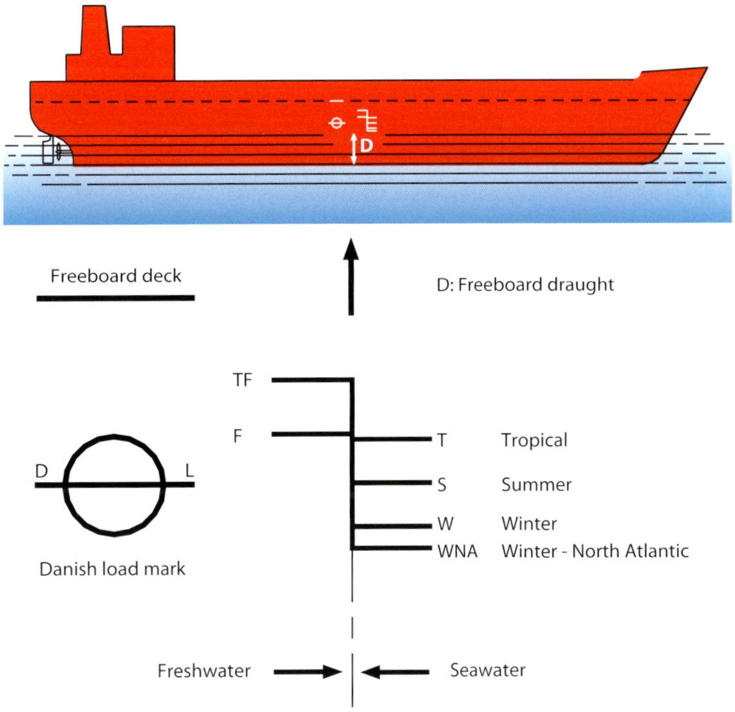

▲

The load lines of a ship.

There are load lines for sailing in fresh water and seawater, in tropical conditions and for sailing in summer and winter times. According to the international freeboard regulations, the summer freeboard draught is equal to the 'Scantling draught', which is the draught when dimensioning the hull.

The winter freeboard draught is lower than summer's, as there is a greater risk of encountering adverse weather. However, the freeboard draught in tropical seas is somewhat higher than the summer freeboard draught.

19.4 Ship dimensions

19.4.1 Displacement and loading capacity

When a ship is loaded to a given waterline, displacement equals the mass of water displaced by the ship.

So displacement equals the full weight of the laden ship; seawater has a specific mass of 1025 kg per m^3.

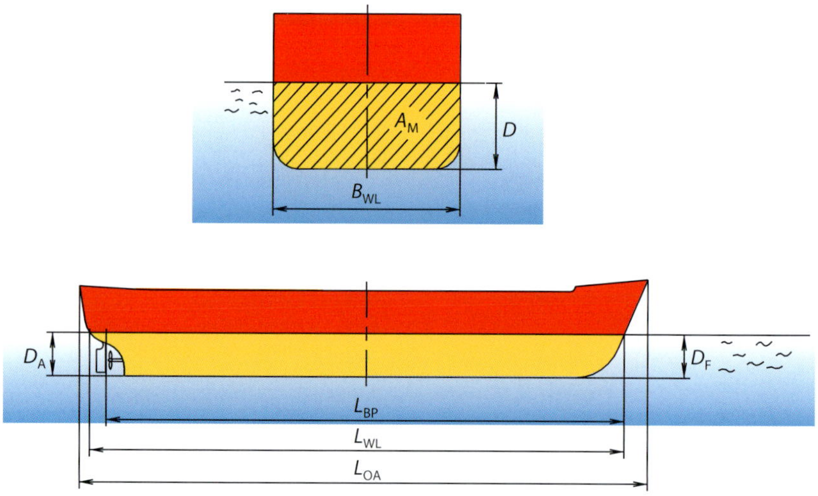

▶

Displacement and loading capacity.

L_{BP} Length between perpendiculars
L_{WL} Length on the waterline
L_{OA} Length overall
B_{WL} Breadth on the waterline
D Draught: $\frac{1}{2}(D_F + D_A)$
D_F Draught foreward
D_A Draught aft
A_M midship section area

Displacement includes the weight of the ship itself and its loading capacity. Loading capacity is often expressed in tons deadweight. Deadweight also includes fuel and other necessities such as lubricants and water required for ship propulsion.
Therefore, deadweight constitutes the difference between displacement at a given moment and the ship's own weight, expressed in metric tons.

Ship type	dwt/light weight ratio	Displ./dwt ratio
Tanker and Bulk carrier	6	1.17
Container ship	2.5-3.0	1.33-1.4

▲

The relationship between a ship's deadweight tonnage, lightweight and displacement.

The lightweight of a ship is not usually used to indicate the size of a ship. Deadweight, based on the loading capacity of the ship, including fuel, lubricating oil and other supplies required for ship operation provides a good indication of the size (read: loading capacity) of the ship.

19.4.2 Deadweight = displacement = ship's total weight

Displacement may also be expressed as the ship's displacement volume. SI unit is m^3.

19.4.3 Gross register tons/Net Register Tons

Ship's volume can also be expressed in Gross Register Tonnage (GRT) and Net Register Tonnage (NRT). A gross register ton equals 100 cubic feet or 2.83 cubic metres.
These values indicate the total internal volume of the ship according to certain complex rules that apply to said values and are often used to calculate commercial fees such as port and canal dues.

19.5 Hull shape

The part of the ship located below the waterline is most relevant in ship propulsion.
The ship measurements found below are used for the design draught. The selection of the design draught depends on load, weather conditions and operating conditions. The degree to which the ship is loaded is of importance: is the ship generally lightly- or fully loaded? In general, the most common draught between full-load displacement and ballast displacement is used.

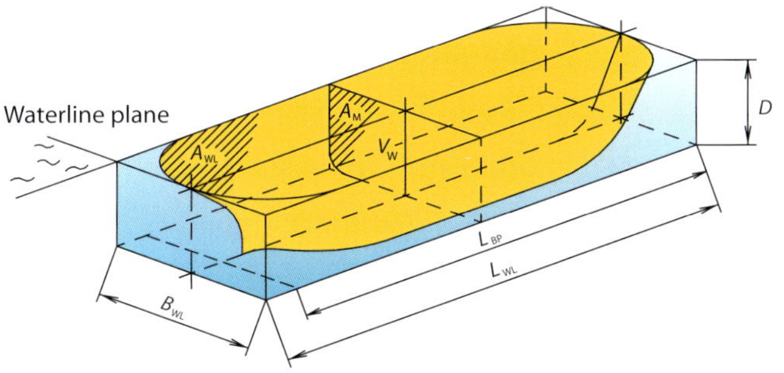

Volume of displacement: V_W

Waterline area: A_{WL}

Block coefficient, L_{WL} based: $C_B = \dfrac{V_W}{L_{WL} \times B_{WL} \times D}$

Midship section coefficient: $C_M = \dfrac{A_M}{B_{WL} \times D}$

Longitudinal prismatic coefficient: $C_P = \dfrac{V_W}{A_M \times L_{WL}}$

Waterplane area coefficient: $C_{WL} = \dfrac{A_{WL}}{L_{WL} \times B_{WL}}$

▲
Hull designs

Ship's lengths

Generally, **the overall length (L_{OA})**, the total length of the ship, does not affect the calculation of the hull's water resistance.
The length on the waterline (L_{WL}) and the length on the load lines (L_{PP}) are factors used for the calculation.
The length between perpendiculars (L_{PP}) is the length between the first load line, usually a vertical line through the stem's intersection with the waterline, and the aft load line, which normally coincides with the axis of the rudder.
Generally, this length is slightly shorter than the waterline length and is expressed as:
$L_{BP} = 0.97 \times L_{wl}$.

Draught D

The draught of the ship (*D*) is defined as the vertical distance from the waterline to the point on the hull, which is most, submerged. The draughts, foreward and aft, are usually identical when the ship is loaded.

Breadth on the waterline B_{WL}

This is the largest breadth on the waterline.

Block coefficient C_B

Various shape coefficients are used to denote the shape of the hull.

Ship type	Block coefficient C_B	Approximate ship speed *V* in knots
Lighter	0.90	5 – 10
Bulk carrier	0.80 – 0.85	12 – 17
Tanker	0.80 – 0.85	12 – 16
General cargo	0.55 – 0.75	13 – 22
Container ship	0.50 – 0.70	14 – 26
Ferry boat	0.50 – 0.70	15 – 26

▲
Block coefficient C_B.

The most important of these coefficients is the block coefficient C_B, which is the ratio of displacement volume and V_W the volume of a block with dimensions: length on the waterline, the breadth on the waterline and the draught according to the formula:

$$C_B = \frac{V_W}{L_{WL} \times B_{WL} \times D}$$

In this instance, the block coefficient refers to the length on the waterline.
However, ship manufacturers often use a block coefficient, which is based on the length between perpendiculars, thus producing a slightly larger block coefficient, as the length on the perpendiculars is slightly smaller than the length on the waterline.

$$C_B = \frac{V_W}{L_{BP} \times B_{WL} \times D}$$

A small block coefficient means less resistance and consequently potentially higher speeds.

19.5.1 Waterplane coefficient C_{WL}

The waterplane coefficient is the ratio of the waterline area A_{WL} and the product of the length on the waterline and the largest width of the ship.

$$C_{WL} = \frac{A_{WL}}{L_{WL} \times B_{WL}}$$

In general, the waterplane coefficient is approximately 0.10 higher than the block coefficient or:

$$C_{WL} = C_B + 0.10$$

The difference is slightly larger for fast ships with low block coefficients, where the stern is partially submerged and therefore part of the water-plane.

19.5.2 Midship coefficient C_M

The midship coefficient C_M is the ratio of the submerged midship section A_M (amidships between the fore and aft perpendiculars) and the product of the ship's width B_{WL} and the draught D.

$$C_M = \frac{A_M}{B_{WL} \times D}$$

For bulk carriers and tankers this coefficient is approximately 0.98 to 0.99 and for container ships 0.97 to 0.98.

19.5.3 Longitudinal prismatic coefficient C_P

Longitudinal prismatic coefficient C_P is the ratio of the volume of displacement V_W and the product of the mid-ship section area A_M and the length to the waterline L_{WL}.

$$C_P = \frac{V_W}{A_M \times L_{WL}} = \frac{V_W}{C_M \times B_{WL} \times D \times L_{WL}} = \frac{C_B}{C_M}$$

The C_P is not an independent form coefficient, but entirely dependent on the block coefficient C_B and the midship coefficient C_M.

19.5.4 Longitudinal centre of buoyancy LCB

The longitudinal centre of buoyancy (LCB) is the longitudinal distance from a point of reference to the centre of the fore and aft perpendiculars.
The distance is usually established as part of the length between the perpendiculars and is positive if the reference point is situated in the first half of the length between the perpendiculars and negative when situated in the last half.
For fast ships such as container ships, the LCB is negative, while slow-speed ships such as tankers and bulk carriers usually have a positive LCB.
The LCB is usually between – 3% and + 3%.

19.5.5 Fineness-ratio C_{LW}

The length/displacement ratio or Fineness ratio (C_{LW}) is the ratio of the waterline of the ship L_{WL} and the length of a cube with a volume that is equal to the volume displaced:

$$C_{LW} = \frac{L_{WL}}{3\sqrt{V_W}}$$

19.6 Ship's resistance *R*

A ship must move efficiently through the water and it is therefore imperative to diminish the ship's resistance to the water. The force of ship's resistance opposes the propulsion force.
The calculation of this resistance R plays a crucial part in the selection process of a propeller and consequently the propulsion engine.

19.6.1 General

The ship's resistance is partially influenced by the ship's speed, displacement and the hull shape.
The total resistance R_T consists of various resistances R, divided into three groups:

1 frictional resistance;
2 residual resistance;
3 air resistance.

The effects of the frictional and residual resistances depend on the shape of the hull and draught; the air-resistance effects are dependent on the surface area of the ship above the waterline.
In container ships, or heavily laden ships transporting large objects on deck, the air resistance is high.

◄ A loaded heavy-cargo ship has significant air resistance.

This increases when operating in head winds.

Water with a velocity, v and a density, ρ has a dynamic pressure of: $\frac{1}{2} \times \rho \times v^2$ (Bernoulli's principle). If water comes to a complete standstill as the result of an object (in this instance a hull) moving through it, the drag load of the water on the hull is equal to the dynamic pressure exerted on the hull, which results in a dynamic force on the hull, namely pressure × surface area.

This fact forms the basis for calculating and/or measuring the source resistances, R, of a ship's hull, using a dimensionless resistance coefficient C_ρ as related to the force K, which is dependent on the dynamic water pressure at a certain speed v on a surface area which in turn is equal to the hull's wet surface A_S. The rudder surface is also included in the wet surface area.
The general formulas for calculating ship's resistances are:

reference force = $K = \frac{1}{2} \times \rho \times v^2 \times A_S$ and

source resistance: $R = C \times K$

Based on many model-resistance tests in towing tanks and using the corresponding dimensionless hull data, methods have been developed to calculate the required resistance coefficients C, and consequently the corresponding source resistances R. In practice, the calculation of the ship's resistance for a particular ship design is performed in a towing tank using corresponding model and provides accurate data.

19.6.2 Frictional resistance R_F

The frictional resistance R_F of the ship's hull is dependent on the wet surface-area A_S of the hull and the specific frictional coefficient R_F.
The resistance increases as a result of fouling, the growth of algae, weeds and barnacles on the ship's hull. Many anti-fouling painting systems on the market limit or prevent fouling of the hull. These often contain tributyl tin compounds (TBT), which are extremely toxic.
IMO prohibited these paint systems for new constructions as of the first of January 2003, and a general prohibition was implemented on the first of January 2008.
TBT alternatives are available, such as copper-based anti-fouling paints. However, these appear to be less effective.

When a ship moves through the water, the frictional resistance increases to approximately the square root of the ship's velocity.

The frictional resistance of the ship in the water is approximately 70 to 90% of the total resistance. This applies to large low-speed ships, such as bulk carriers and oil tankers (deep draught).
For large high-speed cruise ships of relatively small draught, this can be less than 40%.

The frictional resistance is formulated by:

$R_F = C_F \times K.$

▲
Frictional resistance R_F. High-speed, large container ships have propulsive power outputs of up to approximately 70,000 kW.

With velocities up to 25 knots, more power is required. Container ships are always loaded; containers are added and removed, but these ships rarely sail unloaded. The propeller load and therefore the engine load are usually identical.

19.6.3 Residual resistance R_R

The residual resistance R_R comprises the wave- and eddy resistance. Wave resistance refers to the energy loss caused by waves produced by the ship travelling through the water, whereas eddy resistance that leads to power loss is caused by the separation of the water surface by a hull resulting in swirls and eddies. These eddies are prevalent at the stern.
Wave resistance at low speeds is proportional to the square root of the velocity. However, it increases rapidly at higher speeds.

This essentially means that a speed restriction is created, the so-called the maximum hull velocity. A further increase of the diesel-engine output above this velocity will not result in increased ship speed! **All the extra power is converted into wave energy!**

▼
The 'Berge Stahl' with a load capacity of more than 300,000 tonnes and a draught of 19 metres is one of the largest bulk carriers in the world. Large tankers and bulk carriers have low operational speeds. In a loaded condition, 90% of the hull is below the waterline.

To achieve high speeds with these displacements, and breadths reaching from 30 to 70 metres would require huge power outputs.

▼
The picture shows an empty tanker in ballast, the bow and the rudder are partially protruding from the water. The propeller operates just below the water surface.

The residual resistance amounts to approximately 8 to 25% of the total resistance for low-speed ships.
For high-speed ships, this can rise from 40 to 60% of the total resistance.
Shallow water also has a significant effect on the residual resistance, as it is more difficult for the water displaced by the ship to flow astern. (Resistance of the displaced water volume over the ground).

The specific frictional resistance is defined as follows:

$R_R = C_R \times K.$

19.6.4 Air resistance R_A

In principle, the air resistance in calm water is proportional to the square of the ship's velocity and proportional to the cross-sectional surface area of the ship above the waterline.
Normally, air resistance is approximately 2% of the total resistance.
For container ships carrying high loads in a head wind, air resistance can rise to 10% of the total resistance. This also applies for large passenger liners, and in particular, heavily laden freighters.

Similar to the aforementioned resistances, air resistance is expressed as: $R_A - C_A \times K$, but can occasionally be based on 90% of the dynamic air pressure with a velocity v. In this case, the formula is as follows:

$R_A = 0.90 \times ½ \times \rho_{AIR} \times v^2 \times A_{AIR}$,

where ρ_{AIR} constitutes the specific air mass and A_{AIR}, the cross-sectional surface area of the ship above the waterline.

19.6.5 Towing resistance R_T and the effective (towing) power P_E

The total towing resistance R_T is calculated as follows: $R_T = R_E + R_A$.
Therefore, the corresponding effective (towing) power, P_E, required to move the ship or allow it to be towed through the water at a velocity v is:

$P_E = v \times R_T.$

However, the power supplied to the propeller, P_D, required to move the ship at velocity v, is slightly higher. This is due to the flow conditions around the revolving propeller and propeller efficiency itself, which will be discussed later in this chapter.

19.6.6 General total ship resistance

If the frictional resistance is divided into wave- and eddy resistances, as described earlier, the distribution of the total towing resistance R_T can be expressed in percentages (see figure). This is applicable to both low-and high-speed ships.

19.6.7 Ship's resistance increase during operation

When the ship is in service, the paint coatings on the hull will gradually degrade. Steel corrosion will slowly increase; weeds and barnacles will attach themselves to the hull. Adverse weather conditions, perhaps in conjunction with a badly laden ship, can cause deformation of the bottom plates of the ship. This becomes evident when the beams, which look like ribs, protrude from the ship's hull. The hull becomes fouled and its surface is no longer smooth, which consequently results in a frictional resistance increase. The propeller surface can become rough and fouled.

◄ Air resistance R_A. In relation to their displacement, passenger ships, cruise ships and ferries have tall superstructures and are therefore susceptible to (head) winds.

▶

Total ship's resistance in general.

The right column shows low-speed ships, such as bulk carriers and tankers, **the left column** applies to cruise ships and ferries. Container ships may sometimes be placed in this column or between the two columns.
The main difference between both columns is, as previously mentioned, the wave resistance.
In conclusion, all resistances are proportional to the square of the ship's velocity, but for higher speeds, the wave resistance increases more rapidly and is therefore a larger portion of the total resistance.

Type of resistance	% of R_T High speed ship	% of R_T Low speed ship
R_F = Friction	45	90
R_W = Wave	40	5
R_E = Eddy	5	3
R_A = Air	10	2

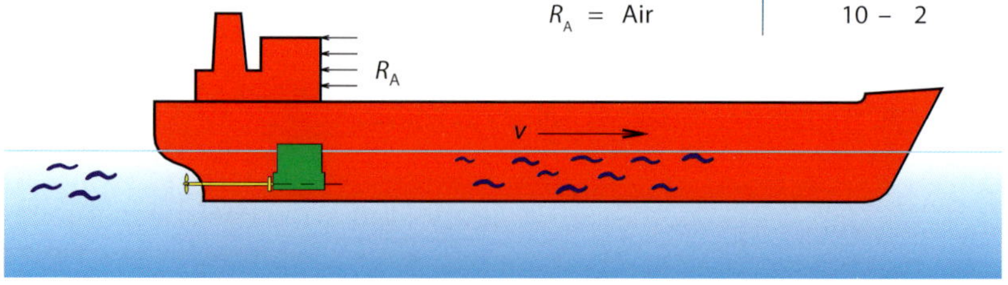

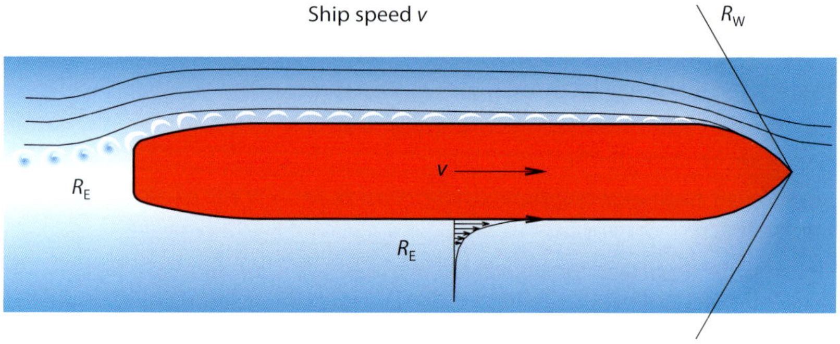

▶

This tendency is clearly visible for a 600 TEU container ship, originally designed for a service speed of 15 knots.

Without any alterations to the hull design, it was requested that the speed for a sister ship of similar size be increased to approximately 17.6 knots. However, this would result in a relatively high wave resistance and would require **twice** as much propulsion power!

A further increase of the propulsion power would only produce a slight increase in the ship's velocity as the largest part of the additional propulsive power is converted into wave energy.
The ship has reached the maximum hull velocity valid for the hull design and only produces a so-called 'wave wall'. The shape of the ship's hull will have to be modified to obtain the required speed increase.
In 2008, there was a tendency to reduce slightly the service velocity of ships in order to reduce fuel costs.
Since 2009, a large number of container ships ran at a reduced speed. This is referred to as slow steaming.

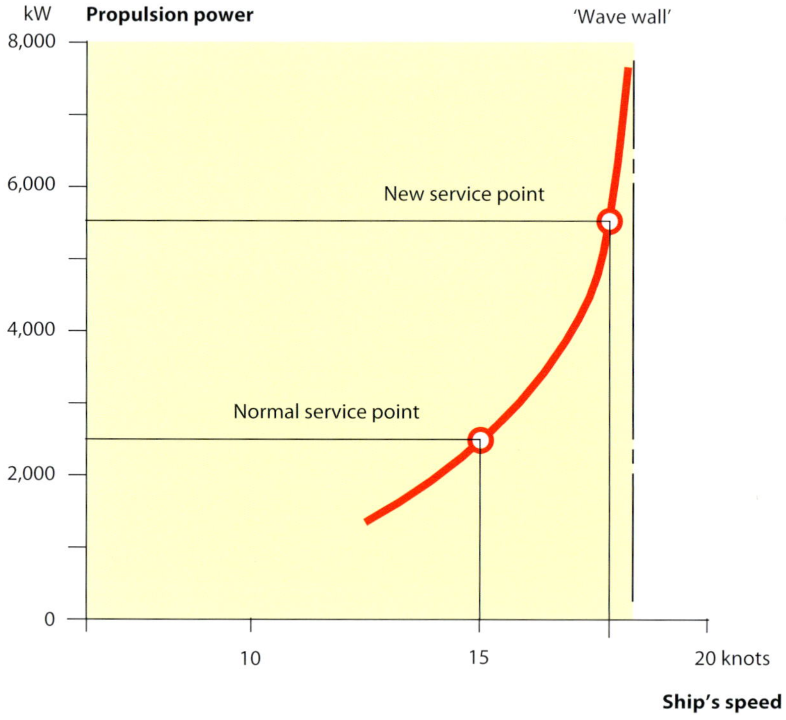

Power and speed relationship for a 600 TEU container ship

Estimates of average increase in resistance for ships navigating the main routes	
North Atlantic route, navigation westward	25 – 35%
North Atlantic route, navigation eastward	20 – 25%
Europe – Australia	20 – 25%
Europe – East Asia	20 – 25%
The Pacific routes	20 – 30%

▲

Obviously, resistance is increased by (heavy) seas, wind conditions and currents.

In comparison to a ship's resistance in calm weather conditions, heavy seas and head winds can increase the total ship's resistance by 50 to 100%.

On North Atlantic routes, the first percentage of a 50% increase corresponds to the summer period and the second percentage of a 100% increase to the winter period.

Data reveals that for a typical 140,000-ton (deadweight) bulk carrier on certain routes, especially on a northern route in the Pacific Ocean between Japan and Canada, the increased resistance when loaded shows extreme increases of 220% with an average increase of 100%.

In calm water, the resistance caused by fouling can result in a total ship's resistance increase of 25 to 50%.

Generally, the larger the ship, the less the increase in resistance.

However, the resistance of large, laden ships with a deep draught, such as bulk carriers and tankers, may alter rapidly with hull fouling below the waterline.

In Summary

In principle and theory, the resistance increase in heavy weather is proportional to:

- the amount of adverse wind and current conditions;
- the size of the waves.

It is difficult to distinguish between these two factors.

19.7 Screw propulsion

The traditional method of marine propulsion is with a propeller. This has been the most widely used method of propulsion for most ships for over a hundred years. Sometimes two and sporadically three propellers are fitted.

The required thrust T, for the propeller enables the ship to move at a velocity, v and this is usually higher than the generated towing resistance R_T.

As described earlier, this divergence is caused by differences in the water-flow between propulsion and towing.

19.8 Propeller types

Propellers can be divided into two main groups:

1 fixed-pitch propellers (**FPP**);

2 controllable-pitch propellers (**CPP**). These are also known as **AP** (adjustable pitch) and **VP** (variable pitch) propellers.

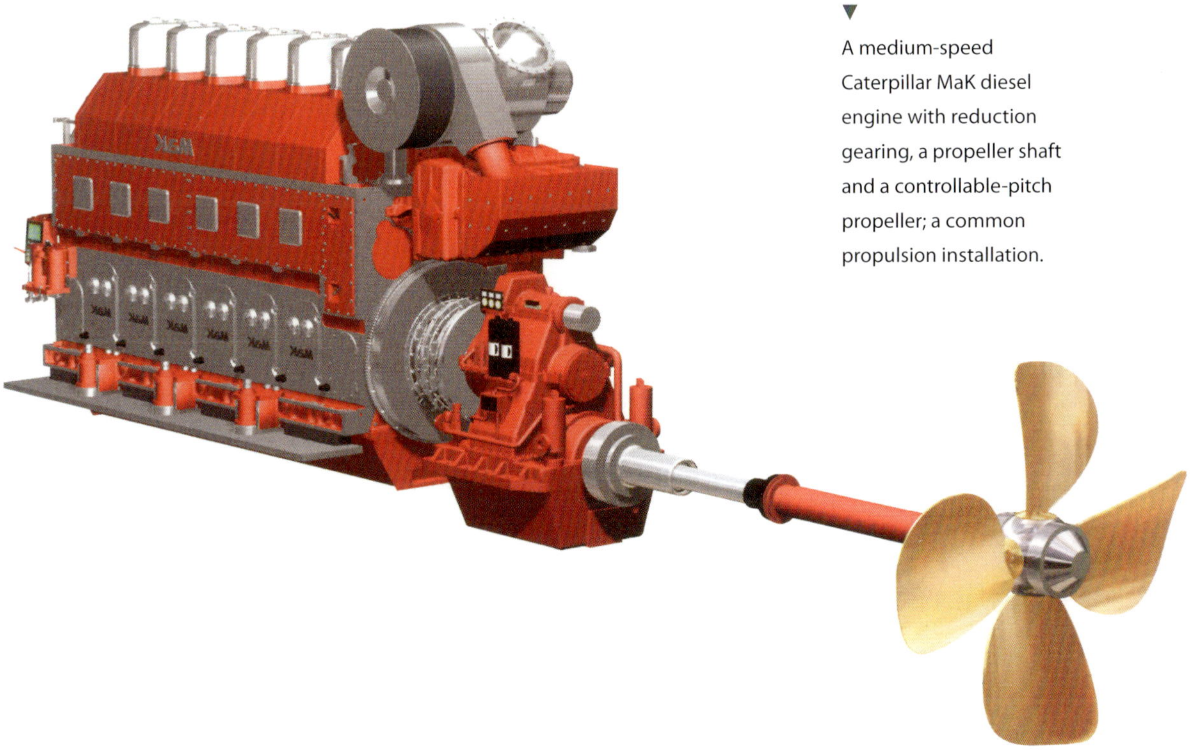

▼

A medium-speed Caterpillar MaK diesel engine with reduction gearing, a propeller shaft and a controllable-pitch propeller; a common propulsion installation.

Velocities

Ship's speed	: v
Arriving water velocity to propeller (Speed of advance of propeller)	: v_A
Effective wake velocity	: $v_W = v - v_A$
Wake fraction coefficient	: $w = \frac{v - v_A}{v}$

Forces

Towing resistance	: R_T
Thrust force	: T
Thrust deduction fraction	: $F = T - R_T$
Thrust deduction coefficient	: $t = \frac{T - R_T}{T}$

Power

Effective (towing) power	: $P_E = R_T \times v$
Thrust power delivered by the propeller to water	: $P_T = \frac{P_E}{\eta_H}$
Power delivered to propeller	: $P_D = \frac{P_T}{\eta_B}$
Brake power of main engine	: $P_B = \frac{P_D}{\eta_S}$

Efficiencies

Hull efficiency	: $\eta_H = \frac{1 - t}{1 - w}$
Relative rotative efficiency	: η_R
Propeller efficiency – open water	: η_O
Propeller efficiency – behind hull	: $\eta_B = \eta_O \times \eta_R$
Propulsive efficiency	: $\eta_O = \eta_H \times \eta_B$
Shaft efficiency	: η_S
Total efficiency	: η_T

$$\eta_T = \frac{P_E}{P_B} = \frac{P_E}{P_T} \times \frac{P_T}{P_D} \times \frac{P_D}{P_B} = \eta_H \times \eta_B \times \eta_S = \eta_H \times \eta_O \times \eta_R \times \eta_S$$

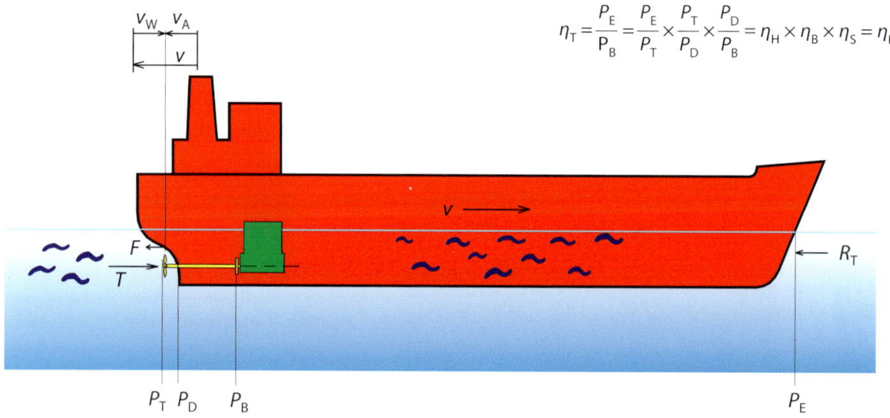

▲

The propulsion of a ship theory.

►

Winter in the Baltic sea.

The resistance that a ship experiences with ice floes is not included in this chapter.

19.8.1 Fixed-pitch propellers

FP-propellers are cast in one piece and usually made of a very strong copper alloy. The position of the blades and therefore the propeller pitch is fixed and cannot be modified. When operating in, for instance, poor weather conditions, propeller performance properties such as thrust will only alter in combination with the supplied engine power and propeller speed.
Most ships not requiring sophisticated manoeuvring capabilities such as, container ships, bulk carriers and tankers are equipped with FP-propellers. These ships generally travel great distances on the high seas and do not berth frequently. They can safely enter ports to load and discharge aided by their bow thrusters (container ships) and/or by tugs.

19.8.2 Controllable-pitch propellers

Controllable-pitch propellers have a relatively large hub as this space houses the hydraulic mechanism that controls the pitch (angle) of the adjustable blades. The price of a controllable-pitch propeller can be three to four times that of a fixed-pitch propeller. Furthermore, the propeller efficiency is somewhat lower due to the relatively larger hub.
The propeller is also more vulnerable in service than a fixed-pitch propeller.

Controllable-pitch propellers are often used in ships that make frequent manoeuvres. Ferries, container-feeder ships, dredgers, Ro-Ro ships and a great many other, slightly smaller ships use CPP-propellers.

Another important advantage is that a directly reversible engine is no longer required. Thus, the diesel-engine design remains simple and it operates at a high, fixed rpm, rapidly producing the desired power for the propeller during manoeuvring. So manoeuvring can take place in a safe, swift manner.

For further details regarding propellers, propeller shafts and seals, see Chapter 31, *Propellers*.

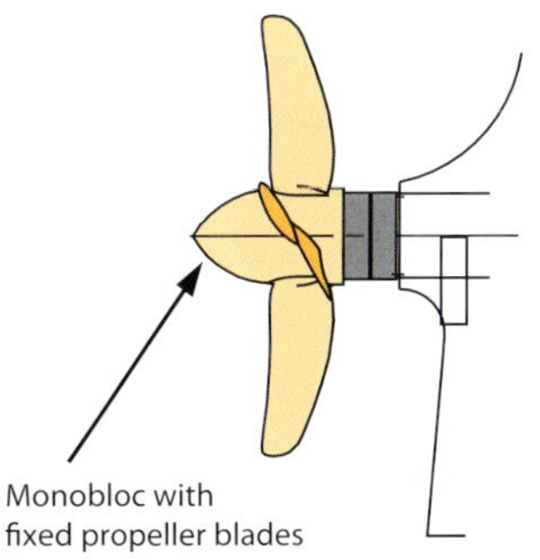

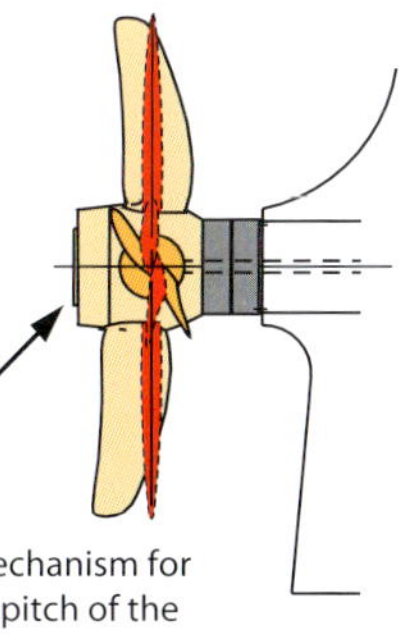

▲ Propeller types.

19.9 Flow conditions around the propeller

19.9.1 Wake coefficient *w*

When the ship moves through water, the friction of the hull creates a 'friction layer' of water on the surface of the hull. In this friction layer the water velocity at the surface of the hull is equal to that of the ship, but is reduced as the water distances itself from the hull surface until the water velocity is zero. This takes place at a certain distance from the hull, which is equal to the outer surface of the friction layer.

◄ A large six-bladed fixed-pitch propeller for a large container ship (Wärtsilä Propulsion).

▲

The wake fraction behind a large container ship; the 'Norasia Enterprise' at an operational speed of 26 knots.

The enormous displacement of the ship's hull caused by the high speed creates a 'pit' of about 1 metre behind the ship. As shown here, the water cannot immediately fill up the pit from the sides. The eddies in the water generated by the rotating propeller are also clearly visible.

Knots for ship's velocities knot = 1 nautical mile per hour = 1852 metres per hour

Therefore explaining why water near the propeller has a wake velocity, v_W in the same direction as the ship's velocity v. This means that the speed of the arriving water, v_A, at the propeller, given as the average velocity over the propeller's wake area v_W, **is lower** than the ship's velocity v.

Therefore, the effective wake velocity at the propeller is equal to $v_W = v - v_A$ and is expressed as a dimensionless number by means of the wake coefficient w. The generally applied wake coefficient as defined by Taylor is as follows:

$$w = \frac{v_W}{v} = \frac{v - v_A}{v}$$

The value of the wake coefficient is largely dependent on the shape of the hull, but also on the location of the propeller and its size. It has a significant effect on the propeller efficiency.

The propeller diameter or rather the ratio of the propeller diameter d and the ship's length L_{WL} somewhat affects the wake fraction coefficient as $\frac{d}{L_{WL}}$ and gives a rough indication of the way in which the propeller performs in the wake field of the hull. The higher the ratio d to L_{WL}, the lower the wake-fraction coefficient w.

Obviously, the wake-fraction coefficient w increases when the hull is fouled.

For single-screw ships the wake-fraction factor w is usually in the 0.20 to 0.45 region, with a corresponding flow velocity to the propeller of 0.80 to 0.55 of the ship's velocity v. The larger the block coefficient, the larger the wake-fraction coefficient.

On twin-screw vessels with a conventionally shaped stern, the propellers are normally positioned outside the 'friction layer', which results in a significantly lower wake-fraction coefficient w.

A high wake-fraction coefficient increases the risk of propeller cavitation, as the water speed around the propeller is often irregular and heterogeneous.

A more homogeneous wake field for the propeller, involving a higher speed of advance of the propeller v_A, is at times required and can be obtained in various ways, for instance by placing the propellers in nozzles. Nevertheless, the preferred method is to design the stern in such a way that an optimum wake fraction is achieved.

19.9.2 Thrust-deduction coefficient *t*

Screw rotation draws the water in front of it astern towards the propeller. This results in an extra resistance on the hull, called 'augment of resistance' or, in relation to the total required thrust force T on the propeller, 'thrust-deduction fraction' F. This means that the thrust force T on the propeller has to overcome the ship's resistance R_T as well as this 'loss of thrust' F.

The thrust-deduction fraction F is expressed as a dimensionless number.

The thrust-deduction fraction t is defined as:

$$t = \frac{F}{T} = \frac{T - R_T}{T}$$

This thrust-deduction coefficient t is calculated with calculation models established by research carried out on various hull shapes.

Generally, the thrust-deduction coefficient increases as the wake fraction coefficient w increases.

The shape of the hull plays a key role; a bulbous bow can decrease the thrust-deduction coefficient at, for instance, low speeds.

The thrust-deduction coefficient t for a single-screw ship is usually in the region of 0.12 to 0.30 as a ship with a large block coefficient has a large thrust-deduction coefficient.

For twin-screw vessels with a conventional stern, the thrust-deduction coefficient t is much lower as the propellers draw water further away from the hull.

19.9.3 Efficiencies

The hull efficiency η_H is defined as the ratio between effective (towing) power $P_E = R_T \times v$ and the thrust power which the propeller delivers to the water $P_T = T \times v_A$, from which follows that:

$$\eta_H = \frac{P_E}{P_T} = \frac{R_T \times v}{T \times v_A} = \frac{R_T/T}{v_A/v} = \frac{1 - t}{1 - w}$$

For a single-screw vessel, the hull efficiency η_H is normally in the region of 1.1 to 1.4 with high values for ships with high block coefficients.
For twin-screw ships with a conventional stern, the hull efficiency η_H is approximately 0.95 to 1.05, again with the highest values for ships with high block coefficients.

19.9.4 Open-water propeller efficiency η_O

The propeller efficiency η_O **is related to working in open water**, which is to say, the propeller operates in a homogeneous wake field with no hull in front of it. The propeller efficiency is dependent on the speed of advance v_A, the thrust force T, the rate of revolution of the propeller per time unit (minute), the propeller diameter and the design of the propeller.
Considerations are:
- the number of blades;
- the surface area of the propeller blades in relationship to the total circular surface area of the propeller circle;
- the pitch/diameter ratio, which will be discussed later in this chapter.

The propeller efficiency η_O may vary from 0.35 to 0.75, with high values for fast-running propellers.

As a rule of thumb, it can be stated that roughly a third of all the energy supplied to the engine in the form of fuel is used to provide the ship with a particular speed.

19.9.5 Relative rotative efficiency η_R of the propeller

The actual velocity of water flowing to the propeller behind the hull is not constant and does not reach the circular propeller surface area at the correct angle, but has a rotational flow. This explains the discrepancy between a propeller operating in open water and a propeller operating behind a ship's hull. This difference is called the relative rotative efficiency η_R.

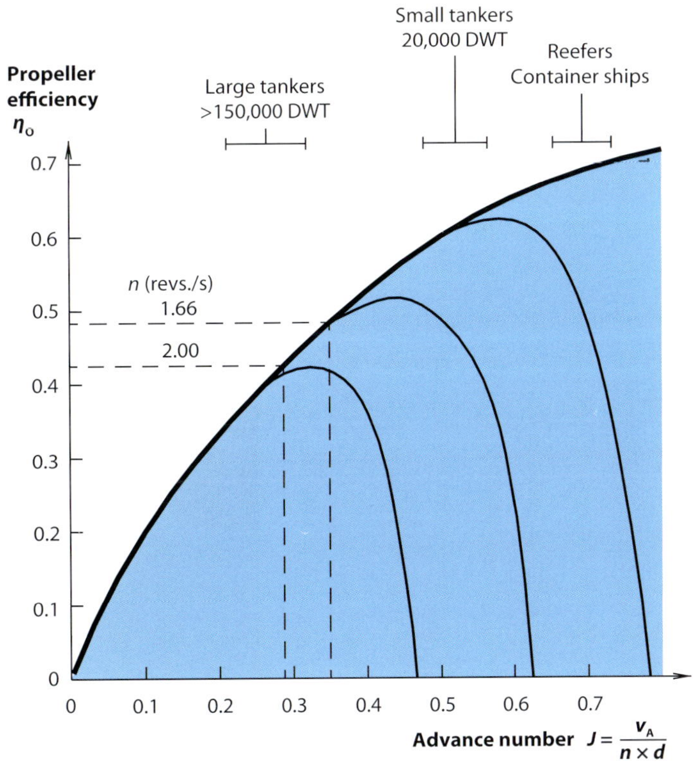

▲
The potential propeller efficiency η_O in relation to the speed of advance v_A.

J is the 'advance number' of the propeller and is dimensionless.

Furthermore, the rate of revolution per second; this is clearly a table for fixed-pitch propellers of large ships, directly driven by two-stroke crosshead engines.
Therefore, the efficiency of a propeller varies from 45 to 70%. This means that at a total engine efficiency of 50% the total propulsive efficiency is much lower.
Let us assume that the total engine efficiency is 50% and the propeller efficiency 65%.
Then, the total propulsive efficiency of the fuel (100%) is $0.5 \times 0.65 = 0.3250$ or 32.5%.

▶
A picture of a large container ship with a speed of 27 knots.

The bow wave, 1, is relatively small. The propeller water, 2, with the wake fraction can be clearly seen in the picture. This container ship of 5500 TEU produces fewer waves at sea than a cutter at full speed!

The relative rotative efficiency η_R on single propeller vessels lies between 1.0 and 1.07. In other words, the rotating motion of the water has a positive effect.
The relative rotative efficiency for two propellers lies slightly below 1, approximately 0.98.
Combined with w, the wake fraction factor, and t, the thrust deduction coefficient, the relative rotative efficiency η_R is often used to compare the model tank tests with the theory.

19.9.6 Propeller efficiency η_B operating astern

The ratio between the thrust power P_T, which is delivered to the water by the propeller, and the power P_D, which is delivered to the propeller by the propulsion engine is indicated by the propeller efficiency η_B, and defined as follows:

$$\eta_B = \frac{P_T}{P_D} \times \eta_O \times \eta_R$$

19.9.7 Propulsive efficiency η_D

The propulsive efficiency η_D, not to be confused with the propeller efficiency η_O in open water, is equal to the ratio between the effective (towing-) power P_E and the required power delivered to the propeller or:

$$\eta_D = \frac{P_E}{P_D} = \frac{P_E}{P_T} \times \frac{P_T}{P_D}$$

$$\eta_D = \eta_H \times \eta_B = \eta_H \times \eta_O \times \eta_R$$

As can be seen the propulsive efficiency η_D is equal to the product of the hull efficiency η_H, the open-water propeller efficiency η_O and the relative rotative efficiency η_R of the propeller, in which the latter is less important.
In view of this, it can be inferred that a ship's hull with a high wake-fraction coefficient w, and consequently a high hull efficiency η_H provides the best conditions for the highest propulsive efficiency η_D.
However, the open-water propeller efficiency η_O is also dependent on the speed of advance v_A which increases as the wake-fraction coefficient w rises, and therefore often the opposite effect is achieved.

Generally, the best propulsive efficiency is achieved when the propeller operates in a homogeneous wake field.

19.9.8 Shaft efficiency η_S

The shaft efficiency η_S is, amongst others, dependent on the alignment and the lubrication of the shaft bearings and on the reduction gear, if any.

The shaft efficiency η_S is equal to the ratio of the power P_D transferred to the propeller and the shaft power P_B delivered by the main engine.

$$\eta_S = \frac{P_D}{P_B}$$

The shaft efficiency normally amounts to about 0.985, but may vary from 0.96 to 0.995.

19.9.9 Total propulsive efficiency η_T

The total propulsive efficiency η_T is equal to the ratio between effective (towing) power P_E and the brake power supplied to the propulsion engine η_B and is expressed as follows:

$$\eta_T = \frac{P_E}{P_B} = \frac{P_E}{P_D} \times \frac{P_D}{P_B}$$

$$\eta_T = \eta_D \times \eta_S = \eta_H \times \eta_O \times \eta_R \times \eta_S$$

For strength and manufacturing purposes, the propeller diameter will seldom exceed 10 metres and a power output is in the region of 90,000 kW. These values were achieved in the 'Emma Maersk' in 2006!
The largest propeller diameter (2011) is 11 metres and the propeller has four blades!

19.10 Propeller dimensions

Propeller diameter *d*.

The largest propeller diameter is normally selected in order to achieve the largest possible propulsive efficiency η_D. However, a number of conditions must be considered.
For instance: the shape of the hull at the stern of the ship may vary greatly and is predominantly dependent on the type of ship, the ship design and the required clearance between the tips of the propeller blades and the ship's hull. This is closely related to the propeller type.

For bulk carriers and tankers, which often sail in ballast conditions, a request for a fully immersed propeller is an option.
Container ships do not require a maximum size as they seldom sail in ballast conditions.
The above-mentioned factors attest to the fact that it is relatively difficult to calculate an exact diameter for the propeller. However, the rule of thumb with regard to propeller size is as follows:

The ratio between the propeller diameter and the draught of the ship can be described as follows: a large propeller diameter automatically means a low speed of rotation *n*.

For bulk carriers and tankers, the following applies:

$d/D <$ approximately 0.65.

For container ships:

$d/D <$ approximately 0.74.

▼

Propac; a Wärtsilä propulsion system.

Shown here, in-line engines with 6, 8 and 9 cylinders and bores of 200 and 260 millimetres.
Included: the nozzle-propeller systems with all data.

Propac ST selection table

Engine type		6L20		8L20		9L20		6L26		8L26		9L26	
MCR engine power CPP	kW	1080		1440		1620		2025		2600		2925	
	HP	1469		1958		2203		2754		3536		3978	
MCR engine power FPP	kW	1080		1440		1620		1950		2480		2790	
	HP	1469		1958		2203		2652		3373		3794	
MCR engine speed	rpm	1000		1000		1000		1000		1000		1000	
Thruster type		200		225		225		250		275		300	
Propeller diameter	mm	1900	2100	2100	2300	2100	2300	2400	2600	2600	2800	2800	3000
Propeller speed	rpm	318	318	248	248	274	274	257	257	245	245	215	215
Bollard pull with CPP													
in 19A nozzle	tonne	35	37	46	49	50	53	63	66	79	82	90	93
in HR nozzle	tonne	39	40	51	53	55	57	69	71	86	89	99	102
Bollard pull with FPP													
in 19A nozzle	tonne	35	36	46	49	50	52	62	64	76	79	87	90
in HR nozzle	tonne	38	39	51	53	55	57	67	69	83	85	95	98
MCD-type for steerable thrusters with FPP													
LD type		3000-3	3000-3	3000-4	3000-4	3000-4	3000-4	3000-5	3000-5	3000-6	3000-6	not available ->	
HD type		3000-3	3000-3	3000-6	3000-6	3000-7	3000-7	3000-7	3000-7	3000-8	3000-8	not available ->	

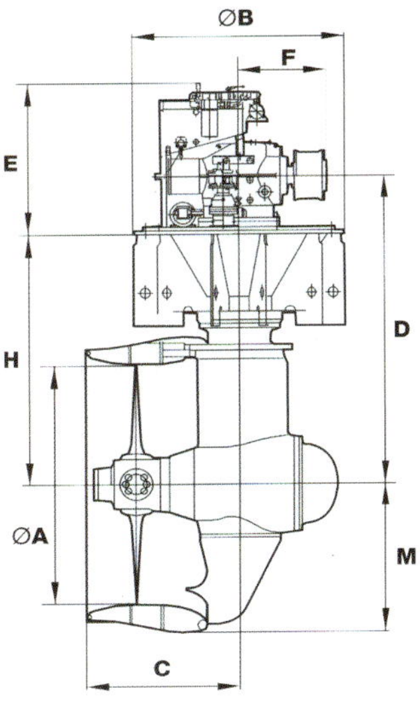

Propac ST main dimensions

Thruster type FS/CS	A mm	B mm	C mm	D mm	E mm	F mm	H mm	M mm
175	1600	1600	1200	2100	910	610	1743	985
	1800		1230	2200			1843	1109
200	1900	1900	1300	2500	1200	740	2048	1179
	2100		1350	2600			2148	1303
225	2100	2100	1400	2630	1210	740	2160	1299
	2300		1450	2830			2360	1423
250	2400	2100	1525	3100	1435	840	2512	1485
	2600		1575	3200			2612	1614
275	2600	2850	1665	3500	1465	877	2890	1609
	2800		1765	3620			3010	1731
300	2800	2850	1770	3700	1465	877	3090	1735
	3000		1870	3700			3090	1858

▶

The water should advance towards the propeller as smoothly as possible, so without eddies. The propeller rotates completely 'freely' under the cut-away section of the ship.

▶

This graph represents the ideal propeller diameter, here from 0.3 metres (small ships, yachts) to 3 metres (larger ships, fishing craft).

horizontal: propeller speed from 100 to 5000 rpm

vertical: power in kW to 2000 kW

slanted line: This represents the ideal propeller diameter from, 0.3 metres (small ships, yachts) up to 3 metres (larger ships, fishing craft). The larger the propeller diameter, the higher the propeller efficiency. The smaller the propeller diameter, the lower the propeller efficiency.

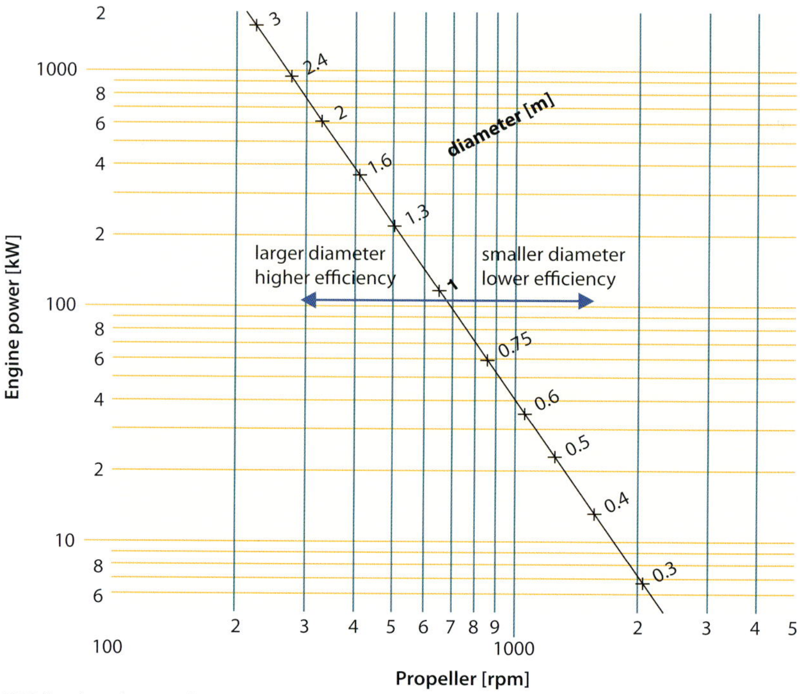

CIPS fixed-pitch propeller sizes

▶

A graph from MAN Diesel showing the optimum ratio of the propeller diameter (vertical) to the engine power (horizontal) and the speed (top right).

Example: At a propeller diameter of 6000 mm (6 metres) and a speed of 100 rpm, a shaft power of 7000 kW is required. An optimum propeller efficiency of approximately 70% is obtained.

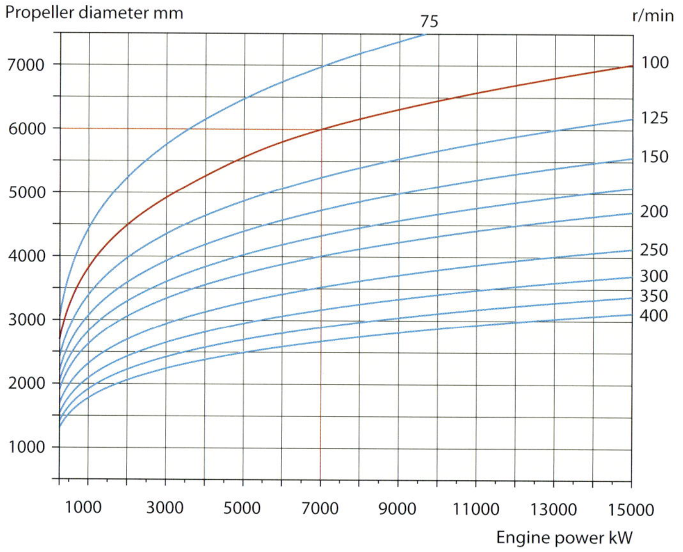

19.10.1 Number of propeller blades

Propellers can be manufactured with two, three, four, five or six blades. The lower the number of blades; the higher the propeller efficiency. For reasons of strength, two or three blades do not suffice for heavily loaded propellers.
Large ships, which are propelled by two-stroke crosshead engines generally, use four-bladed propellers. Large container ships with speeds reaching 25 knots and large propulsion engines sometimes require five- or six-bladed propellers.
Due to vibrations, some propellers are not used in certain instances. Propellers with a different number of blades are selected in order to avoid generating natural frequencies in the ship's superstructure.

19.10.2 Disk-area coefficient

The disk-area coefficient, in older literature referred to as the expanded-blade area ratio, is the ratio between the surface of the propeller blade and its disk area. A factor of 0.55 is considered adequate.

For traditional four-bladed propellers, the disk-area coefficient is insignificant, as a value in excess of 0.55 will result in extra resistance on the propeller in the water and therefore result in a minimal increase in propeller efficiency.

Heavy propeller loads are found on large ships, they often have five- or six-bladed propellers and higher disk-area coefficients. On naval vessels they can be as high as 1.2 (frigates).

▼
The disk-area coefficient is high, about 0.7.

19.10.3 Pitch/diameter ratio of the propeller *p/d*

The pitch/diameter ratio *p*/*d* expresses the ratio between the pitch of the propeller *p* and the propeller diameter *d*.
The pitch *p* is the distance that the propeller travels through the water per revolution, if there is no 'slip' (see next section).
As the pitch may vary along the length of the blade (the blade's radius), the ratio is normally related to the pitch at 0.7 of the diameter (0.7 *r*), where $r = \frac{1}{2} d$, the radius of the propeller.
In order to obtain optimum propeller efficiency at a given propeller diameter, an optimum pitch/diameter ratio should be found, which in turn should correspond to a certain design rate of revolution.
If, for instance, a propeller is designed for a low revolution rate, the pitch/diameter ratio should be increased and vice versa for efficiency.
If a low design rate of revolution is required and providing that the draught of the ship allows this, it is possible to opt for a larger propeller diameter, which will effect an increase of the propulsive efficiency.

▼
A simple pitch measurement for a fixed-pitch propeller for a cutter.

Using a measuring pin, the pitch on each radius of the propeller can be determined. In fact, the drop of the measuring pin at a given angle is measured.
The pitch of a fixed-pitch propeller is often measured at 0.7 of the radius.

19.10.4 Propeller coefficients J, K_T and K_Q

Propeller theory is based on models, but in order to enable a general use of the theory, a number of dimensionless propeller coefficients have been introduced, in relation to:

- the propeller diameter d;
- the rate of revolution n;
- the specific mass of the water ρ.

The three most important coefficients are discussed below.

The advance number of the propeller J is, as mentioned earlier, a dimensionless expression of the propeller velocity of advance v_A.

$$J = \frac{v_A}{n \times d}$$

The thrust coefficient is expressed dimensionless as follows:

$$K_T = \frac{T}{\rho \times n^2 \times d^4}.$$

And the propeller torque as:

$$Q = \frac{P_D}{2 \times \pi \times n} \qquad K_Q = \frac{Q}{\rho \times n^2 \times d^5}.$$

The open-water propeller efficiency η_O can be calculated from the aforementioned coefficients, because as previously mentioned, the propeller efficiency η_O is as follows:

$$\eta_O = \frac{P_T}{P_D} = \frac{T \times V_A}{Q \times 2 \times \pi \times n} = \frac{K_T}{K_Q} \times \frac{J}{2 \times \pi}.$$

With the help of special and complex propeller diagrams, which, amongst others, include J, K_T and K_Q curves, it is possible to calculate/find the propeller's sizes, efficiency, thrust, and shaft power.

19.10.5 Manufacturing accuracy for propellers

ISO 484-1:1981 (CE)		
Class	Manufacturing accuracy	Mean pitch for propeller
S	Very high accuracy	+/– 0.5%
I	High accuracy	+/– 0.75%
II	Medium accuracy	+/– 1.00%
III	Wide tolerances	+/– 3.00%

▲

Prior to manufacturing a propeller, the accuracy-class standard must be established.

Example ISO 484-1:1981 (CE) which has four different accuracy classes.
Each class describes, amongst others, the maximum tolerance of the average pitch of the manufactured propeller, including the maximum tolerance of the corresponding propeller speed (the rate of revolution).
The propeller price is dependent on the chosen accuracy class, with the lowest price for class III.
However, it is inadvisable to opt for class III as the tolerances (+/– 3%) are too high.
The tolerance of the pitch should not exceed +/– 1.0%.
The accuracy of manufacturing the propeller corresponds to the **accuracy** of the propeller speed (rate of revolution) tolerance and is +/– 1.0%.
Here the total accuracy of the propeller has a tolerance of +/– 2.0%.
This also corresponds with the operating conditions in heavy weather.

19.10.6 Influence of propeller diameter and the pitch/diameter ratio on the propulsive efficiency η_D

As mentioned earlier, the highest possible propulsive efficiency at a certain required velocity of the ship is achieved by using the largest possible propeller diameter d, in conjunction with the corresponding optimum pitch/diameter ratio p/d.

▶

An example for an 80,000 dwt crude-oil tanker, with a service speed of 14.5 knots and a maximum propeller diameter of 7.2 metres.

According to the **blue line**, a propeller with a maximum diameter of 7.2 metres has an optimum pitch/diameter ratio of 0.7 and the lowest possible shaft power of 8820 KW at a propeller speed of 100 r/min.

If the pitch for this diameter is altered, the propulsive efficiency is decreased, or the necessary shaft power will increase, see the **red curve**.

The **blue line** shows that when a larger propeller diameter is feasible, the required shaft power is decreased to 8690 KW at 94 r/min; in other words, the larger the propeller, the lower the optimum rate of revolution of the propeller.

The **red curve** shows that with regard to propulsion, it is always advantageous to use the largest possible propeller diameter even though the propeller speed could be too low at the optimum pitch/diameter ratio. This with relation to the required engine speed.

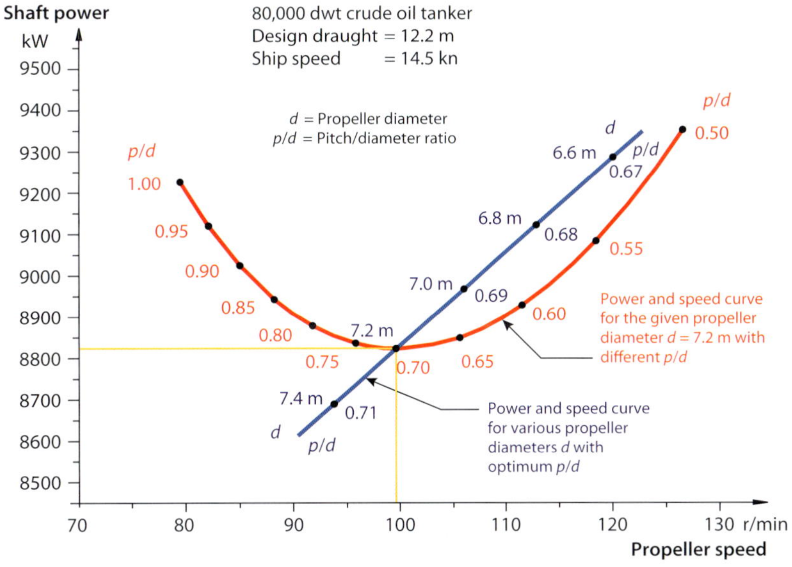

So, by choosing a slightly lower pitch/diameter ratio, comparable to the optimum ratio, the propeller and consequently the engine speed will increase and only require a minimal power increase.

19.11 Operating conditions of a propeller

19.11.1 Slip ratio S

▶

The slip S.

If the propeller had no slip in the water, or if the propeller did not thrust the water backwards, the propeller would move forward with a velocity of $v = p \times n$ in which n is the rate of revolution of the propeller and p the pitch of the propeller.

Imagine a similar situation with the propeller moving in **cork (a solid mass)**; then the slip would be zero and the propeller would move in relation to the cork with a velocity of $v = p \times n$.

However, as water is a liquid and moves backwards, the apparent propeller speed forward decreases with the slip and is equal to the ship's velocity v, and the apparent slip can be expressed as: $S_A = p \times n - v$.

The apparent slip S_A, which is dimensionless, is defined as follows:

$$S_A = \frac{p \times n - v}{p \times n} = 1 - \frac{v}{p \times n}$$

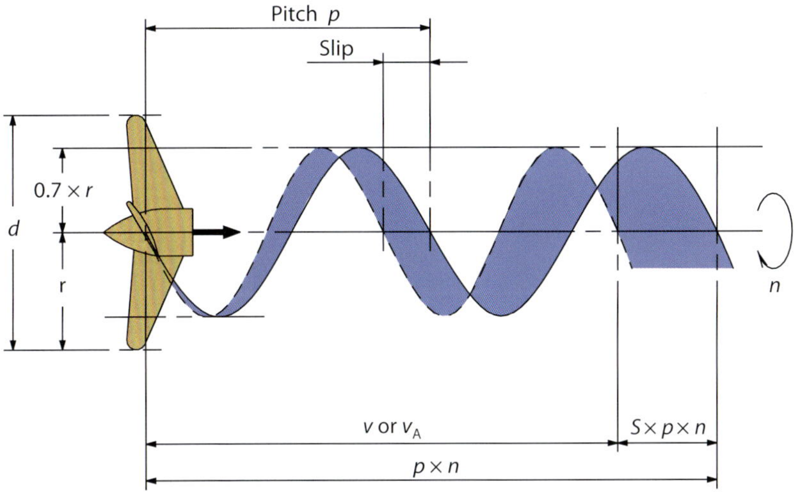

The apparent slip ratio: $S_A = \frac{p \times n - v}{p \times n} = 1 - \frac{v}{p \times n}$

The real slip ratio: $S_R = \frac{p \times n - v_A}{p \times n} = 1 - \frac{v_A}{p \times n}$

▶

A rotating 'propeller' in a solid mass (in this instance, cork).

The 'propeller' speed is equal to the distance travelled $v = p \times n$.
There is no slip.

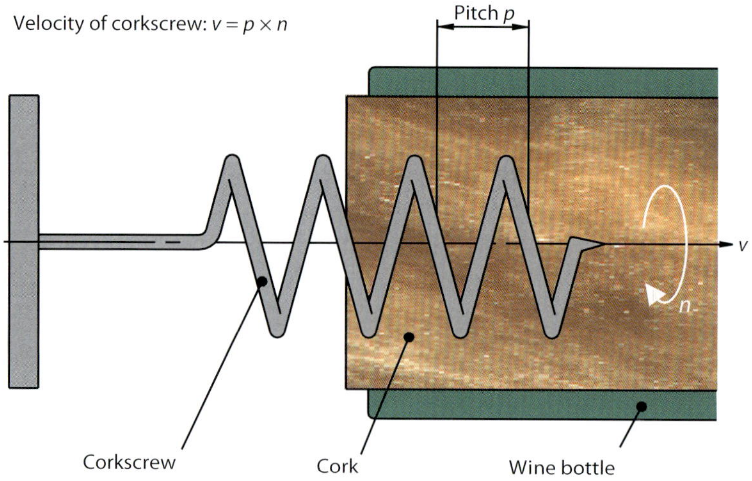

The apparent slip, which is calculated by the crew, provides valuable information as it gives the propeller load in varying operating conditions.
The apparent slip increases when:
- the ship sails into the wind and/or waves;
- the ship sails in shallow waters;
- the ship's hull is fouled;
- the ship accelerates.

Under increased resistance, this means that the propeller velocity has to increase to maintain the required ship's speed.

19.11.2 Real slip S_R

The real slip S_R is greater than the apparent slip S_A due to the fact that the actual speed of advance v_A of the propeller is smaller than the ship's velocity v.
The real S_R, which provides a more accurate picture of the propeller's function, is:

$$S_R = 1 - \frac{v_A}{p \times n} = 1 - \frac{v \times (1 - w)}{p \times n}$$

If a ship is moored (at quay trials) and the propulsion engine operates at full load, the ship's velocity v equals 0 and the real slip 1.0. The slip is sometimes given in percentages.

19.11.3 Propeller law in general

The resistance R for low ship's speeds is proportional to the square of the ship's velocity: $R = c \times v^2$, where c is constant. The required power P is therefore proportional to the ship's velocity v to the power of three.

$P = R \times v = c \times v^3$

For a ship with a fixed-pitch propeller, the ship speed v is proportional to the size and the rate of revolution, so:

$\boldsymbol{P = c \times n^3}$

which precisely expresses the propeller law which states: 'the required engine power output delivered to the propeller is proportional to the rate of revolution to the third power'.

Measurements demonstrate that the relation between the power output and the engine speed for certain weather conditions is fairly congruent, whereas the relation between the power output and the ship's velocity is often higher than the v to the power of three.
A reasonable relationship to be used for estimations in normal ship's velocities could be as follows:

For large high-speed container ships:
velocity ± 22 – 26 knots.

$P = c \times v^{4.5}$

For medium sized, medium-speed ships such as feeder-container ships, RoRo ships etc.:
velocity ± 16 – 20 knots.

$P = c \times v^{4.0}$

For low-speed ships such as tankers and bulk carriers and small feeder-container ships:
velocity ± 12 – 14 knots.

$P = c \times v^{3.5}$

19.11.4 Propeller law for a propeller running under heavy conditions

Obviously, the propeller law can only be applied to identical operating conditions.
If, for instance, after a certain amount of time the hull is fouled and consequently has an increased water resistance, the wake field differs from the wake field of a 'clean' hull during for example, the trial runs.
A ship with a fouled hull is subject to extra resistance in the water, which effects an increased propeller load at an identical engine output. The rate of revolution will therefore be lower.
The propeller law is now applied to a different 'heavier' propeller curve than that of a 'clean' hull.
Similar considerations apply to ships labouring in heavy seas, in strong counter-currents, in stormy weather and high waves. There will also be a heavier propeller operation in tail wind and high waves than in calm weather operation.

19.11.5 Ballast conditions

When a ships sails in ballast, which is to say, with less displacement, the propeller law applies to a 'lighter' curve; at a similar power output the propeller speed will increase slightly.

The propeller law is predominantly applied to fixed-pitch propeller ships running at partial load.
Diesel manufacturers' project guides and engine lay-out and load diagrams specify the engine's curves for both light- and heavy-running conditions.
These diagrams use logarithmic scales and straight lines.

▼
A load diagram for a four-stroke medium-speed Wärtsilä 38B diesel engine operating on HFO.
Propeller type: fixed-pitch propeller.

vertical: cylinder power in kW.
horizontal: rate of revolution per minute (rpm):
- minimum rpm: 200;
- maximum rpm: 600.

MCR: at 600 revolutions and 725 kW per cylinder.
normal service power: at 85% of MCR at 600 revolutions and 625 kW per cylinder.
vertical hatched area: operating area for the propeller load on the engine.
propeller hatched area: area in which the governor is activated.
top: mechanical fuel stop.

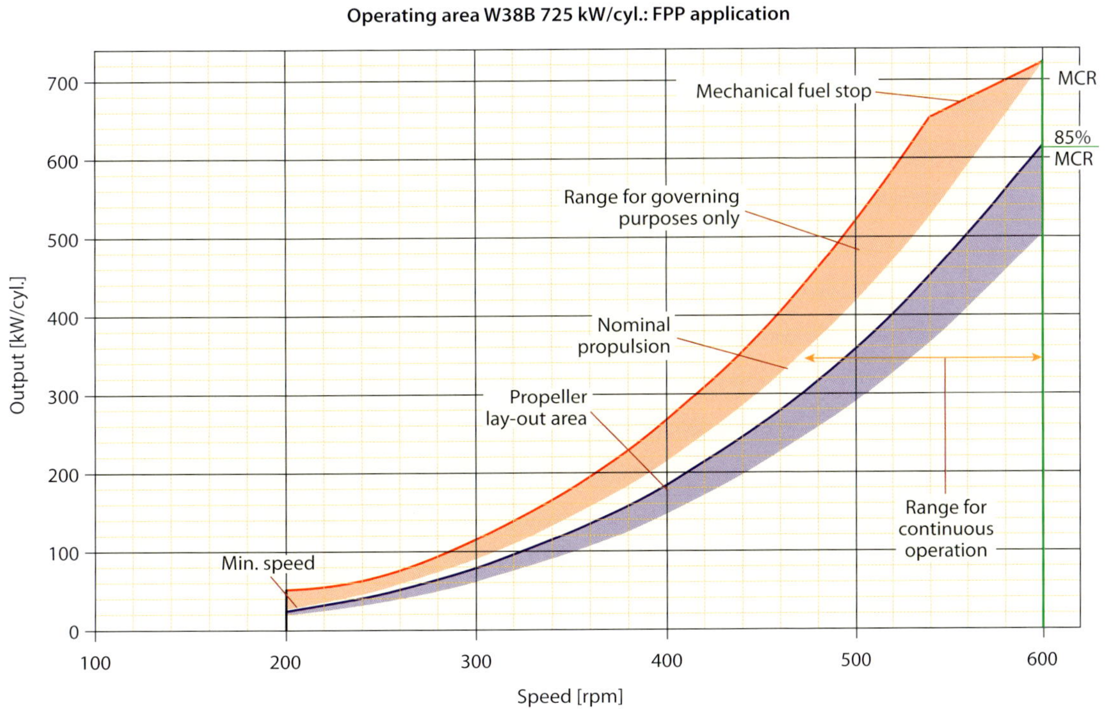

▶

A load diagram for a four-stroke medium-speed Wärtsilä 38 B diesel engine operating on HFO. Propeller type: controllable-pitch propeller.

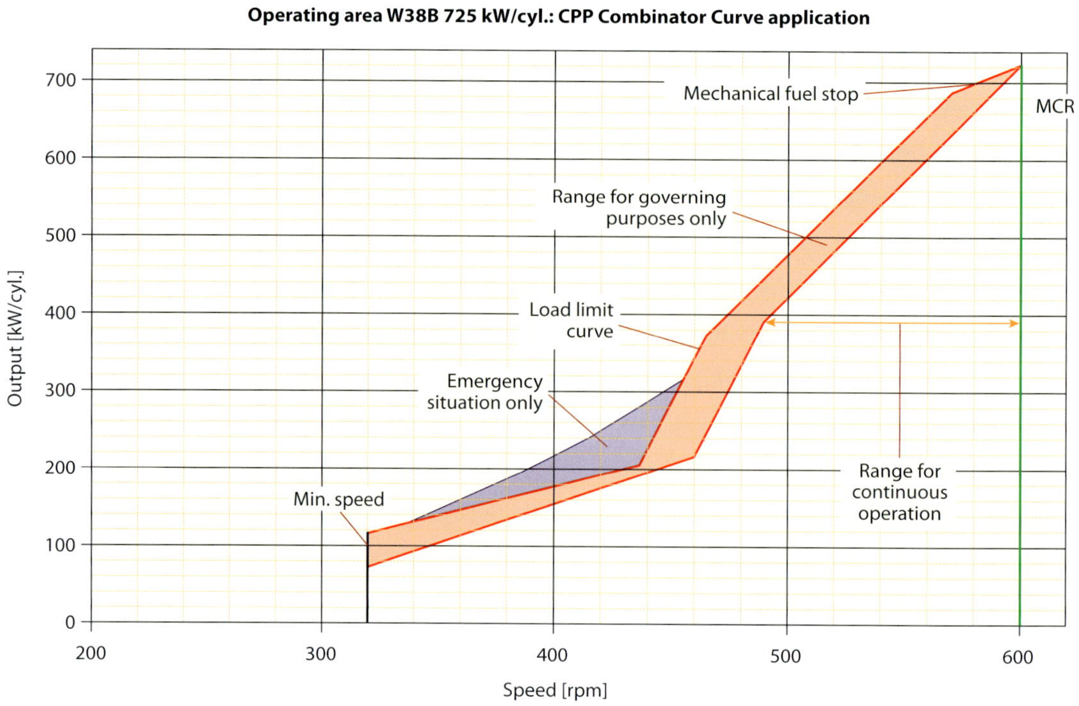

19.11.6 General propeller performance at increased ship resistance

Heavy running

C Extremely bad weather	**6%**
B Average weather	**3%**
A Extremely good weather	**0%**

Clean hull and draught D
D_{mean} = 6.50 m
D_F = 5.25 m
D_A = 7.75 m

Source: Lloyd's Register

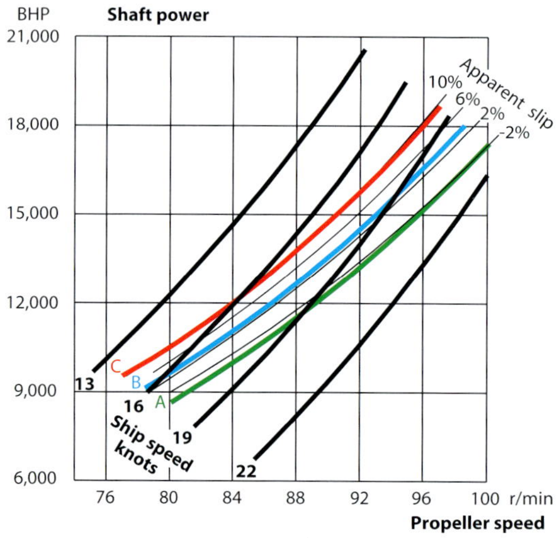

▲

Heavy seas and adverse winds and seas.

Resistance resulting from heavy seas can cause the propeller to run 7 tot 8% heavier than in calm seas, or at the same power output to the propeller, the rate of revolution of the propeller may be reduced by 7 to 8%.

This chart applies to a medium-sized container ship.

horizontal: the rate of revolution of the propeller per minute
vertical: the shaft power BHP (brake horse power)
Brake Horse Power: 736 Watt
draught mean: D_{mean} = 6.50 m
draught forward: D_F = 5.25 m
draught aft: D_A = 7.75 m
green line A = extremely good weather
blue line B = normal weather
red line C = extremely bad weather

Bottom left of the figure: the ship's velocity in knots (knots = 1 nautical mile per hour = 1,852 metres).

Top right of the figure: the apparent slip in percentages.

This data is gathered over a period of a year and only includes the influences of weather conditions!

These measurements have been reduced to three average weather types and show an average 'heavy' running of 6%, in reality, slightly more.

In order to prevent slamming of the ship and consequent damage to the stern, the ship's velocity is reduced to a more acceptable level. If not, the bow of the ship could be damaged and the propeller speed would increase the moment the head of the ship is submerged in a wave while the stern of the ship, including the propeller, would lift out of the water.

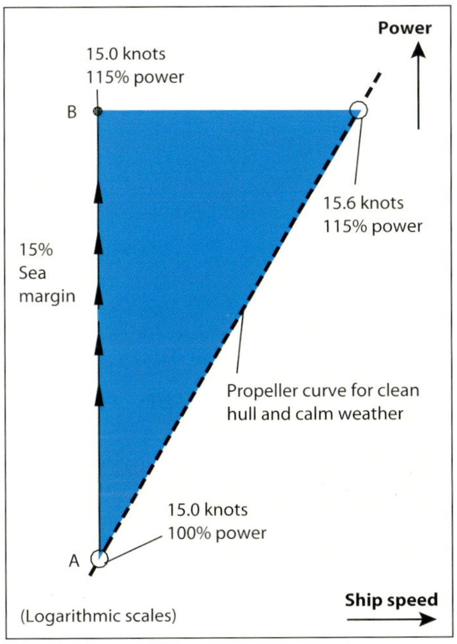

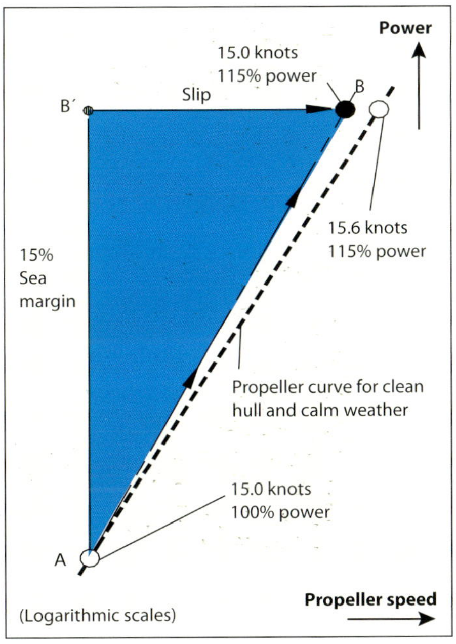

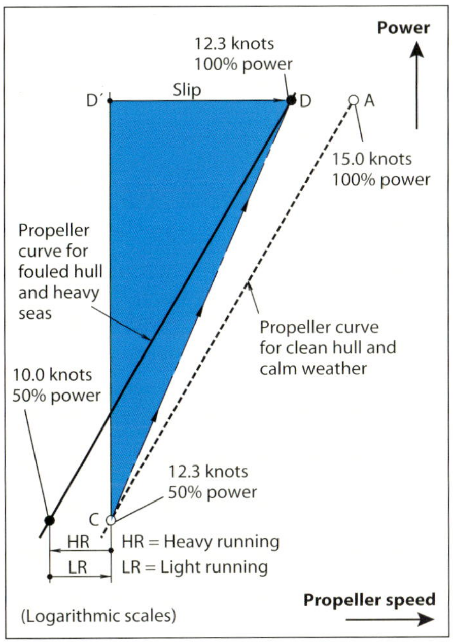

▲

The difference between the curves for a light- and heavy running propeller is clearly visible in these charts.

Here the ship's velocity is 15 knots and the delivered engine power 100%. This applies to a clean hull just after dry-docking and in calm water. With 15% extra engine power, the corresponding velocity increases from 15.0 to 15.6 knots.

As described earlier and corresponding to the calm weather conditions, an extra margin, the so-called 'sea margin', is established for poor weather conditions such as wind, waves and currents. This is the extra power margin that can be used for bad weather conditions.

So most propulsion marine engines usually operate at 85% of their power output. The 15% margin serves to bridge the increased resistance.

In this example, the ship's velocity at 100% of the engine power is 15.0 knots, at point A. At a speed of 12.3 knots with a clean hull and in calm weather, only 50% of the engine power is required, point C at the bottom of the figure.

figure left: horizontal: the ship's velocity in knots; vertical: the power supplied to the propeller in percentages.

If the resistance increases to a level that requires an additional 15% power, the operating point shifts from A to B.

figure middle: horizontal: the propeller speed in knots; vertical: the power delivered to the propeller in percentages.

One would initially think that point A would move to B as with fixed-pitch propellers the propeller moves through the water at the same speed. If the propeller were to rotate through a solid mass, this would be accurate. However, water is not solid and will yield, and the propeller will have slip, which increases at a higher thrust, produced by a higher resistance of the ship. This is why point A moves to point B, which in fact is close to the propeller line of the left figure.

Operational point B is now positioned on the propeller line, which runs heavier than the propeller line of the left figure in calm water.

In some cases, for instance when the hull is fouled and the ship sails in heavy seas and head winds, the resistance can increase significantly and extra, up to 100% or more, power is required. See the example in the right figure.

In this example, the velocity of the ship at 100% engine power is 15.0 knots, at point A. At a velocity of 12.3 knots with a clean hull and in calm water, only 50% of the engine power is required at point C, bottom of the figure.

In heavy conditions similar to those mentioned previously, the ship could achieve a maximum speed of 12.3 knots at a 100% engine-power output.

The operational point now shifts from point A to D. Operational point D is now shifted to the left, relatively far from point A, which implies heavy running.

Situations such as these should be meticulously calculated when designing the propeller.

A propeller with bent blade tips is more suitable for these heavy loads than a standard propeller as the propeller is more capable of absorbing the higher torque in heavy conditions.

► In waves and head winds, the resistance of the sailing ship increases significantly. The engine's power output is considerably increased to maintain the same speed.

► Another example of a reefer during trials.

horizontal: the engine/propeller rpm at a maximum shaft power of 13,000 kW and an rpm of 105.

vertical: the engine power in percentages.

weather conditions: wind velocity 2m/sec, wave height 4 metres.

Line 1 is the standard engine/propeller line for normal speed at a normal load at maximum 100% shaft power and an rpm of 100 (point SMCR).

The **red line** applies to 'heavy running', a slight head wind and waves. The speed lies between 20.5 and 21.1 knots. Here the required power is 85 to 90% and the rpm 98 to 99%.

The **black line** on the right applies to tail wind; the speed increases to 21.1 and 22.3 knots, depending on the power output and the rpm.

At 21.1 knots, the rpm is 98% and the power output 82%.

Near the red line and at a speed of 21.1 knots this is approximately 98.5% and 86% and therefore more power (read fuel) will be required.

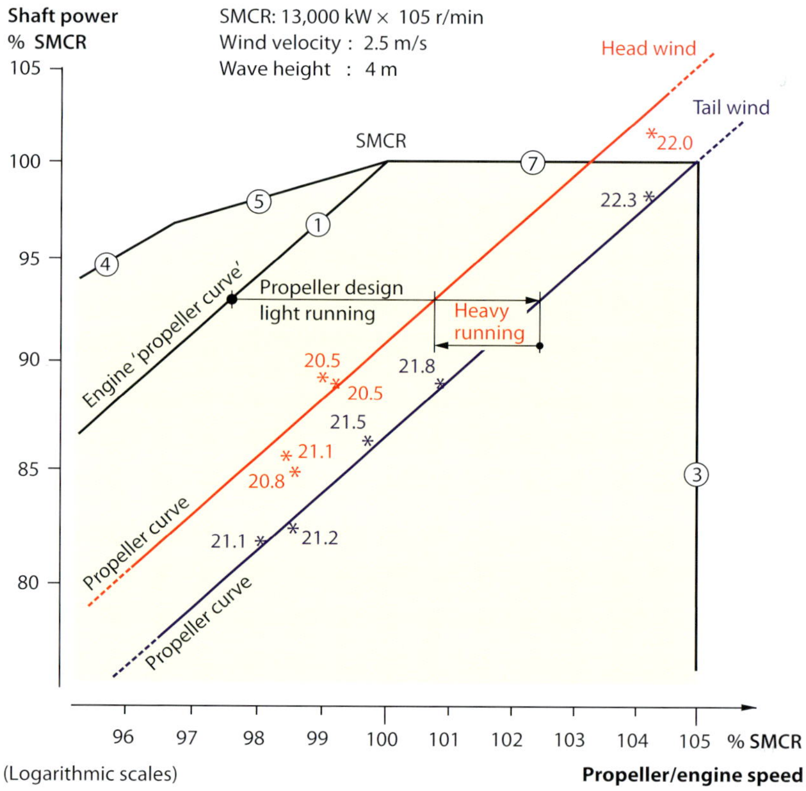

19.12 Increasing ship speed

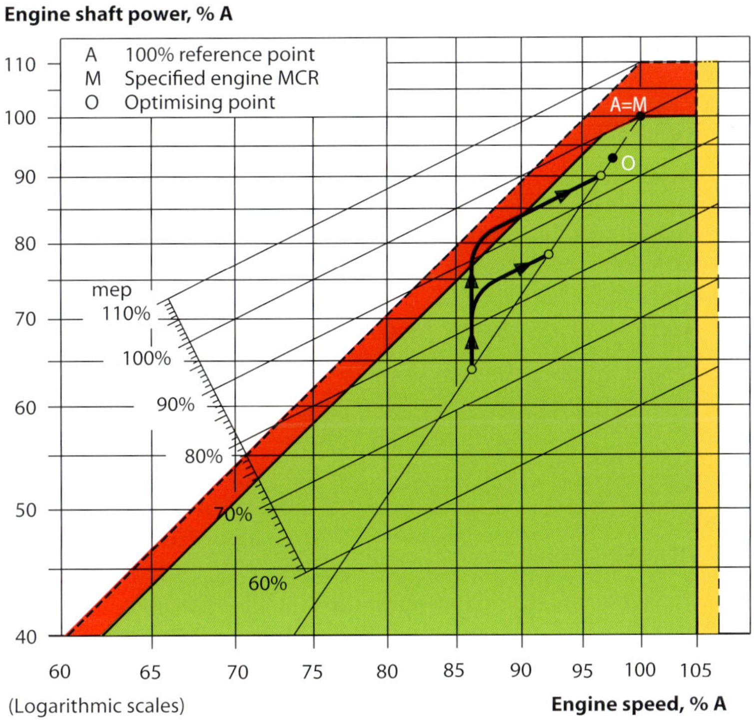

▲

Increasing the ship's speed.

vertical: shaft power in percentages

horizontal: engine rpm

green area: normal engine load

yellow area: maximum rpm for short periods

red area: engine over-load

m.e.p. 110-60%: mean effective pressure in percentages of a maximum 100%

point A = M: 100% power, 100% engine speed, 100% mean effective pressure

Slanted lines – acceleration of the ship – the top line traverses the over-load area temporarily.

When the ship accelerates, the propeller can be subjected to an even larger load than during free sailing.

The engine power, which at that time is higher than during free sailing, is high and it may take some time for the propeller rate to achieve a new and higher revolution rate.

19.12.1 Shallow waters

Sailing in shallow waters may cause an increase in the ship's resistance, thus subjecting the propeller to a larger load than would be the case during free sailing in open water.

19.12.2 Influence of displacement

A laden ship may carry 10% more or less than the average load. This, of course, affects the ship's resistance and the required propeller power, but only has a minor effect on the propeller curve.
When the ship sails in ballast, the displacement volume can be considerably less when compared to the loaded condition, and the propeller curve could, for instance, be 2% 'lighter' at an identical engine power. The speed would then increase by 2%.

19.13 Parameters causing heavy running conditions

The following examples give an indication of the risks involved in sailing with a heavily loaded propeller in poor weather conditions.

19.13.1 Relatively small ships (less than 70,000 dwt)

Relatively small ships (less than 70,000 dwt) such as reefers and medium-sized container ships are more vulnerable than larger ships, such as large tankers and container ships, because the latter types are large in relation to the wave size.

19.13.2 Small ships (length between perpendiculars equal to or less than 135 m and approximately 20,000 dwt)

Small ships have a low directional stability and consequently require frequent rudder corrections, which increase the ship's resistance.

19.13.3 High-speed ships

High-speed ships are more sensitive to waves than low-speed ships, because waves have a higher reaction force to high-speed ships than to low-speed ships.

19.13.4 Ships with a flat stern

Ships with a flat (truncated) stern are more likely to be slowed down by waves than ships with a sharp stern. Similarly, a sharp stem will cut the waves more efficiently and therefore reduce the heavy running of the propeller.

19.13.5 Fouling of the hull and the propeller

Fouling of the hull and the propeller, including corrosion of the hull, will also increase the resistance and therefore the propeller torque.
The propeller should be polished each time the ship is dry docked (even when floating in a dry dock). This particularly applies to the propeller tips, as the speed and therefore the frictional force are at their highest here. Using anti-fouling paints prevents the growth of living organisms such as seaweeds and barnacles.

19.13.6 Ship acceleration

Accelerating the ship's velocity will increase the propeller torque and therefore temporarily give a heavy-running propeller.

19.13.7 Sailing in shallow waters

Sailing in shallow waters increases hull resistance and reduces the directional stability of the ship.

19.13.8 Ships with a skewed propeller

Ships with a skewed propeller are able to absorb a higher torque under heavy running conditions.

19.14 Manoeuvring speeds

Below a certain ship speed, the so-called manoeuvrability becomes increasingly difficult, as the water velocity arriving at the rudder is too low. It is rather difficult to indicate the exact value for the minimum manoeuvrability as the water velocity arriving at the rudder is dependent on the slip stream of the propeller.
Often manoeuvring speeds of 3.5 to 4.5 knots are observed.
According to the propeller law, low engine power is required, but this is, of course, dependent on the weather conditions which can cause an increased resistance and therefore more propulsion power is required.

19.15 Direction of propeller rotation: lateral forces

The blades of a propeller of a ship afloat have better grip in their bottom position than in their top position.
The shallower the water the larger the resulting lateral force, which is the case during port manoeuvring or in shallow waters.

Therefore, a clockwise-turning propeller will push the ship in starboard direction and push the bow of the ship towards port side. This is counteracted by the rudder.

So, when reversing with a fixed-pitch propeller, the lateral-force direction is reversed. The stern of the ship will move to port and the bow to starboard.

The actual reason for a clockwise-turning propeller to thrust the stern towards starboard is the fact that the top part of the water flow around the propeller rotates and strikes the aft side of the ship's hull.

19.16 Engine layouts and load diagrams

As mentioned earlier, the effective brake power P_B is proportional to the mean effective pressure (mep) p_e and the engine speed (rate of revolution) n.
In the engine formula of a certain type of engine, p_e and n are variables, whereas the others

$(\frac{\pi}{4} \times D^2 \times S \times \frac{1}{a} \times Z)$ are constant and referred to as C.

$P_B = C \times p_e \times n$ and for a constant mean effective pressure $P_B = C \times n^1$
As previously mentioned, the shaft power with a fixed-pitch propeller, may in accordance with the propeller law be expressed as:

$P_B = C \times n^3$ (the propeller law).

Therefore, for the examples shown above, the brake power can be expressed as a function of the speed n to the power output of $P_B = C \times n^i$.

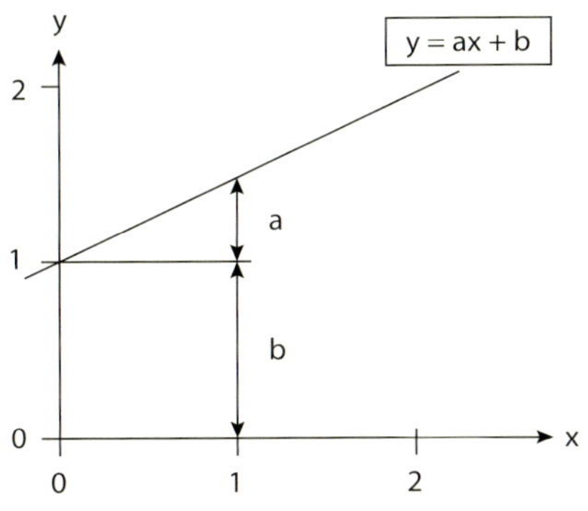

A. Straight lines in linear scales

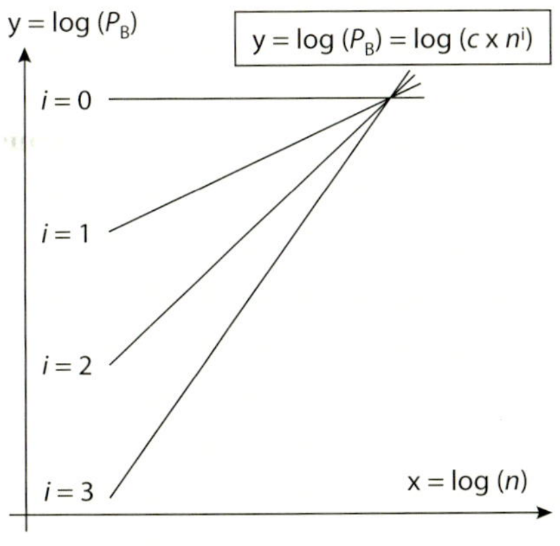

P_B = engine brake power
c = constant
n = engine speed

$$P_B = c \times n^i \Rightarrow \begin{cases} \log(P_B) = i \times \log(n) + \log(c) \\ y \quad = \quad ax \quad + \quad b \end{cases}$$

B. Power function curves in logarithmic scales

◀
The relation between the linear functions with linear scales and power functions with logarithmic scales. These are generally used by large engine manufacturers.

19.17 Propulsion- and engine characteristics

19.17.1 Fouled hull

After a certain sailing time, the hull and propeller resistance will increase due to fouling and corrosion. Consequently, the ship's velocity will decrease if the propulsion power remains at the same level.
If, however, the engine delivers more power to the propeller, the operating conditions of the propeller change to what is called 'heavy running'.

Furthermore, modern, highly efficient ships have a relatively high velocity and very smooth hull- and propeller surface areas (at sea trials). This means that the inevitable build-up of surface roughness on hull and propeller during regular service causes the propeller to rotate more heavily, in comparison to ships already constructed with a rougher hull surface.

19.17.2 Heavy-weather conditions and sea margin used for engine layout

If, at the same time, the weather is poor with head winds, the ship's resistance increases significantly, this leads to even heavier running.
In determining the required engine power, it is common practice to add an extra power margin, the so-called sea margin, which is a standard 15% of the propeller design power P_D.
However, for large container ships margins of 20 to 30% are sometimes used.

▼
With the anemometer 'True Wind', the wind speed can be established accurately. 42.5 knots! 80 km per hour.

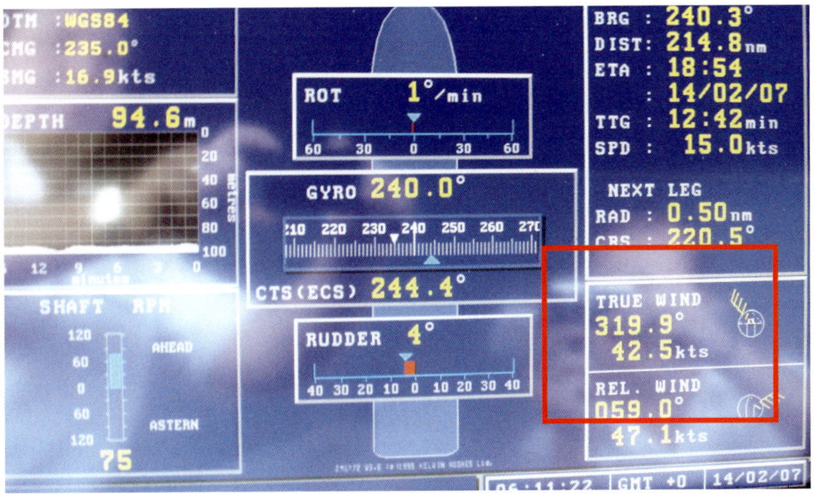

▶

A storm as seen from the bridge of a large container ship.

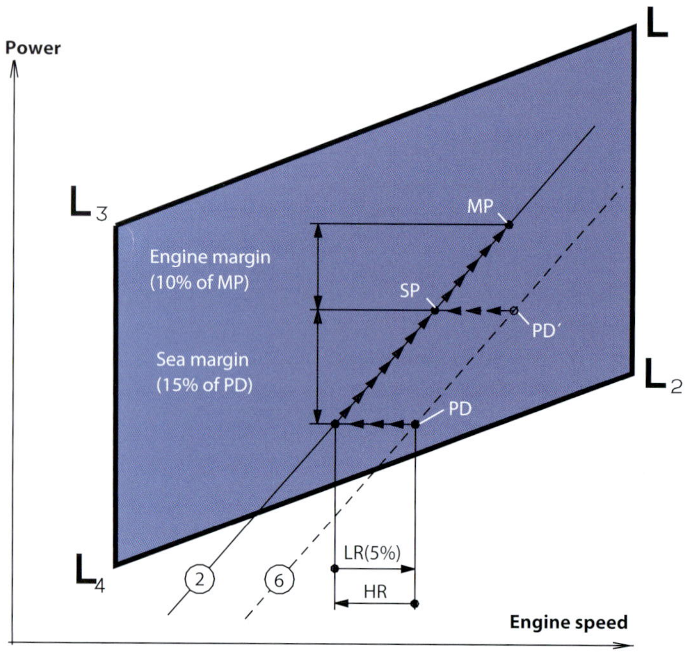

Heavy seas and 'sea margin' in the graph. The propeller design point is PD.

② Heavy propeller curve – fouled hull and heavy weather
⑥ Light propeller curve – clean hull and calm weather

MP: Specified propulsion point
SP: Service propulsion point
PD: Propeller design point
PD´: Alternative propeller design point
LR: Light running factor
HR: Heavy running

Usually, estimations of the required propeller power and the speed are based on theoretical calculations for a loaded ship and often based on tank tests executed in a hydraulics laboratory.
The combination of the speed and the required power output is referred to as the design point, PD, of the ship's propeller. In the figure for a light running propeller, this is straight line no 6.
Some shipyards and/or propeller manufacturers use a design point PD that also partially or entirely incorporates the sea margin (power surplus).
When establishing the required engine speed for a propulsion engine, it is advisable, in comparison to line 6, clean hull, propeller and calm weather, to select the heavier propeller line 2.
Line 2 has 3 to 7% fewer revolutions.
On average a 5% difference is acceptable.
Note that the chosen 'sea margin' does not signify that a heavier propeller line was selected.

The SP point = the maximum service propulsion point for a sailing ship.

The combination of the speed and the engine power for a heavier running propeller and a sea margin is referred to as the maximum service propulsion point SP. This applies to a fouled hull and heavy-weather conditions.
The 'heavy' propeller line 2 will generally be applied as the basis for the engine characteristics during free sailing and the propeller line for a clean hull and calm weather, line 6, for a 'light' running propeller, LR.

19.18 Electronic governors with load limitation

To protect the diesel engine against thermal- and mechanical overload, the accepted electronic governors include two limiter functions.

19.18.1 Torque limiter

The torque limiter ensures that the approved limitation lines of the load diagram are adhered to at all times.
The torque limiter algorithm compares the calculated fuel-pump index (fuel amount) and the actual measured engine speed with a standard reference B limiter curve, which gives the maximum allowable fuel-pump index at a given engine speed. If the calculated fuel-pump index is above this line, the actual fuel-pump index is reduced accordingly.
The standard reference limiter curve should be adjusted to correspond to the limitation lines of the load diagram.

19.18.2 Scavenging-air pressure limiter

The scavenging-air pressure limiter prevents the engine from being over-fuelled in relation to the amount of air supplied by turbochargers while accelerating. This is mainly the case during rapid load fluctuations such as in manoeuvring.
The scavenging-air pressure limiter algorithm compares the calculated fuel-pump index and the measured scavenging-air pressure with a standard reference limiter curve that gives the maximum fuel-pump index at a certain scavenging-air pressure.
If the calculated fuel-pump index is above this limiter line, the fuel-pump index is reduced accordingly.
The reference limiter curve should be adjusted so there is a sufficient air supply for a good combustion process.

19.18.3 Recommendations

Continuous operation within a certain time limit is **only** allowed in the area limited by the lines 4, 5, 7 and 3 of the load diagram.
For fixed-pitch propellers that operate in calm weather conditions with a loaded ship and a clean hull, the propeller/engine may run along or close to the propeller-design line 6.
After a certain running time, the ship's hull and the propeller grow increasingly rough by fouling and contamination, thus causing the propeller to run more heavily. The propeller line will shift to the left from line 6 to line 2. Therefore, more propulsion power will be required to maintain the ship's velocity.
If the propeller runs heavily in calm weather, this could indicate that the hull requires cleaning and possibly the propeller wants polishing.
The area between lines 4 and 1 is available for operation in shallow waters, for acceleration and if required, for non-steady operation without any time limit.

19.19 Use of diagrams

Examples

The following are four different examples; three examples on fixed-pitch propellers and one example based on a controllable-pitch propeller. They illustrate the flexibility of these diagrams.
In this aspect, the choice of the optimum point O has an important influence.

19.19.1 Example 1: Normal operating conditions without a shaft generator

M: Specified MCR of Engine
S: Continuous service rating of engine
O: Optimising point of engine
A: Reference point of load diagram

Point A of load diagram
Line 1: Propeller curve through optimising point (O)
Line 7: Constant power line through specified MCR (M)
Point A: Intersection between lines 1 and 7

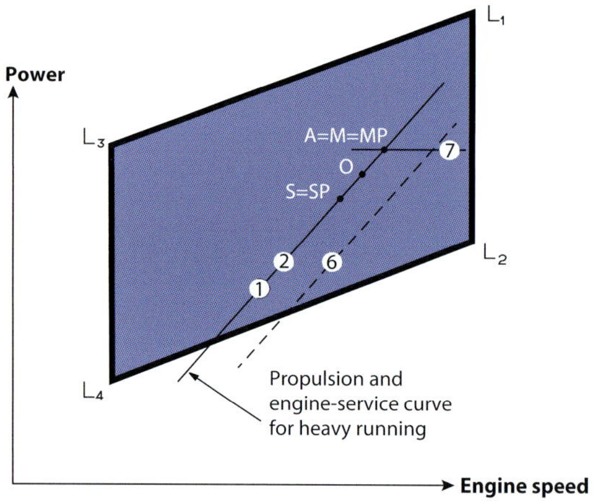

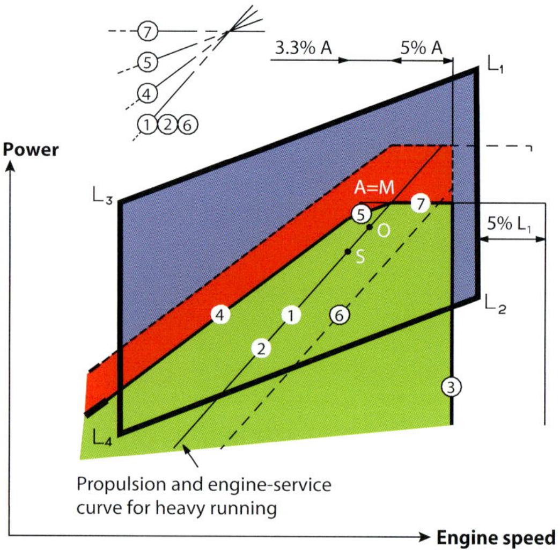

▲
Example 1: normal operation without shaft generator.

Normally, the optimising point O and the engine line 1 are placed on line 2 (for heavy running).

Point A can be found between propeller line 1 (2) and the line of the continuous rate line 7, through M. In this instance, point A is equal to point M.

When point A has been located, the layout diagram such as in the figure above can be drawn and the current load limitations of the diesel engine can be found.

19.19.2 Example 2: Special load conditions without a shaft generator

M: Specified MCR of Engine
S: Continuous service rating of engine
O: Optimising point of engine
A: Reference point of load diagram

Point A of load diagram
Line 1: Propeller curve through optimising point (O)
Line 7: Constant power line through specified MCR (M)
Point A: Intersection between lines 1 and 7

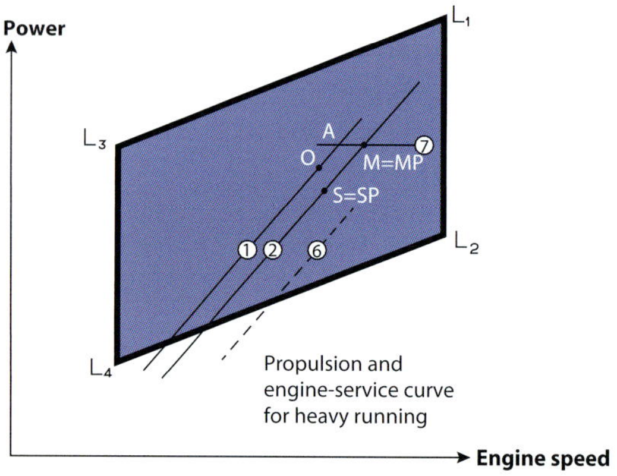

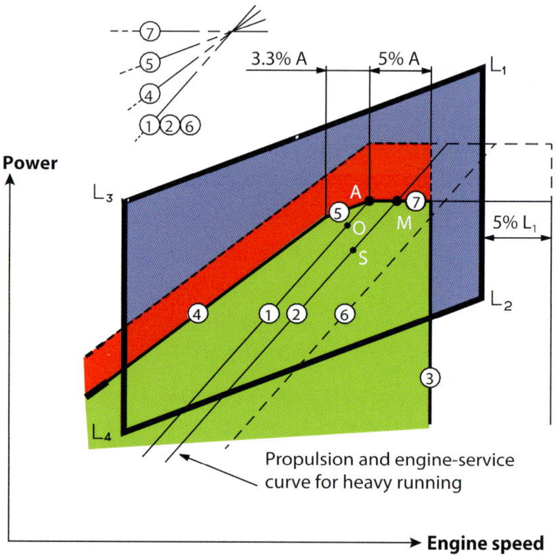

▲
Example 2: Special load situations without shaft generator.

The propeller can be subjected to a heavier load when the ship accelerates than during full load. The same applies when the ship is subjected to extra resistances, for instance when sailing in heavy winds with high wave resistances.

In both instances, the engine-service point moves to the left of the normal service line as the propeller will run heavier.

To avoid exceeding limitation line 4 of the load diagram to the left, it is sometimes necessary to reduce the speed and/or the power output.

It is recommended to move the lay-out diagram further right, when it is expected that the ship will sail on a route where poor weather conditions are a common occurrence.

The latter is performed by shifting the optimising point O and consequently propeller line 1 to the left.

However, this will result in a fuel-cost increase due to a slight increase in fuel consumption.

An example is shown in both graphs. The top figure shows, compared to the normal situation in the first example, that limitation line 4 has been shifted to the left, thus creating more space between the lines 2 and 4; here a light load factor has been applied.

19.19.3 Example 3: Normal operating conditions with a shaft generator

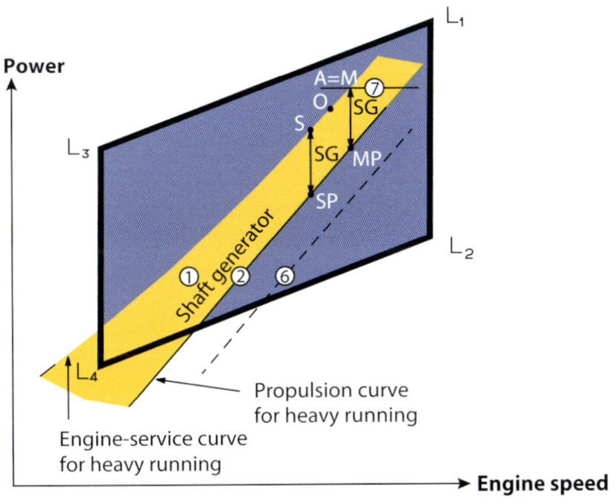

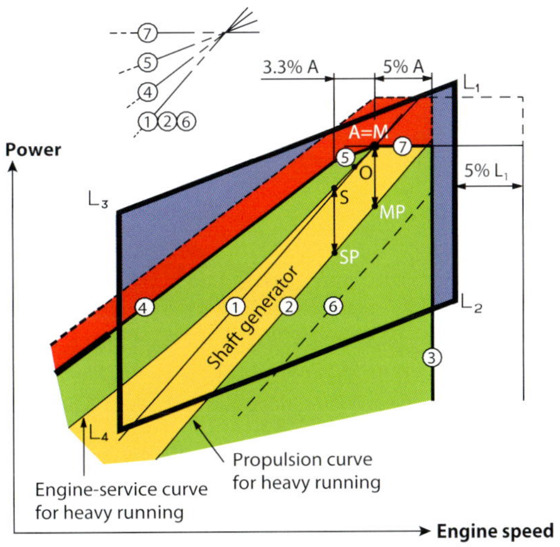

M: Specified MCR of Engine
S: Continuous service rating of engine
O: Optimising point of engine
A: Reference point of load diagram

Point A of load diagram
Line 1: Propeller curve through optimising point (O)
Line 7: Constant power line through specified MCR (M)
Point A: Intersection between lines 1 and 7

▲
Example 3: A standard case, here however, with a shaft generator. In the figure, the engine line has been drawn for heavy running, including the extra power required for the shaft generator.

In this instance, a shaft generator has been installed and therefore the power output delivered to the propeller is reduced by the power supplied to the shaft generator.

The optimising point O and the engine line 1 are then found in a similar fashion as in example 1.

19.19.4 Example 4: A special case, with a shaft generator

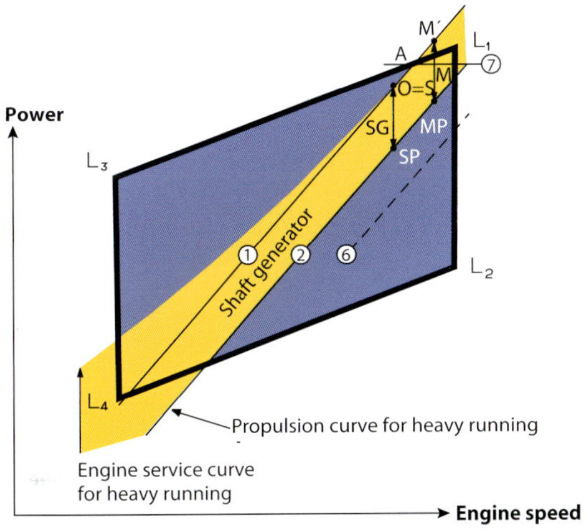

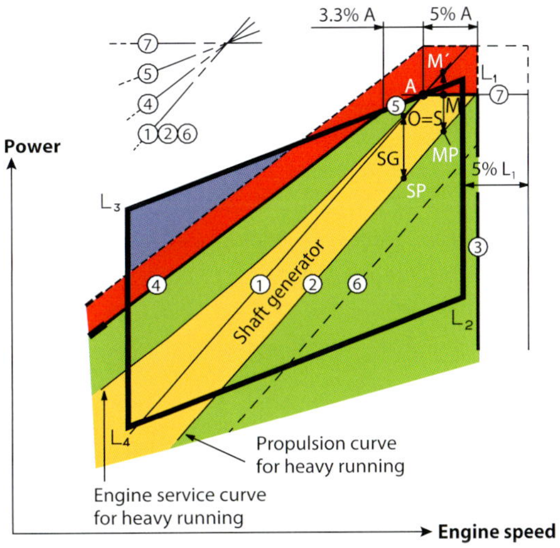

M: Specified MCR of Engine
S: Continuous service rating of engine
O: Optimising point of engine
A: Reference point of load diagram

Point A and M of load diagram
Line 1: Propeller curve through optimising point (O)
Point A: Intersection between lines 1 and line L1 – L3
Point M: Located on constant power line 7 through point A and at MP's speed

▲
Example 4: A special case with a shaft generator.

Also in this special case a shaft generator is installed, but unlike example 3, the MCR-point MP is placed at the top of the diagram This means that the MCR of the engine, point M, will be placed outside the top of the diagram.
This problem could be resolved by choosing a diesel engine with an extra cylinder (for instance a nine-cylinder instead of an eight-cylinder in-line engine). Another and cheaper solution is to reduce the electrical power to the shaft generator when the ship is sailing at full power output. This is achieved by using a diesel genset.

When the latter solution is chosen, the required MCR power of the engine can be reduced from point M[1] to M as is shown in the figure.
Situations such as these seldom occur as ships rarely have a normal operating speed with high engine power outputs such as these.
In the example, optimising point O is chosen equal to point S, and so line 1 is found.
Point A, having the highest possible power can be found at the intersection of line L1-L3 and the complete load diagram has been drawn in the right figure. Point M is found on line 7 at MP's speed.

19.19.5 Example 5: A controllable-pitch propeller, with or without a shaft generator

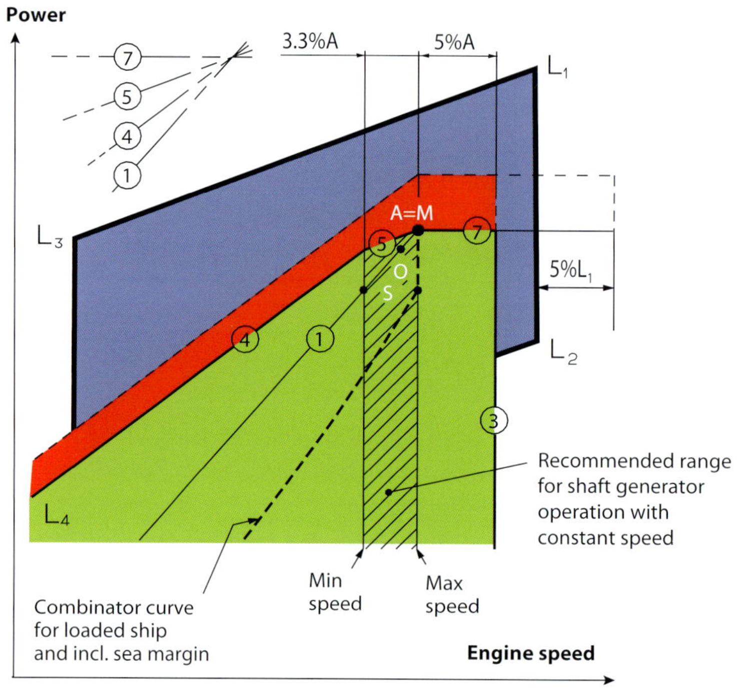

▲

Example 5: A controllable-pitch propeller, with or without a shaft generator. Most ships have a controllable-pitch propeller.

Load chart, without a shaft generator

If a ship is fitted with controllable-pitch propellers (CPP) the combinator line (of the propeller with optimum efficiency) is usually chosen for a loaded ship including a sea margin.
At a given propeller speed and pitch the CP propeller runs heavily in poor weather conditions.
Therefore, it is advisable to select the line for a lightly loaded propeller, such as the dotted line, ensuring a larger margin between the normal line and the lines for heavy-weather conditions such as 4 and 5.

Load chart, with a shaft generator

The hatched area indicates the recommended speed range between 100% and 96.7% of the maximum continuous rating (MCR) for an engine with a shaft generator operating at a constant speed.
The service point S can be situated at any point in the hatched area.
The combinator line of a loaded ship including a sea margin still has an ample margin indicated by lines 4 and 5.

19.20 Summarising the effects of the various types of resistance on engine operation

A brief summary of the effects of the various types of ship's resistance on a fixed-pitch propeller and the propulsion engine is shown by the diagram below.

PD: Propeller design point, clean hull and calm weather

Continuous service rating for propulsion with a power equal to 90% specified MCR, based on:

S0: Clean hull and calm weather, loaded ship

S1: Clean hull and calm weather, ballast (trial)

S2: Clean hull and 15% sea margin, loaded ship

SP: Fouled hull and heavy weather, loaded ship

S3: Very heavy sea and wave resistance

Engine shaft power % of A

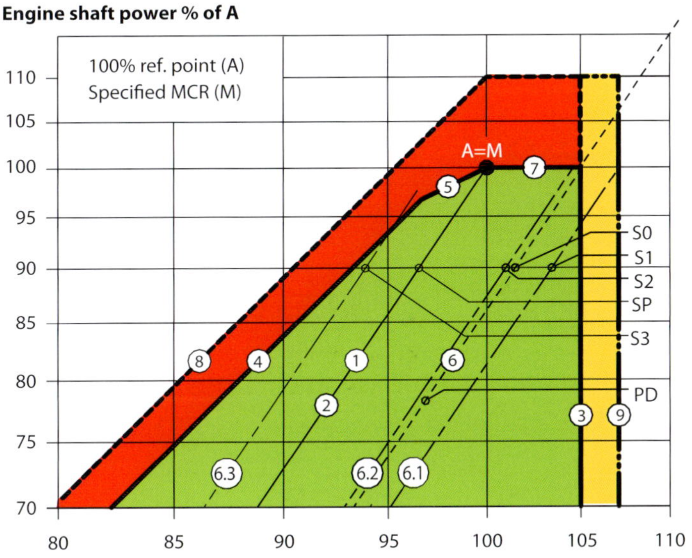

Engine speed, % of A

Line 1: Propeller curve through point A=M, lay-out curve for engine

Line 2: Heavy propeller curve, fouled hull and heavy weather, loaded ship

Line 6: Light propeller curve, clean hull and calm weather, loaded ship, lay-out curve for propeller

Line 6.1: Propeller curve, clean hull and calm weather, ballast (trial)

Line 6.2: Propeller curve, clean hull and 15% sea margin, loaded ship

Line 6.3: Propeller curve, very heavy sea and wave resistance

▲

The influence of the various resistances the ship is subject to is illustrated by means of corresponding service points for propulsion with an identical propulsion power, based on the propeller-design point PD, plus 15% additional power.

Propeller-design point PD

As previously mentioned the propeller is normally designed for a certain ship velocity *v* valid for a loaded ship with a clean, smooth hull in calm weather.

The corresponding engine speed and power are indicated by point PD on the propeller line 6 of the load diagram.

Increased ship speed, point SO

If the engine power increases by, for instance, 15% and the ship has a smooth hull and operates in calm weather, point SO, the ship velocity *v* and the engine speed *n* will increase in accordance with the propeller law (more or less valid for normal speeds).

Point SO is placed on the same propeller line as point PD.

$$v_{SO} = v \times \sqrt[3.5]{1.15} = 1.041 \times v$$

$$n_{SO} = n \times \sqrt[3.0]{1.15} = 1.048 \times n$$

Normal speed ranges with clean hulls and 15% sea margin, point S2

Conversely, still operating with a loaded ship and clean hull, but now with additional resistance due to heavy seas, extra power of, for instance, 15% is required to maintain the ship velocity *v* (15% sea margin).

If the velocity of a ship $v_{S2} = v$ and the propeller has no slip, the engine speed will remain constant.

However, as the water abaft the propeller gives way, the propeller does slip, the engine speed will increase and the service point S2 will be placed on the propeller curve 6.2 close to SO on the propeller curve 6.

The propeller curve 6.2 could possibly indicate a slightly heavier loaded propeller than curve 6 (approximately 0.5%).

Depending on the type of ship and its dimensions, the heavy running of the propeller and therefore the factor will be 0.5% higher or lower.

For resistances that correspond to about 30% power increase (30% sea margin) the corresponding relative heavy-operating factor will be in the region of 1%.

S=PD : Propeller-design point incl. sea margins, and continuous service rating of engine
Line 1 : Propeller curve for lay-out of engine
Line 1 . Combinator curve for propeller design, clean hull and 15% sea margin, loaded ship
Line 6.1 : Light combinator curve, fouled hull and calm weather, loaded ship
Line 2 : Heavy combinator curve, fouled hull and heavy weather, loaded ship
Line 2.1 : Very heavy combinator curve, very heavy sea and wave resistance

Engine shaft power % of A

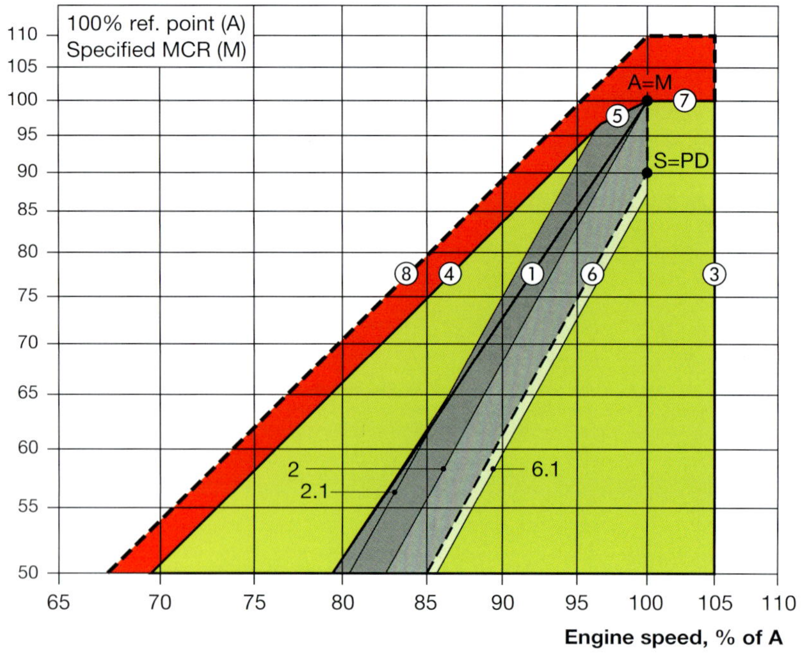

Sailing with a fouled hull and in poor weather conditions, point SP

When, after a certain operational time, the ship's hull is fouled and its surface consequently rougher, the wake field will alter in relation to that of a smooth ship (clean hull).
A ship with a fouled hull experiences more resistance, which results in a modified wake field and therefore a heavier rotating propeller than would be expected by deteriorating weather conditions alone.
The propeller curve will shift to the left, see propeller curve 2 in the load diagram. This propeller curve indicates that the hull is fouled and the ship operates in poor weather conditions. The propeller/engine load has increased approximately 5% in relation to propeller curve 6.
In order to safeguard the air supply for the diesel engine's combustion process, which determines the limit of the torque and the engine speed, it is common practice to ensure that the engine and the turbocharger match each other according to the load on propeller curve 1 of the load diagram, equal to the 'heavier' propeller curve 2.
As opposed to point S2, point SP is normally used to determine the engine lay-out by referring this rating to 'average operating' power at 90% of the specific MCR-point, which matches a 10% margin with the maximum engine power.

In other words, in the example the propeller-design curve is approximately 5% 'lighter' than the propeller curve that is applied to the lay-out of the main engine.

Operating in heavy seas with extremely high waves, point S3

When operating in these conditions the propeller can run between 7 and 8% (or more) 'heavier' than in calm conditions. Consequently, at the same engine/propeller power, the rate of revolution is decreased by 7 to 8%.
For a propeller power equal to 90% of MCR, point S3 in the load diagram shows an example of this operating condition.

In some instances when operating in heavy headwinds, the heavy running is often such that the service point is placed left of line 4. In such instances, when the ship begins to pitch and the propeller periodically surfaces, it is recommended to decelerate in order to avoid damage to the ship and the propulsion installation.

Accelerating and operating in shallow waters

When the ship accelerates and the propeller is working heavier than during normal operating speed, the effect on the propeller is at times equal to the operating conditions at point S3.
The accelerating effects are sometimes even greater.
The same applies to operating in shallow waters.

Trials at open sea point S1

Normally, the curve 6 for a clean hull will be equal to the trial-trip propeller curve. However, as the ship is seldom in loaded condition during the trials and usually operates in ballast, the propeller curve at 6.1 has a somewhat lighter load than curve 6.
Point S1 in the diagram provides an example of the power supplied to the propeller at 90% of the MCR.
It may be necessary to exceed the maximum engine speed of the propeller to 107%, so curve 9 in the diagram, during the trials to evaluate how the ship operates at 100% power.

19.21 Comments

In practice, ship's resistance is compared to the outcome of the model tests in a laboratory, for instance Marin, Wageningen, The Netherlands. The results from the towing tests are also used for optimising the propeller- and hull design. When the required engine power, including the margins and the propeller speed is known, the correct engine can be selected with an engine selection programme developed by the engine manufacturer. It is even possible to state the requirements via internet, so the manufacturer can assess the client's demands in advance.

In view of this, the relation between the ship and the propulsion engine is extremely important, as well as the position of the load diagram, which is to say the choice of the engine-layout diagram in relation to the operational curve in order to achieve an optimum propulsion engine. In order to avoid overloading of the main engine during excessive operating conditions it is advisable to use an electronic governor with load control.

For a main engine that also drives a shaft generator for generating electricity for the ship, the interaction between the ship and the propulsion engine is complex. Nevertheless, in practice problems very seldom occur.

19.22 Future improvements in propulsion efficiency and reduction of CO_2

Source: A summery of a study by MAN Diesel, the largest engine manufacturer for medium- and low speed- diesel engines in the world.

Generally, the propulsion efficiency in ships is low. Four-stroke trunk-piston engines and two-stroke crosshead diesel engines have a total efficiency of just below 50% to just over 50% for larger cylinder diameters.

The efficiency of fixed- and adjustable propellers depends on type, size, rpm and the number of blades, ship's shape and various other factors. It lies usually between 50 and 70%.

Therefore, the total propulsion efficiency in a ship lies between 25 and 35%. Applying the rule of thumb, this means that a mere third of the energy released during combustion is actually used for ship propulsion.

19.22.1 IMO-Energy Efficiency Design Index

IMO regulations stipulate that the permissible CO_2 emissions for ships between 2012 and 2018 will have to be reduced to 70% of the current permissible emission (2012) in relation to the 100% index of new ships under construction today. Reduction in the CO_2 emissions is roughly directly proportional to a reduction in fuel consumption. The manufacturers' objective is to achieve a 30% decrease in fuel consumption for future voyages under normal conditions. Clearly, this is no easy task. In the last 100 years, diesel-engine efficiency has only gradually increased from +/- 25 to +/- 50%. In order to realise a 30% decrease in fuel consumption in less than 10 years, attention must be paid not only to diesel engine capabilities but also:

- Fuel consumption at a lower output per hour, per nautical mile or per freight ton.
- Fuel consumption with a change in propeller dimensions, that is, a larger diameter.
- Fuel consumption by changing the number of propeller blades.

The next example is based on the following assumptions:

1. The propulsion engines are those of a two-stroke crosshead type and in conformance with the IMO Tier II emission regulations even if decreased fuel consumption implies a reduction in NO_x emissions.

2. Moreover, all ships have a fixed-pitch propeller (FPP) and each propulsion engine is directly connected to the propeller and for this reason have an identical rpm.

MAN Diesel has studied three types of ships.

Type 1:
A 75,000 dwt Panamax Product Tanker with an operational speed of 15.1 knots.

Three engine types:
A MAN Diesel 5S60MC-C8 at normal power output (MCR).
A MAN Diesel 6S60MC-C8 at reduced power output.
A MAN Diesel 6S60ME-C8.

Type 2
A 4,500 TEU Panamax Container Ship at reduced cruising speed.

Propulsion engine MAN Diesel 6S80ME-C9 and a MAN Diesel 6K80ME-C9 compared to a MAN Diesel 8K90MC-C6 and a MAN Diesel 9K90MC-C6.
- Effects of the reduced speed.
- Effects of adjusting the number of blades.

Type 3
An 8,000 TEU Post-Panamax Container Ship at reduced speed.

A MAN Diesel 9S90ME-C8 with reduced power output compared to a MAN Diesel 10K98ME-7 and a MAN Diesel 12K98ME-C7.

19.22.2 Propeller and main engine: important data

In general one can say that the larger the propeller diameter, the greater the propeller efficiency. It has also been established that a larger propeller diameter requires a lower rpm in order to achieve optimum speed of both the propeller and propeller diameter.

- Two-stroke crosshead-engine efficiency is in part dependent on the maximum combustion pressure and the mean effective pressure ratio. The higher the ratio, the higher the engine efficiency and the lower the fuel consumption (b_e).

- Furthermore, the higher the stroke-diameter ratio of a two-stroke crosshead engine, the higher the engine efficiency. This means that engine types with a high stroke diameter ratio (super-long stroke), such as a MAN Diesel S80ME-C9, has a slightly higher efficiency than a short-stroke engine such as a MAN Diesel K80ME-C9.

- Electronically controlled engines have better capabilities for engine adjustment during operation than engines equipped with a camshaft. This means that the MAN Diesel ME/ME-C type engines have relatively higher efficiency when running under low NO_x IMO Tier II rules than the traditional MC/MC-C type engines.

If the design speed of a ship is reduced, both the required engine power and the rpm decrease, which in turn may affect the aforementioned propeller- and main-engine particulars.

19.22.3 Summing up the above

Propeller
A larger propeller diameter produces:
- A slight increase in propeller efficiency.
- An increase in optimal rpm.

Reducing the number of propeller blades produces:
- A slight increase in propeller efficiency.
- An increase in the optimal rpm.

Example: Reducing the number of propeller blades from 6 to 5 should achieve a 10% rpm increase for optimum propeller efficiency.

Main Engine
An increase in the p_{max} and p_{mep} ratio results in:

- Higher efficiency, also at reduced power.
- An electronically controlled engine, type ME has a marginally higher efficiency and more control over the NO_x emissions than an engine equipped with a camshaft.

19.22.4 Example 1

Main ship particulars assumed:

Scantling draught	m	14.2
Design draught	m	12.6
Length overall	m	228.6
Length between pp	m	219.0
Breadth	m	32.2
Sea margin	%	15
Engine margin	%	10
Design ship speed	kn	15.1
Type of propeller		FPP
No. of propeller blades		4
Propeller diameter	m	target

◄

This example shows that by derating the main engine, considerable fuel efficiency can be achieved if a four-bladed propeller with a large diameter is used.

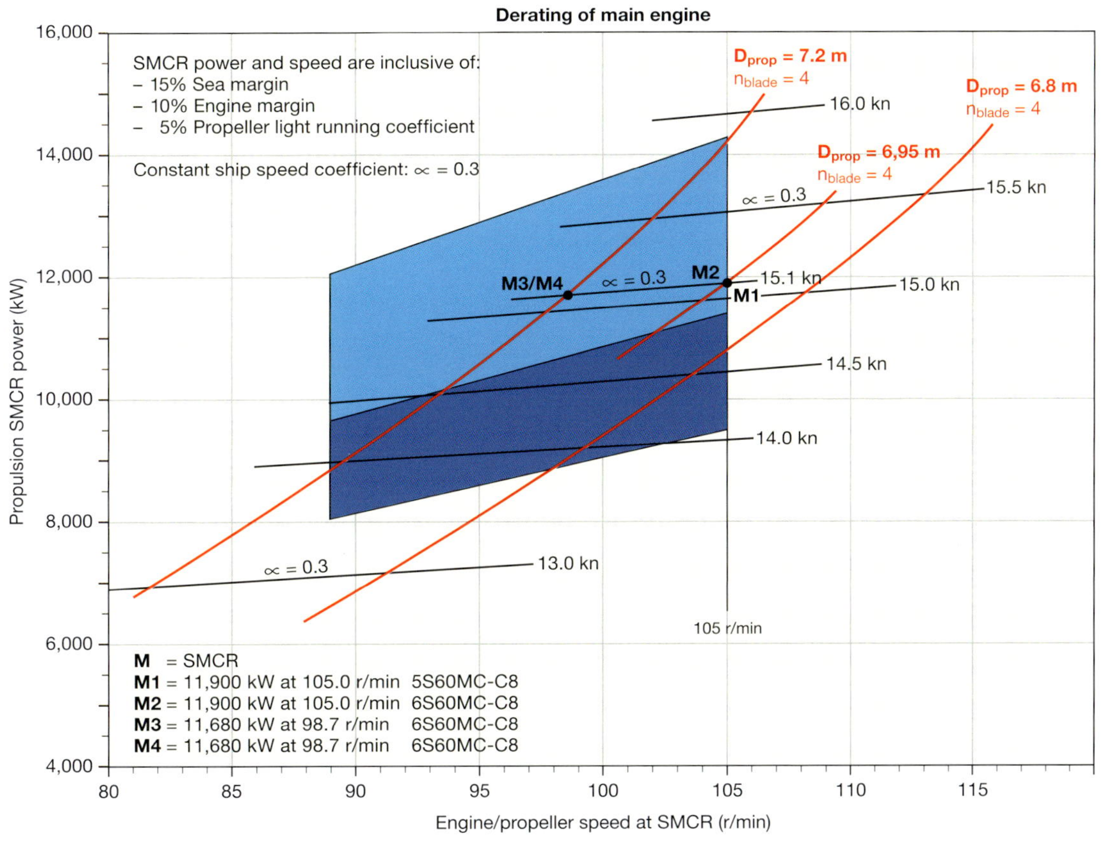

▲

Various main engine and propeller graphs and potential propulsion SMCR power outputs (M1, M2, M3 and M4) for a 75,000 dwt Panamax product tanker running at a similar speeds of 15.1 knots.

This diagram shows the lay-out diagrams of the 5 and 6 S60MC-C8/ME-C8 type engines including the SMCR points, M1, M2, M3 and M4 at 15.1 knots with the propeller curves of the three different propeller diameters, 6.8 m, 6.95 m and 7.2 m, each with four propeller blades, respectively.

Standard example

Option 1: the standard ship's main engine, a MAN Diesel, type 5S60MC-08, a SMCR= M2 = 11,900 kW at 105.0 r/min and a designed ship speed of 15.1 knots, runs at a normal speed.

Derating

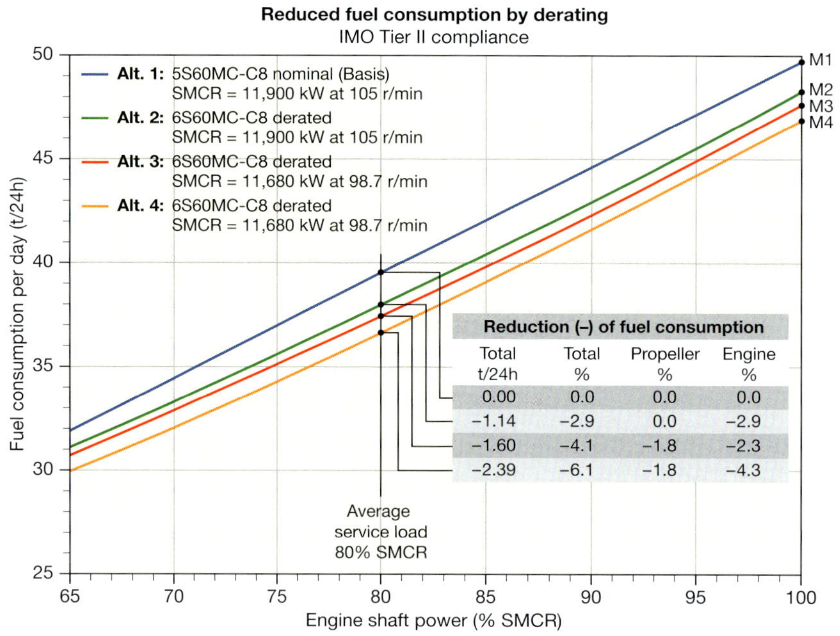

▲

The MAN Diesel type 5S60MC-C8 with a SMCR = M2 = M1, Alt. 2 with one additional cylinder allows for derating to the same SMCR point as the MAN Diesel type 5S60MC-C8 with a SMCR = M2 = M1 operating at full power. This reduces the fuel consumption by 2.9 % at a SMCR of 80%.

A larger propeller diameter

Also, if the stern of the ship can be adjusted (example 3), a propeller can be installed with a (larger) diameter of 7.2 metres and a corresponding SMCR = M3 = 11,680 kW at 98.7 revs/min comparable with a MAN Diesel type 6S60MC-C8 with reduced engine power.

Reduced fuel consumption per day or per voyage

In example 3, the reduced fuel consumption of 4.1% is achieved by the combined improvement of propeller- and main engine efficiency.

Main engine MAN Diesel type 6S60ME-C8

A further decrease in fuel consumption is obtained by installing an electronically controlled main engine (example 4) with an identical SMCR = M4 = M3. The diagram shows the overall reduced fuel consumption of 6.1% effected with a 6S60ME-C8, with an additional reduction of 2% compared to the MAN Diesel type 6S60MC-C8. This can be explained by the fact that the ME-C type has a higher efficiency than the MC-C, IMO NOx Tier II type, which makes it easier for this type to meet the IMO Tier II requirements for NO_x emission.

19.22.5 Example 2

A 4,500 TEU Panamax Container Vessel

Based on a ship where propeller diameter is unchanged. This example clearly shows the reduction in fuel consumption by reducing the original ship speed of 24.7 knots. The calculation concentrates on the influence that the number of propeller blades and the corresponding influence of the selected engine type that enables the design speed of the ship, 22 knots, to be achieved. Apart from the data of the selected main engine type, the ship's particulars are as follows:

Main ship particulars assumed:

Scantling draught	m	13.3
Design draught	m	12.0
Length overall	m	286
Length between pp	m	271
Breadth	m	32.2
Sea margin	%	15
Engine margin	%	10
Type of propeller		FPP
Propeller diameter	m	8.3
No. of propeller blades		target
Design ship speed	kn	target

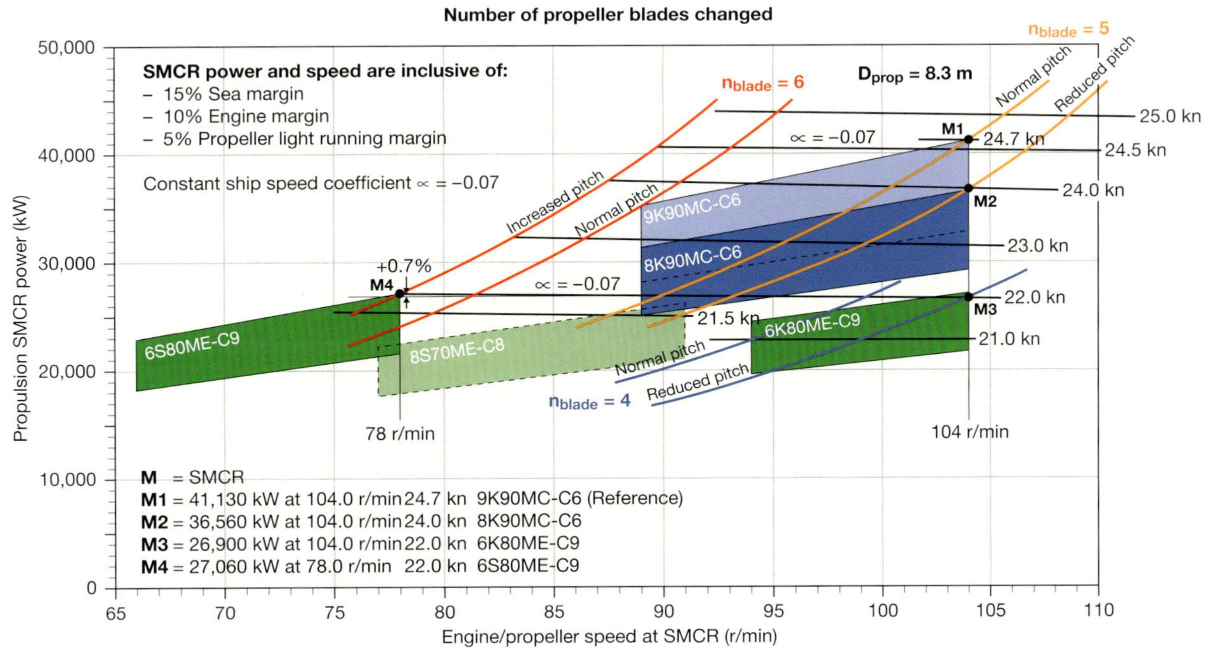

▲
Various types of main engines and their propeller curves and potential SMCR points (M1, M2, M3 and M4) for a 4,500 Panamax Container Ship with varying design ship speeds.

Five-bladed propeller

A MAN Diesel type 9K90 MC-C6 with a SMCR = M1 = 41,130 kW set at nominal propulsion power of 104.0 r/min, a ship speed of 24.7 knots and a five-blade propeller as shown in the graph. The optimum propeller curve with a normal pitch with five blades crossing M1 indicates a corresponding SMCR power and line M for the main engine for reduced ship speeds.

Point M2 = 36,560 kW at 104.0 r/min applies to a MAN Diesel main engine, type 8K90MC-C6 with a propeller curve for reduced pitch and five propeller blades, which potentially can achieve the design speed of 24.0 knots.

The required SMCR point at 22.0 knots is approximately 26,800 kW at 90 r/min.
The lay-out diagram of a MAN Diesel 8S70ME-C8 with L1 = 26,160 kW, a speed of 91,0 r/min and five propeller blades shows that the maximum design ship-speed for this engine type is 21.8 knots.

Four-bladed Propeller

When the number of blades is reduced from 5 to 4, the corresponding optimum SMCR propeller curve at a normal pitch shifts to the right at a speed increase of approximately 10% and is represented by a corresponding SMCR pitch curve at reduced pitch. The last curve shows a ship speed of 22.0 knots, a SMCR point M3 = 26,900 kW at 104.0 r/min.
This point is situated in the top corner of the lay-out graph of a MAN Diesel type 6K80ME-C9 main engine.

Six-bladed Propeller

The corresponding SMCR = point M4 = 27,060 kW at 78 r/min for a ship speed of 22 knots at an increased propeller pitch, this time with six propeller blades and therefore a reduced propeller speed.
Point M4 is equal to the nominal SMRC point of the MAN Diesel type 6S80ME-C9 main engine.

Reduced fuel consumption per day

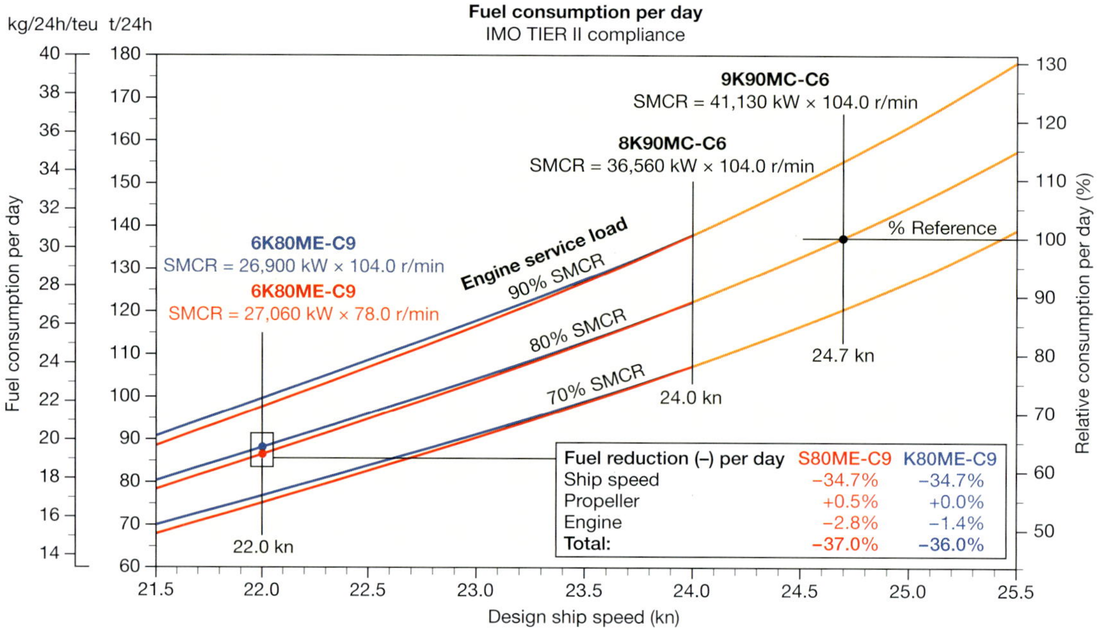

▲

Reduced fuel consumption per day.

Fuel consumption per day for the aforementioned alternative main engine is calculated in compliance with the IMO Tier II emission regulations.
The findings as shown in the graph are related to the design ship speed for engine service loads of 70%, 80% and 90% SMCR. Using 24.7 knots as a reference and a SMCR of 80%, the curves demonstrate that the potential fuel reduction per day can be considerable.

At a ship speed of 22.0 knots a reduction of approximately 37% can be achieved for a MAN Diesel type 6S80ME-C9 main engine.

The super-long stroke MAN Diesel type 6S80ME-C9 main engine with higher efficiency effects a higher reduction in fuel consumption than the short-stroke MAN Diesel type 6K80ME-C9 main engine.

Reduced fuel consumption per voyage

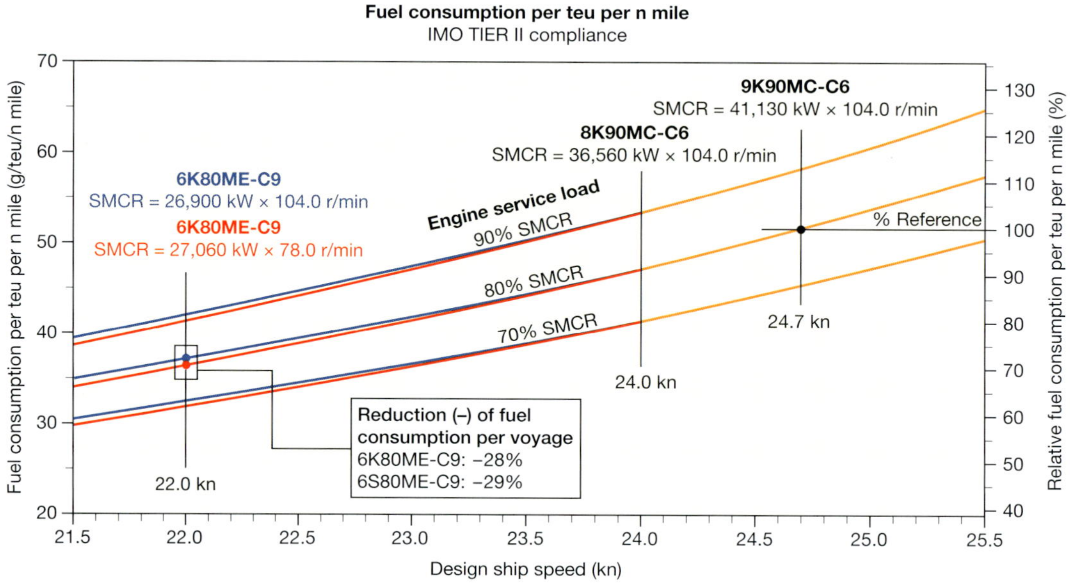

▶

Reduced fuel consumption per voyage.

The graph shows the fuel consumption per nautical mile as well as the fuel consumption per voyage at a reduced power output. The results are significant. If the ships speed is reduced from 24.7 to 22.0 knots, the overall decrease in fuel consumption per voyage is 28% for a MAN Diesel type 6K80ME-C9 and 29% for A MAN Diesel type 6S80ME-C9.

19.22.6 Example 3

An 8,000 TEU Panamax Container Ship

Standard, a six-bladed propeller, shown here with a varying number of propeller diameters. This example shows the quantity of fuel saved by reducing the ship's speed. The calculation is aimed at showing the impact of the propeller diameter at a reduced ship speed and the subsequent choice of main-engine type.

Main ship particulars assumed:

Scantling draught	m	14.5
Design draught	m	13.0
Length overall	m	323
Length between pp	m	308
Breadth	m	42.8
Sea margin	%	15
Engine margin	%	10
Type of propeller		FPP
No. of propeller blades		6
Propeller diameter	m	target
Design ship speed	kn	target

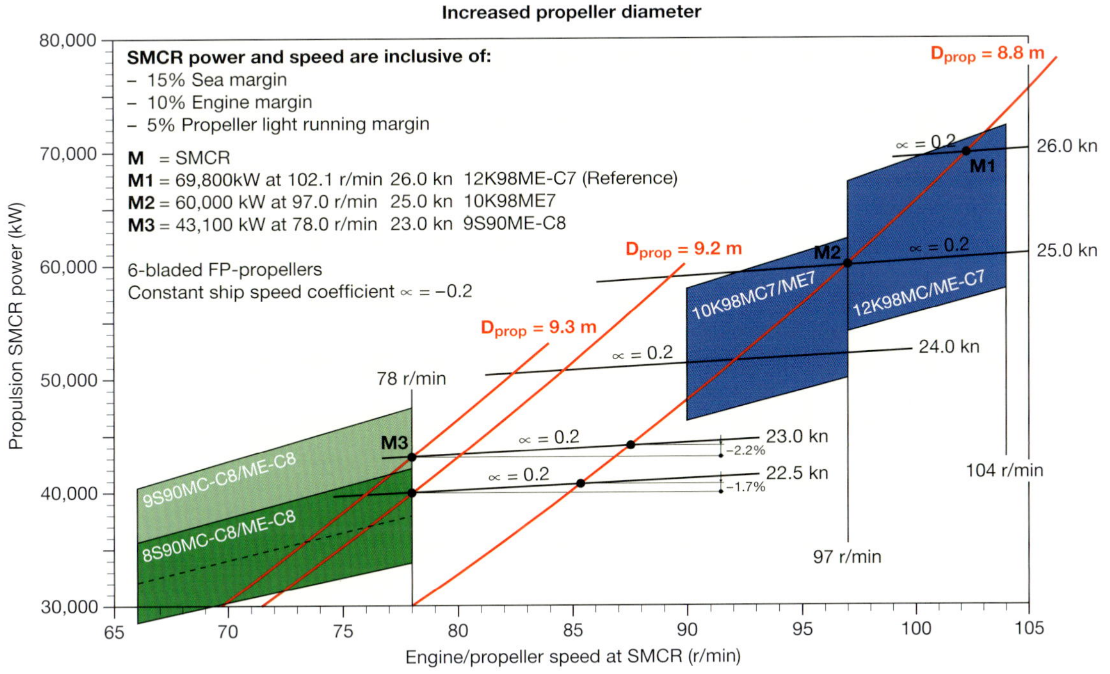

▲

Various main engine and propeller curves and the SMCR possibilities M1, M2 and M3 for an 8,000 TEU Post-Panamax container ship at varying design speeds.

Propeller diameter: 8.8 metres.

The MAN Diesel type 1298ME-C7 with SMCR = M1 = 69,800 kW at 102.1 r/min at reduced power is the standard reference point. The design ship speed is 26.0 knots and the six-blade propeller has a diameter of 8.8 metres.

SMCR = M2 = 60,000 kW at a reduced ship speed of 25.0 knots and an unchanged propeller diameter. In this instance a MAN Diesel type 1098ME-C7 main engine running at 97 r/min will suffice.

A further reduction in the ship speed to 23.0 knots and an unchanged propeller diameter of 8.8 metres requires a SMCR of approximately 44,100 kW at 87.5 r/min.

An increased propeller diameter of 9.3 metres.

At a ship speed of 23.0 knots, a propeller diameter of 9.3 metres and 71.5 % of the design draught (virtually the maximum) the SMCR power and the speed will fall to M3 = 43,100 kW at 78 r/min. Adjusting the propeller diameter corresponds to a ship speed coefficient of $\lambda = 0.2$

$$\lambda = \frac{\ln(43{,}100\text{ kW} / 44{,}100\text{kW})}{\ln(78.0\text{ omw/min} / 87.5\text{ omw/min}} = 0{,}2$$

SMCR point M3 equals the design ship speed of 23.0 knots and can be achieved with a derated MAN Diesel type 9S90ME-C8 main engine.

Reduced fuel consumption per day

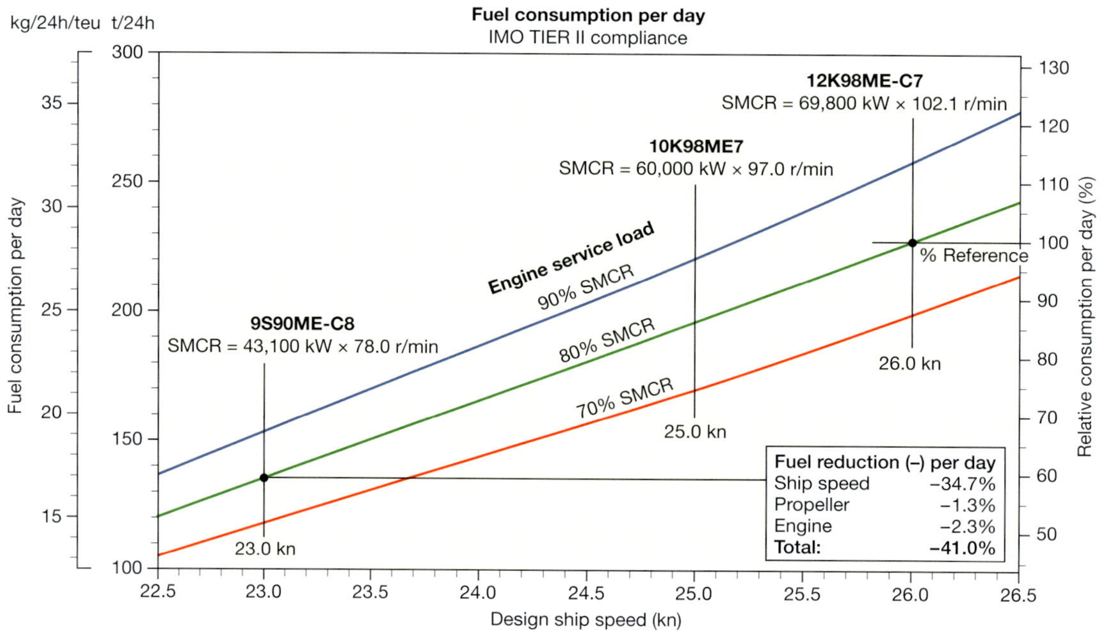

▲

Reduced fuel consumption per day.

The fuel consumption per day for all three main engines is calculated in conformance with the IMO Tier II emission regulations.

The graph shows the findings for ship speeds of 70%, 80% and 90% in relation to the design ship speed.

Having a ship speed of 26.0 knots as a point of reference and at 80% SMCR, the fuel consumption curve shows a potential decrease in the daily fuel consumption of 41% if the MAN Diesel type 12K98ME-C7 and 26 knots is replaced by a MAN Diesel type 9S90ME-C8 and 23 knots.

This reduction is due to:

1 A ship speed reduction of 37.4%
2 An increase in propeller efficiency of 1.3%
3 An increase in engine efficiency of 2.3% due to the super-long stroke MAN Diesel type 9S90ME-C8 main engine as opposed to the short-stroke MAN Diesel type 12K98ME-C7 main engine.

Reduced fuel consumption per voyage

▶

Reduced fuel consumption per voyage.

The graph shows the fuel consumption per nautical mile as well as the reduced fuel consumption per voyage.

Decreasing the ship speed from 26 to 23 knots effects a reduction in fuel consumption of 33% per voyage!

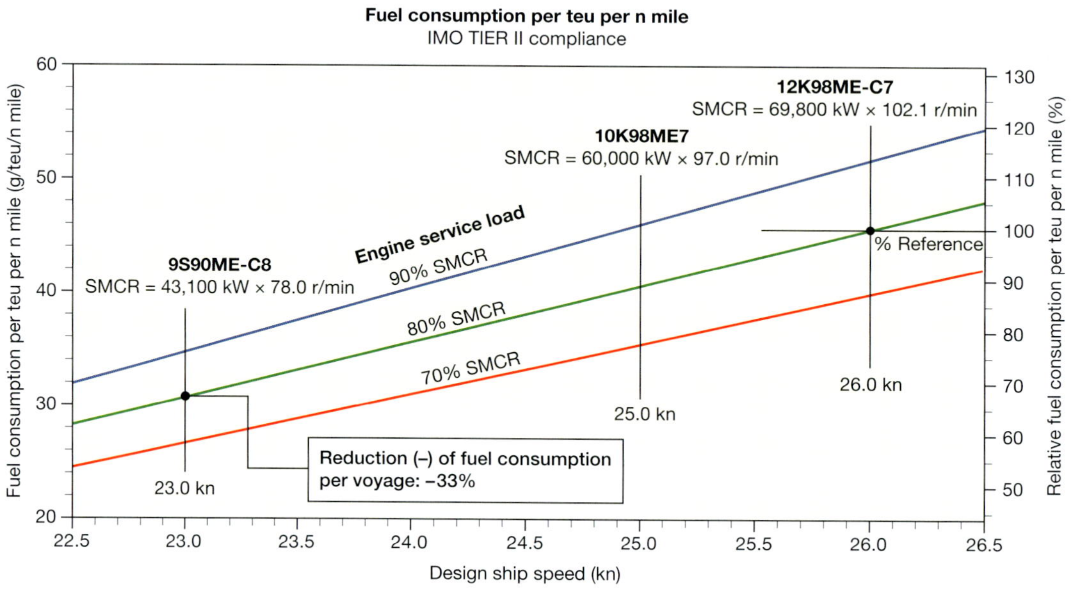

19.22.7 Summary

Future political requirements for reducing CO_2 emissions for sea-going vessels provide an opportunity for interesting, but also costly and drastic developments on board ships, such as amongst others, heat-recovery systems. Here an important point of interest is the design of the stern of the ship.

For instance, at low speed together with either an increased propeller diameter and/or a changed number of propeller blades, the opportunity to select an engine type which normally would not be used in large, high-speed container ships is available.

Super-long stroke engines with a marginally higher efficiency that are commonly used in tankers, such as the MAN Diesel type S80 and S90, will in the future be used in large container ships. This will lead to a decrease in fuel consumption of 30% per voyage and consequently a reduction in CO_2 emissions of 30%, in comparison to the engines presently used in large container ships. Furthermore, using liquefied natural gas (LNG) instead of heavy oil achieves a CO_2 reduction of 23% dependent on the chemical composition of the LNG.

19.22.8 Rule of thumb for achieving optimum efficiency

In general, the largest ships, the largest diesel engines, the largest propellers and the largest electric gensets result in the highest efficiency!

◀

Larger ships require relatively less propulsion power to maintain a certain speed than smaller ships of a similar type.

The 'Colombo Express' from Hapag Lloyd (2005), 8749 TEU, 34,500 kW, was succeeded by the considerably larger 'Emma Maersk (2006) of 13,800 TEU, 83,400 kW. These will soon be followed by 18,000 TEU ships of the 'Triple E-class' by Maersk Line with a shaft power of 63,300 kW and two propellers (2013). The energy costs per teu and consequently the CO_2 emissions will therefore continue to decrease.

▼

Larger propellers running at lower speeds produce higher efficiencies than smaller propellers running at higher speeds. Shown here, a propeller with a diameter of 8.5 metres and maximum speed of approximately 110 r/min. This is a propeller for a 8,600 TEU container ship.

▼

Larger engines with a large stroke volume per cylinder have higher efficiencies than smaller engines with cylinders and a lower stroke volume. Shown here, a large two-stroke crosshead engine driving a large disc generator in a diesel power plant.

>CH 20

Transmission gears, flexible couplings, vibration dampers, shafting and shaft generator drives

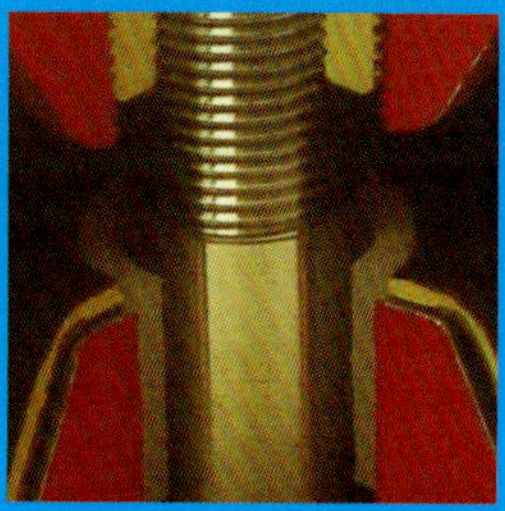

The Wärtsilä 46C engines.

The Propulsion arrangement of a ship from Wagenborg Shipping, the 'Oranjeborg'. Two six-cylinder Wärtsilä 6L46C medium-speed engines drive one adjustable-pitch propeller. Renk reduction gearing controls the correct propeller speed and shaft generator drive, shown top, at the back of the photo.

20.1 Introduction

Transmitting the engine's power output to the pump-, compressor- or electric genset drives requires the use of a number of facilities, such as couplings and gear transmissions.
Often, the engine parts connected to the drives have different speeds, such as shaft generators and propellers. As opposed to the aforementioned drives, gensets in a diesel-power plant or on board ship are equipped with one fixed, specific speed determined by the generator type.

20.2 Diesel-engine arrangements

The following arrangements occur frequently:

A **Ship propulsion** with a low-speed two-stroke crosshead engine.

B **Ship propulsion** with one or multiple medium-speed four-stroke trunk-piston engines.

C **Ship propulsion** with one or multiple high-speed four-stroke diesel engines.

D **Diesel-electric propulsion** with medium-speed or high-speed four-stroke diesel engines.

E **Ship propulsion with water jets** using high-speed four-stroke diesel engines.

F **Diesel-power plants** with both high-speed and medium-speed four-stroke diesel engines and low-speed two-stroke crosshead engines.

G **Special drives** such as those for pumps in the dredging industry, compressors and pumps for use in the gas and oil industry using high-speed- or medium-speed four-stroke diesel engines.

20.2.1 A Ship propulsion with low-speed two-stroke crosshead engines

The engine drives directly a fixed-pitch propeller is directly reversible and runs on heavy-fuel oil (HFO).

This type of propulsion is usually found in large container ships, oil tankers and bulk carriers.
Speed: The low engine speeds of these category IV engines with cylinder bores from approximately 500 mm are exceedingly suitable to drive directly a fixed-pitch propeller. The propeller efficiency is high at low engine speeds of between ± 60 to ± 120 revolutions per minute.
Engines with smaller cylinder diameters and therefore a higher speed up to approximately 200 revolutions per minute are not often directly reversible and are equipped with a transmission gear positioned halfway between the crankshaft and the adjustable-pitch propeller.

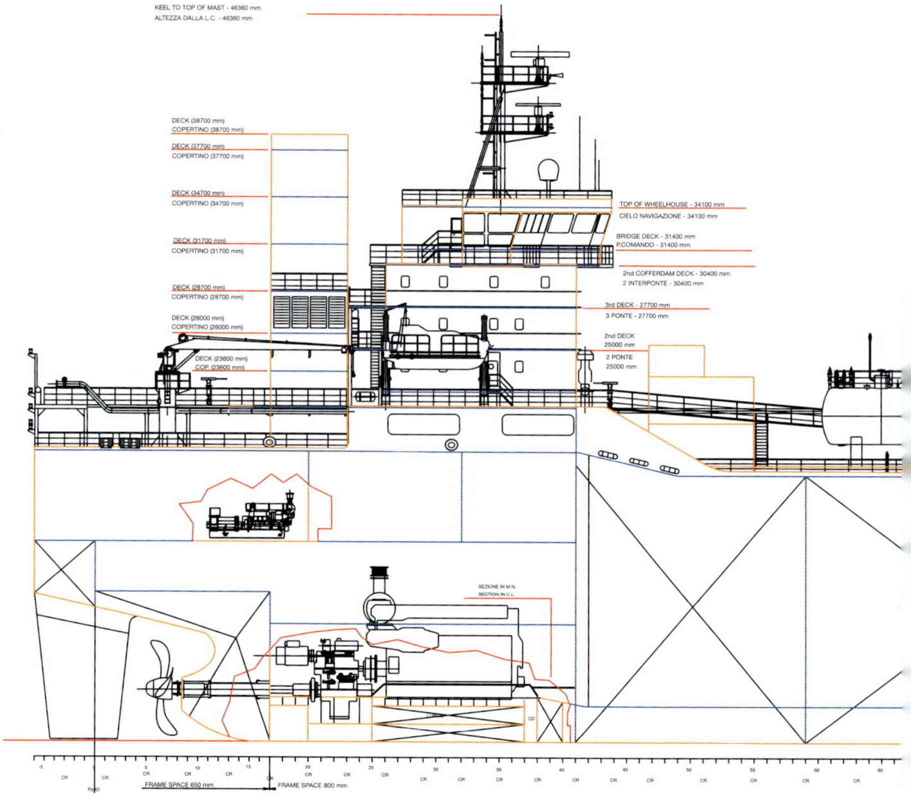

▶
Ship propulsion with a large medium-speed four-stroke engine.

The diesel engine is connected to the propeller shaft via reduction gearing.
The pitch of the propeller is adjustable, so the engine can rotate in one direction and at a fixed speed. A shaft generator is coupled to the gearbox. This allows electricity to be generated using heavy fuel oil, so the auxiliary engines, one deck up, do not have to operate on diesel oil, which is almost twice as expensive.

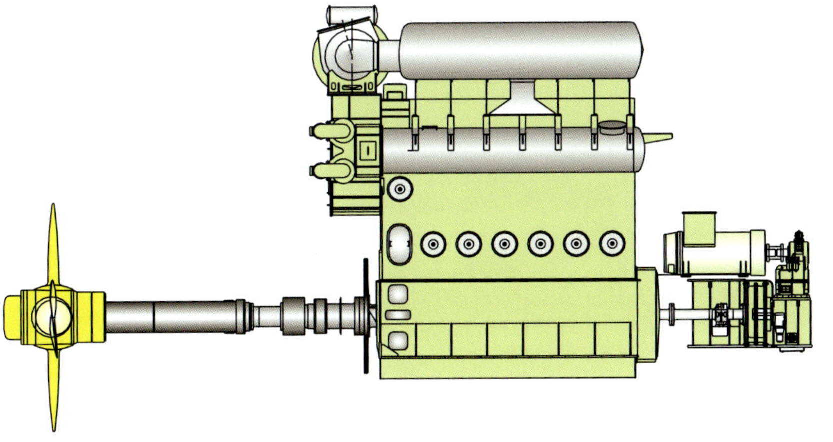

◄ A six-cylinder two-stroke crosshead engine driving a controllable-pitch propeller.

At the other side of the propulsion engine a shaft generator is coupled to the engine.

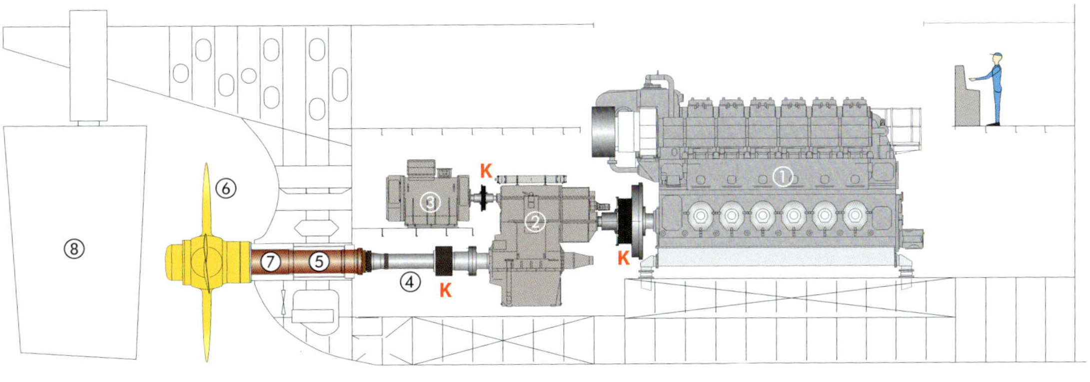

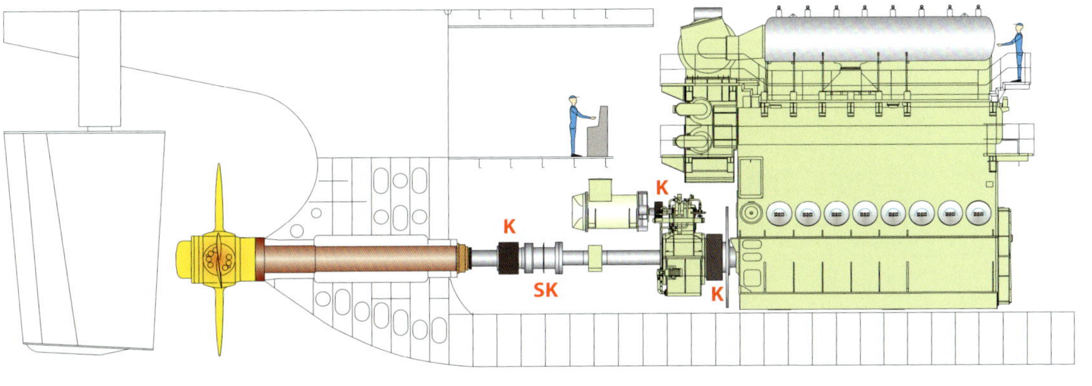

◄ Top picture: ship propulsion with a medium-speed four-stroke diesel engine running on HFO.

The engine has one rotational direction and drives the controllable-pitch propeller and shaft generator.

1 *four-stroke diesel engine*
2 *gearbox*
3 *shaft generator*
4 *intermediate shaft*
5 *propeller shaft*
6 *controllable-pitch propeller*
7 *stern tube*
8 *rudder*

K *flexible couplings*
SK *clutch*

Bottom picture: Ship propulsion with a low-speed two-stroke crosshead engine running on HFO.

Shaft generators

A number of propulsion units are equipped with a shaft generator. When at sea, the shaft generator generates sufficient electric power for all the electric systems on the ship, so the auxiliary engines can be shut down. In this way, the number of running hours of the auxiliary engines is limited as are the corresponding maintenance hours and expenditure.

The total efficiency for power generation increases marginally as the propulsion engine has a higher efficiency than the smaller auxiliary engines. The shaft generator is installed between the engine and the propeller shaft, and on the propeller drive in smaller crosshead engines with a transmission gear to the propeller. This connection is often referred to as Power Take Off (PTO).

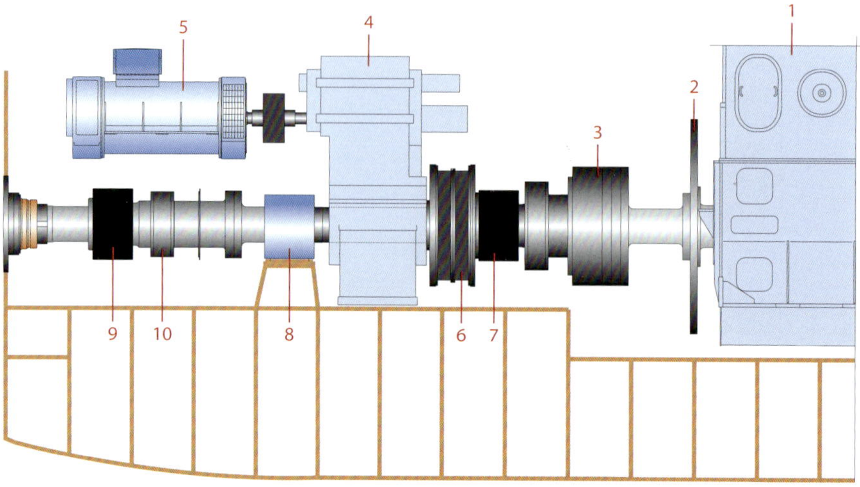

▲

A drive for the propeller shaft and shaft generator in a two-stroke crosshead-engine.

Viewed from the engine

1 *main engine with fixed speed*
2 *flywheel/turning wheel*
3 *conical bolt clutch*
4 *two-speed gearing*
5 *shaft generator*
6 *flexible coupling*
7 *hydraulic clutch*
8 *intermediate bearing*
9 *hydraulic clutch*
10 *oil-distribution ring for controllable-pitch propeller*

Options

I The main engine is running free of the main shaft. Shaft generator and propeller are stopped, for instance, during trials, after repairs or prior to departure.

II The main engine drives the shaft generator and the propeller is stopped.

III The main engine drives the shaft generator as well as the propeller.

VI The shaft generator is switched to the electromotor configuration and drives the propeller with power generated by the auxiliary gensets. This is an emergency operation mode during engine malfunction, the so-called 'getting home' device. This is also used when the vessel must be moved along the quay during main-engine repairs.

▲

The shaft generator on a very large container ship from 6800 TEU.

The rotor of this 3.5 MW generator is an integral part of the shafting between the large two-stroke 68,000 kW crosshead engines and the propeller shaft. The cooling-air ventilators are positioned on top of the shaft generator.

1 *flywheel/turning wheel main engine*
2 *shaft generator*
3 *cooling-air ventilator*

▲

The shaft generator (1) above the propeller shaft on a feeder-container ship.

The space at the back of the engine room near the propeller is rather cramped.

The largest container ships commissioned in 2010 are still equipped with one propulsion engine and one fixed-pitch propeller.

Drives with a fixed-pitch propeller

Most large ships are equipped with the above-mentioned propulsion system.

Advantages

- **The total propulsion efficiency is high in comparison to other systems.** With a diesel-engine efficiency of 50% and a propeller efficiency of 75%, the total propulsion efficiency is therefore: $0.5 \times 0.75 = 0.375$ or 37.5%. All other systems have a lower efficiency.
- **A simple installation.** A large crosshead engine has relatively few moving parts. When using the two-stroke principle, the power output generated is 1.6 times that of four-stroke engines with an identical cylinder bore and engine speed.
- **The lower speed of larger two-stroke crosshead engines is particularly suitable for direct gear drives driving the large fixed-pitch propellers.**

20.2.2 B Ship propulsion with one or several medium-speed four-stroke trunk-piston engines

The engine is non-reversible and often has a fixed engine speed. The fuel is either HFO or MDO.

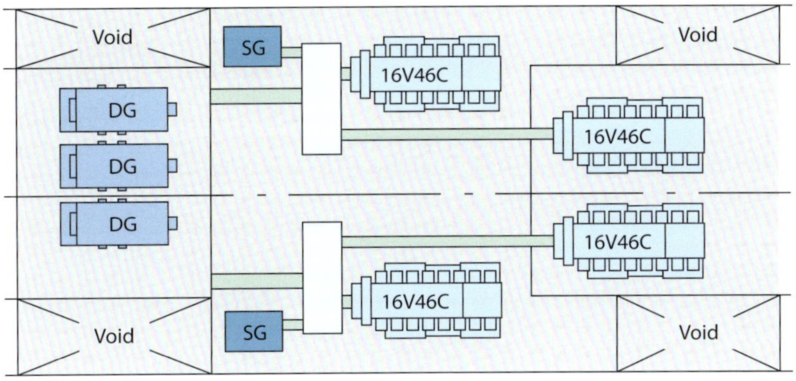

▲
A top view of a propulsion plant with medium-speed four-stroke engines driving twin propellers.

The four sixteen-cylinder Wärtsilä 46 V-engines drive the two adjustable-pitch propellers via reduction gearing. Each reduction gearing has a Power Take Off-shaft with the shaft generator, SG, mounted on top.
In a separate auxiliary engine room another three diesel gensets, DG, are installed.

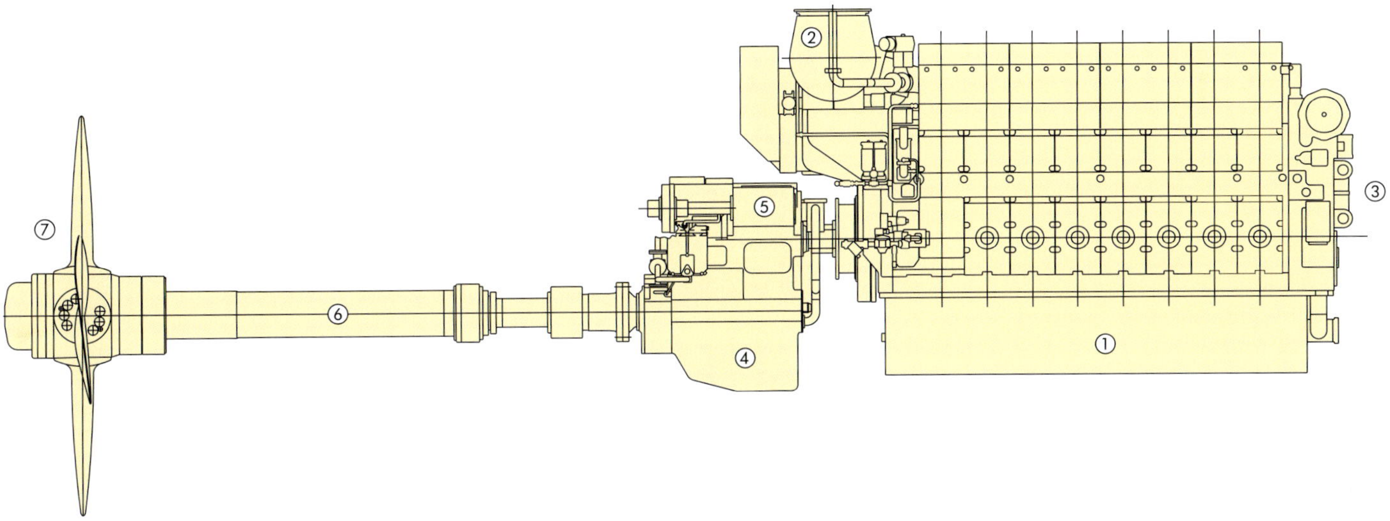

▲
A standard arrangement for a four-stroke medium-speed propulsion engine.

1 *diesel engine*
2 *supercharger with intercooler*
3 *engine-driven cooling-water pumps*
4 *reduction gearing*
5 *PTO-section*
6 *propeller shaft*
7 *adjustable-pitch propeller*

Diesel-mechanical (HFO)
Installed power: 55,980 kW

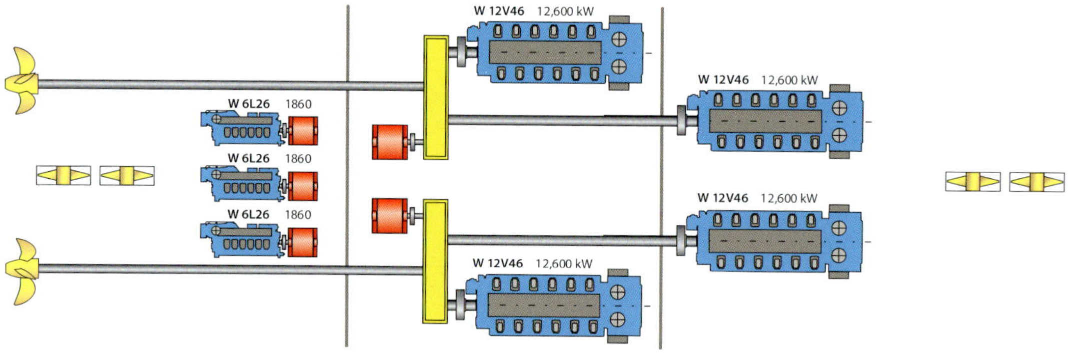

▲

A large propulsion plant with medium-speed four-stroke diesel engines running on HFO, design Wärtsilä.

Four twelve-cylinder Wärtsilä 46 diesel engines in V-shape with a total shaft power of 50,400 KW. The three Wärtsilä 26 six-cylinder in-line engines supply a generator power of 5580 kW. The ship is equipped with two electric bow- and stern thrusters.

▶

Installation of one of the four large MAN Diesel four-stroke medium-speed diesel engines in the engine-room of a large cruise ship. This photograph is taken in the hall of the Jos Meijer shipyard in Papenburg, Germany.

In view of the limited space on these ships, low four-stroke diesel engines have been chosen as opposed to the high two-stroke crosshead engines. Approximately 80% of the volume of the ship is required for the 'hotel' with the remainder is for all the technical installations.

▶

The 'Norwegian Dawn' under construction in Papenburg, Germany. The cruise ship is in dry-dock ready to be 'set afloat'.

This type of propulsion is often seen in smaller ships, such as container feeders, smaller tankers for oil, gas or chemical transport, passenger ships and ferries. The shaft generator often is driven from the gear transmission. The overall height for this type of propulsion is much lower when compared to the crosshead engine, and is therefore suitable for the low overhead space found in engine rooms of large ferries and passenger ships.

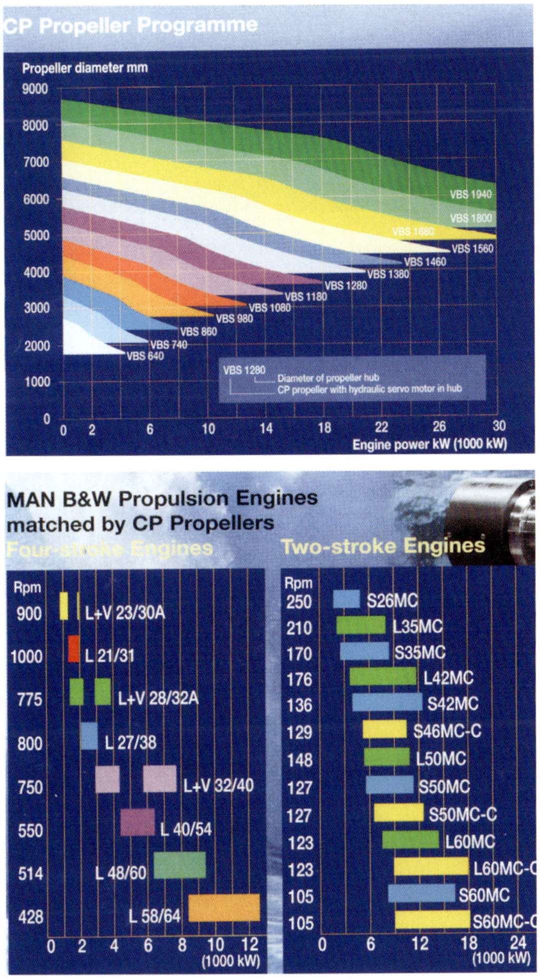

▲

The MAN Diesel program for propulsion engines with controllable-pitch propellers.

In the four-stroke medium-speed diesel engines, the speed varies between 900 and 428 rpm, the cylinder diameter from 230 to 580 mm and the shaft power from 1000 to 12,000 kW.

In low-speed two-stroke crosshead diesel engines the speed varies between 250 and 105 rpm, the cylinder diameter from 260 to 600 mm and the shaft power from 1500 to 18,000 kW.

Larger crosshead engines are directly reversible and have a fixed-pitch propeller. The controllable-pitch propeller diameter varies from 1800 to 8600 mm.

A coupling section often is added to the gear box, so the propeller can be stopped when the diesel engine and shaft generator are still running. All the engines in category III come standard with a gear box, shaft generator, shafting and a controllable-pitch propeller.

20.2.3 C Ship propulsion with one or several high-speed four-stroke diesel engines

These engines operate on MDO. These systems often are found in smaller ships and inland navigation. The complete system comprises the diesel engine, a gear transmission, a coupling section, a PTO for the shaft generator, shafting and adjustable-pitch propeller.

▲

A gear box aft for a four-stroke high-speed MTU-diesel engine.

1 *reduction gearing*
2 *shaft to propeller*
3 *shaft to shaft generator*
4 *lubricating-oil pump*
5 *lubricating-oil cooler*
6 *lubricating-oil filter*
7 *flexile coupling between engine and gear box*

▶
A four-stroke high-speed Caterpillar diesel engine for driving a large yacht.

The reduction gearing with lubricating-oil pump, lubricating-oil cooler and lubricating-oil filter attached to the engine.

Often two engines with two propellers are arranged for extra power generation. This system is also used in luxurious motor yachts.

Photo: Franco Pace

▲
The world's largest private yacht, the 'Athena', has two four-stroke high-speed Caterpillar type 3516 B main engines, each has 1492 kW power output at a speed of 1600 rpm.

There are three gensets of 290 kW each. The twin screws have a diameter of 1.6 metres and a speed of 440 rpm. A maximum of 103,000 litres diesel fuel is in its bunkers.

Photo: Hans Westerink

▲
The engine room in the 'Athena'.

20.2.4 D Diesel-electric propulsion with medium-speed- or high-speed four-stroke diesel engines

In the first instance, they operate on HFO and in the second on MDO.
In this system, the diesel engines drive gensets, which supply power to the electromotors that drive the propeller. This system often is applied today. It is used in passenger ships, ferries and in some self-propelled drilling platforms.
The diesel gensets can be installed in the most convenient location on the ship or vessel, as they do not have to be in-line with the propeller shaft as in traditional arrangements.

Diesel-electric (HFO)
Installed power: 53,100 kW

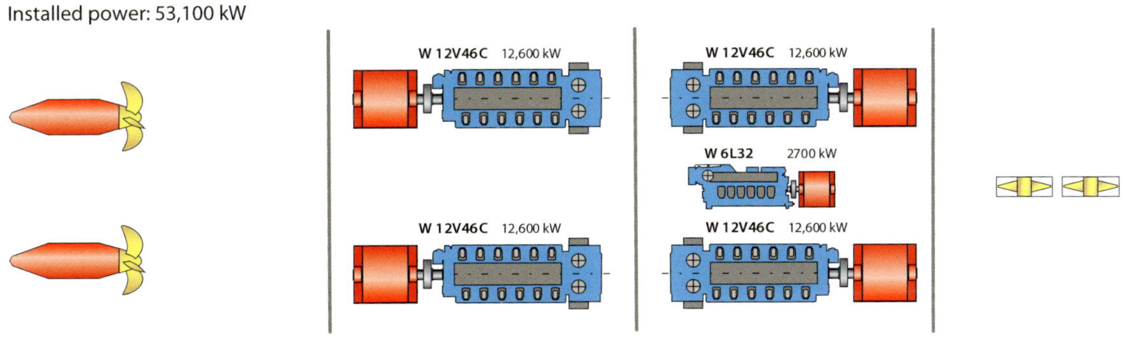

▲
A diesel-electric propulsion plant with four medium-speed four-stroke diesel engines running on HFO.

These plants are often installed on large cruise ships; the propellers are driven by electric motors. Depending on the required propulsion power and electric power, one or more diesel generators may be deployed.

▼
The diesel-electric propulsion plant for a large passenger ship, the 'Queen Mary II'.

Besides four diesel gensets running on HFO, a light-weight and compact gas-turbine genset running on MDO is installed on the upper deck. The power output of this genset is 75% of the total power output of the three diesel gensets.

Diesel power:	33.6 MW
GT power:	25.0 MW
Total installed power:	**58.6 MW**

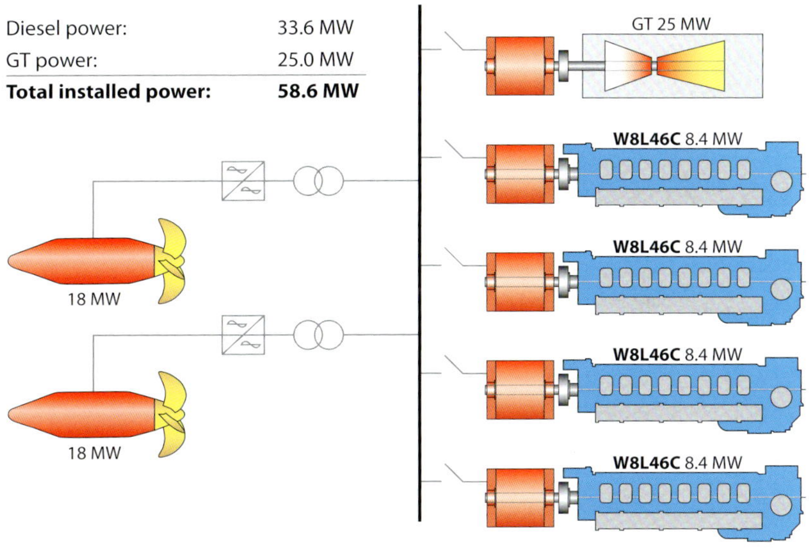

Advantages

- **The diesel gensets operate at a constantly high rpm**, so diesel-engine contamination due to, amongst others, low engine speeds, is reduced at a low load.
- **At low speeds, a section of the diesel genset can be switched off.** Consequently, these have fewer operating hours and therefore require less maintenance.
- **Driving the propellers using electromotors is not dependent of the diesel engines.** With adjustable propellers or frequency-controlled electric-drive motors, any type of propulsion is possible.
- **The generated electric power is also available for energy supply to the ship**, for instance on large passenger ships, ferries, self-propelled platforms and drilling ships.
- **In diesel-electric propulsion, it is also possible to use a gas-turbine genset alongside the diesel genset.** With this approach, a large amount of power is generated quickly with limited size and weight. Only MDO can be used. The exhaust-gas emissions of these gas turbines are significantly lower than those of diesel engines: ideal for 'sight seeing' when sailing in ecologically sensitive areas. Well-known examples are the 'Jewel of Norway' and the 'Queen Mary II'.
- **The total propulsion efficiency** is sometimes higher than that of traditional propulsion when PODS are used.

20.2.5 E Ship propulsion with water jets, using high-speed four-stroke diesel engines

A water jet rather than propellers propels a small, specific group of fast ships. The diesel engine drives a pump, which draws in water at high speed and subsequently pumps this overboard. This so-called water jet can be steered in any direction thus rendering a rudder redundant.

Water jets are used often in high-speed passenger transport such as catamarans, pilot boats and motor yachts.

▶

Water jets driven by high-speed four-stroke diesel engines are the ideal combination for the propulsion of fast, light catamarans for passenger- and patrol craft.

▶

The controllable-discharge nozzle of a water jet.

The amount and direction of the water can be accurately controlled for optimum manoeuvrability of fast ships.

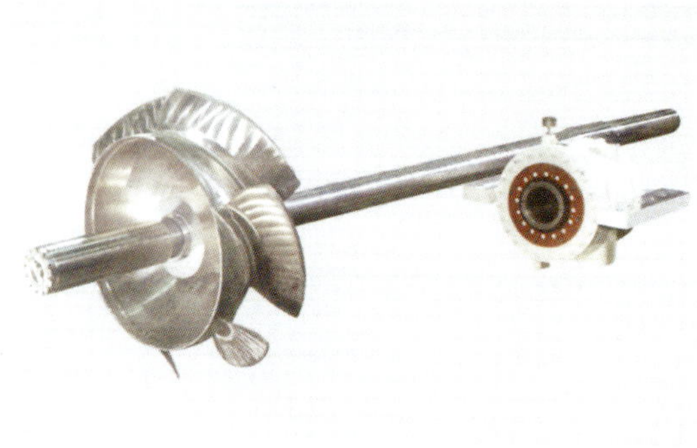

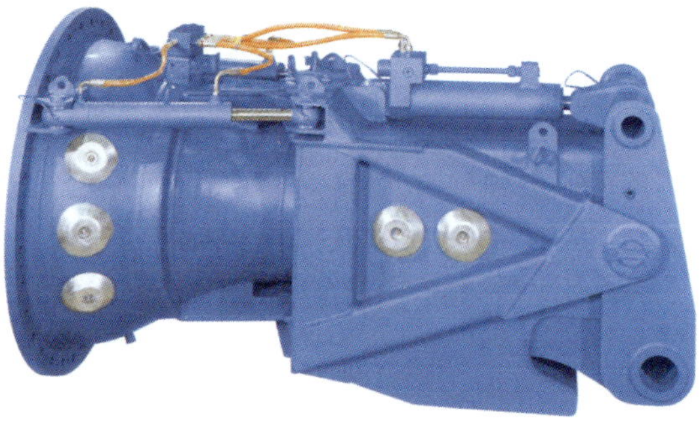

▶

A water jet consists of a central pump section and an adjustable nozzle, shown right. The pump is driven from the left by the diesel engine via a gear box.

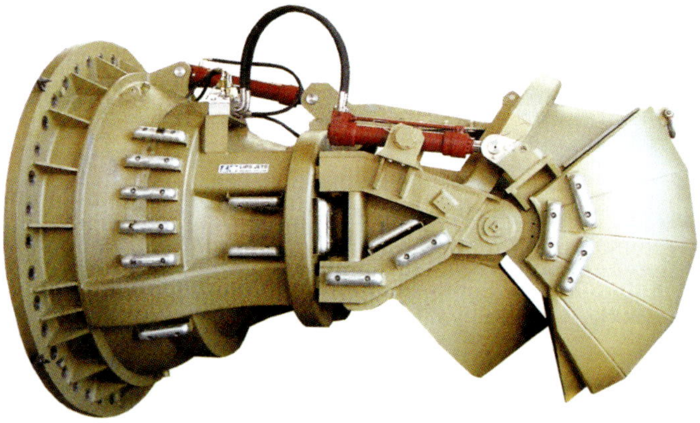

▶

A very large water-jet propulsion plant driven by a gas turbine on a luxury yacht which can achieve a speed of 40 knots.

Left and right of the water jet, not visible underneath the ship are two propellers driven by high-speed four-stroke V-engines.

1 *location of the propellers*
2 *water-jet impeller*
3 *nozzle*

◄

One of the two water jets on the passenger catamaran 'Koegelwieck' from Doeksen Shipping Company, Harlingen, The Netherlands.

1 suction section: the pump is not equipped with a suction box to prevent cavitation.
2 impeller section off the stern bulkhead.
3 cover for inspection and removal of dirt. The impeller and shaft are sturdy, so relatively heavy objects that are drawn in do not cause damage.

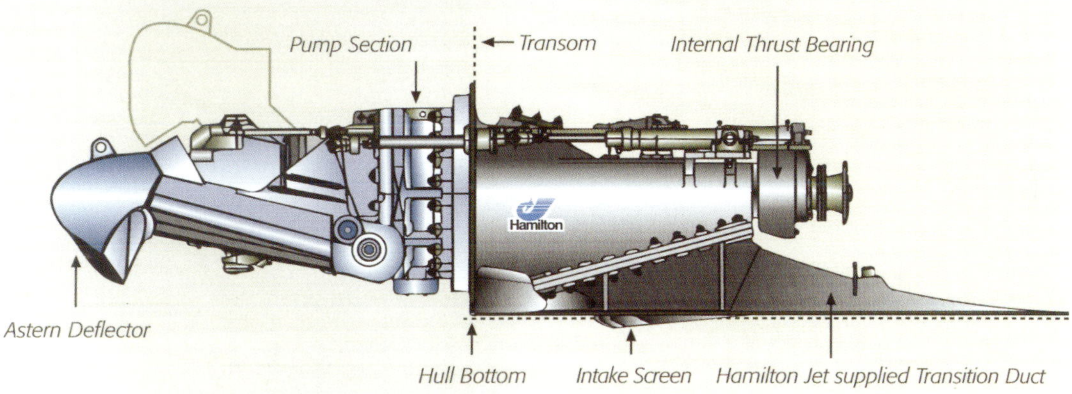

▲

A Hamilton water jet.

The under-keel water is vertically drawn in and discharged on the left. With the 'Astern Deflector' the water-jet power is directed ahead, so that the vessel can sail astern. The pump section is situated just aft of the ship's stern.

20.2.6 F Diesel-power plants with both high-speed- and medium-speed four-stroke diesel engines, as well as low-speed two- stroke crosshead engines. Engine categories I to IV included

The engine operates on MDO and HFO.
Most diesel-power plants are equipped with **medium-speed four-stroke diesel engines** running on HFO. **This is often the most effective arrangement from an economic point of view.**
At smaller capacities, or when no HFO is available, **high-speed four-stroke diesel engines** running on MDO are brought on-line.
Semi-permanent plants and mobile gensets in containers also are found in this category.
Large two-stroke crosshead engines are used often for the larger diesel-power plants both on shore and at sea.

▼

A diesel-power plant with three medium-speed four-stroke engines in V-arrangement from manufacturer Wärtsilä.

Right: the generators (1) and left: the diesel engines (2).

▲
A diesel-power plant disc generator driven by a large low-speed two-stroke crosshead engine.

At 120 rpm or two revolutions per second and a frequency of 50 Hz the number of pole pairs 25 is $n = \frac{f}{p}$

n = revolutions per second

f = frequency in Hertz

p = number of pole pairs

So, this **12 metre** diameter disc generator has **50** poles! This requires a great deal of space and a large generator diameter!

▲
A large 3.5 MW diesel genset on a container ship.

Genset drives

Gear transmission is not used.

A flexible coupling between the diesel engine and the generator usually suffices.

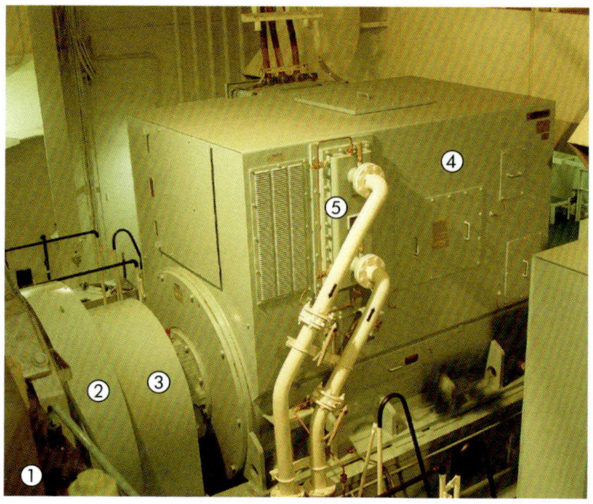

▲
A similar 3.5 MW genset on a container ship.

1 *medium-speed four-stroke diesel engine*
2 *flywheel/turning wheel*
3 *flexible coupling, not visible*
4 *electric generator*
5 *closed cooling-air system: the air is cooled by water. Dust and dirt particles cannot enter this generator!*

▼
Two examples of a flexible coupling between a large disc generator and a large two-stroke crosshead engine.

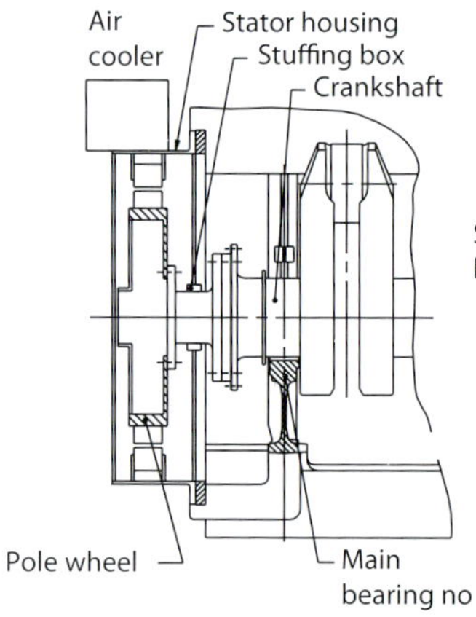

Standard engine, with direct mounted generator (DMG/CFE)

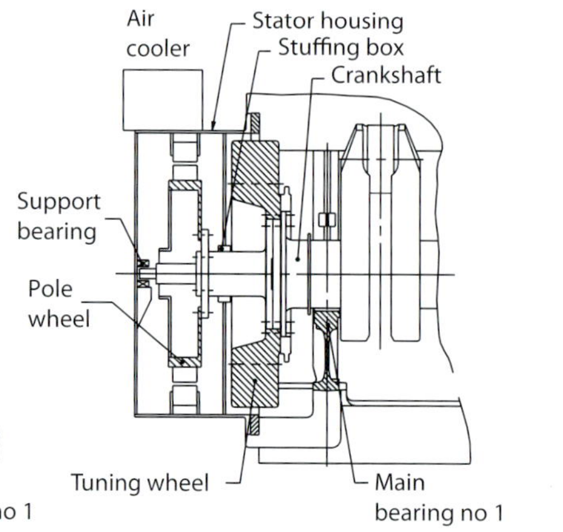

Standard engine, with direct mounted generator and tuning wheel

▶
A simple flexible coupling used between the diesel engine and generator.

The rubber sections of the left shaft flange fit the holes on the right shaft flange. After approximately 30,000 running hours the rubber parts are renewed.

20.2.7 G Special drives, for example, pumps in the dredging industry, compressors and pumps for use in the gas and oil industry using high-speed- or medium-speed four-stroke diesel engines. Engine categories I, II and III

The engine operates on HFO and MDO.
In the dredging industry, either a diesel engine or an electro-motor directly drives large dredging pumps. The larger dredgers use the latter. A high-speed four-stroke diesel engine drives smaller dredgers usually with gear transmissions.

◄ Large dredgers require a lot of power for driving the dredging pumps.

◄ Drive of a large dredging pump on a dredger.

The diesel engines drive the pumps placed in a separate pump room; there is a watertight bulkhead between the engine room and pump room. A flexible coupling (not visible) is located between the diesel engine and the intermediate shaft.

1 *engine-room bulkhead*
2 *intermediate shaft*
3 *pump shaft*
4 *oil-lubricated pump bearing*
5 *pump casing*
6 *pump casing cover for inspection /repairs of the impeller*

◄ The same dredger viewed from the engine-room bulkhead.

The firm SKF Marine Service Centre, Ridderkerk, the Netherlands, is aligning the intermediate shaft to the pump.

In the process industry, diesel engines drive oil pumps or gas compressors. Due to increasingly strict emission regulations, the use of 'Dual Fuel' or gas engines is preferred.
The drive often uses a gear transmission. There is always a flexible coupling present.

20.3 Gear transmission

If gear transmission is used, the propeller speed can be selected independent from that of the diesel-engine crankshaft.
Clearly, the optimal speed is chosen for both diesel engine and propeller respectively, thus optimising the total propulsion efficiency.

Gear transmission is also known as transmission gearing. The ratio of engine speed to the propeller shaft speed is greater than 1.

It often lies between 2 and 4.

$$\text{Reduction} = \frac{\text{engine speed}}{\text{propeller-shaft speed}} > 1 \text{ often 2 to 8.}$$

In shaft generators, this is usually the reverse. The P.T.O.-shaft frequently has a speed that, for instance, exceeds the speed of the medium-speed diesel engine.

For instance, the shaft generator in this type of installation has an rpm of 1500 and the diesel engine an rpm of 750.

The speed of the diesel increases by a factor $\frac{1500}{750} = 2$ via the gearbox.

20.3.1 Examples of the possible uses for gear transmissions

▶

A single reduction gearing with top left, the shaft to the diesel engine and bottom right, the propeller shaft drive.

On top of the reduction gearing, the lubricating-oil system with cooler and filters.

◀

A reduction gearing attached to a four-stroke high-speed diesel engine.

1 *diesel engine*
2 *reduction gearing*
3 *PTO-shaft.*
4 *shaft to propeller*
5 *lubricating-oil cooler*
6 *lubricating-oil filter*
7 *lubricating-oil pump*
8 *vibration damper integrated in the bedplate below the gearbox*

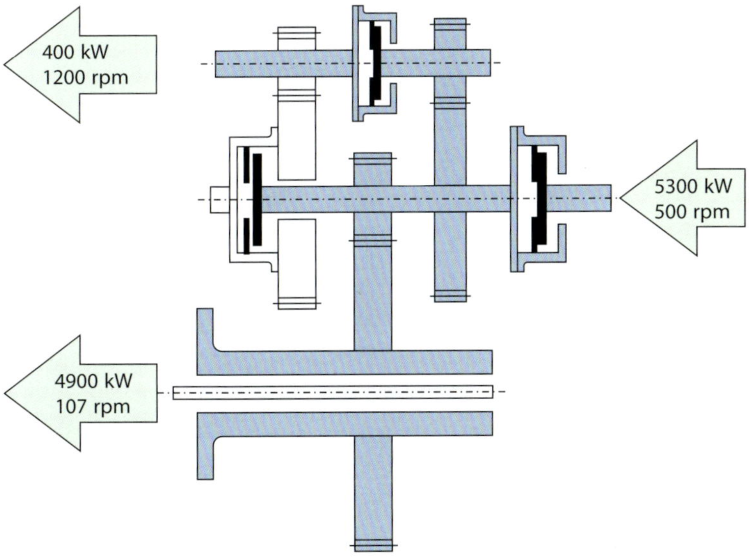

◄ Switch position I: Normal running at sea. Nominal output.

The main engine supplies 5300 kW at 500 rpm and drives the controllable-pitch propeller with a speed of 107 rpm.
At a speed of 1200 rpm, the propeller requires 4900 kW and the shaft generator 400 kW, to supply electricity for the ship's net.

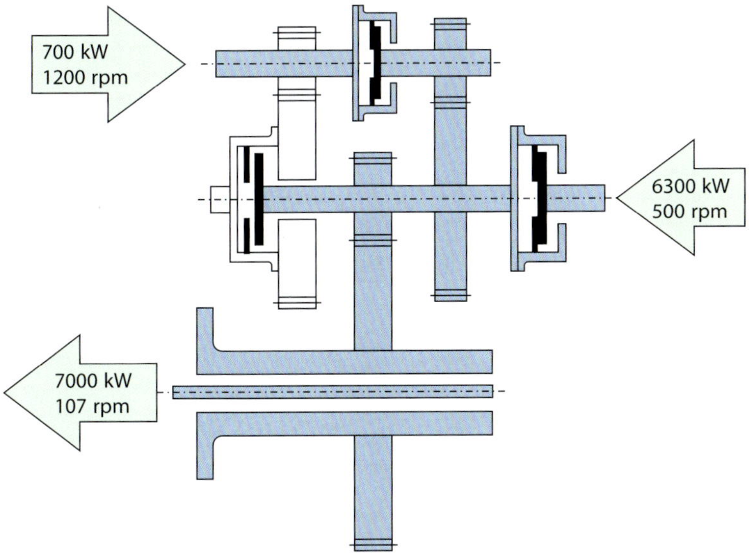

◄ Switch position II: Extra power at sea.

The main engine supplies its full power output of 6300 kW. The shaft generator is switched to the electromotor and supplies 700 kW. The electromotor draws power from the ship's electricity net and together with the main engine drives the controllable-pitch propeller with a maximum power output of 7000 kW.

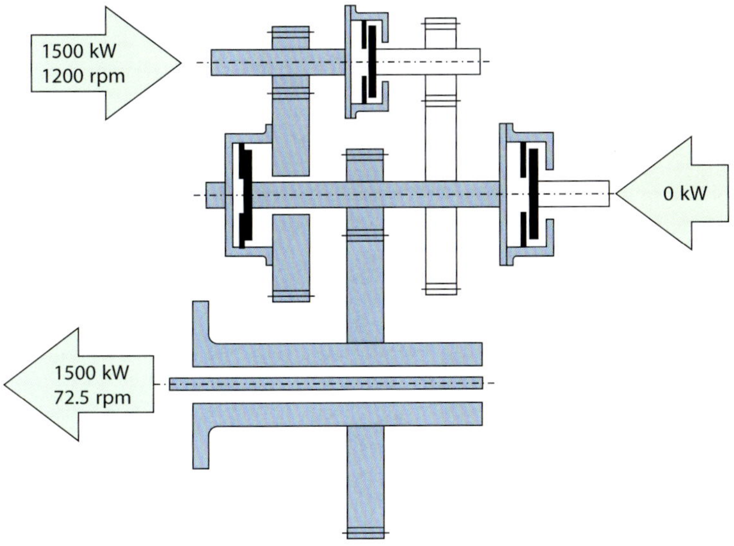

◄ Switch position III: Emergency operation at sea or in port during shifting alongside the quay.

The main engine is stopped and repairs are being done at sea. Major maintenance can be performed in dry-dock.
The switchable shaft generator/electromotor now supplies the maximum capacity of 1200 kW to the controllable-pitch propeller. This switch is referred to as the 'getting home' device.
Admittedly, the speed is low; however, one can proceed to the nearest port for repairs to the main engine. In port the ship can shift its berth.

▶

A cut-away section of a reduction gearing.

1 *engine flange*
2 *gear shaft from the crankshaft*
3 *gear shaft to the propeller shaft*
4 *helical gears*
5 *needle bearings*
6 *lubricating-oil cooler*

20.4 Constructions and designs of gear transmissions

20.4.1 One input shaft from the diesel engine and one output shaft to the propeller.

This is known as a single-gear transmission.

▶

Single reduction gearing with PTO.

1 *diesel drive*
2 *shaft-generator drive*
3 *propeller drive*
4 *clutch*
5 *lubricating-oil sump tank*
6 *lubricating-oil pump*
7 *lubricating-oil distribution unit*
8 *thrust bearing*

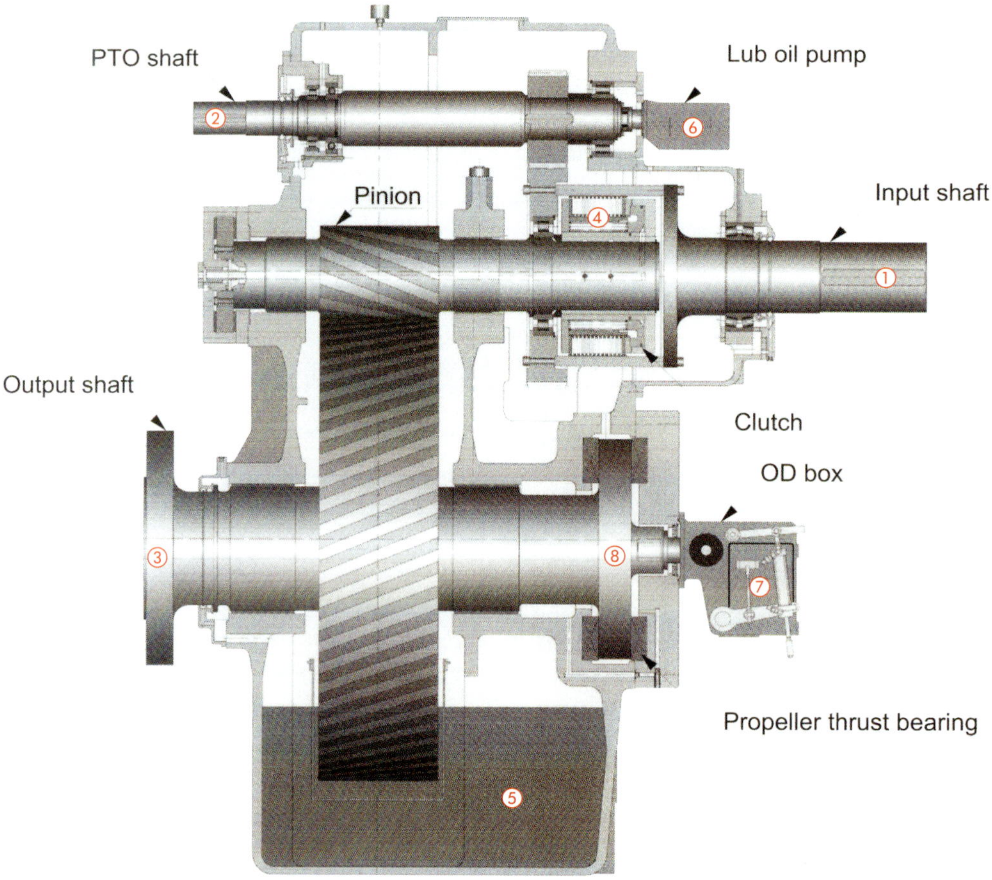

20.4.2 Two input shafts from two diesel engines and one output shaft to the propeller

Significant for the construction of multiple-gear transmissions is the mutual centre-to-centre distance of both diesel engines. If the centre-to-centre distance is large, intermediate shafts are used.

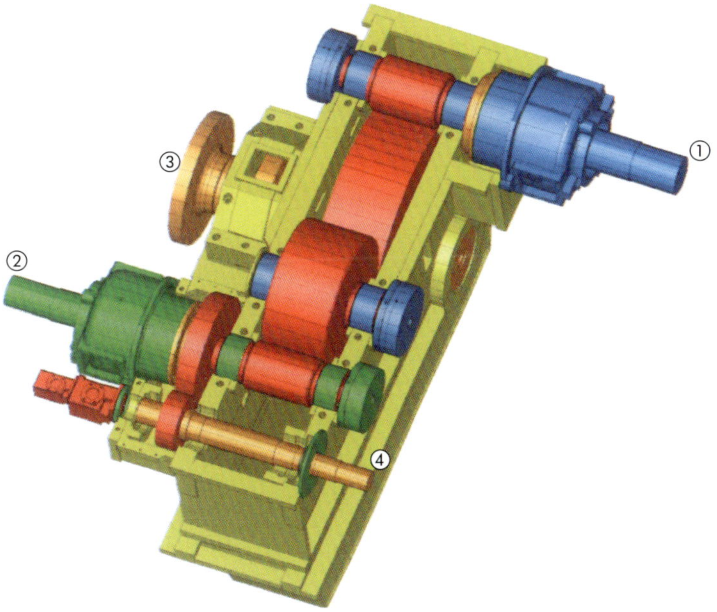

►

Double reduction gearing.

Both diesel engines drive the propeller, one on each side.

1 *diesel engine in forward engine room*
2 *diesel engine in aft engine room*
3 *shaft to the shafting/propeller*
4 *PTO-shaft for driving the shaft generator as well as driving the propeller with the same shaft generator/electromotor going dead slow.*

This arrangement with two propulsion engines in separate engine rooms is obligatory for large cruise ships. If one engine room floods or on fire, 50% of the propulsion power can still be supplied from the other engine room.

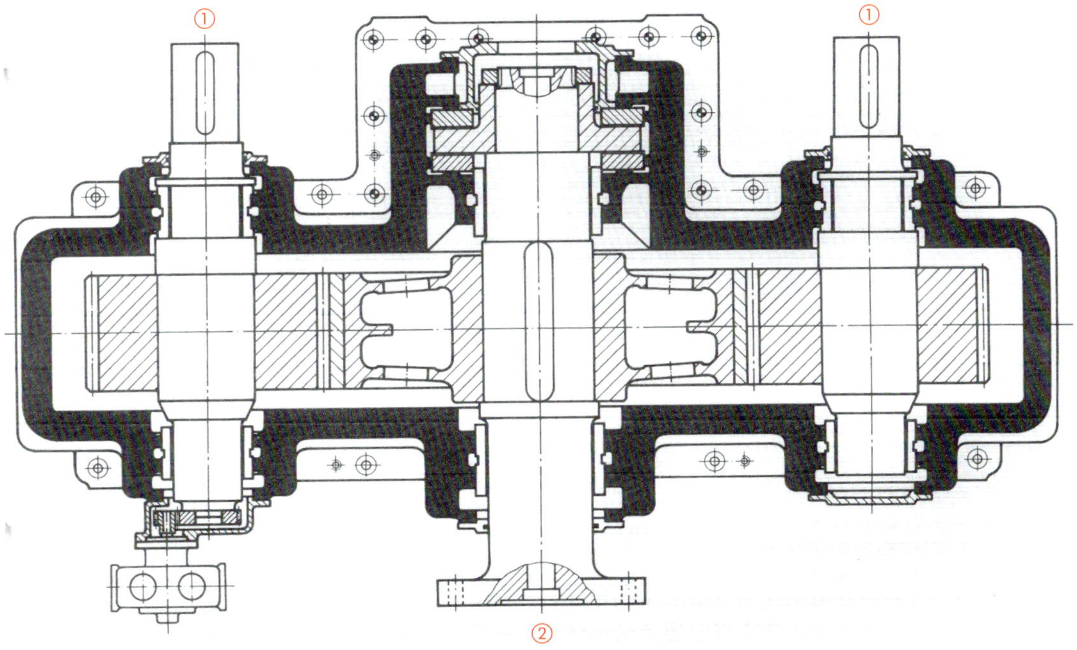

▲

Top: double reduction gearing for two engines, and bottom: a shaft to the propeller.

The gear toothing is straight cut. Bottom left; the lubricating-oil pump. The outgoing pinion wheel is hollow to save weight. Above the pinion wheel; the thrust bearing for absorption of axial forces generated by the propeller thrust.

20.5 Position of the input and output shaft

20.5.1 Input and output shafts on the same side as that of the gear transmission

This is known as a U-arrangement. This construction method is seldom applied.

20.5.2 One shaft on either side of the gearing

This is known as the Z-arrangement. The input and output shafts run parallel to each other. Normally, the centre lines are positioned above each other, but occasionally they are placed alongside.

Designs

The following photos show various gearing designs.

▶ Various versions of reduction gearing.

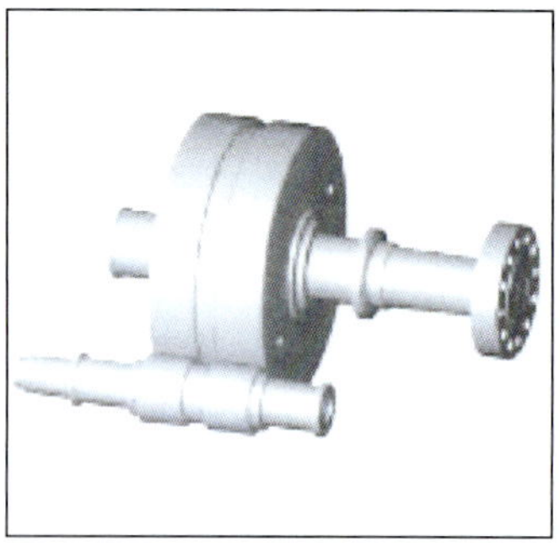

Double helical teeth

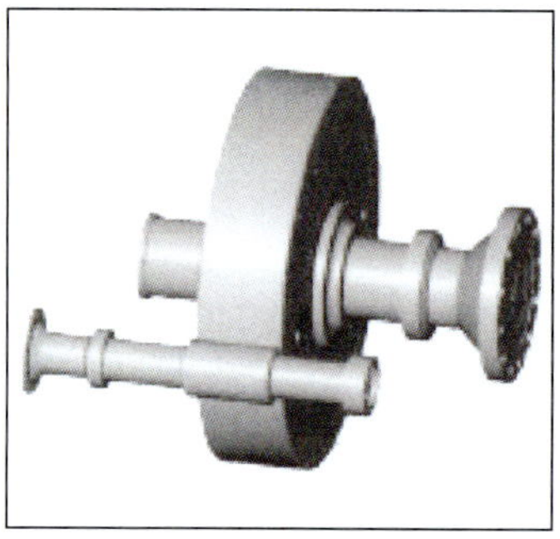

Single helical teeth

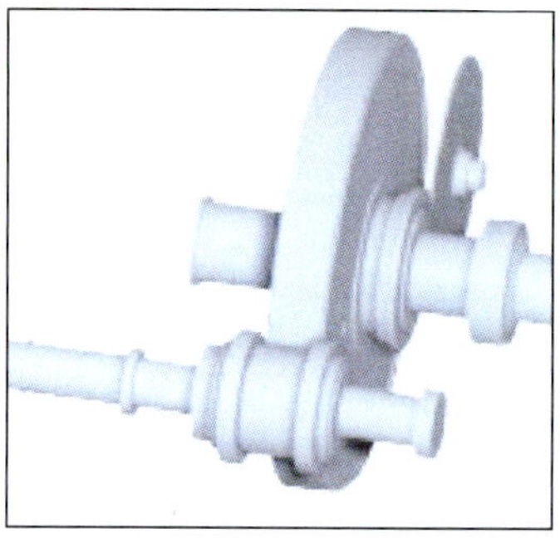

Single helical teeth with thrust collar

Twin in/single out

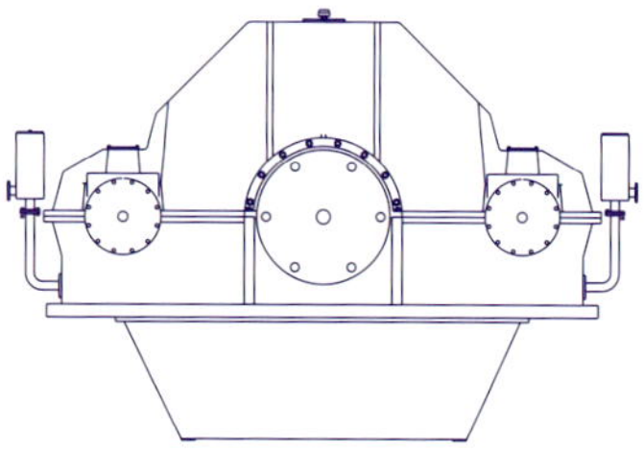

ACG 700TS – ACG 1700TS
2 × 400 kW – 8900 kW

Single PTO/PTI

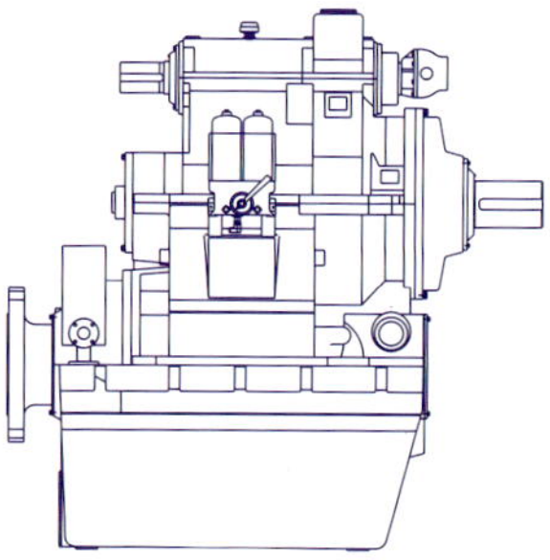

1 or 2 speed
100 kW – 3000 kW

Multiple PTO

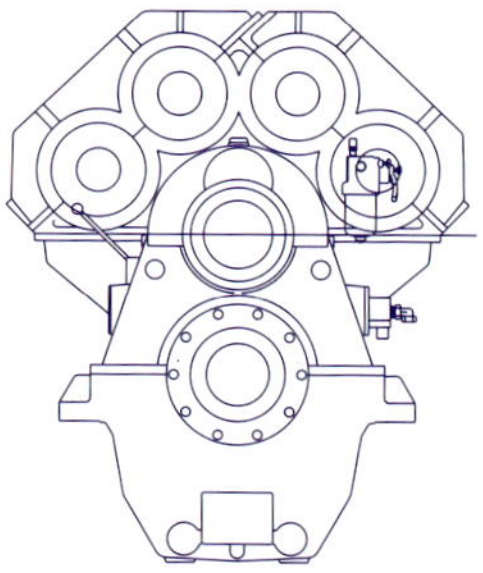

1 to 8 units
100 kW – 3000 kW

▲ Various versions of number and position of the driving shafts.

Left: two driving shafts from the engines; a central shaft for driving the propeller.
Centre: single-reduction gearing with bottom left the driving shaft to the propeller. Centre right, the driving shaft from the diesel engine and top left the drive for the shaft generator/PTO/PTI.

PTO: Power Take Off shaft for driving the shaft generator
PT I: Power Take In shaft for driving the propeller with the shaft generator-electromotor.
Right: reduction gearing and diesel engine together with several PTO shafts for driving the shaft generators, pumps and other equipment.

20.6 Types of teeth

We distinguish three shapes:

1 straight teeth or spur teeth;
2 helical teeth;
3 teeth in V-shape.

The toothed flank of these cogwheels usually has an involute shape. An involute describes the curve traced by a point on a taut, inextensible string as it unwinds in a circle.

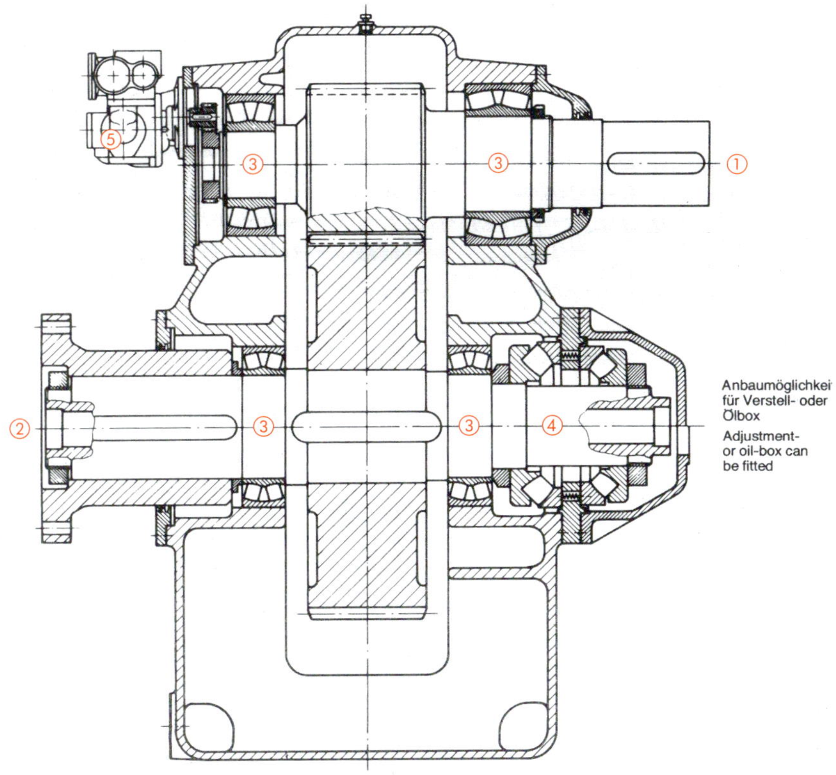

▶

A single transmission with spur gears.

1 *driving shaft of the diesel engine, the so-called inlet shaft*
2 *driving shaft towards the propeller, the outlet shaft*
3 *bearing*
4 *thrust bearing*
5 *lubricating-oil pump*

▲

Double helical teeth.

The space between both rows of helical teeth is clearly visible. This is necessary as the teeth must be ground and polished. Therefore the reduction gearing is fairly long. The axial forces generated by the helical teeth neutralise each other completely.

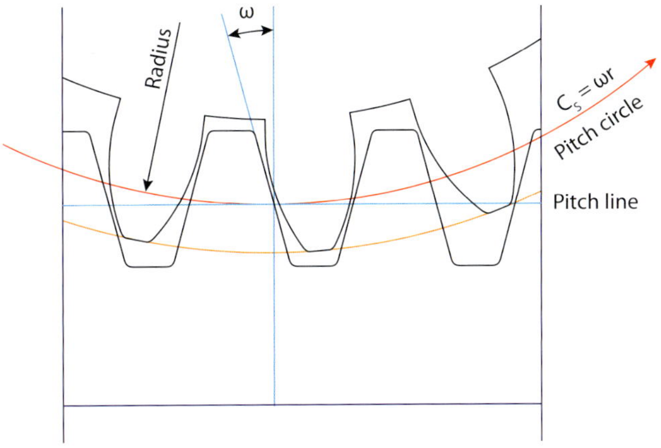

▲

Involute teeth.

In practice, in involute gearing, virtually all the **teeth profiles** are involutes of a circle. These are usually called involute teeth. The force line (or Line of Action) runs along a tangent common to both base circles. The points (of contact) move along the stationary force-vector 'string' as if it were being unwound by the rotating circle. The involute is entirely determined by the radius of the base circle, so all the involutes derived from the base circle are congruent.

A large gear transmission with double helical teeth.

The gear wheels are hollow to reduce the mass.

20.6.1 Comments

Helical gear teeth produce axial forces, which must be absorbed by the axial bearings.
Advantages of helical teeth: more teeth are in contact simultaneously with each other, which gives a smooth operation. With double toothing, the axial forces are eliminated by counter forces. This toothing is very wide, due to the necessary opening between both gear rings in connection with processes, such as polishing.

In this type of toothing, the pinion and the wheel are at a fixed axial position from each other. Therefore, both the pinion and the wheel must be able to move simultaneously. This is made possible with the use of a coupling between the crankshaft and the drive shaft (pinion).

20.7 Gear transmission for shaft generators

Four-stroke- or two-stroke engines drive shaft generators and are occasionally equipped with a special kind of gearing, the so-called **epicyclical transmission.**
This category takes very little space that is imperative for, for instance, a four-stroke propulsion engine installed aft. Here the available space is limited due to the space already used for the gearbox between the engine and the propeller.

It is designed with a solar wheel, a stationary planet carrier and an output shaft with ring wheel.

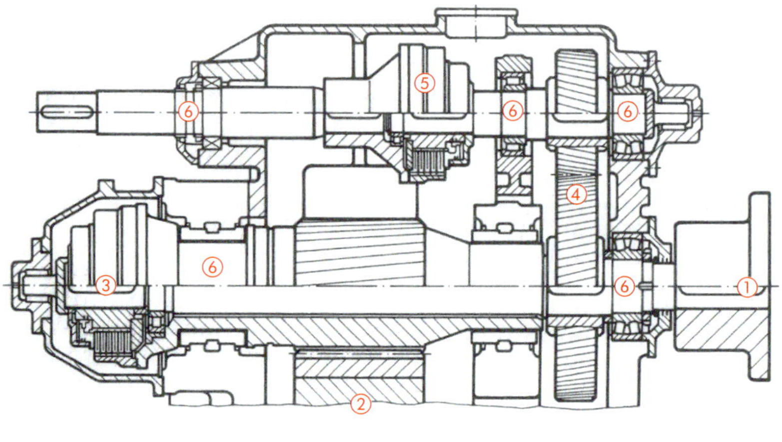

Reduction gearing for a shaft generator.

1 *drive from the diesel engine*
2 *drive towards the controllable-pitch propeller*
3 *clutch to the controllable-pitch propeller*
4 *drive to shaft generator*
5 *clutch to shaft generator*
6 *shaft bearings*

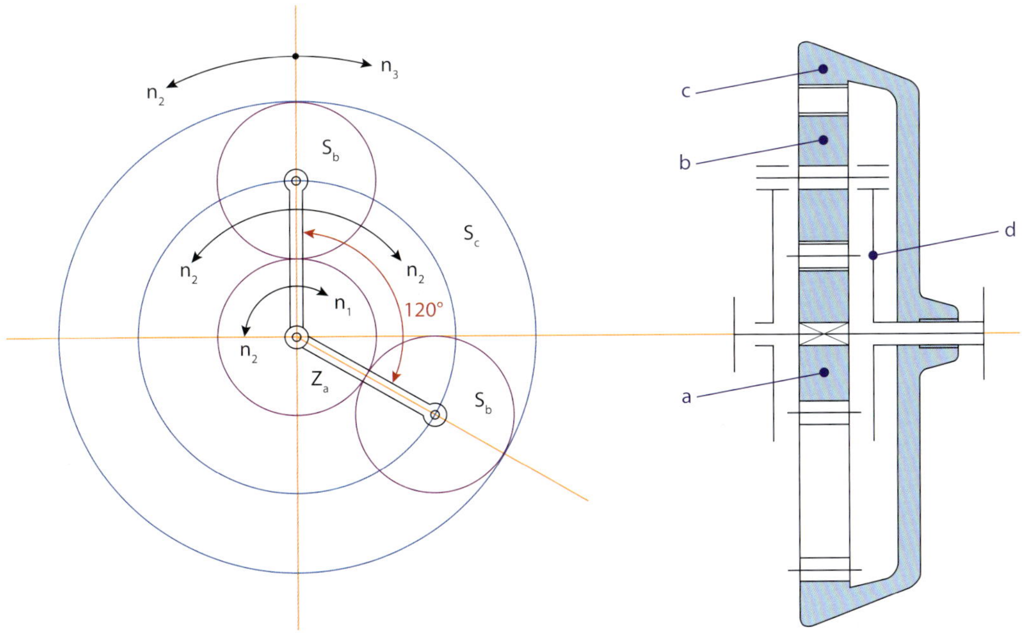

◀ The principle of an epicyclical reduction gearing.

a *solar wheel*
b *planet wheel*
c *ring wheel*
d *shaft support*

Two types are commonly used.

I The driving shaft is connected to the solar wheel, the stationary planet carrier and the ring wheel.

II The driving shaft is connected to the solar wheel, the stationary ring wheel and the output shaft with a planet carrier.

In the first arrangement, the circumferential velocity of all the wheels is identical and the rotation speeds of the wheels are inversely proportional to their diameter. This is known as the star type.

The transmission ratio i is equal to:

$$i = \frac{n_1}{n_3} = \frac{d_r}{d_s},$$

where:

n_1 = rotation speed of the solar wheel
n_3 = rotation speed of the ring wheel
n_2 = rotation speed of the planet wheel
d_s = diameter of the solar wheel
d_r = diameter ring wheel
d_p = diameter planet wheel

Planet and ring wheels rotate in the opposite direction to that of the solar wheel.

In the second arrangement, the ring wheel is fixed and the solar wheel driven by the input shaft; the planet carrier drives the output shaft. This is known as the planetary type.

The rotation speeds now differ from the first instance:

Solar wheel $n_{s^1} = n_1 + n_3$
Planet carrier $n_2 = n_3$
Ring wheel $n_{r^1} = n_3 + n_3 = 0$

In this instance the transmission ratio is:

$$i^1 = \frac{n_{s^1}}{n_2} = \frac{n_1 + n_3}{n_3} = 1 + \frac{n_1}{n_3} \text{ also } i^1 = 1 + \frac{d_r}{d_s}.$$

In the second example, the input and output shafts rotate in the same direction.

There is also a third type, the so-called solar type. Here the solar wheel is fixed permanently.

The transmission ratio then equals $\frac{n_1}{n_2}$ or $\frac{n_s}{n_p}$.

▶

Shaft generator for a two-stroke MAN Diesel crosshead engine using an epicyclical Renk transmission.

The toothed wheel in the centre is called the solar wheel. Surrounding it are three planet wheels and the outer row of teeth is called the ring wheel. An hydraulic-oil pump is driven from the diesel engine driving shaft.

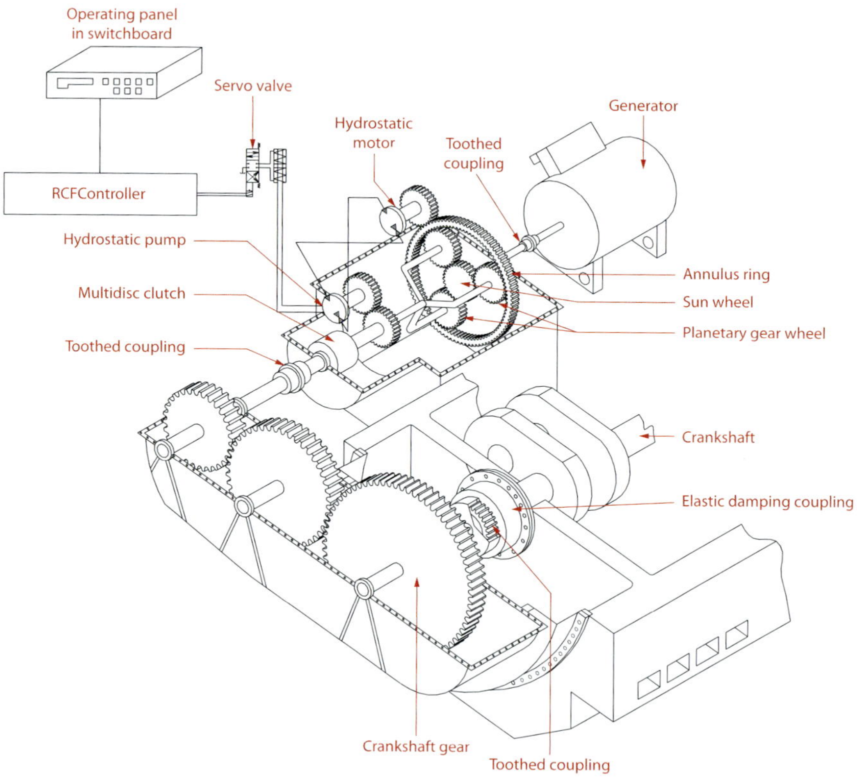

▶

A combination of normal reduction gearing, right, with epicyclical reduction gearing, centre.

1 *first gear wheel, engine speed*
2 *second gear wheel, engine speed slightly increased*
3 *third gear wheel, engine speed higher*
4 *flexible coupling*
5 *coupling flange*
6 *ring gear wheel*
7 *propeller shaft system*
8 *planet wheel*
9 *solar wheel*
10 *coupling flange*
11 *reduction gearing*
12 *hydraulic pump*
13 *hydraulic pipes*
14 *hydraulic motor*
15 *reduction gearing*
16 *coupling flange*
17 *plate coupling*

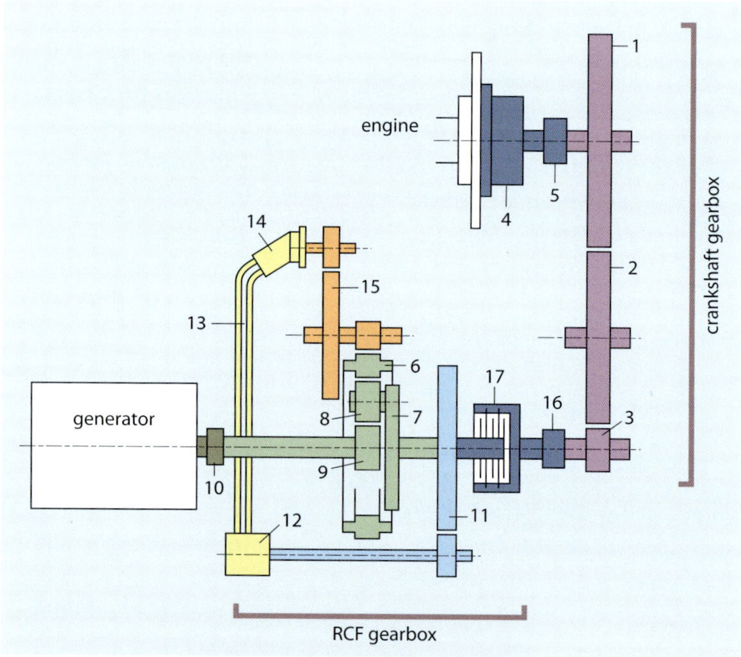

▶

The lubrication of gear wheels is usually done with a piping system where the lubricating oil is sprayed onto the toothed wheels through small perforations.

1 *lubricating-oil supply pipe*
2 *upper bearing shell, the bearing caps have been removed*
3 *helical teeth with intermediate spaces*

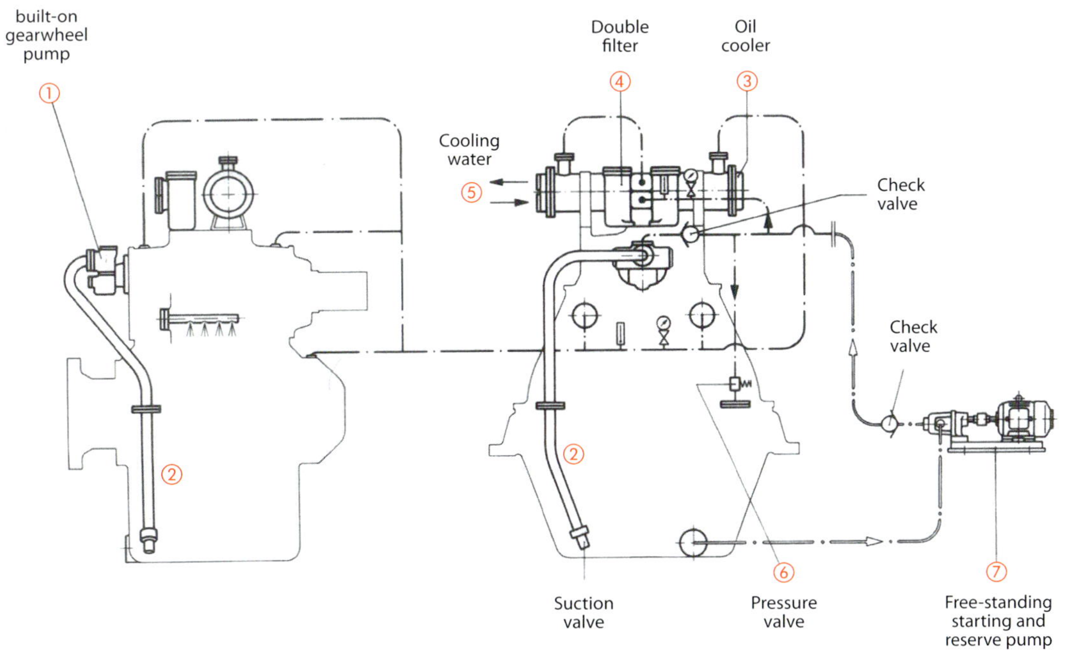

◀

The lubricating-oil system of a single reduction gearing.

1 *built-on gearwheel pump*
2 *supply line with non-return valve*
3 *lubricating-oil coolers*
4 *duplex lubricating-oil filter*
5 *cooling water*
6 *pressure valve*
7 *electrically driven spare lubricating-oil pump*

Wärtsilä reduction gears – Output range

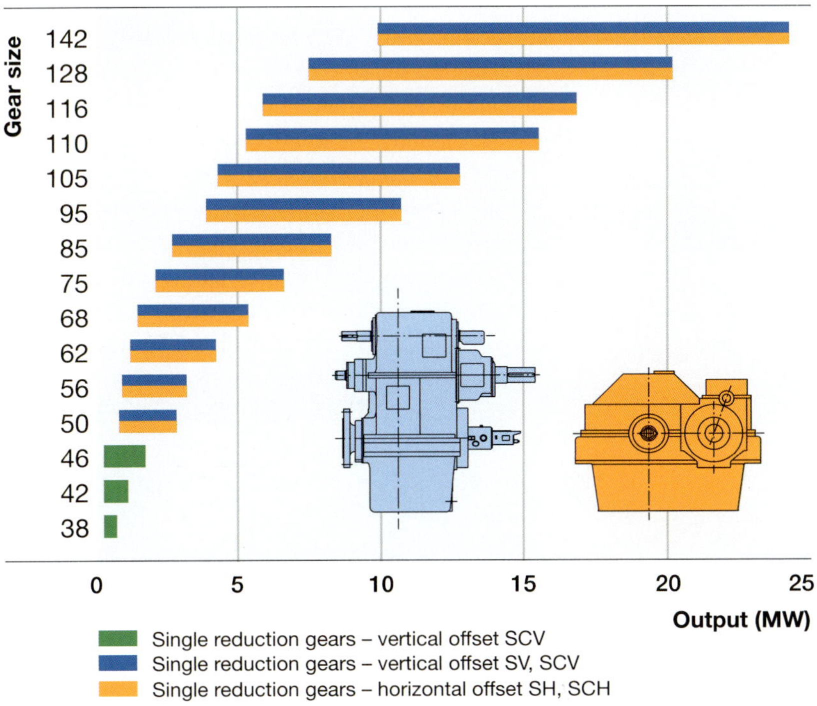

Wärtsilä TCH gear range of twin input-single output gears

Gear size	Engine crankshaft offset (mm)*
TCH190	1 900
TCH240	2 400
TCH250	2 500
TCH270	2 700
TCH290	2 900
TCH300	3 000
TCH320	3 200
TCH330	3 300
TCH340	3 400
TCH350	3 500
TCH370	3 700
TCH380	3 800

* For the TCH type of gears the main parameter is the required horizontal offset between engine crankshafts. However, also the engine power and speed, and propeller speed are dimentioning parameters.

▲

Wärtsilä reduction gearing program.

left figure

vertical: reduction gearing size

horizontal: maximum shaft power

right figure

A schedule with the various types of gear boxes with double diesel-engine drives.

The crankshaft diameter reaches 3800 millimetres per diesel engine.

The power output rises to approximately 25 MW; these are the largest four-stroke medium-speed diesel engines. Higher outputs are usually generated by low-speed two-stroke crosshead engines.

20.8 Couplings

Couplings can be divided in elastic- and flexible couplings. Both are used frequently.

20.8.1 Elastic couplings

Their objective is to reduce torsional vibrations.

Also, see Chapter17, *Vibrations and balancing.*

With the use of an elastic material, plate springs and/or hydraulic damping, the vibrations between, for instance, the engine and the reduction gearing, and the reduction gearing and propeller, and the engine and the generator are reduced.
Damping ultimately achieves a reduction of the torsional vibrations in the crankshaft. Due to increasing power densities in diesel engines, vibration damping has become increasingly important.

▲
Three types of flexible couplings from manufacturer Centa.

Left, the flexible section can be clearly seen. The large holes in the coupling disc are for ventilation.

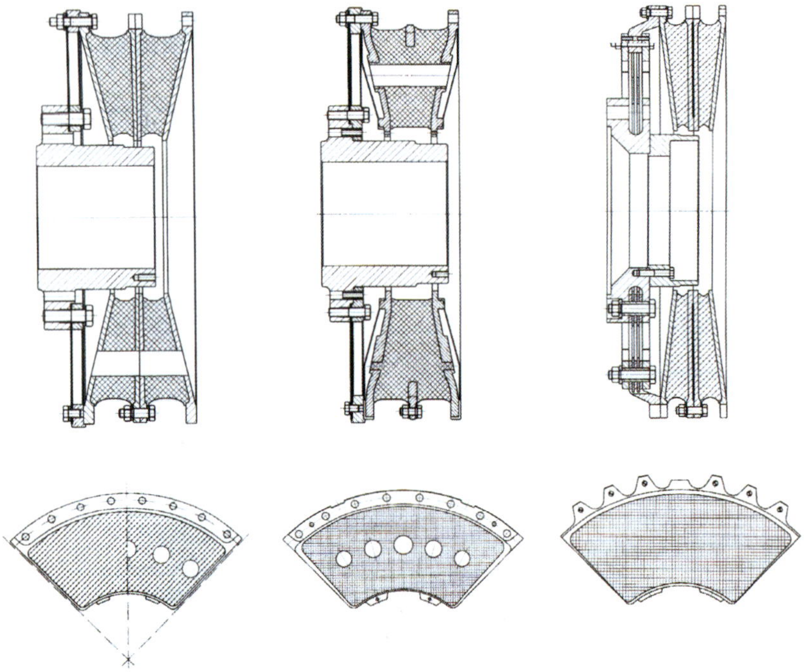

▲
The various types of Vulcan flexible couplings.

Coupling 1: standard coupling with or without axial ventilation holes.
Coupling 2: version with wide flexible segments for maximum radial alignment deviations.
Coupling 3: version with narrow flexible elements for minimum power loss.
The manufacturers of flexible couplings have an elaborate delivery package as flexible coupling requirements may vary considerably.

The working principle is that a coupling with a minor torsional stiffness is placed between the engine and the gearbox. Both fixed parts on the shafts acquire a certain angular rotation in relation to each other due to the flexible section between them. In this manner, the rotation speed of the crankshaft, which produces vibrations begins to **resonate** well below the normal engine speed range.
During start and acceleration and also when stationary, **gear clatter** is possible**.** Gear clatter may occur during torsional vibrations, where the torque moment is periodically negative. This, for instance, could be the case in situations where the engine speed and consequently the power output of the diesel engine is reduced and the propeller is, as it were, driven by the ship. Here the other toothed flank of the cog is temporarily loaded.
There are flexible frictional couplings that can prevent this problem by starting and stopping the engine with a disengaged coupling. The coupling is only put in service at sufficiently high engine speeds when gear clatter no longer occurs.

Examples of elastic couplings

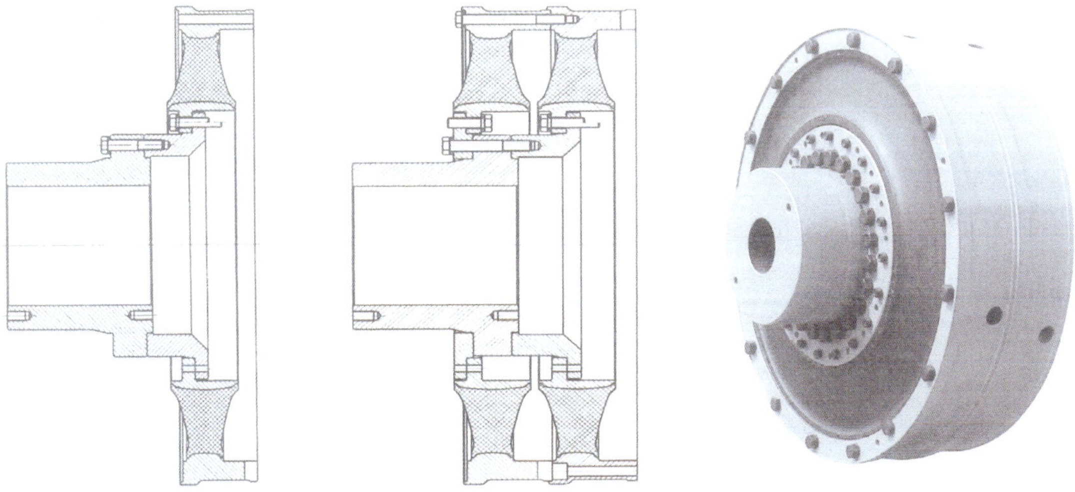

◄ A narrow and a wide coupling.

Clearly visible in the cross-section is that the left coupling is considerably less elastic than the centre coupling, which consists of two flexible rubber parts.

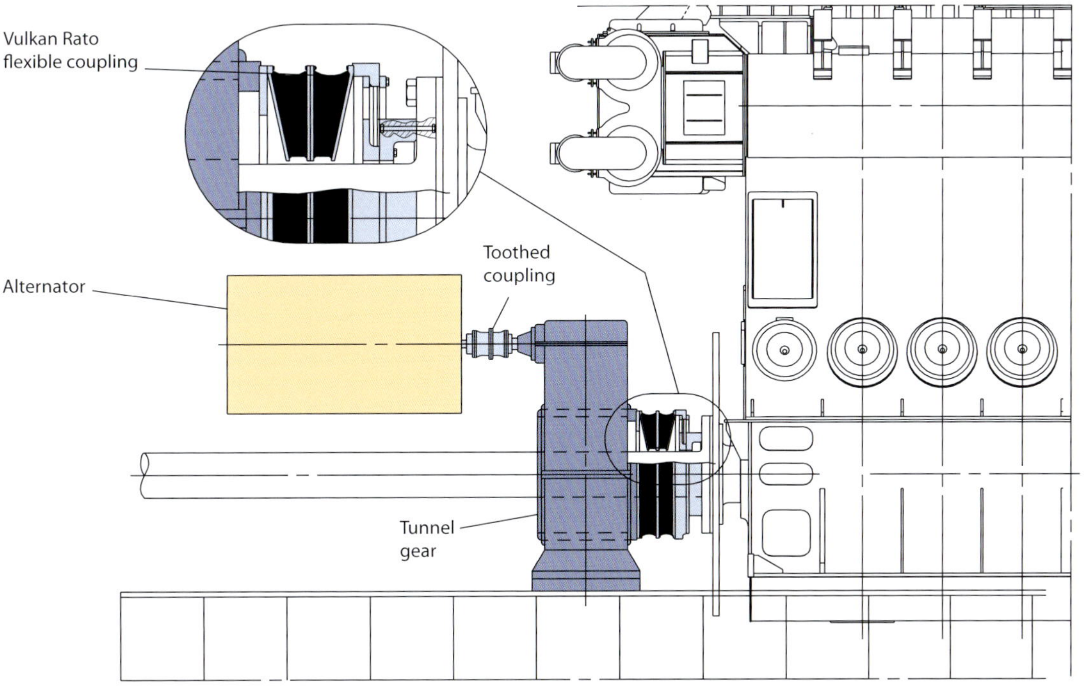

◄ The flexible coupling between a low-speed two-stroke crosshead engine of MAN Diesel and the propeller shafting.

The shaft generator drive has a toothed sliding coupling. In this manner, the axial expansion of the reduction gear transmission that occurs between the stationary and operating temperatures is compensated.

◄ A flexible Centa coupling with two segments placed behind each other for maximum flexibility.

Note the large ventilation holes for discharging the heat generated in the flexible elements that can lead to deformation.

20.8.2 Examples of flexible couplings

▶

A flexible Centa coupling, right; on the left a clutch coupling which runs on air.

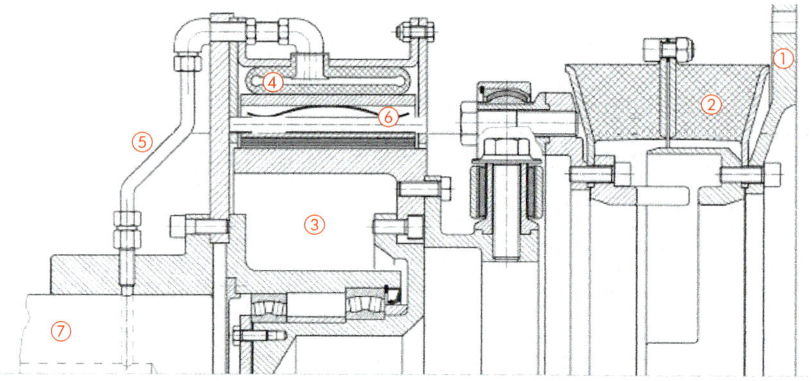

The coupling is achieved by inflating the elastic space (gland).

1 *flange attached to the diesel engine*
2 *flexible segments*
3 *clutch*
4 *completely circular elastic space (gland)*
5 *air pipe*
6 *spring plate*
7 *driving shaft for the controllable-pitch propeller*

▶

A Centa flexible coupling combined with a Wichita friction coupling working on compressed air.

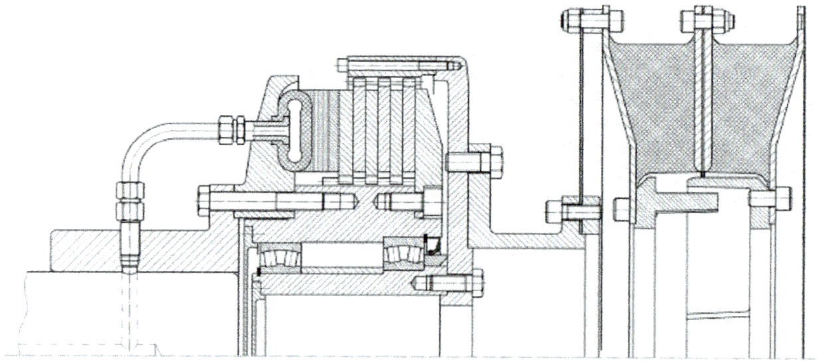

Left: the discs which fix one part to the other half of the shaft and the other part to the other half of the shaft. The discs are pressed together by expansion of the inflatable tube, and the engine shaft drives the shaft (left) from the right towards the propeller. When the propeller shaft is uncoupled, the friction plates are not in contact with each other.

Right: the double flexible coupling.

▶

A cross-section of a reduction gearing with a hydraulically controlled friction coupling for coupling/uncoupling the propeller. Manufacturer MAN Diesel, system Alpha.

1 *diesel-engine drive shaft*
2 *propeller shaft*
3 *electric generator shaft*
4 *hydraulic section for operating the controllable pitch propeller*
5 *lubricating-oil drain tank*
6 *friction coupling*
7 *lubricating-oil pump*
8 *thrust block*

yellow: lubricating oil for the reduction gearing

red/green: hydraulic oil for operating the controllable-pitch propeller

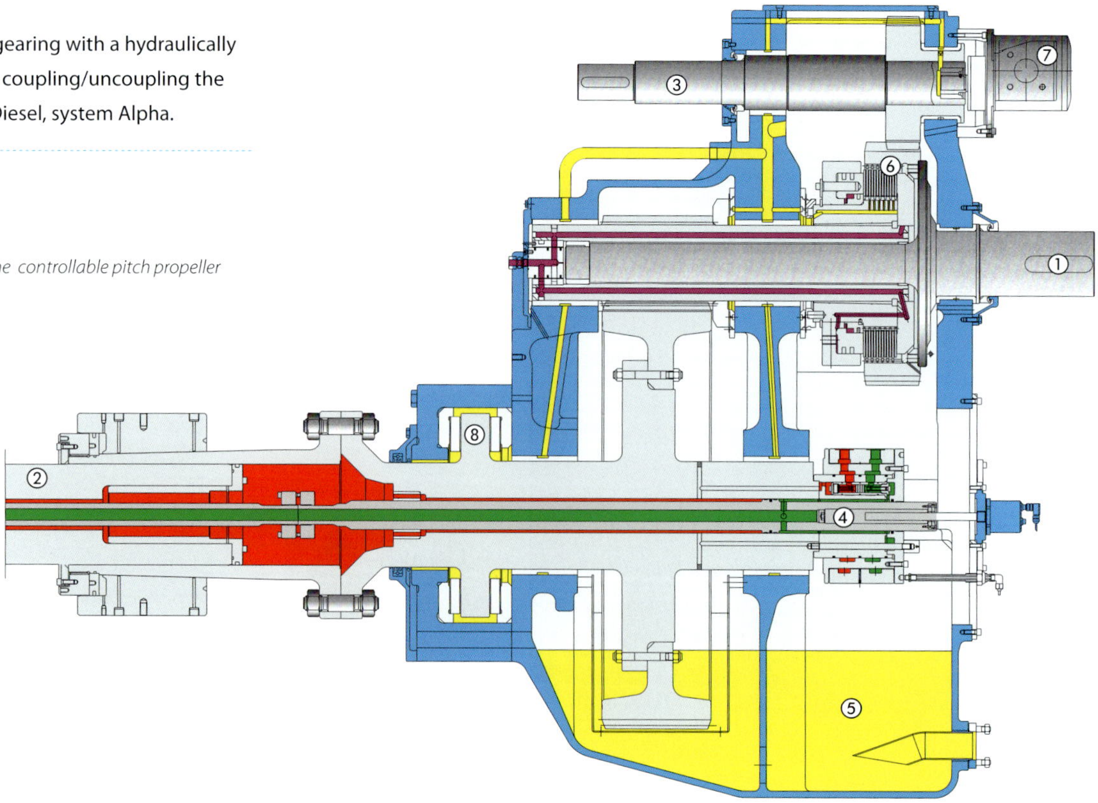

◀ A mechanical coupling from MAN Diesel. Type conical coupling bolts.

◀ The conical coupling bolts of the mechanical coupling from MAN Diesel.

With coupled shafts, the conical ends of the bolts of one shaft slide into the holes of the other shaft. **This can only take place when the shafts are stationary.**

20.9 Torsional-vibration dampers

It is crucial that the normal rotation speed does not coincide with the critical rotation speed as this could produce unacceptable vibrational stresses in the crankshaft.

This can be prevented in the following ways.

1 The correct selection of the cylinder ignition sequence.
2 Adjusting the mass inertia moments.
3 Adjusting the stiffness of certain parts of the shaft.
4 Mounting an elastic coupling between the diesel engine and the gearbox.
5 Installing torsional-vibration dampers, the torsional vibration then can be reduced to an acceptable level, if points 1 to 3 are insufficient.

See Chapter 17, *Vibrations and balancing* for points 1 to 3.

Torsional vibration dampers

These are used in both two-stroke crosshead engines and four-stroke trunk-piston engines. They are mounted on the free end of the crankshaft and in some engines on the free end of the camshaft.

There are two types of vibration dampers.

20.9.1 Energy-absorbing vibration dampers

Here vibrational energy is converted into heat by liquid or dry friction.

20.9.2 Dynamic vibration dampers

These elastic systems have a vibrational frequency that can reduce the torsional vibration in the crankshaft.

20.9.3 General: two-stroke crosshead engines

In larger engines, the fixed-pitch propeller is directly driven by the diesel engine. A simple intermediate shaft suffices as opposed to four-stroke engines, which drive an adjustable-pitch propeller with a flexible coupling and a gearbox. Also in two-stroke engines, the power density increases, thus increasing the torsional vibrations. The crankshaft, the intermediate shafts and the propeller shaft can be protected against these vibrations by placing a vibration damper on the free end of the crankshaft.

Examples

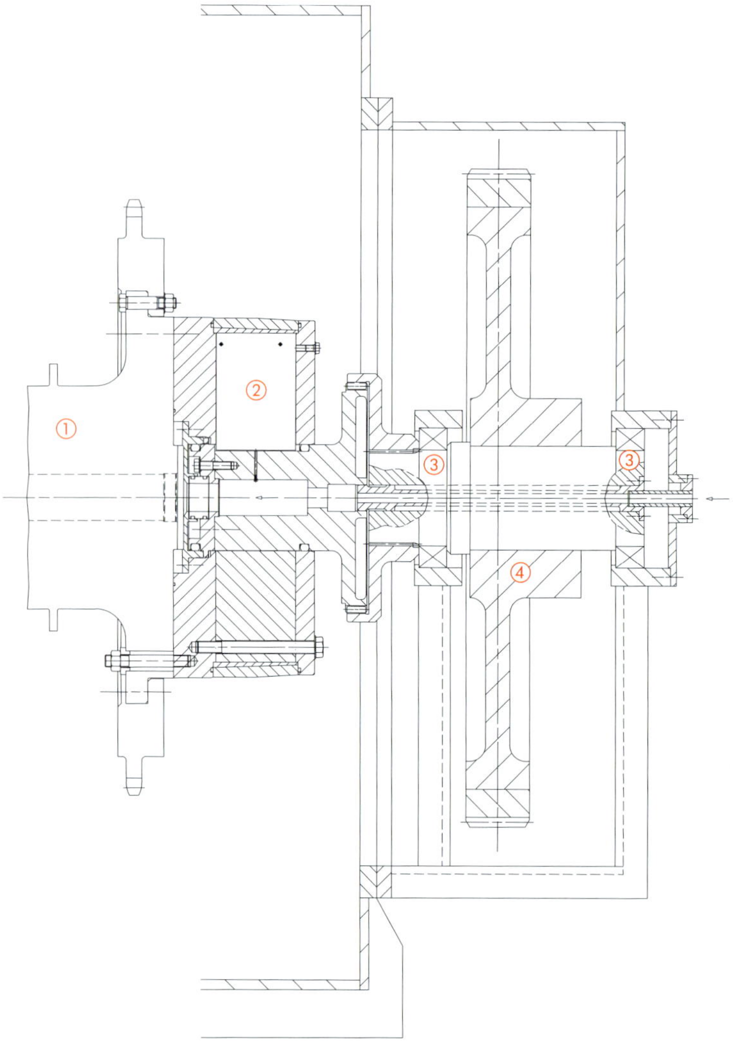

▲ A Geislinger vibration damper with a counterweight for low-speed two-stroke crosshead engines.

1 *crankshaft connection*
2 *vibration damping with spring plate elements*
3 *bearings*
4 *counterweight*

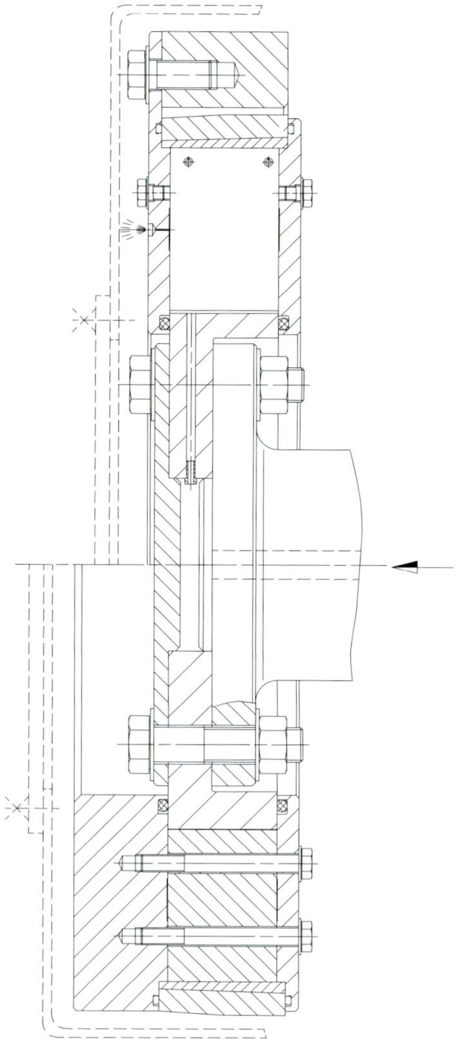

▲ A Geislinger vibration damper with extra mass for low-speed two-stroke crosshead engines.

20.9.4 General: four-stroke engines

In four-stroke engines, the adjustable-pitch propeller is driven by the engine via reduction gearing. Due to the increase in power density, the torsional vibrations have a higher value. Here too, a vibration damper is placed at the free end of the crankshaft and sometimes on the camshaft.

▼

A Geislinger vibration damper with conical bolt attachment for medium-speed four-stroke engines.

1 *diesel-engine shaft*
2 *vibration damper*
3 *spring plate for damping*
4 *protective cap*

The advantage of this kind of attachment is that there is no flange on the crankshaft with a bolt connection to the flange of the vibration damper. The flange is fitted to the crankshaft hydraulically.

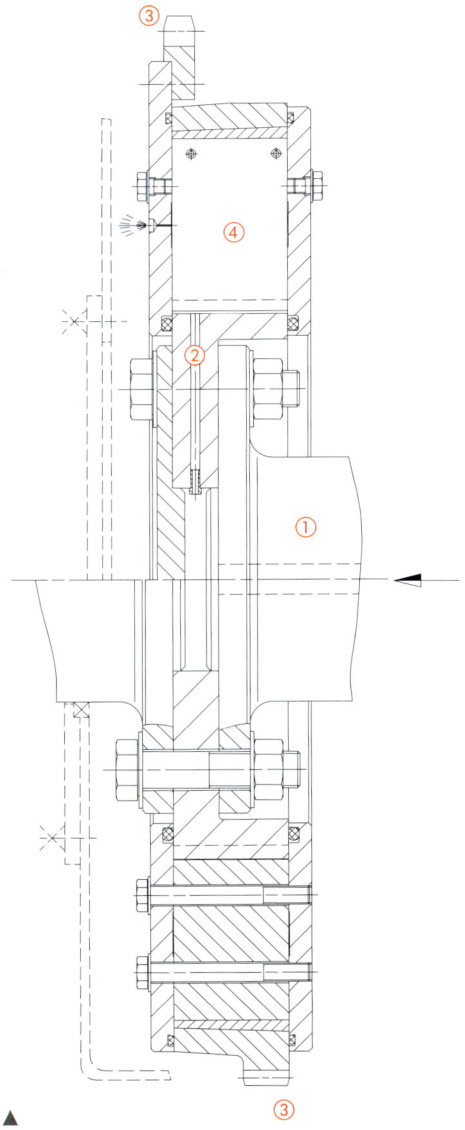

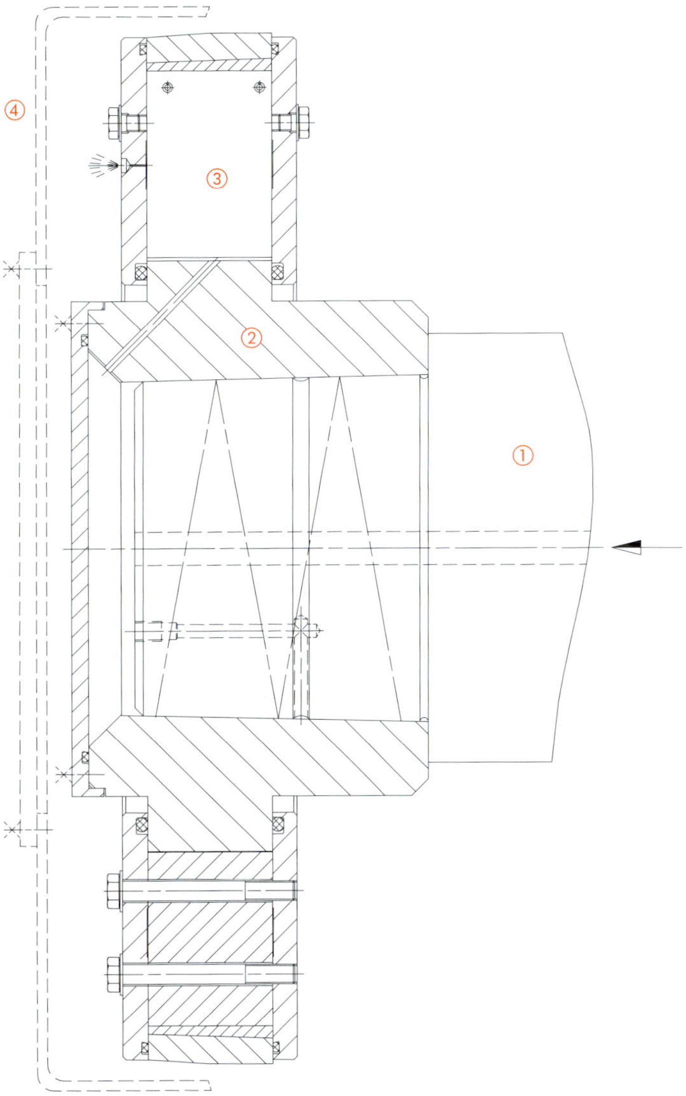

▲

A Geislinger vibration damper with an optional pump drive for medium-speed four-stroke engines.

1 *diesel-engine shaft*
2 *vibration damper*
3 *gearwheel for driving pumps*
4 *spring plates for damping*

▶
The vibration damper on the 'blind' side of the crankshaft of a high-speed four-stroke MTU V-engine.

1 *vibration damper*

▶
The vibration damper on the crankshaft and the camshaft of a four-stroke high-speed Cummins diesel engine.

1 *crankshaft*
2 *power reduction of the crankshaft*
3 *'blind' end of the crankshaft*
4 *vibration damper*
5 *vibration damper on the camshaft*

▲
The torsional-vibration damper on a Deutz high-speed four-stroke marine diesel engine.

The engine is placed on four flexible vibration dampers.

1 *torsional-vibration damper*
2 *flexible vibration damper*

▶
The vibration damper on the 'blind' side of the crankshaft for a high-speed four-stroke Caterpillar V-engine.

Most vibration dampers are not visible as they are protected by a cap.

20.10 Engine shafts

Engine shafts are connections between the diesel engine and the propeller. In two-stroke-crosshead engines with large cylinder diameters, this is a direct drive with crankshaft, intermediate shafts and a propeller shaft. A shaft generator may be placed in the intermediate shafts.

▶

The shafting between main engine (in this instance a two-stroke crosshead engine) and propeller, in the foreground.

1 *main engine*
2 *flywheel*
3 *shaft*

▶

The shafting between main engine and propeller, as seen from the main engine.

1 *intermediate shaft*
2 *bearing block*
3 *shaft coupling*
4 *towards main engine*
5 *towards propeller*
6 *shafting 'tunnel'*

Four-stroke engines have reduction gearing mounted on the crankshaft and then the intermediate shafts to the propeller shaft. The gearing usually has a shaft generator attached to it.

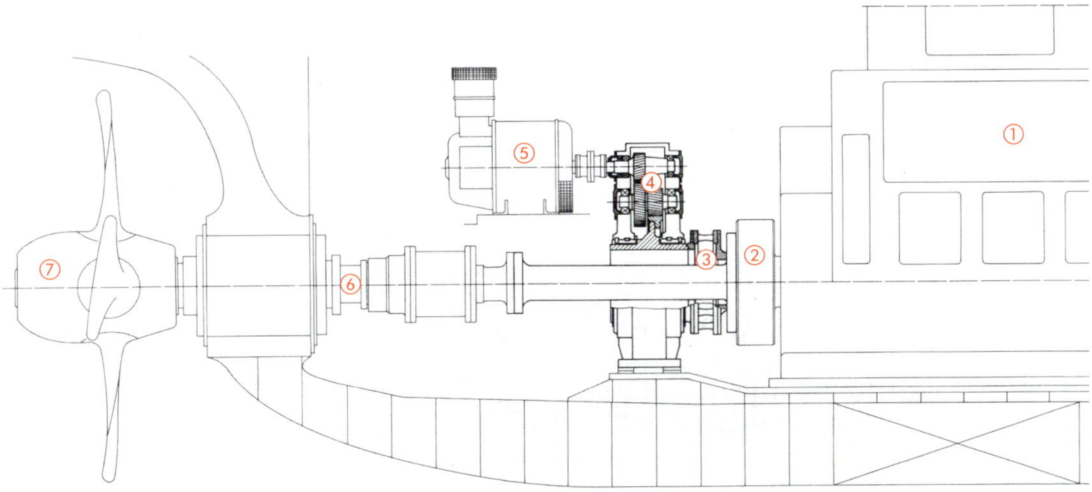

◀

The arrangement of the main engine in relation to the controllable-pitch propeller.

1 *medium-speed four-stroke engine, one rotational direction*
2 *flywheel*
3 *flexible coupling*
4 *reduction gearing for shaft generator*
5 *shaft generator*
6 *propeller shaft*
7 *controllable-pitch propeller*

▲
Reduction gearing with shaft generator on a feeder-container ship from 820 TEU.

1 *main engine*
2 *flexible coupling protective casing*
3 *flywheel protective casing*
4 *gearbox*
5 *lubricating-oil cooler*
6 *shaft generator*
7 *coupling between the reduction gearbox and shaft generator*

The forged-iron shaft, dependent on the length of the propeller shaft, consists of one section; in four-stroke engines positioned aft, or of several sections. The latter is often found in four-stroke engines for ferries and passenger ships, or in two-stroke crosshead engines for container ships, bulk carriers and oil tankers.

20.10.1 Bearings

The intermediate shafts are supported by sleeve bearings, which are located in the tunnel-bearing block. These are often only equipped with a large lower-bearing shell as they bear the weight of the shaft.

20.10.2 Couplings

These are usually **flange couplings**, which are bolted together with coupling bolts. Apart from fixed flanges, there are also **coupling flanges that can be dismantled**. Moreover, a number of manufacturers install **hydraulically fixed couplings**.

▲
The lubricating oil in the shaft tunnel is cooled by a fresh water cooling coil.

1 *bearing block*

20.10.3 Flexible shafting

Flexible shafting is used frequently for specific ships. They are manufactured from composite materials with reinforced carbon compounds. The shafts are lightweight, very strong, have a small expansion coefficient, dampen sound and are corrosion-resistant and flexible.

These carbon-fibre shafts are used for light, fast ships such as catamarans, lifeboats and marine craft.

▼
A speciality: A large ocean-going salvage tug with a two-stroke low-speed crosshead engine from MAN Diesel. The intermediate shafts to the propeller shaft are clearly visible.

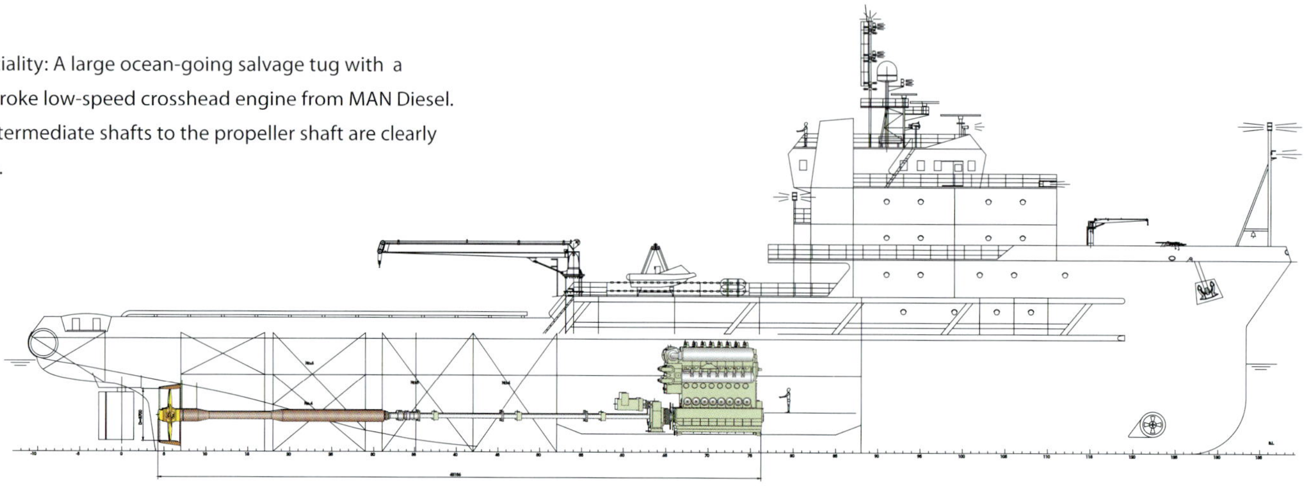

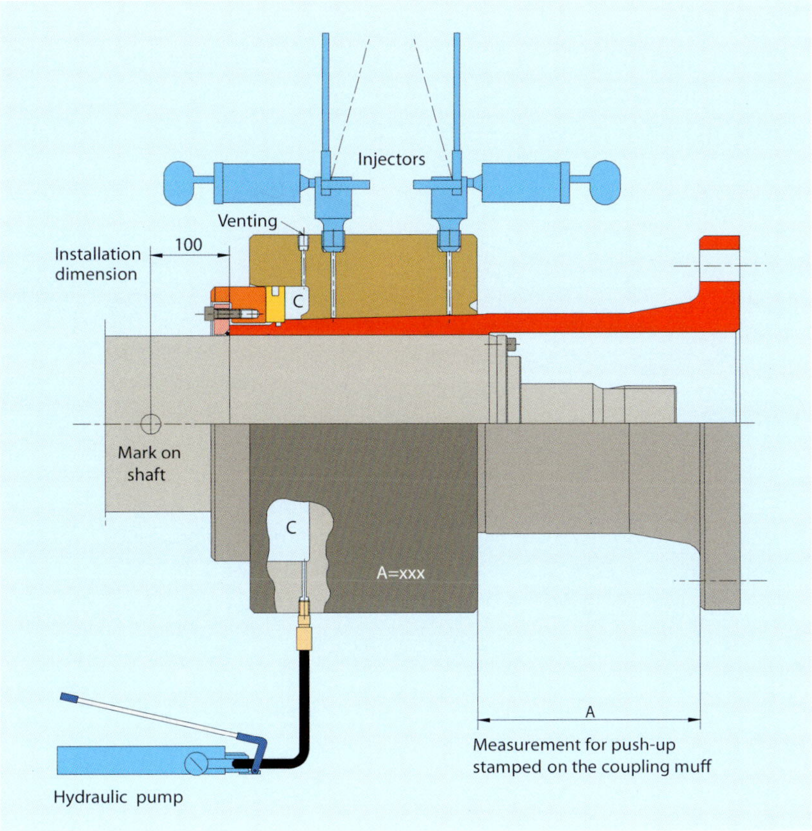

◄

A hydraulically affixed shaft flange.

The required oil pressure is generated using a hand-operated pump.

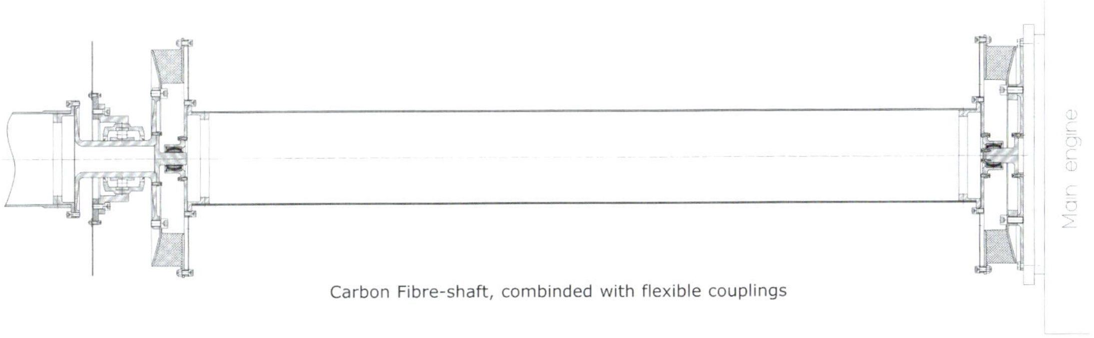

◄

A carbon-fibre flexible shaft combined on both sides with Centa flexible couplings, this is often used for catamarans and other high-speed craft.

The arrangement is very flexible; this is required for the shifting hull sections between the bedplate and the stern tube.

▼

An example of the arrangement with flexible carbon-fibre propeller shafts of Centa shown here in blue.

Between the diesel engines and the gearboxes are two flexible couplings which largely neutralise the torsional vibrations.

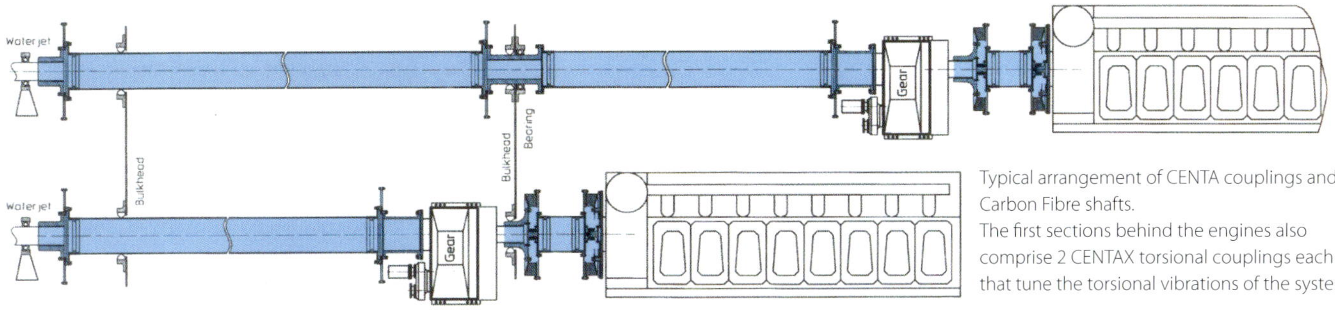

Typical arrangement of CENTA couplings and Carbon Fibre shafts.
The first sections behind the engines also comprise 2 CENTAX torsional couplings each that tune the torsional vibrations of the system.

20.11 Examples of complete systems with diesel engines, reduction gearing, shafting, couplings and vibration dampers

For additional information, see Chapter 17, Vibrations and balancing and Chapter 19, Ship propulsion.

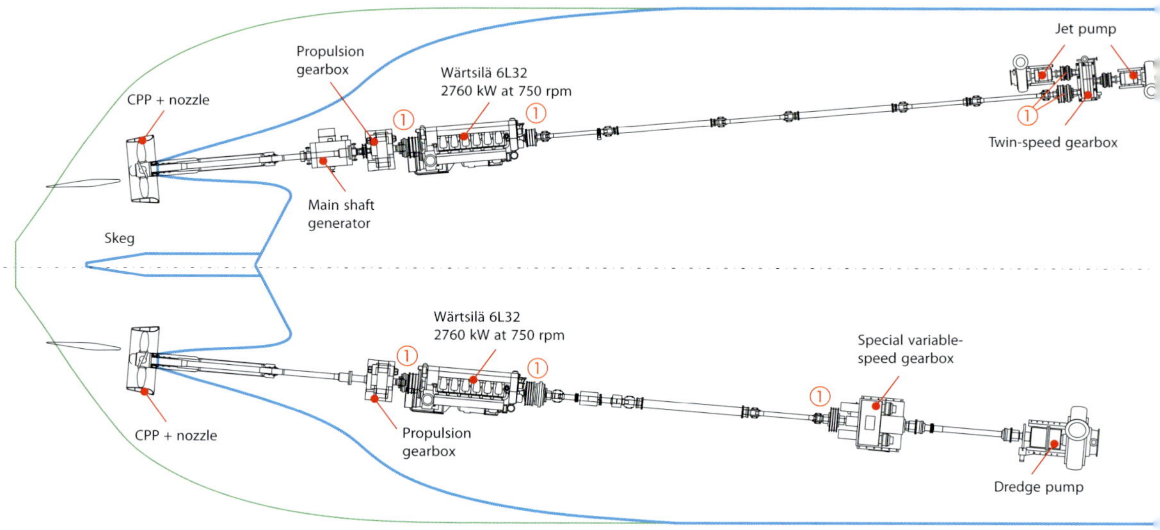

▲

The shafting on a dredger, design Wärtsilä.

Both main engines drive the controllable-pitch propellers in the nozzles via a flexible coupling, reduction gearing and propeller shaft. On the other side of the crankshaft, the port main engine drives two high-pressure water pumps for flushing the hold. On both sides of the lengthy shafting and on both output shafts of the reduction gearing, flexible couplings are installed. On the outgoing reduction gearing to the propeller, a shaft generator has been installed. The starboard main engine drives the dredging pump via reduction gearing with adjustable speed and various in-built flexible couplings.

1 *flexible couplings*

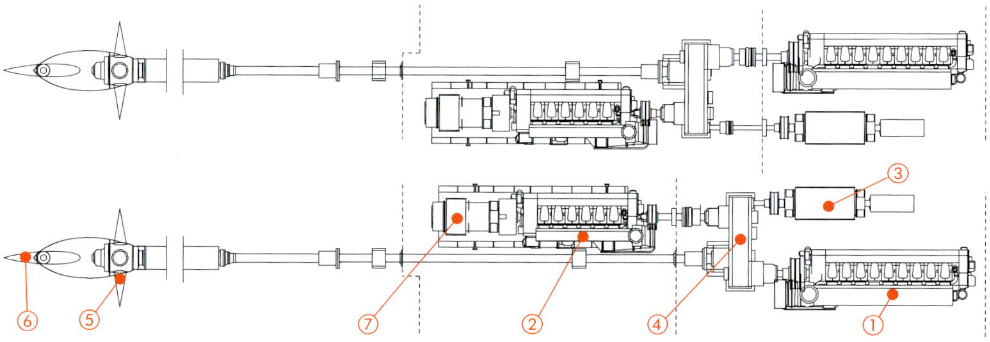

▲

A propulsion system with two main engines and two gensets.

1 *main engine*
2 *genset*
3 *shaft generator*
4 *reduction gearing*
5 *controllable-pitch propeller*
6 *rudder*
7 *generator*

Switch I: Both main engines drive the controllable-pitch propellers via reduction gearing and generate electricity with the shaft generators. Normal speed.
Switch II: Both auxiliary engines generate electricity with their own generator and drive the controllable-pitch propellers at low speed. The main engines are stopped.
Switch III: The four engines drive both controllable-pitch propellers at maximum speed. Electricity may be generated with the shaft generators and/or with the auxiliary generators.
More scenarios are of course optional.
Flexible couplings: All the engines are equipped with flexible couplings on the output shaft, as are the shaft generators and the auxiliary generators.

20.12 Vibration dampers below the engine block or frame

Many four-stroke high-speed- and medium-speed diesel engines in categories I, II and III have vibration dampers below the engine block or frame, apart from the previously discussed flexible couplings to the propeller or the generator, the vibration damper on the blind side of the crankshaft and sometimes on the camshaft, vibration dampers below the engine block or frame. This prevents transfer of engine vibrations to the bedplate and surroundings.
This system requires flexible connections for all the lines, such as fuel, starting air, coolant and exhaust gases. During starting and stopping, the diesel engine moves on the flexible vibration dampers to such an extent that the lines would be torn off if no damping took place.
Especially in small high-speed four-stroke engines in categories I and II all the parts are installed in a flexible manner.

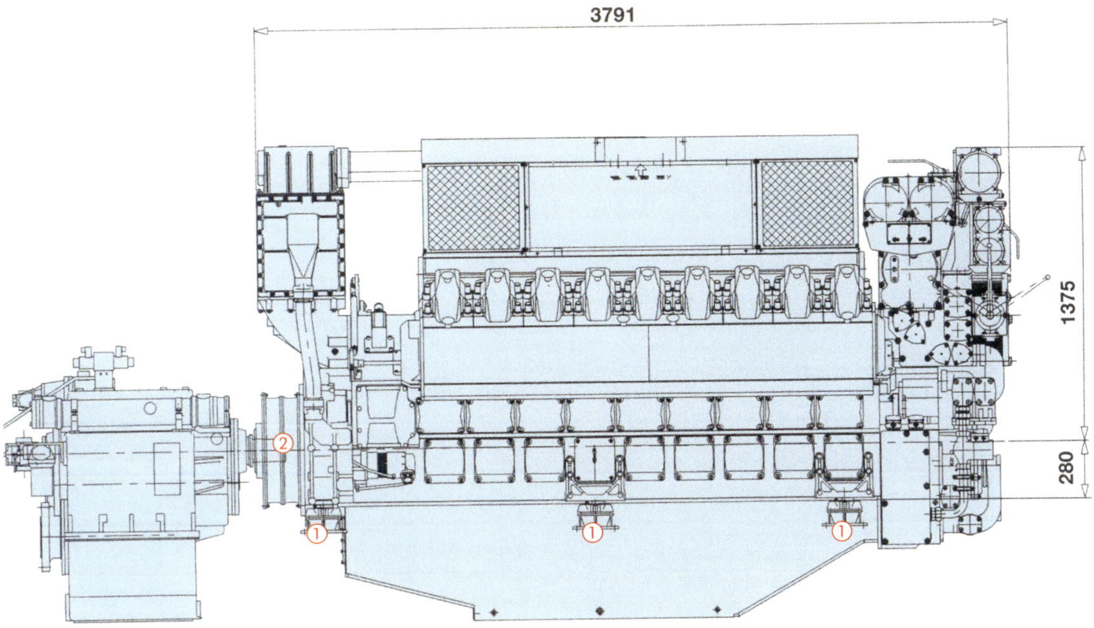

◄
This four-stroke high-speed eighteen-cylinder V-engine with flexibly arranged reduction gearing is fitted with six vibration dampers below the engine block and a flexible coupling between the engine and the reduction gearing.

1 *vibration damper*
2 *flexible coupling*

▲
Vibration dampers (1) below a two-cylinder high-speed four-stroke diesel engine for yacht propulsion.

Left; the built-on reduction gearing with built-in reversing clutch. The lever for the reversing clutch is clearly visible.

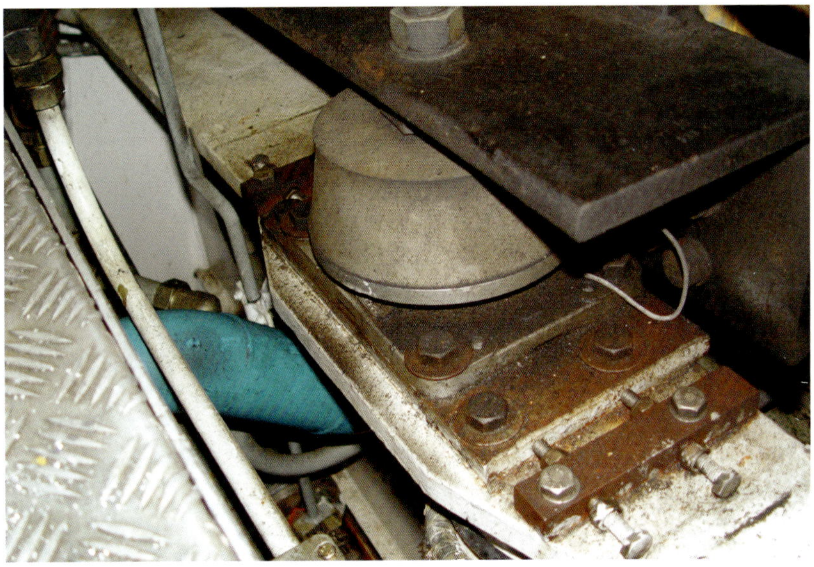

▲
A vibration damper below a propulsion engine.

Note the adjustable base below the vibration damper.

▶

Vibration dampers below the frame of a gas-engine generator set.

Both the gas engine and the generator are fixed rigidly to a mutual frame. A flexible coupling (not visible) has been placed between the engine and the generator.

1 *vibration damper under the frame*
2 *vibration damper on the crankshaft*

20.12.1 Selection and the number of vibration dampers below the diesel engine

The manufacturers concerned have a very elaborate delivery program for each type of engine, genset and pump.

▶

Various types of Rubber Design vibration dampers.

▶

A large Rubber Design vibration damper for heavy engines.

The damping is done with heavy steel compression springs.

▲

A cut-away section of a vibration damper with in the centre the elastic rubber section from Rubber Design, Heerjansdam, The Netherlands.

20.13 Examples of engine arrangements with flexible vibration dampers

20.13.1 Propulsion systems

1 **There are many engine alignments with flexible vibration dampers below the engine block in four-stroke high-speed diesel engines in category I.** This prevents the transmission of vibrations (generated by the engine) to the hull of the ship, in small yachts, working boats and cutters. The bulk of the vibrations is absorbed by the flexible dampers. For most small engines, four dampers, fitted at the corners of the engine block or bedplate, suffice.

2 Often vibration dampers are placed below the engine frame and/or the gear box of **the larger four-stroke high-speed diesel engines in category II**. There are many potential combinations for a propeller driven via a gear box by a high-speed diesel-engine.

- 01 Engine flexibly arranged and gear box semi-flexible.
- 02 Engine and gear box bolted together and arranged flexibly, without thrust absorption.
- 03 Engine and gear box bolted together and arranged flexibly, with thrust absorption.
- 04 Engine flexibly arranged and the gear box semi-flexible in a V-drive.
- 05 Engine and gear box bolted together and flexibly supported with a V-alignment.

3 In larger four-stroke medium-speed diesel engines in category III, less flexible dampers are placed below the frames. Generally, the alignment is fixed.

Also, see Chapter 30, *Bedplates and aligning engines, gearboxes, shafting, propeller shafts and gensets.*

20.13.2 Gensets

Diesel engines that drive gensets often are also equipped with flexible vibration dampers, which are placed below the engine block and the generator. The generator in small gensets is usually fixed to the diesel engine. The underlying frame is often equipped with vibration dampers. The coupling between the engine and the generator is, of course, flexible, but the generator flange is bolted to the engine flange.
Occasionally, the engine is equipped with flexible vibration dampers; the generator is rigidly arranged and driven via a flexible coupling.

20.14 Flexible connections to the diesel engine

Flexible vibration dampers below the engine allow the entire diesel engine to move during starting and stopping. This may vary from several millimetres to several centimetres. With normal engine loads, the production of some motion cannot be avoided.
All connections to the engine such as those for the fuel, coolant, starting-air and exhaust-gas manifolds, as well as cables for electricity, steering and automation must be fitted in such a manner that distortion, damage or tearing is prevented. By placing flexible connections in these lines and cables, the transmission of vibrations generated by the engine to the rest of the installation is prevented.

▼
Left: vibration dampers; in the centre: flexible cooling-water connecting sections and top right: flexible exhaust-gas connecting sections.

▶
A large medium-speed four-stroke diesel engine for the 'Elysee', a container-feeder vessel by J.R. Shipping, Harlingen, The Netherlands.

Man diesel Type L58/ 64, 9-cylinder, 12.6 MW. The engine alignment is flexible. The by-pass valve intermediate the inlet-air manifold and the exhaust-gas pipe is provided with a compensator.

▶
Engine frame mounted on the bedplate and intermediate the flexible connection (1). One of four blockers (2) at the far end of the engine frame, which serve to keep the engine in place in case of a collision or when the ship runs aground.

◀
The flexible part of the engine mounts. Their thickness is frequently measured to establish the degree of depression. The material is oil proof.

▲
Two spare exhaust-gas pipe compensators on board a large container ship.

◀
Flexible hoses for fuel and cooling water are often installed.

Here an example of a small two-cylinder Volvo-Penta high-speed diesel engine for a cutter. Engine category I.

20.15 Double spring alignment: A new design

Diesel engines tend to cause considerable vibrations in the ship. Apart from causing fatigue cracks, these vibrations often produce noise problems in the accommodation aboard the ship. These problems can be diminished by isolating the vibrations.

Rubber Design specialise in vibration damping and are experts in the area of sound reduction. However, the client's requirements regarding construction noise levels generated by gensets were stricter than any other project ever executed in the past.

20.15.1 Double-layer vibration dampers

The standard solution to reduce noise levels is to place the engines on a single layer of vibration dampers. In this particular instance this proved to be insufficient and did not meet the client's requirements. Therefore a double layer of vibration dampers (type RD3-series and RD2-series) provided with a certain interlayer mass was applied. This ultimately resulted in a construction noise level reduction in a cylinder band with an ignition frequency of 117 dB to approximately 55 dB. This was well below the client's requirements.

20.15.2 Expertise, experience and methods

Since the achieved noise reduction of around 56 dB on average was the maximum attainable for a double-layer alignment, it was of the essence to draw up an accurate design for natural frequencies and impedance of the interlayer mass and the location of the natural frequencies of the alignment.

In order to accomplish this, Rubber Design performed the following procedures:

Step 1: establish the position and type of vibration damper based on the load distribution and natural frequencies of the alignment.
Step 2: design the interlayer mass based on 3D-CAD- and finite-element package.
Step 3: calculate the construction-noise transmission of the engine to the bedplate.

It is obvious that nothing was left to chance in the design procedure and that various methods were applied. The 'Finite-Elements Package' provided support in developing an accurate model of the interlayer mass on which the gensets are placed; this eventually resulted in an interlayer mass with favourable natural frequencies and a high impedance. Moreover, the engine frame is modelled and calculated on natural frequencies, strength and impedance.

▶
Diesel-generator alignment with a double-layer of vibration dampers. Rubber Design, Heerjansdam, The Netherlands.

20.15.3 Simulation

The subsequent testing of the entire system not only required entering the models from the finite-element package into a simulation program, but also the properties of the vibration dampers applied. These properties were established throughout the years, thus making a comprehensive database holding all specific properties of vibration dampers available to Rubber Design. The company has comparable databases of the properties of engines and bedplates. The best option is found by optimising the position of the vibration dampers and the geometrics of the intermediate mass in the simulation program.

An independent inspection body measured the vibration levels at a maximum speed in a test stand. Their findings were sufficiently positive to release the double-spring alignment to be mounted in a luxurious yacht.

▲
A modern, high model vibration damper suitable for double-layer application.

▼
In a flexible alignment of the engine, as in this large fourstroke medium-speed diesel engine, the 9L 58/64 MAN Diesel, all other connections to the engine that are fixed must be fitted by means of flexible couplings.

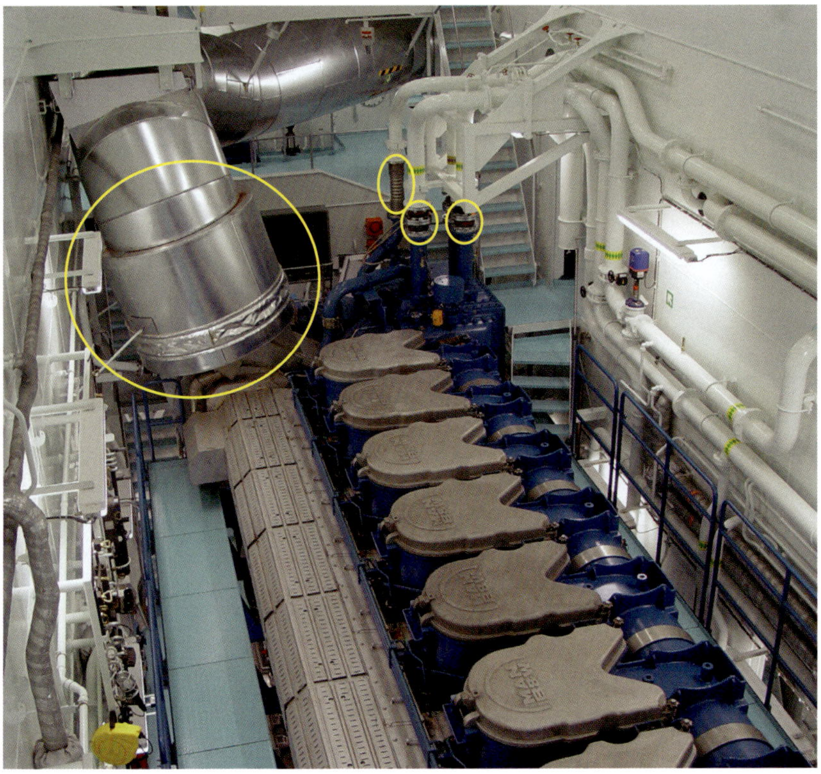

>CH21

Diesel-engine manufacturers

Manufacture of diesel engines in engine category II, high-speed four-stroke engines with shaft power of 100 to 10,000 kW. Fuel: diesel oil (MDO).
Shown here is an overview of a production hall for the BR 2000 series at MTU in Friedrichshafen.

Mounting cylinder liners in a bank for an MTU BR 8000 V-engine.

21.1 Introduction

From the nineteenth century when internal combustion engines could be mass-manufactured, progressively more engineering works started business. Along with their core business of manufacturing machines, they also set up engine-manufacturing production lines. These companies were already manufacturing steam boilers, steam engines and, for example, cannons so the facilities for engine manufacture were present, such as a foundry and a forge; the roughly manufactured parts could be finished with simple planing machines, milling machines and/or lathes.

The work was mostly manual labour. These engine factories were often established near navigable waters, so ships could berth in front of the premises for repairs.

The larger engines in categories II, III and IV were manufactured worldwide.

Very large engines in categories III and IV used to be and still are manufactured throughout the world. Where ship propulsion is concerned, very large low-speed crosshead engines are normally manufactured at the same location as the shipyard. A famous example is Hyundai, the world's largest shipyard/engine factory in South Korea.

21.1.1 Number of diesel engines manufactured on a yearly basis

The total number of diesel engines manufactured annually worldwide lies between 20 and 25 million (2011).

Obviously, the number of small diesel engines manufactured for, amongst others, agricultural machines, excavators, trucks and such, is the largest segment of this market, in excess of 90%.

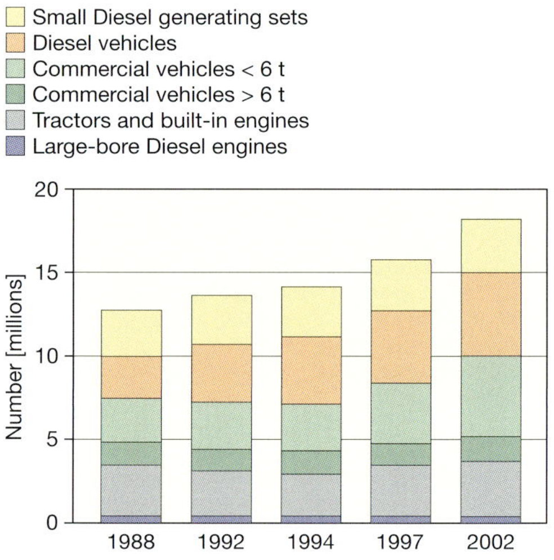

▲

A chart showing the number of diesel engines manufactured in the course of a twenty-five year period.

The number of 'large' diesel engines, dark blue in this overview from Bosch, represents approximately 300,000 pieces.

▶

Large and heavy two-stroke crosshead engines are usually manufactured at the yard where the ship is built. Hyundai, South Korea.

Front left: a smaller MAN diesel engine with an engine frame cast in one piece. Behind it, a MAN diesel engine with a frame made of sheet steel. In the foreground a crankshaft bed for a crosshead engine.

The figures lie below ten thousand for large diesel engines in Category II (1000 kW). For the largest engines in category II, at the most several thousands per annum.
In category III, these figures are in the region of several thousands per annum and in category IV, several hundreds per annum.
In 2007, many more engines were manufactured, in comparison to ten years earlier when engine manufacturers had to shut down or downsize due to extreme overcapacity in the market. In 2012, the market is slowly recovering after the recession years of 2008 en 2009.

21.1.2 Licence holders

Situation in 2010 – World-wide production, countries where engines are manufactured

Diesel-engine manufacturers in categories III and IV issue licences allowing others to build their engines. For instance, MAN Diesel has over 20 licence holders.

Ship propulsion with diesel engines, market share 2006

South Korea	31%
Japan	26%
Europe	23%
China	12%
Other Asian countries	4%
USA and Canada	2%
Rest	2%

Source: Douglas – Westwood Ltd, Canterbury, England.

21.2 Engine categories

The divisions are based on actual experience with using diesel engines.

21.2.1 Characteristics

Fuel

Only medium-speed four-stroke and low-speed two-stroke diesel engines run on heavy fuel oil, HFO.

Operating principle

The majority of diesel engines works on the four-stroke cycle principle, and falls in the group of high-speed engines with over 960 revolutions per minute.
A minority, the larger four-stroke engines with a cylinder diameter above 200 mm fall in the category, medium-speed engines and are suitable for running on heavy fuel oil.

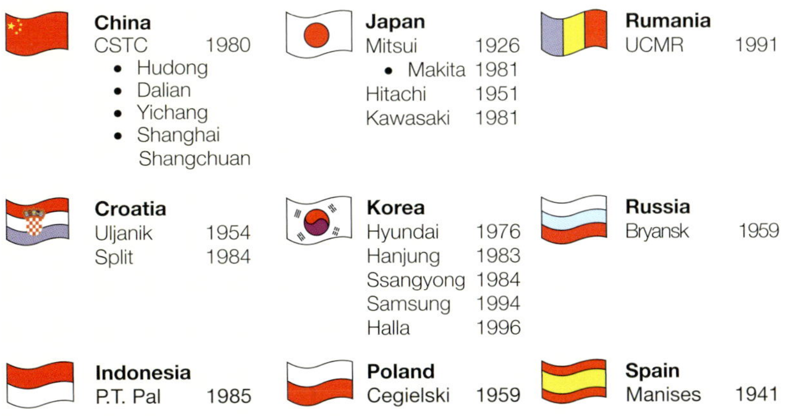

▲ This illustration shows the long period of activity for MAN Diesel licensees.

Capacity classification

A large number of small four-stroke diesel engines fall in category I, up to 100 kW.
They are often used for propulsion of small ships, such a motorboats and yachts, drives for pumps, gensets, anchor winches and building equipment.

- A majority of the larger high-speed four-stroke diesel engines fall in category II, 100 to 10,000 kW.
  They are widely used. The largest diesel engines in this category have a power output of several megawatts are commonly used in the propulsion of fast vessels, catamarans and yachts. The fuel is diesel oil: MDO.
- A smaller number of four-stroke engines fall in category III, medium-speed engines with engine speeds of approximately 960 revolutions per minute and usually operate on HFO.
- The largest diesel engines fall in category IV, low-speed two-stroke crosshead engines with a power output to almost 100,000 kW. They have a maximum engine speed of approximately 240 revs/min and operate on HFO.

General comment:

This classification allows a better insight into engine characteristics. Therefore, it is possible that in the overlapping area between two categories there are other possibilities.

Example

Some factories manufacture engines that can achieve 1000 to 1200 revolutions per minute and categorise them as medium-speed engines.

The Japanese engine factory Mitsubishi was established in 1926!

▶

The four engine categories.

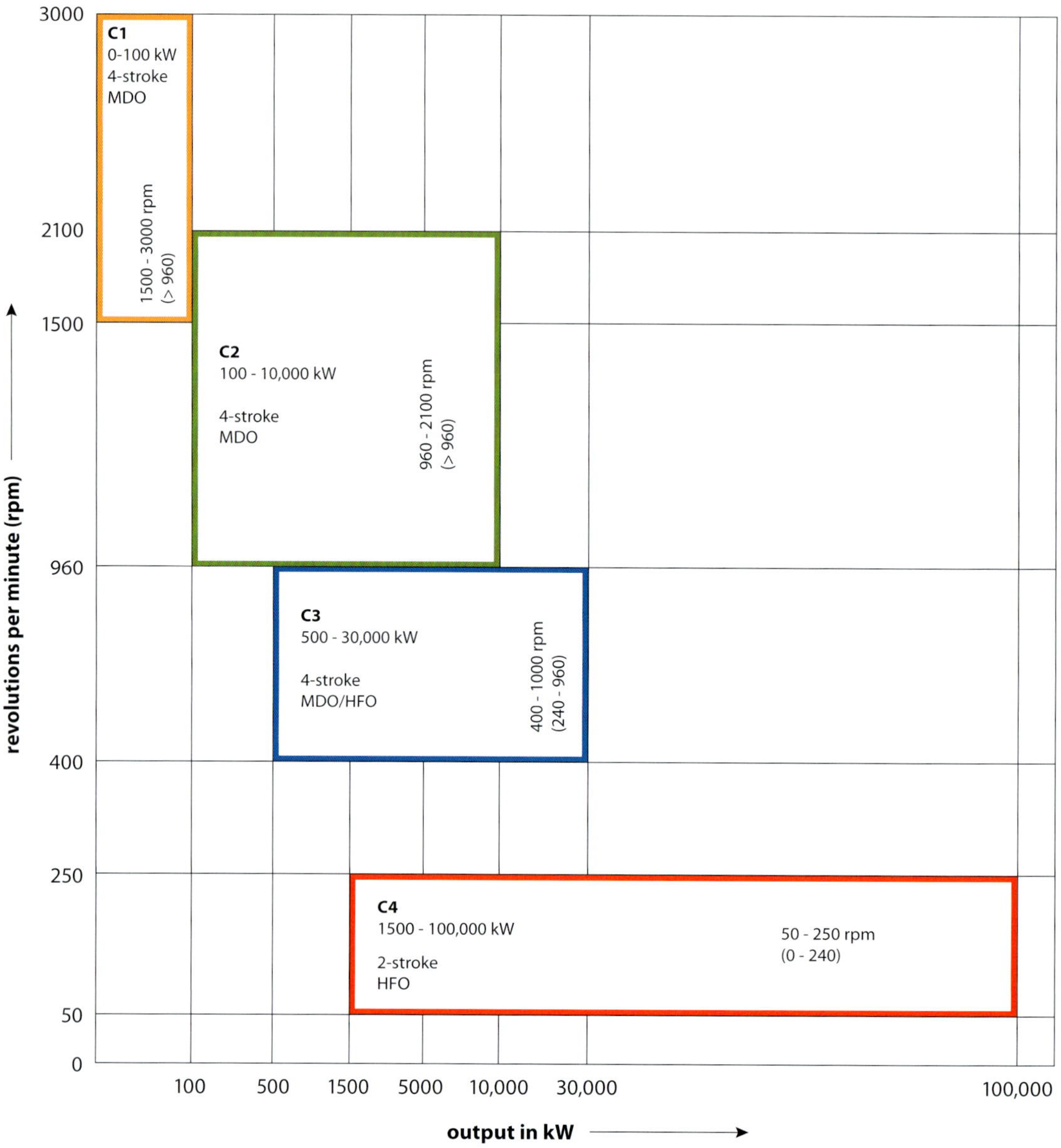

21.3 Engine manufacturers

The two main suppliers MAN Diesel and Wärtsilä produce (defined in shaft power) the majority of engines in Categories III and IV with Caterpillar, number 3 in the smaller capacities, categories I and particularly II.

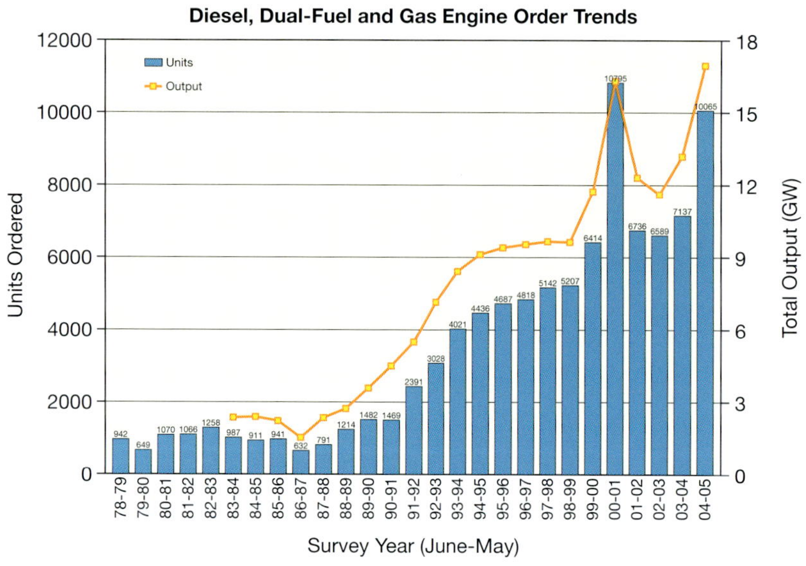

◀

A chart showing the number of diesel, dual-fuel and gas engines ordered between 1978 and 2005.

In this chart all engines from 500 kW shaft power upwards are listed, so categories II, III and IV.

The total number of units lies in the region of 10,000 per year. The average power output is a maximum of approximately 1600 kW (2000–2001).

DIESEL, DUAL-FUEL & GAS ENGINE ORDERS, June 2004 – May 2005																					
Output Range (MW)	Units Ordered	Total Engine Ouput (MW)	Type of Generating Service (Units)			Fuel (Units)				Western Europe	Eastern Europe & Russia	Middle East	Far East	Southeast Asia/ Australia	Central Asia	North Africa	Central W., E., & S. Africa	North America	Central America	South America
			Stand-by	Peak-ing	Contin-uous	Diesel Fuel	Heavy Fuel	Dual-Fuel	Nat. Gas											
1.01 – 2.0	8662	11 465	6174	230	2258	7612	22	0	1028	1421	267	721	1800	870	874	14	55	2138	411	91
2.01 – 3.5	992	2320	616	12	363	751	25	1	214	182	54	45	93	33	33	0	34	485	20	12
3.51 – 5.0	86	362	9	3	74	40	6	0	40	21	8	6	19	7	4	5	0	9	7	0
5.01 – 7.5	136	796	9	7	120	29	33	4	70	14	5	8	36	13	18	0	26	15	1	0
7.51 – 10	137	1155	3	2	132	16	57	0	64	15	3	27	10	14	19	11	11	15	5	7
10.01 – 15	29	354	0	0	29	0	29	0	0	3	0	15	4	0	3	0	1	0	3	0
15.01 – 20	23	411	0	0	23	0	20	3	0	1	0	6	0	0	0	0	3	0	13	0
20.01 – 30	0	0	0	0	0	0	0	0	0	0	0	0	0	0	0	0	0	0	0	0
30.01 & above	1	43	0	0	1	0	1	0	0	0	0	0	0	0	0	0	0	0	1	0
Subtotals	**10 065**	**16 906**	**6811**	**254**	**3000**	**8448**	**193**	**8**	**1416**	**1657**	**337**	**828**	**1962**	**937**	**951**	**30**	**130**	**2662**	**461**	**110**
0.50 – 1.0	12 439	8307	9144	348	2947	11 711	6	6	716	2397	124	1177	2733	1126	1235	4	82	2923	393	245
Totals	**22 504**	**25 213**	**15 955**	**602**	**5947**	**20 159**	**199**	**14**	**2132**	**4054**	**461**	**2005**	**4695**	**2063**	**2186**	**34**	**212**	**5585**	**854**	**355**

▲

An overview from 'Diesel Publications' showing the production to order for engines between June 2004 and May 2005.

Here; the minimum power output is 1000 kW, so these engines fall in category II.

Total number ordered 10,065, of which 8448 are diesel engines.

Most of the engines have a relatively small power output of 1 to 2 MW.

DIESEL, DUAL-FUEL & GAS ENGINE ORDERS, June 2005 – May 2006																					
Output Range (MW)	Units Ordered	Total Engine Ouput (MW)	Type of Generating Service (Units)			Fuel (Units)				Western Europe	Eastern Europe & Russia	Middle East	Far East	Southeast Asia/ Australia	Central Asia	North Africa	Central W., E., & S. Africa	North America	Central America	South America
			Stand-by	Peak-ing	Contin-uous	Diesel Fuel	Heavy Fuel	Dual-Fuel	Nat. Gas											
0.50 – 1.0	17 614	11 840	11 315	1490	4809	16 608	13	12	981	3735	312	2378	2026	1341	2163	5	344	4466	570	274
1.01 – 2.0	11 257	15 403	6298	1280	3679	9765	285	0	1207	2430	180	1016	1610	1181	616	11	182	2827	1075	129
2.01 – 3.5	1778	4226	810	155	813	1269	209	1	298	433	60	150	77	91	29	1	35	659	220	23
3.51 – 5.0	147	574	4	1	142	43	67	3	34	35	9	3	13	24	6	1	5	3	47	1
5.01 – 7.5	129	758	4	10	115	35	15	4	75	22	6	10	51	15	10	4	3	1	5	3
7.51 – 10	151	1303	0	0	151	1	70	0	80	4	58	30	4	5	11	0	4	0	34	1
10.01 – 15	14	167	2	0	12	4	10	0	0	2	0	10	0	0	0	0	0	0	2	0
15.01 – 20	60	1067	1	0	59	13	43	4	0	13	0	9	0	2	1	0	8	0	12	15
20.01 – 30	0	0	0	0	0	0	0	0	0	0	0	0	0	0	0	0	0	0	0	0
30.01 & above	1	43	0	0	1	0	1	0	0	0	0	0	0	0	1	0	0	0	0	0
Totals	**31 151**	**35 381**	**18 434**	**2936**	**9781**	**27 751**	**713**	**24**	**2675**	**6674**	**625**	**3608**	**3781**	**2659**	**2837**	**22**	**581**	**7956**	**1965**	**446**

▲

An overview from 'Diesel Publications' showing the production to order for engines between June 2005 and May 2006.

Here; the minimum power output is 500 kW, so these engines fall in category II.

Total number ordered 31,151, of which 27,751 are diesel engines.

Most engines have a relatively small power output of 0.5 to 2 MW.

This was the first year that engines with a power output of 500 kW as opposed to 1000 kW were included, which meant a significant increase in the number of engines ordered.

The number of engine manufacturers worldwide has dwindled over the last years for various reasons. Today, there are over sixty diesel-engine manufacturers worldwide; approximately half are active in the international market.

A number of reasons for reduction in engine-manufacturers:

- **Due to overcapacity in the past and heavy competition.**
- **Due to increasingly strict regulations for exhaust-gas emissions.** This requires large investments to optimise engine designs that can meet present and future requirements.
- **The customers have stricter requirements.** Important focal points are:
  - capital cost;
  - fuel and lubricating-oil consumption;
  - weight and size in relation to capacity, the so-called power density;
  - fast service and availability of parts worldwide;
  - lifetime of parts.

21.3.1 Mass manufacture for shipping

When designing propulsion installations, the engine type and manufacturer are usually determined before starting the design, as a different engine type will not normally fit the requirements and therefore increases costs. Manufacturing costs are kept to a minimum to compete worldwide among other shipyards.

21.4 Number of engine manufacturers worldwide

21.4.1 Diesel engine manufacturers

There are currently in 2011 more than 60 engine manufacturers operating worldwide. Many will be local and have a relatively small number of engines types. Only a handful of motor manufacturers operate worldwide such as MAN Diesel, Wärtsilä, Caterpillar MaK, Cummins, Hyundai and some others.
There is an overview posted for the year 2011 at the end of this chapter for engines above 500 kW shaft power. This list relates to marine propulsion and diesel power plants: The focus of this book!

21.4.2 Category I: The small high-speed four-stroke diesel engines up to 100 kW shaft power running on MDO

The majority of manufacturers produce these engines. Several dozens for the world market with a larger number producing for the local markets. Large engine manufacturers for categories I and II, such as Caterpillar and Cummins together produce hundreds of thousands to millions of engines annually.
These engines are often used in trucks, buses, earth-moving equipment and cranes.

▶ Manufacturing Perkins diesel engines, category II in Peterborough, England.

Fuel Type: D: Diesel or Heavy Fuel; DF: Dual Fuel; SI: Spark Ingnited. **Number of Cylinders & Configuration:** L: In-Line; V: V-Type; H: Horizontal; O: Opposed

Fuel Type	Engine Model	Bore (mm)	Stroke (mm)	Displacement (liters/cylinder)	Number of Cylinders & Configuration	Output per Cylinder Range (kW/cyl)		Rated Speed Range (r/min)		Maximum Brake Mean Effective Pressure (bar)		Output Range Low (kW)	Output Range High (kW)	Ratings	Application	Emissions level
						min	max	min	max	min	max					
D	4M 42	102	105	0.9	4L	9.7	12.9	1800	3150	5.7	7.5	38.6	51.5		OH	EU Stage 3a
D	3M 41	102	105	0.9	3L	10.4	12.3	1800	3000	5.7	8.1	31.1	36.8		OH	EU Stage 3a
D	2M 41	102	105	0.9	2L	8.2	13.2	1500	3000	6.1	7.6	16.4	26.3		OH	EU Stage 3a
D	4L 42	102	105	0.9	4L	9.4	11.5	1850	3150	5.1	7.1	37.5	46.1		OH	EU Stage 3a
D	3L 41	102	105	0.9	3L	9.3	12.2	1800	3150	5.4	7.3	28	36.7		OH	EU Stage 3a
D	2L 41	102	105	0.9	2L	9.4	12.2	1800	3150	5.4	7.3	18.7	24.4		OH	EU Stage 3a
D	4M 42	102	105	0.9	4L	9.7	12.9	1800	3150	5.7	7.5	38.6	51.5	ISO 3046-1/B	OH	EPA Tier 4i
D	3M 41	102	105	0.9	3L	10.4	12.3	1800	3000	5.7	8.1	31.1	36.8	ISO 3046-1/B	OH	EPA Tier 4i
D	4L 42	102	105	0.9	4L	9.4	11.5	1850	3150	5.1	7.1	37.5	46.1	ISO 3046-1/B	OH	EPA Tier 4i
D	4L 41	102	105	0.9	4L	7.5	12.2	1500	3150	5.4	7	30	48.8	ISO 3046-1/B	OH	EU Stage 2
D	3L 41	102	105	0.9	3L	9.3	12.2	1800	3150	5.4	7.3	28	36.7	ISO 3046-1/B	OH	EPA Tier 4i
D	2L 41	102	105	0.9	2L	9.4	12.2	1800	3150	5.4	7.3	18.7	24.4	ISO 3046-1/B	OH	EPA Tier 4i
D	2 G 40	92	75	0.5	2L	3.7	7.8	1500	3600	5.2	5.9	7.4	15.6	ISO 3046-1/B	OH	
D	1 D 90 V	104	85	0.7	1H	6.4	11.2	1500	3000	6.2	7.1	6.4	11.2	ISO 3046-1/B	OH	
D	1 D 81 C	100	85	0.7	1L	5.4	9.6	1500	3000	5.8	6.5	5.4	9.6	ISO 3046-1/B	OH	
D	1 D 42 C	90	70	0.4	1L	2.7	5.2	1500	3600	3.9	4.9	2.7	5.2	ISO 3046-1/B	OH	
D	1 D 42 C	90	70	0.4	1L	3.7	5.2	2000	3300	4.2	5	3.7	5.2	ISO 3046-1/B	OH	EPA Tier 4i
D	1 D 90	104	85	0.7	1L	6.4	11.2	1500	3000	6.2	7.1	6.4	11.2	ISO 3046-1/B	OH	
D	1 D 81	100	85	0.7	1L	5.5	10.1	1500	3600	5	6.6	5.5	10.1	ISO 3046-1/B	OH	
D	1 D 50	97	70	0.5	1L	3.7	7.5	1500	3600	4.8	5.7	3.7	7.5	ISO 3046-1/B	OH	
D	1 D 42	90	70	0.4	1H	3.3	7	1500	3600	5.2	5.9	3.3	7	ISO 3046-1/B	OH	
D	1 D 42	90	70	0.4	1L	4.7	6.8	2000	3300	5.6	6.3	4.7	6.8	ISO 3046-1/B	OH	EPA Tier 4i
D	1 B 50 V	93	76	0.5	1H	3.9	7.9	1500	3600	5.1	6	3.9	790	ISO 3046-1/B	OH	
D	1 B 40 V	88	76	0.5	1H	3.5	7.3	1500	3600	5.3	6.1	3.5	7.3	ISO 3046-1/B	OH	
D	1 B 30 V	80	69	0.3	1H	2.3	5	1500	3600	4.8	5.3	2.3	5	ISO 3046-1/B	OH	
D	1 B 40	88	76	0.5	1L	3.5	7.3	1500	3600	5.3	6.1	3.5	7.3	ISO 3046-1/B	OH	
D	1 B 30	80	69	0.3	1L	2.3	5	1500	3600	4.8	5.3	2.3	5	ISO 3046-1/B	OH	
D	1 B 27	74	65	0.3	1L	1.6	4	1500	3600	4.8	4.6	1.6	4	ISO 3046-1/B	OH	
D	1 B 20	69	65	0.2	1L	1.5	3.4	1500	3600	4.7	4.9	1.5	3.4	ISO 3046-1/B	OH	
D	1 B 50 V	93	76	0.5	1H	5.3	8	2000	3600	5.1	6.2	5.3	8		OH	EPA Tier 4i
D	1 D 50	97	70	0.5	1L	5.2	7.5	2000	3300	5.3	6	5.2	7.5		OH	EPA Tier 4i
D	1 D 81	100	85	0.7	1L	8.3	10.1	2300	3000	6.1	6.5	8.3	10.1		OH	EPA Tier 4i
D	1 D 90	104	85	0.7	1L	9.2	11.2	2300	3000	6.2	6.6	9.2	11.2		OH	EPA Tier 4i
D	1 D 81 C	100	85	0.7	1L	8	9.6	2300	3000	5.8	6.3	8	9.6		OH	EPA Tier 4i
D	1 D 90 V	104	85	0.7	1H	9.2	11.2	2300	3000	6.2	6.6	9.2	11.2		OH	EPA Tier 4i
D	2 G 40	92	75	0.5	2L	5.3	7.4	2000	3000	5.9	6.3	10.5	14.7		OH	EPA Tier 4i
D	2L 41	102	105	0.9	2L	7.5	12.2	1500	3000	5.7	7	15	24.4		OH	EU Stage 2
D	3L 41	102	105	0.9	3L	6.8	12.2	1500	3000	5.7	7.1	22.9	36.7		OH	
D	4L 41	102	105	0.9	4L	8.8	9	1800	1850	6.8	6.8	35.2	36.1		OH	EPA Tier 4i
D	2M 41	102	105	0.9	2L	8.2	13.1	1500	3000	6.1	7.6	16.4	26.3		OH	EPA Tier 4i
D	3M 41	102	105	0.9	3L	8.3	13.3	1500	3000	6.2	7.8	25	39.8		OH	EU Stage 2
D	4M 41	102	105	0.9	4L	8.5	13.3	1500	3000	6.2	7.9	34	53.1		OH	EU Stage 2
D	1 B 20	69	65	0.2	1L	2.1	3.4	2000	3600	4.7	5.2	2.1	3.4		OH	EPA Tier 4i
D	1 B 27	74	65	0.3	1L	2.5	4	2000	3600	4.8	5.4	2.5	4		OH	EPA Tier 4i
D	1 B 30	80	69	0.3	1L	3.4	5	2000	3600	4.8	5.9	3.4	5		OH	EPA Tier 4i
D	1 B 40	88	76	0.5	1L	4.9	7.3	2000	3600	5.3	6.4	4.9	7.3		OH	EPA Tier 4i
D	1 B 50	93	76	0.5	1L	5.3	7.6	2000	3000	5.9	6.2	5.3	7.6		OH	EPA Tier 4i
D	1 B 50	93	76	0.5	1L	3.9	7.6	1500	3000	5.9	6	3.9	7.6		OH	
D	1 B 20 V	69	65	0.2	1H	1.4	3.4	1500	3600	4.6	4.7	1.4	3.4		OH	
D	1 B 30 V	80	69	0.3	1H	3.4	5	2000	3600	4.8	5.9	3.4	5		OH	EPA Tier 4i
D	1 B 40 V	88	76	0.5	1H	4.9	7.3	2000	3600	5.3	6.4	4.9	7.3		OH	EPA Tier 4i

Source: Diesel Publications

▲

The Hatz GmbH & Co. KG diesel-engine programme, category I.

Fuel Type	Engine Model	Bore (mm)	Stroke (mm)	Displacement (liters/cylinder)	Number of Cylinders & Configuration	Output per Cylinder Range (kW/cyl)		Rated Speed Range (r/min)		Maximum Brake Mean Effective Pressure (bar)		Output Range Low (kW)	Output Range High (kW)	Ratings	Application	Emissions level
						min	max	min	max	min	max					
D	QSK19	159	159	3.2	6L	56	99.5	1800	2000			336	597		OH	EPA Tier 3
D	B3.3	95	115	0.8	4L	11.3	15.8	2200	2600			45	63		OH	EPA Tier 4i
SI	C Gas Plus	114	135	1.4	6L	16.5	34.8	2400	2400			99	209		ON	Euro 4
SI	B Gas Plus	102	120	1	6L	24.2	28.7	2800	2800			145	172		ON	EPA 2007
D	SBE6.7	107	124	1.1	6L	25.2	37.3	2300	2500			151	224		ON	Euro 5
SI	ISL G 2007	114	145	1.5	6L	31	39.8	2200	2200			186	239		ON	EPA 2007
SI	ISB5.9 G	102	120	1	6L	24.2	28.7	2800	2800			145	172		ON	Euro 5
SI	C Gas Plus	114	135	1.4	6L	16.5	34.8	2400	2400			99	209		ON	Euro 3
SI	B Gas International	102	120	1	6L	18.7	28.7	2500	2800			112	172		ON	Euro 3
D	ISF3.8	102	115	1	4L	17.5	20.8	2600	2600			105	125		ON	Euro 5
D	ISF2.8	94	100	0.7	4L	13.3	20	3200	3600			80	120		ON	Euro 5
D	ISLe	114	145	1.5	6L	34.8	49.7	2100	2100			209	298		ON	Euro 5
D	ISBE4.5	107	124	1.1	4L	26	38.8	2300	2500			104	155		ON	Euro 5
D	ISB6.7	107	124	1.1	6L	24.8	40.3	2600	2800			149	242		ON	EPA 2010
D	ISC8.3	114	135	1.4	6L	31	45.3	2200	2200			186	272		ON	EPA 2010
D	ISL9	114	145	1.5	6L	42.8	47.2	2100	2200			257	283		ON	EPA 2010
D	ISX11.9	130	150	2	6L	38.5	52.8	2100	2100			231	317		ON	EPA 2010
D	ISX15	137	169	2.5	6L	49.7	74.5	2000	2100			298	447		ON	Euro 5
SI	B LPG Plus	102	120	1	6L	19.3	24.2	2500	2800			116	145		ON	EPA 2007
SI	B Gas Plus	102	120	1	6L	24.2	28.7	2800	2800			145	172		ON	Euro 5
SI	ISL G	114	145	1.5	6L	31	39.8	2100	2200			186	239		ON	EPA 2010
D	B3.3	95	115	0.8	4L	11.3	15.8	2200	2600			45	63		OH	EPA Tier 3
D	QSB3.3	95	115	0.8	4L	15	20.5	2200	2600			60	82		OH	EPA Tier 3
D	QSB4.5	107	124	1.1	4L	20.5	31.8	1800	2500			82	127		OH	EPA Tier 3
D	QSB6.7	107	124	1.1	6L	16.5	34.2	1800	2500			99	205		OH	EPA Tier 3
D	QSC	114	144.5	1.4	6L	25.5	37.8	1800	2200			153	227		OH	EPA Tier 3
D	QSL	114	145	1.5	6L	34.8	45.3	1800	2200			209	272		OH	EPA Tier 3
D	QSM	125	147	1.8	6L	36	49.7	1800	2100			216	298		OH	EPA Tier 3
D	QSX	137	169	2.5	6L	46.7	82.7	1800	2100			280	496		OH	EPA Tier 3
D	QSK19	159	159	3.2	6L	56	99.5	1800	2000			336	597		OH	EPA Tier 2
D	QSK23	170	170	3.8	6L	94.5	118	1800	2100			567	708		OH	EPA Tier 2
D	QST30	140	165	5	6V	94.5	149.2	1800	2100			567	895		OH	EPA Tier 2
D	QSK38	159	159	3.2	12V	57.2	78.3	1200	1800			686	940		OH	EPA Tier 2
D	QSK45	159	190	3.8	12V	74.6	124.3	1800	1900			895	1491		OH	
D	QSK50	159	159	3.1	16V	66.6	116.5	1300	1900			1066	1864		OH	EPA Tier 2
D	QSK60	159	190	3.9	16V	83.1	132.8	1800	1900			1329	2125		OH	EPA Tier 2
D	QSK78	170	190	4.3	18V	145	145	1900	1900			2610	2610		OH	

Fuel Type: D: Diesel or Heavy Fuel; DF: Dual Fuel; SI: Spark Ingnited. **Number of Cylinders & Configuration:** L: In-Line; V: V-Type; H: Horizontal; O: Opposed

Source: Diesel Publications

▲

The Cummins Inc. diesel-engine programme. Categories I and II

▶

Small pleasure yachts usually have diesel engines with power outputs between 25 and 75 kW, category I.

21.4.3 Category II: Larger high-speed four-stroke diesel engines from 100 to 10,000 kW shaft power running on MDO

There are at present a few dozen manufacturers active in the smaller-capacities market. For engines over 1000 kW, the number of manufacturers decreases significantly and for over 3000 kW there are only a few manufacturers serving this market.

Caterpillar Inc. Engine Products Division

One of the world's largest engine manufacturers for categories I, II and III is the American company Caterpillar. Since their takeover of the German engine manufacturer MaK in Kiel, they have also incorporated medium-speed four-stroke diesel engines running on heavy oil, Category III, in their programme. The production of engines in category II is huge; hundreds of thousands of 'yellow Cats' find their way to the customers each year. There are 49 factories active in the USA and 58 in the rest of the world. An impressive number.

MTU Friedrichshafen GmbH

The German manufacturer of high-speed diesel engines running on MDO is MTU in Friedrichshafen at the Boden Sea. It produces in categories I and II, highly-loaded diesel engines with, in the larger engines, a mean effective pressure of 27 to 32.6 bars. This is very high. These large V-engines with a shaft power of 8200 kW (series 8000), a mean piston speed of 12.1 m/sec. and a mean effective pressure of 27 bars have a high load parameter of 326 bar m/sec. This is the highest of all engines in category II.

Rolls-Royce

The English producer Rolls-Royce manufactures a small number of engine types in category II with a high power output. A maximum power output of 8000 kW and a load parameter of 10 × 24.9 = 249 bar m/sec. is respectable. They are built near Bergen, Norway.

◄ Pistons for Detroit Diesel category II engines, ready to be assembled. Notice the two piston parts. The piston-ring section (1) and section for the absorption of the lateral forces (2).

Fuel Type	Engine Model	Bore (mm)	Stroke (mm)	Displacement (liters/cylinder)	Number of Cylinders & Configuration	Output per Cylinder Range (kW/cyl)		Rated Speed Range (r/min)		Maximum Brake Mean Effective Pressure (bar)		Output Range Low (kW)	Output Range High (kW)	Ratings	Application	Emissions level
						min	max	min	max	min	max					
D	C18 ACERT	145	183	18.1	6L			1800	1900			447	571		OH	EU Stage 3b
D	C15 ACERT	137	171	15.2	6L			1800	2100			354	433		OH	EU Stage 3b
D	C13 ACERT	130	157	12.5	6L			1800	2100			287	354		OH	EU Stage 3b
D	C9.3 ACERT	115	149	9.3	6L			1800	2200			224	261		OH	EU Stage 3b
D	C7.1 ACERT	105	135	7	6L			2200	2200			130	225		OH	EU Stage 3b
D	C6.6 ACERT	105	127	6.6	6L			2200	2200			89	129.5		OH	EU Stage 3b
D	C4.4 ACERT	105	127	4.4	4L			1800	2400			61.5	129.5		OH	EU Stage 3b
D	C3.4B	94	120	3.4	4L			2200	2500			45	86		OH	EU Stage 3b
D	C18 ACERT	145	183	18.1	6L			1800	2100			429	597		OH	EU Stage 3a
D	C15 ACERT	137.2	171.4	15.2	6L			1800	2100			328	444		OH	EU Stage 3a
D	C13 ACERT	130	157	12.5	6L			1800	2100			287	388		OH	EU Stage 3a
D	C11 ACERT	130	140	11.1	6L			1800	2100			242	336		OH	EU Stage 3a
D	C9 ACERT	112	149	8.8	6L			1800	2200			205	280		OH	EU Stage 3a
D	C7 ACERT	110	127	7.2	6L			1800	2200			168	224		OH	EU Stage 3a
D	C6.6 ACERT	105	127	6.6	6L			1800	2500			89	205		OH	EU Stage 3a
D	C4.4 ACERT	105	127	4.4	4L			2200	2200			61.5	106		OH	EU Stage 3a
D	C4.4	105	127	4.4	4L			2200	2400			54	83		OH	EU Stage 3a
D	C3.4	94	120	3.3	4L			2500	2500			47	62		OH	EU Stage 3a
D	C3.4	94	120	3.3	4L			2500	2500			47	62		OH	EPA Tier 3
D	C3.4	94	120	3.3	4L			2500	2500			47	62		OH	EU Stage 3b
D	C2.2	84	100	2.2	4L			2200	3000			31	49.3		OH	EU Stage 3b
D	C1.7	84	100	1.7	3L			2400	2600			23.6	26.1		OH	EU Stage 3b
D	C1.6	77	81	1.5	4L			2800	3000			24.6	26.5		OH	EU Stage 3b
D	C1.5	84	90	1.5	3L			2200	3000			20.9	30		OH	EU Stage 3b
D	C1.1	77	81	1.1	3L			2200	3400			13.7	21		OH	EU Stage 3b
D	C0.7	67	72	0.7	3L			2800	3600			12.2	15.3		OH	EU Stage 3b
D	C0.5	67	72	0.5	2L			2800	3600			8.2	10.2		OH	EU Stage 3b
D	C32 ACERT	145	162	32.1	12V			1800	2100			708	895		OH	EPA Tier 4i
D	C27 ACERT	137.2	152.4	27	12V			1800	2100			597	783		OH	EPA Tier 4i
D	C18 ACERT	145	183	18.1	6L			1800	1900			447	571		OH	EPA Tier 4i
D	C15 ACERT	137	171	15.2	6L			1800	2100			354	433		OH	EPA Tier 4i
D	C13 ACERT	130	157	12.5	6L			1800	2100			287	354		OH	EPA Tier 4i
D	C9.3 ACERT	115	149	9.3	6L			1800	2200			224	261		OH	EPA Tier 4i
D	C7.1 ACERT	105	135	7	6L			2200	2200			130	225		OH	EPA Tier 4i
D	C6.6 ACERT	105	127	6.6	6L			2200	2200			89	129.5		OH	EPA Tier 4i
D	C4.4 ACERT	105	127	4.4	4L			1800	2400			61.5	129.5		OH	EPA Tier 4i
D	C3.4B	94	120	3.4	4L			2200	2500			45	86		OH	EPA Tier 4i
D	3508B	170	190	34.5	8V			1800	1800			746	820		OH	
D	C1.7	84	100	1.7	3L			2400	2600			23.6	26.1		OH	EPA Tier 4i
D	C1.6	77	81	1.5	4L			2800	3000			24.6	26.5		OH	EPA Tier 4i
D	C1.5	84	90	1.5	3L			2200	3000			20.9	30		OH	EPA Tier 4i
D	C1.1	77	81	1.1	3L			2200	3400			13.7	21		OH	EPA Tier 4i
D	C0.7	67	72	0.7	3L			2800	3600			12.2	15.3		OH	EPA Tier 4i
D	C0.5	67	72	0.5	2L			2800	3600			8.2	10.2		OH	EPA Tier 4i
SI	3300 Family	121	152	10.5				1800	1800			71	164		OH	
SI	3400 Family	137	165	27				1400	1800			131	475		OH	

Fuel Type: D: Diesel or Heavy Fuel; DF: Dual Fuel; SI: Spark Ingnited. **Number of Cylinders & Configuration:** L: In-Line; V: V-Type; H: Horizontal; O: Opposed

▲

The Caterpillar Inc. Engine Products Division diesel-engine programme. Categories II and III.

Fuel Type	Engine Model	Bore (mm)	Stroke (mm)	Displacement (liters/cylinder)	Number of Cylinders & Configuration	Output per Cylinder Range (kW/cyl)		Rated Speed Range (r/min)		Maximum Brake Mean Effective Pressure (bar)		Output Range Low (kW)	Output Range High (kW)	Ratings	Application	Emmissions level
						min	max	min	max	min	max					
SI	3500 Family	170	190	86.3				1200	1400			391	1286		OH	
SI	3600 Family	300	300	339.2				1000	1000			1324	3531		OH	
D	C27 ACERT	137.7	152.4	27	12V			1800	2100			597	858		OH	
D	C32 ACERT	145	162	32.1	12V			1800	2100			708	1007		OH	
D	3508	170	190	34.5	8V			1200	1800			507	746		OH	
D	3508B	170	190	34.5	8V			1800	1800			746	820		OH	
D	3512	170	190	51.8	12V			1200	1800			761	1119		OH	
D	3512B	170	190	51.8	12V			1800	1800			1119	1231		OH	
D	3512B	170	190	51.8	12V			1800	1800			1119	1231		OH	
D	3516	170	190	69	16V			1200	1800			1011	1492		OH	
D	3516B	170	190	69	16V			1800	1925			1492	1566		OH	
D	3606	280	300	110.8	6L			750	1000			1490	1850		OH	
D	3608	280	300	147.5	8L			750	1000			1980	2460		OH	
D	3612	280	300	221.7	12V			750	1000			2980	3996		OH	
D	3616	280	300	295.5	16V			750	1000			3960	5310		OH	
D	C2.2	84	100	2.2	4L			2200	3000			31	49.3		OH	EPA Tier 4i
D	C3.4	94	120	3.3	4L			2500	2500			47	62		OH	EPA Tier 4i
D	C4.4	105	127	4.4	4L			2200	2400			54	83		OH	EPA Tier 3
D	C4.4 ACERT	105	127	4.4	4L			2200	2200			61.5	106		OH	EPA Tier 3
D	C6.6 ACERT	105	127	6.6	6L			1800	2500			89	205		OH	EPA Tier 3
D	C7 ACERT	110	127	7.2	6L			1800	2000			168	224		OH	EPA Tier 3
D	C9 ACERT	112	149	8.8	6L			1800	2200			205	280		OH	EPA Tier 3
D	C11 ACERT	130	140	11.1	6L			1800	2100			242	336		OH	EPA Tier 3
D	C13 ACERT	130	157	12.5	6L			1800	2100			287	388		OH	EPA Tier 3
D	C15 ACERT	137.2	171.4	15.2	6L			1800	2100			328	444		OH	EPA Tier 3
D	C18 ACERT	145	183	18.1	6L			1800	2100			429	597		OH	EPA Tier 3

Source: Diesel Publications

▲

The Caterpillar Inc. Engine Products Division diesel-engine programme, (continued). Categories II and III.

▼

Large luxury motor yachts usually have category II diesel engines for propulsion, which operate on diesel oil. The power output ranges between 1000 and 3000 kW. In addition, of course, a genset is necessary.

▶

The diesel engine is an ideal propulsion engine for a variety of yachts. It is reliable, economic, and as it runs on diesel oil, the fire risk is minimal.

Fuel Type	Engine Model	Bore (mm)	Stroke (mm)	Displacement (liters/cylinder)	Number of Cylinders & Configuration	Output per Cylinder Range (kW/cyl) min	Output per Cylinder Range (kW/cyl) max	Rated Speed Range (r/min) min	Rated Speed Range (r/min) max	Maximum Brake Mean Effective Pressure (bar) min	Maximum Brake Mean Effective Pressure (bar) max	Output Range Low (kW)	Output Range High (kW)	Ratings	Application	Emmissions level
D	956	230	230	9.6	12V	313	313		1500		32.6	3750	6250			
D	595	190	210	6	16V	270	270	1750	1800		30.2	3240	4320			
D	595	190	210	6	12V	270	270	1750	1800		30.2	3240	4320			
D	1600	122	150	1.8	12V	41	48	1500	1600			249	576			
D	1600	122	150	1.8	10V	41	48	1500	1600			249	576			
D	1600	122	155	1.8	8V	41	48	1500	1600			249	576			
D	1600	122	150	1.8	6L	41	48	1500	1600			249	576			
D	900	106	136	1.2	4L	24	42		2200		14.5	95	250			
D	2000	130	150	2	18V	38	72	1500	2350		23.6	452	1310			
D	2000	135	156	2.2	8V	90	121	2250	2450		24.6	700	1939			
D	2000	135	156	2.2	10V	90	121	2250	2450		24.6	720	1939			
D	2000	135	156	2.2	12V	90	121	2250	2450		24.6	720	1939			
D	2000	135	156	2.2	16V	90	121	2250	2450		24.6	720	1939			
D	396	165	185	4	8V	125	134	1900	2000		20.4	1000	2150			
D	396	165	185	4	12V	125	134	1900	2000		20.4	1000	2150			
D	396	165	185	4	16V	125	134	1900	2000		20.4	1000	2150			
D	4000	165	190	4.1	8V	87	132	1500	2100		23.9	700	2105			
D	4000	165	190	4.1	12V	87	132	1500	2100		23.9	700	2105			
D	4000	165	190	4.1	16V	87	132	1500	2100		23.9	700	2105			
D	4000	170	190	4.3	12V	160	215	1970	2100		27.6	1920	4300			
D	4000	170	190	4.3	16V	160	215	1970	2100		27.6	1920	4300			
D	4000	170	190	4.3	20V	160	215	1970	2100		27.6	1920	4300			
D	4000	165	210	4.5	20V		136		1800		22.3		2720			
D	4000	170	210	4.8	12V	63	150	1500	1800		19.6	760	3000			
D	4000	170	210	4.8	16V	63	150	1500	1800		19.6	760	3000			
D	4000	170	210	4.8	20V	63	150	1500	1800		19.6	760	3000			
D	956	230	230	9.6	16V	313	313		1500		32.6	3750	6250			
D	956	230	230	9.6	20V	313	313		1500		32.6	3750	6250			
D	1163	230	280	11.6	12V	370	433	1156	1300		29.4	5200	7400			
D	1163	230	280	11.6	16V	370	433	1150	1300		29.4	5200	7400			
D	1163	230	280	11.6	20V	370	423	1150	1300		29.4	5200	7400			
D	8000	265	315	17.4	20V	360	455	1150	1300		27	7200	9100			
D	700	94	100	0.7	3L	10	37	1500	3800		16.8	21	257			
D	700	94	100	0.7	4L	10	37	1500	3800		16.8	21	257			
D	700	94	100	0.7	6L	10	37	1500	3800		16.8	21	257			
D	700	94	107	0.7	4L	12	17	2300	2600		10.4	47	120			
D	700	94	107	0.7	6L	12	17	2300	2600		10.4	47	120			
D	SUN	105	115	1	2L	8	17	1500	2300		10.9	17	75			
D	SUN	105	115	1	3L	8	17	1500	2300		10.9	17	75			
D	SUN	105	115	1	4L	8	17	1500	2300		10.9	17	75			
D	900	102	130	2	4L	19	34		2200		11.4	75	205			
D	900	102	130	2	6L	19	34		2200		11.4	75	205			
D	900	106	136	1.2	6L	24	42		2200		14.5	220	250			

Fuel Type: D: Diesel or Heavy Fuel; DF: Dual Fuel; SI: Spark Ingnited. **Number of Cylinders & Configuration:** L: In-Line; V: V-Type; H: Horizontal; O: Opposed

▲

The MTU Friedrichshafen GmbH diesel-engine programme. Categories I and II.

Fuel Type	Engine Model	Bore (mm)	Stroke (mm)	Displacement (liters/cylinder)	Number of Cylinders & Configuration	Output per Cylinder Range (kW/cyl)		Rated Speed Range (r/min)		Maximum Brake Mean Effective Pressure (bar)		Output Range Low (kW)	Output Range High (kW)	Ratings	Application	Emmissions level
						min	max	min	max	min	max					
D	460	128	166	2.1	6L	37	62		1800		18.8	220	375			
D	500	130	150	2	6V	43	60		1800		16	260	480			
D	500	130	150	2	8V	43	60		1800		16	260	480			
D	S 40E	117	119	1.3	6L	21	31		2200		16.3	130	186			
D	S 60	130	160	2.1	6L	37	62	2100	2200		15.2	224	373			
D	S 60	133	168	2.3	6L	37	103	1800	2300		23.1	224	615			
D	1800	128	166	2.1	6H	52	65		1800		17.2	315	390			
D	2000	130	150	2	12V	38	72	1500	2350		23.6	452	1310			
D	2000	130	150	2	16V	38	72	1500	2350		23.6	452	1310			

Source: Diesel Publications

▲

The MTU Friedrichshafen GmbH diesel-engine programme (continued). Categories I and II.

▲

The 'Seeadler' of the German coast guard with MTU diesel engines for propulsion, category II.

Fuel Type	Engine Model	Bore (mm)	Stroke (mm)	Displacement (liters/cylinder)	Number of Cylinders & Configuration	Output per Cylinder Range (kW/cyl)		Rated Speed Range (r/min)		Maximum Brake Mean Effective Pressure (bar)		Output Range Low (kW)	Output Range High (kW)	Ratings	Application	Emmissions level
						min	max	min	max	min	max					
SI	C26:33L AG	260	330		9L		270		1000		17	1490	2230			
SI	C26:33L AG	260	330		8L		270		1000		17	1490	2230			
SI	C26:33L AG	260	330		6L		270		1000		17	1490	2230			
SI	C26:33L AG	260	330		9L		243		900		17	1340	2010			
SI	C26:33L AG	260	330		8L		243		900		17	1340	2010			
SI	C26:33L AG	260	330		6L		243		900		17	1340	2010			
SI	C26:33AG	260	330		9L		270		1000		18.5	1460	2430			
SI	C26:33AG	260	330		8L		270		1000		18.5	1460	2430			
SI	C26:33AG	260	330		6L		270		1000		18.5	1460	2430			
SI	C26:33PG	260	330		9L		270		1000		18.5	1460	2430			
SI	C26:33PG	260	330		8L		270		1000		18.5	1460	2430			
SI	C26:33PG	260	330		6L		270		1000		18.5	1460	2430			
SI	B32:40PG	320	400		6L		400		750		20	2410	3620			
SI	B32:40PG	320	400		8L		400		750		20	2410	3620			
SI	B32:40PG	320	400		9L		400		750		20	2410	3620			
SI	B32:40AG	320	400		6L		400		750		20	2320	3620			
SI	B32:40AG	320	400		8L		400		750		20	2320	3620			
SI	B32:40AG	320	400		9L		400		750		20	2320	3620			
SI	B35:40PG	350	400		12V		438		750		18.2	5250	7000			
SI	B35:40PG	350	400		16V		438		750		18.2	5250	7000			
SI	B35:40AG	350	400		12V		438		750		18.2	5040	7000			
SI	B35:40AG	350	400		16V		438		750		18.2	5040	7000			
D	B32: 40L	320	400		6L		480		720		22.4	2595	3885			
D	B32: 40L	320	400		8L		480		720		22.4	2595	3885			
D	B32: 40L	320	400		9L		480		720		22.4	2595	3885			
D	B32: 40L	320	400		6L		500		750		22.4	2700	4045			
D	B32: 40L	320	400		8L		500		750		22.4	2700	4045			
D	B32: 40L	320	400		9L		500		750		22.4	2700	4045			
D	B32: 40V	320	400		12V		500		720		22.4	5185	6910			
D	B32: 40V	320	400		16V		500		720		22.4	5185	6910			
D	B32: 40V	320	400		12V		500		750		22.4	5400	7200			
D	B32: 40V	320	400		16V		500		750		22.4	5400	7200			
D	C25: 33L	250	330		6L		240		720		22.3	1295	1945			
D	C25: 33L	250	330		8L		240		720		22.3	1295	1945			
D	C25: 33L	250	330		9L		240		720		22.3	1295	1945			
D	C25: 33L	250	330		6L		250		750		22.3	1350	2025			
D	C25: 33L	250	330		8L		250		750		22.3	1350	2025			
D	C25: 33L	250	330		9L		250		750		22.3	1350	2025			
D	C25: 33L	250	330		6L		290		900		21.5	1565	2090			
D	C25: 33L	250	330		8L		290		900		21.5	1565	2090			
D	C25: 33L	250	330		9L		290		900		21.5	1565	2090			
D	C25: 33L	250	330		6L		300		1000		20.8	1620	2435			
D	C25: 33L	250	330		8L		300		1000		20.8	1620	2435			
D	C25: 33L	250	330		9L		300		1000		20.8	1620	2435			

Fuel Type: D: Diesel or Heavy Fuel; DF: Dual Fuel; SI: Spark Ingnited. **Number of Cylinders & Configuration:** L: In-Line; V: V-Type; H: Horizontal; O: Opposed

▲

The Rolls-Royce diesel-engine programme. Category III.

Fuel Type	Engine Model	Bore (mm)	Stroke (mm)	Displacement (liters/cylinder)	Number of Cylinders & Configuration	Output per Cylinder Range (kW/cyl)		Rated Speed Range (r/min)		Maximum Brake Mean Effective Pressure (bar)		Output Range Low (kW)	Output Range High (kW)	Ratings	Application	Emmissions level
						min	max	min	max	min	max					
SI	B35:40V AG	350	400		12		420		720		18.2	5040	8400			
SI	B35:40V AG	350	400		16		420		720		18.2	5040	8400			
SI	B35:40V AG	350	400		20		420		720		18.2	5040	8400			
SI	B35:40V AG	350	400		12		438		750		18.2	5250	8750			
SI	B35:40V AG	350	400		16		438		750		18.2	5250	8750			
SI	B35:40V AG	350	400		20		438		750		18.2	5250	8750			
SI	B32:40L AG	320	400		6L		387		720		18	2090	3120			
SI	B32:40L AG	320	400		8L		387		720		18	2090	3120			
SI	B32:40L AG	320	400		9L		387		720		18	2090	3120			
SI	B32:40L AG	320	400		6L		401		750		18	2170	3620			
SI	B32:40L AG	320	400		8L		401		750		18	2170	3620			
SI	B32:40L AG	320	400		9L		401		750		18	2170	3620			
D	B32: 40A	320	400		6L		500		750		24.9	2880	4500			
D	B32: 40A	320	400		8L		500		750		24.9	2880	4500			
D	B32: 40A	320	400		9L		500		750		24.9	2880	4500			
D	B32: 40A	320	400		12V		500		750		24.9	5765	8000			
D	B32: 40A	320	400		16V		500		750		24.9	5765	8000			
D	B32: 40P	320	400		6L		500		750		24.9	3000	4500			
D	B32: 40P	320	400		8L		500		750		24.9	3000	4500			
D	B32: 40P	320	400		9L		500		750		24.9	3000	4500			
D	B32: 40P	320	400		12V		500		750		24.9	6000	8000			
D	B32: 40P	320	400		16V		500		750		24.9	6000	8000			
D	C25: 33A	250	330		6L		250		750		24.7	1440	2250			
D	C25: 33A	250	330		8L		250		750		24.7	1440	2250			
D	C25: 33A	250	330		9L		250		750		24.7	1440	2250			
D	C25: 33A	250	330		6L		330		1000		26.4	1920	3000			
D	C25: 33A	250	330		8L		330		1000		26.4	1920	3000			
D	C25: 33A	250	330		9L		330		1000		26.4	1920	3000			
D	C25:33P	250	330		6L		250		750		24.7	1440	2250			
D	C25:33P	250	330		8L		250		750		24.7	1440	2250			
D	C25:33P	250	330		9L		250		750		24.7	1440	2250			
D	C25: 33P	250	330		6L		330		1000		26.4	1920	3000			
D	C25: 33P	250	330		8L		330		1000		26.4	1920	3000			
D	C25: 33P	250	330		9L		330		1000		26.4	1920	3000			

Source: Diesel Publications

▲

The Rolls-Royce diesel-engine programme (continued). Category III.

▶

Placing two ABC medium-speed in-line six-cylinder engines in a large barge. Rear left and right (starboard and port) the two silencers. Behind the engine, just visible, are the two gearboxes to the propeller shaft. MDO fuel.

21.4.4 Category III: The large medium-speed four-stroke diesel engines of 500 to 30,000 kW shaft power running on HFO

Here the number of engine manufacturers is limited to a maximum of ten. Roughly fifty percent are active worldwide.

Caterpillar Inc. (MaK)

The American engine manufacturer Caterpillar also produces diesel engines in category III since their take-over of the German engine manufacturer MaK. An engine series with four cylinder-diameters can be delivered by this supplier of mainly medium-speed diesel engines running on HFO. The largest engine, the sixteen-cylinder M43 V-engine with 16,000 kW has a load parameter of 10.45 × 27.1 = 283.2 bar m/sec. This is reasonably high.

Daihatsu Diesel Co., Ltd.

The Japanese engine manufacturer Daihatsu builds engines in categories II and III. The largest diesel engine, the sixteen-cylinder V-engine DK-32 has a shaft power of 6325 kW and a moderate load parameter of 9.2 x 22.5 = 207 bar m/sec.

MAN Diesel AG

The world's largest engine manufacturer for category IV engines is unquestionably the German-Danish combination MAN B&W, since 2007 known as MAN Diesel AG; it is also one of the principal manufacturers for category III engines. Many category III engines are manufactured in Augsburg, Germany. Larger numbers are manufactured under license in Asia. The load parameter of a popular diesel engine for use in cruise ships, the V48/60B, a twelve-cylinder V-engine with a shaft power of 21,600 kW, is 10.3 × 24.8 = 255 bar m/sec. The L58/64 in-line engine, with the largest cylinder diameter, has a cylinder-power output of 1310 kW.

S.E.M.T. Pielstick

The French engine factory builds four-stroke trunk-piston engines in categories II and III. The slightly lower capacity of the MDO-type is more often used in naval ships and the largest engines running on HFO are used in diesel-power plants. In 2006 MAN Diesel AG acquired full ownership of this engine manufacturer.

Wärtsilä Corporation

The Finnish engine manufacturer Wärtsilä is the market leader in category III, the medium-speed four-stroke engines running on HFO as well as Dual Fuel engines and Spark Gas engines (Otto-engines with a large cylinder diameter). After taking over numerous European engine factories, a large number of engines with a maximum cylinder bore of 64 cm were built in Finland, Italy, France, Korea and China. The largest four-stroke diesel engine in the world has a cylinder power output of 2150 kW (!). A desired diesel engine, the latest 46 F, has a load parameter of 11.6 × 25.9 = 300.4 bar m/sec. This is a high value for a medium-speed four-stroke engine running on HFO.

ABC Belgian Corporation

The Belgian engine manufacturer in Ghent produces medium-speed in-line- and v-type diesel engines. Dual-fuel engines are also available.

Fuel Type	Engine Model	Bore (mm)	Stroke (mm)	Displacement (liters/cylinder)	Number of Cylinders & Configuration	Output per Cylinder Range (kW/cyl)		Rated Speed Range (r/min)		Maximum Brake Mean Effective Pressure (bar)		Output Range Low (kW)	Output Range High (kW)	Ratings	Application	Emmissions level
						min	max	min	max	min	max					
D	MaK M 20 C	200	300	9.5	6, 8, 9L	170	190	900	1000	24.1	24.2	1020	1710			
D	MaK M 25	255	400	20.5	6, 8, 9L	290	330	720	750	23.5	25.8	1800	2970			
D	MaK M 32 C	320	480	38.7	6, 8, 9L	480	500		600	24.9	25.9	2880	4500			
D	MaK M 32 C	320	420	33.8	12, 16V	480	500	720	750		23.7	5760	8000			
D	MaK M 43 C	430	610	88.5	6, 7, 8, 9L	900	1000	500	514	23.7	27.1	5400	9000			
D	MaK M 43 C	430	610	88.5	12, 16V	900	1000	500	514	23.7	27.1	10800	16000			

Source: Diesel Publications

▲

The Caterpillar Inc. (MaK) diesel-engine programme. Category III.

Fuel Type: D: Diesel or Heavy Fuel; DF: Dual Fuel; SI: Spark Ingnited. **Number of Cylinders & Configuration:** L: In-Line; V: V-Type; H: Horizontal; O: Opposed

Fuel Type	Engine Model	Bore (mm)	Stroke (mm)	Displacement (liters/cylinder)	Number of Cylinders & Configuration	Output per Cylinder Range (kW/cyl)		Rated Speed Range (r/min)		Maximum Brake Mean Effective Pressure (bar)		Output Range Low (kW)	Output Range High (kW)	Ratings	Application	Emmissions level
						min	max	min	max	min	max					
D	DC-17	170	270	6.1	5L		101.7		1000		22.1	480	610			
D	DC-17	170	270	6.1	6L		101.7		1000		22.1	480	610			
D	DC-32	320	400	32.2	6L		480		750		23.3	2905	3844			
D	DC-32	320	400	32.2	8L		480		750		23.3	2905	3844			
D	DK-20	200	300	9.4	3L		160		900		22.6	480	1360			
D	DK-20	200	300	9.4	5L		160		900		22.6	480	1360			
D	DK-20	200	300	9.4	6L		160		900		22.6	480	1360			
D	DK-20	200	300	9.4	8L		160		900		22.6	480	1360			
D	DK-26	260	380	20.2	5L		267		750		22	1280	1710			
D	DK-26	260	380	20.2	6L		267		750		22	1280	1710			
D	DK-28	280	390	24	6L		329		750		22.1	1900	5000			
D	DK-28	280	390	24	8L		329		750		22.1	1900	5000			
D	DK-28	280	390	24	16V		329		750		22.1	1900	5000			
D	DK-32B	320	360	29	12V		368		750		21.2		4413			
D	DK-32C	320	390	31.4	6L		395		750		20.3	2445	3162			
D	DK-32C	320	390	31.4	8L		395		750		20.3	2445	3162			
D	DK-36	360	480	48.9	6L		551		600		22.6	3310	4413			
D	DK-36	360	480	48.9	8L		551		600		22.6	3310	4413			
D	DK-36	360	460	46.8	12V		527		600		22.5		6325			
D	DL-16A	165	210	4.5	6L		73.5		1200		16.4	350	530			
D	M2	120	150	1.7	6L		27.5		1800		10.8	73	165			
D	M3	140	160	2.5	6L		44		1800		11.9	198	264			
D	M5	145	160	2.6	6L		51.3		1800		13	220	308			

Source: Diesel Publications

▲

The Daihatsu Diesel Co. Ltd diesel-engine programme. Category III.

▼

A feeder container ship, usually driven by a medium-speed four-stroke diesel engine running on heavy oil, category III.

Fuel Type	Engine Model	Bore (mm)	Stroke (mm)	Displacement (liters/cylinder)	Number of Cylinders & Configuration	Output per Cylinder Range (kW/cyl)		Rated Speed Range (r/min)		Maximum Brake Mean Effective Pressure (bar)		Output Range Low (kW)	Output Range High (kW)	Ratings	Application	Emmissions level
						min	max	min	max	min	max					
D	DX	242	320	14.7	6		110		750		12	276	662			
DF	DXD	242	320	14.7	6		100		750		12	600	600			
D	DZC	256	310	16	6		234		1000		18.4	810	3750			
D	DZC	256	310	16	8		234		1000		18.4	810	3750			
D	DZC	256	310	16	12		234		1000		18.4	810	3750			
D	DZC	256	310	16	16		234		1000		18.4	810	3750			
DF	DZD	256	310	15.5	6		166		1000		13.6	720	2670			
DF	DZD	256	310	15.5	8		166		1000		13.6	720	2670			
DF	DZD	256	310	15.5	12		166		1000		13.6	720	2670			
DF	DZD	256	310	15.5	16		166		1000		13.6	720	2670			

Source: Diesel Publications

▲

The ABC (Anglo Belgian Corporation) diesel-engine programme. Category III.

Fuel Type	Engine Model	Bore (mm)	Stroke (mm)	Displacement (liters/cylinder)	Number of Cylinders & Configuration	Output per Cylinder Range (kW/cyl)		Rated Speed Range (r/min)		Maximum Brake Mean Effective Pressure (bar)		Output Range Low (kW)	Output Range High (kW)	Ratings	Application	Emmissions level
						min	max	min	max	min	max					
D	L58/64	580	640	169	6-9L	1310	1400	400	428		23.2	7860	12600			
D	L+V48/60B	480	600	108.5	6-9L; 12-18V		1200	500	514	25.8	26.5	7200	21600			
D	L48/60	480	600	108.5	9L		1050	500	514	22.6	23.2		9450			
D	L40/54	400	540	67.8	6-9L	700	720	500	550	23.1	24.8	4200	6480			
D	V40/50	400	500	62.8	12-16, 18, 20V		750		600		23.9	9000	15000			
DF	L+V32/40DF	320	400	32.2	6-9L; 12-18V	385	400	720	750		19.9	2310	7200			
D	L+V 32/40	320	400	32.2	6-10L; 12-20V		500	720	750	24.9	25.9	3000	10000			
D	L+V28/32A	280	320	19.7	6-9L; 12, 16V		245		775		19.3	1470	3920			
D	L28/32H	280	320	19.7	5-9L;	210	220	720	750	17.8	17.9	1050	1980			
D	L27/38	270	380	21.7	5-9L		340		800		23.5	2040	3060			
D	L+V23/30A	225	300	11.9	6, 8L; 12V		160		900		17.9	960	1920			
D	L23/30A-E	225	300	11.9	6L		133		825		16.3		800			
D	L23/30H	225	300	11.9	5-8L	130	160	720	900		17.9	650	1280			
D	L21/31	210	310	10.7	5-9L		215	900	1000		24.1	1290	1935			
D	L16/24	160	240	4.8	5-9L	90	100	1000	1200	20.7	22.4	450	900			

Source: Diesel Publications

▲

The MAN Diesel AG diesel-engine programme. Category III.

Fuel Type: D: Diesel or Heavy Fuel; DF: Dual Fuel; SI: Spark Ingnited. **Number of Cylinders & Configuration:** L: In-Line; V: V-Type; H: Horizontal; O: Opposed

Fuel Type	Engine Model	Bore (mm)	Stroke (mm)	Displacement (liters/cylinder)	Number of Cylinders & Configuration	Output per Cylinder Range (kW/cyl)		Rated Speed Range (r/min)		Maximum Brake Mean Effective Pressure (bar)		Output Range Low (kW)	Output Range High (kW)	Ratings	Application	Emmissions level
						min	max	min	max	min	max					
D	PA5	255	270	13.8	4L	210	220	900	1000		19.2	840	3960			
D	PA5	255	270	13.8	18V	210	220	900	1000		19.2	840	3960			
DF	PA5	255	270	13.8	4L	132	147	900	1000		12.8	528	2646			
DF	PA5	255	270	13.8	18V	132	147	900	1000		12.8	528	2646			
D	PA6	280	290	17.9	6L	272	325	900	1050		21.8	1630	6500			
D	PA6	280	290	17.9	20V	272	325	900	1050		21.8	1630	6500			
D	PA6 B	280	330	20.3	6L	325	405	900	1050		22.8	1950	8100			
D	PA6 B	280	330	20.3	20V	325	405	900	1050		22.8	1950	8100			
D	PA6 B STC	280	330	20.3	12V		405		1050		22.8	4860	8100			
D	PA6 B STC	280	330	20.3	20V		405		1050		22.8	4860	8100			
D	PA6 CL	280	350	21.6	6L		295	720	750		21.9	1770	5880			
D	PA6 CL	280	350	21.6	20V		295	720	750		21.9	1770	5880			
D	PA6 STC	280	290	17.9	12V		324		1050		20.7	3880	5180			
D	PA6 STC	280	290	17.9	16V		324		1050		20.7	3880	5180			
D	PC2.6	400	460	57.8	6L		550	500	520		22	3300	9900			
D	PC2.6	400	460	57.8	18V		550	500	520		22	3300	9900			
D	PC2.6 B	400	500	62.8	12V		750		600		23.9	9000	15000			
D	PC2.6 B	400	500	62.8	20V		750		600		23.9	9000	15000			
D	PC4.2	570	620	158.2	10V		1215	400	429		21.5	12150	21870			
D	PC4.2	570	620	158.2	18V		1215	400	429		21.5	12150	21870			
D	PC4.2 B	570	660	168.4	10V	1300	1400	400	430		23.3	13000	28000			
D	PC4.2 B	570	660	168.4	20V	1300	1400	400	430		23.3	13000	28000			
D	PC40	570	750	191.4	5L		1325		375		22.2	6625	13250			
D	PC40	570	750	191.4	10L		1325		375		22.2	6625	13250			
D	PA4-185 VG	185	210	5.6	6L	98	123	1200	1500		17.4	590	2215			
D	PA4-185 VG	185	210	5.6	18V	98	123	1200	1500		17.4	590	2215			
DF	PA4-185 VG	185	210	5.6	6L	98	123	1200	1500		17.4	590	2215			
DF	PA4-185 VG	185	210	5.6	18V	98	123	1200	1500		17.4	590	2215			
SI	PA4-185 VG	185	210	5.6	6L	98	123	1200	1500		17.4	590	2215			
SI	PA4-185 VG	185	210	5.6	18V	98	123	1200	1500		17.4	590	2215			
D	PA4-200 VG	200	210	6.6	8V	132	166	1200	1500		20	1060	2650			
D	PA4-200 VG	200	210	6.6	16V	132	166	1200	1500		20	1060	2650			

Source: Diesel Publications

▲

The S.E.M.T. Pielstick diesel-engine programme.
A subsidiary of MAN Diesel AG. Category III.

Fuel Type	Engine Model	Bore (mm)	Stroke (mm)	Displacement (liters/cylinder)	Number of Cylinders & Configuration	Output per Cylinder Range (kW/cyl)		Rated Speed Range (r/min)		Maximum Brake Mean Effective Pressure (bar)		Output Range Low (kW)	Output Range High (kW)	Ratings	Application	Emmissions level
						min	max	min	max	min	max					
DF	Wärtsilä 34DF	340	400	36.3	16V		450		750		19.8	2700	7200			EPA Tier 2
DF	Wärtsilä 34DF	340	400	36.3	12V		450		750		19.8	2700	7200			EPA Tier 2
DF	Wärtsilä 34DF	340	400	36.3	9L		450		750		19.8	2700	7200			EPA Tier 2
DF	Wärtsilä 34DF	340	400	36.3	6L		450		750		19.8	2700	7200			EPA Tier 2
DF	Wärtsilä 20DF	200	280	8.8	9L		176		1200		20	1056	1584			EPA Tier 2
DF	Wärtsilä 20DF	200	280	8.8	8L		176		1200		20	1056	1584			EPA Tier 2
DF	Wärtsilä 20DF	200	280	8.8	6L		176		1200		20	1056	1584			EPA Tier 2
DF	Wärtsilä 50DF	500	580	113.9	18V		950		514		20	17100	17100			

▲

The Wärtsilä Corporation diesel-engine programme. Category III.

Fuel Type	Engine Model	Bore (mm)	Stroke (mm)	Displacement (liters/cylinder)	Number of Cylinders & Configuration	Output per Cylinder Range (kW/cyl)		Rated Speed Range (r/min)		Maximum Brake Mean Effective Pressure (bar)		Output Range Low (kW)	Output Range High (kW)	Ratings	Application	Emmissions level
						min	max	min	max	min	max					
DF	Wärtsilä 32GD	320	400	32.2	20V		460		750		20	2760	9200			
D	Wärtsilä 20	200	280	8.8	4L		200		1000		28	800	1800			
D	Wärtsilä 20	200	280	8.8	6L		200		1000		28	800	1800			
D	Wärtsilä 20	200	280	8.8	8L		200		1000		28	800	1800			
D	Wärtsilä 20	200	280	8.8	9L		200		1000		28	800	1800			
D	Wärtsilä 26	260	320	17	6L		340		1000		25.5	2040	3060			EPA Tier 2
D	Wärtsilä 26	260	320	17	8L		340		1000		25.5	2040	3060			EPA Tier 2
D	Wärtsilä 26	260	320	17	9L		340		1000		25.5	2040	3060			EPA Tier 2
D	Wärtsilä 26	260	320	17	12V		340		1000		25.5	4060	5440			EPA Tier 2
D	Wärtsilä 26	260	320	17	16V		340		1000		25.5	4060	5440			EPA Tier 2
D	Wärtsilä 32	320	400	32.2	6L		500		750		24.9	3000	4500			EPA Tier 2
D	Wärtsilä 32	320	400	32.2	7L		500		750		24.9	3000	4500			EPA Tier 2
D	Wärtsilä 32	320	400	32.2	8L		500		750		24.9	3000	4500			EPA Tier 2
D	Wärtsilä 32	320	400	32.2	9L		500		750		24.9	3000	4500			EPA Tier 2
D	Wärtsilä 32	320	400	32.2	12V		500		750		24.9	6000	9000			EPA Tier 2
D	Wärtsilä 32	320	400	32.2	16V		500		750		24.9	6000	9000			EPA Tier 2
D	Wärtsilä 32	320	400	32.2	18V		500		750		24.9	6000	9000			EPA Tier 2
DF	Wärtsilä 32GD	320	400	32.2	6L		460		750		20	2760	9200			
DF	Wärtsilä 32GD	320	400	32.2	9L		460		750		20	2760	9200			
DF	Wärtsilä 32GD	320	400	32.2	12V		460		750		20	2760	9200			
DF	Wärtsilä 32GD	320	400	32.2	16V		460		750		20	2760	9200			
DF	Wärtsilä 34DF	340	400	36.3	6L		450		750		19.8	2700	4050			EPA Tier 2
DF	Wärtsilä 34DF	340	400	36.3	9L		450		750		19.8	5400	9000			EPA Tier 2
DF	Wärtsilä 34DF	340	400	36.3	20V		450		750		19.8	5400	9000			EPA Tier 2
DF	Wärtsilä 34DF	340	400	36.3	16V		450		750		19.8	5400	9000			EPA Tier 2
SI	Wärtsilä 34SG	340	400	36.3	9L	450	500		750		19.8	4050	10000			
SI	Wärtsilä 34SG	340	400	36.3	16V	450	500		750		19.8	4050	10000			
SI	Wärtsilä 34SG	340	400	36.3	20V	450	500		750		19.8	4050	10000			
D	Wärtsilä 38	380	475	53.9	6L		725		600		26.9	4350	6525			EPA Tier 2
D	Wärtsilä 38	380	475	53.9	8L		725		600		26.9	4350	6525			EPA Tier 2
D	Wärtsilä 38	380	475	53.9	9L		725		600		26.9	4350	6525			EPA Tier 2
D	Wärtsilä 38	380	475	53.9	12V		725		600		26.9	8700	11600			EPA Tier 2
D	Wärtsilä 38	380	475	53.9	16V		725		600		26.9	8700	11600			EPA Tier 2
D	Wärtsilä 46	460	580	96.4	18L		975		514		24.3	11700	17550			EPA Tier 2
D	Wärtsilä 46	460	580	96.4	12V		1000		514		24.9	12000	16000			EPA Tier 2
D	Wärtsilä 46	460	580	96.4	16V		1000		514		24.9	12000	16000			EPA Tier 2
D	Wärtsilä 46F	460	580	96.4	6L		1200		600		24.9	7200	10800			EPA Tier 2
D	Wärtsilä 46F	460	580	96.4	7L		1200		600		24.9	7200	10800			EPA Tier 2
D	Wärtsilä 46F	460	580	96.4	8L		1200		600		24.9	7200	10800			EPA Tier 2
D	Wärtsilä 46F	460	580	96.4	9L		1200		600		24.9	7200	10800			EPA Tier 2
D	Wärtsilä 46F	460	580		12V		1200		600		24.9	14400	19200			EPA Tier 2
DF	Wärtsilä 50DF	500	580	113.9	6L		975		514		20	5850	16940			EPA Tier 2
DF	Wärtsilä 50DF	500	580	113.9	8L		975		514		20	5850	16940			EPA Tier 2
DF	Wärtsilä 50DF	500	580	113.9	9L		975		514		20	5850	16940			EPA Tier 2
DF	Wärtsilä 50DF	500	580	113.9	12V		975		514		20	5850	16940			EPA Tier 2
DF	Wärtsilä 50DF	500	580	113.9	16V		975		514		20	5850	16940			EPA Tier 2
DF	Wärtsilä 50DF	500	580	113.9	18V		975		514		20	5850	17100			EPA Tier 2
D	Wärtsilä 64	640	900	290	6L		2150		333.3		27.2	12900	17200			EPA Tier 2
D	Wärtsilä 64	640	900	290	7L		2150		333.3		27.2	12900	17200			EPA Tier 2
D	Wärtsilä 64	640	900	290	8L		2150		333.3		27.2	12900	17200			EPA Tier 2

Fuel Type: D: Diesel or Heavy Fuel; DF: Dual Fuel; SI: Spark Ingnited. **Number of Cylinders & Configuration:** L: In-Line; V: V-Type; H: Horizontal; O: Opposed

Source: Diesel Publications

▲
The Wärtsilä Corporation diesel-engine programme (continued). Category III.

21.4.5 Category IV: The large low-speed two-stroke crosshead engines of 1500 to 100,000 kW shaft power running on HFO

These are only manufactured by three engine manufacturers: MAN Diesel, Wärtsilä and Mitsubishi. The ratios are startling:

± 89% MAN Diesel;
± 8% Wärtsilä;
± 3% Mitsubishi.

Obviously, the percentages fluctuate somewhat around these averages.

MAN Diesel AG

The world's largest producer of two-stroke crosshead engines with a market share of ± 89%, they manufacture a large number of engine types with a cylinder diameter from 26 to 108 centimetres. Here the power output rises from 1080 to an incredible 97,300 kW, approximating the magical boundary of 100,000 kW per engine. These 'Cathedrals of the Oceans' weigh approximately 2800 tons and consume approximately 300 tons of HFO each day. The ME-series has electrically controlled fuel injection and the MC-series, a traditional fuel injection.

Mitsubishi Heavy Industries, Ltd.

The Japanese producer of two-stroke crosshead engines Mitsubishi builds engines with cylinder diameters of 33 to 85 centimetres. The worldwide share of 3% in the two-stroke crosshead-engine market indicates that many of these engines are bought by the Japanese market for oil tankers, bulk carriers and fast container ships. At present, they are manufacturing two-stroke crosshead engines together with Wärtsilä.

Wärtsilä Corporation

The two-stroke crosshead engine's programme of Wärtsilä Sulzer, RT-Flex, indicates the use of the common-rail fuel system while RTA indicates engines with the traditional injection systems. Wärtsilä manufactures the originally Sulzer two-stroke crosshead engines. Sulzer's original factory was situated in Winterthur, Switzerland. The large crosshead engines used to be transported by road and subsequently by ship to their final destination. Today, all engines are built by license holders throughout the world, often at the locations where the ships are built. Wärtsilä manufacture a large number of engines types with a cylinder diameter from 35 to 96 centimetres.

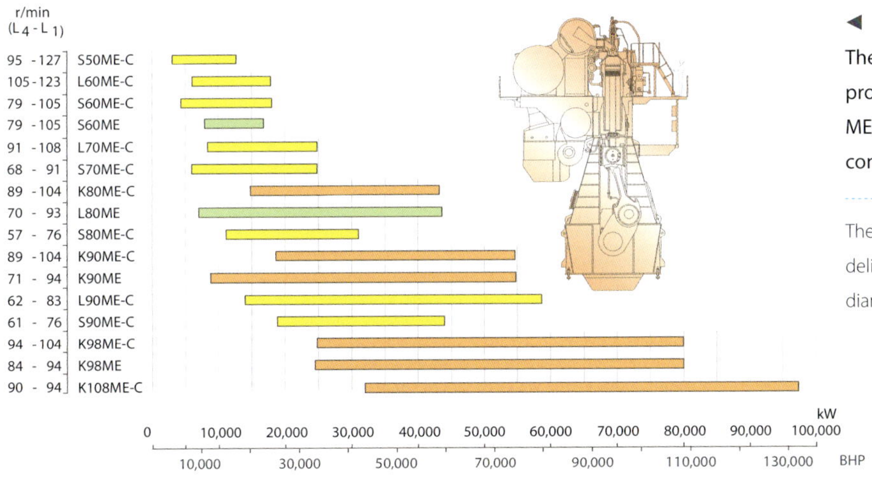

◄ The MAN Diesel engine programme for the ME-series, with electrically controlled fuel injection.

The ME-series can also be delivered in smaller cylinder diameters. See page 120.

◄ Large container ships, tankers and bulk carriers are usually propelled by the 'Cathedrals of the Oceans', large two-stroke crosshead engines, category IV.

Shown here the largest container ship in 2007, the 'Emma Maersk' departing from the shipyard at Fünen, Denmark.

Fuel Type	Engine Model	Bore (mm)	Stroke (mm)	Displacement (liters/cylinder)	Number of Cylinders & Configuration	Output per Cylinder Range (kW/cyl)		Rated Speed Range (r/min)		Maximum Brake Mean Effective Pressure (bar)		Output Range Low (kW)	Output Range High (kW)	Ratings	Application	Emmissions level
						min	max	min	max	min	max					
D	S35MC7	350	1400	135	12L	505	740	147	173	15.3	19.1	2525	8880			
D	S35ME-B9	350	1550	149	5L	595	870	142	167	16.8	21	2975	6960			
D	S35ME-B9	350	1550	149	6L	595	870	142	167	16.8	21	2975	6960			
D	S35ME-B9	350	1550	149	7L	595	870	142	167	16.8	21	2975	6960			
D	S35ME-B9	350	1550	149	8L	595	870	142	167	16.8	21	2975	6960			
D	L35MC6	350	1050	101	5L	440	650	178	210	14.7	18.4	2200	7800			
D	L35MC6	350	1050	101	6L	440	650	178	210	14.7	18.4	2200	7800			
D	L35MC6	350	1050	101	7L	440	650	178	210	14.7	18.4	2200	7800			
D	L35MC6	350	1050	101	8L	440	650	178	210	14.7	18.4	2200	7800			
D	L35MC6	350	1050	101	9L	440	650	178	210	14.7	18.4	2200	7800			
D	L35MC6	350	1050	101	10L	440	650	178	210	14.7	18.4	2200	7800			
D	L35MC6	350	1050	101	11L	440	650	178	210	14.7	18.4	2200	7800			
D	L35MC6	350	1050	101	12L	440	650	178	210	14.7	18.4	2200	7800			
D	S26MC6	260	980	52	5L	270	400	212	250	14.8	18.5	1350	4800			
D	S26MC6	260	980	52	6L	270	400	212	250	14.8	18.5	1350	4800			
D	S26MC6	260	980	52	7L	270	400	212	250	14.8	18.5	1350	4800			
D	S26MC6	260	980	52	8L	270	400	212	250	14.8	18.5	1350	4800			
D	S26MC6	260	980	52	9L	270	400	212	250	14.8	18.5	1350	4800			
D	S26MC6	260	980	52	10L	270	400	212	250	14.8	18.5	1350	4800			
D	S26MC6	260	980	52	11L	270	400	212	250	14.8	18.5	1350	4800			
D	S26MC6	260	980	52	12L	270	400	212	250	14.8	18.5	1350	4800			
D	S35MC7	350	1400	135	11L	505	740	147	173	15.3	19.1	2525	8880			
D	S35MC7	350	1400	135	5L	505	740	147	173	15.3	19.1	2525	8880			
D	S35MC7	350	1400	135	6L	505	740	147	173	15.3	19.1	2525	8880			
D	S35MC7	350	1400	135	7L	505	740	147	173	15.3	19.1	2525	8880			
D	S35MC7	350	1400	135	8L	505	740	147	173	15.3	19.1	2525	8880			
D	S35MC7	350	1400	135	9L	505	740	147	173	15.3	19.1	2525	8880			
D	S35MC7	350	1400	135	10L	505	740	147	173	15.3	19.1	2525	8880			
D	S50ME-B9	500	2214	435	7L	1210	1780	99	117	16.8	21	6050	16020			
D	S50ME-B9	500	2214	435	8L	1210	1780	99	117	16.8	21	6050	16020			
D	S50ME-B9	500	2214	435	9L	1210	1780	99	117	16.8	21	6050	16020			
D	S50ME-B8	500	2000	393	5L	1130	1660	108	127	16	20	5650	14940			
D	S50ME-B8	500	2000	393	6L	1130	1660	108	127	16	20	5650	14940			
D	S50ME-B8	500	2000	393	7L	1130	1660	108	127	16	20	5650	14940			
D	S50ME-B8	500	2000	393	8L	1130	1660	108	127	16	20	5650	14940			
D	S50ME-B8	500	2000	393	9L	1130	1660	108	127	16	20	5650	14940			
D	S46MC-C8	460	1932	321	5L	940	1380	110	129	16	20	4700	11040			
D	S46MC-C8	460	1932	321	6L	940	1380	110	129	16	20	4700	11040			
D	S46MC-C8	460	1932	321	7L	940	1380	110	129	16	20	4700	11040			
D	S46MC-C8	460	1932	321	8L	940	1380	110	129	16	20	4700	11040			
D	S46MC-C7	460	1932	321	5L	880	1310	108	129	15.2	19	4400	10480			
D	S46MC-C7	460	1932	321	6L	880	1310	108	129	15.2	19	4400	10480			
D	S46MC-C7	460	1932	321	7L	880	1310	108	129	15.2	19	4400	10480			
D	S46MC-C7	460	1932	321	8L	880	1310	108	129	15.2	19	4400	10480			
D	S46ME-B8	460	1932	321	5L	940	1380	110	129	16	20	4700	11040			
D	S46ME-B8	460	1932	321	6L	940	1380	110	129	16	20	4700	11040			
D	S46ME-B8	460	1932	321	7L	940	1380	110	129	16	20	4700	11040			
D	S46ME-B8	460	1932	321	8L	940	1380	110	129	16	20	4700	11040			

Fuel Type: D: Diesel or Heavy Fuel; DF: Dual Fuel; SI: Spark Ingnited. **Number of Cylinders & Configuration:** L: In-Line; V: V-Type; H: Horizontal; O: Opposed

▲

The MAN Diesel AG diesel-engine programme. Category IV.

Fuel Type	Engine Model	Bore (mm)	Stroke (mm)	Displacement (liters/cylinder)	Number of Cylinders & Configuration	Output per Cylinder Range (kW/cyl)		Rated Speed Range (r/min)		Maximum Brake Mean Effective Pressure (bar)		Output Range Low (kW)	Output Range High (kW)	Ratings	Application	Emmissions level
						min	max	min	max	min	max					
D	S42MC7	420	1764	244	5L	730	1080	115	136	15.6	19.5	3650	12960			
D	S42MC7	420	1764	244	6L	730	1080	115	136	15.6	19.5	3650	12960			
D	S42MC7	420	1764	244	7L	730	1080	115	136	15.6	19.5	3650	12960			
D	S42MC7	420	1764	244	8L	730	1080	115	136	15.6	19.5	3650	12960			
D	S42MC7	420	1764	244	9L	730	1080	115	136	15.6	19.5	3650	12960			
D	S42MC7	420	1764	244	10L	730	1080	115	136	15.6	19.5	3650	12960			
D	S42MC7	420	1764	244	11L	730	1080	115	136	15.6	19.5	3650	12960			
D	S42MC7	420	1764	244	12L	730	1080	115	136	15.6	19.5	3650	12960			
D	S40ME-B9	400	1770	222	5L	770	1135	124	146	16.8	21	3850	9080			
D	S40ME-B9	400	1770	222	6L	770	1135	124	146	16.8	21	3850	9080			
D	S40ME-B9	400	1770	222	7L	770	1135	124	146	16.8	21	3850	9080			
D	S40ME-B9	400	1770	222	8L	770	1135	124	146	16.8	21	3850	9080			
D	S50ME-C8	500	2000	393	8L	1130	1660	108	127	16	20	5650	14940			
D	S50ME-C8	500	2000	393	9L	1130	1660	108	127	16	20	5650	14940			
D	S50MC-C8	500	2000	393	5L	1130	1660	108	127	16	20	5650	14940			
D	S50MC-C8	500	2000	393	6L	1130	1660	108	127	16	20	5650	14940			
D	S50MC-C8	500	2000	393	7L	1130	1660	108	127	16	20	5650	14940			
D	S50MC-C8	500	2000	393	8L	1130	1660	108	127	16	20	5650	14940			
D	S50MC-C8	500	2000	393	9L	1130	1660	108	127	16	20	5650	14940			
D	S50ME-C7	500	2000	393	5L	950	1580	95	127	15.2	19	4750	14220			
D	S50ME-C7	500	2000	393	6L	950	1580	95	127	15.2	19	4750	14220			
D	S50ME-C7	500	2000	393	7L	950	1580	95	127	15.2	19	4750	14220			
D	S50ME-C7	500	2000	393	8L	950	1580	95	127	15.2	19	4750	14220			
D	S50ME-C7	500	2000	393	9L	950	1580	95	127	15.2	19	4750	14220			
D	S50MC-C7	500	2000	393	5L	950	1580	95	127	15.2	19	4750	14220			
D	S50MC-C7	500	2000	393	6L	950	1580	95	127	15.2	19	4750	14220			
D	S50MC-C7	500	2000	393	7L	950	1580	95	127	15.2	19	4750	14220			
D	S50MC-C7	500	2000	393	8L	950	1580	95	127	15.2	19	4750	14220			
D	S50MC-C7	500	2000	393	9L	950	1580	95	127	15.2	19	4750	14220			
D	S50MC6	500	1910	375	5L	860	1430	95	127	14.4	18	4300	11440			
D	S50MC6	500	1910	375	6L	860	1430	95	127	14.4	18	4300	11440			
D	S50MC6	500	1910	375	7L	860	1430	95	127	14.4	18	4300	11440			
D	S50MC6	500	1910	375	8L	860	1430	95	127	14.4	18	4300	11440			
D	S50ME-B9	500	2214	435	5L	1210	1780	99	117	16.8	21	6050	16020			
D	S50ME-B9	500	2214	435	6L	1210	1780	99	117	16.8	21	6050	16020			
D	S60MC6	600	2292	648	7L	1230	2040	79	105	14.4	18	6150	16320			
D	S60MC6	600	2292	648	8L	1230	2040	79	105	14.4	18	6150	16320			
D	L60ME-C8	600	2022	572	5L	1600	2340	105	123	16	20	8000	21060			
D	L60ME-C8	600	2022	572	6L	1600	2340	105	123	16	20	8000	21060			
D	L60ME-C8	600	2022	572	7L	1600	2340	105	123	16	20	8000	21060			
D	L60ME-C8	600	2022	572	8L	1600	2340	105	123	16	20	8000	21060			
D	L60ME-C8	600	2022	572	9L	1600	2340	105	123	16	20	8000	21060			
D	L60MC-C8	600	2022	572	5L	1600	2340	105	123	16	20	8000	21060			
D	L60MC-C8	600	2022	572	6L	1600	2340	105	123	16	20	8000	21060			
D	L60MC-C8	600	2022	572	7L	1600	2340	105	123	16	20	8000	21060			
D	L60MC-C8	600	2022	572	8L	1600	2340	105	123	16	20	8000	21060			
D	L60MC-C8	600	2022	572	9L	1600	2340	105	123	16	20	8000	21060			
D	L60ME-C7	600	2022	572	5L	1520	2230	105	123	15.2	19	7600	20070			

▲

The MAN Diesel AG diesel-engine programme (continued). Category IV.

Fuel Type	Engine Model	Bore (mm)	Stroke (mm)	Displacement (liters/cylinder)	Number of Cylinders & Configuration	Output per Cylinder Range (kW/cyl)		Rated Speed Range (r/min)		Maximum Brake Mean Effective Pressure (bar)		Output Range Low (kW)	Output Range High (kW)	Ratings	Application	Emmissions level
						min	max	min	max	min	max					
D	L60ME-C7	600	2022	572	6L	1520	2230	105	123	15.2	19	7600	20070			
D	L60ME-C7	600	2022	572	7L	1520	2230	105	123	15.2	19	7600	20070			
D	L60ME-C7	600	2022	572	8L	1520	2230	105	123	15.2	19	7600	20070			
D	L60ME-C7	600	2022	572	9L	1520	2230	105	123	15.2	19	7600	20070			
D	L60MC-C7	600	2022	572	5L	1520	2230	105	123	15.2	19	7600	20070			
D	L60MC-C7	600	2022	572	6L	1520	2230	105	123	15.2	19	7600	20070			
D	L60MC-C7	600	2022	572	7L	1520	2230	105	123	15.2	19	7600	20070			
D	L60MC-C7	600	2022	572	8L	1520	2230	105	123	15.2	19	7600	20070			
D	L60MC-C7	600	2022	572	9L	1520	2230	105	123	15.2	19	7600	20070			
D	S50ME-C8	500	2000	393	5L	1130	1660	108	127	16	20	5650	14940			
D	S50ME-C8	500	2000	393	6L	1130	1660	108	127	16	20	5650	14940			
D	S50ME-C8	500	2000	393	7L	1130	1660	108	127	16	20	5650	14940			
DF	S65ME-GI8	650	2730	906	6L	1960	2870	81	95	16	20	9800	22960			
DF	S65ME-GI8	650	2730	906	7L	1960	2870	81	95	16	20	9800	22960			
DF	S65ME-GI8	650	2730	906	8L	1960	2870	81	95	16	20	9800	22960			
D	S60ME-C8	600	2400	679	5L	1610	2380	89	105	16	20	8050	19040			
D	S60ME-C8	600	2400	679	6L	1610	2380	89	105	16	20	8050	19040			
D	S60ME-C8	600	2400	679	7L	1610	2380	89	105	16	20	8050	19040			
D	S60ME-C8	600	2400	679	8L	1610	2380	89	105	16	20	8050	19040			
DF	S60ME-GI8	600	2400	679	5L	1610	2380	89	105	16	20	8050	19040			
DF	S60ME-GI8	600	2400	679	6L	1610	2380	89	105	16	20	8050	19040			
DF	S60ME-GI8	600	2400	679	7L	1610	2380	89	105	16	20	8050	19040			
DF	S60ME-GI8	600	2400	679	8L	1610	2380	89	105	16	20	8050	19040			
D	S60ME-B8	600	2400	679	5L	1610	2380	89	105	16	20	8050	19040			
D	S60ME-B8	600	2400	679	6L	1610	2380	89	105	16	20	8050	19040			
D	S60ME-B8	600	2400	679	7L	1610	2380	89	105	16	20	8050	19040			
D	S60ME-B8	600	2400	679	8L	1610	2380	89	105	16	20	8050	19040			
D	S60ME-C7	600	2400	679	5L	1360	2260	79	105	15.2	19	6800	18080			
D	S60ME-C7	600	2400	679	6L	1360	2260	79	105	15.2	19	6800	18080			
D	S60ME-C7	600	2400	679	7L	1360	2260	79	105	15.2	19	6800	18080			
D	S60ME-C7	600	2400	679	8L	1360	2260	79	105	15.2	19	6800	18080			
D	S60MC-C8	600	2400	679	5L	1610	2380	89	105	16	20	8050	19040			
D	S60MC-C8	600	2400	679	6L	1610	2380	89	105	16	20	8050	19040			
D	S60MC-C8	600	2400	679	7L	1610	2380	89	105	16	20	8050	19040			
D	S60MC-C8	600	2400	679	8L	1610	2380	89	105	16	20	8050	19040			
D	S60MC-C7	600	2400	679	5L	1360	2260	79	105	15.2	19	6800	18080			
D	S60MC-C7	600	2400	679	6L	1360	2260	79	105	15.2	19	6800	18080			
D	S60MC-C7	600	2400	679	7L	1360	2260	79	105	15.2	19	6800	18080			
D	S60MC-C7	600	2400	679	8L	1360	2260	79	105	15.2	19	6800	18080			
D	S60MC6	600	2292	648	5L	1230	2040	79	105	14.4	18	6150	16320			
D	S60MC6	600	2292	648	6L	1230	2040	79	105	14.4	18	6150	16320			
D	L70ME-C8	700	2360	908	7L	2200	3270	91	108	16	20	11000	26160			
D	L70ME-C8	700	2360	908	8L	2200	3270	91	108	16	20	11000	26160			
D	L70MC-C8	700	2360	908	5L	2200	3270	91	108	16	20	11000	26160			
D	L70MC-C8	700	2360	908	6L	2200	3270	91	108	16	20	11000	26160			
D	L70MC-C8	700	2360	908	7L	2200	3270	91	108	16	20	11000	26160			
D	L70MC-C8	700	2360	908	8L	2200	3270	91	108	16	20	11000	26160			
D	L70ME-C7	700	2360	908	5L	2090	3110	91	108	15.2	19	10450	24880			

Fuel Type: D: Diesel or Heavy Fuel; DF: Dual Fuel; SI: Spark Ingnited. **Number of Cylinders & Configuration:** L: In-Line; V: V-Type; H: Horizontal; O: Opposed

▲

The MAN Diesel AG diesel-engine programme (continued). Category IV.

Fuel Type	Engine Model	Bore (mm)	Stroke (mm)	Displacement (liters/cylinder)	Number of Cylinders & Configuration	Output per Cylinder Range (kW/cyl)		Rated Speed Range (r/min)		Maximum Brake Mean Effective Pressure (bar)		Output Range Low (kW)	Output Range High (kW)	Ratings	Application	Emmissions level
						min	max	min	max	min	max					
D	L70ME-C7	700	2360	908	6L	2090	3110	91	108	15.2	19	10450	24880			
D	L70ME-C7	700	2360	908	7L	2090	3110	91	108	15.2	19	10450	24880			
D	L70ME-C7	700	2360	908	8L	2090	3110	91	108	15.2	19	10450	24880			
D	L70MC-C7	700	2360	908	5L	2090	3110	91	108	15.2	19	10450	24880			
D	L70MC-C7	700	2360	908	6L	2090	3110	91	108	15.2	19	10450	24880			
D	L70MC-C7	700	2360	908	7L	2090	3110	91	108	15.2	19	10450	24880			
D	L70MC-C7	700	2360	908	8L	2090	3110	91	108	15.2	19	10450	24880			
D	S65ME-C8	650	2730	906	5L	1960	2870	81	95	16	20	9800	22960			
D	S65ME-C8	650	2730	906	6L	1960	2870	81	95	16	20	9800	22960			
D	S65ME-C8	650	2730	906	7L	1960	2870	81	95	16	20	9800	22960			
D	S65ME-C8	650	2730	906	8L	1960	2870	81	95	16	20	9800	22960			
DF	S65ME-GI8	650	2730	906	5L	1960	2870	81	95	16	20	9800	22960			
D	K80MC-C6	800	2300	1156	6L	2470	3610	89	104	14.4	18	14820	43320			
D	K80MC-C6	800	2300	1156	7L	2470	3610	89	104	14.4	18	14820	43320			
D	K80MC-C6	800	2300	1156	8L	2470	3610	89	104	14.4	18	14820	43320			
D	K80MC-C6	800	2300	1156	9L	2470	3610	89	104	14.4	18	14820	43320			
D	K80MC-C6	800	2300	1156	10L	2470	3610	89	104	14.4	18	14820	43320			
D	K80MC-C6	800	2300	1156	11L	2470	3610	89	104	14.4	18	14820	43320			
D	K80MC-C6	800	2300	1156	12L	2470	3610	89	104	14.4	18	14820	43320			
D	S70ME-C8	700	2800	1078	5L	2210	3270	77	91	16	20	11050	26160			
D	S70ME-C8	700	2800	1078	6L	2210	3270	77	91	16	20	11050	26160			
D	S70ME-C8	700	2800	1078	7L	2210	3270	77	91	16	20	11050	26160			
D	S70ME-C8	700	2800	1078	8L	2210	3270	77	91	16	20	11050	26160			
DF	S70ME-GI8	700	2800	1078	5L	2210	3270	77	91	16	20	11050	26160			
DF	S70ME-GI8	700	2800	1078	6L	2210	3270	77	91	16	20	11050	26160			
DF	S70ME-GI8	700	2800	1078	7L	2210	3270	77	91	16	20	11050	26160			
DF	S70ME-GI8	700	2800	1078	8L	2210	3270	77	91	16	20	11050	26160			
D	S70ME-C7	700	2800	1078	5L	1860	3110	68	91	15.2	19	9300	24880			
D	S70ME-C7	700	2800	1078	6L	1860	3110	68	91	15.2	19	9300	24880			
D	S70ME-C7	700	2800	1078	7L	1860	3110	68	91	15.2	19	9300	24880			
D	S70ME-C7	700	2800	1078	8L	1860	3110	68	91	15.2	19	9300	24880			
D	S70MC-C8	700	2800	1078	5L	2210	3270	77	91	16	20	11050	26160			
D	S70MC-C8	700	2800	1078	6L	2210	3270	77	91	16	20	11050	26160			
D	S70MC-C8	700	2800	1078	7L	2210	3270	77	91	16	20	11050	26160			
D	S70MC-C8	700	2800	1078	8L	2210	3270	77	91	16	20	11050	26160			
D	S70MC-C7	700	2800	1078	5L	1860	3110	68	91	15.2	19	9300	24880			
D	S70MC-C7	700	2800	1078	6L	1860	3110	68	91	15.2	19	9300	24880			
D	S70MC-C7	700	2800	1078	7L	1860	3110	68	91	15.2	19	9300	24880			
D	S70MC-C7	700	2800	1078	8L	1860	3110	68	91	15.2	19	9300	24880			
D	S70MC6	700	2674	1029	5L	1680	2810	68	91	14.4	18	8400	22480			
D	S70MC6	700	2674	1029	6L	1680	2810	68	91	14.4	18	8400	22480			
D	S70MC6	700	2674	1029	7L	1680	2810	68	91	14.4	18	8400	22480			
D	S70MC6	700	2674	1029	8L	1680	2810	68	91	14.4	18	8400	22480			
D	L70ME-C8	700	2360	908	5L	2200	3270	91	108	16	20	11000	26160			
D	L70ME-C8	700	2360	908	6L	2200	3270	91	108	16	20	11000	26160			
D	S80MC-C8	800	3200	1609	8L	2830	4180	66	78	16	20	16980	33440			
D	S80ME-C7	800	3200	1609	6L	2320	3880	57	76	15.2	19	13920	31040			
D	S80ME-C7	800	3200	1609	7L	2320	3880	57	76	15.2	19	13920	31040			

▲
The MAN Diesel AG diesel-engine programme (continued). Category IV.

Fuel Type	Engine Model	Bore (mm)	Stroke (mm)	Displacement (liters/cylinder)	Number of Cylinders & Configuration	Output per Cylinder Range (kW/cyl)		Rated Speed Range (r/min)		Maximum Brake Mean Effective Pressure (bar)		Output Range Low (kW)	Output Range High (kW)	Ratings	Application	Emmissions level
						min	max	min	max	min	max					
D	S80ME-C7	800	3200	1609	8L	2320	3880	57	76	15.2	19	13920	31040			
D	S80MC-C7	800	3200	1609	6L	2320	3880	57	76	15.2	19	13920	31040			
D	S80MC-C7	800	3200	1609	7L	2320	3880	57	76	15.2	19	13920	31040			
D	S80MC-C7	800	3200	1609	8L	2320	3880	57	76	15.2	19	13920	31040			
D	S80MC6	800	3056	1536	5L	2180	3640	59	79	14.4	18	10900	29120			
D	S80MC6	800	3056	1536	6L	2180	3640	59	79	14.4	18	10900	29120			
D	S80MC6	800	3056	1536	7L	2180	3640	59	79	14.4	18	10900	29120			
D	S80MC6	800	3056	1536	8L	2180	3640	59	79	14.4	18	10900	29120			
D	K80ME-C9	800	2600	1307	6L	3280	4530	94	104	16	20	19680	54360			
D	K80ME-C9	800	2600	1307	7L	3280	4530	94	104	16	20	19680	54360			
D	K80ME-C9	800	2600	1307	8L	3280	4530	94	104	16	20	19680	54360			
D	K80ME-C9	800	2600	1307	9L	3280	4530	94	104	16	20	19680	54360			
D	K80ME-C9	800	2600	1307	10L	3280	4530	94	104	16	20	19680	54360			
D	K80ME-C9	800	2600	1307	11L	3280	4530	94	104	16	20	19680	54360			
D	K80ME-C9	800	2600	1307	12L	3280	4530	94	104	16	20	19680	54360			
D	K80MC-C9	800	2600	1307	6L	3280	4530	94	104	16	20	19680	54360			
D	K80MC-C9	800	2600	1307	7L	3280	4530	94	104	16	20	19680	54360			
D	K80MC-C9	800	2600	1307	8L	3280	4530	94	104	16	20	19680	54360			
D	K80MC-C9	800	2600	1307	9L	3280	4530	94	104	16	20	19680	54360			
D	K80MC-C9	800	2600	1307	10L	3280	4530	94	104	16	20	19680	54360			
D	K80MC-C9	800	2600	1307	11L	3280	4530	94	104	16	20	19680	54360			
D	K80MC-C9	800	2600	1307	12L	3280	4530	94	104	16	20	19680	54360			
D	K80ME-C6	800	2300	1156	6L	2470	3610	89	104	14.4	18	14820	43320			
D	K80ME-C6	800	2300	1156	7L	2470	3610	89	104	14.4	18	14820	43320			
D	K80ME-C6	800	2300	1156	8L	2470	3610	89	104	14.4	18	14820	43320			
D	K80ME-C6	800	2300	1156	9L	2470	3610	89	104	14.4	18	14820	43320			
D	K80ME-C6	800	2300	1156	10L	2470	3610	89	104	14.4	18	14820	43320			
D	K80ME-C6	800	2300	1156	11L	2470	3610	89	104	14.4	18	14820	43320			
D	K80ME-C6	800	2300	1156	12L	2470	3610	89	104	14.4	18	14820	43320			
D	K90ME-C9	900	2600	1654	9L	4150	5730	94	104	16	20	24900	68760			
D	K90ME-C9	900	2600	1654	10L	4150	5730	94	104	16	20	24900	68760			
D	K90ME-C9	900	2600	1654	11L	4150	5730	94	104	16	20	24900	68760			
D	K90ME-C9	900	2600	1654	12L	4150	5730	94	104	16	20	24900	68760			
D	K90MC-C9	900	2600	1654	6L	4150	5730	94	104	16	20	24900	68760			
D	K90MC-C9	900	2600	1654	7L	4150	5730	94	104	16	20	24900	68760			
D	K90MC-C9	900	2600	1654	8L	4150	5730	94	104	16	20	24900	68760			
D	K90MC-C9	900	2600	1654	9L	4150	5730	94	104	16	20	24900	68760			
D	K90MC-C9	900	2600	1654	10L	4150	5730	94	104	16	20	24900	68760			
D	K90MC-C9	900	2600	1654	11L	4150	5730	94	104	16	20	24900	68760			
D	K90MC-C9	900	2600	1654	12L	4150	5730	94	104	16	20	24900	68760			
D	S80ME-C9	800	3450	1734	6L	3050	4510	66	78	16	20	18300	40590			
D	S80ME-C9	800	3450	1734	7L	3050	4510	66	78	16	20	18300	40590			
D	S80ME-C9	800	3450	1734	8L	3050	4510	66	78	16	20	18300	40590			
D	S80ME-C9	800	3450	1734	9L	3050	4510	66	78	16	20	18300	40590			
D	S80MC-C9	800	3450	1734	6L	3050	4510	66	78	16	20	18300	40590			
D	S80MC-C9	800	3450	1734	7L	3050	4510	66	78	16	20	18300	40590			
D	S80MC-C9	800	3450	1734	8L	3050	4510	66	78	16	20	18300	40590			
D	S80MC-C9	800	3450	1734	9L	3050	4510	66	78	16	20	18300	40590			

▲

The MAN Diesel AG diesel-engine programme (continued). Category IV.

Fuel Type: D: Diesel or Heavy Fuel; DF: Dual Fuel; SI: Spark Ingnited. **Number of Cylinders & Configuration:** L: In-Line; V: V-Type; H: Horizontal; O: Opposed

Fuel Type	Engine Model	Bore (mm)	Stroke (mm)	Displacement (liters/cylinder)	Number of Cylinders & Configuration	Output per Cylinder Range (kW/cyl)		Rated Speed Range (r/min)		Maximum Brake Mean Effective Pressure (bar)		Output Range Low (kW)	Output Range High (kW)	Ratings	Application	Emmissions level
						min	max	min	max	min	max					
D	S80ME-C8	800	3200	1609	6L	2830	4180	66	78	16	20	16980	33440			
D	S80ME-C8	800	3200	1609	7L	2830	4180	66	78	16	20	16980	33440			
D	S80ME-C8	800	3200	1609	8L	2830	4180	66	78	16	20	16980	33440			
D	S80MC-C8	800	3200	1609	6L	2830	4180	66	78	16	20	16980	33440			
D	S80MC-C8	800	3200	1609	7L	2830	4180	66	78	16	20	16980	33440			
D	S90ME-C7	900	3188	2028	6L	3140	4890	61	76	15.2	19	18840	44010			
D	S90ME-C7	900	3188	2028	7L	3140	4890	61	76	15.2	19	18840	44010			
D	S90ME-C7	900	3188	2028	8L	3140	4890	61	76	15.2	19	18840	44010			
D	S90ME-C7	900	3188	2028	9L	3140	4890	61	76	15.2	19	18840	44010			
D	S90MC-C7	900	3188	2028	6L	3140	4890	61	76	15.2	19	18840	44010			
D	S90MC-C7	900	3188	2028	7L	3140	4890	61	76	15.2	19	18840	44010			
D	S90MC-C7	900	3188	2028	8L	3140	4890	61	76	15.2	19	18840	44010			
D	S90MC-C7	900	3188	2028	9L	3140	4890	61	76	15.2	19	18840	44010			
D	K90ME9	900	2870	1826	6L	4090	5720	84	94	16	20	24540	68640			
D	K90ME9	900	2870	1826	7L	4090	5720	84	94	16	20	24540	68640			
D	K90ME9	900	2870	1826	8L	4090	5720	84	94	16	20	24540	68640			
D	K90ME9	900	2870	1826	9L	4090	5720	84	94	16	20	24540	68640			
D	K90ME9	900	2870	1826	10L	4090	5720	84	94	16	20	24540	68640			
D	K90ME9	900	2870	1826	11L	4090	5720	84	94	16	20	24540	68640			
D	K90ME9	900	2870	1826	12L	4090	5720	84	94	16	20	24540	68640			
D	K90MC9	900	2870	1826	6L	4090	5720	84	94	16	20	24540	68640			
D	K90MC9	900	2870	1826	7L	4090	5720	84	94	16	20	24540	68640			
D	K90MC9	900	2870	1826	8L	4090	5720	84	94	16	20	24540	68640			
D	K90MC9	900	2870	1826	9L	4090	5720	84	94	16	20	24540	68640			
D	K90MC9	900	2870	1826	10L	4090	5720	84	94	16	20	24540	68640			
D	K90MC9	900	2870	1826	11L	4090	5720	84	94	16	20	24540	68640			
D	K90MC9	900	2870	1826	12L	4090	5720	84	94	16	20	24540	68640			
D	K90MC-C6	900	2300	1463	6L	3130	4570	89	104	14.4	18	18780	54840			
D	K90MC-C6	900	2300	1463	7L	3130	4570	89	104	14.4	18	18780	54840			
D	K90MC-C6	900	2300	1463	8L	3130	4570	89	104	14.4	18	18780	54840			
D	K90MC-C6	900	2300	1463	9L	3130	4570	89	104	14.4	18	18780	54840			
D	K90MC-C6	900	2300	1463	10L	3130	4570	89	104	14.4	18	18780	54840			
D	K90MC-C6	900	2300	1463	11L	3130	4570	89	104	14.4	18	18780	54840			
D	K90MC-C6	900	2300	1463	12L	3130	4570	89	104	14.4	18	18780	54840			
D	K90ME-C9	900	2600	1654	6L	4150	5730	94	104	16	20	24900	68760			
D	K90ME-C9	900	2600	1654	7L	4150	5730	94	104	16	20	24900	68760			
D	K90ME-C9	900	2600	1654	8L	4150	5730	94	104	16	20	24900	68760			
D	K98MC7	980	2660	2007	10L	4630	6230	90	97	15.4	19.2	27780	87220			
D	K98MC7	980	2660	2007	11L	4630	6230	90	97	15.4	19.2	27780	87220			
D	K98MC7	980	2660	2007	12L	4630	6230	90	97	15.4	19.2	27780	87220			
D	K98MC7	980	2660	2007	13L	4630	6230	90	97	15.4	19.2	27780	87220			
D	K98MC7	980	2660	2007	14L	4630	6230	90	97	15.4	19.2	27780	87220			
D	K98MC6	980	2660	2007	6L	4100	5720	84	94	14.6	18.2	24600	80080			
D	K98MC6	980	2660	2007	7L	4100	5720	84	94	14.6	18.2	24600	80080			
D	K98MC6	980	2660	2007	8L	4100	5720	84	94	14.6	18.2	24600	80080			
D	K98MC6	980	2660	2007	9L	4100	5720	84	94	14.6	18.2	24600	80080			
D	K98MC6	980	2660	2007	10L	4100	5720	84	94	14.6	18.2	24600	80080			
D	K98MC6	980	2660	2007	11L	4100	5720	84	94	14.6	18.2	24600	80080			

▲

The MAN Diesel AG diesel-engine programme (continued). Category IV.

Fuel Type	Engine Model	Bore (mm)	Stroke (mm)	Displacement (liters/cylinder)	Number of Cylinders & Configuration	Output per Cylinder Range (kW/cyl)		Rated Speed Range (r/min)		Maximum Brake Mean Effective Pressure (bar)		Output Range Low (kW)	Output Range High (kW)	Ratings	Application	Emmissions level
						min	max	min	max	min	max					
D	K98MC6	980	2660	2007	12L	4100	5720	84	94	14.6	18.2	24600	80080			
D	K98MC6	980	2660	2007	13L	4100	5720	84	94	14.6	18.2	24600	80080			
D	K98MC6	980	2660	2007	14L	4100	5720	84	94	14.6	18.2	24600	80080			
D	S90ME-C8	900	3188	2028	6L	3570	5270	66	78	16	20	21420	47430			
D	S90ME-C8	900	3188	2028	7L	3570	5270	66	78	16	20	21420	47430			
D	S90ME-C8	900	3188	2028	8L	3570	5270	66	78	16	20	21420	47430			
D	S90ME-C8	900	3188	2028	9L	3570	5270	66	78	16	20	21420	47430			
D	S90MC-C8	900	3188	2028	6L	3570	5270	66	78	16	20	21420	47430			
D	S90MC-C8	900	3188	2028	7L	3570	5270	66	78	16	20	21420	47430			
D	S90MC-C8	900	3188	2028	8L	3570	5270	66	78	16	20	21420	47430			
D	S90MC-C8	900	3188	2028	9L	3570	5270	66	78	16	20	21420	47430			
D	K98MC-C7	980	2400	1811	12L	4510	6020	97	104	15.4	19.2	27060	84280			
D	K98MC-C7	980	2400	1811	13L	4510	6020	97	104	15.4	19.2	27060	84280			
D	K98MC-C7	980	2400	1811	14L	4510	6020	97	104	15.4	19.2	27060	84280			
D	K98ME-C6	980	2400	1811	6L	4140	5710	94	104	14.6	18.2	24840	79940			
D	K98ME-C6	980	2400	1811	7L	4140	5710	94	104	14.6	18.2	24840	79940			
D	K98ME-C6	980	2400	1811	8L	4140	5710	94	104	14.6	18.2	24840	79940			
D	K98ME-C6	980	2400	1811	9L	4140	5710	94	104	14.6	18.2	24840	79940			
D	K98ME-C6	980	2400	1811	10L	4140	5710	94	104	14.6	18.2	24840	79940			
D	K98ME-C6	980	2400	1811	11L	4140	5710	94	104	14.6	18.2	24840	79940			
D	K98ME-C6	980	2400	1811	12L	4140	5710	94	104	14.6	18.2	24840	79940			
D	K98ME-C6	980	2400	1811	13L	4140	5710	94	104	14.6	18.2	24840	79940			
D	K98ME-C6	980	2400	1811	14L	4140	5710	94	104	14.6	18.2	24840	79940			
D	K98MC-C6	980	2400	1811	6L	4140	5710	94	104	14.6	18.2	24840	79940			
D	K98MC-C6	980	2400	1811	7L	4140	5710	94	104	14.6	18.2	24840	79940			
D	K98MC-C6	980	2400	1811	8L	4140	5710	94	104	14.6	18.2	24840	79940			
D	K98MC-C6	980	2400	1811	9L	4140	5710	94	104	14.6	18.2	24840	79940			
D	K98MC-C6	980	2400	1811	10L	4140	5710	94	104	14.6	18.2	24840	79940			
D	K98MC-C6	980	2400	1811	11L	4140	5710	94	104	14.6	18.2	24840	79940			
D	K98MC-C6	980	2400	1811	12L	4140	5710	94	104	14.6	18.2	24840	79940			
D	K98MC-C6	980	2400	1811	13L	4140	5710	94	104	14.6	18.2	24840	79940			
D	K98MC-C6	980	2400	1811	14L	4140	5710	94	104	14.6	18.2	24840	79940			
D	K98MC7	980	2660	2007	6L	4630	6230	90	97	15.4	19.2	27780	87220			
D	K98MC7	980	2660	2007	7L	4630	6230	90	97	15.4	19.2	27780	87220			
D	K98MC7	980	2660	2007	8L	4630	6230	90	97	15.4	19.2	27780	87220			
D	K98MC7	980	2660	2007	9L	4630	6230	90	97	15.4	19.2	27780	87220			
D	K98ME7	980	2660	2007	10L	4630	6230	90	97	15.4	19.2	27780	87220			
D	K98ME7	980	2660	2007	11L	4630	6230	90	97	15.4	19.2	27780	87220			
D	K98ME7	980	2660	2007	12L	4630	6230	90	97	15.4	19.2	27780	87220			
D	K98ME7	980	2660	2007	13L	4630	6230	90	97	15.4	19.2	27780	87220			
D	K98ME7	980	2660	2007	14L	4630	6230	90	97	15.4	19.2	27780	87220			
D	K98ME6	980	2660	2007	6L	4100	5720	84	94	14.6	18.2	24600	80080			
D	K98ME6	980	2660	2007	7L	4100	5720	84	94	14.6	18.2	24600	80080			
D	K98ME6	980	2660	2007	8L	4100	5720	84	94	14.6	18.2	24600	80080			
D	K98ME6	980	2660	2007	9L	4100	5720	84	94	14.6	18.2	24600	80080			
D	K98ME6	980	2660	2007	10L	4100	5720	84	94	14.6	18.2	24600	80080			
D	K98ME6	980	2660	2007	11L	4100	5720	84	94	14.6	18.2	24600	80080			
D	K98ME6	980	2660	2007	12L	4100	5720	84	94	14.6	18.2	24600	80080			

▲

The MAN Diesel AG diesel-engine programme (continued). Category IV.

Fuel Type: D: Diesel or Heavy Fuel; DF: Dual Fuel; SI: Spark Ingnited. **Number of Cylinders & Configuration:** L: In-Line; V: V-Type; H: Horizontal; O: Opposed

Fuel Type	Engine Model	Bore (mm)	Stroke (mm)	Displacement (liters/cylinder)	Number of Cylinders & Configuration	Output per Cylinder Range (kW/cyl)		Rated Speed Range (r/min)		Maximum Brake Mean Effective Pressure (bar)		Output Range Low (kW)	Output Range High (kW)	Ratings	Application	Emmissions level
						min	max	min	max	min	max					
D	K98ME6	980	2660	2007	13L	4100	5720	84	94	14.6	18.2	24600	80080			
D	K98ME6	980	2660	2007	14L	4100	5720	84	94	14.6	18.2	24600	80080			
D	K98ME-C7	980	2400	1811	6L	4510	6020	97	104	15.4	19.2	27060	84280			
D	K98ME-C7	980	2400	1811	7L	4510	6020	97	104	15.4	19.2	27060	84280			
D	K98ME-C7	980	2400	1811	8L	4510	6020	97	104	15.4	19.2	27060	84280			
D	K98ME-C7	980	2400	1811	9L	4510	6020	97	104	15.4	19.2	27060	84280			
D	K98ME-C7	980	2400	1811	10L	4510	6020	97	104	15.4	19.2	27060	84280			
D	K98ME-C7	980	2400	1811	11L	4510	6020	97	104	15.4	19.2	27060	84280			
D	K98ME-C7	980	2400	1811	12L	4510	6020	97	104	15.4	19.2	27060	84280			
D	K98ME-C7	980	2400	1811	13L	4510	6020	97	104	15.4	19.2	27060	84280			
D	K98ME-C7	980	2400	1811	14L	4510	6020	97	104	15.4	19.2	27060	84280			
D	K98MC-C7	980	2400	1811	6L	4510	6020	97	104	15.4	19.2	27060	84280			
D	K98MC-C7	980	2400	1811	7L	4510	6020	97	104	15.4	19.2	27060	84280			
D	K98MC-C7	980	2400	1811	8L	4510	6020	97	104	15.4	19.2	27060	84280			
D	K98MC-C7	980	2400	1811	9L	4510	6020	97	104	15.4	19.2	27060	84280			
D	K98MC-C7	980	2400	1811	10L	4510	6020	97	104	15.4	19.2	27060	84280			
D	K98MC-C7	980	2400	1811	11L	4510	6020	97	104	15.4	19.2	27060	84280			
D	K98ME7	980	2660	2007	6L	4630	6230	90	97	15.4	19.2	27780	87220			
D	K98ME7	980	2660	2007	7L	4630	6230	90	97	15.4	19.2	27780	87220			
D	K98ME7	980	2660	2007	8L	4630	6230	90	97	15.4	19.2	27780	87220			
D	K98ME7	980	2660	2007	9L	4630	6230	90	97	15.4	19.2	27780	87220			

Source: Diesel Publications

▲

The MAN Diesel AG diesel-engine programme (continued). Category IV.

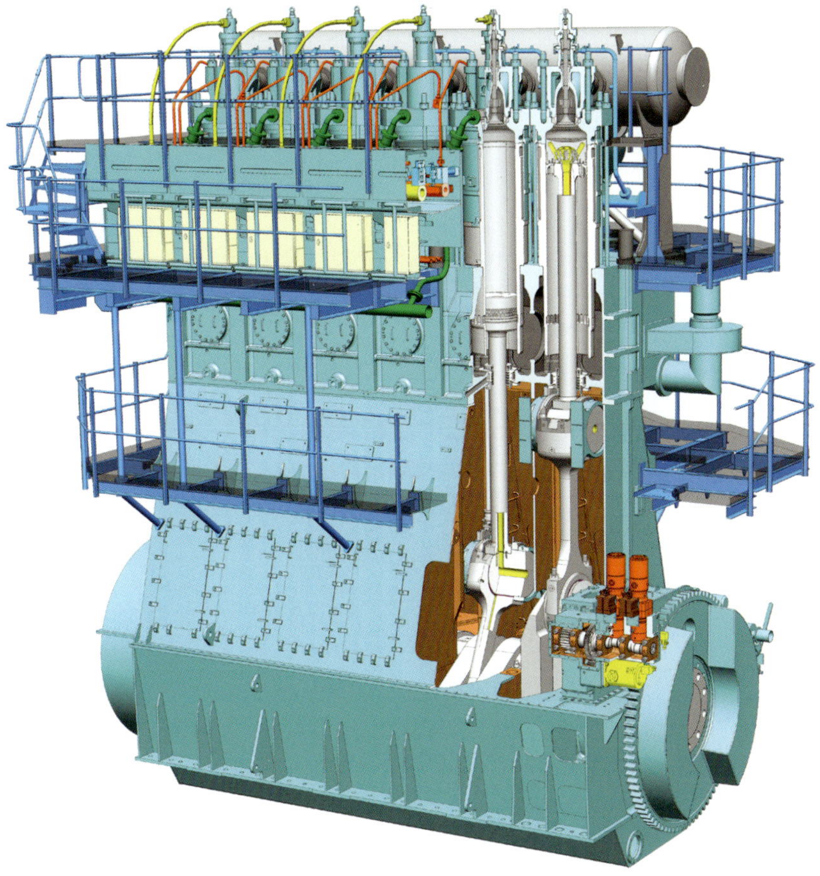

◄

A six-cylinder two-stroke crosshead engine from Wärtsilä. Type 6RT-Flex50. See page 129.

Fuel Type	Engine Model	Bore (mm)	Stroke (mm)	Displacement (liters/cylinder)	Number of Cylinders & Configuration	Output per Cylinder Range (kW/cyl)		Rated Speed Range (r/min)		Maximum Brake Mean Effective Pressure (bar)		Output Range Low (kW)	Output Range High (kW)	Ratings	Application	Emmissions level
						min	max	min	max	min	max					
D	UEC33LSII	330	1050	90	4-8L		566	162	215		18	1230	4530			
D	UEC37LA	370	880	95	4-8L		520	158	210		15.6	1120	4160			
D	UEC37LSII	370	1290	139	5-8L		772	140	186		18	2095	6180			
D	UEC43LSII	430	1500	218	4-8L		1050	120	160		18	2280	8400			
D	UEC45LA	450	1350	215	4-8L		890	119	158		15.6	1920	7120			
D	UEC45-LSE	450	1840	293	5-8L		1245	111	130		19.6	4250	9960			
D	UEC50LSII	500	1950	383	4-8L		1445	95	127		17	3120	11560			
D	UEC50LSE	500	2400	403	5-8L		1620	99	124		19.5	5175	12960			
D	UEC52LA	520	1600	340	4-8L		1180	100	133		15.6	2560	9440			
D	UEC52LS	520	1850	393	4-8L		1330	90	120		16.9	2880	10640			
D	UEC52LSE	520	2000	425	4-8L		1705	95	127		19	3700	20460			
D	UEC60LA	600	1900	537	4-8L		1550	83	110		15.7	3360	12400			
D	UEC60LS	600	2200	622	4-8L		1770	75	100		17	3800	14160			
D	UEC60LSII	600	2300	650	4-8L		2045	79	105		17	4440	16360			
D	UEC60LSE	600	2400	2713	5-8L		2255	90	105		19	7650	18040			
D	UEC68LSE	680	2690	977	5-8L		2940	81	95		19	10050	23520			
D	UEC75LSII	750	2800	1237	4-9L		2940	63	84		17	6380	26460			
D	UEC85LSII	850	3150	1787	5-9L		3860	54	76		17	9900	34740			
D	UEC85LSC	850	2360	1339	5-12L		3900	76	102		17	10575	46800			

Source: Diesel Publications

▲

The Mitsubishi Heavy Industries, Ltd. diesel-engine programme. Category IV.

►

The Wärtsilä Sulzer engine programme for the RTA-series, with camshaft and traditional fuel injection.

This program is regularly changed due to designs for new types of engines for some ships. Engine types also disappear after a certain period (considerable number of years). In recent years many new designs have been brought onto the market, primarily to reduce fuel consumption and harmful emissions.

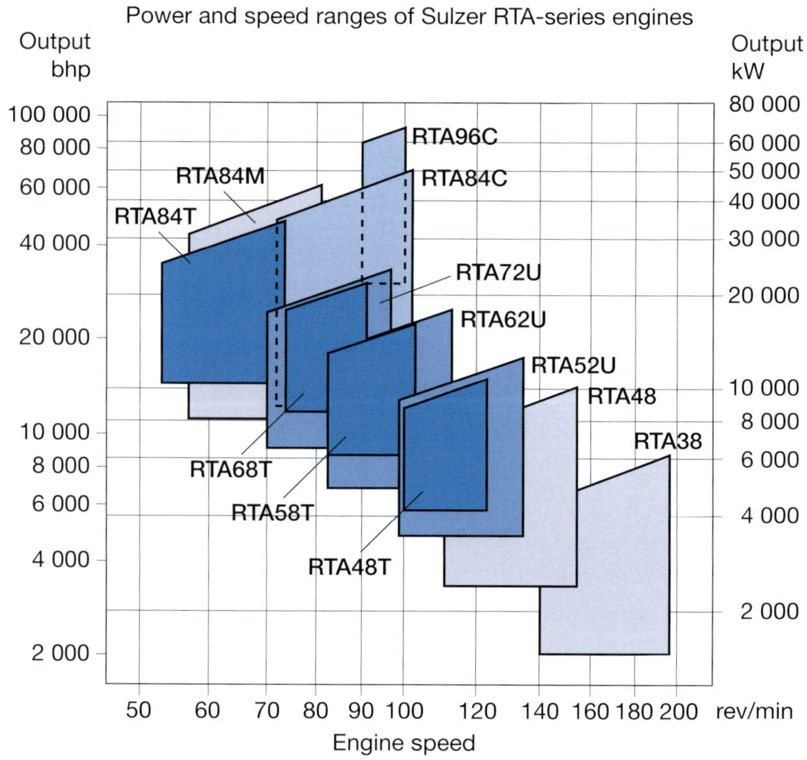

Fuel Type	Engine Model	Bore (mm)	Stroke (mm)	Displacement (liters/cylinder)	Number of Cylinders & Configuration	Output per Cylinder Range (kW/cyl)		Rated Speed Range (r/min)		Maximum Brake Mean Effective Pressure (bar)		Output Range Low (kW)	Output Range High (kW)	Ratings	Application	Emissions level
						min	max	min	max	min	max					
D	Wärtsilä 46DF	460	580	96.4	16V		1200		600		24.9	14400	19200			
D	Wärtsilä 46DF	460	580	96.4	14V		1200		600		24.9	14400	19200			
D	Wärtsilä RT-flex50-D	500	2050	402.5	5L	1220	1745	99	124		21	6100	13960	IMO		EPA Tier 2
D	Wärtsilä RT-flex50-D	500	2050	402.5	6L	1220	1745	99	124		21	6100	13960	IMO		EPA Tier 2
D	Wärtsilä RT-flex50-D	500	2050	402.5	7L	1220	1745	99	124		21	6100	13960	IMO		EPA Tier 2
D	Wärtsilä RT-flex50-D	500	2050	402.5	8L	1220	1745	99	124		21	6100	13960	IMO		EPA Tier 2
D	Wärtsilä RTA48T-D	480	2000	362	5L	1020	1455	102	127		19	5100	11640	IMO		EPA Tier 2
D	Wärtsilä RTA48T-D	480	2000	362	6L	1020	1455	102	127		19	5100	11640	IMO		EPA Tier 2
D	Wärtsilä RTA48T-D	480	2000	362	7L	1020	1455	102	127		19	5100	11640	IMO		EPA Tier 2
D	Wärtsilä RTA48T-D	480	2000	362	8L	1020	1455	102	127		19	5100	11640	IMO		EPA Tier 2
D	Wärtsilä RT-flex48T-D	480	2000	362	5L	1020	1455	102	127		19	5100	11640	IMO		EPA Tier 2
D	Wärtsilä RT-flex48T-D	480	2000	362	6L	1020	1455	102	127		19	5100	11640	IMO		EPA Tier 2
D	Wärtsilä RT-flex48T-D	480	2000	362	7L	1020	1455	102	127		19	5100	11640	IMO		EPA Tier 2
D	Wärtsilä RT-flex48T-D	480	2000	362	8L	1020	1455	102	127		19	5100	11640	IMO		EPA Tier 2
D	Wärtsilä RT-flex40	400	1770	222	5L	910	1135	124	146		21	4550	9080	IMO		EPA Tier 2
D	Wärtsilä RT-flex40	400	1770	222	6L	910	1135	124	146		21	4550	9080	IMO		EPA Tier 2
D	Wärtsilä RT-flex40	400	1770	222	7L	910	1135	124	146		21	4550	9080	IMO		EPA Tier 2
D	Wärtsilä RT-flex40	400	1770	222	8L	910	1135	124	146		21	4550	9080	IMO		EPA Tier 2
D	Wärtsilä RT-flex35	350	1550	149	5L	695	870	142	167		21	3475	6960	IMO		EPA Tier 2
D	Wärtsilä RT-flex35	350	1550	149	6L	695	870	142	167		21	3475	6960	IMO		EPA Tier 2
D	Wärtsilä RT-flex35	350	1550	149	7L	695	870	142	167		21	3475	6960	IMO		EPA Tier 2
D	Wärtsilä RT-flex35	350	1550	149	8L	695	870	142	167		21	3475	6960	IMO		EPA Tier 2
D	Wärtsilä RT-flex84T-D	840	3150	1746	8L	2940	4200	61	76		19	14700	37800	IMO		EPA Tier 2
D	Wärtsilä RT-flex84T-D	840	3150	1746	9L	2940	4200	61	76		19	14700	37800	IMO		EPA Tier 2
D	Wärtsilä RTA84T-D	840	3150	1746	5L	2940	4200	61	76		19	14700	37800	IMO		EPA Tier 2
D	Wärtsilä RTA84T-D	840	3150	1746	6L	2940	4200	61	76		19	14700	37800	IMO		EPA Tier 2
D	Wärtsilä RTA84T-D	840	3150	1746	7L	2940	4200	61	76		19	14700	37800	IMO		EPA Tier 2
D	Wärtsilä RTA84T-D	840	3150	1746	8L	2940	4200	61	76		19	14700	37800	IMO		EPA Tier 2
D	Wärtsilä RTA84T-D	840	3150	1746	9L	2940	4200	61	76		19	14700	37800	IMO		EPA Tier 2
D	Wärtsilä RT-flex82T	820	3375	1782	6L	3620	4520	68	80		20	21720	40680	IMO		EPA Tier 2
D	Wärtsilä RT-flex82T	820	3375	1782	7L	3620	4520	68	80		20	21720	40680	IMO		EPA Tier 2
D	Wärtsilä RT-flex82T	820	3375	1782	8L	3620	4520	68	80		20	21720	40680	IMO		EPA Tier 2
D	Wärtsilä RT-flex82T	820	3375	1782	9L	3620	4520	68	80		20	21720	40680	IMO		EPA Tier 2
	Wärtsilä RTA82T	820	3375	1782	6L	3620	4520	68	80		20	21720	40680	IMO		EPA Tier 2
	Wärtsilä RTA82T	820	3375	1782	7L	3620	4520	68	80		20	21720	40680	IMO		EPA Tier 2
	Wärtsilä RTA82T	820	3375	1782	8L	3620	4520	68	80		20	21720	40680	IMO		EPA Tier 2
	Wärtsilä RTA82T	820	3375	1782	9L	3620	4520	68	80		20	21720	40680	IMO		EPA Tier 2
D	Wärtsilä RT-flex82C	820	2646	1397	6L	3620	4520	87	102		20	21720	54240	IMO		EPA Tier 2
D	Wärtsilä RT-flex82C	820	2646	1397	7L	3620	4520	87	102		20	21720	54240	IMO		EPA Tier 2
D	Wärtsilä RT-flex82C	820	2646	1397	8L	3620	4520	87	102		20	21720	54240	IMO		EPA Tier 2
D	Wärtsilä RT-flex82C	820	2646	1397	9L	3620	4520	87	102		20	21720	54240	IMO		EPA Tier 2
D	Wärtsilä RT-flex82C	820	2646	1397	10L	3620	4520	87	102		20	21720	54240	IMO		EPA Tier 2
D	Wärtsilä RT-flex82C	820	2646	1397	11L	3620	4520	87	102		20	21720	54240	IMO		EPA Tier 2
D	Wärtsilä RT-flex82C	820	2646	1397	12L	3620	4520	87	102		20	21720	54240	IMO		EPA Tier 2
D	Wärtsilä RTA82C	820	2646	1397	6L	3620	4520	87	102		20	21720	54240	IMO		EPA Tier 2
D	Wärtsilä RTA82C	820	2646	1397	7L	3620	4520	87	102		20	21720	54240	IMO		EPA Tier 2
D	Wärtsilä RTA82C	820	2646	1397	8L	3620	4520	87	102		20	21720	54240	IMO		EPA Tier 2
D	Wärtsilä RTA82C	820	2646	1397	9L	3620	4520	87	102		20	21720	54240	IMO		EPA Tier 2

▲
The Wärtsilä Corporation diesel-engine programme. Category IV.

Fuel Type: D: Diesel or Heavy Fuel; DF: Dual Fuel; SI: Spark Ingnited. **Number of Cylinders & Configuration:** L: In-Line; V: V-Type; H: Horizontal; O: Opposed

Fuel Type	Engine Model	Bore (mm)	Stroke (mm)	Displacement (liters/cylinder)	Number of Cylinders & Configuration	Output per Cylinder Range (kW/cyl)		Rated Speed Range (r/min)		Maximum Brake Mean Effective Pressure (bar)		Output Range Low (kW)	Output Range High (kW)	Ratings	Application	Emmissions level
						min	max	min	max	min	max					
D	Wärtsilä RTA82C	820	2646	1397	10L	3620	4520	87	102		20	21720	54240	IMO		EPA Tier 2
D	Wärtsilä RTA82C	820	2646	1397	11L	3620	4520	87	102		20	21720	54240	IMO		EPA Tier 2
D	Wärtsilä RTA82C	820	2646	1397	12L	3620	4520	87	102		20	21720	54240	IMO		EPA Tier 2
D	Wärtsilä RT-flex68-D	680	2720	988	5L	2150	3130	76	95		20	10950	25040	IMO		EPA Tier 2
D	Wärtsilä RT-flex68-D	680	2720	988	6L	2150	3130	76	95		20	10950	25040	IMO		EPA Tier 2
D	Wärtsilä RT-flex68-D	680	2720	988	7L	2150	3130	76	95		20	10950	25040	IMO		EPA Tier 2
D	Wärtsilä RT-flex68-D	680	2720	988	8L	2150	3130	76	95		20	10950	25040	IMO		EPA Tier 2
D	Wärtsilä RT-flex68-B	680	2720	988	5L	2150	3130	76	95		20	10950	25040	IMO		EPA Tier 2
D	Wärtsilä RT-flex68-B	680	2720	988	6L	2150	3130	76	95		20	10950	25040	IMO		EPA Tier 2
D	Wärtsilä RT-flex68-B	680	2720	988	7L	2150	3130	76	95		20	10950	25040	IMO		EPA Tier 2
D	Wärtsilä RT-flex68-B	680	2720	987.8	8L	2150	3130	76	95		20	10950	25040	IMO		EPA Tier 2
D	Wärtsilä RT-flex60C-B	600	2250	636	5L	1650	2420	91	114		20	8450	21780	IMO		EPA Tier 2
D	Wärtsilä RT-flex60C-B	600	2250	636	6L	1650	2420	91	114		20	8450	21780	IMO		EPA Tier 2
D	Wärtsilä RT-flex60C-B	600	2250	636	7L	1650	2420	91	114		20	8450	21780	IMO		EPA Tier 2
D	Wärtsilä RT-flex60C-B	600	2250	636	8L	1650	2420	91	114		20	8450	21780	IMO		EPA Tier 2
D	Wärtsilä RT-flex60C-B	600	2250	636	9L	1650	2420	91	114		20	8450	21780	IMO		EPA Tier 2
D	Wärtsilä RT-flex58T-D	580	2416	638	5L	1530	2260	84	105		20.2	7900	18080	IMO		EPA Tier 2
D	Wärtsilä RT-flex58T-D	580	2416	638	6L	1530	2260	84	105		20.2	7900	18080	IMO		EPA Tier 2
D	Wärtsilä RT-flex58T-D	580	2416	638	7L	1530	2260	84	105		20.2	7900	18080	IMO		EPA Tier 2
D	Wärtsilä RT-flex58T-D	580	2416	638	8L	1530	2260	84	105		20.2	7900	18080	IMO		EPA Tier 2
D	Wärtsilä RTA58T-D	580	2416	638	5L	1530	2260	84	105		20	7900	18080	IMO		EPA Tier 2
D	Wärtsilä RTA58T-D	580	2416	638	6L	1530	2260	84	105		20.2	7900	18080	IMO		EPA Tier 2
D	Wärtsilä RTA58T-D	580	2416	638.3	7L	1530	2260	84	105		20.2	7900	18080	IMO		EPA Tier 2
D	Wärtsilä RTA58T-D	580	2416	638	8L	1530	2260	84	105		20.2	7900	18080	IMO		EPA Tier 2
D	Wärtsilä RT-flex96C	960	2500	1810	6L	4000	5720	92	102		18.6	24000	80080	IMO		EPA Tier 2
D	Wärtsilä RT-flex96C	960	2500	1810	7L	4000	5720	92	102		18.6	24000	80080	IMO		EPA Tier 2
D	Wärtsilä RT-flex96C	960	2500	1810	8L	4000	5720	92	102		18.6	24000	80080	IMO		EPA Tier 2
D	Wärtsilä RT-flex96C	960	2500	1810	9L	4000	5720	92	102		18.6	24000	80080	IMO		EPA Tier 2
D	Wärtsilä RT-flex96C	960	2500	1810	10L	4000	5720	92	102		18.6	24000	80080	IMO		EPA Tier 2
D	Wärtsilä RT-flex96C	960	2500	1810	11L	4000	5720	92	102		18.6	24000	80080	IMO		EPA Tier 2
D	Wärtsilä RT-flex96C	960	2500	1810	12L	4000	5720	92	102		18.6	24000	80080	IMO		EPA Tier 2
D	Wärtsilä RT-flex96C	960	2500	1810	13L	4000	5720	92	102		18.6	24000	80080	IMO		EPA Tier 2
D	Wärtsilä RT-flex96C	960	2500	1810	14L	4000	5720	92	102		18.6	24000	80080	IMO		EPA Tier 2
D	Wärtsilä RTA96C	960	2500	1810	6L	4000	5720	92	102		18.6	24000	80080	IMO		EPA Tier 2
D	Wärtsilä RTA96C	960	2500	1810	7L	4000	5720	92	102		18.6	24000	80080	IMO		EPA Tier 2
D	Wärtsilä RTA96C	960	2500	1810	8L	4000	5720	92	102		18.6	24000	80080	IMO		EPA Tier 2
D	Wärtsilä RTA96C	960	2500	1810	9L	4000	5720	92	102		18.6	24000	80080	IMO		EPA Tier 2
D	Wärtsilä RTA96C	960	2500	1810	10L	4000	5720	92	102		18.6	24000	80080	IMO		EPA Tier 2
D	Wärtsilä RTA96C	960	2500	1810	11L	4000	5720	92	102		18.6	24000	80080	IMO		EPA Tier 2
D	Wärtsilä RTA96C	960	2500	1810	12L	4000	5720	92	102		18.6	24000	80080	IMO		EPA Tier 2
D	Wärtsilä RTA96C	960	2500	1810	13L	4000	5720	92	102		18.6	24000	80080	IMO		EPA Tier 2
D	Wärtsilä RTA96C	960	2500	1810	14L	4000	5720	92	102		18.6	24000	80080	IMO		EPA Tier 2
D	Wärtsilä RT-flex84T-D	840	3150	176	5L	2940	4200	61	76		19	14700	37800	IMO		EPA Tier 2
D	Wärtsilä RT-flex84T-D	840	3150	1746	6L	2940	4200	61	76		19	14700	37800	IMO		EPA Tier 2
D	Wärtsilä RT-flex84T-D	840	3150	1740	7L	2940	4200	61	76		19	14700	37800	IMO		EPA Tier 2

Fuel Type: D: Diesel or Heavy Fuel; DF: Dual Fuel; SI: Spark Ingnited. **Number of Cylinders & Configuration:** L: In-Line; V: V-Type; H: Horizontal; O: Opposed

Source: Diesel Publications

▲ The Wärtsilä Corporation diesel-engine programme (continued). Category IV.

21.5 Diesel engines for ship propulsion and diesel-power plants

21.5.1 Numbers

Obviously, the number of small engines manufactured in particular for categories I and II is enormous. Every year, hundreds of thousands of engines are manufactured for various vessels and other drives by a large number of engine manufacturers. These total approximately 15 million diesel engines per annum.
In surveys of engines for shipping and diesel-power plants, the number of engines manufactured annually is 'counted' from a certain bottom limit, that of shaft power; often 1000 or 2000 kW.
Obviously, when a large two-stroke low-speed crosshead engine with fourteen cylinders in the largest diameter is ordered, it attracts more attention in the industry than when a much smaller four-stroke medium-speed diesel engine is ordered.

21.5.2 Striking aspects of engine manufacturing

1. Major engine manufacturers produce a large number of engine types in terms of cylinder diameters, speed and power output. In this manner, they endeavour to 'cover' their market.
2. Engines in category I often start with a single cylinder and move on to a six-cylinder in-line arrangement.
3. Engines in category II often start with a three- or four-cylinder in-line arrangement and move to sixteen-, eighteen- or twenty-cylinder V-arrangements.
4. Engines in category III often start with a four-cylinder in-line arrangement and move on to an eighteen-cylinder V-arrangement.
5. Engines in category IV usually start with a four-cylinder in-line arrangement and at present are produced with up to fourteen-cylinder in-line arrangements.
6. Especially for category III with the limited number of engine manufacturers, it is noticeable that they often have engine programmes for the same cylinder diameter. Other manufacturers have 200, 250 and 320 mm cylinder diameters.
7. The mean effective pressures and the mean piston speeds are increasing due to keen competition between the manufacturers. Consequently the load parameter also increases!
8. Very few new engine manufacturers enter the market. However, the number of manufacturers that have closed their businesses or merged with others continues to increase. After a merger, certain types of engines are only maintained if they are economically viable and after several years the brand name disappears and an increasing number of renowned international and national trade-marks end up in the history books.
9. **Fuel used.** As a result of increasingly strict regulations regarding maximum emission levels for pollutants in exhaust gases and a growing realisation that supply of exploitable crude oil is finite, engine manufacturers and consumers are attempting to find 'alternative' fuels. The diesel process where small amounts of liquid fuel are combined with gaseous fuel, the so-called **diesel/gas process**, is for instance in development. The emissions are often lower in comparison with those of a regular diesel engine. The usage of bio-fuels, such as olive oil, palm oil, rape oil and other vegetable oils is also being developed. These fuels work **CO_2 neutral**, which means that they emit the same amount of CO_2 to the air during combustion as they absorb from the atmosphere during their growth.

Also see chapter 29, *New developments in the fuel industry*.

21.6 Future of the diesel-engine industry

Diesel-engine performances have to meet high requirements.
The legal regulations regarding emission levels for pollutants are being applied to more areas in the world and will become more stringent in the future.
Users of diesel engines demand an optimum service network and a growing number of these consumers operate worldwide. This means that engine manufacturers must maintain an extensive service organisation. This is only economically viable for large manufacturers.
The development costs of new engine types are extremely high. These costs are made to:

a. reduce fuel consumption;
b. reduce lubricating oil consumption;
c. create a higher mean effective pressure;
d. produce a higher mean piston speed;
e. create a higher load parameter ($c \times d$);
f. produce longer life for the engine parts;
g. produce lower emissions;
h. be suitable for alternative fuels;
i. a lower specific weight;
j. devise simple construction and maintenance-friendly disassembly and assembly methods.

There are numerous other points with which an engine manufacturer may stay ahead of the competition!
Due to above-mentioned reasons, many engine manufacturers are either economically or due to organisational issues, no longer capable of developing new engines and discontinue certain engine series, merge or simply close their business. Over fifty per cent of the engine manufacturers discontinued their businesses between 1950 and 2010.

Alternative methods for energy generation other than by diesel engines are:

- fuel cells;
- gas turbines;
- hydrogen;
- kite sailing.

These alternatives could become successful and are dependent on numerous factors, so very little can be said about the future developments at this time.

▶ Container ships.

Large, fast ships such as this container ship from SAF Marine, now Maersk Line always have large two-stroke crosshead engines, speeds of 100-120 revs/min, with a single 5- or 6-bladed screw for propulsion, propulsion in the easiest and most effective manner. Engine category IV, cylinder diameter 800 mm and greater.

◀ The super container ship, 'Emma Maersk'.

In 2007, the E-series of the world's largest shipping company, Denmark's Maersk, was the first container ship with container capacity of over 10,000 TEU, TEU = twenty-foot equivalent unit, (a standard container is 20 feet). These ships were the first since the use of two-stroke crosshead engines over twelve cylinders in-line, a fourteen-cylinder engine.

In February 2011, the next generation of super container vessels were ordered by Maersk.

Data:	Triple E-class.
Dimensions:	*Length*: 400 metres *Height*: 73 metres *Beam*: 59 metres *Draft*: 14.5 metres
Capacity:	18,000 TEU
Deadweight:	displacement 165,000 metric tonnes.
Max. speed:	23 knots.
Transport costs per container:	costs are 26% lower than the present super container ships that can transport 13,100 TEU.
CO_2 emissions:	20% lower per container than the 13,100 TEU class.
Propulsion:	two ultra-long two-stroke crosshead engines each with its own propeller.
Power output:	*New Triple E*: 65-70 MW. *Emma Maersk*: 80 MW, single engine.
Revolutions per minute:	80 rpm.
2 main engines:	Manufacturer: MAN Diesel.
Type:	Man Diesel G series, two engines with a total shaft power of 63 MW, ultra-long stroke resulting in a lower rpm requiring a larger propeller area.
Extra:	the exhaust gases are partially used to generate electricity with an exhaust-gas boiler and a steam-turbine genset.
Screw diameters:	*New Triple* E: 9.8 metres with four blades. *Emma Maersk*: 9.6 metres with six blades.
Propulsion efficiency:	The lowest possible rpm with the largest possible propeller area allow the highest propulsion efficiency with lowest fuel consumption.

Each ship will sail 115,000 sea miles yearly, approximately 4.5 times around the world.
The 10 ships will be launched between 2013 and 2015.

Triple E stands for: Economy, Energy and Environment.

1 **Economy**, transporting more containers while using less energy. Due to its size, energy consumption compared to the Emma Maersk class is reduced.
2 **Energy**, increased energy efficiency. The ship is designed for slower speeds and has a heat recovery system.
3 **Environment**, reduction of harmful emissions, such as CO_2 emissions.
Energy consumption per container is expected to be 50% lower. The materials used in its construction can also be fully and responsibly recycled.

Maersk has chosen for two two-stroke crosshead engines with two large screws, the 'twin skeg' principle. A slower operating speed compared to the Emma Maersk class of 23 knots instead of 25 knots translates to a fuel savings of 19%.
The 'twin skeg' principle allows for an extra 4% fuel savings compared to a single screw.

▶

The LNG tanker Maersk Ras Laffan.

Anno 2011, more and more tankers to transport LPG (Liquefied Petroleum Gas) and LNG (Liquefied Nitrogen Gas) are being built. They are mostly propelled by four-stroke or two-stroke 'Dual-Fuel' engines.

Using the cargo gas together with a small amount of liquid fuel as an ignition source for the engine, these propulsion installations have very low harmful emissions.

The latter is a very important fact, now in 2012, and in the near future.

▶

The passenger ship, the 'Zuiderdam' of the Holland America Line.

These 'floating hotels' with thousands of passengers use a great deal of energy for their accommodation next to the energy required for propulsion.

This energy is generated by large medium-speed four-stroke diesel engines in category III, fuel HFO.

▶

The water and power plant of Aruba, the WEB.

The water and electricity company uses together with the older steam power plant, four-stroke medium-speed diesel engines, engine category III.

Besides the existing engines from Wärtsilä, the Vasaa and Wärtsilä 32, four new gensets with Wärtsilä 12V46 engines will be installed in 2011.

◀

The three-masted Athena, one of the largest private sailing yachts in the world.

Yachts are frequently fitted with propulsion and generator engines of engine categories I and II, the highspeed four-stroke diesel engine, fuel MDO, Marine Diesel Oil.

◀

Dual-fuel engines.

A sector that is rapidly developing is that of engines using the dual-fuel principle. Both the four-stroke medium-speed engines, category III, as the two-stroke low-speed crosshead engines, category IV, are used.
Here's a MAN Diesel two-stroke crosshead engine for an LNG tanker. They are also increasingly common in diesel-power plants.

◀

Power plants all over the world.

Transportation of the many engines used in power plants is often spectacular. Using low-loaders, ships, trains and shown here, a cantilever system, the engines finally reach their destination.

▶

The development of MAK four-stroke diesel engines between 1950 and 2010.

The continuous development of this principle for a higher efficiency has been a laborious and expensive process since the first test of a diesel engine by Rudolf Diesel at MAN in Augsburg, Germany. Customers demand ever increasing power outputs from the same stroke volume, therefore the mean effective pressure must increase and with this the speed.
Occasionally remarkable steps are made, such as, supercharging with turbochargers driven by the exhaust gases, a better piston ring packet with lubrication or the antipolishing ring.
Another step that will be taken in 2011 is an even greater charging pressure, this is achieved by two turbochargers in series with the four-stroke engine.

Left: the piston of a Krupp-Diesel with a cylinder diameter of 571 mm weighing 300 kg. Cylinder power output, 300 kW. Material: one-piece forged aluminium alloy.
Right: the piston of a Caterpillar MaK 25 with a cylinder diameter of 255 mm weighing 46 kg. Cylinder power output, 300 kW. Material: forged steel alloy piston crown with cast iron skirt. Two parts.

21.7 Overview of diesel engines ordered in 2010

▶

Overview of diesel engines above 0.5 MW ordered world-wide in 2010

An insight into the production share in percentages for the large manufacturers, MAN Diesel, Wärtsilä, Caterpillar MaK, HHI Himsen and others. This overview has 9 segments, from I to IX.

I Marine Medium Speed Propulsion > 0.5 MW

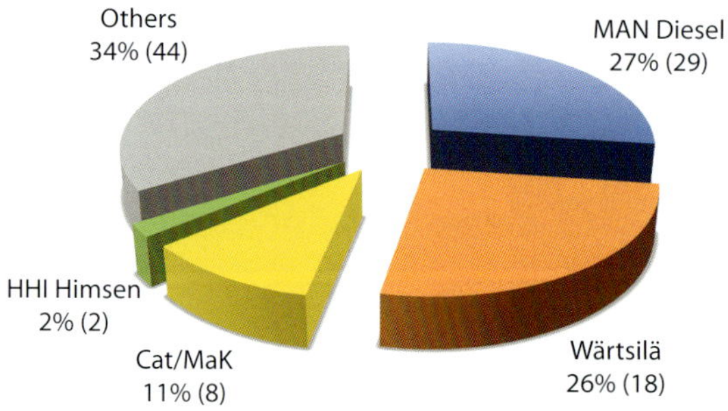

100% = 4,235 MW (2009: 3,011 MW, +41%)

II Marine Medium Speed Auxiliary > 0.5 MW

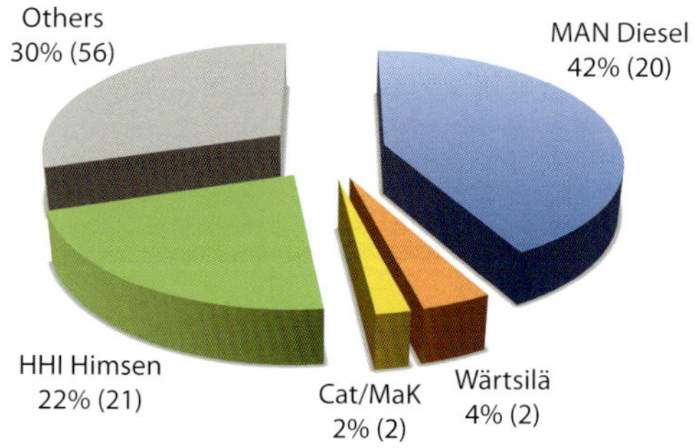

100% = 4,298 MW (2009: 2,650 MW, +62%)

III Marine Medium Speed Propulsion & Auxiliary > 0.5 MW

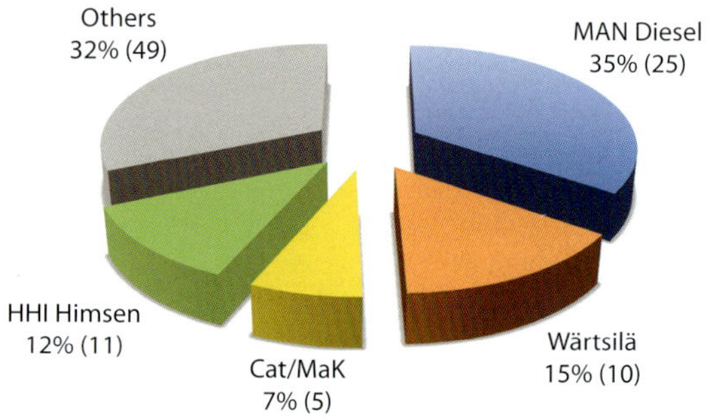

100% = 8,528 MW (2009: 5,661 MW, +51%)

IV Power Plant Medium Speed Diesel/HFO > 0.5 MW

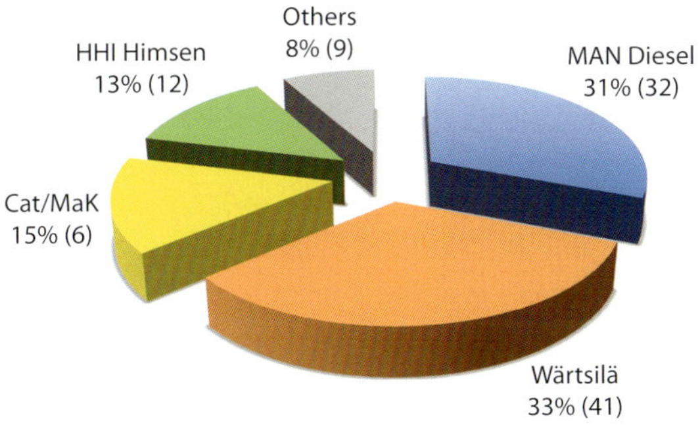

100% = 5,451 MW (2009: 2,875 MW, +90%)

V Power Plant Medium Speed Gas/DF > 0.5 MW

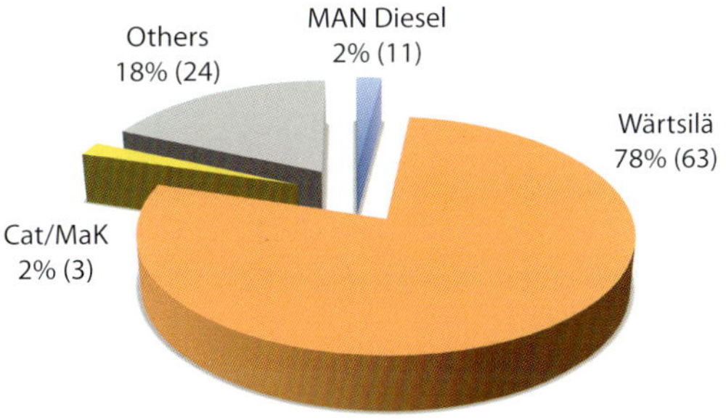

100% = 1,767 MW (2009: 1,277 MW, +38%)

VI Power Plant Medium Speed Liquid Biofuel > 0.5 MW

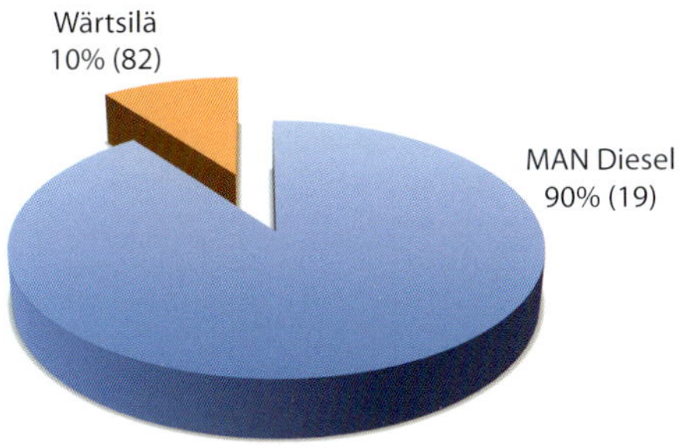

100% = 9 MW (2009: 43 MW, –79%)

VII Power Plant Medium Speed Total: Diesel; Gas/DF, LBF > 0.5 MW

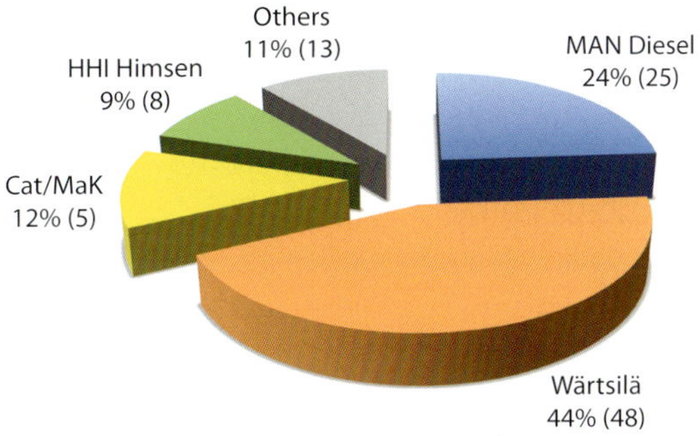

100% = 7,227 MW (2009: 4,195 MW, +72%)

Conclusion:

In the marine propulsion market, there are only a few leading companies that have a considerable share in practically all sectors.

Wärtsilä is clearly the market leader in four-stroke medium-speed engines, followed by MAN Diesel and others.

MAN Diesel is the largest producer of two-stroke crosshead engines, Wärtsilä and Mitsubishi have a much smaller market share.

MAN diesel has the largest market share in the diesel engine industry for ship propulsion and diesel engines.

VIII Medium Speed > 0.5 MW Propulsion, Auxiliary, Power Plant

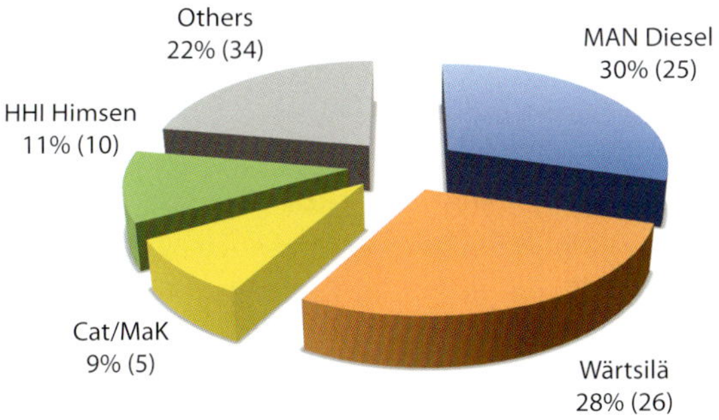

100% = 15,755 MW (2009: 9,855 MW, +60%)

IX Marine Low Speed Propulsion

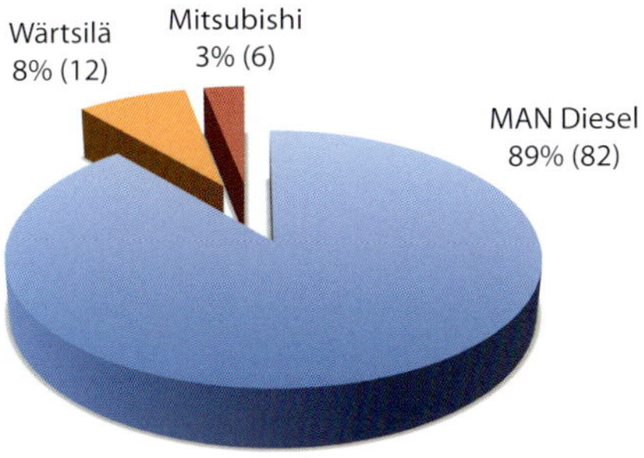

100% = 18,100 MW (2009: 6,600 MW, +175%)

I Ship propulsion above 0.5 MW
Medium-speed diesel engines, category III, HFO.

II Ship auxiliary above 0.5 MW
Medium-speed diesel engines, category III, HFO.

III Ship propulsion and auxiliary above 0.5 MW
Medium-speed diesel engines, category III, HFO.

IV Power plants above 0.5 MW
Medium-speed diesel engines, category III, HFO.

V Power plants above 0.5 MW
Medium-speed diesel engines, category III, gas/dual fuel.

VI Power plants above 0.5 MW
Medium-speed diesel engines, category III, liquid biofuel.

VII Power plants above 0.5 MW
Medium-speed diesel engines, category III, total overview: diesel, gas/dual fuel, liquid biofuel.

VIII Medium-speed four-stroke diesel engines above 0.5 MW
Category III, total overview: propulsion, genset and power plant.

IX Ship propulsion
Low-speed two-stroke crosshead main engines, category IV, HFO.

21.8 Complete list of diesel engine manufacturers, March 2011

Contents Diesel Publications.

AGCO Sisu Power (7 Diesel Engines)
Akasaka Diesels Ltd. (3 Diesel Engines)
Anglo Belgian Corporation (10 Diesel Engines)
Briggs & Stratton Commercial Power (5 Diesel Engines)
Burmeister & Wain Scandinavian Contractor A/S (11 Diesel Engines)
Caterpillar Inc. Caterpillar Marine Power Systems
Cummins Inc. Power Systems Division (72 Diesel Engines)
CJSC Transmasholding/PJSC Kolemensky Zavod (9 Diesel Engines)
Compagnia Tecnica Motori S.p.A.
CRM Spa Motori Marini (2 Diesel Engines)
Cummins Inc. (37 Diesel Engines)
Daihatsu Diesel Mfg. Co, Ltd. (23 Diesel Engines)
DEUTZ AG (53 Diesel Engines)
Diesel United, Ltd. (19 Diesel Engines)
Doosan Infracore Co, Ltd. Engine & Material BG (77 Diesel Engines)
Electro-Motive Diesel, Inc. (4 Diesel Engines)
Fairbanks Morse Engine (97 Diesel Engines)
Fiat Powertrain Technologies (94 Diesel Engines)
Greaves Cotton Limited (69 Diesel Engines)
Greaves Farymann Diesel GmbH. (8 Diesel Engines)
Guascor. S.A.U. (46 Diesel Engines)
H. Cegielski, Poznan S.A. (291 Diesel Engines)
Hanshin Diesel Works Ltd. (10 Diesel Engines)
Hino Motors , FPT. (10 Diesel Engines)
Hitachi Zosen Corporation (9 Diesel Engines)
Isotta Fraschini Motori SpA. (7 Diesel Engines)
Isuzu Motors, Ltd. (9 Diesel Engines)
JCB Power Systems Ltd. (1 Diesel Engine)
John Deere Power Systems (60 Diesel Engines)
JSC ZVEZDA (39 Diesel Engines)
Kawasaki Heavy Industries Ltd.
Kubota Corporation (22 Diesel Engines)
Liebherr Machines Bulle SA (8 Diesel Engines)
Lister Petter Ltd. (30 Diesel Engines)
Lombardini Srl. (34 Diesel Engines)
Lovol Engines (44 Diesel Engines)
MAN Diesel SE (391 Diesel Engines)
MAN Truck & Bus AG (47 Diesel Engines)
Marine Diesel Sweden AB (11 Diesel Engines)
Mitsui Engineering & Shipbuilding Co, Ltd. (11 Diesel Engines)
Motorenfabrik Hatz GmbH & Co, KG (51 Diesel Engines)
MTU Friedrichshafen GmbH. (52 Diesel Engines)
Navistar International Corporation (50 Diesel Engines)
Niigata Power Systems Co., Ltd. (34 Diesel Engines)
OMT Officine Meccaniche Torino SpA
Perkins Engine Company Limited (33 Diesel Engines)
RUMO, JSC (22 Diesel Engines)
S.E.M.T. Pielstick, a brand of the MAN Diesel Group (32 Diesel Engines)
Scania (10 Diesel Engines)
Seatek (5 Diesel Engines)
STEYR MOTORS GmbH. (35 Diesel Engines)
Tedom (2 Diesel Engines)
Tianjin Lovol Engines Co., Ltd. (46 Diesel Engines)
VM Motori (42 Diesel Engines)
Volkswagen AG (8 Diesel Engines)
Volvo Penta (22 Diesel Engines)
Wärtsilä Corporation (151 Diesel Engines)
Yamaha Motor Co, Ltd. (5 Diesel Engines)
Yanmar Co, Ltd. (55 Diesel Engines)

◄

Preparing to tow near an oil field off the coast of Nigeria.

Both the large tow-boat and the platform are equipped with large four-stroke medium-speed diesel engines running on HFO, category III.

◄

The 'Berlin Express' from Hapag Lloyd approaches the container harbour at Hamburg, Germany.

These ships are equipped with all the engine categories:

- life-boat engines, category I;
- emergency generator, category II;
- gensets, category III;
- propulsion engine, category IV.

◄

A large ship suitable for all kinds of sea rescues.

Equipped with MAN Diesel engines, category III.

>CH22

Engine emissions

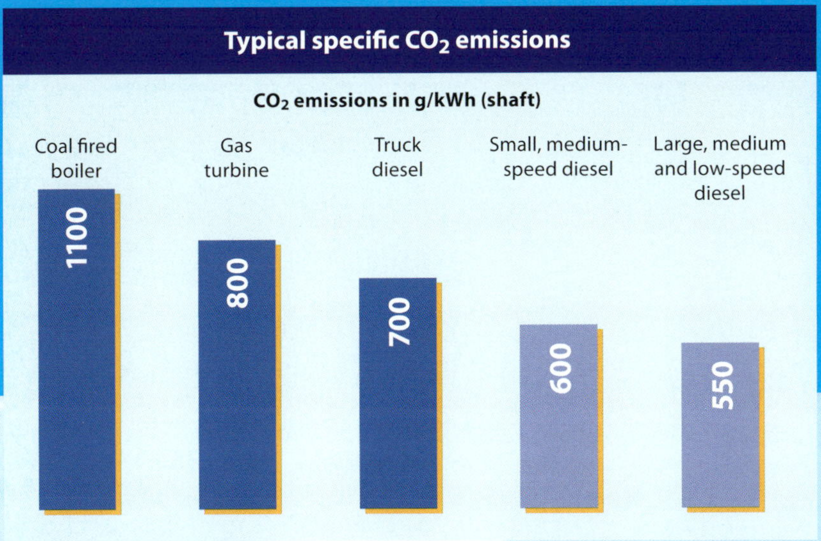

Carbon-dioxide emissions in grams per kWh from various energy production units.

In many parts of the world, electricity is generated using large diesel engines. In Honduras, where heavy-fuel oil is half the price of diesel oil, this is the fuel of choice. Pollutants from heavy-fuel oil are higher than those of diesel oil. Due to increasingly strict pollutant-emission requirements, engine manufacturers are using new technologies, such as, common-rail systems and exhaust-gas cleaning. Ultimately, the price per generated kWh is the deciding factor in determining which fuels and systems are selected in order to comply with emission regulations.

▶

Many human activities have an impact on the environment.

Cruise ships sailing to beautiful regions in the world pollute the environment. The emission of pollutants must therefore be restricted as much as possible. Increasingly strict legislation ensures that this is a major point when designing an engine.

22.1 Introducing 'the fossil fuel' society

Society has been transformed from around 1750, the beginning of the industrial revolution in Western Europe. With the introduction of the steam boiler and steam engine, continuous power was available for the first time in history, independent of wind, water or animals.
However, this required a constant fuel supply. At first coal was used as a fuel for the steam boiler.
In later technical developments, steam turbines were used on a large scale.
Especially after 1850, when the Otto internal-combustion engine was developed, and just before 1900 when the Diesel internal-combustion engine was introduced, liquid and gaseous fossil fuels were being increasingly used.
Around 1950, the diesel engine was used more frequently for ship propulsion and soon new diesel engines suitable for operation with relatively inexpensive fuel oil (without excessive problems), were introduced. Initially, this fuel oil was only used for steam boilers. Originally this HFO was used in the large two-stroke crosshead engines and has in the past 25 years, been introduced for medium-speed diesel engines.

Use of fossil fuels

The use of fossil fuels has dramatically risen over the last hundred years and after ±1950, the increase has been explosive.
In 2011 the demand for energy, in countries such as China and India in particular as well as the other emerging Asian countries, has increased significantly and the price of a barrel of heavy oil, 380 cSt has risen to 700 USD per ton weight.

Dependence on fossil fuels

This is in 2012, extreme. Apart from its use in the generation of energy for an increasing number of sectors, a large number of products are manufactured from crude oil. Among these are synthetics for building materials and clothing,

▶

A future vision for sources of energy generation to 2100.

The use of coal remains constant. The use of crude oil decreases slightly. The use of natural gas increases significantly. The use of nuclear energy increases as well as the use of waterpower and other energy sources, such as wind and sun. It can be concluded that although the energy sources change somewhat, the total amount of energy consumption will double. This has enormous consequences for the environment and therefore our existence.

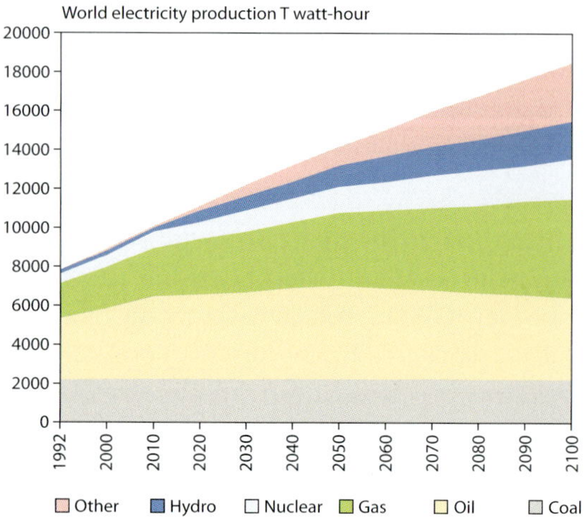

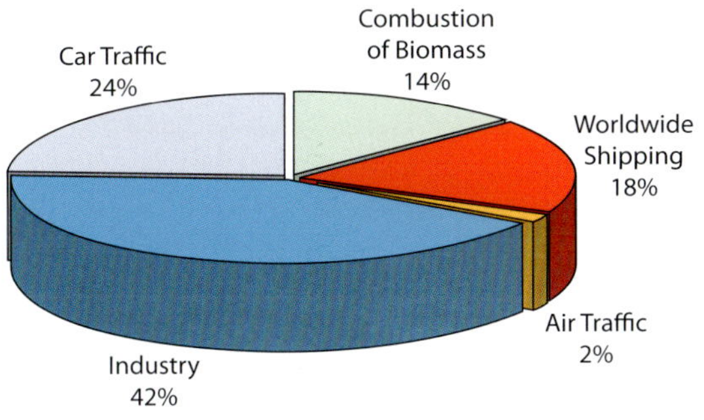

▲

The global energy consumption in 2000. The emission of pollutants is indicated for each consumer group.

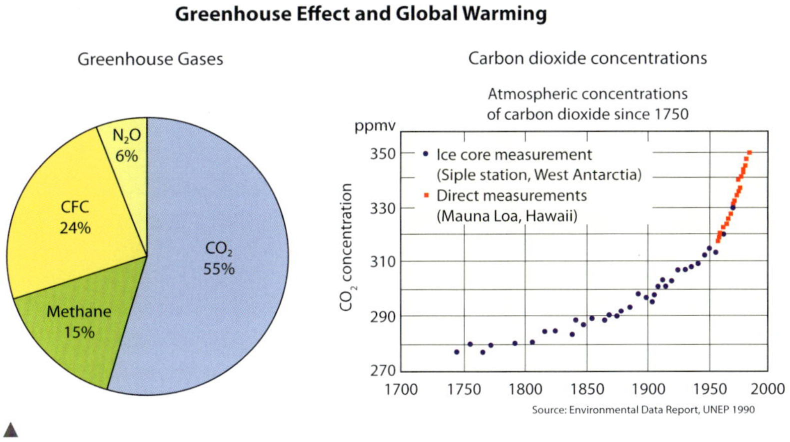

▲

This graph shows how various gases contribute to the greenhouse effect.

Note the increase starting from approximately 1750 (first industrial revolution) and the increase around 1950 (mechanisation/electrification throughout society). The graph after the year 2000 also shows an ominously sharp incline.

fertilisers and pesticides for agricultural use as well as an infinite number of raw materials and semi-finished products in the chemical industry.
Crude oil dependence is universal and comprehensive.
In fact, modern society is built on the availability of a continuous oil supply!

Environmental effects

Effects of the large-scale usage of fossil fuels on the environment have long been known and researched in detail throughout the years. It has been established that fossil-fuel combustion weighs heavily on our environment.
Topics such as the greenhouse effect, acid deposition, harmful emissions and a reduced habitat quality for flora and fauna are daily recurring conversational topics.

22.2 Exhaust-gas composition

A modern two-stroke crosshead engine from Wärtsilä is used as an example.
Wärtsilä builds a series of crosshead engines with cylinder diameters from 480 to 960 millimetres, with power outputs between 7000 and 76,000 KW.
Heavy-fuel oil is used as the fuel. It has a maximum viscosity of 730 cSt at 50 °C; ISO 8217, category ISO-F-RMK55. This is heavy-fuel oil with a sulphur content of ± 3.5%.

Also, see Chapter 8, *Fuels, fuel-line systems and cleaning fuel.*

In the combustion-process overview, air, fuel and lubricating oil are added as follows to the engine.

▼

An overview of the combustion process in a two-stroke crosshead engine running on heavy fuel.

Note the ratio between the amount of the fuel and air supplied to the engine. Here the ratio is 1: 45.6. The oxygen content (O_2) shows that there is still a considerable amount of oxygen in the exhaust gas. In the combustion of 7.8 kilograms air, the oxygen content is reduced from 1.638 kilograms to 1.1 kilogram. The consumption during the combustion process is 1.638 – 1.1 = 0.538 kilogram per kW. The excess air (Lambda) of this engine is $\frac{1.638}{0.538} = 3.04$.

Reminder: air comprises approximately 21% oxygen and 79% nitrogen. 21% of 7.8 kilograms air is 1.638 kilograms oxygen.

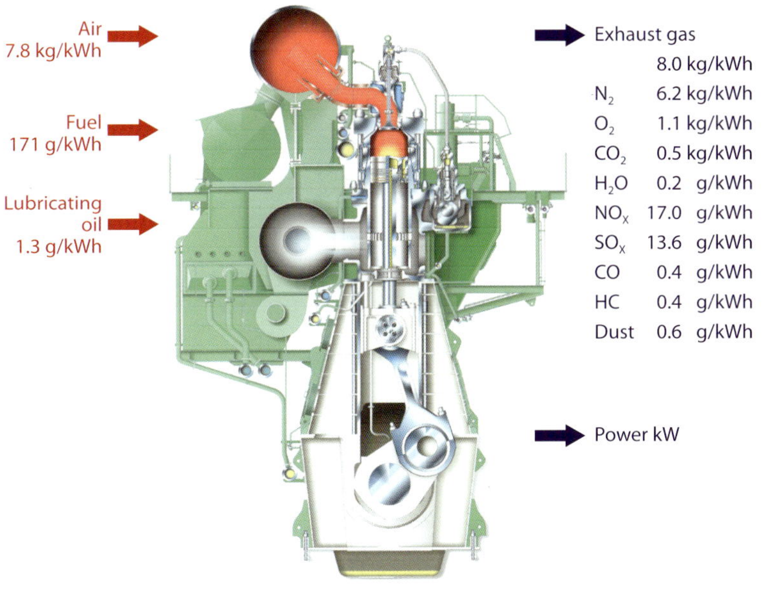

22.2.1 Air

Air consists predominantly of:
- **nitrogen,** N_2, approximately 78% volume;
- **oxygen,** O_2, approximately 21% volume;
- **residual gases,** amongst which are Argon, CO_2 and other inert gases, ± 1% volume.

Comment
The greenhouse effect gas, CO_2 in air is approximately 0.03% volume!

The amount of air supplied to this engine is approximately three times the amount of air required for a complete chemical combustion of the fuel, or: the excess air is 3.

Also, see Chapter 8, *Fuels, fuel-line systems and cleaning of fuel.*

7.8 kilograms of air are supplied per kWh.

22.2.2 Fuel

For this engine, heavy-fuel oil that consists almost entirely of:
- **hydrocarbon compounds**, CH, approximately 97% volume;
- **sulphur**, S, approximately 3% volume.

Heavy-fuel oil, of course, also contains small amounts of other substances.

22.2.3 Cylinder-lubricating oil

Used lubricating oil combusts. In a two-stroke crosshead engine, cylinder lubrication is separate to that of the engine-lubricating system. The engine's lubricating oil almost entirely combusts and is discharged in the exhaust gases. A smaller part flows to the scavenging-air space and is drained.

Also, see Chapter 11, *Lubrication of engines.*

This consists of:
- **hydrocarbon compounds,** CH, approximately 97% volume;
- **calcium,** Ca, approximately 2.5% volume;
- **sulphur**, S, approximately 0.5% volume.

The three above-mentioned substances, air, fuel and lubricating oil are definitive for the combustion process.

In the Wärtsilä overview, 7.8 kilograms of air, 171 grams of fuel and 1.3 grams of cylinder-lubricating oil are supplied per kW hour (kWh), approximately 8 kilograms in total.
Subsequently, 8 kilograms of exhaust gas are formed with the following composition:

– nitrogen	N_2	6.2 kg/kWh
– oxygen	O_2	1.1 kg/kWh
– carbon dioxide	CO_2	0.5 kg/kWh
– water	H_2O	0.2 g/kWh
– nitrogen oxides	NO_X	17.0 g/kWh
– sulphur oxides	SO_X	13.6 g/kWh
– carbon monoxide	CO	0.4 g/kWh
– hydrocarbon compounds	HC	0.4 g/kWh
– dust particles		0.6 g/kWh

An example for the volume percentages for two-stroke crosshead engines type MC from engine manufacturer MAN Diesel.

▶
In this figure, the gas volume percentages are clearly shown.

Similar to fuel, the cylinder-lubricating oil comprises mainly of hydrocarbon chains. The composition of noxious substances in the exhaust gases is shown in p.p.m. (parts per million).

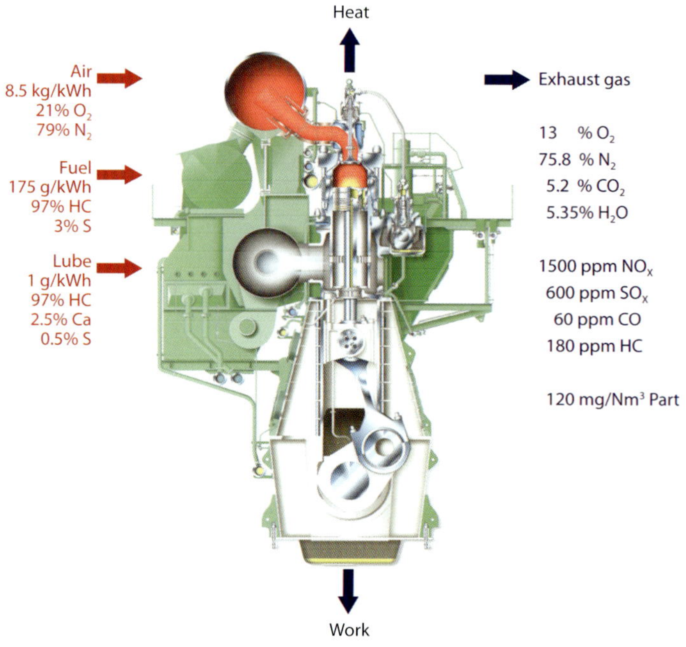

Striking features:

- The largest part of the nitrogen is, as in air, found in the exhaust gases. Essentially, nitrogen is an inert gas and does not normally combust with other substances. Under certain circumstances, for instance in a diesel engine, nitrogen oxide will react with oxygen. This requires process temperatures of over 1200 °C and these frequently occur, often up to 1800 °C!
- The exhaust gases still contain oxygen, also referred to as excess air.
- Water vapour is produced from the hydrogen component in the fuel.

22.2.4 Exhaust-gas composition

1 Oxygen

Diesel-engine exhaust gases always contain oxygen. Oxygen is required for the combustion process, but the air is also used for the scavenging and cooling processes in the engine. Therefore, due to excess air in diesel engines, exhaust gases always contain oxygen, which also indicates the magnitude of the excess air. The oxygen content varies from approximately 13 to 16%. The excess air factor is often 3 or more.

2 Nitrogen

As in air, nitrogen is the main component of exhaust gases. It is inert and therefore does not react with other substances.

3 Carbon dioxide

The carbon in the fuel reacts with the oxygen. This produces carbon dioxide at perfect combustion. This gas appears to contribute to the so-called greenhouse effect. For diesel engines, it is only possible to reduce this emission with the use of light fuels, which have smaller quantities of carbon atoms and consequently more hydrogen atoms, and to develop engines with a higher total efficiency and therefore lower fuel consumption. Maintenance and manner of operation also offer (limited) possibilities to reduce this emission.

Also, see Chapter 27, *Maintenance and repairs* and Chapter 25, *Operational management and automation.*

4 Water vapour

Hydrogen in fuels reacts with oxygen to form water vapour.

5 Carbon monoxide

This is produced during imperfect combustion of the fuel. In principle this occurs when too little oxygen is available for perfect combustion, for instance, close to the walls of the combustion chamber. Poor fuel and air mixture may also be a cause. Carbon monoxide can react with oxygen and this occurs very easily in the engine. It is a very toxic gas for both man and fauna. **In a closed space, carbon monoxide can kill instantaneously when inhaled.**

6 Sulphur dioxide

This is produced with perfect combustion and is a reaction of sulphur with oxygen. In addition to sulphur dioxide SO_2, sulphur trioxide SO_3 is also formed. Sulphur dioxides are the main compounds in acid rain and have a destructive effect on flora and fauna and human respiratory systems. They also attack building materials, in particular soft limestone. The effects of sulphur-oxide compounds are negligible at sea. Seawater is slightly alkaline and therefore they are easily neutralised. One litre of seawater can neutralise approximately

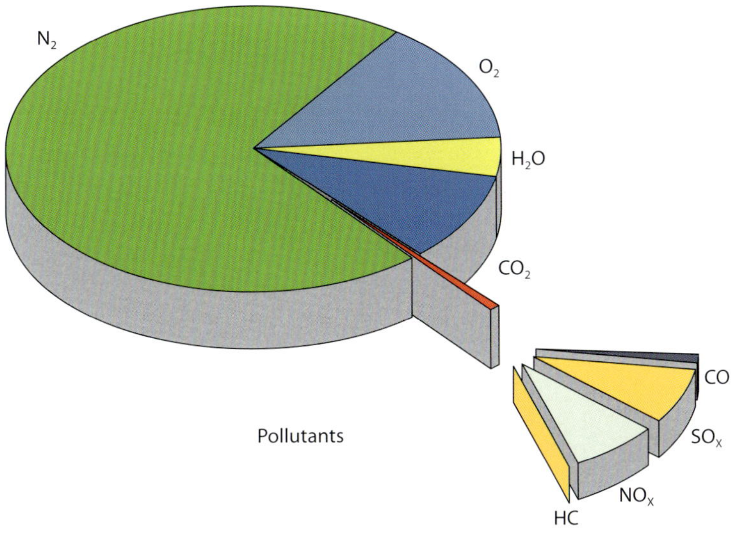

◄ In this figure the composition of exhaust gases.

N_2 = nitrogen
O_2 = oxygen
H_2O = water (steam)
CO_2 = carbon dioxide
SO_X = sulphur oxide
HC = hydrocarbon

300 milligrams of sulphur. Furthermore, sulphur dioxide is not carried long distances but relatively quickly precipitated into the sea. A greater negative effect could occur within 100 kilometres from the coast. However, measurements taken in the busy English Channel have demonstrated that the contribution of shipping to air pollution in the English Channel region is only 5%. The remainder is caused by sulphur oxide emissions from shore-based producers, such as electric-power plants and chemical factories. Sulphur dioxide emissions fall under the IMO, the International Maritime Organisation, regulations.

7 Nitrogen oxides

These are produced during the combustion process at high temperatures between nitrogen and oxygen. Fuel contains small amounts of nitrogen. Heavy fuels contain more nitrogen than light fuels.

Nitrogen oxides are compounded under the following circumstances:

01 **Sufficient oxygen available.** This is always the case in diesel engines. The excess air factor is often more than two or even three.

02 **A temperature in the combustion process in excess of 1200 °C.** In the hottest part of the process, this is normally the case. Temperatures from 1400 °C to 1800 °C are no exception.

03 **Time in the combustion process** so that nitrogen oxides are formed at this high a temperature and in an oxygen-rich environment.

Comment

The total combustion process including expansion is; for two-stroke engines approximately 120 crank degrees and; for four-stroke engines about 140 crank degrees.

▼

The generation of nitrogen oxides.

Horizontal: the temperature in Kelvin of the combustion process and vertical: the time at which these high temperatures occur.

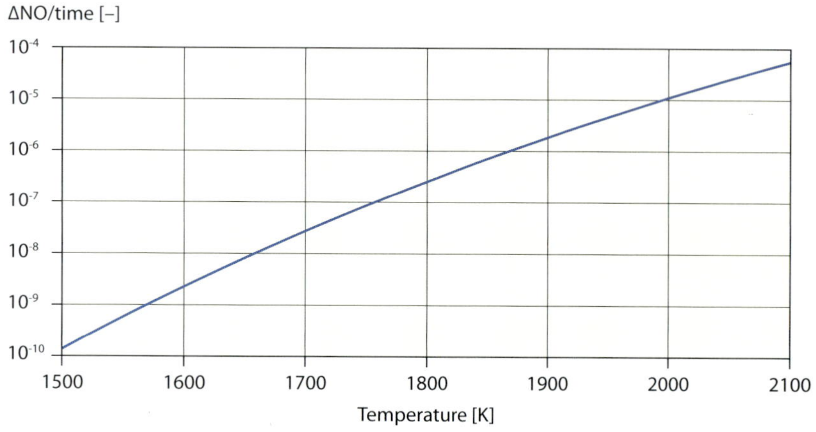

Example 1

Assumption: A two-stroke crosshead engine has an rpm of 94.

The total combustion process including expansion is: the number of passed-through crank degrees per second is the number of revolutions per second × 360 crank degrees:

$\frac{94}{60} \times 360 = 564$ crank degrees per second.

Then the combustion process takes

$\frac{120}{564} =$ **0.212 second,** so approximately $\frac{2}{10}$ second.

Example 2

Assumption: A four-stroke diesel engine has an rpm of 720.

The number of passed-through crank degrees per second:

$\frac{720}{60} \times 360 = 4320$ crank degrees per second.

Then the total combustion process including expansion takes $\frac{140}{4320}$ seconds = **0.032 second,** so approximately $\frac{3}{100}$ second.

▼

A diesel-power station in China for the generation of electricity.

Cheap heavy fuel oil is becoming increasingly restricted in its use all over the world. The use of sulphur-poor heavy fuel oil or even banning the use of heavy fuel oil is occurring more often. Relatively clean natural gas will be used more often as an engine fuel in the future. This is already taking place to a significant scale in engines with the 'Dual Fuel' (DF) principle, a combination of diesel oil and gas, and the 'spark gas' (SG) principle, a gas engine.

A diesel-power plant with a number of medium-speed four-stroke V-engines from manufacturer, ABC in Ghent, Belgium. They run on heavy fuel oil.

Example 3

A small high-speed four-stroke engine with an engine speed of 2100 revolutions per minute:
The number of passed-through crank degrees per second:

$$\frac{2100}{60} \times 360 = 12{,}600 \text{ crank degrees per second.}$$

Then the total combustion process including expansion takes:

$$\frac{140}{12{,}600} = \mathbf{0.011\ second} \text{ or } \frac{1}{100} \text{ second.}$$

In conclusion: Combustion **process:**

I	**low-speed engine**	94 rev/min.	0.212 sec.
II	**medium-speed engine**	720 rev/min.	0.032 sec.
III	**high-speed engine**	2100 rev/min.	0.011 sec.

From the above, it is shown that a low-speed engine has more time for the formation of nitrogen oxides at high process temperatures than medium- or high-speed engines.
So high-speed engines by definition emit far less nitrogen oxide, not only because the fuel used, diesel oil, contains less nitrogen.
The higher the rpm of an engine, the fewer nitrogen oxides it produces.
The IMO regulations assume a certain engine speed and a standard emission weight in grams NO_X per produced kWh.

22.2.5 Hydrocarbon compounds

These are often partially combusted or uncombusted hydrocarbons from fuel oil or lubricating oil. These are very diverse compounds with virtually any

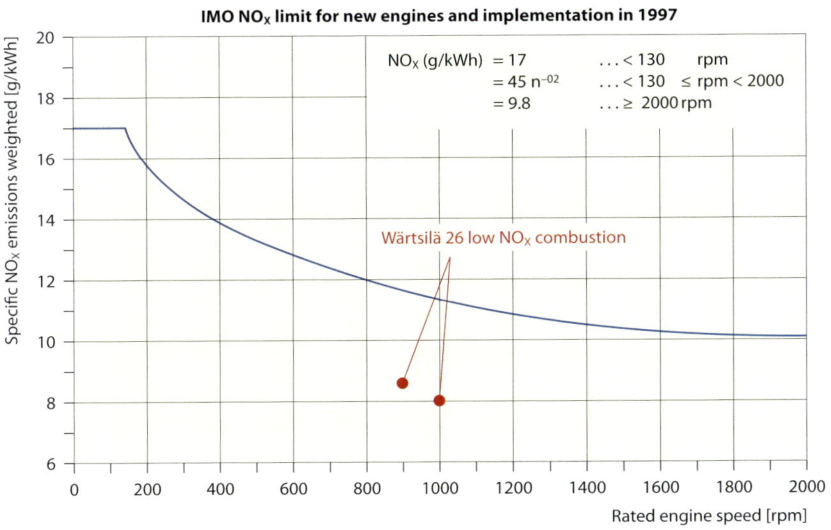

A well-known graph.

Horizontal: the nominal engine speed of the diesel engines and vertical: the nitrogen oxide emissions in grams per kilowatt-hour. For slow-speed engines, the maximum emission is 17 grams per kWh. Medium-speed engines may discharge between 17 and 12 grams per kWh and fast-speed engines between 12 and 9.8 grams per kWh, dependent on the speed. Every engine manufacturer today, builds engines with low nitrogen oxide emissions, the 'Low NO_x Combustion'. Shown is an example of a Wärtsilä 26 engine.

possible chemical combination of carbon, hydrogen, nitrogen, oxygen and sulphur, albeit in extremely low concentrations. The bonds these compounds will form during the combustion process are hard to predict, as are their effects on the environment and human and animal health.
Generally, hydrocarbon compounds are a result of imperfect combustion and in fact largely depend on the parameters of the combustion process, such as velocity, temperature and oxygen content. Well-tuned engines ensure that these HC-compounds are kept as low as possible.

Solids

These are a complex mixture of organic and inorganic substances, amongst others carbon, ash particles, heavy metals, precipitated sulphur oxides, water, corrosion particles and a variety of partially combusted hydrocarbons from the fuel and the lubricating oil. The diameter of these substances is less than $\frac{1}{1000}$ millimetres. They easily be carried to the outside air.
Inhalation of these exhaust gases can be detrimental to health. Naturally, fuel quality plays an important role.

Smoke, visible smoke

Smoke is of course visible and can be quantified by the following methods:

- Bosch smoke number;
- SAE smoke number;
- Bacharach (ASTM) smoke number.

Smoke mainly consists of carbon particles, also referred to as soot.

▲
Almost all diesel engines produce visible smoke.

This is mostly the case with a reduced power output or with quick power changes as with this container ship passing through the locks at Kiel in Germany.

▲
A ferry arriving in a harbour.

Clearly shown is the visible smoke, consisting of soot particles (carbon) and fine dust (metal particles).

▶
A large container ship approaching Hamburg terminal.

At a reduced power output, no smoke is visible. In designing new engines types, engineers try to create the smokeless engine.

Today, engine manufacturers are developing means in which 'visible smoke' can be reduced.

Black smoke is the result of imperfect combustion. Grey smoke may indicate the presence of water in the exhaust gases.

When 100 kilograms of fuel is burnt, approximately 120 kilograms of water is produced with an additional 20 kilograms of water from the scavenging-air supply.
Smoke visibility is also dependent on atmospheric conditions.

Yellow smoke indicates the presence of sulphur in the fuel.
If heavy-fuel oil contains above 2% of sulphur, this is clearly visible.

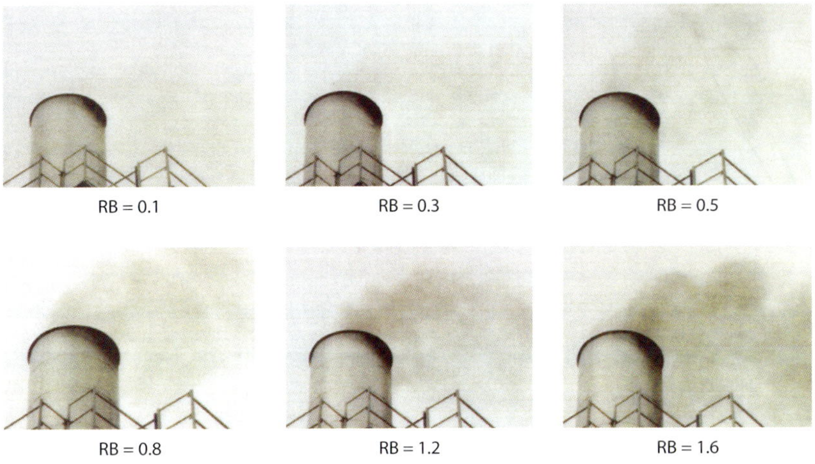

▲
Visible smoke can be defined in smoke numbers.

Shown: the Bosch method. Using these pictures, smoke emissions can be determined.

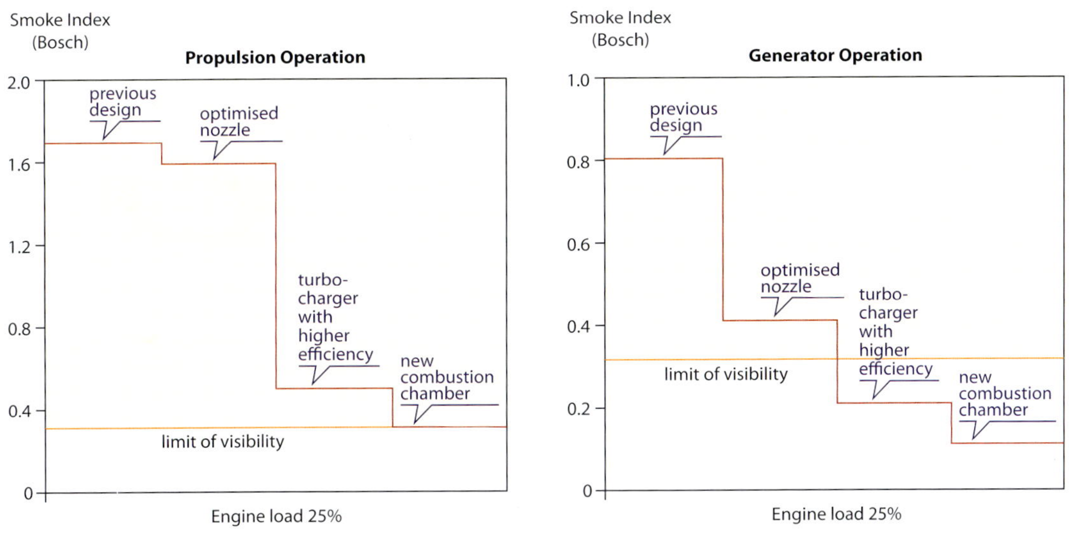

▲
Two examples of the Bosch smoke index.

Left: an example for a propulsion engine and right: an engine for a genset.
The regular load of the generator and the constant high operational speed ensure low values. The irregular load of the propulsion engine at reduced speed has at a low load of 25%, a much higher soot emission. The high constant speed of the generator set allows a better combustion because of the higher process temperatures.

▼
The difference between conventional fuel injection and the common-rail system of Wärtsilä.

horizontal: engine load between 0 and 100%
vertical left: smoke index (FSN is Fuel Smoke Number)
vertical right: permissible visible smoke dependent on the engine power output in MW.

Using modern fuel injection methods such as the common rail, smoke emissions are convincingly reduced between zero load and 80% load. At full load, there is almost no difference.

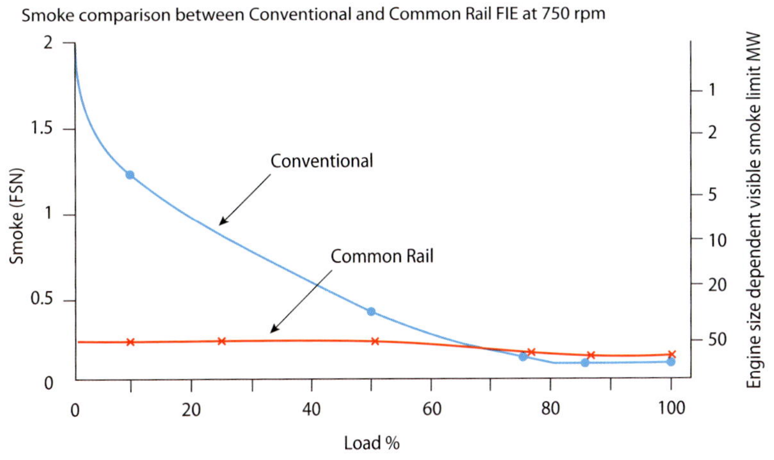

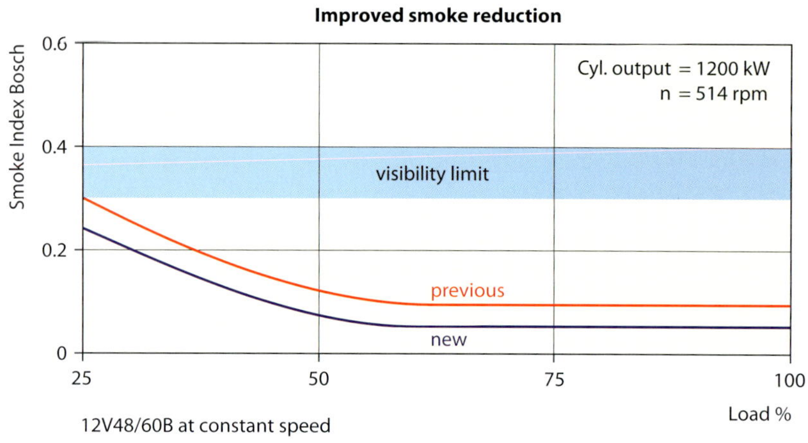

▲

Combining the design of new combustion technologies with existing engine types can significantly reduce smoke emissions.

The visible smoke limit for this four-stroke-diesel engine from MAN Diesel has been established at 0.4 on the Bosch smoke index. The old (previous) design is represented by the red line and the new design by the green line. The engine has a constant speed at full load with a cylinder power output of 1200 kilowatt at 514 revolutions per minute. The Bosch smoke index is below 0.1. The blue area designates the smoke visibility limits.

22.2.6 An example of smoke reduction

In response to universal concern regarding environmental issues, engine manufacturer MAN Diesel started a programme in 1996 to reduce smoke in exhaust gases, especially at low engine loads. The following measures were taken:

- **Pre-heating of the scavenging air.** This method involved compensating the reduced process temperatures caused by the injection of less fuel. This takes place at loads below 20%.
- **Installing partial-load injector nozzles.** This nozzle increases the injection pressure improving the fuel distribution throughout the combustion chamber and fuel-droplet size decreases.
- **A by-pass or 'waste-gate' mounted on the turbocharger.**

Overall, a turbocharger is most efficient at heavy engine loads. The efficiency can be 'advanced' at lower loads by adjusting the charger and diffuser of the air compressor. The scavenging air pressure will also often increase allowing a more effective scavenging of the cylinder. If this engine runs at full load, the scavenging-air pressure will be too high. Therefore, a by-pass or waste-gate valve is placed in the exhaust-gas supply of the exhaust-gas turbine.

▶

Adjusting the amount of air supplied to the engine is important at variable loads.

Left, the exhaust gas by-pass so that the amount of charge air is limited at full load. Right, the amount of charge air is controlled by the quantity of charge air allowed through the by-pass. Both methods have the same goal: the optimal amount of air at all engine loads so that optimum combustion can occur.

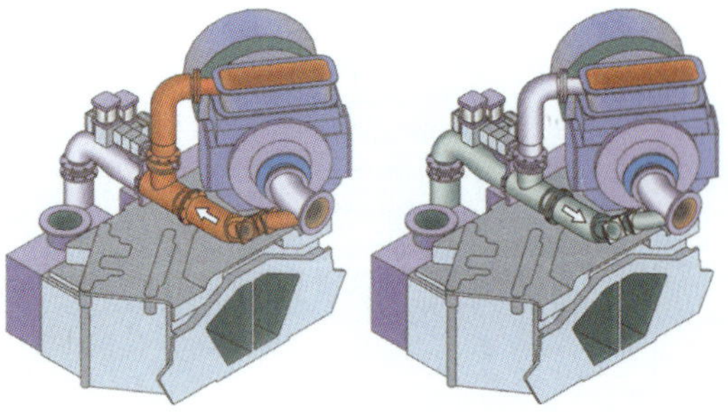

▶

An example with a mechanism on both the inlet manifold and the exhaust-gas manifold so that a 'waste-gate' is mounted on both systems.

Both valves are adjustable. Furthermore, a by-pass valve has been placed between the air compressor and the exhaust-gas turbine that controls the amount of air to the engine.

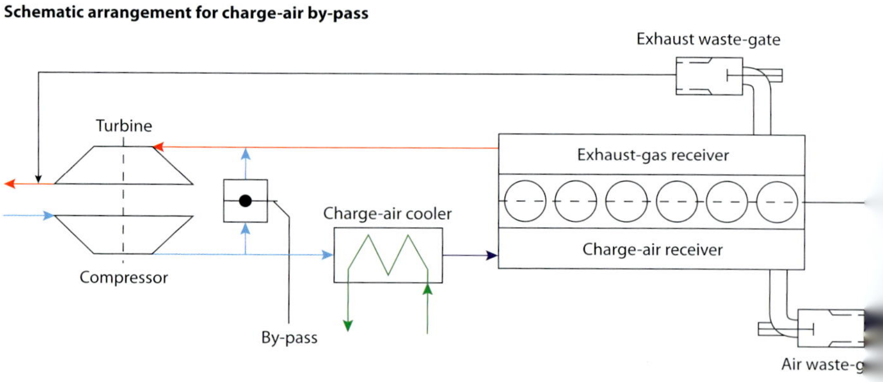

Fuel emulsions

The main objective of a fuel-water mixture (fuel emulsion) is the reduction of nitrogen oxides in the exhaust gases. At low loads, visible smoke is reduced as the fuel-water emulsion is injected at higher pressures.
Placing a **bypass valve** between the scavenging-air receiver and the exhaust-gas manifold is also an option. This valve is opened at low engine loads. Due to reduced air resistance, the turbocharger will revolve faster, producing a higher flow efficiency effecting a higher airflow through the engine.

Auxiliary charger

For a standard turbocharger, the efficiency at no load or low loads lies far below the maximum value. In this instance, an electrically driven auxiliary charger can be installed to reduce significantly visible smoke.

Fuel quality

This is the main influence on the combustion-process quality and therefore smoke production. At times, fuels with different specifications are bunkered. This specification change can cause a considerable increase in smoke production. It is then advisable to order the original specified fuel for the next delivery.

22.3 Units of contamination

The substance load is often indicated in kilograms or tons per hour or year. Presently, grams of contamination per generated unit kilowatt-hour, is frequently used when referring to energy generation.
Simply: contamination is denoted in g/kWh.
The unit, parts per million or **ppm is also used.**

Exhaust-gas emissions must be specified at a certain oxygen concentration, for example 5, 13, or 15% O_2. This is often not related to the actual oxygen percentage found in air in the diesel engine in question. It is, however, easy to convert, for example p.p.m. in volume to g/kWh by applying the following formula (ISO 8178 part 1).

$$X_{corr.} = X_{ref.} \frac{20.95\% - O_{2\ corr.}}{20.95\% - O_{2\ ref.}}$$

$X_{corr.}$ = actual gas concentration
$X_{ref.}$ = reference oxygen concentration

22.3.1 Pollutant-emission regulations with respect to diesel engines

In the final decades of the last century, growing environmental concern led to regulations for limiting pollutants in exhaust gases of, amongst others, diesel engines.
Universal regulations have been established for shipping. This prevents individual countries or continents drawing up their own regulations. They were initiated by the United Nations.
The IMO has recorded these in Annex IV of Marpol 73/78, which for the first time in history stipulates the regulations for maximum exhaust-gas emissions. The annex concerns the limitation of sulphur- and nitrogen-oxide emissions and has been compulsory since 2000 for all newly built engines with a shaft power over 130 kW.
This ratification of the NO_X and SO_X limitations became effective when more than fifteen flag-states, members of the IMO with a tonnage of at least 50% of the total world tonnage, signed this agreement.

These agreements entail:

- IMO regulation for the maximum sulphur- and nitrogen-oxide emissions;
- optimising diesel-engine designs in order to comply with the IMO regulations;
- maintain a 'technical' bookkeeping;
- terms such as 'single engine', 'engine group' and 'engine family';
- comments/instructions from engine manufacturers;
- comments/instructions from shipping companies.

Sulphur oxides

Fuel sulphur content limited to a maximum of 4.5% mass. In areas with local limitations, 1.5% mass sulphur.

Nitrogen oxides

The maximum limits are based on the rpm for diesel engines with a shaft power of over 130 kW. This rpm graph is only applicable for ships built after the 1st of January 2000, or to engines that have undergone extensive modifications.
Test methodologies to check emissions at certain engine loads have also been established.
For ships built before 2000 that have not undergone extensive engine modifications or substitutions, the fuel sulphur limit is applicable.
In short, each diesel engine with a shaft power in excess of 130 kW must have an Emission Certificate. The country under whose flag the ship sails or will

▶

The IMO directives for regulating the emission of the noxious substances, NO_X and SO_X.

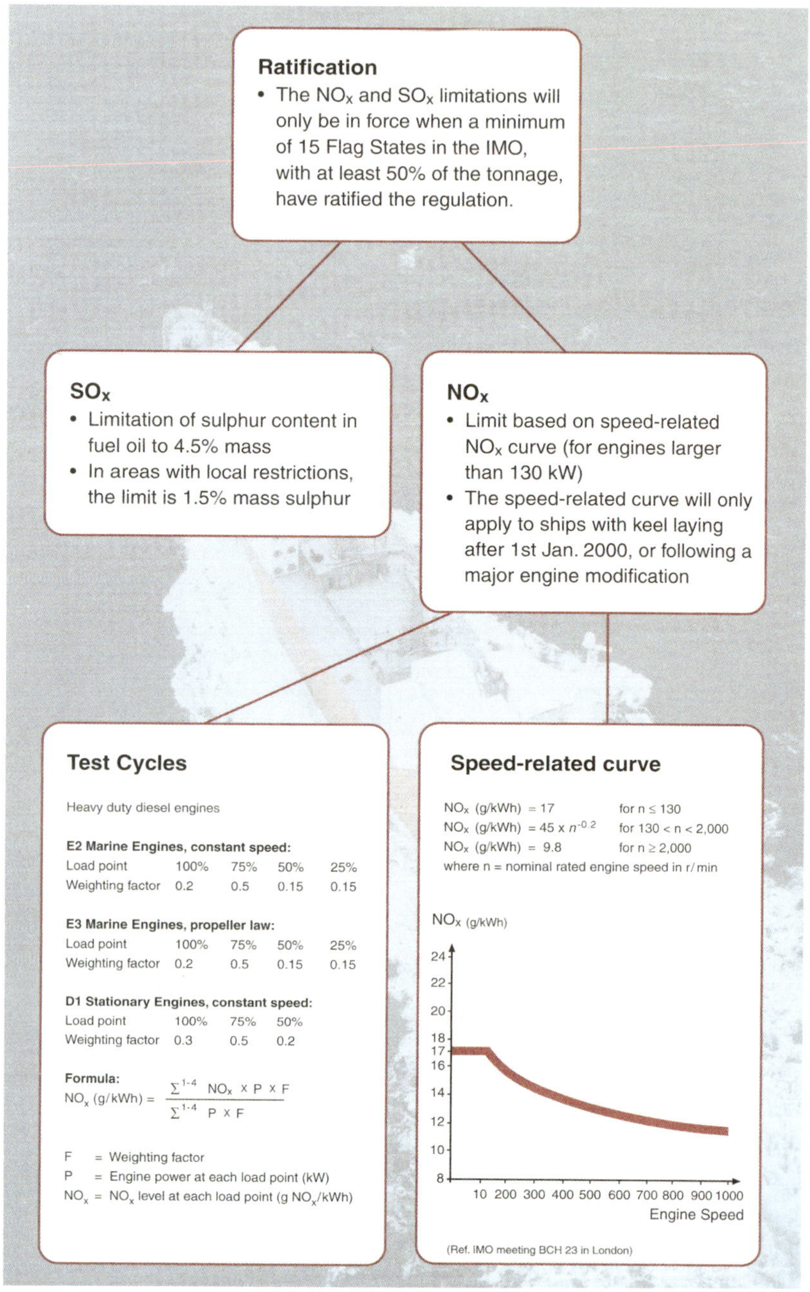

sail is responsible for the inspection and certification of the diesel engine. So worldwide, each engine has an **'EIAPP': Engine International Air Pollution Prevention Certificate.**

Apart from the technical data, **'The Technical File'** should also state which engine parts could influence the NO_X emission values.
After the engine has been approved, these parts may not be replaced by other parts unless the substitutes have been tested and approved.

22.3.2 Additional legal regulations for diesel engines

There are regulations already in force with regard to diesel engines for road- and inland-shipping transport. These will become increasingly stringent in different phases.

The **Euro norms for road transport show** that the NO_X emission requirements have become increasingly strict between 2000 and 2012.

Other regulations, such as those stipulated by the **World Bank, Det Norske Veritas** or **Lloyd's,** underline that people worldwide are taking nitrogen-oxide emissions increasingly seriously.
Proprietors of **diesel-power plants,** predominantly built in developing countries, are obliged to comply with the World Bank emission requirements if they wish to borrow money.

In **Scandinavia**, there are local regulations that apply in coastal waters, for instance, fjords.

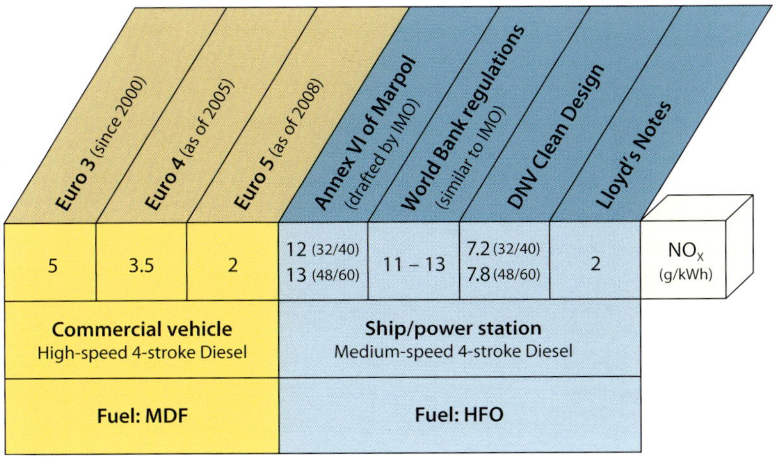

◀ The Euro-standards for road transport compared to the standards for ship propulsion and diesel-power plants.

22.3.3 Important data for the sulphur content in ship fuels

2005	May 19 Aug. 11	IMO EU	MARPOL Annex VI enters into force EU Sulphur Directive 1999/32 as amended by 2005/33 enters into force
2006	May 19 Aug. 11	IMO EU EU	Baltic SECA enters into force Baltic SECA enforced by EU directive 2005/33 Max. 1.5% sulphur for passenger ships to/from EU ports (2005/33)
2007	Jan. 1 Aug. 11 Nov. 22	CARB EU IMO	Max 0.5% sulphur within 24 miles of California shore North Sea and English channel SECA enters into force North Sea and English channel SECA enters into force
2010	Jan. 1	EU EU CARB	Max. 0.1% sulphur bunker fuel in use at EU berths Max. 0.1% sulphur in all EU inland waterways Max. 0.1% sulphur within 24 miles of California shore
2012	Jan. 1	EU	Max 0.1% sulphur fuel in use by Greek ferries at Greek ports

Maximum sulphur content in ship fuels in the European Community

Fuel	S-%	When
Marine diesel oils	1.5	Aug. 11, 2006
Marine gas oils	0.1	Jan. 1, 2010

Maximum sulphur content in ship fuels per ship type in the European community following the EU directive 2005/33/EC

Ship type	Area	%	When
All	Baltic SECA	1.5	Aug. 11, 2006
All	North Sea + English Channel SECA	1.5	Aug. 11, 2007
All	All EU ports	0.1	Jan. 1, 2010
Passenger ships	All EU	1.5	Aug. 11, 2006
Inland waterway vessels	All EU inland waterways	0.1	Jan. 1, 2010

Combination of rules for fuels according to the Marpol Annex IV and the new EU directives

When	Ship type	Area	%	Act
May 19, 2006	All	Baltic SECA	1.5	Marpol
Aug. 11, 2006	All	Baltic SECA	1.5	EU
Aug. 11, 2006	Passenger ships	All EU	1.5	EU
Aug. 11, 2007	All	NS+E C SECA	1.5	EU
Nov. 22, 2007	All	NS+E C SECA	1.5	Marpol
Jan. 1, 2010	All	All EU ports	0.1	EU
Jan. 1, 2010	Inland waterway vessels	All EU inland waterways	0.1	EU
Jan. 1, 2012	16 Greek ferries	Greek ports	0.1	EU

▲
A diesel-power plant in Kenya using heavy fuel oil.

In countries as these, the rules for the emission of noxious substances are often inadequate or not present. When new power stations are now built, far more attention is paid to the environment and bankers such as the World Bank have their own requirements. Environmental measures cost a lot of money and therefore are often only available for the richer industrial countries.

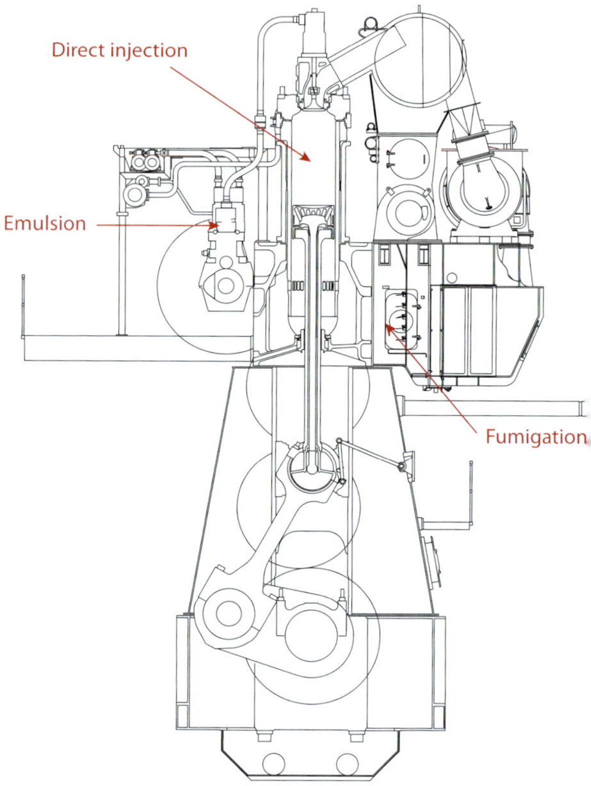

▲
Three possible ways to influence the combustion process for large crosshead engines using water, ensuring that maximum combustion temperatures decrease and therefore producing less nitrogen oxides.

method 1: mixing water with fuel to obtain an emulsion (emulsion)
method 2: humidification of the charge air (fumigation)
method 3: direct injection of water in the combustion space (direct injection)

22.4 Methods for the reduction of exhaust-gas emissions

We distinguish between **primary** and **secondary** methods.

Primary methods deal with the combustion space and are therefore referred to as combustion-technique improvements.
Secondary methods deal with the exhaust gas and are referred to as exhaust-gas treatment.

22.5 Primary methods

22.5.1 Nitrogen oxides – NO_X

The quantity of nitrogen oxides produced is dependent on the gas temperature in the combustion chamber. The higher the gas temperature the larger the quantity of NO_X formed. Therefore, a lower gas temperature offers a solution. However, total efficiency is somewhat reduced by these lower process temperatures.
In the following figure, all the methods aim at nitrogen-oxide reduction.

▶
Methods to reduce nitrogen oxides.

green: decrease in the amount of nitrogen oxides in the exhaust gases
red: increase of the specific fuel consumption per measure

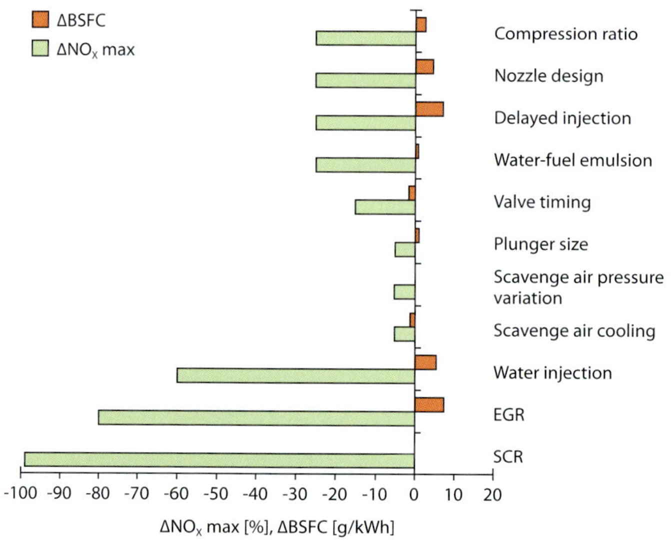

Important issues are:
- scavenging-air temperature;
- Miller method for supercharging;
- turbocharger cooling;
- scavenging-air pressure;
- valve timing;
- compression ratio;
- delayed fuel injection;
- injector-nozzle design;
- fuel-plunger dimensions;
- exhaust-gas recirculation;
- addition of water.

22.5.1.1 Decreasing the scavenging-air temperature and using the Miller-principle

Both entail the reduction of air temperature at the beginning of the compression stroke. Every 3 °C reduction in air temperature produces a nitrogen reduction of approximately 1%. One cannot normally cool below the dew point, as the water vapour will condense inhibiting a temperature reduction. In tropical conditions, the dew point is approximately 32 °C.

A simple NO_X reduction can be achieved by applying the Miller-process for four-stroke engines. In the **Miller-process, the inlet valve is closed before**

Engine-room ventilation (4V69E8169)

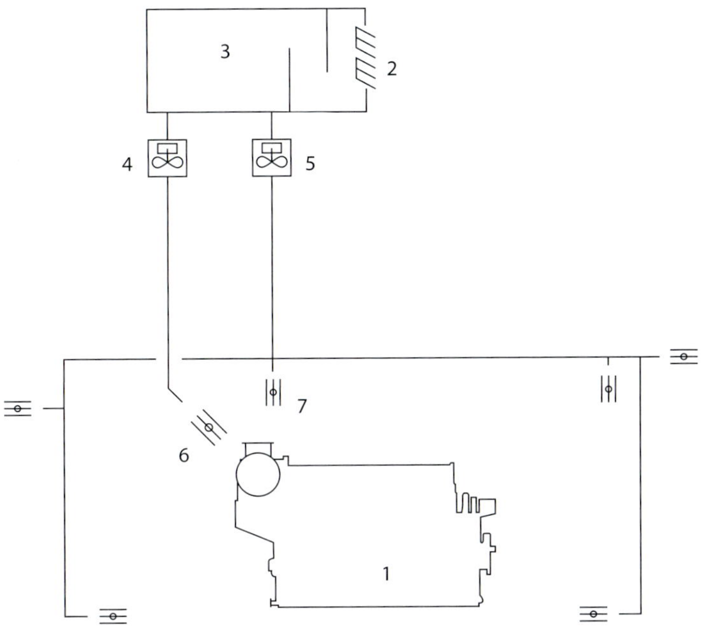

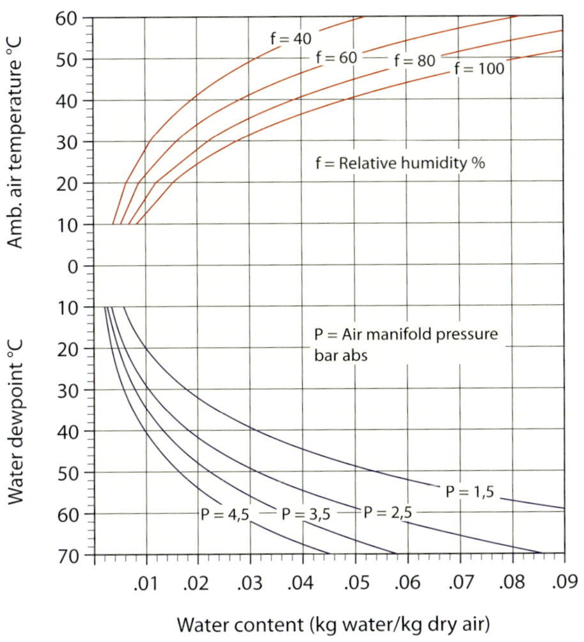

◀

Upper picture: Schematic arrangement for normal engine-room ventilation on a ship.

1 *diesel engine*
2 *suction louver*
3 *water trap*
4 *combustion-air fan*
5 *engine-room ventilation fan*
6 *flap*
7 *outlets with flaps*

Lower picture: Graph of the condensation of steam in air coolers.

Example

- At an ambient air temperature of 35 °C and a relative humidity of 80%, the content of water in the air is 0.029 kg water/kg dry air. If the air manifold pressure (receiver pressure) under these conditions is 2.5 bar (= 3.5 bar absolute), the dew point will be 55 °C. If the air temperature in the air manifold is only 45 °C, the air can only contain 0,018 kg water per kg dry air. The difference, 0.011 kg (0.029 – 0.018) will precipitate as condensed water.
- In an engine using 5 kilograms of air per second, according to the graph, 5 x 0.011 × 3600 = 198 kilograms water per hour is discharged. This is almost 5000 litres per day!

the piston has reached bottom dead centre.
The expansion of the scavenging air in the closed cylinder leads to a lower scavenging-air temperature!
The reduction of nitrogen-oxide emissions is remarkable: a reduction of 15 to 20% can be achieved.

22.5.1.2 Adjusting the scavenging-air pressure and valve timing

If excess air is reduced from, for instance, 2.2 to 1.8, nitrogen-oxide reduction becomes noticeable, as there is less oxygen in the combustion space to react with nitrogen.
Additionally, the exhaust valve is closed earlier and the compression rate increased keeping a constant ratio of combustion pressure to compression pressure.
The reduction of the nitrogen oxides can be as high as 15%. However, fuel consumption increases slightly.

22.5.1.3 Delayed fuel injection

An important factor in the production of NO_X is post-compression of the combusted gases. When fuel and air combust, the temperatures peak.
If the combusted gases are then further compressed before the piston reaches top dead centre, a considerable amount of extra NO_X will be produced.
This can only be resolved by delaying the fuel injection. This is in fact the best method to reduce nitrogen-oxide emissions.

In principle, delayed fuel injection warrants a lower maximum combustion pressure and slightly lowers engine efficiency. In the two-stroke engines from Wärtsilä Sulzer RTA, a delayed fuel injection of two crank degrees produces a reduction in the maximum combustion-pressure of 10 bars, which in turn effects a 10% reduction in the production of NO_X and a fuel-consumption increase of 3 grams per kWh.
Delayed fuel injection can achieve a maximum NO_X reduction of 25%.

22.5.1.4 Increasing the compression ratio

The loss of engine efficiency caused by the fuel-injection delay can be compensated by increasing the compression ratio.
In this way, the production of NO_X can be reduced by 2% and the air consumption will only increase by 1%.

▶
The production of nitrogen oxides in relation to the crankshaft position with three different fuel-injection methods.

Delayed injection method and a high compression ratio give the lowest emission of nitrogen oxides.

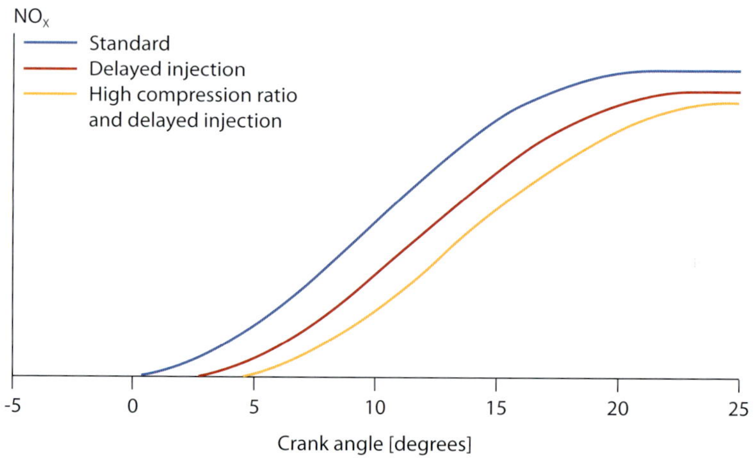

▶
The decrease in engine efficiency caused by delayed fuel injection can be compensated by increasing the compression ratio.

Therefore, NO_X production can be decreased by 2% and the specific air consumption only increases by 1%.

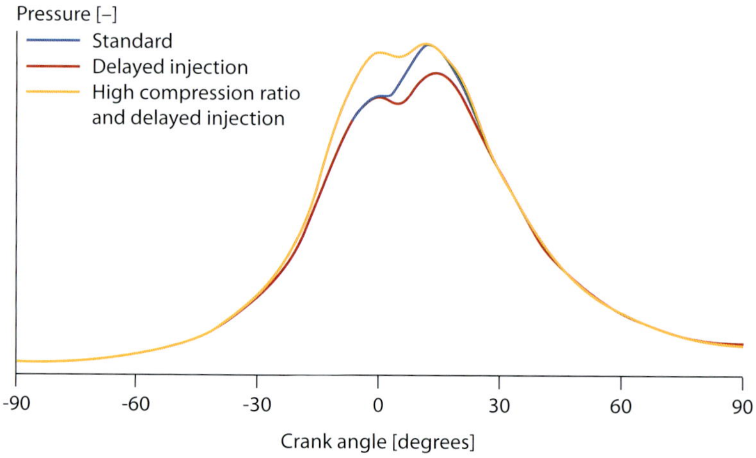

22.5.1.5 Adjusting the injector nozzles

The injector-nozzle design significantly affects combustion.
Nozzle-aperture diameter determines the injected-fuel velocity and consequently the depth of penetration of the fuel droplets in the high-pressure compressed air. This in turn is a determinant in the quality of the evaporation process, turbulence, mixing and fuel combustion.

In a four-stroke engine, the fuel is injected centrally into the relatively flat combustion chamber, which is related to the relatively small stroke-diameter ratio of 1.1 to 1.2.
In a two-stroke engine, the fuel is injected from the external ring of the combustion chamber via two or three nozzles, positioned in a circle around the rim. This can influence the combustion process considerably. Therefore, 'Low NO_X' nozzles have been designed which result in slightly lower nitrogen-oxide emissions.

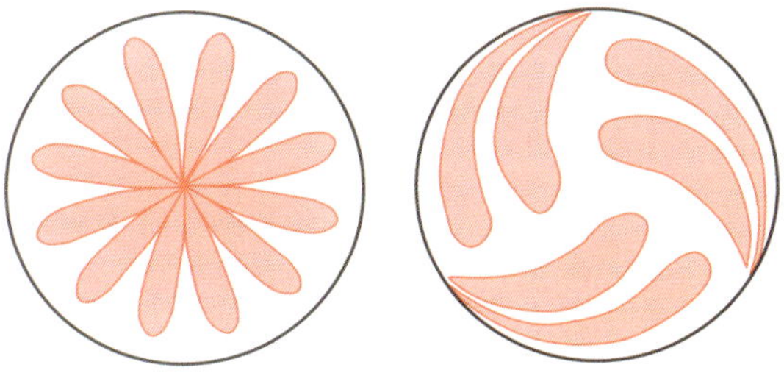

▲
Fuel-Injection patterns.

Left: the injection pattern for a four-stroke engine shown from the heart of the cylinder and right: the injection pattern for a two-stroke crosshead engine from the cylinder periphery with three injectors.

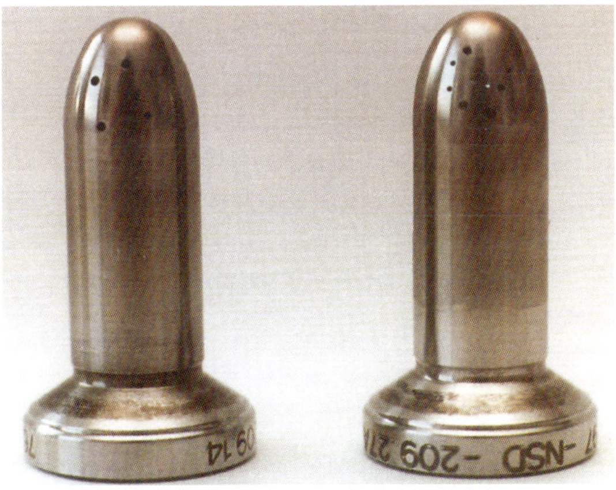

▲
Left: a standard injector nozzle for a two-stroke crosshead engine and right the Low NO_X nozzle.

22.5.1.6 Modifying fuel-injection pressure and timing

Fuel injection can be adjusted by modifying the diameter of the injector-pump plunger or the nozzle-aperture surface area.
Both parameters influence the fuel-spray penetration, fuel-spray pattern in the cylinder and the turbulence generated in the combustion chamber.
Research has demonstrated that the effect on the NO_X production is disappointing. In the most favourable situation, a 5% reduction of the NO_X production can be achieved when prolonging the injection time by 10%.

22.5.1.7 Addition of water

A significant reduction of nitrogen oxides can be achieved by the addition of water to the combustion process. This can be done in one of three ways:

1. direct water injection in the combustion chamber;
2. atomised water in the scavenging air, thus increasing the scavenging-air moisture content;
3. addition of water to the fuel, producing a fuel-water emulsion.

Engine designers have justified concerns regarding the addition of water to the scavenging air. Mist separators extract excess moisture from the entering scavenging air in order to avoid damage of the lubricating-oil film on the cylinder liner. Moreover, water droplets affect the lubricating-oil film rather than the moist air itself. In practice, the system for increasing the relative humidity is not used and on the contrary, moisture is deliberately extracted from the scavenging air.
To achieve a satisfying result by the addition of water, it must be injected in the combustion chamber at the correct time and place. This is the area where most NO_X is formed.
It is imperative that the water quality is acceptable. Generally, water is used that has been produced by the seawater evaporator heated by the HT cooling-water system of the engine.

▶

A fuel injector with two nozzles, left: the water nozzle and right: the fuel nozzle.

Direct injection of water in the cylinder

The water is injected by a completely independent injection system and monitored by an electronic control system. This allows huge amounts of water to be injected without compromising the fuel injection to the engine, which would decrease power output. Moreover, the water-injection timing does not have to coincide with the moment of fuel injection. Water injection takes place via the standard fuel injector, which is an integrated water and fuel injector with two separate nozzles, a fuel nozzle and one for the water.

In two-stroke engines, results of a 50% NO_X reduction have been measured when injecting 140 grams of water per kWh.

For a crosshead engine of 30,000 kW this means that per hour $\frac{30{,}000 \times 140}{1000}$ 4200 kilograms of water must be injected. A large storage tank and a good seawater evaporator are an absolute necessity!

The water is injected 60 crank degrees before TDC, far before the start of fuel injection. This commences approximately 20 crank degrees before TDC.

The figure shows an overview of nitrogen-oxide reduction with various methods.

Therefore, there is no water mist in the scavenging air.

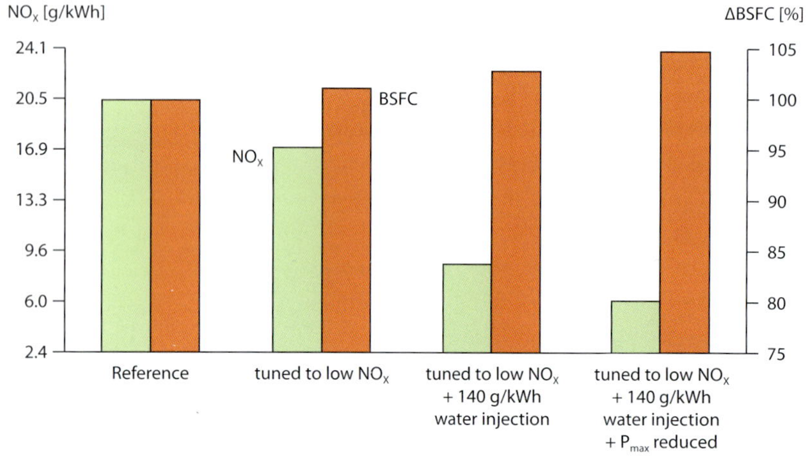

▲

In this graph, another example showing the methods used for reducing nitrogen-oxide emissions.

horizontal: three methods in respect to the standard engine. Far left in the graph, the reference

vertical: emission of NO_X in grams per kilowatt-hour

The green column represents the NO_X levels; the red column represents the increase in specific fuel consumption (BSFC) in percentages.

Conclusion: A large decrease in nitrogen oxides results in a visible increase in the fuel consumption.

Adding water to fuel oil

This is a well-known technique in the reduction of NO_X emissions and can reduce NO_X emissions up to 50%.

This is shown in the figure. However, fuel consumption may increase slightly.

As the fuel pump supplies extra water to the cylinder, the maximum amount of water supplied is dependent on the pump capacity. If the pump capacity is insufficient, water can displace the fuel, which will lead to a power-output decrease!

It is therefore possible that the camshaft-driven fuel-injection system requires adjustments.

It follows that the injector nozzle must also be reconfigured, as the amount of liquid pumped through it will increase. This results in more or larger apertures in the nozzle.

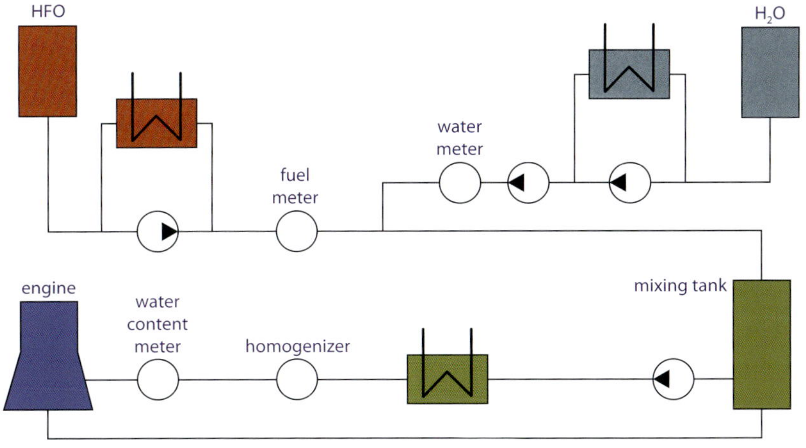

An installation for the supply of water before the engine.

22.5.1.8 Exhaust-gas recirculation

This is a very efficient method for the reduction of NO_X emissions. It is generally used in engines for road transport.

This system is based on three factors.

1. The oxygen content in the exhaust gases which is reduced by the combustion process, causes, when re-circulated, an increase in the amount of gas that requires heating before combustion. This results in a decrease in combustion temperatures, which in turn leads to a reduction in the formation of NO_X. The average process temperature is not affected if the recirculating gas is cooled.
   Engine efficiency does not decline when the oxygen content in the combustion process is decreased.
2. The reduced oxygen content lowers the reactions between the nitrogen and oxygen, which forms nitrogen oxide (NO), and therefore reduces the NO_X emissions.
3. Inert components, such as water and carbon dioxide, which recirculate to the cylinder, are reheated during combustion. Water and carbon dioxide absorb more heat than air. This leads to an increase in the total heat capacity of the gas that is compressed and consequently levels local peak temperatures in the process. This, naturally, reduces the production of nitrogen oxides.

This system is not used in two-stroke crosshead engines, the exhaust-gas pressures before the exhaust-gas turbine of the turbocharger are always lower than the scavenging-air pressure after the air compressor of the turbocharger due to cylinder scavenging. Moreover, the exhaust gas of engines running on heavy-fuel oil is too polluted and would therefore contaminate the engine too quickly.
In fact, the recirculation method can only be applied to four-stroke engines using relatively clean fuels, such as light diesel oil, petrol and natural gas and is frequently used in Otto engines.

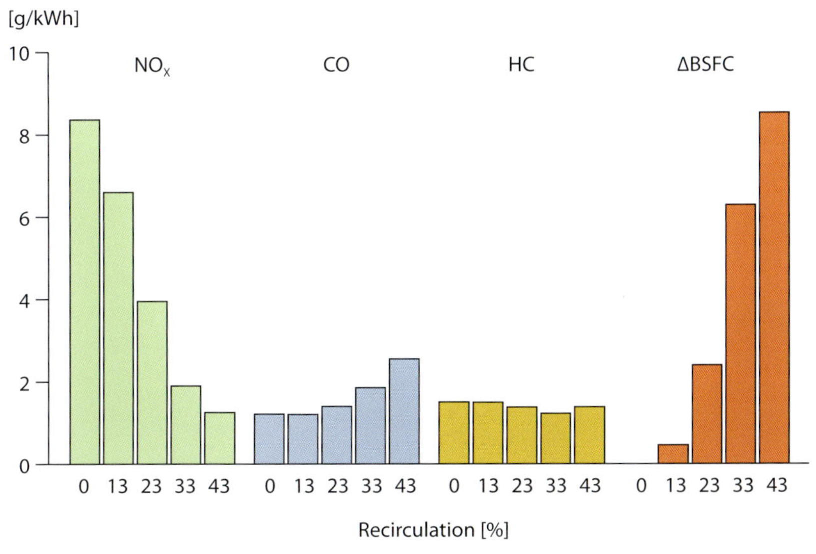

A graph clearly showing the effect of partial recirculation of exhaust gases.

green: NO_X decrease at different recirculation percentages
blue: carbon monoxide increase
yellow: hydrocarbon decrease
red: specific fuel consumption increase

▼

Decreasing the sac volume under the injection needle can reduce the emission of uncombusted hydrocarbons.

1 *sac volume (space below the injector needle)*

22.5.2 Hydrocarbon particles and other dirt particles

Due to improvements in the diesel process aimed at achieving low specific fuel consumption, uncombusted hydrocarbon particles have virtually disappeared.
This means that combustion is practically 100%.
The hydrocarbons in the exhaust gases occur mainly due to lubricating-oil consumption and fuel evaporation in the apertures of the fuel-injector nozzle.
As the current lubricating-oil consumption already is very low, reducing the hydrocarbon content can only be achieved by diminishing the volume of the injector-nozzle space after the injector needle.

22.6 Secondary methods

After all the primary methods have been exhausted and emission norms are still not attained, the exhaust gases can be post-treated.
The following methods have been in long-term use.

22.6.1 Catalytic converters

Oxidation of hydrocarbon compounds and reduction of nitrogen oxides.

22.6.2 Thermal reactors

For hydrocarbon compounds and carbon monoxide.

22.6.3 Systems and filters for the interception of fine particles

Methods used are, amongst others, electrostatic- and dry filters.

22.6.4 Selective catalytic reduction

In professional literature referred to as SCR.
For diesel engines, this is the only method that well established, as opposed to Otto engines where several systems work satisfactorily.
This method involves the mixture of exhaust gases with ammonia, which subsequently flows through a labyrinth that has a specially coated surface area at temperatures between 300 and 400 °C.
Here NO_X is converted to N_2 and H_2O.

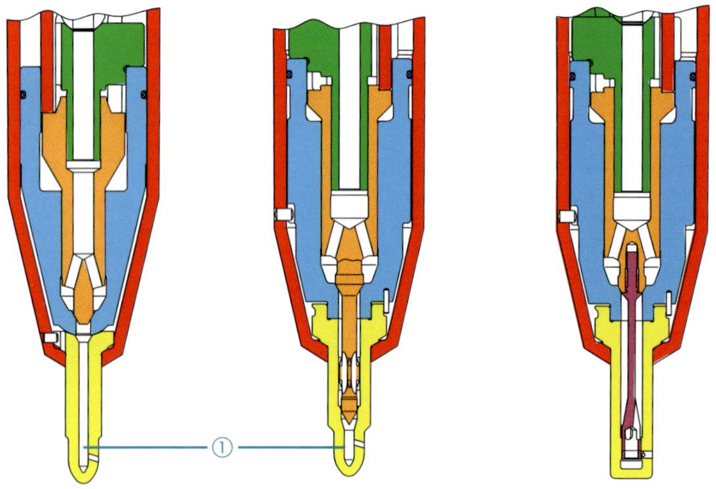

Conventional fuel valve
Sac volume 1690 mm³

Mini-sac valve
Sac volume 520 mm³

Slide-type fuel valve
Sac volume 0 mm³

▶

In the near future, progressively more diesel-power plants will use exhaust-gas cleaning in order to meet the increasingly stringent requirements concerning the emission of noxious substances.

A power station using heavy fuel oil. This is clearly seen in the exhaust gases.

The nitrogen oxides are converted into the harmless compound nitrogen and water vapour in accordance with the following chemical reaction:

$4NO + 4NH_3 + O_2 \rightarrow 4N_2 + 6H_2O$

$6NO_2 + 8NH_3 \rightarrow 7N_2 + 12\,H_2O$

NO and NO_2 combine to form NO_x. NH_3 is ammonia (urea solution).
Oxygen is necessary for this reaction and there is sufficient oxygen in exhaust gases.

The temperature at which this reaction occurs is important. Excessively high exhaust-gas temperatures combust ammonia before it can react with NO and NO_2. Low exhaust-gas temperatures produce a partial reaction and the formation of ammonium sulphates, which damage the catalyst.

The amount of ammonia used in the system depends on the amount of nitrogen oxides produced by the engine, which in turn is dependent on the engine load. A process computer controls the entire system and immediately responds to changes, for instance, more engine power means the injection of extra ammonia.

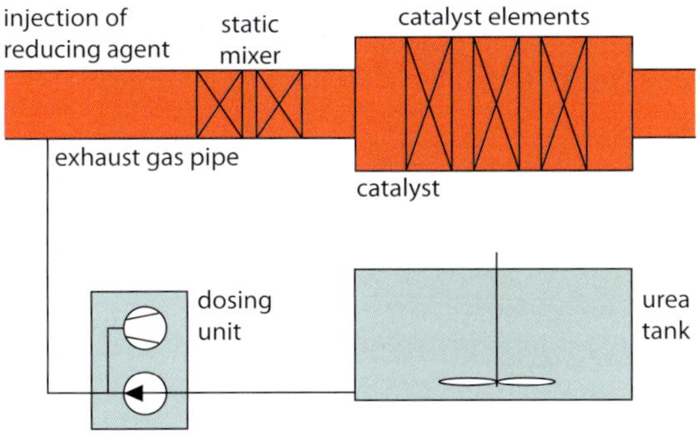

▲
Cleansing the exhaust gases by the injection of ammonia to decrease the nitrogen content.

The amount of nitrogen oxide that can be removed from the exhaust gases is dependent on the amount of ammonia injected.
A very small amount of ammonia is not used and flows with the exhaust gases into the system. It, for instance, may flow through an exhaust-gas boiler. When too much ammonia is injected, the unused ammonia may react with sulphur trioxide (SO_3) found in the exhaust gases and subsequently contaminate the heating surface of the waste-heat boiler with ammonium sulphates.

Ammonia can be:
- liquid;
- water-free under pressure;
- mixed with water under atmospheric pressure;
- urea, a dry product and dissolved in water before use.

▼
A large two-stroke crosshead engine supplied with a complete system for exhaust-gas cleaning.

At present, this system is seldom applied in diesel power plants. On ships, this system is never used, as the space required is too vast.

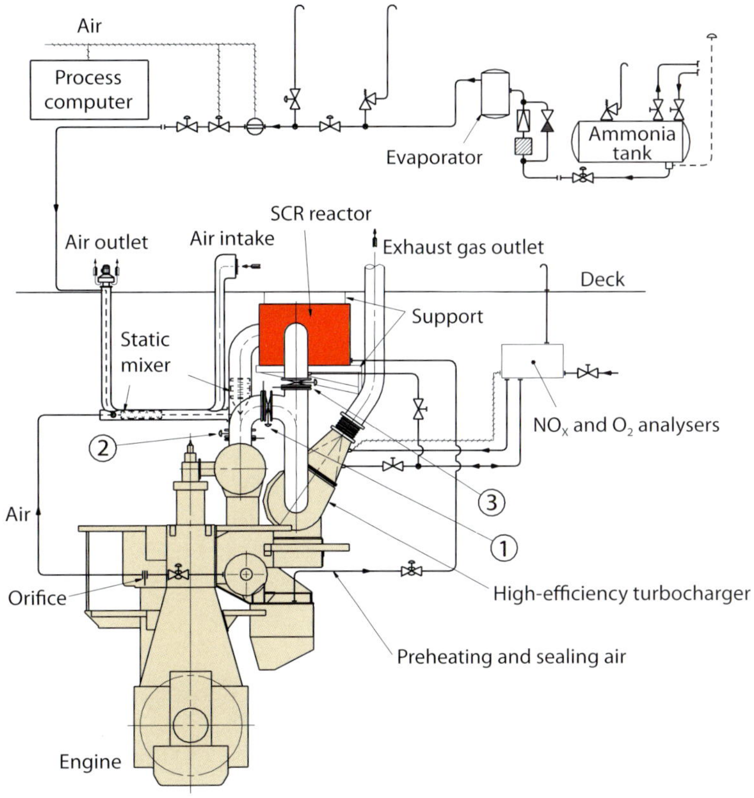

Engine manufacturers are increasingly putting systems on the market that significantly reduce the amount of emissions. An optimum combination of fuel injection with the common rail system, which is particularly effective at partial load and the Miller system, which is an earlier closing of the exhaust valve are now standard in new engines. These are called the primary cleaning methods: in the combustion chamber.

In these systems, nitrogen oxides can be reduced by 80 to 90%.

▶
A turbocharger with adjustable blades (1) in the diffuser section.

The amount of air can be carefully adjusted dependent on the optimum air requirement of the engine using this mechanism. This is known as VTA (Variable Turbine Area).

▼
Cleansing flue gases.

Using an ammonia solution, nitrogen oxides are converted to nitrogen and water (vapour). In the background, the catalysator (1) that catalyses the reaction.

◄

Ammonia, which is a urea solution, is sprayed into the flue gases. The partitions enable a good mixture of urea solution and flue gases.

▼

Urea is widely used to convert nitrogen oxides to nitrogen and water (vapour).

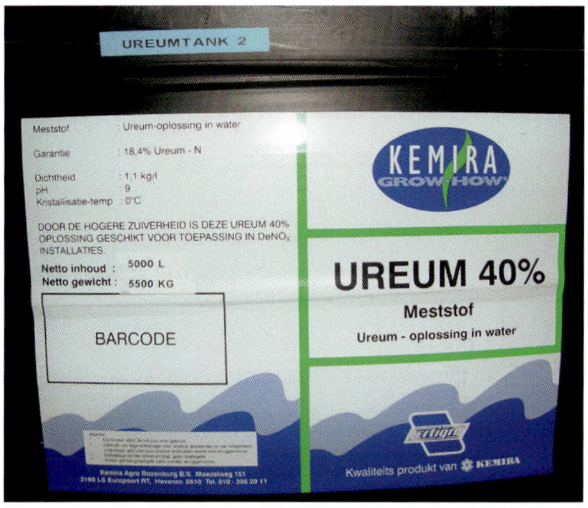

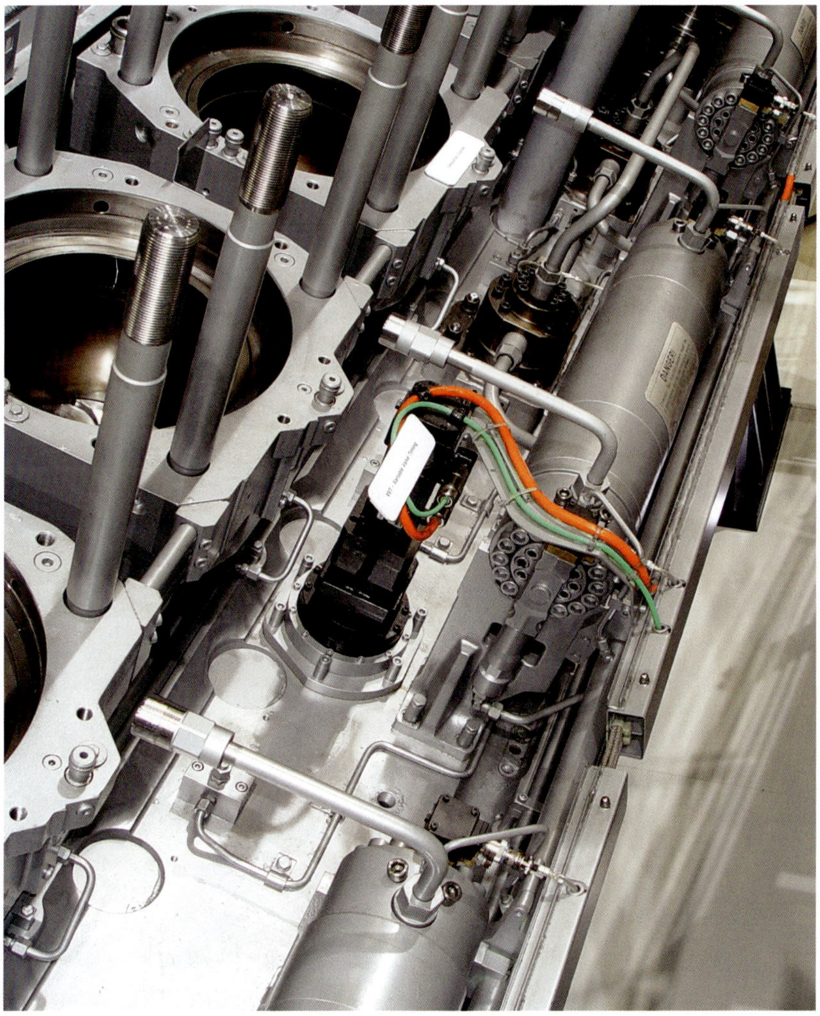

▲

Urea dissolved in water can simply and without risk be stored in plastic containers.

◄

In this MAN Diesel engine, type 32/44, the whole 'package' is used to reduce amongst others, the emission of nitrogen oxides and soot particles.

In the background, the common rail system enabling reduced production of visible smoke. Nitrogen oxides are drastically reduced by advancing the inlet-air valve closure and cleansing the flue gases

▲

An SCR (Selective Catalytic Reduction) installation in a diesel-power plant in Zurich, Switzerland.

Note the dimensions of the complete upper installation (1 – sound damper) and behind (2 – SCR) the diesel engine generator set.

▼

A delayed fuel injection ensures lower process temperatures and therefore lower nitrogen content in the exhaust gases.

horizontal: position of the crankshaft TDC = top dead centre

For conventional injection, needle lift is 35 crank degrees. Needle lift at delayed injection is 28 crank degrees.

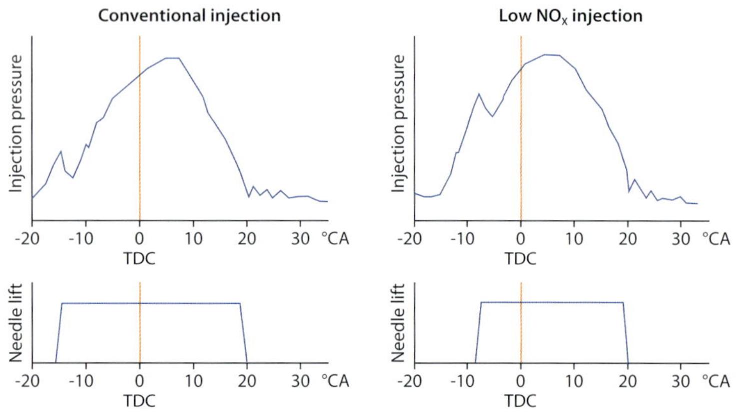

Engine manufacturer Wärtsilä claims a significant nitrogen-oxide reduction in a Wärtsilä Vaasa 32 using:

- delayed fuel injection: 20 to 30%;
- direct water injection: 50%;
- exhaust-gas recirculation: 50%;
- SCR-systems: 80 to 90%.

22.7 Reduction of sulphur oxides in exhaust gases

Reducing sulphur oxides is technically simple and is used in diesel-power plants. However, this has not been translated for use on board ships, as desulphurisation plants are very large and costly.

It is easier and less expensive for ships to reduce the sulphur content in fuel than decrease the amount of sulphur oxides in exhaust gases. Sulphur oxides cannot be removed by catalysts. They can only be washed out.

▼

A project for the supply of district heating in Vaasa, Finland.

Two twelve-cylinder medium-speed four-stroke V-engines with a cylinder bore of 46 and 64 centimetres generate electricity for the grid using an electrical generator. The heat from the exhaust gases is used by two exhaust-gas boilers to produce steam for a steam turbine. The spent steam of the steam turbine condenses in the heat exchanger for city heating; a very large piping network throughout the city provides buildings with warmth. The exhaust gases are cleaned in the NO_X- and the SO_X systems. The normal heat discharge of both engines is used as much as possible for heating purposes. Shown here, the heating of the water supply to the boilers.

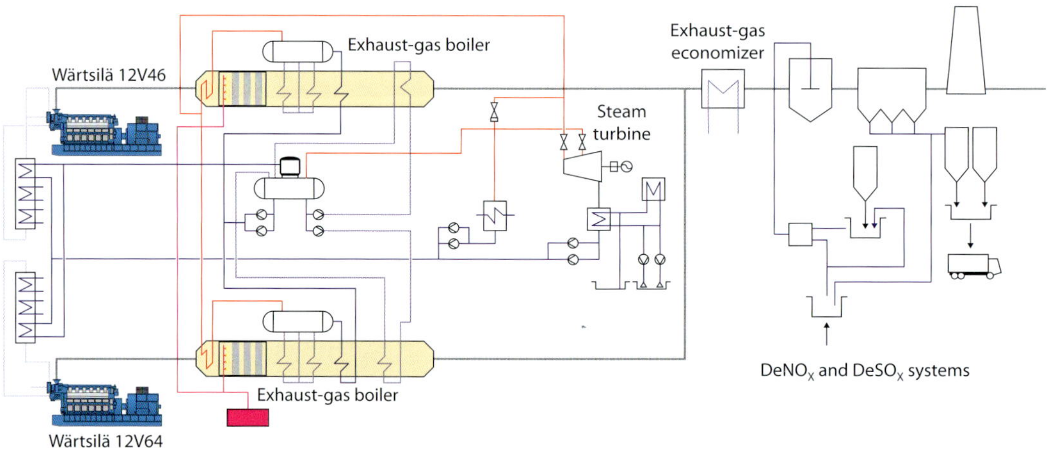

This is usually accomplished with alkaline solutions, such as calcium carbonate, sodium carbonate or caustic soda.

The reaction is as follows:

$Ca\,CO_3 + SO_3 \rightarrow CA\,SO_4 + CO_2$

$Ca\,CO_3 + SO_2 + ½\,O_2 \rightarrow Ca\,SO_4 + CO_2$

In this instance, calcium carbonate is used as a detergent and reacts with the SO_2/SO_3 to form **gypsum** (calcium sulphate).

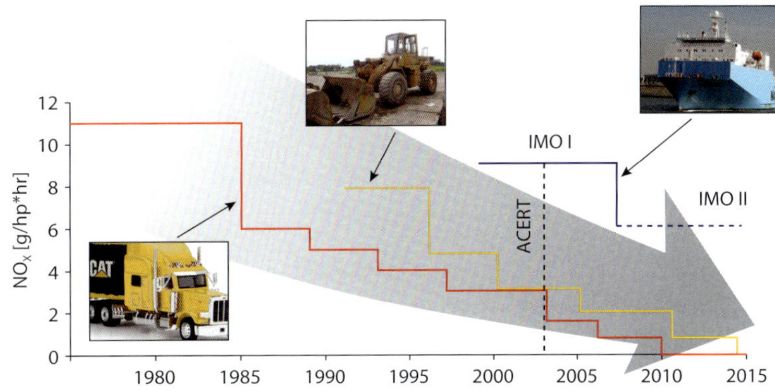

▲
The 'Acert' technology of Caterpillar MaK.

22.8 Removal of fine particles from exhaust gases

These minuscule dust particles often have diameters from 0.1 to 1.0 microns and cannot be removed in a cyclone, as opposed to sand grains found in diesel-power plant air-inlet systems in arid areas.
Electrostatic filters are generally used. In this system, the minuscule particles are ionised and adhere to the electrodes. Dust particles can then be washed away. The installations, however, are too bulky for ships and are therefore not used.
Tests have shown that a combination of a cyclone and an electrostatic filter remove over 60% of fine particles.

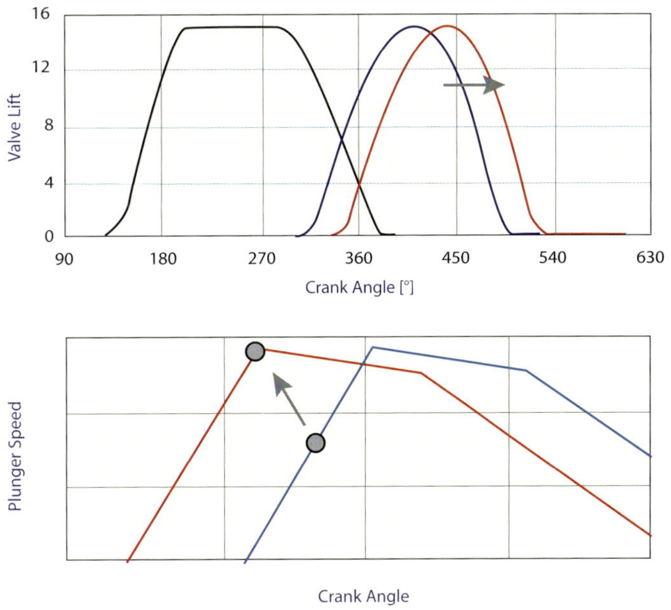

▲
The 'Flex Cam' technology of Caterpillar.

22.9 Examples of techniques engine manufacturers apply to reduce emissions

22.9.1 Engine manufacturer Caterpillar

Caterpillar developed for its medium-speed Caterpillar MaK heavy-fuel engines, the so-called ACERT-technology.

This technology is characterised by three aspects:

1 A reliable system when using heavy-fuel oil.
2 The highest engine efficiency in this power-output range compared with the competition.
3 The lowest engine emissions with the simplest possible system.

This is achieved with **Flex Cam-technology.** Here the inlet-valve timing and the fuel injection is adjusted to diesel-engine load.
With a mechanical adjustment device, the timing of the inlet valve and the fuel pump can be controlled. An actuator adjusts the driving gear.

This system is abbreviated: FCT = Flex Cam Technology.

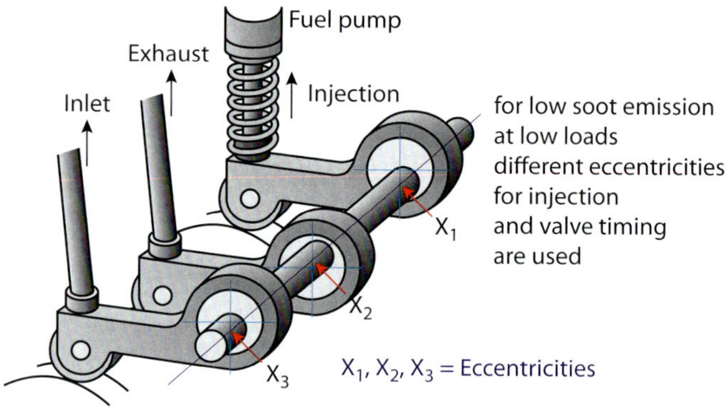

▲

The mechanical valve and the fuel pump adjustments are made possible in Caterpillar MaK diesel engines by the variable eccentricities X_1, X_2 and X_3.

This first step entails the synergy between a flexible fuel system and an advanced air-supply system so attaining an optimum engine design.
The fuel injection retains an extremely high pressure throughout the entire operational range while the amount of fuel and the timing of opening the inlet-valve are load-controlled. A lever system controls the injection timing and pressure, and the inlet-valve timing using an actuator.
The valve timing changes at partial loads, thus allowing effective compression to rise and consequently the maintenance of optimum combustion.
Furthermore, by modifying the fuel cam, the fuel-injection pressure increases. This method involves the fuel being sprayed in a series of finely timed microbursts in the engine-load range where preventing the production of smoke is difficult.

The **second step** has the following objectives: to reduce emission of exhaust gases below the current IMO regulations and create a technology that complies with the emission requirements in vulnerable areas. Improve the mechanical adjustment system with additional built-in flexibility.

Other engine manufacturers are also improving the combustion process with the following successes:
- reduced fuel consumption;
- reduced soot-particle emissions;
- lower nitrogen-oxide emissions.

The graphs at the bottom of this page show improvements for medium-speed four-stroke diesel engines operating on heavy fuel from MAN Diesel, type 32/40.

New fuel-injection techniques provide a welcome contribution, such as the common rail, which is extensively discussed in Chapter 9.

22.9.2 Engine manufacturer Wärtsilä

Engine manufacturer Wärtsilä has also taken various measures to reduce exhaust-gas pollutant emissions for its four- and two-stroke diesel engines.

An advantage of medium-speed diesel engines is that they can operate on poor quality residues from the refining process, an inexpensive fuel source. The main disadvantage is that the pollutant emissions mean that extensive costly installations are required to limit these emissions. **Ultimately, selecting fuel is a complex matter in which the financial aspect is the most important subject.**

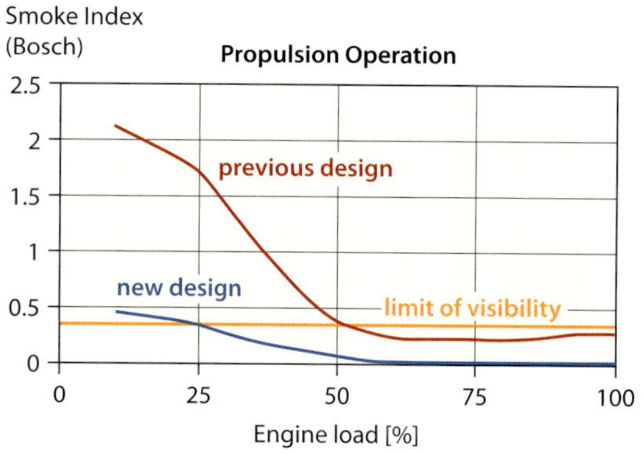

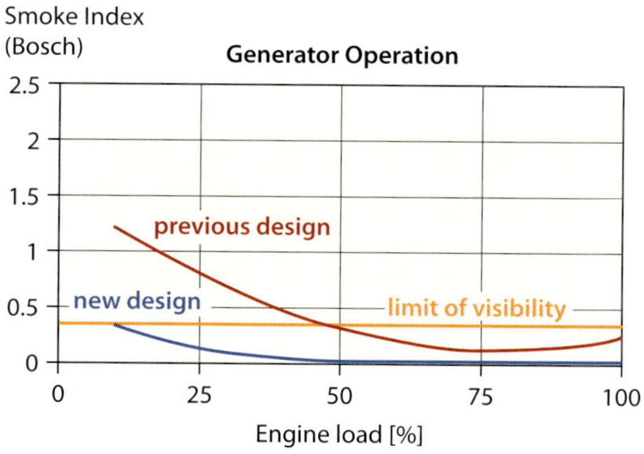

▲

In the graphs above, the decrease of smoke using new combustion technologies for MAN Diesel, type 32/40 diesel engines category III.

Some examples

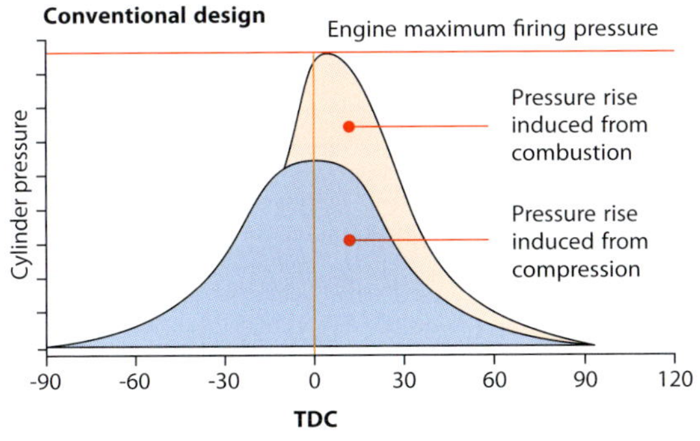

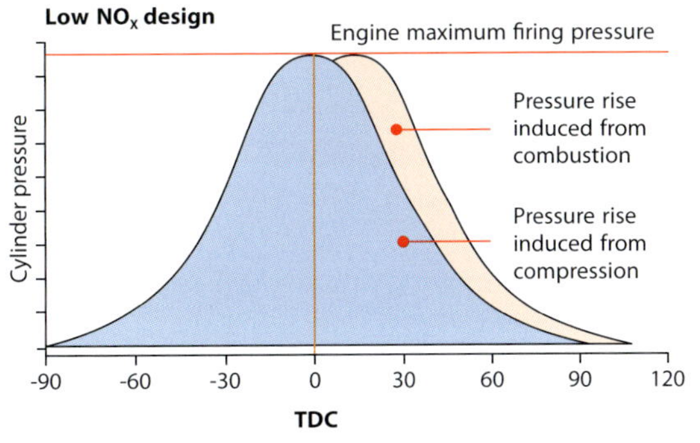

◀ A very high compression ratio ensures a decrease in the production of nitrogen oxides.

Example 1: Extremely high compression ratio

This requires that the final compression pressure is equal to the final combustion pressure. Therefore, the final compression temperature is considerably higher and the fuel combusts rapidly. The so-called ignition delay, the time lapse between the beginning of fuel injection into the cylinder and fuel ignition is shortened. There is, then, more time for the fuel to completely combust. As a result, the fuel injection can be delayed slightly and combustion will occur at the most propitious time. The high peak temperatures in the combustion process are slightly lower, resulting in less nitrogen oxide in the exhaust gases.

Example 2: Water injection into the cylinder

As mentioned previously, this method consists of cooling the combustion process by decreasing the peak temperatures. This in turn leads to a decrease of nitrogen oxides in the exhaust gases.

▼ A technical packet for direct water injection in Wärtsilä diesel engines.

Important: The injected water must be very pure. The fuel injector has two functions: the injection of fuel and water. The water-injection pressure is supplied by its own high-pressure water pump. The water flow is checked at every cylinder as a safety measure. A control unit provides the controls.

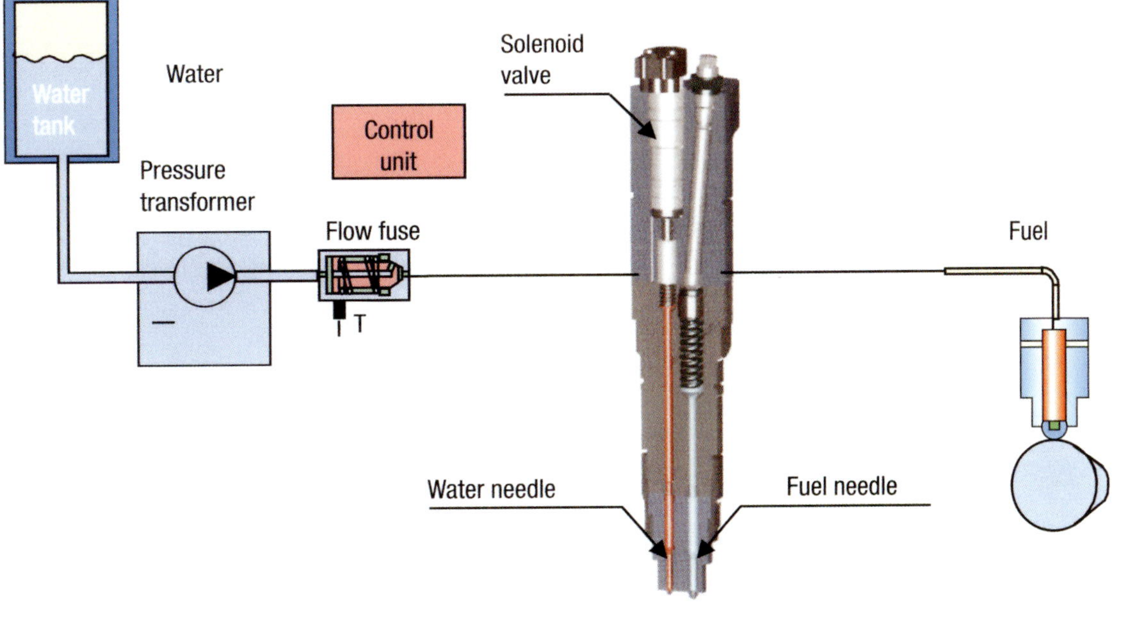

Example 3: Conversion of nitrogen oxides in exhaust gases using the 'Selective Catalytic Reduction' method, or in short, placing a catalyst in the exhaust-gas line

All engine manufacturers apply this method whenever a radical nitrogen-oxide reduction is required.

Future developments due to increasingly stringent regulations with regard to diesel-engine emissions

Designing 'cleaner' engines means that engine manufacturers must invest time and money. A large number have to or will discontinue their businesses or merge. Consequently, the number of industrial diesel engine manufacturers continues to dwindle.
The use of very heavy and contaminated fuels in certain areas will be further restricted and a diesel-oil fuel with a low sulphur content will be chosen.
The use of gas turbines running on relatively light diesel oil will increase for certain types of ships, such as cruise ships.
If heavy-fuel oil and therefore 'inexpensive' fuel is used, the secondary systems for cleaning exhaust gases will be used more frequently. In future, this will also apply to diesel-power plants.
The use of natural gas and other gases that have much lower carbon dioxide and nitrogen-oxide emissions will be used more often. The construction of natural-gas networks is on the increase in many countries. Gaseous fuels can be used for gas engines working on the Otto principle and for dual-fuel engines, which use gaseous fuel and small amounts of diesel oil.

Also, see Chapter 8, *Fuels, fuel-line systems and cleaning fuels.*

Total engine efficiency can still be increased. The efficiencies in the various engine categories rise from 30% to over 50%. Efficiency increments save fuel, which in turn will effect the reduction of the hazardous emissions per kWh-generated power.

Total efficiencies 2011

Category I	0 – 100 kW	Maximum 35%
Category II	100 – 10,000 kW	Maximum 44%
Category III	500 – 30,000 kW	Maximum 51%
Category IV	1500 – 100,000 kW	Maximum 54%

22.10 Actions taken in large modern sea-going vessels

22.10.1 Tanks for sulphur-free fuel

In the last few years, ships have been built with a separate fuel tank for sulphur-free heavy-fuel oil. The moment a sea-going ship is in port, the fuel supply can be changed to this slightly more environmentally friendly fuel.

22.10.2 Electric quayside connection

This connection is standard in sea-going ships so that they can connect to the local grid while in port or drydock, during repairs and inoperative periods. The available power, however, is small in comparison to the power required for loading and discharging, and maintaining the correct temperature of refrigerated containers.
Since 2008, large container ships with heavy shore connections have been built, ensuring sufficient electricity for all operations in harbour.
This means that all the diesel engines on board will be stopped, so no pollutants are emitted. This was first used in Los Angeles. It is known as the Alternative Maritime Power Programme (AMP).
However, building an electrical shore connection is costly and requires considerable on-board investments.

22.11 New techniques: the Miller process

The principle of the Miller-process is that in four-stroke engines, the inlet valve closes earlier during the inlet stroke, approximately time wise 20% before the end of the inlet stroke. The air in the cylinder will expand and consequently cool. Therefore, the compression stroke commences at a lower temperature, approximately 15 °C lower.
The result is that the final compression temperature is about 50 degrees lower and the final combustion temperature is therefore lower.
It is clear that at these consequently lower combustion temperatures, fewer nitrogen oxides form.
The NO_X reduction amounts to approximately 30%.

22.11.1 Adjustments

Turbocharger

To provide the cylinder with a sufficient amount of air, a modified turbocharger with a delivery pressure of 5 bars is required.
This can be achieved using a two-stage turbocharger system.

Camshaft

Starting the engine with Miller timing system is difficult; moreover a large amount of soot is produced at partial load because the supercharging pressure is too low.
An adjustable camshaft allows adjustments to be made to valve closure and fuel injection during ignition and at partial load.
The Miller timing is then switched off so the engine operates normally and the soot emissions are reduced.

Tests with a Caterpillar MaK 7M43C have shown that the system can also be installed in existing engines. The costs in this case amount to 15 to 25% of a new engine.
The engine is then IMO phase II-norm compliant.

Example 1

In Rostock, Germany, the MaK 7M43C main engine of the chemical tanker Fure West (144 × 21 metres, 16,000 dwt) was converted to a Low-Emission Engine (LEE). It is the first ship with a heavy-fuel engine that complies with the future IMO phase II-emission norms. The Fure West is owned by the Swedish Furetank Rederi.
The NO_X-emission at present amounts to 8.3 grams per kWh and the soot emission is less than 0.5 FSN (Filter Smoke Number).

According to Caterpillar Engines in Kiel, the engine achieves low emissions without a fuel-consumption increase.
The most important modification made to reduce the NO_X emission is the increase of the compression ratio. Ten years ago, the standard compression ratio was 1:11 or 12.
To meet the IMO I emission requirements, the ratio had already been reduced to 1:14 or 15 and IMO II regulations require compression ratios in the region 1:17.
The second important modification is the Miller cycle: An adjustment of the inlet-valve timing effects a cooler combustion process.
In order to achieve the IMO I norm, a relatively small Miller effect of five per cent was required. However, to achieve IMO II norms, a Miller effect of twenty per cent is required.
To attain this Miller effect, the inlet valve must close when the piston still has to move downwards a further twenty per cent. This ensures that the inlet air locked in the cylinder, expands and further cools. Therefore combustion is less hot, which reduces the NO_X production. In order to supply a sufficient amount of air within the short time that the valves are open, an adjusted turbocharger is required. The inlet air must be supplied at a pressure of five bars. Combining the higher compression ratios with the Miller effect, the NO_X emission is reduced by approximately thirty per cent while efficiency remains the same.

22.11.2 Camshaft technique

The main drawback of the Miller timing is difficult start-up and the engine produces a great deal of soot when operating at partial load, because the turbocharger pressure is too low to provide the cylinders with sufficient air.
To prevent this, the LEE-system (Low Emission Engine) works with an adjustable camshaft. This 'Flexible Camshaft Technology' (FCT) allows valve-closing times and fuel injection during start and at partial loads to be adjusted. Then, the Miller timing is switched off, the engine operates normally, and soot emissions are reduced. By accelerating the fuel injection and increasing the injection pressure, combustion is improved, and soot emission is also further reduced.
According to MaK, the flexible-camshaft technique reduces soot emissions by 75 per cent, while the operational performance of the engine is also improved.
Essentially, the FCT-system comprises a modified camshaft with a pneumatically controlled actuator and Programmable Logic Controller (PLC).
Retrofitting and certifying the engine took four days. During these days, the engineers adjusted the compression ratios and the camshaft.
A higher compression ratio leads to a smaller combustion chamber as the piston is slightly raised, consequently the anti-polishing rings had to be replaced by shorter rings.
Parts of the valve drive, high-pressure fuel pumps and the injector nozzles were also replaced.
ABB specialists modified the turbocharger to meet the higher requirements. After the retrofit, trials were held and the engine was certified by Germanischer Lloyd.

The engine has operated on this new system for several years (2012).
All existing Mak M20C, M25C, M32C and M43C engines can be adapted to meet the future IMO II-standards.
The retrofit costs amount to 15 to 25 per cent of new engine costs.

Example 2

The Wärtsilä Sulzer four-stroke medium-speed engines of the ZA40 type with a 400 mm bore and a 480 mm stroke and the ZA40 A type with a 400 mm bore and a 560 mm stroke have been tested for important improvements regarding emission reduction.

Two systems were installed for trials.

The dry system
An improved piston, injector nozzle, anti-polishing ring and a variable inlet-valve timing system.

The wet system
This is based on water and heavy-fuel emulsions and is referred to as a micro-emulsion.

In the 'dry system', the Miller principle is incorporated with the so-called Variable Inlet Valve Closing or VIC-system.

The system works using a hydraulic system between the cam roller and the push rod.

▶
The mode of operation of the Miller principle for Wärtsilä Sulzer ZA 40 engines.

At conventional full engine load, the throttle valve B is open, the oil can flow freely, and the push rod moves according to the cam and cam height.
At low loads, the throttle valve is closed by a servo-air system with a small valve, which closes the inlet valve. Next, the inlet valve is opened in the normal way since the throttle valve is opened by the increase in oil pressure and the inlet valve closes at the correct time, namely after reaching bottom position.
VIC is built on the engine and no external mechanical system is necessary.
A simple automation system is required for servo air.
This can be placed in existing engines.

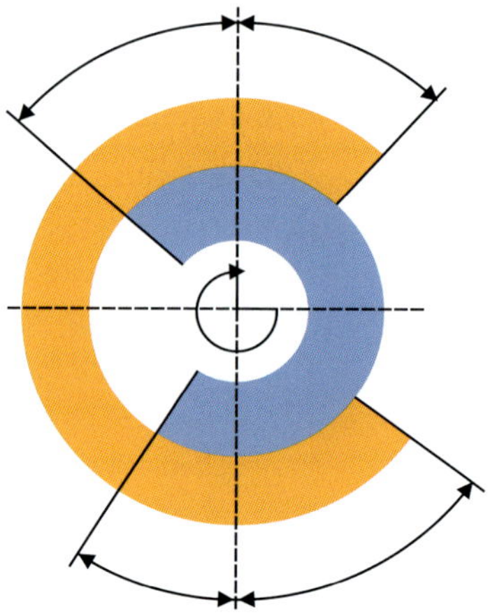

Delayed timing
for improved smoke behaviour at low load

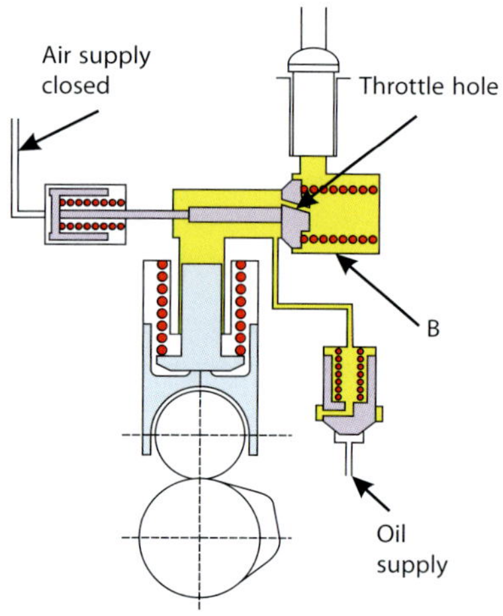

▶
Soot emissions of a standard engine compared to the 'dry system' of Wärtsilä for Sulzer ZA 40 medium-speed engines running on HFO.

Clearly shown are the large differences between zero load to approximately 70%.
After this, there is no improvement!

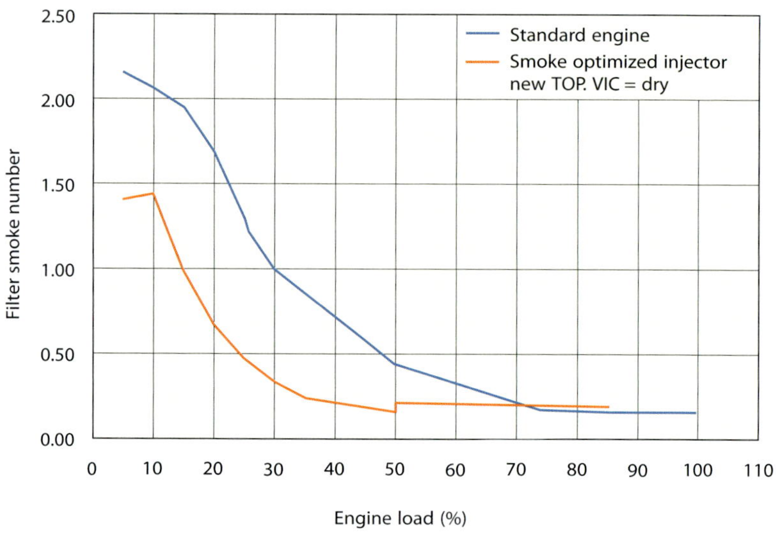

Example 3

The Wärtsilä 32 series.
In this series, the objective was to reduce nitrogenoxide emissions while retaining a high efficiency.
This is difficult from a physics point of view as, in theory; an internal combustion engine achieves high efficiency with a high compression ratio and the highest possible maximum combustion temperature. Carnot shows this.

$$\text{Efficiency Carnot} = \frac{T_1 - T_2}{T_2}$$

Where T_1 is the highest temperature in the process, that is, the maximum combustion temperature, with T_2 being the lowest temperature, the exhaust-gas temperature.

Improvements.
For the Wärtsilä 32 engines, this entailed the following.
- High compression ratio.
- New fuel-injection technique.
- New injector nozzle.
- New valve timing; the Miller principle.
- Increased turbocharger pressure ratio.
- The SPEX charging-air system for both in-line and V-engines.

▶
Wärtsilä 32 performances.

Clearly shown is that the fuel consumption is lowest in the 'economy' type and the NO_X emissions are lowest in the 'emission' type.

Red line: The maximum World Bank/IMO limit, 970 ppm at 15% O_2.
Green line: The maximum nitrogen-oxides emission in India, 710 ppm at 15% O_2.
Purple line: The NO_X for various 32 types.
Blue line: The development of the specific fuel consumption for various 32 types.

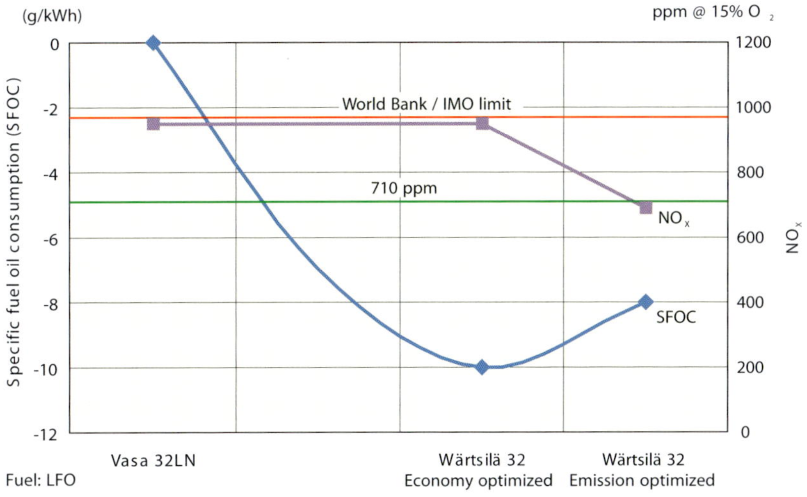

▶
A comparison of the traditional valve timing and Miller valve timing.

Horizontal:	The number of crank degrees.
Vertical:	
Left:	The pre-charge pressure in bars.
Right:	The air temperature.

Red dotted line:	cylinder temperatures, traditional valve timing.
Black dotted line:	cylinder pressures, traditional valve timing.
Red line:	cylinder temperatures, Miller valve timing.
Black line:	cylinder temperatures, Miller valve timing.

The graph shows that the pressure in the cylinder in the Miller system is higher than that of traditional valve timing. Obtaining this high pressure requires approximately a 0.6 bar higher pre-charge pressure.

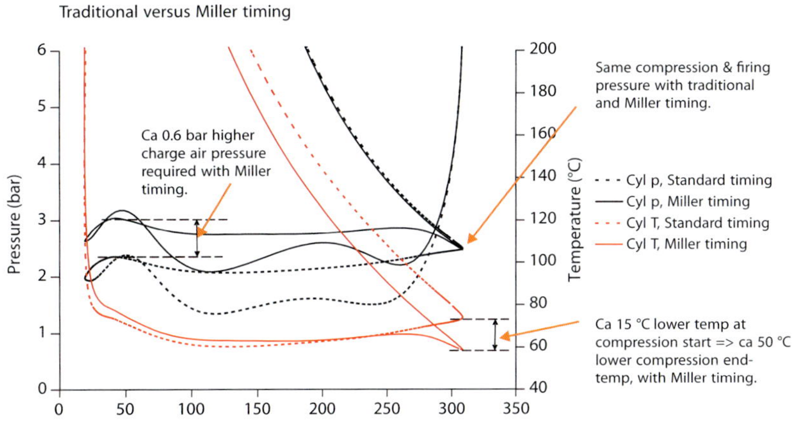

It is also shown that the initial compression temperature for the Miller system is lower than that of the traditional system. This results in a 50 °C lower final compression temperature.

▶
The difference in the valve timing of the inlet valve between the traditional system and the Miller system.

Horizontal: The number of crank degrees.
TDC – Top Dead Centre.
BDC – Bottom Dead Centre.

The graph shows that with the Miller system, the inlet valve closes earlier, just before the bottom dead centre (BDC) of the piston.

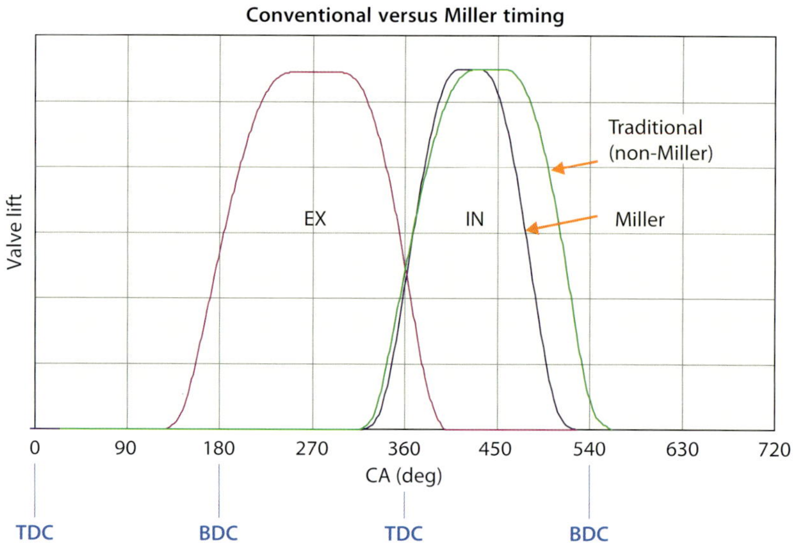

22.12 New developments: the Hercules project

This is an extensive project with the objective: further optimisation of the modern internal combustion engine.
A large number of companies and institutes are involved in this project; each has their own expertise in the engine field. Collectively the participants have approximately 80% of global turnover in the medium-speed four-stroke and low-speed two-stroke crosshead engines sectors.
The project is supported by the European Commission and the Swiss authorities.

Hercules is the abbreviation for **H**igh **E**fficiency **R**&D on **C**ombustion with **U**ltra **L**ow **E**missions for **S**hips.

A few project objectives

- Drastic reduction of pollutant emissions and dirt particles in exhaust gases.
- Improvements in the total efficiency and reliability of the engine. This reduces specific fuel consumption and at the same time decreases the carbon-dioxide emissions. Maintenance costs are also reduced. This project should be realized between 2010 and 2020.

It is obvious that some technical improvements to the
engine must be realized, such as:

- Extremely high cylinder pressures.
- Improved combustion.
- Special turbochargers with several pressure stages.
- A diesel engine with an exhaust-gas boiler producing steam for the steam-turbine genset.
- Emission-reduction systems with water in the combustion chamber.
- Emission-reduction systems with exhaust-gas recirculation.
- Emission-reduction systems using post-treatment outside the engine.
- Reduction of the driving-gear friction in the engine.
- A diesel engine with elaborate electronic controls for optimal combustion-process control.

Hercules Vision	Year 2010	Year 2020
Reduction of fuel consumption and CO_2 emissions	-3%	-5%
Reduction of NO_x (Relative to IMO 2000 standard)	-30%	-60%
Reduction of other emission components (PM, HC)	-20%	-40%
Improvement in engine reliability	+20%	+40%
Reduction of time to market	-15%	-25%
Reduction in lifecycle cost	-10%	-20%

◄ The Hercules project goals to 2020.

Hercules Objective	BAT-IS (2003)	Year 2007 Targets	Year 2010 Targets
Reduction of fuel consumption and CO_2 emissions	2-stroke: 170 g/kWh 4-stroke: 175 g/kWh	-1%	-3%
Reduction of NO_x (Relative to IMO 2000 standard)	IMO 2000 limits (g/kWh) 17 — N < 130 rpm 45 x (rpm)-0.2 — 130 < N < 2000 rpm 9.8 — N > 2000 rpm	-20%	-30%
Reduction of other emission components (PM, HC)	< No limits for marine engines > Visible smoke limit FSN 1.1 Opacity 20%	-5%	-20%
Improvement in engine reliability	18,000 hours to overhaul of major components	+10%	+20%
Reduction of time to market	5 Years	-10%	-15%
Reduction in lifecycle cost – Initial cost – Fuel/lub-oil cost – Maintenance	< Costs depend on engine size >	 0% -1% -4%	 -1% -3% -6%

◄ The improvements to 2010.

Overall, it can be said that these improvements have now been achieved, (2011).

Potential Exploitation Items		
Workpackage	**Item**	**Time**
WP1: Extreme design parameters	Engine components for extreme output operation (pistons, rings, bearings)	M
	Extreme value engine	L
WP 2: Advanced Combustion Concepts	Combustion models	S
	Chemical kinetics models	S
	Full cycle simulation tools	S-M
WP 3: Multistage/Intelligent turbocharging	Variable geometry turbocharger	M
	Power take-in, take-out systems (Integration motor/generator/turbocharger)	M
	Multistage intercooled turbocharger	M
WP4: Turbo-compound/hot engine	Composite structures for hot-engine	L
	Engine compounding systems and components (boilers, TG, TCS)	S
WP6: Emission reduction methods (internal-water)	Direct water injection system	M
	Inlet air humidification system	S
	Control systems for above	S
WP7: Emission reduction methods (internal-Exh. Gas)	Exhaust gas recirculation system	M
	PM measuring techniques	S
WP8: Emissions after treatment	In-service emissions monitoring system	S
	Non-Thermal Plasma Technology	L
	Wet Scrubber Technology	M
	Select-cylinder emission measurement technology	M
WP9: Reduced friction engine	Low friction engine components (liner, pistons, rings, bearings, injection)	M
	In-service monitoring system for cylinder and lub feed rate adjustment	S
	Low friction engine	M
WP11 Adaptive engine	Onboard engine electronics	S-M

Timespan: Short = S, Medium = M, Long = L

◄ The ten items chosen for specialised studies by separate groups.

>CH23

Calculating fuel- and lubricating-oil consumption

Fuel- and lubricating-oil consumption of large engines with a substantial number of operating hours forms the most important cost factor. The fuel- and lubricating-oil consumption of this two-stroke MAN Diesel MC crosshead engine in a large container ship is huge and costs millions of Euros each year.

The annual fuel- and lubrication costs for this luxury yacht with almost no operating hours are negligible.

23.1 Introduction

The costs for fuel- and lubricating-oil are high. The fuel- and lubricating-oil costs for ships or diesel-driven power plants constitute between 40 and 70% of the total exploitation costs. Obviously, this is dependent on, for instance, the number of operating hours, engine category, and depreciation of the plant, personnel expenses, maintenance, and insurance.

Naturally, harbour dues, towing and pilotage costs are also important. Costs for ships sailing through the Suez or Panama canals amount to € 100,000 or more (one-way only!).

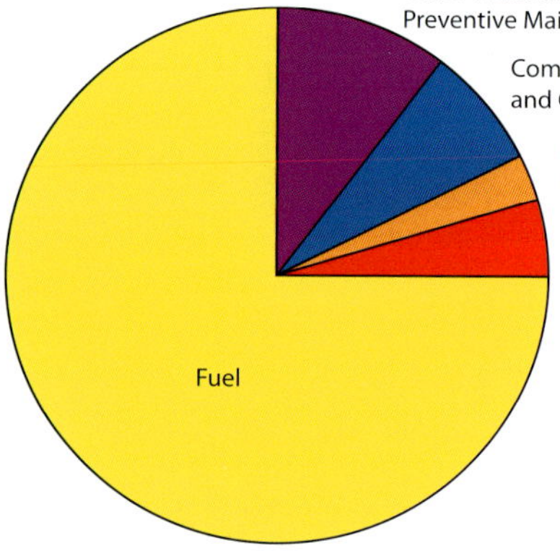

▲

An overview from Caterpillar.

The total running costs of a diesel engine in its life span of between 10 and 25 years.
Here fuel costs alone amount to 75%!

▼

Shipping costs.

There are many cost factors in shipping:
- interest and amortisation on purchase costs;
- insurance;
- maintenance and repairs;
- quay and pilot dues;
- fuel- and lubricating oil;
- canal and locks dues, such as the Suez Canal and Panama Canal;
- crew;
- loading/discharging.

One cost factor is considerable: the cost of fuel and lubricants. This is the largest cost factor for ships that travel long distances, between 40 and 70% of all costs.

23.2 Diesel-engine efficiency

Depending on engine type and the cylinder volume, the efficiency varies from ± 25 to 50%.
Small four-stroke high-speed diesel engines in category I have an efficiency of over 30%, and the large two-stroke low-speed crosshead engines have an efficiency of over 50%.

Efficiency is equal to: $\frac{\text{Shaft power}}{\text{Fuel power}}$.

23.3 Shaft power in kW or MW

Shaft power can be measured with a torsion meter placed on the output shaft. It can also be measured with the engine formula, where the mean induced pressure is measured by an indicator.

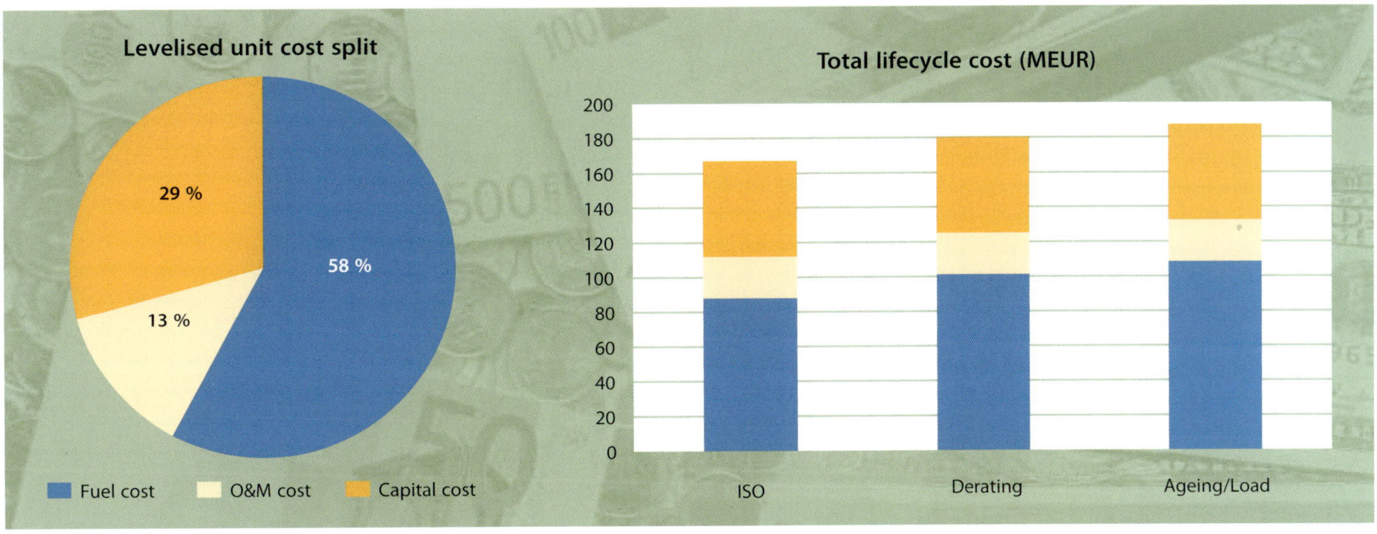

▲

An overview from Wärtsilä.

The total costs of a diesel-power plant. The fuel costs are 58%. O&M cost is operation and maintenance costs.

At present, this is done by an electronic measurement system where all the data from the engine formula are directly processed to an effective cylinder output and consequently the shaft power for all the cylinders, so the entire engine.
In practice, these mean induced pressures are measured for engines in categories III and IV.

The fuel power is equal to: fuel mass in kg/sec × calorific value of the fuel in kJ/kg.

$$\text{Fuel power} = \frac{\text{kg}}{\text{sec}} \times \frac{\text{kJ}}{\text{kg}} = \frac{\text{kJ}}{\text{sec}} = \text{kW.}$$

The amount of energy supplied to the engine in fuel form is always set at 100% and therefore the shaft power always plays a part in this fuel power.

23.4 Specific fuel consumption

At present, this is the standard unit for fuel consumption and is indicated in grams per kW hour, or **g/kWh.**

Some values

The specific fuel consumption varies from approximately 240 to 160 grams per kWh at 100% load (Maximum Continuous Rating, MCR), depending on size and type of engine.
The calorific value of HFO is set at 42,700 kJ/kg and that of MDO at 43,500 kJ/kg.
Manufacturers of large engines usually provide a calorific value of 42,700 kJ/kg for HFO. This is in accordance with ISO 3046-1 'lower calorific value without engine-driven pumps'. The tolerance is +/- 5% and is valid at an 85% load for propulsion engines according to the propeller law. Obviously, in practice, the calorific value stated on the bunkering bill or the fuel specification should be used.

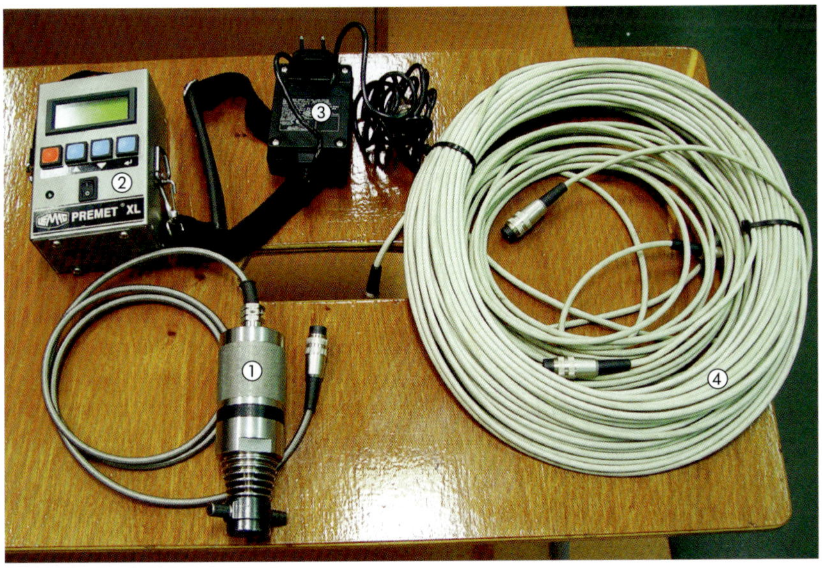

◄

A Lemag electronic pressure sensor, the Premet XL.

1 *pressure sensor*
2 *data storage*
3 *electric-power plug*
4 *pc data cable*

Example 1: Diesel engine category IV – MAN Diesel K98 MC. ± 80% of MCR

b_e = 160 g/kWh. HFO, Fixed screw.

$$\text{efficiency} = \frac{\text{shaft power}}{\text{fuel power}}$$

$$\text{fuel power} = 0.160 \times 42{,}700$$
$$= 6832 \text{ kJ/hour}$$

$$\text{per second} = \frac{6832}{3600} = 1.97 \text{ kJ/sec} = 1.897 \text{ kW}$$

$$\text{therefore, efficiency } \frac{1}{1.897} = 0.5271 \text{ or } 52.71\%$$

Example 2: Diesel engine category III – Wärtsilä 38 B

HFO. 85% of MCR. Adjustable-pitch propeller

b_e = 173 g/kWh.

fuel power =

$$\frac{0.173 \times 42{,}700}{3600} = 2.051 \text{ kJ/sec} = 2.051 \text{ kW}$$

$$\text{Therefore efficiency } \frac{1}{2.051} = 0.4875 \text{ or } 48.75\%$$

Example 3: Diesel engine category II

MDO, calorific value 43,500 kJ/kg

b_e = 190 g/kWh, genset at 100% MCR

fuel power =

$$\frac{0.190 \times 43{,}500}{3600} = 2.2958 \text{ kW}$$

$$\text{therefore, efficiency } \frac{1}{2.2958} = 0.4355 \text{ or } 43.55\%$$

Example 4: Diesel engine category I

MDO, calorific value 43,500 kJ/kg

b_e = 230 g/kWh. Genset at 100% MCR

fuel power =

$$\frac{0.230 \times 43{,}500}{3600} = 2.7791 \text{ kW}$$

$$\text{therefore, efficiency } \frac{1}{2.7791} = 0.3598 \text{ or } 35.98\%$$

In these examples, the efficiency varies from ± 36 to 53%. Note: this applies to a new engine operating under optimal circumstances!

For a diesel engine with a large number of running hours and consequently increased wear of the parts, the efficiency may be several per cent lower! Due to:

- deficiencies in the fuel injection and consequently combustion;
- inefficient turbocharger;
- ineffective intercooler;
- lower compression due to wear of the piston rings and cylinder liner.

Specific fuel consumption and engine load

The lowest specific fuel consumption is not usually achieved at full-load, but at approximately 85% of full-load.

23.5 Fuel consumption

Fuel gauges in systems usually give fuel quantity in litres and m³. To convert from litres to kilograms, the specific mass and the fuel specifications must be known. The specific mass varies from 950 to 1010 kg per m³ at 15 °C for HFO and 840 to 920 for MDO.

$$\text{Fuel in kilograms} = \frac{\text{Fuel in litres}}{\text{Specific mass of the fuel}}$$

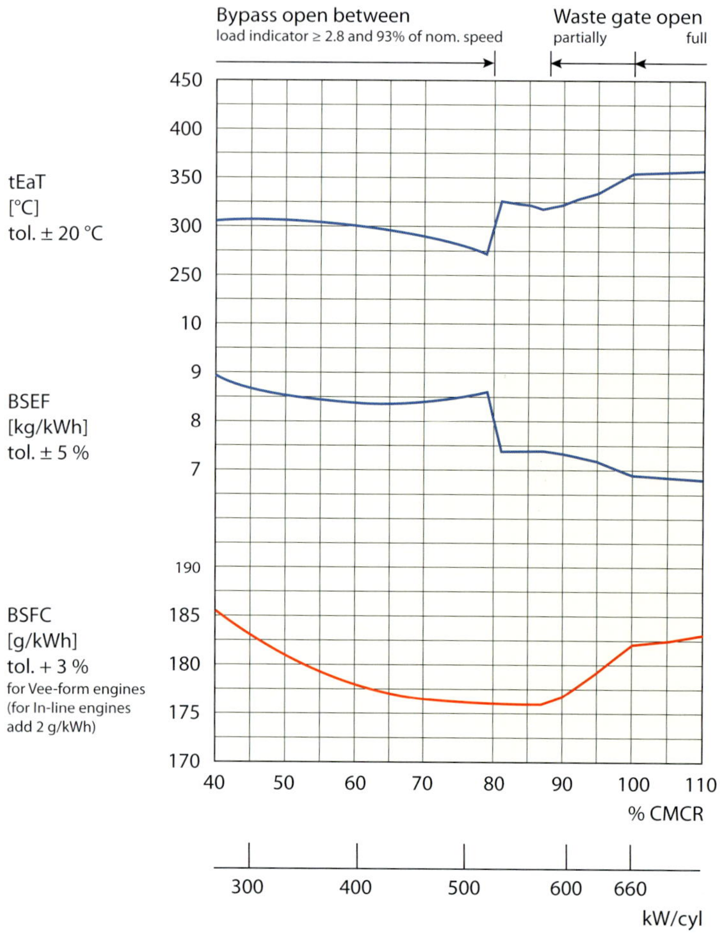

◀ The specific fuel consumption of a Sulzer ZA40S diesel engine, cylinder power capacity 660 kW at 100% MCR, category III.

Between 40 and 100% MCR, the specific fuel consumption varies from 177.5 grams per kWh at 75% MCR to 185 grams per kWh at 100% MCR.

23.6 Fuel consumption for engines in diesel power-plants

Electro-generators have a high efficiency of between 95 and 98%. The higher the power output of the generator, the higher the efficiency.
Dependent on generator load, the fuel consumption is directly proportional to the engine load. The specific fuel consumption seldom varies with varying engine loads. For example, see the curve in the chart on the following page for the Sulzer ZA 40 S diesel engine

23.6.1 Calculating the fuel consumption and the specific fuel consumption for diesel power-plants

Example

An engine drives a three-phase generator, which feeds an extensive electricity network via a step-up transformer; the voltage generated is transformed to the higher voltage of the power grid. For instance, when 6600-volt is stepped up to 100,000-volt mains voltage, frequency 50 Hz. then:
fuel consumption per hour =

$$\frac{\text{supply gauge reading} - \text{return gauge reading}}{\text{number of hours measured}}$$

The results are denoted either in litres or in m^3 per hour. At the same time the electricity production measured at the main switchboard =

$$\frac{\text{final kWh} - \text{initial kWh}}{\text{number of hours measured}}$$

From this calculation, the specific fuel consumption can be calculated (b_e). It is important to note that the specific mass of the fuel must be given at the temperature that is measured at the fuel gauge. Specific mass of the bunker fuel is known (often measured at 15 °C). This specific mass must be converted for the fuel temperature at the gauges.

$$sm_2 = sm_1 - \Delta t \times \alpha$$

sm_2 = specific mass during measuring
sm_1 = specific mass from bunkering note
Δt = temperature difference
α = expansion coefficient of the fuel

The amount of fuel per second can now be calculated in grams per kWh.

Example

A power plant consumes 200,000 kg HFO per day. There are five diesel generators with a power output of 8000 kW. The electrical efficiency of the generator is 98%. What is the specific fuel consumption in g/kWh?

Solution

The shaft power of one diesel engine is:

$$\frac{8000}{0.98} = 8163 \text{ kW.}$$

The fuel consumption (b_e) per hour per engine is:

$$\frac{200{,}000}{5 \times 24} = 1667 \text{ kg.}$$

The specific fuel consumption (b_e) is:

$$\frac{1.667{,}000}{8163} = 204 \text{ grams per kWh.}$$

23.7 Fuel consumption for propulsion diesel engines

The fuel consumption per unit time or the specific fuel consumption is not the only important factor, but also the fuel consumption per nautical mile, or tonnage cargo per nautical mile or per kilometre (inland).
Engines in categories III and IV often have electronic measuring devices, which are connected to the indicator cock of each cylinder. The measuring device gives the average induced pressure at a certain rpm.
The mechanical efficiency of the engine is more or less fixed: usually 90% for four-stroke engines and 95% for two-stroke engines. Therefore the mean effective pressure can also be calculated:

$$p_{\text{eff.}} = p_{\text{ind.}} \times \eta_{\text{mech.}}$$

The shaft power can now be calculated using the engine formula and measured rpm. The fuel consumption per unit time can easily be calculated if both an input and an output gauge have been incorporated in the fuel-supply system. One fuel gauge is also sufficient if placed before the fuel-supply system.
The specific fuel consumption is then simply calculated as shown in the previous example.
It can be difficult to measure the mechanical efficiency accurately. However, using a torsion gauge together with the fuel consumption, it can be calculated exactly.

Note

It is advisable to perform regular calculations to either check if the results diverge from previous calculations or the protocol data provided by engine manufacturers. Data provided by engine manufacturers are gathered from trials, often under ideal conditions and may have a maximum deviation of 5% according to the project guides for the respective engine.

The fuel consumption per nautical mile or voyage. It is important to know that the fuel consumption per unit of the distance travelled is proportional to speed to the third power.

23.7.1 The most economical speed

According to the propeller law, the ship's resistance and therefore the required thrust is proportionate to the square of the velocity. The following comparisons pertain to displacement ships that remain well below the hull velocity.

Thrust is force times the velocity, and the velocity of a ship is more or less proportional to the propeller and engine speeds, which means that the required thrust is the velocity cubed.

Therefore, as seen in chapter 19, Ship propulsion, the graph has a curve to the third power.

Note

A minimum speed reduction effects a significant decrease in fuel costs.

Example

Speed reduction to 70% of maximum. The fuel consumption according to the third power curve is considerably reduced:

$0.7^3 = 0.343$ or 34.3%. Therefore, for an identical time unit, the fuel consumption has decreased significantly.

▼

A fuel-consumption curve.

The angle of inclination, OC, is a measure of the fuel consumption per distance. The lowest fuel consumption per nautical mile is shown by the curve to the third power, point L.

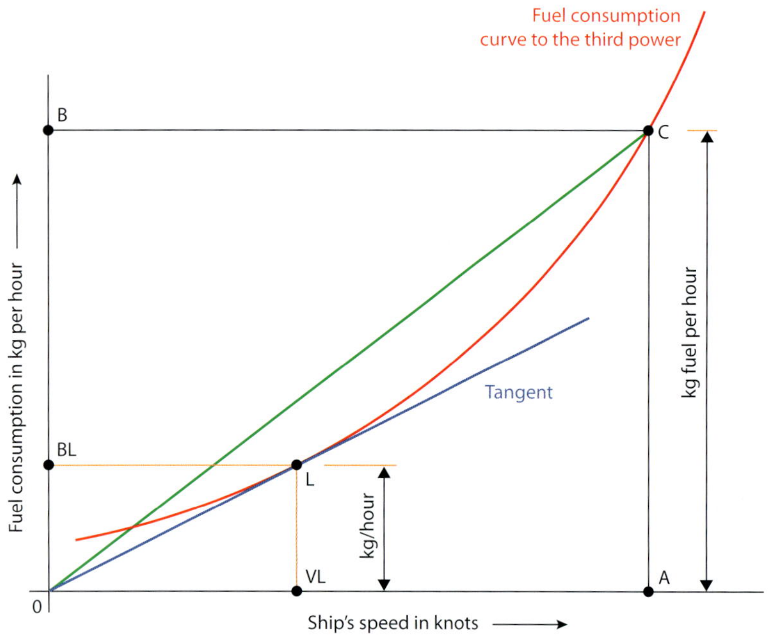

A = Service speed at 100% MCR
B = Fuel consumption at 100% MCR
C = Service point at 100% MCR

OC = The angle of inclination
L = Lowest fuel consumption in kg/hour/nautical mile

▼

The required engine-power output for a certain speed.

Sometimes, increasing the output and therefore the propeller speed does not necessarily increase the ship's speed: the maximum hull velocity has been reached.

At 15 nautical miles per hour, the output is merely 2500 kW; it subsequently increases rapidly per extra nautical mile per hour.

When increasing the engine speed, the velocity and required output increase proportionately; therefore, the fuel consumption increases to the third power.

When reducing the engine speed, the velocity and required output decrease proportionately; therefore, the fuel consumption decreases to the third power.

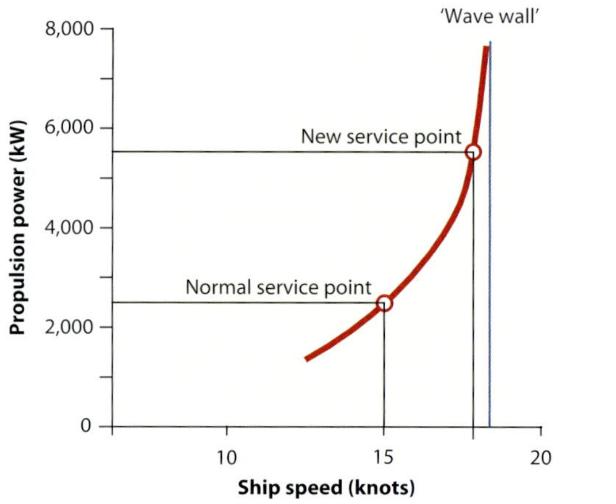

Power and speed relationship for a 600 TEU container ship

23.7.2 Consumption for a certain route

In the example, speed is reduced by 30%. This means that it takes $\frac{100}{70}$ or 1.43 times longer to travel a certain route.
Ultimately, the fuel consumption for the route is equal to $1.43 \times 0.343 = 0.49$ or 49%.
The fuel consumption for the entire route is therefore halved at a velocity reduction of 30% and a 43% longer voyage time.

The speed at which the fuel consumption for a certain distance is lowest can be read from the graphs available for each specific propulsion plant.

Minimum consumption is the point where the tangent meets the curve to the third power.
This tangent has a certain angle with the horizontal: the slope.

23.7.3 Modern calculation programmes

The propulsion engine data, the fixed-pitch or the controllable-pitch propeller data and various other data concerning the ship as well as the weather conditions en route allow the expected speed and fuel consumption to be calculated using special programmes. This in conjunction with current satellite navigation that accurately provides the distance covered making the calculations increasingly precise.

23.7.4 Sailing schedules

Ships receive a sailing schedule from their shore-based organisations, giving the distances to be covered between ports together with some extra leeway allowing for increased and decreased speeds. This allows an average speed to be calculated which using the ship's slip table gives a certain rpm.

23.7.5 Weather routeing

The SPOS system of Meteo Consult, the Netherlands, provides the weather forecast for a certain voyage. Onboard, by calculating alternatives based on, for instance, maximum swell, the master can optimize voyage planning according to weather conditions.
Maersk Lines works with the VES-system, Voyage Efficiency System.
This system calculates the most economic ship's speed and takes the currents, tides, route information and port-related data into account.
This system also allows sharing of ocean-current data between ships sailing the same route. The users exchange information regarding their positions and speed so allowing the strength and direction of currents to be determined.
These statistics collected by the ships are processed and the results are then made available to other shipping. When new information is received by the system about the route of a certain ship, the results of the recalculations are transmitted to that ship, so an alternative route and/or speed can be chosen.
Apart from the most favourable speed and propeller speed, the estimated expected fuel consumption is calculated and relayed.
The VES-system gives the amount of fuel that was consumed for each former voyage, the distance from one port to the next, the fuel quantity to be ordered and the amount of fuel in shore storage. Vescos, the central Databank in Copenhagen, stores all this data.
Virtually all variables are taken into account, for instance, calm-sea routes or shallow-water routes.
The VES system also supplies:
- harbour information;
- safety instructions;
- port details;
- information for masters and officers.

23.8 Lubricating-oil consumption and specific lubricating-oil consumption

The lubricating-oil consumption of an engine is also important. Not only for the lubricating-oil costs but also as above-average lubricating-oil consumption is indicative of engine wear. Increased lubricating-oil consumption may be caused by:
- **worn piston rings**
  Too much lubricating oil remains on the cylinder liner where it burns and contaminates the exhaust-gas system via the exhaust valve. Pollutant emissions increase.
- **worn valve stems and exhaust valves in particular**
  This results in lubricating-oil leakage to the exhaust system.
- **leakage along the seals** of the valve covers, crankcase covers and such.
- when a **lubricating-oil separator** is connected to the system, oil may be lost due to a poorly functioning separator. The sludge tank used to collect the soiled discharge from the lubricating-oil system should therefore be checked regularly.

- **leaking lubricating-oil cooler**
  This results in lubricating oil leaking into the cooling-water system. In order to prevent cooling-water leakage to the lubricating-oil system, which would cause severe damage to the engine, the lubricating-oil pressure must always be higher than that of the cooling-water system.

23.8.1 Specific lubricating-oil consumption

This is denoted in grams per kWh; litres per hour at full load is also frequently used.

For four-stroke trunk-piston engines specific lubricating-oil consumption is between 0.6 and 1.0 gram per kWh with a remark that the actual lubricating-oil consumption is dependent on operating conditions. An extra consumption of 5% is standard.

There are two types of consumption **for two-stroke crosshead engines**:

1 cylinder-lubricating oil
This varies from approximately 0.8 to 1.4 grams per kWh and denotes pure consumption.
A part is expelled by the exhaust gases to the environment or is burnt and deposited in the exhaust-gas system, comprising the turbochargers, the waste-heat boiler, if any, and the silencers.
The other part seeps along the cylinder liner around the piston rod into the scavenging-air duct and is drained. This part is regularly checked by taking measurements of the quantity. The cylinder lubricating-oil supply to the cylinder lubricator is also often measured in order to establish the amount of cylinder-lubricating oil lost along with the exhaust gases. Remember: the quantity of cylinder-lubricating oil supplied to the cylinder = amount of drained cylinder-lubricating oil + the amount of cylinder-lubricating oil lost in the exhaust gases.

2 driving-gear lubricating oil
This consumption expressed in litres per day at full load (MCR) is extremely low as this system is entirely separate from the combustion section.

A MAN Diesel K98MC engine uses between 7.5 and 11.0 litres of lubricating oil per day. The power output varies from, for a six-cylinder engine, 34,320 kW to a twelve-cylinder engine, 68,640 kW.

In grams, at a specific lubricating-oil mass of 950 kg per m^3, this is respectively 7.125 to 11.020 litres:

Per kWh this is approximately $\frac{7125}{34,320} = 0.2$ to 0.3 gram. A small amount.
Maximum consumption occurs with leakage at the crankcase covers or at other seals. A certain amount of oil is wasted during shooting of the lubricating-oil separator, despite first discharging the lubricating oil in the centrifuge bowl before shooting.

23.8.2 Choosing between a four-stroke and a two-stroke engine

Cylinder lubricating-oil consumption is one of the decisive factors when choosing between a four-stroke and two-stroke engine. This is on average 15 to 35% per kWh higher for two-stroke engines than the lubricating-oil consumption for four-stroke engines.

23.9 Measuring fuel consumption

It is important to obtain an accurate measurement of engine-fuel consumption, not only from a financial viewpoint, but also to establish the specific fuel consumption for comparison with previous calculations. This indicates whether engine efficiency has been reduced as a result of, for instance, a poorly functioning engine. This could be caused for instance by insufficient fuel atomisation or a contaminated turbo blower providing less air, etcetera.

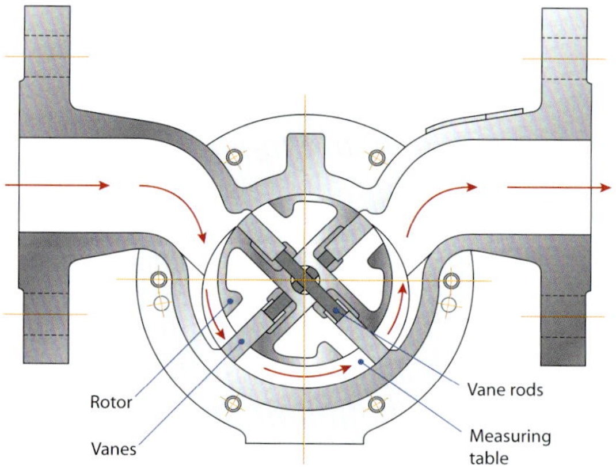

▲
A cross-section of a VAF fuel gauge. The rotating rotor and its four well-sealed vanes must be accurately mounted in the casing.

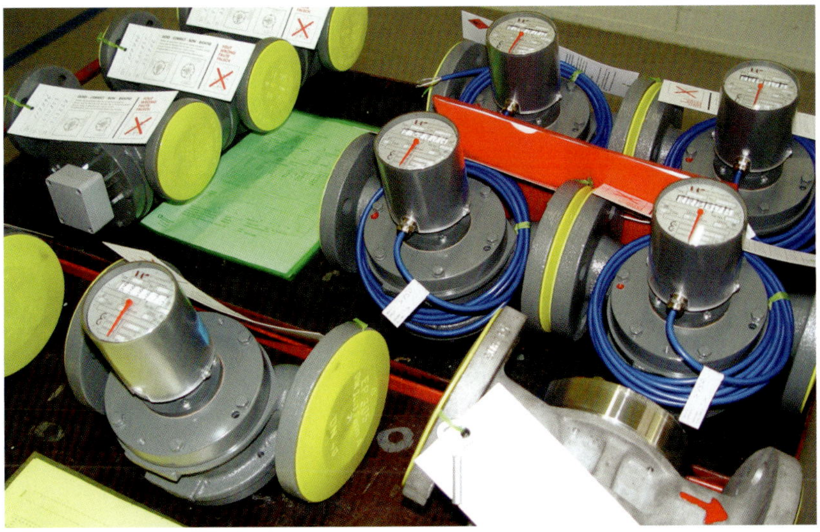

◀
Different types of fuel gauges ready for shipping. VAF Instruments, Dordrecht, The Netherlands.

◀
A dismantled VAF fuel gauge.

1 *the casing*
2 *the rotor*
3 *the end covers*
4 *the vanes*

Fuel-gauge construction

Most fuel gauges consist of a moveable rotor inside a casing. The number of revolutions of the rotor can be measured. The rotor is mounted in the casing in such a way that there is very little clearance between the rotor vanes and the inside of the casing. The inflowing fuel rotates the rotor and continues its path towards the engine.

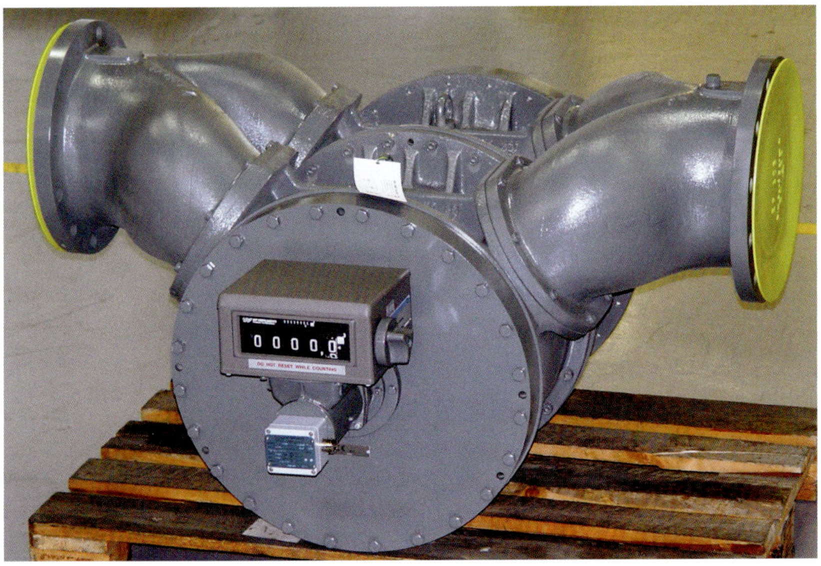

◀
A large fuel gauge. This is for example, installed in a supply line on a bunker boat delivering fuel to ships.

▲
The casing. Note the manner in which the fuel flows through the partitions.

The shape of this patented casing ensures that the measurement is very accurate, even at minimal flow.

◀
The rotor of this large fuel gauge.

The gauge is equipped with a pulse transmitter, which reads the amount of fuel flowing through it and a temperature transmitter that can be read remotely.
A filter with a mesh size detailed by the manufacturer should be fitted before the fuel gauge.

Measuring fuel consumption with several quantity gauges.

Fuel systems are often equipped with inlet and outlet lines. Small engines in categories I and II often have simple systems with only a fuel-supply system.

In heavy-fuel oil systems, where the supplied fuel is maintained at a certain temperature, the temperature of the return fuel is often considerably lower. This means that the fuel-temperature difference must be taken into account when calculating the correct fuel consumption.

An example:

A diesel engine with HFO consumes 5 m^3 per hour and has a fuel circulation of 20 m^3 per hour, the temperature difference between the fuel supply and return is 25 °C.

When the temperature difference is not compensated, the deviation is 2.5% of the amount of circulating fuel. In this instance, 2.5% of 20m^3 is 0.5 m^3 or 500 litres!

The measurement deviation in fuel consumption is:

$$\frac{500}{5000} \times 100\% = 10\%$$

Temperature difference = $\Delta T = T - T_{ret}$ (°C)

Expansion = $j_{tot} = j \times \Delta T$ (%)

Corrected fuel amount is:

$$Q_{cor} = \frac{Q_{measurement}}{1 + \frac{j_{tot}}{100}}$$

Here T_{ret} = 15 °C
$Q_{measurement}$ = measured amount
j_{tot} = expansion coefficient in %

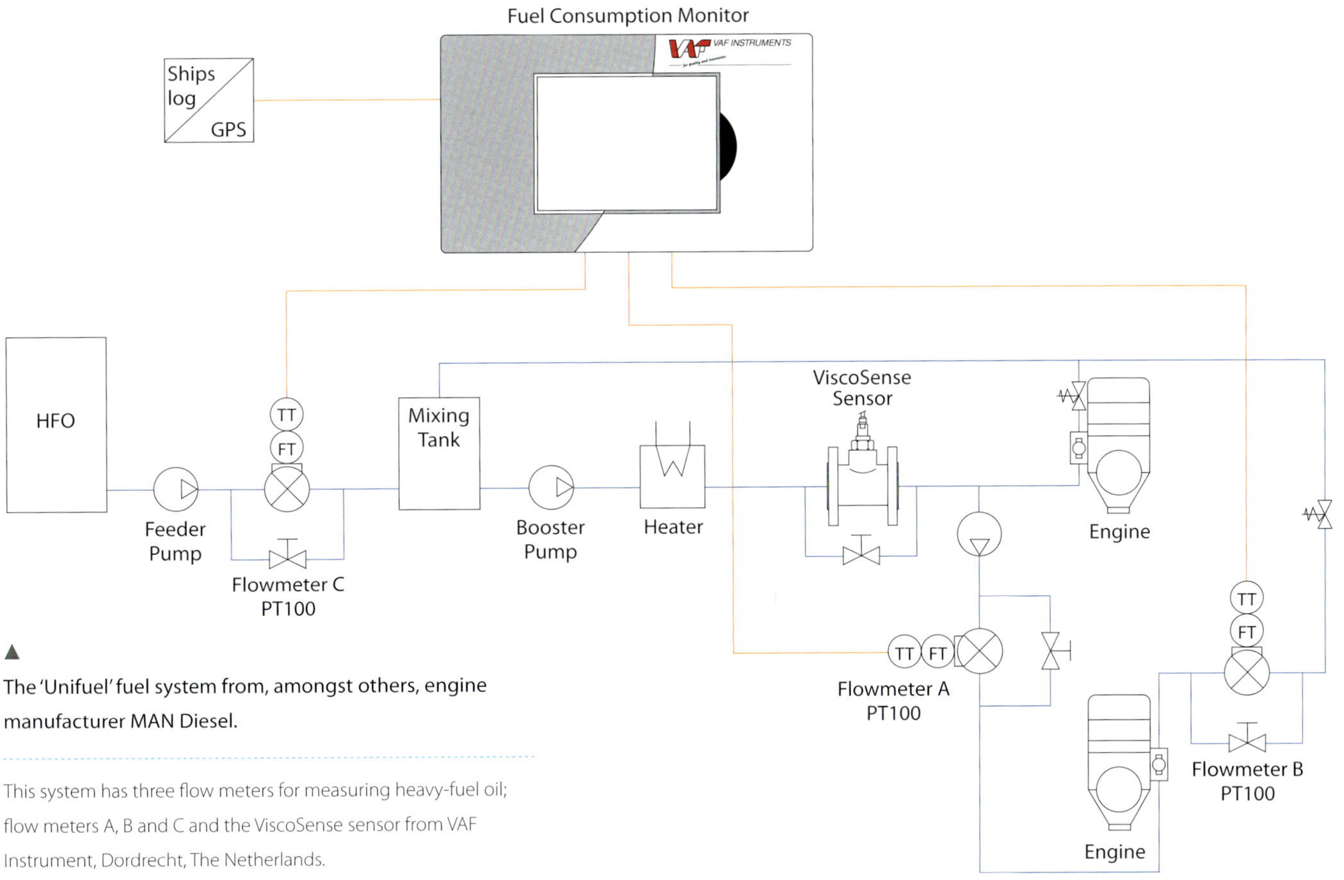

▲

The 'Unifuel' fuel system from, amongst others, engine manufacturer MAN Diesel.

This system has three flow meters for measuring heavy-fuel oil; flow meters A, B and C and the ViscoSense sensor from VAF Instrument, Dordrecht, The Netherlands.

Example of economic sailing.

Fuel measurements in the 'Unifuel' systems.
In this system, one fuel-supply system is used for the main engine and auxiliary engines. This requires three fuel gauges: one for the total amount of fuel supplied, one for fuel supply to the auxiliary engine(s) and one return gauge for the auxiliary engines.

Consumption main engine = total amount of fuel supplied – (supply auxiliary engines – return auxiliary engines).

There is often a temperature difference: 145 – 70 = 75 °C between the fuel supplied (cold section) and the fuel supplied after the heat exchanger (hot section). Therefore, all three measurements must be meticulously adjusted with respect to temperature to avoid large deviations in actual fuel consumption.

As the fuel temperature fluctuates slightly during operation, temperature corrections should be made constantly. In this way an accurate measurement of the fuel consumption for all the engines supplied with fuel via the 'Unifuel' system is found.

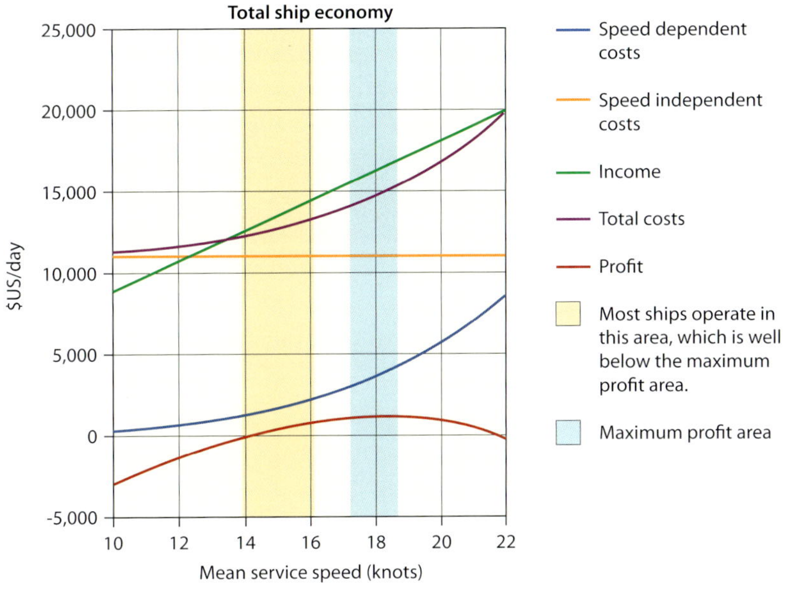

▲

An example of the costs for a certain ship per day in $ U.S.

23.10 Fuel consumption measured in trials

Measuring the fuel consumption, and consequently the specific fuel consumption accurately is best done during the mandatory tests for acceptance of the engine.
Here the engine is loaded with a water brake.
The tests are performed in accordance with regulations, and the results are a solid basis for the calculations and comparisons performed when the engine is operating in a ship or diesel power-plant.

It must be noted that the engine load in a diesel-power plant, an electric generator can be measured very accurately.
This is not the case in ship propulsion.

Also, see Chapter 19, *Ship propulsion* and Chapter 32, *Regulations for propulsion engines, classification, repair and damage.*

▶ One of the largest diesel engines being tested; the 'blue' water brake is being used as a load.
This is a Wärtsilä Sulzer RT-Flex96C.

Note the size in relationship to the engineers!

The price development of crude oil in $ U.S. per barrel between 1950 and January 2008.

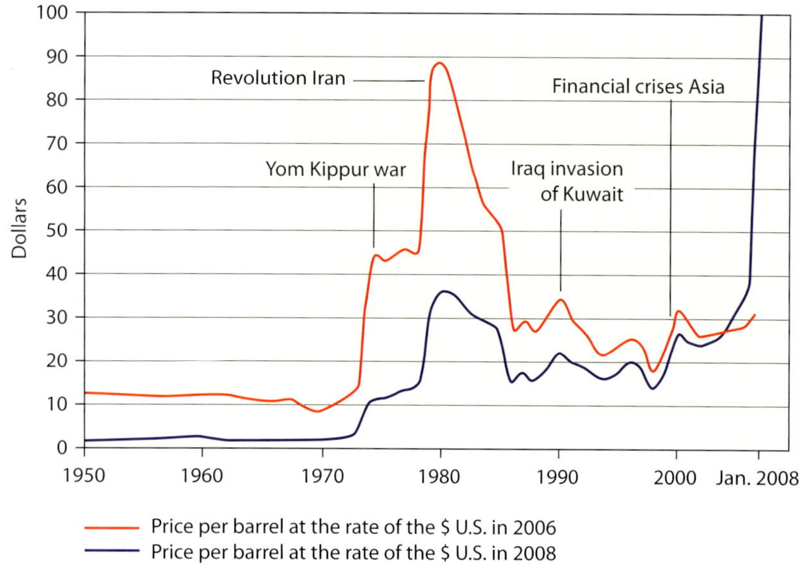

23.11 Oil-price development

The price of crude oil is and has traditionally been calculated in U.S. dollars per barrel (159 litres).
The barrel price was below twenty dollars in the 1970s and is presently (2012) over one hundred and twenty U.S. dollars!
This price is dependent not only on supply and demand but also on political and financial circumstances.

◀ The price development of crude oil in $ U.S. per barrel (159 litres) between 1950 and January 2008.

23.12 Sailing at reduced engine power or 'slow steaming'

Source: A summary of a study by Maersk Lines, the largest container shipping company in the world.

23.12.1 Introduction

Due to fuel prices reaching record levels in 2011, fuel consumption is once again a hot topic. The current fuel prices fluctuate from 690 USD per ton HFO to 990 USD per ton for MDO (2012).

A further point of interest is reducing exhaustgas emissions, in particular CO_2, carbon dioxide. Obviously, fuel consumption can be cut drastically when the ship's speed is reduced and less power is required. Reducing the speed of a high-speed, large container ship from 25 to 21 knots achieves a 50% power reduction! However, this is only feasible provided the engine can run at reduced power for an extended period without causing additional pollution and engine wear.

Two important questions should be addressed in the case of reduced ship speeds:

1. Is the propulsion engine capable of operating at reduced power for extended periods?
2. What are the fuel-consumption rates at low loads?

Therefore, prior to building these ships one should first establish if the ship's design should meet the existing high-speed requirements or if one should opt for lower speeds with a relatively smaller propulsion engine and lower engine capacity.

So far (2012), ship owners continue to opt for high speeds and ample propulsion power, allowing them to adjust the speed to future requirements. Naturally, the IMO Tier II emission requirements for newly built ships that were implemented on 01-01-2011also play a role. These render the presence of electronically-controlled fuel systems such as the common rail indispensable. All three engine manufacturers of two-stroke crosshead engines, MAN Diesel, Wärtsilä Sulzer and Mitsubishi have such systems. Tier II encompasses a reduction of NO_X emissions from the present 17 g/kWh to 14.5 g/kWh. This can only be executed if the specific fuel consumption is reduced by 2 g/kWh (electronically controlled engines). Conventional propulsion engines such as the MAN Diesel MC and the Wärtsilä Sulzer RTA's require a reduction of 4 g/kWh.

The 2009/2010 recession in, amongst others, container haulage has shown that large container ships need to reduce their speed drastically from 22 - 24 knots to 12 - 13 knots. This results in an engine load of 10 to 15% of the full load. Two-stroke crosshead engines are generally designed for engine loads between 60 and 100%. Therefore, depending on the manufacturer, special attention should be paid to the engines.

MAN ME type	electronically controlled.
Wärtsilä RT Flex	electronically controlled.
MAN MC/C type	mechanically controlled.
Wärtsilä RTA types	mechanically controlled.
Mitsubishi UEC type	mechanically controlled.

▶

The required propulsion power relative to the ship speed.

Clearly shown is that with a slight decrease in speed, the propulsion power required and therefore fuel consumption drastically decrease. The present large container ships that were all designed for high speeds, often more than 20 knots, now undertake voyages that last for 9 weeks instead of the earlier 8 weeks as their speed has been drastically reduced.

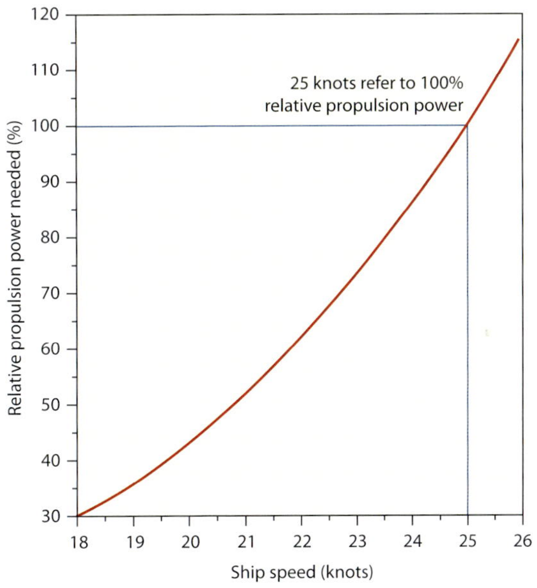

The first two engine types do not require special attention. According to the manufacturers they can readily achieve 10% of the Maximum Continuous Rating (MCR).

However, manufacturers do ask for special attention to be paid for the last three engine types. The recommendations are as follows:

1 The fuel valves/injectors require proper maintenance and inspection in accordance with the stipulated number of operating hours, or, depending on their operating mode, in a shorter time span. MAN engines require slide valves.
2 Cleaning turbochargers. To prevent soiling on the exhaust-gas side, the engine should operate at full speed at regular intervals. This should be executed as stipulated by the turbocharger manufacturer and the exhaust-gas turbine should be cleaned with soft materials such as walnut shells.
3 In- and decreasing the engine power should be performed according to the time table drawn up by the manufacturer. Generally, it takes over an hour for the engine to reach full power and half an hour to reduce power.
4 The high-temperature cooling-water system (HTC) should be kept at the highest allowable temperature (85-95 °C) to minimise lowtemperature corrosion of the cylinder liners.
5 The fuel viscosity should be kept low (between 12 and 17 cSt). The fuel temperature at the highpressure fuel pumps should not exceed 150 °C.
6 Cylinder lubrication should remain below 1.2 grams per kWh at a constant load. The doses may only be increased in the case of power shifts due to, for instance, manoeuvring, or if the cylinder liner is rapidly deteriorating. Contamination as a result of excessive lubrication causes further contamination of the scavenging-air space, which should be prevented at all times. It is recommended that an electronically controlled cylinder-lubrication system is installed (Pulse Jet or Alpha).
7 Regular scavenging-air space inspections should be performed to check scavenging-air ports and piston rings as well as preventing ring sticking, soot, dirt, non-combusted fuel and sludge. These inspections should occur more often than those at normal engine loads in order to avoid scavenging-air fires.
8 If the electric auxiliary turbochargers are continually on stand-by, the return valves in the scavenging-air line must be checked for wear and tear at regular intervals.
9 The exhaust-gas pipe, the exhaust valves and the turbocharger ring absorbers should be inspected regularly and soot build-up should be removed.
10 Electrically driven turbochargers that run continuously, experience more wear than in their original design calculations. After all they were designed to operate solely during manoeuvring. Therefore, it is imperative that open bearings be lubricated at regular intervals.
11 The exhaust-gas temperatures before and after the turbochargers should remain within alarm limits. It may be necessary to avoid engine loads just over the auxiliary turbo-charger shutdown.
12 The exhaust-gas boiler should be checked regularly and cleaned, if required, to prevent potential fires which, at worst, may destroy the boiler.

23.12.2 An example of slow steaming

Engine: Wärtsilä two stroke crosshead 10RTA96C.

– Speed, nautical mile.	23.5	12.5
– Power, kW	39,305	5,366
– Fuel consumption, kg/hr.	7,152	1,195
– Fuel consumption per nautical mile.	719.09	112.52
– Spec. fuel consumption, gr/kW/hr.	180.74	212.15
– Number of rev. min.	92	45
– Scav. air pressure, bar	1.7	0.0
– Rev. per minute. Turbo chargers	8995/ 8521/ 8545	2316/ 2120/ 2158
– Exhaust gas temp, mean, after cylinders, °C.	374	232

Slow steaming

$$\text{Service speed} = \frac{12.5}{23.5} \times 100\% = \mathbf{53.2\%}$$

$$\text{Fuel consumption} = \frac{1{,}195}{7{,}152} \times 100\% = \mathbf{16.7\%}$$

$$\text{Spec. fuel consumption} = \frac{212.15}{180.74} \times 100\% = \mathbf{117.4\%}$$

$$\text{Fuel consumption per nm} = \frac{112.52}{719.09} \times 100\% = \mathbf{15{,}6\%}$$

23.12.3 Cleaning the exhaust-gas turbine of turbochargers at a 50% engine load

Dry cleaning of the turbine is only effective at a minimum engine load of 75%; lower loads gave insufficient results, increased fuel costs and had the added risk of causing additional wear and tear of the cylinder lining, the so-called 'scuffing'.

Despite these recommendations, it remains imperative that inspections take place at regular intervals in order to identify deviations and take the measures required.

Engine Load	Action	Load during cleaning	Interval Days
10 – 50%	Turbine dry cleaning	50%	1 – 2
50 – 75%	Turbine dry cleaning	Actual Load	1 – 2
75 – 90%	Turbine dry cleaning	Actual Load	2 – 4
10 – 90%	Turbine water cleaning	20 – 35% (0.3 – 0.6 bar Pscav)	2 – 20
10 – 50%	Compressor, water cleaning	50% (MAN & ABB) Maximum 6000 RPM (MHI)	1 – 2
50 – 90%	Compressor, water cleaning	Actual Load (MAN & ABB) Maximum 6000 RPM (MHI)	1 – 15

◀ Procedures for cleaning turbochargers.

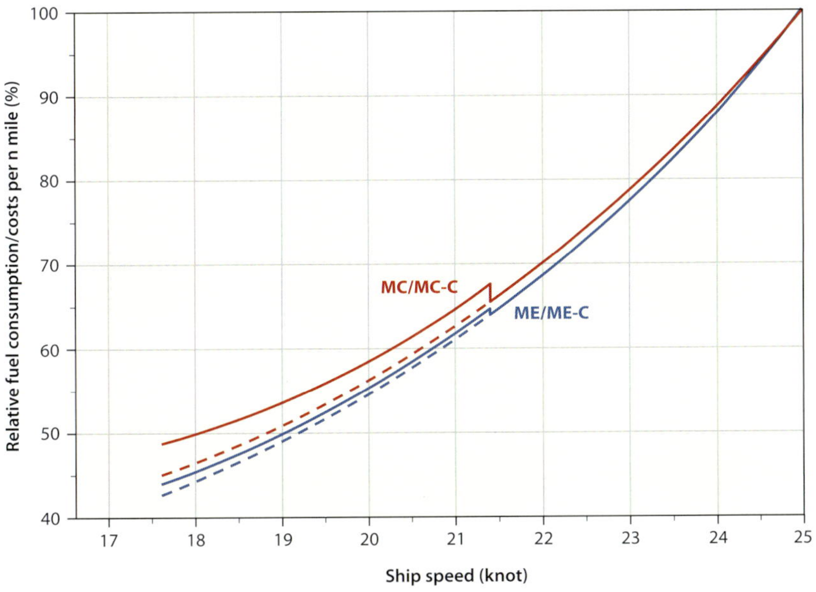

◀ Decreased fuel consumption at low propulsion power for large container ships with MAN Diesel MC/MC-C and ME/ME-C two-stroke crosshead engines.

At partial loads, the traditional designs MC/MC-C consume more fuel.

◀ A slide valve in a MAN Diesel MC/C two-stroke crosshead engine.

This has been specially designed for slow steaming.

▶

A cylinder lubricating-oil device. It is advisable to adjust the capacity electronically when changing the engine power output.

▶

Inspection of the piston, piston rings and cylinder liner can be done from the scavenging-air space through the scavenging ports. Shown here an extremely dirty piston crown above the piston ring packet.

▶

A glance through the scavenging ports to the cylinder liner. Using the flash on a digital camera, very clear photos can be taken. Information such as the cylinder number, date, number of operating hours and determinations can then be saved together with the photograph in a file.

Parameters	Engine power	SFOC	Fuel consumpt.	Operating time	Fuel consumpt.	Ship speed	Sailed distance	Fuel consumpt. per n mile	Fuel costs per n mile	Relative fuel costs per n mile
Mode	% SMCR	g/kWh	t/24h	h/week	t/week	knot	n mile/week	kg/n mile	USD/n mile	%
Standard	90.0	167.5	263.3	168	1843.3	25.0	4200	439	219	100.0
12K98MC-C6										
Low	75.0[1]	165.1	216.2	14	126.1	75.0	Included in low load operation assuming no ship speed increase			
Load	30.0	174.0	91.2	+ 154	+ 585.0	30.0				
runn.				168	711.1	18.5[2]	3108	229	114	52.1
12K98ME-C6										
Low	75.0[1]	163.6	214.4	2	17.9	23.0	Included in low load operation assuming no ship speed increase			
Load	30.0	172.8	90.6	+ 166	+ 626.4	18.5				
runn.				168	644.3	18.5[2]	3108	207	103.6	47.3

1. Including temporary load increase of up to 75% SMCR at low load operation. Ref. service letter SL07-480/SBE, June 2007
2. During 75% running load, the ship will often sail a longer curved/circular path instead of straight ahead. Therefore, no ship speed increase is assumed when calculating sailed distance (even if a longer sailed distance is assumed there will still be significant fuel cost savings)
   - SFOC refers to LCV = 42,700 kJ/kg
   - Fuel consumption refers to LCV = 40,200 kJ/kg
   - Fuel price used is 500 USD/t

▲

An example of reduced fuel consumption at low propulsion power for large container ships operating with a MAN Diesel 12K98MC-C6 and a 12K98ME-C6, power output, SMCR-68,529 kW at 104 rpm.

Man Diesel two-stroke diesel engines are a good example of the possibilities of sailing at reduced engine power. The ME/ ME-C engine series can operate at four different programmes installed in the software of the ME engine. The mechanical engineer on board can decide which programme to use.

In principle each individual program can be programmed to meet the requirements of the ship owner, but the programme must be in conformance with the IMO requirements for NO_x emissions.
MAN Diesel has standard programmes for minimal fuel consumption and NO_x emissions.
Licence builders have opted for supplying engines with a programme for fuel economy mode with the NO_x emission programme being optional. The NO_x emission programme is of interest for shipping companies operating in areas where financial compensation can be obtained for low NO_x and SO_x emission investments.
The general description for Low Load Operation with MC/MC-C engines also applies to ME/ME-C engines.

This entails that MC/MC-C engines running at low-load operation should run at 75% of the maximum engine power or Maximum Continuous Rating (MCR) for one hour every twelve hours, to reduce engine contamination.

This is the equivalent of 14 hours per week.
ME/ME-C engines have to run at 75% MCR two hours a week; a significant reduction in comparison to the traditional MC/MC-C engines!
This is also evident in the table above showing measurements for a MAN Diesel 12K98MC-C6 and the MAN Diesel 12K 98ME-C6 engines. The fuel cost savings can amount to as much as 10% of the overall fuel costs!

23.12.4 Low-load operation in MAN Diesel MC/MC-C engines

Slide valves

Today it is standard procedure to fit slide valves with nozzles. The absence of additional volume below the atomizer allows improved combustion at every load.

Deactivating fuel pumps

Below 40% MCR some of the fuel pumps may be switched off putting extra load on the operational cylinders which consequently results in an improved and cleaner combustion process. The engine continues to run efficiently at speeds of 20-25% of the nominal speed.

Note

The maximum permissible speed for an engine with a number of discontinued fuel pumps is calculated by MAN Diesel on a case by case basis, as the vibrational behaviours in the crankshaft must be taken into account. Occasionally, MAN Diesel engines require the fitting of turbochargers with an increased capacity.

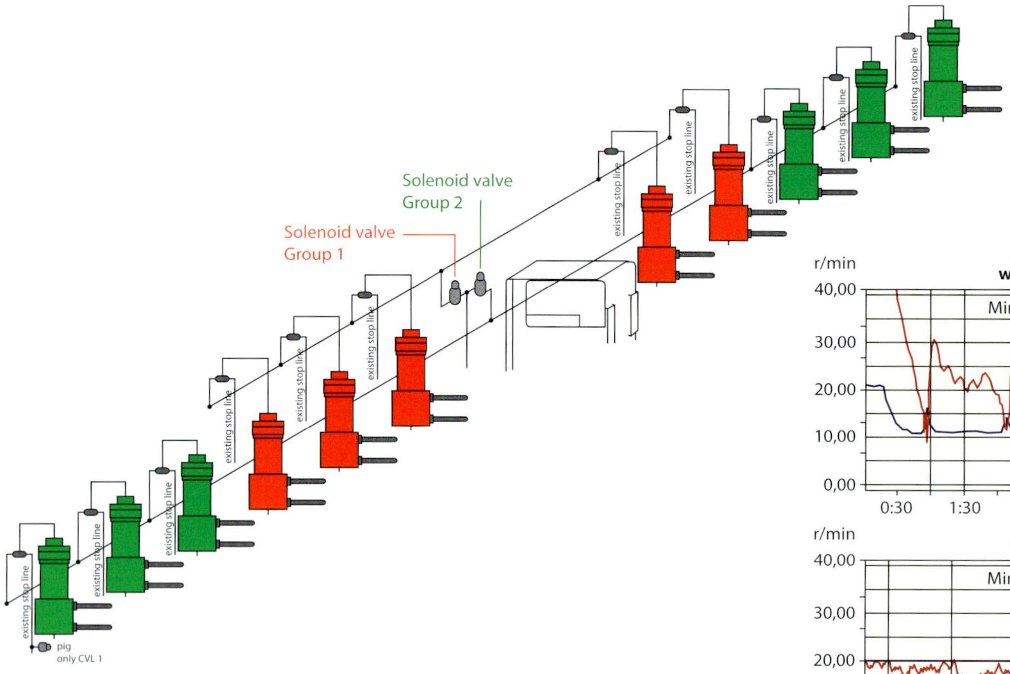

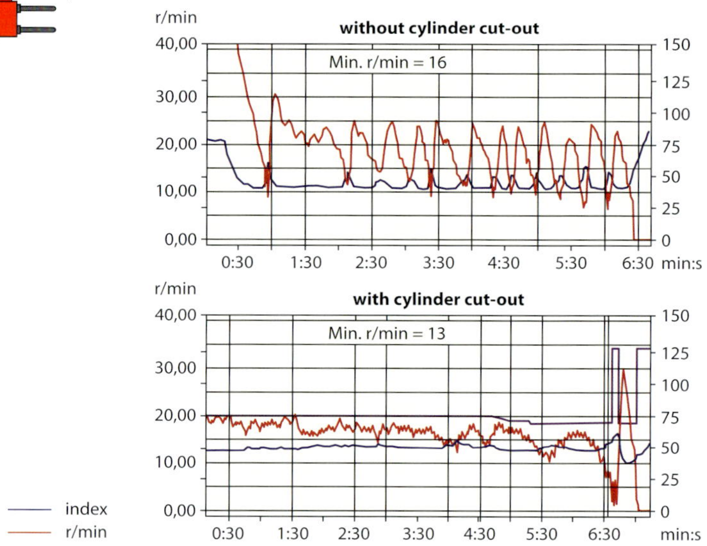

▲

The system for shutting down the high-pressure fuel pumps. In this example, a maximum of five cylinders can be shut down. The system is used below 40% MCR.

The engine can operate in a stable condition up to 20-25% of nominal rpm.

The maximum rpm must be calculated per engine by specialists at MAN Diesel, because of possible vibrations in the crankcase.

◀

If the turbochargers are in continuous operation, it is advisable to pay extra attention to the bearings and connections. They are not designed for this purpose.

1 *electro-motor*
2 *ventilator*
3 *scavenging-air duct*

◀ Shown here, a steam driven exhaust-gas boiler, which becomes very contaminated at a reduced power output.

▼ It is easy to inspect the piston, piston rings and cylinder liner from the scavenging-air space. Using good lighting and controlling the turning engine remotely from the scavenging-air ducting, every cylinder can be evaluated by taking digital photographs saved to a file.

▲ The scavenging-air inlet of a turbocharger, foreground. At reduced power output, the exhaust-gas turbine must be cleaned regularly with water.

▶
Before inspection, the entire scavenging-air duct can be cleaned.
In this case, the used cylinderlubricating oil does not completely drain.
It is advisable to install an extra drainage.

▼
Piston ring inspection. The rings are clean, whole and loose.
The ports are clean. Well maintained.

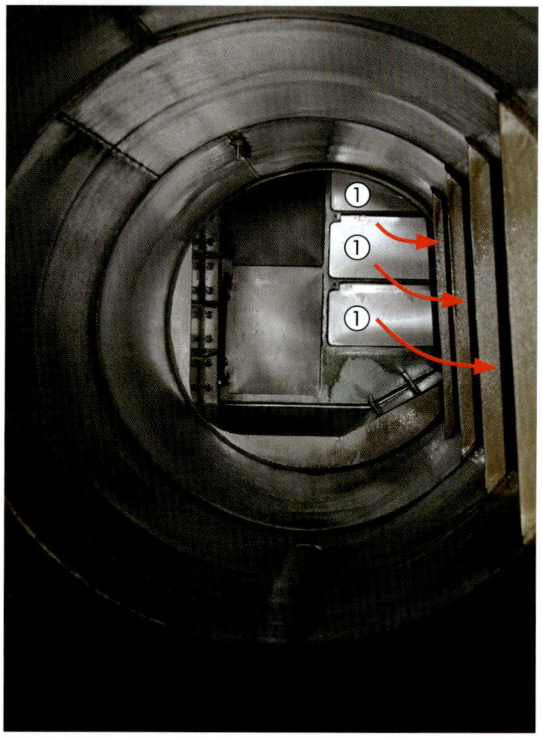

▲
The non-return valves in an electrically driven auxiliary turbocharger must be checked regularly.

1 *non-return valve*

The red arrows indicate the airflow direction to the inlet or scavenging ports.

◀
An example of a very heavily damaged steam driven exhaust-gas boiler.
The pipes are partially burned.

An example of a heavily damaged steam exhaust-gas boiler

Most likely cause: soot fire. In exceptional cases possible hydrogen fire. Due to engine contamination, the flue-gas side of the boiler has been soiled with soot particles. The soot fire with extremely high temperatures caused a leakage in the water/steam section.

When water is broken down in oxygen and hydrogen, which only occurs at extremely high temperatures of over 1368 °C, a so-called hydrogen fire occurs. The entire steel construction is rapidly melted down by these high temperatures and the fire subsequently spreads to the rest of the engine room. However, it is also feasible that the fire was caused by a heavily contaminated exhaust-gas boiler. During this so-called soot fire, oil and soot combined with sufficient air (oxygen) can cause considerable damage.

Often part of the pipe cluster is cooled inadequately due to failure or switching-off of the feed-water pump. The main engine must operate at full power as the relatively cool exhaust gases (approximately 350 °C) cool the boiler at these intense temperatures. The boiler should be cleaned at regular intervals (preferably every 2 months). Monthly cleaning is certainly required in the case of complex boilers with economisers, dampers and superheaters. Pressure differences before and after the boiler give a good indication when the boiler should be cleaned.

▲

An example of a very heavily damaged steam driven exhaust-gas boiler. A pipe plate after a boiler fire.

▲

An example of a very heavily damaged steam driven exhaust-gas boiler. Debris from the steam boiler.

23.12.5 Reduction of specific fuel consumption by reduced power at an unchanged speed

A MCR point is chosen below the nominal MCR point in the diagram, down the vertical line. The speed is unchanged. This system had been used for many years and was the manner used to reduce power in most engines. The advantage of this system being that the specific fuel consumption was reduced and the disadvantage was (not in this instance) that less power was generated.

Derating an engine may at times be complicated. However, one advantage is that the MCR point can be re-adjusted to the original nominal MCR 1 point L1 when fuel prices drop and there is an increased demand for higher speeds. When choosing turbochargers and air coolers this should be taken into account.

▼

Example: If the power output of a MAN Diesel 10K98ME is required but a MAN Diesel, type 11K98ME with reduced power output is chosen, the specific fuel consumption is reduced from 162.5 g/kWh to 159.5 g/kWh. A reduction of 3 g/kWh.

The purchase of the extra cylinder is of course an extra expense, but the return on investment is 3 to 4 years with the present fuel prices. If a MAN Diesel 12k98ME7 is chosen, the specific fuel consumption can be reduced to 157 g/kWh, almost halved!

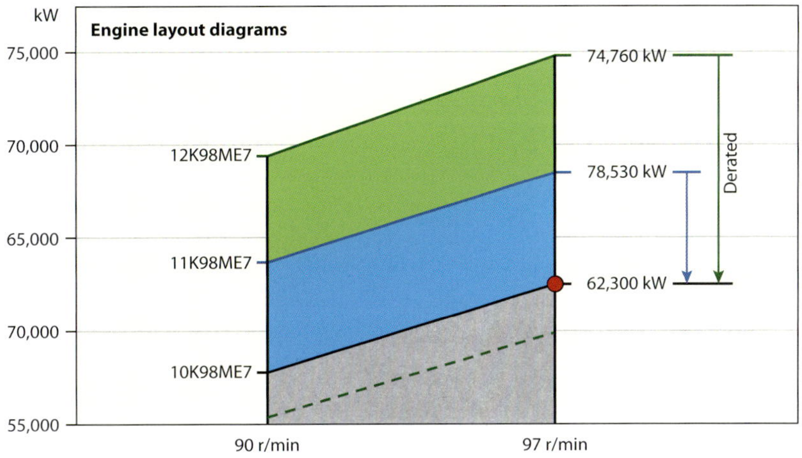

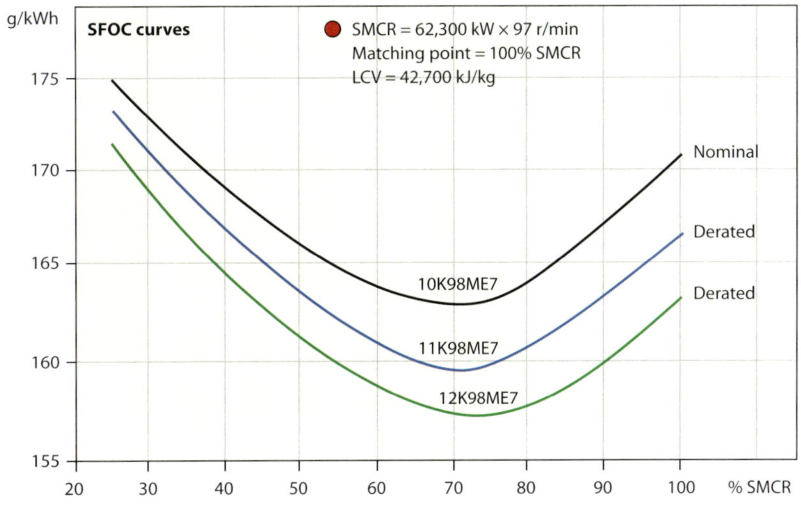

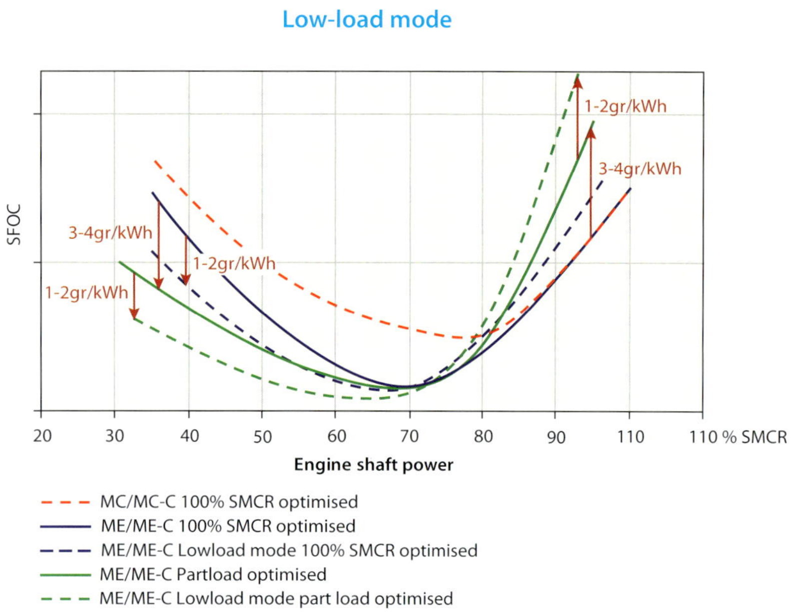

▲

A simple low- and partial-load power output optimization of the MAN Diesel ME/ME-C engine is obtained by measures such as adjusting the exhaust-valve openings and the fuel-injection timing for specific low- and partial load power outputs.

The reduction of the specific fuel consumption is approximately 1 to 2 g/kWh in economy- and low-load mode. Changing from economy- to low-load mode is as simple as pressing a button on the operating panel.

When the engine is operated in low load and has been optimized for economy mode, the total specific fuel consumption reduction is the sum of the reduction due to partial-load optimization and the low-load power output optimization. In this case, 4 - 6 g/kWh reduction compared to 100% SMCR.

23.13 VAF Instruments T-sense torque and torsion measurement of the propeller shaft

23.13.1 Principle of operation

The T-sense (Torque) system for torque measurement can be fitted on the propeller shaft or intermediate shaft. A shaft loaded by torque will cause a slight distortion of the shaft surface. A LED (Light Emitting Diode) and an optical eye are mounted on the rings that are fixed to the rotating shaft.

▲

A LED and optical sensor.

In the event that the beam of the LED moves in the X or Y direction, the optical eye perceives this movement.

The optical eye is capable of registering movement of 25 nanometres, which is equal to 0.000025 mm. A very small value!

▶

A complete measuring system on the shaft.

The part that rotates with the shaft is clamped on the shaft. The rotor part has a thick insulation layer to avoid outside influences such as varying temperatures as much as possible.

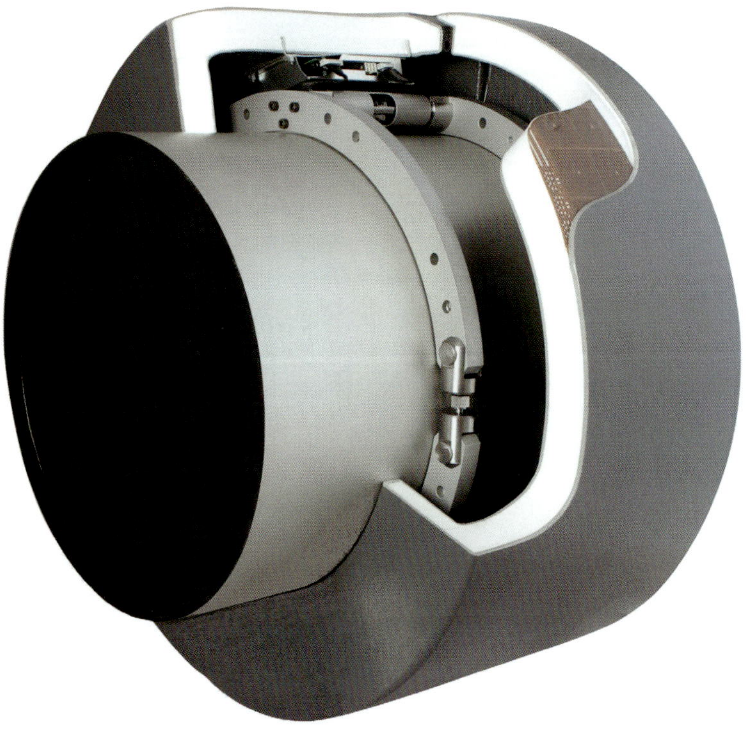

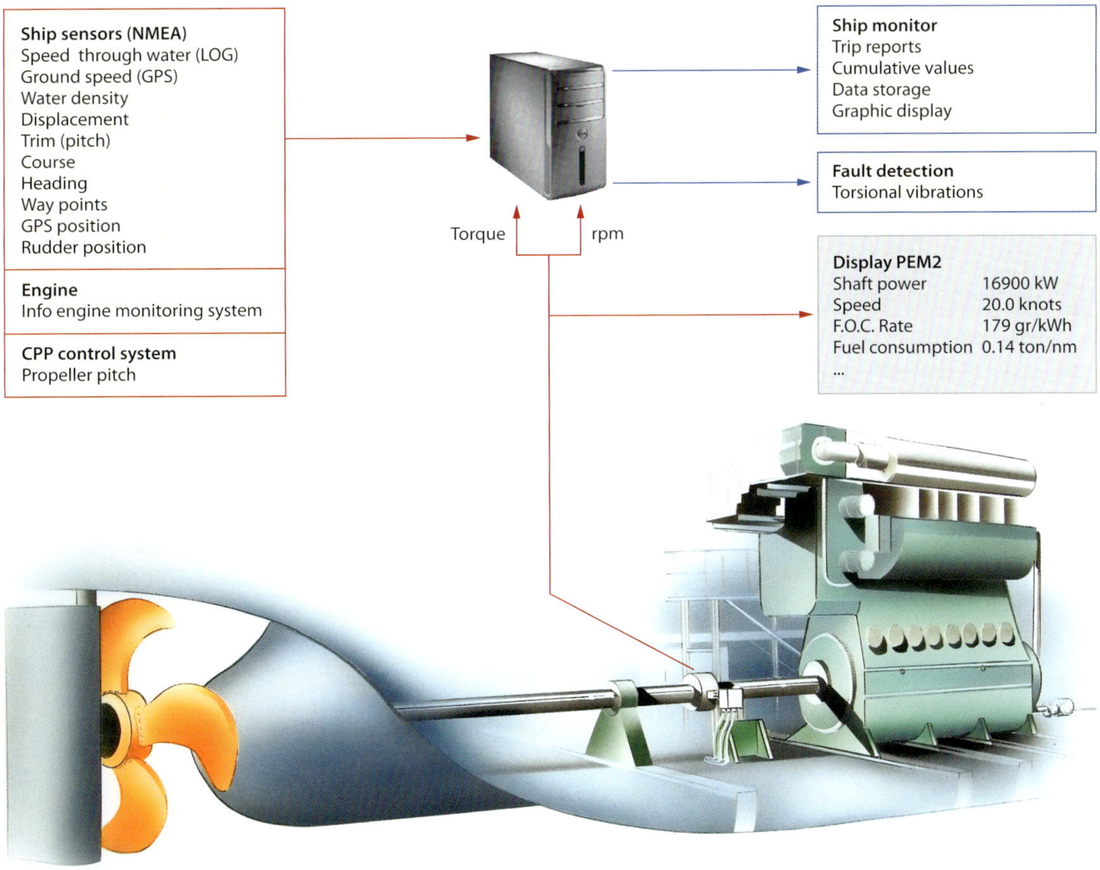

◄ A torque measuring system on the propeller shaft.

The parts attached to the rotating shaft are housed in a splash-proof casing, fixed to the circumference of the rotating shaft. The system functions contactless, the measured values are continuously transferred from the rotating shaft to the stationary housing using Bluetooth.

Energy supply from the stator section to the rotating shaft occurs by induction.

The stator section holds the primary coils for the energy transfer, a Bluetooth receiver and a control module with digital and analogue outputs.

23.13.2 Bluetooth applications

The important applications are sending files between devices such as computers and mobile phones, sending a document or image from a computer to a printer, sending keystrokes from a keyboard to a computer, sending images from a scanner to a computer and sending sound bites to, for instance, wireless headphones.

23.13.3 Security

As radio signals can be picked up by all receivers in the vicinity of the Bluetooth devices, the Bluetooth basic protocol is supported by authentication and encryption. Authentication is realised by means of a secret key supplied to both devices. The protocol allows for one device to authenticate another. If the authentication is successful the connection can be encrypted.

▼ A complete torque measuring system.

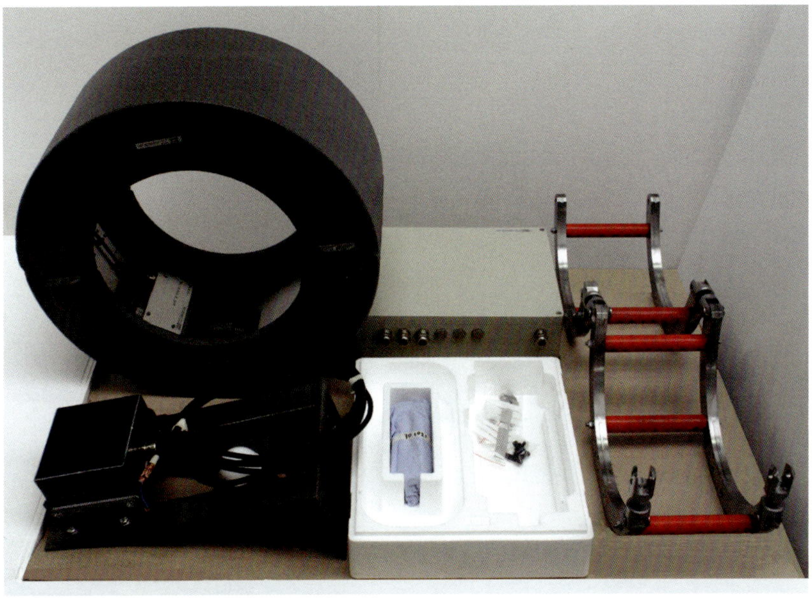

The stator can be connected to the 'Propulsion Efficiency Monitor' (PEM2). The PEM2 touch screen is easy to operate and continually supplies up-todate values for the torque, speed and power output. Moreover, the average values of the torque, speed, power and total power output can be read from the screen displaying propulsion data.

23.13.4 Torque signal and energy transfer

Close to the rotating shaft a coil is installed which holds the primary winding for transferring the induced electronic energy to the rotating shaft equipped with the measuring instruments. The alternating current is then converted and feeds the measuring system on the rotating shaft. An amplifier and an AD (Analogue to Digital) converter send the outgoing signal of the optical eye directly to the Bluetooth transmitter. The completely shielded, wireless 2.4 Giga Hertz signal is transmitted from the rotating shaft to the stationary section where the Bluetooth receiver is situated.

The Bluetooth signal received is the input for the control module and is converted into a RS-485 Modbus signal that can be sent to the Propulsion Efficiency Monitor where it is subsequently used. The control module also has additional connections for RS-485 Modbus and analogue 4-20 mA output signals, which can be transmitted to the ship's network with its measuring- and control systems.

23.13.5 Speed signal

The speed is measured by a gravity sensor located on the rotating shaft. The outgoing signal generated by this sensor is transmitted to the electronics on the stator via the Bluetooth transmitter.

23.13.6 Propulsion Efficiency Monitor, PEM2

The PEM2 system including touch screen continuously displays the torque, speed, shaft power and other selected data. Depending on the selected system, the fuel quantity and -temperature can be

▶
Looking at the T-sense measuring system from another angle.

Orange crosshatched part in the illustration: the stator spool induces electrical energy for the rotating secondary spool and supplies energy to the measuring system.
The spool, which supplies energy to the rotating section, is mounted next to the shaft on a pedestal. See drawing.

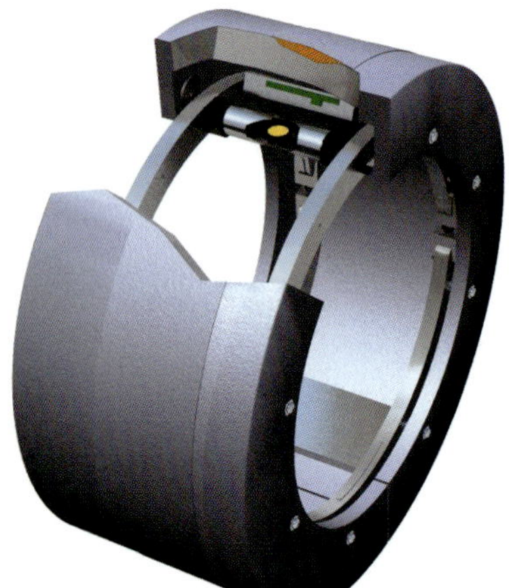

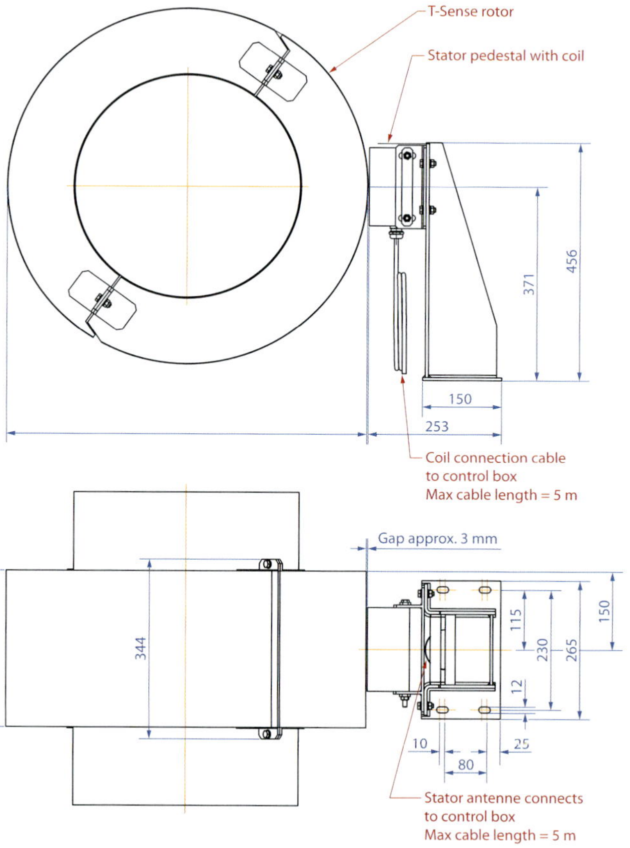

▶
Left: the LED arm.
Right: the arm with the optical eye.

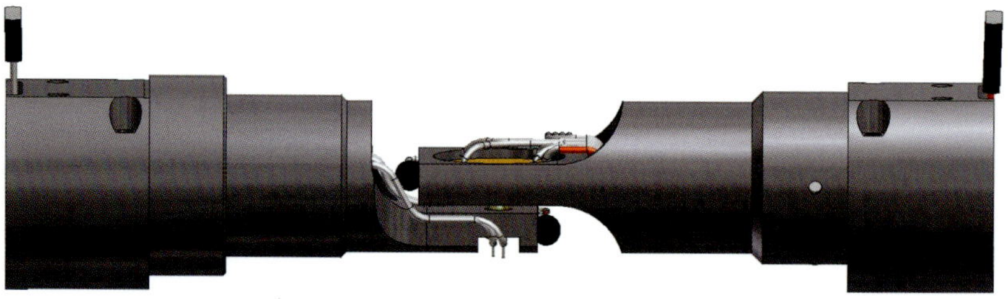

an input and be displayed on the screen. The fuel consumption measured by the PT2 flow meter can be converted to fuel consumption in kilograms per hour using temperature compensation. Displaying fuel consumption in kilograms per hour or kilograms per kWh constitutes the most accurate method to determine the ship's engine efficiency. Specific Fuel Oil Consumption (SFOC) is a figure utilised by all ship-engine manufacturers as an indicator of engine efficiency. In conjunction with the data entered from the ship's log or GPS, the PEM2-system displays the specific fuel consumption in kilograms per nautical mile sailed. This fuel consumption is a reliable indicator of the ship's efficiency.

The T-Sense signal assists in finding the lowest specific fuel consumption at an optimum point in the power diagram when the ship's speed and time of arrival are not important. The T-Sense also helps to compute the most favourable speed where fuel consumption is kept to a minimum. Importantly, for most hull shapes a 10% speed increase results in an increased fuel consumption of 33%!
Additionally, a standard ship with 1,000 kW of shaft power uses approximately 4000 litres of fuel daily.

23.13.7 Monitoring fuel consumption

In large propulsion plants it is crucial to obtain information regarding performance and propulsion power in relation to fuel consumption. Fuel consumption can be measured by installing flow meters and PT100 temperature gauges in the supplyand discharge sections of the fuel pipes. The PEM2-system will calculate the fuel consumption corrected by the various inlet- and outlet temperatures of the fuel. The system will subsequently display the fuel consumption per kW shaft power or per nautical mile travelled, provided the ship's log is connected to the system.

23.13.8 VAF T-Sense follow-up

The crew can process this data and determine the most economical setting for the propulsion engine, the trim or the pitch. The potential changes are displayed on the screen.

23.13.9 Additional options

The standard system supplies generated torque, shaft speed and shaft power using a torque sensor. The system can be expanded with the following modules:

- Maximum torque safety.
- Energy consumption per nautical mile or per kWh.
- Power calculations.
- Propeller shaft vibration analyses for frequencies up to 50 Hz.

These additional optional features require that a number of extra signals are connected to the system, such as the fuel quantity, fuel temperature, ship's log data, GPS system or the shaft-generator data.

23.13.10 Accuracy

The overall measuring error for the system is below 0.5%.

23.13.11 Advantages of the T-Sense torque measuring system

- Suitable for all shaft diameters.
- No measuring errors during its operational life.
- The transmission of the measurements as well as the energy supply occurs without the use of contacts, so no wear and tear or soiling.
- Vibrational measurement of the propeller shaft up to frequencies of 50 Hz.
- No maintenance.
- Solid construction.

23.13.12 New developments in TT-Sense torque and thrust gauge

This optical sensor is capable of measuring very small deviations in the axial direction of the shaft resulting from the thrust exerted on the shaft by the propeller. These values are measured at full load in the order of 100 microns (μm) for torsion and 5 microns (μm) for thrust, with values being considerably lower at low loads. By using a set of four yokes equipped with contra weights, the effect of the mechanical distortion is negligible.
To avoid expansion of the various materials, the rings and windings are made of the same material as that of the shaft. Furthermore, they are thermally isolated from the environment by a protective cap. In order to compensate for potential temperature differences, a number of temperature sensors are mounted, which compensate for deviations in the measurements using a calculation program.
The TT-Sense helps ship-owners to determine the ship's resistance and propeller efficiency very precisely.

>CH24

Auxiliary systems: fuel- and lubricating-oil separators

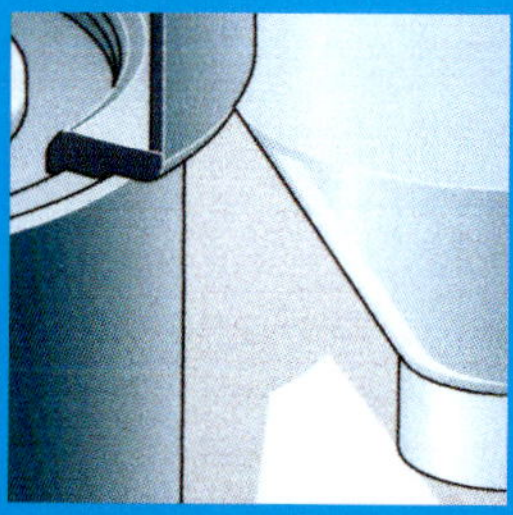

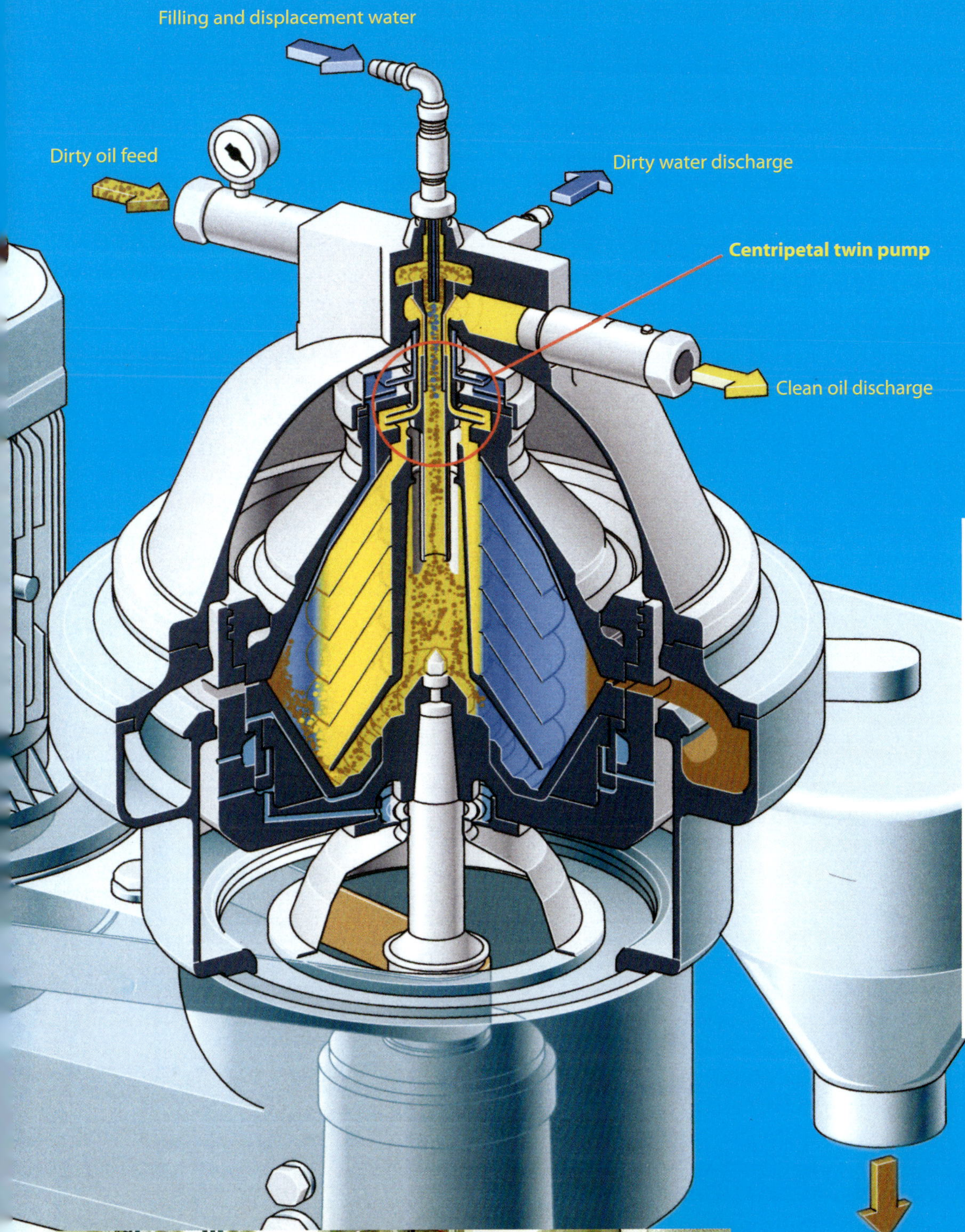

It is necessary that large diesel-engine plants on ships and in power plants have centrifuges or centrifugal separators installed to clean fuel and lubricating oil. The use of a modern automatic centrifuge is imperative given today's inferior heavy-fuel oil quality. Shown is a cut-away Westfalia separator.

A 'booster unit' with an automatic back-flushing fine-filter.

24.1 Introduction

Cleaning fuels and lubricating oils is crucial for optimal and long-life operation of diesel engines. Fuels, especially inexpensive HFO, contain various contaminants that are detrimental to the fuel system and cause problems when injected in the combustion space.

Lubricating oil is the 'blood' of the engine and serves many purposes. The quality of the circulating lubricating oil is vital. Many combustion and corrosion products are absorbed by or carried in the lubricating-oil system and must be removed in time to prevent severe damage to the engine.

24.2 Fuels

The following points are important for 'contaminant free' fuel oil:

- Ensure that the fuel tanks are clean and have no condensation.
- A drain at the lowest point of the tank is only useful when periodically used.
- Bunker fuel at well-known firms with a regular turn-over.
- Fuel has to comply with minimum requirements. This is usually officially organised when bunkering large quantities. However, when refuelling small amounts as with for instance yachts, it is assumed that 'things are all right'.
- In wintertime or in cold regions additives are added to fuels to prevent the paraffins in the fuel from coagulating. This would clog the fuel filters and stop the engine.
- From the smallest engines to the largest crosshead engines, proper fuel filters are crucial. They must be regularly changed, or manually or automatically back-flushed; this is dependent on engine size.
- Small diesel-engine plants often have a simple water separator and a magnet, to which the iron particles adhere.
- Duplex filters are also useful in small engines. One filter is operational while the other filter is cleaned for use when the operational filter is contaminated.
- In larger installations, settling tanks are also used. In settling tanks, water and impurities are separated from the fuel oil and drained off. In heavy-fuel oil systems, the contents of the settling tanks are heated to a temperature at which the specific mass difference between water and fuel is greatest, thus facilitating the separation.

Nevertheless, this does not suffice for the cleaning of heavier fuels such as HFO. Adequate cleaning is achieved with the use of centrifuges or separators; the centrifugal force is a decisive factor.

See also Chapter 8, *Fuels, fuel-line systems and cleaning fuels.*

24.3 The principle of centrifugal separators

When solid or liquid particles are carried in a solution that is under the influence of a centrifugal force, they will assume a constant speed after some time.

This phenomenon is known as separation speed, also referred to as Stoke's Law:

$$v_g = d^2 \left(\frac{sm_d - sm_v}{18 \times \eta} \right) \times g$$

v_g = velocity in metres per second
d = diameter of a particle in metres
sm_d = specific mass for a particle in kg per m^3
sm_v = specific mass of the liquid in kg per m^3
g = gravitational acceleration in metres per sec^2
η = viscosity in 10^{-3} per metres second

From the formula, the following assumptions can be made:

- the separation of solid particles from the liquid is dependent on the specific mass of both the particles and the liquid;
- the larger the difference between the specific masses, the more complete the separation;
- the lower the specific mass of the liquid, the greater the separation.

24.4 Separation in a settling tank

This is limited as specific mass difference alone aids separation. In the following assumption, a particle is 8 microns = 8 x 10^{-6} metres.

The specific mass difference $sm_d - sm_V$ = 50 kg per m^3.

The viscosity $\eta = 10^{-3}$ kg per metre second in accordance with Stoke's Law:

v_g = 1,75 x 10^{-6} metres per second.

This velocity is very low! Fuel must therefore remain in the settling tank for a significant time in order to achieve a basic separation. Fuel, especially heavy-fuel oil, is heated to a certain temperature in order to create the largest possible specific mass differences and so improve the separation.

24.5 Separation with a centrifugal separator

'Stokes law' reads: $\frac{d^2 \times \Delta p}{18\eta} \times g$.

The acceleration g is constant, as the centrifuge rotates at a constant speed.

The force exerted on a particle in the centrifuge is related to the engine speed (the angular velocity) and the radius to the centre of the centrifuge.

The larger the centrifuge's diameter and the higher the engine speed, the greater the force exerted on the particle:

$a = r \times \omega^2$

a = accelerating force
r = radius to the centre
ω = angular velocity

In accordance with Stoke's Law:

$$g = r \times \omega^2$$

$$v = \frac{d^2 \times \Delta p}{18\eta} \times r \times \omega^2$$

$$v = v_g = \frac{r \times \omega^2}{g} \times v_g \times Z$$

$$Z = \frac{r \times \omega^2}{g} = \text{accelerating force}$$

Example

In the settling tank, v_g was 1.75 x 10^{-6} metres per second. Assuming that in the centrifuge:
r = 0.2 metre. rpm n = 5000.

$$\omega = \frac{\pi \times n}{30} = \frac{\pi \times 5000}{30}$$

$$Z = \frac{r \times \omega^2}{g} = \frac{0.2 \times \pi^2 \times 5000^2}{30^2 \times 9.81} = 5600$$

$$v = v_g \times Z = 1.75 \times 10^{-6} \times 5.6 \times 10^{-3} = 0.98 \times 10^{-2}$$

Therefore, the ratio of the separation force in the centrifuge to that in the settling tank is:

$$\text{ratio} = \frac{0.98 \times 10^{-2}}{1.75 \times 10^{-6}} = 5600.$$

On average, the separation force of centrifuges is 6000 times larger than that in settling tanks!

Adequately functioning centrifuges are imperative in properly cleaning fuels and lubricating oils in larger installations/diesel engines. Naturally, extremely fine pressure filters, either automatically back-flushing or manual, are also used in both instances.

24.6 Types of separators

Essentially, there are two types of separators:

1. Clarifiers: for the separation of solid particles from the fuel or lubricating oil.
2. Purifiers: for the separation of solid particles and water from the fuel or lubricating oil.

▼

A typical clarifier of Alfa Laval.

The untreated oil is supplied by a fixed pipe to the centre of the bowl.
This untreated oil flows into the rapidly rotating bowl or drum.
The dirt particles are separated and collected in the peripheral space, the sludge space.
The cleaned oil floats to the surface in the drum and is discharged at 2.
The bottom of the bowl periodically drops. After the bowl is emptied, the bowl is high-pressure cleaned with water.
Subsequently, the bowl supplied with water from 3 is closed.

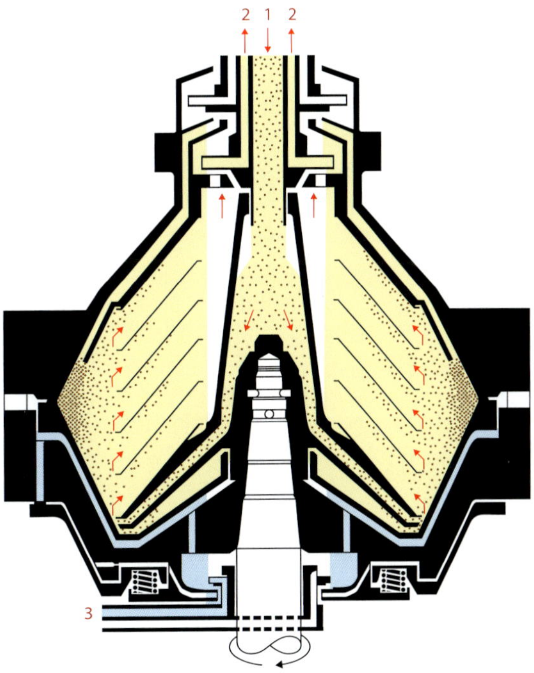

24.6.1 Interface

The interface is crucial for the operation of the centrifuge, as an equilibrium must be created between the various liquids in the rotating bowl.

▶

A cross-section of half a separator bowl showing the radius for liquids.

r_L = radius to clean-oil discharge
r_H = radius to water discharge
r_S = radius to interface with water

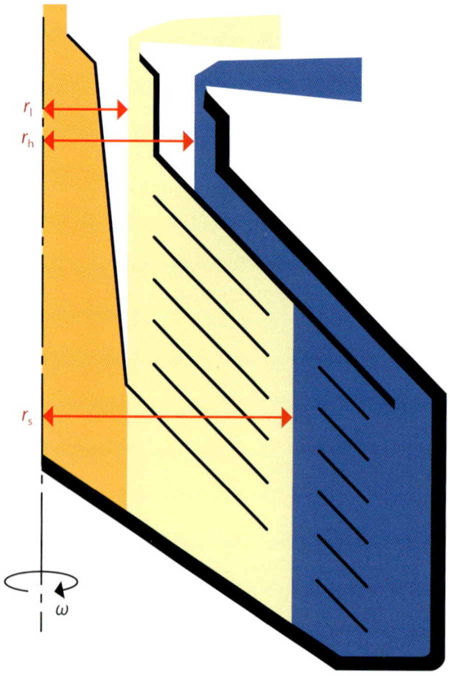

Here the radius between the bowl and the various liquid interfaces and the specific masses of the liquids are the decisive factors.
The liquid pressure in the bowl is expressed as follows:

$$p = \frac{\omega^2}{2} \times \rho \times r^2 - r_0^2.$$

p = liquid pressure
ω = angular velocity
ρ = specific mass
r_o = inner radius of the liquid
r = outer radius of the liquid

$$p_s = \frac{\omega^2}{2} \times \rho_1 (r_s^2 - r_o^2) = \frac{\omega^2}{2} \times \rho_H (r_s^2 - r_h^2) \text{ of}$$

$$\rho_1 (r_s^2 - r_1^2) = \rho_H (r_s^2 - r_h^2)$$

▼

The Purifier of Alfa Laval.

1 *contaminated-oil supply*
2 *clean-oil discharge*
3 *water discharge*
4 *water supply*

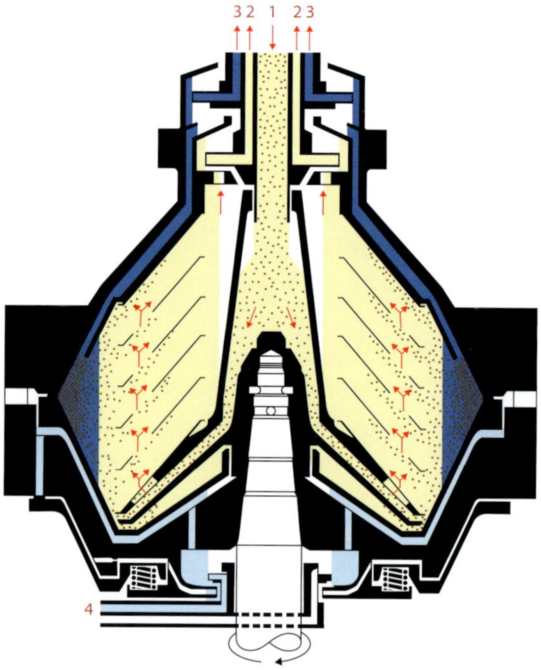

24.6.2 Purifier centrifuge

Hereby, both debris and water is separated.

The dividing line between the liquids is called the interface.
All the major modern centrifuges are cleaned automatically. This entails closing the supply of the contaminated fuel-oil and water so that the drum contents can be discharged; there is no oil present. This is known as 'shooting'. The bottom of the drum drops and the high water pressure forces the debris through the slits in the outer drum and is discharged, the drum is then closed and the supply of water and contaminated fuel-oil resumes. After most of the water is discharged (3), the equilibrium (interface) is restored.

Limits for conventional purifiers

- Fuels with a specific mass of maximum 991 kg per m³ at 15 °C.
- Optimal performance is dependent on the correct specific-gravity disc, which corresponds to the average density, the viscosity, the desired capacity, and the temperature of the fuel oil.

- Checking and mounting of the specific-gravity disc is time-consuming and often done by 'trial and error'. If the incorrect gravity disc is used, the water seal breaks and fuel oil is discharged via the water output to the contaminated-oil and sludge tank.
- A second purifier placed in series is required to ensure that the fuel is properly cleaned. This **clarifier** works as a kind of safety net. The impurities not yet separated in the purifier are removed from the fuel with this step. Here the maximum specific mass of 991 kg per m^3 at 15 °C also applies.

24.6.3 Clarifiers

Limits for conventional clarifiers

The main problems with the cleaning of fuels, irrespective of density, are fuel-oil losses and water-discharge limitations. When a centrifuge has been arranged as a clarifier, during shooting/cleaning of the clarifier, no water is supplied to displace the oil and only water and impurities are removed, so all the fuel oil present is discharged.

This centrifuge type, clarifier, does not remove water because the water discharge is closed and can only be removed with the sludge during shooting.

Placed in series with a purifier (water and dirt) and a subsequent clarifier (remaining dirt), it is a useful type of separator.

24.7 New separators by Alfa Laval

New types of separators have been developed to clean fuels with a high specific mass.

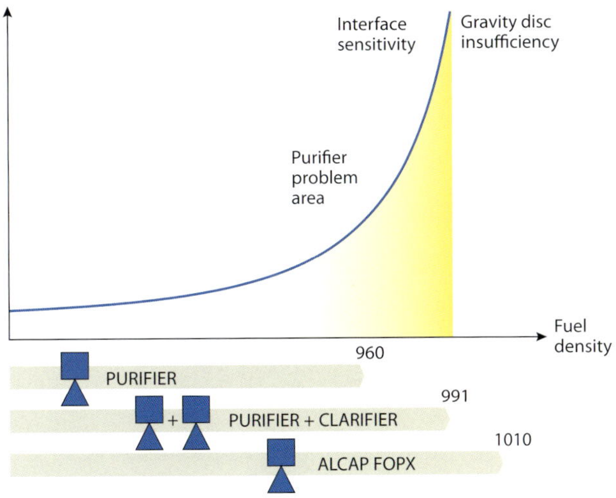

▲

The working areas of various Alfa Laval centrifugal separators.

horizontal: specific density of the fuel in kg per m^3.

vertical: sensitivity of the interface.

reduced functioning of the gravity disc.

The purifier functions adequately to approximately 960 kg per m^3.

The combination of purifier and clarifier is sufficient for approximately 991 kg per m^3.

The new Alcap FOPX of Alfa Laval functions to approximately 1010 kg per m^3.

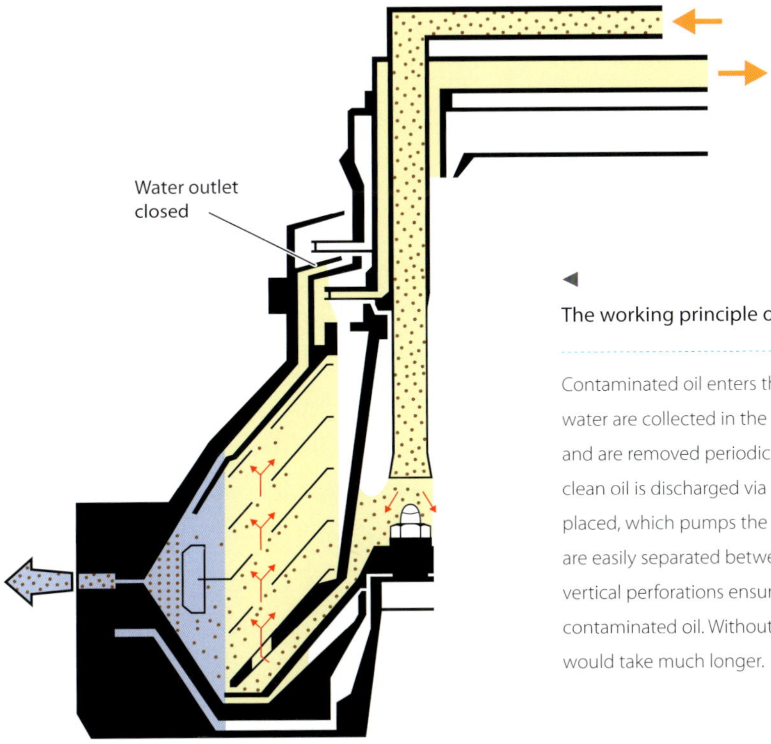

◀

The working principle of a conventional clarifier.

Contaminated oil enters the centre of the bowl. The dirt and the water are collected in the sludge space (periphery) of the bowl and are removed periodically by dropping the inner bottom. The clean oil is discharged via a disc stack. Here a stationary rotor is placed, which pumps the oil from the separator. The dirt particles are easily separated between the conical discs in the bowl. The vertical perforations ensure a proper distribution of the contaminated oil. Without these discs in the disc stack, separation would take much longer.

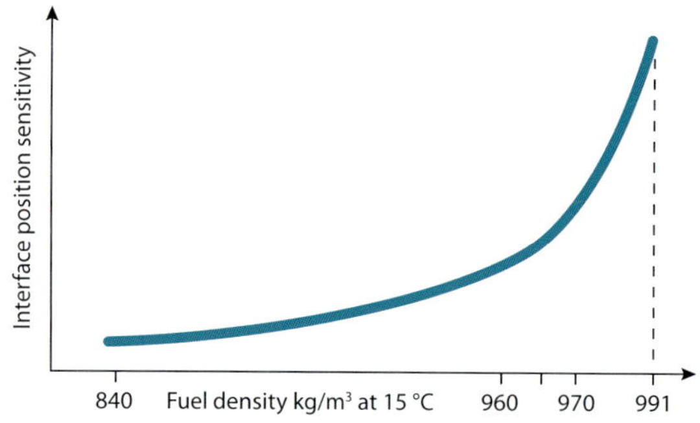

▲ A graph indicating interface position's sensitivity.

As shown, interface sensitivity (deteriorating operation) increases significantly over 960 kg per m³.

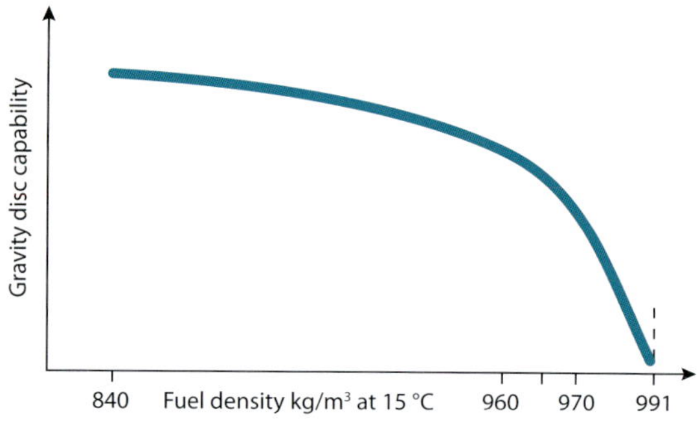

▲ A graph for the capacity of the gravity disc.

As shown, the capacity of the gravity disc decreases significantly above an approximate specific gravity of 950 kg per m³.

24.7.1 'Alcap'-technology by Alfa Laval

Contaminated pre-heated fuel oil is fed continuously to the 'S' type separator, which essentially functions as a clarifier. Clean fuel oil is continuously pumped away. When separated water approaches the disc stack, traces of water start to escape with the cleaned oil. This minor increase in water content of the cleaned oil is detected by the transducer MT 50, which is installed in the clean-oil outlet. Sludge and water accumulate at the bowl periphery. When separated water reaches the dirty disc stacks, traces of water escape with the clean-oil flow to the clean-oil outlet.

A measuring device MT 50 detects the water content in the fuel. At an unacceptable increase in water content, the clarifier is triggered to initiate an automatic discharge of the water via a water drain controlled by a magnetic valve.
When shooting the sludge – discharging the dirt – the contaminated-fuel supply is closed and the fuel remaining in the bowl is displaced by the supply of water and discharged to the clean-fuel supply. The separator is cleaned at great force.
So in principle, this centrifuge is a clarifier with an automatic water drain!

▼ The Alcap technology I.

Low free-water in the fuel. The water discharge opens automatically as soon as the water content carried in the clean oil is above a preset limit.

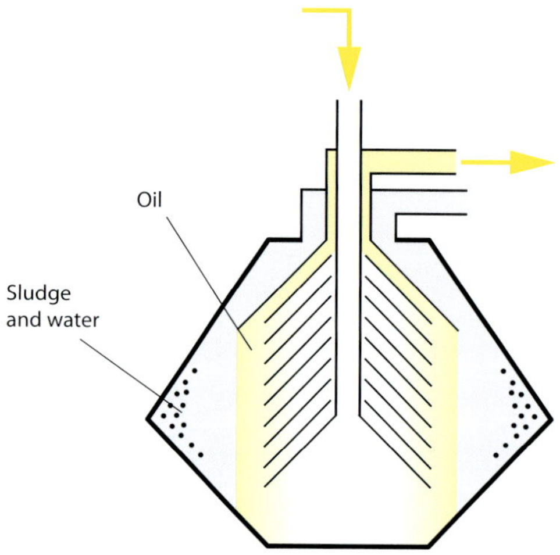

▼ The Alcap technology II.

Higher free-water content in the fuel. The measuring device detects more 'free' water in the clean fuel and opens the water discharge via the process controller.

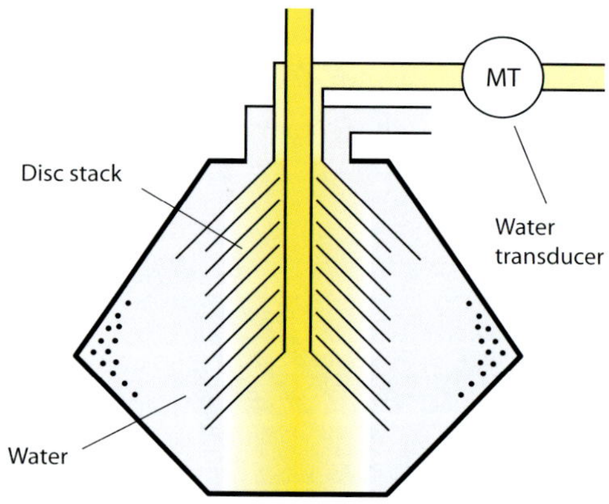

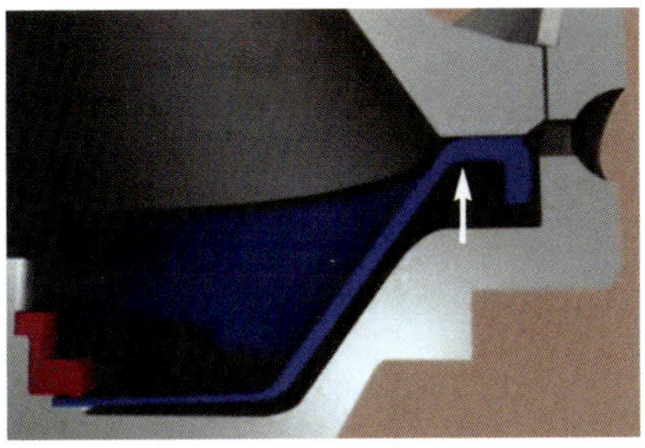

STEP ONE: The CentriShoot discharge slide is fixed at the centre. During separation it blocks the discharge ports.

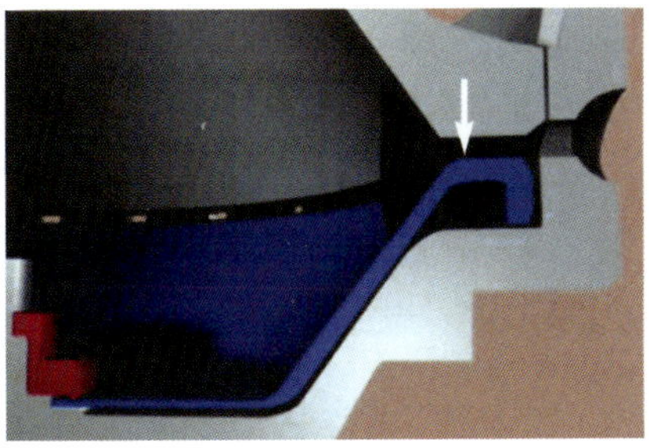

STEP TWO: During sludge discharge, the edge of the slide flexes downward, exposing the ports.

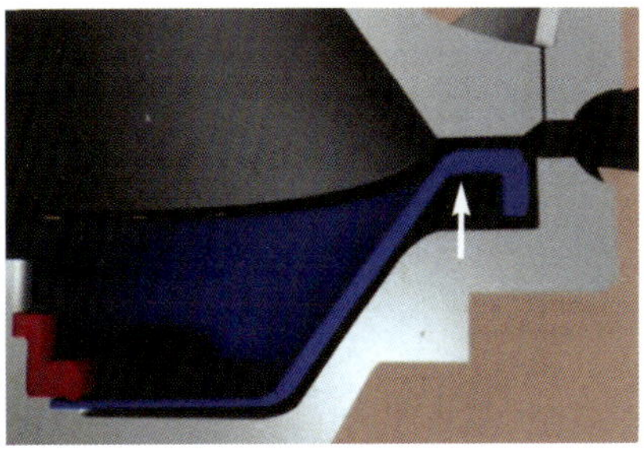

STEP THREE: After discharge, the slide moves gently back into position, closing the ports. Closing is done hydraulically, without any springs.

▲

Opening and closing the bowl by moving the flexible inner bottom.

Normally, the water pressure, maintained by the centrifugal force, keeps the bottom closed in top position. When opened, the bottom drops slightly.

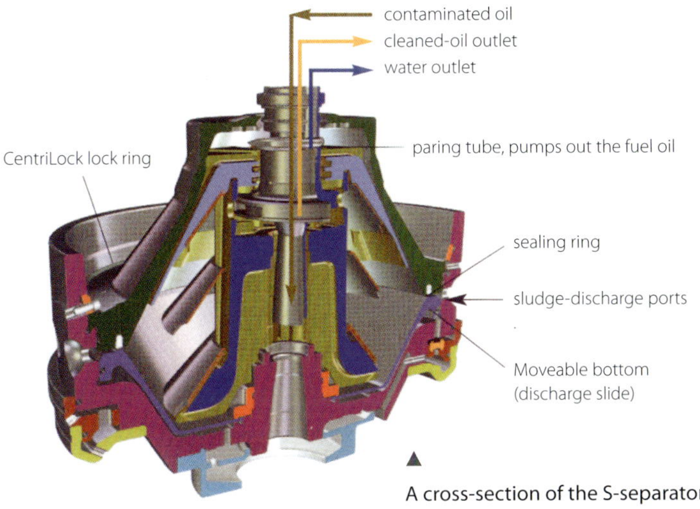

▲

A cross-section of the S-separator.

▼

Construction of an S-separator of Alfa Laval.

1 *electromotor*
2 *belt*
3 *bearing with special springs for a flexible drive*
4 *bowl*
5 *casing*
6 *sludge and water discharge*

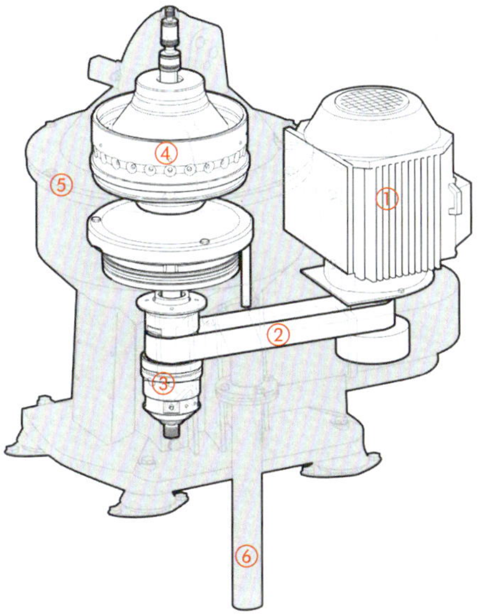

▼

The Alcap technology III.

The water is discharged.

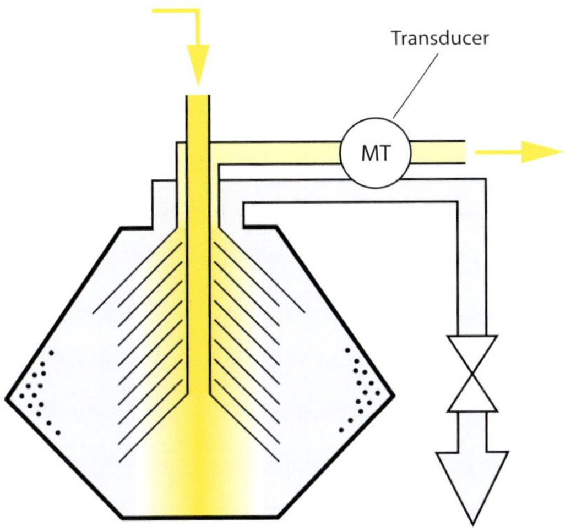

▼

The Alcap technology IV.

'Shooting' of the separator to remove the solids. The contaminated fuel is returned, water displaces the remaining fuel and the bowl opens which causes the sludge to be discharged.

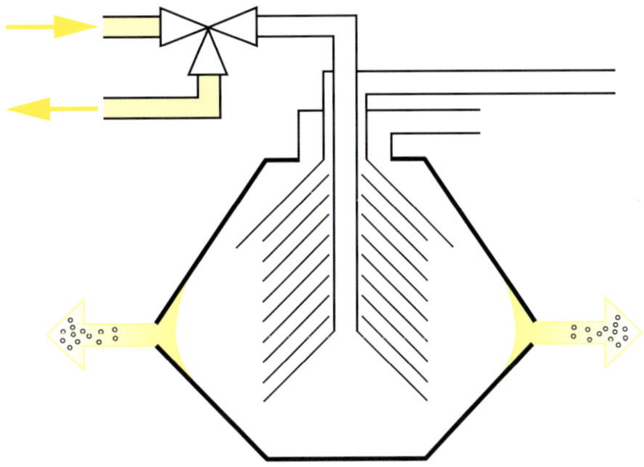

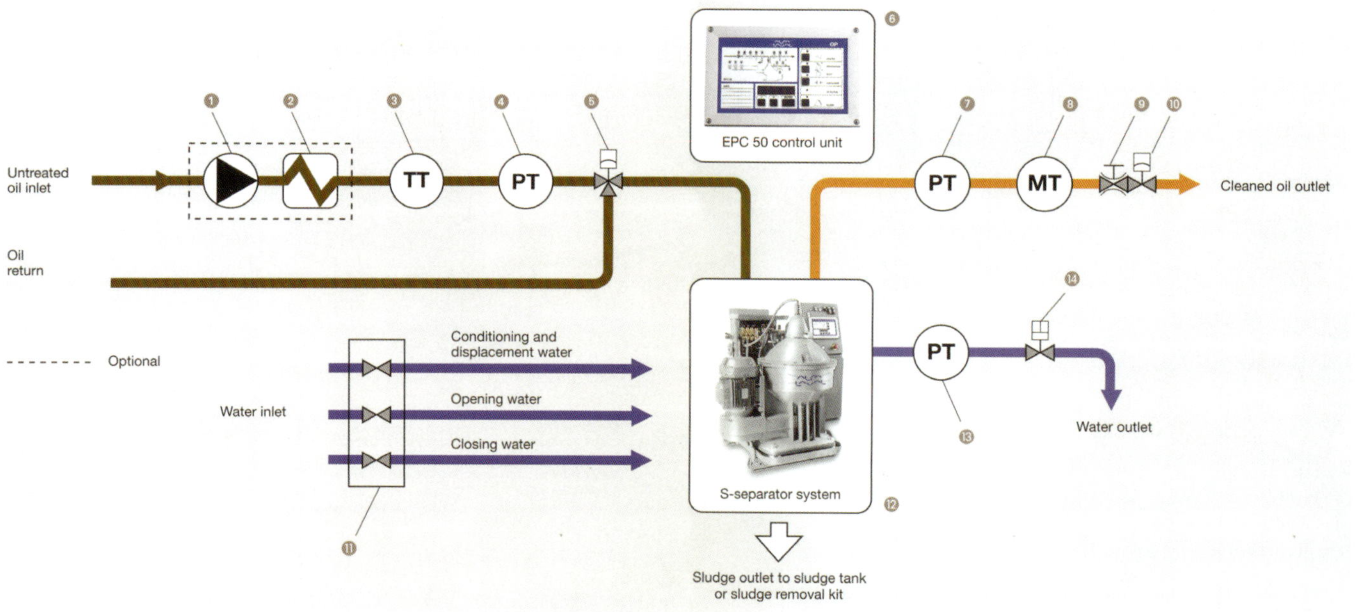

▲

The complete flow diagram of an Alfa Laval S-separator.

1 *Feed pump for contaminated fuel.*
2 *Pre-heater, pre-heats the fuel to the separator temperature; for heavy oil often just below 100 °C (the boiling point of water), approximately 98 °C.*
3 *Temperature transmitter measures the temperature and signals the process-control unit.*
4 *Fuel-pressure transmitter measures the fuel-supply pressure and signals the process-control unit.*
5 *Pneumatically controlled three-way valve controls the contaminated fuel to the separator or to the recirculation pump line.*
6 *Process-control unit regulates the separator.*
7 *Pressure transmitter, measures the discharge pressure of the cleaned fuel and signals the process-control unit.*
8 *Transducer continuously monitors the water content in the cleaned fuel. Ensures that the separator is functioning correctly. A water surplus, shoot!*
9 *Control valve, regulates the discharge pressure of the clean fuel.*
10 *Pneumatically controlled shut-off valve. Closes the clean-fuel outlet.*
11 *Solenoid valve for water discharge which regulates the supply of separator water and conditioning water in the drum/bowl.*
12 *Separator, removes the solids and water from the fuel.*
13 *Water-pressure transmitter measures the water pressure in the drain outlet and signals the control unit.*
14 *Drain valve, magnet valve, opens when the water is drained from the separator.*

▼

The new lock rings of the Alfa Laval separator: a simple flexible steel strip, fixed with hexagonal socket screws.

A sledge hammer to obtain the correct mark and a special ring spanner belong to the past.

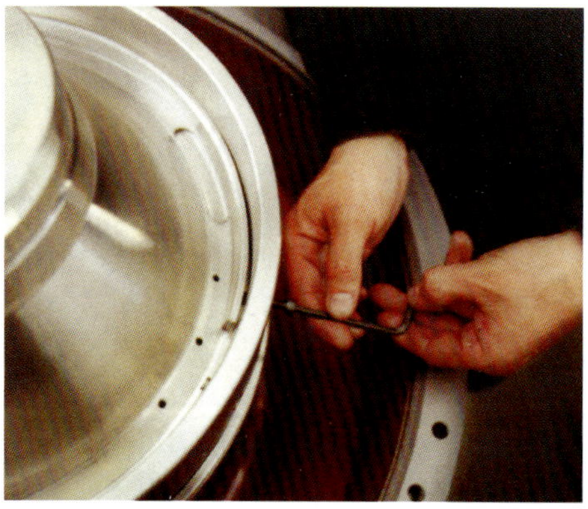

CentriLock can only be removed with an Allen key. No sledge hammer is necessary.

CentriLock lifts out and snaps in easily – without any threads to wear.

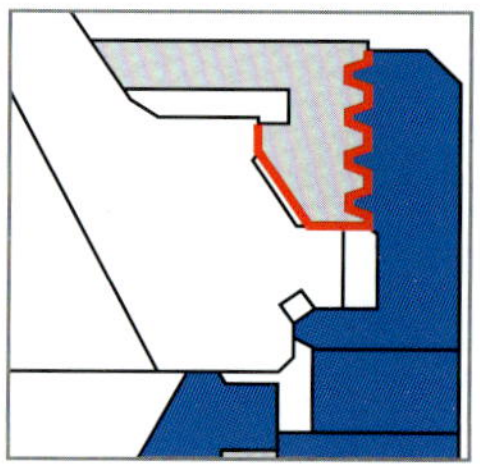

Threaded lock rings must be removed with a sledge hammer. Over time, metal-to-metal wear between bowl and lock ring, leads to expensive bowl repair or replacement.

24.8 Effects of separators

At present, heavy fuels contain residues of catalysts used during the refining process, the so-called cat fines.
These aluminium oxides can cause severe damage to a fuel-injection system, such as, the high-pressure fuel pumps and injectors. They can also damage the remainder of the engine via the combustion space. **These must be removed from the fuel by the separator as efficiently as possible.**

24.8.1 Separation-enhancing factors

The following factors optimise separation:
- fuel temperature in the separator;
- viscosity in the separator;
- feed rate of the separator;
- use of the disc stack;
- specific mass of fuel and water;
- size and number of the solid particles in the 'untreated' fuel;
- amount of water in 'untreated' fuel;
- chemical properties of fuel oil.

24.8.2 Rule of thumb: the lower the capacity the cleaner the fuel!

This is why most separators have a far higher capacity than the maximum fuel consumption. Separators always only operate at a percentage of their maximum capacity.

24.8.3 Cat-fines content

According to the ISO 8217 regulations, the maximum cat-fines content may amount to 80 parts per million (ppm). However, engine manufacturers require, if possible, a maximum content of 15 ppm. This makes it necessary to reduce the cat-fines content in the fuel from 80 to 15 ppm. Separators are capable of separating particles as small as 2 microns.

24.9 Examples of cleaning systems for lubricating oil, fuel, sludge and bilge water

On board ships and in diesel-power plants, there are two main suppliers of separators: Alfa Laval, a Swedish firm with its well-known blue machinery, and Westfalia, a German firm with its well-known green machinery.
Both supply an elaborate program of systems and machinery for treatment of lubricating oil, fuels, sludge and bilge water.

Here a number of examples of Westfalia.

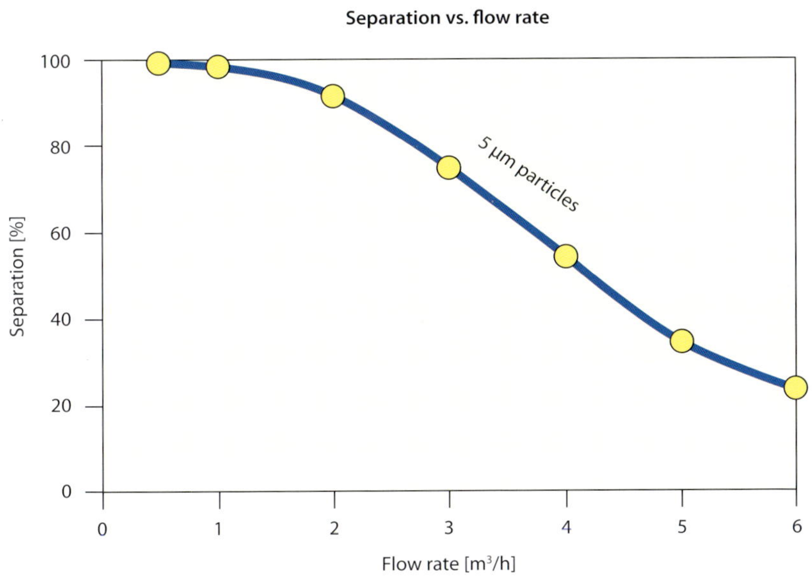

◄ Separation of aluminium and silicon particles, size 5 microns.

horizontal: flow in m^3 per hour.
vertical: efficiency of the separation in percentage.
It is clear that separation decreases as the flow increases. This is logical as the fuel remains in the bowl for a shorter time span, so the separator cannot remove all the small particles.

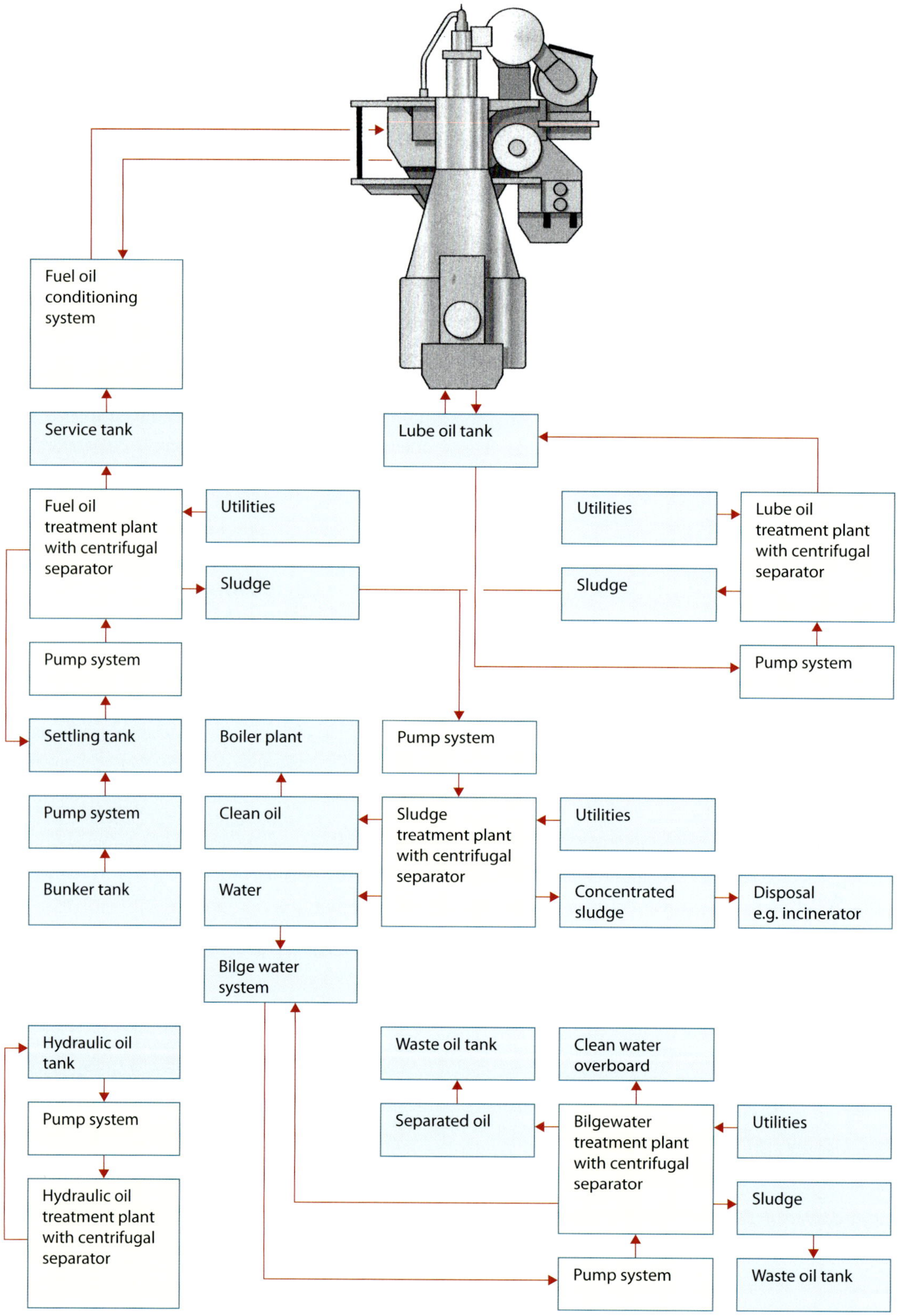

▲

An overview of all the systems surrounding a two-stroke crosshead engine.

In large plants each system has a separator: shown here for heavy-fuel oil, lubricating oil, sludge and bilge water. Hydraulic systems may also be equipped with a separator. This occurs mainly in very large hydraulic installations and seldom on board ship.

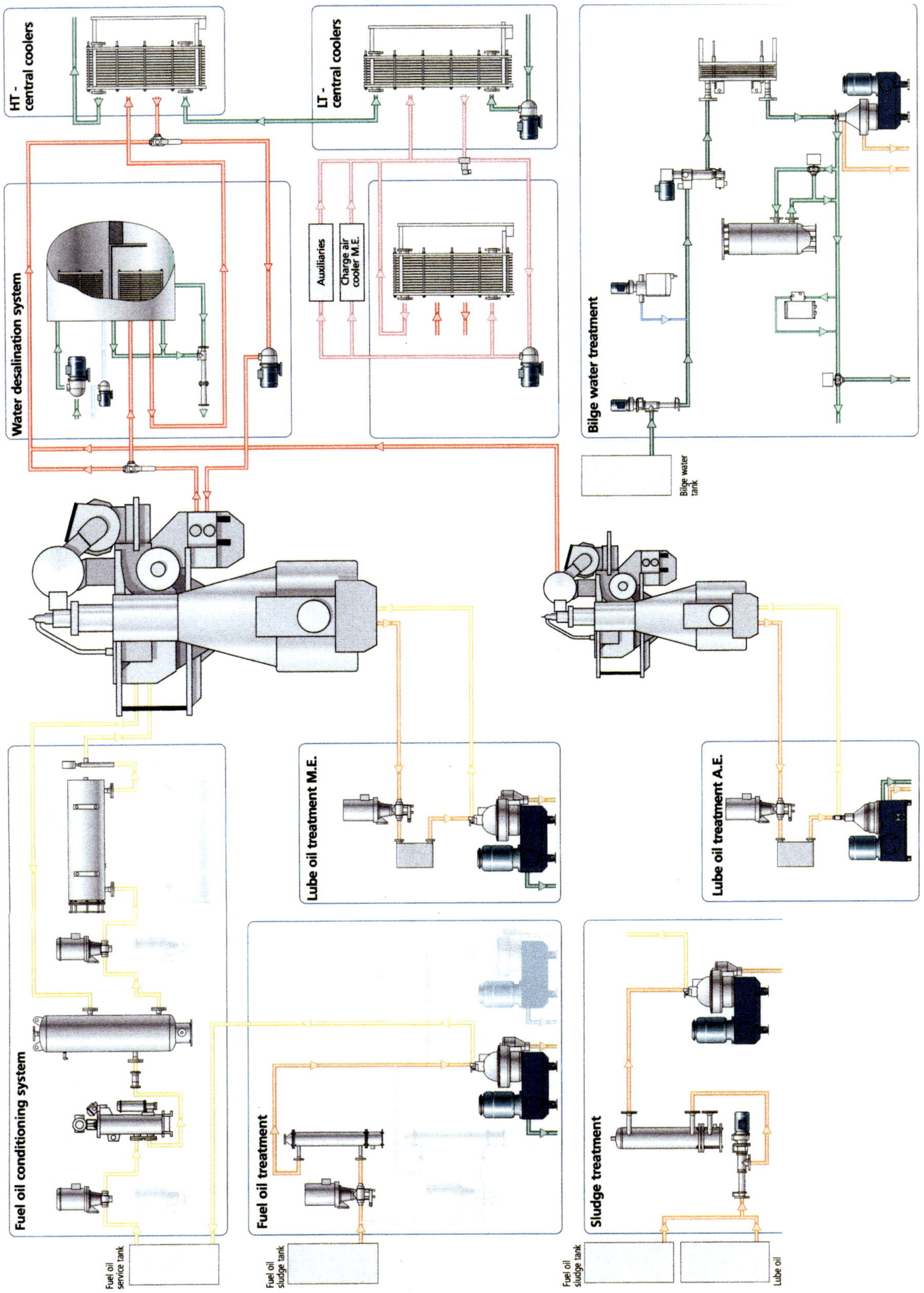

▲
An overview of the piping with all the essential equipment.

Apart from cleaning heavy oil, lubricating oil, sludge and bilge water for both the main engine (ME) and auxiliary engines (AE), the HT and LT fresh-water cooling systems are also incorporated as well as a sea-water evaporator, which is cooled with HT cooling water.

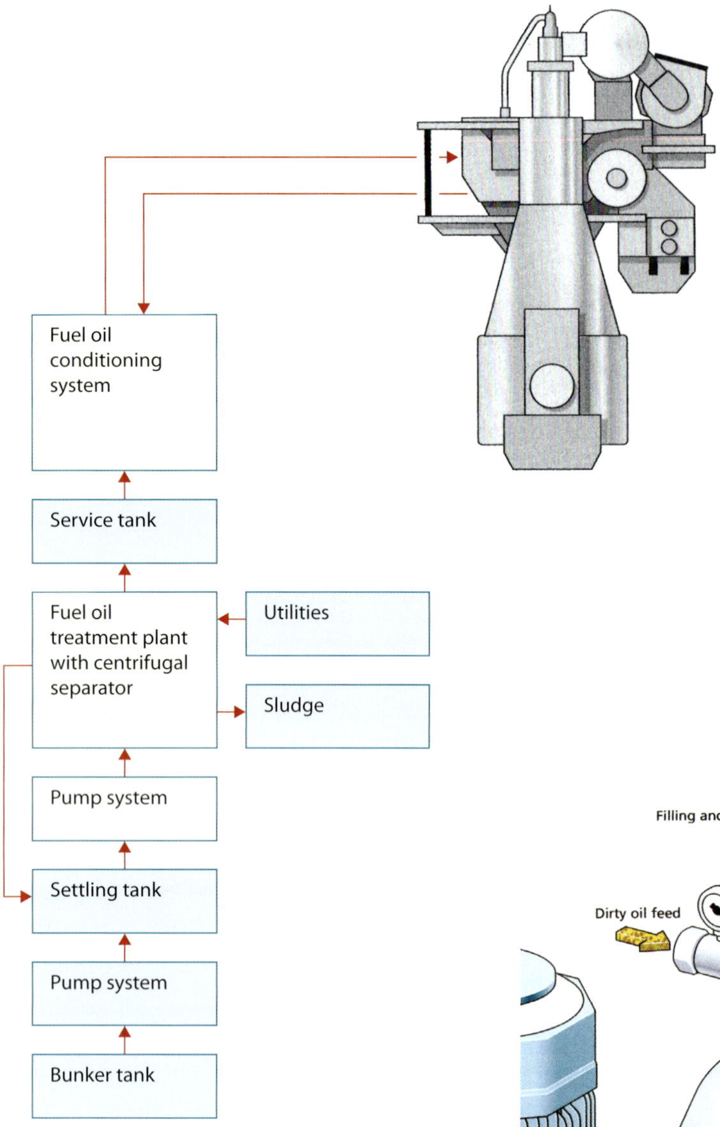

◄

A flow diagram for the cleaning of heavy oil for the main engine.

▼

A simple system with a separator for a small plant operating on MDO.

The separator must be manually cleaned at regular intervals. There is no fuel pre-heater.

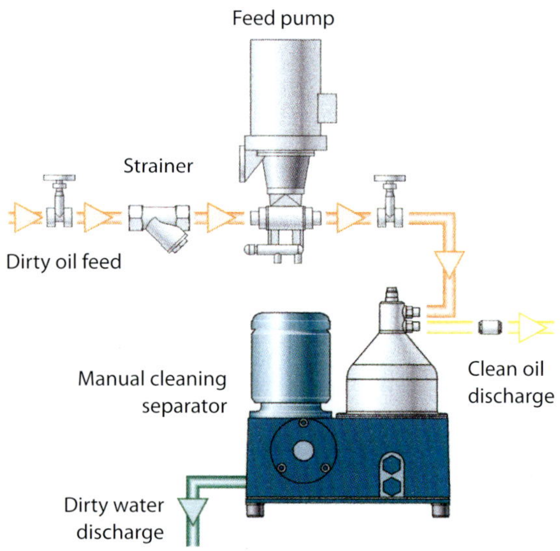

▼

A cross-section of a self-cleaning fuel and lubricating-oil separator of Westfalia.

Bottom right: sludge discharge. The contaminated fuel or dirty lubricating oil is supplied centrally along the vertical centre line of the bowl.

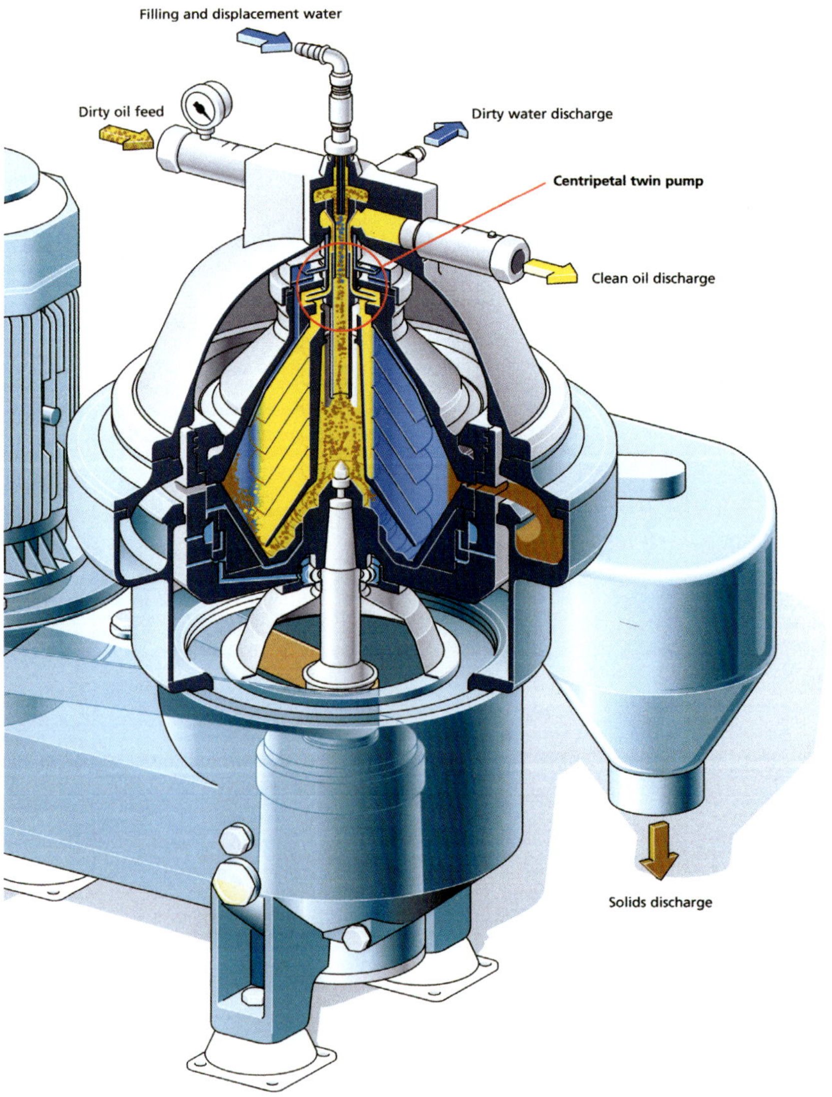

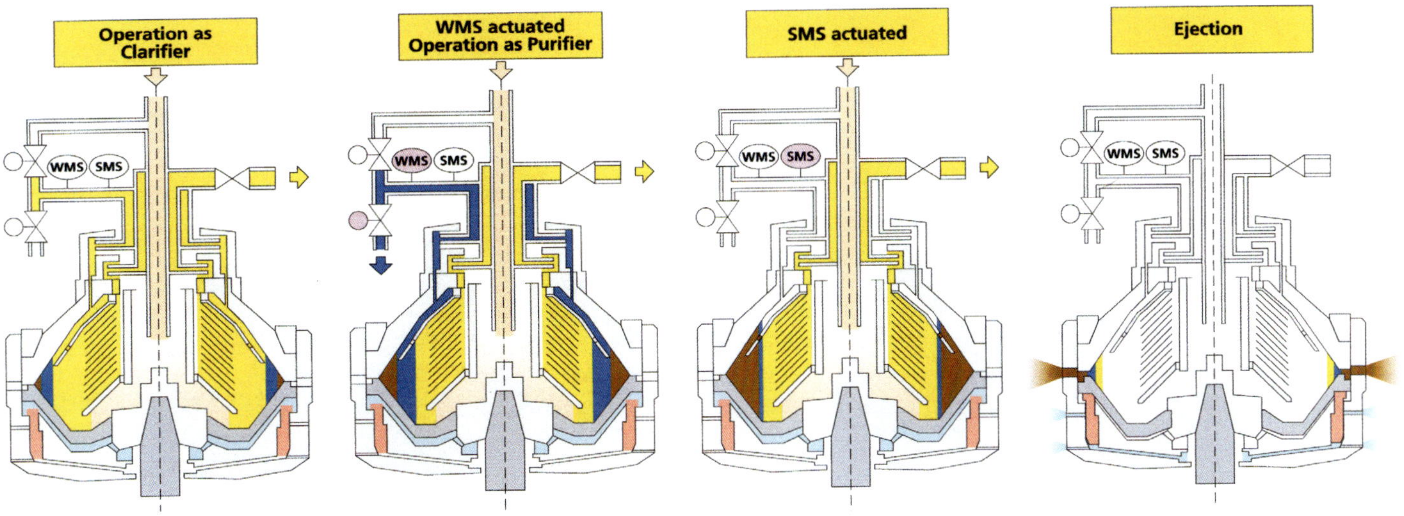

Features:

- UNITROL® system in one stage WMS and SMS
- Treatment of fuel oils up to 1.01 kg/l
- Regulating rings are not required
- High separation efficiency even in critical situations
- Automatic adjustment as clarifier or purifier
- Automatic bowl ejection
- Highly concentrated solids
- Continuous processing
- Self-cleaning effect of the bowl and the disc stack by total ejection
- Versatility of application
- Safe operation
- Belt drive
- Low noise level.

▲

The working principle of the Clarifier and Purifier of Westfalia.

left: operation of the Westfalia Clarifier, which separates water and solids periodically by shooting.

right: operation of the same separator as a Purifier which separates solids periodically and water continuously.

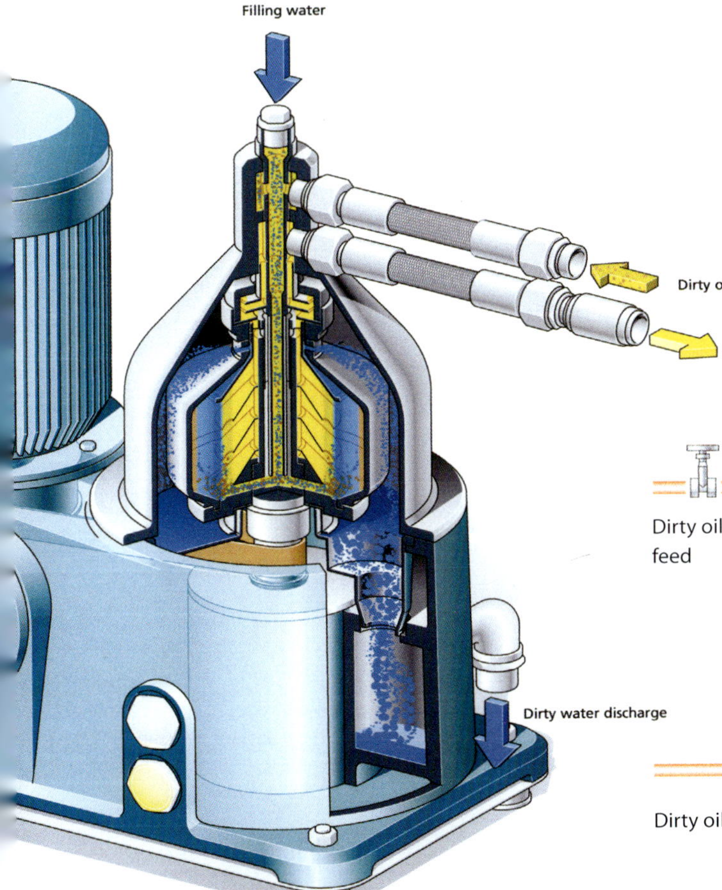

▲

A cross-section of a simple separator.

MDO usually has few contaminants; the separator can remain operational for several days before cleaning is required. Experience will show when cleaning is required.

▼

A simple system including a fuel pre-heater.

The correct oil-separation temperature can now also be maintained in, for instance, arctic circumstances.

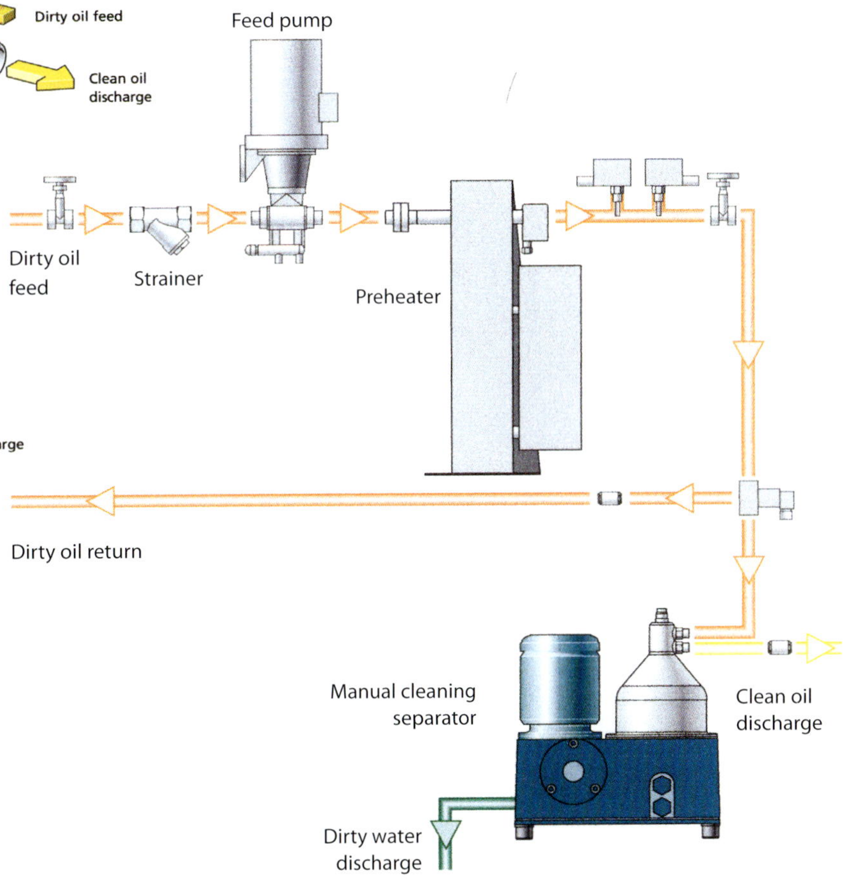

▶
A complete system with pre-heater and a self-cleaning separator.

This version is common for both MDO and HFO.

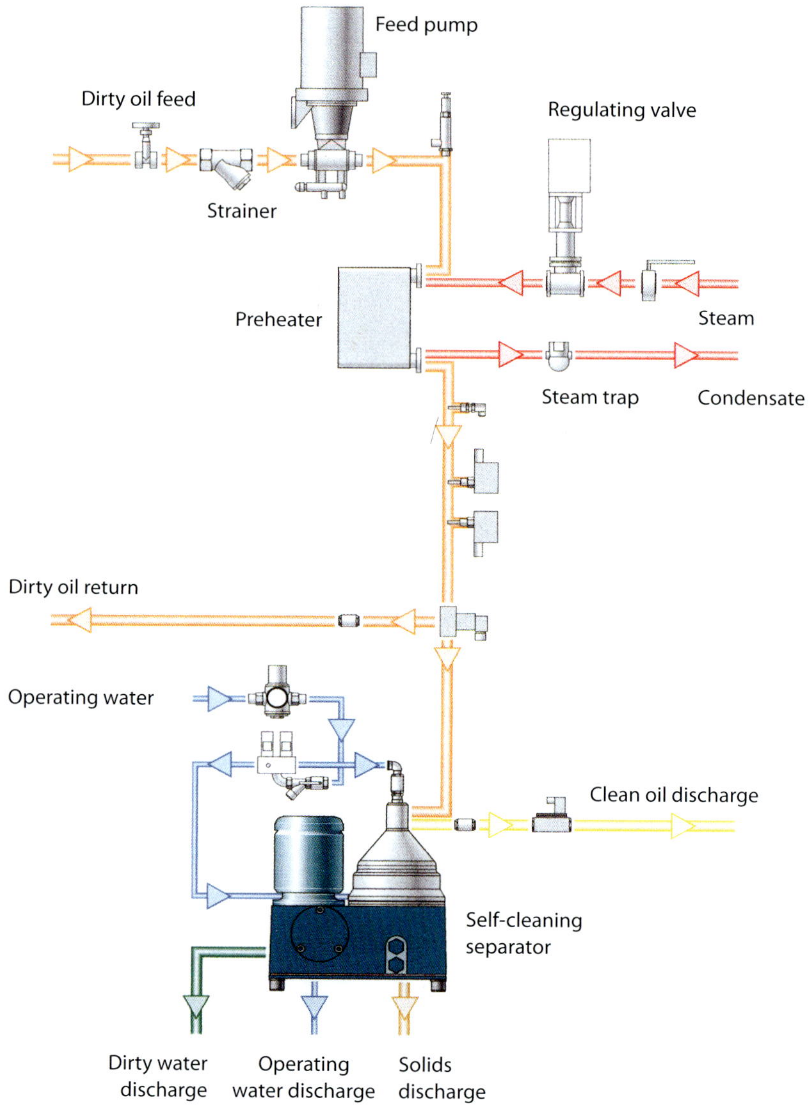

▼
A similar system with a self-cleaning separator equipped with a Unitrol system.

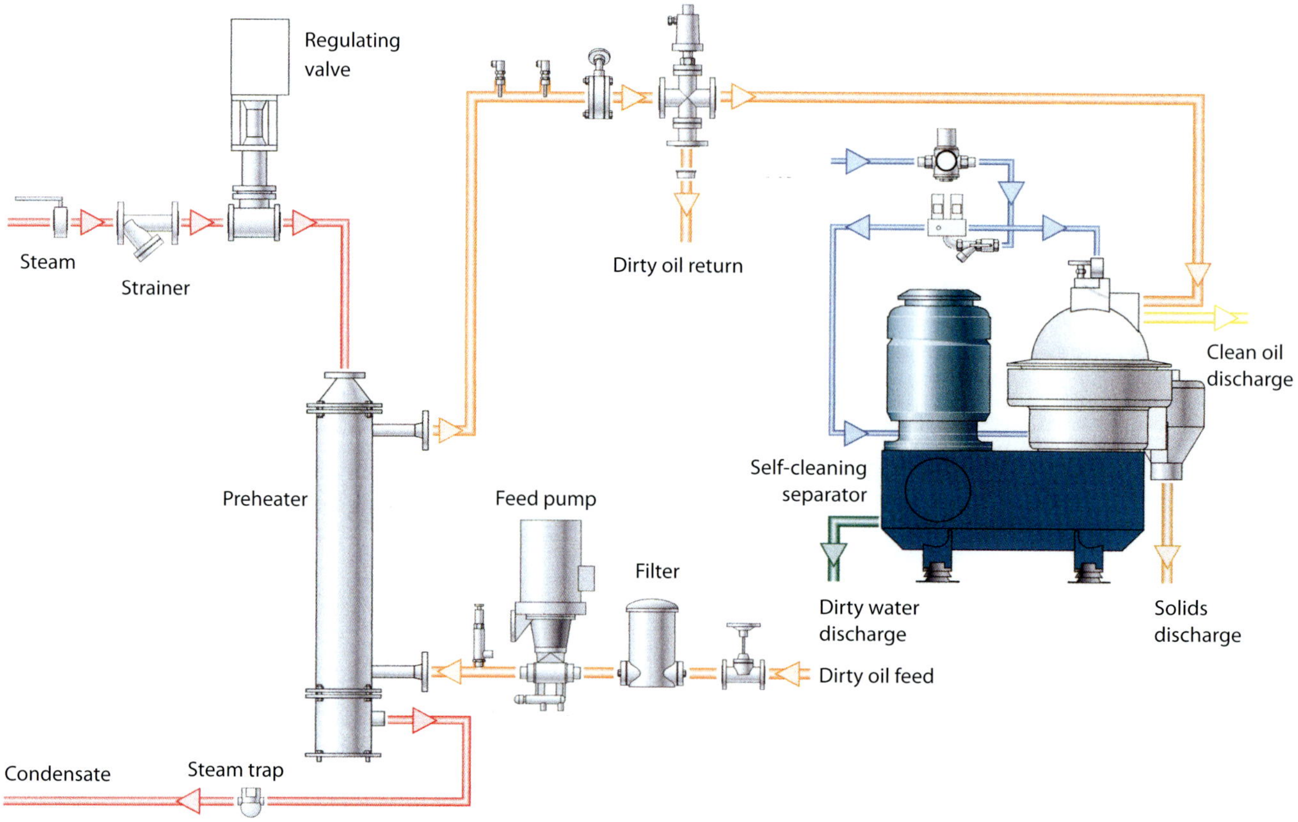

24.9.1 UNITROL system of Westfalia

This system constantly measures the water content in the clean-oil output and the amount of sludge in the sludge tank.

This makes specific-gravity discs superfluous and excludes human operating errors. Naturally, Westfalia also uses abbreviations such as:

WMS: Water Monitoring System;

SMS: Sludge Space Monitoring System.

▲

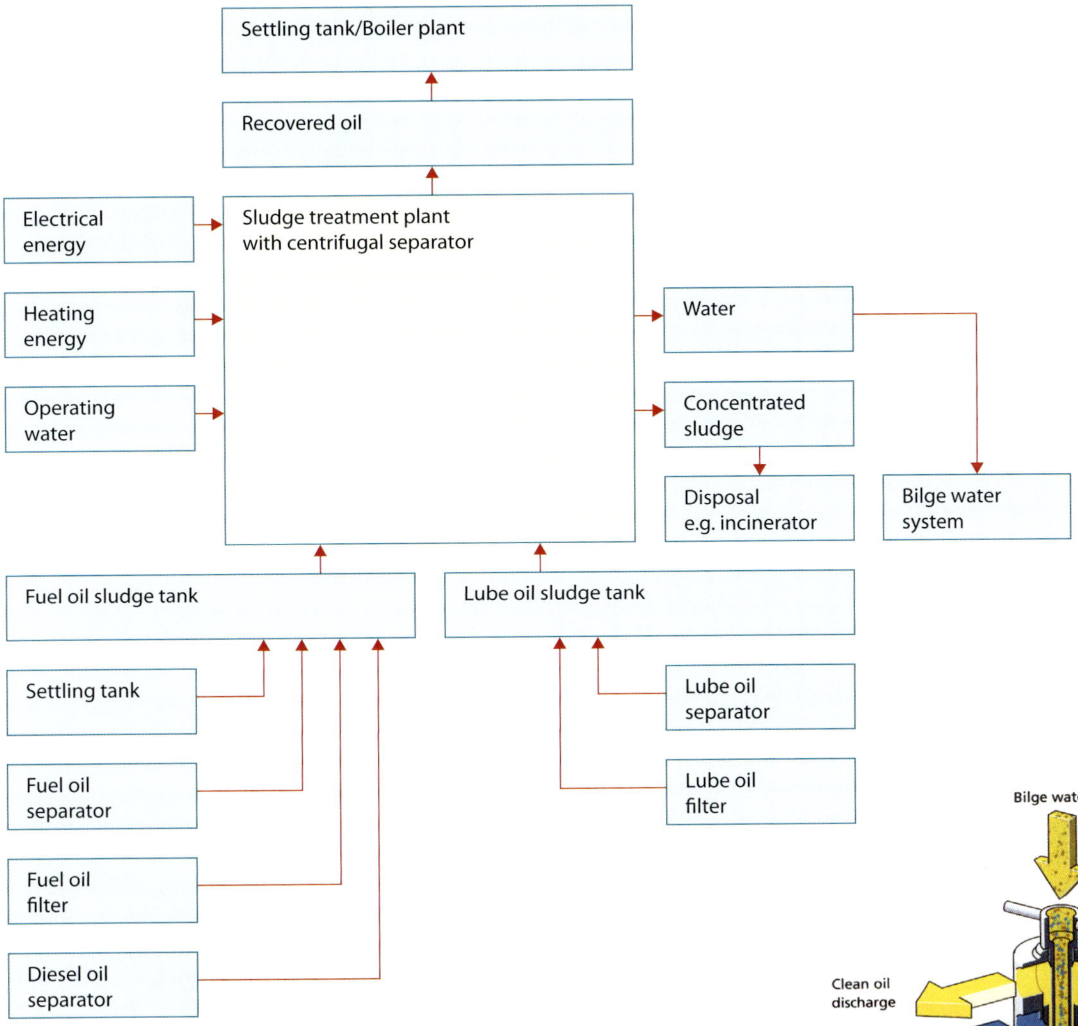

◀

A flow diagram of the complete sludge treatment.

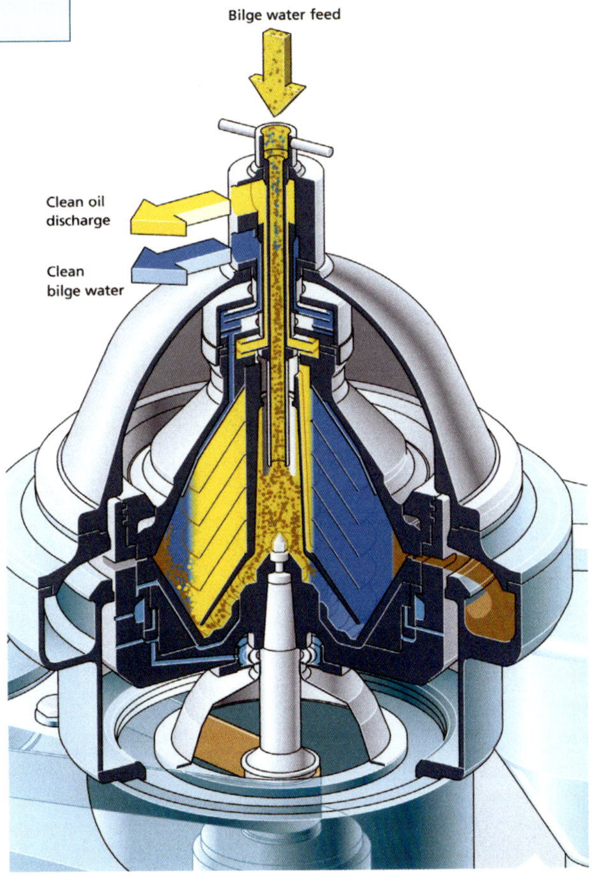

▶

A bilge-water separator, virtually identical to the fuel and lubricating-oil separators.

Bilge water: this is water mixed with the dirt and oil that gathers on the bottom of the ship's engine room. The double bottom often has drain tanks in which the bilge water collects.

It is prohibited to discharge bilge water directly overboard (IMO). Therefore, bilge water is treated. A bilge-water separator and a strainer remove as much dirt and oil as possible.

A measuring device in the over-board pressure line prevents bilge water with too high an oil content, measured in ppm, being pumped overboard. In this case, a three-way valve returns the water to the bilge.

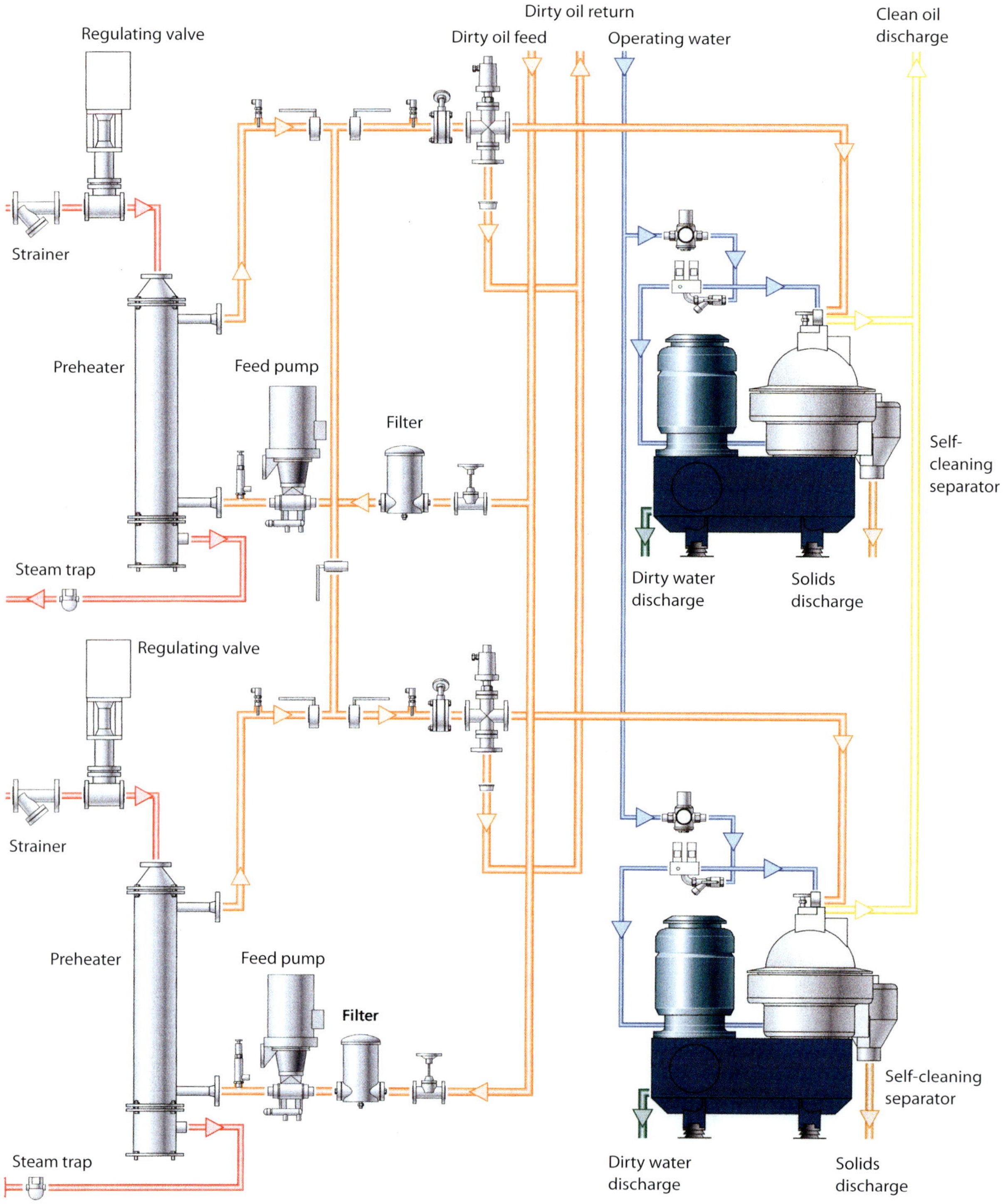

▲

Larger installations with marine diesel engines often have several heavy-fuel oil separators placed in series in order to increase operational reliability.

An arrangement of two separators in series, which operate on the same supply line to ensure a purified oil discharge.

Sludge treatment

It is increasingly important to concentrate the sludge of various centrifugal separators because:

- Increasingly often sludge must be disposed of on shore: it is prohibited to pump it overboard.
  There are often costs involved in sludge disposal;
- part of the sludge can be re-used as fuel, which is cost-saving.

The residues after post-treatment are:

- **water**; this is pumped to the bilge-water cleaning system;
- concentrated sludge; this must be disposed of on shore by certified firms;
- dirt and such; this can be burnt in an **incinerator**.

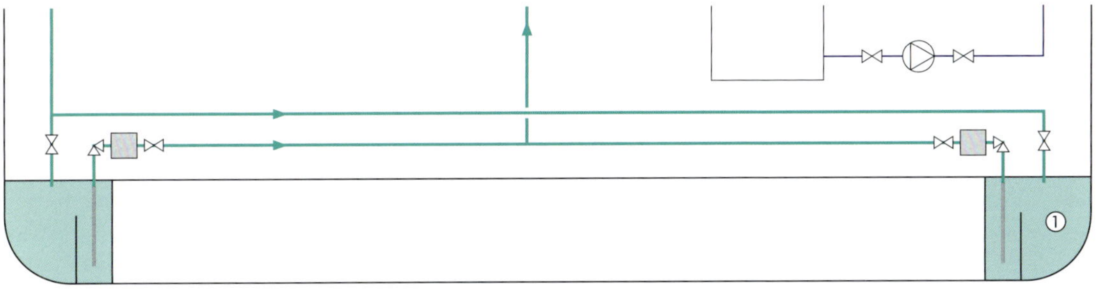

Cross-section of a ship.

The bilges on port and starboard side in the engine room of this ship.

1 *bilge*

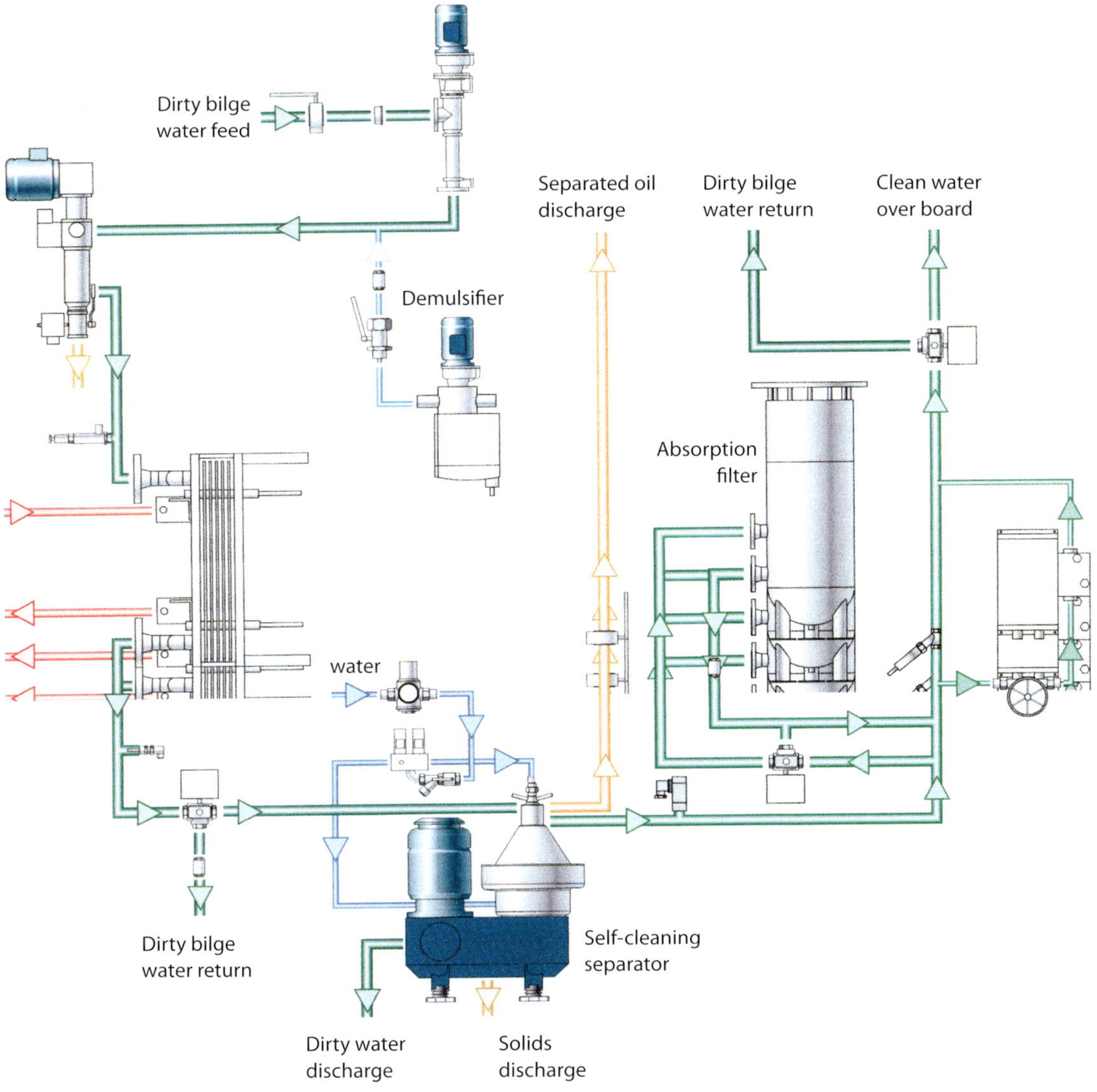

A complete bilge-water cleaning installation.

For adequate separation the bilge water must have a temperature at which the specific mass difference between water and oil is at its highest.

A special filter removes the last oil particles from the water.

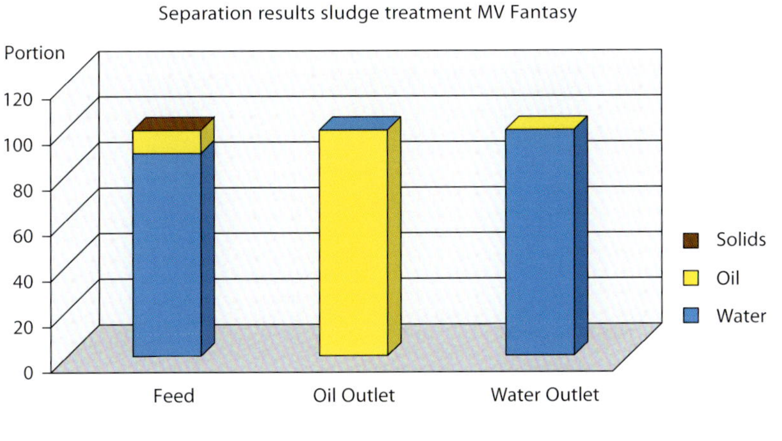

A graph of the effects of a bilge-water cleaning installation.

left: contaminated bilge water with 90% water and 10% oil

centre: oil outlet which consists of 99.6% oil and only 0.4% water

right: water outlet with 99.5% water and 0.5% oil

The 0.5% oil percentage is cleaned in an absorption filter, so the quality of the 'over-board water' is compliant with all legal requirements.

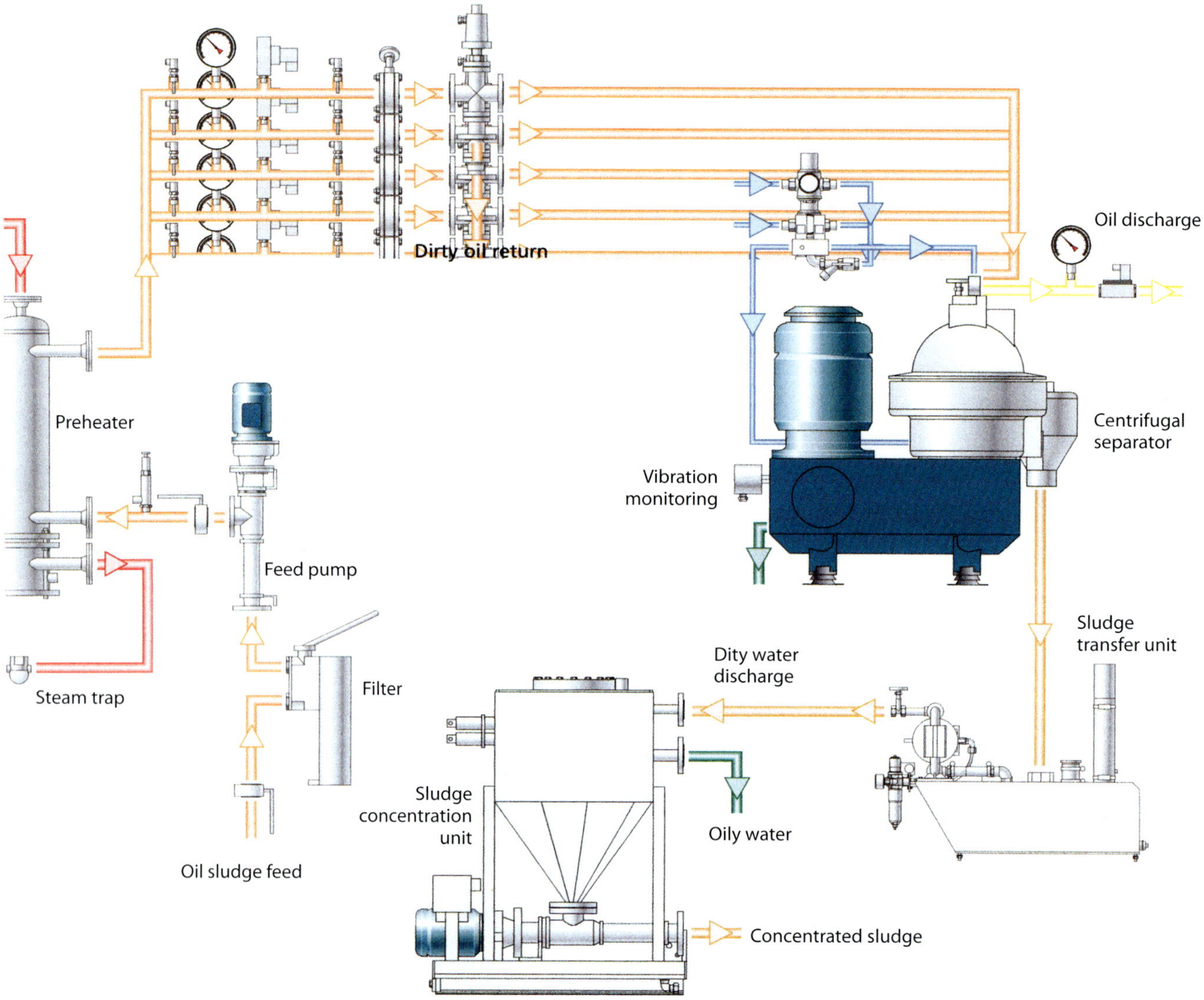

▲

Schema of a sludge-cleaning system for in the near future.

The sludge from the centrifuge is treated further so that the mass is as small as possible. Ultimately, only concentrated sludge remains.

> CH 25

Operational management and automation

A modern 'technical office' on a large ship is no longer referred to as a control room. The remote control is situated on the bridge. Only an emergency control device directly connected to the propulsion engine is in the engine room. All engine-room activities can be monitored and adjusted in an office off the engine room using monitors, amongst others. Data can be sent via satellite to, for instance, the engine manufacturer who can then assess the engine information and discuss it with the marine engineers.

The emergency control of a two-stroke crosshead engine, type Wärtsilä Sulzer 12RTA96C.

The control room of a modern diesel-power plant.

25.1 Introduction

Today, an increasing number of engines in all categories are operated and monitored remotely. With modern communication technologies and the general use of computers, the diesel-engine industry has changed dramatically over the last 25 years. It was common practice to have personnel on location when engines were operational. However, today the reverse applies: there is no one in situ.
At night, the engine rooms of virtually all modern ships are unmanned. This applies to small coasters as well as to the largest container ships.
There are some exceptions, for example passenger vessels; legal requirements stipulate that personnel must be present at all times in the engine room.
The largest container ships with power outputs of 100,000 kW and a displacement of 70,000 tons can sail the oceans at 25 knots with one officer on the bridge, assisted by one crewmember.
Relatively few accidents occur considering the number of sea-going and inland vessels plying the globe's waters at any one time.

25.2 Automation of diesel engines

1 Diesel-engine data

All relevant information is displayed on the monitor and saved in a database making it possible to refer to the history, which is extremely important when problems occur.
The data can be sent anywhere, for instance, to the engine manufacturer or the ship's proprietor. The large engine manufacturers have a special service department who can continuously provide the relevant comparative data. This means that assessments can be made from a distance. This also applies to diesel-power plants.

2 Operating the engine

This is remotely controlled: the engine load and speed can be adjusted or shut down. This may occur in various ways, either mechanically, hydraulically, pneumatically, electrically, electronically or even with radio-controls.

3 Engine alarms

All-important pressures, temperatures, loads, speeds, liquid levels and such are measured. The measuring gauges are fitted with alarms, often firstly a pre-alarm and if there is no response a second alarm automatically interrupts the process, if necessary.
In a small sloop engine, only a few lamps will flash and a buzzer is activated. However, in a large marine engine, an elaborate and programmable procedure is put into action. The alarms in large installations are not restricted to the engine, but all the auxiliary systems, such as those for cooling water, lubricating oil, fuel, refrigeration equipment and starting air. The groups have been categorised so the nature of the problem can be speedily detected.

25.3 Examples of automation systems

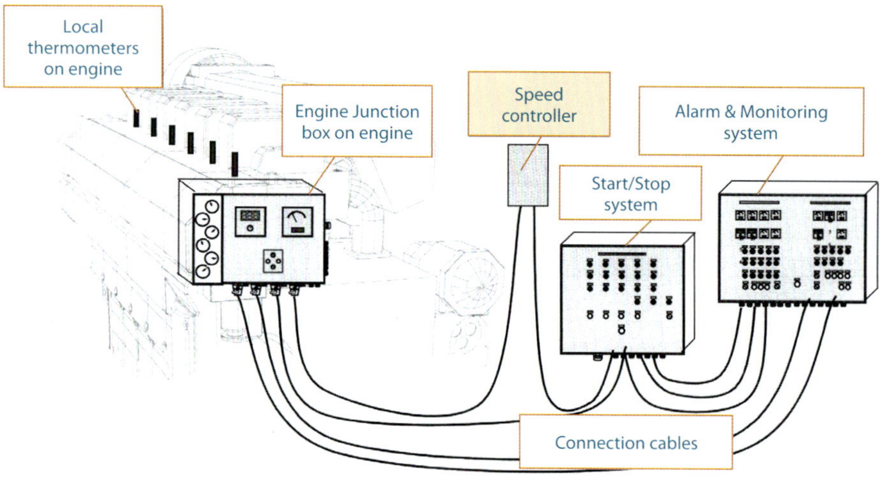

◄
Standard automation system.

The system that Wärtsilä can supply for a W 26 in-line diesel engine is:

- various types of meters mounted on the engine, such as temperature, pressure and speed;
- standard junction boxes on the engine with important values which can be read on the spot;
- remote-controlled starting and stopping system;
- remote-controlled alarm system;
- speed-control system.

The engine is also provided with an 'Overspeed' module. Naturally, the actuator that regulates the fuel supply to the engine is also mounted on the engine. The speed-control and start/stop system are not mounted on the engine but on the bridge. In an emergency, the engine can be shut down in the engine room.

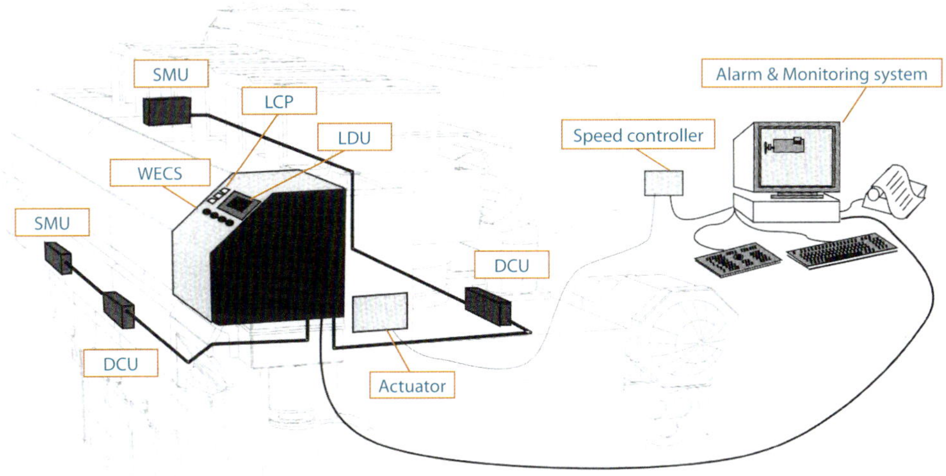

◄ The Wärtsilä WECS system.

This automation system for engines registers all the data, and monitors and ensures the safe operation of the engine.

Monitoring and alarm system

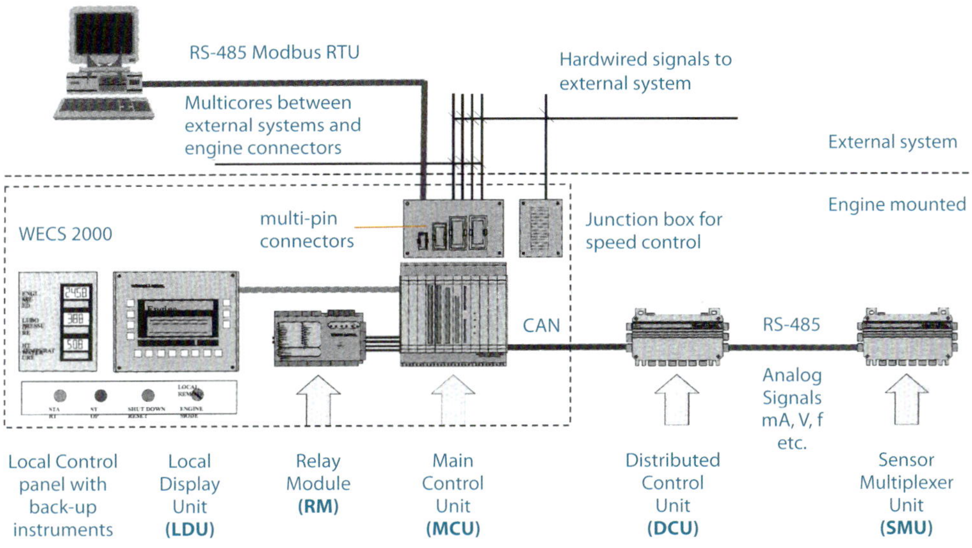

Wärtsilä uses the WECS system: the 'Wärtsilä Engine Control System'

This system has the following functions:

- measurement of analogue and digital signals;
- measurement of both the engine- and turbocharger speeds;
- engine protection;
- start/stop the engine;
- start-blocking;
- automatic switch off;
- load adjustment;
- arming sensor alarms;
- important parameters can be read;
- exchange data with other systems via MODBUS communication, such as general alarm systems and safety systems.

Abbreviations used by the WECS 2000 system

DCU's: Distributed Controls Units, a sensor interface based on microprocessors.

SMU's: Sensor Multiplexer Units, a sensor interface based on microprocessors.

MCU: Main Control Unit, central main control unit.

RM: Relay Module, a safety back-up system. This is required by many classification societies.

LDU: Local Display Unit, a monitor on location. Often vibration-free suspended from the engine.

LCP: Local Control Panel, an on-site panel with important data concerning the engine.

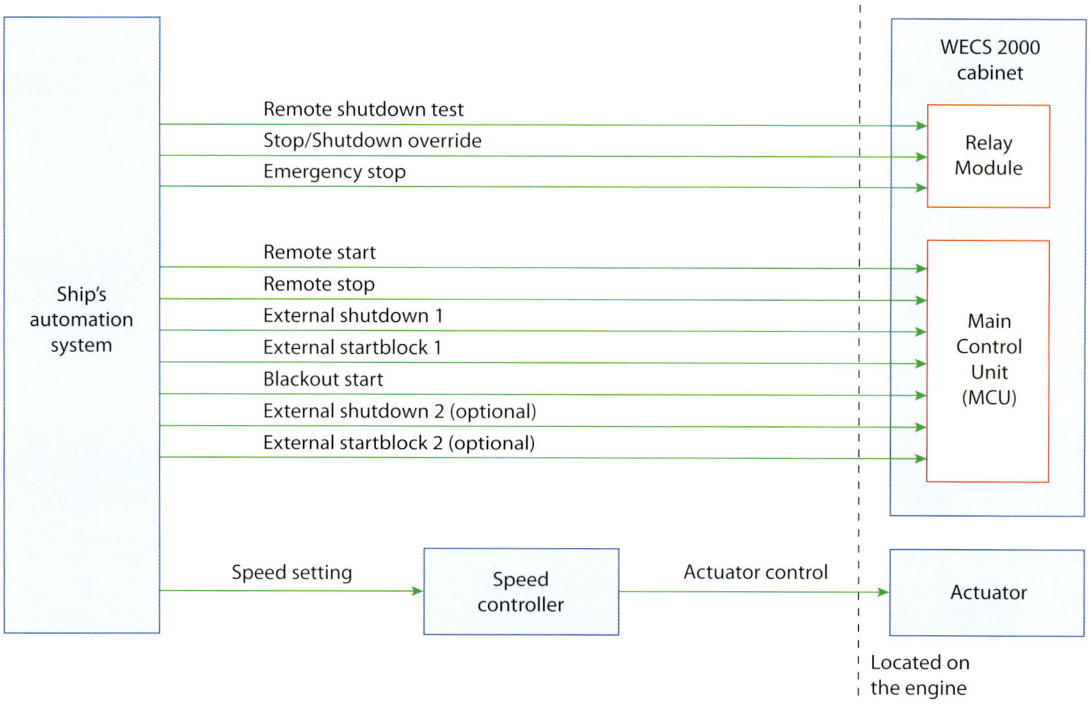

▲
The safety system in the WECS 2000.

There are five elements:

1 *start*
2 *stop*
3 *start blocking*
4 *emergency stop*
5 *executing load reductions*

Ad 1: Start

Starting can either be remotely controlled or at the engine. After an emergency stop the reset button must be pressed.

'Black-out start'. This is an emergency start, which is selected despite the fact that starting may cause damage to the engine. Only when the turning engine is in operation will the engine not start. The pre-lubrication operation can be by-passed for five minutes.

Emergency engine start. This can be done at the engine.

Ad 2: Stop

Normal stop.

Emergency stop. These can be mounted in various places.

Ad 3: Start-blocking

Blocking the start.

The start is blocked under the following circumstances:

- engine is operational;
- low pre-lubricating oil level of the turbocharger;
- turning engine is on stand-by;
- stop lever in stop position;
- local start control is set on remote control, which blocks the start at the engine;
- emergency stop is active;
- external start blocking.

Ad 4: Emergency stop

Similar to a normal stop, an emergency stop is executed by activating two solenoid valves. An emergency stop must be reset prior to re-start.

Ad 5: Performing load reductions

There are many options, such as:

- high lubricating-oil temperature;
- high coolant temperature;
- high bearing temperature;
- high exhaust-gas temperatures.

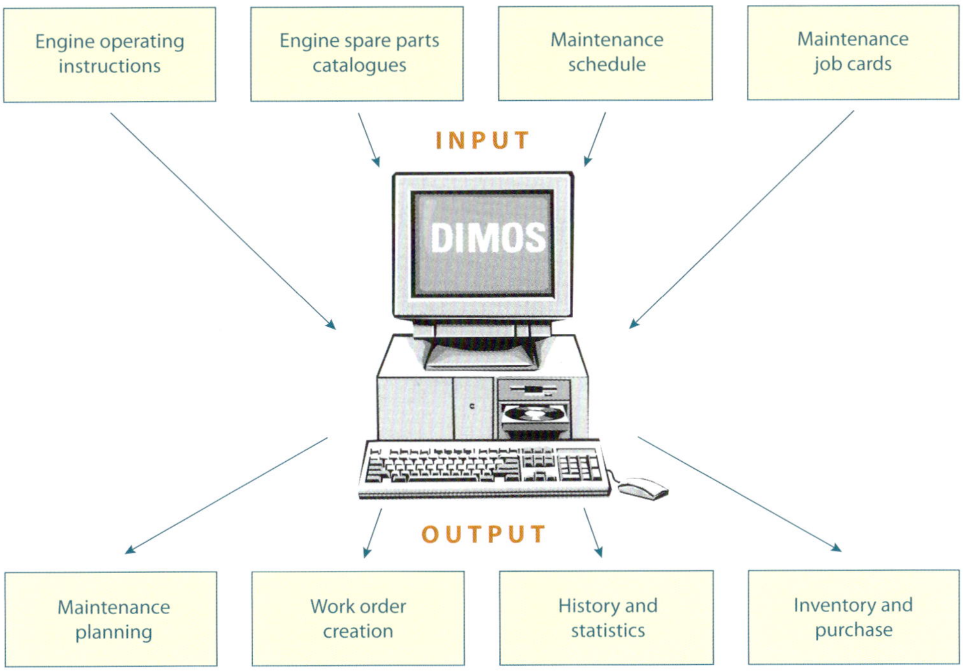

▲

The diesel management system, DIMOS from Caterpillar.

A complete system for maintenance, spare-parts control management, statistics and expense monitoring.
All conceivable information is saved and processed during operation. Operating hours, preventative maintenance actions and breakdowns are categorised and stored here.

25.4 Operational management

For the operational management of diesel engines on ships and in power plants, it is important that the data from the installations can be stored in computer systems.
When only a handwritten engine journal and lists of the operating-hours for all the instruments were once available, one can now save all the relevant data and process it for use in the maintenance system, and spare-parts storage for instance, or the management of a propulsion installation, and power generation on ships or on-shore.

25.5 Complete systems for diesel engines, some examples

All the major engine manufacturers have developed programs, which cover monitoring systems for the entire engine operating data.

25.5.1 Caterpillar MaK

Caterpillar MaK has developed the DICARE-system.
According to Caterpillar MaK, this system has the following advantages:
- early discovery of wear and tear;
- optimal operational control as all the alarms caused by data deviating from normal conditions are made clearly visible;
- reduction in maintenance costs by using predictive-trend analysis;
- longer life of parts by comparing current and required values;
- information regarding the maintenance status of the entire engine;
- allows personnel to anticipate the planning of maintenance and spare-part ordering.

The engine management system at Caterpillar is known as MaK DIMOS. This is a computer-supported maintenance and spare-parts management programme.

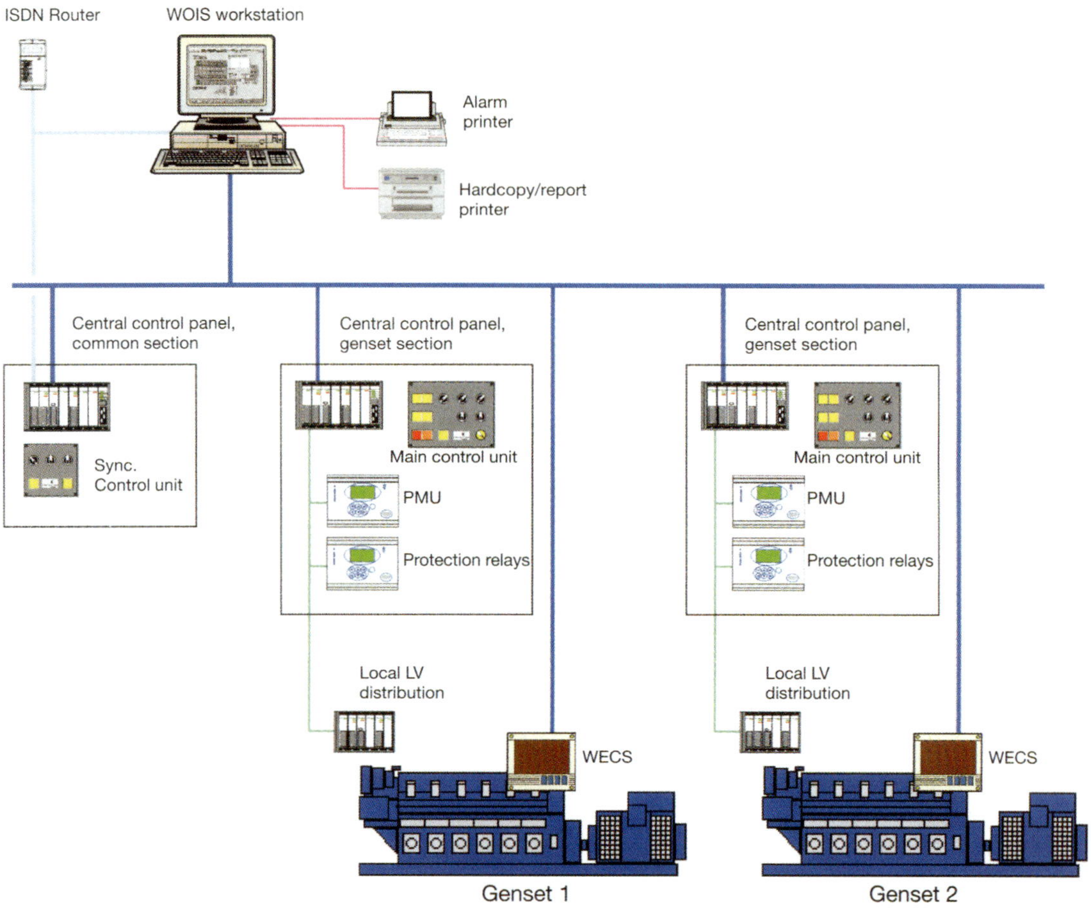

▲

A complete system for Wärtsilä generators.

A pc and monitor is attached to every diesel genset, showing all the data collected.

In the Wärtsilä Operators Interface System (WOIS) found in the control room of for instance, a diesel-power plant (or the generator of a ship), all the operating details are stored and alarms processed.

This WOIS-system can be connected to various control systems that are available in the industry.

25.5.2 Wärtsilä's Faks system

Faks (Fault Avoidance Knowledge System) is an elaborate software system for monitoring and fault avoidance for large engines. The system retrieves various values from the monitoring systems of the engine, and analyses them for possible failures and shows where the failure occurs. This is not 100% watertight, but gives an indication of what the most likely fault is.

The system:

- safeguards the engine by using early warning systems;
- performs measurement inspections;
- monitors the maintenance status;
- performs research using the knowledge base;
- communicates with other web-enabled applications.

For example, special sensors measure the exhaust-valve seat temperature, the temperature of the cylinder liner near the combustion space and the surface temperature of the main bearing, along with many other standard temperatures.

Faks is the result of years of expertise, the analysis of engine data, faultfinding and their solutions. All the data can be sent to the service department of the engine manufacturer via modern communication systems. Subsequently, the engine manufacturers can either on request, or by contractual obligations perform fault analysis and provide advice to prevent or resolve potential or actual problems.

25.5.3 CoCos system from MAN Diesel

CoCos is the abbreviation for: Computer Controlled Surveillance. According to the manufacturer, this system has the following advantages:

- increase the availability and the reliability of diesel engines;
- considerable reduction in operating costs;
- effective planning of engine maintenance;
- easy and simple identification of spare parts;
- integrated stock control.
-

The CoCos system consists of four parts:

1 CoCos-EDS – Engine Diagnostics System

An engine diagnostics system that simplifies decision-making on board, at the diesel-power plant or in the office:

- it assists in making decisions regarding measures that are to be taken;
- it improves the availability and reliability of diesel engines;
- it reduces operating costs and losses due to engine failure.

These main objectives are attained by:

- collecting, monitoring and storing operating data;
- analysis of the stored data;
- timely detection of irregularities.

2 CoCos-MPS – Maintenance Planning System

An all-in-one maintenance system, which ensures the timely execution of preventative diesel-engine maintenance, and maintenance of auxiliary equipment (in the proximity of the engine). The system works in conjunction with CoCos-SPC and CoCos-SPO.

This allows easy access to the number and type of new parts required for maintenance, the quantity of parts in stock and the quantity of parts that must be ordered.

3 CoCos-SPC Spare-Part Catalogue

The spare-part catalogue is a good system for tracking the quantity of parts in stock in an orderly manner. The detailed records are available digitally including all the drawings and diagrams.

Main objectives of the system are:

- to chart the engine parts and other equipment in a simple fashion
- to provide extensive information with respect to these parts.

4 CoCos-SPO – Stock-control and spare-part procurement

This effectively sets up the stock administration and registration of the spare parts. This system provides accurate updated details of the location and history of components including spare parts in stock for the engine or power plant. This system is elaborate and, of course, entirely digital.

The main objectives:

- provide assistance in handling the spare-part stock;
- provide up-to-date information regarding current stock;
- forecast spare-part availability for future overhauls;
- provide assistance in spare-part procurement.

If this CoCos system is used correctly, it provides all the necessary information so that an accurate overview regarding stock control of spare parts is obtained.

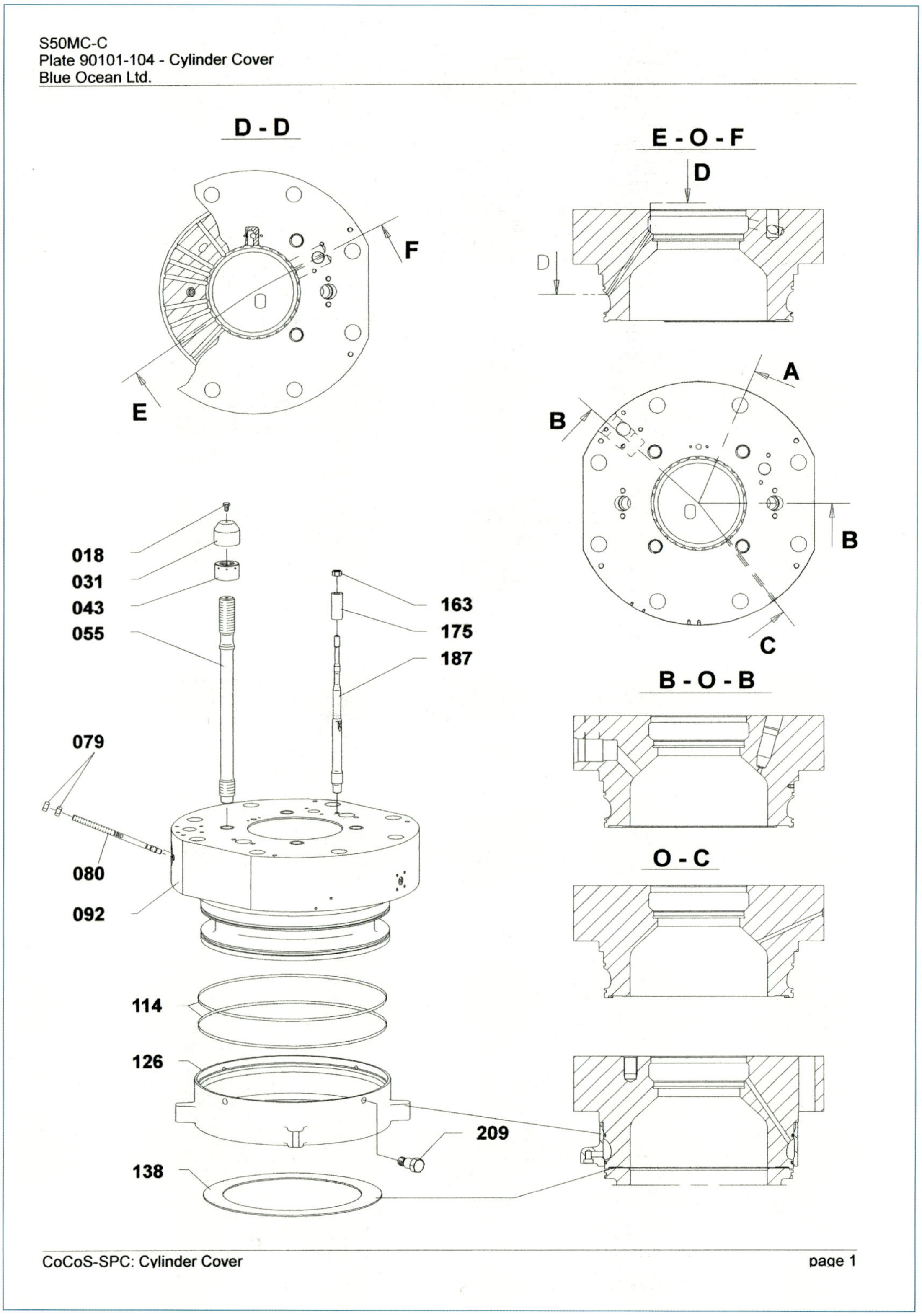

▲

The beginnings of the CoCos system.

All the parts are numbered. Shown here, the cylinder head of a two-stroke crosshead engine type MAN Diesel S50MC-C. If one clicks on the numbers, the details of the part are listed according to the assembly/sub-assembly, including the tools required for assembly and dismounting. Stock-control numbers can also be obtained. Ultimately, a very complete system for the operation of diesel engines.

Comment

Naturally, it is important to include all modifications of each part into the CoCos system. Once again, man is the weakest link!

>H 26

Reconditioning engines and their parts

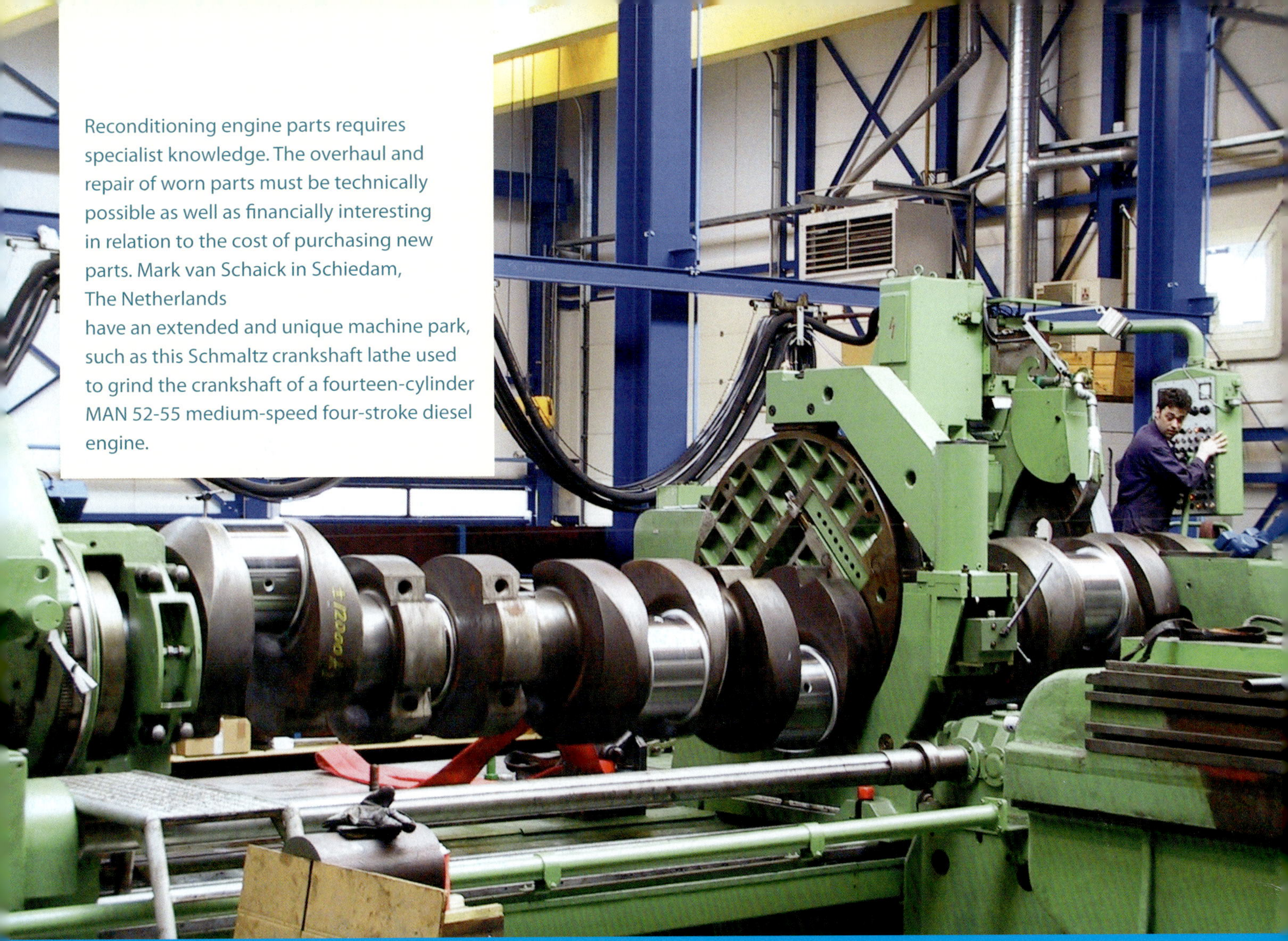

Reconditioning engine parts requires specialist knowledge. The overhaul and repair of worn parts must be technically possible as well as financially interesting in relation to the cost of purchasing new parts. Mark van Schaick in Schiedam, The Netherlands
have an extended and unique machine park, such as this Schmaltz crankshaft lathe used to grind the crankshaft of a fourteen-cylinder MAN 52-55 medium-speed four-stroke diesel engine.

Keeping a two-stroke crosshead-engine piston warm before rewelding.
At the MAN Diesel AG Reconditioning company in Hamburg.

26.1 Introduction

Mechanical, thermal and chemical factors cause wear in engines. The worn parts must be either be replaced or **reconditioned** within a certain time span.

Several examples of parts that require replacement are:

- cylinder heads, especially the exhaust valves;
- pistons and piston rings;
- cylinder liners;
- shafts and shaft bearings;
- supercharging groups;
- pumps;
- governors;
- starting motors;
- fuel pumps and injectors.

▲

An exhaust valve affected by high-temperature corrosion, HTC.

◄

Worn parts.

After a certain number of operating hours, or after having established that a particular part no longer functions normally, it must be inspected.

Shown here, a disassembled turbocharger.

▲

A filthy turbocharger casing.

▲

After cleaning, soiled exhaust valves are checked for wear, hair cracks and size.

◄

Exhaust-valve seats are heavily loaded.

AFK HAARS = scrapped after a hair-crack test. Hair-cracks result in valve leakages or fractures.

◄

Various engine parts requiring maintenance. This is referred to as 'overhauling' or 'reconditioning'.

◄

In connecting rods, bores in the big end of the crankpin and the eye for the piston pin often require truing.

▲ If repairs are postponed for too long, an exhaust valve may start to leak severely, the so-called blow-through.

▲ A microscopic view of an exhaust valve on the combustion-chamber side, which is seriously affected by high-temperature corrosion.

Engine manufacturers supply specifications concerning the normal operating life for the various parts, and the engine proprietors or operators put together a maintenance plan based on this information.
Regular maintenance is performed at the most favourable time in shipping: during dry-docking, surveys or while the ship is in port.
Diesel-power plant maintenance is executed preferably when there is a reduced electricity demand, for instance, at night or in weekends.

In the case of failures or damage, action must be taken immediately.
In the discussion regarding reconditioning of engine parts, the following subjects arise.

Reconditioning:

- cylinder heads including inlet and exhaust valves;
- pistons and piston rings;
- cylinder liners;
- crankshafts and bearings;
- camshafts and bearings;
- geared wheels;
- fuel pumps and injectors;
- turbochargers or supercharging groups;
- governors;
- engine blocks;
- coolers;
- pumps;
- connecting rods and piston rods.

Four-stroke engines will be first discussed and subsequently two-stroke crosshead engines.

► The lifetime of parts may vary considerably. Fuel quality is a deciding factor.

Component	Time between overhauls		Expected lifetime [h]	
Fuel quality	HFO	MDO	HFO	MDO
Piston	12 000 – 20 000	20 000 – 24 000	24 000 – 40 000	40 000 – 48 000
Piston rings	12 000 – 20 000	20 000 – 24 000	12 000 – 20 000	20 000 – 24 000
Cylinder liner	12 000 – 20 000	20 000 – 24 000	60 000 – 100 000	60 000 – 100 000
Cylinder head	12 000 – 20 000	20 000 – 24 000	60 000 – 100 000	60 000 – 100 000
Inlet valve	12 000 – 20 000	20 000 – 24 000	24 000 – 40 000	40 000 – 48 000
Exhaust valve	12 000 – 20 000	20 000 – 24 000	12 000 – 20 000	24 000 – 32 000
Injection valve nozzle	2000	2000	4000 – 8000	8000
Injection pump	16 000	16 000	16 000 – 24 000	32 000
Main bearing	16 000 – 20 000	16 000 – 20 000	32 000 – 40 000	32 000 – 40 000
Big end bearing	12 000 – 20 000	20 000 – 24 000	12 000 – 20 000	20 000 – 24 000

26.2 Four-stroke engines

26.2.1 Cylinder heads with inlet and exhaust valves

Valves

Perfect fuel combustion (100%) is purely theoretical. In practical terms, fuel combustion is approximately 96 to 98%. This means that a small amount of the fuel either does not burn or does not completely burn.

Uncombusted carbon particles are hard and cause damage in the combustion space.

Other heavy-oil components can produce high-temperature corrosion, (HTC). The **Vanadium** in fuel is chemically bound with sodium from (sea) water and forms a sticky slag, which can affect the oxide coating on the valves.

Also, see Chapter 8, *Fuels, fuel-line systems and cleaning fuels.*

Moreover, the slag is oxygen-rich and therefore forms an extra oxide coating. This process repeats itself continuously in an aggressive, rapid fashion. Special valve materials may be able to prevent this process.

Also, see Chapter 7, *Use of materials for diesel engines.*

An rpm of 1500 comprises 750 complete processes in a four-stroke diesel engine, one every second revolution. Therefore, valves are opened and closed 750 times! With a life expectancy for valves of 15,000 operating hours, this amounts to 750 × 60 × 15,000 = 750,000,000 times!

▲
Cylinder heads after disassembly of the valves and valve drives, ready for cleaning.

Key points for exhaust valves

Valve seat: pitting due to the impact of **solid particles**. Blow-through, wear and tear.
Seat in the head: ditto, pay attention to cracks.
Valve disc: corrosion of material by HTC, usually visible through porous surface material.
Valve stem: damage to the surface; crumbling of the chromium top layer, if present; top of the valve stem: crater forming due to valve drives.

Key points for inlet valves

Obviously, these do not experience as high a pressure as the exhaust valves. The airflow is cool and uncontaminated.
Valve seat: check for pitting and wear.
Seat: ditto, pay attention to cracks.
Valve stem: check for wear, pitting and damage to the top of the stem.

▼
High-temperature corrosion frequently occurs in engines running on HFO.

The valve material exhibits craters and cracks.

▼
Two cleaned multi-cylinder heads for Caterpillar engines.

▶
A conveyor washing machine for cleaning engine parts at Engine factory Bolier, Dordrecht, The Netherlands.

Carbon deposits and other solids are effectively removed.

Procedure for reconditioning valves

- Cleanse in a special washing machine, tank or other device with a carbon- and degreasing agent.
- If required, scrape off the hardened layers of blast-clean them in a blasting box with an abrasive.
- Check for surface damage, burns and cracks.
- Check the quality of the sealing surface. Often a thin layer of valve material has peeled off; this occurs in, for instance, rewelded valves.

▼
A dry-blast cabinet for cleaning parts.

1 *entry hatch*
2 *viewing window*
3 *gloves for handling the parts inside the cabinet*

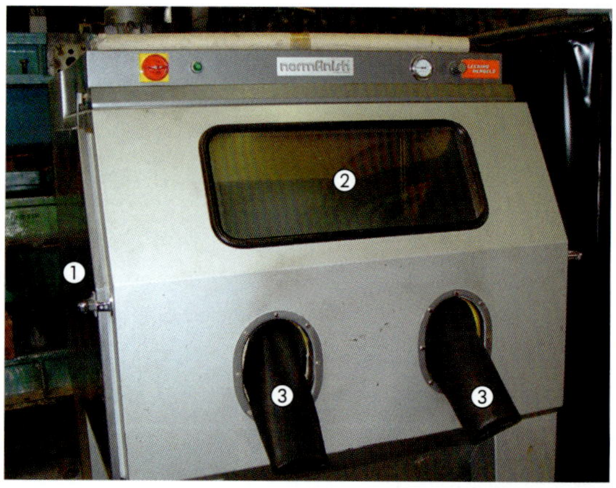

▼
After cleaning, inlet and exhaust valves are inspected for hair-cracks and the dimensions are checked.

▶
Installing inlet and exhaust valves in reconditioned cylinder heads for Caterpillar MaK 32 medium-speed heavy-fuel engines.

The valves are hand polished in their new seats.

Trueing up the valve seat

Essentially, two methods are used.

1 **Grinding the seat with a grinding machine.**

▲

A 'Hunger' valve-grinding machine.

The valve seat is ground at the correct angle until it is smooth and clean.

2 **Turning the seat with a chisel housed in a cutting device.**

▲

The valve seat is reconditioned using a chisel using this machine from 'Hunger'.

Shown here, a seat is overhauled using a rotary cutting machine with a chisel mounted on the new valve guide that used as the centre line.

◄

Grinding a valve seat using a grinding stone mounted on a special grinding machine.

The grinding angle is important. The accuracy of the grinding angle is crucial. Grinding takes place using degrees and minutes where one minute is 1/60th of a degree!

▶
A valve-seat grinding machine from Chris-Marine.

Small grinding stones are used.

Trueing up the valve-seat casing

The machining is similar to that used for the valve seats:

1 grinding
2 turning using a small chisel.

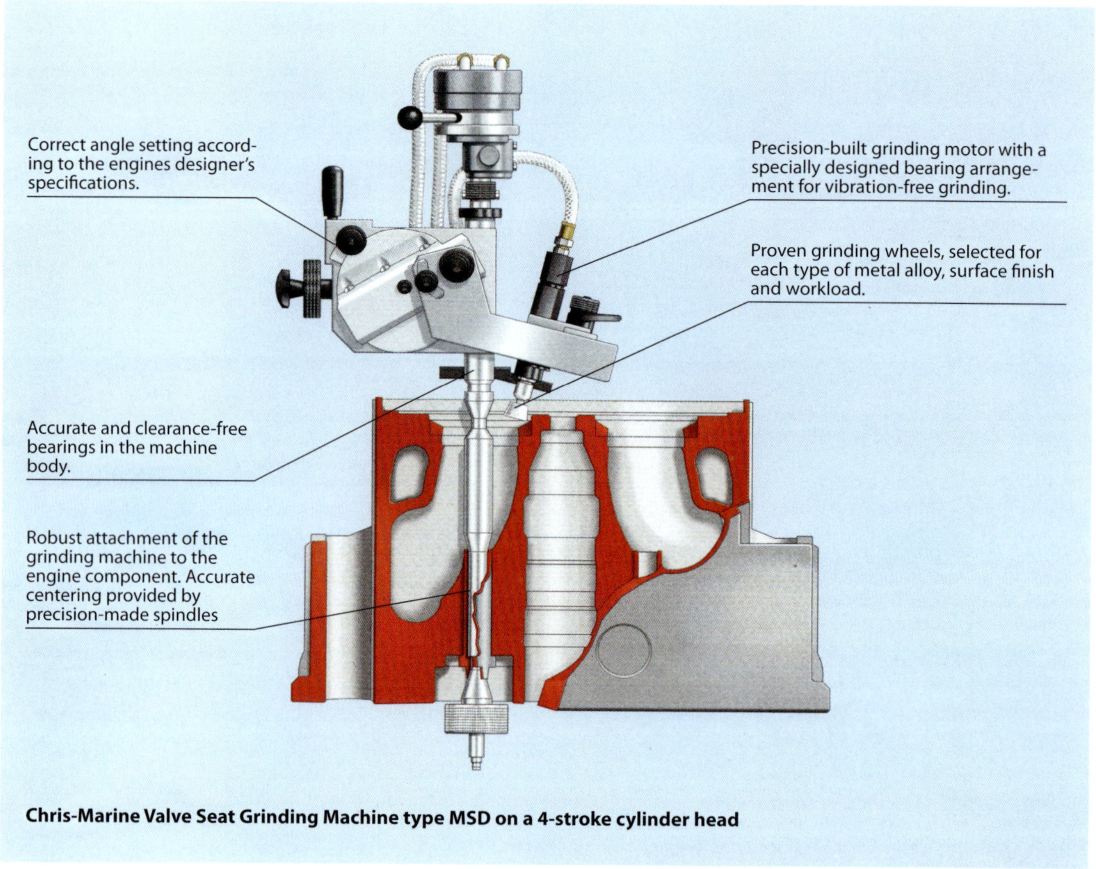

Chris-Marine Valve Seat Grinding Machine type MSD on a 4-stroke cylinder head

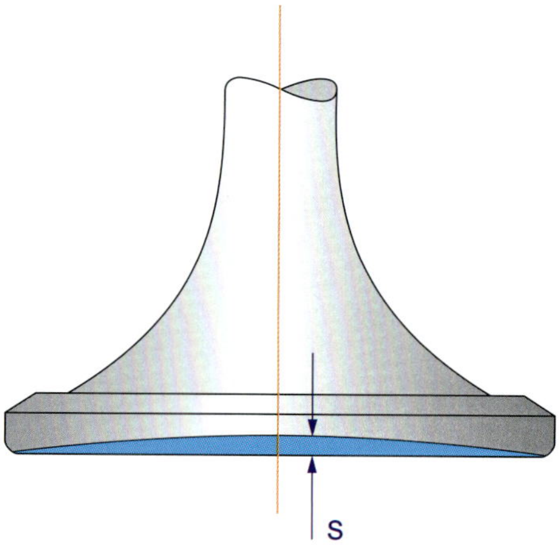

▲
S is the maximum burn depth for a valve before scrapping.

▲
Cylinder heads after thorough reconditioning, ready to be mounted on an engine.

The cardboard envelope contains a complete packing set. Each head is provided with a data label.

Drives

Obviously, a general overhaul of the cylinder head includes inspection of the valve drives and the valve-lever mechanisms.

Procedure

- Degrease the parts.
- Inspect for damage, cracks and pitting.
- Measure the pivot clearances.
- Repair excessive clearances.

Bushes of, amongst others, valve levers are usually renewed.
Valve springs are checked for fractures and the relaxed spring length is measured. If the springs have become too short due to material fatigue, they are scrapped and replaced.

Push rods must be straight. Rolling them along a level surface shows if they are bent.

Cylinder heads

Due to increased pressures in the cylinder, the mechanical load on the cylinder heads is constantly increasing. Most four-stroke engines with individual cylinder heads have **four head-bolts,** which in large cylinder bores are hydraulically tensioned. The cylinder-liner seal is heavily loaded necessitating a precise fit of the surfaces.
They can be machined in various ways. The cylinder head on the cooling-water side must also be cleaned and meticulously inspected.
The heads are submerged in a cleansing tank, for the removal of deposits. Prior to inspection, they are rinsed with clean water for the removal of residual solvents.
The freeze plugs are re-fitted.

The cylinder-head flame plate is checked for burning, corrosion and cracks.

◄ Valve springs and the entire valve movement require thorough inspection for damage and corrosion.

◄ Cylinder heads of a Caterpillar MaK engine under construction.

1 *freeze plugs*

▶

The cylinder-head flame plate.

1 *inlet valves*
2 *exhaust valves*
3 *bore for starting-air valve*
4 *bore of fuel injector*
5 *bore of relief safety valve*
6 *bore for indicator cock*

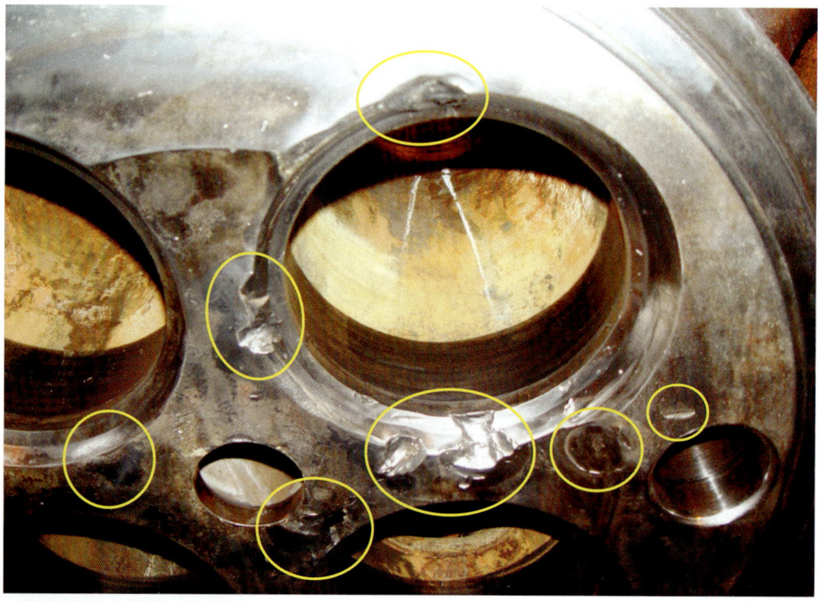

▶

A damaged cylinder head.

Clearly visible is a piece of material, probably part of an exhaust valve, lodged between the cylinder head and the piston.

The cooling-water spaces of the heads are 'pressure tested' with clean tap water to check for leaks through covers or freeze plugs or caused by tearing.

◀

Pressure testing the cooling-water space containing the cylinder liners for a newly built medium-speed four-stroke HFO engine.

The yellow glands are used to place the cylinder liner under the same pressure the cylinder heads will exert on them when assembled. For 24 hours, the cooling-water pressure is checked for pressure loss. Above a certain pressure-reduction percentage, it can be concluded that leaks have occurred.

◀
Valve seats are cooled in liquid nitrogen to a temperature of 180 °C below zero to facilitate their assembly in the cylinder heads.

When the seats are at ambient temperature, they are firmly fixed. Top right; the starting-air supply bore. In the centre of the head; the fuel-injector bore.

Valve seats are usually nitrogen-cooled and then fitted. When the seat is heated, it expands to fit snugly onto the cylinder head.

▲
An insulated liquid-nitrogen drum.

◀
Levelling a cylinder head with a portable planing machine.

In multiple- or multi-cylinder heads, the sealing surface is often refaced so achieving a true seal on the block.
This is done using a grinding stone on a surface-planing table.

The safety valve and indicator cock, if present, must be removed and cleaned. The sealing surfaces must be inspected and re-assembled after renewing gaskets and refacing the sealing areas. Safety valves must be tested; they must open at the correct pressure.

26.2.2 Pistons and piston rings

These parts are particularly heavily loaded in diesel engines.
Together they must seal the cylinder liner at increasingly higher pressures.
Furthermore, the piston must transfer the pressure in the combustion space to the crank-connecting rod mechanism.
Piston- and piston-ring wear is caused by:
- abrasive combustion products and particles;
- metal-on-metal contact of the piston rings, piston-ring grooves and the cylinder liner. This mainly occurs at the beginning of the combustion stroke.
- corrosion, chemical damage at low engine loads; low temperature corrosion occurs when sulphuric acid is formed. Sulphur is found in heavy fuel.

Light-metal pistons in smaller engines are cleaned and inspected. When the total number of approved operating hours has been reached, they are replaced together with the piston rings.

Pistons of larger engines are dismantled, cleaned and inspected. If the dimensions are correct and the remaining operating hours is still adequate, both parts are re-used.
The three-part piston-ring package is usually replaced.

If the top piston-ring groove is chromium-plated, it will require a new chromium coating. During inspection of the piston, it is imperative that the inside is adequately cleaned and checked. Cooling the piston with lubricating oil is vital.

26.2.3 Cylinder liners

Pistons, piston rings and cylinder liners all have a certain 'service life': the number of operating hours between major overhauls they should achieve without failure.
Abrasion by combustion products and **chemical damage** brought about by low-temperature corrosion found at low engine loads are the main causes of wear.

▲
Damage to engine parts can be considerable.

Shown a piston (1) of a four-stroke trunk-piston engine, this has seized in the cylinder. The camshaft (2) is bent. A connecting-rod bolt probably broke allowing it to move freely in the crankcase, and damaging the engine.

▼
Pistons manufactured in two parts, such as those found in medium-speed engines, are taken apart, cleaned, measured and, if necessary, repaired.

Foreground: pistons with a dismantled crown.
Background: the piston crowns have been disassembled.

1 *piston skirt*
2 *piston crown*
3 *bolt holes for attaching the crown/skirt*
4 *cooling-oil drilling for piston-crown cooling*

▼
New piston, assembled from two parts.

▲
Cylinder liners prior to honing.

◀
Honing-stone holder with honing stones.

1 *honing stone*
2 *holder*

Procedure

- Decarbonise, degrease and clean the entire liner.
- Inspect rim at the top of the cylinder.
- Inspect the honing pattern on the cylinder wall.
- Inspect the sealing areas.
- Check the cooling bores.
- Measure the liner diameter at various heights and at 90°.
- Wear of the rim.
- Plateau-honing of the cylinder liner.
- Measure the liner diameter, see above.

Glazing

After a certain operating time, cylinder liners can show signs of glazing. The honing pattern completely disappears and the surface area shows signs of being polished by carbon particles produced during the combustion process. Honing can remove this glazing completely.

There are many premises regarding proper honing, subjects such as honing patterns, honing times, honing liquids and honing movements are just a few of the many.
Generally, plateau honing is performed after coarse honing, creating a polished honing pattern. Therefore 'running in' the liner surface is no longer necessary.

▼
The honing pattern after honing in a used cylinder liner.

◀
Honing gives a precise finish of the cylinder-liner running surface.

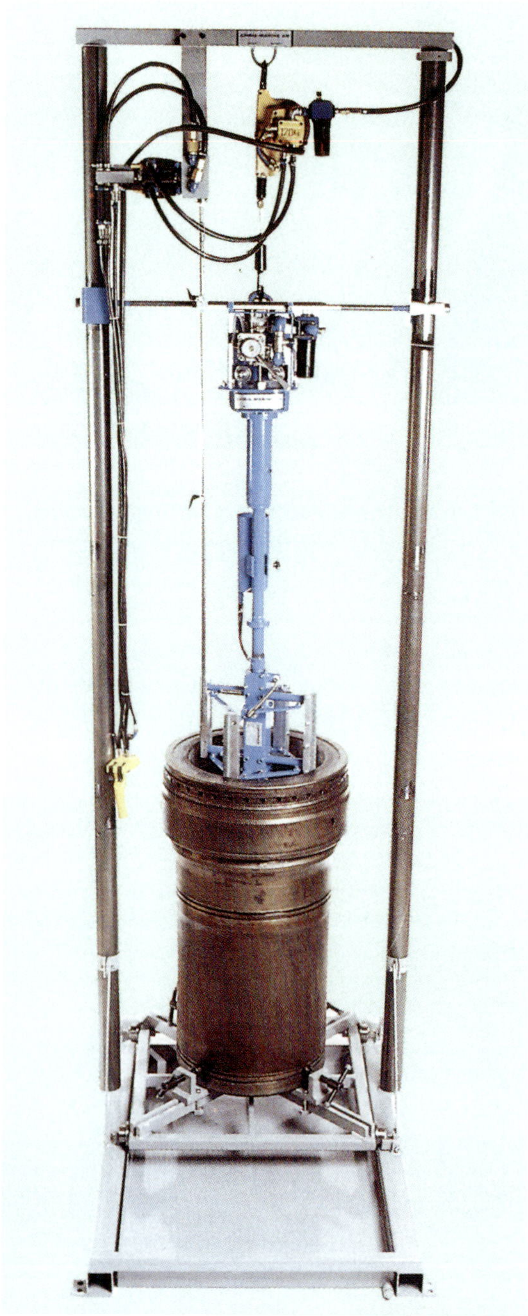

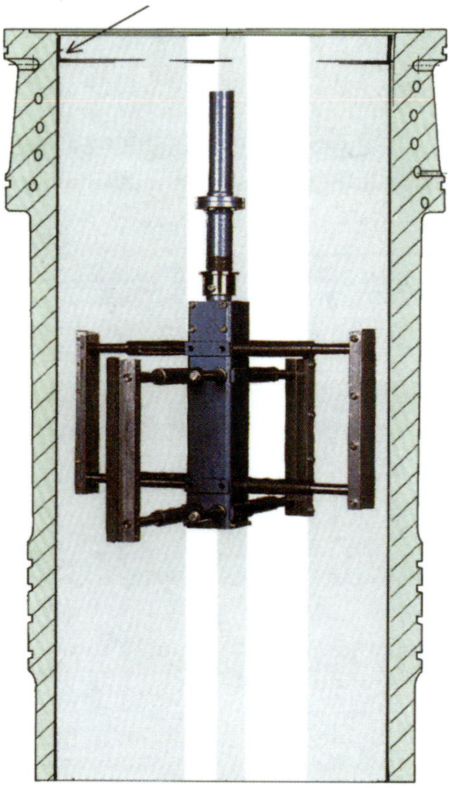

▲

Cylinder honing set with four honing stones. Cylinder-honing sets for larger cylinder diameters may have as many as eight honing stones giving a true 'round' honing, also for larger cylinder diameters.

◄

This mobile honing machine allows cylinder liners to be honed without disassembling the engine. It can also remove glazing.

▼

Honing a cylinder liner in the engine.

Several features of correct honing (Mahle)

1 **Uniformity of the honing pattern**
In general, the cylinder wall must be shiny, uniform and true.

2 **Honing groove shape**
Cutting edges must be clean and sharp, uniform groove width with a maximum width of 0.1 mm.

3 **Honing groove position**
Uniform cutting pattern for the reciprocating stroke.

4 **Honing angle**
Must lie between 30 and 60 degrees.

Used liner

The overall condition of a liner may not be apparent until after a few honing strokes.

After 3 minutes of basic honing

Soon the surface exhibits marks, ovalty and imperfections.

After 8 minutes of basic honing

The full extent of material to be removed becomes clear.

After 14 minutes of basic honing

The imperfections are beginning to disappear.

After 22 minutes of basic honing

Roundness has now been restored and surface marks removed.

After 3 minutes of plateau honing

The plateau honing has removed the coarse peaks, providing a good carrying surface for the piston rings.

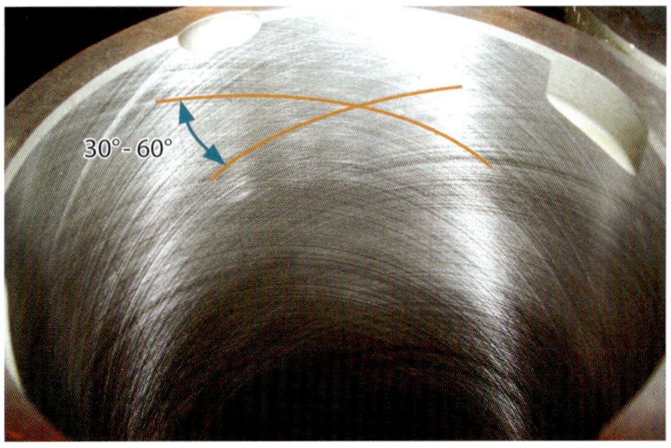

▲

Normal honing angles lie between 30 and 60 degrees.

◀

An overview of cylinder-liner honing, the procedure followed by Chris Marine, the so-called plateau honing.

The entire procedure of transforming a 'glazed' cylinder liner to an optimal running surface takes approximately one hour.

▼

Plateau honing; the ideal carrying surface is created in three phases.

left picture: worn and almost completely glazed surface.
centre picture: roughly honed surface with deep grooves and high peaks (R_{max}) that are still too large.
right picture: the third finish, hones down the peaks and creates an ideal honing pattern, giving an ideal carrying surface for the piston and the piston rings, yet leaving sufficient groove depth and width for the lubricating oil.

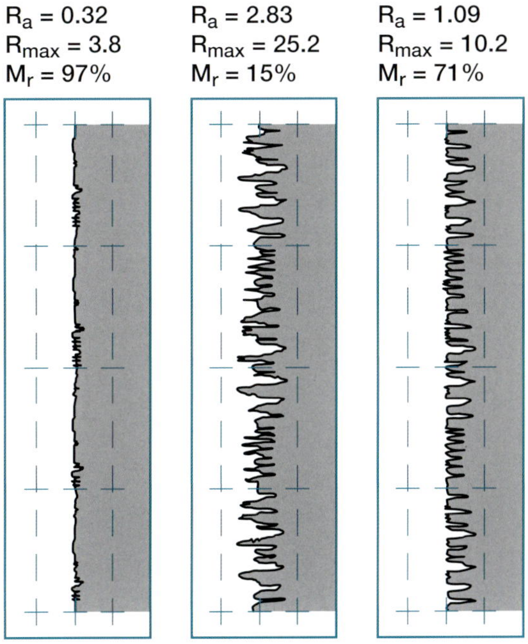

5 **Honing pattern**
Sharply cut, neither blocked nor flattened and no burring.

6 **Plateau-honing**
By pre-honing as well as post-honing.

7 **Roughness of the honing pattern.**
The guide values for diesel engines are
R_z = 4 – 8 µm (microns)
R_a = 0.8 – 1.2 µm (microns)

TR–value	plateau-honing pattern	A	1.5 µm
		B	83%
	normal honing pattern	A	2.5 µm
		B	91%

R_k = 0.5 – 1.3 µm	(Core Roughness)
R_{pk} < 0.3 µm	(Peak Roughness)
R_{vk} = 1.0 – 4.0 µm	(Valley Roughness)
R_a < 1.0 µm	(Average Roughness)
R_z = 4 – 8 µm	(Peak/Valley Roughness)

Examples of correct cylinder-liner honing methods from Chris Marine.

C-M Result = Chris Marine.

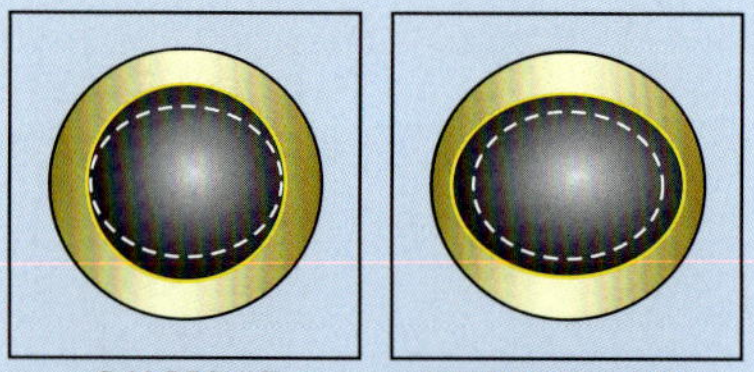

Oval cylinder liners are a common problem that should be regularly dealt with. Round cylinder liners can be achieved by using machines that have:

- adequately designed 4 or 8-armed honing head with a rack and pinion system
- a totally reliable mechanical stone pressure adjustment system

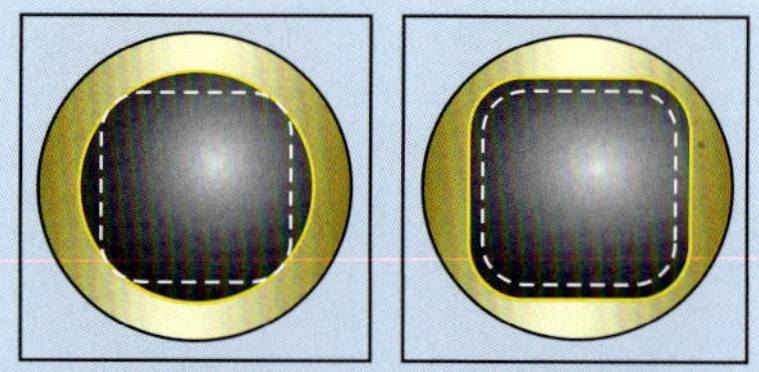

Under certain circumstances, a four-leafed clover wear can occur in a cylinder liner. To restore good roundness requires a machine that has:

- adequately designed 8-armed honing head with a rack and pinion system
- a totally reliable mechanical stone pressure adjustment system

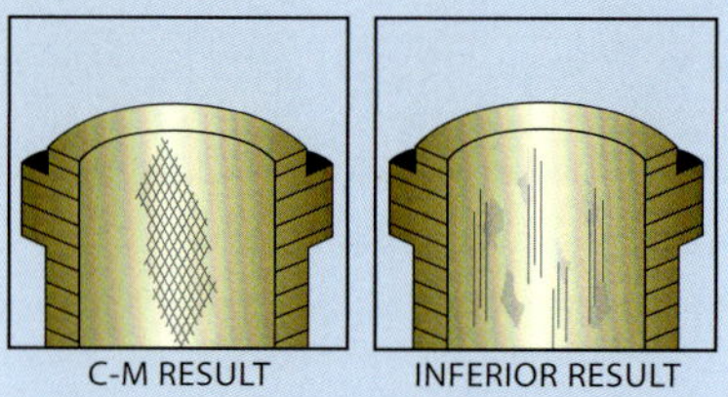

Scuffing marks may be caused by worn piston rings, carbon deposits or fuel-borne contaminants. Such a condition can be difficult to improve and therefore requires a high capacity machine that:

- comes with a variety of tested and proven honing stones
- has a suitable stroke length to fit the cylinder liners
- provides for effective operator control of the honing work

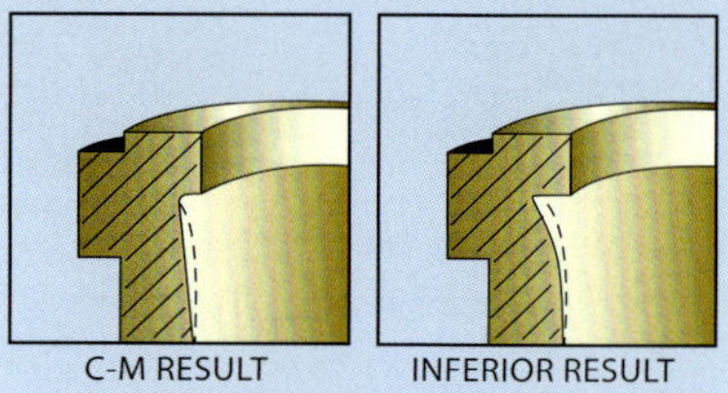

As cylinder liners usually wear at the top sooner than the bottom, a trumpet shape can sometimes occur. This needs to be removed leaving a long transition that reduces the load exerted on both the piston and the piston rings. A cylindric or slightly tapered geometry can best be achieved with a machine that:

- is specifically designed for demanding cylinder liner maintenance work
- provides for the correct honing methods for different kinds of cylinder liners

Inspection methods

1 **Visual inspection, finger nail and coin tests**
These are frequently employed. Experienced technicians can give assessments in combination with other tests. These methods are inadequate and indefinable.

2 **Fax-film**
An imprint is made of the cylinder surface using a plastic cellulose acetate film of approximately 0.2 millimetres thick. This imprint is magnified which allows for precision assessments of the honing features as detailed in points 2, 3, 4, 5 and 6.
Procedure: Clean the test surface with acetone (a degreaser), re-treat with acetone and press the sheet of fax-film on the surface until the acetone has evaporated (10 to 15 seconds). Remove the film, and place the imprint of the cylinder carrying-surface under a microscope or a microfiche reader and assess.

3 **Surface measurement**
The cylinder surface is probed by an electronic device and the roughness (in R_t, R_z or R_a) is measured according to a certain scale. Furthermore, it is advisable to use a profilometer, which registers the roughness on a profile diagram. Preferably, a roughness measurement according to Abbott's bearing curve.

Important aspects for the achievement of good results in the surface processing for cast-iron cylinders

- Only use suitable machines and accessories.
- Take note of the following when tensioning, tensioning regulations drawn up by the engine factory, levelling, equal tension forces and prevent twisting of the part.
- Precision drilling to approximately 0.04-0.006 mm, dependent on the diameter.
- Hone for a minimum of 0.03 mm material removal, dependent on the diameter. This is

required as the groove depth and the resulting hard zone can reach this depth during precision drilling.
- Do not hone dry.
- Hone cylinders throughout the entire surface so that the thickened rim is completely removed.
- Use the prescribed honing oil (correct viscosity) for the honing stones and material. Cool, if required.
- The honing oil must be sprayed under pressure against the cylinder wall. Material and honing-stone residues must be flushed away immediately so that they are not pressed into the cylinder wall or honing-pattern grooves.
- Re-used honing oil must be thoroughly filtered. Filters must be cleaned or replaced on a regular basis. Honing oil must be refreshed regularly.
- Use ceramic honing stones, if possible; these allow a sharper cut than diamond stones as they have self-sharpening properties.
- When selecting honing stones, the hardness of the cylinder material must be taken into account.
- Rule of thumb: the harder the material, the softer the honing stones.
- Perform a visual inspection after honing: all the cylinders should have a uniform shine and a glossy-silver finish.
- After honing, clean the cylinder block and cylinder liners thoroughly.

Machining

Cylindrical shape finishes.

- **Taper**
  Stroke is too long.
  Stroke is set too high.
  Honing stones and guides are not trued up.
  Honing stroke on the wide side is too high and too low on the narrow side.

- **Bell mouth (top and bottom too wide).**
  Stroke too long.
  Honing stroke on both sides too small.

- **Barrel (top and bottom narrower than in the centre)**
  Stroke too short.
  Honing stroke on both sides too small.

◄ A worn cylinder liner. The thickened rim is clearly visible.

◄ Assembly of a re-conditioned diesel engine. The engine has new bearings, cylinder liners, pistons and re-conditioned cylinder heads.

◄ The cylinder liner of a l arge four-stroke engine for repair.

▲ In used cylinder liners, a coating can be applied by spraying a material on the liner above the highest position of the piston. This has the same effect as an anti-polishing ring.

1 *carrying surface of the cylinder liner*
2 *coating*

26.2.4 Crankshafts and bearings

▲
New crankshafts made from a single forging ready for their final finishing, such as the bolt holes for the counterweights. MAN Diesel, Hamburg, Germany.

In four-stroke engines, the engine-driving gear consists of the crankshaft, the connecting rod and the piston.

The crankshaft drives the camshaft via the geared wheels, which in turn drive the fuel pumps and inlet and exhaust valves.
In many four-stroke engines, the main lubricating-oil pump, fresh-water pump and untreated cooling-water pumps are also driven from the crankshaft and the camshaft.
The camshaft of a four-stroke engine rotates at half the speed of the crankshaft, it makes one revolution for every two revolutions of the crankshaft.
The entire process takes 720 crank degrees.

If the engine is started using starting-air systems connected to each cylinder, additional control-cams are mounted on the camshaft.
Today, an increasing number of larger engines are started via a starter on the flywheel.

Also, see Chapter 14, *Starting systems of diesel engines.*

Together with the engine frame, the crankshaft is the most expensive part of the engine.
Crankshaft damage necessitates elaborate and expensive repairs. In large engines, damage to, for instance, the journals can sometimes be repaired without dismantling the crankshaft. This is also referred to as **in situ reconditioning.**
Crankshaft damage or damage to the drive gearing often involves severe damage to the entire engine.

▶
The drive gearing of an engine, such as the crankshaft, gears, the camshaft, connecting rods and pistons, must be precisely installed in the engine.

The centre lines of the crankshaft and the camshaft must run parallel. The centre line of the pistons must be perpendicular to the centre line of the crankshaft. Gear alignment is important.

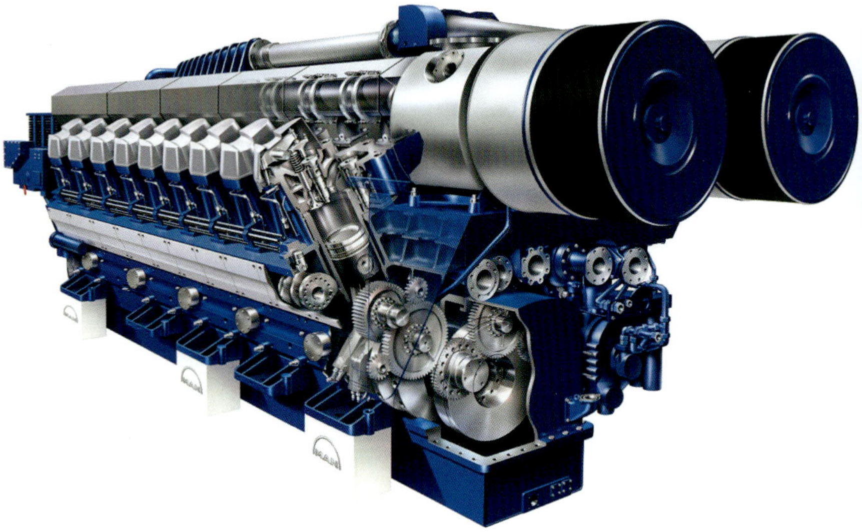

◄

Damage in the crankcase usually means damage to the crankshaft itself. Shown here, a total loss case.

▲

The carrying surface of this crankshaft journal is damaged. It is possible to grind this journal to the next possible 'undersize'.

▼

A large crankshaft lathe at Mark van Schaick, Schiedam, The Netherlands.

Classification agencies have very detailed regulations with respect to material use, size and manufacturing methods for crankshafts.

Also, see Chapter 32, *Regulations for propulsion engines, classification, repair and damage.*

►

Damaged crankshafts for re-conditioning.

Usually, the crankshaft journals or crankpins are damaged. After a thorough inspection, which, consists of measuring, surface hardness check and a hair-crack test amongst others, the shaft is ground to an undersize. 'Thicker' bearing shells must be used.

26.2.4.1 Common crankshaft damage.

1 **Overheating**
Mostly a result of inadequate lubrication between the crankshaft journal or crankpin and the bearing. All crankshafts that overheat are bent. Generally, crankshaft beds or engine blocks are rigid and the crankshaft is 'limp'. The lubricating-oil pressure in the bearing keeps it 'afloat', that is, the crankshaft is supported by the oil film in the bearing so it is not in contact with the bearing material. If the bed or block as well as the crankshaft are too rigid, the rotating crankshaft will not be supported by the lubricating-oil film resulting in wear and increased friction.

2 **Fractured connecting-rod bolts**
By overloading or poor assembly of the connecting-rod bolts, the connecting rod can become detached from the crankshaft and cause severe damage.
Connecting-rod bolts can lose their strength if re-used too often. The technical manual should always be consulted in order to check how often the bolts might be used!

3 **Dilution of lubricating oil**
The lubricating properties of lubricating oil can deteriorate considerably due to, for instance, either ageing of the lubricating oil or fuel or coolant leaks.
Diluted lubricating oil results in a reduction of the hydrodynamic lubrication in the bearings resulting in the shaft coming into contact with the bearing.

26.2.4.2 Crankshaft hardness

The hardness for 'soft' crankshafts is measured according to the Brinell scale. For 'hard' crankshafts, the Rockwell C scale is used.
For spray coatings, such as chromium, the Vickers scale is used, the base coat of thin coatings is often also measured, thereby giving incorrect measurements!
Large crankshafts for engines in categories III and IV are generally 'soft'. The material hardness is approximately 150 to180 Brinell for C 45 crankshafts and 220 to 260 Brinell for 42 Cr crankshafts.
Smaller crankshafts in engine categories I and II are usually 'hard'. They have been surface-hardened. The hardness is approximately 45 to 65 Rockwell C.
Soft crankshafts (categories III and IV) can become 'hard' when the crankpin or the crankshaft journal overheats. The hardness can rise from 200 to 600 Brinell.

Warming a heated crankshaft. Sometimes an overheated journal is re-heated in an attempt to reduce the hardness. This is performed in situ with electric 'heat strips'. The results vary, but the original hardness is not usually achieved and the hardness remains high. If the engine supplier accepts the acquired hardness, approval from the classification agency usually follows. This method avoids the time-consuming and expensive dismantling of the engine to remove the crankshaft.

Hard crankshafts (categories I and II) turn 'soft' when the crankpin or the journals overheat.
Flame hardening is a surface-hardening process applied to the journal.
Method: The material is heated with a high-temperature gas flame to approximately 600 °C followed by quenching using jets of water resulting in rapid cooling. The structure of the material surface is transformed and acquires a significant increase in hardness.

▼
Changing the crankshaft in a diesel engine onboard.

This is an expensive and time-consuming procedure.
In four-stroke engines, the entire block must be hydraulically supported, so the crankshaft can be 'dropped' out from below the engine. All partitions or doors must be removed so the crankshaft can be transported out of the engine room.
Often a large hole is cut in the hull closest to the engine room using cutting torches.

Surface hardening can also be achieved with a nitrogen process: induction hardening. A hard plated chrome layer with a maximum thickness of approximately one millimetre can also be applied.

Tearing of the journals often occurs because of material fatigue, torque and overloading. Microscopic tears deepen resulting in crankshaft fractures and over-all damage to the engine. Inferior vibration dampers may be the root cause of the problem. Vibration dampers should be regularly inspected.
The filling found in dampers usually contains a highly viscous silicone fluid (jelly or oil). It is therefore advisable to test the vibration dampers. A special testing device is used that introduces vibrations into the coupling. The extent to which the vibrations are damped indicates how effective the vibration damper is.

Heat-induced cracks are generally superficial and can be removed by polishing on a (crankshaft) grinding machine. If the crack cannot be removed by polishing or grinding, the crankshaft is scrapped.

The material used in crankshafts can always be traced via a database. The hardness is particularly important. The undamaged section always has the initial base hardness.

Renewing bearing shells
It is imperative that the selected bearing is manufactured from the same material as that of the original shells. Bearing shells have standard 'oversizes'.
The engineer operating a crankshaft-grinding machine knows precisely how much material can be removed in order to obtain the first oversize. Only OEM-shells should be used
(OEM = Original Equipment Manufacturer).

Straightening/trueing up an overheated crankshaft
Crankshafts with a crankpin or main journal that has overheated are always bent. They can be straightened hydraulically or by hammering around the radius with hammering tools and a heavy hammer.

26.2.4.3 Grinding the crankshaft

The crankshaft is positioned at 'zero' in the centre of the grinding machine.
Depending on the size of the crankshaft, stays are placed underneath the journals to prevent crankshafts from bending under their own weight. They also absorb the pressure exerted by the grinding process.

Grinding the crankshaft starts from the centre axis to the left and then right and then to the middle.
In the centre of the crankshaft, the value given on the dial gauges must be 0.000 mm!

▼ Grinding a large crankshaft at Mark van Schaick, Schiedam, The Netherlands.

1 *supports*
2 *grinding stone*

Clearance of journals and crankpins

Rule of thumb: $\frac{1}{100}$ mm or 0.001 times the diameter clearance per 10 mm shaft diameter.

A shaft with a diameter of 250 mm:
the clearance of the bearings may not exceed
$\frac{250}{10} \times \frac{1}{100} = \frac{25}{100}$ mm.

26.2.4.4 Crankpin wear

Crankpins and crankpin bearings are heavily loaded. Hydrodynamic lubrication is not optimal as the forces fluctuate in both size and direction. This often causes boundary lubrication where insufficient lubricant-film thickness allows surface contact between pins and bearings. Localised wear occurs resulting in an oval-shaped shaft. This occurs more often in a certain type of engine from certain engine manufacturers.

◄ A large crankpin for a medium-speed diesel engine.

An attempt to straighten the crankshaft by hammering has only succeeded partially. This crankshaft is irreparable!

▶

The crankpin must be true.

Measuring the roundness is important after a large number of operating hours. The pin often assumes an oval shape, which is then ground down to a smaller diameter.

▲

After repairs, the crankshaft must be approved according to the classification bureau regulations. This is referred to as the 'release' of the crankshaft.

An absolutely round shape with a smooth surface area can be obtained by grinding the shaft.
The main- and crankpin journals of the crankshafts are manufactured with a certain oversize and may be ground-down several times without dimensioning problems. Obviously, oversized (thicker) bearing shells must be used.
Standard sizes are normally used so the shaft can be ground to a certain diameter (undersize).

After this machining process, a protocol is drawn up with the new dimensions of the crankpins and the journals. This practice is referred to as 'acceptance': the crankshaft meets classification agency standards.

Grooves in the running surface of crankpins and journals

▶

A damaged crankshaft on a crankshaft lathe, a type of grinding machine.

The grinding stones (1) have a large radius and varying hardness and grit sizes.

▲

A worn crankpin.

These grooves are often a result of overheating, caused in part by insufficient lubricating oil. Moreover, dirt particles in the lubricating oil may have come into contact with the shaft, for instance, via a damaged filter or as a result of certain activities that have contaminated the cooling system.
If the grooves are shallow, the shaft or pin can be ground.

▼

Grinding a small crankshaft.

26.2.4.5 Modification of the material structure

Overheating and groove formation cause a temporary temperature increase in the material. The carrying surfaces often turn blue. This usually means that the carrying-surface temperature has risen to approximately 700 °C.
This changes the structure of the material.

Unalloyed carbon-steel crankshafts harden.

Local soft areas occur in alloyed hardened crankshafts.

These phenomena cause a tension increase in the surface layers of the material. This can result in the formation of hairline cracks, ultimately resulting in a shaft fracture.
These 'areas' can be located simply by performing a hardness test on location. Dependent on the difference between the hardness of the areas and the hardness of the shaft, the decision is made as to whether the shaft or crankpin should be ground to a certain undersize so that there is no or very little discrepancy between the harnesses. Clearly, the shaft must be ground to below the minimum prescribed shaft diameter.

26.2.4.6 Cracking of crankshafts

▲
Check the journals for hairline cracks.

The shaft can often be repaired by grinding if the cracks are superficial.
If the cracks are too deep, the shaft is written off as a total loss.
The crack depth can be established using a small grinding stone.
Some second-rate firms weld the cracks.
This is obviously pointless and the shaft is then scrapped!

▲
A magnetic field is used to detect superficial cracking.

A coil is placed on the shaft and this generates a magnetic field. A contrast liquid is applied to the shaft and the surface is inspected with a lamp. A crack can be discerned by an interruption in the colour of the shaft. This process is called 'Magnafluxing'.

Crankshaft cracks are a result of material overload. The excessive load may be produced by torsional stress, generated by torque and torsional vibrations. Under normal operating conditions, this need not be a problem. Often torsional vibrations are at the core of the problem. When the engine speed approximates the number of natural vibrations of the crankshaft, the torsional stress can increase to a point where the crankshaft cracks.
These fractures are generally easily recognisable. The cracked surface is often at an angle of approximately 45° to the centre line of the crankshaft.
Extremely large bending stress may also cause overall damage to the crankshaft when the crankshaft is not aligned.
The most important causes are:
- wear of the main shaft bearings;
- loading of the ship;
- distortion of the engine bedplate due to extreme weather conditions or the ship running aground;
- incorrect bedplate design in diesel-power plants and power barges;
- temperature influences.

26.2.4.7 Damage to the crankshaft by a seized piston

If piston rings seize in the cylinder because of insufficient cooling or poor lubrication, severe damage often follows.

The piston crown seizes at the top of the cylinder liner. The piston skirt with the piston pin is torn from the piston crown, and the moving connecting rod severely damages the frame and the crankshaft. The connecting rod is often detached from the crankpin when the connecting-rod bolts break and the counterweight bolts on the crankshaft detach. All these fractured parts cause denting in the crankshaft. These are irreparable resulting in the crankshaft being declared a total loss.
If the seized piston damages the frame irreparably, the engine is a total loss.

26.2.4.8 Damage caused by counterweights detaching from the crankshaft

When the engine speed far exceeds the maximum allowable speed, the counterweights may detach from the crankshaft because the bolts attaching the counterweights to the crankshaft break. In this case, the maximum allowable bolt tensile stress is exceeded. Total damage to the crankshaft follows.

26.2.4.9 Crankshaft damage as a result of incorrect mounting of the main bearing shells

This can occur when technicians do not know how to mount the bearing shells or where special attention is required.
Key points: the bearing shells must be clean and show no signs of damage. During fitting, grease or oil must not be applied to the back of the shell to facilitate insertion of the bearing shell. The thickness of this grease or oil layer is often greater than the clearance between the shaft and the bearing shell! When installing the bottom bearing, the bolts for the bearing-cap with the bottom bearing shell must be equally tensioned. In accordance with the regulations of the manufacturer, the vertical and horizontal bearing-fixing bolts can then be attached.

Furthermore, it is important to know if a crankshaft is equipped with the originally fitted bearings or if one or several oversized bearings have been used after crankshaft repairs.

▼
The bolts of the counterweights must be tightened and secured with the utmost precision. Detached counterweights can cause irreparable damage to the crankshaft.

It can be determined if the crankshaft has undergone an 'overspeed' even when the counterweights are still attached to the crankshaft properly.
After removal of the counterweights, the crankshaft faces and the counterweight faces exhibit smooth friction spots. This is caused by stretching of the fixing bolts to the extent that the faces of the crankshaft and the counterweight are no longer in contact, thus allowing for micro-movements independent of each other.

26.2.4.10 Acid damage to the journals and crankpins

Sulphur found in fuel can form sulphur dioxide in the combustion space and can move past the piston rings and be found in the crankcase if maintenance is not performed regularly, for instance, lubricating-oil- cleaning and changing.
Subsequently sulphuric acid may be formed with water vapour. Under good maintenance conditions, the acid is neutralised by an alkaline additive in the lubricating oil and therefore is not damaging to the crankshaft.

See Chapter 11, *Lubrication of engines.*

If no or little attention is paid to the quality of the lubricating oil, for instance, failure to sample and test the lubricating oil, the alkaline additive could be exhausted, thus allowing sulphuric acid to damage the crankshaft. Black stains or pitting will appear on the journals, which will damage the running surface. **Corrosion of the crankshaft may also occur when there is leakage of cooling water or coolant, although this is less aggressive in comparison to sulphuric-acid corrosion.**
This problem is averted by taking samples at regular intervals!

▲
An in-line engine frame 'on its head'. The crankshaft counterweights must still be fitted.

Damaged bearings

Bearings in an engine have a short life span as they are heavily loaded; corrosive and mechanical wear in particular are significant.
In four-stroke trunk-piston engines, the quality of the fuel plays a decisive role.

26.2.4.11 Key points for an extended operating time

These are:
- suitable lubricating-oil treatment, such as filtration, separation and sampling;
- preventing external contamination of the engine during maintenance;
- capping lubricating-oil drillings in crankpins during repairs, and capping all openings leading to the lubricating-oil system;
- pre-lubrication of the engine ensuring all vital engine parts are properly lubricated prior to start;
- check lubrication during operation by monitoring lubricating-oil pressures, lubricating-oil temperatures and the lubricating-oil level in the crankcase. Monitoring lubricating-oil consumption is important as this indicates the state of engine maintenance. Inadequately functioning piston rings and exhaust-valve stems whose clearances are outside normal parameters lead to an increase in lubricating-oil consumption.
- Obviously, the operating life of bearings is largely dependent on operating conditions and varies from manufacturer and engine type.

Modern **trimetal bearings,** mainly used in both high-speed and medium-speed four-stroke engines, are not reconditioned.
They are usually replaced with a new bearing after a certain number of operating hours.
The total thickness of the bearing shell can be measured in order to establish the extent of wear.

After a certain number of operating hours, the lead-bronze base coat or the emergency-running coating gradually emerges. This means that the actual friction material, an aluminium alloy, has been worn away. The criterion for scrapping a bearing is when a certain percentage of the lead-bronze surface is visible.

26.2.5 Camshafts and bearings

The cams on the camshaft are heavily loaded when driving the high-pressure fuel pumps and inlet and exhaust valves.

▲
A camshaft with, from left to right, the fuel cam, the inlet cam, and the exhaust cam.

▶
A finished engine block as seen from the front. The location of the crankshaft, the camshaft, the scavenging-air ducting, and the internal pipes is clearly visible.

▶
An older type camshaft for a four-stroke high-speed diesel engine.

This is made from a single forging, as opposed to modern camshafts, which are usually manufactured in sections.

The cams are often damaged by grooves and indentations.
Small camshafts manufactured from a single forging are usually replaced.
In large camshafts, which consist of separate sections for each cylinder, only the damaged section will be replaced.

In separate hydraulically mounted cams, the damaged cam can simply be replaced.

▲
Re-welding steel engine parts is a meticulous and complex procedure.

The quality of reconditioned engine parts does not always match new engine parts. Shown here, a section of a camshaft.

On occasion, cams are rewelded using a wear-resistant material. The original dimensions are then restored by grinding.

Camshaft bearings are replaced by new ones when scrapped.

26.2.6 Gears

Gears are subject to wear.
The shaft bearings of the gears are of primary importance.
When the valve clearance is too large, the gearwheels mesh differently causing wear.
The carrying surface must be checked for damage and wear.
This normally only occurs with insufficient gearwheel lubrication, contaminated lubricating oil or an irregular load on the teeth.
Generally, when reduction gears are correctly aligned, the wheels mesh smoothly and few problems can be expected.

◀

The camshaft driven by the crankshaft via gears.

This is a MAN four-stroke medium-speed diesel engine with a separate camshaft for the fuel pump, and inlet- and exhaust valves.

1 *fuel cam*

26.2.7 Fuel pumps and injectors

Fuel pumps

Pressures and temperatures in the fuel pumps can rise considerably.

Peak fuel pressures of 1000 to 2000 bar and fuel temperatures of 140 °C are common when running on HFO.

Extreme pressures also occur when running on diesel oil; however, the fuel temperature is significantly lower, 15 to 30 °C.

The camshaft drive with fuel cam, cam and plunger are also heavily loaded.

◀

Fuel-pump casing in the reconditioning workshop.

▲

Honing a fuel-pump cylinder at MAN Prime Serv, Schiedam, The Netherlands.

During the honing process, very little material is removed. The running surface is finished so that the seal between the fuel-pump plunger and the cylinder is restored.

▶

Honing the pump cylinder.

1 *honing machine with honing stones (not visible)*
2 *pump cylinder*
3 *fasteners*
4 *mechanical rotating drive, the reciprocating movement is done by hand*

▶

A variety of honing tools and honing stones.

▶

Reconditioned fuel pumps.

All the openings are sealed to keep out moisture and dirt. Each pump is labelled.

Procedure

The fuel pump is disassembled and cleaned in a washing machine.
Visual inspection is performed on parts subject to wear, such as plungers, casings, pressure valves, cavitation bolts and the driving mechanism.
The dimensions of, in particular, the plunger and cylinder are measured.

▶

Fuel pump and parts.

1 *worn fuel plunger*
2 *reconditioned fuel plunger*
3 *cylinders or 'barrels'*
4 *reconditioned fuel pump*

▲

Plungers and barrels (cylinders).

▲

A polishing machine for trueing up parts for atomisers and fuel pumps.

The parts are trued up by the applying a fine sanding paste to the vibrating plate.

◀

Disassembled fuel pumps, cleaned and ready for inspection.

1 *plunger*
2 *barrel*
3 *pump casing*
4 *gear rack*
5 *springs*
6 *buffer plate*
7 *adjustment sleeve*
8 *cap with locking gasket*

Injectors

Adequate fuel-injector performance is essential for a perfect combustion process in the cylinder. The injectors must atomise the fuel at the correct pressure.

See also Chapter 9, *Fuel-injection systems*.

Injectors must be removed from the cylinder head after a prescribed number of operating hours and tested by an injector-testing device. These are simple manual test pumps, or mechanically driven test devices used in specialised firms.
During the test, the opening pressure and fuel atomization is checked, including possible injector leaks after the needle valve closes.

▲
Fuel atomisers require regular and meticulous maintenance. They are essential for optimum engine operation.

▼
Injectors for large diesel engines. The contact faces have been buffed, a fine-sanding process.

The latter is indicative of poor sealing of the needle valve.
Furthermore, the injector casing and connection to the fuel line is inspected for leaks.

When the injectors are inspected by specialised companies, the entire injector is disassembled; the various parts are cleaned, measured, repaired or renewed.
During inspection, the following is checked:

- the sealing surface of the injector needle;
- the length of the relaxed spring;
- the shape and dimensions of the injector perforations;
- the wear of the plunger and the casing;
- the pressure valve or the residual pressure valve, if present;
- the shape of the nozzle and possible corrosion. High temperatures may cause distortion of the nozzle.

After assembly, the injectors are tested.
The spray patterns are also checked by placing special paper or cardboard underneath the nozzle.

◀
The nozzles. Every diesel engine has its own nozzle type.

◀

Fuel pumps for all the Wärtsilä four-stroke engines. The injectors are tested at varying speeds and loads.

The pumps are mounted on an actuating machine attached to a camshaft, which measures injector performances at varying speeds.

▼

An injector-test machine.

1 *injector*
2 *fuel-supply line*
3 *test machine*

Warning: Always stay clear of an operational injector. The speed of the fuel jet is such that accidents can occur!

Comment

A regular and meticulous inspection of the injectors is important for optimal operation of diesel engines.

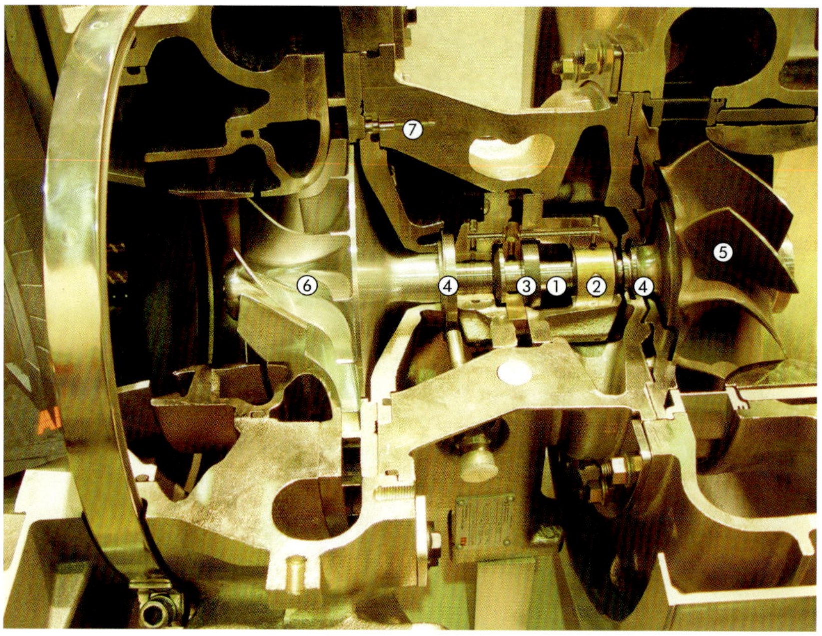

▶
A cut-away section of a turbocharger.

1 *turbine shaft*
2 *bearing on the exhaust-gas side*
3 *thrust bearing*
4 *seal*
5 *exhaust-gas turbine*
6 *centrifugal compressor*
7 *turbine casing*

▶
Blades of an exhaust-gas turbine for a turbocharger that are irreparably damaged. This is most likely caused by fragmented exhaust valves.

26.2.8 Turbochargers or supercharging groups

Supercharging groups are essential in the operation of the diesel engine.
Sufficient air supply for complete fuel combustion, cylinder scavenging and cooling of hot engine parts in the combustion space, are important for efficient engine operation and for its operating life. The turbocharger must supply sufficient air at a certain pressure.

Optimal turbocharger performance requires the following:
The degree of contamination and damage to the exhaust-gas turbine.

▶
Damaged exhaust-gas rotor, possibly caused by fractured piston rings.

A piston-ring catcher is sometimes placed before the exhaust-gas turbine inlet (two-stroke).

Combustion products such as carbon deposits may settle on the blades and interfere with the flow of hot exhaust gases, thus causing the charger speed and consequently the capacity and the supply pressure to decrease: the engine receives less air.

26.2.8.1 Contamination and damage to the air compressor

Obviously, this occurs less frequently. The air-inlet filter captures virtually all solid particles and the air temperature on the suction and delivery sides is relatively low. The diffuser, the stationary section of the air compressor that converts the high air velocity generated by the impeller to pressure, is usually in good condition.

26.2.8.2 Rotor-shaft bearings

The speed of extremely large chargers is high, approximately 4000 to 6000 rpm. Very small chargers for high-speed diesel engines can reach speeds from 60,000 to 120,000 rpm.
The pump impeller and the turbine wheel are attached to a rotating charger shaft. Correct balancing of this shaft is essential for the operating life of both bearings.
Therefore, the turbocharger is always dynamically balanced after turbocharger maintenance.
Both bearings are disassembled, cleaned and all the dimensions measured. The rotating parts are replaced.

▼

Appropriate and regular maintenance is essential for diesel-engine performance.

This air-inlet filter for a turbocharger is heavily contaminated. It impedes the air supply to the engine-.

◀

A complete rotor ready to be used.

▲

A contaminated exhaust-gas turbine reduces the capacity of the turbocharger.

▼

A cross-section of an ABB VTC 254 turbocharger. Note the large air filter with sound damping, to the left in the picture.

▲ Bearing casings are cleaned, inspected and measured, and then re-assembled and mounted with new bearings.

▲ All the moving parts of the turbine-rotor bearings, such as ball bearings, spring packings and bushes are always replaced by new parts with major turbocharger overhaul.

▼ Each rotor is dynamically balanced at approximately 15% of nominal speed. Subsequently as shown here, the other parts are assembled.

▼ After arrival at the reconditioning workshop, the turbochargers are taken apart, cleaned, and measured.

► A new compressor-impeller is heated and then shrunk to the compressor shaft. ABB in Rotterdam, The Netherlands.

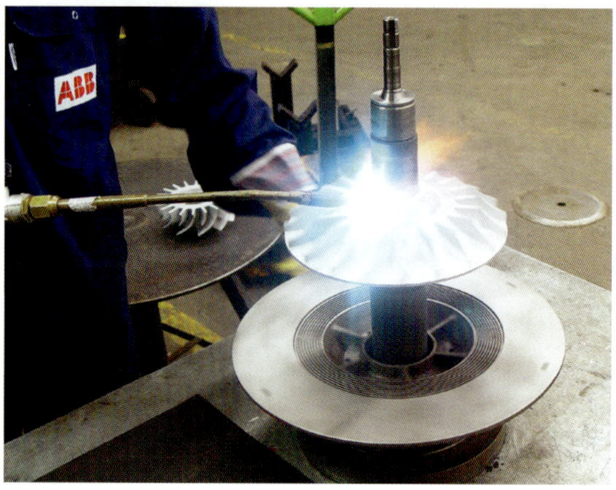

▶

A new design of a bearing-casing.

The connection between the outer ring, attached to the casing and the ball-bearing support ring is flexible. This is a self-aligning bearing.

◀

Exhaust-gas turbines for the larger turbochargers are equipped with individual blades; these are attached to the turbine-wheel rim using blade feet.

1 *turbine shaft*
2 *labyrinth seal*
3 *turbine wheel*
4 *turbine blade*
5 *lacing wire to dampen vibrations*

Casing

The materials used in the manufacture of the exhaust-gas section are exposed to exhaust-gas temperatures from 300 °C to 600 °C. Until recently, the exhaust-gas section was double-walled and large chargers were cooled with cooling water circulating through this section. This is still common practice in some new types of chargers and, naturally, in older types. **Corrosion and contamination** of the casing are the most frequently occurring problems. There is also the possibility of tearing at high temperatures, particularly of uncooled parts or turbine casings.

▲

Turbocharger casings after painting. TurboNed in Zwijndrecht, The Netherlands.

▲
A blade-attachment system for the exhaust-gas turbine on the turbine wheel. This is called the 'Christmas tree configuration'.

▲
Polishing the bearing position on a spray-coated built-up shaft section.

▶
Spare impellers and nozzle rings in a stock room.

1 *impellers*
2 *nozzle rings*

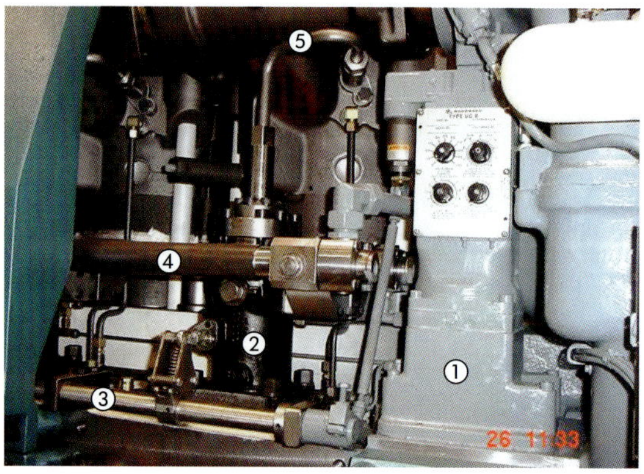

▶
A small but vital part of a diesel genset is the engine-speed governor.

1 *governor*
2 *fuel pump*
3 *fuel-adjusting spindle*
4 *supply-line for the fuel pump*
5 *delivery-line for the fuel pump*

26.2.9 Governors

Minor maintenance to governors is performed when they are mounted on the engine.
This maintenance comprises refilling or changing the lubricating oil, lubricating the shaft system to the fuel pumps and inspection of the governor during operation.
Reconditioning is performed by specialised companies, such as Woodward and Europa.
The governor is disassembled, cleaned and inspected. Worn parts are replaced and the governor is tested subsequent to assembly.

See chapter 15, *Speed control.*

◄

The governor is essential for speed adjustment.

A defective or poorly operating governor results in engine over-speed, which can cause engine damage. Proper maintenance and the correct settings are important. The mechanical connections between the governor and the fuel pump, a lever system, must be able to move freely.

1 *fuel-block pump*
2 *drive from the crankshaft*
3 *emergency stop*
4 *adjusting spindle for the fuel-block pump*

26.2.10 Engine blocks

The engine block or frame of an engine may vary with respect to construction and size from a small, compact, light-metal cast block to a large, heavy cast-iron block or a built-up frame such as those in large two-stroke crosshead engines.
Generally, the following damage occurs in engine blocks.
Seized pistons resulting in fracturing of the crank-connecting rod mechanism.
This often produces severe damage to the engine block, frequently leading to scrapping, especially with four-stroke engines.
The block is irreparable.

◄

A fractured crankshaft and engine block.

When a small dredger sunk, water was sucked into the engine above the pistons causing the crankshaft to break.

◄

A fractured cylinder liner due to a seized piston.

Insufficient lubrication of the piston is the cause here.

Cooling-water corrosion

This occurs in small engines with 'wet' cylinder liners as well as in larger engines. Most modern four-stroke medium-speed diesel engines have a **dry block**. Here the cooling water flows through the thickened upper rim of the cylinder liner protruding from the block.

▼

Cooling-water corrosion of the cylinder liner. This is pit corrosion, a localized and deep corrosion.

▶

A heavily oxidized cylinder liner. This is caused by inadequate cooling-water treatment.

Damaged alignment edges

These are often dented or corroded if two engine parts micro-move against one another, such as cylinder liners in the engine block. In blocks without protruding cylinder liners, this may also be the case with a cylinder head and engine block.
The supporting edge of the frame, for instance, may be damaged at the same height as the cylinder liners.

This occurs regularly in large built-up frames for two-stroke crosshead engines, especially near the bolted connections.

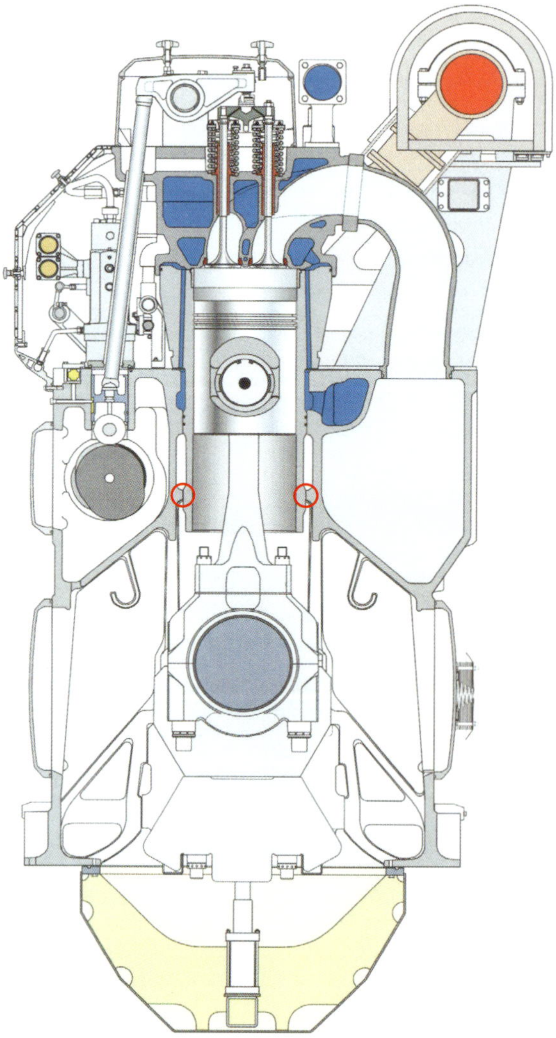

▶

A cross-section of a Wärtsilä 46 diesel engine, category III.

○ *Supports*

◀
The crankshaft must be aligned to prevent bending or fracturing of the shaft. The shaft centre-line is constantly checked, from casting to assembly in the diesel engine.

Misalignment of the crankshaft and bearings

After a certain time, the crankshaft may no longer be aligned due to deformation of the engine block. This may be because a seized part, such as, a crankshaft that seized due to an overheated bearing. Furthermore, problems related to the attachment of the engine block to the bedplate or the bedplate itself may cause the misalignment of the frame and consequently the crankshaft.

Tears in the cast frame in four-stroke engines

This occurs at excessively high material temperatures or with mechanical overloading of the material. Rapid temperature fluctuations and exceeding the maximum operating temperatures may produce tearing in a cast block.

Problem solving

A damaged cast-iron block can sometimes be repaired, provided no essential parts have been damaged. This is referred to as 'meta-locking', a method where the 'fragments' are put back together with stainless steel clips.
Corrosion of the block is indicative of inadequate cooling-water treatment.

If the corrosion is superficial and shallow, cleaning and intensive cooling-water treatment is sufficient. Many cooling-water treatment methods leave a film on the cooled part, therefore protecting the parts from, for instance, oxygen found in the cooling water. The oxygen is chemically bound by additives in the treatment of cooling-water.

See Chapter 10, *Cooling diesel engines.*

Damaged alignment edges in the block can be trued up by machining processes such as boring, milling, grinding and sanding. In large engines, the alignment edges in blocks can be trued up in

▼
'Meta-lock', a method of putting fragments together using stainless-steel clips.

▶
Special attention must be given to the position of the centre line of the crankshaft. This is valid for new engines, inspection during operation and during major engine repairs.

▲
Trimming the top alignment edges of an engine block with a reaming machine.

situ with special equipment. When performing machining operations, dimensions are altered due to the removal of material. At present, alignment edges of older parts are repaired with a synthetic resin.

In smaller engine blocks, misalignment of the crankshaft may be fixed with line boring: the block is machined with line-boring equipment in a specialized workshop or in situ. During this process, material is removed from the block and the mounted lower bearing cap to the next undersize, a standard dimension for larger bearing shells.

Tearing of the cast frame

Generally, this cannot be repaired. In cases of serious tearing, scrapping of the block must be considered. Occasionally the start of the tear is 'drilled' to prevent further tearing. This is done by drilling a hole in the direction of the crack.

▼
Line boring. The cylinder block has been turned 180°.

1 *engine frame*
2 *bearing cap*
3 *pre-tensioned bearing-cap bolts*
4 *driving shaft of line-boring machine*
5 *cutting section with small chisel near the bearing cap, barely visible*

▼
An installation that allows mounted bearing caps to be made in an oversize so that the oversized bearing shells, and ultimately the crankshaft can be aligned.

1 *engine block*
2 *mounted bearing caps*
3 *line-boring tool*
4 *drive*

◄ Reconditioning a connecting rod. Line boring the connecting-rod big end/crankpin (in the background) and (in the foreground), the line boring of the connecting-rod head/piston pin.

26.2.11 Connecting rods

A common problem is oval shaped crankpin bearings.

This excessive ovality may cause damage to the crankshaft.

Inspection

- The connecting rod is cleaned.
- Visual inspection of the parts.
- Hair-crack test.
- Thereafter, the connecting rod with cap is mounted and the bolts are tightened to the normal starting tension.
- Then the crankpin-bearing bore is measured in various positions.
- The pitch and position of the teeth are measured.
- The top angle of the teeth is found.
- The position of the locating piston pin is measured.
- The straightness of the connecting rod is checked.
- The distance between the centre line of the piston pin and crankpin is measured.
- The total weight of the rod is measured for balancing purposes.

There are two repair methods

Connecting rod without cracks

- The teeth are skimmed on a milling machine.
- Both toothed sections of the connecting-rod parts are fitted again and mating faces checked with blue paste, bolts are tightened in accordance to the engine-manufacturers' instructions.
- Subsequently, the crankpin-bearing bore is machined to its original dimensions using a rotating chisel.
- The complete connecting rod is labelled and lubricated to prevent corrosion.

Connecting rod with cracks

- The cracked serrations are machined until the cracks have disappeared.
- After pre-heating the connecting rod in an oven, the machined section is rebuilt with an alloy certified by the manufacturer and according to a strict welding procedure. The rod is then immediately stress-relieved by gradually cooling. The final process is that all teeth and bores will be machined to their original dimensions

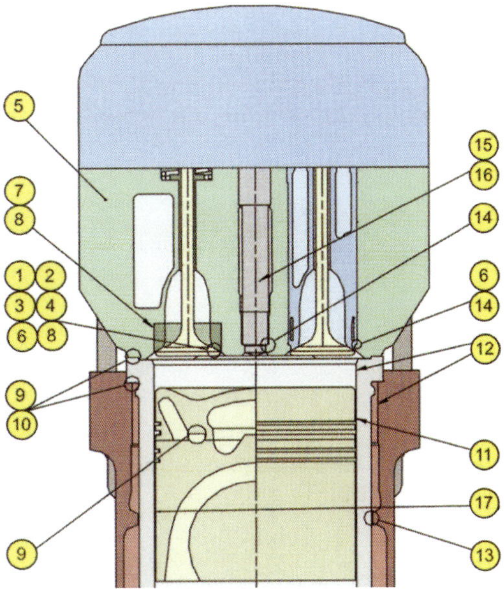

◄ All the faces that require inspection and possibly trueing up during reconditioning a four-stroke engine.

▶
Two-stroke crosshead engines have various parts composed of steel or steel alloys, for instance, the A-frame, scavenging-air space, drive gearing, piston, cylinder cover and exhaust valve.
Many of these parts can be reconditioned by welding.

▲
Top right: a disassembled cylinder cover (1) is cleaned for inspection.

26.3 Two-stroke engines

General

The pistons and the cylinder covers are manufactured from forged steel or cast steel, which can be welded. Therefore, the surface can be rewelded after removal of the damaged material. This is also valid for the exhaust valve.

26.3.1 Cylinder covers

The section of the cylinder cover, which forms the top of the combustion chamber, is heavily loaded thermally. Crack formation is often a result of this thermal loading.

▶
Reconditioned parts are supplied with a certain code, for tracing the reconditioning company.

Visible cracks are often found in the section adjacent to the central exhaust valve.
The exhaust-valve casing containing the valve seat in the cylinder cover is also heavily loaded. The seat is frequently damaged to the extent that the entire valve seat requires replacing.

▼
Several cracks in an older type cylinder cover with inlet- and exhaust ports in a crosshead engine.

The injector bore is positioned in the centre of the cylinder, surrounded by large starting-air and safety valve bores. This cylinder cover can be reconditioned.

▼
Cracks in a cylinder cover. These are known as thermal cracks.

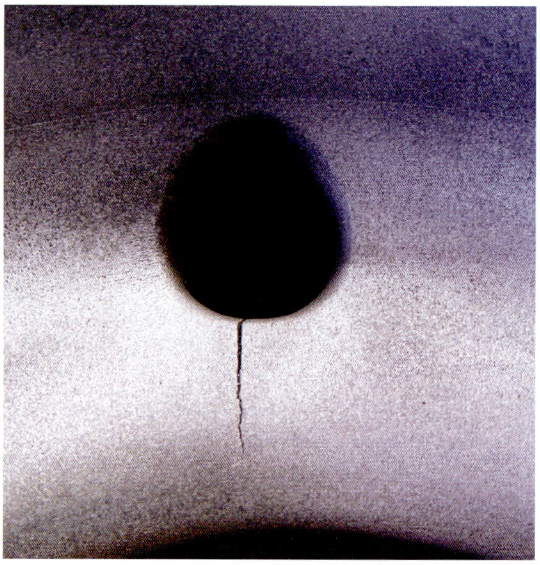

▲

Machining a cylinder cover.

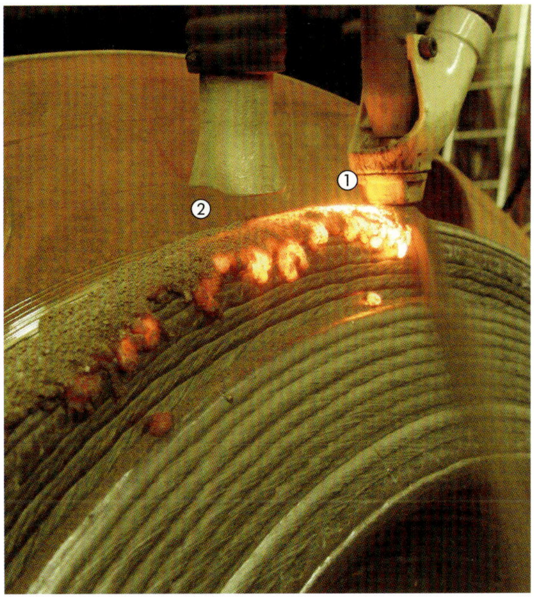

▲

Rewelding a cylinder cover by means of fully automatic arc welding. Note: the edge closest to the exhaust valve has been damaged.

1 *welds*
2 *powder suction*

Procedure cylinder cover

After cleaning in a washing machine, the cover is inspected for:

- hair-cracks in the thermally loaded parts;
- irregularities of the alignment edge with the cylinder liner;
- contamination and cracking in the cooling-water ducts;
- impurities in: the sealing faces of the exhaust-valve casing, injectors, starting-air valve, the safety valve and the indicator cock.

After careful analysis of the above, the decision is made whether it is technically possible and economically viable to recondition the cylinder cover.

Reconditioning

- **Machine** all the cracks by grinding until they have disappeared.
- **Weld the cracks according to the 'Plasma Transfer Arc'** process.
- **Annealing** in an oven at a temperature of approximately 620 °C for 24 hours.
- **Polishing or metal removal** of all welding irregularities to the original dimensions.
- **Trueing up all alignment edges** of the various parts.
- **Registering all data** and numbering all the parts of the reconditioned cylinder cover.
- **Anti-corrosion treatment** and preparation for transport.

▶

Welding methods for various types of cylinder covers.

The hatched or black section of the cylinder head is first removed and then rewelded and machined so that the original size is restored.

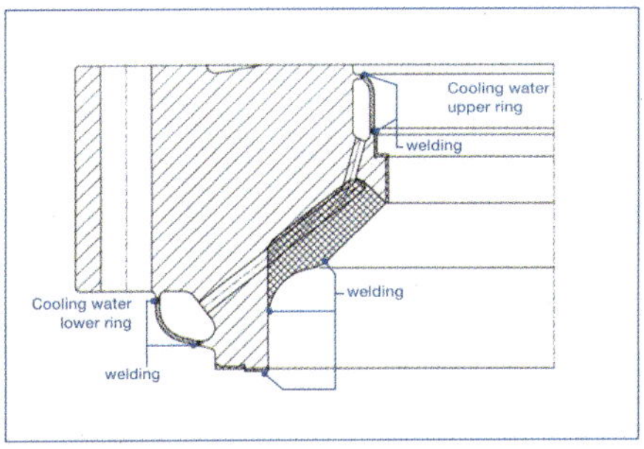

Cylinder cover B&W L GBE

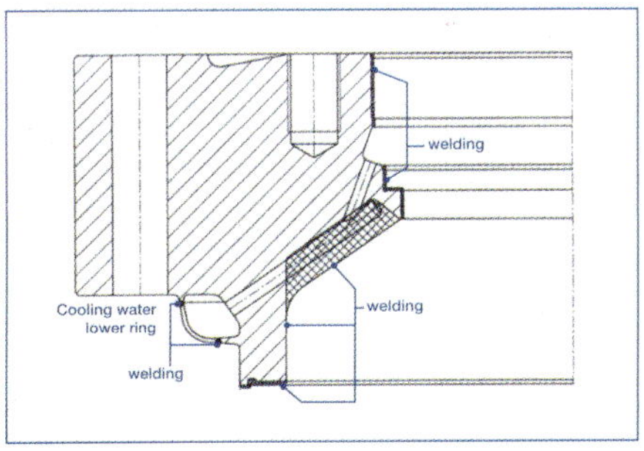

Cylinder cover B&W L MC

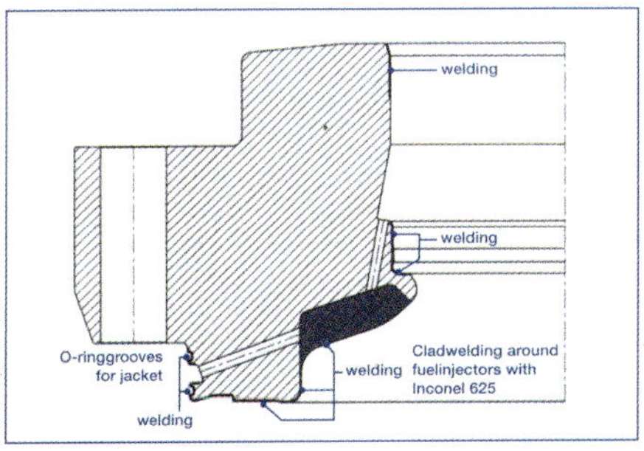

Cylinder cover Sulzer RTA standard repair method

▶
With a large gas flame, the slowly turning cylinder cover is kept warm for further machining. This is important for adequate adhesion.

▶
After the piston has been drawn, it is cleaned, inspected and measured.

When the piston is severely damaged, it cannot be replaced and will be reconditioned on shore

1 *piston with ring package*
2 *piston rod*
3 *spare cylinder liner*

◀
Finishing a reconditioned cylinder cover, grinding the faces, trueing up the bores and threaded holes followed by a visual inspection. Finally, if so requested by the customer, the various parts are assembled.
Van West Holland, IJmuiden, The Netherlands.

26.3.2 Pistons and piston rings

▼

New pistons with piston rods and the piston-rod stuffing box, ready to be assembled in a crosshead engine.

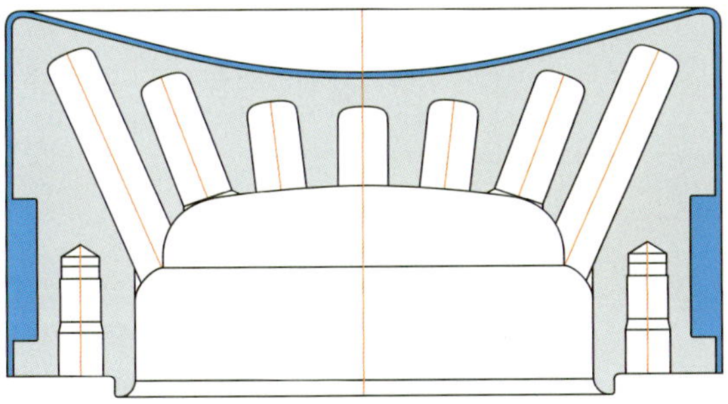

▲

If the piston-ring grooves are worn to the extent that they must be scrapped, rewelding is an option.

The piston-ring grooves and rims are removed, and then the space is filled by arc welding. The piston-ring grooves are restored to their original dimensions.

Blue = rewelding

▲

Piston crowns for a crosshead engine for a MAN Diesel MC and a Wärtsilä Sulzer RTA.

Notice the different cooling arrangements. After cleaning, they are inspected and a reconditioning report is drawn up.

▼

A piston, which has been inadequately cooled by the lubricating oil circulating in the piston.

The entire centre section of the bottom of the piston has torn to such an extent that a conical shaped 'cover' has been formed, only the overpressure above the piston kept it in its place.

During inspection at the reconditioning company, one tap of the hammer was sufficient to remove the 'cover'!

▼

The piston crown for a two-stroke crosshead engine.

This damage was caused by a leaking fuel injector. The piston bottom has been torn in a star shape.

Piston crowns

As a result of high mechanical- and thermal loading, the piston crowns containing the piston rings require replacement or reconditioning after a certain number of operating hours.

►

A disassembled piston prior to cleaning.

The carbon deposit above the piston-ring package is clearly visible. All the piston rings are intact and able to move.

►►

Arc welding the bottom of a large piston. The side has been removed and will be rewelded. Subsequently, new piston-ring grooves are machined.

►

A piston after a large number of operating hours.

The piston crown shows signs of carbon deposits. The piston rings are in good condition and can move freely. There is no carbon build-up behind the piston rings.

▲

The rewelding is finished. Now the piston is machined to its original size on a lathe.

The following is often established:

- **Internal crack formation.** These cracks must be located, removed, and rewelded.
- **Cracks in the piston crown.** These may be rebuilt using a special rewelding process.

►

Arc welding the bottom of a large piston.

This is an automatic process.

◄

The piston crown has been deformed to such an extent during rewelding that the connecting-bolt holes are no longer aligned.

This is the reason that they are filled in when the piston is rewelded and then re-bored.

▲

Drilled cooling holes in a piston crown that with tears, marked in chalk.

▲

Deep craters formed by corrosion of the piston material.

▼

A piston crown showing signs of burning where fuel injection occurs.

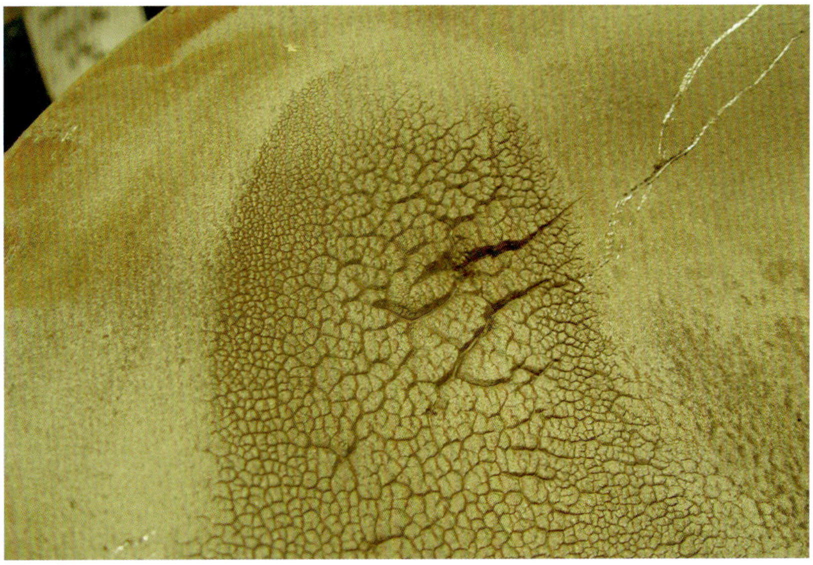

- **Crown burning.** All damage and cracks must be removed; the crown must subsequently be rebuilt to the original height using a special rewelding process. A special protection coating may be applied if required.
- **Severe crown burning and cracking.** All cracks deeper than 10 millimetres must be located and removed, the crown must be rebuilt to its original height and profile. In the latest types of crosshead engines, crown burning can be so severe that it must be replaced.

▲
Measuring piston-crown burn with a piston-crown mould.

▲
Reconditioned pistons for two-stroke crosshead engines.

The piston-ring grooves are usually worn to such an extent that the complete surface must be removed, rewelded and machined. As the bottom of the piston is usually partially burnt away, this is also removed, rewelded and machined. This gives a completely rewelded piston!

▼
A piston crown and measuring tool used to measure the extent of crown burn.

A for a drawn piston
B for a piston in the engine, the measuring mould is suspended from the inlet ports

A

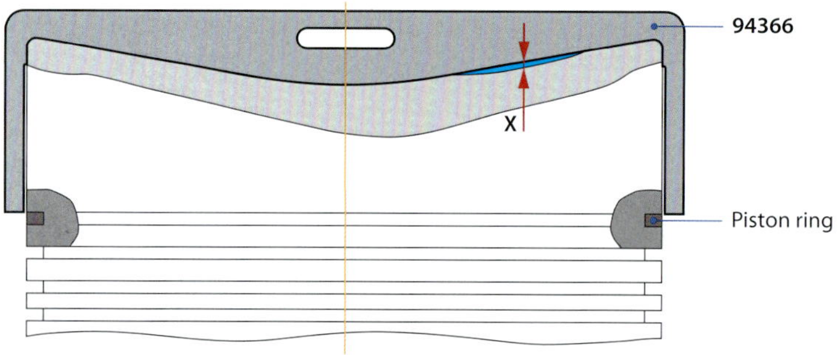

B

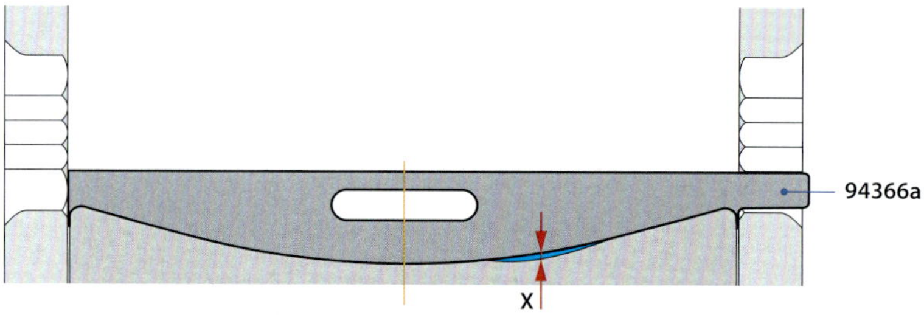

▼
After arc welding, the piston is machined to its original dimensions.

▲
Cylinder covers ready for repair at Wärtsilä, Kruiningen, The Netherlands. The covers are manufactured from forged steel and almost solid.

26.3.3 Cylinder liners

Obviously, cast-iron cylinder liners wear out during usage (see four-stroke engines). Inlet ports are fitted to the bottom of the cylinder liner. All the large two-stroke crosshead engines have cylinder-lubrication drillings in the liner wall and lubricating-oil grooves in the carrying surface.
The upper thickened rim has drillings for cooling-water circulation.
Conditions in the combustion chamber cause considerable wear and tear of the cylinder liners.

Key points

26.3.3.1 Wear of the carrying surface

- Normal frictional wear of the piston rings. Boundary lubrication is often inevitable.
- Abrasive wear by hard carbon and ash particles.
- Corrosion, especially from sulphur found in fuel. The latter is referred to as low-temperature corrosion, (LTC).

At present, the sulphur content of HFO lies between 2 and 3.5%.

See Chapter 8. *Fuels, fuel-line systems and cleaning fuels.*

▼

Modern cylinder liners with inlet ports.

- Glazing of the carrying surface because of 'polishing' by abrasive particles, which are produced from wear and during the combustion process.
- Contamination of cylinder-lubricating oil supply ducts. This mainly occurs due to problems with cylinder-lubricating devices or an inferior quality of the cylinder-lubricating oil.
- Contamination of the cooling-water drillings, and/or spaces, and corrosion of the ducts.
- Damage to the alignment edges of the cylinder block with the cylinder head.

▲

A spare cylinder liner.

◄

The cylinder-lubrication distribution on the circumference of the cylinder liner.

26.3.3.2 Procedures

After cleaning the cylinder liner, its geometry is measured.

▲

Measuring cylinder-liner wear.

The diameter of the cylinder liner is measured using a calliper gauge. This is done side to side at various heights. One person takes the measurements and a second person records all the values. These are compared with the original values of the liner. A decision is then taken as to whether the liner can remain in place until the next overhaul or if the liner must be replaced.

If a dimension is outside the prescribed limits or if the wear and tear is such that after replacement of the liner, the number of remaining operating hours is too small, the liner is scrapped.
If the dimensions are within the prescribed limits, the following procedure is performed:

- **rehoning** of the carrying surface. Firstly, the rim of the liner is removed by grinding;
- **cleaning** of the cooling-water ducts in the thickened upper rim;
- **machining** the liner-alignment edges with the block and cylinder head;
- **trueing** up the cylinder-lubricating oil bores.

26.3.4 Crankshafts and bearings

Large two-stroke crosshead engines have heavy, bulky crankshafts.
A crankshaft from a Wärtsilä Sulzer FLEX, a fourteen-cylinder diesel engine with a bore of 960 millimetres and a stroke of 2500 millimetres is 27 metres long and weighs 155 tons.
Damage to the crankshaft for this type of engine requires extensive and expensive repairs, which can take several months. The entire engine frame must be lifted hydraulically and the scrapped crankshaft removed through a hole in the hull.

▼

A large damaged crankshaft is removed from the engine and replaced by a new crankshaft. This procedure takes more than two months.

▼

A crosshead engine with a cylinder diameter of 500 millimetres, a MAN Diesel, in manufacture.

Note the large diameter of the journals and the very small distance between the crank webs. In order to reduce weight, the crankpin has been hollowed out.

1 *crankpin*
2 *crankshaft journal*
3 *web*

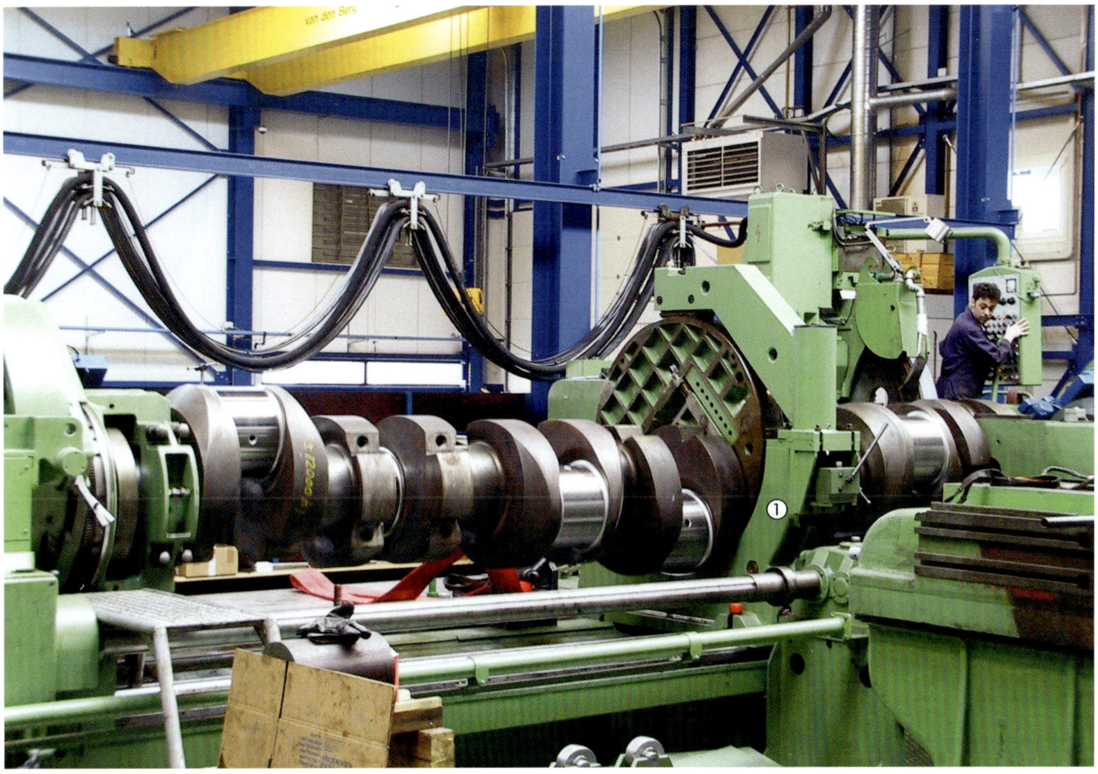

◄
A very large crankshaft lathe.

A very large crankshaft lathe that has a distance of 12,000 millimetres between the centres, large enough to grind nearly all the crankshafts for four-stroke engines. Mark van Schaick, Schiedam, The Netherlands. Note the large supports (chucks), which prevent the crankshaft from sagging.

1 *chuck*

The new crankshaft must be carefully hoisted into the engine room and positioned in the 'crankshaft bed'.

For damaged crankpins and journals, an in-place technological approach is often used: the in-place rounding and finishing by grinding the journals. This is done by various companies worldwide. The damaged crankshaft is stationary while the grinding device is in operation. Removal of several tenths of millimetres of material can take more than half a day.

◄
A new method for reconditioning a crankpin in-situ using a manufacturing machine from the company, Goltens in Spijkenisse, The Netherlands. This is known as 'in-situ' reconditioning of crankpins.

1 crankpin dummy
2 manufacturing machine with:
   - guides on the crankpin radius
   - chisel holder with cutting tool
   - adjustment
3 manufacturing machine is chain-driven by an electromotor

Operation: The machine runs on the radius between the crankpin and the crank web. After trueing up in relation to the heart line of the crankpin, the crankpin is machined at high-speed. 5 to 15 mm of material is removed in three to four days. Using the older method, this could take several weeks! Another advantage is that there is less fine residue in the crankcase. The clamps for the chisel are relatively easy to remove. In this procedure, the crankshaft is stationary and the machine rotates around the crankpin.

▶
In the crankcase, also known as 'carter', of a large crosshead engine.

1 *the A-frame columns*
2 *crankshaft*
3 *crank web*
4 *connecting-rod big end*
5 *main bearing cap*
6 *bearing-cap thrust bolt*
7 *steps towards the crosshead*
8 *main lubricating-oil supply to the main bearing*
9 *crosshead*
10 *crosshead guide*
11 *hydraulic jack*

▲
The hydraulically tensioned cams on the camshaft.

▲
Manual finish of a camshaft bearing.

▶
Modern exhaust valves often have a long operating life. The use of material such as 'Nimonac' results in an operating life of 40,000 hours or more.

▲
This exhaust-valve seat has been partially worn away by high-temperature corrosion.

The exhaust-valve casing containing the valve seat, situated in the cylinder head is also heavily loaded. The seat is often damaged to the extent that the entire valve seat must be replaced.

▲

Exhaust valves during reconditioning.

▲

A cylinder cover. In the middle, place for the central exhaust valve.

front left: the bore for one of the two injectors

It is important that all the sealing faces close properly. They are usually machined with special equipment, which can be delivered with the engine.

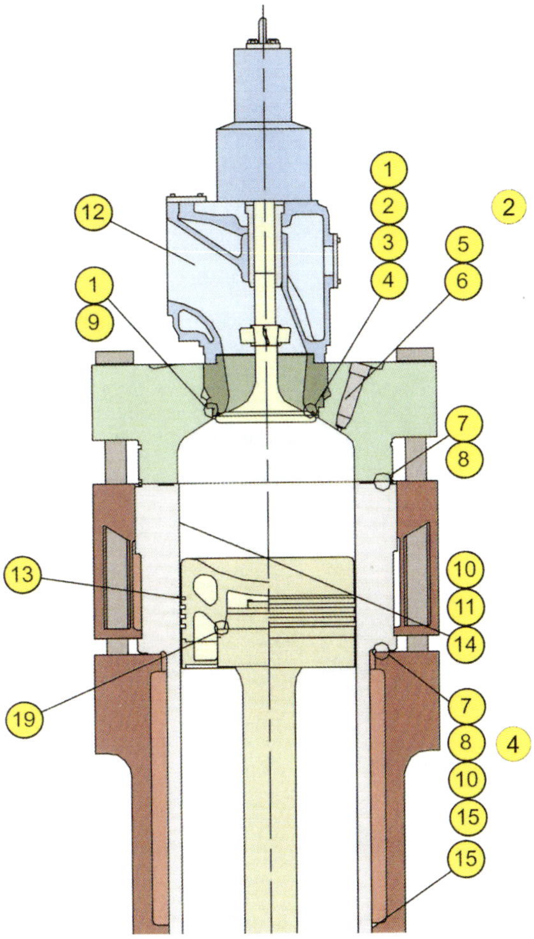

◀

All the sealing faces are checked and, if required, trued up during reconditioning of a two-stroke crosshead engine.

Welded A-frames

A-frames manufactured from steel plate are light and strong. However, they can sometimes cause cracking. This occasionally occurs in engines from all of the three engine manufacturers.

The power play in these large, up to 32-metre engines, is enormous. Usually, the dimensions are measured in trials to investigate the behaviour of these welded constructions.

Therefore, regular checks for cracks are imperative.

Most problems can be solved by grinding away the cracks and rewelding.

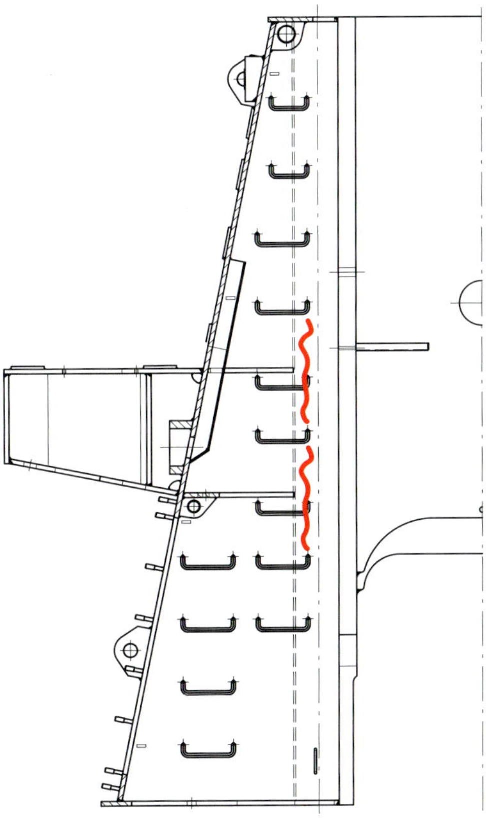

◀

Tears in the welded A-frame in a two-stroke crosshead engine.

Tears can also occur in welded A-frames for two-stroke crosshead engines due to high material stresses. These problems can often be resolved on location to the client's satisfaction. Long-term testing is often too costly and time-consuming. Most bugs are solved in the 'field', so at the customer.

26.3.5 Examples of damage to four-stroke- and two-stroke engines

▶
A dirty turbocharger, exhaust-gas section.

Clearly, the air capacity is enormously reduced and the engine contaminated.

▶
A seized piston in the cylinder liner.

▶▶
A torn piston bottom in a two-stroke crosshead engine.

▶
An attempt to weld a cooling-water jacket.

In this case, the cooling-water jacket should be replaced as quickly as possible.

◄◄
An overheated and distorted cam.

◄
A large exhaust valve for a two-stroke crosshead engine.

Fragments from the broken valve can cause enormous damage to the combustion space and in the turbocharger.

▲
A split piston.

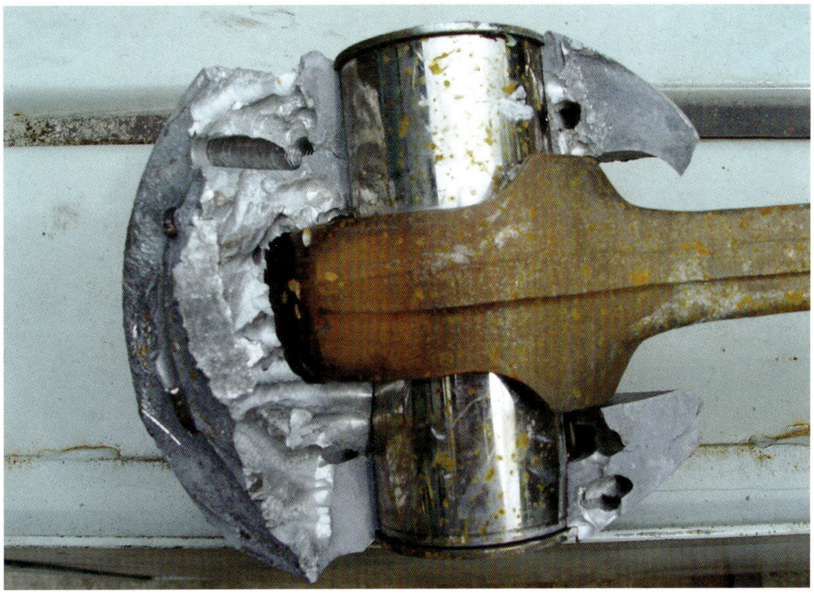

▲
A damaged piston with connecting rod.

◄
A damaged engine block.

The cast iron block is severely damaged. The engine block can be reconditioned, if the block has not torn and new cylinder liners can be inserted.

▶

A damaged crankshaft.

The damage is so severe that it will be scrapped. Note the blue mark on the crank pin; this is the most probable position where was overheated.

▶▼

Line-boring the crankshaft bearings in an engine block.

▶

A 'folded' connecting-rod big end caused by poor lubrication.

The discolouration and distortion shows that extremely high material temperatures were reached most probably caused by poor lubrication of the crank pin.

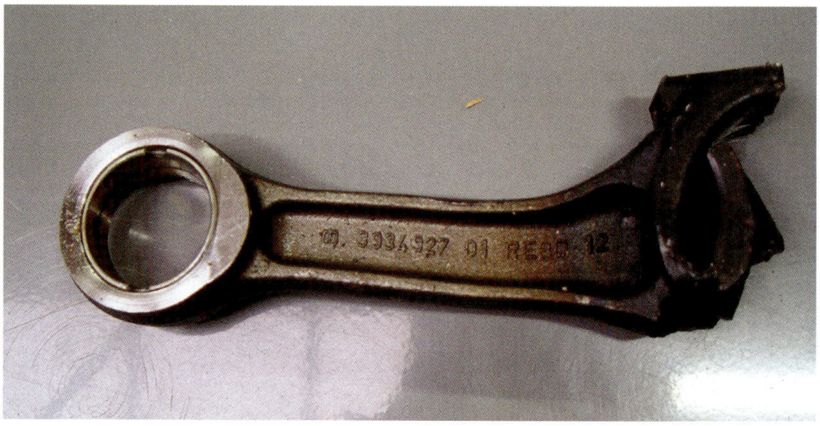

▶

A broken piston; these fractures usually occurs at the drillings through the piston pin.

◄◄

A connecting rod has come loose and perforated the block.

◄

Damaged blades in the exhaust-gas section of a turbocharger.

◄

A corrosion hole in the cylinder liner, viewed from the cylinder.

▲

A hole in the casing of the exhaust-gas section in a turbocharger. Today, turbochargers are manufactured in such a manner that the fragments remain inside the casing.

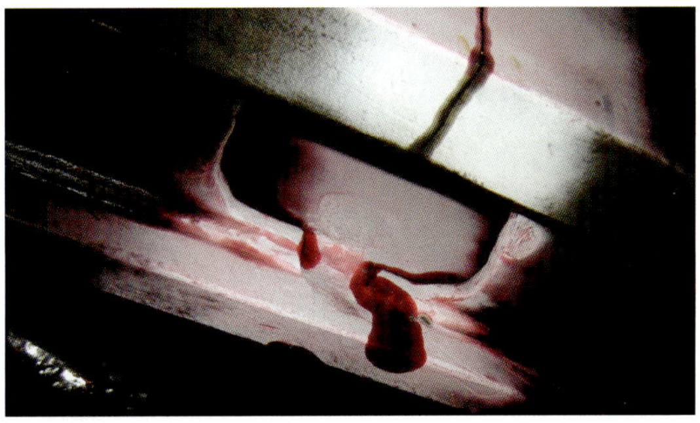

◄

Cracking in various places in a welded steel A-frame for a two-stroke crosshead engine.

In this case, the engine is operational. If the cracks become larger, they will often be repaired.

▶

Tearing on the inside of the piston crown in a two-stroke crosshead engine.

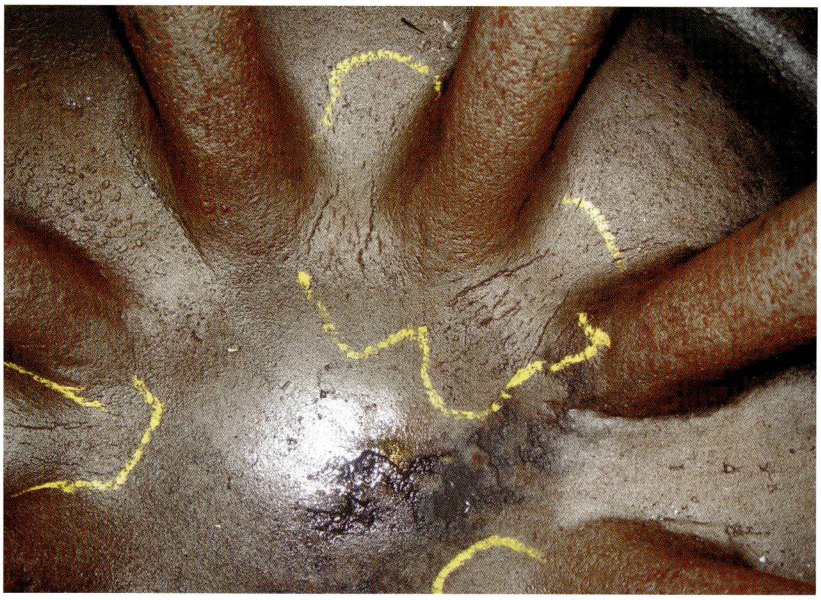

▲

A heavily worn crosshead-pin bearing in a two-stroke crosshead engine.

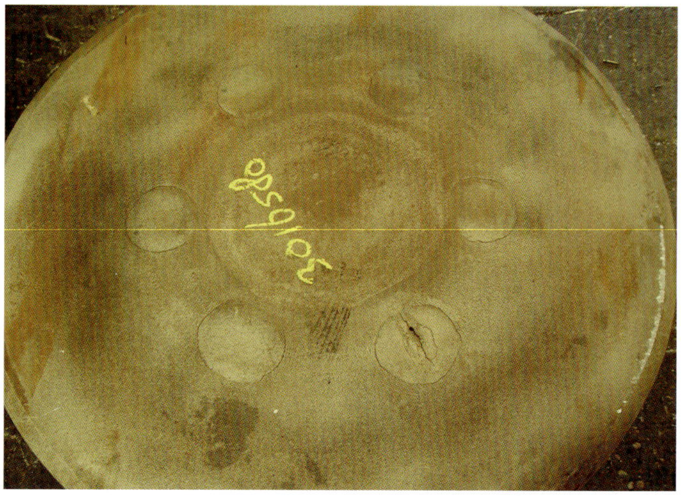

▲

Tearing at the top of the piston crown in the same type of two-stroke crosshead engine.

▶

A heavily damaged crankshaft caused by a 'loose' connecting rod. This crankshaft cannot be repaired.

◀
Tearing in the cooling ducts on the inside of the piston crown in a two-stroke crosshead engine.

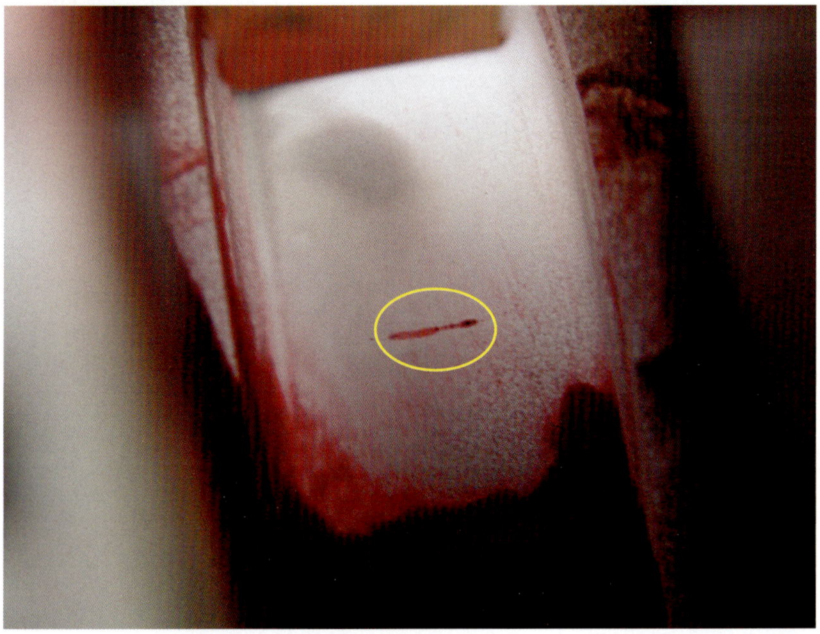

◀
A visible hairline crack in a crankpin seen during inspection.

The depth of the fracture can be determined by grinding down the surface around the hairline crack. This crankshaft will be scrapped if the crack is too deep for the largest (thickest) bearing shell.

◀
A damaged cylinder liner. The piston has not yet seized!

▶

Measuring cylinder-liner wear.

The diameter of the cylinder liner is measured using a calliper gauge. This is done side to side at various heights. One person takes the measurements and a second person records all the values. These are compared to the original values of the liner. A decision is then taken as to whether the liner can remain in place until the next overhaul or if the liner must be replaced.

▶

A cracked block in a four-stroke engine.

When the cracks in the upper edge of the cylinder block were detected, the cast iron was ground down to such an extent that the crack disappeared. Eventually, this block was scrapped.

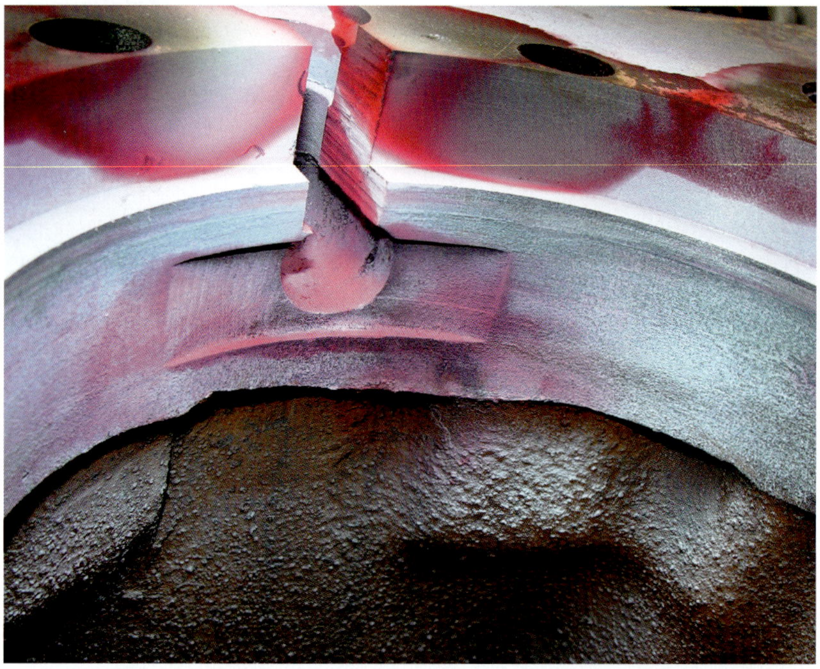

▲

A broken crankshaft.

The crankpin broke near the drilling for the lubricating-oil. Probable cause: An extremely high load on the crankshaft caused by a seized piston.

▶

A gudgeon pin in a two-stroke crosshead engine.

Insufficient lubrication has damaged the pin. Measuring the hardness can show whether the pin had overheated, a pin subjected to extreme heat shows increased hardness.

◄

A Rottler machine for trueing up valve seats.

This machine is installed at Noord-Hollandse Motoren Revisie Company, Winkel, The Netherlands. It can machine worn seats quickly and effectively.

◄

This exhaust valve for a two-stroke engine is badly damaged and has been scrapped.

◄

Over-all damage in a medium-speed four-stroke diesel engine.

After a piston seized, the connecting rod with gudgeon pin broke away from the piston and damaged the engine block to such an extent that it had to be scrapped. The connecting rod is still connected to the gudgeon pin.

▶
Repairing a fuel pump for a Caterpillar diesel engine at Pon Power in Papendrecht, The Netherlands. The correct measurements and settings are important.

▲
The process.

▶
A Rottler machine for line-boring engine blocks for high-speed four-stroke diesel engines from Caterpillar and Cummins.

Line boring is far less time-consuming than traditional processes. Shown here, at the premises of Mark van Schaick in Schiedam, The Netherlands.

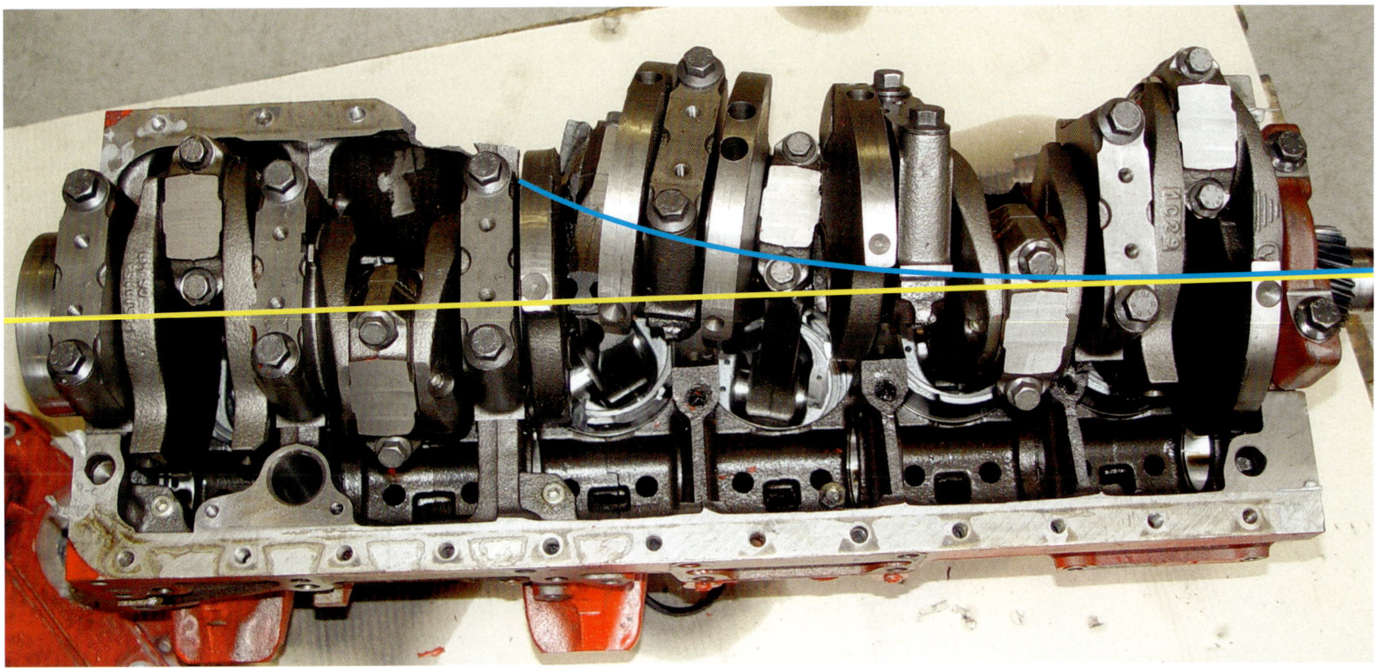

centre line normal crankshaft

centre line broken crankshaft

▲

A broken crankshaft in a damaged block.

This diesel engine drove a water pump on a dredger. The engine most probably suddenly drew in a large amount of water.

◀

Machining a light-metal block for a small diesel engine.

Noord Hollandse Revisie, Winkel, The Netherlands.

▼

A broken exhaust valve caused severe damage to the flame plates in a cylinder head.

▶

A fractured exhaust valve has pierced this piston in a four-stroke engine.

▶

The valve seat in an exhaust valve for a two-stroke crosshead engine.

This seat is severely heat damaged. The decision as to repair or scrap will be taken after testing for cracking.

▶

Arc welding visualised.

Right, the gas supply and the electrode. Left, the powder suction. The slag to the left is manually removed with a hammer.

1 *Electrode wire supply*
2 *Gas supply*
3 *Suction of the powder sealant.*
4 *Hard slag layer.*

Shown here, a piston crown is being welded.

◀

An oven used to anneal the welded parts. Shown here; two piston crowns.

After heating, the welded parts are allowed to cool gradually. By gradual cooling, the stresses introduced in the welding process are mostly relaxed. Wärtsilä, Kruiningen, The Netherlands.

◀

Testing a vibration damper for a diesel engine, category III.

The vibration damper is mounted on a revolving table driven at a certain speed. A measuring instrument is connected to the table that can measure the vibrations in the shaft. The vibration damper is then hit with a hammer; the measuring instrument shows the gradation of damping of the vibration damper.
Koedood Dieselservice, Hendrik Ido Ambacht, The Netherlands.

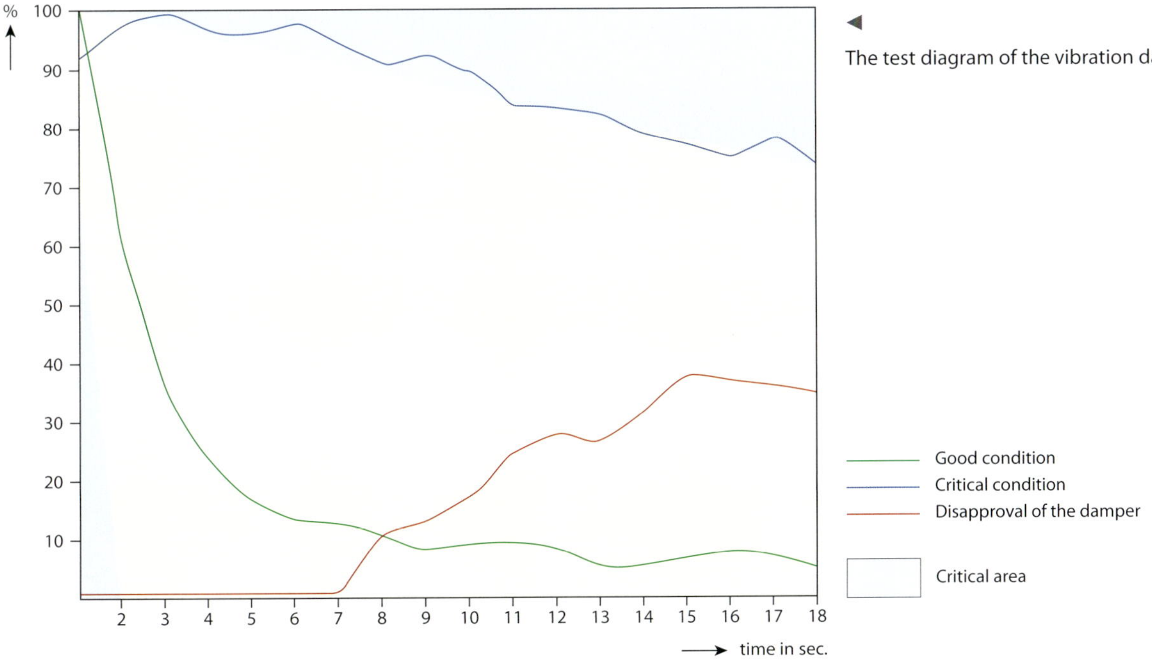

◀

The test diagram of the vibration damper.

>CH 27

Maintenance and repairs

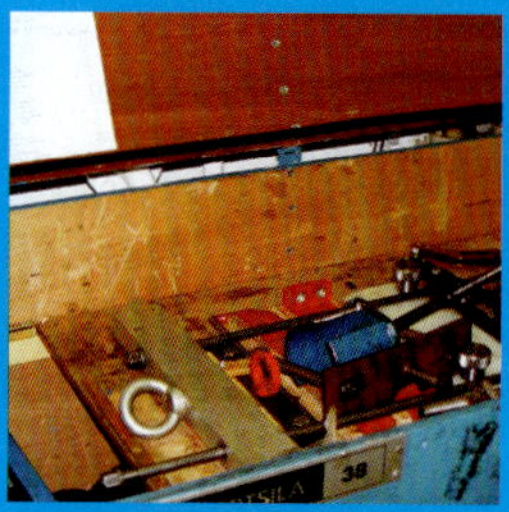

Expertise is required for the repair and maintenance of modern, highly charged diesel engines.
As shown here, a good inventory-management system, the correct tools and especially a competent and enthusiastic crew are required to keep the propulsion installation of the 'Oranjeborg' from Wagenborg Shipping in good condition.

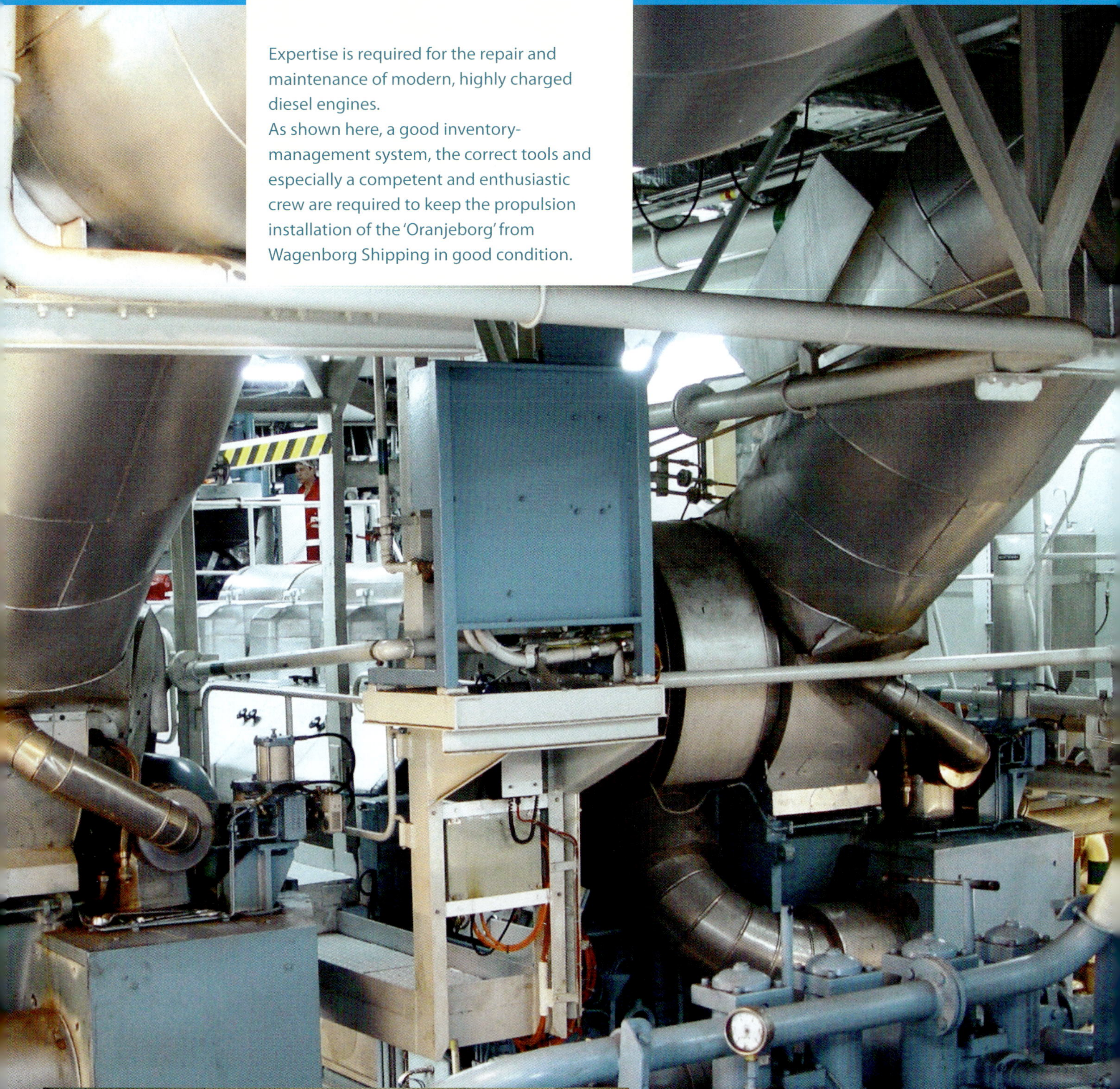

The engine-control room with left, the main switchboard and in the middle, amongst others, the remote control and monitoring of the main engines.

27.1 Introduction

Engines in operation wear and require maintenance to remain in good condition.
Obviously, engines that are faulty or poorly maintained cause problems.
Diesel engines for shipping and electrical-power stations have to be in an excellent condition. They must be capable of running at full load for most of the year.
Nowadays, ships have short periods in dock; there is barely time to perform major maintenance. If required, engine-repair and -maintenance companies are hired that either with or without the (now small) crew, perform the activities.
It is of utmost importance that engine manufacturers maintain a worldwide service network so that they can provide a good service standard for their clients. Quick delivery of spare parts is also essential.
Major ship maintenance is scheduled, if possible, to coincide with the planned dockings.
In certain circumstances, more time is available for the maintenance of diesel-generated power stations, but here there is demand for electricity at a full-load and ever-increasing operating hours.
The mean percentage for the operating hours in a good quality power plant at full-load is between 90 and 95%. Sometimes higher: per year between 7500 and 8500 operating hours! In addition, the maintenance must be first-rate and well organised.

27.2 Types of maintenance

There are in general terms three types of maintenance.

1 **Preventative maintenance**
This is planned maintenance aimed at preventing breakdowns and failures. Often, parts that are not worn or in need of replacement are exchanged in accordance with a schedule.

2 **Corrective maintenance**
This is maintenance performed to restore an item to its working order when it has failed or worn out.

3 **Condition-based maintenance**
This is maintenance based on the state-of-repair at a certain time, combined with operational experience. To prevent failures, the spare parts are removed, repaired or replaced after a predetermined number of operating hours. At times, it can become apparent that the number of operating hours can either be increased or decreased. This is a matter of experience.
Engine suppliers often provide spare-part lists and information regarding the maximum number of operating hours. The prescribed number of operating hours is often chosen so maintenance activities for different items can take place at the same time.

27.3 Instruction manuals/ Maintenance manuals

These must be thoroughly read in order to have a good understanding of the operation, maintenance, repairs, and settings of diesel engines. **All too often instruction manuals are not consulted before starting various activities.**
Often, the smallest details are decisive in executing successful maintenance.

It is advisable to follow the procedure below.

1 Read the instruction manuals and maintenance procedures carefully. This includes the drawings.
2 Find out which special tools are required and collect them.
3 Determine the required period for maintenance.
4 Discuss the activities with workers and exchange information.
5 Ensure the working environment is clean and well lit.
6 Use approved hoisting equipment and well-maintained tools.
7 Check the oil in the hydraulic tensioning equipment and ensure that the various hoses are not damaged. Make sure that the couplings are clean.
8 Avoid getting dirt in the engine when open.
9 Store all disassembled spare parts neatly; preferably on white cardboard, use clean cloths and place heavy parts on wood, never on steel. In the latter case, cylinder-head covers may be damaged.
10 Hoist heavy parts whenever possible; do not lift them. **Be careful with your back!**
11 Most engines have special attachments to facilitate disassembly/assembly, use and keep them in good condition.
12 Work methodically and do not rush. In the latter case, all too often mistakes are made.
13 Discuss problems with a colleague: two heads are better than one.
14 Modern communication allows easy access to the service organisation of engine suppliers. Use them for advice and consultation; after all, they are there for you!

	HFO	MDF	HFO	MDF
	Time between overhauls (h)	Time between overhauls (h)	Expected comp. lifetime (h)	Expected comp. lifetime (h)
Main bearing	12000	16000	36000	48000
Big end bearing	12000	16000	24000	32000
Gudgeon pin bearing	12000	16000	48000	48000
Camshaft bearing bush	16000	16000	32000	32000
Camshaft intermed. gear bearing	16000	16000	32000	32000
Balancing shaft bearing, 4L20	12000	16000	24000	32000
Cylinder head	12000	16000		
Inlet valve	12000	16000	36000	32000
Inlet valve seat	12000	16000	36000	32000
Exhaust valve	12000	16000	24000	32000
Exhaust valve seat	12000	16000	36000	32000
Valve guide, EX	12000	16000	24000	32000
Valve guide, IN	12000	16000	36000	48000
Piston Crown	12000	16000	24000	48000
Piston rings	12000	16000	12000	16000
Cylinder liner	12000	16000	48000	64000
Antipolishing ring	12000	16000	24000	32000
Connecting rod	12000	16000		
Connecting rod screws	12000	16000	24000	32000
Valve tappet and roller			24000	32000
Injection pump tappet and roller			24000	32000
Injection element	12000	16000	24000	32000
Injection valve	6000	8000		
Injection nozzle	6000	8000	6000	8000
Water pump shaft seal	12000	12000	12000	12000
Water pump bearing			24000	24000
Turbocharger	24000	24000		
Governor	12000	12000		
Vibration damper	Acc. to manuf.	Acc. to manuf.		

◄ The number of operating hours between two overhauls and the expected lifetime. Shown here, a Wärtsilä 20 diesel engine, Category III.

Clearly shown is that the service lifetime of a diesel engine running on HFO is significantly shorter than that of a diesel engine running on MDO. Approximately 25% less!
Many spare parts must be replaced in the third overhaul.
When planning the (HFO) number of operating hours, companies will attempt to execute an optimum effective overhaul.
An injector overhaul is planned after 6000 operating hours, and then after 12,000 hours. Most spare parts at 12,000 operating hours are set for overhauling, with the exception of the camshaft (16,000) and the turbocharger (24,000). On average, these types of engines operate between 6000 and 7000 hours per year.

15 Note all the activities done and record all measurements in a journal.
16 Many engine manufacturers have their maintenance programs computerized. All known engine manufacturers have these programs. Advice from the factory can also be obtained via a remote connection when using these systems.

▼ Tools stored in a well-arranged system are a primary requirement to perform the work effectively.

▶ Special tools, such as these for a Wärtsilä 38 diesel engine Category III, are delivered with the engine in a special 'toolbox' with detailed instructions, either on plasticized cards or digitally for use with a laptop computer.

27.4 Engine maintenance

Division of the activities

In the past, many of these activities were carried out by the crew.

- There were sufficient skilled engineers.
- There was sufficient time; periods of inactivity were common and long.
- Electronics were not available.
- Communication with the engine factory was difficult.
- The operating life of some spare parts was so short, requiring frequent repair. Notorious examples are the exhaust valves.

Nowadays maintenance is performed in a very different manner.

- There are few engineers on location.
- There is very little time, ships dock for a very short time.
- Modern engines work with numerous electronic systems. Working with these systems requires specialised knowledge.
- The operating life of spare parts is much longer.
- At present, worn spare parts are removed and replaced by overhauled or new spare parts. On board or in a diesel-power station, very few repairs are carried out. Replacement is the remedy.
- Shipping companies often replace all the components in a propulsion engine that wear with overhauled or new components. This is done either in port or during a major overhaul. After two to three years, dependent on the number of operating hours, these parts are simply replaced. This type of major maintenance can be performed in a couple of days to one week.
- Only minor maintenance, such as checks or adjustments, changing fuel- and lubricating-oil filters or testing/exchanging injectors, is still carried out by the crew.
- The revision of cylinder heads, turbocharger overhaul and for example, testing and repairing fuel pumps and injectors is done by specialised companies.

Reminder

One should never forget that these systems do not solve breakdowns in the middle of an ocean. One should therefore always ensure that the crew/engineers are capable of 'bringing the ship home'. **If engineers only work as operators, the risk of 'being adrift' is considerable!**

27.4.1 Hydraulic tensioning tools

Many engines in category II and almost all in categories III and IV are currently supplied with hydraulic bolt-tensioning systems. This is often the case in the following:

- cylinder-head bolts;
- connecting-rod bolts;
- horizontal and vertical main-bearing bolts;
- fixing bolts for large counterweights on the crankshaft;
- in two-stroke engines, bolts for the exhaust-valve casing;
- in two-stroke engines, the tie rods;
- in two-stroke engines fixing the piston rod on the crosshead;
- fixing the crosshead-bearing covers;

◄
Good maintenance of the technical installation and therefore the diesel engines is of utmost importance for safe sailing, especially in bad weather conditions. An enthusiastic, motivated and well-trained crew is essential for every shipping company.

- foundation bolts for all the larger engines;
- cams on camshafts.

Other fastenings that are assembled/disassembled with hydraulic tools are, amongst others, coupling sleeves between shaft-tunnel parts and attaching the propeller to the propeller shaft.

Advantages of hydraulic tools

- The bolt is stretched within normal elasticity limits using a hydraulic jack. The correct oil pressure applies the exact elastic tension.
- The cylinder-shaped hydraulic bolt is simply hand-tightened with a small rod. Large force or heavy tools are not necessary.
- The bolt is not under a torsion load and therefore does not loosen. The bolt does not have to be tightened with a large force in the tap holes, hand tightening is sufficient.
- Tightening or loosening can be carried out in stages allowing the forces to remain stable on the bolts during assembly or disassembly.
- Hydraulic tensioning demands little time and requires less muscle. The hydraulic lubricating-oil pressure is supplied via a manual plunger pump and in larger systems by pneumatic or electric high-pressure pumps.

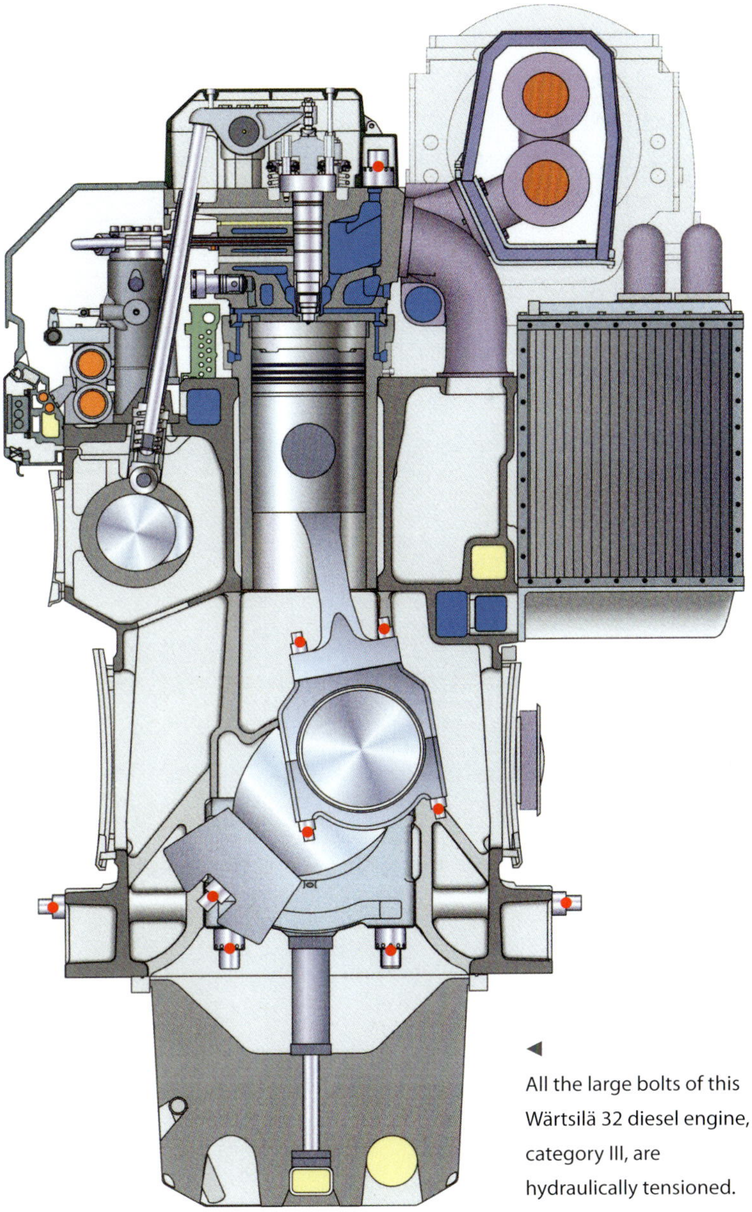

◄
All the large bolts of this Wärtsilä 32 diesel engine, category III, are hydraulically tensioned.

▶

The hydraulic jacks installed on the two connecting-rod bolts and the vertical main-bearing bolts of a Wärtsilä 20 diesel engine, category III, HFO.

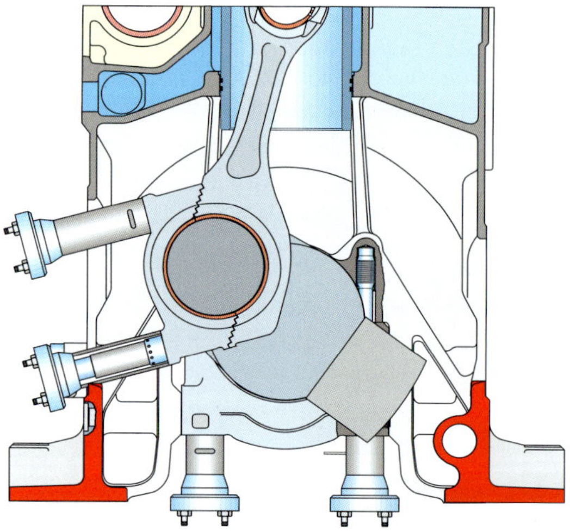

▶

The four hydraulic jacks installed on the cylinder head of a Wärtsilä 20 diesel engine.

The piping, the manometer, and the pipe (foreground) of the hand pump have already been mounted.

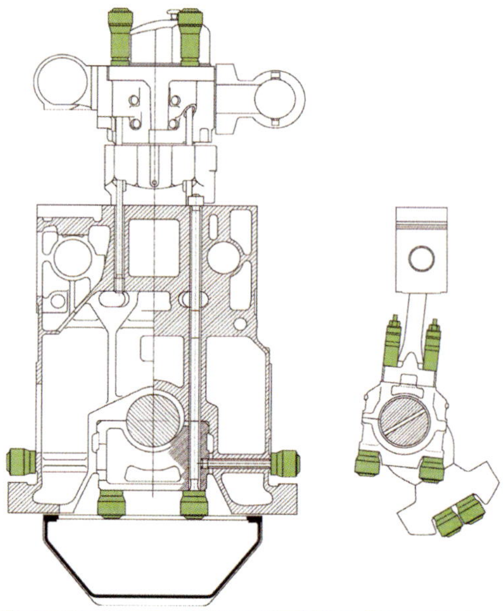

Application of the hydraulic tools for the L32/40 type

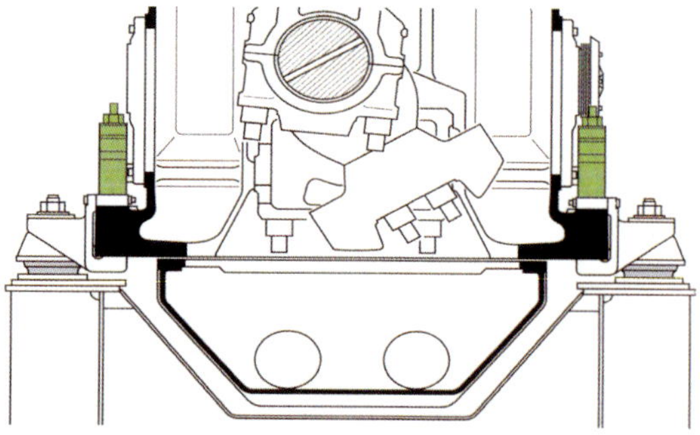

Application of the hydraulic tools on the bearing screws in the case of resilient support

▲

The bolted connections of a MAN Diesel L32/40 diesel engine, category III.

The hydraulic jacks have been placed on all the hydraulically tensioned bolts (green).

▶

'Loosening' the connecting-rod big end from the crankpin.

1. *Place accessory between connecting-rod big end and bearing cap.*
2. *Loosen connecting-rod bolts.*
3. *Remove the nuts.*
4. *Hoist the piston out so that the crankpin-bearing shells are released.*

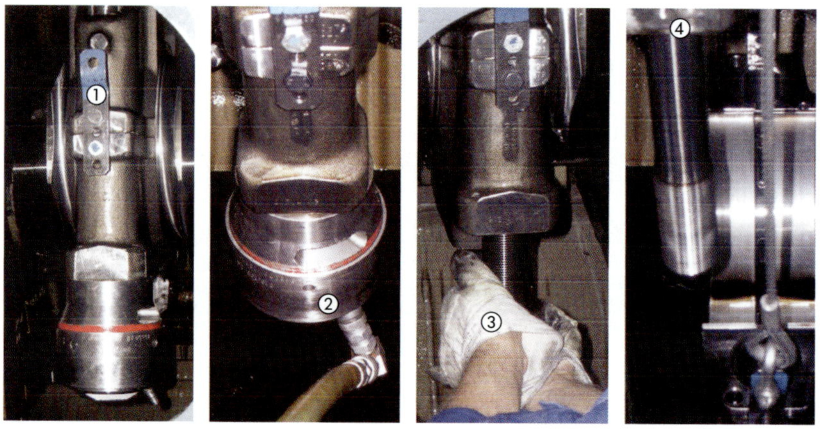

Fore-end view of a S65ME-C

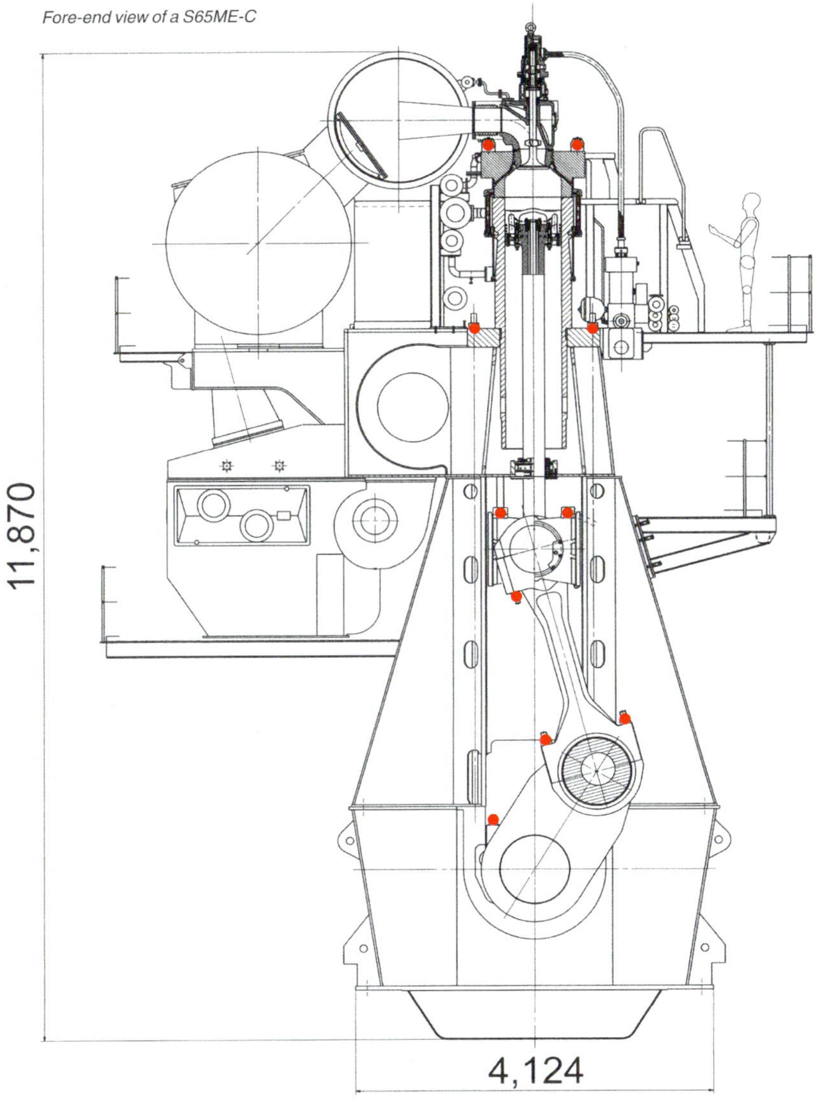

◄

Also with this new type of two-stroke crosshead engine, a MAN Diesel 7S65ME-C, all the larger bolts are hydraulically tensioned.

▼

A machine for hydraulic 'threading' of engine bolts for diesel engines.

The two rollers are hydraulically moved towards each other and pressure their threaded sections between the clamped blank bolt giving the required thread profile. Even though the blank bolt is not pre-heated, after time it must be cooled to dissipate the the resulting distortion heat. There is no bolt shown in this picture. Between the rollers, the cooling-oil pipe is shown.

▲

Hydraulic tensioning device for the cylinder head for a Wärtsilä Sulzer RTA 84 two-stroke crosshead engine, category IV.

All the tools are moved using an overhead crane.

▲

The four hydraulic jacks with hoisting equipment for a Wärtsilä 46 diesel engine, category III.

27.4.2 Bolted connections

Many of the bolts used for engine gears, as well as for cylinder heads and main bearings, are supplied with a rolled thread as opposed to a cut thread.
By 'rolling', that is, the hydraulic deformation of the material, the steel-grain structure is not distorted and the bolt is stronger than a bolt with a cut thread.

▲
Hydraulic bolts with rolled thread.

The conical nut can only be turned with a steel pin.

27.4.3 Grinding and honing tools

For many marine engines and engines for power plants, specialist tools are available for maintenance of engines.

27.4.4 Grinding and cutting tools

These are mainly used for the repair of valve seats in the cylinder head or for the seats of the valves themselves. One manufacturer of these tools may grind the seats while another may use a cutting tool, a small chisel.

Honing tools

It is common procedure to have mobile honing tools available in the proximity of the engine. These are used to provide the cylinder liner with a good honing pattern after the piston removal.
Smooth areas, 'glazing', are removed ('deglazing').
It should be noted that liner wear is minimal, so it can remain in place for the full operating life of the piston.
Honing extends the operating life of the cylinders.

27.4.5 Turbochargers, fuel pumps, injectors and governors

In general, the maintenance of these parts entails either exchanging or reconditioning during planned major overhauls; progressively less maintenance is being performed on location. Maintenance is mostly carried out by specialised companies.

27.4.6 Piston maintenance

In general, the vessel's own staff will carry out maintenance of the pistons. They are 'drawn', inspected and cleaned. Afterwards, the important dimensions are checked to ensure that the piston can be re-used or if it should be replaced. This also applies to the gudgeon pin, the piston-pin bushes, and the piston rings. For two-stroke crosshead engines, the piston rod and the piston-rod stuffing box.

27.5 Maintenance for small engines, category I

This diesel-engine category contains numerous (normally) lightly loaded engines with few operating hours. This is mostly the case for pleasure craft, such as yachts and motorboats.

Maintenance

It is of utmost importance to read carefully the instruction manual. The long winter stops often lead to more damage caused by frost, moisture, and disuse than wear and tear caused by hundreds of engine operating hours in the summer season!

For adequate maintenance, the next points are of importance:

- Read the instruction manual thoroughly, all normal activities are mentioned.
- Ensure that fuel, lubricating oil and cooling water/coolant is of good quality.

Fuel

A **fuel tank** should be left full in the winter stop. This prevents build-up of condensate and eliminates the water contamination of the fuel.
Use a good fuel filter and an oil-water separator.
Fill the fuel tank with diesel oil at tank stations that have a regular supply-and-demand network.

Lubricating oil

Lubricating oil rarely obtains the number of authorised operating hours. Changing the couple of litres of lubricating oil every season is advisable as the water and dirt in the oil are disposed of at the same time. Always renew the lubricating-oil filter.

Coolant

In a closed cooling system, use only the best coolant. Coolant must be kept at the correct level and exchanged according to the instructions.
In an open cooling system or the secondary cooling system (untreated-water system), the 'sea inlet filter' must be cleaned regularly. Check, if possible, the coolant discharge.

Leaks

Stop fuel, lubricating-oil and cooling water/coolant leaks as quickly as possible; keep the outside of the engine clean and dry.

Small maintenance

This is often limited to the following points:

- Checking the valve clearance.
- Checking the fuel, lubricating-oil and coolant systems, see above.
- Checking the starting system including starting motor and accumulator.
- Checking the dynamo, often belt-driven (correct tension).
- Checking the rotor of the coolant pump(s) and cleaning the algal filter.
- Checking the vibration dampers, the flexible couplings, and the propeller-shaft seals.
- Checking the lubricating-oil level in the high-pressure fuel pump.
- Checking the coolant pressure in the closed system. Checking the coolant for frost protection.

Testing the main engine

- Check that the engine is operating in a regular loop; if not; this is often an indication of improper combustion.
- Check pressures and temperatures.
- Check visible smoke in the exhaust gas after the engine is warm.
- Check vibrations.
- Check leaks.
- Check to see if the dynamo and accumulator are working.

27.5.1 Major maintenance

This is mostly carried out by specialised firms or the service department of the engine manufacturer.

27.5.2 Some maintenance activities

- **Reconditioning and cleaning the cylinder heads:**
  - cylinder-head levelling;
  - replaning the valve seats in the head.
- **Reconditioning the injector(s):**
  - clean the injector nozzle (outer side);
  - test the opening pressure and the atomisation;
  - check for fuel leaks.
- **Honing the cylinder liner:**
  When the cylinder liner shows some wear, honing with an optimal crosshatch pattern will extend the operating life of the liner. Serious wear or a loose cylinder liner in the block requires replacing the liner.
- **Piston reconditioning and cleaning:**
  - renew piston rings, if required;
  - measure the gudgeon-pin bush dimensions and renew, if required.
- **Check connecting-rod bearings and renew if required.**
- **Check turbocharger. If required, allow reconditioning to be done by a specialised firm.**
- **Replace all packings, if required, for example, valve caps and crankcase doors.**
  Replace the copper rings of the injector and fuel lines.
- **Crankshafts have a long life as do the main bearings.**
  Only with serious wear must they be disassembled. The crankshaft must be ground and over-dimensioned main bearings and crankpin bearings must be adjusted.
- **Check governor and lubricating-oil piping:**
  - change the lubricating oil.
- **High-pressure fuel pump.**
  Change the lubricating oil regularly. This lubricating oil is often very thin! This points towards possible fuel leaks, which are not a major problem if the lubricating oil is regularly changed.
- **Crankcase.**
  Clean the crankcase well with lint-free cloth. Remove all small impurities. Carefully check all the nooks and crannies with a lamp.

27.6 Examples of maintenance for engines, category III

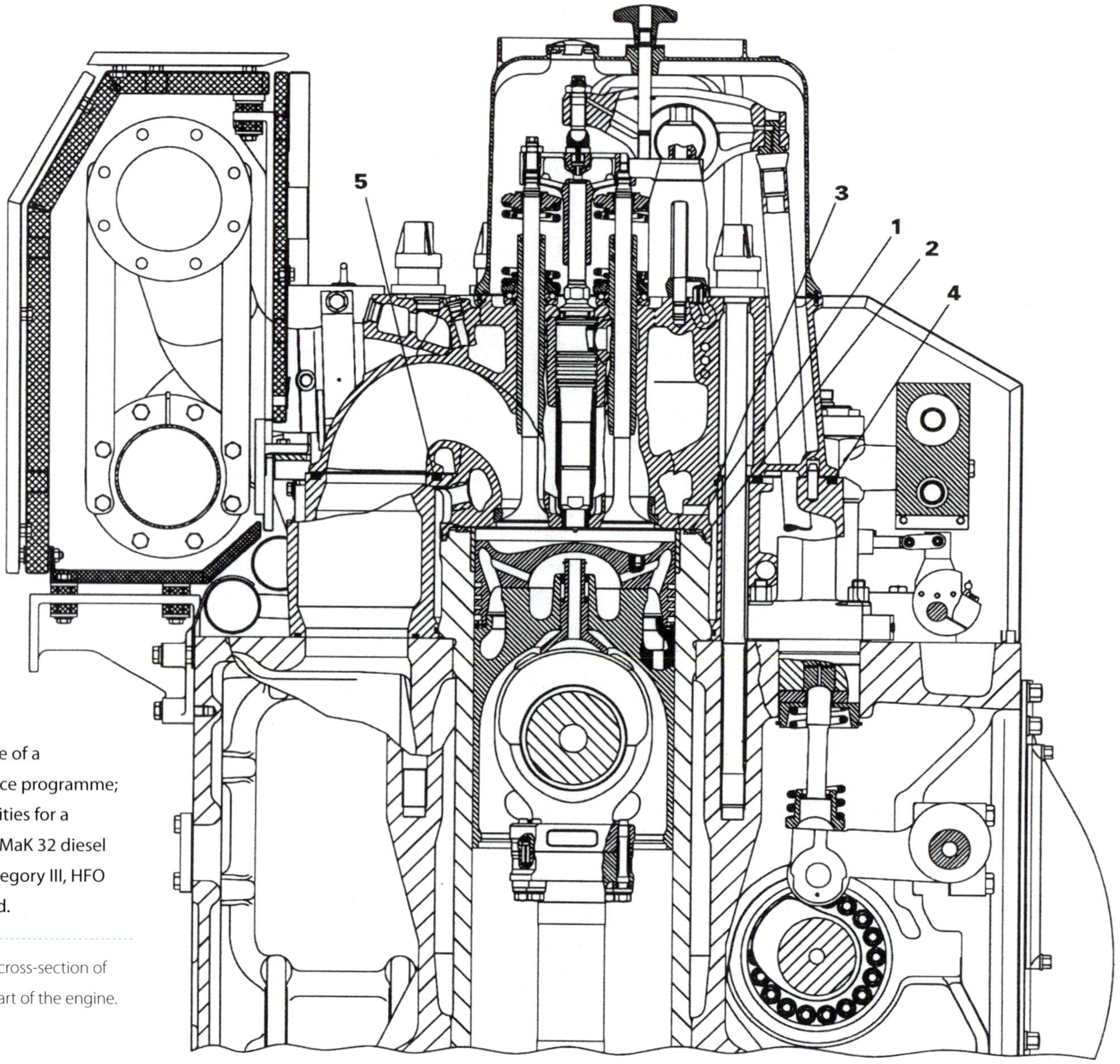

▶ An example of a maintenance programme; some activities for a Caterpillar MaK 32 diesel engine, category III, HFO are detailed.

Shown, the cross-section of the upper part of the engine.

27.6.1 Crankshaft clock gauging

Ensure that the heart line of the crankshaft is in line.
Goal: avoid material fatigue in the crankshaft: this could result in crankshaft breakage.
Cause: wear of the main bearing or deformation of the engine frame (large engines).
Method crankshaft clock gauging: basically, a dial gauge is placed in two standard centre holes between the crankshaft webs. The position of the crankshaft-heart line is obtained by setting the dial gauge to zero for every cylinder in a certain piston position. Then progressively per cylinder, measurements are taken in four different positions. The dial gauge 'results' must always be within certain values.

Crankshaft clock gauging is performed for the following objectives.
- As a maintenance check.
- After a collision.
- If there is a suspicion that the engine foundation is deformed.
- With bearing damage.
- Before and after dry-docking.

M32

Operating hours	Vessel	Date
Engine Type	Engine no.	Engineer/Service Station

Engine cold ☐

Shut down for ☐ hours

Preheated ☐ °C

Indication of Dial Gauge

down ☐ up ☐

press

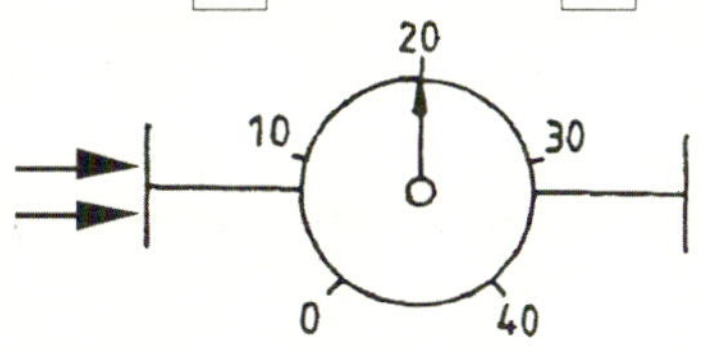

Draft forward:

Draft aft:

Measuring point	Cylinder 1	2	3	4	5	6	7	8	9
(1)	20	20	20	20	20	20	20	20	20
(2)									
(3)									
(4)									
(5)									
△ max									

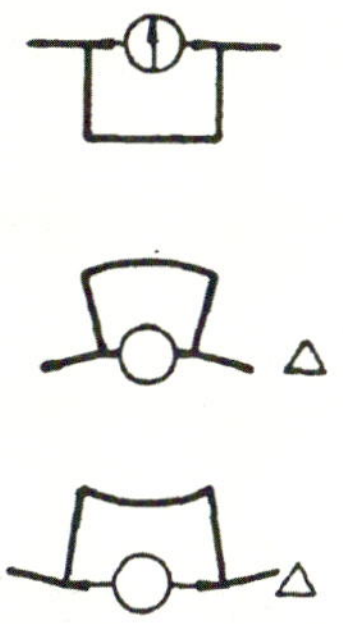

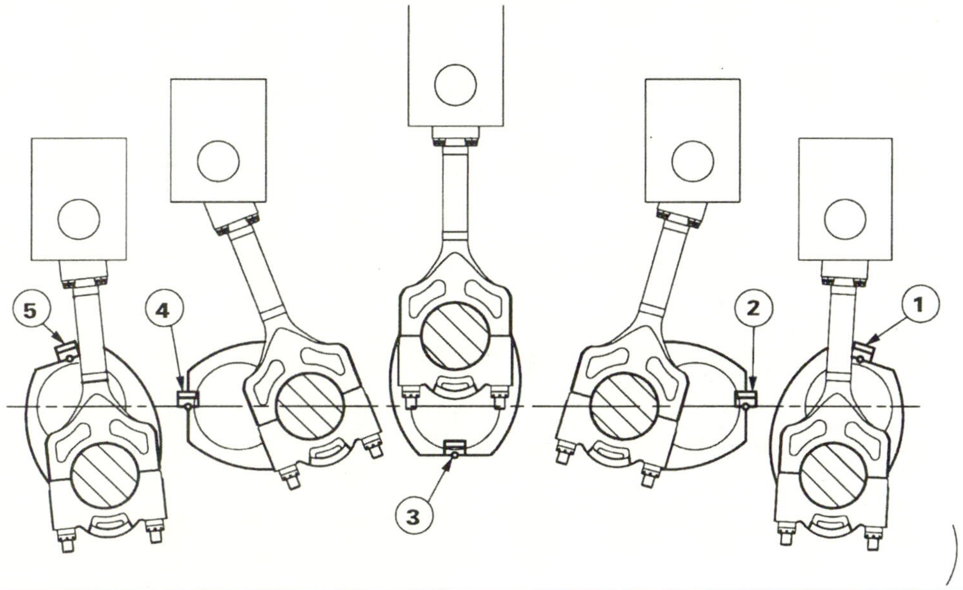

◀ A measuring table for the crankshaft position of a Caterpillar MaK 32 engine. Shown, a nine-cylinder engine.

negative: smaller distance between the crank webs

positive: larger distance between the crank webs

In the lower drawing, the measuring positions are clearly shown.

As the connecting rod is in the way, measurements are taken once on the one side, position 5, and once on the other side, position 1.

◀ There are also electronic crankshaft-deflection devices that store the measured values in a file.

27.6.2 Valve-clearance check

Checking the clearances in the drive mechanism of the inlet and exhaust valves.
Goal: Valve clearance ensures that the valves are **completely** closed when this is required for the process. The clearance is often between 0.2 and 1.2 millimetres dependent on the type and size of the engine.

The exhaust-valve clearances are usually larger; the exhaust valve is usually warmer than the inlet valve and therefore expands more.
There are different methods for cold and hot engines.
Method: A feeler gauge is used to measure the clearance between both drive parts.

▶
Measuring the valve clearance for a Caterpillar MaK 32 diesel engine.

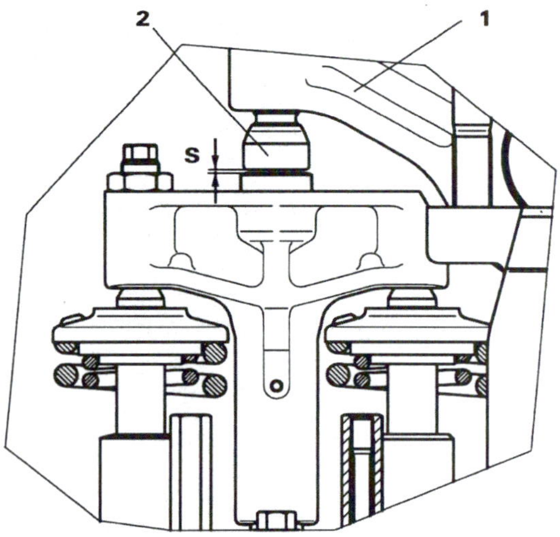

Fig. 1

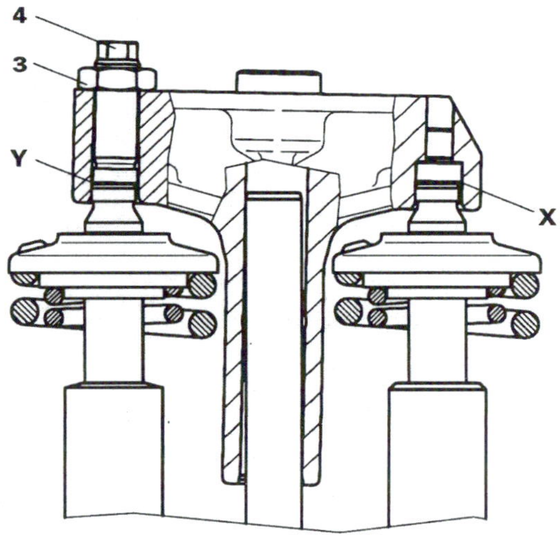

Fig. 2

In figure 1:

1 *valve-rocker arm*
2 *valve-adjusting bolt*
S *valve clearance*

Using a feeler gauge, the clearance S can be checked. The feeler gauge must move with reasonable difficultly.

In figure 2:

3 *counter nut*
4 *setting bolt*

If the clearance is not correct, the counter nut is loosened and using the setting bolt, the correct clearance can be set.
When tightening the counter nut, ensure that the setting bolt does not rotate. Always recheck the clearance after any adjustments! Always press both valve yokes down so that there is no clearance.

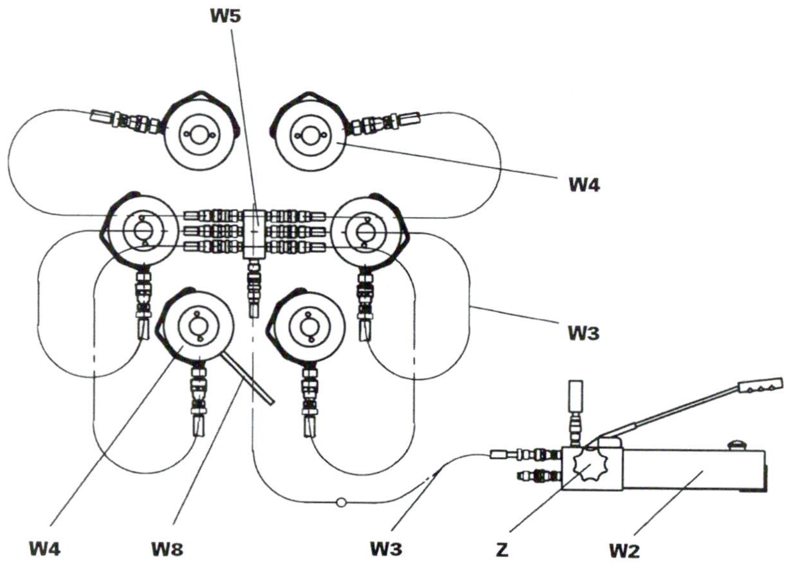

27.6.3 Cylinder-head dismantling

Remove the cylinder head.
Goal: Check the cylinder head, such as, inlet- and exhaust valves and seats, coolant spaces, contamination of the bottom side of the head, fuel injectors, and etcetera. Dismantle the cylinder head further, so the piston and liner can be dismantled.

◀
An hydraulic jack set ('spider') for the six cylinder-head bolts of a Caterpillar MaK 32 diesel engine.

In this case, the oil pressure is supplied using a hand pump. In stages, the pressure over all six hydraulic jacks can be increased until the 'hydraulic' nuts on the head bolts are released so that the cylinder head can be removed.

Method: Small engines in categories I and II have bolted connections, which can be loosened manually. Cylinders with a cylinder diameter from 200 millimetres in the categories II and III have hydraulically tensioned bolts and hydraulic jacks are used.

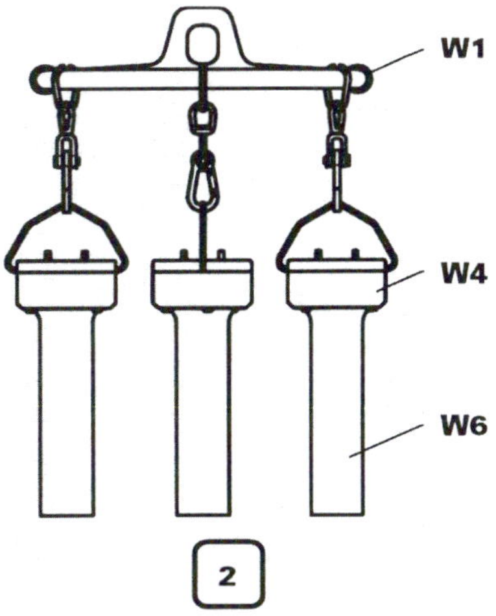

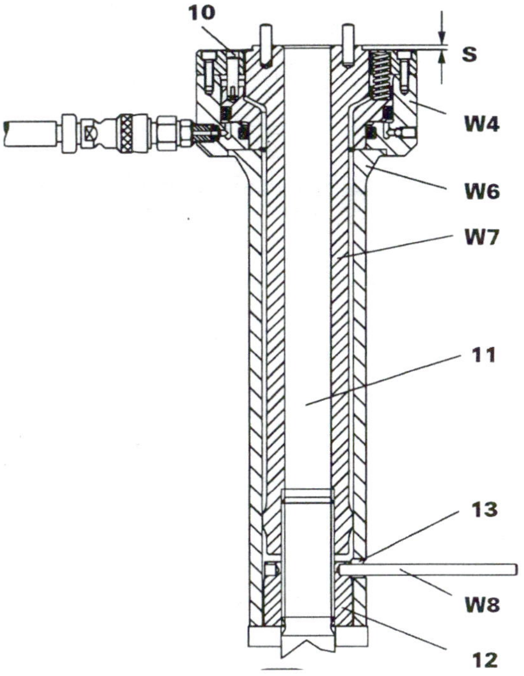

▶

Upper drawing: the hydraulic jacks hanging on a spreader and hoisted above the cylinder with an overhead crane. Lower drawing: a cross-section of the relatively long hydraulic jack for the cylinder-head bolts.

The oil is supplied from the upper left. Outside pipe, W6, is on the cylinder head. Inside pipe, W7, is screwed onto the rolled thread of the cylinder-head bolt and due to the high oil pressure, is moved slightly upwards. As a result, the bolts are slightly stretched and the cylinder-head nut 12, with pin W8, can be rotated upwards ('eight holes') so that when the oil pressure is removed, they remain free of the cylinder head. The head can then be removed.

Working pressure of the oil system is 850 bars.

13 *grooves in outer pipe*
10 *venting*
S *measured distance of the screwed inside pipe, W7*

Inlet and exhaust valves

These are cleaned and measured, the welded valve seats are then ground to the correct angle.

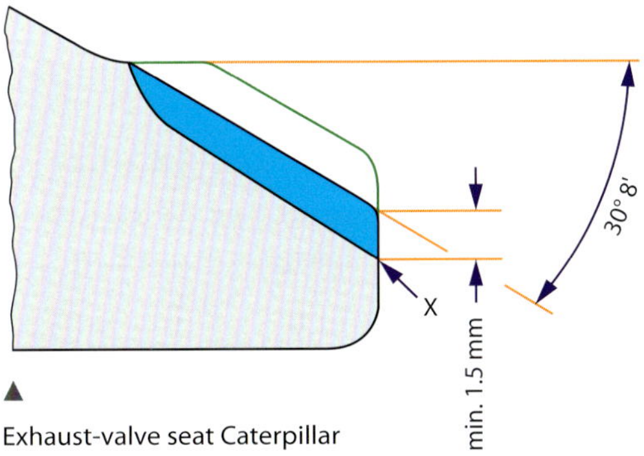

▲

Exhaust-valve seat Caterpillar MaK 32.

The welded section must have a minimal height of 1.5 mm after grinding. The angle must be 30 degrees 8 minutes.

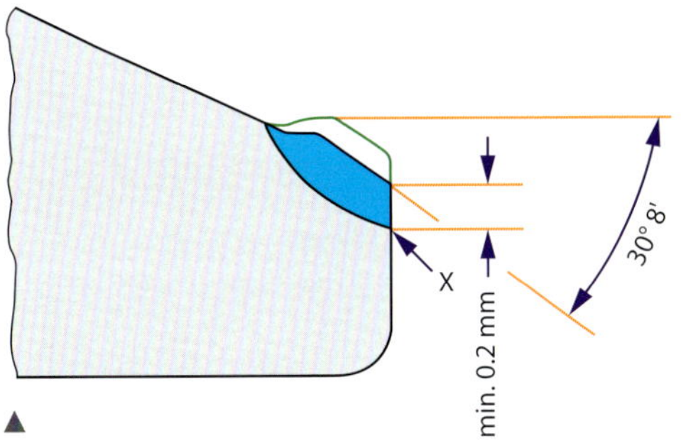

▲

Inlet-valve seat Caterpillar MaK 32.

The welded section must have a minimal height of 0.2 mm after grinding. The angle must be 30 degrees 8 minutes.

▶

A measuring table for the valves of a Caterpillar MaK 32 diesel engine.

Amongst others, measurements are taken:

- if the valve stem is bent: A-B;
- to find the thickness S, the burn section caused by high-temperature corrosion. S is a maximum of 1.5 mm.

M32

- when the deviation from the straightnesss does **not exceed** the following values (**Fig. 1**).

Measuring range "**s**" = **390 mm**

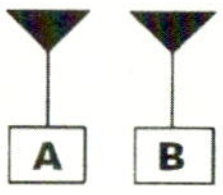

= contact points for test device

- when the radial runouts of the valve cone do not exceed the following values:

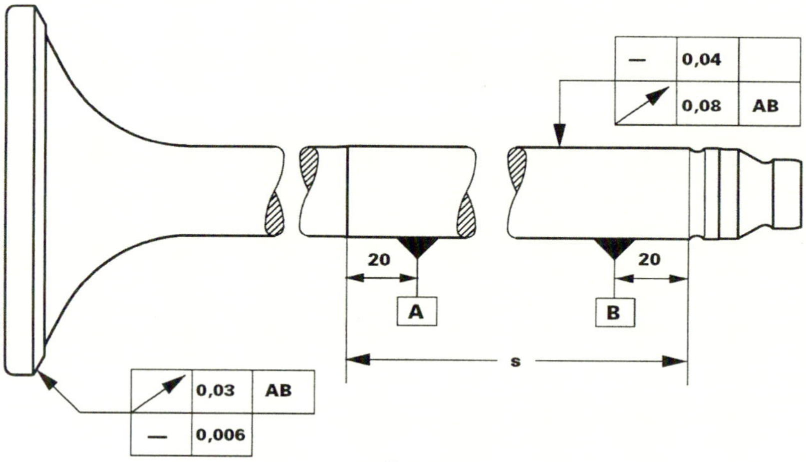

Fig. 1

1.2 The valves cannot be reused in the following cases

- the valve cone face is damaged (cracks, blowholes)
- the wear caused by corrosive material exceeds **> 2 %** of the valve head diameter (compare with new valve)
- concave burns "**s**" caused by high temperature corrosion on the underside of the valve head exceeds **> 1.5 mm** (**Fig. 2**)
- the underside of the valve head indicates severe pitting (formation of so-called "paving stones")
- corrosion pits and mechanical damage in zone of the valve stem, the stem transition and valve head
- excess of limit dimension for refinishing.

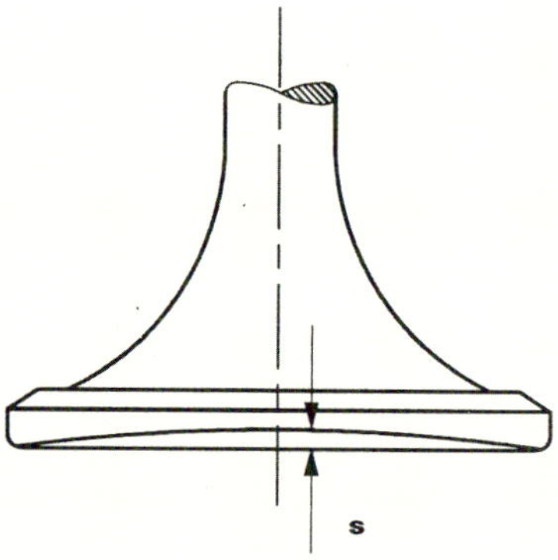

Fig. 2

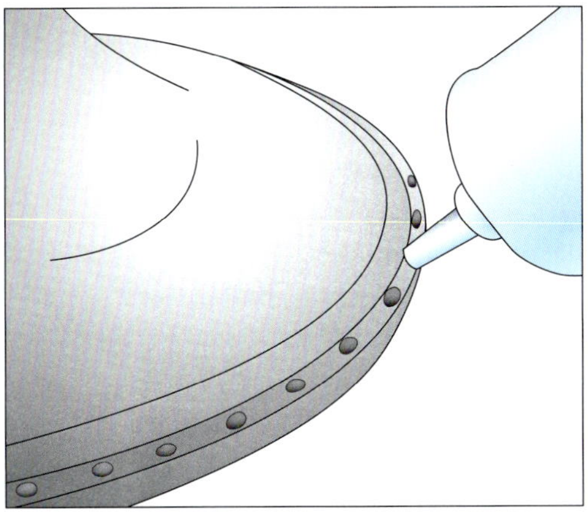

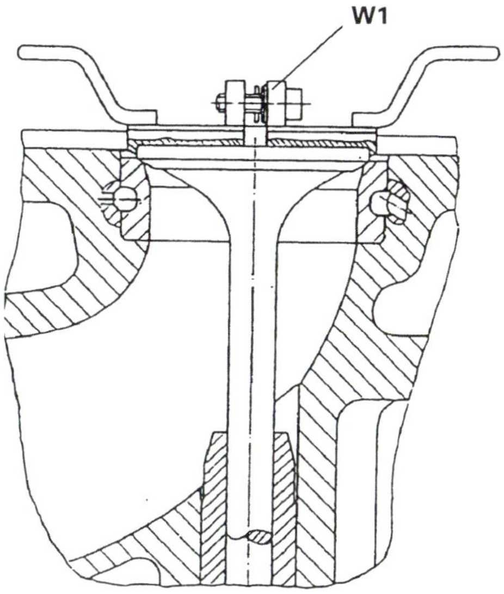

▲

Checking the valve-seat surface.

- Sand lightly by hand.
- This is done with a very fine grinding compound (Dp 30/10 – 15 microns).
- After application of the grinding compound, a solvent is sprayed over the paste to improve the polishing process.

The handgrip is placed on the valve. The valve stem is oiled and positioned in the valve guide. After this, the valve seat is lightly polished for a very short time. The contact surface of the valve with the seat is between 30 to 50% of the seat width.

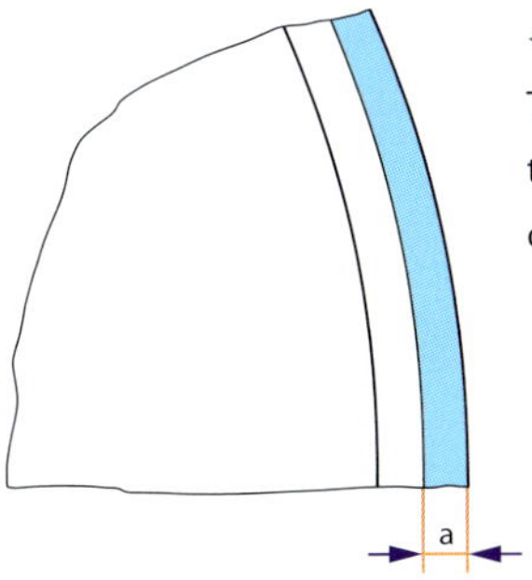

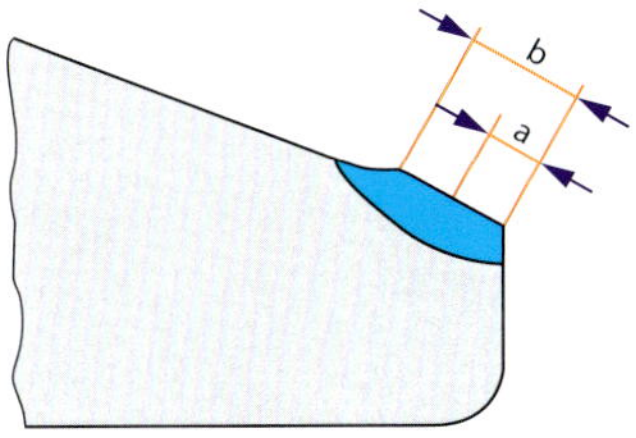

◄

The polished outer ring of the valve seat is a maximum of 50% of the total width.

27.6.4 Checking the bearing shells

Goal: Detect damage and wear.
Comment: 85% of all bearing damage is caused by contaminated lubricating oil, solid contaminants and water!

A bearing shell for a medium-speed Caterpillar MaK 32 contains two layers: a 10 mm-thick steel support layer and a 0.7 mm to 1.1 mm thick running layer consisting of a lead/bronze or aluminium alloy.

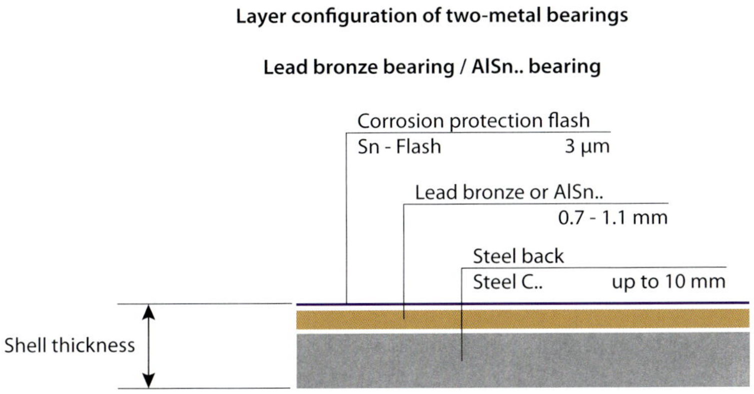

◄

The arrangement of the bearing shells for a Caterpillar MaK 32 diesel engine.

The steel shell itself is approximately 10 mm. The lead bronze or aluminium tin layer is between 0.7 and 1.1 mm. The anti-corrosion layer is very thin: 0.003 mm.

Method

Bearing replacement is required:

- when there are deep scratches on the running layer;
- when due to wear, the minimum layer thickness is reached;
- in special cases, only with Al-Sn bearing shells.

▲

The load-bearing area is distributed equally over the entire running layer. The dirt in the lubricating oil has had no negative influence on the bearing reliability. The bearing shell can be re-used.

▲

The load-bearing area is distributed equally over the entire layer. The bearing shell can be re-used.

▲

The load-bearing area is distributed equally over the entire layer. The circumferential scratches are caused by very fine solids, such as metal particles and/or sand in the lubricating oil.

Given that the depth of these scratches cannot be measured or felt, the bearing shell can be re-used. If the number, the depth and the size of the scratches negatively affect the lubricating-oil film build-up, then **both** bearing shells must be renewed.

▲

The running layer shows a heavily overloaded bearing with fretting corrosion in the aluminium alloy.

This bearing shell cannot be re-used. **Both** bearing shells must be replaced. The cause of this damage must be investigated and resolved.

▲

The load-bearing area shows damage of the bearing in a large surface area, in certain parts, fretting corrosion.

This bearing shell cannot be re-used, renew both bearing shells. The cause of this damage must be investigated and resolved. Check the other bearing shells.

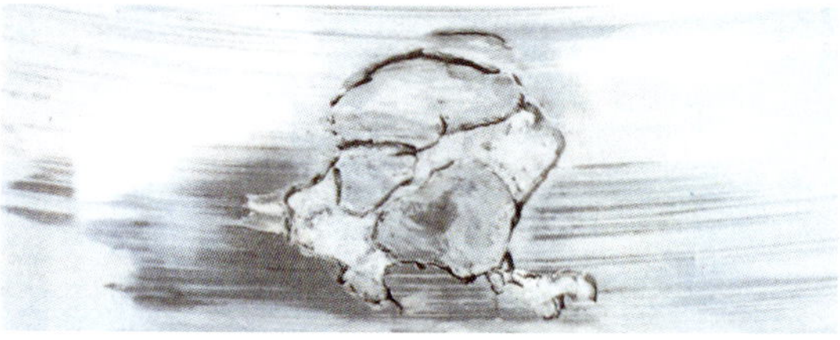

▲

The heavily loaded section of the bearing shows the stripping of bearing material, probably by overloading.

Stripped bearing material can be found in the area. Clean! This bearing shell cannot be re-used; both bearing shells must be renewed.

The cause of this damage must be investigated and resolved.

Check the other bearing shells.

27.6.5 Piston 'withdrawal'

Goal: Check the condition of piston, piston rings, cylinder liner, gudgeon pin, connecting rod, connecting-rod bush, and connecting-rod bearings.
Method: Loosen the connecting-rod bolts at the connecting-rod division when the piston is at top position and attached to an overhead crane with a hoist strap.
The anti-polishing ring is then removed with a special tool.
Carbon deposition on the cylinder liner is also removed.
If the above two steps are not executed, the (sharp) piston rings could remain behind!
After this, the piston rings are removed from the piston using piston-ring pliers and the piston crown is dismantled. The inside of the piston crown can now be inspected for contamination.

Comment

When the lubricating oil used for cooling the piston overheats, the internal piston-surface area can become caked with carbon residues. This impedes the heat transfer from the piston crown to the lubricating oil, causing overheating of the piston. The pressure shocks occurring in the combustion process can load the piston crown to such an extent that the piston crown fails, causing huge damage to the piston and its surroundings. All these parts must be cleaned.
The surface areas checked for damage and all the important dimensions are measured.
All rejected parts, such as piston rings, are replaced.

As a matter of course, the piston pin and the connecting-rod bushes are measured to locate possible wear. The rejected connecting-rod bushes are removed in the engine factory and with nitrogen (shrinkage), a new bush is attached to the connecting-rod eye. This can also be done with CO_2 'snow'. All the bolted connections of the piston and connecting-rod eye are tightened with an approved torque spanner.

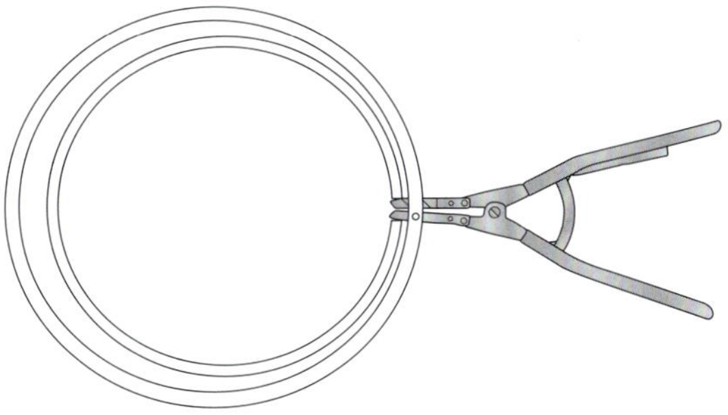

◄ Piston-ring pliers used for the simple removal of piston rings. The risk that the rings will break when using these pliers is small.

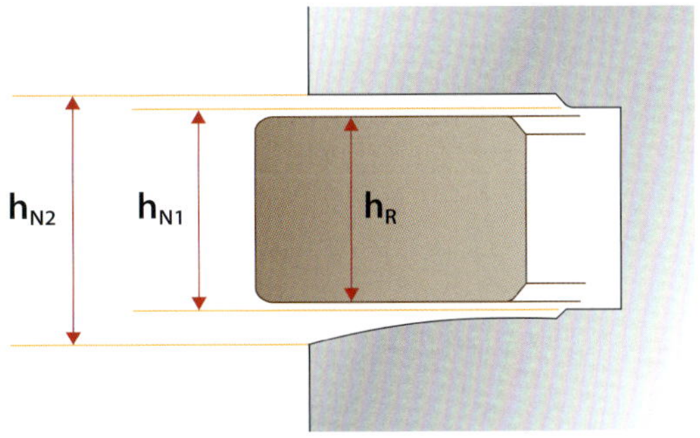

◄ Measuring the piston-ring groove in two places and the thickness of the piston ring.

Groove	Ring thickness h_R Nominal mm	Groove width h_{N2} Wear limit mm	Limit $h_{N2} - h_R$ mm
1	8	8.45	0.5
2, 3	6	6.45	0.5
4	10	10.2	0.3

◄ A measuring table for the piston-ring grooves and the ring height for all four rings.

27.6.6 Cylinder liner

Goal: Check the condition of the cylinder liner, such as wear, crosshatch pattern, cracks, and scratches.
Method: Both, when installed in the block or disassembled, the diameter of the liner can be measured at different heights, in the length and in width (fore and aft direction as well as abeam). Mostly wear is found on the upper side.

► Measuring the cylinder diameter in a two-stroke crosshead engine, a Wärtsilä Sulzer RTA96C.

The diameter of the cylinder liner is measured in various places to determine the wear. To position the calliper gauge correctly, use is made of steel strips with holes suspended at various heights in the liner. See next figure.

With a certain amount of wear, the cylinder liner will be considered as faulty and replaced. Although the cylinder liner may still be in working order, one should realise that it cannot be replaced until the next overhaul.

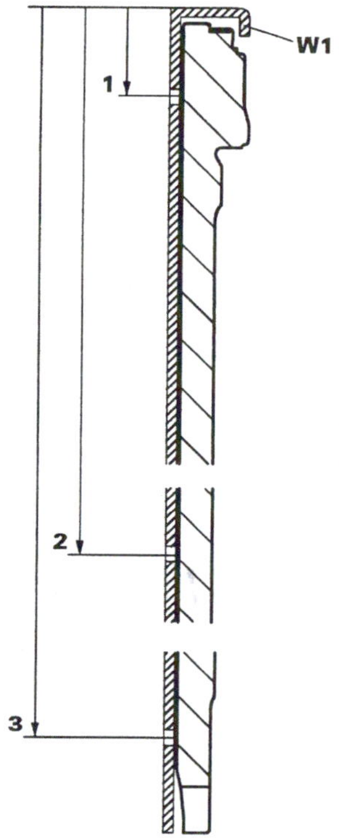

◀ Two measuring strips are placed in the cylinder so that the wear can be measured at three different heights by placing the calliper gauge in both holes.
The measuring strips must be mounted exactly opposite each other in the cylinder.

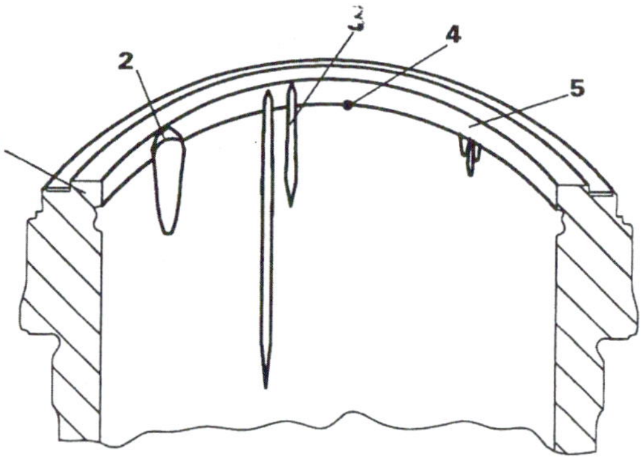

▶ Different types of scratches in the running layer of the cylinder liner. The number and their maximum dimensions are defined for all scratches, regarding possible rejection of the cylinder liner.

27.7 Some examples of maintenance for large two-stroke crosshead engines, category IV

27.7.1 Wärtsilä Sulzer RTA96C (Engine category IV)

In this very large engine, hardly any parts can be lifted by hand, the weight of the spare parts are imposing. Some of these very heavy parts are given below as an example.

bottom-half six-cylinder crankcase	118,000 kg
main bearing shell	184-234 kg
bare A-frame	94,000 kg
tie rod	2210 kg
tie-rod nut	41 kg
cylinder liner	9006 kg
cylinder cover complete	10,442 kg
exhaust valve complete with casing	2909 kg
exhaust valve	236 kg
crankshaft six-cylinder	186,000 kg
crankshaft twelve-cylinder two parts	359,700 kg
fuel cam	253 kg

This entails the following:

- Almost all the parts must be moved using an overhead crane with a hoist capability of 10 to 15 tonnes.
- All the large bolted connections are hydraulically tensioned.

Thread diameters in millimetres

foundation bolts	M 64
anchor foundation bolt	M 180 × 6
cylinder-head bolts	M 110 × 6
exhaust-valve casing	M 110 × 6 T
piston rod, on crosshead	M 110 × 6

▲ The running surfaces in the crankcase of a large two-stroke crosshead engine. Slipping during these activities can be life-threatening.

- The requisite hydraulic jacks are very heavy and moved by a crane.
- Manual inspection is inadequate; only by measuring can one ascertain if a part is in the correct place.
- Working with heavy parts requires a lot of attention, safety is very important. The possibility of permanent injury is always present!
- The crankcase and the scavenging-air space are large enough to walk around in; especially the crankcase, covered in lubricating oil, can be very slippery. Permanent stairs facilitate climbing.

Examples of maintenance schedules

Study spare-part drawings, which contain all the relevant data. Today this is done digitally. All the spare parts are represented in blue in both longitudinal- and cross-section. Clicking on them provides detailed spare-part drawings.

SULZER RTA96C Maintenance **0380–1/A1**

Maintenance Schedule
Inspection and Overhaul Intervals (Guidelines)

Component	Work to be carried out	Group	Intervals
Group 0			
Lubricating oil . . .	– Laboratory analysis	0900–1	3'000 op. hours (= operating hours)
Cooling water . . .	– Check concentration of inhibitor (as per supplier's instructions)		
Group 1			
Bedplate	– Check pre-tension of foundation bolts . . . . (first time after 1'500 operating hours)	1112–1	every 2 years
Main bearing	– Check pre-tension of waisted stud, if necessary re-tension (first time after 1 year)	1132–1	15'000 ÷ 20'000 op. hours
	– Remove main bearing upper half for inspection	1132–2	acc. to class. society
	– Remove bottom bearing shell for inspection	1132–1	as required
Thrust bearing . .	– Check axial and radial clearance	1203–1	6'000 ÷ 8'000 op. hours
	– Check bottom drain for free passage	1203–1	6'000 ÷ 8'000 op. hours
	– Remove thrust bearing pads for inspection . .	1224–1	acc. to class. society
Tie rod	– Check pre-tension, if necessary re-tension . . (first time after 1 year)	1903–1	every 4 years
Group 2			
Cylinder liner	– Establish wear in bore (in fitted condition) . . .	2124–1	at every piston removal
	– Remove cylinder liner	2124–2	as required
	– Replace O-rings	2124–2	at every removal
	– Water guide jacket, replace O-rings	2124–2	at every removal
	– Grind off wear ridge in bore	2124–3	at every piston removal
	– Recondition scavenge ports	2124–3	as required
	– Reshape lubricating grooves	2124–3	as required
Lubricating quill and accumulator	– Check function and tightness	2136–1	at every piston removal
	– Replace O-rings	2136–1	at every liner removal
	– Check diaphragm of accumulator	2136–1	15'000 ÷ 20'000 op. hours

◀ A small section of the maintenance schedule for the largest two-stroke crosshead engine from Wärtsilä, the 'Sulzer' RTA96C.

Activities are often combined. Clicking the blue code numbers in the digital programme provides detailed information on the part. There are six pages in total.

Checking the main-bearing shells

The bearing-cap bolts are loosened hydraulically. The bearing cap is hoisted from the crankcase onto a crane trolley outside the engine using a hoisting point in the crankcase. After this, the upper main-bearing shell is hoisted out of the crankcase.

Subsequently the bottom shell, with the use of an attachment and the turning motor, is rotated around the crankshaft so it lies on top of the crankshaft, and can now be hoisted out of the crankcase.

Now, the shells are inspected and the thicknesses measured in various places. If the shells are approved, the procedure above is reversed.

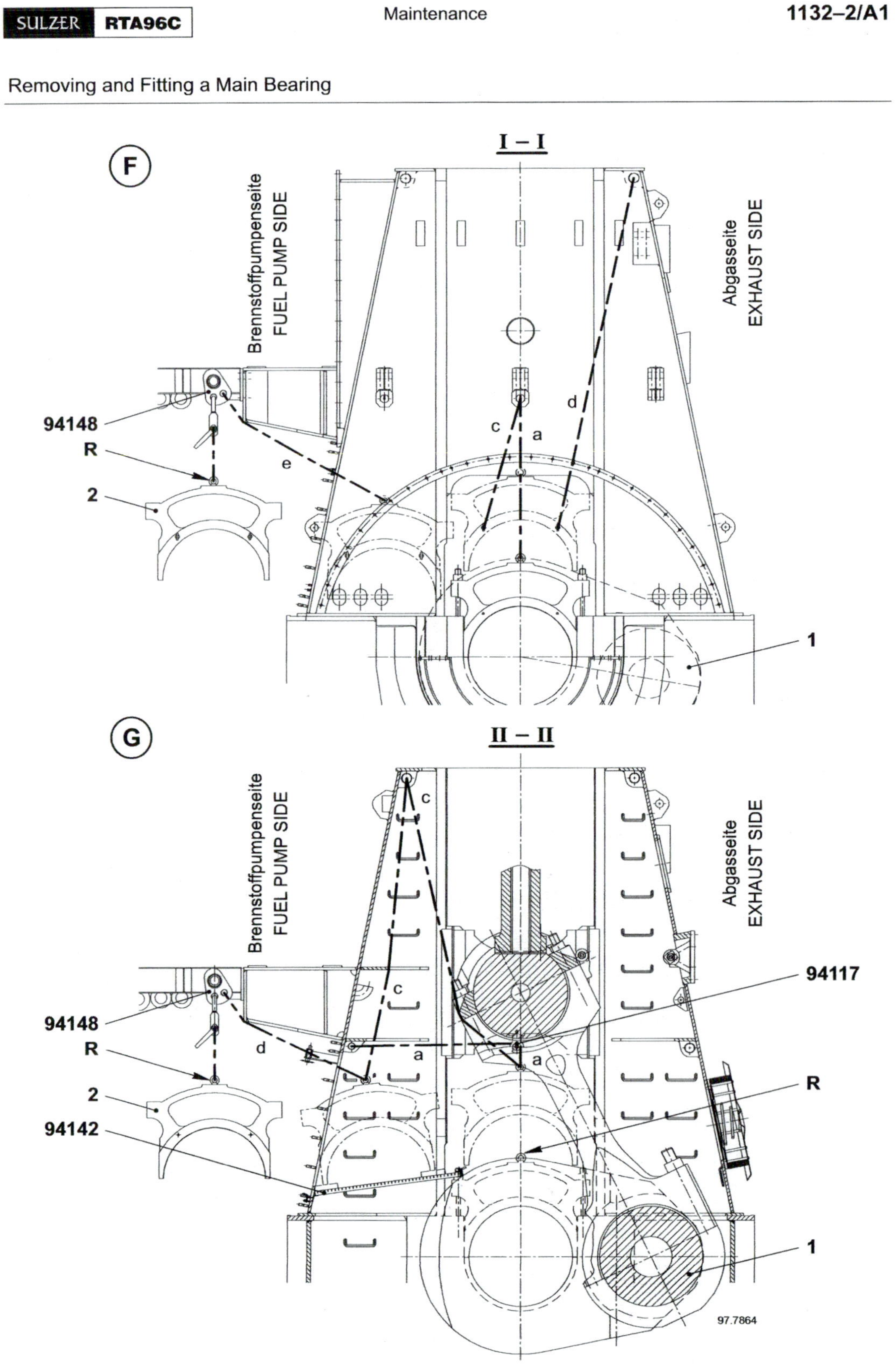

▶ Hoisting the bearing cap of the main bearing out of the crankcase.

Various hoisting points can be found inside the crankcase; outside the crankcase, hoist beams and a trolley.

Adjusting and checking the VIT-system of the high-pressure fuel pumps

This extensive procedure involves three dial gauges being placed on the suction valve, the overflow valve, and the high-pressure plunger.

The start of injection, the end of injection and the effective plunger stroke are measured. In this procedure, the camshaft must be progressively re-positioned by turning the engine.

SULZER RTA96C | Maintenance | 5512–1/A1

Setting and Checking the Control with Load-dependent V.I.T. (Variable Injection Timing)

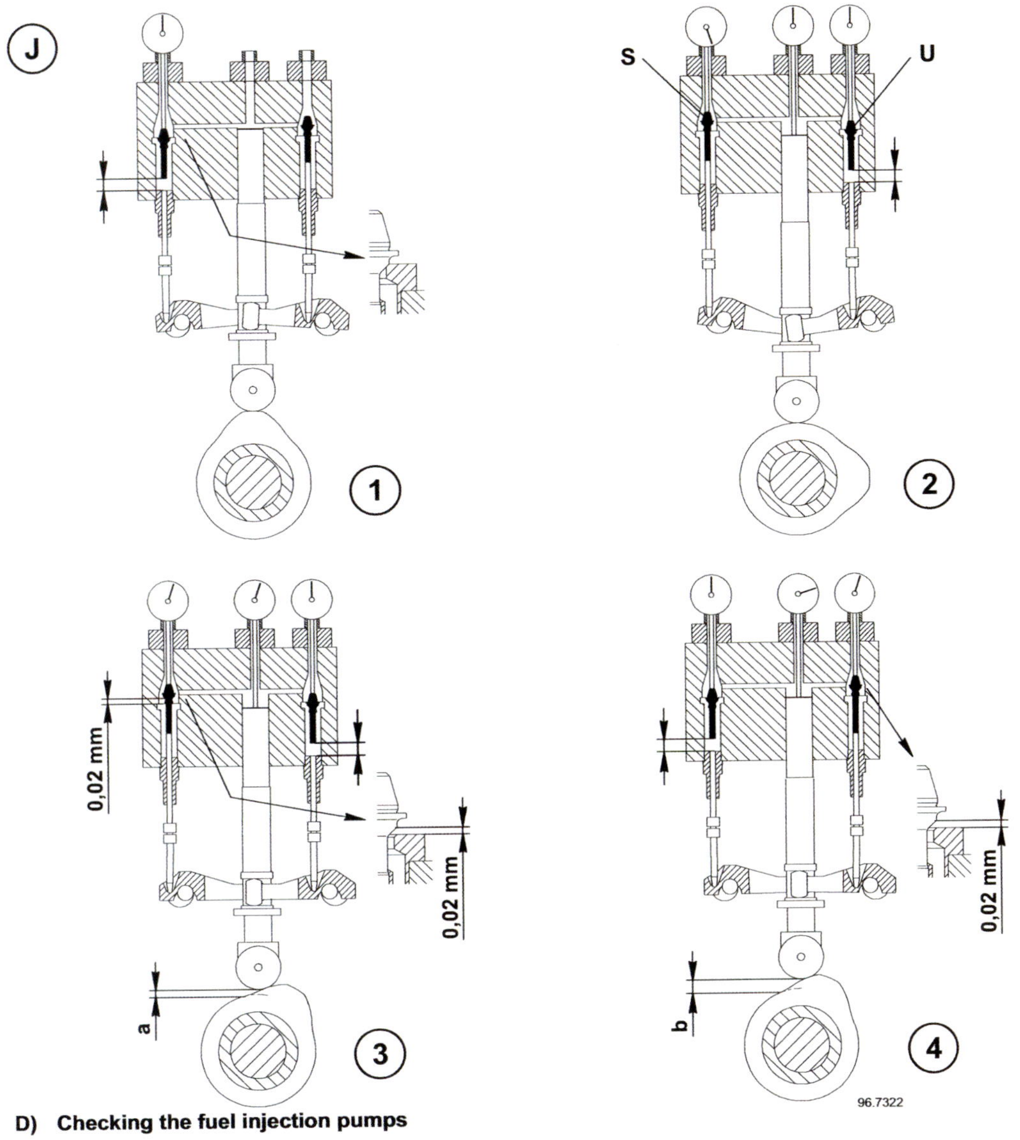

◄ Checking and adjusting the fuel pumps.

An extensive digital manual ensures that the procedure can be easily followed.

D) Checking the fuel injection pumps

The same setting values as given for the injection pump setting must be used.
Carry out preparations in accordance with section B).

Checking the begin of injection, end of injection and of the effective plunger stroke

The description below applies to **AHEAD rotation.**

1. Turn the engine until the cam roller of the fuel pump to be checked rests on the peak of the cam.
   Fit the dial gauge with about 1mm pre-tension over the now closed suction valve (S) and set it to '0' (see Illustr. 'J', Fig. 1).
2. Turn engine ASTERN, until the cam roller rests on the base circle of the cam.
   Fit the dial gauges over the plunger as well as over the now closed spill valve (U) and adjust to '0' with a slight pretension (see Illustr. 'J' Fig. 2).

Measuring crankshaft deflection

Crankshaft deflection is measured at least once a year. When doing so, the ship must be afloat. Crankshaft deflection must be measured as soon as possible after a ship has run aground.
After changing the main-bearing shells, the deflections have to be measured if possible, within a period of 100 operating hours.

The measurements are performed for five crankshaft positions. From a start position – the same for every cylinder – the dial gauge is set to zero and then read for every position. The readings can be either positive or negative (larger of smaller).
For very large crankshafts, the deviations are tenths of a millimetre. The maximum deviation for crankshafts in these engines is 0.66 mm!

▶
Measuring crankshaft deflections.

3103–1/A1 Maintenance **SULZER RTA96C**

Measuring Crank Deflection

The difference (Δa) between the indicated values at B.D.C. and T.D.C. shows the amount of crank deflection during one revolution (see Fig. 'B').
All influences which cause a temporary deformation of the hull and/or the engine affect the crank deflection.

These are for example:
- loaded condition of the ship
- engine cold or at service temperature
- differing air/water temperatures
- strong sunshine

The values of Δa given below are valid for all conditions independent of outside influences.

Where values are measured which lie over the maximum permissible limits, the cause has to be found and the necessary remedial measures taken (defective main bearing, engine support altered due to hull deformation, loose holding-down bolts, defective shaftline bearings, etc.).

When measuring the crank deflection, ensure that the crank journals are fully seated in their bearings.

Ⓑ

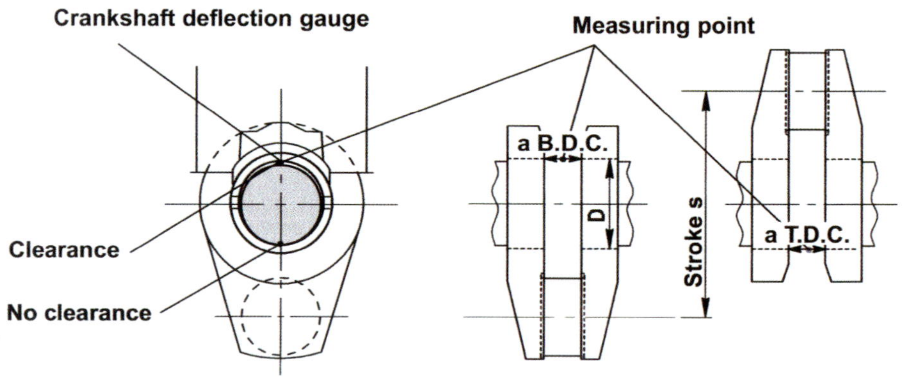

Difference	**Δ a = a T.D.C. – a B.D.C.**
Good values	**Δ a = 0 ÷ 0.33 mm**
Values still admissible	**Δ a = 0.33 ÷ 0.66 mm**

◀
The different positions when measuring the crankshaft deflections.

SULZER **RTA96C** Maintenance **3103–1/A1**

Crankshaft
Measuring Crank Deflection

Tools:

1 Dial gauge 94305

In order to ascertain whether the axes of the crank journals deviate from the theoretical shaft axis, the crank deflections must be measured. When doing so the ship must be floating freely.
Under normal circumstances, it is sufficient to measure the crank deflections once a year. They must, however, be measured as soon as possible in case the ship has grounded or after replacing the main bearing shells. The deflections have to be measured again approx. 100 service hours after replacing the bearing shells.

Should signs of damage of the main bearings be found during inspection of the crankcase, the crank deflections should also be measured.

Measuring is carried out using a crankshaft dial gauge (tool 94305) which, for this purpose, is inserted between the crank webs of the crank to be measured. When turning the crankshaft, the change in distance between the crank webs can be read from the dial gauge as this indicates any opening or closing up. The smaller the variations the better the position of the shaft.

Measuring procedure (indicator valves must be open)

With the running gear in place, the crank to be measured has to be turned towards B.D.C. until the dial gauge can be fitted next to the connecting rod at the position indicated. Pretension the dial gauge slightly and set it to "0" (see Fig. 'A').

Turn the crankshaft AHEAD with the turning gear and read off and note the readings shown on the dial gauge at the crank positions 90° before T.D.C., at T.D.C., 90° after T.D.C. and before B.D.C. (dial gauge still next to the connecting rod). The last value serves as a check. If the correct procedure has been followed, this should be back at nearly "0".

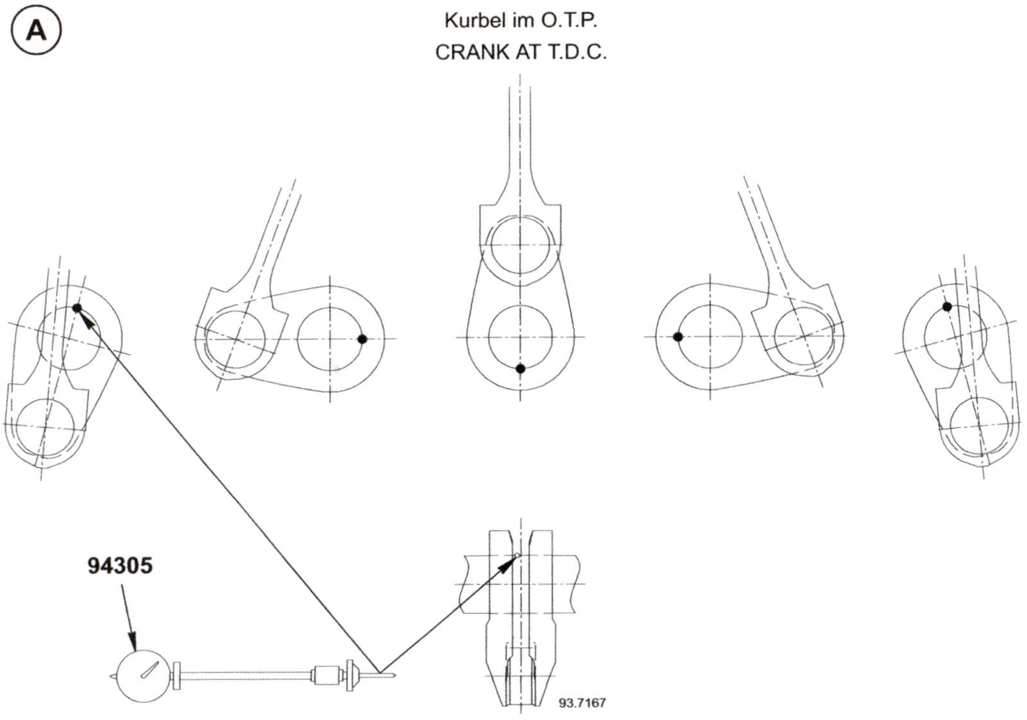

Dismantling or assembling an SKF-shaft coupling of the camshaft

The camshaft of this engine comprises various parts. The shaft couplings and cams are attached hydraulically. When repairing or changing a cam, part of the shaft is first removed from the cam casing so that the cams can be dismantled.

▶ Dismantling and assembling an SKF-shaft coupling of the camshaft.

The blue codes provide information about the tools (digital info).

SULZER **RTA96C** Maintenance **4203–4/A1**

Camshaft

Removing and Fitting of SKF Shaft Coupling

Tools:

1	HP oil pump	94931
1	SKF hand oil pump	94931a
1	Pressure gauge	94932
1	HP hose	94935
1	HP hose	94935a
1	Hydraulic unit	94942

Key to Illustrations:

1	Camshaft	P	Fit gap
2	Inner bush	R	Ring space
3	Coupling sleeve	V	Relief valve
4	Locking plate		
5	Valve screw	HPC	High Pressure Connection
6	Screw		
7	Nut	LPC	Low Pressure Connection
8	Seal ring		

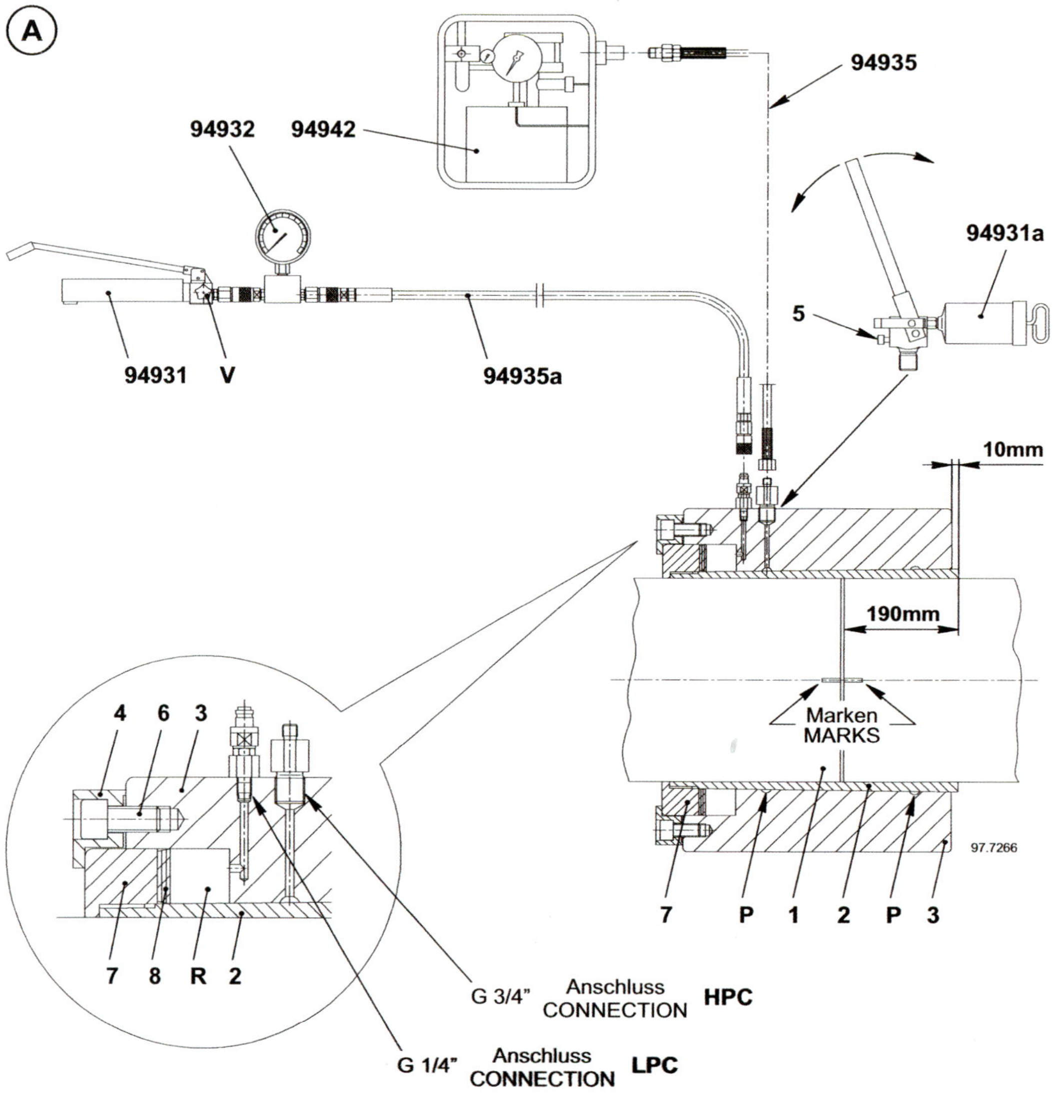

◄ Reconditioning a damaged fuel cam for a Wärtsilä Sulzer 10RTA96C engine.

1 *camshaft*
2 *cam*
3 *fuel pump*

◄ Removing a section of the camshaft.

Of course, many auxiliary tools are required, such as guiding beams, hoists and hydraulic tools.

◄ The hydraulic loosening of the SKF-shaft coupling. Every cylinder has its own removable camshaft section.

Checking the foundation bolts

In order to keep the various parts of a crosshead engine, the crankshaft bed, the A-frame, and the cylinder beam together, very heavily hydraulically tensioned foundation bolts are used.
These are tensioned using very large hydraulic jacks.

Working with these hydraulic jacks is extensively described in instruction manuals, the so-called 'Work procedures'. All the necessary detailed drawings can be found in 'electronic instruction manuals' by a simple click on the blue-highlighted text in the drawings.

▶ The hydraulic jack for the largest bolt in the engine industry: the tie rods for the two-stroke crosshead engine: 180 millimetres!

Tools:

2	Pre-tensioning jacks	94180
1	Pressure gauge	94932a
1	Connection piece	94934a
1	HP hose	94935
2	HP hoses	94935a
1	Hydraulic unit	94942

Key to Illustrations:

1	Cylinder jacket
2	Intermediate ring
3	Upper tie rod nut
4	Cylinder
5	Piston
6	Sealing ring
7	Round bar
8	Pin
9	Vent screw
10	Tie rod
11	Screw
12	Protection cap
13	Opening for solvent
14	Clamp screw
15	Column
16	Bedplate
17	O-ring
18	Lower backing ring
19	Lower tie rod nut
20	Bush
AN	Drain slot
K	Slot
L	Thread protrusion
S	Measuring place

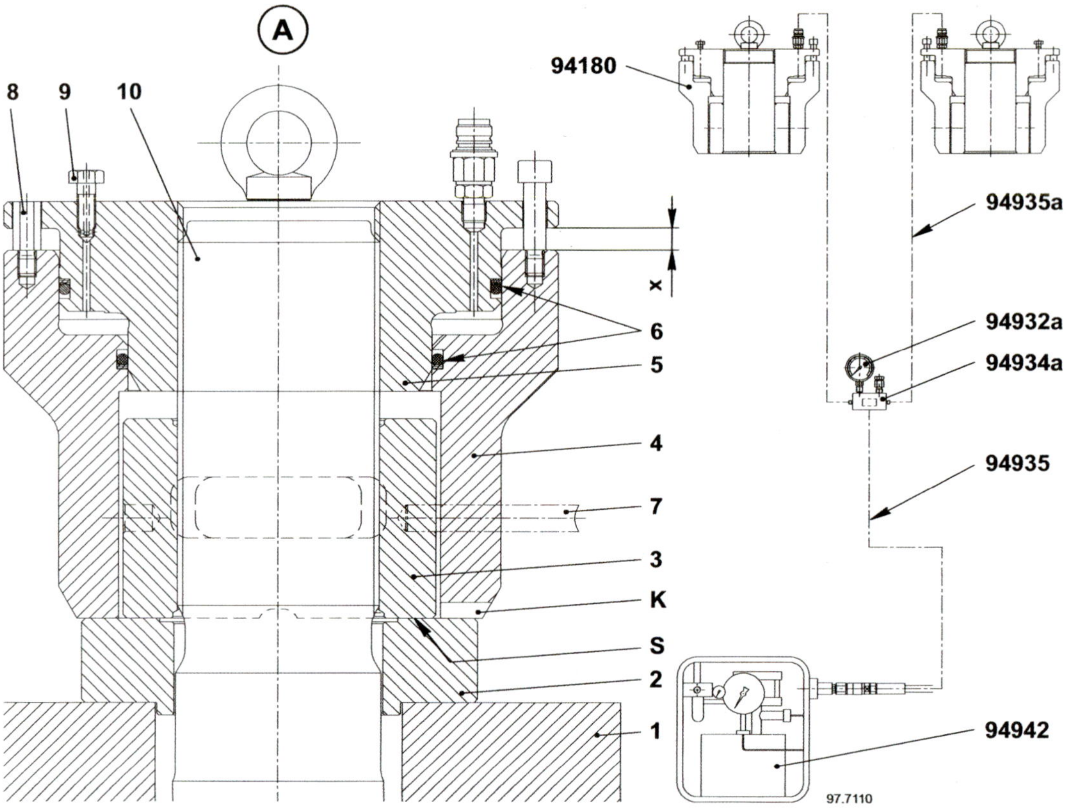

◀
Good maintenance depends for a large part on motivated and specialised personnel that, in the short time in which diesel engines are stopped, can perform the maintenance and repairs competently, quickly and successfully.

◀
The construction of a M32 line engine at Caterpillar MaK in Kiel, Germany.

This is a medium-speed diesel engine, fuel HFO.
Note the covering of the crankcase vents. Special 'Adhesive foil lids' are used to prevent debris entering the crankcase.

▶

Pulling the piston, shown here, a Wärtsilä 38, a medium-speed four-stroke propulsion engine.

Note the notches in the piston skirt for weight savings.

▶▶

Pulling the piston liner.

Clearly shown here, the thickened edge that can be seen protruding above the engine.

▼

A cylinder cover of a two-stroke crosshead engine being placed on the cylinder-head landing.

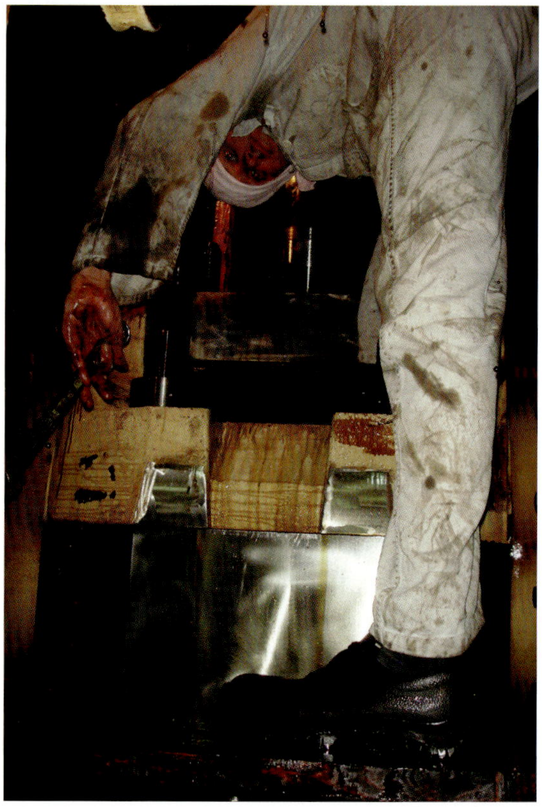

▲

Dismantling a crosshead in a large two-stroke engine.

The whole gearbox is extremely slippery because of the finish and lubricating oil. Extreme caution is advised. Fixed steps, aluminium ladders and scaffolding as well as a safety harness should be used.

27.8 Excessive wear or 'scuffing' of the cylinder liners in a two-stroke crosshead engine

27.8.1 Introduction

Source: A summary of a study by Maersk Lines, the largest container shipping company in the world.

Excessive wear of cylinder liners, pistons and piston rings is a regularly occurring problem in two-stroke crosshead engines.

This is partly due to the following:

- The material quality from which the cylinder liners and piston rings are manufactured.
- The amount and location of the dosage of cylinder lubrication.
- The quality of the cylinder lubricating-oil.
- The manner in which the liner is honed.
- The type of piston rings for instance the profile, the surface treatment such as chrome plating or ceramic coating.
- The size and finish of the piston ring grooves and the surface treatment.
- The engine load.
- The fuel quality.
- The clearance of the various parts in relation to each other such as liner, piston, piston rings and piston ring grooves.
- The amount of moisture in the scavenging air.

As shown in Chapter 26, '*Reconditioning diesel engines and their parts*', this is a rather complex procedure.
Engineers often have divergent ideas as to how honing should be performed. However, today a number of standard procedures are generally accepted.

27.8.2 Working on a cylinder liner in the engine after a certain number of operating hours

Removing the wear edge in the cylinder liner

After many operating hours or excessive wear of the liner, some residue remains right above the top ring when the piston is in top position, whereas the surface below has worn away: the so-called wear edges. In order to draw the piston, the edge must be removed; otherwise the piston is caught behind the edge. Grinding away the edge must be performed with great precision.

Just 5 mm above the edge should be ground away. This creates a space in which the carbon deposit is contained, which prevents it from reaching the cylinder liner.

This avoids wearing of the liner. If the wear edge is not removed in time, a **newly** fitted top ring could potentially crack allowing hot gases to destroy part of the lubricating film and subsequently damaging the second ring. In time, hot gases could pass through the entire spring package. This would result in a heavily damaged piston and a badly worn cylinder liner.

Grinding the wear edge

A wear edge is often removed using an angle grinder. As it is difficult to keep this tool in the correct position, some material **below** the wear edge may inadvertently be removed. This affects the seal between piston and liner, increasing the chances of a blow-out.
It is preferable to use a special tool able to remove the wear edge with great precision, such as a edge grinder from Chris-Marine.

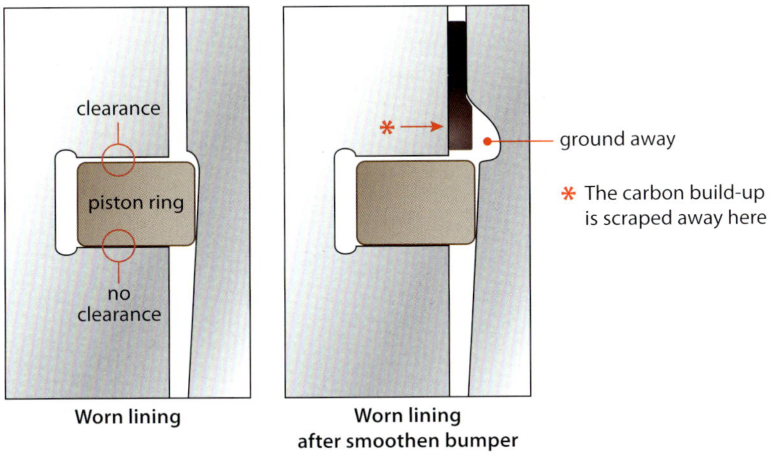

◄ Left; a top piston ring and cylinder liner after a considerable number of operating hours. The spring and the liner are worn just above the top position of the piston ring where the liner does not wear, creating a 'fender', landing surface.
Right; the fender has been ground away creating a space to hold the carbon, preventing deposition in the piston ring package.

▶
A 'fender' grinding machine from Chris-Marine for accurate grinding/lathing of the landing surfaces.

In about 15 minutes the fender is ground away.

Checking the cylinder liner

It is advisable to inspect the running surface of the cylinder liner thoroughly, prior to drawing the piston.

- Is there a sufficient amount of cylinder lubricating-oil throughout the liner?
- Is the cylinder liner still reasonably round?

This is easily ascertained by applying talcum powder to the liner running surface and subsequently turning the engine.

Lack of contact between the piston rings and the cylinder liner can easily be detected as the talcum powder remains on the cylinder liner.

A cylinder liner with a fine honing pattern and correct internal dimensions need not be removed. However, if the liner shows very smooth patches, the liners require re-honing, provided they have not already worn out.
These patches could also be the result of semi-burnt lubricating oil and/or semi-burnt fuel droplets or water mist droplets. It is imperative to check the cylinder-liner running surface even though it can be difficult to establish the cause of the wear spots.

27.8.3 Three types of cylinder wear

Abrasive wear

This is easy to establish: vertical scratches appear on the running surface of the cylinder liner. The wear is caused by abrasive particles in dirty fuel. These so-called 'cat-fines' are remains of catalysts utilised in refining crude-oil, desert sand or grit from sand-blasting. However, abrasive wear could also result from adhesive wear products wearing the liner and the piston rings.

Adhesive wear

This type of wear occurs, for various reasons, when the lubricating-oil film dissolves causing metal-to-metal contact between the piston ring and the cylinder liner. In some instances this phenomenon

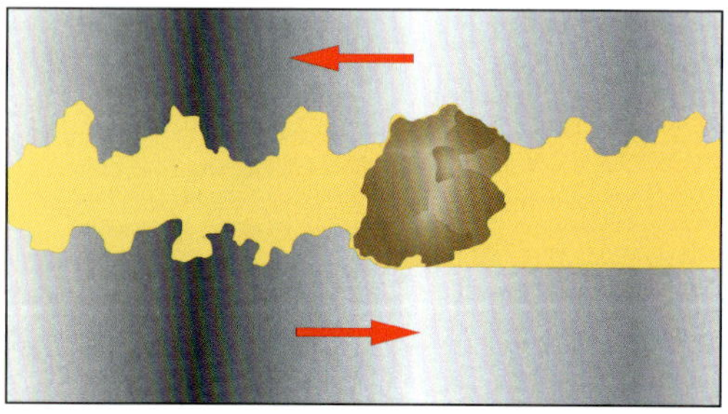

▲
Abrasive wear.

The dimensions of the dark dirt particle are greater than the distance between the moving surfaces relative to each other and create vertical scratches on the running surface of the cylinder liner.

▶
Abrasive wear.

The vertical scratches on the cylinder liner are clearly visible.

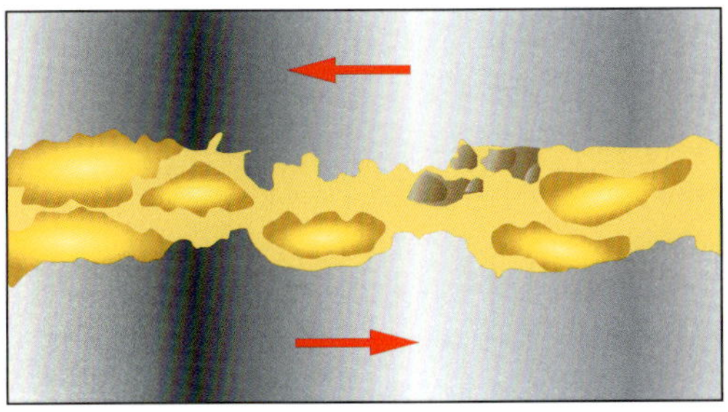

▲

Adhesive wear.

Some particles 'stick' on the surface when the lubricating-oil film disappears, this may result excessive wear or 'scuffing'.

▲

Corrosive wear.

The sulphuric acid produced erodes the material.

Red: sulphuric acid.

is referred to as 'scuffing'. The degree of wear can be considerable and occur in a very short space of time. Scuffing may also result from poor water drainage in the scavenging-air coolers. This is clearly visible through a magnifying glass with a magnification of 8-10.

Corrosive wear

Corrosive wear of the running surface of the cylinder liner is caused by sulphuric acid. This occurs at temperatures below the dew point (about 160° C) combined with low oil film thickness. 'Controlled' and limited corrosive wear enhances the bonding of the cylinder lubricating-oil. The surface of the cylinder liner appears slightly speckled with numerous tiny indentations.

Abrasive and corrosive wear

These two types of wear often go hand in hand. Corroded particles drawn from the running surface by the piston movement cause abrasive wear.

27.8.4 Clover leafing

Clover leafing is a phenomenon where the running surface of the cylinder liner shows signs of wear in areas where the cylinder lubricating-oil is relatively far removed from the supply points. It is the result of an insufficient supply of cylinder lubricating-oil combined with the TBN dip on the running surface. Most of the wear is located between the lubricating points where the lubricating-oil supply and therefore neutralisation of sulphuric acid is less than in areas nearer to the lubricating points. These cylinder liners must be honed until the 'clover' structure has been removed. Chromed piston rings are especially vulnerable to this phenomenon as the chrome layer can disintegrate. The damaged ring surface will cause scuffing or, in some instances, fracturing of the ring.

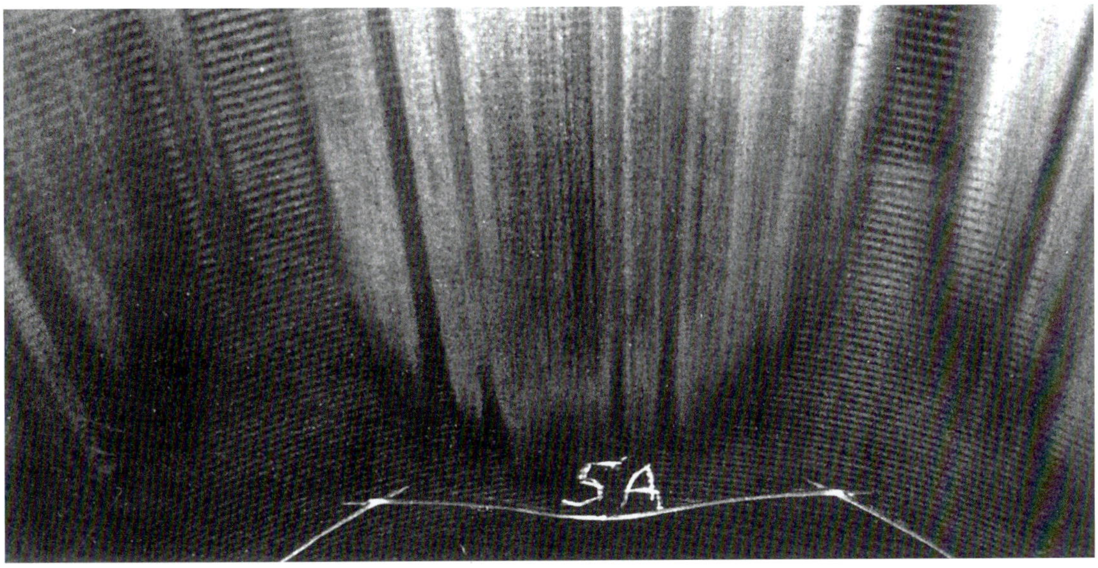

◄

Clover Leaf pattern wear, a form of wear which is regularly found between the supply points of the cylinder lubricating-oil.

▲
If liners run for a while with broken or stuck piston rings, there is a risk that the running surface will be thermally overloaded and that cracks will appear in the upper part of the cylinder liner.

27.8.5 Hammering

Horizontal edges and hammer profiles at the top of the cylinder liner result from broken piston rings beating against the running surface. Hammer profiles produced by parts of the exhaust valve or the piston rings have varying patterns. The cylinder liner can still be used if the scuffing is superficial. Honing is a reliable method for checking the surface and provides it with a proper pattern. When fuel injectors are fitted incorrectly, or in poor condition, thermal tearing can also occur in the top section of the cylinder liner.

27.8.6 Honing of the cylinder

The flow decision chart for honing can be consulted to establish if honing is required. The time required for honing can be as much as 8 hours!
Only in exceptional cases is honing is performed on board. It is usually performed at a revision shop on shore, where further knowledge of honing is available.

Checking the piston and the piston rings

Prior to cleaning the piston and fitting new piston rings, the exterior of the piston should be closely examined.

Precipitation on the piston crown

Precipitation on the piston crown or on the section above the rings provides information regarding the degree of cylinder lubrication. A layer of carbon may have accumulated to the side of the piston, above the piston rings. This layer affects the lubricating-oil film on the cylinder liner. The layer is either formed by an imbalance between the alkaline additives in the lubricating oil and the sulphuric acid formed during the combustion process and/or the result of poor fuel atomisation, leading to the formation of various small carbon particles which adhere to the piston surface.
It could also be caused by fuel leakage.

Too high a dose of lubricating oil is both expensive and may produce an adverse effect.

Dirt and carbon deposits are also often found on the piston ring surface and in the piston ring grooves. Poor combustion and piston ring blow-out add to carbon formation.
Carbon build up generally occurs in ring grooves and behind the piston rings. In four-stroke engines this often results in 'ring sticking', which at times leads to piston seizure. In two-stroke crosshead engines this phenomenon is seen less frequently, although it does cause additional wear.
It is imperative that the rings have free movement. Furthermore, piston ring grooves should always be kept clean.

Scuffing generally starts at the bottom of the cylinder liner as this has the smallest diameter.

Burning of the piston crowns

This is a common occurrence and results from inadequate fuel injection. It may also be caused by the lubricating-oil cooling of the piston crown. Due to internal contamination the piston crown is insufficiently cooled, which causes the surface to burn. Piston rings often fracture when they are lodged to the grooves. Potential problems are easier to detect by examining the piston surface rather the cylinder liner. The type of wear can be determined by using a magnifying glass with an 8-10 magnification.

Clearance of the piston ring in the groove

This is essential and should always be checked during maintenance. Overall, the clearance is measured to ensure that it is not excessive. Too little clearance hinders the ring tension on the cylinder liner, which means that the ring is not firmly pressed against the cylinder liner.

27.8.7 Piston rings

Due to rapidly increasing combustion pressures over the past twenty years, designers of two-stroke crosshead engines have been required to significantly improve the quality of piston ring technology and the materials applied in order to acquire an acceptable degree of wear.

▼

A flow chart showing the decision making process for the necessity of honing of the cylinder liner.

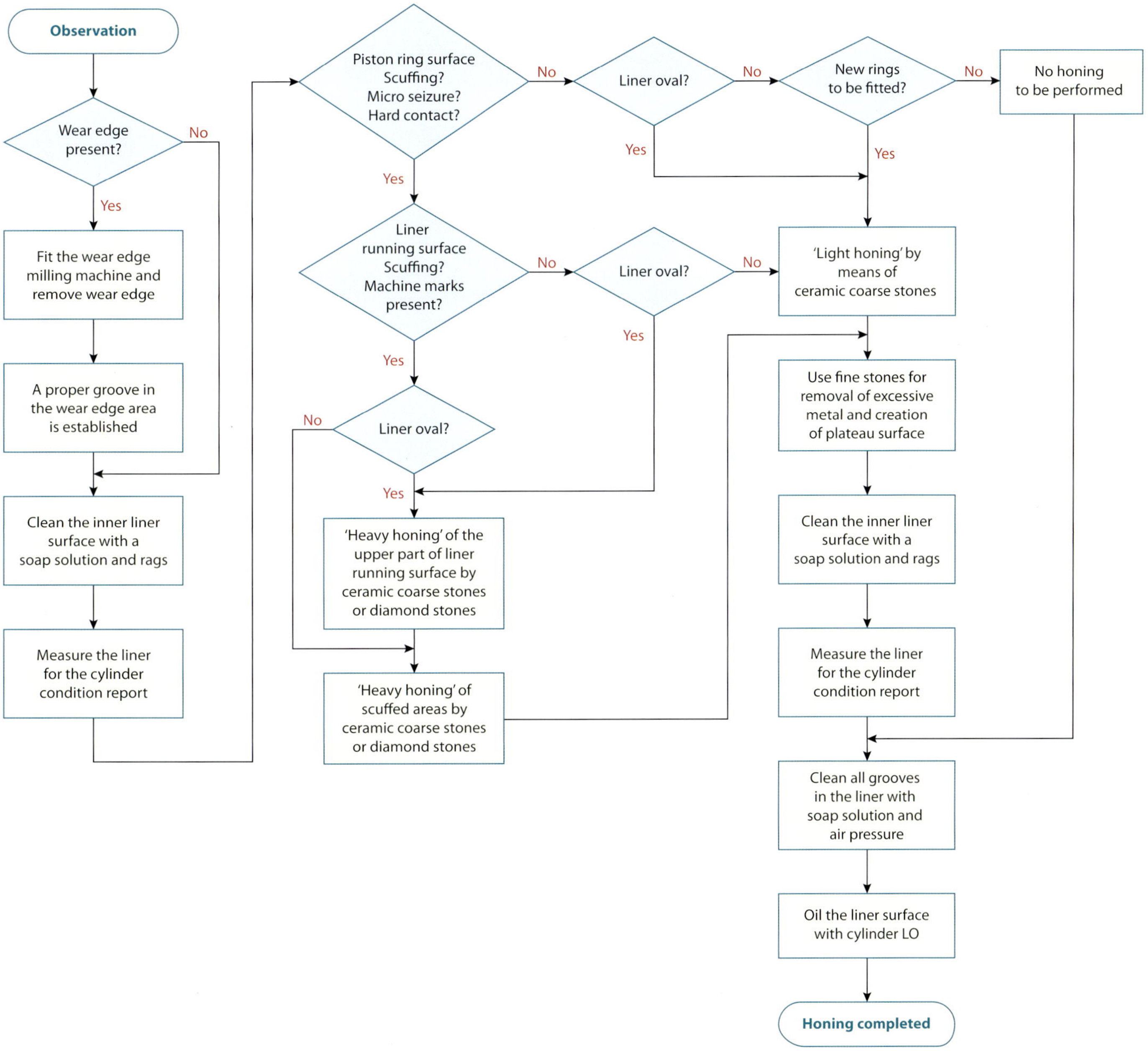

►

A 'pulled' piston, the piston is not very dirty; all piston rings are loose and not broken. The running surface of the piston rings is smooth so that they run **entirely** on the running surface of the cylinder liner. A perfect piston!

►

A piston with a little more contamination. After cleaning and measuring the piston rings, piston ring grooves and the diameter of the cylinder liner, it can be decided whether the piston can be re-installed in the engine.

►

This piston crown of a two-stroke crosshead engine is severely worn. After the dry blasting, it is clearly visible that the rings are broken or stuck. At the left, the piston material is so hot that it melted and the piston ring has been moved! The piston can be reconditioned, inspected and re-installed in the engine.

27.8.8 Materials

Although simple cast iron piston rings are still used today, they are usually not suitable for carrying high loads. Particularly the top piston ring and increased cylinder diameters create problems.

▲
A typical example of a conventional piston ring. Depending on the microstructure, it is possible to increase the wear resistance with so-called hard iron-carbides in the perlitic crystal lattice.

▲
Blunt Edge graphite structure, manufacturer's name: Daros RM 5. This has a better structure than the perlitic structure; it is stronger, tougher and conducts heat better.

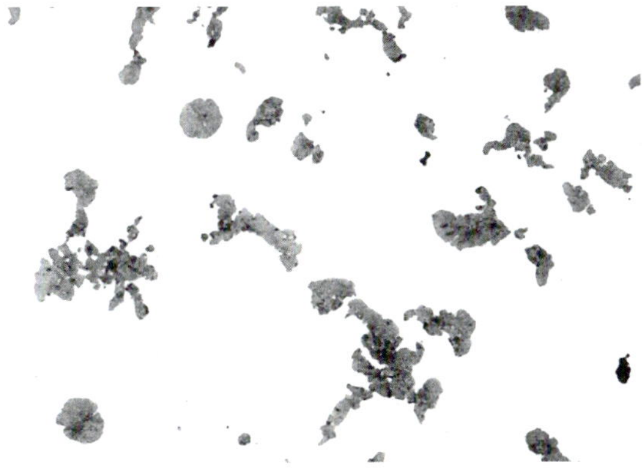

▲
Vermicular iron, manufacturer's name: Daros Darcast. Engine manufacturers often choose the material in consultation with the piston ring manufacturer.

In recent years this has increasingly led to the development of advanced piston ring coatings and piston ring profiles. These developments went hand in hand with an increase in development- and manufacturing-costs and while in theory the extended life of the improved piston rings should easily compensate for these additional costs, this has not been the case.

Material for a standard piston ring

There are three variants:

- Laminar cast iron;
- Fine perlitic cast iron;
- Nodular cast iron.

The cast iron structure is enhanced by alloys to obtain the required properties such as material strength, microstructure, thermal stability, thermal conductivity and wear resistance.
Fine perlitic cast iron, often provided with a coating, is commonly used for the top piston ring.
Laminar cast iron is used for the lower piston rings which are also often coated. Nodular iron is currently **not** used.

27.8.9 Coatings

Applying a special coating in heavily loaded engines is beneficial as it either enhances the running-in of the ring to the cylinder liner or reduces wear and therefore extends the life of the rings.

Electrolytic hard chrome piston rings are widely used for medium-speed four-stroke- and high-speed diesel engines, for heavy oil, diesel oil or gas oil. Mitsubishi was the first manufacturer of two-stroke crosshead diesel engines to use these rings as their standard top ring in UEC75LSII engines.

However, one of the drawbacks of these chrome piston rings is that they are vulnerable to high temperatures. Blow out causes the coating to detach, which subsequently affects the pre-tensioning of the ring which consequently loses its seal.

However, if the material temperature of a chrome piston ring remains below 230 °C, it remains the best option where wear resistance and low friction coefficient are concerned. At temperatures over 250 °C it rapidly loses its strength /hardness, which reduces its quality significantly.

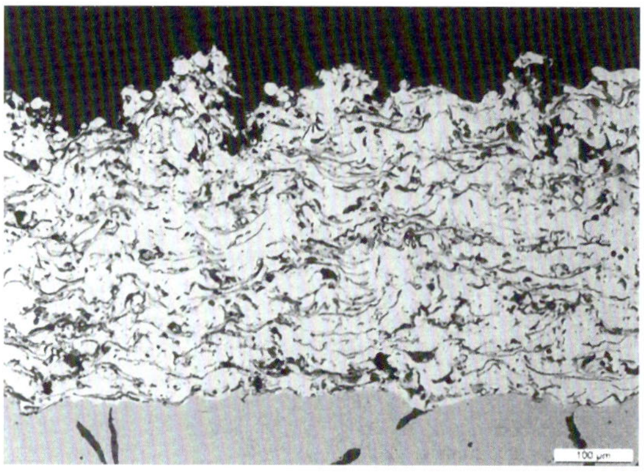

▲ A piston ring with a ceramic layer.

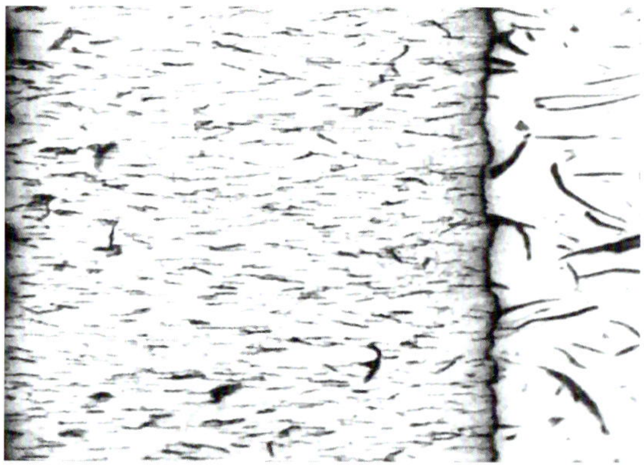

▲ A piston ring with an electrolytic plated hard chromium layer.

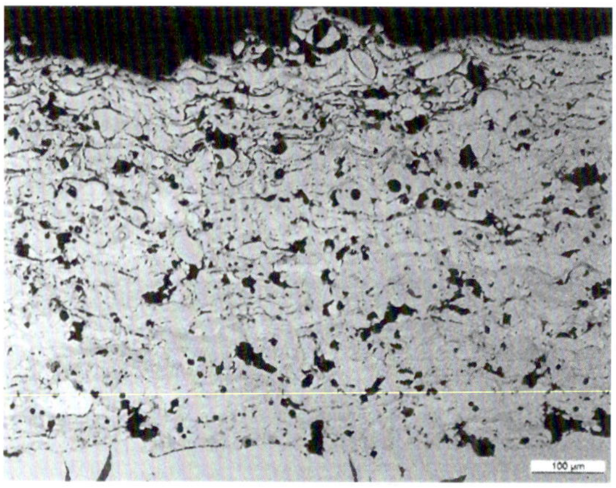

▲ A piston ring with an aluminium layer.

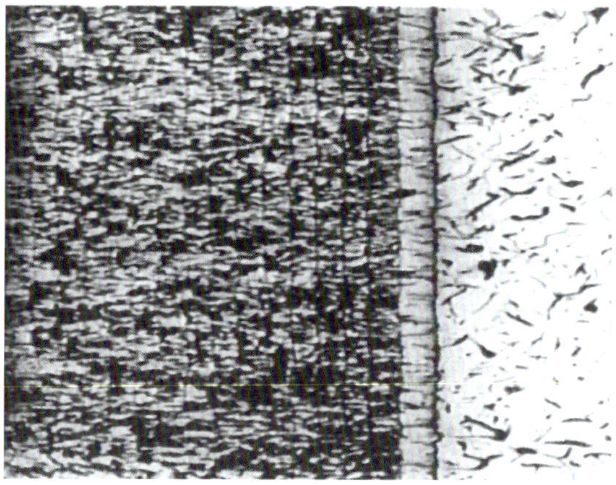

▲ A piston ring with a chrome-ceramic coating used in the modern Wärtsilä two-stroke crosshead engines.

Plasma ceramic coating

This consists of a mixture of carbides and metal components and has a hardness of approximately 1,500 HV and 250 HV (Vickers pyramid number). The 'Cermet', ceramic metal is sprayed onto the pivoting piston ring and has much better material temperature tolerance than both electrolytic and ceramic chrome, usually in the order of 500 °C, before it is permanently damaged. Moreover, the material has an increased scuffing resistance. In fact, it is virtually indestructible. One of the disadvantages of plasma coating is that it can cause extremely high cylinder liner wear and not only during 'running in'. In some instances, 'running in' can not be completed as the piston rings and the cylinder liner need to be replaced due to excessive wear. Both cast iron piston rings and plasma coated piston rings can be supplied with a copper or aluminium coating, also referred to as RIAS, with a profiled running surface. This surface allows for a rapid and safe 'running-in' process.

Below a list of the most common piston rings and the material used by the three manufacturers of two-stroke crosshead engines.

Note: MAN Diesel engines with large borings have CPR (constant pressure relief) pistons in the upper piston ring groove. Both the bottom parts of this ring as well as the second ring are provided with a chrome coating.

MAN	Götze	Daros	NPR	Riken	Material Composition	Coating	Cylinder liner Material
Uballoy		Uballoy	Uballoy	Rik45	Gray cast iron	None or Alu-coat	Tarkalloy
CF5	LP8	RM5	Uballoy-S	Rik47	Copper-molybdenum alloyed, flake graphite	None, Alu-coat or Cermet + Alu-coat	Tarkalloy or Tarkalloy-C
CF4		RM4			Vanadium and copper-molybdenum alloyed flake graphite	None or Alu-coat	**PVA or Tarkalloy-C***
		RVK**	Tarkalloy G	Rik29VMD	Alloyed high strength compact graphite	Cermet + running-in coating	Tarkalloy-C
CV1	VP6	RVK-C	Tarkalloy G	Rik29VMD	Alloyed high strength compact graphite with wear reducing carbides	Alu-coat or Cermet + Alu-coat	Tarkalloy-C

Cermet coating is a plasma sprayed ceramic metal alloy having low wear properties and high scuffing resistance.
Alu-coat (golden appearance) running-in coating with a soft Cu matrix including hard aluminium oxide phase.
* C8 and RM4 piston rings are used in special cases (old standard).
** RVK = Darcast

▲

MAN Diesel two-stroke crosshead engines. Types MC and ME. Preferred piston rings.

Note: The larger cylinder piston diameters have CPR piston rings in the top piston-ring groove and chromed surfaces at the bottom of the ring. The second piston ring also has chromed surfaces on at the bottom of the ring.

▼

Wärtsilä two-stroke crosshead engines. Types RTA and RT Flex. Preferred piston rings.

Engine Type	Piston ring position	Daros	Götze	Riken	Material Composition	Cylinder liner Manufacturer
RTA/RT-flex96C	1	Darcast PM2 BF	SCP1CC20	Rik29Mo	GGV, Chrome Ceramic	**
RTA/RT-flex96C	2	RM4TF	SCP2CC20	Rik29Mo	GG spec. III, Chrome Ceramic	**
RTA/RT-flex96C	3	RM4TF	SCP2CC20	Rik47	GG spec. III, Chrome Ceramic	**
RTA/RT-flex96C	4	RM4TF	SCP2CC20	Rik47	GG spec. III, Chrome Ceramic	**
RTA84C/CU	1	Darcast	SCP1RC20	Rik29	GGV, RIAS	**
RTA84C/CU	2	RM4	SCP1RC20	Rik47	GG spec. III, RIAS	**
RTA84C/CU	3	RM4	SCP1RC20	Rik47	GG spec. III, RIAS	**
RTA84C/CU	4	RM4	SCP1RC20	Rik47	GG spec. III, RIAS	**
RTA84C/CU	5	RM4	SCP1RC20	Rik47	GG spec. III, RIAS	**
RTA/RT-flex84T	1	Daroseal***	SCP1CC20	Rik29	GGV, Chrome Ceramic alt. RIAS	**
RTA/RT-flex84T	2	RM4***	SCS1RC*	Rik47	GG spec. III, RIAS alt. Chrome Ceramic	**
RTA/RT-flex84T	3	RM4***	SCP1RC20	Rik47	GG spec. III, RIAS alt. Chrome Ceramic	**
RTA/RT-flex84T	4	RM4***	SCP1RC20	Rik47	GG spec. III, RIAS alt. Chrome Ceramic	**
RTA62U	1	Darcast	SCP1CC17	Rik29	GGV, Chrome Ceramic alt. RIAS	**
RTA62U	2	RM4	SCP1RC13	Rik47	GG spec. III, RIAS	**
RTA62U	3	RM4	SCP1RC13	Rik47	GG spec. III, RIAS	**
RTA62U	4	RM4	SCP1RC13	Rik47	GG spec. III, RIAS	**
RT-flex58T	1	Darcast	SCP1CC16	Rik29	GGV, Chrome Ceramic	**
RT-flex58T	2	RM4	SCP1RC16	Rik47	GG spec. III, RIAS	**
RT-flex58T	3	RM4	SCP1RC16	Rik47	GG spec. III, RIAS	**
RT-flex58T	4	RM4	SCP1RC16	Rik47	GG spec. III, RIAS	**

☐ WHC Standard Specification
☐ Designation of compatible or alternative piston rings
* Valid for engines fitted with SIPWA only (SIPWA piston rings are uncoated)
** Typically HHI, CAPE, or ToaKoki, or anyone fulfilling spec. GJL-250 according DIN 1561 ISO 185
*** New spec. under construction

▶

Mitsubishi two-stroke crosshead engines. Types UEC and MHI/RTA96C.

Engine Type	Piston ring position	Riken	NPR	Material Composition	Cylinder liner Manufacturer
UEC50LSII	1	Rik29Cr		Cr-plated vermicular graphite	Toa Koki
UEC50LSII	2	Rik47		Cu-Mo alloyed flake graphite	Toa Koki
UEC50LSII	3	Rik45		Plain grey cast iron	Toa Koki
UEC50LSII	4	Rik45		Plain grey cast iron	Toa Koki
UEC60LS	1	Rik47Cr		Cr-plated Cu-Mo alloyed flake graphite	Toa Koki
UEC60LS	2	Rik47Cr		Cr-plated Cu-Mo alloyed flake graphite	Toa Koki
UEC60LS	3	Rik47		Cu-Mo alloyed flake graphite	Toa Koki
UEC60LS	4	Rik47		Cu-Mo alloyed flake graphite	Toa Koki
UEC 60HA & 52/125H	1		Uballoy S	Cu-Mo alloyed flake graphite	Toa Koki
UEC 60HA & 52/125H	2		Uballoy S	Cu-Mo alloyed flake graphite	Toa Koki
UEC 60HA & 52/125H	3		Uballoy C	Plain grey cast iron	Toa Koki
UEC 60HA & 52/125H	4		Uballoy C	Plain grey cast iron	Toa Koki
UEC 60HA & 52/125H	5		Uballoy C	Plain grey cast iron	Toa Koki
MHI/RTA96C	1	Rik29Mo		Mo-coated vermicular graphite	Toa Koki
MHI/RTA96C	2	Rik29Mo		Mo-coated vermicular graphite	Toa Koki
MHI/RTA96C	3	Rik47		Cu-Mo alloyed flake graphite	Toa Koki
MHI/RTA96C	4	Rik47		Cu-Mo alloyed flake graphite	Toa Koki

27.8.10 Piston ring design

Today, all two-stroke crosshead piston engines have profiled cross sections, usually just the upper ring, but occasionally all rings, depending on the engine design. This is due to the increased ring height and the subsequently increased radius of the curvature of the ring's running surface. This is the simplest manner in which to obtain a sufficient oil-film thickness between the piston ring and the cylinder liner for a given cylinder pressure. An increase in the viscosity of the lubricating oil by, for instance, reducing the liner temperature, can also increase the oil-film thickness.

The introduction of 'tall' piston rings poses a problem during the 'running-in' process as a, say, 30% increase in the height of a ring produces a factor 2 increase in the distortion resistance. This affects the piston ring adaptability to the geometric conditions at varying engine loads, temperatures and piston positions.

Shipping company A.P. Møller-Maersk introduced the so-called 'minutes piston rings' or 'conical rings' in large two stroke crosshead engines towards the end of the 1980's to improve the 'running-in' process. Shortly thereafter, MAN Diesel also introduced the conical piston rings as their new piston rings.

▶

Two different types of piston rings with evidently different diameters.

Left: A 'barrel' type piston ring with rounded corners, a barrel shape.

Right: A tapered piston ring.

Upper drawings: Not loaded, the engine stopped.

Lower drawings: Loaded, the engine running.

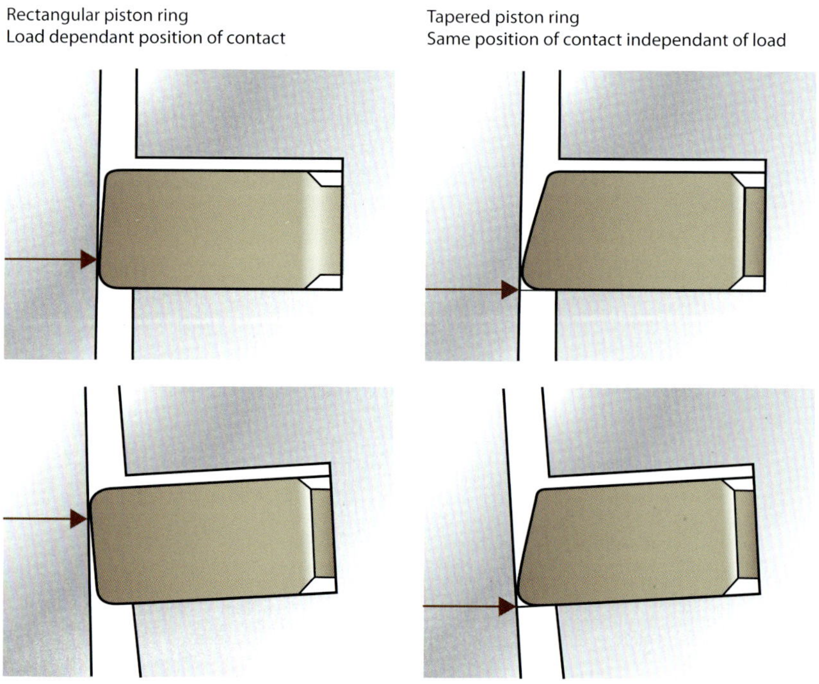

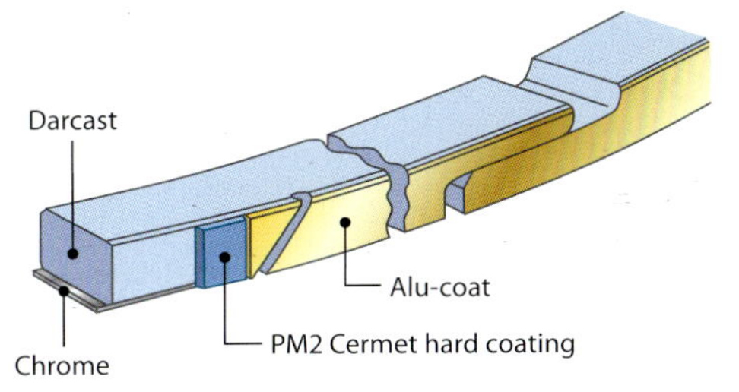

▲
A piston ring with clearly visible, one of its four slanted ducts, a 'Controlled Pressure Relief' type.

Clearly, a taller and hence more rigid piston ring requires a more conical profile than the standard shorter ring.

It should be noted that a new Wärtsilä cylinder liner has a surface roughness comparative to a 'run-in' liner. The underlying idea is to reduce the 'running-in' process and the consequent problems.

It is of the utmost importance that the aforementioned conically shaped piston rings are fitted with the marked 'top' side up. In the absence of a top mark, the top is provided with the name of the piston ring manufacturer!

Controlled Pressure Relief piston rings

The running surface of this type of piston ring has been supplied with four grooves. This way the heat is more evenly distributed over the cylinder liner, which allows for a steady pressure reduction over the ring package. This reduces the wear of the piston rings. A wear resistant plasma coating is required to protect the groove during operation as it prevents carbon deposit formation. Otherwise the ring could be rendered completely gas tight.

27.8.11 Inspection via inlet ports

Inspections via the inlet ports are imperative for monitoring the condition of the cylinders. It is advisable to perform an inspection every two to four weeks.
This inspection is certainly recommended after a long crossing.

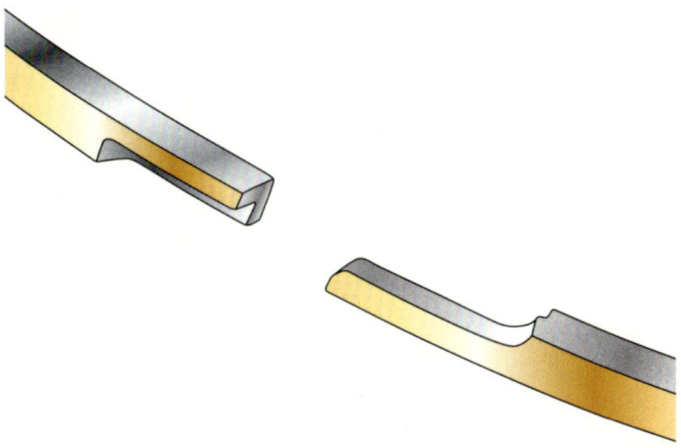

▲
A piston ring with a lock hole: Type; airtight. Ring with an aluminium bearing-surface.

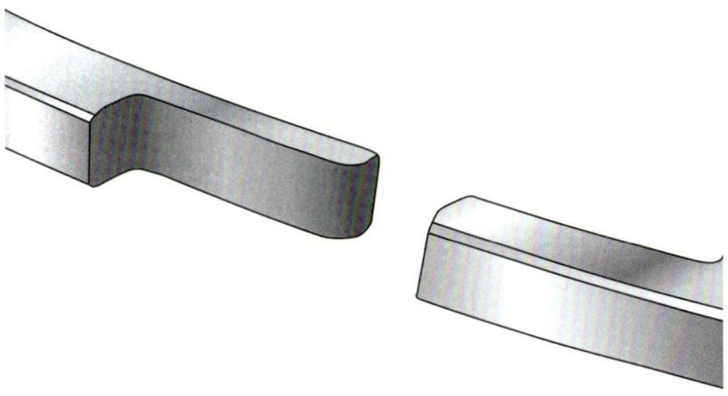

▲
A piston ring with a lock hole: Type; 'Daroseal'.

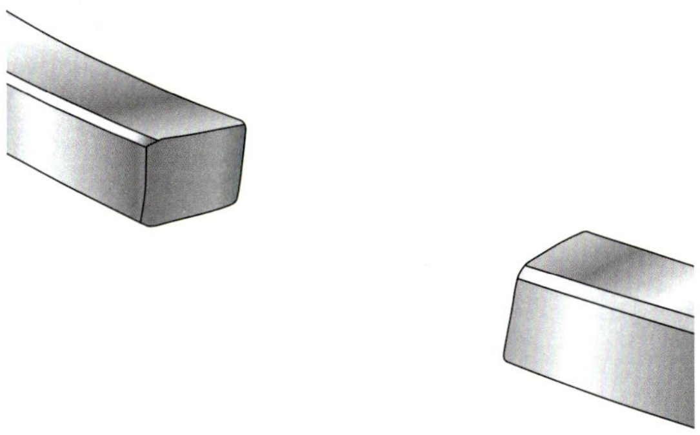

▲
A piston ring with a lock hole: Straight ring type.

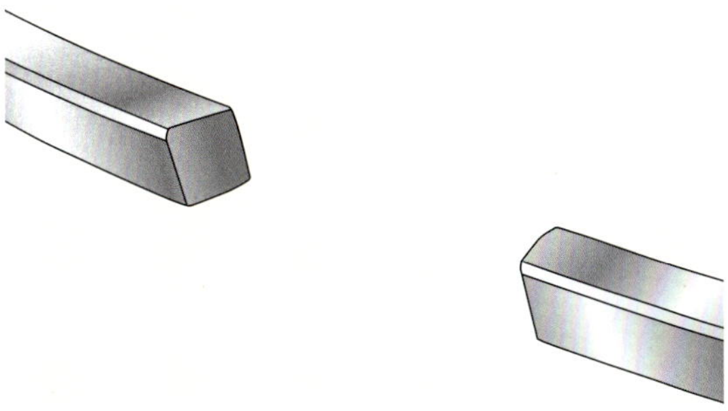

▲
A piston ring with lock hole: Type; oblique rings.

▲
A good picture of the wear process and pollution of piston rings, piston ring grooves, piston and bearing surface of the cylinder liner can be obtained via the scavenging-air receiver through the scavenging air ports.

▲
The image through the scavenging-air ports. The wear can be tracked with the use of photos.

27.8.12 The inspection

This can be performed in two ways:
Through the covers in the cam shaft section.
Through the scavenging air space. The latter is preferable.

The piston should be inspected in the following locations:
- Top side of the piston.
- Piston ring package.
- Piston skirt.
- When the piston is in top dead centre position in order to inspect the cylinder liner.

When checking the piston rings special attention must be paid to:

- Pointed edges or burrs at periphery of the piston rings.
- Measuring the piston ring-to-groove clearance after a considerable number of operating hours.
- Fractured rings.

▶
Piston ring wear.

Vertical: The increase of the slot opening in millimetres.
Horizontal: The wear of the piston ring in millimetres.
New ring: Gap, 6 millimetres.
Worn ring: Gap, 21 millimetres.
Wear ring: (21-6)/2π = 2.4 millimetres.

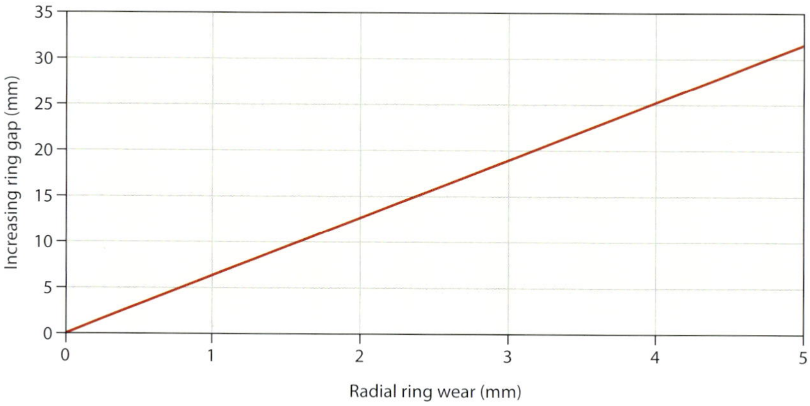

▲
In a two-stroke crosshead engine, an inspection can be carried out from the scavenging-air beam through the scavenging-air port. These piston rings are in good condition, clean, well run in with more than enough lubricating oil.

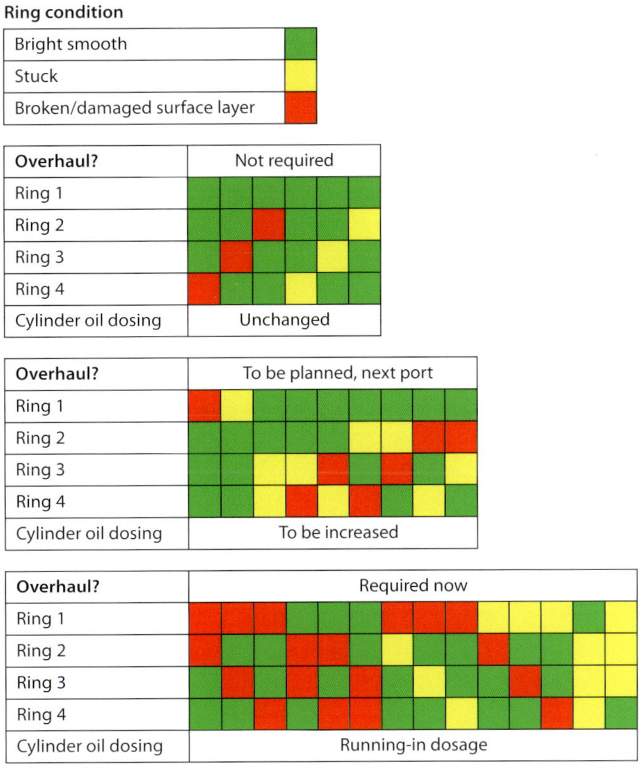

▲
This graph shows the inspections performed via the scavenging-air ports and recommendations on the steps to be taken.

27.8.13 'Running-in' the conical rings

This can simply be established since the lower section of the ring rubs against the liner, whereas the top section can move freely.

This can be established after 18 to 24 operating hours.

27.8.14 Evaluation of scavenging-air ports inspections

After photographing all rings and charting each ring of every cylinder an action plan can be drawn up.

Additional information, such as carbon deposits on the top section of the piston crown, behind and in between the piston rings, the running surface of the cylinder liner and potential detachment of ceramic- or chrome coatings on the piston rings may help when drawing-up such a plan.

27.8.15 A few examples of damage, wear and soiling

The optimal cylinder

- Good quality piston rings are not fractured, have smooth running surfaces and no pointed edges.
- There is no carbon build-up in the grooves behind the piston rings and there is a limited amount of carbon deposition above and in between the piston rings.
- The entire surface is provided with a layer of lubricating oil.
- The inlet ports are still dirt-free.
- The cylinder liner shows no significant signs of local or overall wear when the piston is in BDC.
- The pattern of a wave-cut surface liner is still intact.
- The cylinder liner shows no corrosive wear as a result from sulphuric acid formation.
- The liner surface has not been hammered by fractured piston rings or by parts of the exhaust valve.
- The upper section of the cylinder liner bears no thermal fractures.
- No tearing of the liner surface due to a poor condition of the piston ring package.

- No scuffing of the liner surface due to dirt behind the piston rings in the vicinity of the inlet ports.
- Neither the piston crown, nor the section above the piston rings or the piston rings themselves have been burnt as a result of a defective fuel injector.
- The piston grooves have not been burnt by heavy leakage of hot gases along the piston rings (heavy blow out).
- The grooves show no thermal tearing resulting from heavy leakage of the piston ring.

Further focal points:

- One often finds that pistons are overly dry and stuck in the ring groove (sticking). To resolve this issue, the lubricating-oil dosage should be increased and the cylinder monitored closely. The piston should be drawn for maintenance or replacement in the next port.
- The running surface of a functioning piston ring is smooth and lightly coloured. When a ring is wedged in a groove it has a much darker colour.
- Piston rings supplied with a chrome ceramic coating have excellent running properties, but do require an even surface and optimal lubrication conditions. If the running surface of a piston shows irregular patches, the piston should be replaced immediately as the remainder of the coating will peel off rapidly.
- A piston ring with an aluminium coating often wears at the inlet ports: here the ring's ends are pushed rather hard into the ports. In these areas the aluminium coating may start to detach and is further peeled away. Inspections should be performed at regular intervals. Replacement of the piston rings should be considered only if the coating detachment is significant and is not limited to the ends.

▶

A clearly worn cylinder liner with vertical scratches. Here, double vertical rows with cylinder lubricating-oil supply channels.

▼

A photo taken from the top of the liner.

At the time of taking this picture the cylinder liner was being tested. The cylinder lubricating-oil runs straight down here (1).

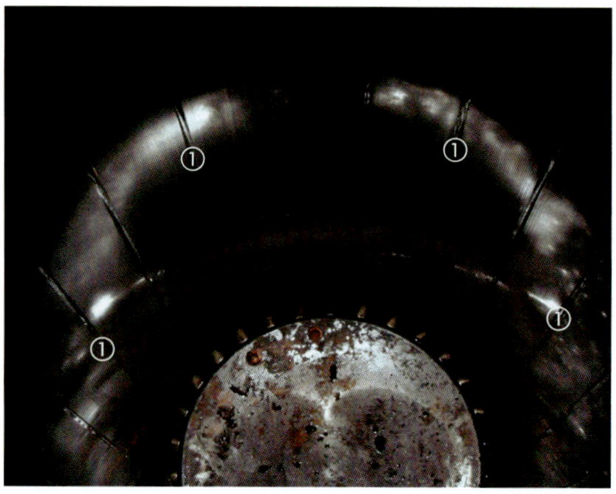

▼

Inspection of a two-stroke crosshead engine.

Digital cameras are commonly used nowadays for the inspection of a two-stroke crosshead engine to ascertain the condition of the piston crown, piston rings, piston skirt, piston rod and cylinder-liner wall. It is important to include/use:

- A 'flash' which allows a much clearer picture, particularly of scratches and stains;
- Indicate the position and direction from where the photo was taken;
- Record the cylinder number and
- Record the date and number of operating hours.

With good lighting and the remote control of the turning motor, an image can be taken from any desired piston position.

1

2

3

4

5

6

7

8

9

10

11

12

13

14

15

16

27.8.16 Checking the maintenance of cylinder liners

Two reasons:

1 To avoid sudden engine failure.
2 Keep the engine operational in the most cost-effective manner.

Traditional measuring

Measuring the cylinder liner and rings over given periods provide values that are difficult to diagnose. However, it is the most reliable method to acquire insight into the maintenance of a cylinder. It is imperative that the mechanical engineer literally be inside the cylinder liner while these measurements are taken, so the running surfaces can be closely examined. No measurement could ever replace the eyes, nose and finger tips.

Engine manufacturers take at least 10 measurements to obtain a good insight into the degree of wear. These tests are performed both longitudinally and athwartships. Measuring is ideally executed under test conditions. On board a ship, measuring the first ring in BDC, the fourth ring in BDC, and at ⅓ of the downward stroke of the piston suffices.

Obviously, the results of these measurements need to be recorded. In most engines, the piston ring performance determines the time in between two maintenance periods. Particularly, in case of fractured top piston rings, maintenance is easily performed. **Piston rings are and will continue to be the most heavily loaded engine parts and despite the best technology piston rings may still break.**

The duplicate method from 'Replica's'

Originally developed some 30 years ago by Rolls Royce aviation engineers, the replica technique is a unique method of non-destructive research and analysis.

The so-called Repliset from Struers is a rapidly drying method which takes between 1 and 20 minutes, depending on the material temperature. A two component silicon-rubber mixture with a 0.1 micron resolution can provide a very accurate 3-D representation of the microstructure of the liner surface in the engine.

The Replica technique is often used in power stations and in the aviation-, chemical- and offshore industry. It was developed for advanced laboratory analyses, both optically and using SEM (scanning electron microscopes).

Over the past 10 years A.P. Möller Maersk has applied this technique successfully using a simple and inexpensive Struers microscope with a magnification of 100.

The reason is that the requirements on the most important parameters which significantly affect the cylinder liner condition, such as hard graphite particles, catalytic products, corrosive and abrasive wear are relatively low.

The running surface of the cylinder liner should be cleaned thoroughly prior to applying the test tape with the replica compound on the back of the tape to the running surface.

Occasionally, the running surface is polished with a diamond paste and subsequently etched in order to bring out the optimum microstructure. Generally, this is only carried out in a foundry or a processing shop for evidence and documentation of the correct microstructure and in particular for checking for hard graphite particles, distribution and the type of graphite.

Analysis of the drained lubricating oil in the scavenging air pipe

In recent years, various propulsion engines have been fitted with measuring devices on the lubricating-oil drain in the scavenging air pipe for each individual cylinder. This way, a lubricating-oil analysis can be obtained showing, for instance, the correlation between the amount of sulphur in the fuel and the amount of lubricating oil supplied to a certain cylinder and the wear of the cylinder liner.

There are various options:

Check for a typical wear of the cylinder liner by establishing the cast iron contents and TBN.

- An engine 'footprint' indicates if the manufacturer's instructions concerning the sulphur related lubricating-oil dosage corresponds to the number of iron particles in the lubricating-oil samples at various loads.

- **Type of wear: abrasive or corrosive.** Here the TBN number is correlated to the iron content. A high TBN number for the drained lubricating-oil samples is indicative of a corrosive layer, which means that the type of wear is abrasive (also referred to as 'scuffing').

It is important to know the extent to which the drained lubricating oil has been diluted with crank-case oil. Each engine has its own specific characteristics and the degree of dilution depends on the manner in which the piston-ring stuffing boxes are maintained. However, in practice the dilution can be disregarded when comparing cylinders.
Erroneous measurements may occur when cylinders are affected by one cylinder showing excessive wear. This may occur in Wärtsilä Sulzer engines, which have openings in the frame between the cylinders. When testing Wärtsilä engines, the three adjoining cylinders should also always be measured. The ensuing test results must be carefully analysed before any conclusions can be drawn.
This type of contamination does not occur in MAN Diesel and Mitsubishi engines.

Temperature measurements of the cylinder liner

Cylinder liner temperature measurements have been taken for all three engine types since the beginning of the 1990's. Applying this method may prevent the perishing of the liner. Liner temperature measuring in conjunction with the analyses of drained cylinder lubricating-oil provides a good insight into cylinder-liner wear.

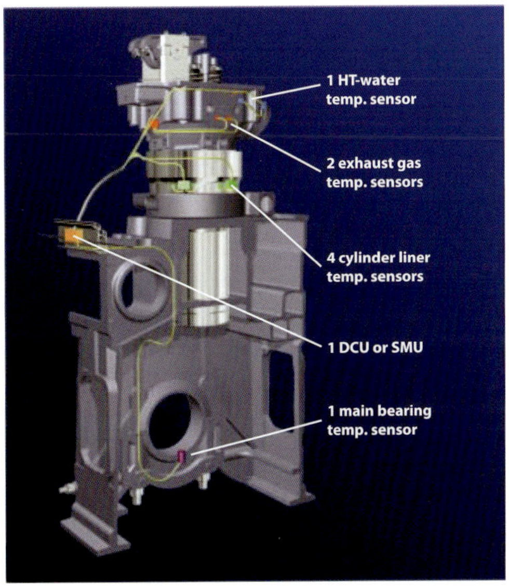

◄ The temperature measurements for a Wärtsilä medium-speed four-stroke engine operating with heavy fuel oil.

Measurements of the H.T. coolant temperature, exhaust-gas temperatures after the exhaust valve (2×), material temperatures of the cylinder liners (4×) and material temperature of the main bearing shell. All data is sent via the DCU or SMU to the engine control system.

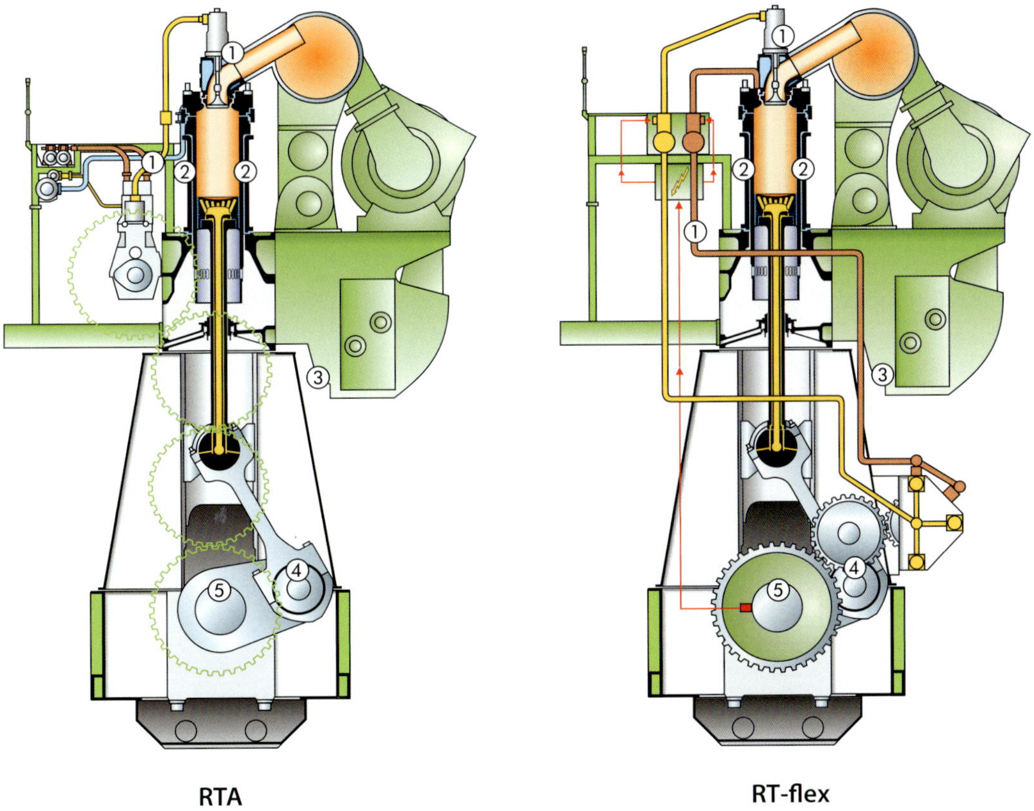

◄ Measured values of the Mapex system from Wärtsilä.

1 *cylinder cooling-water inlet and outlet temperatures*
2 *liner wall temperature*
3 *scavenging-air temperature*
4 *engine load*
5 *engine speed*

27.8.17 Other causes of cylinder-liner wear

The air supply

This may be heavily contaminated with dust, sand and water. It is of the utmost importance that dust and sand are prevented from entering the cylinder liner by filters fitted on the turbochargers, Especially in North Africa, the Red Sea- and the Persian Gulf area, dust- and sand storms can carry very fine dust particles over long distances. Although the dust appears powdery, it is hard and abrasive. Furthermore, the air may contain airborne iron oxide, in particular in the region north of Oman where the mountain soil holds large amounts of iron oxide. This can be seen in the iron contents of the drained cylinder lubricating-oil.
The fine dust particles produce a tremendous abrasive action in the cylinder, thus causing excessive wear of the piston rings and the cylinder liner.

Therefore, special attention must be paid to the filters on the turbochargers!

Air coolers, water mist catchers and draining

Air coolers require proper maintenance and regular cleaning to sustain a high-cooling effect and ensure low scavenging-air temperatures. Water mist catchers and draining systems are equally important. Automatic draining systems should function properly and consistently and the drains should be checked regularly for clogging and dirt. An orifice could be entirely plugged with sludge.

Engine performance

One should avoid exceeding the maximum material temperature of the cylinder liner as this would upset the thermal stability of the lubricating oil: the lubricating oil will burn!

The thermal load of the engine parts will increase in the course of time, but can be positively influenced by proper maintenance.

1. Turbochargers significantly affect the condition of the cylinder if their efficiency is reduced and the thermal load of the engine increases. It is of the essence that the exhaust-gas turbine is kept clean and the blades and clearances match the instructions.
2. Air cooler performance will gradually diminish if the heat exchangers in the cooling water system are defective and the material temperatures of the cylinder liners will consequently increase.
3. Fractured of seriously worn piston rings will effect a 10 ° to 15 °C material temperature increase of the cylinder liners and will require immediate replacement.

Delayed combustion should be restored to test conditions

Some mechanical engineers keep the maximum combustion pressure below that of the test condition. This, combined with the effects of the VIT systems can delay the fuel injection considerably and will therefore delay the combustion process.
In the pV-chart this will create a 'dromedary' effect. In modern engines operating conform the IMO NO_X requirements, a similar effect can be seen at loads over 90% MCR.
This means that a larger section of the cylinder liner running surface is exposed to the flames generated by the combustion process, resulting in a deterioration of the cylinder lubricating-oil and causing increased wear, seized and fractured piston rings.
Generally, the maximum combustion pressure for all loads should comply with the manufacturers' specifications.

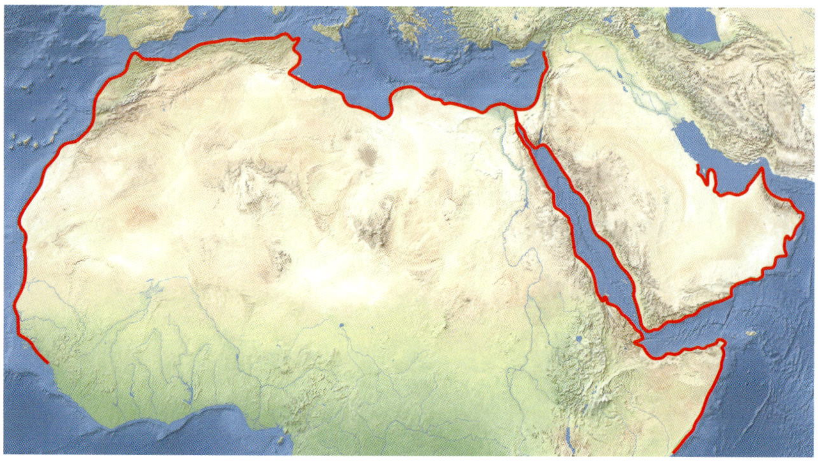

▶ Part of North Africa, the area around the Red Sea and part of the Persian Gulf where dust storms frequently cause problems for diesel engines!

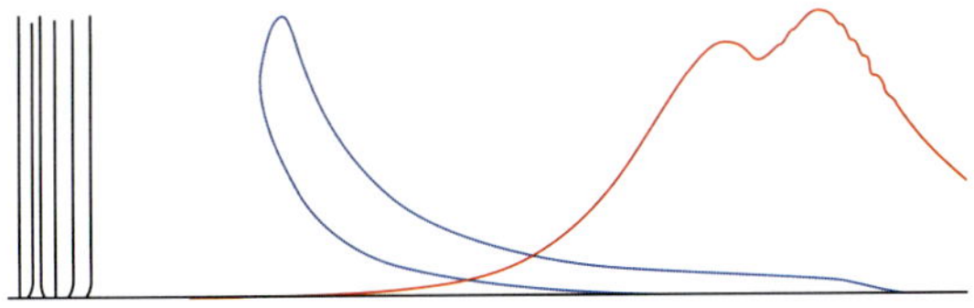

▲
A *pV*-diagram for late fuel injection. On the left, vertical peaks, the maximum combustion pressures (black). Thereafter a 'banana' (blue) and finally a 90° shifted curve (red).

27.8.18 Fuel and fuel systems

Today's poor fuel quality poses problems which were formerly unknown.
Three fuel parameters are crucial for the condition of the liner running surface:

- The sulphur content.
- The CCAI.
- The number of catalytic products, in particular, aluminium and silicon.

The sulphur content of the fuel does cause any problems provided that the lubricating dosage has been accurately modified. However, one must bear in mind that the sulphur content affects the degree of wear.

Problems arise when mechanical engineers are required to frequently change the type of fuel due to demands from local authorities. This means that either the cylinder lubricating-oil dosage and/or the TBN-number must be modified to the lowest TBN-number to acquire a stable oil film.

An elevated CCAI number causes a delayed combustion. The implications were discussed earlier.

The problems associated with catalytic products, the so-called 'cat fines' occur in all heavy oil systems. Refinery residue such as hard particles consisting of aluminium, silicon and oxygen will be transferred to the cylinder liners and piston rings through fuel injection in the cylinder. The negative effects of cat fines can only be countered by the usage of well-functioning fuel centrifuges.

A well-functioning centrifuge is indispensible even if the cat fine content falls well below the maximum permissible value!

Remarkably, fuel with low sulphur content often contains a high percentage of cat fines. This is due to the fact that low sulphur content is obtained by blending, which may produce any percentage of cat fines. So a malfunctioning fuel centrifuge can cause significant and rapid wear of the piston rings and cylinder liner.

▼
Sudden cylinder liner wear due to a poorly functioning centrifuge system.

After approximately 13,000 hours, wear on all cylinders suddenly appears. The green coloured cylinder which had been renewed has the same problem!
After solving the centrifuge problem, the wear decreases immediately.

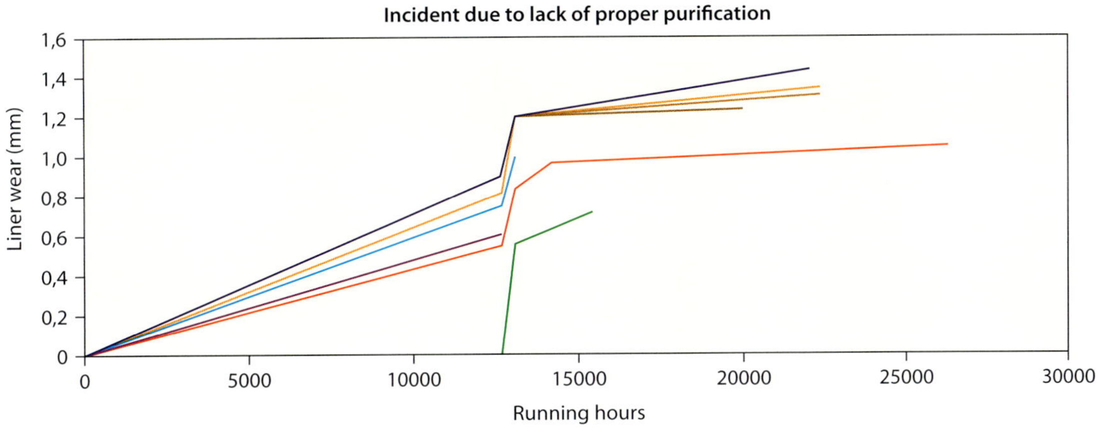

Standard regulations for cylinder lubricating-oil

Standard lubricating oils supplied by large oil companies have all been approved by the engine manufacturers. However, this is not to say that they all have identical properties to prevent wearing and scuffing of piston rings and cylinder liners, although no lubricating-oil company produces lubricating oil too inferior to use. Some low-priced lubricating oils are only suitable for low-load engines.

The composition of the additives also varies. The cylinder liner is predominantly affected by corrosive wear resulting from sulphur forming sulphuric acid. This acidic compound can be neutralised with alkaline components, usually calcium salts. This additive to neutralise the acid in the lubricating oil is called the Total Base Number or TBN.

Engine manufacturers' list of requirements:

Wärtsilä Sulzer:
TBN 70
Over 1.5% S standard recommended dosage.
From 1.5% S to 3.0% S minimum recommended dosage.
TBN 40
Over 1.5% S standard recommended dosage.
From 1.5% S to 3.0% S increased dosage.

MAN Diesel:
TBN 70
Below 2.0% S standard recommended dosage.
From 2.0 S to 1.5% S minimum recommended dosage.
TBN 40
Over 1.5% S standard recommended dosage.

Apart from the TBN number, thermal stability against oxidation forms an important factor in assessing the amount of deposition that may occur when using a certain type of cylinder lubricating-oil. Through oxidation, compounds are formed which adhere to the piston surface and are a magnet for other substances, such as soot and various other particles, which form a solid, thick coating on the piston crown and on the sides of the piston between the ring grooves. This can eventually cause a decrease in the lubricating-oil film on the running surface of the liner. Furthermore, the lubricating properties of the oil residue on the running surface are reduced, thus creating scuffing.

Obviously, the temperature of the cylinder liner is the most important variable for checking the thermal stability of the lubricating oil. Not all engines are designed with identical clearances, and generally engines with small cylinder diameters have larger clearances than engines with large diameters. The engineer can control the clearance to an extent by performing quality maintenance as stipulated in the manuals.

Excessive wear or scuffing can be avoided by sustaining a good lubricating film between the liner running- surface, the piston rings and the piston skirt.

Therefore it is imperative that the various lubricating methods be studied in order to acquire insight into their effectiveness and obtain optimum lubrication at a minimum dosage.

Numerous measurements have demonstrated that the traditional lubricating systems are far from ideal. Sample analyses from scavenging-air drains show that only a minor amount of the lubricating-oil supply is used during each piston stroke. This means that either the lubricating-oil distribution is inefficient or that the localized lubricating-oil dosage is excessive. It is usually a combination of these two factors.

The conclusion applies to all types of two-stroke crosshead engines from MAN Diesel, Wärtsilä Sulzer and Mitsubishi.

Essentially, the lubricating oil is distributed over the liner running surface in two different manners:

1. By injecting the lubricating oil via the supply channel into the liner grooves.
2. By injecting the lubricating oil directly onto the running surface, in the Wärtsilä 'Pulse-Jet' system or indirectly, the 'SIP' system.

Irrespective of the method, the piston rings obviously perform most of the vertical and horizontal distribution of the lubricating oil.

When the liner running surface has been supplied with grooves, the zigzag lubricating method is most effective. This type of cylinder liner has various advantages over those only supplied with supply channels or horizontal grooves.

A most interesting experiment is the so-called 6k pocket liner, which has 6000 small, horizontal oil chambers and has been tested in Wärtsilä Sulzer RT 96C engines for several years. The design's starting point is that the running surface is constantly covered with a lubrication film, irrespective of the moment of supply. This partially applies to 'Wave-Cut' cylinder liners as the lubricating oil found in the machined troughs of a MAN Diesel K90MC engine corresponds to the amount that is supplied at 120 rev/min and at full power.

If the cylinders temporarily have to operate with either seized or fractured piston rings, the running surface can become thermally overloaded causing tearing of the top section of the cylinder liner where the piston ring package is situated when the piston is in TDC. The larger the cylinder diameter of the engine, the greater the risk. These cylinder liners cannot be repaired and will have to be replaced.

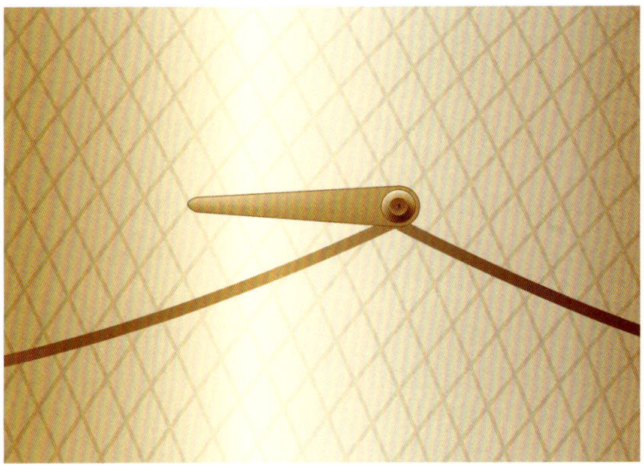

▲

A cylinder liner from Wärtsilä Sulzer, deeply honed with Pulse Jet cylinder lubrication and zigzag grooves.

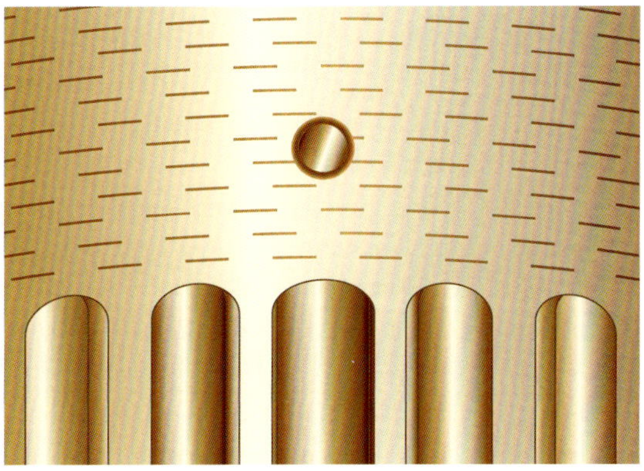

▲

A cylinder liner from Wärtsilä Sulzer, with 6000 pockets and cylinder lubrication from a single point.

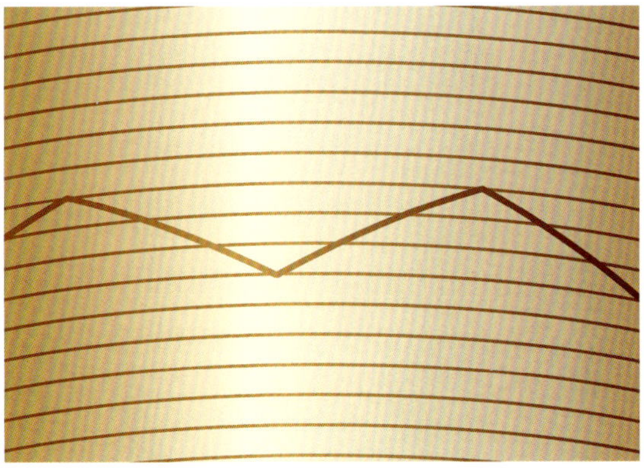

▲

A cylinder liner from MAN Diesel, 'Wave Cut' with Alpha cylinder lubrication and zigzag grooves.

▲

A cylinder liner from Wärtsilä Sulzer/Mitsubishi, 'Wave Cut, SIP cylinder lubrication without dividing grooves, pockets and cylinder lubrication from a single point.

27.9 Trouble shooting excessive cylinder liner wear

The numbers in the chart are discussed below.

1. The wear edge at the top of the cylinder liner just above the first piston ring should be removed with a special cutter. This grinding method requires accuracy, so the risk of the newly fitted piston ring hammering against the wear edge is minimised. If not, the piston ring may break causing gasses to escape to the second ring (blow by). This could damage the entire ring package.
2. The liner running-surface will almost certainly be damaged if the wear edge is not removed with a special machine. Consequently, the new piston-ring seals cannot be checked and the ensuing leakage will cause broken and seized piston rings.
3. In case the new piston rings are mounted with a standard tool, they are likely to fracture either at the ends or in the middle after a few operating hours.
4. Condensation or cooling-water leakage in the air coolers must be drained; otherwise the water droplets entering the cylinder will damage the lubricating film on the cylinder liner and piston rings. This will effect increased wear and probably scuffing, starting at the bottom of the cylinder liner, predominantly at the front- and port side of the liner. Therefore, the air-cooler drains should be inspected at regular intervals. In addition, the air coolers should be opened at least once a year to check the drain internally as it often congested with dirt and rust.

5. Severe piston-skirt scuffing may distribute wear material throughout the cylinder and damage the piston rings. Scuffing may also cause localised damage of the cylinder liner, which produces scuffing and/or wear of the piston rings. The piston skirts are usually somewhat worn, however, with no direct implications for the piston rings and cylinder liner.
6. In case of severe piston-skirt scuffing, the piston alignment should be checked. Abrupt scuffing in older engines is generally caused by loose guide ways.
7. Excessive deposition on the piston crown may damage the oil film on the cylinder liner, causing scuffing of the liner and piston rings. The sulphur content of the fuel should be checked and the lubricating-oil dosage should conform to the guidelines for sulphur content. Should the dosage be in excess of the specifications, then it should be reduced. Deposit formation on the piston crown may also result from inadequate fuel injection. Therefore, the injectors/fuel valves should be checked regularly.
8. If the piston rings have not been run in properly they are not gastight. This may cause blow-by and subsequent seizure or fracturing of the piston rings. Conical or rounded piston rings enhance the running-in process. The engine manufacturer provides ample information with regard to this process.
9. Insufficient dosage of lubricating oil produces scuffing, resulting in seized and/or broken piston rings. The correct lubricating-oil doses are clearly stipulated in the manufacturer's operating manual.
10 The time of dosage should be in accordance with the manufacturer's instructions. An incorrect time-setting of the lubricating-oil dosage will cause scuffing, blow by and seized and/or fractured piston rings.

11. The piston-ring grooves must be cleaned thoroughly before mounting the piston rings. This should be performed with a special tool: a piston ring groove cleaner. If the grooves are not cleaned, the piston ring will eventually cease to move in the ring grooves and either cause seizure or fracturing of the piston rings. Scuffing may

also occur at the liner's smallest diameter, usually just above the inlet ports.

12. Piston-ring groove wear should be inspected at regular intervals. Excessive wear causes increased wear of the cylinder liner and may lead to seized and fractured piston rings. Insufficient ring to piston clearance will damage the rings and they consequently break.
13. If the liner wear is in excess of 0.5% of the diameter, it is difficult keep the cylinder in good condition. It is, therefore, advisable to replace the cylinder liner rather than continually overhauling the piston as the piston rings are either seized and/or fractured. Scuffing will often occur.
14. Oval cylinder liners or clover leafing will also cause fracturing and scuffing. Check the roundness of the cylinder liner by using talcum powder and, if required, send the cylinder liner to shore for honing.
15. Fuel-valves/injectors must be inspected regularly and in accordance with manual guidelines. Heavily worn injectors cause delayed combustion. As a consequence, fuel drops may reach the liner's running surface and damage the lubrication film resulting in scuffing, blow out and seized and/or fractured piston rings. If the fuel-valves/injectors are defective, the liner should be checked for thermal tearing.
16. Fuel analysis provides information regarding the CCAI (The Calculated Carbon Aromaticity Index). If the CCAI exceeds 900, combustion may be delayed, thus affecting the lubrication film on the liner running surface. This results in scuffing, blow by seized and/or fractured piston rings. In this instance, the lubricating-oil dosage must be increased.
17. Running the engine at full power (maximum of 85% MCR) and a simultaneous high scavenging- and inlet-temperature may cause thermal overload of the engine. When the engine load is 85% and the scavenging- and inlet-temperatures total 75 °C, the thermal load of the engine is 100%. Obviously, this affects the condition of the cylinder, leaving little room for errors. In extreme cases this will cause scuffing and seized and/or fractured piston rings. Moreover, the high thermal load forms a catalyst for other problems. The coinciding high cooling-water temperature exacerbates the problems. The problem may be solved by reducing the temperatures especially that of the scavenging air. However, the cooling-water temperature should not fall below 70 °C. The lubricating-oil dosage can be temporarily increased by 25%.
18. If the cooling-water temperature is in excess of 85 °C, scuffing may occur at high engine loads. Long-term low engine power may produce corrosive wear.
19. The air-inlet filter of the engine room and that mounted on the turbochargers serve to prevent water, dust and sand from entering the engine. Dust and sand have abrasive properties, which accelerate the wear of the piston rings and the cylinder liner. Therefore, the air-inlet filter should be inspected at regular intervals.
20. The best method to remove abrasive particles from the fuel is to align the centrifuges in series. Setting the engine at low load capacity can also be beneficial.

21. Check the fuel filter at regular intervals.
22. If the catalytic product content, aluminium- and silicon oxides, approximates or exceeds the maximum permissible value of the fuel specifications, extra attention should be paid to centrifuging the fuel. A low-capacity setting is useful.
23. In case the cylinder liner is honed without first removing the piston, one should be aware of the grindings produced by the honing process. This should be intercepted and prevented from entering the piston and onto the liner as they cause severe wear. Lubricating-oil drenched cloths covering the cylinder wall catch much of the grindings.
24. Sensors measuring the cylinder-liner temperature indicate whether there are signs of scuffing wear. Reduce the load of the cylinder concerned by decreasing the fuel supply to the cylinder.

Trouble shooting Chart for solving liner wear problems.

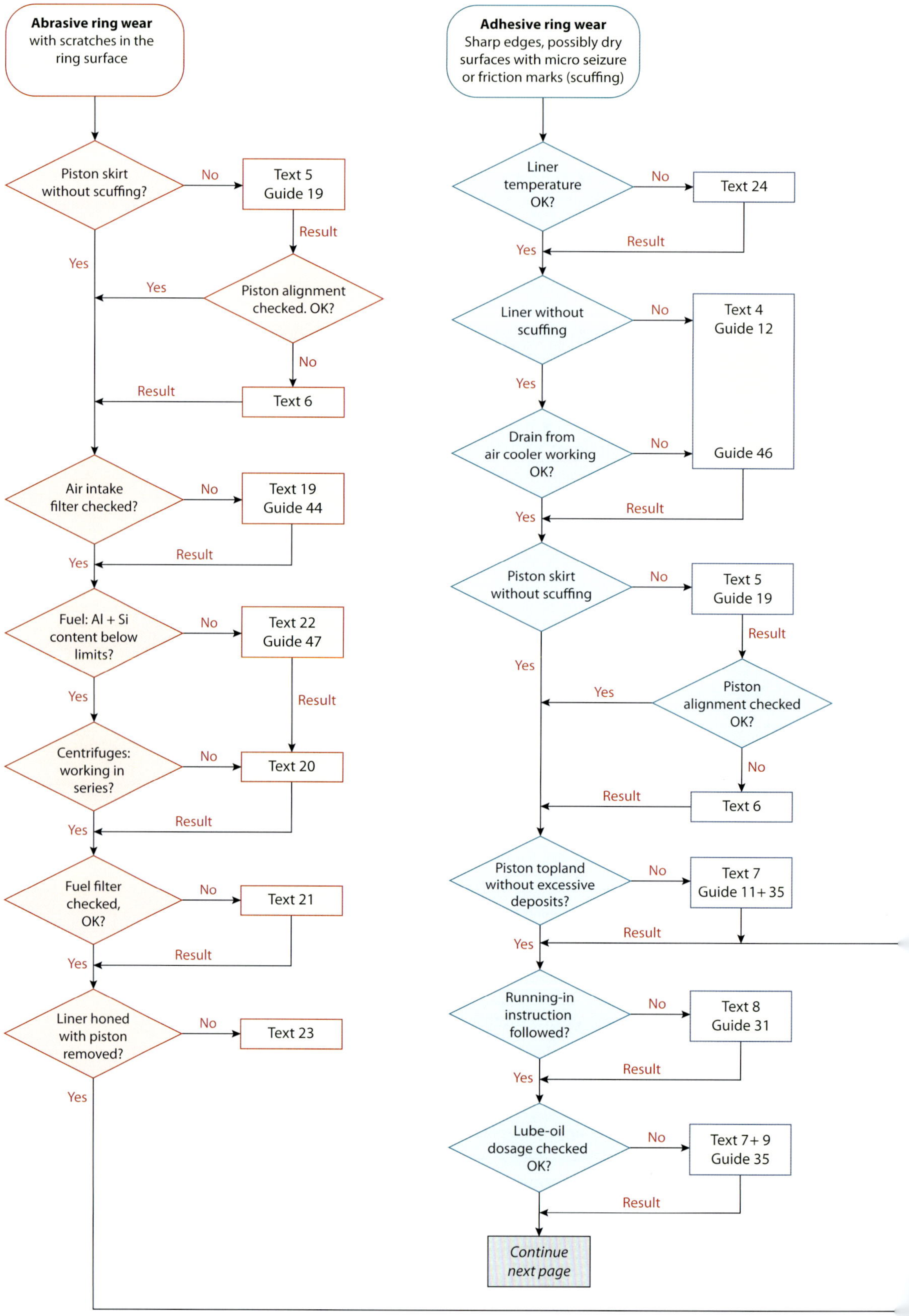

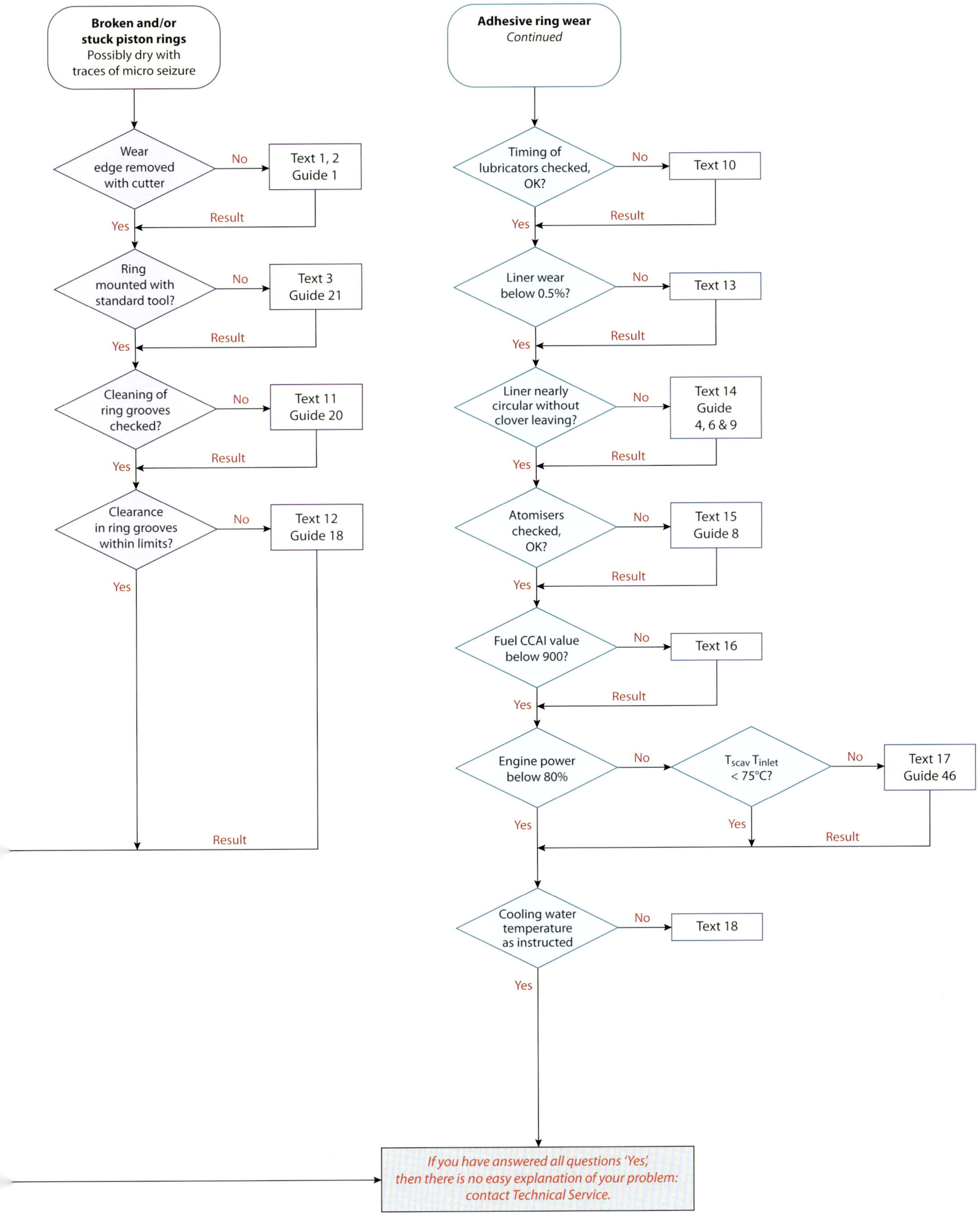
Broken and/or stuck piston rings
Possibly dry with traces of micro seizure
Wear edge removed with cutter
No
Text 1, 2 Guide 1
Result
Yes
Ring mounted with standard tool?
No
Text 3 Guide 21
Result
Yes
Cleaning of ring grooves checked?
No
Text 11 Guide 20
Result
Yes
Clearance in ring grooves within limits?
No
Text 12 Guide 18
Yes
Result
Adhesive ring wear
Continued
Timing of lubricators checked, OK?
No
Text 10
Result
Yes
Liner wear below 0.5%?
No
Text 13
Result
Yes
Liner nearly circular without clover leaving?
No
Text 14 Guide 4, 6 & 9
Result
Yes
Atomisers checked, OK?
No
Text 15 Guide 8
Result
Yes
Fuel CCAI value below 900?
No
Text 16
Result
Yes
Engine power below 80%
No
T_{scav} T_{inlet} < 75°C?
No
Text 17 Guide 46
Yes
Result
Yes
Cooling water temperature as instructed
No
Text 18
Yes
If you have answered all questions 'Yes', then there is no easy explanation of your problem: contact Technical Service.

27.10 Damage report Wärtsilä Vasa 4R32

27.10.1 Introduction

Damage to the driving gear and the engine block of a Wärtsilä Vasa 4R32 medium-speed four-stroke diesel engine running on HFO 380 cSt resulted in a major overhaul.

27.10.2 Cause

Broken fixing bolts between the piston crown and the piston skirt. Possibly caused by a cracked or otherwise damaged contact surface between the crown and the skirt, this resulted in piston-crown seizure. The piston skirt was also damaged and the connecting rod was moving freely in the crankshaft. This caused damage to the engine block, the crankshaft counter weight and both balance weights. The valve discs of the in- and outlet valves broke off and the fuel injectors were damaged. Eventually, all parts of one cylinder were rejected.

27.10.3 Explanation

According to the Wärtsilä manufacturer, the engine type Vase 32 with an average load below 75% and using HFO requires major maintenance after 20,000 operating hours, such as:
- Inspection of engine parts,
- Replacement of parts,
- Trueing up parts, such as valves.
- Measuring parts.

This applies to:

1. Connecting rods
- Replace the crank-pin bearing.
- Measure the boring of the connecting-rod big end.
- Replace the nuts and studs of the connecting-rod bolts after 24,000 operating hours. Inspect and measure the ovality of the piston-pin bushing.

2. Cylinder head
- Clean and inspect. Inspect cooling-water section. If required, regrind or replace valve seats.
- Inspect valve rotators.
- Check valve movement.
- Replace o-rings in the valve guides.
- Replace the o-rings below the cylinder head every second overhaul.

3. Cylinder liner
- Measure diameter according to guidelines.
- Replace cylinder liner if the maximum measuring values are exceeded.
- Hone cylinder liner, if required.
- Inspect anti-polishing ring.
- Turn anti-polishing ring every 20,000 hours.

4. Cylinder liner waterside
- Draw one of the cylinder liners.
- Draw all liners and clean the cooling water sections in the block if the deposition thickness is in excess of 1 mm,
- Replace o-rings at the bottom of the liner retaining-edge every 20,000 hours.

5. Pistons
- Draw, open, inspect and clean pistons.
- Measure ring-groove height according to guidelines.
- Inspect the retaining ring of the piston pin.
- Replace all piston rings.

Obviously, the list is much longer!

27.10.4 Maintenance

The manufacturer's guidelines demonstrate that engine maintenance had been performed recently. The maintenance had been performed by the crew, only certain sections of the instruction manual had been consulted; the contents of the service letters were unknown. The maintenance was poorly documented; measurements and findings were not processed in the maintenance system.
Often bolt connections are periodically checked for hair-line cracks. Connecting-rod bolts and piston-skirt and -crown connecting bolts require meticulous monitoring and testing (crack test/ dye check) conform the engine manufacturer's guidelines. Bolt replacement should also be performed in accordance with the specifications. Fractures have major implications.

Below, the engine damage documented with photographs.

▲

The results of damage.

Top left, the piston crown.
Middle and right: buckets filled with fragments of cast iron from the piston crown and cylinder liner.
Top left; the gudgeon.
Top right; the crankpin bearings.

▲

The cylinder head viewed from below.

All four valve flanges are broken. The flame plate of the head is badly damaged. The cooling channels, as a cross from the centre outward, are visible!

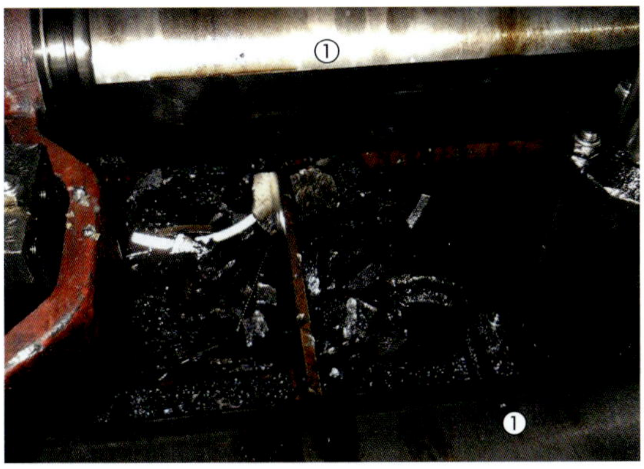

▲

Fragments of a piston skirt, cylinder liner and the engine block in the crankcase.

Above and below, the counterweights (1) which are absolutely necessary in this four-cylinder engine to compensate the secondary free forces.

◄

Part of the piston crown and above, the cylinder liner collar.

◄

The distorted crankpin bearing-shells.

◄◄

The 'small end' of the connecting rod.

The forged steel connecting rod is very strong and when it detaches from the piston, it destroys everything in his path. The rod itself was badly bent.

▲

The crankpin immediately after the damage.

Heavy oil leakage is visible on the pen mainly from the disassembly of the parts. The entire lubrication system including crankshaft were flushed with a special flushing oil to remove all the dirt.

▲

The cleaned crankpin.

The running surface is slightly damaged, but still good to clean.

▶

The probable cause of the damage.

The broken bolts used to attach the piston to the piston crown skirt. It is of vital importance during the inspection of the disassembled piston to check carefully for irregular alignment, damages or other irregularities of the contact surfaces between the piston crown and piston skirt. This is described in detail in the operating manual supplied with the engine, for bolts; there are also clear instructions for: inspections, measuring, testing for cracks, bolt replacements, replacing parts such as bushings, nuts, washers and other specific details.

▶

The counterweights.

The counterweights are quite 'cold deformed'. Since these are made of forged steel, they never break! However removing a counterweight took, because of the deformation, more than 40 man hours!

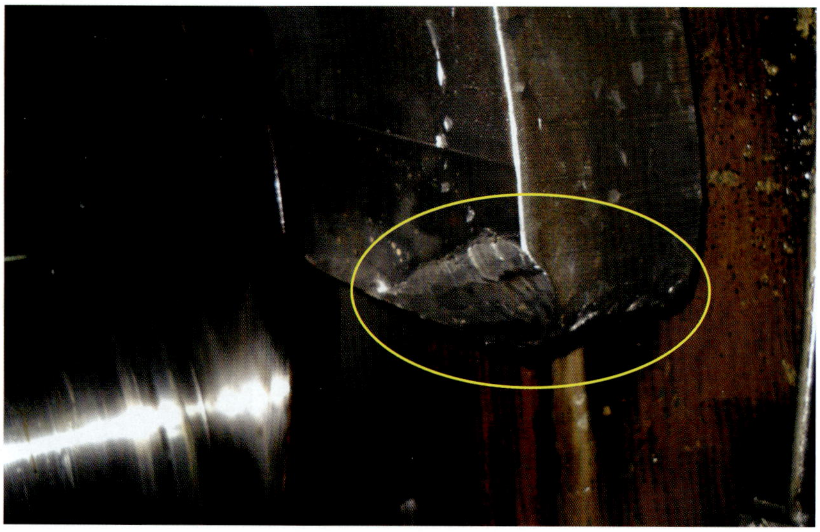

▲

The beaten flat spray nozzle.

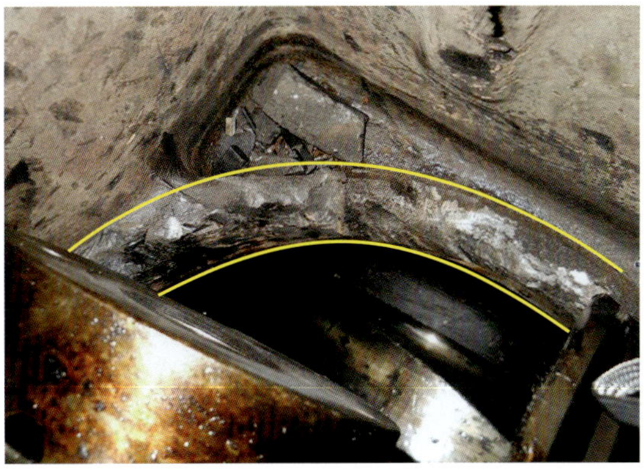

▲

The major damage.

The support for the bottom edge of the cylinder liner in the cast iron block was severely damaged. Fortunately, no cracks were detected in the engine block. This would be a reason for rejecting the block. Here, after a number of operations, a new support ring must be manufactured, a fairly complicated repair.

◄

A slightly damaged counterweight.

This does not need to be replaced. Often the rough edges are slightly re-cut.

◄

The nozzle segment for the impeller of the exhaust-gas turbine of a turbocharger.

Many pieces of broken material have severely damaged the turbine via the exhaust pipe.

▼

The impeller of the exhaust-gas turbine is significantly damaged.

The engine manufacturer and the classification society play an important role in assessing this type of damage. Repairs must be performed in compliance with the manufacturer's specifications, who present the ship-repair plan to the classification society. The latter will enforce a so-called CoC (condition of class) and repairs will have to be completed within a certain time frame. A reference to the repair specification is included in the CoC. If a CoC is not observed, a ship may lose its class.

Step 1
Both the damage report and the repair plan are drawn up by the engine manufacturer. These include repair details, the welding method, the welding electrodes used, welding surveys, the required certificates and the manufacturer's instructions for welding repairs.

Step 2
The damage report and the repair plan are assessed by the classification society and approved.

Step 3
The repairs are performed under supervision of the manufacturer and generally carried out by their own personnel.

Step 4
Following the repairs, the engine is test run under the supervision of a surveyor from the classification society and the CoC is lifted.

27.10.5 Repair operation details

- Removal of all parts, debris and dirt.
- Boring out the bottom of the retaining edge to create a ledge on which the stainless steel sleeve can rest.
- Manufacturing a stainless steel sleeve with a diameter allowing for a solid crimped connection with the engine block after it has been cooled with nitrogen.
- Honing of the crank pin using a mobile honing machine.
- Overhauling, cleaning and replacement of the other three cylinders, including the driving gear.
- Inspection of the inside of the engine.
- Cleaning the entire inside of the engine.
- Checking the position of the crankshaft: the crankshaft clocking.
- Replacement of all damaged cylinder parts, such as: cylinder head, cylinder liner, piston, connecting rod, bearings and packing.
- Scavenging with a scavenging oil of all lubricating-oil pipes, filters, pumps and ducts.
- Refilling the sump with fresh lubricating-oil and filling the entire system.
- Checking lubrication of all driving gears with the pre-lubricating pump with opened crankcase- and camshaft doors. Obviously, while the engine is idle.
- Test running the engine in compliance with the manufacturer's specifications, including the load (generating operation).

▶
The movable boring machine for turning the lower support ring.

The machine is aligned to the centre axis of the upper surface of the engine in the centre axis of the cylinder. The crankshaft axis direction is then determined perpendicular to this point. The machine is fixed with four cylinder-head bolts.

◀

The line-boring of the lower support edge.

The machine is supported at the bottom of the engine with four threaded rods against the cooling-water wall of the engine. By measuring and adjusting the threaded supports, the machine is placed absolutely perpendicular to the block. The chisel (1) runs at a speed of about 30 rev/min.

▲

Trueing up of the crankpin.

The crankshaft is stationary and the rotating part with the honing stones of the honing machine is pneumatically driven by a chain.

1 *Air connection*
2 *Drive motor*
3 *Chain*
4 *Honing machine*
5 *Crankpin*

▲

Honing.

The honing machine 'runs' on the side against the radius of the journal and crank web.

▲

The crankpin journal is honed.

The honing stone experiences a slight vibration because of the air-driven cylinder. Gas oil is used for the lubrication and flushing of the crankpin and honing stones.

◀

Fixing the new lower support edge.

The supporting edge is turned and the stainless steel ring is fitted after being nitrogen cooled. After the ring has assumed the same temperature as the environment and therefore the block, the fit is then tight when the dimensions of the block and ring are then correct.
On the top a malleable steel 'Belzona' is applied for additional support and finish. The corresponding repair kit includes a malleable steel degreaser to attach the parts properly. The ring approaches the specifications of the supporting edge. Any slight ovality is absorbed by the O-rings on liner so that it is air tight.

27.11 Sleeve bearings, theory and damage

Source: A summary of a study by the Miba Bearing Group, a manufacturer of engine bearing applications and bearings.

27.11.1 Sleeve bearing

Since ancient times, bearings have constituted an indispensable part of a rotating shaft. Today, sleeve bearings are widely used, also in the driving gears of combustion engines, such as:

- The main bearings and crankpin bearings of the crankshaft.
- The bearing bushings of the camshaft.
- The bearings in the thrust blocks of the crankshaft.
- The bushing of valve levers.

The diesel engine's greatest load is located at the crankpin- and main shaft bearings. However, the piston-pin bushings are also heavily loaded.

27.11.2 The function of a sleeve bearing

Sleeve bearings serve to bear and support the rotating or moving parts of the engine. This means that there is direct contact between the shaft and the bearing, which causes friction and consequently heat development and wear. The friction must be kept to a minimum to achieve an acceptable bearing life. Friction is reduced by lubrication. Ideally, the sliding surfaces of the sleeve bearing and the shaft surface are completed separated by a thin oil film. This is referred to as fluid film lubrication. Due to internal oil-film friction there may be some heat development. This must be discharged by cooling the lubricating oil. The Stribeck curve shows the relation of the friction coefficient to shaft speed. As mentioned in Chapter 11, Lubrication of engines, there are three types of friction:

I-Boundary lubrication,
II-Fluid film lubrication,
III-Hydrostatic lubrication.

The latter uses a separate high-pressure supply pump and will therefore not be discussed.

▼

Left figure: Downtime of the shaft in the sleeve bearing. The two parts touch each other.
I-Boundary lubrication, middle figure: the parts occasionally come into contact with each other in spite of the lubricating-oil film between the rotating shaft and the stationary bearing.
The high 'peaks' of the 'rough' surface of the bearing shell and the axis come into contact with each other more or less, depending on the degree of boundary lubrication.
II- Full film lubrication, right figure: the rotating shaft and the stationary bearing are consistently separated by a very thin lubricating-oil film. The ideal situation: there is no wear on either part.

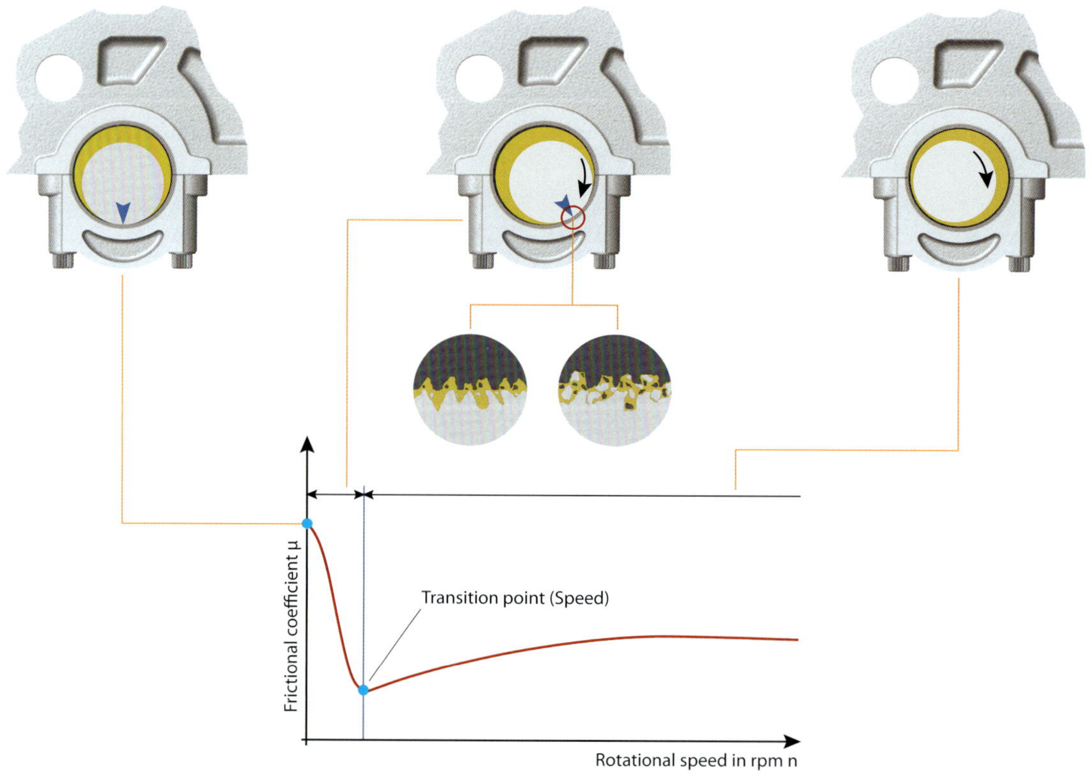

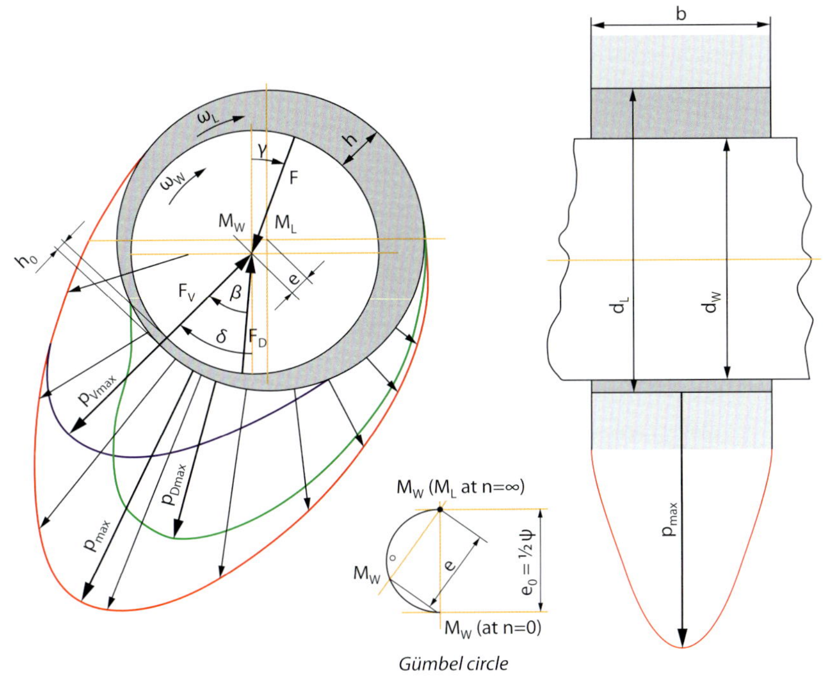

b	Bearing width [mm]
d_L	Bearing diameter [mm]
d_W	Shaft diameter [mm]
e	Excentricity
F	Bearing load [N]
F_D, F_V	Bearing load portions caused by rotation and squeeze effect [N]
h	Lubricant film thickness [mm]
h_0	Minimum lubricant film thickness [mm]
p_{max}	Maximum lubricant film pressure [MPa]
p_{Dmax}, p_{Vmax}	Maximum lubricant film Pressure caused by rotation and squeeze effect [MPa]
β	Angular position of h_0, relative to F_D
δ	Angular position of h_0
γ	Angular position of bearing load F
ω_L	Angular velocity of bearing
ω_W	Angular velocity of shaft
Ψ	Relative clearance $(\frac{d_L - d_W}{d_L})$

▲

The hydrodynamic effect of bearing.

To minimize friction, the bearing is always designed for full-film lubrication. Many engineers have made these calculations, such as Reynolds, Sommerfeld, Gürkel and many others.

- The image set a rotating shaft in a plain bearing.
- The shaft rotates clockwise.
- The space between the rotating shaft and the bearing is known as the lubricating-oil clearance and is filled with lubricating oil. Clearance is h_0.
- The rotating shaft squeezes the lubricating-oil, on the shaft, into the gap and so decreasing the clearance. Through this wedge, a pressure build up is created the shaft 'floats' on the accumulated lubricating-oil pressure.

  In most cases, the lubricating-oil pressure is axially oriented and lifts the shaft. The forces exerted on the shaft such as the weight of the shaft and forces from the combustion process are completely compensated by the oil pressure in the lubricating film. The shaft drops slightly when the forces on the shaft increase; causing the wedge to increase and therefore the oil pressure. It is once again balanced! The centre axis of the rotating shaft moves in comparison to the centre axis of the bearing because the shaft rotates. Depending on the operating conditions, the centre axis of the shaft moves along a curved line, the circle of Gümbel. For an infinitely high speed, the centre axis of the shaft will in theory coincide with the centre axis of the bearing.

The small figure in the middle shows that the minimum lubricating-oil film thickness depends on the size of the bearing, speed, bearing load, viscosity of the lubricant and the bearing clearance. The lubricant film is thicker with a decreasing load, increased viscosity of the lubricating oil and higher speeds. Yet there are no general rules of thumb for the clearance in the bearings because the various parameters also influence each other. Parameters include the bearing load, the temperature rise of the lubricant due to friction, the decrease in the amount flowing lubricant and therefore the cooling capacity, the increasing bearing clearance due to wear and also changes in the primary shape and dimensions of the bearing.

The following values can be used as reference for the minimum clearances:

- Crankpin bearings, 0.06% of the crankpin diameter, dw;
- Main bearings, 0.075%.

The corresponding maximum tolerances are calculated according the above values.

Experience is obviously paramount; engine manufacturers always mention these tolerances in their manuals.

▶
The main bearings in a small diesel engine, category I.

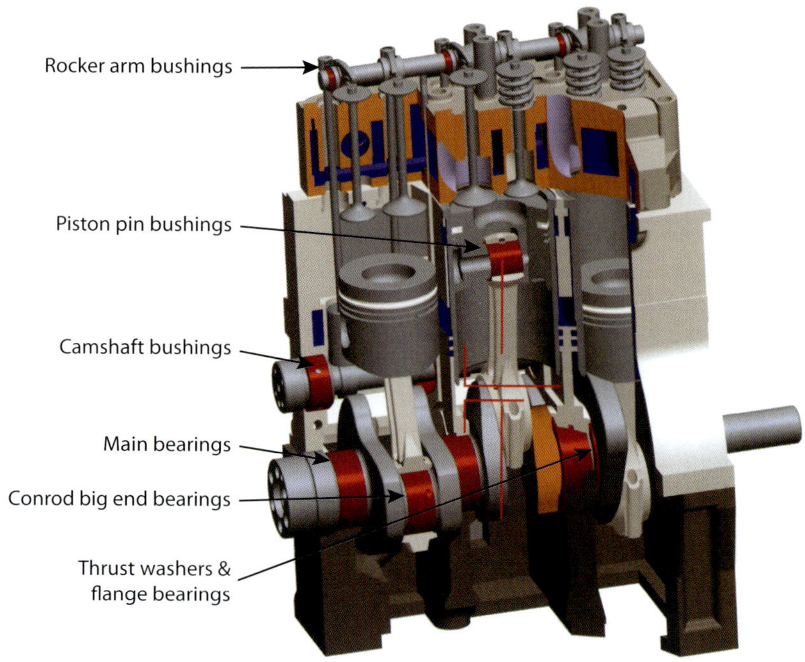

▶
The bearing loads.

The load on the main bearings of the crankpin is due to the gas forces on the piston and the acceleration and deceleration forces of the gearing.
The direction and size of these bearing forces change continuously.
In the figure, the dynamic load on the bearings of a four-stroke engine.

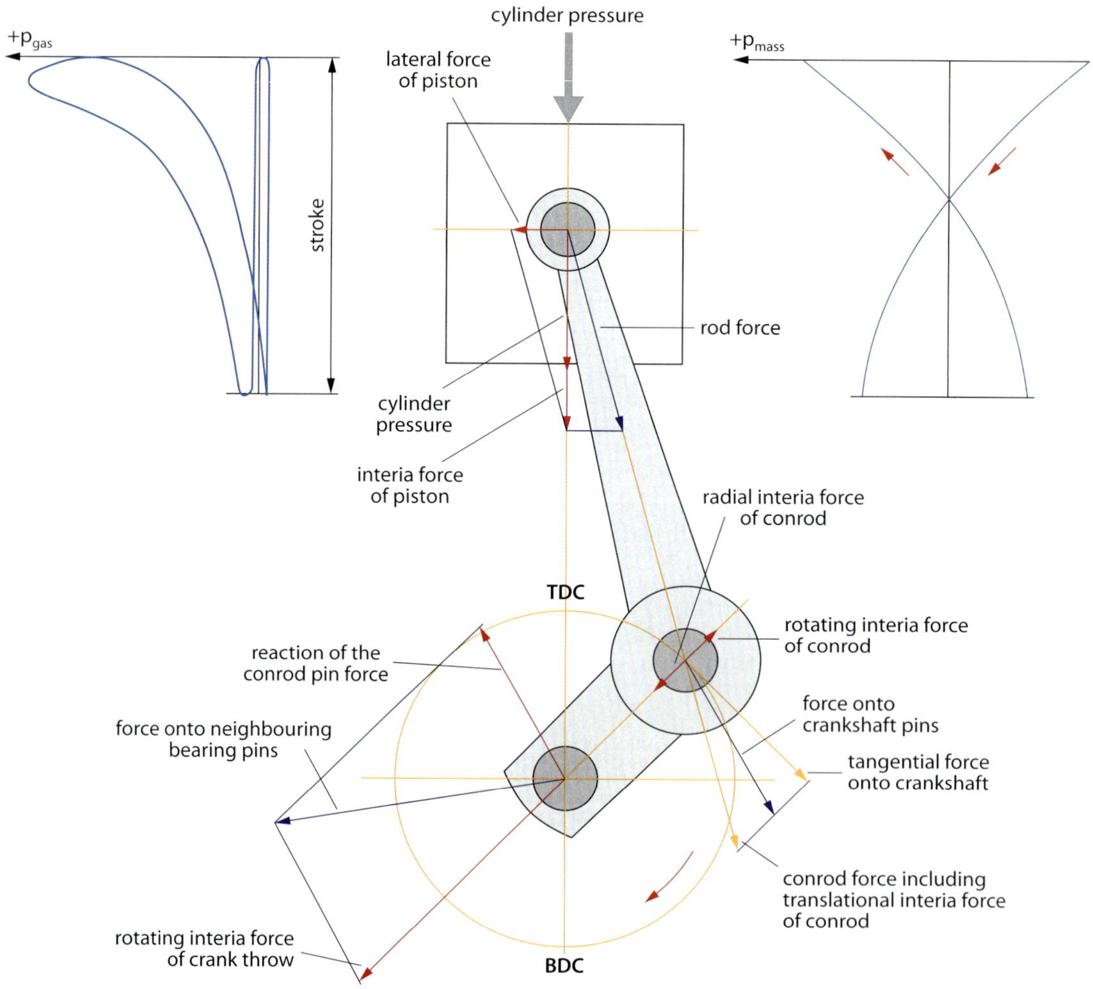

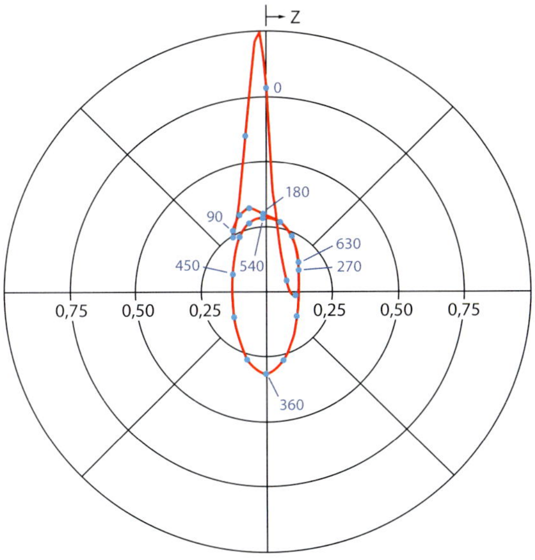

◀

A polar diagram of the forces on the connecting rod bearing for a four-stroke diesel engine.

The closed red curve shows the power curve. The blue values represent the crank angles. When the crank is in its top position during combustion, the greatest force is exerted.

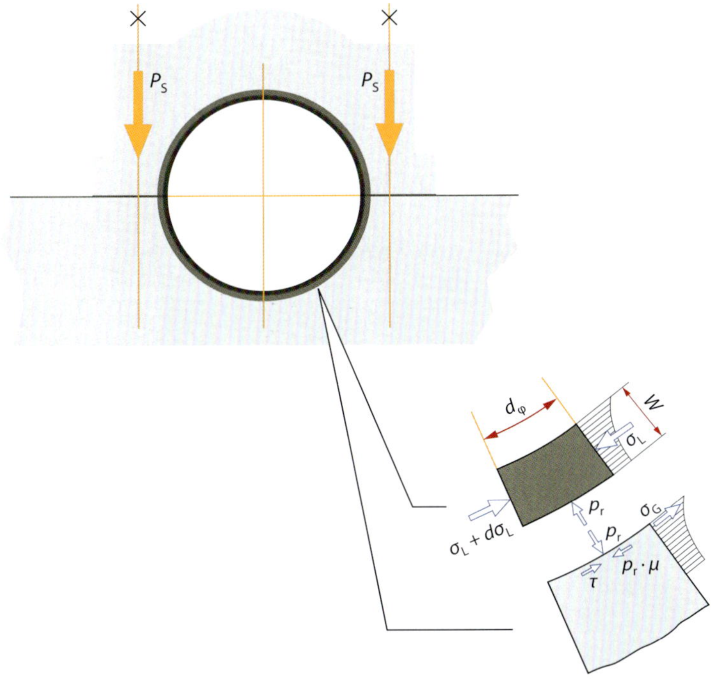

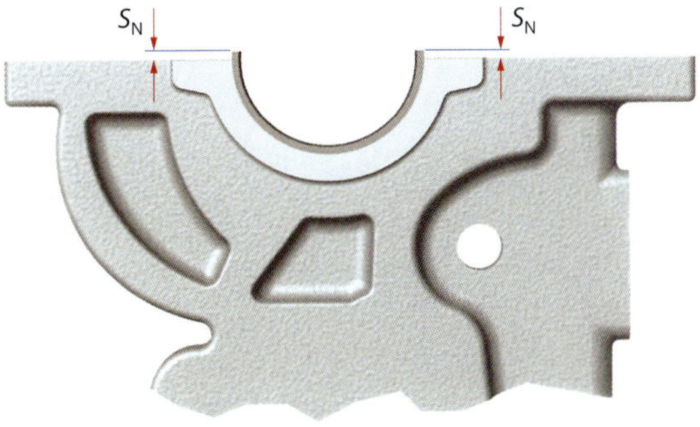

p_r = radial holding pressure [MPa]
P_S = bolt force [N]
S_N = crush height [mm]
W = bearing wall thickness [mm]
d_φ = angle of section cutted out
μ = friction coefficient
σ_L = circumferential pressure in the bearing [MPa]
σ_G = circumferential pressure in the housing [MPa]
τ = tangential stress by friction [MPa]

▲

Adjustment and tensioning of bearing shells.

It is extremely important that the two bearing shells are fitted properly in the bearing housing and are correctly loaded.
Good contact with the bearing housing is important: if there is too much clearance, the shells tend to rotate along with the shaft, of course, this leads to immediate damage!
In sleeve bearing designs, a large number of formulas are used to create the most ideal design. Amongst others, the complete circumference of a sleeve bearing is greater than the product of the diameter of the bearing housing multiplied by a factor of 2 of S_N, the clearance required to load the bearing correctly.
When designing bearing clearance under varying loads during operation and when starting, field experience and a host of other information play an important role.

Due to the increasing load on the bearings over the years, the direct result of the greatly increased load on the engine, it is becoming increasingly complicated to design a good bearing.

Note:
For the engineer in field, it is of paramount importance to pay attention to the maintenance of the gearing, the bearings and their rotating shafts.
The shaft surface (rough, scratched, out of centre) is especially important.
Always measure when replacing the shaft bearings!
Lubricating oil remains a major focus.

▶

The movement of a crankpin bearing.

Rotating loaded bearings continuously change the thickness of the oil film and oil pressure.
For each revolution of the shaft, the position of the centre line can be calculated using a special program 'Pin Orbital Path'.
In this case for bearing manufacturer Miba, Laakischen, Austria, the method is based on Holland-Lang with the numbers of Sommerfeld consistent to Butenschon. This method is widely used for the design of the main-, crankpin- and camshaft bearings.
The red line in the figure shows the movement of a crankpin bearing.
The crankpin changes place not only in its circumference, axially, but also radially! The closer the crankpin comes to the bearing, the greater the load capacity of the oil film.
Normal values for the thinnest lubricant film, h_{omin} are 1-5 micro-millimetres, a small percentage of the bearing clearance.
This oil film is often much thinner than the size of impurities such as wear particles carried in the lubricating oil.

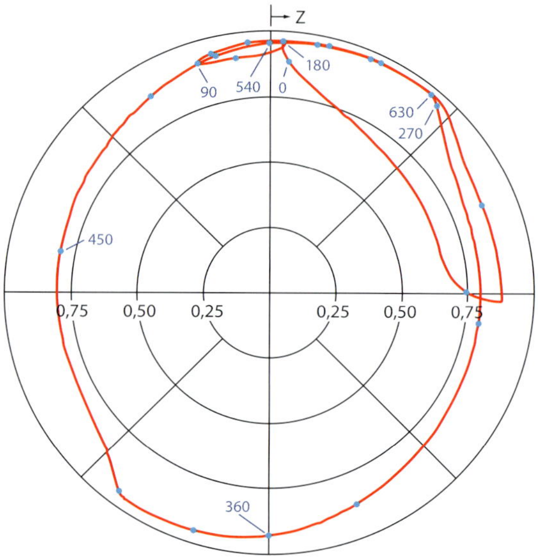

▶

Theoretical calculations.

Theoretical calculations for bearings and bearing research have provided a great deal of information about the right design for sliding bearings.
Calculations of the hydrodynamic behaviour combined with the deformation of the bearing shows how a bearing behaves under load.
The figure a standard crankpin bearing is shown with left, part of the connecting rod.
The oil grooves are clearly marked.

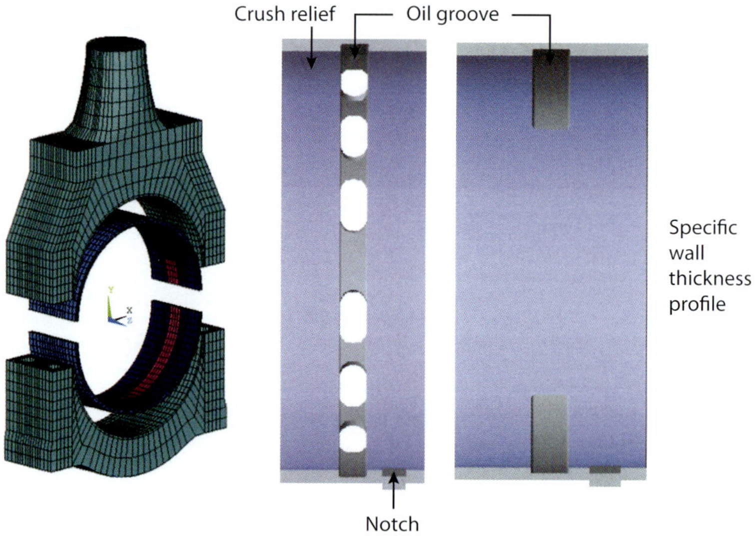

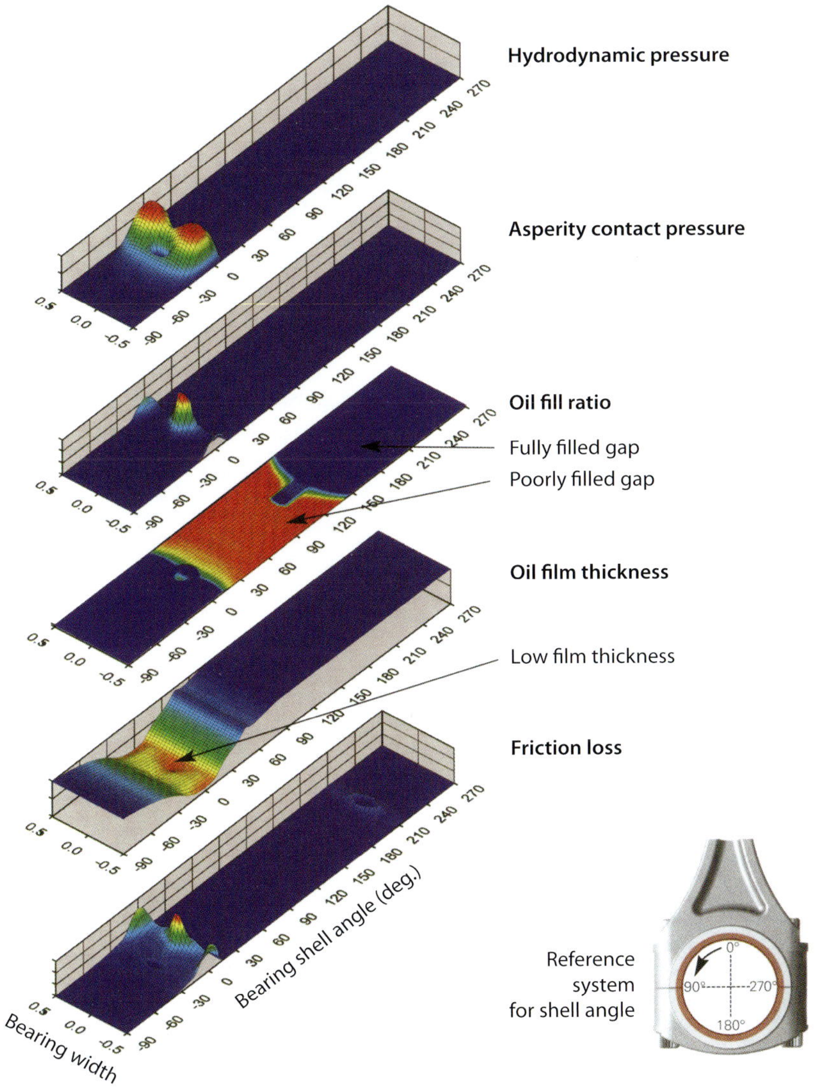

◄ The load of a sleeve bearing.

This figure displays the characteristics of a sleeve bearing, such as:

- *Hydrodynamic pressure:* The hydrodynamic pressure build-up. The highest pressure, the so-called peak pressure is shown in red. The numbers indicate the position of the crankshaft. 0 degrees is the top position. At top to just after the top, the oil pressure is at its highest.
- *Asperity contact pressure:* Contact pressure with impurities.
- *Oil fill ratio:* Extent to which the oil film is present.
- *Fully filled gap:* Sufficient oil film present.
- *Poorly filled gap:* Insufficient oil film present. Wear possible!
- *Oil film thickness:* Thickness of the oil film.
- *Low film thickness:* Location of the thinnest oil film.
- *Friction loss:* Friction losses.

The image clearly shows that the greatest load of slide bearings for engines is found in the middle of the lower shell (main bearing) or upper bearing (crankpin bearing) and occurs when the load on the piston, at the beginning of the power stroke, is greatest.
In images of bearing shells, the wear is usually easy to see!

27.11.3 Worn and damaged sleeve bearings

It is important to establish the cause of severe wear or damage. The following information is vital:

- The number of operating hours.
- The operating conditions.
- Lubricating-oil supply/refreshing/examination.
- The engine load.
- Other damaged engine parts.

If required, the bearing shell supplier should be consulted.

1 Running surface scratches in the direction of rotation

▼

Image

Often local embedding of foreign particles.

Causes:

- Dirty oil.
- Contamination of the lubricating-oil system after repairs to the engine.
- Malfunctioning of the lubricating-oil filters and / or centrifuges.

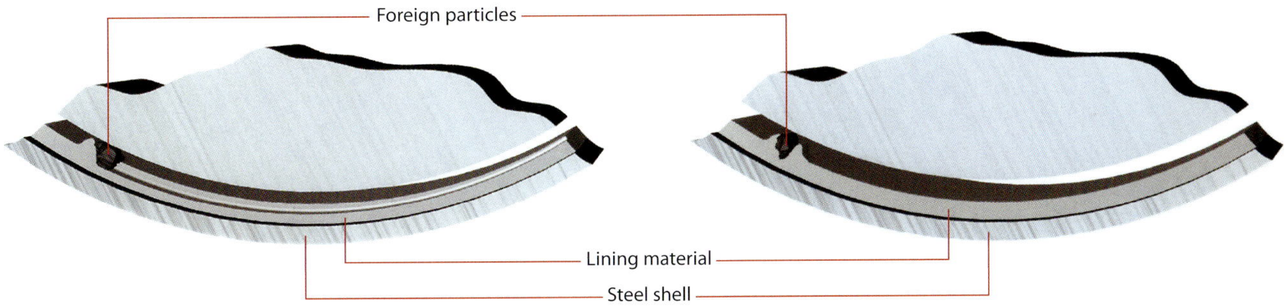

2 Local overload of a bearing shell

a

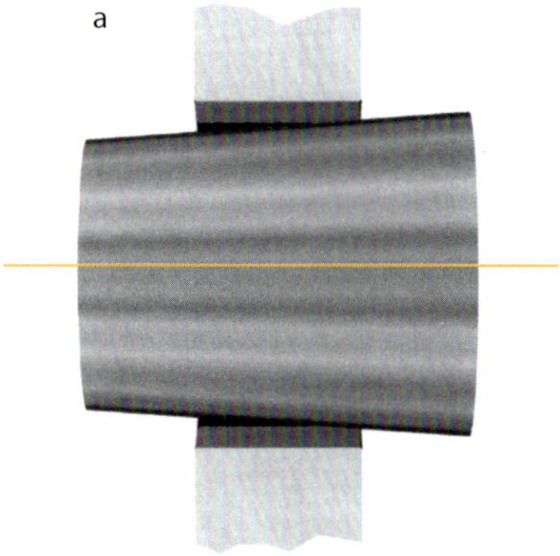

c

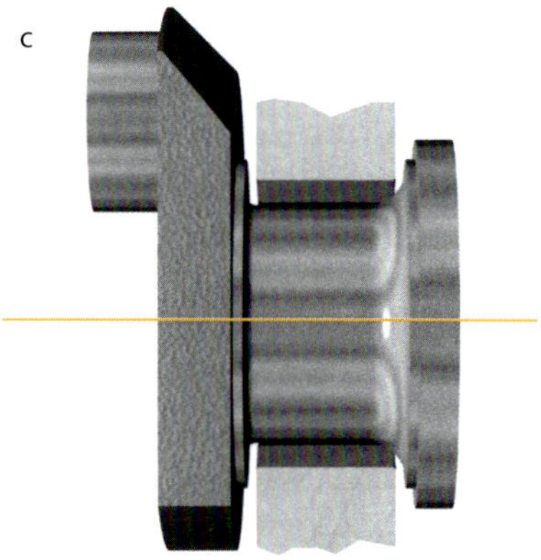

b

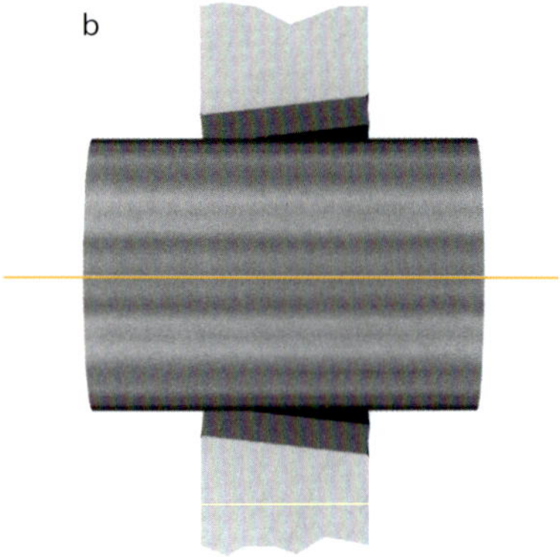

d

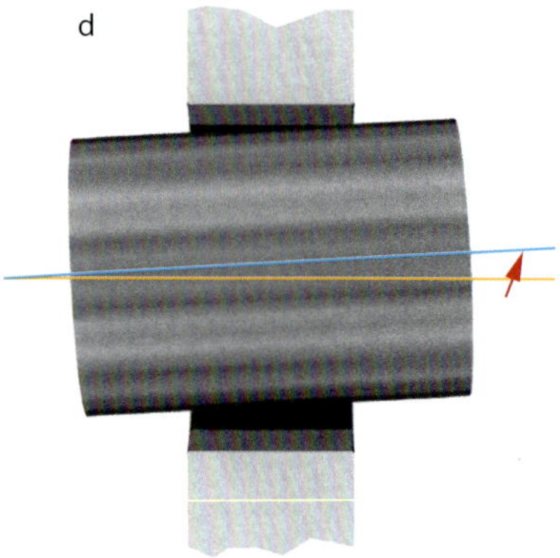

▲

Image

This is often local, the running surface is worn and very often lighter in colour.

Causes:

- Conical (and therefore incorrect) grinding of the shaft pin or crankpin.
- Conical shape of the housing where the bearing shells are installed.

Example:

Figure a. Conical turned connecting rod foot, the so-called 'big eye' in which the crankpin bearings, the most heavily loaded bearings in the engine are placed.

Figure b. Crankshaft is not properly aligned. Control by means of crankshaft deflections.

Figure c. Abnormal movements of the rod, often 'waving' in the radial direction.

Figure d. Crankshaft imbalance: eccentricity.

4 Diagonal wear of the bearing shells

a

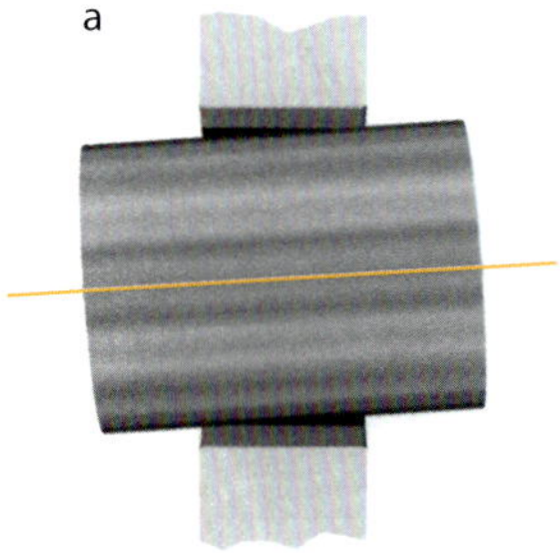

b

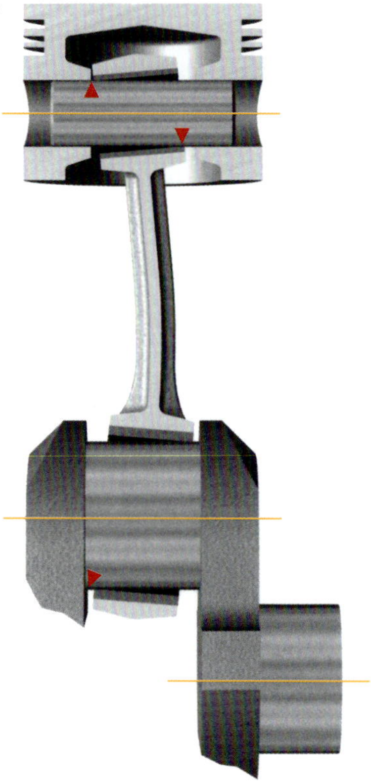

▲

Image

Shown, the wear in the centre axis direction of the shaft on one side of the lower shell and on the other side of the upper shell.

Causes:

Figure a. Misalignment of the shaft and/or the centre axis of the bearing housing.

Misalignment of the bearing housing.

Figure b. Bent or twisted rod.

3 Overload in the centre of the bearing shell

a

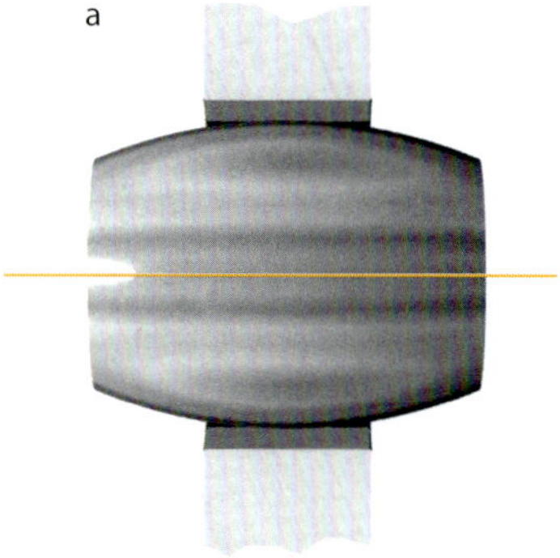

b

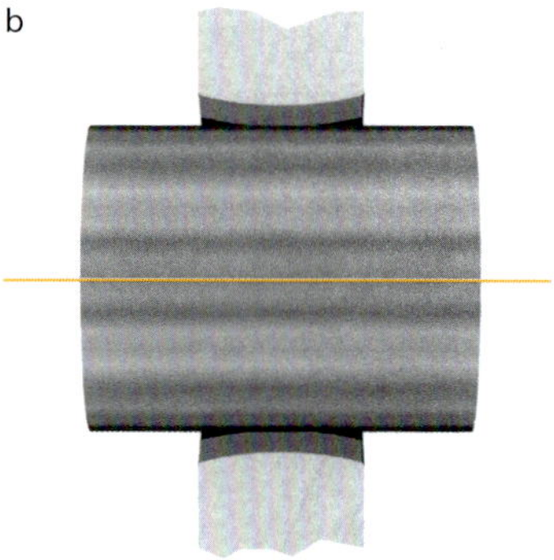

◄

Image

In the middle of the shell, excessive wear is visible, possibly even cracks in the running surface due to overloading.

Furthermore, 'glazing' of the friction layer is visible.

Causes:

- *See Figure a.* barrel polished journal.
- *See Figure b.* 'Concave' bored bearing caps.
- Lack of lubrication.

5 Local load at the extremities of the bearing shells

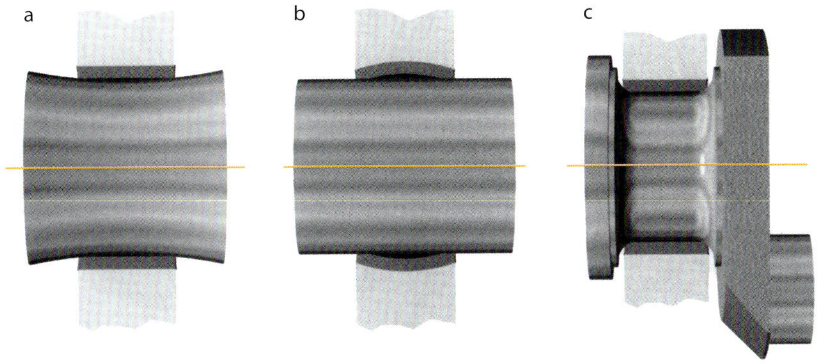

◀

Image

Severe wear on both ends of the bearing shells.

Causes:

- Concave ground shaft journal.
- Convex ground shaft journal.
- Radii of the corners of the crank webs are too large.
- Radial clearances in the crankshaft are too large.
- 'Waving' of the connecting rod.

6 A 'high' oval boring in, for instance, the connecting rod big end

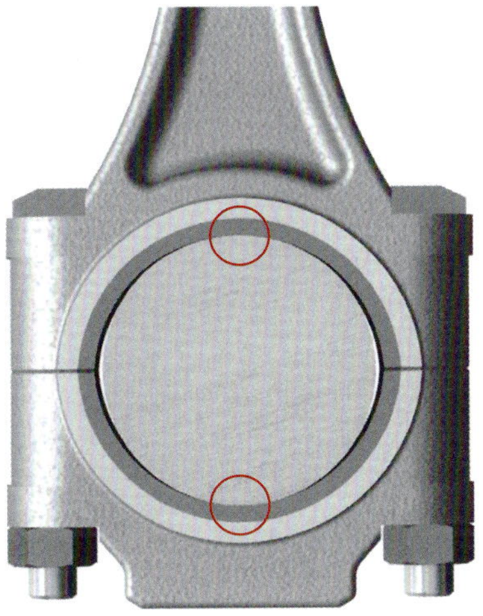

◀

Image

Very heavy local wear in the middle of the two bearing shells and 'glazing' of the friction layer. Sometimes on both bearing shells.

Causes:

- Excessive deformation of the bearing housing.
- Connecting rod or main-bearing bolts are not securely tightened after repairs or maintenance.

7 A 'wide' oval boring in, for instance, the connecting rod big end

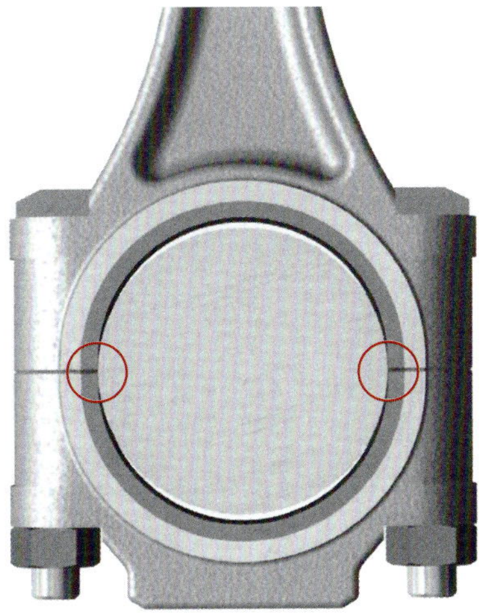

◀

Image

Very severe wear in the middle of the bearing shells, sometimes with glazing of the running surface. Occurs in loaded and unloaded bearing shells.

Causes:

- Exceptional plastic deformation of the bearing housing or connecting rod foot.
- Bolts not properly tightened after an overhaul of the drilling.

8 Wear in the heavily loaded section of the bearing shells

Image

Overall wear of the entire bearing surface, especially in the heavily loaded midsection.

Cause:

- Normal wear resulting from a large number of operating hours.
- Inadequate lubrication produces boundary lubrication between the shaft and the bearing shell.
- Increased roughness of the shaft surface.
- Radial oscillation of the shaft.

9 Wear in the transverse direction of the crankshaft centreline

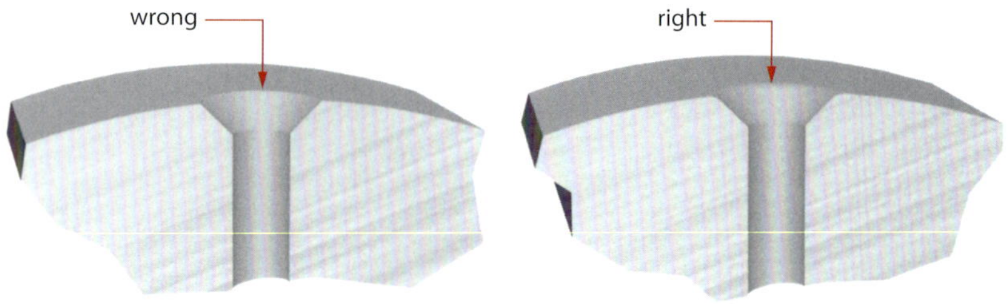

▲

Image

Local wear at the location of the lubricating-oil hole in the crankshaft.

Cause:

- Insufficient rounding of the lubricating-oil hole.

10 Wear of the bottom crankpin bearing shell

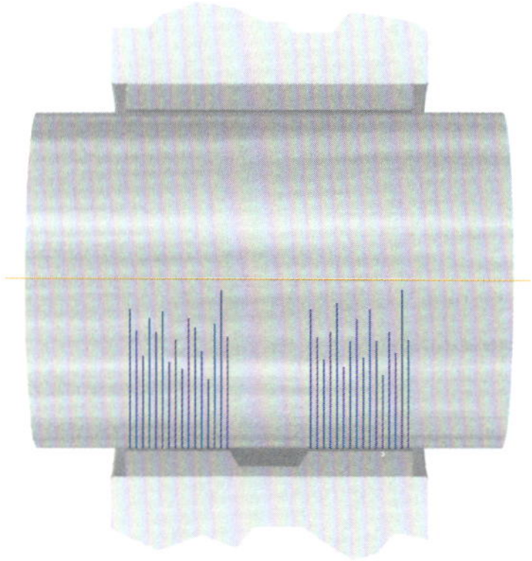

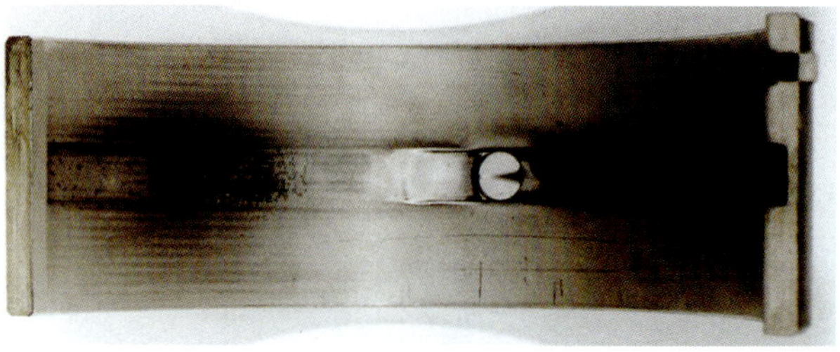

◄

Image

Local wear with lots of scratches on the underside of the crankpin. In some cases, fatigue of the friction layer which results in their detachment.

Cause:

The bottom of the crankpin is carried locally and comes into direct contact with the bearing surface. In the middle, the crank pin is no longer carried. This causes strain on the rest of the crankpin. Result: scratching.

11 The sides of the bearing shells are not 'run-in'

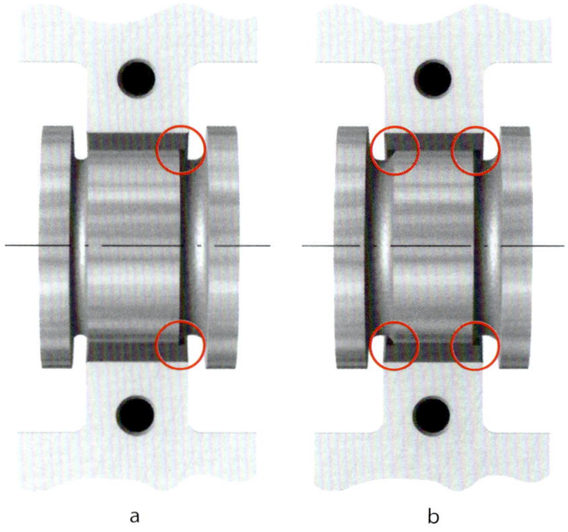

▲

Image

No wear along the edges of the bearing shells in the centre axis of the crankshaft or crankpin.

Cause:

- Badly ground journals, finishing too narrow.

12 Bearing-shell corrosion

Image

Rough, porous or smooth running surface with dark patches.
Often, the running surface has largely disappeared, revealing divergently coloured transition zones. In extreme cases, wearing of lead-bronze emergency running-surface.

Causes:

- Aggressive chemical additives in the lubricating oil.
- Lubricating oil contaminated with coolants and anti-freeze or acidic additives.
- Combustion residue in lubricating oil. For example: Sulphuric acid due to combusting heavy oil in engine.
- Exceeding the maximum stability of the lubricating oil.
- A high-water content in the lubricating oil.

13 Erosion resulting from cavitation

Generally, cavitation is caused by local lubricating-oil pressure drops. This may produce cavities ("bubbles") as oil boils at relatively low pressures. These cavities implode once they enter a high-pressure section and may subsequently cause a pressure wave. Large forces are released with the collapse of a cavity, which may cause serious damage to the running surface of the bearing shells. This may result in pitting. Cavitation in bearings are usually the result of low oil-pressure at high lubricating-oil temperatures, by liquids in the lubricating oil with low boiling points such as fuel, and also due to vibrations.

Image

Long-stretched erosion of the bearing shells with ragged lubricating-oil grooves.

Cause:

Turbulence in the lubricating-oil supply stream, in particular in areas where the flow direction is opposed to the rotation direction of the shaft.

14 Cavitation as a result of lubricating-oil discharge at a hole in the bearing

Image

The erosion appears as mushroom-shaped mark near the oil hole. Occasionally, the edges of the lubricating-oil grooves are also affected.

Cause:

Turbulence in the lubricating oil due to a poor transition from the groove to the bearing running-surface.

15 Cavitation as a result of fluctuating lubricating-oil pressure

Image

Oval or round erosion in the rotation direction. The erosion is generally found near the main bearings.

Cause:

Pressure waves in the oil boring in the crankshaft due to directional changes of the oil flow.

16 Cavitation as a result of oscillation

Image

Pitting erosion with round patches on the running surface, usually in the vicinity of the connecting-rod bearings. Main bearings often show oblong patching in the running surface of the top bearing shells.

Cause:

- The connecting-rod bearing is out of true due to bearing load.
- Oscillation due to rapid, cyclic movements of the crankpin.

17 Running-surface fatigue

▼

Image

Cracks or a network of cracks in the running surface.

Cause:

- Local overloading caused by high bearing loads, often in conjunction with high material temperatures.
- Load on the bearing edge or increased boundary lubrication, often caused by an oval hole in the bearing housing or connecting-rod foot.

1

2

3

18 Fatigue of all the bearing material, the running surface (anti-friction layer) and emergency running-layer (mostly bronze)

▲

Image

Cracks or a network of cracks, stone-shaped impurities in the bearing material.
Pieces of the bearing material have broken off.

Causes:

- Local overloading due to very high bearing loads.
- Local overloading due to uneven loads on the bearing.

Damage sequence

- Firstly, small tears in the running surface of the bearing occur.
- The cracks in the steel bearing shell increase in depth and width.
- Finally, pieces of the bearing material break off.

19 Fretting corrosion of the back of the bearing shells

Image

Dark, rough sections at the back of the bearing housing. The steel surface has already been affected.

Causes:

- Excessive clearance of the bearing shells in relation to the bearing housing or the connecting-rod big end. Usually, the bearing housing- or big-end borings are too wide.
- The main bearing- or the connecting-rod bolts have not been fixed according to the specifications.

20 Welding spots on the back of the bearing shells

Image

Fissures at the back of the bearing shells and the bearing housing or the big end, usually visible at the joints of the bearing shells.

Causes:

- Excessive clearance of the bearing shells in relation to the bearing housing or the big end. Usually, the bearing housing- or big end borings are too wide.
- The main bearing- or the connecting-rod bolts have not been fixed according to the specifications.

21 Fracturing of the steel bearing shell

Image

Fractured steel bearing shell.
The fractured surface is indicative of a sudden fracture. Occasionally, the damage indicates material battering.

Causes:

- The fretting of material and/or welding spots produced indentations.
- Loose bearing shells.
- Incorrect fit of the bearings in the bearing housing or big end.

27.11.4 Incorrect mounting

22 Bearing shell is misaligned

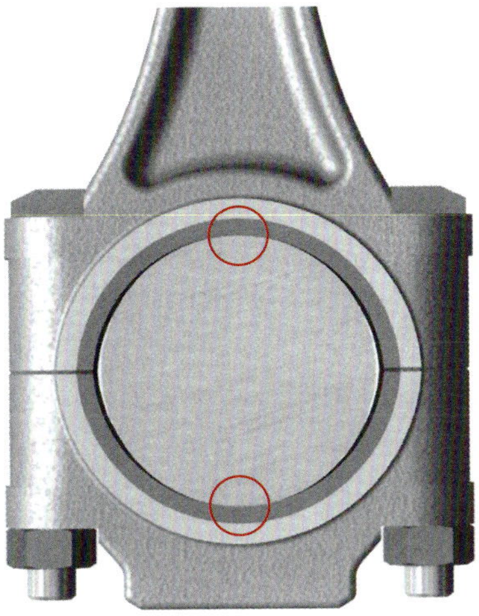

◀

Image

Very severe abrasions at the connection between the upper and lower bearing shells on the diagonally opposite surfaces of both bearings.

Causes:

- Improper bearings.
- Bearing shell rotated 180°.
- Damaged tabs on the back of a bearing shell.
- Bearing-cap or connecting-rod bolts have been incorrectly fixed.

23 Incorrect placing of oil holes in a bearing shell

Image

Overheated running surface, bearing shell is no longer functional.
Imprint of lubricating hole at the back of the bearing shell.

Cause:

- Incorrect mounting of the bearing shell impeding the lubricating-oil supply to the bearing.

24 Damages bearing shell due to dowel pin

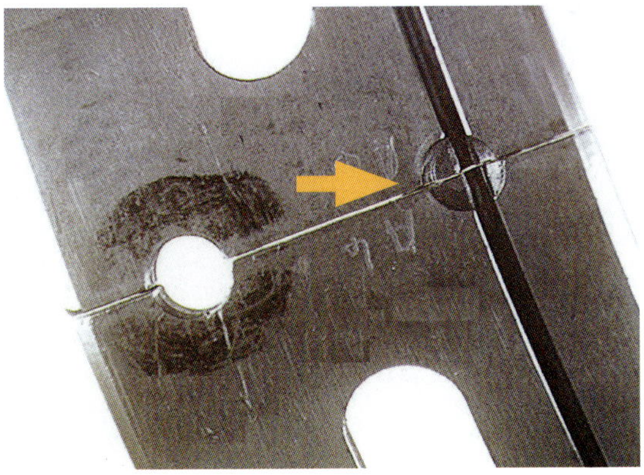

◀

Image

Serious damage to the running surface. The imprint of the pin is clearly visible next to the hole on the back of the lower bearing shell.

Cause:

- Dowel pin too long, remnants in the pin hole.
- Dowel pin misaligned.

27.11.5 Bearing types for various purposes. Some examples

▶

A type of bearing for large four- and two-stroke engines.

Engine categories III and IV.

The steel back has a carbon percentage of 0.13% and a manganese percentage of 0.6%.

The running surface is an aluminium (60%) tin (40%) alloy.

The bonding layer is 99.5% aluminium required for good adhesion.

The specific load is 25 MPa, which is moderate!

The maximum sliding speed of the shaft relative to the sleeve bearing is 162 m/sec. This is very high!

Steel-backed AlSn40 Bimetal Bearing

Cross-section

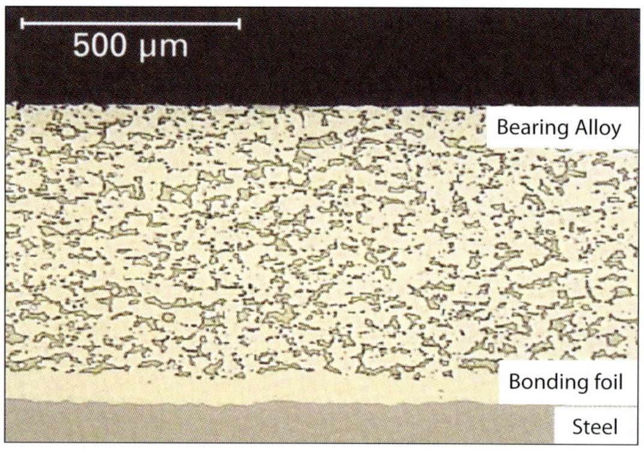

Application

Typical application

Large 4-stroke and 2-stroke engines, compressor and medium speed engines

Additional Qualifications

High corrosion resistance

Chemical Composition

	Chemical composition (% by mass)			
Bearing Alloy	Al		Sn	Cu
	rem.		40	0,6
Bonding Foil	Al 99,5 according DIN 1712/part 3			
Steel Back	C	Mn	P	S
max.	0,13	0,60	0,045	0,050

Hardness

	Hardness (HB)	
Bearing Alloy	min.	max.
	28	35
Steel back	min. 139	

Operating Limits

Short term temperature limit (°C)	120
Long term temperature limit (°C)	
Operation at design point	100
Max. sliding speed (m/s)	162
Specific unit load (MPa)	25

Performance Characteristics

Wear resistance	+
Fatigue life	0
Emergency runnig capability	+
Cavitation resistance	0

Legend: 0 = Standard; + = Improved; ++ = Excellent

▶

A type of bearing for heavily loaded four-stroke engines.

Engine categories II and III.

The steel back has a carbon percentage of 0.25% and a manganese percentage of 0.7%.

The coating on the running surface is an alloy of aluminium (79%), tin (20%) and copper (1%).

The running surface is a copper (75.5%), lead (22%) and tin (2.5%) alloy.

The specific load is 100 MPa, which is very high!

The maximum sliding speed of the shaft relative to the sleeve bearing is 22 m/sec. This is low!

Steel-backed Leaded Bronze Trimetal Bearing (Sputter Overlay)

Cross-section

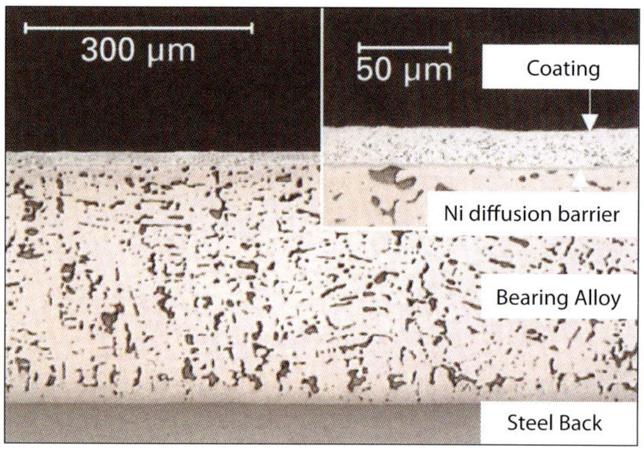

Application

Typical application

High performance truck and medium speed engines

Additional Qualifications

Excellent corrosion resistance

Chemical Composition

	Chemical composition (% by mass)			
Coating	Al		Sn	Cu
	rem.		20	1,0
Bearing Alloy	Cu		Pb	Sn
	rem.		22	2,5
Steel Back	C	Mn	P	S
max.	0,25	0,70	0,045	0,050

Hardness

	Hardness (HB)	
Bearing Alloy	min.	max.
	70	100
Steel back	min. 100	

Operating Limits

Short term temperature limit (°C)	190
Long term temperature limit (°C)	
Operation at design point	165
Max. sliding speed (m/s)	22
Specific unit load (MPa)	100

Performance Characteristics

Wear resistance	++
Fatigue life	++
Emergency runnig capability	0
Cavitation resistance	++

Legend: 0 = Standard; + = Improved; ++ = Excellent

Steel-backed Leaded Babbit Bimetal Bearing

Cross-section

Chemical Composition

	Chemical composition (% by mass)			
Bearing Alloy	Pb	Sb	Sn	Cd
	rem.	14,0	9,0	1,0
Steel Back	C	Mn	P	S
max.	0,25	0,60	0,045	0,050

Hardness

	Hardness (HB)	
Bearing Alloy	min.	max.
	22	–
Steel back	min. 100	

Application

Typical application

Main- and conrod bearings,
conrod small end bushings in compressors
and heavy fuel engines

Additional Qualifications

Excellent conformability and embeddability

Operating Limits

Short term temperature limit (°C)	120
Long term temperature limit (°C) *Operation at design point*	120
Max. sliding speed (m/s)	14
Specific unit load (MPa)	14

Performance Characteristics

Wear resistance	0
Fatigue life	0
Emergency runnig capability	++
Cavitation resistance	0

Legend: 0 = Standard; + = Improved; ++ = Excellent

◄ A type of bearing for the main connecting-rod bearings and piston pin bushings for heavily loaded four-stroke engines running on heavy fuel oil. Engine categories II and III.

The steel back has a carbon percentage of 0.25% and a manganese percentage of 0.6%.
The running surface is a copper (77%), lead (14%) and tin (9%) alloy.
The specific load is 14 MPa, which is low!
The maximum sliding speed of the shaft relative to the sleeve bearing is 14 m/sec. This is very low!

27.11.6 The three types of bearing shells

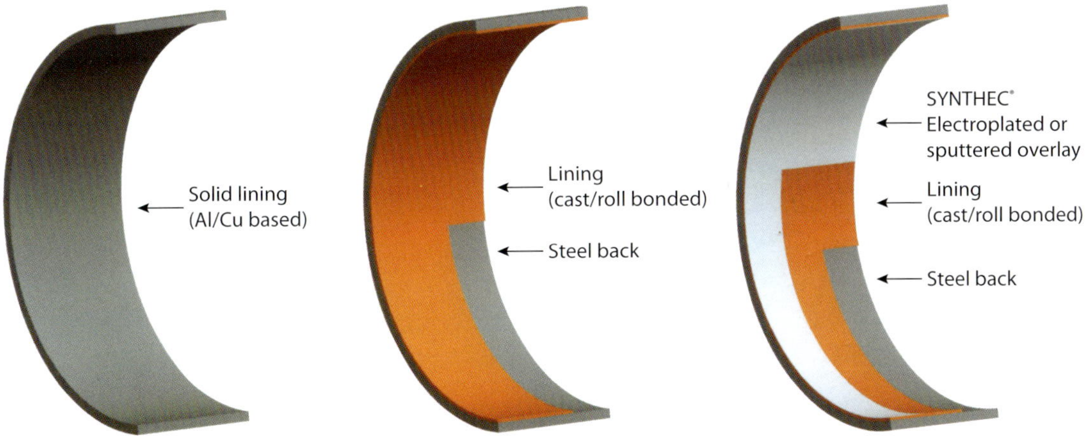

◄ The three types of bearing shells.

Left, a single piece.
A one-piece bearing, an aluminium copper alloy.

Middle, bi-metal.
A built up bearing shell, a steel back and a running surface comprising a friction material, such as an aluminium tin alloy.

Right, tri-metal.
A built up bearing shell with a steel back, an intermediate layer consisting of a copper alloy, which also functions as an emergency running surface and a sputtered running surface.
This usually consists of an aluminium tin alloy.

27.11.7 The names of all plain-bearings parts

▶

Names of all the parts of a bearing. These are important when drawing up a damage report.

Bearing shells and thrust washer (fig. 1)

1 *Free spread*
2 *Lug (tang) width*
3 *Lug (tang) position*
4 *Lug (tang) relief*
5 *Bearing width*
6 *Partial oil groove*
7 *Parting line relief*
8 *Bearing internal diameter*
9 *Countersunk oil hole*
10 *Bearing back*
11 *Wall thickness*
12 *Outside edge*
13 *Inside edge*
14 *Lug (tang) length*
15 *Joint face*
16 *Oil pocket*
17 *Lug (tang) length*
18 *Thrust washer thickness*
19 *Running surface*
20 *Thrust face*
21 *Oil Groove*
22 *Locating lug (tang)*
23 *Bearing length*
24 *Distance between flanges*
25 *Locating notch*
26 *Relief on thrust face*
27 *Releif*
28 *Flange thickness*
29 *Lubrication slot*
30 *Relief on thrust face*
31 *Oil groove*
32 *Flange diameter*

Fig. 1

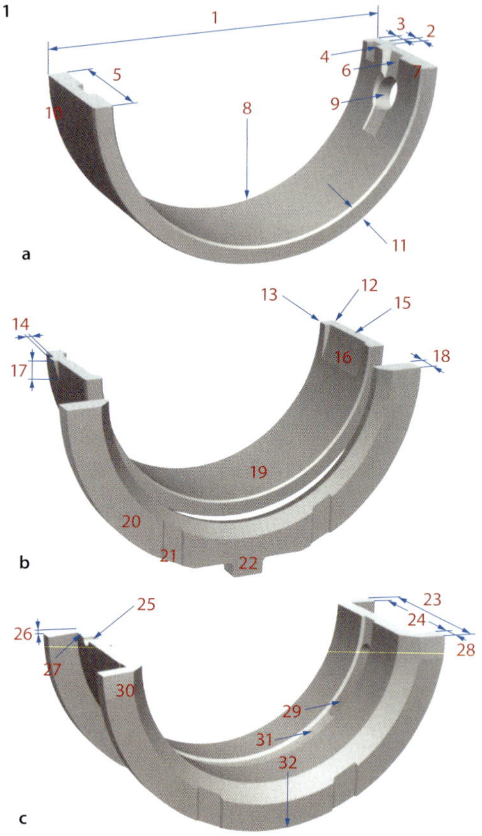

Bushings (fig. 2)

1 *Oil groove*
2 *Oil hole*
3 *Inside diameter*
4 *Outside diameter*
5 *Length*
6 *Mating faces*
7 *Clinch butt*
8 *Oil groove*
9 *Location notch*
10 *Axial oil groove*

Fig. 2

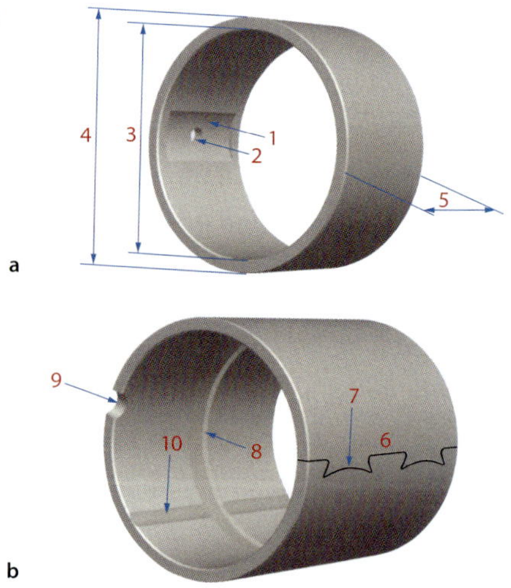

27.11.8 Guide to determine the cause of bearing damage (checklist)

◀ A checklist to determine causes of slide bearing damage.

<table>
<tr><th>Function</th><th>Characteristics</th><th colspan="2">Parameters to check</th></tr>
<tr><td rowspan="8">External conditions</td><td rowspan="8"></td><td colspan="2">Operating conditions – particular load cases or applications</td></tr>
<tr><td colspan="2">Breakage of adjacent parts</td></tr>
<tr><td colspan="2">Failure of piston</td></tr>
<tr><td colspan="2">Piston pin</td></tr>
<tr><td colspan="2">Piston pin bushing</td></tr>
<tr><td colspan="2">Engine design, vibrations</td></tr>
<tr><td colspan="2">Failure of lubrication – reason?</td></tr>
<tr><td colspan="2">Assembly/dirt in the system</td></tr>
<tr><td rowspan="13">Lubricant</td><td rowspan="6">Viscosity</td><td colspan="2">Oil analysis, additives, oil aging</td></tr>
<tr><td colspan="2">Single or multiple viscosity oil</td></tr>
<tr><td colspan="2">Viscosity</td></tr>
<tr><td colspan="2">Water content</td></tr>
<tr><td colspan="2">Foaming</td></tr>
<tr><td colspan="2">Oil treatment</td></tr>
<tr><td>Dilution
Contamination</td><td colspan="2">Running surface and housing, transitions,
evaluation of bearing set (conrod, main)</td></tr>
<tr><td rowspan="4">Cooling performance</td><td rowspan="2">Hot operation condition</td><td>Standard temperature</td></tr>
<tr><td>Temperature under specific operating conditions</td></tr>
<tr><td rowspan="2">Overheating</td><td>Temperature measurement before failure</td></tr>
<tr><td>Metallurgical analysis of the components</td></tr>
<tr><td rowspan="2">Mixed friction condition</td><td colspan="2">Running in/before stop/lubrication in advance</td></tr>
<tr><td colspan="2">Specific appearance of the bearing</td></tr>
<tr><td rowspan="13">Bearing/load</td><td rowspan="4">Bearing design</td><td colspan="2">Specific surface load[1]</td></tr>
<tr><td colspan="2">Pin orbital path/limits[1]</td></tr>
<tr><td colspan="2">(Engine's own) vibrations</td></tr>
<tr><td colspan="2">Comparable engines</td></tr>
<tr><td rowspan="2">Installation situation</td><td colspan="2">Latest data from assembly test</td></tr>
<tr><td colspan="2">Available geometric measurements</td></tr>
<tr><td rowspan="2">Housing</td><td colspan="2">Installation situation</td></tr>
<tr><td colspan="2">Condition of bearing back</td></tr>
<tr><td rowspan="3">Bearing</td><td colspan="2">Corrosion</td></tr>
<tr><td colspan="2">Cavitation[2]</td></tr>
<tr><td colspan="2">Metallurgy</td></tr>
<tr><td rowspan="2">Pin</td><td colspan="2">Cavitation[2]</td></tr>
<tr><td colspan="2">Metallurgy</td></tr>
<tr><td rowspan="6">Quality</td><td rowspan="2">Housing</td><td colspan="2">Manufacturing records</td></tr>
<tr><td colspan="2">Measurements of geometry, surface roughness, etc.</td></tr>
<tr><td rowspan="2">Bearing</td><td colspan="2">Manufacturing records</td></tr>
<tr><td colspan="2">Detailed geometry, including spread, crush height, wall thickness,
bearing clearance, back contact</td></tr>
<tr><td rowspan="2">Pin</td><td colspan="2">Manufacturing records</td></tr>
<tr><td colspan="2">Measurements of geometry, surface roughness, etc.</td></tr>
</table>

1: Relevant for the design and testing of new engines.

2: Individual types of cavitation are not considered here in detail.

>CH 28

Casting, forging and welding engine parts

Casting small engine parts.

A cast for a very large MAN Diesel four-stroke V-engine, category III being hoisted out of the casting pit. There is still a long road ahead before this unfinished block is ready for the assembly of the engine!

28.1 Introduction

From the very first trials with Otto- and Diesel-engines, most parts, such as the cylinder block, the cylinder head and various other components, were made of cast iron.
This was a common manufacturing method in the middle of the nineteenth century. One hundred years earlier, steam engines were produced in a similar fashion from cast iron.
Casting was already a common craft around 1750 for weapons, such as cannons!
Many of the present foundries that manufacture engine parts have a very long history, which often goes back centuries.

28.2 Cast-iron parts for diesel engines

1 **Engine blocks**
These have been cast from the very beginning of the engine industry. Only in large two-stroke crosshead engines, are the bottom half of the crankcase and the A-frame manufactured from welded steel plates.
In the past, this was also the case for two-stroke engines from Brons and Bolnes. All four-stroke engines have as a rule, a cast engine block.

2 **Cylinder heads**
Customarily, these are cast.

► The Caterpillar MaK foundry, at Kiel, Germany.

▼ The MAN Diesel AG foundry in Augsburg, Germany.

3 **Pistons**
Many pistons are still cast. Moreover, pistons that consist of two parts often have a cast-piston skirt.

4 **Other components**
Components such as the air-intake ducts, integrated pumps, oil sumps, rocker arms, air coolers, and governor seats are frequently cast.

5 **Gears**
Due to ever-increasing interaction forces in the engine, fewer connecting rods and crankshafts are manufactured from cast iron. These cast-iron parts are now only found in small, often older engines.

▲
The topside of an electric oven. The liquid cast iron is visible; temperatures are reached between 1000 °C and 1050 °C.

28.3 Advantages of cast engine-parts

Often engine parts such as the block and the cylinder head have a very complex shape. At present, the air supply to the cylinders and the lubricating-oil supply lines are placed inside these parts whenever possible. Both the block and cylinder head have shapes that are difficult to manufacture by welding steel plates.

For material properties, see Chapter 7, *Use of materials for diesel engines*.

28.4 Foundries

Many engine manufacturers have their own foundry, such as that at MAN Diesel AG in Augsburg, Germany, where most of the photographic material in this chapter was taken.
There are foundries that not only cast engine parts for engine manufacturers, but also for other industries.
The very large shipyards in the Far East often build complete ships and manufacture all the casts for their four-stroke- and two-stroke engines on location.
Parts that are forged, such as crankshafts, connecting rods, pistons and cylinder heads (two-stroke) are also manufactured on location.

▲
High-tensile scrap iron, such as waste steel from the manufacture of beams, profiles and/or metal sheeting used for ship manufacture is used as a base material for cast iron.

Waste from the thin metal sheeting used in the auto-industry is also used. Every foundry has its own preferences. There are some waste-sheeting materials containing undesirable substances, such as copper.

28.5 Casting process

In a smelting furnace, often an induction oven, relatively clean waste metals, such as steel used for shipbuilding and metal sheeting used in the auto-industry, are melted with the required alloys.

The steel turns liquefies at a temperature of approximately 1050 °C. For every type of cast iron, a different raw-material composition is required. Removal of impurities from the molten steel is of the utmost importance.

▶
The smelting furnace at MAN Diesel AG, Augsburg, Germany. Shown here, just before casting ladles are filled.

▲
Transporting a casting ladle using an overhead crane through the casting hall. Prepared mould-cast assemblies ready for casting are in various places.

The moment that a fully loaded oven reaches the required temperature, the molten cast iron is kept at this temperature and immediately distributed equally over a number of large casting ladles; this is dependent on the size and number of the casts.

The casting ladles are insulated against the very hot liquid cast iron; this is done using special heat resistant firebricks. These casting ladles are moved through the casting hall to every casting location with an overhead crane.

28.5.1 Casting moulds, the principle

▼
Removing the slag from the smelting process.

With a special rake, all impurities that float to the surface are removed.

▲
Adding alloys.

Special alloys are added during the filling of the crucibles to improve the properties of cast iron. MAN B & W, Frederikshavn, Denmark (2009).

▲

The casting ladle.

Several ladles filled from the furnace are used to fill all the assemblies at once.

▲

A series of assemblies ready for casting.

The assemblies are ready to be filled via the funnels on the left. The vent tubes can be seen to the right in the middle of the picture.

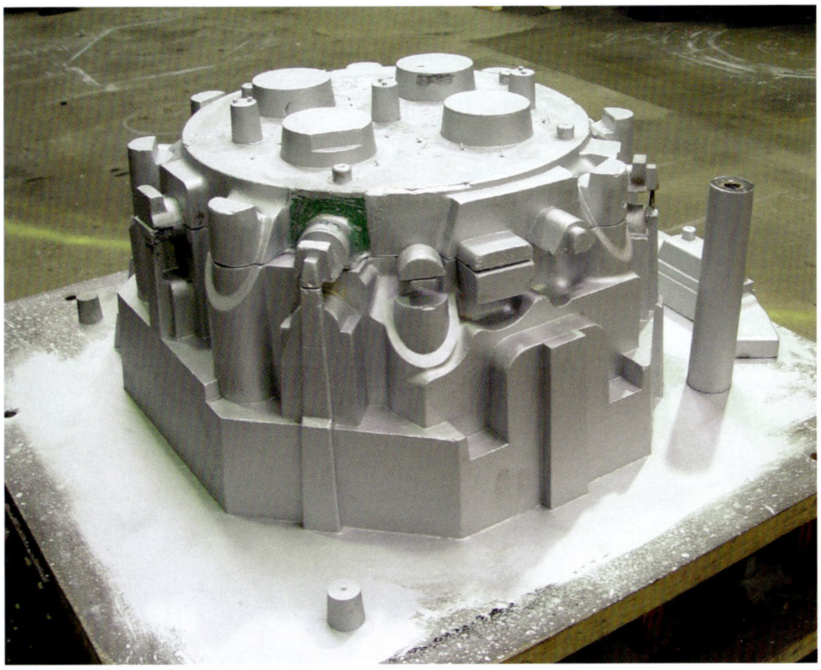

▲

A cured mould consisting of two parts of a cylinder.

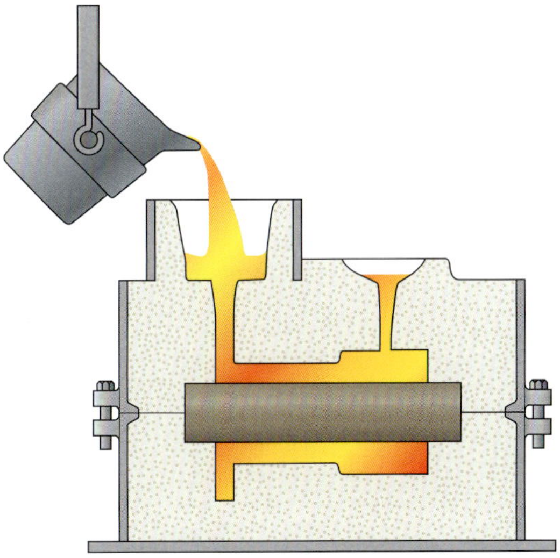

▲

The principle of casting with a horizontally partitioned setup.

The core like the surface consists of cured casting sand. Top right, the air can escape while to the left while the cast is being filled.

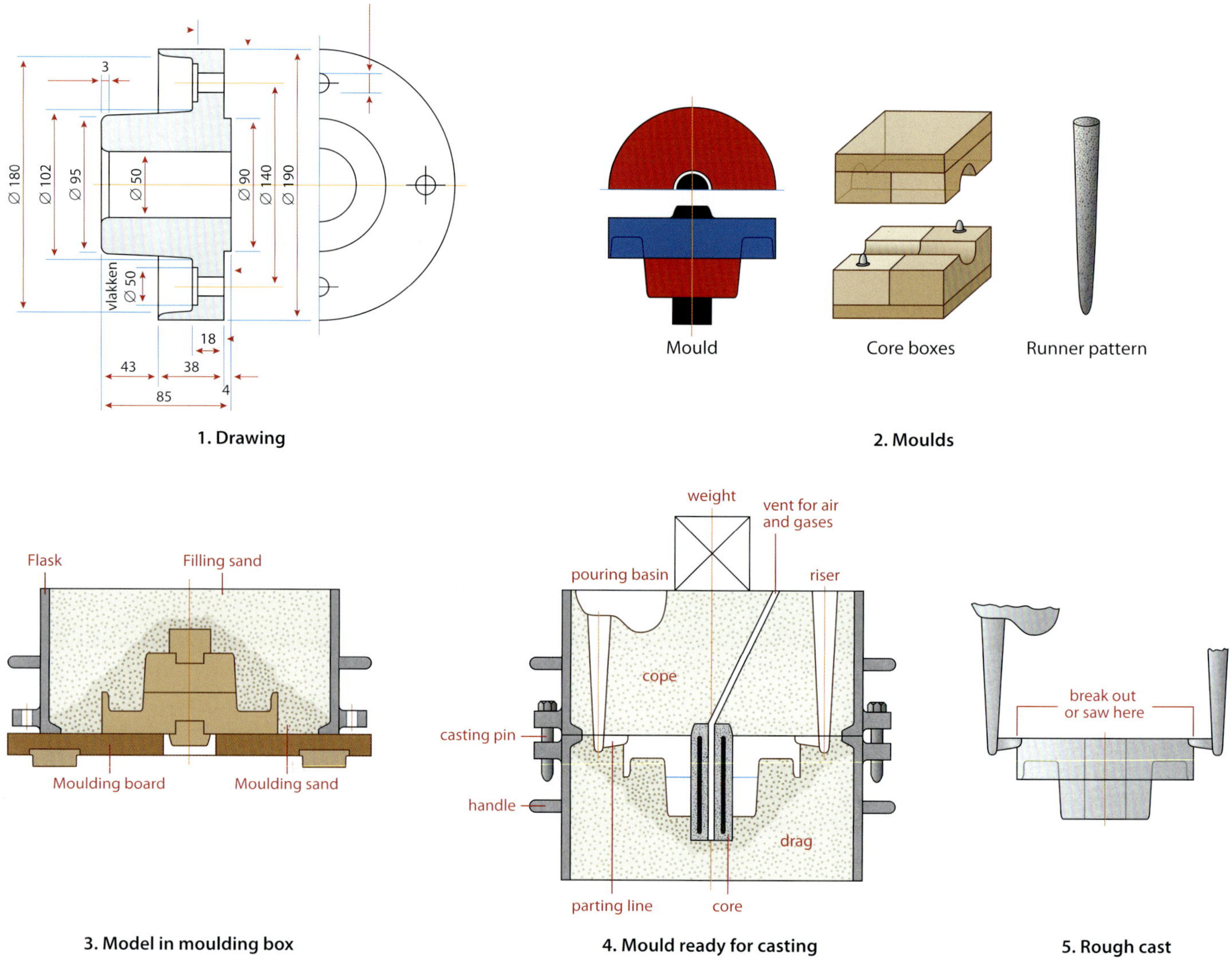

▲

The principle of the manufacture of castings.

1 *The drawing of a to be cast coupling flange.*
2 *A wood model of the sand mould:*
 – *the core box to make the runner;*
 – *a model for both channels in the cope so that the sand mould can be manufactured.*
3 *The sand mould is made in a moulding box.*
4 *The casting assembly ready for casting.*
5 *The raw casting before machining, the two channels are often removed manually.*

▶

A principle of the assembly for casting an engine block without paying too much attention to details.

1 *free space in the casting basement*
2 *foundation casting hall*
3 *steel casting cabinet*
4 *support under/next to cast*
5 *liquid cast iron feed*
6 *risers/air duct*
7 *hardened casting sand*
8 *machined surfaces*

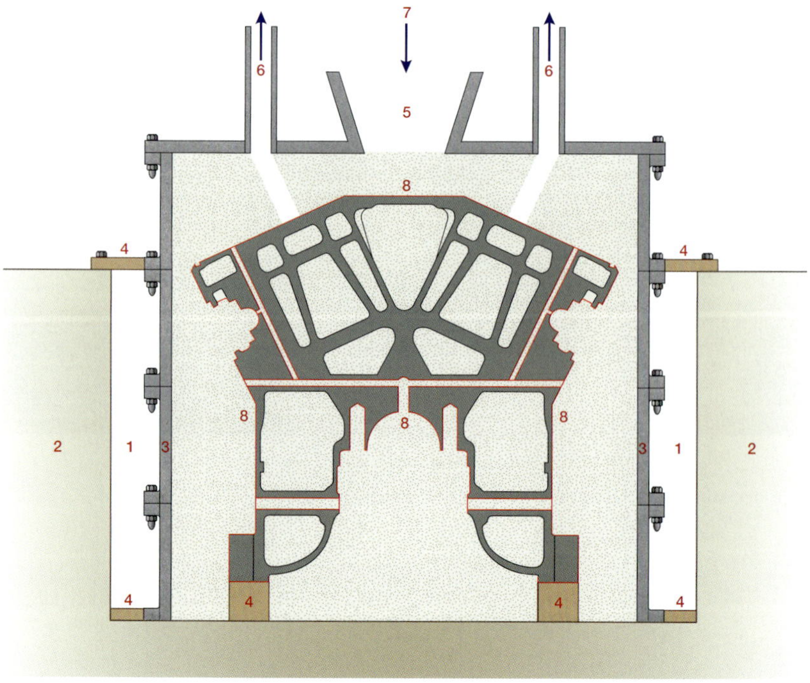

◄
A mould-casting assembly in a casting pit. The square openings are casting mouths so that this large engine block can be filled quickly.

28.6 Casting location

Large casts, such as engine blocks are cast in the floor of a casting pit. Square steel moulds on the floor, built up in layers are used for smaller casts.

28.7 Moulds

Throughout the centuries, casting has used moulds manufactured from casting sand and a special substance that ensures that the complex and sometimes very thin shapes harden so that they can be transported either manually or by using a crane. Casting dies are necessary so that the moulds can be manufactured. These used to be made of wood but are now manufactured from wood as well as plastics.

▼
A mould-casting assembly in a moulding box. In order to take shrinkage of the cast iron into account during cooling, a part of the funnel is also filled with the liquid cast iron.

28.8 Filling the casting dies

Small complex moulds are made by hand. The casting dies are filled by hand with casting sand and intermittently compacted ('stamped down') and re-filled. When the die has been completely filled, the casting sand is left to harden so that it can be used under normal operating conditions.
Casting dies are partitioned in different ways so that they can be taken apart easily.
Larger moulds are filled using filling machines that vibrate and therefore compact the casting sand at the same time.

▼
The manual process for manufacture of smaller moulds for engine parts.

Moulding sand is often used as a base material. There are two categories: clay-bound sand for small castings or chemically bound sand for large castings.

▶
Filling machines for the larger moulds.

In order to ensure that the casting sand completely fills the moulds, the casting sand is 'stamped down'.
Shown here are partitioned wooden casting dies.

28.9 Mould assembly

28.9.1 Engine blocks

1 Almost all the space in the block must be filled with the prepared moulds, not only inside the block but also on the outside.
2 Most blocks are cast in a square casting pit built into the work floor or in large steel moulding boxes. The space between the floor, sides and the top of the block must also be filled with moulds.
3 In order to obtain a uniform material structure, casting must take place fast. The largest four-stroke engine blocks are cast in 100 seconds. The weight of the cast-iron block is 100 tonnes. A casting speed of 1 ton per second! It is therefore necessary to have large casting openings, so that the molten iron can be poured swiftly, filling the mould from 'below', while the air moves rapidly upwards!
4 It is also of the utmost importance that there are no air inclusions in the casting! This can be detrimental, especially near the crankshaft, where very large loads take place. This often results in the rejection of the block and consequently the loss of € 100,000.00, the production costs!

▶
Casting moulds for the upper part of the engine block, here the two cylinder blocks are fitted after casting and removal of the casting sand.

◄
One must ensure that the construction for large mould-casting assemblies does not 'float'. This is prevented by the use of heavy weights in large mould-casting assemblies and screwed bars for the smaller moulding boxes.

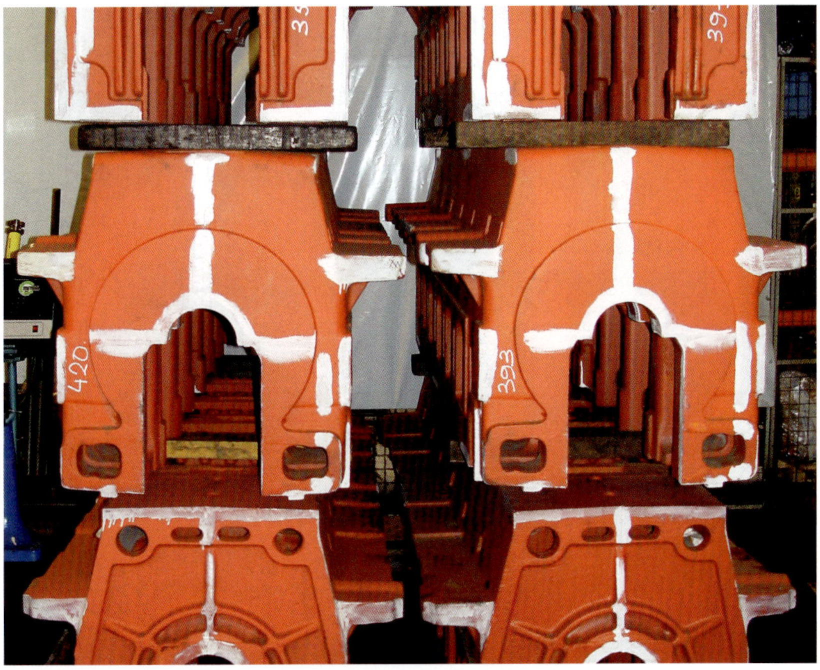

◄
Engine blocks for ABC in-line engines ready for machining.

The engines blocks are cleaned after casting, finished and ready for machining. The blocks are paint-coated in order to reveal potential irregularities. White marks indicate the areas where measurements ought to be taken prior to machining.

◄
Part of the engine block in a sand mould, ready to be placed in the casting pit. Air ducting for venting is clearly shown in the location where the crankshaft will later be mounted.

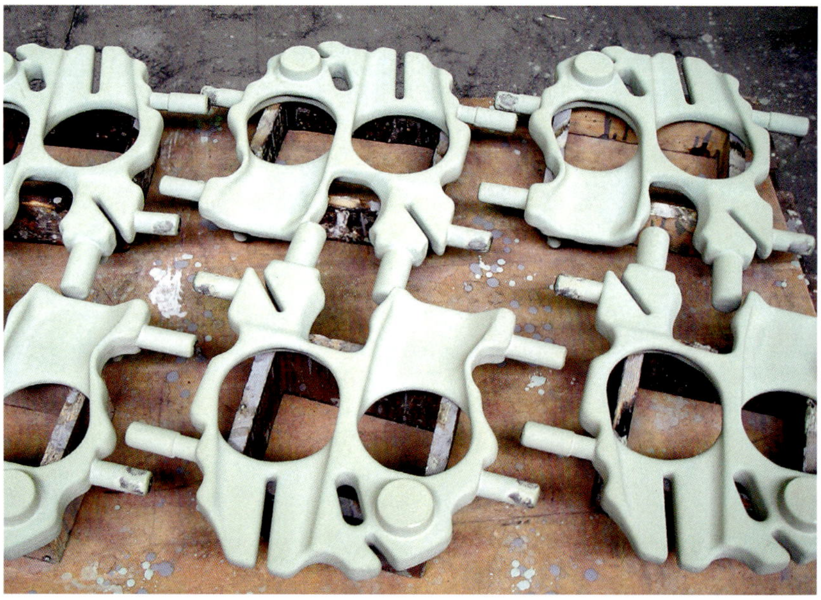

▶
Casting moulds as part of the cylinder heads. The casting sand is hardened by adding of binding agents allowing all the casting moulds to be prepared in a single casting arrangement

All that remains of the sand in a casting arrangement after casting is open space: air!

▲
A series of casting dies, suspended by hoists. They will later be used to build the complete cast.

▲
Casting. The caster is completely protected from liquid cast-iron spatters.

28.9.2 Smaller parts, such as cylinder heads

Exactly the same procedure is followed for these and similar parts, albeit less complicated in scope and process.

▶
The upper lid is still to be placed on this mould-casting assembly. Before casting, every cast follows a very precise procedure. Rejection of the cast is expensive.

28.9.3 Serial production

The lower crankshaft-bearing caps are cast in one die containing, for instance, six pieces.

▶

Castings ready to be filled with liquid cast iron at 1300 degrees Celsius.

Smaller engine parts mostly cast in series. Left the funnel, the stoppers are installed to prevent pollution of the casting. The pipes on the right allow the air to escape. Also shrinkage can be compensated by the iron in these pipes.

◀

Part of a four stroke engine after casting. Note the razor-sharp edges created by the clearances between the various sand moulds.

◀

Cylinder heads for four-stroke engines immediately after casting.

In order to ensure that the component will have the proper dimensions after the various machining processes, the superfluous material must be removed so that measurements can be taken.

After casting, the casts must be allowed to cool for several days. Casts must not cool too quickly as they can crack.
Outside the casting pit or moulding boxes, small flames can be seen for days on the coatings that are applied by immersion (small moulds) or lubrication (large moulds).
After dismantling the mould-casting assembly, the now black casting sand is collected, cleaned, and re-used.

▶
Unfinished engine parts after casting.

▶
Casting engine components in series.

▶
A V-engine block ready for finishing.

28.10 Cleaning the castings

As shown in the images, casts must undergo various processes before being machined at a different location.

Procedure

- An initial superficial inspection is performed to determine whether the casting has been successful.
- The casting sand is removed.
- Protruding 'icicles and ice sheets', remnants of molten iron that has flowed out of the splits and air-vent drillings are removed by hand using hammers, chisels and grinding stones.

▲ 'Icicles and ice sheets'. The removal of all prominences is heavy and dirty work requiring a great deal of experience.

◀ Finishing the cast, a seven-cylinder in-line engine block.

▼ A cylinder head being finished. This is done with pick hammers, chisels and for the final finish, grinding stones and brushes.

◀ Just cast. The liquid cast iron is visible in the pouring basins. The non-stick coating is on fire in some places.

The block is then de-burred using de-burring discs until all the sharp edges and protuberances have been removed. The block is then 'preserved' (cast iron scarcely rusts) with a coating of varnish or paint.

28.11 Casting stresses

Some engine manufacturers place their cast engine blocks in storage for many months. In doing so, the stresses formed during casting decrease.
At the Wärtsilä factory in Vaasa, Finland, they place the cast blocks for the W20 engine outside in the snow during the winter months to reduce material stresses.

28.12 Checking for air inclusions and damage

This is carried out before machining, if possible. Of course, certain discrepancies are more serious.

28.13 Dimensional checks

It is of utmost importance that the final engine block can be machined from the roughcast block.
On the other hand, if too much material must be removed, this can be expensive due to the extra machining hours required.
Engine blocks delivered to customers are re-measured on location to guarantee that they have the correct dimensions.

►
A finished and painted casting, a block for a V-engine. The block is measured before machining.

▼
A block for a large V-engine, cleaned, painted, measured and ready for machining.

▲
Ultrasonic inspection at a location where the engine will be heavily loaded, specifically around the crankshaft.

28.14 Operations at the machining factory

With a minimum removal of material, that is, material that falls 'outside the dimensions', the requisite part is manufactured from the roughcast at the machining factory.
A rule of thumb: 15% of the cast material must be removed before the part is ready for use.
Moreover, it is of utmost importance that no air or other impurities are incorporated. The modern heavily loaded parts, such as the engine block, can be rejected for these inclusions.

◄ Before a cast is machined, it is inspected for damage, shown here at the edge of the oil sump, serious or not.

◄ An in-line- and a V-block in storage before machining at Wärtsilä, Zwolle, The Netherlands (2004).

◄ An unfinished in-line-engine block is being aligned on the processing machine. Shown here, at Caterpillar MaK, Kiel, Germany.

▲
The block for a nine-cylinder Caterpillar MaK in-line engine ready for machining after being cleaned, painted, re-measured, and checked for air inclusions.

►
Cylinder heads.

►
Final fine-machining of the cast-iron block for a four-stroke medium-speed diesel engine, V type. MRC Schiedam, The Netherlands.

◄ Moulds in storage with, amongst others, an engine block in the centre left of the photo.

◄ Wooden casting moulds with inside, the casting sand. Note the casting-sand finish on the upper surface.

▼ Working on unfinished four-stroke engine blocks at MAN Diesel AG, Augsburg, Germany.

Engine blocks for in-line and V-engines.
They are finished, cleaned, re-measured and ready for machining at Wärtsilä in Zwolle, The Netherlands (2004).

Line boring the crankshaft of a Caterpillar MaK in-line engine at Kiel, Germany.
Firstly, the bearing caps are hydraulically fixed at the normal pre-load.
Afterwards, the boring for all the crankshaft bearings is completed in one operation.

A machine for machining engine blocks.

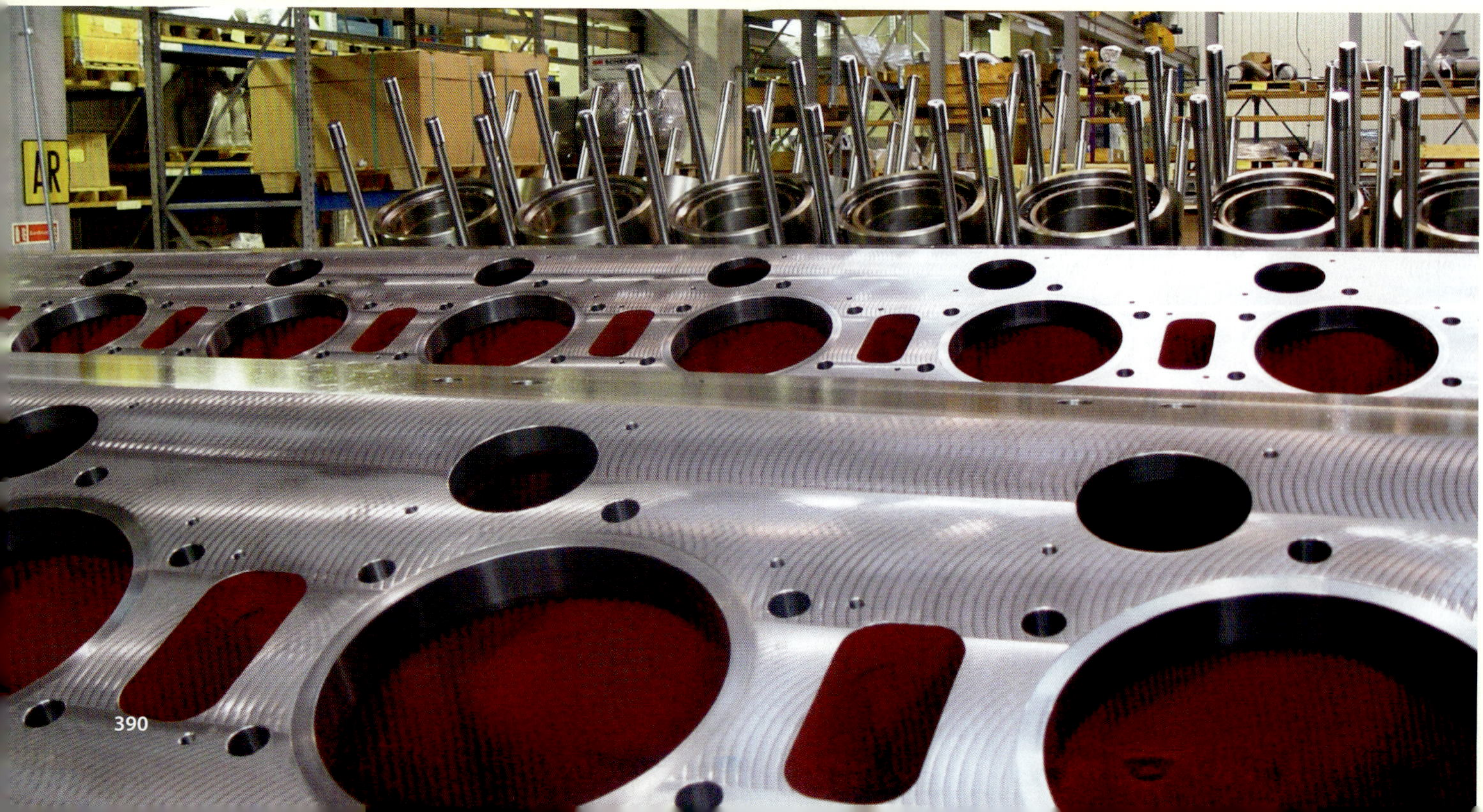

►

Processing a block for a V-engine at MAN Diesel, Augsburg, Germany.

▲

A V-engine block with the bearing caps already mounted in relation to the line bores.
Left and right in the block, the bores for both camshafts.

◄

Foreground: a finished V-engine block.

▼

Washing in the washing machine is an important process executed before the block is re-measured and assembled. All impurities, such as filings and other material, must be removed.

Shown here, the block is placed on a trolley-cart.

Many washing machines now have a horizontal turning shaft with a flange to which the engine flange can be bolted.

This means that the block can rotate through 360°, and so gravity assists in material removal.

A final inspection is performed.

Using special tools and mirrors, the very last and minute particles can be removed.

Leaving any material in the block can cause major damage to an engine!

▲

The new washing machine for cleaning machined engines for Caterpillar Mak at Kiel, Germany.

All blocks of this Caterpillar Mak series can be placed in these pivoting frames.

By allowing the block to rotate slowly in the wash street in the background, whilst being cleaning gravity assists. This facilitate easy and complete removal of all the machining material from the block.

Cleaning and rotation is completely automatic. The cleaning fluid is filtered and reused.

▼

The construction of a four-stroke medium-speed V-engine at MAN Diesel in Frederikshavn, Denmark.

After carefully cleaning the engine, it is extremely important that dirt does not enter the engine during its assembly.

Clean and cover with plastic wrap. It is of paramount importance that parts that will be used are inspected before assembly! (2010)

▲

Deburring of forgings.

The upper and lower dies specific to this crankshaft are placed on a heavy hydraulic press.

The shaft is pressed into the dies with huge forces of thousands of tons; the result is the rough crankshaft.

1 *the forging*
2 *the cutting plate*
3 *the stamp*
4 *hydraulic press*

28.15 Forging crankshafts

28.15.1 Introduction

Crankshafts are the most heavily loaded parts in diesel engines, subjected to torsional and bending stresses. Crankshaft manufacturing must meet high requirements due to the fluctuating loads on every crank.

The forces on the crankshaft are impressive. Even the smallest Category I engines, there are forces of hundreds of Newton being exerted on the crankpin during combustion.

The combustion forces exerted on the piston and consequently the crankshaft vary from 0.5 tons up to as much as 1,100 tons in the various engine categories!

Manufacturing the largest crankshafts for medium-speed diesel engines, category III is very expensive.

The crankshaft is forged from a solid piece of the approved material. Large crankshafts are forged per crank. This is called 'throw by throw' or 'Continuous Grain Flow', as opposed to the manufacturing of smaller crankshafts, which are forged 'in one throw'. This is called closed die forging and is also a 'continuous grain flow' method. This is followed by very meticulous machining and the shaft must be ultimately approved by a Classification agency.

▼

Roughly forged crankshafts stored outside before machining. Every crankshaft undergoes approximately **45** machining steps.

The total production time from rough rod form to a ready too assemble crankshaft is on average, **six** months.

▼

A large used crankshaft for a four-stroke in-line engine at Mark van Schaick, Schiedam, The Netherlands for reconditioning.

▲
In the engine factory of Alfing Kessler, Aalen, Germany.

In the foreground; the steel shafts, which after heating in an oven are stamped in a hydraulic press to form the rough crankshaft. Also in the picture, a number of crankshafts ready for inspection. They will be subsequently machined to the correct dimensions.

28.15.2 Crankshafts requirements

High requirements are set for:
- The material used;
- The crankshaft design;
- The calculation of the maximum tensions inherent in the material, such as the nominal tension, resistance to metal fatigue and the ensuing acceptable maximum values;
- The manufacture of the lubricating oil ducts in the crankshaft;
- The finish and surface hardness of the journals;
- The shrink fit allowances of the crank web and crankshaft in two-stroke crosshead engines.

28.15.3 Material used

A	Manganese – carbon-steel	Casting	400-550 N/mm^2
B	Manganese – carbon-steel	Forging	400-600 N/mm^2
C	Manganese – carbon-steel	Forging	maximum 700 N/mm^2
D	Alloyed steel	Casting	maximum 700 N/mm^2
E	Alloyed steel	Forging	maximum 1.000 N/mm^2
F	Spherical or nodular graphite steel	Casting	370-800 N/mm^2

A number of steel alloys with a certain tensile strength can be considered.

28.15.4 Manufacturing crankshafts

Casting
Small lightly loaded crankshafts are often cast, for instance, air-cooling and industrial air compressors and small lightly loaded engines in Category I.
The liquid material is poured into moulds, where it immediately assumes the rough shape of the required crankshaft.
The final shape is achieved by subsequent machining processes, such as turning, milling, drilling and other machining operations.
As the required manufacturing equipment is relatively simple, these crankshafts are comparatively inexpensive.
The granular structure of the material resembles sand; this is noticeable in crankshaft fractures.
Casting has fewer positive properties than forging. The material has a lower tensile strength and therefore is more prone to fractures.

Forging
Forging large crankshafts requires large heavy hydraulic presses. Moreover, subsequent elaborate machining is required. The base material, a steel shaft with the correct composition, is heated to almost its melting point (1100 °C to 1500 °C) and then placed between dies and stamped into the required shape.
This requires presses that can exert pressure forces of thousands of tons!
Stamping produces a continuous grain flow; continuous grain flow gives the crankshaft better fatigue resistance.

This continuous grain flow is due to the reduced space between the molecules of the material; the molecules are more firmly joined together. Subsequent to the forging, the material is stress-relieved using heat-treating methods. The crankshafts are then machined extensively.
The hydraulic presses used for forging are very expensive. Moreover, each type of crankshaft requires a specific mould.
Companies producing crankshafts or cranks (for two-stroke engines) for engine manufacturers in Categories II, III and IV, who do not manufacture their own crankshafts, have a series of crankshaft dies in storage for each type with respect to cylinder bore, in-line or V-shape and the number of cylinders. This is very expensive!

Forged crankshafts are stronger and more expensive than cast crankshafts. The price of a forged crankshaft for small craft in Category I is two or three times higher than that for a cast crankshaft. Large crankshafts are always forged.

28.15.5 Material requirements and tests by Classification bureaus

There are many requirements, such as, the material composition, the heat treatment, the mechanical and non-mechanical tests; such as hardness-, tensile strength-, magnetic- and ultrasonic test methods.
As the engines discussed here seldom have cast crankshafts, only the manufacturing of forged-steel crankshafts will be taken into account.

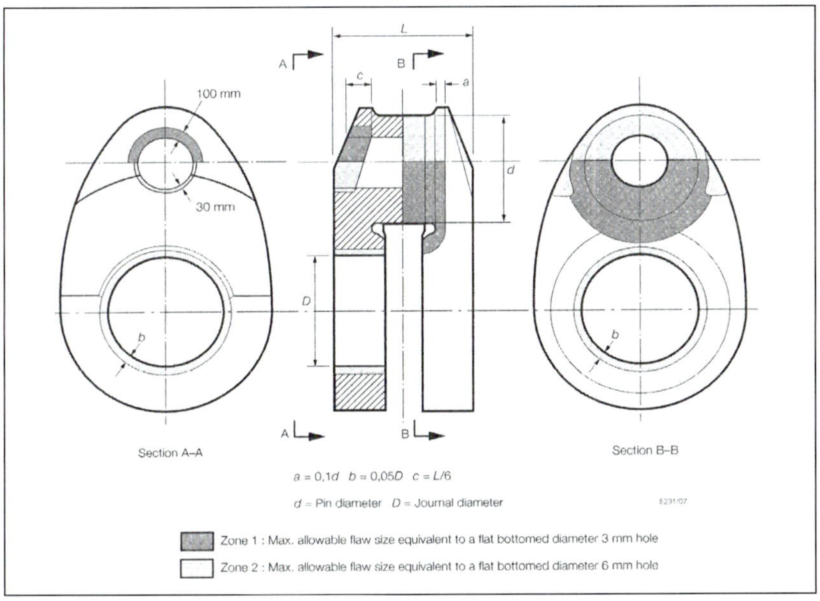

◀ The approved inspection of the crankshaft by Lloyd's Register using the ultrasonic method. The requirements that apply to crankshafts are very stringent.

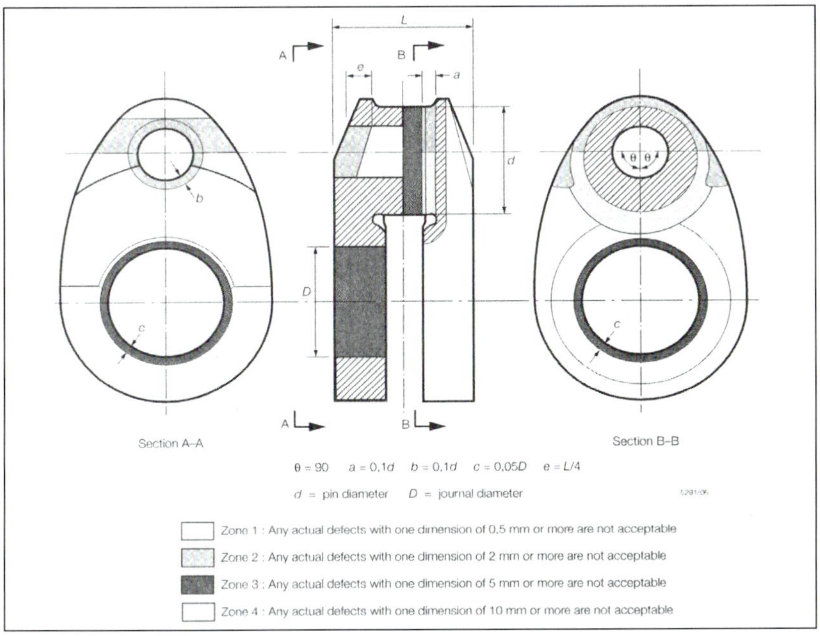

◀ The approved inspection of the crankshaft by Lloyd's Register using the magnetic-iron particle method.

In order to achieve a sturdy crankshaft, the dimensions of the crankshaft parts must be in a certain proportion to each other.

▶ Rough forged crankshafts with an alloyed steel shaft in the centre, the base material.

The white marker paint is applied in order to determine if the final crankshaft can be manufactured from the material.

The procedure.

- Provide the forge with a drawing of the crankshaft and material specifications, approved by the Classification bureau. The drawings must be made accessible in 3D so that all possible loads on the crankshaft can be established and calculated!
- Order a steel shaft with the correct dimensions and material composition.
- Prepare the hydraulic crankshaft press with the correct dies for the crankshaft.
- Heat the shaft in a gas or electric oven until the correct temperature is achieved.
- Place the heated shaft in the press.
- The shaft is usually pre-processed in a separate machine: pre-forging.
- Press the entire crankshaft in a series of forgings or, for smaller crankshafts, in a single throw.

To ensure that the forged steel maintains the correct temperature, the rough crankshaft is removed from the press with a forklift three times and re-heated in the oven to a temperature of 900 °C. The hydraulic pressing is carried out 25 times per cycle, so a rough crankshaft is completed in 75 pressings! This is for closed die forging, not for throw-by-throw forging.

- Larger crankshafts are forged per crank: throw-by-throw pressing.
  One crankshaft section of the rough shaft is heated to 900 °C and subsequently forged in a series of throws.
  Here the cranks are also re-heated. For the following crank, the shaft is placed in the correct position for that crank.
- Heat treatment techniques are performed, such as annealing and tempering.
- Machining the crankshaft to its final shape and dimensions.
- Post-treatments, such as hardening of the journals for crankshafts in Categories II, (generally), and III (occasionally).

▶ For this cast crankshaft, the roughly cast shape (the crank webs with the fitted single-throw counterweights) is clearly visible.

Other engine manufacturers require cleanly machined crankshafts.

◀

A roughly forged crankshaft being placed in the oven.

The partially forged crankshaft is re-heated to 900 °C.

◀

In the hammer forging (closed die forging) the forging gets forged in one piece.

◀

Forged crankshafts undergo heat treatments, such as annealing, for stress relief.

The parts are placed in a gas-heated oven for a certain amount of time, (in this case, approximately 11 hours), prior to hardening. This is the time required for the centre of the crankshaft to obtain the same temperature as the outer parts of the crankshaft.

▶

Quenching and tempering during the forging process.

This process is installed in order to receive the mechanical properties as requested.

▲

A finished crankshaft for an eight-cylinder in-line engine, Category III. The crankshaft is ready for transport to the engine manufacturer.

These final finishes are followed by a number of treatments. The crankshaft is washed in a special washing machine and transported to a dust-free room. Here, a final inspection takes place and the crankshaft is tested by a Classification bureau and marked.
Most crankshafts are delivered to the client in corrosion-proof packaging. The packaging is generally a strong case, with a dust-proof carton lining and with sufficient supports to prevent bending of the shaft during transport.

▶

A new crankshaft is placed in a washing machine to remove the final dirt particles before the crankshaft is placed in the engine block. Caterpillar MaK, Kiel, Germany.

◀

A forged crankshaft for a Category II engine and a connecting rod with a high division for a Category III engine.

This connecting rod is forged in a single throw after which the 'small eye' and the 'big end' are milled.
It is then finished and both divisions are made.

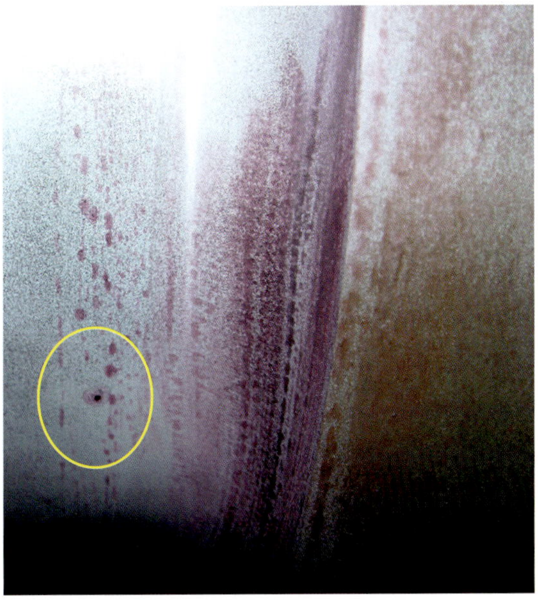

▲

Hair-crack inspection.

A small hole in the surface area of the shaft becomes visible after removal of the red layer with a clean cloth and a solvent.
This is usually indicative of inexpert reconditioning methods, such as the rewelding of a damaged journal.

▲

When reconditioning, such as the grinding of journals, a hair-crack test is always performed.

The method using a white and red spray shows possible cracking of the journal after it has been cleaned; it is known as the fluorescent test.

◀

The cranks of large two-stroke crosshead engines, category IV, are forged in a single throw and then the crankshaft parts are inserted in the holes in the crank webs and fixed using the shrink-fit method.

28.16 Forging gear parts for two-stroke crosshead engines

28.16.1 Introduction

Engines with large cylinder diameters have tremendous forces being exerted on the piston, therefore gearing for two-stroke crosshead engines are also forged as is the gearing for four-stroke engines.
The forces at the maximum combustion pressure are more than 1,000 tons for the largest cylinder diameters from Wärtsilä, the RTA 96 engine, and MAN Diesel, the MC 98 engine.
Only forged gear parts with large dimensions can withstand these enormous forces.
Forged parts have the following advantages:

- The naturally occurring bonds in the molecules and the crystal structure are not affected by forging.
- These bonds become even stronger with forging as the molecules are 'pressed together'. This increases the resistance to metal fatigue.
- The constant firing and cooling of the white-hot steel has another advantage: It improves the properties of the steel such as the tensile strength and the impurities shift to the outside and can be removed in the machining process. In particular, the negative element phosphorus and sulphate compounds.

Cylinder covers, piston crowns, piston rods and connecting rods are always forged out of one-piece of base material, steel.
Of course the exhaust valves are also forged from one piece and all the cranks of the crankshaft.
Successive cranks are attached to the journals by 'shrink-fit' connections.
All forgings are made from large cast-steel basic shapes: the ingots (blocks). They are also known as billets.
These ingots are cast in an octagon shape for forging heavy parts such as crankshafts and other gearing for two-stroke crosshead engines.
Steel bars have a square cross-section whereas sheets have a rectangular cross-section.

▶
The manufacture of ingots/billets.

The liquid steel is poured into a special mould, which is placed in a separate insulated chamber so that the ingot can be cooled very slowly.

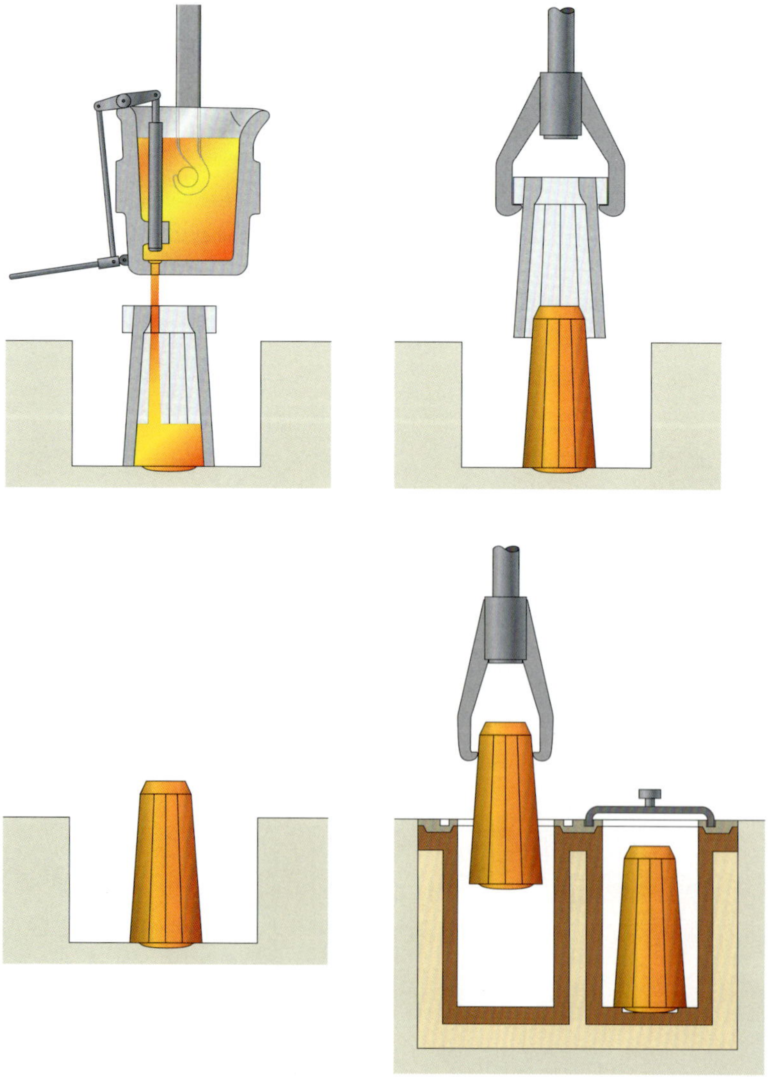

Forging these large parts is done using hydraulic presses that can exert very large forces on the work piece, which is heated to the correct forging temperature. These large presses can exert forces of more than 10,000 tons.

Forging processes are:

Drawing out, extending the length of the material at the expense of its cross-section.

Swaging, rotating the piece during the press strokes. This is the case for amongst others, the connecting rod between the big-eye and small-eye.

Upsetting is increasing the cross-sectional area thereby reducing the length of the forging. This is the case when forging a cylinder cover.

Flattening, this is the flattening of the forging giving it a greater width and reduced thickness. Swaging takes place before flattening.

Impression die forging is used when part of a forging has a much smaller cross-section than the other part.

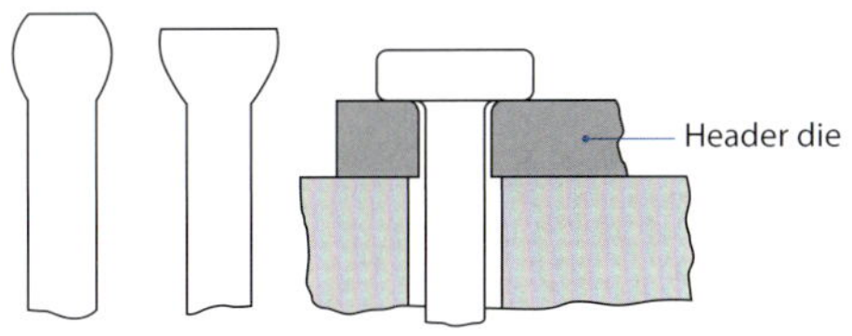

▲ Upsetting

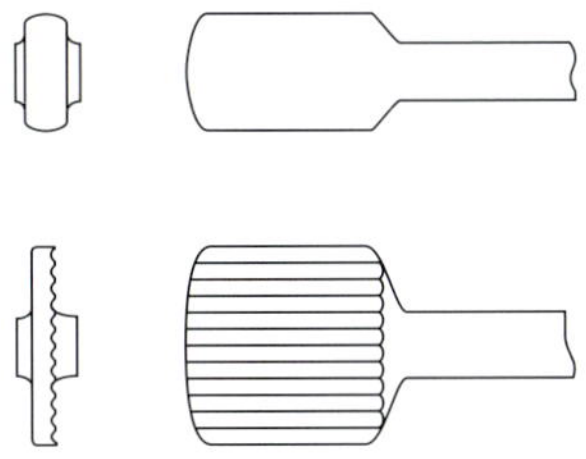

▲ Flattening

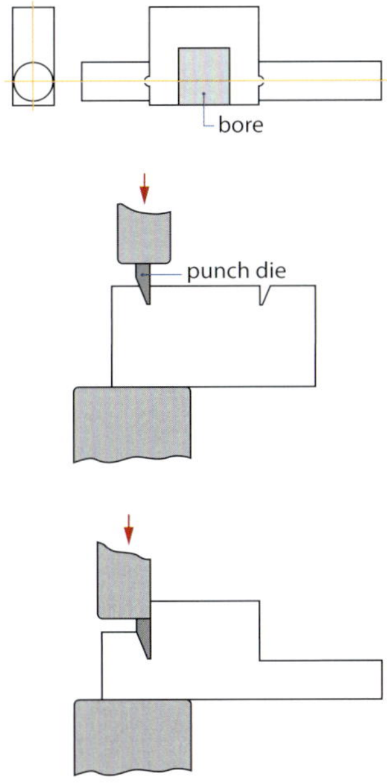

▲ Impression die forging

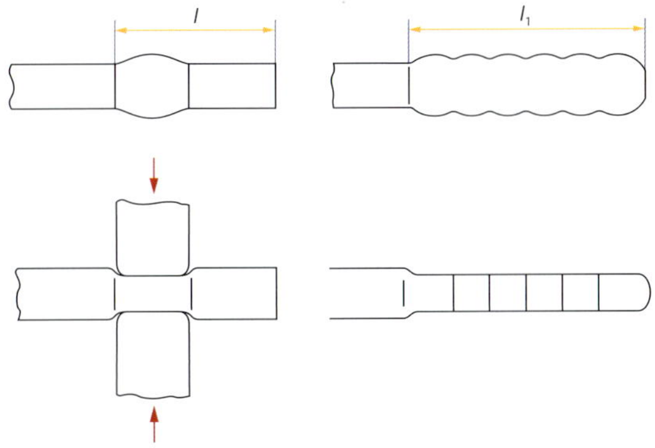

▲ Drawing-out

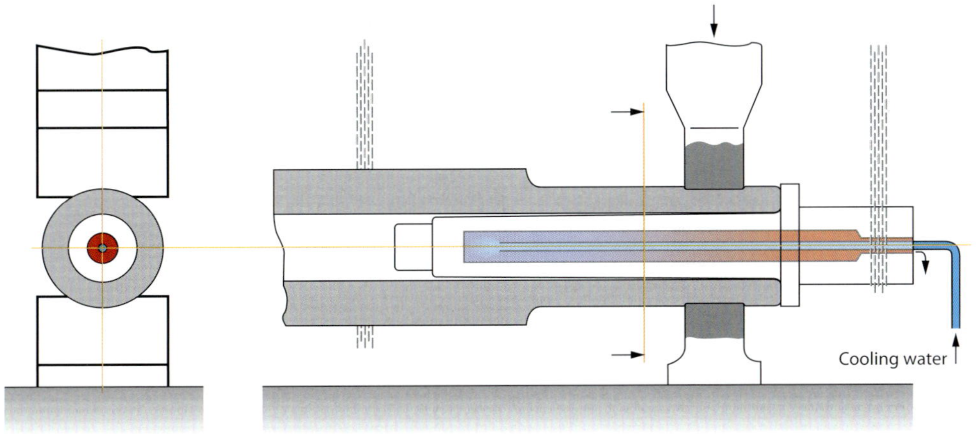

◄ Swaging

Roll forging is used in the manufacture of crankshafts in one piece.

Closed die forging is used when parts of the forging must have a circular cross-section.

Cleaving is a forging process in which a hot chisel is used. This process is used to manufacture the connecting rod big end.

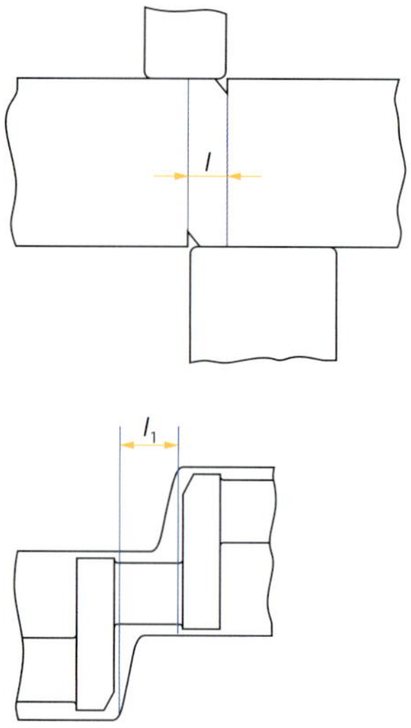

▲ Rolling

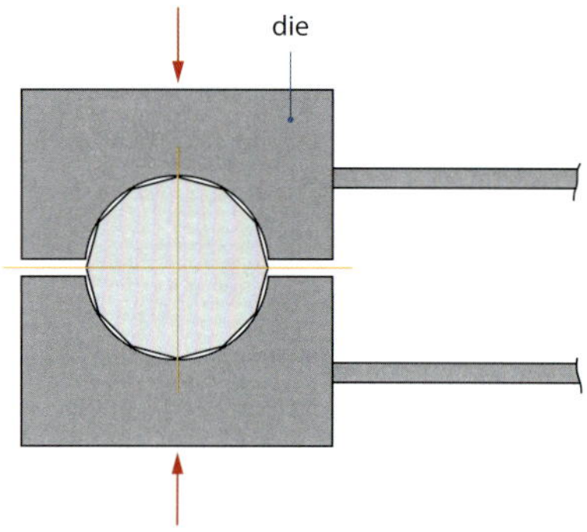

▲ Closed die

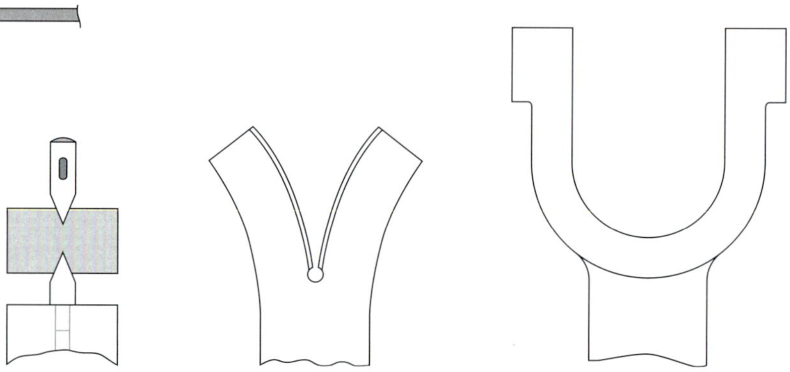

▶ Cleaving

▼ The raw material obtained directly from the steelworks.

These 'ingots' are available in all sizes.
The cylinder cover, piston crown, the crosshead, the piston rod and connecting rod are forged from the ingots.

▼ The material composition is stated on the ingot.

This is determined by the engine manufacturer and must be maintained by the parts manufacturer. Towards the back and on the right of the image are large pieces of raw material.
The shrinkage cavities on the head of the ingot, formed during cooling, have already been removed.

◀

The ingots are sliced into parts so that they can be used as raw material for forging cylinder covers.

◀

A forging is removed from the gas-fired furnace so that it can be forged into a certain form using a hydraulic press.

The high temperatures, often exceeding 1,100 °C are important for forging because:

- At high temperatures, the drawing point of the metal is low therefore the force required to deform the metal is small.
- At high temperatures, the elongation of a metal is high.
- At high temperatures, strain hardening of the metal does not occur because of the operation.

Due to the deformation of the material during forging, heat is generated in the material thus negating the necessity of reheating the forging.

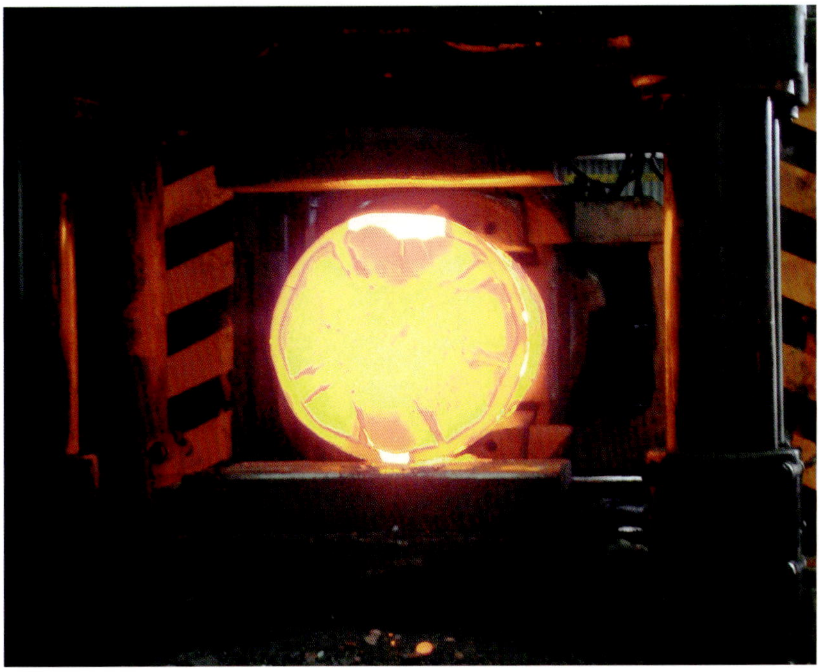

◀

Forging a large connecting rod.

The entire connecting rod is forged from one piece of material. The hydraulic press exerts a driving force between 8,000 and 12,000 tons!
In this phase the material is stretched, the diameter is smaller.

▶
Swaging.

This photo clearly shows that forging lengthens the forging. This is known as swaging.

▶
The forging should be reheated regularly. Shown here, the rough shape of the connecting rod is already visible.

A smaller diameter is forged for the section between the 'fork' and 'foot' of the connecting rod.This is called impression die forging.

▶
The connecting rod has been forged.

The planes which must be measured are painted white. If the dimensions are found to be correct, machining of the rod can start.

▶
Connecting rods, partially machined.

◀
Rough and partially machined forged piston rods.

To obtain rounded bars, closed die forging is used.

◀
Pistons complete with piston rod and piston rod stuffing boxes.

▲
Machining a piston rod in a lathe.

The rough shape of forging is clearly visible at the two ends.

▶

Forging a cylinder cover: first stage.

Place the forging in the forging press. Shown here, a cylinder cover is being manufactured. In order to obtain the proper shape, the mould (1) is very important.
Obviously, every type of engine has its own mould.

▶

Forging a cylinder cover: second stage.

Compared with the first stage, it can be clearly seen that the rough forging has been flattened. Eventually the forging will have a rather flat shape.
This forging process called upsetting.

▲

These ingots were forged into the rough shape required for the cylinder covers. Now the machining process can start.

▲

The moulds used in forging cylinder covers.

Each type of engine has its own mould. For example:
MC 50, a MAN Diesel two-stroke crosshead engine;
RT96C, a Wärtsilä two-stroke crosshead engine.

◄

The machining of the rough cylinder cover is ready: the hole in the middle is for the central exhaust-valve housing.

The large holes are for the cylinder cover bolts.

◄

Forged piston crowns.

As with crankshafts is the "grain flow" very important here. The force lines "run" along the shape of the crown!

◄

A finished piston crown for a Wärtsilä RTA engine.

The cooling holes for the lubricating-oil injectors are clearly visible.

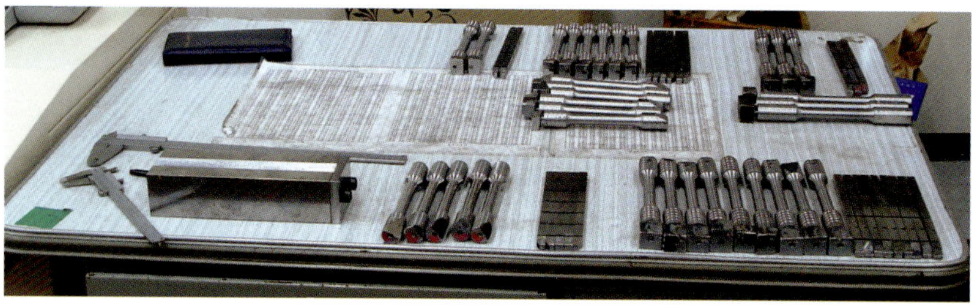

▲

Test bars for testing the tensile strength.

▶

Regular material testing is performed. Here a tensile testing machine.

28.17 Building a two-stroke crosshead engine A-frame

Small two-stroke crosshead engines frames, such as the MAN, MC 26 and 35 are cast, (sometimes partially), whereas all larger engines have welded steel A-frames. It is a light-weight and solid construction applied by all three crosshead-engine manufacturers.
Welding stress is partially relieved by heat treatment. In virtually all engines tearing of the A-frame occurs during operation. Generally, the cracks run along the weld, however, they may also occur across the construction. They can be repaired simply by drilling a hole at the end of each crack, grinding the fissure and re-welding it. Occasionally reinforcement is used, which has an adverse effect as it increases the possibility of crack formation.
The main-bearing seats are usually manufactured from thick-walled cast steel. They absorb the great forces released during the combustion process.
As cast steel can be welded, they are entirely integrated in the A-frame.

▶

A cast steel bearing support for a rigid and strong structure under the main bearing of a crankshaft.

Carbon steel can be welded which allows easy mounting of the steel frame. It can be clearly seen in this photo where liquid steel is supplied (1) and where the air has escaped (2).

◀ Welding the steel frame.

The lightweight and rigid construction is ideal for building the A-frame. Welds create a great deal of tension in the frame. The heat treatment afterglow releases these tensions. Often, an insulated cap with electric spirals on the inside is placed over a large part. These provide the correct, accurate and adjustable material temperatures for stress relieving the welded A-frame.

◀ To be able to carry out the welding operations, a frame can be tilted.

This allows for the proper angle and in a good working position.

◀ Automatic welder.

Long welds are often carried out by by arc welding. Short welds in difficult positions are welded by hand.

▶

The lower bed ('bedplate') or sump.

The cast steel bearing bracket (1) is welded to the frame (2).

Cast steel bearing seats are very stiff and ideal for the very large varying combustion forces. Cast steel has good welding properties, which allows for easy mounting of the steel A-frame.

▼

A complete A-frame of a twelve-cylinder engine.

28.18 Forging crankshafts, 2009 report

Alfing Kessler (Aalen, Germany), one of world's largest crankshaft suppliers.

Crankshafts constitute a basic and costly part of the diesel engine. They convert the up- and down movement of the piston into a revolving motion via the connecting rod. Crankshafts have to meet high requirements, such as:

- The crankshaft has to be sufficiently powerful to convert the massive forces exerted on the crankshaft by the connecting rod into a rotating movement.
- The material should have good tensile strength, especially in view of the considerable torsional stress.
- The material should be easy to machine giving the numerous machining processes.
- The crank journals and the crankpin journals require top-quality finishes.
- All four-stroke crankshafts are manufactured from a single piece, which entails a complex and time-consuming forging process. All larger four-stroke crankshafts are forged throw by throw.

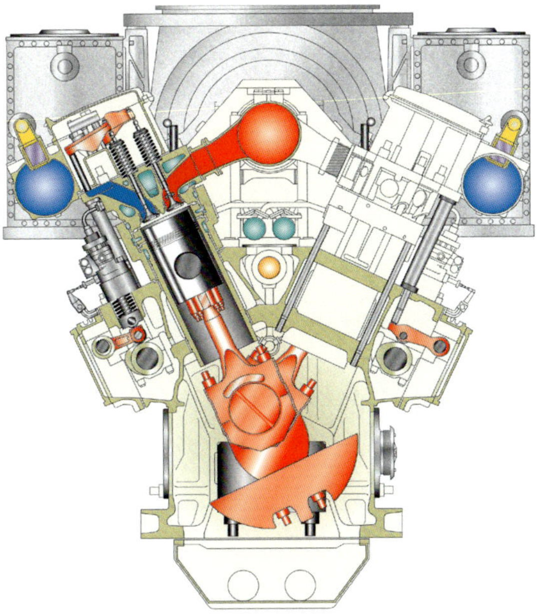

▲
The forged steel crankshaft, the heart of the diesel engine, is the most essential and also most expensive part of a diesel engine.

◄
A crank shaft in the assembly shop. The cog wheel for driving the gearing for, among others, the camshaft and most of the counterweights has already been mounted.

▶

The simplified principle of forging a crankshaft.

a The rough round or square rod: the base material.
b The rough forged shape of the crankshaft.
c The forged crankshaft in the final machining process.
d The complete crankshaft after all machining processes.
e The first forging.
f The second forging.
g The third forging.

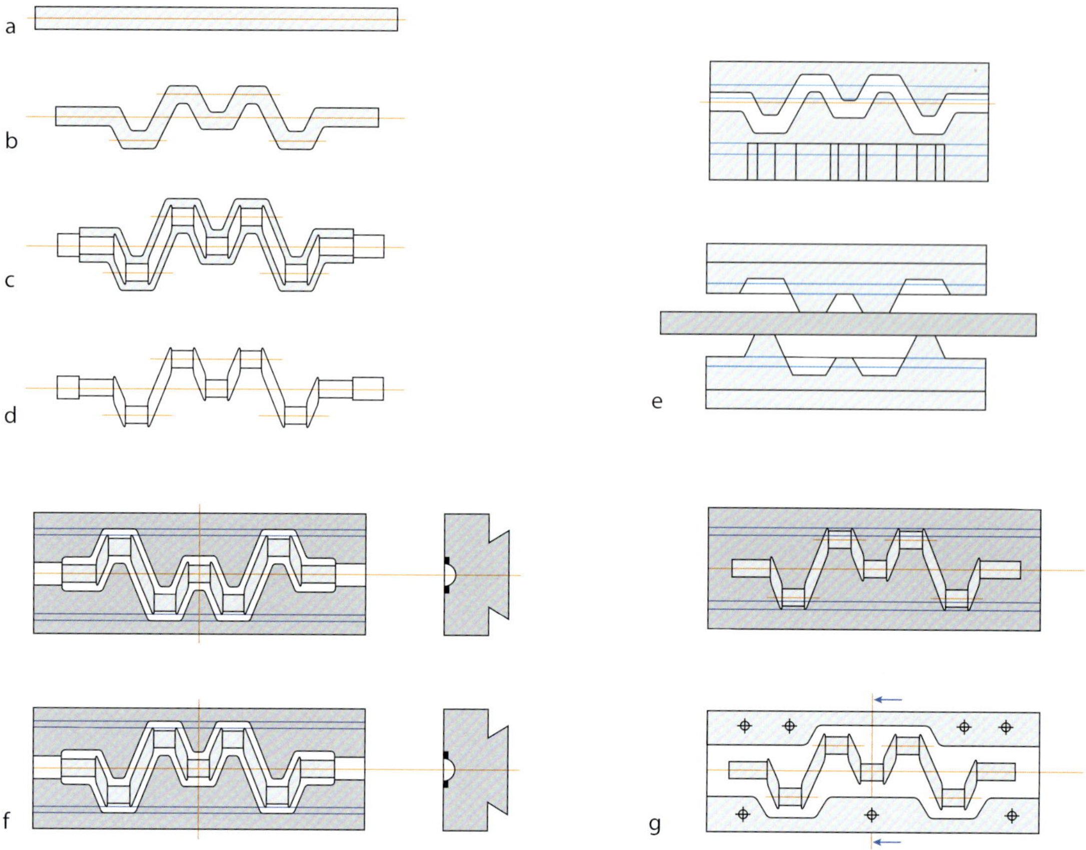

▶

The basic raw material: round or rectangular steel rods.

The client, generally the engine manufacturer, determines the composition of the material.

▶

The so-called pre-machining of the raw base material.

In the background, a large lathe. In the foreground, the rod for a seven-cylinder diesel engine.

◄
The lower mould for forging the crankshaft. The material is very hard so that distortion cannot occur.

Machining such a hard material takes time.

◄
The complete upper and lower moulds. The moulds have holes (1) so that they can be positioned correctly relative to each other.

▼
The new hall at the firm Alfing Kessler in Aalen (Germany) in 2009. The gas-fired ovens are operational. A forklift is removing a large crankshaft from the oven. Only the last part is heated, a forged flange will be attached here.

▶
The oven is open. Again, only the last part of the crankshaft is heated.
A flange will be 'pressed' in the hydraulic press from this bar shape.

▶
A series of six crankshafts undergo the same processes, so that the moulds can remain in the hydraulic press. This is much faster than finishing one crank shaft at a time.

▶
Steel moulds for the various processes in the manufacture of large crankshafts that are forged 'throw by throw'.

◄

Every type of crankshaft has its own series of moulds.

This means that when several customers who each have their own type of engine construction, a large number of moulds must be in stock! Many thousands of moulds are stocked in the larger crankshafts forges. This is very expensive, since all moulds are made of a very hard and durable material.

The moulds must be solid to prevent deviations and must be able to be used often.

◄

Hydraulic presses for forging crankshafts can exert from 5,000 to 10,000 tons of pressure force!

◄

The forging of the flange on the crankshaft for power transmission. By holding the crankshaft in the press and exerting two opposing forces horizontally in the longitudinal axis, the flange is formed.

1 *flange.*
2 *hot crank.*
3 *force in the centre axis of the crankshaft.*

▲

The shaft is clamped.

▲

The shaft is pressed together using very large forces. Here, the distance between the restraints is reduced whereas the diameter increases.

▲

The coupling flange is formed.

▲

The crankshaft is removed from the hydraulic press.

▶

Roughly forged crankshafts of the same type, ready for further processing.

1 *crankshaft.*
2 *crankshaft web.*
3 *crankpin.*

By forging the crankshaft out of one-piece of steel, the force lines in the material run along the shape of the crankshaft.

Red – force lines natural grain-flow.

In forging smaller crankshafts in a complete mould, part of the base material is pressed between the upper and lower moulds. This excess material is later removed by a hydraulic press.

Circle: The transition between the lower and upper moulds is visible by the sharp edge.

To prevent adhesion of the very hot forging material to the moulds, a layer of sawdust is applied the inner surface of the moulds. This is burnt off during forging.

The excess material removed. This is recycled by the steel supplier for the production of new base material.

▶

After forging, certain parts of the crankshaft are painted so that the dimensions can be recorded. They are used to check whether the crankshaft can be manufactured from the raw material.

▶

The first machining operations, such as determining the centre line and the rough turning of the crankshaft, are carried out.

▲

Heat treatment to reduce for the material tensions that occur as a result of forging.

This is called 'stress-relieving'. The crankshafts are heated in an oven to ± 850 °C and then set aside so that they can cool slowly. Most of the forging stresses are relieved.

▲

A heat treatment to harden the surface. This makes the shafts (crankshaft and crankpin) harder and more wear resistant.

The crankshaft is again heated to ± 850 °C and then rapidly cooled in an oil bath. This changes the composition of the outer skin to a depth of ± 2 mm so that the material is much harder and therefore more durable.

◄

Smaller cranks often undergo heat treatment 'surface hardening'. Here a series of surface-hardened crankshafts waiting for further processing.

◄

The crankshaft taps, crank pins and the sides of the crank webs are machined roughly. Most of the material is removed here. This is followed by a precise finish of the shaft and to a lesser degree, the crank webs.

◄

The rough milling of the end of the crankshaft at the crankshaft flange location.

Here the journals are not machined to truly round, but to 16- or 32-sided using an external router. Importantly, most of the excess material is quickly removed in this process.

▶ Machining the crankshaft in a special crankshaft bench. Note the many supports to ensure that the shaft is kept in true.

▼ A series of fully automatic CNC machines that carry out the many operations on the crankshaft.

▲

The setted crankshaft in the CNC machine, the supports ensure that the crankshaft remains true during the operation.

◀

The custom grinding of the crankshaft and crankpin to their true dimensions. The machine can be moved horizontally so that the crankshaft can be positioned both crankshaft axis and crankpin axis.

Horizontal red arrow - movement.

1 *grindstone.*
2 *crankshaft.*
3 *supports.*

▶

A grinding bench with manually adjustable supports (1). Each journal is supported.

▶

After all the machining processes, very small burrs are removed by hand and roundings applied. Also, all the machining processes are checked, amongst other defects.

▶

The rotating crankshaft is polished in a wooden mould with very fine emery cloth for a true smooth finish to the crankshaft and crankpin. Each shaft has a special wooden mould.

28.18.1 Material analysis of crankshafts: steel's natural grain-flow

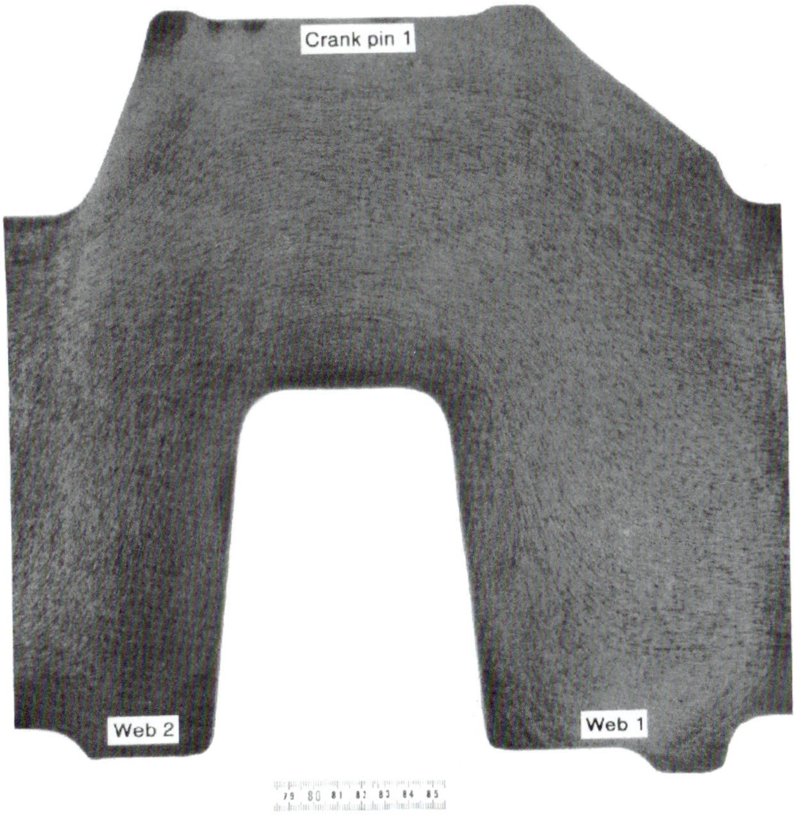

◀

A slice of a crankshaft crank, which has been machined, polished and finally undergone an acid etching treatment. Alfing Kessler in Aalen, Germany.

This allows the structure of the material to clearly emerge. The lines of force (grain flow) run along the shape of the forged crank. It can also be seen that the forged steel structure on the outside of the crank is closer together. This is caused by the enormous forces exerted on the forging by the steel moulds during forging.

Forged parts have the following advantages.

The natural bonds of the steel molecules and the crystal structure remain intact during forging. The natural bonds are stronger: the molecules are pressed tightly together during forging. This increases the fatigue resistance.

The constant white-hot firing of the steel followed by the cooling processes have another advantage: it improves the steel properties such as tensile strength and the impurities move to the outside of the forging and can then be removed in the machining processes. This is relevant especially for the harmful elements such as phosphorus and sulphates.

▼

A surface hardness test at the crankpin location with the depth of the hardened layer given.

The crankshaft is hardened by the induction method. The hardening depth varies from 9.2 to 19.7 mm.

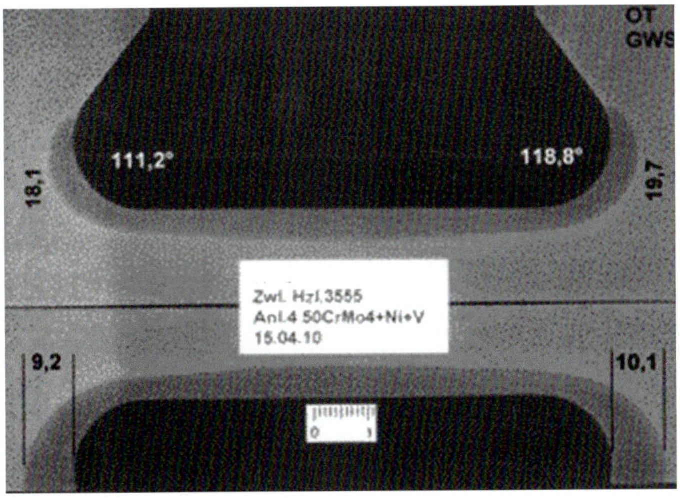

▼

Crack Examination using ultrasound.

The crack is approximately 31 cm from the crank web and has a length of approximately 9millimeters.

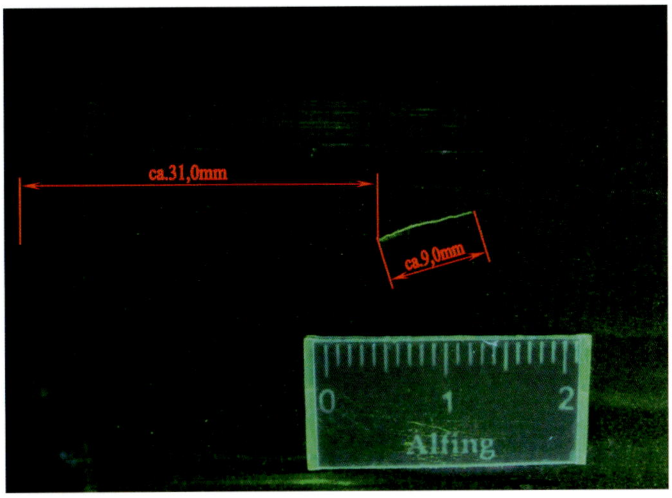

▶

The crankshaft is guided through an electromagnetic coil in order to check for hairline cracks.

▶

Any hairline cracks can also be detected using the ultrasonic method.

▶

Crankshaft measuring assembly.

It is essential that the crankshaft have the correct dimensions after all the machining operations have been completed. All dimensions are accurately measured and recorded; these are exclusive to this particular crankshaft. Measurements must be recorded for every crankshaft even for two identical engines with the same crankshaft. These crankshafts always have very small deviations with respect to each other!

◄

Once all the processing and inspections have been carries out, the crankshaft is wrapped in a plastic foil so that oxygen cannot come into contact with the crankshaft. This prevents rust, which can damage the running surfaces.

◄

A very rigid wooden frame is made so ensuring that the crankshaft is not deformed during transport is (1).

◄

Finally, the crankshafts are shipped in specially constructed crates, in which the crankshaft is firmly supported so that it remains aligned. The wood used to construct the crates must now meet different requirements: for example, it may not contain insects or woodworm.

28.19 Manufacturing pistons

A piston design is rather complex: pistons have to meet various requirements.

- Must withstand the large forces generated by the combustion-process pressures. Pressures from 30 bars up to 250 bars, depending on the engine category, translate into forces of several tens of kilonewtons to over 1,000 tons in two-stroke crosshead engines.
- Must resist material temperatures of as much as 350 °C to 450 °C. Therefore, the dimensions have to be geared to compensate for a given expansion coefficient.
- In four-stroke engines, the lateral forces produced by the crank-/connecting-rod mechanism must be transferred to the cylinder liner by the piston skirt.
- The piston, in conjunction with the piston rings, lubricating oil and cylinder liner, should provide optimum sealing for a large number of operating hours.
- The forces exerted on the piston must be transferred to the crankshaft via the piston pin and the connecting rod.
- The piston material must be impervious to aggressive and abrasive substances produced by the combustion process.
- Piston distortion as a result from the combustion process should be minimal.
- The piston parts should be effectively cooled by lubricating oil to avoid overheating of the materials.
- The movement of the piston rings in the ring grooves should cause minimum wear.
- The friction between the piston ring and piston-ring grooves and cylinder liner should be minimal. The largest friction losses in a diesel engine are the friction losses due to the piston rings.

▲
Pistons for four stroke diesel engines have a long piston skirt to absorb the lateral forces. These forces are led via the piston skirt to the cylinder liner and absorbed by the engine block.

▶
The piston rings, piston, connecting rod and crankshaft are the heart of each diesel engine: the drive gears bring about a reciprocating movement of the piston and a rotational movement of the crankshaft.

This has been the case from the invention of the steam engine in ± 1700. Shown here, an assembly of the complete gearing in front of the Mahle headquarters in Stuttgart, Germany.

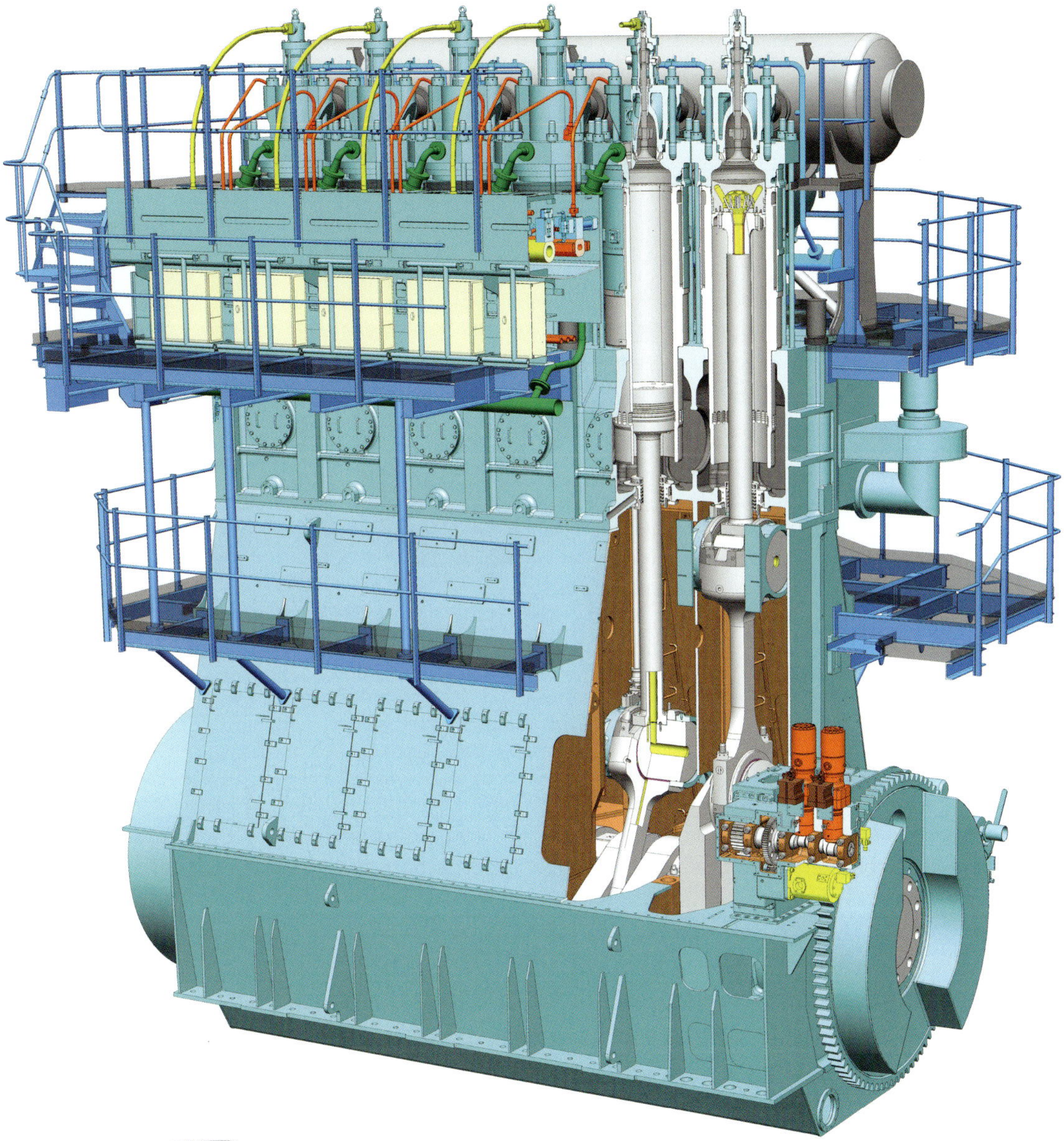

◀
Two-stroke crosshead diesel engines have short pistons. This is possible because the lateral forces generated by the crank rod mechanism are absorbed by the crosshead, the slide shoes and bars and then the engine frame. Not by the piston skirt as with four stroke engines!

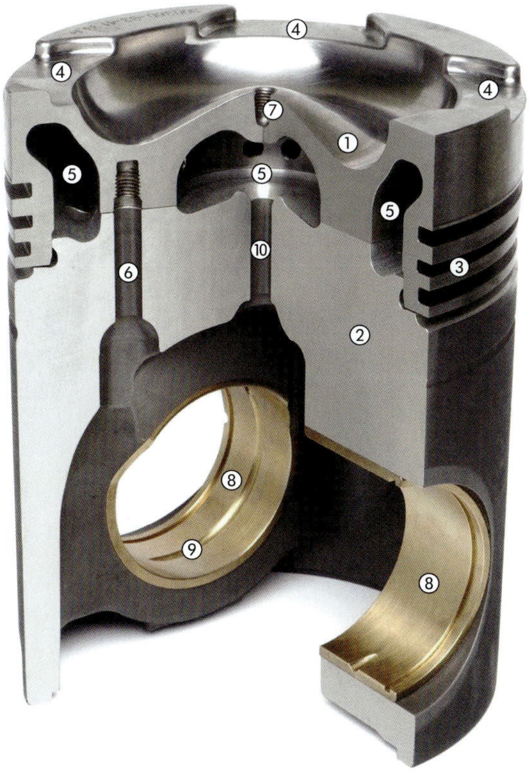

◀
A piston manufactured by Mahle in Markgroningen, Germany. The piston comprises two parts: the forged steel crown and forged aluminium skirt to give enough clearance for the values when the piston is in top dead centre (intake stroke).

All the piston rings are mounted on the piston crown. The piston crown has value pockets for the inlet and exhaust valves. Note the cooling channels of the cooling lubricant.

1 *piston crown*
2 *piston skirt*
3 *piston-ring grooves*
4 *pockets for the inlet and exhaust valves*
5 *cooling galleries*
6 *fixing bolt piston crown/piston skirt*
7 *tap hole for hoisting the piston*
8 *bronze bushes in the piston pin bore*
9 *oil groove for cooling oil inlet*
10 *drain channel for cooling oil*

▶

Various types of pistons, manufactured by Mahle.

In principle, the smaller pistons, engine category I, are made of light metals, an aluminium alloy.
Many pistons in category II engines are also made of light metals. The larger the diameter of the piston, the greater the ratio of the forces exerted on the piston. Often, because the compression and combustion pressures are much higher: currently more than 200 bars!
The mean piston speed will also increase, so the acceleration and deceleration forces on the piston.
In larger medium-speed engines, the piston comprises two parts: the forged steel piston crown and the nodulair cast iron or forged aluminium or forged steel piston skirt.
Pistons in two-stroke crosshead engines are now all made of forged steel.

28.19.1 New techniques

New techniques are being developed continually to process the enormous forces exerted on the piston with increased combustion pressures.

Mahle has developed the Ferrocomp® piston, in which the piston skirt is manufactured from forged steel. Apart from the fact that this type of piston has higher durability than its light-metal and cast-iron counterparts, it has better wear resistance. This increases piston life.

Furthermore, light-metal piston skirts have been used in the last years. These are forged and some are welded together (Electron Beam Welding).

▼

The development of materials and construction of Mahle pistons.

Figure left:

Vertical; the combustion pressures up to 300 bars (in the future).
In the figure: The various types of pistons with also the specific and patented Mahle pistons®.

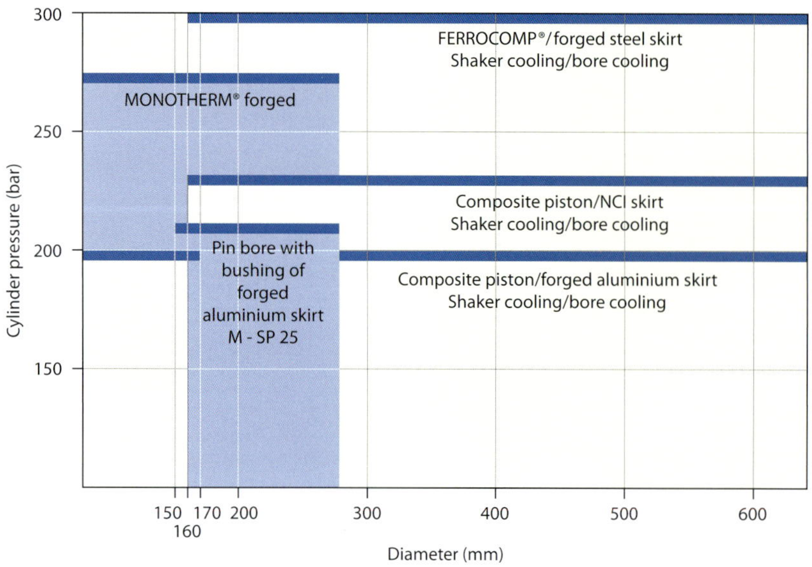

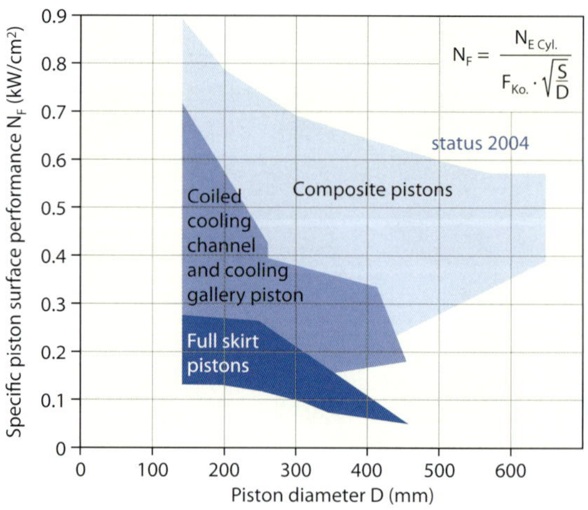

◀
The storage of roughly processed, forged piston crowns.

The piston crowns are made in separate forges, as are piston rods and crankshafts.

◀
Machining piston skirts.

◀
Measuring a piston crown.

▶

The forged piston crowns after the last machining process.

Note the low position of the piston rings relative to the top of the piston.

▶

Measuring a piston skirt.

The dimensions are very important: the interplay of forces are in fact very large and complicated. All the dimensions of the piston skirt are measured and recorded using a 3D probe.

▶

Roughly forged piston crowns.

Forged piston crowns are much more resistant to high combustion pressures and temperatures than cast-iron piston crowns.

◄
Piston skirts before final machining.

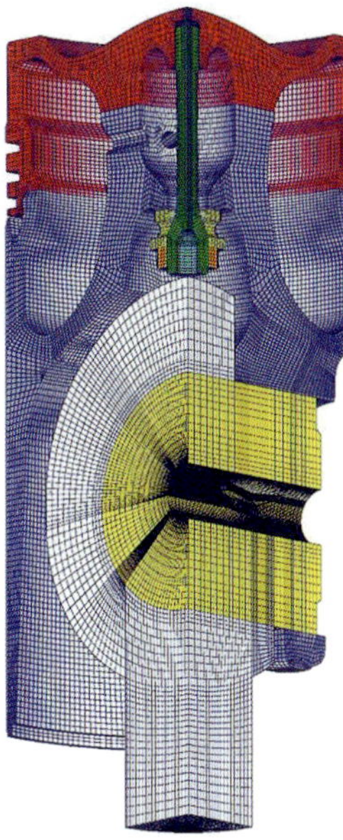

◄
Using the finite element method, the forces on the piston, piston pin and connecting rod can be calculated in the design phase.

Red: the piston crown, blue: the piston skirt, yellow: the piston pin and white: the connecting rod.
In this manner, the piston design is verified and optimised before doing any hardware.

▲
The automatic processing of the piston crowns.

◄
The automatic forging of piston crowns using a robot: in the past, this was done by a blacksmith!

28.19.2 Electron Beam Welding

This is a welding process in which two components are joined using large amounts of energy, supplied by a beam of electrons. This takes place under high-vacuum/ low-pressure conditions.

- Aluminium alloys can be welded up to a depth of 200 mm.
- The advantage of the vacuum-welding process is the lack of atmosphere during welding in the melting bath. Consequently, the aluminium parts can be joined simply using cold welding.
- The width of the electron beam varies from 0.1 to 0.01 mm, thus allowing for a very accurate weld.
- This welding method is distinguished by a narrow and deep weld.
- The entire welding process is automated and performed under high-vacuum conditions.
- During the welding process, electron beams are fired at the work piece at 3/4 of the speed of light. To protect the operators from the radiation emitted during this process, welding machines are equipped with lead shielding.

Advantages:

- Single weld for thick-walled joints.
- Optimum welding conditions due to the vacuum.
- Minimal distortion.
- Minimal contamination of the weld.
- Narrow weld.
- Only partial heating of the material.
- Parts consisting of different materials can be joined.
- No filling material required for weld.

Disadvantages: The following problems may occur:

- Tearing.
- Insufficient fusion of both components.
- Porous weld.

▶

The principle of electron beam welding.

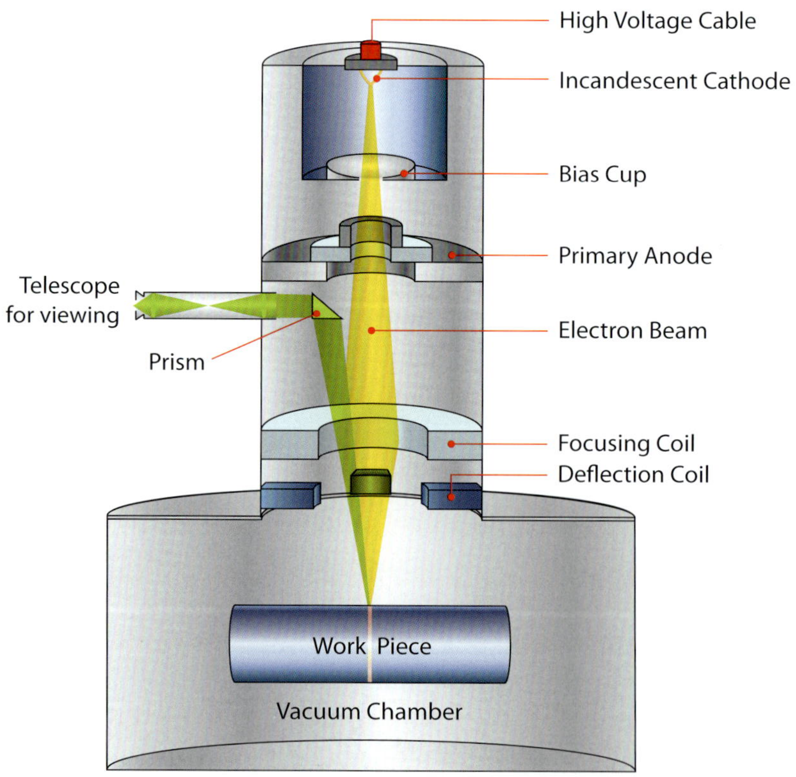

◀

Aluminium pistons for high-speed diesel and gas engines are often of welded design; where a lower piston body of high strength forged aluminium is joined by welding with an upper casted part of aluminium that contains a cooling gallery as well as a ring for the tap ring.

◀

Here shown the arrangement for welding in the vacuum chamber.

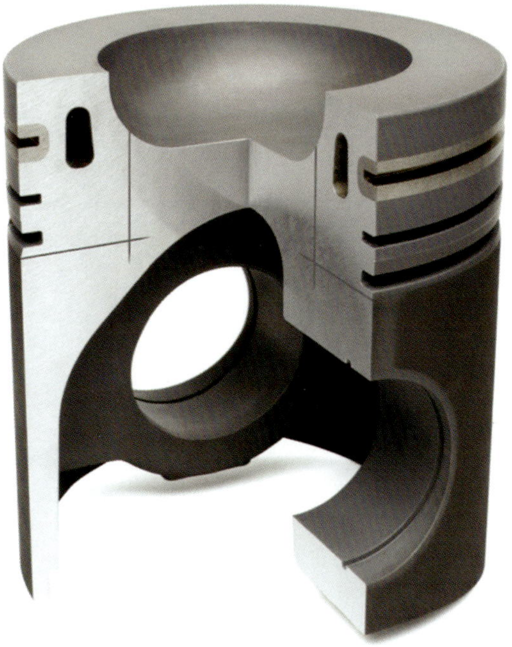

◀

A Mahle piston with oil cooling gallery, insert for the top piston ring. The piston consists of two parts. Welded by the electron beam method.

►
A cast iron piston skirt during magnetic particle crack detection.

▼
The inside of a piston crown being machined.

▲
A multi axis drilling machine does all the oil drillings in one setup.

▼
Materials research is essential to ensure quality of engine components.

Material testing is essential in manufacturing diesel-engine components. The material requirements for moving parts such as the piston, piston pins, connecting rods and crankshafts are particularly high, as breakage could result in total engine loss.

Three types of material testing:

- Standard material testing of the base material. This is performed prior to machining.
- Material testing during manufacturing.
- Material testing following manufacturing. Generally, this consists of periodically sampling certain engine parts.

28.20 Manufacture of inlet and outlet valves

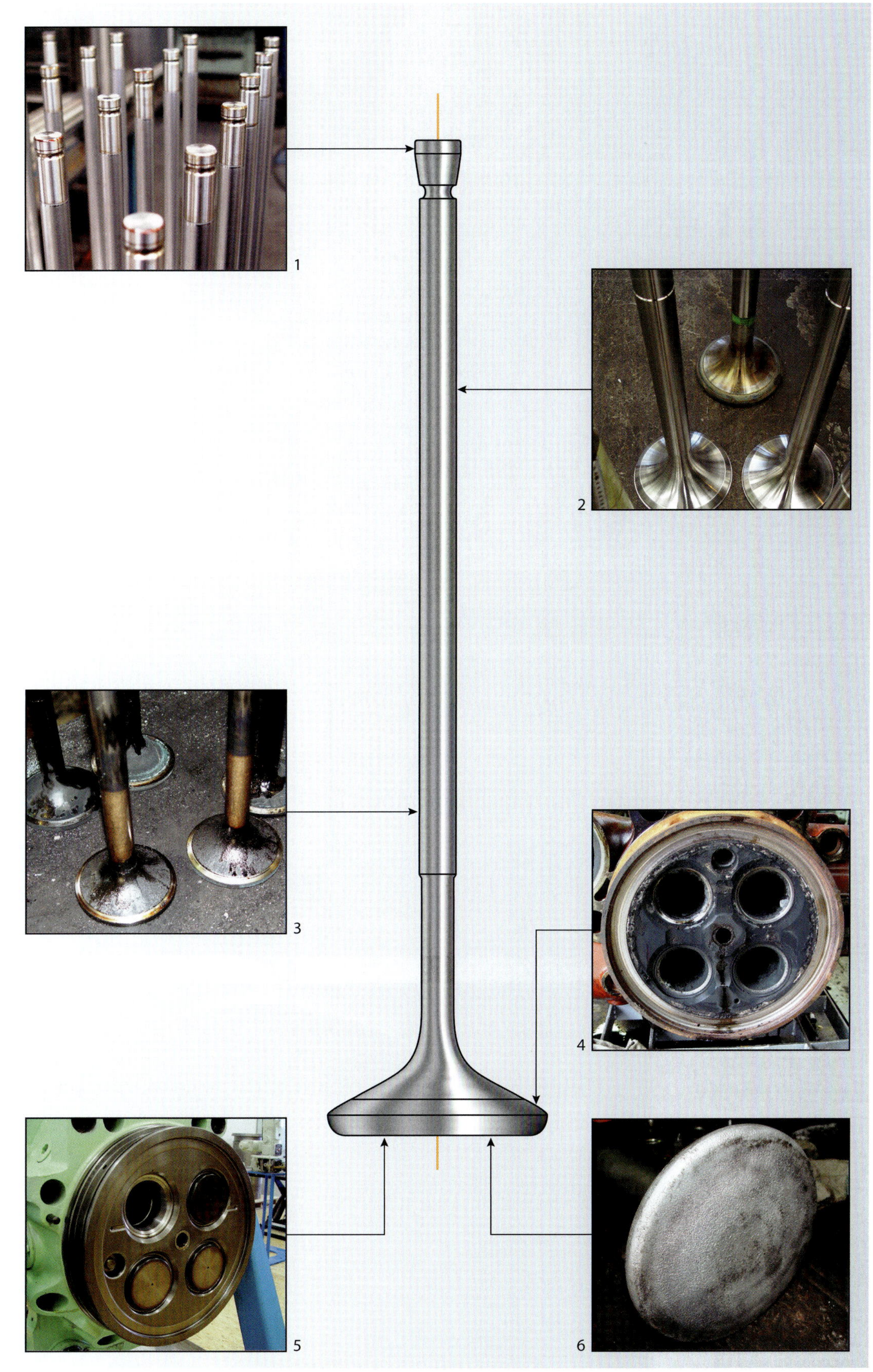

▲
View of a valve.
Exhaust valve.

1. *The end of the valve stem is hardened to prevent deformations (pitting). Despite the smallest possible valve clearance, the rocker arm strikes the head through the adjusting bolt when the valve is opened.*
2. *The valve stem must be true in the valve guide. The running surface should be very durable. The clearance between valve stem and valve guide must be such that there is always a thin film of lubricant present between the two components. Furthermore, the centre-axis of the valve stem and the centre-axis of the valve guide must be aligned in true. Finally, the valve must seat itself cleanly around the valve seat!*
3. *The most wear takes place on this part of the valve stem. The material temperature is high here due to the exhaust gasses, and the lubricating-oil film can be affected by exhaust fumes and dirt from the combustion process.*
4. *The valve should seal at all times when closed. The high exhaust-gas temperatures cause the material temperature of the valve and seat to reach up to 550 °C. Furthermore, hard carbon particles and wear particles (combustion process) between the valve and seat can cause damage to the sealing surface. The valve striking the seat is a common phenomenon that ultimately can cause the valve to leak.*
5. *The valve must have adequate resistance to deformation because of the high pulsating pressures above 200 bars found in four-stroke medium-speed diesel engines. If not, the valve would not close tightly on the seat.*
6. *When using heavy fuels, high temperature corrosion can occur. It can damage the valve very quickly.*

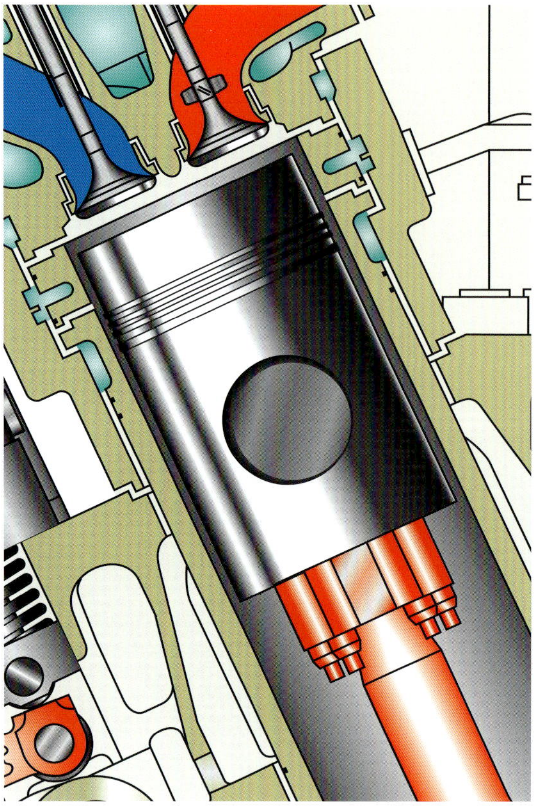

▶
Inlet and exhaust valves allow the gas exchange in the cylinder for the combustion process.

Note the small space between the valves and piston crown at top dead centre!

▲
A modern cylinder head with two inlet and two exhaust valves for optimum gas exchange.

The exhaust valve is especially heavily loaded. The maximum material temperature is more than 700 °C and during the combustion process, products are produced that can cause chemical corrosion such as high temperature corrosion, HTC. Contamination and wear are caused by the unburned carbon from the fuel.

▶
Two cylinder heads in a Caterpillar high-speed four-stroke diesel engine.
All modern four-stroke engines have four valves in the head. In particular, the exhaust valves wear out quickly. Since this diesel engine runs on diesel oil, the valve wear is slight in comparison to the wear and tear caused by heavy fuel-oil used in categories III and IV.

▶
The two inlet and exhaust valves of a Caterpillar Mak four-stroke medium-speed diesel engine operating on heavy fuel-oil. The service life of this engine compared to engines operating on diesel is almost halved!

28.20.1 Manufacturing valves

The alignment is followed by elaborate machining processes consisting of up to thirty steps.

These are executed by specialised firms. Most of the following photos were taken at Maerkisches Werk in Halver, Germany.

Procedure

- Saw the selected bars to size.
- Upsetting process creating a tulip shape by means of a conduction method.
- Forging process with a die under the forging hammer.
- Attach the upper part of the stem using friction welding.
- Straighten valve and stem.
- Machining, rough-turning.
- Weld seat with stellite alloy.
- Weld valve seat.
- Fine-turn valve.
- Heat treatment to improved material properties.
- Harden valve-stem end.
- Test hardness.
- Machining, turn to size.
- Polish valves.
- Ultrasonic testing of the material.
- Material survey/sampling.
- Chrome plating, if required.
- Final check for hair-line cracking using ultraviolet light.
- Packing of valve in polystyrene boxes treated with an anti-corrosive.

◀ Bars that are cut to size and ready for the first step: upsetting.

Inlet and exhaust valves are forged from steel bars. The composition of the steel including the alloys of chromium, nickel and silicon is determined by the client, usually the engine manufacturer. Naturally, the thermal, chemical and mechanical loads on exhaust valves are much higher than inlet valves; therefore exhaust valves are always alloyed. The bars are cut to size.

◀ Electro Upsetting.

The bar is tensioned before being heated by an electric current. The bar is heated to the required temperature of, for example, 1200 °C so that the hydraulic press can forge the tulip shape. Actually, this is called swaging: thickening of the material by forging.

▲

The complete electro-upsetting machine.

1 *Tensioning.*
2 *Bar to be forged.*
3 *Hydraulic press, both above and below.*
4 *Heating the valve with a current loop between 1 and 4.*

▲

Forging the valve.

The rough tulip shape is now placed in a mould in the forging press so that the true shape of the valve can be forged. With a press exerting a pressure of ± 1,200 tons, the valve is forged in one movement into shape. The forging press hits the rough shape valve and turns the valve in a very rapid movement into its correct shape.

▶

The rough valve shape is forged to its correct shape for further machining.

▲

The forged valve can now be removed with a pair of large tongs.

▲

Removing the valve from the forging press is done manually.

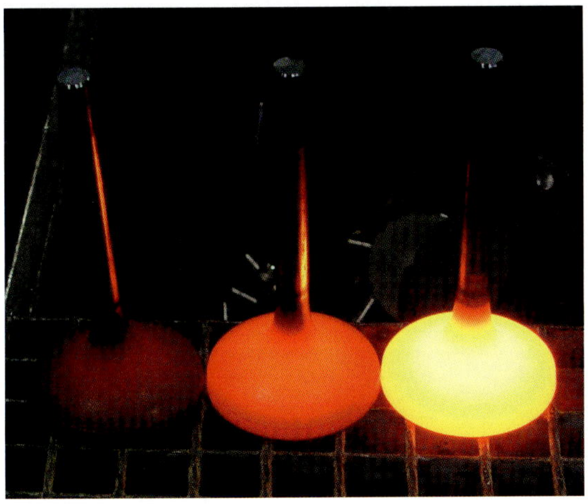

◀

Cooling of the valves.

Three valves cooling after forging. Note the short valve stem. The section of the stem that runs in the valve guide has yet to be attached; this is performed using friction welding.

▼

Valve production at Caterpillar Mak.

Roughly forged valves. They are manufactured at a separate, external foundry.

▼

Friction welding, Märkisches Werk GmbH in Halver, Germany.

The upper part of the stem is attached to the valve by friction welding. The upper part of the stem should have good running properties and be wear resistant. The stem end is hardened so that deformations caused by the valve drive striking the tip can be avoided. The lower part of the stem with the valve head must be able to withstand very high temperatures, corrosive acids and the impact of hard particles such as carbon and other contaminants.

▶
Valve production at Caterpillar Mak.

A series of roughly forged valves during the rough machining process at the engine plant of Caterpillar Mak in Kiel, Germany.

▶
Valve production at Caterpillar Mak.

Rough machined valves, arranged in a buffer/carousel for automated finishing.

▶
Valve production at Caterpillar Mak.

The triangular granules are made from a special synthetic material, which is very hard.

▶
Valve production at Caterpillar Mak, Kiel, Germany.

The surfaces of the roughly-turned valves are compacted in a process using triangular plastic granules. This makes the surface harder and more wear resistant. The valves are placed in holders and moved through these granules, which compact the surface.

Example of a trajectory of the valve stem in guide.

Valve-disc diameter 40 mm.
Lift valve ¼ x D = 10 mm.
Number of revolutions four-stroke 960 rev/min.
Valve life 18,000 hours (HFO).

Valve stem trajectory:
Number of engine revolutions $\frac{960}{60}$ = 16 per second.

Work process every second revolution.
Therefore, valve moves 8 times per second.

Trajectory per second 2 × 8 × 0.01 metre = 0.16 metres.
This means 0.16 × 3600 × 18,000 = 10,368,000 metres or 10,368 km per 18,000 hours.
During this period, the seat must make a gas seal with the valve disc 8 × 3600 × 18,000 = 51,840,000 times!
This is a considerable accomplishment considering the circumstances, such as, high pressures and temperatures, chemical corrosion, contamination resulting from the combustion process and a moderate lubrication of the stem.

◀ Straightening the valve stem.

After friction welding, the valve stems are straightened so that the centre line of the valve stem is almost in line with the shaft. Then, a special machine straightens the stems before machining can take place. The stem is bent past its elasticity boundary so creating a lasting change in the stem.

◀ Heat Treatment.

A series of valves ready to be annealed in an electric furnace. As they are then cooled very slowly, almost all the material stresses tensions disappear.

◀ Heat Treatment.

The swaging, forging and induction hardening process cause many stresses in the material. These are largely removed by annealing. This is carried out in a valve holder here at a temperature of 800 °C. The valves are annealed in the holder and put to one side to slowly cool, or they are oil cooled. By the use of different temperatures and times, the material properties can be established.

▶
Machining.

The machining processes are carried out after forging and friction welding. After pre-turning the stem, the valve head is roughly turned. The groove where the hard stellite layer will be welded can be clearly seen. This groove is formed during forging.

▲
Welding the stellite layer.

This is a very hard and heat wear resistant material widely used in exhaust valves. It has very good properties:
- Very hard surface: pitting by striking carbon particles is almost impossible.
- Resistant to high material temperatures.

▲
Stellite welding in the picture.

The good adhesion and weld quality are essential for a long operating life of the valve.

▶
Annealing.

Welding the stellite layer creates heat stresses in the valve material. Heating and then slowly cooling largely removes these stresses.

▼
Induction heating has occurred. Cooling the end of the valve stem tip quickly with water creates a very hard and durable tip.

▲
A series of valves showing the welded stellite layers.

▼
An exhaust valve before and after the welding the stellite layer on the valve seat. The discoloration occurred during welding.

▶

Apply contact fluid.

After application of the contact fluid, the entire stellite layer is checked with a probe. Every valve is checked.

▶

Measurement of the value.

At the final inspection, the valves are measured using cameras. This data is recorded and immediately processed.

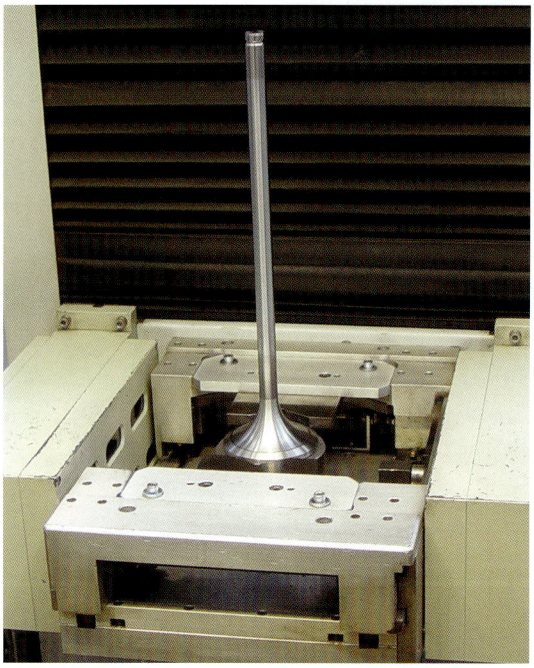

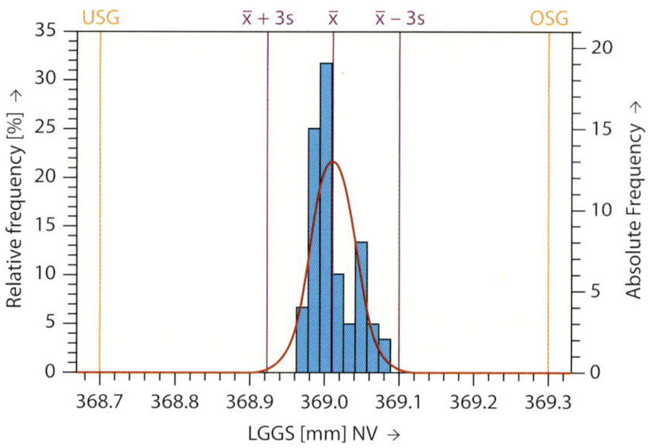

▲

Measurement results.

The measurement results are displayed on a screen. If the values are too high, the valve is rejected.

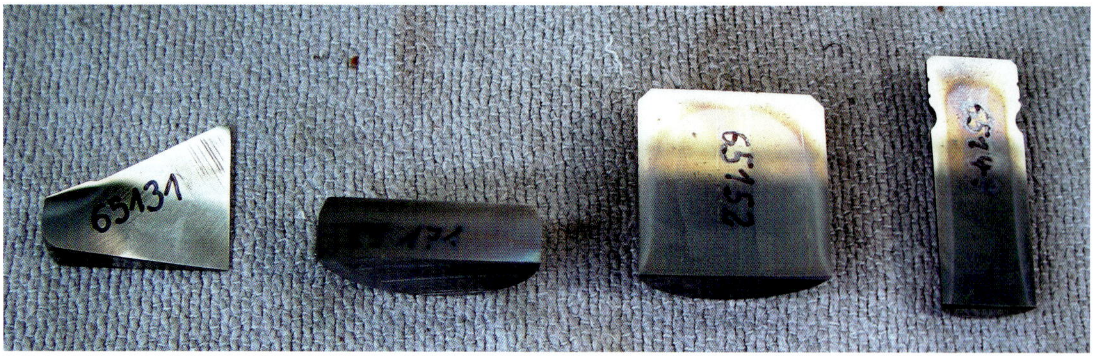

▲

Controlling the material quality.

A valve is taken from every batch in order to control the material properties.

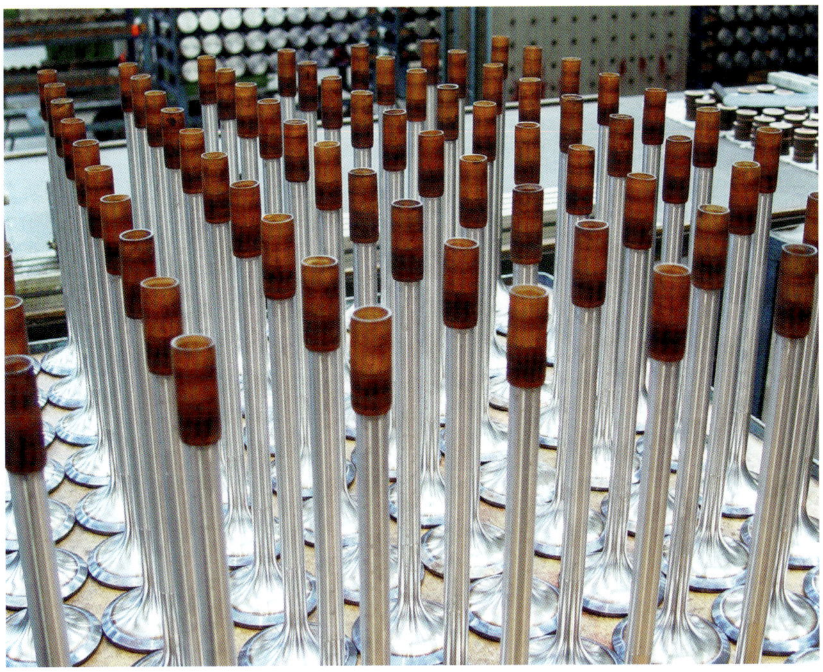

◀

Chromiem plating the valves.

At the customer's request, the valves are chromed. The sections that should not be plated are protected by pieces of flexible plastic tubing, as shown here, the valve-stem end.

◀

Valve holders for chromiem plating.

The valves are placed in these holders so that they can be cleaned in immersion baths. This ensures the adhesion of the chromiem plating. After cleaning, they are immersed in the chromiem plating bath. The thickness of the chrome layer is dependant on, among other things, the time the valves are immersed in the bath.

◀

The various immersion baths for cleaning the valves.

▶

Polishing the valves.

Valves are always polished for their appearance.

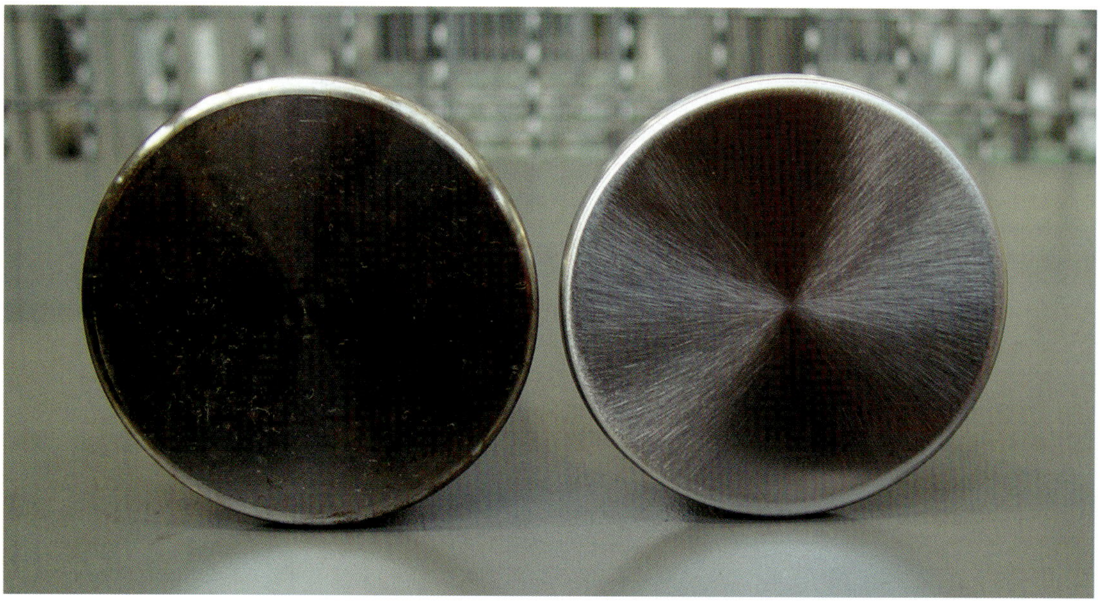

▶

The detection of cracks.

The detection of cracks in the welded stellite coating on the valve seat. The cracks are visible in ultraviolet light. This valve is therefore certainly rejected!

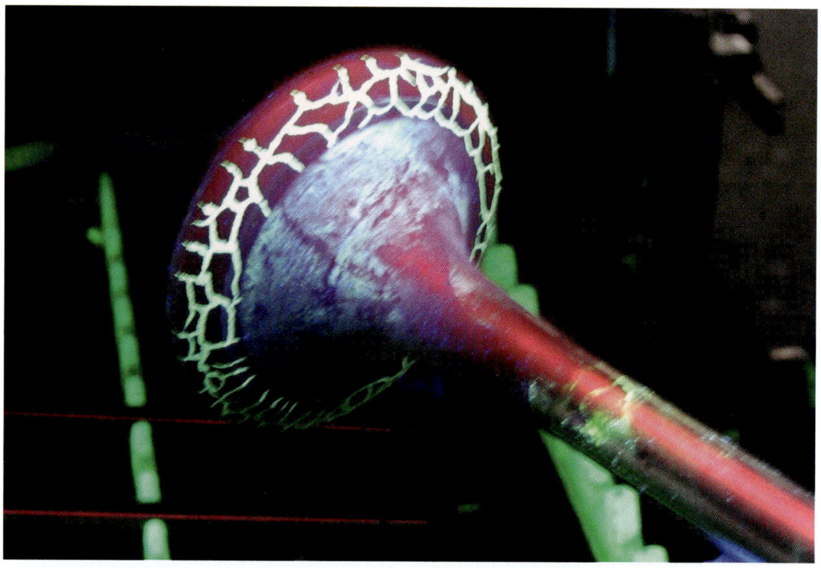

▶

Rejected valve.

An irregularity in the valve disc is clearly shown here under an ultraviolet light source. This valve is rejected.

28.21 Manufacturing cylinder liners

Both four-stroke- and two-stroke diesel-engines cylinder liners must meet strict requirements, such as:

- Be sufficiently strong to absorb the pulsating, extremely high pressures of up to 250 bars produced by the combustion process.
- In four-stroke diesel engines, be sufficiently strong to absorb the lateral forces generated by the force exerted on the connecting-rod crank mechanism by the piston. The horizontal forces the piston skirt exerts on the cylinder liner and therefore the engine block are of a similar order and may amount to approximately 2000 kN. or 200 tons in large four-stroke engines with cylinder diameters of 400 mm or over.
- Have good running properties as the cylinder liner should be able to supply a proper seal in conjunction with the piston, piston rings and the intermediate lubricating-oil coating of approximately 0.04 mm.
- Have the correct dimensions, concerning both the inner diameter and the outer measurements of the cylinder liner.
- Allow for applying good honing patterns on the inside of the cylinder liner.
- Allow for machining processes, such as boring cooling-water ducts.

28.21.1 Centrifugal casting of cylinder liners

This method is applied widely. The material quality using the centrifugal casting method is superior in comparison to the traditional sand-casting method. Generally, cylinder liners in four-stroke diesel engines are cast using this technique, as are smaller cylinder liners in two-stroke crosshead engines. In larger two-stroke crosshead cylinder liners, the masses of the rotating parts, such as the moulds partially filled with iron, are too large.

The illustrations made on the occasion of a visit by the author at Jürgensen, in Sörup, Germany, a large manufacturer of cylinder liners for, among others, high-speed- and medium-speed diesel- and gas engines.

▼

Cylinder liners.

Cylinder liners should meet very high standards as the diesel engine must operate problem free for many thousand hours.

◀

Cylinder liners.

Cylinder liners are essential parts of the engine as they form together with the cylinder head and piston, the combustion chamber, with its high pressures and temperatures and residual products from the combustion process, which can cause contamination and wear.

28.21.2 The material utilised

Generally, cylinder liners are manufactured from fine, perlitic cast iron allowing for excellent running properties. The base material comprises residual steel from shipyards and construction firms, the plating industry, for instance, manufacture of automobiles, and residual material from the company itself.
In order to acquire the correct properties for the cylinder liner, materials such as chromium, titanium, nickle, copper and other alloys are added.

The required composition of the cast iron is supplied by the client, usually engine manufacturers such as Wärtsilä, MAN Diesel, Caterpillar Mak, MTU or Rolls Royce.
The composition of the cast iron varies for each manufacturer and may depend on the engine type.

▲

Ingots.

One of the basic raw materials for the production of cylinder liners, pieces of cast iron, called ingots.

▲

Basic raw materials.

Excess steel from shipyards and engineering works, a good basis for cast iron.

▲

Basic raw materials.

Also chips from the machining of cylinder liners are a good raw material.

►

Adding alloys.

Alloys are added to many high-quality castings. They create new material properties, such as a higher tensile strength and wear resistance.

►►

The alloy elements are precisely weighed and placed together with the base material in the electrical furnace.

28.21.3 The melting process

The base materials are heated to approximately 1300 °C in an electro-oven. The slag is skimmed and once the centrifugal cast machines are assembled with the moulds suitable for the cylinder liner concerned, the casting ladles are filled with a liquid cast iron that has a low viscosity. To achieve high-quality casting, the number of kilograms of material required for each liner is predetermined. A sample is taken from each liner in order to ascertain the composition and material properties quickly and automatically.

28.21.4 Casting

The cylinder shaped liners are especially suitable for centrifugal casting. This method produces a (homogenous) structure superior to that produced by sand casting. The cylindrical cast-steel mould is placed in the electromotor-driven casting machine and attains high speeds. Due to centrifugal forces the liquid cast iron is pressed against inner side of the casting mould. The low viscous, liquid cast iron is poured using a funnel, which can be easily moved into the casting mould.

The advantages of this method:

- No core.
- The mould can be utilised numerous times.
- High-quality casting.

A special mould and end cover are manufactured for each liner type.

Cooling and storing roughly-cast liners

The castings are extracted from the mould and stored in calm air to cool to ambient temperature. The roughly-cast liners are stored inside a hall or outside.

◀ The slag, a collective name for impurities must be removed.

◀ The electric furnace, the heart of the foundry.

The cover of the furnace has been raised. In the foreground, some ladles.

▶

The cast-steel cylindrical moulds.

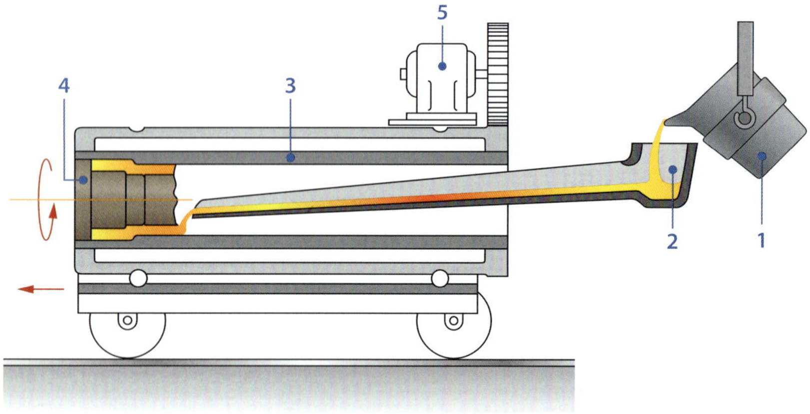

▲

The basic principle of the centrifugal casting of cylinder liners.

1 *ladle with liquid iron, temperature ± 1300 °C.*
2 *pour spout, the caster moves during the casting process.*
3 *mould, clamped in the casting machine.*
4 *end cover, is in reality also found to the right with a large hole for the funnel.*
5 *casting machine drive.*

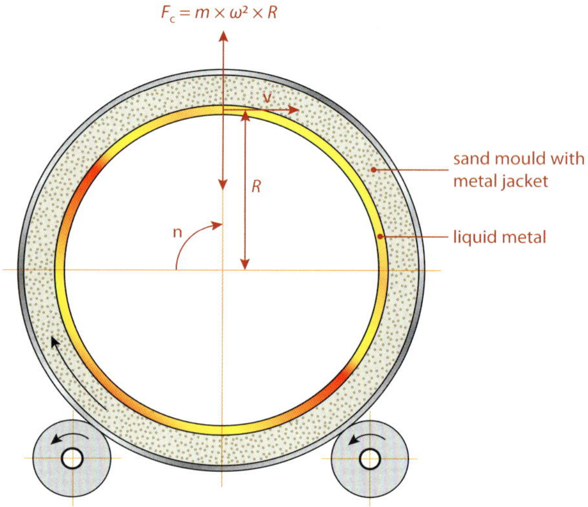

▲

Principle of centrifugal casting.

The low viscosity iron is thrown outwards due to the large centrifugal force. This creates a high quality and homogeneous material composition.

▶

The cylindrical cast-iron end covers for the cylindrical moulds.

Separate cylindrical moulds and end covers should be made for each type of cylinder liner.

◀

The casting of the cylinder liner.

A ladle ready to pour liquid iron for casting a cylinder liner, the ladle is suspended from a weighing scale and is filled to the correct weight for this cylinder liner, 204 kilograms.

◀

Casting the cylinder liner.

The casting of the cylinder liner, the correct flow rate is important. The speed of the casting is performed manually: a matter of considerable experience.

▼

Pouring the thin liquid iron into the funnel of the casting machine.

Due to the centrifugal force, the liquid cast-iron is thrown very fast and at a great force against the outer wall of the mould, so creating the symmetrical cylinder shape of the roughly-cast cylinder liner.

▼

The casting of the cylinder liner.

Casting machines are electrically driven and have very high rotational speeds. The speed is dependant on the size of the cylinder liner.

Lower speeds are kept for large cylinder liners as the centrifugal force depends primarily on the radius of the rotating apparatus.

$F_c = m \times \omega^2 \times R$

F_c = force in Newton

m = mass in kilograms

ω = angular velocity in m/sec.

R = radius in meters.

▲

Test specimen.

During casting, a test specimen is cast. This is used to analyse the material properties.

▲

Cooling after casting.

It takes a long time for cast iron to cool. This is intentional as fewer stresses are then found in the casting.

▲

The rough castings.

The cooled cylindrical mould with the rough casting after cooling. The end covers with the central hole have been removed.

▲

The rough castings.

The rough cast-iron cylinder liners prior to machining.

►

Pre-machining cylinder liners for four-stroke diesel engines.

◀

Unfinished cylinder liners.

Unfinished cylinder liners in the 'waiting room' for further processing. Often a whole series of the same cylinder liners are finished in a single course of action.

◀

Design.

The final design is already visible in these roughly machined cylinder liners.

◀

Sand-casted cylinder liners for smaller two-stroke crosshead engines.

The roughness of the casting is clearly shown: The liner-wall thickness varies considerably! Sand-casted cylinder liners have a low quality compare to centrifugal casted cylinder liners.

▶
The finished surface of a cylinder liner of MAN Diesel that deviates from the normal honing, called 'wave cut' must be machined on a special way.

28.21.5 Rough machining

Firstly, most of the superfluous material from the outer flange is removed from the roughly-cast liners. Next, a material sample is taken from the outside of the liner flange. Subsequently after approving the quality of the material properties, the fine machining can be started.

28.21.6 Annealing

Prior to finish machining the pre-machined cylinder liner, the liners are annealed in an electro-oven. After heating to approximately 600 °C, the liners are cooled gradually, thus relieving casting stress.

▶
Rough cylinder liners with a thickened upper collar.

▶
Complete cylinder-units with the complete cylinder head, cylinder liner, connecting rod and piston.

28.21.7 Honing the cylinder liner

Subsequent to turning the liner to its custom size with tolerances under 0.01 mm, the cylinder liner running surface is honed: the mesh of shallow grooves retains the lubrication oil required to create a thin lubrication film between the piston and piston rings, and cylinder.

▲ Roughly machining the inside of the cylinder liner.

1 *cylinder liner*
2 *tensioners*
3 *cutting tool/shaft*

◄ A cylinder liner for a diesel engine with bore cooling and an anti-polishing ring during processing.

◄ A setup with a large number of CNC machines for precision finishing.

After the cylinder liner is clamped to the work table, these machines carry out many of the operations fully automatically.

▶

Stored parts such as anchor bolts, cylinder head bolts, a complete connecting rod and a cylinder liner for a MAN four-stroke medium-speed diesel engine, type 9L58/64, power output 12.6 MW. M.V. Elyssée, J.R. Shipping, the Netherlands.

▶

A honing machine hones the cylinder liner using a special operating program.

1 *Cylinder liner*
2 *Clamps for holding the liner*
3 *Shaft honing machine*

▶

A honing machine.

1 *Shaft honing machine*
2 *Honing stone*
3 *Nozzle lubrication*

▼

The honing pattern on the cylinder wall.

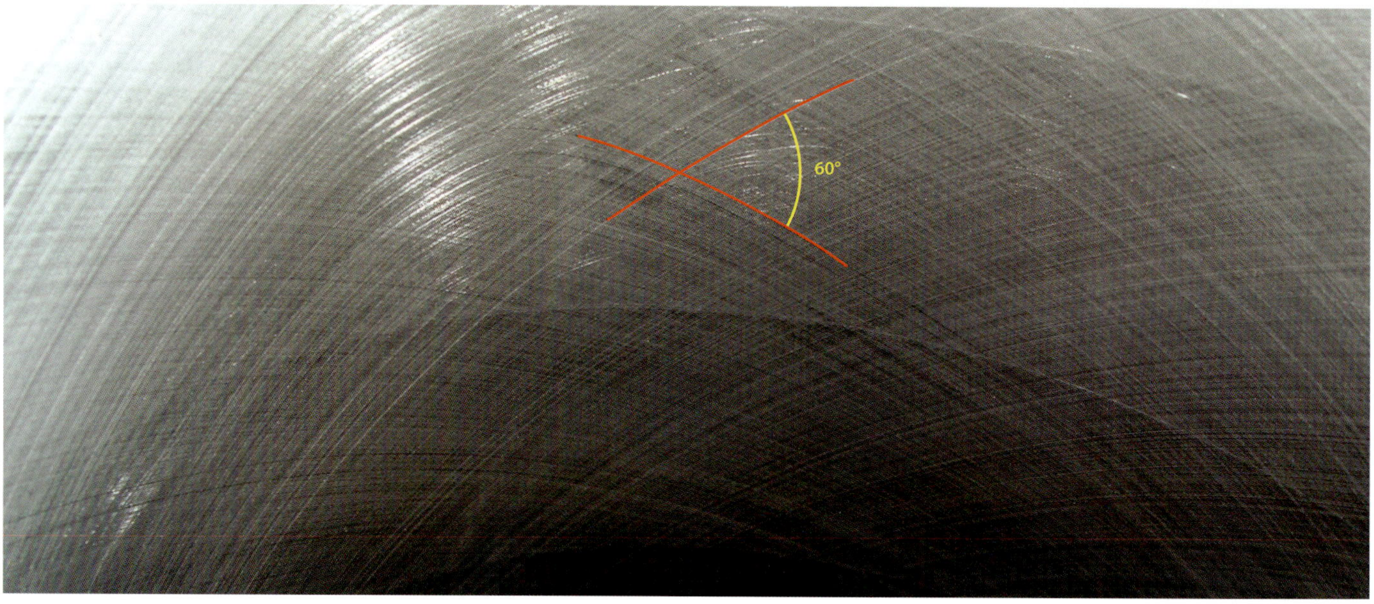

28.21.8 Quality Control

Subsequently, a detailed and accurate control of the dimensions and finishing follows. The cylinder liners are then given an anti-corrosion treatment and stored for transport to the customer.

▶

Quality control.

The dimensions and finishes are all checked in this conditioned room before the liners are released for shipment to the customer.

◀◀

A large number of cylinder liners ready to be packaged before shipment.

◀

A large warehouse for the storage of new cylinder liners.

A stock position is sometimes built at regular customers requests.

◀◀

Pulling the piston.

When servicing diesel engines, the cylinder liner is often measured after the piston has been 'pulled'.
Shown here, pulling the piston on the 'The Pride of Rotterdam'.

◀

Pulling cylinder liners.

The cylinder liner is 'pulled' when it has been rejected due to excessive wear. It is imperative that the spare liners on board are suitable as replacements. This will avoid unnecessary delays. Spare parts should be in good condition at all times!

>CH 29

New fuel developments

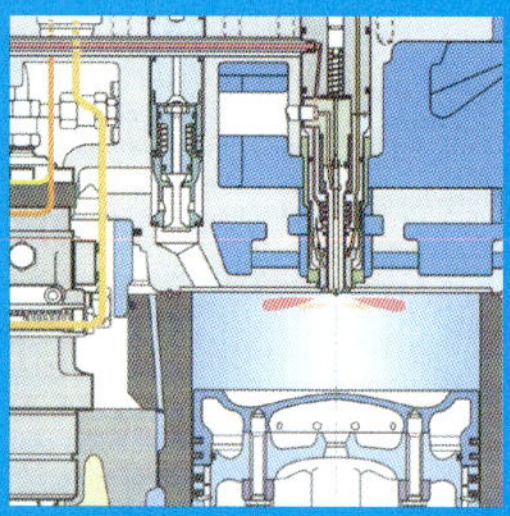

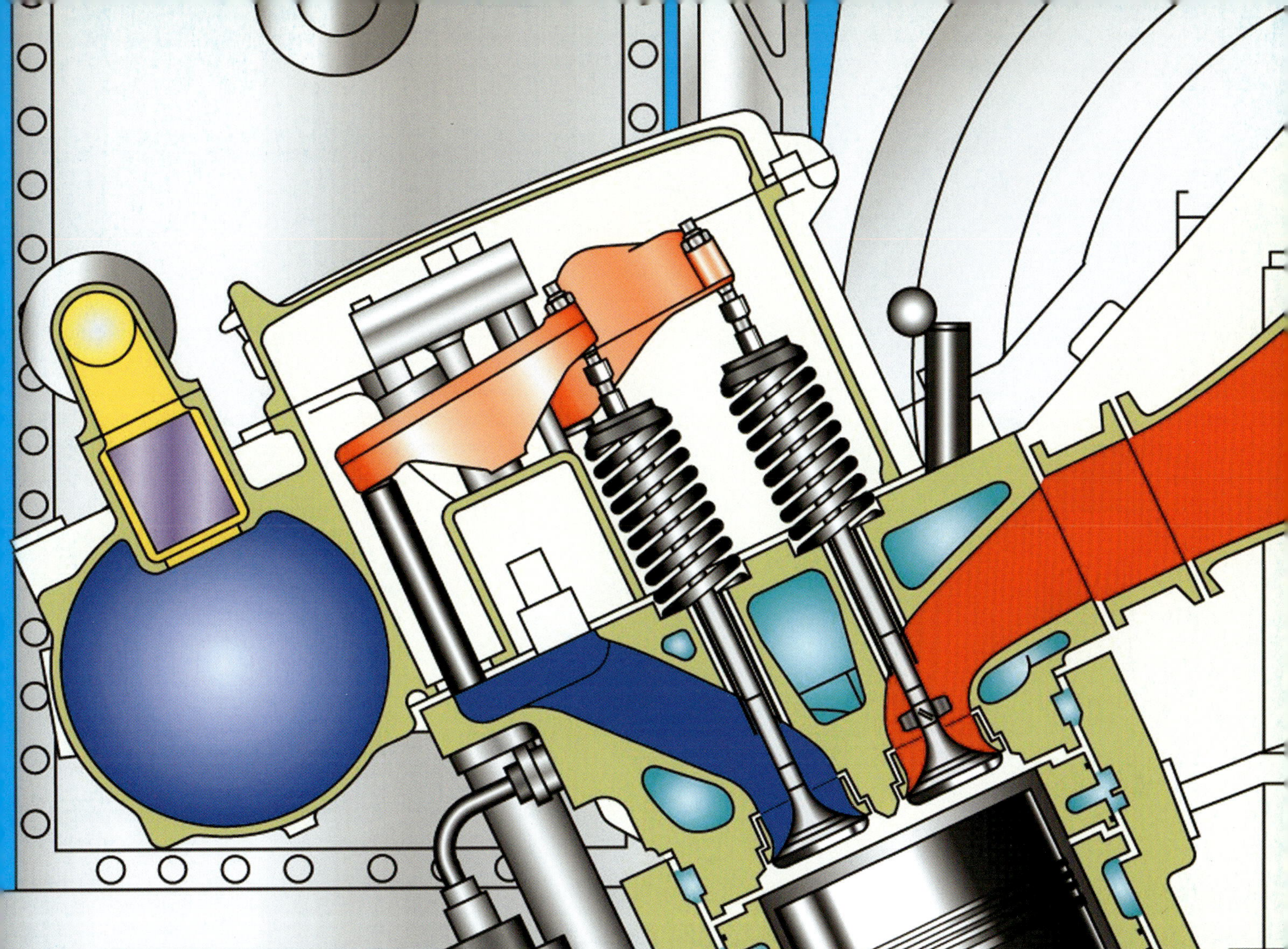

A modern LNG tanker, driven by dual-fuel engines.

One of the new diesel engine developments is the enormous growth in the number of gas tankers equipped with dual-fuel engines currently being manufactured.
An original fuel pump with a special injector is used, which enables combustion of this compressed mixture.
This is a mixture of air and gas, drawn from the ship's storage tanks.
Due to this, the temperature of the cargo (liquid gas) is sufficiently low and therefore emission of noxious gases significantly reduced.
Often less than 5% liquid fuel (heavy-fuel oil) is supplied. The remainder is gas.

29.1 Introduction

At present, the search for fuels that effect a significant reduction in the pollutant emissions continues, this is due to demands to meet environmental requirements regarding the emission of noxious substances in the exhaust gases of diesel engines.

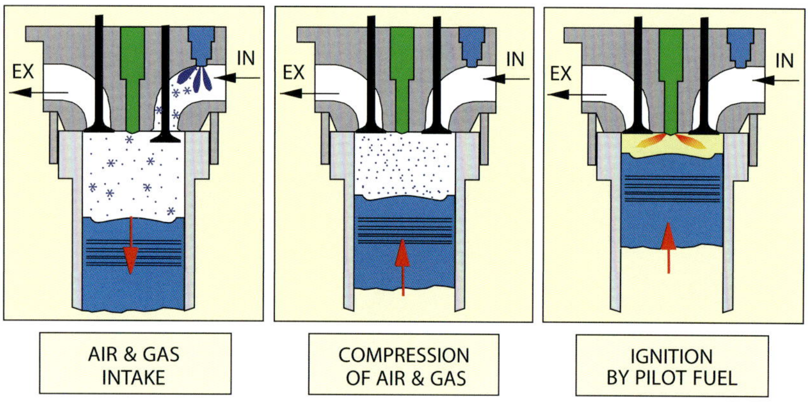

▲

The dual-fuel engine principle. The diesel process is retained.

left: inlet stroke

As the piston moves downwards, a mixture of air and gas is drawn into the cylinder. The gas is injected in the inlet channel via a solenoid valve.

middle: compression stroke

As the piston moves upwards, the mixture is compressed. The mixture is poor, which means that excess air volume is large (approximately 2). Due to this, the mixture does not ignite spontaneously.

right: just before the top position of the piston, the fuel is injected into the cylinder under high pressure. This ignites immediately causing the air/gas mixture to ignite. Afterwards, the normal process takes place.

29.2 Use of combination fuels

29.2.1 Dual-Fuel (DF)

In this process, a very small percentage of between 1 and 10% liquid heavy fuel or diesel oil is injected. During the intake stroke, gas is admitted just before the inlet valves. The liquid fuel ignites the air/gas mixture.

The advantage is that the exhaust gases are cleaner than when heavy-fuel oil or diesel oil are combusted. Gaseous fuel contains fewer carbon atoms than a liquid fuel therefore reducing CO_2 production, the emission of soot particles and the sulphur content is significantly decreased. The contamination of the engine decreases.

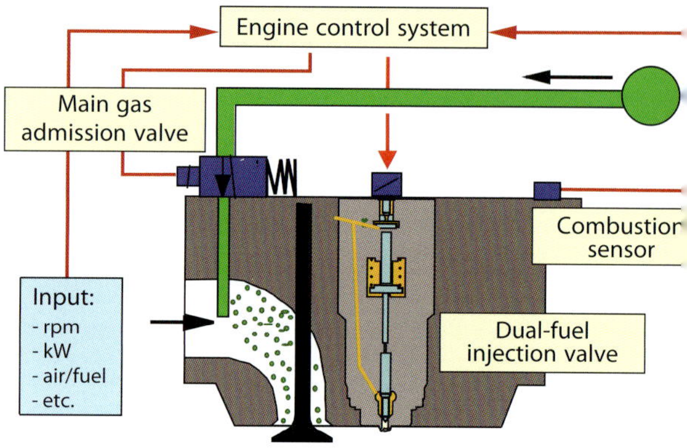

▲

The dual-fuel principle.

The engine can automatically change over from heavy-fuel oil to gas below an 80% load. This takes approximately one minute. When there is an interruption in the gas supply, the engine automatically changes over to heavy-fuel oil.

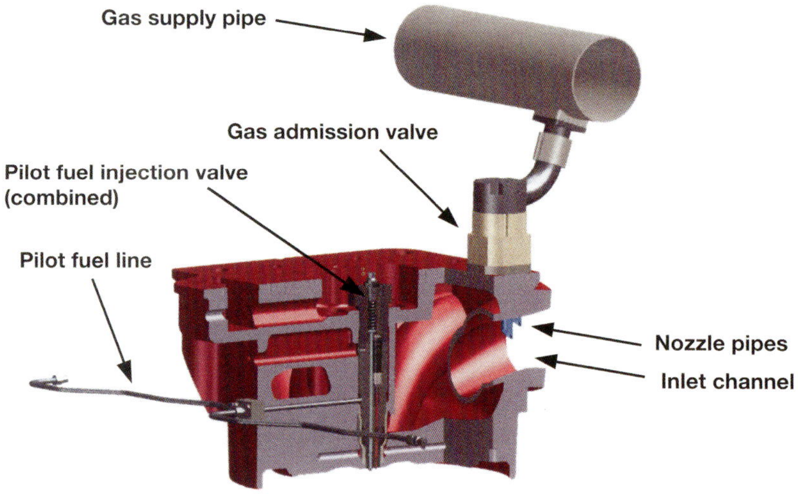

◀

The dual-fuel system principle.

The one fuel, gas, is admitted just before the inlet valves during the induction stroke. Gas and air are intensively mixed. The air factor is between 1.5 and 2.0. This is known as a 'poor' mixture. 50 to 100% more air volume is present than required to fully burn all the fuel chemically. At the end of the compression stroke, the very small amount of injected fuel ignites immediately followed by the rapid ignition of the air/gas mixture. The 'Pilot'-injector replaces the traditional sparkplug. The process continues according to the diesel principle, which has a higher efficiency than the Otto-principle.

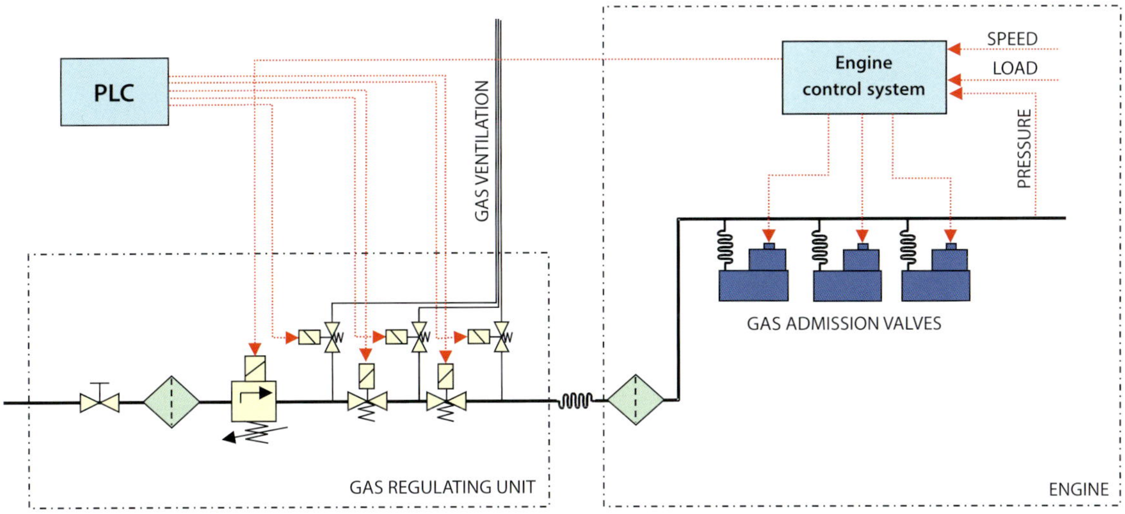

▲

The gaseous fuel system.

The yellow section, the 'gas valve unit', ensures that gas is safely supplied to the engine under all conditions.

A manual shut-off valve, a filter, a pressure-regulating device and two solenoid valves are standard. With engine failure and/or activation of the emergency switch, both solenoid valves quickly closed and the engine shuts down.

The 'Engine Control System' controls the solenoid valves that inject the gas in both air ducts before both inlet valves.

▼

The liquid heavy-fuel oil system for this dual-fuel diesel engine.

The fuel before the injectors comprises 1% heavy-fuel oil and 99% gas supplied by a 'Pilot' high-pressure pump unit that via a common-rail system provides the injectors with sufficient fuel.

With the transition to 100% heavy-fuel oil use, the large high-pressure fuel pumps provide sufficient fuel.

The injectors have two injector nozzles, the pilot and main injector.

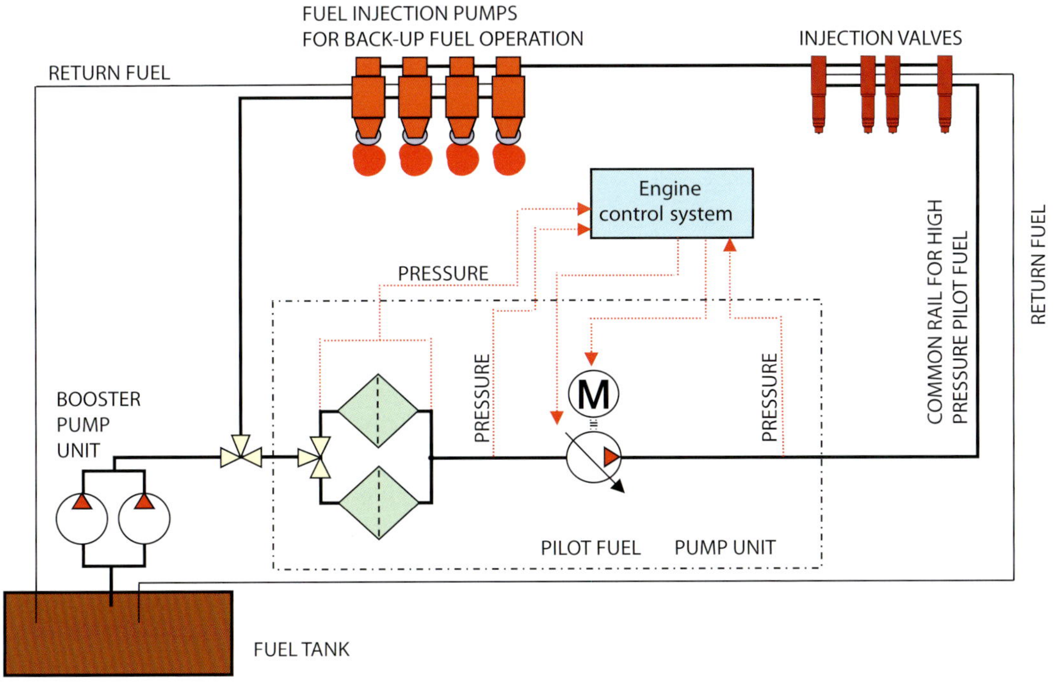

▲

Until approximately 2004, Liquid Natural Gas (LNG) tankers were propelled by steam consisting of a high-pressure water-tube boiler, a steam turbine and a condenser unit.

The fuel is supplied from tanks in which the liquid gas is kept at a temperature of -180 °C. The evaporating liquid fuel is constantly boiling during the voyage and therefore removing the warmth from the LNG tanks, keeping the LNG at a constant low temperature. Per voyage, dependent on the time taken, between 2 and 10% of the cargo is 'boiled off'.
The steam boiler can be fired with heavy-fuel oil when failures occur. At present, the orders for dual-fuel engines continue to experience tremendous growth; all the LNG-tankers ordered in 2006 were supplied with these engines.

Most LNG-tankers are presently supplied with a propulsion engine that mainly operates on the gas from the cargo. In order to keep the cargo (liquid gas) at a sufficiently low temperature, a very small part of the cargo evaporates. This evaporation requires heat, drawn out of the cargo. This method keeps the liquid gas at the desired low temperature of **-180 °C** and gaseous fuel becomes available for propulsion. Previously, this gas was also used as fuel for the steam boiler of the steam-turbine propulsion installation. The efficiency of such propulsion installations is much lower than that of the dual-fuel engine.

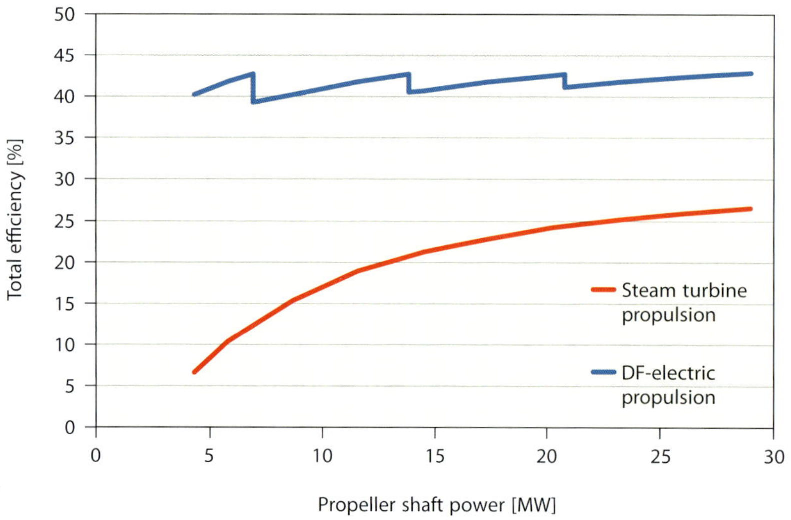

◄

The difference in propulsion efficiency of conventional steam turbines in relation to the dual-fuel systems in diesel engines.

For this tanker, dual-fuel diesel engines were chosen which drive electric generators. At a reduced speed, one or more gensets can be shut down. Due to this, the propulsion efficiency remains high, also at partial loads. The propulsion efficiency of a steam turbine at full load is significantly lower than that of dual-fuel engines. At partial load, the efficiency drops quickly.

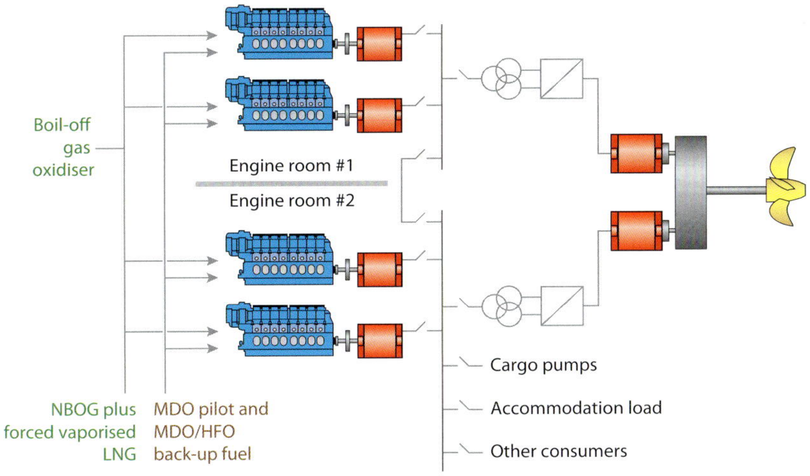

An example of dual-fuel propulsion with electro-motors.

The four DF diesel engines drive generators, which supply the electricity net on board. The electromotors are supplied with energy from this net via a transformer; the speed is controlled by a frequency control.
The electromotors are reversible; therefore a fixed-pitch propeller is adequate.

Dependent on the distance of the voyage, propulsion uses approximately 10% of the cargo. When a tanker is not carrying cargo, the main engine can run on 100% heavy-fuel oil via the main injector. The separate 'pilot' injector (nozzle) is then switched off.

29.2.2 Gas-diesel (GD)

This is similar to the Dual-Fuel principle: the supply of modest amounts of liquid fuel, often less than 1% of the fuel consumption at full load and a substantial amount of gaseous fuel.

During the intake stroke, only air is drawn in and compressed. Just before the TDC position, a small amount of liquid fuel is injected, immediately followed by the injection of the gaseous fuel. If no gaseous fuel is present, the engine can run on liquid fuel. A disadvantage of this **gas/diesel** system in relation to the **dual-fuel** system is that the gas must be compressed to a very high pressure, approximately 350 bars, so it can be injected at the end of the compression stroke. This requires a considerable amount of power for gas compression.

The same type of propulsion system of a Wärtsilä design for a very fast roll-on roll-off ferry.

Note the second propeller drive. This POD can be operated in any direction for maximum manoeuvrability.

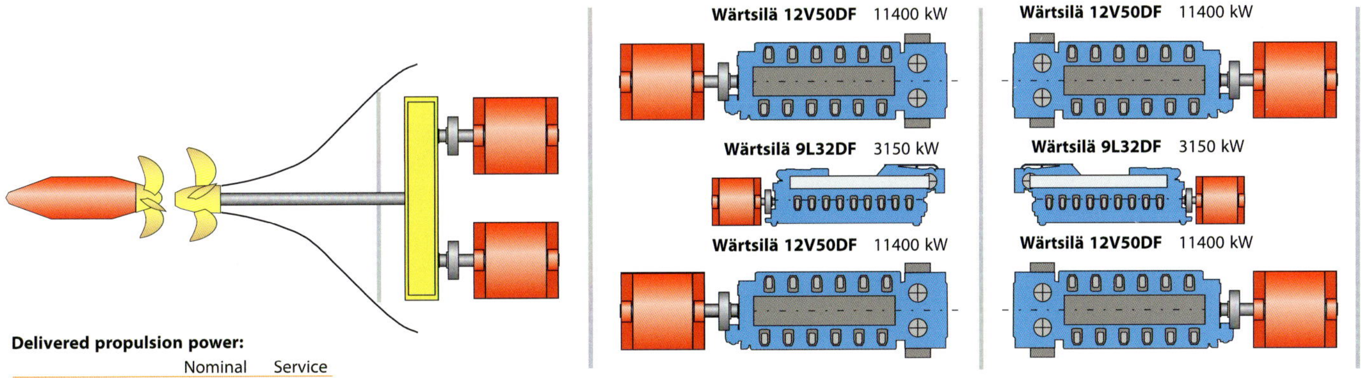

Delivered propulsion power:

	Nominal	Service
Electric pod	17 MW 40 %	17 MW 41 %
LIPS CP propeller	25 MW 65 %	24 MW 62 %
Total shaft power	**42 MW**	**41 MW**

Installed engine power:

Mechanical transmission	0 MW
Electrical power generation	51.9 MW
Total installed power	**51.9 MW**

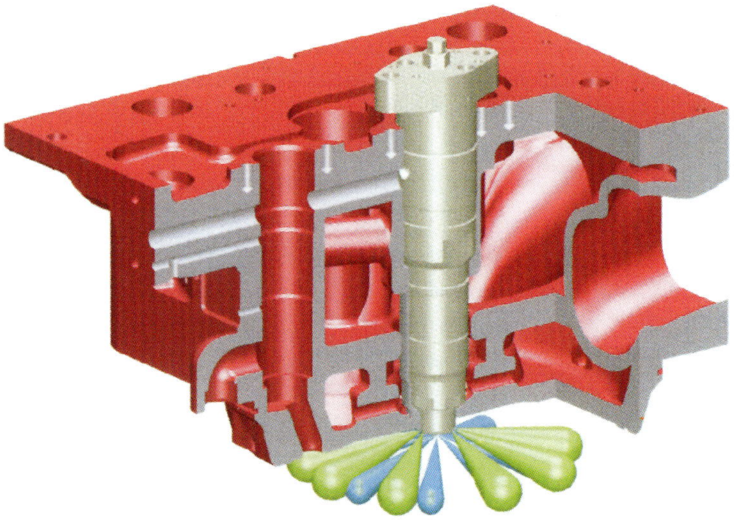

▲

The gas-diesel (GD) engine principle.

blue: heavy-fuel oil

green: gaseous fuel

▼

The gas-diesel (GD)-engine operating principle.

white: air intake

blue: compressed air

red: heavy-fuel oil

green: gaseous fuel

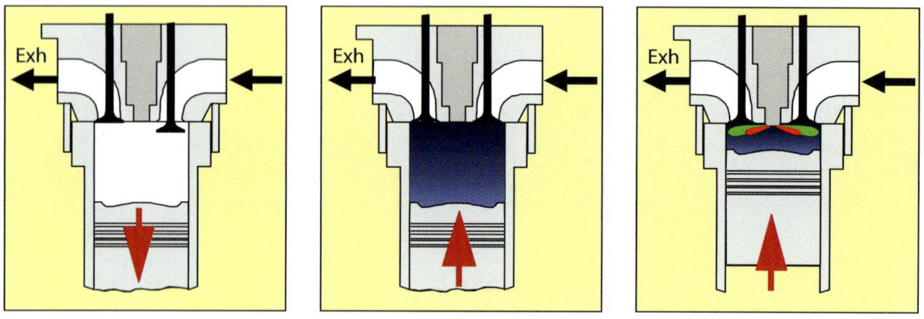

29.2.3 Biofuels, such as olive oil, palm oil, rapeseed oil and other vegetable oils

In approximately 1900, Rudolf Diesel demonstrated a diesel engine that could run on peanut oil. In 2012, the interest in biofuels is huge. This is the result of the following two items.

1 **The increasingly stringent requirements regarding noxious emissions in the exhaust gases, amongst which, the toxic greenhouse gas CO_2, carbon dioxide.**
Biofuels absorb CO_2 from the atmosphere during their growth process; the CO_2 is converted to vegetable matter. This is known as photosynthesis. Photosynthesis is the conversion of, amongst others, CO_2, extracted from the atmosphere as a base material and light as an energy source to (energy containing) carbohydrates and O_2, oxygen, which is released into the atmosphere. This is known as 'CO_2 neutral'. Therefore, when biofuels are burnt, no extra CO_2 is emitted, as is the case with fossil fuels.

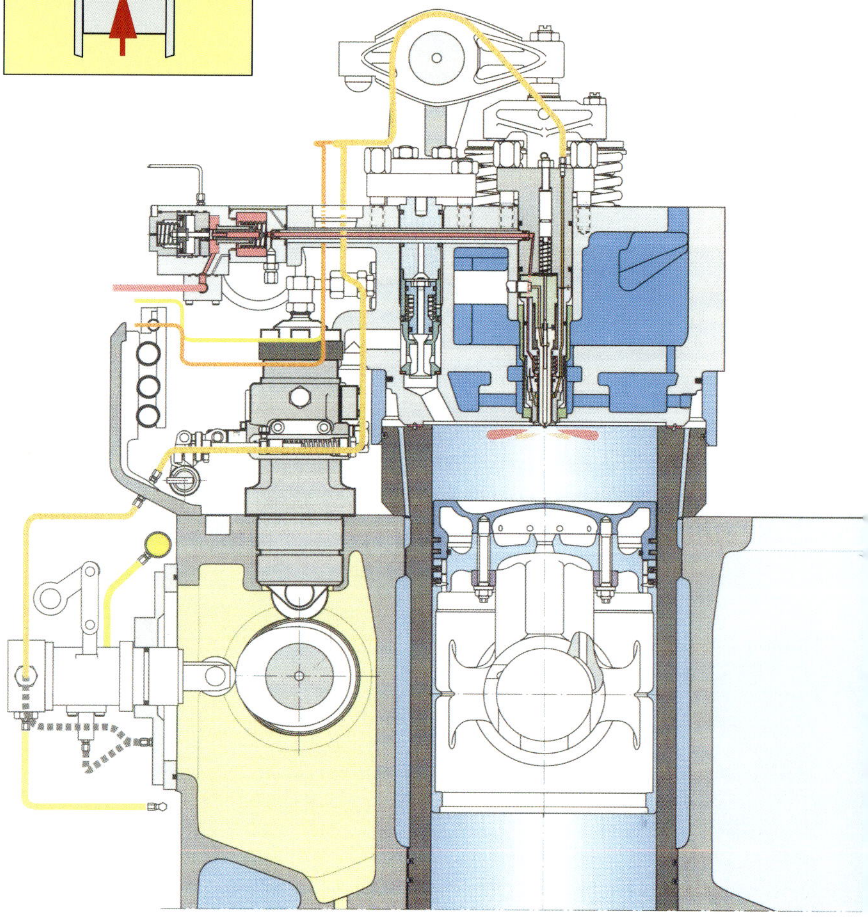

▶

A cross-section of a Wärtsilä 32 GD engine.

In principle, gas-diesel engines are manufactured in the same manner as diesel engines, only the fuel systems differ. Note the horizontally positioned pilot-fuel pump on the side of the camshaft. Above left; the (red) gas supply is shown. The heavy-fuel oil pump, above right of the camshaft, circulates the fuel without the main injector being activated.

◄

CO_2 fertilization, another way of reducing carbon dioxide emissions.

The purified exhaust gases from gas engines are transported with large plastic pipes to the greenhouses, where the natural CO_2 content of the air (0.0375%) is increased, allowing many plants, such as roses and peppers to grow faster. In this way, less CO_2 from fossil fuels is expelled into the atmosphere.

◄

Peppers are suitable for CO_2 fertilization.

2 **Finitude of fossil fuels**

Due to an explosive growth in the use of energy, the demand for mineral fuels has increased exponentially. Countries such as China, India, Brazil and Indonesia require an ever-increasing amount of energy for their industries. Due to this, the worldwide demand is increasing, as is the fuel price. It is becoming increasingly clear how much large industrial countries are dependent on oil-producing countries. The economic and political need for the development of alternative fuel sources is increasing and in the search for alternatives, subjects such as 'less dependent on fossil fuels' are ever more important.

Some biofuel properties

The most important biofuels are olive oil, palm stearine and refined palm oil. Rapeseed oil is at present being investigated in Western Europe. Some properties:

- The pour point is high.
- A large variation in ash content.
- A large variation in acid number.
- A low calorific value.
- Very low sulphur content.

The following measures are necessary for the use of these fuels.

- Fuel systems: the storage tanks must be heated and the fuel pre-heated before injection.
- Fuels with high ash and acid content, which cause contamination and corrosion, must not be used.

▶
The properties and composition of some vegetable fuels in comparison to diesel oil (LFO) and heavy-fuel oil (HFO).

Note, the high sulphur content and high acid number of olive oil!
Refined palm oil is at present the most frequently used vegetable oil being tested.
LT = lower than.
Acid number LT3 lower than 3 mg KOH/g.

Bio-oil's physical & chemical property	Unit	LFO	HFO	Olive oil	Palm stearine	Refined palm oil
Density 15°C	kg/m³	864	993	910.7	921	915.8
Viscosity50176C	mm²/s	3.0	622	12.9	32.5	30.9
M. carbon residue	% m/m	< 0.5	0.6	0.0	0.20	0.15
Water	% v/v	0.0	0.6	0.5	0.1	0.1
Sediment total	% m/m	0.0	0.7	0.01	0.01	0.01
Ash	% m/m	0.01	0.082	0.20	0.01	0.01
Vanadium	mg/kg	LT1	370	LT1	LT1	LT1
Nickel	mg/kg	LT1	55	LT1	LT1	LT1
Sodium	mg/kg	LT1	30	594	LT1	LT1
Calcium	mg/kg	LT1	3	8	1	LT1
Phosphorus	mg/kg	LT1	N.A.	15	2	LT1
Sulphur	% m/m	0.7	2.3	0.05	0.05	0.05
Pour point	°C	-15	15	21	39	18
Acid number	mg KOH/g	LT1	LT3	136	0.08	0.12
Net calorific value	MJ/kg	42.6	40.1	36.55	36.49	36.77

▶
Noxious emissions from some vegetable fuels in comparison to diesel oil and heavy-fuel oil.

Once again, olive oil is the least viable.

Emission	Unit	LFO	HFO	Refined palm oil	Palm stearine	Olive oil
NO_X	ppm, dry 15% O_2	860	970	990	990	1040
CO	ppm, dry 15% O_2	40	40	50	50	50
THC (as CH_4)	ppm, dry 15% O_2	170	80	LT30	LT30	LT30
SO_2	ppm, dry 15% O_2	118	463	< 2	< 2	< 2
CO_2	vol %	5.3	5.46	0.0	0.0	0.0
Particles	mg/Nm³ at 15% O_2	30	40-60	LT10	LT10	70

- The fuel-supply and high-pressure systems must have larger dimensions as they transport more fuel, due to the lower calorific value of biofuels in comparison with diesel and heavy-fuel oils.
- Biofuels can vary in composition. This is dependent on the refining process and the source of the vegetable matter supplied for the production of biofuels. It is therefore advisable to obtain the biological matter from the food industry, where stringent requirements for quality and composition of their raw materials are in place.

Waste oils: Used oils, such as used cooking oils will play a part in the search for alternative fuel sources for diesel engines in the near future.

▶
This large LNG tanker has two two-stroke crosshead engines from MAN Diesel, each drives a propeller.

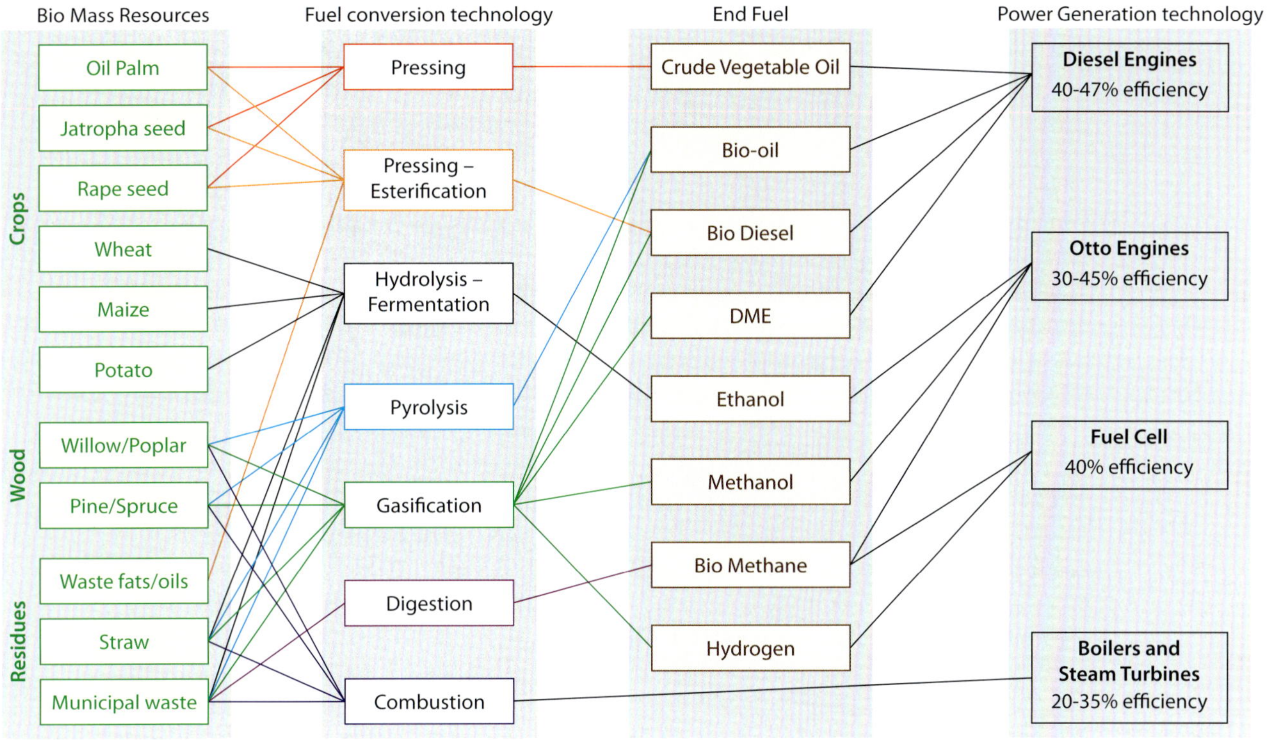

▲
Uses of alternative fuels for amongst others, internal combustion engines according to engine manufacturer Wärtsilä.

Four possible liquid fuels are available for **diesel engines**: Vegetable/plant oils, bio-oils, biodiesel and DME (Dimethylether or methoxymethane).

Vegetable/plant oils

Palm oil:
A good fuel. However huge plantations are required, often in areas where rain forests and other large forested areas such as, swamp forests leading to deforestation. Slowly but surely, palm oil is becoming a highly controversial alternative fuel.

Jatropha seed:
Jatropha oil, a biofuel, can be extracted from the seed. The oil is poisonous. At present there are high expectations for its use as a biofuel.

Rape oil:
Extracted form rape seed.

These three oils are extracted by pressing and are made suitable for use as fuel for diesel engines.

Bio-oils
Willow, poplar, pine, spruce, straw and municipal waste.
Process by pyrolyse or gasification.

Pyrolysis is the process of heating the biomass in the absence (or under reduction) of air.

Biodiesel
The raw material can consist of palm oil, jatropha oil, rape oil or used fats and oils.
At the moment used cooking oils are popular.
Biodiesel is obtained by pressing/cold extraction and transesterification; a reaction with methanol. This produces the 'diesel fuel' with glycerine, the by product from transesterifed biodiesel. Biodiesel can be reasonably simply blended with 'fossil fuels' up to approximately 5% content. Biodiesel can also be manufactured from wood based raw materials and waste cooking oils with the ***gasification*** process.

Gasification is a process that converts carbonaceous materials at high temperatures with a controlled amount of oxygen and/or steam into syngas (synthetic gas) and liquid carbon compounds, such as biodiesel!

DME (Dimethylether)
DME can also be produced from wood based raw materials and waste cooking oils.

General: The degree to which these alternative fuels can be used for diesel engines is dependant on many factors. These can be technical or financial; the environment is becoming an increasingly important factor.
Time will only tell which alternative fuel will play a major role in the industry. Rudolf Diesel ran his first engine on peanut oil!

29.3 Example of the propulsion of a large LNG tanker

A large LNG tanker propelled by a two-stroke crosshead engine.

Gas is a relatively 'clean' fuel and therefore there is worldwide a rising demand for this product.
Where transporting natural gas in pipelines is not technically possible or economical, it can be transported by specially designed LNG tankers.
LNG (Liquid Natural Gas) has at atmospheric pressure, $\frac{1}{600}$ the volume of natural gas.
'Natural' gas means naturally occurring gas; it is not produced in a refinery.
As expected, LNG must undergo extensive processing prior to transport and use as a fuel.
LNG is an odourless, colourless liquid with a specific mass of 450 kg per m^3, 45% of the specific mass of water, which is 1000 kg per m^3.
LNG is primarily methane. The amount of methane and other compounds is dependent on the oil fields / natural gas fields.

LNG is always transported 'cold' at atmospheric pressure; the boiling point is **-163 °C.**
The LNG is kept cold during transport by the removal of methane gas produced in the 'boil off', which is diverted to the engines and used as a fuel for ship propulsion.

The LNG tanks are normally never completely emptied when the ship is sailing because a certain amount of LNG is required to keep the tanks at this low temperature.
The liquid is 'kept cold' during storage according to a well-known physics principle; as the vapour boils off, there is a phase change from liquid to gas (at -163 °C). Heat for this phase change comes from the environment so cooling the remaining liquid.

Only a relatively small amount of boil-off is required to maintain temperature because the cargo tank insulation is very efficient. This phenomenon is also called auto-refrigeration.
There are two main types of gas tanks used today in LNG carriers. They are normally; the spherical (Moss) type or the membrane type.

29.3.1 The spherical or 'Moss' type storage tank

The gas tankers are constructed according to the double-hull concept, including the bottom. Furthermore, for LNG storage, the cargo tanks must be installed separately in the ship's holds, and may not be part of the ship's structure.

29.3.2 The membrane type

Tanks with the membrane system are rectangular, fully integrated into the hull and rely on the strength of the ship's hull. The stainless steel walls of the tank-membrane system are a very thin barrier (0.7-1.5 mm) supported by the insulation. In these tanks, only a relatively small amount of steel has to be cooled.

At present, LNG tankers are mostly being ordered with membrane tanks, because of the relatively higher utilisation of their hull volume for the cargo.

▶
One of the two-stroke crosshead engines used for the propulsion of a large LNG tanker, a MAN Diesel 6S70ME-C engine with a shaft power of 18,660 kW.

The fuel is heavy-fuel oil. These engines can be equipped with the GI system, the gas-injection system.

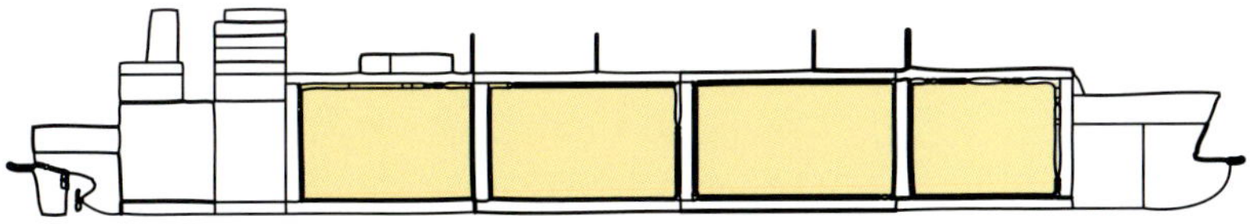

138,000 m^3 LNG carrier of the membrane type

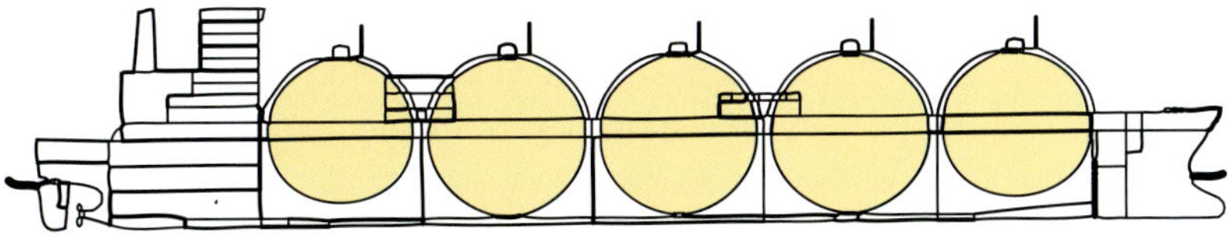

138,000 m^3 LNG carrier of the spherical (Moss) type

◄ The two containment systems used in LNG tankers.

Above, the membrane type and below the spherical 'Moss' type.

29.3.3 LNG tanker classes

There are five LNG tanker classes, ranging from small up to Q-max.

LNG carrier classes	Dimensions	Ship size - LNG capacity
Small	B: up to 40 m L_{OA}: up to 250 m	up to 90,000 m^3
Small Conventional	B: 41 - 49 m L_{OA}: 270 - 298 m	120,000 - 149,999 m^3
Large Conventional	T_{des}: up to 12.0 m B: 43 - 46 m L_{OA}: 285 - 295 m	150,000 - 180,000 m^3
Q-flex	T_{des}: up to 12.0 m B: approx. 50 m L_{OA}: approx. 315 m	200,000 - 220,000 m^3
Q-max	T_{des}: up to 12.0 m B: 53 - 55 m L_{OA}: approx. 345 m	more than 260,000 m^3
Examples of special LNG carrier sub-classes: Med-max (Mediterranean maximum size) about 75,000 m^3 Atlantic-max (Atlantic sea maximum size) about 165,000 m^3		

◄ The size classes for LNG tankers.

LNG tankers, as oil tankers and ore carriers, are very wide; this means that they have a slow sailing speed.
Twenty knots is the maximum speed for the largest ships; Q max.

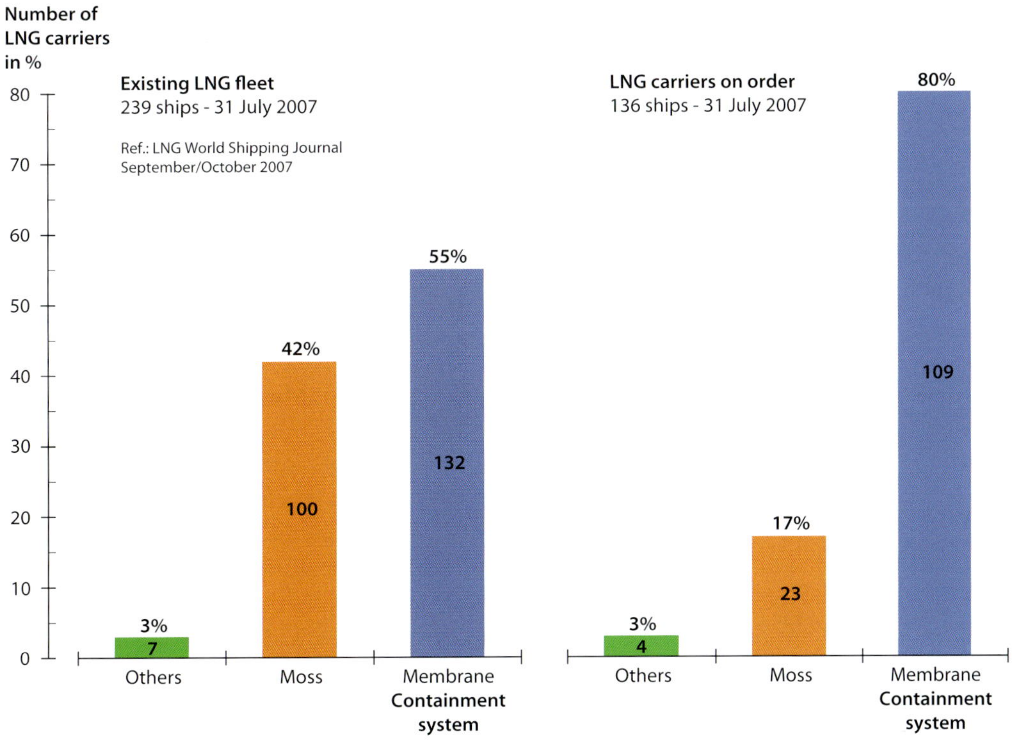

◄ The number of existing LNG tankers in service per 31 July 2007, left, and the number LNG tankers on order per 31 July 2007, right.

Clearly shown is that most of the LNG tankers will be equipped with a membrane containment system.

Recently, there has been a dramatic increase in the size of the LNG carriers being ordered.
Due to this, extra propulsion-power output is required.
The shift from steam-turbine propulsion with a low efficiency to the internal combustion engine with a much higher efficiency gives not only new prospective for four-stoke engines but also larger power outputs for the two-stroke crosshead engine.

The LNG tankers on order at shipyards as of July 2007 show a clear trend.
Besides the conventional steam-turbine propulsion installations (40%), there are a large number fitted with diesel engines using the dual-fuel systems (59%).

29.3.4 The fuel system for LNG tankers with two-stroke crosshead engines

This is known at MAN-Diesel as the ME–GI system.
ME electronically controlled liquid fuel.
GI gas-injection system.

The fuel supply system consists of a heavy-fuel oil- and an LNG supply system.
Obviously, the supply of this HFO system is also available in the ME version.

29.3.5 The LNG-supply system: the principle

In order to be able to inject gas in the cylinder during the compression stroke, a fuel-gas compressor is required that can supply the LNG gas at a gas injection pressure of 150 to 265 bars from the atmospheric low tank pressure.
The compressor must be able to supply different types of gases with a controllable capacity at different suction pressures and temperatures.
The supply pressure must also be variable.
In addition, the compressor must also be able to compress gas for the gas-flaring installation when propulsion is not required.

Note: the inlet-gas temperature is -160 °C!
A total of five compression stages are provided and the compressor is directly driven by an electric motor.

29.3.6 The ME-GI system for the engine

The engines are identical to the existing ME version. The only difference is that the ME-GI version uses gas as the major fuel.

29.3.7 LNG gas supply

The gas is supplied to the cylinder using double-wall piping and a gas-valve on the slightly modified cylinder cover. The gas-control block has an internal accumulator to level pressure fluctuations. A gas injection valve is installed in the cylinder cover for controlling the amount of injected gas and injection timing.

In addition to modifications to the engine/cylinder head, the following new systems have been installed:
- **a ventilation system**, for venting the space between the inner and outer pipe of the double-walled piping system;
- **a sealing-oil system**, delivering sealing oil to the gas valves separating the servo oil and the gas oil;
- **an inert gas system**, which enables purging of the gas system with inert gas.

Additionally, there is a control and safety system consisting of a hydrocarbon analyser for checking the hydrocarbon content of the air between the double-wall gas pipes.

All failures in the gas-fuel system result in an immediate changeover to the HFO system or engine shutdown.
The changeover to fuel-oil mode is always done without any power loss to the engine.

The gas-injection valve is controlled from the existing hydraulic system in the ME diesel engines.

Dual-fuel operation requires the injection of not only fuel, but also gas in the combustion space.
Therefore, there are two fuel and two gas injectors in the cylinder head.

29.3.8 The fuel supply during operation

One of the advantages of the ME-GI version is its fuel flexibility which benefits an LNG carrier. At the start of a laden voyage, the natural 'boil-off gas' holds a large amount of nitrogen and the heat value is low. If the boil-off gas is pressurized, it can consist of both methane and propane, and the heat value is high.

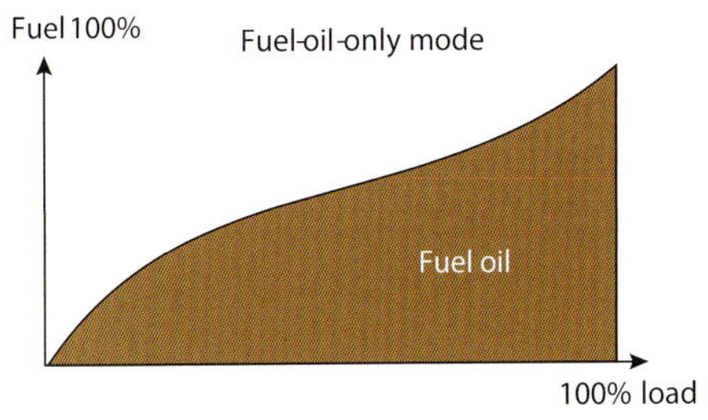

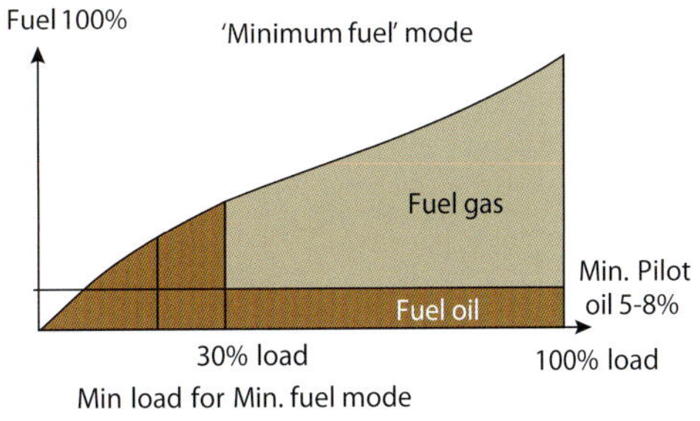

▲

The possible types of fuel supply for MAN Diesel ME-G1 two-stroke crosshead engines in LNG tankers.

Left: Fuel oil only at every load.

Right: The 'dual fuel' system.

To 30% load, only liquid fuel is supplied.

From 30 to 100% load, 5 to 8% liquid fuel is supplied.

The remaining fuel, dependent on the load, is gas.

At full load, maximum 95% gas and minimum 5% liquid fuel (MDO of HFO).

A two-stroke, high-pressure gas injection engine is able to burn these different fuels without a drop in thermal efficiency.

There are two possible types of fuel supply:

- The fuel-oil mode, well-known from the ME engine. The engine is considered 'gas safe'. If a failure in the gas system occurs, it will result in a gas shutdown and will be secured.
- The minimum fuel mode is developed for gas operation, and it can only be started manually by an operator on the Gas Main Operating Panel in the control room.

The minimum fuel-oil supply is between 5 and 8% at full engine load dependent on the fuel quality. Both MDO and HFO can be used.

MAN Diesel is not able to guarantee a stable gas- and pilot-oil combustion at an engine load below 30% of full load, and in that case the engine returns to Fuel-oil only mode.

29.3.9 Cylinder-lubricating oil

Gaseous fuels have similar properties to fuels with low sulphur contents. Therefore, TBN 40 cylinder-lubricating oil is advised.

◄

The cylinder cover for a MAN Diesel two-stroke crosshead engine.

A small change in the cover and the mounted gas control block allow a diesel engine to be converted for use in LNG-tankers. Left in the photo: the mounted gas control unit.

>CH 30

Bedplates and engine alignments, gearboxes, shafts, propeller shafts and generators

When a new diesel engine is installed in an existing ship, the existing bedplate is modified by a specialised company.
Shown here, the installation of a new engine at the Breko Company in Papendrecht, The Netherlands.
This is referred to as repowering.

A new engine with lower specific fuel consumption and lower pollutant emissions standing ready in the yard.

30.1 Introduction

Diesel engines for ship propulsion and as well as in diesel gensets for power plants are usually mounted on welded sheet-steel bedplates. Diesel engines can be rigidly or flexibly mounted on their bedplates.

30.2 Ship propulsion

The bedplate of a propulsion engine must meet the following requirements.

- The bedplate must be able to support the weight of the engine without distortion to the bedplate or hull. The weight of the engine can vary, in category I, some tens of kilograms, in engine category IV up to nearly 3000 tons!
- The engine bedplate must be able to absorb and transmit the fluctuating forces formed during engine operation.
- The bedplate must be designed and manufactured in such a way that the engine does not shift when the ship runs aground or is involved in a collision. Therefore, the engine is often supplied with end stoppers.
- If the thrust block is installed on the bedplate or in the gearbox at the rear of the engine, the bedplate must be able to transmit these fluctuating forces to the ships structure.

30.3 Construction of the bedplate, engine category IV

Larger ships usually have double bottoms; these can be used for the storage of fuels, ballast- and drinking water. In order to achieve this, the double bottom is divided into various compartments, both longitudinally and transversely.
Around the propulsion engine, there is a separate space with double bottom, often surrounded by a 'cofferdam'.
The double bottom below the engine is reinforced with girders and often forms an integral part of the engine bedplate.
The plates are also usually thicker. At the locations where the longitudinal holding-down bolts are placed, the horizontal strip is thicker and ribbed.

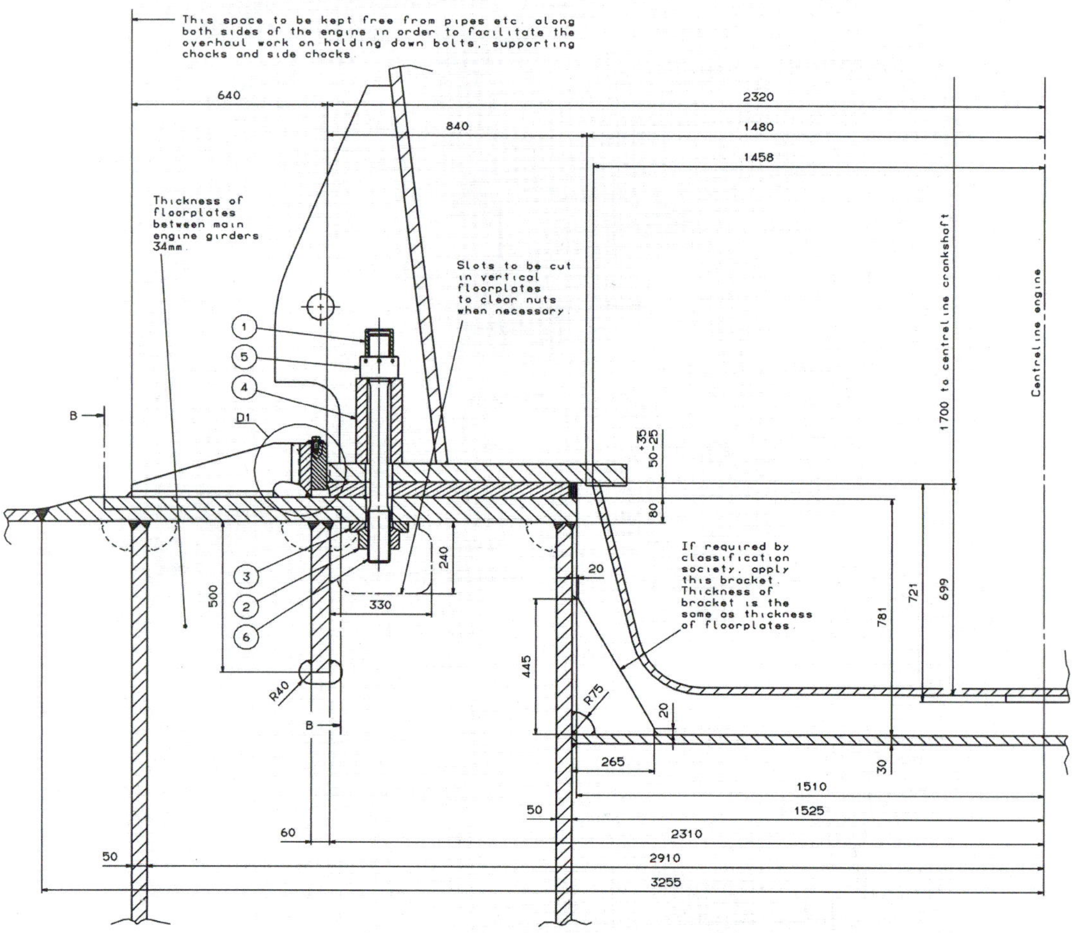

▶
Detailed drawing of the bedplate for a two-stroke crosshead engine from MAN Diesel ME.

The bedplate of a MAN Diesel ME engine with an epoxy resin filler below the welded engine flange. Take note of the long elastic holding-down bolt.

1 *protecting cap*
2 *spherical nut*
3 *spherical washer*
4 *distance pipe*
5 *round hydraulic nut*
6 *holding-down bolt*

◄ The hydraulically tensioned holding-down bolts for a large two-stroke crosshead engine with side chocks

1 *engine frame*
2 *hydraulic holding-down bolts*
3 *side chocks*

▼ The hydraulic unit with two cylinders.

The controls are represented by the double slanted lines on the electric wiring.

'Top bracing'

The engine is equipped with mechanical or hydraulic top bracing to absorb the lateral forces. These forces can be significant.

Also, see Chapter 17, *Vibrations and balancing.*

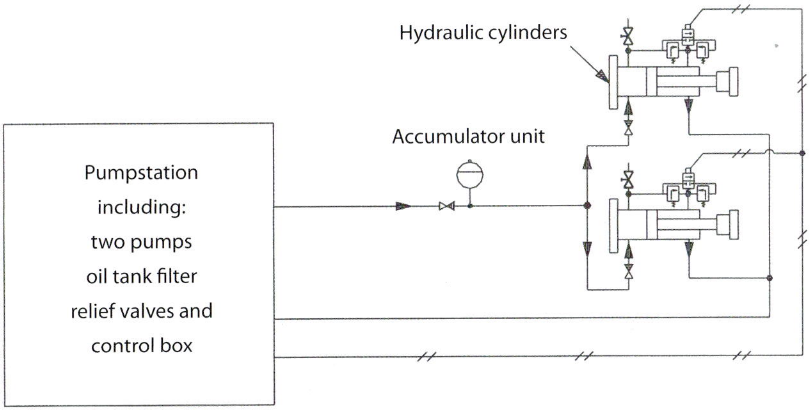

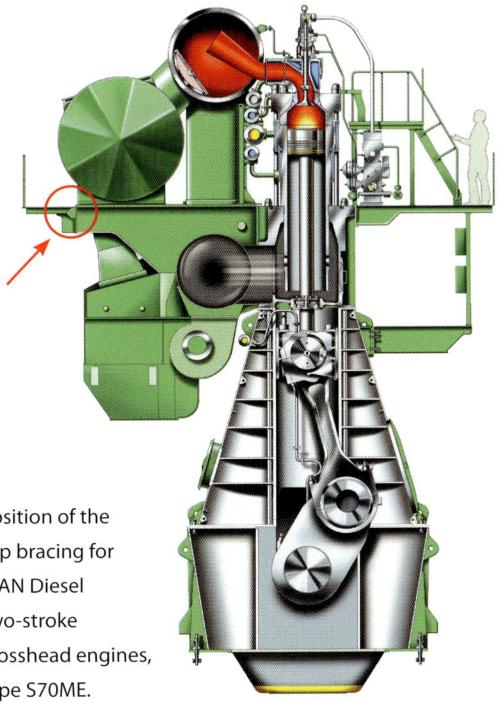

▲ Position of the top bracing for MAN Diesel two-stroke crosshead engines, type S70ME.

Force per mechanical top bracing and minimum horizontal rigidity at attachment to the hull

Engine type	Force per bracing in kN	Minimum horizontal rigidity in MN/m
K108ME-C	available on request	
K98ME	248	230
K98ME-C/GI	248	230
S90ME-C	209	210
K90ME/GI	209	210
K90ME-C	209	210
S80ME	available on request	
S80ME-C	165	190
K80ME-C	165	190
S70ME-C/GI	126	170
L70ME-C	126	170
S65ME-C/GI	available on request	
S60ME-C/GI	93	140
L60ME-C	93	140
S50ME-C	64	120

◄ Mechanical top bracing.

The forces per top bracing at attachment to the hull for MAN Diesel ME two-stroke crosshead engines. The forces per top bracing vary from 6.4 to 24.8 tons!

▶

A cross sectional view of hydraulic bracing for matching engines.

This system consists of two to four hydraulic cylinders, two accumulators and hydraulic pump with control unit. The locations on the hull are reinforced where the bracings are fitted. The hydraulic cylinder maintains a constant force between engine and hull. A relief valve ensures that the maximum allowable force is not exceeded.

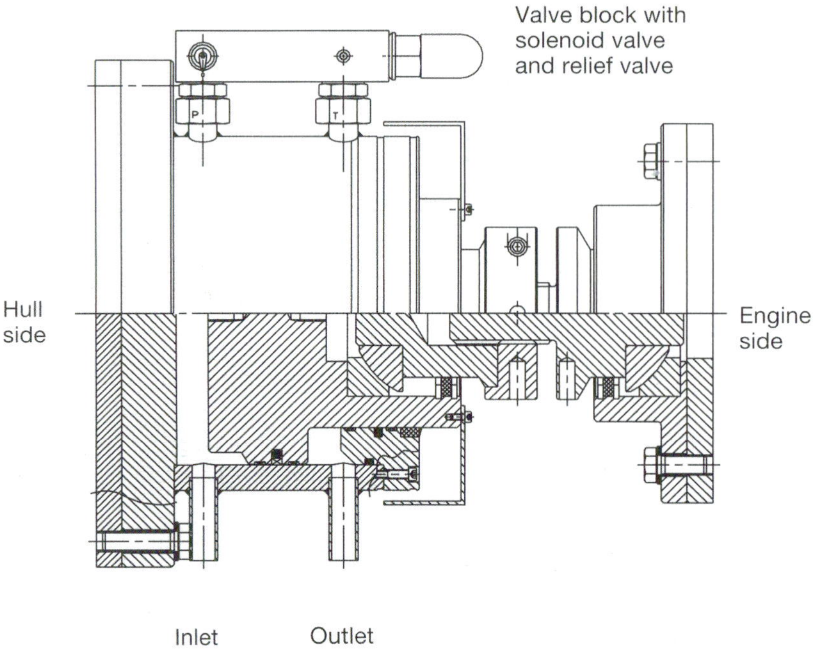

▶

A certain type of top bracing in a Wärtsilä Sulzer RTA84 two-stroke crosshead engine.

The use of top bracing is dependent on the number of cylinders. For instance, a five-cylinder engine usually requires top bracing, occasionally longitudinal bracing is required! The X- and H-vibrational modes of a seven-cylinder engine are low and top bracing is not required (Project Manual RT-Flex60C).

▶

A type of top bracing in a Wärtsilä Sulzer RTA96 two-stroke crosshead engine.

30.4 Examples of rigid mounting, engine category III

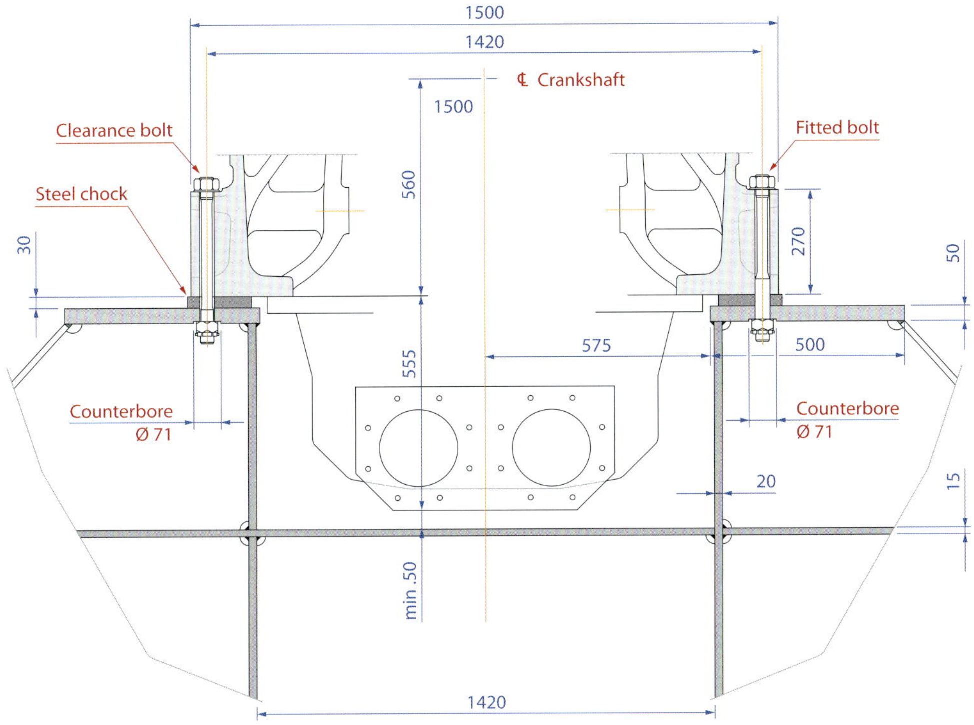

▲

The bedplate for a Wärtsilä 38B diesel engine.

The engine is rigidly mounted on steel or resin chocks (epoxy resin), or flexibly mounted on rubber elements.

The bedplate and double bottom are as stiff as possible in all directions in order to absorb the dynamic forces generated by the engine, reduction gearing and thrust block.

The steel strips below the engine are often tapered to an inclination of 1 to 100.

The engine manufacturer provides the dimensions of the wedge-type steel chocks. The chock material may be cast iron or steel.

The surfaces supporting the chocks are often mechanically levelled giving a large contact face.

At least 80% of the chock's bearing surface must be achieved.

The clearance holes in chock and top-plate strip must be at least 2 mm larger than the bolt diameter, except when equipped with fitted bolts.

All engines are equipped with side supports (chocks). These are fitted before the alignment of the engine. The steel wedges between the engine block and the side supports must be fitted as soon as the engine and bedplate have reached their operating temperatures.

The holding-down bolts are usually provided with a hydraulic nut at the top and a conventionally secured hexagon lock nut at the bottom. It is recommended that the bolts be manufactured from high-tensile strength steel, for instance, 42 CR MO 4 TQ + T alloyed steel or comparable. The contact faces of the nuts on the seat of the engine block (top) and in the bedplate (bottom) must be counter-bored to prevent additional bending stresses in the bolts.

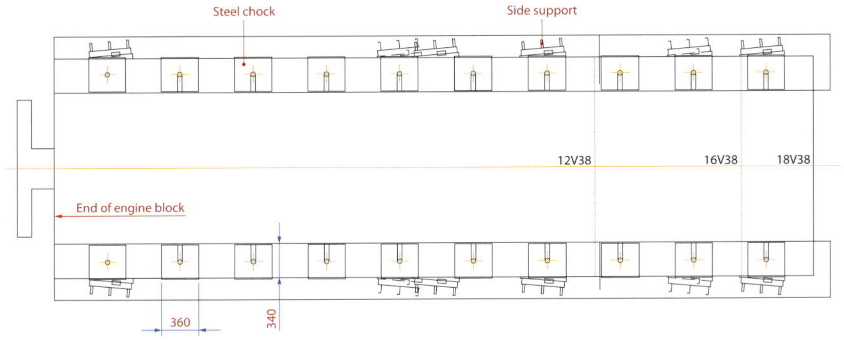

◀

Wärtsilä 38B V-engine.

The bedplate viewed from the top with ten holding-down bolts on each side, steel chocks and six side supports on each side.

▶
A holding-down bolt for a Wärtsilä 38B in-line engine.

Holding-down bolts have to meet high standards

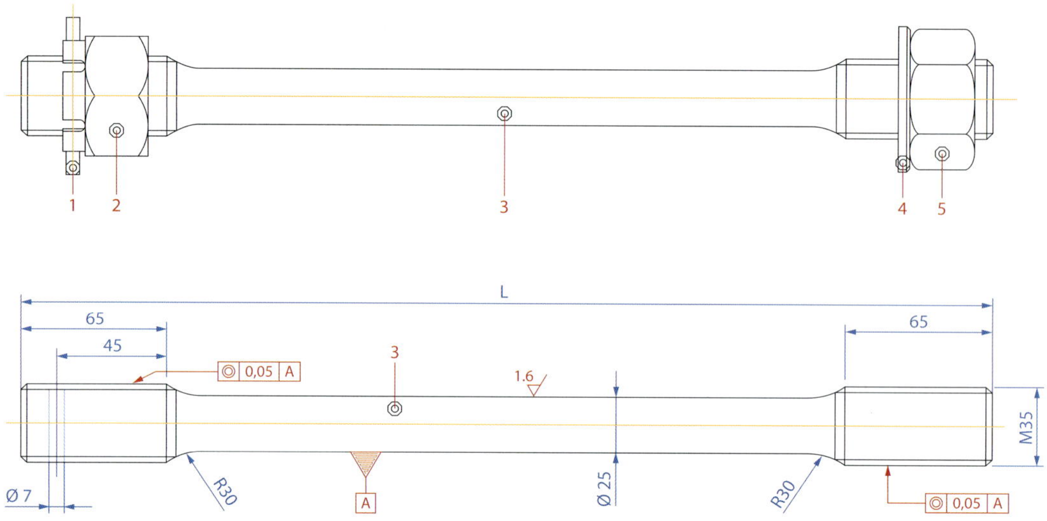

3
Bolt material: 12CrMo1(TQ+T) according EN 100083–1
Thread tolerance bolt according DIN 267
Thread bolt according DIN 267
Thread product class: A according DIN 267

L = topplate + filling + 370

Item description
01 Split pin (6.3 × 63 – St) DIN 91
02 Castle nut DIN 935–M36–8
03 M36 space tap bolt
04 Washer DIN 125 A–37–St
05 Hexagon nut ISO 4032–M36–8 (nut height 0.80D)

Mounting the engine on the bedplate

Hydraulically tensioned holding-down bolts are used for large engines in categories III and IV. These have a large bolt elongation (strain) which improves the protection against loosening of the nuts and breakage.
The bolts are extremely large.

▶
Installing the engine block for a medium-speed four-stroke diesel engine, used to drive a generator in a large container ship.

1 *welded steel-plate lower frame bolted to the reinforced double bottom*
2 *hydraulically tensioned holding-down bolts*
3 *engine block*

30.5 Flexible mounting of propulsion engines, engine categories I, II and III

This is often applied in categories I and II and occasionally in category III engines.
All two-stroke crosshead engines in category IV are rigidly mounted. Most other engines are flexibly mounted to reduce the transmission of sound and vibrations produced by the engine to the hull and consequently the ship's structure including the accommodation. A diesel engine produces low-frequency vibrations and high-frequency noise in the ship's structure.

Also, see Chapter 16, *Noise, origin and damping* and Chapter 17, *Vibrations and Balancing.*

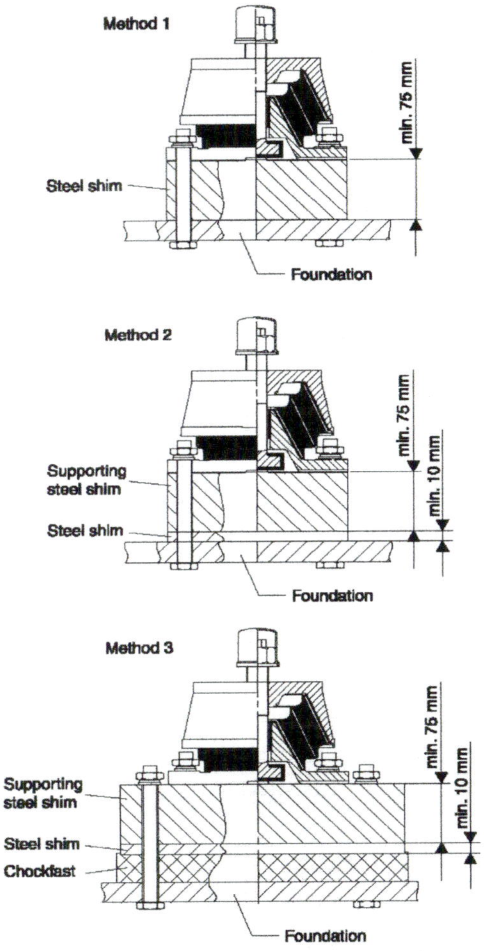

▲

The flexible mounting options for the MAN Diesel L16/24 diesel engines used for power generation.

Shown a series of three conical flexible rubbers. The compression of the rubber elements by the weight of the engine lies between 5 and 11 millimetres.

▲

Various flexible elements from manufacturer, Rubber Design, Heerjansdam, The Netherlands.

There are special flexible elements available for each type of engine and arrangement.

▲

A flexible rubber element below a large Caterpillar MaK 43-L propulsion engine.

The total surface of the rubber mounting for this engine which weighs 110,000 kg is 2 m^2. Engine category III.

◄

A side-support bolt for the same Caterpillar MaK 43-L propulsion engine.

The engine can be bolted with end-chock bolts if the rubber elements are too compacted.

In engine category II, flexible arrangements are often used, not only for the propulsion engine, but also for the gearbox.
In propulsion engines, the engine is frequently mounted on flexible elements. The engine and the generator in gensets are rigidly mounted on the bedplate and the bedplate is mounted on flexible elements.

In engine category I, flexible arrangements are often chosen to improve comfort on dinghies and yachts. The propeller thrust is usually absorbed by a thrust bearing.

▲

The arrangement of two diesel-power aggregates, engine category II.

Scania diesel engines drive the electric generators on a feeder-containership.
The engine and the generator are mounted on flexible elements.

▲

A small diesel engine (Deutz) for a boat or small vessel, engine category I.

System: Vetus, Schiedam, The Netherlands.
The flexible elements ensure that transmissions of vibrations to the hull are minimal.

1 *flexible elements*

▶

The flexible arrangement (1) of a genset from Mastervolt.

Obviously, a vibration- and soundproof arrangement is a prerequisite.
Mastervolt, Drachten,
The Netherlands.

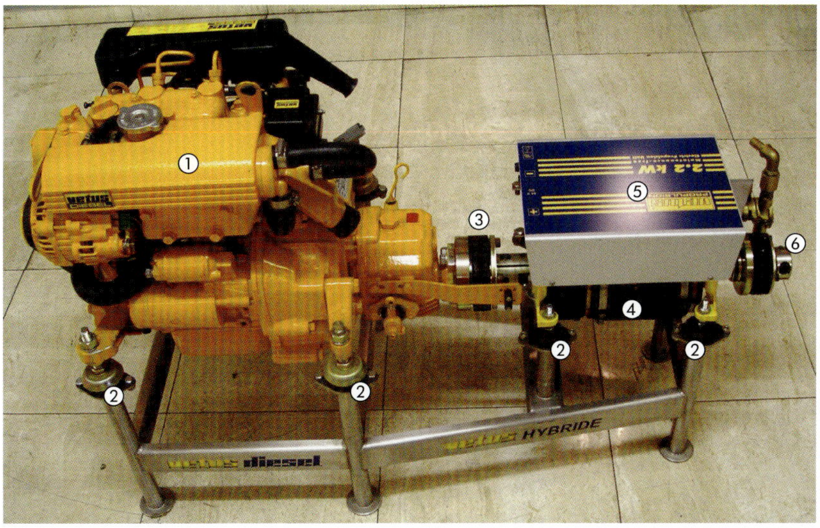

◄

A hybrid arrangement from Vetus, Schiedam, The Netherlands.

A small, flexibly arranged diesel engine, category I, drives a flexibly arranged generator. A battery is placed on top of the generator.
A flexible coupling between the diesel engine and the electric generator/electromotor.

1 *diesel engine, category I*
2 *flexible elements*
3 *flexible coupling between the engine and the generator/ electromotor*
4 *generator/electromotor*
5 *battery*
6 *flexible coupling between the electromotor and propeller*

30.6 Engine alignment

The alignment of propulsion systems with engines is crucial and vital in preventing damage to the engine system.

Objective:

- To prevent damage to shafts, bearings, couplings, gearboxes and similar components.
- To reduce the vibrational levels in the engines, bedplates and their surroundings.
- To reduce the noise generated by these vibrations.

30.6.1 Alignment of stern tube, gearbox and four-stroke diesel engine during construction of a new ship

30.6.1.1 Standard method

When a ship is under contruction, the components are, when possible, placed in their final installation position.
During the construction of the engine room, the engines, gearboxes and generators are placed on their bedplates.
The available space on both sides of these components is evenly distributed.
The engine is positioned as described previously and then the gearbox is installed to the custom build and provisionally aligned.

◄

Line boring a stern tube.

This is firstly aligned to the best possible alignment and then welded in place. The line-boring machine is then aligned, so that the stern tube can be drilled to the correct size.

Once the gearbox is aligned, the shaft generator can be installed in relation to the gearbox. Until now, the space has been evenly distributed, dependent on the accuracy required in each of the sections.
When the rear section, containing the stern tube and brackets is installed, the position is determined and checked using 3D-measuring methods and/or lasers during the entire welding process. The reference for positioning the rear section is the outlet flange on the gearbox.

After welding the rear section, a reference line is established to determine how the various bearings must be positioned. The positioning of the bearings is determined with a propeller-shaft alignment calculation during the engineering phase of the ship's manufacture. After positioning the bearings, the installation of the propeller shaft can continue. The propeller-shaft flange is the reference for the final alignment of the outgoing gearbox shaft.
In the aforementioned calculation, a so-called GAP/SAG value is obtained. The gearbox is aligned within a tolerance of +/– 0.05 mm of this GAP/SAG value. GAP is radial deflection, SAG is axial deflection.

After mounting the gearbox, the alignment of the engine can begin. The engine flywheel is aligned in relation to the input shaft of the gearbox. For this alignment, a GAP/SAG value is also calculated in advance. The engine is aligned within a tolerance of +/– 0.05 mm of this GAP/SAG-value.
Subsequently the crankshaft deflection is measured. The combination of the GAP/SAG values and the measured crankshaft deflection determines if the engine is correctly aligned. When correct, mounting and fastening of the engine can be completed.

▶
When aligning shafts, a GAP-SAG analysis is always performed.

GAP: Opening, deviation of the neutral shaft
SAG: Sag, deviation from the horizontal centre axis.

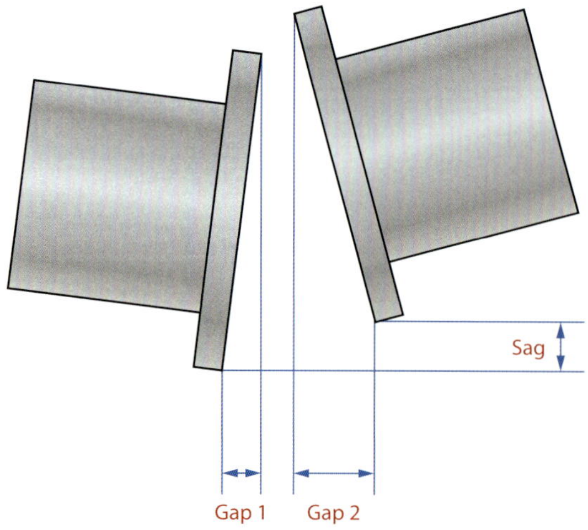

Positive Sag
Negative Gap 1
Positive Gap 2
Gap = –Gap 1 + Gap 2

Once the engine has been aligned, the shaft generator can be aligned with the same GAP/SAG-principle.
When the installation is complete and all the components are connected, the factual bearing load at each bearing position can be established by weighing the shaft in combination with strain-gauge measurements.
The values measured must correspond with the values calculated in the engineering phase.

Alignment of an existing engine that is rigidly mounted

There may be various reasons for the re-alignment of the engine and gearbox. The flywheel is aligned in relation to the input shaft of the gearbox. A GAP/SAG value is calculated in advance. The engine is aligned within a tolerance of +/– 0.05 mm of the calculated GAP/SAG value. The crankshaft deflection is then measured. The GAP/SAG values in conjunction with the measured crankshaft deflection ultimately determine if the engine is correctly aligned. If this is the case, the main engine can be mounted and affixed to the bedplate. The alignment and the crankshaft deflection are re-checked.
After the classification agency approval, the flexible coupling is installed.

Alignment of an existing engine with flexible mounts

Elaborate installation instructions must be provided by the engine supplier for the alignment of an engine on flexible elements.
These instructions detail the process of engine alignment. These instructions can vary per flexible element type.
The flexible elements must be first be loaded for 48 hours after installation in order to obtain the correct measurements. The elements must be levelled with respect to the front and rear within a certain tolerance (an indication: within 1 mm). Compression of the various elements is also restricted (an indication: within 2 mm).
It is recommended that the engine be filled with water and oil to achieve an optimal alignment. Once the components are aligned, the flexible elements can be fixed.

Some data with regard to damage ensuing from inaccurate alignment

Damage as a result from incorrect alignment may vary from damage to the white metal in the sleeve bearings, breakage of the couplings, and generation of severe vibrations to crankshaft breakage.
The bearings for the different components are monitored with temperature sensors. As soon as the temperature rises, an alarm signal is given. Vibration and/or temperature increases in the engine can be indicative of an alignment problem.
It is self-explanatory that not all vibrations or temperature-related problems are created by incorrect alignment.

30.6.1.2 New methodology

ShaftDesigner: Software for shaft alignment and vibration calculations

ShaftDesigner was developed by Intellectual Maritime Technologies in the Ukraine in cooperation with SKF Marine Service Centre B.V. in Ridderkerk, The Netherlands, for part of their service division in order to validate the theoretical and practical measurements.
One of this system's greatest advantages is that the theory can be cross-validated en vice versa. This can be done using the Reverse Engineering module.
The software calculates the theoretical values for a jack-load, which can then be used for comparison to the measurements done during alignment of the installation. The reverse is also possible, actual jack-load measurements taken can be entered into the software and this data can then be used to calculate the actual alignment and which modifications should be made to correct the alignment of the installation.
Working with this software gives clarity as the 3D graphical environment allows for quick and easy visual checks after values have been entered.

ShaftDesigner is the first software package that works with a central base model for the calculation of both alignment and vibrations. Previously, these calculations were done in separate programmes and the values had to be re-entered for different conditions, a laborious procedure!

This is useful software for the design, construction and maintenance of ships.

Design of a ship

In the design phase, ShaftDesigner assists in calculating the optimal positions of all components, including the main engines, gearboxes, bearings, propellers and generators on the shaft line. The easy drag and drop function of ShaftDesigner allows a relatively fast and easy placement of all components.
In the engineering stage, ShaftDesigner provides a useful tool to calculate an optimal alignment and good bearing-load distribution. Also, in this design stage, it is possible to calculate if a certain lay-out has the potential to create harmful vibrations. As ShaftDesigner can identify these issues in the ship's design phase, possible time-consuming and costly (re-)alignment problems and vibrational problems can be avoided during trials. In the design phase, the optimal placement of components can be calculated. Using several criteria, such as, acceptable bearing loads and offsets between the shaft and bearings, an optimized alignment is determined.

Construction of a ship

If ShaftDesigner was not used during the design phase, the use of this software can still be used in the construction phase and give significant benefits. Designers benefit from the drag and drop capabilities, ship builders profit from the fact that correct measurements can be entered into the software, giving an accurate representation of the

◄ A basic 3D model of the propeller shaft line.

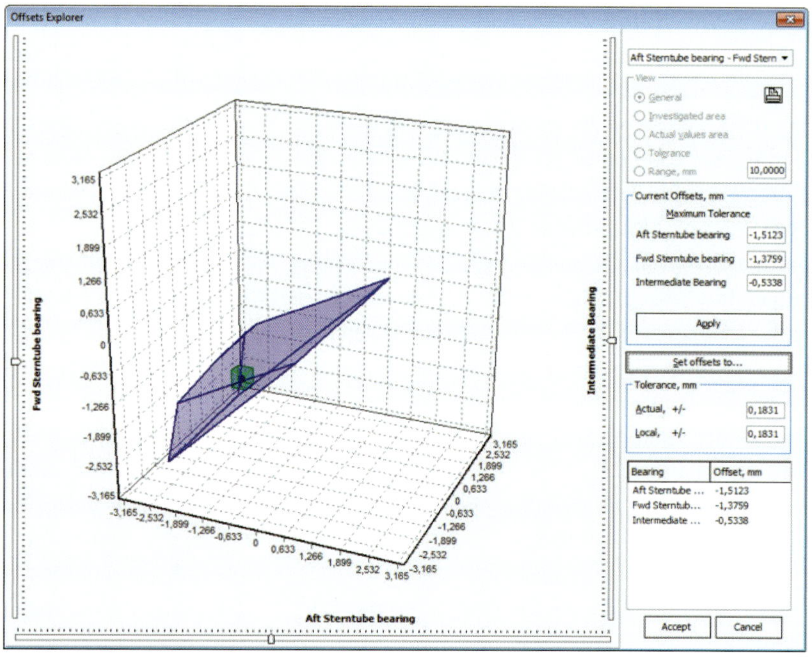

▶

A 3D representation of the allowed layer deviations/ offsets during alignment optimization.

entire shaft line. This is automatically updated whenever a new entry is made. ShaftDesigner supports various shaft alignment techniques guaranteeing the shaft alignment. These techniques include jack-loads, GAP/SAG and strain gauge alignments. These can be used at various stages of the entire installation giving an optimized shaft line.

Maintenance and repair phase

ShaftDesigner is also convenient for maintenance and repair purposes. Creating a model of the shaft line is an easy way to subsequently calculate the alignment and bearing loads. Stress loads of bearings can be determined and thus checked for wear on board, avoiding problems during operation. This system can be used to verify shaft alignment after a ship is grounded or other damage. This can be done on board by measuring the alignment and entering these values into the system as a comparison to the theoretical model. The shaft alignment module differentiates itself by having various advanced functions, such as the possibility to optimize the positioning of bearings, the simultaneous analysis of multiple operating conditions, bush calculations with hydrodynamic lubrication, and automatic calculation of the propeller load and gearbox loading in various conditions.

Additional to these functions, the Shaft Alignment module has reverse-engineering capabilities making it possible to calculate the positioning and loads for the bearings. Additional modules allow various vibration analyses to be performed. There are different modules for the calculation of bending vibrations, axial vibrations such as those found in thrust bearings and torsional vibrations of the entire shaft line. ShaftDesigner makes it easy to control the theoretical alignment. Jack-load and strain gauge measurements taken on board can simply be entered into the software. These values are then used for an optimal alignment. Thanks to the 3D graphics, it is easy to find the cause of the alignment problems and to calculate how these can be solved optimally.

▼

A 3D representation of the pressure distribution of a bearing with hydrodynamic lubrication.

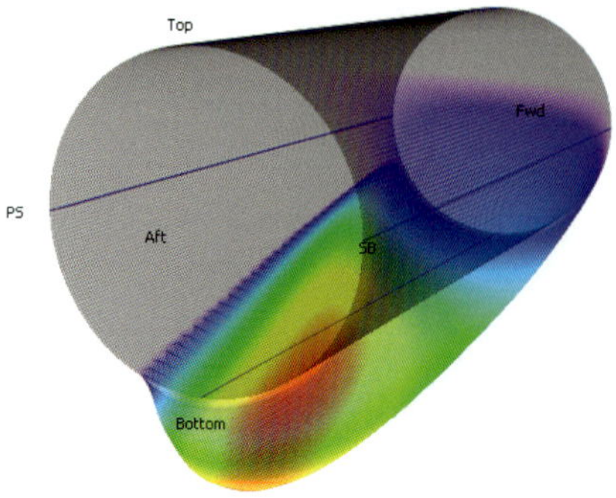

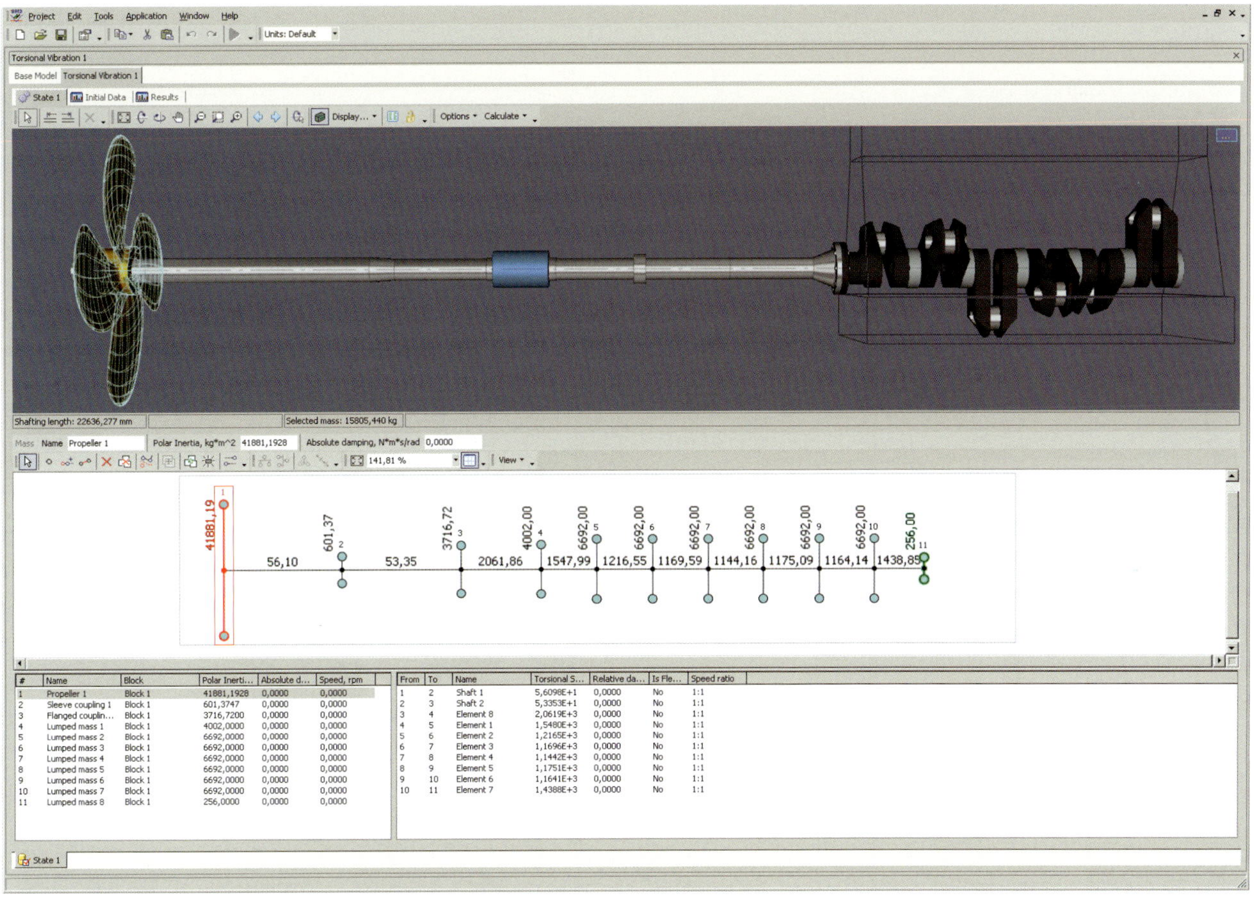

▲

This system can generate calculation models for torsional vibration in the crankshaft, intermediate shaft and propeller shaft. Shown here, for a six-cylinder two-stroke crosshead engine.

Important properties:

- All values entered can be checked visually in a 3D graphical environment in which it is possible to drag and drop all components. These components can be viewed from any angle in detail.
- Models can simulate different operating conditions from ballast to laden, from 'in dock' to 'at full power'.
- All calculations are taken from one base model with the advantage that values only have to be entered once.
- Any changes to the base model are automatically updated in all modules.
- The Graphical user interface can be customized to show the items one requires, making it multi-functional and allowing a clear overview.
- Calculations are done in horizontal as well as vertical plane simultaneously making the results easier to analyse.
- The results of the different calculations are compiled in detailed reports that can be customized so that all information needed by the user is represented. To facilitate analysis of the reports, the software automatically generates graphical outputs of the results. All reports are created as an XML document so they can easily be exported to various formats, such as html, Microsoft Word or Windows Web Archive.

▼

A simulation for hydrodynamic lubrication and for a bearing in a propeller shaft tube and the representative lubricating-oil pressure.

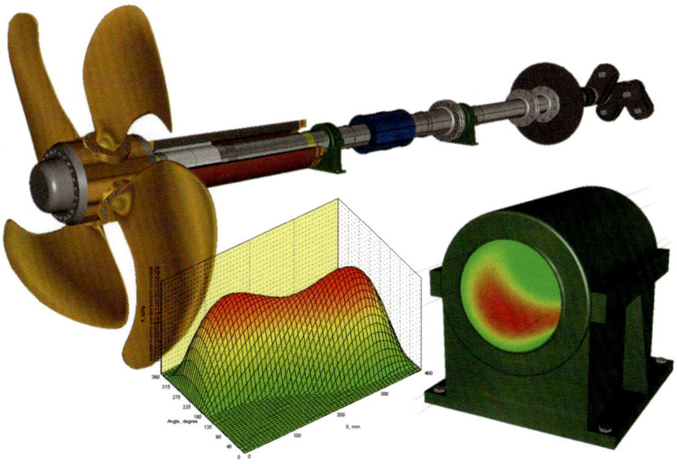

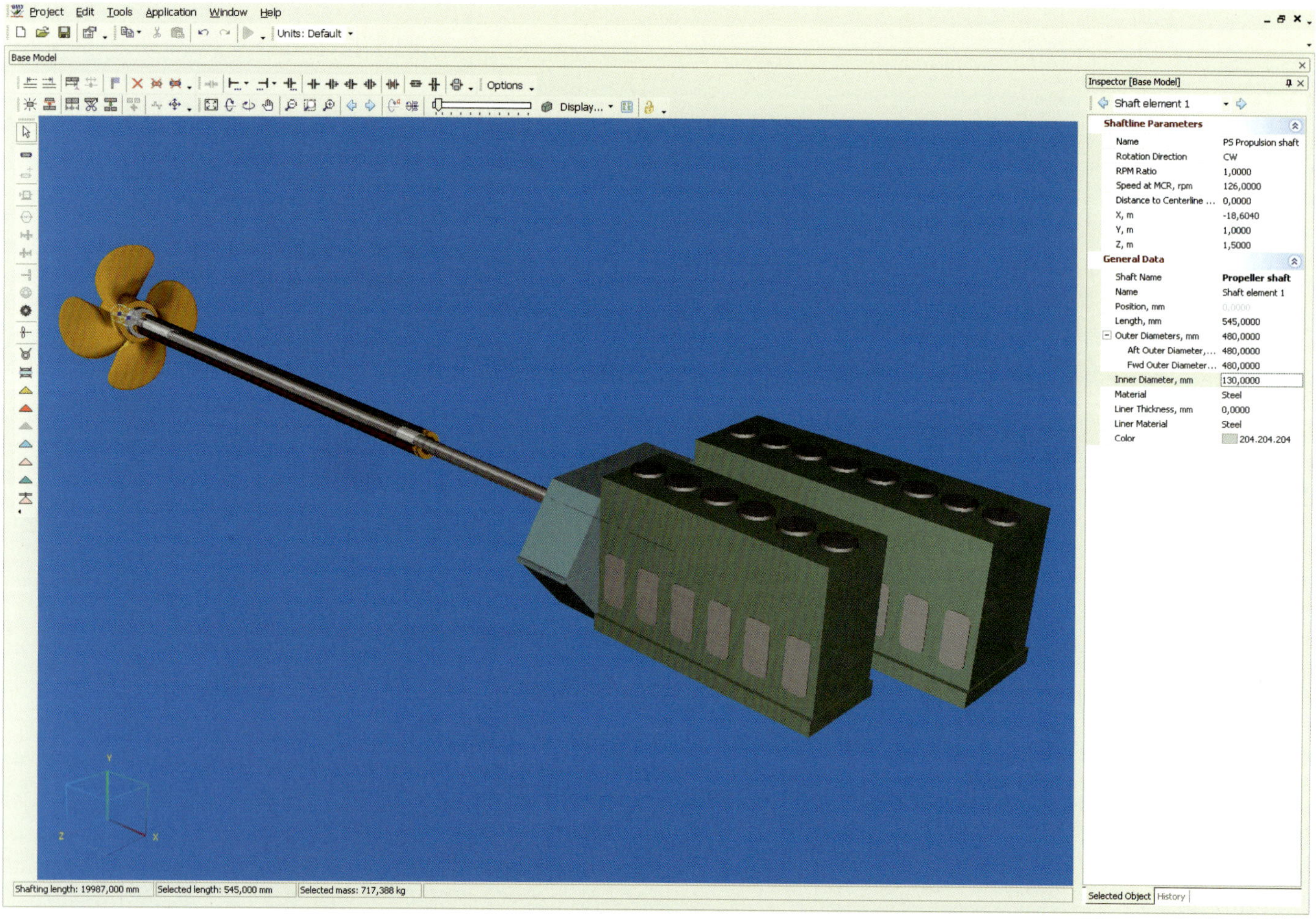

▲

Model for double engine installation that drives a propeller via reduction gears, which has a lower rpm than that of both engines.

▶

The lubricating-oil pressure in the stern tube. The lubricating-oil pressure is given in MPa. The maximum lubricating-oil pressure here is approximately 1.7 MPa = 17 bars.

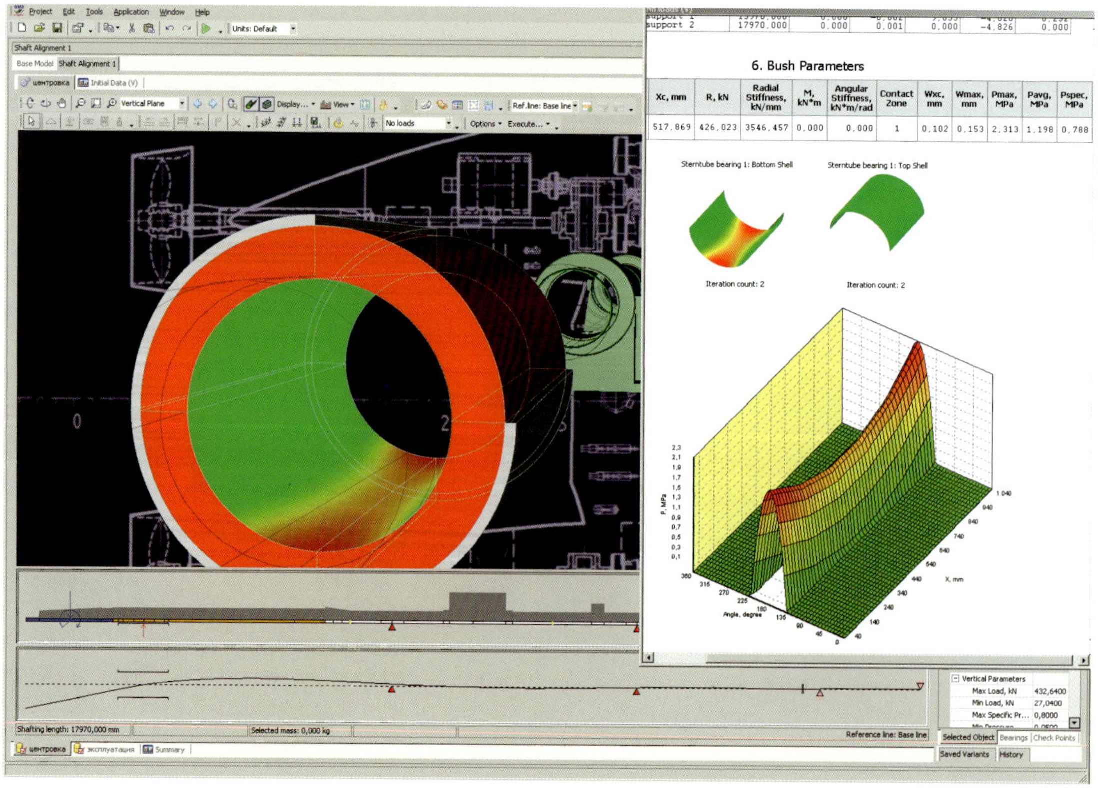

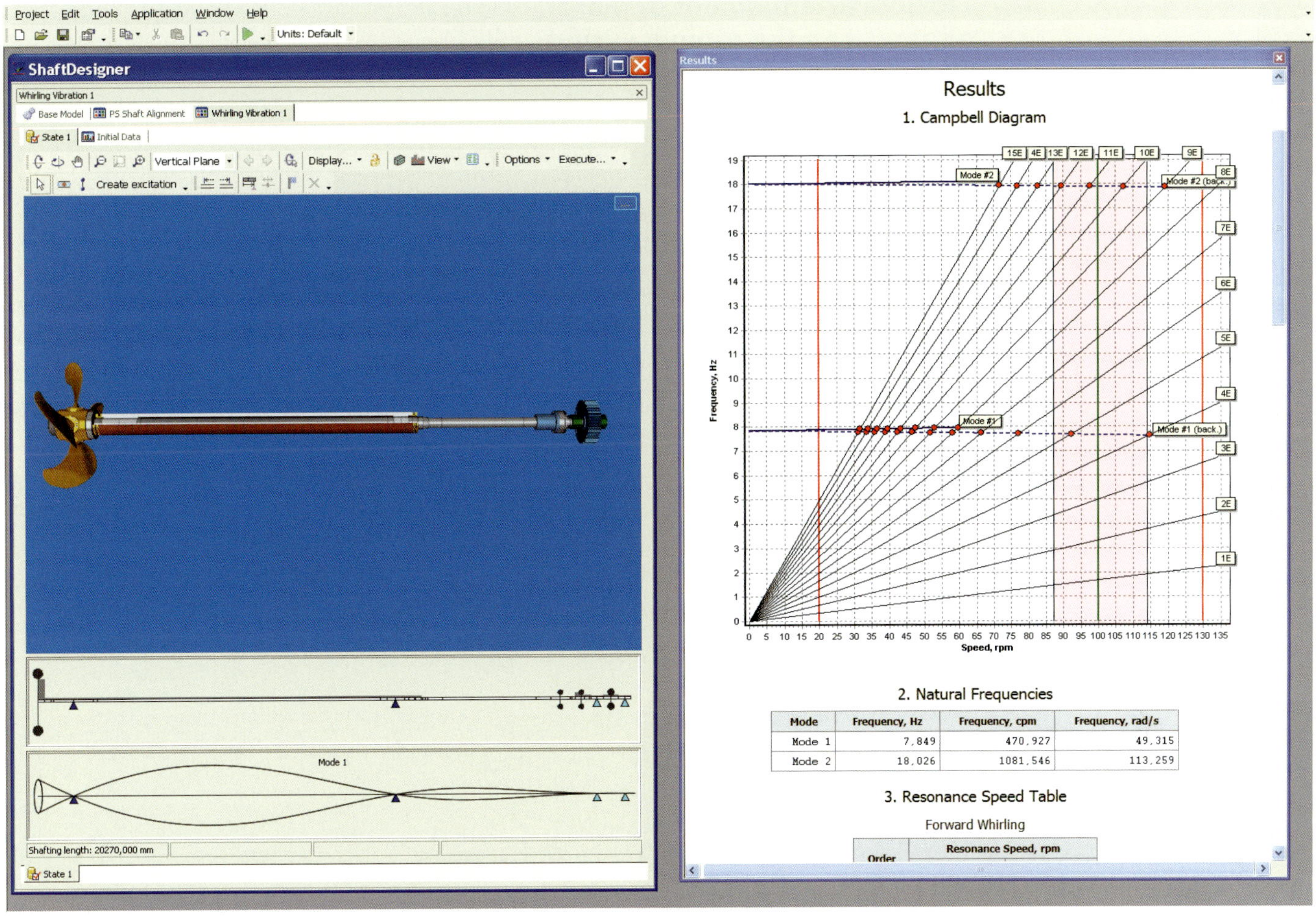

▲

Left above: An example of a propulsion system with the propeller shaft in the stern tube and to the right; a gear wheel of the reduction gearing.

Left below: The propeller with four support points and four masses.

Far below: Bending of the shaft.

Right: A Campbell diagram showing how the torsional vibrations deviate for different shaft speeds.

The red points show where possible resonances can occur.

30.6.1.3 Key considerations

- Ship structures distort during operation.
- Components 'grow' during operation.
- These above-mentioned factors influence the alignment.
- There are many factors that influence alignment with flexible mountings.
- Performing regular alignment check-ups extends engine life.
- Generally, flexible couplings have better absorption capacity than the component bearings between which they are mounted.
- Flexible couplings are used as torsional-vibration compensators, not for compensating alignment errors.
- Economising during construction does not often compensate for the cost of repairs during operation. The factual alignment of the propeller shaft can be precisely determined using a combination of laser measurements, shaft-weight and strain-gauge measurements.
- Rapid re-adjustment of the components after an alignment check saves money.

30.7 Flexible arrangement of diesel engines and the piping, wiring and other equipment that are attached to the engine

Large four-stroke trunk-piston engines are also often mounted on flexible rubber strips to limit vibration- and sound transmissions to the hull, and therefore the ship's accommodation.
Modern feeder-containerships that have the propulsion engine and accommodation aft are sensitive to vibrations and noise nuisances.
The high engine-power output, the slim stern, the light hull construction and the tall narrow superstructure with the rotating propeller below are not conducive to a quiet and vibration-free environment!
This means that the various pipelines attached to the diesel engine must be supplied with flexible connections.

Piping and fittings around the diesel engine

The exhaust-gas pipes in a rigidly mounted diesel engine have flexible compensators in the piping; this allows them to move freely during fluctuating temperatures. Vibration-related problems in diesel engines often cause tearing behind the turbocharger, towards the sound damper. In smaller and flexibly arranged engines or engines, which have a bedplate mounted on flexible elements, all pipes and wiring (ducting) connected to the engine and/or moving bedplate must be taken into account. There are a large number of techniques available to prevent damage or tearing for the fuel, lubricating oil, starting air, exhaust gases and cooling water systems. It is advisable to check periodically the deflection in a certain point on the engine in relation to a fixed point in a flexibly arranged diesel engine.
If, for instance, the fuel injection for a six-cylinder in-line engine functions inadequately, this can usually be found by measuring the deflection.

▶ The stern of a feeder-container ship.

The accommodation is located astern. The propulsion engine is situated directly below the rear containers. The propeller is situated directly below the accommodation. This construction is conducive to vibrations.

▶ Two spare exhaust-gas piping compensators for a two-stroke crosshead engine.

30.8 Mounting methods for propulsion engines and other components with critical alignments

The following filling methods in the alignment of diesel engines, gearboxes and generators are applied:
- with steel chocks;
- with steel shims;
- with a resin chock, often epoxy resin such as Epocast 36;
- with adjustable chocks.

30.8.1 Steel chocks

Advantages
- Relatively inexpensive.
- Easy to install.

Disadvantages
- It is prohibited to place them below propulsion engines.
- They do not fill the space sufficiently.
- Too many contact points are created.

30.8.2 Epoxy resin such as Epocast 36

Advantages
- Provides all-over support, both for the bedplate and the engine flange.
- Cheaper than steel chocks and less time-consuming.
- Reduces high-frequency vibrations.

Disadvantages
- Time-consuming and more expensive compared to adjustable chocks.
- Hardening of the epoxy resin takes relatively long; depending on the ambient temperature, 24 to 48 hours.
- The method is permanent. The engine cannot be re-positioned.
- Special packaging material is required for air transport.
- Installation must be performed by specialists.

30.8.3 Adjustable chocks

Advantages
- In height adjustable.
- Self-levelling, therefore they can accommodate angular differences in mounting surfaces and can compensate errors.
- Adjustable during the life cycle of the machinery.
- Accurate mounting for engine alignment is simple and quick.
- Expensive milling of the bedplates is not required.
- No extra work required for the bedplates.
- There are adjustable chocks available in all sizes and shapes.
- Minimal height for confined installation spaces is 20 mm (dependent on the type).

◄ Compensators for, amongst others, cooling-water pipelines and exhaust-gas piping required for the flexible arrangement of diesel engines.

▶

A cross section of an adjustable chock, brand 'Vibracon'.

The chock is self-levelling and therefore can accommodate angular differences in both contact faces.

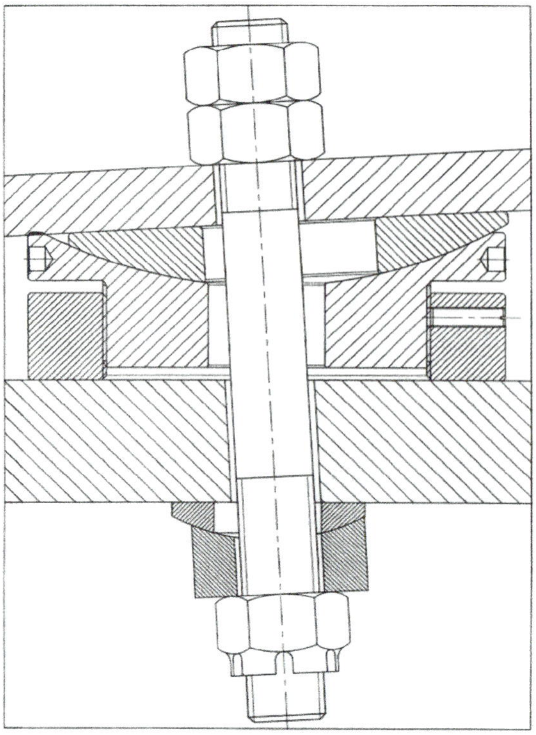

▶

An adjustable chock from 'Vibracon'.

The height is adjustable and a spherical washer compensates for the angular upper face.

▶

Generator and diesel engine installed on a fixed bedplate and supplied with adjustable chocks from Vibracon.

▶

A gearbox mounted on adjustable chocks.

▲ Adjusting the chocks underneath the diesel engine.

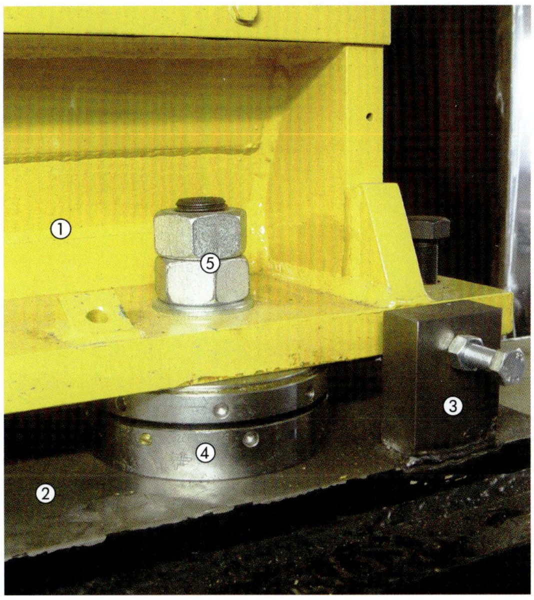

▲ An adjustable chock.

1 *engine frame*
2 *bedplate*
3 *adjusting bolts for alignment*
4 *adjustable chock*
5 *holding-down bolt*

◀ Vibracon Low Profile chocks below Stork gearboxes and high-pressure pumps driven by a Caterpillar 3512 diesel engine, Category II.

◀ An electric generator (yellow) between a Yanmar diesel engine and gearbox with reverse clutch.

The power output at a voltage of 230 volt and frequency of 50 Hz is 3 kW; ideal for a moderately sized yacht.

▼ The design of a Vibracon Low Profile adjustable element.

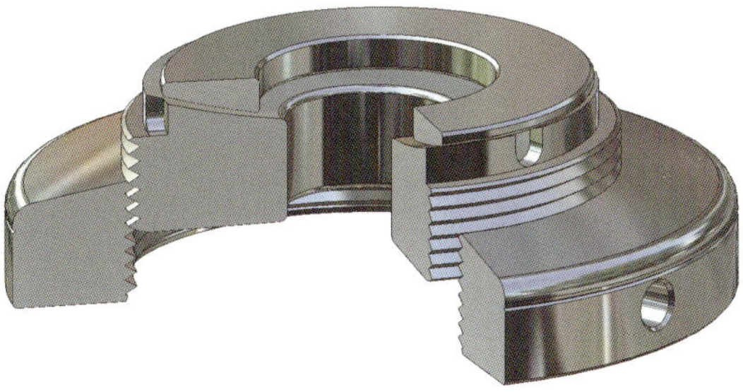

>CH 31

Propellers

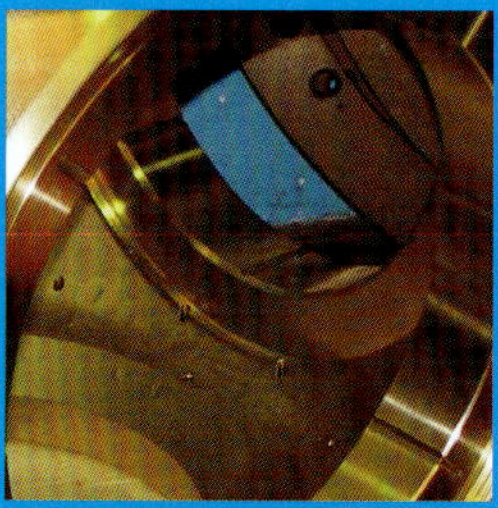

A large controllable-pitch propeller.

One of world's largest propellers weighing 85 tons;
the fixed-pitch, six-bladed propeller of a large container ship

31.1 Introduction

Most marine propellers are manufactured of copper alloys. However, for inland shipping, small ships and yachts, propellers are also made of cast iron, aluminium alloys, stainless steel and numerous synthetic compounds.

31.1.1 Large fixed-pitch propellers

Large fixed-pitch propellers are cast in a mould and then machined.
Apart from the accurate finish of the propeller hub, a smooth finish of the blades is also essential; the propeller must be able to have the lowest possible drag when rotating through water.

Small propellers can be cast in large series and sometimes only weigh several kilograms. Larger propellers are individually cast and weigh up to 140,000 kg!

▲
Fixed-pitch propellers and propeller blades being cast.

A large foundry at the Hyundai engine works and shipyard, South Korea. The unfinished fixed-pitch propeller sometimes weighing 100 tons must be cast rapidly in a single casting in order to obtain a homogeneous structure.

▶
Propeller blades for controllable-pitch propellers at the MAN Diesel engine works in Frederikshavn, Denmark. They also manufacture two-stroke crosshead engines with cylinder bores up to and including 500 mm, as well as four-stroke engines up to 2011. The blades are of the 'high-skew design' and highly polished.

▶
The magnified surface of a propeller blade.

The smoother the surface, the less the resistance of the rotating propeller through the water.

▲
Machining a large six-bladed fixed-pitch propeller for a container ship.

Note the approximate 4 metre supports for the propeller blades. These prevent bending during the machining processes.

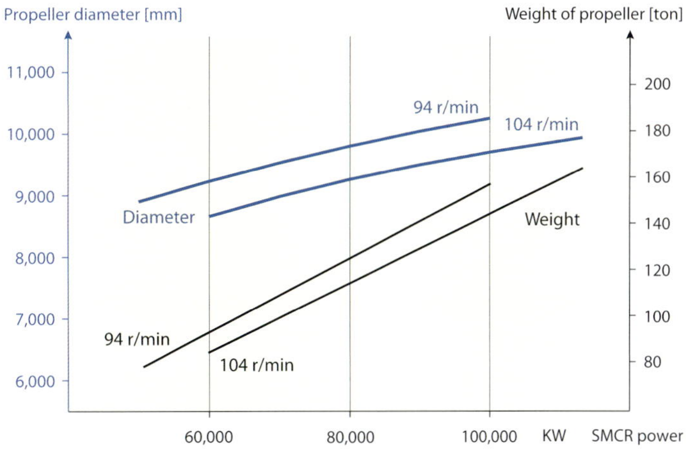

▲
A table for propeller sizes.

horizontal: shaft power of the engine in kW

Example: At 80,000 kW shaft power and an rpm of 94, the propeller diameter is approximately 9700 mm and it weighs 130 tons!

31.1.2 Accuracy of the finish

It is important that the hub and the blades are finished with the utmost precision. This is required to ensure optimal operation, as well as preventing the generation of additional vibrations, which would negatively affect the propulsion of the ship.

The mechanical properties of Cunial	
Mass density	7650 kg/m^3
E-modulus (at 20 °C)	121000 N/mm^2
Poisson's ratio	0.33
Yield stress	min. 250 N/mm^2
Tensile strength	min. 650 N/mm^2
Elongation (longitudinal)	min. 18%
Coefficient of thermal expansion (from 20 to 100 °C)	16-10^{-6} K^{-1}

◀
The mechanical properties of the propeller material 'Cunial'.

31.1.3 Materials used in the manufacture of large propellers

Material requirements for propellers:
- corrosion resistant;
- sturdy ;
- ease of casting;
- easily machined.

A well-known alloy is 'Cunial', an alloy with copper, nickel, iron and manganese. 'Cunial' represents the chemical symbols for the elements: Cu-copper, Ni-nickel and Al-aluminium.

The raw material for large propellers is often 'old' coins, essentially a copper alloy.

Material	NiAl	CrNi
0.2 proof stress	min. 250 N/mm^2	min. 380 N/mm^2
Tensile strength	590 – 780 N/mm^2	600 – 790 N/mm^2
Elongation	min. 16%	min. 19%
Impact strength Charpy V notch	30 J	21 J
Brinell Hardness	min. 150	240 – 300

◀
Comparison of Cunial 625 with other propeller Cunial 625 and materials,13/4 chromium-nickel-steel.

31.2 Fixed-pitch propellers

Advantages

- Simple and relatively inexpensive to manufacture.
- Very strong.
- High efficiency due to relatively small hub.
- During docking, the propeller is often stationary and therefore cannot be damaged by obstacles. It is also unlikely that a hawser will become entangled in the propeller.

▶
A fixed-pitch propeller fitted hydraulically to the hub. A key seat as locking device is no longer required.

Disadvantages

- The propeller must be driven in **two rotational directions**:
  1. either the propulsion engine must be reversible (large plants) or
  2. a reverse gear must be mounted between the propulsion engine and the propeller (smaller plants).
- In extreme conditions, such as heavy-weather conditions with head-on winds and high swells, the propeller can become heavily loaded.
- During manoeuvring with large plants, the maximum speed is generally the 'manoeuvring' speed, approximately 80% of the full-load speed.

31.2.1 Use of fixed-pitch propellers

Fixed-pitch propellers are often used in the large power-output category IV engines, as well as smaller power-output engines in categories I and II.

Engine category IV

Large container ships, bulk carriers and oil tankers are usually supplied with fixed-pitch propellers.
The number of times annually that manoeuvring is required for these vessels is limited, and with the use of bow and stern thrusters (container ships) and tugs (container ships, in particular bulk carriers and oil tankers), manoeuvring when arriving or departing from port is facilitated.
Due to the size of the ships in this category, adverse weather conditions hardly affect the hull and consequently the engine power output.
The engines are always low-speed two-stroke crosshead engines with such a low speed that reduction gearing is not required.
All the engines are directly reversible.

Also, see Chapter 14, *Starting systems of diesel engines.*

▶
The tunnel shaft between the two-stroke crosshead engine (background) and the propeller shaft (not visible in the foreground).

A reversible diesel engine allows a high-efficiency fixed-pitch propeller to be installed. A simple and common propulsion system for large ships.

A change in the rotational direction can be done quickly using a free wheel, which disengages the propeller from the drive shaft. For very small vessels, such as yachts and cabin cruisers, this is standard. Vessels that often change direction use a similar system, as it is user-friendly and simple to operate.

▲

A relatively small two-stroke crosshead engine from MAN Diesel, type ME.

This MC 50 can drive a fixed-pitch propeller and is directly reversible. However, a controllable-pitch propeller is often installed. This engine has a common-rail system.

◄

A small four-stroke high-speed diesel engine ('Vetus') with an angle gear set through the yacht's hull.

The reverse coupling is located directly behind the engine. The propeller is a foldable propeller type: during sailing, it folds back due to the flow of the water over it providing minimum drag when sailing.

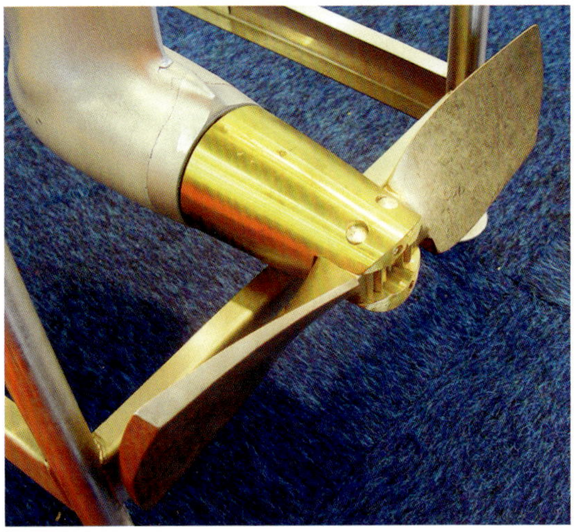

◄

A foldable propeller in operational setting. The rotating blades fold outwards due to the centrifugal forces.

Engine categories I and II

These four-stroke high-speed diesel engines running on MDO usually have a fixed-pitch propeller with a reverse gear to change the rotational direction of the propeller.

◄

A propulsion installation with a four-stroke high-speed Caterpillar diesel engine running on MDO. Used for inland navigation.

Reduction gearing with a reverse coupling connected to a fixed-pitch propeller is installed at the rear of the engine (back).

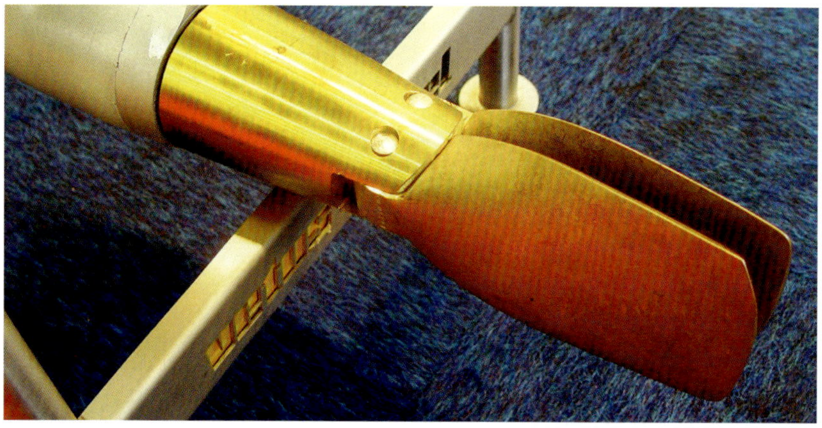

▶ The propeller shaft is stationary. The yacht sails. The propeller blades fold back. There is minimum drag from the propeller.

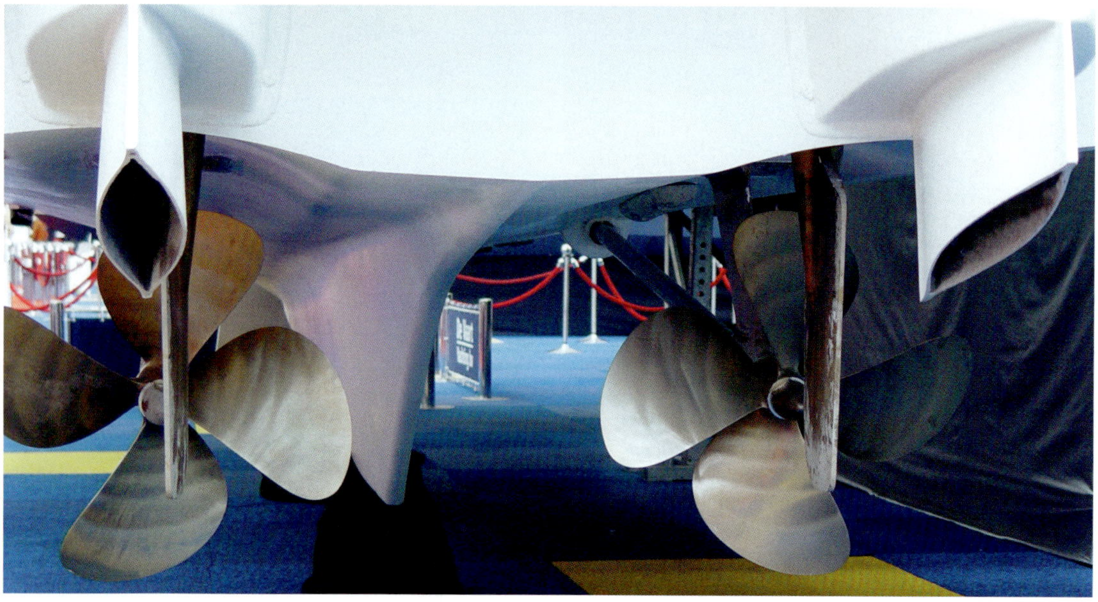

▶ A yacht with two fixed-pitch propellers.

Both four-stroke high-speed diesel engines are installed with reduction gearing and reverse gearing.

31.3 Use of controllable-pitch propellers

The position of the propeller blades can be gradually modified so that with the same rotational direction of the propeller, it can be steered from full speed ahead via the 'zero feathering position' to full speed astern. (Its operation is explained later in this chapter).

Advantages of controllable-pitch propellers

Controllable-pitch propellers are widely used and have many advantages.

- It allows for rapid manoeuvring, from full speed ahead, to full speed astern; the propulsion engine continues to run in one direction at the same nominal speed, and can therefore rapidly be heavily loaded.
- Every speed can be achieved without having to stop the engine.
- The average efficiency for ships with alternating loads, such as tugs, fishing boats and dredgers can be improved by adjusting the position of the propeller blades to the load ('pitch mode').
- With the correct reduction gearing, the engine speed can be reduced to the propeller speed which has the highest efficiency. This is effected by placing an electric-shaft generator drive on the reduction gearing.
- It is possible to substitute a damaged propeller blade under water, if required.

Disadvantages of controllable-pitch propellers

- Propeller efficiency is lower than that for fixed-pitch propellers due to the relatively large hub, which contains the mechanism for adjusting the blade pitch and the bearings.
- They are more prone to damage by objects in the water than fixed-pitch propellers.
- On average they are three to four times more expensive than fixed-pitch propellers.
- The hydraulic system can fail and requires maintenance.

◀
A four-bladed controllable-pitch propeller. Note the shape of the blade: the 'skew' model.

This minimizes the forces on the actuating mechanism inside the propeller hub and decreases cavitation.

31.3.1 Use of controllable-pitch propellers

These are used in three engine categories.

Engine category II

For the larger four-stroke high-speed diesel engines running on MDO, they are mostly used for engines with higher power outputs.
Is this is not the case; a reverse gear is installed in the reduction gearing.

Engine category III

The controllable-pitch propeller is extensively used in this category for four-stroke medium-speed diesel engines running on HFO. Furthermore, a large number of feeder-container ships, chemical tankers, cargo carriers, ferries and passenger ships are equipped with controllable-pitch propellers.

▼
The most common propulsion plant with controllable-pitch propeller: a medium-speed diesel engine with reduction gearing and shaft generator.

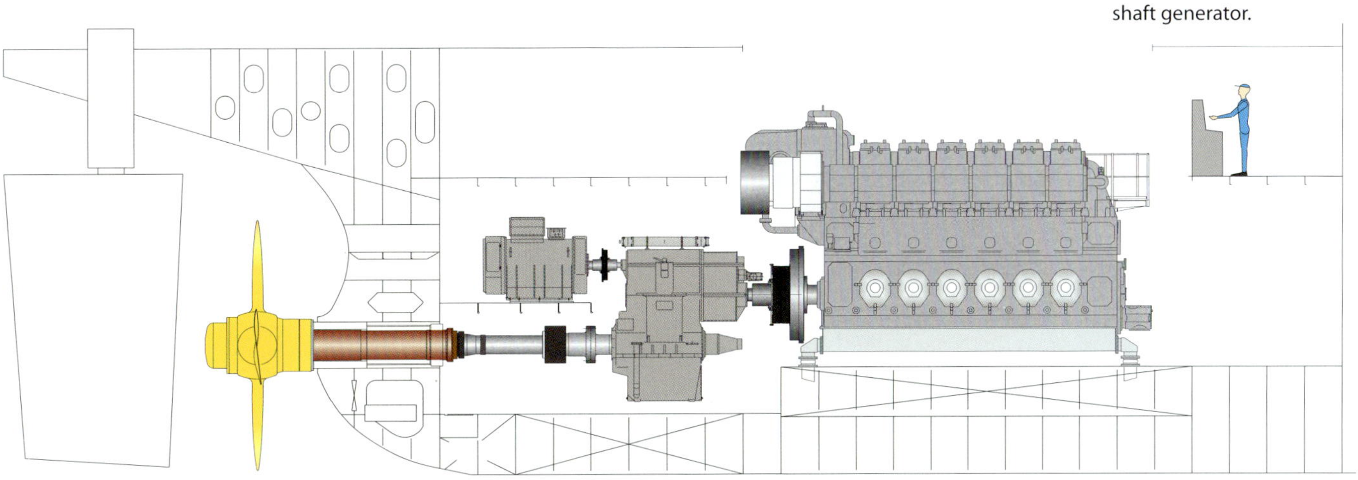

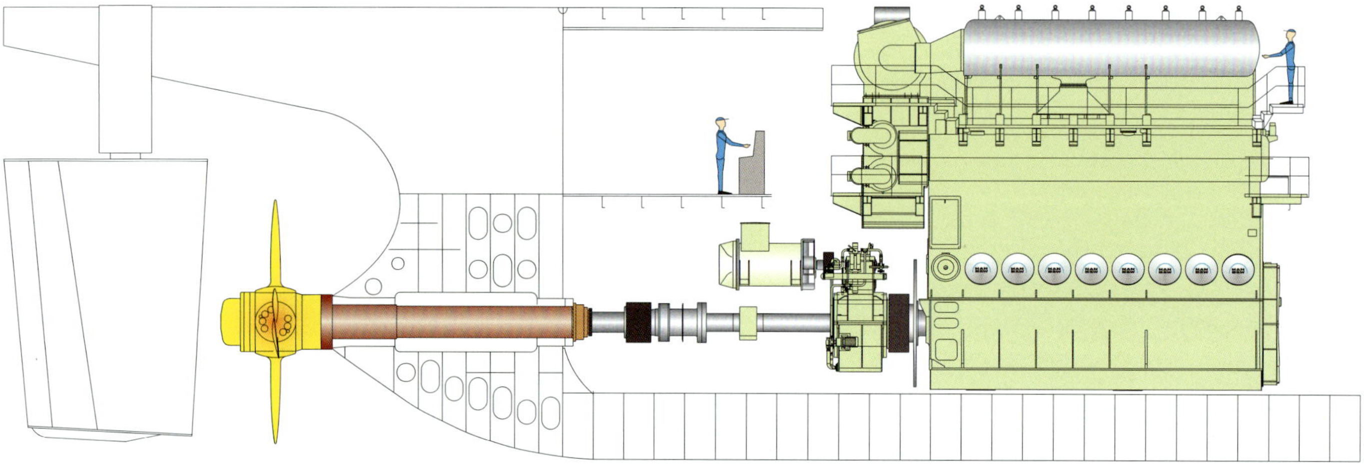

▲

A smaller type of two-stroke crosshead engine with shaft generator and controllable-pitch propeller. Reverse gearing, common in these types of engines, is not required.

Engine category IV

In this category for two-stroke low-speed crosshead engines with small bores, complete propulsion systems with intermediate shaft, shaft-generator drive, propeller shaft and a controllable-pitch propeller are often installed.

Engine manufacturer MAN Diesel supplies these systems up to and including a 700 mm cylinder diameter.

31.3.2 Propeller shapes

For each ship type, propellers are designed to provide the best efficiency, produce low noise and generate minimal vibrations. The propeller shape is a decisive factor. **The propeller is custom-made for every ship!**

Several examples

Example 1: Large container ships

These types of ships travel huge distances each year, have large engine-power outputs and the fuel costs are a major component of operational costs, total propulsion efficiency is of utmost importance. A large two-stroke crosshead engine with a total efficiency of 52% and large diameter propeller with a total efficiency of 75% can achieve a total efficiency of $0.52 \times 0.75 = 0.39$ or 39%. This is **exceptionally high!** Therefore, 39% of the supplied fuel (100%) is eventually used to propel the ship at a certain speed. In all other propulsion systems, this is always lower and at times significantly lower.

Example 2: Ferries

Modern ferries must have low noise and vibration levels. Propellers are specially designed to achieve this. Manoeuvring is of paramount importance.

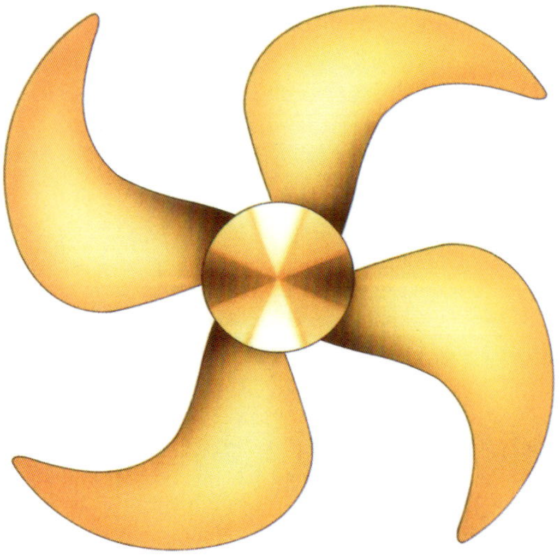

▲

The four-bladed fixed-pitch propeller of a large container ship. 'Skew' model blades.

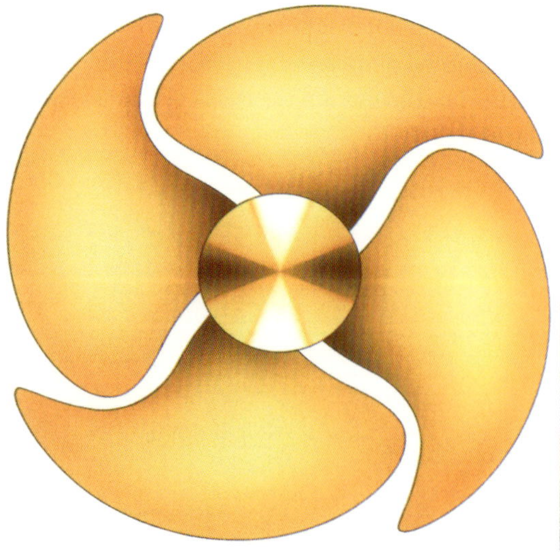

▲

The controllable-pitch propeller for ferries.

Example 3: Pontoons

Pontoons, which travel relatively short distances, often have two propellers port and starboard for efficient mooring and departure, thus allowing for the rapid transportation of passengers and vehicles. When both propellers are set in a position in which they do not provide propulsion, they must be placed in such a position that minimizes resistance.
This also applies to ships with two propellers, where one propeller is regularly taken out of commission.

Example 4: Reversible propeller for passenger ships

These are especially designed for maximum 'comfort', that is: low noise and vibration levels. The blade shape is referred to as the 'skew'-shape. The blades are bent in the rotational direction.

Example 5: Propellers for icebreakers

These are designed to comply with the highest ice classes stipulated by various classification societies. Furthermore, they must be able to withstand very high tensile forces ('bollard pull').

Example 6: Propellers for anchor-handling ships, tugboats and tenders

They are able to supply high tensile forces and also have good manoeuvrability even under extreme weather conditions.
A high torque is essential and is achieved with either a fixed tunnel thruster or a steerable thruster.

▲
The controllable-pitch propeller for pontoons.

▲
The controllable-pitch propeller for passenger vessels.

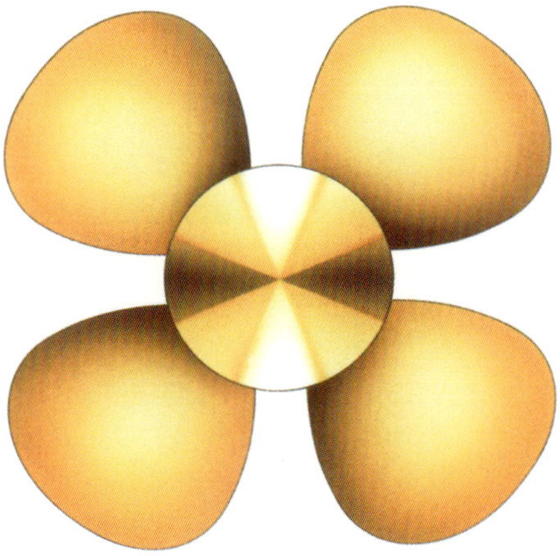

▲
A controllable-pitch propeller for icebreakers.

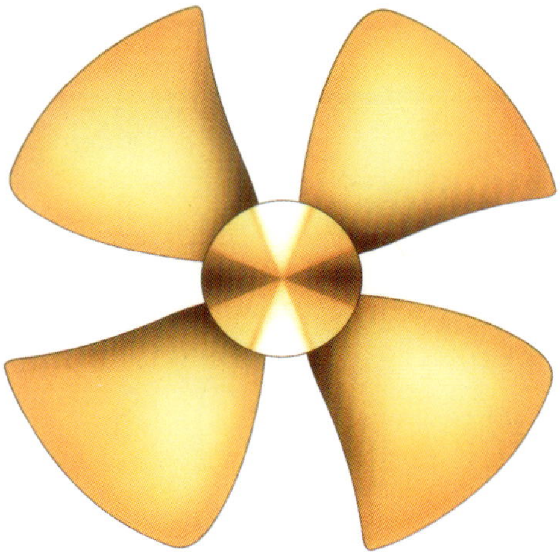

▲
The controllable-pitch propeller for various vessels, such as anchor-handling ships, tugboats and tenders.

31.4 Fixed-pitch propellers, construction

31.4.1 Traditional attachment

The traditional attachment of the propeller on the tapered end of the propeller shaft is with a keyed joint comprising a key and locking nut.
The taper is approximately 1:12, so the hub can be firmly affixed to the shaft. The key does not absorb the thrust, but is/was intended as security. When complications arise, it must absorb the entire engine torque.

▶ Disassembling a fixed-pitch propeller from a nozzle. The hammer wrench is in position.

The taper between the hub and the propeller shaft must be able to absorb at least twice the engine torque.
The keyed joint is in fact undesirable for the construction of the taper or conical connection. In practice, hair cracks and fractures may occur in this location. In small propellers of, for instance, pleasure craft, attention must be paid to this issue.

▼ A fixed-pitch, six-bladed propeller weighing 80 tons for the 6800 TEU Southampton class, a large container ship. The raw material used for manufacture is 'old coins'.

31.4.2 Propeller fixing without a key, a key-less joint

The key-less assembly is common for large propellers. The torque of this conical connection should be at least three times that of the diesel engine as this construction lacks the guarantee of a keyed joint. Large propellers are always hydraulically fitted to the shaft. This requires a meticulous working method. As security a locking nut, protected from the corrosive seawater by a cap, is hydraulically attached to a bolt at the shaft end.

▲ Five-bladed propeller, without key.

▼ A four-bladed controllable-pitch propeller.

31.5 Controllable-pitch propellers, construction

The hydraulic actuating device for the blades in the larger controllable-pitch propellers is housed in the hub.

◀

The cast and then machined propeller hub prior to installation of the parts, such as the pitch-actuating mechanism and propeller blades.

▲

A propeller hub attached to the propeller shaft.

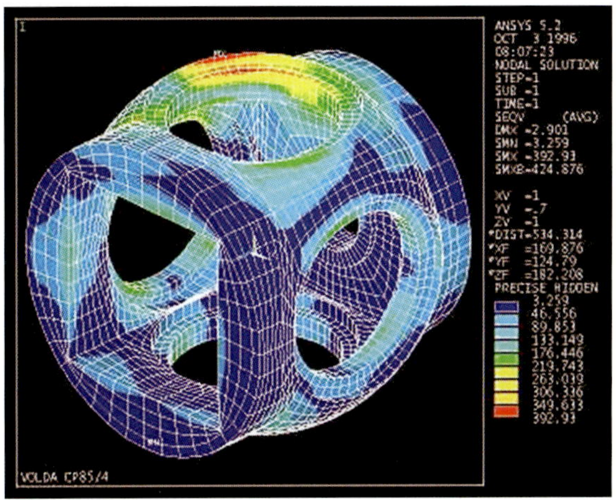

▲

A propeller hub must be very strong.

A 'finite-element method' can be used to calculate the load on every hub element.
red: highest load.
dark blue: lowest load.

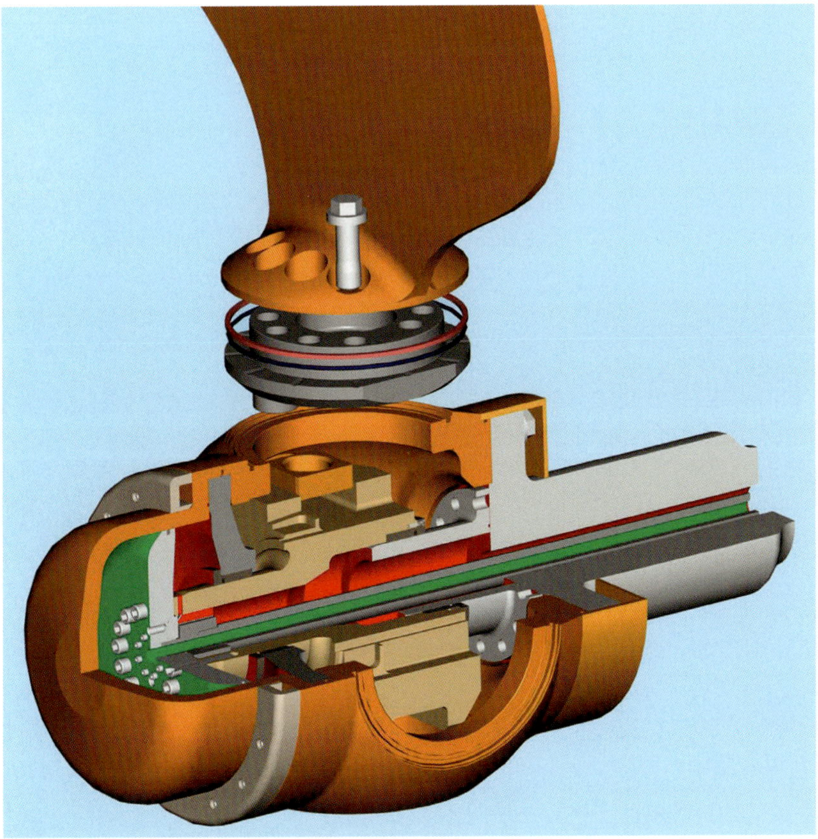

◀

A cut-away section of a propeller hub showing the pitch adjustment mechanism.

The oil supply pipe and the adjustment piston on the pressure side are coloured green.
The oil drainage area is coloured red.
The pitch adjustment mechanism is coloured light yellow.

▶

A cross-section of a Wärtsilä Lips controllable-pitch propeller.

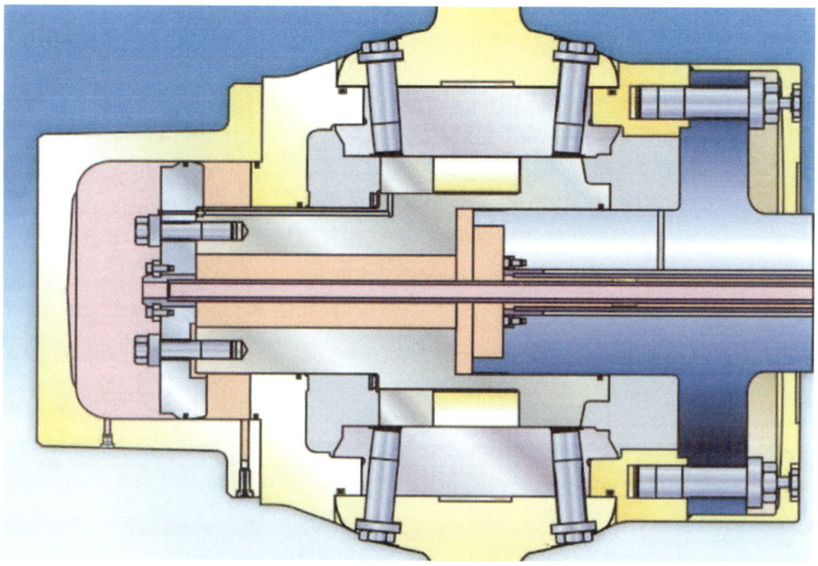

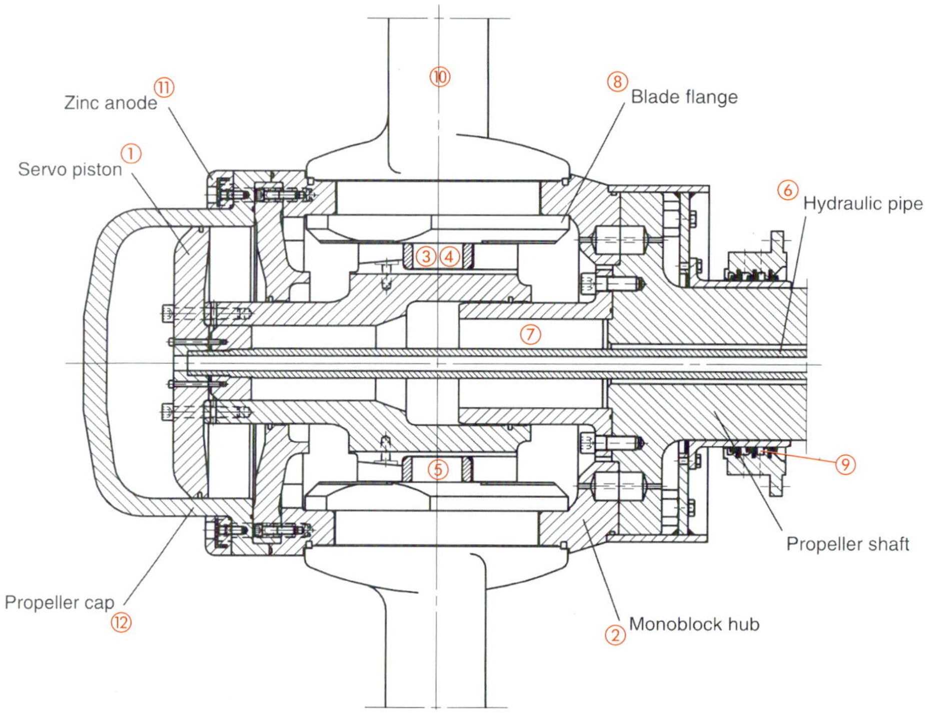

▲

A cross-section of the propeller hub.

1 *The servo-piston is displaced with high-pressure hydraulic oil through a hollow tube inside the propeller shaft. This allows precision-pitch adjustment.*
2 *The hub casing ('mono-block hub') is bolted to the flange of the propeller shaft.*
3 *The servo-piston moves an eccentric pin of the propeller blade with a 'slot-crosshead construction'.*
4 *Slot-crosshead construction.*
5 *Eccentric pin.*
6 *High-pressure hydraulic-oil supply line.*
7 *Hydraulic-oil return line.*
8 *Blade flange.*
9 *Rear-shaft seal.*
10 *Propeller blade.*
11 *Zinc anode: this is the so-called sacrificial anode, which prevents seawater corrosion. The zinc anode must be regularly substituted.*
12 *Propeller cap.*

Explanation

The hydraulic-oil system for controlling the 'servo'-piston is completely isolated from the oil system for the propeller blades. Only the blade seals are subjected to gravitational oil pressure. The blade sealing rings are made of Teflon.

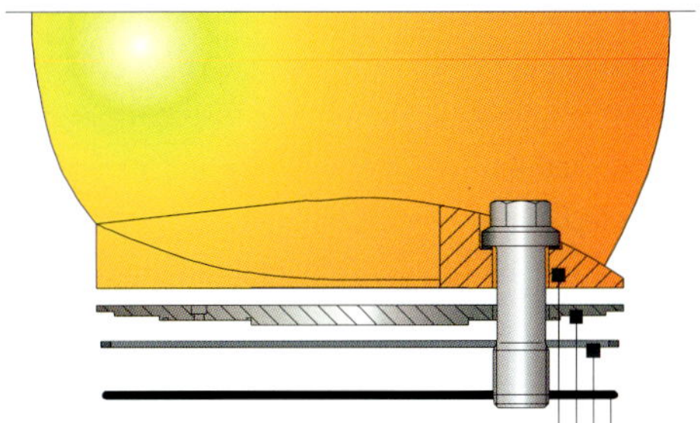

▲

The insertion of an extra intermediate flange below the blade foot allows underwater replacement of the blades.

The propeller-blade flange is lubricated via a separate drilling.

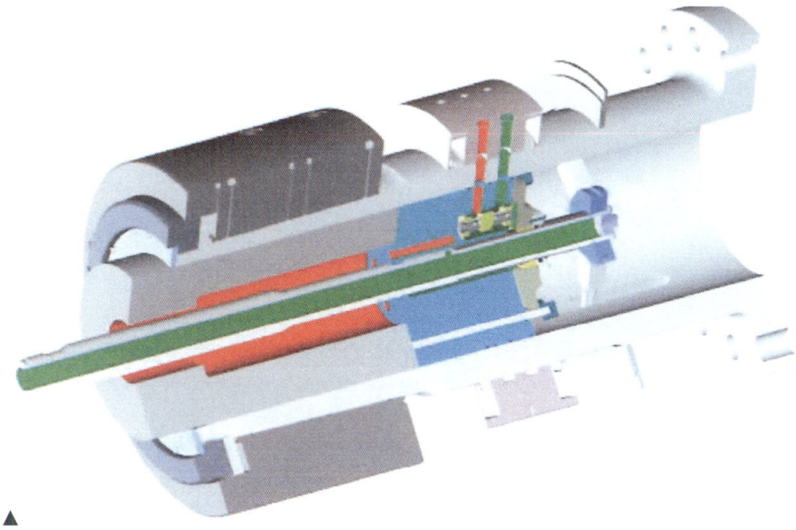

▲

The same coupling flange with the oil distribution and pitch indicator (in colour).

black: hydraulic coupling bush.
light blue: oil distribution.
green: hydraulic supply oil.
red: hydraulic return oil.
dark blue: seal.

Oil supply and drainage

Two different systems are used.
For propulsion systems with low-speed two-stroke-crosshead engines, the oil-distribution box is positioned on the shaft.

Propulsion systems for medium-speed four-stroke engines equipped with reduction gearing, the oil-distribution box is positioned in front of the reduction gearing or on the shaft.

▶

The oil distribution is integrated in the reduction gearing on the right, in the propeller shaft extension.

Above, the engine drive with coupling.
The top shaft is for driving the shaft generator.
Oil for emergency adjustment:
green: oil supply.
red: return oil.
purple: oil for the coupling.
yellow: lubricating oil for bearing lubrication.

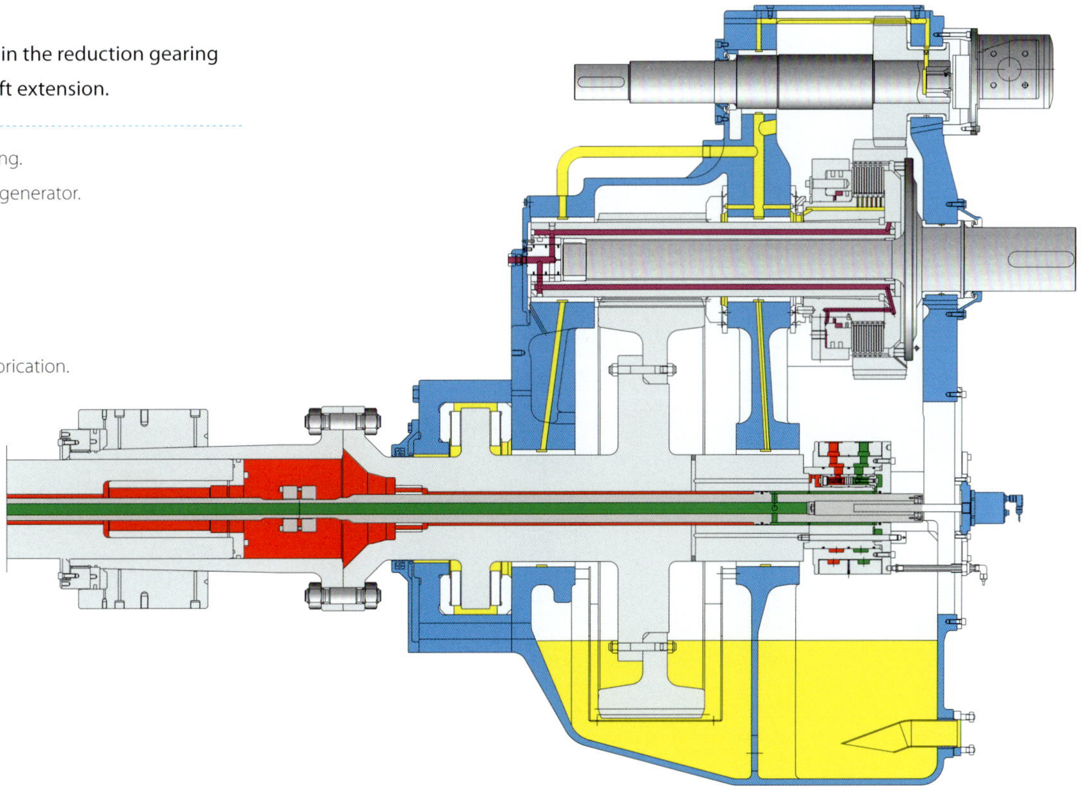

Loss of lubricating-oil pressure
When the plugs are removed, the pitch of the propeller can be manually set in the required position using a hand pump. The closed oil system maintains the selected position.

Oil distribution placed in the reduction gearing
Propulsion plants equipped with reduction gearing between the engine and the propeller have an OD-ring fitted in front of the reduction gearing. An electric-pitch gauge is mounted in the ring.

Hydraulic-oil system
The system consists of a hydraulic unit ('Hydra Pack'), a coupling flange with an electrical-pitch feedback box and an oil-distribution ring.
In the reduction-geared version, this piston-operated system is integrated in the gearbox. An intermediate shaft is installed between the piston and the pitch-control device for the blades located in the hub,
In case of an emergency, the pitch can be adjusted in the gearbox.

▶

A complete system for a controllable-pitch propeller from MAN Diesel.

The hydraulic power unit controls the hydraulic-oil pressure for pitch adjustment. The static pressure tanks ensure that the propeller-shaft seals 'forward' and 'astern' are kept under pressure. This arrangement is used in the direct drive of controllable-pitch propellers by a small low-speed two-stroke crosshead engine.

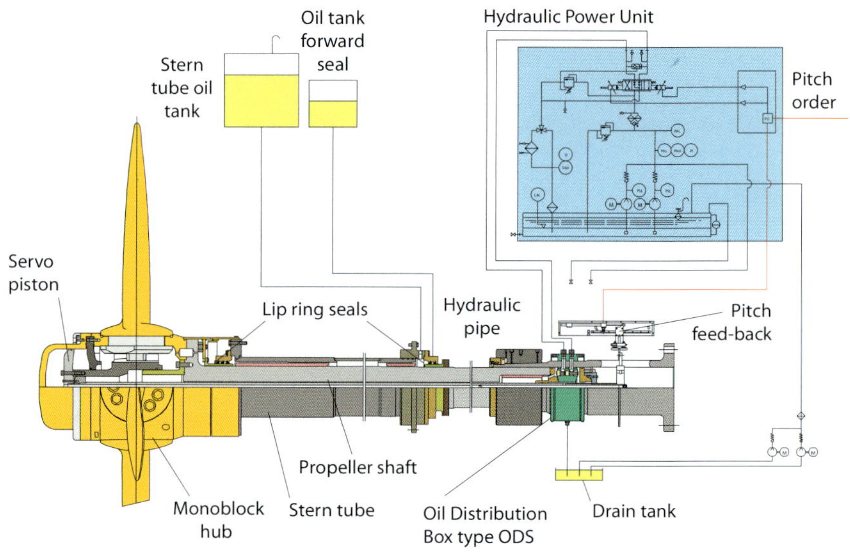

▶

A similar system for a controllable-pitch propeller. However, the oil distribution (green) is placed in the reduction gearing. This is typical for a four-stroke medium-speed diesel engine.

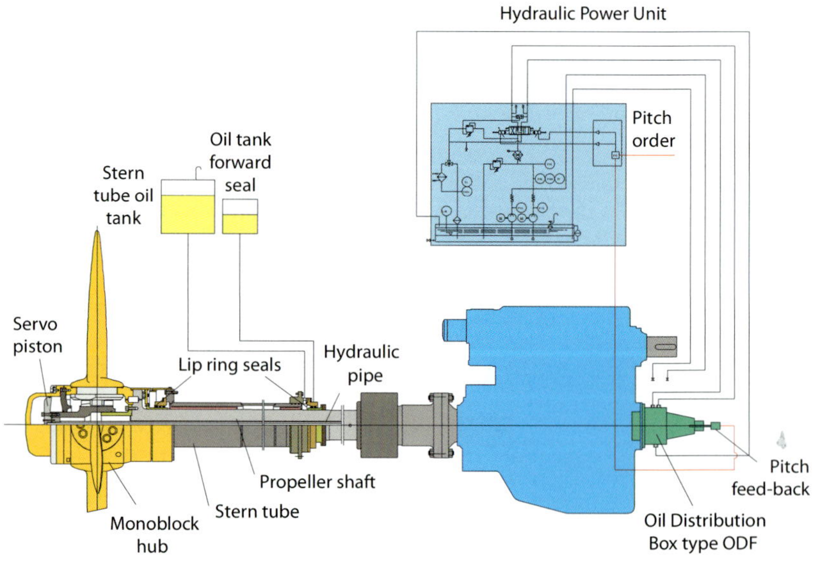

Lubricating systems

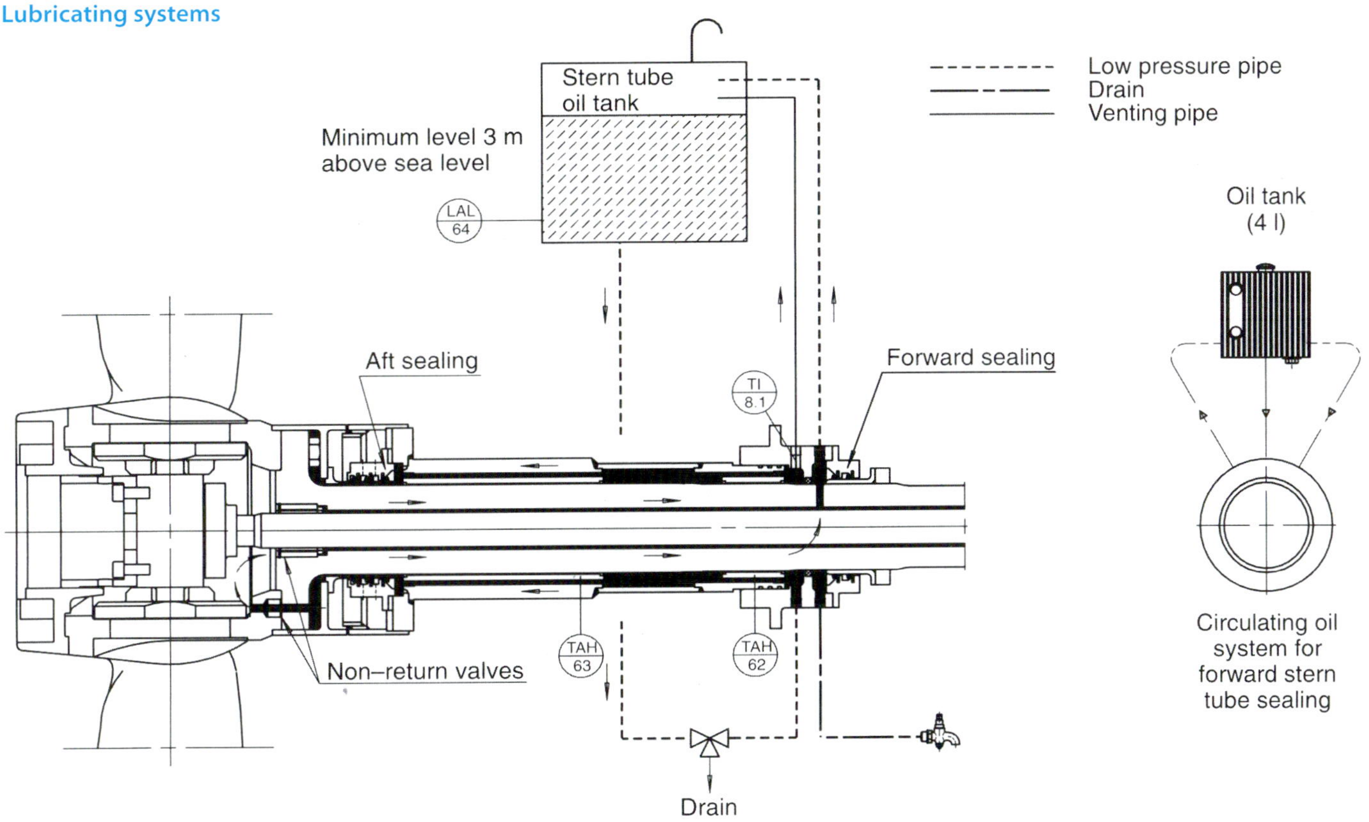

▲
The lubricating system.

The stern tube and controllable-pitch propeller have a common lubricating-oil system.

In order to prevent seawater penetration, a lubricating-oil system is kept under static pressure by placing an elevated gravity tank at least 3 metres above sea level, in accordance with the instructions from the manufacturer of the stern-tube seals.

Pitch changes create a pumping effect in the hub, resulting in a lubricating-oil circulation throughout the system, including the elevated oil tank.

Non-return valves in the propeller hub and pitch-control rod ensure that the oil flow to the hub also flows to the stern-tube journal bearings and continues along the chromium steel stern tube to the adjustable-pitch mechanism.

The return oil flows back to the storage tank via the pitch-control rod.

The propeller hub is fitted with two plugs for draining and venting during docking.

The pitch-control rod is lubricated with grease where the intermediate shafts are fitted.

31.6 Propeller shaft and coupling flange

The propeller shaft is manufactured of forged steel and then stress relieved. Material Forged Steel S 45P. The shaft is machined finished and hollow-bored for positioning the pitch-control rod or pipes for the supply and drainage to the servo-piston.

In large propulsion plants with low-speed two-stroke crosshead engines, bearings are fitted at regular intervals along the shaft according to the formula:

$$L = 450\sqrt{\text{shaft diameter (mm)}}$$

L = maximum bearing distance
rpm < 350 rpm / min.

Coupling flange

▶

The hydraulic-coupling flange.

A coupling flange consisting of two parts is clamped to the propeller using oil pressure. A lubricating-oil pressure in excess of 2000 bars is squeezed between the bush and the coupling flange using injectors. By increasing the oil pressure in the annular space C with the manual-hydraulic pump, the bush is pushed up the conical section of the shaft in stages. Longitudinal placing of the coupling flange as well as final push-up of the bush is marked on the shaft and the bush.

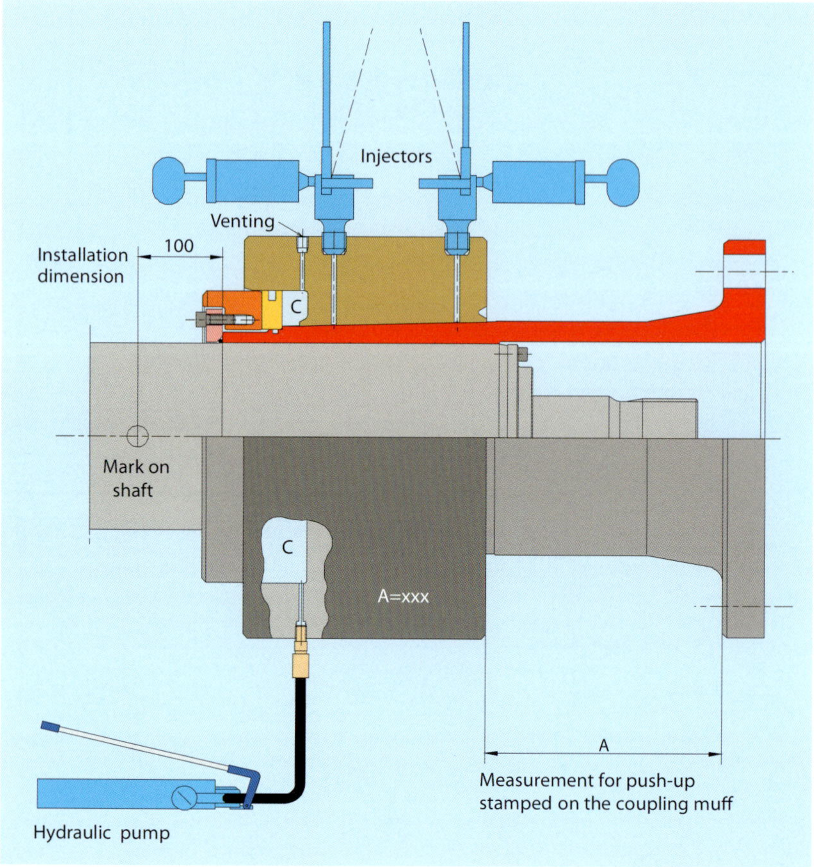

31.7 Stern tube

▶

The standard stern tube.

The standard stern tube is fitted aft and bolted to the propeller-frame boss. The front section of the stern tube has an oil box, bolted to the stern-frame boss. This allows thermal expansion/ contraction of the stern tube and facilitates the installation of the stern tube. Close tolerances of the stern tube are no longer required, and the stern tube can have a potential machining tolerance of 5 mm. MAN Diesel, Frederikshavn, Denmark.

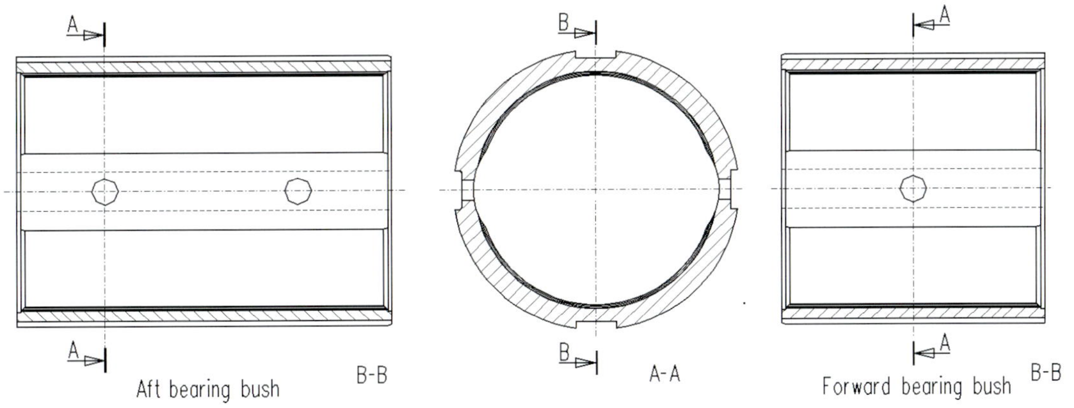

◄ An example of a stern-tube bearing from IHC Lagersmit B.V., The Netherlands.

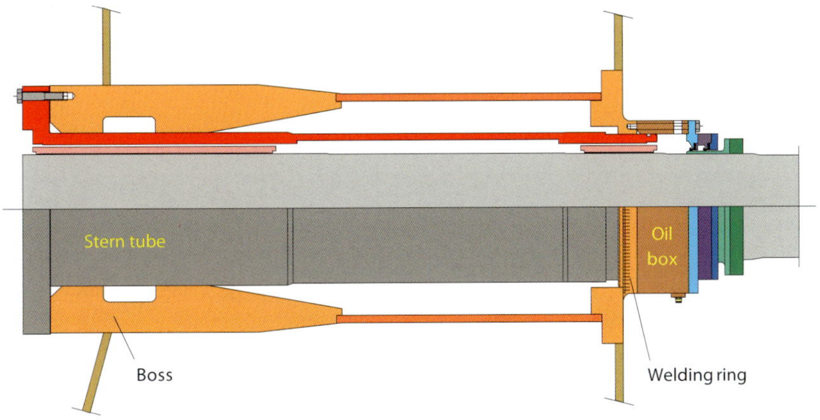

◄ Stern-tube bushes.

The stern tube is fitted with white metal front and rear bushings (pink).

If required, temperature sensors can be fitted.

The stern tube and the oil space can be treated with epoxy resin.

31.8 Stern-tube seals

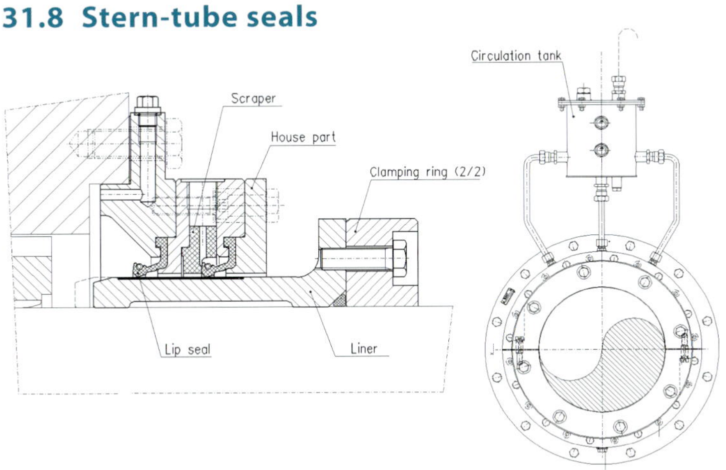

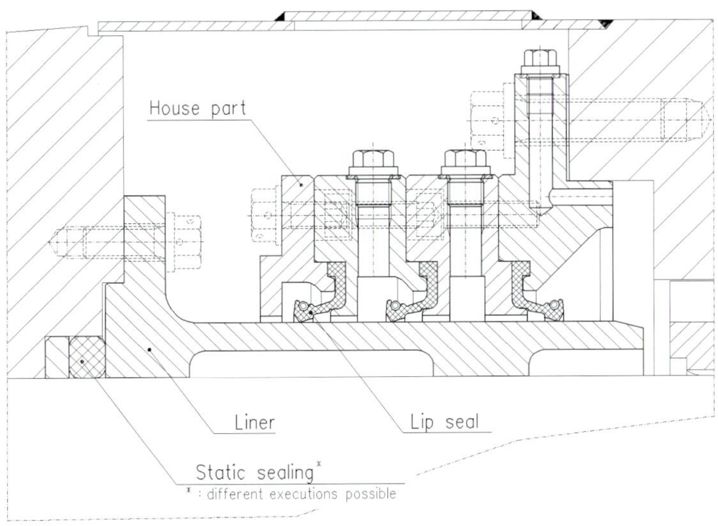

◄ Propeller shafts are standard with forward and aft seals. Here, an example from IHC Lagersmit B.V., Kinderdijk, The Netherlands.

There are numerous types of seals. Their most important property is the prevention of (sea) water and dirt particles from entering the aft seal. Equally important is that leakage of lubricating oil into the (sea) water is kept to a minimum.

Today there are also seals that work without lubricating oil (large plants) or grease (small plants).

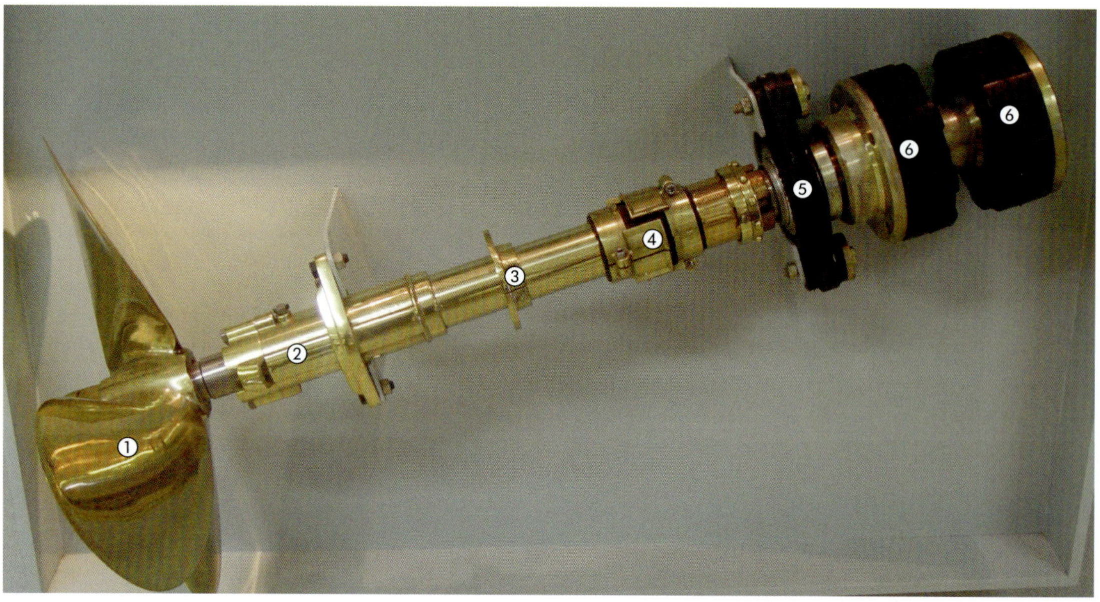

▶ A fixed-pitch propeller with stern tube and seals for small yachts.

1 *fixed-pitch propeller*
2 *stern tube*
3 *shaft seal*
4 *cardan coupling*
5 *thrust block for absorbing the thrust*
6 *flexible coupling towards the engine*

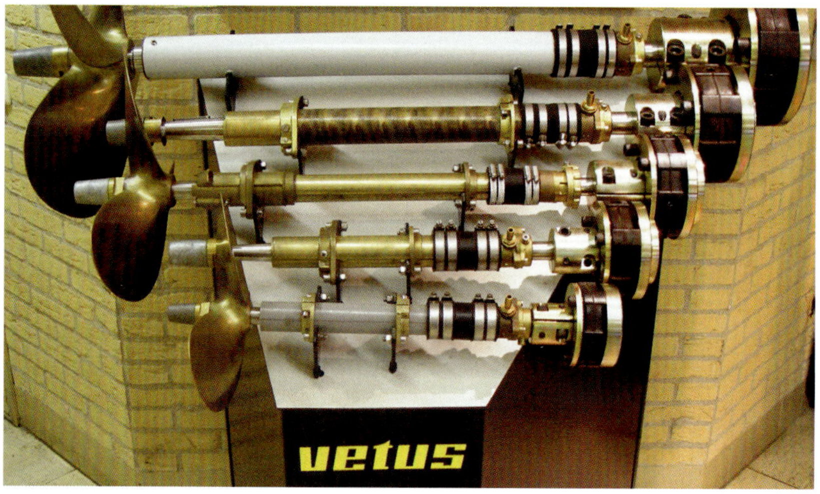

▶ Various propeller systems with their corresponding shafts.

31.8.1 Dry seals, pollution-free shaft seals

This system ensures that no lubricating oil from the shaft seals can contaminate the environment.

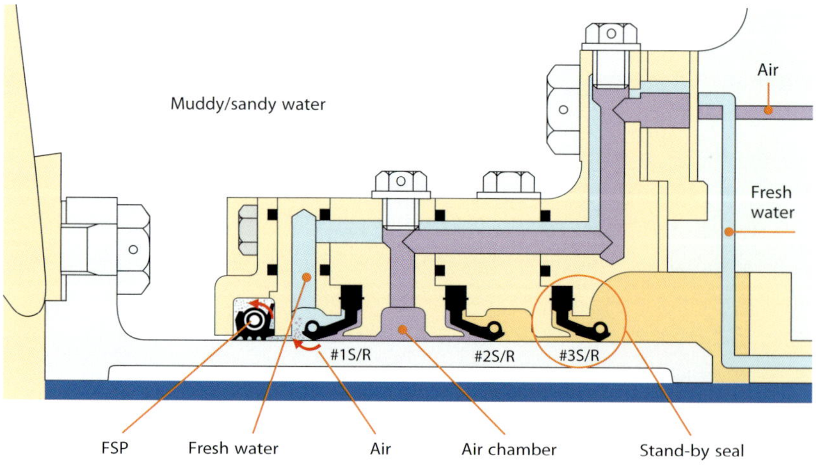

▶ A complete shaft seal where lubricating oil is not used.

Fresh water, light blue, is fed into the far left outer seal and ensures that no sand or mud can enter the seal.
There is an air chamber in the middle of the seal, which is kept under pressure. To the right of the air chamber, another two seals are installed. The seal on the far right is a stand-by seal in case of failure of the 2S/R seal.

31.9 Hydraulic bolts

A complete plant can be fitted with hydraulically tensioned bolts, making assembly/disassembly of the propeller shaft, the intermediate shafts and the flange to the flywheel of the (two-stroke) engine swift, simple and meticulous.

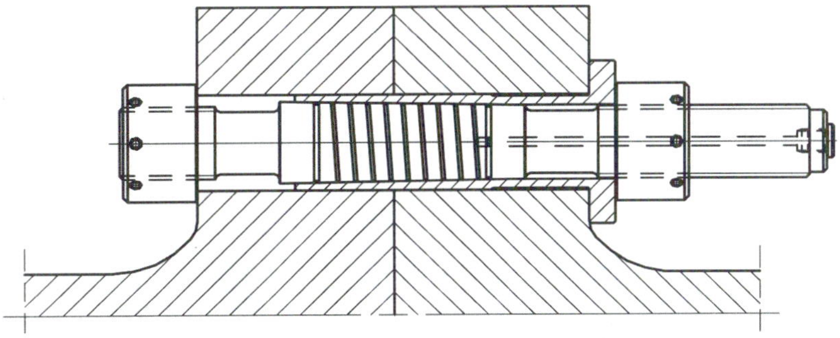

▲
A hydraulic connecting bolt.

The correct pressure can be applied simply and accurately to the flanges.
As the bolts are pre-tensioned by an elastic stretch process, they never 'loosen' of their own accord.

31.10 Material for controllable-pitch propellers

ISO, the International Standard Organisation, has introduced a series of standards for the manufacture of propellers (ISO 484).
The accuracy class can be selected by the customer.

Propeller blades are manufactured of a nickel-aluminium-bronze (Ni, Al)-alloy or a chromium-nickel-stainless steel (Cr, Ni)-alloy.

Class	Manufacturing accuracy
S	Very high accuracy
I	High accuracy
II	Medium accuracy
III	Wide tolerances

▲
The accuracy classes for manufacturing propellers.

If the customer does not specify a class, the propeller is manufactured according to class 1.

Ice class		C		1A*	
Material		NiAl	CrNi	NiAl	CrNi
Thickness r/R = 0.35	mm	132	146	169	187
Thickness r/R = 0.60	mm	71	78	90	100
Thickness r/R = 1.00	mm	0	0	15	13
Blade weight	kg	729	877	952	1053

▲
The graph shows the properties of both materials.

Both materials have high resistance to **cavitation-erosion**.
NiAl has better fatigue characteristics in a corrosive environment than CrNi.
Propeller blades are extensively exposed to periodic varying stresses. Consequently, resistance to fatigue is critical.
According to classification societies, blades made of CrNi must have a 10% higher thickness than a propeller blade made of NiAl.
The thicker the blade, the lower the propeller efficiency!
It is usually recommended that propeller blades be manufactured of NiAl.

31.11 Propeller design

Hydrodynamic design

Propeller blades are computer designed. The basic data is derived from advanced hydrodynamic theories, practical experience and numerous model tests at various hydrodynamic institutes.
Each blade is designed for a certain hull and adapted to the operating conditions of the ship. The total propeller efficiency, suppressing noise levels and vibrations are prime design objectives.

Propeller efficiency is predominantly determined by propeller diameter and the corresponding optimum speed. To a lesser extent, the blade surface area, the pitch and the thickness gradient of the propeller blade. These also affect the propeller efficiency.
The blade surface is determined according to the requirements for minimum cavitation, noise- and vibration levels.
To keep cavitation to a minimum, the pitch is reduced towards the tip of the blade.

Example 1

What is the tip speed of a propeller with a diameter of 8000 mm and a speed of 82 rpm?

$$v_{tip} = 2 \times \pi \times R \times n$$

$$= 2 \times 3.14 \times 4 \times \frac{82}{60} = 34.2 \text{ m/sec.}$$

Speed in kilometres per hour =

$$34.2 \times \frac{3600}{1000} = 123 \text{ km per hour.}$$

Example 2: Small yachts

How high is the tip speed of a propeller with a diameter of 300 mm and a speed of 900 rpm?

$$v_{tip} = 2 \times \pi \times R \times n$$

$$= 2 \times 3.14 \times 0.15 \times \frac{900}{60} = 14.3 \text{ m/sec.}$$

Speed in kilometres per hour =

$$14.3 \times \frac{3600}{1000} = 50.8 \text{ km per hour.}$$

The **thickness gradient** of the propeller blade must match the requirements of the Classification societies for ‘unskewed’ propellers.

▶ Part of a rudder showing the effects of cavitation.

Note the pitting caused by cavitation from the propeller.

31.12 Cavitation

Cavitation is caused by a localized decrease in the vapour pressure near the propeller.
At this reduced vapour pressure, water boils causing vapour bubbles. When these vapour bubbles reach other regions near the propeller, the shell of the ship or parts of the rudder with a higher pressure, they implode. This is accompanied by so much energy release that severe damage may be caused to the surface material close to the imploding bubbles.

Cavitation types

There are three main types of cavitation in terms of their origin and position on the propeller blades.

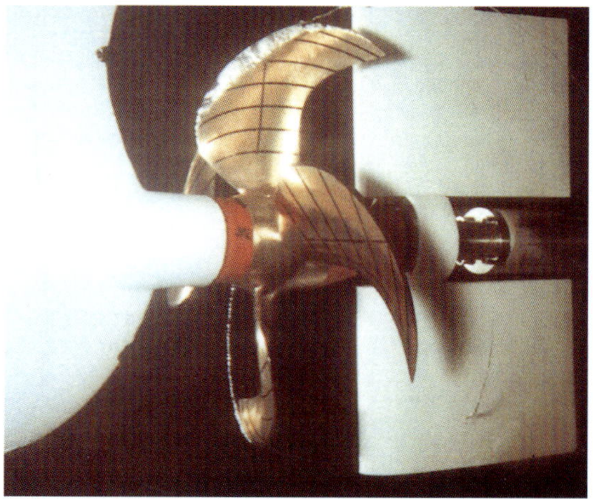

▲ Research into cavitation is particularly important in the design of a propeller for a particular hull.

Not only can severe damage to the propeller, hull and rudder be prevented but noise generation can also be determined in a fluid laboratory.

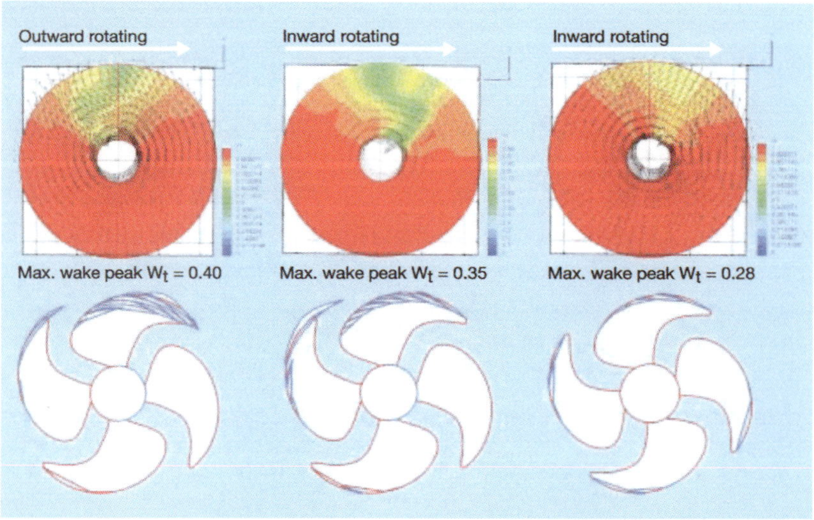

▶ When testing propeller designs, much research is done to minimize the extent of cavitation.

The right propeller has a much smaller surface area where cavitation can occur than the two examples on the left.
The wake peaks of these two examples are higher.
The homogeneous and uninterrupted supply of sufficient water is a prerequisite for minimizing cavitation.

31.12.1 'Sheet' cavitation on the 'suction side' of the blade

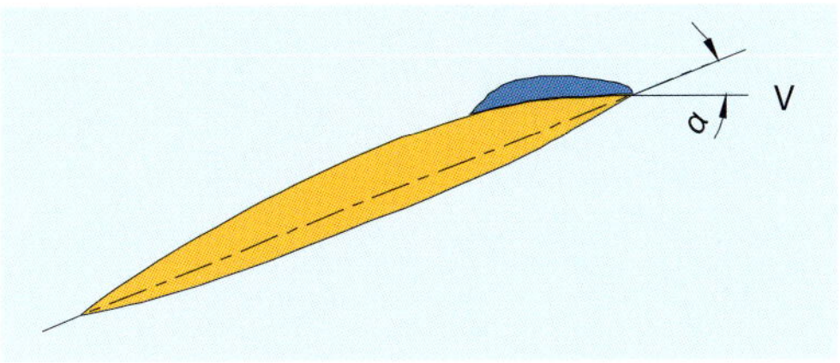

▲
Sheet cavitation on the 'suction side' of the propeller blade. Sheet cavitation occurs at the leading edge of the propeller blade due to a low-pressure peak in this area.

The sound and vibration levels will be low if the cavitation is limited and the clearance to the hull is sufficient.
If cavitation extends over half of the radius, it can increase significantly. This can lead to erosion of the propeller-blade material and must therefore be prevented.

31.12.3 'Sheet' cavitation on the pressure side of the blade

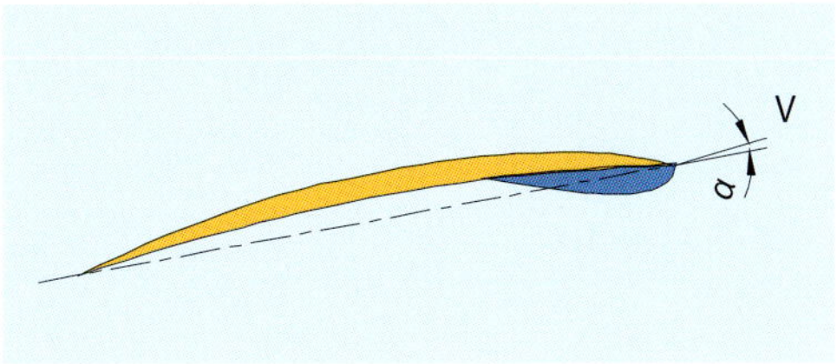

▲
Sheet cavitation on the pressure side of the blade.

This form of cavitation is similar to 'sheet' cavitation on the suction side. However, the generated vapour bubbles tend to implode on the blade surface due to the increased pressure.
This is likely to produce erosion. Therefore, the blades should be designed without any pressure-side cavitation.

'Sheet' cavitation at the tip of the propeller blade may develop into a tip vortex cavitation of the propeller blade and often extends over the centreline of the propeller.
If the tip vortex spreads to the rudder, it may cause material erosion.

31.12.2 Bubble cavitation

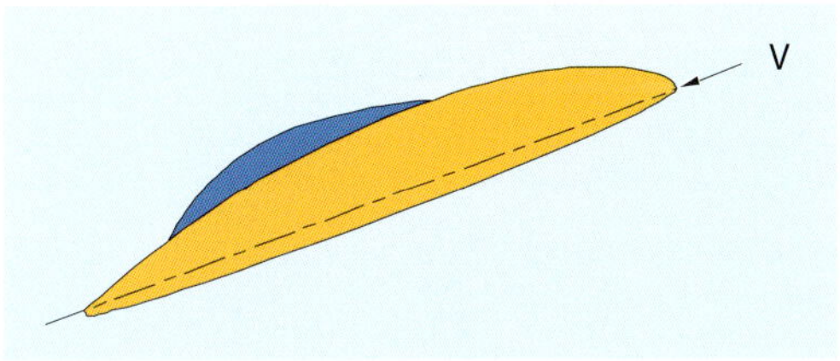

▲
Bubble cavitation.

If the propeller is overloaded and the blade surface area is too small to deliver the required thrust, the middle section on the suction side of the centre of the propeller blade will be subject to cavitation. This type of cavitation can be extensive and lead to erosion. When designing a particular propeller, this must be avoided.

31.12.4 Graphs of specific fuel consumption and towing force

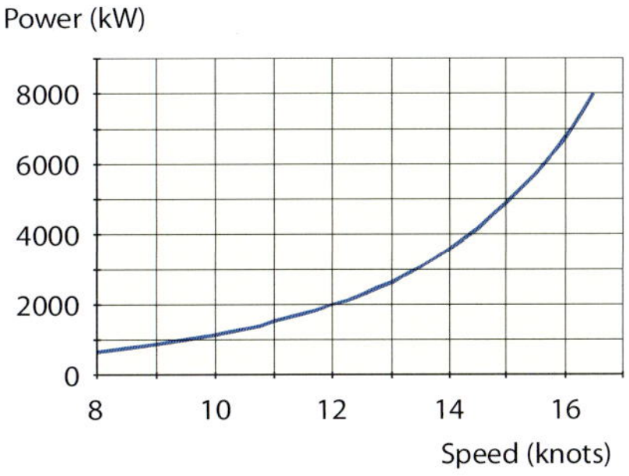

◀
A graph depicting the required power output in kW (vertical) and the speed in knots (horizontal).

At 12 knots, a power output of 2000 kW is required and at 14 knots, 4000 kW is required. Increasing the speed by 2 knots requires twice the engine output and therefore doubles the fuel consumption!

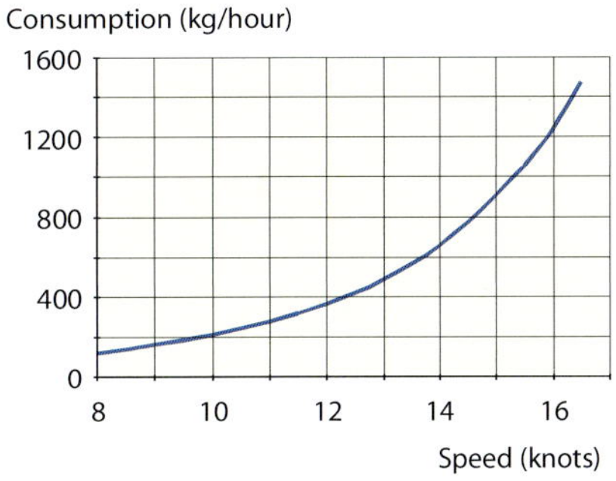

▲
A graph showing fuel consumption in kilograms per hour (vertical) and the ship's speed in knots (horizontal).

Clearly, fuel-consumption increase is significantly higher than the increase in speed. Fuel consumption is 200 kg per hour at a speed of 10 knots. This is doubled to 400 kg per hour at 12 knots and at 16 knots; it is 1200 kg per hour. Therefore, an increase in speed of six knots requires 1200 kg per hour of fuel, not the expected 200 kg of fuel. This amounts to six times the expected fuel consumption to achieve a speed increase of six knots.

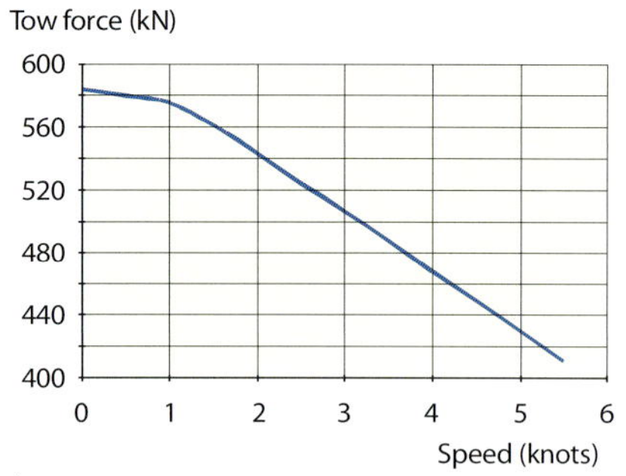

▲
The graph shows the towing force in kN (vertical) and the speed in knots (horizontal). Increasing the speed effects a significant decrease in towing force.

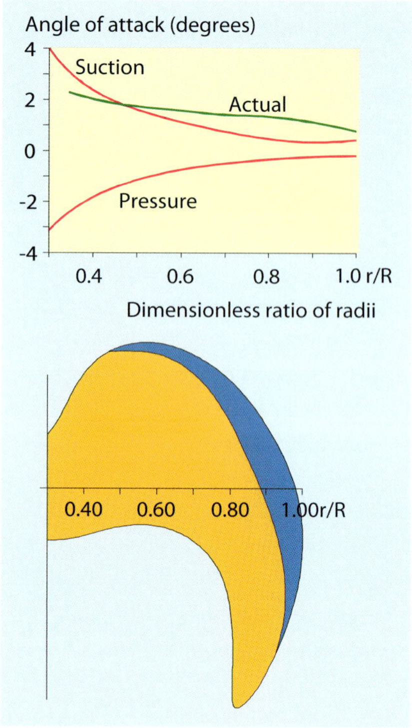

▲
A modern propeller, the 'skew' model. Cavitation can occur in the blue area under unfavourable conditions.

31.13 Calculating the propeller-blade model

For all conditions in which propeller can operate, and for all pitch angles of the blades, the flow of water behind the corresponding hull is calculated. The extent of cavitation is evaluated with respect to the degree of noise and vibration.

31.13.1 'High skew' design of a propeller

A special design is used to improve cavitation suppression by reducing induced-pressure fluctuations.
The blade is hereby skewed in the rotational direction. The vibrations generated by a 'skewed' propeller blade are reduced to less than 30% of those generated by an 'unskewed' design.

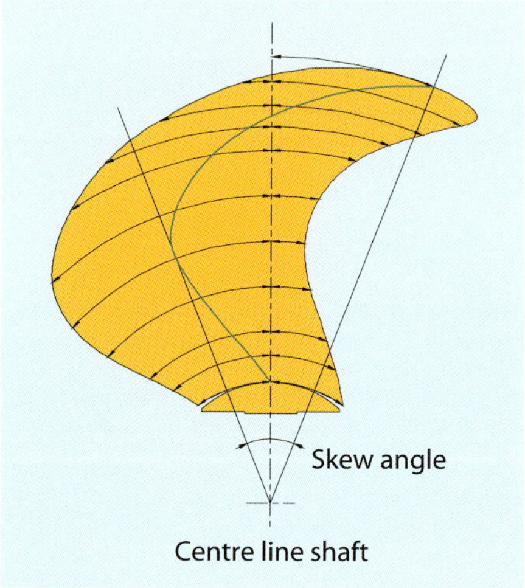

▲
The 'high-skew' design.

As this type of propeller does not affect propeller efficiency, it is used in ferries and cruise ships where low vibration levels are a requirement.
Today the 'skew'-type propeller is bent in two directions, both rotational and contra-rotational. The blades of this kind of propeller can be balanced so that the smallest possible force occurs in the adjustable-pitch mechanism.
A finite-element analysis is carried out for determining the optimum 'skew'.

▼
'High-skew' propeller blades ready for transportation.

31.14 Examples of other propulsion systems with controllable-pitch propellers

Apart from the traditional propulsion systems with propellers, specifically the propulsion engine with behind it the reduction gearing, coupling, intermediate shafts, propeller shaft, propeller along with many versions, there are many other variations that deviate from the standard pattern: in line with propulsion engine, shafts and propeller.

31.14.1 Example 1: Diesel-electric propulsion

By driving propellers with electromotors with the required electrical power generated by diesel engine driven generators, the location of the propellers and diesel engines may deviate from the traditional arrangement.

▼

Diesel-electric propulsion from Wärtsilä. Shown here, an LNG-tanker with Dual Fuel (DF) engines.

Each of the four main gensets, twelve-cylinder V-engines type 12V50DF supply 11,400 kW and both auxiliary genset, nine-cylinder in-line engines type 9L32DF, supply 3150 kW each. Therefore, the total installed generating capacity is 51,900 kW. This amount of electrical power can drive two electromotors of 12.5 MW each, driving a controllable-pitch Lips propeller. An electrically driven POD serves as a contra-rotating propeller at normal speeds, and as an excellent manoeuvring propeller when entering the LNG terminals.

Electric thruster

Here, the propulsion unit, an electromotor, is situated on the outside of the ship's hull. This system is known by its brand name 'Azipod'.
The electro-motor with a variable speed is installed in a separate pod below the ship. There are versions that are fixed and also those that can rotate 360°.
The propeller can be mounted before or behind the pod making it a tow or thrust version.
The ability to rotate the pods renders rudders superfluous as the vessel has excellent and rapid manoeuvring characteristics.

Today, the above-mentioned system is increasingly used for passenger ships, vessels with a dynamic-positioning system, and ferries.

Several advantages

- **Both the propeller system and the diesel gensets can be installed in the most favourable location.** It should be remembered that the 'hotel' section on passenger vessels takes up over 80% of the total space available on the ship.
- **Depending on the required propulsion output, diesel-driven gensets can be placed off-line.** In this manner, the diesel engines are always heavily loaded and emissions are kept to a minimum and a higher efficiency is achieved.
- **In general, larger ships have medium-speed four-stroke engines, suitable for running on HFO.** This is relatively inexpensive and consequently exploitation costs are reduced.

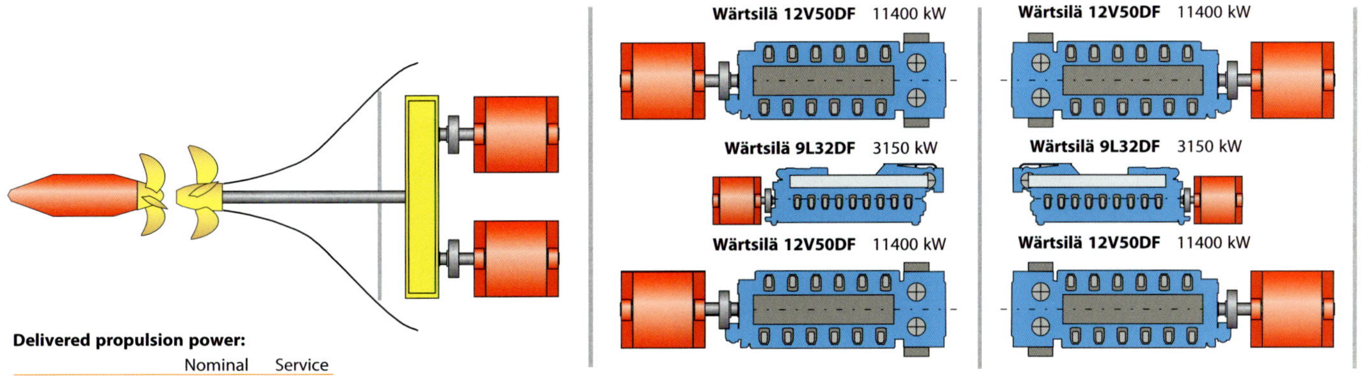

Delivered propulsion power:

	Nominal	Service
Electric pod	17 MW 40 %	17 MW 41 %
LIPS CP propeller	25 MW 65 %	24 MW 62 %
Total shaft power	**42 MW**	**41 MW**

Installed engine power:

Mechanical transmission	0 MW
Electrical power generation	51.9 MW
Total installed power	**51.9 MW**

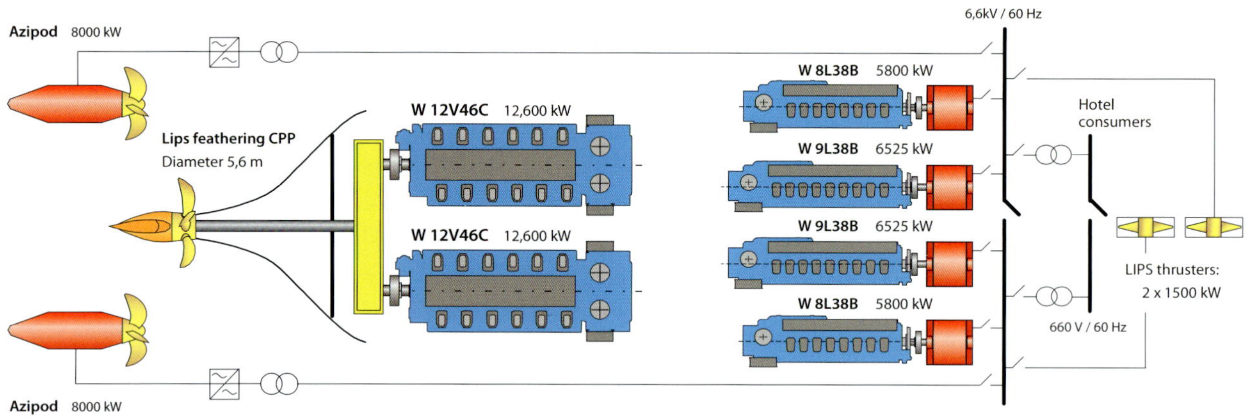

▲

An alternative arrangement of Wärtsilä design.

Both main engines, type 12V46C, drive one controllable-pitch Lips propeller with a diameter of 5.6 metres. Total shaft power is 2 x 12,600 kW.

The four gensets with 8L38B and 9L38B engines, total output 24,650 kW, are available for driving both 'Azipods' of 6000 kW each. The voltage of this network is 6600 volts at 60 Hz. Both Lips bow thrusters of 1500 kW each are also on this grid. The normal ship's grid is supplied with a voltage of 660 volts at 60Hz via two transformers.

This arrangement allows various means of propulsion:

a the main engines and auxiliary engines with three propellers.
b only with the Azipods.
c one main engine driving the controllable-pitch propeller.
d portion of the auxiliary engines supplying power to the Azipods.

– **Sometimes, a larger genset driven by a gas turbine (fuel MDO) is installed in the larger propulsion plants.** It can be put on-line if the diesel gensets fail or when the electricity consumption of the hotel, for instance, the air conditioning, increases significantly. A well-known example is the 'Queen Mary II'.
– **Excellent manoeuvrability.** Generally, tugboats are not necessary and therefore not required by the port authorities.

31.14.2 Example 2: Compact thruster

The propeller is driven by diesel engines via cardan shafts. The engine power-output is transmitted to the propeller via two bevel gears. There are many possibilities.

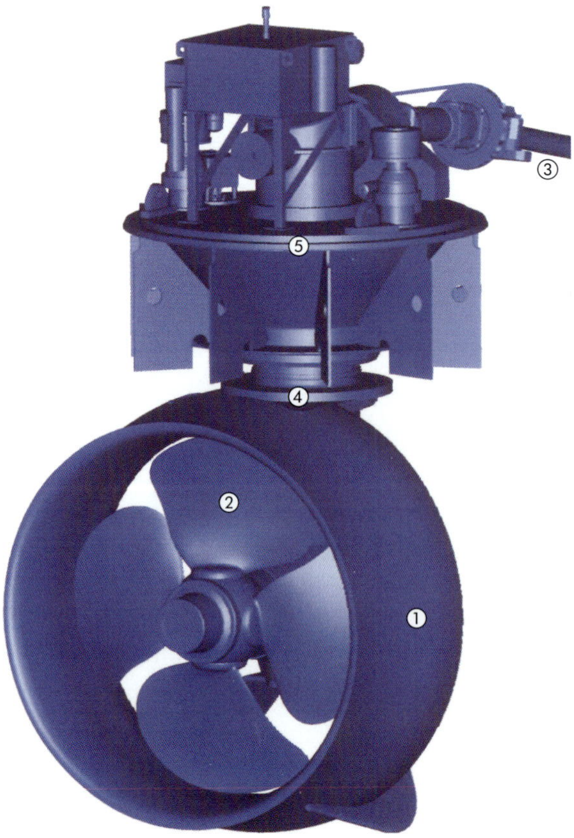

◄

A Lips compact thruster.

This thruster can rotate through 360 degrees and therefore has excellent manoeuvrability making a rudder redundant. The thruster is driven by the diesel engine via long cardan shafts.

1 *nozzle*
2 *controllable-pitch propeller*
3 *drive from the diesel engine*
4 *hull*
5 *rotating disc*

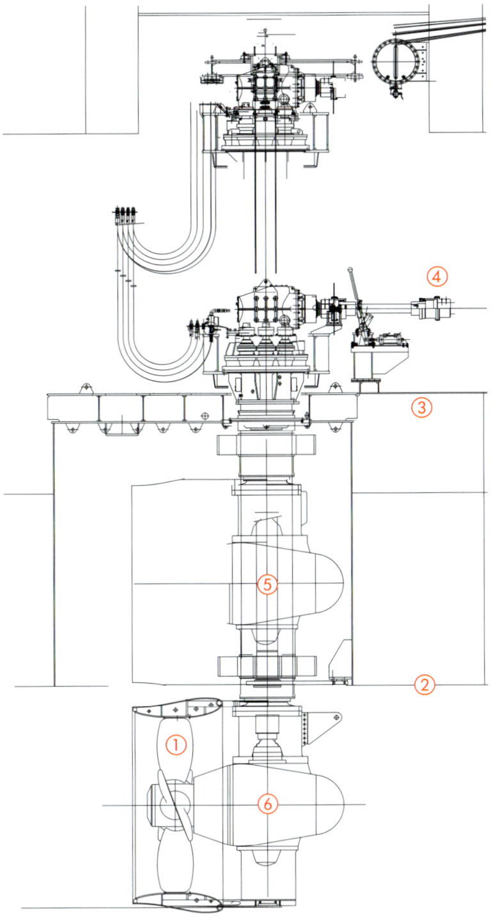

▲
A retractable thruster system in which the thrusters can be completely retracted into the hull.

In sailing position, the thruster is lowered outside the hull.

1 *propeller*
2 *bottom of the ship*
3 *hull*
4 *engine-drive shaft*
5 *high position inside the hull*
6 *low position outside the hull*

Fixed systems

This allows the propeller to move vertically, depending on the ship's draught. In an unloaded/ empty vessel, the propeller will always be immersed with acceptable propeller efficiency.

A retractable system

These are often only used at sea.

Bow and stern thrusters

These are arranged in a tunnel located in the forward superstructure or astern to generate lateral thrust. This can be used to manoeuvre or to hold a position.

31.14.3 Example 3: Propellers in nozzles

These are widely used. The purpose of these nozzles is to increase the thrust by the water flowing around the nozzle section. When the speed of the water flow inside the nozzle exceeds that of the outside flow, a pressure difference is created, which produces horizontal thrust and therefore a propulsive force. In addition, it often reduces the noise and vibration levels generated by the propeller. The local pressure differences decrease due to the homogeneous flow of the water through the nozzle. Therefore, cavitation is reduced.
A nozzle can be designed for any type of ship or propeller. Nozzles are predominantly used on smaller ships, for instance inland shipping, tugs, tenders and fishing vessels.

Propulsion types of smaller vessels

These diverge considerably from types for the larger inland shipping and ocean navigation.
Clients often wish to install the propulsion engine in a different place, or require electric- or hydraulic-driven propellers.

31.14.4 Example 4: Electric propulsion

This is limited to use in small craft such as dinghies.

Advantages

- whisper propulsion
- no emissions when running
- easy installation with a simple reversible electromotor

▼
There are numerous ways of propulsion for small vessels. Shown here, an electric drive of a fixed-pitch propeller. The engine is reversible and can operate at any speed.

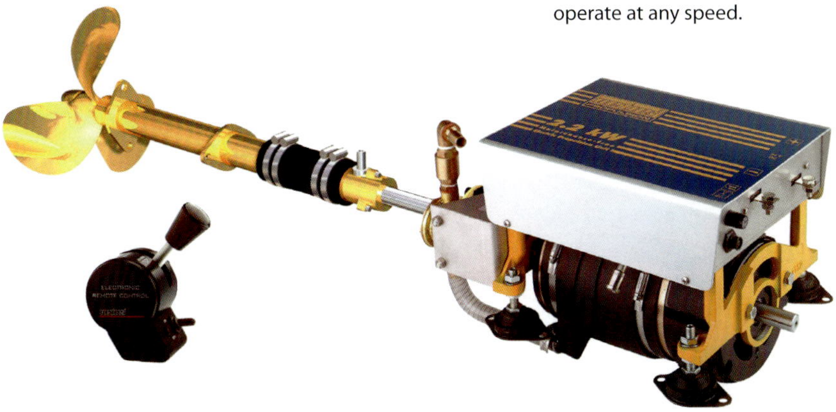

Disadvantages

- Limited range.
- Storing electricity in batteries has a low efficiency. The electricity required from the public grid with respect to fuel supplied and generated by the electric-power plants has a low efficiency of approximately 40%. An additional 50% of this is lost with loading and storage. Therefore a mere 40% – 20% = 20% of the original fuel energy remains for driving the propeller.
- Consequently, the total propulsion efficiency from the fuel supplied to the power plant at a propeller efficiency of 40% is at most 8%. A small dinghy with a small diesel engine with an efficiency of 25% does not exceed 15%.

31.14.5 Example 5: Hydraulic propulsion

This is also limited to use in small motor vessels for recreational purposes and for water-based companies. The hydraulic unit consists of a diesel engine, which drives an oil pump.
Here the propeller is driven by a hydraulic engine. Furthermore, various other systems can be driven by this system, such as bow and stern thrusters, an anchor winch or a light crane.

Advantages.

- The diesel-driven hydraulic set can be placed in the most favourable place in the vessel. The diesel engine operates at a constant speed and can therefore be loaded quickly via the hydraulic pump.
- The auxiliary power does not require a separate battery.

Disadvantages

The purchase price of a hydraulic installation is high and its propulsion efficiency is slightly lower.

31.14.6 Example 6: Propulsion with a diesel engine and custom intermediate shaft system to the propeller shaft

Here the engine is not placed in line with the propeller shaft but, for instance, astern. The propeller is driven via a so-called 'Z drive'.

An advantage is that the diesel engine can be installed in the most favourable location on the boat. This system is common in high-speed motor boats.

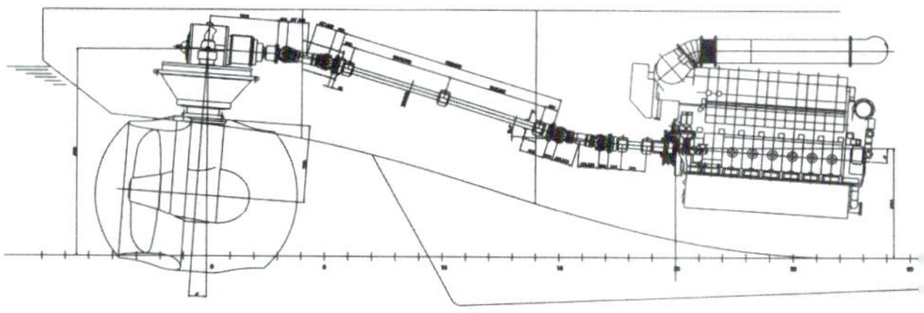

▲

Propulsion with a diesel engine and an intermediate-shafting system to the propeller shaft.

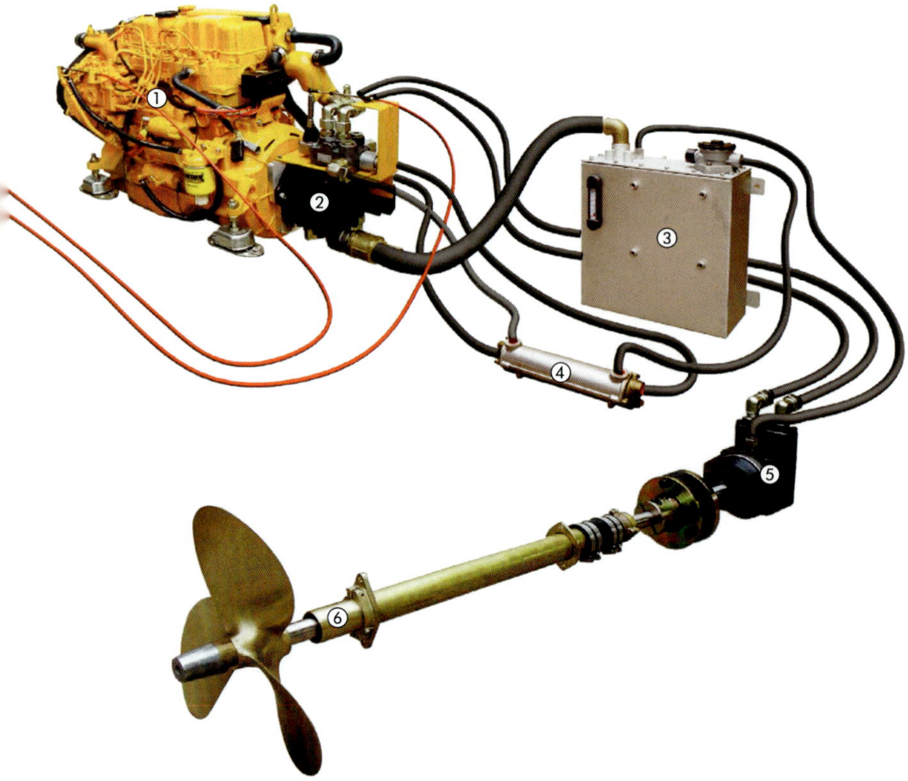

◄

A diesel-driven hydraulic propeller drive.

1 *diesel engine*
2 *hydraulic pump*
3 *hydraulic-oil storage tank and buffer tank*
4 *oil cooler*
5 *hydraulic motor*
6 *propeller shaft with propeller*

The components are connected by flexible hydraulic hoses. The diesel engine can be positioned in an ideal place when opting for this system. The hydraulic system can also be used for other purposes, such as an anchor winch.

31.14.7 Example of a large propeller

These propellers are used for very large and fast container ships.
They are fixed-pitch propellers often with six blades.
The propeller diameter can increase to up to 9 metres and the weight up to 100 tons!

◄ A large fixed-pitch six-bladed propeller.

Note the large amount of free space around the propeller, allowing for the free flow of water around the propeller. Nevertheless, cavitation will be caused by the rudder! See inside the yellow circles.

◄ Removing this type of propeller requires at least 10 engineers!

>CH32

Regulations for propulsion engines, classification, repair and damage

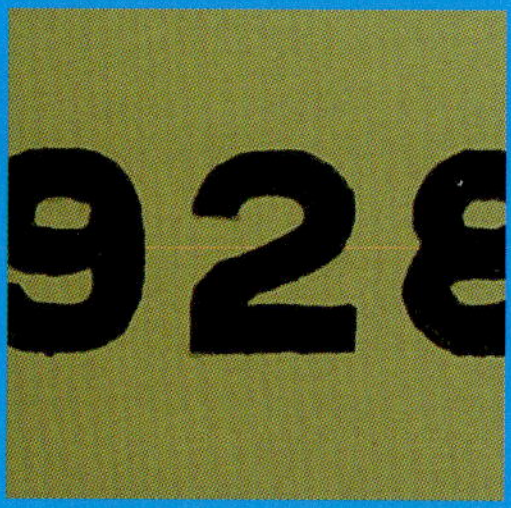

Each propulsion engine is tested on a test bed according to strict regulations. Shown, a large two-stroke crosshead engine, a Wärtsilä Sulzer RT-Flex96 with a shaft power of 72,360 kW at a speed of 102 rpm.

If a ship is in dry-dock for either maintenance or, in this case, for repairs, regular inspection and overhaul can also be done.

32.1 Introduction

Many regulations are in force pertaining to the technical installations on ships. Adherence to these rules and regulations ensures constructive compliance with the law and with the stipulations of insurance companies. The principle is the safe operation of commercial shipping with respect to crew, ship, cargo and environment.
Many countries used to have their own regulations, all more or less with the intention of achieving the above goals, but all differing slightly. As shipping is perhaps the most international of the world's industries, there is a need for international standards to regulate shipping – which should be adopted and accepted by all.

32.2 The IMO: International Maritime Organization

The IMO (International Maritime Organization) has 167 Member States and 3 Associate Members, representing a worldwide range of shipping interests in 2008.
The International Maritime Organization, based in London, UK, embodies the coordination of governments and shipping in maritime-technical issues, in order to improve maritime and environmental safety. The IMO is a specialized agency of the United Nations.
The concept of the IMO was developed after the MS Titanic disaster, the ship that 'could not sink' but sank on her maiden voyage. Until then, every country had its own regulations with respect to design, construction and equipping of ships. In 1948 at an international conference in Geneva, the IMCO (**Intergovernmental Maritime Consultative Organisation**) was formally established. In 1982, the name IMCO was changed to IMO, International Maritime Organization.

The first major treaty was the establishment of a new international convention on safety, **SOLAS** (Safety Of Life At Sea). Through the years, SOLAS has changed, adapted and brought up-to-date by the inclusion of modern views, scientific development and new technological possibilities.

Safety was the main objective of the IMO, however, pollution control being put high on the agenda after the accident with the Torrey Canyon on 18th March 1967, when 119,000 tons of Kuwait crude was spilled into the sea. . In the years following, many procedures were started to prevent recurrences of major oil spills and to reduce the effects. This resulted in 1973 in the International Convention for the Prevention of Pollution from Ships, which in 1978 was changed in the **MARPOL** protocol. Additionally, conventions were also drawn up to establish liability and compensation when accidents occur.

Safety at sea is largely determined by communication. The establishment of the **International Mobile Satellite Organization (IMSO)** in the nineteen-seventies made communication via radio and VHF maritime radio-telephony possible. In 1992, the next step was taken with **GMDSS**, the **Global Maritime Distress and Safety System.** All ships are required to be equipped with satellite emergency-position indicating radio beacons. This ensures that in emergencies, an automatic emergency signal is sent that can be localised by satellite.

IMO, following the 9/11 terrorist attacks in the US in 2001, has been highly alarmed with terrorism. The fight to prevent and suppress these acts has resulted in the **International Ship and Port Facility Security Code**, where a risk analysis must be performed to determine what security measures are appropriate in harbours to prevent not only terrorism, but also human trafficking.

32.2.1 Rules and regulations

Below are a number of mandatory regulations adopted by the IMO.

Maritime security

- 1966, LL, International Convention on Load Lines. The Load Lines Convention contains detailed regulations on the assignment of the freeboard (the vertical distance between the top of the hull and the waterline) and the specific load-line limitations for different types of ships. It also sets down the required construction integrity of a ship so that it can safely carry cargoes.
- 1971, STP, Special Trade Passenger Ships Agreement.
- 1973, Protocol on Space Requirements for Special Trade Passenger Ships.
- 1972, COLREG, International Regulations for Preventing Collisions at Sea. A set of traffic rules for shipping that can be compared with road traffic regulations.
- 1972, CSC, International Convention for Safe Containers. Regulations that set out procedures regarding the safety of freight containers

transported by sea: size, strength, handling, storing and stacking.

- 1974, SOLAS, International Convention for the Safety of Life at Sea. Regulations which specify minimum standards for the construction, equipment and operation of ships, compatible with their safety for those on board.
- 1979, SAR, International Convention on Maritime Search and Rescue. The system covers a search and rescue operation – who does what, where and under what conditions.
- 1976, INMARSAT, Convention on the International Maritime Satellite Organization. The operational requirements for a satellite communication system for maritime purposes.
- 1977, SFV, The Torremolinos International Convention for the Safety of Fishing Vessels.
- 1978, STCW, International Convention on Standards of Training, Certification and Watch-keeping for Seafarers. Regulations establishing basic requirements for training, certification and watch-keeping for seafarers at an international level.
- 1995, STCW-F, International Convention on Standards of Training, Certification and Watch keeping for Fishing Vessel Personnel. Regulations to improve the training, certification and watch-keeping standards for crews on fishing vessels.

Marine pollution

- 1969, International Convention Relating to Intervention on the High Seas in Cases of Oil Pollution Casualties.
- 1972, LDC, Convention on the Prevention of Marine Pollution by Dumping of Wastes and Other Matter.
- 1978, (MARPOL 73/78) International Convention for the Prevention of Pollution from Ships. The prevention of accidental or operational pollution of the marine environment by ships.
- 1990, OPRC, International Convention on Oil Pollution Preparedness, Response and Co-operation. Measures for dealing with pollution incidents, either nationally or in co-operation with other countries.
- 2000, HNS Protocol, Protocol on Preparedness, Response and Co-operation as to Pollution Incidents by Hazardous and Noxious Substances, 2000.
- 2001, AFS, International Convention on the Control of Harmful Anti-fouling Systems on Ships.
- 2004, International Convention for the Control and Management of Ships, Ballast Water and Sediments.

Liability and compensation

- 1969, CLC, International Convention on Civil Liability for Oil Pollution Damage. These are arrangements concerning liability, especially when disasters occur in international waters.
- 1971, FUND, International Convention on the Establishment of an International Fund for Compensation for Oil Pollution Damage.
- 1971, Convention relating to Civil Liability in the Field of Maritime Carriage of Nuclear Material.
- 1974, PAL, Athens Convention relating to the Carriage of Passengers and their Luggage by Sea.
- 1976, LLMC, Convention on Limitation of Liability for Maritime Claims.
- 1996, HNS, International Convention on Liability and Compensation for Damage in Connection with the Carriage of Hazardous and Noxious Substances by Sea.
- 2001, BOPD, International Convention on Civil Liability for Bunker Oil Pollution Damage.

◀

A cargo-ship docked for overhaul, repairs and periodic inspection by a classification society.

Many precisely defined regulations are in force that oversee the technical installations on ships. The underlying principle is that merchant vessels sail safely with respect to crew, ship, cargo and environment.

Since most seafaring countries already had highly varying rules and regulations pertaining to shipping, as shipping is indubitably a purely international industry, it was deemed better to have uniform international regulations.

Other

- 1965, FAL, Convention on Facilitation of International Maritime Traffic.
- 1969, International Convention on Tonnage Measurement of Ships.
- 1988, SUA, Convention for the Suppression of Unlawful Acts Against the Safety of Maritime Navigation.
- 1989, International Convention on Salvage.

32.2.2 The MSC, the Marine Safety Committee

The legislature of the IMO is The Assembly comprising committees and sub-committees that deal with different issues.
Issues concerning safety are dealt with by the MSC (Maritime Safety Committee).
The MEPC, the Marine Environment Protection Committee, was established to handle environmental issues such as oil pollution of the sea, air pollution by the emission of noxious substances or the dumping of industrial and other waste at sea.
Over the years, two important conventions have been adopted:

- **SOLAS** – Safety Of Life At Sea.
- **Marpol** – Marine Pollution.

For both conventions, many rules have been drawn up. These must be ratified by a sufficient number of member states before the convention is adopted. It then enters into force and is implemented by all member states.

▶ Every sea-going vessel has a clearly visible IMO number in the engine room and on the deck.

32.2.3 Considerations for diesel engines

For diesel engines, the fuel used and the emission of noxious substances is of the utmost importance. Naturally, the discharge of oil and oily wastes at sea is bound by very strict regulations.
Seagoing vessels are required to have certain certificates on board that establish that they meet the requirements of IMO.

The compulsory equipment on board must be approved by the flag states and/or by one of the classification societies.

The following four certificates are in use:

For SOLAS:

1 Cargo Ship Safety Construction Certificate.
2 Cargo Ship Safety Equipment Certificate.
3 Cargo Ship Safety Radio Certificate.
4 Cargo Ship Safety Certificate, this combines 1, 2 and 3.

For Marpol:
This is expressed in The International Oil Pollution Certificate 9I.O.P.C.
This applies to, amongst others, oil tankers and cargo ships above 400 GT.
In detail, the following parts are covered:

Annex I: normal oil, such as, fuel and lubricating oil. Sludge and bilge water also belong to this annex.
Annex II: noxious liquids and chemicals.
Annex III: packed substances.
Annex IV: sewage.
Annex V: garbage.
Annex VI: air pollution, the exhaust gases of diesel engines.
Annex VII: ballast water.

32.3 Classification societies

They are active worldwide in the survey and certification of systems, assets and facilities to improve quality, increase safety and protect environment.
Many classification societies have more than one discipline, of which shipping is an important one.

32.3.1 Shipping

Their activities in the shipping field mainly comprise 'classifying' ships.
Classification societies draw up standards for the quality and the reliability with respect to design, construction and operation of ships.
Besides carrying out classification surveys and statutory inspections, compulsory under international shipping conventions and regulations, a classification society gives ship-owners and operators insight in the risks they face and helps them to improve business performance.

32.3.2 Classification

A classification society establishes the regulations regarding the construction of a certain type of ship. The society approves the drawings, gives advice for possible modifications and verifies that the construction of the ship and all the technical installations are conform its standards.
After completion of the ship, the ship-owner will receive a **Certificate of Classification, for the hull and the machinery.**
An Operating **Certificate** is also issued that is valid for 5 years. Every year, the certificate must be endorsed when the annual survey has been satisfactorily completed.
In order to provide this service, classification societies maintain a worldwide network of surveyors.
A Class certificate allows the vessel to be insured.

The society surveys the technical condition of the ship, the safety and the living conditions on board. Flag states allow these activities to be carried out by the classification societies.
Ships carry not only class certificates but also comply with statutory regulations.
Both are issued by the societies.

32.3.3 Acceptance by a flag state

A company can have a sea-going vessel registered in a particular country, the flag state. The flag state accepts the ship as a member of its fleet.
The authorities of a country allow the shipping company to sail under their jurisdiction at a certain (financial) tariff.
These tariffs vary considerably per country/flag state.
The flag state's name must be visibly displayed on the stern of the ship.
The certificate of registration, the international tonnage certificate, is issued by the flag state or by the classification company on behalf of the flag state.
The official details of the ship are listed in the tonnage certificate (also important for, amongst others, harbour dues).

32.4 Periodic inspections of the diesel engine and its parts

The condition of the engine parts must be checked regularly. Classification societies have strict regulations regarding the inspection periods and activities.
In view of the brief docking times of modern ships these 'surveys' are predominantly performed during dry-docking and other periods when the ship is out of service.
Therefore, timely contact between the shipping company/proprietor and the classification society is of utmost importance.

◄ On the stern of an approved vessel, the name of the vessel and the port of registry must be displayed. It is then clear to see under whose flag the vessel is sailing.

▶
During dry-docking, as shown here due to damage, many activities can be carried out to the propulsion plant, without extra delays and therefore extra expenditure.

32.4.1 Regulations for reconditioned parts

▶
Stamped code in a cylinder liner.

Class Germanischer Lloyd.
Pressure 7 bar.
A repair performed in 1995.
The 1995 year logo of GL.

Worn parts are often reconditioned.

Also, see Chapter 26, *Reconditioning engines and their parts.*

After repair, the part must be re-tested and therefore should be 'made available' to the classification society.
The reconditioning company makes an appointment for inspection with a surveyor from the society.
Often one (new constructions) or several (reconditioned) registration numbers are marked on the part by one or several societies.

▶
Testing the cooling-water spaces of newly installed cylinders. Left, the foundry code. This code displays all the manufacturing data for the engine block.

▲

Stamped codes on a crankshaft web.

After a number of repairs are done to the web, the 'fuller' it becomes.

▲

Stamped codes on the connecting-rod big end.

The material composition is also displayed, 42 CR M0 4.

Those who are aware of all the numbers and marks can easily verify which parts have been reconditioned and/or tested!

Vessels are constructed in accordance with stipulations from classification societies.
The classification society approves drawings, details and material types. Every detail is clearly registered.
During construction of the vessel, the classification society monitors the activities and ensures that the vessel is constructed in accordance with the approved drawings.
When construction is complete, all the systems are checked and tested.
The final check of the vessel is a test at sea, a sea trial.
The entire construction of the vessel, all the tests and ultimately the trials are performed in accordance with a specific class.
The classification society supplies the vessel with a certificate – the Certificate of Class for the vessel and machinery.
An operating certificate, valid for five years, is also issued. This certificate must be re-approved every year after an annual inspection.
The classification societies are all members of **IACS, the International Association of Classification Societies.**
This association has its headquarters in London and is the umbrella organisation of the ten largest societies.

For a complete classification, the classification society checks the vessel from the drawing table, the beginning of construction, the skeleton, the installations and the engines.
The classification society subsequently follows the entire life cycle of a vessel.
All vessels must be surveyed in order to comply with the requirements of seaworthiness.
The flag states are accountable. They usually outsource the actual testing to classification societies.

32.4.2 Classification societies

The following societies are active worldwide:

ABS – American Bureau of Shipping
BV – Bureau Veritas
CCS – China Classification Society
DNV – Det Norske Veritas
GL – Germanischer Lloyd
KR – Korean Register of Shipping
LR – Lloyd's Register
NKK – Nippon Kayi Kyokai
RINA – Registro Italiano Navale
RS – Russian Maritime Register of Shipping

The regulations of, for instance, **Lloyd's Register, comprise seven parts**:

Part 1: Rules/Regulations.
General regulations, the actual classification, periodic inspection.

Part 2: Regulations for the manufacture, for the testing and the classification of materials.
These tests are performed at foundries, iron-works and component manufacturers.

Part 3: Ship's structure.
Basis structure of the skeleton, longitudinal strength, bow and stern constructions.

Part 4: Ship's structure (ship types).
Skeleton-construction requirements for the carcass per ship type, such as towing boats, ferries, oil tankers and container ships.

Part 5: Main and auxiliary engines.
Including propeller-shaft alignment and vibrations, piping for oil tankers and steering gear.
This part is especially important for us; it provides a detailed description of the manner in which the propulsion engine must be constructed.

Part 6: Control, electrical systems, cooling systems and fire prevention.

Part 7: Remaining vessel types and plants.

32.5 Examples: Germanischer Lloyd

The following examples are a small part of the regulations.
It is not our intention to give a complete overview.
The societies have comprehensive regulations.
'Rules for Classification and Construction Ship Technology/Seagoing Ships/Machinery Installations'.
Key points are the acquisition of an idea of what these regulations mean. This overview is not complete!

▶
The general regulations for the set-up, selection and arrangement of machinery with respect to the maximum angle of inclination.
Furthermore, the water temperature, the air conditions and other ambient conditions in spaces, enclosed spaces and on deck.

Table 1.1 Inclinations

<table>
<tr><th rowspan="3">Installations, components</th><th colspan="4">Angle of inclination [°][2]</th></tr>
<tr><th colspan="2">Athwartship</th><th colspan="2">Fore-and aft</th></tr>
<tr><th>static</th><th>dynamic</th><th>static</th><th>dynamic</th></tr>
<tr><td>Main and auxiliary machinery</td><td>15</td><td>22,5</td><td>5[4]</td><td>7,5</td></tr>
<tr><td>Ship's safety equipment e.g. emergency power installations, emergency fire pumps and their drives</td><td rowspan="2">22,5[3]</td><td rowspan="2">22,5[3]</td><td rowspan="2">10</td><td rowspan="2">10</td></tr>
<tr><td>Switchgear, electrical and electronic appliances[1] and remote-control systems</td></tr>
</table>

1 Up to and angle of inclination of 45° no undesired switching operations or functional changes may occur.
2 athwartships and fore- and aft-inclinations may occur simultaneously.
3 On ships for the carriage of liquefied gases and chemicals the emergency power supply must also remain operational with the ship flooded to a final athwartships inclination up to a maximum of 30°.
4 Where the length of the ship exceeds 100 m, the fore-and-aft static angle of inclination may be taken as 500/L degrees.

Table 1.2 Water temperature

Coolant	Temperature [°C]
Seawater	+ 32[1]
Charge air coolant inlet to charge air cooler	+ 32[1]

1 GL may approve lower water temperatures for ships operating only in special geographical areas.

Table 1.3 Air temperature
at atmospheric pressure = 1000 mbar
and relative humidity = 60 %

<table>
<tr><th>Installations, components</th><th>Location, arrangement</th><th>Temperature range [°C]</th></tr>
<tr><td rowspan="4">Machinery and electrical installations[1]</td><td>in enclosed spaces</td><td>0 to 45[2]</td></tr>
<tr><td>on machinery components, boilers</td><td rowspan="2">According to specific local conditions</td></tr>
<tr><td>in spaces subject to higher or lower temperatures</td></tr>
<tr><td>on the open deck</td><td>- 25 to + 45</td></tr>
</table>

1 Electronic appliances shall be designed and tested to ensure trouble free operation even at a constant air temperature of + 55 °C.
2 The Society may approve lower air temperatures for ships designed only for service in particular geographical areas.

Table 1.4 Other ambient conditions

<table>
<tr><th>Location</th><th>Condition</th></tr>
<tr><td rowspan="3">In all spaces</td><td>Ability to withstand oil vapour and salt-laden air</td></tr>
<tr><td>Trouble-free operation within the temperature ranges stated in Table 1.3, and with a relative humidity up to 100 % at a reference temperature of 45 °C</td></tr>
<tr><td>Tolerance to condensation is assumed</td></tr>
<tr><td>In specially protected control rooms</td><td>80 % relative humidity at a reference temperature of 45 °C</td></tr>
<tr><td>On the open deck</td><td>Ability to withstand temporary flooding with seawater and salt-laden spray</td></tr>
</table>

32.5.1 General regulations and instructions

These comprise general subjects, related to, amongst others:

- The propulsion plants and the auxiliary devices required to warrant the ship's safety during operation.
- The required drawings and all necessary information.
- The atmospheric conditions such as air temperature, humidity and air pressure.
- Allowable vibrations.
- The design of the installation, in particular with regard to the stresses and loads, material types, accessibility of the engine room, operation of the propulsion plant and manoeuvring.
- Fuels used. The flash point, for instance, may not drop below 60 °C and for auxiliary engines must be equal to or higher than 43 °C.
- Additional regulations for measuring instruments, such as pressure gauges, lighting, bilge spaces/pipes, ventilation, enclosure of rotating parts, alarms and communication.
- All the equipment that is part of the propulsion plant, such as diesel generators, pumps, starting-air systems, turbochargers and ventilation.

32.5.2 Internal combustion engines and starting-air compressors

Today, there are individual regulations for dual-fuel engines for, amongst others, LNG tankers. These are separate from the diesel-engine regulations.

◄ Special regulations apply for 'dual fuel' engines, such as on this modern LNG tanker.

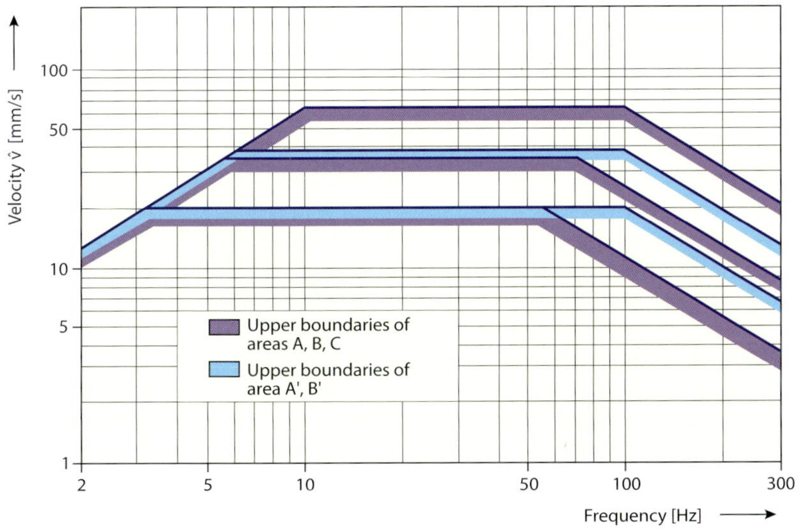

◄ The maximum vibrations permissible in certain spaces.

Table 1.5 Numerical definition of the area boundaries shown in figure above

Areas	A	B	C	A'	B'
ŝ [mm]	< 1	< 1	< 1	< 1	< 1
v̂ [mm/s]	< 20	< 35	< 63	< 20	< 40
v_{eff} [mm/s]	< 14	< 25	< 45	< 14	< 28
â [9,81 m/s^2]	< 0,7	< 1,6	< 4	< 1,3	< 2,6

32.5.3 Important regulations

To calculate the power output of the engine, the following atmospheric conditions are applied:

Air pressure	1000 millibar
Inlet-air temperature	45 °C
Relative humidity of the air	60%
Seawater temperature	32 °C

32.5.4 Standard nominal power

The use of a graph is required.

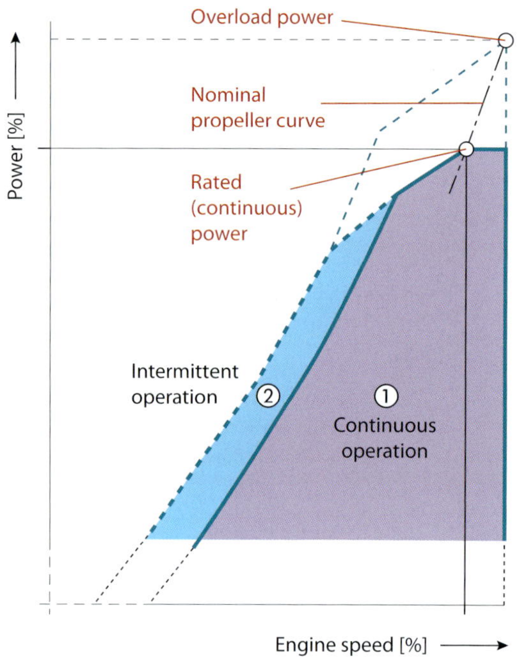

A power diagram for a diesel engine driving a propeller.

1 Continuous operation.
2 Intermittent operation.

It should be possible to exceed the standard nominal power with 10% for as long as one hour.

- After trials, an engine overload should not be possible. Therefore, the fuel supply must be restricted to 100% standard nominal power.
- This rule does not apply for diesel generators.
- However, the rule does apply for dual-fuel engines.
- Accessibility. This is an important topic for engineers who work on engines.
- All the activities, such as the changing of components, must be able to be performed without any problems.
- Electric components and measuring instruments, which operate and control the engine, must meet certain requirements. This applies, for instance, to computer systems.
- If an electrical/electronic system fails, the propulsion engine should not stop or undergo a drastic change in the power output. This also applies to speed regulators such as governors.

According to the regulations, all engine parts must be accessible for inspection, repair and maintenance.

In practice, this is sometimes very difficult to realise. Often, the costs of the alterations are too high.
Shown here, two service engineers from Wärtsilä removing a cover from the 'hot box'. The upper side of this Wärtsilä 46 is easily accessible. Note, the IMO number on the engine-room bulkhead.

32.5.5 Documents of approval

The engine manufacturer must provide the classification society with drawings and data for each engine, such as those listed in the table below.

Manufacturers operating under licence are also required to present these documents; these must include a statement by the licenser when modifications have taken place.

◀ A table of all parts and drawings that must be submitted for approval.

Table 2.1 Documents for approval

Ser. No.	A/R	Description	Quantity	Remarks (see below)
1	R	Detail required on GL forms F 144 and F 144/1 when applying for approval of an internal combustion engines	3	
2	R	Engine tansverse cross-section	3	
3	R	Engine longitudinal section	3	
4		Bedplate and crankcase		
	R	– cast	1	
	A	– welded, with welding details and instructions	3	9
5	R	Thrust bearing assembly	3	3
		Thrust bearing bedplate		
6	R	– cast	1	
	A	– welded, with welding details and instructions	3	9
		Frame/framebox		
7	R	– cast	1	1
	A	– welded, with welding details and instructions	3	1, 9
8	R	Tie rod	1	
9	R	Cylinder cover/head, assembly	1	
10	R	Cylinder liner	1	
11	A	Crankshaft for each number of cylinder, with data sheets for calculation of crankshafts	3	
12	A	Crankshaft assembly, for each number of cylinders	3	
13	A	Thrust shaft of intermediate shaft (if integral with engines)	3	
14	A	Shaft coupling bolts	3	
15	R	Counterweights including fastening bolts	3	
16	R	Connecting rod, details	3	
17	R	Connection rod, assembly	3	
18	R	Crosshead assembly	3	2
19	R	Piston rod assembly	3	2
20	R	Piston assembly	1	
21	R	Camshaft drive, assembly	1	
22	A	Material specifications of main parts with information on non-destructive material tests and pressure tests	3	8
23	R	Arrangement of foundation (for main engines only)	3	
24	A	Schematic layout or other equivalent documents of starting air system	3	6
25	A	Schematic layout or other equivalent documents of fuel oil system	3	6
26	A	Schematic layout or other equivalent documents of lubricating oil system	3	6
27	A	Schematic layout or other equivalent documents of cooling water system	3	6
28	A	Schematic diagram of engine control and safety system	3	6
29	A	Schematic diagram of electronic components and systems	1	
30	R	Shieldin and insulation of exhaust pipes, assembly	1	
31	A	Shieldin of high-pressure fuel pipes, assembly	3	4
32	A	Arrangement of crankcase explosion relief valves	3	5
33	R	Operation and service manuals	1	7
34	A	Schematic layout or other equivalent documents of hydraulic system (for valve lift) on the engine	3	
35	A	Type test progam and type test report	1	
36	A	High pressure parts for fuel oil injection system	3	10

1 only for one cylinder.
2 only necessary if sufficient details are not shown on the transverse cross section and longitudinal section.
3 if integral with engine and not integrated in the bedplate.
4 for all engines.
5 only for engines with a bore > 200 mm, or a crankcase volume ≥ 0,6 m^3
6 and the systems, where this is supplied by the engine manufacturer. If engines incorporate electronic control systems a failure mode and effect analysis (FMEA) is to be submitted to demonstrate that failure of an electronic control system will not result in the loss of essential services for the operation of the engine and that operation of the engines will not be lost or degraded beyond an acceptable performance criteria of the engine.
7 operation and service manuals are to contain maintenance requirements (servicing and repair) including details of any special tools and gauges that are to be used with their fitting/settings together with any test requirements on completion of maintenance.
8 for comparison with GL requirements for material. NDT and pressure testing as applicable.
9 The weld procedure specification is to include details of pre and post weld heat treatment, welding consumables and fit-up conditions.
10 The documentation has to contain specifications of pressures, pipe dimensions and materials.
A for approval
R for reference

32.5.6 Definition of the diesel-engine type

The type specification of an internal-combustion engine is defined by the following data:
- Type number of the manufacturer.
- Cylinder diameter.
- Stroke.
- Fuel-injection method.
- Fuel used.
- Operating principle: four stroke or two-stroke.
- Scavenging system: natural aspiration or supercharging.
- Nominal cylinder power at nominal speed and the maximum mean effective pressure.
- Method of supercharging: intermittend pressure system or constant-pressure system.
- Charging-air cooling system: type of intercooler.
- Cylinder arrangement: in-line or V.

For additional engine components such as turbochargers, heat exchangers, engine-driven pumps and other machinery, the manufacturer must provide the classification society with the required particulars.

32.5.7 Calculations for the crankshaft

The most important and yet most vulnerable part of the engine is the crankshaft.
The load on crankshafts is high and damaged crankshafts can result in an engine total loss.

Repairing crankshafts is costly, as they usually have to be removed from the engine room. This is in itself expensive. Consider disassembly, assembly and the time the vessel is non-operational!

There are many considerations.

- Shrink joints of built-up crankshafts for two-stroke crosshead engines must be designed in accordance with certain regulations.
- The bolts used to connect power-end flange couplings must be the specifically prescribed fitted bolts.

32.6 Materials for diesel engines

The mechanical properties of materials used for engine parts must meet strict requirements.

The following individually designed parts must be submitted for material tests:
- The crankshaft.
- The crankshaft-coupling flange for main power transmission.
- The crankshaft-coupling bolts.
- Pistons or piston crowns made of steel, cast steel or nodular cast iron.
- Piston rods.
- Connecting rods including bearing caps.
- Crossheads.

▶
Very strict regulations are applicable for crankshafts due to their high loads.

- Cylinder liners made of steel or cast iron.
- Cylinder covers made of steel or cast iron.
- Welded crankshaft bedplate and the bearing transverse girders made of forged or cast steel.
- Welded A-frames and crankcases.
- Welded fitting caps.
- Tie rods.
- Bolts and studs for cylinder covers, crossheads, main bearings and connecting-rod bearings.
- Camshaft-drive gear wheels and chain wheels for, amongst others, some two-stroke crosshead engines, made of steel or cast iron.
- Dependent on the cylinder diameter and therefore the engine size, certain parts are submitted for material testing.
- Above a cylinder diameter of 400 mm, **all** the parts mentioned above must be submitted for material testing.

Table 2.2 Material tests

Cylinder bore	Parts to be tested*
≤ 300 mm	1 - 6 - 10 - 11 - 12 - 13
> 300 ≤ 400 mm	1 - 6 - 8 - 9 - 10 - 11 - 12 - 13 - 14
> 400 mm	all parts
* numbered according to the list under D.2.1 above	

▲

The number of parts that must be subjected to material testing is dependent on the cylinder diameter of the engine. Above 400 millimetres, all parts must be submitted for testing.

- In addition, material tests must be carried out on pipes and parts of the starting-air system and other pressure systems.
- Materials for air coolers must be delivered with test reports from the manufacturer.
- In individually manufactured engines, non-destructive material tests must be performed.
- Especially the drive gearing and other components subjected to heavy loads, such as cylinder covers, are subjected to special testing.

32.7 Tests and trials

32.7.1 Inspection during engine construction

If the engine manufacturer is approved as a 'Supplier of mass-produced engines' by a classification society, special regulations apply for this manufacturer.

▼

A table of approved materials and type of test certificate.

Table 2.3 Approved materials and type of test certificate

Approved materials	Society's Rules*	Components	Test certificate†		
			A	B	C
Forged steel $R_m \geq 360$ N/mm²	Section 3, C.	Crankshafts	X	–	–
		Connecting rods	X	–	–
		Pistons rods	X[3]	X[4]	–
		Crossheads	X[3]	X[4]	–
		Pistons and piston crowns	X[3]	X[4]	–
		Cylinder covers/heads	X	–	–
		Camshaft drive wheels	X[3]	X[4]	–
Rolled or forged steel rounds $R_m \geq 360$ N/mm²	Section 3, C.	Tie rods	X	–	–
		Bolts and studs	X[1]	X[2]	–
Special grade cast steel $R_m \geq 440$ N/mm²	Section 4, C.	Throws and webs of build-up crankshafts	X	–	–
Special grade forged steel $R_m \geq 440$ N/mm²	Section 3, C.				
Cast steel	Section 4, C.	Bearing transverse girders (weldable)	X	–	–
		Pistons and piston crowns	X[3]	X[4]	–
		Cylinder covers/heads	X[1]	X[2]	–
		Camshaft drive wheels	X[3]	X[4]	–
Nodular cast iron, preferably ferritic grades $R_m \geq 370$ N/mm²	Section 5, B.	Engine blocks	–	X[1]	–
		Bedplates	–	X[1]	–
		Cylinder blocks	–	X[1]	–
		Pistons and piston crowns	X[3]	X[4]	–
		Cylinder covers/heads	–	X[1]	–
		Flywheels	–	X[1]	–
		Valve bodies	–	X[1]	–
Lamellar cast iron $R_m \geq 200$ N/mm²	Section 5, C.	Engine blocks	–	–	X
		Bedplates	–	–	X
		Cylinder blocks	–	–	X
		Cylinder liners	–	–	X
		Cylinder covers/heads	–	–	X
		Flywheels	–	–	X
Shipbuilding steel, all GL grades for plate thickness ≤ 35 mm	Section 1. B.	welded bedplates	X	–	–
Shipbuilding steel, GL grade B for plate thickness > 35 mm		welded frames	X	–	–
		Welded housings	X	–	–
Structural steel, unalloyed, for welded assemblies	Section 1, F.				

* All details refer to the GL Rules II – Materials and Welding, Part I – Metallic Materials, Chapter 2 - Steel and Iron Materials

† Test certificates are to be issued in accordance with GL Rules II – Materials and Welding, Part I – Metallic Materials, Chapter 1 – Principles and Test Procedures – Section 1, H. with the following abbreviations:
A: GL Material Certificate. B: Manufacturer Inspection Certificate. C: Manufacturer Test Report

1 only for cylinder bores > 300 mm
2 for cylinder bores ≤ 300 mm
3 only for cylinder bores > 400 mm
4 for cylinder bores ≤ 400 mm

▶

The compulsory magnetic particle tests and ultrasonic tests are applicable for a number of engine parts at certain cylinder diameters.

Table 2.4 Magnetic particle tests

Cylinder bore	Parts to be tested*
≤ 400 mm	1 - 2 - 3 - 4 - 5
> 400 mm	all parts
* numbered according to the list under D.2.2 above	

Table 2.5 Ultrasonic tests

Cylinder bore	Parts to be tested*
≤ 400 mm	1 - 2 - 3 - 4 - 7 - 10
> 400 mm	1 - 2 - 3 - 4 - 5 - 6 - 7 - 10 - 11
* numbered according to the list under D.2.2 above	

▶ A number of parts must also be pressure tested.

Table 2.6 Pressure tests[1]

Component		Test pressure, p_p [bar][2]
Cylinder cover, cooling water space[3]		7
Cylinder liner, over whole length of cooling water space[5]		7
Cylinder jacket, cooling water space		4, at least $1,5 \cdot p_{e,zul}$
Exhaust valve, cooling water space		4, at least $1,5 \cdot p_{e,zul}$
Piston, cooling water space (after assembly with piston rod, if applicable)		7
Fuel injection system	pump body, pressure side	$1,5 \cdot p_{e,zul}$ or $p_{e,zul} + 300$ (whichever is less)
	Valves	$1,5 \cdot p_{e,zul}$ or $p_{e,zul} + 300$ (whichever is less)
	Pipes	$1,5 \cdot p_{e,zul}$ or $p_{e,zul} + 300$ (whichever is less)
Hydraulic system	High pressure piping for hydraulic drive of exhaust gas valves	$1,5 \cdot p_{e,zul}$
Exhaust gas turbocharger, cooling water space		4, at least $1,5 \cdot p_{e,zul}$
Exhaust gas line, cooling water space		4, at least $1,5 \cdot p_{e,zul}$
Coolers, both sides[4]		4, at least $1,5 \cdot p_{e,zul}$
Engine-driven pumps (oil, water, fuel and bilge pumps)		4, at least $1,5 \cdot p_{e,zul}$
Starting and control air system		$1,5 \cdot p_{e,zul}$ before installation

1 In general, items are to be tested by hydraulic pressure as indicated in the Table. Where design or testing features may require modification of these test requirements, special consideration will be given.
2 $P_{e,zul}$ [bar] = maximum working pressure in the part concerned.
3 For forged steel cylinder covers test methods other than pressure testing may be accepted e.g. suitable non-destructive examination and dimensional control exactly recorded.
4 Charge air coolers need only be tested on the water side.
5 For centrifugally cast cylinder liners, the pressure test can be replaced by a crack test.

32.7.2 Pressure tests

Certain parts of the engine are subjected to pressure testing.

32.7.3 Type-approval testing

Diesel engines used on board ships, must be type-tested in the presence of a classification society's surveyor.

32.7.4 Conditions for testing certain engine types

The engine must be tested in accordance with the specific engine properties.
The inspections and measurements performed by the engine manufacturer for reliable continuous operation are submitted to the classification society.

32.7.5 Scope of type-approval testing

The type-approval test comprises three parts:

A Internal test.
Tests under operating conditions where the operating data of the engine, such as pressures, temperatures, power output and speed are collected.

B Type test.
The tests take place in the presence of a surveyor from a classification society.

C Part inspection.
After the test run, certain parts of the engine must be submitted to the classification society for inspection. This entails the disassembly of these parts.

32.7.6 Load during the test

In normal cases, the engine load is 25%, 50%, 75%, 100% and 110% of the maximum rated power.

a. This is performed along the nominal (theoretical) propeller curve and/or at constant speed in propulsion engines.
b. At rated speed with a constant governor setting for continuous operation in a genset.

For engines supplied with supercharged air by a turbocharger, the achievable power output in the case of turbocharger damage is indicated as follows:

- Engines with one turbocharger, if the rotor is blocked or removed.
- Engines with two of more turbochargers, if the damaged turbocharger is turned off.

B test: Type test

This test is carried out in the presence of a surveyor from a classification society.
Deviations from this program require the agreement of the classification society.

▶

This two-stroke crosshead engine is ready to be installed in the ship (docked in the background). It has been subjected to intensive testing in accordance with the stipulations at the factory.

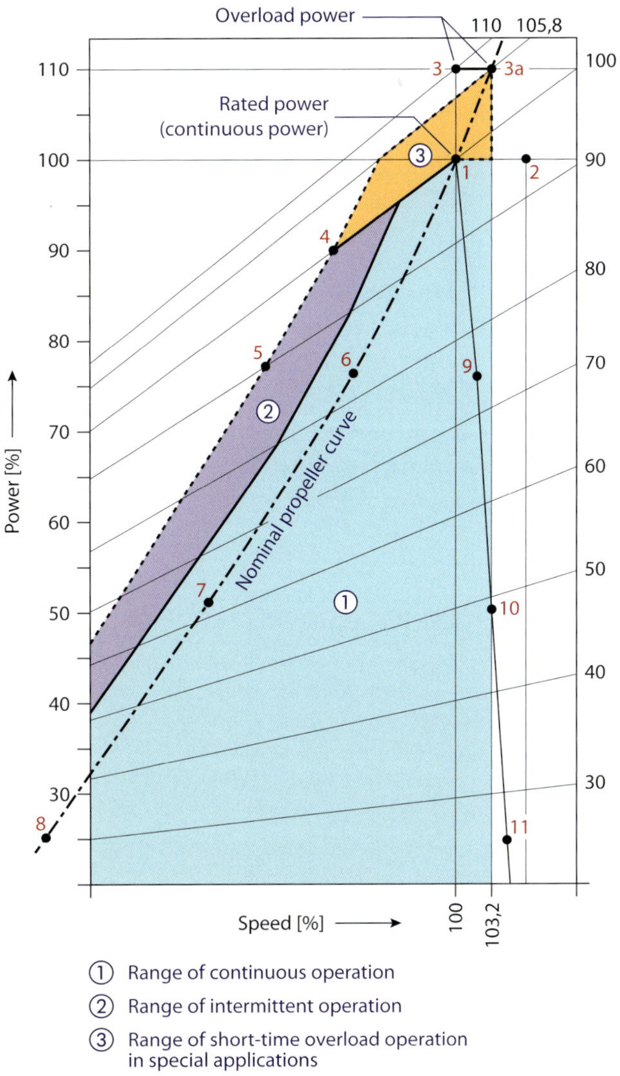

▶

An engine-load diagram.

Horizontal: the speed
Vertical left: the power output
Vertical right: the torque

- The data measured and recorded when testing the engine at various load points must include all the parameters necessary for an assessment.
- The operating time per load point depends on the engine size and on the time required for collection of the operating values. The measurements shall in every case only be taken after a steady-state condition is achieved.
- Normally, an operating time of half an hour per load point is kept.
- At 100% output (rated power), an operating time of half an hour is required. At least two sets of readings must be taken with a 1 hour interval in each case.

The nominal maximum continuous power, 100% power.

Rated power/continuous power.

The operating point of 100% output at maximum allowable speed must correspond to load point 2 during the test.

Maximum permissible torque.

The maximum permissible torque normally results at 110% output at 100% speed corresponding to load point 3.

The minimum permissible speed for intermittent operation has to be adjusted:

- at 100% torque corresponding to load point 4.
- at 90% torque corresponding to load point 5.

Partial load.

For partial-load operations of 75%, 50% and 25% of the rated power with speeds corresponding to the maximum nominal propeller curve, load points 6, 7 and 8 are held. Proceeding from the nominal speed at a constant governor setting, the operating points have to be adjusted corresponding to load points 9, 10 and 11.

Emergency operation for problems with turbochargers

The maximum achievable power can be attained:

- At a speed conforming to normal propeller curve.
- With a constant governor setting for rated speed.

Functional tests

These must be performed as follows:

- Ascertaining the lowest speed according to the nominal propeller curve.
- Starting tests for non-reversible and/or reversible engines.
- Governor tests.
- Test of the safety system, especially for 'overspeed' and failure of the lubricating-oil pressure.

C test: Component inspection

Immediately after testing, the parts from one cylinder for in-line engines and two cylinders from V-engines must be presented for inspection as follows:

- Piston, dismantled.
- Crosshead bearing, dismantled.
- Crankpin and main bearing, dismantled.
- Cylinder liner may remain in the engine block.
- Cylinder head/cover, all the valves disassembled.
- Camshaft, camshaft drive and crankcase, with the covers removed.
- If the classification society deems it necessary, further dismantling of the engine may be required for the inspection of other parts.

After this, the classification society draws up a test report and if this is favourable, a **'Type Approval Certificate'** is issued.

32.8 Testing mass-produced engines

For engines with a cylinder bore smaller than or equal to 300 mm, numerous conditions apply.

Some important points:
If the power of a certain type of engine increases by more than 10%, a new test program is required. This frequently occurs in practice. When introducing new types of engines, the power is usually increased.

32.9 Shipboard trials

After the conclusion of the testing program prescribed by the engine manufacturer, the vessels undergo further trials in open water, both stationary and sailing.
The final trials are known as 'the sea trials'.

▶
After construction of the vessel has been completed at the wharf, the final procedure before the vessel is delivered to her new owner takes place: the sea trials.

32.10 Some important points

32.10.1 Main-engine propulsion with a fixed propeller

The tests must be carried out as follows:

a. At maximum permissible speed during at least four hours and an engine speed that is usual for normal cruising power for at least two hours.
b. At a speed at 103% of the maximum allowable speed for 30 minutes when the engine adjustments allow this.
c. Determining the minimum speed at load to determine when the engine stalls.
d. Starting/stopping and reversing manoeuvres.
e. Reversing the direction of the propeller rotation at a speed of at least 70% for 10 minutes.
f. Testing the monitoring and safety systems.

32.10.2 Main-engine propulsion systems with a controllable-pitch propeller or reverse gearing

Controllable-pitch propellers must be tested at various propeller pitches.
In combined systems for speed and a certain pitch, a combination curve must be plotted and verified by measurements.

32.10.3 Main engines that drive generators, such as those for diesel-electric propulsion

The tests must be performed at nominal maximum speed with a constant governor setting under the following conditions:

a. 100% nominal maximum power for at least four hours and at normal continuous cruising power for at least two hours.
b. At 110% of maximum power for 30 minutes.
c. At a minimum speed of at least 70% of the maximum speed for 10 minutes when reversing the direction of propeller rotation during sea trials.
d. Starting/stopping procedures.
e. Testing the monitoring and safety systems.

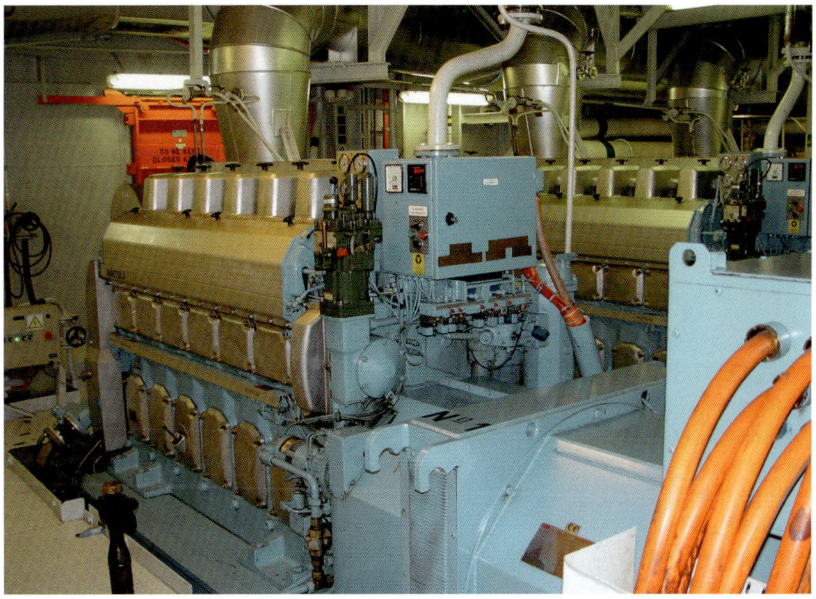

▲ Diesel engines for generators also undergo extensive testing.

32.10.4 Diesel-engine driven auxiliaries and electric generators

These engines must be subjected to an operational test for at least four hours.
During the entire test, the set must operate at maximum permissible power.
The test must establish that the set is capable of operating at 110% of its normal maximum power, and for diesel gensets, the amount of time required for the actuation of the overload protection system.

32.10.5 Fuel type

If the main and auxiliary engines can run on heavy oil or another type of fuel oil, this must be demonstrated by a test.
The program for the sea trials can be extended for special operating conditions, such as towing and fishing.

32.10.6 Earthing

It is necessary to check the limits. These are specified by the engine manufacturers, the potential difference between the crankshaft and propeller shaft to the hull must not be exceeded during operation.
Appropriate earthing devices must ensure that this cannot occur. Furthermore, the potential difference must be monitored and alarmed.

▶
A slip-ring construction on the propeller shaft prevents too large a potential difference between crankshaft, propeller shaft and hull.
This prevents damage to the engine, especially the crankshaft.

32.10.7 Safety devices

Here many regulations are applicable, for instance:

32.10.8 Speed control and protection against overspeed

Some comments:
- Each diesel engine not used to drive an electric generator must be protected with a governor, which is set so that the engine speed cannot exceed the normal maximum speed by more than 15%.
- Each main engine with a shaft power of 220 kW or more, which can be declutched in operation or drives a controllable-pitch propeller, must be equipped with an independent overspeed protection. This must ensure that the normal maximum speed cannot be exceeded by more than 20%.

32.10.9 Engines driving gensets

Here strict regulations apply, also in view of the damage that can be caused to electrical-power consumers.
Each engine or emergency diesel-power unit must be supplied with a governor, which ensures that the frequency fluctuations in the electrical network do not exceed plus or minus 10% of the normal frequency, with a recovery time to the steady-state frequency conditions of a maximum of five seconds, when the maximum electrical load is switched on or off.
If a certain load equal to the maximum load is switched off, a speed deviation of 10% of the nominal

▶
Strict requirements also apply for governors.

◀
A safety valve on the cylinder cover for a two-stroke crosshead engine. This opens at 40% above the normal maximum combustion pressure.

Serious engine damage can occur, especially with a temporary engine shutdown.
A leaking injector can result in a very high peak pressure. Cooling-water leaks can cause a great deal of damage to the piston and connecting-rod assembly in the cylinder head. This is known as water hammer.

maximum speed is allowed. It should be noted that the speed safety of + 15% must not be activated.
Apart from the normal governor, each diesel engine with a power of 220 kW of more must be equipped with an **overspeed protection device**, operating independently of the normal governor.
The protection device must prevent normal maximum speed exceeding 15%.
There are many regulations for maintaining a constant frequency and voltage of the electrical network.
When increasing the load in two stages, this is allowed from 0 to 50% and subsequently from 50 to 100%.
Here, the frequency is not allowed to deviate too much from the normal frequency, with a maximum deviation of 10% and not for more than five seconds.
Gensets installed in 'stand-by' systems must operate on full load within 30 seconds after the start, also when the engine is cold, and therefore without heated cooling water.
For emergency diesel gensets, this is 45 seconds.

32.11 Regulations for propulsion engines

To ensure that the speed of the engine is controllable in all circumstances, the following requirements apply:

a. The governor system has an independent 'back-up' system.
b. There is an extra governor system that can be manually controlled with a separate, protected power supply.
c. The engine has a hand-operated fuel control, suitable for manoeuvring.

32.11.1 Safety system against excessive pressures in the combustion space

All the cylinders of engines with a bore larger than 230 mm must be fitted with a safety system that is actuated when the pressure exceeds 40% of the normal maximum combustion pressure.

32.11.2 Crankcase airing

This is not allowed.

32.11.3 Crankcase venting

The openings should be as small as possible.
In systems for extracting lubricating-oil vapours or, for instance, monitoring the oil-vapour concentration in crankcase mist detectors, the under-pressure in the crankcase may not exceed 2.5 millibar.
The vent pipes of two or more engines may not be combined.
In two-stroke engines, the oil vapour from the crankcase should not be transported to the scavenging-air receiver or the air-inlet receiver of the engine.

32.11.4 Safety valves on crankcases

Each type must be provided with a certificate. Safety valves for the crankcase must be fitted on each engine with a cylinder bore larger than 200 mm or a crankcase volume equal to or larger than 0.6 m^3.

All the other separate spaces in the crankcase, such as, the gear- or chain casings that drive the camshaft or similar drives, must be fitted with their own safety valves if the volume of these spaces exceeds 0.6 m^3.

Engines with a cylinder bore larger than 200 millimetres and equal to or smaller than 250 millimetres must be fitted with at least one safety valve at each end of the crankcase.

An engine with more than eight cylinders must be fitted with an extra safety valve in the centre of the crankcase.

Engines with a cylinder bore larger than 250 mm and smaller than or equal to 300 mm must be fitted with at least one safety valve placed near every second crank web with a minimum of two.

Engines with a cylinder bore larger than 300 millimetres must be equipped with at least one safety valve per cylinder. Each safety valve must at a minimum have a free cross-sectional area of 45 cm^2.

The total free cross-sectional area of the safety valve on a crankcase in an engine should not be less than 115 cm^2 per m^3 crankcase volume.

▶ A relief valve.

Prescribed for a gear-casing section with a capacity of 0.6 m^3 or more.
Prescribed for engines with a cylinder diameter greater than 200 mm.

▶ Relief valves are required for every crank of a two-stroke crosshead engine with cylinder diameters larger than 300 mm.

◄ Clearly shown, the exiting pressure wave is directed towards the floor when the relief valves open. This makes it as safe as possible for people in the vicinity.

Notes about operation

The safety valve must be of the rapidly 'open and close' type.
In normal operation, they must be oil-tight when closed and prevent air from entering the crankcase.
The gas flow must be deflected by a baffling plate in order to guarantee the safety of persons standing nearby when an actual explosion occurs.
The gas flow is often directed downwards.

32.11.5 Warning sign

A warning sign indicating that the crankcase covers or doors should not be opened immediately after stoppage, but only after a certain cooling-down period, must be mounted on both sides of the engine.

32.11.6 Oil-mist detection system

Engines with a cylinder bore larger than 300 millimetres or with a maximum power output of 2250 kW or higher must be equipped with an oil-mist detection system.
One should be able to read this at a safe distance from the engine.

◄ An oil-mist detection system is mandatory on diesel engines with a cylinder diameter of more than 300 millimetres or a power output of 2250 kW or higher.

▼ The compulsory 'warning sign' mounted on the engine next to or on the crankcase cover.

▶ The compulsory relief valves for the scavenging-air space of a two-stroke crosshead engine.

32.11.7 Safety devices in the starting-air system

Many starting-air systems work with pressures of 30 bars. This may be dangerous in certain circumstances.
A separate check valve must be placed in the starting-air line of every engine.
Engines with a cylinder bore larger than 230 millimetres must be supplied with a flame arrestor under the following conditions:

a. On directly reversible engines, at the starting-air valve on each cylinder.
b. On non-reversible engines, at the supply line of the main starting-air duct for the engine.

▲ The flame arrestors in the supply line for the starting-air valve of a two-stroke crosshead engine from manufacturer Wärtsilä, RT-Flex96C.

32.11.8 Safety devices in the lubricating-oil system

Each engine with a maximum continuous power of 220 kW or more must be fitted with safety devices, which automatically shut down the engine in the event of a failure in the lubricating-oil supply.
This is not required for engines driving emergency gensets or emergency fire pumps. These engines require a lubricating-oil pressure alarm.

32.11.9 Safety devices in inlet-air manifolds

Inlet-air manifolds connected with the cylinders must be fitted with safety valves similar to those fitted to the crankcase.

32.11.10 Fuel systems

Many regulations are applicable.

- Flanged fuel-line connections may only to be fitted with metal sealing surfaces or a similar approved design.
- Feed and return fuel-line systems of the engine must be designed in such a way that no excessively high pressures can occur in the system.
- All the parts of the system must be sufficiently strong to withstand the maximum pressures that occur in the system.
- Fuel-oil lines should not be placed too close to high-temperature systems. Steam pipelines, exhaust manifolds, silencers and other equipment must be adequately insulated.

Shielding

Regardless of the intended use and location of internal combustion engines, all external fuel-injection lines between the high-pressure fuel pumps and injectors must be double-walled.

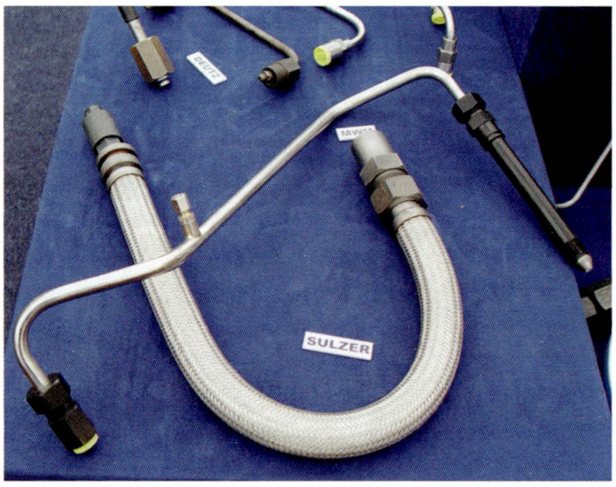

▲ The compulsory double-walled high-pressure fuel line, between the high-pressure fuel pump and high-pressure injector.

The fuel must be:

- Safe enough to be collected.
- Drained without an over-pressure.
- Efficiently monitored and alarmed.
- If pressure variations larger than 20 bars occur in feed- and return-fuel lines, these lines must also be double-walled.
- The double-walled high-pressure fuel lines must be of permanent assembly.

▲

The double-walled high-pressure fuel- and lubricating-oil pipes for opening the exhaust valves and the three injectors in a two-stroke crosshead engine.

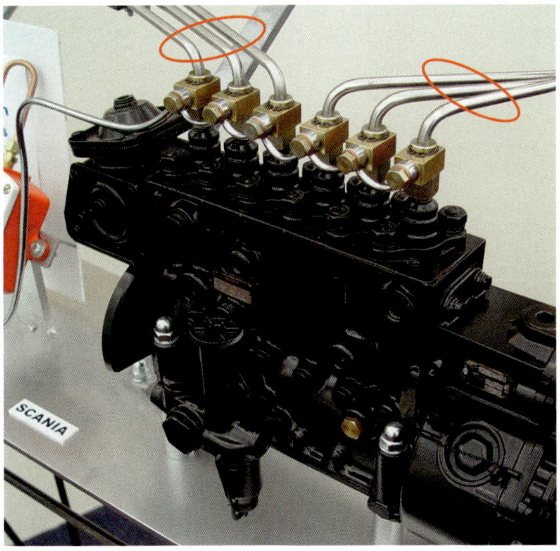

▲

A fuel pump for a six-cylinder Scania diesel engine.

The high-pressure supply lines to the injectors are double-walled.

32.11.11 Fuel-oil emulsions

Engines running on fuel emulsions or other liquid fuels must be able to convert to normal fuel-oil treatment systems in cases of failure.

32.11.12 Filters for fuel- or lubricating-oil systems

Filters directly mounted on the engine should not be located in the immediate proximity of rotating or hot parts. If this is not feasible, the rotating or hot parts must be shielded.

◀

A readily accessible heavy-oil fuel filter. The heavy oil can be easily led from the one to the other filter element; it is also easy to see when the filter is dirty.

32.11.13 Duplex filters

- Filters must be provided with a drip tray.
- Filters must be able to be safely switched off and drained for cleaning purposes.
- It should be clear which filter is on-line in the system.
- The off-line filter must be installed so that it can be cleaned while the engine is running.

◀

Strict requirements also apply for these fuel filters.

- It should also be feasible to vent the filter and depressurise the off-line filter.

32.11.14 The lubricating-oil system

- The sumps must be able to be refilled during engine operation.
- One should be able to drain the sump.
- Combining the oil-drainage lines from the sumps of two or more engines is not allowed.
- Main lubricating-oil pumps driven by the engine must be designed to maintain the supply of lubricating oil under all operating conditions.
- Main engines, which drive main lubricating-oil pumps, must be equipped with independently-driven stand-by lubricating-oil pumps.
- Cylinder lubricating-oil systems for the engine provided with electronically controlled dosing systems must be approved by a classification society.

▼
Cylinder lubricating-oil systems for the engine must be approved by a classification society, as they are supplied with electronically controlled dosing systems.

32.11.15 Cooling systems

Similar to the lubricating-oil pumps, cooling-water pumps must have enough capacity to supply a sufficient amount of cooling water for all operating conditions.
Main engines that drive main cooling-water pumps must be equipped with independently driven stand-by pumps.
If cooling air is obtained from the engine room, the design of the cooling system must be based on an engine-room temperature of at least 45 °C.

32.11.16 Turbocharger systems

Turbochargers must not reach critical speed ranges under any operating conditions.
The supply of lubricating oil must also be sufficient during starting-up and shutting-down the turbochargers.

Even at low engine speeds, the main engines must be supplied with a sufficient amount of turbocharged air to ensure a reliable operation.

▲
The lubricating-oil system of an ABB turbocharger.

The turbine shaft drives the lubricating-oil pump that lubricates the bearings. Therefore, the lubricating oil in the system is separate from the rest of the engine and remains clean. Look at the sight glass!
When the turbocharger is operational, the bearings are always well lubricated under all conditions.

▲
An extra lubricating-oil tank supplies the lubricant for this turbocharger's bearings when the engine stops, until the turbocharger stops.

▲
The electrically driven auxiliary charger for a two-stroke crosshead engine.

◀
The steel sheet protection of the exhaust-gas line in this propulsion engine may attain a maximum surface temperature of 220 °C.

If required, two-stroke crosshead engines must be fitted with directly- or indirectly driven scavenging-air pumps.
In the event of a turbocharger failure, the main engines must be able to maintain emergency operation.

32.11.17 Scavenging-air cooling

Means are to be provided to ensure that the temperature of the scavenging air is within the temperature range specified by the engine manufacturer.
The air-inlet pipes of engines with air coolers must be supplied with good drainage points.

32.11.18 Fire-extinguishing systems on scavenging-air lines

The scavenging-air line for two-stroke crosshead engines must be supplied with an approved fire-extinguishing system that operates independently of the engine-room's fire-extinguishing system.

32.11.19 Exhaust-gas lines

Exhaust-gas lines must be insulated and/or cooled to prevent the surface temperature exceeding 220 °C.
The insulation material must be fireproof.

32.11.20 Starting-air systems

Many regulations also apply.
- There must be at least two starting-air compressors for the main engines.
- At least one of the two starting-air compressors should operate independently of the main engine, and should provide at least 50% of the total compressor capacity.
- The volume of the starting-air reservoirs should be such that they can be filled in one hour to the legally required capacity from atmospheric pressure to the maximum permitted pressure.
- If the main engine is started with starting air, the available amount of starting air must be distributed over two starting-air reservoirs of similar capacity.
- The total capacity of the starting-air reservoirs must be of sufficient capacity in order to achieve twelve sequential starts of the engine in ahead and astern direction in the case with reversible engines, without re-charging. A minimum of six starts is prescribed for non-reversible engines driving a controllable-pitch propeller or they must be equipped with another device, which allows the engine to start without a load.
- For auxiliary engines with a starting-air system, this should be at least three starts.

▶
Two large starting-air reservoirs for a twelve-cylinder main engine, a Wärtsilä 12RT-Flex96C.

The engine must start with these reservoirs 6 times ahead and 6 times astern **without** assistance from the air compressors.

▲
All the necessary equipment for an emergency-diesel generator set must be in the same room as the generator itself. Shown here, the diesel-oil daily supply tank.

32.11.21 Electric starting systems

Here a number of regulations also apply.

- Main engines that are electrically started must be equipped with two separate starting batteries, which cannot be operated in parallel.
- Each starting battery must be able to start the engine from a cold condition.
- The total capacity of the starting batteries must be sufficient so that within half an hour, the previously described starts can be done without having to re-charge the starting batteries.

32.11.22 Auxiliary engines

- At least three starts.
- Starting batteries should only be used to start the engine and its monitoring installations.

32.11.23 Starting emergency-diesel generators

- Emergency-diesel generators should be able to start at temperatures as low as 0°C.
- If it is not possible to start at these low temperatures, the generator must be supplied with heating.
- The genset must be operational in all weather- and sea-going conditions.
- The starting, charging and energy-storage equipment, such as the battery and the diesel-oil tank should be installed in the location of the emergency-diesel generator.

32.11.24 Local operating-control system

There are a number of requirements that apply to local operating control.

▶
The emergency manoeuvring position for a two-stroke crosshead engine, a Wärtsilä 10RTA96C.

In circle: starting- and reversing levers.

Table 2.7 Alarms and indicators

Description	Propulsion engines	Auxiliary engines	Emergency engines
speed/direction of rotation	I		
engine overspeed[5]	A, S	A, S	A, S
lubricating oil pressure at engine inlet	I, L, S	I, L, S	I, L
lubricating oil temperature at engine inlet	I, H	I[5],H[5]	I[5],H[5]
fuel oil pressure at engine inlet	I	I	
fuel oil temperature at engine inlet[1]	I	I	
fuel oil leakage from high pressure pipes	A	A	A
cylinder cooling water pressure or flow at engine inlet	I, L	I, L[4]	I[5], L[5]
cylinder cooling water temperature at engine outlet	I, H	I, H	I, H
piston coolant pressure at engine inlet	I, L		
piston coolant temperature at engine outlet	I, H		
charge air pressure at cylinder inlet	I		
charge air temperature at charge air cooler inlet	I		
charge air temperature at charge air cooler outlet	I, H		
starting air pressure	I, L		
control air pressure	I, L		
exhaust gas temperature[2]	I, H[3]		
oil mist concentration in crankcase or alternative monitoring system[7, 8]	I, H[6]	I, H[6]	I, H[6]

1 for engines running on heavy fuel oil only
2 where ever the dimensions permit, at each cylinder outlet and at the turbo charger inlet and outlet
3 at turbo charger outlet only
4 cooling water pressure or flow
5 only for an engine output ≥ 220 kW
6 for engines having an output > 2250 kW or a cylinder bore > 300 mm
7 alternative methods of monitoring may be approved by GL
8 an engine shutdown may be provided where necessary

I: Indicator
A: Alarm
H: Alarm for upper limit
L: Alarm for lower limit
S: Shutdown

◀ Table of compulsory alarms and indicators that must be able to be read in the emergency-start manoeuvring mode.

32.11.25 Main engine

- A locally-situated system that can operate and monitor the engine must be fitted.
- If reduction gearing and controllable-pitch propeller systems have been fitted, the above-mentioned control system must be able to enable emergency operation.
- Critical speed must be indicated by red sectors on the speed gauges.

▲ The emergency manoeuvring mode for a four-stroke trunk-piston engine, a nine-cylinder Caterpillar Mak 43 engine.

Arrow: toothed and reverse lever.

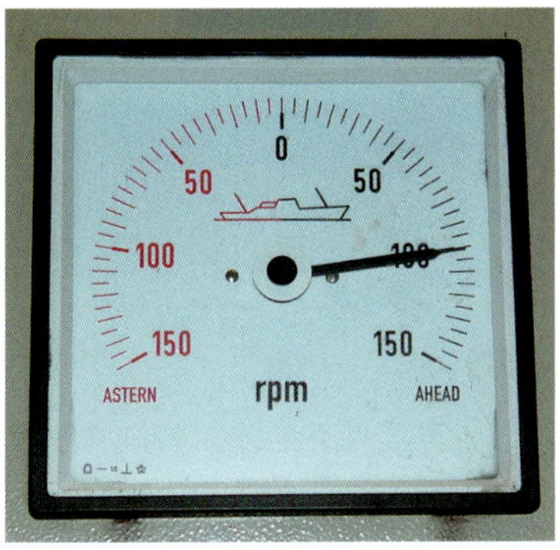

◀ Critical speeds must be indicated by red sectors on the speedometers. This engine does not have a critical speed.

32.12 Engine alignment

Special regulations have been drawn up.

- The crankshaft alignment must be checked each time an engine is re-aligned on its bedplate.
- This is done by **clock gauging** of the crankshaft. The crank-web deflection of every cylinder must be measured.

In order to compare later crankshaft deflections, note should be taken of the loading and draught of the vessel. The conditions of the engine when the deflection was measured: cold, preheated or at operating temperature.

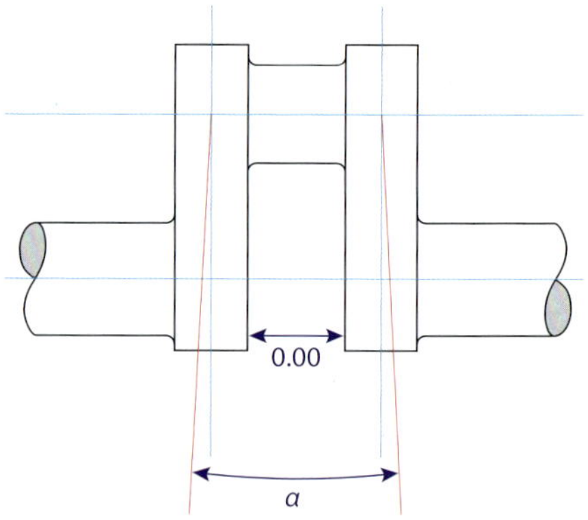

▲

When the crankshaft rotates, if the crank webs do not have the original values between the (engine manufacturer) stamped centres, they have a deviation known as the angular rotation of the webs. This is known as 'crankshaft clock gauging' or 'a deflection measurement'. It is obvious that the larger a crank the larger measured value at the same deflection (angle α) than a smaller crank.

▼

A chart showing clearly how large the maximum crankshaft deflections may be.

Horizontal: the distance r_o indicating where the measurement must be taken.

Vertical: the deflection result Δa

In the figure are lines for trunk-piston engines and crosshead engines.

Both lower lines are the recommended values after initial engine installation.

Both upper lines are the advised maximum crankshaft deflection.

The larger r_o is, the larger the crankshaft diameter and the larger the deflection may be.

Guidelines are in the process of being established for the angular rotation that a crank may deviate from the ideal line. The same maximum angular rotation has a larger absolute value with a larger crank.

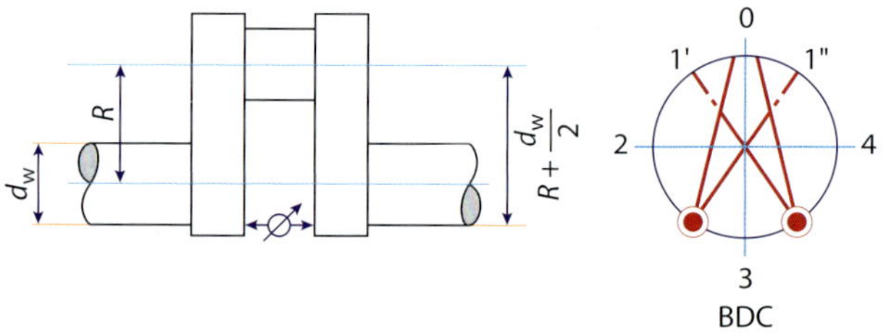

▲

The basic standard measurements for defining the crankshaft-deflection value, r_o.

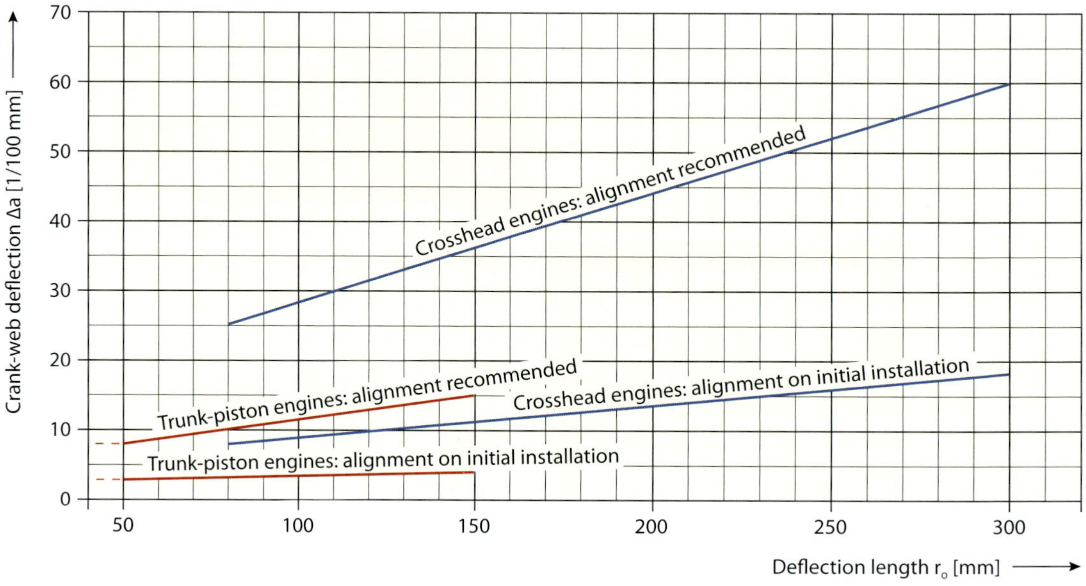

Remarks for clock gauging of the crankshaft and measuring the distance between the crank webs of each cylinder

The measurements should be performed at a certain distance from the heart line of the crankshaft

that is: $R + \frac{d_w}{2}$.

In virtually all crankshafts, both webs of the crank are provided with a centre hole suitable for placing the crankshaft clock gauge.
In normal crankshaft clock-gauging, the connecting rod remains attached to the crankpin. Therefore, the values 1' and 1" are an average (see the figure on the previous page).
The location of the centre holes is dependent on the type of crankshaft.
One can distinguish four types: A, B, C and D (see the figure on the right).

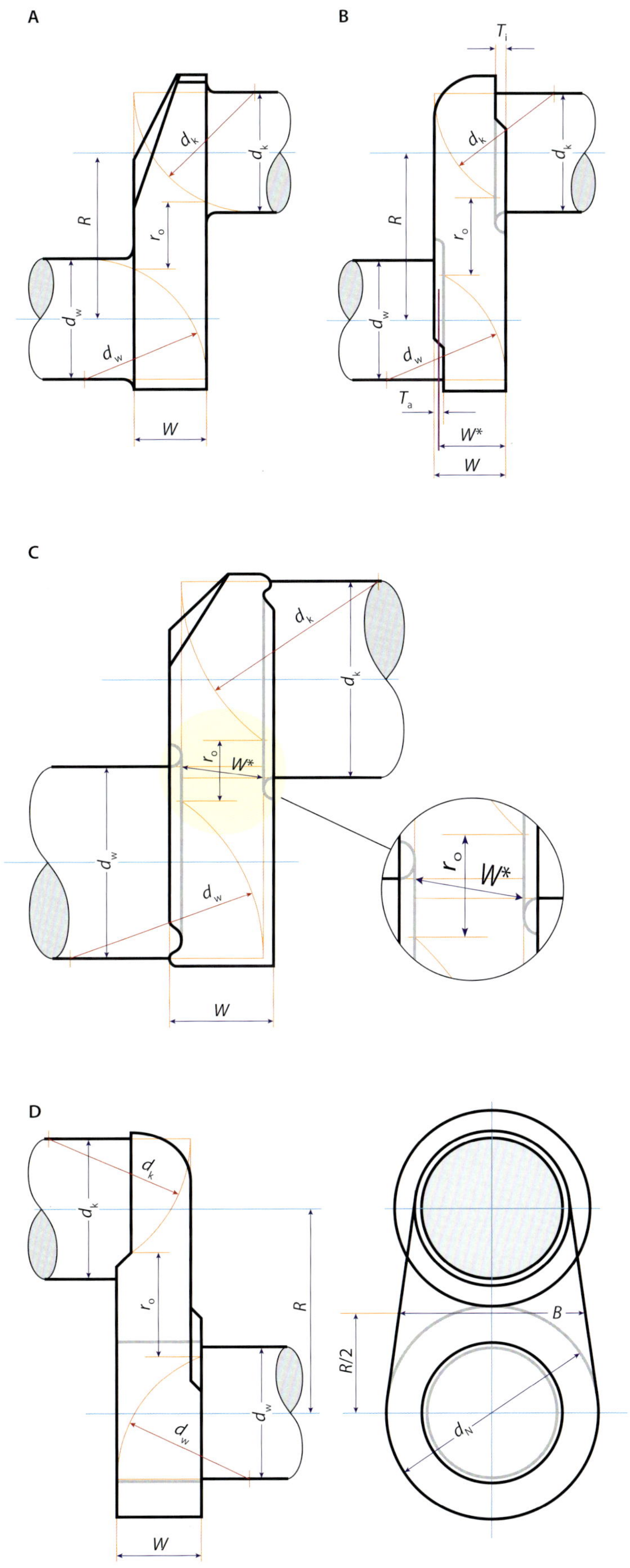

▶

The standard measurements for defining the crankshaft deflection values, r_o, for four different types of crankshafts.

They are all forged steel crankshafts.
Types A, B and C are manufactured in a single throw forging and used for four-stroke trunk-piston engines.
The type D crankshaft is manufactured by joining together single 'throws' consisting of two webs and the crankpin and used for two-stroke crosshead engines.

Type A: standard crankshaft, negative crankpin/crankshaft-formula solution.
Type B: with milled sections T_i and T_a
Type C: positive crankpin/crankshaft-formula solution.

32.13 The standard measurements and calculations to determine the crankshaft deflection measurement ro for four different types of crankshafts

32.13.1 Calculating the deflection length ro according to Germanischer Lloyd

There are separate calculations for shafts forged in a single throw forging, A, B and C and for a crankshaft manufactured by joining single throws together D.

The symbols applied are:

R = crank radius/radius (mm).
H = piston stroke (2R) (mm).
d_k = crankpin diameter (mm).
d_w = crankshaft diameter (mm).
W = width of the crank web (mm).
B = width of the crank web at distance $\frac{R}{2}$ (mm).
T_i = depth of milled section in the crank web at the crankpin side (mm).
T_a = depth of the milled section in the crank web at the crankshaft side (mm).
S = crankpin/bearing overlap (mm).
$S = \frac{(d_k + d_w)}{2} - R$

At a negative crankpin/bearing overlap ($S < o$) the deflection length r_o is in accordance with figure A and is calculated as follows:

$$r_o = 0.5(H + d_k + d_w) - W\left(\sqrt{\frac{2d_k}{W}} - 1 + \sqrt{\frac{2d_w}{W}} - 1\right)$$

In the other cases; B, C and D the formula must be adjusted constantly.
Apart from the material, the shape, construction and detailing of the crankshaft, much attention is paid to the exact location of the deflection measurement.

All are forged-iron crankshafts.
Types A, B and C are forged by single throw forging and usually manufactured for four-stroke trunk-piston engines.

Type A: with a negative result is:

$S < o$ in de formule $S = \frac{(d_k + d_w)}{2} - R$

$$r_o = 0.5(H + d_k + d_w) - W\left(\sqrt{\frac{2d_k}{W}} - 1 + \sqrt{\frac{2d_w}{W}} - 1\right)$$

When sections of a crank web have been milled away, the so-called 'web undercut', W in the formula must be modified to:

$$W^1 = W - \frac{(T_i + T_a)}{2}.$$

This is crankshaft type B.

For crankshaft type D, d_w in the formula must be substituted by $d_w^{\ 1}$:

$$d_w^{\ 1} = \frac{1}{3}(d_n - d_w) + d_w$$

For crankshaft type C, with a positive crankpin/crankshaft overlap ≥ 0 in the formula $s = \frac{(d_k + d_w)}{2} - R$, the W value in the formula must be replaced by W^1.

$$W^1 + \sqrt{(W - T_i - T_a)^2} + [0.5(d_k + d_w - H)]^2$$

From the abovementioned formulas, it can be inferred that the position of the crankshaft measurement for each engine was calculated earlier. Clearly, an engine manufacturer applies the deflection centres when finishing the crankshaft rather than leaving it to the mechanic to guess where the centres should be located.

32.14 Procedure for reconditioning parts

Many engine parts repaired after wear has been detected must be re-tested and approved. The following amongst others:

- Crankshafts
- Connecting rods
- Engine blocks
- Cylinder covers
- Pistons

The procedure is often as follows:

- The parts are disassembled and sent to a specialised reconditioning company.
- The parts are cleaned, inspected, measured and the material hardness of the journals is measured for, amongst others, a crankshaft.
- Advice is given to the client regarding the repair. It is possible that the advice is to repair the part; however, with serious wear or damage, it can be recommended that the part should be replaced.

▲
The installation of a new sleeve bearing in the connecting-rod eye is part of the overhaul of engine parts.

Shown here, the new sleeve bearing is cooled using liquid nitrogen so that the sleeve can be easily fitted in the connecting-rod eye. If both parts are at ambient temperature, the sleeve will not fit into the eye.
Wearing special 'cold resistant' gloves is necessary; the material temperatures are below -100 °C!

▲
An overheated crankpin from a medium-speed four-stroke engine.

The question is whether this type of damage can be repaired effectively.
Together with the classification society and the specialised repair company, an overhaul plan is prepared. The final result must guarantee that the surface hardness of the overheated crankpin does not deviate excessively from the surface hardness of the other crankpins.

The repair advice will also include the cost of the repair.
After the repair has been approved by the client, the part will be repaired according to the specifications of the engine manufacturer.
This means for damaged crankshaft journals, that the crankpin and/or crankshaft is ground to a certain **undersize**, and then fitted with **oversized** bearing shells.
In order to achieve this, engine manufacturers manufacture new crankshafts in such a manner that the shafts are over-dimensioned so that grinding is possible.
Therefore, the entire repair is performed according to the manufacturers' specifications.
After reconditioning/repair, the classification society for the particular engine, is invited to perform a survey.
When approved, the part is given a number code and signs by the classification society.
These are stamped. Surveyors of a classification society also have their own personal number, which is stamped into the part.

32.15 New parts

These are supplied with certificates by classification societies.
The reason is that often after purchase by the client, the engine manufacturers do not yet know to which class the engine and its spare parts belong.
Some engine manufacturers have **six** classification societies visit!

32.16 Special cases of wear and damage to engine parts

It can occur in very special cases that with damage to, for instance an engine block or a crankshaft, the classification society will be asked to partake in the discussions regarding the repair. An example is, whether the crankshaft of an engine in a vessel should be disassembled and removed, or if the disassembly of an engine block in a diesel-power station should be considered.

In both cases, the disassembly and then assembly cost a great deal of time and money. During the repair, the vessel or a part of the diesel power station is non-operational. A stationary ship costs a huge amount of money!

▶

A severely damaged cast-iron engine block.

The question remains: Is the repair technically possible and will the engine then be reliable again? How long does the customer have use this engine?

▲

A damaged two-stroke crosshead-engine guide.

The guide is removed from the A-frame and inspected on the engine-room floor.

The white metal running surface has separated from the guide.

This guide must be repaired.

▶

Inspection of this camshaft drive for a fuel pump shows that both running surfaces are still in a good state of repair.

The running surfaces are smooth and have no scratches or damage.

32.16.1 Engine blocks and connecting rods

Occasionally, after notification by a reconditioning company, a representative of the classification society will visit for an inspection. When the notification states that the parts will be reconditioned according to the specifications of the engine manufacturer and the parties are known to each other, often no checking visit takes place.

◀

The specifications from the engine manufacturer must also be strictly adhered to when line-boring commissioned engines.

◀

Damage to crankshafts is costly; when in repair at a specialised repair company, the crankshaft must be disassembled and transported.

The complete engine block must be placed on hydraulic jacks. The rejected crankshaft must be replaced; the costs can reach hundreds of thousands Euros; especially for category III engines. Severe damage to the crankshaft of two-stroke crosshead engines results in even higher costs. The ship is out of operation for a long time!

32.17 Damages to the engine or engine parts

A large number of parties are involved:

- An expert from the company, which carries the insurance of the engine.
- A surveyor from the classification society.
- The engine supplier and/or manufacturer.
- The owner/user of the engine.

Usually, these parties meet on the ship or at the diesel-power station for deliberations.

32.18 Damage

Insurance companies, that insure diesel engines, draw remarkable conclusions regarding damage to engines.

32.18.1 Engine damage, all engine types

Claims type	Number	Total cost (USD)	Avg. Cost (USD)
Turbocharger	84 (36,2%)	16.983.474 (24,4%)	202.184
Crankshaft, Connecting rod	23 (9,9%)	13.949.870 (20,0%)	606.516
Cylinder liner	17 (7,3%)	4.267.795 (6,1%)	251.047
Entablature, Staybolts	17 (7,3%)	3.505.803 (5,0%)	206.224
Journal, Bearing	15 (6,5%)	6.653.302 (9,5%)	443.553
Fuelpump, Gears	12 (5,2%)	3.161.929 (4,5%)	263.494
Camshaft, Coupling	10 (4,3%)	3.804.377 (5,5%)	380.438
Piston, Pistonrod	9 (3,9%)	2.702.420 (3,9%)	300.269

◀

Noticeable is the high rates of damage to turbochargers. A crucial part for modern highly-loaded diesel engines. The crankshaft, connecting rod and cylinder liner follow.

32.18.2 Engine damage, low-speed two-stroke crosshead engines

Claims type	Number	Total cost (USD)	Avg. Cost (USD)
Turbocharger	63 (42,6%)	14.057.056 (40,7%)	223.128
Entablature, Staybolts	17 (11,5%)	3.505.803 (10,2%)	206.224
Cylinder liner	15 (10,1%)	3.810.363 (11,0%)	254.024
Journal, Bearing	9 (6,1%)	2.916.201 (8,4%)	324.022
Piston, Pistonrod	7 (4,7%)	1.609.588 (4,7%)	229.941

◀

The turbocharger has the highest rate of damages for the low-speed two-stroke crosshead engines, followed by the stay bolts, cylinder liners, journal bearings, pistons and piston rods.

32.18.3 Engine damage, medium-speed four-stroke engines

Claims type	Number	Total cost (USD)	Avg. Cost (USD)
Turbocharger	21 (25,6%)	2.926.417 (8,9%)	139.353
Crankshaft, Connecting rod	21 (25,6%)	13.593.961 (41,5%)	647.331
Camshaft, Coupling	8 (9,8%)	3.451.850 (10,5%)	431.481
Journal, Bearing	6 (7,3%)	3.737.102 (11,4%)	622.850
Fuelpump, Gears	5 (6,1%)	765.436 (2,3%)	153.087

◀

For the four-stroke medium-speed diesel engines, the rates of damage to turbochargers and crankshafts with connecting rods are of an equal number; a remarkable shift to the drive gearing.
Gearing is the result of the high thermal- and mechanical loads of these engines types. See the load numbers in Chapter 5.

The graph below gives clarity regarding the type of propulsion.

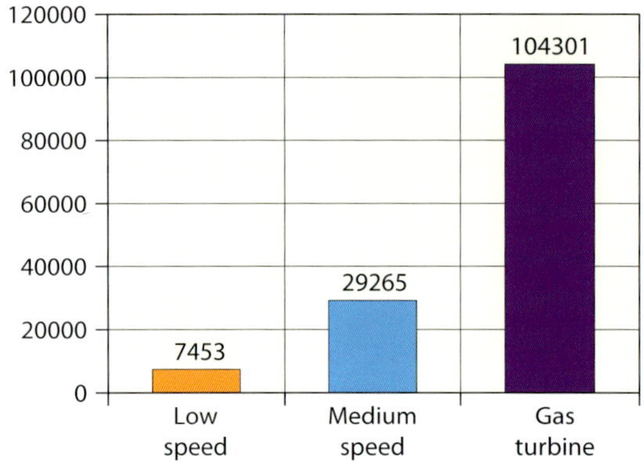

▲
Conclusion:
The amount of damage to low-speed two-stroke crosshead engines is by far the lowest; almost four times lower than the damage to medium-speed four-stroke engines and fourteen times lower than the damage to gas turbines!

32.18.4 Some general remarks

1 Damage to propulsion plants are costly.
2 Mainly medium-speed four-stroke diesel engines have high risks with respect to damages and therefore costs. They only represent 20% of the total number of insured engines, but 50% of the total costs of damages.
3 Damage to turbochargers is the largest loss item in diesel engines.
4 The most costly damage is to the crankshaft and connecting rod; especially for four-stroke medium-speed engines.
5 Damage to the crankshaft of a low-speed two-stroke crosshead engine forms the major part in the total amount of claims paid out for damage. Think about the disassembly and removal of the damaged crankshaft and the installation of a new crankshaft. A new crankshaft is a very expensive engine part!
6 The total loss of a medium-speed four-stroke engine, where the crankshaft and engine block are beyond repair is also part of damages that insurance companies must pay out.

▶
Damage to turbochargers is the most common occurring damage, for both four-stroke medium-speed and two-stroke crosshead engines.

32.19 Examples of certificates

▼

A certificate for 'oil engines' and therefore for diesel engines.

The certificate is for a Mak 8M32C propulsion engine for a tanker built at Volharding Shipyards B.V., Hoogezand, The Netherlands.

CERTIFIED COPY

Certificate no: **HAM 0740246/1**
Page 1 of 1

Lloyd's Register

Certificate For Oil Engines (Quality Assurance)

Office	Work's order number
Kiel	**261 956**
Manufacturer	Purchaser
Caterpillar Motoren GmbH & Co. KG	**Pritchard Gordon Ltd., Slaugham, UK**
Name and address of works	Purchaser's order number
Caterpillar Motoren GmbH & Co. KG **Falckensteiner Strasse 2** **24159 Kiel** **Germany**	**Tanker** Intended service **Main engine** Intended for ship/yard/engine number **Volharding Shipyards B.V. Hoogezand, Yard No. 603**

Particulars

Engine serial number	Number of cylinders	Diameter of cylinder (mm)
38809	**8**	**320**
Manufacturers designation	Stroke (mm)	
'MaK' 8M32C	**480**	
Shaft power kW and RPM (continuous rating)	Method of cooling	
3840 kW at 600	**Water**	

Design Appraisal Document no. and date
LR EMEA Hamburg, Plan Approcal Centre ref. HMD 11808-03/3 dated 10 November 2006

Testing
The engine has been run on a dynamometer and has satisfied the agreed test schedule

Identification Marks (Serial No. Control No. and Date)

QAM **042 LRS HAM 0740246 (entablature stamped)**

Manufacturer's Statement
This is to certify that the oil engine described above has been constructed and tested with satisfactory results in accordance with the current Rules and Regulations. A copy of the applicable Design Appraisal Document is attached.

Signed on behalf of manufacturer

Signature	Date
Name in block capitals	Position in company
G. HERRMANN / D. DITTRICH / Dr. GALLUN	**Quality Supervisor Department CQ3**

This certificate is issued by the manufacturer in accordance with the arrangements authorised by Lloyd's Register EMEA in Quality Assurance approval

Certificate number QAM **042/A1**

I certify that these arrangements are being kept under review by regular and systematic auditing of the approved Manufacturing and quality control procedures

Date
18 October 2007

Lloyd's Register EMEA
Signature
Kiel Office
H. PURCZ
Surveyor to Lloyd's Register EMEA

A member of the Lloyd's Register Group

This engine will be accepted for fitting in ships classed or intended to be classed with Lloyd's Register subject to satisfactory installation under the usual conditions of survey and testing, and where required by the rules, approval of vibration characteristics.

Form QA018 (2004.12)

LLOYD'S REGISTER GROUP PAPER (01/2005)

▼
Report on the main engine, an internal combustion engine with all the important engine data, such as, the cylinder diameter, the stroke, the shaft power, the mean indicated pressure and the maximum combustion pressure.
The data includes information regarding the torsional-vibration damper, the crankcase volume, the number of relief valves, the type of governor, the cylinder-cover material, piston crowns and oil-mist detector.

Report no: **HAM 0740246/2**
Page 1 of 3

Lloyd's Register

Report on Main Internal Combustion Reciprocating Engine

Ship's Name	Port	IMO number
-		
Place of Survey, if different from above		Office
		Kiel
Number of visits in shops	First Date	Last date
1	**17 October 2007**	**17 October 2007**
Ship built by		Yard number
Volharding Shipyards B.V., Hoogezand, The Netherlands		**603**
Engine made by		Year and month
Caterpillar Motoren GmbH & Co. KG, Falckensteiner Strasse 2, 24159 Kiel, Germany.		**2007/10**
Engine make and type		Engine Serial No
'MaK' 8M32C		**38809**
State if cylinders in V or other special formation	Number of engines	Number of crankshafts
In line	**1**	**1**
2 or 4 stroke cycle	Maximum kW	Number of cylinders, each engine
4	**3840**	**8**
Bore	Service kW	At rpm
320 mm	**3840**	**600**
Stroke	Corresponding M.I.P.	Maximum cylinder pressure
480 mm	**27.2 bar**	**191 bar**

Is engine of crosshead or trunk piston type? **Trunk piston type**
Are relief valves fitted to each cylinder? **Yes**

Is welded construction used for:
Bedplate? **No** Frames? **No** Entablature? **No**

Cooling medium for:
Cylinders? **F.W.** Pistons? **L.O.** Fuel Valves? **L.O.**

Must engine be removed for overhaul of main bearings, etc.? **No**
T.V.C. approval letter date and reference number: -

If torsional vibration damper or detuner fitted, state type, e.g. spring viscous etc. **Spring type**
State where fitted: **Crankshaft free end**

Total internal volume of crankcase: **7.7 m³**
Number and total area of crankcase explosion relief devices: **8 = 1144 cm²**

Are relief devices fitted to scavenge/supercharger manifold?	Is engine reversible?	How is engine started?
No	**No**	**Compressed air**

Type of governor fitted: **UG-40 DI**
Material of: Cylinder covers **Cast iron** Piston crowns **Forged steel**

Are flame guards or traps fitted to:
Crankcase relief devices? **Yes** Starting air pipes at cylinder starting air valves? **Yes**

Is an approved type crankcase oil mist detector fitted?
Schaller VN 115/87

Flywheel Shaft

Separate, integral with crank or thrust shaft
Integral with crankshaft

Material	Approved tensile strength	Actual tensile strength
Integral with crankshaft	**As for crankshaft**	**As for crankshaft**
Diameter	Flywheel weight	Flywheel diameter
As for crankshaft	**1405 kg**	**1439 mm**

Thrust Shaft

Separate, integral with crank or thrust shaft
Separate

Material	Approved tensile strength	Actual tensile strength	Diameter adjacent to collar
N/A	**N/A**	**N/A**	**N/A**

NOTICE: The particulars in this report are to be given as fully and as clearly as possible. Where the answer is "NO" or "NONE", state. Ticks and other signs of a doubtful meaning are not to be used. Wording not applicable to be cancelled.

Form 2011 (2004.11)

LLOYD'S REGISTER GROUP PAPER (01/2005)

▼

Further details of the same engine with the crankshaft data, an important part.
Documentation on the composition of material used, the manufacturing method, the tensile strength and pump data.

Report no: **HAM 0740246/2**
Page 2 of 3

Two Stroke Engines Only

Number and type of scavenge pumps of blowers each engine and how driven **N/A**	
For E.G. blowers state makers name and type **N/A**	Where exhaust gas driven blowers only are fitted, can engine operate with one out of action? **N/A**
State any emergency arrangements fitted **N/A**	Is the exhaust discharged through ports in the cylinders or through valves in the cylinder covers **N/A**
Are undersides of pistons used as scavenge pumps? **N/A**	Scavenge air pressure at full power. **N/A**

Four Stroke Engines Only

Is engine supercharged? **Yes**	Number of exhaust has driven supercharge blowers per engine **1**
Are undersides of pistons used as supercharge pumps? **No**	Supercharge pressure at full power **2,937 bar**

All Engines

Number of blower/supercharger air coolers each engine **1**	Number of blower/supercharger oil coolers each engine **None**

Crankshaft

Number of main journals **9**	Are main bearings of ball or roller type? **No**	Distance between inner edges of bearings in way of cranks **407 mm**
Distance between centre lines of side rods opposed piston engines **N/A**	Built, semi-built or solid crankshaft **Solid crankshaft**	
Diameter: Journals **300 mm**	Centre crank pins **280 mm**	Side crank pins **N/A**
Breadth of webs at mid throw **420 mm**	Axial thickness of webs **No. 1 web = 130 mm, remainder = 101 mm**	
If webs shrunk, state radial thickness round eyeholes **N/A**	Nominal shrinkage allowance if dowel pins are not fitted **N/A**	

Material (state whether cast or forged)

Pins	Webs	Journals
50CrMo4+Ni+V	**50CrMo4+Ni+V**	**50CrMo4+Ni+V**

If forged, state method
Continuous grain flow forged

Approved Tensile Strength

Pins	Webs	Journals
860 N/mm²	**860 N/mm²**	**860 N/mm²**

Actual Tensile Strength

Pins	Webs	Journals
900 N/mm²	**900 N/mm²**	**900 N/mm²**

Material of coupling bolts
42 CrMo 4 V

Main Engine Driven Pumps each engine.

State number and purpose of each type of pump (including fuel injection and fuel surcharge), type of drive, and for bilge pumps capacity at normal rpm
1 L.O. pressure pump (gear drive), 2 F.W. circulation pumps (gear drive), 8 F.O. injection pumps (cam drive)

Form 2011 (2004.11)

LLOYD'S REGISTER GROUP PAPER (01/2005)

▼

Identification marks on forgings and castings.

Report no: **HAM 0740246/2**
Page 3 of 3

Air Compressors
State number and how connected and whether can be declutched
No air compressor attached to the engine

Identification Marks of important forgings and castings (Shafting Certificates to be forwarded)
Crankshaft
QA023 STU 0700164, Heat No. 71220, Serial No. G 993 325,
Thrust/flywheel shaft
N/A
Is a detailed list of certificates attached to the report stating item, manufacturer, port of issue, certificate number, and identification markings?
Yes

Declaration
Declaration to be signed by engine builders

To the best of our knowledge this machinery has been soundly constructed in conformity with the Rules and Regulations, and the foregoing particulars of main engines are correct.

Signature – Manufacturer	Name in BLOCK CAPITALS	Date
i.A.	**C. MOHR**	25.10.07

Date and Port of Approval of Plans
State if general approval
General Approval
Crankshaft
LR EMEA Hamburg, Plan Approcal Centre ref. HMD 11808-03/3 dated 10 November 2006
Thrust/flywheel shaft
N/A
Air Receivers
N/A
A previous similar case was for (name)
N/A
Eng. Number
N/A
Port and report number
N/A

The machinery reported above has been built under Special Survey in accordance with the Rules, approved plans and Secretary's letters, examined, running on the test bed and found satisfactory. The materials and workmanship are good, the spare gear required by the Rules has been supplied and the machinery is eligible in my opinion to be fitted in a classed ship.

Issued at **Kiel** on **18 October 2007**

Lloyd's Register EMEA
Kiel Office
H. PURCZ
Surveyor to Lloyd's Register EMEA
Lloyd's Register
A member of the Lloyd's Register Group

Form 2011 (2004.11)

LLOYD'S REGISTER GROUP PAPER (01/2005)

▼
Certificate for forged steel crankshafts.
All the important data is shown. The identification number, method of manufacture and material tests.

Certificate no: **STU 0700164**
Page 1 of 1

Lloyd's Register

Certificate for Forged Steel Crankshafts (Quality Assurance)

Office
Stuttgart

Manufacturer
Maschinenfabrik Alfing Kessler GmbH, Aalen

Name and address of works
Maschinenfabrik Alfing Kessler GmbH
Auguste-Kessler-Str. 20
73433 Aalen

Work's order number
14560/16

Purchase order number
58369

Purchaser
Caterpillar Motoren GmbH & Co. KG, Kiel

Client
Manufacturer

Particulars

Crankshaft designation
One (1) cgf-forged, Type: 8 M 32-500

Plan number
1.94.7-25.10.01-04"d/d", Weight: 4400 kg (each)

Type and grade of steel
50CrMo4+Ni+V, quenched and tempered

Design appraisal Document no. and date
HMD 11808-3 Issue 1 dated 09.02.2004

Number of crankshafts
Forging No(s): G 993 325

Number of throws
8-throw

Stage of manufacture **finished machined**

Forging Process
continous grain flow forged

Non-Destructive Examination

Magnetic Particle
yes

Ultrasonic
yes

Identification

Cast number
71 220

LR **QA023, STU 0700164**
Forging No(s): see above

Test number
N/A

Date
N/A

This is to certify that the forgings detailed above have been inspected and tested with satisfactory results and conform in all respects to the specification and order.

Signed on behalf of (Manufacturer)
Maschinenfabrik Alfing Kessler GmbH, Aalen

Remarks
N/A

Signature – Manufacturer
J. Schmidt

Status
Quality Manager

Date
24 August 2007

This certificate is issued by the manufacturer in accordance with the arrangements authorised by **Lloyd's Register EMEA** in Quality Assurance Approval Certificate Number **QA023**

I certify these arrangements are being kept under review by regular and systematic auditing of the approved manufacturing and quality control procedures.

Remarks
For details of Mechanical Properties - see attached Works Test Certificate 38068 dated 26.06.07

Signature
O. Fuchs
Surveyor to Lloyd's **Register EMEA**
A member of the Lloyd's Register Group
Date **26 September 2007**

The crankshafts will be accepted for engines, compressors or other items of machinery which are required to be constructed in accordance with the Rules and Regulations for the classifications of ships.

Form 1281 (2006.04)

LLOYD'S REGISTER GROUP PAPER (01/2005)

▼
Inspection report for a crankshaft from the engineering works, Alfing in Aalen, Germany.

MASCHINENFABRIK ALFING KESSLER GMBH
73433 AALEN - WASSERALFINGEN

ALFING
MASCHINENFABRIK
ALFING KESSLER GMBH
AUTOMOTIVE CRANKSHAFTS
LARGE CRANKSHAFTS
HARDENING EQUIPMENT

Inspection - Certificate
acccording to EN 10204 3.1

No. 60580

Subject	Type	Drawing-No.	Material
1 Crankshaft, 8 throw , finished maschined	8 M 32-500	1.94.7-25.10.01-04 "d/d" 263 250 125	50CrMo4+Ni+V +QT quenched and tempere
Purchase	**Order No.**	**Alfing-Reference-No.**	**Inspection Society**
Caterpillar Motoren GmbH & Co.KG Kiel	58369 v. 19.06.2006	14560/16 HfH-4101 4010922	LR

The above-mentioned products have been manufactured and tested in accordance with the order and, where appropriate, with the stipulated specifications.
The results recorded are documented on the form/forms

Mech. Eigenschaften Nr. 38068.
crankshaft no. G 993 325.

Which as appendix / appendices to this covering sheet forms / form part of the inspection certificate in accordance with EN 10204.

Remarks:

Ultraschall-Prüfung am Schmiedeprodukt entsprechend Vorschrift AA 10/02/03/A durchgeführt.
Befund: Innerhalb der Vorschrift

Magnetische Rißprüfung gemäß Vorschrift AA 10/03/27/B durchgeführt.
Befund: Ohne Anzeigen.

We hereby certify that the results recorded for the above-mentioned products correspond to the customer's requirements or have been approved by the customer.

Maschinenfabrik Alfing Kessler GmbH
Quality Assurance Dept.

(Work's Inspector)
(J. Schmidt)

Date
24.08.2007
QWA - Schö

Enclosure(s)

▼
Mechanical properties of the crankshaft drawn up by the engineering works, Alfing.

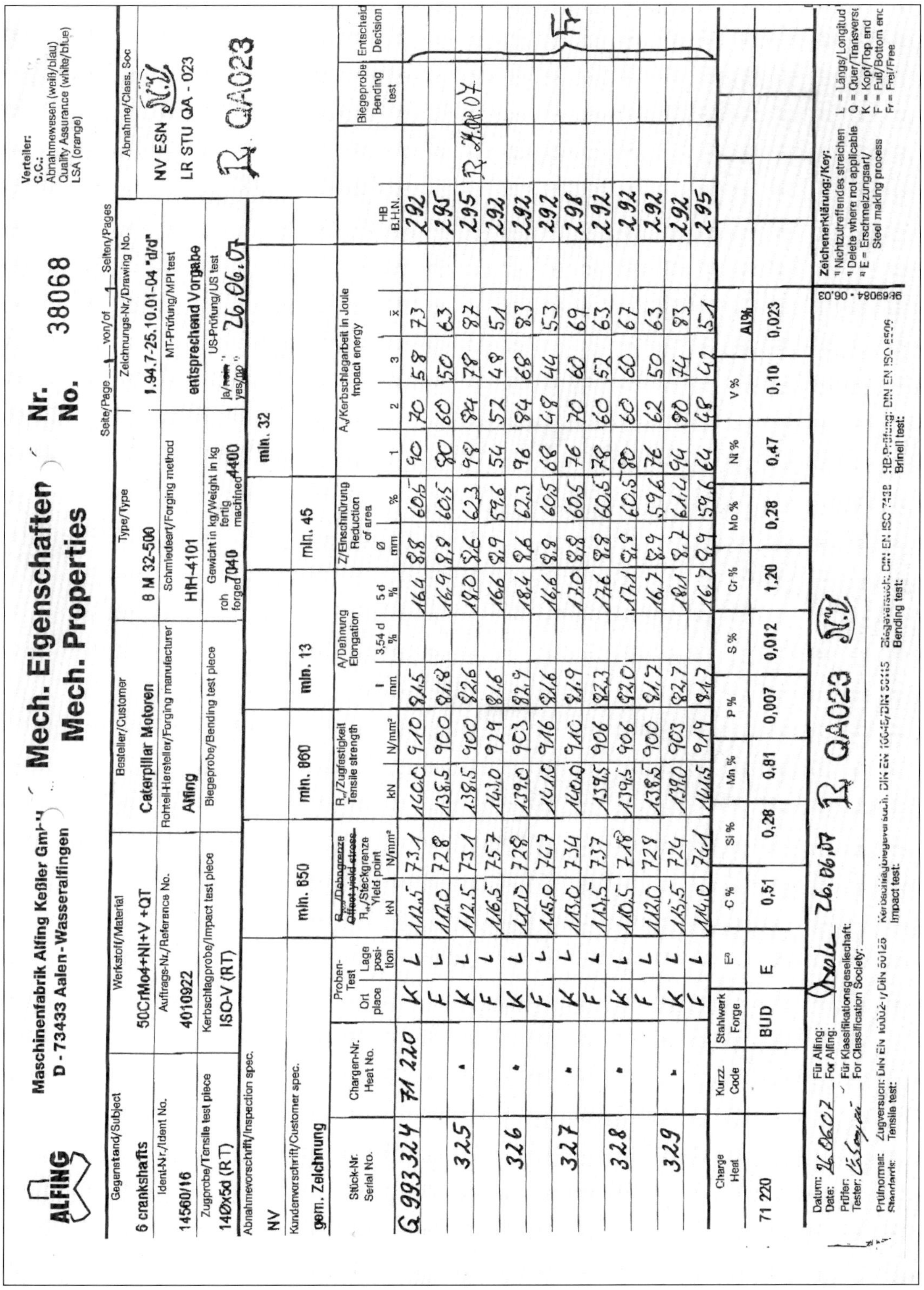

ALFING — Maschinenfabrik Alfing Keßler GmbH, D - 73433 Aalen-Wasseralfingen

Mech. Eigenschaften / Mech. Properties — Nr./No. 38068

Verteiler: C.C.: Abnahmewesen (weiß/blau), Quality Assurance (white/blue), LSA (orange)

Seite/Page 1 von/of 1 Seiten/Pages

Gegenstand/Subject	Werkstoff/Material	Besteller/Customer	Type/Type	Zeichnungs-Nr./Drawing No.	Abnahme/Class. Soc
6 crankshafts	50CrMo4+Ni+V +QT	Caterpillar Motoren	8 M 32-500	1.94.7-25.10.01-04 "d/d"	NV ESN N.V.
Ident-Nr./Ident No.: 14560/16	Auftrags-Nr./Reference No.: 4010922	Rohteil-Hersteller/Forging manufacturer: Alfing	Schmiedeart/Forging method: HH-4101	MT-Prüfung/MPI test: entsprechend Vorgabe	LR STU QA - 023
Zugprobe/Tensile test piece: 140x5d (RT)	Kerbschlagprobe/Impact test piece: ISO-V (RT)	Biegeprobe/Bending test piece	Gewicht in kg/Weight in kg: roh/forged 7040, fertig/machined 4400	US-Prüfung/US test: ja/yes 26.06.07	R QA023

Abnahmevorschrift/Inspection spec.: NV
Kundenvorschrift/Customer spec.: gem. Zeichnung

Stück-Nr. Serial No.	Chargen-Nr. Heat No.	Ort place	Lage position	R_{eH}/Streckgrenze Yield point kN	N/mm²	R_m/Zugfestigkeit Tensile strength kN	N/mm²	A/Dehnung Elongation l mm	3,54 d %	5 d %	Z/Einschnürung Reduction of area Ø mm	%	A_v/Kerbschlagarbeit in Joule Impact energy 1	2	3	x̄	HB B.H.N.	Biegeprobe Bending test	Entscheid Decision
Requirement					min. 850		min. 860			min. 13		min. 45				min. 32			
G 993324	71 220	K	L	112,5	731	140,0	910	91,5		16,4	8,8	60,5	90	70	58	73	292		Fr
		F	L	112,0	728	138,5	900	91,8		16,9	8,9	60,5	80	60	50	63	295		
325	•	K	L	112,5	731	138,5	900	92,6		18,0	8,6	62,3	98	84	78	87	295	R 24.08.07	
		F	L	116,5	757	143,0	929	91,6		16,6	8,9	59,6	54	52	48	51	292		
326	•	K	L	112,0	728	139,0	903	92,9		18,4	8,6	62,3	96	84	68	83	292		
		F	L	115,0	747	141,0	916	91,6		16,6	8,8	60,5	68	48	44	53	292		
327	•	K	L	113,0	734	140,0	910	91,9		17,0	8,8	60,5	76	70	60	69	298		
		F	L	113,5	737	139,5	906	92,3		17,6	8,8	60,5	78	60	52	63	292		
328	•	K	L	110,5	718	139,5	906	92,0		17,1	8,8	60,5	80	60	60	67	292		
		F	L	112,0	728	138,5	900	91,7		16,7	8,9	59,6	76	62	50	63	292		
329	•	K	L	111,5	724	139,0	903	92,7		18,1	8,7	61,4	94	80	74	83	292		
		F	L	114,0	741	141,5	919	91,7		16,7	8,9	59,6	64	48	42	51	295		

Charge Heat	Kurzz. Code	Stahlwerk Forge	E³	C %	Si %	Mn %	P %	S %	Cr %	Mo %	Ni %	V %	Al%
71 220		BUD	E	0,51	0,28	0,81	0,007	0,012	1,20	0,28	0,47	0,10	0,023

Datum/Date: 26.06.07 — Für Alfing/For Alfing: 26.06.07 — R QA023 N.V.
Prüfer/Tester: — Für Klassifikationsgesellschaft/For Classification Society:

Prüfnormen/Standards: Zugversuch/Tensile test: DIN EN 10002-1/DIN 50125; Kerbschlagbiegeversuch/Impact test: DIN EN 10045/DIN 50115; Biegeversuch/Bending test: DIN EN ISO 7438; HB-Prüfung/Brinell test: DIN EN ISO 6506

Zeichenerklärung/Key: ¹) Nichtzutreffendes streichen / Delete where not applicable; ²) E = Erschmelzungsart/Steel making process; L = Längs/Longitud.; Q = Quer/Transverse; K = Kopf/Top end; F = Fuß/Bottom end; Fr = Frei/Free

98/69084 · 06.03

▼

The crankshaft dimensions drawn up by the engineering works, Alfing.

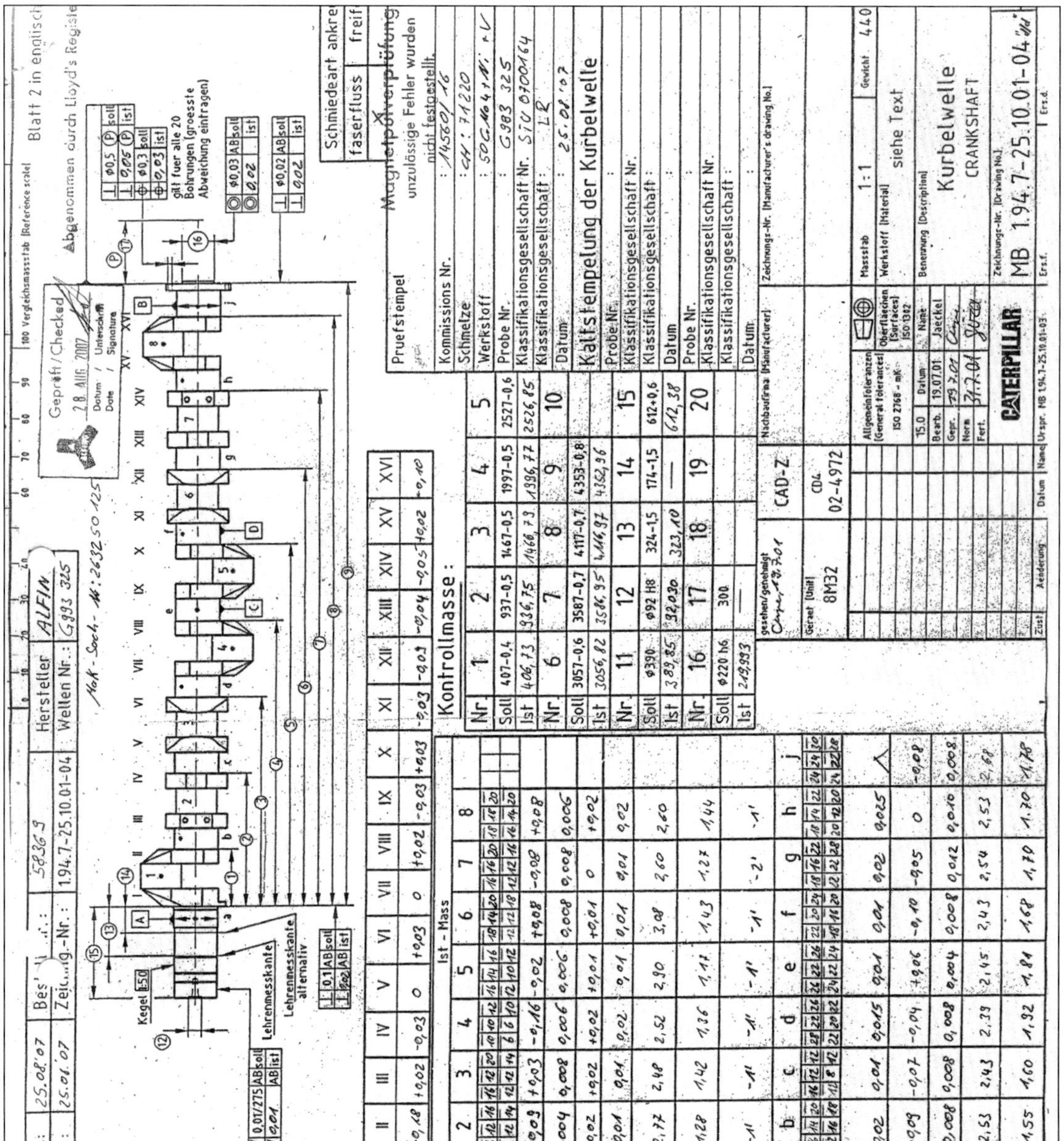

▼
A report from Lloyd's Register regarding the reduction gearing of a main engine.
Manufacturer, Renk AG in Rheine, Germany.

Report no: **DTM 0562069/1**
Page 1 of 3

Lloyd's Register

Report on Main Engine Reduction Gearing

Ship's Name

IMO number

Port
Dortmund

Place of Survey, if different from above
Rheine

Office
Dortmund

Number of visits in shops
2

First date
29 June 2006

Last date
26 September 2006

Ship built by

Yard number
NB 101

Gearing made by
Renk AG, Rheine / Germany

Gear number
857936

Year and month
2006/09

Number of sets and description of gearing, including reversing arrangements, and clutches, if any. State type of bearings.
Marine Reduction Gearbox Type RSV-630 with PTO. Center distance between the input/output shafts 630 mm. Input shafts and output shafts mounted on sleeve bearings and the PTO shaft are running in roller bearing.

Diagrammatic Sketch Showing Arrangement of Gearing

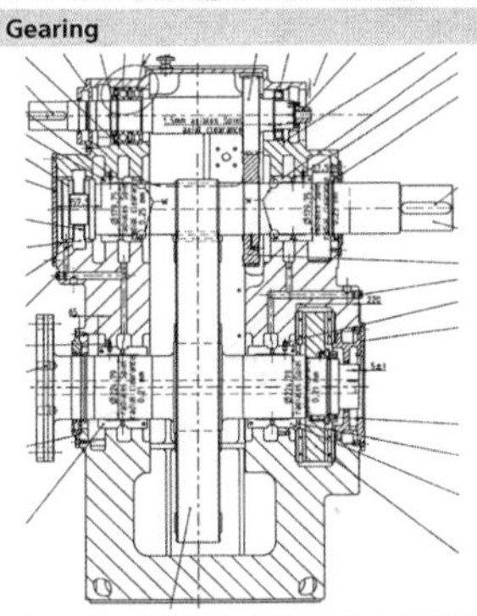

Approved maximum total kW each set
1520 KW

Corresponding rpm of main wheel
169,5 rpm

State ice class notation, if applicable
no

If single helical, state type and position of gear thrust bearing
see sketch or approved drawing of gear

Have all pinions, flexible coupling sleeves and service attachments been dynamically balanced?
no

State if wheels dynamically or statically balanced
PTO shaft dynamically balanced

Are bodies of cast or welded construction? State material.
solid forged

How are bodies connected to shafts?
shrink fitted

If rims separate, state how secured
no

If rims shrunk, has shrinkage allowance been checked and found as approved?
n.a.

Number and diameter of radial or dowels fitted
n.a.

Which gears were cut under controlled temperature conditions
Yes

How were teeth cut
ground

Machine used for finishing process

Maker's name
Niles / Pfauter Kapp

Serial number

If teeth surface hardened, state method
case hardened

Is gearcase of cast or welded construction
cast

Note: The particulars in this report are to be given as fully and clearly as possible. Where the answer is "NO" or "NONE", say so. Ticks and other signs of doubtful meaning are not to be used.

Form 2016 (2005.04)

▼

Extra data for the same gear box.

Report no: **DTM 0562069/1**
Page 2 of 3

If welded construction, has it been stress relieved

Have means been supplied for verifying that gearcase is free from distortion when secured in ship?

n.a.

Has gearing been run light loaded in shop, and tooth contact found satisfactory? State max. rpm reached

light load run main stage at 300 KW, PTO at 300 KW

What is the backlash? (State whether measured circumferentially or normal to teeth and if no clearance bearings)

main item 5040/5020: 0,30 mm; PTO item 5060/5100: 0,57 mm (pitch circle)

	Primary H.P.	Primary L.P.	Secondary H.P.	Secondary L.P.
Pinions				
Max. kW to be delivered to primary pinions	**item 5040: 1520 KW**		**item 5100: 372 KW**	
Corresponding rpm.	**1000 rpm**		**1507,7 rpm**	
Number of teeth	**20**		**65**	
Total width of gearface parallel to axis	**150 mm**		**55 mm**	
Width of gap between helices	--		--	
Approved pinion material tensile strength	**1080 N/mm²**		**1080 N/mm²**	
Actual pinion material tensile strength	**1334 N/mm²**		**1287 N/mm²**	
Quillshafts				
Approved tensile strength	--		--	
Actual tensile strength	--		--	
Flexible Couplings				
Type of coupling	--		--	
Approved driving member tensile strength	--		--	
Actual driving member tensile strength	--		--	
Approved driven member tensile strength	--		--	
Actual driven member tensile strength	--		--	
Material of sleeve	--		--	
Approved sleeve tensile strength	--		--	
Actual sleeve tensile strength	--		--	

Wheels	Primary H.P.	Primary L.P.	Main Wheel
Number of teeth	**item 5020: 118**	**item 5060: 98**	
Approved rim material tensile strength	**1080 N/mm²**	**1080 N/mm²**	
Actual rim material tensile strength	**1208 N/mm²**	**1283 N/mm²**	
Approved shaft material tensile strength	**item 5010: 600 N/mm²**	--	
Actual shaft material tensile strength	**715 N/mm²**	--	

Form 2016 (2005.04)

▼
Representatives from the manufacturer and the surveyor of Lloyd's Register must sign the report.

Report no: **DTM 0562069/1**
Page 3 of 3

Declaration to be Signed by Gearmakers

To the best of our knowledge this reduction gearing has been soundly constructed in conformity with the Rules and Regulations, and the foregoing particulars of reduction gearing are correct.

Signature	Name in BLOCK CAPITALS	Date
[signature]	**Dirk Kremkow**	**19 October 2006**

Date and port of approval plans
LLoyd's Register EMEA, Hamburg, Ref.: HMD 16496-06 Issue no. 0 dated 02 January 2006

A previous similar case was for (name)

Port and Report Number — Gear number

Declaration to be Completed and Signed by Surveyor

The reduction gearing reported above has been built under Special Survey in accordance with the Rules, approved plans and Secretary's letters. The materials and workmanship are good; the spare gear required by the Rules has been supplied and the gearing is eligible, in my opinion, to be fitted in a classed ship. A detailed list of certificates is attached to the report stating item, manufacturer, port of issue and certificate number and identification marking.

Issue Date: **03 November 2006**

Issue Location: **Dortmund**

[Stamp: Lloyd's Register EMEA, K. Allermann, Business Centre Central, Dortmund, Lloyd's Register]

K. Allermann
Surveyor to Lloyd's Register EMEA
A member of the Lloyd's Register Group

Declaration to be Completed and Signed by Surveyor at Port of Installation

The above reduction gearing has been fitted on board the MV/SS at in a fit and proper manner and found satisfactory when tested on under full power conditions for hours and when examined subsequently.

Date:

Location:

Surveyor to

A member of the Lloyd's Register Group

Form 2016 (2005.04)

LLOYD'S REGISTER GROUP PAPER (01/2005)

▼

The material properties for all the parts of the gear box are tested.
In this case, a pinion shaft in a gear seating.
The composition of the material and the mechanical properties are of utmost importance.
The heat treatment is also important, in this case case-hardening.

Prüfbescheinigung nach EN 10204 - Werkstoffprüfung
Inspection document acc. to EN 10204 - Material testing

RENK

Vorgangsdaten / Process data

RENK Auftrags-Nr. / RENK order no.	Produkt-Type / Product type	Herstellerzeichen / Manufacturer's sign	Ext. Abn.-Organisation / Ext. Insp. Organisation	Abnahme-Zertifikat-Nr. / Certificate no.
857936	**RSV-630**	RENK	**LRS**	
Positions-Nr. / Position no.	Material-Nr. / Material no.	Zeichnungs-Nr. / Drawing no.	Werkstoff-Norm / Material standard	Werkstoff / Material
5040	**W90662796253**	**0796253/0**	**TN70484**	**18CrNiMo7-6+A**

Anzahl / Quantity	Benennung/Denomination	Wärmebehandlung/Heat treatment
1	**RITZELWELLE M.RADSITZ / PINION SHAFT WITH GEAR SEATING**	**Einsatzhärten / case-hardening**

Anforderungen / Specifications	Rohabmessung / Rough dimension	Gewicht [kg] / Weight	Stahllieferant / Steel supplier	Stahlerzeuger / Steel maker
TN70481	**Ø 230**	**241,2**	**BGH- Siegen**	**BGH- Siegen**

Werkszeugnis 2.2 nach EN 10204 über chemische Zusammensetzung
Test report 2.2 acc. to EN 10204 - Chemical composition

Schmelzen-Nr. / Heat no. **760305**	C %	Si %	Mn %	P %	S %	Cr %	Mo %	Ni %	Al %	V %
Sollwerte / Theoretical values min.	0,15	-	0,40	-	-	1,50	0,25	1,40	-	-
max.	0,19	0,30	0,60	0,010	0,005	1,80	0,35	1,70	0,040	-
Istwerte[1] / Actual values[1]	**0,18**	**0,24**	**0,50**	**0,006**	**0,001**	**1,66**	**0,33**	**1,56**	**0,033**	**-**

[1] Chemische Zusammensetzung gemäß Lieferantenangaben / Chemical composition as indicated by the supplier

Abnahmeprüfzeugnis 3.1 nach EN 10204 über mechanisch-technologische Eigenschaften
Inspection certificate 3.1 acc. to EN 10204 - Mechanical-technological properties

	Zugversuch/Tensile test gemäß/acc. to EN 10002-1				Kerbschlagbiegeversuch/Impact test gemäß/acc. to EN 10045-1			
Proben-Nr.[2] / Specimen no.[2] **1866**; Lage[3] / Location.[3] **L**	Streck-grenze / Yield point $R_{p\,0,2}$ N/mm²	Zugfestigkeit / Tensile strenght R_m N/mm²	Bruch-dehnung / Elongation (l = 5 d) %	Einschnürung / Reduction of area %	Kerbschlagarbeit / Impact work ISO-V Joule			Prüf-temp. / Temp. °C
Sollwerte / Theoretical values	min. 785	1080-1500	min. 11	min. 40	min. 40			RT
Istwerte / Actual values	**1032**	**1334**	**13,4**	**60,8**	**95**	**91**	**73**	**RT**

[2] Art der Probe: B nach DIN 50125 / Specimen type: B acc. to DIN 50125 [3] L = längs/longitudinal; T = tangential/tangential; Q = quer/transverse

Bemerkungen:
Remarks:

Prüfergebnisse/Test results:		Die gestellten Anforderungen sind erfüllt / The requirements have been complied with.		
20.01.2006	**Ostholthoff**	**20.01.2006**	Weiffenbach	
Prüfdatum / Date of test	RENK-Prüfer / RENK Operator	Datum / Date	RENK-Abnahmebeauftragter / RENK Inspection representative	Externer Abnehmer / External Inspector

RENK AG Werk Rheine Qualitäts-Management

Diese elektronisch erstellte Prüfbescheinigung ist ohne Unterschrift gültig / This electronically issued inspection document is valid without signature.

RENK Aktiengesellschaft · Werk Rheine
Qualitätsmanagement / Quality Management
Postfach 1953 · 48409 Rheine
Tel.: +49 5971 790-0 · Fax: +49 5971 790-208

DIN ISO 9001 DNV Certified Company Seit 1989

Werkstoff_de_en / 2005-06 Fo.1020

The companies mentioned below all have contributed to the realisation of this book.
This has been made possible by allowing interviews, consenting to the use of photographs, providing abundant information or by correcting texts.
For this we would like to express our sincere gratitude!

Company	Location
A.M.W. Marine	Hendrik-Ido-Ambacht
ABB B.V.	Rotterdam
ABB Turbochargers	Baden, Switzerland
ABC, Anglo Belgium Corporation	Ghent, Belgium
Alfa Laval Benelux B.V.	Breda
Arhus Maskinmesterskole	Arhus, Denmark
Breko Reparatie B.V.	Papendrecht
Brinkman & Niemeijer Motoren B.V.	Twello
Caterpillar Mak	Kiel, Germany
Caterpillar Marine Power Systems	Hamburg, Germany
Caterpillar Motoren GmbH	Kiel, Germany
CENTA Nederland B.V.	Stellendam
Chalmers Maritime Academy	Kalmar, Sweden
Chris Marine	Malmö, Sweden
Cummins Holland B.V.	Dordrecht
DBR	Hardinxveld
Det Norske Veritas B.V.	Groningen
Diesel Engines Online B.V.	Barendrecht
Diesel Publications	Waukesha, USA
Discom B.V.	Alblasserdam
Easy Connections B.V.	Dordrecht
Eefting Engineering	Kolham
Fredericia Maskinmesterskole	Fredericia, Denmark
G.M.S. Instruments B.V.	Rotterdam
Geislinger GmbH	Salzburg, Austria
Genpower B.V.	Groot-Ammers
Germanischer Lloyd/Benelux	Schiedam
Goltens Dubai	Dubai, UAE
Goltens Rotterdam B.V.	Spijkenisse
Hatz Nederland B.V.	Culemborg
HRP Europe B.V.	Krimpen a/d Lek
IHC Holland	Kinderdijk
IHC Lagersmit B.V.	Kinderdijk
JR SHIPPING B.V.	Harlingen
Kemper en Van Twist Diesel B.V.	Dordrecht
Koedood Dieselservice B.V.	Hendrik-Ido-Ambacht
L.P. Koster & Zn. B.V.	Rotterdam
Ludwig Hunger	München, Germany
M. Jürgensen GmbH & Co KG	Sörup, Germany
Machinefabriek BOLIER	Dordrecht
Machinefabriek G. Olthof N.V.	Capelle a/d IJssel
Machinenfabrik Alfing Kessler GmbH	Aalen, Germany
Maersk Line	Copenhagen, Denmark
Maerkisches Werk GmbH	Halver, Duitsland
Mahle GmbH	Markgroningen, Germany
Maine Maritime Academy	Pleasant St. Castine, USA

MAK Nederland B.V.	Dordrecht
MAN Diesel A/G	Augsburg, Germany
MAN Diesel A/G	Hamburg, Germany
MAN Diesel A/S Alpha	Frederikshavn, Denmark
MAN Prime Serve	Schiedam
MAN Diesel Prime Serv Academy	Copenhagen, Denmark
Maritieme Academie Willem Barentsz	West-Terschelling
Mark van Schaick B.V.	Schiedam
Massachusetts Maritime Academy	Buzzards Bay, USA
Mastervolt B.V.	Drachten
Merford	Gorinchem
Miba Bearing Group	Laakirchen, Austria
Mobil Oil B.V.	Breda
M.R.C.	Schiedam
MTI Holland B.V.	Kinderdijk
MTU Detroit Diesel Benelux B.V.	Dordrecht
Nicoverken Holland B.V.	Schiedam
Noordhollandse Motorenrevisie B.V.	Winkel
Pasman Motoren & Aggregaten	Baak
Perkins	Peterborough, UK
Pon Power B.V.	Papendrecht
Rederij Doeksen	Harlingen
Royal Huisman Shipyard	Vollenhove
Rubber Design B.V.	Heerjansdam
Ruysch Technisch Handelsbureau	Zutphen
Sandfirden Technics	Den Oever
SKF Marine Service Centre	Ridderkerk
The Vestfold Academy	Göteborg, Sweden
Tognum, MTU Friedrickshafen GmbH	Friedrickshafen, Germany
Topec B.V.	Papendrecht
Total Nederland N.V.	's-Gravenhage
Turbo Ned	Zwijndrecht
Turbo's Hoet Nederland B.V.	Baarn
Turner B.V.	Hoofddorp
VAF Instruments B.V.	Dordrecht
Van Dinteren Technische Handelsonderneming	Tiel
Van West-Holland B.V.	IJmuiden
Vereniging Importeurs Verbrandingsmotoren	Zoetermeer
Vetus B.V.	Schiedam
Wagenborg Shipping B.V.	Delfzijl
Wärtsilä Nederland B.V.	Kruiningen
Wärtsilä Nederland B.V.	Schiedam
Wärtsilä Nederland B.V.	Zwolle
Wärtsilä Switzerland Ltd.	Winterthur, Switzerland
Waterschap Groot Salland	Zwolle
Westfalia Separator Nederland B.V.	Cuijk
Woodward Governor Nederland B.V.	Hoofddorp